nature
Encyclopedia of the Human Genome

nature
Encyclopedia of the
Human Genome

Volume 5

Editor in Chief

David N Cooper
University of Wales College of Medicine

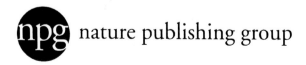
npg nature publishing group

London, New York and Tokyo

Published by
Nature Publishing Group, 2003
The Macmillan Building, 4 Crinan Street, London N1 9XW, UK

Associated companies and representatives throughout the world

www.nature.com
ISBN: 0-333-80386-8
345 Park Avenue South, New York, NY 10010, USA

British Library Cataloguing in Publication Data

A catalog record for this book is available from the British Library

Library of Congress Cataloging in Publication Data

A catalog record for this book is available from the Library of Congress

Typeset by Newgen Imaging Systems (P) Ltd, Chennai, India
Printed and bound by Clowes, Beccles, UK

Editorial and Production Staff

Project Managers
Michael Calais
Amy Lockyer
Mark Mones

Editorial Assistant
Fozia Khan

Production Controllers
Fenella Cooke
Daniel Price
Darren Smith

Copy Editors/Proofreaders
Anne Ashford
Anne Brown
Mary Carpenter
Andrew Colborne
Lucy Evans
Anna Hodson
Bridget Johnson
Allan Masson
Jessica Prokup
Carrie Walker
Alison Woodhouse
Mark Yawdoszyn

Additional Proofreading
Macmillan India

Indexer
Merrall-Ross International

Glossary Editor
Eleanor Lawrence

Managing Editor
David Atkins

Commissioning Editor and Publisher
Sean Pidgeon

Contents of Volume 5

Contents of Volume 5

Contents of Volume 5

Topical Outline of Articles

The entries in the *Nature Encyclopedia of the Human Genome* were conceived according to the conceptual categories listed below. Some entries are listed more than once because the conceptual categories are not mutually exclusive. The entries themselves appear alphabetically in the main body of the work.

Structural Genomics

Genome Organization
Chromosome 1
Chromosome 2
Chromosome 3
Chromosome 4
Chromosome 5
Chromosome 6
Chromosome 7
Chromosome 8
Chromosome 9
Chromosome 10
Chromosome 11
Chromosome 12
Chromosome 13
Chromosome 14
Chromosome 15
Chromosome 16
Chromosome 17
Chromosome 18
Chromosome 19
Chromosome 20
Chromosome 21
Chromosome 22
Chromosome X
Chromosome Y
Genetic Variation: Polymorphisms and Mutations
Genome Organization in Vertebrates
Genome Size
Single Nucleotide Polymorphism (SNP)

Gene Structure and Organization
Chromosomes 21 and 22: Comparisons
DNA Structure
Exonic Splicing Enhancers
Gene Structure and Organization
mRNA Untranslated Regions (UTRs)
3 UTRs and Regulation
5 UTRs and Regulation

Gene Types
Duchenne Muscular Dystrophy (*DMD*) Gene
GAS5 Gene
Genes: Types
Intronless Genes
Nonprotein-coding Genes
Protein Coding
Structural Proteins: Genes
Titin (*TTN*) Gene

Regulatory Sequences
Alternative Promoters: Duchenne Muscular Dystrophy (*DMD*) Gene
CpG Islands and Methylation
Enhancers
Locus Control Regions (LCRs)
Negative Regulatory Elements (NREs)
Promoters

RNA Processing
RNA Processing
Splice Sites

Gene Families
γ-Aminobutyric Acid (GABA) Receptors
Apolipoprotein Gene Structure and Function
Chaperones, Chaperonins and Heat Shock Proteins (HSPs)
Crystallins
Cystatins
Cytochrome P450 (*CYP*) Gene Superfamily
Gene Families
Heat Shock Proteins (HSPs): Structure, Function and Genetics
Histones
Hox Genes: Embryonic Development
Immunoglobulin Genes
Integrins
Interferons
Keratins and Keratin Diseases
Major Histocompatibility Complex (MHC) Genes
Nuclear Receptor Genes
Olfactory Receptors
Ovalbumin Serpins
Serine Proteases
Zinc-finger Genes

Repetitive Sequence Elements
Centromeric Sequences and Sequence Structures
Chromosome-specific Repeats (Low-copy Repeats)
Long Interspersed Nuclear Elements (LINEs)
Megasatellite DNA
Microsatellites
Minisatellites
Retroviral Repeat Sequences
Short Interspersed Elements (SINEs)
Simple Repeats
Telomeric and Subtelomeric Repeat Sequences
Transposons
Trinucleotide Repeat Expansions: Mechanisms and Disease Associations

Mitochondrial Genome
Mitochondrial DNA: Fate of the Paternal Mitochondrial Genome
Mitochondrial DNA Polymorphisms
Mitochondrial Genome
Mitochondrial Proteome: Origin

Functional Genomics

Gene Expression
Gene Expression Networks
mRNA Stability and the Control of Gene Expression
Promoter Haplotypes and Gene Expression
Transcription Factors

RNA: Function at the Genome Level
Alternative Promoter Usage
Chromatin Structure and Domains
Clustering of Highly Expressed Genes in the Human Genome
DNA Coiling and Unwinding
DNA Helicases
DNA Polymerases: Eukaryotic
DNA Recombination
DNA Repair
DNA Replication
DNA Replication Fidelity
DNA Replication Origins
Histone Acetylation: Long-range Patterns in the Genome
Matrix-associated Regions (MARs) and Scaffold Attachment Regions (SARs)
Mitochondrial DNA Repair in Mammals
Telomeres: Protection and Maintenance
Topoisomerases
Transcription-coupled DNA Repair

RNA: Function at the Transcriptome Level
Alternative Processing: Neuronal Nitric Oxide Synthase
Ectopic Transcription
mRNA Editing
mRNA Export
mRNA Localization: Mechanisms
mRNA Stability and the Control of Gene Expression
mRNA Turnover
mRNA: Intranuclear Transport
Noncoding RNAs: A Regulatory Role?
Nucleolar Dominance
Nucleolus: Structure and Function
Posttranscriptional Processing
RNA-binding Proteins: Regulation of mRNA Splicing, Export and Decay
RNA Polymerases and the Eukaryotic Transcription Machinery
snoRNPs
Spliceosome
Splicing of pre-mRNA
SR Proteins
Tissue-specific Locus Control: Structure and Function

Transcriptional Regulation: Coordination
Translation Initiation: Molecular
  Mechanisms in Eukaryotes
Trans Splicing
tRNA

Epigenetics
DNA Demethylation
DNA Methylation and Histone
  Acetylation
DNA Methylation and Mutation
DNA Methylation: Enzymology
DNA Methylation in Development
Genetic Conflict and Imprinting
Genomic Imprinting at the Transcriptional
  Level
Methylated DNA-binding Proteins

Developmental Genetics
Hox Genes: Embryonic Development
Germ Plasm and the Molecular
  Determinants of Germ Cell Fate
Human Developmental Molecular
  Genetics
Mapping Gene Function in the Embryo
Skeletogenesis: Genetics
Tooth Morphogenesis and Patterning:
  Molecular Genetics

Genetics of Physiological Systems
Absolute Pitch: Genetics
Circadian Rhythm Genetics
Complement System and Fc Receptors:
  Genetics
Handedness, Left/Right: Genetics
Longevity: Genetics
Skin Pigmentation: Genetics

Analytical Techniques
Animal Models
Comparative Genomic Hybridization in
  the Study of Human Disease
Cre-*lox* Inducible Gene Targeting
DNA Chip Revolution
DNA Chips and Microarrays
Expression Analysis *In Vitro*
Expression Analysis *In Vivo*
Expression Analysis *In Vivo*: Cell Systems
Expression Studies
Functional Complementation
Fusion Proteins as Research Tools
Gene Targeting by Homologous
  Recombination
Infectomics: Study of Response to Infection
  using Microarrays
In Vitro Mutagenesis
Microarrays and Single Nucleotide
  Polymorphism (SNP) Genotyping
Microarrays in Disease Diagnosis and
  Prognosis
Microarrays in Drug Discovery and
  Development
Microarrays in Toxicological Research
Microarrays: Use in Gene Identification
Microarrays: Use in Mutation Detection
Minigenes
Mouse Genetics as a Research Tool
Mutation Detection
RNA Interference (RNAi) and MicroRNAs
Yeast Two-hybrid System and Related
  Methodology

Chromosome Structure and Function

Chromosome Structure and Organization
Blocks of Limited Haplotype Diversity
Cell Cycle: Chromosomal Organization
Centromeres
Chromatin in the Cell Nucleus: Higher-
  order Organization
Chromosomal Bands and Sequence
  Features
Chromosome
Chromosome Structures: Visualization
Chromosomes 21 and 22: Gene Density
Chromosomes and Chromatin
Chromosomes: Higher-order Organization
Epigenetic Factors and Chromosome
  Organization
Gene Distribution on Human Chromosomes
Heterochromatin: Constitutive
Isochores
L Isochore Map: Gene-poor Isochores
Microdeletions and Microduplications:
  Mechanism
Telomere

Chromosomes during Cell Division
Cell Cycle Control: Molecular Interaction
  Map
Chromosomes during Cell Division
Kinetochore: Structure, Function and
  Evolution
Meiosis
Meiosis and Mitosis: Molecular Control of
  Chromosome Separation
Mitosis
Mitosis: Chromosomal Rearrangements
Mitosis: Chromosome Segregation and
  Stability
Synchronous and Asynchronous
  Replication
Telomerase: Structure and Function

Chromosome Analysis and Identification
Banding Techniques
Chromosome Analysis and Identification
Chromosome Preparation
Cytogenetic and Physical Chromosomal
  Maps: Integration
Digital Image Analysis
Far-field Light Microscopy
Flow-sorted Chromosomes
Fluorescence *In Situ* Hybridization (FISH)
  Techniques
Fluorescence Microscopy
Genetic and Physical Map Correlation
Karyotype Interpretation

Sex Chromosomes
Chromosome X: General Features
Chromosome Y: General and Special
  Features
Male Sex Determination: Genetics
Mammalian Sex Chromosome Evolution
Sex Chromosomes
X and Y Chromosomes: Homologous
  Regions
X-chromosome Inactivation

Chromosome Evolution in Mammals
Chromosomal Rearrangements in Primates
Chromosome Numbers in Mammals

Chromosome Rearrangement Patterns in
  Mammalian Evolution
Chromosomes in Mammals: Diversity and
  Evolution
Comparative Chromosome Mapping:
  Rodent Models
Comparative Cytogenetics
Comparative Cytogenetics
  Technologies
Microdeletions and Microduplications:
  Mechanism

Evolution and Comparative Genomics

Evolution of the Human Genome
Bacterial DNA in the Human Genome
Evolutionary History of the Human
  Genome
GC-rich Isochores: Origin
Mitochondrial Genome: Evolution
Mitochondrial Non-Mendelian
  Inheritance: Evolutionary Origin and
  Consequences
Mitochondrial Origins of Human Nuclear
  Genes and DNA Sequences
Mitochondrial Proteome: Origin
Polyploid Origin of the Human
  Genome

Comparative Genomics
Caenorhabditis elegans Genome
  Project
Codon Usage
Coevolution: Molecular
Comparative Genomics
Evolution: Convergent and Parallel
  Evolution
Fugu: The Pufferfish Model Genome
Genetic Code: Evolution
Great Apes and Humans: Genetic
  Differences
Homologous, Orthologous and
  Paralogous Genes
Human and Chimpanzee Nucleotide
  Diversity
Karyotype Evolution
Mammalian Phylogeny
Orthologs, Paralogs and Xenologs in
  Human and Other Genomes
Primates: Phylogenetics
Speciation
Species and Speciation

Evolution of Gene Structure
Alternative Splicing: Evolution
Cystic Fibrosis Transmembrane
  Conductance Regulator Sequences:
  Comparative Analysis
Exons and Protein Modules
Exons: Insertion and Deletion during
  Evolution
Exons: Shuffling
Gene Structure: Evolution
Introns: Movements
Introns: Phase Compatibility
Phylogenetic Footprinting
Primate Evolution: Gene Loss and
  Inactivation
Promoters: Evolution
Pseudoexons

Behavioral and Psychiatric Genetics

Mathematical and Population Genetics

Repetitive Elements: Detection
RNA Gene Prediction

Macromolecule Structure Prediction
Protein Homology Modeling
RNA Secondary Structure Prediction
RNA Tertiary Structure Prediction:
    Computational Techniques

Ethical Legal and Social Issues

Genetic Counseling and Clinical Genetics
Clinical Genetics and Genetic Counseling
    Professionals: Attitudes to Contentious
    Issues
Clinical Genetic Services in the United
    Kingdom
Code of Ethical Principles for Genetics
    Professionals
Genetic Carrier Testing
Genetic Counseling
Genetic Counseling: Communication
Genetic Counseling: Consanguinity
Genetic Counseling: Consanguinity and
    Cultural Expectations
Genetic Counseling Consultations:
    Uncertainty
Genetic Counseling for Muslim Families of
    Pakistani and Bangladeshi Origin in
    Britain
Genetic Counseling: Impact on the Family
    System
Genetic Counseling: Nondirectiveness
Genetic Counseling Profession in Europe
Genetic Counseling: Psychological Issues
Genetic Counseling: Psychological Models
    in Research and Practice
Genetic Counseling Services: Outcomes
Genetic Education of Primary Care Health
    Professionals in Britain
Genetic Information and the Family in
    Japan
Genetic Registers
Genetic Risk
Genetic Testing of Children
Huntington Disease: Predictive Genetic
    Testing
Nondirectiveness
Predictive Genetic Testing
Predictive Genetic Testing: Psychological
    Impact

Genetic Screening
Carrier Screening for Inherited
    Hemoglobin Disorders in Cyprus and the
    United Kingdom
Carrier Screening of Adolescents in
    Montreal
Down Syndrome: Antenatal Screening
Familial Breast Cancer: Genetic Testing
Genetic Screening Programs
Genetic Screening: Facilitating Informed
    Choices
Genetic Susceptibility
Informed Consent and Multiplex
    Screening
Newborn Screening Programs
Polygenic Inheritance and Genetic
    Susceptibility Screening

Population Carrier Screening:
    Psychological Impact
Pregnancy Termination for Fetal
    Abnormality: Psychosocial
    Consequences
Reproductive Genetic Screening: A Public
    Health Perspective from the United
    Kingdom

Genes, Research and Industry
Animal Rights: Animals in Genetics Research
Celera Genomics: The Race for the Human
    Genome Sequence
Children in Genetic Research
Cloning of Animals in Genetic Research:
    Ethical and Religious Issues
Commercialization of Human Genetic
    Research
deCODE: A Genealogical Approach to
    Human Genetics in Iceland
deCODE and Iceland: A Critique
DNA Technology: A Critical European
    Perspective
ELSI Research Program of the National
    Human Genome Research Institute
    (NHGRI)
Gene Therapy: Expectations and Results
Gene Therapy: Motivations for Research
Genetic Age: A Vision
Human Genome Project as a Social
    Enterprise
Human Genome Project, HUGO and
    Future Health Care
Informed Consent in Human Genetic
    Research
Ownership of Genetic Material and
    Information
Patenting of Genes: A Personal View
Patenting of Genes: Discoveries or
    Inventions?
Patent Issues in Biotechnology
Stored Genetic Material: Use in Research

Human Genetics, Ethics and the Law
China: The Maternal and Infant Health
    Care Law
Community Consent for Genetic Research
Criminal Responsibility and Genetics
Data Protection Legislation
DNA Fingerprinting, Paternity Testing and
    Relationship (Immigration) Analysis
Gene Therapy: Ethics and Regulation
Genetic Disability and Legal Action:
    Wrongful Birth, Wrongful Life
Genetic Testing of Children: Capacity of
    Children to Consent
Genetic Testing of Children: Parental
    Requests
Helsinki Declaration
Human Cloning: Legal Aspects
Informed Consent
Informed Consent: Ethical and Legal Issues
Interests of the Future Child
In Vitro Fertilization: Regulation
Privacy and Genetic Information
Privacy: Confidentiality and Responsibility

*Insurance, Employment and Human
    Genetics*
Discrimination in Insurance: Experience in
    the United States

Genetic Factors in Life Insurance: Actuarial
    Basis
Genetic Information, Genetic Testing and
    Employment
Health Care and Health Insurance in the
    United States
Insurance and Genetic Information
Insurance and Human Genetics:
    Approaches to Regulation

Disability and Genetics
Conceptualization of the Body in
    'Disability'
Disability and Genetics: A Disability Rights
    Perspective
Disability, Human Rights and
    Contemporary Genetics
Disability: Diagnostic Labeling
Disability: Philosophical Issues
Disability: Stigma and Discrimination
Disability: Western Theories
Mentally Handicapped in Britain: Sexuality
    and Procreation
Quality of Life: Human Worth Reduced to
    Measures of Ability

*Human Genetics and Reproductive
    Technologies*
Autonomy and Responsibility in
    Reproductive Genetics
Dolly and Polly
Feminist Perspectives on Human Genetics
    and Reproductive Technologies
Gamete Donation and 'Race'
Genetic Enhancement
Genetic Enhancement: The Role of Parents
Genetics and the Control of Human
    Reproduction
Human Cloning
Human Cloning: Arguments Against
Human Cloning: Arguments For
In Vitro Fertilization
In Vitro Fertilization: Regulation
Preimplantation Genetic Diagnosis: Ethical
    Aspects
Reproductive Choice
Reprogenetics: Visions of the Future
Sex Selection

Human Genetics and Society
Art and Genetics
Behavioral Phenotypes: Goals and
    Methods
China: The Maternal and Infant Health
    Care Law
Citizens' Jury on Genetic Testing for
    Common Disorders
Confronting Genetic Disease:
    Psychological Issues
Deaf Community and Genetics
Ethnic Inequalities in Health
Gene as a Cultural Icon
Genetic Futures and the Media
Genetic Information and the Family in
    Japan
Genetic Risk: Social Construction
Genetic Services: Access
Geneticization: Debates and Controversies
Genetics in Contemporary Germany
Health Care Ethics: The Four Principles
Heredity and the Novel
Heredity: Lay Understanding

History

Renal Carcinoma and von Hippel–Lindau Disease

Patrick H Maxwell, *Imperial College of Science, Technology and Medicine, London, UK*

Von Hippel–Lindau disease is a rare autosomal dominant condition with a high risk of renal carcinoma. The underlying tumor suppressor gene is also mutated in most sporadic clear cell renal carcinomas, and has given insight into how cells detect and respond to changes in the availability of oxygen.

Intermediate article

Introduction

Renal cell carcinoma and the insights provided by the relatively uncommon hereditary cancer syndrome, von Hippel–Lindau disease, are topical for several reasons. First, renal cell carcinoma is a common cause of cancer death, killing nearly 12 000 individuals in the United States in 2000. Second, von Hippel–Lindau (VHL) disease, although rare, is a potentially devastating familial condition in which screening for clinical manifestations offers great benefit. Third, the identification of the von Hippel–Lindau syndrome (*VHL*) tumor suppressor gene in 1993 represented a breakthrough in understanding both VHL disease and the commonest form of renal cell carcinoma. Fourth, the VHL gene product has emerged as a key regulator of cellular responses to changes in oxygen tension, thus lying at the center of an evolutionarily conserved system that regulates diverse processes including angiogenesis and cellular metabolism.

Classification of Renal Carcinoma

Renal cell carcinoma originates from the tubular epithelial cells of the renal cortex. Other tumors arising in the kidney are Wilms tumor (nephroblastoma) in children, and transitional cell carcinoma of the renal pelvis. Renal cell carcinoma is subclassified on the basis of histological characteristics. The commonest form accounting for about 80% of cancers is clear cell renal cell carcinoma (CCRCC), followed by papillary (15%), chromophobic (5%), oncocytic (<1%) and collecting duct tumors (very rare). This article focuses on the commonest type of tumor, CCRCC.

Clinical Aspects of Clear Cell Renal Cell Carcinoma

The incidence of renal carcinoma is increasing steadily. Some of this is due to a higher likelihood of diagnosis, which relates to increased use of magnetic resonance imaging (MRI), computed tomography (CT) and ultrasound. However, population-based studies suggest that underlying incidence is increasing. Significant risk factors for developing CCRCC include male sex, cigarette smoking, obesity, hypertension, diuretic treatment and having a first-degree relative with the disease. Presentation in a small proportion of patients is with pain in the flank and/or hematuria. The great majority of cases are found when imaging studies are being carried out for nonspecific abdominal complaints. Prognosis mainly depends on clinical stage at the time of presentation: about 30% of patients will have metastases at presentation, and these individuals have a median survival of less than 1 year. Common sites of metastasis are the lung, bone, liver and brain. In as many as 30% of patients with larger tumors and apparently limited disease, metastases will become evident after nephrectomy.

The mainstay of treatment is surgery. The historical treatment for early stage disease has been radical nephrectomy, in which the whole kidney, together with surrounding fat and the adrenal gland, is removed and regional lymph nodes are dissected. Although not subjected to a direct comparison, apparently satisfactory results have been obtained by a more conservative approach to smaller tumors (<4 cm) in which only the portion of the kidney containing the tumor is removed. Even in advanced disease, surgery may be appropriate since it can improve symptoms. Surgical removal of a solitary metastasis can be curative. On some occasions, secondary tumors will regress once the primary tumor is removed. Unfortunately, CCRCC is relatively resistant to chemotherapy and radiotherapy. Cytokine treatment with interleukin-2 and/or interferon-α2a induces a response in some patients.

There are promising recent reports using cell-based therapies – nonmyeloablative allogeneic peripheral blood stem cell transplantation and vaccination with tumor cell–dendritic cell hybrids. In studies of fewer than 20 patients with metastatic disease these treatments have been curative in about 15% of patients (Childs et al., 2000; Kugler et al., 2000).

von Hippel–Lindau Disease

As with cancers of other organs, notably breast and colon, major insights into the genetic events underlying CCRCC have come from infrequent kindreds with a very high risk of disease. In VHL disease, which is transmitted as an autosomal dominant trait, individuals have a lifetime risk of CCRCC of around 70%. VHL patients will often have multiple tumors with first presentation at a younger age than nonfamilial CCRCC. In addition to tumors, patients typically develop multiple renal cysts. The incidence of VHL disease is approximately 1 per 36 000 live births. Other clinically important manifestations of VHL disease include retinal hemangioblastoma, central nervous system hemangioblastoma and pheochromocytoma. Patients may also develop nonsecreting islet cell tumors of the pancreas, endolymphatic sac tumors (which can result in deafness), epididymal papillary cystadenoma (men) and cysts of the uterine broad ligament (women). Clinical management of VHL disease involves regular screening for retinal hemangioblastoma, CCRCC and pheochromocytoma. Early treatment of these lesions prevents blindness, reduces the risk of metastatic CCRCC and aims to prevent serious consequences of pheochromocytoma. At the moment (mid-2002), presymptomatic treatment of cerebellar hemangioblastoma is not recommended since there is no evidence for a favorable effect, and treatment before symptoms appear could be harmful. Lesions in the kidney can be safely observed until they reach a diameter of 3 cm. At this size they are removed surgically, along with any other smaller tumors present in the same kidney. An important advance has been the use of kidney-sparing operations, allowing renal replacement therapy to be avoided for as long as possible.

von Hippel–Lindau Syndrome (*VHL*) gene

A positional cloning approach led to isolation of the VHL gene in 1993 (Latif et al., 1993). Essentially, all individuals with VHL disease bear a mutation in the VHL gene, which is situated at 3p25. In the tumors and in the renal cysts, function of the remaining normal VHL allele is lost through mutation, deletion or methylation. Importantly, the great majority of nonfamilial CCRCC also have inactivation of both copies of the VHL gene. When a VHL gene is reintroduced into CCRCC cell lines, tumor growth in nude mice is suppressed (Iliopoulos et al., 1995). Thus VHL is a classical tumor suppressor gene that conforms to Knudson's two-hit hypothesis. The fact that it suppresses growth of fully transformed cells suggests a 'gatekeeper' rather than a 'caretaker' role. Apart from sporadic CCRCC and hemangioblastoma, mutations in VHL are most unusual in other cancers. This suggests that in other cancers, VHL loss-of-function would not confer a selective advantage.

The VHL gene encodes 213 amino acids, giving rise to a protein that migrates with an apparent molecular weight of 24–30 kDa on sodium dodecyl sulfate polyacrylamide gel electrophoresis (SDS-PAGE). An internal methionine at codon 54 provides an alternative translational initiation site, giving rise to an isoform that migrates on polyacrylamide gels with an apparent molecular weight of about 19 kDa. Several lines of evidence indicate that this shorter isoform incorporates the key biological functions:

- the shorter isoform suppresses tumor growth in *in vivo* assays;
- mutations in the first 53 amino acids are not found in VHL disease or sporadic CCRCC;
- the sequence 5′ to the internal methionine codon is not conserved in rodents or *Caenorhabditis elegans*.

Function of the VHL protein in hypoxia-inducible factor-1 regulation

The VHL protein is associated with several other proteins – elongin B, elongin C, Cul2 and Rbx1. The overall complex is an E3 ubiquitin ligase enzyme, with extensive sequence and structural similarities to the Skp1-Cdc53/Cul2-F box class of ubiquitin ligases, with elongin C having similarities to Skp1 and VHL acting like an F box protein (Kamura et al., 1999; Stebbins et al., 1999). Recently, it has become clear that VHL has a critical role in cellular responses to oxygen, being required for oxygen-regulated destruction of two specific transcription factors – the alpha subunits of hypoxia-inducible factor-1 (HIF-1), HIF-1α and HIF-2α (Maxwell et al., 1999). This is mediated by a direct interaction between VHL and the oxygen-dependent destruction domain of the HIF alpha subunit, which leads to ubiquitylation and destruction of the HIF alpha subunit. Crucially, capture is regulated; it does not occur when the oxygen tension is low. Consequently, in hypoxic cells, HIF alpha subunits rapidly accumulate and an active HIF-1 complex forms. This results in altered transcription of genes involved in glucose uptake, glycolysis, matrix metabolism, vasodilatation, angiogenesis and decisions concerning proliferation/apoptosis.

The basis of the oxygen dependence of the VHL/HIF alpha interaction has been discovered. Capture requires hydroxylation of specific prolyl residues in the HIF alpha subunit. Two such hydroxylation sites have been defined, namely Pro402 and Pro564 of HIF-1α. The hydroxylation reaction requires oxygen and is carried out by a newly recognized family of three prolyl hydroxylase enzymes encoded by the genes *EGLN1, 2, 3* which are homologs of *Caenorhabditis elegans eg19* (Bruick and McKnight, 2001; Epstein *et al.*, 2001). These enzymes belong to the extended family of 2-oxoglutarate-dependent dioxygenases, which includes the prolyl hydroxylases that modify procollagen. The VHL complex is therefore the first example of an E3 ligase where capture and destruction of the target depends on regulated enzymatic oxidation of the target. The HIF prolyl hydroxylase–HIF–VHL system is highly conserved in evolution, being present in *C. elegans*.

Comparison of CCRCC cell lines that are defective for VHL function with derivatives that stably reexpress a functional *VHL* gene has been very useful in understanding the function(s) of this protein. In cells lacking VHL, HIF alpha subunits are stable even in the presence of oxygen. CCRCC cells that lack VHL consequently have a fully activated HIF system in normoxia, resulting in a high constitutive level of production of angiogenic growth factor and glycolysis. These characteristics are corrected by reexpression of VHL. A reasonable question is whether the strict requirement of VHL for HIF-1 regulation is a special characteristic of CCRCC cell lines, perhaps accounting for the tissue specificity of the hereditary syndrome. The evidence suggests that this is not the case since VHL is also required for inactivation of HIF-1 in normoxia in *C. elegans*, Chinese hamster ovary cells and mouse hepatocytes.

Other functions of the VHL protein

Besides the role of VHL in capture of the HIF alpha oxygen-dependent destruction domain, VHL has been reported to have other functions; these are as yet incompletely understood and include the following:

- VHL is implicated in other oxygen-regulated processes; examples are the regulation of transactivator recruitment by HIF-1 and the regulation of the stability of the messenger ribonucleic acid (mRNA) of certain oxygen-responsive genes.
- VHL has been reported to be physically associated with fibronectin in cells, and to be necessary for proper assembly of a fibronectin matrix.
- VHL is necessary for cell cycle exit on serum withdrawal, and VHL loss predisposes to apoptosis under certain circumstances.
- VHL has been reported to interact with the transcription factor Sp1 and some protein kinase C isoforms.

Effects of VHL status on gene expression

Stable introduction of a wild-type *VHL* gene in CCRCC cells alters the expression of a large number of genes, a substantial number of which are direct targets of HIF-1. Since HIF-1 alters expression of other transcription factors, expression of targets of these will also be influenced indirectly by HIF-1 activation. Some specific examples of genes influenced by VHL status are those for the angiogenic growth factor VEGF, transforming growth factors alpha and beta 1, glycolytic enzymes and glucose transporters and the transcription factor DEC1. Carbonic anhydrase IX (*CA9*) is a VHL/HIF-1 target gene of particular interest (Ivanov *et al.*, 1998). Before it was known to be encoded by a VHL target gene, this protein was identified as a tumor-specific antigen to which monoclonal antibodies had been generated. It is expressed on the cell membrane and is stable, providing a sensitive marker for HIF activation when it occurs either through local hypoxia or loss of function of *VHL*. Antibodies to carbonic anhydrase IX may therefore be useful for imaging and therapy in CCRCC.

Genotype–Phenotype Correlations in von Hippel–Lindau-related Disease

An extensive number of different mutations have been detected in kindreds with VHL disease and in sporadic CCRCC. Consistent with the importance of interactions with elongin C and HIF alpha subunits, clusters of mutations affect the surface region of the molecule at these sites.

VHL kindreds can be classified according to the risk of CCRCC and pheochromocytoma, as shown in **Table 1**. All mutations that are associated with hemangioblastoma (i.e. type I, IIA and IIB disease) that have been tested to date prevent VHL from regulating HIF alpha subunits in CCRCC cell lines, consistent with HIF activation being central to this angiogenic phenotype. In contrast, mutations associated with pheochromocytoma alone do not appear to influence the capacity for HIF-1 regulation, suggesting that some other function of VHL is affected by these mutations.

Table 1 Classification of VHL kindreds on the basis of tumor risk

Category	Risk of pheochromocytoma	Risk of hemangioblastoma	Risk of renal cell carcinoma
1	Low	High	High
2A	High	High	Low
2B	High	High	High
2C	Yes	No	No

High risk: tumor type observed in over 50% affected individuals. Low risk: tumor type observed in less than 5% of affected individuals. Yes: tumor type observed in all affected individuals. No: tumor type not observed in affected individuals.

Other Genetic Insights into Renal Carcinoma

In addition to loss of VHL function in CCRCC, there is frequent loss of heterozygosity in two other regions of 3p: 3p12–14.2 (containing the fragile histidine triad (*FHIT*) gene) and 3p21–22 (containing the *RASSF1A* candidate tumor suppressor gene). Analogous to the situation with CCRCC and VHL disease, an important insight into papillary renal cell carcinoma has come from families with hereditary papillary renal carcinoma. The majority have been shown to have mutations in the met proto-oncogene (*MET*). *MET* encodes the receptor for hepatocyte growth factor, and the inherited mutations result in a constitutively active receptor. *MET* mutations have also been found in a proportion (around 15%) of sporadic cases of papillary renal cell carcinoma.

References

Bruick RK and McKnight SL (2001) A conserved family of prolyl-4-hydroxylases that modify HIF. *Science* **294**: 1337–1340.

Childs R, Chernoff A, Contentin N, *et al.* (2000) Regression of metastatic renal-cell carcinoma after nonmyeloablative allogeneic peripheral-blood stem-cell transplantation. *New England Journal of Medicine* **343**: 750–758.

Epstein ACR, Gleadle JM, McNeill LA, *et al.* (2001) *C. elegans* EGL-9 and mammalian homologues define a family of dioxygenases that regulate HIF by prolyl hydroxylation. *Cell* **107**: 43–54.

Iliopoulos O, Kibel A, Gray S and Kaelin Jr WG (1995) Tumour supression by the human von Hippel–Lindau gene product. *Nature Medicine* **1**: 822–826.

Ivanov SV, Kuzmin I, Wei M-H, *et al.* (1998) Down-regulation of transmembrane carbonic anhydrases in renal cell carcinoma cell lines by wild-type von Hippel–Lindau transgenes. *Proceedings of the National Academy of Sciences of the United States of America* **95**: 12 596–12 601.

Kamura T, Koepp DM, Conrad MN, *et al.* (1999) Rbx1, a component of the VHL tumor suppressor complex and SCF ubiquitin ligase. *Science* **284**: 657–661.

Kugler A, Stuhler G, Walden P, *et al.* (2000) Regression of human metastatic renal cell carcinoma after vaccination with tumor cell–dendritic cell hybrids. *Nature Medicine* **6**: 332–336.

Latif F, Tory K, Gnarra J, *et al.* (1993) Identification of the von Hippel–Lindau disease tumor suppressor gene. *Science* **260**: 1317–1320.

Maxwell PH, Wiesener MS, Chang G-W, *et al.* (1999) The tumour suppressor protein VHL targets hypoxia-inducible factors for oxygen-dependent proteolysis. *Nature* **399**: 271–275.

Stebbins CE, Kaelin Jr WG and Pavletich NP (1999) Structure of the VHL–elonginC–elonginB complex: implications for VHL tumor suppressor function. *Science* **284**: 455–461.

Further Reading

Godley PA and Taylor M (2001) Renal cell carcinoma. *Current Opinion in Oncology* **13**: 199–203.

Karumanchi SA, Merchan J and Sukhatme VP (2002) Renal cancer: molecular mechanisms and newer therapeutic options. *Current Opinion in Nephrology and Hypertension* **11**: 37–42.

Knudson AG (2000) Chasing the cancer demon. *Annual Review of Genetics* **34**: 1–19.

Kondo K and Kaelin Jr WG (2001) The von Hippel–Lindau tumor suppressor gene. *Experimental Cell Research* **264**: 117–125.

Maxwell PH, Pugh CW and Ratcliffe PJ (2001) Activation of the HIF pathway in cancer. *Current Opinion in Genetics and Development* **11**: 293–299.

Motzer RJ, Bander NH and Nanus DM (1996) Renal-cell carcinoma. *New England Journal of Medicine* **335**: 865–875.

Phillips JL, Pavlovich CP, Walther M, Ried T and Linehan WM (2001) The genetic basis of renal epithelial tumors: advances in research and its impact on prognosis and therapy. *Current Opinion in Urology* **11**: 463–469.

Semenza GL (2000) HIF-1 and human disease: one highly involved factor. *Genes and Development* **14**: 1983–1991.

Web Links

Carbonic anhydrase IX (*CA9*); Locus ID: 768. LocusLink:
http://www.ncbi.nlm.nih.gov/LocusLink/LocRpt.cgi?l = 768

Fragile histidine triad gene (*FHIT*); Locus ID: 2272. LocusLink:
http://www.ncbi.nlm.nih.gov/LocusLink/LocRpt.cgi?l = 2272

Met proto-oncogene (hepatocyte growth factor receptor) (*MET*); Locus ID: 4233. LocusLink:
http://www.ncbi.nlm.nih.gov/LocusLink/LocRpt.cgi?l = 4233

Von Hippel–Lindau Syndrome (*VHL*); Locus ID: 7428. LocusLink:
http://www.ncbi.nlm.nih.gov/LocusLink/LocRpt.cgi?l = 7428

Carbonic anhydrase IX (*CA9*); MIM number: 603179. OMIM:
http://www.ncbi.nlm.nih.gov/htbin-post/Omim/dispmim?603179

Fragile histidine triad gene (*FHIT*); MIM number: 601153. OMIM:
http://www.ncbi.nlm.nih.gov/htbin-post/Omim/dispmim?601153

Met proto-oncogene (hepatocyte growth factor receptor) (*MET*); MIM number: 164860. OMIM:
http://www.ncbi.nlm.nih.gov/htbin-post/Omim/dispmim?164860

Von Hippel–Lindau Syndrome (*VHL*); MIM number: 193300. OMIM:
http://www.ncbi.nlm.nih.gov/htbin-post/Omim/dispmim?193300

Repetitive Elements and Human Disorders

Mohammed El-Sawy, *Tulane University Health Sciences Center, New Orleans, Louisiana, USA*

Prescott Deininger, *Tulane University Health Sciences Center, New Orleans, Louisiana, USA*

Repetitive elements comprise over 50% of the mammalian genome, the majority of these being made up of mobile elements. Although most of the mobile elements are inactive, a number are still actively amplifying. This results in a significant level of insertional mutagenesis. Mobile elements contribute even more to aberrant recombination events that produce germ-line mutations, as well as probably playing a major role in genetic instability in cancer. Other repetitive elements, such as chromosomal duplications and microsatellites (particularly triplet repeats), also lead to genetic damage.

Human Mobile Elements

Mobile elements fall into several major classes: deoxyribonucleic acid (DNA)-based transposable elements, autonomous retrotransposons of both the long terminal repeat (LTR) and non-LTR types, and nonautonomous retrotransposons. All of these elements can contribute to human disease through insertional mutagenesis events. In addition, as multiple copies of these elements have built up in the genome, the potential for unequal, homologous recombination has increased, leading to disease through either genetic deletion or duplication (**Figure 1**).

The DNA transposons have the ability to encode a transposase and transpose through a DNA intermediate. The most prevalent members of this class are mariner elements that comprise about 1–2% of the human genome (Lander *et al.*, 2001). Although there is some evidence for the mobility of mariner in recent evolutionary time, we are not aware of any reports of human disease caused by insertion, and only limited examples of potential recombination caused by these elements have been reported (Reiter *et al.*, 1999).

Autonomous retrotransposons are mobilized via a ribonucleic acid (RNA) intermediate that is reverse transcribed by an enzyme encoded by the element and reintegrated into the genome. They fall into two subclasses: those that contain LTRs and those that lack LTRs. LTR-containing retrotransposons resemble retroviruses in structure but lack a functional envelope gene. This class of elements contributes significantly to genetic damage in the mouse genome but minimally in the human genome (Kazazian and Moran, 1998). The most abundant members of this class in humans are the human endogenous retroviruses (HERVs), which comprise about 8% of the genome (Lander *et al.*, 2001). There have been a number of suggestions that HERVs may cause a

number of human diseases by inducing autoimmunity (Urnovitz and Murphy, 1996).

Non-LTR retrotransposons in humans are mainly long interspersed nuclear element-1 (LINE-1 or L1) elements, which comprise about 17% of the genome, with a copy number of over 800 000. Only about 60 copies maintain functional copies of both open-reading frames with amplification capability (Kazazian and Moran, 1998). It is estimated that there is one new L1 insertion somewhere in the genome in every 50–100 human births. (*See* Long Interspersed Nuclear Elements (LINEs).)

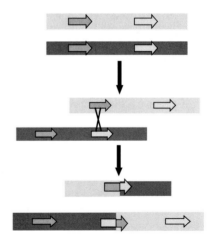

Figure 1 Schematic of unequal homologous recombination between related elements. The dark bars represent genome sequence and the arrows represent related mobile, or repetitive, elements found in that genome. These dispersed, repetitive elements provide the opportunity for mispairing between different copies of the same element, resulting in an unequal, homologous recombination. This recombination can result in either a deletion of the sequences between the elements or a duplication of the sequences in the same region. This same process may also occur between elements on different chromosomes, resulting in a chromosomal translocation.

The nonautonomous retrotransposons are represented primarily by Alu elements in the human genome. Alu elements make an approximately 300 base RNA polymerase III-derived transcript that has no protein coding capacity. It has been proposed that they rely on the L1 retrotransposition machinery (Boeke, 1997). There are over 1 million Alus in the human genome, with about 2000 Alus inserted recently enough in evolution to be specific to the human genome (Deininger and Batzer, 1999). This also represents about one insertion in every 100 human births. Like L1 elements, very few Alu elements are capable of amplification. (*See* Retrosequences and Evolution of Alu Elements.)

There are several other members of each of the various families of mobile elements present in the human genome. Some, like L2 and Mer (autonomous and nonautonomous retrotransposons respectively), are present at high copy numbers, but appear not to have actively amplified in recent times (Lander *et al.*, 2001).

L1 Insertions and Human Disease

Newly inserted L1 elements can cause human disease, both by disrupting reading frames and by interfering with RNA splicing (Kazazian and Moran, 1998). **Table 1** is a compilation of known diseases caused by L1 insertion. This frequency of insertion has been reported to be equivalent to about 0.1% of human germ-line disease. It is interesting that several of the genes have been disrupted multiple, independent times. This suggests that some genes are much more prone to disruption than others. Particularly striking is the observation that the majority of cases of disease (the first three examples in **Table 1**) represent disruption of genes on the X chromosome. This may reflect ascertainment biases because X-linked diseases are essentially dominant in males, allowing the disease state may be manifested with only a single mutant allele. However, diseases from Alu insertions do not show this same strong X-chromosome bias. The X chromosome is also enriched for L1 insertions

(Bailey *et al.*, 2000), with approximately 30% of the chromosome being L1.

Alu Insertions and Disease

Alu elements are normally located throughout the genome and in almost any location within a gene except those where they would disrupt the function of essential genes (Deininger and Batzer, 1999). They have some preference for the more gene-rich regions (Lander *et al.*, 2001). Their small size allows Alus to be relatively well tolerated in introns and between genes. They tend to disrupt function and cause disease when they disrupt exons or land in an intron in a way that alters splicing of the gene. **Table 2** is a list of diseases that are caused by Alu insertions. Alu elements contribute to approximately 0.1% of human genetic diseases (Deininger and Batzer, 1999). Although some genes have much higher Alu content than others, in general Alus appear to contribute to a much broader range of diseases than do L1 elements. This would suggest that Alu elements have a more random level of insertion in the genome.

Mobile Elements and Recombination

Even if their insertion does not cause any immediate damage to the genome, multiple copies of closely related sequences located in the genome can provide opportunities for unequal homologous recombination, leading to duplications, deletions or even chromosomal translocations (**Figure 1**). There are limited examples of L1 and mariner elements potentially contributing to unequal homologous recombination; Alu elements appear to be the dominant elements in these types of rearrangements. **Table 3** contains a list

Table 1 L1 insertions and human disease

Locus	Disease
3× *F8*	Hemophilia A
3× *DMD*	Dystrophinopathies
RP2	X-linked retinitis pigmentosa
2× *CYBB*	Chronic granulomatous disease
HBB	Thalassemia-beta
APC	Colon cancer
FCMD	Fukuyama-type congenital muscular dystrophy

Loci that have been independently inactivated multiple times are preceded by the number of events (e.g. 3×).

Table 2 Alu insertions and disease

Locus	Disease
BRCA2	Breast cancer
MLVI2	Associated with leukemia
NF1	Neurofibromatosis type 1
APC	Hereditary desmoid disease
BtK	X-linked agammaglobulinemia
IL2RG	XSCID
Cholinesterase	Angioedema
CASR	Hypocalciuric hypercalcemia
SERPING1	Complement deficiency
2× factor IX	Hemophilia B
2× *FGFR2*	Apert syndrome
GK	Glycerol kinase deficiency
EYA1	Bronchio-oto-renal syndrome
PBGD	Acute intermittent porphyria

Loci that have been independently inactivated twice are preceded by 2×.

Table 3 Alu/Alu recombination and disease

Locus	Disease
8× *LDLR*	Hypercholesterolemia
5× α-Globin (*HBA*)	α-Thalassemia
5× G1-inhibitor (*SERPING1*)	Angioneurotic edema
2× *PLOD*	Ehlers–Danlos syndrome type VI
DMD	Duchenne muscular dystrophy
ADA	Severe combined immunodeficiency
APOB	Hypo-betalipoproteinemia
INSR	Insulin-independent diabetes mellitus
GLA	Fabry disease
HPRT	Lesch–Nyhan syndrome
ITGAZB	Glanzmann thrombasthenia
PHKA2	Glycogen storage disease VIII
GALNS	MPS IVA
SERPINC1	Thrombophilia
XY	XX male
HEXA	Tay–Sachs disease
C3	C3 deficiency
HEXB	Sandhoff disease
2× *BRCA1*	Familial breast cancer
FAA	Fanconi anemia
HMOX-1	Heme oxygenase-1 deficiency
NCF1	Chronic granulomatous disease
RB1	Gliomas
RBL1	B-cell lymphoma
MLH1	HNPCC
EWSR1	Ewing sarcoma
10× *MLL*	AML

Loci that have been independently inactivated multiple times are preceded by the number of events (e.g. 8×).
MPS IVA, mucopolysaccharidosis type IVA; C3, complement component 3; HNPCC, hereditary nonpolyposis colorectal cancer; AML, acute myelogenous leukemia.

of many of the examples of diseases caused by Alu/Alu recombinations that have been reported. We estimate that this causes at least 0.3% of human germ-line diseases (Deininger and Batzer, 1999). These findings are likely to have a significant bias based on the genes that have been most extensively studied to date, as well as the methods used for studying them. Some methods, like Southern blots, are likely preferentially to detect insertion and recombination events over point mutations, while polymerase chain reaction-based techniques are more likely to overlook these events. Although a very broad range of genes are influenced by Alu/Alu recombination events, a few genes, such as *LDLR*, *SERPING1* and *MLL*, appear to be very susceptible to these events. Although a high density of Alu elements in these genes may be a contributing factor, there are many genes with high densities of Alu elements that do not seem to be very prone to rearrangement. It may be properties of the chromosomal regions, the specific Alus or other molecular properties of these genes that contribute to the high rate of Alu/Alu recombination.

A previous report suggests that a specific segment of the Alu element serves as a chi-like sequence, resulting in high levels of recombination in the left portion of Alu (Rudiger *et al.*, 1995). In order to test this hypothesis further, we have mapped the approximate recombination sites in the diseases shown in **Table 3** (**Figure 2**). Although the low-density lipoprotein receptor recombinations (which dominated earlier studies) do seem to show some specificity, in general

Alu/Alu recombination junctions

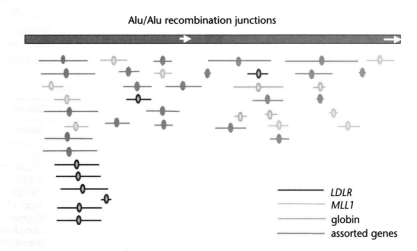

LDLR
MLL1
globin
assorted genes

Figure 2 [*Figure is also reproduced in color section.*] Mapping recombination junctions in Alu/Alu unequal homologous recombination events. Alu/Alu recombination, as shown in **Figure 1**, can result in a crossover occurring at any point in the Alu element. Slight differences in the Alu elements can be used to map the site of crossover approximately. The dark bar at the top represents the approximately 300 bp Alu sequence. The white arrows within the bar mark the A-rich regions found at the downstream end of both halves of the dimer-like structure. The approximate locations of the recombination junctions for the Alu/Alu recombination events shown in **Table 3** are shown as lines with a dot to show the center position. Some of these are color-coded for the genes with the largest numbers of such events. The low-density lipoprotein receptor recombination events seem heavily biased to the left end, near the Alu promoter region.

the recombination sites are spread fairly evenly across the Alu map. Although the Alus in some genes may show preferential sites for recombination, this does not appear to be a specific property of Alus.

Alu elements have also been suggested as playing a role in illegitimate recombination events (Rudiger et al., 1995). Although illegitimate recombination events frequently occur in or near Alu elements, it is difficult to assess whether the Alu element actually contributed to the recombination. Because Alus represent 11% of the human genome, they should be involved in a large number of recombination events by chance. However, studies using Alu constructs in an inverted orientation (Gebow et al., 2000) have shown that they may cause more recombination through this illegitimate mechanism than through unequal homologous recombination. Thus, the full extent of mobile elements in chromosomal rearrangements, insertions and deletion is still to be determined.

Mobile Elements and Cancer

Tables 1–3 include a number of examples in which mobile elements have contributed to various cancers. However, very few of these cases are known to result from somatic insertion or rearrangement. There are few data on the impact of somatic changes caused by mobile elements in humans. We do not know, for instance, whether cellular transformation leads to higher frequency of insertion of Alu or L1 elements. However, mutations in p53 have been shown to lead to tremendous increases in unequal homologous recombination, specifically between Alu elements (Gebow et al., 2000). Thus, it is likely that recombination between mobile elements is a much more frequent cause of damage in cancer cell genomes than the rates we have described in germ-line mutations and may be responsible for a high proportion of the loss of heterozygosity seen in tumor cells.

Microsatellites

Microsatellites have a tandem duplication of many copies of short segments of DNA (Bowater and Wells, 2000). These segments are typically 2–4 base pairs (bp) long, present as ten to thousands of copies. However, tandem repetitions of much longer lengths are also found dispersed throughout chromosomes. There are hundreds of thousands of microsatellites in the human genome, making up several per cent of the mass of the genome. A large proportion of these elements are derived from the A-rich tails at the 3′ end of the retroelements. Microsatellites have been found to be relatively unstable over an evolutionary time period. This has made them very useful markers of genetic

diversity in the human population. However, in the presence of certain predisposing mutations, they can be extremely unstable; for example, mutations in mismatch repair genes can lead to instability of microsatellite sequences throughout the genome and correlate strongly with tumorigenicity in some tumors, such as colon cancer. In some cases, expansion of microsatellites within a gene may lead to gene defects. However, the instability of microsatellite sequences associated with many tumors may only be an indirect measure of other types of genetic instability associated with the defective repair genes. (*See* Microsatellites.)

Triplet repeats, microsatellites with specific three-base sequences tandemly repeated, are an unusual class of microsatellite that has been associated with a wide range of diseases, primarily neurological disease (Bowater and Wells, 2000). When specific triplet repeat sequences reach a critical size, they become remarkably unstable. This leads to a phenomenon called 'anticipation' in which the disease can become worse with each generation. Furthermore, triplets can amplify from modest lengths to thousands of copies in different somatic tissues within a single individual. Triplet repeats appear to have specific structural features that lead to this instability, while most microsatellites seem to amplify through slippage of DNA polymerase or, in some cases, through unequal recombination. (*See* Trinucleotide Repeat Expansions: Disorders.)

Chromosomal Duplicons

In addition to the high copy number elements, like microsatellites and mobile elements described above, the human genome also has a number of low copy number sequences created through recombination processes, which then provide opportunities to create further recombination and chromosomal translocations. Of particular note is the tendency of sequences near the chromosomal centromeres to be prone to these types of rearrangements (Eichler, 1998). A number of these pericentromeric regions contribute to frequent, recurring rearrangements leading to disease. Prader–Willi syndrome, Charcot–Marie–Tooth disease and Smith–Magenis syndrome represent common examples of these genomic diseases. The exact nature of the instability of these regions is not well understood, but seems likely to represent some basic property of those chromosomal locations that fosters the recombination process.

See also
Long Interspersed Nuclear Elements (LINEs)
Microsatellites
Retrosequences and Evolution of Alu Elements

Retrotransposition and Human Disorders
Transposons

References

Bailey JA, Carrel L, Chakravarti A and Eichler EE (2000) From the cover: molecular evidence for a relationship between LINE-1 elements and X chromosome inactivation: the Lyon repeat hypothesis. *Proceedings of the National Academy of Sciences of the United States of America* **97**: 6634–6639.

Boeke JD (1997) LINEs and Alus – the polyA connection. *Nature Genetics* **16**: 6–7.

Bowater RP and Wells RD (2000) The intrinsically unstable life of DNA triplet repeats associated with human hereditary disorders. *Progress in Nucleic Acid Research and Molecular Biology* **66**: 159–202.

Deininger PL and Batzer MA (1999) Alu repeats and human disease. *Molecular Genetics and Metabolism* **67**: 183–193.

Eichler EE (1998) Masquerading repeats: paralogous pitfalls of the human genome. *Genome Research* **8**: 758–762.

Gebow D, Miselis N and Liber HL (2000) Homologous and nonhomologous recombination resulting in deletion: effects of p53 status, microhomology, and repetitive DNA length and orientation. *Molecular and Cellular Biology* **20**: 4028–4035.

Kazazian Jr HH and Moran JV (1998) The impact of L1 retrotransposons on the human genome. *Nature Genetics* **19**: 19–24.

Lander ES, Linton LM, Birren B, *et al.* (2001) Initial sequencing and analyses of the human genome. *Nature* **409**: 860–921.

Reiter LT, Liehr T, Rautenstrauss B, Robertson HM and Lupski JR (1999) Localization of mariner DNA transposons in the human genome by PRINS. *Genome Research* **9**: 839–843.

Rudiger NS, Gregersen N and Kielland-Brandt MC (1995) One short well conserved region of Alu-sequences is involved in human gene rearrangements and has homology with prokaryotic chi. *Nucleic Acids Research* **23**: 256–260.

Urnovitz HB and Murphy WH (1996) Human endogenous retroviruses: nature, occurrence, and clinical implications in human disease. *Clinical Microbiology Reviews* **9**: 72–99.

Further Reading

Deininger P, Batzer M, Hutchison IC and Edgell M (1992) Master genes in mammalian repetitive DNA amplification. *Trends in Genetics* **8**: 307–312.

Craig NL, Craigie R, Gellert M and Lambowitz AM (eds.) (2002) *Mobile DNA II*. Washington, DC: American Society for Microbiology.

Hu X and Worton RG (1992) Partial gene duplication as a cause of human disease. *Human Mutation* **1**: 3–12.

Ji Y, Eichler EE, Schwartz S and Nicholls RD (2000) Structure of chromosomal duplicons and their role in mediating human genomic disorders. *Genome Research* **10**: 597–610.

Kazazian HHJ (1998) Mobile elements and disease. *Current Opinion in Genetics and Development* **8**: 343–350.

Maraia R (ed.) (1995) *The Impact of Short, Interspersed Elements (SINEs) on the Host Genome*. Georgetown, TX: RG Landes.

Moran JV, Holmes SE, Naas TP, *et al.* (1996) High frequency retrotransposition in cultured mammalian cells. *Cell* **87**: 917–927.

Schmid CW (1996) Alu structure, organization, evolution, significance and function of one-tenth of human DNA. *Progress in Nucleic Acid Research and Molecular Biology* **53**: 283–319.

Urnovitz HB and Murphy WH (1996) Human endogenous retroviruses: nature, occurrence, and clinical implications in human disease. *Clinical Microbiology Reviews* **9**: 72–99.

Repetitive Elements: Detection

Intermediate article

Jerzy Jurka, Genetic Information Research Institute, Mountain View, California, USA

Vladimir V Kapitonov, Genetic Information Research Institute, Mountain View, California, USA

Arian FA Smit, The Institute for Systems Biology, Seattle, Washington, USA

Article contents

- Overview and Definitions
- Reference Collections of Human Repeats
- Identification, Classification and Analysis of Repetitive DNA

Multicopy, or repetitive, deoxyribonucleic acid (DNA) is routinely being detected and analyzed by computer-assisted comparison of genomic DNA with representative reference databases of repeats.

Overview and Definitions

The term 'repetitive DNA' is usually applied to genomic sequences that are present in multiple copies but have no clearly assigned biological function. From a biological and computational point of view it is important to make a basic distinction between two types of repetitive DNA, generally referred to as either 'simple' or 'complex' repeats. They are often associated with each other in the genomic DNA.

The most abundant and thoroughly studied are tandemly repeated 'simple sequence repeats' (SSRs), typically referred to as micro- and minisatellites (**Figure 1**). Microsatellites are defined as SSRs with an arbitrary basic unit length of anywhere from 1 to 6 base pairs (bp). Minisatellites are SSRs with a basic

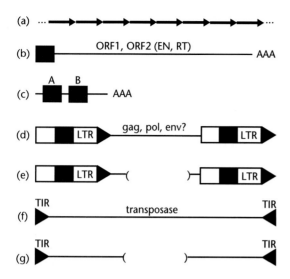

Figure 1 Basic prototypes of human repetitive DNA:
(a) tandemly repeated DNA (minisatellites, microsatellites, centromeric and telomeric repeats); (b) LINE retro(trans)posons; (c) SINE retro(trans)posons; (d, e) autonomous and nonautonomous endogenous retroviral elements; (f, g) autonomous and nonautonomous DNA transposons. ORF1 and ORF2 denote open reading frames 1 and 2 in the human L1 (LINE1) element. ORF2 encodes an enzyme with endonuclease (EN) and reverse transcriptase (RT) activities. LTR: long terminal repeat; TIR: terminal inverted repeat.

elements' (TEs). There are two major categories of TEs in the human genome: retroelements and DNA transposons (see **Figure 1**). Retroelements (long interspersed nuclear elements (LINEs), short interspersed nuclear elements (SINEs) and retrovirus-like elements) reproduce via reverse transcription, which is followed by integration into the host DNA. DNA transposons are capable of integrating themselves to, and excising themselves from, the host genome, thus taking advantage of the host replication through this 'cut-and-paste' mechanism. The reproduction of TEs is very ineffective and most of them are 'dead on arrival', that is, they cannot reproduce any further due to incomplete duplication, lethal mutations or suppression of transcription by the host. The integration sites of TEs are usually short stretches of DNA, typically undergoing duplication in most retroelements and DNA transposons. Only a small fraction of mobile retroelements and DNA transposons are 'autonomous', that is, encoding active enzymes involved in their proliferation. The majority of mobile TEs are 'nonautonomous', that is, their proliferation is determined by the enzymatic machinery of their autonomous relatives. The proliferation of active TEs can be, and probably is, tightly controlled by the host and most TEs proliferate in bursts rather than continuously, leaving a discrete record of families and subfamilies in the host genome. Each (sub)family is derived from a separate active TE (a source gene), which eventually becomes extinct or, in rare instances, recruited by the host. The biological role of TE-derived genes is not yet known.

In analogy to a paleontological record, the deposits of repeats represent a 'genomic fossil record', which preserves valuable information about the history and evolutionary impact of TEs. Copies of TEs deposited in the human genome during the last 200 million years or so are detectable as repetitive DNA. Much older copies are no longer recognizable due to 'erosion' by mutations, but based on indirect evidence it is believed that the proliferation of TEs continued for a very long time and that they possibly even predate the origin of eukaryotic systems.

To understand the significance of the genomic fossil record, it is important to proceed with the reconstruction of TEs from their genomic copies by building consensus sequences and biological characterization (Jurka, 1998). Inactive copies of TEs are usually mutated and are often incomplete. Such copies, once integrated into the genome, mutate at approximately neutral rates with the exception of CpG dinucleotides. A significant fraction of point mutations can be 'reversed' by building consensus sequences based on multiple alignment of individual repeats. CpG doublets mutate to TpG or CpA at a rate about an order of magnitude higher than the neutral rate due to the

unit length of 15 bp or more. SSRs with a basic unit length of 7–14 bp fall into a 'twilight zone', as they are listed in either category or often left out from the scientific literature. The upper limit for minisatellite unit size is not well defined. Micro- and minisatellites are often polymorphic and the biological mechanisms behind this phenomenon are thought to be different for the two groups. Owing to the polymorphism and relatively uniform distribution in chromosomal DNA, micro- and minisatellites are among the most informative genetic markers. Other types of SSRs include 'cryptically simple repeats', which result from the reshuffling of a limited number of DNA sequence motifs in various orientations, and 'low-complexity repeats', which are usually derived from other simple repeats, although their periodic character may be obscured by mutations.

A separate category of tandem repeats is represented by satellite and telomeric repeats. Unlike micro- and minisatellites, these repeats are confined to well-defined chromosomal regions. Satellites are primarily found in the centromeric regions of chromosomes whereas telomeric repeats occupy chromosomal ends, or telomeres, and are maintained by specialized reverse transcriptases.

'Complex repeats' from the human genome are primarily represented by defective copies, or 'pseudogenes', derived from biologically active 'transposable methylation of cytosine followed by spontaneous conversion of the resulting 5-methylcytosine to thymine by deamination. Pending sequence context, the homologous CpA and TpG doublets do not always align with each other and the alignment often converges to either CpA or TpG in the resulting consensus sequence. Therefore, the potential CpG spots must be evaluated separately by visual inspection.

Individual repeats are more similar to their consensus sequence than to each other. This can be seen from the following approximate formula (Jurka, 1994, 1998) relating average similarity of individual repeats aligned to their consensus (y) with average similarity between the repeats (x):

$$y = \frac{1 + \sqrt{12x - 3}}{4}$$

One can calculate, for example, that repeats that are 40–60% similar to each other will be around 58.5–76.2% similar to their quality consensus sequence. Therefore, reconstructing TEs, especially the old ones, is important for detecting diverged elements. The majority of human reference repeats described below are consensus sequences.

Reference Collections of Human Repeats

The first reference collection of human repetitive elements, containing 53 separate entries, was established in 1992 (Jurka *et al.*, 1992). It became known as 'Repbase' and then in 1997 as 'Repbase Update', abbreviated to RU (Jurka, 2000). RU is the sole reference collection used in annotating the human genome (Lander *et al.*, 2001) and is increasingly being used in other genome projects.

As of December 2002, RU contained over 3100 unique entries from diverse eukaryotic species, of which over 620 represent repeats identified in the human genome. Each entry describes a consensus sequence or a representative sequence of a family or subfamily of repeats. The description includes a unique locus name, definition, date of creation, date of the latest update, keywords and basic credits to individual contributors or references to the original scientific literature. Many repetitive families and subfamilies published in RU are original contributions unreported anywhere else. To illustrate, approximately 300 entries describing human repeats have been added to RU since 1998, virtually all of them unpublished anywhere else. Therefore, RU plays a central role as both a database and an electronic journal.

Almost every human repetitive element deposited in RU has been assigned to one of several well-defined

biological categories. In **Table 1**, locus names of non-LTR retroelements are listed in sections 1 and 2 as SINEs and LINEs; retrovirus-like retroelements including long terminal repeats (LTRs) are listed in sections 3 and 4 and DNA transposons in section 5. The remaining sections cover minisatellites, satellites, composite repeats representing a mosaic sequence partially reducible to known elements, and repeats still to be classified (section 9). For quick classification purposes, the 620 or so original locus names have been reduced to 171 theme names listed in **Table 1**. For example, MER105 in section 5 indicates a single DNA transposon family, while MER106* represents the MER106 and 106B subfamilies listed in RU. Similarly, with the exception of MER41I and MER4BI listed in section 3b, all names starting with 'MER4' followed by an asterisk indicate LTRs (see section 4). Obviously, only the youngest subfamilies are species specific and only some so-called 'human repeats' are present exclusively in the human genome. Most are shared with nonhuman primates. Many are shared with other mammalian and even nonmammalian species due to a common evolutionary heritage and occasional horizontal transfer. However, for practical purposes, the RU collection is being divided into a growing number of sections such as human (humrep.ref), rodent (rodrep.ref), plant (plnrep.ref), *Caenorhabditis elegans* (celrep.ref), etc., to facilitate species-oriented genome studies. RU is also prepared in the software-specific version used as a reference file by the RepeatMasker program (see Web Links). In general, it may be advisable to separate or combine different parts of RU depending on specific search strategies and biological needs.

Identification, Classification and Analysis of Repetitive DNA

Routine identification of repetitive DNA is a highly automated process, which has been evolving since the early 1990s, and was first systematically described in 1994 (Jurka, 1994). The core of the identification procedure is the comparison of a query sequence from an input file against a 'gold standard' represented by a reference collection of repeats (see **Figure 2**). The reference collection of complex human repeats from RU contains around 1 Mb of representative sequences that represent over 1200 Mb of discernible repetitive DNA interspersed throughout the genome. This DNA can generate nonspecific sequence similarities that may obscure the analysis of homologous proteins or other regions of interest. Furthermore, DNA fragments containing repetitive DNA can produce overly broad DNA–DNA hybridization

Table 1 Major categories of human repetitive families from Repbase Update and their family/subfamily loci names

Major categories	Subcategories	Family/subfamily loci names
1. LINEs		IN25, L1*; L2A, KER2, L3, CR1_HS
2. SINEs		Alu*, FLA*, HAL1*, L2B, MIR, MIR3, SVA, SVA2
3. Retroviruses and retrovirus-like elements (internal sequences only)	(a) MaLRs	MLT1R, MLT1AR, MLT1CR, MLT1FR, MSTAR, THE1BR
	(b) Other retroviruses	ERVL, HARLEQUIN, HERV*, HRES1, HUERS-P*, LOR1I, MER4I, MER4BI, MER21I, MER31I, MER41I, MER50I, MER51I, MER52AI, MER57I, MER57A_I, MER61I, MER65I, MER66I, MER70I, MER83AI, MER83BI, MER84I, MER89I, MER110I, PABL_AI, PABL_BI, PRIMA4_I, PRIMA41
4. Long terminal repeats (LTRs)[a]	(a) MaLR LTRs	MLT1*, MST*, THE1*
	(b) Other retrovirus LTRs	HARLEQUINLTR, LTR*, LOR1, MLT2*, MER4*, MER9, MER11*, MER21*, MER31*, MER34*, MER39*, MER41*, MER48, MER49, MER50*, MER51*, MER52*, MER54*, MER57*, MER61*, MER65*, MER66*, MER67*, MER68*, MER70*, MER72*, MER73, MER74*, MER76, MER77, MER83*, MER84, MER87, MER88, MER89, MER90, MER92*, MER93*, MER95, MER101*, MER110*, PABL_A, PABL_B, PRIMA4_LTR, PTR5
5. DNA transposons		BLACKJACK, CHARLIE*, CHESHIRE*, GOLEM*, HSMAR*, HSTC2, LOOPER, MADE1, MARNA, MER1*, MER2*, MER3, MER5*, MER6*, MER8, MER20*, MER28, MER30*, MER33, MER44*, MER45*, MER53, MER63*, MER69*, MER75, MER80, MER81, MER82, MER85, MER91*, MER94, MER96*, MER97*, MER99, MER103, MER104*, MER105, MER106*, MER107, MER113, MER115, MER116, MER117, MER119, ORSL, PMER1, RICKSHA*, TIGGER*, ZAPHOD, ZOMBI*
6. Minisatellites		IVR, R66
7. Satellites		ALR*, BSR, CER, (GGAAT)n, GSAT*, HSATI*, LSAU, MER22, MER122, MSR1, REP522, SAR, SATR*, SN5, TAR1
8. Composite		MER120
9. Unclassified		HIR, MER35, MER109, MER112, MER121

* Indicates multiple families/subfamilies. For example, MER97* represents three subfamilies listed in RU as MER97A, B and C. LTR* stands for 127 different families and subfamilies of long terminal repeats.
[a]This list does not include the internal sequence names listed above.

signals or nonspecific DNA amplification if used as primers in a polymerase chain reaction (PCR). Therefore, identification of repetitive DNA has become an essential routine associated with sequence analysis and probe/primer design. More importantly, repetitive elements carry invaluable information applicable to a wide variety of biological problems ranging from phylogenetic and population studies to analysis of chromosomal structure, stability and mechanisms of evolution.

In most cases, analysis of repetitive DNA begins with the identification of SSRs. They can be detected by alignment to known categories of simple repeats listed in RU as a separate file named 'simple.ref'. Alternatively, their nonrandom base composition can be used as a basis for detection (Wootton and

Federhen, 1996). Some cryptically simple repeats can best be untangled using the SMPL program (Milosavljevic and Jurka, 1993) run as part of a routine repeat analysis by the CENSOR server (see

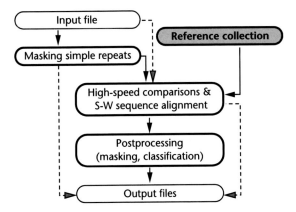

Figure 2 General scheme for computer-assisted identification of repetitive DNA. Full and broken arrows indicate major and alternative steps in the process respectively. The first step is identifying and masking simple sequence repeats (SSRs) using any variety of programs described in the text, or by sequence alignment against a reference collection of simple repeats. This is followed by identification of complex repeats, by aligning the masked sequence against the respective reference collection. There are several types of possible output file: a list of alignments against the reference sequences, an input file with masked simple or complex repeats, or both. Output files listing repeat location and other characteristics can be organized in a form of map similar to that in **Table 2**. S-W: Smith–Waterman algorithm.

Web Links). Another aspect of computer-assisted SSR studies is an effort devoted to sequence-based DNA structure prediction, driven by biological interest (Baldi and Baisnee, 2000).

During routine analyses of repetitive DNA, it is advisable to mask all identifiable SSRs in the query sequence prior to searches for complex repeats in order to prevent nonspecific matches between ubiquitous simple repeats (**Figure 2**). Masking usually means replacing each base in an SSR or other chosen sequence region by an arbitrary letter, for example 'N', so that the length of the masked sequence remains unchanged. The masked query sequence can then be compared against the reference collection of complex repeats. The standard tool for high-speed comparison and masking of complex repeats is the XBLAST program (see Web Links) (Claverie, 1994). However, more sensitive analyses implemented in repeat annotation programs, such as CENSOR (see Web Links) (Jurka *et al.*, 1996) and RepeatMasker (see Web Links), are based on the Smith–Waterman algorithm. These programs are more successful in detecting old repeats (such as MIR3, LINE3, etc. in **Table 1**), and are more accurate in classifying repetitive DNA into distinct subfamilies of repeats as described in RU. The major limitation of both programs is their inadequate speed to keep up with growing DNA sequence databases. One way in which acceleration can be achieved is by running the programs on dedicated hardware or multiple processors. Another approach focuses on

Table 2 Sample map of repetitive elements and its interpretation

GB acc.	Beg.	End	RU locus	Beg.	End	O	S
AF036876	22217	23766	TIGGER1	1	1764	d	0.87
AF036876	23767	24245	LOR1	1	495	d	0.83
AF036876	24638	25995	MER51I	901	2218	d	0.77
AF036876	25996	26198	MER66I	3775	3976	d	0.74
AF036876	26256	26792	MER51I	2219	2766	d	0.75
AF036876	26801	26931	(TGTA)$_n$	4	134	d	0.67
AF036876	27067	27251	MER51I	2745	2942	d	0.69
AF036876	28136	28423	AluSz	1	290	c	0.85
AF036876	28425	29814	ZOMBI	995	2623	c	0.86
AF036876	30231	30497	MER4I	3798	4053	d	0.80
AF036876	30668	30809	MER66I	4318	4459	d	0.87
AF036876	31815	32288	LOR1	1	494	d	0.83
AF036876	32289	32915	TIGGER1	1761	2406	d	0.90

TIGGER1	LOR1	LOR1I	AluSz & ZOMBI	LOR1I	LOR1	TIGGER1

Columns from left to right: (1) GenBank accession number of the query sequence; (2, 3) beginning and end coordinates of the identified repeats; (4) the names of the reference repeats as used in the Repbase Update; (5, 6) coordinates relative to the reference sequence; (7) orientation: direct (d) and complementary (c); (8) per cent of sequence identity to the reference. The bar below the table provides the current interpretation of the map as consisting of a retrovirus (LOR1I) flanked by LTRs (LOR1) and inserted in TIGGER1. AluSz and ZOMBI are inserted in the internal portion of the retrovirus.

developing acceleration software currently targeted at RepeatMasker (Bedell *et al.*, 2000).

Typical output files from either of the last two programs include the query sequence with all repeats masked, sequence alignments for visual inspection if needed and a so-called 'map of repeats' summarizing biologically relevant information about the identified repetitive DNA. Maps of repeats are equivalent to repeat annotation in databases such as GenBank and they can be very informative about individual repeats. A sample of a map obtained using the CENSOR server is shown in **Table 2**. It shows a Tigger1 DNA transposon split around positions 1761–1764 into two consequential fragments containing a number of repetitive elements apparently inserted as one block. Patchy similarities between the insert and different endogenous retroviruses (MER51I, MER66I and MER4I; see **Table 1**) permitted its classification as a new endogenous retrovirus flanked by two LTRs (LOR1) previously known only as 'low copy repeats'. The internal region of the new retrovirus was subsequently entered to RU as 'LOR1I'. Nonviral elements, AluSz and ZOMBI (see **Table 1**), were most likely inserted into the virus at some point. The microsatellite $(TGTA)_n$ is a likely part of the retrovirus, possibly expanded after the integration took place.

Genomic structure and evolutionary history cannot be understood without detailed studies of repetitive DNA. Detection and basic analysis of repetitive DNA outlined here is only a first step in this direction.

See also

Long Interspersed Nuclear Elements (LINEs)
Microsatellites
Minisatellites
Short Interspersed Elements (SINEs)
Transposable Elements: Evolution

References

Baldi P and Baisnee P-F (2000) Sequence analysis by additive scales: DNA structure for sequences and repeats of all lengths. *Bioinformatics* **16**: 865–889.

Bedell JA, Korf I and Gish W (2000) Maskeraid: a performance enhancement to RepeatMasker. *Bioinformatics* **16**: 1040–1041.

Claverie JM (1994) Large scale sequence analysis. In: Adams MD, Fields C and Venter JC (eds.) *Automated DNA Sequencing and Analysis*, pp. 267–279. San Diego, CA: Academic Press.

Jurka J (1994) Approaches to identification and analysis of interspersed repetitive DNA sequences. In: Adams MD,

Fields C and Venter JC (eds.) *Automated DNA Sequencing and Analysis*, pp. 294–298. San Diego, CA: Academic Press.

Jurka J (1998) Repeats in genomic DNA: mining and meaning. *Current Opinion in Structural Biology* **8**: 333–337.

Jurka J (2000) Repbase Update: a database and an electronic journal of repetitive elements. *Trends in Genetics* **16**: 418–420.

Jurka J, Klonowski P, Dagman V and Pelton P (1996) CENSOR – a program for identification and elimination of repetitive elements from DNA sequences. *Computers and Chemistry* **20**: 119–121.

Jurka J, Walichiewicz J and Milosavljevic A (1992) Prototypic sequences for human repetitive DNA. *Journal of Molecular Evolution* **35**: 286–291.

Lander ES, Linton LM, Birren B, *et al.* (2001) Initial sequencing and analysis of the human genome. *Nature* **409**: 860–921.

Milosavljevic A and Jurka J (1993) Discovering simple DNA sequences by the algorithmic significance method. *Computer Applications in Biosciences* **9**: 407–411.

Wootton JC and Federhen S (1996) Analysis of compositionally biased regions in sequence databases. *Methods in Enzymology* **266**: 554–571.

Further Reading

Brosius J (1999) Genomes were forged by massive bombardments with retroelements and retrosequences. *Genetica* **107**: 209–238.

Jurka J (1995) Human repetitive elements. In: Meyers RA (ed.) *Molecular Biology and Biotechnology. A Comprehensive Desk Reference*, pp. 438–441. New York, NY: VCH.

Milosavljevic A (1998) Repeat analysis. In: Spurr NK, Young BD and Bryant SP (eds.) *ICRF Handbook of Genome Analysis*, pp. 617–628. Oxford, UK: Blackwell Science.

Prak ET and Kazazian HHJ (2000) Mobile elements and the human genome. *Nature Reviews Genetics* **1**: 134–144.

Smit AFA (1996) The origin of interspersed repeats in the human genome. *Current Opinion in Genetics and Development* **6**: 743–748.

Smit AFA (1999) Interspersed repeats and other mementos of transposable elements in mammalian genomes. *Current Opinion in Genetics and Development* **9**: 657–663.

Web Links

Genetic Information Research Institute. CENSOR server screens DNA sequences for simple and interspersed repeats using the current version of Repbase Update (Jurka, 2000)
http://www.girinst.org

Repeat Masker Server at the University of Washington. RepeatMasker2 screens DNA sequences for interspersed repeats and low complexity DNA sequences using Repbase Update (Jurka, 2000)
http://repeatmasker.genome.washington.edu

XBLAST program. It masks regions similar to query sequences based on output from blast
ftp://ftp.ncbi.nlm.nih.gov/pub/jmc/xblast

Representational Difference Analysis

Peter C Groot, *Utrecht University, Utrecht, The Netherlands*

Bernard A van Oost, *Utrecht University, Utrecht, The Netherlands*

Representational difference analysis (RDA) is a highly sensitive technique for the identification of differences between complex deoxyribonucleic acid samples. RDA is based on polymerase chain reaction amplification in combination with multiple rounds of subtractive hybridization and can be used for the identification of unknown pathogens, genetic lesions in cancer and polymorphic markers linked to a trait without the use of a preexisting map, as well as for the identification of differentially expressed genes.

Advanced article

Article contents

- Introduction
- How Representational Difference Analysis Works
- Applications
- Future of the RDA Technique

Introduction

Subtractive hybridization can be used to isolate deoxyribonucleic acid (DNA) fragments present in one DNA sample but absent in another. The basic idea is simple: one DNA sample, called the 'tester', is mixed with a large excess of another DNA sample, the 'driver', melted and hybridized again. After hybridisation, double-stranded tester-DNA molecules are recovered. Due to the large excess of driver DNA in the hybridization mixture, tester-DNA molecules will preferentially hybridize with driver-DNA counterpart molecules and not with the tester-DNA counterparts. Molecules that are present in the tester and not in the driver escape this competition, and a strong enrichment of these molecules will occur.

Despite the fact that the classic subtraction has been used successfully in several instances, it has always been very demanding technically. Very large amounts of starting materials are required, because only very small amounts of the desired product(s) remain at the end of a subtractive hybridization and because several rounds of subtraction are usually required to obtain a sufficiently pure difference product.

This has changed since the application of the polymerase chain reaction (PCR), which has made it possible to start with much smaller amounts of DNA and to use DNA amplification to recover the difference product after subtractive hybridization. Lisitsyn *et al.* (1993) published this new and much improved method for the isolation of differences between two DNA samples which they call 'representational difference analysis' (RDA). (*See* Nucleic Acid Hybridization.)

How Representational Difference Analysis Works

RDA consists of several steps (see **Figure 1**). The first step is the generation of so-called representations or amplicons. This is done by digestion of a DNA sample with a restriction enzyme. The digestion will contain a large number of restriction fragments, all of different lengths. These fragments are ligated to specific linker primers and PCR-amplified using a single primer which recognizes the linker. Each fragment can thus be amplified by the same primer. During amplification, larger fragments will amplify less efficiently than smaller fragments and in practice the larger fragments will get lost. Very small amplicon fragments, less than 150–200 bp, are removed by either agarose gel electrophoresis or by size-exclusion chromatography. What remains then is a random fraction of molecules in a size range of approximately 200–700 bp, which is ideal for subtractive hybridization. The representations have a much smaller complexity than the DNAs they came from, but are truly representative of the DNA they were made from. Very small amounts of starting material suffice for the generation of a large number of amplicons.

The second step is to generate a tester- and a driver-DNA sample from the amplicons. The driver is made by simply removing the linker. Tester DNA is made by removing the linker and replacing it by a second linker, with a sequence that is different from the first one. This is one of the crucial steps in RDA, because the change of linkers makes it possible to only amplify tester DNA, even in the presence of a large excess of driver DNA. Without changing the adapters this would not be possible.

The linkers used for amplicon amplification are a 12-mer and a 24-mer. Both primers are not phosphorylated and the primers are designed in such a way that during the ligation only the 24-mer will be joined at its 3′ end to the DNA. The 12-mer is only necessary for the DNA duplex formation required for the ligation. After ligation it can be removed easily by heating and precipitating the ligated DNA. What is then left are amplicon DNA molecules with single-stranded 24-bp 5′ extensions. It is only possible to amplify these DNA

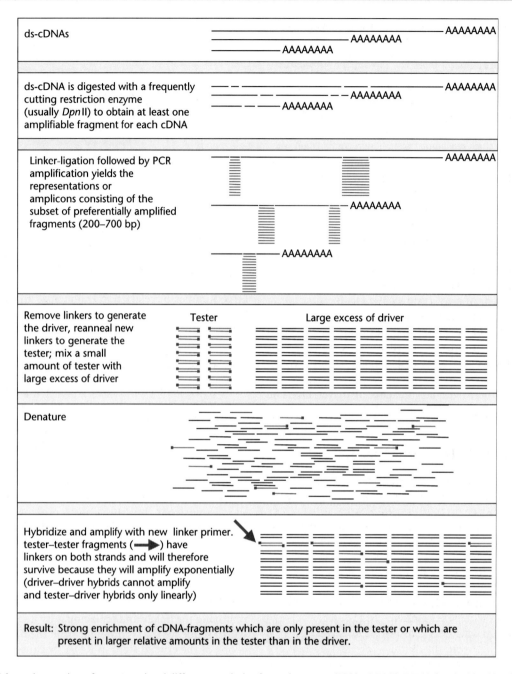

ds-cDNAs

ds-cDNA is digested with a frequently cutting restriction enzyme (usually *Dpn*II) to obtain at least one amplifiable fragment for each cDNA

Linker-ligation followed by PCR amplification yields the representations or amplicons consisting of the subset of preferentially amplified fragments (200–700 bp)

Remove linkers to generate the driver, reanneal new linkers to generate the tester; mix a small amount of tester with large excess of driver

Tester   Large excess of driver

Denature

Hybridize and amplify with new linker primer. tester–tester fragments (➤) have linkers on both strands and will therefore survive because they will amplify exponentially (driver–driver hybrids cannot amplify and tester–driver hybrids only linearly)

Result: Strong enrichment of cDNA-fragments which are only present in the tester or which are present in larger relative amounts in the tester than in the driver.

Figure 1 Schematic overview of representational difference analysis of complementary DNAs (cDNA-RDA) for the identification of differentially expressed genes. ds: double-stranded.

molecules with the 24-bp linker primer after these extensions are filled in by the Taq polymerase during the first step of PCR. The important point here is that, after hybridization of a single-stranded DNA tester and a single-stranded driver-DNA molecule, a double-stranded DNA molecule will form with a 5′ extension present on only one strand (the tester-DNA strand). PCR of such a molecule will only lead to linear amplification, whereas molecules consisting of two tester-DNA strands will amplify exponentially. Driver-only DNA molecules will of course not amplify at all, since they do not carry the linker primer. This difference in amplification is the basis of the very specific amplification of only tester-DNA molecules.

For subtractive hybridization, a small amount of tester DNA, usually approximately 40 ng, is mixed with a 100-fold excess, 40 μg, of driver DNA. All this DNA is purified and concentrated by ethanol precipitation into a very small volume of approximately 3.5 μL and left to hybridize for approximately

20 h at 65°C. Under these conditions, even DNA molecules that are present at very low concentrations can hybridize again. Just as with classic subtraction, the driver DNA will very strongly inhibit the formation of DNA duplexes consisting of two tester strands. Tester molecules with no counterpart present in the driver will rehybridize efficiently and give rise to a very strong enrichment of these molecules, but tester molecules that are present in larger relative amounts in the tester than in the driver will also be enriched. This whole process is repeated 1–3 times, leading to very strong enrichment of DNA fragments, which are present at higher levels in the tester than in the driver. The so-called kinetic enrichment contributes strongly to this. This is the phenomenon that the speed at which single-stranded DNA fragments reanneal is proportional to the square of the concentration. In most cases, after two RDA cycles the amplicon library is skewed in favor of only a few molecules. In order to get a more comprehensive overview of the difference product, the RDA procedure can be repeated with the prominent amplicons added to the driver DNA as shown in **Figure 2** (Groot and van Oost, 1998).

Applications

DNA deletion/DNA insertion analysis

The RDA technique has been applied with great success to find differences in genomic DNA between different organisms or cell types. For example, using genomic DNA of the male as tester and genomic DNA of the female DNA as driver, Y chromosome-specific sequences can be identified (Navin *et al.*, 1996). More salient examples are the analyses of deletions in genomic DNA in tumor material (Lisitsyn *et al.*, 1995), which have contributed to the discovery of the tumor suppressor genes *BRCA2* (Schutte *et al.*, 1995) and *PTEN* (Li *et al.*, 1997) among others. Also genomic differences between bacterial strains differing in pathogenicity or viral integrations in chromosomal DNA have been detected by the RDA technique.

Identification of differentially expressed genes

In 1994, Hubank and Schatz reported the application of the RDA technique for the comparison of two complementary DNA (cDNA) populations (cDNA-RDA). This has led to numerous publications to identify genes expressed under different environmental conditions and to identify target genes for activator or suppressor proteins.

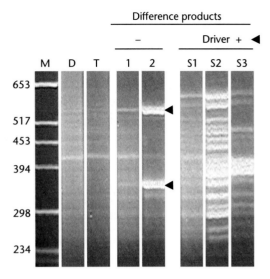

Figure 2 Effect of suppression of previously identified products in representational difference analysis (RDA). Ethidium bromide staining of an agarose gel loaded with driver (D) and tester (T) DNA amplicons and the difference products obtained after one and two rounds of subtractive hybridization respectively. The dramatic effect of RDA is clear: the difference product 2 consists mainly of two major products (indicated by triangles), derived from genes highly expressed in the cells from which the tester was derived only. A disadvantage of this very high level of purification of these two gene fragments is the fact that many other differentially expressed genes are underrepresented or absent in the difference product 2. However, cDNA-RDA offers the unique possibility to suppress the reappearance of previously identified products by adding those to the driver in a new RDA experiment. Difference products obtained after 1 (S1), 2 (S2) and 3 (S3) rounds of subtractive hybridization are shown; the two previously identified products are completely absent, and many additional differentially expressed genes can be identified this way (for details, see Groot and van Oost (1998)).

Positional cloning

When the tester and driver samples are selected on the basis of their genetic relationship, the technique is called genetically directed representational difference analysis (GDRDA). The usefulness of this technique was first demonstrated with congenic mouse lines (Lisitsyn *et al.*, 1994). A striking example of the power of the GDRDA technique is the positional syntenic cloning of the mammalian circadian mutation *tau*. This was accomplished by RDA linkage analysis in the hamster without an actual genetic map (Lowrey *et al.*, 2000). (*See* DNA Cloning; Gene Mapping and Positional Cloning.)

Future of the RDA Technique

Since the seminal publication by Lisitsyn *et al.* (1993), more than 200 papers dealing with the optimization and applications of the RDA technique have appeared. It can be expected that gene arrays will take over the RDA technique for organisms for which the complete genome sequence is available. On the other hand, the time-consuming analysis of the difference products can be speeded up by using gene arrays. For the so-called 'orphan' genomes for which the complete genome sequence is still far off, the RDA technique will continue to be a very useful tool.

See also
Expression Studies

References

Groot PC and van Oost BA (1998) Identification of fragments of human transcripts from a defined chromosomal region: representational difference analysis of somatic cell hybrids. *Nucleic Acids Research* 26(19): 4476–4481.

Hubank M and Schatz DG (1994) Identifying differences in mRNA expression by representational difference analysis of cDNA. *Nucleic Acids Research* 22: 5640–5648.

Li J, Yen C, Liaw D, *et al.* (1997) *PTEN*, a putative protein tyrosine phosphatase gene mutated in human brain, breast, and prostate cancer. *Science* 275: 1943–1947.

Lisitsyn N, Lisitsyn N and Wigler M (1993) Cloning the differences between two complex genomes. *Science* 259(5097): 946–951.

Lisitsyn NA, Lisitsina NM, Dalbagni G, *et al.* (1995) Comparative genomic analysis of tumors: detection of DNA losses and amplification. *Proceedings of the National Academy of Sciences of the United States of America* 92: 151–155.

Lisitsyn NA, Segre JA, Kusumi K, *et al.* (1994) Direct isolation of polymorphic markers linked to a trait by genetically directed representational difference analysis. *Nature Genetics* 6: 57–63.

Lowrey PL, Shimomura K, Antoch MP, *et al.* (2000) Positional syntenic cloning and functional characterization of the mammalian circadian mutation *tau*. *Science* 288: 483–491.

Navin A, Prekeris R, Lisitsyn NA, *et al.* (1996) Mouse Y-specific repeats isolated by whole chromosome representational difference analysis. *Genomics* 36: 349–353.

Schutte M, da Costa LT, Hahn SA, *et al.* (1995) Identification by representational difference analysis of a homozygous deletion in pancreatic carcinoma that lies within the *BRCA2* region.

Proceedings of the National Academy of Sciences of the United States of America 92: 5950–5954.

Further Reading

Everts RE, Versteeg SA, Renier C, *et al.* (2000) Isolation of DNA markers informative in purebred dog families by genomic representational difference analysis (gRDA). *Mammalian Genome: Official Journal of the International Mammalian Genome Society* 11: 741–747.

Lisitsyn NA (1995) Representational difference analysis: finding the differences between genomes. *Trends in Genetics* 11: 303–307.

Michiels L, van Leuven F, van den Oord JJ, de Wolf-Peeters C and Delabie J (1998) Representational difference analysis using minute quantities of DNA. *Nucleic Acids Research* 26: 3608–3610.

Muller K, Heller H and Doerfler W (2001) Foreign DNA integration: genome-wide perturbations of methylation and transcription in the recipient genomes. *Journal of Biological Chemistry* 276: 14271–14278.

Sagerström CG, Sun BI and Sive HL (1997) Subtractive cloning: past, present, and future. *Annual Review of Biochemistry* 66: 751–783.

Simons JN, Pilot-Matias TJ, Leary TP, *et al.* (1995) Identification of two flavivirus-like genomes in the GB hepatitis agent. *Proceedings of the National Academy of Sciences of the United States of America* 92: 3401–3405.

Sornson MW and Rosenfeld MG (1999) Genetically directed representational difference analysis-based positional cloning and Northern analysis of candidate genes. *Methods in Enzymology* 306: 67–89.

Toyota M, Canzian F, Ushijima T, *et al.* (1996) A rat genetic map constructed by representational difference analysis markers with suitability for large-scale typing. *Proceedings of the National Academy of Sciences of the United States of America* 93: 3914–3919.

Wallrapp C and Gress TM (2001) Isolation of differentially expressed genes by representational difference analysis. *Methods in Molecular Biology* 175: 279–294.

Welford S, Gregg J, Chen E, *et al.* (1998) Detection of differentially expressed genes in primary tumor tissues using representational differences analysis coupled to microarray hybridization. *Nucleic Acids Research* 26: 3059–3065.

Reproductive Choice

John Harris, *University of Manchester, Manchester, UK*

Assisted reproductive technologies, including preimplantation genetic diagnosis, may be used to influence nonmedical traits. When may this be appropriate? What potential problems are there with such applications of the technology?

Intermediate article

Article contents

- What is Medical Technology For?
- The Presumption in Favor of Liberty
- Reproductive Liberty
- Is there such 'Good and Sufficient Cause' to Deny Access to Artificial Reproductive Technologies?
- Preimplantation Genetic Diagnosis

What is Medical Technology For?

Our question concerns the ethics of using assisted reproductive technologies (ART) to influence nonmedical traits; but this immediately raises another question: when and why is it appropriate to use ART to influence medical traits or for therapeutic purposes?

If we know why this might be appropriate then we might have some idea about nontherapeutic uses.

Medical technology, that is the use of technology for medical and therapeutic purposes, is morally justified by the good that it will do, by the importance of that good, and by the fact that it is desired by those whom it benefits or (where patients cannot request or consent to its use) is in their best interests. The urgency of the use of medical technology is proportional to the magnitude of the good that it will do and the use of the technology will be ethical if it will do the good that the patients want or when patients cannot request it, its use will be ethical if it does good and its use is in the patient's best interests. Medical goods are often important because they protect life, ameliorate pain or suffering, restore mobility and so on, all very important goods. However, note that there is an ambiguity in what we mean when we say that the use of technology is justified by the good that it does. There is ambiguity between the question of whether the deployment of public resources for the achievement of the particular good is justified and the question of whether individuals are justified in accessing or others are justified in offering the technology. The use of public resources may be justified if the technology does a good that ought to be done, although pressure on resources will always influence what is actually funded. However, the individuals will be justified in accessing the technology and others justified in providing it if it does no harm or no significant harm, even if there are no moral imperatives for doing that harmless or marginally harmful thing.

The use of medical technology in plastic surgery is a good example here. If the surgery is necessary to prevent or mitigate suffering, then it will be justified both to do it and to spend public resources on it. However, if it is purely a matter of personal preference ('I'd like a smaller nose – but I am not made miserable by my existing one'), then again the deployment of the technology, while not required by morality and while not in the public interest, is not wrongful in any way. However, if the individual wants the slogan 'Death to all Americans' indelibly burnt into her forehead, then although only mildly harmful to herself it is immoral for independent reasons, for example, that it constitutes incitement to racial hatred, is against the public interest and arguably harmful or is a danger to others.

Note again, that questions of scarce resources aside, the provision of services that require technology is not really a moral issue unless some harm can be shown to result from their provision. There is nothing special about 'medical' technology save that it is assumed to partake of the general justifications and 'prestige' of the practice of medicine more generally. Without those general justifications it requires its own. And here the question of with whom the burden of proof lies assumes importance: who has to justify the use of technology and to whom must it be justified?

The Presumption in Favor of Liberty

In the United Kingdom and the United States and in many other democracies, there is a presumption in favor of liberty. This means that the burden of justifying their actions falls on those who would deny liberty and not on those who would exercise it. This idea is constitutive of liberalism and hence of all liberal democracies. Indeed I believe that it is part of all societies that treat people as equals, and hence of all democracies properly so-called; but to argue for that proposition is beyond the scope of this article. If this is right, the presumption must be in favor of the liberty to access ART unless good and sufficient reasons can be shown against so doing. However, suppose this presumption is not accepted. Can anything else be said about the liberty to access ART which might give credibility to a presumption of liberty?

Reproductive Liberty

When people express their choices about procreation, even in unusual or idiosyncratic ways, they are claiming an ancient, if only recently firmly established, right of a very fundamental sort. This right is found in all the principal conventions on human rights. It has been expressed as the right to marry and found a family or as the right to privacy and to respect for family life (European Convention on Human Rights, 1953; International Covenant of Civil and Political Rights, 1976; United Nations, 1978).

Some see the right or entitlement to reproductive liberty as derived from the right to reproduce *per se*, others as derivative of other important rights or freedoms. (I defended this liberty, albeit in a somewhat convoluted form, in Harris (1992); for a more explicit and more elegant defence see Robertson (1994).)

There is no consensus as to the nature and scope of this right; however, it is clear that it must apply to more than conventional sexual reproduction and that it includes a range of values and liberties that normal sexual reproduction embodies or subserves. For example, John Robertson outlining his understanding of this right suggests:

> The moral right to reproduce is respected because of the centrality of reproduction to personal identity, meaning and dignity. This importance makes the liberty to procreate an important moral right, both for an ethic of individual autonomy and for the ethics of community or family that view the purpose of marriage and sexual union as the reproduction and rearing of offspring. Because of this importance the right to reproduce is widely recognised as a prima facie moral

right that cannot be limited except for very good reason. (Robertson, 1994)

Ronald Dworkin has defined reproductive liberty or procreative autonomy as 'a right to control their own role in procreation unless the state has a compelling reason for denying them that control'.

The right of procreative autonomy has an important place ... in Western political culture more generally. The most important feature of that culture is a belief in individual human dignity: that people have the moral right – and the moral responsibility – to confront the most fundamental questions about the meaning and value of their own lives for themselves, answering to their own consciences and convictions ... The principle of procreative autonomy, in a broad sense, is embedded in any genuinely democratic culture. (Dworkin, 1993)

Dworkin's and Robertson's accounts both center on respect for autonomy and for the values that underlie the importance attached to procreation. These values see procreation and founding a family as involving the freedom to choose one's own lifestyle and express, through actions as well as through words, the deeply held beliefs and the morality that families share and seek to pass on to future generations.

The freedom to pass on one's genes is widely perceived to be an important value; it is natural to see this freedom as a plausible dimension of reproductive liberty, not least because so many people and agencies have been attracted by the idea of the special nature of genes and have linked the procreative imperative to the genetic imperative. Whether or not this suggestion is ultimately persuasive, it is surely not possible to dismiss the choices about reproduction and access to the relevant technologies that constitute the point of claiming reproductive liberty as a simple and idle exercise of preference. Reproductive choices, whether or not they prove to be protected by a right to procreative liberty or autonomy, have without doubt a claim to be taken seriously as moral claims. As such they may not simply be dismissed wherever and whenever a voting majority can be assembled against them. Those who seek to deny the moral claims of others (as opposed, possibly, to the exercise of their idle preferences) must show good and sufficient cause.

Is there such 'Good and Sufficient Cause' to Deny Access to Artificial Reproductive Technologies?

Without listing them all, the main technologies currently available are, first, technologies that generate an embryo: *in vitro* fertilization (IVF) and intracytoplasmic sperm injection (ICSI) and cell nuclear transfer (reproductive cloning). Then there are technologies that allow embryo selection such as IVF, preimplantation genetic diagnosis (PIGD) and various methods of testing *in utero*. There are genetic tests available prior to conception: carrier testing and the like. And finally in the further future there may be genetic manipulation available to select for nonmedical traits.

It is difficult to deal briefly with all the possible uses of such technologies to influence nonmedical traits, but perhaps a general observation can indicate a possible approach. If we can identify states of affairs (nonmedical traits) that would be morally problematic *of themselves*, we might be secure in judgments as to which of such states of affairs or traits it would be morally problematic deliberately to produce. The answer seems to be only those traits that would be harmful to the individual produced or harmful to others. Thus it would not be a morally problematic event if a boy rather than a girl were produced (or vice versa), and it would not be morally problematic if a child with a particular skin color, hair color, eye color or a range of useful abilities – sporting prowess, musical talent, intelligence, etc. – were to be born or created. It could not be said that children with any of these features would be born in a harmed condition or at any disadvantage whatsoever, neither would it be plausible to claim that they would be in any way harmful or dangerous to others. If it is not wrong to wish for a bonny, bouncing, brown-eyed, intelligent baby girl with athletic potential and musical ability, in virtue of what might it be wrong to use technology to play fairy godmother to oneself and grant the wish that was father to the thought?

There is of course room for claims that the choice of such features would not be in the public interest, but this raises the question of just how something harmless or even positively beneficial *in itself* might become contrary to the public interest when deliberately chosen. It is sometimes claimed that those who would produce children with particular nonmedical traits would have burdensome expectations of those children – expectations that would in fact harm the children (Harris, 2000; Holm, 1998). Notice here that such expectations are not peculiar to the use of technology to select for particular traits, but are a feature of all burdensome parental expectations. If we object to such expectations we must take care to demonstrate their unreasonableness and protect children from them. Among such expectations would have to be included those concerning religious observance, career path, values and occupation.

What objections could there then be (which do not refer to the expectations that parents might have that children would use these traits in particular ways) to the creation of nonmedical traits? There are other sorts of objections that might be made, but such objections are of necessity more 'remote' and more speculative than objections that concern the welfare of the child to be born or the welfare of others. 'Remote' in the sense that they must appeal to abstract notions like 'the public interest' or 'public acceptability', not harm to the individuals concerned; and speculative in that they may refer to possible harmful effects of patterns of choice of nonmedical traits on society or baseless speculation about the harmful effects on the individual that knowledge that they were selected to have such traits may have. I say 'baseless speculation' not because it might not be proved right in the event, but baseless in the sense that there is no prospective reason to expect harmful rather than beneficial effects to be the ones that actually occur.

Some people have been attracted to the idea of talking as if the disabled are simply differently abled and not in any way harmed. Deafness is often taken as a test case here (Journal of Medical Ethics, 2001). Insofar as it is plausible to believe that deafness is simply a different way of experiencing the world, but by no means a harm or disadvantage, then of course the deaf are not suffering from any disability. However is it plausible to believe any such thing? Would the following statement be plausible: 'I have just accidentally deafened your child, it was quite painless and no harm was done so you needn't be concerned or upset!'? Or suppose a hospital were to say to a pregnant mother 'Unless we give you a drug your fetus will become deaf, since the drug costs £5 and there is no harm in being deaf we see no reason to fund this treatment'. But there is harm in being deaf and we can state what it is.

Imagine a child whose deafness could have been successfully treated, saying the following to the parents who denied the treatment: 'I could have enjoyed Mozart and Beethoven and dance music and the sound of the wind in the trees and the waves on the shore, I could have heard the beauty of the spoken word and in my turn spoken fluently but for your deliberate denial'. In response it might be suggested that 'One may acknowledge the joy that (these things) bring others without insisting that the inability to perceive them is a harm or a deficit. After all, many persons are 'deaf' to the pleasures of classical music (or jazz, or reggae, or rap, etc.) and yet none assume their limits of comprehension reflect a deficit or harm'. But to be 'deaf' to the pleasures of classical musical is to be deaf in inverted commas, not really deaf. Musical taste can be educated, but not so hearing for the profoundly deaf.

Deafness is a harmed condition properly so-called but equally to say so does not imply that the deaf do not have lives that are thoroughly worth living and are the moral equals of anyone. This may be seen if we consider the issue not of disabilities but of enhancements. Suppose some embryos had a genetic condition that conferred complete immunity to many major diseases – HIV/AIDS, cancer and heart disease for example – coupled with increased longevity. We would, it seems to me, have moral reasons to prefer to implant such embryos given the opportunity of choice. However, such a decision would not imply that normal embryos had lives that were not worth living or were of poor or problematic quality. If I would prefer to confer these advantages on any future children that I may have, I am not implying that people like me, constituted as they are, have lives that are not worth living or that are of poor quality nor that they are in any the sense lesser or inferior as persons.

Preimplantation Genetic Diagnosis

Finally we should note the use of PIGD in selection. In addressing the ethics of PIGD, a comparison is sometimes made with abortion. The purpose of such a comparison is to suggest that PIGD must be justified using criteria comparably stringent with those required to justify abortion and that in particular a woman's motives for requesting PIGD must be minutely scrutinized. However, this comparison is fallacious. The fallacy involved in the comparison is that a decision to abort must, in most jurisdictions, be justified and comply with the law, whereas a decision not to implant embryos requires no legal or moral justification whatsoever. The decision not to implant *in vitro* embryos is within the unfettered discretion of any woman.

A woman cannot be forced to implant any embryos and she can also usually determine the fate of those embryos, requiring them to be destroyed or donating them for research or to another woman for IVF. How might it be justifiable to deny a woman information on her *in vitro* embryo, which would enable her to make a rational and informed decision as to what to do with it? Suppose she were denied tests that she sought, and implanted an embryo that turned into a damaged child. However slight the damage, the mother (and according to some, also the child) would have moral and perhaps also legal grounds for complaint that the damage was due to the denial of access to information.

See also
Genetic Enhancement
Human Cloning
In Vitro Fertilization
Preimplantation Genetic Diagnosis: Ethical Aspects
Reprogenetics: Visions of the Future

References

Brownlie I (ed.) (1971) *Basic Documents on Human Rights*, pp. 211–232, 338–363. Oxford, UK: Clarendon Press.

Dworkin R (1993) *Life's Dominion*, pp. 166–167. London, UK: HarperCollins.

Harris J (1992) *Wonderwoman & Superman: The Ethics of Human Biotechnology*, chapters 2 and 3. Oxford, UK: Oxford University Press.

Harris J (2000) The welfare of the child. *Health Care Analysis* **8**(1).

Holm S (1998) A life in shadows. *Cambridge Quarterly of Healthcare Ethics* **7**: 160–162.

Journal of Medical Ethics (2001) Symposium: equality and disability. *Journal of Medical Ethics* **27**(6): 370–392.

Robertson JA (1994) *Children of Choice*. Princeton, NJ: Princeton University Press.

United Nations (1978) *Universal Declaration of Human Rights*, Article 16. New York, NY: United Nations.

Further Reading

Burley JC (ed.) (1999) *The Genetic Revolution and Human Rights: The Amnesty Lectures 1998*. Oxford, UK: Oxford University Press.

Glover J (1984) *What Sort of People Should There Be?* Harmondsworth, UK: Pelican.

Harris J (1998) *Clones Genes and Immortality*. Oxford, UK: Oxford University Press.

Harris J (2001) One principle and three fallacies of disability studies. *Journal of Medical Ethics* **27**(6): 383–388.

Harris J and Holm S (eds.) (1998) *The Future of Human Reproduction*. Oxford, UK: Clarendon Press.

Reproductive Genetic Screening: A Public Health Perspective from the United Kingdom

Layla Al-Jader, *University of Wales College of Medicine, Cardiff, UK*

Public health has several roles in reproductive genetic screening. It should emphasize the voluntary nature of these programs and ensure that people are adequately informed before making decisions and supported afterward. It must ensure that a central coordinating body will monitor the implementation and the outcome of these programs. Public health must also ensure that the definition of 'success' or 'failure' of a program is not based on too narrow monetary calculations.

Advanced article

Article contents

- Introduction
- Background
- Role of Public Health Medicine
- Methods of Evaluating the Benefits of Antenatal Screening
- Factors Influencing Uptake of Screening
- Conclusion

Introduction

There is a general confusion over the distinction between the linked but separate concepts of screening and diagnosis. An antenatal screening test identifies an increased likelihood of a fetal abnormality in an apparently normal pregnancy. A diagnostic test confirms or refutes the existence of fetal abnormality in those at increased risk.

Screening, in general, has important differences from clinical practice because the health service is targeting apparently healthy people, offering to help individuals to make better-informed choices about their health. However, screening has potential adverse consequences and it is important to have realistic expectations of such programs. (*See* Genetic Screening Programs.)

The National Screening Committee of the Department of Health in the United Kingdom (NSC) signaled the need for a changed approach such that individuals are offered a choice that is based on an appreciation of risks and benefits. The types of reproductive genetic screening programs available are listed in **Table 1**.

Background

Analysis of the National Down's Syndrome Cytogenetic Register 1989–1997, which holds records of 10 651 cases from England and Wales (Alberman, 2002), shows that the proportion diagnosed prenatally in that period rose from 30% to 53% (Mutton *et al.*, 1998). Ultrasound indications for referrals to prenatal diagnosis also rose during that period, especially referrals made on the basis of an increase in nuchal thickening. The rise in Down syndrome cytogenetics registrations is thought to be because of the demographic trend towards starting families later in life, the high proportion of older mothers from postwar birth registration bulge, and the increase in early prenatal diagnoses of trisomy 21 in fetuses leading to terminations of pregnancy that would have previously miscarried spontaneously. All this tended to stabilize the annual number of affected births, in spite of the increase in the number diagnosed prenatally.

Table 1 Approaches to reproductive genetic screening

Type of disease	Screening approach and/or reasons for screening
Congenital malformation including aneuploidy	**Maternal infections** for rubella virus, which is the main recognized infectious cause of congenital malformations. Definitive prenatal diagnosis of affected fetuses is possible by measuring the appropriate antibodies in fetal blood. A maternal screening blood test is carried out early in pregnancy to detect susceptibility to rubella infection.
	Maternal age for chromosomal abnormalities including Down syndrome.
	Maternal serum screening for Down syndrome, neural tube defects and a few other chromosomal abnormalities, which constitute about half the abnormal results. Maternal serum screening for Down syndrome involves biochemical tests such as α-fetoprotein, human chorionic gonadotrophin and unconjugated estriol, either alone or in combination, that have variable detection and false-positive rates.
	Ultrasound anomaly scanning for detecting some syndromes and structural abnormalities of the cardiovascular, central nervous system including neural tube defects, renal, gastrointestinal and skeletal systems.
Inherited diseases	**Cystic fibrosis**, the most common inherited disorder among Caucasians. Only available through the National Health Service in Edinburgh, Scotland.
	Family history of a recessive genetic condition, such as cystic fibrosis, inborn errors of metabolism, and so on.
	Selection by ethnic group, including hemoglobin disorders such as thalassemia in women of Mediterranean or Asian origin, sickle cell disorders in women of African or Afro-Caribbean origin, and Tay–Sachs disease in the Ashkenazi Jewish population.

Role of Public Health Medicine

The role of public health medicine in reproductive genetic screening includes primary prevention, secondary prevention, evaluating the screening program and setting quality standards, and informing public policy on genetic screening services.

Primary prevention

By epidemiological analysis of specific malformations through national and regional malformation registers, public health surveillance should be able to identify malformation clusters and epidemics. Setting up epidemiological studies to test theories for the cause of these anomalies is another role for public health medicine and epidemiology. These disciplines advise the government on public health policy on primary prevention to reduce the incidence of congenital anomalies.

The main example of primary prevention is that of folic acid supplements to reduce the incidence of neural tube defects. The current recommendation is that all women of childbearing age ingest 0.40 mg of folic acid per day to reduce the occurrence of neural tube defects (Czeizel, 1995), with a higher dose recommended for those at an increased risk.

Secondary prevention

There is technology available to test for several hundred genetic disorders. Public health medicine can contribute to the debate about what tests should be available by assessing the needs for screening services. The principal objective of genetic screening in pregnancy is to allow the widest possible range of informed choice to women at risk of having children with an abnormality. It also provides reassurance, avoids the birth of seriously affected children through selective abortion, and ensures optimal treatment of affected infants through early diagnosis.

Evaluating the screening program and setting quality standards

Public health physicians are key professionals that are involved in evaluating screening programs. In 1993 in a south Wales district in the United Kingdom, a survey of health professionals about Down syndrome screening found that obstetric units had introduced a universal screening policy with no written policy or staff training (Al-Jader et al., 2000). As part of service evaluation, a qualitative semistructured survey of pregnant women was also carried out. Most women were unaware that screening tests were voluntary. Half were not well-informed to make decisions. All women wanted to be offered the choice of whether to be screened. Only five out of 100 women refused screening; these women tended to be better educated and of higher social class. As women became more affluent and knowledgeable about screening and its limitation, the chance of opting out of screening was higher than otherwise. This trend is confirmed by other studies (A. Summers, personal communication) and has implications for the clarification of our primary public health goal: is it to reduce the burden of abnormalities through selective abortions or to give women informed choices?

In late 1990s, a clear policy of offering informed choice to people undergoing screening has emerged. Stone and Stewart (1996) argued that the two most frequently cited objectives of screening for a recessive carrier state are to reduce the incidence of the disorder and to inform individuals and couples at risk of their reproductive choices. They claimed that the latter aim represents a model shift in the philosophy of screening in that no preventive principle is involved. Instead, information is regarded as worthwhile in itself, regardless of outcome. Stone and Stewart argued that the benefits arising from the information generated in the course of genetic carrier screening, for cystic fibrosis as an example, cannot be presumed just by asserting a 'right to know' ethical imperative. They draw attention to the danger that a combination of technical capability, professional zeal and consumer demand will override currently accepted screening principles. If this were to happen, they argued, future efforts to subject screening programs to rational evaluation could be undermined. (*See* Autonomy and Responsibility in Reproductive Genetics.)

However, reproductive genetic screening programs are different from other screening programs such as those for early signs of toxemia of pregnancy and gestational diabetes, for which a more directive approach is taken. A different nondirective approach is required in screening for genetic and congenital abnormalities, because if it were aimed at reducing the incidence of disorders then it would inevitably come into conflict with people who are opposed to termination of pregnancy. Such screening programs are also different because the results may have implications for other family members, who may wish to know the outcome; there is the concept of identifying healthy carriers of recessive disorders, such as cystic fibrosis, who may pass on the gene to the next generation, and there are possible future implications for employment and insurance. Perhaps the most important difference between genetic antenatal screening and other screening programs is the possible decision to terminate life with all its ethical and moral implications. What is acceptable for one family may be abhorrent to others. In a civilized society, treating and caring for severely handicapped children can go hand in hand with supporting those who exercise their informed choice not to bring a severely handicapped child into the world. (*See* Genetic Counseling: Nondirectiveness; Nondirectiveness; Reprogenetics: Visions of the Future.)

Assessment of a genetic screening program should incorporate the evaluation of ethical, social, cultural and political dimensions. Active promotion by the medical profession and regulation by the political decision-makers may leave little room for learning about the social and cultural meaning and for the public acceptability of serum screening – as in the context of Dutch maternity care. (*See* Genetics and the Control of Human Reproduction.)

The benefits of screening for Down syndrome in a socioeconomically deprived inner city area in Birmingham, which has a high ethnic minority population, were judged to be small by researchers. They found that the uptake of amniocentesis after a high-risk result of serum screening was 35% among Asian women compared with 67% among Caucasian women. When analyzing possible contributing factors for this difference, we should be clear of what constitutes 'benefits' of such a screening program.

Informing public policy on genetic screening services

Screening services for Down syndrome throughout the United Kingdom often have no clear line of managerial responsibility and no clear mechanism for modifying policy or service monitoring. The NSC is reviewing this service to produce a strategy for England. Similar reviews are being carried out in Wales and Scotland. Until the new infrastructure is in place, the NSC has recommended in 2001 that all pregnant women, irrespective of age, should be offered second trimester serum screening. The test used should comprise at least a double test, but it would be desirable for laboratories to move to triple or quadruple tests when possible. For these tests the cutoff level for increased risk of Down syndrome should usually be around 1 in 250 at term. Using ultrasound dating this would be expected to yield a detection rate of about 60% or better for a 5% false-positive rate. They are continuing to review and amend this guidance as appropriate. A network of regional coordinators is being set up across the country to ensure proper management and coordination of local programs. The *Health Technology Assessment* report on antenatal screening for Down syndrome (Wald *et al.*, 1998) recommended that there should be a written policy, improved staff education and training, specific funding for screening and line management responsibility, but preservation of local commitments.

A public health report detailing women's experiences of the service and making recommendations in 1997 in south Wales led to the establishment of a professional advisory group to review the service, recommend a model of care, and decide on quality standards for audit and review. A model of screening midwife specialist in every maternity unit was developed; this specialist was appropriately trained and responsible for audit and coordination of the service together with a named obstetrician (Al-Jader and Hopkins, 2000). Such a model is being adopted for the all-Wales strategy and other regions in the United Kingdom.

Methods of Evaluating the Benefits of Antenatal Screening

Methods of evaluating the benefits include averted costs, quality-adjusted life years, and willingness to pay. Some researchers have suggested an alternative method that assumes that screening can be regarded as an investment in improved information, and that the benefits of screening can be measured by the value placed on this information. The information is valued by presenting individuals with standard gamble questions framed in terms of prenatal diagnostic choice (Shackley and Cairns, 1996). Choosing a screening protocol requires a trade-off between a desired detection rate and an acceptable false-positive rate. Various modeling programs have been developed on the basis of a range of assumptions to optimize the societal net benefit (Beazoglou et al., 1998). The policies used must also consider the freedom of individuals in decision-making at each step of the prenatal diagnosis pathway.

The effects, safety and cost-effectiveness have been compared for nine strategies currently available for screening for Down syndrome in the United Kingdom. There are four screening strategies that are cost-effective and efficient. Nuchal thickening measurement is apparently the cheapest procedure, but the quadruple test, the first trimester combined test and the integrated tests represent the best options in terms of effectiveness, cost-effectiveness and safety. Mathematical modeling, based on a systematic review of antenatal screening for Down syndrome, was used in the comparison. Some argued that screening policy should not be driven by modeling without evidence from controlled trials that confirm the presumed benefits of the various policies that are achieved in practice.

Early screening, which means counseling women before coming to hospitals, increases the demand for chorion villus sampling, with its associated higher miscarriage rate, higher cost and restricted availability. Moving from the double test to the quadruple test may prove effective; although there will be an increased laboratory cost, the quadruple test is not dependent on a skilled and equipment-based technique, such as nuchal thickening measurement. This is especially significant when the additional problem of radiographer shortages nationally is realized. All studies of Down syndrome screening have had a relatively small number of Down syndrome pregnancies – not enough to provide the required level of statistical significance and overcome the problem of wide confidence intervals.

Factors Influencing Uptake of Screening

Knowledge of the condition

Researchers found that racial groups other than Caucasians had a poorer understanding of Down syndrome. The factors that affected knowledge of Down syndrome included quality of spoken English, knowing an affected child, parity and religion. The most significant factor that affected acceptance of screening was the woman's knowledge of Down syndrome. To increase uptake of the screening test for Down syndrome, the authors recommended that better education and counseling are needed for women who attend antenatal care. (*See* Down Syndrome.)

A randomized controlled trial (RCT) was carried out in six geographically and demographically diverse sites in Ontario. It was found that written patient information could contribute in an important way to patient knowledge. Limitations were found for subgroups (women under 25 and those not speaking English at home; Glazier et al., 1997).

An RCT was carried out to assess the effect of a Down syndrome screening video on test uptake, knowledge and psychological stress. Although the video had no effect on the rate of screening uptake, it did increase knowledge of screening for both groups; those who received the video before booking and those who did not. There were no significant differences between the groups in either specific worries about abnormalities in the baby or general anxiety (Hewison et al., 2001).

Another RCT was carried out in Aberdeen, Scotland, to compare the effectiveness of a touch screen system with an information leaflet. The touch screen seemed to convey no benefit over well-prepared leaflets in improving understanding of prenatal testing among the pregnant women. It did, however, seem to reduce anxiety and may be most effective for providing information to selected women (Graham et al., 2000).

A systematic review has been made of trials of interventions aimed at increasing the uptake of screening. Two trials also assessed decision-making: neither found that interventions such as videos, information leaflets with decision trees, or touch screen computers conveyed any additional benefits over well-prepared leaflets (Jepson et al., 2001).

Humanitarian and religious beliefs

Women who have strong religious or humanitarian beliefs against terminating an affected fetus almost invariably choose not to be screened. Some of these women may still want to know if their fetus is affected in order to prepare themselves for the birth of their affected baby. In south Wales in the United Kingdom, two surveys were carried out to explore parents' attitudes to screening, antenatal diagnosis and termination of an affected fetus. Although the numbers were small, both studies showed similar findings. The earlier survey of mothers of cystic fibrosis children found that after antenatal diagnosis, 52% of mothers would abort a cystic fibrosis fetus, whereas 24% would not (Al-Jader et al., 1990). Similar attitudes were found 10 years later when pregnant women were surveyed regarding termination of a Down syndrome fetus: 56% of women said they would consider termination, whereas 21% would not (Al-Jader et al., 2000).

Although the earlier survey on cystic fibrosis showed no correlation between the social classes of the parents and their views on terminating an affected fetus, there were relatively more parents from social classes 2 and 3 who would refuse termination. By contrast, the survey on Down syndrome showed that those who would refuse to terminate an affected pregnancy were from both ends of the social class spectrum (i.e. a U-shaped curve).

How the service is organized

Staff who are well-trained and given sufficient time to talk with each expectant mother are a prerequisite for an informed choice. A study of women in the postnatal period found that with an 'opt-out' or routine system of screening, high rates of uptake were achieved at the expense of informed choice for some participants. The factors that most often influenced choice were fetus health, maternal age and knowledge of the consequences of screening. Another study found that women's attitudes to prenatal screening are very dependent on how the prenatal screening program is already organized in their local area.

Conclusion

For many years, public information about screening has been aimed at achieving high uptake. By giving information that emphasizes only the positive aspects of screening, the autonomy of individuals is ignored. If we adopt instead an approach that makes explicit the limitations and adverse effects, then a different set of problems will be encountered. Those most likely to be deterred from accepting screening may be the most socially disadvantaged. (*See* Autonomy and Responsibility in Reproductive Genetics.)

There will be a cost in terms of the staff time required to explain screening more fully to participants,

and cost-effectiveness could be reduced if uptake falls so low as to make services barely viable. What makes this matter even more critical is that the cost of staff training in delivering appropriate counseling is not taken into account. The cost of training staff was not included in a 1999 *Health Technology Assessment* report, which recommended the introduction of antenatal screening for cystic fibrosis (Murray et al., 1999).

See also
Clinical Genetic Services in the United Kingdom
Genetic Screening Programs
Public Health Genetics

References

Alberman E (2002) The National Down Syndrome Cytogenetic Register (NDSCR). *Journal of Medical Screening* **9**(3): 97–98.

Al-Jader LN, Goodchild MC, Ryley HC and Harper PS (1990) Attitudes of parents of CF children towards neonatal screening and antenatal diagnosis. *Journal of Clinical Genetics* **38**: 460–465.

Al-Jader LN and Hopkins S (2000) Development of an antenatal screening programme for congenital abnormalities in a South Wales district; lessons learnt for clinical governance. *Community Genetics* **3**: 31–37.

Al-Jader LN, and Parry-Langdon N and Smith RJW (2000) Survey of attitudes of pregnant women towards Down's syndrome screening. *Prenatal Diagnosis* **20**: 23–29.

Beazoglou T, Heffley D, Kyriopoulos J, Vintzileos A and Benn P (1998) Economic evaluation of prenatal screening for Down's syndrome in the USA. *Prenatal Diagnosis* **18**(12): 1241–1252.

Czeizel AE (1995) Nutritional supplementation and prevention of congenital abnormalities. *Current Opinion in Obstetrics and Gynaecology* **7**(2): 88–94.

Glazier R, Goel V, Holzapfel S, et al. (1997) Written patient information about triple-marker screening: a randomized, controlled trial. *Obstetrics & Gynaecology* **90**(5): 769–774.

Graham W, Smith P, Kamal A, et al. (2000) Randomised controlled trial comparing effectiveness of touch screen system with leaflet for providing women with information on prenatal tests. *British Medical Journal* **320**(7228): 155–160.

Hewison J, Cuckle H, Baillie C, et al. (2001) Use of videotapes for viewing at home to inform choice in Down's syndrome screening: a randomised controlled trial. *Prenatal Diagnosis* **21**(2): 146–149.

Jepson RG, Forbes CA, Sowden AJ and Lewis RA (2001) Increasing informed uptake and non-uptake of screening: evidence from systematic review. *Health Expectations* **4**(2): 116–130.

Murray J, Cuckle H, Taylor G, Littlewood J and Hewison J (1999) Screening for cystic fibrosis. *Health Technology Assessment* **3**: 8.

Mutton D, Ide RG and Alberman E (1998) Trends in prenatal screening for and diagnosis of Down's syndrome: England and Wales, 1989–97. *British Medical Journal* **317**: 922–923.

Shackley P and Cairns J (1996) Evaluating the benefits of antenatal screening: an alternative approach. *Health Policy* **36**(2): 103–115.

Stone DH and Stewart S (1996) Screening and the new genetics; a public health perspective on the ethical debate. *Journal of Public Health Medicine* **18**(1): 3–5.

Wald JW, Kennard A, Hackshaw A and McGuire A (1998) Antenatal screening for Down's syndrome. *Health Technology Assessment* **2**(1): 99.

Further Reading

Acheson D (1988) *Public Health in England – The Report of the Committee of Enquiry into the Future Development of the Public*

Health Function, Cmd 289. London: Her Majesty's Stationary Office.

Al-Jader LN (1996) *Antenatal Screening for Down Syndrome*, part II. London, UK: Faculty of Public Health Medicine of the Royal Colleges of Physicians.

Chapple J (1994) *Screening Issues – The Public Health Aspect. Prenatal Diagnosis, the Human Side*. Abramsky L and Chapple J (eds.) London, UK: Chapman & Hall.

Chilaka VN, Konje JC, Stewart CR, Narayan H and Taylor DJ (2001) Knowledge of Down's syndrome in pregnant women from different ethnic groups. *Prenatal Diagnosis* **21**(3): 159–164.

Cochrane AL and Holland WW (1971) Validation of screening procedures. *British Medical Bulletin* **27**: 3.

Crang-Svalenius E, Dykes AK and Jorgensen C (1998) Factors influencing informed choice of prenatal diagnosis: women's feelings and attitudes. *Fetal Diagnosis* **13**(1): 53–61.

Ford C, Moore AJ, Jordan PA, *et al.* (1999) The value of screening for Down's syndrome in a socioeconomically deprived area with high ethnic population. *British Journal of Obstetrics & Gynaecology* **105**(8): 855–859.

Gilbert RE, Augood C, Gupta R, *et al.* (2001) Screening for Down's syndrome: effects, safety, and cost effectiveness of first and second trimester strategies. *British Medical Journal* **323**: 423.

Gray JAM (1996) *Dimensions and Definitions of Screening*. Milton Keynes, UK: NHS Executive Anglia and Oxford, Research and Development Directorate.

Holland WW and Stewart S (1990) *Screening in Healthcare: Benefit or Loss*. The Nuffield Provincial Hospitals Trust, UK.

Howe D (2001) Modeling does not reflect reality. *British Medical Journal* **323** (7310): 423.

Jorgensen FS (1996) Attitude of pregnant women to prenatal screening. *Ugeskrift for Laeger* **158**(39): 5447–5452.

Raffle AE (2001) Information about screening – is it to achieve high uptake or to ensure informed choice? *Health in Expectancy* **4**(2): 92–98.

Reynolds TM (2000) Down's syndrome screening: a controversial test, with more controversy to come! *Journal of Clinical Pathology* **53**(12): 893–898.

Sackett DL and Holland WW (1975) Controversy in the detection of disease. *Lancet* **2**: 357–359.

Searle J (1997) Routine antenatal screening: not a case of informed choice. *Australian and New Zealand Journal of Public Health* **21**(3): 268–274.

Stemerding D and van Berkel D (2001) Maternal serum screening, political decision-making and social learning. *Health Policy* **56**(2): 111–125.

Wald NJ (ed.) (1984) *Antenatal and Neonatal Screening*. Oxford, UK: Oxford University Press.

Wilson JMG and Jungner G (1968) *Principles and Practice of Screening for Disease*, Public Health Paper No. 34. Geneva: WHO.

Web Links

The National Screening Committee Library. This is a library of literature published by the UK National Screening Committee of the Department of Health. It advises ministers on matters related to screening programmes
http://www.doh.gov.uk/nsc/library/lib_ind.htm

Public Health Genetics Unit. The UK National Health Service funds the Public Health Genetics Unit. Its Web site provides news and information about advances in genetics and their impact on public health and the prevention of disease
http://www.medinfo.cam.ac.uk/phgu

Reprogenetics: Visions of the Future

Regine Kollek, *University of Hamburg, Hamburg, Germany*

As humans acquire the capacity to modify their germ-line DNA for purposes of enhancement, one can imagine a variety of different scenarios for the future. Some of these possible futures may have an important social context and an influence on how genetic enhancements may be socially valued.

Intermediate article

Article contents

- Background
- Context
- Health Purposes: The Slippery Slope
- Enhancing Looks: Preprogrammed Disappointment
- Controlling Behavior: Society Strikes Back
- Modesty and Wisdom: Let Future Generations Decide

Background

The term reprogenetics characterizes the conceptual and practical convergence of two lines of medical and scientific developments: human genetics, spurred by the theoretical and technical progress of the Human Genome Project, and artificial reproductive technologies, through which human embryos can be created *in vitro* and therefore are accessible to further interventions.

Existing technologies which are applied to introduce inheritable genetic modifications in such embryos still have many flaws and therefore may lead to physical problems in the future individual. Embryos or fetuses with abnormal features or developments following germ-line alterations would probably be discarded or aborted; this practice might lead to the view that such interventions are irresponsible and unethical. Furthermore, it is quite possible that the complexities and dynamics of the human genome and of genotype–environment interactions will preclude anything but the manipulation of the most simple phenotypes in a controlled manner. For the sake of argument, however, it is assumed that germ-line engineering procedures in humans are no more risky than natural conception. Parents could then choose to enhance their embryos with genes that they themselves do not carry.

If this could be done safely, what kind of possible futures might be imagined for the carriers of modified genes, enhanced traits or capabilities, and for the society they live in?

Context

Several problems arise when we attempt to assess future consequences of germ-line alterations. The first is that enhancement can mean different things in different interpretative and social contexts (Juengst, 1998). Second, benefits or harm resulting from interventions aimed at enhancing features or traits cannot be defined in absolute terms. Their interpretation depends to a large extent on the social or cultural contexts shaping the norms according to which human features or abilities are valued. Although these norms are generally characterized by considerable stability, they are also prone to change over time. Therefore, features of the human body or mind which are highly valued today might be irrelevant or even have negative social connotations in future societies.

The third problem is that we can hardly foresee how the effects of the broad implementation of germ-line alterations will interact with society and possibly change it. Reasonable guesses about future developments can therefore only be made by presuming that future societies will not differ much from current societies in their main characteristics. Scenarios developed against the background of liberal Western societies would not be coherent or meaningful in the context of paternalistic, authoritarian cultures or even dictatorships. Based on these assumptions it seems reasonable to assume that the application of future reprogenetics will be shaped by a plurality of ethical standpoints, by a scarcity of resources, by commercialism and by globalization (Norgren, 1998).

Furthermore, modern industrialized societies are characterized by a high degree of individualism. Old belief systems and guiding norms are losing ground and being replaced by pluralism with respect to religion or secular normative views. Autonomy and freedom of choice are among the most highly regarded liberties.

Lifestyles are no longer bound to conventional roles but multiply as societies become more and more heterogeneous by education, permeability of traditional social structures, migration, mobility and so on. These developments pose quite a challenge to the individual. The lack of orientation and social embeddedness resulting from these changes needs to be balanced. In this situation, private and personal relationships become more important. Since large families have become increasingly rare, relationships between different generations are emphasized as strongly as the perceived dependency and feeling of mutual responsibility between members of vertical family lines. These developments are promoted by increasing life expectancy in Western industrialized societies.

Reprogenetics now promises to provide the means to stabilize and strengthen such family lines. It expands control not only over the process of procreation but also to future generations, providing an element of security and orientation in an insecure and ambiguous world (Kollek, 2000). It not only allows couples to choose freely the time for procreation and the number of children, but also the health status, sex, and – at least the assumption here – behavioral traits and other phenotypes of their offspring, independent of their own genetic endowment. Reprogenetics therefore seems to match perfectly individualized lifestyles characterized by the need and the desire to plan and structure one's own life course rationally and in accordance with the requirements of modern society. The question is whether reprogenetics can keep its promise of security and control or whether old uncertainties will simply be replaced by new ones.

Health Purposes: The Slippery Slope

It is likely that germ-line interventions will first be carried out in order to prevent severe inherited disorders. Imagine the extremely rare case of both partners being homozygous carriers of the same recessive mutation, leading inevitably to a pathological condition in every child. If such a couple insists on having a healthy child of its own, germ-line therapy may be regarded as the method of choice.

Once it has been proven that germ-line alterations for therapeutic purposes are safe – at least in the individual directly manipulated – other applications will soon be considered, some of which will inevitably cross the line between treatment and enhancement, a line which is difficult to draw (Parens, 1998). In the context of prevention, the introduction of genes that confer resistance against widespread infectious diseases such as AIDS or offer protection against cancer or enhance the activity of detoxifying genes could be defended. Germ-line immunization to protect against severe diseases could be offered to groups or communities at risk. However, since germ-line interventions still require one or more cycles of *in vitro* fertilization (IVF), including hormone treatment, health insurers may soon be forced to refuse coverage for such treatments because too many people apply for it.

Only individuals or families with considerable financial means would then be able to afford such treatment, which might dramatically widen the gap in health status and life expectation between the more and the less well off (Silver, 1997), resulting in increased social tensions. In countries with public retirement insurance, well-off people living significantly longer than average may even be accused of exploiting the social system and living at the expense of others. To ensure equality of opportunity, some people may then claim a right to become enhanced and force the state to grant equal access to such treatments to everyone.

Financial restrictions and/or cultural preferences may also lead to unequal gender distribution of benefits from germ-line interventions. If a couple can afford genetic enhancement of one child only, they may invest in a boy rather than in a girl. This may especially be true if birth rates are controlled, or in cultures preferring a boy for the first-born child.

Enhancing Looks: Preprogrammed Disappointment

As a result of a lack of consensus on reproductive matters in current and future democratic societies, and thanks to the support of proponents of a radical reproductive autonomy view (Robertson, 1994) and the effects of market forces, inheritable genetic modifications may not only be offered for health-related purposes, but also for cosmetic reasons. Some people would consider altering their skin color or body shape and size. Since favorable physical attributes are regarded as one of the preconditions for social and economic success, a woman undergoing IVF treatment because of fertility problems may also be interested in treatments enhancing the physical attributes of her future child.

Given predictable outcomes, this may finally increase the number of good-looking men and women as compared to current standards. Will this make them happier or even more successful? There are grounds for doubt. Such developments could, for example, reinforce social prejudice by promoting stereotypes of race and gender and supporting questionable standards for normalcy. Offering people opportunities to choose the phenotype of a child may therefore result in psychosocial pathologies, including deeper class and racial divisions in society (Krimsky, 2000).

But even more benign scenarios may have ambiguous outcomes. For instance, features conferred by genetic enhancement follow the ideals of others and thus differ from cosmetic surgery, where people decide for themselves. Children quite often do not have the same ideas about beauty as their parents and may be very dissatisfied with parents' choices. The resulting frustration may be even more profound than it would be if one's physical appearance was solely the result of natural processes. In the latter case, no one could be held responsible whereas in the case of disliked physical enhancements, parents or grandparents can be blamed for their decisions.

Ideals of physical attractiveness may also change with time. It is not unlikely that fashion and the media will some day become bored with interchangeable beauties and replace them with models considered more attractive, because they are characterized by natural heterogeneity in shape, size and proportions and endowed with interesting deviations from previous norms. In that case, cosmetic enhancement will prove to have been, at best, a waste of money. At worst, it will have resulted in a group of unsatisfied beauties who wish they had been born as naturals or who might even sue their parents for denying them a genetic outfit that is the product of chance. Living with the knowledge that one's own hereditary factors have been programmed therefore may restrict an individual's right to an open future and undermine the essentially symmetrical relations between free and equal human beings (Habermas, 2003).

Controlling Behavior: Society Strikes Back

Violence can be imagined as a growing problem in future societies. Since governments and influential social groups in general are averse to social changes, because they fear the loss of privileges, they may provide massive funding for research in neurogenetics and behavioral genetics. Despite substantial drawbacks – many promising results cannot be replicated in later research – geneticists finally may succeed in convincing politicians that the diverse activities labeled under the umbrella term 'violence' – from rape to terrorism – are all manifestations of an individual's genetically determined aggression, and not regular human reactions to social and economic oppression.

Trials will be initiated involving twinned embryos from couples undergoing IVF treatment, in order to test the hypothesis of neurogenetic determinism. Couples will volunteer for this experiment because they are eager to support science and to promote societal evolution. In each twin-pair, one individual will have his or her neurometabolic outfit genetically altered to reduce aggressive tendencies. After several years of monitoring, however, it may become evident that behavioral outcomes do not correlate with genetic interventions. For example, many of the nonmanipulated twins do not exhibit violent behavior, even under adverse circumstances, and many of those manipulated unexpectedly do. Interestingly, among those twins who carry germ-line alterations and exhibit less aggressive behavior, fewer individuals may prove to be successful in business or medicine, presumably because they lack the aggressive approach which is needed to survive in such competitive environments.

Children who realize that they have been intentionally manipulated by their parents to suit their interests and the interests of society may experience psychological distress. They may feel that their destiny has been arbitrarily preprogrammed. Some will deliberately choose to behave in contrast to the expected outcome of genetic modification.

Neuroscientists and behavioral geneticists will finally agree with psychologists and sociologists that the expression of violence and how it is valued depends, to a large extent, on the social context. The same violent act may be socially desirable or condemned; a soldier shooting a suspected terrorist may receive a medal or be charged with murder (Rose, 1998). Since results of germ-line interventions aimed at controlling personality traits are so much amenable to societal values, there is a growing consensus that socially complex interactive processes can neither be reduced to the properties of individual neurometabolic patterns or genes, nor manipulated on the level of genetics with controlled outcomes. After several decades of experimentation, the practice of genetically interfering with behavioral traits will be abandoned.

Modesty and Wisdom: Let Future Generations Decide

One could also imagine a completely different scenario. In the coming 10 or 20 years, not only will knowledge about the human genome increase dramatically, but also our understanding of the complexities of genotype–phenotype relationships. We will learn how the expression of genes is modulated by other genes and environmental factors. Various neuroscientific disciplines will show how personal and social parameters retroact on the expression of genes and behaviors. Although the technology and skills needed to perform genetic modifications with high precision will have been developed, it will also have become evident that the resulting social outcomes will not be predictable. Thanks to this insight, researchers and specialists

in molecular and reproductive medicine will become increasingly reluctant to interfere with the human germ-line, especially for the purpose of enhancement. A call for an international ban on germ-line interventions will enjoy massive support in many countries on the part of a broad spectrum of societal groups as well as scientists. Such an agreement will be signed and become internationally effective before the rights of individuals yet to be born have been violated.

See also

Eugenics: Contemporary Echoes
Genetic Enhancement
Genetic Enhancement: The Role of Parents
Genetics and the Control of Human Reproduction
Reproductive Choice

References

Habermas J (2003) *The Future of Human Nature*. Cambridge, UK: Polity Press.

Juengst ET (1998) What does enhancement mean? In: Parens E (ed.) *Enhancing Human Traits: Ethical and Social Implications*, pp. 29–47. Washington, DC: Georgetown University Press.

Kollek R (2000) Technicalization of human procreation and social living conditions. In: Haker H and Beyleveld D (eds.) *Ethics in Genetics in Human Procreation*, pp. 131–152. Aldershot, UK: Ashgate.

Krimsky S (2000) The psychosocial limits on human germ-line modifications. In: Stock G and Campbell J (eds.) *Engineering the Human Germ Line: An Exploration of the Science and Ethics of Altering the Genes We Pass to Our Children*, pp. 104–107. New York, NY: Oxford University Press.

Norgren A (1998) Reprogenetics policy: three kinds of models. *Community Genetics* **1**: 61–70.

Parens E (1998) Is better always good? The enhancement project. In: Parens E (ed.) *Enhancing Human Traits: Ethical and Social Implications*, pp. 1–28. Washington, DC: Georgetown University Press.

Robertson JA (1994) *Children of Choice: Freedom and the New Reproductive Technologies*. Princeton, NJ: Princeton University Press.

Rose SPR (1998) Neurogenetic determinism and the new euphenics. *British Medical Journal* **317**: 1707–1708.

Silver LM (1997) *Remaking Eden: Cloning and Beyond in a Brave New World*. New York, NY: Avon Books.

Further Reading

Billings P, Hubbard R and Newman SA (1999) Human germ-line gene modification: a dissent. *Lancet* **353**: 1873–1875.

Buchanan A, Brock DW, Daniels N and Wikler D (2000) *From Chance to Choice: Genetics and Justice*. Cambridge, UK: Cambridge University Press.

Haker H and Beyleveld D (eds.) (2000) *The Ethics of Genetics in Human Procreation*. Aldershot, UK: Ashgate.

Kitcher P (1996) *The Lives to Come: The Genetic Revolution and Human Possibilities*. London, UK: Penguin Books.

Parens E (ed.) (1998) *Enhancing Human Traits: Ethical and Social Implications*. Washington, DC: Georgetown University Press.

Shapiro MH (200) Human enhancement uses of biotechnology, policy, technological enhancement and human equality. In: Murray TH and Mehlman MJ (eds.) *Encyclopaedia of Ethical, Legal and Policy Issues in Biotechnology*. vol. 1, pp. 527–548. New York, NY: John Wiley & Sons.

Stock G and Campbell J (eds.) (2000) *Engineering the Human Germ Line: An Exploration of the Science and Ethics of Altering the Genes We Pass to Our Children.* New York, NY: Oxford University Press.

Walters LR and Palmer JG (1997) *The Ethics of Human Gene Therapy.* New York, NY: Oxford University Press.

Web Links

Frankel MS and Chapman AR (2000) Human inheritable genetic modifications: assessing scientific, ethical, religious, and policy issues. Prepared by the American Association for the Advancement of Science
http://www.aaas.org/spp/dspp/sfrl/germ line/main.htm

Nuffield Council of Bioethics (2002) Genetics and human behaviour: the ethical context. Published by the Nuffield Council of Bioethics, London
http://www.nuffieldbioethics.org

Restriction Enzymes

Mala Mani, *Johns Hopkins University, Baltimore, Maryland, USA*

Karthikeyan Kandavelou, *Johns Hopkins University, Baltimore, Maryland, USA*

Joy Wu, *Johns Hopkins University, Baltimore, Maryland, USA*

Srinivasan Chandrasegaran, *Johns Hopkins University, Baltimore, Maryland, USA*

Advanced article

Article contents

- Introduction
- Role and Nature of Restriction Enzymes
- Frequency of Sites in Human Deoxyribonucleic Acid
- Complete Digestion of Deoxyribonucleic Acid
- 'Star' Activity
- Effect of Methylation of Deoxyribonucleic Acid
- Partial Digestion of Deoxyribonucleic Acid
- Digestion of Deoxyribonucleic Acid Embedded in Agarose
- Summary

Restriction enzymes have played a key role in the completion of the initial sequencing and analysis of the first draft of the human genome. They were important tools in the cloning, analysis and sequencing of the various deoxyribonucleic acid fragments as well as in the construction of a physical map of the human genome.

Introduction

Restriction enzymes, also known as restriction endonucleases, are molecular scissors that recognize specific sequences in deoxyribonucleic acid (DNA) and cleave within or adjacent to these sites. Their discovery over 30 years ago ushered in the revolution of recombinant DNA technology in molecular biology. In 1978, Werner Arber, Hamilton Smith and Daniel Nathans were awarded the Nobel prize for medicine for their work on restriction enzymes. (*See* Recombinant DNA.)

Role and Nature of Restriction Enzymes

Restriction enzymes are endonucleases that recognize specific DNA sequences and make double-stranded cuts (New England Biolabs, 2000–2001; Ausubel *et al.*, 2001). They require Mg^{2+} as a cofactor. The common type II enzymes recognize specific DNA sequences with a dyad axis of symmetry, called palindromes, and cleave within or adjacent to these sites. The phosphodiester bond cleavage results in 3' hydroxy and 5' phosphate termini. Some enzymes cleave at the axis of symmetry to yield 'flush' or 'blunt' ends, while others make staggered cuts to yield single-stranded 3' or 5' ends known as 'cohesive' termini (**Table 1**). DNA fragments with complementary ends may be ligated to each other by using DNA ligases to produce recombinant DNA molecules. The restriction enzymes have

Table 1 Examples of type II and type IIs restriction endonucleases and methylases

R–M system	Recognition site/cleavage site ($\downarrow$) of the restriction endonuclease	Recognition site/methylation site (m) of the corresponding modification enzyme or methylase
*Alu*I	5' AG$\downarrow$CT 3' 3' TC$\uparrow$GA 5'	5' AG^{m}CT 3' 3' T^{m}CGA 5'
*Bam*HI	5' G$\downarrow$GATCC 3' 3' CCTAG$\uparrow$G 5'	5' GGAT^{m}CC 3' 3' C^{m}CTAGG 5'
*Eco*RI	5' G$\downarrow$AATTC 3' 3' CTTAA$\uparrow$G 5'	5' GA^{m}ATTC 3' 3' CTT^{m}AAG 5'
*Pst*I	5' CTGCA$\downarrow$G 3' 3' G$\uparrow$ACGTC 5'	5' CTGC^{m}AG 3' 3' G^{m}ACGTC 5'
*Fok*I[a]	5' GGATG(N$_9$)$\downarrow$ 3' 3' CCTAC(N$_{13}$)$\uparrow$ 5'	5' GG^{m}ATG 3' 3' CCT^{m}AC 5'

[a]Type IIs R–M system.

made it possible to purify homogeneous fragments of defined length by molecular cloning, which are then used as substrates in a wide variety of other experiments. The specific cut sites provide unique molecular landmarks for obtaining a physical map of DNA. Thus, restriction enzymes have proven to be the most essential tools for analyzing and manipulating DNA. There are also numerous enzymes that recognize an asymmetric sequence and cleave a short distance from the sequence. These are termed type IIs enzymes. These enzymes do not recognize any specific sequence at the cleavage site. (*See* Genome Map; Restriction Fragment Length Polymorphism (RFLP).)

Type II enzymes are homodimers, with subunits of 25–50 kDa. Since most type II enzymes recognize palindromic sequences with a twofold rotational symmetry, it was expected that the two enzyme subunits arranged symmetrically bind the recognition sites. Crystal structures of several enzyme–cognate binding site complexes have shown this to be true (Newman *et al.*, 1994). Since the type IIs enzymes bind an asymmetric sequence, they were believed to bind DNA as a monomer. The crystal structure of the *Fok*I–DNA complex has shown this to be correct (Wah *et al.*, 1997). Furthermore, the type IIs enzymes appear to contain two separate protein domains: one responsible for DNA sequence recognition and the other for DNA cleavage (Li *et al.*, 1992). The modular nature of *Fok*I restriction endonuclease has made it possible to construct chimeric nucleases with novel sequence specificities by linking other DNA-binding proteins to the *Fok*I cleavage domain (Kim *et al.*, 1996). These engineered chimeric nucleases have been used to stimulate homologous recombination through targeted cleavage in frogs' eggs (Bibikova *et al.*, 2001). In contrast, the DNA recognition and DNA cleavage functions of type II enzymes overlap each other and are not separable. (*See* DNA Recombination; DNA Repair; Gene Therapy; Gene Therapy: Motivations for Research; Transgenic Mice.)

Screening of thousands of bacteria and culture collections from around the world has yielded nearly 3000 restriction enzymes, exhibiting over 200 different sequence specificities (New England Biolabs, 2000–2001). *Chlorella* viruses that infect the unicellular eukaryotic green algae have been shown to encode many restriction enzymes. (*See* Genomic DNA: Purification; Polymerase Chain Reaction (PCR): Design and Optimization of Reactions.)

The biological function of restriction enzymes appears to be the protection of host cells from foreign DNA. This is consistent with the observation that most of the enzymes tested in their natural hosts restrict (i.e. cleave) infecting viral or plasmid DNA.

Thus, restriction enzymes prevent infecting foreign DNA from successfully replicating within the host cell.

How do the host organisms that make restriction enzymes protect their own DNA from cleavage? This is achieved by making a corresponding modification enzyme, also known as DNA methyltransferase, that protects their own DNA from cleavage by the restriction enzymes. The methyltransferases recognize the same DNA sequence as the corresponding restriction enzyme, but instead of cleaving DNA, they methylate one of the bases within their recognition sites in each DNA strand (Cheng and Roberts, 2001). The restriction enzyme and its cognate modification methyltransferase together form a restriction–modification (R–M) system. The modification methyltransferases require only *S*-adenosyl methionine (SAM) for their activity. They transfer the methyl group from SAM to either a cytosine or adenine base within the recognition sequence. This methylation renders the site insensitive to cleavage by the corresponding restriction endonuclease. Three types of methylations are used to provide protection against the cognate restriction enzyme: N_6 methyladenine, N_4 methylcytosine and C_5 methylcytosine. (*See* DNA Methylation in Development; Methylated DNA-binding Proteins; Methylation-mediated Transcriptional Silencing in Tumorigenesis.)

Frequency of Sites in Human Deoxyribonucleic Acid

The G+C content of the human genome is approximately 41%. From the draft human genome sequence, it is clear that the local G + C content shows substantial variation from its genome-wide average of 41%. There are huge regions (> 10 megabases (Mb)) with a G + C content far from average. The distal 48 Mb of chromosome 1p has an average G+C content of 47.1% while chromosome 13 has a 40 Mb region with only 36% G + C content (International Human Genome Sequencing Consortium, 2001). Furthermore, the CpG dinucleotide is about five times rarer than expected from the G + C content. Also, most of the CpG dinucleotides are methylated on the cytosine base. However, the human genome contains many 'CpG islands' in which CpG dinucleotides are not methylated and these occur at a frequency closer to that predicted by the local G + C content. The CpG islands are of particular interest because many occur at 5′ ends of genes, that is, the regulatory regions of genes. (*See* GC-rich Isochores: Origin; L Isochore Map: Gene-poor Isochores; Human Genome: Draft Sequence.)

These two factors, the G+C content and CpG dinucleotide distribution, affect the frequency of restriction sites in human DNA. Restriction enzyme recognition sequences that contain CpG are very rare in human DNA and in mammalian genomes in general (New England Biolabs, 2000–2001). Since most CpG sequences are methylated in human DNA, almost all the enzymes with CpG in their recognition sequence cannot cleave if CpG is methylated. However, cleavage will occur at sites where CpG is not methylated. The average fragment sizes that result from these digestions are therefore quite large. (*See* CpG Islands and Methylation; CpG Dinucleotides and Human Disorders; DNA Demethylation.)

Complete Digestion of Deoxyribonucleic Acid

The efficiency of restriction endonuclease cleavage is dependent on the purity of the DNA and buffer conditions like pH and salt concentration. Impurities like protein, phenol, chloroform, ethylenediamine-tetraacetic acid (EDTA), etc., found in some DNA preparations, inhibit the cleavage activity of restriction endonucleases (New England Biolabs, 2000–2001; Ausubel *et al.*, 2001). Typically, Tris (2-amino-2-hydroxymethylpropane-1,3-diol) Cl buffer is used to maintain optimal pH for enzyme activity. Some enzymes are very sensitive to the concentration of K^+ or Na^+ ions. Others are active over a wide range of ionic strengths. For each enzyme the optimal reaction condition and buffer solution are provided by the manufacturer. Recommended assay buffers in some cases differ widely in terms of pH and specific ion requirements. Upon complete digestion, the cleavage products should be present in equimolar amounts, and the band intensity as visualized by staining with ethidium bromide should be proportional to the fragment length (Ausubel *et al.*, 2001). Bands that appear weaker than expected for the fragment length are probably due to incomplete digestion; bands that appear stronger than expected for the fragment length are probably due to two or more fragments of similar size. The supercoiled forms of plasmid DNA often require more units of enzyme for complete digestion than linear DNA. (*See* Gel Electrophoresis.)

'Star' Activity

Some restriction enzymes relax their recognition sequence specificity under nonoptimal reaction conditions. These include high glycerol concentrations, high enzyme concentration, high pH or low ionic strength (New England Biolabs, 2000–2001). In these instances the enzyme cleaves sequences which are similar but not identical to their defined recognition sequence. This altered specificity is called the 'star' activity. For example, *Eco*RI normally cleaves at the canonical site G↓AATTC. Under high pH and low ionic strength, *Eco*RI cleaves the sequence N↓AATTN. Recent studies indicate that *Eco*RI star activity results in the cleavage of any site that differs from the canonical recognition site by a single base substitution (New England Biolabs, 2000–2001). This 'star' activity is not usually observed under the optimal buffer conditions recommended by the manufacturer.

Effect of Methylation of Deoxyribonucleic Acid

As indicated earlier, restriction enzyme cleavage is blocked when the recognition sequence is methylated by the cognate methylase. Methylation at other bases or overlapping sites can inhibit cleavage, or reduce the rate and extent of cleavage (New England Biolabs, 2000–2001). The rate of cleavage may also be affected by the DNA sequence flanking the recognition site, even more so by the methylation of the DNA sequence flanking the recognition site (New England Biolabs, 2000–2001).

Many laboratory strains of *Escherichia coli* have been shown to contain three site-specific DNA methylases, namely Dam methylase, Dcm methylase and *Eco*KI methylase (**Table 2**). Dam methylase transfers a methyl group from SAM to the N_6 position of the adenine residue in the sequence GATC; the Dcm methylase methylates the internal cytosine residues in the sequences CCAGG and CCTGG at the C_5 position while *Eco*KI methylase modifies adenine residues in the sequences AAC(N_6)GTGC and GCAC(N_6)GTT. Some or all of the sites for a restriction endonuclease may be refractory to cleavage due to overlapping methylation sites when isolated from strains

Table 2 Site-specific DNA methylases that are present in most commonly used laboratory strains of *E. coli*

DNA methylase	Target site(s)/methylation site (m)
Dam	5′ G^{m}ATC 3′
Dcm	5′ C^{m}CAGG 3′
	5′ C^{m}CTGG 3′
M. *Eco*KI	5′ A^{m}AC(N_6)GTGC 3′
	5′ GC^{m}AC(N_6)GTT 3′

expressing the Dam or Dcm methylases. Almost all cloning strains are Dam$^+$ Dcm$^+$ and many are M$^+$ EcoKI (New England Biolabs, 2000–2001).

Partial Digestion of Deoxyribonucleic Acid

Partial or incomplete digestion of DNA indicates that the enzyme has cleaved at only a subset of restriction sites. Under optimal conditions for enzyme activity, partial digestion of DNA suggests that the substrate is not very pure, that the specific activity of the enzyme is low, or that overlapping methylation sites are present. The last is particularly true of human DNA, in which most but not all CpG dinucleotide sequences are methylated.

Under some circumstances, like restriction mapping or cloning DNA segments that contain the cloning sites internally within the segments, it is useful to produce partial digestion of DNA (Ausubel et al., 2001). A physical map of DNA entails construction of an accurate map of recognition sites where restriction endonucleases cleave DNA. This is also commonly referred to as restriction mapping, and it is based upon the cleavage of DNA at specific sites with restriction enzymes, followed by the determination of the length of the DNA fragments by gel electrophoresis. The map is built after digesting the DNA of interest with a variety of enzymes. An alternative method of restriction mapping utilizes partial endonuclease digestions (Ausubel et al., 2001). In this case, the DNA fragment of interest is radiolabeled at one of its two ends. It is gel purified and subjected to partial cleavage by a restriction endonuclease. The resulting products are then analyzed by gel electrophoresis, which enables one to define the distance of restriction enzyme sites from the labeled end. This technique for partial digestion of DNA fragments is relatively simple and rapid; it is ideally suited to fine structure restriction mapping. The analysis is also rather straightforward and is not complicated by the presence of many cleavage sites for a given enzyme. (See Restriction Mapping.)

Digestion of Deoxyribonucleic Acid Embedded in Agarose

Restriction endonuclease digestion of agarose-embedded DNA (also referred to as agarose plugs) has proven to be a valuable tool in the physical mapping of chromosomes. The resolution of large DNA molecules up to 1 Mb in length is possible using pulse field gel electrophoresis. Handling and manipulation of large DNA molecules of megabase lengths in solution is extremely difficult. The mechanical shear forces lead to random double-strand breaks. DNA can be embedded in agarose gel matrix to avoid the fragmentation of molecules of this size. To generate agarose plugs of genomic DNA, intact cells are first immobilized in agarose (New England Biolabs, 2000–2001). Their cell walls are then disrupted to remove the cellular proteins. The resulting agarose-embedded DNA can be manipulated and analyzed in the same way as the naked DNA in solution. Most restriction enzymes cleave DNA embedded in agarose (New England Biolabs, 2000–2001). However, diffusion into agarose plugs is slow; therefore, higher concentrations of enzymes and longer incubation times are required for complete DNA digestion as compared with solution. Highly purified restriction endonucleases devoid of any contaminating nucleases are critical for these experiments (New England Biolabs, 2000–2001). (See Chromosome 22: Sequencing.)

Summary

Restriction endonucleases have played an important role in the production and analysis of the draft human genome. Two approaches, namely 'hierarchical shotgun sequencing' (International Human Genome Sequencing Consortium, 2001), and 'genome-wide shotgun sequencing' (Venter et al., 2001), were used to produce the initial sequencing and analysis of the human genome by the International Human Genome Sequencing Consortium and Celera groups respectively. The former method, also referred to as the clone-based approach (International Human Genome Sequencing Consortium, 2001), involved breaking up the human genome into DNA segments of about 150 000 bp by partial digestion using site-specific restriction endonucleases. The large segments were cloned into bacterial artificial chromosomes (BACs). Individual BAC clones were sheared into smaller fragments, then cloned and sequenced. Thus, the public project's sequencing strategy involved producing a physical map of the human genome and then painting the sequence on it. Celera's proprietary sequencing data were not available to be included in such a draft. The latter approach employed by Celera (Venter et al., 2001) involved preparation of high-quality plasmid libraries of the human genome in a variety of insert sizes: 2, 10 and 50 kb. Pairs of sequence reads were obtained from both ends of each insert. Powerful computers were then used to order and assemble the draft sequences of the human genome. This approach also included all the sequences from the freely available public database to produce the draft genome. Thus, restriction enzymes have not only played an important role in the cloning and sequencing of various DNA segments but also in the production of a physical map of the human genome. (*See* Genome Sequencing; Gene Therapy: Technology; Shotgunning the Human Genome: A Personal View.)

Acknowledgement

The work in Professor Chandrasegaran's laboratory is supported by a grant (GM53923) from the National Institutes of Health, USA.

See also
Restriction Mapping

References

Ausubel FN, Brent R, Kingston RE, *et al.* (eds.) (2001) Enzymatic manipulation of DNA and RNA. In: *Current Protocols in Molecular Biology*, vol. 1, chap. 3, pp. 3.0.1–3.19.8, New York, NY: John Wiley.

Bibikova M, Carrol D, Segal DJ, *et al.* (2001) Stimulation of homologous recombination through targeted cleavage by chimeric nucleases. *Molecular and Cellular Biology* 21: 289–297.

Cheng X and Roberts RJ (2001) AdoMet-dependent methylation, DNA methyltransferases and base flipping. *Nucleic Acids Research* 29: 3784–3795.

International Human Genome Sequencing Consortium (2001) Initial sequencing and analysis of the human genome. *Nature* 409: 860–921.

Kim Y-G, Cha J and Chandrasegaran S (1996) Hybrid restriction enzymes: zinc finger fusions to *Fok*I cleavage domain. *Proceedings of the National Academy of Sciences of the United States of America* 93: 1156–1160.

Li L, Wu LP and Chandrasegaran S (1992) Functional domains in *Fok*I restriction endonuclease. *Proceedings of the National Academy of Sciences of the United States of America* 89: 4275–4279.

New England Biolabs (2000–2001) *New England Biolabs Catalog and Technical Reference, 2000–2001*. Beverly, MA: New England Biolabs. (See the appendix for detailed information about restriction–modification enzymes.)

Newman M, Strzelecka T, Dornor LF, Schildkraut I and Aggarwal AK (1994) Structure of restriction endonuclease *Bam*HI and its relationship to *Eco*RI. *Nature* 368: 660–664.

Venter JC, Adams MD, Myers EW, *et al.* (2001) The sequence of the human genome. *Science* 291: 1304–1351.

Wah DA, Hirsch JA, Dorner LF, Schildkraut I and Aggarwal AK (1997) Structure of the multimodular endonuclease *Fok*I bound to DNA. *Nature* 388: 97–100.

Further Reading

Aggarwal AK and Wah DA (1998) Novel site-specific DNA endonuclease. *Current Opinions in Structural Biology* 8(1): 19–25.

Chandrasegaran S (2001) Restriction enzymes. *Encyclopedia of Life Sciences*, http://www.els.net. London, UK: Nature Publishing Group.

Chandrasegaran S and Reddy S (2001) Ligation: theory and practice. *Encyclopedia of Life Sciences*, http://www.els.net. London, UK: Nature Publishing Group.

Chandrasegaran S and Smith J (1999) Chimeric restriction enzymes: what is next? *Biological Chemistry* 380: 841–848.

Gormley NA, Watson MA and Halford SE (2001) Bacterial restriction–modification systems. *Encyclopedia of Life Sciences*, http://www.els.net. London, UK: Nature Publishing Group.

Klug A (1999) Zinc finger peptides for the regulation of gene expression. *Journal of Molecular Biology* 293: 215–218.

Pabo CO, Peisach E and Grant A (2001) Design and selection of novel Cys2 His2 zinc finger proteins. *Annual Review of Biochemistry* 70: 313–340.

Pingoud A and Jeltsch A (2001) Structure and function of type II restriction endonuclease. *Nucleic Acids Research* 29: 3705–3727.

Roberts RJ and Cheng X (1998) Base flipping. *Annual Review of Biochemistry* 67: 181–198.

Roberts RJ and Hulford SE (1993) Type II restriction endonuclease. In: Linn SM, Lloyd RS and Roberts RJ (eds.) *Nuclease*, pp. 35–38. Cold Spring Harbor, NY: Cold Spring Harbor Laboratory Press.

Restriction Fragment Length Polymorphism (RFLP)

Paul H Dear, *Medical Research Council Laboratory of Molecular Biology, Cambridge, UK*

Introductory article

A restriction fragment length polymorphism is a fragment produced by restriction enzyme digestion of genomic deoxyribonucleic acid, whose length is variable because of a sequence polymorphism that affects the presence or location of a target site for the stated restriction enzyme.

Although of no intrinsic interest (unless the sequence polymorphism happens to lie within a gene or other functional sequence and affects its action), restriction fragment length polymorphisms (RFLPs) are valuable in genetic linkage mapping, because the pattern of inheritance of the distinct alleles can be observed experimentally. If two RFLPs show closely similar patterns of inheritance (i.e. if they cosegregate often), then they can be inferred to lie close together on the same chromosome.

RFLPs were used extensively as genetic linkage markers, particularly in the human genome in the 1980s, but have now been largely superseded by other types of variable sequence (such as microsatellites), which are more highly polymorphic and more easily scored using techniques such as the polymerase chain reaction. (*See* Polymerase Chain Reaction (PCR).)

See also
Genetic Linkage Mapping
Genome Mapping
Microsatellites

Restriction Mapping

Burkhard Tümmler, *Medizinische Hochschule Hannover, Hannover, Germany*

Frauke Mekus, *Medizinische Hochschule Hannover, Hannover, Germany*

Deoxyribonucleic acid is partially or completely digested with one or more restriction endonucleases. The generated fragments are separated by size, and then ordered by combinatorial analysis to generate a physical map.

Advanced article

Article contents
- Basic Principles
- Applications
- Methodology

Basic Principles

Restriction mapping is a generally applicable approach to characterize deoxyribonucleic acid (DNA) molecules by the position of cleavage sites for restriction endonucleases. Thereby the specificity of restriction enzymes is exploited to cleave within or adjacent to an oligonucleotide recognition sequence. Depending on the oligonucleotide usage of the target DNA and the length of the recognition sequence of the enzyme, the DNA is digested into fragments ranging from a 100 base pairs (bp) to more than a 1 000 000 bp in length. The restricted DNA is separated by size and the fragments are subsequently visualized by DNA stain or hybridization with sequence specific probes. Depending on the size and the topology of the DNA molecule and the required resolution of the restriction map, the order of fragments is ascertained from the combinatorial analysis of one or more complete and/or partial digestions with one or more restriction enzymes. The map positions of the restriction sites constitute a physical map of the DNA molecule. The assignment of genes to restriction fragments converts the physical map into a genetic map.

If the sequence of the DNA of interest is known, restriction maps are generated *in silico* by straightforward search for the recognition sequence of the enzyme. However, in most cases restriction mapping is the first step to analyze a DNA molecule of unknown sequence, which is followed by more thorough sequencing or functional analyses.

Applications

Ordered restriction maps provide precise distances between the physical landmarks of restriction sites. Applications range from the characterization of single gene loci or small DNA molecules such as plasmids, phages, cloned or segments amplified by the polymerase chain reaction (PCR) to the construction of whole-genome maps, or high-throughput ordering of large numbers of recombinant clones. Restriction maps facilitate contig formation for gene mapping and for sequencing efforts and can assist in preliminary characterization of genetic alterations such as insertions, deletions and inversions. However, independent confirmation is required to exclude other explanations for such alterations, such as differential methylation or point mutations, which create or destroy recognition sites. (*See* Genome Map; Genome Mapping; Polymerase Chain Reaction (PCR).)

Methodology

The optimum strategy depends on the total size of the DNA target, the size of the region to be mapped, and on the frequency and spacing of restriction sites in the region of interest. Restriction fragment patterns of PCR-amplified DNA fragments, phages, plasmids, recombinant derivatives thereof, and of small genomes (e.g. viruses, bacteria or lower eukaryotes) can be visualized directly by gel staining. With larger genomes, the complex mixture of fragments is blotted on to membranes after electrophoresis, and hybridized with suitable probes to detect fragments of interest. (*See* Genome Map: Resolution; Nucleic Acid Hybridization.)

Purification and separation of DNA

For short-range mapping of restriction fragments with a maximal size of 30 kb, DNA of any source is purified from aqueous solution and the restricted DNA is separated by continuous field agarose gel electrophoresis. Long-range restriction mapping requires the preparation of unsheared intact DNA. Cells are embedded in agarose and then treated with detergents and enzymes which lyse the cell wall and allow proteins and other molecules to diffuse out. The long DNA molecules remain trapped in the agarose and protected by it. Thereafter the DNA is digested with a rare-cutting restriction endonuclease and the generated macrorestriction fragments are separated by pulsed-field gel electrophoresis (PFGE). (*See* Gel Electrophoresis; Genomic DNA: Purification; Kilobase Pair (kbp); Megabase Pair (Mbp).)

Choice of restriction endonuclease

DNA molecules from a few kilobases up to 100 kb in length are the most common substrates for restriction mapping. One should first test the inexpensive restriction endonucleases with 6 base pair (pb) recognition sequences such as *Eco*RI, *Hind*III, *Bam*H1 or *Pst*I. Small DNA fragments of 1 kb or less may be digested with frequently cleaving restriction endonucleases with a four-base recognition sequence. However, restriction mapping with these enzymes is rather expensive and in most cases sequencing will be the more informative, rapid and cost-effective alternative.(*See* Restriction Enzymes.)

Enzymes that recognize sequences longer than 6 bp are potentially useful for long-range restriction mapping, as they tend to cut infrequently. Useful criteria for the selection of enzymes are the GC content, codon usage, degree of methylation, nearest-neighbor data of dinucleotide frequencies and nucleotide sequence data. The methylation-sensitive restriction enzymes with two or more CpG dinucleotides in their recognition sequence such as *Not*I or *Bss*HII are particularly useful for long-range mapping of mammalian DNA. The dinucleotide CpG is fivefold more rare in mammalian

genomes than would be expected from the GC content alone. With the exception of the nonmethylated 'CpG islands' associated with the 5′ end of housekeeping genes, most CpG dinucleotides are methylated at the cytidine residue. Therefore, these enzymes with GC-rich recognition sequences will predominantly cleave in the islands at the 5′ end of genes and generate on the average fragments of several hundred kilobase pairs in size. (*See* CpG Islands and Methylation; DNA Methylation and Mutation; DNA Methylation: Enzymology.)

Restriction mapping strategies

Mapping of DNA molecules up to 100 kb in size

Complete digestion

The DNA of interest is completely digested with several restriction endonucleases, either individually or in combination. Single and double digests are separated in parallel by gel electrophoresis and stained with ethidium bromide. Using DNA fragments of known lengths for comparison, the lengths of all restriction fragments are calculated. The map is built up from the least complex to the most complex fragment patterns. As a complementary approach, complete single or double digestions are hybridized with gene probes or gel-purified restriction fragments obtained by cleavage with another enzyme.

Partial digestion

The linear DNA fragment is labeled at one of its ends and then partially cleaved by a restriction endonuclease. The lengths of the gel-separated labeled fragments indicate the distance of restriction sites from the labeled end. Alternatively, end-labeling may be substituted by hybridization with a fragment end probe. The latter technique is particularly useful for restriction mapping of cloned inserts with the conserved vector ends as probes.

Restriction mapping in complex genomes

Restriction maps of loci or chromosomal regions can be assembled by Southern blot hybridizations of gel-separated partial, single, double or triple restriction digestions. High-resolution maps of DNA segments in genomes smaller than 10 Mb in size are rapidly constructed by Smith–Birnstiel mapping: genomic DNA is completely digested with a rare-cutting restriction endonuclease and then subjected to partial digestion kinetics with a frequently cleaving restriction enzyme. Gel-separated fragments are visualized by hybridization with a rare-cutter fragment end probe. The partial-digestion fragment lengths define the distance of frequent-cutter sites from the rare-cutter site.

Two-dimensional genome mapping

The smaller genomes of bacteria and lower eukaryotes can be mapped by two-dimensional PFGE without any need for supplementary genetic data. The fragment order is established by two strategies: in the case of 'partial–complete mapping', a partial restriction digest is first separated by PFGE in one dimension, then redigested to completion with the same enzyme, and subsequently resolved in the second orthogonal dimension. In the case of 'reciprocal digest mapping', a complete restriction digest with enzyme A is separated in the first dimension, the gel lane is excised and the fragments it contains are redigested with enzyme B, and then separated in the second orthogonal direction. On a separate series of gels, the order of restriction digestions is reversed.

Critical parameters

Insufficient spatial resolution of fragments, the differential susceptibility of recognition sites to cleavage and the close proximity of recognition sites may lead to fragment patterns that are difficult to interpret. Fragments of similar length will give rise to 'comigrating' fragments, which can be identified by their increased signal intensity and band broadening. If the sum of fragment lengths of a complete digestion is larger than expected, some of the bands probably arose from incomplete cleavage. Bands that are fainter than expected for their molecular weight most likely represent partial digestion fragments. If fewer bands than expected are seen in double or triple digestions, some tiny fragments may have run out of the gel or got lost by diffusion. Restriction maps deduced from partial restriction digestions may be incomplete because of an unfavorable combination of 'hard-to-cut' and 'easy-to-cut' sites. In summary, although the principles of restriction mapping are straightforward, the evaluation of the primary data, particularly of a two-dimensional gel, often represents a challenging task.

See also

Restriction Enzymes
Two-dimensional Gel Electrophoresis

Further Reading

Bautsch W, Römling U, Schmidt KD, et al. (1997) Long-range restriction mapping of genomic DNA. In: Dear PH (ed.) *Genome Mapping – A Practical Approach*, pp. 281–313. Oxford, UK: IRL/ Oxford University Press.

Boseley PG, Moss T and Birnstiel ML (1980) 5′ Labelling and poly(dA) tailing. *Methods in Enzymology* **65**: 478–494.

Danna AJ (1980) Determination of fragment order through partial digests and multiple enzyme digests. *Methods in Enzymology* **65**: 449–467.

Römling U, Fislage R and Tümmler B (1996) Macrorestriction mapping and analysis of bacterial genomes. In: Birren B and Lai E (eds.) *Nonmammalian Genome Analysis*, pp. 165–195. San Diego, CA: Academic Press.

Retinitis Pigmentosa

Qing Wang, *The Cleveland Clinic Foundation, Cleveland, Ohio, USA*

Qiuyun Chen, *The Cleveland Clinic Foundation, Cleveland, Ohio, USA*

Intermediate article

Article contents

- Introduction
- Autosomal Dominant RP
- Autosomal Recessive RP
- X-linked RP
- Digenic RP
- Conclusions

Retinitis pigmentosa is a heterogeneous group of retinal dystrophies that are characterized by photoreceptor cell degeneration, night blindness, a gradual loss of peripheral visual fields and eventual loss of central vision. Several genes responsible for the development of retinitis pigmentosa have now been identified and cloned, and these discoveries have defined the genetic pathways for pathogenesis of retinitis pigmentosa.

Introduction

Retinitis pigmentosa (RP) refers to the 'bone spicule' pigmentation that occurs in the retina in some forms of blindness (Donders, 1857). This typical bone spicule-like pigmentation pattern observed in individuals affected with RP is shown in **Figure 1**. Affected individuals suffer from night blindness, constriction and a gradual loss of peripheral visual field (tunnel vision), followed by eventual loss of central vision. Individuals with advanced disease show very small or nondetectable electroretinograms (Heckenlively *et al.*, 1988).

RP affects 1 in 4000 people and is responsible for the visual handicap of 1.5 million individuals worldwide (Haim *et al.*, 1992). The disease shows high genetic heterogeneity, with various inheritance modes

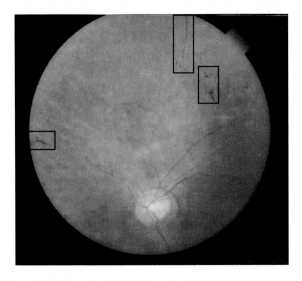

Figure 1 Photograph of the fundus of an individual affected with retinitis pigmentosa who carries the Glu341X mutation in the rhodopsin gene (Zhao *et al.*, 2001). Note the typical bone-spicule pigmentation pattern (boxed). (Figure kindly provided by S. Xiong.)

including autosomal dominant, autosomal recessive, X-linked and digenic forms (Wang *et al.*, 2001). Autosomal dominant RP (adRP) and autosomal recessive RP (arRP) are estimated to each account for about 20% of RP cases, whereas X-linked RP (xlRP) accounts for 10% of cases. Roughly 50% of RP cases are simplex RP, which represents either sporadic RP or arRP. Digenic RP (dRP) is rare. The molecular genetics of RP is summarized below. (For references, see Wang *et al.* (2001) or specific citations in the text.)

Autosomal Dominant RP

HPRP3. The adRP-linked gene on chromosome 1p13–q21 (locus *RP18*) has been identified as the human PRP3 gene (*HPRP3*) (Chakarova *et al.*, 2002). *HPRP3* encodes a protein of 682 amino acids with homology to yeast protein factor PRP3, which is a component of the U4/U6 small nuclear ribonucleoprotein (snRNP) particle involved in precursor messenger ribonucleic acid (pre-mRNA) splicing.

RHO. The *rhodopsin* (*RHO*) gene on chromosome 3q21–24 was the first RP gene to be identified. *RHO* encodes opsin (the apoprotein of rhodopsin), which is the light-sensing molecule in the rod photoreceptor cell. Opsin and the chromophore 11-*cis*-retinal then form rhodopsin in the outer segments of rod photo-receptors. Rhodopsin can activate a G protein (transducin), which initiates the phototransduction cascade. Most *RHO* mutations cause adRP, but a few mutations cause arRP or congenital stationary night blindness.

RDS. Mutations in the *retinal degeneration, slow* (*RDS*) gene on chromosome 6p21 cause adRP. *RDS* encodes a protein that is expressed highly in both rod and cone photoreceptors. RDS protein, which is called peripherin because it localizes to the rim region of outer segment disks, functions to create the bend at the disk rim and to connect the disk to the cytoskeleton at the disk rim. Most *RDS* mutations cause adRP, but some mutations also cause progressive macular degeneration, dRP, pattern dystrophy of retina, vitelliform macular dystrophy, butterfly-shaped pigmentary macular dystrophy and foveomacular dystrophy.

RP1. Mutations in the *retinitis pigmentosa 1* (*RP1*) gene cause the adRP variant linked to chromosome 8p11–21. Mutations in *RP1* have been identified in roughly 4% of unrelated individuals with adRP in North America. *RP1* is expressed specifically in the retina. Studies with *in situ* hybridization have shown that in the retina, *RP1* is expressed in photoreceptor cell bodies and inner segments. *RP1* encodes a protein of 2156 amino acids. The function of the RP1 protein is not clear, but its amino (*N*)-terminus shares homology with the protein kinase doublecortin.

RGR. The *retinal G protein coupled receptor* (*RGR*) gene of the retinal pigment epithelium (RPE) encodes a rhodopsin homolog and has been found exclusively in retinal pigment epithelium and Müller cells. RGR binds all-*trans*-retinal, which is then converted to 11-*cis*-retinal in rhodopsin by photons of light. An insertion of one nucleotide in codon G275 near the 3′ end of the coding region of *RGR* has been identified in a small RP family affected with adRP.

ROM1. The *retinal outer segment membrane protein 1* (*ROM1*) gene on chromosome 11q13 encodes a retinal membrane protein of 37.3 kDa that is localized to the rod photoreceptor outer segments. *ROM1* forms a heterotetramer and larger oligomers at the margins of the rod outer segment disks by interacting with peripherin. *ROM1* mutations have been identified in families with adRP, but the definitive pathogenic nature of these mutations is in question. Studies with mice lacking *ROM1* suggest that *ROM1* is important in regulating disk morphogenesis and the viability of rod photoreceptors – results that are consistent with the possibility that *ROM1* mutations may cause adRP.

NRL. The *neural retina leucine zipper* (*NRL*) gene is responsible for adRP linked to chromosome 14q11.1–q11.2 and encodes a transcription factor of the leucine zipper family that regulates expression of rhodopsin. The NRL protein is expressed in retinal neurons and shares homology with the deoxyribonucleic acid (DNA)-binding domain of the *MAF* oncogene product.

PRPC8. The adRP gene on chromosome 17p13.3 (*RP13*) has been identified as the *PRPF8* gene (McKie *et al.*, 2001). *PRPF8* encodes a large protein of 2335 amino acids with homology to the yeast pre-mRNA splicing factor PRP8. This splicing factor is the core component of the U5 snRNP required for forming the U4/U6 · U5 tri-snRNP involved in spliceosome assembly and pre-mRNA processing.

FSCN2. The *fascin homolog 2, actin-bundling protein, retinal (FSCN2)* gene encodes fascin, a photoreceptor-specific protein of 516 amino acids with actin-binding activity (Wada *et al.*, 2001). Fascin is a member of the family of actin-binding proteins and may be important in the actin-based structures of the connecting cilium plasma membrane and formation of the photoreceptor disk. A one-nucleotide deletion predicted to generate a nonfunctional fascin has been identified in four Japanese families affected with adRP.

CRX. Mutations in *cone–rod homeobox* (*CRX*) are associated with adRP, as well as cone–rod dystrophy and Leber congenital amaurosis. *CRX* encodes a protein of 299 amino acids that is homologous to the human OTX1 and OTX2 homeodomain proteins. CRX is a retina-specific transcription factor that regulates the expression of several photoreceptor-specific genes, including rhodopsin, rod transducin α subunit, cone arrestin, recoverin and green/red cone opsin. CRX and NRL show transcriptional synergy in transcriptional activation of the *RHO* promoter.

PRPF31. The adRP gene on chromosome 19q13.4 (locus *RP11*) has been identified as the *PRP31 pre-mRNA processing factor 31 homolog (PRPF31)* gene that encodes a protein of 499 amino acids with homology to the yeast pre-mRNA splicing factor PRP31 (Makarova *et al.*, 2002). Like HPRP3, PRPF31 is a component of the U4/U6 snRNP particle involved in pre-mRNA splicing.

Other adRP genes. Recently, the genes for RP9 and RP10 have been identified as *PAP1* (OMIM *607331) and *IMPDH1* (OMIM *146690), respectively. *PAP1* encodes a protein that interacts with the PIM1 kinase (encoded by the *PIM1* protooncogenes), and is phosphorylated by PIM1 kinase. *IMPDH1* is the gene for IMP (inosine-5′-monophosphate) dehydrogenase-1, which catalyzes the rate-limiting step of *de novo* guanine nucleotide synthesis. Guanine nucleotide synthesis is important for DNA and RNA synthesis, cellular growth, differentiation and apoptosis. The chromosomal location of another adRP gene has been mapped by linkage analysis, locus *RP17* (17q22), but the responsible gene remains to be cloned or identified (**Table 1**).

Autosomal Recessive RP

RPE65. Mutations in *retinal pigment epithelium-specific protein* (65 kDa) (*RPE65*) cause arRP linked to chromosome 1p31, as well as early-onset severe

Table 1 Genes involved in autosomal dominant retinitis pigmentosa

	Locus	Location	Gene	Protein/ predicted function
1	*RP18*	1p13–q23	*HPRP3*	Pre-mRNA splicing factor
2	*RP4/RP5*	3q21–q24	*RHO*	Rhodopsin
3	*RP7*	6p21.1–cen	*RDS*	Peripherin
4	*RP1*	8p11–q21	*RP1*	Oxygen-regulated photoreceptor gene, homologous to doublecortin
5	*RGR*	10q23	*RGR*	RPE-retinal G-protein-coupled receptor, a rhodopsin homolog
6	*ROM1*	11q13	*ROM1*	Retinal outer segment membrane protein 1
7	*RP27*	14q11.1–q11.2	*NRL*	Transcription factor of the leucine zipper family
8	*RP13*	17p13.3	*PRPF8*	Pre-mRNA splicing factor
9	*FSCN2*	17q25	*FSCN2*	Member of the actin-binding protein family
10	*CRX*	19q13.3	*CRX*	Transcription factor
11	*RP11*	19q13.4	*PRPF31*	Pre-mRNA splicing factor
12	*RP9*	7p15.1–p13	*PAP1*	PIM1 kinase-associated protein
13	*RP10*	7q31–35	*IMPDH1*	Guanine nucleotide synthesis
14	*RP17*	17q22		?

rod–cone dystrophy and Leber congenital amaurosis (a severe childhood-onset retinal dystrophy that affects rods and cones simultaneously). *RPE65* encodes a protein of 61 kDa in the RPE. The RPE65 protein is important in the RPE/photoreceptor vitamin A cycle and is essential for the re-isomerization of all-*trans*-retinol in the visual cycle.

ABCA4. The gene for the chromosome 1p13–p21-linked arRP locus has been identified as *ATP-binding cassette, sub-family A (ABC1), member 4 (ABCA4)*, which encodes a retina-specific ATP-binding cassette transporter. *ABCA4* mutations have been also found in individuals with autosomal recessive Stargardt disease, fundus flavimaculatus (an autosomal recessive disorder that leads to macular degeneration in adulthood) and cone–rod dystrophy. Mutations in *ABCA4* have been also implicated in the pathogenesis of age-related macular degeneration, although some researchers argue that allelic variation in *ABCA4* is not associated with this disorder.

CRB1. The gene on the *RP12* locus on chromosome 1q31–32.1 was identified as *crumbs homolog 1*

(*CRB1*), which encodes an extracellular protein homologous to the *Drosophila* protein Crumbs. CRB protein localizes to the inner segment of photoreceptors and may form a molecular scaffold that is important for photoreceptor morphogenesis (Pellikka *et al.*, 2002).

USH2A. Usher syndrome 2A (*USH2A*) on chromosome 1q41 encodes usherin, which contains epidermal growth factor and fibronectin type III motifs that are most commonly found in proteins of the basal lamina and extracellular matrix. Although *USH2A* mutations cause a mild form of autosomal recessive Usher syndrome (type IIA) characterized by both RP and hearing loss, a homozygous missense mutation (Cys759Phe) has been identified in 4.5% of 224 individuals with arRP without hearing loss.

MERTK. c-Mer proto-oncogene tyrosine kinase (*MERTK*) encodes a novel tyrosine kinase (c-Mer) of 984 amino acids that is highly homologous to the chicken retroviral oncogene product v-Ryk. Mutations in *MERTK* have been found in individuals affected with various retinal dystrophies including arRP.

SAG. A homozygous one-nucleotide deletion (1147delA) in codon 309 of *S-antigen; retina and pineal gland* (*SAG*) has been identified in individuals with Oguchi disease, a rare autosomal recessive congenital stationary night blindness with a golden or gray–white discoloration of the fundus. The same deletion has been identified in two individuals affected with arRP. *SAG* encodes S-antigen (also known as arrestin), a soluble rod photoreceptor protein of 45 kDa with 405 amino acids that is involved in the recovery phase of phototransduction.

RHO. A few *RHO* mutations have been found to cause arRP.

PDE6A, PDE6B. Phosphodiesterase 6A, cGMP-specific, rod, alpha (*PDE6A*) and *phosphodiesterase 6A, cGMP-specific, rod, beta* (*PDE6B*) encode, respectively, the α and β subunit of cGMP phosphodiesterase, which is involved in the retinal rod phototransduction cascade. Mutations in both *PDE6A* and *PDE6B* have been identified in individuals with arRP.

CNGA1. The *cyclic nucleotide gated channel alpha 1* (*CNGA1*) gene encodes the α subunit of the cGMP-gated cation channel that is located on the rod outer segment plasma membrane and is the final component of the phototransduction cascade. Mutations in *CNGA1* have been identified in families with arRP.

TULP1. The arRP gene on 6p21.3 has been identified as *tubby-like protein 1* (*TULP1*), a gene that encodes a protein with homology to murine tubby. *TULP1* is expressed specifically in the retina, but its function is not clear.

RGR. A homozygous missense mutation in *RGR* (Ser66Arg) has been reported in a family affected with arRP.

NR2E3. The arRP gene on chromosome 15q22–q24 was identified as *nuclear receptor subfamily 2, group E, member 3* (*NR2E3*), which encodes a photoreceptor-cell-specific nuclear receptor with a DNA-binding domain followed by a putative ligand-binding domain. Mutations in *NR2E3* have been also identified in individuals with an autosomal recessive retinopathy, in which affected individuals have increased sensitivity to perception of blue light (a process mediated by the S (short wavelength, blue) cone photoreceptors), which is also known as enhanced S-cone syndrome.

RLBP1. The arRP gene on chromosome 15q26 was identified as the *retinaldehyde binding protein 1* (*RLBP1*) gene, which encodes the cellular retinaldehyde-binding protein CRALBP – a water-soluble protein of 36 kDa that carries 11-*cis*-retinaldehyde or 11-*cis*-retinal as physiological ligands. Mutations in *RLBP1* can cause arRP by disrupting vitamin A metabolism in the RPE and Müller cells of the retina. The *RLBP1* gene is not expressed in photoreceptors but is expressed highly in the RPE and Müller cells of the neural retina.

CNGB1. A mutation in the *cyclic nucleotide gated channel beta 1* (*CNGB1*) gene has been identified in a consanguineous French family affected with arRP (Bareil *et al.*, 2001). *CNGB1* encodes the β subunit of the rod cGMP-gated channel involved in phototransduction.

Other arRP genes. The chromosomal locations of six additional genes linked to arRP have been mapped by linkage analysis, *RP28* (chromosome 2q11–p15), *RP26* (2q31–33), *RP29* (4q32–34), *RP25* (6q14–21), *RP16* (14q11) and *RP22* (16p12.1–12.3), but the responsible genes remain to be cloned or identified (**Table 2**).

X-linked RP

RPGR. The gene on the *RP3* locus on chromosome Xp11.4–21.2 was identified as *retinitis pigmentosa GTPase regulator* (*RPGR*), which encodes a protein of 90 kDa that is homologous to the regulator of chromosome condensation (RCC1) – the guanine-nucleotide-exchange factor of the Ras-like GTPase Ran. Mutations in *RPGR* account for 70% of xlRP cases. A retina-specific exon (orf15) of *RPGR* that encodes 567 amino acids at the carboxy (*C*)-terminus of RPGR seems to be a mutation hot spot. Mutations in *RPGR* disrupt the interactions between RPGR and its interacting proteins, including the δ subunit of rod cGMP phosphodiesterase and RPGRIP (a novel retinal protein with two coiled-coil domains at the *N*-terminus). Studies with RPGR-deficient mice suggest that RPGR may maintain the polarized protein distribution across the connecting cilium by facilitating directional transport or restricting redistribution.

Table 2 Genes involved in autosomal recessive retinitis pigmentosa

	Locus	Location	Gene	Protein/predicted function
1	*RP20*	1p31	*RPE65*	Involved in retinal recycling
2	*RP19*	1p21–p13	*ABCA4*	Retina-specific ATP-binding cassette transporter
3	*RP12*	1q31–q32.1	*CRB1*	Homologous to *Drosophila crumbs*
4	*USH2A*	1q41	*USH2A*	Protein with laminin EGF and fibronectin type III domains
5	*MERTK*	2q14.1	*MERTK*	c-*mer* proto-oncogene encoding a receptor tyrosine kinase
6	*SAG*	2q37.1	*SAG*	S-antigen (arrestin)
7	*RHO*	3q21–24	*RHO*	Rhodopsin
8	*PDE6B*	4p16.3	*PDE6B*	Phosphodiesterase β subunit
9	*CNGA1*	4p14–q13	*CNGA1*	Cyclic nucleotide-gated channel α subunit
10	*PDE6A*	5q31.2–q34	*PDE6A*	Phosphodiesterase α subunit
11	*RP14*	6p21.3	*TULP1*	Human homolog of mouse *Tub*
12	*RGR*	10q23	*RGR*	RPE-retinal G-protein-coupled receptor, a rhodopsin homolog
13	*NR2E3*	15q23	*NR2E3*	Nuclear receptor subfamily 2 group E3
14	*RLBP1*	15q26	*RLBP1*	Cellular retinaldehyde-binding protein
15	*CNGB1*	16q13–q21	*CNGB1*	β subunit of the rod cGMP-gated channel
16	*RP28*	2q11–p15		?
17	*RP26*	2q31–q33		?
18	*RP29*	4q32–q34		?
19	*RP25*	6q14–q21		?
20	*RP16*	14q11		?
21	*RP22*	16p12.1–p12.3		?

Table 3 Genes involved in X-linked retinitis pigmentosa

Locus	Location	Gene	Protein/predicted function
RP3 (including *RP15*)	Xp11.4–p21.1	*RP3* or *RPGR*	Homologous to the guanine-nucleotide exchange factor
RP2	Xp11.2–p11.4	*RP2*	Homologous to cofactor C involved in β-tubulin folding
RP23	Xp22		?
RP6	Xp21.3–p21.2		?
RP24	Xq26–27		?

RP2. The *retinitis pigmentosa 2* (*RP2*) gene on chromosome Xp22.13–22.11 was identified in 1998. *RP2* encodes a novel protein of 350 amino acids with homology to human cofactor C, which is involved in the last step of β-tubulin folding. Both *RP2* mRNA and RP2 protein are ubiquitously expressed in human tissues, and RP2 protein is localized on plasma membranes. Mutations in the *RP2* gene have been identified in 18% of individuals with xlRP.

Other xlRP genes. The chromosomal locations of three additional genes linked to xlRP have been mapped by linkage analysis, *RP23* on Xp22, *RP6* on Xp21.2–21.3 and *RP24* on Xq26–q27, but the genes remain to be identified (**Table 3**).

Digenic RP

ROM1/RDS. A combination of a heterozygous mutation in *ROM1* on chromosome 11q13 and a heterozygous mutations in *RDS* on 6p21.1–cen causes dRP. The *RDS* mutation that causes dRP is a heterozygous Leu185Pro substitution, and the *ROM1* mutations include a heterozygous one-nucleotide insertion at codon 80 and another heterozygous one-nucleotide insertion at codon 114. Only individuals with double heterozygous mutations, one in *ROM1* and the other in *RDS*, develop dRP. This is the first molecular documentation that a human disease is caused by the interaction of two different genes (the 'two-locus' mechanism).

Table 4 Functional grouping of genes involved in retinitis pigmentosa

Functional group	Genes
Visual cascade[a]	RHO, PDE6A, PDE6B, CNGA1, CNGB1, SAG
Visual cycle[b]	RPE65, RLBP1, RGR, ABCA4
Structural proteins	RDS, ROM1
Transcriptional factors	NRL, CRX, NR2E3
Extracellular proteins	CRB1, USH2A
Protein kinases and their target proteins	MERTK, RP1, PAP1
Splicing factors	PRP3, PRPF8, PRPF31
Cytoskeletal proteins	RP2, FSCN2
Cellular growth, differentiation, apoptosis	IMPDH1
Proteins with unknown function	RPGR, TULP1

[a]The visual cascade is a process consisting of the uptake of photons of visible light by photoreceptors, conversion into a neuronal signal, and eventual perception of the signal by the brain as sight.
[b]The visual cycle is a metabolic pathway that recycles and transforms all-*trans*-retinol (vitamin A) to form the chromophone of rhodopsin, 11-*cis*-retinaldehyde.

Conclusions

So far, 28 distinct RP genes have been cloned or identified, and additional RP genes remain to be identified (**Table 4**). Molecular genetics studies have made genetic testing possible in some cases of RP and may lead to the development of effective treatments for individuals affected with RP. The known RP genes can be classified into several functional groups (**Table 4**). Genes involved in the visual cascade (phototransduction) include *RHO*, *PDE6A*, *PDE6B*, *CNGA1*, *CNGB1* and *SAG*; *RPE65*, *RLBP1*, *RGR* and *ABCA4* are involved in the visual cycle. RP genes also encode structural proteins (peripherin, ROM1), transcriptional factors (NRL, CRX, NR2E3), extracellular matrix proteins (CRB1, USH2A), protein kinases (MERTK, RP1), pre-mRNA splicing factors (HPRP3, PRPF8, PRPF31), cytoskeletal proteins (RP2, FSCN2), cellular growth, differentiation, and apoptosis (IMP dehydrogenase 1), and proteins with undefined functions (RPGR, TULP1).

Individuals affected with dysfunctional visual cycle proteins may be the first group of people to benefit from gene-specific therapies (e.g. by modulating vitamin A metabolism). High-dose vitamin A has been reported to have a modest slowing of disease progression in some individuals with RP. Strategies that can inhibit a major lipofuscin fluorophore (A2-E) accumulation in RPE cells may become potential therapeutic treatments for individuals with *ABCA4* mutations. Those with *RPE65* and *RLBP1* mutations may benefit from therapeutic trials that focus on manipulation of the vitamin A cycle, for example strategies to increase 11-*cis*-retinyl esters in individuals with *RPE65* mutations.

See also

Eye Disorders: Hereditary

References

Bareil C, Hamel CP, Delague V, et al. (2001) Segregation of a mutation in CNGB1 encoding the beta-subunit of the rod cGMP-gated channel in a family with autosomal recessive retinitis pigmentosa. *Human Genetics* **108**: 328–334.

Chakarova CF, Hims MM, Bolz H, et al. (2002) Mutations in HPRP3, a third member of pre-mRNA splicing factor genes, implicated in autosomal dominant retinitis pigmentosa. *Human Molecular Genetics* **11**: 87–92.

Donders F (1857) Beitraege sur pathologischen anatomie des auges. 2. Pigmentbildung in der netzhaut. *Archives of Ophthalmology* **3**: 139–165.

Haim M, Holm NV and Rosenberg T (1992) Prevalence of retinitis pigmentosa and allied disorders in Denmark. I. Main results. *Acta Ophthalmologica (Copenhagen)* **70**: 178–186.

Heckenlively JR, Yoser SL, Friedman LH and Oversier JJ (1988) Clinical findings and common symptoms in retinitis pigmentosa. *American Journal of Ophthalmology* **105**: 504–511.

McKie AB, McHale JC, Keen TJ, et al. (2001) Mutations in the pre-mRNA splicing factor gene PRPC8 in autosomal dominant retinitis pigmentosa (RP13). *Human Molecular Genetics* **10**: 1555–1562.

Makarova OV, Makarov EM, Liu S, Vornlocher HP and Luhrmann R (2002) Protein 61K, encoded by a gene (PRPF31) linked to autosomal dominant retinitis pigmentosa, is required for U4/U6 · U5 tri-snRNP formation and pre-mRNA splicing. *EMBO Journal* **21**: 1148–1157.

Pellikka M, Tanentzapf G, Pinto M, et al. (2002) Crumbs, the Drosophila homologue of human CRB1/RP12, is essential for photoreceptor morphogenesis. *Nature* **416**: 143–149.

Wada Y, Abe T, Takeshita T, et al. (2001) Mutation of human retinal fascin gene (FSCN2) causes autosomal dominant retinitis pigmentosa. *Investigative Ophthalmology and Visual Science* **42**: 2395–2400.

Wang Q, Chen Q, Zhao K, et al. (2001) Update on the molecular genetics of retinitis pigmentosa. *Ophthalmic Genetics* **22**: 133–154.

Zhao K, Xiong S, Wang L, et al. (2001) Novel rhodopsin mutation in a Chinese family with autosomal dominant retinitis pigmentosa. *Ophthalmic Genetics* **22**: 155–162.

Further Reading

Blackshaw S, Fraioli RE, Furukawa T and Cepko C (2001) Comprehensive analysis of photoreceptor gene expression and the identification of candidate retinal disease genes. *Cell* **107**: 579–589.

Berson EL (1994) Retinitis pigmentosa and allied diseases. In: Albert DM and Jakobiec FA (eds.) *Principles and Practice of Ophthalmology: Clinical Practice*, pp. 1214–1237. Philadelphia, PA: WB Saunders.

Inglehearn CF (1988) Molecular genetics of human retinal dystrophies. *Eye* **12**: 571–579.

Phelan JK and Bok D (2000) A brief review of retinitis pigmentosa and the identified retinitis pigmentosa genes. *Molecular Vision* **6**: 116–124.

Rattner A, Sun H and Nathans J (1999) Molecular genetics of human retinal disease. *Annual Review of Genetics* **33**: 89–131.

van Soest S, Westerveld A, de Jong PT, *et al.* (1999) Retinitis pigmentosa: defined from a molecular point of view. *Survey of Ophthalmology* **43**: 321–334.

Web Links

Retinal Information Network
http://www.sph.uth.tmc.edu/Retnet

Retinal degeneration, slow (*RDS*); LocusID: 5961. LocusLink:
http://www.ncbi.nlm.nih.gov/LocusLink/LocRpt.cgi?l = 5961

Rhodopsin (*RHO*); LocusID: 6010. LocusLink:
http://www.ncbi.nlm.nih.gov/LocusLink/LocRpt.cgi?l = 6010

Retinal outer segment membrane protein 1 (*ROM1*); LocusID: 6094. LocusLink:
http://www.ncbi.nlm.nih.gov/LocusLink/LocRpt.cgi?l = 6094

Retinal pigment epithelium-specific protein (65kD) (*RPE65*); LocusID: 6121. LocusLink:
http://www.ncbi.nlm.nih.gov/LocusLink/LocRpt.cgi?l = 6121

Retinitis pigmentosa GTPase regulator (*RPGR*); LocusID: 6103. LocusLink:
http://www.ncbi.nlm.nih.gov/LocusLink/LocRpt.cgi?l = 6103

Retinal degeneration, slow (*RDS*); MIM number: 179605. OMIM:
http://www.ncbi.nlm.nih.gov/htbin-post/Omim/dispmim?179605

Rhodopsin (*RHO*); MIM number: 180380. OMIM:
http://www.ncbi.nlm.nih.gov/htbin-post/Omim/dispmim?180380

Retinal outer segment membrane protein 1 (*ROM1*); MIM number: 180721. OMIM:
http://www.ncbi.nlm.nih.gov/htbin-post/Omim/dispmim?180721

Retinal pigment epithelium-specific protein (65kD) (*RPE65*); MIM number: 180869. OMIM:
http://www.ncbi.nlm.nih.gov/htbin-post/Omim/dispmim?180869

Retinitis pigmentosa GTPase regulator (*RPGR*); MIM number: 312610. OMIM:
http://www.ncbi.nlm.nih.gov/htbin-post/Omim/dispmim?312610

Retinoblastoma

Intermediate article

Dietmar Rudolf Lohmann, *Universität Essen, Essen, Germany*

Brenda L Gallie, *Princess Margaret University, Toronto, Canada*

Article contents

- Clinical Presentation
- Genetics of Retinoblastoma

Retinoblastoma is a malignant tumor that originates from the developing retina. Mutations in both alleles of the retinoblastoma gene are a prerequisite for the development of this tumor. Germ-line mutations in the retinoblastoma gene cause heritable predisposition to retinoblastoma, a Mendelian trait that shows autosomal dominant inheritance.

Clinical Presentation

Retinoblastoma is a malignant tumor that originates from the developing retina. The estimated incidence is between one in 15 000 and one in 20 000 live births. Diagnosis is based on clinical signs and symptoms, and is usually made in children under the age of 5 years. The first presenting sign is most often a white pupillary reflex (leucocoria). Strabismus is the second most common sign and may accompany or precede leucocoria. The diagnosis of retinoblastoma is usually established by examining the fundus of the eye using indirect ophthalmoscopy. Additional diagnostic tools such as computed tomography, magnetic resonance imaging and ultrasonography may be required for differential diagnosis and staging. If tumor material has been obtained, histopathology can confirm the diagnosis of retinoblastoma.

Most patients (about 60%) have retinoblastoma in one eye only (unilateral retinoblastoma), and occasionally multiple tumor foci can be found (unilateral multifocal retinoblastoma). Patients with unilateral retinoblastoma most often have sporadic disease, that is, no other case of retinoblastoma has been noted in their family. In about 40% of patients, both eyes are affected (bilateral retinoblastoma), usually with more than one focus per eye (bilateral multifocal retinoblastoma). In children with bilateral retinoblastoma, diagnosis is made earlier than in children who develop retinoblastoma in only one eye (median age at diagnosis 11 and 22 months respectively). In only 10% of patients is there a family history of retinoblastoma (familial retinoblastoma). However, examination of the retina in all first-degree relatives of retinoblastoma patients is required to identify retinal scars or quiescent tumors (retinomas) because these lesions also indicate familial disease.

The treatment of retinoblastoma depends on tumor stage, the number of tumor foci (unifocal, unilateral multifocal or bilateral disease), the localization and size of the tumor(s) within the eye, the presence of vitreous seeding and the age of the child. Treatment options include enucleation, external-beam radiation, cryotherapy, photocoagulation, brachytherapy with episcleral plaques and systemic chemotherapy combined with local therapy. If tumor cells have not yet invaded the extraocular tissues, treatment is successful in most patients. Metastasizing retinoblastoma is fatal in most patients. Following successful treatment, children require frequent follow-up examinations for the early detection of new intraocular tumors.

Patients with bilateral retinoblastoma have an increased risk of specific neoplasms outside the eye (second tumors). The spectrum of second tumors includes osteogenic sarcoma, soft tissue sarcoma and malignant melanoma. The risk for a second tumor developing is increased in patients who have received external beam radiation for treatment of bilateral retinoblastoma.

Genetics of Retinoblastoma

The development of retinoblastoma is initiated by two mutations (two-hit model) that impair the function of both alleles of the retinoblastoma gene (*RB1*). The origin of the first of the two oncogenic mutations is important in terms of the genetic risk to offspring and the clinical presentation of the retinoblastoma.

In patients with nonhereditary retinoblastoma, the first oncogenic mutation has occurred in a somatic cell, and the mutant allele is not present in any of the patient's germ-line cells. Development of a tumor focus is initiated by loss of the normal allele (second mutation). With rare exceptions, patients with nonhereditary retinoblastoma have unilateral retinoblastoma. About 90% of patients with isolated unilateral tumors, and in addition a few patients with isolated bilateral tumors, have nonhereditary retinoblastoma.

Retinoblastoma can be inherited as an autosomal dominant trait (hereditary retinoblastoma) that is caused by germ-line mutations in the *RB1* gene. Patients who are heterozygous for a germ-line mutation are predisposed to retinoblastoma because just one second *RB1* gene mutation in a retinal progenitor cell is sufficient to initiate the development of a single tumor focus. In most patients with isolated bilateral retinoblastoma, the predisposing *RB1* gene mutation has occurred *de novo*. Typically, patients with hered-

itary retinoblastoma have multiple tumor foci in both eyes. Phenotypic expression, however, is dependent on the nature of the predisposing mutation. Specific *RB1* gene mutations are associated with milder expression (unilateral retinoblastoma) or incomplete penetrance in families (see below). In addition to patients with familial retinoblastoma and almost all patients with isolated bilateral retinoblastoma, some patients with isolated unilateral retinoblastoma also have hereditary retinoblastoma.

In some patients, the first oncogenic *RB1* gene mutation has occurred during embryonic development. In these patients, the mutation is not present in all cells (mutational mosaicism), and clinical presentation as well as the transmissibility of the retinoblastoma will depend on the number and the types of cell that carry the oncogenic mutation. It is reasonable to assume that mutational mosaicism is relatively common among patients with isolated unilateral retinoblastoma. Available data suggest that about 8% of patients with isolated bilateral retinoblastoma carry mosaicism.

Although mutations in both alleles of the *RB1* gene are a prerequisite for the development of retinoblastoma, there is reason to believe that additional genetic alterations are required for progression to a malignant tumor. Retinoma, which is a nonmalignant tumor that is identified in some carriers of oncogenic *RB1* gene mutations, might be regarded as a precursor lesion, and the malignant transformation of such a tumor has been observed. Moreover, investigation of retinoblastoma by cytogenetic analysis or comparative genomic hybridization (CGH) has revealed recurrent genetic alterations, including isochromosome of the short arm of chromosome 6 (i(6p)), in about 60% of tumors.

Retinoblastoma gene

The retinoblastoma gene (*RB1*) is located on chromosome 13q14. It consists of 27 exons that are scattered over 183 kb of genomic sequence (**Figure 1a**). At its 5′ end, the *RB1* gene has a CpG island that is normally unmethylated. The promoter region contains binding motifs for transcription factors Spl and ATF but no TATA or CAAT elements. In tissues investigated so far, the gene is transcribed into a 4.7 kb messenger ribonucleic acid (mRNA) with no convincing evidence for alternative splicing. The open reading frame, which starts in exon 1 and is terminated in exon 27, has 2.7 kb and is followed by a 2-kb untranslated region (**Figure 1b**). Homologs of the human *RB1* gene have been identified in a wide variety of organisms and show a high level of sequence similarity in translated regions. The part of the gene that encodes the domains

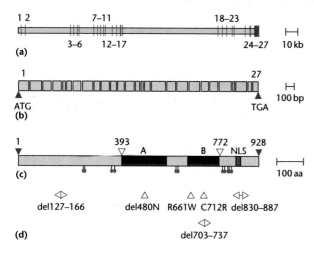

Figure 1 Organization of the retinoblastoma (*RB1*) gene and protein, pRB, and location of mutations associated with incomplete penetrance. (a) Genomic organization of the *RB1* gene. (b) Organization of the 27 exons containing the open reading frame. (c) Structure of the pRb protein. A and B: A/B pocket domain; NLS: nuclear localization signal; filled circles: phosphorylation sites. (d) Location of missense substitutions and in-frame deletions known to be associated with incomplete penetrance and mild expressivity.

for the A/B pocket (see below) has a homolog also in higher plants (*mat3*).

Retinoblastoma protein

The *RB1* gene encodes the retinoblastoma protein (pRb), a 928 amino acid nuclear phosphoprotein that migrates at 110 kDa in SDS-PAGE when hypophosphorylated (**Figure 1c**). pRb is part of a small family of nuclear proteins that includes p107 and p130. These proteins share significant sequence similarity in two discontinuous areas that constitute the A/B pocket (pocket proteins). Conditional on the phosphorylation status at multiple serine and threonine residues in other regions of the protein, the A/B pocket can bind to members of the E2F family of transcription factors as well as to transforming proteins of deoxyribonucleic acid tumor viruses (e.g. adenovirus E1a, simian virus 40 T antigen and human papillomavirus E7) and endogenous nuclear proteins that contain the LxCxE peptide motif (such as histone deacetylases 1 and 2). The *C*-terminal region of pRb contains a nuclear localization signal and a cyclin–cyclin-dependent kinase (cdk) interaction motif that enables it to be recognized and phosphorylated by cyclin–cdk complexes. The *C*-terminal region can also bind to the nuclear c-Abl tyrosine kinase and to MDM2, which are proteins with oncogenic properties.

The role of pRb that is understood best is its function as a gatekeeper that negatively regulates progression through G_1 phase of the cell cycle. During the G_1 phase of the cell cycle, pRb is hypophosphorylated. Unphosphorylated Rb can bind E2F and cause a repression of E2F-mediated transcription. Beginning in late G_1 and continuing into the M phase, pRb is phosphorylated by G_1 cdks. Upon phosphorylation of pRb, E2F is released and promotes the expression of genes that are required for cell division. Consequently, pRb controls cell cycle phase transition by transcriptional repression. In addition to phosphorylation, cell cycle-dependent acetylation has been found to control pRb function. Acetylation hinders the phosphorylation of pRb and enhances binding to the MDM2 oncoprotein. Besides cell cycle regulation, pRb probably has a variety of roles including control of apoptosis and stimulation of differentiation.

Spectrum of *RB1* gene mutations

Cytogenetic aberrations

Conventional cytogenetic analysis of peripheral blood lymphocytes shows deletions and rearrangements involving 13q14 in about 8% of patients with bilateral and up to 5% of patients with sporadic unilateral retinoblastoma. Large interstitial deletions are often associated with facial dysmorphism and developmental delay (13q deletion syndrome).

Large deletions

About 17% of patients with bilateral or familial retinoblastoma have subcytogenetic deletions of parts of the *RB1* gene. Until now, no evidence for recurrent rearrangements or hot spots of deletion breakpoints has emerged.

Small-length mutations

Deletions or insertions of one or a few base pairs are identified in about 30% of patients with bilateral or familial retinoblastoma. These mutations are found in all coding regions and splice sites except the 3′-terminal exons. A few sites in the *RB1* gene that contain repetitive sequence motifs show a higher mutation frequency. Most small-length mutations result in a premature termination codon because of either a frameshift or a disruption of splice signals. In-frame deletions are rare and are almost restricted to the regions that code for the A/B pocket.

Single-base substitutions

Single-base substitutions are found in about 50% of patients with hereditary retinoblastoma and have been identified in all coding regions and splice sites except the two 3′-terminal exons. In addition, transcription factor binding sites upstream of the start-ATG can be disrupted by base substitutions. Most single-base substitutions are nonsense mutations. Among these, recurrent CpG transitions at 12 of the 15 CGA codons within the open reading frame are most frequent. In addition, the CpG dinucleotide contained in the 5′ splice site of intron 12 (AACgta to AACata) is a target of recurrent transitions. About 10% of reported substitutions result in missense changes and most of them affect the A/B pocket. A relatively frequent missense mutation is a CpG transition that causes an R661W in exon 20.

Oncogenic mutations in retinoblastoma

With two important exceptions, the spectrum of oncogenic mutations in retinoblastomas corresponds to that observed in constitutional cells. In about 65% of tumors from patients that are heterozygous for an *RB1* gene mutation, the normal allele is lost because of chromosomal mechanisms that also result in a loss of heterozygosity at polymorphic loci (LOH). Most of these tumors are homozygous for the mutant allele and show complete or partial isodisomy because of chromosomal nondisjunction and mitotic recombination respectively. Hypermethylation of the CpG-rich island at the 5′ end of the *RB1* gene, which is normally unmethylated, is observed in about 10% of retinoblastomas and results in transcriptional silencing and thus loss of function.

Genotype–phenotype associations

Heterozygous carriers of an oncogenic *RB1* gene mutation can show variable phenotypic expression. This is to be expected because the second mutation that initiates tumor formation is a chance event. Analysis of phenotypic variation has shown that, within most families, the proportion of mutation carriers with bilateral, unilateral and no tumors complies with the ratios that are expected if the second mutation events follow a Poisson distribution. However, penetrance and expressivity can vary between families. A significant proportion of interfamilial variance of penetrance and expressivity can be explained by allelic heterogeneity.

In most families with retinoblastoma, penetrance is complete (100%) and, with rare exceptions, all mutation carriers who have inherited the mutant allele show bilateral retinoblastoma (**Figure 2a, 2b**). Typically, retinoblastoma predisposition in these families is caused by mutant alleles with premature termination codons or deletions of significant parts or the whole *RB1* gene. In RNA from constitutional cells of heterozygous mutation carriers, nonsense transcripts are less abundant than transcripts from the normal

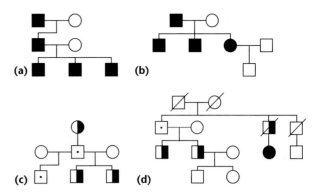

Figure 2 Pedigrees with retinoblastoma. Filled symbols: bilateral retinoblastoma; half-filled symbols: unilateral retinoblastoma; symbols marked with a dot: heterozygous carriers. (a,b) Families with complete penetrance associated with a nonsense (R579X) and a frameshift mutation (651X) respectively. (c,d) Families with incomplete penetrance and mild expressivity associated with the recurrent missense mutation R661W.

allele. This indicates that mutant *RB1* transcripts can be subject to nonsense-mediated decay and may explain why oncogenic alleles with premature termination codons in any of exons 2–25 show a similar phenotypic expression.

In a few families, penetrance is incomplete, and, between families, observed values may vary from 20% to 80% (low-penetrance retinoblastoma; **Figure 2c, 2d**). In most of these families, incomplete penetrance is accompanied by mild expressivity (unilateral retinoblastoma). Families with incomplete penetrance and mild expressivity have mutant alleles that result in reduced levels of structurally normal transcript (promoter mutations) or result in only partial loss of function because of missense or in-frame alterations (**Figure 1d**). Most of these mutations involve the regions that encode the A/B pocket of pRb.

Carriers of an oncogenic *RB1* gene mutation also show variable phenotypic expression with regard to second tumors. Although a higher incidence of second tumors in patients who were exposed to external beam radiation for the treatment of retinoblastoma indicates that environmental factors are important, there might be an influence of genetic variation. A link between specific oncogenic *RB1* gene mutations and a higher risk for second tumors has not been established. Available data suggest that carriers of mutations associated with low penetrance might have a lower incidence of second tumors compared to carriers of null mutations.

See also

Caretakers and Gatekeepers
Tumor Suppressor Genes

Further Reading

DiCiommo D, Gallie BL and Bremner R (2000) Retinoblastoma: the disease, gene and protein provide critical leads to understand cancer. *Seminars in Cancer Biology* **10**: 255–269.

Gallie BL, Campbell C, Devlin H, Duckett A and Squire JA (1999) Developmental basis of retinal-specific induction of cancer by RB mutation. *Cancer Research* **59**: 1731s–1735s.

Harbour JW and Dean DC (2000) Rb function in cell-cycle regulation and apoptosis. *Nature Cell Biology* **2**: E65–E67.

Kaelin Jr WG (1999) Functions of the retinoblastoma protein. *BioEssays* **21**: 950–958.

Knudson AG (2001) Two genetic hits (more or less) to cancer. *Nature Reviews Cancer* **1**: 157–162.

Lohmann DR (1999) RB1 gene mutations in retinoblastoma. *Human Mutation* **14**: 283–238.

Lohmann DR, Gerick M, Brandt B, *et al.* (1997) Constitutional RB1-gene mutations in patients with isolated unilateral retinoblastoma. *American Journal of Human Genetics* **61**: 282–294.

Sherr CJ (2001) The INK4a/ARF network in tumour suppression. *Nature Reviews Molecular and Cell Biology* **2**: 731–737.

Sippel KC, Fraioli RE, Smith GD, *et al.* (1998) Frequency of somatic and germ-line mosaicism in retinoblastoma: implications for genetic counseling. *American Journal of Human Genetics* **62**: 610–619.

Zheng L and Lee WH (2001) The retinoblastoma gene: a prototypic and multifunctional tumor suppressor. *Experimental Cell Research* **264**: 2–18.

Web Links

Retinoblastoma 1 (including osteosarcoma) (RB1); Locus ID 5925. LocusLink:
http://www.ncbi.nlm.nih.gov/LocusLink/LocRpt.cgi? = 5925

Retinoblastoma 1 (including osteosarcoma) (RB1); MIM number: 180200. OMIM:
http://www.ncbi.nlm.nih.gov/htbin-post/Omim/dispmim?180200

Retrosequences and Evolution of Alu Elements

Astrid M Roy-Engel, *Tulane University Health Sciences Center, New Orleans, Louisiana, USA*

Mark A Batzer, *Louisiana State University, Baton Rouge, Louisiana, USA*

Prescott Deininger, *Tulane University Health Sciences Center and Ochsner Medical Foundation, New Orleans, Louisiana, USA*

Intermediate article

Article contents

- Retrosequence: Features and Classification
- Human Mobile Retrosequences
- Origin of Alu Elements
- Evolutionary Expansion of Alu Elements
- Alu Elements may be L1 Parasites
- Evolutionary Impact of Alu Elements on the Human Genome

Retrosequences make up almost half of the human genome. They insert in new locations via a ribonucleic acid (RNA) intermediate and can amplify to over a million copies, as is the case for Alu elements. Although almost any transcribed RNA may undergo occasional reverse transcription and insertion into the genome, most retrosequences are limited to a few types of retroelements with high retrotransposition efficiencies. Despite these high copy numbers, the vast majority of retroelements are defective and therefore the number of active copies is generally very low; thus, the evolution of these elements follows the evolution of these few active elements with the remainder representing 'fossils' of the active elements at any given evolutionary time.

Retrosequence: Features and Classification

A retrosequence is defined as a deoxyribonucleic acid (DNA) sequence that amplifies itself in the genome via a ribonucleic acid (RNA) intermediate, but does not have an infectious form. Several viruses, like retroviruses, caulimoviruses and hepadnaviruses, use reverse transcription of RNA as a step in their cycle, but are not considered retrosequences. However, some of the viruses are related to retrosequences. Retroviruses are thought to have evolved from long terminal repeat (LTR) retrotransposons (Malik *et al.*, 2000), and they share several features in common (**Figure 1**). There are several types of retrosequences: LTR retrotransposons, non-LTR retrotransposons (LINEs (long interspersed nuclear elements), nonautonomous retroposons (short interspersed nuclear elements; SINEs) and retropseudogenes (Weiner *et al.*, 1986) (see **Figure 1**). Some of the retrosequences code for all the enzymes and proteins they require for amplification. In contrast, the nonautonomous elements (SINEs and retropseudogenes) need to 'parasitize' the factors they require from external sources.

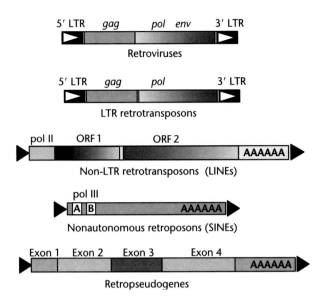

Figure 1 Basic schematics of classes of retrosequences. Retroviruses are flanked by long terminal repeat (LTR) sequences (black boxes with white arrows) that contain a strong promoter and open reading frames (ORFs) that code for three essential genes, *gag*, *pol* and *env*, although they may have other genes as well. LTR retrotransposons are also flanked by LTRs and contain the equivalent of *gag* and *pol* genes. The non-LTR retrotransposons or LINE-like elements contain an RNA polymerase II promoter, two ORFs and a variable poly(A) tail. In the human LINE, L1, ORF 1 codes for an RNA-binding protein, while ORF 2 codes for a protein with endonuclease and reverse transcriptase activity. The nonautonomous retroposons or SINEs have an internal RNA polymerase III (pol III) promoter boxes A and B flanked at the 3' end by a variable A-rich tail. Retropseudogenes, or processed pseudogenes, arise from reverse transcription of spliced messenger RNAs of transcribed genes. They are characterized by an absence of a 5' promoter and introns, and the presence of flanking direct (solid arrows) repeats and a poly(A) segment. Diagram is not drawn to scale.

Human Mobile Retrosequences

The human genome contains all types of retrosequences, which account for about 40% of its mass, with a substantial part being LINEs and SINEs (Lander *et al.*, 2001). LTR retrotransposons make up 8% of the human genome, including MaLR, Mer4 and several families of endogenous retroviruses like HERV-K elements. There are two major forms of non-LTR retrotransposons found in the human genome. The LINE family L2 accounts for about 3–4% of the human genome, while L1 accounts for about 16%. Estimates indicate that the human genome contains from 23 000 to 33 000 retropseudogenes (Goncalves *et al.*, 2000). The nonautonomous, non-LTR retrotransposons fall into the general class termed SINEs. Although there are different types of SINEs (like MIRs), Alu elements are the most abundant in the human genome. Over one million copies of Alu elements are present in the human genome, contributing to approximately 11% of its mass.

Origin of Alu Elements

Alu elements are dimeric molecules, composed of two nonidentical units or arms joined in the middle by an A-rich region (Weiner *et al.*, 1986). Two phases are observed in Alu evolution: an ancient monomeric period with the origins of the progenitor sequences leading to the Alu family (**Figure 2**), and the evolutionarily more recent period (discussed below) involving the amplification of dimeric sequences (**Figure 3**). Alu elements are proposed to have originated from a partial deletion of a pseudogene of the 7SL RNA gene, an integral part of the signal recognition particle (SRP) involved in protein secretion (Weiner *et al.*, 1986). It is suggested that an Alu monomer arose through the deletion of the central 60% of an approximately 300 base pair (bp) long 7SL RNA molecule. This first old Alu-like monomeric family, or fossil Alu monomer (FAM), possibly arose early in mammalian radiation (Labuda and Zietkiewicz, 1994). Because the FAM still contains the internal polymerase III (pol III) promoter of the 7SL, it has the potential to transcribe RNA and amplify. Amplification and subsequent evolution of the FAM family generated the free left Alu monomer (FLA) and the free right Alu monomer (FRA) families of monomeric elements. These two families vary in sequence, where the Alu left arm is characterized by the absence of 31 bp present in FRA. Two subfamilies of FLA (A1 and C1) and one subfamily of FRA elements are present at modest copy numbers in the human genome. Detection of human-specific subfamilies suggests that both Alu monomer progenitor sequences were active after the emergence of humans. It is thought that the fusion of an FLA and FRA monomer gave rise to the progenitor of the Alu family.

Evolutionary Expansion of Alu Elements

Alu elements appear to be present only in primates, although there are some related elements in other organisms. Alu started to amplify about 65 million years ago (mya) with peak amplification between 60 and 35 mya (**Figure 3**). Because the non-LTR retrotransposons do not have mechanisms for specific removal of elements, this has resulted in a steady increase in copy number. Alus created during the peak

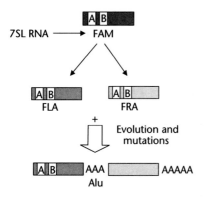

Figure 2 Origin of Alu elements. Alu elements are thought to have arisen from a processed 7SL RNA giving rise to the ancestral element: fossil Alu monomer (FAM). FAM evolved to the free left Alu monomer (FLA) and the free right Alu monomer (FRA) families with sequence variations between each other. The first progenitor of the Alu dimeric family possibly arose through the fusion of FLA and FRA monomers. The Alu left monomer still retains the internal polymerase III promoter with the A and B boxes. In Alu, the monomers are separated by an A-rich region, and the 3′ flank contains a poly(A) tract.

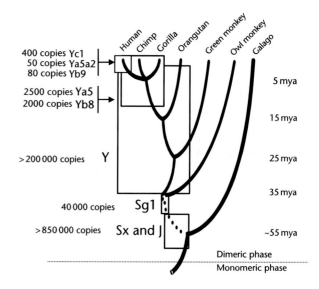

Figure 3 Evolutionary tree of the Alu subfamilies and amplification rates throughout the primate radiation. The old Alu subfamilies J, Sx and Sg1 were most active around 35 to 60 million years ago (mya) (indicated at the right) giving rise to the majority of the Alu elements present today in the human genome. Boxed areas represent the potential period of maximum activity for each Alu subfamily. The amplification rate of Alu decreased with evolutionary time as observed by the reduction of the copy numbers (indicated at the left). Currently the young Alu subfamilies (Y, Ya5, Ya8, Yb8, Yb9, Ya5a2 and Yc1) contribute to all known polymorphisms in the human genome.

amplification period were primarily members of the J and Sx subfamilies. The current rate of amplification of approximately one Alu insertion in every 200 births is about two orders of magnitude slower than that peak rate. The accepted model of Alu amplification proposes that only a few retroposition competent or 'master/source' Alu elements generated the copies present in the human genome (Deininger *et al.*, 1992). Thus, the rate is very dependent on the activity of the specific master elements. Any mutations in the master element would be observed in the copies, thus creating subfamilies of Alu elements reflecting the active master elements during any period of primate evolution. During more recent primate evolution, a series of smaller subfamilies derived from the Y subfamily (**Figure 3**) have been the only elements to contribute to the increase in copy number. Several of these subfamilies of young Alu elements inserted after the human radiation from the African apes are therefore not found in nonhuman primates. These young subfamilies are the only ones currently active in the human genome and can result in genetic polymorphism between individuals (Deininger and Batzer, 1999).

Each Alu element reflects the specific sequence of the master element that created it. However, once inserted, the copies begin to accumulate changes. Alu master elements are rich in CpG dinucleotides. In the copies, these positions are subject to methylation and mutation at high rates and therefore are lost relatively rapidly. The copies also gradually accumulate random mutations at a neutral rate, suggesting no selective pressure on the individual elements. Therefore, divergence from the consensus correlates well with the

insertion time of elements. This evolutionary pattern is somewhat confused by a limited number of gene conversions that can occur between elements and alter both subfamily features and divergence level.

Alu Elements may be L1 Parasites

The evolution of mobile elements, particularly nonautonomous elements such as Alu, is intricately tied to their mechanism of evolution. Several lines of evidence suggest that both Alu elements and processed pseudogenes are dependent on the endonuclease and reverse transcriptase activities provided by the L1 elements (Boeke, 1997). This is consistent with the observation that L1 and Alu share a similar evolutionary amplification pattern, although L1 elements are somewhat older. L1 elements also have a subfamily evolutionary history similar to that of Alu, suggesting that the amplification of L1 may be dominated by a small number of elements and the vast majority are also nonfunctional (Deininger *et al.*, 1992). It may be possible that the Alu subfamilies have evolved in order to efficiently parasitize the L1 retroposition apparatus as the L1 elements have evolved. In addition, genome analysis suggests that there were earlier SINE (MIR) and LINE (L2) families with a similar relationship, prior to the primate expansion (Smit, 1999). This suggests that this interrelated amplification of LINE and SINE families may be a common theme in genomic evolution.

L1 elements have been found to be preferentially located in A + T-rich regions, while Alu elements are located in G + C-rich genomic regions (Lander *et al.*, 2001). This seems inconsistent with their sharing of a common retrotransposition machinery. However, the youngest inserts of both types of elements show little bias relative to their enrichment in different genomic regions. This suggests that both types of elements may insert at relatively similar frequencies throughout the genome, but while the L1 insertions are preferentially lost from the gene-rich G + C-rich regions, the Alu elements are selectively lost from the A + T-rich regions. The relative depletion of Alus from A + T-rich regions is difficult to understand, but may represent some undefined selection for Alus in the G + C-rich regions (Lander *et al.*, 2001), or alternatively a selective loss mechanism (i.e. selective recombination or chromosomal bias against stability of Alus) in the A + T-rich regions).

Evolutionary Impact of Alu Elements on the Human Genome

There are over one million copies of Alu elements spread across the entire human genome (Lander *et al.*, 2001). The amplification of Alu elements has had both a positive and negative impact. Alu has had a negative effect on the genome both through insertional mutagenesis and Alu–Alu recombination. These two mechanisms account for 0.1% and 0.3% respectively of human genetic disorders (Deininger and Batzer, 1999) (for updates see the Website of the Deininger Laboratory of Selfish DNA, listed in Web Links). Alu–Alu recombination has caused disease both at the germ-line (genetic) and somatic (causing cancer) levels. Some of these unequal homologous recombinations occur in a higher frequency in specific genes, such as the low-density lipoprotein receptor (familial hypercholesterolemia) (*LDLR*) gene and the myeloid/lymphoid or mixed-lineage leukemia (trithorax homolog, *Drosophila*) (*MLL*) for acute myelogenous leukemia (Deininger and Batzer, 1999). It seems likely that the negative selection on an organism caused by inter-element recombination is the primary mechanism for limiting amplification rates of elements in populations.

Alu elements have also had a positive impact on the human genome. Recombination between Alu elements can result in deletion or duplication of sequences that may allow for duplication of exons and formation of

new protein variants. Some of the recombinations occurring between different chromosomes may have led to alterations involved in speciation. Insertion of an Alu can introduce a CpG island in a new location, possibly contributing to the evolution of gene expression and imprinting (Schmid, 1996). In addition, Alu elements have several half sites for steroid hormone receptors. There are some cases where an insertion of these sites through an Alu near a gene has altered gene expression. Overall, Alu elements have played an important role in the shaping of the human genome to what it is today, and continue to play a major role in human genetic instability.

See also
Long Interspersed Nuclear Elements (LINEs)
Short Interspersed Elements (SINEs)
Transposable Elements: Evolution
Transposons

References

Boeke JD (1997) LINEs and Alus – the polyA connection. *Nature Genetics* **16**: 6–7.

Deininger PL and Batzer MA (1999) Alu repeats and human disease. *Molecular Genetics and Metabolism* **67**: 183–193.

Deininger P, Batzer M, Hutchison IC and Edgell M (1992) Master genes in mammalian repetitive DNA amplification. *Trends in Genetics* **8**: 307–312.

Goncalves I, Duret L and Mouchiroud D (2000) Nature and structure of human genes that generate retropseudogenes. *Genome Research* **10**: 672–678.

Labuda D and Zietkiewicz E (1994) Evolution of secondary structure in the family of 7SL-like RNAs. *Journal of Molecular Evolution* **39**: 506–518.

Lander ES, Linton LM, Birren B, *et al.* (2001) Initial sequencing and analysis of the human genome. International Human Genome Sequencing Consortium. *Nature* **409**: 860–921.

Malik HS, Henikoff S and Eickbush TH (2000) Poised for contagion: evolutionary origins of the infectious abilities of invertebrate retroviruses. *Genome Research* **10**: 1307–1318.

Schmid CW (1996) Alu: structure, origin, evolution, significance and function of one-tenth of human DNA. *Progress in Nucleic Acid Research in Molecular Biology* **53**: 283–319.

Smit AF (1999) Interspersed repeats and other mementos of transposable elements in mammalian genomes. *Current Opinion in Genetics and Development* **9**: 657–663.

Weiner A, Deininger P and Efstradiatis A (1986) The reverse flow of genetic information: pseudogenes and transposable elements derived from nonviral cellular RNA. *Annual Review of Biochemistry* **55**: 631–661.

Further Reading

Batzer MA, Deininger PL, Hellmann-Blumberg U, *et al.* (1996) Standardized nomenclature for *Alu* repeats. *Journal of Molecular Evolution* **42**: 3–6.

Deininger P and Batzer MA (1993) Evolution of retroposons. In: Heckht MK, *et al.* (eds.) *Evolutionary Biology*, pp. 157–196. New York, NY: Plenum.

Deininger P and Batzer M (1995) SINE master genes and population biology. In: Maraia R (ed.) *The Impact of Short, Interspersed Elements (SINEs) on the Host Genome*, pp. 43–60. Georgetown, TX: RG Landes.

Deininger P and Roy-Engel A (2002) Mobile elements in animal and plant genomes. In: Craig NL, Craigie R, Gellert M and Lambowitz A (eds.) *Mobile DNA II*, pp. 1074–1092. Washington, DC: American Society for Microbiology.

Kass DH (2001) Impact of SINEs and LINEs on the mammalian genome. *Current Genomics* **2**: 199–219.

Kapitonov V and Jurka J (1996) The age of Alu subfamilies. *Journal of Molecular Evolution* **42**: 59–65.

Malik HS, Burke WD and Eickbush TH (1999) The age and evolution of non-LTR retrotransposable elements. *Molecular Biology and Evolution* **16**: 793–805.

Schmid CW (1998) Does SINE evolution preclude Alu function? *Nucleic Acids Research* **26**: 4541–4550.

Web Links

The Deininger Laboratory of Selfish DNA, includes updated information on Alu and L1 elements
http://129.81.225.52/

Low density lipoprotein receptor (familial hypercholesterolemia) (*LDLR*); Locus ID: 3949. Locus Link:
http://www.ncbi.nlm.nih.gov/LocusLink/LocRpt.cgi?l = 3949

Myeloid/lymphoid or mixed-lineage leukemia (trithorax homolog, Drosophila) (*MLL*); Locus ID: 4297. LocusLink:
http://www.ncbi.nlm.nih.gov/LocusLink/LocRpt.cgi?l = 4297

Low density lipoprotein receptor (familial hypercholesterolemia) (*LDLR*); MIM number: 606945. OMIM:
http://www.ncbi.nlm.nih.gov/htbin-post/Omim/dispmim?606945

Myeloid/lymphoid or mixed-lineage leukemia (trithorax homolog, Drosophila) (*MLL*); MIM number: 159555. OMIM:
http://www.ncbi.nlm.nih.gov/htbin-post/Omim/dispmim?159555

Retrotransposition and Human Disorders

Eric M Ostertag, *University of Pennsylvania School of Medicine, Philadelphia, Pennsylvania, USA*

Haig H Kazazian Jr, *University of Pennsylvania School of Medicine, Philadelphia, Pennsylvania, USA*

Retrotransposition is a 'copy-and-paste' mechanism whereby a retrotransposable element is copied from one genomic location and inserted into another genomic location, using a ribonucleic acid intermediate.

Intermediate article

Article contents

- Introduction
- Insertional Mutagenesis
- Unequal Homologous Recombination
- Frequency of Disease Caused by Retrotransposons

Introduction

Retrotransposons are a class of transposable element that are extremely prevalent in humans, constituting approximately one-third of the genome (International Human Genome Sequencing Consortium, 2001). Retrotransposons are classified as autonomous or nonautonomous based upon whether they encode proteins required for their retrotransposition. In the human genome, the most abundant retrotransposons are the autonomous L1 element and the nonautonomous Alu element. Many scientists consider retrotransposons to be molecular parasites that have expanded selfishly into the genome. By expanding into the genome over evolutionary time, it is clear that retrotransposons have played a substantial role in determining the structure of the genome, and the possibility exists that they may indeed serve some beneficial function for the host. However, their ability to occasionally cause genetic disease is irrefutable. The main mechanisms by which transposable elements cause disease are insertional mutagenesis and unequal homologous recombination. (*See* Short Interspersed Elements (SINEs); Telomeric and Subtelomeric Repeat Sequences.)

Insertional Mutagenesis

Insertional mutagenesis occurs when a retrotransposon inserts into or near a gene and thereby abolishes or alters the gene's expression or results in the production of a mutant protein. Most of the retrotransposons in the human genome have acquired mutations sufficient to inactivate them and therefore they cannot cause disease by insertional mutagenesis. However, some retrotransposons remain active. In 1988, the first examples of human disease caused by insertional mutagenesis were reported (Kazazian *et al.*, 1988). Two independent insertions of L1 elements had inserted *de novo* into the factor VIII gene (coagulation factor VIII, procoagulant component (hemophilia A)

(*F8*)) of hemophilia A patients. There are now 13 known recent or *de novo* L1 insertions that have resulted in independent cases of human disease.

Mechanism of retrotransposition

As autonomous retrotransposons, L1 elements encode the proteins required to promote their own mobilization. These elements produce a bicistronic ribonucleic acid (RNA), a single transcript with two open reading frames (ORFs). The first ORF encodes an RNA-binding protein and the second ORF encodes a protein with endonuclease and reverse transcriptase activity. While it is clear that both the ORF1 and ORF2 proteins are required for autonomous retrotransposition, it is unknown if additional host proteins are also required (Moran *et al.*, 1996).

The mechanism of human L1 retrotransposition remains largely undetermined. However, the study of these elements in a cultured cell assay has provided some clues and additional inferences can be made based upon the study of similar retrotransposons from other organisms (**Figure 1**). The L1 element is first transcribed from an internal promoter, probably by RNA polymerase II. The message is next translated in the cytoplasm by an unknown mechanism to produce ORF1 and ORF2 protein. After gaining access to the nucleus, the transcript is reverse transcribed and reintegrated into the genome. It is thought that reintegration occurs by a process called target primed reverse transcription (TPRT). During TPRT, the endonuclease domain of the ORF2 protein cleaves genomic deoxyribonucleic acid (DNA), creating a structure that the reverse transcriptase domain of the ORF2 protein can use to prime reverse transcription of the RNA (Luan *et al.*, 1993) (**Figure 2**). The retrotransposition process results in a DNA copy of the L1 element that produced the RNA transcript.

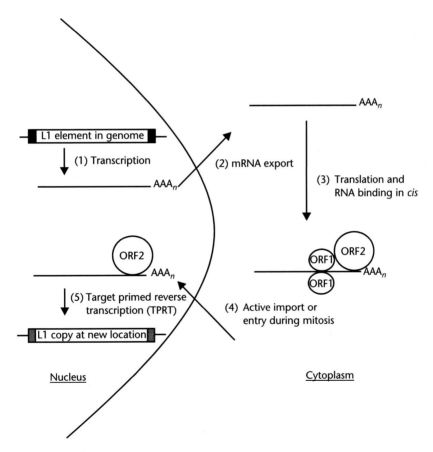

Figure 1 Simplified schematic of the proposed mechanism of L1 retrotransposition. (1) A full-length active L1 element in the genome is first transcribed using an internal promoter. (2) The L1 transcript is exported to the nucleus. (3) The ORF1 and ORF2 proteins are translated and preferentially bind the RNA molecule that encoded them (*cis* preference). (4) The L1 RNA and associated protein(s) return to the nucleus by active transport or entry during nuclear membrane breakdown at mitosis. (5) The L1 RNA is reverse transcribed and integrated into the genome by target primed reverse transcription (TPRT). The process depicted results in a DNA copy of the original L1 element at a new genomic location. Note that the target site duplications flanking the original L1 element (represented by black rectangles) will differ from the target site duplications flanking the L1 copy at a new genomic location (represented by shaded rectangles). The new L1 copy also often differs from the original by truncating or rearranging during the retrotransposition process.

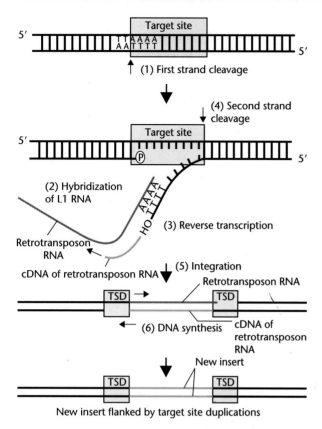

Figure 2 Target primed reverse transcription (TPRT). The L1 retrotransposon is thought to integrate by TPRT. (1) During L1 TPRT, the retrotransposon's endonuclease cleaves one strand of genomic DNA at its target site (rectangle), producing a 3' hydroxyl (OH) at the nick. (2) The retrotransposon RNA hybridizes at the nick. (3) The retrotransposon's reverse transcriptase uses the free 3' OH to prime reverse transcription. Reverse transcription proceeds, producing a cDNA of the retrotransposon RNA. (4) The endonuclease cleaves the second DNA strand of the target site to produce a staggered break. (5) The cDNA inserts into the break by an unknown mechanism. (6) Removal of RNA and completion of DNA synthesis produces a complete insertion flanked by target site duplications (TSDs).

Reintegration often creates a 7–20 base pair direct repeat of the endonuclease target site on each end of the inserted L1, called the target site duplication (TSD). The endonuclease domain of the L1 ORF2 protein displays a weak target site preference such that L1 insertions occur relatively randomly throughout the genome. However, a target site preference for the AA|TTTT hexanucleotide and minor variants is sufficiently strong such that the TSD can be used as a genetic signature of an insertion effected by the L1 endonuclease (Jurka, 1997).

In addition to the mutations caused by insertion of L1 elements, insertions of nonautonomous retrotransposons that are either causative or associated with disease include at least 18 insertions of Alu elements (Deininger and Batzer, 1999) and two insertions of an element termed SVA (Kobayashi et al., 1998; Rohrer et al., 1999). Recently inserted nonautonomous elements are flanked by TSDs that resemble those created by the L1 endonuclease. Therefore, it appears that all reports of de novo insertional mutagenesis by transposable elements in the human genome are the result of the L1 ORF2 protein working in cis to retrotranspose its own transcript or in trans to retrotranspose transcripts of nonautonomous retrotransposons. Recent studies of L1 retrotransposition in cultured cells have demonstrated that the L1 proteins show a strong preference for retrotransposing the transcript that encoded them (Esnault et al., 2000; Wei et al., 2001). Apparently, some nonautonomous retrotransposons have a mechanism to undermine the strong cis preference.

Mechanism of disease

Nine of the 13 reported disease-causing L1 insertions are inserted directly into a gene exon. Examples include two independent insertions into the factor VIII gene (F8) causing hemophilia A, four independent insertions into the dystrophin gene (dystrophin (muscular dystrophy, Duchenne and Becker types) (DMD)) causing Duchenne muscular dystrophy or X-linked dilated cardiomyopathy, an insertion into the cytochrome b-245, beta polypeptide (chronic granulomatous disease) (CYBB) gene causing chronic granulomatous disease, an insertion into the factor IX gene (coagulation factor IX (plasma thromboplastic component, Christmas disease, hemophilia B) (F9)) causing hemophilia B, and an insertion into the adenomatosis polyposis coli (APC) gene causing colon cancer. In most of these cases, the mechanism of disease is presumably the introduction of nonsense codons into the coding sequence or the skipping of the disrupted exon during splicing. For example, one of the dystrophin insertions results in exon skipping and subsequent out-of-frame translation. However, in the case of the L1 insertion causing X-linked dilated cardiomyopathy, the insertion was into the 5' untranslated region of the muscle exon 1 of the DMD gene. The insertion did not disrupt the dystrophin reading frame, but likely affected transcription or transcript stability. The disease-causing Alu insertions are also frequently the result of direct insertions into a gene exon. Examples include, but are not limited to, independent insertions into the F9 gene causing hemophilia B, an insertion into the eyes absent homolog 1 (Drosophila) (EYA1) gene causing brachiootorenal syndrome, an insertion into the fibroblast growth factor receptor 2 (bacteria-expressed kinase, keratinocyte growth factor receptor, craniofacial dysostosis 1, Crouzon syndrome, Pfeiffer syndrome, Jackson–Weiss syndrome) (FGFR2) gene causing Apert syndrome, and an insertion into the hydroxymethylbilane synthase (*HMBS*) gene causing acute intermittent porphyria.

The four known L1 inserts into gene introns cause disease by introducing alternative splice sites that result in an improperly spliced transcript, by decreasing transcription of the gene or by decreasing the stability of the primary transcript. For example, insertion of an L1 into intron 5 of the *CYBB* gene of a patient with chronic granulomatous disease and an L1 insertion into intron 7 of the Fukuyama type congenital muscular dystrophy (fukutin) (*FCMD*) gene of two related Fukuyama-type congenital muscular dystrophy patients both resulted in heterogeneous splicing, while insertions into intron 2 of the β-globin gene (hemoglobin, beta (*HBB*)) of a β-thalassemia patient and intron 1 of the retinitis pigmentosa 2 (X-linked recessive) (*RP2*) gene of a retinitis pigmentosa patient caused low or absent messenger RNA (mRNA) levels without evidence of aberrant splicing. Likewise, Alu elements can cause disease when inserting into gene introns. Examples include an insertion into the neurofibromin 1 (neurofibromatosis, von Recklinghausen disease, Watson disease) (*NF1*) gene of a neurofibromatosis patient resulting in exon skipping and out-of-frame translation, and an insertion into the *FGFR2* gene of a patient with Apert syndrome also resulting in exon skipping. Apert syndrome is a dominant gain-of-function disorder and in this case the Alu insertion resulted in ectopic expression of the mutant protein. Occasionally, an innocuous retrotransposon insertion into an intron can result in disease over time by accumulating mutations that create potential splice sites. An ancient Alu element in intron 3 of the ornithine aminotransferase (gyrate atrophy) (*OAT*) gene had acquired a single base mutation creating a donor splice site that resulted in aberrant splicing of the *OAT* mRNA (Mitchell *et al.*, 1991). A patient with gyrate atrophy of the choroid and retina inherited two copies of this mutation from consanguineous parents and produced almost no normal mRNA.

All but one of the L1 insertions occurred either prefertilization in the germ line or postfertilization and very early during embryogenesis. However, the insertion into the *APC* gene was a somatic event occurring in a colon cancer patient. The *APC* insertion was present in the cancerous cells but not in the surrounding normal cells. The pattern of retrotransposition corresponds well with the known activity of the mammalian L1 promoter. All studies to date have demonstrated L1 transcripts present only in germ-line cells and certain transformed cells but not in normal somatic tissues. As Alu elements are thought to use the L1 retrotransposition machinery for their mobility, it is not surprising that the Alu insertions also occurred in the germ line or early in development. One Alu insertion into the *MLVI2* locus causing an association with leukemia may represent a somatic event although normal tissue was not available for comparison. If L1 is able to retrotranspose in cancerous cells as in the case of the *APC* insertion, it is plausible that Alu elements can be mobilized in cancerous cells also.

Ten of the 13 reported disease-causing L1 insertions are into genes on the X chromosome. This is probably due in part to a selection bias for discovering recent insertions in hemizygous genes. As an example, consider genes that cause recessive disorders when mutated. A new L1 insertion into a gene on the X chromosome will manifest phenotypically in the first male offspring who inherits the mutation because he will have no wild-type copy of the gene, while a new L1 insertion into a gene on one of the autosomes will be phenotypically silent for both males and females unless the gene on their other autosome is also mutated. The X chromosome does contain a higher density of L1 elements than the autosomes, which could reflect an insertion preference for the X chromosome. However, the increased density of L1 on the X chromosome may also be a postinsertion selection bias such as a slower rate of loss of L1 elements from the X chromosome because of reduced recombination rates or positive selection because of possible participation of L1 elements in X-chromosome inactivation.

For a table summarizing the retrotransposon insertions resulting in human disease, visit the Website at the University of Pennsylvania Health System (see Web Links).

Unequal Homologous Recombination

Human disease can also occur when two similar transposable elements undergo unequal homologous recombination, thereby causing a deletion or duplication. It is no surprise that the transposable elements which are most abundant in the genome, the L1 and Alu elements, are responsible for all reported cases of human disease caused by unequal homologous recombination of transposable elements. There have been three reported examples of disease caused by L1 elements, one producing a partial deletion of the paired collagen type IV genes collagen, type IV, alpha 5 (Alport syndrome) (*COL4A5*) and collagen, type IV, alpha 6 (*COL4A6*), one producing a partial deletion of the human beta subunit of the phosphorylase kinase gene (*PHKB*), and one producing a partial deletion of the ataxia telangiectasia mutated (includes complementation groups A, C and D) (*ATM*) gene (Gatti, personal communication; Burwinkel and Kilimann, 1998; Segal *et al.*, 1999). These L1-mediated deletions resulted in Alport syndrome with associated diffuse leiomyomatosis, glycogen storage disease type two and ataxia telangiectasia respectively. A recent survey of disease-producing mutations caused by unequal homologous recombination of Alu elements reported 49 independent mutations occurring in both the germ-line and somatic cells (Deininger and Batzer, 1999).

It is unclear why Alu elements participate in a much greater number of unequal homologous recombination events than the L1 elements do, especially considering that L1 elements are on average longer than Alu inserts, providing longer stretches of sequence identity for recombination, and also make up a greater percentage of the genome by mass. Several possible explanations have been proposed to explain this discrepancy. First, Alu elements may contain sequences that make them more recombinogenic; however, this was not the finding in at least one experiment using cultured cells. Second, the average genomic distance between two L1 elements is greater than the average distance between two Alu elements, making any L1/L1 unequal recombination event both less likely to occur and more likely to result in a lethal mutation. Third, L1 elements tend to reside in more AT-rich DNA, while Alu elements reside in GC-rich DNA. As Alu elements are thought to use the L1 proteins for integration, this distribution is somewhat perplexing and may represent a postintegration selection bias. In any case, the fact that L1 elements tend to reside in the AT-rich, gene-poor DNA may indicate that L1/L1 unequal homologous recombination events do occur more frequently than suspected, but do not usually result in deletions of gene sequences.

Frequency of Disease Caused by Retrotransposons

Most of the several hundred thousand copies of L1 in the human genome are inactive and unable to cause insertional mutagenesis. L1 elements tend to truncate or rearrange during insertion (11 of the 13 disease-causing L1 insertions truncated during integration), resulting in an inactive copy of the progenitor element. Over time, L1 elements also accumulate spontaneous mutations that may inactivate them. Inactive L1 elements that can produce RNA are not retrotransposed in *trans* by active L1 elements because of the *cis* preference of the L1 proteins. Only full-length elements with ORFs are capable of insertional mutagenesis. Interestingly, 12 of the 13 reported recent disease-causing L1 insertions contain a short diagnostic nucleotide sequence indicating that they are members of a small active subfamily of L1 elements termed the Ta subfamily. Current estimates indicate that the average human diploid genome contains

between 40 and 70 active L1 elements (International Human Genome Sequencing Consortium, 2001; Sassaman *et al.*, 1997). Similarly, the *de novo* Alu element insertions are members of closely related subfamilies termed Ya5, Yb8 and Alu Y, indicating that only a small number of the Alu elements in the genome are currently capable of insertional mutagenesis (Deininger and Batzer, 1999).

The frequency of insertional mutagenesis caused by transposable elements has not yet been estimated experimentally. However, estimates can be calculated by taking the number of new mutations caused by insertion of transposable elements and dividing by the total number of characterized human mutations. Although arguments can be made that such calculations may be biased on either the high side or the low side, calculations by several groups indicate that L1 and Alu insertions each contribute to approximately 0.1% of human disease (Kazazian, 1999; Deininger and Batzer, 1999). Similar calculations indicate that approximately 1 in 4 to 1 in 100 people will have a new transposable element inserted somewhere in their genome, a fraction of which will insert into and mutate genes. An additional 0.3% or more of human disease is caused by unequal homologous recombination of retrotransposable elements. Taken together, these data indicate that retrotransposable elements continue to be a notable cause of human disease.

See also
Long Interspersed Nuclear Elements (LINEs)
Repetitive Elements and Human Disorders
Retrosequences and Evolution of Alu Elements
Transposons

References

Burwinkel B and Kilimann MW (1998) Unequal homologous recombination between LINE-1 elements as a mutational mechanism in human genetic disease. *Journal of Molecular Biology* 277: 513–517.

Deininger PL and Batzer MA (1999) Alu repeats and human disease. *Molecular Genetics and Metabolism* 67: 183–193.

Esnault C, Maestre J and Heidmann T (2000) Human LINE retrotransposons generate processed pseudogenes. *Nature Genetics* 24: 363–367.

International Human Genome Sequencing Consortium (2001) The human genome: initial sequencing and analysis. *Nature* 409: 860–921.

Jurka J (1997) Sequence patterns indicate an enzymatic involvement in integration of mammalian retroposons. *Proceedings of the National Academy of Sciences of the United States of America* 94: 1872–1877.

Kazazian Jr HH (1999) An estimated frequency of endogenous insertional mutations in humans. *Nature Genetics* 22: 130.

Kazazian Jr HH, Wong C, Youssoufian H, *et al.* (1988) Haemophilia A resulting from *de novo* insertion of L1 sequences represents a novel mechanism for mutation in man. *Nature* 332: 164–166.

Kobayashi K, Nakahori Y, Miyake M, *et al.* (1998) An ancient retrotransposal insertion causes Fukuyama-type congenital muscular dystrophy. *Nature* 394: 388–392.

Luan DD, Korman MH, Jakubczak JL and Eickbush TH (1993) Reverse transcription of R2Bm RNA is primed by a nick at the chromosomal target site: a mechanism for non-LTR retrotransposition. *Cell* **72**: 595–605.

Mitchell GA, Labuda D, Fontaine G, *et al.* (1991) Splice mediated insertion of an Alu sequence inactivates ornithine δ-aminotransferase: a role for Alu elements in human mutation. *Proceedings of the National Academy of Sciences of the United States of America* **88**: 815–819.

Moran JV, Holmes SE, Naas TP, *et al.* (1996) High frequency retrotransposition in cultured mammalian cells. *Cell* **87**: 917–927.

Rohrer J, Minegishi Y, Richter D, Eguiguren J and Conley ME (1999) Unusual mutations in *Btk*: an insertion, a duplication, an inversion, and four large deletions. *Clinical Immunology* **90**: 28–37.

Sassaman DM, Dombroski BA, Moran JV, *et al.* (1997) Many human L1 elements are capable of retrotransposition. *Nature Genetics* **16**: 37–43.

Segal Y, Peissel B, Renieri A, *et al.* (1999) LINE-1 elements at the sites of molecular rearrangements in Alport syndrome-diffuse leiomyomatosis. *American Journal of Human Genetics* **64**: 62–69.

Wei W, Gilbert N, Ooi SL, *et al.* (2001) Human L1 retrotransposition: *cis*-preference vs. *trans*-complementation. *Molecular and Cellular Biology* **21**: 1429–1439.

Furano AV (2000) The biological properties and evolutionary dynamics of mammalian LINE-1 retrotransposons. *Progress in Nucleic Acid Research and Molecular Biology* **64**: 255–294.

Kazazian Jr HH (2000) L1 retrotransposons shape the mammalian genome. *Science* **289**: 1152–1153.

Kazazian Jr HH and Moran JV (1998) The impact of L1 retrotransposons on the human genome. *Nature Genetics* **19**: 19–24.

Luan DD and Eickbush TH (1996) Downstream 28S gene sequences on the RNA template affect the choice of primer and the accuracy of initiation by the R2 reverse transcriptase. *Molecular and Cellular Biology* **16**: 4726–4734.

Makalowski W, Mitchell GA and Labuda D (1994) Alu sequences in the coding regions of mRNA: a source of protein variability. *Trends in Genetics* **10**: 188–193.

Maraia RJ (1995) Alu elements as a source of genomic variation: deleterious effects and evolutionary novelties. In: Maraia RJ (ed.) *The Impact of Short Interspersed Elements (SINEs) on the Host Genome*, pp. 1–24. Georgetown, TX: Landes Bioscience.

Miki Y (1998) Retrotransposal integration of mobile genetic elements in human diseases. *Journal of Human Genetics* **43**: 77–84.

Moran JV and Gilbert N (2001) Mammalian LINE-1 retrotransposons and related elements. In: Craig N (ed.) *Mobile DNA*, pp. 836–869. Washington, DC: American Society for Microbiology.

Further Reading

Deragon JM and Capy P (2000) Impact of transposable elements on the human genome. *Annals of Medicine* **32**: 264–273.

Web Links

University of Pennsylvania Health System
http://www.med.upenn.edu/genetics/labs/kazazian/human.html

Retroviral Repeat Sequences

Intermediate article

Dixie L Mager, *British Columbia Cancer Agency, Vancouver, Canada*

Patrik Medstrand, *Lund University, Lund, Sweden*

Retroviral repeats are genomic DNA segments that have structural or sequence similarity to the integrated (proviral) forms of vertebrate retroviruses. Such sequences are termed endogenous retrovirus-like elements or, more commonly, endogenous retroviruses.

Article contents
• Origin and Sequence
• Diversity and Classification
• Age of Retroviral Repeats in the Genome
• Frequency and Distribution
• Possible Biological Relevance

Origin and Sequence

It is generally accepted that retroviral repeats are the remnants of ancient germ-line infections of exogenous retroviruses that became fixed in the species. It is also possible that some retrovirus-like repeats did not originate from infectious entities but instead are ancient retrotransposons. Regardless of their origin, these sequences now make up a normal part of the human genome and are inherited in a Mendelian fashion.

Segments of deoxyribonucleic acid (DNA) are categorized as 'retroviral' if they have some features in common with the prototypical proviral structure

shown in **Figure 1**. A typical provirus has two identical long terminal repeats (LTRs), contains *gag*, *pol* and *env* genes, and is flanked by an integration-site target duplication of 4–6 base pairs (bp). Structurally intact retroviral elements (ERVs) resemble full-length proviruses, with two LTRs bordering internal regions that have similarity to the *gag*, *pol* and *env* genes of known retroviruses and are typically 8–9 kilobases (kb).

Different human ERVs are classified into families or groups on the basis of sequence similarity to infectious retroviruses (Wilkinson *et al.*, 1994; Andersson *et al.*, 1999; Tristem, 2000); however, most retroviral repeat sequences in the human genome

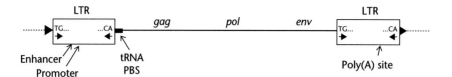

Figure 1 Prototypical structure of an integrated retrovirus or provirus. The identical LTRs, shown as open boxes, flank an internal region containing the three genes of simple retroviruses. *gag* encodes structural proteins, *pol* encodes RNase H, reverse transcriptase and endonuclease, and *env* encodes the viral envelope coat protein that recognizes a cell surface receptor and enables the virus to infect the cell. Noninfectious LTR-containing retrotransposons have a similar structure but lack the *env* gene. LTRs typically begin with TG and end with CA, which are part of a short inverted repeat shown as arrows within the open boxes. Each LTR also contains a transcriptional enhancer, promoter and polyadenylation signal. The tRNA primer binding site (PBS) is a sequence of 18–21 bp that is complementary to the 3′ end of a specific tRNA. This site binds a tRNA molecule that is necessary to prime viral reverse transcription. Cellular DNA is shown as a dotted line and the cellular sequence of 4–6 bp that is duplicated on integration is shown as two thick arrows bordering the provirus. A solitary LTR, created by recombination between two LTRs, is also bordered by a direct repeat of 4–6 bp. The diagram is not drawn to scale.

resemble solitary LTRs, which are presumably derived from recombination events between the 5′ and 3′ LTRs of an intact proviral element. This presumption has been proved for some genomic loci that contain a solitary LTR in humans but contain the progenitor intact retroviral element in other primates (Wilkinson *et al.*, 1994). Solitary endogenous LTRs typically vary from 300 to 1000 bp, which is similar to the size range of LTRs from different classes of exogenous retroviruses. In addition, a group of anonymous DNA repeats is considered to be 'LTR-like' if members contain promoter and polyadenylation motifs, end in a TG... / ...CA inverted repeat, and are flanked by a short target site duplication, even if none of the sequences is associated with internal gene-like regions. (*See* DNA Recombination; Transposable Elements: Evolution.)

Most human ERV sequences (HERVs) are not coding-competent because either they are partially deleted or they have accumulated single nucleotide changes through time that disrupt the open reading frames (ORFs) of the retroviral genes. However, some members of the HERV-K group contain intact ORFs and have been shown to encode retrovirus-like gag and pol proteins and a small protein, c-orf, that is functionally homologous to the *rev*-encoded protein of human immunodeficiency virus (HIV; Löwer *et al.*, 1996). At least one HERV-W element encodes an env-like protein (see below) as does the single-copy element HERV-R. HERV-K elements also encode immature retrovirus-like particles, which have been detected in placenta and in some testicular tumor cell lines, but there is no evidence that HERV-K or any other endogenous retrovirus in humans retains the capacity to produce infectious virus.

Diversity and Classification

More than 200 different endogenous retroviral or LTR-like groups of elements have been identified

through a systematic cataloging of human repeats contained in the Repbase database (Jurka, 2000), but relatively few of these have been analyzed in any depth. Different HERV groups have commonly been named by using the single-letter amino acid code for the transfer ribonucleic acid (tRNA) that is complementary to the retroviral primer-binding site (PBS) located just internal to the 5′ LTR (**Figure 1**). For example, HERV-K and HERV-W denote specific retroviral groups that have lysine and tryptophan tRNA PBSs, respectively. This naming scheme is not universal, however, and it is in need of revision because different groups with the same PBS have been characterized (**Table 1**).

Table 1 contains a list of HERV groups that have some full-length members. Some of these groups have been reasonably well studied, but information on many others is limited to their entry in RepBase. HERV groups generally fall into three classes. Class I groups share internal sequence similarity with the type C mammalian retroviruses, now termed gamma-retroviruses (the prototype being mouse Moloney leukemia virus; MLV). Class II groups share similarity to type B or beta-retroviruses (typified by type B mouse mammary tumor virus; MMTV). A third broad class is sometimes denoted as class III and has limited similarity to spumaretroviruses.

Figure 2 shows the relationship of DNA sequences from the reverse transcriptase domain of *pol* between selected exogenous retroviruses and all HERVs that have sufficient internal sequence for comparison. The clustering of HERVs into the three classes is notable, as is the diversity of sequences. Some HERV groups are related in regions of *pol*, but are justifiably defined as distinct groups because they have unique LTRs with little, if any, similarity to the LTRs of any other group. Databases for HERV sequences include the National Center for Biotechnology Information (NCBI) Taxonomy Browser and the Human Endogenous Retroviruses Database (see Web Links).

Table 1 Human endogenous retroviral groups

Common name[a]	Repbase name		Other names	PBS[b]	Copies[c]
	Internal	LTR			
Class I					
HERV-I	HERVI	LTR10	RTVL-I, HERV-FTD	I	250 (1000)
HERV-ADP	HERVP71A_I	LTR71A		P	40 (300)
HERV-HS49C23	MER57I	MER57		L	200 (1000)
MER110	MER110I	MER110,110A		?	20 (60)
HERV-Rb	PABL_BI	PABL_A, PABL_B		R/L	8 (1000)
HERV-E	HERVE	LTR2,2B,2C	4-1	E	250 (1000)
HERV-R	HERV3	LTR4	ERV-3	R/L	100 (125)
RRHERV-I	HERV15I	LTR15		I	40 (250)
S71	HERVS71	LTR6A,B		T	80 (400)
PRIMA41	PRIMA41	MER41A,41B,41C,41D		?	40 (3000)
HERV-Z69907	MER66I	MER66A,66B		R	50 (1500)
HERV-FRD	MER50I	MER50		?	50 (2000)
MER84	MER84I	MER84		I	25 (300)
HERV-W	HERV17	LTR17	MSRV	W/R	40 (1100)
HERV30	HERV30I	LTR30		R/L	35 (120)
ERV9	HERV9	LTR12,PTR5		R	300 (5000)
MER52A	MER52AI	MER52A,52B,52C		P	200 (1800)
HERV-P	HUERS-P3	LTR9	HuERS-P3	R/G	200 (1000)
HERV-XA34	HERVFH21	LTR21A	HERV-F(XA-34)	F	30 (40)
HERV-H	HERVH	LTR7	RTVL-H, RGH	H	1000 (1000)
HERV46	HERV46I	LTR46		F	40 (200)
HERV-Fb	HERVH48I	MER48		F	60 (100)
HERV-F	HERVFH19I	LTR19		F	45 (550)
Class II					
HML-1	HERVK14I	LTR14A,14B	NMWV6	K	70 (350)
HERV-K(HML-2)	HERVK	LTR5	HERV-K,HERV-K10, HTDV,NMWV1	K	60 (2500)
HML-3	HERVK9I	MER9	HERV-K70A,HERV76, HERV50,NMWV5	K/Y	150 (700)
HML-4	HERVK13I	LTR13	ERV-MLN,HERV-K(T47D)	K	10 (800)
HML-5	HERVK22I	LTR22,22A,22B	NMWV2	I/M	100 (600)
HERV-K(HML-6)	HERVK3I	LTR3,3B	NMWV4	K	50 (400)
HML-7	HERVK11D1	MER11D	NMWV7	K	20 (140)
HML-8	HERVK11I	MER11A,11B,11C	NMWV3	?	60 (600)
HML-9	–	–	NMWV9	W	10 (40)
HERV-K(C4)	HERVKC4,	LTR14	HML-10	K	10 (100)
HERVK14C	HERVK14CI	LTR14C		K	15 (120)
Class III					
HERV-L	HERVL,ERVL	MLT2A1,2A2,MLT2B1,2B2		L	200 (6000)
HERV16	HERV16	LTR16,16A,16A1,16C,16D		L	15 (25)
HERV-S	HERV18	LTR18,18B		S	50 (150)
HERV57	HERV57	LTR57		S	30 (200)
MER70	MER70I	MER70A,70B		S	40 (200)
HERV52	HERV52	LTR52		S	10 (50)
HERVL74	HERVL74	MER74,74A,74B,74C		S	25 (200)
HERVL40	HERVL40	LTR40A,40B,40C		S	10 (300)

[a]Published or Repbase designation.

[b]Single-letter amino acid code.

[c]Estimated copy number per haploid genome of internal sequences and solitary LTRs in parentheses. Number of elements were determined by BLAST searches of human DNA sequences larger than 100 kb. Copy numbers for the whole genome were estimated from the 25% of the human genome sequence that was available in the nonredundant database at the NCBI in August 2001.

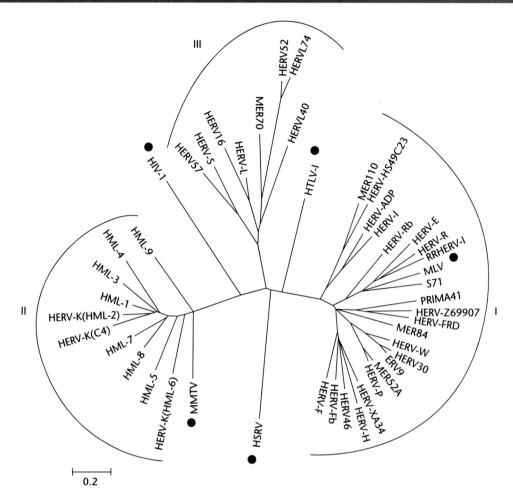

Figure 2 Relationship of exogenous and human endogenous retroviruses. The tree depicts the relationship between selected exogenous retroviruses and entries in Repbase or previously reported HERV sequences in a 550-bp region encoding the reverse transcriptase of the retroviral polymerase. Sequences fall into one of three (I–III) classes. Exogenous retroviruses are indicated with a filled circle. HIV: human immunodeficiency virus; HSRV: human spumavirus; HTLV: human T-cell leukemia virus. The branch length corresponds to the sequence distance indicted below the class II sequences. (This tree was constructed using programs CLUSTALW, and DNADIST and KITSCH of PHYLIP.)

Not listed in **Table 1** is the highly repetitive mammalian apparent LTR retrotransposon (MaLR) superfamily of repeats (Smit, 1993). Full-length MaLR elements (including transposon-like human element (THE-1) repeats) have two LTRs but their internal regions of ~1.5 kb lack similarity to *pol* genes. MaLRs are considered as retroviral or retrotransposon-like because of their structure and because some members have similarity to the *gag* region of HERV-L, which suggests that MaLRs and HERV-L elements may share a common origin. Many other retrovirus-like sequences reported in Repbase are partial elements or are a chimera of several types of repeat and so are difficult to classify.

An additional complication to cataloging retroviral repeats is that concerted evolution or recombination between elements has occurred repeatedly throughout evolution (Johnson and Coffin, 1999; Costas and Naveira, 2000). Furthermore, for many HERV groups, most members that retain both LTRs are still partially deleted, which can hamper classification efforts. There are also over 100 groups of solitary LTR-like repeats for which internal retroviral regions do not exist or have not been identified. (*See* Transposons.)

Age of Retroviral Repeats in the Genome

The endogenous retroviral repeats present in the human genome have also been found in primates, which indicates that these elements are not specific to humans but have been integrated during the course of primate evolution. All groups that have been analyzed are present in higher primates, including Old World monkeys, but most of them are not detectable in New World monkeys. These findings indicate that most retroviral groups of an exogenous origin first entered the germ line and became fixed in primate common ancestors roughly 30 million years (Myr) ago, which is the estimated time of divergence between the New World and Old World branches. HERV-L and MaLR elements, however, can be traced back 100–150 Myr (Smit, 1993; Bénit *et al.*, 1999). (*See* Evolutionary History of the Human Genome; Primates: Phylogenetics.)

Several studies have shown that ERV subpopulations amplified or spread throughout the genome at different times in evolution. For example, at least nine distinct subtypes of HERV-K LTR that vary in age and/or integration time can be distinguished. The youngest of these was amplified after the divergence of humans and chimpanzees, less than 5 Myr ago (Medstrand and Mager, 1998). Distinct periods of amplification have also been noted for other ERV groups, such as ERV-9 (Costas and Naveira, 2000) and HERV-H (Wilkinson *et al.*, 1994; Anderssen *et al.*, 1997), but all integrations seem to be older than the youngest subpopulations of HERV-K sequences. **Figure 3** depicts the approximate time line of amplifications of these three HERV groups during primate evolution. Evidence suggests that the genomic spread or amplification of ERV elements occurred primarily through intracellular retrotransposition as opposed to repeated infections, but the latter mechanism cannot be ruled out. Notably, in some cases, amplification of a partially deleted ERV has occurred in preference to its full-length progenitor (Wilkinson *et al.*, 1994). Although most retroviral repeats are fixed in humans, insertional polymorphisms of full-length HERV-K elements have been reported, which raises the possibility that members of this family are still actively retrotransposing (Turner *et al.*, 2001).

One striking feature evident from **Figure 2** is the lack of similarity of any HERV to the known human exogenous retroviruses HIV, human T-cell leukemia virus 1 (HTLV-1) and human spumavirus (HSRV). In fact, the variety of endogenous retroviral elements reflects the wide spectrum of infectious retroviruses that have bombarded the genome in the evolutionary past but have since become extinct as infectious entities. It should also be emphasized that the genomic retroviral elements that exist today represent only a small fraction of total germ-line integration events that have occurred during primate evolution, namely those that were not detrimental to the host and that also became fixed in the genome of common ancestors. (*See* Fixation Probabilities and Times.)

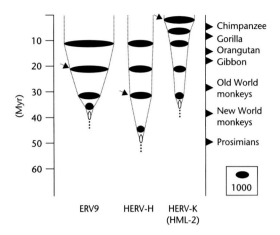

Figure 3 Examples of expansion patterns of three HERV groups during primate evolution. Unfilled ovals represent the approximate time of the initial fixation of the HERV element in the primate lineage. Filled ovals indicate approximate windows in time of different expansions of the elements, where the widths of the oval correspond roughly to the total copy number of the HERVs and the solitary LTRs at the time indicated on the left. An arrow indicates a major burst or expansion resulting in a significantly higher copy number of the elements. The left axis indicates the time in millions of years (Myr) from the present. The right axis indicates the approximate time that different lineages diverged from humans. The data shown are derived mainly from Anderssen *et al.* (1997), Medstrand and Mager (1998) and Costas and Naveira (2000).

Frequency and Distribution

Together, LTR and ERV-like repeats comprise 8% of the human genome. Copy numbers of individual HERV groups vary widely, with many solitary LTR groups being present in several thousand copies (**Table 1**). MaLR elements are more numerous than the ERVs in humans, with estimated copy numbers of over 150 000 for full-length elements and solitary LTRs together. In general, the frequency of solitary LTRs in a given group greatly outweighs that of full-length elements, which suggests that a solitary LTR is genetically more stable and/or less deleterious than a full-length element. The rate of homologous recombination between repeats is known to decrease markedly with increasing sequence divergence. Therefore, because the 5′ and 3′ LTRs of an integrated provirus are initially identical and then diverge with time, it is likely that LTR–LTR recombination occurs relatively soon after integration. (*See* Isochores.)

Some studies using *in situ* hybridization and somatic cell hybrids with single human chromosomes suggest that ERV integration patterns might not be completely random. Particular regions, notably within the major histocompatibility complex (MHC), seem to have a higher number of ERV/LTR elements than would be expected by chance, but this may be accounted for, at least in part, by large DNA duplications rather than by distinct integration events. It has also been shown that retroviral repeats tend to be underrepresented in regions of high guanine plus cytosine (G + C) content. We have found that various retroviral-like repeats are distributed differently with respect to surrounding (G + C) content in the human genome. Elements of older age are found more frequently in regions of low (G + C) compared to young elements of the same retroviral-like class (Medstrand et al., 2002). The same study also found that retroviral LTRs typically are excluded from gene regions and in regions close to genes suggesting that retroviral repeats may frequently interfere with gene transcription. (*See* Isochores.)

Possible Biological Relevance

Most human LTR-like or retroviral repeats presumably have no direct functional role in the cell and are therefore typically considered 'junk' or parasitic DNA. However, it is difficult to imagine that such a significant 'retroviral load' has not affected the organization and evolution of the human genome. For example, it is possible that retroviral repeats within the MHC have contributed to the rapid diversification of this region and to associated differences in susceptibility to disease. There has also been much speculation that some elements may themselves contribute to certain pathogenic conditions, particularly autoimmune diseases or cancer. It is certainly well documented that endogenous retroviruses in other species, such as mice and chickens, can be pathogenic.

Although transcript levels of endogenous retroviruses can increase in certain cancers or other diseases, there is little definitive evidence for a causative role of an ERV or related sequence in any human disease. Unlike in mice where close to 10% of mutations are estimated to be due to endogenous retroviral insertions, no mutations caused by insertions of retroviral elements have been found in humans. Repeats, particularly Alu elements, can also cause disease by acting as targets for illegitimate recombination that leads to deletions, inversions or duplications. A few reports suggest that MaLR elements may promote recombination, and there is an example of HERV–HERV recombination that deletes a gene involved in spermatogenesis, but this mechanism is not thought to be a frequent cause of human mutations. (*See* Repetitive Elements and Human Disorders; Retrotransposition and Human Disorders.)

Several nonpathogenic roles have been postulated for human endogenous retroviruses. In mice, endogenous retroviral sequences can confer protection against exogenous retrovirus infection, and it is possible that some ERVs have had a similar function

during human evolution. Indeed, the endogenization of a retrovirus may well lead to the demise of its infectious counterpart. There is no direct evidence for such a mechanism in humans, but it is intriguing that no active human exogenous virus with significant similarity to an ERV has been identified (**Figure 2**).

Another postulated role is that an ERV or ERVs may function in normal development of the placenta. Retroviral envelope proteins naturally have fusiogenic properties and can induce the formation of syncytia between cells – a process that is similar to the formation of syncytiotrophoblasts in the placenta. Env proteins encoded by at least two retroviral groups, HERV-R (or ERV-3) and HERV-W, are highly expressed in syncytiotrophoblasts, and studies have shown that an HERV-W *env* protein, termed syncytin, promotes syncytia formation in cell lines (Mi *et al.*, 2000). It is therefore possible that a retroviral protein may have an important role in the placenta, but this remains to be demonstrated definitively.

Because retroviral LTRs naturally contain transcriptional promoter, enhancer and polyadenylation signals (**Figure 1**), it is possible that some retroviral repeats have evolved to influence the expression of adjacent genes, and many examples of such a phenomenon have been found in humans and other species (Brosius, 1999). One well-documented example involves the human salivary amylase gene cluster where it has been shown that HERV-E sequences contribute to parotid-specific expression. An HERV-E element also influences expression of the *pleiotrophin* (*PTN*) gene in placenta, and an HERV-K sequence provides part of the coding region for one form of the *leptin receptor* (*LEPR*). Other examples of human genes influenced by HERVs or LTRs include *HERV-H LTR-associating 2* (*HHLA2*), a member of the immunoglobulin gene superfamily that uses an HERV-H LTR polyadenylation signal; *zinc-finger protein 80* (*ZNF80*), which is promoted by a solitary ERV-9 LTR; and *apolipoprotein C-I* (*APOC1*) and *endothelin receptor type B* (*EDNRB*), both of which use an HERV-E LTR as an alternative promoter (Medstrand *et al.*, 2001). Given the large number of LTR-like elements in the genome, it is likely that many of these ancient sequences have assumed gene regulatory functions. (*See* Alternative Promoter Usage.)

See also
Repetitive Elements: Detection

References

Anderssen S, Sjottem E, Svineng G and Johansen T (1997) Comparative analyses of LTRs of the ERV-H family of primate-specific retrovirus-like elements isolated from marmoset, African green monkey and man. *Virology* **234**: 14–30.

Andersson ML, Lindeskog M, Medstrand P, *et al.* (1999) Diversity of human endogenous retrovirus class II-like sequences. *Journal of General Virology* **80**: 255–260.

Bénit L, Lallemand JP, Casella JF, Philippe H and Heidmann T (1999) ERV-L elements: a family of endogenous retrovirus-like elements active throughout the evolution of mammals. *Journal of Virology* **73**: 3301–3308.

Brosius J (1999) Genomes were forged by massive bombardments with retroelements and retrosequences. *Genetica* **107**: 209–238.

Costas J and Naveira H (2000) Evolutionary history of the human endogenous retrovirus family ERV9. *Molecular Biology and Evolution* **17**: 320–330.

Johnson WE and Coffin JM (1999) Constructing primate phylogenies from ancient retrovirus sequences. *Proceedings of the National Academy of Sciences of the United States of America* **96**: 10254–10260.

Jurka J (2000) Repbase update: a database and an electronic journal of repetitive elements. *Trends in Genetics* **16**: 418–420.

Löwer R, Löwer J and Kurth R (1996) The viruses in all of us: characteristics and biological significance of human endogenous retrovirus sequences. *Proceedings of the National Academy of Sciences of the United States of America* **93**: 5177–5184.

Medstrand P, Landry JR and Mager DL (2001) Long terminal repeats are used as alternative promoters for the endothelin B receptor and apolipoprotein C1 genes in humans. *Journal of Biological Chemistry* **276**: 1896–1903.

Medstrand P and Mager DL (1998) Human-specific integrations of the HERV-K endogenous retrovirus family. *Journal of Virology* **72**: 9782–9787.

Medstrand P, Van De Lagemaat LN and Mager DL (2002) Retroelement distributions in the human genome: variations associated with age and proximity to genes. *Genome Research* **12**: 1483–1495.

Mi S, Lee X, Li XP, *et al.* (2000) Syncytin is a captive retroviral envelope protein involved in human placental morphogenesis. *Nature* **403**: 785–789.

Smit AF (1993) Identification of a new, abundant superfamily of mammalian LTR-transposons. *Nucleic Acids Research* **21**: 1863–1872.

Tristem M (2000) Identification and characterization of novel human endogenous retrovirus families by phylogenetic screening of the human genome mapping project database. *Journal of Virology* **74**: 3715–3730.

Turner G, Barbulescu M, Su M, *et al.* (2001) Insertional polymorphisms of full-length endogenous retroviruses in humans. *Current Biology* **11**: 1531–1535.

Wilkinson DA, Mager DL and Leong JC (1994) Endogenous human retroviruses. In: Levy JA (ed.) *The Retroviridae*, vol. 3, pp. 465–535. New York, NY: Plenum Press.

Further Reading

Andersson G, Svensson AC, Setterblad N and Rask L (1998) Retroelements in the human MHC class II region. *Trends in Genetics* **14**: 109–114.

Bock M and Stoye JP (2000) Endogenous retroviruses and the human germline. *Current Opinion in Genetics and Development* **10**: 651–655.

Boeke JD and Stoye JP (1997) Retrotransposons, endogenous retroviruses, and the evolution of retroelements. In: Coffin JM, Hughes SH and Varmus HE (eds.) *Retroviruses*, pp. 343–435. Cold Spring Harbor, NY: Cold Spring Harbor Laboratory Press.

Boese A, Sauter M, Galli U, *et al.* (2000) Human endogenous retrovirus protein cORF supports cell transformation and associates with the promyelocytic leukemia zinc finger protein. *Oncogene* **19**: 4328–4336.

Dawkins R, Leelayuwat C, Gaudieri S, *et al.* (1999) Genomics of the major histocompatibility complex: haplotypes, duplication, retroviruses and disease. *Immunological Reviews* **167**: 275–304.

Hughes JF and Coffin JM (2001) Evidence for genomic rearrangements mediated by human endogenous retroviruses during primate evolution. *Nature Genetics* **29**: 487–489.

Löwer R (1999) The pathogenic potential of endogenous retroviruses: facts and fantasies. *Trends in Microbiology* **7**: 350–356.

Sverdlov ED (2000) Retroviruses and primate evolution. *BioEssays* **22**: 161–171.

Towers G, Bock M, Martin S, *et al.* (2000) A conserved mechanism of retrovirus restriction in mammals. *Proceedings of the National Academy of Sciences of the United States of America* **97**: 12 295–12 299.

Whiteclaw E and Martin DIK (2001) Retrotransposons as epigenetic mediators of phenotypic variation in mammals. *Nature Genetics* **27**: 361–365.

Web Links

Human Endogenous Retroviruses Database
http://herv.img.cas.cz/
National Center for Biotechnology Information (NCBI)
http://www.ncbi.nlm.nih.gov/
NCBI BLAST
http://www.ncbi.nlm.nih.gov/BLAST/
NCBI Taxonomy Browser
http://www.ncbi.nlm.nih.gov/htbin-post/Taxonomy/wgetorg?
mode = Undef&id = 11632&lvl = 3&genome = 1&srchmode = 1
Repbase Update
http://www.girinst.org/Repbase_Update.html

Retroviral Vectors in Gene Therapy

A Dusty Miller, *Fred Hutchinson Cancer Research Center, Seattle, Washington, USA*

Retroviruses have a positive-strand nonsegmented RNA genome that undergoes reverse transcription to a double-stranded DNA form prior to random integration into the genome of a target cell in a form called the provirus. The provirus is transcribed into mRNAs that encode the viral proteins, and these proteins package the full-length genomic mRNA into virions to complete the virus life cycle. Retroviral vectors are retroviruses engineered to transfer and integrate specific DNA sequences into the genomes of target cells, and are often designed to be replication-defective to avoid further spread after the initial transfer event. Gene transfer and expression by a replication-defective retroviral vector is referred to as transduction to differentiate this process from infection by a typical virus that leads to virus production and spread.

Introduction

An early model for the development of retroviral vectors was the discovery that many oncogenic retroviruses are actually mixtures of replication-defective retroviruses carrying oncogenic foreign genes with replication-competent or 'helper' retroviruses that promote spread of the oncogenic virus. This arrangement suggested that any foreign deoxyribonucleic acid (DNA) could be incorporated into a retrovirus to generate a vector capable of transferring and integrating the DNA into the genomes of target cells. The ability of retroviruses to integrate into the host cell genome while accurately preserving the viral genomic structure is a particularly useful feature of this gene transfer system. Early studies of retroviral vectors used helper virus to promote vector transfer, but this resulted in virus replication in target cells and the potential for vector spread. This problem was solved by the development of retrovirus packaging cells that are engineered to produce all of the viral proteins required to package retrovirus genomes into infectious virions, but that do not produce replication-competent virus. These developments produced a vector system that has found many applications for gene transfer in cultured cells and in animals, and for human gene therapy.

Retrovirus Biology

Retroviruses can be divided into three major groups: the oncoretroviruses (oncogenic retroviruses), the lentiviruses (slow retroviruses) and the spumaviruses (foamy viruses). Retroviral vectors have been developed based on viruses from each of these groups. All of these viruses are defined by the requirement of reverse transcription of the viral plus-strand RNA genome into double-stranded DNA, which is then integrated into the host cell genome by the viral integrase, a part of the Pol gene product. Basic proteins required by these viruses for replication include the Gag or internal structural proteins, the Pol or enzymatic proteins, including reverse transcriptase, integrase and a protease required for cleavage of the viral polyprotein precursors, and the Env proteins that reside on the virion surface and promote virus binding and entry into cells. In addition to these proteins, the lentiviruses and spumaviruses require accessory proteins that modulate various viral and cellular functions involved in virus gene expression and interaction with the host animal.

Major features that distinguish vectors derived from these different virus groups include the ability of the vector to transduce nondividing cells, with the oncogenic retroviruses generally being unable to transduce nondividing cells, the lentiviral vectors being able to transduce nondividing cells and the foamy virus vectors having intermediate properties. The pathogenicity of the parental viruses and the possibility of production of similar viruses during vector production is an important concern for human gene therapy applications, with the oncoretroviruses being able to cause cancer in some species, the lentiviruses able to cause severe immunodeficiency and death in humans and other animals, and the spumaviruses being benign and not linked to any disease in humans or animals. The size of the viruses, and thus the capacity for insertion of foreign DNA, also varies, with oncoretroviral vectors capable of accommodating around 7 kb of foreign DNA, lentiviral vectors around 8 kb and spumaviral vectors around 10 kb. The oncoretroviral vectors have been extensively studied in cell culture, in animals and in human gene therapy trials; the lentiviral vectors have been developed into a usable vector system relatively recently and human use should be permitted in the near future; and the spumaviral vectors are the least well developed and have not yet been used in humans.

Design of Retroviral Vectors for Gene Transfer

Vectors based on oncogenic retroviruses were developed first and are widely used. These viruses require only Gag–Pol and Env proteins for replication, and the majority of the viral genome can be replaced with foreign DNA to generate a retroviral vector. **Figure 1** illustrates a replication-competent oncoretrovirus and the derivatives of this virus that are used to produce retroviral vectors. This figure is based on *Moloney murine leukemia virus* (MoMLV), one of the most thoroughly studied viruses for vector design, but applies in general to all retroviral vector systems.

The MoMLV proviral DNA is transcribed into two messenger ribonucleic acids (mRNAs) that encode the Gag–Pol proteins and a sliced RNA that encodes the Env proteins (**Figure 1a**). The majority of the protein translated from the genomic RNA is Gag protein, but infrequent translation through a stop codon at the end of the Gag coding region results in the production of a small amount of Gag–Pol protein. The typical strategy for making viral proteins for packaging of retroviral vectors involves expressing the Gag–Pol and Env proteins from two expression plasmids using nonretroviral transcriptional signals to reduce the possibility of recombination to yield replication-competent virus (**Figure 1b**). As much viral sequence is removed as possible, including retroviral packaging, reverse transcription and integration signals.

Retroviral vectors contain the minimum amount of viral DNA to promote encapsidation of the transcribed RNA into virions and integration of the reverse-transcribed DNA into the target cell genome while minimizing the presence of viral coding regions (**Figure 1c**). Thus all of the Env and most of the Gag–Pol coding regions can be removed. Only the terminal regions of the wild-type virus, including the long terminal repeats (LTRs) and adjacent sequences, need be retained. Many vectors are available with cloning sites for insertion of foreign DNA, for example LXL (**Figure 1c**). The LASN vector is typical of vectors that encode more than one protein (**Figure 1c**), and was the vector used in the first human gene therapy trial (Blaese *et al.*, 1995). In this vector, the therapeutic adenosine deaminase (*ADA*) gene is expressed by using the strong promoter and enhancers

in the retroviral LTR, and the bacterial neomycin phosphotransferase (*neo*) gene is expressed by using an internal *simian virus-40* promoter.

Several strategies are available to produce retroviral vector virions from the plasmids containing Gag–Pol, Env and retroviral vector. All involve introducing these DNAs into eukaryotic cells where the proteins and vector RNA are made and assembled into virions which bud from the cells into the culture medium where they are harvested. Retroviral vectors can be produced following transient introduction of all of the plasmids into cells by using calcium phosphate–DNA coprecipitation, liposomes, electroporation or other techniques. In these procedures, the cells are treated with DNA on the first day, the culture medium is replaced the next day and virus is harvested for several days at 12- to 24-h intervals 2 days after DNA introduction. Advantages of this technique are the rapidity of vector production and the possibility of producing vectors containing toxic genes before the cells producing the vector die. A disadvantage is the relative inhomogeneity of the vector produced which reflects vector rearrangement following plasmid introduction (Miller *et al.*, 1988; Lynch *et al.*, 1993).

Vectors can also be produced by generating stable cell lines that produce the Gag–Pol and Env proteins and the vector genomic RNA. A wide range of packaging cell lines are available that produce the retroviral Gag–Pol and Env proteins. These cells are made by introduction of the Gag–Pol and Env expression plasmids into cells and selection of subclones for high-level production of the proteins. Introduction of vector DNA into these cells results in the production of vector virions. Typically, the vector DNA is stably introduced into the packaging cells to generate stable vector-producing cell lines from which vector virions can be harvested virtually indefinitely. In addition, packaging cell lines can be transduced with vector virions derived by transient transfection of other packaging cells to generate clonal cell lines with single vector integrants (Miller *et al.*, 1993). These integrants can be further characterized by Southern analysis and sequencing to confirm their integrity. Vector virions produced from such lines are homogeneous and are preferred for many applications including human gene therapy. Thus, while this technique for vector production is more time-consuming than transient production techniques, the virus produced is more homogeneous and can be produced continuously without the need for additional transfection procedures.

A useful feature of retroviral vectors is the large range of receptors that can be targeted using different Env proteins. Not only is the range of receptors targeted by naturally occurring retroviruses large, but Env proteins from other virus families have been

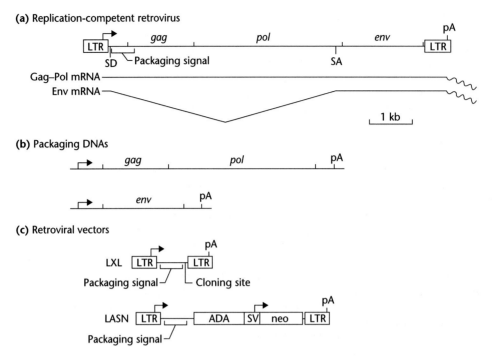

(a) Replication-competent retrovirus

(b) Packaging DNAs

(c) Retroviral vectors

Figure 1 The structure of a simple replication-competent retrovirus and derivative components used to produce retroviral vectors. (a) The DNA provirus structure of a simple oncoretrovirus, *Moloney murine leukemia virus*. RNAs transcribed from the provirus are shown at the bottom. (b) Packaging DNAs encode the viral proteins required for virus replication and are introduced into eukaryotic cells to make retrovirus packaging cells. (c) LXL is a minimal retroviral vector with all elements required for efficient transfer and a cloning site for insertion of genes or other sequence elements. LASN is an example of a retrovirus vector that encodes human adenosine deaminase and bacterial neomycin phosphotransferase. This was the vector used in the first human gene therapy trial to treat severe combined immunodeficiency due to lack of adenosine deaminase (see text). LTR: retroviral long terminal repeat; SD: splice donor; SA: splice acceptor; pA: polyadenylation signal; ADA: adenosine deaminase; SV: *simian virus-40* promoter and enhancers; neo: bacterial neomycin phosphotransferase cDNA; Wavy lines: polyA sequences. Diagrams are drawn to scale and the scale marker indicates 1 kb length.

successfully incorporated into retrovirus virions to further extend the range of receptors that can be targeted. A good example of the latter is provided by the Env protein from *vesicular stomatitis virus* (VSV), a rhabdovirus with a single negative-strand RNA genome. Incorporation of this Env protein into a retroviral vector in place of normal retrovirus Env proteins promotes broad species tropism of the vector and also stabilizes the vector against the denaturing effects of centrifugation used to concentrate the virus and filtration used to remove cellular contaminants.

Lentiviral Vectors

There is increasing interest in lentiviral vectors due to their ability to transduce nondividing cells (Vigna and Naldini, 2000). Thus while standard retroviral vectors (derived from oncoretroviruses) appear unable to enter the cell nucleus except during mitosis, several lentiviral proteins facilitate passage of the virus capsid through nuclear pores and into the nucleus of nondividing cells. In addition, animal studies have clearly shown the benefit of lentiviral vectors for

transduction of stationary or slowly dividing cells *in vivo* in comparison to standard retroviral vectors, indicating that lentiviral vectors should be particularly useful for human gene therapy involving direct administration of the vector. Vectors have been derived from *human immunodeficiency virus type 1* (HIV-1) and from lentiviruses of cats (*feline immunodeficiency virus*), horses (*equine infectious anemia virus*) and other species. Vectors from nonprimate lentiviruses are expected to be safer for use in humans since these viruses are not known to cause disease in humans.

Safety Issues

A primary concern with the use of retroviral vectors for human gene therapy is the potential for generation of replication-competent virus, also called helper virus, during vector production, since all of the components required for such an event are present in vector-producing cells. While the dangers to humans of producing a virus like HIV during vector production are clear, the possible dangers of other recombinant viruses are less clear. For example, current methods for production of lentiviral vectors based on HIV cannot generate wild-type HIV because the HIV *env* gene and most of the HIV accessory genes are not present in the production process. Typically, the HIV Env protein is replaced with that of VSV. One could imagine recombination events leading to a virus with *gag–pol* genes from HIV and an *env* gene from VSV, but whether such a virus could replicate or cause disease in humans is unclear. Standard retrovirus vectors based on murine leukemia viruses might be contaminated with replication-competent *murine leukemia virus* (MLV), but these viruses have not been associated with disease in humans. Indeed, the prototype oncoretrovirus MoMLV does not cause disease when injected into adult mice and must be injected into immunologically immature newborn mice to induce leukemia. However, under extreme conditions involving high-titer contamination of a vector preparation with replication-competent amphotropic retrovirus, infection of monkey bone marrow cells *ex vivo* with large amounts of this virus, and transplantation of the infected cells into lethally irradiated recipients did result in the production of leukemia in almost half of the treated animals (Donahue *et al.*, 1992).

The approach taken to minimize the possibility of production of replication-competent virus has been to separate the virus components on different plasmids such that multiple recombination events would be necessary to generate replication-competent virus, to minimize or eliminate regions of homologous overlap between these elements and to introduce the components into cells in different transfection steps to avoid the high rates of recombination that can occur during simultaneous transfection of plasmids (Miller *et al.*, 1993). These strategies have led to packaging cell lines that are capable of producing liters of high-titer vector stocks in the absence of replicating virus. In contrast, early packaging cell lines that relied only on deletion of the retroviral packaging signal to cripple the parental virus would often produce helper virus after introduction of a retroviral vector, presumably by a single homologous recombination event to restore an intact 5′ end to the virus. Experience has shown that current generations of retroviral vectors and packaging cells are suitable for human use and these vectors have been used in more than 100 early-stage clinical trials.

Another concern regarding the use of retroviral vectors for human gene therapy is the possibility of oncogene activation following the semirandom integration of the vector into the target cell genome. However, many gene transfer techniques lead to integration of the genes into the cell genome, and for long-term gene expression, especially in the progeny of transduced cells, it is advantageous that the vector integrate. Replication-competent retroviruses can cause cancer in animals by insertional activation of cellular oncogenes, but this is in the context of continuous virus replication and reinfection of cells such that many cells are involved and individual cells can suffer multiple integration events. While many animal studies and gene therapy trials have been performed by using retroviral vectors without evidence of oncogene activation, very recent results in a trial involving transduction of hematopoietic stem cells with a retroviral vector encoding the γc cytokine receptor subunit (see below) led to a T-cell leukemia in one of 11 treated patients (see Office of Biotechnology Activities in Web Links). This leukemia was associated with transcriptional activation of the *LMO2* proto-oncogene due to integration of the retroviral vector into this gene. More experience will be required to evaluate the frequency of such events, and whether some aspects of the disease treated or a pre-existing cancer susceptibility of individual patients contributes to cancer induction by vector insertion into the genome.

Host immune responses against retroviral vectors are also a concern for human gene therapy. Since retroviral vectors are typically designed such that no viral proteins are expressed by the vector, there should be no antiviral immune response against transduced cells. It is important to eliminate foreign selectable marker genes from therapeutic vectors to prevent immune response against these foreign proteins. For example, the bacterial hygromycin phosphotransferase gene has been included in retroviral vectors to allow selection of transduced cells using the antibiotic hygromycin, but immune responses against this protein have been documented (Riddell *et al.*, 1996). Lastly, one expects that repeat infusion of a retroviral vector will elicit neutralizing antibodies against the vector and reduce or eliminate transduction by subsequent vector doses, but this problem is typical of all virus vectors.

Clinical Applications

The main application of retroviral vectors based on oncoretroviruses has been to transduce replicating cells *ex vivo* prior to reintroduction of the cells into humans. The ability of the vector to integrate precisely with respect to the vector genome and persist as part of the cellular genome during cell replication is appealing for modification of cells that can expand in number, such as T cells or hematopoietic stem cells. Indeed, the first human gene therapy trial involved the *ex vivo* transduction of T cells with a retroviral vector encoding adenosine deaminase and reintroduction into patients to correct an immunodeficiency caused by lack of this enzyme (Blaese *et al.*, 1995). The trial was a partial success and resulted in clinical improvement, but the one patient who most clearly benefited still required infusions of adenosine deaminase protein to treat her disease. More recently, an immunodeficiency due to a defect in the γc cytokine receptor subunit of several interleukin receptors was completely corrected in affected patients by using a retroviral vector encoding the γc protein to transduce hematopoietic stem cells *ex vivo* followed by reinfusion of the cells (Cavazzana-Calvo *et al.*, 2000). This represents the first unqualified success for gene therapy, and represents the effective treatment of an otherwise incurable disease. It remains to be seen if the effects will lead to a normal lifespan for these patients, but initial results after over 2 years are very encouraging (Hacein-Bey-Abina *et al.*, 2002).

While lentiviral vectors had not been used in human clinical trials as of 2002, these vectors have the advantage that they can transduce nondividing cells *in vivo*, unlike the retroviral vectors based on oncoretroviruses. Thus, relatively efficient transduction of brain and liver cells in animals has been documented, indicating the potential utility of these vectors for the treatment of human disease.

See also

Autoimmune Diseases: Gene Therapy
Gene Therapy: Technology
Hemophilias: Gene Therapy
Immunodeficiency Syndromes: Gene Therapy
Lentiviral Vectors in Gene Therapy

References

Blaese RM, Culver KW, Miller AD, *et al.* (1995) T lymphocyte-directed gene therapy for ADA-SCID: initial trial results after 4 years. *Science* **270**: 475–480.

Cavazzana-Calvo M, Hacein-Bey S, de Saint Basile G, *et al.* (2000) Gene therapy of human severe combined immunodeficiency (SCID)-X1 disease. *Science* **288**: 669–672.

Donahue RE, Kessler SW, Bodine D, *et al.* (1992) Helper virus induced T cell lymphoma in nonhuman primates after retroviral mediated gene transfer. *Journal of Experimental Medicine* **176**: 1125–1135.

Hacein-Bey-Abina S, Le Deist F, Carlier F, *et al.* (2002) Sustained correction of X-linked severe combined immunodeficiency by *ex vivo* gene therapy. *New England Journal of Medicine* **346**: 1185–1193.

Lynch CM, Israel DI, Kaufman RJ and Miller AD (1993) Sequences in the coding region of clotting factor VIII act as dominant inhibitors of RNA accumulation and protein production. *Human Gene Therapy* **4**: 259–272.

Miller AD, Bender MA, Harris EAS, Kaleko M and Gelinas RE (1988) Design of retroviral vectors for transfer and expression of the human β-globin gene. *Journal of Virology* **62**: 4337–4345.

Miller AD, Miller DG, Garcia JV and Lynch CM (1993) Use of retroviral vectors for gene transfer and expression. *Methods in Enzymology* **217**: 581–599.

Riddell SR, Elliott M, Lewinsohn DA, *et al.* (1996) T-cell mediated rejection of gene-modified HIV-specific cytotoxic T lymphocytes in HIV-infected patients. *Nature Medicine* **2**: 216–223.

Vigna E and Naldini L (2000) Lentiviral vectors: excellent tools for experimental gene transfer and promising candidates for gene therapy. *Journal of Gene Medicine* **2**: 308–316.

Further Reading

Coffin JM, Hughes SH and Varmus HE (eds.) (1997) *Retroviruses*. New York, NY: Cold Spring Harbor Laboratory Press.

Miller DG, Adam MA and Miller AD (1990) Gene transfer by retrovirus occurs only in cells that are actively replicating at the time of infection. *Molecular and Cellular Biology* **10**: 4239–4242.

Web Links

Office of Biotechnology Activities: National Institutes of Health. National Institutes of Health Recombinant DNA Advisory Committes (RAC)
http://www.webconferences.com/_nihoba/4–6_december_02.html

Adenosine deaminase (*ADA*); Locus ID: 100. LocusLink:
http://www.ncbi.nlm.nih.gov/LocusLink/LocRpt.cgi?l = 100

Adenosine deaminase (*ADA*); MIM number: 102700. OMIM:
http://www.ncbi.nlm.nih.gov/htbin-post/Omim/dispmim?102700

Rett Syndrome

Eric P Hoffman, *Research Center for Genetic Medicine, Children's National Medical Center, Washington, DC, USA*

Linda Moses, *Research Center for Genetic Medicine, Children's National Medical Center, Washington, DC, USA*

Cheryl AG Scacheri, *Research Center for Genetic Medicine, Children's National Medical Center, Washington, DC, USA*

Kristen C Hoffbuhr, *Research Center for Genetic Medicine, Children's National Medical Center, Washington, DC, USA*

Rett syndrome is a neurological disease of early postnatal brain growth found almost exclusively in girls. Typically girls with Rett syndrome show developmental regression including loss of communication and motor skills, stereotypic hand movements and a deceleration of head growth. The defective gene, methyl CpG binding protein 2 (Rett syndrome) (*MECP2*) is normally involved in the transcription silencing of genes (turning genes 'off'): in Rett syndrome, the MECP2 protein does not work properly in approximately half of the patient's cells (heterozygous), leading to the aberrant expression of a cascade of other genes, normally regulated by MECP2.

Clinical Features

In 1966 Andreas Rett, an Austrian physician, first described a neurological syndrome which presented in young girls as profound regression in developmental milestones – loss of communication and motor skills following a period of seemingly normal early development. The developmental regression was accompanied by a loss of purposeful hand use, stereotypical and repetitive hand movements (hand mouthing, hand wringing and hand washing), autistic-like behavior and a failure of brain growth as measured by a deceleration in head circumference. Rett syndrome, not widely recognized in North America until a report published by Hagberg and colleagues in the early 1980s, is now thought to be one of the most common causes of mental retardation in girls, with an incidence estimated to be 1 in 10 000 to 1 in 22 000 female births.

The progression of Rett syndrome typically follows four clinical stages (**Table 1**). Girls with Rett syndrome show an apparently normal prenatal and postnatal period until 6–18 months of age. Developmental regression can occur between 1 and 4 years of age and may be sudden, lasting only several weeks to months, or can be incremental, lasting several years. The characteristic hand movements, loss of communication skills and the failure of head growth become apparent during this period. Girls may also develop seizures, breathing abnormalities and a wide-based shaky gait. After regression, the clinical course becomes more stable. Girls may regain some speech ability and interest in communication and may seem to become less autistic-like. However, seizures and scoliosis may develop or become more pronounced. The fourth clinical stage is marked by reduced mobility and motor deterioration. Patients with Rett syndrome may survive into adulthood and middle age but several studies have shown a lower survival rate in the 25–40 age group for women with Rett syndrome (69%) compared with the general population (97%).

The clinical stages describe the typical presentation of Rett syndrome; however, 'atypical' Rett syndrome is relatively common. Atypical Rett syndrome can often be described as an incomplete clinical presentation of Rett syndrome such as a later and more gradual onset of regression, or girls who have some purposeful hand movements or preserved speech. Included in this category are girls who develop seizures early in infancy but otherwise show a typical presentation.

Management

Treatment for Rett syndrome is mainly symptomatic and focuses on improving the quality of life for girls with Rett syndrome. Traditional physical and occupational therapies are often used to maintain or improve motor ability. Several innovative therapies including water therapy, music therapy and horseback riding all have been found to have a calming effect on

Table 1 Clinical stages of Rett syndrome

1. Early onset stage: birth until 6–18 months	Normal/near-normal perinatal and postnatal period
	Normal head circumference at birth
	Slow achievement of developmental milestones, especially motor skills
	Low muscle tone
	Nonspecific hand movements
2. Rapid destructive stage: 1–4 years of age	Loss of acquired skills such as learned words and purposeful hand movements
	Development of stereotypic hand movements – hand to mouth, hand washing, hand wringing behavior
	Deceleration of head circumference at 3 months to 4 years of age
	Seizures, breathing abnormalities
	Wide-based, shaky gait
	Teeth grinding, chewing and swallowing problems
	Screaming sessions, sleep disturbances
3. Plateau stage: late childhood (2–10 years of age)	Increased interest in communication, longer attention span
	Fewer problems with crying and irritability
	Muscles become hypertonic, starting at extremities
	Seizures and scoliosis may develop/worsen
4. Late motor deterioration: >10 years of age	Reduced mobility, motor deterioration
	Increased muscle rigidity and weakness, worsened scoliosis
	Hand movements may improve

Adapted from The Rett Syndrome Diagnostic Criteria Work Group (1988) Diagnostic criteria for Rett syndrome. *Annals of Neurology* **23**: 425–428.

patients. Arm braces are used to reduce stereotypic hand motions, and have been shown to increase both relaxation and concentration in girls with Rett syndrome.

Pathological Findings

Both clinical and pathological findings in Rett syndrome are symptomatic of a neurodevelopmental process involving a disruption in early brain development during a period of active growth. Consistent with this model are the striking reductions (14–34%) in both brain weight and size found in affected girls compared with age-matched controls. Certain areas of the brain show greater involvement, including the cerebral cortex, midbrain, corpus callosum, basal ganglia and cerebellum. These areas of the brain have been shown to be important for higher thinking, motor coordination and the automatic processing of movement. At the cellular level, dendrites (antenna-like structures of the neurons which receive signals from other neurons) are smaller, have fewer spines (communication sites), and exhibit reduced branching and different orientation patterns. (*See* Chromosome X; Linkage Analysis.)

Genetics

Early observations that Rett syndrome almost exclusively affected girls and was rarely diagnosed in males sparked speculation that Rett syndrome was caused by a gene on the X chromosome. In an X-linked dominant model, girls with Rett syndrome would be heterozygous for the defective gene (for example, one normal X chromosome and the other X chromosome with the defective gene) and males, who only have one X chromosome, would not survive the gene defect. Affected girls usually do not reproduce, otherwise half of their daughters would be expected to be affected. Genetic studies were hampered for many years by the very few instances of multiplex, familial cases, presumably because most cases represented new mutations. Less than 1% of Rett syndrome cases show any type of familial inheritance; the vast majority are single, isolated cases (sporadic). Traditional methods geneticists use to identify disease genes rely on the comparison of genetic markers between affected and unaffected family members. Because Rett syndrome was rarely diagnosed in more than one member of a family, such linkage analysis techniques proved difficult to apply to Rett syndrome. (*See* Chromosome X; Linkage Analysis.)

Significant progress toward localizing the gene causing Rett syndrome occurred with reports of several multiplex families, including a large Brazilian family with three girls diagnosed with Rett syndrome. The three affected sisters and their mother in the Brazilian family all shared a single region on the X chromosome, the Xq28 region, which was not inherited by their three healthy sisters. Statistical analysis of this family and of other published familial cases indicated that the likelihood of the Xq28 region associating with the disease trait simply by chance approached odds of 1000 to 1 (a logarithmic odds ratio (lod) score of 2.9). Within a year, scientists at the laboratories of Huda Zogbi at Baylor College of Medicine and Uta Franke at Stanford University found the methyl CpG binding protein 2 (Rett syndrome) (*MECP2*) gene in Xq28 to harbor mutations causing Rett syndrome. The MECP2 protein is involved in the regulation of other genes by recognizing and binding methylated CpG dinucleotides and recruiting other proteins that are actively involved in turning genes 'off', the process termed 'transcription silencing'. (*See* Lod Score.)

In mammalian cells, methylation of cytosine residues in CpG dinucleotides is one of the most ubiquitous mechanisms associated with transcriptional silencing. MECP2 belongs to a group of five proteins which all bind methylated CpG dinucleotides. Once MECP2 binds CpG residues, the transcriptional repressor Sin3a and histone deacetylase (HDAC1 and HDAC2) are recruited to the MECP2–deoxyribonucleic acid (DNA) complex (**Figure 1**). Studies have shown that HDAC alters the charge of histones by removing acetyl moieties. Chromatin becomes more compacted due to the presence of deacetylated histones, and this in turn prevents the binding of transcriptional activator proteins to the promoters of genes, making it impossible to turn genes 'on'. (*See* DNA Methylation and Histone Acetylation; Histones.)

In Rett syndrome, the MECP2 protein is lacking in half of the patient's cells, leading to a disruption of this silencing process. More than 75 different mutations in *MECP2* have been described in patients with both classical and atypical Rett syndrome. Most *MECP2* mutations are predicted to result in a loss of function of the MECP2 protein because conserved regions of the protein, such as the methyl-binding domain or the transcriptional repression domain, are disrupted by the mutations (**Figure 2**). Studies have shown that *MECP2* mutations cause most cases (65–80%) of typical forms of Rett syndrome. However, girls with atypical, incomplete phenotypes and familial cases of Rett syndrome have shown a lower incidence of mutations (<45%). Despite the large number of mutations identified in *MECP2*, eight 'common' mutations account for the majority (65%) of

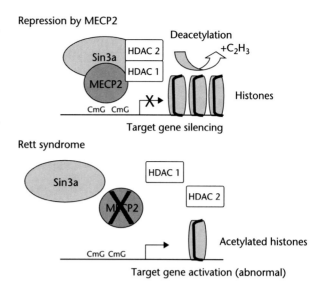

Figure 1 Role of MECP2 in transcription silencing and Rett syndrome. Representative diagram of MECP2-mediated silencing of genes (top panel). MECP2 binds methylcytosines (CmG) and recruits transcriptional repressor Sin3a and histone deacetylase HDAC1 and HDAC2 to the MECP2–DNA complex. HDAC1 and HDAC2 remove acetyl moieties from histones. Deacetylated histones, which are more negatively charged, cause the chromatin to become more compacted, leading to the silencing of genes. As shown by the lower diagram, MECP2 does not function properly in Rett syndrome, leading to the expression of genes normally regulated by MECP2.

mutations found in girls with Rett syndrome (**Figure 2**). These common *MECP2* mutations are typical of *de novo* (new) mutations, where spontaneous deamination of methylated cytosine residues (^{m7}C) leads to a specific type of nucleotide mutation (C→T transition).

While the diagnosis of Rett syndrome is typically restricted to females, rare male patients with a diverse spectrum of clinical symptoms have been found to harbor *MECP2* mutations. In several cases, the hemizygous boys' clinical symptoms were considerably more severe than those of heterozygous girls with the same *MECP2* mutation. These findings have shown that the previous assumption of male lethality (before the discovery of the gene) is not necessarily true.

Genetic Counseling

MECP2 mutations are only very rarely inherited. The vast majority of cases of Rett syndrome represent new mutation events which occurred in either parent's reproductive cells (egg or sperm). The empirical risk of recurrence of Rett syndrome for a couple who already have a daughter with Rett syndrome is 0.5% or 1/200. Couples can have a second child with Rett syndrome if either parent is a gonadal mosaic (parent has a

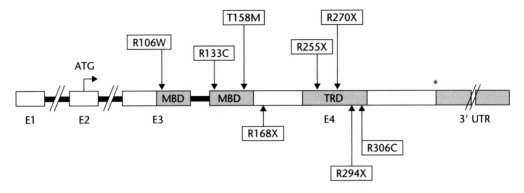

Figure 2 Schematic diagrams of the *MECP2* gene and common mutations in Rett syndrome. The *MECP2* gene consists of four exons (E1–E4), three introns and a large (> 8.5 kb) 3′ untranslated region (3′ UTR). The two highly conserved regions of the gene are the methyl-binding domain (MBD) and the transcriptional repression domain (TRD).

population of reproductive cells that have the *MECP2* mutation) or the mother is an asymptomatic carrier of an *MECP2* mutation. In the latter scenario, asymptomatic carriers have been found to show skewed X inactivation, with the large majority of cells having the normal X active. In these cases where the mother is an asymptomatic carrier of an *MECP2* mutation, the risk of recurrence of Rett syndrome is 50%.

In females, one X chromosome in every cell is turned off by a process called X-chromosome inactivation (XCI). Generally, this process is random: roughly equal numbers of cells have either the maternal X chromosome 'on' (the X chromosome inherited from the mother) or the paternal X chromosome 'on'. Nonrandom XCI, which occurs in 2–5% of females, can protect females who are *MECP2* mutation carriers from having symptoms of Rett syndrome by 'turning off' the X chromosome with the abnormal *MECP2* gene in most or all of the cells in the body. Although this is rare, several families have been reported in which the mother of a girl with Rett syndrome is a 'silent' carrier of an *MECP2* mutation. (*See* X-chromosome Inactivation.)

Conclusion

Future research into Rett syndrome will focus on the new challenges brought by the discovery of the gene and on advancing our knowledge of the pathophysiology of Rett syndrome. The molecular consequences for cells lacking MECP2 are not known. An MECP2 defect would most likely lead to aberrant expression of genes that are normally regulated by MECP2, resulting in the inappropriate expression of these genes. Emerging technologies such as gene chips where approximately 40 000 human genes and ESTs (expressed sequence tags) can be screened simultaneously will provide a powerful approach toward identifying the downstream targets of

MECP2 regulation. Ultimately, dissecting this process of gene silencing and uncovering genes regulated by MECP2 will lead to a better understanding of postnatal brain development and how dysfunction in these genes during brain development cause Rett syndrome. (*See* Expressed-sequence Tag (EST); Microarrays: Use in Gene Identification.)

See also

Epigenetics: Influence on Behavioral Disorders
Histone Acetylation and Disease

Further Reading

Amir RE, Van den Veyver IB, *et al.* (1999) Rett syndrome is caused by mutations in X-linked MECP2, encoding methyl-CpG-binding protein 2. *Nature Genetics* **23**: 185–188.

Armstrong DD (1997) Review of Rett syndrome. *Journal of Neuropathology and Experimental Neurology* **56**: 843–849.

Bird AP and Wolffe AP (1999) Methylation-induced repression – belts, braces, and chromatin. *Cell* **99**: 451–454.

Budden SS (1995) Management of Rett syndrome: a ten year experience. *Neuropediatrics* **26**: 75–77.

Dragich J, Houwink-Manville I and Schanen C (2000) Rett syndrome: a surprising result of mutation in *MECP2*. *Human Molecular Genetics* **9**: 2365–2375.

Hagberg B, Aicardi J, Dias K and Ramos O (1983) A progressive syndrome of autism, dementia, ataxia, and loss of purposeful hand use in girls: Rett's syndrome: report of 35 cases. *Annals of Neurology* **14**: 471–479.

Hagberg BA and Skjeldal OH (1994) Rett variants: a suggested model for inclusion criteria. *Pediatric Neurology* **11**: 5–11.

Hoffbuhr K, Devaney JM, LaFleur B, *et al.* (2001) MeCP2 mutations in children with and without the phenotype of Rett syndrome. *Neurology* **56**: 1486–1495.

Hunter K (1999) *The Rett Syndrome Handbook*. Clinton, MD: International Rett Syndrome Association.

Naidu S (1997) Rett syndrome: a disorder affecting early brain growth. *Annals of Neurology* **42**: 3–10.

Razin A (1998) CpG methylation, chromatin structure and gene silencing – a three-way connection. *EMBO Journal* **17**: 4905–4908.

Rett A (1977) Cerebral atrophy associated with hyperammonaemia. In: Vinken P, Bruyn G and Klawans H (eds.) *Handbook of Clinical Neurology*, pp. 305–329. Amsterdam: Elsevier Science.

The Rett Syndrome Diagnostic Criteria Work Group (1988) Diagnostic criteria for Rett syndrome. *Annals of Neurology* **23**: 425–428.

Sirianni N, Naidu S, Pereira J, Pillotto RF and Hoffman EP (1998) Rett syndrome: confirmation of X-linked dominant inheritance, and localization of the gene to Xq28. *American Journal of Human Genetics* **63**: 1552–1558.

Van den Veyver IB and Zoghbi HY (2000) Methyl-CpG-binding protein 2 mutations in Rett syndrome. *Current Opinion in Genetics and Development* **10**: 275–279.

Web Links

Methyl CpG binding protein 2 (Rett syndrome) (*MECP2*); Locus ID: 4204. LocusLink:
http://www.ncbi.nlm.nih.gov/LocusLink/LocRpt.cgi?l = 4204

Methyl CpG binding protein 2 (Rett syndrome) (*MECP2*); MIM number: 300005. OMIM:
http://www.ncbi.nlm.nih.gov/htbin-post/Omim/dispmim?300005

RFLP

See Restriction Fragment Length Polymorphism (RFLP)

Rheumatoid Arthritis: Genetics

Peter K Gregersen, *North Shore Long Island Jewish Research Institute, Manhasset, New York, USA*

Rheumatoid arthritis is a heterogeneous inflammatory disorder of unknown cause with features suggestive of autoimmunity. The genetic components are complex and appear to overlap with other autoimmune diseases.

Intermediate article

Article contents

- Introduction
- The Clinical Syndrome
- Disease Mechanisms and Pathogenesis
- Familial Aggregation
- Human Leukocyte Antigen Genes and Rheumatoid Arthritis
- Outside the Human Leukocyte Antigen Region: Susceptibility Genes
- Candidate Genes and Rheumatoid Arthritis
- Animal Models
- Conclusion

Introduction

Rheumatoid arthritis (RA) is the most common form of inflammatory polyarthritis in humans. This type of arthritis is termed 'inflammatory' because the arthritic joints often exhibit the classic signs of inflammation, including redness, swelling, warmth and tenderness. It is called a 'polyarthritis' because it generally involves many joints in the body, with a characteristic pattern. RA is just one of many inflammatory, metabolic or infectious diseases that can manifest themselves as arthritis. Many of these diseases are rare and can be distinguished easily from RA. The most common form of arthritis, osteoarthritis or 'degenerative' arthritis, is also distinct from RA, and is usually not accompanied by prominent signs of inflammation. Nevertheless, RA may not be a single disease and can show overlapping features with other forms of inflammatory arthritis. It is diagnosed on the basis of a constellation of symptoms and physical findings. There are currently no specific laboratory tests for the disease. Therefore, it is likely that RA is a heterogeneous disorder, and this presents major difficulties for geneticists who are trying to identify the specific genes involved.

The Clinical Syndrome

The typical patient with RA is a middle-aged woman who begins to develop pain, swelling and redness in the small joints of the hands, accompanied by a prominent sensation of stiffness when arising in the morning. Women are affected three times as often as men. The mean age of onset is in the mid- to late 40s, although adolescents and young adults can also be affected, and onset later in life is not unusual. Although symmetrical involvement of the hands is the classic finding, it is common for RA to involve many other joints, including elbows, shoulders, neck, hips, knees, ankles and feet. Interestingly, and for reasons that are completely unknown, RA does not involve the small joints at the very ends of the fingers (the distal interphalangeal, DIP, joints). This is in contrast to the inflammatory polyarthritis that can occur in the setting of psoriasis, in which involvement of the DIP joints is a typical finding.

Over time, the chronic inflammation of RA often leads to destruction of the joint cartilage and bone, and this may result in major disability for the patient (Pincus, 1995). Hand function can be compromised, making it difficult for patients to carry out the activities of daily living such as buttoning a shirt or opening a door. Shoulder involvement may compromise the ability to brush one's hair. Involvement of the hips, knees and feet can lead to loss of mobility. In extreme untreated cases, RA may leave a patient wheelchair- or bed-bound. Fortunately, modern treatment approaches make this a rare event.

In addition to the major effects on joints, RA can also be accompanied by 'extra-articular' manifestations. These include the development of subcutaneous nodules, most often around the elbows. These rheumatoid nodules are painless and appear intermittently. More serious extra-articular manifestations include lung disease, heart disease, spleen enlargement with low white blood cell counts and generalized blood vessel inflammation. With the exception of rheumatoid nodules, these extra-articular features are not common.

Disease Mechanisms and Pathogenesis

RA is generally classified as an autoimmune disease. The term 'autoimmune' implies that part of the disease mechanism involves a failure of the immune system to distinguish 'self' from 'nonself'. In the case of RA, it is assumed that the immune system inappropriately attacks joint tissues. While autoantibodies against joint constituents such as collagen can be found in RA, the more typical autoantibodies are directed against antigens in the cell nucleus or against subgroups of antibodies themselves. This latter type of autoantibody is called 'rheumatoid factor'. Despite its name, rheumatoid factor is not specific for RA and can be found in a variety of other autoimmune and infectious diseases. More recently, autoantibodies against other self-proteins have been described in RA (Baeten et al., 2001). Overall, however, these autoimmune phenomena do not explain fully why the disease mainly involves the joints.

In terms of causation, the autoimmune aspect of RA could be either primary or contributory. Nevertheless, it is still evident that activation of the immune system plays an important role in RA, particularly in the area of cytokines. Cytokines are proteins released

by immune or inflammatory cells that control the inflammatory response. A number of cytokines are elevated in the joints of RA patients, including the cytokine tumor necrosis factor-alpha (TNFα). A major advance in the treatment of RA has been the development of drugs that inhibit TNFα (Moreland, 2001). This is a strong indication, along with studies of animal models of RA, that cytokines are a key mechanism by which the chronic inflammation in RA is maintained.

The inflammation of RA begins in the tissue that surrounds the joint space, known as the synovium. The synovium normally consists of a thin lining layer of cells. In RA, the synovium becomes markedly thickened due to the proliferation of the lining cells themselves, and the invasion of immune and inflammatory cells from the blood. There is also marked production of new blood vessels in the synovium. Autoantibodies and cytokines are produced within the thickened and inflamed synovium, and the synovial tissue typically invades the surrounding bone and cartilage of the joint. Bone and cartilage destruction is a characteristic feature of RA and is the most important clinical manifestation. The mechanisms for this have begun to be understood only recently, but it appears that specific cells and cytokines derived from the synovial tissue may orchestrate the bone and cartilage destruction (Gravallese and Goldring, 2000). A genetic tendency to progress from transient joint inflammation to chronic invasive synovitis may be an important inherited characteristic of patients with RA.

It should be apparent from this brief summary that RA cannot be precisely defined, either clinically or in terms of disease mechanism. RA also has clinical features that overlap with other autoimmune diseases and, indeed, other autoimmune diseases occur more frequently in patients with RA or their family members. These include autoimmune thyroid disease, juvenile diabetes and Sjogren syndrome. Thus it is likely that the genetic basis of what we now call 'RA' is quite heterogeneous and complex.

Familial Aggregation

Like other autoimmune disorders, RA exhibits a modest tendency to run in families. Approximately 3–4% of the siblings of RA patients will also develop RA, compared with a background population prevalence of 0.25–1.0%. (There is considerable uncertainty about the actual population prevalence of RA because of the difficulty of defining the disease accurately in large population surveys.) Overall, the data suggest that siblings of RA patients are 5–10 times more likely to get the disease than the general population (Seldin et al., 1999). It is also very likely that shared genes, rather than shared environment, are the major reason for this increased risk to siblings.

Human Leukocyte Antigen Genes and Rheumatoid Arthritis

It has been known since the late 1970s that genes within the human leukocyte antigen (HLA) complex on chromosome 6 confer susceptibility to RA (Stastny, 1978). The current view is that a group of alleles at the *major histocompatibility complex, class II, DR beta 1* (*HLA-DRB1*) locus are responsible for this. These alleles include *HLA-DRB1*0401*, *HLA-DRB1*0404* and *HLA-DRB1*0101*. The *HLA-DRB1* gene encodes a cell-surface protein that is essential for normal immune function, and normal variants of this protein can regulate the strength and specificity of the immune response. Indeed, all of the *HLA-DRB1* alleles associated with RA are 'normal' alleles that occur at substantial frequencies in the general population. For example, the *HLA-DRB1*0401* allele is found in approximately 40% of White patients with RA, in contrast to approximately 10% of the normal White population. Thus, *HLA-DRB1*0401* is common, and most people who inherit this allele never develop RA.

Why do multiple different *HLA-DRB1* alleles confer risk for RA? Most of the *HLA-DRB1* alleles associated with RA share a structural feature that has been termed the 'shared epitope' (Gorman and Criswell, 2002). This may be the element that is responsible for the HLA associations with RA; however, the exact role of the shared epitope has not been established. In addition, there is evidence that there may be other genes within the HLA complex, aside from *HLA-DRB1*, that independently confer susceptibility to RA. The role of HLA genes in RA is far from resolved.

Outside the Human Leukocyte Antigen Region: Susceptibility Genes

It has been estimated that the HLA region on chromosome 6 accounts for only 30–50% of the genetic risk for RA. Data suggest that there are a number of other genes, perhaps 10 or more, which also contribute to genetic risk for RA. Individually, these genes appear to have much smaller effects than the HLA-linked genes. As of early 2002, none of these other genes have been definitively localized, much less identified. The general approach has been to look for regions of the genome that are shared among affected individuals in families with RA, usually looking at affected sibling pairs. Many hundreds of such sibling pairs have been screened for evidence of allele sharing across the entire genome.

Despite the current lack of definitive information, one interesting aspect of these genome screens is that some chromosomal regions appear to be implicated in several different autoimmune diseases, in addition to RA (Jawaheer and Gregersen, 2002). Thus, systemic lupus erythematosus (SLE) and insulin-dependent (type 1) diabetes may have susceptibility genes in common with RA. This fits with the observation that these diseases may also run together in the same families. The general consensus in the scientific community is that autoimmune diseases as a group are likely to share overlapping sets of genes; these genes may confer risk for specific autoimmune diseases when they are present in particular combinations.

Candidate Genes and Rheumatoid Arthritis

In addition to screening the entire genome for RA susceptibility, many investigators have attempted to make intelligent guesses as to which genes could be involved. These genes are then examined for polymorphisms that might be associated with RA. Usually this is done using a case–control study, in which the prevalence of selected polymorphisms in a candidate gene is studied in populations of patients and matched controls. Indeed, it was this exploratory approach that led to the original discovery of the association of HLA genes with RA (Stastny, 1978). Nevertheless, this approach to gene identification is likely to fail for any particular candidate, since there are many thousands of genes that could reasonably be implicated, and it is unlikely that more than 10 or 20 are actually involved in the disease.

A number of candidate cytokine genes have been examined in RA, including the *tumor necrosis factor* (*TNF superfamily, member 2*) (*TNF*) gene within the HLA complex on chromosome 6. It appears likely that genetic variation in *TNF*, or in a gene nearby, confers some risk for RA or influences disease outcome (Mulcahy *et al.*, 1996). Similar findings have been made for other cytokine genes (Yamada *et al.*, 2001); however, these findings need to be confirmed and further refined to ascertain whether these genes are actually directly involved in RA.

Animal Models

It is possible to induce an arthritis that resembles human RA by immunizing genetically susceptible strains of rodents with collagen, bacterial products or various other substances (Joe *et al.*, 1999). By performing genetic crosses between different strains of mice or rats, it is now possible to map the genes that control susceptibility to these experimental forms of arthritis. There is a question about how closely these animal models resemble the human disease. However, the study of these animal models can provide new insights into disease mechanisms (Benoist and Mathis, 2000). Thus, as with animal models for other autoimmune diseases, it is quite likely that this experimental approach will aid in the identification of the human genes for these illnesses.

Conclusion

RA has a complex etiology, with considerable heterogeneity at both the clinical and genetic levels. The ultimate definition of the genes involved will require the study of large numbers of well-defined patients and their families, using both linkage and association methods. It is likely that some of the genes involved in RA susceptibility also play a role in other forms of autoimmunity. The complexity of these genetic factors probably reflects the highly complex organization of the underlying biological systems involved in immunity and inflammation.

References

Baeten D, Peene I, Union A, *et al.* (2001) Specific presence of intracellular citrullinated proteins in rheumatoid arthritis synovium: relevance to antifilaggrin autoantibodies. *Arthritis and Rheumatism* **44**: 2255–2262.

Benoist C and Mathis D (2000) A revival of the B cell paradigm for rheumatoid arthritis pathogenesis? *Arthritis Research* **2**: 90–94.

Gorman JD and Criswell LA (2002) The shared epitope and severity of rheumatoid arthritis. *Rheumatic Diseases Clinics of North America* **28**: 59–78.

Gravallese EM and Goldring SR (2000) Cellular mechanisms and the role of cytokines in bone erosions in rheumatoid arthritis. *Arthritis and Rheumatism* **43**: 2143–2151.

Jawaheer D and Gregersen PK (2002) Rheumatoid arthritis. The genetic components. *Rheumatic Diseases Clinics of North America* **28**: 1–15.

Joe B, Griffiths MM, Remmers EF and Wilder RL (1999) Animal models of rheumatoid arthritis and related inflammation. *Current Rheumatology Reports* **1**: 139–148.

Moreland LW (2001) Potential biologic agents for treating rheumatoid arthritis. *Rheumatic Diseases Clinics of North America* **27**: 445–491.

Mulcahy B, Waldron-Lynch F, McDermott MF, *et al.* (1996) Genetic variability in the tumor necrosis factor-lymphotoxin region influences susceptibility to rheumatoid arthritis. *American Journal of Human Genetics* **59**: 676–683.

Pincus T (1995) Assessment of long-term outcomes of rheumatoid arthritis: how choices of measures and study designs may lead to apparently different conclusions. *Rheumatic Diseases Clinics of North America* **21**: 619–654.

Seldin MF, Amos CI, Ward R and Gregersen PK (1999) The genetics revolution and the assault on rheumatoid arthritis. *Arthritis and Rheumatism* **42**: 1071–1079.

Stastny P (1978) Association of the B-cell alloantigen DRw4 with rheumatoid arthritis. *New England Journal of Medicine* **298**: 869–871.

Yamada R, Tanaka T, Unoki M, *et al.* (2001) Association between a single-nucleotide polymorphism in the promoter of the human interleukin-3 gene and rheumatoid arthritis in Japanese patients, and maximum-likelihood estimation of combinatorial effect that two genetic loci have on susceptibility to the disease. *American Journal of Human Genetics* **68**: 674–685.

Further Reading

Klippel JH, Weyand CM and Wortmann RL (eds.) (1997) *Primer on the Rheumatic Diseases*, 11th edn. Atlanta, GA: Arthritis Foundation Press.

Weyand CM and Goronzy JJ (2000) Rheumatoid arthritis. In: Lahita RG, Chiorazzi N and Reeves WH (eds.) *Textbook of Autoimmune Diseases*, pp. 573–594. Philadelphia, PA: Lippincott Williams & Wilkins.

Web Links

HLA-DRB1 (major histocompatibility complex, class II, DR beta 1); Locus ID: 3123. LocusLink:
http://www.ncbi.nlm.nih.gov/LocusLink/LocRpt.cgi?=3123

TNF (tumor necrosis factor (TNF superfamily, member 2)); Locus ID: 7124. LocusLink:
http://www.ncbi.nlm.nih.gov/LocusLink/LocRpt.cgi?=7124

HLA-DRB1 (major histocompatibility complex, class II, DR beta 1); MIM number: 142857. OMIM:
http://www.ncbi.nlm.nih.gov/htbin-post/Omim/dispmim?142857

TNF (tumor necrosis factor (TNF superfamily, member 2)); MIM number: 191160. OMIM:
http://www.ncbi.nlm.nih.gov/htbin-post/Omim/dispmim?191160

Ribosomes and Ribosomal Proteins

Naoya Kenmochi, *Miyazaki Medical College, Miyazaki, Japan*

As the catalyst for protein synthesis, the omnipresent ribosome is essential to all organisms and is composed, in mammalian cells, of four ribosomal ribonucleic acid species and 79 different ribosomal proteins. As many as several hundred copies of the genes encoding the RNA constituents form clusters within the human genome; by contrast, the genes encoding the proteins are widely dispersed, though they are functionally related and coordinately expressed.

Number and Diversity in This Class

The ribosome is a massive structure, with components organized into two subunits designated by their sedimentation coefficients: the smaller is 40S in eukaryotes, the larger 60S. In mammalian cells, the 40S subunit contains a single molecule of ribosomal ribonucleic acid rRNA (18S) and 32 different ribosomal proteins (RPs), while the 60S subunit contains three molecules of rRNA (28S, 5.8S and 5S) and 47 different RPs (Wool *et al.*, 1996). With the exception of two proteins, each of these components is present as a single copy within the ribosome. There are about 4×10^6 ribosomes per cell, and these RNAs and proteins constitute 80% of all cellular RNA and 5–10% of cellular protein. (*See* Nucleolus: Structure and Function.)

Ribosomal components have been conserved to a significant degree throughout evolution. For instance, the amino acid sequences of RPs are nearly identical among mammals, and one can find a close homolog in yeast for all but one of the mammalian proteins. Even eubacteria and archebacteria possess orthologs of human RPs, though the number of proteins expressed by members of these kingdoms is much smaller than in eukaryotes. The sequences of the small and large rRNAs from thousands of organisms are now known (see The Ribosomal Database Project in Web Links). Although the sizes and primary nucleotide sequences of these rRNAs vary considerably, the folded stem–loop structures, including the functional domains, are comparable in all organisms. The ribosome has thus retained its overall structure and its fundamental function as the cell's protein-synthesizing machinery over a span of more than 2 billion years. (*See* Phylogenetics.)

Table 1 shows the ribosomal components of several species whose genomes have been sequenced entirely. Note that the RNA constituents in eubacteria and archebacteria are smaller than those in eukaryotes, as is the number of proteins. In *Escherichia coli*, for

Table 1 Numbers of ribosomal components in six selected organisms

	RNA	Protein
Eukaryotes		
Human	4	79
Drosophila melanogaster	4	79
Caenorhabditis elegans	4	79
Saccharomyces cerevisiae	4	78
Archea		
Methanococcus jannaschii	3	63
Bacteria		
Escherichia coli	3	54

instance, the small (30S) subunit consists of the 16S rRNA and 21 proteins; the large (50S) subunit contains the 23S rRNA, the 5S rRNA and 33 proteins. (*See Caenorhabditis elegans* Genome Project; Comparative Genomics.)

Aspects of Structure and Function

Recent X-ray crystallographic analysis of the *Haloarcula marismortui* 50S subunit, the *Thermus thermophilus* 30S subunit, and the complete structure of the *T. thermophilus* 70S ribosome in the presence of messenger RNA (mRNA) and transfer RNA (tRNA) bound in the A (aminoacyl), P (peptidyl) and E (exit) sites (Yusupov *et al.*, 2001) has increased our understanding of the structural basis of ribosome function. Indeed, we are now able to trace the path taken by mRNA within the ribosome at a resolution of 7 Å. These observations have led to the conclusion that the ribosome is a ribozyme in which rRNAs function as catalytic elements, and RPs serve as structural units, organizing the RNA into the appropriate configuration.

It has proved more difficult to resolve the fine structure of eukaryotic ribosomes, though the essential elements involved in protein synthesis would be expected to have features similar to those in all organisms. There is, however, some information available on the functional interactions between various components of mammalian ribosomes. Most extensively studied is the guanosine triphosphatase (GTPase) associated RNA domain, which binds a complex composed of ribosomal P proteins (P1, P2 and P0) and L12. This region probably serves as a binding site for elongation factor EF-2 and, during protein synthesis, drives translocation of peptidyl tRNA from the A site to the P site through GTP hydrolysis. Furthermore, the finding that a eukaryotic P protein complex and L12 also bind to the *E. coli* GTPase-associated domain suggests that the folded structure of this domain is conserved in all organisms and is essential for its function (Uchiumi *et al.*, 1999). (*See* tRNA.)

Characteristics of Ribosomal Proteins

An average human RP has a molecular weight of 18 877, contains 167 amino acids (**Table 2**), is highly basic and has an isoelectric point (p*I*) of 10.63. Although these proteins function together, their amino acid sequences are dissimilar. Four RPs (S6 and the three P proteins) are subject to phosphorylation, which in the case of S6 varies in response to various stimuli; phosphorylation of the P proteins, by contrast, is unaffected by changes in physiological conditions (Wool *et al.*, 1996). It has long been argued that S6 phosphorylation can affect the translation efficiency of certain classes of mRNA, although no completely convincing evidence of that has yet been presented. Unlike most, three RPs are formed by cleavage from a larger hybrid protein: S27a and L40 are the carboxyl extensions of ubiquitin, and S30 is the carboxyl extension of a ubiquitin-like protein. The function of the ubiquitin moiety remains unknown.

Ribosomal proteins may also have extraribosomal functions (Wool, 1996). In mammals and *Drosophila melanogaster*, for example, S3 functions as both a ribosomal protein and an endonuclease; in *Drosophila*, S6 functions as a tumor suppressor in the hematopoietic system, and S2 functions in oogenesis; and in humans, L5 forms a protein complex with Mdm2 and p53, and L22 forms a ribonucleoprotein (RNP) complex with Epstein–Barr virus-encoded RNA. To account for their extraribosomal functions, Wool (1996) suggested that, during the course of evolution, proteins with diverse functions must have been recruited to the ribosome in order to stabilize rRNA or enhance translation, but the recruited proteins must have also retained their ancestral functions.

Gene Distribution

Multiple copies of the genes encoding rRNAs are clustered together within the human genome. The 28S, 18S and 5.8S rRNAs are generated by processing a single 45S precursor derived from tandemly repeated (~250 times) gene arrays, comprising five clusters of ~50 tandem repeats located on the short arms of chromosomes 13, 14, 15, 21 and 22. The 5S rRNA derives from gene clusters of several hundred tandem repeats on the long arm of chromosome 1. (*See* Chromosome-specific Repeats (Low-copy Repeats); Non-protein-coding Genes.)

Unlike rRNAs, RPs are typically encoded by a single gene; however, each of these genes generates a large number (10–20 copies) of silent, processed pseudogenes at sites dispersed throughout the genome. All the functional genes encoding the 79 RPs have been mapped to the chromosomes (Uechi *et al.*, 2001) (**Table 2**). Though functionally related and coordinately expressed, RP genes are widely dispersed – both sex chromosomes and 20 of the 22 autosomes (all but chromosomes 7 and 21) carry one or more RP genes. Only the notable presence of 13 RP genes on chromosome 19 is at odds with their seemingly random distribution throughout the human genome. Even when located on the same chromosome, these genes are widely separated from one another: except in the case of the ribosomal protein L13a gene (*RPL13A*) and ribosomal protein S11 gene (*RPS11*), which have an interval of only 4.6 kilobases (kb) between them. Consequently, the clustering of RP genes into operons, which is a characteristic feature of eubacteria and archebacteria, is not apparent in humans, nor is it seen in other eukaryotes such as *D. melanogaster*, *Caenorhabditis elegans* and *Saccharomyces cerevisiae*. (*See* Chromosomal Bands and Sequence Features; Gene Clustering in Eukaryotes; Pseudogenes and their Evolution.)

Corresponding to the four RP genes present on the X chromosome (*RPS4X*, *RPL10*, *RPL36A* and *RPL39*) are second functional copies on the Y chromosome and on chromosomes 3 and 14 (**Table 2**). Those on the autosomes are probably generated by retrotransposition of the X-located genes and therefore contain no introns in their coding regions. By contrast, *RPS4Y* present on the Y chromosome does contain introns inserted at positions that correspond to their insertion points in the X-located copy. Although the function of the second genes remains unclear, they might be required to maintain the gene dosage so as to provide

Table 2 Human ribosomal proteins

Protein	No. of residues[a]	Gene Size (kb)	No. of exons	Chromosomal location	Accession no.	Locus ID[c]
Small subunit						
Sa	295	5.7	7	3	U43901	3921
S2	293	2.8	7	16	AC005363	6187
S3	243	6.2	7	11	AB061838	6188
S3a	264	5.0	6	4	Z83334	6189
S4	263	4.6	7	X, Y	AF041428	6191
S5	204	7.5	6	19	AC012313	6193
S6	249	4.0	6	9	X67309	6194
S7	194	5.6	7	2	Z25749	6201
S8	208	3.2	6	1	X67247	6202
S9	194	6.8	5	19	AB061839	6203
S10	165	8.6	6	6	AL138726	6204
S11	158	3.3	5	19	AB028893	6205
S12	132	3.0	6	6	AB061840	6206
S13	151	3.3	6	11	D88010	6207
S14	151	5.4	5	5	M13934	6208
S15	145	2.1	4	19	M32405	6209
S15a	130	7.4	5	16	AC020716	6210
S16	146	2.7	5	19	AB061841	6217
S17	135	3.7	5	15	M18000	6218
S18	152	4.4	6	6	AL031228	6222
S19	145	11.2	6	19	AC010616	6223
S20	119	1.4	4	8	AB061842	6224
S21	83	1.4	6	20	AB061843	6227
S23	143	2.3	4	5	AC005406	6228
S24	133	4.5	5	10	U12202	6229
S25	125	2.6	5	11	AB061844	6230
S26	115	2.0	4	12	U41448	6231
S27	84	1.4	4	1	AB061845	6232
S27a[b]	80	2.9	6	2	AB062071	6233
S28	69	0.9	4	19	AB061846	6234
S29	56	2.8	3	14	AB061847	6235
S30[b]	59	1.5	5	11	X65921	2197
Large subunit						
L3	403	6.7	10	22	AL022326	6122
L4	427	5.5	10	15	AB061820	6124
L5	297	9.9	8	1	AL162740	6125
L6	288	4.4	7	12	AB042820	6128
L7	248	3.0	7	8	AC111149	6129
L7a	266	3.2	8	9	X52138	6130
L8	257	2.6	6	8	AB061821	6132
L9	192	4.8	8	4	U09954	6133
L10	214	2.5	7	X, 14	M81806	6134
L10a	217	2.4	6	6	AL022721	4736
L11	178	4.6	6	1	AL451000	6135
L12	165	3.7	7	9	AL445222	6136
L13	211	2.6	6	16	AC092123	6137
L13a	203	4.3	8	19	AB028893	23521
L14	220	5.0	6	3	AB061822	9045
L15	204	2.4	4	3	AB061823	6138
L17	184	4.0	7	18	AB061824	6139
L18	188	3.9	7	19	AB061825	6141
L18a	176	3.4	5	19	AC005796	6142
L19	196	4.4	6	17	AC004408	6143
L21	160	5.0	6	13	AB061826	6144
L22	128	13.0	4	1	AL031847	6146
L23	140	3.6	5	17	AB061827	9349
L23a	156	3.0	5	17	AF001689	6147
L24	157	5.6	6	3	AB061828	6152
L26	145	5.7	4	17	AB061829	6154
L27	136	4.5	5	17	AC055866	6155
L27a	148	3.1	5	11	AB020236	6157

Table 2 Continued

| Protein | No. of residues[a] | Gene | | | | |
		Size (kb)	No. of exons	Chromosomal location	Accession no.	Locus ID[c]
L28	137	2.4	5	19	AC020922	6158
L29	159	2.3	3	3	AC097636	6159
L30	115	3.8	5	8	AP003352	6156
L31	125	4.1	5	2	AB061830	6160
L32	135	5.5	4	3	AB061831	6161
L34	117	4.7	5	4	AB061832	6164
L35	123	4.1	4	9	AL354928	11224
L35a	110	–	–	3	–	6165
L36	105	1.4	4	19	AB061833	25873
L36a	106	2.9	5	X, 14	U78027	6173
L37	97	2.8	4	5	AB061834	6167
L37a	92	2.6	4	2	AC073321	6168
L38	70	6.2	5	17	AC100786	6169
L39	51	5.1	3	X, 3	AB061835	6170
L40[b]	52	3.4	5	19	X56997	7311
L41	25	1.2	4	12	AB010874	6171
P0	317	4.4	8	12	AC004263	6175
P1	114	2.7	4	15	AB061836	6176
P2	115	2.9	5	11	AB061837	6181

[a]Deduced from the coding sequence of complementary DNA, which includes the first methionine.
[b]Fusion protein with ubiquitin or a ubiquitin-like protein. Gene size and the number of exons include the ubiquitin moiety.
[c]Entry IDs in LocusLink (see LocusLink in Web Links).

equimolar amounts of proteins to the ribosome. It is believed that a full complement of RPs is required to assemble a functional ribosome and to sustain normal cell development. In *D. melanogaster*, for example, haploinsufficiency of any one of the RP genes results in an abnormal phenotype (Lambertsson, 1998). (**See** Haploinsufficiency; Mammalian Sex Chromosome Evolution; Retrotransposition and Human Disorders; X-chromosome Inactivation.)

Gene Structure

Human RP genes are fairly small; most are less than 5 kb in length (Yoshihama *et al.*, 2002) (**Table 2**). The average gene size from the transcription start site is 4.1 kb, with *RPL22* being the largest (13 kb long) and *RPS28* the smallest (only 0.9 kb). Each gene contains from 3 to 10 exons, with 5.6 being the average. The first exon is small, an average of 44 base pairs (bp); the others are 124 bp in length, which is comparable to the genome-wide average of 145 bp. The 5' and 3' noncoding regions are, respectively, 42 and 57 bp in length, making them significantly smaller than the genome averages (300 and 770 bp respectively). The translation initiator ATG is located either in the first or second exon, in most cases near the splice junction of the first intron, while the stop codon is usually in the last exon (all but *RPS3*, *RPS25*, *RPS28* and *RPL9*). The average length of the coding sequences is 502 bp.

The average GC content of an entire RP gene is 49%, and that of the promoter region (-250 to $+250$ bp) is 61%, which is substantially higher than the genome average of 41%. The promoter regions also contain CpG islands. (**See** CpG Islands and Methylation; Gene Structure and Organization.)

A number of small nucleolar RNA (snoRNA) genes have been found within the introns of the RP genes. Indeed, a large number of distinct snoRNAs have been identified in the nucleoli of eukaryotic cells, and in vertebrates most are encoded by introns of other genes. These RNAs function as guides for RNA ribose methylation and pseudouridylation, mostly of rRNA and snRNA. To date, 106 methylations and 91 pseudouridylations have been identified in human rRNA, about half of which were tentatively assigned to known snoRNAs. Among these, 38 snoRNA genes (including putative ones) have been identified within introns of 26 RP genes, accounting for about one-third of the known snoRNAs (Yoshihama *et al.*, 2002). (**See** *Gas5* Gene; Noncoding RNAs: A Regulatory Role?; Nonprotein-coding Genes; Nucleolus: Structure and Function; snoRNPs.)

Regulation of Gene Expression

Regulated coproduction of the ribosomal components may be achieved by actions at virtually any level of gene expression. It has been suggested, for example, that *trans*-acting regulatory mechanisms affecting both transcription and translation play key roles in coordinating production of RPs in mammals, and the contributions of other mechanisms, such as feedback regulation, await further investigation.

The 5′ ends of RP mRNAs in vertebrates contain a unique 5′ terminal oligopyrimidine tract (5′ TOP). This structural motif comprises the core of the translational *cis*-regulatory elements of these mRNAs. Translational control of these mRNAs, which has been extensively studied in mammalian cells and *Xenopus laevis* (Meyuhas *et al.*, 1996), is thought to be governed by growth and nutritional stimuli, but although several 5′ TOP-binding proteins have been identified, no direct relationship between the binding activity of these proteins and the growth-dependent translational efficiency of RP mRNAs has been demonstrated so far. (*See* mRNA Stability and the Control of Gene Expression; 5′ UTRs and Regulation.)

On the other hand, experiments using DNA array technology have shown that the transcription of RP mRNAs in yeast is strictly regulated in a manner responsive to changes in growth conditions. Common regulatory factors, such as Rap1 and Abf1, are involved in most yeast RP gene transcriptions. Likewise, three mouse RP genes (*Rpl30*, *Rpl32* and *Rps16*) with promoters having similar architectures – five or more elements over a 200 bp region – have been found to bind a variety of *trans*-acting factors (Hariharan *et al.*, 1989). In fact, systematic analysis of the expression data (transcriptome) from 64 human tissues (see BodyMap in Web Links) suggests that expression of this class of genes is regulated to a substantial degree through modulation of transcription. (*See* DNA Chips and Microarrays; Promoters; Transcriptional Regulation: Coordination.)

Despite the central role played by the ribosome in the growth and development of organisms, the effects of ribosomal mutations, particularly their role in human disease, have been largely ignored. One recent study, however, showed that the gene encoding RPS19 is mutated in patients with Diamond–Blackfan anemia (Draptchinskaia *et al.*, 1999), a rare form of chronic anemia characterized by absent or low levels of erythroid precursors in the bone marrow, implicating RP gene defects in human disease. Moreover, transcriptome analysis showed that the expression patterns of 14 RP genes in the brains of Ts65Dn mice, a model of Down syndrome, are significantly different from those in normal mice (see Division of Medical Genetics, University of Geneva in Web Links), implicating abnormal ribosome biogenesis in the development and maintenance of Down syndrome phenotypes. (*See* Down Syndrome.)

Mitochondrial Ribosomes

Mitochondria have their own translation system, including ribosomes that are distinct from their cytoplasmic counterparts. In mammalian cells, mitochondrial ribosomes comprise two rRNAs and about 80 RPs and are responsible for the production of 13 proteins essential for oxidative phosphorylation. The small (28S) subunit is made up of one rRNA molecule (12S) and an estimated 31 RPs, while the large (39S) subunit is made up of one rRNA molecule (16S) and an estimated 48 RPs. Mitochondria are thought to be descendants of ancient bacteria that became incorporated into the preeukaryotic cell through endosymbiosis. During the course of evolution, however, the RNA constituents became smaller in both size and number, while the proteins increased in number. The genes encoding mitochondrial rRNAs still reside in the mitochondrial genome. On the other hand, mitochondrial RPs (MRPs) are encoded in the nuclear genome where their genes are widely dispersed in a manner similar to their cytoplasmic counterparts. MRPs are thus synthesized in the cytoplasm and imported into the mitochondria where they are assembled into the ribosome. (*See* Mitochondrial Genome; Mitochondrial Genome: Evolution; Mitochondrial Origins of Human Nuclear Genes and DNA Sequences.)

See also
Nucleolus: Structure and Function
Translation Initiation: Molecular Mechanisms in Eukaryotes

References

Draptchinskaia N, Gustavsson P, Andersson B, *et al.* (1999) The gene encoding ribosomal protein S19 is mutated in Diamond–Blackfan anaemia. *Nature Genetics* **21**: 169–175.

Hariharan N, Kelley DE and Perry RP (1989) Equipotent mouse ribosomal protein promoters have a similar architecture that includes internal sequence elements. *Genes and Development* **3**: 1789–1800.

Lambertsson A (1998) The *Minute* genes in *Drosophila* and their molecular functions. *Advances in Genetics* **38**: 69–134.

Meyuhas O, Avni D and Shama S (1996) Translational control of ribosomal protein mRNAs in eukaryotes. In: Hershey JWB, Mathews MB and Sonenberg N (eds.) *Translational Control*, pp. 363–388. Cold Spring Harbor, NY: Cold Spring Harbor Laboratory Press.

Uchiumi T, Hori K, Nomura T and Hachimori A (1999) Replacement of L7/L12.L10 protein complex in *Escherichia coli* ribosomes with the eukaryotic counterpart changes the specificity of elongation factor binding. *Journal of Biological Chemistry* **274**: 27578–27582.

Uechi T, Tanaka T and Kenmochi N (2001) A complete map of the human ribosomal protein genes: assignment of 80 genes to the cytogenetic map and implications for human disorders. *Genomics* **72**: 223–230.

Wool IG (1996) Extraribosomal functions of ribosomal proteins. *Trends in Biochemical Sciences* **21**: 164–165.

Wool IG, Chan YL and Glück A (1996) Mammalian ribosomes: the structure and the evolution of the proteins. In: Hershey JWB, Mathews MB and Sonenberg N (eds.) *Translational Control*, pp. 685–732. Cold Spring Harbor, NY: Cold Spring Harbor Laboratory Press.

Yoshihama M, Uechi T, Asakawa S, *et al.* (2002) The human ribosomal protein genes: sequencing and comparative analysis of 73 genes. *Genome Research* **12**: 379–390.

Yusupov MM, Yusupova GZ, Baucom A, *et al.* (2001) Crystal structure of the ribosome at 5.5 Å resolution. *Science* **292**: 883–896.

Further Reading

Amaldi F, Camacho-Vanegas O, Cardinall B, *et al.* (1995) Structure and expression of ribosomal protein genes in *Xenopus laevis*. *Biochemistry and Cell Biology* **73**: 969–977.

Brown PO and Botstein D (1999) Exploring the new world of the genome with DNA microarrays. *Nature Genetics* **21**: 33s–37s.

Chrast R, Scott HS, Papasavvas MP, *et al.* (2000) The mouse brain transcriptome by SAGE: differences in gene expression between P30 brains of the partial trisomy 16 mouse model of Down syndrome (Ts65Dn) and normals. *Genome Research* **10**: 2006–2021.

Fisher EM, Beer-Romero P, Brown LG, *et al.* (1990) Homologous ribosomal protein genes on the human X and Y chromosomes: escape from X inactivation and possible implications for Turner syndrome. *Cell* **63**: 1205–1218.

Higa S, Yoshihama M, Tanaka T and Kenmochi N (1999) Gene organization and sequence of the region containing the ribosomal protein genes *RPL13A* and *RPS11* in the human genome and conserved features in the mouse genome. *Gene* **240**: 371–377.

Kawamoto S, Yoshii J, Mizuno K, *et al.* (2000) BodyMap: a collection of 3' ESTs for analysis of human gene expression information. *Genome Research* **10**: 1817–1827.

Kenmochi N, Kawaguchi T, Rozen S, *et al.* (1998) A map of 75 human ribosomal protein genes. *Genome Research* **8**: 509–523.

Kenmochi N, Suzuki T, Uechi T, *et al.* (2001) The human mitochondrial ribosomal protein genes: mapping of 54 genes to the chromosomes and implications for human disorders. *Genomics* **77**: 65–70.

Smith CM and Steitz JA (1997) Sno storm in the nucleolus: new roles for myriad small RNPs. *Cell* **89**: 669–672.

Uechi T, Maeda N, Tanaka T and Kenmochi N (2002) Functional second genes generated by retrotransposition of the X-linked ribosomal protein genes. *Nucleic Acids Research* **30**: 5369–5375.

Warner JR (1999) The economics of ribosome biosynthesis in yeast. *Trends in Biochemical Sciences* **24**: 437–440.

Zhang Z, Harrison P and Gerstein M (2002) Identification and analysis of over 2000 ribosomal protein pseudogenes in the human genome. *Genome Research* **12**: 1466–1482.

Web Links

BodyMap. A databank of expression information of human and mouse genes, novel or known, in various tissues or cell types and various timings
http://bodymap.ims.u-tokyo.ac.jp

Division of Medical Genetics, University of Geneva. Differential gene expression in trisomy Ts65Dn mouse model of Down syndrome
http://medgen.unige.ch/research/Ts65Dn/

LocusLink. Provides a single query interface to curated sequence and descriptive information about genetic loci.
http://www.ncbi.nlm.nih.gov/LocusLink/

Ribosomal Protein Gene database (RPG). A curated database that provides data related ribosomal protein genes, including sequences, map positions, and comparative data between orthologs.
http://ribosome.miyazaki-med.ac.jp

The Ribosomal Database Project. A curated database that provides data, programs and services related to ribosomal RNA sequences.
http://rdp.cme.msu.edu/html/

Ribozymes

See Antisense and Ribozymes

RNA-binding Proteins: Regulation of mRNA Splicing, Export and Decay

Jayanthi P Gudikote, *University of Texas MD Anderson Cancer Center, Houston, Texas, USA*

Miles F Wilkinson, *University of Texas MD Anderson Cancer Center, Houston, Texas, USA,*

Mammalian messenger ribonucleic acid (mRNA)-binding proteins regulate gene expression at several steps including pre-mRNA processing, nucleocytoplasmic export and degradation.

Introduction

Precursor ribonucleic acid (RNA) molecules are transcribed from deoxyribonucleic acid (DNA) templates in the nucleus by RNA polymerases. These precursor RNA molecules then undergo maturation to form functional RNAs. For example, precursors for messenger RNAs (pre-mRNAs) typically go through a series of posttranscriptional processing steps, including 5′ capping, splicing, 3′ end cleavage and polyadenylation. Specific proteins associate with these mRNA molecules at every step of their life, including when they are first transcribed, when they undergo maturation, when they are exported to the cytoplasm, and when they are translated and later degraded. Many of these proteins play essential roles in directing and regulating these steps. In this article we will focus mainly on RNA-binding proteins that participate in and control mammalian mRNA splicing, export and decay. Excellent reviews that describe the RNA-binding proteins involved in other aspects of mRNA metabolism, as well as those involved in the processing, export and decay of other RNAs, including transfer RNAs (tRNAs) and ribosomal RNAs (rRNAs), can be found elsewhere (see Further Reading).

RNA-binding Proteins Involved in Pre-mRNA Processing

Eukaryotic protein-coding genes are transcribed into pre-mRNA molecules in the nucleus by RNA polymerase II. Nascent pre-mRNAs emerging from the transcriptional apparatus are coated with abundant RNA-binding proteins called heterogeneous ribonuclear proteins (hnRNPs). At least 20 major hnRNP proteins, A1 to U, have been identified, many of which are thought to stabilize newly synthesized mRNAs. They also participate in a wide variety of other processes, including transcriptional regulation, pre-mRNA 3′ end processing, mRNA localization, translation and mRNA turnover (Shyu and Wilkinson, 2000). hnRNPs may also play a role in regulating pre-mRNA splicing and export, which will be discussed later.

RNA splicing

Most metazoan pre-mRNA transcripts contain protein-coding sequences (contained within exons) that are interrupted by noncoding sequences (known as introns). The 5′ and 3′ exons also contain untranslated regions (UTRs) that are located upstream (5′ UTR) and downstream (3′ UTR) of the start and stop codons respectively. Introns are excised from the adjacent exons in the nucleus by a process called RNA splicing prior to the transport of mRNA to the cytoplasm. The pre-mRNA splicing reaction is carried out by spliceosomes – multicomponent ribonucleoprotein complexes containing five small nuclear (sn) RNAs: U1, U2, U4, U5 and U6 (Kramer, 1996). snRNAs associate with several proteins to form sn ribonucleoprotein (snRNP) particles that bind to pre-mRNA as integral parts of the splicing machinery. Spliceosome assembly is directed, in part, by the RNA sequences at the exon–intron junctions (splice sites). Three intronic sequence elements are important in establishing the high fidelity of splice-site recognition by the spliceosome: the 5′ splice site, the branchpoint sequence (BPS) and the pyrimidine tract/AG dinucleotide at the 3′ splice site. Pre-mRNA splicing occurs via a two-step mechanism. In the first step, the 5′ end of the intron is joined to an adenosine residue located within the BPS to form a branched intermediate called an intron lariat. In the second step, the two exons are ligated together and the intron lariat is released (**Figure 1**). The splicing machinery recognizes the correct 5′ (GU) and 3′ (AG) splice sites based on their proximity to exons. The binding of U1 snRNP to the 5′ splice site, branchpoint binding protein (BBP) to the BPS and U2 auxiliary factor (the U2AF 35/65 kDa heterodimer) to the pyrimidine tract and the 3′ YAG initiates spliceosome assembly and formation of the early (E) spliceosome complex (**Figure 1**). This complex commits the pre-mRNA to the splicing pathway. This E complex is converted to an A complex when the U2 snRNP binds to the branchpoint. Subsequently, U4, U5 and U6 snRNPs enter the spliceosome as a tri-snRNP particle to form the B complex. Finally a massive rearrangement occurs, in which U6 snRNP replaces U1 snRNP at the 5′ splice site, U6 snRNP and U2 snRNP interact, U5 snRNP bridges the splice sites, and U1 snRNP and U4 snRNP become destabilized to form the catalytically active C complex (see Graveley, 2000 and the references therein).

Many RNA-binding proteins facilitate spliceosome assembly on pre-mRNAs. These include members of the serine/arginine-rich (SR) protein family, which are required at several different steps of spliceosome assembly (Graveley, 2000). SR proteins contain a *C*-terminal RS domain rich in alternating arginine and serine residues, required for interactions with other RNA-binding proteins that contain RS domains, and one or two *N*-terminal RNA recognition motifs (RRMs) that function in sequence-specific RNA binding. At least 10 mammalian SR proteins have been well characterized: SRp20, SC35, SRp46, SRp54, SRp30c, SF2/ASF, SRp40, SRp55, SRp75 and 9G8. Other factors that participate in setting up the splicing apparatus are SR-related proteins that contain RS domains but are structurally and functionally distinct from the SR family proteins; for example the U2 auxiliary factors U2AF35 and U2AF65, Tra2α and β, SRm 160 and hPrp16 (Graveley, 2000). These SR-related proteins and classical SR proteins work together to recruit U1 snRNP to the 5′ splice site and U2 snRNP to the branchpoint by functioning as bridging factors. They

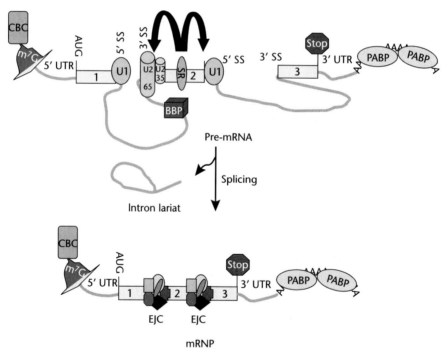

Figure 1 Pre-mRNA splicing. Mammalian pre-mRNAs contain a 5′ cap (m⁷G), a 5′ untranslated region (5′ UTR), coding regions (numbered 1, 2 and 3) interrupted by noncoding regions called introns, a 3′ untranslated region (3′ UTR) and a poly(A) tail. Cap binding protein (CBC) and poly(A)-binding protein (PABP) interact with the 5′ cap and 3′ poly(A) tail, respectively, where they serve to protect the mRNA from decay. The exons are joined together by a complex splicing process initiated by binding of U1 snRNP (U1) to the 5′ splice site (5′ SS). This is followed by the binding of the branchpoint binding protein (BBP) to the branchpoint, and U2 auxiliary factors U2AF 65 (U2 65) and U2AF 35 (U2 35) to the polypyrimidine tract just upstream of the 3′ splice site (3′ SS). SR proteins (SR) are RNA-binding proteins that are thought to help recruit these splicing factors to pre-mRNA. Other molecules involved in RNA splicing described in the text are not shown here. The exon junction complex (EJC) is a multiprotein complex deposited near exon–exon junctions after RNA splicing. It appears to facilitate mRNA export and play a critical role in the NMD RNA surveillance pathway. The positions of the start (AUG) and stop (stop sign) codons are marked.

can also recruit the U4/U6/U5 tri-snRNP particle to pre-mRNA.

RNA-binding proteins that regulate splicing

RNA-binding proteins are not only essential for the basic mechanism of RNA splicing but they can also regulate the decisions made by the splicing apparatus. In other words, they can regulate alternative splicing, one of the important mechanisms by which protein diversity is generated in vertebrates (Black, 2000). Alternative pre-mRNA splicing is a process in which multiple mRNAs are generated from the same pre-mRNA by the differential joining of 5′ and 3′ splice sites. Regulatory proteins can trigger alternative splicing by binding directly to the 5′ or 3′ splice sites, or to other pre-mRNA sequences called exonic or intronic splicing enhancers (ESEs or ISEs) and silencers (ESSs or ISSs). These enhancers and silencers stimulate or repress splice-site selection respectively. ESEs act by attracting SR proteins, which in turn enhance the utilization of the adjacent splice sites by forming networks of interactions with each other and with SR-related proteins. For example, SR proteins bound to ESEs can promote splicing by facilitating the binding of U2AF-65 kDa to the polypyrimidine tract through an interaction mediated by U2AF-35 kDa (**Figure 1**) (Blencowe, 2000). Tissue-specific splicing is imparted by the splicing factor neurooncological ventral antigen-1 (NOVA-1), which stimulates inclusion of specific exons in neurons by binding to the ISEs adjacent to the included exons in target pre-mRNAs. In contrast, polypyrimidine tract binding protein (PTB, also known as hnRNP I) represses exon inclusion by directly interfering with general splicing factors binding to the pyrimidine tract (Wagner and Garcia-Blanco, 2001). hnRNP A1 is another example of a splicing repressor. For example, it represses the splicing of the human immunodeficiency virus by binding to an ESS in HIV pre-mRNAs, allowing export of partially spliced and unspliced HIV mRNAs encoding essential viral proteins.

RNA-binding Proteins that Promote mRNA Export

Although pre-mRNA transcription and processing are important early steps in gene expression, the expressed form of mRNA, the protein, can be generated only after the mRNA is translated. Hence, mRNAs that are properly spliced must be exported to the cytoplasm. Most mRNAs are exported to the cytoplasm as an mRNA–RNP complex (mRNP). Several proteins in these mRNPs have been implicated as playing a role in

mRNA export, although few have clearly defined roles. In this section, we will cover RNA-binding proteins that appear to play a role in mammalian mRNA export.

Export of mRNAs requires three classes of factors: the adaptor proteins that bind directly to the mRNA, the receptor proteins that recognize and bind the adaptor proteins and the nuclear pore complex (NPC) components that mediate transport across the nuclear membrane (Reed and Hurt, 2002). NPCs are huge macromolecular assemblies spanning the nuclear envelope that contain a central channel lined with proteins called nucleoporins that interact with the transport receptor proteins (**Figure 2**). The different classes of RNAs, including tRNAs, rRNAs, snRNAs and mRNAs, are exported through distinct pathways directed by the signals present on these RNAs and/or the different receptor proteins that bind to them. Export of most RNA cargos requires transport receptors called karyopherins, also known as importins/exportins. Ran guanosine triphosphatase (GTPase), a member of the Ras family of small GTPases, regulates the interaction between the RNA cargos and karyopherins. Although tRNAs, snRNAs and rRNAs use the Ran-GTP pathway for RNA export, most mRNAs do not (Reed and Hurt, 2002; Zenklusen and Stutz, 2001) and so the Ran-GTP pathway will not be covered here.

REF/TAP-mediated mRNA export

Nuclear export of mRNAs is only just beginning to be understood and is expected to be far more complex than the export of other RNA molecules. One aspect of this complexity is that mRNA export is coupled to pre-mRNA splicing. Studies done in the *Xenopus laevis* oocyte show that mature mRNAs generated by splicing are exported more efficiently than those that are not spliced. In addition, several lines of evidence suggest that the RNA export factor REF (also known as Yra 1 and Aly) functionally links pre-mRNA splicing to mRNA export (Zenklusen and Stutz, 2001; Reed and Hurt, 2002). REF colocalizes with pre-mRNA splicing factors in nuclear speckles and is recruited to mRNAs during splicing and then is deposited near exon–exon junctions after splicing where it serves to promote export. REF accomplishes this by recruiting the mRNA export receptor TAP (also known as Mex67p). This TAP/REF pathway, which unlike other RNA transport pathways does not require Ran-GTP, appears to be the major system that shuttles mRNAs to the cytoplasm.

REF has a central RNA-binding domain (RBD), which suggests that it functions in mRNA export by directly binding to RNA and then recruiting TAP. However, the interaction of REF with RNA is

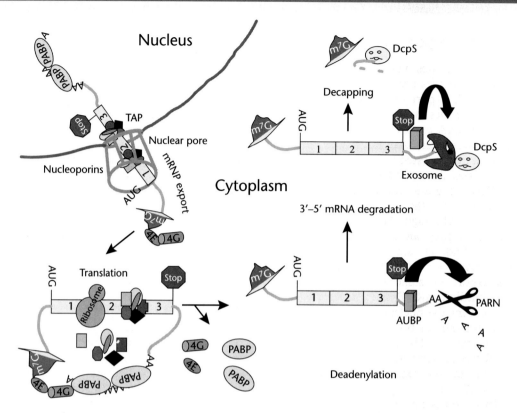

Figure 2 Nuclear mRNA export and mRNA decay in the cytoplasm. Nucleocytoplasmic export of mRNA requires its interaction with mRNA-binding proteins (collectively known as an mRNA–protein complex (mRNP)), which in turn interact with nucleoporins lining the nuclear pore. TAP, a component of the EJC, assists in the movement of mRNP through the nuclear pore. Evidence suggests that the first round of translation ejects the EJC and reorganizes the mRNP in a manner that brings the 5′ and the 3′ ends of the mRNA together to form a circular mRNP complex (this first round of translation is pictured as occurring in the cytoplasm but it may also occur in the nucleus). RNA circularizaton allows for more efficient translation by ribosomes. The poly(A) tail of cytoplasmic mRNAs is degraded by poly(A)-specific ribonucleases (e.g. PARN) at a rate dependent on intrinsic features of the mRNA and specific RNA-binding proteins that bind to the 3′ UTR. One class of RNA-binding proteins that regulate deadenylation is ARE-binding proteins (AUBPs). AUBPs can also trigger mRNA decay by recruiting the exosome, a multiprotein complex that has multiple 3′–5′ nucleases that degrade the body of RNAs. Decapping enzymes, including those associated with the exosome (e.g. DcpS) degrade the 5′ cap.

not mediated through the RBD but instead through its N- and C-terminal regions, neither of which has known RNA-binding motifs. Thus, whether REF binds directly to RNA or instead interacts with RNA using adaptor proteins is not known. The N- and C-termini of REF also serve as two separate binding sites for TAP. TAP has an RBD that is thought to allow it to bind directly to RNA. TAP cross-links to poly(A)$^+$ RNA *in vitro* and its RBD is known to be required for it to bind to the constitutive transport element (CTE) essential for TAP-mediated export of some retroviral mRNAs. However, because TAP exhibits low-affinity interactions with RNA, it is likely that TAP interacts with RNA through protein–protein interactions (such as by binding to REF) as well as protein–RNA interactions. TAP forms a heterodimer with a small protein subunit, p15 (NXT); together TAP and p15 mediate the translocation of mRNAs through the

nuclear pore by sequential interactions with the phenylalanine–glycine (FG) repeats in the nucleoporins of the NPC (Zenklusen and Stutz, 2001; Reed and Hurt, 2002).

Recent studies have indicated that the spliceosomal component UAP56 (also known as Sub2) also plays a role in coupling splicing and mRNA export. UAP56 is a DEAD-box family member possessing ATP-dependent helicase activity required for the export of spliced mRNAs. It appears to function in mRNA export by recruiting REF to mRNA during splicing. Although essential for export, UAP56 is only involved in the initial steps of transport as it is released after binding of the TAP–p15 heterodimer to REF. UAP56 also stimulates the transport of mRNAs derived from intronless genes (that do not undergo RNA splicing) by an apparently REF-independent mechanism (Reed and Hurt, 2002).

Alternative mRNA export pathways

Recent evidence indicates that many other RNA-binding protein molecules besides TAP and REF are involved in mRNA export (Reed and Hurt, 2002). Some of these molecules may participate in the TAP/REF-mediated mRNA export pathway, while others probably function in alternative mRNA export pathways. hnRNPs were suggested to play a role in mRNA export when it was discovered that some hnRNPs, including hnRNP A1 and K, can shuttle with mRNAs from the nucleus to the cytoplasm (Shyu and Wilkinson, 2000). Although there is no proof that shuttling hnRNPs play an active role in mRNA export, several findings are consistent with this possibility, including the finding that excess hnRNP A1 inhibits export of some specific mRNAs. In contrast, nonshuttling hnRNPs may have the opposite function: to prevent the nuclear export of the RNA molecules that they remain bound to, including spliced introns and aberrant mRNAs.

SR and SR-like proteins have also been suggested to play a role in mRNA export. Like some hnRNP proteins, many SR-containing proteins can shuttle out of the nucleus, consistent with such a role. The shuttling SR proteins Srp20 and 9G8 were recently shown to promote the export of the intronless mRNA encoding histone H2a by binding to a 22-nucleotide *cis* element. In contrast, the shuttling SR-like proteins RNPS1 and SRm160 bind to exonic sequences from spliced mRNAs and remain bound to these sequences after export, consistent with a role in export, although whether they are instead merely passive exported passengers is not known. A *Saccharomyces cerevisiae* RNA-binding protein related to mammalian SR proteins, Nlp3p, has been shown to be essential for mRNA export. Interestingly, experiments have suggested that Nlp3p is linked with factors bound to the 5′ and 3′ ends of mRNAs, suggesting that Nlp3 may regulate mRNA export indirectly by participating in a monitoring system that analyzes the processing and packing of mRNPs.

Some specialized mRNAs may use Ran-dependent mechanisms for export. The unusually unstable c-*fos* mRNA is exported by two distinct pathways, both involving Ran-dependent receptors of the karyopherin superfamily. Some mRNAs in the testes may also be exported by a Ran-dependent pathway, as the testes express a novel form of Tap (NXF3) that lacks the *C*-terminal domain required for the export by other Tap family members but instead contains a nuclear export signal known to be bound by karyopherins. The possibility that testicular cells use an unusual nuclear export pathway is perhaps not surprising, as many testes genes are regulated by novel transcriptional mechanisms and their transcripts undergo novel alternative splicing and polyadenylation events.

RNA-binding Proteins Controlling mRNA Stability

Mature mRNAs exported to the cytoplasm typically undergo several rounds of translation before they are degraded. mRNA decay is a major control point in gene expression, as it ensures that mRNAs do not build up in the cell, directing the synthesis of unnecessary proteins. Furthermore, developmental regulation of mRNA stability is an important means to ensure that a given mRNA is available for translation only at the appropriate point of development and in the appropriate cell type. Situations in which it is particularly advantageous to control mRNA stability are those in which the encoded protein has the potential to engender uncontrolled growth or acute toxic effects. Thus, mRNAs encoding oncoproteins such as c-*fos* and c-*myc* are highly unstable unless a proliferative signal is sent to the cell. Cytokine mRNAs are highly unstable unless activation signals are received by lymphocytes, as this ensures that the potentially toxic systemic effects of cytokines are avoided unless an infection has occurred. In this section, we will first discuss RNA-binding proteins that protect mRNAs from decay and the pathways that overcome this protection to degrade mRNAs. We will then go over examples of RNA-binding proteins that control the rate of this decay. Finally, we will discuss a recently discovered RNA-binding complex that appears to play a role in an RNA surveillance pathway that degrades aberrant mRNAs.

RNA decay pathways

mRNAs are protected from decay by several different RNA-binding proteins. The poly(A) tail on mRNAs is bound by poly(A)-binding protein (PABP), which serves to protect the 3′ terminus of the mRNA from 3′ to 5′ exonucleases. Likewise, the 5′ cap structure is bound by eIF-4E (4E in **Figure 2**), a component of the translation initiation complex, which protects the 5′ terminus of the mRNA from 5′ to 3′ exonucleases. Interestingly, eIF-4E binds to the translation initiation factor eIF-4G, which in turn binds to PABP. This discovery led to the proposal that these proteins direct the 5′ and 3′ ends of mRNA to interact with each other, allowing the formation of RNA circles (**Figure 2**). Such RNA circularization, which has been shown by microscopy to occur normally, permits efficient translation, as ribosomes terminating translation at the 3′ end of an mRNA could quickly scan downstream to find the 5′ end and reinitiate another round of translation.

This circularization of RNAs also dictates how mRNAs are degraded. In *S. cerevisiae*, the major pathway of mRNA decay is initiated by removal of the poly(A) tail, which liberates PABP, causing the RNA circle to open. This event is followed by removal of the 5′ cap, which then leaves the body of the mRNA susceptible to rapid 5′–3′ exonuclease attack (Tucker and Parker, 2000). As in yeast, mammalian mRNAs are also degraded predominantly by a deadenylation-dependent decay pathway. Deadenylation is typically the rate-limiting step and current data suggest that it is accomplished in part by the poly(A)-specific ribonuclease PARN, a 3′–5′ exonuclease (**Figure 2**). However, unlike yeast, mammalian cells appear to degrade the body of mRNAs primarily by the action of 3′–5′ exonucleases, rather than 5′–3′ exonucleases, based on *in vitro* and to a lesser extent *in vivo* evidence. This 3′ to 5′ decay is mediated by the cytoplasmic exosome, a highly conserved complex of 10 or more 3′–5′ exonucleases. The cap structure left after exosome-mediated decay is then degraded by DcpS, a 'scavenger' decapping enzyme associated with the exosome (**Figure 2**) (van Hoof and Parker, 2002).

Although most mammalian mRNAs appear to be degraded by a deadenylation-dependent pathway, some instead may be degraded by a pathway initiated by endonucleolytic cleavage (Guhaniyogi and Brewer, 2001). Examples of mRNAs degraded by this endonuclease-initiated pathway are those encoding insulin-like growth factors and the transferrin receptor. The RNA-binding proteins that may regulate this pathway have not been defined.

Cis-acting elements and *trans*-acting factors that dictate the decay of normal mRNAs

The rate at which an mRNA is degraded is dictated by *cis*-acting elements within the mRNA and protein factors that bind to these elements. *Cis* elements regulating RNA stability are found in the 5′ UTR, the protein-coding region and the 3′ UTR. The mechanisms by which these *cis* elements act have not yet been precisely defined but progress is being made. An important class of destabilizing *cis* elements is AU-rich elements (AREs) (Chen and Shyu, 1995). AREs typically contain one to multiple copies of the pentamer AUUUA and are found in the 3′ UTR of a wide variety of short-lived mRNAs, including those encoding cytokines, proto-oncoproteins and growth factors. Several ARE-binding proteins (AUBPs) have been characterized. Some AUBPs destabilize ARE-bearing mRNAs, while other AUBPs stabilize these mRNAs. An example of the former is AUF1 (also

known as hnRNP D), which promotes ARE-directed decay as part of a multisubunit complex that includes eIF-4G, PABP and the heatshock proteins hsp/hsc 70. Recently, tristetraprolin (TTP), the prototype of a family of zinc-finger proteins that have two copies of the unusual Cys-Cys-Cys-His (CCCH) domain, was also found to destabilize ARE-bearing mRNAs. AUBPs that stabilize ARE-bearing mRNAs include Hel-N1, HuC, HuD and HuR. The precise mechanisms by which AUBPs modulate the stability of ARE-bearing mRNAs are not known, but recent evidence suggests that one mechanism by which AUBPs can function is by recruiting the exosome (van Hoof and Parker, 2002).

Other mRNA-destabilizing elements are those found in the coding region. These coding region destabilizers (called coding region determinants (CDRs)) have been best characterized in c-*fos*, c-*myc* and β-tubulin mRNA. For example, two distinct destabilizing elements (CRD-1 and CRD-2) have been identified in the coding region of c-*fos* mRNA. CRD-1, the major element, is a well-characterized 320-nucleotide (nt) purine-rich segment that binds to a multiprotein complex containing at least five proteins: PABP, AUF1, PAIP-1 (a PABP-interacting protein), Unr (a purine-rich RNA-binding protein) and NSAP1 (an hnRNP R-like protein). It has been proposed that these proteins form a bridge between CRD-1 and the poly(A) tail. According to this model, ribosome transit across CRD-1 disrupts or reorganizes the complex, leading to rapid RNA deadenylation and decay (Grosset *et al.*, 2000).

RNA-binding factors involved in the degradation of aberrant mRNAs

Another set of mRNAs targeted for rapid mRNA decay are aberrant transcripts harboring premature termination codons (PTCs). PTCs are generated by nonsense mutations, frameshift mutations, splicing errors and transcriptional errors. Transcripts harboring PTCs are degraded by an RNA-surveillance pathway known as nonsense-mediated decay (NMD) (Mendell and Dietz, 2001; Wilkinson and Shyu 2002). Two signals in an RNA are usually required to trigger the NMD response in mammalian cells: a stop codon and a downstream intron. The second signal, the downstream intron, does not appear to act directly to trigger RNA decay; rather it acts indirectly by attracting RNA-binding proteins that serve in this capacity. A good candidate to mediate this is an RNA-binding complex that is left behind near exon–exon junctions after RNA splicing. This complex, dubbed the exon junction complex (EJC), is deposited 20–24 nt upstream of exon–exon junctions after RNA splicing

(Lykke-Andersen, 2001). Components of the EJC include the spliceosome-associated proteins SRM160 and RNPS1, the nuclear export factors REF and TAP, the small nucleocytoplasmic shuttling protein Y14, and the essential NMD proteins UPF2 and UPF3. Because the EJC contains components involved in splicing, export and NMD, this has led to a model that links these three events in a stepwise manner. According to this model, the deposition of the EJCs after RNA splicing drives the export of spliced mRNAs. After export, the EJCs are stripped off by ribosomes during the first round of translation. However, in the case of mRNAs with PTCs, translation terminates before all of the EJCs have been stripped off, causing the posttranslational surveillance machinery (that has been suggested to form after translation termination) to come in contact with one or more EJCs, thereby triggering mRNA decay. Although it is not known precisely how NMD is triggered, it has been suggested to occur as a result of the essential NMD protein UPF1 bound to the posttranslational surveillance machinery coming in contact with UPF2 in the EJC (Lykke-Andersen, 2001). The location in which NMD occurs has been controversial; the increasing evidence for nuclear translation suggests it may occur in the nucleus rather than the cytoplasm. This has led to alternative models for how the EJC directs NMD and RNA transport (Wilkinson and Shyu, 2002).

Summary

In this article we have described how RNA-binding proteins play crucial roles in mammalian mRNA splicing, mRNA export and mRNA stability. We showed that while some RNA-binding proteins are essential for these processes to occur, other RNA-binding proteins regulate these processes in a tissue- or context-specific manner. Increasing evidence indicates that many RNA-binding proteins are multi-functional proteins that regulate many events (Shyu and Wilkinson, 2000). For example, some RNA-binding proteins can regulate both RNA processing and RNA stability. Others can regulate both transcription and translation. This underscores the emerging view that the many steps of gene expression, beginning with transcription and RNA maturation in the nucleus and ending with translation and mRNA decay in the cytoplasm, are biochemically and spatially linked, allowing gene expression to operate more efficiently. We are only just beginning to understand the complex communication networks between RNA-binding proteins that allow this elaborate interconnected pathway to function.

See also

See also
mRNA Export
mRNA Stability and the Control of Gene Expression
RNA Processing

References

Black D (2000) Protein diversity from alternative splicing: a challenge for bioinformatics and post genome biology. Cell 103: 367–370.

Blencowe BJ (2000) Exonic splicing enhancers: mechanism of action, diversity and role in human genetic diseases. Trends in Biochemical Science 25: 106–110.

Chen C-YA and Shyu A-B (1995) AU-rich elements: characterization and importance in mRNA degradation. Trends in Biochemical Science 20: 465–470.

Graveley BR (2000) Sorting out the complexity of SR protein functions. RNA 6: 1197–1211.

Grosset C, Chen C-YA, Xu N, et al. (2000) A mechanism for translationally coupled mRNA turnover: interaction between the poly(A) tail and a c-fos RNA coding determinant via a protein complex. Cell 103: 29–40.

Guhaniyogi J and Brewer G (2001) Regulation of mRNA stability in mammalian cells. Gene 265: 11–23.

van Hoof A and Parker R (2002) Messenger RNA degradation: beginning at the end. Current Biology 12: R285–R287.

Kramer A (1996) The structure and function of proteins involved in mammalian pre-mRNA splicing. Annual Review of Biochemistry 65: 367–409.

Lykke-Andersen J (2001) mRNA quality control: making the message for life or death. Current Biology 11: R88–R91.

Mendell JT and Dietz HC (2001) When the message goes away: disease-producing mutations that influence mRNA content and performance. Cell 107: 411–414.

Reed R and Hurt E (2002) A conserved mRNA export machinery coupled to pre-mRNA splicing. Cell 108: 523–531.

Shyu A-B and Wilkinson MF (2000) The double lives of shuttling mRNA binding proteins. Cell 102: 135–138.

Tucker M and Parker R (2000) Mechanisms and control of mRNA decapping in Saccharomyces cerevisiae. Annual Review of Biochemistry 69: 571–595.

Wagner EJ and Garcia-Blanco MA (2001) Polypyrimidine tract binding protein antagonizes exon definition. Molecular and Cellular Biology 21: 3281–3288.

Wilkinson MF and Shyu A-B (2002) RNA surveillance by nuclear scanning? Nature Cell Biology 6: E144–E147.

Zenklusen D and Stutz F (2001) Nuclear export of mRNA. FEBS Letters 498: 150–156.

Further Reading

Lafontaine DL and Tollervey D (2001) The function and synthesis of ribosomes. Nature Reviews Molecular and Cell Biology 7: 514–520.

Maniatis T and Tasic B (2002) Alternative pre-mRNA splicing and proteome expansion in metazoans. Nature 418: 236–243.

Nagai K and Mattaj IW (1994) RNA–protein interactions in the splicing snRNPs. In: Nagai K and Mattaj IW (eds.) RNA–Protein Interactions, pp. 150–177. Oxford, UK: IRL Press at Oxford University Press.

Sachs AB, Sarnow P and Hentze MW (1997) Starting at the beginning, middle, and end: translation initiation in eukaryotes. Cell 89: 831–838.

Sachs AB and Varani G (2000) Eukaryotic translation initiation: there are (at least) two sides to every story. Nature Structural Biology 5: 356–361.

Orphanides G and Reinberg D (2002) A unified theory of gene expression. *Cell* **108**: 439–451.

Wagner E and Lykke-Andersen J (2002) mRNA surveillance: the perfect persist. *Journal of Cell Science* **115**: 3033–3038.

Wilusz CJ, Wang W and Peltz SW (2001) Curbing the nonsense: the activation and regulation of mRNA surveillance. *Genes and Development* **15**: 2781–2785.

Wolin SL and Matera AG (1999) The trials and travels of tRNA. *Genes and Development* **13**: 1–10.

RNA Editing and Human Disorders

Jonatha M Gott, *Case Western Reserve University, Cleveland, Ohio, USA*

Small, targeted changes at the ribonucleic acid level can have profound effects on gene expression. Such programmed alterations can be controlled in a tissue-specific or developmentally regulated fashion, providing an alternative means of augmenting the information encoded within the genome.

Intermediate article

Article contents

- Introduction to RNA Editing
- Editing in Human Cells
- Effects of RNA Editing Defects in Model Systems
- RNA Editing in Human Disorders
- RNA Editing in Human Pathogens
- Potential Connections to Other Cellular Processes
- Implications in the Postgenomic Era

Introduction to RNA Editing

The term ribonucleic acid (RNA) editing was initially coined by Benne and colleagues to describe the insertion of nonencoded uridine residues into messenger RNAs (mRNAs) in trypanosome kinetoplasts. Since that time the definition has expanded to include other examples of site-specific alterations of RNA sequences (but excluding processes such as RNA splicing and polyadenylation). These include both base changes and insertion/deletion of nucleotides in a range of organisms, including humans and other mammals, viruses, marsupials, parasites, plants, *Drosophila*, slime molds and a variety of single-cell organisms (Smith *et al.*, 1997). Most editing events have very defined effects on gene expression, leading to the creation of open-reading frames (via both frameshifts and the generation or elimination of translational start and stop codons), changes in encoded amino acids and splice site utilization, alterations in RNA secondary structures and even changes in transfer RNA (tRNA) identity (Gott and Emeson, 2000). Thus, RNA editing is a potent means of generating genetic diversity.

Editing in Human Cells

Two forms of RNA editing, cytidine to uridine (C to U) and adenosine to inosine (A to I) changes, have been shown to occur in humans (reviewed in Keegan *et al.*, 2001). Both of these processes are hydrolytic deamination reactions that affect bases within RNA, resulting in the replacement of the amino groups of A

and C with keto groups. Substitution of this functional group alters the base-pairing potential of the base: C becomes a U, which pairs with A rather than G, and A becomes I, which pairs with C rather than U (**Figure 1**). The highly specific base alterations found in human mRNAs have significant consequences in the tissues in which they occur, generally resulting in the production of two or more proteins from a single gene. The relative levels of individual isoforms are often subject to regulation via modulation of the extent of editing in different cell types.

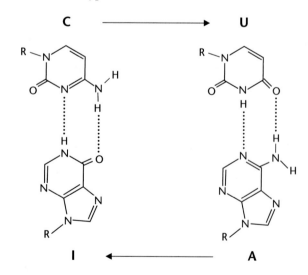

Figure 1 Changes in base-pairing upon conversion of cytidine (C) to uridine (U) or adenosine (A) to inosine (I).

C to U editing

The first example of RNA editing in mammalian cells was discovered during the characterization of the transcripts from the apolipoprotein B (apoB) gene (reviewed in Davidson and Shelness, 2000). A single C to U substitution within the apoB mRNA changes a glutamine codon (CAA) to a stop codon (UAA) via a process that is regulated both developmentally and hormonally. In humans, this form of editing occurs in the intestine, resulting in tissue-specific expression of the shorter form of the protein, APOB48, a component of chylomicrons. The longer form of the protein, APOB100, is expressed in the liver, where it is assembled into very low-density lipoproteins, which are then processed to low-density lipoproteins. The C-terminal domain of the APOB100 protein, which is absent in APOB48, contains both the low-density lipoprotein receptor-binding domain and the attachment site of apoliprotein (a), giving the two proteins markedly different properties. Other, less well-characterized targets of C to U editing include the tumor suppressor gene *NF1*, the tumor susceptibility gene *WT1* and the putative translational repressor *NAT1*.

Targeting the single, highly specific C to U change within the ~14 000 nucleotide apoB mRNA is dependent upon an 11 nucleotide sequence, termed the mooring sequence, just downstream of the editing site, as well as other efficiency elements that lie both upstream and downstream of the site. These elements are thought to form a partially double-stranded structure that is recognized by editing factors. Two proteins are required to carry out the C to U change within the apoB mRNA *in vitro*: APOBEC1 (apoB mRNA editing enzyme catalytic polypeptide 1) and ACF (APOBEC1 complementing factor) (reviewed in Chester *et al.*, 2000). Other proteins that enhance or inhibit apoB mRNA editing or are homologous to APOBEC1 have also been identified. These proteins may serve to regulate the levels of editing *in vivo* and/or participate in editing of additional targets of C to U changes. One of the human APOBEC1 homologs, activation-induced cytidine deaminase (AID), is expressed only in a subset of B cells and has been linked to somatic hypermutation and immunoglobulin class-switching, leading to speculation that editing might play a critical role in B-cell differentiation and the generation of antibody diversity (Longacre and Storb, 2000).

A to I editing

Most of the characterized A to I changes described to date fall within transcripts specific to the central nervous system (CNS) (Reenan, 2001). Many of these mRNAs contain multiple A to I targets, and because editing is usually not complete, the end result of editing is the production of multiple protein isoforms that vary by one or a few amino acids. These small changes can have profound effects on permeability of ion channels, signal transduction via G protein coupling and drug interactions.

A to I changes are carried out by proteins called ADARs (adenosine deaminases that act on RNA). Three ADAR genes have been identified in the human genome, but because ADAR mRNAs are subject to both alternative splicing and A to I editing, there are also multiple ADAR isoforms, some of which have altered specificity and/or localization. The expression of additional ADAR isoforms is induced by interferon via activation of alternative promoters. ADAR proteins act on largely double-stranded regions of RNA, formed by base-pairing between exon target sequences and an editing site complementary sequence within an adjacent intron. Given that two of the three human ADARs are expressed in non-CNS tissues and inosine-containing RNAs are found in many cell types, additional ADAR substrates are likely to emerge.

Effects of RNA Editing Defects in Model Systems

The physiological impact of RNA editing has been largely explored through the use of model systems (**Tables 1** and **2**). Increased levels of apoB transcripts have been observed in genetically obese Zuker rats and after ethanol feeding, accompanied by elevated levels of triglycerides. Overexpression of APOBEC1 in mice and rabbits results in liver dysplasia and hepatocarcinomas accompanied by aberrant 'hyperediting' of NAT1 (and potentially other) transcripts. And although the APOBEC1 homolog AID has not been shown to be involved in editing of a particular target, the unexpected finding that deletion of the AID gene both eliminates class switch recombination (CSR) and abrogates somatic hypermutation in mice suggests that RNA editing may influence some mechanistic step common to these two processes.

The importance of A to I editing has been demonstrated *in vivo* through manipulation of the ADAR genes and some of their known targets, the G-protein-coupled serotonin receptor, 5-HT$_{2c}$ and the glutamate-gated ion channels (**Table 2**). Comparisons of 5-HT$_{2c}$ receptors generated by differential editing demonstrated differences in G-protein coupling when expressed in NIH3T3 cells. A gene knockout of the single ADAR isoform in *Drosophila* resulted in significant neurological phenotypes, including coordination and mating defects, tremors, temperature-sensitive paralysis and progressive neurodegeneration. Mouse

Table 1 Alterations in C to U editing and apoB mRNA editing enzyme catalytic polypeptide 1 (APOBEC1) and activation-induced cytidine deaminase (AID) expression in model systems

Experimental system	Observation	Reference
Overexpression of APOBEC1 in mice and rabbits	Liver dysplasia, hepatocellular carcinoma, NAT1 mRNA editing	Yamanaka *et al.* (1995, 1997)
APOBEC1 knockout mice	apoB editing eliminated	Hirano *et al.* (1996)
	HDL deficiency	Nakamuta *et al.* (1996)
Genetically obese Zuker rats	Increased apoB mRNA editing	Phung *et al.* (1996)
Ethanol-treated rats	Increased apoB mRNA editing	Lau *et al.* (1995)
	Elevated triglycerides	
AID$^{-/-}$ mice	Failure to undergo class switch recombination, severely reduced somatic hypermutation	Muramatsu *et al.* (2000)
Overexpression of AID in murine lymphoma cells	Augmented class switching in the absence of stimulation	Muramatsu *et al.* (2000)

AID: activation-induced cytidine deaminase; HDL: high-density lipoproteins; mRNA: messenger RNA; NAT1: novel APOBEC target 1.

Table 2 Effects of altered A to I editing in model systems

Mutation/defect	Observation	Reference
5-HT$_{2c}$ receptor isoforms (NIH3T3 cells)	Altered receptor-mediated G-protein activation	Price *et al.* (2001)
Mutated GluR-6 Q/R site (mouse)	Increased seizure susceptibility	Vissel *et al.* (2001)
Mutated GluR-2 Q/R site (mouse)	Early-onset epilepsy, early death, neurological defects	Brusa *et al.* (1995) Feldmeyer *et al.* (1999)
ADAR2-null mutant (mouse)	Seizure-prone, early death	Higuchi *et al.* (2000)
ADAR-null (*Drosophila*)	Behavior defects (coordination, mating, etc.), neurodegeneration	Palladino *et al.* (2000)
ADAR1-chimeras (mouse)	Embryonic lethal with erythropoiesis defects	Wang *et al.* (2000)

ADAR: adenosine deaminase that acts on RNA.

ADAR1 null mutants display an embryonic-lethal phenotype, probably caused by defects in embryonic erythropoiesis. Studies in mice also indicate that the most important target of ADAR2 is the GluR2 Q/R site. Expression of Glu2 that is unedited at the Q/R site (i.e. containing a glutamine (Q) rather than an arginine (R) at this position) leads to increased Ca^{2+} permeability, epileptic seizures and early death. ADAR2$^{-/-}$ mice are prone to seizures and die shortly after birth, with phenotypes similar to those of GluR2 mutants. These phenotypes can be rescued by constitutive expression of GluR2 containing arginine (R) at the Q/R editing site. Thus, editing plays a critical role in normal gene expression, signal transduction, homeostasis and development.

RNA Editing in Human Disorders

What little is known about the effects of altering editing in humans is generally consistent with observations in model systems (**Table 3**). For instance, editing at the GluR2 Q/R site is decreased in malignant gliomas and, like mice deficient in editing at this site, patients with malignant gliomas are prone to epileptic seizures. Similarly, consistent with the mouse studies mentioned above, patients with an autosomal recessive form of hyper-IgM syndrome (HIGM2) characterized by severe deficiencies in immunoglobin CSR and hypermutation all contained mutations in the AID gene. Several of these mutations localized to the putative deaminase domain, while the others were predicted to lead to the production of truncated AID proteins.

Changes in APOBEC-1 expression patterns and elevated NF1 editing have been reported in tumor cells. Altered levels of A to I editing have also been observed in patients with amyotrophic lateral sclerosis (GluR2), epileptics (GluR5 and GluR6) and suicide victims (5-HT2c receptors). Conflicting results have been reported for patients with Alzheimer disease, Huntington disease and schizophrenia, with some

Table 3 RNA editing and human disorders

System/subject	Defect	Reference
Peripheral nerve sheath tumors	Editing of NF1 mRNA, abnormal APOBEC1 mRNA expression	Mukhopadhyay *et al.* (2002)
Colon cancer cells	Alternatively spliced form of APOBEC1 mRNA	Lee *et al.* (1998)
Neurofibromatosis type 1 (NF1)	Elevated NF1 mRNA editing	Cappione *et al.* (1997)
Hyper-IgM syndrome (HIGM2)	AID deficiency, absence of class switch recombination and hypermutation	Revy *et al.* (2000)
Suicide victims	Elevated editing at $5HT_{2c}R$ A site	Niswender *et al.* (2001)
Malignant brain tumors	Decreased editing at GluR2 Q/R site	Maas *et al.* (2001)
	Decreased activity of ADAR2	
Amyotrophic lateral sclerosis (ALS)	Decreased editing of GluR2 mRNA	Takuma *et al.* (1999)
Epileptic patients	Altered editing of GluR5, GluR6	Kortenbruck *et al.* (2001)
Acute myeloid leukemia	Hyperediting of hematopoietic cell phosphatase (PTPN6) transcripts	Beghini *et al.* (2000)

ADAR: adenosine deaminase that acts on RNA; AID: activation-induced cytidine deaminase; APOBEC1: apoB mRNA editing enzyme catalytic polypeptide 1; NF1: neurofibromatosis type 1.

studies reporting minor changes in editing levels while others see no significant differences in affected regions of the brain.

Hyperediting of both mammalian and viral transcripts has also been implicated in human disease (Keegan *et al.*, 2001). The tyrosine phosphatase PTPN6 downregulates a number of signaling cascades through direct dephosphorylation of receptor tyrosine kinases. Analysis of PTPN6 transcripts isolated from blasts from acute myeloid leukemia patients demonstrated that these RNAs contained multiple A to G conversions, one of which affected the putative branchpoint adenosine within the third intron. This mutation led to the retention of intron 3, a change predicted to lead to the production of an inactive, truncated protein. Measles matrix RNAs isolated from the brains of victims of subacute sclerosing panencephalitis and measles inclusion body encephalitis also contain very high levels of transitions (A to G in the genome, U to C in the antigenome). These and other instances of biased hypermutations (Bass, 1997) are thought to be the result of hyperediting of transiently double-stranded RNAs by ADAR enzymes.

RNA Editing in Human Pathogens

RNA editing is also an obligatory step in expression of genes in a variety of human pathogens, including both viruses and unicellular parasites. Editing was first discovered in *Trypanosoma brucei*, which causes African sleeping sickness, and also occurs in the related kinetoplastid protozoa that cause Chagas disease and leishmaniasis (Alfonso *et al.*, 1997). Mitochondrial transcripts in these organisms are subject to both insertion and deletion of uridines,

leading to the creation of translatable open-reading frames not found in the mitochondrial genome. This process is the result of a series of cleavage–ligation reactions which are directed by small RNAs called guide RNAs (gRNAs). In the most extreme examples of kinetoplastid editing, greater than 50% of the transcript is created posttranscriptionally (reviewed in Stuart *et al.*, 1997).

Viral transcripts are also frequently edited. The subviral human pathogen hepatitis delta virus (HDV) exacerbates hepatitis B infections, causing severe acute hepatitis and often progressive chronic disease. Production of the form of the HDV antigen required for packaging requires A to I editing of the HDV antigenome RNA (Bass, 1997). Hemorrhagic fever viruses (such as Ebola and Marburg viruses) as well as the paramyxoviruses (including parainfluenza, respiratory syncytial, meningeal, measles and mumps viruses) utilize editing to produce multiple proteins from a single region of their genomes (Smith *et al.*, 1997). Programmed insertion of additional residues at a single editing site results in the creation of translational frameshifts, leading to the production of proteins having a common *N*-terminus but differing in their *C*-terminal regions. Changes have also been reported in HIV transcripts, but the mechanism by which these alterations are effected has not yet been characterized.

Potential Connections to Other Cellular Processes

The use of intron sequences to target editing sites indicates that A to I editing must occur prior to splicing. However, other interconnections between editing and splicing are also likely. ADAR2 edits its own message, creating an alternative splice site that leads to the production of an inactive isoform. Interestingly, GluR2 pre-mRNAs accumulate in ADAR2 mutant mice, suggesting that editing at the Q/R site might be required for efficient splicing. Based on the observation that not all editing events occur within coding sequences, editing may also affect RNA stability, transport and localization. In addition, links to RNA interference have been proposed, given the similarity between substrates. Finally, the recent discovery of brain-specific snoRNAs, one of which may target an A to I editing site for methylation, has suggested a possible regulatory interplay between RNA editing and modification.

Implications in the Postgenomic Era

Clearly, editing adds another twist to the problem of gene identification. It is well established that alternative splicing is a major source of variation in the human proteome, but the contribution of RNA editing is just beginning to emerge, and is likely to be substantial; for example, partial editing at five different sites within the 5-HT$_{2c}$ receptor mRNA is known to lead to the production of at least 12 isoforms of the serotonin receptor from a single gene. These combinatorial possibilities, coupled with the potential for temporal and spatial regulation, make editing a powerful means of generating diversity beyond what is encoded in the genome.

See also

mRNA Editing
Posttranscriptional Processing
RNA Processing
RNA Processing and Human Disorders

References

Alfonso JD, Thiemann O and Simpson L (1997) The mechanism of U insertion/deletion RNA editing in kinetoplastid mitochondria. *Nucleic Acids Research* 25: 3751–3759.

Bass BL (1997) RNA editing and hypermutation by adenosine deamination. *Trends in Biochemical Sciences* 22: 157–162.

Beghini A, Ripamonti CB, Peterlongo P, et al. (2000) RNA hyperediting and alternative splicing of hematopoietic cell phosphatase (PTPN6) gene in acute myeloid leukemia. *Human Molecular Genetics* 9: 2297–2304.

Brusa R, Zimmermann F, Koh DS, et al. (1995) Early-onset epilepsy and postnatal lethality associated with an editing-deficient GluR-B allele in mice. *Science* 270: 1677–1680.

Cappione AJ, French BL and Skuse GR (1997) A potential role for NF1 mRNA editing in the pathogenesis of NF1 tumors. *American Journal of Human Genetics* 60: 305–312.

Chester A, Scott J, Anant S and Navaratnam N (2000) RNA editing: cytidine to uridine conversion in apolipoprotein B mRNA. *Biochimica et Biophysica Acta* 1494: 1–13.

Davidson NO and Shelness GS (2000) Apolipoprotein B: mRNA editing, lipoprotein assembly and presecretory degradation. *Annual Review of Nutrition* 20: 169–193.

Feldmeyer D, Kask K, Brusa R, et al. (1999) Neurological dysfunctions in mice expressing different levels of the Q/R site-unedited AMPAR subunit GluR-B. *Nature Neuroscience* 2: 57–64.

Gott JM and Emeson RB (2000) Functions and mechanisms of RNA editing. *Annual Review of Genetics* 34: 499–531.

Higuchi M, Maas S, Single FN, et al. (2000) Point mutation in an AMPA receptor gene rescues lethality in mice deficient in the RNA-editing enzyme ADAR2. *Nature* 406: 78–81.

Hirano K, Young SG, Farese RV, et al. (1996) Targeted disruption of the mouse apobec-1 gene abolishes apolipoprotein B mRNA editing and eliminates apolipoprotein B48. *Journal of Biological Chemistry* 271: 9887–9890.

Keegan LP, Gallo A and O'Connell MA (2001) The many roles of an RNA editor. *Nature Reviews Genetics* 2: 869–878.

Kortenbruck G, Berger E, Speckmann EJ and Musshoff U (2001) RNA editing at the Q/R site for the glutamate receptor subunits GLUR2, GLUR5, and GLUR6 in hippocampus and temporal cortex from epileptic patients. *Neurobiology of Disease* 8: 459–468.

Lau PP, Cahill DJ, Zhu HJ and Chan L (1995) Ethanol modulates apolipoprotein B mRNA editing in the rat. *Journal of Lipid Research* 36: 2069–2078.

Lee RM, Hirano K, Anant S, Baunoch D and Davidson NO (1998) An alternatively spliced form of apobec-1 messenger RNA is overexpressed in human colon cancer. *Gastroenterology* 115: 1096–1103.

Longacre A and Storb U (2000) A novel cytidine deaminase affects antibody diversity. *Cell* 102: 541–544.

Maas S, Patt S, Schrey M and Rich A (2001) Underediting of glutamate receptor GluR-B mRNA in malignant gliomas. *Proceedings of the National Academy of Sciences of the United States of America* 98: 14 687–14 692.

Mukhopadhyay D, Anant S, Lee RM, et al. (2002) C→U editing of neurofibromatosis 1 mRNA occurs in tumors that express both the type II transcript and apobec-1, the catalytic subunit of the apolipoprotein B mRNA-editing enzyme. *American Journal of Human Genetics* 70: 38–50.

Muramatsu M, Kinoshita K, Fagarasan S, et al. (2000) Class switch recombination and hypermutation require activation-induced cytidine deaminase (AID), a potential RNA editing enzyme. *Cell* 102: 553–563.

Nakamuta M, Chang BHJ, Zsigmond E, et al. (1996) Complete phenotypic characterization of apobec-1 knockout mice with a wild-type genetic background and a human apolipoprotein B transgenic: Background, and restoration of apolipoprotein B mRNA editing by somatic gene transfer of apobec-1. *Journal of Biological Chemistry* 271: 25 981–25 988.

Niswender CM, Herrick-Davis K, Dilley GE, et al. (2001) RNA editing of the human serotonin 5-HT2C receptor. Alterations in suicide and implications for serotonergic pharmacotherapy. *Neuropsychopharmacology* 24: 478–491.

Palladino MJ, Keegan LP, O'Connell MA and Reenan RA (2000) A-to-I pre-mRNA editing in Drosophila is primarily involved in adult nervous system function and integrity. *Cell* 102: 437–449.

Phung TL, Sowden MP, Sparks JD, Sparks CE and Smith HC (1996) Regulation of hepatic apolipoprotein B RNA editing in the genetically obese Zucker rat. *Metabolism* 45: 1056–1058.

Price RD, Weiner DM, Chang MS and Sanders-Bush E (2001) RNA editing of the human serotonin 5-HT2C receptor alters receptor-mediated activation of G13 protein. *Journal of Biological Chemistry* 276: 44 663–44 668.

Reenan RA (2001) The RNA world meets behavior: A→I pre-mRNA editing in animals. *Trends in Genetics* **17**: 53–56.

Revy P, Muto T, Levy Y, *et al.* (2000) Activation-induced cytidine deaminase (AID) deficiency causes the autosomal recessive form of the hyper-IgM syndrome (HIGM2). *Cell* **102**: 565–575.

Smith HC, Gott JM and Hanson MR (1997) A guide to RNA editing. *RNA* **3**: 1105–1123.

Stuart K, Allen TE, Heidmann S and Seiwert SD (1997) RNA editing in kinetoplastid protozoa. *Microbiology and Molecular Biology Reviews* **61**: 105–120.

Takuma H, Kwak S, Yoshizawa T and Kanazawa I (1999) Reduction of GluR2 RNA editing, a molecular change that increases calcium influx through AMPA receptors, selective in the spinal ventral gray of patients with amyotrophic lateral sclerosis. *Annals of Neurology* **46**: 806–815.

Vissel B, Royle GA, Christie BR, *et al.* (2001) The role of RNA editing of kainate receptors in synaptic plasticity and seizures. *Neuron* **29**: 217–227.

Wang Q, Khillan J, Gadue P and Nishikura K (2000) Requirement of the RNA editing deaminase ADAR1 gene for embryonic erythropoiesis. *Science* **290**: 1765–1768.

Yamanaka S, Balestra ME, Ferrell LD, *et al.* (1995) Apolipoprotein B mRNA-editing protein induces hepatocellular carcinoma and dysplasia in transgenic animals. *Proceedings of the National Academy of Sciences of the United States of America* **92**: 8483–8487.

Yamanaka S, Poksay KS, Arnold KS and Innerarity TL (1997) A novel translational repressor mRNA is edited extensively in livers containing tumors caused by the transgene expression of the apoB mRNA-editing enzyme. *Genes and Development* **11**: 321–333.

Further Reading

Bass B (ed.) (2001) *RNA Editing*. Oxford, UK: Oxford University Press.

Bass BL (2000) Double-stranded RNA as a template for gene silencing. *Cell* **101**: 235–238.

Cavaille J, Buiting K, Kiefmann M, *et al.* (2000) Identification of brain-specific and imprinted small nucleolar RNA genes exhibiting an unusual genomic organization. *Proceedings of the National Academy of Sciences of the United States of America* **97**: 14311–14316.

Gerber AP and Keller W (2001) RNA editing by base deamination: more enzymes, more targets, new mysteries. *Trends in Biochemical Sciences* **26**: 376–384.

Grosjean H and Benne R (eds.) (1998) *Modification and Editing of RNA*. Washington, DC: ASM Press.

Hammond SM, Caudy AA and Hannon GJ (2001) Post-transcriptional gene silencing by double-stranded RNA. *Nature Reviews Genetics* **2**: 110–119.

Hausmann S, Garcin D, Delenda C and Kolakofsky D (1999) The versatility of paramyxovirus RNA polymerase stuttering. *Journal of Virology* **73**: 5568–5576.

Kinoshita K and Honjo T (2001) Linking class-switch recombination with somatic hypermutation. *Nature Reviews Molecular and Cellular Biology* **2**: 493–503.

Sanchez A, Trappier SG, Mahy BWJ, Peters CJ and Nichol ST (1996) The virion glycoproteins of Ebola virus are encoded in two reading frames and are expressed through transcriptional editing. *Proceedings of the National Academy of Sciences of the United States of America* **93**: 3602–3607.

RNA Gene Prediction

Stephen R Holbrook, *Lawerence Berkeley National Laboratory, Berkeley, California, USA*

Richard J Carter, *Lawerence Berkeley National Laboratory, Berkeley, California, USA*

Richard F Meraz, *Lawerence Berkeley National Laboratory, Berkeley, California, USA*

Intermediate article

Article contents

- Introduction
- Computational Approaches
- Conclusion

The rapid growth of sequenced genomes and the discovery of numerous novel RNA species has made the development of methods for the computational identification of genes encoding functional RNA a high priority. Recent developments of algorithms for RNA gene prediction are based on diverse criteria, including location of promoters and terminators, sequence conservation among related genomes, RNA base-pairing and nucleotide composition.

Introduction

The promise of genome sequencing has been to reveal the complete set of genes through which the cell performs its functions, as well as regulatory elements that control their expression levels. To a great extent this goal has been reached for protein genes. Powerful computer programs such as GLIMMER, GRAIL and GeneMark have been developed for protein gene discovery and are routinely applied to genomic sequences to identify all open reading frames (ORFs) as potential protein genes. Sequence similarity is used to infer function and confirm gene identity. Hypothetical and unknown proteins are assigned as the products of the remaining genes that lack sequence homology to proteins of known function. The presence of sequence homologs in other organisms confirms the identification of hypothetical proteins even though their function is not assigned. (*See* Hidden Markov Models.)

However, despite the recent rapid expansion in our understanding of the function of ribonucleic acid (RNA) in biological systems, until recently no analogous programs have been developed for the identification of genes encoding functional RNAs (fRNAs) in genomic sequences. This omission was due to both the conceptual underappreciation of the number and significance of RNA genes in genomic sequences and technical difficulties due to lack of signals within and around RNA genes in comparison to protein genes. (*See* Gene Structure and Organization; Noncoding RNAs: A Regulatory Role?)

Figure 1 shows these differences for prokaryotic genes. As shown, protein genes incorporate several signals that can be used in their recognition. These include the consensus ribosome binding (Shine–Dalgarno) site, the appropriately spaced start and stop codons, and a triplet-encoded ORF based on the genetic code and corresponding to the organism's codon usage frequencies. On the other hand, RNA genes contain none of these signals.

Computational prediction of fRNAs in genomic sequences would allow experimental testing of expression levels, functional assay by deletion or mutagenesis, and structural analysis. These untranslated fRNAs have also been referred to as noncoding (ncRNA), small RNAs (sRNA, smRNA), untranslated and small nonmessenger (snmRNA). Here they are referred to as fRNAs.

Two recent experimental studies have identified an unexpectedly large and diverse population of expressed and presumably fRNA molecules. First, an expressed sequence tag (EST)-based experimental technique was used to generate complementary deoxyribonucleic acid (cDNA) libraries from the small fraction (50–500 nucleotides) of total RNA isolated from mouse brain (Huttenhofer *et al.*, 2001). This

approach resulted in the identification of a total of 201 different expressed RNA species potentially encoding fRNA species. Of these RNAs, 113 were identified as small nucleolar RNAs (snoRNAs), guiding modification of ribosomal, snRNA or other RNA modifications. Most of the remainder consisted of novel RNAs of unknown function. (*See* Expressed-sequence Tag (EST).)

Several groups have used a range of biochemical techniques, including cDNA cloning and cloning from selected fractions of total RNA to discover the presence of micro-RNAs (miRNAs; for a review see Ruvkin (2001)). These miRNAs act in a variety of roles, including development and regulation, and were identified in *Caenorhabditis elegans*, *Drosophila melanogaster* and HeLa cells. It is expected that these miRNAs modulate the translation or stability of messenger RNAs (mRNAs; Lau *et al.*, 2001). (*See* DNA Cloning.)

Computational Approaches

Identification of well-characterized RNA species

Many fRNAs are shared among organisms, such as transfer RNA (tRNA) and ribosomal RNA (rRNA), ribonuclease P (RNase P) RNA, snoRNAs (eukaryotes and archaea) and tmRNAs (bacteria). These known RNAs can usually be identified in genomes by sequence and/or structure similarity and conservation. Sophisticated programs such as tRNAscan-SE and Snoscan utilize conserved sequence and/or structure patterns, covariance models and stochastic context-free grammars (SCFGs; Eddy and Durbin, 1994) to accurately and automatically find tRNAs (Lowe and Eddy, 1997) and snoRNAs (Lowe and Eddy, 1999; Omer *et al.*, 2000) in genomes. These methods were developed to find new members of well-characterized RNA types, but are not applicable to the identification of novel or poorly characterized RNA species. (*See* RNA Secondary Structure Prediction.)

Formal Bayesian probabilistic models have been introduced as tools to identify complicated consensus features in biological sequences. Hidden Markov models (HMM) are probably the best known of these approaches. Another class of model, the covariance model, is able to capture both primary consensus and secondary structure information through the use of SCFGs. Much like sequence profiles, covariance models are constructed from multiple sequence alignments.

In the tRNAscan-SE program (Lowe and Eddy, 1997), sequences are searched against a given

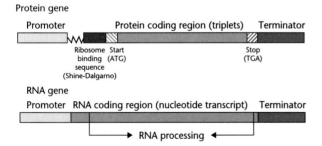

Figure 1 Comparison of RNA and protein genes in prokaryotes. In the RNA gene, there are no ribosome binding sites, no start or stop codons, and no triplet code.

covariance model using a three-dimensional dynamic programming algorithm, similar to a Smith–Waterman alignment but including base-pairing terms also. RNA covariance models have the advantages of high sensitivity, high specificity, and general applicability to any RNA sequence family of interest. Using these general tools, the search for a tRNA takes three steps. Firstly, the DNA is screened for the presence of a short, intergenic promoter sequence that is found in the T and D arms of tRNA. This is followed by a search for stem–loop structures in the location of the promoter. The second step involves calculating a log-odds score for conserved sequences and the distance between. The final stage parses the output from these programs and undertakes a probabilistic search for tRNA. A secondary structure prediction of any putative matches will reveal the presence of an anticodon region. (*See* Dynamic Programming; Smith–Waterman Algorithm.)

Snoscan (Lowe and Eddy, 1999) is a program that recognizes methylation guide snoRNAs in archaeal genomes. The program identifies six components characteristic of the class of fRNA: box D; box C; a region of sequence complementary to rRNA; box D9 if the rRNA complementary region is not directly adjacent to box D; the predicted methylation site within the rRNA based on the complementary region; and the terminal stem base pairings, if present. The program also takes into account the relative distance between identified features within the snoRNA, information that is useful in reducing the rate of false positives.

Identification of novel fRNAs

Recently, a number of studies have been undertaken to find novel RNA genes using gene boundary prediction (Olivas *et al.*, 1997; Argaman *et al.*, 2001), comparative genomics (Argaman *et al.*, 2001; Wassarman *et al.*, 2001), a combination of comparative sequence analysis and probabilistic models of nucleotide mutation bias in regions of conserved secondary structure (Rivas *et al.*, 2001), and contrasting sequence and structural patterns between known RNA genes and noncoding sequences within genomes (Carter *et al.*, 2001). The various approaches to computational identification of novel RNA genes are summarized in **Table 1** and described in the following sections.

Gene boundary prediction

The prediction of potential RNA promoters and terminators has been used to narrow the search for potential RNAs. Parker and coworkers used a genomics guided-search technique to find novel RNA genes (Olivas *et al.*, 1997). Two strategies were used in this study. First, strong RNA polymerase III sites were identified by sequence, and transcripts from these sites were probed experimentally. Second, large gaps between predicted ORFs were analyzed for RNA expression. The first method identified a new, but nonessential, 170-nucleotide noncoding RNA, and the second method found 15 RNA transcripts, one of which appeared to be a snoRNA. While this approach was laborious and not comprehensive, it did show the presence of previously unidentified RNAs in the yeast genome. Argaman *et al.* (2001) used promoter and terminator prediction methods combined with comparative genomics to predict potential RNA coding regions in *Escherichia coli*.

Currently, the polymerase binding sites of RNA promoters and terminator sites are not unambiguously predicted in bacteria, archaea or eukaryotes. In addition, due to the variable distance between promoters and start sites, RNA processing, and the presence of introns in eukaryotes, these signals alone

Table 1 Computational approaches for the identification of novel RNA genes

Basis	Method	Organism	Number predicted	Reference
Boundary prediction (promoter–terminator patterns)	CSA	Yeast	16	Olivas *et al.* (1997)
		E. coli	24	Argaman *et al.* (2001)
Comparative genomics – sequence conservation	CSA	*E. coli*	23	Wassarman *et al.* (2001)
			24	Argaman *et al.* (2001)
Stability of RNA (GC content)	NCS	Archaea	NA	Schattner (2002)
			>200	Omer *et al.* (2000)
Nucleotide mutation bias in conserved secondary structure	SCFG/HMM	*E. coli*	275	Rivas *et al.* (2001)
Comparison of known RNAs to noncoding non-conserved intergenic sequences	NN SVM	Archaea *E. coli*	370	Carter *et al.* (2001)

NN: neural networks; SVM: support vector machines; SCFG: stochastic context-free grammars; HMM: hidden Markov models; CSA: comparative sequence analysis (BlastN); NCS: nucleotide composition statistics.

do not define the fRNA very well. Progress in this area, combined with other approaches, will, however, be important in improving predictive methods. For example, an improved algorithm for prediction of rho-independent terminators in prokaryotes has been reported recently (Lesnik *et al.*, 2001).

Comparative genomics

Wassarman *et al.* (2001) have used a comparative genomics study to identify 19 novel RNA genes in *E. coli*. They examined intergenic sequences of more than 180 nucleotides (nt) and then carried out a BLAST search against a set of bacterial genomes. Those that had a high degree of conservation over at least 80 nt were further examined. After screening to remove those that were possibly ORFs or promoters, the remaining sequences were tested using microarray data and traditional biochemical methods. The authors speculate that the comparative genomics approach may be easily applied to other organisms, although it is noted that a high degree of conservation does not necessarily infer the presence of an RNA gene, but may instead be a protein binding site, control element, insertion sequence or even a small ORF. (*See* BLAST Algorithm.)

Another comparative genomics study of *E. coli* (Argaman *et al.*, 2001) has examined intergenic sequences and additionally attempted to identify sites of transcription initiation or termination within them. If the distance between the initiation and termination sites was between 50 and 500 nucleotides then a BLAST search of the sequence was undertaken using three bacterial genomes for comparison. This produced 24 potential RNA genes, of which 14 were biochemically verified. It is hypothesized that, provided the genomic and sequence features characteristic of an fRNA can be defined in an explicit manner, then an algorithmic approach can be used to find fRNA in higher organisms.

An analysis of the *E. coli* genome has been undertaken using a program called QRNA (Rivas *et al.*, 2001) that combines comparative genomics and probabilistic methods. The program requires as input a set of sequence alignments from closely related organisms and then classifies the sequence as an fRNA, a protein coding sequence or an uncorrelated sequence region. The hypothesis is that conserved sequence regions will show a pattern of mutation more consistent with a probabilistic model of covariation of nucleotides within base-paired secondary structure rather than an analogous model of conservation of an encoded amino acid sequence (protein coding) or the null model of uncorrelated, position-independent mutation. While this method is entirely general, requiring no prior knowledge about the

fRNAs of the respective genomes, it is limited to detecting fRNAs with conserved intramolecular structure. The method was used to identify 275 putative fRNAs in *E. coli*, a number of which have been verified with biochemical methods (Rivas *et al.*, 2001). Testing indicated a true-positive prediction accuracy of approximately 85% for classification of RNAs in *E. coli*.

Compositional and structural characterization

Defining the features characteristic of fRNAs has led to a great deal of recent research. A computational approach to the identification of RNA genes has been applied to archaeal genomes (Omer *et al.*, 2000; Schattner, 2002) based on the increased $G + C$ content of RNA genes in hyperthermophilic organisms such as *Methanococcus jannaschii* that have an AT-rich overall genome composition. While this method appears to have predictive value, its use is restricted to organisms with AT-rich genome compositions. (*See* Sequence Complexity and Composition.)

Carter *et al.* used machine-learning methods to predict the presence of novel RNA genes in the *E. coli* genome and several other bacterial and archaeal genomes. In order to discriminate fRNA from background genomic sequence, a number of neural networks were trained. The first neural network was trained on the single and dinucleotide composition of known fRNAs genes and a dataset of presumed nonfunctional genomic sequence. A second network was trained using 'structural motifs'; these included the well-known sequence motifs UNCG, GNRA and CUYG found in RNA tetraloops, the AAR subsequence of the tetraloop receptor motif, and the DNA sequence CTAG, which occurs rarely in bacterial protein genes and noncoding regions compared with RNA genes. The final parameter of this set was the calculated free energy of folding. When a third voting-network was trained on the combined results of the previous networks, it achieved an overall predictive accuracy of over 90% in bacterial genomes, over 95% in hyperthermophilic archaeal genomes, and was successfully able to predict a number of recently identified fRNAs that were not included in the training sets. (*See* Neural Networks.)

Conclusion

It is evident from the work of a number of laboratories as of the year 2002 that RNA gene prediction is an area of intense research and that the techniques are improving substantially. It is hoped that RNA gene prediction can become as reliable and have the same level of confidence as genomic protein prediction methods, and that new biological pathways, incorporating novel regulatory, catalytic or structural RNAs, will be identified.

Much of this recent work has concentrated on bacterial genomes, especially *E. coli*. This is mainly because of the wealth of available information about that genome, both in terms of sequence and biological function. A number of groups have estimated that there are probably a further 300 RNA genes to be found in *E. coli*. These same techniques are now being applied to higher organisms, especially the human genome, which may contain thousands of novel RNA genes waiting to be discovered.

See also
Gene Feature Identification

References

Argaman L, Hershberg R, Vogel J, *et al.* (2001) Novel small RNA-encoding genes in the intergenic regions of *Escherichia coli*. *Current Biology* **11**: 941–950.

Carter RJ, Dubchak I and Holbrook SR (2001) A computational approach to identify genes for functional RNAs in genomic sequences. *Nucleic Acids Research* **29**(19): 3928–3938.

Eddy SR and Durbin R (1994) RNA sequence analysis using covariance models. *Nucleic Acids Research* **22**: 2079–2088.

Huttenhofer A, Kiefmann M, Meier-Ewert S, *et al.* (2001) RNomics: an experimental approach that identifies 201 candidates for novel, small, non-messenger RNAs in mouse. *EMBO Journal* **20**: 2943–2953.

Lau NC, Lim LP, Weinstein EG and Bartel DP (2001) An abundant class of tiny RNAs with probable regulatory roles in *Caenorhabditis elegans*. *Science* **294**: 858–862.

Lesnik EA, Sampath R, Levene HB, *et al.* (2001) Prediction of rho-independent transcriptional terminators in *Escherichia coli*. *Nucleic Acids Research* **29**: 3583–3594.

Lowe T and Eddy SR (1997) tRNAscan-SE: a program for improved detection of transfer RNA genes in genomic sequence. *Nucleic Acids Research* **25**: 955–964.

Lowe TM and Eddy SR (1999) A computational screen for methylation guide snoRNAs in yeast. *Science* **283**: 1168–1171.

Olivas WM, Muhlrad D and Parker R (1997) Analysis of the yeast genome: identification of new non-coding and small ORF-containing RNAs. *Nucleic Acids Research* **25**: 4619–4625.

Omer AD, Lowe TM, Russell AG, *et al.* (2000) Homologues of snoRNAs in Archaea. *Science* **288**: 517–522.

Rivas E, Klein RJ, Jones TA and Eddy SR (2001) Computational identification of noncoding RNAs in *E. coli* by comparative genomics. *Current Biology* **11**: 1369–1373.

Ruvkin G (2001) Glimpses of a tiny RNA world. *Science* **294**: 797–799.

Schattner P (2002) Searching for RNA genes using base-composition statistics. *Nucleic Acids Research* **30**: 2076–2082.

Wassarman KM, Repoila F, Rosenow C, Storz G and Gottesman S (2001) Identification of novel small RNAs using comparative genomics and microarrays. *Genes and Development* **15**: 1637–1651.

Further Reading

Chen S *et al.* (2002) A bioinformatics based approach to discover small RNA genes in the *Escherichia coli* genome. *BioSystems* **65**: 157–177.

Eddy SR (2001) Non-coding RNA genes and the modern RNA world. *Nature Reviews Genetics* **2**: 919–929.

Eddy SR (2002) Computational genomics of noncoding RNA genes. *Cell* **109**: 137–140.

Laslett D, Canback B and Andersson S (2002) BRUCE: a program for the detection of transfer-messenger RNA genes in nucleotide sequences. *Nucleic Acids Research* **30**: 3449–3453.

Regalia M, Rosenblad MA and Samuelsson T (2002) Prediction of signal recognition particle RNA genes. *Nucleic Acids Research* **30**: 3368–3377.

Storz G (2002) An expanding universe of noncoding RNAs. *Science* **296**: 1260–1263.

Szymanski M and Barciszewski J (2002) Beyond the proteome: non-coding regulatory RNAs. *Genome Biology* 1–8.

RNA Interference (RNAi) and MicroRNAs

Gregory J Hannon, *Cold Spring Harbor Laboratory, Cold Spring Harbor, New York, USA*

Intermediate article

Article contents

- Discovery of RNAi
- A Flexible Silencing Machinery Operates at Multiple Regulatory Levels
- Mechanism of Posttranscriptional Gene Silencing
- Biological Functions of RNAi
- RNAi – Will it Deliver the Genome?
- Perspective

In 1990, an experiment designed to alter floral pigmentation in *Petunia* sowed the seeds of what has since become a major new field of biology. Efforts to understand the mechanisms that underlie double-stranded RNA-induced (dsRNA) gene silencing are now bearing fruit of many varieties. It is clear that a conserved biological response to dsRNA, known variously as RNA interference (RNAi) or posttranscriptional gene silencing, mediates resistance to both endogenous parasitic and exogenous pathogenic nucleic acids, such as transposons and RNA viruses, and regulates the expression of protein-coding genes. In addition, RNAi has been cultivated as a means to experimentally manipulate gene expression. In the near future, the use of RNAi to probe gene function at a whole-genome scale is likely to yield a rich harvest, not only providing insights into basic biological processes but also the tools to identify more rapidly therapeutic targets for numerous human diseases.

Discovery of RNAi

Ribonucleic acid interference (RNAi), *per se*, was first discovered in *Caenorhabditis elegans* as a response to double-stranded RNA (dsRNA), which resulted in sequence-specific gene silencing (Fire *et al.*, 1998). Silencing by dsRNAs had a number of remarkable properties. RNAi could be provoked by injection of dsRNA into the *C. elegans* gonad or by introduction of dsRNA through feeding either dsRNA itself or bacteria engineered to express dsRNA (Timmons and Fire, 1998). Exposure of a parental animal to quantities of dsRNA that amount to only a few molecules per cell triggered gene silencing throughout the treated animal (systemic silencing) and in its F_1 progeny.

The discovery that dsRNAs induced sequence-specific silencing in worms caused a coalescence from which emerged the notion that a number of previously characterized homology-dependent gene silencing mechanisms might share a common biological root. Several years previously, Richard Jorgensen had been engineering transgenic petunias with the goal of altering pigmentation. Introduction of exogenous transgenes had far from the expected effect of deepening color. Instead, flowers showed variegated pigmentation, with some lacking pigment altogether (van der Krol *et al.*, 1990; Napoli and Jorgensen, 1990) (and reviewed in Jorgensen, 1990). This result indicated that not only were the transgenes themselves inactive but also that these deoxyribonucleic acids (DNAs) somehow affected expression of the endogenous loci, and thus this phenomenon was termed c-suppression. In parallel, a number of laboratories, including David Baulcombe's group (e.g. Angell and

Baulcombe, 1997; Ruiz *et al.*, 1998) Dougherty and colleagues (Dougherty *et al.*, 1994) and Grill and colleagues (Kumagai *et al.*, 1995), found that plants responded to RNA viruses by targeting viral RNAs for destruction. Notably, silencing of endogenous genes could also be triggered by inclusion of homologous sequences in a virus replicon.

What is now clear in retrospect is that both transgene arrays and replicating RNA viruses generate dsRNA. In fact, in plant systems dsRNAs that are introduced from exogenous sources or that are transcribed from engineered inverted repeats are potent inducers of gene silencing (reviewed in Bernstein *et al.*, 2001b). Cosuppression phenomena are not restricted to plants. In fact, similar outcomes of transgenesis experiments have been noted in unicellular organisms, such as *Neurospora*, and in metazoans, such as *Drosophila*, *C. elegans* and mammals (e.g. Fire *et al.*, 1991; Romano and Macino, 1992; Pal-Bhadra *et al.*, 1997; Dernburg *et al.*, 2000).

A Flexible Silencing Machinery Operates at Multiple Regulatory Levels

In *C. elegans*, initial observations were consistent with dsRNA-induced silencing operating at the post-transcriptional level. Exposure to dsRNAs resulted in loss of corresponding messenger RNAs (mRNAs), and promoter and intronic sequences were largely ineffective as silencing triggers (Fire *et al.*, 1998). A posttranscriptional mode was also consistent with data from plant systems in which exposure to dsRNA (de Carvalho et al., 1992), for example in the form of an RNA virus, triggered depletion of mRNA sequences without an apparent effect on the rate of promoter firing (e.g. Jones et al., 2001). Indeed, viral transcripts themselves were targeted, despite the fact that these were generated in the cytoplasm by transcription of RNA genomes (e.g. Ruiz et al., 1998). These studies led to the notion that RNAi enforced silencing by inducing degradation of homologous mRNAs, and this hypothesis has been validated by biochemical analysis of the silencing pathways. However, the RNAi machinery affects gene expression through other mechanisms, as well. In plants, exposure to dsRNA induces genomic methylation of sequences homologous to the silencing trigger (e.g. Wassenegger et al., 1994). If the trigger shares sequence with a promoter, the targeted gene can become silenced at the transcriptional level (Mette et al., 2000). Hints that this regulatory mode extends to other organisms were initially scarce, but recent studies have suggested that the RNAi machinery may also interact with the genome in Drosophila, C. elegans and fungi (Pal-Bhadra et al., 1997, 2002; Tabara et al., 1999; Dudley et al., 2002; Volpe et al., 2002). Finally, in C. elegans, endogenously encoded components of the RNAi machinery, for example lin-4, operate at the level of protein synthesis (Wightman et al., 1993). Although translational control by dsRNA has not been definitively established in other systems, the conservation of let-7 and related RNAs (Reinhart et al., 2000) suggests that this regulatory mode is a common if not universal mechanism through which RNAi pathways may control the expression of cellular genes (**Figure 1**).

Mechanism of Posttranscriptional Gene Silencing

Our present understanding of the mechanisms underlying dsRNA-induced gene silencing comes largely from genetic studies in C. elegans and plants and from biochemical studies of Drosophila extracts. Silencing can be divided into two biochemically separable steps: initiation in which dsRNAs are cleaved to generate ~22 nucleotide (nt) RNAs, known as small interfering RNAs (siRNAs), and the effector step in which silencing is carried out.

The initiation step

The model outlined in **Figure 2** implies that siRNAs are generated as nucleolytic products of the dsRNA silencing trigger. Support for this notion emerged first from studies of Drosophila embryo extracts, which contained an activity capable of processing long

dsRNA substrates into ~22 nt fragments (Zamore et al., 2000). Detailed analysis of these RNAs suggested they were double-stranded in nature and contained $5'$ phosphorylated termini (Zamore et al., 2000; Elbashir et al., 2001a). One group of RNases, the RNaseIII family, displays specificity for dsRNAs and generates such termini. These observations formed the basis for the identification of the enzyme that catalyzed this process. Dicer is a member of the RNaseIII family of enzymes, which contains dual catalytic domains and additional terminal domains, including a helicase homology and a PAZ motif (Bernstein et al., 2001a). This protein was found to process dsRNA into siRNAs and was therefore proposed to initiate RNAi (Bernstein et al., 2001a). This family, now named the Dicer enzymes, is evolutionarily conserved, and proteins from Drosophila, Arabidopsis, Spodoptera frugiperda, tobacco, C. elegans, mammals and Neurospora have all been shown to specifically recognize dsRNA and to process that RNA into siRNAs that are of a characteristic size for the species from which the enzyme is derived (Bernstein et al., 2001a; Ketting et al., 2001). Genetic evidence has also emerged from C. elegans and from Arabidopsis that is consistent with Dicer acting in the RNAi pathway. For example, Dicer is required for RNAi in the C. elegans germ line (Grishok et al., 2001; Ketting et al., 2001; Knight and Bass, 2001).

The effector step

In the Drosophila system, RNAi is enforced by an effector nuclease complex that recognizes and destroys target mRNAs. RNA-induced silencing complex, RISC, is a multicomponent enzyme with both essential RNA and protein components. This has been purified from Drosophila S2 cells as a ~500 kDa RNP with characteristics that differ slightly from the enzyme formed in embryo extracts (Hammond et al., 2000, 2001a).

RISC from S2 cells consistently copurifies with a member of the Argonaute gene family, Ago2 (Hammond et al., 2001a). Argonaute proteins were first identified in Arabidopsis as mutants that produced altered leaf morphology (Bohmert et al., 1998). It is now clear that Argonaute proteins form a large evolutionarily conserved gene family with representatives in most eukaryotic genomes, with the possible exception of Saccharomyces cerevisiae (reviewed in Hammond et al., 2001b). These proteins are characterized by the presence of two homology regions, the PAZ domain and the Piwi domain, the latter of which is unique to this group of proteins. The PAZ domain also appears in one other group of proteins, the Dicer family, and has been proposed to play a role in the assembly of silencing complexes (Bernstein et al., 2001a).

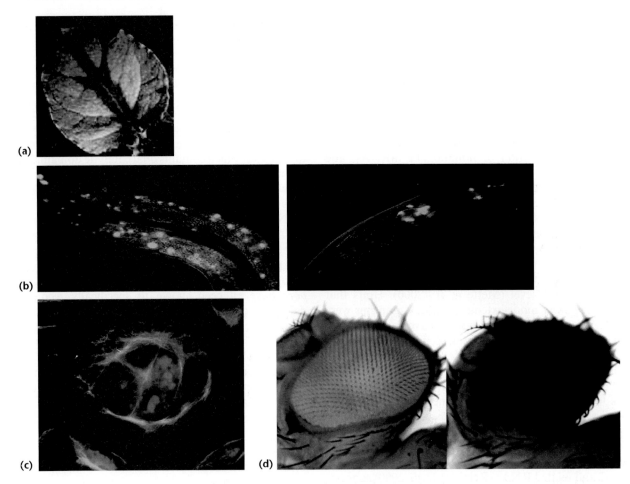

Figure 1 [*Figure is also reproduced in color section.*] Double-stranded RNA (dsRNA) is a potent inducer of gene silencing. dsRNA can be introduced experimentally to silence target genes of interest. In plants, silencing can be triggered, for example, by engineered RNA viruses or by inverted repeat transgenes. In worms, silencing can be triggered by injection or feeding of dsRNA. In both these systems, silencing is systemic and spreads throughout the organism. (a) A silencing signal moving from the veins into leaf tissue. Green is GFP fluorescence and red is chlorophyll fluorescence that is seen upon silencing of the GFP transgene. Similarly, in (b), a *C. elegans* has been engineered to express GFP in nuclei. Animals on the right have been treated with a control dsRNA, while those on the left have been exposed to GFP dsRNA. Some neuronal nuclei remain fluorescent, correlating with low expression of a protein required for systemic RNAi (Winston *et al.*, 2002). In (c), HeLa cells have been treated with an ORC6 siRNA and stained for tubulin (green) and DNA (red). Depletion of ORC6 results in accumulation of multinucleated cells. Stable silencing can also be induced by expression of dsRNA as hairpins or snap-back RNAs. In (d), adult *Drosophila* express a hairpin homologous to the white gene (left), which results in unpigmented eyes as compared to wild type (right).

Recently, RISC has been shown to contain additional RNA-binding proteins. Among these are vasa intronic gene (VIG) and the *Drosophila* homolog of the human fragile X mental retardation protein (FMRP) (Caudy *et al.*, 2002). The latter observation raises the intriguing possibility that defects in the regulation of endogenous gene expression by RNAi might contribute to human disease.

The RISC complex from human cells also contains members of the Argonaute gene family (Mourelatos *et al.*, 2002; Martinez *et al.*, 2002) and has been shown to cleave substrates as directed by either exogenous dsRNA triggers (e.g. siRNAs) or upon incubation with substrates that are 100% complementary to endogenous miRNAs (see below) (Martinez *et al.*, 2002; Hutvagner and Zamore, 2002). In addition, the RISC complex from human cells contains Gemin3 and Gemin4 (Mourelatos *et al.*, 2002; Hutvagner and Zamore, 2002).

Biological Functions of RNAi

Since target identification depends upon Watson–Crick base-pairing interactions, the RNAi machinery has the potential to be a flexible but exquisitely specific silencing apparatus. Thus, this regulatory paradigm may have been adapted and adopted for numerous cellular

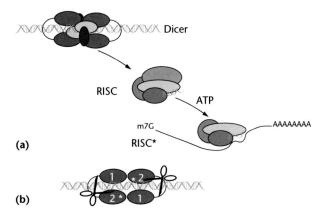

(a)

(b)

Figure 2 Dicer and RNA-induced silencing complex (RISC)
(a) RNAi is initiated by the Dicer enzyme, which processes dsRNA
into ∼22 nt siRNAs (Bernstein *et al.*, 2001a). Based upon the known
mechanisms for RNaseIII enzymes, Dicer is thought to work as a
dimeric enzyme. Cleavage into precisely sized fragments is
determined by the fact that one of the active sites in each Dicer
protein is defective, shifting the periodicity of cleavage from
∼9–11 nt for bacterial RNaseIII to ∼22 nt for Dicer family
members (Blaszczyk *et al.*, 2001). The siRNAs are incorporated into
a multicomponent nuclease, RISC, which uses the siRNA as a guide
to substrate selection (Hammond *et al.*, 2000). Recent reports
suggest that RISC must be activated from a latent form, containing a
double-stranded siRNA to an active form, RISC*, via unwinding of
siRNAs (Nykanen *et al.*, 2001). (b) Diagrammatic representation of
Dicer binding and cleaving dsRNA. Mutations in the second
RNaseIII domain inactivate the central catalytic sites, resulting in
cleavage at 22 nt intervals.

functions. For example, in plants, RNAi forms the
basis of virally induced gene silencing (VIGS), leading
to the proposal that RNAi might play an important
role in pathogen resistance. An elegant proof of this
hypothesis comes from the genetic links between
virulence and RNAi pathways (e.g. Mourrain *et al.*,
2000; Voinnet *et al.*, 2000 and reviewed in Baulcombe,
1999). For example, many plant viruses encode suppressors of posttranscriptional gene silencing (PTGS),
which are essential for pathogenesis, and these virulence determinants can be rendered irrelevant by host
mutations in silencing pathways. RNAi has also been
linked to the control of endogenous parasitic nucleic
acids. In *C. elegans*, some RNAi-deficient strains are
also 'mutators' due to increased movement of endogenous transposons (Tabara *et al.*, 1999; Ketting *et al.*,
1999). In many systems, transposons are silenced by
their packaging into heterochromatin (reviewed in
Martienssen and Colot, 2001). Thus, it is tempting to
speculate, particularly in the light of the results
described above, that RNAi may stabilize the genome
by sequestering repetitive sequences, such as mobile
genetic elements, both preventing transposition and
making repetitive elements unavailable for homologous
recombination events that would lead to chromosomal

translocations. However, it remains to be determined
whether RNAi pathways regulate transposons
through effects at the genomic level or by posttranscriptionally targeting mRNAs (e.g. those encoding
transposases) that are required for transposition.

A role for RNAi pathways in the normal regulation
of endogenous, protein-coding genes was originally
suggested through the analysis of plants and animals
containing mutant proteins that are linked to RNAi
pathways. For example, mutations in the *argonaute1*
gene of *Arabidopsis* cause pleiotropic developmental
abnormalities that are consistent with alterations
in stem cell fate determination (Bohmert *et al.*, 1998).
A hypomorphic mutation in *carpel factory*, an
Arabidopsis Dicer homolog, causes similar defects in
leaf development and also overproliferation of floral
meristems (Jacobsen *et al.*, 1999). Mutations in
Argonaute family members in *Drosophila* also impact
normal development. In particular, mutations in
argonaute-1 have drastic effects on neuronal development (Kataoka *et al.*, 2001), and *piwi* mutants have
defects in both germ-line stem cell proliferation and
maintenance (Cox *et al.*, 1998).

A possible mechanism underlying the regulation of
endogenous genes by the RNAi machinery emerged
from the study of *C. elegans* containing mutations in
their single Dicer gene, *DCR-1*. Unlike most other
RNAi-deficient worm mutants, *dcr-1* animals were
neither normal nor fertile. In fact, mutation of Dicer
induced a number of phenotypic alterations in addition
to its effect on RNAi (Ketting *et al.*, 2001; Knight and
Bass, 2001; Grishok *et al.*, 2001; Hutvagner *et al.*,
2001). Specifically, Dicer mutant animals showed
alterations in developmental timing similar to those
observed in *let-7* and *lin-4* mutant animals.

The *lin-4* gene was originally identified by Ambros
and colleagues as a mutant that affects larval
transitions (Lee *et al.*, 1993). Later, *let-7* was isolated
by Ruvkun and colleagues as a similar heterochronic
mutant (Reinhart *et al.*, 2000). These loci encode small
RNAs, which are synthesized as ∼70 nt precursors but
which are posttranscriptionally processed to an
∼21 nt mature form. Both genetic and biochemical
studies in *C. elegans* and human cells have indicated
that maturation of these RNAs is catalyzed by Dicer
(Ketting *et al.*, 2001; Knight and Bass, 2001; Grishok
et al., 2001; Hutvagner *et al.*, 2001).

The small temporal RNAs (stRNAs) encoded by
let-7 and *lin-4* are negative regulators of target protein-
coding genes, as might be expected if stRNAs trigger
RNAi. However, stRNAs do not trigger mRNA
degradation. Instead, they regulate expression at the
translational level (Olsen and Ambros, 1999; Slack
et al., 2000). This raised the possibility that the only
link between stRNAs and RNAi was at the level of
the initiator enzyme, Dicer. However, Mello and

colleagues demonstrated a requirement for Argonaute family proteins (i.e. Alg-1 and Alg-2) in both stRNA biogenesis and stRNA-mediated suppression (Grishok *et al.*, 2001). This observation has led to a model in which the effector complexes containing siRNAs and stRNAs are closely related but provokes questions regarding how these related complexes regulate expression via distinct mechanisms (**Figure 3**). Of note, neither lin-4 nor let-7 forms a perfect duplex with its cognate target (e.g. Ha *et al.*, 1996). Thus, in one possible model an essentially identical RISC complex is formed containing either siRNAs or stRNAs. In the former case, cleavage is dependent upon perfect complementarity. In the latter case, cleavage does not occur, but the complex blocks ribosome transit and arrests elongation. In support of this notion, miRNAs can direct cleavage of completely complementary substrates in both human cell extracts and in plants (Ketting *et al.*, 2001; Llave *et al.*, 2002).

An alternative possibility is that stRNAs and siRNAs are discriminated in cells and enter related but distinct complexes. Complexes containing siRNAs target substrates for degradation while those containing stRNAs are specialized for translational regulation. Consistent with the latter model is the observation that siRNAs or exogenously supplied hairpin RNAs that contain single mismatches with their substrates fail to repress, rather than simply shifting their regulatory mode to translational inhibition (Elbashir *et al.*, 2001a, b; Paddison *et al.*, 2002).

If the latter model is correct, one may now think of RISC as a flexible platform upon which different regulatory modules may be superimposed. The core complex would be responsible for receiving the small RNA from Dicer and using this as a guide to identify its homologous substrate. Depending upon the nature of the signal (e.g. its structure, its localization, etc.), different effector functions could join the core. In the case of RNAi, nucleases would be incorporated into RISC, whereas in the case of stRNA-mediated regulation, translational repressors would join the complex. Transcriptional silencing could be accomplished by the inclusion of chromatin remodeling factors, and one could imagine numerous other adaptations for control of RNA biogenesis and function.

Whether or not RISC is a flexible regulator becomes particularly important in the light of recent findings that let-7 and lin-4 are archetypes of a large class of endogenously encoded small RNAs. More than one hundred of these, known collectively as microRNAs or miRNAs, have now been identified in *Drosophila*, *C. elegans* and mammals (Lagos-Quintana *et al.*, 2001; Lau *et al.*, 2001; Lee and Ambros, 2001; Mourelatos *et al.*, 2002), and such searches are far from saturated. While the functions of these novel miRNAs are unknown, their prevalence provokes the hypothesis that RNAi-related mechanisms may play pervasive roles in the control of cellular gene expression. In this regard, a number of miRNAs from *Drosophila* are partially complementary to two sequences, the K box and the Brd box, that mediate

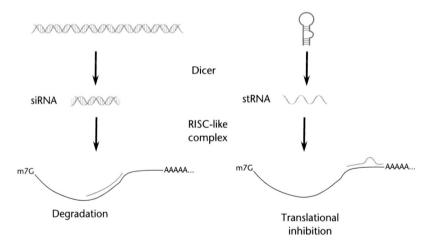

Figure 3 Small interfering RNAs versus small temporal RNAs. siRNAs are produced by Dicer from dsRNA silencing triggers as dsRNAs of ~21–23 nt in length. Characteristic of RNaseIII products, these have 2 nt 3′ overhangs and 5′ phosphorylated termini. In order to work with maximum efficiency, siRNAs must have perfect complementarity to their mRNA target (with the exception of the two terminal nucleotides, which contribute only marginally to recognition). stRNAs, such as lin-4 and let-7, are transcribed from the genome as hairpin precursors. These are also processed by Dicer, but in this case, only one strand accumulates. Notably, neither lin-4 nor let-7 shows perfect complementarity to its targets. In addition, stRNAs regulate targets at the level of translation rather than RNA degradation. It remains unclear whether the difference in regulatory mode results from a difference in substrate recognition or from incorporation of siRNAs and stRNAs into distinct regulatory complexes.

posttranscriptional regulation of numerous mRNAs (Lai, 2002).

RNAi – Will it Deliver the Genome?

It was clear from the outset that RNAi would evolve as a powerful tool for probing gene function. In *C. elegans*, strategies for testing the functions of individual genes have transformed into tools that now extend to analysis of nearly every one of the worm's predicted ~19 000 genes. Similar strategies are being pursued in plant systems and in other organisms, as well. Recently, Tuschl and colleagues, showed that siRNAs themselves could be used to induce potent and sequence-specific silencing in many types of mammalian cells (Elbashir *et al.*, 2001b). It is presumed that these small RNAs, essentially chemically synthesized

mimics of Dicer products, are incorporated into RISC and target cognate substrates for degradation. The siRNAs fail to induce nonspecific dsRNA responses, such as PKR, because they fail to meet the minimum size required to activate these pathways (Clarke and Mathews, 1995).

Despite their tremendous utility, siRNAs have one drawback. Their effects are, by definition, transient, since mammals apparently lack the mechanisms that amplify silencing as exist in worms and plants. Recently, a number of laboratories (Elbashir *et al.*, 2001b; Brummelkamp *et al.*, 2002; Sui *et al.*, 2002) have shown that short hairpin RNAs (shRNAs), modeled on the miRNA paradigm, can be transformed into a tool for experimentally manipulating gene expression. Of note, these small hairpin RNAs can be expressed *in vivo* from RNA polymerase III promoters to induce stable suppression in mammalian somatic cells.

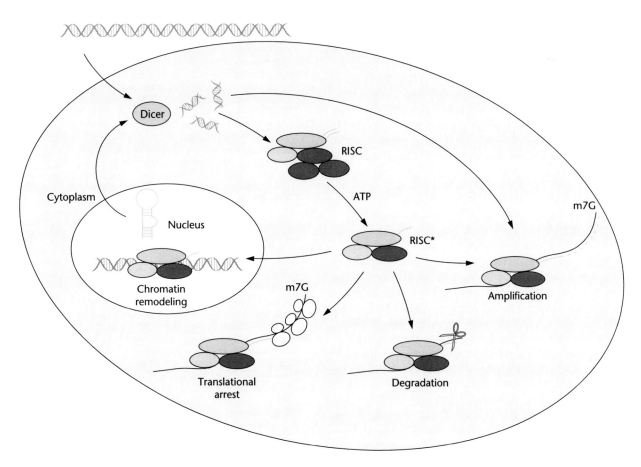

Figure 4 Model for the mechanism of RNAi. Silencing triggers in the form of dsRNA may be presented in the cell as synthetic RNAs, replicating viruses or may be transcribed from nuclear genes. These are recognized and processed into siRNAs by Dicer. The duplex siRNAs are passed to RISC, and the complex becomes activated by unwinding of the duplex. Activated RISC complexes can regulate gene expression at many levels. Almost certainly, such complexes act by promoting RNA degradation and translational inhibition. However, similar complexes probably also target chromatin remodeling. Amplification of the silencing signal in plants may be accomplished by siRNAs priming RNA-dependent RNA polymerase (RdRP)-dependent synthesis of new dsRNA. This could be accomplished by RISC-mediated delivery of an RdRP or by incorporation of the siRNA into a distinct, RdRP-containing complex.

The ability to encode stable triggers of the RNAi pathway builds upon the utility of siRNAs in several key ways. Experimental observation of phenotypes can now be recorded over long time spans. Stably engineered cells can be assayed either *in vitro* or *in vivo*, testing, perhaps, the angiogenic or metastatic potential of tumor cells in xenograft models. Similar strategies may prove efficacious in the construction of transgenic animals in which RNAi is used to create hypomorphic alleles in a rapid and genetically dominant fashion. Coupled with the possible use of inducible RNA polymerase III promoters (Meissner *et al.*, 2001; Ohkawa and Taira, 2000), this could generate a powerful approach akin to the collection of tissue-specific Gal4-drivers in *Drosophila*. Finally, shRNA expression strategies may be combined with existing high-efficiency gene delivery vehicles to create *bona fide* RNAi-based therapeutics. In this regard, we have already validated the delivery of shRNAs from replication-deficient retroviruses and envision numerous possible applications for *ex vivo* manipulation of stem cells based upon this paradigm. For example, the potential exists to engineer a patient's own bone marrow stem cells to resist HIV infection by suppressing expression of either the HIV RNA itself or of receptors necessary for HIV infection (e.g. CCR5). Ultimately, the exquisite specificity of the RNAi machinery may make it possible to distinguish and specifically silence a disease-causing mutant allele, such as an activated oncogene, without affecting a normal allele.

Perspective

Over the past several years, a new regulatory paradigm in biology has emerged in which cells respond to the presence of dsRNA by silencing homologous genes (**Figure 4**). This response can be triggered in many different ways, ranging from experimental introduction of synthetic silencing triggers to transcription of endogenous RNAs, which regulate gene expression. We are only beginning to see both the mechanistic complexity of this process and its biological ramifications. RNAi has already begun to revolutionize experimental biology in organisms ranging from unicellular protozoans to mammals. RNAi has been applied on the whole-genome scale in *C. elegans* and this goal is also being pursued in mammalian systems. This will permit for the first time in mammalian cells the performance of large-scale, loss-of-function genetic screens and rapid tests for genetic interactions. Resources such as these hold tremendous promise for unleashing the potential that currently lies dormant in sequenced genomes.

Acknowledgement

This work was supported in part by a grant from the NIH (RO1-GM62534).

See also

Caenorhabditis elegans as an Experimental Organism

References

Angell SM and Baulcombe DC (1997) Consistent gene silencing in transgenic plants expressing a replicating potato virus X RNA. *EMBO Journal* **16**: 3675–3684.

Baulcombe D (1999) Viruses and gene silencing in plants. *Archives of Virology Supplement* **15**: 189–201.

Bernstein E, Caudy AA, Hammond SM and Hannon GJ (2001a) Role for a bidentate ribonuclease in the initiation step of RNA interference. *Nature* **409**: 363–366.

Bernstein E, Denli AM and Hannon GJ (2001b) The rest is silence. *RNA* **7**: 1509–1521.

Blaszczyk J, *et al.* (2001) Crystallographic and modeling studies of RNase III suggest a mechanism for double-stranded RNA cleavage. *Structure* **9**: 1225–1236.

Bohmert K, *et al.* (1998) AGO1 defines a novel locus of *Arabidopsis* controlling leaf development. *EMBO Journal* **17**: 170–180.

Brummelkamp TR, Bernards R and Agami RA (2002) System for stable expression of short interfering RNAs in mammalian cells. *Science* **21**: 21.

de Carvalho F, *et al.* (1992) Suppression of beta-1,3-glucanase transgene expression in homozygous plants. *EMBO Journal* **11**: 2595–2602.

Caudy AA, Myers M, Hannon GJ and Hammond SM (2002) Fragile X-related protein and VIG associate with the RNA interference machinery. *Genes and Development* **16**: 2491–2496.

Clarke PA and Mathews MB (1995) Interactions between the double-stranded RNA binding motif and RNA: definition of the binding site for the interferon-induced protein kinase DAI (PKR) on adenovirus VA RNA. *RNA* **1**: 7–20.

Cox DN, *et al.* (1998) A novel class of evolutionarily conserved genes defined by piwi are essential for stem cell self-renewal. *Genes and Development* **12**: 3715–3727.

Dernburg AF, Zalevsky J, Colaiacovo MP and Villeneuve AM (2000) Transgene-mediated cosuppression in the *C. elegans* germ line. *Genes and Development* **14**: 1578–1583.

Dougherty WG, *et al.* (1994) RNA-mediated virus resistance in transgenic plants: exploitation of a cellular pathway possibly involved in RNA degradation. *Molecular Plant–Microbe Interactions* **7**: 544–552.

Dudley NR, Labbe JC and Goldstein B (2002) Using RNA interference to identify genes required for RNA interference. *Proceedings of the National Academy of Sciences of the United States of America* **99**: 4191–4196.

Elbashir SM, Martinez J, Patkaniowska A, Lendeckel W and Tuschl T (2001a) Functional anatomy of siRNAs for mediating efficient RNAi in *Drosophila melanogaster* embryo lysate. *EMBO Journal* **20**: 6877–6888.

Elbashir SM, *et al.* (2001b) Duplexes of 21-nucleotide RNAs mediate RNA interference in cultured mammalian cells. *Nature* **411**: 494–498.

Fire A, Albertson D, Harrison SW and Moerman DG (1991) Production of antisense RNA leads to effective and specific inhibition of gene expression in *C. elegans* muscle. *Development* **113**: 503–514.

Fire A, *et al.* (1998) Potent and specific genetic interference by double-stranded RNA in *Caenorhabditis elegans*. *Nature* **391**: 806–811.

Grishok A, et al. (2001) Genes and mechanisms related to RNA interference regulate expression of the small temporal RNAs that control C. elegans developmental timing. *Cell* **106**: 23–34.

Ha I, Wightman B and Ruvkun G (1996) A bulged lin-4/lin-14 RNA duplex is sufficient for *Caenorhabditis elegans* lin-14 temporal gradient formation. *Genes and Development* **10**: 3041–3050.

Hammond SM, Bernstein E, Beach D and Hannon GJ (2000) An RNA-directed nuclease mediates post-transcriptional gene silencing in *Drosophila* cells. *Nature* **404**: 293–296.

Hammond SM, Boettcher S, Caudy AA, Kobayashi R and Hannon GJ (2001a) Argonaute2, a link between genetic and biochemical analyses of RNAi. *Science* **293**: 1146–1150.

Hammond SM, Caudy AA and Hannon GJ (2001b) Post-transcriptional gene silencing by double-stranded RNA. *Nature Reviews Genetics* **2**: 110–119.

Hutvagner G and Zamore PD (2002) A microRNA in a multiple-turnover RNAi enzyme complex. *Science* **297**: 2056–2060.

Hutvagner G, et al. (2001) A cellular function for the RNA-interference enzyme Dicer in the maturation of the let-7 small temporal RNA. *Science* **293**: 834–838.

Jacobsen SE, Running MP and Meyerowitz EM (1999) Disruption of a RNA helicase/RNAse III gene in *Arabidopsis* causes unregulated cell division in floral meristems. *Development* **126**: 5231–5243.

Jones L, Ratcliff F and Baulcombe DC (2001) RNA-directed transcriptional gene silencing in plants can be inherited independently of the RNA trigger and requires Met1 for maintenance. *Current Biology* **11**: 747–757.

Jorgensen R (1990) Altered gene expression in plants due to trans interactions between homologous genes. *Trends in Biotechnology* **8**: 340–344.

Kataoka Y, Takeichi M and Uemura T (2001) Developmental roles and molecular characterization of a Drosophila homologue of *Arabidopsis* Argonaute1, the founder of a novel gene superfamily. *Genes to Cells* **6**: 313–325.

Ketting RF, Haverkamp TH, van Luenen HG and Plasterk RH (1999) Mut-7 of *C. elegans*, required for transposon silencing and RNA interference, is a homolog of Werner syndrome helicase and RNaseD. *Cell* **99**: 133–141.

Ketting RF, et al. (2001) Dicer functions in RNA interference and in synthesis of small RNA involved in developmental timing in *C. elegans*. *Genes and Development* **15**: 2654–2659.

Knight SW and Bass BL (2001) A role for the RNase III enzyme DCR-1 in RNA interference and germ line development in *Caenorhabditis elegans*. *Science* **293**: 2269–2271.

van der Krol AR, Mur LA, de Lange P, Mol JN and Stuitje AR (1990) Inhibition of flower pigmentation by antisense CHS genes: promoter and minimal sequence requirements for the antisense effect. *Plant Molecular Biology* **14**: 457–466.

Kumagai MH, et al. (1995) Cytoplasmic inhibition of carotenoid biosynthesis with virus-derived RNA. *Proceedings of the National Academy of Sciences of the United States of America* **92**: 1679–1683.

Lagos-Quintana M, Rauhut R, Lendeckel W and Tuschl T (2001) Identification of novel genes coding for small expressed RNAs. *Science* **294**: 853–858.

Lai EC (2002) Micro RNAs are complementary to 3′ UTR sequence motifs that mediate negative post-transcriptional regulation. *Nature Genetics* **30**: 363–364.

Lau NC, Lim LP, Weinstein EG and Bartel DP (2001) An abundant class of tiny RNAs with probable regulatory roles in *Caenorhabditis elegans*. *Science* **294**: 858–862.

Lee RC and Ambros V (2001) An extensive class of small RNAs in *Caenorhabditis elegans*. *Science* **294**: 862–864.

Lee RC, Feinbaum RL and Ambros V (1993) The *C. elegans* heterochronic gene lin-4 encodes small RNAs with antisense complementarity to lin-14. *Cell* **75**: 843–854.

Llave C, Xie Z, Kasschau KD and Carrington JC (2002) Cleavage of Scarecrow-like mRNA targets directed by a class of Arabidopsis miRNA. *Science* **297**: 2053–2056.

Martienssen RA and Colot V (2001) DNA methylation and epigenetic inheritance in plants and filamentous fungi. *Science* **293**: 1070–1074.

Martinez J, Patkaniowska A, Urlaub H, Luhrmann R and Tuschl T (2002) Single-stranded antisense siRNAs guide target RNA cleavage in RNAi. *Cell* **110**: 563.

Meissner W, Rothfels H, Schafer B and Seifart K (2001) Development of an inducible pol III transcription system essentially requiring a mutated form of the TATA-binding protein. *Nucleic Acids Research* **29**: 1672–1682.

Mette MF, Aufsatz W, van der Winden J, Matzke MA and Matzke AJ (2000) Transcriptional silencing and promoter methylation triggered by double-stranded RNA. *EMBO Journal* **19**: 5194–5201.

Mourelatos Z, et al. (2002) miRNPs: a novel class of ribonucleo-proteins containing numerous microRNAs. *Genes and Development* **16**: 720–728.

Mourrain P, et al. (2000) *Arabidopsis* SGS2 and SGS3 genes are required for posttranscriptional gene silencing and natural virus resistance. *Cell* **101**: 533–542.

Napoli CA and Jorgensen R (1990) Introduction of a chimeric chalcone synthetase gene in petunia results in reversible cosuppression of homologous genes in trans. *Plant Cell* **2**: 279–289.

Nykanen A, Haley B and Zamore PD (2001) ATP requirements and small interfering RNA structure in the RNA interference pathway. *Cell* **107**: 309–321.

Ohkawa J and Taira K (2000) Control of the functional activity of an antisense RNA by a tetracycline-responsive derivative of the human U6 snRNA promoter. *Human Gene Therapy* **11**: 577–585.

Olsen PH and Ambros V (1999) The lin-4 regulatory RNA controls developmental timing in *Caenorhabditis elegans* by blocking LIN-14 protein synthesis after the initiation of translation. *Developmental Biology* **216**: 671–680.

Paddison PJ, Caudy AA, Bernstein E, Hannon GJ and Conklin DS (2002) Short hairpin RNAs (shRNAs) induce sequence-specific silencing in mammalian cells. *Genes and Development* **16**: 948–958.

Pal-Bhadra M, Bhadra U and Birchler JA (1997) Cosuppression in *Drosophila*: gene silencing of alcohol dehydrogenase by white-Adh transgenes is polycomb dependent. *Cell* **90**: 479–490.

Pal-Bhadra M, Bhadra U and Birchler JA (2002) RNAi-related mechanisms affect both transcriptional and posttran-scriptional transgene silencing in *Drosophila*. *Molecular Cell* **9**: 315–327.

Reinhart BJ, et al. (2000) The 21-nucleotide let-7 RNA regulates developmental timing in *Caenorhabditis elegans*. *Nature* **403**: 901–906.

Romano N and Macino G (1992) Quelling: transient inactivation of gene expression in *Neurospora crassa* by transformation with homologous sequences. *Molecular Microbiology* **6**: 3343–3353.

Ruiz MT, Voinnet O and Baulcombe DC (1998) Initiation and maintenance of virus-induced gene silencing. *Plant Cell* **10**: 937–946.

Slack FJ, et al. (2000) The lin-41 RBCC gene acts in the C. elegans heterochronic pathway between the let-7 regulatory RNA and the LIN-29 transcription factor. *Molecular Cell* **5**: 659–669.

Sui G, et al. (2002) A DNA vector-based RNAi technology to suppress gene expression in mammalian cells. *Proceedings of the National Academy of Sciences of the United States of America* **99**: 5515–5520.

Tabara H, et al. (1999) The rde-1 gene, RNA interference, and transposon silencing in C. elegans. *Cell* **99**: 123–132.

Timmons L and Fire A (1998) Specific interference by ingested dsRNA. *Nature* **395**: 854.

Voinnet O, Lederer C and Baulcombe DC (2000) A viral movement protein prevents spread of the gene silencing signal in *Nicotiana benthamiana*. *Cell* **103**: 157–167.

Volpe TA, *et al.* (2002) Regulation of heterochromatic silencing and histone H3 lysine-9 methylation by RNAi. *Science* **297**: 1833–1837.

Wassenegger M, Heimes S, Riedel L and Sanger HL (1994) RNA-directed *de novo* methylation of genomic sequences in plants. *Cell* **76**: 567–576.

Wightman B, Ha I and Ruvkun G (1993) Posttranscriptional regulation of the heterochronic gene lin-14 by lin-4 mediates temporal pattern formation in *C. elegans*. *Cell* **75**: 855–862.

Winston WM, Molodowitch C and Hunter CP (2002) Systemic RNAi in *C. elegans* requires the putative transmembrane protein SID-1. *Science* **295**: 2456–2459.

Zamore PD, Tuschl T, Sharp PA and Bartel DP (2000) RNAi: double-stranded RNA directs the ATP-dependent cleavage of mRNA at 21 to 23 nucleotide intervals. *Cell* **101**: 25–33.

RNA Polymerases and the Eukaryotic Transcription Machinery

Nancy A Woychik, *University of Medicine and Dentistry of New Jersey, Piscataway, New Jersey, USA*

Intermediate article

Article contents

- Introduction
- Human RNA Polymerases
- RNA Polymerase II
- RNA Polymerases, the Transcription Machinery and Human Diseases

RNA polymerase and its associated proteins play a central role in human gene regulation. This transcription machinery is poised to sense and respond to the multiplicity of cellular signals designed to customize synthesis of both the amount and type of RNA based on the specialized needs of cells comprising human tissues.

Introduction

In eukaryotes, three deoxyribonucleic acid (DNA)-dependent ribonucleic acid (RNA) polymerases (RNAPs) catalyze the synthesis of RNA encoding different sets of genes. RNAP I transcribes the precursors for most ribosomal RNAs (rRNAs), RNAP II transcribes pre-messenger RNA (pre-mRNA) and most small nuclear (snRNAs) that assemble into splicing particles, and RNAP III transcribes genes for transfer RNAs (tRNAs), 5S rRNA, and other small, stable RNAs.

Human and all other eukaryotic RNAPs not only perform the generic functions ascribed to all RNAPs – initiation, elongation and termination of RNA synthesis – but their activity during the entire process of RNA synthesis appears to be regulated by transcription factors. This regulation of enzyme activity is especially intricate for RNAP II, reflected by the myriad of transcription factors that have a hand in this process. However, even more layers of complexity are added by the use of chromatin remodeling, histone acetylation and deacetylation, and histone methylation (Narlikar *et al.*, 2002). Therefore, transcription factors and modulation of chromatin structure work coordinately to impart the exacting and intricate gene regulation crucial for the development and normal function of the human body. (*See* Transcription Factors.)

As with all eukaryotes, the physiological template for all genes transcribed by the three human RNAPs is chromatin. Chromatin is DNA that is folded and compacted through the action of histones that form nucleosomes and more condensed higher-order structures. The packing and unpacking of regions of DNA containing promoters and genes is regulated by a number of posttranslational covalent modifications to the histone *N*-terminal exposed tails. Histone acetylation and methylation are the best studied, but other modifications (e.g. ubiquitination, poly(ADP-ribosylation) and phosphorylation) also modify histone tails (Eberharter and Becker, 2002; Kuzmichev and Reinberg, 2001; Rice and Allis, 2001; Roth *et al.*, 2001; Zhang and Reinberg, 2001). In general, acetylation of histones is associated with transcriptional activation, since it results in unpacking of the DNA by altering the charge of the lysine targets (which weakens the DNA–nucleosome interaction and opens up the promoter region for binding of the transcription machinery). In contrast, repression (or 'silencing') of transcription is associated with more compacted, deacetylated histones. Like acetylation, methylation of histones is also associated with modulation of histone structure and function. A second regulatory event involving chromatin – known as chromatin remodeling – is also invoked in eukaryotic cells (Peterson, 2002). In this case, the position of the nucleosomes on DNA is modulated by adenosine triphosphate (ATP)-dependent chromatin remodeling protein complexes. (*See* Histone Acetylation and Disease; Methylation-mediated Transcriptional Silencing in Tumorigenesis.)

In total, human genes are regulated by a large number of transcription factors and transcription factor complexes (whose activities are regulated by a number of signaling pathways), alterations in chromatin structure through covalent modifications, or alterations in chromatin position. The action of RNAP II is manipulated (directly or indirectly) as a consequence of these regulatory events.

Human RNA Polymerases

The most highly characterized eukaryotic RNAPs are derived from the brewer's yeast *Saccharomyces cerevisiae*, a genetically and biochemically tractable model system whose entire transcriptional machinery is strikingly similar to that in human cells. Yeast RNAP I is composed of 14 subunits, RNAP II of 12 subunits and RNAP III of 17 subunits (Geiduschek and Kassavetis, 2001; Huang and Maraia, 2001; Paule and White, 2000; Woychik, 1998). Human RNAPs I and III are not as well characterized as RNAP II at the subunit level (Grummt, 1999). Human RNAP II has 12 subunits that are conserved in sequence and function with a corresponding yeast subunit. In fact, in many instances, individual human RNAP II subunits can function in place of their yeast counterparts *in vivo*. **Figure 1** demonstrates the functional relationships between RNAP II in yeast and human compared with bacterial RNAP. The essential β, β′ and α bacterial RNAP subunits, as well as the nonessential ω subunit, have counterparts in all eukaryotes. Not only are human RNAP II, yeast RNAP II and bacterial RNAP related, but all three classes of RNAP in humans and other eukaryotes share substantial sequence and functional similarities. For example, the two largest subunits of human RNAPs I, II and III are all related in sequence and function to the bacterial β and β′ subunits. Two other subunits in RNAPs I and III are related to Rpb3 and Rpb11 and the bacterial α_2 subunit pair. Finally, one subunit (Rpb6) is identical in all three RNAPs and is the functional counterpart of the bacterial ω subunit. In summary, all known eukaryotic RNAPs are strikingly similar in their subunit content, amino acid sequence and function. Therefore, information gained from studies using model systems that are easy to

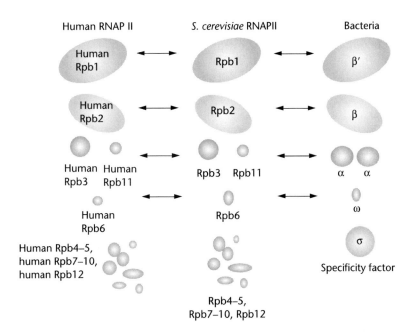

Figure 1 Evolutionary conservation of ribonucleic acid polymerases (RNAPs). Relative subunit sizes are approximate; 'human' represents all metazoan RNAP subunits, yeast (*S. cerevisiae*) represents lower eukaryotic RNAP subunits. The bacterial subunit composition represents that used under normal culture conditions. The eukaryotic subunit nomenclature originated from yeast RNAP.

manipulate genetically and biochemically (such as yeast) can be extrapolated to human RNAPs.

RNA Polymerase II

Since RNAP II catalyzes the synthesis of all protein-encoding mRNAs, it is subject to the most regulation, and consequently aberrations in processes related to mRNA synthesis can lead to cancer and a number of diseases. In order to understand the cause of human diseases related to RNAP II, it is necessary to provide some background about the key features of this enzyme.

RNA polymerase II C-terminal domain

Human RNAP II is composed of 12 subunits. All of the largest subunits of RNAP II (i.e. human Rpb1 counterparts) in eukaryotes have a C-terminal domain (CTD) with an unusual heptapeptide consensus sequence Tyr–Ser–Pro–Thr–Ser–Pro–Ser (Corden, 1990; **Figure 2**). This domain is absent in RNAPs from Archaea, vaccinia and other mammalian viruses, most protozoa, and RNAP I and III. The CTD length varies among eukaryotes, and the number of heptapeptide repeats increases with genome complexity, with humans (and mammals in general) possessing the most repeats in the tail domain. Although consensus variability exists, certain amino acids within the CTD (especially tyrosines) are less variant. The CTD plays an essential role in transcription. Removal of the entire CTD is lethal in yeast cells, and removal of about half of the CTD results in a number of growth and transcriptional abnormalities. In human B cells,

exclusive expression of a form of human Rpb1 lacking most of the CTD causes global transcription defects.

CTD phosphorylation

The CTD is highly phosphorylated, predominantly at the serine residues. At least four protein kinases are known to be involved in CTD phosphorylation (Oelgeschlager, 2002; Prelich, 2002). Although the enlistment of multiple CTD kinases might seem redundant, surprisingly, each of the four known kinases appear to have very specialized and distinct regulatory roles. For example, some CTD kinases enhance transcription, while others are involved in transcription repression.

The phosphorylation state of the CTD influences transcription initiation and elongation. The CTD is unphosphorylated when RNAP II associates with the transcription initiation complex. However, the CTD then undergoes extensive phosphorylation during the transition from initiation to elongation, and the level of phosphorylation of the elongating RNAP appears to be promoter specific.

Current models suggest that phosphorylation of the CTD causes a disruption in interactions with 'general transcription factors' required for stable preinitiation complex formation, allowing release of the elongating enzyme. The CTD and its phosphorylation state also influence RNA capping, splicing, poly(A) addition and degradation of the largest RNAP II subunit (Hirose and Manley, 2000; Proudfoot et al., 2002).

A CTD phosphatase has also been identified and characterized in human and yeast cells. Along with CTD kinases, this enzyme plays an equally important role in vivo; it stimulates RNAP II elongation and

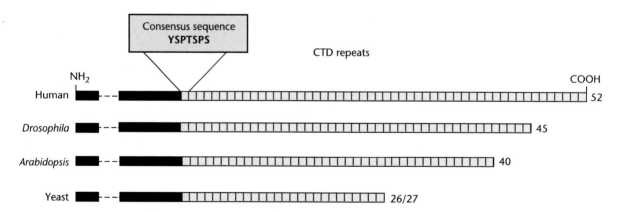

Figure 2 RNA polymerase II C-terminal domain (CTD). Each square box represents one CTD heptapeptide; the filled rectangles (not drawn to scale) represent RNAP II large subunit amino acids in front of the CTD. The consensus sequence shown represents the most prevalent repeated sequence in the overall CTD. A one- or two-amino-acid deviation from the consensus sequence is also common. The length of the yeast CTD shows strain variability: it can comprise either 26 or 27 repeats.

allows recycling of the polymerase back to initiation complex. Therefore, modulation of RNAP II activity through CTD phosphorylation states is one of the most critical signaling steps in the regulation of mRNA synthesis.

RNA polymerase II structure

Tremendous insight has been gained by not only using classic genetic and biochemical approaches for studying RNAP subunits in model systems such as yeast, but also studying the atomic structure of the enzyme using X-ray crystallography (Geiduschek and Bartlett, 2000). The laboratory of Roger Kornberg at Stanford University has been a pioneer in the study and interpretation of the structure of eukaryotic RNAP II; recently they have solved the structure of yeast RNAP II at 2.8 Å resolution and an RNAP II elongating complex at 3.3 Å (Cramer et al., 2001; Gnatt et al., 2001). Since the yeast and human enzymes are very similar in both amino acid sequence and function, the structural parallels are likely to be strong. In fact, the family of DNA-dependent RNAPs across all species is generally well conserved, so much so that striking structural similarities even exist between bacterial RNAP and the eukaryotic yeast RNAP II.

The overall shape of the enzyme has been described as a 'crab claw'. Another key feature of the enzyme, evident in both the low-resolution and the refined 2.8 Å RNAP II crystal structure, is the 25 Å diameter DNA-binding cleft where RNAP II clamps onto the DNA template. This 25 Å nucleic acid-binding cleft is also present in other RNAPs such as bacterial RNAP, bacteriophage T7 RNAP, human immunodeficiency virus (HIV)-1 reverse transcriptase and even in DNA polymerases. Therefore, human RNAP II is expected to possess this feature as well. Also, as with their orthologs in yeast and bacteria, human Rpb1 and human Rpb2 undoubtedly form the two jaws of the cleft. A combination of other structural studies has also allowed for the placement of the majority of individual subunits within the overall structure of the yeast enzyme. As with the two large subunits, the overall placement of the other subunits within the human structure is expected to mirror yeast RNAP II.

Human RNA polymerases and mediator

Unlike bacterial RNAPs, which can adroitly transcribe naked DNA to RNA, human and all other eukaryotic enzymes require accessory proteins (general transcription factors) in order to synthesize RNA. Although a combination of purified RNAP II and recombinant general transcription factors TFIID, TFIIB, TFIIE, TFIIF and TFIIH can support low levels of promoter-specific transcription in a test tube (basal transcription), many more proteins are required for the enhanced and regulated expression of genes in living cells. The complex coordinating and signaling processes necessary to effectively regulate gene expression in humans requires the close association of the RNAP II enzyme with distinct sets of accessory transcription factors, collectively referred to as Mediator (Malik and Roeder, 2000; Myers and Kornberg, 2000; Boube et al., 2002).

Mediator functions as a global 'coactivator' – it is required for activation by transcription factors that bind DNA in a site-specific manner but itself does not possess the ability to bind specific DNA sequences. Consistent with the regulatory role played by Mediator, the RNAP II–Mediator interaction is reversible. Therefore, Mediator enables activator or repressor proteins to modulate RNAP II activity and in some cases target information toward chromatin. Mediator also stimulates basal transcription in vitro and the protein kinase component of TFIIH. The entire RNAP II transcription machinery includes the multisubunit RNAP, Mediator and the five general transcription factors (TFIID, TFIIH, TFIIE, TFIIF and TFIIB). All of the components of this machinery appear to work together to coordinate mRNA synthesis in living cells (**Figure 3**).

The Mediator complex can be purified from cell extracts using a combination of biochemical and immunological techniques. Not surprisingly, the yeast RNAP II Mediator is well characterized, while that from human and other metazoan cells is larger and its composition more controversial. A snapshot of the components of yeast Mediator and an example of one well-characterized human Mediator complex is shown in **Figure 4**.

Interestingly, the exact composition of this megadalton human Mediator complex appears to depend upon the rationale behind its isolation and purification. For example, one Mediator complex called 'thyroid receptor-associated protein' (TRAP) was isolated based on its ability to enhance thyroid hormone receptor function, while another Mediator complex referred to as 'cofactor required for Sp1' (CRSP) was isolated as a complex necessary for the activation by the human transcription factor Sp1. Mediator complexes isolated using independent approaches are sometimes identical – such as the TRAP and 'Srb- and Med-containing cofactor complex' (SMCC) complexes shown in **Figure 4** – or instead have only certain subunits in common. The identification of subunits common to different Mediator preparations indicates that human Mediator is modular. Other variables such as cell type and growth conditions may also influence the composition of the purified Mediator complex. Current proposals view the common subunits as a 'core Mediator', while the

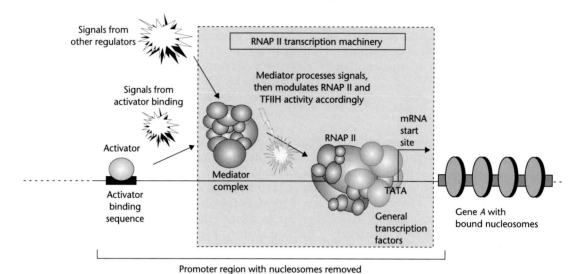

Figure 3 RNAP II transcription machinery. All sizes are approximate. Subunit and transcription factor placement in the respective complexes is not intended to reflect known interactions. Adapted from Woychik and Hampsey (2002).

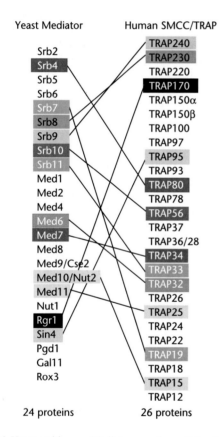

Figure 4 Yeast and human Mediator complexes. Subunits of yeast Mediator are grouped and listed by name, not by apparent molecular weight. The example used for human Mediator is the Srb- and Med-containing cofactor complex–thyroid receptor-associated protein complex (SMCC/TRAP); subunit names reflect apparent molecular weight. Evolutionary conservation exists between twelve subunits in the two complexes; each related pair is linked by a line (subunit similarities adapted from Malik and Roeder (2000); Boube *et al.* (2002)).

remaining Mediator subunits impart the functional specificity required to perform the given task at hand. Therefore, the modular nature of Mediator may allow cells to custom-design Mediator complexes according to constantly fluctuating needs of mature and developing human cells. Finally, certain proteins of the Mediator are evolutionarily conserved while others appear to be metazoa-specific. For example, about half of the subunits of yeast Mediator have counterparts in the human SMCC/TRAP complex (**Figure 4**).

RNA Polymerases, the Transcription Machinery and Human Diseases

The precise firing and extinguishing of gene expression at specific times, at the right intensity and in certain tissues, is crucial to all aspects of human life, from normal embryonic development to maintenance of health. Not surprising, as more and more players involved in regulated mRNA transcription are disclosed, scientists are gradually linking mutations in these proteins to an array of human diseases. In fact, most oncogenes act by altering gene expression. (*See* Transcription Factors and Human Disorders.)

Table 1 highlights examples of maladies caused by aberrant RNAP II mRNA synthesis due to mutations: (1) in components of the RNAP II transcription machinery, (2) in a transcription factor that regulates the RNAP II transcription machinery or (3) resulting in altered chromatin structure and position. These mutants ultimately result in either alterations in a step in the transcription process (i.e. elongation) or enhancement of gene expression by releasing the repressive

Table 1 Transcription and human disease

Disease	Mutated target	Normal target protein function	Comments
Cancer			
Von Hippel–Lindau (VHL) heredity cancer	*VHL* tumor suppressor	Regulates growth and differentiation in kidney cells	VHL has a negative effect on the elongin complex that stimulates elongation
Wilms (WT; kidney) tumors	*WT1* tumor suppressor	Transcriptional regulator of kidney differentiation	
Malignant rhabdoid tumors	Human SNF5 (*SMARCB1*)	Member of a chromatin remodeling complex	Mutations appear to be loss of function
Familial retinoblastoma (Rb) and many other tumors	*RB1* tumor suppressor	Interacts with HDAC[a]-associated protein; antagonizes activation of transcription by E2F family of RNAP II transcription factors; represses RNAP III transcription; regulates RNAP I transcription	Mutant protein cannot bind to E2F or HDAC complex
Breast and ovarian-specific tumors	BRCA1 tumor suppressor	Interacts with HDAC-associated proteins	
Multiple cancers	*TP53* tumor suppressor	Interacts with HDAC-associated proteins	
Glioblastomas, colorectal cancers, breast cancer	CBP/p300 tumor suppressor (*PCAF*)	Interacts with HDAC-associated proteins	Association to these cancers found upon loss-of-heterozygosity analysis for CBP
Leukemia	MOZ[c] (*RUNXBP2*) and TIF2 (*NCOA2*)	HATs[b]	Inversions of chromosome 8 lead to MOZ–TIF2 fusion protein
Acute myeloid leukemia (AML)	MOZ (*RUNXBP2*) and CBP (*PCAF*)	MOZ-HAT; CBP interacts with HDAC-associated proteins	Translocation results in MOZ–CBP fusion protein
	MLL	MLL interacts with human SNF5 (member of a chromatin remodeling complex)	Translocation leads to MLL–ELL fusion isolated from AML patients
	ELL and ELL2	ELL and ELL2 stimulate elongation rate; ELL inhibits RNAP II initiation	
Mixed lineage leukemia (MLL)	MLL and CBP/p300	MLL interacts with human SNF5; CBP/p300 intracts with HDAC-associated proteins	Translocation results in MLL fusion to either CBP or p300
Dermatomyositis and cancer	NuRD	Chromatin remodeling complex that also possesses HDAC activity	One component of NuRD complex closely related to metastasis-associated factor MTA1, whose increased expression level correlates with metastatic potential; a second component of NuRD is an autoimmune antigen for dermatomyositis (15–30% of these patients also develop cancer)
Certain breast and ovarian cancers	AIB1 (ACTR)	HAT	Normal gene is overexpressed
Breast cancer	Human HP1	Repressor protein	Downregulated in breast cancer cells that display an invasive, metastatic phenotype
Other			
Human immunodeficiency virus (HIV)	Tat	Transciptional activator	Reverses host cell repression of elongation; Tat recruits elongation factor P-TEFb, P-TEFb phosphorylates the CTD leading to stimulated transcription elongation of the HIV-1 long terminal repeat
Xeroderma pigmentosum	XP-A to -G and variant	XP-D and XP-B are TFIIH subunits	Associated with defective nucleotide excision repair
Trichthiodystrophy	XP-B, XP-D and TTD-A	XP-D and XP-B are TFIIH subunits	Associated with defective nucleotide excision repair
Cockayne syndrome A and B	CS-A, CS-B, XP-B	XP-D and XP-B are TFIIH subunits	Associated with defective nucleotide excision repair

Table 1 Continued

Disease	Mutated target	Normal target protein function	Comments
	XP-D and XP-G	CS proteins only involved in transcription-coupled nucleotide excision repair	
Rubinstein–Taybi syndrome	CBP (*CREBBP*)	Interacts with HDAC-associated proteins	Patients have lost one copy of the CBP gene
Coffin–Lowry syndrome (CLS)	RSK2 (*RP56KA31*)	Protein kinase	Cell lines from CLS patients have defective histone H3 phosphorylation
Holt–Oram syndrome	Tbx-5 (*TBX5*)	Heart-specific transcription factor	
Atrial–septal defect	Nkx2-5 (*NKX2-5*)	Heart-specific transcription factor	

[a]Histone deacetylase.
[b]Histone acetylase.
[c]Monocytic leukemia zinc-finger protein.

action of chromatin (by remodeling nucleosomes or altering the acetylation pattern of the core histones). In some cases, the precise mechanism linking altered transcription to the disease has even been established. Because transcriptional aberrations are the foundation of so many diseases and disorders, the sustained intensity of research in this area of biology is certain to provide more valuable avenues for drug and gene therapy treatments for many debilitating or fatal disorders.

See also

Transcriptional Regulation: Coordination
Transcription-coupled DNA Repair

References

Boube M, Joulia L, Cribbs DL, Bourbon H-M (2002) Evidence for a Mediator of RNA polymerase II transcriptional regulation conserved from yeast to man. *Cell* **110**: 143–151.

Corden JL (1990) Tails of RNA polymerase II. *Trends in Biochemical Sciences* **15**: 383–387.

Cramer P, Bushnell DA and Kornberg RD (2001) Structural basis of transcription: RNA polymerase II at 2.8-angstrom resolution. *Science* **292**: 1863–1876.

Eberharter A and Becker PB (2002) Histone acetylation: a switch between repressive and permissive chromatin: second in review series on chromatin dynamics. *EMBO Reports* **3**: 224–229.

Geiduschek EP and Bartlett MS (2000) Engines of gene expression. *Nature Structural Biology* **7**: 437–439.

Geiduschek EP and Kassavetis GA (2001) The RNA polymerase III transcription apparatus. *Journal of Molecular Biology* **310**: 1–26.

Gnatt AL, Cramer P, Fu J, Bushnell DA and Kornberg RD (2001) Structural basis of transcription: an RNA polymerase II elongation complex at 3.3-angstrom resolution. *Science* **292**: 1876–1882.

Grummt I (1999) Regulation of mammalian ribosomal gene transcription by RNA polymerase I. *Progress in Nucleic Acids Research and Molecular Biology* **62**: 109–154.

Hirose Y and Manley JL (2000) RNA polymerase II and the integration of nuclear events. *Genes and Development* **14**: 1415–1429.

Huang Y and Maraia RJ (2001) Comparison of the RNA polymerase III transcription machinery in *Schizosaccharomyces pombe*, *Saccharomyces cerevisiae* and human. *Nucleic Acids Research* **29**: 2675–2690.

Kuzmichev A and Reinberg D (2001) Role of histone deacetylase complexes in the regulation of chromatin metabolism. *Current Topics in Microbiology and Immunology* **254**: 35–58.

Malik S and Roeder RG (2000) Transcriptional regulation through Mediator-like coactivators in yeast and metazoan cells. *Trends in Biochemical Sciences* **25**: 277–283.

Myers LC and Kornberg RD (2000) Mediator of transcriptional regulation. *Annual Review of Biochemistry* **69**: 729–749.

Narlikar GJ, Fan HY and Kingston RE (2002) Cooperation between complexes that regulate chromatin structure and transcription. *Cell* **108**: 475–487.

Oelgeschlager T (2002) Regulation of RNA polymerase II activity by CTD phosphorylation and cell cycle control. *Journal of Cellular Physiology* **190**: 160–169.

Paule MR and White RJ (2000) Survey and summary: transcription by RNA polymerases I and III. *Nucleic Acids Research* **28**: 1283–1298.

Peterson CL (2002) Chromatin remodeling enzymes: taming the machines: third in review series on chromatin dynamics. *EMBO Reports* **3**: 319–322.

Prelich G (2002) RNA polymerase II carboxy-terminal domain kinases: emerging clues to their function. *Eukaryotic Cell* **1**: 153–162.

Proudfoot NJ, Furger A and Dye MJ (2002) Integrating mRNA processing with transcription. *Cell* **108**: 501–512.

Rice JC and Allis CD (2001) Histone methylation versus histone acetylation: new insights into epigenetic regulation. *Current Opinion in Cell Biology* **13**: 263–273.

Roth SY, Denu JM and Allis CD (2001) Histone acetyltransferases. *Annual Review of Biochemistry* **70**: 81–120.

Woychik NA (1998) Fractions to functions: RNA polymerase II thirty years later. *Cold Spring Harbor Symposia on Quantitative Biology* **63**: 311–317.

Woychik NA and Hampsey M (2002) The RNA polymerase II machinery: structure illuminates function. *Cell* **108**: 453–463.

Zhang Y and Reinberg D (2001) Transcription regulation by histone methylation: interplay between different covalent modifications of the core histone tails. *Genes and Development* **15**: 2343–2360.

Further Reading

Buratowski S (2000) Snapshots of RNA polymerase II transcription initiation. *Current Opinion in Cell Biology* **12**: 320–325.

Conaway JW and Conaway RC (1999) Transcription elongation and human disease. *Annual Review of Biochemistry* **68**: 301–319.

Conaway JW, Shilatifard A, Dvir A and Conaway RC (2000) Control of elongation by RNA polymerase II. *Trends in Biochemical Sciences* **25**: 375–380.

Hampsey M (1998) Molecular genetics of the RNA polymerase II general transcriptional machinery. *Microbiology and Molecular Biology Reviews* **62**: 465–503.

Hampsey M and Reinberg D (1999) RNA polymerase II as a control panel for multiple coactivator complexes. *Current Opinion in Genetics and Development* **9**: 132–139.

Lee TI and Young RA (2000) Transcription of eukaryotic protein-coding genes. *Annual Review of Genetics* **34**: 77–137.

Meisterernst M (2002) Transcription. Mediator meets Morpheus. *Science* **295**: 984–985.

Orphanides G and Reinberg D (2000) RNA polymerase II elongation through chromatin. *Nature* **407**: 471–475.

Orphanides G and Reinberg D (2002) A unified theory of gene expression. *Cell* **108**: 439–451.

RNA Processing

Introductory article

Christopher WJ Smith, *University of Cambridge, Cambridge, UK*

ADJ Scadden, *University of Cambridge, Cambridge, UK*

Eukaryotic RNA transcripts undergo multiple processing reactions before becoming mature functional RNA products. Processing reactions involve base and sugar modifications as well as the removal of noncoding sequences (introns). RNA processing is not only necessary for proper gene expression, but also allows individual genes to produce multiple protein isoforms by alternative splicing and RNA editing.

Introduction

According to Francis Crick's 'Central Dogma' of molecular biology, information transfer during gene expression can be summarized in the following way:

$$DNA \rightarrow RNA \rightarrow Protein$$

An RNA copy of the DNA gene is made by the process of transcription, and the information in this is then translated into protein. This simple and helpful formulation hides a number of complexities. Of the three eukaryotic RNA polymerases, only RNA polymerase II produces messenger RNA (mRNA) products that are templates for protein translation. All three polymerases produce RNAs that are themselves functional and do not serve as templates for protein synthesis. Moreover, all primary RNA transcripts – protein-coding or not – undergo extensive processing reactions before the RNA is fit for its cellular function. The precise processing reactions depend upon the class of RNA, and the polymerase that transcribes it.

RNA processing reactions include modifications to the bases (e.g. methylation, deamination) or ribose groups (e.g. $2'$-O-methylation), trimming of the $5'$ and $3'$ ends by exonuclease enzymes, and splicing reactions in which internal segments of RNA are removed and the flanking sequences are ligated together. Many of these processing reactions are dependent upon small RNAs that must themselves be heavily processed before becoming functional. In the case of the protein-coding mRNAs, the generation of a continuous open reading frame that can be correctly decoded during translation requires the removal from the pre-mRNA transcript of the noncoding sequences (introns) that lie between the coding sequences (exons). This process of pre-mRNA splicing is not only an essential step in gene expression, but it also provides a key regulatory step. In many human genes – probably the majority – the process of alternative splicing allows the production of more than one protein from a single gene.

Types of Modification

RNA processing reactions can be classified according to whether they affect the backbone phosphodiester bonds, the ribose sugars, or the bases. Phosphodiester bonds can be broken or new ones formed via hydrolysis, condensation or concerted transesterification reactions (as occurs in pre-mRNA splicing).

The ribose groups can be methylated at the 2′ position to yield 2′-*O*-methyl nucleosides (**Figure 1a**). This may affect higher-order RNA structure due to the replacement of a hydrogen bond donor by a hydrophobic group, and it also creates resistance to many nucleases. Base modifications (examples in **Figure 1b**) are far more diverse and include methylations (e.g. ribothymidine, 7-methyl guanosine), rotation of the base so that the glycosidic bond is formed at a different base position (pseudouridine (ψ)), reduction of double bonds (dihydrouridine), substitution of oxygen by sulfur (e.g. 4-thiouridine, 2-thiocytidine) and hydrolytic deamination (cytidine (C) to uridine (U) and adenosine (A) to inosine (I)). An important distinction between the various base modifications is whether they affect the hydrogen-bonding groups involved in Watson–Crick base-pairing (shown in dark gray (H-bond donors) and light gray (H-bond acceptors) in **Figure 1b**). For instance, C to U deamination can change the information content of mRNA dramatically by changing a CAA glutamine codon to a UAA translation termination codon. In contrast, many base modifications do not alter Watson–Crick base-pairing. For example, ψ and ribothymidine have the same Watson–Crick hydrogen-bonding groups as uridine and can base pair normally with A. However, the additional amine group or methyl substituents can affect tertiary RNA contacts important for higher-order structure. The additional N1 hydrogen bond donor of ψ is particularly useful in establishing tertiary contacts, and may explain why ψ is such a common modified base in transfer RNA (tRNA), ribosomal RNA (rRNA) and small nuclear RNAs (snRNAs). Unlike the other uridine derivatives the dihydrouridine base is non-planar and does not base-pair well. Even though it

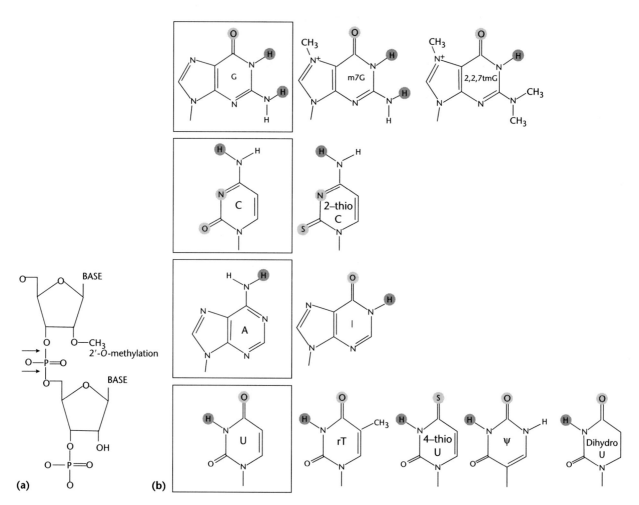

Figure 1 Types of RNA processing and modifications. The phosphodiester backbone can be cleaved (arrows) to leave either 5′ or 3′ phosphate groups. The ribose sugars can be methylated on the 2′ positions. The four canonical bases (G, C, A, U) can be modified in various ways (examples shown on the right). Watson–Crick hydrogen-bonding groups are indicated by dark gray (donor) or light gray (acceptor) circles. Replacement of oxygen by sulfur results in a weaker hydrogen bond acceptor. m7G: 7-methyl guanosine; 2,2,7 tmG: 2,2,7-trimethyl guanosine; I: inosine; rT: ribothymidine; ψ: pseudouridine.

maintains all the Watson–Crick hydrogen-bonding groups, it tends to act as a helix breaker.

Processing of Different Classes of RNA

Messenger RNA

Pre-mRNA is synthesized by RNA polymerase II in the nucleoplasm and undergoes several processing reactions before becoming mature mRNA. All mRNAs acquire a cap structure at the 5′ end, and the vast majority are cleaved at the 3′ end, following which a poly (A) tail is added. In addition, with few exceptions, most human genes contain noncoding introns, which are transcribed into the pre-mRNA but which must be removed by the process of pre-mRNA splicing. Finally, some mRNAs undergo editing processes, in which specific internal bases are modified to change the coding capacity of the mRNA. All of these processes occur cotranscriptionally while the RNA is still being synthesized. Indeed many of the pre-mRNA processing factors associate directly with RNA polymerase II via the C-terminal domain of its large subunit.

Capping

The cap is added cotranscriptionally to the 5′ end of pre-mRNAs, but is not templated within the DNA. It consists of an 'inverted' 7-methylated GTP group involving an unusual 5′-ppp-5′ linkage with the first templated nucleotide (**Figure 2**). It serves several functions:

- confers stability against 5′ exonucleases;
- activates splicing of 5′ proximal introns;
- activates translation.

The cap is added in the following steps:

1. RNA terminal phosphatase (RTPase) removes the terminal phosphate of the 5′ triphosphate group of the nascent RNA.
2. RNA guanylyl transferase (RGTase) adds GMP to the 5′ end using GTP as the donor.
3. RNA (guanine-7) methyl transferase adds a methyl group to the N7 position of the cap guanosine, using S-adenosyl methionine as a donor, to produce Cap 0.
4. Methyl groups can be added sequentially to the 2′ groups of the first and second ribose groups (Cap 1 and 2 respectively). If the first templated base is A, it can be methylated on the N6 position, but only if it has first been 2′-O-methylated.

The specificity of capping is primarily determined by the RNA polymerase. Human RTPase and RGTase activities are both associated with a single polypeptide that binds to the C-terminal domain of RNA polymerase II's large subunit. In addition, only RNAs with a di- or triphosphate at the 5′ end are substrates for capping. Therefore products of RNA polymerase I and III, or of endonuclease cleavage, are not capped.

Splicing

With a few exceptions (e.g. histones, heatshock proteins, interferons), most human genes contain introns, noncoding sequences that interrupt the sequences that will appear in the mature mRNA (exons). The process of pre-mRNA splicing removes the introns and ligates the exons to generate translatable mRNA to be exported to the cytoplasm. Introns vary enormously in length; the human average is 3.3 kilobases with an absolute minimum size of ∼60

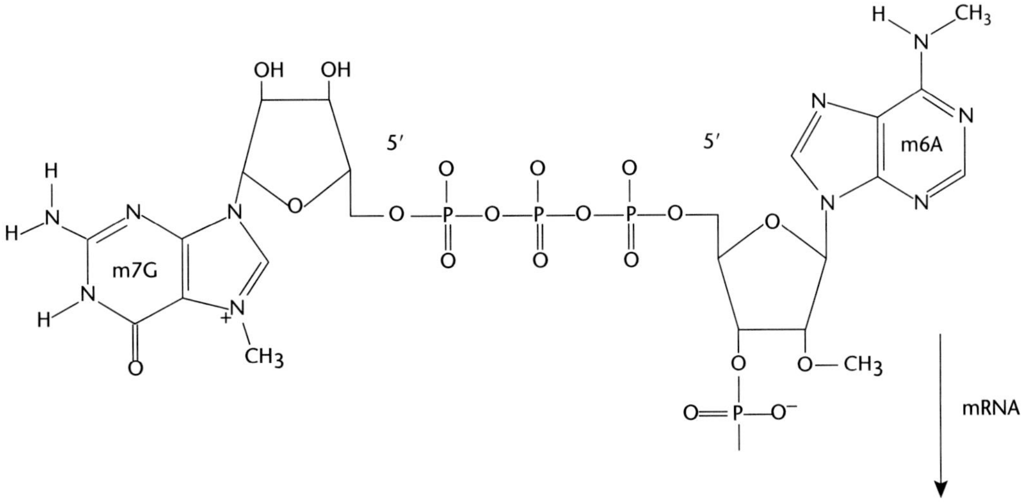

Figure 2 7-Methyl guanosine inverted cap structure. The cap is shown with a 2′-O-methyl group on the first templated base, which has also been methylated at the adenine N6 position.

bases. There appears to be little restriction on maximum length; many introns are tens of kilobases, and some are hundreds of kilobases long. Likewise, the number of introns varies widely from a few introns up to 177 in the human titin gene. RNA splicing is clearly essential to generate functional mRNAs. However, it also has additional roles:

- The process of alternative splicing (see below) allows the regulated production of more than one protein from a single gene.
- It assists in the subsequent export of mRNAs to the cytoplasm.
- It helps to distinguish functional RNAs from those with premature stop codons (via the process of nonsense mediated decay).
- Long introns may be important as a timing device for gene expression by introducing a significant delay between initiation of transcription and production of functional protein. (*See* Splicing of pre-mRNA.)

Mechanism of splicing

Introns have short consensus sequences at their boundaries, including the almost invariant GU and AG dinucleotides at their 5′ and 3′ ends respectively and the A at the 'branchpoint' (**Figure 3**). The 5′ splice site (sometimes called the 'donor' site) has a 9 nucleotide (nt) consensus (A/C)AG|GURAGU, where R denotes a purine and the vertical line represents the exon–intron boundary. The consensus sequence is complementary to a conserved single-stranded region in the abundant U1 (snRNA), and this base-pairing interaction is critical in 5′ splice site recognition. (*See* Splice Sites; Splicing of pre-mRNA.)

The 3′ splice site (sometimes called the 'acceptor' site) has the consensus CAG|G and is preceded by a region enriched in pyrimidines (the polypyrimidine tract). Upstream of the polypyrimidine tract is the 'branchpoint' sequence with consensus YNYURAY, where R is a purine, Y is a pyrimidine, N any nucleotide and A is the branchpoint residue (see below). In yeast, there is a much tighter version of this consensus: UACUAAC. The yeast branchpoint sequence is exactly complementary with a conserved single-stranded region of U2 snRNA, allowing for a single mismatch in which the branchpoint A is bulged out. This U2 snRNA-branchpoint interaction is also important in splicing.

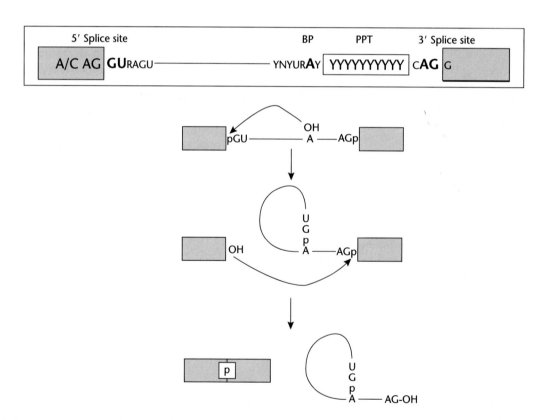

Figure 3 Consensus sequences and reaction of pre-mRNA splicing. Consensus splice site sequences. Exons are shown as boxes and the intron as a line. The invariant GU and AG dinucleotides at the ends of the intron and the branchpoint A are shown in bold. R: purine; Y: pyrimidine; BP: branchpoint; PPT: polypyrimidine tract. Splicing occurs by two successive transesterification reactions. In step 1, the 2′ OH of the BP attacks the phosphate at the 5′ splice site, forming the reaction intermediates. In step 2, the 3′ OH of the 5′ exon attacks the 3′ splice site forming the spliced mRNA (left) and the intron, which is released as a loop or 'lariat' structure.

Other than these limited degenerate consensus sequences, minimal distances are required between the branchpoint and both the 5′ and 3′ splice sites, leading to the minimal intron size of ∼60 nt. There are also elements known as 'splicing enhancers', which can be located in either exons or introns and which activate adjacent splice sites. Splicing enhancers have relatively degenerate sequences, typically purine rich, and act as binding sites for activating proteins. (*See* Exonic Splicing Enhancers.)

Chemically, the splicing reaction proceeds via two transesterification reactions involving groups at the branchpoint, 5′ and 3′ splice sites (**Figure 3**). First, the 2′-OH group of the branchpoint A makes a nucleophilic attack on the phosphate between the 5′ exon and intron, leading to cleavage of the phosphodiester bond between the exon and intron and formation of a new 2′–5′ phosphodiester between the branchpoint A and the G at the 5′ end of the intron. The intermediates of the reaction are therefore the 5′ exon, now with a free 3′-OH group, and the intron–3′ exon, with the intron in a looped or 'lariat' configuration. The second splicing step involves a similar nucleophilic attack by the 3′-OH at the end of the 5′ exon upon the phosphate at the intron–3′ exon junction. The products are the spliced exons and the intron, still in the lariat form.

The splicing reaction only occurs after the consensus splice site elements have been recognized by various splicing factors leading to assembly of a 'spliceosome'. The spliceosome is a large ribonucleoprotein complex containing 50–100 proteins and five snRNA components (U1, 2, 4, 5 and 6). The snRNAs are each contained within preformed small nuclear ribonucleoprotein (snRNP) complexes (U1, U2, U4/6 and U5), containing several snRNPs, some of which are common to all the spliceosomal snRNPs. U1 and U2 snRNAs are responsible for recognition of the 5′ splice site and branchpoint respectively by base pairing. In the fully assembled spliceosome, U2 and U6 snRNAs help to form a network of RNA–RNA interactions that bring together the reactive groups in the pre-mRNA, which are distantly separated in the primary RNA sequence. They also form the catalytic core of the spliceosome, in part by providing specific sites for binding the Mg^{2+} ions important for catalysis of splicing. Another important group of splicing factors are RNA helicases, which are involved in remodeling the intricate RNA–RNA and RNA–protein interactions during spliceosome assembly. For example, base pairing of U1 snRNA to the 5′ splice site gets disrupted and replaced by an interaction between the 5′ splice site and U6 snRNA. Such changes are driven by the RNA helicases using ATP hydrolysis as an energy source. (*See* Spliceosome.)

ATAC introns and the minor U11/U12 spliceosome

A very small number of introns have splice sites and branchpoints with different consensus sequences. In particular, these introns typically have AU and AC dinucleotides at their 5′ and 3′ ends (and so are often referred to as ATAC introns). They are spliced by a low-abundance spliceosome, which contains the same U5 snRNP as the major spliceosome, but in which U1, U2, U4 and U6 snRNPs are replaced by the related U11, U12, $U4_{atac}$ and $U6_{atac}$ respectively. Both major and minor spliceosome introns can be found in the same genes.

3′ end processing

Termination of transcription by RNA polymerase II does not usually occur at fixed positions. Instead it commonly transcribes far beyond the 3′ end of the mature mRNA. The 3′ end of mRNAs is generated by a specific endonuclease cleavage followed by non-templated addition of ∼250 adenosines to form the poly(A) tail. The poly(A) tail confers resistance to 3′ exonucleases, and is necessary for efficient translation initiation. Both stability and translational efficiency can be regulated by the poly(A) tail length, and subsequent lengthening or shortening of the poly(A) tail in the cytoplasm is commonly used as a regulatory strategy (for example during early development in *Xenopus* embryos).

The poly(A) tail is added in two steps (**Figure 4**). First the RNA is cleaved by an endonuclease, 15–30 nt downstream of a highly conserved hexanucleotide sequence AAUAAA. Poly(A) polymerase then synthesizes the poly A tail using ATP as a substrate. The 3′ end processing reaction requires at least two sequences in the RNA. The AAUAAA element provides a binding site for cleavage polyadenylation specificity factor (CPSF), while a region downstream

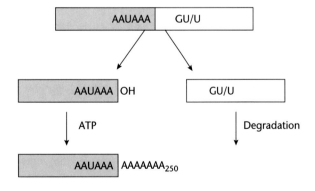

Figure 4 Pre-mRNA 3′ processing. 3′ end processing occurs in two steps. First the pre-mRNA is cleaved and then a tail of ∼250 A residues is added. The reactions depend upon the highly conserved AAUAAA sequence and a GU/U-rich sequence, which flank the site of cleavage.

of the cleavage site enriched in U or G and U residues is necessary for binding of cleavage stimulation factor (CStF). Both CPSF and CStF contain multiple protein subunits, but various additional cleavage factors (CFs) are required for the cleavage reaction. Poly(A) polymerase alone synthesizes very long poly A tails, but the presence of poly(A)-binding protein II (PAB II) limits the length of the tail to the physiological length of ~250 nt.

Histone 3′ ends

Replication-dependent histone mRNAs provide an exception to the general rule for mRNA 3′ end processing. They are cleaved by an unidentified endonuclease, but no poly(A) tail is added. The cleavage reaction requires two RNA elements. The first is a short stem–loop structure (a hairpin) just upstream of the cleavage site, which is bound by stem–loop-binding protein (SLBP). The second element is a purine-rich element about 10–17 nt downstream of the cleavage site. This is recognized by base pairing with U7 snRNA, which is promoted by SLBP, and provides a reference point for the site of cleavage. SLBP remains bound to the histone mRNA and plays roles in export to the cytoplasm, translation and mRNA stability.

mRNA editing

A small number of mRNA transcripts contain some nucleotides that differ from those encoded by the gene. These are due to base modifications that alter the coding capacity of the RNA. The best characterized types of mRNA editing events in mammalian systems are C to U and A to I editing, which occur by a chemically similar deamination mechanism (**Figure 5a**), but which are specified in distinct ways. (*See* mRNA Editing; RNA Editing and Human Disorders.)

A to I editing

Adenosine can be converted to inosine by hydrolytic deamination at the N6 position (**Figure 5a**). Since I is recognized by the translation machinery as G, this can change the amino acid encoded by the edited codon, although it cannot create new stop codons. For example, in GluRB, an ion channel mRNA, a CAG glutamine codon is edited to CIG, which is recognized as a CGG arginine codon. All RNAs identified to undergo A to I editing are expressed in the central nervous system (e.g. glutamate-gated ion channel receptors (GluR) and serotonin (5-HT$_{2C}$) receptors). In at least one case, A to I editing has also been shown to create a new 3′ splice site by changing an AA dinucleotide to AI, which is recognized as a 3′ splice site AG by the splicing machinery.

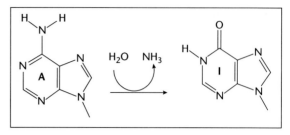

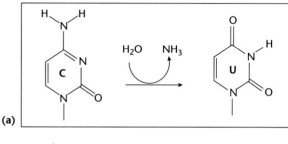

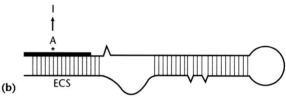

Figure 5 mRNA editing. (a) A to I (inosine) and C to U editing occur by chemically identical hydrolytic deamination reactions at the N6 of adenosine and N4 of cytosine. (b) A to I editing is specified by intramolecular secondary structure in which an intronic 'exon complementary sequence' (ECS) base pairs with the exon containing the A to be edited. The exon–ECS interaction is stabilized by additional base pairing within the intron.

A to I editing is carried out by adenosine deaminases that act on RNA (ADARs). In humans there are three members of this gene family – ADAR (formerly *ADAR1*), *ADARB1* (formerly *ADAR2*) *ADAR3* – and alternative splicing gives rise to different isoforms, which show some variation in activity. ADARs contain discrete deaminase domains and double-stranded RNA (dsRNA)-binding domains. They catalyze the hydrolytic deamination of A to I in specific pre-mRNAs as well as in long perfect double-stranded RNA duplexes, where they carry out hyperediting of up to 50% of the adenosines.

Specific pre-mRNA A to I editing relies on RNA secondary structure rather than the primary RNA sequence. Specific editing occurs within a short intramolecular RNA duplex that comprises the exon sequences flanking the edited site and a short 'exon complementary sequence' (ECS) within the downstream intron (**Figure 5b**). Point mutations that disrupt Watson–Crick base-pairing within this duplex abolish editing, while compensatory mutations that restore base-pairing reactivate editing. Additional RNA secondary structure within the intron is important for correctly aligning the ECS with the exon sequence.

C to U editing

C to U editing occurs in apoB mRNA. ApoB-100, encoded by the unedited mRNA in liver, is a major component of low-density lipoprotein (LDL) and very low-density lipoprotein (VLDL) particles. In the intestine, editing of apoB mRNA at a single cytidine residue (C^{6666}) to U changes a glutamine codon (CAA) to a stop codon (UAA). The resultant truncated protein (apoB-48) lacks the *C*-terminal LDL receptor binding domain of apoB-100 and is a component of chylomicrons.

C to U editing is catalyzed by APOBEC-1 (apoB mRNA editing enzyme catalytic polypeptide 1), a cytidine deaminase, which in humans is expressed only in the intestine. APOBEC-1 has weak RNA-binding properties and at least one auxiliary factor, ACF, which binds APOBEC-1 and ApoB mRNA, is also required. In contrast to A to I editing, the primary sequence of apoB mRNA is important for editing. The minimal RNA required for specific editing comprises a 5′ regulator element (immediately upstream of C^{6666}), a 4 nt spacer element following the edited site, and an 11 nt 'mooring sequence' immediately downstream. The mooring site is necessary for high-affinity binding by ACF, which docks APOBEC-1 to the mRNA for specific editing at C^{6666}. ApoB is the only identified substrate for APOBEC1, but it is possible that other substrates remain to be identified.

Ribosomal RNA

The four rRNAs form the core of the ribosomes, the protein synthesis machines of the cell. While they used to be thought of as a passive 'scaffold' upon which the more important ribosomal proteins assembled, it is now clear that the rRNAs themselves form the catalytic active sites of the ribosome, with the ribosomal proteins playing the supporting roles. The 45S rRNA precursor contains the sequences of mature 18S, 5.8S and 28S rRNAs (5′ to 3′) with additional internal and external transcribed spacer sequences (ITS and ETS) between the mature rRNAs and at the extreme 5′ and 3′ ends respectively (**Figure 6**). It is transcribed by RNA polymerase I in the nucleolus from multicopy rDNA genes, is heavily modified by multiple 2′-O-methylations, ψ modifications and base methylations and processed by a series of endonuclease cleavages and exonuclease trimming steps. The base and ribose modifications are concentrated within the highly conserved core regions of mature rRNA that are involved in ribosome function. All the rRNA processing occurs in the nucleolus, concurrent with association of the ribosomal proteins with the maturing rRNA. The 5S rRNA is transcribed by RNA polymerase III and undergoes very little

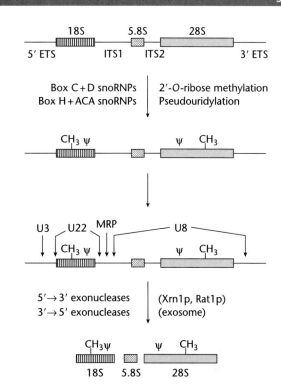

Figure 6 Pre-rRNA processing. 18S, 5.8S and 28S rRNAs are synthesized as a 45S pre-rRNA precursor with 5′ and 3′ external transcribed spacers (ETS) and internal transcribed spacers (ITS). Extensive ribose 2′-O-methylations (CH₃) and pseudouridylations (ψ) are carried out by Box C/D and Box H/ACA snoRNPs. Subsequent endonuclease cleavages are also guided by snoRNPs, including U3, 8 and 22 and RNase MRP. Finally, exonuclease trimming events produce the mature rRNAs.

processing other than removal of a short 3′ trailer sequence. (**See** rRNA Genes: Evolution.)

Modifications are specified by snoRNAs

The initial processing event is cleavage downstream of 28S rRNA by an endonuclease homologous to *E. coli* RNase III (Rnt1p in yeast). This is rapidly followed by ~110 ribose 2′-O-methylations and nearly 100 pseudouridylations. These modification reactions are specified by multiple small nucleolar RNAs (snoRNAs), which are associated with proteins in snoRNPs. Two major classes of snoRNAs are defined according to the presence of particular conserved sequence and structural elements: box C/D and box H/ACA (**Figure 7**). Both classes of snoRNA also contain sequences complementary to the specific site of modification in the rRNA. The role of the snoRNAs is to bring the appropriate modifying enzyme to specific sites for modification. (**See** snoRNPs.)

Box C/D snoRNAs specify sites of 2′-O-methylation. Each one has one or two extended regions (10–21 bases) that are perfectly complementary to regions of

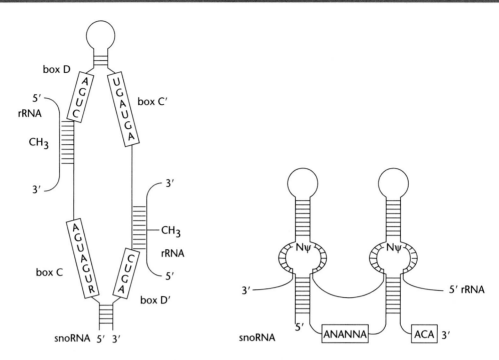

Figure 7 Pre-rRNA modifications are specified by snoRNA : pre-rRNA base-pairing.

rRNA to be methylated (**Figure 7**). Box D is positioned 5 bp from the ribose to be methylated. Methylation of the target ribose appears to be catalyzed by fibrillarin, a protein subunit of the box C/D snoRNPs.

Box H/ACA snoRNAs each specify one or two sites of ψ formation. Isomerization of U to ψ involves rotation of the base about the N3–C6 axis (**Figure 1**). Consistent with this, each snoRNA has a region that base-pairs perfectly with the rRNA, except that the U to be modified is bulged out of the intermolecular helix (**Figure 7**). The base isomerization is catalyzed by an snoRNP-associated pseudouridine synthase, Cbf5p/dyskerin.

Nuclease processing

Removal of the internal and external transcribed spacer elements (ITS and ETS) involves a series of endonuclease cleavages, many of which are followed by exonuclease 'trimming'. Some of the endonuclease events are guided by base-pairing between pre-rRNA and snoRNAs. For instance, base-pairing of U3 snoRNA with sequences in both the 5′ ETS and 18S rRNA is essential for cleavage in the 5′ ETS. A later cleavage, just upstream of 5.8S rRNA, is carried out by RNase MRP, an enzyme with one RNA subunit and nine protein subunits. In yeast, eight of the nine protein subunits are shared with RNase P, the RNP enzyme involved in pre-tRNA 5′ processing (see below).

The 5′ exonuclease trimming steps in yeast are carried out by Rat1p and Xrn1p, both of which are also involved in processing and turnover of other classes of RNA. The 3′ exonuclease steps are carried out by a remarkable complex called the exosome, the nuclear form of which contains 11 subunits, which have either been demonstrated to have 3′ exonuclease activity or which, on the basis of homology, are predicted to have this activity. In the nucleus the exosome is also involved in processing snRNAs and snoRNAs and in turnover of pre-mRNA. A cytoplasmic form of the exosome, lacking one subunit, is involved in mRNA turnover. The 3′ exonucleases of the exosome can be distinguished biochemically according to whether they catalyze a hydrolysis or phosphorolysis reaction and by the processive or distributive nature of their reactions. Interestingly, although exosome subunits are very active individually, the intact exosome complex is relatively inactive. Complex formation may keep the exosome constituent components inactive until a specific substrate is encountered, at which point the appropriate enzymes are activated. This would help to prevent indiscriminate degradation of RNA.

rRNA processing also requires a large number of RNA helicases (16 RNA helicase family members are necessary in yeast). It is likely that these enzymes help to drive particular changes in conformation of the maturing rRNAs in the assembling ribosomal subunits.

Transfer RNA

Although tRNAs are only 70–80 nt in size, they undergo more intensive processing than any other RNA (summarized in **Figure 8**). They are transcribed by RNA polymerase III as precursor molecules with 5′ leader and 3′ trailer sequences that must be removed. The essential CCA sequence is subsequently added to the 3′ end of tRNA by a nucleotidyltransferase. Pre-tRNA molecules also undergo a series of base modification steps by a variety of enzymes that specifically recognize the tRNA. Some pre-tRNAs also contain an intron that must be removed to generate mature tRNAs. The order of maturation events does not seem to be fixed and can be influenced by the concentration of the precursors. Although they are transcribed in the nucleoplasm, it appears that tRNA processing occurs in the nucleolus before the mature tRNAs are exported to the cytoplasm. (*See* tRNA.)

Base and sugar modifications

tRNAs contain numerous base and ribose 2′-*O*-methylations. Various unusual bases are produced by modification of standard bases following their incorporation into the tRNA. The modifications may be simple, such as methylation, or complex, such as rearrangement of the purine ring. Base modifications occur throughout the tRNA molecule, including the anticodon loop. Some modifications are constant features of all tRNA molecules, for example the dihydrouridine residues in the dihydrouridine arm and the pseudouridines in the TψX sequence. On the 3′ side of the anticodon, there is always a modified purine, although its identity varies. Some of the more common modifications of tRNA are shown in **Figure 1**. For example, ribothymidine, dihydrouridine, pseudouridine and 4-thiouridine are all common derivatives of U. A base modification important in position 1 of the anticodon is the hydrolytic deamination of adenosine to inosine. Inosine's ability to form wobble pairs with several bases (A, C, U) reduces the number of tRNAs needed to recognize all possible codons. This modification is carried out by enzymes called adenosine deaminases that act on tRNA (ADATs), which have homology with ADARs (see section on mRNA editing). Some purine modifications can be rather complex. For example, wyosine is a guanosine derivative with an additional ring fused with the purine ring, and this has a long carbon chain, which can have additional groups added. A large number of enzymes are involved in base modifications in tRNA,

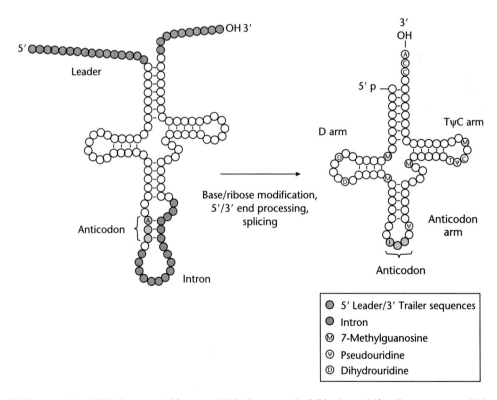

Figure 8 Pre-tRNA processing. tRNA is generated from pre-tRNA by removal of 5′ leader and 3′ trailer sequences, addition of the 3′ end CCA, splicing, ribose methylations and various base modifications, including invariant pseudouridines and dihydrouridines, and frequent A to I modification at the first position of the anticodon.

although the mechanism by which specificity is achieved is unknown. Unlike the rRNA (above) and snRNAs (below), none of the modifications appears to be specified by snoRNAs. It is possible that a single enzyme is not responsible for all modifications of a specific type (e.g. U to ψ), but that different enzymes are involved.

tRNA splicing

Many eukaryotes contain introns in their tRNAs, typified by the well-characterized yeast tRNA introns (14–60 nt in length). These introns, which are unrelated to pre-mRNA introns, are invariably found in the anticodon loop of tRNA, one nucleotide downstream of the anticodon. The secondary structure of a pre-tRNA only differs from the mature tRNA in the anticodon region, where part of the intron base pairs with the anticodon loop (**Figure 8**).

Mechanism of tRNA splicing

Unlike the introns found in pre-mRNA, tRNA introns do not have conserved sequences at the 5′ and 3′ ends to mark the intronic boundaries. Instead, the secondary structure of the tRNA is important in determining the points of intron excision. The site of cleavage at the exon–intron boundaries is determined by measuring from a reference point in the mature domain of the tRNA. In addition, a conserved intronic purine residue 3 nt from the 3′ splice site must pair with a pyrimidine in the anticodon loop 6 nt upstream of the 5′ splice site.

tRNA splicing requires three enzymes – a site-specific endonuclease, a tRNA ligase and a phosphotransferase. In the first step a highly conserved endonuclease cleaves the tRNA at both the exon–intron boundaries. This cleavage event yields the excised linear intron and two half tRNA molecules, which have 2′–3′ cyclic phosphate and 5′-OH termini. The two half tRNA molecules remain together by virtue of the tRNA structure. The tRNA ligase now joins the two half molecules in a multistep reaction. Initially, both termini undergo modification. The cyclic 2′–3′ phosphate is opened to give a 2′-phosphate and 3′-OH. The 5′-OH becomes phosphorylated. tRNA ligase then ligates the two ends together in an ATP-dependent reaction to give a 5′–3′ phosphodiester bond. The 2′-phosphate at the splice junction is subsequently removed to an NAD acceptor by a phosphotransferase enzyme.

5′ end processing

Removal of the 5′ leader sequence to generate mature tRNA is catalyzed by RNase P, an RNP enzyme related to RNase MRP (see section on rRNA processing). Nuclear RNase P in humans consists of at least 10 protein subunits associated with the H1 RNA subunit. While the catalytic center of eubacterial RNase P is contained entirely within the RNA component (indeed it was the first naturally occurring ribozyme to be discovered), the eukaryotic RNA subunit is incapable of cleaving pre-tRNA in the absence of protein.

Mechanism of 5′ end processing

The cleavage of tRNA by RNase P occurs in three consecutive steps – binding the precursor tRNA, hydrolyzing the scissile phosphodiester bond to generate a 3′-OH and a 5′ phosphate, and dissociation of mature tRNA and 5′ leader product. RNase P activity absolutely requires divalent metal ions (preferably Mg^{2+}).

All tRNAs in a cell are processed by the same RNase P, which therefore must be able to recognize features common to all tRNA species. Various pieces of evidence suggest that it is the tertiary tRNA structure that provides the determinants necessary for recognition rather than the leader sequence itself. It is likely that a ruler mechanism is used to ensure that cleavage occurs at the correct position at the 5′ end, where some features of the mature tRNA are used as a reference point. In contrast to prokaryotic RNase P, the CCA triplet at the 3′ end of the tRNA is not required for processing of the 5′ end.

3′ end processing

Trimming of the 3′ end

The 3′ trailer sequence must be removed to enable addition of the terminal CCA sequence, which is necessary for tRNA aminoacylation. The reactions involved in 3′ end processing are complex and differ from organism to organism, although they usually involve endonucleases. A detailed investigation of 3′ end processing enzymes in *Saccharomyces cerevisiae* suggests that endonucleases are exclusively used for 3′ end processing *in vivo*. The protein Lhp1p, the yeast homolog of the human La protein, which binds to all polymerase III transcripts, is essential for this reaction. This protein is thought to stabilize the tRNA structure and thus enable endonuclease cleavage. In the absence of Lhp1p 3′ end maturation occurs by a 'back-up' mechanism involving exonuclease cleavage.

Addition of the CCA at the 3′ end

The essential CCA is present at the 3′ end of all mature tRNAs. The tRNA is charged with its cognate amino acid at the 3′ end by a specific aminoacyl-tRNA synthetase and the CCA triplet is critical for transferring the amino acid to the growing polypeptide chain during protein synthesis. While it is encoded in some bacterial tRNA genes, most organisms rely on posttranscriptional addition of CCA. The enzyme responsible for catalyzing the addition of the CCA at the 3′-OH terminus of tRNA is the tRNA nucleotidyltransferase or CCA adding enzyme. The CCA adding enzyme adds nucleotides in a primer-dependent, template-independent manner to tRNA substrates lacking one, two or three of the CCA nucleotides. Specificity is achieved by RNA–protein interactions between the CCA adding enzyme and various nucleotides in the upper portion of tRNA (acceptor and TψC arms). The CCA adding enzyme has a single catalytic site that is responsible for addition of both C and A residues. During addition of the CCA triplet the tRNA nucleotidyltransferase and tRNA form a dynamic RNP structure with a single nucleotide binding pocket. The growing 3′ end of the tRNA continually refolds to release the binding pocket of the enzyme for interaction with the next nucleotide to be added, in a process known as collaborative templating.

Small nuclear RNAs

U1, 2, 4 and 5 spliceosomal snRNAs are transcribed by RNA polymerase II, while U6 is transcribed by RNA polymerase III. Before they can carry out their roles in processing of pre-mRNAs (see above), the snRNAs, like the rRNAs, are first processed and assembled into RNP particles. Processing at the 3′ end differs from that of Pol II pre-mRNA transcripts. The snRNA genes have defined transcription termination signals, and snRNAs do not need a poly(A) tail. They get cleaved shortly beyond the position of the mature 3′ end, and are subsequently processed by 3′ exonuclease trimming to the mature 3′ end by the action of the exosome (see section on rRNA processing). Like other Pol II transcripts, snRNAs cotranscriptionally acquire the 7 methyl GpppG cap structure at the 5′ end. This is subsequently further modified in the cytoplasm to a 2,2,7-trimethyl guanosine cap (**Figure 1**), which is characteristic of these snRNAs and is necessary for their reimport to the nucleus after full assembly into snRNPs. Upon reimport, the snRNPs travel to specialized structures known as the Cajal bodies where numerous 2′-O-methylation and pseudouridylation modifications occur. At least some of the modifications are carried out by a mechanism similar to the rRNA modifications. The small RNAs involved have been referred to as small Cajal body RNAs (scaRNAs). ScaRNAs with sequences complementary to U1, 2, 4 and 5 snRNAs have been identified. U85 scaRNA has both C/D and H/ACA boxes and directs both 2′-O-methylation and pseudouridylation of U5 snRNA. In the case of U2 snRNA, these modifications are known to be essential for function.

U6 (and U6$_{atac}$) is transcribed by Pol III and does not have the 7-methylguanosine cap. Instead it retains the 5′ triphosphate end from the initiating nucleotide, and the terminal γ phosphate becomes methylated. At its 3′ end, U6 snRNA is modified to have a cyclic 2′–3′ phosphate group. It has eight 2′-O-methyl groups and three pseudouridines, which are introduced in the nucleolus by box C/D and H/ACA snoRNPs.

Small nucleolar RNAs

snoRNAs and scaRNAs comprise a large class of small RNAs localized to the nucleolus and Cajal bodies respectively and are involved in rRNA and snRNA processing (see above). Some snoRNAs are synthesized by RNA polymerase II from independent promoters, and acquire the same 2,2,7-trimethyl guanosine cap as snRNAs. All vertebrate box C/D and H/ACA snoRNAs have unmodified 5′ monophosphates and are produced by posttranscriptional processing from larger transcripts. In fact, many snoRNAs are economically contained within introns of other genes. Many of these host genes encode proteins that play some role in ribosome synthesis or function, for example ribosomal proteins or translation factors. An intriguing subset of snoRNA host genes does not produce any protein products; although the genes are transcribed to produce polyadenylated mRNAs, they do not contain any long open reading frames and are short-lived. In stark contrast to normal genes, it is the nonconserved exons that can be considered 'junk' in these genes, while the introns encode functional products. Two classes of intronic snoRNAs have been identified. Some snoRNAs are synthesized by debranching of the excised intron followed by exonuclease trimming. Other snoRNAs are generated by endonucleolytic cleavage of the pre-mRNA within the intron, in a process mutually exclusive with splicing. Processing intermediates are produced that are promptly trimmed to the mature ends of the snoRNA. Following production of mature snoRNAs, snoRNP-specific complexes that protect the snoRNA termini from further exonucleolytic digestion are formed. (*See* snoRNPs.)

Other small RNAs

In addition to the major classes of RNA discussed above, there are numerous other functional RNAs in the cell, including the RNA subunits of RNase P, RNase MRP, signal recognition particle and telomerase. Many of these RNAs undergo processing events similar to those described above. Recently, a large number of 'micro-RNAs' (miRNAs) of ∼21–22 nt have been identified in many species. For example *lin-4* and *let-7* regulate developmental timing in *Caenorhabditis elegans*. They negatively regulate translation of target mRNAs by base-pairing with a complementary sequence in the 3′ untranslated region. Mature *let-7* RNA (21 nt) is produced from a precursor RNA stem–loop structure (∼77 nt) by an RNase III enzyme. Homologs of the *let-7* gene that encodes the *let-7* precursor RNA, and also the mature 21 nt *let-7* RNA have been found in a wide range of animal species including humans. The roles of *let-7* RNA and of the numerous other human miRNAs are as yet unknown.

Processing of double-stranded RNA

Although many of the important cellular RNAs are highly structured, it is rare for them to contain stretches of more than ∼10 uninterrupted base pairs. In contrast, it is not uncommon for long perfect dsRNA duplexes to be formed during viral infections, either as dsRNA replication intermediates or due to overlapping transcription using opposite strands of the DNA as templates. Cells have therefore evolved mechanisms to recognize perfect dsRNA as a foreign molecule. Two of these mechanisms involve modification or processing of the dsRNA.

Hyperediting

The same ADAR enzymes that catalyze specific pre-mRNA editing can convert up to 50% of As to Is in long dsRNA duplexes. This introduces multiple missense mutations into the RNA, and it also alters the structure since multiple A–U Watson–Crick base pairs are replaced by I·U wobble pairs. Such hyperedited RNA can either be retained in the nucleus or specifically degraded.

RNA interference

Long dsRNA can also provoke the phenomenon of RNA interference (RNAi) in which endogenous mRNAs with sequence identity to the dsRNA are degraded. This occurs by a two-step process. First, in a step mechanistically related to the formation of miRNAs, the dsRNA is processed to 21 bp duplexes by an RNase III type enzyme (known as Dicer). These small interfering RNAs (siRNAs) are then incorporated into a second ribonuclease complex where they are used as templates to identify mRNAs to be degraded. While the phenomenon of RNAi has been best characterized in *C. elegans* and *Drosophila*, at least part of the pathway operates in mammalian cells, but is sometimes obscured by other responses to dsRNA. The RNAi response can be harnessed experimentally as a very convenient method to switch genes off.

Alternative pre-mRNA processing and editing generate diversity in the proteome

Many of the RNA processing events described above are simple obligatory steps required to produce a functional RNA. However, some pre-mRNA processing events can be carried out by alternative pathways, giving rise to more than one possible mature mRNA and hence more than one protein from a single gene. For instance, RNA editing is not a required step in the expression of most genes but can allow production of functionally distinct protein isoforms from some genes. A more widespread phenomenon is alternative splicing, in which the different pairs of splice sites can be joined by the splicing machinery (**Figure 9a**). Each simple alternative splicing event allows a binary choice. The switch between production of different isoforms is commonly regulated in a cell-type specific or developmental fashion. (*See* Alternative Promoter Usage; Alternative Promoters: Duchenne Muscular Dystrophy (*DMD*) Gene; Alternative Splicing: Evolution.)

Alternative splicing can have various effects upon protein function:

- It can alter only the untranslated sequences (usually the 5′UTR) so that the actual protein produced is identical, but its translation or the stability of its mRNA may be regulated differentially.
- It can lead to the introduction of premature stop codons so that one splicing pattern leads to functional protein while the other pattern does not. In these cases, alternative splicing acts as an on/off switch for gene expression.
- Most commonly, different isoforms of a protein share common sequences (and hence functional domains), but vary only in specific regions. This can often allow the production of antagonistic protein isoforms. For instance, transcription factors can be produced as antagonistic isoforms that contain the same DNA-binding domain, but only one of which has a transcriptional activation domain. Likewise, at many levels of the apoptotic pathway, antagonistic pairs of isoforms are produced, which are either pro- or antiapoptotic. Another common outcome is that the same functional protein can be delivered to different cellular (or extracellular) locations dictated by alternatively spliced targeting signals.
- Completely different products can be produced. For instance, the INK4a gene produces two distinct inhibitors of cyclin-dependent kinases by alternative splicing of 5′ exons, leading to use of alternative reading frames across the rest of the mRNA. (*See* 3′ UTRs and Regulation; 5′ UTRs and Regulation; Nonsense-mediated mRNA Decay.)

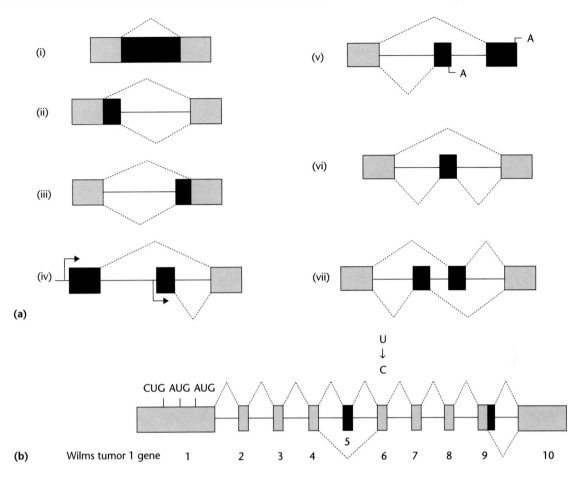

Figure 9 Generation of proteomic complexity by alternative pre-mRNA splicing and editing. (a) Types of simple alternative splicing event. In each case, constitutively spliced exon sequences are in gray and alternatively spliced sequences in black. Splicing patterns are indicated by diagonal dashed lines. (i) Retained intron; (ii) competing 5′ splice sites; (iii) competing 3′ splice sites; (iv) alternative promoters and 5′ end exons; (v) alternative 3′ end exons and poly(A) sites; (vi) cassette exon; (vii) mutually exclusive exons. (b) Combinations of these basic events allow individual genes to generate multiple isoforms. A combination of three translation start codons in exon 1, a U to C editing event in exon 6, alternative splicing of exon 5, and competing 5′ splice sites on exon 9 allows the Wilms tumor 1 gene to generate up to 24 isoforms. Misregulation of the exon 9 alternative splice is the basis of Frasier syndrome, a urogenital developmental defect.

While many genes produce only a small number of isoforms, combinations of alternative splicing events can give rise to very large numbers of isoforms. The most spectacular example is the *Dscam* (Down syndrome cell adhesion molecule) gene of *Drosophila*. This gene has 115 exons of which 95 are alternatively spliced. DSCAM is a cell-surface adhesion molecule with a number of immunoglobulin-like domains, and is involved in axon guidance in the developing brain. Two mutually exclusive exons encode the membrane spanning domain, while arrays of 12, 33 and 48 exons encode variants of Ig domains. Within each array, only one exon can be selected for any individual mRNA. However, there is no apparent coupling between exon selection from each of the arrays. The total number of DSCAM isoforms that can be generated is therefore 38 016 ($2 \times 12 \times 33 \times 48$), which exceeds the number of genes in *Drosophila* by 2–3 fold.

While alternative splicing has long been recognized as a crucial mechanism for regulating gene expression, it is only recently that its widespread role has been noted. In 1993, Phil Sharp estimated that about 5% of vertebrate genes may be alternatively spliced. Mass EST/cDNA and genome sequencing projects have subsequently revealed that its true prevalence is an order of magnitude higher. Recent genome-wide analyses of alternative splicing give conservative estimates that 30–50% of human genes are alternatively spliced. This helps to close the gap between the relative modest number of genes in the human genome and the much higher number of expressed proteins.

RNA editing is another important generator of functionally distinct protein isoforms, particularly in the nervous system. Various receptors and ion channels expressed in the brain are edited. The serotonin (5-HT_{2C}) receptors, thought to be involved in a number of psychiatric and behavioral disorders, have up to 12 isoforms generated by A to I editing. Editing occurs at five sites affecting three amino acids within an intracellular loop involved in coupling to the downstream G-protein. These isoforms are expressed in different regions of the brain, suggesting that differentially edited receptors may serve distinct biological functions in the regions in which they are expressed. AMPA receptors, which are involved in fast synaptic neurotransmission in the brain, undergo editing at the Q/R site, which results in converting a CGA (glutamine) codon to a CIG (arginine) codon. This amino acid substitution causes the Ca^{2+} permeability of the channel to be reduced. Inosine is found at levels of 1 in 17 000 bases in RNA isolated from rat brain, suggesting that on average 1 in 11 mRNAs are edited by ADARs. If each of these inosine residues correlated with a biologically relevant codon change, A to I editing would play a major role in the regulation of gene expression. However, it is possible that a significant proportion of these inosine residues may be the result of hyperediting of few transcripts rather than specific editing of many mRNAs.

The combinatorial output from some genes is further enhanced when alternative splicing and RNA editing combine with other mechanisms for generating isoform diversity. For example, a combination of alternative splicing of cassette exon 5 and competing 5' splice sites on exon 9, a U to C editing event in exon 6, and three alternative translation initiation codons in exon 1, allows the Wilms tumor 1 gene (*WT1*) to produce up to 24 protein isoforms (**Figure 9b**). The importance of alternative mRNA processing is underlined by the many genetic diseases that can arise when processing goes wrong. For example, alternative selection of the two 5' splice sites of exon 10 of *WT1* leads to optional inclusion of a KTS tripeptide (lysine–threonine–serine). Isoforms lacking the KTS tripeptide act as transcriptional regulators, while isoforms containing the KTS tripeptide function with the splicing machinery. The importance of this apparently subtle change is underlined by the fact that point mutations in the downstream 5' splice site of exon 10 resulting in a switch toward production of the KTS isoform are sufficient to cause a urogenital developmental defect known as Frasier syndrome. (*See* Alternative Splicing: Evolution; mRNA Editing; Wilms Tumor.)

See also

mRNA Editing
RNA Interference (RNAi) and MicroRNAs
snoRNPs
Splicing of pre-mRNA
tRNA

Further Reading

Bernardi G (1995) The human genome: organization and evolutionary history. *Annual Review of Genetics* **29**: 445–476.

Burge CB, Tuschl T and Sharp A (1999) Splicing of precursors to mRNAs by the spliceosome. In: Gestetland RF, Cech TR and Atkins JF (eds.) *The RNA World*, 2nd edn, pp. 525–560. New York, NY: Cold Spring Harbor Laboratory Press.

Graveley BR (2001) Alternative splicing: increasing diversity in the proteomic world. *Trends in Genetics* **17**: 100–107.

Keegan LP, Gallo A and O'Connell MA (2001) The many roles of an RNA editor. *Nature Reviews Genetics* **2**: 869–878.

Kiss T (2001) Small nucleolar RNA-guided post-transcriptional modification of cellular RNAs. *EMBO Journal* **20**: 3617–3622.

Proudfoot N (2000) Connecting transcription to messenger RNA processing. *Trends in Biochemical Science* **25**: 290–293.

Smith CWJ and Valcárcel J (2000) Alternative pre-mRNA splicing: the logic of combinatorial control. *Trends in Biochemical Science* **25**: 381–388.

Venema J and Tollervey D (1999) Ribosome synthesis in *Saccharomyces cerevisiae*. *Annual Review of Genetics* **33**: 261–311.

Wolin SL and Matera AG (1999) The trials and travels of tRNA. *Genes and Development* **13**: 1–10.

Yu Y-T, Scharl EC, Smith CM and Steitz JA (1999) The growing world of small nuclear ribonucleoproteins. In: Gestetland RF, Cech TR and Atkins JF (eds.) *The RNA World*, 2nd edn, pp. 487–524. New York, NY: Cold Spring Harbor Laboratory Press.

RNA Processing and Human Disorders

Fabrice Lejeune, *University of Rochester, Rochester, New York, USA*

Lynne E Maquat, *University of Rochester, Rochester, New York, USA*

Errors in messenger ribonucleic acid biogenesis at any one of a number of steps often result in human disease because of the failure to synthesize functional protein.

Introduction

Expression of the approximately 30 000 human genes that produce protein requires that the encoded precursor messenger ribonucleic acid (pre-mRNA), which is collinear with the gene, undergoes processing to form mRNA, which is generally not collinear with the gene. Pre-mRNA processing involves a concerted and often mechanistically interconnected series of steps that occur in the nucleus and include, but are not limited to, (i) splicing to remove intervening sequences (introns), (ii) cleavage and polyadenylation to mature the 3′ end and, occasionally, (iii) editing, which converts one nucleotide to another or, possibly, changes the number of nucleotides (**Figure 1**). After mRNA is fully processed, it is exported to the cytoplasm, where it is translated to produce protein and, ultimately, degraded. Defects in any step can result in inefficient gene expression. Defective RNAs are often degraded by one of a number of mechanisms that ensure the quality of mRNA function. This article describes a few instances where a defect in splicing, 3′ end formation or editing is associated with human disease (**Table 1**).

Pre-mRNA Splicing

Intron removal by pre-mRNA splicing is generally required for the production of an uninterrupted translational reading frame. If splicing is inefficient or inaccurate, then mRNA will encode a structurally and, often, functionally aberrant protein; for example, retention of all or part of an intron caused by a mutation that weakens a splice site often results in

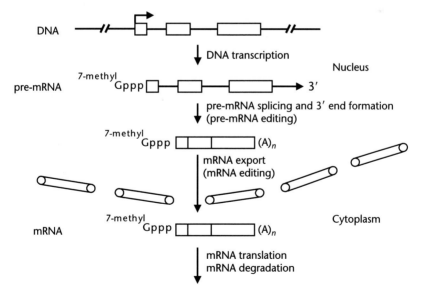

Figure 1 Overview of the steps involved in human gene expression. In the nucleus, genes are transcribed to form pre-mRNA. Pre-mRNA, which is characterized by a $^{7\text{-methyl}}$GpppG cap structure that is added cotranscriptionally, is then processed to mRNA. Processing includes (i) splicing, that is the ligation of exons (boxes) and the concomitant removal of intervening sequences or introns (lines between boxes), and (ii) 3′ end formation, that is endonucleolytic cleavage followed by the addition of poly(A) ((A)$_n$) to the upstream cleavage product. Depending on the particular transcript and type of editing, editing can occur either before or after splicing. Finally, processed mRNA is exported to the cytoplasm, where it is translated and, ultimately, degraded. The translation initiation codon usually resides within the first exon, and the translation termination codon usually resides within the last exon.

Table 1 Summary of causes of defective RNA processing and their resulting effects

Gene	Defect	Consequence
HBB	T-to-G transversion in intron 2	Creation of splice site leads to intron retention and decrease of synthesis of full-length β-globin protein (β-thalassemia)
CD45	C-to-G transversion in exon 4	Destruction of exonic splicing silencer promotes exon 4 inclusion and, possibly, the synthesis of less active protein as a result of increased dimerization (muscular dystrophy)
BRCA1	G-to-T transversion in exon 18	Destruction of exonic splicing enhancer leads to exon 18 skipping and abnormal protein sequence (some cancers)
TAU	Mutation in hairpin structure	Constitutive accessibility of 5' splice site of intron 10 leads to overexpression of the Tau protein isoform containing four microtubule-binding repeats (fronto-temporal dementia with parkinsonism)
HBA2	Mutation in polyadenylation sequence	Decrease in α1-globin gene transcription and decrease in stability of α2-globin transcripts lead to decreased levels of α1-globin and α2-globin proteins (α-thalassemia)
NF1	Upregulation of editing	Creation of nonsense codon in exon 23-1 results in a decreased level of full-length NF1 protein (von Recklinghausen neurofibromatosis)
Serotonin receptor	Misregulation of editing	Modification of serotonin receptor sequence in the second intracellular loop reduces the interaction with G-protein (suicide susceptibility and lack of response to antipsychotic drug therapies)
Protein tyrosine phosphatase	Upregulation of editing	Destruction of intron 3 branchpoint leads to retention of intron 3 in mRNA. This abnormal mRNA is found in the bone marrow of acute myeloid leukemia patients

expansion of the translational reading frame to include intronic sequences and one of three scenarios: (i) generation of an intron-derived nonsense codon in frame with the upstream reading frame; (ii) a shift in the reading frame that generates a downstream nonsense codon or (iii) no effect on the reading frame so that translation terminates at the normal site (**Figure 2**). As another example, deletion of part of an exon caused by a mutation that activates a cryptic splice site often results in deletion of the translational reading frame and one of two scenarios: (i) a shift in the reading frame that generates a downstream nonsense codon or (ii) no effect on the reading frame so that translation terminates at the normal site (**Figure 2**). An estimated one-third of inherited human diseases are attributable to aberrations in pre-mRNA splicing. The relatively high frequency of splicing aberrations has been explained by the remarkable prevalence of *cis*-residing pre-mRNA sequences that influence either constitutive or alternative splice site choice. The majority of human genes encode pre-mRNAs that harbor an average of seven to eight introns. For an estimated 30–60% of these genes, the encoded pre-mRNA is alternatively spliced, often in a cell-specific manner, to generate different mRNAs that encode different proteins. Mutations within *cis*-residing sequences that alter the use of constitutive or alternative splice sites, or create new sites, have the potential to disrupt the proper translational reading

frame and cause disease. (*See* Anxiety Disorders; Splicing of pre-mRNA.)

Constitutive as well as alternative splice site usage depends in part on splice site sequence. The 5' splice site consensus is AG/GURAGU (where / denotes the exon/intron boundary). The 3' splice site consists of three elements: the branchpoint (YNYURAC), a polypyrimidine tract and the 3' splice site itself (YAG/N). Mutation of a splice site can preclude efficient splice site recognition by the splicing machinery and, as a consequence, result in retention of all or part of the affected intron. Alternatively, mutation of either an exon or an intron can create an ectopic splice site or activate a cryptic splice site; for example, a T-to-G mutation within intron 2 of the human β-globin gene creates a 5' splice site ($GAT_{705}/GTAAGA \rightarrow GAG_{705}/GTAAGA$) and activates a cryptic 3' splice site (UCUCUUUCUUUCAG/G) upstream of the mutation, leading to retention of part of the intron, a reduced level of hemoglobin and the disease β-thalassemia (Gorman *et al.*, 2000 and references therein) (**Figure 3**). (*See* Globin Genes: Polymorphic Variants and Mutations; Hemoglobin Disorders: Gene Therapy; Splice Sites; Thalassemias.)

Both constitutive and alternative splice site usage generally also depends on one or more *cis*-acting regulatory elements that reside outside of the splice site. These regulatory elements, some of which have been shown to bind members of the serine- and

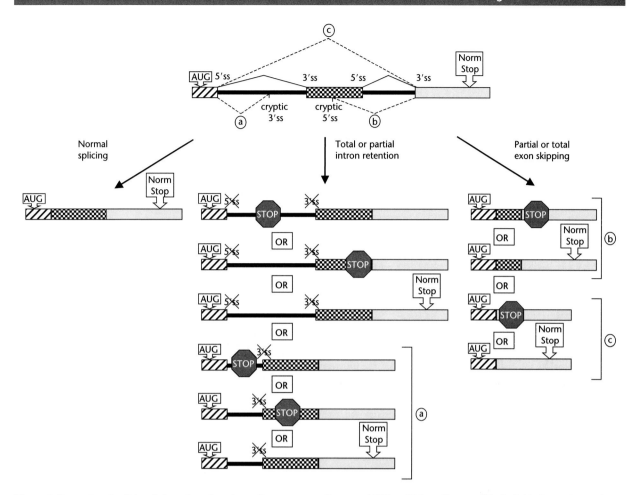

Figure 2 Examples of splicing defects that alter the coding sequence of a gene. AUG and Norm Stop specify the initiation codon and the normal termination codon, respectively, while STOP signs specify premature termination codons. Patterned boxes and thick lines represent exons and introns respectively. Thin and dashed lines connect normally used and abnormally used (cryptic) 5′ and 3′ splice sites (ss) respectively. Intron retention can result in an intron-derived premature termination codon, a shift in the reading frame that generates a downstream premature termination codon or an insertion in the reading frame. Exon skipping can result in a shift in the reading frame that generates a downstream premature termination codon or deletion of the reading frame.

arginine-rich protein family, usually consist of 6–8 nucleotides (nt) and include four types: (i) exonic splicing enhancers; (ii) intronic splicing enhancers; (iii) exonic splicing silencers and (iv) intronic splicing silencers. The existence of exonic regulatory elements indicates that exon sequences are functionally constrained, not only by requirements imposed by the encoded protein and codon usage but also by requirements imposed to ensure proper pre-mRNA splicing. Likewise, the existence of intronic regulatory elements indicates that intron sequences are functionally constrained, not only at 5′ and 3′ splice sites but also at internal sites. An analysis of sequence databases that had been performed before the prevalence of splicing regulatory elements was understood revealed that approximately 15% of point mutations associated with genetic diseases affect splicing, mostly by disrupting splice sites, creating ectopic splice sites or

activating cryptic splice sites. In reality, this percentage is probably much higher. (*See* SR Proteins.)

The *CD45/PTPRC* gene encodes a transmembrane tyrosine phosphatase that is expressed on all nucleated hematopoietic cells. A C-to-G mutation within exon 4 of the gene probably contributes to some forms of muscular dystrophy because it is more prevalent in patients than in healthy controls and cosegregates with the disease in at least four independent families. This mutation does not change the amino acid specified by the affected codon but inactivates an exonic splicing silencer (Lynch and Weiss, 2001 and references therein). *CD45* mRNA consists of 33 exons, and exon 4 is the most tightly regulated of three alternatively spliced exons. Exon 4 is included within the mRNA of naïve T cells but removed from the mRNA of activated and memory T cells. Inactivation of the exon 4 splicing silencer activates the associated 5′

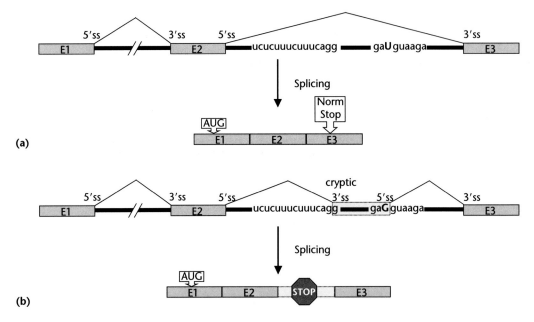

Figure 3 Intronic mutation can create a new splice site within β-globin pre-mRNA. Splicing of exon 1 (E1), exon 2 (E2) and exon 3 (E3) of a normal (a) or mutated (b) β-globin pre-mRNA that harbors a U-to-G transversion (bold capital letters) within the second of two introns (thick bold lines). In the case of the mutated pre-mRNA, the transversion is recognized as a 5′ splice site and is paired with a cryptic 3′ splice site that resides upstream. As a consequence, an intronic region (dotted box) that harbors a premature termination codon is retained in the spliced product so that it encodes abnormal β-globin protein.

splice site which, in turn, leads to exon 4 inclusion in the *CD45* mRNA of activated T cells and monocytes. Inclusion has been hypothesized to decrease CD45 protein function by increasing protein dimerization.

As another example, a G-to-T transversion that generates a nonsense codon within the constitutively spliced exon 18 of the breast cancer susceptibility gene *BRCA1* results in inappropriate exon 18 skipping by inactivating an exonic splicing enhancer (Liu *et al.*, 2001). This transversion has been found in a family with a history of breast and ovarian cancers and in five unrelated cancer cases.

Changes in pre-mRNA secondary structure can also alter splice site usage. The microtubule binding protein Tau, which promotes the assembly and stabilization of microtubule tracks, derives from a neuronal pre-mRNA that contains alternative exons 2, 3 and 10. Several intronic mutations residing downstream of exon 10 have been shown to activate exon 10 inclusion. Inclusion increases the level of Tau protein harboring the fourth microtubule binding repeat and results in frontotemporal dementia with parkinsonism linked to chromosome 17, an autosomal dominant pathology characterized by brain atrophy accompanied by neuronal cell death, gliosis and formation of intraneuronal deposits containing Tau protein (Jiang *et al.*, 2000). The intronic mutations destabilize a stem–loop around the exon 10–intron 10 junction that normally precludes recognition of the 5′ splice site and

splicing of exon 10 to exon 11. Poorly defined *trans*-acting factors have been proposed to destabilize the stem–loop structure under specific conditions that allow for exon 10 retention within mRNA. (**See** Parkinson Disease.)

Notably, aberrant splicing that generates a premature termination codon that derives from either exonic or intronic sequences usually results in degradation of the product mRNA by a process called nonsense-mediated mRNA decay. Generally, this type of mRNA decay is elicited when the premature termination codon resides more than 50–55 nt upstream of a splicing-generated exon–exon junction. Splicing-generated exon–exon junctions are generally characterized by a complex of proteins deposited 20–24 nt upstream as a consequence of the splicing reaction. It is known that one more of these proteins recruit other proteins that function in nonsense-mediated decay after translation terminates prematurely. (**See** Nonsense-mediated mRNA Decay.)

Pre-mRNA 3′ End Formation

Pre-mRNA 3′ ends are matured depending on the presence of a consensus AAUAAA polyadenylation sequence. This consensus sequence directs endonucleolytic cleavage 10–30 nt downstream and, subsequently, the addition of approximately 200 A

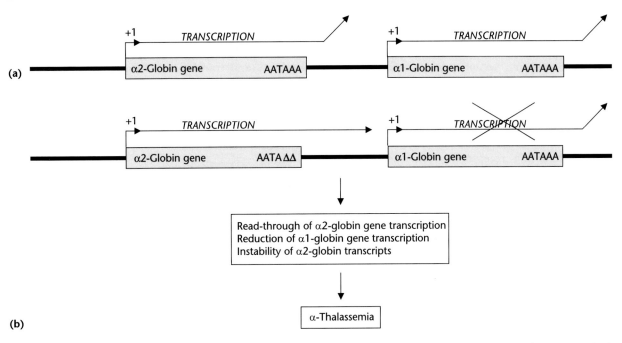

Figure 4 Deletion within the polyadenylation sequence of the α2-globin gene can lead to decreased expression of both α2- and α1-globin genes. 3′ end formation of a normal (a) or mutated α2-globin gene (b) that lacks the last two base pairs of the polyadenylation sequence (bold ΔΔ). In the case of the mutated gene, 3′ end formation is inefficient, causing transcription termination to be inefficient. As a consequence, transcription of the downstream α1-globin gene is compromised, and the α2-globin transcript is unstable.

nucleotides to the 3′ end of the upstream cleavage product. Polyadenylation not only releases pre-mRNA from its DNA template, but also increases transcript stability and translation. A deletion that removes the last two nucleotides of the consensus sequence exemplifies one of four polyadenylation sequence mutations within the human α2-globin gene. These mutations not only downregulate the α2-globin gene but also interfere with transcription of the downstream α1-globin gene, resulting in the disease α-thalassemia (Giordano *et al.*, 1999 and references therein; **Figure 4**).

Occasionally, a gene will have multiple polyadenylation sequences that alter the composition of the 3′ untranslated region of the encoded mRNA so as to create RNA diversity, if not protein diversity, from a single gene. Mutations that create or destroy an alternative polyadenylation sequence could create disease considering the potential for changes in the amounts or types of protein isoforms produced; for example, use of the second of the two polyadenylation sequences in the gene encoding amyloid precursor protein produces mRNA containing a *cis*-acting translational activator (Mbella *et al.*, 2000). Therefore, at least in theory, mutations that enhance use of the second polyadenylation sequence could promote Alzheimer disease. (*See* Alzheimer Disease.)

Pre-mRNA or mRNA Editing

Editing is a cotranscriptional or posttranscriptional process in which one or more nucleotides within pre-mRNA or mRNA are modified. The four certified types of RNA editing include: (i) A deamination to I; (ii) C deamination to U; (iii) U conversion to C and (iv) U conversion to A. Like alternative splicing and alternative 3′ end formation, editing provides a means for increasing the types of proteins produced from a single gene. Tumor suppressor genes that undergo editing include *NF1*, which encodes neurofibromin. Defective neurofibromin that no longer suppresses mitogenic signaling results in von Recklinghausen neurofibromatosis. C-to-U editing of *NF1* RNA introduces a premature termination codon that results in neurofibromin lacking the GTPase activating domain and, supposedly, lacking tumor suppressor function based on correlations between the degrees of editing and tumor progression (Cappione *et al.*, 1997). (*See* mRNA Editing; RNA Editing and Human Disorders.)

Pharmacological studies point to a role for serotonin in a number of psychiatric disorders. Serotonin receptor RNA undergoes A-to-I editing to variable extents at five sites, and editing changes the coding potential of the translational reading frame. Notably, patients who committed suicide exhibited significantly more editing at the first of these sites. Also, the efficacy of antipsychotic drugs varies with variations in the pattern of editing. Therefore, information about the status of serotonin receptor RNA editing may be critical in choosing appropriate therapies for certain psychiatric disorders (Niswender *et al.*, 2001).

RNA encoding protein tyrosine phosphatase, which modulates the growth and function of hematopoietic cells by downregulating a broad spectrum of growth-promoting receptors, is subject to double-stranded A-to-G editing (i.e. editing dependent on RNA secondary structure) mainly at the putative branchpoint sequence of intron 3 (Beghini *et al.*, 2000). Experiments using HeLa cell extracts active in splicing demonstrated that A-to-I editing produces mRNA that retains intron 3, which harbors a nonsense codon in frame with the upstream translational reading frame. The encoded protein is predicted to be nonfunctional. Intron-containing protein tyrosine phosphatase mRNA was detected at significantly higher levels in the bone marrow of acute myeloid leukemia patients relative to either normal individuals or patients in remission, indicating that editing contributes to leukemogenesis.

Another type of editing that results in +1 frameshifting as a consequence of deletions of two nucleotides has been proposed based on deposits of aberrant β amyloid precursor protein and ubiquitin-B protein in the cerebral cortex of Alzheimer and Down syndrome patients. These proteins derive from mRNAs harboring dinucleotide deletions within GAGAG or, less frequently, CUCU motifs (van Leeuwen *et al.*, 1998). Because the deletions are not evident in the corresponding genes, they are thought to reflect editing events. (*See* Down Syndrome; Down Syndrome: Antenatal Screening.)

Conclusions

Proper RNA processing is required for proper gene expression, that is, a normal level of functional protein. It follows that mutations that disrupt proper processing can cause disease by a number of mechanisms. Mutations that weaken a splice site by altering the splice site *per se*, an exonic splicing enhancer or an intronic splicing enhancer generally result in inefficient processing and, often, RNA degradation in the nucleus. Mutations that strengthen or create a splice site directly or indirectly by weakening an exonic or intronic splicing inhibitor generally result in inaccurate processing and, often, spliced RNA that harbors a premature termination codon and is subject to nonsense-mediated decay. Mutations that result in inefficient 3′ end formation downregulate the affected gene, presumably by triggering nuclear RNA degra-

dation, while mutations that alter the site of 3′ end formation can alter gene expression in ways that reflect effects of the altered 3′ untranslated region on mRNA stability and translation. Mutations that alter RNA editing can influence gene expression by changing sequences that influence RNA metabolism, the encoded protein or both.

See also
Alternative Promoter Usage
mRNA Editing
RNA Editing and Human Disorders
RNA Processing
Splicing of pre-mRNA

References

Beghini A, Ripamonti CB, Peterlongo P, *et al.* (2000) RNA hyperediting and alternative splicing of hematopoietic cell phosphatase (PTPN6) gene in acute myeloid leukemia. *Human Molecular Genetics* **9**: 2297–2304.

Cappione AJ, French BL and Skuse GR (1997) A potential role for NF1 mRNA editing in the pathogenesis of NF1 tumors. *American Journal of Human Genetics* **60**: 305–312.

Giordano PC, Harteveld CL, Bok LA, *et al.* (1999) A complex haemoglobinopathy diagnosis in a family with both beta zero- and alpha(zero/+)-thalassemia homozygosity. *European Journal of Human Genetics* **7**: 163–168.

Gorman L, Mercatante DR and Kole R (2000) Restoration of correct splicing of thalassemic β-globin pre-mRNA by modified U1 snRNAs. *Journal of Biological Chemistry* **275**: 35914–35919.

Jiang Z, Cote J, Kwon JM, Goate AM and Wu JY (2000) Aberrant splicing of tau pre-mRNA caused by intronic mutations associated with the inherited dementia frontotemporal dementia with parkinsonism linked to chromosome 17. *Molecular and Cellular Biology* **20**: 4036–4048.

van Leeuwen FW, de Kleijn DP, van den Hurk HH, *et al.* (1998) Frameshift mutants of beta amyloid precursor protein and ubiquitin-B in Alzheimer's and Down patients. *Science* **279**: 242–247.

Liu H-X, Cargegni L, Zhang MQ and Krainer AR (2001) A mechanism for exon skipping caused by nonsense or missense mutations in BRCA1 and other genes. *Nature Genetics* **27**: 55–58.

Lynch KW and Weiss A (2001) A CD45 polymorphism associated with multiple sclerosis disrupts an exonic splicing silencer. *Journal of Biological Chemistry* **276**: 24341–24347.

Mbella EG, Bertrand S, Huez G and Octave JN (2000) A GG nucleotide sequence of the 3′ untranslated region of amyloid precursor protein mRNA plays a key role in the regulation of translation and the binding of proteins. *Molecular and Cellular Biology* **20**: 4572–4579.

Niswender CM, Cerrick-Davis K, Dilley GE, *et al.* (2001) RNA editing of the human serotonin 5-ht(2c) receptor, alterations in suicide and implications for serotonergic pharmacotherapy. *Neuropsychopharmacology* **24**: 478–491.

Further Reading

Cartegni L, Chew SL and Krainer AR (2002) Listening to silence and understanding nonsense; exonic mutations that affect splicing. *Nature Reviews Genetics* **3**: 285–298.

Cooper TA and Mattox W (1997) The regulation of splice-site selection, and its role in human disease. *American Journal of Human Genetics* **61**: 259–266.

Keegan LP, Gallo A and O'Connell MA (2001) The many roles of an RNA editor. *Nature Reviews* **2**: 869–878.

Maquat LE and Carmichael GG (2001) Quality control of mRNA function. *Cell* **104**: 173–176.

Web Links

Breast cancer 1, early onset (*BRCA1*); LocusID: 672. LocusLink:
http://www.ncbi.nlm.nih.gov/LocusLink/LocRpt.cgi?l = 672

Neurofibromin 1 (*NF1*); LocusID: 4763. LocusLink:
http://www.ncbi.nlm.nih.gov/LocusLink/LocRpt.cgi?l = 4763

Protein tyrosine phosphatase, receptor type, C (*PTPRC*); LocusID: 5788. LocusLink:
http://www.ncbi.nlm.nih.gov/LocusLink/LocRpt.cgi?l = 5788

Breast cancer 1, early onset (*BRCA1*); MIM number: 113705. OMIM:
http://www.ncbi.nlm.nih.gov/htbin-post/Omim/dispmim?113705

Neurofibromin 1 (*NF1*); MIM number: 162200. OMIM:
http://www.ncbi.nlm.nih.gov/htbin-post/Omim/dispmim?162200

Protein tyrosine phosphatase, receptor type, C (*PTPRC*); MIM number: 151460. OMIM:
http://www.ncbi.nlm.nih.gov/htbin-post/Omim/dispmim?151460

RNA Secondary Structure Prediction

Ivo L Hofacker, *University of Vienna, Vienna, Austria*

Functional RNA molecules tend to fold into evolutionarily well-conserved structures. On the level of secondary structures, such folding can be predicted by a variety of algorithms from the nucleotide sequence. The predicted structures can help to identify and compare functional RNAs.

Intermediate article

Article contents

- Ribonucleic Acid Secondary Structures
- Secondary Structure Representations
- Energetics of Ribonucleic Acid Secondary Structures
- Structure Prediction by Energy Minimization
- Suboptimal Folding and Pair Probabilities
- Structure Prediction using Sequence Covariation
- Well-defined Regions and Reliability
- Available Programs and Web Services

Ribonucleic Acid Secondary Structures

Single-stranded nucleic acid molecules can form double helical regions by folding back on themselves, resulting in a pattern of helices connected by single-stranded regions called secondary structure. This secondary structure is believed to form first during folding and then act as a scaffold for the formation of the three-dimensional tertiary structure. Functional ribonucleic acid (RNA) molecules (as opposed to pure coding sequences) usually have characteristic secondary structures that are prerequisites for their function and highly conserved in evolution. While prediction of tertiary structure remains an elusive goal, secondary structure prediction has become a routine tool in the analysis of RNA function. (*See* RNA Tertiary Structure Prediction: Computational Techniques.)

For RNA, the double helical regions will consist almost exclusively of Watson–Crick C–G and A–U pairs as well as G–U wobble pairs. All other combinations of pairing nucleotides, called *noncanonical* pairs, are neglected in secondary structure prediction, although they do occur especially in tertiary structure motifs (Leontis *et al.*, 2002).

A secondary structure is thus primarily a list of base pairs. To ensure the structure is feasible, a valid secondary structure should fulfill the following constraints:

- A base may participate in at most one base pair.
- Paired bases must be separated by at least three bases.
- There are no pseudoknots, that is, there cannot be two base pairs (i, j) and (k, l) with $i < k < j < l$.

The first condition excludes tertiary structure motifs such as base triplets and G-quartets; the second takes into account the fact that the RNA backbone cannot bend too sharply.

The last condition (somewhat arbitrarily) classifies pseudoknots as tertiary structure motifs. This is done in part because most dynamic programming algorithms cannot deal with pseudoknots. However, including pseudoknots entails other complications, since most hypothetical structures that violate the third condition in the list would also be sterically impossible. Furthermore, little is known about the energetics of pseudoknots, except for some data on H-type pseudoknots (Gultyaev *et al.*, 1999). Pseudoknots should therefore be regarded as a first step toward prediction of RNA tertiary structure.

Secondary Structure Representations

Secondary structures are most commonly presented as graphs with each vertex representing a nucleotide and edges connecting consecutive nucleotides and base pairs. The result is an outer planar graph and thus can always be drawn without crossing edges. Nevertheless, secondary structure drawings of larger structures tend to get messy, and finding a good layout can be an art.

A quick overview of large structures is conveniently obtained from the *mountain representation*. In the mountain representation, a single secondary structure is represented in a two-dimensional graph, in which the *x* coordinate is the position *k* of a nucleotide in the sequence and the *y* coordinate the number *m(k)* of base pairs that enclose nucleotide *k*. The mountain representation allows for a easy visual comparison of secondary structures.

Finally, secondary structures can be compactly stored in a string consisting of dots and matching brackets. For each unpaired positions we place a dot '.' at the corresponding position of the string, and for each pair (i, j), $i < j$ an opening bracket '(' at position *i* and a closing bracket ')' at *j*. **Figure 1** shows examples of these representations.

Energetics of Ribonucleic Acid Secondary Structures

Secondary structures can be uniquely decomposed into loops, characterized by the number of unpaired nucleotides in the loop (loop size) and the number of helices emerging from the loop (degree) (**Figure 2**). Two consecutive stacked base pairs thus form a loop of degree 2 and size 0.

The standard energy model is based on this loop decomposition and assumes that the free energy of a structure can be obtained as the sum over the free energies of its constituent loops:

$$E(S) = \sum_{l \in S} E(l)$$

Because the energy contribution of a pair in the middle of a helix depends only on the following and the previous pair, such energy rules have been termed 'nearest-neighbor' rules.

The largest stabilizing energy terms are the stacked pairs, which include both hydrogen bond and stacking energies. A single stacked pair can stabilize the structure by more than $3 \, \text{kcal mol}^{-1}$. Loop energies in general consist of a size-dependent entropic term describing loss of conformational freedom and a sequence-dependent term describing mainly stacking interactions of unpaired nucleotides adjacent to pairs. Additionally, there are several empirical rules for important structure motifs such as tetraloops and coaxial stacking.

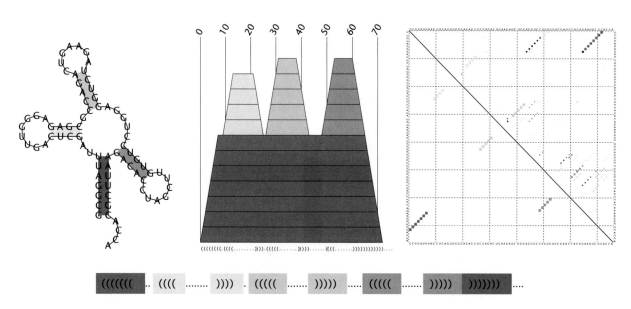

Figure 1 [*Figure is also reproduced in color section.*] Transfer ribonucleic acid clover leaf structure as a secondary structure graph, mountain plot, dot plot and in bracket notation.

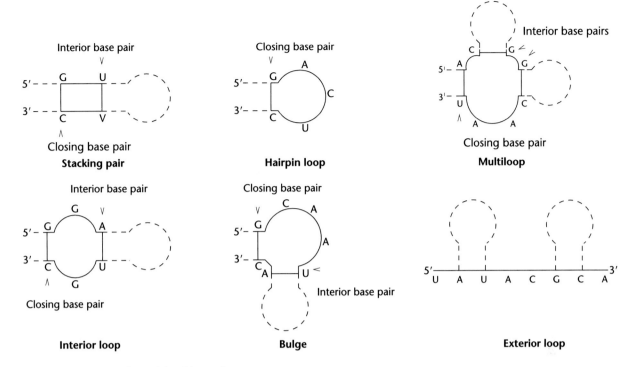

Figure 2 Loop types in ribonucleic acid secondary structures.

While the most important free energy parameters have been measured experimentally, others are still estimates. In particular, the first multiloop energies were measured only recently (Diamond *et al.*, 2001). A compilation of the current energy parameters for RNA was published by Mathews *et al.* (1999) and is available for download from the Turner Group site (see Web Links). Corresponding parameters for deoxyribonucleic acid (DNA) folding were published by SantaLucia (1998).

The energies of secondary structure formation are large compared with those of tertiary structure interactions. This is the reason why one can successfully predict secondary structures independent of tertiary structure, although there have been reports of exceptions where tertiary structure formation causes some secondary structure rearrangements (Wu and Tinoco, 1998).

Structure Prediction by Energy Minimization

Given an energy model, the simplest approach to structure prediction is to determine the optimal structure with respect to its free energy. Since the number of possible structures increases exponentially with the sequence length n, exhaustive enumeration of all structures quickly becomes unfeasible. Fortunately, the additive form of the energy model allows a more efficient approach: each base pair (i, j) divides a structure into two independent parts (inside and outside of the pair). In an optimal structure, both parts must again be optimal and can of course be further subdivided into smaller and smaller substructures. This observation was the starting point for the design of algorithms that compute the optimal structure recursively via dynamic programming (Zuker and Sankoff, 1984). Because the algorithm has to compute and store the optimal energy for each interval $[i \ldots j]$ of the sequence, memory requirements grow as the square of the sequence length, while computation time grows as its cube. (*See* Dynamic Programming.)

In the simplest case, the dynamic programming algorithm returns a single optimal solution, the minimum free energy (mfe) structure. This is unsatisfactory because inaccuracies in the energy parameters will lead to errors in the predicted structure, and also because significantly different structures may be needed to represent the molecule in thermodynamic equilibrium.

Suboptimal Folding and Pair Probabilities

The most common strategy for generating additional suboptimal structures is the algorithm of Zuker (1989), which considers for each possible base pair the best structure containing that pair. The number of structures in the output is further reduced by considering only structures within some energy interval of the mfe and filtering out structures that are too similar to others. The method usually returns a shortlist of possible foldings that form a representative sample. Occasionally, however, important alternatives will be missed.

A more rigorous approach is the computation of the partition function and base-pairing probabilities using McCaskill's algorithm (McCaskill, 1990). For every possible base pair (i, j), the algorithm yields the probability p_{ij} that the base pair will be formed, that is, the sum of the probabilities of all structures containing that pair. The partition function can also be used to calculate heat capacities and thus characterize melting transitions.

Base-pair probabilities can be nicely represented in so-called dot plots (**Figure 1**). On a two-dimensional grid indexed by i and j, we plot for each pair (i, j) a square with area p_{ij}. Similarly, Zuker's suboptimal folding algorithm can be used for energy dot plots, where instead of the probability, the best possible energy in structures containing (i, j) is plotted. For the novice, dot plots tend to be harder to interpret than a small list of alternative structures but provide an excellent overview of possible foldings.

The complete suboptimal folding algorithm of Wuchty *et al.* (1999) can generate *all* suboptimal structures in a predefined energy range above the mfe. For small molecules, it can be illuminating to look at the exhaustive list of structural possibilities. For larger molecules, the information quickly becomes overwhelming. With further postprocessing, however, these data can be used for detailed analysis of an RNA energy landscape (Flamm *et al.*, 2000).

This is important since transitions between secondary structures often face huge energy barriers. In extreme cases, an RNA molecule may not reach thermodynamic equilibrium within its lifetime. In such cases, better predictions may be achieved through explicit simulation of the folding kinetics (Gultyaev *et al.*, 1995; Morgan and Higgs, 1998), with the added advantage that pseudoknots can be included in the simulation.

Recently, Rivas and Eddy (1999) have shown that treating pseudoknots is possible in dynamic programming algorithms as well. While their algorithm is too costly to be practical, simple H-type pseudoknots can probably be handled reasonably fast.

Structure Prediction using Sequence Covariation

If several sequences are known to fold into (almost) the same structure, their common structure can be inferred from sequence covariation, typically measured as mutual information between two columns of a multiple sequence alignment. If enough sequences with a reliable alignment are available, as for ribosomal RNAs, these phylogenetic methods produce excellent predictions, including even some tertiary interactions (Gutell *et al.*, 1992).

Recently, a number of methods have appeared that combine thermodynamic prediction with covariation analysis (Hofacker and Stadler, 1999; Juan and Wilson, 1999; Lück *et al.*, 1999). These methods achieve accurate predictions with only a few related sequences and can also be used to detect conserved functional structure motifs.

Prediction accuracy

Structures determined by covariation methods are also used as a yardstick to measure the accuracy of single-sequence predictions. On a test set containing some 43 000 base pairs, the latest energy parameters predicted about 70% of pairs correctly (Mathews *et al.*, 1999). Even in unfortunate cases, where a predicted mfe structure may have fewer than 30% correct pairs, good structures are found in the vicinity of the mfe structure by suboptimal folding.

Often a small number of constraints can improve prediction accuracy dramatically. Most folding programs allow the specification of constraints, such as specifying positions as (un)paired. Such constraints can be obtained without too much effort from chemical probing experiments.

Well-defined Regions and Reliability

Pair-probability end-energy dot plots can also give a good visual impression of the quality of prediction and well-defined regions. A dot plot cluttered by many alternatives may indicate structural flexibility but also makes the prediction less reliable. Well-defined structures are likely to be correctly predicted, since they will be robust with respect to small variations of the energy parameters.

Several quantitative measures of well-definedness are being used. In the simplest case, the well-definedness of the prediction can be quantified by the difference between the mfe and the free energy of the best suboptimal structure. A more robust measure is the difference between the mfe and the ensemble free energy $G = -RT \ln(Q)$, where Q is the partition function. The latter is equivalent to the probability of the mfe structure in the ensemble, given by Boltzmann's law $p(\text{mfe}) = \exp\ (-E_{\min}/RT)/Q = \exp [-(E_{\min} - G)/RT]$.

Even more useful are positionwise measures that help to identify credible parts of the prediction. One can, for example, compute from the pair probabilities the positional entropy:

$$S_k = -\sum_i p_{ik} \ln p_{ik}$$

where p_{ii} is defined as the probability that i does not pair $p_{ii} = 1 - \sum_{j \neq i} p_{ij}$. A useful application of such measures is to the annotation of structure drawings (Zuker and Jacobson, 1998). (*See* Information Theories in Molecular Biology and Genomics.)

Available Programs and Web Services

Users who need to do structure predictions only occasionally will find excellent web services (see Web Links), such as Michael Zuker's mfold server. The mfold program for Unix machines is available from the same site, free for academic use, an older version of `mfold` is included in the commercial GCG package. David Mathews' `RNAstructure` program is a re-implementation of `mfold` for PCs running Windows and is freely available in binary form.

The Vienna RNA Package includes software for secondary structure prediction and analysis, including calculation of partition functions and pair probabilities, as well as complete suboptimal folding, prediction of consensus structures and sequence design. The software is meant to be easily extensible for interested programmers and is freely available as C code; a fold server can be found at the same web address.

See also
RNA Tertiary Structure Prediction: Computational Techniques

References

Diamond JM, Turner DH and Mathews DH (2001) Thermodynamics of three-way multibranch loops in RNA. *Biochemistry* **40**: 6971–6981.

Flamm C, Fontana W, Hofacker IL and Schuster P (2000) RNA folding at elementary step resolution. *RNA* **6**: 325–338.

Gultyaev AP, van Batenburg FHD and Pleij CWA (1995) The computer simulation of RNA folding pathways using a genetic algorithm. *Journal of Molecular Biology* **250**: 37–51.

Gultyaev AP, van Batenburg FHD and Pleij CWA (1999) An approximation of loop free energy values of RNA H-pseudoknots. *RNA* **5**: 609–617.

Gutell RR, Power A, Hertz GZ, Putz EJ and Stromo GD (1992) Identifying constraints on the higher-order structure of RNA: continued development and application of comparative sequence analysis methods. *Nucleic Acids Research* **20**(21): 5785–5795.

Hofacker IL and Stadler PF (1999) Automatic detection of conserved base pairing patterns in RNA virus genomes. *Computers & Chemistry* **23**: 401–414.

Juan V and Wilson C (1999) RNA secondary structure prediction based on free energy and phylogenetic analysis. *Journal of Molecular Biology* **289**(4): 935–947.

Lück R, Graf S and Steger G (1999) Construct: a tool for thermodynamic controlled prediction of conserved secondary structure. *Nucleic Acids Research* **27**: 4208–4217.

Leontis NB, Stombaugh J and Westhof E (2002) The non-Watson-Crick base pairs and their associated isostericity matrices. *Nucleic Acids Research* **30**(16): 3497–3531.

Mathews D, Sabina J, Zucker M and Turner H (1999) Expanded sequence dependence of thermodynamic parameters provides robust prediction of RNA secondary structure. *Journal of Molecular Biology* **288**: 911–940.

McCaskill JS (1990) The equilibrium partition function and base pair binding probabilities for RNA secondary structure. *Biopolymers* **29**: 1105–1119.

Morgan SR and Higgs PG (1998) Barrier heights between groundstates in a model of RNA secondary structure. *Journal of Physics A: Mathematical and General* **31**: 3153–3170.

Rivas E and Eddy SR (1999) A dynamic programming algorithm for RNA structure prediction including pseudoknots. *Journal of Molecular Biology* **285**: 2053–2068.

SantaLucia Jr J (1998) A unified view of polymer, dumbbell, and oligonucleotide DNA nearest-neighbor thermodynamics. *Proceedings of the National Academy of Sciences of the United States of America* **95**: 1460–1465.

Wu M and Tinoco I (1998) RNA folding causes secondary structure rearrangement. *Proceedings of the National Academy of Sciences of the United States of America* **95**: 11 555–11 560.

Wuchty S, Fontana W, Hofacker IL and Schuster P (1999) Complete suboptimal folding of RNA and the stability of secondary structures. *Biopolymers* **49**: 145–165.

Zuker M (1989) On finding all suboptimal foldings of an RNA molecule. *Science* **244**: 48–52.

Zuker M and Jacobson AB (1998) Using reliability information to annotate RNA secondary structures. *RNA* **4**: 669–679.

Zuker M and Sankoff D (1984) RNA secondary structures and their prediction. *Bulletin of Mathematical Biology* **46**(4): 591–621.

Further Reading

Flamm C, Hofacker IL and Stadler PF (1999) RNA *in silico*: the computational biology of RNA secondary structures. *Advances in Complex Systems* **2**: 65–90.

Higgs PG (2000) RNA secondary structure: physical and computational aspects. *Quarterly Review of Biophysics* **33**: 199–253.

Schuster P, Stadler PF and Renner A (1997) RNA structures and folding: from conventional to new issues in structure predictions. *Current Opinion in Structural Biology* **7**: 229–235.

Schuster P, Fontana W, Stadler PF and Hofacker IL (1994) From sequences to shapes and back: a case study in RNA secondary structures. *Proceedings of the Royal Society of London, Series B: Biological Sciences* **255**: 279–284.

Tinoco Jr I and Bustamante C (1999) How RNA folds. *Journal of Molecular Biology* **293**: 271–281.

Zuker M (2000) Calculating nucleic acid secondary structure. *Current Opinion in Structural Biology* **10**: 303–310.

Web Links

Mathews' RNA structure program
    http://rna.chem.rochester.edu/RNAstructure.html
Turner Group. The Turner Group home page offers the RNAstructure program for download, as well as a compilation of current RNA energy parameters.
    http://rna.chem.rochester.edu/index.html

Vienna RNA package. The Vienna RNA page offers documentation and source codes for the Vienna RNA software package as well as several web fold servers.
    http://www.tbi.univie.ac.at/RNA/
Zuker's mfold server. The Zuker Group site hosts the popular mfold server and contains links to much additional information on RNA structure prediction and thermodynamics
    http://bioinfo.rpi.edu/~zukerm/rna/

RNA Tertiary Structure Prediction: Computational Techniques

François Major, *Université de Montréal, Montréal, Canada*

Patrick Gendron, *Université de Montréal, Montréal, Canada*

Ribonucleic acid (RNA) tertiary structure prediction is an activity that consists of inferring complete sets of atomic coordinates of RNAs, in Euclidean space, on the basis of observation, knowledge or construct. The principal goal of prediction is to obtain precise RNA tertiary structures and, thus, to reduce the costs of the discovery process by offering to scientists new possibilities to design efficient pinpoint experiments to decipher RNA function.

> **Advanced article**
>
> **Article contents**
> - Introduction
> - Hydrogen Bonds and Secondary Structure Elements
> - Computational Techniques
> - Applications

Introduction

Ribonucleic acid (RNA) molecules are flexible polymers that fold into a large diversity of three-dimensional (3D) shapes. The tertiary or 3D structure of an RNA is a set of atomic coordinates or positions $(x\ y\ z)$ in Euclidean space, that is one position for each atom composing the RNA (see **Figure 1a**). Tertiary structures can be determined from high-resolution data obtained from physical experimental methods such as X-ray crystallography and nuclear magnetic resonance (NMR). Structure prediction is an alternative approach, which consists of inferring tertiary structures from primary and secondary structures and lower-resolution data that can be derived computationally from primary and secondary structures or obtained experimentally (cf. chemical and enzymatic probing, mutagenesis, chemical modifications, etc.).

Structure prediction is a two-step iterative activity involving observation and 3D interpretation. Observation consists of elaborating and performing experiments, and collecting or computing tertiary structural data from sequence and secondary structure. Interpretation consists of building tertiary structures from observation. The iterative aspect comes from the fact that one of the goals of building tertiary structures is to help in the design of concise experiments whose results are usually fed back into 3D construction. Tertiary structure prediction can reveal aspects of RNA structure and function that, sometimes, cannot be captured using experimental determination methods. Therefore, tertiary structure prediction is applied to RNAs of biological interest, whether experimental tertiary structures are available or not.

Hydrogen Bonds and Secondary Structure Elements

Nucleotides contain hydrogen donor and acceptor groups, whose juxtaposition in 3D space results in the formation of hydrogen bonds and hydrogen-bonding patterns. The most frequent types are the Watson–Crick patterns forming the A–U and G–C base pairs, which involves N–H...N and N–H...O hydrogen bonds, and the wobble hydrogen-bonding pattern between G and U bases. Two hydrogen bonds can bifurcate between one acceptor, for instance C=O, and

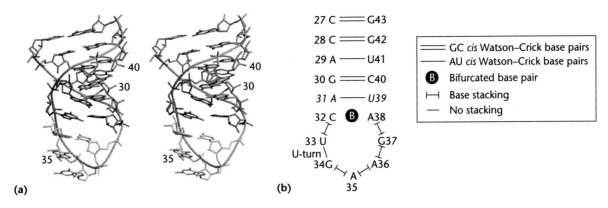

Figure 1 [*Figure is also reproduced in color section.*] Yeast tRNA^{Phe} anticodon stem–loop. Cylinders are drawn between covalently bonded atoms. Hydrogen atoms are not shown. (a) 3D X-ray crystal structure. The stem is shown in blue. The U-turn motif is shown in yellow. (b) Secondary structure using a newly proposed annotation by Leontis and Westhof. Regular font indicates C3′_endo puckers, italic font indicates C2′_exo pucker and all nucleotides have antiglycosyl torsions.

two donor atoms on its partner, for instance two N–H groups. The secondary structure of an RNA is a set of ordered base pairs (i, j) over its sequence S, such that $1 \bullet i < j \bullet n$. Given two pairs (i, j) and (i', j'), then either:

$i < j < i' < j'$ (base pair (i, j) precedes base pair (i', j'))

$i < i' < j' < j$ (base pair (i, j) includes base pair (i', j'))

$i < i' < j < j'$ (the base pairs form a pseudo-knot)

A consecutive stretch of one or more Watson–Crick base pairs is called a stem, and forms, in the tertiary structure, a regular double-helical conformation, similar to the double helix of double-stranded DNA. A single-stranded region is also called a loop. When a loop connects the terminal base pair of a stem it is called a hairpin loop, whereas when it connects two or more distinct stems it is called a multibranched loop. A stem–loop is composed of a stem and its hairpin loop. A stem contains a bulge if one or several nucleotides in one of its strands interrupt it. In the tertiary structure, single bulged nucleotides can either swing-out or stack-in the stem helix. Internal loops are two-branched loops that interrupt stems, and are composed of unpaired nucleotides in both strands. **Figure 2** illustrates the above elements.

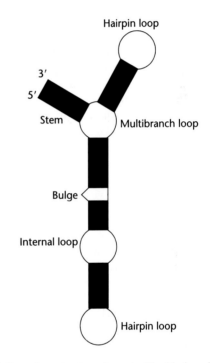

Figure 2 Secondary structure elements. The black regions correspond to double-helical stems.

Computational Techniques

Computational techniques to predict the formation of stems, composed of the isosteric G–C and A–U Watson–Crick, and G∘U wobble base pairs, are referred to as secondary structure prediction techniques, an important step in tertiary structure prediction. Secondary structure prediction can be made by the use of combinatorial or dynamic programming algorithms that generate the most likely sets of stems according to their contribution to free energy.

The combinatorial algorithm first computes the exhaustive list of all possible stems, and then generates secondary structures based on subsets of this list. Each stem is assigned a score, and various algorithms to choose the optimal subset of stems can be applied. Because of the large number of possible subsets, virtually only probabilistic approaches work in practice. An alternative is to request a user to select the stems.

Zuker *et al.* (1999) implemented the dynamic programming algorithm in their computer program *mfold*, which does not consider pseudo-knots. Rivas and her colleagues recently implemented a dynamic programming algorithm that considers pseudo-knots in her computer program *PKNOTS* (Rivas and Eddy, 1999).

Despite great efforts to improve secondary structure prediction from a single sequence, comparative sequence analysis is the most productive method in practice. For instance, using comparative sequence analysis, Gutell predicted most base pairs that were observed in the X-ray crystal structure of the 23S rRNA subunit. Computational techniques that can be used to align multiple RNA sequences (and thus to predict secondary structure) include those based on stochastic context-free grammars from Haussler's group, and similarly by Eddy and Durbin, and genetic algorithms, as implemented in *RNAGA* from Maizel's group.

When the secondary structure has been determined, and enough structural data have been accumulated, 3D interpretation can take place. Computational techniques for projecting structural data in 3D are usually defined by three components: a data structure, a conformational search space and an inference engine.

Structural graph

The structural graph is a general data structure that can be used to store all types of RNA structural data. It has been implemented as an internal data structure by Major and co-workers in the *MC-Sym* RNA tertiary structure building computer program (Major *et al.*, 1991; Gendron *et al.*, 2001).

A structural graph, $G = (V, E)$, is a graph composed of a set of vertices, V, representing the nucleotides, and a set of edges, E, representing the internucleotide spatial relations. The representation employed in **Figure 1b** contains the information of the structural graph of the yeast tRNA^{Phe} anticodon stem–loop (Leontis and Westhof, 1998). When available, structural data about the exposure to the solvent of specific chemical groups, nucleotide conformations, as well as more precise geometric data, such as intranucleotide atomic distances, angles and torsions, can also be introduced in the vertices of structural graphs. The tertiary structure of RNAs is highly constrained by base pairing, but also base stacking, which involves dipole–dipole and dipole-induced dipole interactions (London dispersion), and hydrophobic forces, which from a thermodynamic viewpoint are weaker interactions than hydrogen bonds (Saenger, 1984).

Conformational search space

An RNA conformational search space (CSS) is a set of tertiary structures that we are interested in exploring. The CSS is defined by a series of parameters. The product of the numbers of allowed values per parameter defines the *size* of the CSS. As an example, consider the CSS defined by the seven torsion angles of nucleotides, where each angle is assigned one of 360 degrees. The size of such a CSS for an RNA of N nucleotides is thus 360^{7N}. In tertiary structure, the flexibility of nucleotide conformations is mostly conferred by seven torsion angles: six in the backbone (α, β, γ, δ, ε and ζ) and one around the glycosyl bond linking the ribose sugar and nitrogen base (χ) (Saenger, 1984).

Operators that modify the value of each parameter relate the tertiary structures of a CSS. In the above example, the operator that modifies the ζ torsion relates in the CSS two structures that differ by their values of the ζ torsion. In searching a CSS, we have a particular *goal* in mind, which describes what is searched for. Here, the goal corresponds to a subset of tertiary structures that satisfy the constraints described in the structural graph. Most search methods proceed by systematically applying a set of operators and verifying whether the resulting tertiary structures are elements of the goal. When it is impossible to perform an exhaustive search over the CSS, probabilistic methods, such as Monte Carlo or simulated annealing, can be applied.

A metric over a CSS allows us to compute some measure of the value of a given structure. The best example of such a metric is the model potential energy function defined for molecular mechanics and dynamics simulations. This metric can direct the search based on the assumption that applying the operators to a structure estimated to be closer to the goal will lead to it more rapidly than applying operators to more distant structures. For instance, the model potential energy function serves as an evaluation function, which assigns a value to each structure according to its potential energy. In practice, however, partly due to the local minima problem, the model potential energy function is rather employed to refine tertiary structures of the goal. Search methods guided by metrics are called heuristics, and are based on algorithms supposed to perform well in practice, although they provide no guarantees they will find the goal.

Inference engine

An inference engine searches the CSS for the goal. Computational techniques for predicting RNA tertiary structure differ in what information from the structural graph they use and how they use it, how they define the CSS, and what algorithm they employ to find the goal. An example is the backtracking algorithm implemented in the *MC-Sym* computer program, which generates tertiary structures consistent with most types of RNA structural data. The *MC-Sym* engine backtracks on systematic assignments of rigid base conformations, base pairs and base stacking examples, as extracted for instance from the tertiary structures deposited in databases such as the Protein Data Bank (see Web Links). Resulting structures can be refined, and made thermodynamically sound, with the use of the *AMBER* or *CHARMM* packages and force fields.

Applications

Large RNAs

Westhof and co-workers have developed 3D computer interactive modeling tools for building large RNAs. In collaboration with Michel, they built a well-received tertiary structure of a group I intron using comparative sequence analysis. Massire developed an interactive tool, called *MANIP*, based on those developed by Westhof, and used it to build the tertiary structure of the RNase P RNA using cross-linking data, as shown in **Figure 3** (Massire *et al.*, 1998).

The use of cross-linking agents produces structural data that are less precise than covariation analysis, but allows one to approximate the distance between specific nucleotides, and to confirm experimentally predicted base pairs. Harvey's group used similar cross-linking data to derive a different tertiary structure of the RNase P RNA (Malhotra *et al.*, 1994; Haas *et al.*, 1994). The method Harvey and his colleagues developed is based on molecular mechanics. They progressively use models from one to five points per nucleotide that are submitted to molecular mechanics using a special potential energy function composed of constraint terms. Their program was called *YAMMP* for 'yet another molecular mechanics potentials'. The approximate tertiary structures generated by *YAMMP* can be converted in all-atom structures using *MC-Sym*, as in the prediction of the 16S rRNA decoding region tertiary structure by Wollenzien and co-workers.

Case and Macke have developed a system in which several tertiary structure computational tools are embedded in a computer language called *NAB*. The language allows one to combine rigid body transformations and distance geometry to create candidate tertiary structures, which can then be refined using *AMBER*. *NAB* includes *YAMMP*, the generalized Born model for solvation effects, and the *AVS* visualization system.

Cross-linking data can also be used in conjunction with the program *ERNA-3D*, developed by Mueller and Brimacombe, to derive tertiary structure of large RNAs. *ERNA-3D* allows one to interactively move and orient double helices. Using *ERNA-3D*, Mueller and Brimacombe derived a new model for the tertiary structure of the *E. coli* 16S rRNA using a 3D electron microscopic map at 20 Å resolution.

Leadzyme

The tertiary structure prediction of small internal loops, as the one found in the lead-activated ribozyme (**Figure 4**), can be made by applying a protocol based on the intersection of conformational space, developed in Major's laboratory (Major *et al.*, 1998). First, a series of hypotheses concerning the base pairing interactions among the nitrogen bases of the internal loop were made, and, for each one, tertiary structures of the wild-type leadzyme were generated using *MC-Sym*. Variants obtained by chemical modifications and *in vitro* selection were threaded into each hypothetical tertiary structure. Fuzzy logic was applied to select the most plausible structures, which were further analyzed. Structural features from the best candidate tertiary structures were challenged by more concise chemical modifications, the results of which were fed back into tertiary structure prediction. Although the computer program to generate tertiary structures was *MC-Sym*, in fact any tertiary structure generator could have been used in this protocol.

Hairpin ribozyme

Burke and colleagues generated tertiary structures of the hairpin ribozyme. A tertiary structure predicted by Westhof and co-workers was used in conjunction with recent structural data obtained from Burke's laboratory. Westhof's prediction was reversed-engineered to an *MC-Sym* input script that became the starting point for Burke and co-workers' tertiary prediction. The input script was modified to incorporate the new structural data, and produced structures showing a base triple at the center of the cleavage site. Experimental investigations suggested by this discovery demonstrated the presence of the triple, and allowed the research team to define the tertiary structure of the catalytic core (see **Figure 5**). Refined tertiary structures of the catalytic core led to the development of a model of the catalytic activity of the hairpin ribozyme (Pinard *et al.*, 2001). The X-ray crystal structure shows similar features to the prediction, but represents a different catalytic state.

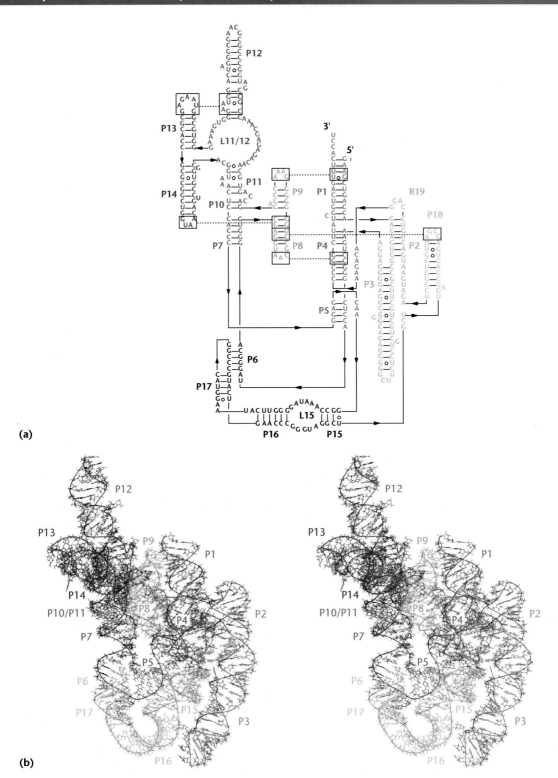

Figure 3 [*Figure is also reproduced in color section.*] RNase P RNA from Haas and co-workers. (a) Secondary structure. Straight lines indicate *cis* Watson–Crick base pairs. Dots represent GU wobble base pairs. Arrows indicate tertiary base pairs revealed by cross-linking data. (b) Stereoview of the tertiary structure.

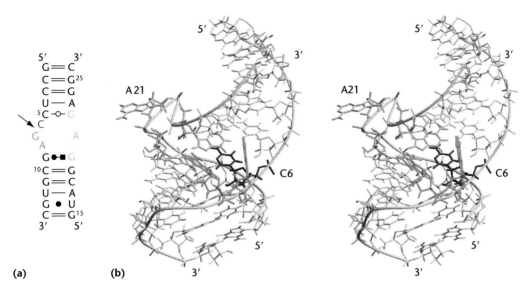

Figure 4 [*Figure is also reproduced in color section.*] Lead-activated ribozyme. (a) Secondary structure. The symbols defined in **Figure 1b** were used. The empty circle indicates a *trans* Watson–Crick base pair. The filled circle followed by the filled box indicates a *cis* base pair with Watson–Crick and Hoogsteen interacting edges. The single filled circle indicates a GU wobble base pair. The arrow indicates the cleavage site. (b) Stereoview of the tertiary structure. The stems flanking the internal loop are shown in gray. The nucleotides in the internal loop are colored: C6 in red, G7 and A8 in green, and G20, A21 and G22 in yellow.

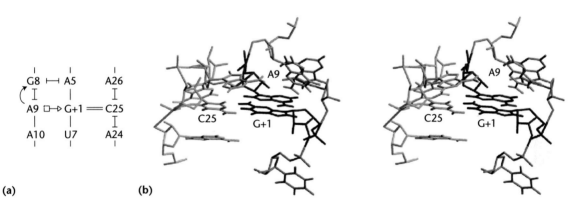

Figure 5 [*Figure is also reproduced in color section.*] Catalytic core of the hairpin ribozyme. (a) Secondary structure. The symbols defined in **Figure 1b** were used. The empty square followed by the arrow indicates a *trans* base pair with Hoogsteen and sugar interacting edges. The curved arrow indicates a direction change in the backbone. (b) Stereoview of the tertiary structure. A8–A10 are shown in blue, A24–A26 are shown in green and B5–B7 (substrate) are shown in red. G+1 (substrate), A9 and C25 form a base triple.

See also
RNA Secondary Structure Prediction

References

Gendron P, Lemieux and Major F (2001) Quantitative analysis of nucleic acid three-dimensional structures. *Journal of Molecular Biology* **308**(5): 919–936.

Haas ES, Brown JW, Pitulle C and Pace NR (1994) Further perspective on the catalytic core and secondary structure of ribonuclease P RNA. *Proceedings of the National Academy of Sciences of the United States of America* **91**: 2527–2531.

Leontis NB and Westhof E (1998) Conserved geometrical base-pairing patterns in RNA. *Quarterly Reviews of Biophysics* **31**: 399.

Major F, Lemieux S and Ftouhi M (1998) Computer RNA three-dimensional modeling from low-resolution data and multiple-sequence information. In: Leontis NB and Santa Lucia J (eds.) *Molecular Modeling of Nucleic Acids*, pp. 394–404. Washington, DC: American Chemical Society Books.

Major F, Turcotte M, Gautheret D, *et al.* (1991) The combination of symbolic and numerical computation for three-dimensional modeling of RNA. *Science* **253**(5025): 1255–1260.

Malhotra A, Tan RK and Harvey SC (1994) Modeling large RNAs and ribonucleoprotein particles using molecular mechanics techniques. *Biophysical Journal* **66**: 1777.

Massire C, Jaeger L and Westhof E (1998) Derivation of the three-dimensional architecture of bacterial ribonuclease P RNAs from comparative sequence analysis. *Journal of Molecular Biology* **279**(4): 773–793.

Pinard R, Hampel K, Heckman JE, *et al.* (2001) Functional involvement of G8 in the hairpin ribozyme cleavage mechanism. *EMBO Journal* 20(22): 6434–6442.

Rivas E and Eddy SR (1999) A dynamic programming algorithm for RNA structure prediction including pseudoknots. *Journal of Molecular Biology* 285: 2053–2068.

Saenger W (1984) *Principles of Nucleic Acid Structure.* New York, NY: Springer-Verlag.

Zuker M, Mathews DH and Turner DH (1999) Algorithms and thermodynamics for RNA secondary structure prediction: a practical guide. In: Barciszewski and Clark BFC (eds.) *RNA Biochemistry and Biotechnology.* NATO ASI Series. Dorchrecht: Kluwer Academic Publishers.

Further Reading

Cedergren RJ and Major F (1998) Modeling the tertiary structure of RNA. In: Simons RW and Grunberg-Manago M (eds.) *RNA Structure and Function.* New York, NY: Cold Spring Harbor Laboratory Press.

Lemieux S, Chartrand P, Cedergen R and Major F (1998) Modeling active RNA structures using the intersection of conformational space: application to the lead-activated ribozyme. *RNA* 4: 739–749.

Major F, Turcotte M, Gautheret D, *et al.* (1991) The combination of symbolic and numerical computation for three-dimensional modeling of RNA. *Science* 253: 1255.

Michel F and Costa M (1998) Inferring RNA structure by phylogenetic and genetic analyses. In: Simons RW and Grunberg-Manago M (eds.) *RNA Structure and Function.* New York, NY: Cold Spring Harbor Laboratory Press.

Michel F, Costa M, Massire C and Westhof E (2000) Modelling RNA tertiary structure from patterns of sequence variation. *Methods in Enzymology* 317: 491.

Mueller F and Brimacombe R (1997) A new model for the three-dimensional folding of *E. coli* 16S ribosomal RNA: I. Fitting the RNA to a 3D electron microscopic map at 20 angstroms. *Journal of Molecular Biology* 271: 524.

Web Links

Protein Data Bank. A database that contains three-dimensional coordinates of proteins, DNA and RNA structures, as determined mainly by experimental, but also theoretical methods http://www.rcsb.org/

RNA Viral Vectors in Gene Therapy

Stephen J Russell, *Mayo Clinic, Rochester, Minnesota, USA*

Genes can be delivered into mammalian cells using synthetic nonviral vectors or recombinant viral vectors derived from DNA virsuses, retroviruses and, more recently, RNA viruses.

Intermediate article

Article contents
• Introduction
• Picornavirus Vectors (*Poliovirus*)
• Rhabdovirus Vectors (Vesicular Stomatitis Viruses)
• Paramyxovirus Vectors
• Influenza Virus
• Alphavirus Vectors

Introduction

For the field of gene therapy, the ribonucleic acid (RNA) viruses should be viewed as a vast and largely untapped resource. During the 1980s and 1990s, the spotlights for vector development efforts fell predominantly on deoxyribonucleic acid (DNA) viruses (e.g. adenoviruses, herpesviruses, adenoassociated viruses), and on retroviruses, which are converted to DNA inside the cell. However, throughout this time, technologies for the generation of recombinant RNA viruses carrying foreign gene segments have been well established, and RNA vectors with a variety of novel and potentially exploitable features are rapidly emerging on the gene therapy scene.

The RNA viruses comprise an extraordinarily diverse collection of distinct virus families and genera, including many well-established or evolving human and animal pathogens. The viruses themselves may be small or large, enveloped or nonenveloped, and their genomes may be positive- or negative-sense, single- or double-stranded, segmented or nonsegmented. Available systems for the rescue of plus-strand RNA viruses are relatively simple because they can be generated from messenger RNA (mRNA). Rescue systems for negative-strand RNA viruses are more complex because the minimal infectious unit is the ribonucleoprotein–polymerase complex. Replication-competent RNA virus vectors have potential utility both as oncolytic viruses mediating selective tumor destruction and for vaccination against infectious agents or tumor antigens.

Because of space limitations, this article discusses emerging RNA virus vectors from only five virus families (Picornaviridae, Rhabdoviridae, Paramyxo-viridae, Orthomyxoviridae and the alphaviruses (Toga-viridae)). These particular vectors have been selected on the basis that they have been developed to a relatively advanced level. However, it should be borne in mind that virtually any virus can be used for gene transfer.

Picornavirus Vectors (*Poliovirus*)

Introduction

Poliovirus is the best known and most thoroughly characterized member of the Picornaviridae family, one of the largest and most important families of human and agricultural pathogens, comprising five genera: rhinoviruses (e.g. common cold), enteroviruses (e.g. *Poliovirus*), aphthoviruses (e.g. foot and mouth disease), cardioviruses and hepatoviruses (e.g. hepatitis A). They are very small (30 nm), spherical, nonenveloped viruses containing a single positive-strand RNA genome.

Virus biology

Genome

The single-stranded positive-sense RNA genome contains approximately 7500 nucleotides and is covalently linked at its 5′ end to the VPg protein and is polyadenylated at its 3′ end. The 5′ noncoding region contains an internal ribosome entry site (IRES) that precedes a single, long, open reading frame encoding a large precursor polypeptide that is processed to form about 12 individual viral proteins. The polyprotein is divided into three regions encoding the viral capsid proteins (P1) or the proteins involved in protein processing ($2A^{pro}$, $3C^{pro}$, $3CD^{pro}$) and genome replication ($2B$, $2C$, $3AB$, $3D^{VPg}$, $3CD^{pro}$ and $3D^{pro}$).

Life cycle

Binding to the CD155 cellular *Poliovirus* receptor triggers structural changes in the capsid that lead to uncoating and intracellular delivery of the viral RNA. The 5′ terminal protein is removed from the RNA which is then translated, producing a polyprotein that is nascently and posttranslationally cleaved to release the individual viral proteins. The viral RNA polymerase copies the plus strand to generate full-length minus strands, which in turn act as templates for progeny plus-strand RNAs, which are either translated or packaged into newly formed virions. Virus release occurs by cell lysis. One important aspect of *Poliovirus* replication is that 2 h after viral infection, polyribosomes are disrupted and translation of host messenger RNAs is shut off.

Use for gene transfer

The live attenuated Sabin vaccine strain of *Poliovirus* has been used extensively for human vaccination and induces long-lasting protective immunity after oral administration. Recombinant polioviruses, expressing foreign antigens, may therefore provide a convenient vaccine vector to engender mucosal immunity. *Poliovirus* was the first animal RNA virus to be rescued from an infectious DNA clone. However, the recombinant viruses are most easily rescued by transfecting HeLa cells with *in vitro* transcribed RNA from recombinant DNA clones.

Several strategies have been used for engineering *Poliovirus* vectors. Small antigenic epitopes can be inserted into one of the capsid proteins, foreign sequences can be inserted at various positions into the viral polyprotein flanked by *Poliovirus* protease recognition sites, bicistronic *Poliovirus* vectors can be constructed by inserting the foreign transgene as a second IRES-driven expression cassette, and *Poliovirus* minireplicons can be constructed by replacing the capsid coding sequences with a foreign sequence.

More recently, the IRES of a pathogenic *Poliovirus* was replaced by the IRES from human rhinovirus type 2 to generate a recombinant virus that was shown to be potently oncolytic for glioma xenografts, but greatly attenuated for neurovirulence.

Longevity/magnitude of gene expression

Through the combined mechanisms of shut-off of cellular protein synthesis, inhibition of transport of cellular glycoproteins and the proteolytic digestion of transcription factors, *Poliovirus* vectors are potently cytotoxic. The expression profile from these vectors is therefore very short-lived, but very high level.

Safety features

Poliomyelitis is the most serious human disease caused by any of the picornaviruses, but the full-blown paralytic disease occurs in less than 1% of infected individuals. Moreover, *Poliovirus* vectors are typically derived from the live attenuated Sabin vaccine strain of the virus. Routine childhood vaccination against polioviruses has resulted in a very high prevalence of seropositivity to the virus that needs to be considered when contemplating the use of *Poliovirus* vectors in human subjects. Intracellular homologous recombination between enterovirus genomes can readily occur, leading to the generation of interspecies hybrids that may have novel host range properties.

Clinical applications

Poliovirus vectors, expressing foreign antigens, may provide a convenient vaccine vector to engender mucosal immunity against important pathogens such as human immunodeficiency virus (HIV). Given the favorable safety profile and other characteristics of the Sabin vaccine strain, many investigators have attempted to adapt the virus for the expression of various foreign viral antigens. However, clinical testing of recombinant *Poliovirus* vaccines has not yet been undertaken. Also, potential applications are limited by the low capacity of these vectors for foreign sequences and by vector instability, leading to rapid deletion of foreign sequences. In addition to their promise as recombinant vaccines, oncolytic polioviruses are likely to prove beneficial to the treatment of glioma. Clinical trials are awaited.

Rhabdovirus Vectors (Vesicular Stomatitis Viruses)

Introduction

Vesicular stomatitis viruses (VSV) belong to the genus Vesiculovirus of the family Rhabdoviridae, the simplest of the nonsegmented, negative-strand RNA viruses. These viruses are widely distributed in nature, and infect not only mammals but also the insects from which they are transmitted. The virions are characteristically bullet-shaped (180×75 nm) particles comprising a helically packed ribonucleoprotein core and an outer envelope that carries 1200 copies of the viral glycoprotein G, forming about 400 trimeric spikes. There are approximately 50 copies of the viral polymerase in each particle.

Virus biology

Genome

The 11.2 kilobase (kb) negative-sense RNA genome of VSV is neither capped nor polyadenylated and contains five genes designated N-P-M-G-L ($3'$ to $5'$). Each mRNA begins with the sequence Cap–AACAG and ends with the sequence UAUG–poly(A), and there are only two intergenic nucleotides between adjacent genes. The $3'$ leader and $5'$ trailer sequences contain all signals required for genome replication and encapsidation.

Life cycle

After binding to its (relatively ubiquitous) receptor, which has been identified as phosphatidylserine,

endocytosis of the virus leads to a low pH-triggered conformational rearrangement of the surface glycoprotein (G) that leads to membrane fusion and intracellular release of the nucleoprotein core. The viral genes are next transcribed by the viral polymerase phosphoprotein complex (L–P3) with 20–30% attenuation of transcription at each gene junction, leading to the accumulation of newly synthesized N, P, M, G and L proteins in decreasing abundance.

Genome replication begins only after sufficient nucleocapsid (N) and phosphoprotein (P) has accumulated. Initially, the negative strand is copied to produce full-length positive-strand antigenomes, which are then (more efficiently) copied to generate a much greater abundance of progeny negative-strand genomes. Progeny genomes are encapsidated during their synthesis into ribonucleoprotein cores, which then condense with the matrix (M) protein at the plasma membrane. The M protein interacts with the cytoplasmic tail of the membrane-anchored G protein and drives the process of budding and release of bullet-shaped virus progeny from the cell surface.

Use for gene transfer

The rescue of negative-strand RNA viruses from cloned DNA is complicated because the transcription and replication of their genomes occurs only in the context of a correctly assembled ribonucleoprotein complex comprising the negative-sense genome and the viral N, P and L proteins. Recovery of VSV and other nonsegmented negative-strand RNA viruses is therefore achieved by transfecting the following components into a rescue cell line to achieve cytoplasmic transcription of the viral genome in the presence of N, P and L proteins:

- A plasmid (or virus) coding for the bacteriophage T7 RNA polymerase.
- Plasmids coding for the N, P and L proteins under T7 promoter control, and preceded by an internal ribosome entry site to facilitate cap-independent translation.
- A plasmid encoding the full-length antigenome under T7 promoter control followed by a delta T ribozyme to ensure correct genome termination.

Infectious VSV is released into the supernatants of cells (typically BHK cells) transfected with these components. Foreign genes, flanked by the appropriate transcriptional start and stop signals, can be inserted into the VSV genome upstream or downstream of each of the viral genes. The foreign genes participate in the transcriptional gradient of the viral genome, attenuating transcription of downstream genes, and being attenuated by upstream genes.

However, VSV recombinants expressing foreign genes are able to replicate almost as well as wild-type VSV. Replication-defective VSV recombinants, lacking the G protein coding sequences, have been successfully rescued on cells that provide the VSV G protein in *trans*.

Longevity and magnitude of gene expression

Host protein synthesis is suppressed in VSV-infected cells, and virally expressed proteins can represent as much as 5% of total cell protein 7 h after infection. However, expression is short-lived because the virus is cytotoxic, largely due to the expression of M protein, which depolymerizes actin, tubulin and vimentin, leading to cell rounding and detachment within a few hours of infection.

Safety features

VSV infection of cattle, horses and swine is associated with significant but nonfatal disease, manifesting as vesicular lesions around the mouth, hooves and teeth (easily confused with foot and mouth disease). Human infection is rare, and usually asymptomatic, although febrile illness, including a single case of encephalitis in a 3-year-old boy has been reported. Replication-defective VSV-derived vaccine vectors, lacking the G protein, were completely attenuated for pathogenesis in a mouse model, and may therefore be more attractive candidates for human vaccination.

Although subject to point mutations, recombinant VSV genomes remain tightly encapsidated in ribonucleoprotein complexes during their replication in the cytoplasm and do not therefore readily undergo homologous or nonhomologous recombination.

Because of the low prevalence of antibodies to VSV in the human population and their ability to infect mucosal surfaces, VSVs expressing foreign viral glycoproteins may be effective vaccine vectors. Mucosal immunization with recombinant VSVs expressing influenza hemagglutinin has been shown to noninvasively stimulate mucosal and systemic immunity. The vectors also induce high-titer neutralizing antibodies against the VSV G protein. VSVs expressing G proteins of different serotypes are therefore required when repeat dosing is necessary, for example to boost the response to a viral antigen.

VSVs are exquisitely sensitive to the antiviral activity of interferons, and therefore replicate more efficiently in abnormal cells (e.g. cancer cells) that are defective in their antiproliferative and antiviral responses to alpha and beta interferon.

Clinical applications

VSV-derived vaccine vectors encoding foreign antigens are being developed for mucosal vaccination against HIV. In an intriguing development of this vector system, a noncytolytic rhabdoviral vector, expressing a *Rabies virus* neutralizing antibody, was recently reported. Such a vector might be a useful alternative to passive immunization for postexposure prophylaxis of rabies. The VSVG protein was inserted into the vector in place of rabies G to allow promiscuous infection of mammalian cells. Finally, replication-competent VSV has potential as an oncolytic agent for the treatment of a variety of human malignancies that are defective in their antiviral responses to interferon. However, to mid-2002 VSV vectors had not been tested in the clinic.

Paramyxovirus Vectors

Introduction

The Paramyxoviridae are enveloped, negative-stranded RNA viruses that replicate entirely in the cytoplasm. The defining characteristic of the Paramyxoviridae is the presence of an F protein that mediates virus cell fusion at neutral pH. Many well-known pathogens of animals and humans belong to this family of viruses (e.g. *Measles virus*, *Mumps virus*, parainfluenza, respiratory syncytial virus, *Newcastle disease virus*, *Canine distemper virus*, *Rinderpest virus*), but several of these diseases have been controlled through the judicious and widespread use of live attenuated vaccines. Reverse genetic systems are available for many different paramyxoviruses, but the most promising systems currently available for gene transfer applications are those based on *Sendai virus* and *Measles virus*.

Virus biology

Genome

Viruses of the family Paramyxoviridae contain nonsegmented, single-stranded, negative-sense RNA genomes ranging in size from 15 to 19 kb with 6–10 genes. The N protein is tightly associated with the viral nucleic acid (one copy every six nucleotides) to form a long, helical nucleocapsid, which in turn associates with the viral P and L proteins forming a functional polymerase complex. Polymerase-mediated genome replication and gene transcription are regulated by well-characterized nucleic acid signals at either end of the viral genome and at the intergenic boundaries.

Life cycle

Attachment via the H (e.g. *Measles virus*) or HN (e.g. *Sendai virus*) attachment protein triggers a conformational rearrangement of the F protein leading to fusion peptide exposure, membrane fusion and cytoplasmic release of the viral ribonucleoprotein (vRNP). Proteolytic activation of the F protein precursor, most commonly mediated by the Golgi protease furin, is a prerequisite for proper fusion triggering and function. Paramyxoviruses with furin-resistant monobasic F protein cleavage signals are not infectious until activated by extracellular trypsin-like proteases and therefore have greatly reduced virulence.

Viral mRNAs are transcribed from the negative-sense genome. Transcripts are polyadenylated by reiterative copying of a poly(U) signal at the end of each gene. Most of the polymerase complexes then cross the three-nucleotide intergenic boundary to initiate transcription of the next gene. Because this process is not 100% efficient, there is a gradient of decreasing transcriptional efficiency across the viral genome, the N gene being expressed most abundantly and the L (polymerase) gene least abundantly.

In addition to viral mRNA synthesis, the polymerase complex also mediates genome replication. After sufficient N protein has accumulated in the cytoplasm, transcripts initiating at the 3′ end of the viral genome are cotranscriptionally encapsidated, and the polymerase ignores all poly(A) signals to generate a full-length antigenome that provides the template for the synthesis of progeny-negative-strand RNP complexes which associate with the P–L polymerase complex prior to their assembly into progeny viruses.

Virus assembly occurs at internal membranes and at the plasma membrane, coordinated by M, the matrix protein. M drives virus assembly and budding through its combined interaction with the cytoplasmic tails of newly synthesized viral membrane glycoproteins and with the progeny nucleocapsids.

Use for gene transfer

As for other nonsegmented negative-strand RNA viruses, rescue of paramyxoviruses from cloned DNA is achieved by providing the N, P and L proteins in *trans* (from expression plasmids) to encapsidate and replicate the full-length antigenome which is expressed from a T7 promoter and terminates at a T7 transcriptional terminator downstream of a hepatitis delta ribozyme which accurately trims the 3′ end of the antigenome. The bacteriophage T7 RNA polymerase is provided in *trans* by *Vacciniavirus* T7 or by an additional expression plasmid.

Provided they are flanked by the appropriate transcriptional regulatory signals, foreign genes can be inserted upstream or downstream of any one of the paramyxovirus genes. Because of the helical nature of the nucleocapsid and the pleomorphic, nonicosahedral envelope, recombinant paramyxoviruses can easily accept additional genes without compromising virus assembly. For example, up to 5 kb of foreign sequence has been inserted into the genome of a replication-competent recombinant *Measles virus*.

As of mid-2002, replication-defective paramyxovirus vectors devoid of viral coding sequences had not been described. However, several groups are working toward the generation of high-capacity, helper-dependent paramyxovirus vectors.

Longevity/magnitude of gene expression

The level of expression of a foreign gene inserted into a paramyxovirus genome is governed by the transcriptional gradient across the genome, and is therefore determined by the precise site of insertion. Genes inserted upstream of N, the first viral gene, are expressed at extremely high levels and typically do not impair virus replication because they do not distort the transcriptional gradient across the remainder of the viral genome.

The longevity of gene expression from replication-competent paramyxovirus vectors is restricted both by the viral cytopathic effect and by the host immune response to virus-infected cells. Many paramyxoviruses fuse cultured cells into multinucleated syncytia, which eventually die, often by apoptosis. *In vivo*, infected cells may also be eliminated by the host immune response before the virus has caused significant cellular damage. Gene expression usually peaks within 48–72 h and is undetectable 2 weeks after transduction.

Safety features

As mentioned previously, the family Paramyxoviridae includes a number of devastating human and animal pathogens. However, for several of these pathogens, live attenuated forms of the viruses have been widely and safely applied for human and/or veterinary vaccination. Vaccine strain paramyxoviruses therefore provide a convenient platform from which to develop replication-competent paramyxovirus vectors that can be safely applied *in vivo*.

As a general rule, recombinant paramyxoviruses are very stable during serial tissue culture passage. Their remarkable genetic stability is largely a consequence of the mode of replication whereby progeny genomes are tightly bound in ribonucleoprotein (RNP) complexes and are therefore not available for homologous or nonhomologous recombination with cellular RNA. Also, since replication of the RNA genome is entirely cytoplasmic, paramyxoviruses have no capacity for integration or damage to cellular DNA.

Preexisting humoral or cell-mediated immunity to *Measles virus* and *Mumps virus* is widespread as a consequence of global vaccination programs. Preexisting immunity is expected to curtail expression of measles and mumps vectors, but also provide a safety firewall to prevent accidental spread of the vector or a recombinant in the human population. For *Sendai virus* and other nonhuman paramyxovirus vectors, expression may be prolonged, but the risk of uncontrolled spread in the human population is higher.

Clinical use

Paramyxovirus vectors have not been clinically tested. However, the potential for clinical benefit is considerable. Live attenuated measles and mumps vaccines are routinely used in childhood vaccination programs throughout the world, and are therefore ideal platforms from which to express additional viral antigens (e.g. from HIV or hepatitis C) to expand the protective effect of existing vaccine programs.

In addition, certain attenuated paramyxoviruses (e.g. *Newcastle disease virus* and *Measles virus*) have been shown to be potently oncolytic against human tumors, and may therefore have potential as novel anticancer therapeutics. To demonstrate the feasibility of *Measles virus* retargeting, growth factors and single-chain antibodies have been displayed on the surface of recombinant measles viruses by fusing them to the *C*-terminus of the H glycoprotein. Additional modifications to the viral genome are being explored to further enhance the performance of attenuated measles viruses as oncolytic agents.

Because of the broad and varied tropisms of the different paramyxoviruses, the potential applications for high-capacity paramyxovirus vectors, devoid of viral genes, will be very far-reaching. However, in mid-2002 the production of such vectors was still awaited.

Influenza Virus

Introduction

The Orthomyxoviridae are roughly spherical enveloped viruses with a negative-sense single-stranded RNA genome that is divided into multiple segments. Best known for their epidemic spread in human populations, influenza viruses can also infect pigs, horses, seals and a wide variety of birds. Indeed, aquatic birds (e.g. ducks) are considered to be a most important reservoir from which new human influenza viruses periodically emerge. Uniquely among the RNA viruses, orthomyxovirus genomes are transcribed and replicated in the nucleus of the infected cell.

Virus biology

Genome

The 13 500 nucleotide *Influenza A virus* genome is divided into eight segments ranging from 890 to 2341 nucleotides, each one coding for one or two viral proteins. Each segment is encapsidated by the nuclear protein NP and is associated with the three virally encoded proteins, PB2, PB1 and PA, that comprise the polymerase complex to form an RNP structure. Segments one to three code for the transcriptase complex (PB2, PB1, PA); segment five codes for the nuclear capsid protein NP; and segments four and six code for the hemagglutinin (HA) and neuraminidase (NA) protein. Transcripts from segment seven and eight are partially spliced and thereby code for two proteins each. Segment seven codes for the matrix protein M1 and for the M2 ion channel protein, an important component of the virus coat. Segment eight codes for the nonstructural protein NS1 and for a second protein named NS2.

Life cycle

HA-mediated binding to sialic acid is followed by endocytosis, endosomal acidification and low pH triggered rearrangement of the HA trimer. This leads to insertion of the previously buried HA fusion peptide in the target cell membrane, and subsequent membrane fusion. Also triggered by low endosomal pH, the RNP core dissociates from the M1 matrix protein and is actively transported into the nucleus where viral mRNA synthesis begins.

Viral mRNA synthesis is initiated by capped RNA primers which the viral polymerase cleaves from cellular mRNAs (cap snatching) and is terminated by reiterative copying of a stretch of uridines close to the 5′ end of the viral RNA (vRNA).

Viral genome replication commences only after sufficient free NP protein has accumulated in the nucleus, whereupon full-length vRNA is copied by the viral transcriptase complex to form a positive-strand complementary RNA (cRNA), which provides the template for progeny vRNA synthesis. In contrast to viral mRNAs, all vRNAs and cRNAs are encapsidated in RNP complexes. The matrix (M1) and NS2 proteins associate with vRNPs and mediate their export from the nucleus to the plasma membrane where they associate with, and bud from, membrane rafts containing the viral membrane proteins HA, NA and N2. The mechanism that ensures incorporation of at least one copy of each of the eight viral genomic segments has not yet been elucidated.

Use for gene transfer

The rescue of infectious influenza virus entirely from cloned DNA was first reported by two independent research groups in 1999. The eight vRNAs are transcribed from eight plasmids, each one coding for a single vRNA. In the most popular system, vRNA transcription is controlled by pol I promoter and terminator signals, thereby ensuring that the 5' and 3' termini are neither capped nor polyadenylated. Additional pol II promoter-driven expression plasmids, coding for the viral polymerase complex and nucleoprotein, are cotransfected to ensure correct vRNA encapsidation, transcription and replication. Expression plasmids coding for the remaining viral proteins are cotransfected to enhance the efficiency of virus rescue.

In a modification of this system, the RNA pol I driven constructs are flanked by a pol II promoter and poly(A) signal such that both vRNA and mRNA are generated from the same template with the consequence that the viruses can be generated from only eight plasmids.

Foreign genes can be expressed from recombinant influenza viruses in different ways. For example, peptide coding sequences can be inserted into surface loops of the HA protein, and full-length protein can be linked to influenza virus genes and expressed from an inserted internal ribosome entry site or cleaved from an influenza protein at an inserted protease site. Foreign genes can also be expressed from inserted RNA segments by generating influenza viruses with nine or more segments. Replication defective influenza viruses can be generated by expressing all of the viral structural proteins from a set of nine expression plasmids along with an influenza-like vRNA coding for the gene of interest. Titers in the region of 10^4 per ml have been achieved using this system, and further increases in the efficiency are anticipated.

Longevity/magnitude of gene expression

Influenza-mediated gene expression is very high level but relatively short lived because the virus is potently cytotoxic. Influenza virus cytotoxicity is the consequence of several viral activities. For example, the virus cannibalizes the cap structures from cellular mRNAs, inhibits cellular mRNA polyadenylation and splicing, inhibits the cellular interferon response, and efficiently shuts off host/cell protein synthesis within a few hours of infection.

Safety features

Influenza viruses cause a highly contagious acute respiratory illness. Epidemics are frequent and can be devastating if a new viral strain with increased virulence emerges; for example, the influenza pandemic of 1918–1919 killed between 20 and 40 million people. Evolution of new pandemic influenza virus strains is thought to occur through the reassortment of RNA segments between different strains of influenza virus, primarily in the aquatic bird population.

Needless to say, the strains of influenza virus used for gene transfer are safe and nonpathogenic, but there is currently no safeguard to prevent the engineered gene segments contained in these vectors from reassorting into pathogenic influenza viruses circulating in the human population.

Humoral immunity to influenza virus is specific to the HA and NA of the infecting strain, and is relatively short lived. Also, there are 15 distinct HA subtypes and nine distinct NA subtypes from which influenza-based vectors can be constructed to circumvent preexisting immunity. However, certain pathogenic NA/HA combinations should be avoided.

Clinical applications

Whilst engineered influenza viruses have not been tested in the clinic, live attenuated influenza viruses are currently (2002) being tested in clinical vaccine trials for protection against influenza. Replication-competent influenza virus vectors, expressing foreign viral proteins or tumor antigens, may therefore provide a useful platform for the development of live attenuated virus vaccines. Also, because of their ability to abortively infect dendritic cells, replication-competent influenza virus vectors provide a highly efficient gene delivery system to generate human dendritic cells expressing tumor-associated antigens for stimulation of antitumor immunity.

Recombinant influenza viruses may also have potential as oncolytic agents for the treatment of human malignancy. Influenza viruses lacking the NS1 protein are readily inhibited by PKR, a cellular enzyme that is activated by virally induced interferon. NS1-deleted influenza viruses are therefore nonpathogenic because they are readily inhibited by interferon. PKR activity is suppressed in certain tumor cells having an activated Ras pathway, and these tumors therefore remain susceptible to destruction by the otherwise harmless virus.

Alphavirus Vectors

Introduction

The 27 known alphaviruses belong to the Togaviridae family. They are simple enveloped viruses comprising a plus-strand RNA genome in an icosahedral capsid surrounded by a lipid envelope containing 80 trimeric glycoprotein spikes.

Virus biology

Genome

The 12 kb positive-sense alphavirus genome is capped and polyadenylated and is organized into two regions: the 5' two-thirds encodes the four nonstructural proteins (nsP1–4) and the 3' third codes for the viral structural proteins (capsid and envelope glycoproteins E1 and E2).

Life cycle

Virus attachment is mediated through the E2 glycoprotein and leads to endocytosis via coated pits, which become acidified endosomes in which fusion is triggered through the dissociation of E1 from E2 followed by the trimerization of E1. After membrane fusion, nucleocapsid is released into the cytoplasm and uncoated. Nonstructural proteins are then translated from the genomic RNA, which also serves as a template for minus-strand synthesis. The minus strand is the template for new genomic RNA and for the 26S subgenomic RNA that codes for the viral structural proteins. Both nonstructural and structural proteins are cleaved from larger polyprotein precursors. NsP4 is the viral polymerase whose activity is regulated by nsP1, 2 and 3.

Assembly and budding of alphaviruses from vertebrate cells occurs at the plasma membrane. Efficient encapsidation of genomic RNA is mediated by a packaging signal in the nsP1 coding region of the genomic RNA which is recognized by the capsid protein during nucleocapsid assembly.

Use for gene transfer

Alphaviruses can be rescued from cloned DNA by *in vitro* transcription and subsequent RNA transfection of mammalian cells. Replication-competent alphavirus vectors have been generated by insertion of heterologous amino acid sequences into the envelope glycoproteins such that they are displayed in the surface of the virus. Alternatively, foreign genes can be expressed from a second subgenomic RNA promoter inserted close to the 5' end of the viral genome.

Defective alphavirus vectors (derived e.g. from *Sindbis virus* and *Semliki Forest virus*) are generated by replacing the viral structural genes with foreign transgenes. The vector genome retains the viral nonstructural genes and *cis*-acting regulatory elements and is therefore self-replicating (a replicon). Packaging of vector genomes is achieved by providing the viral structural proteins in *trans*, either transiently by cotransfection of a defective helper genome or stably by means of a packaging cell line.

Longevity/magnitude of gene expression

Alphavirus vectors characteristically give rise to very high level but transient gene expression. The vectors encode the viral ns proteins which are potently cytotoxic (and immunogenic). However, certain mutations in the ns proteins have recently been shown to abrogate cytotoxicity, advancing the prospect of noncytotoxic alphavirus vectors more suitable for longer-term gene expression.

Safety features

Alphavirus replicons, administered as RNA transcribed *in vitro* from DNA copies of vector genomes, are safe because they lack viral structural genes and do not persist. This contrasts with alphavirus packaging systems in which homologous recombination, leading to the generation of replication-competent helper virus contamination, is inevitable.

Alphaviruses are transmitted by mosquitoes and may lead to encephalitis or polyarthritis, but most often do not cause identifiable disease. *Sindbis virus* and *Semliki Forest virus* disseminate rapidly in newborn but not adult mice, leading to fatal encephalitis. In humans, *Sindbis virus* causes a mild polyarthritis and *Semliki Forest virus* has rarely been associated with encephalitis.

Clinical applications

Alphavirus vectors have not yet (2002) been tested in the clinic. *Semliki Forest virus* replicons encoding heterologous viral antigens have shown promise in preclinical vaccination studies when administered as nucleic acid by intramuscular injection and should be safe for human application. Recombinant alphavirus particles may be of value for vaccination, for cytoreductive gene therapy or for oncolytic virotherapy, but there are significant outstanding safety and manufacturing issues that need to be addressed before clinical trials are conducted.

Acknowledgements

Dr Stephen J. Russell is supported by the Siebens Foundation, Eisenberg Foundation, Multiple Myeloma Research Foundation, and NIH grants CA83181-03 and 1PO1 HL66958-01A1.

See also

Gene Therapy: Technology

Further Reading

Evans DJ (1999) Reverse genetics of picornaviruses. *Advances in Virus Research* **53**: 209–228.

Griffin D (2001) Alphaviruses. In: Knipe D and Howley P (eds.) *Field's Virology*, pp. 917–962. Philadelphia, PA: Lippincott, Williams & Wilkins.

Lamb R and Kolakofsky D (2001) Paramyxoviridae: the viruses and their replication. In: Knipe D and Howley P (eds.) *Field's Virology*, pp. 1305–1340. Philadelphia, PA: Lippincott, Williams & Wilkins.

Lamb R and Krug R (2001) Orthomyxoviridae: the viruses and their replication. In: Knipe D and Howley P (eds.) *Field's Virology*, pp. 1487–1531. Philadelphia, PA: Lippincott, Williams & Wilkins.

Nagai Y (1999) Paramyxovirus replication and pathogenesis. Reverse genetics transforms understanding. *Reviews in Medical Virology* **9**: 83–99.

Neumann G and Kawaoka Y (2001) Reverse genetics of influenza virus. *Virology* **287**: 243–250.

Racaniello V (2001) Picornaviridae: the viruses and their replication. In: Knipe D and Howley P (eds.) *Field's Virology*, pp. 685–722. Philadelphia, PA: Lippincott, Williams & Wilkins.

Roberts A, Buonocore L, Price R, Forman J and Rose JK (1999) Attenuated vesicular stomatitis viruses as vaccine vectors. *Journal of Virology* **73**: 3723–3732.

Rose J and Whitt M (2001) Rhabdoviridae: the viruses and their replication. In: Knipe D and Howley P (eds.) *Field's Virology*, pp. 1221–1244. Philadelphia, PA: Lippincott, Williams & Wilkins.

Schlesinger S (2000) Alphavirus expression vectors. *Advances in Virus Research* **55**: 565–577.

Robotics and Automation in Molecular Genetics

Trevor Hawkins, US DOE Joint Genome Institute, Walnut Creek, California, USA

Chris Elkin, US DOE Joint Genome Institute, Walnut Creek, California, USA

Martin Pollard, US DOE Joint Genome Institute, Walnut Creek, California, USA

Robotic platforms and automation can be applied to laboratory molecular biology protocols to achieve high-throughput sample processing for DNA sequencing, genomics, proteomics and other biological applications.

Introductory article

Article contents
• Introduction
• Sample Processing and Preparation
• DNA Sequencing and Detection Devices
• Conclusion

Introduction

Automation has been the backbone of the Human Genome Project and the foundation of the biotechnology and genomics industry. In our quest to turn laboratory methods into industrial-scale processes, we have turned largely to robotic solutions in tandem with new biochemical approaches. There were many changes made during the last few years of the twentieth century in the types of automation used, as new methods were developed and the scale of the processes increased. It is also the case that the availability of automation has changed dramatically since the late 1990s as dozens of companies have started to supply custom and off-the-shelf systems. (*See* Pharmacogenetics and Pharmacogenomics.)

Sample Processing and Preparation

Sample-processing automation in genomics is designed to assist in the processing of samples prior to loading in deoxyribonucleic acid (DNA)-sequencing instruments. This generally includes the following process steps: creating a library of cloned DNA from the genome, isolating and purifying the DNA, and attaching the labeled fluorescent tags to the DNA prior to electrophoresis and detection. During the course of the Human Genome Project, a great deal of commercial automation has been developed to address these most common process needs. Some of the instruments will only perform a single, specialized task, while others are flexible, programmable systems that can be programmed to perform multiple steps under user control. Industry standards for microtiter plates have resulted in instruments from all manufacturers being generally compatible with each other. While all genomic sequencing strategies share common types of steps, the specific protocol details at each genome center or company require instruments to be individually configurable to some degree. The common major automation platforms in use today are described in the following sections.

DNA shearing

DNA fragments to be used to create subclone libraries can be created using enzymatic, ultrasonic or mechanical shearing methods. One automated mechanical shearing instrument, the GeneMachines HydroShear, shears DNA via hydrodynamic shearing by pumping the DNA sample through an orifice plate. As a pump pushes the DNA through a small hole in the plate, the velocity of the sample is increased locally to stretch and fragment the DNA. The velocity of the pump and the size of the orifice tune the average DNA fragment size. Typically, genomic or large DNA is sheared into fragments averaging 2000 base pairs (pb) which are then end-repaired and used to clone into subclone vectors to produce libraries.

Guided single colony or plaque picking

Colony picking was one of the first sample-processing tasks to be automated, dating back to systems developed by academic groups in the early 1980s. All colony pickers operate on the same basic design; randomly arrayed colonies or plaques are presented to the picker on agar plates. A CCD camera imaging system is then used to guide the picking needles by creating a digital image of the colony locations and converting these to mechanical picking coordinates. The needles are then moved to the physical coordinates and a small number of cells picked onto the needle tips. On the destination microtiter plate, either 96 or 384 wells are filled with growth medium, and then inoculated with the colonies. Finally, the picking needles are cleaned, using combinations of ultrasonic baths, mechanical scrubbing and heat drying, to be reused in the next picking cycle. The imaging software selects colonies based on user-defined criteria used to select single, isolated colonies. Throughputs of these devices vary depending on colony density and imaging criteria, the length of the tip-cleaning procedure, and whether the machines are manually loaded or are equipped with autoloaders. Throughputs should be evaluated based on the number of output samples per hour and can range from 1000 to 4000 in most currently available systems. Examples of systems currently available are Genetix Qbot and Qpix, Genemachines Gel-2-Well, and Genomic Solutions Flexys.

Thermal cyclers

Thermal cyclers run programmed temperature cycling of samples for processes such as polymerase chain reaction (PCR) and cycle-sequencing reactions. The most popular thermal cyclers operate using Peltier effect devices that can heat or cool electrically such as the Perkin Elmer Geneamp 9700, and the MJ Research Tetrad thermal cycler. The typical operating range for these thermal cyclers is 4–100°C, with ramp times in the range or 1–$2°C S^{-1}$. The instruments can be programmed for any sequence of temperatures, ramp speeds and dwell times. These thermal cyclers require the samples to be processed in thin-walled microtiter plates with either 96- or 384-well plates to ensure good, rapid heat transfer from the temperature reservoir and the sample, using heated lids to prevent evaporation of the sample. Water-based instruments are also available for some applications, though they usually only have three temperature reservoirs and are typically used for high-throughput, highly specialized applications. For use of small reaction volumes, less than 5 µL, capillary tubes can be used in place of the thin-walled microtiter plates. These capillary tubes can be heated and cooled very fast and, in the case of the Idaho Technology RapidCycler, can provide very rapid cycle times and significant reductions in reaction volumes.

Automated pipetting and dispensing

Pipetting and dispensing are some of the most common automated processes in genomics. Pipetting and dispensing instruments are available in 1-, 2-, 4-, 8-, 96- and 384-channel systems, that is, 384 samples are moved together in the 384-channel system. Genomics applications generally use standard microtiter plates such as 96 or 384 wells, although even 1536-well plates are now being introduced. Typical pipette systems such as the Tecan Genesis RSP, Packard Instruments Multiprobe and PlateTrak, Matrix PlateMatePlus and Robbins Hydra Microdispenser almost always use some form of syringe mechanism to aspirate and dispense liquids. Instruments are often available in fixed-tip arrays of 8, 96 or 384 channels for dispensing or pipetting. All channels operate in unison in these systems. Some pipette systems are designed with individually programmable pipettes that can randomly access individual wells of plates. This feature is especially useful in rearraying specific wells from many plates into a single working plate.

Pipetting robot systems are available in a variety of designs. Some systems have stationary pipetting heads and move the plates to the pipetting heads using grippers, plate shuttles or conveyor belts. Others move the heads within the work envelope to access an array of plates spread out on the work surface; while some systems have integral plate-moving mechanisms and can be integrated with external robot arms.

Liquid-handling systems are generally designed to operate in the 2 to 200 µL volume range. Successful applications are very much dependent on pipetting technique and accuracy requirements. The pipetting of

80 µL of growth medium for use on the colony and plaque pickers does not generally require high accuracy. However, the addition of sequence chemistry reagents at 1–2 µL does require high accuracy (better than 5% variance) and high reliability (many hundreds of dispensing steps per day). Each application therefore requires decisions to be made about the speed and approach made. Dispensing volumes of less than 2 µL requires careful techniques and modifications to approaches. It is relatively difficult to dispense 1 µL of liquid into the dry well of a standard polypropylene microtiter plate, since liquid viscosity, nozzle material, plastic plate material and speed requirements all factor into protocol development. Future systems are being developed using ink-jet style technologies to dispense nanoliter volumes into specialized plates and reaction chambers. With these types of systems come additional issues such as evaporation, cross-contamination and accuracy.

Plate-moving systems

Plate-moving systems, as defined here, consist of a mechanical arm and gripper or a conveyor belt integrated with several peripheral modules. In recent years, the use of such plate systems has been very popular, for example the dedicated track systems used at the Whitehead Institute and the articulated robotic arm systems employed at the US DOE Joint Genome Institute. Robot arms can be as simple as a cylindrical robot, with just vertical and rotational motion plus a gripper, to sophisticated articulated arms, with as many as six axes of motion such as those developed by CRS robotics. Robot arms come in many configurations, such as polar, Cartesian (x,y,z), gantry and cantilever for general applications, as well as custom designs for specialized applications. Conveyor belts are well-known inexpensive systems for moving samples between peripherals in a prescribed sequence and have been used to great effect by Packard Biosciences Incorporated. Grippers can be designed to grasp almost any type of object and can be controlled with motors or pneumatics. An essential element of any robot system is the control software. The software must allow the user to teach the robot where and how to move to each point in the system. The software must also operate each of the peripheral devices such as pipettes, plate washers and plate shakers, and place plates in static stations such as incubators or magnet stations. An important requirement of the software is to control the arm and peripherals so that several operations can proceed in parallel to increase throughput.

There are several advantages to robot systems. Increased productivity can be achieved because an operator can operate several robots and often still have time to perform other laboratory tasks during a batch process. Training costs are reduced because it is often easier to train an operator to use the robot than to perform the protocol manually. Protocol consistency is improved due to the use of an invariant control program, which can report and log errors.

Integrated robot systems have a high initial cost and are most suited for high-throughput production processes. However, since plates are moved from module to module, it is often the case that methods can be worked out manually with the modules and then automated with the arm or belt system.

DNA Sequencing and Detection Devices

Modern sequencing instruments have seen major advancements since their introduction over a decade ago. Based on a manual method devised by Frederick Sanger in 1974, these instruments now produce over 60 000 bases of sequence information in just 2 h. While planned upgrades such as the 384-capillary systems will increase this capacity fourfold, several new technologies may provide this data in just minutes or seconds.

Initial slab-gel sequencing instruments

Modern versions of Frederick Sanger's original technique generate fluorescently labeled DNA fragments that are read in an automated fashion with electronic data capture. All possible fragment sizes from 25 bases up to ~1000 bases are produced by the sequencing chemistry, with each fragment containing a single fluorescent label that corresponds to the last DNA letter of that fragment. By size-fractionating these fragments and matching the color to the size, the original DNA sequence can easily be discerned.

The first automated DNA sequencers, such as the Applied Biosystems 373 and Amersham Pharmacia ALF, were introduced in the early 1990s as slab-gel electrophoresis instruments that utilized a thin (400 µm), cross-linked polyacrylamide gel sandwiched between two glass plates. Dye-labeled sequence fragments, initially primer labeled and now terminator labeled, were manually loaded to the top of the vertical gel assembly and an electric field was applied. As the negatively charged DNA fragments migrated through the gel, they were size-fractionated by the polyacrylamide gel. Near the bottom of the gel assembly, separated fragments were automatically excited with a scanning argon laser and detected by a CCD camera. While these instruments automated several steps, slab-gel set-up and breakdown were still manual and quite laborious. These issues were addressed in the newer capillary systems.

Capillary sequencers

Modern capillary systems are currently the most robust and fastest sequencing systems available. These instruments utilize the same basic electrophoresis and detection technology as slab-gel systems, with several added advantages. First, each sample is separated and analyzed in an individual capillary. This eliminates sample-tracking issues, sample bleed-over and gel-uniformity problems associated with a slab gel. Also heat dissipation from a capillary is much more efficient due to its higher surface area-to-volume ratio, and this allows for higher run voltages that separate DNA fragments significantly faster than slab-gel systems.

The introduction of commercial capillary instruments such as the Applied Biosystems 3700 and the Amersham Pharmacia Biotech MegaBACE 1000 has automated two key steps: gel preparation and sample loading. Slab-gel set-up and breakdown were replaced with a linear polyacrylamide gel (LPA) matrix in the case of the MegaBACE and a cross-linked polymer (POP6) in the case of the ABI 3700 that could be forced into and out of the capillary. These replicable gels reduced technician-related errors, since the new gel required no polymerization and was automatically cleaned out of the capillaries as new gel was injected into the capillaries. Electrokinetic sample injection was another key automation development. With slab-gel instruments, samples were hand-pipetted on top of the gels, leading to misloading and contamination of samples. Electrokinetic injection automated sample loading by passing a negative current through the liquid sample, forcing the negatively charged DNA fragments to migrate into the capillary.

Electrokinetic injection and the separation matrices do have drawbacks that are a significant source of variability in sequencing quality. During injection, smaller molecules diffuse faster through the sample than larger molecules. Small charged molecules like salt and dye effectively compete with labeled DNA fragments for injection into the capillary. This lowers the signal strength by inhibiting injection of labeled DNA fragments. Another issue is the injection of large DNA molecules. LPA contains pores that are easily clogged by large DNA molecules such as plasmid and genomic DNA. Both of these issues place very high requirements on front-end sample purity and consistency.

A typical capillary instrument produces about 30 000 bases in 1 h, a fivefold increase over the original slab sequencers. Further increases of up to 20-fold will be attained using higher-density arrays and new

gel formulations. One high-density system is a 384-capillary sequencer that utilizes the same optical and electrophoresis technologies employed in previous systems. Scientists are bundling four times as many capillaries into the MegaBACE sequencer, as well as upgrading the optical system to maintain scanning speeds. Whereas it took 9 years to develop the first commercial 96-capillary sequencer, the MegaBACE 4000 DNA sequencer was completed in just 2 years.

Future sequencing systems

Future sequencing instrument designs will increasingly incorporate more of the upstream processes for total system integration. While extension of electrophoresis technologies such as microchannel devices promises only a fourfold increase in capacity over capillary systems, they offer the ability to etch an entire front-end process onto a single device. This will result in increased process stability and the capability to work with subnanoliter volumes.

Theoretical technologies such as nanopores and atomic force microscopy could provide the throughputs required for the total genomic analysis of many species. Estimated sequencing speeds of up to 10 000 bases per second could sequence a human genome in a single day. The following is an introduction to these technologies, and interested readers are referred to the Further Reading section.

Microchannel devices

Several research groups first described the use of microchannel devices for DNA sequencing in the early 1990s. These integrated microfluidic devices are similar to capillary instruments except that samples pass through enclosed channels etched into circular, 8 in diameter glass or plastic wafers. Advantages include the potential for a total analysis system (known as 'laboratory on a chip'), shorter separation times, inexpensive fabrication and miniaturization for lower-sample volumes, and increased array density.

To compete with capillary arrays, 96 or more channels must be etched into each chip. One limiting issue is that single-base resolution requires 20 cm long separation channels, while current manufacturing technology limits wafer diameters to 20 cm. Since the channels can only use half the diameter to eliminate channel crossing, other solutions are required. Scientists at the University of California, Berkeley, are investigating the use of turns to effectively increase the path length of the channels. Further miniaturization of channel space is also possible with new injector designs that greatly improve base resolution. Samples are injected as a small thin plug across the main electrophoresis channel instead of being directly injected into the capillary. With 'cross-injection', the sample enters the main electrophoresis channel from a sample channel and then flows out of the main channel into a waste channel. Impurities flowing ahead of and behind the sample plug are eliminated, resulting in a partial purification of the sample based on mobility.

Cross-injection also generates tighter sample plugs. During a direct injection, the sample plug becomes wider as the injection time progresses. In cross-injection, the sample plug injects with a steady-state flow, resulting in a sample plug of less than 1 μm. Compared with direct injection, cross-injection plugs can reduce the required channel lengths below 20 cm and still provide resolution comparable with 40 cm long capillaries.

Another favorable attribute of microchannel devices is their superior heat dissipation, which allows for even higher run voltages and faster separation times. The large thermal mass of the wafer allows for excellent heat dissipation from the small channels. Microchannel devices typically run at 300 $V\,cm^{-1}$, while capillary systems operate at less than 140 $V\,cm^{-1}$.

Sample detection in microchannel devices (**Figure 1**) is very similar to capillary devices. Samples are electrophoresed toward the center of the wafer, where they are excited with an argon laser beam. The laser is focused through a rotating microscope objective that illuminates each channel about 3 times a second. The

resulting fluorescence is passed back through the objective and is directed to various beam splitters and filters. This light is then detected by one of four photomultipliers, which sends the information to a computer for analysis.

In the first decade of the twenty-first century, we may see microchannel devices and high-density capillary arrays dominate high-throughput sequencing. However, these systems are limited to an increase in throughput of about fourfold over the MegaBACE 4000. More radical approaches that eliminate the time-consuming electrophoresis and sample-preparation steps are needed to reach the ultimate goals of the genome initiative. The following sections briefly review two of the most promising technologies; interested readers are directed to the more comprehensive reviews cited in Further Reading.

Nanopore technology

The future of high-speed chemical analysis resides with novel subnanotechnologies combined with single-molecule detection. These methods involve the difficult tasks of discerning structures less than a nanometer in diameter and the handling of femtoliter volumes. As these fields advance, most of the sample preparation and cost associated with DNA sequencing will disappear.

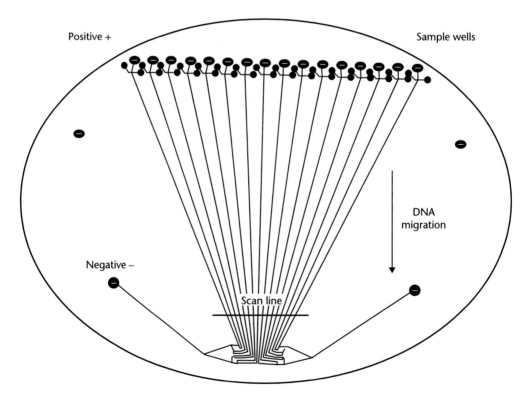

Figure 1 Microchannel electrophoresis chip used for separation and detection of labeled deoxyribonucleic acid (DNA) sequencing fragments. This 16-channel chip was fabricated at Molecular Dynamics, Sunnyvale, CA, by Dr Stevan Jovanovich.

Nanopore technology or single-channel current measurements is one of the newest methods being applied to DNA sequencing. It is based on the study of membrane ion channels that are formed by proteins in lipid membranes. By attaching a 1 μm diameter glass pipette around the protein channel and placing electrodes on either side of the membrane, one can characterize the flow of ions through the channel by measuring the current flow.

The idea is to place DNA in the glass tip and measure the changes in ion current flow as the DNA transverses the pore. Measurements of current fluctuations in time intervals equal to a single base's translocation time may reveal the base's identity. Most of the current research uses the protein α-hemolysin and focuses on proof of principle and physical characterizations. α-Hemolysin is a predictable membrane ion channel protein with picoamp resolution. Its limiting pore diameter of 1.5 nm just allows single-stranded DNA to experience some drag as it flows through the pore.

Recent studies have detected differences between poly(A) and poly(C) strands that are 100 bases long. However, single-base resolution is required for a feasible sequencing technology. Further studies have revealed that α-hemolysin's 5 nm long pore interacts with about seven DNA bases at one time. Since the current measurements would average all seven base interactions, single-base resolution may not be possible with this pore. Modern current measuring equipment is also not accurate enough to measure the current fluctuations associated with single-base interactions. While nanopore technology seems feasible, commercial instruments are not yet available.

Atomic force microscopy

Another interesting approach is the use of atomic force microscopy (AFM) to directly read the sequence of a DNA strand. AFM uses a flexible cantilever attached to a nanosize tip that is scanned over the sample surface. As the tip moves across the surface, the cantilever deflection is monitored by a laser that controls the tip height. This deflection angle is used to generate a topological map.

Using AFM to measure the force required to unzip DNA has given encouraging sequencing results. One research group attached one end of a double-stranded DNA to the AFM tip and unzipped the DNA one base at a time by moving the tip. They found that A–T pairs required 10 pN and G–C pairs required 20 pN to unzip. Using a similar technique with glass needles, researchers have resolved 500 bp in 25 s. This technology may prove to be a valuable resource to specialized sequencing or genotyping applications.

Conclusion

The development of new technologies to automate, integrate and generally simplify the field of genomics and biotechnology is an ongoing process. Most large public genome centers are retooling their production lines and automation every 18–24 months to keep up with the improvements. Despite this rapid change, the sequencing phase of the Human Genome Project has only been the start point of applying engineering and factory-like principles to bioprocesses. As the next phases of the genome project lead us to molecular medicine and point-of-care diagnostics, the automation of genomic analysis will become even more central to the future of this field. The initial academic prototype development has grown into a vibrant support industry that is now developing technologies and making these available as discovery systems to academic and commercial sources, then will follow the merging and integration of many of the technologies described here. The concept of a single device to rapidly read out genetic information such as a DNA sequence from a blood sample or other material is believable in the first decades of the twenty-first century.

See also

Genome Sequencing
Industrialization of Proteomics: Scaling-up Proteomics Processes
Molecular Entry Point: Strategies in Proteomics
Structural Proteomics: Large-scale Studies

Further Reading

Deamer DW and Akeson M (2000) Nanopores and nucleic acids: prospects for ultrarapid sequencing. *Trends in Biotechnology* **18**: 147–151.

Dolnik V, Liu S and Jovanovich S (2000) Capillary electrophoresis on microchips. *Electrophoresis* **21**: 41–54.

Kheterpal I, Scherer JR, Clark SM, *et al.* (1996) DNA sequencing using a four-color confocal fluorescence capillary array scanner. *Electrophoresis* **17**: 1852–1859.

Marziali A and Akeson M (2001) New DNA sequencing methods. *Annual Review of Biomedical Engineering* **3**: 195–223.

Meldrum D (2000) Automation for genomics, part II: sequencers, microarrays, and future trends. *Genome Research* **10**: 1288–1303.

Medintz IL, Paegel BM and Mathies RA (2001) Microfabricated capillary array electrophoresis DNA analysis systems. *Journal of Chromatography A* **924**: 265–270.

Paegel BM, Hutt LD, Simpson PC and Mathies RA (2000) Turn geometry for minimizing band broadening in microfabricated capillary electrophoresis channels. *Analytical Chemistry* **72**: 3030–3037.

Rief M, Clausen-Schaumann H and Gaub HE (1999) Sequence-dependent mechanics of single DNA molecules. *Nature Structural Biology* **6**: 346–349.

Schmalzing D, Koutny L, Salas-Solano O, *et al.* (1999) Recent developments in DNA sequencing by capillary and microdevice electrophoresis. *Electrophoresis* **20**: 3066–3077.

Zhou H, Miller AW, Sosic Z, *et al.* (2000) DNA sequencing up to 1300 bases in two hours by capillary electrophoresis with mixed replaceable linear polyacrylamide solutions. *Analytical Chemistry* **72**: 1045–1052.

Roma (Gypsies): Genetic Studies

Luba Kalaydjieva, *Western Australian Institute for Medical Research, Perth, Australia*

Bharti Morar, *Edith Cowan University, Perth, Australia*

The Roma, also known as Gypsies, are a transnational founder population, resembling a mosaic of socially and genetically divergent groups. Common origins from a small group of related founders result in sharing of maternal and paternal lineages and of ancient disease-causing mutations across endogamous Romani groups in different countries. Genetic differentiation between groups, with an impact on genetic epidemiology, is the product of bottleneck events, random genetic drift and differential admixture, and correlates with the migrational history of the Roma in Europe.

Intermediate article

History and Cultural Anthropology

The Roma are a people of 12–15 million, scattered across the world, with 8–10 million resident in Europe. Their arrival in Byzantium and early settling in the Balkans is dated to the eleventh to twelfth century AD. A small initial population size is suggested by the Ottoman Empire Tax Registry from the fifteenth century (300–400 years after the arrival of the Roma), estimating their overall number in the Balkan provinces at about 17 000. The routes of the Roma diverged after arrival in the Balkans: while the majority remained there, some (now known as Vlax Roma) crossed the Danube river into Wallachia (present-day Romania) and were confined to slavery until the middle of the nineteenth century, while still others continued their journey west in small groups, reaching most parts of Europe by the end of the fifteenth century. Superimposed on these initial migratory movements are three large recent migration waves: out of present-day Romania in the end of the nineteenth century; out of Yugoslavia in the middle of the twentieth century; and out of most Eastern European countries in the last decade of the twentieth century.

The Roma have no written history, and their origins have been the subject of legends and of linguistic studies. The similarities between Romanes (the language of the Roma) and languages spoken in the Indian subcontinent have led to the hypothesis of Indian origins, now universally accepted by social scientists. Some researchers suggest that Romanes developed after the exodus from India as the product of close interaction between ethnically mixed speakers of diverse languages and dialects. The idea of the proto-Roma as an ethnically mixed population is supported by the traditional organization of Romani society, still in place today, which closely resembles the endogamous professional *jatis* of India. Its complex mosaic structure is based on the Romani group, whose identity is the compound product of traditions and customs, organs of self-rule, trades (often long abandoned but still reflected in ethnonyms), language and dialects related to history of migrations and religion. Romani groups are separated by various, sometimes complicated, rules of endogamy, generally proscribing intermarriage between groups as much as intermarriage with non-Roma. While the social history of Romani groups is unclear, different sources point to their historically small size: early European records consistently describe traveling Gypsies as groups of 30–300 people, and the 1893 Austro-Hungarian census data on Transylvania suggest that individual Romani groups at the time numbered around 2000 individuals each.

Genetic Data on Origins and Population Structure

A comprehensive list of references on population and medical genetic studies can be found in a recent review (Kalaydjieva *et al.*, 2001b).

The issue of the Indian origins has been addressed in numerous population-genetic studies, conducted in different European countries over the last 70 years. Despite limitations imposed by the use of 'classical' markers (mainly blood groups) and single-locus comparisons, and the deficiencies of the information available on populations in the Indian subcontinent, these investigations have shown that overall the Roma are genetically closer to Indians than to their European neighbors. Cross-section sampling and disregard for the traditional group organization of the Roma has not allowed an insight into internal genetic structure, but has generally suggested that differentiation does exist, with greater genetic distances between Romani populations of different countries than between the autochthonous populations of these countries.

Strong support for the common Asian origins of divergent endogamous Romani groups comes from a study of Y-chromosome and mitochondrial deoxyribonucleic acid (mtDNA) markers of different mutability, used to examine different time depths of the history of male and female Romani lineages (Gresham et al., 2001). Two-thirds of 252 Romani males carry Y chromosomes of Asian descent (**Figure 1a**). The major Y-chromosome haplogroup, VI-68, shared by 45% of Roma, occurs at low frequency in India and Central Asia, while haplogroup VI-56 (13% in the Roma) is present in Pakistan, Central Asia and the Middle East (nomenclature and global distribution from Underhill et al., 2000). Some additional Asian haplogroups are found in the Roma at minor frequencies, and most remaining Y chromosomes belong to common European haplogroups. Female lineages display greater

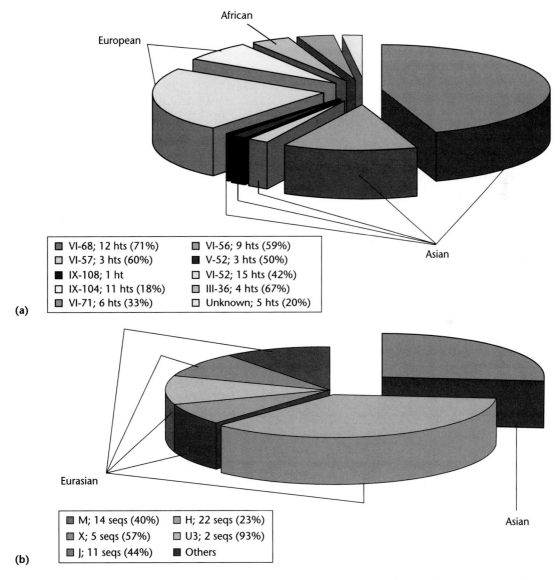

(a)

VI-68; 12 hts (71%)   VI-56; 9 hts (59%)
VI-57; 3 hts (60%)   V-52; 3 hts (50%)
IX-108; 1 ht   VI-52; 15 hts (42%)
IX-104; 11 hts (18%)   III-36; 4 hts (67%)
VI-71; 6 hts (33%)   Unknown; 5 hts (20%)

(b)

M; 14 seqs (40%)   H; 22 seqs (23%)
X; 5 seqs (57%)   U3; 2 seqs (93%)
J; 11 seqs (44%)   Others

Figure 1 [*Figure is also reproduced in color section.*] (a) Distribution of Y-chromosome haplogroups in 252 male Roma from different endogamous groups. The haplogroups are defined by unique event polymorphisms as described by Underhill et al. (2000). Shaded in gray are haplogroups whose geographic association is not known. The number of haplotypes (defined by the alleles of seven STR loci) along each haplogroup is given next to the haplogroup. The figures in parentheses refer to the frequency of the most common STR haplotype within that haplogroup. (b) Distribution of mtDNA haplogroups in 275 Roma from different endogamous groups. The haplogroups are defined by restriction fragment length polymorphisms in the mitochondrial DNA. The number of different hypervariable segment 1 (HVS1) sequences within each haplogroup is given next to the haplogroup. The figures in parentheses refer to the frequency of the most common HVS1 sequence variant within that haplogroup. The category labeled 'Others' consists of haplogroups I, N, T, U1, U5, U(K) and W.

variety (**Figure 1b**), similar to other populations, yet about one-quarter of 275 Roma display Asian mtDNA haplogroup M (Torroni *et al.*, 1994). The origins of three additional haplogroups (H, U3 and X), together accounting for over 50% of Romani mtDNA, are unclear. They are widely represented in Eurasia and could be present in the Roma as a result of European admixture. However, the most common Romani sequences of these haplogroups have not been identified previously in Europe, and might also be founding lineages of Indian origin or early admixture on the route to Europe.

The major founding paternal and maternal lineages of Asian origin are shared between all Romani groups (**Figure 2**), pointing unambiguously to common ancestry. Moreover, the limited internal diversification of these lineages in the Roma, in sharp contrast to their overall diversity in non-Romani Asians, is compatible with a small group of related founders splitting from an ethically defined population, such as a distinct caste or tribal group.

The genetic evidence is thus different from the expectations based on cultural anthropology and on some linguistic data, and suggests that the population fissions, spawning numerous Romani groups, occurred after the exodus from India. However, despite their common origins and the recent nature of the splits, Romani groups have diverged genetically and show substantial differences in the frequencies of Y-chromosome and mtDNA lineages (Gresham *et al.*, 2001) (**Figure 2**). This differentiation is the likely product of

two major factors: (a) random genetic drift related to the small population size of the proto-Roma and the even smaller number of founders of individual Romani groups; and (b) different levels and sources of admixture. Strong drift effects can explain the wide variation in Y-chromosome and mtDNA lineages, for example the frequency of haplogroup VI-68 ranging from <10% to >80% in different Romani groups, while differential admixture from surrounding European populations is illustrated by haplogroups VI-52 and IX-104, whose occurrence in the Roma reflects their reported clinical distribution in Europe. An additional factor that merits further investigation is nomadism. By limiting population growth and gene inflow, a long history of nomadism may have played a role in shaping the genetic profile of some Romani groups, such as the Monteni (Kalaydjieva *et al.*, 2001a) where Nei's unbiased estimator of diversity for the highly mutable minisatellite MSY1 is 0.711 ± 0.114, significantly lower than in any other European founder population.

Based on the available data, the early separation into three migrational groups within Europe (early settlers in the Balkans, Vlax Roma, and migrants to northern and western Europe) is the best predictor of genetic affinities among the numerous social anthropological criteria for the classification of Romani groups (Gresham *et al.*, 2001). Current residence and ethnonyms reflecting traditional trades and religion are of no genetic relevance.

Further differentiation within the migrational grouping is suggested by a study of three endogamous

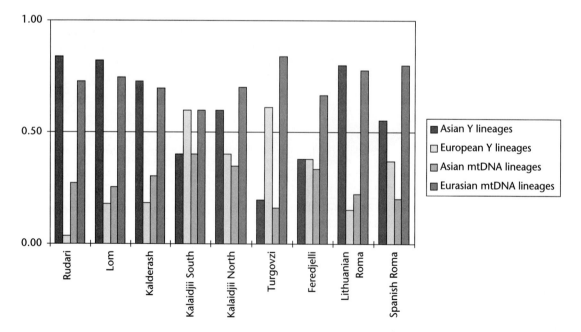

Figure 2 [*Figure is also reproduced in color section.*] Distribution of geographically localized Y-chromosome and mtDNA haplogroups in different endogamous Roma groups. Color coding: Asian Y chromosome haplogroups: red; Asian mtDNA haplogroups: orange; European Y chromosome haplogroups: yellow; Eurasian mtDNA haplogroups: green.

groups of Vlax Roma (Kalaydjieva *et al.*, 2001a), with time to the most recent common male ancestor estimated at 400–500 years, consistent with the early period of slavery in Wallachia. The evidence of internal stratification, provided by the fast mutating Y-chromosome minisatellite MSY1, needs confirmation on a larger sample size.

In summary, the extant Romani population can be described as a mosaic of founder subpopulations, the compound product of superimposed migrations, genetic drift and admixture. In terms of social history and its impact on genetic structure, Romani groups resemble the Ashkenazi Jewish communities dispersed throughout Eastern Europe, prior to their amalgamation in the ghettos of the big cities. The same process of amalgamation is now beginning for the Roma.

Mendelian Disorders in the Roma

The pattern of single-gene disorders has been defined by the same factors that have shaped population genetic structure.

Admixture from surrounding populations has introduced common mutations causing cystic fibrosis, phenylketonuria, medium-chain acyl-coenzyme A dehydrogenase deficiency and Wilson disease in some Romani groups. An intriguing example of admixture is provided by the *BRCA1* mutation 185delAG, occurring in Spanish Gypsies on the same conserved haplotype as in Jewish chromosomes (Diez *et al.*, 1998). The epidemiology and molecular basis of such 'imported' disorders is related to that in the surrounding majority and to the time and number of admixture events.

The information on private genetic disorders and mutations of the Roma is still partial and unsystematic, a situation very different from long-studied founder populations, such as the Ashkenazi Jews and the Finns. To date, 11 single-gene disorders caused by founder mutations have been described (**Table 1**). These include three novel entities: hereditary motor and sensory neuropathies types Lom (HMSN-L) and Russe (HMSN-R), and the congenital cataracts facial dysmorphism neuropathy (CCFDN) syndrome, recently shown to be genetically identical to a subtype of the Marinesco–Sjögren syndrome (Merlini *et al.*, 2002). Data in the literature (references in Kalaydjieva *et al.*, 2001b) point to other autosomal recessive disorders, including rare malformation syndromes such as Bowen–Conradi, Jarcho–Levin, Meckel, Smith–Lemli–Opitz and Fraser, which may be caused by still unidentified private mutations in the Roma.

The age of the mutation, its frequency in the ancestral population and random drift are among the major factors defining current epidemiology and molecular genetic characteristics. While international data are limited, the systematic information collected in Bulgaria (Tournev I, DSci thesis) allows some general conclusions.

Mutations, whose origins predate the population splits of the Roma, tend to occur in numerous groups with high frequency. Examples include congenital myasthenia, galactokinase deficiency and HMSN-L (**Figure 3**), with overall carrier rates among the Roma in Bulgaria in the 3–6% range. In our experience, the usual size of the conserved region of complete homozygosity surrounding such founder mutations is around 100–200 kb, precluding the use of homozygosity mapping even in single highly inbred kindreds. However, refined mapping is greatly facilitated by the independent historical recombinations and haplotype divergence in different affected Romani groups.

Other mutations occur in specific groups and may be altogether absent in others living in close geographic

Table 1 Mendelian disorders of the Roma caused by private founder mutations

Disorder[a]	OMIM	Inheritance	Map location	Gene	Mutation
Congenital myasthenia	254210	AR	17p13	*CHRNE*	1267delG
Hereditary motor and sensory neuropathy-Lom	601455	AR	8q24	*NDRG1*	R148X
Galactokinase deficiency	230200	AR	17q24	*GK1*	P28T
Limb-girdle muscular dystrophy type 2C	253700	AR	13q12	*SGCG*	C283Y
Primary congenital glaucoma	231300	AR	2p21	*CYP1B1*	E387K
Glanzmann thrombasthenia	273800	AR	17q21	*ITGA2B*	IVS15DS, G-A+1
von Willebrand disease type 3[b]	277480	AR	12p13.3	*VWF*	Q1311X
Spinal muscular atrophy[c]	253300	AR	5q13	*SMN*	3 different defects
Hereditary motor and sensory neuropathy-Russe	605285	AR	10q23		
Congenital cataracts facial dysmorphism neuropathy	604168	AR	18qter		
Polycystic kidney disease	173900	AD	4q21–q23	*PKD2*	R306X

[a]References for most disorders and mutations listed in the table can be found in Kalaydjieva *et al.* (2001b).
[b]Casana *et al.* (2000).
[c]Jordanova *et al.* (2002). Romani SMA chromosomes shared a common founder haplotype; however, they carried three different molecular defects, suggesting an ancestral allele particularly prone to rearrangements.

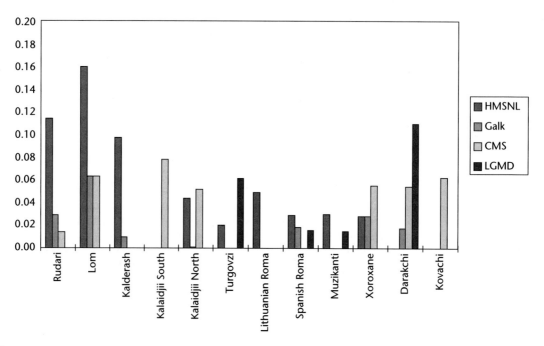

Figure 3 [*Figure is also reproduced in color section.*] Carrier rates of four founder mutations (R148X in *NDRG1*, 1267delG in *CHRNE*, P28T in *GK1* and C283Y in *SGCG*) among 800 Roma from different endogamous groups from Bulgaria, Lithuania and Spain. These mutations are responsible for the autosomal recessive disorders hereditary motor and sensory neuropathy type Lom (HMSNL), congenital myasthenia (CMS), galactokinase deficiency (Galk) and limb-girdle muscular dystrophy 2C (LGMD) respectively.

proximity. Such clustering may result from drift effects or from the recent origin of the mutation. The C283Y mutation in *SGCG*, causing limb-girdle muscular dystrophy type 2C, is an example of extreme drift effects. C283Y occurs in Romani patients in Western Europe and has been estimated to predate the exodus from India by at least 1000 years (Piccolo *et al.*, 1996), yet extensive studies in Bulgaria have shown its very limited distribution, confined to a small subset of the early Balkan settlers (**Figure 3**). Another example of limited distribution, likely to be due to the recent origin of the mutation, is provided by CCFDN, which occurs exclusively in a specific group of Vlax Roma. While such differences are obvious within a single country, the characteristics of the Roma as a transnational founder population mean that the same mutations occur in patients from different parts of the world. Since its identification in Bulgaria, CCFDN has been diagnosed in individuals from the same Romani group residing in Romania, Hungary, Germany, Italy and the United States.

The current changes in the traditional demography of Romani minorities mean that geneticists and medical practitioners in many countries will encounter patients with previously unknown or very rare disorders. A recent example is provided by the Berlin screening program, with no galactokinase deficiency detected among 260 000 tested in the period

1978–1991, and an 'epidemic' of cases since the beginning of the wars in former Yugoslavia, all from Bosnian refugee families and homozygous for P28T (Reich *et al.*, 2002).

Analysis of four founder mutations (R148X in *NDRG1*, 1267delG in *CHRNE*, P28T in *GK1* and C283Y in *SGCG*) among 700 Roma from Bulgaria has shown that one in seven individuals is a carrier. The carrier rate for all Romani founder mutations will obviously be higher. These figures are by no means exceptional and fall in the same range as the carrier rates for hemoglobinopathies in Mediterranean populations. The health implications for the marginalized Romani minorities are serious and need to be addressed in a collaborative international effort.

See also

Peopling of India: Insights from Genetics
Population History of Europe: Genetics

References

Casana P, Martinez F, Haya S, *et al.* (2000) Q1311X: a novel nonsense mutation of putative ancient origin in the von Willebrand factor gene. *British Journal of Haematology* **111**: 552–555.

Diez O, Domenech M, Alonso MC, *et al.* (1998) Identification of the 185delAG BRCA1 mutation in a Spanish Gypsy population. *Human Genetics* **103**: 707–708.

Gresham D, Morar B, Underhill PA, *et al.* (2001) Origins and divergence of the Roma (Gypsies). *American Journal of Human Genetics* **69**: 1314–1331.

Jordanova A, Kargaci V, Kremensky I, *et al.* (2002) Spinal muscular atrophy among the Roma (Gypsies) in Bulgaria and Hungary. *Neuromuscular Disorders* **12**: 378–385.

Kalaydjieva L, Calafell F, Jobling MA, *et al.* (2001a) Patterns of inter- and intra-group genetic diversity in the Vlax Roma as revealed by Y chromosome and mitochondrial DNA lineages. *European Journal of Human Genetics* **9**: 97–104.

Kalaydjieva L, Gresham D and Calafell F (2001b) Genetic studies of the Roma (Gypsies): a review. *BiomedCentral Medical Genetics* **2**: 5–18.

Merlini L, Gooding R, Lochmuller H, *et al.* (2002) Genetic identity of Marinesco–Sjogren/myoglobinuria and CCFDN syndromes. *Neurology* **58**: 231–236.

Piccolo F, Jeanpierre M, Leturcq F, *et al.* (1996) A founder mutation in the γ-sarcoglycan gene of Gypsies possibly predating their migration out of India. *Human Molecular Genetics* **5**: 2019–2022.

Reich S, Hennermann J, Vetter B, *et al.* (2002) An unexpectedly high frequency of hypergalactosemia in an immigrant Bosnian population revealed by newborn screening. *Pediatric Research* **51**: 598–601.

Torroni A, Miller JA, Moore LG, *et al.* (1994) Mitochondrial DNA analysis in Tibet: implications for the origin of the Tibetan populations and its adaptation to high altitude. *American Journal of Physical Anthropology* **93**: 189–199.

Underhill PA, Shen P, Lin AA, *et al.* (2000) Y chromosome sequence variation and the history of human populations. *Nature Genetics* **26**: 358–361.

Further Reading

Fraser A (1992) *The Gypsies*. Oxford, UK: Blackwell Publishers.

Liégeois J-P (1994) *Roma, Gypsies, Travellers*. Strasbourg: Council of Europe Press.

Hancock I (1987) *The Pariah Syndrome*. Ann Arbor, MI: Karoma Publishers Inc.

Marushiakova E and Popov V (1997) Gypsies (Roma) in Bulgaria. *Studien zur Tsiganologie und Folkloristik*. Frankfurt am Main: Peter Lang.

Romanov Family

Suni M Edson, *Armed Forces DNA Identification Laboratory, Rockville, Maryland, USA*

The bodies of the last Russian royal family were recovered in 1991. Genetic markers have had an impact on their lives, deaths and identification.

Intermediate article

Article contents

- The Tsarina and Rasputin
- The Deaths
- The Remains and Identification

The Tsarina and Rasputin

There are many genetic traits in the European royal families that are interesting for their own sake but not useful for identification. The most significant of these is the X-linked recessive trait hemophilia. A carrier female most commonly passes down X-linked recessive traits, as the traits are generally deleterious and males expressing them usually do not live to a reproductive age. Hemophilia is characterized by the inability of the blood to clot and until the mid-1960s was largely untreatable.

Within the family of the last tsar of Russia, Tsarveich Alexei, the youngest child of Nicholas and Alexandra, was a hemophilic male. Alexandra was a carrier of this trait, having acquired it from her mother, Alice, daughter of Queen Victoria of England. Victoria was a carrier of the disease, and passed it to three of her nine children, who went on to marry into other royal families of Europe including those of Russia and Spain (Stevens, 1999).

At the beginning of the twentieth century, there was no way to treat the Tsarevich. He regularly hemorrhaged and was bed-ridden. The happiness that the birth of an heir had brought to the Romanovs was quickly diminished by his illness. Alexandra was desperate for a cure and searched among a number of mystics and physicians before finding Grigory Efimovich Rasputin.

Rasputin was largely considered to be a fraud who changed the course of the Romanov dynasty. While it is true that he was a drunkard and womanizer, it was through Alexei's illness that he became a great force behind the last of the Romanovs. Alexandra truly believed that Rasputin could heal her son, for Alexei improved daily after meeting the 'elder'. Alexandra and Nicholas kept Rasputin close to their home to 'heal' Alexei on a regular basis. Whether or not he actually did is irrelevant. The Tsarina believed and that gave Rasputin remarkable power (Radzinsky, 2000).

Through her unbending faith in him, Rasputin played on the Tsarina's fears and desires and managed to change the face of the Russian Duma (the Russian 'congress') and Synod (the ruling body of the Russian

Church), and alienated both Alexandra's and Nicholas' families. The Tsar and his immediate family were left with very little support by the time the Russian revolution came.

Rasputin had prophesied that his death would foretell the fall of the Romanovs. He was murdered in 1916; the last Tsar of Russia and his family in 1918.

The Deaths

In the summer of 1918, the family of Tsar Nicholas II of Russia vanished along with four servants and the family dog. The official statement from the ruling Soviet government was that Nicholas had been executed, and his family had been allowed to flee to the Americas or England. For 8 years the Soviets did not contradict these statements. As no bodies had been discovered, it was easy to deny their existence. However, in 1926, Stalin allowed the publication of a text stating that Nicholas and his family had been executed in 1918. But by then, the myth that the last ruling Romanov family was still alive in hiding had become truth to millions.

Their survival was a romantic ideal that colored the search for and recovery of the remains. Those that did search were assured of censure or jail if their inquiries were not discreet. The search for and eventual location of the remains are detailed in Massie (1995) and will not be discussed here.

Suffice it to say that the remains of the last Tsar and his family were located in 1979 and reburied, as the political atmosphere of the country was not conducive to the revelation. With the advent of *perestroika* and Glasnost, the finders felt assured of their personal safety and in 1991 the remains were unearthed.

The Remains and Identification

What was recovered in 1991 were the skeletal remains of nine individuals. The difficulty here is that there were eleven people murdered in the early morning of 17 July 1918: the former royal family of Tsar Nicholas II, Tsarina Alexandra, the Tsarevich Alexei, and the four grand duchesses, Olga, Maria, Tatiana and Anastasia; and their remaining servants. The skeletons were reassembled and studied by both Russian anthropologists and a team of scientists from the United States. Agreement could not be reached over the identity of the individuals represented. Thus, the Russian chief medical examiner, Valdislav Plaskin and Dr Pavel Ivanov of the Englehardt Institute of Molecular Biology of the Russian Academy of Sciences in Moscow proposed sending the remains to Dr Peter Gill at the Molecular Research Centre of the Home Office Forensic Sciences Service (FSS) in the UK for deoxyribonucleic acid (DNA) identification.

Two different types of DNA can be used in the identification of remains: nuclear and mitochondrial. Nuclear DNA (nucDNA) is inherited from both parents and could be used to determine whether the remains are part of a family group. It is found in the nucleus of the cell, and is what is typically thought of when DNA is discussed.

Mitochondrial DNA (mtDNA) is a small circular genome found in the mitochondria of the cell. It is inherited only through the maternal line and can be traced through a number of generations virtually unchanged. mtDNA exists in large copy numbers, thus making it very appropriate for use in identifying ancient remains, in which the nucDNA, existing in only two copies per cell, is often highly degraded and insufficient for identification (Holland *et al.*, 1993).

Total genomic DNA was extracted from powdered bone that had been treated with extreme care. The amount of DNA in bone samples tends to be very low, and the DNA from the scientists handling the samples can easily contaminate the extracts and either overwhelm the ancient DNA or blend with it to form a hybrid or mixed sequence. Very small amounts of DNA were recovered using standard techniques. This was sufficient to attempt nucDNA testing through a technique called short tandem repeat (STR) analysis.

Short tandem repeat analysis uses polymerase chain reaction to amplify a series of loci, the combination of which is typically a unique identifier for an individual; for example, an individual can be A,B at a given locus: the A was received from one parent, the B from the other. A series of such loci studied for any given person is virtually unique. These loci can also be used to identify children if the parents' sequences are known. If two of the six adult skeletons found were the parents of the adolescent remains, then the childrens' STR pattern should be a combination of the two.

STR analysis was conducted using five different STR loci and a test for amelogenin, which permits sex determination. Gill's testing showed that five of the nine skeletons uncovered were indeed part of a family group and the three sets of adolescent remains were female. But were they the Romanov family? This question could not be answered with nucDNA.

It was then that the researchers turned to mtDNA. There are two regions within mtDNA that are typically studied for identification purposes: hypervariable regions one and two. Gill and his team amplified DNA from the presumed Romanov remains in these two regions. The four female remains produced identical results. These were compared to a mtDNA sequence generated from Prince Philip, the husband of Queen Elizabeth II of England and the maternal grandnephew of Alexandra. The results were identical, indicating that these were indeed the remains of Tsarina Alexandra and three of the grand duchesses. It was presumed that the related male skeleton was indeed that of Nicholas, but in view of the controversy already surrounding the remains, Gill wanted to be sure.

The FSS located two living maternal relatives of Nicholas who were willing to donate samples. The first was Xenia Sfiris, the great-granddaughter of Nicholas's sister, Grand Duchess Xenia and ironically the granddaughter of one of Rasputin's murderers, Prince Felix Yusupov. The second was Lord James Carnegie, traced back through Nicholas's grandmother and six generations of the royal European family tree. Mrs Sfiris and Lord Carnegie matched each other exactly. However, when compared to the sample presumed to be from Nicholas, there was a single mismatch. Upon very close examination of the data, it was determined that Nicholas' sample actually had two bases at the mismatched position (Ivanov et al., 1996). As no other positions within the sample contained a similar overlay of bases, Gill concluded that this was a true heteroplasmy and not a contaminant.

At that time, heteroplasmies (true mixtures of mtDNA sequence types within one individual) were considered very rare. Currently, while not common, they are regularly seen in mtDNA analysis. Gill and his team calculated the probabilities that such a mixture might occur in an individual and determined a 98.5% probability that this was indeed Nicholas II (Gill et al., 1994). The Russian authorities refused to accept this result and withheld burial of the bodies until such time as further testing could be undertaken.

A more closely related maternal relative was needed to confirm the heteroplasmy. The Russian government gave permission for exhumation of the body of Georgij Romanov, Nicholas's brother, who had died in 1899 of tuberculosis. Ivanov took a sample from Georgij's remains and a fresh cutting from those of the 'Tsar' and delivered them to the Armed Forces DNA Identification Laboratory (AFDIL) in 1995. Not only did the Tsar's remains match the results generated by Gill and the FSS, but those of Georgij expressed the same heteroplasmy (Ivanov et al., 1996).

Given the combination of nucDNA, mtDNA and the anthropological evidence obtained, there is now no reasonable doubt that the recovered remains were those of Tsar Nicholas II, Tsarina Alexandra and three of the grand duchesses. Anthropologists are not in agreement as to whether the missing daughter is Maria or Anastasia (the two youngest), and it is beyond the scope of DNA analysis without a direct reference.

Disclaimer

The opinions and assertions contained herein are those of the author and do not necessarily reflect those of the United States Department of Defense or the United States Department of the Army.

See also
Ancient DNA: Recovery and Analysis

References

Gill P, Ivanov PL, Kimpton C, et al. (1994) Identification of the remains of the Romanov family by DNA analysis. *Nature Genetics* **6**: 130–135.

Holland MM, Fisher DL, Mitchell LG, et al. (1993) Mitochondrial DNA sequence analysis of human skeletal remains: identification of remains from the Vietnam war. *Journal of Forensic Sciences* **38**(3): 542–553.

Ivanov PL, Wadhams MJ, Roby RK, et al. (1996) Mitochondrial DNA sequence heteroplasmy in the Grand Duke of Russia Georgij Romanov establishes the authenticity of the remains of Tsar Nicholas II. *Nature Genetics* **12**(4): 417–420.

Massie RK (1995) *The Romanovs: The Final Chapter*. New York, NY: Random House.

Radzinsky E (2000) *The Rasputin File*. New York, NY: Anchor Books.

Stevens RF (1999) The history of haemophilia in the royal families of Europe. *British Journal of Haematology* **105**: 25–32.

Further Reading

Ivanov PL (1998) The expert identification of the remains of the imperial family by means of molecular genetic verification of genealogical relations. [In Russian.] *Sudebno-Meditsinskaia Ekspertiza (Moskva)* **41**(4): 30–47.

Stoneking M, Melton T, Nott J, et al. (1995) Establishing the identity of Anna Anderson Manahan. *Nature Genetics* **9**: 9–10.

rRNA Genes: Evolution

Iris L Gonzalez, *Al duPont Hospital for Children, Wilmington, Delaware, USA*

James E Sylvester, *Nemours Children's Clinic, Jacksonville, Florida, USA*

The head-to-tail tandem arrangement of the rRNA gene repeats leads to their concerted evolution on both homologous and nonhomologous chromosomes. But not all parts of the rDNA unit are homogenized to the same extent.

Introduction

This article concentrates on the genes for the 18S, 5.8S and 28S ribosomal ribonucleic acids (rRNAs). These RNAs are encoded in 300–400 copies of 43-kilobase (kb) ribosomal deoxyribonucleic acid (rDNA) units that are tandemly arranged, head to tail, on the short arms of the five pairs of acrocentric chromosomes. Each rDNA repeat is transcribed into a 13-kb primary transcript, which is subsequently processed to release the 18S, 5.8S and 28S rRNAs (1870, 156 and ∼ 5034 bases respectively). The transcribed regions are separated by a 30-kb intergenic spacer (IGS; **Figure 1**). (**See** Ribosomes and Ribosomal Proteins.)

The 18S rRNA is incorporated into the small ribosomal subunit, whereas the 28S and 5.8S are part of the large subunit along with 5S rRNA (120 bases). The genes for 5S rRNA are also tandemly repeated and are located on chromosome 1; however, the evolution of these genes is not described here. We discuss two aspects of the evolution of rRNAs and their genes: first, how the structure of the encoded rRNAs has evolved; and second, how this multigene family evolves.

Evolution of rRNA Structure

Ribosomal RNAs are a very ancient feature of cellular organisms and carry out catalytic functions during messenger RNA (mRNA) translation by the ribosome. The function of rRNA depends on secondary and tertiary structures that have been conserved over the eons, although the primary nucleotide sequences have diverged markedly between kingdoms and phyla. Conservation of structure was achieved by compensatory base changes that maintained intramolecular base pairing in the rRNAs. Thus, the basic structural modules or domains that are present in the prokaryotic and archaebacterial rRNAs are also contained in the much longer human 18S and 28S rRNAs (Raue *et al.*, 1988).

These functional domains are called the 'conserved' regions and are separated by GC-rich segments called 'variable regions' or 'expansion segments', which make up as much as 50% of 28S rRNA. The origin of variable regions is unclear (Gerbi, 1986; Gray and Schnare, 1996). It is possible that variable regions are remnants of spacers that separated cotranscribed primordial rRNA coding modules (the present-day

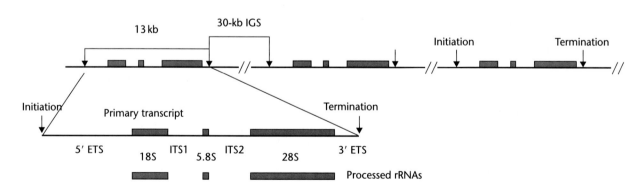

Figure 1 Production of rRNAs from rDNA. Transcription from the rDNA repeats (top) produces primary transcript (middle), which then yields the processed rRNAs (bottom). ETS: external transcribed spacer; IGS: intergenic spacer; ITS: internal transcribed spacer.

conserved domains), which have been retained in the eukaryotic lineages but were eliminated from the more economical prokaryotic lineages as the modules merged into larger genes. Evidence for this hypothesis comes from the existence of modular rRNAs (rRNAs in pieces) in protistan cytoplasmic ribosomes and in algal mitochondrial ribosomes. Alternatively, variable regions could be late acquisitions, that is, sequences that have been inserted into simple prokaryotic-like molecules.

It is striking that both the variable regions and the internal transcribed spacers (ITSs) and external transcribed spacers (ETSs) have similar high G + C compositions and can maintain their secondary structures under the denaturing conditions that were used in early electron microscopy studies of the primary rRNA; melting curves of heteroduplexes of rRNAs and genomic DNA also predicted that the sequences between the 18S and 28S coding regions would have a high G + C content. The shared sequence and structural features of variable regions and transcribed spacers do suggest a common descent (Gray and Schnare, 1996). These are the only GC-rich segments in the whole rDNA molecule – neither the conserved regions nor the IGS have this composition, and the myth that rDNA has a high G + C content should be laid to rest. The variation of the more rapidly evolving variable regions and transcribed spacers arises largely from replication slippage in simple sequence motifs but also includes base substitutions.

The conserved regions have maintained the functionally important secondary and tertiary structures by compensatory base changes as they evolved. In fact, comparative studies of prokaryotic and eukaryotic rRNAs, combined with biochemical studies, have helped to elucidate these structures. As their name implies, conserved regions show almost no variation within a species or between closely related species (0% divergence between human, chimpanzee and gorilla, and only 0.2% between human and orangutan), and very low divergence among mammals (about 0.6% between human and mouse).

Using rRNAs for Phylogenetic Reconstruction

The universality of rRNAs, the multiple copies of rRNA genes, and the large amount of rRNA in cells make these molecules easy to clone and sequence (rRNA can also be sequenced directly). These features have made rRNA attractive for phylogenetic studies,

which are carried out in two ways: the slowly evolving conserved region sequences are used for phylogenetic inference between distantly related organisms, which helps to establish relationships between major lineages and phyla (Hillis and Dixon, 1991); and the fast evolving variable regions, ITSs and ETSs are used for phylogenetic studies of closely related taxa, such as primates, and are also used extensively for taxonomic identification. (*See* Hominids: Molecular Phylogenetics; Primates: Phylogenetics.)

Evolution of rRNA Genes

The rDNA shown in **Figure 1** consists of 30–40 tandemly arranged repeats of the 43 kb unit on each acrocentric chromosome p arm, as shown by electron microscopy studies (Wellauer and Dawid, 1979). With hundreds of genes as targets for mutation, one would expect variation to accumulate because selection must be relaxed (e.g. people with Robertsonian fusions lack two clusters of rDNA and show no detrimental effect, so that one could imagine that 60–80 inactive genes might have no effect).

Variation is indeed present. It was noted initially in electron microscopy measurements that showed differences in the length of DNA between the transcribed regions. Subsequently, Southern blots allowed characterization of restriction-site polymorphisms in the transcribed region and the IGS, and of length variation in the IGS (Arnheim et al., 1980). Although some of these studies were done with total genomic DNA, some were done with rodent–human somatic cell hybrids containing single acrocentric chromosomes, which showed that variants could be shared among nonhomologous chromosomes and that a single chromosome could carry more than one variant. Several clones derived from single individuals were found to show intraspecies sequence variation in the 28S coding region, and S1 nuclease studies showed that these variants are actually present in cellular rRNA. Finally, after the advent of polymerase chain reaction it was possible to clone and sequence many fragments from along the whole rDNA derived from individual acrocentric chromosomes, and the extent of variation of different parts of rDNA could be compared (Gonzalez and Sylvester, 2001). These studies and others have shown that there is low variation in transcribed and potential regulatory regions of rDNA and much higher divergence in the IGS.

In view of the variability that was being found, the homogeneity among the hundreds of rDNA copies distributed into 10 rDNA clusters was striking. Early work clearly showed that rDNA is subject to concerted evolution (Arnheim *et al.*, 1980). This process operates on tandemly arranged repeated genes and results in much less divergence among the multiple copies than if the copies were allowed to evolve independently. Concerted evolution occurs by the mechanisms of unequal homologous exchanges and gene conversion, and promotes two opposite results: it can eliminate new mutations and is therefore responsible for stability and uniformity of the gene copies – even those located on nonhomologous chromosomes; and it can aid the rapid spread and fixation of mutations among the multiple gene copies. How fast variants can spread is shown by finding increases in the number of variant copies from one generation to the next in human pedigrees. (*See* Concerted Evolution; Globin Genes: Evolution.)

The picture that has emerged indicates that although rDNA units evolve in concert, not all parts of rDNA evolve equally in concert (Arnheim *et al.*, 1980; Gerbi, 1986; Gonzalez and Sylvester, 2001). The transcribed and regulatory regions (which must feel the pressure of selection), and also the sequences distal to rDNA on the acrocentric p arms, evolve rather slowly and seem to be exchanged frequently and/or corrected intrachromosomally and among homologs and nonhomologs (Gonzalez and Sylvester, 2001). For example, variation was found to be 0% in a conserved region, of 28S, and 0.12–0.13% in a variable region, in ITS1 and in a region upstream of transcription initiation that contained p53-binding sites. These values contrast with a divergence of 2.5–6.7% between certain fragments in the IGS.

The IGS evolves much faster, as would be predicted from its high content of simple sequence and from the numerous retroposons present (mostly *Alu* elements). Length variation in the IGS has been shown to result from unequal homologous exchanges between sets of tandemly repeated sequence blocks (La Volpe *et al.*, 1985; Sylvester *et al.*, 1989); it does not seem to arise from recombination among paralogous Alu elements, as they have entered the rDNA at various times and their sequences are too divergent (15–26%) to align properly. (*See* Retrosequences and Evolution of Alu Elements; Short Interspersed Elements (SINEs).)

IGS sequence variation was characterized by obtaining multiple clones of a specific fragment that is present only once per rDNA unit. Unexpectedly, the sequences of these clones could be classified into five classes that show low intraclass divergence (0–1.3%) and high interclass divergence (2.5–6.7%; Gonzalez and Sylvester, 2001). Concerted evolution among the members of a class must account for their low divergence. But 1–3 IGS fragment classes can coexist in a single rDNA array. This alters our picture of

rDNA concerted evolution, as it suggests that IGSs are preferentially homogenized intrachromosomally through sequence exchanges or gene conversions among adjacent rDNA units to form and maintain these sequence classes; less frequent interchromosomal exchanges then spread these IGS classes to the other acrocentrics.

How can concerted evolution result simultaneously in stringently conserved regions and in less conserved regions? First, DNA repair along the rDNA is probably not uniform, as judged by conflicting results from various studies. Second, the close proximity of rDNAs and their topological arrangement in the nucleolus may differentially expose various segments to recombination factors such as nucleolin and topoisomerase I (TOPOI). Third, there are various localized recombination-promoting sequences in rDNA, such as repeats of CT and CTTT (triplex DNA), TG (Z-DNA) and consensus sequences for TOPOI binding in the IGS, and also G- and GC-rich sequences that are characteristic of recombination hot spots in the transcribed region. Some of the recombination-promoting sequences may favor intrachromosomal exchanges, whereas others may work equally well within and between chromosomes. This notion is supported by studies of *Drosophila* rDNA that show that interchromosomal homogenization of IGS sequences is due to the presence of TOPOI-binding sites, which results in sharing of IGS sequences on the X and Y chromosomes. But *Drosophila* ITS1 sequences tend to homogenize only within X or Y chromosome (Polanco *et al.*, 1998). Last, there is the differential methylation status of transcribed (hypomethylated) and nontranscribed (hypermethylated) parts of rDNA, where hypomethylation is thought to increase recombination frequency (hypomethylation is thought to contribute to the concerted evolution in the human RNU2 system). (*See* DNA Recombination; DNA Repair; GC-rich Isochores: Origin; Nucleolus: Structure and Function; Topoisomerases.)

In conclusion, rRNA genes on the acrocentric chromosomes evolve in concert, but not uniformly along the length of the rDNA repeat. In addition, there is as much exchange among homologous as among nonhomologous chromosomes, such that no chromosome-specific variants have been found, even in 8.5 kb of adjacent sequences on the telomeric side of rDNA.

See also

Concerted Evolution
Nucleolus: Structure and Function
Ribosomes and Ribosomal Proteins

References

Arnheim A, Krystal M, Schmickel R, *et al.* (1980) Molecular evidence for genetic exchanges among ribosomal genes on nonhomologous chromosomes in man and apes. *Proceedings of the National Academy of Sciences of the United States of America* **77**: 7323–7327.

Gerbi SA (1986) The evolution of eukaryotic ribosomal DNA. *BioSystems* **19**: 247–258.

Gonzalez IL and Sylvester JE (2001) Human rDNA: evolutionary patterns within the genes and tandem arrays derived from multiple chromosomes. *Genomics* **73**: 255–263.

Gray MW and Schnare MN (1996) Evolution of rRNA organization. In: Zimmerman RA and Dahlberg AE (eds.) *Ribosomal RNA: Structure, Evolution, Processing, and Function in Protein Synthesis*, pp. 49–69. Boca Raton, FL: CRC Press.

Hillis DM and Dixon MT (1991) Ribosomal DNA: molecular evolution and phylogenetic inference. *Quarterly Review of Biology* **66**: 411–453.

La Volpe A, Simeone A, D'Esposito M, *et al.* (1985) Molecular analysis of the heterogeneity region of the human ribosomal spacer. *Journal of Molecular Biology* **183**: 213–223.

Polanco C, Gonzalez AI, de la Fuente A and Dover GA (1998) Multigene family of ribosomal DNA in *Drosophila melanogaster* reveals contrasting patterns of homogenization for IGS and ITS spacer regions: a possible mechanism to resolve this paradox. *Genetics* **149**: 243–256.

Raue HA, Klootwijk J and Musters W (1988) Evolutionary conservation of structure and function of high molecular weight ribosomal RNA. *Progress in Biophysics and Molecular Biology* **51**: 77–129.

Sylvester JE, Petersen R and Schmickel R (1989) Human ribosomal DNA: novel sequence organization in 4.5 kilobases upstream of the promoter. *Gene* **84**: 193–196.

Wellauer PK and Dawid IB (1979) Isolation and sequence organization of human ribosomal DNA. *Journal of Molecular Biology* **128**: 289–303.

Further Reading

Dover G (1986) Molecular drive in multigene families: how biological novelties arise, spread and are assimilated. *Trends in Genetics* **2**: 159–165.

Dover GA, Linares AR, Bowen T and Hancock JM (1993) Detection and quantification of concerted evolution and molecular drive. *Methods in Enzymology* **224**: 525–541.

Gonzalez IL and Sylvester JE (1995) Complete sequence of the 43-kb human ribosomal DNA repeat: analysis of the intergenic spacer. *Genomics* **27**: 320–328.

Kuick R, Asakawa J-I, Neel JV, *et al.* (1996) Studies of the inheritance of human ribosomal DNA variants detected in two-dimensional separations of genomic restriction fragments. *Genetics* **144**: 307–316.

Maden BEH, Dent CL, Farrell TE, *et al.* (1987) Clones of human ribosomal DNA containing the complete 18S-rRNA and 28S-rRNA genes. *Biochemical Journal* **246**: 517–527.

Nitta I, Kamada Y, Ueda T and Watanabe K (1998) Reconstitution of peptide bond formation with *Escherichia coli* 23S ribosomal RNA domains. *Science* **281**: 666–669.

Noller HF (1984) Structure of ribosomal RNA. *Annual Reviews in Biochemistry* **53**: 119–162.

Schlötterer C and Tautz D (1994) Chromosomal homogeneity of *Drosophila* ribosomal DNA arrays suggest intrachromosomal exchanges drive concerted evolution. *Current Biology* **4**: 777–783.

Wellauer PK, Dawid IB, Kelley DE and Perry RP (1974) Secondary structure maps of ribosomal RNA. II. Processing of mouse L-cell ribosomal RNA and variations in the processing pathway. *Journal of Molecular Biology* **89**: 397–407.

Worton RG, Sutherland J, Sylvester JE, *et al.* (1988) Human ribosomal RNA genes: orientation of the tandem array and conservation of the 5′ end. *Science* **239**: 64–68.

Saccharomyces

See Yeast as a Model for Human Diseases

Sample Size Requirements

Michael Knapp, *University of Bonn, Bonn, Germany*

Sample size requirements in the context of linkage analysis between a trait and genetic markers refer to the calculation of the sample size required for a linkage study to guarantee a certain power of the study against an assumed inheritance model. The power against an assumed inheritance model is the probability of correctly declaring a linkage when in fact the assumed inheritance model is true and the trait and marker loci are linked.

Factors Influencing the Power of a Linkage Study

The power to detect linkage depends on a variety of factors. The central difficulty, especially in the context of genetically complex diseases, lies in the specification of the trait characteristics or inheritance model. In most applications, a single-locus, diallelic disease model is assumed, but even this simplified approach requires the specification of four parameters (disease allele frequency and penetrance for each of three genotypes at the disease locus). The marker characteristics include the number of alleles at the marker locus and their frequencies, which determine the informativeness of the marker locus. The genetic distance between marker and disease locus is described by the recombination fraction θ. The types of family (e.g. nuclear families or extended pedigrees) in the sample, as well as the specific statistical test used to decide on linkage, together with the type I error rate, will influence the power of the study.

Methods for Power Calculations

Three approaches for power calculations can be distinguished. The power of a statistical test can sometimes be calculated exactly but more often, because of the complexity of the calculations, requires some form of approximation. This may be via simulation methods or through approximation of the test statistic under the alternative hypothesis.

Firstly, provided that the sample space is sufficiently small, it can be possible to calculate the exact power of a statistical test for linkage by summing up the probabilities (under the alternative of interest) of samples that lead to the rejection of the null hypothesis.

Secondly, power calculations may be performed by an approximation of the distribution of the test statistic under the alternative of interest. A power estimate is then obtained by calculating the tail probabilities of this distribution.

The third approach for power calculation is based on simulations. Having specified the alternative of interest, a certain number of replicates of the sample are generated on the basis of a random number generator. Each replicate is then analyzed by the statistical test for which the power has to be estimated. The relative frequency of replications leading to the rejection of the null hypothesis of no linkage is an estimate of the power of this statistical test against the alternative under which the samples were generated.

The following sections describe examples for each of these approaches.

Power Calculations for Affected Sib Pair Tests: Exact Calculation

The affected sib pair method is a popular approach to linkage analysis, especially in the context of genetically complex diseases. A sample for the affected sib pair method consists of n nuclear families, each family consisting of two affected sibs and their parents. All members in all families are typed at the marker locus. The assumption that the marker is completely informative assures that, for each sib pair, the number of marker alleles that are identical by descent (ibd) can be determined unambiguously. Let n_i ($i = 0, 1, 2$) denote the observed number of sib pairs sharing i marker alleles. Then, (n_0, n_1, n_2) is a realization of a trinomial distributed random variable (N_0, N_1, N_2) with parameters n and (p_0, p_1, p_2). In the case of no linkage,

$\mathbf{p} = (p_0, p_1, p_2)$ becomes $\mathbf{p}^0 = (1/4, 1/2, 1/4)$. Different statistical tests have been proposed to decide between hypotheses H_0: $\mathbf{p} = \mathbf{p}^0$ and H_1: $\mathbf{p} \neq \mathbf{p}^0$. In the following, the focus is on the so-called mean test (Blackwelder and Elston, 1985), which rejects the null hypothesis if the total number $T(n_0, n_1, n_2) := 2n_2 + n_1$ of shared alleles in the sample is sufficiently large. Under the null hypothesis, T is binomially distributed $B(2n, 1/2)$. Therefore, the critical value $c(n, \alpha)$ of the mean test corresponding to a type I error rate α is given by

$$c(n, \alpha) := \min\{c : \sum_{k=c}^{2n} \binom{2n}{k} (1/2)^{2n} \leq \alpha\}$$

For an exact calculation of the power of the mean test against a specific alternative, the first step is to calculate the ibd distribution $\mathbf{p}^* = (p_0^*, p_1^*, p_2^*)$ induced by this alternative. Suarez et al. (1978) provided formulae that express $\mathbf{p}^*$ in terms of disease allele frequency and penetrances for a general diallelic single-locus disease model. Risch (1990) parameterized $\mathbf{p}^*$ by using relative recurrence risks for the offspring and sib of affected individuals. The second step is to sum up the probabilities of samples leading to a rejection of H_0, that is,

$$\sum_{\{(n_0, n_1, n_2): 2n_2 + n_1 \geq c(n,\alpha)\}} n! \prod_{i=0}^{2} \frac{(p_i^*)^{n_i}}{n_i!}$$

which can equivalently be expressed by a weighted sum of tail probabilities of binomial distributions, that is,

$$\sum_{n_1=0}^{n} \binom{n}{n_1} (p_1^*)^{n_1} (1 - p_1^*)^{n-n_1}$$

$$\sum_{n_2 \geq (c(n,\alpha)-n_1)/2} \binom{n-n_1}{n_2} \left(\frac{p_2^*}{1-p_1^*}\right)^{n_2} \left(1 - \frac{p_2^*}{1-p_1^*}\right)^{n-n_1-n_2}$$

It is straightforward to incorporate this latter formula into a computer program. The resulting value is the exact power of the mean test under the assumption of a completely informative marker and can be considered an upper bound for the power of an incompletely informative marker.

Power Calculations for the Transmission/Disequilibrium Test: Approximate Calculation

The transmission/disequilibrium test (TDT) introduced by Spielman et al. (1993) is a linkage test that exploits the presence of linkage disequilibrium between alleles at the marker and the disease locus. The TDT can be applied to nuclear families with an arbitrary number of affected children. As shown by Risch and Merikangas (1996), the TDT can be substantially more powerful than affected sib pair tests, provided that strong linkage disequilibrium is present.

Assuming a diallelic marker locus with alleles A and B, the TDT compares the number of transmissions of allele A with the number of transmissions of allele B by heterozygous parents to their affected children. Even in its simplest situation, given by a sample consisting of nuclear families with a single affected child in each family, six essentially different types of families have to be distinguished. The probabilities of these family types can be calculated for an assumed inheritance model. However, the number of different samples consisting of n families is $\binom{n+5}{5}$ (Feller, 1970), which for $n = 100$ is already approximately 9.66×10^7. Therefore, an exact power calculation by summing up the probabilities of samples leading to the rejection of H_0 will be possible only for very small sizes of n.

The test statistic of the TDT is a function of multinomially distributed random variables. By replacing this multinomial distribution with a multidimensional normal distribution, the distribution of the square root of the test statistic of the TDT can be approximated by a normal distribution. The expectation and variance of this approximating normal distribution depend on the alternative of interest but can easily be calculated numerically, as has been shown by Knapp (1999). Whereas Knapp (1999) considered only a simplified situation by assuming that there is complete linkage disequilibrium between the alleles at the marker and the disease locus, and that marker and disease locus are completely linked, both of these assumptions can be relaxed. An SAS (SAS Institute Inc., 1990) macro that computes power approximations for the TDT under general alternatives can be obtained via the internet (see Web Links).

Power Calculations for Parametric Linkage Analysis: Simulation Method

For specified trait and marker characteristics and an assumed recombination fraction, it is straightforward to simulate pedigree data (i.e. marker genotypes and trait phenotypes) based only on pedigree structure. For example, the program SIMULATE, described by Terwilliger et al. (1993; see Web Links) can be used for this purpose. This kind of simulation would be appropriate if random families were sampled, but in almost all linkage studies, the families are selected for the presence of disease in at least one family member. A similar situation occurs when disease phenotypes in all members of all families are already available and, prior to undertaking the typing of marker genotypes, it has to be decided whether the collected pedigrees provide sufficient information to demonstrate linkage. In both scenarios, the simulations have to be consistent with partially or completely observed trait phenotypes. Therefore, the simulations have to sample marker genotypes from the conditional distribution of the genotypes given trait phenotypes. Ploughman and Boehnke (1989) described an algorithm for this purpose and implemented their method in the program SIMLINK (see Web Links). A similar approach to simulate marker genotypes conditional on disease phenotypes has been implemented in the SLINK program (Weeks et al., 1990; see Web Links). Both programs (see Web Links) provide the user with information on the proportion of replicates in which the maximum lod score exceeds some threshold, which is an estimate of the power of the linkage study.

See also

Linkage Analysis
Linkage Disequilibrium
Parametric and Nonparametric Linkage Analysis
Relatives-based Tests of Association

References

Blackwelder WC and Elston RC (1985) A comparison of sib-pair linkage tests for disease susceptibility loci. *Genetic Epidemiology* **2**: 85–97.

Feller W (1970) *An Introduction to Probability Theory and its Applications*, 3rd edn. New York, NY: Wiley.

Knapp M (1999) A note on power approximations for the transmission/disequilibrium test. *American Journal of Human Genetics* **64**: 1177–1185.

Ploughman LM and Boehnke M (1989) Estimating the power of a proposed linkage study for a complex genetic trait. *American Journal of Human Genetics* **44**: 543–551.

Risch N (1990) Linkage strategies for genetically complex traits. II. The power of affected relative pairs. *American Journal of Human Genetics* **46**: 229–241.

Risch N and Merikangas K (1996) The future of genetic studies of complex human diseases. *Science* **273**: 1516–1517.

SAS Institute Inc. (1990) SAS Language: Reference, version 6, 1st edn. Cary, NC: SAS Institute.

Spielman RS, McGinnis RE and Ewens WJ (1993) Transmission test for linkage disequilibrium: the insulin gene region and insulin-dependent diabetes mellitus (IDDM). *American Journal of Human Genetics* **52**: 502–516.

Suarez BK, Rice J and Reich T (1978) The generalized sib pair IBD distribution: its use in the detection of linkage. *Annals of Human Genetics* **42**: 87–94.

Terwilliger JD, Speer M and Ott J (1993) Chromosome-based method of rapid computer simulation in human genetic linkage analysis. *Genetic Epidemiology* **10**: 217–224.

Weeks DE, Ott J and Lathrop GM (1990) SLINK: a general simulation program for linkage analysis. *American Journal of Human Genetics* **47**: A204.

Further Reading

Chen WM and Deng HW (2001) A general and accurate approach for computing the statistical power of the transmission disequilibrium test for complex disease genes. *Genetic Epidemiology* **21**: 53–67.

Holmans P (2001) Nonparametric linkage. In: Balding DJ, Bishop M and Cannings C (eds.) *Handbook of Statistical Genetics*, pp. 541–563. Chichester, UK: Wiley.

Holmans P and Clayton D (1995) Efficiency of typing unaffected relatives in an affected-sib-pair linkage study with single-locus and multiple tightly linked markers. *American Journal of Human Genetics* **57**: 1221–1232.

Krawczak M (2001) ASP – a simulation-based power calculator for genetic linkage studies of qualitative traits, using sib-pairs. *Human Genetics* **109**: 675–677.

Risch N (2000) Searching for genetic determinants in the new millennium. *Nature* **405**: 847–856.

Thompson EA (2001) Linkage analysis. In: Balding DJ, Bishop M and Cannings C (eds.) *Handbook of Statistical Genetics*, pp. 541–563. Chichester, UK: Wiley.

Weeks DE and Lathrop GM (1995) Polygenic disease: methods for mapping complex disease traits. *Trends in Genetics* **11**: 513–519.

Web Links

SAS macro to compute power approximations for the TDT
http://www.uni-bonn.de/~umt70e/soft.htm
SIMLINK program
http://www.sph.umich.edu/statgen/boehnke/simlink.html
SIMULATE program
ftp://linkage.rockefeller.edu/software/simulate
SLINK program
ftp://linkage.rockefeller.edu/software/slink

Sanger, Frederick

Noel G Coley, *The Open University, Milton Keynes, UK*

Frederick Sanger (1918–) is a British biochemist who determined the sequence of amino acid residues in insulin and devised the dideoxy (Sanger) method of sequencing DNA.

Frederick Sanger was born in Rendcombe, Gloucestershire, on 13 August 1918 and educated at Bryanston School and St John's College, Cambridge, from where he graduated as a Bachelor of Arts in Natural Sciences in 1939 and a Doctor of Philosophy in 1943. He held a Beit Memorial Fellowship for medical research (1944–1951) and then worked for the Medical Research Council (MRC) in Cambridge until his retirement in 1983. He was awarded the Nobel Chemistry Prize in 1958 for his work on the molecular structure of insulin, and shared a second Nobel prize with Paul Berg and Walter Gilbert in 1980 for contributions to determining the base sequences in nucleic acids. Sanger is one of the very few scientists to have been awarded two Nobel prizes.

Sanger began his study of the insulin molecule with A. C. Chibnall's discovery of its two polypeptide chains that each terminate in a free α-amino acid group. He found that 2,4-dinitro-1-fluorobenzene (Sanger's reagent) combined with these amino acids and that one chain ended with phenylalanine and the other with glycine. The chains were linked by disulfide bonds between cystine residues, which Sanger oxidized with performic acid to separate the insulin chains. He fragmented each polypeptide chain by mild acid hydrolysis and identified the amino ends of each fragment, together with its composition using a new electrophoresis separation procedure learned on a visit in 1947 to Arne Tiselius at Uppsala, Sweden. Then, by using proteinase enzymes that split proteins after specific amino acids, he determined the order of the amino acids in each chain. He found that the glycyl chain contained 21 amino acids and the phenylalanyl chain had 30. The fragments were difficult to separate, but, by combining paper chromatography and ionophoresis techniques, he constructed ever-longer sequences and by 1953 he had determined the order of the 51 amino acids in both chains. Only the positions of the three sulfur bridges in the insulin molecule remained to be found and, by carefully analyzing residues from the double-chain molecule, Sanger located all three, thus completing the structure of the insulin molecule from cow. He subsequently found minor differences in the amino acid sequences in insulins from the pig, sheep, horse and whale. Sanger's work, for which he received the 1958 Nobel prize for

Chemistry, showed that the molecular structure of a complex biochemical molecule can be determined given a clear enunciation of the specific problems, innovatory laboratory techniques and dogged persistence. (*See* Insulin-dependent Diabetes Mellitus (IDDM): Identifying the Disease-causing Gene at the *IDDM11* Locus.)

Around 1958, Sanger became interested in nucleic acids – biological macromolecules with base sequences that determine the information carried by the genes. In 1962 he joined the group of Max Perutz in the new MRC Laboratory of Molecular Biology in Cambridge, where Francis Crick encouraged him to seek a rapid method for nucleic acid sequencing. To attack this problem Sanger used radioactive phosphorus (^{32}P) as a label and partial polymer degradation methods similar to those that he had used for insulin. Enzymes capable of cutting RNA at specific points had been discovered. Ribonuclease T_1, which cuts the molecule after guanine, was digested with an RNA strand and the fractions were separated using two-dimensional ionophoresis. The products were then detected by autoradiography. The isolated oligonucleotides were treated further with ribonucleases, and the resulting small fragments were separated by two-dimensional chromatography. (*See* Capillary Electrophoresis; Perutz, Max Ferdinand.)

In the mid-1960s, Sanger began work on single-stranded DNA using the same methods, but he found them too cumbersome for sequences of more than 50 nucleotides. For DNA containing hundreds or thousands of nucleotides, a quicker method was needed. In his first attempt to achieve this, Sanger used DNA polymerase to copy single-stranded DNA molecules that incorporated radiolabeled ribonucleotides. The new DNA strand had weak links that could be broken by digestion with alkali wherever the ribonucleotide was inserted. By this method sections of DNA containing about 80 nucleotides could be sequenced, but, because even the smallest DNA genomes contain several thousand nucleotides, still more powerful methods were required.

Sanger observed that when a low concentration of a labeled nucleotide was used in the process, DNA fragments of various sizes were synthesized, each of which stopped before reaching the nucleotide in short supply. This suggested a new method of fractionating on the basis of chain length. The relative sizes of the chains would be related to the location in the DNA of the depleted nucleotide. This technique was described as the 'plus and minus' method. Later it was found that incorporating dideoxy nucleoside triphosphates into growing DNA chains halted growth. By using dideoxy nucleotides of each of the four DNA bases in separate reactions, segments of DNA of varying length were produced each of which ended with a specific base. The segments were separated using electrophoresis on polyacrylamide gels and their positions were revealed on an autoradiograph. By applying the four samples adjacent to each other on a gel, the exact sequence of the DNA could be read directly by eye from an autoradiograph. Quicker and more accurate than the plus and minus method, this technique was used in 1975 to complete the DNA sequence of the virus ϕ X174, which contains 5386 nucleotides. The sequencing showed overlapping genes in regions of the viral DNA coding for two genes. (*See* Gel Electrophoresis; Sequence Accuracy and Verification; Sequence Finishing; Two-dimensional Gel Electrophoresis.)

The methods based on dideoxy sequencing all required a single strand of DNA, which causes no difficulty for some viruses; however, most DNA is double-stranded and Sanger went on to devise techniques for separating the strands and, more importantly, for cloning DNA into single-strand-producing viruses. This method resulted in the sequencing of DNA in bovine and human mitochondria, the latter with a sequence of 16 569 nucleotides. In his early work, Sanger showed that certain sequences of nucleotides encode the production of particular proteins, and he pioneered methods that validated the genetic code. His study of human mitochondrial DNA also revealed a difference between the mitochondrial genetic code and that of other biological systems, which showed that the code is not universal as had been thought previously.

(*See* Celera Genomics: The Race for the Human Genome Sequence; DNA Cloning; Evolutionary History of the Human Genome; Mitochondrial Genome; Mitochondrial Genome: Evolution.)

Sanger's researches stemmed from his conviction that knowledge of sequences in biological molecules could contribute significantly to our understanding of living matter. In devising techniques for identifying the exact sequence of nucleotides in DNA, he provided a valuable tool for producing useful proteins, such as interferon and human hormones, by the manipulation of the DNA molecule. He was named a Companion of Honour in 1981 and a member of the Order of Merit in 1986. (*See* Gonadotropin Hormones: Disorders.)

See also
Genome Sequencing
Perutz, Max Ferdinand

Further Reading

Anderson S, Bankier AT, Barrell BG, *et al.* (1981) Sequencing and organisation of the human mitochondrial genome. *Nature* **290**: 457–465.

Barker GR (1993) In: James LK (ed.) *Nobel Laureates in Chemistry 1901–1992*, pp. 406 – 411, 633–638. Washington DC: American Chemical Society and the Chemical Heritage Foundation.

Burke HM (1990) In: Magill FN (ed.) *The Nobel Winners, Chemistry 1901–1989*, 3 vols; vol. 2, pp. 683–690. Pasadena, CA, Englewood Cliffs, NJ: Salem Press.

Daintith J, Mitchells S, Gjertsen D and Tootile E (eds.) (1994) *Biographical Dictionary of Scientists*, 2nd edn. Bristol: Institute of Physics.

Sanger F (1964) The chemistry of insulin. *Nobel Lectures, Chemistry*, 1942–1962, pp. 544–556. Amsterdam: Elsevier.

Sanger F (1975) Nucleotide sequences in DNA. *Proceedings of the Royal Society of London B* **191**: 317–333.

Sanger F (1981) Determination of nucleotide sequences in DNA. *Les Prix Nobel en 1980*, pp. 143–159. Stockholm: Almqvist & Wiksell International.

Silverstein A and Silverstein V (1969) *Frederick Sanger; The Man who Mapped out a Chemical of Life*. New York, NY: John Day.

Smith R (1990) In: Magill FN (ed.) *The Nobel Winners, Chemistry 1901–1989*, 3 vols; vol. 3, pp. 1049–1057. Pasadena, CA, Englewood Cliffs, NJ: Salem Press.

SARs

See Matrix-associated Regions (MARs) and Scaffold Attachment Regions (SARs)

Schizophrenia and Bipolar Disorder: Linkage on Chromosomes 5 and 11

Michel Maziade, *Laval University, Sainte-Foy, Quebec, Canada*

Roberta Palmour, *McGill University, Montreal, Quebec, Canada*

Daniel Phaneuf, *Laval University, Sainte-Foy, Quebec, Canada*

Chantal Mérette, *Laval University, Sainte-Foy, Quebec, Canada*

Marc-André Roy, *Laval University, Sainte-Foy, Quebec, Canada*

Two premature claims of linkage on chromosomes 5 and 11 for schizophrenia (SZ) and bipolar affective disorder (BP) were reported in *Nature* in the late 1980s. Although the findings failed to be replicated, accumulated knowledge from the results of the first generation of molecular genetic studies of SZ and BP sets the basis for a promising second generation of studies.

Introductory article

Article contents

- Introduction
- Linkage Studies as a Starting Point to Detect Susceptibility Genes
- A Well-grounded Start with False Expectations
- Methodological Pitfalls to Avoid in the Second Generation of Molecular Genetic Studies
- Conclusion

Introduction

Science moves forward through trial and error and demands thorough and replicable methods. By recognizing its failures, it has contributed historically to new ways of thinking. Now and then, mistaken goals or observations in one field have led to great medical discoveries in another. For example, chlorpromazine, the first effective antipsychotic, was a failure as an antihistamine, but revolutionized brain pharmacology. The brief history of research into the genetics of major psychiatric disorders such as schizophrenia (SZ) and bipolar affective disorder (BP) has already been plagued (or blessed) by such inadvertent errors.

Linkage Studies as a Starting Point to Detect Susceptibility Genes

After successes with previously intractable disorders such as Huntington disease, the application of linkage strategies, as a first step to identifying the susceptibility genes for heritable diseases, generated much enthusiasm in the late 1980s. There followed more than a decade of disappointing results for psychiatry. However, the negative findings of the first generation of molecular genetic studies of SZ and BP have led to new conceptual and methodological approaches to studying the complex mechanisms underlying these common diseases. It is thus essential to analyze the reasons why previous claims of strong linkage findings, reported in very prestigious journals, provided shaky starts. The priority is to avoid similar pitfalls at the

beginning of the twenty-first century, during the second generation of molecular genetic studies of major psychoses.

Logic of linkage studies as an efficient tool to locate susceptibility genes

Linkage analysis is based on meiotic recombination and on the probabilistic estimation of the distance between a marker, the location of which is known on a chromosome, and the putative disease susceptibility gene. Given that the rate of recombination (θ) increases with the physical distance between two chromosomal loci, linkage is declared between the chromosomal locus and a disease gene if the estimated distance is short. This step is followed by the molecular search for the disease gene at this locus. The 'lod score' is the measurement used in linkage analysis that assesses the level of evidence of the location of a gene, based on the maximization of a likelihood function over θ. In other terms, lod score analyses examine this linkage phenomenon within a sample of multigenerational families densely affected by the disease. It inspects the cosegregation of an allelic marker, the chromosomal location of which is known, with the affected status of the family members. By following the segregation of marker alleles from the affected parents to the offspring, chromosomal loci in which affected offspring inherit one marker allele, contrary to the unaffected, can be circumscribed. Such a strategy has been successful in detecting disease genes or mutations for several classical Mendelian genetic illnesses and for some complex polygenic disorders.

A Well-grounded Start with False Expectations

Linkage claim for schizophrenia in the late 1980s

In one of the first linkage studies of SZ, Sherrington *et al.* reported, in 1988, a linkage signal that the authors interpreted as strong evidence for an SZ susceptibility locus on the long arm of chromosome 5. They proposed it as the 'first concrete evidence for a genetic basis for SZ'. Sherrington *et al.* had data from seven families of UK and Icelandic origin which yielded an lod score of 6.49 in 5q11–q13. Moreover, a *Lancet* report published in the same year by Bassett *et al.* indicated a chromosomal abnormality at 5q11–q13 in a Chinese family affected by SZ. Candidate genes such as the glucocorticoid receptor gene and the 5-hydroxytryptamine-1A receptor gene were already known to be localized at 5q11–q13. Given the scientific enthusiasm of the late 1980s regarding the genetics of psychiatric disorders, as well as the high prevalence of this devastating brain disorder, the story was broadcast worldwide, before any replication studies. Subsequent studies did not replicate the chromosomal region, and the combined reanalysis of published data ruled out straightforward linkage heterogeneity. Different transmission models and different disease definitions were used in tens of subsequent genome scans and replication studies for SZ, which made a final consensus difficult.

Premature claim of linkage for bipolar affective disorder in the 1980s

In 1987, there was also a widely publicized *Nature* report by Egeland *et al.* of linkage in Amish pedigrees densely affected with multigenerational BP. The data appeared to provide strong support for a linkage (lod: 3.0–4.0) between affective disorder and a locus at the tip of chromosome 11p. The authors proposed a gene, perhaps tyrosine hydroxylase, in the region between the *v-Ha-ras Harvey rat sarcoma viral oncogene homolog* (*HRAS*; MIM number: 190020; formerly *HRAS1*) or *insulin* (*INS*; MIM number: 176730) genes region as the site of the susceptibility gene for BP. Dopamine, the metabolic product of tyrosine hydroxylase, was already implicated in the development of major psychiatric disorders, so there was widespread excitement over the report. Two years later, the authors published more data, in which the evidence of linkage vanished. Similar to the story of Sherrington *et al.*, several reports followed, but failed to replicate

the 11p finding for BP. As for the SZ studies, the replication in studies for BP used different populations, methods and phenotype definitions.

Scientific and social reasons for premature claims of linkage

The initial enthusiasm for the reports was fueled by the valid linkage findings for simple Mendelian disorders obtained since 1984. Also, the extensive and time-consuming linkage studies of Sherrington *et al.* and Egeland *et al.* represented the state of the art and were highly credible. Such studies, reported in the late 1980s, had to be conceptualized methodologically in the late 1970s/early 1980s, and scientists had to utilize concepts based on available knowledge. It took the failures of the first generation of molecular genetic studies of major psychiatric disorders to increase researchers' awareness of the different complexity levels of these disorders. Third, unsurprisingly, scientists were highly motivated by the knowledge that candidate genes expressed in the central nervous system (CNS) were located very close to 5q11–q13 or 11p15–p14, which gave credence to the findings. Finally, the glamour of the newly available molecular technologies undoubtedly promoted the elated attitudes of the 1980s. This happened despite the warning of several prestigious genetic epidemiologists such as Ming Tsuang and Irving Gottesman that the process was likely to be long and arduous due to the complexities involved.

Short period of disenchantment

These first hasty claims of linkage were followed by a decade of molecular genetic studies scanning the genome, using both parametric and nonparametric linkage statistics, which left scientists with a bulk of frustrating and contradictory results. Few, if any, of the findings resisted the statistical criteria for significant linkage. However, the knowledge accumulated from these disconcerting efforts has called for a second generation of studies to search for the disease genes. Because of genetic epidemiology, we now have consistent evidence that: (i) SZ and BP have a definite genetic component that is complex, polygenic and may involve epistatic interaction between loci; and (ii) a component of this complexity resides at the level of phenotype, which needs considerable redefinition. Indeed, the first generation of molecular genetic studies used variable and less than optimal diagnostic categories, which was one reason why replication studies of positive findings were difficult to interpret.

Methodological Pitfalls to Avoid in the Second Generation of Molecular Genetic Studies

The first generation of molecular genetic studies of SZ and BP did not deliver a major psychosis gene. Yet it has yielded major scientific conclusions to ground the second generation of studies. Conceptually, the idea that one major gene can account for either SZ or BP can be rejected, leading the new studies to look for several gene effects with complex interactions. Methodologically, molecular studies must address this complexity at three main levels.

Power of samples resides in validity of phenotype as well as in size

A seminal idea to emerge from the former studies is that statistical power to detect genes in complex traits will not come primarily from the size of the samples, but rather from the accuracy with which scientists are able to define the phenotype under scrutiny. A consensus now exists that a major level of complexity for SZ and BP resides in the uncertainty of the phenotype definition as used in former linkage and association studies. A crucial prerequisite is, of course, that stringent blindness on a study must be maintained as clinicians collect pedigrees and diagnostic data, and laboratories generate genotypes. This was not always the case in pioneer studies. Scientists will have to establish and maintain blindness; this requires a close collaborative relationship among the research clinicians, the clinical network, and the basic scientists in statistics and molecular genetics.

The diagnostic categories used in the 1980s were necessary but not sufficient to be validly related to the genotype. Moreover, these diagnoses may reflect too distal or variable effects of the genes that underlie susceptibility to the two major psychoses. Hence, the use of dimensional phenotypes (or clusters of symptoms) and neurocognitive deficits related to the illness may be genetically less complex, that is, be influenced by fewer genes than diagnostic categories and thus be easier to elucidate. Also, more interest is now devoted to endophenotypes, that is, neurobiological markers related to SZ or BP that may be more proximal to the effect of the disease mutations. Eye tracking and P50 auditory evoked potential are promising candidates for such a strategy in SZ.

Also, the variable expression of SZ and BP needs to be taken into account in linkage analysis. Anticipation is an example of a dynamic genetic mechanism that may underlie such variability. Anticipation is an increase in illness severity from one generation to the next, a phenomenon related to expanding trinucleotide repeats. Such dynamic mutations have already been identified for other CNS disorders such as Huntington disease.

Heterogeneity of schizophrenia and bipolar affective disorder

Another major challenge is that SZ, as currently defined, does not constitute a single illness that can be explained by a unitary genetic mechanism. Instead, SZ seems made of several etiologically different entities. The same observation applies for BP. The resolution of this genetic heterogeneity calls for further work on phenotype definition and classification to delineate genetically homogeneous or distinct subsets of patients and families. This in turn will increase power to detect the susceptibility genes with the developing sophisticated molecular technologies. Such efforts have already been successful in the mapping of genes for other heterogeneous complex disorders such as Alzheimer disease and type 1 diabetes.

Moreover, there is increasing evidence that some susceptibility loci might be specific to either SZ or BP, whereas others would be common to both. Such knowledge must be taken into account in the coming generation of genetic studies, because it is essential for the detection and understanding of the intricate additive or interactive effects of the numerous susceptibility genes causing SZ and BP.

Independent replication and complex statistical modeling

Statistical standards have been better delineated and improved for the next series of molecular genetic studies. Indeed, most scientists now adopt Lander and Kruglyak's criteria: (i) the linkage evidence in at least one report must be significant using genome-wide statistical thresholds; and (ii) confirmation of linkage findings by at least one independent group remains the standard for a valid finding. But more importantly, the implications of the true mode of transmission of SZ must not be brushed aside and need to be considered in each step of the overall design and analytic strategy, from the sampling to the final statistical testing. Since most genetically distinct subgroups of SZ and BP probably depend upon the interactive and coactive effects of multiple genes and nongenetic factors, new multilocus statistical analyses urgently need to be developed to fit the complex reality of the family data, as well as new analytic models developed to address the reality that continuous traits are associated with the illness.

Conclusion

Giving attention to the former methodological issues is essential to avoid the embarrassments of the past. New nonparametric and parametric complex methods of analyses, advances in high-throughput genotyping, the development of microarrays and the emerging maps of single-nucleotide polymorphism will revolutionize the genetic studies of complex diseases such as SZ and BP. Although microarray experiments can determine expression levels of thousands of genes, they will not give automatic information about higher-order relationships or hidden patterns among genes or among samples. Brain circuitry and plasticity is much more than static patterns of gene expression. Emphasis must be placed not only on uncovering static patterns of gene expression that appear to be associated with SZ or BP, but also on discovering the developmental dynamics of these patterns. This is the type of information that will ultimately produce the knowledge required to design curative and preventive treatments.

High-performance molecular technologies are necessary but not sufficient to detect the genes that will matter in the mechanistic cascade of events leading to SZ or BP. We are now facing the same peril that plagued the first generation of molecular genetic studies that, unfortunately, focused too much on molecular technologies while neglecting essential methods imposed by the complex genetics and the clinical epidemiology of SZ and BP. Scientists in the 1980s expected too much from the sole power of restriction fragment length polymorphism and new technologies in genotyping, forgetting the complexities of modeling and of phenotypic uncertainty. Indeed, the danger now resides in a similar glamorous attraction, among basic scientists, to rely on the sole apparent power of new technologies, such as DNA microarrays, as a tool to analyze brain functioning in both illness and health. The potential of the microarray-based expression analysis, if not combined with strategies addressing brain tissue complexity, or with careful sample collections, or with proper phenotype characterization, may just replay the past and end up in disappointment instead of enthusiasm. It is imaginable that in the coming generation of studies, scientists, rather than computer-assisted, high-throughput

technologies alone, will remain the main contributors of new ideas to tackle the complex mechanisms involved in SZ and BP.

See also
Genetic Susceptibility
Linkage and Association Studies: Replication
Psychiatric Disorders: The Search for Genes
Schizophrenia: Molecular Genetics
Susceptibility Genes: Detection

Further Reading

Berrettini W (1998) Progress and pitfalls: bipolar molecular linkage studies. *Journal of Affective Disorders* 50: 287–297.
Cook Jr EH (2000) Genetics of psychiatric disorders: where have we been and where are we going? *American Journal of Psychiatry* 157: 1039–1040.
Gottesman II (1991) Schizophrenia genesis. The origins of madness. In: Atkinson RC, Lindzey G and Thompson RF (eds.) *Books in Psychology*, p. 296. New York, NY: WF Freeman and Company.
Gottesman II and Moldin S (1997) Schizophrenia genetics at the millennium: cautious optimism. *Clinical Genetics* 52: 404–407.
Lander E and Kruglyak L (1995) Genetic dissection of complex traits: guidelines for interpreting and reporting linkage results. *Nature Genetics* 11: 241–247.
Maziade M, Roy MA and Rouillard E, *et al.* (2001) A search for specific and common susceptibility loci for schizophrenia and bipolar disorder: a linkage study of 13 target chromosomes. *Molecular Psychiatry* 6: 684–693.
Riley BP and McGuffin P (2000) Linkage and associated studies of schizophrenia. *American Journal of Medical Genetics* 97: 23–44.
Tsuang MT and Faraone SV (2000) The frustrating search for schizophrenia genes. *American Journal of Medical Genetics* 97: 1–3.
Tsuang MT, Stone WS and Faraone SV (2000) Toward reformulating the diagnosis of schizophrenia. *American Journal of Psychiatry* 157: 1041–1050.
Watson SJ, Meng F, Thompson RC and Akil H (2000) The 'chip' as a specific genetic tool. *Biological Psychiatry* 48: 1147–1156.

Web Links

v-Ha-ras Harvey rat sarcoma viral oncogene homolog (*HRAS*); LocusID: 3265. Locus Link:
http://www.ncbi.nlm.nih.gov/LocusLink/LocRpt.cgi?l=3265
Insulin (*INS*); LocusID: 3630. Locus Link:
http://www.ncbi.nlm.nih.gov/LocusLink/LocRpt.cgi?l=3630
v-Ha-ras Harvey rat sarcoma viral oncogene homolog (*HRAS*); MIM number: 190020. OMIM:
http://www3.ncbi.nlm.nih.gov/htbin-post/Omim/dispmim?190020
Insulin (*INS*); MIM number: 176730. OMIM:
http://www3.ncbi.nlm.nih.gov/htbin-post/Omim/dispmim?176730

Schizophrenia: Molecular Genetics

Michael J Owen, *University of Wales College of Medicine, Cardiff, UK*

Molecular genetic studies of schizophrenia have used both linkage and association approaches. Both types of study have resulted in conflicting findings, and unequivocal evidence for linkage or association is as yet lacking. However, systematic linkage studies have revealed several chromosomal areas of potential interest, and candidate gene association studies have suggested that variation within the *HTR2A* and *DRD3* genes might confer susceptibility.

Introduction

It is clear from family, twin and adoption studies that there is an important genetic contribution to schizophrenia. However, studies on the recurrence risk in various classes of relative allow us to exclude the possibility that schizophrenia is a single-gene disorder or collection of single-gene disorders even when incomplete penetrance is taken into account. Rather, the mode of transmission, like that of other complex disorders, is complex and non-Mendelian (McGue and Gottesman, 1989). The commonest mode of transmission is probably oligogenic or polygenic, or a mixture of the two. However, the number of susceptibility loci, the disease risk conferred by each locus and the degree of interaction between loci all remain unknown. Risch (1990) has calculated that the data for recurrence risks in the relatives of probands with schizophrenia are incompatible with the existence of a single locus of $\lambda s > 3$ and, unless extreme epistasis exists, models with two or three loci of λs less than or equal to 2 are more plausible. These calculations are based upon a homogeneous population, and under a model of heterogeneity, it is quite possible that genes of larger effect are operating in some subpopulations of patients. In spite of these uncertainties and the difficulties that ensue, schizophrenia has seemed to many a compelling candidate for molecular genetic studies using both linkage and association approaches.

Linkage Studies

The majority of systematic linkage studies of schizophrenia have focused upon large, multiply affected pedigrees for analysis. Initially it was hoped that such families, or at least a proportion of them, are segregating alleles that are sufficiently common and of sufficiently large effect to be unequivocally detected and replicated using linkage analysis. This approach initially produced positive findings (Sherrington *et al.*,

1998), but unfortunately these could not be replicated and are likely to be false positives, due largely to a combination of multiple testing and the use of statistical methodology and significance levels derived from work on single-gene disorders.

Many subsequent studies have been reported but have failed as yet to produce unequivocal, replicated demonstrations of linkage. However, modest evidence for several regions has been reported in more than one data set. Areas implicated for which supportive data have also been obtained from large international collaborative studies include chromosomes 22q11–12, 6p24–22, 8p22–21 and 6q. There are also a number of other promising areas of putative linkage, which have not received convincing support from international consortia. These include 13q14.1–q32, 5q21–q31 and 10p15–p11. Other regions that are currently being investigated in collaborative studies include 1q21–q22 and 18p22–21. However, it should be noted that in every case there are negative as well as positive findings; and, in only two cases, those of chromosomes 13q13q14.1–q32 and 1q21–q22, did any single study achieve genome-wide significance at $P < 0.05$.

These positive findings contrast with those from a large, systematic search for linkage using a sample of 196 affected sibling pairs (ASPs) drawn typically from small nuclear families rather than extended pedigrees (Williams *et al.*, 1999). The results of simulation studies suggested that the power of this study was more than 0.95 to detect a susceptibility locus of $\lambda s = 3$ with a genome-wide significance of 0.05, but only 0.70 to detect a locus of $\lambda s = 2$. This study yielded no evidence for linkage approaching a genome-wide significance of 0.05.

The findings from linkage studies of schizophrenia to date demonstrate several features that are to be expected in the search for genes for complex traits. First, no finding replicates in all data sets. Second, levels of statistical significance are unconvincing and estimated effect sizes are usually modest. Third, chromosomal regions of interest are typically broad (often more than 20–30 cM).

At the present time, therefore, the linkage literature supports predictions from genetic epidemiological studies: it is highly unlikely that a commonly occurring locus of effect size $\lambda s > 3$ exists, but there is evidence implicating a number of regions, which is consistent with the existence of some susceptibility alleles of moderate effect ($\lambda s = 1.5$–3). Moreover, encouraging results in several chromosomal regions suggest that less common alleles of larger effect may be segregating in some large, multiply affected families.

Linkage methods in sample sizes that are realistically achievable can detect smaller genetic effects than the studies to date. For example, it is possible to detect alleles with $\lambda s = 1.5$–3 in a sample of 600–800 ASPs. Priority should now be given to collecting such samples using robust clinical methodology that is comparable across all interested research groups. However, if liability to schizophrenia is entirely due to the operation of many genes of small effect, then even these large-scale studies will be unsuccessful.

Candidate Gene Association Studies

Once genes of smaller effect than $\lambda s = 1.5$ are sought, the number of affected family members required becomes prohibitively large. For this reason, many researchers have sought to take advantage of the potential of candidate gene association studies to identify such loci. Although a potentially powerful means of identifying genes of small effect, association studies are not without their problems. Moreover, for a complex and poorly understood disorder such as schizophrenia, the choice of candidate genes is limited largely by the imagination and resources of the researcher. This problem places a stringent burden of statistical proof on positive results.

Most candidate gene studies have been based upon neuropharmacological studies, suggesting that abnormalities in monoamine neurotransmission, in particular dopaminergic and serotonergic systems, play a role in the etiology of schizophrenia. Overall, the results in this extensive literature are disappointing, but it should be noted that the sample sizes in many of the older studies would now generally be regarded as inadequate, particularly in view of the fact that the polymorphic markers in question did not in themselves represent functional variants and few genes have been systematically screened even for common functional variants. However, there have been more promising reports of candidate gene associations. Although the effect sizes are small, in no case has a stringent burden of proof been met, and it is possible that these are false positives or the result of association with a confounding variable such as severity or a comorbid syndrome.

Serotonin 5HT2a receptor gene (*HTR2A*)

A large European consortium consisting of seven centers and involving 571 patients and 639 controls replicated a finding from a previous small Japanese study showing an association between schizophrenia and the C allele of the T102C polymorphism in *HTR2A* (*5-hydroxytryptamine (serotonin) receptor 2A*; MIM number: 182135; Williams *et al.*, 1996). While many other studies have followed with mixed results, a recent meta-analysis of all available data, including over 3000 subjects, supports the original finding ($P = 0.0009$), and this does not appear to have been owing to publication bias (Williams *et al.*, 1997).

Since this meta-analysis was undertaken, a few further negative reports have followed, but none has approached the sample sizes required. If we assume homogeneity and if the association is true, the putative odds ratio (OR) for the C allele can be expected to be around 1.2 in any replication sample. Sample sizes of 1000 subjects are then required for 80% power to detect an effect of this size even at a relaxed criterion of $\alpha = 0.05$. Thus, the negative studies are effectively meaningless, but it is also true that the evidence for association even in the meta-analysis ($P = 0.0009$) is not definitive if genome-wide significance levels are required.

If the association is real, it is unlikely that T102C is the susceptibility variant, because this nucleotide change does not alter the predicted amino acid sequence of the receptor protein, nor is it in a region of obvious significance for regulating gene expression. T102C is in complete linkage disequilibrium (LD) with a polymorphism in the promoter region of this gene, but there is as yet no evidence that this has a functional effect either (Spurlock *et al.*, 1998). Recent evidence of polymorphic monoallelic expression of the *5HT2A* gene points to the possible existence of sequence variation elsewhere that influences gene expression (Bunzel *et al.*, 1998), and this may be the true susceptibility variant.

Dopamine receptor D3 gene (*DRD3*)

Association has been reported between schizophrenia and homozygosity for a Ser9Gly polymorphism in exon 1 of the *dopamine receptor D3 gene* (*DRD3*; MIM number: 126451; Crocq *et al.*, 1992). As with the *5HT2A* association, the results have now been confirmed in several independent samples, including one family-based study (Williams *et al.*, 1998), but several negative studies have also been reported. Meta-analysis of data from over 5000 individuals revealed a small (OR 1.23) but significant ($P = 0.0002$) association between homozygosity at Ser9Gly and schizophrenia (Williams *et al.*, 1998). Again this association could not easily be ascribed to selective publication (Williams *et al.*, 1998). At present then, the status of the *DRD3* finding is similar to that of *5HT2A*; that is, the balance of evidence at present favors association, but the null hypothesis still cannot be confidently rejected. Those wishing to replicate or reject these findings should bear in mind that in order to obtain power of more than 0.80 to detect an effect of this size at a criterion of $\alpha = 0.05$, a sample of 1500 cases and 1500 controls will be required. So far, no other polymorphisms have been found that might explain the putative D3 association, but several new polymorphisms have been identified in previously unknown exons 5′ to the exon referred to here as exon 1. These are currently being tested to establish whether variants in this region in LD with the Ser9Gly polymorphism provide a more functionally plausible explanation of the association with schizophrenia.

Future Directions

It is hoped that advances, as in other common diseases, might come eventually through genome-wide association studies using single-nucleotide polymorphism (SNPs) and new methods of genotyping and statistical analysis. However, the era of genome-wide association studies is not yet at hand. Instead, studies in the next few years should probably focus either upon intense analysis of regions implicated by linkage studies or upon the analysis of SNPs from the coding and regulatory regions in a wide range of functional and positional candidate genes. Preferably, complete functional systems should be dissected by a combination of the application of sensitive methods for mutation detection, followed by association studies in appropriately sized samples. We should also use our knowledge of functional pathways to make predictions about probable epistasis. However, given our ignorance of pathophysiology, the expectation should be that most reported associations will be false and will only be resolved by replication in large, well-characterized samples. Successful application of these methods requires access to large, well-characterized patient samples, and collection of these is a priority at the present time.

We also need to focus research on the development and refinement of phenotypic measures and biological markers, which might simplify the task of finding genes. Perhaps, genetic validity could be improved by focusing upon aspects of clinical variation such as symptom profiles, or by identifying biological markers that predict degree of genetic risk or define more homogeneous subgroups. However, it seems unlikely that this will provide a rapid solution to the problem. First, we will need to ensure that the measures used are stable and determine the extent to which they are affected by state. Second, to be of use in gene mapping, such measures will have to be practically applied to a sufficient number of families or unrelated patients. Third, we will need to ensure that the traits identified are highly heritable, which will itself require a return to classic genetic epidemiology and model fitting.

Finally, it is possible that new genes and pathways might be implicated by emerging transcriptomic and proteomic approaches. The problem here is that human studies will be hampered by the many confounding variables associated with postmortem studies of brain, while animal studies suffer from the difficulties inherent in extrapolating from animal behavior to a complex human psychiatric disorder such as schizophrenia.

See also

Developmental Psychopathology

Psychoses

Schizophrenia and Bipolar Disorder: Linkage on Chromosomes 5 and 11

Schizophrenia Spectrum Disorders

Velocardiofacial Syndrome (VCFS) and Schizophrenia

References

Bunzel R, Blumcke I, Cichon S, *et al.* (1998) Polymorphic imprinting of the serotonin-2A (5-HT2A) receptor gene in human adult brain. *Molecular Brain Research* 59: 90–92.

Crocq MA, Mant R, Asherson P, *et al.* (1992) Association between schizophrenia and homozygosity at the dopamine D3 receptor gene. *Journal of Medical Genetics* 29: 858–860.

McGue M and Gottesman II (1989) A single dominant gene still cannot account for the transmission of schizophrenia. *Archives of General Psychiatry* 46: 478–479.

Risch N (1990) Linkage strategies for genetically complex traits. 2. The power of affected relative pairs. *American Journal of Human Genetics* 46: 229–241.

Sherrington R, Brynjolfsson J, Petursson H, *et al.* (1998) Localization of a susceptibility locus for schizophrenia on chromosome-5. *Nature* 336: 164–167.

Spurlock G, Heils A, Holmans P, *et al.* (1998) A family based association study of *T102C* polymorphism in 5HT2A and schizophrenia plus identification of new polymorphisms in the promoter. *Molecular Psychiatry* 3: 42–49.

Williams J, McGuffin P, Nothen M and Owen MJ (1997) The EMASS Collaborative Group meta-analysis of association between the 5-HT2a receptor T102C polymorphism and schizophrenia. *Lancet* 349: 1221.

Williams J, Spurlock G, Holmans P and Mant R (1998) A meta-analysis and transmission disequilibrium study of association between the dopamine D3 receptor gene and schizophrenia. *Molecular Psychiatry* 3: 141–149.

Williams J, Spurlock G, McGuffin P and Mallet J (1996) Association between schizophrenia and *T102C* polymorphism of the 5-hydroxytryptamine type 2A-receptor gene. *Lancet* 347: 1294–1296.

Williams NM, Rees MI, Holmans P and Norton N (1999) A two-stage genome scan for schizophrenia susceptibility genes in 196 affected sibling pairs. *Human Molecular Genetics* 8: 1729–1739.

Further Reading

Baron M (2001) Genetics of schizophrenia and the new millennium: progress and pitfalls. *American Journal of Human Genetics* **68**: 299–312.

Gottesman II (1991) *Schizophrenia Genesis: The Origins of Madness.* New York, NY: WH Freeman.

McGuffin P, Owen MJ and Farmer AE (1995) Genetic basis of schizophrenia. *Lancet* **346**: 678–682.

O'Donovan MC and Owen MJ (1999) Candidate gene association studies of schizophrenia. *American Journal of Human Genetics* **65**: 587–592.

Owen MJ, Cardno AG and O'Donovan MC (2000) Psychiatric genetics: back to the future. *Molecular Psychiatry* **5**: 22–31.

Owen MJ and O'Donovan MC (2002) Schizophrenia. In: Plomin R, DeFries JC, Craig IC and McGuffin P (eds.) *Behavioral Genetics in a Postgenomics World.* Washington DC: APA Books.

Sherrington R, Rogaev EI, Liang Y, *et al.* (1995) Cloning of a gene bearing missense mutations in early-onset familial Alzheimer's disease. *Nature* **375**: 754–760.

Web Links

5-hydroxytryptamine (serotonin) receptor 2A (*HTR2A*); MIM number: 182135. OMIM:
http://www3.ncbi.nlm.nih.gov/htbin-post/Omim/dispmim?182135

Dopamine receptor D3 (*DRD3*); MIM number: 126451; OMIM:
http://www3.ncbi.nlm.nih.gov/htbin-post/Omim/dispmim?126451

5-hydroxytryptamine (serotonin) receptor 2A (*HTR2A*); Locus ID: 3356; LocusLink:
http://www.ncbi.nlm.nih.gov/LocusLink/LocRpt.cgi?l=3356

Dopamine receptor D3 (*DRD3*); Locus ID: 1814; LocusLink:
http://www.ncbi.nlm.nih.gov/LocusLink/LocRpt.cgi?l=1814

Schizophrenia Spectrum Disorders

Michael F Pogue-Geile, *University of Pittsburgh, Pittsburgh, Pennsylvania, USA*

The history, importance, and definition of schizophrenia spectrum disorders are described. Findings regarding the nature of the familial association between spectrum disorders and schizophrenia are reviewed and evaluated. Evidence does suggest that spectrum disorders may be genetically correlated with schizophrenia and thus potentially useful in efforts at identifying genes for schizophrenia liability.

Intermediate article

Article contents

- Scope
- Conceptual Background
- Importance
- History
- Association between Schizotypal Personality Disorder and Schizophrenia
- Conclusions

Scope

Before proceeding to consider the relevant literature, it will be useful to delimit and discuss the concept of 'spectrum disorders' more generally. First, the present summary will focus on spectrum *diagnoses* and will not include other potential spectrum characteristics that are not part of the official diagnostic nomenclature, such as quantitative abnormalities of cognitive or other psychological function. This discussion will further focus even more specifically on only those diagnoses that have come to be included as part of the Personality Disorders section (Axis II) of the American Psychiatric Association Diagnostic and Statistical Manual (DSM) (APA, 1994) and the International Classification of Diseases (ICD) (WHO, 1999). Psychotic disorders that may also be related to schizophrenia (e.g. schizoaffective disorder) but that are considered as part of the DSM Axis I will not be considered here (Kendler, 2000).

Conceptual Background

The term 'spectrum' implies several distinguishable but related classes, such as the different colors of the

visible light spectrum. There can be different bases for deciding 'relatedness'. In psychopathology and psychiatry, two general bases that have been used are relatedness either at an observational, phenotypic level or at a causal, etiological level. Some uses of the term spectrum disorders have emphasized observed similarity between diagnoses in signs and symptoms. For example, schizophrenia spectrum disorders historically have sometimes included diagnoses that are phenotypically similar to schizophrenia, but perhaps less severe symptomatically (i.e. schizophrenia-like) (Kendler, 1985). In contrast, this summary will focus on spectrum disorders that may be related etiologically to schizophrenia, regardless of their observed similarity to schizophrenia signs and symptom.

An important goal for research on schizophrenia has been to develop diagnoses that have some causes in common with schizophrenia. This work has almost exclusively sought resemblance in genetic but not environmental causes, and as a result this will also be the focus here. In all such studies, the presumption is that although spectrum diagnoses may have some genetic causes in common with schizophrenia, they are not etiologically identical with schizophrenia (or otherwise presumably they would be schizophrenia). Spectrum diagnoses thus differ from schizophrenia itself either in environmental exposures and/or some genetic effects. Implicit in this framework is a related developmental presumption, namely that spectrum diagnoses do not always develop into schizophrenia itself at some later time (i.e. perhaps due to later exposure to relevant environmental risk factors). If they always developed into schizophrenia, then they would be useful predictors but etiologically identical to schizophrenia. Current classification approaches allow for the possibility that spectrum diagnoses in some individuals may precede later schizophrenia, but in these cases the diagnosis is changed to schizophrenia. The aim of most research on spectrum diagnoses is thus to identify diagnoses that are genetically related to schizophrenia, but do not always precede the later onset of schizophrenia.

Importance

Why is it important to identify or develop diagnoses that share some genetic causes in common with schizophrenia? There are several motivations for such research. First, many attempts to identify specific genetic loci contributing to the causes of schizophrenia utilize genetic linkage methods that often consist of investigating the correlation within families between allelic variation at known genetic marker loci and schizophrenia diagnoses among family members. In such traditional linkage studies of complete families, relatives who actually do share alleles with the schizophrenia proband at the marker locus but who are not diagnosed as schizophrenic may be considered as 'false negatives' and serve to reduce any observed correlation between marker and diagnosis. In these designs, such 'false negatives' reduce the power of the study to detect linkage. Because only about 48% of the monozygotic (MZ) co-twins (who share 100% of their genotype) of schizophrenia probands are themselves diagnosed as schizophrenic, it seems very likely that such 'false negatives' exist (Gottesman and Shields, 1982; Gottesman and Bertelsen, 1989). Not all genetic linkage study designs are susceptible to 'false negatives' however, but those that are, not because they include only relatives diagnosed as schizophrenic (Weeks and Lange, 1988), do so at a cost in statistical power. Therefore, any valid identification of nonschizophrenic relatives who have some genetic liability to schizophrenia should be helpful in genetic linkage studies attempting to locate genes for the disorder. A second reason for the importance of this

enterprise is that such nonschizophrenic, but genetically liable, individuals may provide insights into the pathology of schizophrenia that is uncontaminated by medication and other iatrogenic effects and may also yield information relevant to the prediction of schizophrenia onset. (*See* Schizophrenia: Molecular Genetics; Linkage Analysis; Genetic Epidemiology of Complex Traits; Psychiatric Disorders: The Search for Genes.)

History

The attributes of relatives of patients have been remarked upon even from the first proposed description of what is now called schizophrenia (Ingraham, 1995). Emil Kraepelin (1919/1971) in his classic treatise describing *dementia praecox* commented upon the sometimes eccentric personalities of the siblings of patients and hypothesized that they should be regarded as 'latent schizophrenias and therefore essentially the same as the principal malady' (p. 234). Eugen Blueler (1911/1950), who coined the term 'schizophrenia', also proposed a diagnosis of 'latent' schizophrenia. Kendler (1985) summarized the informal observations of these and other early workers and concluded that most commented on the following characteristics of some relatives of schizophrenia patients: eccentric–odd, irritable–unreasonable, socially isolated, aloof and suspicious. Meehl (1962), influenced by Rado (1956), contributed importantly to the theoretical conceptualization of spectrum diagnoses with his hypothesis that schizophrenia was generally caused by a single major gene that always produced central nervous system abnormalities, which he termed 'schizotaxia', that in turn always produced personality abnormalities, termed 'schizotypy'. Only a subset of schizotypes who were exposed to particular environmental experiences would then become schizophrenic and those without such exposures would remain schizotypal. In this model, a hypothesized genetic relationship between spectrum diagnoses (e.g. schizotypy) and schizophrenia was formalized. The next major development in the conceptualization of spectrum diagnoses was provided by Kety, Rosenthal and Wender in their adoption study of schizophrenia in Denmark in which they not only diagnosed schizophrenia but also systematically assessed relatives for 'borderline schizophrenia' in an attempt to include what they termed the entire 'schizophrenia spectrum of disorders' (Kety *et al.*, 1968, p. 353). The case records of borderline schizophrenia diagnoses from this study (including both biological relatives of schizophrenia adoptees and other subjects) were abstracted by Spitzer *et al.* (1979) and formed the definition of the new diagnosis, schizotypal personality disorder (SPD), which was introduced into Axis II of the DSM-III (APA, 1980) and the ICD-9. The current DSM IV definition of SPD (APA, 1994) requires the presence of five or more of the following symptoms and signs:

1. ideas of reference;
2. odd beliefs;
3. unusual perceptual experiences;
4. odd thinking and speech;
5. suspiciousness;
6. inappropriate or constricted affect;
7. odd behavior;
8. lack of close friends; and
9. excessive social anxiety.

These signs bear a resemblance to the criteria used for the diagnosis of schizophrenia but in milder, nonpsychotic form and include abnormalities in the three general areas of thinking, emotions and social interactions.

Association between Schizotypal Personality Disorder and Schizophrenia

What evidence is there that SPD is genetically correlated with schizophrenia? First, we will briefly summarize nuclear family studies bearing on this question, only considering studies utilizing DSM III or later diagnoses of SPD based on structured interviews and including either nonpatient or patient control groups. Although other personality disorder diagnoses have also been examined among relatives of schizophrenia probands, especially paranoid and schizoid diagnoses, which comprise subsets of the symptoms defining SPD (i.e. schizoid/asociality and paranoid/suspiciousness), they will not be considered here due to space limitations. Nine nuclear family studies meet these criteria (**Table 1**).

Several points are notable about these studies. First, in all cases, DSM SPD is more common among relatives of index schizophrenic probands than among relatives of nonpatient control probands, with relative risks ranging between 1.1 and 7.0, which provides evidence of a familial correlation between SPD and schizophrenia. However, the variation in rates across studies is marked, ranging among index relatives from a high of 19% to 1%. Although it seems clear that DSM SPD is familially associated with schizophrenia, most studies would suggest that its risk is unlikely to be greater than 5% among relatives of schizophrenia patients. Four studies have also addressed the question of the specificity of schizotypal diagnoses to schizophrenia by including a control group of the relatives of nonpsychotic affective disordered probands. SPD still tends to aggregate more in families of schizophrenia patients than in families of affective probands (relative risks from 7.0 to 1.6), although there may also be some slight increased risk for SPD among families of affective patients compared to relatives of nonpatient controls (relative risks from 2.8 to 1.6). (*See* Adoption Studies.)

From these nuclear family studies, it is unclear whether this familial correlation is due to the genetic variation shared among family members or perhaps due to the experiences that they share. Adoption studies allow these causes to be separated because in the absence of relevant selective placement, adoptees share only genetic resemblance with their biological relatives. Three independent adoption studies have investigated the genetic association between schizophrenia and SPD (**Table 2**).

Table 1 Rates of DSM schizotypal personality disorder among relatives of schizophrenia, affective disorder and nonpatient probands

Study	Index relatives (%)[a]	Affective control relatives (%)[b]	Nonpatient control relatives (%)[c]
Baron *et al.* (1985)	14.6	na	2.1
Frangos *et al.* (1985)	1.0	na	0.3
Coryell *et al.* (1988)	2.8	na	2.5
Gershon *et al.* (1988)	1.5	na	0.0
Onstad *et al.* (1991)	7.0	0.0	na
Parnas *et al.* (1993)	19.0	na	5.0
Kendler *et al.* (1993)	6.9	2.3	1.4
Maier *et al.* (1994)	2.1	0.7	0.3
Erlenmeyer-Kimling *et al.* (1995)	4.5	2.8	0.0

[a]Rate of SPD among relatives of schizophrenia probands.
[b]Rate of SPD among relatives of nonpsychotic affective disorder probands.
[c]Rate of SPD among relatives of nonpatient probands.

Table 2 Rates of DSM schizotypal personality disorder among reared-apart biological relatives of schizophrenia and nonpatient control probands

Study	Index relatives (%)[a]	Control relatives (%)[b]
Lowing et al. (1983)	15.4	7.7
Kendler et al. (1994)	13.2	3.7
Tienari et al. (2000)	2.4	0.0

[a]Rate of SPD among reared-apart biological relatives of schizophrenia probands.
[b]Rate of SPD among reared-apart biological relatives of nonpatient control probands.

As can be seen, SPD occurs more frequently among reared-apart biological relatives of schizophrenia probands than control relatives (relative risk from 3.6 to 2.0), suggesting that SPD is genetically associated with schizophrenia. However, again there is a wide range of estimates of risk among index relatives, ranging from 15.4% to 2.4%. (*See* Twin Studies.)

Twins studies, in which phenotypic resemblance is compared between twin pairs who share 100% of their genetic variation (MZ twins) and 'control' pairs who show on average 50% (dizygotic, DZ twins), are a second method that can detect genetic effects independent of shared family environmental effects. Genetic effects are implicated to the extent that MZ twins show greater observed resemblance than DZ twins. Although investigation of nonpsychotic personality traits has a long history in twin studies of schizophrenia (Gottesman et al., 1976), there have been few studies of DSM SPD among the MZ and DZ co-twins of schizophrenia patients. In a small study in Norway, Torgersen et al. (1993) found 3/15 (20%) of MZ co-twins of schizophrenia probands to meet criteria for DSM SPD compared to 4/27 (15%) of DZ co-twins. This similarity in rates for MZ and DZ co-twins implicates shared family environmental effects rather than genetic effects, but the sample size is too small for any firm inferences. (*See* Schizophremia: Molecular Genetics.)

Conclusions

In conclusion, studies generally agree that DSM SPD diagnosis is familially associated with schizophrenia and furthermore, based on the adoption studies, that it shares genetic effects with schizophrenia. However, these studies also raise questions concerning substantial across-study variation in rates, specificity to schizophrenia, and whether DSM SPD is any more familially associated with schizophrenia than is schizophrenia itself. In addition, several other important questions remain concerning the overall schizophrenia spectrum concept. Although the focus of some studies, (Kendler et al., 1995a), it is currently unclear which individual symptoms of SPD may be most familially associated with schizophrenia and furthermore whether there are other signs and symptoms outside of the DSM SPD diagnosis that should also be included in a spectrum that might make it more useful. Perhaps even more importantly for gene mapping efforts in schizophrenia is the specific nature of the genetic association between SPD and schizophrenia. It is currently not known with any certainty whether SPD represents a milder threshold on the same quantitative dimension of liability as schizophrenia (Kendler et al., 1995b) or whether SPD itself might be multidimensional with different dimensions being associated with separate quantitative trait loci (Venables and Bailes, 1994). These issues have potentially important consequences for the usefulness of the schizophrenia spectrum concept for schizophrenia gene mapping studies. Although, schizophrenia spectrum diagnoses of various sorts have been included in genetic linkage studies from early on (Sherrington et al., 1988), the importance of their contribution to these efforts to date is uncertain (Baron, 2001).

See also

Developmental Psychopathology
Psychoses
Schizophrenia: Molecular Genetics

References

American Psychiatric Association (1980) *Diagnostic and Statistical Manual of Mental disorders*, 3rd edn. Washington DC: American Psychiatric Association Press.

American Psychiatric Association (1994) *Diagnostic and Statistical Manual of Mental Disorders*, 4th edn. Washington DC: American Psychiatric Association Press.

Baron M (2001) Genetics of schizophrenia and the new millennium: progress and pitfalls. *American Journal of Human Genetics* **68**: 299–312.

Baron M, Gruen R, Rainer J, et al. (1985) A family study of schizophrenic and normal control probands: implications for the spectrum concept of schizophrenia. *American Journal of Psychiatry* **142**: 447–455.

Bleuler E (1911/1950) *Dementia Praecox or the Group of Schizophrenias*. New York, NY: International Universities Press.

Coryell W and Zimmerman M (1988) The heritability of schizophrenia and schizoaffective disorders. *Archives of General Psychiatry* **45**: 323–327.

Erlenmeyer-Kimling L, Squires-Wheeler E, Adamo UH, et al. (1995) The New York high-risk project: psychoses and cluster A personality disorders in offspring of schizophrenic parents at 23 years follow up. *Archives of General Psychiatry* **52**: 857–865.

Frangos E, Athanassenas G, Tsitourides S, Katsanou N and Alexandrakou P (1985) Prevalence of DSM III schizophrenia among the first-degree relatives of schizophrenic probands. *Acta Psychiatrica Scandinavica* **72**: 382–386.

Gershon ES, DeLisi LE, Hamovit J, et al. (1988) A controlled family study of chronic psychoses. *Archives of General Psychiatry* **45**: 328–336.

Gottesman II and Bertelsen A (1989) Confirming unexpressed genotypes for schizophrenia. *Archives of General Psychiatry* **46**: 867–872.

Gottesman II and Shields J (1982) *Schizophrenia: The Epigenetic Puzzle.* New York, NY: Cambridge University Press.

Gottesman II, Shields J and Heston LL (1976) Characteristics of the twins of schizophrenics as fallible indicators of schizoidia. *Acta Geneticae Medicae et Gemellologiae* **25**: 225–236.

Ingraham LJ (1995) Family-genetic research and schizotypal personality. In: Raine A, Lencz T and Mednick SA (eds.) *Schizotypal Personality*, pp. 19–42. New York, NY: Cambridge University Press.

Kendler KS (1985) Diagnostic approaches to schizotypal personality disorder: a historical perspective. *Schizophrenia Bulletin* **11**: 538–555.

Kendler KS (2000) Schizophrenia genetics. In: Sadock BJ and Sadock VA (eds.) *Comprehensive Textbook of Psychiatry*, pp. 1147–1159. Philadelphia, PA: Lippincott.

Kendler KS, Gruenberg AM and Kinney DK (1994) Independent diagnoses of adoptees and relatives as defined by DSM-III in the provincial and national samples of the Danish adoption study of schizophrenia. *Archives of General Psychiatry* **51**: 456–468.

Kendler KS, McGuire M, Gruenberg AM and Walsh D (1995a) Schizotypal symptoms and signs in the Roscommon family study. *Archives of General Psychiatry* **52**: 296–303.

Kendler KS, Neale MC and Walsh D (1995b) Evaluating the spectrum concept of schizophrenia in the Roscommon family study. *American Journal of Psychiatry* **152**: 749–754.

Kendler K, McGuire M, Gruenberg AM, et al. (1993) The Roscommon family study: III Schizophrenia-related personality disorders in relatives. *Archives of General Psychiatry* **50**: 781–788.

Kety SS, Rosenthal D, Wender PH and Schulsinger F (1968) The types and prevalence of mental illness in the biological and adoptive families of adopted schizophrenics. In: Rosenthal D and Kety SS (eds.) *The Transmission of Schizophrenia*, pp. 345–362. New York, NY: Pergamon Press.

Kraepelin E (1919/1971) *Dementia Praecox and Paraphrenia.* New York, NY: Krieger RE Publishing.

Lowing PA, Mirsky AF and Pereira R (1983) The inheritance of schizophrenia spectrum disorders: a reanalysis of the Danish adoptee study data. *American Journal of Psychiatry* **140**: 1167–1171.

Maier W, Lichtermann D, Minges J and Heun R (1994) Personality disorders among the relatives of schizophrenia patients. *Schizophrenia Bulletin* **20**: 481–493.

Meehl PE (1962) Schizotaxia, schizotypy, schizophrenia. *American Psychologist* **17**: 827–838.

Onstad S, Skre I, Edvardsen J, Torgersen S and Kringlen E (1991) Mental disorders of first-degree relatives of schizophrenics. *Acta Psychiatrica Scandinavica* **83**: 463–467.

Parnas J, Cannon TD, Jacobsen B, et al. (1993) Lifetime DSM-III-R diagnostic outcomes in the offspring of schizophrenic mothers. *Archives of General Psychiatry* **50**: 707–714.

Rado S (1956) *Psychoanalysis of Behavior.* New York, NY: Grune and Stratton.

Sherrington R, Brynjolfsson J, Petursson H, et al. (1988) Localization of a susceptibility locus for schizophrenia on chromosome 5. *Nature* **336**: 164–170.

Spitzer RL, Endicott J and Gibbon M (1979) Crossing the border into borderline personality and borderline schizophrenia. *Archives of General Psychiatry* **36**: 17–24.

Tienari P, Wynne LC, Moring J, et al. (2000) Finnish adoptive family study: sample selection and adoptee DSM-III-R diagnoses. *Acta Psychiatrica Scandinavica* **101**: 433–443.

Torgersen S, Onstad S, Skre I, Edvardsen J and Kringlen E (1993) 'True' schizotypal personality disorder: a study of co-twins and relatives of schizophrenic probands. *American Journal of Psychiatry* **150**: 1661–1667.

Venables PH and Bailes K (1994) The structure of schizotypy, its relationship to subdiagnoses of schizophrenia and to sex and age. *British Journal of Clinical Psychology* **33**: 277–294.

Weeks DE and Lange K (1988) The affected-pedigree-member method of linkage analysis. *American Journal of Human Genetics* **42**: 315–326.

World Health Organization (1999) *Hospital and Payor International Classification of Diseases, 9th revision (Clinical Modification)*, 5th edn. Salt Lake City, UT: Medicode.

SCID

See Severe Combined Immune Deficiency (SCID): Genetics

Secular Humanism

Intermediate article

H Tristram Engelhardt Jr, *Rice University, Houston, Texas, USA*

Lisa M Rasmussen, *Rice University, Houston, Texas, USA and University of Alabama, Birmingham, Alabama, USA*

Secular humanism is an attempt to offer a common moral vision based on a nonreligious conception of what is most truly 'human'.

Article contents

- Introduction: Historical Roots
- Genetic Choice in the Face of Moral Pluralism and Guided by Secular Moral Understandings
- Bioethics, Secular Humanism and Health care Policy in Pluralist Societies

Introduction: Historical Roots

Health care decision-making in a secular, nonreligious society presupposes a moral understanding able to guide choices in the absence of appeals to God or to a particular moral vision. In a secular pluralist society, health care policy requires a moral *lingua franca*, a general moral perspective that can transcend particular moral and religious commitments. Secular humanism has aspired to provide this common moral vision. Indeed, bioethics as it is generally understood and practiced in the United States claims to offer a regional portrayal of the general morality. Secular bioethics in being grounded in the common human condition aspires to be a secular humanism. Secular humanism joins two concepts: an understanding of religious neutrality with assumptions regarding the ability to draw moral guidance from human nature and the human condition. In seeking to inform choices in public morality, policy and law in terms open to all on the basis of a shared morality, secular humanism and bioethics run the risk of purchasing universality at the price of content or content at the price of universality.

History of secular humanism

Although secular moral understandings can be discerned in both Greece and Rome prior to the Christian era (one might think of Protagoras of Abdera, who saw the human as the Criterion for all moral choices), contemporary accounts developed primarily over against Christian views. Secular humanism emerged out of Enlightenment reactions against Clericism and the religious establishment, as well as on behalf of developing a view of the world based on human experience and reason. Secular humanism has roots in the thought of individuals such as David Hume (1711–1776), Immanuel Kant (1724–1804) and the French philosophes. One must also include such thinkers as Anthony Collins (1676–1729) with *A Discourse of Free-thinking* (Collins, 1978 [1713]) and Thomas Paine (1737–1809) with *The Age of Reason* (Paine, 1993 [1794]). Secular moral understandings came to have a marked impact on public policy through the French Revolution (1789–1799) and the legal reforms of Napoleon (1804), as well as the central European Secularization in 1803, which involved a massive transfer of the responsibility for welfare and education from the hands of the Roman Catholic church to secular authorities. These changes took place in the wake of wide-reaching shifts of political power and monastery properties from religious to nonreligious authority. The salience of secular moral accounts is connected to specific historical developments in thought and political structure that marginalized religious moral reflection.

Emergence of a secular humanist account of morality

Humanism as the source of the content for a secular morality draws its roots from the Renaissance, which sought to recapture Greco-Roman celebrations of scholarly refinement, personal development, human grace and immanently understandable values. That concern with human flourishing and the properly

human was reemphasized in subsequent movements such as the Second Humanism at the beginning of the nineteenth century, the New Humanism at the end of that century and the Third Humanism in the first part of the twentieth century. The *humanum* as the intellectual, esthetic and moral realization of that which is most truly human was regarded as a source for guiding science, technology and public policy, and in the case of the reformer of American medical education, Abraham Flexner (1866–1959), for medicine as well (Flexner, 1910, 1928).

Secular interpretations of morality gained ground in the latter part of the nineteenth century because of particular movements aimed at further marginalizing the influence of clerical power and religious moral assumptions. These included the French laicist movement of Jules Ferry (1832–1893) and the secularism movement of George Jacob Holyoake (1817–1906), who attempted to reshape morality and education in nonreligious terms (Reclus, 1947; Holyoake, 1851, 1871, 1896). Significant also was the development out of Unitarian Universalist churches and Reform Judaism of a cluster of groups focused on the articulation of secular communities with secular moral understandings such as the ethical culture movement of Felix Adler (1851–1933) (Kraut, 1979). These various groups analogous to nonreligious churches have in general come under the umbrella of such associations as the Ethical Humanists, the International Humanist and Ethical Union, the American Humanist Association and the British Humanist Association. Of significant influence was the Humanist Manifesto (A Humanist Manifesto, 1933), which was signed by John Dewey (1859–1952). The subsequent Humanist Manifesto of 1973 (Humanist Manifesto II, 1973), Secular Humanist Declaration (A Secular Humanist Manifesto, 1980) and other manifestoes carried on the tradition of attempting to frame a basis for moral and public policy choices free of a religious grounding. Numerous terms have been coined to characterize such a moral understanding, including Oliver L. Reiser's 'scientific humanism' (1940) and Sidney Hook's 'democratic humanism' (1980) (Reiser, 1940; Hook, 1980; Engelhardt, 1991).

Secular humanism in bioethics

It is against that broad spectrum of attempts to frame a nonreligiously available account of morality that one must place the remarkable development in the 1960s and 1970s of both the medical humanities and bioethics. Bioethics in particular emerged as a field to supply moral experts in the face of the marginalization of religious moral advisers and the deprofessionalization of medicine in the face of US Supreme Court holdings that transformed medicine from a guild to a trade. In the wake of these changes, the hope was to secure from the medical humanities in general and from bioethics in particular an account of morality to guide moral and public policy choices regarding medicine and the biomedical sciences. Even though there was little explicit connection with the secular humanist movements of the late nineteenth and early twentieth centuries, concern with the medical humanities developed in America from a recognition that society was becoming morally pluralist and *de jure* detached from the Christian viewpoints that had framed health care policy and law. With respect to the latter, one would need to underscore the Supreme Court holdings regarding contraception and abortion such as *Griswold versus Connecticut* (1965), *Eisenstadt versus Baird* (1972) and *Roe versus Wade* (1973). The court's decisions placed health care choices in these areas in the hands of patients.

The focus on individual autonomy and decision-making was accentuated by developments in informed consent, following the District of Columbia decision of *Canterbury versus Spence* (1972), which enshrined a 'reasonable and prudent person' test for health care choices. That is, *Canterbury* court advocated, as a standard of disclosure for informed consent, whatever a 'reasonable and prudent person' would want to know about a particular medical treatment. In a society that had chosen not to ground its moral choices in religious commitments and in the face of numerous competing secular moral understandings, consent and rights to privacy became central. As a result, autonomous decision-making expressed in informed consent became the lynchpin of secular bioethics. Although there was a recognition of the importance to pursuing the good, avoiding harm and acting justly, there was also a recognition of the multiple understandings of benefit, harm and justice.

Genetic Choice in the Face of Moral Pluralism and Guided by Secular Moral Understandings

Secular moral reflection and bioethical analysis have not delivered an uncontroversial moral perspective. Often the arguments consist of appeals to a secular 'sacredness' of human nature that delineates what we ought and ought not to be permitted to change – which by extension forbids tampering with the human genome. However, the success of such arguments depends upon a dubious connection between the specific makeup of the human genome and the broader existential question of what it means to be human. After all, in secular terms the genome is a result of millennia of contingent events, including random

mutation, cosmic and terrestrial catastrophes, evolutionary pressures and genetic drift. In the face of numerous accounts of that which is truly human, secular morality has by default placed its focus on the principle of permission (Engelhardt, 1996, 2000). The result has been a significant departure from traditional Christian and Judaic medical moral commitments. In particular, secular moral choices in the area of genetics have placed special accent not on a particular content-rich account of proper conduct but instead on procedural moral approaches. Thus, the focus is on clearly articulating the benefits and harms of particular genetic interventions with a view to allowing individual moral agents to make more effective choices. For example, secular morality may suggest that we conduct significant scientific research before considering genetic engineering, ensuring that the possible benefits of a particular genetic intervention outweigh risks. It also implies that we should work to eliminate fatal genetic diseases prior to engineering humans to be taller, smarter or faster.

As a result of the focus on the centrality of moral agency, given a rejection of notions of the sanctity of life and the impropriety of playing God – and in the face of a plurality of understandings of human flourishing – secular moral understandings have in general accepted the moral propriety of prenatal screening and abortion, preimplantation selection of embryos and other technologically mediated interventions designed to maximize the choices of parents regarding the health of their future children. As parents have come to have fewer children, the concern to have fully healthy children has increased. With the prospect of genetic technologies such as cloning, germ-line genetic engineering and gene therapies, the choices available to such parents may increase substantially. In the absence of a single moral vision that binds all people, it will be difficult to articulate any prohibitions in principle against such interventions to which parents or the single reproducing person consents. Instead, if an uncontroversial focus can be identified, it is the consent of the actual moral agents involved in genetic interventions, as well as the realization of the greatest balance of benefits over harms. But since the ranking of benefits and harms is itself a matter of dispute, the choice of a proper ranking has again placed centrally moral agents and their agreement.

Bioethics, Secular Humanism and Health Care Policy in Pluralist Societies

The contemporary, sociopolitical condition of most Western democracies embraces an actively religiously neutral perspective from which to resolve the controversial moral decisions they face. Salient among the decisions are those guided by the new genetic sciences and technologies. Increased genetic knowledge has motivated selective abortion, put people at jeopardy of having their job and insurance possibilities limited because of their health risks, and opened the challenge of how to direct human evolution that will increasingly be under personal control through genetic engineering. Although bioethics for the most part has not recognized its connection to secular humanist reflection, it is in fact an outgrowth of the aspiration to offer a morality that transcends religious commitments and that is grounded in human nature and the human condition.

Although the hope has been to disclose a particular, normative, content-rich view of appropriate moral choices, secular moral understandings of bioethics have by default succeeded precisely where they have not relied on an appeal to a putative canonical moral content (i.e. one that is uniquely normatively governing) but instead on the permission of moral agents and procedures for creating moral answers.

See also

Bioethics: Institutionalization of
Christianity and Genetics

References

A Humanist Manifesto (1933) *The New Humanist* **6**(3): 1–5.
A Secular Humanist Declaration (1980) *Free Inquiry* **1**: 3–7.
Canterbury versus Spence, 464 F. 2nd. 772, 791 (D.C. Cir.) (1972).
Collins A (1978[1713]) *A Discourse of Free-thinking*. New York, NY: Garland Publishers.
Eisenstadt versus Baird, 405 U.S. 438, 92 S. Ct. 1029, 31 L.Ed. 2nd 349 (1972).
Engelhardt Jr HT (1991) *Bioethics and Secular Humanism: The Search for a Common Morality*. London, UK: SCM Press/Philadelphia, PA: Trinity Press International.
Engelhardt Jr HT (1996) *Foundations of Bioethics*. 2nd edn. Oxford, UK: Oxford University Press.
Engelhardt Jr HT (2000) *The Foundations of Christian Bioethics*. Lisse, The Netherlands: Swets & Zeitlinger Publishers.
Flexner A (1910) *Medical Education in the United States and Canada: A Report to the Carnegie Foundation for the Advancement of Teaching*, Bulletin no. 4. New York, NY: Carnegie Foundation.
Flexner A (ed.) (1928) Philology in the technical sense is science, not humanism. *The Burden of Humanism*. Oxford, UK: Clarendon Press.
Griswold versus Connecticut, 381 U.S. 479, 85 S. Ct. 178, 14 L.Ed. 2nd 510 (1965).
Holyoake G (1851) *The History of the Last Trial by Jury for Atheism in England*. London, UK: James Watson.
Holyoake G (1871) *The Principles of Secularism*. London, UK: Austin & Co.
Holyoake G (1896) *The Origin and Nature of Secularism*. London, UK: Watts & Co.
Hook S (1980) The ground we stand on: democratic humanism. *Free Inquiry* **1**: 8–10.
Humanist Manifesto II (1973) *The Humanist* **23**(5): 4–9.
Kraut B (1979) *From Reform Judaism to Ethical Culture: The Religious Evolution of Felix Adler*. Cincinnati, OH: Hebrew Union College Press.
Paine T (1993[1794]) *The Age of Reason*. Avenel, NJ: Gramercy Books.
Reclus M (1947) *Jules Ferry*. Paris: Flammarion.
Reiser OL (1940) *The Promise of Scientific Humanism*. New York, NY: Oskar Piest.
Roe versus Wade, 410 U.S. 113 (1973).

Further Reading

D'Agostino M (1995) Reason and rationality: the core doctrines of secular humanism. *Free Inquiry* **15**(1): 47–50.
Grunbaum A (1992) In defense of secular humanism. *Free Inquiry* **12**(4): 30–39.
Hoeveler Jr JD (1977) *The New Humanism*. Charlottesville, VA: University Press of Virginia.
Huxley J (1961) *The Humanist Frame*. New York, NY: Harper.
Kurtz P (1983) *In Defense of Secular Humanism*. Buffalo, NY: Prometheus Books.
Kurtz P (1985) *Humanist Manifestos I and II*. Buffalo, NY: Prometheus Books.
Kurtz P (1988) *Forbidden Fruit: The Ethics of Humanism*. Buffalo, NY: Prometheus Books.
Pellegrino ED (1979) *Humanism and the Physician*. Knoxville, TN: University of Tennessee Press.
Potter CF (1930) *Humanism: A New Religion*. New York, NY: Simon and Schuster.
Schiller FCS (1969) *Humanism: Philosophical Essays*. Freeport, NY: Books for Libraries Press. [1st edn, 1903.]

Segmental Duplications and Genetic Disease

Beverly S Emanuel, *The Children's Hospital of Philadelphia, Philadelphia, Pennsylvania, USA*

Tamim H Shaikh, *The Children's Hospital of Philadelphia, Philadelphia, Pennsylvania, USA*

More than 50% of our genome is made up of repetitive DNA elements, which are classified into different categories based on their size, copy number, mechanisms of dispersal and several other factors. Segmental duplications represent one such class of repetitive element. Recently, several genetic diseases have been shown to be the result of chromosomal rearrangements mediated by segmental duplications.

Intermediate article

Article contents

- Introduction
- Genomic Disorders Mediated by Segmental Duplications
- Organization and Structure of Segmental Duplications
- Chromosomal Rearrangements Mediated by Segmental Duplications
- Mechanisms for Segmental Duplication-mediated Chromosomal Rearrangements
- Evolution of Segmental Duplications
- Conclusions

Introduction

The ongoing analysis of the human genome sequence has suggested that more than 50% of our genome is made up of repetitive deoxyribonucleic acid (DNA) elements, which are classified into different categories based on their size, copy number, mechanisms of dispersal, etc. (IHGSC, 2001). Segmental duplications (SDs) represent one such class of low-copy-number repetitive DNA element found in certain regions of the human genome. SDs can range in size from 1 to 400 kilobases (kb) and share 96–98% sequence identity (Shaikh *et al.*, 2001; IHGSC, 2001). Most recent estimates suggest that about 5% of our genome is composed of SDs belonging to two separate classes, interchromosomal and intrachromosomal (IHGSC, 2001). The interchromosomal class is duplicated on nonhomologous chromosomes and many of them localize to the pericentromeric and subtelomeric regions of human chromosomes (Eichler *et al.*, 1996, 1997; Regnier *et al.*, 1997; Trask *et al.*, 1998a, b). The intrachromosomal duplications, also referred to as region- or chromosome-specific low-copy repeats, are typically found on a single chromosome or within a single chromosomal band (Mazzarella and Schlessinger, 1998; Lupski, 1998a; Ji *et al.*, 2000; Emanuel and Shaikh, 2001; Stankiewicz and Lupski, 2002a). Many of these intrachromosomal SDs have been implicated in mediating chromosomal rearrangements associated with genetic diseases (Ji *et al.*, 2000; Shaffer and Lupski, 2000; Emanuel and Shaikh, 2001).

The chromosomal rearrangements mediated by SDs include deletions, interstitial duplications, translocations, inversions and small marker chromosomes. Misalignment followed by recombination between nonallelic SDs on homologous chromosomes has been proposed to give rise to these rearrangements. The resulting disruption of a gene, or alternatively an altered copy number of a gene(s), leads to genetic disease or disorder. Since a change at the genomic level is involved, these disorders have also been referred to as genomic disorders (Lupski, 1998a).

Genomic Disorders Mediated by Segmental Duplications

A few of the better known genomic disorders that are mediated by SDs include Charcot–Marie–Tooth disease type 1A (CMT1A)/hereditary neuropathy with liability to pressure palsies (HNPP) on 17p11.2 (Chance *et al.*, 1994; Lupski, 1998b), Prader–Willi syndrome (PWS)/Angelman syndrome (AS) on 15q11–q13 (Christian *et al.*, 1999; Amos-Landgraf *et al.*, 1999), neurofibromatosis type 1 (NF1) on 17q11.2 (Dorschner *et al.*, 2000), Smith–Magenis syndrome (SMS)/duplication 17p11.2 (Chen *et al.*, 1997; Potocki *et al.*, 2000), Williams–Beuren syndrome (WBS) on 7q11.23 (Perez-Jurado *et al.*, 1998; Peoples *et al.*, 2000), DiGeorge syndrome (DGS)/velocardiofacial syndrome (VCFS) and cat-eye syndrome (CES) on 22q11 (McTaggart *et al.*, 1998; Edelmann *et al.*, 1999a, b; Shaikh *et al.*, 2000). The list of SD-mediated genomic disorders has been continually expanding (Emanuel and Shaikh, 2001; Stankiewicz and Lupski, 2002a) and it is generally accepted that more SDs will be identified after the completion of the human genome sequence (IHGSC, 2001; Eichler, 2001). Thus it is likely that more instances of SDs will be shown to be responsible for a variety of genomic disorders.

Organization and Structure of Segmental Duplications

Segmental duplications are either simple in structure or contain a complex arrangement of duplicated modules (Ji *et al.*, 2000; Emanuel and Shaikh, 2001). The CMT1A-REP element, which mediates the interstitial duplication/deletion associated with CMT1A/HNPP respectively, is an example of a simple SD. CMT1A-REP is 24 kb in size and is present in two copies that flank the duplicated/deleted region on either side (Reiter *et al.*, 1996, 1997). The two copies of CMT1A-REP share 98.7% nucleotide sequence identity (Reiter *et al.*, 1996). Other disorders mediated by simple SDs include X-linked ichthyosis (Yen *et al.*, 1990; Li *et al.*, 1992) and hemophilia A (Naylor *et al.*, 1995, 1996).

Genomic disorders mediated by SDs that are large and complex in structure include SMS (Chen *et al.*, 1997; Potocki *et al.*, 2000), PWS/AS (Christian *et al.*, 1999; Landgraff *et al.*, 1999), WBS (Perez-Jurado *et al.*, 1998; Peoples *et al.*, 2000), NF1 (Dorschner *et al.*, 2000) and DGS/VCFS (Edelmann *et al.*, 1999b; Shaikh *et al.*, 2000). The complex SDs range in size from 100 to 500 kb and consist of multiple smaller, duplicated modules arranged in complex configurations. Therefore, different copies of the complex SDs may differ from each other in size, content and organization of the duplicated modules within them. The modules that are shared between any two given copies of these SDs share 95–98% nucleotide sequence identity. This complex architecture of SDs makes them excellent substrates for misalignment, followed by aberrant recombination leading to chromosomal rearrangements.

Segmental duplications may contain fragments that were originally derived from the same chromosome or from other chromosomes (Ji *et al.*, 2000; Emanuel and Shaikh, 2001). These fragments may include truncated gene segments/pseudogenes like the SDs on 17p11.2 (Potocki *et al.*, 2000; Park *et al.*, 2002), 7q11.23 (Pujana *et al.*, 2001) and 22q11 (Shaikh *et al.*, 2000). The SDs on 15q11–q13 are composed to a large extent of duplications of the hect domain and RLD 2 (*HERC2*) gene (Ji *et al.*, 1999). Similarly, the SDs that mediate the NF1 deletions on 17q11.2 contain at least four ESTs and expressed SH3-domain GRB2-like pseudogene 1 and 2 (*SH3GL1, SH3GL2*) pseudogenes (Dorschner *et al.*, 2000). SDs may also contain potentially recombinogenic sequences such as the palindromic AT-rich repeats (PATRR) and variable number tandem repeat (VNTR) sequences within the SDs on 22q11 (Shaikh *et al.*, 2000, 2001) or the ψ-like sequences, which in *Escherichia coli* are hot spots for recombination, within the SD on 17q11.2 (Lopez-Correa *et al.*, 2001). However, there is as yet no direct evidence for the involvement of any particular sequence in the rearrangements associated with the above chromosomal regions.

Chromosomal Rearrangements Mediated by Segmental Duplications

Segmental duplications are known to mediate recurrent chromosomal rearrangements. The recurrent rearrangements are site specific and include deletions, interstitial duplications, translocations, inversions and small marker chromosomes. SD-mediated microdeletions include the 1.9 megabase (Mb) deletion on Xp22 associated with X-linked ichthyosis (Ballabio and Andria, 1992); a 1.6 Mb deletion in 7q11.23 associated with WBS (Perez-Jurado *et al.*, 1998); the approximately 4 Mb deletion of 15q11–q13 (del(15)(q11–q13)) associated with PWS/AS (Christian *et al.*, 1999; Amos-Landgraf *et al.*, 1999); the 1.5 Mb deletion on 17q11.2 associated with NF1 in a minority (2–13%) of patients (Dorschner *et al.*, 2000) and the deletions in 22q11.2 associated with DGS/VCFS (Edelmann *et al.*, 1999a, b; Shaikh *et al.*, 2000, 2001).

In several SD-mediated rearrangements both reciprocal products of the unequal crossovers, a duplication and a deletion, leading to distinct phenotypic disorders have been observed. These include the reciprocal rearrangements on 17p12 associated with CMT1A and HNPP (Lupski, 1998b). The duplication of a 1.5 Mb region on 17p12 which contains the peripheral myelin protein 22 (*PMP22*) gene leads to CMT1A and the reciprocal deletion of the same 1.5 Mb fragment leads to HNPP (Lupski, 1998b). A similar scenario has been demonstrated for the SMS deletion and the reciprocal interstitial duplication of the same region of 17p11.2 that leads to a phenotype that is distinct from SMS (Chen *et al.*, 1997; Potocki *et al.*, 2000).

SDs can also mediate inversions like the one that disrupts the factor VIII gene on Xq28 leading to hemophilia A (Lakich *et al.*, 1993). Other rearrangements mediated by SDs include the inverted duplications of 15q11–q13 (inv dup (15)) (Wandstrat *et al.*, 1998; Wandstrat and Schwartz, 2000) and the inverted duplication of proximal 22q11, which when present as a supernumerary chromosome leads to CES (McTaggart *et al.*, 1998).

Evidence suggests that SDs are not only involved in rearrangements associated with genomic disorders, but also play an important role in generating genomic variation within the human population. The SDs that cause the ~1.6 Mb microdeletion associated with WBS also mediate a polymorphic inversion in about one-third of the parents of WBS patients (Osborne *et al.*, 2001). Another, perhaps more striking, example comes from the the genomic region on Xq28 surrounding the emerin (Emery–Dreifuss muscular dystrophy) (*EMD*) gene, which is flanked by two large SDs. An inversion polymorphism mediated by the flanking SDs was observed in 33% of normal females (Small *et al.*, 1997). Thus, in this instance, the inversion represents a benign, population-based variant mediated by the presence of SDs. Similar polymorphisms have been reported on chromosome 8p in regions flanked by SDs containing the olfactory receptor gene superfamily (Giglio *et al.*, 2001). The duplication of 15q24–q26 associated with panic and phobic disorders is also present in 7% of control, normal individuals (Gratacos *et al.*, 2001).

Mechanisms for Segmental Duplication-mediated Chromosomal Rearrangements

The sequence analysis of SDs has demonstrated 96–99% sequence identity between duplicated copies over their entire length. Based on this high level of sequence identity, several different models have been proposed to explain chromosomal rearrangements mediated by SDs (Emanuel and Shaikh, 2001; Stankiewicz and Lupski, 2002a). Interchromosomal rearrangements mediated by misaligned SDs on homologous chromosomes have been proposed to explain some of the deletions associated with DGS/VCFS, WBS, PWS/AS and most of the duplications/deletions associated with SMS/dup 17p11.2 and CMT1A/HNPP (reviewed in Emanuel and Shaikh (2001)). Intrachromosomal recombination between two SDs on the same chromosome has been used to explain the deletions associated with NF1, a few deletions associated with HNPP and some deletions associated with DGS/VCFS. It can also explain the paracentric inversions observed in the factor VIII gene that lead to hemophilia A (Lakich *et al.*, 1993), the polymorphic but benign inversion found close to the *EMD* gene (Small *et al.*, 1997) and the inversions that disrupt the iduronate 2-sulfatase (Hunter syndrome) (*IDS*) gene in Hunter syndrome (Bondeson *et al.*, 1995). Similar, inter- and intrachromosomal recombination events between SDs have also been proposed to explain the formation of bisatellited supernumerary marker chromosomes (reviewed in Emanuel and Shaikh (2001)).

The presence of unstable palindromic sequences in some SDs has led to the suggestion that such sequences may be directly involved in SD-mediated rearrangements. It has been proposed that palindromic sequences lead to the formation, at physiological temperature, of hairpins or cruciforms that may be sensitive to nucleases. Thus, the hairpin nicking activity may lead to double-stranded breaks (DSBs) in the chromosome at the site of the palindrome within the SD. These DSBs could further facilitate the rearrangements associated with SDs. This underlying mechanism has been proposed to explain the recurrent t(11;22) translocation as palindromic sequences have been reported at the t(11;22) breakpoints on both chromosomes 11 and 22 (Kurahashi *et al.*, 2000a, b). Large palindromes have also been reported within the SDs in Yq11.2 that flank the ~3.5 Mb region containing the azoospermia factor c (*AZFC*) gene, deleted in infertile males (Kuroda-Kawaguchi *et al.*, 2001).

Evolution of Segmental Duplications

Segmental duplications appear to be limited to the genomes of primates. Most chromosome-specific SDs that have been examined appear to have occurred during the past 35 million years of primate evolution (IHGSC, 2001). This estimate of their evolutionary age is based on the phylogenetic analysis of several SDs in the genomes of various primates and rodents. Molecular analysis of the SDs associated with CMT1A and SMS reveals that these SDs are not present in rodents (Stankiewicz and Lupski, 2002b; Park *et al.*, 2002). Comparative sequence analysis of human 22q11.2 and the syntenic region of the mouse genome has revealed no evidence for the presence of the SDs associated with DGS/VCFS in the mouse genome (Lund *et al.*, 1999; Shaikh, unpublished data). These data strongly suggest that the SDs that mediate genomic disorders are not present in the rodent genome.

Further, the analysis of SDs in humans and nonhuman primates suggests that there are differences in the copy number and location of the SDs on the chromosomes of representative members of the higher primates. Thus, the second copy of the SD associated with CMT1A is human and chimpanzee specific, suggesting that the duplication occurred after the divergence of the gorilla and chimpanzee about 7 million years ago (Kiyosawa and Chance, 1996). Similarly evidence suggests that the SDs that mediate the rearrangements leading to PWS/AS, SMS and DGS/VCFS may have originated and amplified over the past 35–50 million years (Christian *et al.*, 1999; Locke *et al.*, 2001; Park *et al.*, 2002; Shaikh *et al.*, 2001). Thus, SDs associated with genomic disorders have originated relatively recently during the evolution of the primate genome.

Furthermore, further evidence suggests that SDs may have played a role in the evolution of the primate chromosomes. Comparative analysis of human chromosome 19 and the syntenic region of the mouse genome has revealed the presence of duplicated sequences at the sites of the evolutionary breakpoints between the mouse and human chromosomes

(Dehal *et al.*, 2001). Similarly, duplications have been identified at or close to the breakpoints of evolutionary rearrangements on various human chromosomes when compared with the chromosomes of evolutionarily older primates (Nickerson and Nelson, 1998; Valero *et al.*, 2000). Therefore, the evolution and amplification of SDs in the primate genome may be directly related to their role in primate chromosome evolution.

Conclusions

Genetic change, besides being a major driving force for evolution, is also a source of human disease. SDs represent an underappreciated source of genetic change due to their ability to serve as substrates for aberrant genomic rearrangements. Data strongly suggest that SDs create genomic instability by predisposing certain chromosomes or chromosomal bands to rearrangements. SDs may also generate variation within the human population by causing rearrangements that may be phenotypically benign. It is possible that a few of these SD-mediated variant genome architectures may cause predisposition to secondary rearrangements that can further lead to genetic disease. Although all of the SD-mediated rearrangements identified so far (late 2002) have been involved in constitutional genomic disorders, they may also occur in somatic tissue. It remains to be seen if SDs will have a role in the rearrangements associated with certain cancers. Thus, the role of SDs in human disease appears to have been underestimated. We predict that as more SDs are identified in the human genome sequence, more genetic diseases will be found to be associated with them and vice versa.

See also
Chromosome-specific Repeats (Low-copy Repeats)

References

Ballabio A and Andria G (1992) Deletions and translocations involving the distal short arm of the human X chromosome: review and hypotheses. *Human Molecular Genetics* **1**: 221–227.

Bondeson ML, Dahl N, Malmgren H, *et al.* (1995) Inversion of the IDS gene resulting from recombination with IDS-related sequences is a common cause of the Hunter syndrome. *Human Molecular Genetics* **4**(4): 615–621.

Chance PF, Abbas N, Lensch MW, *et al.* (1994) Two autosomal dominant neuropathies result from reciprocal DNA duplication/deletion of a region on chromosome 17. *Human Molecular Genetics* **3**: 223–228.

Chen KS, Manian P, Koeuth T, *et al.* (1997) Homologous recombination of a flanking repeat gene cluster is a mechanism for a common contiguous gene deletion syndrome. *Nature Genetics* **17**: 154–163.

Christian SL, Fantes JA, Mewborn SK, Huang B and Ledbetter DH (1999) Large genomic duplicons map to sites of instability in Prader–Willi/Angelman syndrome chromosome region (15q11–q13). *Human Molecular Genetics* **8**(6): 1025–1037.

Dehal P, Predki P, Olsen AS, *et al.* (2001) Human chromosome 19 and related regions in mouse: conservative and lineage-specific evolution. *Science* **293**(5527): 104–111.

Dorschner MO, Sybert VP, Weaver M, Pletcher BA and Stephens K (2000) NF1 microdeletion breakpoints are clustered at flanking repetitive sequences. *Human Molecular Genetics* **9**(1): 35–46.

Edelmann L, Pandita RK and Morrow BE (1999a) Low-copy repeats mediate the common 3-Mb deletion in patients with velo-cardio-facial syndrome. *American Journal of Human Genetics* **64**: 1076–1086.

Edelmann L, Pandita RK, Spiteri E, *et al.* (1999b) A common molecular basis for rearrangement disorders on chromosome 22q11. *Human Molecular Genetics* **8**(7): 1157–1167.

Eichler EE (2001) Recent duplication, domain accretion and the dynamic mutation of the human genome. *Trends in Genetics* **17**: 661–669.

Eichler EE, Budarf ML, Rocchi M, *et al.* (1997) Interchromosomal duplications of the adrenoleukodystrophy locus: a phenomenon of pericentromeric plasticity. *Human Molecular Genetics* **6**(7): 991–1002.

Eichler EE, Lu F, Shen Y, *et al.* (1996) Duplication of a gene rich cluster between 16p11.1 and Xq28: a novel pericentromeric-directed mechanism for paralogous genome evolution. *Human Molecular Genetics* **5**(7): 899–912.

Emanuel BS and Shaikh TH (2001) Segmental duplications: an 'expanding' role in genomic instability and disease. *Nature Reviews Genetics* **2**: 791–800.

Giglio S, Broman KW, Matsumoto N, *et al.* (2001) Olfactory receptor-gene clusters, genomic-inversion polymorphisms, and common chromosome rearrangements. *American Journal of Human Genetics* **68**(4): 874–883.

Gratacos M, Nadal M, Martin-Santos R, *et al.* (2001) A polymorphic genomic duplication on human chromosome 15 is a susceptibility factor for panic and phobic disorders. *Cell* **106**(3): 367–379.

IHGSC (International Human Genome Sequencing Consortium) (2001) Initial sequencing and analysis of the human genome. *Nature* **409**: 860–921.

Ji Y, Eichler EE, Schwartz S and Nicholls RD (2000) Structure of chromosomal duplicons and their role in mediating human genomic disorders. *Genome Research* **10**(5): 597–610.

Ji Y, Walkowicz MJ, Buiting K, *et al.* (1999) The ancestral gene for transcribed, low-copy repeats in the Prader–Willi/Angelman region encodes a large protein implicated in protein trafficking, which is deficient in mice with neuromuscular and spermiogenic abnormalities. *Human Molecular Genetics* **8**(3): 533–542.

Kiyosawa H and Chance PF (1996) Primate origin of the CMT1A-REP repeat and analysis of a putative transposon-associated recombinational hot spot. *Human Molecular Genetics* **8**: 745–753.

Kurahashi H, Shaikh TH, Hu P, *et al.* (2000a) Regions of genomic instability on 22q11 and 11q23 as the etiology for the recurrent constitutional t(11;22). *Human Molecular Genetics* **9**: 1665–1670.

Kurahashi H, Shaikh TH, Zackai EH, *et al.* (2000b) Tightly clustered 11q23 and 22q11 breakpoints permit PCR-based detection of the recurrent constitutional t(11;22). *American Journal of Human Genetics* **67**(3): 763–768.

Kuroda-Kawaguchi T, Skaletsky H, Brown LG, *et al.* (2001) The AZFc region of the Y chromosome features massive palindromes and uniform recurrent deletions in infertile men. *Nature Genetics* **29**(3): 279–286.

Lakich D, Kazazian HH, Antonarakis SE and Gitschier J (1993) Inversions disrupting the factor VIII gene are a common cause of severe haemophilia A. *Nature Genetics* **5**: 236–241.

Landgraf JM, Ji Y, Gottlieb W, *et al.* (1999) Chromosome breakage in the Prader–Willi and Angelman syndromes involves recombination between large, transcribed repeats at proximal and distal breakpoints. *American Journal of Human Genetics* **65**(2): 370–386.

Li XM, Yen PH and Shapiro LJ (1992) Characterization of a low copy repetitive element S232 involved in the generation of frequent deletions of the distal short arm of the human X chromosome. *Nucleic Acids Research* **20**(5): 1117–1122.

Locke DP, Yavor AM, Lehoczky J, *et al.* (2001) Structure and evolution of genomic duplication in 15q11-q13 [abstract]. *American Journal of Human Genetics* **69**: 17.

Lopez-Correa C, Dorschner M, Brems H, *et al.* (2001) Recombination hot spot in NF1 microdeletion patients. *Human Molecular Genetics* **10**(13): 1387–1392.

Lund J, Roe B, Chen F, *et al.* (1999) Sequence-ready physical map of the mouse chromosome 16 region with conserved synteny to the human velocardiofacial syndrome region on 22q11.2. *Mammalian Genome* **10**(5): 438–443.

Lupski JR (1998a) Genomic disorders: structural features of the genome can lead to DNA rearrangements and human disease traits. *Trends in Genetics* **14**(10): 417–422.

Lupski JR (1998b) Charcot–Marie–Tooth disease: lessons in genetic mechanisms. *Molecular Medicine* **4**: 3–11.

Mazzarella R and Schlessinger D (1998) Pathological consequences of sequence duplications in the human genome. *Genome Research* **8**(10): 1007–1021.

McTaggart KE, Budarf ML, Driscoll DA, *et al.* (1998) Cat eye syndrome chromosome breakpoint clustering: identification of two intervals also associated with 22q11 deletion syndrome breakpoints. *Cytogenetics and Cell Genetics* **81**: 222–228.

Naylor JA, Buck D, Green P, *et al.* (1995) Investigation of the factor VIII intron 22 repeated region (int22h) and the associated inversion junctions. *Human Molecular Genetics* **4**(7): 1217–1224.

Naylor JA, Nicholson P, Goodeve A, *et al.* (1996) A novel DNA inversion causing severe hemophilia A. *Blood* **87**(8): 3255–3261.

Nickerson E and Nelson DL (1998) Molecular definition of pericentric inversion breakpoints occurring during the evolution of humans and chimpanzees. *Genomics* **50**(3): 368–372.

Osborne LR, Li M, Pober B, *et al.* (2001) A 1.5 million-base pair inversion polymorphism in families with Williams–Beuren syndrome. *Nature Genetics* **29**(3): 321–325.

Park SS, Stankiewicz P, Bi W, *et al.* (2002) Structure and evolution of the Smith-magenis syndrome repeat gene clusters, SMS-REPs. *Genome Research* **12**(5): 729–738.

Peoples R, Franke Y, Wang YK, *et al.* (2000) A physical map, including a BAC/PAC clone contig, of the Williams–Beuren syndrome – deletion region at 7q11.23. *American Journal of Human Genetics* **66**(1): 47–68.

Perez-Jurado LA, Wang YK, Peoples R, *et al.* (1998) A duplicated gene in the breakpoint regions of the 7q11.23 Williams–Beuren syndrome deletion encodes the initiator binding protein TFII-I and BAP-135, a phosphorylation target of BTK. *Human Molecular Genetics* **7**: 325–334.

Potocki L, Chen KS, Park SS, *et al.* (2000) Molecular mechanism for duplication 17p11.2 – the homologous recombination reciprocal of the Smith–Magenis microdeletion. *Nature Genetics* **24**(1): 84–87.

Pujana MA, Nadal M, Gratacos M, *et al.* (2001) Additional complexity on human chromosome 15q: identification of a set of newly recognized duplicons (LCR15) on 15q11–q13, 15q24, and 15q26. *Genome Research* **11**(1): 98–111.

Regnier V, Meddeb M, Lecointre G, *et al.* (1997) Emergence and scattering of multiple neurofibromatosis (NF-1)-related sequences during hominoid evolution suggest a process of pericentromeric interchromosomal transposition. *Human Molecular Genetics* **6**: 9–16.

Reiter LT, Murakami T, Koeuth T, Gibbs RA and Lupski JR (1997) The human COX10 gene is disrupted during homologous recombination between the 24 kb proximal and distal CMT1A-REPs. *Human Molecular Genetics* **6**(9): 1595–1603.

Reiter LT, Murakami T, Koeuth T, *et al.* (1996) A recombination hotspot responsible for two inherited peripheral neuropathies is located near a mariner transposon-like element. *Nature Genetics* **12**: 288–297.

Shaffer LG and Lupski JR (2000) Molecular mechanisms for constitutional chromosomal rearrangements in humans. *Annual Review of Genetics* **34**: 297–329.

Shaikh TH, Kurahashi H and Emanuel BS (2001) Evolutionarily conserved duplications in 22q11 mediate deletions, duplications, translocations and genomic instability. *Genetics in Medicine* **3**(1): 6–13.

Shaikh TH, Kurahashi H, Saitta SC, *et al.* (2000) Chromosome 22-specific low copy repeats and the 22q11.2 deletion syndrome: Genomic organization and deletion endpoint analysis. *Human Molecular Genetics* **9**: 489–501.

Small K, Iber J and Warren ST (1997) Emerin deletion reveals common X-chromosome inversion mediated by inverted repeats. *Nature Genetics* **16**: 96–99.

Stankiewicz P and Lupski JR (2002a) Genome architecture, rearrangements and genomic disorders. *Trends in Genetics* **18**: 74–80.

Stankiewicz P and Lupski JR (2002b) Molecular-evolutionary mechanisms for genomic disorders. *Current Opinion in Genetics and Development* **12**(3): 312–319.

Trask B, Friedman B, Martin-Gallardo A, *et al.* (1998a) Members of the olfactory receptor gene family are contained in large blocks of DNA duplicated polymorphically near the ends of human chromosomes. *Human Molecular Genetics* **7**(1): 13–26.

Trask BJ, Massa H, Brand-Arpon V, *et al.* (1998b) Large multi-chromosomal duplications encompass many members of the olfactory receptor gene family in the human genome. *Human Molecular Genetics* **7**(13): 2007–2020.

Valero MC, de Luis O, Cruces J and Perez Jurado LA (2000) Fine-scale comparative mapping of the human 7q11.23 region and the orthologous region on mouse chromosome 5G: the low-copy repeats that flank the Williams–Beuren syndrome deletion arose at breakpoint sites of an evolutionary inversion. *Genomics* **69**(1): 1–13.

Wandstrat AE, Leana-Cox J, Jenkins L and Schwartz S (1998) Molecular cytogenetic evidence for a common breakpoint in the largest inverted duplications of chromosome 15. *American Journal of Human Genetics* **62**(4): 925–936.

Wandstrat AE and Schwartz S (2000) Isolation and molecular analysis of inv dup(15) and construction of a physical map of a common breakpoint in order to elucidate their mechanism of formation. *Chromosoma* **109**(7): 498–505.

Yen PH, Li XM, Tsai SP, *et al.* (1990) Frequent deletions of the human X chromosome distal short arm result from recombination between low copy repetitive elements. *Cell* **61**(4): 603–610.

Further Reading

Bailey JA, Gu Z, Clark RA, *et al.* (2002) Recent segmental duplications in the human GENOME. *Science* **297**: 1003–1007.

Web Links

Emerin (Emery-Dreifuss muscular dystrophy) (*EMD*); Locus ID: 2010. LocusLink:
http://www.ncbi.nlm.nih.gov/LocusLink/LocRpt.cgi?l = 2010

Hect domain and RLD 2 (*HERC2*); Locus ID: 8924. LocusLink:
   http://www.ncbi.nlm.nih.gov/LocusLink/LocRpt.cgi?l = 8924
Iduronate 2-sulfatase (Hunter syndrome) (*IDS*); Locus ID: 3423.
   LocusLink:
   http://www.ncbi.nlm.nih.gov/LocusLink/LocRpt.cgi?l = 3423
Peripheral myelin protein 22 (*PMP22*); Locus ID: 5376. LocusLink:
   http://www.ncbi.nlm.nih.gov/LocusLink/LocRpt.cgi?l = 5376
Emerin (Emery-Dreifuss muscular dystrophy) (*EMD*); MIM
   number: 300384. OMIM:
   http://www.ncbi.nlm.nih.gov/htbin-post/Omim/
   dispmim?300384

Hect domain and RLD 2 (*HERC2*); MIM number: 605837. OMIM:
   http://www.ncbi.nlm.nih.gov/htbin-post/Omim/
   dispmim?605837
Iduronate 2-sulfatase (Hunter syndrome) (*IDS*); MIM number:
   309900. OMIM:
   http://www.ncbi.nlm.nih.gov/htbin-post/Omim/
   dispmim?309900
Peripheral myelin protein 22 (*PMP22*); MIM number: 601097.
   OMIM:
   http://www.ncbi.nlm.nih.gov/htbin-post/Omim/
   dispmim?601097

Segregation Analysis Software

Advanced article

Christian Stricker, *Applied Genetics Network, Altendorf, Switzerland*
Rohan L Fernando, *Iowa State University, Ames, Iowa, USA*

Article contents

- Introduction
- Mendel, Version 4.1.1
- SAGE
- PAP, Revision 5

Segregation analysis software is used to test modes of inheritance and to provide parameter estimates for genetic models.

Introduction

Elston (1981) among others defined segregation analysis as fitting Mendelian segregation ratios to family data. Thus, it is not necessary to distinguish between segregation and linkage analysis since both comply with this definition, as long as data are recorded on families. However, the focus of segregation analysis is usually on inferring the mode of inheritance of a certain trait, whereas in linkage analysis marker data are used to assign such putative loci to specific chromosomal areas.

Traditionally, analyzing single-locus recessive or codominant diseases assuming complete penetrance and no phenocopies was probably the first area where segregation analysis was applied. When these simple models were extended to account for marker information, that is, the cosegregation between the putative disease locus and an observed genetic marker locus, it was still considered segregation analysis by some scientists (see e.g. Ott (1999)) while others referred to it as linkage analysis. However, for more complex modes of inheritance and thus more complex genetic models, adding marker information to a segregation model for a trait was no longer considered segregation analysis, it was rather referred to as linkage analysis. There are three ways a mode of inheritance is considered complex: either there is more than a single locus underlying the trait with intra- and/or interallelic interactions to be considered, or there is a nontrivial environmental influence on the expression of the trait,

or both. In all three cases, the penetrance function is not simple, irrespective of the phenotypic distribution of the trait. Hence, the phenotypic distribution of a trait has little implication with respect to the underlying mode of inheritance. The advantage of considering marker and trait data jointly is twofold. First, using informative markers, chromosomal segments can be traced through a pedigree. This adds to the power to determine the mode of inheritance for linked trait loci. Second, markers can be used to determine the location of a trait locus. Thus, if the intention of adding markers is to increase the statistical power to infer the mode of inheritance of a trait, then some scientists prefer to retain the name segregation analysis. In the latter case, when the intention is to determine the location of the trait locus, the name linkage analysis is usually used.

The important differences between a marker and a putative trait locus are that Mendelian inheritance at a marker locus is more readily accepted and that a marker locus tends to have a simpler penetrance function. This is the main reason why modeling the cosegregation of a marker with the trait is of interest. Due to inconsistent definitions of the term segregation analysis, and considering that the concepts of modeling the inheritance of a marker and a trait locus are not necessarily different, Elston's definition given above for segregation analysis seems appropriate. According to this definition, linkage analysis and segregation analysis are synonymous.

In the remainder of this article we will discuss segregation analysis software capable of fitting segregation ratios to phenotypic trait data only. Software for the joint analysis of marker and trait data is considered elsewhere. A useful list of segregation and linkage analysis software can be found at the website of the Laboratory of Statistical Genetics at Rockefeller University (see Web Links). Existing software is constantly being revised and extended as well as new software being released. Due to this and the limited space for this article, we will present three typical software packages for fitting segregation ratios to trait data in families. (*See* Quantitative Trait Loci (QTL) Mapping Software.)

Mendel, Version 4.1.1

The package can be downloaded via the list of statistical software packages at the Rockefeller site or from the UCLA Human Genetics website (see web Links). The program is free of charge. The program documentation is short and concise, likely requiring the interested user to read additional material. The code is in Fortran 95. The program runs in a line-oriented mode and will run on all operating systems for which an appropriate Fortran 95 compiler is available. Some adaptations may be required, depending on the compiler used. The program uses the Elston–Stewart algorithm (Elston and Stewart, 1971) or the Langer–Green algorithm (Lander and Green, 1987) to calculate the likelihood of models involving a small number of loci or pedigree members, respectively, and simple penetrace functions. Mendel is a powerful package for the analysis of phenotypic data in pedigrees for simple genetic models, its capabilities are not exhausted by segregation analysis only.

SAGE

There are currently (late 2002) two versions of SAGE (Statistical Analysis for Genetic Epidemiology) available. The latest officially released version is 3.1, version 4.0 is in the beta-testing phase. The software package is not free of charge, except for the beta-testing versions. The license charges are based on a yearly basis and include program as well as theoretical and practical support for specific data analyses. The ANSCI C source code is not publicly available. The package is rigorously documented. SAGE is a very powerful package, classical segregation analyses being only one of the various options within model-based and model-free analysis of trait and marker data in

pedigrees. The principal investigator connected to the SAGE package is Robert Elston, one of the founders of modern segregation analysis (Elston and Stewart, 1971). Information about the package, its pricing as well as the download areas for manuals and beta versions can be found at the SAGE Homepage (see Web Links). SAGE may be run in a line-oriented batch mode, or alternatively it features a graphical user interface. Compiled programs are available for most operating systems. If only trait data without marker information are to be analyzed, the program uses the Elston–Stewart algorithm (Elston and Stewart, 1971) to calculate the likelihood of models involving a small number of loci. The pedigree structure is required to be simple, that is, the pedigree cannot contain loops. Mixed oligogenic and polygenic inheritance can be handled via regressive models (Bonney, 1984, 1986). Various penetrance functions allow one to handle discrete and continuous phenotypes. Fitting of covariables and ascertainment correction is available. There is a preprocessing program in SAGE generating appropriate input files. Alternatively, this task may be handled via the graphical user interface. The segregation model is specified by a parameter file. In summary, using SAGE for segregation analysis may serve as a good starting point for more profound analyses when additional information such as marker genotypes on some individuals of the pedigrees studied becomes available. The excellent scientific support behind SAGE is a great advantage of this package.

PAP, Revision 5

The most recent (as of early 2003) release of PAP, Revision 5, was mainly written by Sandy Hasstedt and may be obtained from hasstedt.genetics.utah.edu (see Web Links). Alternatively, PAP integrates various model-based and model-free approaches to analyze pedigrees and nuclear families (marker and trait data). Thus, analyzing trait data only will not exhaust its capabilities. The program uses the Elston–Stewart algorithm (Elston and Stewart, 1971) to calculate the likelihood of oligogenic models for pedigree data. For segregation analysis, it uses the generalization of Cannings et al. (1978) to allow for arbitrary complex pedigree structure. Mixed oligogenic and polygenic inheritance can be accommodated by additionally assuming a normally distributed polygenic component. For such mixed inheritance models, the likelihood needs to be approximated for efficient calculation in larger pedigrees (Hasstedt, 1991). Segregation models may include an arbitrary number of loci and alleles. However, memory and CPU time requirements increase exponentially with the number of loops in the pedigree and the number of genotypes considered. PAP has various penetrance functions and transmission options implemented, and can thus handle discrete or continuous phenotypes. Covariables can be accommodated in the analysis and correcting for ascertainment is possible. PAP expects pedigree data in two input files, a pedigree and a file containing phenotypic records. A format file links this information to the file specifying the model. This latter file may be created interactively by stepping through on-screen menus. All files may alternatively be created using a text editor. Information in these files, is in fixed format. To properly supply these files a minimal understanding of format statements in Fortran 77 is required. In summary, using PAP for segregation analysis may serve as a good starting point for more profound analyses when additional information such as marker genotypes on some individuals becomes available.

See also
Linkage Analysis

References

Bonney GE (1984) On the statistical determination of major gene mechanisms in continuous human traits: regressive models. *American Journal of Medical Genetics* **18**: 731–749.

Bonney GE (1986) Regressive logistic models for familial disease and other binary traits. *Biometrics* **42**: 611–625.

Cannings C, Thompson EA and Skolnick MA (1978) Probability functions on complex pedigrees. *Advances in Applied Probability* **10**: 26–61.

Elston RC (1981) Segregation analysis. In: Harris H and Hirschhorn K (eds.) *Advances in Human Genetics*, vol. II, pp. 63–120. New York, NY: Plenum Publishing.

Elston RC and Stewart J (1971) A general model for the analysis of pedigree data. *Human Heredity* **21**: 523–542.

Hasstedt SJ (1991) A variance components/major locus likelihood approximation on quantitative data. *Genetic Epidemiology* **8**: 113–125.

Lander ES and Green P (1987) construction of multilocus genetic maps in humans, *Proceedings of the National Academy of Sciences of the United States of America* **84**: 2363–2367.

Ott J (1999) *Analysis of Human Genetic Linkage*, 3rd edn. Baltimore, MD: Johns Hopkins University Press.

Further Reading

Morton NE and McLean CJ (1974) Analysis of family resemblence III. Complex segregation analysis of quantitative traits. *American Journal of Human Genetics* **26**: 489–503.

Web Links

Laboratory of Statistical Genetics at Rockefeller University. Web resources of genetic linkage analysis
http://linkage.rockefeller.edu

HGAR: the Human Genetic Analysis Resource (SAGE home page)
http://darwin.cwru.edu/octane/sage/sage.php

PAP, revision 5
http://hasstedt.genetics.utah.edu

Mendel 4.1.1
http://www.genetics.ucla.edu/software/

Selective and Structural Constraints

Austin L Hughes, *University of South Carolina, South Carolina, USA*

The neutral theory of molecular evolution predicts that the rate of evolution will be low in portions of genes and of proteins that are functionally important. This insight is the basis for bioinformatics strategies that use sequence information to make functional predictions.

Advanced article

Article contents

- Mutation and Molecular Evolution
- Synonymous and Nonsynonymous Substitution
- Constraints on Amino Acid Sequence Evolution
- Conclusion

Mutation and Molecular Evolution

Kimura's neutral theory of molecular evolution (Kimura, 1977, 1983) provides elegant mathematical expressions for predicting patterns of molecular evolution. One of the simplest but most widely applicable results of this theory is shown in the following equation:

$$K = u_t f_0 \qquad (1)$$

In eqn [1], K is the rate of selectively neutral molecular evolution per site per generation; u_t is the total mutation rate per generation; and f_0 is the fraction of mutations that are selectively neutral. Although the fact is not widely recognized by experimental biologists, eqn [1] is one of the most fundamental equations in modern biology, and it provides the theoretical basis for many of the computational analyses of macromolecular sequence data that are essential to the emerging field of bioinformatics.

Given a constant or nearly constant mutation rate throughout a gene or genomic region, eqn [1] implies that the rate at which nucleotide substitutions will accumulate over time is proportional to the fraction of mutations that are selectively neutral. Thus, there will be a high rate of nucleotide substitution in genomic regions where all or nearly all possible mutations are neutral, that is, when f_0 is equal to 1, or nearly equal to 1. On the other hand, when f_0 is equal to zero or is very low, the rate of nucleotide substitution will be zero or nearly zero.

The main factor that will cause f_0 to be very low at a given site is natural selection against mutations occurring at that site. In other words, f_0 will be low for a given site or set of sites when all or most possible mutations at those sites are selectively deleterious. Such selectively deleterious mutations will be quickly eliminated from the population by natural selection. The type of natural selection that eliminates deleterious mutations from a population is known as 'purifying selection', in contrast to positive Darwinian selection, which leads to the fixation of selectively advantageous mutations. The neutral theory predicts that purifying selection has been a much more prevalent factor in shaping the evolution of genomes than has positive Darwinian selection, which is expected to be a relatively rare phenomenon (Kimura 1983; Hughes 1999).

Equation [1] gives rise to the idea of a selective constraint. A genomic region subject to a high degree of selective constraint is simply a region in which f_0 is low. Similarly a region with a low degree of selective constraint is one in which f_0 is high. The degree of selective constraint is equivalent to the strength of purifying selection – or selection against harmful mutations – operating on a specific site or set of sites (including a region within a gene, an entire gene or a section of the genome containing more than one gene).

Synonymous and Nonsynonymous Substitution

In coding regions, there is expected to be a difference in selective constraint between synonymous sites (sites at which a mutation does not change the amino acid encoded) and nonsynonymous sites (sites at which a mutation will cause an amino acid change). Equation [1] leads to the prediction that, in most coding regions, the rate of synonymous nucleotide substitution per synonymous site (designated d_S) will exceed the rate of nonsynonymous nucleotide substitution per nonsynonymous site (designated d_N). When these quantities have been calculated for samples of genes from a wide variety of organisms, the results have shown that d_S exceeds d_N in coding regions of the great majority of genes (Li *et al.*, 1985; Duret and Mouchiroud, 2000; Jordan *et al.*, 2002).

For example, Hughes *et al.* (2002) compared 3298 putatively orthologous genes between two genotypes of the bacterium *Mycobacterium tuberculosis*. Mean d_S for all loci was 0.000328 ± 0.000022 (standard error). Mean d_N for all loci was 0.000206 ± 0.00011. The difference between these means is highly significant. Such results support the predictions of eqn [1] and thus the neutral theory (Kimura, 1977).

The fact that d_S exceeds d_N in most protein-coding genes makes exceptions to this typical pattern especially interesting. When positive Darwinian selection occurs, it can cause a reversal of the usual relationship between d_S and d_N (Hughes and Nei 1988; Hughes 1999). Comparison of the rates of synonymous and nonsynonymous nucleotide substitution can thus tell us whether a gene or gene region is subject to purifying selection or positive selection.

Constraints on Amino Acid Sequence Evolution

In most proteins, not all amino acid residue positions evolve at the same rate. Certain amino acids are crucial to the function of the protein, and mutations causing amino acid changes at these positions will be eliminated by purifying selection. Other amino acids may be relatively unimportant functionally, and mutations causing changes at these positions may be selectively neutral. Over evolutionary time, the amino acids that are important to the function of the protein are the least likely to be changed. Such positions are said to be evolutionarily conserved. Conserved positions are said to be subject to functional constraint if changes at these positions are selected against because they would be deleterious to the function of the protein. The evolutionary conservation of functionally important amino acid residues is a consequence of eqn [1] that has proved useful to biologists because it provides a link between the rate of molecular evolution and protein function.

Figure 1 illustrates conservation of functionally important amino acid residues in the case of pepsin A and pepsin C from human and chicken. Pepsin A and pepsin C belong to the protein family known as aspartic proteinases; these are proteolytic enzymes

```
Human pepsin C     TVTYEPMAYMDAAYFGEISIGTPPQNFLVLF DTGSSNLWVPSVYCQSQACTSHSRFNPSE
Chicken pepsin C   TAYEPLANNMDMSYYGEISIGTPPQNFLVLF DTGSSNLWVPSTLCQSQACANHNEFDPNE
Chicken pepsin A   ESYEPMTNYMDASYYGTISIGTPQQDFSVIF DTGSSNLWVPSIYCKSSACSNHKRFDPSK
Human pepsin A     VDEQPLENYLDMEYFGTIGIGTPAQDFTVVF DTGSSNLWVPSVYCSSLACTNHNRFNPED

Human pepsin C     SSTYSTNGQTFSLQYGSGSLTGFFGYDTLTVQSIQVPNQEFGLSENEPGTNFVYAQFDGI
Chicken pepsin C   SSTFSTQDEFFSLQYGSGSLTGIFGFDTVTIQGISITNQEFGLSETEPGTSFLYSPFDGI
Chicken pepsin A   SSTYVSTNETVYIAYGTGSMSGILGYDTVAVSSIDVQNQIFGLSETEPGSFFYYCNFDGI
Human pepsin A     SSTYQSTSETVSITYGTGSMTGILGYDTVQVGGISDTNQIFGLSETEPGSFLYYAPFDGI

Human pepsin C     MGLAYPALSVDEATTAMQGMVQEGALTSPVFSVYLSNQQGSSGGAVVFGGVDSSLYTGQI
Chicken pepsin C   LGLAFPSISAGGATTVMQKMLQENLLDFPVFSFYLSGQEGSQGGELVFGGVDPNLYTGQI
Chicken pepsin A   LGLAFPSISSSGATPVFDNMMSQHLVAQDLFSVYLS-KDGETGSFVLFGGIDPNYTTKGI
Human pepsin A     LGLAYPSISSSGATPVFDNIWNQGLVSQDLFSVYLS-ADDQSGSVVIFGGIDSSYYTGSL

Human pepsin C     YWAPVTQELYWQIGIEEFLIGGQASGWCSEGCQAIV DTG TSLLTVPQQYMSALLQATGAQ
Chicken pepsin C   TWTPVTQTTYWQIGIEDFAVGGQSSGWCSQGCQGIV DTG TSLLTVPNQVFTELMQYIGAQ
Chicken pepsin A   YWVPLSAETYWQITMDRVTVGN-KYVACFFTCQAIV DTG TSLLVMPQGAYNRIIKDLGVS
Human pepsin A     NWVPVTVEGYWQITVDSITMNG-EAIACAEGCQAIV DTG TSLLTGPTSPIANIQSDIGAS

Human pepsin C     EDEYGQFLVNCNSIQNLPSLTFIINGVEFPLPPSSYILSNNG-YCTVGVEPTYLSSQNGQ
Chicken pepsin C   ADDSGQYVASCSNIEYMPTITFVISGTSFPLPPSAYMLQSNSDYCTVGIESTYLPSQTGQ
Chicken pepsin A   --SDGE--ISCDDISKLPDVTFHINGHAFTLPASAYVLNEDG-SCMLGFENMGTPTELGE
Human pepsin A     ENSDGDMVVSCSAISSLPDIVFTINGVQYPVPPSAYILQSEG-SCISGFQGMNLPTESGE

Human pepsin C     PLWILGDVFLRSYYSVYDLGNNRVGFATAA
Chicken pepsin C   PLWILGDVFLRVYYSIYDMGNNQVGFATAV
Chicken pepsin A   -QWILGDVFIREYYVIFDRANNKVGLSPLS
Human pepsin A     -LWILGDVFIRQYFTVFDRANNQVGLAPVA
```

Figure 1 Alignment (Thompson et al., 1994) of the mature polypeptide sequences of four vertebrate aspartic proteinases. The conserved active site residues are boxed.

characterized by two aspartic acid residues in the active site. Each aspartic proteinase consists of two similar domains (the N-terminal and C-terminal domains), and the three-dimensional structure of the molecule brings together portions of both domains to form the active site of the enzyme in which substrate polypeptides are bound (Sielecki et al., 1990). In most known aspractic proteinases, the active site includes the motif aspartic acid-threonine-glycine (DTG) in each of the two domains (**Figure 1**). A number of other residues around the active site residues are also conserved in human and chicken pepsin A and pepsin C (**Figure 1**). Even if we did not know which residues are important in the active site of the aspartic proteinases shown in **Figure 1**, the strong sequence conservation in the region of the two DTG motifs would point to these regions as possible candidates for functionally important regions of these proteins.

Computer programs that align nucleotide or amino acid sequences are widely used in modern biology. Sequence alignments can be used to identify amino acid residues that have been conserved over the evolutionary history of a set of related sequences. In the absence of functional or structural information about a given protein family, conservation of a given amino acid residue or set of residues provides evidence that these residues are important for the protein's function. The theoretical prediction that a low rate of

molecular evolution should characterize functionally important sites allows biologists to use alignment as a tool to make predictions about function that can subsequently be tested by experimental methods.

Conclusion

The theoretical prediction that the rate of molecular evolution is highest at sites that are the least important functionally and lowest at the sites of greatest functional importance has many applications in the analysis of biological sequences and structures. It provides the basis for quantitative analysis of the effects of natural selection at the molecular level and for using sequence comparisons to yield information regarding protein function.

See also
Evolution: Neutralist View
Purifying Selection: Action on Silent Sites
Synonymous and Nonsynonymous Rates

References

Duret L and Mouchiroud D (2000) Determinants of substitution rates in mammalian genes: expression pattern affects selection intensity but not mutation rate. *Molecular Biology and Evolution* **17**: 68–74.

Hughes AL (1999) *Adaptive Evolution of Genes and Genomes*. New York, NY: Oxford University Press.

Hughes AL, Friedman R and Murray M (2002) Genome-wide pattern of synonymous nucleotide substitution in two complete genomes of Mycobacterium tuberculosis. *Emerging Infectious Diseases* **8**: 1342–1346.

Hughes AL and Nei M (1988) Pattern of nucleotide substitution at MHC class I loci reveals overdominant selection. *Nature* **335**: 167–170.

Jordan IK, Rogozin IB, Wolf YI and Koonin EV (2002) Microevolutionary genomics of bacteria. *Theoretical Population Genetics* **61**: 435–444.

Kimura M (1977) Preponderance of synonymous changes as evidence for the neutral theory of molecular evolution. *Nature* **267**: 275–276.

Kimura M (1983) *The Neutral Theory of Molecular Evolution*. Cambridge, UK: Cambridge University Press.

Li W-H, Wu C-I and Luo C-C (1985) A new method for estimating synonymous and nonsynonymous rates of nucleotide substitu-

tion considering the relative likelihood of nucleotide and codon changes. *Molecular Biology and Evolution* **2**: 150–174.

Sielecki AR, Fedorov AA, Boodhoo A, Andreeva NS and James NG (1990) Molecular and crystal structures of monoclinic porcine pepsin refined at 1.8 Å resolution. *Journal of Molecular Biology* **214**: 143–170.

Thompson JD, Higgins DG and Gibson TJ (1994) CLUSTALW: improving the sensitivity of progressive multiple sequence alignment through sequence weighting, position-specific gap penalties and weight matrix choice. *Nucleic Acids Research* **22**: 4673–4680.

Further Reading

Kimura M (1986) DNA and the neutral theory. *Philosophical Transactions of the Royal Society of London* **B312**: 343–354.

Li W-H (1997) *Molecular Evolution*. Sunderland MA: Sinauer.

Mount DW (2001) *Bioinformatics: Sequence and Genome Analysis*. Cold Spring Harbor, NY: Cold Spring Harbor Laboratory Press.

Sequence Accuracy and Verification

Stephan Beck, *The Sanger Centre, Cambridge, UK*

Analyses involving deoxyribonucleic acid sequences have to consider three main parameters concerning accuracy: sequence quality, sequence contiguity and sequence fidelity. Here, sequence quality defines the probability of error for any base-call, contiguity defines the completeness and correctness of the assembly of subsequences and fidelity defines the correctness of the genomic representation of the assembly.

Advanced article

Article contents

- Introduction
- Quality
- Contiguity
- Fidelity
- Outlook
- Cautionary Note

Introduction

Like all experimentally derived data, deoxyribonucleic acid (DNA) sequences contain errors. The causes of such errors have been well documented in several studies (reviewed by Abola *et al.* (2000)) and early on led to the suggestion that sequence quality should be monitored (Beck, 1993). This is particularly the case for the human genome sequence which has been generated by multiple centers and different approaches. All contributors to the public-domain Human Genome Project (HGP) agreed in 1997 – an agreement known as the 2nd Bermuda Agreement (see Web Links) – that the finished sequence should have an accuracy of 99.99% or better, which is equal to no more than one error per 10 000 bases.

Quality

Estimating base-call quality was probably the easiest and therefore the first parameter in respect to sequence

accuracy to become completely automated and widely accepted. Using log-transformed probabilities to detect errors close to the 0% level, the PHRED algorithm computes quality values ($Q = -10 \times \log10(p)$) for each base-call where p is an estimated error probability calculated from various parameters of the sequence trace from which the base has been called (Ewing *et al.*, 1998; Ewing and Green, 1998). Quality (Q) values are assigned on a scale from 1 (lowest) to 99 (highest), and for the HGP it was agreed that only base-calls of PHRED $Q20$ or better should be used for calculating sequence coverage. A $Q20$ base-call has a probability $1/100$ or 1% of being incorrect (a $Q30$ base-call has a probability of $1/1000$ or 0.1%, etc.). For verification, each base is sequenced on average four times for 'draft' quality and about six to eight times for 'finished' quality sequence. By November 2000, about 30% of the human genome sequence was of finished quality and 70% of unfinished quality. Sequences of 'finished' quality have furthermore been subjected to visual inspection using powerful sequence editors such as GAP (Bonfield *et al*., 1995) and CONSED (Gordon *et al*., 1998) to identify and eliminate ambiguities and gaps. Any annotation associated with finished sequence submissions will also be quality checked by the corresponding EMBL/GENBANK/DDBJ databases (e.g. see NCBI News (Web Links)). Preliminary calculations of the first assembly of the human genome suggest an average base-call accuracy of greater than 99.5% for draft sequence and well above 99.99% for finished sequence. These calculations indicate that the achieved quality of the human genome sequence is well above the expected accuracy. In order to verify these predictions, random spot-checking was and still is regularly carried out involving resequencing where necessary (Felsenfeld *et al*., 1999; see also National Human Genome Research Institute (NHGRI) Standard for Quality of Human Genomic Sequence in Web Links). In addition, there are efforts in progress to make all individual human reads including their original trace and quality data available via a central trace archive (see the Ensembl Trace Server in Web Links). (*See* Sequence Finishing.)

Contiguity

Sequence contiguity is achieved by assembly of overlapping shotgun and finishing reads to form a single contig sequence. For the human genome, the bulk of the sequence was assembled with a program called PHRAP (see Web Links). Among many other features, PHRAP utilizes entire sequence reads (not just the trimmed high-quality parts) and a combination of user-supplied and internally computed information on data quality (including the above PHRED Q values) to improve the accuracy of the assembly, particularly in the presence of repeats. It constructs a contig sequence as a mosaic of the highest quality parts of reads (rather than a consensus) and provides extensive information about the resulting assembly including PHRAP quality scores for such contig sequences. These are the actual quality scores that are submitted to the EMBL/GENBANK/DDBJ databases for all unfinished sequences. In brief, PHRAP uses the highest PHRED quality value at a given position and adjusts it by base with confirmation from another read (either from the opposite-strand or by different sequencing chemistry). Therefore, PHRAP quality scores are the products of error probabilities for two independent reads. The example in **Figure 1** shows the PHRAP quality scores of a typical human genome 'draft' sequence as available from the EMBL database. Each number represents a base and the string of 100 zeros (0) represents a gap in the sequence. The average quality scores are lowest at contig ends, where usually there is low read coverage, but quickly rise to the minimum 'finished' quality (greater than or

```
HTG_QSCORE:AL449425.1
>AL449425.1 Phrap Quality (Length:70088, Min: 1, Max: 99)
   ... 99 99 99 99 99 99 99 99 99 99 99 99 99 99 99 99 99 99 99 99 99 99 99
99 99 99 99 99 99 99 99 99 99 99 96 95 90 90 87 90 87 90 83 79 68 69 78 90 90
85 88 70 70 70 60 60 56 56 56 60 60 65 60 60 49 49 56 56 56 56 60 56 49 44
44 49 55 47 52 44 42 42 46 53 55 63 63 63 63 63 63 47 47 44 44 44 47 63 63
63 63 63 63 63 63 63 63 63 63 63 53 53 49 49 49 52 53 58 58 58 53 63 63 63
63 63 63 58 58 58 58 58 58 63 63 63 63 49 53 55 58 53 44 53 53 44 44 44 47
52 52 52 58 63 63 63 63 53 53 49 49 49 49 49 49 49 49 63 63 63 63 63 63 47
40 53 39 39 39 38 39 31 18 26 36 44 47 47 47 47 47 47 40 41 41 46 46 47 47
47 47 47 47 47 47 47 41 20 36 32 39 39 39 43 43 53 51 39 39 18 34 28 19 17
24 20 20 12  0  0  0  0  0  0  0  0  0  0  0  0  0  0  0  0  0  0  0  0  0  0
 0  0  0  0  0  0  0  0  0  0  0  0  0  0  0  0  0  0  0  0  0  0  0  0  0  0
 0  0  0  0  0  0  0  0  0  0  0  0  0  0  0  0  0  0  0  0  0  0  0  0  0  0
 0  0  0  0  0  0  0  0  0  0  0  0  0  0  0  0  0  0  0  0  0  0  0  0  0  0
 0  0  0  0  4 15 16 19 25 13  8  8  4  6  8 18 24 27 27 22 22 22 22 22 22
26 40 37 35 22 22 22 22 23 22 32 22 22 22 22 20 22 22 24 25 25 29 29 35 31
31 35 35 35 39 37 46 34 34 34 34 34 39 39 35 35 34 34 34 34 34 34 35 35 35
35 35 35 31 35 37 46 46 42 40 35 35 35 35 35 35 35 35 35 35 35 35 35 35 28
28 29 35 33 45 45 45 45 40 40 40 35 28 34 34 34 34 35 42 35 35 35 28 28 29
31 31 31 31 37 45 35 35 35 41 32 35 39 48 48 47 49 43 49 40 41 46 52 52 58
69 73 73 77 75 80 79 72 72 83 83 81 68 73 70 74 76 77 73 74 76 75 62 54 58
60 57 57 71 82 84 77 85 81 81 81 78 73 78 69 75 71 75 80 94 90 83 79 85
89 99 99 99 94 74 71 74 81 74 74 73 77 84 88 88 78 73 73 73 85 75 82 79 71
71 75 93 90 99 99 94 89 80 75 59 69 69 73 78 78 86 86 92 84 79 72 80 89 80
90 96 91 81 93 84 99 99 99 99 99 94 99 99 99 91 91 91 91 99 95 88 88 88 88
96 94 99 84 79 84 90 99 99 99 99 91 91 80 87 83 87 94 99 99 98 98 98 99 88
94 99 99 96 92 99 99 99 99 99 99 99 99 99 99 99 93 94 96 99 99 99 99 99 ...
```

Figure 1 The PHRAP quality scores of a typical human genome 'draft' sequence as available from the EMBL database.

equal to $Q30$) and above (see Genome Sequencing Center in Web Links).

For the HGP, two levels of sequence contiguity were attempted as illustrated in **Figure 2**. The first level was to achieve sequence contiguity of the individual clones in the mapped tile path for each chromosome. For unfinished clones, this involved ordering and orienting sequence subcontigs based on computational localization of cloning vector end and read-pair sequences. Further contig ordering and orientation, and to some degree verification, was achieved by alignment with matching complementary DNA (cDNA) and expressed-sequence tag (EST) sequences. Gaps interrupting the sequence contiguity of some unfinished clones were sized by polymerase chain reaction (PCR) or restriction digest and found to be on average 0.5–1 kb in size. For finished clones, the sequence contiguity was experimentally verified by comparison of *in vitro* and *in silico* restriction digests of the sequence assembly, for example, using the CONFIRM program (see The Wellcome Trust Sanger Institute: software in Web Links).

The second level was to achieve sequence contiguity along each chromosomal tile path. On the clone level (for finished sequences only), this was accomplished by determining the exact overlap position with its left and right neighbors and by annotating this information in the corresponding EMBL/GENBANK/DDBJ database entry as shown in the example in **Figure 3**.

In the rare cases where no sequence overlap could be confirmed, gaps were sized by DNA fiber fluorescent *in situ* hybridization (FISH). At the time of publication only 10 and 11 such gaps were present in chromosomes 21 and 22 respectively. With only 10–150 kb per gap, these two chromosomes can be considered as 'essentially' finished meaning that more than 95% of the euchromatic sequence is present and only less than 5% of currently unclonable sequence is missing. At this stage, the HGP did not aim to sequence heterochromatic regions (e.g. centromeres, telomeres and tandem repeats on short arms of acrocentric chromosomes). A first approximation to achieve chromosomal sequence contiguity including finished and unfinished sequences was generated using a BLAST-based sequence ordering and orientation algorithm (GigAssembler). The resulting human genome 'working draft' sequence (see UCSC Human Genome Project Working Draft in Web Links) provided the basis for the very first completely automated and nonredundant annotation (see Project Ensembl Ensembl Genome Browser in Web Links) of the human genome. However, until all sequences have been finished and verified as described above, the 'working draft' will contain gaps, *in silico* translocations and over-collapsed regions owing to repeats or chromosomal duplications.

Fidelity

In order to guarantee the highest possible sequence fidelity, the HGP implemented several safeguards. The DNA for the construction of clone libraries was from anonymous male or female individuals and was

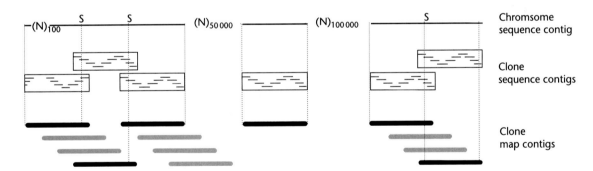

Figure 2 Levels of sequence contiguity. $(N)_{100}$ indicates sequence gap in the clone assembly, $(N)_{50\,000}$ indicates a bridged sequence gap in the chromosome assembly and $(N)_{100\,000}$ indicates an unbridged sequence gap in the chromosome assembly. S indicates switch points between clone sequences in the chromosome assembly. Switch points are chosen arbitrarily within the middle sections of overlapping clone sequences.

```
AC    AL121938
CC    The true right end of clone RP1-16705 is at 156054 in this sequence.
CC    The true left end of clone RP11-735G4 is at 36645 in this sequence.
CC    The true right end of clone RP1-84N20 is at 100 in this sequence.
```

Figure 3 EMBL/GENBANK/DDBJ database entry for the draft sequence shown in **Figure 1**.

isolated from fresh blood to avoid rearrangements often observed in established cell lines. Only the most stable cloning systems such as P1-derived artificial chromosomes (PACs) and bacterial artificial chromosomes (BACs) were used for the bulk of the libraries, and the DNA was digested with different restriction enzymes for different libraries to minimize local over/underrepresentation. Redundant restriction finger-printing and FISH analysis further ensured maximum fidelity of any clone selected for the sequencing tile path. At this point it should be noted that humans are heterozygous and therefore most sequence differences seen within as well as between individuals originate from natural polymorphism. In fact, the identification of one type of natural polymorphism, single nucleotide polymorphism (SNP), has already been greatly helped by the mandatory inclusion of quality scores in the database submissions of all 'unfinished' sequences.

Outlook

So, how much confidence can we have in the accuracy of DNA sequences deposited in public databases and in the results of their subsequent analyses? Based on the safeguards and quality measures described here, we can be very confident about finished sequences and fairly confident about unfinished sequences which will, of course, all be finished in time. As mentioned above, the quality of finished sequences will continue to be rigorously monitored, and plans are under discussion to set up a central DNA sequence quality control center (Felsenfeld et al., 1999; see also National Human Genome Research Institute (NHGRI) Standard for Quality of Human Genomic Sequence in Web Links). In addition, a Committee on Data for Science and Technology of the International Council of Scientific Unions (CODATA) has been established to act as 'ombudsman' for users of public data including DNA sequences. CODATA's role is to advise, to be consulted and to warn (Abola et al., 2000).

What about the analysis involving unfinished sequences – how do software tools cope with these imperfect data? The most frequently used analysis tools for DNA sequences are similarity search programs such as BLAST. Most programs (including BLAST) rely on dynamic matching algorithms and cope well with substitutions, insertions and deletions. Some BLAST servers such as the Wellcome Trust Sanger Institute Human Blast Server (see Web Links) even color-code matches to unfinished sequences according to their PHRED/PHRAP quality scores. Another frequently carried out analysis is ab initio gene prediction using programs such as GENSCAN (Burge and Karlin, 1997). For unfinished sequences, the main problem seems to be fragmentation of genes

into different contigs. Current estimates are that on average each human gene sequenced in 'draft' form is present in 1.8 different 'draft' contigs. This means that genes predicted from 'draft' sequence will often represent incomplete fragments of genes. A secondary effect is the lower sequence quality of the 'draft' which results in a higher frequency of frameshifts, causing a few more exons to be missed or mispredicted. In short, sensitivity is more affected than specificity, but new programs such as GENOMESCAN look promising to address some of these issues (Yeh et al., 2001).

Cautionary Note

The criteria discussed above are valid only for sequences generated by the International HGP Consortium. As of November 2000, no private-domain human genome sequence data or quality criteria were available for review.

See also
Sequence Assembly
Sequence Finishing

References

Abola EE, Bairoch A, Barker WC, et al. (2000) Quality control in databanks for molecular biology. *BioEssays* **22**: 1024–1034.

Beck S (1993) Accuracy of DNA sequencing: should the sequence quality be monitored? *DNA Sequence* **4**: 215–217.

Bonfield JK, Smith KF and Staden R (1995) A new DNA sequence assembly program. *Nucleic Acids Research* **23**: 4992–4999.

Burge C and Karlin S (1997) Prediction of complete gene structures in human genomic DNA. *Journal of Molecular Biology* **268**: 78–94.

Ewing B and Green P (1998) Base-calling of automated sequencer traces using phred II. Error probabilities. *Genome Research* **8**: 186–194.

Ewing B, Hillier L, Wendl MC and Green P (1998) Base-calling of automated sequencer traces using phred I. *Genome Research* **8**: 175–185.

Felsenfeld A, Peterson J, Schloss J and Guyer M (1999) Assessing the quality of the DNA sequence from the Human Genome Project. *Genome Research* **9**: 1–4.

Gordon D, Abajian C and Green P (1998) Consed: a graphical tool for sequence finishing. *Genome Research* **8**: 195–202.

Yeh RF, Lim LP and Burge CB (2001) Computational inference of homologous gene structures in the human genome. *Genome Research* **11**: 803–816.

Further Reading

Altschul SF, Gish W, Miller W, Myers EW and Lipman DJ (1990) Basic local alignment search tool. *Journal of Molecular Biology* **215**: 403–410.

Dunham I, Shimizu N, Roe BA, et al. (1999) The DNA sequence of human chromosome 22. *Nature* **402**: 489–495.

Gregory SG, Howell GR and Bentley DR (1997) Genome mapping by fluorescent fingerprinting. *Genome Research* **7**: 1162–1168.

Hattori M, Fujiyama A, Taylor TD, *et al.* (2000) The DNA sequence of human chromosome 21. *Nature* **405**: 311–319.

Marra MA, Kucaba TA, Dietrich NL, *et al.* (1997) High throughput fingerprint analysis of large-insert clones. *Genome Research* **7**: 1072–1084.

Mullikin JC, Hunt SE, Cole CG, *et al.* (2000) An SNP map of human chromosome 22. *Nature* **407**: 516–520.

Osoegawa K, Mammoser AG, Wu C, *et al.* (2001) A bacterial artificial chromosome library for sequencing the complete human genome. *Genome Research* **11**: 483–496.

International Human Genome Sequencing Consortium (2001) Initial sequencing and analysis of the human genome. *Nature* **409**: 860–921.

Web Links

Ensembl Trace Server
http://trace.ensembl.org/

Genome Sequencing Center. International Finishing Standards for the Human Genome Project (Version September 7, 2001)
http://genome.wustl.edu/gsc/Overview/finrules/hgfinrules.html

National Human Genome Research Institute (NHGRI). NHGRI Standard for Quality of Human Genomic Sequence
http://www.nhgri.nih.gov:80/Grant_info/Funding/Statements/RFA/quality_standard.html

National Center for Biotechnology Information: NCBI News
http://www.ncbi.nlm.nih.gov/Web/Newsltr/feb98.html#GenBank

Project Ensembl. Ensembl Genome Browser
http://www.ensembl.org/

Summary of the Report of the Second International Strategy Meeting on Human Genome Sequencing Bermuda, 27th February–2nd March 1997 sponsored by the Wellcome Trust
http://www.gene.ucl.ac.uk/hugo/bermuda2.htm

The Phred/Phrap/Consed System home page
http://www.phrap.org

The Wellcome Trust Sanger Institute Human Blast Server
http://www.sanger.ac.uk/HGP/blast_server.shtml

The Wellcome Trust Sanger Institute: software
http://www.sanger.ac.uk/Software/

UCSC Human Genome Project Working Draft
http://genome.ucsc.edu/

Sequence Alignment

Intermediate article

Article contents

- Introduction
- What is an Alignment?
- What is the Best Alignment?
- Alignment Algorithms

Julie D Thompson, *Institut de Génétique et de Biologie Moléculaire et Cellulaire, Strasbourg, France*

Olivier Poch, *Institut de Génétique et de Biologie Moléculaire et Cellulaire, Strasbourg, France*

Definition sequence comparison or alignment is the cornerstone of bioinformatics, providing the basis for sequence database searching, three-dimensional structure modeling and evolutionary studies. A sequence alignment shows how a set of sequences may be related by identifying and arranging in columns the structurally and functionally equivalent residues common to all the sequences.

Introduction

Sequence comparison or alignment plays a fundamental role in most areas of modern molecular biology, from shaping our basic conceptions of life and its evolutionary processes, to providing the foundation for the new biotechnology industry. The currently accepted universal tree of life, in which the living world is divided into three domains (Bacteria, Archaea and Eukarya), was constructed from comparative analyses of ribosomal RNA sequences. However, this view, in which life evolved from simple prokaryotes to eukaryotes, has been somewhat shaken by studies of the complete sequences of several microbial genomes, which discovered widespread evidence of significant

horizontal gene transfers between diverse organisms. Comparisons of the complete sequences of the more closely related genomes have revealed that far from being static, genomic DNA is actually highly plastic, being subject to recombinations, rearrangements, insertions and deletions. (**See** Comparative Genomics; DNA Recombination; Ribosomes and Ribosomal Proteins.)

In addition to comparing whole-genome sequences, the genome sequencing projects are also attempting to determine and identify the individual genes encoded by the genome. Sequence alignments provide a powerful way to compare novel sequences with previously characterized genes available in the sequence databases, in order to identify potential homologs, that is, sequences that have evolved from a common ancestor. Generally, homologous proteins share the same three-dimensional (3D) structure and have similar functions, active sites or binding domains. In most genome annotation projects, the standard strategy to determine the function of a novel protein is, therefore, to search the sequence databases for homologs and to transfer the structural/functional annotation from the known to the unknown protein. (*See* Gene Structure and Organization; Homologous, Orthologous and Paralogous Genes; Protein Characterization in Proteomics.)

A combination of sequential and structural information has been shown to increase the accuracy of structure prediction, both 2D and 3D, relative to the exclusive use of sequence or structure. The comparison of homologous sequences provides one of the most efficient methods of modeling the 3D structure of a protein. Protein threading techniques, that rely on matching 3D information predicted from the query sequence with corresponding features of a known structure, can also be improved by incorporating sequence information. (*See* Protein Homology Modeling; Protein Structure Prediction and Databases.)

Sequence alignments on the genomic scale are also having a widespread effect in the pharmaceutical industry, providing an opportunity to identify the proteins associated with a particular disease, which are therefore potential drug targets. Recent advances in the computational analyses of enzyme structures and functions have improved the strategies used to modify enzyme specificities and mechanisms by site-directed mutagenesis, and to engineer biocatalysts through molecular reassembly. The analysis of genomes of extremophile microorganisms has led to the identification of many enzymes showing activity and stability at extremes of temperature, pH, pressure and salinity, many of which have potential for industrial and biotechnological applications. The potential impact on protein structure prediction, biology, protein engineering and medicine is enormous. (*See* Drug Metabolic Enzymes: Genetic Polymorphisms; Gene Targeting by Homologous Recombination; Protein Targeting.)

What is an Alignment?

There exist two main categories of sequence alignment: pairwise alignment (or the alignment of two sequences) and multiple alignment. Pairwise alignments are most commonly used in database search programs such as BLAST and FASTA in order to detect homologs of a novel sequence. Multiple alignments, containing from three to several hundred sequences, are more computationally complex than pairwise alignments and, in general, simultaneous alignment of more than a few sequences is rarely attempted. Instead a series of pairwise alignments are performed and amalgamated into a multiple alignment. Nevertheless, multiple alignments have the advantage of providing an overall view of the family, thus helping to decipher the evolutionary history of the protein family. Multiple sequence alignments are useful in identifying conserved patterns in protein families, which may not be evident from pairwise alignments. They are also used in the determination of domain organization, to help predict protein secondary/tertiary structure and in phylogenetic studies. (*See* Multiple Alignment; Similarity Search.)

The purpose of any sequence alignment, whether pairwise or multiple, is to show how a set of sequences may be related, in terms of conserved residues, substitutions, insertion–deletion events ('indels'). In the most general terms, an alignment represents a set of sequences using a single-letter code for each amino acid (for protein sequences) or nucleotide (for DNA/RNA sequences). Structurally and functionally equivalent residues are aligned either in rows or, more usually in columns (**Figure 1**). When the sequences are of different lengths, insertion–deletion events are postulated to explain the variation, and gap characters are introduced into the alignment. Sequence alignments can be further divided into global alignments that align the

```
            :.*     .  .*     *           *:  *      .. * *       *:* *.:          .  *
SCX6_TITSE  -regypadskgckitcflta-agycntectlk--kgssgycawp-----acycyglpesvkiwtsetnk-c
SCX1_CENNO  -kdgylvdakgckkncyklgkndycnrecrmkhrggsygycygf-----gcyceglsdstptwplp-nktcsgk
SIX2_LEIQU  --dgyirkrdgcklsclfg--negcnkecksy--ggsygycwtw---glacwceglpd-ektwksetn-tcg
SCX1_TITSE  -kdgypveydncayicwnyd-naycdklckdk--kadsgycywv---hilcycyglpdseptktn--gk-cksgkk
SCXA_BUTEU  vrdgyiaddkdcayfcgr---naycdeeckk---gaesgkcwyagqygnacwcyklpdwvpikqkvsgk-cn
SCXA_LEIQH  vrdayiaknyncvyecfr---daycnelctkn--gassgycqwagkygnacwcyalpdnvpirvp--gk-chrk
```

Figure 1 Alignment of six scorpion toxin proteins. Conserved positions are shown in bold. Gaps are represented by dashes between the letter strings. The secondary structure elements of the scorpion *Leiurus quinquestriatus hebraeus* protein (SCXA_LEIQH) are shown below the alignment. Right arrow: β-sheet; coil: α-helix.

complete sequences and local alignments that identify only the most similar segments or sequence patterns (motifs). While global alignment algorithms produce more accurate alignments for proteins of similar length, local alignment algorithms are better at identifying similar regions within sequences when the sequences are not related over their entire length. (*See* Global Alignment; Sequence Similarity.)

What is the Best Alignment?

For any two sequences, there are an exponential number of potential alignments with gaps. Therefore, it is critical to be able to distinguish 'good' alignments from 'bad' ones. A good alignment is one that corresponds to the biologically correct alignment, accurately reflecting the evolutionary, structural and functional relationships between the sequences. Sequence alignment programs have, until recently, used only the primary sequence information to reconstruct these complex relationships. In order to find the best alignment, most alignment programs assign a similarity score to all possible alignments and try to maximize this score. These alignment scores, also known as objective functions, are generally based on scores for aligning single residues with penalties for introducing gaps into the sequences. While these scores are generally adequate for the alignment of relatively well-conserved sequences, it is clear that more elaborate scoring schemes will be required for the highly complex proteins detected by today's advanced database searching methods.

Scoring matrices

Most alignment programs make comparisons between pairs of bases or amino acids by looking up a value in a scoring matrix. The matrix contains a score for the match quality of every possible pair of residues (**Figure 2**). The simplest way to score an alignment is to count the number of identical residues that are aligned. When the sequences to be aligned are closely related, this will usually find approximately the correct solution. For more divergent sequences sharing less than 25–30% identity, however, the scores given to nonidentical residues become critically important.

More sophisticated scoring schemes exist for both DNA and protein sequences and generally take the form of a matrix defining the score for aligning each pair of residues. For alignments of nucleotide sequences, the simplest scoring matrix would assign the same score to a match of the four classes of bases, ACGT, and 0 for any mismatch. However, transitions (substitution of A–G or C–T) happen much more frequently than transversions (substitution of A–T or G–C), and it is often desirable to score these

	A	R	N	D	C	Q	E	G	H	I	L	K	M	F	P	S	T	W	Y	V	B	Z	X
A	2	-2	0	0	-2	0	0	1	-1	-1	-2	-1	-1	-3	1	1	1	-6	-3	0	0	0	0
R	-2	6	0	-1	-4	1	-1	-3	2	-2	-3	3	0	-4	0	0	-1	2	-4	-2	-1	0	-1
N	0	0	2	2	-4	1	1	0	2	-2	-3	1	-2	-3	0	1	0	-4	-2	-2	2	1	0
D	0	-1	2	4	-5	2	3	1	1	-2	-4	0	-3	-6	-1	0	0	-7	-4	-2	3	3	-1
C	-2	-4	-4	-5	12	-5	-5	-3	-3	-2	-6	-5	-5	-4	-3	0	-2	-8	0	-2	-4	-5	-3
Q	0	1	1	2	-5	4	2	-1	3	-2	-2	1	-1	-5	0	-1	-1	-5	-4	-2	1	3	-1
E	0	-1	1	3	-5	2	4	0	1	-2	-3	0	-2	-5	-1	0	0	-7	-4	-2	3	3	-1
G	1	-3	0	1	-3	-1	0	5	-2	-3	-4	-2	-3	-5	0	1	0	-7	-5	-1	0	0	-1
H	-1	2	2	1	-3	3	1	-2	6	-2	-2	0	-2	-2	0	-1	-1	-3	0	-2	1	2	-1
I	-1	-2	-2	-2	-2	-2	-2	-3	-2	5	2	-2	2	1	-2	-1	0	-5	-1	4	-2	-2	-1
L	-2	-3	-3	-4	-6	-2	-3	-4	-2	2	6	-3	4	2	-3	-3	-2	-2	-1	2	-3	-3	-1
K	-1	3	1	0	-5	1	0	-2	0	-2	-3	5	0	-5	-1	0	0	-3	-4	-2	1	0	-1
M	-1	0	-2	-3	-5	-1	-2	-3	-2	2	4	0	6	0	-2	-2	-1	-4	-2	2	-2	-2	-1
F	-3	-4	-3	-6	-4	-5	-5	-5	-2	1	2	-5	0	9	-5	-3	-3	0	7	-1	-4	-5	-2
P	1	0	0	-1	-3	0	-1	0	0	-2	-3	-1	-2	-5	6	1	0	-6	-5	-1	-1	0	-1
S	1	0	1	0	0	-1	0	1	-1	-1	-3	0	-2	-3	1	2	1	-2	-3	-1	0	0	0
T	1	-1	0	0	-2	-1	0	0	-1	0	-2	0	-1	-3	0	1	3	-5	-3	0	0	-1	0
W	-6	2	-4	-7	-8	-5	-7	-7	-3	-5	-2	-3	-4	0	-6	-2	-5	17	0	-6	-5	-6	-4
Y	-3	-4	-2	-4	0	-4	-4	-5	0	-1	-1	-4	-2	7	-5	-3	-3	0	10	-2	-3	-4	-2
V	0	-2	-2	-2	-2	-2	-2	-1	-2	4	2	-2	2	-1	-1	-1	0	-6	-2	4	-2	-2	-1
B	0	-1	2	3	-4	1	3	0	1	-2	-3	1	-2	-4	-1	0	0	-5	-3	-2	2	2	-1
Z	0	0	1	3	-5	3	3	0	2	-2	-3	0	-2	-5	0	0	-1	-6	-4	-2	2	3	-1
X	0	-1	0	-1	-3	-1	-1	-1	-1	-1	-1	-1	-1	-2	-1	0	0	-4	-2	-1	-1	-1	-1

Figure 2 PAM250 matrix. Substitution scores for amino acids.

substitutions differently. More complex matrices also exist in which matches between ambiguous nucleotides are given values whenever there is any overlap in the sets of nucleotides represented by the two symbols being compared. For protein sequence comparisons, scoring matrices generally take into account the biochemical similarities between residues and/or the relative frequencies with which each amino acid is substituted by another.

The most widely used scoring matrices are known as the PAM (point accepted mutation) matrices (Dayhoff *et al.*, 1978). The original PAM1 matrix construct was based on the mutations observed in a large number of alignments of closely related sequences. A series of matrices was then extrapolated from the PAM1. The matrices range from strict ones, useful for comparing very closely related sequences, to very 'soft' ones that are used to compare very divergent sequences. For example, the PAM250 matrix corresponds to an evolutionary distance of 250%, or approximately 80% residue divergence. Other matrices have been derived directly from either sequence-based or structure-based alignments, such as the Blosum matrices, which are based on the observed residue substitutions in aligned sequence segments from the Blocks database. The proteins in the database are clustered at different percentage identities to produce a series of matrices. For example, the Blosum-62 matrix is based on alignment blocks in which all the sequences share at least 62% residue identity. Other more specialized matrices have been developed, for example, for specific secondary structure elements (helices, β-sheets), or for the comparison of particular types of proteins such as transmembrane proteins. (*See* Substitution Matrices.)

Gap schemes

In addition to assigning scores for residue matches and mismatches, most alignment scoring schemes in use today calculate a cost for the insertion of gaps in the sequences. One of the first gap scoring schemes for the alignment of two sequences charged a fixed penalty for each residue in either sequence aligned with a gap in the other. Under this system, the cost of a gap is proportional to its length. Alignment algorithms implementing such length-proportional-gap penalties are efficient; however, the resulting alignments often contain a large number of short indels that are not biologically meaningful. To address this problem, linear or 'affine' gap costs are used that define a gap insertion or 'gap opening' penalty in addition to the length-dependent or 'gap extension' penalty. Thus, a smaller number of long gaps is favored over many short ones. Fortunately, algorithms using affine gap costs are only slightly more complex than those using length-proportional gap penalties, requiring only a constant factor more space and time.

Again, more complex schemes have been developed, such as 'concave' gap costs or position-specific gap penalties. Most of these are attempts to mimic the biological processes or constraints that are thought to regulate the evolution of DNA or protein sequences. (*See* Exons: Insertion and Deletion during Evolution; Gross Insertions and Microinsertions in Evolution.)

Alignment statistics

An important aspect of sequence alignment is to establish how meaningful a given alignment is. It is always possible to construct an alignment between a set of sequences, even if they are unrelated. The problem is to determine the level of similarity required to infer that the sequences are homologous. A simple rule-of-thumb for protein sequences states that if two sequences share more than 25% identity over more than 100 residues, then the two sequences can be assumed to be homologous. However, many proteins sharing less than 25% residue identity, said to be in the 'twilight zone' (Doolittle, 1986), do still have very similar structures. The measure of the percentage identity or similarity of the sequences is generally not sensitive enough to distinguish between alignments of related and unrelated sequences. (*See* Evolutionary Distance; Gene Families: Formation and Evolution; Gene Structure: Evolution; Protein Structure.)

Much work has been done on the significance of both ungapped and gapped pairwise local alignments (Altschul and Gish, 1996; Pearson, 1998), although the statistics of global alignments or alignments of more than two sequences are far less well understood. The aim of the statistical analysis is to estimate the probability of finding by 'chance' at least one alignment that scores as high as or greater than the given alignment. For ungapped local alignments, these probabilities or *P*-values may be derived analytically.

For alignments with gaps, empirical estimates are used, based on the scores obtained during a database search, or from randomly generated sequences.

For database search programs, the significance of an alignment between the query sequence and a database sequence is often expressed in terms of expect- or *E*-values, which specifies the number of matches with a given score that are expected to occur by chance in a search of a database. An *E*-value of zero, with a given score, would indicate that no matches with this score are expected purely by chance. (*See* Alignment: Statistical Significance.)

Alignment Algorithms

Pairwise alignments

The comparison or alignment of biological sequences began in the early 1970s, with the first dynamic programming algorithm for the global (or full-length) alignment of two sequences (Needleman and Wunsch, 1970). The optimal local alignment between a pair of sequences, in which only the highest scoring subsegments of the two sequences are aligned, involves a simple modification to the Needleman–Wunsch method (Smith and Waterman, 1981). (*See* Smith–Waterman Algorithm.)

Dynamic programming is a rigorous mathematical technique that is guaranteed to find the maximal scoring alignment for any two sequences. It does this by constructing a 2D alignment matrix or path graph of partial alignment scores (**Figure 3**). Each position in

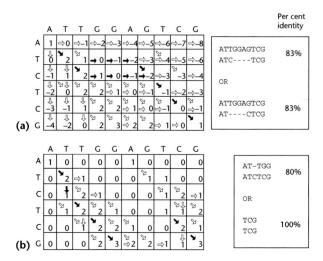

Figure 3 Dynamic programming alignment matrices for global (a) and local (b) alignments of two DNA sequences. Per cent identity scores for each alignment are calculated by dividing the number of identical residues aligned by the total number of residues aligned.

the matrix contains the score for the best partial alignment that ends at that position. The best scoring partial alignment will be extended to subsequent positions in the matrix, either by aligning one residue from each sequence, or by inserting a gap into one or other of the sequences. In this way, all possible alignments are considered and the final alignment is thus the best-scoring alignment possible. The optimal global alignment score is given in the bottom, right-hand corner of the alignment matrix, while the optimal local alignment score is defined as the highest-scoring position anywhere in the alignment matrix. (*See* Dynamic Programming.)

Heuristic methods

A different approach to the local alignment problem involves the use of heuristics or 'approximate' methods, which do not guarantee an optimal alignment solution but are less time-consuming than the rigorous dynamic programming techniques. These approximate alignment algorithms are used in programs such as FASTA (Pearson and Lipman, 1988) and BLAST (Altschul *et al.*, 1990) to search the protein and DNA sequence databases for homologs of a target sequence. The general approach involves comparing the target or 'query' sequence to all the sequences in a specified database in a pairwise fashion. Each comparison is given a score reflecting the degree of similarity between the query and the sequence being compared. The higher the score, the greater the degree of similarity. The similarity is measured and shown by aligning the two sequences. (*See* BLAST Algorithm; FASTA Algorithm.)

The heuristics used involve finding patches of regional similarity, rather than trying to find the best alignment between the entire query and an entire database sequence. FASTA uses a two-step pairwise alignment algorithm. The first step consists of a search for exactly matching strings or 'words' that are common to both sequences. This is done in order to identify regions in a 2D table similar to that shown for the dynamic programming algorithm above, and are likely to correspond to highly similar segments shared by the two sequences. These regions will consist of a diagonal or a few closely spaced diagonals in the table, which have a high number of word matches between the sequences. The second step involves a Smith–Waterman local alignment centered on these regions. The speed-up achieved by a FASTA alignment relative to a full Smith–Waterman alignment is due to the restriction of the dynamic programming algorithm to only the high-scoring regions. In the BLAST program, the first step also involves a word-based heuristic, similar to that of FASTA. However, the high-scoring segments found are then extended in both forward and backward directions to generate an alignment that

continues until the sequence ends, or the alignment becomes nonsignificant. In both FASTA and BLAST, in addition to the alignment scores, the significance of each alignment is computed as a *P*-value or an *E*-value (see the section Alignment statistics).

Dot plots

A dot plot is a powerful, visual method for comparing two complete sequences that provides a global view of all possible regions of similarity between the two sequences. Dot plot programs often provide an inter-active environment in which the user can select signi-ficant sequence segments in order to guide the final alignment.

In the dot plot in **Figure 4**, the *X* and *Y* axes of the plot correspond to the two sequences to be compared. The dots represent all the possible matches of identical residues in the two sequences. Any region of similar sequence appears as a diagonal row of dots. Isolated dots not on the diagonal represent random matches, which are probably not related to any significant alignment.

Visualization of matching regions may be improved by filtering out these random matches using a sliding window calculation. Instead of comparing single sequence positions in the two sequences, the average score in a window of adjacent positions is calculated, and a dot is printed only if the score for the window is above a certain average score. Scoring matrices such as the PAM or Blosum matrices may be used instead of residue identities.

Dot plots are particularly valuable for finding repeats or inversions in protein and DNA sequences, and for predicting regions in RNA that are self-complementary and that, therefore, might form a double-stranded region or secondary structure. For an excellent description of the dot plot method, see States and Boguski (1991).

Multiple sequence alignment

The first formal algorithm for multiple sequence alignment (Sankoff, 1975) was developed as a direct extension of the pairwise dynamic programming algo-rithm. However, the optimal multiple alignment of more than a few sequences (more than 10) remains impractical due to the intensive computer resources required, despite some recent space and time improve-ments. Therefore, in order to multiply align larger sets of sequences, most programs in use today employ some kind of heuristic approach to reduce the problem to a reasonable size.

Traditionally, the most popular method has been the progressive alignment procedure (Feng and Doolittle, 1987). A multiple sequence alignment is

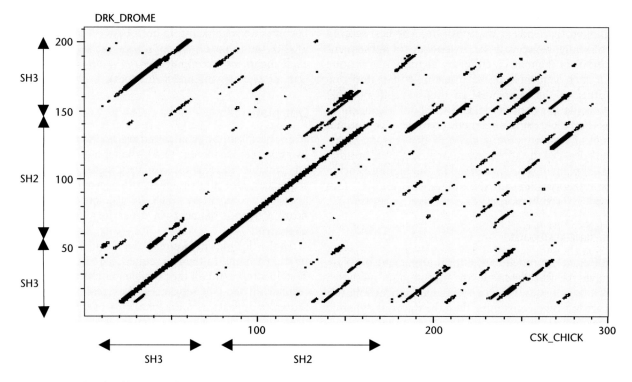

Figure 4 Dot plot of a chicken tyrosine-protein kinase protein (CSK_CHICK) compared to a *Drosophila* SH2–SH3 adaptor protein (DRK_DROME).

built up gradually using a series of pairwise alignments. The two closest sequences are aligned first and then larger and larger sets of sequences are merged, until all the sequences are included in the multiple alignment. Programs implementing the progressive multiple alignment method may use either a local or a global alignment algorithm.

One of the most widely used progressive multiple alignment programs is CLUSTAL W (Thompson *et al.*, 1994). More recently, algorithms other than dynamic programming have been exploited in the search for more accurate multiple alignments in a wider variety of situations. New developments include the use of hidden markov models (HMMs), genetic algorithms, segment-to-segment alignments, Gibbs sampling or iterative refinement techniques. (*See* Hidden Markov Models.)

See also
Alignment: Statistical Significance
BLAST Algorithm
FASTA Algorithm
Global Alignment
Multiple Alignment

References

Altschul SF and Gish W (1996) Local alignment statistics. *Methods in Enzymology* **266**: 460–480.

Altschul SF, Gish W, Miller W, Myers EW and Lipman DJ (1990) Basic local alignment search tool. *Journal of Molecular Biology* **215**: 403–410.

Dayhoff M, Schwartz RM and Orcutt BC (1978) A model of evolutionary change in proteins. *Atlas of Protein Sequence and Structure*, vol. 5(supplement 3), pp. 345–358. Silver Springs, MD: National Biomedical Research Foundation.

Doolittle RF (1986) *Of Urfs and Orfs: Primer on How to Analyze Derived Amino Acid Sequences*. Mill Valley, CA: University Science Books.

Feng DF and Doolittle RF (1987) Progressive sequence alignment as a prerequisite to correct phylogenetic trees. *Journal of Molecular Evolution* **25**: 351–360.

Needleman SB and Wunsch CD (1970) A general method applicable to the search for similarities in the amino acid sequence of two proteins. *Journal of Molecular Biology* **48**: 443–453.

Pearson WR (1998) Empirical statistical estimates for sequence similarity searches. *Journal of Molecular Biology* **276**: 71–84.

Pearson WR and Lipman DJ (1988) Improved tools for biological sequence comparison. *Proceedings of the National Academy of Sciences of the United States of America* **85**: 2444–2448.

Sankoff D (1975) Minimal mutation trees of sequences. *SIAM Journal of Applied Mathematics* **78**: 35–42.

Smith TF and Waterman MS (1981) Identification of common molecular subsequences. *Journal of Molecular Biology* **215**: 403–410.

States DJ and Boguski MS (1991) Similarity and homology. In: Gribskov M and Devereux J (eds.) *Sequence Analysis Primer*, pp. 92–124. NewYork, NY: Stockton Press.

Thompson JD, Higgins DG and Gibson TJ (1994) CLUSTAL W: improving the sensitivity of progressive multiple sequence alignment through sequence weighting, position-specific gap penalties and matrix choice. *Nucleic Acids Research* **22**: 4673–4680.

Further Reading

Altschul SF (1991) Amino acid substitution matrices from an information theoretic perspective. *Journal of Molecular Biology* **219**: 555–565.

Apostolico A and Giancarlo R (1998) Sequence alignment in molecular biology. *Journal of Computational Biology* **5**: 173–196.

Benner SA, Cohen MA and Gonnet GH (1993) Empirical and structural models for insertions and deletions in the divergent evolution of proteins. *Journal of Molecular Biology* **229**: 1065–1082.

Durbin R, Eddy S, Krogh A and Mitchison G (1999) Pairwise alignment. In: Durbin R (ed.) *Biological Sequence Analysis: Probabilistic Models of Proteins and Nucleic Acids*, pp. 12–45. Cambridge, UK: Cambridge University Press.

Gotoh O (1999) Multiple sequence alignment: algorithms and applications. *Advanced Biophysics* **36**: 159–206.

Henikoff S and Henikoff JG (1993) Performance evaluation of amino acid substitution matrices. *Proteins* **17**: 49–61.

Henikoff S (1994) Comparative sequence analysis: finding genes. In: Smith DW (ed.) *Biocomputing, Informatics and Genome Projects*, pp. 87–117. New York, NY: Academic Press.

Smith TF (1999) The art of matchmaking: sequence alignment methods and their structural implications. *Structure with Folding and Design* **7**: R7–R12.

Vogt G, Etzold T and Argos P (1995) An assessment of amino acid exchange matrices in aligning protein sequences: the twilight zone revisited. *Journal of Molecular Biology* **249**: 816–831.

Waterman MS (1995) Dynamic programming alignment of two sequences. In: Michael SW (ed.) *Introduction to Computational Biology: Maps, Sequences and Genomes*, pp. 183–232. London, UK: Chapman & Hall/CRC Press.

Yona G and Brenner SE (2000) Comparison of protein sequences and practical database searching. In: Higgins DG and Taylor WR (eds.) *Bioinformatics: Sequence, Structure and Databanks. A Practical Approach*, pp. 167–190. Oxford, UK: Oxford University Press.

Web Links

Blocks database
http://www.blocks.fhcrc.org/

Sequence Assembly

Joao Meidanis, *University of Campinas, Campinas, Brazil*

Any shotgun DNA sequencing effort involves a first phase in which short, contiguous samples of a long DNA molecule are generated. These samples must be assembled afterwards to yield the original sequence in a process called sequence assembly.

Intermediate article

Article contents

- Introduction
- Cleaning the Data
- Output of Assembly
- Pure Shotgun
- Shotgun with Mated Pairs
- Software Tools

Introduction

The advent of fast deoxyribonucleic acid (DNA) sequencing technologies (Maxam and Gilbert, 1977; Sanger *et al.*, 1977) was a landmark in genomic science. Despite all improvements since their inception, however, the techniques still have one crucial limitation: only contiguous stretches of at most 1000 nucleotide residues (bases) can be read directly in one sequencing reaction. To determine the sequence of bases of longer molecules one has to clone smaller pieces of the molecule, sequence them and assemble the sequence of the original molecule on the basis of the sequences of the clones.

Chromosome walking or primer walking is one strategy that can be used to uncover longer molecules. Starting at one extremity of the DNA molecule, successive reads (fragments) are sequenced, each one beginning near the end of the previous one. A new primer has to be used for each new fragment. Another strategy, termed shotgun, consists of generating several of randomly positioned clones along the target

DNA molecule, either by random shearing (the preferred method) or by restriction enzymes (inferior because it is harder to achieve the necessary randomness). The actual number of clones generated depends on the desired coverage, which is defined as the total length of all fragments (clones) divided by the length of the target molecule. Typically, at least a coverage of 5–6 is used, but values of 10 or more are not uncommon. (*See* DNA: Mechanical Breakage; Primer Walking; Restriction Enzymes.)

For tasks that require more than a few walking steps, the shotgun strategy is likely to be faster because a universal primer can be used, although it is more demanding in terms of total sequencing effort. In addition, only the shotgun strategy requires sequence assembly, because in the walking strategy it is straightforward to position the next fragment with respect to the portion already sequenced.

Whereas it admittedly does not pay to sequence short molecules by shotgun, there is controversy about whether there is an upper limit on the size of the DNA that can be sequenced by this method. The debate over the public versus private sequencing of the human genome is deeply related to this issue. Some think that very long and complex (in terms of repeats and other structures) DNA molecules cannot be sequenced completely by shotgun, even if mated pairs (see below) are used, and that the process should involve necessarily an intermediate step in which large clones are obtained from the original molecule and positioned within it by physical mapping techniques, before each one is sequenced by shotgun. Others advocate that even large and complex genomes such as the human can be sequenced using the shotgun strategy, as long as reliable paired end information is available. (*See* Genome Mapping.)

We outline here the main issues involved in sequence assembly, both with and without mated pairs, keeping in mind that it always implies the use of the shotgun strategy for generation of the primary (or raw) data, and that in principle it can be applied to a target molecule of arbitrary size.

Cleaning the Data

In the early days of DNA sequencing, human operators had to read the patterns off electrophoresis films and write down the sequence of bases in the computer. Today this is done automatically by sequencing machines that output a file with the sequence. This process, termed base-calling, is subject to error in both cases. For a successful assembly, the frequency of base-calling errors must be kept under control. Base-calling software Phred (Ewing *et al.*, 1998; Ewing and Green, 1998) achieves this control by assigning to each base a 'quality value' that is related to the probability that the base is wrong. With this information, reads can be trimmed and overlaps can be determined more reliably. It is probably not an exaggeration to say that the introduction of quality values revolutionized assembly technology.

Trimming is the process by which low-quality portions at the beginning and end of reads are removed or masked before assembly. It is hard to overstress the importance of good trimming, which may easily make the difference between closing a genome and not closing it.

Another source of error that affects assembly is contamination. In the process of cloning and sequencing the clones, several vectors are used. Each of these vectors is a virus, bacterium or other microorganism whose DNA receives the cloned DNA. As a result, the raw sequence obtained directly from the sequencing

procedure is generally a chimera consisting partly of vector DNA and partly of cloned DNA. It is necessary to remove or mask the vector DNA. This is usually an easy task because the whole vector DNA is known in advance.

Output of Assembly

Ideally the result of sequence assembly is the correct sequence of bases of the target DNA molecule. Although this can be achieved for relatively short targets (e.g. up to a few tens of thousand bases), in general the output of assembly is a collection of one or more contigs, in other words, representations of contiguous regions in the target. If mated pairs are used (see below), then the output is one or more scaffolds, that is, groups of contigs linked by mated pairs.

There are several reasons why assembly does not always return one single sequence, including lack of coverage in certain regions. Some regions in the target DNA may contain genes that are deleterious to the cloning hosts, and therefore no clones are generated that cover these regions. Alternatively, secondary structure arrangements in certain regions may prevent the sequencing reactions from occurring normally in these regions, again yielding no clones there. Repeated DNA stretches in the target, either contiguous (tandem) or separated by other DNA, may also confuse the assembler into generating contigs in more quantity than is necessary.

Pure Shotgun

In this section we assume a shotgun strategy in which no mated pair information is available. The traditional approach for assembly in this case includes identifying overlapping fragments, constructing long chains of overlapping fragments that will give rise to contigs, and assigning a consensus sequence to each contig. A completely different method has also been proposed – the Eulerian method – which bypasses these steps and reduces the problem to the computation of an Euler tour (Pevzner *et al.*, 2001).

Overlaps

To verify whether two fragments overlap or not, a type of comparison that does not charge for gaps in the termini of fragments must be used (**Figure 1**). This kind of alignment has been termed semiglobal (Setubal and Meidanis, 1997), because it is intermediate between global alignment and local alignment. It is global in the sense that the scored region extends both to the left

```
GATAGATGCTCGT-GTCGA---------------
---------CGTAGTCTAGATAGTGCTGATAGAT
```

Figure 1 A semiglobal alignment, which is used to detect overlap between fragments.

and to the right up to the point where at least one of the fragments ends. And it is local in the sense that only part of each sequence is used for scoring.

The classical dynamic programing algorithm for global alignment can be readily adapted for semiglobal alignment. The differences are that the first row and column are initialized with zeros and that the maximum value along the last row and column indicates the best score. (*See* Dynamic Programming.)

Semiglobal alignment can also detect whether a fragment is contained in another fragment. Such contained fragments are removed from the layout construction phase but are added in later, when they can help in consensus building.

Overlap graph

The information about overlapping fragments can be summarized in a graph structure, where each fragment (and each reverse complement of a fragment) is a node and a directed edge *a* to *b* means that the end of *a* overlaps with the beginning of *b*. Every directed path in this graph corresponds to a contig. In principle, the problem of sequence assembly according to this approach is to find a collection of disjoint paths covering all fragments in the graph. But this path-covering problem is hard for general graphs, that is, it is difficult in the sense that no fast (polynomial time) algorithm is known for it and there is little hope of finding one.

One of the greatest complications that assemblers face are the repeats. A repeated region in the target DNA molecule causes cycles in the overlap graph. These cycles are responsible for the difficulty of the problem, because finding the better path cover in acyclic graphs can be solved in polynomial time.

Above we said that paths in the overlap graph correspond to contigs. In fact, the step of transforming a path in the overlap graph into a contig amounts to a form of multiple alignment, because the various fragments must be aligned with respect to one another to allow consensus computation. The consensus is computed position by position, and in each position the consensus base is a result of all fragment bases (or spaces) aligned at that position. The consensus base should reflect the most probable alternative in the target DNA, given the fragment sequences as aligned.

An assembly that results in many contigs generally means that we have the sequence of various contiguous

parts of the target molecule but not the intervening regions. These intervening regions are called gaps. In a circular genome, the number of gaps is equal to the number of contigs, whereas in a linear genome the number of gaps is equal to the number of contigs minus one.

Asymptotics of assembly

Lander and Waterman (1988) and Roach (1995) have developed mathematical models that predict the number of contigs, the size of contigs and other parameters given the number of fragments, minimum detectable overlap (in per cent of fragment length) and size of the original molecule. The projections are useful to check whether the sampling of clones is truly random, and can also be used to determine the point in time when a shift from random sampling to directed sequencing is advisable to close gaps and finish the genome. (*See* Statistical Methods in Physical Mapping.)

Eulerian approach

A totally different approach to overlaps has been proposed (Pevzner *et al.*, 2001). In this approach, which could be called the Eulerian approach, all fragments are error-corrected with respect to one another, and then all contiguous pieces of exactly *l* letters from all fragments are used to construct a different kind of overlap graph. In this graph, the assembly problem can be translated into finding a collection of disjoint paths that cover all edges, instead of covering all vertices. It turns out that this slightly different formulation admits a polynomial time algorithm, thus providing hope that larger data sets can be handled.

Shotgun with Mated Pairs

To help cope with repeats and other problems, two reads coming from opposite ends of the same clone, known as mated pairs, are often used to advantage. They help assembly because the approximate distance between them is known. If one read of a mated pair falls into one contig and the other read falls into a second contig, this establishes a relative positioning of the two contigs, even if the actual sequence between them is still unknown (it corresponds to the middle of the clone). Groups of contigs linked in this fashion constitute a scaffold.

Gaps that are spanned by a clone in a scaffold are called virtual gaps, because they can be closed by completely sequencing the spanning clone. In contrast, real gaps are those for which no further information is

Table 1 Software packages used in sequence assembly

Software	Source
Staden Package	http://www.mrc-lmb.cam.ac.uk/pubseq
TIGR Assembler	http://www.tigr.org/softlab/assembler
Phred/Phrap/Consed System	http://www.phrap.org
CAP3	http://genome.cs.mtu.edu/sas.html
CAP4 (now Paracel GenomeAssembler)	http://www.paracel.com/products/pga.html
Celera's Assembler	Not available outside Celera
Arachne	http://www.genome.wi.mit.edu/wga
EULER	http://nbcr.sdsc.edu/euler

available, that is, neither their size nor the contigs that flank it are known.

Software Tools

Dozens of software tools have been created to carry out fragment assembly. Here we mention some of the most representative; the relevant websites are given in **Table 1**.

Rodger Staden's package (Staden *et al.*, 2001) has been the dominating choice for many years, including the 1980s and part of the 1990s. The package includes a complete system for project management, and other tools for sequence analysis that are not necessarily related directly to assembly. It still is widely used today.

The Institute for Genomic Research (TIGR) developed the TIGR assembler (Sutton *et al.*, 1995) to deal with their numerous genome projects, mainly of bacterial origin. This program was among the first to use the information about mated pairs to guide assembly, and gave the world the first publicly available bacterial genome in 1995.

The group led by Phil Green developed the suite Phred, Phrap and Consed, which are, respectively, a base-caller with quality assessment, an assembler and a contig visualization tool (Ewing and Green, 1998; Ewing *et al.*, 1998; Gordon *et al.*, 1998). Among the first software to use quality values for base-calling, trimming, overlap detection and consensus determination, the 'quality values' tool rapidly spread over a large user community, including many research groups involved in the public effort toward sequencing the human genome.

Xiaoqiu Huang began writing assemblers in the early 1990s, and his program CAP2 was the best in a 1995 Implementation Challenge Workshop organized by Rutgers University Center for Discrete Mathematics and Theoretical Computer Science (DIMACS). Improved versions of this tool are CAP3 (Huang and Madan, 1999) and CAP4 (a commercial product).

To assemble very large, complex eukaryotic genomes using a whole-genome shotgun approach with mated pairs, Celera Corporation built their own assembler under the leadership of Gene Myers. As of mid-2001, the software has not been released to the public in any form, but has been used to assemble the *Drosophila melanogaster* (Myers *et al.*, 2000) and human genomes. A new software tool called Arachne that will deal with mated pairs has been released by a team from the Whitehead Institute (Batzoglou *et al.*, 2002).

The software EULER, developed by Pavel Pevzner and colleagues, uses the Eulerian approach described above.

See also
Bioinformatics
Genome Sequencing
Global Alignment
Sequence Accuracy and Verification
Sequence Finishing

References

Batzoglou S, Jaffe DB, Stanley K, *et al.* (2002) ARACHNE: a whole-genome shotgun assembler. *Genome Research* **12**: 177–189.

Ewing B and Green P (1998) Base-calling of automated sequencer traces using phred: II. Error probabilities. *Genome Research* **8**: 186–194.

Ewing B, Hillier L, Wendl MC and Green P (1998) Base-calling of automated sequencer traces using phred: I. Accuracy assessment. *Genome Research* **8**: 175–185.

Gordon D, Abajian C and Green P (1998) Consed: a graphical tool for sequence finishing. *Genome Research* **8**: 195–202.

Huang X and Madan A (1999) CAP3: a DNA sequence assembly program. *Genome Research* **9**: 868–877.

Lander E and Waterman MS (1988) Genomic mapping by fingerprinting random clones: a mathematical analysis. *Genomics* **2**: 231–239.

Maxam AM and Gilbert W (1977) A new method for sequencing DNA. *Proceedings of the National Academy of Sciences of the United States of America* **74**: 560–564.

Myers EW, Sutton GG, Delcher AL, *et al.* (2000) A whole-genome assembly of *Drosophila*. *Science* **287**: 2196–2204.

Pevzner PA, Tang H and Waterman MS (2001) An Eulerian path approach to DNA fragment assembly. *Proceedings of the National Academy of Sciences of the United States of America* **98**: 9748–9753.

Roach JC (1995) Random subcloning. *Genome Research* **5**: 464–473.

Sanger F, Nicklen S and Coulson AR (1977) DNA sequencing with chain terminating inhibitors. *Proceedings of the National Academy of Sciences of the United States of America* **74**: 5463–5467.

Setubal JC and Meidanis J (1997) *Introduction to Computational Molecular Biology*. Boston, MA: PWS.

Staden R, Judge DP and Bonfield JK (2001) Sequence assembly and finishing methods. In: Baxevanis DA and Ouellette BFF (eds.) *Bioinformatics: A Practical Guide to the Analysis of Genes and Proteins*, 2nd edn. New York, NY: John Wiley & Sons.

Sutton G, White O, Adams M and Kerlavage A (1995) TIGR Assembler: a new tool for assembling large shotgun sequencing projects. *Genome Science and Technology* **1**: 9–19.

Further Reading

Pevzner P (2001) *Computational Molecular Biology: An Algorithmic Approach*, chaps 4 and 5, pp. 93–132. Cambridge, MA: MIT Press.

Setubal JC and Meidanis J (1997) *Introduction to Computational Molecular Biology*, chap. 4, pp. 47–104. Boston, MA: PWS.

Waterman MS (1995) *Introduction to Computational Biology: Maps, Sequences, and Genomes*, chap. 7, pp. 183–252. Cambridge, UK: Chapman & Hall.

Sequence Complexity and Composition

Andrzej K Konopka, *BioLingua Research Inc., Gaithersburg, Maryland, USA*

Local compositional complexity is a numerical measure of repetitiveness of sequences of symbols from a finite alphabet. Highly repetitive sequences are considered simple, whereas highly nonrepetitive sequences are considered complex.

Intermediate article

Article contents

- Introduction
- Alphabets
- Local Compositional Complexity
- Periodic Patchy Complexity
- Complexity Classes and Surprisal
- Ancillary Topics Inspired by Sequence Complexity

Introduction

There are over 30 different definitions of complexity in modern science (Horgan, 1995) and all of them are far from being satisfactory in biology (Mikulecky, 2001). In this article, conventions are followed from computational biology and bioinformatics in which biopolymers (nucleic acids and proteins) are represented in the form of sequences of symbols from finite alphabets. In these conventions (Konopka and Owens, 1990a, Konopka, 1994) the term local compositional complexity is related to the concept of algorithmic complexity (Kolmogorov, 1965; Solomonoff, 1964) in a sense that repetitive sequences over a given finite alphabet A are considered 'simple' and nonrepetitive sequences are considered 'complex'. Random (i.e. patternless) sequences are considered maximally complex (Chaitin, 1966).

The numerical value of compositional complexity of a string of symbols depends on both the choice of alphabet and the frequencies with which specific letters are used. But it does not depend on anything else and therefore constitutes an excellent means of sequence data reduction. Applications of compositional complexity to sequence analysis include functionally or structurally relevant segmenting of nucleotide and protein sequences, genome sequence annotation, and finding new functionally relevant properties through studies of large collections of functionally equivalent sequence data.

Alphabets

Nucleic acids and proteins are represented by contiguous sequences of symbols of nucleotides and amino acid residues respectively. The set of symbols used for this primary representation constitutes an elementary alphabet.

The most natural elementary alphabet for nucleic acids is $E_1 = \{A, C, G, T \text{ (or U)}\}$, where the letters stand for adenine, cytosine, guanine and thymine (or uracil) nucleotides respectively. Other alphabets that are used in studies of nucleic acids include:

$E_2 = \{K, M\}$, where K is either guanine or thymine (uracil in ribonucleic acid; RNA) and M is either adenine or cytosine;

$E_3 = \{R, Y\}$, where R is a purine (adenine or guanine) and Y is pyrimidine (either cytosine or thymine);

$E_4 = \{S, W\}$, where S is either cytosine or guanine and W is either adenine or thymine (or uracil in RNA).

Given a fixed integer n we can define the n-extension E^n of an elementary alphabet as a set of all n-grams (strings of length n; n-tuples) of symbols from E. The terms n-extension of E and n-gram alphabet over E mean the same thing.

Example 1

A 2-gram alphabet over E_1 is a set of 16 dinucleotides, $E_1^2 = \{A, C, G, T\}^2 = \{AA, AC, AG, AT, CA, CC, CG, CT, GA, GC, GG, GT, TA, TC, TG, TT\}$. Similarly a 3-extension of E_2 is a set of eight trinucleotides, $E_2^3 = \{RRR, RRY, RYR, RYY, YRR, YRY, YYR, YYY\}$.

One can also consider alphabets of symbols that represent secondary and tertiary structures of biopolymers, physicochemical properties of identifiable regions of nucleic acids and proteins, and even symbols for biological functions of such regions. Because these alphabets require knowledge of properties other than biopolymer sequence alone, it is impractical to use them in the capacity of elementary alphabets. Instead, they can be used as secondary alphabets superimposed on biopolymer sequences that are represented in an elementary alphabet.

Local Compositional Complexity

Quantitative measures of compositional complexity (and simplicity) are based on a concept of deviation from discrete uniform frequency distribution (Konopka, 1994, 1997). Before we can define such measures we need to formalize the sequence composition.

For a string S of length L (L-gram) of letters from a given alphabet A, we construct the composition vector:

$$\tilde{\mathbf{n}} = (\mu_1, \mu_2, \ldots, \mu_N) \qquad (1)$$

where N is the number of letters in the alphabet A (size A), and $\mu_1, \mu_2, \ldots, \mu_N$ are the numbers of occurrences of consecutive letters of A in S. The sum of components of eqn [2] equals L (i.e. $\mu_1 + \mu_2 + \cdots + \mu_N = L$).

To normalize the composition vector (i.e. make sum of components equal to 1), we define 'relative frequencies' of the letters:

$$f_j = \mu_j / L \qquad (2)$$

where L is again the length of string S and each of μ_j's is a frequency (number of occurrences) of the jth letter from the alphabet A of size N.

The relative frequencies vector has the form

$$\tilde{\mathbf{n}}' = (f_1, f_2, \ldots, f_N) \qquad (3)$$

where each of f_j's is the relative frequency of the jth letter. The sum of relative frequencies in eqn [3] equals 1 (i.e. $f_1 + f_2 + \cdots + f_N = 1$), as it should.

One measure of compositional complexity (simplicity) is a square of the norm of the relative frequencies vector (the scalar product of the vector with itself):

$$\text{Norm}^2(\tilde{\mathbf{n}}') = \sum f_i^2 \qquad (4)$$

where the summation is from $i = 1$ to $i = N$.

The minimum possible value of the 'compositional simplicity' in eqn [4] is $1/N$, whereas the maximum equals 1.

Another measure of compositional complexity is Shannon entropy (Shannon, 1948):

$$H_N(\tilde{\mathbf{n}}') = -\sum f_i \log_N f_i \qquad (5)$$

where f_i and N are the same as above, and the summation is from $i = 1$ to $i = N$. The minimum value of N-ary entropy in eqn [5] is 0, and the maximum is 1 (which is why N, the size of the alphabet, is chosen to be the basis of the log function).

Example 2

Table 1 lists five different octanucleotides that are expressed in the familiar mononucleotide alphabet $E_1 = \{A, C, G, \text{and } T \text{ (or U)}\}$, together with the relative frequencies, f_j's, of the mononucleotides. Nucleotides A, C, G and T correspond to the index $j = 1, 2, 3$ and 4 respectively. L is the total number of mononucleotides in the string (eight in this example).

One can see immediately that oligonucleotide 3 should have the lowest complexity and that the other four are quite complex. Indeed, inspection of the relative frequencies indicates that oligonucleotides 1, 2 and 4 should be maximally complex because the relative frequencies of mononucleotides are equal to each other in these cases. Oligonucleotide 5 is complex but not maximally complex. These intuitions are confirmed quantitatively by inspection of the simplicity (expressed by a norm^2 measure) and the complexity (expressed by a quaternary (4-ary) entropy measure).

It should be noted that oligonucleotide 4 is maximally complex in the mononucleotide alphabet,

Table 1 Composition and complexity of selected octanucleotides in an {A, C, G, T} alphabet

Sequence ($L = 8$)	f_A	f_C	f_G	f_T	Norm^2 of $\tilde{\mathbf{n}}'$ (simplicity)	H_4 of $\tilde{\mathbf{n}}'$ (complexity)
(1) GCATACGT	1/4	1/4	1/4	1/4	0.25000 minimum	1.00000 maximum
(2) ATGCTACG	1/4	1/4	1/4	1/4	0.25000 minimum	1.00000 maximum
(3) AAAAAAAA	1	0	0	0	1.00000 maximum	0.00000 minimum
(4) ACGTACGT	1/4	1/4	1/4	1/4	0.25000 minimum	1.00000 maximum
(5) ACGATAGC	3/8	1/4	1/4	1/8	0.28125	0.95284

NATURE ENCYCLOPEDIA OF THE HUMAN GENOME / ©2003 Macmillan Publishers Ltd, Nature Publishing Group / www.ehgonline.net

but as a tandem repeat of the tetranucleotide ACGT, it should display minimum complexity in the four-nucleotide alphabet. This fact illustrates the importance of the alphabet that is used to determine compositional complexity.

Modified compositional complexity

In practical applications one often uses functions of $H_N(\tilde{\mathbf{n}})$ instead of $H_N(\tilde{\mathbf{n}})$ itself. One such function, modified compositional complexity (MCC), has been defined (Konopka, 1990; Konopka and Owens, 1990b) as

$$\mathrm{MCC} = H_N(\tilde{\mathbf{n}}')^{\mathrm{PAI}} \qquad (6)$$

where P^{AI}, the periodic asymmetry index, is a ratio of two-base to three-base sequence periodicities. It tends to be close to 1 for exons of protein-coding genes and significantly greater than 1 for introns and intergenic spacers (Konopka, 1990).

Complexity charts for sequence segmenting and annotation

Compositional complexity chart is a function that assigns a complexity value to every position in a given (long) sequence. In practice one needs to calculate complexity in a window moving along the sequence. The parameters of the chart include window size, W, and the window step, s (the number of nucleotides by which window is moved to make the subsequent window).

Examples of applications of compositional complexity charts to practical and qualitative sequence segmenting and approximate annotation are described in **Figure 1**.

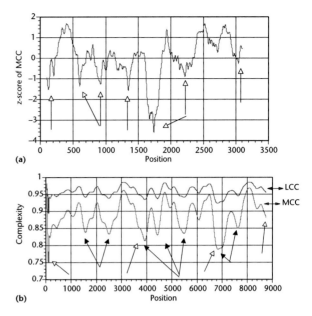

Figure 1 Examples of complexity charts used for DNA sequence segmenting and approximate functional annotation. Both charts were generated with a window width, W, of 200 nucleotides moving one nucleotide at a time (window step, $s = 1$). The accuracy of correctly annotated positions is too low (100 nucleotides) to be useful for exact gene structure determination, but it is clear that compositional complexity is correlated with gene structure. (a) Modified compositional complexity chart (z-score of MCC) for the region analogous to the α-operon of *Escherichia coli* in halophilic archaea, *Halobacterium halobium*. Arrows with transparent points indicate probable intergenic regions between protein-coding sequences. (b) Local compositional complexity and modified compositional complexity in chicken ovalbumin gene X. Arrows with filled points indicate probable positions of introns. Arrows with transparent points show the false-positive indications of intergenic spacers. (Determining the number of different genes in putative protein-coding regions is a serious problem that plagues all computer-assisted gene prediction methods. Compositional complexity chart methods face this problem as well.)

Periodic Patchy Complexity

Strings of symbols for determining the composition vector $\tilde{\mathbf{n}}$ can be generated by selecting noncontiguous nucleotides from a given (long) sequence. As above, we can calculate compositional complexity (simplicity) of strings of L symbols over an alphabet that contains N letters. But now the strings are obtained by taking every kth nucleotide, $k = 1, 2, 3, \ldots$, and so on, from the original nucleotide sequence. 'Patchy oligonucleotides' defined in this way are characterized by the 'regular patchiness' (period) k and block length L.

For example, normal (i.e. nonpatchy) hexanucleotides are characterized by $k = 0$ and $L = 6$, whereas hexanucleotides obtained by concatenating (six times) every other nucleotide in a given sequence are described by $k = 1$ and $L = 6$. Similarly, octanucleotides generated by taking (eight times) every third nucleotide from the original sequence are characterized by $k = 2$ and $L = 8$, whereas normal octanucleotides have a period of $k = 0$.

Example 3

Let us consider the sequence ACGTGTGTACAGG-TACGA. The consecutive patchy dinucleotides with patchiness $k = 1$ are AG, GG, AA and GA. The leftmost patchy tetranucleotide with patchiness $k = 2$ is ATGC. The consecutive dinucleotides with patchiness $k = 3$ are AG and another AG.

We refer to complexity calculated using composition vectors of patchy oligonucleotides as 'periodic patchy complexity' (or just 'patchy complexity'). Of course the normal (i.e. nonpatchy) local compositional complexity is a special case of patchy complexity (patchiness $k = 0$).

Complexity Classes and Surprisal

For a class of L-grams that share the same value of compositional complexity, we can define the 'surprisal' (or log[odds ratio]):

$$S(\mathbf{C}) = \log_N[P_{observed}(\mathbf{C})/P_{expected}(\mathbf{C})] \qquad (7)$$

where N is the size of the alphabet and $\mathbf{C}$ is compositional complexity. $P_{observed}(\mathbf{C})$ is the observed probability of the set of all L-grams that have the same complexity $\mathbf{C}$. In 'Bernoulli text', N letters of the alphabet occur independent of each other and with equal probabilities (i.e. equal to $1/N$). As a consequence, each of the L-grams has the same a priori probability as each of the remaining N^L-1 strings of length L. Therefore, the a priori probability $P_{expected}(\mathbf{C})$ for any compositional complexity value $\mathbf{C}$ is simply the number of different L-grams that share that value divided by the number of possible L-grams N^L.

Complexity classes of short oligonucleotides ($L = 1–8$) in four- and two-letter alphabets are listed in **Tables 2** and **3** respectively. The numbers of occupancies for every class are given as well.

It can be seen that even for Bernoulli text and a fixed L, the complexity classes are not equiprobable. Combinatorial methods to determine the number of occupancies of complexity classes are precise but too intricate to be discussed here. Their description can be found in Salamon and Konopka (1992), Salamon et al. (1993), as well as in Wootton and Federhen (1993).

Surprisal–complexity relationship

All collections of functionally equivalent nucleotide sequences studied so far show a significant (at a confidence level of 5% and better) linear correlation between surprisal and complexity for short oligonucleotides (L up to 20 nucleotides). The slope of the linear regression line is negative in all cases studied (Konopka and Owens, 1990a; Salamon and Konopka, 1992). This means that naturally occurring simple oligonucleotides tend to be overrepresented, whereas

Table 2 Complexity classes for short oligonucleotides in a four-letter alphabet, $E_1 = \{A, C, G, T \text{ or } U\}$

Complexity class (no. of classes)	No. of L-grams	$P_{expected}$ ($\bar{\mathbf{n}}$)
Mononucleotides (1)		
Complexity $= 0.0000$	No. of 1-grams $= 4$	$4/4$, $P_0 = 1.0000$
Dinucleotides (2)		
Complexity $= 0.5000$	No. of 2-grams $= 12$	$12/16$, $P_0 = 0.7500$
Complexity $= 0.0000$	No. of 2-grams $= 4$	$4/16$, $P_0 = 0.2500$
Trinucleotides (3)		
Complexity $= 0.7925$	No. of 3-grams $= 24$	$24/64$, $P_0 = 0.3750$
Complexity $= 0.4591$	No. of 3-grams $= 36$	$36/64$, $P_0 = 0.5625$
Complexity $= 0.0000$	No. of 3-grams $= 4$	$4/64$, $P_0 = 0.0625$
Tetranucleotides (5)		
Complexity $= 1.0000$	No. of 4-grams $= 24$	$24/256$, $P_0 = 0.0938$
Complexity $= 0.7500$	No. of 4-grams $= 144$	$144/256$, $P_0 = 0.5625$
Complexity $= 0.5000$	No. of 4-grams $= 36$	$36/256$, $P_0 = 0.1406$
Complexity $= 0.4056$	No. of 4-grams $= 48$	$48/256$, $P_0 = 0.1875$
Complexity $= 0.0000$	No. of 4-grams $= 4$	$4/256$, $P_0 = 0.0156$
Pentanucleotides (6)		
Complexity $= 0.9610$	No. of 5-grams $= 240$	$240/1024$, $P_0 = 0.2344$
Complexity $= 0.7610$	No. of 5-grams $= 360$	$360/1024$, $P_0 = 0.3516$
Complexity $= 0.6855$	No. of 5-grams $= 240$	$240/1024$, $P_0 = 0.2344$
Complexity $= 0.4855$	No. of 5-grams $= 120$	$120/1024$, $P_0 = 0.1172$
Complexity $= 0.3610$	No. of 5-grams $= 60$	$60/1024$, $P_0 = 0.0586$
Complexity $= 0.0000$	No. of 5-grams $= 4$	$4/1024$, $P_0 = 0.0039$
Hexanucleotides (9)		
Complexity $= 0.9591$	No. of 6-grams $= 1080$	$1080/4096$, $P_0 = 0.2637$
Complexity $= 0.8962$	No. of 6-grams $= 4$	$480/4096$, $P_0 = 0.1172$

Table 2 Continued

Complexity class (no. of classes)	No. of L-grams	$P_{expected}$ ($\tilde{n}$)
Complexity = 0.7925	No. of 6-grams = 360	360/4096, $P_0 = 0.0879$
Complexity = 0.7296	No. of 6-grams = 1440	1440/4096, $P_0 = 0.3516$
Complexity = 0.6258	No. of 6-grams = 360	360/4096, $P_0 = 0.0879$
Complexity = 0.5000	No. of 6-grams = 120	120/4096, $P_0 = 0.0293$
Complexity = 0.4591	No. of 6-grams = 180	180/4096, $P_0 = 0.0439$
Complexity = 0.3250	No. of 6-grams = 72	72/4096, $P_0 = 0.0176$
Complexity = 0.0000	No. of 6-grams = 4	4/4096, $P_0 = 0.0010$
Heptanucleotides (11)		
Complexity = 0.9751	No. of 7-grams = 2520	2520/16 384, $P_0 = 0.1538$
Complexity = 0.9212	No. of 7-grams = 5040	5040/16 384, $P_0 = 0.3076$
Complexity = 0.8322	No. of 7-grams = 840	840/16 384, $P_0 = 0.0513$
Complexity = 0.7783	No. of 7-grams = 2520	2520/16 384, $P_0 = 0.1538$
Complexity = 0.7244	No. of 7-grams = 1680	1680/16 384, $P_0 = 0.1025$
Complexity = 0.6894	No. of 7-grams = 2520	2520/16 384, $P_0 = 0.1538$
Complexity = 0.5744	No. of 7-grams = 504	504/16 384, $P_0 = 0.0308$
Complexity = 0.4926	No. of 7-grams = 420	420/16 384, $P_0 = 0.0256$
Complexity = 0.4316	No. of 7-grams = 252	252/16 384, $P_0 = 0.0154$
Complexity = 0.2958	No. of 7-grams = 84	84/16 384, $P_0 = 0.0051$
Complexity = 0.0000	No. of 7-grams = 4	4/16 384, $P_0 = 0.0002$
Octanucleotides (15)		
Complexity = 1.0000	No. of 8-grams = 2520	2520/65 536, $P_0 = 0.0385$
Complexity = 0.9528	No. of 8-grams = 20 160	20 160/65 536, $P_0 = 0.3076$
Complexity = 0.9056	No. of 8-grams = 6720	6720/65 536, $P_0 = 0.1025$
Complexity = 0.8750	No. of 8-grams = 10 080	10 080/65 536, $P_0 = 0.1538$
Complexity = 0.7806	No. of 8-grams = 6720	6720/65 536, $P_0 = 0.1025$
Complexity = 0.7744	No. of 8-grams = 1344	1344/65 536, $P_0 = 0.0205$
Complexity = 0.7500	No. of 8-grams = 5040	5040/65 536, $P_0 = 0.0769$
Complexity = 0.7028	No. of 8-grams = 6720	6720/65 536, $P_0 = 0.1025$
Complexity = 0.6494	No. of 8-grams = 4032	4032/65 536, $P_0 = 0.0615$
Complexity = 0.5306	No. of 8-grams = 672	672/65 536, $P_0 = 0.0103$
Complexity = 0.5000	No. of 8-grams = 420	420/65 536, $P_0 = 0.0064$
Complexity = 0.4772	No. of 8-grams = 672	672/65 536, $P_0 = 0.0103$
Complexity = 0.4056	No. of 8-grams = 336	336/65 536, $P_0 = 0.0051$
Complexity = 0.2718	No. of 8-grams = 96	96/65 536, $P_0 = 0.0015$
Complexity = 0.0000	No. of 8-grams = 4	4/65 536, $P_0 = 0.0001$

Table 3 Complexity classes for short oligonucleotides in two-letter alphabets {K, M}, {R, Y} and {S, W}

Complexity class (no. of classes)	No. of L-grams	$P_{expected}$ ($\tilde{n}$)
Mononucleotides (1)		
Complexity = 0.0000	No. of 1-grams = 2	2/4, $P_0 = 1.0000$
Dinucleotides (2)		
Complexity = 1.0000	No. of 2-grams = 2	2/4, $P_0 = 0.5000$
Complexity = 0.0000	No. of 2-grams = 2	2/4, $P_0 = 0.5000$
Trinucleotides (2)		
Complexity = 0.9183	No. of 3-grams = 6	6/8, $P_0 = 0.7500$
Complexity = 0.0000	No. of 3-grams = 2	2/8, $P_0 = 0.2500$

Table 3 Continued

Complexity class (no. of classes)	No. of L-grams	$P_{expected}$ ($\bar{n}$)
Tetranucleotides (3)		
Complexity = 1.0000	No. of 4-grams = 6	6/16, $P_0 = 0.3750$
Complexity = 0.8113	No. of 4-grams = 8	8/16, $P_0 = 0.5000$
Complexity = 0.0000	No. of 4-grams = 2	2/16, $P_0 = 0.1250$
Pentanucleotides (3)		
Complexity = 0.9710	No. of 5-grams = 20	20/32, $P_0 = 0.6250$
Complexity = 0.7219	No. of 5-grams = 10	10/32, $P_0 = 0.3125$
Complexity = 0.0000	No. of 5-grams = 2	2/32, $P_0 = 0.0625$
Hexanucleotides (4)		
Complexity = 1.0000	No. of 6-grams = 20	20/64, $P_0 = 0.3125$
Complexity = 0.9183	No. of 6-grams = 30	30/64, $P_0 = 0.4688$
Complexity = 0.6500	No. of 6-grams = 12	12/64, $P_0 = 0.1875$
Complexity = 0.0000	No. of 6-grams = 2	2/64, $P_0 = 0.0312$
Heptanucleotides (4)		
Complexity = 0.9852	No. of 7-grams = 70	70/128, $P_0 = 0.5469$
Complexity = 0.8631	No. of 7-grams = 42	42/128, $P_0 = 0.3281$
Complexity = 0.5917	No. of 7-grams = 14	14/128, $P_0 = 0.1094$
Complexity = 0.0000	No. of 7-grams = 2	2/128, $P_0 = 0.0156$
Octanucleotides (5)		
Complexity = 1.0000	No. of 8-grams = 70	70/256, $P_0 = 0.2734$
Complexity = 0.9544	No. of 8-grams = 112	112/256, $P_0 = 0.4375$
Complexity = 0.8113	No. of 8-grams = 56	56/256, $P_0 = 0.2188$
Complexity = 0.5436	No. of 8-grams = 16	16/256, $P_0 = 0.0625$
Complexity = 0.0000	No. of 8-grams = 2	2/256, $P_0 = 0.0078$

complex oligonucleotides tend to be underrepresented. This can be interpreted as a maximum entropy relationship constrained by mean complexity (Salomon and Konopka, 1992) but other interpretations are possible. For protein sequences and for long n-grams the properties of the relation between complexity and surprisal are more complicated. Readers interested in intricacies of surprisal versus complexity relationship for proteins and for long n-grams should consult Wootton and Federhen (1993) and Salomon *et al.* (1993).

Figure 2 illustrates the following (now well-established) computational–experimental properties of compositional complexity in naturally occurring nucleotide sequences:

- Sequences that correspond to different putative functional domains show systematically different slope values (Konopka, 1990; Konopka and Owens, 1990b) of surprisal versus complexity regression line.
- For every value of patchiness k, a roughly linear relationship between complexity of patchy L-nucleotides and surprisal is observed. The slope values are negative in most cases studied (see Konopka (1997) for more details).

- The larger the block length L (for all values of patchiness k), the more negative the slope of surprisal versus complexity relationship.

Ancillary Topics Inspired by Sequence Complexity

Database searches

Simple (i.e. repetitive) sequences occur in nature frequently enough to be taken into account while designing sequence database search programs. The reason for this special attention is the inordinately large number of 'hits' while searching simple long sequences for the occurrence of their simple subsequences.

Example 4

Let us consider two 'long' sequences: (1) acgtacgtacgtAAAAAAAAcgtacgta and (2) cgcgatatgtgtACGTACGTatcgatgcg. A search for the 4-gram AAAA in the first sequence will reveal five hits, whereas a search for the 4-gram ACGT in the second sequence will reveal only two hits. The reason for the large number of hits in simple sequences is the self-overlap capacity of their (also simple) subsequences. One can place five overlapping tetranucleotides AAAA in the octanucleotide AAAAAAAA because of the high selfoverlap capacity of AAAA.

Of course, the longer a simple sequence in the database is, the more marked the effect of overrepresenting the number of (short) subsequence matches will be. The best way to avoid mistaken conclusions from database searches is to use a properly designed measure of statistical significance for number of matches (hits). Of course, such well-designed measures of significance take into account the subsequence's self-overlap capacity in addition to the

number of possible subsequences of a given length. (Methods of calculating variance for self-overlapping n-grams are part of frequency count analyses that should be discussed elsewhere, for example, see Konopka (1997).)

A less desirable but often effective way of handling the problem is to mask simple sequences in database entries. Most (if not all) of the existing sequence database search tools still rely on this second approach as a temporary solution.

Biological complexity versus compositional complexity

The successful sequencing of over 60 genomes so far (including the human genome) puts issues raised by

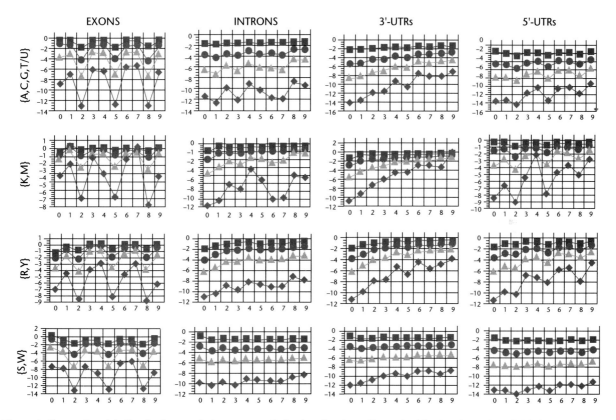

Figure 2 Slopes of straight line fits for surprisal versus complexity data for short oligonucleotides (regular and patchy) in large samples of human exons of confirmed protein-coding genes, introns, 3' untranslated regions (UTRs) and 5' UTRs of these genes. The x coordinate of each plot in the 'matrix' represents patchiness ($k = 0, 1, \ldots, 9$). The y coordinate represents slope values of surprisal versus complexity regression line. (All slope values are significant at a confidence level of 5% or better.) In every figure panel, block lengths $L = 5, 9, 13$ and 20 correspond to the top to bottom lines respectively. Figure panels in the 'exons' column of the matrix show that exons display clear three-base periodicity of occurrence of short oligonucleotides at all levels of patchiness, in all four alphabets. Comparison of 'introns' and '3' UTRs' columns shows that the complexity-related properties of introns and 3' UTRs are remarkably similar in most cases. This explains known difficulties with determining number of protein-coding genes in computationally predicted 'coding regions'. The only significant differences (and precious for practical purposes of gene identification) can be found for 20-grams in {A, C, G, T}, {K, M} and {R, Y} alphabets. Comparison of 'exons' and '5' UTRs' columns also shows that complexity-related properties of exons and 5' UTRs are similar enough to cause problems with computational identification of 5' ends of protein-coding genes. Comparison of figure panels in the bottom right and the bottom left corners of the matrix suggests that using statistics of 20-grams in the {S, W} alphabet should help to correctly identify 5'-UTRs correctly.

'systems biology' in the forefront of biology in particular and science in general. One such re-emerging issue is a definition of complexity that could pertain to a description of living things (or life itself) in both functional and structural terms.

It is clear that local compositional complexity and other variants of algorithmic complexity pertain solely to structural descriptions of sequences of symbols. Thus, the compositional complexity discussed here has nothing to do with biological complexity despite sharing the word 'complexity' in the name.

One can imagine that in the future 'sequence complexity' will be redefined such that it will pertain to functional descriptions of living things; however, it is not yet clear whether such redefinition could be useful in sequence analysis. Perhaps for the time being one should just be aware of the fact that our naming conventions may be subject to change in the future.

See also

Biomolecular Sequence Analysis: Pattern Acquisition and Frequency Counts
Systems Biology: Genomics Aspects

References

Chaitin GJ (1966) On the length of programs for computing finite binary sequences. *Journal of the ACM* **13**: 547–569.

Horgan J (1995) From complexity to perplexity. *Scientific American* June: 104–109.

Kolmogorov AN (1965) Three approaches to the definition of the concept 'Quantity of Information'. *Problems of Information Transmission* (Russian) **1**: 3–11.

Konopka AK (1990) Towards mapping functional domains in indiscriminantly sequenced nucleic acids: a computational approach. In: Sarma RH and Sarma MH (eds.) *Structure and Methods – Human Genome Initiative and DNA Recombination*, vol. 1, pp. 113–125. Guiderland, NY: Adenine Press.

Konopka AK (1994) Sequences and codes: fundamentals of biomolecular cryptology. In: Smith D (ed.) *Biocomputing: Informatics and Genome Projects*, pp. 119–174. San Diego, CA: Academic Press.

Konopka AK (1997) Theoretical molecular biology. In: Meyers RA (ed.) *Encyclopedia of Molecular Biology and Molecular Medicine*, vol. 6. pp. 37–53. Weinheim: VCH Publishers.

Konopka AK and Owens J (1990a) Non-contiguous patterns and compositional complexity of nucleic acid sequences. In: Bell GI and Marr TG (eds.) *Computers and DNA*, pp. 147–155. Redwood City, CA: Addison-Wesley Longman.

Konopka AK and Owens J (1990b) Complexity charts can be used to map functional domains in DNA. *Gene Analysis Techniques and Applications* **7**: 35–38.

Mikulecky DC (2001) The emergence of complexity: science coming of age or science growing old? *Computers and Chemistry* **25**: 341–348.

Salamon P and Konopka AK (1992) A maximum entropy principle for distribution of local complexity in naturally occurring nucleotide sequences. *Computers and Chemistry* **16**(2): 117–124.

Salamon P, Wootton JC, Konopka AK and Hansen LK (1993) On the robustness of maximum entropy relationships for complexity distributions of nucleotide sequences. *Computers and Chemistry* **17**(2): 135–148.

Shannon CE (1948) A mathematical theory of communication. *Bell System Technical Journal* **27**: 379–423, 623–656.

Solomonoff RJ (1964) A formal theory of inductive inference. *Information and Control* **7**: 224–254.

Wootton JC and Federhen S (1993) Statistics of local complexity in amino acid sequences and sequence databases. *Computers and Chemistry* **17**(2): 149–163.

Further Reading

Bell GI and Torney DC (1993) Repetitive DNA sequences: some considerations for simple sequence repeats. *Computers and Chemistry* **17**(2): 185–190.

Britten RJ and Kohne DE (1968) Repeated sequences in DNA. *Science* **161**: 529–540.

Konopka AK (1993) Plausible classification codes and local compositional complexity of nNucleotide sequences. In: Lim HA, Fickett JW, Cantor CR and Robbins RJ (eds.) *The Second International Conference on Bioinformatics, Supercomputing, and Complex Genome Analysis*, pp. 6987. New York: World Scientific Publishing.

Rosen R (2000) *Essays on Life Itself*. New York: Columbia University Press.

Shannon CE (1951) Prediction and entropy of printed English. *Bell System Technical Journal* **30**: 50–64.

Tautz D, Trick M and Dover GA (1986) Cryptic simplicity in DNA is a major source of genetic variation. *Nature* **322**: 652–656.

Wootton JC (1997) Simple sequences of protein and DNA. In: Bishop MJ and Rawlings CJ (eds.) *DNA and Protein Sequence Analysis*, pp. 169–183. Oxford: IRL Press.

Sequence Finishing

Darren Grafham, *The Sanger Centre, Wellcome Trust Genome Campus, Hinxton, UK*

David Willey, *The Sanger Centre, Wellcome Trust Genome Campus, Hinxton, UK*

Sequence finishing utilizes a variety of techniques to produce a highly accurate and complete DNA sequence.

Shotgun to Finishing

The most successful strategy for large-scale production of deoxyribonucleic acid (DNA) sequence, whether for clone-based or whole genome projects, has proved to be 'shotgun' sequencing (Anderson *et al.*, 1982; Deininger, 1983). In this approach, each 'unit' of DNA to be sequenced is sonicated or sheared to generate random discreet DNA fragments that are cloned into appropriate sequencing vectors. There is a hierarchy of scale involved, from whole genome to large insert clones and thence downward through subclones and finally individual reads. This article focuses on large insert bacterial clones and how to finish these within a larger project such as the Human Genome Project. Single-stranded (M13) or double-stranded (plasmid) vectors may be used for 1–2 kilobase DNA inserts or 1–10 kilobase inserts respectively. Sequence spanning 500–1000 bases of one or both ends of the inserts is then collected and assembled into overlapping sets or contigs. There are two important limitations to this approach statistical and biological (Chissoe *et al.*, 1997). The acquisition of novel data from random subclones is limited by the law of diminishing returns. Generally it is not efficient or cost-effective to collect data beyond a redundancy of eight-fold, which provides sequence coverage of greater than 95%. Inevitably, there will be gaps remaining in the assembled sequence. More importantly, the distribution of clones in the subclone libraries is always nonrandom. Some regions will be over-represented and others under-represented to the point of absence. In addition, the sequence of some regions will be inadequate because structural elements in the sequence have a detrimental effect on the sequencing reaction (premature termination for example). In contrast to shotgun sequencing, which is amenable to automation and scaling to a very large extent limited only by sequencing hardware, the process of closing the remaining gaps in assembled sequence, solving problems in difficult regions and resolving ambiguities, which is known as 'finishing', must generally be undertaken by skilled individuals. Finishing is the task of taking all the sequence contigs and assembling them into a single contig (contiguous sequence), which represents the original clone or genome that was shotgunned. Specific reactions are selected to make the clone contiguous and to have no ambiguities (accuracy of >99.99%). This article describes some of the approaches that have been widely used to complete DNA sequences.

Visualization of Sequence Data

Sequence assembly was performed using a program called 'PHRAP' (see Web Links), which utilizes the entire sequence reads to build an assembly. Once shotgun-sequencing projects have completed the assembly phase, they will be in a number of pieces (contigs) with some uncertainties in the sequence. This is the point at which directed finishing is employed. To be able to perform such a task finishers need to have available a method of viewing and manipulating the data. Several viewing tools are available but two packages GAP4 (Staden *et al.*, 2000) and CONSED (Gordon *et al.*, 1998) are most widely used. These two tools have advanced finishing by removing most of the manual editing and utilizing the quality values that are generated for each sequenced base (Ewing *et al.*, 1998). Typically, a bacterial clone of size 100–300 kbp is in 4–5 contigs after full shotgun. Each contig is made up from the sequence of multiple subclones and GAP4 or CONSED displays the position of every subclone within the contig as well as the junction of the clone sequence and cloning vector (**Figure 1**). (*See* Sequence Accuracy and Verification.)

Gap Classification

Sequence gaps can be classified as spanned or unspanned, according to whether or not the gap between two contigs is flanked by forward and reverse

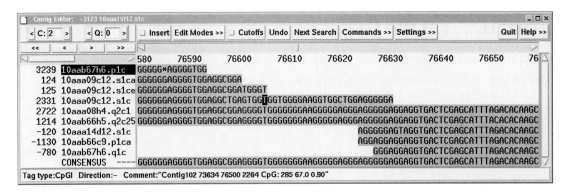

Figure 1 Gap4 screenshot showing subclones within a contig.

Table 1 Types of gaps encountered and available solutions

Gap classification	Reason for occurrence	Solutions
Normal-spanned	No sequence coverage in initial shotgun	Oligo-walking Short-insert libraries Transposon libraries
Sequence structure-spanned	GC-rich or repetitive sequence can form DNA structures that terminate sequencing	dGTP Sequence enhancers Short-insert libraries Transposon libraries
Partially unspanned	M13 projects or sequence failure in initial shotgun	Re-sequencing Oligo-walking Clone PCR
Fully unspanned	No representation in initial shotgun	Clone PCR Oligo-screening

sequences of a subclone. If the shotgun sequence has been generated from all M13 subclones, then all gaps are unspanned since only one shotgun sequence read is generated per subclone. Sequence can be generated from the other end of the subclone but requires a polymerase chain reaction (PCR) to generate a sequencable product. This is the first step of gap closure in all M13 shotgun projects that enter finishing. If the shotgun sequence has been generated from a mixture of plasmid and M13 subclones, or all plasmid, then a mixture of gaps will exist. As the proportion of sequence from plasmid subclones increases, more gaps become spanned with the possibility that all are spanned. In this case, the sequence is said to be ordered and orientated. **Table 1** summarizes the classes of gaps encountered, their reason for occurring and the probable solutions. (*See* Sequence Accuracy and Verification.)

Spanned Gaps

Spanned gaps can be subdivided further into normal and sequence structure gaps.

Normal-spanned gaps

Normal-spanned gaps have no problems other than that they were not covered by sequence data from the original random shotgun. Utilizing the spanning subclone always solves these, most frequently by re-sequencing using a specific oligonucleotide primer instead of the standard M13 forward and plasmid reverse primers. The oligonucleotide primer defines the start site to generate sequence of the desired region of the subclone. This type of sequencing is referred to as oligo-walking and differs from shotgun sequencing protocols only by the substitution of the sequencing primer. This is the simplest way to obtain sequence across this type of gap, but may fail, especially if the gap turns out to contain DNA with secondary sequence structure.

Spanned gaps with sequence structure

These gaps occur when the DNA in the subclone forms a secondary structure and prohibits accurate sequencing through the region. The most common form of secondary structure is the hairpin loop, which often occurs in GC-rich sequence, or between the poly A tails of Alu repeats. These are visualized in the database when the sequencing of a subclone terminates unexpectedly or sequencing of multiple subclones terminates within a few bases of each other. These structures are not always visible but failure to oligo-walk using normal chemistry is a strong indication that a spanned gap contains a sequence structure. Depending on the strength of the bonding in the structures, there are at least two commercial products available that can be used independently or in combination, to facilitate sequence reactions across such regions before the need for using short-insert libraries (SILs) or transposon libraries. They both work with M13, plasmid or custom oligonucleotides. The first is an alternative sequence chemistry that utilizes modified bases to help reduce formation of secondary structures. The second is a group of sequencing 'enhancers', which are added to the sequence mix to reduce the stability of secondary structures.

The utilization of SILs is one particularly useful method for sequencing through problematic secondary structures and permitting the determination of complete sequence data (McMurray et al., 1998). A single plasmid clone spanning the problem region is broken up into very small DNA fragments (50–1000 bp) by extensive sonication. Defined size fractions are isolated (typically 100–300, 300–500 and 500–800 bp) and then re-cloned into both plasmid and/or M13 vectors for sequencing. The fragmentation of the original subclone breaks up any secondary structure and permits the assembly of the subclone sequence. SILs in pUC resolve most problems (secondary structure, repeat sequences,) while M13 SILS are highly successful for GC-rich regions of the genome.

Transposon insertion methods may also be used to generate new priming sites within a spanning plasmid clone and have been utilized for sequencing through problematic sequence regions, particularly repeat sequences (Devine et al., 1997). A number of commercial kits are now available for the generation of transposon insertion libraries.

Unspanned Gaps

These can also be divided into two subgroups, partially and totally unspanned gaps.

Partially unspanned gaps

These arise when one of the pair of sequencing reactions fails or is not carried out. Re-sequencing whichever direction failed should be attempted. If this is successful, the gap is now spanned and additional finishing techniques can be employed. If the reaction fails again, custom oligonucleotides will be used to walk into the gap. If the end of the subclone is reached before the gap is closed, then it becomes an unspanned gap. Without a knowledge of the subclone spanning the gap, it is not advisable to use short-insert or transposon libraries.

Fully unspanned gaps

This type of gap is the most problematic, especially if the BAC clone is not in two contigs, since it creates multiple orientation possibilities. There are two options for solving this type of gap. PCR may be attempted on the BAC DNA, and the reaction aims to amplify a small piece of sequence (typically 500–2000 bp) from the whole clone (which may be as big as 300 000 bp) or a whole genome. The PCR reaction utilizes two unique 17–23mer oligonucleotides (longer can be employed to improve specificity but are more costly) in complementary orientations, designed from the two contigs believed to span the gap (GAP4 or CONSED are best used to check the sequences within the clone) avoiding known repetitive elements. If the oligonucleotides are not unique, then multiple PCR products that are unlikely to produce high-quality sequence data may result. The single PCR product is then sequenced using the two PCR oligonucleotides as sequence primers. If the sequence to close the gap extends into it from two different contigs, the gap is now spanned and the PCR product can be treated like a spanning subclone. If multiple PCR products can be accurately sized by restriction digest, then the gap-spanning product can be used for SIL preparation after purification of the band from an agarose gel. If the primary PCR reaction fails to generate a product, then additional attempts may be tried using different oligonucleotides. If the PCR continues to fail, then gap-spanning subclones may be identified by oligo-screening. Screening of the original shotgun subclone library using specific oligonucleotides is a particularly powerful method for the identification of spanning clones that were not sequenced in the original shotgun phase since they constitute low-representation subclones. The oligo-screen is carried out by replica plating the original shotgun library, usually cloned in plasmid, and probed separately with labeled oligonucleotides, which have been designed from high-quality unique sequence from each of the ends of the existing contigs. Plasmid clones, which span each of the sequence gaps, are identified and may be sequenced directly or used for SIL preparation. Oligo-screening is useful for the assembly of clones that exist as multiple isolated contigs.

Summary

With the vast array of techniques now available, sequence data, which are generated by shotgun sequencing, can be finished and made contiguous. Many final checks are made on the integrity of the sequence before submission and, increasingly, automation, bioinformatics and robotics are enhancing the speed, throughput and accuracy of finishing. (*See* Sequence Accuracy and Verification.)

See also

Sequence Accuracy and Verification
Sequence Assembly

References

Anderson S, de Bruin MHL, Coulson AR, *et al.* (1982) Complete sequence of bovine mitochondrial DNA. *Journal of Molecular Biology* **156**: 683–717.

Chissoe SL, Marra AM, Hillier L, *et al.* (1997) Representation of cloned genomic sequence in two sequencing vectors: correlation of DNA sequence and subclone distribution. *Nucleic Acids Research* **25**: 2960–2966.

Deininger PL (1983) Random subcloning of sonicated DNA: application to shotgun DNA sequence analysis. *Analytical Biochemistry* **129**: 216–223.

Devine SE, Chissoe SL, Eby Y, Wilson RK and Boeke JD (1997) A transposon-based strategy for sequencing repetitive DNA in eukaryotic genomes. *Genome Research* **7**: 551–563.

Ewing B, Hillier L, Wendl MC and Green P (1998) Base-calling of automated sequencer traces using phred I. *Genome Research* **8**: 186–194.

Gordon D, Abajian C and Green P (1998) Consed: a graphical tool for sequence finishing. *Genome Research* **8**(3): 195–202.

McMurray AM, Sulston JE and Quail MA (1998) Short-insert libraries as a method of problem solving in genome sequencing. *Genome Research* **8**: 562–566.

Staden R, Beal KF and Bonfield JK (2000) The Staden package, 1998. *Methods in Molecular Biology* **32**: 115–130.

Web Links

The Phred/Phrap/consed system homepage. Page containing documentation relating to the base caller-'Phred' and the DNA sequence assembler 'Phrap'
http://www.phrap.org

Sequence Similarity

Jaap Heringa, *Free University, Amsterdam, The Netherlands*

Sequence similarity is a measure of an empirical relationship between sequences. A common objective of sequence similarity calculations is establishing the likelihood for sequence homology: the chance that sequences have evolved from a common ancestor. A similarity score is therefore aimed to approximate the evolutionary distance between a pair of nucleotide or protein sequences. Many implementations for measuring sequence similarity exist.

> **Intermediate article**
>
> **Article contents**
> - Comparative Sequence Analysis
> - Sequence Alignment
> - Statistics of Alignment Similarity Scores

Comparative Sequence Analysis

Comparative sequence analysis is a common first step in the analysis of sequence–structure–function relationships in protein and nucleotide sequences. In the quest for knowledge about the role of a certain unknown protein in the cellular molecular network, comparing the query sequence with the many sequences in annotated protein sequence databases often leads to useful suggestions regarding the protein's three-dimensional (3D) structure or molecular function. This extrapolation of the properties of sequences in public databases that are identified as 'neighbors' by sequence analysis techniques has arguably led to the putative characterization (annotation) of more sequences than any other single technology during the last three decades. Although progress has been made, the direct prediction of a protein's structure and function is still a major unsolved problem in molecular biology. Since the advent of the genome sequencing projects and concomitant rapid expansion of sequence databases, the method of indirect inference by comparative sequence techniques has only gained in significance. Many current research projects are aiming to improve the sensitivity of sequence comparison techniques, which requires high-performance computing given the current and rapidly growing database sizes.

Sequence Alignment

Although many properties of nucleotide or protein sequences can be used to derive a similarity score, for example, nucleotide or amino acid composition, isoelectric point or molecular weight, the vast majority of sequence similarity calculations presuppose an alignment between two sequences from which a similarity score is inferred. Ideally, the alignment matches the nucleotide or amino acid sequences from either sequence according to their evolutionary descent from a common ancestor, with conserved and corresponding mutated residues at matched positions and inserted/deleted fragments intervening at proper sequence positions. Often, however, evolution has led to very widely diverged sequences such that at the primary sequence level the ancestral ties have become blurred beyond recognition, leading in many cases to biologically incorrect alignments. Another confounding issue is the fact that an increasing number of cases are identified of nonorthologous displacement, where enzymes carrying out an identical function in different organisms belong to entirely different protein families, and thus are not expected to show any sequence similarity. Examples of nonorthologous displacement include ornithine decarboxylase in *Escherichia coli* and *Saccharomyces cerevisiae*, where the isozymes speF and speC are responsible for this function in *E. coli* and share the same structure comprising three domains (ornithine decarboxylase *N*-terminal 'wing' domain, PLP-dependent transferase and ornithine decarboxylase *C*-terminal domain), whereas the corresponding enzyme spe1 in *S. cerevisiae* is a two-domain protein with entirely different domain structures (PLP-binding barrel and alanine racemase-like domain). In general, sequence alignment techniques are aimed at recognizing divergent evolution by mutation, including changes of gene structure by gene fusion or fission. However, the techniques are not able to trace evolutionary cases of horizontal gene transfer or functional displacement of one gene by another within a genome. (*See* Proteins: Mutational Effects in.)

Techniques for pairwise alignment

Many methods for the calculation of sequence alignments have been developed, of which implementations of the dynamic programming (DP) algorithm (Needleman and Wunsch, 1970; Smith and Waterman, 1981) are considered the standard in yielding the most biologically relevant alignments. The DP algorithm requires a *scoring matrix*, which is an evolutionary model in the form of a symmetrical 4 × 4 nucleotide or a 20 × 20 amino acid exchange matrix. Each matrix cell approximates the evolutionary propensity for the mutation of one nucleotide or amino acid type into another. The DP algorithm also relies on the specification of *gap penalties*, which model the relative probabilities for the occurrence of insertion/deletion events. Normally, a gap opening and extension penalty is used for creating a gap and each extension respectively (*affine* gap penalties), so that the chance for an insertion/deletion depends linearly upon the length of the associated fragment. Given an exchange matrix and gap penalty values, which together are commonly called the *scoring scheme*, the DP algorithm is guaranteed to produce the highest scoring alignment of any pair of sequences: the *optimal alignment*. (*See* Substitution Matrices.)

Two types of alignment are generally distinguished: global and local alignment. *Global alignment* (Needleman and Wunsch, 1970) denotes an alignment over the full length of both sequences, which is an appropriate strategy to follow when two sequences are similar or have roughly the same length. However, some sequences may show similarity limited to a motif or a domain only, while the remaining sequence stretches may be essentially unrelated. In such cases, global alignment may well misalign the related fragments, as these become overshadowed by the unrelated sequence portions that the global method attempts to align, possibly leading to a score that would not allow the recognition of any similarity. If not much knowledge about the relationship of two sequences is available, it is usually better to align selected fragments of either sequence. This can be done using the *local alignment* technique (Smith and Waterman, 1981). The first method for local alignment, often referred to as the Smith–Waterman (SW) algorithm, is in fact a minor modification of the DP algorithm for global alignment. The algorithm selects the best-scoring subsequence from each sequence and provides their alignment, thereby disregarding the remaining sequence fragments. Later elaborations of the algorithm include methods to generate a number of suboptimal local alignments in addition to the optimal pairwise alignment (e.g. Waterman and Eggert, 1987). (*See* Global Alignment; Sequence Alignment; Smith–Waterman Algorithm.)

Calculating alignment scores

Since the DP algorithm essentially models the alignment of two sequences as a Markov process, where the amino acid matches are considered independent, the product of the probabilities for each match within an alignment should be taken. Since many of the scoring matrices contain exchange propensities converted to logarithmic values (*log-odds*), the alignment score can be calculated by summing the log-odd values corresponding to matched residues minus appropriate gap penalties:

$$S_{a,b} = \sum_l s(a_i, b_j) - \sum_k N_k \, \mathrm{gp}(k),$$

where the first summation is over the exchange values associated with l matched residues and the second is over each group of gaps of length k, with N_k the number of gaps of length k and $\mathrm{gp}(k)$ the associated gap penalty. In case affine gap penalties are used (see above), $\mathrm{gp}(k) = \mathrm{pi} + k\,\mathrm{pe}$, where pi and pe are the penalties for gap initialization and extension respectively. A consequence of the widely used affine gap penalty scheme is that long gaps required, for example, to span an inserted domain B in aligning a two-domain sequence AC (where A and C represent domains) with a three-domain sequence ABC, are often too costly, so that such sequences become misaligned.

Sequence database searching

A typical application to infer knowledge for a given query sequence is to compare it with all sequences in an annotated sequence database. Unfortunately, the DP algorithm is too slow for repeated searches over large databases, and may take multiple CPU hours for a single query sequence on a standard workstation. Although some special hardware has been designed to accelerate the DP algorithm, this problem has triggered the development of several heuristic algorithms that represent shortcuts to speed up the basic alignment procedure. These include the currently most widely used method to scour sequence databases for homologies, PSI-BLAST (Altschul *et al.*, 1997), an extension of the BLAST technology (Altschul *et al.*, 1990) and FASTA (Pearson and Lipman, 1988), which is another commonly used heuristic for fast sequence comparison. Owing to advances in computational performance, procedures for homology searching have been developed based on more computationally intense formalisms such as the hidden Markov modeling-based tool SAM-T98 (Karplus *et al.*, 1998) and HMMER2 (Eddy, 1998). (*See* BLAST Algorithm; FASTA Algorithm; Hidden Markov Models; Multiple Alignment; Profile Searching.)

A database search can be performed for a nucleotide or amino acid sequence against an annotated database of nucleotide (e.g. EMBL, GenBank, DDBJ) or protein sequences (e.g. SwissProt, PIR, TrEMBL, GenPept, NR-NCBI, NR-ExPasy). Also the GSS, EST, STS or HTGS nucleotide databases can be scrutinized to find homologies, gain insight into expression data or locate a gene on the genome map. (*See* Comparative Genomics.)

The actual pairwise comparison can take place at the nucleotide or peptide level. However, the most effective way to compare sequences is at the peptide level (Pearson, 1996), which requires that nucleotide sequences must first be translated in all six reading frames followed by comparison with each of these conceptual protein sequences. Although mutation, insertion and deletion events take place at the DNA level, there are several reasons why comparing protein sequences can reveal more distant relationships: (1) Many mutations within DNA are synonymous, which means that these do not lead to a change of the corresponding amino acids. As a result of the fact that most evolutionary selection pressure is exerted on protein sequences, synonymous mutations can lead to an overestimation of the sequence divergence if compared at the DNA level. (2) The evolutionary relationships can be more finely expressed using a 20×20 amino acid exchange table than using exchange values among four nucleotides. (3) DNA sequences contain noncoding regions, which should be avoided in homology searches. Note that the latter is still an issue when using DNA translated into protein sequences through a codon table. However, a complication arises when using translated DNA sequences to search at the protein level because frameshifts can occur, leading to stretches of incorrect amino acids in the wrongly transcribed product and possible elongation of sequences due to missed stop codons. On the other hand, frameshifts typically result in stretches of highly unlikely and distant amino acids, which can be used as a signal to trace their occurrence. (*See* Mutation Rates: Evolution.)

Similarity versus homology

Many times the term homologous sequence is used when in fact a sequence should only be referred to as similar to a given reference sequence (May, 2001). Whereas sequence similarity is a quantification of an empirical relationship of sequences expressed using a gradual scale, the term homology denotes an inference in that the presence of a common ancestor between the sequences and hence divergent evolution is assumed, leading to orthologous genes. This means that homology is a qualitative state; that is, a pair of sequences is homologous or not. As protein tertiary structures are more conserved during evolution than their coding sequences, homologous sequences are assumed to share the same protein fold. Although it is possible in theory that two proteins evolve different structures and functions from a common ancestor, this situation cannot be traced so that such proteins are seen as unrelated. However, numerous cases exist of homologous protein families where subfamilies with the same fold have evolved distinct molecular functions. The term homology is often used in practice when two sequences have the same structure or function, although in the case of two sequences sharing a common function this ignores the possibility that the sequences are analogs resulting from convergent evolution, now often referred to as nonorthologous displacement (see above). Unfortunately, it is not straightforward to infer homology from similarity as enormous differences exist between sequence similarities within homologous families. Many protein families of common descent comprise members that share pairwise sequence similarities, which are only gradually higher than those observed between unrelated proteins. This region of uncertainty has been characterized to lie in the range of 15–25% sequence identity (Doolittle, 1981) (see below), and is commonly referred to as the 'twilight zone'. There are even some known examples of homologous proteins with sequence similarities below the randomly expected level given their amino acid composition (Pascarella and Argos, 1992). As a consequence, it is impossible to prove using sequence similarity that two sequences are not homologous. (*See* Protein Homology Modeling.)

The similarity score for two sequences can be calculated from their alignment using the above formula for $S_{a,b}$, such that it depends on the actual scoring matrix and gap penalties used. It has also been calculated as a fraction of a maximal score possible for two sequences using a normalized scoring matrix and by normalizing the raw alignment score by the length of the shorter sequence (Abagyan and Batalov, 1997).

Sequence similarity versus identity

Numerous studies into protein sequence relationships evaluate sequence alignments using a simple binary scheme of matched positions being identical or nonidentical. Sequence identity is normally expressed in the percentage identical residues found in a given alignment, where normalization can be performed using the length of the alignment or the shorter sequence. The scheme is simple and does not rely on an amino acid exchange matrix. However, if two proteins are said to share a given percentage in sequence identity, this is based on a sequence alignment, which will have been almost always constructed using an amino acid exchange matrix and gap penalty values, so that sequence identity cannot be regarded as being independent of sequence similarity. Using sequence identity as a measure, Sander and Schneider (1991) estimated that if two protein sequences are longer than 80 residues, they could relatively safely be assumed to be homologous whenever their sequence identity is 25% or more. Another commonly used notion is that if two sequences share more than 50% sequence identity, their enzymatic function will be the same (Rost, 2002). Contrary to this notion, however, it has been estimated that 70% of pair fragments above

50% sequence identity might not have a completely identical function (Rost, 2002). An example is *Bacillus subtilus* exodeoxyribonuclease (SwissProt code exoa_bacsu) and rat DNA-lyase (SwissProt code ape1_rat), where the sequences share 57% identity over 122 alignment positions, leading to a very significant BLAST *E*-value of 1.6×10^{-96}, but yet fulfill different functions (DNA degradation and repair respectively). Despite its popularity and use in empirical rules as above, the use of sequence identity percentages is not optimal for homology searches (Abagyan and Batalov, 1997). As a result no major sequence comparison methods employ sequence identity scores in deriving statistical significance estimates. (*See* Gene Families.)

Statistics of Alignment Similarity Scores

Sequence alignment methods are essentially pattern search techniques, leading to an alignment with a similarity score even in case of absence of any biological relationship. Although similarity scores of unrelated sequences are essentially random, they can behave like 'real' scores and, for example, like the latter are correlated with the length of the sequences compared. Particularly in the context of database searching, it is important to know what scores can be expected by chance and how scores that deviate from random expectation should be assessed. If within a rigid statistical framework a sequence similarity is deemed statistically significant, this provides confidence in deducing that the sequences involved are in fact biologically related. As a result of the complexities of protein sequence evolution and distant relationships observed in nature, any statistical scheme will invariably lead to situations where a sequence is assessed as unrelated while it is in fact homologous (*false negative*), or the inverse, where a sequence is deemed homologous while it is in fact biologically unrelated (*false positive*). A relatively frequent cause of erroneous transfer of annotation is based on similarity found over relatively short sequence regions and/or similarity based on different domains in multidomain structures (Rost, 2002).

The derivation of a general statistical framework for evaluating the significance of sequence similarity scores has been a major task. However, a rigid framework has not been established for global alignment, and has only partly been completed for local alignment.

Expected values for global similarity scores

Sequence similarity values resulting from global alignments are known to grow linearly with the sequence length (**Figure 1**), although the growth rate has not been determined. Also, the exact distribution of global similarity scores is yet unknown, and only numerical approximations exist, providing only a rough bound on the expected random scores. As the variance of the global similarity score has not been determined either, most applications derive a sense of the score by using shuffled sequences. Shuffled sequences retain the composition of a given real sequence but have a permuted order of nucleotides or amino acids. The distribution of similarity scores over a large number of such shuffled sequences often approximates the shape of the Gaussian distribution, which is therefore taken to represent the underlying random distribution. Using the mean (m) and standard deviation (σ) calculated from such shuffled similarity scores, each real score S can be converted to the z-score using

$$z\text{-score} = (S - m)/\sigma$$

The z-score measures how many standard deviations the score is separated from the mean of the random distribution. In many studies, a z-score > 6 is taken to indicate a significant similarity.

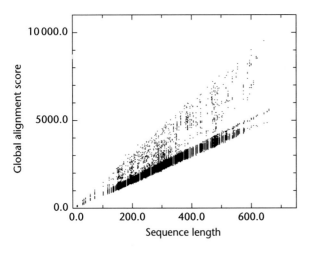

Figure 1 Distribution of 66 066 global similarity scores, derived from pairwise global alignments over an artificial database of sequences derived using a random mutation and insertion/deletion protocol, versus the length of the shortest sequence in each pairwise alignment. The alignments were effected using the PRALINE method (Heringa, 1999, 2002) where the alignment scores were calculated using the BLOSUM62 matrix and gap penalty values of 12 and 1 for gap initiation and extension respectively. A clearly linear lower band of alignment scores of unrelated sequences is visible. The correlation coefficient of the random scores within the lower band is 0.99 while the slope of the regression line is 7.864. Also the higher scores of putatively related sequences above the lower band are correlated: the correlation coefficient is 0.98 and the linear regression line slope is 12.50. Random and real scores were separated by the line $y = 9.334x$.

Expected values for local similarity scores

Statistics of local alignments without gaps

A rigid statistical framework for local alignments without gaps has been derived for protein sequences following the work by Karlin and Altschul (1990), who showed that the optimal local ungapped alignment score growths linearly with the logarithm of the product of sequence lengths of two considered random sequences:

$$S \sim \ln(n \cdot m)/\lambda$$

where n and m are the lengths of two random sequences, and λ is a scaling parameter that depends on the scoring matrix used and the overall distribution of amino acids in the database. Specifically, λ is the unique solution for x in the equation

$$\sum_{i,j} P_i P_j e^{s_{i,j} x} = 1$$

where summation is done over all amino acid pairs, p_i represents the background probability (frequency) of residue type i, and $s_{i,j}$ the scoring matrix.

An important contribution for fast sequence database searching has been the realization (Karlin and Altschul, 1990; Dembo and Karlin, 1991; Dembo *et al.*, 1994) that local similarity scores of ungapped alignments follow the *extreme value distribution* (EVD) (Gumbel, 1958). This distribution is unimodal but not symmetrical like the normal distribution, because the right-hand tail at high scoring values falls off more gradually than the lower tail, reflecting the fact that a best local alignment is associated with a score that is the maximum out of a great number of independent alignments (**Figure 2**).

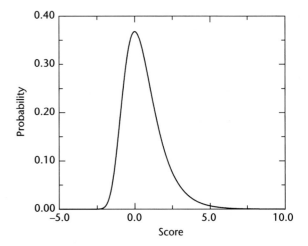

Figure 2 Probability density function for the extreme value distribution resulting from parameter values $\mu = 0$ and $\lambda = 1$, where μ is the characteristic value and λ is the decay constant.

Following the extreme value distribution, the probability of a score S to be larger than a given value x can be calculated as

$$P(S \geq x) = 1 - \exp(-e^{-\lambda(x-\mu)})$$

where $\mu = (\ln Kmn)/\lambda$, and K is a constant that can be estimated from the background amino acid distribution and scoring matrix. (See Altschul and Gish (1996) for a collection of values for λ and K over a set of widely used scoring matrices.) Using the equation for μ, the probability for S becomes

$$P(S \geq x) = 1 - \exp(-Kmne^{-\lambda x}).$$

In practise, the probability $P(S \geq x)$ is estimated using the approximation $1 - \exp(-e^{-x}) \approx e^{-x}$, which is valid for large values of x. This leads to a simplification of the equation for $P(S \geq x)$:

$$P(S \geq x) \approx e^{-\lambda(x-\mu)} = Kmne^{-\lambda x}.$$

The lower the probability for a given threshold value x, the more significant the score S.

Despite the usefulness of the above statistical estimates in recognizing sequence similarity, it should be noted that they do not judge the distribution of similarity along the sequences, which is a crucial aspect in assessing homology, and can correspond to a single domain in a multidomain protein sequence or to a single motif within a domain.

Statistics of local alignments with gaps

Although similarities between sequences can be detected reasonably well using methods that do not allow insertions/deletions in aligned sequences, it is clear that insertion/deletion events play a major role in divergent sequences. This means that accommodating gaps within alignments of distantly related sequences is important for obtaining an accurate measure of similarity. Unfortunately, a rigorous statistical framework as obtained for gapless local alignments has not been conceived for local alignments with gaps. However, although it has not been proven analytically that the distribution of S for gapped alignments can be approximated with the extreme value distribution, there is accumulated evidence that this is the case. For example, for various scoring matrices, gapped alignment similarities have been observed to grow logarithmically with the sequence lengths (Arratia and Waterman, 1994). Other empirical studies have shown it to be likely that the distribution of local gapped similarities follows the extreme value distribution (Smith et al., 1985; Waterman and Vingron, 1994), although an appropriate downward correction for the effective sequence length has been recommended (Altschul and Gish, 1996). The distribution of empirical similarity values can be obtained from unrelated

biological sequences (Pearson, 1998). Fitting of the EVD parameters λ and K (see above) can be performed using a linear regression technique (Pearson, 1998), although the technique is not robust against outliers which can have a marked influence. Maximum likelihood estimation (Mott, 1992; Lawless, 1982) has been shown to be superior for EVD parameter fitting and, for example, is the method used to parameterize the gapped BLAST method (Altschul et al., 1997). However, when low gap penalties are used to generate the alignments, the similarity scores can lose their local character and assume more global behavior, such that the EVD-based probability estimates are not valid anymore (Arratia and Waterman, 1994).

Statistics of database searches

In order to be useful in sequence database searches, the above framework for comparing a pair of random sequences should be adapted to multiple pairwise comparisons. Here, it becomes important to establish the probability for a given query sequence to have a significant similarity with at least one of the database sequences. A *p-value* is the probability of seeing at least one unrelated score S greater than or equal to a given score x in a database search over n sequences. This probability has been demonstrated to follow the Poisson distribution (Waterman and Vingron, 1994):

$$P(x,n) = 1 - e^{-nP(S \geq x)}$$

where n is the number of sequences in the database. In addition to the *p*-value, some database search methods employ the *expectation value* (or *E-value*) of the Poisson distribution, which is defined as the expected number of nonhomologous sequences with scores greater than or equal to a score x in a database of n sequences:

$$E(x,n) = nP(S \geq x).$$

For example, if the *E*-value of a matched database sequence segment is 0.01, then the expected number of random hits with score $S \geq x$ is 0.01, which means that this *E*-value is expected by chance only once in 100 independent searches over the database. However, if the *E*-value of a hit is five, then five fortuitous hits with $S \geq x$ are expected within a single database search, which renders the hit not significant. Database searching is commonly performed using an *E*-value in between 0.1 and 0.001. Low *E*-values decrease the number of false positives in a database search, but increase the number of false negatives such that the sensitivity (see below) of the search is lowered.

Evaluating sequence database searches

A few useful measures are commonly used to measure the accuracy of sequence database search methods over an annotated nonredundant database. The *sensitivity* of a search is defined as

$$\text{Sensitivity} = \text{TP}/(\text{TP} + \text{FN})$$

where TP is the number of true positives and FN the number of false negatives, which reflects the fraction of true hits found relative to the total number of sequences in the database that are homologous to the query. The sensitivity reflects to what extent the method is able to identify distantly related sequences. In many studies this measure is also referred to as *coverage*. The *specificity* (or *selectivity*) is defined as

$$\text{Specificity} = \text{TN}/(\text{FP} + \text{TN})$$

which denotes the fraction of entries correctly excluded as hits, and hence measures the avoidance of unrelated hits. Yet another widely used measure is the *positive predictive value* (PPV), defined as

$$\text{PPV} = \text{TP}/(\text{TP} + \text{FP})$$

which measures the proportion of true homolog within all sequences designated by the search tool as related. In practical database searches, there is a trade-off between sensitivity and specificity: The more the *P*-values or *E*-values are relaxed to allow more distantly related sequences to be found, the more likely it becomes that chance hits infiltrate the search. Moreover, even if a very statistically significant similarity is encountered, problems remain. For example, if high similarity is found over only a portion of the sequences, the sequences may each contain multiple domains and share a single homologous domain only (see above), so that only an aspect of the overall function might be inferred. In iterative homology searches, often carried out using the aforementioned method PSI-BLAST (Altschul *et al.*, 1997), protein sequences containing more than one structural domain can be problematic in that they cause the search to terminate prematurely or lead to an 'explosion' of common domains (George and Heringa, 2002). For example, the occurrence in the query sequence of a common and conserved protein domain such as the tyrosine kinase domain, which is then hit many times in the database, can obscure weaker but also relevant matches to other domain types (George and Heringa, 2002), particularly when the *E*-value is set to only include strong hits. Conversely, when multidomain sequences with the same sequential order of domains as in the query sequence are found initially during an iterative search, homologs with different domain combinations might well be missed due to early convergence of the search. To reduce the chance of including spurious hits, some database search engines, such as PSI-Blast (Altschul *et al.*, 1997), scan query sequences for the presence of so-called *low-complexity regions* comprising biased residue compositions such

as coiled-coil or transmembrane regions (Wooton and Federhen, 1996). These are then excluded from alignment to limit the inclusion of false-positive hits due to database sequence matches with these regions. However, the occurrence of database sequences with low-complexity regions can still cause an explosion of false positives in iterative homology searches (George and Heringa, 2002). Despite recent improvements of search techniques, complications such as above illustrate that automatic biological evaluation of homology searches in genomic pipelines remains elusive. (*See* Bioinformatics: Technical Aspects.)

See also

Alignment: Statistical Significance
Protein Homology Modeling
Sequence Alignment
Similarity Search

References

Abagyan RA and Batalov S (1997) Do aligned sequences share the same fold? *Journal of Molecular Biology* 273: 355–368.

Altschul SF and Gish W (1996) Local alignment statistics. In: Doolittle RF (ed.) *Methods in Enzymology*, vol. 266, pp. 460–480. San Diego, CA: Academic Press.

Altschul SF, Gish W, Miller W, Meyers EW and Lipman DJ (1990) Basic local alignment search tool. *Journal of Molecular Biology* 215: 403–410.

Altschul SF, Madden TL, Schäffer AA, *et al.* (1997) Gapped BLAST and PSI-BLAST: a new generation of protein database search programs. *Nucleic Acids Research* 25: 3389–3402.

Arratia R and Waterman MS (1994) A phase transition for the sore in matching random sequences allowing depletions. *Annals of Applied Probability* 4: 200–225.

Dembo A and Karlin S (1991) Strong limit theorems of empirical functionals for large exceedances of partial sums of i.i.d. variables. *Annals of Probability* 19: 1737.

Dembo A, Karlin S and Zeitouni O (1994) Limit distributions of maximal non-aligned two-sequence segmental score. *Annals of Probability* 22: 2022.

Doolittle RF (1981) Similar amino acid sequences: chance or common ancestry. *Science* 214: 149–159.

Eddy SR (1998) Profile hidden Markov models. *Bioinformatics* 14: 755–763.

George RA and Heringa J (2002) Protein domain identification and improved sequence searching using PSI-BLAST. *Proteins – Structure Function and Genetics* 48: 672–681.

Gumbel EJ (1958) *Statistics of Extremes*. New York, NY: Columbia University Press.

Heringa J (1999) Two strategies for sequence comparison: profile-preprocessed and secondary structure-induced multiple alignment. *Computers and Chemistry* 23: 341–364.

Heringa J (2002) Local weighting schemes for protein multiple sequence alignment. *Computers and Chemistry* 26: 459–477.

Karlin S and Altschul SF (1990) Methods for assessing the statistical significance of molecular sequence features by using general scoring schemes. *Proceedings of the National Academy of Sciences of the United States of America* 87: 2264–2268.

Karplus K, Barrett C and Hughey R (1998) Hidden markov models for detecting remote protein homologies. *Bioinformatics* 14: 846–856.

Lawless JF (1982) *Statistical Models and Methods for Lifetime Data*. pp. 141–202. New York, NY: John Wiley & Sons.

May AC (2001) Related problems. *Nature* **413**: 453.

Mott R (1992) Maximum-likelihood estimation of the statistical distribution of Smith–Waterman local sequence similarity scores. *Bulletin of Mathematical Biology* **54**: 59–75.

Needleman SB and Wunsch CD (1970) A general method applicable to the search for similarities in the amino acid sequence of two proteins. *Journal of Molecular Biology* **48**: 443–453.

Pascarella S and Argos P (1992) A data bank merging related protein structures and sequences. *Protein Engineering* **5**: 121–137.

Pearson WR (1996) Effective protein sequence comparison. In: Doolittle RF (ed.) *Methods in Enzymology*, vol. 266, pp. 227–258. San Diego, CA: Academic Press.

Pearson WR (1998) Empirical statistical estimates for sequence similarity searches. *Journal of Molecular Biology* **276**: 71–84.

Pearson WR and Lipman DJ (1988) Improved tools for biological sequence comparison. *Proceedings of the National Academy of Sciences of the United States of America* **85**: 2444–2448.

Rost B (2002) Enzyme function is less conserved than anticipated. *Journal of Molecular Biology* **318**: 595–608.

Sander C and Schneider R (1991) Database of homology derived protein structures and the structural meaning of sequence alignment. *Proteins – Structure Function and Evolution* **9**: 56–68.

Smith TF and Waterman MS (1981) Identification of common molecular subsequences. *Journal of Molecular Biology* **147**: 195–197.

Smith TF, Waterman MS and Burks C (1985) The statistical distribution of nucleic acid similarities. *Nucleic Acids Research* **13**: 645.

Waterman MS and Eggert M (1987) A new algorithm for best subsequences alignment with applications to the tRNA–rRNA comparisons. *Journal of Molecular Biology* **197**: 723–728.

Waterman MS and Vingron M (1994) Rapid and accurate estimates of statistical significance for sequence data base searches. *Proceedings of the National Academy of Sciences of the United States of America* **91**: 4625.

Wooton JC and Federhen S (1996) Analysis of compositionally biased regions in sequence databases. In: Doolittle RF (ed.) *Methods in Enzymology*, vol. 266, pp. 554–571. San Diego, CA: Academic Press.

Further Reading

Doolittle RF (ed.) (1996) *Methods in Enzymology*. vol. 266, p. 711. San Diego, CA: Academic Press.

Higgins D and Taylor WR (eds.) (2000) *Bioinformatics: Sequence, Structure and Databanks*, p. 249. Oxford, UK: Oxford University Press.

Sequence-tagged Site (STS)

Introductory article

Paul H Dear, *Medical Research Council Laboratory of Molecular Biology, Cambridge, UK*

A sequence-tagged site is simply a short segment of known deoxyribonucleic acid sequence. Generally, it is one for which polymerase chain reaction (PCR) primers have been designed to allow it to be amplified using the PCR.

A sequence-tagged site (STS) is a short (typically 50–500 bp) segment of known DNA sequence from a given genome. In common usage, an STS is a sequence that can be conveniently amplified (and hence detected) using the polymerase chain reaction (PCR) with primers complementary to either end of the sequence.

STSs are most commonly used as 'markers' – identifiable genomic features – in genome mapping. For example, in physical mapping, each of the cloned DNA fragments in a library may be tested by PCR to see if it contains a given STS (if it does, the PCR will yield a product of the expected length, which can be detected by gel electrophoresis); if two clones are found to contain the same STS, it can be deduced that they represent the same (or overlapping) parts of the genome.

An expressed-sequence tag (EST) is an STS that derives from an expressed sequence (gene). An STS that contains, in its middle, a simple, tandemly repeated sequence (such as CACACACACA ...) is commonly called a 'microsatellite', although strictly speaking the microsatellite is only the repeated part of the sequence.

Serine Proteases

Enrico Di Cera, *Washington University School of Medicine, St Louis, Missouri, USA*

Maxwell M Krem, *Washington University School of Medicine, St Louis, Missouri, USA*

Serine proteases are enzymes that use a serine hydroxyl group to cleave other proteins. The human genome encodes serine proteases from four evolutionarily unrelated clans that perform a wide variety of physiologic functions.

Introduction

Serine proteases conduct proteolysis using a serine side chain as the active site nucleophile. These enzymes carry out a diverse array of physiologic functions in organisms ranging from archaea and eubacteria to humans. Serine proteases have been classified into evolutionarily unrelated clans by Barrett and Rawlings (1995). The clans are, in turn, subdivided into families of proteases whose homology can be established statistically. Clans are differentiated according to the order of catalytic residues in the primary sequence, not according to active site structure or chemistry, which is conserved among different clans. The human genome encodes serine proteases of clans PA, SB, SC and SF.

Clan PA: Chymotrypsin-like Serine Proteases

Clan PA (originally named clan SA) contains serine proteases in which the order of the catalytic triad is His-Asp-Ser and related cysteine proteases in which the order of the catalytic triad is His-Asp-Cys. Human proteins of clan PA are exclusively serine proteases of family S1, the chymotrypsin family. The number of human chymotrypsin-like proteases is 118, based on the results of the Human Genome Project. Family S1 features a gamut of important physiologic functions including digestive and degradative processes, blood coagulation, cellular and humoral immunity, fibrinolysis, fertilization and embryonic development (**Table 1**). The proteases in the blood coagulation, fibrinolytic and complement systems are multidomain proteins with auxiliary modules located upstream of the protease domain; these modules help localize the protease domain in space. The most common modules are kringle, epidermal growth factor and sushi domains.

Serine protease domains of clan PA have achieved functional diversity by generating variations on a common fold (Lesk and Fordham, 1996). This fold consists of two open-ended β-barrels with the active site and substrate-binding clefts lying in between

(**Figure 1**). Within the active site, the hydroxyl oxygen of the catalytic Ser is made a more potent nucleophile by the presence of the basic side chain of the catalytic His; positive charge that develops on the His side chain is stabilized by the negatively charged side chain of the active site Asp. This active site composition and geometry, shared with serine proteases of clans SB, SC and SF, indicates a pattern of convergent evolution (**Figure 2**). Variation in substrate recognition is generated in part by differences in the identities of residues that lie within the substrate-binding cleft (Perona and Craik, 1995). The most important residue governing specificity contacts the substrate side chain just upstream of the scissile bond and lies within a buried water-lined pocket formed by β-strands. Other structural features that generate variations in substrate specificity and enzyme activity are Na^+ and Ca^{2+} binding sites, surface insertion loops, and exosites that are allosterically linked to the specificity and active site residues. The majority of this specificity-determining architecture is concentrated within the 50 C-terminal residues of the protease domain, which appear to be the residues that encode physiologic function in family S1 (Krem *et al.*, 1999).

Members of family S1 are scattered throughout the genome, but several clusters of closely related protease genes exist. Tissue kallikreins cluster on chromosome 19q13.3–q13.4; cell-mediated immune system proteases cluster on chromosomes 19p13.3 and 14q11.2; and trypsins cluster on chromosome 7q35. Closely related genes often but do not always cluster: factors VII and X are adjacent on chromosome 13q34, but the other clotting proteases are located on different chromosomes. Several pseudogenes have also been identified for this clan. There are two pseudogenes and one relic gene for trypsin. There are pseudogenes for elastase-I, mannose-binding protein-associated serine protease-1 and complement factor B. The structure of genes encoding family S1 varies. Proteases with auxiliary modules utilize one to two additional, discrete exons to encode these domains, suggesting a role for exon shuffling in the evolution of those proteases. The protease domain is encoded by as

Table 1 Physiologic functions and sites of expression of clan PA serine proteases

Physiologic function	Protease(s)	Site of expression (adult)
Blood coagulation	Factor VII	Liver
	Factor IX	Liver
	Factor X	Liver
	Factor XI	Liver
	Factor XII	Liver
	Plasma kallikrein	Liver
	Protein C	Liver
	Thrombin	Liver
Cell-mediated immunity	Azurocidin[a]	Neutrophils
	Cathepsin G	Neutrophils
	Chymase	Mast cells
	Granzymes	T lymphocytes
	Myeloblastin	Neutrophils
	Neutrophil elastase	Neutrophils
	Tryptases	Mast cells
Complement	Factor B	Liver
	Factor C1r	Liver
	Factor C1s	Liver
	Factor C2	Liver
	Factor D	Adipocytes, macrophage lineages
	Factor I	Liver
	Haptoglobin[a]	Liver
	Mannose-binding protein-associated serine proteases	Liver
Digestion	Chymotrypsin	Pancreas
	Elastases	Pancreas
	Enterokinase	Proximal small intestine
	Trypsins	Pancreas
Fertilization	Acrosin	Sperm
	Prostate-specific antigen	Prostate
Fibrinolysis	Plasmin	Liver
	Tissue plasminogen activator	Endothelium
	Urokinase	Kidney
Growth stimulation and development	Hepatocyte growth factor[a]	Liver
	Hepatocyte growth factor activator	Liver
	Hepsin	Liver
	Neuropsin	Brain
Kallikrein	Tissue kallikreins	Pancreas, kidney
Tissue remodeling	Apolipoprotein A[a]	Liver
	Stratum corneum chymotryptic enzyme	Skin
Unknown function	Neurotrypsin	Brain
	Osteoblast serine protease	Bone
	Pancreatic endopeptidase E	Pancreas
	Prostasin	Prostate
	Protease M	Breast
	Transmembrane protease 2	Small intestine

[a]Mechanism of function is not proteolytic.

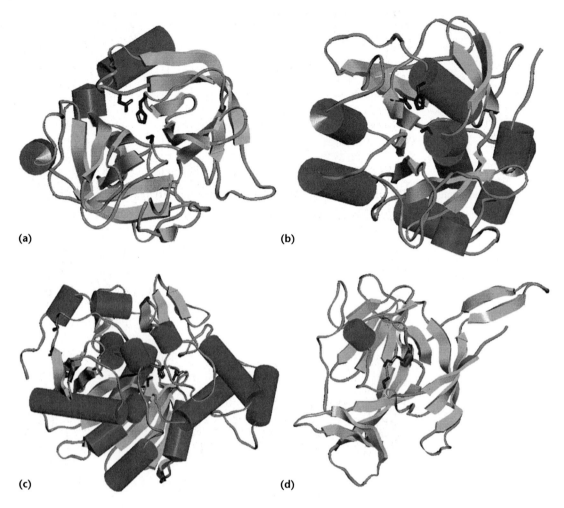

Figure 1 [*Figure is also reproduced in color section.*] Three-dimensional structural representations of serine proteases from clans PA, SB, SC and SF. β-Sheets are represented by broad ribbons and α-helices are represented by cylinders. Catalytic residues are shown in stick representation. Coordinates are from Protein Data Bank entries, indicated in parentheses after each enzyme name. (a) Human chymotrypsin (4CHA) of clan PA. (b) *Bacillus amyloliquefaciens* subtilisin BPN′ (2ST1) of clan SB. (c) Human lysosomal carboxypeptidase A (1IVY) of clan SC. (d) *Escherichia coli* signal peptidase (1B12) of clan SF. Coordinates for human representatives of clans SB and SF were not available at the time of writing.

many as seven exons, but complement factors C1r and C1s utilize just one exon. The final exon of the protease domain shows the least variation and has nearly invariant boundaries; it encodes the 50 C-terminal residues that are crucial for determining specificity and function.

Proteases of family S1 are expressed during development and adulthood. Thrombin, tissue plasminogen activator, urokinase and neuropsin are expressed in developing neural tissues, and other proteases of clan SA are likely to play crucial roles in embryonic development. Adult serine protease expression patterns are summarized in **Table 1**. All of the known family S1 proteases, with the possible exception of hepsin, localize to the extracellular space.

Serine protease activity is regulated on several levels. Myeloblastin is regulated transcriptionally by

a PU.1 promoter sequence. Transcription of chymase and prostate-specific antigen is induced in response to cytokines and steroid hormones respectively. Posttranslational proteolysis is more widely utilized both to activate and deactivate enzymes, particularly among the digestive and clotting proteases; enzymes such as chymotrypsin and factor XII are regulated by autolysis. Enzyme activity is also regulated by the use of serpin and Kunitz-type inhibitors and required cofactors. One additional means of regulation is by environmental pH: chymase is stored in acidic granules and becomes active when released into the neutral extracellular space.

Absence, mutation or impaired regulation of numerous clan SA proteases leads to disease. Tissue damage in pancreatitis, rheumatoid arthritis and emphysema result from the improper release, lack of

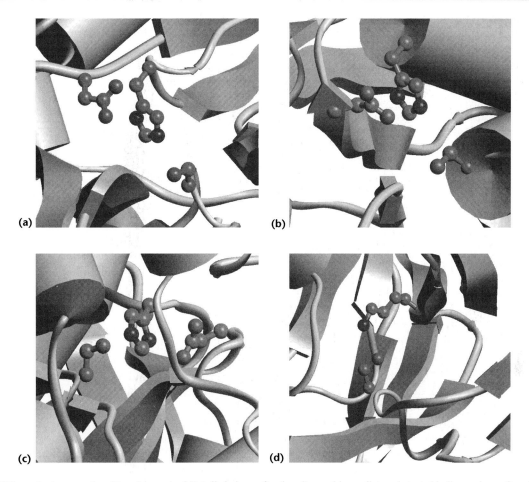

Figure 2 [*Figure is also reproduced in color section.*] Detailed views of active sites and immediate substrate-binding regions of protease structures in **Figure 1**. Catalytic residues are shown in ball-and-stick representation. It can be seen that the orientation of the catalytic triad for clan SC (c) is the mirror-image reverse of that seen for clans PA (a) and SB (b). The Lys seen in the *Escherichia coli* signal peptidase structure would be replaced by His in eukaryotic homologs, which would also have an Asp residue in the catalytic site.

inhibition, or resistance to degradation of digestive enzymes like trypsin and elastase. Mutation of factor IX results in the sex-linked hemophilia B, and deficiency of factor VII results in severe bleeding; excess tissue plasminogen activator can also lead to severe hemorrhage. Tissue kallikrein overexpression is implicated in hypertension, and the kallikrein/kinin system contributes to septic shock and hereditary angioedema. Several serine proteases have roles in autoimmune disease. Myeloblastin is the target of antineutrophil autoantibodies in Wegener granulomatosis, and deficiencies of complement factors C1r, C1s and C2 lead to lupus or lupus-like symptoms. Clan SA serine proteases also contribute to neoplasia. Prostate-specific antigen is a marker for and may stimulate invasiveness of prostatic carcinoma, and hepsin stimulates growth of hepatoma cells. In the fetus, presence of the factor B pseudogene leads to congenital adrenal hyperplasia.

Clan SB: Subtilisin-like Serine Proteases

Clan SB contains protease families in which the order of the catalytic triad is Asp-His-Ser. Eight human proteases are currently classified in family S8, known as the subtilisin family. The human subtilisin-type serine proteases are intracellular or transmembrane proteins primarily concerned with protein processing. The proprotein convertases process propeptides targeted for secretion, including von Willebrand factor and pituitary peptide hormones. Tripeptidyl-peptidase II aids in intracellular protein recycling and also degrades a variety of neuropeptides *in vitro*. Several subtilisins have multidomain structures, with a propeptide upstream of the protease domain and auxiliary modules downstream of the protease domain. The most commonly used modules are

P-domains, cysteine-rich domains and transmembrane domains.

The protease domain is a mixture of α-helix and β-sheet (Wright *et al.*, 1969). The central core of the domain contains a seven-stranded parallel β-sheet surrounded by nine α-helices; a two-stranded antiparallel β-sheet is also located near the *C*-terminus of the catalytic domain (**Figure 1**). The catalytic residues are adjacent to the central β-sheet near the surface of the molecule. Most subtilisin-type proteases prefer substrates with large, hydrophobic groups immediately upstream of the scissile bond, but the proprotein convertases cleave after dibasic sequences (such as Lys-Lys) because their substrate-binding clefts are lined with acidic residues (Perona and Craik, 1995). Substrate specificity in subtilisins is determined largely by the nature of the residues contacting the substrate; however, most subtilisins show broad specificity profiles and are relatively easily engineered with new specificities, unlike the proteases of the chymotrypsin-like family.

Members of family S8 are scattered throughout the genome, except for furin and paired basic amino acid cleaving enzyme (PACE4), which are located together on chromosome 15q25–26. No pseudogenes are known. Those proteases of which the genome structures have been determined, proprotein convertase (PC) 2 and furin, have 12 and 14 exons respectively.

Proteases of clan SB are expressed during development and adulthood. Furin is expressed widely in both embryonic and adult tissues. PC7 is expressed widely during development, but the adult has only lymphoid expression. PACE4 is also widely expressed during development, and is expressed primarily in endocrine and central nervous system cells in adults. PC5 is expressed early in development and in the adult adrenals, gut, Sertoli cells and endothelium. PC1 and PC2 are expressed in adult endocrine and neural cells. Tripeptidyl-peptidase II has been localized to the adult liver and brain.

The proprotein convertases are activated by proteolytic cleavage of the propeptide *N*-terminal to the protease domain. Furin expression is transcriptionally governed by two housekeeping (GC-rich) promoters and one regulated (TATA-containing) promoter. PC1 and PC2 are upregulated by dopamine antagonists and changes in thyroid status, although the mechanism is not established.

Overexpression of subtilisin-type proteases has been implicated in neoplasia. Furin and PACE4 are expressed in tumor cell lines and may activate matrix metalloproteases, aiding in tumor invasion and progression. PC7 is involved in the t(11;14) translocation in high-grade lymphoma. Absence of PC1 has been shown to cause obesity and diabetes.

Clan SC: Serine Proteases with the α/β Hydrolase Fold

Clan SC contains protease families in which the order of the catalytic triad is Ser-Asp-His. Human sequences are known in families S9, S10 and S28. Peptidases of this clan share a common fold with the acetylcholinesterases, lipases and dehalogenases. The eight known proteases of this clan have been localized to the extracellular space, cell membrane and intracellular space. Most participate in the regulation of peptide hormones, with the exception of fibroblast activation protein α (FAP), which is believed to function in tissue remodeling. Several peptidases of this clan are multidomain proteins with cytoplasmic, transmembrane and cysteine-rich domains placed *N*-terminal to the catalytic domain.

Protease domains possess the α/β hydrolase fold (Ollis *et al.*, 1992), which features mostly parallel β-strands connected by α-helices on either surface of the sheet (**Figure 1**). The active site residues reside near the *C*-terminus of the β-sheet and are located at turns between β-strands and α-helices. The spatial orientation of the catalytic triad is a mirror image of that seen in clan SA. Substrate specificity is governed by variations in the length, charge and steric properties of the surface loops formed by the α-helices that link the β-strands. Recognition of different classes of substrates by nonprotease enzymes that share the α/β hydrolase fold is governed by minor alterations in that fold, most notably by the amount of curvature of the central β-sheet.

Members of clan SC are scattered throughout the genome, with the exception of dipeptidyl-peptidase IV and FAP, which are located together on chromosome 2q23–24. No pseudogenes are known. The gene structures of three clan SC members have been determined. From six to nine exons encode the catalytic domain, and in prolyl oligopeptidase the catalytic domain is encoded by two nonadjacent sets of exons (1–3 and 10–15).

Several of the peptidases of clan SC are expressed in nearly all adult tissues, including prolyl oligopeptidase, dipeptidyl-peptidase IV, acylaminoacyl-peptidase, lysosomal carboxypeptidase A and lysosomal Pro-X carboxypeptidase. Of these, all but acylaminoacyl-peptidase show enrichment in the kidney and liver. FAP is expressed in fetal mesenchymal cells but in adults is only expressed in A cells of pancreatic islets, sarcomas and neuroectodermal tumors.

Regulatory strategies vary depending on localization and function. Lysosomal Pro-X carboxypeptidase is activated by proteolysis and has optimum activity under acidic conditions. Lysosomal carboxypeptidase

A acts as a carboxypeptidase at pH 5.0, but is a deamidase and esterase at pH 7.0. FAP, on the other hand, is one of several proteins upregulated as part of a general fibroblast activation program.

Clan SC has not been shown to play an extensive role in disease. Fibroblast activation protein is expressed in tumor stromal fibroblasts, but a pathologic role has not been established. Mutation of lysosomal carboxypeptidase A leads to the lysosomal storage disease galactosialidosis, but loss of catalytic activity is not responsible.

Clan SF: Signal Peptidases

Clan SF contains protease families with a catalytic triad in the sequence Ser-His-Asp (eukaryotes) or a dyad in the sequence Ser-Lys/His (prokaryotes). The human mitochondrial and endoplasmic reticulum (ER) microsomal signal peptidases are found in family S26 and are responsible for cleavage of signal peptides from proteins targeted to the mitochondria (mitochondrial signal peptidase) or the secretory pathway (ER signal peptidase).

The human ER signal peptidase consists of five subunits, all of which are transmembrane proteins (Dalbey *et al.*, 1997). The two catalytic subunits have similar membrane topologies: a cytoplasmic *N*-terminus, a transmembrane domain, and the catalytic domain and *C*-terminus in the lumen. The mitochondrial signal peptidase consists of two subunits, both of which are catalytic and are similar to the ER catalytic subunits in topology. The eukaryotic signal peptidases are distinguished from prokaryotic variants by use of the Ser-His-Asp triad as opposed to a Ser-Lys dyad (VanValkenburgh *et al.*, 1999). The most recent three-dimensional structure available for this family is of the *Escherichia coli* leader peptidase (Paetzel *et al.*, 1998). The structure is primarily β-sheet, with the active site residues on the surface of the protein positioned at the end of a hydrophobic substrate-binding groove (**Figure 1**). Substrate specificities of the signal peptidases appear to be evolutionarily conserved, with Ala preferred upstream of the scissile bond.

The structure and localization of the human signal peptidase genes has not been determined. Tissue expression of the proteins is presumed to be universal, but may be differentially regulated in specific tissues. Regulation of human signal peptidases may include autocatalytic deactivation.

Mutations of the ER signal peptidase that cause inefficient processing of the prevasopressin signal peptide result in central diabetes insipidus. Likewise, mutations in signal peptides of particular secreted proteins also cause disease phenotypes.

See also

α₁-Antitrypsin (AAT) Deficiency
Proteases and Human Disorders
Proteases: Evolution
Serpins: Evolution

References

Barrett AJ and Rawlings ND (1995) Families and clans of serine peptidases. *Archives of Biochemistry and Biophysics* **318**: 247–250.

Dalbey RE, Lively MO, Bron S and van Dijl JM (1997) The chemistry and enzymology of the type I signal peptidases. *Protein Science* **6**: 1129–1138.

Krem MM, Rose T and Di Cera E (1999) The C-terminal sequence encodes function in serine proteases. *Journal of Biological Chemistry* **274**: 28063–28066.

Lesk AM and Fordham WD (1996) Conservation and variability in the structures of serine proteases of the chymotrypsin family. *Journal of Molecular Biology* **258**: 501–537.

Ollis DL, Cheah E, Cygler M, *et al.* (1992) The α/β hydrolase fold. *Protein Engineering* **5**: 197–211.

Paetzel M, Dalbey RE and Strynadka NCJ (1998) Crystal structure of a bacterial signal peptidase in complex with a β-lactam inhibitor. *Nature* **396**: 186–190.

Perona JJ and Craik CS (1995) Structural basis of substrate specificity in serine proteases. *Protein Science* **4**: 337–360.

VanValkenburgh C, Chen XM, Mullins C, Fang H and Green N (1999) The catalytic mechanism of endoplasmic reticulum signal peptidase appears to be distinct from most eubacterial signal peptidases. *Journal of Biological Chemistry* **274**: 11519–11525.

Wright CS, Alden RA and Kraut J (1969) Structure of subtilisin BPN' at 2.5 Å resolution. *Nature* **221**: 235–242.

Further Reading

Barrett AJ, Rawlings ND and Woessner JF (eds.) (1998) *Handbook of Proteolytic Enzymes.* San Diego, CA: Academic Press.

Diaz-Lazcoz Y, Hénaut A, Vigier P and Risler JL (1995) Differential codon usage for conserved amino-acids: evidence that the serine codons TCN were primordial. *Journal of Molecular Biology* **250**: 123–127.

Gorbalenya AE, Donchenko AP, Blinov VM and Koonin EV (1989) Cysteine proteases of positive strand RNA viruses and chymotrypsin-like serine proteases: a distinct protein superfamily with a common structural fold. *FEBS Letters* **243**: 103–114.

Krem MM and Di Cera E (2001) Molecular markers of serine protease evolution. *EMBO Journal* **20**: 3036–3045.

Krem MM and Di Cera E (2002) Evolution of enzyme cascades from embryonic development to blood coagulation. *Trends in Biochemical Sciences* **26**: 67–74.

Perona JJ and Craik CS (1997) Evolutionary divergence of substrate specificity within the chymotrypsin-like serine protease fold. *Journal of Biological Chemistry* **272**: 29 987–29 990.

Rawlings ND and Barrett AJ (1993) Evolutionary families of peptidases. *Biochemical Journal* **290**: 205–218.

Seidah NG, Chrétien M and Day R (1994) The family of subtilisin/kexin like pro-protein and pro-hormone convertases: divergent or shared functions. *Biochimie* **76**: 197–209.

Siezen RJ and Leunissen JAM (1997) Subtilases: the superfamily of subtilisin-like serine proteases. *Protein Science* **6**: 501–523.

Web Links

Protein Data Bank (formerly Brookhaven Protein Data Bank)
http://www.rcsb.org/pdb/

Serpins: Evolution

Intermediate article

Gary A Silverman, *Children's Hospital, Harvard Medical School, Boston, Massachusetts, USA*

David J Askew, *Children's Hospital, Harvard Medical School, Boston, Massachusetts, USA*

James A Irving, *Monash University, Melbourne, Victoria, Australia*

Cliff J Luke, *Children's Hospital, Harvard Medical School, Boston, Massachusetts, USA*

Dion Kaiserman, *Monash University, Melbourne, Victoria, Australia*

Phillip I Bird, *Monash University, Melbourne, Victoria, Australia*

James C Whisstock, *Monash University, Melbourne, Victoria, Australia*

The serpin superfamily of serine (and cysteine) proteinase inhibitors contains over 750 members represented in Bacteria, Archaea and Eukarya. All members of the serpin family share a unique tertiary structure that is critical to both inhibitory and noninhibitory functions. Serpins are involved in a vast array of biological processes, including the regulation of thrombosis, fibrinolysis and inflammation, and host defense.

Serpin Evolution

Members of the serine proteinase inhibitor (serpin) superfamily possess a unique tertiary structure and mechanism of inhibition that distinguishes them from all other classes of active site inhibitors (Silverman *et al.*, 2001). The presence of serpin genes in plants and metazoa, and their absence from *Saccharomyces cerevisiae* and those eubacterial species initially sequenced, led to the notion that serpins appeared sometime before the divergence of plants and animals, roughly 2.7 billion years ago (Ga). However, the increasing amounts of species-specific genomic sequencing data provide additional opportunities to evaluate the size and extent of the serpin superfamily. BLAST (Basic Local Alignment Search Tool) analysis and molecular cloning studies have revealed that serpin family members with a bonafide reactive site loops are more widely distributed then originally suspected (Irving *et al.*, 2000, 2002). Serpin genes are present in all three kingdoms of life, (eu)Bacteria, Archaea and Eukarya. Assuming nonhorizontal gene transfer, this distribution places the emergence of an ancestral serpin gene somewhere between the origins of cellular life (roughly 4.1 Ga) and the divergence between Archaea and Eukarya (about 3.9 Ga). Regardless of the exact timing, these observations suggest that serpins are part of a very ancient antiproteinase defense system. (*See* Ovalbumin Serpins.)

There are now well over 750 serpin sequences deposited in the sequence databases. Phylogenetic analysis reveals strong statistical support for the existence of 16 clades (A–P; **Figure 1**), with humans having representatives in the first 9 (A–I). Many of the remaining serpins are orphans, but these should form additional clades as additional serpin sequences are identified.

Human Serpins

The most recent analysis of the human genome reveals 36 serpin genes (including one pseudogene) that are divided into nine clades (Irving *et al.*, 2000). Clade A (α1-antrypsin) contains 11 members. Eight of these serpin genes map to a cluster at 14q32.1. The others map to Xq22.2 (*SERPINA7*, encoding thyroxine-binding globulin) and 1q42 (*AGT* (previously known as *SERPINA8*), encoding angiotensinogen). All of the serpins of clade a contain typical signal sequences and are secreted into the circulation or the extracellular milieu. *SERPINA1*, *3*, *4*, *5* and *10* all contain functional reactive site loops and inhibit trypsin or chymotrypsin-like proteinases involved in the thrombotic, fibrinolytic or inflammatory cascades, as well as white blood cell granule proteinases. *SERPINA2* and *SERPINA9* harbor what appear to be functional reactive site loops, but no target proteinases have been reported. Two of the clade A serpins (*SERPINA6,7*) are transporters of thyroid and corticosteroid hormones, respectively, whereas an amino (*N*)-terminal proteolytic fragment of AGT increases vascular tone.

The clade B (ov-) serpins comprise the largest group, with 13 members. These serpins cluster into two groups, located at 18q21.3 and 6p25, and are described elsewhere. (*See* Ovalbumin Serpins.)

Currently, both clades C and D contain a single member. *SERPINC1* (encoding antithrombin III) and *SERPIND1* (encoding heparin cofactor II) map to 1q23–25 and 22q21 respectively. SERPINC1 and D1 are regulators of the clotting cascade and neutralize both thrombin and factor Xa.

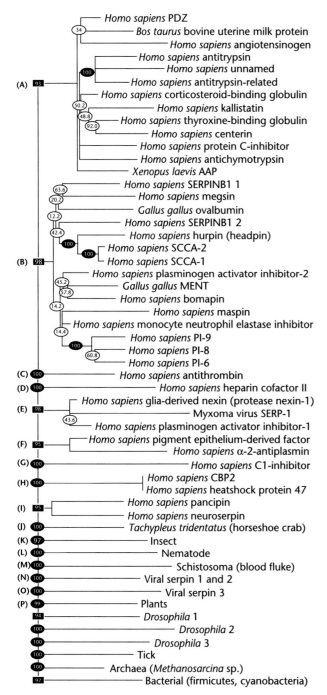

The clade E serpins, *SERPINE1* (encoding plasminogen activator inhibitor type-1) and *SERPINE2* (encoding proteinase nexin 1), map to 7q21.3 and 2q33 respectively. Both of these serpins neutralize thrombin, plasmin, urokinase-type plasminogen activator and tissue-type plasminogen activator.

The two clade F serpins map to 17p12–13. *SERPINF1* (encoding a pigment epithelium-derived factor) has lost its inhibitory function, but serves as both an antiangiogenic and neurotrophic factor. *SERPINF2* (encoding α2-antiplasmin) inactivates plasmin and is an important regulator of the fibrinolytic system.

SERPING1 (encoding C1-esterase inhibitor), which maps to 11q11–q13, is the only member of clade G. SERPING1 regulates inflammatory cascades by neutralizing C1 esterase and kallikrein.

The clade H serpins, *SERPINH1* and *SERPINH2*, map to 11p15 and 11q13.5 respectively. The structure of the reactive site loops of both serpins suggests that they have lost their inhibitory functions. However, both proteins bind collagen and may serve as chaperones. SERPINH1 activity is increased in the heatshock response, but its role in this process is unknown.

Both clade I serpins map to 3p26. *SERPINI1* (encoding neuroserpin) inhibits the plasminogen activators and plasmin. *SERPINI1* has a neuroprotective role in stroke. *SERPINI2* (encoding pancipin) has no known proteinase targets, but can function as an inhibitor of cell motility.

Human serpin mutations and disease

Patients with mutations in either the regulatory elements or coding sequence of serpin genes may present with an abnormal clinical phenotype (**Table 1**). For example, the PiZZ mutation of *SERPINA1* (α1-antitrypsin) facilitates the generation of loop-sheet polymers (Stein and Carrell, 1995). These polymers accumulate in the endoplasmic reticulum of hepatocytes and impair secretion of serpin monomer into the circulation. Consequently, the pulmonary parenchyma is left partially unprotected from a lifelong exposure to neutrophil elastase activity, and hepatocytes die by stress-induced apoptosis. These events result in emphysema and cirrhosis respectively. Mutations of *SERPINI1* (neuroserpin) also lead to polymerization, neuronal cell death and a familial form of presenile dementia.

A mutation of the reactive site loop P1, from a methionine to an arginine, converts *SERPINA1* to an effective inhibitor of thrombin. This gain-of-function mutation (Pittsburgh mutation) results in a lethal bleeding diathesis.

Haploid insufficiency associated with *SERPINC1* and *SERPING1* results in a propensity toward

Figure 1 Multifurcating phylogenic tree showing the relationship between 35 human serpins and other members of the serpin superfamily. The tree was constructed using previously described methods (Irving *et al.*, 2000). Conventional bootstrap values derived from maximum parsimony trees are highlighted by ovals. Hexagons indicate clades identified using the strict consensus method, and rectangles highlight clades identified using the comparison method. Each major clade is labeled A–P, consistent with the nomenclature described (Silverman *et al.*, 2001).

Table 1 Serpins associated with human disease

Gene	Mutation	Disease	Mode of inheritance
SERPINA1 (α1-AT)	ZZ mutation proximal hinge	Cirrhosis, Emphysema	Autosomal recessive
SERPINA1 (α1-AT)	Pittsburgh mutation P1 position	Bleeding	Autosomal dominant
SERPINA3 (α1-ACT)	Multiple	Emphysema	Autosomal recessive
SERPINC1 (ATIII)	Multiple	Venous thrombosis	Autosomal dominant
SERPING1 (C1-inh)	Multiple	Angioedema	Autosomal dominant
SERPINI2 (neuroserpin)	Helix B	Presenile dementia	Autosomal dominant

thrombosis and acute episodes of angioedema respectively. Both of these potentially life-threatening disorders can be corrected by serpin replacement therapy.

Mouse Serpins

In addition to studying the evolution and phylogeny of serpin family members, comparative genomic analysis can be used to identify paralogous and orthologous genes. Comparative genomic studies in humans and other mammals led to the identification of multiple serpin genes that can be sorted into different clades. While certain clade members seem to be conserved between different mammalian species, enormous variability, probably secondary to local duplication events and different selection pressures, have endowed each species with a unique serpin repertoire. For example, the clade A serpins on human chromosome 14 and the clade B serpins on human chromosomes 6 and 18 show varying degrees of gene amplification in the syntenic clusters located on mouse chromosomes 12, 13 and 1 respectively (**Figure 2**). While some of these amplicons contain pseudogenes, there is clear evidence that they also contain increased numbers of functional serpin genes. Why the mouse should have an absolute increase in the number of functional serpin genes relative to those in humans is not known, but may reflect its exposure to a more diverse array of endogenous and/or exogenous proteinase targets.

Targeted deletions of mouse serpins

Targeted deletions of serpin genes in mice are providing new insights into the biological function of human orthologs (**Table 2**). Homozygous loss of Serpinb5 (encoding maspin), Serpinc1 (encoding antithrombin III; Ishiguro et al., 2000), Serpinh1 (also known as Hsp47; Nagai et al., 2000), and possibly

Serpina8 (encoding angiotensinogen; Kim et al., 1995) results in embryonic lethality. Homozygous loss of some serpin genes also leads to structural abnormalities. Animals lacking Serpina5 (encoding a protein C inhibitor) show loss of Sertoli cells and abnormal spermatogenesis (Uhrin et al., 2000). Loss of Serpina8 results in cortical thinning and atrophy of the kidneys as well as arteriolar thickening. The absence of Serpinh1, a chaperone and collagen binding protein, results in abnormal basement membranes, a loss in epithelial cell polarity, a decrease in the number of somites and a delay in neural tube closure. Deficiencies of Serpinc1 and Serpinf2 (encoding α2-antiplasmin) result in massive bleeding and enhanced fibrinolytic potential respectively (Lijnen et al., 1999). Animals with homozygous or heterozygous loss of Serping1 (encoding C1 esterase inhibitor) show increased vascular permeability (Han et al., 2002). Moreover, this effect is eliminated when these mice are crossed with animals lacking the bradykinin type 2 receptor. These data suggest that hereditary angioedema is mediated by the bradykinin (C1 esterase inhibitor also neutralizes kallikrein, which converts kininogen to the vasodilator, bradykinin) system rather than the inhibition of early complement components. Homozygous loss of Serpind1 (encoding heparin cofactor II) results in an increased propensity toward arterial occlusion after endothelial injury (He et al., 2002). Loss of Serpine1 (encoding plasminogen activator inhibitor type 1) and Serpine2 (encoding proteinase nexin 1) leads to a mild hyperfibrinolytic state and increased susceptibility to seizures respectively (Carmeliet et al., 1993; Luthi et al., 1997). Collectively, targeted deletions of mouse orthologs of human genes are yielding valuable insights into the involvement of serpin in development and complex physiologic processes. Crosses between different mutant strains should yield even more information as compensatory overlaps between serpins are eliminated.

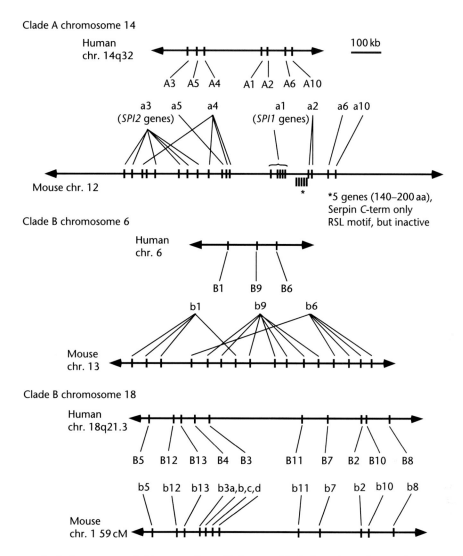

Figure 2 Three mouse serpin clusters are expanded in comparison with humans. Clade A serpins on human chromosome 14 and the two clade B clusters on chromosomes 6 and 18 (top maps) are compared with their mouse counterparts (bottom maps). Synteny is well conserved with expansion of the number of mouse genes in certain regions. C-term only refers to gene fragments that encode for the terminal 200 amino acids only. Although these fragments have an RSL, the absence of N-terminal sequences precludes the synthesis of an active inhibitor. aa: amino acids; cM: centimorgans; kb: kilobases; RSL: reactive site loop.

Prokaryotic Serpins

Recent BLAST analysis of draft or completely sequenced microbial genomes reveals the presence of seven bacterial and five archaeal serpins (**Table 3**; Irving *et al.*, 2002). These serpins show 23–29% similarity with SERPINA1 and contain 38–47 of the 51 conserved residues that are present in more than 70% of all serpin family members. Moreover, all contain a functional hinge region, which predicts that they serve as proteinase inhibitors. Based on the presence of hydrophobic residues (**Table 3**) in the canonical P3–P1 positions, nine of these serpins could

inhibit papain-like cysteine proteinases. Interestingly, no serpins have been detected in the genomes of common human pathogens such as *Pseudomonas aeruginosa*, *Staphylococcus aureus*, *Haemophilus influenzae* and *Neisseria meningitidis*. The appearance of serpins in the genomes of several bacterial species, but not in others, raises questions regarding the origins of these prokaryotic genes. Did some bacterial species acquire eukaryotic serpin genes by horizontal gene transfer, or were serpins present in primordial cellular organisms but selectively lost in certain prokaryotic lineages? By scanning a broader array of prokaryotic genomes for evidence of serpin genes or motifs, by

Table 2 Targeted deletions of mouse orthologs of human serpin genes

Human gene	Phenotype		
	Embryonic lethality	Structural anomalies	Physiological abnormalities
SERPINA5 (PCI)	No	Destruction of Sertoli cell barrier in the testes	Male sterility, abnormal spermatogenesis with malformed sperm
SERPINA8 (ANG)	No, but most die by weaning in one study, not in another	Kidneys show cortical thinning, atrophy and arteriolar thickening	Hypotension, impaired blood–brain barrier function in response to cold injury
SERPINB2 (PAI2)	No	No	None to date
SERPINB3/4 (SCCA)	No	No	?
SERPINB5 (Maspin)	Yes	?	
SERPINB6	No	No	?
SERPINC1 (ATIII)	Yes by E15.5	no	Massive hemorrhage in −/−, increased tendency toward thrombosis in +/−
SERPIND1 (HCII)	No	No	Greater tendency toward arterial thrombosis after endothelial injury
SERPINE1 (PAI1)	No	No	Mild hyperfibrinolytic state; resistance to venous thrombosis but no impairment of hemostasis; loss paradoxically inhibits tumor cell invasion and angiogenesis
SERPINE2 (PN1)	No	No	Increased susceptibility to seizures induced by kainic acid (glutamate agonist), but reduced θ-burst-induced LTP and NMDA receptor-mediated synaptic transmission
SERPINF2 (α2AP)	No	No	Enhanced fibrinolytic potential without overt bleeding
SERPING1 (C1 inh)	No	No	Increased vascular permeability in +/− and −/−
SERPINH1 (HSP47)	Yes by E11.5	Abnormal triple helix formation for several types of collagen (I, IV) leads to disruption of basement membranes, disoriented epithelial cells and ruptured blood vessels; decreased number of somites; delayed neural tube closure	

determining the GC content of serpin genes relative to that of eukaryotes, and by examining how serpin genes are organized within the prokaryotic genome itself (e.g. are serpins part of conserved operons, or are they clustered into islands with the same types of genes?), further insight into the origins of this gene family should be ascertained.

Serpins are metastable in their active conformation and tend to polymerize or denature when incubated at temperatures approaching 65°C. Thus, the presence of an apparently inhibitory-type serpin in the hyperthermophile, *Pyrobaculum aerophilum* (optimal growth at 100°C), must signal the existence of an adaptation that enhances the structural integrity of the serpin backbone without affecting the nimbleness needed to trap its target proteinases. The existence of a disulfide bond between the cysteine residues located at the P1′ position and on strand 3 of β-sheet C might be that adaptation. This linkage could stabilize the serpin backbone without impairing reactive site loop insertion. Structural analysis of the active and cleaved form of this serpin should help determine whether this linkage occurs and whether this is the mechanism whereby a serpin can function at high temperatures.

Serpins in Other Species

Comparative genomic analysis should help elucidate the biological functions of newly discovered as well as previously characterized serpins. For example, the genomes of two Poxviridae genera, *Orthopoxvirus* (variola, vaccinia and rabbitpox) and *Leporipoxvirus* (myxoma), each encode three serpin genes. While none of these serpins is essential for viral growth in culture, they are required for the full host range of infection. SPI-2/CrmA is particularly important, as this serpin may enhance the infectious process by blocking the host's apoptotic pathways or suppressing the

Table 3 Prokaryotic serpins

Species	Reactive site loop					
	P4	P3	P2	P1	P1'	P2'
Bacteria						
Dehalococcoides ethenogenes	Met	Asn	Leu[a]	Thr	Ser	Ala
Anabaena sp. PCC 7120	Met	Val[a]	Ala	Thr	Ser	Leu
Nostoc punctiforme	Ile	Val[a]	Ala	Thr	Ser	Leu
Desulfitobacterium hafniense	Val	Asn	Thr	Thr	Ser	Met
Thermoanaerobacter tengcongensis	Ile	Thr	Ala	Ala	Gly	Ile
Thermobifida fusca	Met	Leu[a]	Leu[a]	Ala	Gly	Ala
Ruminococcus albus	Met	Ile[a]	Thr	Glu	Ala	Ala
Archaea						
Methanosarcina acetivorans 1	Ile	Leu[a]	Glu	Glu	Glu	Ile
Methanosarcina acetivorans 3	Met	Ala	Met	Gly	Val	Ser
Methanosarcina mazei	Met	Thr	Val[a]	Gly	Met	Asp
Methanosarcina acetivorans 2 (possible pseudogene)	Ile	Gly	Ser	Val[a]	Ser	Ser
Pyrobaculum aerophilum	Phe	Lys	Pro	Val[a]	Cys	Ala

[a]Hydrophobic amino acids, which are preferred P2 residues for papain-like cysteine proteinases.

inflammatory response. This observation led to the discovery that serpins can serve as cross-class inhibitors by neutralizing cysteine proteinases of the caspase family.

Model organisms with well-established experimental systems, more tractable to genetic manipulations (e.g. reverse genetics, suppressor screens) than those of higher vertebrates, promise to provide new insights into the role of serpin genes. The genome of *Drosophila melanogaster* contains at least 22 serpin genes. The majority ($n = 18$) of these genes seem to encode inhibitory-type serpins. A loss-of-function mutation in one of the genes (*Spn43Ac*) leads to constitutive activation of the Toll pathway and production of the antifungal peptide drosomycin (Levashina *et al.*, 1999). Spn43Ac functions in the proximal portion of the signaling pathway by regulating the proteolytic processing of the Toll ligand spaetzle. This is the first example of a serpin that participates in the regulation of an innate immune response.

Examination of the *Caenorhabditis elegans* genome has revealed the presence of at least nine serpin genes on chromosome V. Recombinant *srp-1*, *srp-2*, *srp-3*, *srp-6* and *srp-7* encode for true proteinase inhibitors. *Srp-5*, *srp-8*, *srp-9* and *srp-10* may be noninhibitory serpins or pseudogenes. Preliminary analysis of nematodes with serpin gene mutations suggests that this organism should provide additional insights into the role of serpins in innate immunity and homeostatic processes.

Intra- and interspecies genomic comparisons also can be used to identify conserved genomic elements or

locus control regions that coordinate the expression of individual serpin genes or serpin gene clusters respectively. Finally, comparative genomic studies between diverse unicellular organisms such as cyanobacteria and multicellular invertebrates such as *Cynea capillata* should provide critical insights into the role of serpins in cellular metabolism, as well as identifying those structural elements that are crucial for inhibitory and noninhibitory functions.

See also

α₁-Antitrypsin (AAT) Deficiency
Ovalbumin Serpins

References

Carmeliet P, Stassen JM, Schoonjans L, *et al.* (1993) Plasminogen activator inhibitor-1 gene-deficient mice. II. Effects on hemostasis, thrombosis, and thrombolysis. *Journal of Clinical Investigation* **92**: 2756–2760.

Han ED, MacFarlane RC, Mulligan AN, Scafidi J and Davis AE (2002) Increased vascular permeability in C1 inhibitor-deficient mice mediated by the bradykinin type 2 receptor. *Journal of Clinical Investigation* **109**: 1057–1063.

He L, Vicente CP, Westrick RJ, Eitzman DT and Tollefsen DM (2002) Heparin cofactor II inhibits arterial thrombosis after endothelial injury. *Journal of Clinical Investigation* **109**: 213–219.

Irving JA, Pike RN, Lesk AM and Whisstock JC (2000) Phylogeny of the serpin superfamily: implications of patterns of amino acid conservation for structure and function. *Genome Research* **10**: 1845–1864.

Irving JA, Steenbakkers PJM, Lesk AM, *et al.* (2002) Serpins in prokaryotes. *Journal of Molecular Biology* **19**: 1881–1890.

Ishiguro K, Kojima T, Kadomatsu K, *et al.* (2000) Complete antithrombin deficiency in mice results in embryonic lethality. *The Journal of Clinical Investigation* **106**: 873–878.

Kim HS, Krege JH, Kluckman KD, *et al.* (1995) Genetic control of blood pressure and the angiotensinogen locus. *Proceedings of the National Academy of Sciences of the United States of America* **92**: 2735–2739.

Levashina EA, Langley E, Green C, *et al.* (1999) Constitutive activation of toll-mediated antifungal defense in serpin-deficient *Drosophila. Science* **285**: 1917–1919.

Lijnen HR, Okada K, Matsuo O, Collen D and Dewerchin M (1999) α2-Antiplasmin gene deficiency in mice is associated with enhanced fibrinolytic potential without overt bleeding. *Blood* **93**: 2274–2281.

Luthi A, Putten H, Botteri FM, *et al.* (1997) Endogenous serine protease inhibitor modulates epileptic activity and hippocampal long-term potentiation. *Journal of Neuroscience* **17**: 4688–4699.

Nagai N, Hosokawa M, Itohara S, *et al.* (2000) Embryonic lethality of molecular chaperone hsp47 knockout mice is associated with defects in collagen biosynthesis. *Journal of Cell Biology* **150**: 1499–1506.

Silverman GA, Bird PI, Carrell RW, *et al.* (2001) The serpins are an expanding superfamily of structurally similar but functionally diverse proteins: evolution, mechanism of inhibition, novel functions, and a revised nomenclature. *Journal of Biological Chemistry* **276**: 33293–33296.

Stein PE and Carrell RW (1995) What do dysfunctional serpins tell us about molecular mobility and disease? *Nature Structural Biology* **2**: 96–113.

Uhrin P, Dewerchin M, Hilpert M, *et al.* (2000) Disruption of the protein C inhibitor gene results in impaired spermatogenesis and male infertility. *Journal of Clinical Investigation* **106**: 1531–1539.

Further Reading

Atchley WR, Lokot T, Wollenberg K, Dress A and Ragg H (2001) Phylogenetic analyses of amino acid variation in the serpin proteins. *Molecular Biology and Evolution* **18**: 1502–1511.

Davis RL, Shrimpton AE, Holohan PD, *et al.* (1999) Familial dementia caused by polymerization of mutant neuroserpin. *Nature* **401**: 376–379.

Gettins PGW, Patston PA and Olson ST (eds.) (1996) *Serpins: Structure, Function and Biology.* Austin, TX: RG Landes and Chapman & Hall.

Huber R and Carrell RW (1989) Implications of the three-dimensional structure of α1-antitrypsin for structure and function of serpins. *Biochemistry* **28**: 8951–8966.

Marshall CJ (1993) Evolutionary relationships among the serpins. *Philosophical Transactions of the Royal Society of London, Series B: Biological Sciences* **342**: 101–119.

Ragg H, Lokot T, Kamp PB, Atchley WR and Dress A (2001) Vertebrate serpins: construction of a conflict-free phylogeny by combining exon–intron and diagnostic site analyses. *Molecular Biology and Evolution* **18**: 577–584.

Remold-O'Donnell E (1993) The ovalbumin family of serpin proteins. *FEBS Letters* **315**: 105–108.

Web Links

serine (or cysteine) proteinase inhibitor, clade A (α-1 antiproteinase, antitrypsin), member 1 (*SERPINA1*); LocusID: 5265. Locus Link: http://www.ncbi.nlm.nih.gov/LocusLink/LocRpt.cgi?l = 5265

serine (or cysteine) proteinase inhibitor, clade C (antithrombin), member 1 (*SERPINC1*); LocusID: 462. LocusLink: http://www.ncbi.nlm.nih.gov/LocusLink/LocRpt.cgi?l = 462

serine (or cysteine) proteinase inhibitor, clade E (nexin, plasminogen activator inhibitor type 1), member 1 (*SERPINE1*); LocusID: 5054. LocusLink: http://www.ncbi.nlm.nih.gov/LocusLink/LocRpt.cgi?l = 5054

serine (or cysteine) proteinase inhibitor, clade G (C1 inhibitor), member 1 (*SERPING1*); LocusID: 710. LocusLink: http://www.ncbi.nlm.nih.gov/LocusLink/LocRpt.cgi?l = 710

serine (or cysteine) proteinase inhibitor, clade I (neuroserpin), member 1 (*SERPINI1*); LocusID: 5274. LocusLink: http://www.ncbi.nlm.nih.gov/LocusLink/LocRpt.cgi?l = 5274

serine proteinase inhibitor, clade A (*SERPINA1*); MIM number: 107400. OMIM: http://www.ncbi.nlm.nih.gov/htbin-post/Omim/dispmim?107400

serine (or cysteine) proteinase inhibitor, clade C (antithrombin), member 1 (*SERPINC1*); MIM number: 107300. OMIM: http://www.ncbi.nlm.nih.gov/htbin-post/Omim/dispmim?107300

serine (or cysteine) proteinase inhibitor, clade E (nexin, plasminogen activator inhibitor type 1), member 1 (*SERPINE1*); MIM number: 173360. OMIM: http://www.ncbi.nlm.nih.gov/htbin-post/Omim/dispmim?173360

serine (or cysteine) proteinase inhibitor, clade G (C1 inhibitor), member 1 (*SERPING1*); MIM number: 606860. OMIM: http://www.ncbi.nlm.nih.gov/htbin-post/Omim/dispmim?606860

serine (or cysteine) proteinase inhibitor, clade I (neuroserpin), member 1 (*SERPINI1*); MIM number: 602445. OMIM: http://www.ncbi.nlm.nih.gov/htbin-post/Omim/dispmim?602445

Severe Combined Immune Deficiency (SCID): Genetics

A Fischer, *Hôpital Necker, Paris, France*

Severe combined immunodeficiency consists of an array of genetically determined blocks in T lymphocyte development. Characterization of genetic defects contributes to a better understanding of lymphocyte differentiation pathways.

Intermediate article

Article contents

- Introduction
- Purine Metabolism Deficiency – Premature Apoptosis of Lymphocyte Precursors
- γc-dependent Cytokine Receptor Pathway Deficiencies
- V(D)J Recombination Deficiencies
- CD45 Deficiency
- Atypical SCID Phenotypes

Introduction

Severe combined immunodeficiency (SCID) consists of a group of rare inherited diseases characterized by a block in T-lymphocyte development. The overall frequency is estimated to one in 75 000–100 000 births (Buckley *et al.*, 1997). Eight different conditions fulfilling the definition of SCID have been described both at the molecular level and phenotype (**Figure 1**). In addition, a small number of SCID phenotypes have so far not received a molecular definition. Identification of the developmental block in T-lymphocyte differentiation provides information on basic aspects

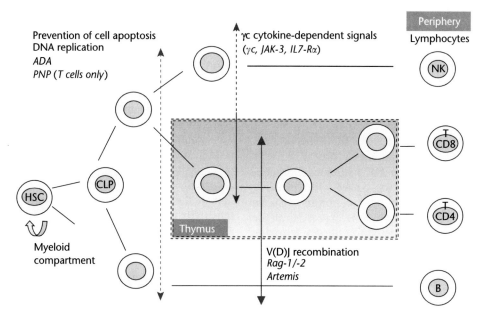

Figure 1 SCID diseases and mechanisms. HSC with self renewal capacity can generate lymphocyte precursors such as putative CLP. The latter give rise to NK, T and B lymphocytes. Arrows indicate block in lymphocyte development caused by indicated gene mutations and resulting in a SCID ADA: adenosine deaminase; PNP: purine nucleoside phosphorylase; Rag-1/-2: recombination activating gene 1/2; IL-7R: interleukin-7 receptor α chain; JAK-3: janus kinase 3; HSC: hematopoietic stem cell; CLP: common lymphoid progenitor; NK: natural killer.

of lymphocyte development as well as the medical consequences, that is, molecular diagnosis, genetic counseling and new therapeutics (Fischer, 2001).

The absence of mature functional T cells, that is, a lack of adaptive immunity, carries devastating clinical consequences caused by multiple infections of different origins: bacterial, mycobacterial, fungal, protozoal and viral organisms can spread and provoke protracted infections or an acute fatal attack within the first year of life. In addition, live vaccines, such as BCG, can also cause a disseminated fatal infection. Circulating maternal T lymphocytes are very frequently detected in the blood of SCID patients. In most cases, the maternal T-cell count is low and there are no clinical consequences. However, maternal T cells can, in some instances, expand (up to several thousands per microliter of blood) and cause immunological attack to the skin, gut and liver. However, only a small number of maternal T-cell clones do expand.

As discussed below, the description of SCID conditions relies on several parameters including immunological phenotype, inheritance pattern, gene identification and mechanism when available (Buckley, 2000). SCID conditions represent intrinsic defects of lymphocyte differentiation and should thus be distinguished from faulty development of T cells caused by environmental defects as observed in the DiGeorge syndrome or the rare nude phenotype in which the thymic epithelial component is missing.

Purine Metabolism Deficiency – Premature Apoptosis of Lymphocyte Precursors

The first SCID condition to be characterized at the molecular level was adenosine deaminase (ADA) deficiency. ADA is an enzyme involved in purine metabolism, and transforms adenosine and deoxyadenosine (dado) into inosine and deoxynosine respectively. Deoxyadenosine is phosphorylated into deoxyadenosine triphosphate (dATP) in immature lymphocytes; accumulation of dATP results in a block in synthesis of deoxyribonucleic acid (DNA) and also induces chromosomal breaks. A complete deficiency of ADA leads to almost total disappearance of T, B and natural killer (NK) lymphocytes. Partial deficiency results in an incomplete block in lymphocyte differentiation. ADA deficiency also leads to primary bone, lung and liver abnormalities. The immunodeficiency resulting from *adenosine deaminase (ADA)* gene mutations can be cured by allogeneic hematopoietic stem cell transplantation (HSCT) but can also be alleviated in part by enzymatic supplementation. Gene therapy is being tested in this setting.

Another defect in purine metabolism, that is, purine nucleoside phosphorylase (PNP) deficiency, also causes T-cell immunodeficiency because of the toxicity of accumulating deoxyguanosine triphosphate.

γc-Dependent Cytokine Receptor Pathway Deficiencies

Lymphocyte precursors do actively proliferate once they receive signals through appropriate cytokine receptors (**Figure 1**). Among them, interleukin-7 (IL-7), produced by both bone marrow stromal cells and thymic epithelial cells, provides survival as well as proliferative signals to T-lymphocyte progenitor cells while IL-15 acts similarly for NK lymphocyte progenitors. IL-7 and IL-15 receptors share with the IL-2, IL-4, IL-9 and IL-21 receptors a common cytokine receptor subunit named γc which is expressed at the surface of lymphocyte precursors. Upon cytokine binding, the tyrosine kinase JAK-3 (janus kinase 3) associates with the cytoplasmic domain if γc is activated and phosphorylates a downstream mediator, that is, the signal transduction activator of the transcription (STAT) 5 molecule. The most common form of SCID consists of a block in development of both T and NK lymphocytes. Its inheritance pattern is either recessive X-linked or autosomal recessive. The X-linked form of SCID (SCID-X1) accounts for more than half of all SCID cases. It is caused by mutations of the gene encoding γc. Multiple distinct mutations have been described, leading to the same phenotype. This observation illustrates the important role of the cytokine-dependent phase of T/NK lymphocyte development. Surprisingly, B-cell differentiation occurs normally despite defective IL-7 signaling. Actually, in most patients the peripheral B-cell counts is even increased. These B lymphocytes are at least partially functional *in vitro* as well as *in vivo* following HSCT in some cases; γc-negative B cells have been shown to undergo class-switch recombination and produce protective antibodies following physiological activation.

Deficiency in JAK-3 accounts for the autosomal recessive form of T(–), NK(–), B(+) SCID demonstrating that γc-dependent signals require JAK-3 activation (Macchi *et al.*, 1995). Recently, six SCID patients were shown to exhibit deleterious mutations in the gene encoding the IL-7 receptor (IL-7R) α chain. IL-7Rα associates with γc to form IL-7R. IL-7Rα deficiency results in a selective block in T-lymphocyte development, demonstrating the crucial role of IL-7 in a very early step of T-cell differentiation (Puel *et al.*, 1998). It is actually unknown whether or not this step precedes thymus homing of precursors.

V(D)J Recombination Deficiencies

A subset of SCID is typically characterized phenotypically by an absence of both mature T and B lymphocytes while NK cells are spared. It accounts for approximately 20% of cases. Inheritance is autosomal recessive. This condition has turned out to be genetically heterogeneous. This SCID condition is reminiscent of the scid mouse model, a natural mutant in which the V(D)J recombination process of the T- and B-cell receptor genes is impaired, due to a deficiency in the DNA-dependent protein kinase activity within the nonhomologous end-joining (NHEJ) repair process. It was therefore reasoned that this form of SCID also results from defective V(D)J recombination.

T(–), B(–) SCID can be divided into two subsets according to cell radiosensitivity. In approximately half of the cases, patients' cells exhibit a normal cell sensitivity to ionizing radiation while in the other half, as in murine scid, cells are abnormally radiosensitive.

The first phenotype is accounted for by deficiencies in recombination activating gene 1 or 2 proteins (Rag-1 or Rag-2) (Schwartz *et al.*, 1996). These elements, which are lymphocyte specific, initiate V(D)J recombination by creating site-specific DNA breaks at recombination signal sequences flanking coding elements of the variable part of T-cell receptors (TCRs) and B-cell receptors (BCRs). Complete defects result in an absence of mature T and B lymphocytes.

The second phenotype, associating defective V(D)J recombination and defective repair of DNA double-strand breaks (dsb), has been shown to be the consequence of mutations of the gene encoding the artemis protein (*DNA cross-link repair 1C* (*PSO2 homolog, S. cerevisiae*) (*DCLRE1C*)) (Moshous *et al.*, 2001). It appears that this ubiquitously expressed protein is involved in the NHEJ process required for both TCR and BCR gene rearrangements and dsb DNA repair. The artemis sequence has some homology with proteins involved in repair of DNA lesions caused by inter-DNA strand bonds, but its exact function remains unknown. Artemis deficiency accounts for the SCID observed in the Athabascan-speaking Native Americans, where it is the most frequent inherited disease (about one in 1000 live births).

CD45 Deficiency

Finally, a CD45 deficiency was found to be responsible in two patients of a SCID phenotype consisting of absence of TCRαβ+ T cells. CD45, a membrane-associated phosphatase, plays a major role in pre-TCR/TCR-induced signaling by removing an inhibitory phosphate in the CD4/CD8 associated src kinase p56 Lck.

Atypical SCID Phenotypes

Residual activities of mutated proteins associated with SCID can attenuate the severity of the phenotype.

This has been known for more than a decade for ADA deficiency. It has also been shown for Rag-1 and Rag-2, as residual activities of either product can be associated with a unique phenotype known as Omenn syndrome (Notarangelo et al., 1999). Omenn syndrome consists of consequences of a massive oligoclonal T-cell expansion. These T cells infiltrate the skin, the gut and sometimes organs, they exhibit a T_{H2} phenotype and cause severe disease with diffuse erythroderma and protracted diarrhea. While it can be understood that residual productive TCR gene rearrangements account for the development of a small number of T-cell clones, it is not clear why such T cells expand and behave as autoimmune clones especially at mucosal sites. There is an overlap between Rag-1/Rag-2 mutations causing absence of T and B lymphocytes and Omenn syndrome, suggesting that additional factor(s) might be involved.

Partial phenotypes have also been described in association with γc or JAK-3 mutations. While in some cases, partial T/NK cell development and function can be explained by γc/JAK-3 residual expression and function, in other instances, the phenotype is not well understood. For instance, in at least one case, a mutation leading to a truncated γc protein unable to bind JAK-3 was found associated with the progressive development over time of partially functioning T cells. Of interest, in one case, a partial T-cell deficiency associated with a deleterious γc mutation actually resulted from a spontaneous reverse mutation event that occurred in a T-cell precursor. For 5 years this child exhibited a mild T-cell lymphopenia with a somewhat diverse repertoire. This observation indicates the tremendous ability of T-cell precursors to proliferate prior to TCR gene rearrangement.

Other characterized genetic defects induce partial deficiencies of T-cell development. There are several examples including CD3γ and ε defects, ZAP-70 kinase deficiency, which, in humans, lead to an impairment in CD8 T-cell development and defective activation of CD4 T cells and expression deficiency of HLA class II molecules which prevents CD4 T-cell development, at least in part. More complex genetic diseases with multiple associated features can also impair T-cell development to some extent, as observed in DNA ligase 4 deficiency and the Nijmegen breakage syndrome caused by NBS/nibrin deficiency. DNA ligase 4 is involved in ligation of the ends of dsb DNA at the end of the NHEJ process. Its partial deficiency is compatible with life but results in microcephaly, a variable T-cell immunodeficiency and susceptibility to cytotoxic drugs. NBS/nibrin is part of a heterotrimeric protein complex with MRE-11 and Rad 50, which bind to DNA dsb and probably act as sensors of the DNA lesion.

Identification of the various genetic defects associated with impairment of T-cell development is of major importance, not only for basic science but also in medicine as it provides accurate tools for disease diagnosis and genetic counseling. In addition, it may pave the way for genetic treatment, as illustrated by the successful gene therapy of SCID-X1 (γc deficiency) (Cavazzana-Calvo et al., 2000).

Finally, the molecular basis of a few SCID phenotypes has not yet been characterized. This is the case of the very rare so-called reticular dysgenesis syndrome characterized not only by partially faulty development of T, NK and B lymphocytes, but also by a block in development of phagocytic cells. Associated deafness has been found in several cases. The yet unknown gene product should play a major role in the early steps of hematopoiesis. In some instances, a pure T-cell deficiency is observed without known etiology. Any gene product with a selective role in early steps of T-cell differentiation could be involved. There are also a number of more complex and somewhat ill-defined partial T-cell deficiencies – also termed combined immunodeficiencies – for which no molecular mechanisms are known. It is likely that further study of these phenotypes will delineate the role of yet unknown proteins in T-cell differentiation and function.

See also
Immunodeficiency Syndromes: Gene Therapy
Immunological Disorders: Hereditary

References

Buckley RH, Schiff RI, Schiff SE, et al. (1997) Human severe combined immunodeficiency: genetic, phenotypic, and functional diversity in one hundred eight infants. *Journal of Pediatrics* **130**: 378–387.

Buckley RH (2000) Advances in immunology: primary immunodeficiency diseases due to defects in lymphocytes. *The New England Journal of Medicine* **343**: 1313–1324.

Cavazzana-Calvo M, Hacein-Bey S, de Saint Basile G, et al. (2000) Gene therapy of human severe combined immunodeficiency (SCID)-X1 disease. *Science* **288**: 669–672.

Fischer A (2001) Primary immunodeficiency diseases: an experimental model for molecular medicine. *The Lancet* **357**: 1863–1869.

Macchi P, Villa A, Gillani S, et al. (1995) Mutations of Jak-3 gene in patients with autosomal severe combined immune deficiency (SCID). *Nature* **377**: 65–68.

Moshous D, Callebaut R, de Chasseval R, et al. (2001) ARTEMIS, a novel DNA double-strand break repair/V(D)J recombination protein is mutated in human severe combined immune deficiency with increased radiosensitivity (RS-SCID). *Cell* **105**: 177–186.

Notarangelo LD, Villa A, Schwarz K (1999) RAG and RAG defects. *Current Opinion in Immunology* **11**: 435–442.

Puel A, Ziegler SF, Buckley RH and Leonard WJ (1998) Defective IL7R expression in T(−)B(+)NK(+) severe combined immunodeficiency. *Nature Genetics* **20**: 394–397.

Schwarz K, Gauss GH, Ludwig L, *et al*. (1996) RAG mutations in human B cell-negative SCID. *Science* **274**: 97–99.

Further Reading

Bassing CH, Swat W, Alt FW (2002) The mechanism and regulation of chromosomal V(D)J recombination. *Cell* **109**: S45–S55.

Fischer A (2001) Primary immunodeficiency diseases: an experimental model for molecular medicine. *Lancet* **357**: 1863–1869.

Leonard WJ (2001) Cytokines and immunodeficiency diseases. *Nature Reviews Immunology* **1**: 200–208.

Reith W, Mach B (2001) The bare lymphocyte syndrome and the regulation of MHC expression. *Annual Review of Immunology* **19**: 331–373.

Rooney S, Sekiguchi J, Zhu C, *et al*. (2002) Leaky scid phenotype associated with defective v(d)j coding end processing in artemis-deficient mice. *Molecular Cell* **10**: 1379–1390.

Villa A, Santagata S, Bozzi F, *et al*. (1998) Partial V(D)J recombination activity leads to Omenn syndrome. *Cell* **93**: 885–896.

Web Links

Adenosine deaminase (*ADA*); Locus ID: 100. LocusLink:
http://www.ncbi.nlm.nih.gov/LocusLink/LocRpt.cgi?l = 100

DNA cross-link repair 1C (PSO2 homolog, *S. cerevisiae*) (*DCLRE1C*); Locus ID: 64421. LocusLink:
http://www.ncbi.nlm.nih.gov/LocusLink/LocRpt.cgi?l = 64421

Adenosine deaminase (*ADA*); MIM number: 102700. OMIM:
http://www.ncbi.nlm.nih.gov/htbin-post/Omim/dispmim?102700

DNA cross-link repair 1C (PSO2 homolog, *S. cerevisiae*) (*DCLRE1C*); MIM number: 605988. OMIM:
http://www.ncbi.nlm.nih.gov/htbin-post/Omim/dispmim?605988

Sex Chromosomes

John C Lucchesi, *Emory University, Atlanta, Georgia, USA*

Sexual reproduction increases the number of different genetic constitutions that are present at each generation and that can be acted upon by natural selection. The genetic information required for the differentiation of fertilized eggs into individuals of different sexes is contained on a special pair of chromosomes: the sex chromosomes.

Introductory article

Article contents

- Introduction
- How and When Were They Discovered?
- How do Sex Chromosomes Determine the Sex of an Individual?
- How are the Sex-determining Genes Distributed on the Sex Chromosomes?
- How did Sex Chromosomes Become Different from One Another and from the Other Chromosomes?
- How Does Dosage Compensation Work?

Introduction

In the majority of animal species (and in some plants), the difference between the two sexes is based on the presence of special chromosomes that are called the 'sex chromosomes'. In diploid species, that is, in organisms that at conception received one set of their chromosomes from each parent, there is usually only one pair of sex chromosomes. All of the other chromosomes that are present in the genetic material are called 'autosomes' to distinguish them from the sex chromosomes. The difference in the sex chromosomes present in males and females is not the same in all groups of animals. In mammals, including humans, and in some types of insects such as flies, females have two sex chromosomes, which are identical in shape and genetic content, called the X chromosomes; males have only one X chromosome and, instead of a second X, they have a sex chromosome, which is different in shape and genetic content, called the Y chromosome. In some insects such as grasshoppers and crickets, and in some roundworms, the only sex chromosome present in males is the X chromosome. In birds and in some insects such as butterflies and moths, the sex chromosome situation is a mirror-image of the situation just described for mammals: males have two identical sex chromosomes and it is the females that have only one of these chromosomes and another sex chromosome that is different. The sex chromosome that is found in both sexes of these species is called the Z chromosome and the other sex chromosome, found only in females, is called the W chromosome.

How and When Were They Discovered?

In the latter part of the nineteenth century, a German scientist named H. Henking studied the formation of the reproductive cells of a particular insect. He noted that, in males, half of the sperm carried a chromosome that was not present in the other half. He called this peculiar chromosome 'x', because he was not sure that it was a chromosome at all. As was shown a few years later by C. E. McClung (1902), the x element was indeed a chromosome. Furthermore, McClung also showed that when an egg was fertilized by a sperm containing the x element, it gave rise to a female, while an egg fertilized by a sperm lacking the element gave rise to a male. He concluded that the x chromosome was the basis for the inheritance of sex in this insect.

How do Sex Chromosomes Determine the Sex of an Individual?

The sex of an organism reflects all of the physical and behavioral characteristics that allow us to recognize a particular individual as a male or a female. These traits are the result of the expression of specific genes. The first conclusion that one can draw from this is that the sex chromosomes are responsible for determining the sex of an individual through the function of all or some of the genes that they carry. There exists a difference in these sex-determining genes between the fertilized eggs that will give rise to females and those that will give rise to males. So, the second conclusion that one can draw is that the sex chromosomes, by virtue of the fact that they are different in the two sexes, must be responsible for the necessary difference in sex-determining genes.

How are the Sex-determining Genes Distributed on the Sex Chromosomes?

There are different types of sex-determining genes and their presence on particular sex chromosomes is different in given groups of organisms. In mammals, the male sex is determined by the presence of a specific gene in the fertilized egg; if this gene is absent, then a fertilized egg will develop into a female. Clearly, the most efficient way of including a gene in some of the fertilized eggs and excluding it from others is to have the gene reside on the sex chromosome that is present in only half of these eggs, the Y chromosome.

In humans sex is determined by the presence or absence of a male-determining gene present on the Y chromosome. This fact was established by examining the chromosomes of two types of abnormal individuals. The first individuals to be examined had Klinefelter syndrome (named after the physician who had first described it). These individuals are sterile males, with varying levels of femaleness in some of their sexual characteristics, such as hair and fat distribution. Their sex chromosome constitution is XXY. The other type of abnormal individuals had a condition called the Turner syndrome. Clearly female, but exhibiting such abnormalities as very poor sexual development, small stature and a characteristic set of facial features, these individuals were found to have an XO sex chromosome constitution. The fact that XX individuals carrying a Y chromosome are predominantly male and XO individuals are predominantly female led to the conclusion that in normal humans it is the presence of the Y chromosome and not the number of X chromosomes that determines sex (**Table 1**).

In the light of the previous statement, how can one explain the determination of sex in organisms where the only sex chromosome present in the male is one X chromosome? (It is customary to represent the sex chromosome constitution of these males as XO to emphasize the absence of a second sex chromosome.) In XO individuals, it is not the presence or absence of a particular gene that is the key factor in determining the sex of the fertilized egg. The difference is in the level of the product of one or more genes that are present on the X chromosome: females with two X chromosomes have twice the number of these genes and twice the amount of gene product that is found in males with a single X chromosome and only one dose of genes. It is this difference in the amount of gene products of specific genes on the X chromosome that determines whether a fertilized egg will develop into a male or a female.

Surprisingly, the difference in the level of product made by genes on the X chromosome in males and females is responsible for sex determination, even in some organisms where the males have both an X and a Y chromosome. This was discovered, for example in fruit flies, by the American geneticist C. B. Bridges, who based his conclusion on the following evidence: Eggs normally carry an X chromosome,

Table 1 Effect of the sex chromosomes on sexual development

Organism	Chromosome constitution	Sex
Fruit flies	XX	Female
	XY	Male
	XXY	Female[a]
	XO	Male[b]
Humans	XX	Female
	XY	Male
	XXY	Male[c]
	XO	Female[d]
Birds	ZW	Female
	ZZ	Male

[a]A fly with this sex chromosome constitution is a normal and fertile female.
[b]A male, normal in all respects except that he is sterile.
[c]A Klinefelter male.
[d]A Turner female.

while sperm carry either an X or a Y chromosome. Fertilization of an egg by an X-bearing sperm will result in a female (XX) and by a Y-bearing sperm will result in a male (XY). Occasionally, during the production of sex cells, errors occur and an inappropriate number of sex chromosomes is included in an occasional egg or sperm. Fertilization of a normal egg by a sperm which by accident contains both an X and a Y chromosome will give rise to an individual with a sex chromosome constitution of XXY. Bridges determined that such an individual is a perfectly normal and fertile female. Similarly, if by chance an egg is produced without any sex chromosomes at all and if this egg is fertilized by an X-bearing sperm, an individual with a sex chromosome constitution of XO would be produced. Bridges noted that such individuals are males, normal in all respects except that they are sterile (**Table 1**). Bridges concluded that in fruit flies the Y chromosome does not carry genetic information that will cause a fertilized egg to develop into a male. The only genetic information carried by this chromosome is necessary for the male to form sperm and, since it is useful only to males, it is not surprising that it is found on a chromosome that is present only in males.

How did Sex Chromosomes Become Different from One Another and from the Other Chromosomes?

In many species, the two members of the pair of chromosomes on which the sex-determining genes are located are identical in appearance. In these species, each member of the pair carries a different form of the sex-determining gene. Different forms of a gene are called 'alleles'; different alleles cause the gene to have different functions. The presence of a particular allele of the sex-determining gene in a fertilized egg leads to one sex, while absence of this allele leads to the other. This is thought to be the most primitive and ancient mechanism of sex determination.

In many other species, one member of the sex chromosome pair is very different from the other, in shape and with respect to the genes that it carries. The changes that eventually result in different sex chromosomes can be limited to a small segment of the chromosome or they can affect the entire chromosome. Ever since the discovery of sex chromosomes and their correlation with sex determination, the causes of these changes have been the object of considerable speculation. It is generally believed that all of the sex chromosome types that are found in modern species were derived from the more primitive

situation of sex determination by a simple difference in the alleles of a gene present on a pair of identical chromosomes. If the two alleles of the sex-determining gene were called A and a, respectively, one sex would be aa and the other Aa. Since the presence of the A allele in a fertilized egg leads to one sex and its absence to the other, the A allele is restricted to one sex. Mutations (changes in the genetic material) that may occur by chance in the vicinity of this allele would also tend to be restricted to the same sex. Since the vast majority of mutations reduce or abolish the normal function of genes, the sex-limited region of the A-bearing chromosome would tend to become more and more dysfunctional. This region or, in some cases, the whole chromosome, except for the sex-determining gene, eventually would degenerate into inactive genetic material. The accumulation of mutations on the chromosome that carries the A allele of the sex-determining gene would be accelerated if the two sex chromosomes were somehow prevented from exchanging material by the process of recombination that normally occurs between the two identical members of chromosome pairs.

This scenario could just as easily explain sex determination in species where the sex chromosomes of females are XX and those of the males are XY. In these species, the Y chromosome carries a male-determining allele; absence of this allele leads by default to female sexual development. In species where females form two types of eggs (with a Z or a W chromosome) and all of the sperm carry a Z chromosome, the allele carried by the W chromosome would be female-determining and the absence of this allele would cause fertilized eggs with ZZ sex chromosomes to develop as males (**Table 1**).

The sex chromosomes of species such as fruit flies, where the Y chromosome is not involved in determining sex, were most likely derived from the same type of primitive system of sex determination based on two different alleles on a pair of identical chromosomes. The only difference is that, in this case, it was not the presence of the A allele that mattered; rather, it was the difference in the dose of the a allele in aa compared with Aa fertilized eggs, leading to a difference in the level of this allele's specific gene product, that determined the sex of the developing fertilized egg. A similar explanation holds for species where the sex chromosome constitution of the males is XO.

In most mammals, although the X and Y chromosomes are different in size, appearance and genetic content, they have a region at one of their tips where the genetic content is the same on the two chromosomes. This region is called the 'pseudoautosomal region', to indicate that the genes that it contains are present in two doses, just as they are on the pairs of autosomes. In humans, the region of genetic similarity is split into two small regions located at the tips of the short and the long arm of the X and Y chromosomes. Recombination, the process that normally results in an exchange of material between the two identical members of chromosome pairs, can and does occur between the X and Y chromosomes in this region. Surprisingly, the sex chromosomes of marsupials do not have a pseudoautosomal region.

At this point it is very important to note that, although they perform a very essential genetic function, sex chromosomes also present a very serious problem to the species where they are found. The Y (or W) chromosome of some species contains genetic information that is present nowhere else in the chromosomes and that determines that a fertilized egg will develop into an adult of a particular sex. The X chromosome of other species contains sex-determining genes leading to different levels of products in XX and XY or XO individuals. But what about all of the genes that were present on the pair of ancestral sex chromosomes, chromosomes that were identical and differed only with respect to the alleles of a single sex-determining gene? All or many of these genes that have nothing to do with sex determination are still located on the modern X chromosomes; therefore, these genes are present in two doses in XX fertilized eggs and in a single dose in XO or XY fertilized eggs. Since the products of these genes are equally important to the development and well-being of both sexes, a difference in their level in XX and XO or XY individuals would be problematic. Due to the presence of a special regulatory mechanism called 'dosage compensation', the level of most of these gene products is the same in the two sexes.

How Does Dosage Compensation Work?

Dosage compensation has been investigated in three groups of organisms: flies, roundworms and mammals. Each group uses a different regulatory mechanism to equalize the expression of sex-linked genes that differ in dose between females and males. In *Drosophila*, where it was first discovered, dosage compensation is achieved by increasing the rate of activity of genes on the single X chromosome of males (with a sex chromosome constitution of XY) to approximately twice that of the genes on each X chromosome of females (XX). In the roundworm *Caenorhabditis elegans*, males are XO, but individuals with a sex chromosome constitution of XX are hermaphrodites, that is individuals that produce both male and female sex cells. In hermaphrodites, the rate of activity of genes on both X chromosomes is

approximately halved, rendering the level of gene products in this sex equivalent to the level in males with a single X chromosome. Much has been learned about the molecules that are involved in achieving these differences in the activity of genes (an increase in activity in *Drosophila* males or a decrease in activity in the case of *Caenorhabditis* intersexes). In both cases, the regulation is accomplished through the activity of groups of proteins that interact with one another and with the genetic material. Some of these proteins are involved in other functions in the cell or are very similar to proteins that are involved in other cellular functions. Directing these complexes to the correct chromosome, that is, the X chromosome, in the appropriate sex relies on additional specific proteins. It also relies on the presence of targets, in the form of specific stretches of the DNA that are found on the X chromosome.

Mammals achieve compensation by inactivating one of the two X chromosomes in all of the cells of females (with the exception of certain cells in the ovaries). So male mammals (XY) have one active dose of genes on the X chromosome, and females (XX), having inactivated one of their two X chromosomes, also have one active dose of these genes. The genes in the pseudoautosomal region, present in equal doses in males and females, escape inactivation. The proteins involved in mammalian X-chromosome inactivation have not been identified to date.

Although the mechanisms responsible for achieving dosage compensation in fruit flies and in mammals are very different, both mechanisms share a similarity in that they involve a component that is not a protein. This component is ribonucleic acid (RNA). There are at least two types of RNA molecules that associate with the dosage-compensation proteins all along the X chromosome of *Drosophila* males. In mammals, a gene that is present on the X chromosome produces an RNA product that decorates the inactive X chromosome but not the active X in the cells of females. In neither case are the roles of these RNA molecules understood.

See also
Chromosome X: General Features
Chromosome Y: General and Special Features
Mammalian Sex Chromosome Evolution
X and Y Chromosomes: Homologous Regions
X-chromosome Inactivation

Further Reading

Graves AM, Wakefield MJ and Toder R (1998) The origin and evolution of the pseudoautosomal region of human sex chromosomes. *Human Molecular Genetics* **7**(13): 1991–1996.

Kelley RL and Kuroda MI (2000) The role of chromosomal RNAs in marking the X for dosage compensation. *Current Opinion in Genetics and Development* **10**(5): 555–561.

McElreavey K and Fellous M (1999) Sex determination and the Y chromosome. *American Journal of Medical Genetics* **89**(4): 176–185.

Meller VH (2000) Dosage compensation: making 1X equal 2X. *Trends in Cell Biology* **10**(2): 54–59.

Meyer BJ (2000) Sex in the worm-counting and compensating X-chromosome dose. *Trends in Genetics* **16**(6): 247–253.

Pannuti A and Lucchesi JC (2000) Recycling to remodel: evolution of dosage-compensation complexes. *Current Opinion in Genetics and Development* **10**(6): 644–650.

Sex Determination: Genetics

See Male Sex Determination: Genetics

Sex: Evolutionary Advantages

Intermediate article

Jukka Jokela, *Swiss Federal Institute of Technology (ETH), Zürich, Switzerland*

Article contents

- Long-term Benefits of Sexual Reproduction
- Short-term Costs of Sexual Reproduction
- Geographic Patterns and Origin of Asexual Reproduction
- Theories and Tests for the Advantage of Sex
- Pluralistic Theory

One of the most intriguing quests in evolutionary biology is to discover why sexual reproduction is so common in nature, and how it is maintained. In theory, sexual reproduction is expected to carry considerable costs compared with asexual reproduction. Competing hypotheses for the advantage of sex range from better resistance to coevolving parasites to efficient clearance of deleterious mutations.

Long-term Benefits of Sexual Reproduction

Sexual reproduction (combining the genes of two parents to produce offspring) has two long-term advantages. First, the process breaks associations between deleterious mutations that are present in each of the parent genomes. Hence sexual reproduction may create genotypes that have a lower mutational load than the parent genomes. Mutation clearance allows sexual populations to avoid the accumulation of deleterious mutations, as a result of 'Muller's ratchet' (Muller, 1964), which would be expected to erode the fitness of asexual lineages over time (Gabriel *et al.*, 1993). The second advantage of sexual reproduction is that recombination allows good genes (and new beneficial mutations) to find worthy company. Reshuffling of the parental genes may create novel gene combinations and allow efficient responses to natural selection (Barton and Charlesworth, 1998). Efficient adaptation to environmental change is considered as the main long-term benefit of sexual reproduction. In fact, asexuals are often referred to as evolutionary dead-ends. It is possibly because of this that exclusively asexual clades are rare; bdelloid rotifers are the only large clade that is suspected to have stayed exclusively asexual over long periods of time.

The problem of sex is not in understanding the long-term benefits. The problem is that, in the short term, sexual reproduction has costs that may easily void the long-term advantages; yet sex seems to prevail in nature.

Short-term Costs of Sexual Reproduction

The cost of recombination

Recombination has a direct cost in organisms that are well adapted to their environment: it breaks apart favorable gene combinations (Barton and Charlesworth, 1998). It is not currently clear how common or important the cost of recombination is compared with other short-term costs of sex, but because the cost of recombination increases with the level of adaptedness, it certainly adds to the mystery.

The cost of males

The most important short-term cost of sex is the cost of male production (Maynard Smith, 1978). The cost of males is best illustrated by a comparison of the reproductive rates of otherwise equal sexual and asexual populations. Females that reproduce asexually do not have to produce males to ensure reproduction, while sexual females need males to reproduce. This has

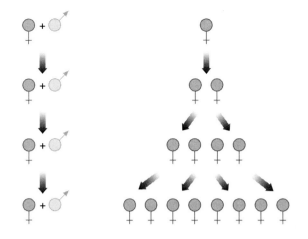

Figure 1 The cost of males. The growth rate of asexual lineage is faster than the growth rate of an otherwise similar sexual lineage. In this case (sex ratio 1 : 1) the growth rate of the asexual population is double that of the sexual population.

important consequences when the sexual and asexual females compete for the same resource. If all else is equal between the two groups (i.e. females have similar life histories, similar ecological enemies, similar habitat preferences and are sympatric), sexual females are predicted to lose in competition to asexual females, simply because males do not contribute to the growth rate of the sexual population (**Figure 1**). Hence, the cost of males refers to the slower growth rate of the sexual population caused by male production. Modelling of the severity of this cost reveals that a sexual population of 10^6 individuals (sex ratio 1 : 1) is expected to succumb to an invasion of a single asexual mutant in less than 50 generations (Lively, 1996). Obviously this is too short a time for the long-term benefits of sex to rescue the population, which is why the maintenance of sex under the pressure of invading asexuals is such an intriguing problem for evolutionary biologists.

Geographic Patterns and Origin of Asexual Reproduction

Geographic parthenogenesis

Asexual reproduction is more common in adverse environments (arid, polar, high altitude) than in the tropics, and more common in freshwater habitats than in the sea (Bell, 1982). These patterns are known as geographic parthenogenesis. The occurrence of geographic parthenogenesis has been taken to suggest that asexual reproduction is favoured in environments where biotic interactions (competition, predation, parasitism) are less important determinants of

community structure. Whatever the true reason for this interesting pattern, a general explanation for the advantage of sex should be compatible with the geographic distribution of sexual reproduction, and should help us to understand what leads to geographic parthenogenesis.

Single or multiple origin of parthenogens?

The easiest explanation for the predominance of sex is that sex is phylogenetically constrained: once it has evolved, it is not easy to revert back to asexuality, and what we see at present may just be a reflection of past events. Comparative studies do not support this view, however (Bell, 1982). In contrast, present-day parthenogens are almost exclusively of sexual origin, and parthenogenesis has evolved independently several times in many groups of organisms. Sex seems to be phylogenetically constrained only in higher vertebrates where parthenogenetic mutants are very rare. The occurrence of occasional parthenogens in stable and predominantly sexual plant and animal populations suggests that sex has to prove itself continuously in the wild, again raising the question of why sex is maintained in nature.

Different types of asexuals

How does the transition to asexuality takes place? Many genetic processes may generate offspring that develop to parthenogenetic females, but the most important distinction is between 'spontaneous' and 'hybrid' parthenogens. Spontaneous parthenogens arise from one sexual population, either through conjugation of female and male gametes that produce a parthenogenetic female, or directly from a viable unfertilized egg that turns into a parthenogenetic female. These parthenogenetic clones may be thought of as random samples of the genotypes present in the original sexual population that have frozen one genotype to a nonrecombining state (Vrijenhoek, 1984).

Hybrid parthenogens, on the other hand, arise through a sexual cross between two close species. Hybridization is common in nature – for example, all asexual vertebrates are hybrids (Radtkey et al., 1995). The difference between spontaneous parthenogens and hybrids is that the former are expected to be phenotypically indistinguishable from the phenotypes of the parental population, while the latter often express intermediate phenotypes to parental species, and may therefore represent novel variants. This is important when hypotheses on the maintenance of sexual reproduction are tested. Spontaneous parthenogens compete directly with the original sexual population, while hybrids may actually occupy a different niche. Hence the coexistence of hybrid parthenogens with sexuals may not present such a challenge to the theory as the coexistence of spontaneous parthenogens with sexuals (Jokela *et al.*, 1997).

Theories and Tests for the Advantage of Sex

Maintenance of sex has attracted much theoretical interest (Kondrashov, 1993). It is generally recognized that a hypothesis that satisfactorily explains the maintenance of sex has to show how the short-term advantage of parthenogens is counterbalanced in such a general way that is applicable to the diverse group of organisms that rely on sexual reproduction. A general theory should also be able to deal with geographic parthenogenesis; the commonness of asexuals in certain environments suggests that ecological processes may be important. A general theory should also be able to deal with ploidy differences between sexuals and parthenogens; clones are often of a higher ploidy than their sexual counterparts. These requirements of generality imply that the hypothesis should work under many different ecological and genetic scenarios: it should be independent of population size, or presence or absence of spatial heterogeneity or temporal variation in the environment, and it should not require a specific genetic system. It has become clear that it is not difficult to generate a model that counterbalances the cost of males in the short term, but it is very difficult to find a model that does not require very specific constraining assumptions (Kondrashov, 1993). Of all the hypotheses with more or less strict assumptions, two seem most promising as an explanation for why sex is so advantageous: the Red Queen hypothesis, and the Mutation Deterministic hypothesis.

How to test the alternative hypotheses?

When looking for an opportunity to test theories on the advantage of sex, it should not be forgotten that the main question is why sexual reproduction predominates in nature, not whether each of the models is formally correct. In fact, it is generally agreed that the principal models are formally correct, and, within the limits of their assumptions, each of them is able to show the advantage of sex (Kondrashov, 1993). The critical question is which of the theories is best, when applied in nature, and whether any of them is sufficient as a sole explanation (West *et al.*, 1999).

From a practical point of view, the advantage of sex is best studied in a system in which asexuals and sexuals coexist in nature, that is, in a system where sex

has to prove itself continuously. Tests with organisms for which parthenogenesis is virtually impossible because of phylogenetic constraints may not be very informative in a search for the general advantage of sex. Likewise, tests with organisms that are confined to stripped-down interactions with other species in controlled laboratory environments are also prone to be generally uninformative. Thorough tests of the theory should incorporate data from both natural populations and laboratory experiments, and are therefore likely to be laborious. Further, the advantage of sex may have to be shown on a case-by-case basis, because only the accumulation of a critical number of case studies will allow one to draw general conclusions about the most likely processes that make sex such a success in nature.

Step one: is all else equal?

The paradox of sex stems from the cost of males. The cost of males depends on the 'all-else-equal' assumption: similarity of the asexual and sexual populations in all other aspects than the reproductive mode. The easiest way to resolve the question of maintenance of sex is to determine whether all else really is equal for the specific system in question (Jokela *et al.*, 1997). If asexuals have inherently lower fitness, whatever the reason, then the cost of males may be counterbalanced, with no further explanations needed. In fact, only a few tests of this assumption have been conducted.

Why would parthenogens have lower fitness than their sexual relatives? A simple answer is that there are many things that may go wrong when a sexual egg suddenly turns into a parthenogenetic female. For example, the switch to parthenogenesis commonly includes a change in the ploidy level, which may be associated with ecological differences between the sexuals and asexuals, as suggested in some studies of hybrids. Hence it is advisable to test how similar are the life history traits of parthenogens and sexuals, how similar is their ecology, and whether the populations are truly sympatric; that is, might the replacement of sexuals by asexuals really take place.

Step two: strong inference?

Having obtained a satisfactory answer to the question regarding the cost of males, it is relevant to contrast the competing hypotheses concerning maintenance of sex. But how might one prove that a particular theory is correct? Philosophy of science suggests that proving theories correct is very difficult. Often the most fruitful way to proceed may be to test the specific assumptions of the theories that, when proven false, would allow these theories to be discarded from the set of alternative explanations. Hence a productive way to proceed may be to ask which of the hypotheses are clearly not applicable. If the set of hypotheses narrows down to few plausible alternatives, effort may be focused to test and contrast these. This is the approach adopted in empirical research programmes addressing the maintenance of sex.

Parasite–host coevolution: the Red Queen hypothesis

The Red Queen hypothesis suggests that coevolving parasites may counterbalance the cost of males, and therefore lead to the maintenance of sex (Jaenike, 1978). The metaphor is borrowed from the familiar children's book *Through the Looking Glass*, by Lewis Carroll.

Imagine a sexual population that is a resource for a specific harmful parasite. Allow a clone to invade the sexual population. Because the clone does not pay the cost of males, its proportion in the population is expected to increase rapidly, threatening the sexuals with rapid extinction. This is where the parasites enter the stage. The commonness of the newly arisen host clone gives a fitness advantage to those parasite genotypes that are able to infect this particular genotype. The parasites now have a readily available resource, and their offspring will also have plenty of opportunities to successfully complete their lifecycle and reproduce. This leads to an overproportional infection of the common clone, which erodes the fitness advantage it gained by not paying the cost of males.

However, there is a time lag in the response of the parasite population. As the fitness advantage of the common host clone has already decayed, the clone is still susceptible to a large proportion of the parasite population, and remains overproportionally infected. Hence, the parasite has checked the increase of the clone in the host population, and further chases the clone down, turning it into a rarity. As the previously common clone becomes rare, the specific parasite strains lose their relative fitness advantage, but again with a time lag. Now the cycle is completed and a new one starts. The clone has regained rare advantage with respect to parasites, starts to increase in frequency because it is not paying the cost of males, and may be tracked by the parasite after a certain time lag. Meanwhile, sex is maintained in the population, because the clone is knocked down before it competes the sexual population to extinction. This type of selection is called time-lagged negative frequency-dependent selection. The parasites represent one scenario of how fluctuating selection may lead to the maintenance of sex.

The Red Queen hypothesis has two main assumptions (Lively, 1996). The parasite has to be highly virulent, and infection has to have a genetic basis (i.e. the host has to have genetic variance for resistance). High virulence in this case means that the parasite has to kill its host, or to prevent the host from reproducing, otherwise it is unlikely that selection is strong enough to rescue sexuals from extinction. Genetic variance for resistance allows the parasite population to adapt to common resistance genotypes through natural selection, a prerequisite for negative frequency-dependent selection to work.

All free-living species that have been examined so far host one or several species of specialist parasites or pathogens. Furthermore, completely parasite- or disease-free host populations are very rare. Parasitism being as common as it is, and genetic variation for resistance also not being rare, it may well be that the Red Queen hypothesis meets the general requirements for a theory that is able to explain why males exist. However, more case studies are needed to evaluate this point fully. Note that these two assumptions, high virulence and genetic variation for resistance, are relatively easy to test even in a natural host–parasite system.

The problem of the Red Queen hypothesis is in the selection for increased host diversity that the parasite exerts. The hypothesis predicts advantage of rare types, but not necessarily an advantage of sex. Under the Red Queen hypothesis, sex is only advantageous if the host population has no other means than recombination to produce genotypic diversity. An alternative is that host populations may be clonally diverse. In simulation studies where a sexual population is invaded by several asexual clones, sex tends to lose rapidly in competition with diverse clonal population, even if the parasites drive the clones through Red Queen-type cyclical dynamics (Lively and Howard, 1994). The sudden emergence of multiple clones is unlikely but, if the extinction probability of clones is not high (Red Queen only drives them to become rare), the emergence of new clones will add to the diversity of the present ones, and the level of clonal diversity will eventually be sufficient for the replacement of the sexuals. To survive, therefore, the Red Queen hypothesis needs an additional process to explain how clonal diversity is prevented from accumulating.

Empirical support for the Red Queen hypothesis

Empirical studies of maintenance sex are unfortunately not as numerous as theoretical studies. Some studies that have addressed the Red Queen hypothesis are reviewed briefly below.

The principal premise of the Red Queen hypothesis, that parasites are able to adapt to the most common host genotypes, was shown in mixed populations of sexual and asexual fish by Lively *et al.* (1990). Mexican topminnows (*Poeciliopsis*) live in pools that often have a mixture of sexuals and a few gynogenetic clones. The fish are parasitized by a subcutaneous trematode that encysts under the skin and produces visible black spots (black-spot disease). The research group found that common fish clones were more infected than the sexuals in two of the three pools studied. The third pool, where the common clone was less infected than the sexual, proved to be quite interesting. The sexual population in that pool was known to be severely inbred as a result of a recent dry period, after which only a few sexual individuals had colonized the pool and founded the population. The introduction of 30 genetically variable sexuals from upstream was sufficient to boost the genetic variation of the sexuals so that the parasite switched to infect the common clone. This transition took only two years.

In this case, however, the parasite was not highly virulent, and the clones most likely were of hybrid origin and known to occupy slightly different niches than the sexual fish (Schenck and Vrijenhoek, 1986). It is therefore unlikely that maintenance of sexuals in these populations was due to parasite-mediated negative frequency-dependent selection. However, the study shows how the parasites are rapidly able to specialize on common resistance genotypes in the host population.

The most detailed studies of various assumptions of the Red Queen hypothesis have been conducted in a common New Zealand freshwater snail, *Potamopyrgus antipodarum*. Populations of *P. antipodarum* commonly have mixtures of diploid sexual and triploid clonal individuals. Snails also harbor a diverse community of trematode parasites that castrate the host snails, effectively sterilizing both males and females. In extensive field surveys of variation in the proportion of sexuals in the populations, Curt Lively found that in lakes where parasites were rare, the clonal snail populations dominated, as predicted by the Red Queen hypothesis (Lively, 1987, 1992). In this series of studies, Lively contrasted three other ecological hypotheses, the Tangled Bank, the Lottery and the Reproductive Assurance hypotheses, with the Red Queen hypothesis by comparing the predictions of the alternative hypotheses, with the ecological distribution of sexuals and asexuals. The Red Queen hypothesis was the only one that survived the test. For example, the proportion of sexuals was better explained by variation in parasite prevalence than by unpredictability of the habitat (prediction of the Lottery hypothesis), or density of the population (prediction of the Reproductive Assurance hypothesis). Furthermore, within similar competition-dominated stable lake environments, clones predominated in lakes where parasites were not common, which goes against the prediction of the Tangled Bank hypothesis. Later studies have also found the predicted association

between parasitism and proportion of sexual individuals between depth-specific habitat zones of a single lake (Jokela and Lively, 1995).

It is important to remember at this point that causal interpretation of correlative evidence may easily be misleading, and that experiments are often needed to examine further the causes behind the observed field correlation. However, insights given by correlative evidence should not be wasted. Correlative evidence may give strong support for the hypothesis if the plausible alternative explanations can be logically excluded or shown to be unlikely. Furthermore, not finding a correlation where there are good grounds to expect one is usually sufficient to cast serious doubt on the relevance of the tested hypothesis. Correlational data are also essential because laboratory experiments addressing specific assumptions of specific theories reveal very little of how relevant the theory actually is in nature.

Further studies in the *Potamopyrgus* system have shown that parasites are locally adapted to their hosts (Lively, 1989; Lively and Jokela, 1996), triploid clones are spontaneously spun off from local sexual populations (Dybdahl and Lively, 1995a) and the clones are heterogeneously infected by the most common parasites (Dybdahl and Lively, 1995b). In the most recent studies, Dybdahl and Lively also describe clonal dynamics in an all-clonal population of these snails that look amazingly similar to parasite-driven Red Queen cycles (Dybdahl and Lively, 1998). In this particular study, Dybdahl and Lively tracked the rapid increase of a clone in a natural population over several years, followed by a decrease after an overproportional infection by a common parasite. Furthermore, rare clones, which had been common in the past couple of years but had recently crashed, proved to be very susceptible to the parasites when compared with a group of other rare clones.

These studies have critically explored the assumptions of the Red Queen hypothesis and have compared alternative ecological and genetic (see below) hypotheses for the maintenance of sex. Studies have come close to the Red Queen hypothesis and have not been able to refute it. Clearly what is needed are similar studies in other systems.

The mutation-deterministic hypothesis

As already mentioned above, sexuals are expected to be better at clearing deleterious mutations than asexuals. Mutation-deterministic (MD) theory reveals the conditions under which better mutation clearing may counterbalance the short-term costs of sex (Kondrashov, 1988). Better clearing is predicted to give a short-term advantage to sexuals if the genomic mutation rate is high (more than one deleterious mutation per genome per generation), and if deleterious mutations follow synergistic epistasis (the negative effect of each new deleterious mutation increases as a function of the number of deleterious mutations already present in the genome). The MD hypothesis makes no assumption about population size, ecology or population biology of the system. The only critical assumptions are high mutation rate and synergistic epistasis between the deleterious mutations.

Empirical support for the MD theory

The MD hypothesis for maintenance of sex has been celebrated because it is easy to falsify; finding a low genomic mutation rate or a lack of synergistic epistasis is sufficient to refute the hypothesis at the species level. Although these assumptions seem to be less easy to address than advocated by theoreticians, tests are accumulating. The genomic mutation rates seem to be higher than one in mammals (Kondrashov, 1988), but in other organisms tested so far this does not seem to be the case (Kibota and Lynch, 1996; Deng and Lynch, 1997; Keightley and Caballero, 1997) – although evidence from *Drosophila* is contradictory (Fry *et al.*, 1999). Experiments designed to detect the synergistic effects of deleterious mutations have been conducted in several different organisms ranging from bacteria to *Potamopyrgus* snails, but have rarely found evidence for the effect (Hartman, 1985; De Visser *et al.*, 1997; Elena and Lenski, 1997; Lively *et al.*, 1998). Hence the accumulating empirical studies suggest that the MD hypothesis may not be a sufficient general explanation for the maintenance of sex. It is expected that advances in techniques of molecular biology will allow the fate of the MD hypothesis to be resolved conclusively in the near future.

Pluralistic Theory

Frustrated with the problems of each single theory, researchers in the field have looked for ways to combine them. The idea is that a combination of some prevailing theories will help us to understand why sex seems so victorious, and overcomes some shortcomings of the present models (West *et al.*, 1999). Howard and Lively (1994) presented one such alternative. Their idea was that Red Queen may be rescued by the accumulation of deleterious mutations in the clones. Using computer simulations, they showed that the accumulation of deleterious mutations speeds up when a clone is driven through a bottleneck by a parasite. This is because Muller's ratchet clicks faster in small populations. The Red Queen combined with the ratchet gives the sexual population a much higher chance of survival even in competition with multiple clones. The logic of the pluralistic approach is

attractive; a single explanation may be beautiful (mutations versus parasites), but two processes may do the trick even more efficiently.

See also
Genetic Load
Male-driven Evolution
Species and Speciation

References

Barton NH and Charlesworth B (1998) Why sex and recombination? *Science* **281**: 1986–1990.

Bell G (1982) *The Masterpiece of Nature: The Evolution and Genetics of Sexuality.* Berkeley, CA: University of California Press.

De Visser JAGM, Hoekstra RF and van Den Ende H (1997) Test of interaction between genetic markers that affect fitness in *Aspergillus niger. Evolution* **51**: 1499–1505.

Deng HW and Lynch M (1997) Inbreeding depression and inferred deleterious mutation parameters in *Daphnia. Genetics* **147**: 147–155.

Dybdahl MF and Lively CM (1995a) Diverse endemic and polyphyletic clones in mixed populations of the freshwater snail, *Potamopyrgus antipodarum. Journal of Evolutionary Biology* **8**: 385–398.

Dybdahl MF and Lively CM (1995b) Host–parasite interactions: infection of common clones in natural populations of a freshwater snail (*Potamopyrgus antipodarum*). *Proceedings of the Royal Society of London* **260**: 99–103.

Dybdahl MF and Lively CM (1998) Host–parasite coevolution: evidence for rare advantage and time-lagged selection in a natural population. *Evolution* **52**: 1057–1066.

Elena SF and Lenski RE (1997) Test of synergistic interactions among deleterious mutations in bacteria. *Nature* **390**: 395–398.

Fry JD, Keightley PD, Heinsohn SL and Nuzhdin SV (1999) New estimates of the rates and effects of mildly deleterious mutation in *Drosophila melanogaster. Proceedings of the National Academy of Sciences of the United States of America* **96**: 574–579.

Gabriel W, Lynch M and Bürger R (1993) Muller's ratchet and mutational meltdowns. *Evolution* **47**: 1744–1757.

Hartman PS (1985) Epistatic interactions of radiation-sensitive (RAD) mutants of *Caenorhabditis elegans. Genetics* **109**: 81–94.

Howard RS and Lively CM (1994) Parasitism, mutation accumulation and the maintenance of sex. *Nature* **367**: 554–557.

Jaenike J (1978) An hypothesis to account for the maintenance of sex in populations. *Evolutionary Theory* **3**: 191–194.

Jokela J and Lively CM (1995) Parasites, sex and early reproduction in a mixed population of freshwater snails. *Evolution* **49**: 1268–1271.

Jokela J, Lively CM, Dybdahl MF and Fox JA (1997) Evidence for a cost of sex in the freshwater snail *Potamopyrgus antipodarum. Ecology* **78**: 452–460.

Keightley PD and Caballero A (1997) Genomic mutation rates for lifetime reproductive output with lifespan in *Caenorhabditis elegans. Proceedings of the National Academy of Sciences of the United States of America* **94**: 3823–3827.

Kibota TT and Lynch M (1996) Estimate of the genomic mutation rate deleterious to overall fitness in *E. coli. Nature* **381**: 694–696.

Kondrashov AS (1988) Deleterious mutations and the evolution of sexual reproduction. *Nature* **336**: 435–440.

Kondrashov AS (1993) Classification of hypotheses on the advantages of amphimixis. *Journal of Heredity* **84**: 372–387.

Lively CM (1987) Evidence from a New Zealand snail for the maintenance of sex by parasitism. *Nature* **328**: 519–521.

Lively CM (1989) Adaptation by parasitic trematode to local populations of its snail host. *Evolution* **43**: 1663–1671.

Lively CM (1992) Parthenogenesis in a freshwater snail: reproductive assurance versus parasitic release. *Evolution* **46**: 907–913.

Lively CM (1996) Host–parasite coevolution and sex: Do interactions between biological enemies maintain genetic variation and cross-fertilization? *BioScience* **46**: 107–114.

Lively CM and Howard RS (1994) Selection by parasites for clonal diversity and mixed mating. *Philosophical Transactions of the Royal Society of London* **346**: 271–281.

Lively CM and Jokela J (1996) Clonal variation for local adaptation in a host-parasite interaction. *Proceedings of the Royal Society of London* **263**: 891–897.

Lively CM, Craddock C and Vrijenhoek RC (1990) Red queen hypothesis supported by parasitism in sexual and clonal fish. *Nature* **344**: 864–866.

Lively CM, Lyons EJ, Peters AD and Jokela J (1998) Environmental stress and the maintenance of sex in a freshwater snail. *Evolution* **52**: 1482–1486.

Maynard Smith J (1978) *The Evolution of Sex.* Cambridge, UK: Cambridge University Press.

Muller HJ (1964) The relation of recombination to mutational advance. *Mutation Research* **1**: 1–9.

Radtkey RR, Donnellan SC, Fisher RN, *et al.* (1995) When species collide: the origin and spread of an asexual species of gecko. *Proceedings of the Royal Society of London* **259**: 145–152.

Schenck RA and Vrijenhoek RC (1986) Spatial and temporal factors affecting coexistence among sexual and clonal forms of *Poeciliopsis. Evolution* **40**: 1060–1070.

Vrijenhoek RC (1984) Ecological differentiation among clones: the frozen niche variation model. In Levin SA (ed.) *Population Biology and Evolution*, pp. 217–231. Berlin: Springer-Verlag.

West SA, Lively CM and Read AF (1999) A pluralist approach to sex and recombination. *Journal of Evolutionary Biology* **12**(6): 1003–1013.

Sex Selection

Dena S Davis, *Cleveland State University, Cleveland, Ohio, USA*

People have always attempted to influence the sex of their children. Until recently, prenatal screening followed by abortion of the undesired fetus was the only effective method, and most people who used that method did it to avoid the birth of girls. Today, preconception sex selection is becoming available, and people in the West are more likely to want to 'balance' their families than to avoid girls. These developments raise new questions about the ethics of sex selection.

Intermediate article

Article contents

- Introduction
- Techniques of Sex Selection
- Motives for Sex Selection
- Legal Issues
- Ethical Issues

Introduction

People have always speculated on what influences the conception and birth of boys or girls; in many cultures, people have attempted to control or influence the outcome of that process. Since the second half of the twentieth century, it has become possible to select the sex of one's offspring by selective abortion; at the beginning of the twenty-first century, techniques for sex selection that occur before implantation, or even before conception, are becoming available. How those techniques are used and the ethical, legal and social implications of that use can only be evaluated within cultural and historical contexts.

Techniques of Sex Selection

Premodern

Folk prescriptions abound for selecting the sex of one's offspring. These include: putting an axe under the bed before intercourse to conceive a boy; hanging the man's overalls on one bedpost or another; eating sweet or sour foods; and placing the bed to face in a particular direction. Hebrew and Greek sages agreed that girls came from the left testicle and boys from the right; French noblemen were advised to tie off or even amputate their left testicles in order to ensure male heirs. In many cultures, men blamed their lack of male heirs on their female partners. Even today, in many parts of the world, it is common for men to divorce or displace wives who do not produce sons.

Abortion

Until recently, the only reasonably sure way of selecting the sex of one's offspring was to ascertain the sex of a fetus *in utero*, by amniocentesis or ultrasound, and to abort the undesired fetus. This remains the most common technique and is commonly used in a number of developing countries.

Preconception sex selection

In the 1970s and 1980s, techniques such as douching and strategic timing of intercourse sought to exploit the differences between X- and Y-bearing sperm to influence the sex of the conceptus. Although a number of Western couples made use of these inexpensive, low-technology strategies, their success was never proved (Davis, 2001). In 1998, researchers in the USA announced a new and more successful technology that sorted the male partner's sperm in the laboratory and then used artificial insemination to impregnate the female partner with sperm carrying the desired trait. Although their technique is not completely effective, it improves the odds for conceiving a girl to more than 90% and is somewhat less successful at conceiving boys (Belkin, 1999).

Preimplantation genetic diagnosis as a concomitant of *in vitro* fertilization (IVF) is another way to ensure a fetus of a desired sex. In this technique, embryos created through IVF are tested for their genetic characteristics, including sex, and only those of the desired sex are implanted in the womb. This technique is so unwieldy and has such a high rate of pregnancy failure that it is unlikely to be used for sex selection alone, in the absence of some other problem.

Motives for Sex Selection

In some cultures, daughters are significant economic burdens, especially if they are destined to leave home and join their husband's family. Sons are therefore an important form of old-age security for their parents. Sons may also contribute to the parents' standing in the community and may be crucial for performing religious rituals that ensure a parent's well-being in the hereafter. Economist Amartya Sen has calculated that, due to feticide, infanticide and simple neglect, there are 100 million missing females around the world (Kristof, 1991). It is not only poor families who seek to avoid daughters. In contexts where people are achieving middle-class status by decreasing their family size, the total number of children desired falls more swiftly than the total number of desired sons, leaving even less 'room' for daughters (Das Gupta and Bhat, 1997).

In North America, there is strong evidence that people no longer prefer boys over girls. Numerous studies show that when people express interest in sex selection, it is to 'balance' their family by seeking to conceive a girl if they already have a boy, or vice versa (Davis, 2001).

Legal Issues

Many countries in both the developed and developing worlds, including some states within India, have laws forbidding abortion for sex selection, or even forbidding the use of screening techniques such as ultrasound if the purpose is to determine the sex of the fetus (in the absence of a sex-related genetic disease). Most of these laws are ineffective, as it is impossible to police a couple's motivation for screening. Further-

more, when a person has a legitimate reason for screening, such as advanced maternal age, it is usually considered her legal right to know all the information that results from the screening, of which fetal sex is inevitably one part.

Ethical Issues

Feminist philosophers find the problem of sex selection to be an enormous challenge. On the one hand, they have sympathy with the women in those countries for whom the birth of too many daughters and not enough sons may spell divorce or even death. They also are staunchly supportive of a woman's right to decide for herself why she should have an abortion. Among genetic counselors in the USA, a 1989 survey found that a majority would perform prenatal diagnosis for the purpose of sex selection or refer the couple to someone who would (Fletcher and Wertz, 1990). On the other hand, as most sex selection around the world is directed against girls and expresses a profound sexism and even misogyny, feminists deplore the practice (Powledge, 1981). (*See* Feminist Perspectives on Human Genetics and Reproductive Technologies.)

For many ethicists, simply the fact that sex selection is accomplished by abortion is enough to make it immoral. A common ethical position is to support abortion for 'serious' reasons such as fetal anomalies or the burden on the family of raising another child; aborting a fetus of the 'wrong' sex is often held up as the most egregious of 'trivial' reasons.

Another argument against sex selection is that it skews the normal gender ratio in a society. In India, for example, a 1991 census found 92.9 females for every 100 males (Balakrishnan, 1994, p. 269). Although one might think that 'market forces' would act to make women more desirable and improve their lives, what happens instead is that women are increasingly valued for their domestic and reproductive capacities alone and find their lives increasingly narrowed (Guttentag and Secord, 1983). On the other hand, if sex selection will happen anyway, through infanticide and neglect, perhaps abortion is the more humane alternative. Even in countries where feticide and infanticide account for the loss of many girls, the most significant decrease in girls occurs between the ages of 1 and 4 years, because boys are favored in receiving nutrition and medical care (Balakrishnan, 1994, p. 269).

Sex selection in North America at the beginning of the twenty-first century presents a very different set of ethical issues. First, with the advent of effective means for preconception sex selection, the argument will shift away from the focus on abortion. Prochoice feminists may then feel more free to criticize sex selection, but other commentators, who relied on abortion as their main argument against sex selection, will have to rethink their position. Second, most people in North America, if they were to use sex selection at all, would probably use it only to balance the numbers of children of each sex in their family (the exceptions appear mainly to be among people who have immigrated from countries where the pressure to produce boys is strong). There is even some evidence that people who avail themselves of preconception selection techniques are more likely to be trying to have girls (Belkin, 1999). Thus, it is no longer obvious that sex selection devalues women or will lead to societal imbalance. In the absence of these arguments, some commentators assert that sex selection is morally neutral (Wertz and Fletcher, 1989).

Other commentators continue to find that sex selection is 'sexist', but they must make this argument in more nuanced terms. Selection can be sexist if it is motivated by rigid notions of gendered behavior, so that parents who want an 'assertive' child will make sure to have a boy, while parents seeking a 'loving' child, or perhaps one who values music over athletics, will want a girl. Parents who believe that only boys can go on fishing trips with their fathers and who make sure that they have at least one boy for that reason will miss out on the discovery that girls can enjoy fishing as well and will perpetuate gender stereotypes by selecting children to meet those expectations (Bayles, 1984; Pogrebin, 1980). Even parents who want a girl so as to oppose gender stereotyping and raise the first woman president of the United States are still seeing her primarily in terms of gender. Some ethicists caution that parents who invest substantial resources in making sure that they have a girl or a boy may be insufficiently attuned to the child's unique characteristics and individual flourishing (Davis, 2001; Ryan 1992).

See also
Clinical Genetics and Genetic Counseling Professionals: Attitudes to Contentious Issues
Genome Sequencing

References

Balakrishnan R (1994) The social context of sex selection and the politics of abortion in India. In: Sen C and Snow R (eds.) *Power and Decision: The Social Control of Reproduction*, Harvard Series on Population and International Health, pp. 267–286. Cambridge, MA: Harvard University Press.

Bayles M (1984) *Reproductive Ethics*. Englewood Cliffs, NJ: Prentice-Hall.

Belkin L (1999) Getting the girl. *New York Times Magazine* 25 July.

Das Gupta M and Bhat MPN (1997) Fertility decline and increased manifestation of sex bias in India. *Population Studies* **51**: 307–315.

Davis D (2001) *Genetic Dilemmas: Reproductive Technology, Parental Choices, and Children's Futures*. New York, NY: Routledge.

Fletcher JC and Wertz DC (1990) Ethics, law, and medical genetics. *Emory Law Journal* **39**: 747–809.

Guttentag M and Secord PF (1983) *Too Many Women? The Sex Ratio Question*. Beverly Hills, CA: Sage Publications.

Kristof ND (1991) Stark data on women: 100 million are missing. *New York Times* 5 November.

Pogrebin L (1980) *Growing Up Free: Raising Your Child in the 80s*. New York, NY: McGraw-Hill.

Powledge T (1981) Unnatural selection: on choosing children's sex. In: Holmes HB, Hoskins BB, Gross M (eds.) *The Custom-made Child? Women-centered Perspectives*, pp. 93–100. Clifton, NJ: Humana Press.

Ryan M (1992) The argument for unlimited procreative liberty: a feminist critique. In: Campbell C (ed.) *What Price Parenthood? Ethics and Assisted Reproduction*, pp. 83–90. Aldershot, UK/ Brookfield, VT: Dartmouth Publishing Company.

Wertz DC and Fletcher JC (1989) Fatal knowledge? Prenatal diagnosis and sex selection. *Hastings Center Report* **19**: 21–27.

Further Reading

Bumiller E (1990) *May You Be the Mother of a Hundred Sons: A Journey Among the Women of India*. New York, NY: Random House.

Robertson JA (2001) Preconception gender selection. *American Journal of Bioethics* **1**: 2–39.

Rothman BK (1986) *The Tentative Pregnancy: Prenatal Diagnosis and the Future of Motherhood*. New York, NY: Viking Press.

Rothman BK (1998) *Genetic Maps and Human Imaginations: The Limits of Science in Understanding Who We are*. New York, NY: WH Norton and Company, Incorporated.

Warren MA (1985) *Gendercide: The Implications of Sex Selection*. Totowa, NJ: Rowman and Allanheld.

Sexual Orientation: Genetics

Intermediate article

J Michael Bailey, *Northwestern University, Evanston, Illinois, USA*

Research into a possible genetic basis for homosexuality has suggested the role of an as yet undiscovered gene localized to the q28 region of the X chromosome. However, this suggestion is controversial for a number of reasons, and the matter is unlikely to be resolved without a considerable amount of additional research.

Sexual orientation is the degree to which one is attracted to the same versus the other sex. The trait is often measured via a seven-point Kinsey scale (Kinsey *et al.*, 1953), with a score of 0 representing complete heterosexuality, a score of 6 representing complete homosexuality and scores of 1–5 representing intermediate degrees of preference. At least 90% of men and women rate themselves 0 on the Kinsey scale, but perhaps 3% of men and 1% of women admit to intense homosexual attractions (Bailey *et al.*, 2000; Laumann *et al.*, 1994). Homosexual orientation is associated with sex atypicality (i.e. femininity in males and masculinity in females), especially during childhood (Bailey and Zucker, 1995).

The origins of variation in sexual orientation and related traits have been controversial for both scientific (Byne, 1994) and social reasons. The scientific controversies have been those associated with nature–nurture questions. The most influential scientific hypothesis in recent years has been the neurohormonal theory (Ellis and Ames, 1987), that homosexual orientation results from atypical action of androgen during prenatal development. That is, the brains of gay men are hypothesized to have been exposed to low levels of androgen and the brains of lesbians to high levels of androgens, compared with heterosexual individuals of the same sex. This hypothesis could help explain why homosexuality is often accompanied by other sex-atypical behavior. Causes of atypical androgen exposure are unknown, but might include genetic variation. Nurture hypotheses have been less well specified, but include parental socialization and social attitudes (Bailey and Dawood, 1998).

Kallmann (1952) conducted the first sizable genetic study of sexual orientation. He ascertained gay twins in the 'homosexual underworld' and correctional/mental institutions of New York City during the late 1940s. Remarkably, 100% of 37 monozygotic (MZ) twin pairs were concordant compared with 15% of 26 dizygotic (DZ) pairs. His study had a number of methodological defects, including its overreliance on (evidently) mentally ill gay men, lack of information on zygosity diagnosis, and especially its anomalously high rate of MZ concordance compared with other studies (Rosenthal, 1970). Still, it is remarkable that despite its promising results, nearly 40 years passed before another large twin study of male homosexuality was attempted (Bailey and Pillard, 1991). In this study twins were recruited via advertisements in gay-oriented publications. It yielded much lower concordance rates for male homosexuality: 52% and 22% for MZ and DZ twins respectively. Adoptive brothers of gay men had a rate of homosexuality of 11%. These results are consistent with substantial, but imperfect, heritability. A similar study was conducted for female homosexuality (Bailey *et al.*, 1993), and it yielded similar results. The rate of homosexuality for sisters of lesbians was 48% for MZ twins, 16% for DZ twins and 6% for adoptive sisters.

Several twin studies have been conducted since then, and in general their results are generally consistent with the following conclusions. First, nearly all studies find higher concordance rates for MZ than for DZ twins. Second, the specific rates vary widely, with the most carefully collected samples yielding rates well below 50% even for MZ twins (Bailey *et al.*, 2000; Kendler *et al.*, 2000). Heritability estimates depend both on concordance rates and on the mode of transmission that is assumed (e.g. polygenic versus monogenic), and both of these facts are unknown. Thus, estimating heritability is premature. Nevertheless, sexual orientation appears to have at least modest heritability in both men and women.

Environmental effects obviously also exist, because MZ twins often have different sexual orientations. One well-established environmental effect on male sexual orientation is fraternal birth order. Specifically, gay men tend to be later born in a series of male births (Blanchard, 2001). MZ twins who differ in sexual orientation recall different patterns of childhood behavior, with the homosexual twin recalling more sex atypicality. Thus, environmental influences appear to act early, and could be prenatal. Two case histories have been reported representing an extreme environmental manipulation. In both cases, a male infant lost his penis due to a surgical accident and was subsequently reared as a girl (Diamond and Sigmundson, 1997; Bradley *et al.*, 1998). In both cases, adult sexual attraction was primarily to women. This suggests that it is difficult to redirect sexual orientation after birth.

Although twin studies can provide evidence for genetic influence, they reveal nothing about the particular genes involved. In 1993, Dean Hamer published a report suggesting that a gene on the X chromosome influenced male sexual orientation (Hamer *et al.*, 1993). Two independent findings supported this conclusion. First, gay men reported more gay male relatives (uncles and first cousins) on their mother's than on their father's side. Second, a sample of 40 gay brothers shared markers on chromosomal segment Xq28 much more often than would be expected by chance. The finding was controversial for a variety of reasons, including the fear that it would lead to a prenatal test for homosexuality. Risch *et al.* (1993) made several scientific criticisms. These included specific aspects of Hamer's methodology as well as scepticism that a gene for homosexuality could be sufficiently common to be detectable in such a small study. Homosexual orientation is associated with markedly reduced reproductive success (Bell and Weinberg, 1978) and thus, genes for homosexuality should be rare.

Hamer's lab replicated the result shortly thereafter (Hu *et al.*, 1995). In this study, they also included data from heterosexual brothers of gay men, showing that heterosexual siblings share Xq28 markers less often than would be expected by chance. The rate of haplotype sharing was somewhat weaker in this study (67%) than in the prior study (83%), but it was still statistically reliable. Furthermore, they also included data from pairs of lesbian sisters, who did not show a tendency to share Xq28 markers.

Scientists tend to prefer independent replications to replications from the same lab. One study failed to replicate Hamer's finding that gay men report more maternal than paternal gay relatives (Bailey *et al.*, 1999). In the only independent linkage study to date Rice *et al.* (1999) found that 52 pairs of gay brothers did not share Xq28 markers more often than would be expected by chance. They engaged Hamer in a technical exchange, in which each side claimed victory. Hamer concluded that after combining all relevant studies, there appears to be linkage between Xq28 markers and male homosexuality. Rice *et al.* argued that results from independent replication attempts had failed to support Hamer's original finding.

What is the current status of Xq28 as a potential site of a gene for male sexual orientation? The available data do not yield a definitive verdict. The studies that have been done are small, and they have produced conflicting results. We will not resolve this controversy by arguing about the data that have been published so far, but by conducting new, larger studies. Unfortunately, it is unclear whether this will happen in the near future. Social controversy has impeded scientific work in this area. This controversy has been fueled by the belief that causal hypotheses about sexual orientation have important social and moral implications (Kallmann, 1952). This belief is mistaken. The value of a trait should be judged by its consequences rather than its causes. Social and political controversies about homosexuality have had an adverse effect on scientific research. Despite keen interest in the topic by scientists and the public, obtaining research funding to study it has been difficult.

See also
Chromosome X: General Features

References

Bailey J and Dawood K (1998) Behavioral genetics, sexual orientation, and the family. In: Patterson CJ and D'Augelli AR (eds.) *Lesbian, Gay, and Bisexual Identities in Families: Psychological Perspectives*, pp. 3–18. New York, NY: Oxford University Press.

Bailey J, Dunne MP and Martin NG (2000) Genetic and environmental influences on sexual orientation and its correlates in an Australian twin sample. *Journal of Personality and Social Psychology* **78**(3): 524–536.

Bailey J and Pillard RC (1991) A genetic study of male sexual orientation. *Archives of General Psychiatry* **48**(12): 1089–1096.

Bailey J, Pillard RC, Dawood K, *et al.* (1999) A family history study of male sexual orientation using three independent samples. *Behavior Genetics* **29**(2): 79–86.

Bailey J, Pillard RC, Neale MC and Agyei Y (1993) Heritable factors influence sexual orientation in women. *Archives of General Psychiatry* **50**: 217–223.

Bailey J and Zucker KJ (1995) Childhood sex-typed behavior and sexual orientation: a conceptual analysis and quantitative review. *Developmental Psychology* **31**(1): 43–55.

Bell A and Weinberg M (1978) *Homosexualities: A Study of Diversity Among Men and Women*. New York, NY: Simon and Schuster.

Blanchard R (2001) Fraternal birth order and the maternal immune hypothesis of male homosexuality. *Hormones and Behavior* **40**(2): 105–114.

Bradley SJ, Oliver GD, Chernick AB and Zucker KJ (1998) Experiment of nurture: ablatio penis at 2 months, sex reassignment at 7 months, and a psychosexual follow-up in young adulthood. *Pediatrics* **102**(1): e9.

Byne W (1994) The biological evidence challenged. *Scientific American* May: 50–55.

Diamond M and Sigmundson HK (1997) Sex reassignment at birth. Long-term review and clinical implications. *Archives of Pediatrics and Adolescent Medicine* **151**(3): 298–304.

Ellis L and Ames MA (1987) Neurohormonal functioning and sexual orientation: a theory of homosexuality–heterosexuality. *Psychological Bulletin* **10**(2): 233–258.

Hamer DH, Hu S, Magnuson VL, Hu N and Pattatucci AML (1993) A linkage between DNA markers on the X chromosome and male sexual orientation. *Science* **261**(5119): 321–327.

Hu S, Pattatucci AML, Patterson C, *et al.* (1995) Linkage between sexual orientation and chromosome Xq28 in males but not in females. *Nature Genetics* **11**: 248–256.

Kallmann FJ (1952) Twin and sibship study of overt male homosexuality. *American Journal of Human Genetics* **4**: 136–146.

Kendler KS, Thornton LM, Gilman SE and Kessler RC (2000) Sexual orientation in a U.S. national sample of twin and nontwin sibling pairs. *American Journal of Psychiatry* **157**(11): 1843–1846.

Kinsey AC, Pomeroy WB, Martin CE and Gebhard PH (1953) *Sexual Behavior in the Human Female*. Philadelphia, PA: WB Saunders.

Laumann EO, Gagnon JH, Michael RT and Michaels S (1994) *The Social Organization of Sexuality: Sexual Practices in the United States*, pp. 283–320. Chigago, IL: University of Chicago Press.

Rice G, Anderson C, Risch N and Ebers G (1999) Male homosexuality: absence of linkage to microsatellite markers at Xq28. *Science* **284**(5414): 665–667.

Risch N, Squires-Wheeler E and Keats BJB (1993) Male sexual orientation and genetic evidence. *Science* **262**: 2063–2065.

Rosenthal D (1970) *Genetic Theory and Abnormal Behavior*. New York, NY: McGraw-Hill.

Short Interspersed Elements (SINEs)

Carl W Schmid, *University of California, Davis, California, USA*

Carol M Rubin, *University of California, Davis, California, USA*

Intermediate article

Article contents

- A Class of Repetitive Mobile Elements
- SINE Evolution
- Mechanism of SINE Retrotransposition
- SINE Mobility and Mutagenic Effects
- SINE Effects
- SINE Function

Short interspersed elements (SINEs) are highly repetitive sequences that retrotranspose into eukaryotic DNA through intermediates transcribed by RNA polymerase III (pol III). In many species, SINEs are a ubiquitously dispersed feature of the whole genome, often constituting a significant ($\sim$ 10%) mass fraction of total DNA. SINEs cause mutations both by their retrotransposition within genes and by unequal recombination, and are widely considered to be examples of 'selfish' or 'parasitic' DNA.

A Class of Repetitive Mobile Elements

The very first genomic studies showed that human deoxyribonucleic acid (DNA), and similarly the DNA of most higher eukaryotes, is largely interspersed with highly repetitive, retrotransposed sequences (Schmid, 1996). Most of these interspersed repeats consist of elements ending in an A-rich 3′ tail surrounded by short direct repeats that result from duplication of the genomic insertion site – a hallmark of their transposition. Unlike viral retrotransposons, these elements do not have flanking long terminal repeats (LTRs). This class of retrotransposed sequences has been subdivided further into long and short interspersed nuclear elements (LINEs and SINEs respectively) – a seemingly arbitrary division according to length that now highlights the different requirements for their retrotransposition (see below).

As the most extensively studied SINE, human Alu repeats (named for a diagnostic restriction site) illustrate features of this class of sequence. Alu repeats share a consensus sequence of about 282 nucleotides (nt), followed by an A-rich 3′ tail that varies in both length and sequence (Schmid,1996,1998). The 1–2 million copies of Alu sequences would correspond to an average genomic spacing about 2000 base pairs (bp) if they were uniformly distributed throughout the genome, and most regions of human DNA contain multiple copies of this sequence. Individual Alu elements are usually flanked by the duplicated genomic target site, which also defines the 5′ end of the sequence. Except for some truncated copies, the 5′ end corresponds precisely to the transcriptional initiation site for ribonucleic acid polymerase III (RNA pol III).

Alu repeats are ancestrally derived from the gene for signal recognition particle (SRP) RNA (also called 7SL RNA) and like that gene contain an internal promoter for transcription directed by pol III.

SRP-RNA-related SINEs are restricted to primates and rodents. In all other species, SINEs, if present (see below), are instead derived from a particular transfer RNA (tRNA; Ohshima *et al.*, 1996; Okada and Ohshima, 1995). Examples of tRNA-derived SINEs are present in cow, silk worm, rice blast fungus and *Arabidopsis*. tRNA homology provides these SINEs with internal pol III promoter elements so that, on their insertion, all SINEs import the promoter that is necessary for their transcription. This internal pol III promoter is therefore probably essential for maintaining the copy number of SINEs.

SINE Evolution

Although anthropoid primates do not have an active tRNA-related SINE, prosimians have both Alu and tRNA-related SINEs (Schmid, 1996, 1998). Similarly, rodents have both an SRP-RNA-related SINE (usually called B1) and a tRNA-related SINE (B2). Although rodent B1 and primate Alu repeats are each homologous to SRP RNA, they are distinct sequences (see **Figure 1**). The 282-nt Alu consensus sequence is a tandem dimer of SRP RNA homologs; the shorter B1 sequence is monomeric. These and the abrupt phylogenetic differences described above raise issues of how and why new SINE families appear and how previously successful families disappear.

Comparisons of orthologous Alu repeats in primates show that some Alu insertions predate even monkey–human divergence and indicate further that Alu elements diverge at the rate expected for unselected sequences (i.e. sequences that are neutral to an organism's fitness) without sequence conversion (Schmid, 1996). On average, human Alu repeat sequences differ by 15–20% owing to the accumulation of mutations that have occurred after their insertions. In addition to random mutations, Alu elements are divisible into

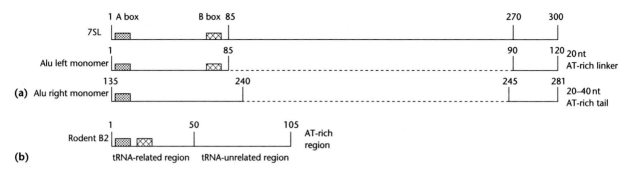

Figure 1 Interspersed repeat structure. (a) The primate Alu left and right monomers are shown relative to their ancestor, 7SL. The two monomers are linked by a short AT-rich region. Only the left monomer contains functional A and B boxes that are essential for RNA polymerase III transcription. The right monomer contains an additional 31 nucleotides of 7SL sequence not found in the left monomer (numbering is approximate). (b) Structure of the tRNA-related rodent B2 SINE. The ancestor of the B2 tRNA-related portion is lysine tRNA. Other species contain SINEs derived from glycine or arginine tRNAs that have similar structures (see Okada and Ohshima, 1995).

distinct subfamilies of different evolutionary age and different degrees of sequence divergence (Schmid, 1996). Each subfamily bears the distinguishing sequence of its actively propagating parent. One evolutionarily young subfamily has been traced back to a founder sequence that is positioned fortuitously near flanking sequences that apparently stimulate its transcription. Simple founder effects explain the appearance of new Alu subfamilies and, by implication, different SINEs in phylogenetic comparisons (Schmid, 1996).

The phylogeny outlined above indicates that the common ancestor of all mammals had tRNA-related SINEs. Very highly divergent fossil tRNA-related SINEs (called MIRs) are present as mouse–human orthologs, which proves their antiquity (Jurka *et al.*, 1995; Smit and Riggs, 1995). This previously successful SINE family became inactive in both mouse and human. Despite the presence of tRNA-related SINEs in insects and fungi, SINEs are apparently absent in two premier genetic models, *Drosophila* and yeast, and presumably have been eliminated from these genomes.

Mechanism of SINE Retrotransposition

Polymerase-III-directed SINE transcripts, like tRNAs and SRP RNA, are not expected to encode proteins; therefore the factors necessary for SINE retrotransposition must be made elsewhere in the genome (i.e. *in trans*). LINEs encode their own reverse transcriptase and other factors required for autonomous retrotransposition. The human LINE-1 endonuclease cleaves at a preferred sequence when opening the genomic

insertion site. The direct repeats that flank Alu elements resemble the LINE target preference, which suggests that LINEs direct their insertion (Jurka, 1997). In addition, tRNA-related SINEs include sequences that are unrelated to the tRNA. In several cases, the tRNA-unrelated portion of an SINE is homologous to an LINE variant that exists or had existed previously in the same lineage (Okada *et al.*, 1997; Okada and Hamada, 1997), which suggests that tRNA primers for LINE reverse transcription became linked to these regions, thereby creating new SINEs. This model suggests further that the genomic maintenance of a particular SINE family depends on the continuing activity of its cognate LINE, the loss of which dooms the SINE to extinction. SINE evolution may be driven totally by LINE evolution (Okada *et al.*, 1997).

SINE Mobility and Mutagenic Effects

As discussed, comparison of orthologous Alu repeats in human and chimpanzee indicates that most human Alu elements were fixed in the ancestral primate genome and maintained during the separate evolutionary histories of the primates. Younger subfamilies that expanded after primate divergence have also been identified (Schmid, 1996). Members of young subfamilies can be recognized because they are less divergent than older Alu elements and are absent in one branch but present in another of two closely related species. In addition, some members of young Alu subfamilies are not fixed in a population, which suggests that they appeared after its radiation.

SINE dimorphisms, corresponding to alleles for the presence and absence of the SINE at a particular locus, are particularly informative for pedigree analysis because the ancestral allele is unambiguously identifiable as the 'empty' locus that predated insertion (Deininger et al., 1997). Ideally, the empty allele contains the unduplicated genomic target site and has the same flanking unique sequences as those surrounding the SINE insertion in the occupied allele. Unequal recombination can and does (see below) prune SINEs from a genome, but these events rearrange the flanking regions and are thus distinct from the ancestral empty site. Alu dimorphisms have already proved useful for determining the relatedness of human populations, and nonhuman SINEs should be equally informative in examining other pedigrees (Stoneking et al., 1997).

De novo germ-line SINE insertions (and possibly even a somatic event) have been observed to cause phenotypic effects in human and rodent (Labuda et al., 1995; Deininger and Batzer, 1999). The frequency of Alu insertions is estimated at 1 in every 200 births, which corresponds to 0.1% of human genetic disorders (Deininger and Batzer, 1999).

Unequal homologous recombination between dispersed SINEs invariably duplicates and deletes intervening regions (Labuda et al., 1995; Schmid, 1996; Deininger and Batzer, 1999). Numerous genetic diseases have been identified as resulting from unequal Alu–Alu recombination, including both intrachromosomal and interchromosomal events. Given the high copy number of many SINE families, the most remarkable aspect of ectopic SINE recombination is perhaps that such events are not more frequent. Perhaps the divergence of SINE members (e.g. 15–20% for Alu elements) interferes with their efficient recombination (Schmid, 1998).

SINE Effects

The dispersion of an abundant sequence family throughout the genome must inevitably affect gene structure and expression (Labuda et al., 1995; Makalowski, 1995). For example, unequal homologous SINE recombination can create new genes by altering and more especially by duplicating existing genes. Such events are more properly regarded as an effect rather than a function of SINEs. There are other such effects. Nonconsensus base substitutions within SINEs can generate binding sites for transcription factors. Because of the accumulation of SINEs in gene-rich regions (Lander et al., 2001), many such sites are located near promoter regions and modulate the expression of neighboring genes. SINEs present in the 3' UTRs of messenger RNAs (mRNAs) provide alternate polyadenylation signals and probably alter

their posttranscriptional regulation. SINEs positioned near the 3' UTR can also supply a termination signal, which results in a truncated protein containing SINE-encoded C-terminal substitutions. Because of their genomic dispersion, SINEs are abundantly represented as pol-II-directed read-through transcripts in precursor mRNAs. Although largely removed during processing, nonconsensus base substitutions within such SINEs can generate cryptic splice sites such that, in some cases, small portions of SINE-derived sequences are spliced into the mature mRNA and supply amino acid substitutions in the protein product.

The genome is sufficiently plastic to be reshaped to some degree by selective pressure. That SINEs are involved in this process is a consequence of their ubiquity and abundance rather than a reason for their existence.

SINE Function

As SINEs have no known function, they possibly have no function (Howard and Sakamoto, 1990). The absence of SINEs in Drosophila and yeast shows that they are not essential, but this does not necessarily mean that SINEs are functionless. Their function might have been lost or is satisfied in some other manner in these organisms. As previously discussed, SINE sequences are not well conserved in evolution, which implies that they are functionless or that the nucleotide sequence per se is not essential for their function. SINEs might function at the level of DNA or through the pol-III-directed transcript that they encode.

The dispersion of SINEs raises the possibility that they have a role in genome structure. Alu repeats account for about a third of the potential sites for modification of human DNA by cytosine methylation (Schmid, 1998). In addition, Alu methylation shows profound tissue-specific differences. Methylation affects both chromatin structure and transcription, which underscores a possible role for Alu and, more generally, SINEs in genome organization.

The regulated pol-III-directed transcription of SINEs is incompletely understood but again most extensively studied for human Alu elements (Schmid, 1998). Alu transcription is subject to many types of control, including repression caused by DNA methylation and chromatin packing, and cis-acting effects of the unique sequences that flank individual Alu elements. Viewed in this perspective, SINEs are members of a multigene family in which each member is, to a degree, regulated uniquely. If this view is correct, then a complete description of their transcriptional regulation will be highly complex.

Despite the abundance of SINEs in many species, their corresponding transcripts are often expressed far less than other pol-III-directed transcripts. This is particularly true for Alu RNA, which is barely detectable in some human cell lines. But various cell stresses transiently increase the abundance of SINE RNA in mammals and silk worm (Schmid, 1998; Kimura et al., 1999). Examples of these stresses include heat shock, ethanol toxicity, exposure to heavy metals, translational inhibition by either cycloheximide or puromycin, and viral infection including human immunodeficiency virus, herpes and adenovirus. The SINE stress response has been observed in both living animals and cultured cells, which shows that these increases are a normal vital response (Li et al., 1999). The SINE stress response occurs with about the same kinetics as the stress-induced increase in heatshock protein mRNAs; thus, SINEs behave like classic cell stress genes. In some instances, the SINE response is even more robust than that of heatshock genes, which indicates that SINE RNA may be a useful bioindicator of stress. Present indications suggest that SINE RNA might serve a regulatory function during stress recovery.

See also
Evolutionarily Conserved Noncoding DNA
Long Interspersed Nuclear Elements (LINEs)
Retrosequences and Evolution of Alu Elements
Transposons

References

Deininger PL and Batzer MA (1999) *Alu* repeats and human disease. *Molecular Genetics and Metabolism* **67**: 183–193.

Deininger PL, Sherry ST, Risch G, et al. (1997) Interspersed repeat insertion polymorphisms for studies of human molecular anthropology. In: Papiha SS, Deka R and Chakraboorty R (eds.) *Genomic Diversity: Applications in Human Population Genetics*, chap. 14, pp. 201–212. New York, NY: Kluwer Academic.

Howard BH and Sakamoto K (1990) *Alu* interspersed repeats: selfish DNA or a functional gene family? *New Biologist* **2**: 759–770.

Jurka J (1997) Sequence patterns indicate an enzymatic involvement in integration of mammalian retroposons. *Proceedings of the National Academy of Sciences of the United States of America* **94**: 1872–1877.

Jurka J, Zietkiewcz E and Labuda D (1995) Ubiquitous mammalian-wide interspersed repeats (MIRs) are molecular fossils from the mesozoic era. *Nucleic Acids Research* **23**: 170–175.

Kimura RH, Choudary PV and Schmid CW (1999) Silk worm Bm1 SINE RNA increases following cellular insults. *Nucleic Acids Research* **27**: 3380–3387.

Labuda D, Zietkiewcz E and Mitchell GA (1995) *Alu* elements as a source of genomic variation: deleterious effects and evolutionary novelties. In: Maraia RJ (ed.) *The Impact of Short Interspersed Elements (SINEs) on the Host Genome*, chap. 1, pp. 1–24. Austin, TX, USA: RG Landes Co.

Lander ES, Linton LM, Birrin B, et al. (2001) Initial sequencing and analysis of the human genome. *Nature* **409**: 860–921.

Li T, Spearow J, Rubin CM and Schmid CW (1999) Physiological stresses increase mouse short interspersed element (SINE) RNA expression *in vivo*. *Gene* **239**: 367–372.

Makalowski W (1995) SINEs as a genomic scrap yard: an essay on genomic evolution. In: Maraia RJ (ed.) *The Impact of Short Interspersed Elements (SINEs) on the Host Genome*, chap. 5, pp. 81–104. Austin, TX, USA: RG Landes Co.

Ohshima K, Hamada M, Terai Y and Okada N (1996) The 3′ ends of tRNA-derived short interspersed repetitive elements are derived from the 3′ ends of long interspersed repetitive elements. *Molecular and Cell Biology* **16**: 3756–3764.

Okada N and Hamada M (1997) The 3′ ends of tRNA-derived SINEs originated from the 3′ ends of LINEs: a new example from the bovine genome. *Journal of Molecular Evolution* **44**(supplement 1): S52–S56.

Okada N, Hamada M, Ogiwara I and Ohshima K (1997) SINEs and LINEs share common 3′ sequences: a review. *Gene* **205**: 229–243.

Okada N and Ohshima K (1995) Evolution of tRNA derived SINEs. In: Maraia RJ (ed.) *The Impact of Short Interspersed Elements (SINEs) on the Host Genome*, chap. 4, pp. 61–80. Austin, Texas, USA: RG Landes Co.

Schmid CW (1996) *Alu*: structure, origin, evolution, significance and function of one-tenth of human DNA. *Progress in Nucleic Acid Research and Molecular Biology* **53**: 283–319.

Schmid CW (1998) Does SINE evolution preclude *Alu* function? *Nucleic Acids Research* **26**: 4541–4550.

Smit AF and Riggs AD (1995) MIRs are classic, tRNA-derived SINEs that amplified before the mammalian radiation. *Nucleic Acids Research* **23**: 98–102.

Stoneking M, Fontius JJ, Clifford SL, et al. (1997) *Alu* insertion polymorphisms and human evolution: evidence for a larger population size in Africa. *Genome Research* **11**: 1061–1071.

Shotgunning the Human Genome: A Personal View

J Craig Venter, *The Center for the Advancement of Genomics, Rockville, Maryland, USA*

Using the whole-genome shotgun method as the primary approach to sequencing large genomes, notably the human genome, Celera has demonstrated the validity of this method, which is now used in genome-sequencing laboratories worldwide for all classes of organisms.

Intermediate article

Article contents

Introduction

The publication of the human genome sequence in the February 16, 2001 issue of *Science* demonstrates the value of the 'shotgun' method of sequencing large genomes. The whole-genome shotgun approach randomly breaks the genome into deoxyribonucleic acid (DNA) fragments whose chemical 'letters' are read by automated DNA sequencers. Then, unique computer algorithms are used to assemble millions of random fragments into a properly ordered genome sequence. The shotgun method eliminates the need to map landmarks throughout the genome prior to sequencing DNA, a laborious and therefore expensive process.

Whole-genome Shotgun Approach

In 1996, the publicly funded Human Genome Project was beginning to shift resources from genome mapping to the more complex sequencing phase when we proposed that the whole-genome shotgun approach be used to finish the job. Our paper in *Nature* (Venter *et al.*, 1996) argued that scientific advances made it feasible to apply the shotgun method to the human genome and complete the project before 2005 under the projected cost of $3 billion. Months earlier, our team at The Institute for Genomic Research (TIGR) had used the shotgun method to decode the bacterium *Haemophilus influenzae*, the largest sequenced genome at the time. We suggested in *Nature* that the use of automated sequencers had clear advantages over the then current strategy.

Our proposal was not well received. Critics said the shotgun method would fail because the human genome is thousands of times larger than that of *H. influenzae* and has long stretches of repetitive DNA not found in bacteria. They said that millions of sequenced but unordered DNA fragments would be impossible to assemble without the landmarks of a physical map.

It was widely predicted that repetitive sequences would confuse the computer programs used in assembly.

In early 1998, PE Biosystems (now Applied Biosystems) invited a team from TIGR to inspect a new, high-throughput DNA analyzer, the ABI PRISM 3700. It was clear that several hundred machines running at full capacity could decode the human genome within approximately 3 years. In June, we announced the plans to form a company, Celera Genomics, and use the shotgun method to sequence the human genome. With a subset of the DNA sequencers, which would eventually read human DNA, Celera and the Berkeley *Drosophila* Genome Project tested the strategy on the fruit fly.

The *Drosophila* genome was sequenced repeatedly 14 times and shred into 3 million fragments. Five hundred letters at the ends of fragments were then sequenced. The sequenced ends (called mate pairs) became markers used by the computer assembly programs, which identified overlapping fragments and used mate pairs to orient the overlaps along contiguous stretches of the genome. The *Drosophila* sequence was published in March 2000 (Adams *et al.*, 2000).

In February 2001, the simultaneous publication of two human genome sequences validated both the data and the shotgun approach (International Human Genome Sequencing Consortium, 2001; Venter *et al.*, 2001). In the months prior to publication, both teams relied on the ABI 3700 analyzer and data generated by shotgun sequencing. Celera reported the order and orientation of more than 95% of the genome with an accuracy greater than 99.6% (Venter *et al.*, 2001). Celera estimates that there are a total of between 26 500 and 30 000 human genes.

See also

References

Adams MD, Celnicker SE, Holt RA, *et al.* (2000) The genome sequence of *Drosophila melanogaster*. *Science* **287**: 2185–2195.

International Human Genome Sequencing Consortium (2001) Initial sequencing and analysis of the human genome. *Nature* **409**: 860–921.

Venter JC, Smith HO and Hood L (1996) A new strategy for genome sequencing. *Nature* **381**: 364–366.

Venter JC, Adams MD, Myers EW, *et al.* (2001) The sequence of the human genome. *Science* **291**: 1304–1351.

Web Links

Celera.com. Here can be found the complete sequence of Celera's Human Reference Genome, as published in *Science*. Researchers can compare this early, highly accurate version of the Celera human genome assembly with that of the public genome project [AB1]

http://public.celera.com/cds/login.cfm

Sickle Cell Disease as a Multifactorial Condition

Martin H Steinberg, *Boston University School of Medicine, Boston, Massachusetts, USA*

The phenotype of sickle cell anemia is heterogeneous, although all patients have the identical sickle mutation. The products of epistatic modifying genes and the sickle hemoglobin gene interact to determine the disease phenotype.

Advanced article
Article contents
• Pathophysiology of Sickle Cell Disease
• Fetal Hemoglobin and Sickle Cell Disease
• Genetic Conditions Known to Affect Sickle Cell Disease
• Red Cell Membrane and Cation Transport
• Endothelium as a Modulator in Sickle Cell Anemia
• Inflammation
• Hemostasis

Pathophysiology of Sickle Cell Disease

Homozygosity for a unique hemoglobin gene *hemoglobin, beta* (*HBB*) mutation ($\beta^{6GAG6GTG; Glu6Val}$; sickle hemoglobin, HbS) causes sickle cell anemia, but the clinical features of this disease are heterogeneous. Deoxy-sickle hemoglobin (deoxyHbS) molecules polymerize and HbS polymer initiates an elaborate pathophysiological cascade, involving: injury of the sickle cell; adhesive interactions among sickle cells, endothelial cells and other blood cells; reperfusion injury; and inflammation. Damaged red cells initiate vaso-occlusion and the vascular pathology of sickle cell disease that injures vital tissues, impairing their function and causing pain and premature death (**Figure 1**).

The spectrum of vaso-occlusive events among patients is variable. Clinical heterogeneity in sickle cell disease, as in other 'single-gene' disorders, must be caused by diversity among genes that shape its intermediate phenotypes. In a 'new' paradigm of sickle cell disease, the phenotype of this single-gene disease is multifactorially determined. Products of epistatic modifier genes are likely to interact to determine the specific pattern of vaso-occlusive phenotypes in a patient (**Figure 1**).

Fetal Hemoglobin and Sickle Cell Disease

Fetal hemoglobin (HbF), the best-known genetic modifier of sickle cell anemia, inhibits the polymerization of HbS, an effect traced to several amino acid differences between the γ-globin chain of HbF and the β-globin chain of HbA (Nagel and Steinberg, 2001). HbF levels are genetically controlled, and usually patients with the highest HbF concentrations have the mildest disease.

Control of HbF expression

Globin gene switching in the β-like globin gene cluster is controlled at cellular and molecular levels. Switching involves the complex interactions of erythroid-specific and ubiquitous *trans*-acting transcription factors with the β-globin gene locus control region (LCR) and other *cis*-acting sequences that act positively or

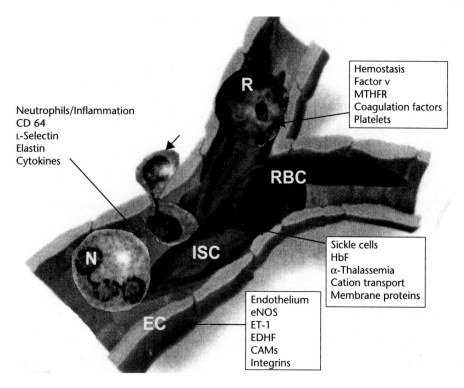

Figure 1 Genes that may modulate the vaso-occlusive complications of sickle cell disease. The actions of different genes expressed in sickle erythrocytes (ISC: irreversibly sickled cells; RBC: sickle discocytes; R: stress reticulocytes), endothelial cells (EC), neutrophils (N) and genes that effect hemostasis may all influence the phenotype of sickle cell disease. Sickle vaso-occlusion involves adhesion of sickle erythrocytes, neutrophils and sickle reticulocytes to each other and to the endothelium and subendothelial tissue. Also involved are platelets (upper arrow), coagulation factors (not shown) and subendothelial elements that might be accessible due to endothelial damage and injury caused by migrating neutrophils (small arrow). MTHFR: methylene tetrahydrofolate reductase; eNOS: endothelial nitric oxide synthase; ET-1: endothelin-1; EDHF: endothelial cell hyperpolarizing factor; CAMs: cell adhesion molecules such as VCAM, ICAM.

negatively in transcriptional regulation. As erythroid progenitors differentiate during embryonic, fetal and adult development, stage-specific transcription programs are activated and deactivated, allowing transcription of embryonic genes in early gestation, fetal genes during most of gestation and early extrauterine life, and adult globin genes afterwards. Normal adults have less than 1% HbF; in sickle cell anemia, HbF levels are increased and may vary by two orders of magnitude.

cis-Acting elements

The HbS mutation occurred at least five times in ancient history against different genetic backgrounds. These backgrounds are defined by haplotypes of the β-globin gene-like cluster. Three haplotypes, Benin, Bantu and Senegal, are found in most African patients with sickle cell anemia, while the Arab-Indian haplotype is common in India (Nagel and Labie, 1989). A C-to-T polymorphism present 5′ to the γ gene in the Senegal and Arab-Indian haplotypes is usually associated with higher HbF levels than other haplotypes. Yet considerable differences in HbF levels are present among individuals with any given haplotype, suggesting that diverse genetic elements simultaneously affect HbF expression.

Located 5′ to the β-globin gene, the LCR comprises a series of DNase-hypersensitive sites that play a critical role in the tissue- and developmental-specific expression of the β-globin-like genes. Each hypersensitive site contains different combinations of conserved binding domains for erythroid-specific and ubiquitous DNA-binding proteins that influence gene transcription, and some hypersensitive sites themselves are polymorphic. The LCR may function as an erythroid cell-specific enhancer, a chromatin-opening domain and an insulator (Li et al., 1999).

Phylogenetic conservation of other noncoding DNA 5′ to the β-globin gene cluster suggests that this region also serves important regulatory functions. The human and mouse β-globin gene clusters show sequences outside the hypersensitive sites of the LCR that are as highly conserved as the LCR. Phylogenetic conservation of some olfactory receptor genes and their flanking sequences 5′ to the LCR suggests that this region may also control expression of β-like globin genes.

Transcription factor binding motifs in proximal and distal promoter elements of the HbF genes may contain point mutations that cause an increased HbF in adults (Wood, 2001). Proteins that bind the CAAT and TATA boxes of the γ-globin gene promoter may act as gene suppressors and promoters, influencing HbF production (Blobel and Weiss, 2001).

trans-Acting factors

trans-Acting factors are the other critical elements regulating expression within the β-like globin gene complex. Erythroid Kruppel-like factor (EKLF) may be the most gene-specific of the known erythroid transcription factors. Interacting with the β-globin gene promoter, EKLF may influence the γ6 to β switch (Donze et al., 1995). Fetal Kruppel-like factors (FKLF, FKLF-2) bind to the CACCC box of the γ-globin gene promoter in vitro and are potential γ-globin gene-specific transcription factors. Quantitative trait loci likely to modulate HbF expression include the F-cell production locus and elements linked to chromosomes 6 and 8. Genetic variation in the cis- and trans-acting factors that regulate HbF expression may account for some disease heterogeneity.

Genetic Conditions Known to Affect Sickle Cell Disease

α-Thalassemia

Chromosome 16 contains two α-globin genes that encode identical proteins. α-Thalassemia is usually caused by the deletion of one or more α-globin genes. More than 30% of the black population has α-thalassemia; however, the loss of a single gene or even two α-globin genes is clinically trivial.

α-Thalassemia affects the phenotype of sickle cell anemia by reducing the HbS concentration (Steinberg and Embury, 1986). As polymerization is concentration-dependent, so α-thalassemia diminishes the polymerization potential of HbS. Clinically, this interaction has a differential effect on the intermediate phenotypes of sickle cell anemia. In sickle cell anemia–α-thalassemia, fewer dense and poorly deformable cells with less HbS polymer raise the hematocrit (packed cell volume, PCV), but, since these cells still contain HbS, blood viscosity is increased. Patients with sickle cell anemia–α-thalassemia have more episodes of pain and more bone disease than do patients with sickle cell anemia alone. Skin ulcers of the leg and retinal vascular disease may be less common in sickle cell anemia–α-thalassemia. In some children with high HbF levels, α-thalassemia appears to preserve splenic function. α-Thalassemia may also reduce the incidence of stroke in children with sickle cell anemia. These observations have no experimentally proven explanation but might result from differences in blood flow and oxygen delivery in dissimilar vascular beds.

In patients with the Arab-Indian haplotype and very high HbF levels, α-thalassemia may be associated with a milder disorder. This salutary effect in sickle cell anemia with exceeding high HbF may be a result of the beneficial effect of high HbF that dominates the increased PCV caused by α-thalassemia.

Glucose-6-phosphatase dehydrogenase deficiency

Glucose-6-phosphatase dehydrogenase (G6PD) deficiency is common in sickle cell anemia. Studies of the phenotype of combined G6PD deficiency and sickle cell anemia have been equivocal. Blood counts are similar in patients with and without G6PD deficiency. The prevalence of G6PD deficiency does not change according to age. In 800 males over the age of 2 years with sickle cell anemia, G6PD deficiency is not associated with differential survival or more pain crises. Under most conditions, little, if any, modulation of the phenotype of sickle cell anemia is caused by coincident G6PD deficiency.

Red Cell Membrane and Cation Transport

Red cell heterogeneity is typical of sickle cell anemia. Cell density and shape depend on the cell's capacity to maintain normal hydration and cation content. Cation transport is modulated by at least three distinct channels.

K^+–Cl^- cotransport is abnormally activated in sickle erythrocytes. When sickle cells are isosmotically swollen or are exposed to acid pH, K^+–Cl^- cotransport promotes K^+ and Cl^- loss with cell dehydration (Brugnara, 2001). Every time sickle erythrocytes are exposed to a hypoxic and/or acidic environment, activation of K^+–Cl^- cotransport is likely to follow, with K^+ loss and dehydration. Ca-dependent movement of K^+ was first described by Gardos in 1958. When sickle erythrocytes are deoxygenated, Ca-dependent dehydration occurs. Deoxygenation increases free intracellular Ca by two- to threefold, values close to the threshold for Gardos channel activation. Deoxygenation of sickle cells is also associated with K⁺ loss and Na⁺ gain, mediated by a diffusion pathway. The inward Na^+ flux and outward K^+ fluxes are balanced, leading to no net changes in ion content and cell volume.

Mutations altering the function of erythrocyte transport channels and cytoskeleton have been described. Genetic variation in cation transporters that alter protein function, and cell homeostasis is also likely to exist and could affect the phenotype of sickle cell disease.

Endothelium as a Modulator in Sickle Cell Anemia

Endothelial cells are central to the pathobiology of sickle cell disease. Endothelial adhesion molecules bind blood cells and proteins. The endothelium helps to control the balance between procoagulant and anti-coagulant properties of the blood and vessel wall. By producing vasoconstrictors and vasodilators, the endothelium modulates vascular tone and contributes to the physiology of inflammation and hemostasis. Many biological modifiers, including interleukins, tumor necrosis factor-alpha (TNFα), thrombin, oxidation, hypoxia and endotoxin, affect the endothelium.

Sickle endothelium may be injured and abnormally activated (Hebbel and Mohandas, 2001). Endothelial activation and damage may be provoked by adherent sickle cells and shear stresses that cause prostacyclin release, oxidant radical formation, inhibition of DNA synthesis, endothelin (ET-1) expression and a disturbance of nitric oxide (NO) biology. As sickle cells perturb the endothelium, vasoconstriction may be favored. Endothelial production of the vasoconstrictor protein ET-1 is increased, while the production of the vasodilator NO may be impaired. Cellular damage enables adhesive interactions between sickle cells and endothelial cells. By assorted attachment mechanisms – perhaps differing by the type of endothelium encountered – the association of sickle and endothelial cells is postulated to delay sufficiently cellular passage so that HbS polymerization, sickling and vaso-occlusion occur before microvasculature transit is complete. Sickle cells stick to the endothelial cell via surface antigens such as CD36 and CD44, integrins, and perhaps other unique features of the sickle erythrocyte membrane. Plasma proteins such as von Willebrand factor, thrombospondin, fibrinogen and fibronectin may be intermediates in the process by which the sickle cell interacts with endothelial cell and extracellular matrix molecules such as laminin, glycoprotein (GPIb), integrins, vascular cell adhesion molecule 1 (VCAM-1) and the Fc receptor.

Table 1 Some genetic polymorphisms possibly associated with modulation of sickle cell disease

Gene	Polymorphism	Linkage
VCAM	Gl238C	CVA protection
VCAM1	Gly1385	CVA
F5	Arg485Lys	CVA, osteonecrosis
TNF	−308Ala	CVA
DBR1	DRB1*0301, DRB1*0302	CVA risk
DBR1	DRB1*1501	CVA protection
MTHFR	Cys677Thr	Osteonecrosis
CBS	68-bp insert at Ile278Thr	CVA protection
AGT	GT repeats	CVA
Nos1	AAT repeats	ACS

Most studies of associations report small numbers of patients and, when more than one study has been done, the results often are conflicting. While suggesting areas for further investigation, none of these studies are authoritative. CVA: cerebrovascular accident; ACS: acute chest syndrome; VCAM: vascular cell adhesion molecule; TNF: tumor necrosis factor; MTHFR: methylene tetrahydrofolate reductase; CBS: cystathione β-synthase; AGT: angiotensinogen; Nos: nitric oxide synthase.

Preliminary evidence exists for genetic variation in molecules affecting the endothelium (**Table 1**). A relationship of a single nucleotide polymorphism (SNP) in *vascular cell adhesion molecule 1* (*VCAM1*) to stroke risk has been reported, but other SNPs, in *intercellular adhesion molecule 1* (*CD54*), *human rhinovirus receptor* (*ICAM1*) and *CD36 antigen* (*collagen type I receptor, thrombospondin receptor*) (*CD36*), are unrelated to risk. SNPs are present in the extracellular domain of *VCAM1* in sickle cell anemia patients but their functional significance is unknown. Genetic variations have been found in vascular endothelial growth factor (VEGF) binding to its receptor, Flt-1.

Inflammation

Neutrophils are increased in patients with sickle cell anemia and are an adverse risk factor for these patients' survival (Platt *et al.*, 1994). Neutrophils of sickle cell anemia patients may be activated, and activation may also accompany acute painful episodes. Neutrophils from sickle cell disease patients are more adherent to endothelial cells than control neutrophils. Increased numbers of neutrophils have been associated with acute chest syndrome and stroke. The role of leukocytes in mediating ischemic injury suggests that differences in neutrophil gene expression or polymorphisms in neutrophil-expressed genes may account for some phenotypic variation in sickle cell disease.

Hemostasis

Thrombosis and hemostasis may play roles in the pathophysiology of sickle cell anemia, so coincidental mutations that favor blood coagulation or thrombosis could influence the phenotype of disease. Polymorphisms in genes for factor V (*F5; coagulation factor V (proaccelerin, labile factor)*), platelet glycoprotein IIIa and 5,10-methylenetetrahydrofolate reductase (*MTHFR; 5,10-methylenetetrahydrofolate reductase (NADPH)*) have been associated with some vaso-occlusive complications (**Table 1**). Most studies are small, and any associations with a phenotype are unconfirmed. Protein C, protein S, heparin cofactor II, thrombin–antithrombin complexes, and resistance to activated protein C have been measured in children being transfused as part of treatment for stroke, children at risk of stroke and untransfused controls, but there are no significant differences among these groups.

See also

Complex Multifactorial Genetic Diseases
Globin Genes: Evolution
Hemoglobin Disorders: Gene Therapy

References

Blobel GA and Weiss MJ (2001) Nuclear factors that regulate erythropoiesis. In: Steinberg MH, Forget BG, Higgs DR and Nagel RL (eds.) *Disorders of Hemoglobin: Genetics, Pathophysiology, and Clinical Management*, pp. 72–94. Cambridge, UK: Cambridge University Press.

Brugnara C (2001) Red cell membrane in sickle cell disease. In: Steinberg MH, Forget BG, Higgs DR and Nagel RL (eds.) *Disorders of Hemoglobin: Genetics, Pathophysiology, and Clinical Management*, pp. 550–576. Cambridge, UK: Cambridge University Press.

Donze D, Townes TM and Bieker JJ (1995) Role of erythroid Kruppel-like factor in human γ6 to β-globin gene switching. *Journal of Biological Chemistry* **270**: 1955–1959.

Hebbel RP and Mohandas N (2001) Cell adhesion and microrheology in sickle cell disease. In: Steinberg MH, Forget BG, Higgs DR and Nagel RL (eds.) *Disorders of Hemoglobin: Genetics, Pathophysiology, and Clinical Management*, pp. 527–549. Cambridge, UK: Cambridge University Press.

Li QL, Harju S and Peterson KR (1999) Locus control regions – coming of age at a decade plus. *Trends in Genetics* **15**: 403–408.

Nagel RL and Labie D (1989) DNA haplotypes and the β^S globin gene. *Progress in Clinical and Biological Research* **316**: 371–393.

Nagel RL and Steinberg MH (2001) Hemoglobins of the embryo and fetus and minor hemoglobins of adults. In: Steinberg MH, Forget BG, Higgs DR and Nagel RL (eds.) *Disorders of Hemoglobin: Genetics, Pathophysiology, and Clinical Management*, pp. 197–230. Cambridge, UK: Cambridge University Press.

Platt OS, Brambilla DJ, Rosse WF, *et al.* (1994) Mortality in sickle cell disease: life expectancy and risk factors for early death. *New England Journal of Medicine* **330**: 1639–1644.

Steinberg MH and Embury SH (1986) Alpha-thalassemia in blacks: genetic and clinical aspects and interactions with the sickle hemoglobin gene. *Blood* **68**: 985–990.

Wood WG (2001) Hereditary persistence of fetal hemoglobin and δβ-thalassemia. In: Steinberg MH, Forget BG, Higgs DR and Nagel RL (eds.) *Disorders of Hemoglobin: Genetics, Pathophysiology, and Clinical Management*, pp. 356–388. Cambridge, UK: Cambridge University Press.

Further Reading

Brugnara C (1997) Erythrocyte membrane transport physiology. *Current Opinion in Hematology* **4**: 122–127.

Bunn HF (1997) Mechanisms of disease – pathogenesis and treatment of sickle cell disease. *New England Journal of Medicine* **337**: 762–769.

Chui DH and Dover GJ (2001) Sickle cell disease: no longer a single-gene disorder. *Current Opinion in Pediatrics* **13**: 22–27.

Fraser P and Grosveld F (1998) Locus control regions, chromatin activation and transcription. *Current Opinion in Cell Biology* **10**: 361–365.

Hebbel RP (1997) Adhesive interactions of sickle erythrocytes with endothelium. *Journal of Clinical Investigation* **99**: 2561–2564.

Li Q, Peterson KR and Stamatoyannopoulos G (1998) Developmental control of ε- and γ-globin genes. *Annals of the New York Academy of Sciences* **850**: 10–17.

Nagel RL (2001) Pleiotropic and epistatic effects in sickle cell anemia. *Current Opinion in Hematology* **8**: 105–110.

Nagel RL and Steinberg MH (2001) Genetics of the β^S gene: origins, epidemiology, and epistasis. In: Steinberg MH, Forget BG, Higgs DR and Nagel RL (eds.) *Disorders of Hemoglobin: Genetics, Pathophysiology, and Clinical Management*, pp. 711–755. Cambridge, UK: Cambridge University Press.

Platt OS (2000) Sickle cell anemia as an inflammatory disease. *Journal of Clinical Investigation* **106**: 337–338.

Steinberg MH and Rodgers GP (2001) Pathophysiology of sickle cell disease: role of cellular and genetic modifiers. *Seminars in Hematology* **38**: 299–306.

Web Links

HBB (hemoglobin, beta); Locus ID: 3043. LocusLink:
http://www.ncbi.nlm.nih.gov/LocusLink/LocRpt.cgi?l = 3043

VCAM1 (vascular cell adhesion molecule 1); Locus ID: 7412. LocusLink:
http://www.ncbi.nlm.nih.gov/LocusLink/LocRpt.cgi?l = 7412

HBB (hemoglobin, beta); MIM number: 141900. OMIM:
http://www.ncbi.nlm.nih.gov/htbin-post/Omim/dispmim?141900

VCAM1 (vascular cell adhesion molecule 1); MIM number: 192225. OMIM:
http://www.ncbi.nlm.nih.gov/htbin-post/Omim/dispmim?192225

Signal Peptides

Gunnar von Heijne, *Stockholm University, Stockholm, Sweden*

Signal peptides are parts of polypeptide chains that are used as 'address labels' for sorting proteins to their correct subcellular destinations.

Overview

The human cell, with its complex ultrastructure, needs to be able to sort proteins from their site of synthesis (most often the cytosol) to their correct subcellular destinations. The sorting information is encoded within the polypeptide chain itself, and one or more distinct parts of the protein – either a linear stretch of amino acids or a 'patch' of residues on the protein's surface – serve as 'address labels' (**Table 1**). The sorting signals are recognized by appropriately located receptors that ensure transport into the proper organelle. The various transport machineries are described in more detail elsewhere in this Encyclopedia. A relatively large number of genetic diseases are correlated with mutations in sorting signals or in components of the transport machineries (Aridor and Hannan, 2000). (*See* Protein Secretory Pathways; Protein Targeting; Protein Transport; Protein: Cotranslational and Posttranslational Modification in Organelles.)

Sorting Signals within the Secretory Pathway

Entry into the secretory pathway is dependent on an N-terminal signal peptide that targets the protein to translocation sites on the endoplasmic reticulum (ER) membrane. Signal peptides are not well conserved in terms of amino acid sequence, but all adhere to a common architecture: a short, positively charged N-terminal segment (the n-region), a central hydrophobic region (the h-region) and a more polar C-terminal segment (the c-region), which contains the recognition site for the signal peptidase enzyme (von Heijne, 1985). The n-region is typically 1–5 residues long, the h-region is 7–12 residues long and the c-region is approximately 5 residues long. In some cases, unusually long n-regions have been found to have additional functions that are expressed only after the signal peptide has been removed by the signal peptidase (Martoglio and Dobberstein, 1998). Theoretical estimates suggest that approximately 10% of all human and yeast proteins are secreted (Drawid and Gerstein, 2000; Emanuelsson *et al.*, 2000).

When the signal peptide emerges from the ribosome, it is recognized by the signal recognition particle (SRP). The SRP has a hydrophobic surface groove into which the signal peptide binds (Batey *et al.*, 2000). The ribosome–SRP–nascent chain complex docks with translocation channels in the ER membrane and the polypeptide is translocated across the ER membrane.

While most proteins only appear transiently in the ER before progressing along the secretory pathway, some are actively retrieved back from the Golgi compartment to the ER. Retrieval depends on a C-terminal-Lys-Asp-Glu-Leu (KDEL) signal that is recognized by a receptor that cycles between the Golgi compartment and the ER.

Lysosomal proteins are diverted from the secretory pathway by virtue of a specific glycan – mannose-6-phosphate – that becomes attached to a 'surface patch' on the protein while it is in transit through the Golgi compartment. The mannose-6-phosphate moiety is recognized by a receptor that cycles between a late compartment in the secretory pathway and lysosomes.

Table 1 A summary of sorting signals

Destination	Description
ER and the secretory pathway	N-terminal positively charged/hydrophobic/polar signal peptide
ER retention	C-terminal-Lys-Asp-Glu-Leu
Lysosome	Mannose-6-phosphate moiety attached to 'surface patch'
Mitochondria	N-terminal amphiphilic α-helix, positive net charge
Nucleus	One or more clusters of positively charged residues
Peroxisomes	PTS1: C-terminal-Ser-Lys-Leu
	PTS2: N-terminal extension

ER: endoplasmic reticulum.

Mitochondrial Targeting Peptides

Mitochondrial targeting peptides need not be N-terminal, though many are. The average length of mitochondrial targeting peptides is approximately 35 residues. The canonical, common structural element is a positively charged, amphiphilic α-helix, typically approximately 10 residues long. The uncharged face of the α-helix binds to a hydrophobic groove on the surface of an outer membrane receptor (Abe *et al.*, 2000). The positively charged residues may be necessary for translocation across both the outer and the inner mitochondrial membranes. After import, the targeting peptide is removed by proteases located in the matrix space. The number of human mitochondrial proteins with internal targeting peptides is not known, although some 10% of all yeast and human proteins have been predicted to be mitochondrial (Drawid and Gerstein, 2000; Emanuelsson *et al.*, 2000).

Proteins destined for the mitochondrial intermembrane space have bipartite sorting signals. A typical *N*-terminal targeting peptide ensures targeting to the mitochondrion and is translocated into the matrix, where it is cleaved off. The targeting peptide is immediately followed by a second sorting signal that is reminiscent of a secretory signal peptide: one or a few positively charged residues, a stretch of nonpolar residues and a signal peptidase cleavage site. It appears that the second sorting signal can act in either of two ways: (i) as a 'stop-transfer' sequence that halts translocation across the inner membrane and leaves the C-terminal parts of the protein in the intermembrane space; or (ii) as a signal peptide that initiates retranslocation of a fully imported protein back across the inner membrane to the intermembrane space. In either case, the second sorting signal is removed by a signal peptidase located in the intermembrane space.

Nuclear Localization Signals

Nuclear localization signals (NLS) are composed of one (monopartite NLS) or two (bipartite NLS) clusters of positively charged residues exposed on the surface of the protein. The monopartite NLS has a short consensus sequence, K(K/R)X(K/R), and binds to a pocket on the surface of the NLS receptor importin-α (Hodel *et al.*, 2001). In bipartite NLS, the monopartite motif is combined with a second small cluster of basic residues 10–12 residues *N*-terminal to the first cluster. The second cluster binds to a distinct pocket on importin-α. A collection of 'NLS motifs' appears to provide a good description of the range of sequence variation (Cokol *et al.*, 2000). Importin-α in turn binds to importin-β, and the whole complex (including the NLS-carrying cargo protein) is imported through nuclear pore complexes into the nucleus, where the complex dissociates. The importin-α and -β subunits are recycled to the cytosol, leaving the cargo protein behind. A similar mechanism

is used for protein export from the nucleus, although the sorting signals are different. Theoretical estimates suggest that as much as one-third of all yeast proteins are routed to the nucleus (Drawid and Gerstein, 2000).

Peroxisomal targeting sequences

There are two distinct types of peroxisomal targeting sequences. The so-called PTS1 type is a C-terminal tripeptide, -Ser-Lys-Leu (or variations thereof), while the PTS2 type is a little-understood *N*-terminal prepeptide.

PTS1 and PTS2 sequences are recognized by distinct cytosolic receptors, although the import pathways then converge onto a common translocation machinery. The PTS1 receptor has a highly specific binding pocket for the Ser-Lys-Leu C-terminal tripeptide (Gatto *et al.*, 2000) and will presumably only bind to proteins where the tripeptide is freely accessible on the surface. Many serious genetic diseases are caused by malfunctions in peroxisomal protein import.

Sorting to Multiple Compartments

Some proteins, although coded by a single gene, are nevertheless found in more than one cellular compartment. At least three different ways in which this can be achieved have been found: 'inefficient' sorting signals, differentially spliced transcripts and multiple initiation codons.

Inefficient sorting signals, as the name implies, do not bind to import receptors with high affinity, and only a fraction of the proteins that are made actually get targeted to an organelle, while the remaining fraction stays in the cytosol.

Differential splicing and multiple initiation codons are two ways to the same end, namely to produce proteins with different *N*-termini from the same gene. Differential splicing is the most versatile strategy, since different 5' exons can encode different sorting signals. The use of multiple initiation codons only allows the production of proteins either containing or lacking an N-terminal sorting signal.

Computer Prediction of Protein-sorting Signals

Methods to predict the subcellular localization of proteins from their amino acid sequence are in general based either on the charateristics of the respective sorting signals already described here or on characteristics of the whole polypeptide chain such as overall amino acid composition. Among the former, the best performance has been obtained with methods that predict signal peptides for the secretory pathway. For such signals, the levels of false-positive and false-negative predictions are both low (less than 10%) and the cleavage sites can be accurately predicted (70–75% of sites are correctly identified). Mitochondrial targeting peptides are somewhat more difficult to predict, and NLs are even worse (Cokol *et al.*, 2000; Emanuelsson *et al.*, 2000).

Methods that use all kinds of available sequence-derived information (chain length, amino acid composition, predicted sorting signals, predicted secondary structure, etc.) to predict subcellular localization have also been developed (Drawid and Gerstein, 2000; Nakai, 2000).

See also
Protein Secretory Pathways
Protein Transport

References

Abe Y, Shodai T, Muto T, *et al.* (2000) Structural basis of presequence recognition by the mitochondrial protein import receptor Tom20. *Cell* **100**: 551–560.

Aridor M and Hannan L (2000) Traffic jam: a compendium of human diseases that affect intracellular transport processes. *Traffic (Copenhagen, Denmark)* **1**: 836–851.

Batey RT, Rambo RP, Lucast L, Rha B and Doudna JA (2000) Crystal structure of the ribonucleoprotein core of the signal recognition particle. *Science* **287**: 1232–1239.

Cokol M, Nair R and Rost B (2000) Finding nuclear localization signals. *EMBO Reports* **1**: 411–415.

Drawid A and Gerstein M (2000) A Bayesian system integrating expression data with sequence patterns for localizing proteins: comprehensive application to the yeast genome. *Journal of Molecular Biology* **301**: 1059–1075.

Emanuelsson O, Nielsen H, Brunak S and von Heijne G (2000) Predicting subcellular localization of proteins based on their N-terminal amino acid sequence. *Journal of Molecular Biology* **300**: 1005–1016.

Gatto Jr GJ, Geisbrecht BV, Gould SJ and Berg JM (2000) Peroxisomal targeting signal-1 recognition by the TPR domains of human *PEX5*. *Nature Structural Biology* **7**: 1091–1095.

von Heijne G (1985) Signal sequences. The limits of variation. *Journal of Molecular Biology* **184**: 99–105.

Hodel M, Corbett A and Hodel A (2001) Dissection of a nuclear localization signal. *The Journal of Biological Chemistry* **276**: 1317–1325.

Martoglio B and Dobberstein B (1998) Signal sequences: more than just greasy peptides. *Trends in Cell Biology* **8**: 410–415.

Nakai K (2000) Protein sorting signals and prediction of subcellular localization. *Advances in Protein Chemistry* **54**: 277–344.

Further Reading

Bayliss R, Corbett A and Stewart M (2000) The molecular mechanism of transport of macromolecules through nuclear pore complexes. *Traffic (Copenhagen, Denmark)* **1**: 448–456.

Fujiki Y (2000) Peroxisome biogenesis and peroxisome biogenesis disorders. *FEBS Letters* **476**: 42–46.

Görlich D and Kutay U (1999) Transport between the cell nucleus and the cytoplasm. *Annual Review of Cell and Developmental Biology* **15**: 607–660.

Johnson AE and Haigh NG (2000) The ER translocon and retrotranslocation: is the shift into reverse manual or automatic? *Cell* **102**: 709–712.

Kornfeld S and Mellman I (1989) The biogenesis of lysosomes. *Annual Review of Cell Biology* **5**: 483–525.

Matlack K, Mothes W and Rapoport T (1998) Protein translocation: tunnel vision. *Cell* **92**: 381–390.

Pelham HR (1996) The dynamic organization of the secretory pathway. *Cell Structure and Function* **21**: 413–419.

Sacksteder KA and Gould SJ (2000) The genetics of peroxisome biogenesis. *Annual Review of Genetics* **34**: 623–652.

Voos W, Martin H, Krimmer T and Pfanner N (1999) Mechanisms of protein translocation into mitochondria. *Biochimica et Biophysica Acta – Biomembranes* **1422**: 235–254.

Similarity Search

William R Pearson, *University of Virginia, Charlottesville, Virginia, USA*

Intermediate article

Article contents

- Introduction
- Similarity and Homology
- Similarity from Sequence Alignments
- Statistical Significance of Sequence Similarity
- DNA versus Protein Sequence Comparison
- Summary

Protein and DNA similarity searches are used to find sequences that are likely to be homologous. Homologous sequences diverged from a common ancestor; when two sequences share statistically significant similarity (much more than would be expected by chance), the most parsimonious explanation for the excess similarity is homology. The most effective similarity searches use protein or translated-DNA : protein comparisons.

Introduction

Searching for sequence similarity is the most widely used method for characterizing genes in newly sequenced genomes. Modern similarity searching programs, such as BLAST and FASTA, are very sensitive (they can identify protein homologs that diverged more than 2500 million years ago (mya)) and very selective; their estimates of statistical significance are very accurate. Statistically significant sequence similarity can be used to infer homology, that is, descent from a common ancestor. Homologous proteins always share significant three-dimensional structural similarity; many homologous proteins have similar functions.

Similarity and Homology

The power of sequence comparison to recognize biologically important relationships was recognized in the early 1980s, with Barker and Dayhoff's discovery of significant similarity between the *src* genes and protein kinases, and Doolittle's identification of surprising similarity between v-*Sis* and platelet derived growth factor. Both of these similarities suggested a clear biological mechanism for the oncogenic potential of the viral genes. Indeed, by early 1985, unexpected protein sequence similarities were becoming commonplace: for example, the relationship between angiotensin precursor and alpha-1-antitrypsin, duplicated domains within the epidermal growth factor precursor (EGFP), and the domains shared by the LDL-receptor and EGFP.

These early discoveries relied on three developments: comprehensive protein sequence databases; computer programs for comparing sequences; and statistical methods for evaluating whether a similarity was likely to arise by chance. By 1983, more than 3000 protein sequences were known; by 2002, comprehensive databases contained sequences from a million different proteins. Moreover, modern sequence comparison algorithms are far more powerful and their statistical methods are more sophisticated. But the fundamental argument remains the same: when two protein or deoxyribonucleic acid (DNA) sequences share much more similarity than is expected by chance, the most parsimonious explanation for the excess similarity is that the two sequences shared a common evolutionary ancestor (Doolittle, 1981). (*See* Protein Databases.)

Sequence similarity searches seek to identify homologous sequences, that is, sequences that share a common ancestor. Homology is inferred from statistically significant sequence or structural similarity; sequences that share statistically significant sequence similarity always have similar three-dimensional structures. An unexpected amount of sequence similarity can have two explanations: either the two sequences arose independently, but converged to share significant similarity because of structural or functional constraints, or the two sequences diverged

from a common ancestor that appeared once. The second explanation is the simpler one, because it assumes that a three-dimensional structure arose only once. When proteins share 75–100% identity over hundreds of amino acids, the single appearance (common ancestry) hypothesis is compelling, but many researchers are more skeptical when two sequences share less than 30% identity. Statistical estimates are more reliable than intuition; proteins sharing 30% identity over 200 amino acids should appear by chance less than once in 10^{15} database searches.

Homologous sequences have diverged from a common ancestor. Examples of the alternative, convergence, are well known, but are limited to protein active sites. Unrelated proteins with convergent functional sites never share statistically significant sequence similarity. The independent emergence of the serine protease catalytic triad active site in the trypsin-like serine proteases and the subtilisins illustrates the difference between divergent evolution from a common ancestor and convergent evolution to a similar function. Outside the catalytic site, the trypsin-like serine proteases and subtilisins have completely different overall three-dimensional structures – trypsins are almost all beta-strand structures, while subtilisins are mostly alpha helical. One might argue that both protein families diverged from a common ancestor, but the number of structural changes required to change one protein's structure into the other would be no different from the number required to build the structure from any other sequence. Indeed, it would be easier to make subtilisin from almost any, mostly alpha helical, protein than to make it from trypsin.

Significant sequence similarity implies homology and common three-dimensional structure, but the converse is not true. While all proteins that share statistically significant sequence similarity can be inferred to be homologous, many proteins that share significant structural similarity, and are thus homologous, do not share significant sequence similarity. Thus, sequence similarity searches can be used to demonstrate that two proteins are related, but searches cannot prove that a sequence is unrelated to all known proteins. The highly conserved eukaryotic actin catalytic domain does not share significant sequence similarity with any prokaryotic protein, yet has strong structural similarity with bacterial HSP70 and sugar kinases. Likewise, while proteins that share significant sequence similarity always have similar structures, their functions may differ; conversely, nonhomologous proteins (e.g. trypsins and subtilisins) sometimes share similar functions.

Similarity from Sequence Alignments

The first sequence alignment programs calculated 'global' sequence similarity (Needleman and Wunsch, 1970); alignments began with the first residue of each sequence and ended with the last. Global similarity scores can be transformed into evolutionary distances that can be used to build phylogenetic trees (sequences that are 100% identical have a distance of 0); as similarity goes down distance goes up. Modern sequence searching programs like FASTA (Pearson and Lipman, 1988) and BLAST (Altschul *et al.*, 1990) measure 'local' sequence similarity, approximating an algorithm first described by Smith and Waterman (1981). In contrast to global similarity scores, which penalize similarities that do not extend to the ends of both sequences, local similarity scores focus on the most similar portions of the two sequences and ignore the rest. Local similarity scores are attractive because they allow conserved domains within proteins to obtain high scores, regardless of the surrounding sequence context, and there is an accurate statistical model for local similarity scores – the extreme value distribution. (*See* BLAST Algorithm; FASTA Algorithm; Global Alignment; Smith–Waterman Algorithm.)

Although sequence similarity is most frequently referred to in terms of per cent identity, it has been known since the late 1970s that more distant evolutionary relationships can be detected by using similarity scores that include a notion of conservative and nonconservative replacements. Thus, scoring matrices that give greater weight to amino acid replacements that require a single codon change are more sensitive than matrices that give the same score to every nonidentity, but the most sensitive scoring systems incorporate empirical measurements of amino acid replacements in related proteins and the likelihood of seeing to amino acid residues align by chance (Altschul, 1991). These scores recognize that leucine–valine–isoleucine, or arginine–lysine–histidine replacements appear frequently in related proteins, while glycine–tryptophan replacements should be very rare. The BLAST and FASTA programs use the BLOSUM series (Henikoff and Henikoff, 1992) of scoring matrices to identify distant sequence relationships, but modern versions of Dayhoff's PAM series of matrices can be used to focus on recent evolutionary events. The relationship between scoring matrices, information content and statistical significance is now well understood (Altschul, 1991), but effective gap penalties must still be determined empirically. The standard matrices and gap penalties used by the BLAST and FASTA programs are very effective in detecting distant evolutionary relationships and pro-

viding accurate statistical estimates. (*See* Substitution Matrices.)

Statistical Significance of Sequence Similarity

The inference of homology is based on statistically significant sequence similarity, that is, similarity that is not expected by chance. Local similarity scores between unrelated protein or DNA sequences are accurately described by the extreme value distribution: $P(S' > x) = 1 - \exp[-\exp(-x)]$. In this formulation, $S' = \lambda S - \log Kmn$, where λ is a factor that scales the raw similarity score S for the scoring matrix used to calculate the similarity, and $\log Kmn$ is a term that corrects for the fact that an alignment between two random long sequences has a better chance of obtaining a high similarity score t than an alignment between two short sequences (m and n are the lengths of the two sequences being compared). For alignments without gaps, λ and K can be calculated analytically (Altschul *et al.*, 1990); for alignments with gaps, they must be estimated empirically. BLAST provides precalculated estimates of λ and K for a variety of scoring matrices and gap penalties; FASTA estimates λ and K from the scores of the unrelated sequences that are calculated in a library search (Pearson, 1998). (*See* Alignment: Statistical Significance.)

The $P(S' > x)$ value describes the probability that two sequences will obtain a score $S' > x$ by chance, in a single sequence alignment. Since a similarity search typically compares a protein with hundreds of thousands to millions of sequences, the single pairwise alignment probability must be adjusted to account for the large number of comparisons performed. Modern versions of BLAST and FASTA both report statistical significance in terms of an expectation or $E()$-value, the expected number of times an alignment score would be obtained by chance in a database search. FASTA calculates the expectation as $E(S' > x) = DP(S' > x)$, where D is the number of sequences in the database (and thus the number of sequence comparisons performed). BLAST uses a slightly different calculation, $E(S' > x) = N/nP(S' > x)$, where n is the length of the library sequence and N is the total number of residues in the library. Thus, the expectation, or $E()$-value, estimates the number of times a similarity score would be expected by chance in a search of a database of size D. $E()$ can range from 0 to D, the size of the database. $E()$-values of <0.001 are considered clearly statistically significant; some investigators use $E() < 0.01$ as a threshold for significance. The highest scoring unrelated sequence in a similarity search should have $E() \sim 1$.

In general, sequence similarity statistics are very accurate – an unrelated (or random) sequence will obtain an $E()$-value <0.01 about once in 100 database searches. The most common cause of inaccurate statistical estimates, that is, $E()$-values <0.001 for unrelated sequences, are low-complexity regions, for example, runs of 'SPSPSPSP'. These regions violate the assumptions of position independence and composition distribution used by the statistical models. The BLAST program routinely masks out (converts to 'XXXX') low-complexity regions with the 'seg' program (Wootton and Federhen, 1993) each time a sequence is searched. FASTA can ignore low-complexity regions in the initial phase of the search, so that they do not obtain high scores with unrelated sequences, and then allow the entire sequence, with low-complexity regions, to be used in the final alignment.

With the exception of low-complexity regions, local regions of biased amino acid composition, for example, the membrane-spanning domains of receptors, channels and other transmembrane proteins, rarely produce statistically significant similarity scores. FASTA calculates its statistical parameters from the distribution of similarity from unrelated sequences in the database (Pearson, 1998); this process corrects for all but the most biased amino acid compositions. The FASTA package provides an alternative method for testing whether a significant similarity score reflects composition bias; the statistical parameters can be estimated from a set of shuffled sequences that preserve length and local composition (Pearson and Lipman, 1988). BLAST uses an alternative approach to correct for composition bias in high-scoring sequences (Altschul *et al.*, 2001).

In many cases it is possible to test whether the statistical significance estimates are accurate by finding the $E()$-value for the highest scoring unrelated sequence, which should be near $E() < 1$. For query sequences that are related to dozens of sequences in the database, it is not difficult to determine the highest scoring sequence that does not share significant similarity with homologs to the query. Alternatively, if the query sequence matches fewer than six sequences with $E() < 0.001$, the statistical significance can be confirmed by comparing the alignment score with the distribution of scores produced from alignments with shuffled sequences.

The dependence of $E()$-value on database size provides a simple strategy for dramatically increasing similarity search sensitivity – one should search the smallest database that is likely to contain the protein of interest. For example, a Smith–Waterman search using mouse glutathione transferase theta1 (SwissProt gtt1_mouse) against the *Escherichia coli* proteome (4289 sequences) identifies four glutathione transferase homologs with $E()$-values <0.001. In a search with the same sequence against the NCBI NR (nonredundant) protein sequence database (1 018 874 sequences), those four sequences have exactly the same similarity scores, but because the database is 250 times larger, expectation values that range from 2.5×10^{-8} to 2.1×10^{-5} in the *E. coli* proteome search become 6.2×10^{-6} to 0.005. Now that complete genome sequences are available for dozens of bacteria and a variety of eukaryotes, searches on specific proteomes can be far more sensitive and efficient than searching comprehensive databases.

DNA versus Protein Sequence Comparison

While searching small, comprehensive databases can improve search sensitivity, the most effective strategy for improving search sensitivity is to use protein, or translated protein, sequence comparison rather than DNA sequence comparison. Protein sequence comparison allows a 5–10-fold greater evolutionary look-back time. As the glutathione transferase example shows, statistically significant protein sequence similarities are often found between eukaryotes and prokaryotes, organisms that last shared a common ancestor more than 2500 mya. In contrast, DNA sequence comparison for protein-coding genes rarely detects sequences that diverged more than 500 mya; for noncoding sequences the look-back time is less than 60 million years. Moreover, protein databases are typically one-tenth the size of DNA databases, providing an additional order of magnitude in statistical significance. The FASTA and BLAST packages provide programs for comparing DNA sequences to protein sequence databases, or vice versa, that allow frameshifts to improve alignment scores, so even when a DNA sequence may have errors, accurate, high-scoring alignments can be calculated. Comparisons of translated DNA and proteins are somewhat less sensitive than protein : protein searches, but both are far more effective than DNA : DNA searches.

The FASTA and BLAST programs have been optimized to identify distantly related proteins and produce accurate statistical estimates. While the focus can be shifted to more recent evolutionary events by using 'shallower' scoring matrices, the statistical significance of most relationships is relatively insensitive to modest changes in gap penalty (e.g. $-10/-2$ to $-12/-2$) or scoring matrix (BLOSUM50 to BLOSUM62). The Smith–Waterman algorithm can sometimes identify significant similarities that BLAST and FASTA miss, but these cases are rare. For higher sensitivity, comparisons are done with position-specific scoring matrices (PSSMs), profiles or hidden

Markov models (HMMs) (Altschul *et al.*, 1997; Park *et al.*, 1998). These methods abstract the evolutionarily and functionally critical positions in a protein sequence and build a scoring matrix that reflects those constraints; regions of the sequence that are highly conserved will have a greater effect on the score than regions that are more variable.

Summary

Modern protein sequence comparison methods are very sensitive and reliable. It is routine to identify homologs that shared a common ancestor billions of years ago. For example, about 30% of *E. coli* proteins share statistically significant sequence similarity with human proteins (2500 million years since a common ancestor), and about 50% of yeast proteins share significant similarity with human proteins after 1000 million years of evolution. The inference of homology from statistically significant sequence similarity ($E()$ < 0.001–0.01) reliably predicts similar three-dimensional structure, and suggests similar function. The effectiveness of similarity searching is likely to improve as more genomes are sequenced, more complete protein sets become available and protein domain families are better characterized.

See also
BLAST Algorithm
FASTA Algorithm
Protein Homology Modeling
Sequence Similarity
Substitution Matrices

References

Altschul SF (1991) Amino acid substitution matrices from an information theoretic perspective. *Journal of Molecular Biology* **219**: 555–565.

Altschul SF, Bundschuh AR, Olsen R and Hwa T (2001) The estimation of statistical parameters for local alignment score distributions. *Nucleic Acids Research* **29**: 351–361.
Altschul SF, Gish W, Miller W, Myers EW and Lipman DJ (1990) A basic local alignment search tool. *Journal of Molecular Biology* **215**: 403–410.
Altschul SF, Madden TL, Schaffer AA, *et al.* (1997) Gapped BLAST and PSI-BLAST: a new generation of protein database search programs. *Nucleic Acids Research* **25**: 3389–3402.
Doolittle RF (1981) Similar amino acid sequences: chance or common ancestry? *Science* **214**: 149–159.
Henikoff S and Henikoff JG (1992) Amino acid substitutions matrices from protein blocks. *Proceedings of the National Academy of Sciences of the United States of America* **89**: 10915–10919.
Needleman S and Wunsch C (1970) A general method applicable to the search for similarities in the amino acid sequences of two proteins. *Journal of Molecular Biology* **48**: 444–453.
Park J, Karplus K, Barrett C, *et al.* (1998) Sequence comparisons using multiple sequences detect three times as many remote homologues as pairwise methods. *Journal of Molecular Biology* **284**: 1201–1210.
Pearson WR (1998) Empirical statistical estimates for sequence similarity searches. *Journal of Molecular Biology* **276**: 71–84.
Pearson WR and Lipman DJ (1988) Improved tools for biological sequence comparison. *Proceedings of the National Academy of Sciences of the United States of America* **85**: 2444–2448.
Smith TF and Waterman MS (1981) Identification of common molecular subsequences. *Journal of Molecular Biology* **147**: 195–197.
Wootton JC and Federhen S (1993) Statistics of local complexity in amino acid sequences and sequence databases. *Computers and Chemistry* **17**: 149–163.

Further Reading

Altschul SF, Boguski MS, Gish W and Wootton JC (1994) Issues in searching molecular sequence databases. *Nature Genetics* **6**: 119–129.
Pearson WR and Wood TC (2001) Statistical significance in biological sequence comparison. In: Balding DJ, Bishop M and Cannings C (eds.) *Handbook of Statistical Genetics*, pp. 39–65. Chichester, UK: Wiley.

Simple Repeats

Christian Schlötterer, *Veterinärmedizinische Universität Wien, Wien, Austria*

Introductory article

Article contents
- Introduction
- Definition
- Mutation Processes
- Evolutionary Perspective
- Applications of Simple Repeat Analysis

Simple sequences (microsatellites) are a ubiquitous component of the genome of higher organisms. Their high mutation rate provides the basis for the successful use of microsatellites as genetic markers.

Introduction

The distribution of the four nucleotides (G, A, T, C) in the human genome is not random, even in nonfunctional regions. One deviation from randomness is the presence of repetitive sequence stretches, which can be visualized easily by a dot matrix. When a given sequence is compared against itself and identical

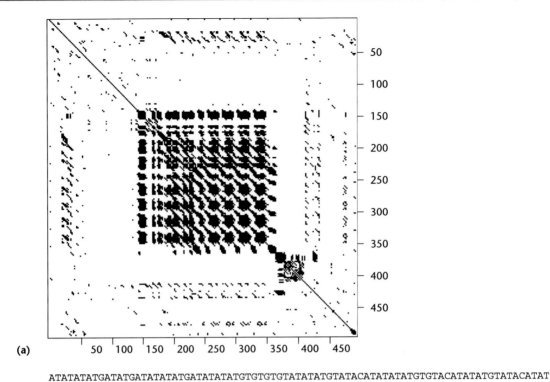

(a)

```
ATATATATGATATGATATATATGATATATATGTGTGTGTATATATGTATACATATATATGTGTACATATATGTATACATAT
\\\\\ \\ • • //\\\\ \\\•••••••••       ••••••••••• •••••• •    ••• • •  •• • • •
```
(b)
```
TATATGTACACACATATATGTATATATATATGTACACACATATATGTATATATATATGTACACACACACACATAGAGAGAG
```

Figure 1 Dot plot of a cryptic simple sequence. (a) Dot plot using the partial human *coagulation factor IX* (*F9*) coagulation factor IX gene. Only identities of at least 75% in an 8-base-pair window are shown. (b) Alignment of a region off the diagonal that shows sequence similarity.

nucleotides are visualized by a black dot, a sequence without regions of self-complementarity will be represented as a diagonal line in a dot plot. **Figure 1** shows a dot plot for a partial sequence of the human *coagulation factor IX* (*F9*) gene. In addition to the expected diagonal, a diamond-shaped area of self-complementarity can be recognized. Closer inspection of the underlying sequence (**Figure 1b**) indicates that several short repetitions of the motif AT are scattered throughout the sequence.

Definition

A deoxyribonucleic acid (DNA) sequence containing such repeats seems to be simpler than a random sequence. Thus, such tandem repeats of short sequence motifs have been called 'simple sequences' or 'simple repeats'. In analogy to other tandemly repeated sequences, such as satellite DNA and minisatellites, simple sequences are often called microsatellites (because the repeat unit consists of only 1–6 base pairs). In addition, the term 'short tandem repeats' is also used frequently in the literature.

Simple sequences can be distinguished by their repeat motif – that is, the length of the repeat unit

Uninterrupted microsatellite	CTCTCTCTCTCTCTCTCTCT
Imperfect microsatellite	CTCTCTCTCTATCTCTCTCTC
Interrupted microsatellite	CTCTCTCTAAACTCTCTCTCT
Compound microsatellite	CTCTCTCTCCACACACACACAC

Figure 2 [*Figure is also reproduced in color section.*] Microsatellite nomenclature. The uninterrupted microsatellite consists of a perfect repetition of CT di-nucleotides, the imperfect microsatellite has a C substituted by a A, the interrupted microsatellite has an insertion of three As, which interrupt the CT repeat, the composite repeat consists of two different juxtaposed repeats.

(mono-, di-, tri- or tetranucleotide repeats) – and the base composition. There are four different dinucleotide repeats, $(AT)_n$, $(GT)_n$, $(GA)_n$ and $(GC)_n$. The most classic example of a microsatellite consists of an uninterrupted sequence of tandem repeats of the same motif (**Figure 2a**). When one or more bases interrupt the repeat array, the microsatellite is called imperfect or interrupted. A juxtaposition of two types of repeat (called composite microsatellites) is also found frequently (**Figure 2**).

The sequence shown in **Figure 1b**, however, does not easily fall into one of the above categories. In this sequence there is more than one type of repeat, but each is short and frequently interrupted. Because the identification of a single repeat type by eye is not obvious, such sequences have been called 'cryptic simple sequences'. A computer algorithm has been developed to obtain an objective measurement of the extent to which a sequence resembles a microsatellite. Various publications using this computer algorithm have demonstrated the widespread occurrence of both cryptic simple sequences and simple sequences in eukaryotic and prokaryotic genomes.

Mutation Processes

Whereas nonrepetitive DNA evolves mainly through the accumulation of base substitutions, simple repeats have their own, specific mutation mechanism called 'DNA replication slippage'. During DNA synthesis, the newly polymerized strand is displaced and re-aligned out of register (**Figure 3**). When DNA synthesis continues, a loop containing one or more unpaired repeat units is generated. Provided that the DNA mismatch machinery does not recognize the loop, the next round of replication will result in a microsatellite sequence, which has either gained or lost repeats. *In vitro* experiments and data from cells

deficient in mismatch repair indicate that the displacement and out of register pairing is a highly frequent phenomenon intrinsic to simple sequences.

Typical microsatellite mutation rates in wild-type cells range from 10^{-6} to 10^{-2} per generation. This wide range of mutation rates is the reflection of various factors that influence microsatellite instability. Probably, the best-characterized factor is the number of repeats in a microsatellite sequence. A positive correlation between microsatellite mutation rate and repeat number has been shown in several studies, but whether the increase in mutation rate is linear or exponential has not been determined as yet. Other factors contributing to microsatellite instability are the length of the repeat unit (mono-, di-, tri- or tetranucleotide repeats), the sequence composition of the repeat (GT, AT, GA) and whether or not the microsatellite stretch is interrupted by base substitutions (**Figure 2**).

Evolutionary Perspective

Even though a functional role has been shown for some microsatellite loci, the genomic distribution of

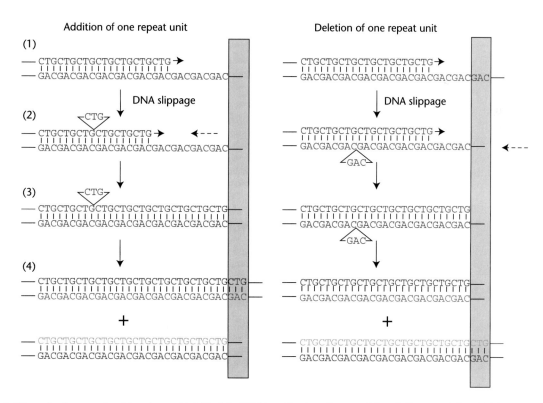

Figure 3 [*Figure is also reproduced in color section.*] Model of DNA replication slippage adding (left) or removing (right) one repeat unit. (1) First round of DNA replication. (2) DNA slippage, causing one repeat unit to loop out; the dashed arrow indicates the direction of DNA slippage. (3) DNA synthesis continues without repair of the loop. (4) Second round of DNA replication leads to the addition or deletion of one repeat unit in one of the strands.

microsatellites remains mostly unclear. Overall, microsatellites are well dispersed over the genome. In coding sequences, mono-, di- and tetranucleotide repeats are underrepresented because slippage mutations would cause a frameshift mutation, which is generally deleterious.

Although the origin of microsatellites is still not understood, an early, but still plausible, hypothesis assumed that short stretches of microsatellite DNA were generated by mutations, which served as templates for DNA replication slippage. Other studies have suggested that not only base substitutions but also short insertions could create microsatellite loci. A completely different mechanism for microsatellite genesis is the derivation of AT-rich microsatellites from a 3′ extension of retrotranscripts, similar to major messenger ribonucleic acid (mRNA) polyadenylation.

Phylogenetic analysis has indicated that microsatellite loci are conserved over several million years. This extraordinary conservation might result, at least in part, from the capacity of DNA replication slippage to remove interruptions in the repeat. By this mechanism an uninterrupted microsatellite could be restored, even when base substitutions occurred during evolution.

Less understood is the observation that most microsatellites are within a certain size range that seems to be species-specific. Early hypotheses for this phenomenon ranged from selection-mediated size constraints to a balance between base substitutions and slippage mutations. Now, however, a size-specific mutation bias of microsatellite alleles has been suggested to explain the genomic size distribution of microsatellites. Short microsatellite alleles have no or a slightly upward mutation bias, whereas long alleles experience downward mutations encompassing several repeat units, resulting in an overall downward mutation bias. Thus, the different mutation spectrum of long and short microsatellite alleles maintains the observed length distribution. Most importantly, the same process seems to apply to different species, except that there is a species-specific size range of microsatellite alleles.

Applications of Simple Repeat Analysis

The intrinsic length variation of microsatellites has rendered them a very popular genetic tool. The combination of polymerase chain reaction with gel electrophoresis has facilitated the discrimination of microsatellite alleles differing by only a single repeat unit.

- For several years, microsatellites were the marker of choice in human genetics, and the isolation and characterization of many genes by positional cloning was dependent on a reliable microsatellite-based map. Subsequently, single nucleotide polymorphisms have become more popular for mapping in humans and genetic model organisms. Nevertheless, the high information content of microsatellite markers will promote further use for genome mapping in nonmodel organisms.
- The popularity of microsatellites in forensics and paternity testing is still unbroken, mainly owing to the high information content of a single microsatellite marker.
- Even though research in the field of population genetics is moving toward multilocus sequence analysis, the high mutation rates of microsatellites will continue to provide important insights into more recent evolutionary timescales.

Acknowledgement

The laboratory of CS is supported through grants of the Fonds zur Förderung der wissenschaftlichen Forschung (FWF).

See also

Further Reading

Ellegren H (2000) Microsatellite mutations in the germline: implications for evolutionary inference. *Trends in Genetics* **16**: 551–558.

Goldstein D and Schlötterer C (1999) *Microsatellites: Evolution and Applications.* Oxford, UK: Oxford University Press.

Schlötterer C (2000) Evolutionary dynamics of microsatellite DNA. *Chromosoma* **109**: 365–371.

Tóth G, Gáspári Z and Jurka J (2000) Microsatellites in different eukaryotic genomes: survey and analysis. *Genome Research* **10**: 967–981.

SINEs

See Short Interspersed Elements (SINEs)

Single-base Mutation

Dan Graur, *Tel Aviv University, Tel Aviv, Israel*

DNA sequences are normally copied exactly during the process of chromosome replication;
however, new sequences are formed if errors in either DNA replication or repair occur.
These errors are called mutations.

Deoxyribonucleic acid (DNA) sequences are normally copied exactly during the process of chromosome replication. Sometimes, however, errors in either DNA replication or DNA repair occur, giving rise to new sequences. These errors are called mutations.

Mutations can occur in either somatic or germ-line cells. Somatic mutations are not inherited, so they can be disregarded in an evolutionary or genetic context. Germ-line mutations are the ultimate source of variation and novelty in evolution. Some organisms (e.g. plants) do not have a sequestered germ line, and therefore the distinction between somatic and germ-line mutations is not absolute.

Mutations may be classified by the length of the DNA sequence affected by the mutational event. If the mutational event affects two or more adjacent nucleotides, then we refer to the event as a *segmental mutation*. Here, we will only deal with mutations affecting a single nucleotide (*single-base* or *point mutations*). Specifically, we deal with *substitution mutations*, that is, the replacement of one nucleotide by another.

Substitution mutations are divided into *transitions* and *transversions*. Transitions are substitution mutations between A and G (purines) or between C and T (pyrimidines). Transversions are substitution mutations between a purine and a pyrimidine. There are four types of transitions (A→G, G→A, C→T and T→C) and eight types of transversions (A→C, A→T, C→A, C→G, T→A, T→C, G→C and G→T).

Substitution mutations occurring in protein-coding regions may also be classified according to their effect on the product of translation, the protein. Because of the degeneracy of all genetic codes, not all single-base mutations affect the sequence of the protein. A substitution mutation is defined as *synonymous* if it changes a codon into another that specifies the same amino acid as the original codon (**Figure 1**). Otherwise, it is *nonsynonymous*. A change in an amino acid due to a nonsynonymous mutation is called a *replacement*. The terms 'synonymous' and '*silent*' mutation are often used interchangeably because, in the great majority of cases, synonymous mutations do not alter the amino acid sequence of a protein and are therefore not detectable at the amino acid level. However, a synonymous mutation may not always be silent. A synonymous mutation may, for instance, create a new splicing site or obliterate an existing one, thus turning an exonic sequence into an intron or vice versa, and causing a different polypeptide to be produced. For example, a synonymous change from the glycine codon GGT to its synonymous codon GGA in codon 25 of the first exon of β globin has been shown to create a new splice junction, resulting in the

	Ile	Cys	Ile	Lys	Ala	Leu	Val	Leu	Leu	Thr
	ATA	TGT	ATA	AAG	GCA	CTG	GTC↓	CTG	TTA	ACA
	ATA	TGT	ATA	AAG	GCA	CTG	GTA	CTG	TTA	ACA
(a)	Ile	Cys	Ile	Lys	Ala	Leu	Val	Leu	Leu	Thr
	Ile	Cys	Ile	Lys	Ala	Asn	Val	Leu	Leu	Thr
	ATA	TGT	ATA	AAG	GCA	AAC	GTC↓	CTG	TTA	ACA
	ATA	TGT	ATA	AAG	GCA	AAC	TTC	CTG	TTA	ACA
(b)	Ile	Cys	Ile	Lys	Ala	Asn	Phe	Leu	Leu	Thr
	Ile	Cys	Ile	Lys	Ala	Asn	Val	Leu	Leu	Thr
	ATA	TGT	ATA	AAG	GCA	AAC	GTC	CTG	TTA	ACA
	ATA	TGT	ATA	TAG↓	GCAAACGTCCTGTTAACA					
(c)	Ile	Cys	Ile	Stop						

Figure 1 Types of substitution mutations in a coding region: (a) synonymous, (b) missense and (c) nonsense. (Reproduced from Graur and Li (2000).).

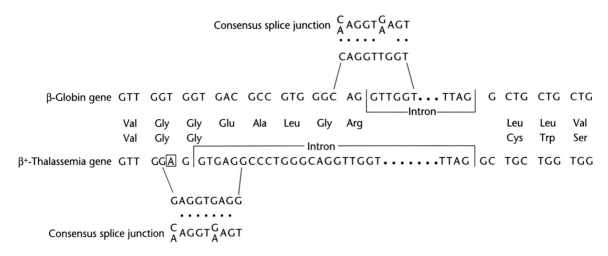

Figure 2 Nucleotide sequences at the borders between exon-1 and intron-I and exon-2 in the β-globin gene from a normal individual and a patient with β⁺-thalassemia, the mutated nucleotide boxed. The leader lines indicate the splicing sites. Each of the splice junctions is compared with the sequence of the consensus splice junction, and dots denote identity of nucleotides between the splice junction and the consensus sequence. Note that the nucleotide substitution in the β⁺-thalassemia gene is synonymous, because both GGT and GGA code for the amino acid glycine. It is not, however, silent, because the activation of the new splicing site in the β⁺-thalassemia gene results in the production of a frameshifted protein. (Reproduced from Graur and Li (2000) after Goldsmith *et al.* (1983).)

production of a frameshifted protein of abnormal length (**Figure 2**). Such a mutation is obviously not 'silent'. Therefore, it is advisable to distinguish between the two terms. Of course, all silent mutations in a protein-coding gene are synonymous.

Nonsynonymous or *amino acid-altering* mutations are further classified into *missense* and *nonsense* mutations (**Figure 1**). A missense mutation changes the affected codon into a codon that specifies a different amino acid from the one previously encoded. A nonsense mutation changes a sense codon into a termination codon, thus prematurely ending the translation process and ultimately resulting in the production of a truncated protein. Codons that can mutate to a termination codon by a single substitution mutation, for example, UGC (Tyr), which can mutate in one step into either UAG or UGA, are called *pretermination codons*.

Readthrough mutations, that is, mutations turning a stop codon into an amino acid-specifying condon, are also known to occur. Such mutations cause the translation to continue beyond the original termination codon until the translation apparatus encounters a new in-frame stop codon. If such a codon does not exist downstream of the mutated stop codon, the translation will continue up to the polyadenylation site.

Each of the sense codons can mutate to nine other codons by means of a single substitution mutation. For example, CCU (Pro) can experience six nonsynonymous changes, to UCU (Ser), ACU (Thr), GCU (Ala), CUU (Leu), CAU (His) or CGU (Arg), and three synonymous changes, to CCC, CCA or CCG. Since the universal genetic code consists of 61

Table 1 Relative frequencies of different types of substitution mutations in a random protein-coding sequence[a]

Mutation	Number	Percentage
Total in all codons	549	100
Synonymous	134	25
Nonsynonymous	415	75
Missense	392	71
Nonsense	23	4
Total in first codon	183	100
Synonymous	8	4
Nonsynonymous	175	96
Missense	166	91
Nonsense	9	5
Total in second codon	183	100
Synonymous	0	0
Nonsynonymous	183	100
Missense	176	96
Nonsense	7	4
Total in third codon	183	100
Synonymous	126	69
Nonsynonymous	57	31
Missense	50	27
Nonsense	7	4

[a]Reproduced from Graur and Li, 2000.

sense codons, there are $61 \times 9 = 549$ possible substitution mutations. If we assume that all possible substitution mutations occur with equal frequencies, and that all codons are equally frequent in coding regions, we can compute the expected proportion of the different types of substitution mutations from the genetic code. These are shown in **Table 1**. Of course, not all mutations occur with equal frequencies, nor is the frequency distribution of codons uniform. However, the results in **Table 1** give us an indication on the buffering capacities of the genetic code against nonsynonymous and nonsense mutations. Because of the structure of the genetic code, synonymous mutations occur mainly at the third position of codons. Indeed, almost 70% of all the possible nucleotide changes at the third position are synonymous. In contrast, all the substitution mutations at the second position of codons are nonsynonymous, as are most nucleotide changes at the first position (96%).

In the vast majority of cases, the exchange of a codon by a synonym, that is, one that codes for the same amino acid, requires only one or at most two synonymous mutations. The only exception to this rule is the exchange of a serine codon belonging to the four-codon family (UCU, UCC, UCA and UCG) by one belonging to the two-codon family (AGU and AGC). Such an event requires two nonsynonymous mutations.

Nucleotide substitution mutations are thought to arise mainly from the mispairing of bases during DNA replication. As part of their formulation of DNA replication, Watson and Crick (1953) suggested that transitions might be due to the formation of purine–pyrimidine mispairs (e.g. A : C), in which one of the bases assumes an unfavored tautomeric form, that is, enol instead of keto in the case of guanine and thymine, or imino instead of amino in the case of adenine and cytosine (**Figure 3**). Topal and Fresco (1976) proposed that purine–purine mispairs can also occur, but pyrimidine–pyrimidine mispairs cannot. The purine–pyrimidine mispairs are $A^* : C$, $A : C^*$, $G^* : T$ and $G : T^*$ and the purine–purine mispairs are $A^* : A$, $A^* : G$, $G^* : A$ and $G^* : G$, in which the asterisk denotes an unfavored tautomeric form. The pathways via which substitution mutations arise are as follows: (1) transitions arise from purine–pyrimidine mispairing and can occur on either strand. For example, the transition $A : T \rightarrow G : C$ can arise from one of four possible mispairs: $A^* : C$, $A : C^*$, $G : T^*$ or $G^* : T$; (2) transversions arise from purine–purine mispairing but can only occur if the purine resides on the template strand. For instance, the transversion $A : T \leftrightarrow T : A$ can arise only from $A^* : A$, where the unfavored tautomer A^* is on the template strand.

Because of the rarity of mutations, the rate of spontaneous mutation is very difficult to determine directly, and at present only a few such estimates exist at the DNA sequence level (e.g. Drake *et al.*, 1998). The rate of mutation, however, can be estimated indirectly by other means. Li *et al.* (1985) and Kondrashov and Crow (1993), for example, estimated the average rate of mutation in mammalian nuclear DNA to be $3–5 \times 10^{-9}$ substitution mutations per nucleotide site per year. The mutation rate, however, varies enormously with genomic region, and in microsatellites, for instance, the rate in humans was

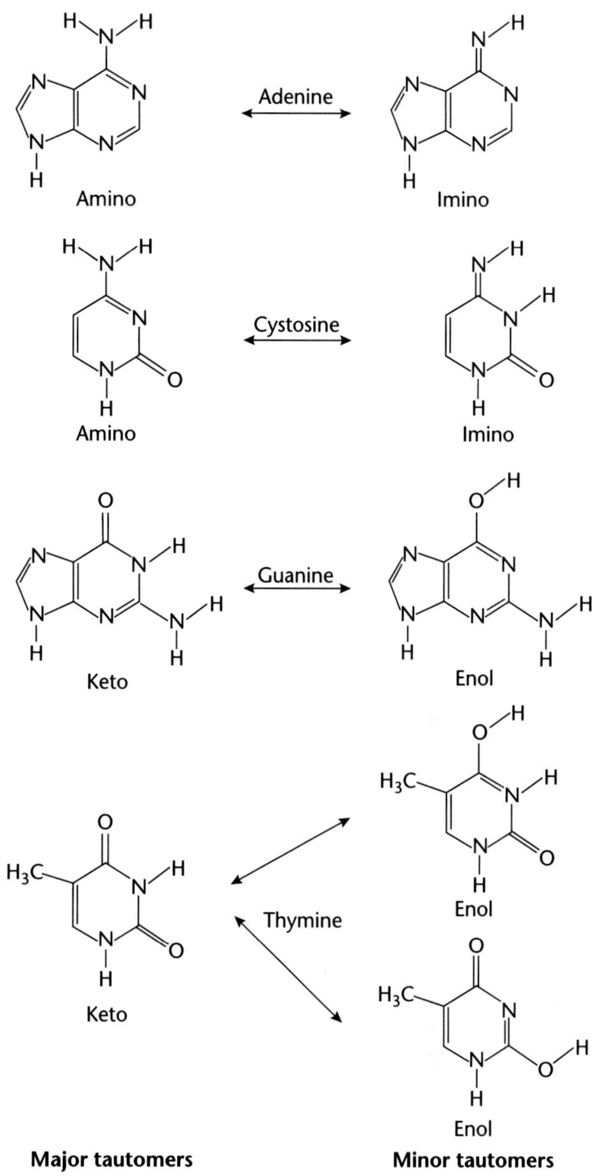

Figure 3 Amin↔imino and ket↔enol tautomerisms. Adenine and cytosine are usually found in the amino form, but rarely assume the imino configuration. Guanine and thymine are usually found in the keto form, but rarely form the enol configuration. Thymine has two enol tautomers. All minor tautomers can assume different rotational forms. (Reproduced from Graur and Li (2000).)

estimated to be over 10^{-3} substitution mutations per nucleotide site per year (Brinkmann *et al.*, 1998).

The rate of mutation in mammalian mitochondria has been estimated to be at least 10 times higher than the average nuclear rate (Brown *et al.*, 1982). Ribonucleic acid viruses have error-prone polymerases (i.e. ones lacking proofreading), and they lack an efficient postreplication repair mechanism for mutational damage. Thus, their rate of mutation is several orders of magnitude higher. Gojobori and Yokoyama (1985), for example, estimated the rates of mutation in the influenza A virus and the Moloney murine sarcoma virus to be of the order of 10^{-2} substitution mutations per nucleotide site per year, that is, approximately 2 million times higher than the rate of mutation in the nuclear DNA of vertebrates. The rate of mutation in the Rous sarcoma virus may be even higher, as nine mutations were detected out of 65 250 replicated nucleotides, that is, a rate of 1.4×10^{-4} substitution mutations per nucleotide site per replication cycle (Leider *et al.*, 1988).

Mutations do not occur randomly throughout the genome. Some regions are more prone to mutate than others, these being called *hot spots* of mutation. One such hot spot is the dinucleotide 5′-CG-3′ (often denoted as CpG), in which the cytosine is frequently methylated in many animal genomes, and is changed to 5′-TG-3′. The dinucleotide 5′-TT-3′ is a hot spot of mutation in prokaryotes but usually not in eukaryotes. In bacteria, regions within the DNA containing short *palindromes* (i.e. sequences that read the same on the complementary strand, such as 5′-GCCGGC-3′, 5′-GGCGCC-3′ and 5′-GGGCCC-3′) were found to be more prone to mutation than other regions.

The direction of mutation is nonrandom. In particular, transitions were found to occur more frequently than transversions. In animal nuclear DNA, transitions were found to account for about 60–70% of all mutations, whereas the proportion of transitions under random mutation is expected to be only 33%. Thus, in nuclear genomes, transitional mutations occur twice as frequently as transversions. In animal mitochondrial genomes, the ratio of transitions to transversions is about 20 transitions to 1 transversion. Some nucleotides are more mutable than others. For example, in the nuclear DNA of mammals, G and C tend to mutate more frequently than A and T.

Mutations are commonly said to occur 'randomly'. However, as we have seen previously, mutations do not occur at random with respect to genomic location, nor do all types of mutations occur with equal frequency. So, what aspect of mutation is random? Mutations are claimed to be random in respect to their effect on the fitness of the organism carrying them. That is, any given mutation is expected to occur with the same frequency under conditions in which this mutation confers an advantage on the organism carrying it, as under conditions in which this mutation confers no advantage or is deleterious. 'It may seem a deplorable imperfection of nature,' said Dobzhansky (1970), 'that mutability is not restricted to changes that enhance the adeptness of their carriers'. Indeed, the issue of whether mutations are random or not with respect to their effects on fitness is periodically debated in the literature, sometimes with fierce intensity (see e.g. Hall, 1990; Lenski and Mittler, 1993; Rosenberg *et al.*, 1994; Sniegowski, 1995).

See also
Mutation Detection
Mutation Nomenclature
Mutation Rate
Mutational Change in Evolution
Mutations in Human Genetic Disease

References

Brinkmann B, Klintschar M, Neuhuber F, Huhne J and Rolf B (1998) Mutation rate in human microsatellites: influence of the structure and length of the tandem repeat. *American Journal of Human Genetics* **62**: 1408–1415.

Brown WM, Prager EM, Wang A and Wilson AC (1982) Mitochondrial DNA sequences of primates: tempo and mode of evolution. *Journal of Molecular Evolution* **18**: 225–239.

Dobzhansky T (1970) *Genetics and the Evolutionary Process*. New York, NY: Columbia University Press.

Drake JW, Charlesworth B, Charlesworth D and Crow JF (1998) Rates of spontaneous mutation. *Genetics* **148**: 1667–1686.

Gojobori T and Yokoyama S (1985) Rates of evolution of the retroviral oncogene of *Moloney murine sarcoma virus* and of its cellular homologues. *Proceedings of the National Academy of Sciences of the United States of America* **82**: 4198–4201.

Goldsmith ME, Humphries RK, Ley T, *et al.* (1983) 'Silent' nucleotide substitution in a β⁺-thalassemia globin gene activates splice site in coding sequence RNA. *Proceedings of the National Academy of Sciences of the United States of America* **80**: 2318–2322.

Graur D and Li W-H (2000) *Fundamentals of Molecular Evolution*, 2nd edn. Sunderland, MA: Sinauer Associates.

Hall BG (1990) Directed evolution of a bacterial operon. *BioEssays* **12**: 551–557.

Kondrashov AS and Crow JF (1993) A molecular approach to estimating human deleterious mutation rate. *Human Mutation* **2**: 229–234.

Leider JM, Palese P and Smith FI (1988) Determination of the mutation rate of a retrovirus. *Journal of Virology* **62**: 3084–3091.

Lenski RE and Mittler JE (1993) The directed mutation controversy and neo-Darwinism. *Science* **259**: 188–194.

Li W-H, Luo C-C and Wu C-I (1985) Evolution of DNA sequences. In: MacIntyre RJ (ed.) *Molecular Evolutionary Genetics*, pp. 1–94. New York, NY: Plenum.

Rosenberg SM, Longerich S, Gee P and Harris RS (1994) Adaptive mutation by deletions in small mononucleotide repeats. *Science* **265**: 405–409.

Sniegowski PD (1995) The origin of adaptive mutants: random or nonrandom? *Journal of Molecular Evolution* **40**: 94–101.

Topal MD and Fresco JR (1976) Complementary base pairing and the origin of substitution mutations. *Nature* **263**: 285–289.

Watson JD and Crick FHC (1953) Genetical implications of the structure of deoxyribonucleic acid. *Nature* **171**: 964–967.

Further Reading

Graur D and Li W-H (2000) *Fundamentals of Molecular Evolution*, 2nd edn. Sunderland, MA: Sinauer Associates.

Griffiths AJF, Miller JH and Suzuki DT (1996) *An Introduction to Genetic Analysis*, 6th edn. New York, NY: Freeman.

Portin P (1993) The concept of the gene: short history and present status. *Quarterly Review of Biology* **68**: 173–223.

Sinden RR, Pearson CE, Potaman VN and Ussery DW (1998) DNA: structure and function. *Advances in Genome Biology* **5A**: 1–141.

Single Nucleotide Polymorphism (SNP)

Anthony J Brookes, *Karolinska Institute, Stockholm, Sweden*

A single nucleotide polymorphism (SNP) is an abundant form of genome sequence variation comprising single base pair alternatives represented at a population frequency of more than 1%. SNPs contribute greatly to the genetic basis of phenotypic variation in ways yet to be fully understood.

Introduction

A single nucleotide polymorphism (SNP) is a fundamental and simple form of sequence variation in deoxyribonucleic acid (DNA). Despite this simplicity, SNPs are highly relevant to a multifarious range of genetics disciplines (Brookes, 1999). An analogy can be drawn with the elemental 1s and 0s ('bits') of the computing world, which are compositionally simple 'numbers' of extreme importance to many fields of science and technology. As the twenty-first century dawns, we are realizing the equivalently wide-ranging and important role that the simple SNP 'letters' (T, C, G, A) play in the realms of biology. This awareness has inspired a tidal wave of activity concerning the nature and exploitation of this form of genome polymorphism. Since their first experimental incarnation as protein variants, and thereafter as restriction fragment length polymorphisms in the 1980s, a truly extensive analysis of SNPs has awaited recent advances in technology plus the maturing Human Genome Project – things that together give us the necessary tools to ask appropriate questions. Particular features of SNPs (abundance, stability and theoretical suitability for automated scoring) can now begin to be exploited in many different contexts. Ongoing activities can be divided simplistically into areas of 'science' (what/why) and 'technology' (how). Due to limitations of space, this article will concentrate on presenting an overview of SNP science with a few general comments about SNP technology.

Basic Facts

SNPs are only one of many forms of sequence variation in the human genome, although at greater than 90% of all observed differences they are by far the most abundant. As indicated by the name, an SNP involves the existence of more than one alternative nucleotide at a single base locus in different copies of a sequence taken from some population. When this entails a coding sequence, the variation would be called a cSNP. Single base insertions/deletions (indels) are not generally included in the definition of an SNP. In principle, two, three or four alternative nucleotides (T, C, G, A) at any particular locus would be classed as an SNP, although in practice the overwhelming majority of SNPs are diallelic. Only four different diallelic SNP types exist: one transition (T/C) and three transversions (T/A, T/G, C/G). Two other substitutions (A/G, A/C) are merely opposite strand representations of two of the above (T/C, T/G), although it may be important to distinguish these in situations where the DNA strands are not equivalent, e.g. transcription studies. Interestingly, the T/C (A/G) variation represents a full two-thirds of all human SNPs, a fact that is partly though not completely explained by the high C to T mutability of CpG dinucleotides.

Another facet of the definition of an SNP is a requirement that the most abundant allele must be represented in the population under test at a frequency of 99% or less. The rare allele(s) will then be present at a total frequency of at least 1%. This is a somewhat arbitrary threshold, and rare alleles below 1% frequency are termed rare variants rather than SNP alleles. Since high-frequency alleles are easier to discover, there is a tendency among experimentalists (and even some theoreticians) to overlook the potential importance of rare variations. However, while often undetected, rare variations actually underlie a great amount of sequence variance in a population. This could be extremely important for considerations of variability in genome function. Overall, the likelihood that any one base will differ between two copies of a sequence chosen randomly from a population is a measure called the 'nucleotide diversity', signified by π (Nei and Li, 1979). Present global estimates for π in human populations range around 0.1–0.2%. But this does not mean that only one or two SNPs typically exist in a thousand bases of the human genome, since more SNPs will be found by sampling more than two chromosome copies. Across a population, four or five

SNPs per 1000 bases is an approximate genome average, implying well over 10 000 000 SNPs across the whole genome. Considerable differences in π have been reported for various coding and noncoding subregions (**Figure 1**), probably indicative of differential evolutionary selective pressures. Interestingly, such data indicate that the 5′ and 3′ untranslated portions of genes are under notably strong evolutionary constraints. Furthermore, 10- to 100-fold differences in π are seen between individual genes or subchromosomal regions. Other factors potentially responsible for the large range of observed π values include chromosomal architecture, differential mutation rates, gene conversion, recombination, structural constraints on ribonucleic acid (RNA), origin of the sampled population and population demographics. The intricacies of these relationships are as yet, however, poorly understood.

Origins of Single Nucleotide Polymorphisms

Although stable over multiple generations, on an evolutionary timescale SNPs are in a dynamic state of flux. The nucleotide mutation rate in humans is about $1/10^8$ to $1/10^9$ per generation, and therefore some 10–100 rare variations (up to five amino acid alterations) are created in each newly conceived individual (Eyre-Walker and Keightley, 1999). Most of these remain extremely rare and are sooner or later lost from the population by random changes (drift) in allele frequency, but some do attain higher representation. This can happen due to drift alone, especially in smaller populations, and positive selection for the new

allele (or nearby sequences) over the old allele will help the transition to occur. The average time required for one or other allele of a neutral variation to become 'fixed' (reach 0% or 100% frequency) is $4N_e$ generations, where N_e is the effective population size (Kimura, 1983), that is, of the order of 1 million years for modern humans. Since modern humans are believed to have radiated from a common African ancestral population only 0.1–0.2 million years ago (Foley, 1998), most variations in the human genome (85%) are today shared between different geographical populations and ethnic groups (although allele frequencies can be different). Ideas of 'race' or major genetic differences between certain groups of humans are therefore scientifically unfounded. Compared with many other species, sequence diversity across all human groups is really not particularly great.

Phenotypic Consequences

A simple and persuasive argument can be made that abundant genome sequence variation (not least SNPs) must underlie genetic influences upon phenotypic variation: the 'common disease–common variant' (CD–CV) model. Human illnesses typically do have a considerable genetic component integral to their etiology, and for common disorders this probably involves many interacting risk factors. Hence, these 'complex diseases' (e.g. mental illness, cardiovascular disease, metabolic disorders, cancer predisposition) are now receiving much research attention focused upon genetic association analysis and the presumed predisposing role played by SNPs. A few common polymorphisms have been conclusively associated with an increased risk for some diseases. Finding these has, however,

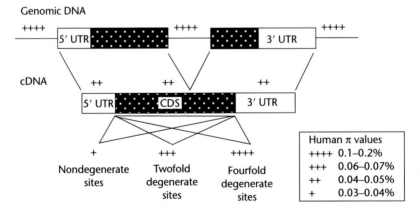

Figure 1 Observed levels of nucleotide diversity (π) in and around human genes. A simple two-exon gene is represented in genomic DNA and processed messenger RNA (mRNA) forms, with various subdomains (including degeneracy classes of coding nucleotides) marked to illustrate the typically observed levels of sequence variation reported across a range of publications. There is, however, a disparity in sequence variability of well over 10-fold between different genes and chromosomal regions. cDNA: complementary DNA; UTR: untranslated region; CDS: coding sequence.

required a great deal of searching, and there have been many false positive (or at least currently unproven) claims along the way. A key problem is that for complex diseases there is at present no way of predicting the intricacy of the etiological architecture. If too many genes (or too many alleles) are involved in influencing the risk of disease, or if the effect of any one allele is too small, then we will probably not be able to detect these risk factors against a background of chance (false) associations and experimental error.

Despite the above, there are good reasons for suggesting that high-frequency polymorphisms might exist with alleles that have differential and substantial effects upon disease risk. Genetic drift, particularly in small or recently bottlenecked populations, could raise pathogenic sequences to appreciable frequencies. Selection for diversity (e.g. cytochrome P-450 gene variants) or balancing selection (e.g. blood group polymorphisms) can likewise increase functionally distinct alleles to substantial frequencies. Risk alleles that cause only late-onset disorders (e.g. Alzheimer disease) would probably escape negative selection acting upon reproductive fitness, and the ideas behind 'compensatory adaptation' propose that alleles that are deleterious late in life may be advantageous earlier in one's existence (Schachter, 1998). Finally, the unknown sophisticated mechanisms by which alleles may interact (e.g. epistatic and multiplicative dependences) plus the everchanging nature of our environment mean that we certainly cannot yet rule out the CV–CD model.

Uses of Single Nucleotide Polymorphisms

Few, if any, genetics disciplines will be untouched by the SNP revolution. Obvious areas include clinical and molecular genetics (Gray *et al.*, 2000), pharmacogenomics, evolutionary and population genetics, anthropology, forensics, functional genomics and structural proteomics. Coding and promoter region SNPs will probably be targeted for initial functional analyses, but all types of SNP will benefit the many other research areas listed above. A description of how these fields might be changed by SNPs is beyond the scope of this article, but one dimension that must be considered is the use of SNPs in clustered groups rather than as isolated entities. First, the polymorphism information content (PIC) available from SNPs can be increased by scoring them as closely linked sets. Second, the phenomenon of linkage disequilibrium (nonrandom assortment of alleles of nearby SNPs) means that one SNP may act as a reasonably effective surrogate for another (which may be part of an unknown pathogenic sequence), thereby aiding genome or regional scanning studies. At present, however, much remains to be defined about the true practical limits of this second proposition. Third, haplotypes (series of alleles at polymorphic loci along a stretch of DNA) not only facilitate a better dissection of the natural history of a region of DNA, but they also probably represent the true functional units that will need to be understood if we are to comprehend phenotypic consequences.

Technologies for Investigating Single Nucleotide Polymorphisms

Most 'hot news' surrounding SNPs at present relates to matters of technology. This is a natural reflection of the need to establish tools before experimental questions can be asked. The two principal areas of activity relate to methods for discovery or scoring of SNPs, and the establishment of large SNP catalogs. On the methods front there are numerous interrelated ideas being developed, each with various advantages and disadvantages (Landegren *et al.*, 1998). None, however, yet approach the level of throughput, low cost, flexibility and robustness that will be needed to fully exploit SNPs in the context of a complex genome. Ironically, the most abundant form of SNP (C/T) is often the most troublesome to score, since in many assay formats this entails detecting a T–G mismatched base pair that is actually quite stable. And despite the appeal of haplotypes, no method can conveniently determine haplotype structures in diploid genomic segments. Additional problems arise due to the presence of many similar sequences within most genomes (e.g. repeat elements, gene paralogs, pseudogenes), which can easily give rise to false SNP predictions and/or compromised genotyping assays. Thus, much remains to be accomplished in the realms of SNP technology.

Many aspects of SNPs and SNP technology are being investigated (Kwok and Gu, 1999). There is an ever-present need to evaluate which SNPs might be the 'best' to catalog (cSNPs versus random SNPs, SNPs with high- or low-frequency alleles, all cSNP versus only nonsynonymous variants, etc.), and to scale and design the discovery procedures accordingly. Present efforts, however, are certain to yield millions of human SNPs within the next few years. The SNP Consortium (TSC) is particularly relevant here, representing an encouraging model of how industry and academia can work together to fund and facilitate the production of impressive primary research tools (over 1 000 000 mapped human SNPs in this case) for unrestricted by all. And if this spirit of sharing is expanded, private sequencing efforts could couple their primary human sequence data to that of the academic community to immediately reveal several million SNPs, thereby removing the discovery bottleneck from the field. To handle the mass of emerging data, a number of SNP databases have been or are being constructed (Lehväslaiho, 2000), mostly based upon some form of open-access model. Good dialog between these exercises is ensuring complete data sharing, harmony of effort and maximizing the availability of data for the community. Together then, the technological challenges are slowly but certainly being met; the only outstanding question is 'what exactly will the SNP revolution reveal?'.

See also

Genetic Variation: Polymorphisms and Mutations
Microarrays and Single Nucleotide Polymorphism (SNP) Genotyping
Polymorphisms: Origins and Maintenance
Single Nucleotide Polymorphisms (SNPs): Identification and Scoring

References

Brookes AJ (1999) The essence of SNPs. *Gene* **234**: 177–186.

Eyre-Walker A and Keightley PD (1999) High genomic deleterious mutation rates in hominids. *Nature* **397**: 344–734.

Foley R (1998) The context of human genetic evolution. *Genome Research* **8**: 339–347.

Gray IC, Campbell DA and Spurr NK (2000) Single nucleotide polymorphisms as tools in human genetics. *Human Molecular Genetics* **9**: 2403–2408.

Kimura M (1983) *The Neutral Theory of Molecular Evolution.* Cambridge, UK: Cambridge University Press.

Kwok P-Y and Gu Z (1999) Single nucleotide polymorphism libraries: why and how are we building them? *Molecular Medicine Today* **5**: 538–543.

Landegren U, Nilsson M and Kwok P-Y (1998) Reading bits of genetic information: methods for single-nucleotide polymorphism analysis. *Genome Research* **8**: 769–776.

Lehväslaiho H (2000) Human sequence variation and mutation databases. *Briefings Bioinformatics* **1**: 161–166.

Nei M and Li W-H (1979) Mathematical model for studying genetic variation in terms of restriction endonucleases. *Proceedings of the National Academy of Sciences of the United States of America* **76**: 5269–5273.

Schachter F (1998) Causes, effects, and constraints in the genetics of human longevity. *American Journal of Human Genetics* **62**: 1008–1014.

Further Reading

Altshuler D, Pollara VJ, Cowles CR, *et al.* (2000) An SNP map of the human genome generated by reduced representation shotgun sequencing. *Nature* **407**: 513–516.

Gould Rothberg BE (2001) Mapping a role for SNPs in drug development. *Nature Biotechnology* **19**: 209–211.

Jorde LB (2000) Linkage disequilibrium and the search for complex disease genes. *Genome Research* **10**: 1435–1444.

Kristensen VN, Kelefiotis D, Kristensen T and Børresen-Dale A-L (2001) High-throughput methods for detection of genetic variation. *Biotechniques* **30**: 318–332.

Kruglyak L (1999) Prospects for whole-genome linkage disequilibrium mapping of common disease genes. *Nature Genetics* **22**: 139–144.

Kruglyak L and Nickerson DA (2001) Variation is the spice of life. *Nature Genetics* **27**: 234–236.

Kwok P-Y and Gu Z (1999) Single nucleotide polymorphism libraries: why and how are we building them. *Molecular Medicine Today* **5**: 538–543.

Przeworski M, Hudson RR and Rienzo AD (2000) Adjusting the focus on human variation. *Trends in Genetics* **16**: 296–302.

Schork NJ, Fallin D and Lanchbury JS (2000) Single nucleotide polymorphisms and the future of genetic epidemiology. *Clinical Genetics* **58**: 250–264.

Weiss KM and Terwilliger JD (2000) How many diseases does it take to map a gene with SNPs? *Nature Genetics* **26**: 151–157.

Single Nucleotide Polymorphisms (SNPs): Identification and Scoring

Pui-Yan Kwok, *University of California, San Francisco, California, USA*

Single nucleotide polymorphisms (SNPs) are the most abundant DNA sequence variations in the human genome. Numerous methods have been developed to identify and score them. A large set of human SNPs have been identified and mapped to the reference human genome sequence. The SNP markers are useful for the study of the genetic basis of many human diseases.

Intermediate article

Article contents

- Introduction
- SNP Identification
- SNP Scoring

Introduction

Identification and scoring of deoxyribonucleic acid (DNA) sequence variations such as single nucleotide polymorphisms (SNPs) are important activities in molecular genetics, with applications in the study of the genetic factors associated with numerous diseases (Collins *et al.*, 1997). Multiple studies have shown that the prevalence of SNPs in the human genome is approximately one in 1000 bp when two chromosomes are compared. Since each human cell contains two sets of chromosomes (one from the mother and one from the father), there are approximately 3 million heterozygous sites in our 3 billion-letter genome at which the chromosomes derived from the two parents differ. SNPs are genetic markers that can be used to distinguish the two sets of chromosomes and allow tracing of the inheritance of the chromosomes from generation to generation.

By following the inheritance pattern of a trait (such as a disease or some other measurable condition) and SNPs either in families or in the population, one can identify the SNPs associated with the trait and map the genes associated with the trait to specific locations in the genome (Risch, 2000). This genetic mapping approach opens up new ways to study traits and diseases where there is little or no biological knowledge. Many of the genetic mapping approaches require a large number of markers, and SNPs are the only class of markers that can satisfy the needs of all genetic mapping strategies. For SNP mapping approaches to work, a large set of SNPs must be identified in the human genome and SNP scoring assays must be developed to study large populations.

SNP Identification

Although the frequency of SNPs in the human genome is reasonably high, searching for unknown polymorphisms is still a tedious and expensive process. There are no efficient ways to compare thousands of base pairs of DNA across multiple individuals. Most commonly used SNP identification methods can only scan approximately 500-bp DNA fragments reliably. To find SNPs in targeted regions, small DNA fragments are amplified from several individuals by the polymerase chain reaction (PCR) and scanned for variants. Although PCR occasionally introduces errors during amplification, these errors are random and are found in a minor fraction of the PCR products formed. Scanning the entire pool of PCR products makes it possible to identify SNPs but not the errors introduced during the PCR process.

DNA sequencing is the most accurate way to identify SNPs. Because DNA sequencing of PCR products is still labor intensive and costly, many groups use a screening method to identify variants followed by DNA sequencing of individual samples that contain variants. The most widely used screening methods include denaturing high-performance liquid chromatography (DHPLC), heteroduplex analysis (HA), single-stranded DNA conformation analysis (SSCA), and chemical or enzymatic cleavage of

mismatched DNA (Kwok and Chen, 1998). DHPLC and HA are SNP scanning techniques based on the difference in speed at which perfectly matching and mismatched double-stranded DNA molecules travel down a column or gel at near-denaturing temperatures or chemical conditions. SSCA is based on single-stranded DNA assuming a tertiary structure dictated by its sequence and a single base difference between two molecules leading to tertiary structures that can be separated on native gel electrophoresis. Chemical or enzymatic cleavage of mismatched DNA relies on chemicals or enzymes that preferentially cut mismatched DNA molecules. Observing smaller DNA fragments after chemical or enzymatic treatment indicates that a variant resides in the DNA being scanned.

The first attempt to identify a large set of SNPs to cover the entire genome was made by scanning thousands of physical landmarks of the human genome called sequence tagged sites (STSs). STSs are unique DNA sequences defined by a PCR assay. The PCR products of these STSs from just a few individuals were scanned by DNA sequencing and several thousand SNPs were identified. A second attempt to identify a large set of SNPs focused on variants found in expressed sequence tags (ESTs). Because there are millions of ESTs being generated by numerous groups using tissues from many different donors, many variants can be found simply by comparing the EST sequences from the same gene loci. Tens of thousands of SNPs have been found by this approach, and a good database of these gene-associated SNPs is to be found at the US National Cancer Institute's Cancer Genome Anatomy Project website (Buetow et al., 1999).

The most comprehensive attempt to identify a large number of SNPs has been made by the International SNP Map Working Group, comprising academic researchers supported by public funding agencies and the SNP Consortium (TSC; a nonprofit organization formed by a number of pharmaceutical companies, high-technology companies and the Wellcome Trust, a British charity). Using several complementary approaches, these groups have identified 1.42 million SNPs that are uniquely mapped to the human genome sequence and are freely available to the public (The International SNP Map Working Group, 2001). By constructing 'reduced representation' libraries of small DNA fragments generated from 24 individuals and sequencing millions of these clones, the SNP Consortium has been able to obtain sequences for the same loci from multiple individuals. Comparing these sequences has led to the discovery of numerous SNPs. In addition, the SNP Consortium has also compared all the sequences with the reference human genome sequence and found additional SNPs.

Researchers at Washington University in St Louis and the Sanger Centre in Cambridge, UK, have compared the DNA sequences from overlapping bacterial artificial chromosome (BAC) clones generated by the Human Genome Project and identified hundreds of thousands of SNPs. A sample has shown that more than 95% of these SNPs are real variants and more than 80% of them are indeed polymorphic in the general population (Marth *et al.*, 2001). In addition to the publicly available SNPs, several commercial concerns have also accumulated millions of SNPs by EST sequencing or whole-genome sequencing (Venter *et al.*, 2001). Because these SNPs are not publicly available, their properties are largely unknown. (**See** Pharmacogenetics and Pharmacogenomics.)

SNP Scoring

SNP scoring (or genotyping) refers to the screening of DNA samples to determine which allele(s) of a known SNP each sample contains. It is a much simpler problem than scanning DNA for unknown polymorphisms. The challenge is to identify a specific base at a polymorphic site in the background of 3 billion base pairs unambiguously. Furthermore, most genetic studies require scoring of numerous SNPs in many individuals, so efficiency and cost are important considerations. Fortunately there are a number of good SNP scoring approaches available through the ingenious use of advances in enzymology and analytical methodologies.

Because no current technology can examine genomic DNA directly and decipher its sequence at specific locations, some kind of amplification is needed. One can amplify the target DNA region by the PCR prior to the genotyping assay or perform the genotyping assay on genomic DNA followed by a signal amplification step. In general, PCR amplification also simplifies the genome and results in an extra level of specificity.

Most SNP scoring methods take advantage of sequence-specific activities of enzymes involved in DNA replication or repair. For example, when extending a primer annealed to a DNA template, DNA polymerases incorporate, very specifically, nucleotides complementary to the target sequence. Similarly, DNA ligases join together two adjacent synthetic oligonucleotides only if they perfectly anneal to the target sequence. If the nucleotides or oligonucleotides are labeled, the products formed in a positive reaction can be visualized and the genotypes inferred. Other SNP scoring methods rely on the ability to distinguish between the physical properties of the genotyping reaction products either in terms of their molecular weight (e.g. as determined by mass

spectrometry) or their mobility through a gel matrix during electrophoresis. Yet other SNP scoring methods are based on differences in the strength of hybridization between matched versus mismatched oligonucleotide probes against the target sequence containing the SNP (Kwok, 2001).

With millions of SNPs and robust SNP scoring methods available for genetic mapping, the stage is set for in-depth genetic analysis of common human diseases. By selecting large enough patient and control populations and analyzing the genotyping data for a large number of appropriately chosen SNPs, there is optimism that the major genetic factors associated with common human diseases can be identified. The SNPs (or mutations) that are predictive of disease susceptibility can be used to make medical decisions and to help individuals prone to certain diseases decide whether to take preventive measures against the diseases (Collins and McKusick, 2001).

See also

Microarrays and Single Nucleotide Polymorphism (SNP) Genotyping
Single Nucleotide Polymorphism (SNP)

References

Buetow KH, Edmonson MN and Cassidy AB (1999) Reliable identification of large numbers of candidate SNPs from public EST data. *Nature Genetics* **21**: 323–325.

Collins FS, Guyer MS and Charkravarti A (1997) Variations on a theme: cataloging human DNA sequence variation. *Science* **278**: 1580–1581.

Collins FS and McKusick VA (2001) Implications of the Human Genome Project for medical science. *Journal of the American Medical Association* **285**: 540–544.

Kwok PY (2001) Methods for genotyping single nucleotide polymorphisms. *Annual Review of Genomics and Human Genetics* **2**: 235–258.

Kwok PY and Chen X (1998) Detection of single nucleotide variations. *Genetic Engineering* **20**: 125–134.

Marth G, Yeh R, Minton M, *et al.* (2001) Single-nucleotide polymorphisms in the public domain: how useful are they? *Nature Genetics* **27**: 371–372.

Risch NJ (2000) Searching for genetic determinants in the new millennium. *Nature* **405**: 847–856.

The International SNP Map Working Group (2001) A map of human genome sequence variation containing 1.42 million single nucleotide polymorphisms. *Nature* **409**: 928–933.

Venter JC, Adams MD, Myers EW, *et al.* (2001) The sequence of the human genome. *Science* **291**: 1304–1351.

Further Reading

Gray IC, Campbell DA and Spurr NK (2000) Single nucleotide polymorphisms as tools in human genetics. *Human Molecular Genetics* **9**: 2403–2408.

Kwok PY (2001) Methods for genotyping single nucleotide polymorphisms. *Annual Review of Genomics and Human Genetics* **2**: 235–258.

Nowotny P, Kwon JM and Goate AM (2001) SNP analysis to dissect human traits. *Current Opinion in Neurobiology* **11**: 637–641.

Schork NJ, Fallin D and Lanchbury JS (2000) Single nucleotide polymorphisms and the future of genetic epidemiology. *Clinical Genetics* **58**: 250–264.

Tsuchihashi Z and Dracopoli NC (2002) Progress in high through-put SNP genotyping methods. *Pharmacogenomics Journal* **2**: 103–110.

Web Links

Cancer Genome Anatomy Project (CGAP). The CGAP website provides researchers with access to all CGAP data and biological resources including genomic data for human and mouse, including expressed sequence tags (ESTs), gene expression patterns, single nucleotide polymorphisms (SNPs), cluster assemblies and cytogenetic information
http://cgap.nci.nih.gov/

Single Nucleotide Polymorphism (SNP). The NCBI single nucleotide polymorphism database contains data on all the publicly available SNPs, with their map positions in the human genome, allele frequencies where available and some haplotype data
http://www.ncbi.nlm.nih.gov/SNP/

The SNP Consortium Ltd. The SNP Consortium (TSC) website contains data on > 1.8 million SNPs discovered and characterized by its member institutions and other publicly funded groups
http://snp.cshl.org/

Skeletal Dysplasias: Genetics

Andrea Superti-Furga, *University of Lausanne, Lausanne, Switzerland*

Sheila Unger, *The Hospital for Sick Children, Toronto, Ontario, Canada*

The skeletal dysplasias and dysostoses are a group of genetic disorders that manifest as disproportionate short stature, deformity and various medical complications. Dysplasias may result from genetic mutations that affect structural components, metabolic pathways, signal transduction mechanisms or gene expression patterns involved in skeletal growth and development.

Intermediate article

Article contents

- Introduction
- Genetic Disorders of the Skeleton
- Overview on Selected Dysplasias and Dysplasia Families

Introduction

The human skeleton consists of more than 200 individual elements. A genetic blueprint determines their shape and their position, with little variation from one individual to another. The skeletal elements are formed as cartilage condensations in the embryo. They undergo considerable growth until adulthood while cartilage tissue is gradually replaced by bone. Once adult size is attained, growth ceases but the tissues continue to be biologically active to ensure maintenance of proper mechanical resistance and joint function. The correct formation of all skeletal elements according to the genetic blueprint is called skeletal morphogenesis. The changes in size and shape that transform the tiny skeletal elements in the embryo into the adult skeleton are skeletal growth. The combination of biological processes by which the skeletal elements maintain their integrity over the life span of the organism is skeletal homeostasis. The remarkable combination of mechanical resistance and biological plasticity of the skeleton is afforded by cartilage and bone, two highly specialized tissues, each with differentiated cells and an array of extracellular macromolecules that interact and assemble to form the extracellular matrix.

Genetic Disorders of the Skeleton

Skeletal dysplasias and dysostoses

When one or more of the processes of skeletal morphogenesis, growth and homeostasis are disturbed, the end result is the disorders called skeletal dysplasias and dysostoses. They are determined by genetic mutations although affected individuals may be sporadic, that is, the only ones to be affected in their families. The manifestations of skeletal dysplasias are manifold, and include visible changes such as short stature, altered body proportions, deformities, changes in shape and number of fingers and toes, or changes in the facial features (Unger, 2002).

Depending on the tissue-specific expression of the affected gene, changes may occur in tissues other than bone and structural tissue, leading to manifestations in organs such as the eye, the ear, the lungs, the brain, the muscles, the kidney, the immune system and others. The degree of severity of skeletal dysplasias is remarkably wide, ranging from isolated, mild shortening of stature of little significance to the individual, to severe skeletal underdevelopment leading to neonatal death or intrauterine demise. In surviving patients, secondary complications of skeletal deformity and manifestations in extra skeletal organs add to the burden of disease.

Incidence

Although individually rare (two among the most common skeletal dysplasias, achondroplasia and osteogenesis imperfecta, have an incidence of approximately one in 15 000 newborns), there are over 200 recognized nosologic entities, so that the collective incidence of skeletal dysplasias has been estimated to be at least one in 5000 newborns.

Skeletal dysplasias versus dysostoses

Dysostoses originate in embryonic, morphogenetic developmental processes. They commonly affect a single bone or a group of bones. The pathogenetic event may be confined to the embryonic period. Dysplasias originate more commonly in structural components or in metabolic pathways of bone and cartilage. Thus, they may affect all bones and lead to a generalized and sometimes progressive disorders of skeletal growth and homeostasis. However, this distinction is gradually losing significance as it is not always possible to draw a clear distinction, as genes responsible for the formation of individual bones in the embryo may be expressed also at much later stages, sometimes even in adult life.

Criteria for phenotypic classification

The many different disorders are distinguished by their peculiar combination of clinical and radiographic features and by mode of inheritance. The *International Nomenclature and Classification of the Osteochondrodysplasias* (Spranger, 1992; Rimoin, 1998; Hall, 2002) lists those skeletal dysplasias and dysostoses meeting criteria to warrant a separate nosologic entity; undoubtedly, many more exist. The nomenclature is revised periodically with the help of new clinical, biochemical and molecular information. The most widely used criteria for a diagnostic approach to the patient are clinical and radiographic features.

Classification of genetic disorders of the skeleton

The complexity of skeletal–genetic phenotypes began to be appreciated almost 50 years ago, leading to the delineation of individual entities from what had

previously been called chondrodystrophy. The criteria used for classification have been clinical features such as growth, body proportions, and, because of the outstanding role of radiography in defining skeletal disease, radiographic criteria (Unger, 2002).

Clinical criteria include early (pre- or perinatal) lethality, prenatal or postnatal onset, proportionate or disproportionate shortening (predominant shortening of the trunk or of the limbs), involvement of the skull and of the hands and fingers. Radiographic criteria refer to the site of skeletal changes as seen on radiographic films: a dysplasia can be spondylar, epiphyseal, metaphyseal, diaphyseal or a combination of these. The skeletal elements may show increased density or reduced density. Although these criteria are of fundamental importance, they serve as landmarks for orientation but are not enough for the differential diagnosis of the great variety of known disorders. Therefore, other characteristic clinical manifestations, as well as specific radiographic signs, are important in making a diagnosis. Data on the mode of genetic transmission as well as biochemical data are also important criteria for classification.

Based on this combination of criteria, over 200 nosologic entities have been distinguished. The *International Nomenclature and Classification of the Osteochondrodysplasias* (1998) has been a valuable instrument in defining existing entities and delineating new ones.

Classification by molecular defect and pathogenetic mechanism

The accumulation of insight into the pathogenesis of skeletal dysplasias and dysostoses has turned the skeletal system into a unique biologic model – hardly another system or organ offers a similarly detailed understanding of its biology and pathobiology. The multitude and variety of genes and proteins involved in the pathogenesis of skeletal dysplasias and dysostoses called for a molecular classification to assist in understanding the pathogenesis and in identifying sequences and cascades of gene expression as well as metabolic pathways involved in skeletal development, and thus to help in pointing at other candidate genes, in designing new diagnostic strategies, and perhaps at finding therapeutic avenues (Superti-Furga, 2001).

The molecular–pathogenetic classification presented below (**Table 1**) has been drafted in analogy to a functional classification of proteins in *Saccharomyces cerevisiae*, *Caenorhabditis elegans* and disease-related human genes and proteins (Jimenez-Sanchez, 2001). It should be useful in elucidating metabolic pathways, signal transduction mechanisms, and cascades of gene regulation relevant to skeletal biology, and thus in making bridges between basic science, animal models and the human clinical setting.

Table 1 Classification of genetic disorders of the skeleton by molecular and pathogenetic criteria

Gene or protein	Mode of inheritance[a]	Clinical phenotype
Group 1: Defects in extracellular structural proteins		
COL1A1, COL1A2 (collagen 1 α1, α2 chain)	AD	Family: osteogenesis imperfecta
COL2A1 (collagen 2 α1 chain)	AD	Family: achondrogenesis 2, hypochondrogenesis, congenital spondyloepiphyseal dysplasia (SEDC), Kniest, Stickler arthro-ophthalmopathy, familial osteoarthritis, other variants
COL9A1, COL9A2, COL9A3 (collagen 9 α1, α2, α3 chain)	AD	Multiple epiphyseal dysplasia (MED; two or more variants)
COL10A1 (collagen 10 α1 chain)	AD	Metaphyseal dysplasia Schmid
COL11A1, COL11A2 (collagen 11 α1, α2 chain)	AR, AD	Otospondylomegaepiphyseal dysplasia (OSMED); Stickler (variant), Marshall syndrome
COMP (cartilage oligomeric matrix protein)	AD	Pseudoachondroplasia, multiple epiphyseal dysplasia (MED, one form)
MATN3 (matrilin-3)	AD	Multiple epiphyseal dysplasia (MED; one variant)
Perlecan (*HSPG2*)	AR	Schwartz–Jampel type 1; dyssegmental dysplasia
Group 2: Defects in metabolic pathways (including enzymes, ion channels and transporters)		
Tissue nonspecific alkaline phosphatase (*ALPL*)	AR, AD	Hypophosphatasia (several forms)
ANKH (pyrophosphate transporter)	AD	Craniometaphyseal dysplasia
DTDST/*SLC26A2* (diastrophic dysplasia sulfate transporter)	AR	Family: achondrogenesis 1B, atelosteogenesis 2, diastrophic dysplasia, recessive multiple epiphyseal dysplasia (rMED)
PAPSS2 (phosphoadenosine-phosphosulfate-synthase 2)	AR	Spondyloepimetaphyseal dysplasia Pakistani type
TCIRG1, osteoblast proton pump subunit (acidification defect)	AR	Severe infantile osteopetrosis
Chloride channel 7 (*CLCN7*)	AR	Severe osteopetrosis
Carbonic anhydrase II (*CA2*)	AR	Osteopetrosis with intracranial calcifications and renal tubular acidosis
Vitamin K-epoxide reductase complex (*GGCX*)	AR	Chondrodysplasia punctata with vitamin K-dependent coagulation defects
MGP (matrix Gla protein)	AR	Keutel syndrome (pulmonary stenosis, brachytelephalangism, cartilage calcifications and short stature)
ARSE (arylsulfatase E)	XLR	X-linked chondrodysplasia punctata (CDPX1)
3-β-Hydroxysteroid-dehydrogenase (*NSDHL*)	XLD	CHILD syndrome
3-β-Hydroxysteroid-Δ(8)Δ(7)-isomerase (*EBP*)	XLD	X-linked chondrodysplasia punctata, Conradi–Hünermann type (CDPX2); CHILD syndrome
PEX7 (peroxisomal receptor/importer)	AR	Rhizomelic chondrodysplasia punctata 1
Dihydroxy-acetonphosphate-acyltransferase, peroxisomal enzyme (*GNPAT*)	AR	Rhizomelic chondrodysplasia punctata 2
Alkyl-dihydroxy-diacetonphosphate synthase (*AGPS*; peroxisomal enzyme)	AR	Rhizomelic chondrodysplasia punctata 3
Group 3: Defects in folding and degradation of macromolecules		
Sedlin (endoplasmic reticulum protein with unknown function) (*SEDL*)	XR	X-linked spondyloepiphyseal dysplasia (SED-XL)
Cathepsin K (lysosomal proteinase) (*CTSK*)	AR	Pycnodysostosis
Lysosomal acid hydrolases and transporters (sulfatase, glycosidase, translocase, etc.)	AR, XLR	Lysosomal storage diseases: mucopolysaccharidoses, oligosaccharidoses, glycoproteinoses (several forms)

Table 1 Continued

Gene or protein	Mode of inheritance[a]	Clinical phenotype
Targeting system of lysosomal enzymes (GlcNAc-1-phosphotransferase) (*GNPTA*)	AR	Mucolipidosis II (I-cell disease), mucolipidosis III
MMP2 (matrix metalloproteinase 2)	AR	Torg type osteolysis (nodulosis arthropathy and osteolysis syndrome)
Group 4: Defects in hormones and signal transduction mechanisms		
25-α-Hydroxycholecalciferol-1-hydroxylase (*CYP27B1*)	AR	Vitamin D-dependent rickets type 1 (VDDR1)
1,25-α-Dihydroxy-vitamin D_3 receptor (*VDR*)	AR	Vitamin D-resistant rickets with end-organ unresponsiveness to vitamin D_3 (VDDR2)
CASR (calcium 'sensor'/receptor)	AD	Neonatal severe hyperparathyroidism with bone disease (if affected fetus in unaffected mother); familial hypocalciuric hypercalcemia
PTH/PTHrP receptor (*PTHR1*)	AD (activating mutations)	Metaphyseal dysplasia Jansen
	AR (inactivating mutation)	Lethal dysplasia Blomstrand
	AD or somatic mutations	Enchondromatosis
GNAS1 (stimulatory Gs alpha protein of adenylate cyclase)	AD	Pseudohypoparathyroidism (Albright hereditary osteodystrophy and several variants) with constitutional haploinsufficiency mutations; McCune–Albright syndrome with somatic mosaicism for activating mutations
PHEX	XL	Hypophosphatemic rickets, X-linked semidominant type (impaired cleavage of FGF23)
FGF23, fibroblasts growth factor 23	AD	Hypophosphatemic rickets, autosomal dominant type (resistance to PEX cleavage)
FGFR1 (fibroblast growth factor receptor 1)	AD	Craniosynostosis syndromes (Pfeiffer, other variants)
FGFR2	AD	Craniosynostosis syndromes (Apert, Crouzon, Pfeiffer; several variants)
FGFR3	AD	Thanatophoric dysplasia, achondroplasia, hypochondroplasia, SADDAN; craniosynostosis syndromes (Crouzon with acanthosis nigricans, Muenke nonsyndromic craniosynostosis)
ROR2 ('orphan receptor tyrosine kinase')	AR	Robinow syndrome
	AD	Brachydactyly type B
TNFRSF11A (receptor activator of nuclear factor κB; RANK)	AD	Familial expansile osteolysis
TGFβ1 (*TGFB1*)	AD	Diaphyseal dysplasia (Camurati–Engelmann)
Cartilage-derived morphogenetic protein 1 (*GDF5*)	AR	Acromesomelic dysplasia Grebe/Hunter–Thompson
	AD	Brachydactyly type C
Noggin ('growth factor', TGF antagonist) (*NOG*)	AD	Multiple synostosis syndrome; synphalangism and hypoacusis syndrome
dll3 (delta-like 3, intercellular signaling) (*DLL3*)	AR	Spondylocostal dysostosis (one form)
IHH (Indian hedgehog signal molecule)	AD	Brachydactyly A1
C7orf2 (orphan receptor)	AR	Acheiropodia
SOST (sclerostin; cystine knot secreted protein)	AR	Sclerosteosis, van Buchem disease
LRP5 (LDL receptor-related protein 5)	AR	Osteoporosis–pseudoglioma syndrome
WISP3 (growth regulator/growth factor)	AR	Progressive pseudorheumatoid dysplasia

Table 1 Continued

Gene or protein	Mode of inheritance[a]	Clinical phenotype
Group 5: Defects in nuclear proteins and transcription factors		
SOX9 (HMG-type DNA binding protein/ transcription factor)	AD	Campomelic dysplasia
Gli3 (zinc-finger gene) (*GLI3*)	AD	Greig cephalopolysyndactyly, polydactyly type A and others, Pallister–Hall syndrome
TRPS1 (zinc-finger gene)	AD	Trichorhinophalangeal syndrome (types 1–3)
EVC2 (leucine-zipper gene)	AR	Chondroectodermal dysplasia (Ellis–van Creveld syndrome)
TWIST (helix–loop–helix transcription factor)	AD	Craniosynostosis Saethre–Chotzen
P63 (p53-related transcription factor) (*TP73L*)	AD	EEC syndrome, Hay–Wells syndrome, limb–mammary syndrome, split hand–split foot malformation (some forms)
CBFA-1 (core binding factor A 1; runt-type transcription factor) (*RUNX2*)	AD	Cleidocranial dysplasia
LMX1B (LIM homeodomain protein)	AD	Nail–patella syndrome
DLX3 (distal-less 3 homeobox gene)	AD	Trichodentoosseous syndrome
HOXD13 (homeobox gene)	AD	Synpolydactyly
MSX2 (homeobox gene)	AD (gain of function)	Craniosynostosis, Boston type
	AD (loss of function)	Parietal foramina
ALX4 (homeobox gene)	AD	Parietal foramina (cranium bifidum)
SHOX (short stature homeobox gene)	Pseudoautosomal	Léri–Weill dyschondrosteosis, idiopathic short stature?
TBX3 (T-box 3, transcription factor)	AD	Ulnar–mammary syndrome
TBX5 (T-box 5, transcription factor)	AD	Holt–Oram syndrome
EIF2AK3 (transcription initiation factor kinase)	AR	Wolcott–Rallison syndrome (neonatal diabetes mellitus and spondyloepiphyseal dysplasia)
NEMO (NFκB essential modulator; kinase activity) (*IKBKG*)	XL	Osteopetrosis, lymphedema, ectodermal dysplasia and immunodeficiency (OLEDAID)
Group 6: Defects in oncogenes and tumor suppressor genes		
EXT1, EXT2 (exostosin-1, exostosin-2; heparan-sulfate polymerases)	AD	Multiple exostoses syndrome type 1, type 2
SH3BP2 (c-*Abl*-binding protein)	AD	Cherubism
Group 7: Defects in RNA and DNA processing and metabolism		
RNase MRP-RNA component	AR	Cartilage–hair hypoplasia (including metaphyseal dysplasia without hypotrichosis)
ADA (adenosine deaminase)	AR	Severe combined immunodeficiency (SCID) with (facultative) metaphyseal changes
SMARCAL1	AR	Schimke immuno-osseous dysplasia

AR: autosomal recessive; XL: X-linked; XLD: X-linked dominant; XLR: X-linked recessive.
Modified from Superti-Furga *et al.* (2001).
[a]AD: autosomal dominant.

Practical significance of biochemical and molecular knowledge

As many as half of the skeletal dysplasia cases likely to be encountered in pediatric and genetics practice today may belong to a molecularly characterized dysplasia family. The identification of biochemical defects and subsequently of the genes and mutations responsible for several skeletal dysplasias has transferred the diagnosis from the clinical–radiographic level to that of the laboratory. Confirmation of a clinical–radiographic diagnosis by biochemical or molecular tests provides a solid basis for information of affected individuals and families, for individual tailoring of therapeutic and prophylactic measures, for personal and genetic counseling, and for prenatal diagnosis when applicable and requested by the family.

Overview on Selected Dysplasias and Dysplasia Families

Fibroblast growth factor receptor 3 (FGFR3) skeletal dysplasia family

Achondroplasia (ACH; MIM 100800) is the most frequent skeletal dysplasia in humans. Achondroplasia, the more severe forms (thanatophoric dysplasia, TD; MIM 187600; and severe achondroplasia with developmental delay and acanthosis nigricans, SADDAN) and the milder form (hypochondroplasia) (HYP; MIM 146000) are all caused by dominant mutations in the *FGFR3* gene which cause activation of the fibroblast growth factor receptor 3 (FGFR3) molecule, a receptor tyrosine kinase. There is a strong paternal age effect, and achondroplasia individuals are likely to be born to elder parents and, in multiplex families, be younger than their unaffected sibs.

Unlike the situation for most other genes responsible for skeletal dysplasia, the great majority of individuals with achondroplasia carry exactly the same point mutation at nucleotide 1138 which leads to the substitution of glycine 380 with arginine. Nucleotide 1138 of *FGFR3* has the highest spontaneous mutation rate known so far in humans. Although the cellular pathogenesis has not been clarified completely, it appears that FGFR3 activation changes the behavior of chondrocytes, shortening the period of proliferation (responsible for bone growth) and anticipating their differentiation to terminal chondrocytes.

Collagen 1 family

Osteogenesis imperfecta (OI; MIM 166200, 166210, 166220, 166240), also known as brittle bone disease, is caused by dominant mutations in the genes *COL1A1* and *COL1A2* which code for the two polypeptide chains of collagen 1, the major collagen type in humans, found predominantly in bone tissue, ligaments, tendons and skin. The severity ranges from perinatally lethal forms with multiple fractures and severe bone deformation at birth to subclinical forms with only a few or even no fractures.

The so-called Sillence classification is useful as clinical stenogram, although the genetic part of the original classification is no longer valid: OI type I is the mildest form, usually not accompanied by overt deformations, and often with normal stature; OI type IV is a moderately severe form, with short stature and mild deformations; OI type III is a severe form with major deformity, markedly short stature and major handicap (patients are usually wheelchair-bound); and OI type II is the perinatally lethal form.

An autosomal recessive form of OI exists but is very rare, accounting for a few per cent of cases at most; its molecular basis is still not known (MIM 259440).

Collagen 2 family and related cartilage-specific collagen disorders

Collagen 2 is the main collagen type in cartilage, and is present also in the vitreous body of the eye and in the internal ear. Mutations in collagen 2 thus affect skeletal growth, joint homeostasis, vision and hearing. There is close homology between the molecular pathogenesis of collagen 1 (producing osteogenesis imperfecta; see above) and that of collagen 2. Mutations in collagen 2 produce a similarly wide phenotypic spectrum including perinatally lethal forms (achondrogenesis type 2 and hypochondrogenesis), severe nonlethal forms (spondyloepiphyseal dysplasia congenita (MIM 183900), and Kniest dysplasia (MIM 156550)) and milder forms manifesting as early-onset arthrosis (Stickler syndrome or hereditary arthro-ophthalmopathy; MIM 108300), familial arthrosis with mild chondrodysplasia).

Mutations in two other collagen genes, *COL11A1* and *COL11A2*, may cause the Stickler phenotype (Stickler type 2 or Marshall syndrome, MIM 184840; Stickler type 3 or mild OSMED; see below). Collagen 11 is expressed in cartilage and participates in the formation of collagen fibrils together with collagens 2 and 9 (see below). Recessive haploinsufficiency mutations in *COL11A2* produce a distinct, severe phenotype called otospondylomegaepiphyseal dysplasia (OSMED; MIM 215150) (short stature, enlarged joints, severe facial hypoplasia, cleft palate, severe sensorineural deafness, collapsible airways with respiratory distress in the newborn).

Multiple epiphyseal dysplasia group

Multiple epiphyseal dysplasia (MED) is a relatively mild, common and heterogeneous disorder. Presenting complaints are waddling gait and/or hip pain in children, and premature osteoarthrosis is common. The diagnosis of bilateral aseptic necrosis of the femoral head (Perthes disease) is often made in children with MED. Stature may be normal or moderately reduced and body proportions are usually unremarkable. Some cases may be asymptomatic (chance observation on X-rays made for other causes). The radiological hallmark is small epiphyses with retarded ossification, particularly evident at hips and knees.

Mutations in at least six genes can produce MED: cartilage oligomeric protein (*COMP*; see below), matrilin-3, the α1, α2 and α3 chains of collagen 9, and the sulfate transporter *DTDST* (see below); there is evidence from linkage studies that more genes for MED must exist. Collagen 9 is expressed in cartilage like collagen 2 (although at lower levels), and is incorporated in collagen fibrils together with collagen 2 and collagen 11. Unlike collagen 2 mutations, the phenotype of *COL9A2* and *COL9A3* mutations appears not to involve eye and ear, while mild myopathy has been associated with *COL9A3* mutations.

Cartilage oligomeric matrix protein

Pseudoachondroplasia (PSACH; MIM 177170) is a condition that superficially resembles achondroplasia because of short stature with short limbs (hence the name), but it has distinctive features. The molecular basis is dominant mutations in the gene coding for cartilage oligomeric protein (*COMP*), a structural component of both cartilage and tendons. Different mutations in the same gene can produce a milder disorder, dominantly inherited multiple epiphyseal dysplasia type 1 (EDM1; MIM 132400; see above).

Lysosomal enzymes and skeletal dysplasias: mucopolysaccharidoses, oligosaccharidoses, mucolipidoses

The prototype of metabolic skeletal dysplasias are the mucopolysaccharidoses. These are lysosomal storage disorders caused by deficient activity of one of several lysosomal hydrolases and sulfatases responsible for the degradation of sulfated glycosaminoglycans, which are abundant in skin and connective tissue, cartilage, and bone, the central nervous system, internal organs, and even blood plasma. Clinical signs include coarsening of facial features, skin thickening, joint limitations, visceromegaly, impaired hearing, corneal clouding, and retarded mental development or mental regression in various combinations.

Within each enzyme deficiency, there is extensive clinical variability. There is usually a generalized skeletal dysplasia known under the somewhat misleading name of dysostosis multiplex. Mucopolysaccharidosis type IVA (Morquio syndrome) is been one of the first short-trunk dysplasias to have been described. The oligosaccharidoses (mannosidosis, fucosidosis, aspartylglucosaminuria) associate neurovisceral storage with skeletal dysplasias.

Mucolipidosis II (I-cell disease) and mucolipidosis III are clinically similar to the mucopolysaccharidoses; they are both caused by a deficiency in *N*-acetyl-glucosaminyl-phosphotransferase, an enzyme responsible for the biosynthesis of the mannose-6-phosphate label which targets enzymes to the lysosome. Although mucopolysaccharidoses, mucolipidoses and oligosaccharidoses are true skeletal dysplasias with short stature and deformities, these disorders are usually placed within the context of lysosomal storage diseases.

Sulfate transporter and the achondrogenesis diastrophic dysplasia family

Proteoglycans in cartilage are highly sulfated and defects both in the degradation of sulfated proteoglycans (mucopolysaccharidoses) (see above) and in the sulfation of proteoglycans (the diastrophic dysplasia sulfate transporter dysplasia family) lead to skeletal dysplasia. Mutations in the gene for the diastrophic dysplasia sulfate transporter (*DTDST*; SLC26A2), a sulfate/chloride antiporter of the cellular membrane, can lead to intracellular sulfate depletion in cells with a high biosynthetic demand for sulfate, such as chondrocytes which synthesize large amounts of sulfated proteoglycans. The phenotypic spectrum associated with such mutations ranges from perinatally lethal forms (achondrogenesis type 1B, MIM 600972; atelosteogenesis type 2, MIM 256050) to the nonlethal forms diastrophic dysplasia and recessively inherited multiple epiphyseal dysplasia (MIM 226900).

Chondrodysplasia punctata group

The finding of irregular, stippled calcification of epiphyses is called chondrodysplasia punctata (CDP) and is found in a wide number of skeletal dysplasias. The characteristic stippling must be looked for at birth and in the first 2 years of age, as physiological progression of epiphyseal calcification may conceal this finding later on.

Two rare variants are peroxisomal disorders presenting with rhizomelic limb shortening in the newborn, accompanied by evidence of severe neurological dysfunction and early lethality. They are caused by deficiency of peroxisomal protein PEX-7p (rhizomelic CDP type 1; MIM 215100) or deficiency in the peroxisomal enzyme acylCoA:dihydroxyacetone acyltransferase (rhizomelic CDP type 2; MIM 222765). Both are inherited as autosomal recessive traits. The X-linked dominant form (CDPX2; Conradi–Hünermann syndrome; MIM 302960) is seen almost exclusively in females, as it is presumably lethal in hemizygous males. There is a phenotypic spectrum ranging from stillborns to asymptomatic. In the classical form, there are punctiform calcifications around the spinal column and the pelvis as well as in the long bone epiphyses, giving moderately reduced stature with mild rhizomelic shortening. The nose is short, and there are oculocutaneous findings such as cataracts, ichthyosis, pigment changes and skin atrophy.

It is important to note that all clinical and radiographic findings can be of variable severity, but can also be asymmetric, probably as a consequence of the X chromosome inactivation pattern. Thus, there may be leg length discrepancy. The molecular basis is mutations in the 3-β-hydroxysteroid-Δ(8)Δ(7)-isomerase gene on Xp11, and a characteristic pattern of steroid metabolites can be seen in tissues from affected females. Yet another defect in cholesterol biosynthesis has been observed in a lethal bone dysplasia known as Greenberg 'moth-eaten' dysplasia (MIM 215140). Finally, epiphyseal stippling is also seen in the Zellweger syndrome (MIM 214100), in mucolipidosis 2 (MIM 252500), as well as in other, still less well-defined conditions that await molecular delineation.

Dyschondrosteosis Léri–Weill

Dyschondrosteosis Léri–Weill (MIM 127300) is a dominantly inherited condition with moderate short stature and moderate mesomelic limb shortening with Madelung deformity of the forearm, that is, ventral subluxation of the carpus from the radius and ulna. The disorder tends to be more pronounced (or more frequently diagnosed) in females. Its molecular basis is heterozygous deletions or point mutations in the short stature-homeobox (SHOX) gene located in the pseudoautosomal region of the X and Y chromosomes. Homeotic genes are transcriptional regulators acting as switches for various developmental programs. Homozygosity or compound heterozygosity for SHOX mutations can occur (both parents having either dyschondrosteosis or short stature) and results in Langer mesomelic dysplasia (MIM 249700), a more severe condition with marked mesomelic shortening of arms and legs.

Ellis–van Creveld syndrome

The Ellis–van Creveld syndrome (EvC; chondroectodermal dysplasia; MIM 225500) is a pleiotropic autosomal recessive disorder combining skeletal changes (narrow thorax, polydactyly) with ectodermal dysplasia (hypoplastic nails, accessory labial frenula, dental anomalies such as hypodontia, cone-shaped teeth) and congenital heart disease atrial septal defect (ASD) or ventricular septal defect (VSD) in approximately one-half of affected individuals). The gene responsible for EvC codes for a protein with leucine-zipper and nuclear localization motifs. The function of the protein and the pathogenesis are as yet undetermined.

Cleidocranial dysplasia

Cleidocranial dysplasia (OMIM 119600) is a relatively common condition with mildly reduced stature of postnatal onset and normal body proportions. The clavicles are hypoplastic (allowing the shoulders to be brought together anteriorly), the skull is incompletely mineralized (giving a characteristic crackled appearance on radiographs with multiple wormian bones)

with frontal bossing, hypertelorism and persistence of fontanelles into adulthood. Anomalies of dentition are common: delayed eruption, and hypo- or aplasia of individual teeth (more rarely supernumerary teeth). This unusual combination of findings is caused by dominant haploinsufficiency mutations in a gene, CBFA-1, which encodes for a transcriptional regulator of osteoblast differentiation and odontogenesis.

Cartilage–hair hypoplasia

Cartilage–hair hypoplasia (MIM 250250) is a recessive disorder first delineated in the Amish population but also frequent among the Finnish population. Generalized metaphyseal dysplasia is combined with fine, sparse hair (hence the name) but also with congenital cellular immune deficiency, anemia, increased susceptibility to cancer, and, rarely, Hirschsprung disease. There is significant variability of all clinical features, the basis of which is not understood. The responsible gene, RMRP, does not code for a protein, but rather for a structural RNA subunit of an RNase complex, RNase MRP. This unusual molecular defect may explain both the multiple symptoms and the clinical variability of the disorder.

Disorders of the vertebrae

Spondylothoracic dysplasia (Jarcho–Levin syndrome; MIM 277300) and spondylocostal dysostosis (MIM 277300 and 122600; but note that the MIM classification may be inaccurate) are characterized by disorganized development of the vertebrae: they are thus homologous to many disorders studied in the mouse and defined as segmentation disorders, as they arise from the early embryonic process of notochord segmentation and somite formation. Lethal and nonlethal forms exist; the main clinical feature is small stature with shortening of the thoracic and lumbar spine.

The severe Jarcho–Levin syndrome has abnormal vertebral segmentation but usually normal ribs that are compressed to give a crab-like appearance of the thorax; inheritance is autosomal recessive, but appears not to be associated with the dll3 gene (see below). The milder spondylocostal dysostoses have changes of vertebrae and ribs (fusion and duplication). Progressive spinal deformity with neurological complications may occur. Many, but not all, instances of spondylocostal dysostosis are caused by recessive mutation in the dll3 gene, while the molecular basis of the dominantly inherited forms is still unclear.

Progressive pseudorheumatoid chondrodysplasia

Progressive pseudorheumatoid chondrodysplasia (also known as spondyloepiphyseal dysplasia tarda with progressive arthropathy; MIM 208230) is a rare autosomal recessive disorder which combines features of a bone dysplasia (platyspondyly, epiphyseal dysplasia) with joint pain and progressive contractures strongly reminiscent of rheumatoid arthritis, although inflammatory signs are minimal or absent and response to anti-inflammatory medicaments is poor. Interest in this rare condition has been stimulated by identification of the responsible gene, *WISP-3*, which encodes a putative homeostatic regulator of cartilage and thus may be implicated in the pathogenesis of more common painful disorders of cartilage.

Asphyxiating thoracic dystrophy

Asphyxiating thoracic dystrophy (ATD; Jeune syndrome; MIM 208500) is an autosomal recessive disorder characterized by a narrow thorax with lung hypoplasia. The pelvic findings (trident acetabular roof, shirt-squared iliac bones) are asymptomatic but fairly constant and the diagnosis of ATD is questionable in their absence. Polydactyly is a frequent (although not obligate) finding. It seems probable that ATD may be related to the lethal short rib–polydactyly syndromes (SRPS).

See also

Skeletogenesis: Genetics

References

Hall CM (2002) International nosology and classification of constitutional disorders of bone (2001). *American Journal of Medical Genetics* **113**: 65–77.

Jimenez-Sanchez G, Childs B and Valle D (2001) The effect of mendelian disease on human health. In: Scriver CR, Beaudet AL, Valle D and Sly WS (eds.) *The Metabolic and Molecular Bases of Inherited Disease*, vol. 1, pp. 167–174. New York, NY: McGraw-Hill.

Rimoin DL (1998) International nomenclature and classification of the osteochondrodysplasias (1997). International Working Group on Constitutional Diseases of Bone. *American Journal of Medical Genetics* **79**: 376–382.

Spranger J (1992) International classification of osteochondrodysplasias. The International Working Group on Constitutional Diseases of Bone. *European Journal of Pediatrics* **151**: 407–415.

Superti-Furga A, Bonafe L and Rimoin DL (2001) Molecular-pathogenetic classification of genetic disorders of the skeleton. *American Journal of Medical Genetics* **106**: 282–293.

Unger SA (2002) A genetic approach to the diagnosis of skeletal dysplasia. *Clinical Orthopaedics and Related Research* 32–38.

Further Reading

Hermanns P and Lee B (2001) Transcriptional dysregulation in skeletal malformation syndromes. *American Journal of Medical Genetics* **106**: 258–271.

McLean W and Olsen BR (2001) Mouse models of abnormal skeletal development and homeostasis. *Trends in Genetics* **17**: S38–S43.

Pourquie O and Kusumi K (2001) When body segmentation goes wrong. *Clinical Genetics* **60**: 409–416.

Spranger J, Brill P and Poznanski A (2002) *Bone Dysplasias – An Atlas of Genetic Disorders of Skeletal Development*. 2nd edn, New York, NY: Oxford University Press.

Taybi H and Lachman RS (1996) *Radiology of Syndromes, Metabolic Disorders, and Skeletal Dysplasias*, 4th edn. St Louis, MO: Mosby-Year Book, Inc.

Wilkie AO (1997) Craniosynostosis: genes and mechanisms. *Human Molecular Genetics* **6**: 1647–1656.

Web Links

Achondroplasia (ACH); LocusID: 2261. LocusLink:
http://www.ncbi.nlm.nih.gov/LocusLink/LocRpt.cgi?l = 2261

Cleidocranial dysplasia; LocusID: 860. LocusLink:
http://www.ncbi.nlm.nih.gov/LocusLink/LocRpt.cgi?l = 860

Osteogenesis imperfecta (OI); LocusID: 1277, 1278, 12843. LocusLink:
http://www.ncbi.nlm.nih.gov/LocusLink/LocRpt.cgi?l = 1277, 1278, 12843

Pseudoachondroplasia (PSACH); LocusID: 1311. LocusLink:
http://www.ncbi.nlm.nih.gov/LocusLink/LocRpt.cgi?l = 1311

Achondroplasia (ACH); MIM number100800. OMIM:
http://www.ncbi.nlm.nih.gov/htbin-post/Omim/dispmim?100800

Cleidocranial dysplasia; MIM number 119600. OMIM:
http://www.ncbi.nlm.nih.gov/htbin-post/Omim/dispmim?119600

Dyschondrosteosis Léri–Weill; MIM number 127300. OMIM:
http://www.ncbi.nlm.nih.gov/htbin-post/Omim/dispmim?127300

Osteogenesis imperfecta (OI); MIM numbers 166200, 166210, 166220, 166240. OMIM:
http://www.ncbi.nlm.nih.gov/htbin-post/Omim/dispmim?166200, 166210, 166220, 166240.

Pseudoachondroplasia (PSACH); MIM number 177170. OMIM:
http://www.ncbi.nlm.nih.gov/htbin-post/Omim/dispmim?177170

Skeletogenesis: Genetics

Pancras CW Hogendoorn, *Leiden University Medical Centre, Leiden, The Netherlands*

Judith VMG Bovée, *Leiden University Medical Centre, Leiden, The Netherlands*

Marcel Karperien, *Leiden University Medical Centre, Leiden, The Netherlands*

Anne-Marie Cleton-Jansen, *Leiden University Medical Centre, Leiden, The Netherlands*

Intermediate article

Skeletogenesis is a tightly regulated process during fetal, postnatal and pubertal growth, finely orchestrated by transcription factors and consecutive signaling pathways and controlled by delicate feedback loops. There is an increasing awareness of the role of these processes during tumorigenesis of primary bone tumors.

Introduction

The formation of bone is a continuous process in vertebrate development, initiating at embryogenesis and persisting after birth with growth, (re)modeling and fracture repair. Several phases of development of the skeleton can be discerned:

1. *Fetal bone growth*: positioning of skeletal elements along the body plan and differentiation of condensed mesenchymal stem cells.
2. *Postnatal growth*: longitudinal bone growth from the growth plate.
3. *Pubertal growth*: temporary increase in longitudinal growth followed by fusion of the growth plate.
4. *Epiphyseal fusion*: causes a complete arrest in longitudinal bone growth; however, bone mass still increases and peaks around the age of 30 years. The skeleton is continuously renewed by a process called bone remodeling, which continues throughout life.

In general terms, the skeleton develops either by intramembranous or endochondral ossification. In intramembranous ossification, osteoblasts differentiate directly from primitive embryonic mesoderm without a cartilaginous intermediary and start to deposit a mineralized bone matrix. In this way mainly so-called woven bone is formed. This type of ossification is confined to the cranial vault, the facial bones and the clavicles. In contrast, most of the skeleton develops by endochondral ossification, a process in which the skeletal elements are preformed in a cartilaginous model. These so-called 'anlagen' arise from multiple condensed, committed areas of the primitive mesenchyme and in due course differentiate into primitive cartilage. These primitive mesenchymal cells are pluripotent and have the ability to differentiate into fibroblasts, chondrocytes, osteoblasts or adipocytes as a result of temporal and sequential expression of specific genes (**Figure 1**). In a tightly orchestrated process, the cartilage is subsequently replaced by bone. (*See* Human Developmental Molecular Genetics.)

Collagen Types and Transition of the Model: Formation of the Growth Plate

Anatomy of the earliest structure

During development of the cartilaginous model, collagens I and III, present in the primitive condensed mesenchyme, are replaced by collagen II and cartilage-specific proteins. This model elongates by proliferation. Subsequently, in the middle of the future diaphysis, chondrocytes start to differentiate and become hypertrophic. This is characterized by typical morphological changes, changes in matrix constituents and finally matrix mineralization. The mineralized cartilage is subsequently invaded by blood vessels. Following primitive vascularization, invasion of osteoclasts, which resorb the mineralized cartilage, and invasion of osteoblastic progenitors occur. Subsequently, the osteoblastic progenitors differentiate into osteoblasts, which in turn start to ossify the remnants of the hypertrophic cartilage. At the same time, signals derived from the hypertrophic chondrocytes induce osteoblastic differentiation in the lateral perichondrium. These osteoblasts begin to form the bone collar or cortex. The region in which this process occurs is called the primary ossification center. During this process, type II and type X collagen, which are characteristic for cartilage, are replaced by type I collagen. As a result, primitive trabeculae are formed, consisting of a cartilage core, which is remodeled into osteoid. This sequence of

NATURE ENCYCLOPEDIA OF THE HUMAN GENOME / ©2003 Macmillan Publishers Ltd, Nature Publishing Group / www.ehgonline.net

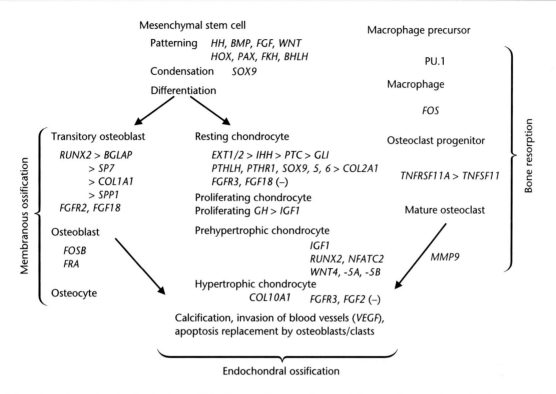

Figure 1 Overview of genes involved in the three cellular lineages that contribute to skeletogenesis: osteoblasts, chondrocytes and osteoclasts. The genes that are involved in the distinct differentiation stages are shown.

events – hypertrophy, vascularization, ossification – moves gradually from the center of the diaphysis toward the epiphyseal ends. Here the development of the growth plate occurs. The growth plate is characterized as the site where both proliferation and hypertrophy of chondrogenic cells take place. It is at this specialized site of the developing bone, juxtaposed to the metaphyses, that bone growth primarily occurs. Later in development, a secondary ossification center is formed at the distal ends of the long bones, and the growth plate becomes entrapped between the epiphysis and the metaphysis.

The growth plate is the committed region of the primitive bone whereby a process of endochondral ossification bone enlargement occurs by interstitial growth of cartilaginous matrix. In humans, this process continues until the end of puberty, when fusion of the epiphyseal plate causes a complete arrest in longitudinal growth. (*See* Vertebrate Evolution: Genes and Development.)

Anatomy of the growth plate

The growth plate or physis has a peculiar microanatomy, which reflects the tight regulation of growth and differentiation at this site (**Figure 2**). The chondrogenic cells are aggregated in orderly, longitudinal columns.

Within these arranged areas, specific morphological regions can be identified, each characterized by the expression of specific markers: the germinal or resting zone, the proliferation zone, the hypertrophic zone, the degeneration zone and the calcification zone. These zones are tightly controlled at the genetic and growth-regulation level.

Cortex formation

The cortex of the bone is formed directly by the periosteal osteoblasts via a process of intramembranous ossification, followed by remodeling into lamellar cortical bone.

Lamellar bone formation

The woven bone, a mesh-like structure found in embryonic bone, is replaced by lamellar bone, the normal bone of the adult skeleton, and characterized by a parallel pattern of the collagen fibers between adjacent portions of bone matrix.

Appendicular skeleton

The long bones in the appendicular skeleton are formed by endochondral ossification. In humans, the

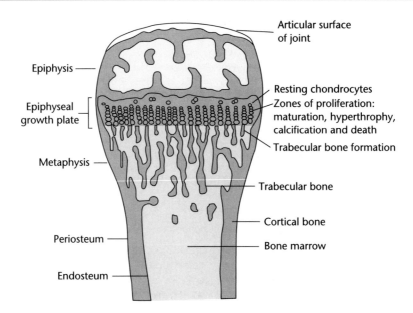

Figure 2 Anatomy of long bones.

limb buds appear during the fourth fetal week. During the initial development of the skeletal elements in the growing limb bud, mesenchymal cells receive patterning information that determines the shape, size and number of mesenchymal condensations. This information is provided by polypeptides of four different families of evolutionarily conserved growth factors: Wnt, hedgehog (Hh), fibroblast growth factor (FGF) and members of the transforming growth factor-β (TGFβ) superfamily (de Crombrugghe *et al.*, 2001). These factors control the expression of series of transcription factors such as members of the Hox, Pax, Forkhead, Homeodomain-containing and Basic helix–loop–helix (bHLH) families which provide detailed information on cell fate. After condensation of the mesenchymal cells, the primitive cartilage is formed. In the further development of the primitive long bones, one polypeptide growth factor appears to play a pivotal role, namely Indian hedgehog (IHh). The primitive cartilage expresses low levels of Ihh. When bone development progresses, Ihh expression becomes restricted to transitional chondrocytes, which have stopped proliferating and have begun to differentiate. Ihh regulates the elongation of the long bones by stimulating chondrocyte proliferation as well as by controlling the pace of hypertrophic differentiation. The latter is accomplished in close concert with the parathyroid hormone-related peptide (PTHrP) in a negative feedback loop. Finally, IHh is responsible for the formation of the bone collar by inducing osteoblast differentiation in the lateral perichondrium, thereby tightly linking chondrogenesis with osteogenesis, which is a typical feature of endochondral bone

formation (St Jacques *et al.*, 1999; Lanske *et al.*, 1996). During further progression of the long bones, the chondrocytes in the growth plate are primarily responsible for longitudinal growth, while the osteoblasts are primarily responsible for the growth in width due to the apposition of matrix on free surfaces of preexisting bone.

Genetic Control of Skeletogenesis

The subsequent processes of either fetal bone growth, postnatal growth or pubertal growth are finely orchestrated by transcription factors, consecutive signaling pathways and controlled by feedback loops or other molecular processes. Many genes play a role in this process, such as transcription factors, growth factors, hormones, receptors, cell-cycle regulators and signal transducers. The curated Gene Ontology Database (see Web Links section) lists 151 genes associated with skeletal development. **Figure 1** gives an overview of some of the major genes that play a role in differentiation and proliferation of the three different cell lineages that are the main players in skeletogenesis and associated with the three processes, that is, osteoblasts in membranous ossification, chondrocytes in endochondral ossification and osteoclasts in bone resorption.

Chondrocyte differentiation

One of the first genes that determines cell fate in skeletogenesis is *SOX9* (*SRY* (*sex determining* region Y)-box 9 (*campomelic dysplasia, autosomal sex-reversal*)), encoding a DNA-binding protein with a high mobility group (HMG) domain. When the mesenchymal stem cells that are dedicated to bone formation are condensing, *SOX9* is expressed and activates the expression of type II collagen, a major, developmentally regulated protein of cartilage. Upstream of *SOX9* are members of the TGFβ superfamily such as *bone morphogenetic protein 2 (BMP2)*, which is regulated by Sonic hedgehog (SHh). Also other BMPs are essential for chondrogenesis. Upon differentiation into chondrocytes, the cells proliferate in response to IHh signaling, which is most probably mediated by Gli, cyclin D and cyclin E. Proliferation is negatively regulated by *FGF18*, through its receptor *FGFR3 (fibroblast growth factor receptor 3 (achondroplasia, thanatophoric dwarfism))*, which inhibits IHh. In postnatal skeletogenesis, longitudinal bone growth continues and is regulated by hormones of the hypothalamus–pituitary–gonadal growth axis, such as growth hormone (GH), estradiol and thyroid hormone. These hormones regulate growth both directly and indirectly via insulin-like growth factor 1 (IGF-1).

The molecules that play a role in chondrocyte maturation are tightly regulated by a negative feedback loop (**Figure 3**). This loop differs in the pre- and postnatal growth plate. IHh, which is secreted by prehypertrophic chondrocytes, upregulates PTHrP

expression via its receptor Ptc (patched). PTHrP is either secreted by cells of the perichondrium prenatally or by prehypertrophic cells postnatally. Binding of PTHrP to type I parathyroid hormone receptor (PTHR1), which is expressed on prehypertrophic chondrocytes, suppresses their differentiation into hypertrophic chondrocytes, thereby inhibiting IHh expression and constituting a negative feedback loop (Lanske *et al.*, 1996). The transcription factor CBFA1/RUNX2 is a positive regulator of chondrocyte maturation. This function is independent of its ability to induce osteoblast differentiation. *EXT1 (exostoses (multiple) 1)* and *EXT2* are genes mutated in hereditary multiple exostoses syndrome, characterized by multiple cartilaginous tumors. These genes are likely also involved in IHh/PTHrP signaling. Nuclear factor of activated T-cells, cytoplasmic 2 (NFATp) is an inhibitor of cartilage growth and differentiation, because NFATp knockout mice have an excess of cartilage in their joints (Ranger *et al.*, 2000). Also WNT genes play a role in negative regulation of chondrogenesis: *WNT4* and *WNT5A* in the perichondrium and *WNT5B* in the prehypertrophic chondrocytes.

Osteoblast differentiation

The transcription factor CBFA1/RUNX2 is expressed in the mesenchymal condensations and at high levels in osteoblasts during embryogenesis. In endochondral

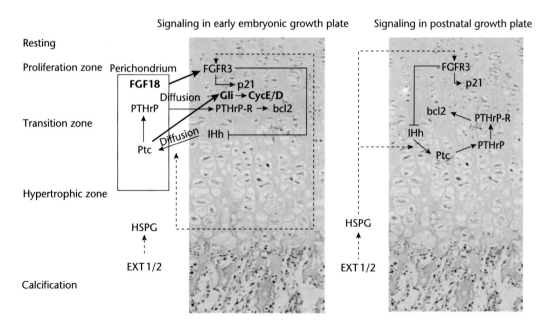

Figure 3 Signaling pathway operative in the early embryonic growth plate and postnatal growth plate respectively. Note substantial differences in growth regulation during development. The paracrine feedback loop involving parathyroid hormone-related peptide (PTHrP) is shown in the embryonic and the postnatal growth plate. Associated molecules are shown in bold. Proteins: CycE/D: cyclin E, -D; EXT1/2: exostosin 1, -2; FGF18: fibroblast growth factor 18; FGFR3: FGF receptor 3; HSPG: heparan sulfate proteoglycan; IHh: Indian hedgehog; Ptc: patched.

bone formation, CBFA1/RUNX2 expression is induced by IHh, which initiates osteoblastogenesis (St Jacques *et al.*, 1999). The control mechanisms for CBFA1/RUNX2 expression during intramembranous ossification are less clear. CBFA1/RUNX2 activity is indispensable for the initiation and progression of osteoblastic differentiation. Many osteoblast-specific genes, such as *collagen, type I, alpha 1* (*COL1A1*), *bone gamma-carboxyglutamate* (*gla*) *protein* (*osteocalcin*) (*BGLAP*), *Sp7 transcription factor* (*SP7* or *OSX*) and *integrin-binding sialoprotein* (*bone sialoprotein, bone sialoprotein II*) (*IBSP*), are induced by CBFA1/RUNX2. Collagen type 1α (Col1A) is the major component of bone matrix and proclaims the differentiation of osteoblasts. Osteocalcin (bone gamma-carboxyglutamic acid protein, BGLAP or BGP) is a small, highly conserved molecule associated with the mineralized matrix of bone and exclusively expressed in osteoblasts. Osterix is a zinc-finger containing transcription factor involved in deposition of bone matrix (Nakashima *et al.*, 2002). Like CBFA1/RUNX2, this transcription factor is indispensable for osteoblast differentiation. Bone sialoprotein (osteopontin) binds tightly to hydroxyapatite and appears to form an integral part of the mineralized matrix.

Two other positive regulators of osteoblast differentiation are, a splice variant of FOSB (*deltaFOSB*) lacking transactivation of repression domains and *FOS-like antigen 1* (*FOSL1*). DeltaFosB is an alternative splice product of FosB and, when overexpressed, results in osteosclerosis by increased osteoblast differentiation. The same phenotype is caused by overexpression of *FOSL1*; however, this gene is not essential for osteoblast differentiation.

In contrast to the inhibiting effect of FGF18 in chondrogenesis, this growth factor has a positive effect on osteoblast differentiation through another receptor, FGFR2 (Liu *et al.*, 2002). The most widely measured osteoblast marker is alkaline phosphatase, a ubiquitous enzyme, which catalyzes the hydrolysis of phosphate esters at an alkaline pH and expressed in the cell membrane of osteoblasts.

Osteoclast differentiation

The first step in osteoclast differentiation is the transition of the macrophage precursor to the mature macrophage and is mediated by PU.1, a transcription factor from the ETS family. PU.1 progressively increases as marrow macrophages assume the osteoclast phenotype. Further commitment to the osteoclast lineage is mediated by c-*fos* and nuclear factor-κB (NF-κB). c-*fos* knockout mice develop osteopetrosis, that is, increased bone mass due to absence of osteoclasts. Committed osteoclasts differentiate into mononuclear osteoclast precursors upon signaling

through the receptor activator of NFκB (RANK) and RANK ligand (RANKL). RANKL expression is regulated by osteoblasts through the transcription factor CBFA1/RUNX2, thereby establishing a tight link between osteoblastogenesis and osteoclastogenesis (Gao *et al.*, 1998). The bioactivity of RANKL is regulated by osteoprotegerin (OPG), which is a soluble decoy receptor for RANKL. The balance between OPG and RANKL determines whether osteoclastogenesis is stimulated or inhibited (Nagai and Sato, 1999). Upon fusion of mononucleated cells, multinucleated osteoclasts are formed. These osteoclasts can actively resorb bone through induction of enzymes involved in acidification (H-ATPase) and degradation (e.g. cathepsin K). This process results in the formation of cavities, which subsequently can be filled with vascular channels and hematopoietic cells. For example, MMP9 is a proteinase secreted by osteoclasts that plays a role in angiogenesis of the growth plate.

Fusion of the growth plate

At the beginning of puberty, a growth spurt is induced. This is followed by cessation of longitudinal bone growth, owing to fusion of the epiphysis and the metaphysis at the end of puberty. Both processes, growth spurt and fusion, are regulated by estrogen, as can be deduced from two types of mutations found in patients with tall stature and osteoporosis, that is, inactivation of either the *estrogen receptor 1 gene* (*ESR1*) or of *CYP19* (*cytochrome P450, subfamily XIX* (*aromatization of androgens*)). *CYP19* encodes aromatase, the enzyme involved in the conversion of androgen into estrogen. In addition to its proliferative activity, estrogen accelerates the programmed senescence of the growth plate, thus causing earlier proliferative exhaustion and consequently earlier epiphyseal fusion (Weise *et al.*, 2001). (*See* Skeletal Dysplasias: Genetics; Limb Development Anomalies: Genetics.)

Genes in Neoplastic Bone Growth

Cartilage-forming tumors

Cartilaginous tumors of bone are mainly located in long bones, developing from endochondral ossification. Peripheral cartilaginous tumors (osteochondroma (benign) and secondary peripheral chondrosarcoma (malignant)) are found in hereditary multiple exostoses (HME; see **Table 1**). The two main syndromes involving central cartilaginous tumors (enchondroma (benign) and central chondrosarcoma

Table 1 Clinical syndromes involving bone tumors

Syndrome	Bone tumors (cases in which the given syndrome is involved)	Other symptoms	Gene locus	Gene (cases in which the gene given is involved)	OMIM
Hereditary multiple exostoses	Multiple osteochondromas (100%) Secondary peripheral chondrosarcoma (1–3%)	Skeletal deformity	8q24 11p11–p12	EXT1 (44–60%) EXT2 (27%)	133700 166000
Ollier disease	Multiple enchondromas (100%) Secondary central chondrosarcoma (35%)	Skeletal deformity, marked unilateral predominance	3p22–p21.1?	PTHR1?	166000
Maffucci syndrome	Multiple enchondromas (100%) Secondary central chondrosarcoma	Vascular lesions	Unknown, nonhereditary	Unknown	166000
Spondyloenchondrodysplasia	Multiple enchondromas	Platyspondyly, short stature, (CNS calcifications?)	Unknown, autosomal recessive	Unknown	271550
Metachondromatosis	Multiple enchondromas, multiple exostoses (pointing to epiphyses)		Unknown, autosomal dominant	Unknown	156250
Li–Fraumeni syndrome	Osteosarcoma	Breast cancer, soft tissue sarcoma, brain tumors, leukemia, adrenocortical carcinoma	17p13.1 22q12.1	TP53 (70%) CHK2	151623
Retinoblastoma	Osteosarcoma	Retinoblastoma	13q14	RB1	180200
Rothmund–Thomson syndrome	Osteosarcoma (32%)	Dermatosis, juvenile cataract, saddle nose, skeletal abnormalities, disturbed hair growth, hypogonadism, gastrointestinal problems	8q24.3	RECQL4	268400
OSLAM syndrome	Osteosarcoma	Limb anomalies, macrocytosis without anemia	Unknown	Unknown	165660
Paget disease of bone	Osteosarcoma (1–10%)	Bone pain, deformity, pathological fracture, hearing loss	6p21.3? 18q21–22 5q35	Unknown RANK? SQSTM1/p62 (46%)	602080
Diaphyseal medullary stenosis with MFH	MFH (35%)	Bone infarcts, cortical growth abnormalities, pathological fracture, painful debilitation	9p22–p21	Unknown	112250

MFH: malignant fibrous histiocytoma.

(malignant)) are Ollier disease and Maffucci syndrome (see **Table 1**). Recent evidence indicates that the factors involved in growth and differentiation within the normal growth plate also play a role in tumorigenesis.

Peripheral cartilaginous tumors

The histology of osteochondroma (benign bony protuberance covered by a cartilage cap) is reminiscent of that of the normal growth plate. Osteochondromas can be solitary sporadic or hereditary multiple, with two genes, *EXT1* and *EXT2,* being cloned so far. The EXT genes are involved in heparan sulfate biosynthesis. Heparan sulfate proteoglycans (HSPGs) are large macromolecules that participate in cell signaling pathways (McCormick *et al.*, 1999). For instance, HSPGs are required for high-affinity interactions of FGF with its receptor (FGFR; Erlebacher *et al.*, 1995). Furthermore, an *EXT1* homolog (tout-velu, *Ttv*) in *Drosophila*, which is also involved in heparan sulfate biosynthesis, is required for diffusion of an important segment polarity protein called Hedgehog (Hh). Hh is a homolog of mammalian IHh, an important molecule in the delicate paracrine feedback loop within the normal growth plate. Indeed, molecules involved in the IHh/PTHrP pathway are mostly absent in osteochondromas (Bovee *et al.*, 2000).

Central cartilaginous tumors

Recently, a case of familial Ollier disease was reported demonstrating a mutation in *PTHR1* (*parathyroid hormone receptor 1*), the receptor for PTHrP. This would indicate that in centrally as well as in peripherally developing cartilaginous tumors, the Ihh/PTHrP signaling pathway is, although at different levels, affected.

Bone-forming tumors

Osteosarcoma, a malignant, bone-forming tumor, can occur in different clinical syndromes (see **Table 1**). The best known are retinoblastoma and Li–Fraumeni syndrome, carrying germ-line mutations in the cell cycle regulators *RB1* (*retinoblastoma 1*) and *TP53* (*tumor protein p53* respectively). In addition, osteosarcoma, sometimes multicentric, is found in 32% of patients with Rothmund–Thomson syndrome (RTS). RTS is a rare autosomal recessive disorder with a rash and bilateral juvenile cataracts, as well as skeletal abnormalities consisting of osteopenia, pathological fractures, dislocations, irregular metaphyses, abnormal trabeculation and stippled ossification of the patella (Wang *et al.*, 2001). A subset is caused by mutations in the DNA helicase gene *RECQL4* (*RecQ protein-like 4*).

In addition, Paget disease of bone (PDB) is a common disorder, characterized by focal increased and disorganized osteoclastic bone resorption, mainly affecting the axial skeleton, leading to bone pain, deformity, pathological fracture and an increased risk of osteosarcoma. PDB can be familial for three loci at 6p21.3, 18q21–22, and 5q35–qter. Familial expansile osteolysis (FEO; MIM number 174810) is a rare, similar disorder demonstrating increased bone remodeling, without involvement of the axial skeleton and so far not associated with an increased risk of osteosarcoma. For FEO, mutations have been described in the *TNFRSF11A* gene (*tumor necrosis factor receptor superfamily, member 11a, activator of NFκB*) at 18q21.2–21.3, a factor essential in osteoclast formation. The short, in-frame insertions within the signal peptide region lead to increased activation of NFκB, possibly leading to increased osteoclast formation. A *TNFRSF11A* gene mutation has also been described in one family with severe early-onset PDB, although other familial and sporadic cases failed to demonstrate mutations. Recently, mutations in another gene associated with the NFκB pathway, *sequestosome 1* (*SQSTM1*) at 5q35–qter were reported to cause 46% of familial cases of PDB (Laurin *et al.*, 2002).

See also
Skeletal Dysplasias: Genetics

References

Bovee JVMG, Van den Broek LJCM, Cleton-Jansen AM and Hogendoorn PCW (2000) Up-regulation of PTHrP and Bcl-2 expression characterizes the progression of osteochondroma toward peripheral chondrosarcoma and is a late event in central chondrosarcoma. *Laboratory Investigation* **80**: 1925–1933.

de Crombrugghe B, Lefebvre V and Nakashima K (2001) Regulatory mechanisms in the pathways of cartilage and bone formation. *Current Opinion in Cell Biology* **13**: 721–727.

Erlebacher A, Filvaroff EH, Gitelman SE and Derynck R (1995) Toward a molecular understanding of skeletal development. *Cell* **80**: 371–378.

Gao YH, Shinki T, Yuasa T, *et al.* (1998) Potential role of cbfa1, an essential transcriptional factor for osteoblast differentiation, in osteoclastogenesis: regulation of mRNA expression of osteoclast differentiation factor (ODF). *Biochemical and Biophysical Research Communications* **252**: 697–702.

St Jacques B, Hammerschmidt M and McMahon AP (1999) Indian hedgehog signaling regulates proliferation and differentiation of chondrocytes and is essential for bone formation. *Genes & Development* **13**: 2072–2086.

Lanske B, Karaplis AC, Lee K, *et al.* (1996) PTH/PTHrP receptor in early development and Indian hedgehog-regulated bone growth. *Science* **273**: 663–666.

Laurin N, Brown JP, Morissette J and Raymond V (2002) Recurrent mutation of the gene encoding sequestosome 1 (SQSTM1/p62) in Paget disease of bone. *American Journal of Human Genetics* **70**: 1582–1588.

Liu Z, Xu J, Colvin JS and Ornitz DM (2002) Coordination of chondrogenesis and osteogenesis by fibroblast growth factor 18. *Genes & Development* **16**: 859–869.

McCormick C, Duncan G and Tufaro F (1999) New perspectives on the molecular basis of hereditary bone tumors. *Molecular Medicine Today* 5: 481–486.

Nagai M and Sato N (1999) Reciprocal gene expression of osteoclastogenesis inhibitory factor and osteoclast differentiation factor regulates osteoclast formation. *Biochemical and Biophysical Research Communications* 257: 719–723.

Nakashima K, Zhou X, Kunkel G, et al. (2002) The novel zinc finger-containing transcription factor osterix is required for osteoblast differentiation and bone formation. *Cell* 108: 17–29.

Ranger AM, Gerstenfeld LC, Wang J, et al. (2000) The nuclear factor of activated T cells (NFAT) transcription factor NFATp (NFATc2) is a repressor of chondrogenesis. *Journal of Experimental Medicine* 191: 9–22.

Wang LL, Levy ML, Lewis RA, et al. (2001) Clinical manifestations in a cohort of 41 Rothmund–Thomson syndrome patients. *American Journal of Medical Genetics* 102: 11–17.

Weise M, De Levi S, Barnes KM, et al. (2001) Effects of estrogen on growth plate senescence and epiphyseal fusion. *Proceedings of the National Academy of Sciences of the United States of America* 98: 6871–6876.

Further Reading

Bi W, Deng JM, Zhang Z, Behringer RR and de Crombrugghe B (1999) Sox9 is required for cartilage formation. *Nature Genetics* 22: 85–89.

Capdevila J and Izpisua Belmonte JC (2001) Patterning mechanisms controlling vertebrate limb development. *Annual Review of Cell and Developmental Biology* 17: 87–132.

Chung UI, Schipani E, McMahon AP and Kronenberg HM (2001) Indian hedgehog couples chondrogenesis to osteogenesis in endochondral bone development. *Journal of Clinical Investigation* 107: 295–304.

Hartmann C and Tabin CJ (2000) Dual roles of Wnt signaling during chondrogenesis in the chicken limb. *Development* 127: 3141–3159.

Jochum W, David JP, Elliott C, et al. (2000) Increased bone formation and osteosclerosis in mice overexpressing the transcription factor Fra-1. *Nature Medicine* 6: 980–984.

Karp SJ, Schipani E, St Jacques B, et al. (2000) Indian hedgehog coordinates endochondral bone growth and morphogenesis via parathyroid hormone related-protein-dependent and -independent pathways. *Development* 127: 543–548.

Karsenty G (1999) The genetic transformation of bone biology. *Genes & Development* 13: 3037–3051.

Le Roith D, Bondy C, Yakar S, Liu JL and Butler A (2001) The somatomedin hypothesis: 2001. *Endocrinology Review* 22: 53–74.

Sabatakos G, Sims NA, Chen J, et al. (2000) Overexpression of DeltaFosB transcription factor(s) increases bone formation and inhibits adipogenesis. *Nature Medicine* 6: 985–990.

The I, Bellaiche Y and Perrimon N (1999) Hedgehog movement is regulated through tout velu-dependant synthesis of a heparan sulfate proteoglycan. *Molecular Cell* 4: 633–639.

Tsumaki N, Nakase T, Miyaji T, et al. (2002) Bone morphogenetic protein signals are required for cartilage formation and differently regulate joint development during skeletogenesis. *Journal of Bone Mineral Research: The Official Journal of the American Society for Bone and Mineral Research* 17: 898–906.

Wuyts W, Van Wesenbeeck L, Morales-Piga A, et al. (2001) Evaluation of the role of *RANK* and *OPG* genes in Paget disease of bone. *Bone* 28: 104–107.

Web Links

Gene Ontology Tools. The website of the Gene Ontology Consortium aims at producing a dynamic controlled vocabulary that can be applied to all organisms even as knowledge of gene and protein roles in cells is accumulating and changing.
http://www.godatabase.org/dev/

COL1A1 (collagen, type I, alpha 1); Locus ID: 1277. LocusLink:
http://www.ncbi.nlm.nih.gov/LocusLink/LocRpt.cgi?l = 1277

COL1A1 (collagen, type I, alpha 1); MIM number: 120150. OMIM:
http://www.ncbi.nlm.nih.gov/-post/Omim/dispmim?120150

ESR1 (estrogen receptor 1); Locus ID: 2099. LocusLink:
http://www.ncbi.nlm.nih.gov/LocusLink/LocRpt.cgi?l = 2099

ESR1 (estrogen receptor 1); MIM number: 133430. OMIM:
http://www.ncbi.nlm.nih.gov/htbin-post/Omim/dispmim?133430

EXT1 (exostoses (multiple) 1); Locus ID: 2131. LocusLink:
http://www.ncbi.nlm.nih.gov/LocusLink/LocRpt.cgi?l = 2131

EXT1 (exostoses (multiple) 1); MIM number: 133700. OMIM:
http://www.ncbi.nlm.nih.gov/htbin-post/Omim/dispmim?133700

FGFR3 (fibroblast growth factor receptor 3 (achondroplasia, thanatophoric dwarfism)); Locus ID: 2261. LocusLink:
http://www.ncbi.nlm.nih.gov/LocusLink/LocRpt.cgi?l = 2261

FGFR3 (fibroblast growth factor receptor 3 (achondroplasia, thanatophoric dwarfism)); MIM number: 134934. OMIM:
http://www.ncbi.nlm.nih.gov/htbin-post/Omim/dispmim?134934

TNFRSF11A (tumor necrosis factor receptor superfamily, member 11a, activator of NFKB); Locus ID: 8792. LocusLink:
http://www.ncbi.nlm.nih.gov/LocusLink/LocRpt.cgi?l = 8792

TNFRSF11A (tumor necrosis factor receptor superfamily, member 11a, activator of NFKB); MIM number: 603499. OMIM:
http://www.ncbi.nlm.nih.gov/htbin-post/Omim/dispmim?603499

Skin: Hereditary Disorders

Laura D Corden, *University of Dundee, Dundee, UK*

WH Irwin McLean, *Our Lady's Hospital for Sick Children, Dublin, Ireland*

Alan D Irvine, *University of Dundee, Dundee, UK*

There are many disorders with skin phenotypes for which the underlying molecular defect is now known. These include blistering diseases, pigmentary disorders, disorders of cornification and those with tumor predisposition.

Introduction

There are approximately in excess of 200 inherited skin diseases where the defective gene has been identified and this number is growing steadily. This article is not therefore a fully comprehensive study of all genetic skin disorders.

Skin Blistering Disorders

Epidermolysis bullosa (EB) is a group of disorders characterized by skin blistering (**Table 1**). There are three main forms: dystrophic, junctional and simplex, classified according to the level in the skin where blistering occurs. In the recessive Hallopeau–Siemens form of dystrophic EB, intradermal blistering leads to chronic, mutilating scarring and a high risk of squamous cell carcinoma. This disorder is caused by mutations in the type VII collagen gene *COL7A1*. Type VII collagen is an essential constituent of the anchoring fibrils. Milder dominant forms of the disease also exist. Junctional EB (JEB) is the most severe subtype. The archetypal form of JEB is the recessive Herlitz form, resulting from the loss of any of the three chains of laminin 5, the main component of anchoring filaments within the cutaneous basement membrane zone. In JEB, blisters occur within the basement membrane. Nonlethal forms (non-Herlitz-JEB; JEB-nH), includes those previously designated generalized atrophic benign EB; GABEB) are less severe and are caused by mutations in type XVII collagen or in laminin-5 genes. A further autosomal recessive subtype of JEB is that caused by mutations in integrins α6 or β4, and this subtype is associated with pyloric atresia (JEB-PA). Epidermolysis bullosa simplex (EBS) encompasses the third group of EB, where blisters occur within the cytoplasm of basal keratinocytes. There are three main autosomal dominant forms: Dowling–Meara; Köbner; and Weber–Cockayne EBS, each caused by mutations in the basal keratins, K5 and K14. In EBS, the phenotypic severity is related to the position of the mutation within the gene. EBS with late-onset muscular dystrophy (EBS-MD) is caused by mutations within plectin, a cytoskeletal associated protein that connects keratin filaments to the hemidesmosomal inner plaque. Plectin also crosslinks cytoskeletal systems in muscle and therefore ablation of expression leads to fragility of the skin and myopathy.

Disorders of Keratinization

Keratin gene defects

Keratins are the type I and II intermediate filament proteins and are generally expressed in pairs, in a differentiation-specific and tissue-specific manner; for example, K5 and K14 are expressed in the basal cell layers of stratified squamous epithelia and K1 and K10 are expressed in suprabasal cells. Mutations in these genes lead to phenotypes of epithelial fragility and/or overproliferation of the tissues expressing the mutated keratin. Mutations have now been found in genes encoding 18 epithelial keratins.

Most keratin disorders exhibit autosomal dominant inheritance, although a few cases of recessive EBS with mutations in *KRT14* have been reported. These recessive kindreds tend to have severe phenotypes as a result of protein ablation. In dominant keratin disorders, missense mutations predominantly occur in the keratin rod domain, leading to structural weakening or complete collapse of the cytoskeleton. As keratins are obligate heteropolymers expressed in specific pairs, mutation of either keratin produces a similar phenotype. Keratinizing disorders caused by keratin gene defects are listed in **Table 2**, with keratin defects of skin appendages listed in **Table 11**.

Table 1 Skin blistering disorders

Disorder	Genes affected	Further information
Dominant dystrophic epidermolysis bullosa (DDEB)	COL7A1	Pro-alpha 1 chain of type VII collagen (chr 3p21) Type VII collagen is a constituent of dermal anchoring fibrils
Recessive dystrophic epidermolysis bullosa Hallopeau–Siemens (RDEB-HS)	COL7A1	Pro-alpha 1 chain of type VII collagen (chr 3p21). Type VII collagen is a constituent of dermal anchoring fibrils
		Most severe form of DEB. Pseudosyndactyly, anemia, growth retardation, ocular and oral lesions, gastrointestinal manifestations and dental caries, are common. High risk of squamous cell carcinoma ($\sim$40% by age 30 years)
Recessive dystrophic epidermolysis bullosa non-Hallopeau–Siemens (RDEB-nHS)	COL7A1	Pro-alpha 1 chain of type VII collagen (chr 3p21). Type VII collagen is a constituent of dermal anchoring fibrils
		Increased risk of squamous cell carcinoma ($\sim$14% by age 30 years). Lower incidence of anemia, growth retardation, gastrointestinal manifestations, dental caries and ocular oral lesions than RDEB-HS
Herlitz junctional EB (JEB-H)	LAMA3, LAMB3, LAMC2	Laminin 5, main component of anchoring filaments (chr 18q11, 1q32, 1q51)
Non-Herlitz junctional EB (JEB-nH)	COL17A1, LAMB3	Alpha 1 chain of collagen XVII (chr 10q24), a transmembrane component of the hemidesmosome
		Beta 3 chain of laminin 5 (chr 1q32)
Junctional EB with pyloric atresia (JEB-PA)	ITGB4, ITGA6	Integrins alpha 6 and beta 4, structural role in hemidesmosome assembly (chr 17q11–qter and chr 2)
Epidermolysis bullosa simplex (EBS)	KRT5, KRT14	Keratin 5 and keratin 14 (chr 12q11–q13 and 17q12–q21)
		Includes three well-recognized variants with increasing severity: EBS–Weber–Cockayne, EBS–Köbner and EBS–Dowling–Meara
EBS with mottled pigmentation (EBS-MP)	KRT5	Keratin 5 (chr 12q11–q13) Association EBS with palmoplantar keratoderma and distinctive mottled pigmentation
EBS with muscular dystrophy (EBS-MD)	PLEC1	Plectin, a cytoskeletal-associated protein of the hemidesmosomal inner plaque (chr 8q24)
EBS–Ogna	PLEC1	Plectin, a cytoskeletal-associated protein of the hemidesmosomal inner plaque (chr 8q24)
		Dominant-acting mutations

Desmosomal proteins

Desmosomes are anchorage points at the plasma membrane of epithelial cells for intermediate filaments. They are composed of proteins from three families: the plakins, including desmoplakin, envoplakin and periplakin; the armadillo family of proteins (plakoglobin and plakophilins 1–3); and the desmosomal cadherins (desmocollins 1–3 and desmogleins 1–3). The protein composition of the desmosome depends upon its position within the stratified tissue.

Dominant mutations in DSP lead to striate palmoplantar keratoderma (SPPK), which is characterized by longitudinal stripes of hyperkeratosis on palms and soles. Recessive mutations in the desmoplakin 1 gene DSP also cause striate keratoderma but accompanied by dilated cardiomyopathy and woolly hair (**Table 3**). Mutations in the plakophilin 1 gene PKP1 which encodes an accessory desmosomal plaque protein, underlie a rare condition termed ectodermal dysplasia/skin fragility syndrome. Naxos disease is another recessive disorder of the desmosome, comprising arrhythmogenic right ventricular cardiomyopathy with palmoplantar keratoderma and woolly hair. Cell adhesion is again affected because of mutations in the JUP gene encoding plakoglobin (McKoy et al., 2000). In all these desmosomal diseases, there is a decrease in keratinocyte–keratinocyte adhesion as a result of aberrant intermediate filament/desmosome

Table 2 Disorders of keratinization: keratin gene defects

Disorder	Genes affected	Further information
BCIE	KRT1, KRT10	Suprabasal keratins 1 and 10 (chr 12q11–q13 and 17q12–q21)
		Skin fragility at birth followed by generalized thickening and scaling
Annular epidermolytic icthyosis variant of BCIE and cyclic ichthyosis	KRT10, KRT1	Suprabasal keratin 10 (chr 17q12–q21) for annular variant; keratin 1 for cyclic icthyosis (chr 12q11–q13).
		Annular plaques with similar appearance and histology to BCIE
BCIE nevus	KRT10, KRT1	Keratin 10 (chr 17q12–q21) and keratin 1 (chr 12q11–q13)
Diffuse nonepidermolytic palmoplantar keratoderma	KRT1	Keratin 1 (chr 12q11–q13)
Epidermolytic palmoplantar keratoderma	KRT9	Keratin 9 (type I keratin; chr 17q12–q21). Keratin specific for palm and sole skin
		Diffuse thickening of the palms and soles with red border. Onset at birth
Focal nonepidermolytic palmoplantar keratoderma	KRT16	Keratin 16 (chr 17q12–q21). Constitutive expression in palm and sole epidermis and other sites
IBS	KRT2E	Epidermal keratin 2e (chr 12q11–q13). Expressed in late suprabasal cells of epidermis
Ichthyosis hystrix (Curth–Macklin)	KRT1	Keratin 1 (chr 12q11–q13)

BCIE: bullous congenital ichthyosiform erythroderma; IBS: ichthyosis bullosa of Siemens.

Table 3 Disorders of keratinization: desmosomal components

Disorder	Genes affected	Further information
Dilated cardiomyopathy with woolly hair and keratoderma	DSP	Desmoplakin 1 (chr 6p2)
		Autosomal recessive with striate palmoplantar keratoderma, left ventricular dilatation and woolly hair. Heterozygotes display no phenotype
Ectodermal dysplasia with skin fragility	PKP1	Plakophilin 1 (chr 1q)
		Skin fragility, nail dystrophy, sparse hair and palmoplantar keratoderma
Naxos disease	JUP	Plakoglobin, component of desmosome and adherens junction (chr 17q21)
Striate palmoplantar keratoderma	DSG1, DSP	Desmoglein 1, a desmosomal cadherin (chr 18q12.1) or desmoplakin 1. Role in attachment of intermediate filaments to the desmosome (chr 6p2)
		Focal palmoplantar keratoderma on the soles of the feet with variable involvement of striate palmoplantar keratoderma on the palms

interaction. Desmosomes are also found in cardiac muscle, explaining the cardiomyopathy seen in some of these disorders.

Defects of calcium pumps

Darier disease is an autosomal dominant disorder with keratotic papules in a sebaceous distribution, sometimes in association with mental illness. Histologically, there is a failure of keratin–desmosome connection. The calcium pump protein SERCA2 (ATP2A2 gene) is mutated in this disease (Sakuntabhai et al., 1999; **Table 4**). Hailey–Hailey disease is caused by mutations in a related calcium pump gene, ATP2C1. Both of these

Table 4 Disorders of keratinization: calcium pump defects

Disorder	Genes affected	Further information
Darier disease	ATP2A2	Sarcoplasmic/endoplasmic reticulum calcium ATPase (SERCA2) (chr 12q24.1)
Hailey–Hailey disease	ATP2C1	Calcium-transporting ATPase (chr 3q21–q24)

disorders reveal the importance of controlling calcium metabolism in the function of stratified squamous epithelia. Desmosome formation is sensitive to calcium levels, which may explain the pathogenesis of these diseases.

Gap junction defects

The connexins are a family of proteins that assemble to form gap junctions between cells (**Table 5**). Connexins are expressed in a tissue-specific manner and mediate the diffusion of ions and metabolites between cells. The epithelia of skin and inner ear are rich in gap junctions, which appear to play a role in the coordination of keratinocyte growth and differentiation. Erythrokeratodermia variabilis is caused by heterozygous mutations in connexin 31 (Richard *et al.*, 1998). The disease is characterized by fixed patches of hyperkeratosis and migratory erythematous areas. Clouston syndrome, characterized by alopecia, nail dystrophy, skin hyperpigmentation and palmoplantar hyperkeratosis, is caused by mutations in connexin 30. Mutations in connexin 26 lead to palmoplantar keratoderma in association with deafness (Vohwinkel syndrome). Specific connexin 26 mutations have recently been linked to keratitis–ichthyosis–deafness syndrome. Palmoplantar keratoderma with deafness has also been shown to be caused by mutation in the mitochondrial serine tRNA gene *MMTS1*.

Metabolic and enzymatic protein defects

Defects in many enzymes lead to skin diseases, including sulfatase deficiency, proteinase defects, disorders of peroxisomal enzymes and enzymes involved in cholesterol biosynthesis (**Table 6**). In X-linked ichthyosis (steroid sulfatase deficiency) mutations in the arylsulfatase C or *STS* gene (in many instances the entire gene is deleted) lead to accumulation of cholesterol sulfate which is thought to be responsible for scale formation and perturbation of barrier function. Netherton and Papillon–Lefèvre syndromes are both caused by defects in proteinase function. Netherton syndrome is characterized by hair shaft abnormality and generalized scaling erythroderma, and is caused by mutations in the

SPINK5 gene, which encodes the serine protease inhibitor LEKT1 (Chavanas *et al.*, 2000). Papillon–Lefèvre syndrome is characterized by palmoplantar keratoderma and premature loss of deciduous and permanent dentition. The disorder is caused by defects in the lysosomal protease gene *CTSC*, encoding cathepsin C which plays an important role in intracellular protein degradation and activation of serine proteinases in immune and inflammatory cells. These defects reveal the importance of protein turnover in normal epidermal differentiation.

Type I lamellar ichthyosis is caused by mutations in the *TGM1* gene, which encodes keratinocyte transglutaminase (Huber *et al.*, 1995). An affected individual will develop large, brown scales over their whole body and palmoplantar hyperkeratosis might appear. Transglutaminase 1 is an enzyme required for crosslinking epidermal proteins during formation of the cornified envelope and when defective, the barrier function of the skin is perturbed.

Rhizomelic chondrodysplasia punctata type I is a peroxisome biogenesis disorder, characterized by foci of calcification in the hyaline cartilage, coronal vertebral clefting, dwarfing, joint contractures, congenital cataract, ichthyosis and severe mental retardation. Skin involvement is found in 27% of affected individuals. Mutations in the gene *PEX7* lead to aberrant peroxisomal protein targeting and accumulation of phytanic acid (**Table 6**). Phytanic acid has been shown to bind the retinoid X receptor (RXR) and may therefore be involved in coordinating cellular metabolism through RXR-dependent signaling pathways. Patients with Refsum disease have mutations in the *PAHX* gene and accumulate phytanic acid as a result of decreased phytanic acid oxidation. Histologically, lipids are shown to accumulate in vacuoles within epidermal cells. X-linked dominant chondrodysplasia punctata (CDPX2) or Conradi–Hünerman syndrome is a second form of chondrodysplasia punctata, which is caused by mutations in the *EBP*

Table 5 Disorders of keratinization: gap junction defects

Disorder	Gene affected	Further information
Clouston syndrome (hidrotic ectodermal dysplasia 2)	*GJB6*	Connexin 30 (chr 13q12). Palmoplantar keratoderma, nail dystrophy and variable alopecia
EKV	*GJB3*	Connexin 31 (chr 1p35.1). Fixed and migratory red plaques of hyperkeratosis
EKV with erythema gyratum repens	*GJB4*	Connexin 30.3 (chr 1p35.1)
Vohwinkel syndrome (classic variant, with deafness)	*GJB2*	Connexin 26 (chr 13q11–q12). Mutilating keratoderma, deafness
Keratitis–ichthyosis deafness	*GJB2*	Connexin 26 (chr 13q11–q12). Keratitis often leading to visual impairment; sensorineural deafness; widespread hyperkeratotic plaques and palmoplantar keratoderma. Specific connexin 26 mutations cause this syndrome

EKV: erythrokeratodermia variabilis.

Table 6 Disorders of keratinization: metabolic and enzymatic protein defects

Disorder	Gene affected	Further information
Chanarin–Dorfman syndrome	*CGI-58*	CGI-58 (chr 3), CGI protein of the esterase/lipase/thioesterase subfamily. Nonlysosomal disorder of neutral lipid metabolism. Mild widespread ichthyosis
CHILD syndrome	*NSDHL, EBP*	3 beta-hydroxysteroid dehydrogenase, an enzyme in cholesterol biosynthesis (chr Xq28). Mouse mutant, *bare patches*
		Mutations also reported in 3 beta hydroxysteroid-delta 8, delta 7 isomerase
Haim–Munk syndrome	*CTSC*	Cathepsin C (see below)
Lamellar ichthyosis (LI1, LI2, LI3)	*TGM1*	LI1, keratinocyte transglutaminase I (chr 14q11.2) cross-linking enzyme in cornified envelope formation
		LI2 (chr 2q33–q35); LI3 (chr 19p12–q12)
Netherton syndrome	*SPINK5*	Serine protease inhibitor LEKT1 (chr 5q32). Atopic diathesis, characteristic ichthyosis in some patients, structural hair defects and growth delay
Papillon–Lefèvre syndrome	*CTSC*	Cathepsin C, a lysosomal protease (chr 11q14.1–14.3). Periodontitis, palmoplantar keratoderma
Refsum disease	*PAHX*	Phytanoyl-CoA hydroxylase (chr 10pter–p11.2), a PTS2 protein
		Autosomal recessive disorder resulting from accumulation of phytanic acid. Characterized by ichthyosis with retinitis pigmentosa, chronic polyneuropathy and cerebellar signs
Rhizomelic chondrodysplasia punctata type I	*PEX7*	Encoding the peroxisomal matrix protein with type 2 peroxisome targeting signal (PTS2) receptor (chr 6q22–q24). Calcification of hyaline cartilage, clefting, dwarfing, joint contractures, congenital cataract, ichthyosis and severe mental retardation
Richner–Hanhart (tyrosine transaminase deficiency)	*TAT*	Tyrosine aminotransferase (chr 16q22). Photophobia, corneal erosions and mental retardation
Sjögren–Larsson syndrome	*FALDH* (also known as *ALDH10* or *ALDH3A2*)	Fatty aldehyde dehydrogenase (chr 17p11.2). Microsomal enzyme that catalyzes medium- and long-chain aliphatic aldehyde oxidation
X-linked icthyosis (steroid sulfatase deficiency)	*STS* (also known as *ARSC*)	Arylsulfatase C gene (chr Xp22.32)
		Affects 1 : 6000–1 : 2000 males, presents by the age of 3 months in most cases. May be associated with hypogonadism and failure to undergo sexual maturation. Associated corneal opacity in some cases. Of the patients 10% have a contiguous gene syndrome
X-linked recessive chondrodysplasia punctata	*ARSE*	Arylsulfatase E gene (chr Xp22.3) resulting in steroid sulfatase deficiency
X-linked dominant chondrodysplasia punctata (CDPX2) or Conradi–Hünermann syndrome	*EBP*	Emopamil binding protein, 3 beta hydroxysteroid-delta 8, delta 7 isomerase. Role in cholesterol biosynthesis (chr Xp11.22–p11.23). Mouse model, *tattered*
		Linear ichthyosis, cataracts, short stature. Occasionally erythroderma at birth with residual follicular atrophoderma and alopecia. Predisposition for involvement of skin flexures (ptychotropism)

CHILD syndrome: congenital hemidysplasia, ichthyosis and limb defects.

gene, encoding the emopamil binding protein, 3 beta hydroxysteroid-delta 8, delta 7 isomerase. This enzyme has a role in cholesterol biosynthesis.

Disorders such as Sjögren–Larsson syndrome and CHILD syndrome are caused by defects in enzymes involved in lipid metabolism and cholesterol biosynthesis. Sjögren–Larsson patients have mental retardation, spasticity and icthyosis. Mutations in the fatty aldehyde dehydrogenase gene (*FALDH*), which is involved in the conversion of fatty alcohol to fatty acid, underlie this disorder (De Laurenzi *et al.*, 1996). CHILD syndrome (congenital hemidysplasia, ichthyosis and limb defects) is caused by mutation in a 3 beta-hydroxysteroid dehydrogenase, an enzyme involved in cholesterol biosynthesis. Lipid lamellae in the stratum corneum are central to the barrier function of the skin. This barrier requires correctly processed ceramides, cholesterol and free fatty acids.

Cornified cell envelope defects

Several cornified cell envelope defects are listed in **Table 7**. The cornified cell envelope is deposited beneath the plasma membrane in terminally differentiating keratinocytes and is composed of highly cross-linked insoluble proteins. Proteins involved in the formation of the cornified envelope include involucrin, envoplakin and periplakin, which are cross-linked by enzymes such as transglutaminase 1 (see **Table 6**) to form a functional scaffold. This scaffold is later reinforced by addition of ceramides and loricrin, which comprises around 80% of the total cornified cell envelope, and small proline-rich proteins. Mutations in the loricrin gene *LOR* lead to loricrin keratoderma (the ichthyotic/Camisa variant of Vohwinkel keratoderma) or progressive symmetrical erythrokeratoderma. This disorder is a severe form of mutilating keratoderma but can be distinguished from classic Vohwinkel keratoderma (connexin 26 mutations) by the absence of deafness.

Secreted protein

Mal de Meleda is caused by a defect in the secreted protein, SLURP-1 (Ly-6/uPAR related protein-1) (**Table 8**). Members of the Ly-6/uPAR superfamily have been implicated in transmembrane signal transduction, cell activation and cell adhesion but the function of SLURP-1 is unclear. The disease is associated with transgressive palmoplantar keratoderma, keratotic skin lesions, perioral erythema, brachydactyly and nail abnormalities.

Pigmentary Disorders

Altered skin pigmentation is a feature of several diseases (**Table 9**). There are several types of oculocutaneous albinism (OCA), which are associated with decreased pigmentation in the skin, hair and eyes. This reduced pigmentation occurs as a result of mutations in genes involved in melanin synthesis (the tyrosinase gene *TYR*, *OCA2* and tyrosinase-related protein gene 1, *TRP-1*) (King *et al.*, 1991). Type I and III OCA are believed to be endoplasmic reticulum retention diseases with a mutation in one melanogenic protein affecting the maturation and stability of the other in the melanogenic pathway. In the Carney complex, there is spotty centro-orofacial pigmentation, in addition to multiple neoplasia. In Peutz–Jeghers syndrome, melanocytic macules are found on the lips, in addition to multiple gastrointestinal hamartomatous polyps and an increased cancer risk. McCune–Albright syndrome is characterized by polyostotic fibrous dysplasia, café-au-lait lesions and endocrine disorders. Hemochromatosis is associated with hypermelanotic pigmentation of the skin and is caused by mutations in the *HFE* gene. The protein encoded by *HFE* is a member of the major histocompatibility complex class 1-like family. Overexpression of *HFE* is thought to decrease the affinity of the transferrin receptor, revealing a link between *HFE* and iron transport. Alterations in this regulatory mechanism are expected to play a role in hereditary hemochromatosis. Mutations have also been found in the gene encoding ferroportin (*SLC11A3*) and the transferrin receptor-2 gene (*TFR2*) in patients with hemochromatosis.

Diseases Involving Skin Malignancy

Muir–Torre syndrome is part of the Lynch cancer family syndrome II (**Table 10**). Patients present with sebaceous skin tumors and low-grade visceral malignancy, often gastrointestinal cancers. In Muir–Torre syndrome, mutations have been found in the *MSH2* and *MLH1* genes. These are both involved in DNA mismatch repair (Kruse *et al.*, 1996). Xeroderma

Table 8 Disorders of keratinization: secreted protein

Disorder	Genes affected	Further information
Mal de Meleda	*ARSB*	Mutilating keratoderma, knuckle pads, koilonychia, subungual hyperkeratosis. Encodes *SLURP-1* (chr 8q24.3)

Table 7 Cornified cell envelope defects

Disorder	Genes affected	Further information
Loricrin keratoderma (Vohwinkel syndrome: ichthyotic/Camisa variant)	*LOR*	Loricrin, component of cornified cell envelope (chr 1q21). Mutilating keratoderma digital constrictions (pseudoainhum), mild ichthyosis, no deafness
Progressive symmetric erythrokeratoderma	*LOR*	Loricrin, component of cornified cell envelope (chr 1q21)

Table 9 Pigmentary disorders

Disorder	Gene affected	Further information
Carney complex	*PRKAR1A*	Protein kinase A regulatory subunit 1-alpha (apparent tumor-suppressor gene), on chr 17q22–24. Centro-orofacial pigmentation, cutaneous and cardiac myxomas. Predisposition to endocrine tumors and Schwannomas
Chédiak–Higashi syndrome	*LYST*	Lysosomal trafficking regulator gene, on 1q42. Hypopigmentation, silver sheen to hair, lymphohistiocytic proliferation in 85% with resultant organ infiltration and anemia, neutropenia and bleeding
Hemochromatosis	*HFE, SLC11A3, TFR2*	Member of major histocompatibility complex class-I-like family (chr 6p21); solute carrier family 11, member A3 (ferroportin) (chr 2q32); transferrin receptor-2 (chr 7q22)
Hermansky–Pudlak syndrome	*HPS1, AP3B1, HPS3, HPS4*	HPS1 (chr 10q23). Oculocutaneous albinism, bleeding and pulmonary fibrosis. Beta 3A subunit of adaptin-3-lysosomal protein (chr 5). HPS3 (chr 3q24). HPS4 gene (chr 22q11.2–q12.2). Melanosomal protein HPS4 is the human ortholog of the mouse *le* (light ear) gene product involved in organelle biogenesis
Incontinentia pigmenti	*NEMO*	Four distinct clinical stages observed: vesicobullous, verrucous, hyperpigmented, hypopigmented/atrophic. High incidence of dental and ocular abnormalities. IKK gamma gene – required for the activation of NFκB transcription factor (chr Xq28)
McCune–Albright syndrome	*GNAS1*	Segmental café-au-lait macules, polyostotic fibrous dysplasia, recurrent fractures, precocious puberty Alpha subunit of the G protein–guanine nucleotide binding regulatory protein (chr 20q13)
Neurofibromatosis 1	*NF1*	Neurofibromin (chr 17q11), tumor-suppressor gene Discussed elsewhere in this encyclopedia
Neurofibromatosis 2	*NF2*	Merlin (also known as schwannomin) (chr 22q11–q13), tumor-suppressor gene. Putative membrane-organizing protein Discussed elsewhere in this encyclopedia
Oculocutaneous albinism type I	*TYR*	Tyrosinase-negative albinism. White hair, nystagmus, photophobia, increased skin cancers Tyrosinase, an enzyme in the melanin biosynthetic pathway (chr 11q14–q21)
Oculocutaneous albinism type II	*OCA2*	Tyrosinase-positive albinism. Yellow-brown hair, multiple ephilides and lentigines and increased skin cancer Human homolog of mouse p locus (chr 15q11–q12)
Oculocutaneous albinism type III	*TRP-1*	Tyrosinase-related protein 1 gene (chr 9p23)
Peutz–Jeghers syndrome	*STK11*	Pigmented macules or periorifacial skin, intestinal polyps, adenocarcinoma. Increased ovarian, breast and pancreatic carcinoma Serine–threonine protein kinase (19p13.3)
Piebaldism	*KIT*	Depigmented patches white forelock, islands of normal pigmentation within these areas. A proto-oncogene, human homolog of v-kit family. Hardy Zuckerman 4 feline sarcoma viral oncogene (chr 4q11–q12)
Waardenburg syndrome type I and III	*PAX3*	White forelock, synophrys, caries, dystopia canthorum, heterochromia irides Paired homeobox DNA-binding protein (chr 2q35)
Waardenburg syndrome type IIA	*MITF*	Involved in melanocyte differentiation (chr 3p14)
Waardenburg syndrome type IV	*SOX10, EDN3, EDNRB*	SOX10, related to testis-determining factor (chr 22q13) Endothelin-3 (chr 20q13) Endothelin-B receptor (chr 13q22)

Table 10 Diseases predisposing to malignancy

Disorder	Gene affected	Further information
Ataxia telangiectasia	*ATM*	Encodes a protein kinase (chr 11q22)
Bloom syndrome	*BLM*	Photodistributed erythema, café-au-lait macules, long narrow face, predisposition to cancer (leukemia, lymphoma, gastrointestinal)
		DNA helicase RecQ protein-like-2 gene, involved in DNA unwinding (chr 15q26)
Cowden syndrome	*PTEN*	Tumor-suppressor gene, phosphatase (chr 10q23)
Dyskeratosis congenita	*DKC1, TERC*	Dystrophic and absent nails, reticulated pigmentation on the neck, trunk and face. Characteristic atrophy, hypopigmentation and telangiectasia. Alopecia and eyebrow thinning. Leukoplakia of the tongue and mucosal surfaces. Fanconi-type pancytopenia. Dental caries. Predisposition to squamous cell carcinoma
		Dyskerin, possible nucleolar and cell cycle function (chr Xq28)
		Telomerase RNA component (chr 3q21)
Familial cutaneous malignant melanoma	*CDKN2A, CDK4*	CDKN2A, known as p16, a cyclin-dependent kinase inhibitor (chr 9p21); CDK4 (chr 12q13) physically interacts with *CDKN2A*, regulating progression through G1 phase of cell cycle
Familial cylindromatosis	*CYLD1*	Tumor-suppressor gene (chr 16q12–q13)
Gardner syndrome	*APC*	Epidermoid cysts, osteomas, gastrointestinal polyps with predisposition to adenocarcinoma
		Hypertrophy of the retinal pigment epithelium
		Adenomatosis polyposis coli gene, cell signaling pathway molecule (5q21–q22)
Gorlin syndrome	*PTCH*	Characteristic facies, skeletal abnormalities, odontogenic cysts. Predisposition to basal cell carcinoma
(Nevoid basal cell carcinoma syndrome)		Human homolog of *Drosophila* segment polarity gene 'patched' (chr 9q22). Likely tumor-suppressor gene
Muir–Torre syndrome	*MSH2, MLH1*	Multiple sebaceous cell carcinomas, adenocarcinoma of the colon. DNA mismatch repair genes *MSH2* (chr 2p22–p21) and *MLH1* (chr 3p21.3)
Rothmond–Thomson syndrome	*RECQL4*	Growth retardation, osteosarcoma. Characteristic poikiloderma
		RecQ helicase protein-like 4 (chr 8q24)
Tuberous sclerosis	*TSC1, TSC2*	*TSC1*, hamartin (chr 9q34), a potential tumor suppressor; TSC2; tuberin (chr 16p13)
		See elsewhere in this encyclopedia
Xeroderma pigmentosum	*XPA*	XPA (chr 9q22.3)
	XPB/ERCC3	XPB (chr 2q21)
Eight complementation groups	*XPC*	XPC (chr 3p25)
	XPD/ERCC2	XPD (chr 19q13)
	XPE/DDB2	XPE (chr 11p12), DNA damage binding protein
	XPF/ERCC4, XPG/ERCC5	XPF (chr 16p13)
	XPV/POLH	XPG (chr 13q)
		DNA polymerase eta (chr 6p21)
		Predisposition to skin carcinoma, extreme sun sensitivity. Some groups have neurological complications

pigmentosum (XP) is an autosomal recessive disorder characterized by extreme sensitivity of the skin to sunlight. Patients have a dramatically increased risk of developing cancer on sun-exposed skin as a result of impairment of the repair of ultraviolet-induced DNA damage. Mutations have been found in genes involved in the nucleotide excision repair pathway. There are eight complementation groups of XP, A–G and V (variant), each caused by mutation in a different repair enzyme. The variant form of XP is caused by mutations in *POLH*, encoding DNA polymerase.

Interestingly, mutation in the *XPB* or *XPD* genes also leads to trichothiodystrophy (TTD), also known as PIBIDS, as patients suffer from Photosensitivity, Ichthyosis, Brittle hair and nails, Impaired intelligence, Decreased fertility and Short stature. Patients with TTD have photosensitivity, but no increase in the prevalence of skin cancers. The *XPB* and *XPD* genes encode two helicase subunits of the transcription/ repair factor TFIIH. It has been suggested that if mutation affects the DNA repair function of TFIIH without affecting transcription, an XP phenotype results, whereas if the mutation affects only transcription, this would produce the TTD phenotype.

Skin Appendage Defects

The ectodermal dysplasias encompass disorders that affect skin appendages (**Table 11**). There are over 170 known ectodermal dysplasias. X-linked hypohidrotic ectodermal dysplasia (ED1) is the most common form.

Table 11 Skin appendage defects

Disorder	Gene affected	Further information
Ankyloblepharon – ectodermal defects – cleft lip/palate syndrome (AEC/Hay–Wells)	*TP73L*	Extensive skin erosions at birth, cleft lip and palate. Recurring scalp infections lead to scarring alopecia
		p63 protein, strong homology to tumor-suppressor protein p53
Autosomal recessive and dominant (hypohidrotic) ectodermal dysplasia	*DL*	Heat intolerance as a result of diminished sweating, sparse brittle hair, hypodontia often with peg-shaped teeth
		Human gene homologous to mouse *downless* (chr 2q11–q13), an ectodysplasin receptor
Congenital (papular) atrichia	*HR*	Total loss of hair after first cycle. Some eyebrow hair may remain. Develop variable numbers of cysts throughout life
		Human gene homologous to mouse *hairless* (chr 8p21)
Ectodermal dysplasia hypohidrotic with immune deficiency	*IKK gamma gene*	Sparse brittle hair, peg-shaped teeth, immunological deficiencies, especially hypogammaglobulinemia
		Encodes *NEMO*, essential for NFκB signaling (chr Xq28). Mutations in same gene also cause incontinentia pigmenti (chr Xq28)
Ectodermal dysplasia (Margarita island type) ED4 or CLPED1	*PVRL1*	Nectin, an immunoglobulin-related transmembrane cell–cell adhesion molecule (chr 11q23–q24). Sparse hair, dental abnormalities, cleft lip/palate, syndactyly and nail changes
Ectrodactyly–ectodermal dysplasia and cleft lip/palate syndrome (EEC)	EEC3 mutations in *TP73L*	EEC1 (chr 7q11.2–q21.3) EEC2 (chr 19)
		EEC3 (chr 3q27), p63 homology to tumor-suppressor gene p53
Monilethrix	*KRTHB1, KRTHB6*	Beaded hairs, excessive weathering of hair. Variable nail defects
		Type II hair keratins (chr 12q13)
Nail patella syndrome	*LMX1B*	Nail defects at birth, palmoplantar hyperhidrosis. Absent or hypoplastic patella and renal failure
		(chr 9q34) Transcription factor of LIM-homeodomain type. Role in dorsal/ventral patterning of vertebrate limb
Pachyonychia congenita type 1	*KRT6A, KRT16*	Keratins 6a and 16 (chr 12q13 and 17q12–q21). Main features are hypertrophic nail dystrophy, oral leukokeratosis and focal keratoderma
Pachyonychia congenita type 2	*KRT6B, KRT17*	Keratins 6b and 17 (chr 12q13 and 17q12–q21)
		Main features are hypertrophic nail dystrophy, multiple pilosebaceous cysts, natal teeth and focal keratoderma
Steatocystoma multiplex	*KRT17*	Type I keratin K17 (chr 17q12–q21)
		Pilosebaceous cysts with little or no nail changes and/or keratoderma
Tricho-dento-osseous syndrome	*DLX3*	Member of distal-less homeobox gene family (chr 17q12–q22). Curly hair, enamel hypoplasia, increased bone density, mainly in skull
Tricho-rhino-phalangeal syndrome types I and III	*TRPS1*	Zinc-finger protein, potential transcription factor (chr 8q24). Sparse hair, finger deformations and broad nose
Tricho-rhino-phalangeal syndrome type II	Deletion incorporating 8q42.11 to 8q24.13	See above
X-linked hypohidrotic ectodermal dysplasia (ED1)	*ED1*	Heat intolerance resulting from diminished sweating, sparse brittle hair, hypodontia often with peg-shaped teeth
		Membrane protein ectodysplasin A involved in epithelial–mesenchymal signaling (chr Xq12–q13). Mouse homolog is *tabby*

Affected males have hypotrichosis, abnormal teeth, absent sweat glands and sometimes mental defects. Mutations have been found in the gene *ED1* encoding the protein ectodysplasin A, a membrane protein found to colocalize with cytoskeletal structures at the lateral and apical surfaces of epithelial cells, and is thought to be involved in communication. Mutations have been found in the *TP63* gene encoding p63 in patients with ectrodactyly–ectodermal dysplasia and cleft lip/palate syndrome (EEC3) and Hay–Wells syndrome (ankyloblepharon-ectodermal defects – cleft lip/palate). These mutations are thought to affect DNA binding capacity of p63 (EEC3) or sites required for protein–protein interaction (Hay–Wells). CLPED1 is another ectodermal dysplasia (cleft lip/palate-ectodermal dysplasia syndrome) known as the Margarita island type. This disease involves cleft lip/palate with hidrotic ectodermal dysplasia and developmental defects of the hands. Mutations have been found in the gene *PVRL1*, which encodes nectin-1, an immunoglobulin-related transmembrane cell–cell adhesion molecule.

Connective Tissue Disorders

Pseudoxanthoma elasticum is a connective tissue defect. Mutations in a transmembrane ATP-binding cassette transporter gene (*ABCC6* or *MRP6*) lead to progressive calcification of elastic fibers in the skin, arteries and the retina (Ringpfeil *et al.*, 2000). Patients suffer from dermal lesions with associated laxity and loss of elasticity, arterial insufficiency and retinal hemorrhages. Ehlers–Danlos syndrome is a heterogeneous group of connective tissue disorders (see **Table 12**). Affected individuals suffer from skin hyperextensibility and fragility, joint hypermobility and bruising. Several of the types result from mutations in collagen genes (*COL1A1*, *COL1A2*, *COL3A1* and *COL5A1*) leading to disruption of collagen fibril packing as a result of abnormal collagen fibrillogenesis.

Marfan syndrome is a connective tissue disorder, manifested in the ocular, skeletal and cardiovascular systems. The skin, lungs and muscle are also involved. The disease shows great clinical variability and has been shown to be caused by mutations in the *FBN1* gene encoding fibrillin, a microfibrillar glycoprotein of the elastic fiber system.

Porphyrias

There are currently five forms of porphyria for which the underlying molecular defect is known: erythropoietic protoporphyria, congenital erythropoietic porphyria, porphyria cutanea tarda, hepatoerythropoietic porphyria, and variegate porphyria (**Table 13**). In variegate porphyria, mutations in the

Table 12 Connective tissue defects

Disorder	Gene affected	Further information
Ehlers–Danlos syndrome, arthrochalasia type (VIIa and b)	*COL1A1*, *COL1A2*	Alpha-1 and -2 chains of type I collagen: COL1A1 (17q21–q22); COL1A2 (chr 7q22)
Ehlers–Danlos syndrome classical type (I/II)	*COL5A1*, *COL5A2*	Alpha-1 and -2 chains of type 5 collagen: COL5A1 (chr 9q34); COL5A2 (chr 2q31)
Ehlers–Danlos syndrome, dermatosparaxis type (VIIC)	*ADAMTS2*	A disintegrin-like and metalloproteinase with thrombospondin type I motif, 2
		Procollagen I N-proteinase (chr 5q23)
Ehlers–Danlos syndrome, hypermobility type (type III)	*COL3A1*	Alpha-1 chain of type 3 collagen (chr 2q31)
Ehlers–Danlos syndrome, kyphoscoliotic type (type VI)	*PLOD*	Gene for lysyl hydroxylase, procollagen lysine, 2-oxoglutarate-5 dioxygenase (chr 1p36)
Ehlers–Danlos progeroid form	*XGPT1*	Gene encoding xylosylprotein 4-beta-galactosyl transferase
Ehlers–Danlos like syndrome caused by tenascin X deficiency	*TNX*	Tenascin X gene mutations (chr 6p21.3)
Ehlers–Danlos syndrome, vascular type (type IV)	*COL3A1*	Alpha-1 chain of type 3 collagen (chr 2q31)
Fabry disease	*GLA*	Alpha-galactosidase A, a lysosomal hydrolase (chr Xq22)
Hereditary angioedema	*C1NH*	C1-esterase inhibitor (chr 11q11–q13), member of a large serine protease inhibitor gene family
Marfan syndrome	*FBN1*	Fibrillin-1, a glycoprotein of the microfibrillar component of the elastic fiber system (chr 15q21.1)
Pseudoxanthoma elasticum	*ABCC6*	Transmembrane ATP-binding cassette transporter (chr 16p13.1)
X-linked cutis laxa	*ATP7A*	Copper-transporting ATPase, alpha peptide (chr Xq12–q13)

Table 13 Porphyria

Disorder	Genes affected	Further information
Congenital erythropoietic porphyria	*URO-synthase*	Uroporphyrinogen III synthase (chr 10q25–q26)
Erythropoietic protoporphyria (EPP)	*FECH*	Ferrochelatase, a mitochondrial enzyme and the final step in the heme synthesis pathway (chr 18q21.3)
Hepatoerythropoietic porphyria	*UROD*	Uroporphyrinogen decarboxylase (chr 1p34)
Porphyria cutanea tarda	*UROD*	Uroporphyrinogen decarboxylase (chr 1p34)
Variegate porphyria	*PPOX*	Protoporphyrinogen oxidase (chr 1q22–q23)

Table 14 Vascular disorders

Disorder	Genes affected	Further information
Hyperkeratotic cutaneous capillary venous malformations (HCCVM)	*CCM1*	Encodes KRIT1, Krev interaction trapped-1 (chr 7q11.2–q21)
Hereditary lymphedema type I	*FLT4*	Vascular endothelial growth factor C receptor (chr 5q35)
Lymphedema-distichiasis syndrome	*MFH1*	chr 16q24.6. Transcription factor. Double row of eyelashes, lymphedema, corneal erosions
Osler–Rendu–Weber disease; hereditary hemorrhagic telangiectasia types I and II (HHT1/2)	*ENG,* *ACVRLK1*	Endoglin, a transforming growth factor-beta binding protein (chr 9q33–q34); activin receptor like kinase 1 gene, member of the serine–threonine kinase receptor family (chr 12q11–q14)
Venous malformations, cutaneous and mucosal (VMCM)	*TEK*	Receptor tyrosine kinase TIE2 (chr 9p21)

Table 15 Miscellaneous skin diseases

Disorder	Gene affected	Further information
Albright hereditary osteodystrophy	*GNAS1*	Guanine nucleotide-binding protein, alpha stimulating activity polypeptide (chr 20q13)
Cerebrotendinous xanthomatosis	*CYP27A1*	Sterol 27-hydroxylase, a cytochrome P450 (chr 2q33–qter)
Chronic granulomatous disease	*CYBB, NCF2,* *CYBA, NCF1*	Glycoprotein components of superoxide generating nicotinamide adenine dinucleotide phosphate (NADPH) oxidase: *CYBB* - p91 phox, (chr Xp21.1); *NCF2* - p67 phox (chr 1q25); *CYBA*-p22 phox (chr 16q24); *NCF1*-p47 phox (chr 7q11)
Cockayne syndrome	*ERCC8,* *ERCC6*	WD-40 repeat protein, involved in DNA excision repair (chr 5, 10q11)
Leprechaunism	*INSR*	Insulin receptor (chr 19p13)
Lipoid proteinosis	*ECM1*	chr 1q21
		Protein name is also ECM1
		Bullae on skin in early life, later development of papules over face, neck and extremities. Infiltration of mucous membranes, including the larynx. Associated with a coarse cry. May have temporal lobe and hippocampal calcification
Menkes syndrome	*ATP7A*	Sparse, brittle hair with shaft abnormalities including pili torti and trichorrexhis nodosa. Failure to thrive. Death by the age of 2–3 years
		Copper-transporting ATPase, alpha peptide (chr Xq12–q13)
Omenn syndrome	*RAG1, RAG2*	Lymphoid-specific proteins Rag-1, Rag-2 (chr 11p13)
Werner syndrome	*RECQL2*	DNA helicase (chr 8p12–p11)
Wiskott–Aldrich syndrome	*WAS*	Wiskott–Aldrich protein containing a GTPase-binding site (chr Xp11)

protoporphyrinogen oxidase gene *PPOX* lead to disruption in the penultimate step of heme biosynthesis. Patients with this form of porphyria may suffer from neurovisceral attacks, hyperpigmentation, hypertrichosis and cutaneous photosensitivity. Congenital erythropoietic porphyria is the most severe form of porphyria and results from a decrease in the activity of an enzyme in the heme biosynthetic pathway, uroporphyrinogen III synthase, encoded by *URO-synthase*. Accumulation of the porphyrin isomers, uroporphyrin I and coproporphyrin I, leads to the phenotype in this form of porphyria. Blistering and scarring in exposed areas may lead to mutilating deformity.

Vascular Disorders

There are many forms of rare inherited vascular malformation syndromes and mutations have been found in genes underlying two of these: venous malformations, cutaneous and mucosal (VMCM) and hyperkeratotic cutaneous capillary-venous malformations associated with cerebral capillary malformation (HCCVM). Mutations have been found in the *TEK* gene in VMCM patients, which encodes the endothelial-cell-specific receptor tyrosine kinase TIE2 (Vikkula *et al.*, 1996). Characterized by crimson-colored, irregularly shaped skin lesions, HCCVM is caused by mutations in the *CCM1* gene, which encodes KRIT1, a protein that interacts with RAP1A (a member of the RAS family of GTPases). Hereditary lymphedema type I is also included under this subheading (**Table 14**).

Miscellaneous Skin Diseases

Menkes syndrome is characterized by hair shaft abnormalities, epidermal hypopigmentation and progressive cerebral degeneration (**Table 15**). It is caused by mutations in *ATP7A* which encodes a copper-transporting ATPase. Wiskott–Aldrich syndrome is an X-linked immunodeficiency characterized by thrombocytopenia, eczema and recurrent infections. Mutations are found within the *WAS* gene, which encodes a protein containing a GTPase-binding site, which interacts specifically with activated CDC42 and is thought to be involved in actin polymerization.

See also
Keratins and Keratin Diseases
Marfan Syndrome
Porphyrias: Genetics

References

Chavanas S, Bodemer C, Rochat A, *et al.* (2000) Mutations in SPINK5, encoding a serine protease inhibitor, cause Netherton syndrome. *Nature Genetics* **25**: 141–142.
De Laurenzi V, Rogers GR, Hamrock DJ, *et al.* (1996) Sjögren–Larsson syndrome is caused by mutations in the fatty aldehyde dehydrogenase gene. *Nature Genetics* **12**: 52–57.
Huber M, Rettler I, Bernasconi K, *et al.* (1995) Mutations of keratinocyte transglutaminase in lamellar ichthyosis. *Science* **267**: 525–528.
King RA, Mentink MM and Oetting WS (1991) Non-random distribution of missense mutations within the human tyrosinase gene in type I (tyrosinase-related) oculocutaneous albinism. *Molecular Biology and Medicine* **8**: 19–29.
Kruse R, Lamberti C, Wang Y, *et al.* (1996) Is the mismatch repair deficient type of Muir–Torre syndrome confined to mutations in the hMSH2 gene? *Human Genetics* **98**: 747–750.
McKoy G, Protonotarios N, Crosby A, *et al.* (2000) Identification of a deletion in plakoglobin in arrhythmogenic right ventricular cardiomyopathy with palmoplantar keratoderma and woolly hair (Naxos disease). *The Lancet* **355**: 2119–2124.
Richard G, Smith LE, Bailey RA, *et al.* (1998) Mutations in the human connexin gene *GJB3* cause erythrokeratodermia variabilis. *Nature Genetics* **20**: 366–369.
Ringpfeil F, Lebwohl MG, Christiano AM and Uitto J (2000) Pseudoxanthoma elasticum: mutations in the MRP6 gene encoding a transmembrane ATP binding cassette (ABC) transporter. *Proceedings of the National Academy of Sciences of the United States of America* **97**: 6001–6006.
Sakuntabhai A, Burge S, Monk S and Hovnanian A (1999) Spectrum of novel ATP2A2 mutations in patients with Darier disease. *Human Molecular Genetics* **8**: 1611–1619.
Vikkula M, Boon LM, Carraway III KL, *et al.* (1996) Vascular dysmorphogenesis caused by an activating mutation in the receptor tyrosine kinase TIE2. *Cell* **87**: 1181–1190.

Further Reading

Arin MJ and Roop DR (2001) Disease model: heritable skin blistering. *Trends in Molecular Medicine* **7**: 422–424.
Fine J-D, Eady RAJ, Bauer EA, *et al.* (2000) *Revised Classification System for Inherited Epidermolysis Bullosa*. Report of the second International Consensus Meeting on Diagnosis and Classification of Epidermolysis Bullosa. *Journal of the American Academy of Dermatology* **42**(6): 1051–1066.
Frank J and Christiano AM (1998) Variegate porphyria: past, present and future. *Skin Pharmacology and Applied Skin Physiology* **11**: 310–320.
Irvine AD and McLean WHI (1999) Human keratin diseases: the increasing spectrum of disease and subtlety of the phenotype–genotype correlation. *British Journal of Dermatology* **140**: 815–828.
Ishida-Yamamoto A, Takahashi H and Iizuka H (1998) Loricrin and human skin diseases: molecular basis of loricrin keratodermas. *Histology and Histopathology* **13**: 819–826.
McGrath JA and Barker JNWN (2001) *Cell Adhesion and Migration in Skin Disease*, pp. 27–55. London, UK: Harwood Academic Publishers.
Mao JR and Bristow J (2001) The Ehlers–Danlos syndrome: on beyond collagens. *Journal of Clinical Investigations* **107**: 1063–1069.
Priolo M and Lagana C (2001) Ectodermal dysplasias: a new clinical–genetic classification. *Journal of Medical Genetics* **38**: 579–585.
Richard G (2000) Connexins: a connection with the skin. *Experimental Dermatology* **9**: 77–96.
van Steeg H and Kraemer KH (1999) Xeroderma pigmentosum and the role of UV-induced DNA damage in skin cancer. *Molecular Medicine Today* **5**: 86–94.

Web Links

Intermediate Filament Mutation Database. By Cassidy AJ, Irvine AD, Lane EB and McLean WHI (2002)
  http://www.interfil.org

Skin Pigmentation: Genetics

Jonathan L Rees, *University of Edinburgh, Edinburgh, UK*

Differences in skin pigmentation are principally due to variation in the amount and type of melanin produced in cutaneous melanocytes. In humans, a large number of Mendelian disorders of pigmentation, including albinism, are now understood at a genetic level. Only one locus, the melanocortin 1 receptor, has been so far identified to contribute to normal variation in skin and hair color, but it is clear that many other loci (yet to be determined) must also play a role.

Intermediate article

Article contents

Introduction

Variation in skin pigmentation between people of different genetic ancestries is one of the most striking human characteristics (Nordlund *et al.*, 1998). Skin color is due to the combination of three different pigments: melanin within the epidermis, hemoglobin in blood vessels within the dermis and dietary derived beta-carotene. The most important of these is melanin, except in instances such as when skin color changes due to blushing or flushing when variation in blood flow is important. Beta-carotene plays a minimal role in determining skin color variation.

Melanin in a complex quinone/indole–quinone derived mixture of biopolymers produced in melanocytes from tyrosine. Melanocytes are dendritic neural crest derived cells that migrate into human epidermis during the first trimester and produce melanin in specialized lysosomal-like organelles called melanosomes, which are then secreted into the surrounding keratinocytes where they undergo further changes. Melanin chemistry is complex and far from completely understood, for a number of reasons: melanin is a complex mixture of polymers, many intermediates are unstable and rapidly autooxidize, and available methods to solubilize melanin alter its primary structure (Ito, 1998). Insight into melanin regulation and production has in large part derived from genetic approaches, especially those based on the resource provided by the mouse fancy (Barsh, 1996).

Variation in human skin color can be accounted for by variation in the total amount of melanin and the types of melanin produced and not differences in the number of melanocytes, which is comparable for all groups. Two broad classes of melanin polymers are recognized: eumelanin, which is brown or black, and the cysteine containing pheomelanin, which is red or yellow. Individuals with dark skin, such as those from much of Africa, have large amounts of eumelanin in their skin and hair, whereas by contrast a red-headed northern European has less eumelanin and proportionally more pheomelanin in their hair and skin. Skin

color is not, however, simply a result of differences in the primary structure of melanin. Melanin is packaged into melanosomes and such melanosomes differ in shape, size and aggregation patterns between persons with different skin colors. Such differences in macromolecular structure would be expected (by way of influencing light scattering) to influence skin color. The genetic and biochemical mechanisms determining these differences are as yet poorly understood.

Evolution of Skin Color

The striking variation in skin color is conventionally explained in terms of the ability of melanin to protect against damage induced by ultraviolet radiation (UVR) (Jablonski and Chaplin, 2000; Nordlund *et al.*, 1998). The ultraviolet part of sunlight is harmful to skin, causing, in the short term, erythema (redness) and burning with possible blister formation and, in the long term, skin cancer. The toxicity of ultraviolet radiation is in part mediated by the direct action of ultraviolet B (UVB) radiation on one of the epidermis' principal chromophores, deoxyribonucleic acid (DNA). Individuals such as albinos who produce no or little melanin, and who live in equatorial regions, are at a major biological disadvantage from burning (and consequently from an inability to go out during much of the day) and skin cancer in their teenage years and early adult life (King *et al.*, 2001). Pigmentation in the form of melanin would be therefore expected to be under strong selective pressure.

In general, persons from more temperate regions have paler skin. A popular explanation for this is that because vitamin D is in part biosynthesized in skin (under the action of UVB) pale skin would have selective advantages in areas where the other source of vitamin D, that is, dietary vitamin D, was poor and where ambient UVB low. By contrast, persons with dark skin would be unable to synthesize as much vitamin D, because their pigment would shield their keratinocytes from UVR, and would be more prone to suffer from rickets, a vitamin D deficiency disease. There is widespread acceptance of this hypothesis (Jablonski and Chaplin, 2000). An alternative hypothesis is that whereas dark skin color may be under strong selective pressure in Africa, when humans moved out of Africa some variation in the genes determining pigmentation was neutral with respect to fitness (see for instance Harding *et al.* (2000)).

Genetic Control of Skin Pigmentation

Human skin pigmentation is genetically complex. In the mouse more than 50 loci are thought to influence coat color (Barsh, 1996). In humans we have few robust estimates of the relative contribution of the number of genes involved or their relative quantitative contribution to normal pigmentation. Indeed, whereas a number of Mendelian disorders (diseases) of skin pigmentation are known, only one locus, the melanocortin 1 receptor (alpha melanocyte stimulating hormone receptor) (*MC1R*), has so far been identified to play a role in normal variation in human pigmentation (Rees, 2000).

Mendelian disorders of skin pigmentation

Albinism

Albinism (King *et al.*, 2001) is one of the archetypal inborn errors of metabolism as described by Archibald Garrod, with a frequency of one in 20 000. Albinism is usually defined as a congenital hypopigmentation of the skin, hair or eyes. When eyes and skin are both involved, it is called oculocutaneous albinism. Ocular manifestations include reduced retinal pigmentation, abnormal decussation of the optic tract, nystagmus and translucent irises. Most definitions require ocular changes to be present as part of the definition of albinism, although the literature is inconsistent on this.

There are a number of different forms of albinism due to mutations in genes important in the production of melanin. All forms of albinism involving the skin have in common that the number of melanocytes is normal but the amount of melanin produced is low or even absent. Many previous classifications of albinism not based on identification of gene products (e.g. partial, incomplete albinism, etc.) are confusing and are now considered inappropriate. Nonetheless the range of phenotypes seen even within mutations at one locus is considerable, and the effects of genetic background (other pigment genes) are important.

Oculocutaneous Albinism Type 1

Oculocutaneous albinism type 1 (OCA1) (MIM 203100; see OMIM in Web Links) is the second most common form of albinism and is due to mutations in the copper-containing enzyme tyrosinase. Tyrosinase is the rate-limiting step in the biosynthesis of melanin catalyzing the hydroxylation of tyrosine to dopa (3,4-dihydroxyphenylalanine) and dopaquinone, as well as other later reactions with the melanin pathway. Tyrosinase shows homology to other enzymes such as tyrosinase-related protein 1, mutations of which cause oculocutaneous albinism type 3 (OCA3), and tyrosine-related protein 2, another key enzyme in melanin biosynthesis. The human tyrosinase locus maps to 11p14–21, and the gene (tyrosinase (oculocutaneous albinism IA) (*TYR*)) contains five exons spanning more than 80 kilobases (kb). Over 100 mutations, including missense, nonsense, frameshift and splice site mutations, have been described.

Tyrosinase-related oculocutaneous albinism is divided into two groups, OCA1A and OCA1B, based on the phenotype, with those in the former group being more severely affected. Classically, patients with OCA1A show a complete absence of melanin in the skin and hair at birth with a failure to tan or to develop pigmented nevi at older ages. By contrast, OCA1B may develop nevi and freckles and may show signs of some melanin production later in life in response to UVR. Some OCA1B patients appear, at least from a cutaneous perspective, to be just very pale-skinned rather than as obvious albinos and may be missed clinically. The difference between OCA1A and OCA1B is one of degree, with the former arising from mutations that are null while those associated with OCA1B leaving the enzyme tyrosinase with some residual activity.

Oculocutaneous Albinism Type 2

Persons with oculocutaneous albinism type 2 (OCA2) (MIM 203200; see OMIM in Web Links) have some pigment of the eyes, skin and hair but do not usually tan. This disorder is due to mutations in the oculocutaneous albinism II (pink-eye dilution homolog, mouse) (*OCA2*) gene, the human homolog of the mouse pink-eyed dilution locus that maps to 15q11.2–q12, a region also associated with two distinct genetic syndromes, Angelman syndrome (MIM 105830; see OMIM in Web Links) and Prader–Willi syndrome (MIM 176270; see OMIM in Web Links). The *OCA2* gene spans 250–600 kb with 25 exons and codes for a product believed to be important in determining the acidic melanosomal pH. The classic phenotype of OCA2 is found in Africans in whom a common deletion accounts for most cases and in whom, as with OCA1B, nevi and freckles may develop. The irises are lightly pigmented.

One particular phenotype, referred to as Brown OCA (BOCA), is also due to mutations of the

OCA2 gene. OCA2 may coexist with Prader–Willi or Angelman syndromes, and pigmentary abnormalities in these latter disorders in the absence of OCA2 may be due to defects in *OCA2* gene function.

Other Forms of OCA

There are a number of other causes of albinism. OCA3 is due to mutations of tyrosinase-related protein 1 (*TYRP1*) causing rufous oculocutaneous albinism (MIM 203290; see OMIM in Web Links) character-ized, at least in African populations, by reddish brown skin, ginger hair and hazel or brown irises. Not all of the usual ocular manifestations of albinism may be present. Other forms of albinism include OCA4 (MIM 606574; see OMIM in Web Links), due to mutations of the human homolog of the mouse underwhite gene; several disorders conceived of as lysosomal disorders (and thereby involving melanosomes), including Hermansky–Pudlak syndrome (MIM 203300; see OMIM in Web Links) characterized by OCA, a bleeding diathesis and ceroid storage defect, which is genetically heterogeneous; and Chediak–Higashi dis-ease (MIM 214500; see OMIM in Web Links), where albinism is combined with a defect in the immune system secondary to mutation in the Chediak–Higashi syndrome 1 (*CHS1*) gene.

Other genetic disorders of pigmentation without albinism

There are many other Mendelian syndromes charac-terized by either general hypopigmentation, such as occurs in the Griscelli syndrome (MIM 214550; see OMIM in Web Links), or focal loss of pigmentation as seen in piebaldism (MIM 172800; see OMIM in Web Links). For further information on these, the reader is directed to specialized texts.

Physiological variation in skin pigmentation: the melanocortin 1 receptor

Most of the Mendelian disorders discussed above were identified based on the presence of mouse mutants, identification of the genetic defect and gene in the mouse, and matching of this information to a rare human phenotype. Attempts to ascribe variation in pigmentation within the normal population to the loci involved in albinism or other pigmentary disorders have been, with one notable exception, unsuccessful. It seems to this author that this may be due to technical factors and a lack of power in describing human pigmentary phenotypes rather than the fact that these loci do not underpin such physiological variation.

The melanocortin 1 receptor (MC1R) was identified in the mouse as the locus underpinning a series of mutants at the extension locus (Cone *et al.*, 1996). Loss-of-function mutants at the mouse *Mc1r* locus have yellow fur characterized by a relative overpro-duction of pheomelanin and a reduction in eumelanin production, whereas dominant gain-of-function mutants were black coated with high eumelanin production (Cone *et al.*, 1996). Humans with red hair and pale skin also show mutations for the *MC1R* gene (Valverde *et al.*, 1995). The *MC1R* gene (16q24.3) codes for a 317 amino acid G-coupled transmembrane receptor expressed on melanocytes, the endogenous ligand for which is a cleavage product of pro-opiomelanocortin (POMC), alpha melanocyte-stimu-lating hormone (α-MSH). Activation of the receptor leads to an increase in intracellular cyclic adenosine monophosphate (cAMP) and a preference for eumelanin over pheomelanin biosynthesis. Agouti signal protein, at least in the mouse, antagonizes the effects of α-MSH as the MC1R with a resulting shift to a pheomelanin phenotype, although a role for agouti in the control of human pigmentation remains under investigation.

Persons with red hair and pale skin are usually homozygous mutants for *MC1R* with three to six sequence variants accounting for most cases (Rees, 2000). Some individuals with red hair appear to have functionally significant changes in only one allele, although changes outside the coding region have been sought but not found. Such red-haired persons are more likely to have strawberry blonde or auburn rather than bright 'carrot' red hair. In some western European populations, such as the United Kingdom and Ireland, over 40% of the population harbor functionally significant changes in *MC1R* on one allele. There is a clear heterozygote effect of the MC1R on tanning, freckle number and cancer risk with hetero-zygotes showing an increased tendency to burn in response to UVR (compared with those with the consensus sequence), a higher freckle count and an elevated risk of melanoma and nonmelanoma skin cancer. Studies of MC1R diversity have shown that most nonsynonymous diversity has been found in populations outside Africa (Harding *et al.*, 2000). One interpretation is that a functional constraint has operated at this locus within Africa but that with movement out of Africa the resulting increase in diversity is compatible with neutral theory. Modeling suggests that extant *MC1R* mutations associated with red hair are around 50 000 years old (Harding *et al.*, 2000).

See also

Albinism: Genetics
Melanoma: Genetics
Skin: Hereditary Disorders

References

Barsh GS (1996) The genetics of pigmentation: from fancy genes to complex traits. *Trends in Genetics* **12**: 299–305.

Cone RD, Lu D, Koppula S, et al. (1996) The melanocortin receptors: agonists, antagonists, and the hormonal control of pigmentation. *Recent Progress in Hormone Research* **51**: 287–317.

Harding RM, Healy E, Ray AJ, et al. (2000) Evidence for variable selective pressures at the human pigmentation locus, MC1R. *American Journal of Human Genetics* **66**: 1351–1361.

Ito S (1998) Advances in chemical analysis of melanins. In: Nordlund JJ, Boissy RE, Hearing VJ, KIng RA and Ortone JP (eds.) *The Pigmentary System: Physiology and Pathophysiology*, pp. 439–450. New York, NY: Oxford University Press.

Jablonski NG and Chaplin G (2000) The evolution of human skin coloration. *Journal of Human Evolution* **39**: 57–106.

King RA, Hearing VJ, Creel DJ and Oetting WS (2001) Albinism. In: Scriver CR, Beaudet AL, Sly WS, et al. (eds.) *The Metabolic and Molecular Bases of Inherited Disease*, pp. 5587–5628. New York, NY: McGraw-Hill.

Nordlund JJ, Boissy RE, Hearing VJ, King RA and Ortonne JP (eds.) (1998) *The Pigmentary System: Physiology and Pathophysiology*. New York, NY: Oxford University Press.

Rees JL (2000) The melanocortin 1 receptor (MC1R): more than just red hair. *Pigment Cell Research* **13**: 135–140.

Valverde P, Healy E, Jackson I, Rees JL and Thody AJ (1995) Variants of the melanocyte-stimulating hormone receptor gene are associated with red hair and fair skin in humans. *Nature Genetics* **11**: 328–330.

Further Reading

Sturm RA, Box NF and Ramsay M (1998) Human pigmentation genetics: the difference is only skin deep. *BioEssays* **20**: 712–721.

Web Links

OMIM (Online Mendelian Inheritance in Man). A database cataloging human genes and genetic disorders
http://www.ncbi.nlm.nih.gov/Omim/

Chediak-Higashi syndrome 1 (*CHS1*); Locus ID: 1130. LocusLink:
http://www.ncbi.nlm.nih.gov/LocusLink/LocRpt.cgi?l = 1130

Melanocortin 1 receptor (alpha melanocyte stimulating hormone receptor) (*MC1R*); Locus ID: 4157. LocusLink:
http://www.ncbi.nlm.nih.gov/LocusLink/LocRpt.cgi?l = 4157

Oculocutaneous albinism II (pink-eye dilution homolog, mouse) (*OCA2*); Locus ID: 4948. LocusLink:
http://www.ncbi.nlm.nih.gov/LocusLink/LocRpt.cgi?l = 4948

Tyrosinase (oculocutaneous albinism IA) (*TYR*); Locus ID: 7299. LocusLink:
http://www.ncbi.nlm.nih.gov/LocusLink/LocRpt.cgi?l = 7299

Tyrosinase-related protein 1 (*TYRP1*); Locus ID: 7306. LocusLink:
http://www.ncbi.nlm.nih.gov/LocusLink/LocRpt.cgi?l = 7306

Chediak-Higashi syndrome 1 (*CHS1*); MIM number: 606897. OMIM:
http://www.ncbi.nlm.nih.gov/htbin-post/Omim/dispmim?606897

Melanocortin 1 receptor (alpha melanocyte stimulating hormone receptor) (*MC1R*); MIM number: 155555. OMIM:
http://www.ncbi.nlm.nih.gov/htbin-post/Omim/dispmim?155555

Oculocutaneous albinism II (pink-eye dilution homolog, mouse) (*OCA2*); MIM number: 203200. OMIM:
http://www.ncbi.nlm.nih.gov/htbin-post/Omim/dispmim?203200

Tyrosinase (oculocutaneous albinism IA) (*TYR*); MIM number: 606933. OMIM:
http://www.ncbi.nlm.nih.gov/htbin-post/Omim/dispmim?606933

Tyrosinase-related protein 1 (*TYRP1*); MIM number: 115501. OMIM:
http://www.ncbi.nlm.nih.gov/htbin-post/Omim/dispmim?115501

Slater, Eliot Trevor Oakeshott

Introductory article

Article contents

- Introduction
- Slater's Prewar Contributions
- The War Years
- Postwar and the First Golden Era of Psychiatric Genetics

Irving I Gottesman, *University of Minnesota, Minneapolis, Minnesota, USA*

Eliot Trevor Oakeshott Slater (1904–1983) was an English psychiatrist and genetic researcher, founding editor of the *British Journal of Psychiatry*, and founder of the first Medical Research Council Psychiatric Genetics Unit.

Introduction

Societal-wide turbulence characterized the functioning of governments, economics, medicine and science in the 1930s, and it would be reflected for years to come in the structure of and attitudes toward the discipline that began to grow at the interface of neuropsychiatry and human genetics. In the 'democratic' German elections of 1932, the Nazi Party won; in the presidential race, Hitler lost to Hindenburg; and then Hitler was appointed chancellor in January 1933 and immediately implemented his dictatorial powers. Soon a 'compulsory' sterilization law was promulgated, with the purpose of 'preventing genetically diseased offspring'. The latter included 200 000 persons with mental retardation and some 80 000 with schizophrenia. Such historical facts frame the fact that Eliot

Slater (1904–1983), widely appreciated as the founder of psychiatric genetics in the UK, received a Rockefeller Foundation Fellowship to study in this field in 1934 and 1935, in Munich and Berlin, primarily in the laboratories of Ernst Rüdin. By 1917, the very first research institute for psychiatry – Deutsche Forschungsanstalt für Psychiatrie – was in place, with Rüdin as head of the Department for Genealogical Demography. Kraepelin had earlier encouraged Rüdin to study 'dementia praecox' (schizophrenia) from a Mendelian viewpoint, resulting in the first major, scientifically sound family study, including the study of adoptees. Preliminary findings were available in 1911, the same year as Kraepelin's rival, Eugen Bleuler (1857–1939), published his book on schizophrenia. (*See* China: The Maternal and Infant Health Care Law; Eugenics; Nazi Scientists.)

Slater joined the staff at the Maudsley Hospital, UK, in October 1931, along with Aubrey Lewis. Lewis was immersed in genetic studies of psychopathology for a definitive chapter in the prescient book edited by C. P. Blacker (1934), *The Chances of Morbid Inheritance*. From the dates in the bibliography, it appears to have been completed early in 1932, before Hitler or the first of the draconian eugenics laws were in place.

In late 1934, Slater began his new life in Munich, replacing Erik Essen-Möller of Sweden.

Slater's Prewar Contributions

Slater arrived in Hitler's Nazi Germany with very mixed emotions. The Nazi law of 1933 for the 'restoration of the professional civil service' forced the dismissal of all Jews and 'half-Jews' from any government position, including universities, research institutes and hospitals. The Nazi law for 'compulsory' sterilization for the prevention of hereditary diseases became effective in 1934. The culpable 'diseases' were congenital feeble-mindedness, schizophrenia, manic-depressive insanity, inherited epilepsy, Huntington chorea, inherited blindness, inherited deafness, severe inherited physical malformation and anyone 'suffering from severe alcoholism'. (*See* Mood Disorders: Molecular; Psychoses; Schizophrenia: Molecular Genetics.)

Word spread and an August 1933 issue of *The Lancet* contained a sharply worded alarm, before the text of the new law was published that September. The author of the warning was Lewis, writing from the vantage point of being a Jew and a genetically informed psychiatrist. In 1934, Lewis made it clear that Rüdin was one of the main authors of this law, and that it was projected that 400 000 citizens, including children older than 10 years, would be suitable for sterilization. Although compulsory, the decision on sterilization in any case was to be taken by the 'nicety' of a nationwide set of 205 Eugenic Courts, consisting of a lawyer, a public health physician and a physician 'versed in eugenics'. With the 'kind permission of Professor Rüdin', whom he only met twice socially in the course of a year, Slater commenced the study of three generations of manic-depressive families, resulting in his first major contribution to psychiatric genetics.

As part of his Fellowship, Slater spent 3 months in Berlin learning basic and *Drosophila* genetics in the laboratories of Timofeeff–Ressovsky. While there, he met Lydia Pasternak, a chemist and sister of the poet–novelist Boris Pasternak. Upon returning to England at the end of 1935, they married, which would have been illegal in Germany as she was Jewish according to the Nuremberg Laws of 1935. In one of the first papers published upon his return (1936), Slater said, echoing Lewis's earlier outrage:

> The Fuehrer directs with a series of ukases. With successive hammer blows, the German citizen is driven into a swastika-shaped hole. The atmosphere of compulsion pervades the whole of his life. The fact that he and his fellow men are now to be selected and bred like a herd of cattle seems to him hardly more distasteful than a hundred other interferences in his daily life... The command now is to breed. (Slater, 1971, p. 292)

Upon his return to the Maudsley Hospital, Slater obtained a grant from the Medical Research Council (MRC) that permitted him to collect data on twins from the 10 London County Council mental hospitals. Research and all other life was disrupted by the outbreak of the Second World War in 1939. The hospital was evacuated, with Lewis in charge of one unit at Mill Hill, while Slater was in command of the other unit at Sutton.

The War Years

Some 20 000 military psychiatric casualties were cared for in Sutton by Slater and his staff. Research continued, and the patients were asked about their twinship and their family histories. Of most importance is the work leading to a general theory of the etiology of neuroses, a wide-ranging diathesis–stressor theory, specifically polygenic on the diathesis side, and multifactorial. Military stress was even quantified into severe, moderate and trifling. See the Further Reading section for more of Slater and his coworkers' accomplishments. The diathesis–stressor model developed to study neurosis was the first to emphasize a polygenic component on the diathesis side of the model. Slater and his brother concluded that the neurotic constitution was a 'useful hypothesis', and that neurochemical investigations in the future were warranted. Innovative yet again, the Slaters used a form of factor analysis, principle components analysis and tetrachoric correlation coefficients in their work. (*See* Heritability Wars.)

Postwar and the First Golden Era of Psychiatric Genetics

The classic twin study of psychotic and neurotic illnesses in twins was restarted in 1947, when Slater had the good fortune to take on James Shields (1918–1978) as his Research Associate. Shields followed up the prewar sample and extended the study to new cases. Speculations about modes of inheritance for any of the disorders studied are conspicuous by their absence; but best lifetime estimates of diagnoses for this unselected sample were made nevertheless. Conclusions were typically modest, but it was clear that the future of the psychiatric genetics of psychoses, neuroses and personality disorders would advance. (*See* Twin Studies; Twin Studies: Software and Algorithms.)

In a masterful but short paper (1958), Slater put forward a monogenic theory of schizophrenia that reads as though it had been written in the 1990s. The Medical Research Council recognized his potential, and Slater was appointed Director of its first Psychiatric Genetics Unit at the Institute of Psychiatry in 1959. Slater took up duties as Editor-in-Chief of the *British Journal of Psychiatry* (*Br J Psychiatry*, 1961–1972; known as the *Journal of Mental Science* before 1963), the importance of which to the history of psychiatric genetics cannot be overlooked. Through his reputation, and mentorship of younger scientists, many papers were attracted to the *Br J Psychiatry*.

Slater was concerned about the issues surrounding etiological heterogeneity. Nongenetic, schizophrenia-like and schizophreniform psychoses attracted his attention in the hope that their identification would leave the field clearer for the identification of the genetic contributors to schizophrenia. In many ways his entire research career involving psychiatric genetics and its underpinnings in psychology and mathematics culminated in the first English language textbook on the genetics of mental disorders. Slater and Cowie's (1971) *The Genetics of Mental Disorders* stands as a launching pad for the modern era of psychiatric genetics.

After Slater's 'retirement' at the age of 65, in 1969, he kept up with many of his scientific interests and managed to earn a PhD in English Literature at Cambridge, UK, in his 76th year. He chose to explore the statistical aspects of word usage in the anonymous sixteenth-century play *King Edward III* and succeeded in attributing authorship to Shakespeare.

See also
Nazi Movement and Eugenics

Further Reading

Proctor RN (1988) *Racial Hygiene – Medicine under the Nazis*. Cambridge, MA: Cambridge University Press.

Roth M and Cowie V (eds.) (1979) *Psychiatry, Genetics and Pathography, A Tribute to Eliot Slater*. London, UK: Gaskell Press.

Shields J and Gottesman II (eds.) (1971) *Man, Mind and Heredity – Selected Papers of Eliot Slater on Psychiatry and Genetics*. Baltimore, MD: Johns Hopkins University Press.

Slater E and Cowie V (1971) *The Genetics of Mental Disorders*. London, UK: Oxford University Press.

Slater E and Shields J (1953) *Psychotic and Neurotic Illnesses in Twins*. Medical Research Council Special Report Series no. 278. London, UK: HMSO.

Smith–Waterman Algorithm

Richard Mott, *University of Oxford, Oxford, UK*

Advanced article

Article contents

- Introduction
- Background and History
- Algorithm
- Extensions and Applications
- Strategies

The Smith–Waterman algorithm is a computer algorithm that finds regions of local similarity between DNA or protein sequences.

Introduction

A central task in bioinformatics is the alignment of pairs of deoxyribonucleic acid (DNA) or protein sequences in order to determine whether they are descended from a common ancestor. Because evolution may have caused the substitution, insertion and deletion of nucleotides or amino acids, the biologically correct sequence alignment – that is, that which reconstructs the true evolutionary history – will match conserved regions, while placing mismatches and gaps wherever mutations have occurred. In general, the alignment will not run from the start to the end of both sequences, but instead will comprise a local similarity. The Smith–Waterman algorithm is used to find these optimal local alignments automatically. The algorithm is important not only because it solves a core task elegantly and efficiently, but also because many other bioinformatics problems, such as predicting genes in genomic DNA, can be solved using a generalization of it, the hidden Markov model (HMM), and because of the impetus it has given to the study of sequence alignment statistics. (*See* Sequence Alignment.)

Background and History

The Smith–Waterman algorithm belongs to a family of dynamic-programming methods that have been repeatedly rediscovered in molecular biology, engineering, computer science, linguistics and statistics. Needleman and Wunsch (1970) introduced dynamic programming to molecular biology in 1970, when they showed how to align two protein sequences from end to end (i.e. perform a global alignment). It was soon realized that a local alignment algorithm was required, because the constraint that the sequences' ends match often produced alignments that were biologically incorrect. Throughout the 1970s, various solutions were proposed, of which Sellers' metric method (Sellers, 1974) is probably the best known. However, all these algorithms are somewhat impracticable. (*See* Dynamic Programming; Global Alignment.)

In 1981, Temple Smith and Michael Waterman (Smith and Waterman, 1981) published an algorithm that finds optimal local sequence alignments. With a refinement of it proposed by Gotoh (1982), this algorithm has become a cornerstone of biological sequence analysis.

Algorithm

The inputs to the algorithm are the DNA or protein sequences to be aligned and a scoring scheme, which determines how mismatches and gaps are to be scored relative to matches. The output is an optimal local alignment that maximizes the alignment score. The choice of scoring scheme strongly affects the optimal alignment, and much research has been devoted to finding scoring schemes that can discriminate between genuine and random similarities. This subject is closely related to the issue of alignment statistical significance. (*See* Alignment: Statistical Significance.)

The scoring scheme comprises:

1. A gap penalty, in which the cost of inserting a gap of k consecutive symbols is some function $g(k)$. The affine gap penalty $g(k) = A + Bk$ is most common, where there is a high cost $A + B$ to start a gap, and a lower marginal cost B to extend it: gaps tend to occur as single evolutionary events rather than to accumulate from many adjacent mutations. Gap penalties that increase more slowly with (or are independent of) gap length are suitable in situations where very long gaps may be expected, for example, the alignment of spliced DNA to genomic DNA, where the presence of very long introns may be suspected.

2. A substitution matrix $\mathbf{M}(a, b)$ for aligning symbols a, b that is positive when the symbols are likely to be related, and negative otherwise. It is a requirement that the average symbol match score should be negative. For DNA sequences, $\mathbf{M}(a, b)$ is usually positive if $a = b$, and negative otherwise. Protein substitution matrices are more complicated, being derived from substitution frequencies observed in sequence alignments that have been validated from protein structure data. In general, amino acids are likely to have a positive substitution score if they have similar chemical properties (e.g. are hydrophobic), because such substitutions tend to preserve protein structure and function, and are therefore favored during evolution. Two commonly used families of matrices are the BLOSUM and PAM series. (*See* Substitution Matrices.)

Every possible alignment of part of one sequence with part of the other can be scored using the substitution matrix and gap penalty. For example, a plausible local alignment of the DNA sequences TTACCGGCCAACT-AA, ACCGTGTCACTAAC is shown in **Figure 1**, which contains nine matches, one mismatch, and two gaps each of length 1. If matches score +2, mismatches −3 and gaps $3 + k$, then the score of this alignment is $9 * 2 - 1 * 3 - 2 * 4 = 7$. Note that this is a local alignment, because the prefix 'tt' has been discarded from the first sequence without penalty, along the suffixes 'a', 'c'. The corresponding global alignment, with score −1, is shown in **Figure 2**.

The objective of the Smith–Waterman algorithm is to find the local alignment(s) with the highest score. In

```
ttACCG-GCCAACTAa
  |||| |*|| |||
  ACCGTGTCA-CTAc
```

Figure 1 Local alignment of the DNA sequences TTACCGGCCAACTAA, ACCGTGTCACTAAC. Aligned portions are shown in upper case and unaligned portions of the sequences are shown in lower case.

```
TTACCG-GCCAACTAA
|||| |*|| |||*
--ACCGTGTCA-CTAC
```

Figure 2 Global alignment for the example shown in **Figure 1**.

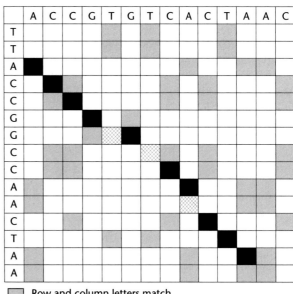

	A	C	C	G	T	G	T	C	A	C	T	A	A	C
T														
T														
A														
C														
C														
G														
G														
C														
C														
A														
A														
C														
T														
A														
A														

☐ Row and column letters match
■ Match
▨ Mismatch or gap

Figure 3 Dot matrix for the optimal local alignment between sequences, showing cells in which the corresponding row and column letters match; cells lying on the optimal alignment that match; and where the path through the cell is a mismatch or gap.

this simple example, it is clear by inspection that we have found a best local alignment for this scoring scheme, but for sequences with thousands of nucleotides or hundreds of amino acids it may no longer be obvious what the optimum is. Furthermore, as the number of local alignments grows exponentially with the sequence lengths, it is clearly impossible to find the optimum by enumerating and scoring all alignments.

Instead, a dynamic-programming algorithm is used to arrange the calculation so that the number of steps required is proportional to the product of the sequence lengths. This is a huge reduction in effort and means that it is possible to search protein databanks in a few minutes on a desktop computer.

Denote the two sequences to be aligned by the strings of symbols $\mathbf{a} = a_1 a_2 a_3 \ldots a_n$, $\mathbf{b} = b_1 b_2 b_3 \ldots b_m$. These symbols will be either the nucleotides A, C, G, T or from the 20-letter amino acid alphabet. Imagine an array of n rows and m columns, with sequence $\mathbf{a}$ arranged vertically down the left-hand side and sequence $\mathbf{b}$ horizontally along the top. There are two key ideas behind dynamic programming. First, every alignment corresponds to a path in this matrix. Where the alignment contains a match or mismatch between a_i and b_j, the path takes a diagonal route through the cell i,j. A gap inserted in $\mathbf{a}$ between a_i and a_{i+1} opposite $b_j \ldots b_k$ corresponds to a vertical path from cell i,j to i,k. Similarly, a gap inserted in $\mathbf{b}$ is represented by a horizontal path. Second, the optimal alignment terminating in the cell i,j must be built from one of the paths passing through the three preceding cells, diagonally $i-1,j-1$, horizontally $i-1,j$ or vertically $i,j-1$. Let $S_{i,j}$ be the score of the optimal alignment terminating at i,j. Let $H_{i,j}$ be the score of the top-scoring alignment to i,j entering the cell horizontally, and $V_{i,j}$ be the best alignment entering vertically. These quantities may be calculated as follows:

$$S_{i,j} = \max\left[0, S_{i-1,j-1} + M(a_i, b_j), H_{i,j}, V_{i,j}\right]$$
$$H_{i,j} = \max_{k<i}\left[0, S_{i-k,j} + g(k)\right]$$
$$V_{i,j} = \max_{k<j}\left[0, S_{i,j-k} + g(k)\right]$$

These are called dynamic-programming recurrence equations. The matrices **S**, **H**, **V** are computed cell by cell, starting from cell 1,1 and ending at m, n, usually row by row.

The score of optimal local alignment can be found by keeping a record of that cell with the overall maximum score. There may be more than one such

cell, implying that there are multiple optimal alignments. The alignment itself can be recovered from its path through the matrix, by keeping a record of each cell's predecessor in the recursion, and then backtracking through the matrix until an edge of the matrix or cell with zero score is reached (see **Figure 3**).

Smith and Waterman's insight was to put the zeros in the above equations; these have the effect of trimming alignments with a negative score back to their positive cores, and ensure that the algorithm does find optimal local alignments. Smith and Waterman define an alignment to be locally optimal if it cannot be extended or shrunk without reducing its score. This is equivalent to saying that a path is locally optimal if no subpath of it has a higher score and if it is not a subpath of a higher-scoring path. To see why the algorithm works, suppose there is an optimal local alignment from cells i,j to k,l. This means that there cannot be a path ending at i,j with a positive score, for if there were, then the alignment could be extended backwards. Similarly, if the alignment could be extended to the right, then k,l would no longer be the global maximum in the matrix, for another cell would exceed it. Finally, if the alignment could be shrunk, then either a prefix or a suffix of the alignment must have a negative score. Now the effect of the zeros in the recursion is to automatically trim away paths with negative score, so, combined with the global maximum rule, we see that the algorithm will find the optimal local alignment.

Gotoh's contribution was to note that if the gap penalty is affine, then the recurrences can be recast in a more efficient form, which reduces the time taken by the algorithm from a quantity proportional to $mn(m + n)$ to mn, i.e. a 100- to 1000-fold saving in a typical sequence comparison:

$$S_{i,j} = \max[0, S_{i-1,j-1} + M(a_i, b_j), H_{i,j}, V_{i,j}]$$

$$H_{i,j} = \max[0, H_{i-1,j} + b, S_{i-1,j} + a + b]$$

$$V_{i,j} = \max[0, V_{i,j-1} + b, S_{i,j-1} + a + b]$$

This is a considerable saving in computer time, and renders the algorithm practical for searching protein databanks. It is also easy to implement on specialist computer hardware which can perform many operations in parallel. **Table 1** lists a number of implementations of the Smith–Waterman algorithm.

Extensions and Applications

There are now many variations on the basic Smith–Waterman algorithm. These include the use of specialized gap penalties for aligning spliced DNA (for instance, a complementary DNA, cDNA) onto genomic DNA, to identify gene structure (Mott, 1997; Florea et al., 1998). Other variants are for aligning a protein sequence to a DNA sequence, allowing for the possibility that the translated reading frame jumps (e.g. due to a sequencing error or the presence of an intron). Algorithms for finding local sequence

Table 1 Implementations of the Smith–Waterman and related algorithms

Program	Web address	Comments
Standard implementations, mainly for protein–protein comparisons		
SSEARCH	ftp://ftp.virginia.edu/pub/fasta	FTP server; part of FASTA. Assesses statistical significance by function-fitting
MPSRCH	http://www.anedabio.com/products/mpsrch/index.html	Fast standard implementation
ARIADNE	http://www.well.ox.ac.uk/rmott/ARIADNE/ariadne.html	Assesses protein statistical significance by a formula; ideal for self-comparisons
Fast implementations using heuristics to accelerate databank searches		
BLAST	http://www.ncbi.nlm.nih.gov/BLAST/	Standard database search package. Family of related programs for DNA and protein searches. Statistical significance assessed by formulas
FASTA	http://www.people.virginia.edu/~wrp/pearson.html	Comprehensive, popular database search software; includes programs for DNA and protein comparisons
CrossMatch	http://www.genome.washington.edu/UWGC/analysistools/Swat.cfm	Optimized for all-versus-all comparisons
Profile-based comparisons		
HMMR	http://hmmer.wustl.edu/	Use hidden Markov models to identify protein similarities with multiple alignments
SAM	http://www.cse.ucsc.edu/research/compbio/sam.html	
ARIADNE	http://www.well.ox.ac.uk/ariadne	
Specialized applications – gene annotation		
EMBOSS: est2genome	http://www.uk.embnet.org/Software/EMBOSS/Apps/est2genome.html	All align a spliced DNA to genomic DNA, skipping long introns if necessary
SIM4	http://globin.cse.psu.edu/	
GenSeqer	http://bioinformatics.iastate.edu/cgi-bin/gs.cgi	
Dynamite	http://www.sanger.ac.uk/Software/Dynamite/	Creates customized dynamic-programming algorithms for specialized applications
Wise2	http://www.sanger.ac.uk/Software/Wise2/	Dynamic-programming algorithms, including DNA–protein comparisons

alignments with nonaffine gap penalties are described by Mott (1999). Recently Arslan *et al.* (2001) have shown how to identify local alignments with maximal per cent homology, rather than maximum score. Several systems such as Dynamite have been developed to help create a dynamic-programming code for novel problems. (*See* Expressed-sequence Tag (EST).)

The most widely used development has been profile alignment. A profile, also known as a position-specific scoring matrix (PSSM), or a profile HMM (Durbin *et al.*, 1999), is a way to represent the variation in sequence conservation typically observed across a multiple alignment of several related sequences. Often certain positions in the multiple alignment are tightly conserved, indicating active sites or other key functional or structural domains, while other parts are highly variable, perhaps corresponding to loops in the protein structure. In the profile, the substitution matrix differs at each position to reflect the level of conservation. Profiles are accommodated very easily in the Smith–Waterman framework. (*See* BLAST Algorithm; Hidden Markov Models; Profile Searching.)

Profiles are generally more sensitive at detecting distantly related sequences than are ordinary sequence comparisons, and this insight has led to the creation of databases of protein multiple alignments, corresponding to conserved protein domains, and their associated profiles (Pfam; SMART). Indeed, over 60% of human genes contain at least one known protein domain.

Strategies

It is generally inefficient to find gene homologies from direct DNA to DNA interspecies sequence comparisons. The redundancy in the genetic code means that even quite closely related genes may exhibit only 60% identity at the DNA sequence level, whereas protein comparisons, which take advantage of substitution matrices that model likely amino acid replacements, are much more sensitive. When a novel gene is discovered, either by *de novo* gene prediction algorithms applied to genomic DNA, or from sequencing a cDNA library, a better characterization is usually obtained by performing the following order of searches:

1. Compare the translated sequence to a library of annotated protein domains such as Pfam or SMART, using profiles or HMMs, or a frameshift-tolerant DNA–protein comparator such as GENEWISE.
2. Perform a PSI-BLAST search of its predicted protein product against a protein sequence database such as NR or SPTREMBL.
3. Perform a Smith–Waterman search using FASTA or BLASTP.

4. Perform a DNA to DNA BLASTN search of genomic DNA, filtered to remove low-complexity sequence and known repeats.

However, DNA–DNA comparison is very useful for discovering conserved short sequences that may have a gene regulation function, e.g. a promotor or transcription-factor binding site. In this context, large regions (tens or hundreds of kilobase pairs long) of genomic DNA from, for instance, human and mouse, and containing genes known to be syntenic are compared. This is an active area of research, and a number of hybrid methods that combine the Smith–Waterman algorithm with fast, dictionary-based matching algorithms have been proposed (Batzoglou *et al.*, 2000). (*See* Comparative Genomics; FASTA Algorithm; Homologous, Orthologous and Paralogous Genes; Promoters.)

During the first decade of the twenty-first century, the number of fully sequenced and annotated genomes will increase dramatically. Almost every gene will become fully characterized in terms of its homologs and paralogs and domain structure. We may find that the central role sequence comparison has played in discovering these relationships is subsumed into databases of precomputed links, and the emphasis is shifted to a synthesis of complementary data such as domain occurrence, metabolic pathways and expression data. (*See* Genetic Databases: Mining.)

See also
Dynamic Programming
Global Alignment
Hidden Markov Models
Sequence Alignment
Substitution Matrices

References

Arslan AN, Egecioglu O and Pevzner PA (2001) A new approach to sequence comparison: normalized sequence alignment. *Bioinformatics* **17**: 327–337.

Batzoglou S, Pachter L, Mesirov JP, Berger B and Lander ES (2000) Human and mouse gene structure: comparative analysis and application to exon prediction. *Genome Research* **10**: 950–958.

Durbin R, Krogh A, Michison G and Eddy S (1999) *Biological Sequence Analysis: Probabilistic Models of Proteins and Nucleic Acids.* Cambridge, UK: Cambridge University Press.

Florea L, Hartzell G, Zhang Z, Rubin GM and Miller W (1998) A computer program for aligning a cDNA sequence with a genomic DNA sequence. *Genome Research* **8**: 967–974.

Gotoh O (1982) An improved algorithm for matching biological sequences. *Journal of Molecular Biology* **162**: 705–708.

Mott R (1997) EST_GENOME: a program to align spliced DNA sequences to unspliced genomic DNA. *Computer Applications in the Biosciences: CABIOS* **13**: 477–478.

Mott R (1999) Local sequence alignments with monotonic gap penalties. *Bioinformatics* **15**: 455–462.

Needleman SB and Wunsch CD (1970) A general method applicable to the search for similarities in the amino acid sequences of two proteins. *Journal of Molecular Biology* **48**: 444–453.

Sellers P (1974) An algorithm for the distance between two finite sequences. *Combinatorial Theory* **16**: 253–258.

Smith TF and Waterman MSW (1981) Identification of common molecular subsequences. *Journal of Molecular Biology* **147**: 195–197.

Further Reading

Waterman MS (1995) *Introduction to Computational Biology Maps, Sequences and Genomes.* Boca Raton, FL: CRC Press.

Web Links

Pfam (protein families database of alignments and HMMs). Updated May 2002
http://www.sanger.ac.uk/Pfam

SMART (simple modular architecture research tool). Updated May 2002
http://smart.embl-heidelberg.de

snoRNPs

Vanda Pogačić , *Friedrich Miescher Institut, Basel, Switzerland*

Witold Filipowicz, *Friedrich Miescher Institut, Basel, Switzerland*

Advanced article

Article contents

- Introduction
- snoRNAs Guiding RNA Modifications
- snoRNAs Involved in rRNA Processing
- Proteins Associated with snoRNAs
- Biogenesis of snoRNAs

snoRNPs are nucleolus-localized ribonucleoprotein particles that comprise small RNAs (snoRNAs) and a few associated proteins. Known snoRNPs can be subdivided into two major classes, C/D and H/ACA, which direct site-specific 2′-O-ribose methylation and pseudouridylation of ribosomal RNA (rRNA) respectively. Some snoRNPs participate in precursor rRNA processing.

Introduction

The nucleoli of eukaryotic cells contain a large population of ribonucleic acid (RNA) molecules of 70–300 nucleotides (nt) called small nucleolar RNAs (snoRNAs). The snoRNAs are involved in the processing and modification of precursor ribosomal RNAs (pre-rRNAs) and in the modification of small spliceosomal RNAs (U-snRNAs). They can be subdivided into two main families, C/D and H/ACA, on the basis of conserved sequence motifs (**Figure 1**). Both snoRNA families function as ribonucleoproteins

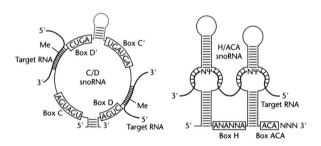

Figure 1 Structure of guide snoRNAs and their interaction with target RNAs. The secondary structures of the C/D-box and H/ACA-box snoRNAs are shown, with the conserved sequence elements and base-paired target RNAs indicated. Most snoRNAs contain one rather than two functional domains. 2′-O-ribose methylation is carried out on the RNA residue that is base-paired with the fifth position upstream from box D (or D′). Pseudouridine (ψ) formation in RNA occurs on the unpaired uridine residue upstream from the H box or the ACA box (usually 14–16 nt from the box). snoRNAs of each class are associated with a set of specific proteins that are not shown.

(snoRNPs) in association with different sets of specific proteins. Most of the C/D-box and H/ACA-box snoRNAs act as guides that direct site-specific 2′-O-ribose methylation and isomerization of uridine to pseudouridine (ψ) respectively; some snoRNAs are also required for different processing steps of pre-rRNA. RNase MRP, an evolutionarily conserved nucleolar RNP, belongs to neither the C/D nor the H/ACA class. It is structurally similar to RNase P and is involved in the endonucleolytic cleavage of pre-rRNA in the yeast *Saccharomyces cerevisiae* and probably also in vertebrates.

Members of the C/D-box family contain short sequence elements, UGAUGA (box C) and CUGA (box D), that are generally located a few nucleotides away from the RNA 5′ and 3′ termini, respectively. The snoRNA ends are usually base-paired, which brings the C and D motifs into close proximity. Many C/D snoRNAs contain additional C- and D-like motifs, termed C′ and D′ boxes, which are located in the central region of the snoRNA molecule. The secondary structure of the C′–D′ region resembles that of the C–D region. The C and D motifs and the terminal stem structure are important for maturation, stability, nucleolar localization and also for the function of snoRNAs.

The H/ACA-box snoRNAs fold into a hairpin–hinge–hairpin–tail structure with the consensus sequence ANANNA (box H, where N is any nucleotide) located in the hinge region and the ACA trinucleotide (box ACA) positioned 3 nt upstream from the 3′ end. The hairpins are generally interrupted by an internal loop, which is important for selecting the modification site. The H and ACA boxes are essential for the processing, accumulation and function of snoRNAs (Bachellerie *et al.*, 2000; Tollervey and Kiss, 1997; Weinstein and Steitz, 1999).

snoRNAs Guiding RNA Modifications

Most known snoRNAs are involved in 2′-*O*-methylation and pseudouridylation of rRNA. Over 100 nucleotides in the rRNA of mammals are methylated on a ribose and nearly 100 are converted to ψ. Because a single snoRNA can guide one or at the most two modifications, 50–100 different snoRNAs of each class are required to specify modifications in rRNA. To date, snoRNAs (or genes encoding snoRNAs) for 91 rRNA ribose methylations and 42 ψ conversions have been identified in mammals.

The C/D-box guide snoRNAs contain sequences of 10–21 nt – positioned immediately upstream of the D or D′ motifs – that are complementary to sequences in the target RNA. The position complementary to the fifth nucleotide from box D (or D′) undergoes 2′-*O*-methylation. In H/ACA-box snoRNAs, short (4–8 nt) sequences complementary to target RNAs are located in single-stranded internal loops present in either one or both long snoRNA hairpins. On hybridization, the complementary sequences form two short helices, which flank the unpaired uridine residue to be isomerized to ψ. This uridine is always positioned 13–15 nt from box H or box ACA (Bachellerie *et al.*, 2000; Tollervey and Kiss, 1997; Weinstein and Steitz, 1999).

Guide snoRNAs can also function in the modification of cellular RNAs other than rRNA. Ribose methylation of several different nucleotides in vertebrate U2, U4 and U6 snRNAs either has been shown or is predicted to be directed by typical C/D-box snoRNAs, and five ψ isomerizations within U2 and U6 snRNAs are probably guided by H/ACA-box snoRNAs (Hüttenhofer *et al.*, 2001; Tycowski *et al.*, 1998). Notably, modification of two adjacent nucleotides in the human U5 snRNA by 2′-*O*-methylation (a cytidine at position 45) and pseudouridylation (a uridine at position 46) is guided by an unusual C/D–H/ACA-box 'hybrid' snoRNA (Jady and Kiss, 2001). This evolutionarily conserved RNA contains the box elements of both classes of snoRNAs and also associates with proteins that are characteristic of each class of snoRNPs.

Many 'orphan' guide-like snoRNAs that show no complementarity to rRNAs and U-snRNAs have been identified (Hüttenhofer *et al.*, 2001). This suggests that other classes of cellular RNA, possibly messenger RNA (mRNA), are also modified by snoRNAs. Some of the orphan C/D and H/ACA-box snoRNAs

are expressed specifically in human and mouse brain. Expression of three of these snoRNAs, all belonging to the C/D subfamily, is paternally imprinted and their genes map to chromosomal regions implicated in the neurogenetic disease Prader–Willi syndrome (Cavaillé *et al.*, 2000). The functions of the imprinted snoRNAs and their relationship, if any, to this disease remain to be established. (*See* Prader–Willi Syndrome and Angelman Syndrome.)

snoRNAs Involved in rRNA Processing

In eukaryotes, 18S, 5.8S and 25–28S rRNAs are cotranscribed into long 35–45S pre-rRNAs. The above-mentioned modifications of rRNA nucleotides, which are specified by guide snoRNAs, are made on pre-rRNAs. After this, the pre-rRNA is processed through a series of complex cleavage and trimming steps that release mature rRNAs. Some of the processing steps require the action of a few snoRNAs, which are referred to as 'processing snoRNAs'. The C/D-box U3 and U14 snoRNAs participate in 18S rRNA processing in vertebrates and yeast, whereas U13 and U22 function in 18S rRNA biogenesis only in vertebrates. The vertebrate C/D-box U8 snoRNA is involved in processing the 5.8S and 28S rRNAs. The vertebrate H/ACA-box snoRNAs U17 and E3 and yeast snoRNAs snR10 and snR30 participate in 18S rRNA synthesis. RNase MRP is involved in the cleavage upstream from 5.8S rRNA. The yeast processing snoRNAs U3, U14, snR10 and snR30 are all essential for growth.

The best-characterized factor to date is the U3 snoRNP. U3 snoRNA base pairs with pre-rRNA in regions upstream of and within the 5′ part of 18S rRNA, and is required for cleavage at sites upstream and downstream of the 18S rRNA region. In yeast, snoRNAs U14, snR10 and snR30, and also many protein *trans*-acting factors, are similarly required for processing at sites flanking the 18S region. This indicates that snoRNPs and additional protein factors all contribute to the formation of the processing complex, by altering the structure of either the pre-mRNA or the nascent pre-ribosomal particle. In addition to U3, processing reactions involving also U8, U13 and U14 snoRNAs require base-pairing interactions between snoRNAs and pre-rRNA (Tollervey and Kiss, 1997; Venema and Tollervey, 1999). (*See* RNA Processing.)

Proteins Associated with snoRNAs

Each class of snoRNA is associated with a distinct set of nucleolar proteins. These proteins have been most extensively studied in *S. cerevisiae*, but the available evidence indicates that all yeast proteins have orthologs in mammals and most probably in other eukaryotes. Four proteins common to all C/D-box snoRNPs have been characterized: fibrillarin (Nop1p in yeast), Nop58 (Nop5p), Nop56p and 15.5 kDa/ Snu13p. In yeast, all these proteins are essential for growth and their depletion or mutation interferes with pre-rRNA processing and 2′-O-methylation. Nop1p/ fibrillarin has primary sequence homology to methyl transferases and may represent an enzyme that catalyzes the methylation reaction. Nop56p and Nop58p are structurally related proteins but their function has not yet been characterized. The 15.5 kDa/ Snu13p, which interacts with the C/D-box motif, is unique in being also a component of the spliceosomal U4/U6·U5 tri-snoRNP. Its binding site in the U4 snRNA is structurally related to the C/D-box motif (Watkins *et al.*, 2000; Weinstein and Steitz, 1999). (*See* Spliceosome; Splicing of pre-mRNA.)

The H/ACA-box snoRNAs also seem to be associated with only four common proteins, in both *S. cerevisiae* and humans (human/mouse nomenclature is given in parentheses): Gar1p (hGAR1), Cbf5p (dyskerin/NAP57), Nhp2p (NHP2) and Nop10p (NOP10) (Pogačić *et al.*, 2000, and references therein). All four proteins are essential for growth in yeast and their depletion inhibits the production of 18S rRNA and causes the pseudouridylation of rRNA. Depletion of Cbf5p, Nhp2p and Nop10p (but not Gar1p) affects the accumulation of H/ACA snoRNAs. Nhp2p contains an RNA-binding domain that is also found in some spliceosomal and ribosomal proteins. Gar1p also shows some RNA-binding potential. Cbf5p has sequence similarity to ψ-synthases that are involved in tRNA modification and thus may represent the ψ-synthase of H/ACA snoRNPs.

Mutations in the gene encoding dyskerin (*dyskeratosis congenita 1*; *DKC1*), the human ortholog of Cbf5p, cause dyskeratosis congenita – a chromosome-X-linked recessive disease that is associated with bone marrow failure, skin defects, chromosome instability and a predisposition to certain types of malignancy. Dyskeratosis congenita is not apparently due to the inhibition of rRNA processing and/or pseudouridylation, but results from the decreased activity of telomerase – an RNP enzyme that is responsible for maintaining chromosome ends. In vertebrates, the telomerase RNA (hTR) contains a 3′ terminal domain that structurally resembles H/ACA snoRNAs, and all four human H/ACA-box snoRNP proteins, including dyskerin, are associated with hTR (Mitchell *et al.*, 1999; Pogačić *et al.*, 2000, and references therein). The telomerase H/ACA domain is required for accumulation of hTR and an *in vivo* function of the enzyme. (*See* RNA Processing and Human Disorders; Telomerase: Structure and Function; Telomeres and Telomerase in Aging and Cancer; Telomeres: Protection and Maintenance.)

At present, there is no evidence that guide snoRNPs contain proteins other than those discussed above. But processing snoRNPs are associated with more proteins. The best-studied particle, U3 snoRNP, contains at least one additional structural WD-40 repeat protein, referred to as Rrp9p in yeast and U3-55k in humans. Like other core snoRNP proteins, Rrp9p is essential for growth. Immunoprecipitation experiments revealed seven additional essential proteins, among them the RNA-helicase-like protein Dhr1p and the RNA 3′-phosphate cyclase-like protein Rcl1p. These proteins may associate with U3 snoRNP only transiently to assist this particle in performing its function during 18S rRNA maturation. RNase MRP in yeast contains nine protein components, and counterparts of most of them are present in the human enzyme. One of the proteins, Snm1p, is specific for RNase MRP; others are shared with RNase P (Van Eenennaam *et al.*, 2000).

Biogenesis of snoRNAs

Biogenesis of snoRNAs in different organisms follows some unusual pathways. In vertebrates, all guide snoRNAs are encoded in introns of genes transcribed by RNA polymerase II. Most of the snoRNA host genes encode proteins that are essential for ribosome biogenesis or function. Notably, some of the hosts do not code for proteins and it is their snoRNA-bearing introns and not exons that are evolutionarily conserved. Intron-encoded snoRNAs are generally processed from spliced-out and debranched introns by exonucleonucleases that trim an excess of the RNA from both ends.

Some of the imprinted genes that encode brain-specific snoRNAs consist of up to 50 tandem repeats, each containing a single snoRNA unit. The genes seem to produce very long transcripts, with snoRNAs processed from the excised introns by the exonucleolytic mechanism. Brain-specific snoRNAs that are expressed from tandemly arranged genes are much more abundant than other guide RNAs. Most vertebrate processing snoRNAs, such as U3 and RNase MRP RNA, are transcribed from independent genes by either RNA polymerase II (U3 RNA) or III (MRP RNA). In plants and yeast, most snoRNAs are transcribed from independent genes as either mono- or polycistronic units. Processing from the polycistronic transcripts is catalyzed by RNase-III-like enzymes and exonucleases. Initial packaging of snoRNAs into RNPs probably takes place in the nucleoplasm, but the final assembly and processing steps may occur in Cajal (coiled) bodies and/or the nucleolus (Bachellerie *et al.*, 2000; Weinstein and Steitz, 1999). (*See* Nucleolus: Structure and Function; RNA Polymerases and the Eukaryotic Transcription Machinery.)

See also

Gas5 Gene
Nonprotein-coding Genes
Nucleolus: Structure and Function
RNA Processing

References

Bachellerie JP, Cavaillé J and Qu LH (2000) Nucleotide modifications of eukaryotic rRNAs: the world of small nucleolar RNA guides revisited. In: Garrett RA, Douthwaite SR, Liljas A, *et al.* (eds.) *The Ribosome: Structure, Function, Antibiotics, and Cellular Interactions*, pp. 191–203. Washington DC: ASM Press.

Cavaillé J, Buiting K, Kiefmann M, *et al.* (2000) From the cover: identification of brain-specific and imprinted small nucleolar RNA genes exhibiting an unusual genomic organization. *Proceedings of the National Academy of Sciences of the United States of America* **97**: 14311–14316.

Hüttenhofer A, Kiefmann M, Meier-Ewert S, *et al.* (2001) RNomics: an experimental approach that identifies 201 candidates for novel, small, non-messenger RNAs in mouse. *EMBO Journal* **20**: 2943–2953.

Jady BE and Kiss T (2001) A small nucleolar guide RNA functions both in 2′-*O*-ribose methylation and pseudouridylation of the U5 spliceosomal RNA. *EMBO Journal* **20**: 541–551.

Mitchell JR, Wood E and Collins K (1999) A telomerase component is defective in the human disease dyskeratosis congenita. *Nature* **402**: 551–555.

Pogačić V, Dragon F and Filipowicz W (2000) Human H/ACA small nucleolar RNPs and telomerase share evolutionarily conserved proteins NHP2 and NOP10. *Molecular and Cellular Biology* **20**: 9028–9040.

Tollervey D and Kiss T (1997) Function and synthesis of small nucleolar RNAs. *Current Opinion in Cell Biology* **9**: 337–342.

Tycowski KT, You ZH, Graham PJ and Steitz JA (1998) Modification of U6 spliceosomal RNA is guided by other small RNAs. *Molecular Cell* **2**: 629–638.

Van Eenennaam H, Jarrous N, Van Venrooij W and Pruijn GJM (2000) Architecture and function of the human endonucleases RNase P and RNase MRP. *IUBMB Life* **49**: 1–8.

Venema J and Tollervey D (1999) Ribosome synthesis in *Saccharomyces cerevisiae*. *Annual Review of Genetics* **33**: 261–311.

Watkins NJ, Segault V, Charpentier B, *et al.* (2000) A common core RNP structure shared between the small nucleolar box C/D RNPs and the spliceosomal U4 snRNP. *Cell* **103**: 457–466.

Weinstein LB and Steitz JA (1999) Guided tours: from precursor snoRNA to functional snoRNP. *Current Opinion in Cell Biology* **11**: 378–384.

Further Reading

Eddy SR (2001) Non-coding RNA genes and the modern RNA world. *Nature Reviews Genetics* **2**: 919–929.

Kiss T (2001) Small nucleolar RNA-guided post-transcriptional modification of cellular RNAs. *EMBO Journal* **20**: 3617–3622.

Lafontaine DL and Tollervey D (1998) Birth of the snoRNPs: the evolution of the modification-guide snoRNAs. *Trends in Biochemical Sciences* **23**: 383–388.

Web Links

Dyskeratosis congenita 1 (*DKC1*); LocusID: 1736. LocusLink: http://www.ncbi.nlm.nih.gov/LocusLink/LocRpt.cgi?l=1736

Dyskeratosis congenita 1 (*DKC1*); MIM number: 300126. OMIM: http://www.ncbi.nlm.nih.gov/htbin-post/Omim/dispmim?300126

SNP

See Single Nucleotide Polymorphism (SNP)

Sociobiology, Evolutionary Psychology, and Genetics

Patrick Bateson, *University of Cambridge, Cambridge, UK*

Advanced article
Article contents
• Rise of Sociobiology
• Rise of Evolutionary Psychology
• Role of Genes
• Conclusion

Much of human behavior, including violence and those activities thought to be characteristic of each sex, has been attributed to the heritage of human evolution. These ideas have stimulated much discussion but have been criticized for playing down the interplay between the developing individual and the environment. In modern times, its clearest expression has been in the writings of the sociobiologists and the evolutionary psychologists.

Rise of Sociobiology

In the 1960s, field studies relating behavior patterns of animals to the social and ecological conditions in which they normally occur led to the enormous popularity and success of behavioral ecology (Krebs and Davies, 1981). A new subject called 'sociobiology' brought to the study of behavior important concepts and methods from population biology, together with some all-embracing claims of its own. The pivotal moment for the growth of sociobiology was the publication of an important book by E. O. Wilson (1975). Imaginations were captured by the way the ideas from evolutionary biology were used. The appeal of evolutionary theory, in which sociobiology was embedded, was that it seemed once again to make a complicated subject manageable. This was particularly true of the gene-centered writings of Richard Dawkins (e.g. Dawkins, 1976), who provided a crutch for understanding the complex dynamics of evolution by attributing intentions to genes.

Individual animals interact with others, have relationships and collectively form societies. Social behavior often seems to involve cooperation and from Charles Darwin onwards this aspect of behavior had continued to tease the theorists. If evolution depended on competition, how could cooperation have evolved? Three types of explanation were offered. The first benefited enormously from the thinking of Bill Hamilton (e.g. Hamilton, 1996). He pointed out that if two parties are related then the benefits of helping a cousin, say, are logically the same as helping a child – although quantitatively less effective in terms of gene propagation. This idea fostered the Dawkins 'selfish gene' approach.

The second explanation was that both parties would benefit from the act of cooperation. Robert Trivers (1985) coined the term 'reciprocal altruism'. An example of mutualism between species is when a large predator fish opens its mouth to cleaner-fish which pick food remnants from between the teeth of the predator without danger to themselves. The third explanation for cooperation is that the individual is part of a group of unrelated individuals that may survive better as an entity than another group as a result of actions taken by the individuals within it. In the face of gene-centered theories, this explanation was widely thought to be implausible, but it may occur if individuals die out more rapidly than groups and immigration between groups is difficult – a not implausible set of circumstances in human tribal societies. Group-selected behavior may also evolve when the actions of the group cannot be subverted by individual free riders (Bateson, 1988).

The gene-centered approach to behavioral biology brought a new look to studies of communication where well-known instances of supposed transfer of information were reinterpreted in terms of selfish manipulation (Krebs and Dawkins, 1984). It also led to re-examination of apparent shared interests of both parents caring for young or the mutually beneficial interactions of parent and offspring (Trivers, 1985). Most adults of most species are able to produce more than one offspring during their lifetime. The charac-

teristics of those offspring that are most successful in leaving descendants will tend to predominate in subsequent generations. Parents who sacrifice too much for one offspring will have fewer descendants.

By the same token, offspring that do less to ensure their own survival than others will have fewer descendants. Such are the broad rules that shape any life span, but just how they look in detail will depend on the species and, within a species, on local conditions. In some species, parental care for the tiny progeny consists of nothing more than providing a small amount of yolk, enough to sustain the offspring until they can feed for themselves. Marine fish such as herring, for instance, produce vast numbers of eggs and sperm, which fuse in the sea. Neither parent provides any care for their progeny. Most fertilized herring eggs consequently die at an early stage in their lives. Other fish produce far fewer fertilized eggs and care for them in a variety of different ways, some keeping them like a mammal in their bodies until the young are born, others gathering the fertilized eggs into their mouths where they are protected.

Of all animal groups, birds and mammals produce the smallest number of young and take the greatest care of those that they do produce. Birds encase their fertilized eggs in a hard shell and eject them from their bodies, whereupon both sexes usually take turns in keeping the developing egg at body temperature until the chick hatches. While it is developing inside the egg the embryonic chick is fed from the yolk, which at the outset is enormous relative to the embryo. After the egg has hatched, both parents usually protect and bring food to the developing young. This has important consequences for the amount of care that is given by the parent and the time taken to become adult. The long period of development that is particularly characteristic of humans, but also true to a lesser extent of most other mammals, is made possible by the protection and the provisioning by the parent. As they prepare for their own eventual reproductive life, children meanwhile have to survive.

The traditional image of parenthood had been one of complete harmony between the mother and her unborn child. However, evolutionary theory in the hands of the sociobiologists cast doubt on this blissful picture. In sexually reproducing species, parents are not genetically identical to their offspring. Consequently, offspring may require more from parents than parents are prepared to give, creating the possibility of a conflict of long-term interests. Trivers (1985) called this 'parent–offspring conflict', a term that refers strictly to a conflict of reproductive interests, not conflict in the sense of overt squabbling (Bateson, 1994). The parent may sacrifice some of the needs of its current offspring for others that it has yet to produce; the offspring maximizes its own chances of survival.

Parent and offspring 'disagree' about how much the offspring should receive. The result of such evolutionary conflicts of interest is sometimes portrayed as a form of arms race, with escalating fetal manipulation of the mother being opposed by ever more sophisticated maternal countermeasures.

However, limits must be encountered in the course of evolution. If the offspring is too aggressive in its demands it will kill its maternal host and, of course, itself. Likewise, if the mother is too mean, her parasitic offspring will not thrive and she might as well have not bred. Moreover, mutually beneficial communication often occurs between parent and offspring so that independence comes at a time that is beneficial to both parties (Bateson, 1994).

Despite the invigorating debates about the function and evolution of social behavior, grand claims were made for the relevance of sociobiology to human behavior. In the final chapter of his seminal book, Wilson (1975) offered biological explanations for much fought-over areas such as the differences between the sexes, homosexuality, xenophobia and religion. Even more provocatively he predicted that before long the social sciences would be incorporated into biology. The zeal of the hard-selling proponents of sociobiology made them unpopular with other academics who felt threatened or insulted by the attempted takeover. Moreover, some sociobiological views have been known to appeal to people with a strong interest in maintaining their power and their customary privileges. These less savory aspects of sociobiology led fierce debates (Segerstråle, 2000; Laland and Brown, 2002).

The subject was deemed by many to have overreached itself, particularly in its claims about the relevance of evolutionary biology to human social behavior. Nonetheless, interest in the links from evolutionary biology to human behavior persisted and flowered in a number of subdisciplines such as human ethology and human behavioral ecology. Many of these studies focused on observables. A view grew up that many of the most interesting legacies of human evolution lay not so much in the surface features of behavior but in the rules underlying their development. The most striking example of this would be language, which differs dramatically from culture to culture but, it is argued, develops according to rules that acquired their present form in the course of human evolution (Pinker, 1994).

Rise of Evolutionary Psychology

Two of the major figures in this development were Tooby and Cosmides (e.g. Tooby and Cosmides, 1992). Taking a lead from philosophical discussions

of the modularity of mind (Fodor, 1983) and from advances in the design of computer architecture, evolutionary psychologists suggested that neural modules in the brain were adapted for specific functions. Famously, the workings of the brain were likened to a Swiss Army knife in which each tool serves a particular job. Each module was thought to generate a unitary group of instinctive behaviour patterns. While some functions such as speech are well described by this style of thought, the concept is loosely used and its generality has been much criticized (Heyes and Huber, 2000).

The idea of a neural module generating behavior suffers from much the same defects as the concept of instinct itself. Apart from its colloquial use, the term instinct has at least eight scientific meanings: present at birth (or at a particular stage of development), not learned, develops before it can be used, unchanged once developed, shared by all members of the species (or at least of the same sex and age), organized into a distinct behavioral system (such as foraging), served by a distinct neural module, adapted during evolution, and differences between individuals are due to genetic differences (Bateson, 2000).

The problem is that one use does not necessarily imply another even though it is often assumed, without evidence, that it does. Behavior that has been probably shaped by Darwinian evolution and appears ready formed without opportunities for learning may be changed in form and the circumstances of expression by subsequent experience. The smile of the human behavior is a good example (Bateson and Martin, 2000).

Despite the weakness of some of its concepts, evolutionary psychology provided a research strategy based on asking the following six questions:

- What problems did humans have to solve in the ancestral environment?
- What were the likely adaptations to that environment?
- What information processing problems had to be solved in evolving such adaptations?
- What are the design features of such cognitive processes likely to be?
- How do the predictions of such models compare with those of other models of cognitive neuroscience?
- What do the models predict about the behavior of humans living in modern conditions?

This general approach has raised doubts about how the characteristics of the so-called environment of evolutionary adaptedness could be ever ascertained. Even so it has led to some interesting studies. For example, sex differences in behavior could be linked to the different evolutionary pressures on the reproduction of human males and human females. Women were expected to rate highly the characteristics of men that would facilitate the care of their offspring. In one major study, the desirable characteristics of individuals of the opposite sex were investigated in more than 10 000 men and women from 37 cultures around the world. Both men and women rated the capacity for love, dependability and emotional stability highly. But big differences were found between men and women in what characteristics were valued highly in a member of the opposite sex. While men rated youth and physical attractiveness more highly than did women, women rated highly the resources held by a man and all the characteristics associated with acquiring such resources, such as health and intelligence. These sex differences were consistent across all cultures (Buss, 1994). Universality does not imply, however, that if conditions changed, the same sex differences would be found. Darwinian mechanisms of evolution could have operated in the past to produce consistent sex differences so long as the rearing and social enviroments on which the differences might depend were stable from one generation to the next.

Role of Genes

The great majority of biologists would agree that the evolution of life has generally involved changes in the genetic composition of the evolving organisms. A second point of consensus is that, if the change over time involves an adaptation to the environment and the postulated process of Darwinian selection has worked at all, then the genes are likely to influence the characteristics of the organism. This is not to exclude the importance of constant features of the environment. The disagreements start over the particular ways in which genes affect the outcome of an individual's development.

Some sociobiologists thought that a straight forward correspondence could be found between the genes and behavior. Wilson (1976) was quite candid about it when replying to reviewers' criticisms that he was naïve about behavioral development in his great book. He argued that most phenotypic traits of parents and offspring are correlated to some extent by virtue of a higher than average possession of the same genes. Therefore, he went on, the exact structure of the nervous system, the developmental pathways and endocrine–behavioral interactions were modules that could be decoupled from the explanatory scheme. The implication was that the development of the individual is merely a complex process by which genes are decoded.

If genes code for structures or behavior patterns, they must bear a straightforward relationship to them. Usually they do not and the correct way to describe what is observed is to state that a genetic difference between two individuals gives rise to a difference in behavior without anything being said about the mode of development. When details are known about the mode of development, the point of this intellectual discipline becomes obvious. For example, people with Kallmann syndrome associated with a lack of sexual interest may differ from others in only one gene. The Kallmann syndrome is caused by damage at one of several specific genetic loci. Cells that are specialized to produce a chemical messenger called gonadotrophin-releasing hormone (GnRH) are formed initially in the nose region of the fetus. Normally the hormone-producing cells would migrate into the brain. As a result of the genetic defect, however, their surface properties are changed and the cells remain dammed up in the nose subsequently causing anosmia. The activated GnRH cells, not being in the right place, do not deliver their hormone to the pituitary gland at the base of the brain. Without this hormonal stimulation, the pituitary gland does not produce the normal levels of two other chemical messengers, luteinizing hormone and follicle-stimulating hormone. Without these hormones in men, the testes do not produce normal levels of the male hormone testosterone. Without normal levels of testosterone, the man shows little sign of normal adult male sexual behavior. As a result, men with Kallmann syndrome who are not given hormone replacement therapy have a reduced libido and are not attracted to either sex. The pathway from gene to behavior is long, complicated and indirect. Each step along the causal pathway requires the products of many genes and has ramifying effects, some of which may be apparent and some not (Pfaff, 1997).

If a gene coding for altruism is proposed, the implication is that the gene represents the behavior and that a one-to-one correspondence will be found between the gene and altruistic behavior. If this were true, as it is for the link between gene and protein, the gene could be properly treated as an absolute unit coding for altruism and insistence on properly referring to differences in genes giving rise to differences in behavior would be mere pedantry. However, in most cases the slippage in meaning is unjustified.

The modern view about development is that the processes involve systems influenced by many different things with properties that are not easily anticipated, even when all the influences are known. Like many artificial systems, developmental processes may be strikingly conditional in character, particularly those in complex organisms. In one set of conditions they proceed in a particular and appropriate direction, in another set they do something different but equally appropriate. In yet others entirely new forms of behavior may be generated. One human example of alternative modes of development, which has been much discussed, is that in poor conditions individuals develop a small body type and a metabolism that is well adapted to those conditions – the so-called thrifty phenotype; in affluent conditions, individuals with identical genotypes develop larger bodies and have biochemical pathways well able to cope with ample food supplies (Bateson, 2001). This conditional dependence on the state of the environment means that the expression of genes and the characteristics that they influence are not inevitable.

An appropriate view of an individual's development is clarified by a culinary metaphor, which copes with the fact that in most cases many factors have been responsible for the detailed specification of behavior. In the baking of a cake, the flour, the eggs, the butter and all the rest react together to form a product that is different from the sum of the parts. The actions of adding ingredients, preparing the mixture and baking all contribute to the final effect. Nobody expects to recognize each of the ingredients and each of the actions involved in cooking as separate components in the finished cake. The biological equivalent of raisins will be found from time to time, but a simple relationship between the developmental determinants and the behavior will be exceptional. The issue then becomes squarely one of understanding the developmental processes.

Conclusion

Applying the biologists' knowledge of social behavior to humans has obvious and sometimes damaging political implications. However, the impact of sociobiology and evolutionary psychology is by no means all on the negative side. Indeed, some of the apparent support for social injustice was based on a muddle about how genes actually work. As this muddle was straightened out and genetic determinism fell away as a serious issue, biological knowledge has served to help the understanding of social issues by showing precisely how human potential is expressed in some conditions and not seen in others. The promise of fruitfully combining the insights of different disciplines is real. Sex differences in humans in terms of biological function bind together and make sense of what is observed without implying that the differences are inevitable, unchangeable or even desirable in a modern context (Hinde, 1984).

It is not necessary to appeal to biology at all, of course, and many would continue to argue that to do so in many cases remains utterly misleading and dangerous. They may well be right so long as people continue to suppose that 'natural' means 'desirable'. However, 'natural' by no means always means 'nasty and selfish' and, in as much as biological arguments are brought into debates about social issues, it is appropriate that the biological value of cooperation, for instance, should be fully appreciated.

See also
Behavior: Role of Genes
Wilson, Edward Osborne

References

Bateson P (1988) The biological evolution of cooperation and trust. In: Gambetta D (ed.) *Trust: Making and Breaking Cooperative Relations*, pp. 14–30. Oxford, UK: Blackwell.

Bateson P (1994) The dynamics of parent–offspring relationships in mammals. *Trends in Ecology and Evolution* 9: 399–403.

Bateson P (2000) Taking the stink out of instinct. In: Rose HRS (ed.) *Alas, Poor Darwin*, pp. 189–207. New York, NY: Harmony.

Bateson P (2001) Fetal experience and good adult design. *International Journal of Epidemiology* 26: 561–570.

Bateson P and Martin P (2000) *Design for a Life: How Behavior Develops*. London, UK: Vintage.

Buss DM (1994) *The Evolution of Desire: Strategies of Human Mating*. New York, NY: Basic Books.

Dawkins R (1976) *The Selfish Gene*. Oxford, UK: Oxford University Press.

Fodor JA (1983) *The Modularity of Mind*. Cambridge, MA: MIT Press.

Hamilton WD (1996) *Narrow Roads of Gene Land*, vol. 1. Basingstoke: WH Freeman.

Heyes C and Huber L (2000) *The Evolution of Cognition*. Cambridge, MA: MIT Press.

Hinde RA (1984) Why do the sexes behave differently in close relationships? *Journal of Social and Personal Relationships* 1: 471–501.

Krebs JR and Davies NB (1981) *Introduction to Behavioral Ecology*. Oxford, UK: Oxford University Press.

Krebs JR and Dawkins R (1984) Animal signals: mind-reading and manipulation. In: Krebs JR and Davies NB (eds.) *Behavioral Ecology*, 2nd edn, pp. 380–402. Oxford, UK: Blackwell.

Laland KN and Brown GR (2002) *Sense and Nonsense: Evolutionary Perspectives on Human Behavior*. Oxford, UK: Oxford University Press.

Pfaff DW (1997) Hormones, genes, and behavior. *Proceedings of the National Academy of Sciences of the United States of America* 94: 14 213–14 216.

Pinker S (1994) *The Language Instinct*. London, UK: Penguin.

Segerstråle U (2000) *Defenders of the Truth: The Sociobiology Debate*. Oxford, UK: Oxford University Press.

Tooby J and Cosmides L (1992) The psychological foundations of culture. In: Barkow JH, Cosmides L and Tooby J (eds.) *The Adapted Mind*, pp. 19–136. New York, NY: Oxford University Press.

Trivers RL (1985) *Social Evolution*. Menlo Park, CA: Benjamin/Cummings.

Wilson EO (1975) *Sociobiology: The New Synthesis*. Cambridge, MA: Harvard University Press.

Wilson EO (1976) Author's reply to multiple review of *Sociobiology*. *Animal Behavior* 24: 716–718.

Further Reading

Barker DJP (1998) *Mothers, Babies and Health in Later Life*. Edinburgh, UK: Churchill Livingstone. [Summary of the evidence for the effects of the fetal environment on the human babies developmental trajectory.]

Bateson P (2001) Where does our behavior come from? *Journal of Biosciences* **26**: 561–570. [A succinct discussion of the origins of human behavior.]

Krebs JR and Davies NB (eds.) (1997) *Behavioral Ecology: An Evolutionary Approach*, 4th edn. Oxford, UK: Blackwell. [A collection of essays on modern approaches to behavioral ecology.]

Oyama S, Griffiths P and Gray R (eds.) (2001) *Cycles of Contingency*. Cambridge, MA: MIT Press. [A modern collection of essays on developmental systems theory, bringing together developmental and evolutionary approaches to behavior.]

Pagel M (2002) *Encyclopedia of Evolution*, 2 vols. Oxford, UK: Oxford University Press. [Excellent collection of articles on all aspects of evolution.]

Somatic Hypermutation of Antigen Receptor Genes: Evolution

Marilyn Diaz, *National Institute of Environmental Health Sciences, Research Triangle Park, North Carolina, USA*

Laurent K Verkoczy, *The Scripps Research Institute, La Jolla, California, USA*

Analysis of the immunoglobulin mutational pattern across species has helped to reveal some of the molecular properties of this mechanism. Strong evidence for the involvement of DNA strand breaks and error-prone DNA polymerases confirms earlier predictions made from phylogenetic data.

Intermediate article

Article contents

- Introduction
- V(D)J Recombination and Diversity of the Preimmune Repertoire
- Immunoglobulin Hypermutation: Enhancing Specificity to Antigen
- Evolution of Immunoglobulin Hypermutation in Adaptive Immunity
- Phylogenetic Studies
- Conclusion

Introduction

In the innate immune system, the molecules responsible for pathogen binding and clearing recognize microorganisms on the basis of the motifs or patterns that are present in their cell surface molecules. However, pathogenic microorganisms are under strong selection to evade immune recognition. This, coupled to the fact that they have short generation times and, in some cases, a high mutation frequency (such as in the case of some viruses), affords them a fast pace of evolutionary change and adaptation. Quick adaptation of pathogens in turn creates selective pressure on the host's innate immune system to evolve novel pattern recognition molecules and adjusted versions of former ones to recognize a larger spectrum of motifs.

This presents animals characterized by complex life histories (such as long life spans, long-generation times and delayed sexual maturity) with two major difficulties: (1) these animals are poor evolutionary competitors against the microorganisms that infect them in terms of responding and modifying the specificity of their recognition molecules; and (2) large-spectrum recognition molecules are potentially cross-reactive with self-antigens. This was undeniably the impetus for the evolution of the vertebrate adaptive immune system, whereby a huge repertoire of highly specific antigen receptor molecules capable of recognizing thousands of antigens is generated somatically and monitored, also by a somatic mechanism, for self-reactivity.

The hallmark molecules of the adaptive immune system are major histocompatability complex (MHC) proteins, T-cell receptors (TCRs) and B-cell receptors (BCRs) and their secreted product, immunoglobulins (Igs). While diversity of MHC molecules is encoded in the genome, the BCR and TCR preimmune repertoire is generated somatically by the mechanism of V(D)J recombination (see below). B cells with a productively rearranged variable domain that have become activated by antigen can then be subjected to a second somatic mechanism of gene alteration known as somatic immunoglobulin hypermutation (Ig hypermutation). This mechanism targets Ig V(D)J regions of activated B cells to fine-tune the specificity of the receptor for recognition and binding of an encountered antigen. We briefly describe V(D)J recombination as an introduction to the generation of the B-cell and T-cell preimmune diversity, and then discuss in depth what is known about Ig hypermutation and its evolution.

V(D)J Recombination and Diversity of the Preimmune Repertoire

BCR and TCR polypeptides consist of two structural domains: a variable domain, which forms the antigen binding site; and one or more constant region domains, which mediate various effector functions. The Ig gene segments encoding the variable domains for the Ig heavy (H) chain or the light (L) chains, kappa (κ) and lambda (λ), are found on separate chromosomes and are assembled during lymphocyte development from variable (V), diversity (D) and joining (J) exons. This assembly is done by a site-specific somatic recombination process that is unique among mammalian systems and is termed V(D)J recombination.

Within the variable domain fall the regions that directly interact with foreign antigen: the complementarity determining regions (CDRs). CDR1 and CDR2 are encoded in the genome by V segments, while CDR3 is generated somatically via the juxtaposition of VDJH or VJL segments by V(D)J recombination. This combinatorial diversity generated by the random generation of many different VJL and VDJH combinations is further increased by the ability of any VH region to pair with any VL region to bind antigen. Random pairing of ~1000 different VL (250×4 possible Vκ $+ 2 \times 3$ possible Vλ) with about 24 000 different VH ($400 \, VH \times 15 \, DH \times 4 \, JH$) results in up to 2.4×10^6 different possible antibody specificities in mice. Actual combinatorial diversity is probably less, because some gene segments rearrange more often than others and some VH–VL pairings bind antigen better than others.

However, further diversity is attained by a process affecting predominantly the CDR3 region called junctional diversity. Junctional diversity results both from the imprecise joining of gene segments via a nonhomologous end-joining process and from the addition of (N) nucleotides by terminal deoxynucleotide transferase (TdT) a homolog of DNA polymerase-β. Junctional diversity is significant, but difficult to quantitate, as it also results in many nonproductive rearrangements, including altering the reading frame of the D elements in the heavy chain. The coupling of combinatorial and junctional diversities, however, has been estimated by some accounts to bring the actual number of preimmune specificities to as many as 10^{10}.

Immunoglobulin Hypermutation: Enhancing Specificity to Antigen

B cells, in addition to V(D)J recombination, undergo at least two more processes whereby their Ig genes are somatically recombined or modified: (1) Ig hypermutation and (2) class switch recombination. Both of these mechanisms are triggered only after the B cell is activated by antigen. During Ig hypermutation,

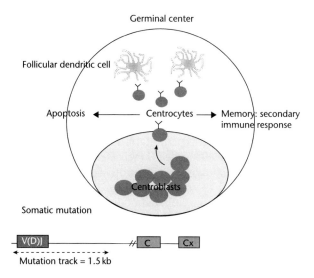

Figure 1 B cells that are activated by antigen form germinal centers in peripheral lymphoid organs such as the spleen and the lymph nodes. There, they divide for 7–21 days and these dividing cells are known as centroblasts. Centroblasts are where somatic hypermutation in V(D)J regions occurs. Subsequently they can differentiate into centrocytes and 'test' their immunoglobulin receptor affinity to antigen in the context of antigen-presenting cells known as follicular dendritic cells. B cells whose receptors have acquired beneficial mutations are selected and differentiate into memory B cells, while the rest, the great majority, undergo apoptosis.

the deoxyribonucleic acid (DNA) encoding the V(D)J region of the Ig receptor is specifically targeted for hypermutation at a frequency that is more than 1 million times higher than the background mutation frequency. This reaction occurs in transient lymphoid structures named germinal centers, which are formed in secondary lymphoid tissues such as the spleen and lymph nodes. In the germinal center, Ig hypermutation is coupled to a cellular machinery that selects for receptors that have obtained mutations that improve affinity to that antigen, while inducing apoptosis in B cells with Ig receptors exhibiting poor affinity to the immunizing antigen (**Figure 1**). This process leads to the formation of memory B cells with Ig receptors exquisitely tuned to recognize a specific foreign antigen, thus greatly contributing to the enhanced response during re-exposure to a pathogen and to the effectivity of vaccination.

Evolution of Immunoglobulin Hypermutation in Adaptive Immunity

Because, in its current state, Ig hypermutation coupled to antigen-driven selection leads to the formation of high-affinity memory B cells, one might speculate that it was this reason, the acquisition of enhanced affinity, that drove the evolution of Ig hypermutation. However, the risks and cost of evolving this process include the potential to generate novel specificities that recognize self-antigen, the potential of mutating nonimmune genes leading to neoplasia, and the metabolic expense of carrying out the reaction itself. Given these potential problems, the benefit to evolving Ig hypermutation was probably great at or near its onset. If in fact Ig hypermutation evolved to enhance affinity to a specific antigen, then the intricate cellular machinery that drives selection for high-affinity variants would have to have been fully available at its onset. Given the unlikelihood of this scenario, it seems more plausible that the original gain from evolving Ig hypermutation was not the enhancement of affinity.

Three potential scenarios that may have driven the evolution of Ig hypermutation are (1) an increase in diversity of the naïve repertoire; (2) mutations self-recognizing receptors away from nonself, and later being recruited and gradually optimized to generate high-affinity receptors to foreign antigen; and (3) an increase in diversity of the repertoire but following exposure to antigen. That Ig hypermutation evolved to generate the naïve, preimmune repertoire is an appealing idea because it does not require an antigen-driven selection machinery to be operative at the time of its onset. However, this scenario appears unlikely given the early inception of V(D)J recombination, and N region addition in phylogeny coupled to the fact that most species studied to date utilize Ig hypermutation to increase affinity to foreign antigen. In addition, the mode of modification by Ig hypermutation, namely through the introduction of single-base substitutions, suggests a mechanism better suited to generate changes for a restricted modulating function (to increase affinity or decrease self-recognition) rather than as a general machinery to augment diversity. The second possible scenario is that Ig hypermutation evolved originally to mutate self-reactive Ig receptors away from self-recognition.

The appeal of this explanation is that, given the generally random nature of mutation introduction (within the context of the Ig V(D)J region), it is easier to evolve a process that 'destroys' a specificity (self) than one that creates a new desired specificity (to an immunizing antigen). On the other hand, such a scenario would suggest that what was originally a positive signal to drive mutation (losing affinity to antigen, in this case, self) later became a negative signal as losing the ability to bind antigen in the current version of Ig hypermutation results in apoptosis. That would require rewiring of the Ig receptor signal transduction pathway. Third, it is possible that Ig hypermutation could have evolved as a way of generating new specificities following exposure to

pathogen, not to increase affinity, but to further diversify the repertoire when most needed, during the course of an infection. However, it is difficult to imagine the gain in diversity would be great in this scenario, because the repertoire generated during V(D)J recombination is massive.

Each of these possibilities predicts that vestiges of the original version of this mechanism may be detectable in extant species with adaptive immunity. As the repertoire of species that utilize Ig hypermutation increases, and by studying ancestral properties of the mechanism, we will soon be able to discern among these scenarios, which will be an important exercise in the study of the evolution of intricate molecular machineries.

Phylogenetic Studies

Ig hypermutation has been described in a variety of species such as sharks, frogs, sheep, camels, humans and mice and, as part of a gene conversion process, in chickens, rabbits, pigs and cows (reviewed by Diaz and Flajnik, 1998). While some differences exist among species, many similarities exist in the mutational pattern across species, suggesting a common mechanism. These include a preponderance to generate base substitutions, a bias in favor of transitions over transversions, and identical hot spots of hypermutation such as the AGC/T sequence (Diaz and Flajnik, 1998). These similarities undeniably reveal ancestral properties of a common mechanism of Ig hypermutation and suggest that, by studying any differences among species, it may be possible to track the footprints of the molecular machinery, because separate evolutionary paths may have optimized different components of the mechanism.

The main cross-species differences in the mutational pattern generated by the Ig mutator include a marked bias to mutate G · C base pairs in shark and frog Ig, a high frequency of mutations occurring in doublets in the nurse shark novel antigen receptor (NAR) V genes, and the occurrence of Ig hypermutation in some species within the context of gene conversion. The implications of these differences are discussed below.

Shark novel antigen receptor

Martin Flajnik and colleagues (Greenberg et al., 1995) discovered that in addition to bona fide Ig, sharks contain a NAR that undergoes extensive Ig hypermutation. The mutational pattern in shark NAR is identical to that of mammalian Ig except for one unique characteristic: 25% of the mutations occur in tandem, particularly in doublets (Diaz et al., 1999). Not one of these doublets was explained by gene conversion events, but instead occurred as part of the normal NAR hypermutational machinery. This is an interesting finding because it suggests that the polymerases involved in hypermutation are capable of frequently extending from a mismatched terminus, a rather difficult task for most polymerases, particularly the high-fidelity DNA polymerases.

Given these findings and the high mutation frequency seen in Ig hypermutation, it is almost unquestionable that the polymerases involved in the process lack proofreading activity and are error-prone. This prompted us to suggest that DNA polymerase-ζ may be an attractive candidate because it lacks proofreading activity and is an efficient mismatch extender. Recent data suggest an involvement of this polymerase in Ig hypermutation, most probably as a mismatch extender (Zan et al., 2001).

G · C bias in shark and frog Ig suggests the involvement of multiple polymerases

Unlike most species, where all nucleotides are roughly targeted equally by the Ig mutator, the horned shark (*Heterodontus francisci*) and frog Ig display a strong bias to mutate G · C base pairs (Wilson et al., 1992; Hinds-Frey et al., 1993). This finding recapitulates what is seen in human B-cell lines that undergo hypermutation of their Ig genes as well as mismatch repair defective mice and humans defective in the error-prone DNA polymerase-η (Rada et al., 1998; Zeng et al., 2001). It now appears to be the case that this G · C bias reveals a two-layered process of Ig hypermutation, an A · T and a G · C layer (Rada et al., 1998), each with different polymerases, both probably sharing a common role for DNA polymerase Zeta as a mismatch extender.

Immunoglobulin hypermutation link to gene conversion

In some species such as chickens, rabbits and pigs, the B-cell repertoire is diversified not only via V(D)J recombination but also via somatic Ig gene conversion (Reynaud et al., 1989; Short et al., 1991). In these species, a few or even single V, D and J genes are utilized to generate a functional V(D)J region, which then serves as an acceptor for gene conversion utilizing any of a number of pseudoV gene segments found in their Ig locus as donor sequences. Interestingly, point mutations with a pattern similar to that of Ig hypermutation are found near the vicinity of the region undergoing gene conversion. In addition, these point mutations tend to fall within the same hot spots seen for Ig hypermutation.

We had previously noted that the sporadic distribution of Ig hypermutation and gene conversion in phylogeny can be best explained by the notion that these two mechanisms are in fact mechanistically linked (Diaz and Flajnik, 1998). This putative mechanistic link now appears to be true as revealed by the following findings: (1) both processes share in common the introduction of DNA strand breaks (Sale and Neuberger, 1998; Papavasiliou and Schatz, 2000); (2) both processes are entirely dependent on the regulatory activity of a novel cytidine deaminase (AID) (Muramatsu et al., 2000; Arakawa et al., 2002); and (3) ablation of two genes involved in homologous recombination, both RAD51 paralogs, in a chicken B-cell line that undergoes constitutive Ig gene conversion induces Ig hypermutation over Ig gene conversion in that cell line (Sale et al., 2001).

These results suggest that both mechanisms share the targeted introduction of DNA strand breaks into the V(D)J region, and their processing via homologous recombination repair. However, the templates utilized for repair differ, as in Ig gene conversion the template is a pseudoV gene, while in the case of Ig hypermutation the template may be a sister chromatid instead (Sale et al., 2001). The fact that the DNA strand breaks in Ig hypermutation are seen in the G2 phase of the cell cycle is consistent with the availability of a sister chromatid for Ig hypermutation (Papavasiliou and Schatz, 2000). It should be noted, however, that studies suggest that the DNA breaks originate from the generation of abasic sites at IgV regions, when AID-generated uracils are removed (Peterson-Mahrt et al., 2002).

Conclusion

The study of the evolutionary aspects of Ig hypermutation has proved to be extremely useful for predicting some of the mechanistic properties of the Ig mutator. A model that is developing on the basis of these and other data suggests that DNA lesions are introduced into the V(D)J region of Ig genes, which are then repaired utilizing one or several error-prone polymerases to fill in the gap. A major gap that remains in our understanding of this elusive mechanism is the nature of the molecules responsible for the targeting and introduction of breaks into the Ig V(D)J region. Given the potentially catastrophic consequences of introducing DNA strand breaks into DNA, it is very likely that the targeting mechanism was in place early on in evolution, and that, once again, phylogenetic approaches may contribute to the elucidation of this aspect of the Ig hypermutation machinery.

See also
DNA Recombination
Immunoglobulin Genes

mRNA Editing
Mutation Rates: Evolution

References

Arakawa H, Hauschild J and Buerstedde JM (2002) Requirement of the activation-induced deaminase (AID) gene for immunoglobulin gene conversion. *Science* **15**: 1301–1306.

Diaz M and Flajnik MF (1998) Evolution of somatic hypermutation and gene conversion in adaptive immunity. *Immunological Reviews* **162**: 13–24.

Diaz M, Velez J, Singh M, Cerny J and Flajnik MF (1999) Mutational pattern of the nurse shark antigen receptor gene (NAR) is similar to that of mammalian Ig genes and to spontaneous mutations in evolution: the translesion synthesis model of somatic hypermutation. *International Immunology* **11**: 825–833.

Greenberg AS, Avila D, Hughes M, *et al.* (1995) A new antigen receptor gene family that undergoes rearrangement and extensive somatic diversification in sharks. *Nature* **374**: 168–173.

Hinds-Frey KR, Nishikata H, Litman RT and Litman GW (1993) Somatic variation precedes extensive diversification of germline sequences and combinatorial joining in the evolution of immunoglobulin heavy chain diversity. *Journal of Experimental Medicine* **178**: 815–824.

Muramatsu M, Kinoshita K, Fagarasan S, *et al.* (2000) Class switch recombination and hypermutation require activation-induced cytidine deaminase (AID), a potential RNA editing enzyme. *Cell* **102**: 553–563.

Papavasiliou FN and Schatz DG (2000) Cell-cycle-regulated DNA double-stranded breaks in somatic hypermutation of immunoglobulin genes. *Nature* **408**: 216.

Petersen-Mahrt SK, Harris RS and Neuberger MS (2002) AID mutates *E. coli* suggesting a DNA deamination mechanism for antibody diversification. *Nature* **418**: 99–103.

Rada C, Ehrenstein MR, Neuberger MS and Milstein C (1998) Hot spot focusing of somatic hypermutation in Msh2-deficient mice suggests two stages of mutational targeting. *Immunity* **9**: 135–141.

Reynaud C-A, Danhan A, Anquez V and Weill J-C (1989) Somatic hyperconversion diversifies the single VH gene of the chicken with a high incidence in the D region. *Cell* **59**: 171–183.

Sale JE, Calandrini DM, Takata M, Takeda S and Neuberger MS (2001) Ablation of XRCC2/3 transforms immunoglobulin V gene conversion into somatic hypermutation. *Nature* **412**: 921–926.

Short JA, Sethupathi P, Zhai SK, and Knight KL (1991) VDJ genes in Vha2 allotype-suppressed rabbits: limited germline Vh gene usage and accumulation of somatic mutations in D regions. *Journal of Immunology* **147**: 4014–4018.

Sale JE and Neuberger MS (1998) TdT-accessible breaks are scattered over the immunoglobulin V domain in a constitutively hypermutating B cell line. *Immunity* **9**: 859–869.

Wilson M, Hsu E, Marcuz A, *et al.* (1992) What limits affinity maturation of antibodies in Xenopus – the rate of somatic mutation or the ability to select mutants? *EMBO Journal* **11**: 4337–4347.

Zan H, Komori A, Li Z, *et al.* (2001) The translesion DNA polymerase zeta plays a major role in Ig and bcl-6 somatic hypermutation. *Immunity* **14**: 643–653.

Zeng X, Winter DB, Kasmer C, *et al.* (2001) DNA polymerase eta is an A–T mutator in somatic hypermutation of immunoglobulin variable genes. *Nature Immunology* **2**: 537–541.

Further Reading

Diaz M and Casali P (2002) Somatic immunoglobulin hypermutation. *Current Opinion in Immunology* **14**: 235–240.

Maizels N (1989) Might gene conversion be the mechanism of somatic hypermutation of mammalian immunoglobulin genes? *Trends in Genetics* **5**: 4–8.

Neuberger MS, Ehrenstein MR, Klix N, *et al.* (1998) Monitoring and interpreting the intrinsic features of somatic hypermutation. *Immunological Reviews* **162**: 107–116.

Storb U, Shen HM, Michael N and Kim N (2001) Somatic hypermutation of immunoglobulin and non-immunoglobulin genes. *Philosophical Transactions of the Royal Society of London, Series B* **356**: 13–19.

Weigert MG, Cesari IM, Yonkovich SJ and Cohn M (1970) Variability in the l light chain sequences of mouse antibody. *Nature* **228**: 1045–1047.

Weill JC and Reynaud CA (1996) Rearrangement/hypermutation/gene conversion: when, where and why? *Immunology Today* **17**: 92–97.

Speciation

Troy E Wood, *Indiana University, Bloomington, Indiana, USA*

Loren H Rieseberg, *Indiana University, Bloomington, Indiana, USA*

Speciation is the formation of two or more new species from one ancestral species.

Introductory article

Article contents

- Introduction
- Are Species Real Biological Entities?
- Species Concepts
- Isolating Mechanisms
- Geography of Speciation
- Genetic Basis of Speciation
- Reinforcement
- Speciation via Hybridization
- Summary

Introduction

Speciation is defined as the formation of two or more new species from one ancestral species. The process of speciation is responsible for the vast amount of biological diversity found today. Estimates of the number of extant (currently existing) species range from 5 million to 30 million. Despite the seemingly limitless number of examples of speciation that nature has provided, we know little about the underlying biological mechanisms of species formation. This lack of knowledge can be attributed largely to the fact that speciation occurs too quickly to be recorded in the fossil record and, with few exceptions, too slowly to be observed in nature or in the laboratory. However, advances in investigative techniques have allowed us, at least in part, to overcome these inherent limitations and make inferences about the processes of species formation.

Are Species Real Biological Entities?

Grouping and naming of organisms is a natural and intuitive process. In fact, it is not hard to imagine that the ability to distinguish between biological forms was essential to the survival of our ancestors. For example, being able to distinguish poisonous from nutritious plants was obviously a critical skill of our forebears. Thus, our inherent desire to partition and name organisms probably has its roots in a need for formal classification of the natural world around us. However, some biologists and philosophers argue that the grouping of organisms into species is an artificial and subjective enterprise. These scientists would argue that 'species' do not represent discrete groups of organisms, but rather points along a continuum of diversity of biological form. In practice, most evolutionary biologists agree that the variation among organisms is discontinuous; that is, species are not simply constructs of the human mind, but rather they represent discrete biological units. For example, consider domestic cats (*Felis domestica*) and mountain lions (*Felis concolor*). Although there is a great deal of variation within domestic cats and within mountain lions, it is unlikely that anyone would misidentify a house cat as a cougar. The variation in house cats does not overlap with the variation in mountain lions. These two species are easily distinguished because they are not connected by intermediate forms.

Species Concepts

As the example above illustrates, certain related species can be readily separated, but in many cases the line between species pairs is blurry. Thus, a precise, operational definition of species is crucial. Of course, the study of speciation is linked intrinsically to how we define species, because these are the entities produced by speciation. Moreover, different definitions of species will lead to different conclusions about pattern and process in species formation. Evolutionary biologists have proposed a diverse, almost innumerable list of species concepts, and the different views expressed in these various formulations have fueled an interminable debate. This ongoing debate emphasizes the fact that species concepts are critical to the study of

evolution and that as yet no one concept is completely satisfactory. Here we focus on the three most instructive concepts and examine the implications each holds for the study of speciation.

Phylogenetic species concept

A phylogenetic species is a group of organisms that share unique morphological or genetic features that make them diagnosably distinct from other groups. These shared features are inferred to be the result of common ancestry. This concept has its roots in the branch of evolutionary biology that focuses on reconstructing the evolutionary history of organisms. The end point of work in this discipline is the phylogenetic tree, a diagram depicting the hypothetical evolutionary relationships among related organisms. At the termini of the branches are phylogenetic species, which cannot be divided further into distinct forms. In this definition, emphasis is placed on the evolutionary history, or parental pattern of ancestry, and not on an ability to interbreed with other groups of organisms. (*See* Gene Trees and Species Trees; Phylogenetics.)

Recognition species concept

According to this concept, a species is the most inclusive population of individual biparental organisms that share a common fertilization system. By focusing on mate recognition, this concept suggests that the forces that produce species are associated with the evolution of mating systems. Consequently, this definition is especially helpful in understanding patterns of diversification in animal taxa that have diverged via sexual selection. More generally, this concept emphasizes characters that are actually subject to natural selection in a population undergoing speciation rather than those that are simply a by-product of divergence.

Biological species concept

Biological species are groups of actually or potentially interbreeding natural populations that are reproductively isolated from other such groups. This concept was developed to explain the absence of intermediates between closely related, yet discontinuous species that co-occur in nature. If two species are unable to interbreed, then the lack of intermediates is no longer puzzling. One of the most useful features of this definition is that it suggests a diagnostic test to distinguish between true species and variants within a species. If two populations cannot interbreed to form viable and fertile progeny, then they represent two distinct species. A weakness of this definition is that many populations never interbreed in nature because they occupy different geographical ranges, hence the *ad hoc* 'potentially'. A more serious problem is the frequent occurrence of morphologically and ecologically distinct species that do interbreed in areas of contact. Furthermore, like the recognition concept, this definition is not applicable to organisms that reproduce asexually.

If we accept the premise of the biological species concept, that species are delimited in terms of reproductive isolation, we can then view speciation as the evolution of mechanisms that prevent gene flow between divergent populations. Thus, the biological species concept provides a logical framework for the study of speciation and, consequently, most speciation theory is based on this concept.

Isolating Mechanisms

Because species are defined as biological entities that are reproductively isolated, it is important to focus on the evolution of barriers to the exchange of genes. Barriers that are based on biological attributes of species are called isolating mechanisms. Isolating mechanisms result from the evolution of morphological, behavioral or chromosomal differences between closely related species. Isolating mechanisms can be divided into premating and postmating barriers, depending on when they act to prevent gene flow (**Table 1**). Most speciation events are thought to result from a combination of two or more types of barrier, with physical barriers (geological or climatic changes that separate formerly contiguous populations) setting the stage for the evolution of isolating mechanisms.

Table 1 Isolating mechanisms

Mechanism	Barriers
Premating barriers	Temporal isolation: populations mate at different times
	Ecological isolation: populations mate in different habitats
	Behavioral isolation: populations rely on different mating cues
	Mechanical isolation: matings unsuccessful due to morphological differences
Postmating barriers	Gametic incompatibility: fertilization limited due to negative interactions between sperm and egg
	Hybrid inviability: hybrid offspring do not reach reproductive maturity
	Hybrid sterility: hybrids sterile due to chromosomal or genic differences that disrupt meiosis

Premating barriers

Temporal isolation

Two closely related species that mate at different times or during different seasons are said to be temporally isolated. Often, temporal isolation prevents gene flow between populations that are fully interfertile. Pink salmon, *Oncorhynchus gorbuscha*, which migrate from salt water to fresh water to spawn, represent an interesting example of how temporal isolation may result in speciation. Because of the timing of reproductive maturation, pink salmon spawn in 2-year cycles. Those individuals that result from a spawn in an even year return to spawn 2 years later, in an even year, while individuals that are produced in odd years return to spawn in an odd year. 'Odd-year' and 'even-year' salmon are thus genetically isolated in time. While odd-year and even-year pink salmon are still considered the same species, studies suggest that the two types are diverging. Forced crosses between odd- and even-year individuals yield hybrids with lowered fitness. Temporal isolation is also an effective barrier to gene flow between two species of pine, *Pinus muricata* and *Pinus radiata*, which are highly interfertile but do not produce hybrids in nature because the anthers and stigmas mature at different times.

Ecological isolation

Ecological isolation, also called habitat or resource isolation, can occur when two species are adapted to different niches. Two species that utilize different resources or occupy different habitats are less likely to encounter each other and, consequently, opportunities for interspecific matings are rare. One of the most convincing examples of ecological isolation at work in speciation involves the threespine stickleback, *Gasterosteus aculeatus*. Different forms of this species are adapted to feeding at different lake depths; one form feeds in open water and the other is a bottom feeder. Although the species status of these forms is unclear, they are morphologically distinct and females prefer to mate with males of the same morphology.

Behavioral isolation

Patterns of behavior during courtship and mating are important in mate recognition systems. If two closely related species have distinct courtship and mating rituals, they will not recognize each other as potential mates, or mating attempts generally will not result in successful copulations. There is evidence that behavioral isolation has played an important role in the diversification of Hawaiian crickets. Species of this group almost always have different calling songs and females of many species can identify conspecific mates, even when heterospecific songs are only slightly different. Although behavioral isolation is thought to be more common in animals than plants, the latter can be isolated by the behavior of their pollinators. In some cases, pollinators display remarkable flower constancy – they preferentially visit one species of flower based on its floral morphology, color or scent. Because the pollinators move only between flowers of the same species, no gene flow between different species occurs.

Mechanical isolation

Certain species are reproductively isolated because of morphological differences that prevent successful interbreeding. Mechanical isolation is common in plant species that rely on animal pollinators to transport gametes between individuals. For example, two species in the snapdragon family, *Pedicularis groenlandica* and *P. attollens*, are both pollinated by bumblebees. Owing to structural differences between the flowers of these two species, the plants deposit and remove pollen from different parts of the bee. Consequently, little or no pollen is exchanged between the two species.

Postmating barriers

Gametic incompatibility

Experimental crosses between related species frequently reveal that mating systems are incompatible because of interactions between egg and sperm or interactions between the female reproductive tract and sperm. For example, heterospecific sperm (or pollen in plants) often suffer a competitive disadvantage. In plants, heterospecific pollen is sometimes unable to germinate on the stigmas of different species, and when it does it may be unable to fertilize ovules. In insects, studies of females serially mated to males of the same and different species show that intraspecific sperm is more likely to fertilize eggs, regardless of mating order. In some fruitfly species, interspecific crosses initiate an immunological response by females that targets and kills sperm of other species.

Hybrid inviability

Progeny of crosses between species frequently show greatly reduced vigor. Because these progeny are unlikely to reproduce, gene flow between the parental species is limited. Developmental abnormalities that prevent gestation or seed production are also a frequent outcome of interspecific crosses.

Hybrid sterility

Sterility barriers are important in maintaining the genetic integrity of very closely related species that have not yet evolved other effective isolating mechanisms. Breeding between recently diverged taxa can

result in viable offspring that reach reproductive maturity. However, these offspring are commonly sterile because of chromosomal or genic differences between the species that disrupt gamete formation. A classic example of this is seen in crosses between female horses and male donkeys. Mules that result from these crosses are completely sterile. (*See* Genetic Conflict and Imprinting.)

Geography of Speciation

Allopatric and peripatric speciation

Populations of the same or related species that occupy nonoverlapping ranges are referred to as allopatric populations. Because allopatric populations are separated by physical barriers (e.g. mountains, water, large expanses of unoccupied land, etc.), they evolve independently, free of the homogenizing effects of gene flow. Under these conditions, populations diverge because they are subject to different selection pressures and undergo unique changes due to genetic drift. Barriers to gene flow can arise easily as a by-product of this divergence, which would be unlikely to occur if the populations were in contact.

Two major types of allopatric speciation are recognized by evolutionary biologists. The first type, called vicariant speciation, is based on geological changes that split formerly connected populations. For example, the uplift that created the spine of mountains that runs down the center of the Baja peninsula also resulted in the formation of the Gulf of California. Since at least the Late Pliocene this sea has isolated populations of many species that now exist on both the peninsula and mainland Mexico. Some of these populations have differentiated enough to be distinguished as separate species. In this example, the same geological processes that split populations also resulted in ecological and climatic shifts that created new selection regimes. The combination of these processes has produced a number of intriguing organisms adapted to environmental conditions unique to Baja California.

The second model of allopatric divergence, peripatric speciation, involves the colonization of habitats that lie along the periphery of a species range. This model is based on the idea that small, founding populations may be genetically distinct because they represent only a small, random sample of the larger source population. The model predicts that subsequent evolutionary changes within these small colonies will result in rapid differentiation or 'genetic revolutions' as selection establishes novel, adaptive gene complexes. Reproductive isolation could result as a secondary effect of these changes. The geographical pattern of morphological differentiation within the New Guinea kingfisher, *Tanysiptera galeata*, is consistent with the predictions of the model. Kingfishers that live on coastal islands are strikingly different from mainland kingfishers, and the variation of plumage phenotypes across these small islands is much greater than the variation seen within the much larger mainland population. However, there is little direct evidence that small population size has played a critical role in speciation in this or other examples.

Sympatric speciation

Sympatric speciation occurs when populations become reproductively isolated in spite of significant amounts of gene flow during the initial stages of divergence. While allopatric speciation inherently allows for differentiation without the constant production of intermediate forms, models of sympatric speciation must meet this challenge head on. These models rely on strong disruptive selection to drive divergence in the presence of gene flow. For example, consider a population that occupies a habitat in which two different resources are distributed in discrete patches. If individuals begin to specialize on one of the two resource types, two distinct forms, which are adapted to utilize the two different resource types, may evolve. In situations where disruptive selection is strong enough to eliminate hybrids between the two morphs, the evolution of reproductive isolation may be possible. Many examples of sympatric speciation through ecological isolation have been reported. For example, many closely related insect species occupy the same range but feed and reproduce on different host plants. In most cases, it is hard to prove that speciation did in fact occur while the incipient species were in sympatry. Purported examples of sympatric speciation can be explained by the alternative view that reproductive isolation evolved in allopatry, and the distribution patterns we see today came about through secondary contact. Some theoretical issues related to sympatric speciation are discussed below in the context of reinforcement.

Genetic Basis of Speciation

If speciation is equated with the evolution of reproductive isolation, the goal of speciation genetics is then to reconstruct the sequence of genetic changes necessary for the development of isolating mechanisms. (*See* Speciation: Genetics.)

Postmating barriers

Much of the theoretical work on speciation genetics is founded on the standard model, which was formulated soon after the synthesis of the theories of natural selection and genetics. This model is based on the evolution of genetic divergence during allopatry. If two populations are separated in space, natural selection causes the populations to evolve along different paths as they adapt to their different environments. Because these populations are isolated and experience different selective regimes, favorable mutations that arise and become fixed are very likely to be different. When the populations come back into contact, the mutated genes may interact unfavorably, thereby causing reproductive isolation. The reason that this model is so powerful is that the mutated genes that contribute to isolation actually confer a fitness advantage to individuals within the population in which they arise. Only when they interact with mutated genes that arose in another isolated population do they cause hybrid incompatibilities. Thus, according to this model, the genetic changes that cause reproductive isolation are a side-effect of natural selection. This model has recently been confirmed in a pair of fruitfly species. In this example, hybrid male sterility appears to be a direct result of strong positive selection on an important developmental gene.

Premating barriers

Unlike postmating barriers, which appear to be a side-effect of adaptive divergence, characters affecting premating isolation are often selected on directly. Thus, the genetics of premating barriers can be viewed as being synonymous with the genetics of adaptation. In the classical view, very small mutations are thought to drive the evolution of adaptations because mutations of large effect are unlikely to be favorable. However, others have argued, based on results from classical crossing studies, that mutations with large morphological effects often contribute to adaptive evolution. This latter viewpoint has received increased support recently because of the recognition that larger mutations, while less likely to be favorable, are more likely to be fixed in the population when advantageous and because mutations with large morphological effects are often detected in genetic mapping studies. For example, genetic analyses of floral isolation between two species of monkeyflowers indicate that 9 of the 12 floral differences studied involve major mutations.

Reinforcement

Reinforcement is one of the most interesting and intensely debated ideas in speciation theory. Reinforcement occurs when premating barriers evolve in sympatry to prevent the production of hybrid offspring. For example, if two incipient species, A and B, come into secondary contact after a period of allopatry, genetic differences that arose during separation may cause hybrid offspring between A and B to have reduced fitness. If a mutation arises that increases within-species mating, individuals that carry this mutation will have increased fitness relative to individuals that mate indiscriminately. Thus, natural selection operates to isolate the incipient species by increasing assortative, or within-species, mating. The term 'reinforcement' is based on the idea that this process 'reinforces' reproductive barriers that evolved in allopatry, and reinforcement is often viewed as a mechanism that serves to complete the process of speciation.

Reinforcement is a popular concept because it describes a scheme in which natural selection plays a direct role in speciation. Some evolutionary biologists would argue that the isolating mechanisms discussed above are not really 'mechanisms' of speciation, because they have not evolved directly to prevent hybridization between divergent populations. Rather, they are thought to be a by-product of allopatric divergence. In contrast, reinforcement is a mechanism of speciation because it evolves directly to prevent gene flow between divergent populations.

Although recent theoretical and empirical work suggests that reinforcement is plausible, unequivocal evidence of reinforcement is lacking. The scarcity of natural examples is not surprising considering the following theoretical constraints on the operation of reinforcement. First, a gene that increases mate discrimination in zones of contact may not be favored in allopatric populations that serve as a source of migrants. Thus, gene flow between the zone of contact and allopatric populations will dampen the effects of selection on the gene that increases mate discrimination. Second, the intensity of selection will decrease as crossmating decreases because nonviable hybrids become increasingly rare. Hence reinforcement is unlikely to result in complete reproductive isolation. Last, for reinforcement to work effectively, it must generate an association between the genes that cause reduced hybrid fitness and the gene for mate discrimination. Because reduced fitness in hybrids is usually based on the effects of many genes, this association is unlikely to be formed. Even if reduced hybrid fitness is caused by only one gene, recombination between this gene and the mate discrimination gene will limit the effects of natural selection. In fact, selection to increase hybrid fitness may operate more effectively than selection against crossmating.

The majority of evidence for reinforcement comes from studies that compare levels of premating isolation between sympatric and allopatric populations of divergent species or subspecies. These studies test the hypothesis that premating isolation should be more pronounced in areas where the ranges of related species overlap. Many studies have revealed this pattern. However, this type of comparison does not provide direct evidence for reinforcement. In most of these studies, the species pairs examined are not able to hybridize. In these cases, increased premating isolation in sympatry may simply be a result of selection to avoid wasting energy on incompatible mating, and not on selection to avoid producing unfit hybrid offspring. In cases where species pairs cannot hybridize, increased premating isolation in sympatry is called reproductive character displacement. The difference between reproductive character displacement and reinforcement is subtle, but the distinction is important. Reproductive character displacement, because it operates between species, is not a mechanism of speciation. Reinforcement, which operates within species, selects for a reduction in gene flow, and is therefore a direct mechanism of speciation. Demonstration of reinforcement requires evidence that gene flow occurs between the populations under comparison and that, in spite of this gene flow, selection is operating to reduce interbreeding.

A recent study meets all of the stringent requirements imposed by opponents of the reinforcement hypothesis. The pied (*Ficedula hypoleuca*) and collared (*F. albicollis*) flycatchers are primarily distributed allopatrically, but come into contact in parts of Central and Eastern Europe. In allopatry, males of the two species have very similar black and white markings, but in sympatric zones, males of the pied flycatcher are brown and collared males have more extreme white markings. Researchers demonstrated that this phenotypic difference plays a role in female mate choice. Females from sympatric areas show a strong preference for males of their own species, with pied females even preferring the brown males over conspecific black and white males. Less than 3% of matings are heterospecific, and the hybrid progeny of these matings are one-third as fit as pure progeny. Hence females that choose conspecific mates accrue a significant reproductive advantage. The conclusion that selection for assortative mating is responsible for the phenotypic differences of sympatric and allopatric males is firmly supported by the data. This study, coupled with theoretical work, has reinvigorated enthusiasm for the reinforcement hypothesis.

Speciation via Hybridization

Geological or climatic causes of isolation are often impermanent. When two populations come back into contact after a period of separation, the evolution of barriers to gene flow may be incomplete. In these cases, hybridization between the two populations will occur. This mixing of genomes can result in the following four outcomes: (1) the populations may merge and evolve as a single species; (2) reinforcement may complete the process of speciation initiated in allopatry; (3) the two populations may retain their genetic identity, with hybrids restricted to zones of contact; (4) new species can arise from the recombination of the differentiated genomes (**Figure 1**). Natural examples of the third scenario are often a focus of study, because they provide windows on the processes of speciation. Here we discuss the fourth outcome: speciation via hybridization.

In some cases, crosses between divergent genomes can result in instant or 'abrupt' speciation. If two divergent populations come into secondary contact, hybrids may be sterile because of chromosomal factors that disrupt meiosis. One mechanism of avoiding such sterility is polyploidy. Polyploid individuals result from genome duplication and are often of hybrid origin. Because polyploids have two identical copies of each chromosome, pairing during meiosis is not affected. However, backcrosses to either parent yield sterile, triploid offspring. Thus polyploid lineages that originate after hybridization are instantly isolated from their parental species and will evolve as distinct entities. Polyploidy, which is often associated with extreme habitats, is very common in ferns and their allies and in flowering plants, suggesting that hybridization has played an important role in the diversification of vascular plants. The fern genus *Ophioglossum* provides an extreme yet interesting example. Members of this genus have chromosome counts ranging from 120 to 630. These elevated chromosome counts are apparently a result of polyploid speciation through hybridization.

Hybridization can also lead to new species that retain the chromosome number of their progenitors. This mode, referred to as homoploid hybrid speciation or recombinational speciation, is believed to occur less frequently than polyploid speciation, because the possibility of backcrosses makes the establishment of reproductive isolation less likely. Like polyploid species, homoploid hybrid derivatives usually possess extreme phenotypes and are adapted to extreme habitats. While the genetic basis of these extreme phenotypes is not fully understood, experimental evidence suggests that homoploid hybrid species combine favorable genes from each parental species to form novel, adaptive genotypes. If these novel genotypes become associated with a distinct mating type, or if individuals with these genotypes become separated spatially or ecologically, the homoploid derivative can establish itself as a distinct species. Although this mode of speciation is believed to be rare, putative examples exist for all major animal and plant groups.

Because hybrid speciation can occur rapidly, investigators have been able to study this type of speciation in the laboratory. Forced crosses between *Gilia malior* and *G. modocensis* yielded highly sterile, homoploid hybrids. This sterility was due to chromosomal differences that disrupt meiosis. Hybrid sterility was ameliorated by artificial selection and continued crossing of the most fertile hybrids. After nine generations, hybrid lines recovered full fertility and were reproductively isolated from the parental species. The hybrid plants were morphologically and chromosomally distinct from the parent species. In a more

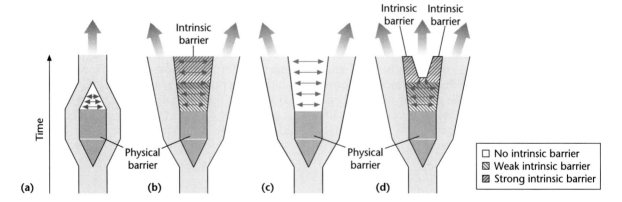

Figure 1 Possible outcomes of contact between populations that were previously isolated by a physical barrier. Broad arrows represent evolutionary lineages (species). Narrow horizontal arrows indicate gene flow between otherwise independently evolving lineages. The shading of the region between diverging lineages depicts the evolution of intrinsic isolating barriers (refer to the key in the figure). (a) Populations merge and evolve as a single species. (b) Reinforcement completes the formation of reproductive isolation initiated in allopatry. (c) Populations retain genetic identity but form hybrid swarms in zones of contact. (d) Recombination of differentiated genomes results in a new, 'hybrid' species.

recent study that employed modern molecular techniques to identify chromosomal blocks, researchers were able to compare the genomic composition of synthetic hybrids of two sunflower species, *Helianthus annuus* and *H. petiolaris*, to that of *H. anomalus*, the supposed natural hybrid derivative of the two species. The molecular data indicated that the genomes of the synthetic and natural hybrids were highly similar, suggesting that hybrid speciation is a deterministic process. In addition, this study revealed that hybridization can result in rapid genomic change.

Summary

Progress in speciation research has been constrained by debates on how to define species. If we assume that reproductive isolation is the most important aspect of species boundaries, we can avoid the semantic quagmire created by definitional debates. Students of speciation who have adopted the biological species concept have made considerable progress in elucidating the evolutionary mechanisms of species formation. Reproductive isolation has been shown to result from a combination of isolating mechanisms that act in concert to preserve the genetic integrity of species. Theoretical models of speciation indicate that isolating mechanisms arise while populations are geographically isolated, although sometimes populations come back into contact before isolation is complete. This can lead to the merging of species through hybridization, the origin of new hybrid derivatives, or the completion of speciation via the reinforcement of preexisting isolating mechanisms. Analyses of the genetic basis of these isolating mechanisms suggest that postmating isolating barriers typically arise as a by-product of adaptive divergence, whereas premating barriers are more likely to be selected on directly. Further study of the genetic basis of isolation will allow us to estimate the relative frequency and tempo of the different modes of species formation.

See also

Evolution: Views of
Speciation: Genetics
Species and Speciation

Further Reading

Avise JC and Wollenberg K (1997) Phylogenetics and the origin of species. *Proceedings of the National Academy of Sciences of the United States of America* **94**: 7748–7755.

Bush GL (1975) Modes of animal speciation. *Annual Review of Ecology and Systematics* **6**: 339–364.

Coyne JA and Orr HA (1998) The evolutionary genetics of speciation. *Philosophical Transactions of the Royal Society B* **353**: 287–305.

Dobzhansky T (1937) *Genetics and the Origin of Species*. New York, NY: Columbia University Press.

Grant VA (1981) *Plant Speciation*. New York, NY: Columbia University Press.

Howard DJ and Berlocher SH (eds.) (1999) *Endless Forms: Species and Speciation*. New York, NY: Oxford University Press.

Mayr E (1942) *Systematics and the Origin of Species*. New York, NY: Columbia University Press.

Otte D and Endler J (eds.) (1989) *Speciation and its Consequences*. Sunderland, MA: Sinauer.

Rice WR and Hostert EE (1993) Laboratory experiments on speciation: what have we learned in 40 years? *Evolution* **47**(6): 1637–1653.

Rieseberg LH (1997) Hybrid origins of plant species. *Annual Review of Ecology and Systematics* **28**: 359–389.

Speciation: Genetics

Intermediate article

Mark C Ungerer, *North Carolina State University, Rayleigh, North Carolina, USA*

Loren H Rieseberg, *Indiana University, Bloomington, Indiana, USA*

New species arise when previously interbreeding populations become isolated by reproductive barriers. These barriers are thought to originate as a by-product of genetic divergence between geographically isolated populations; they may have a complex polygenic basis or be under simple genetic control.

Article contents

- Introduction
- Prezygotic Barriers
- Postzygotic Barriers
- Haldane's Rule
- Hybrid Speciation
- Conclusions

Introduction

One of the greatest challenges to evolutionary biology is understanding the genetic processes that underlie species formation. Although these processes are responsible for the magnificent array of past and present biological diversity, they have been difficult to study because new species often require thousands or millions of generations to form. Moreover, the many different modes of speciation, each with unique features, make the study of speciation the study of multiple processes. (*See* Speciation.)

The most widely employed definition of species is the biological species concept, in which species are defined as 'groups of actually or potentially interbreeding natural populations, which are reproductively isolated from other such groups' (Mayr, 1942). Reproductive isolation can arise through either barriers that act to prevent mating or fertilization (prezygotic barriers) or barriers that reduce the viability or fertility of hybrids (postzygotic barriers). Although the biological species concept may not apply to all organismal groups, it does provide a conceptual and experimental framework for studies of speciation in sexual taxa. In these groups, the genetic basis of speciation can be reduced to the genetic basis of reproductive isolation. By analyzing the genetic architecture of reproductive barriers (numbers, locations, effects and interactions of loci), models for their evolution can be tested and speciational mode and tempo can be inferred. Almost all contemporary research in speciation genetics is conducted within this framework.

Prezygotic Barriers

Theory

Neo-Darwinian theory suggests that prezygotic barriers arise as an incidental by-product of genetic differences that accumulate between geographically isolated populations. This divergence may be driven by natural selection and/or genetic drift. For example, a selective shift in time of reproduction may isolate conspecific populations even if forced crosses are fully compatible. If the populations remain isolated long enough, postzygotic barriers may evolve as well. However, prezygotic factors were responsible for the initial isolation. Because prezygotic isolation often is a by-product of adaptation to divergent biotic or abiotic conditions, the genetic architecture of prezygotic barriers in many instances can be viewed as synonymous with that of adaptations.

The conventional view derived from neo-Darwinian theory holds that adaptation is a gradual process that requires many small genetic changes. However, others have argued that adaptations often have a simple genetic basis, involving few loci of large effect (Gottlieb, 1984). The implications of these differing viewpoints are huge. If adaptations can have a relatively simple genetic basis, prezygotic barriers can arise more easily and speciation can proceed more rapidly. Only recently have experimental methods

become available that allow these alternatives to be distinguished reliably.

Empirical data

Only a handful of well-designed studies have examined the genetic basis of prezygotic barriers. One of the most informative of these analyzed the number and magnitude of chromosomal segments controlling floral isolation between two closely related species of monkeyflowers, *Mimulus lewisii* and *M. cardinalis* (Bradshaw *et al.*, 1995). *M. lewisii* has a floral morphology that attracts bumblebee pollination (pale pink flowers with yellow nectar guides, a wide corolla with petals modified as landing platforms and a small volume of concentrated nectar), whereas *M. cardinalis* has a floral morphology that is more consistent with hummingbird pollination (red petals lacking nectar guides, narrow corolla and a relatively large nectar volume). Although the two species are sympatric over large portions of their geographic distribution and overlap in flowering time, a lack of natural hybrids testifies to the effectiveness of floral isolation as a barrier to reproduction between them. (*See* Quantitative Trait Loci (QTL) Mapping.)

Bradshaw *et al.* used an interspecific F₂ population to analyze the segregation of eight floral traits that contributed to the differences between species. For all traits, at least one chromosomal segment explained at least 25% of the variance. For three of the eight traits, a single segment explained more than 50% of the variance. Although these results were based on relatively small sample sizes, similar results were found in a follow-up study using much larger sample sizes. The authors concluded that the relative simplicity of these genetic differences indicates that speciation may have been a relatively rapid process in this group. They provide a scenario in which only three major mutations affecting pollinator attraction, reward and efficiency would be necessary for a shift from the ancestral bumblebee-pollinated floral syndrome (*M. lewisii*) to one favoring the derived hummingbird-pollinated floral syndrome (*M. cardinalis*).

Another important prezygotic barrier that has been studied is mate discrimination. Courtship and mating behaviors are often based on physical and/or chemical cues in one sex and preferences in the other. For example, male courtship in the *Drosophila melanogaster* clade is influenced by differences in female pheromones (cuticular hydrocarbons) and male preferences. Coyne *et al.* (1994) combined behavioral observations and classical genetic analysis to show that these differences probably contribute to sexual isolation and to reveal their underlying genetic basis. Only one of the four chromosomes in the *Drosophila* genome was significantly associated with phenotypic differences in hybrids, implying a relatively simple genetic basis for this trait.

Although there are few informative studies of prezygotic barriers, the sparse data such studies have provided appear inconsistent with the neo-Darwinian view of gradual adaptive differentiation involving many small genetic changes. Even complex phenotypic shifts, such as floral differentiation in *Mimulus*, can sometimes be accounted for by a few mutations with major effects. These results lend credence to the view that prezygotic isolation can arise rapidly.

Postzygotic Barriers

Theory

Early studies of species differences revealed a puzzling phenomenon: crosses between closely related species often resulted in inviable or sterile hybrids. This phenomenon seemed unlikely because it was thought that these hybrid genotypes represented the most likely 'pathway' through which incipient species would cross *en route* to becoming a species. If one used the analogy of Sewall Wright's adaptive landscape (species occur on adaptive peaks, and hybrids occupy the intervening valley between peaks), it was difficult to understand how, in the face of natural selection, a new species could bud off from its progenitor and become established on a new peak without proceeding through the intervening maladaptive valley. (*See* Wright, Sewall.)

The solution to this problem has become known as the standard model and represents the first genetic theory of speciation (Dobzhansky, 1937). The standard model suggests that reduced fitness in hybrids is due to two or more 'complementary loci' that arise and function well within separate lineages, but that interact negatively when brought together in hybrids. For example, a single diploid population of genotype *aabb* (where *aa* and *bb* represent homozygous alleles at different loci) becomes divided into two populations between which gene flow is inhibited. In one population, allele *a* gives rise to mutant allele *A*, which increases in frequency and eventually fixes. During the transition in which *A* increases in frequency, genotypes *aabb*, *aAbb* and *AAbb* coexist in the population and are fully interfertile. In the second population, allele *b* gives rise to mutant allele *B*, and it too eventually fixes. As in the first population, genotypes *aabb*, *aaBb* and *aaBB* temporarily coexist and are fully interfertile.

However, if these populations expand and come into contact, hybridization between the now divergent races will yield F$_1$ offspring of genotype *AaBb*. This combination of alleles (*A* and *B* in the same individual) may cause a reduction in the fertility or viability of

hybrids. Thus, the success of this model relies on negative interactions between alleles that function perfectly well within their own genetic background. There are variations on this theme. For instance, in a single lineage, mutant allele *A* may arise and fix, followed by the origin and fixation of *B*. The second lineage does not need to change for an incompatibility to arise. This is because the *B* allele may be incompatible with the *a* allele (*A* was already fixed before *B* arose, thus the combination *aB* was never exposed to selection). This model provided a feasible scenario to explain how a new species could occupy a new adaptive peak without first crossing an adaptive valley. (*See* Epistasis.)

The standard model makes several assumptions. First, it assumes that the fixation of alleles *A* and *B* is not due to selection on their contribution toward reproductive isolation, but rather is due to their ecological value or to stochastic processes. Thus, as with prezygotic barriers, reproductive isolation is a by-product of genetic divergence, which is driven in turn by ecological selection and/or genetic drift. A second assumption is that speciation occurs in allopatry; geographic isolation is required in order for speciation to occur. As will be discussed below, allopatry is not required for some modes of speciation (see the section on Hybrid Speciation). Third, the standard model only applies to the evolution of postzygotic barriers. In some organismal groups such as birds and plants, prezygotic barriers appear to arise earlier than postzygotic ones and may play a more critical role in species formation. Owing to the fragility of prezygotic barriers, however, postzygotic isolating factors may provide more permanent barriers to interspecific gene flow.

A recent mathematical formalization of the standard model (Orr, 1995) made a number of predictions regarding the nature of genetic incompatibilities. First, the alleles contributing to reduced fitness in hybrids are predicted to act asymmetrically. For example, in the above model the *A* allele is incompatible with the *aaBB* population but the *a* allele is compatible with the *AAbb* population. This prediction was first noted by Muller (1942) and makes intuitive sense: if the interaction of alleles *A* and *B* causes reduced fitness in hybrids, then allelic variants *a* and *b* must be compatible because their association represents the ancestral genotype *aabb*. A corollary of this prediction is that hybrid incompatibilities should more often result from derived rather than ancestral alleles. Second, hybrid incompatibilities are more likely to evolve if larger numbers of loci and higher-order interactions are considered. As the numbers of substitutions and higher-order interactions increase, so does the number of potential incompatibilities, thereby providing more pathways to speciation. Third, hybrid incompatibilities should evolve progressively faster and become progressively more complex over time (the likelihood that a newly arising mutation in one population will be incompatible with loci in the other population increases with increasing genetic divergence).

Empirical data

Evidence from several taxonomic levels shows that hybrid incompatibilities are due to the action of complementary loci as predicted by the standard model. An excellent example is provided by hybridization experiments between two fish species in the genus *Xiphophorus* (Wittbrodt *et al.*, 1989). Some platyfish (*Xiphophorus maculatus*) have a dominant sex-linked tumor allele that is regulated by an autosomal repressor. The tumor locus is closely linked to a locus responsible for black pigmentation patterns. Within species, individual platyfish that have all three loci (the tumor allele, the repressor and the pigmentation locus) develop melanophore spot patterns but no tumors. Closely related swordtails (*X. helleri*) presumably have none of these loci (or inactive forms) and develop neither spot patterns nor tumors. When platyfish and swordtails are crossed, however, some backcross hybrids develop malignant melanomas. Classical genetic analysis suggests that this is due to the abnormal regulation of the tumor allele in hybrids that lack the repressor.

The formal predictions of the standard model (Orr, 1995) have also been validated empirically. For example, a substantial body of studies indicates that incompatibility loci do act asymmetrically. This is illustrated by hybridization studies between two fruit-fly species, *Drosophila pseudoobscura* and *D. persimilis*, in which heterospecific introgressions that cause hybrid male sterility in one direction fail to do so when the same chromosomal regions are introgressed in the reciprocal direction (Wu and Beckenbach, 1983). Asymmetric hybrid male sterility has been observed in other *Drosophila* crosses (reviewed in Orr, 1997) and probably represents the earliest stages of speciation.

Strong evidence has also been compiled for a polygenic basis for hybrid incompatibilities in many taxa, some with complex interactions. For example, fine-scale introgression studies have identified 15 loci on the X chromosome of *Drosophila simulans* that cause male sterility in a *D. mauritiana* genetic background (Wu *et al.*, 1996). If the distribution of sterility loci across the rest of genome is similar, as many as 120 sterility loci may isolate these species (Wu *et al.*, 1996). Introgression studies in plants have provided similar results. In wild sunflowers, introgression of *Helianthus petiolaris* chromosomes into an *H. annuus* background

revealed 14 segments from seven chromosomes that contribute to reduced hybrid fitness (Rieseberg, 1998). Extrapolating this finding to the rest of the genome suggests that as many as 40 chromosomal segments may negatively affect hybrid fitness. The results of these and other studies (reviewed in Coyne and Orr, 1998) are consistent with predictions of the standard model, but should be interpreted cautiously because it is difficult to differentiate between incompatibilities that arose before and after speciation.

Clinal patterns in natural hybrid zones can also provide clues regarding the genetic architecture of species barriers. This approach employs estimates of the width of the region of reduced viability, dispersal rate, patterns of linkage disequilibrium and strength of selection against hybrids to estimate the number of loci contributing to hybrid inviability. Applying this approach to well-characterized hybrid zones in *Podisma* (Barton and Hewitt, 1981) and *Bombina* (Szymura and Barton, 1991) yields gene number estimates of 50–500 in *Podisma* and 26–88 in *Bombina*. Although these estimates are consistent with the prediction of the standard model that postzygotic isolation is polygenic, they tell us little about the location, effects and interactions of specific loci.

Haldane's Rule

In 1922, J. B. S. Haldane, a founder of the synthetic theory of evolution, made the following observation:

> When in the F_1 offspring of two different animal races one sex is absent, rare, or sterile, that sex is the heterozygous [heterogametic] sex. (Haldane, 1922)

This pattern is referred to as Haldane's rule, and it has been observed in all surveyed animal taxa that have chromosomal sex determination, regardless of which sex is heterogametic. In mammals and *Drosophila*, males are heterogametic and are the afflicted sex, whereas in birds and Lepidoptera, females are heterogametic and are the afflicted sex. Such consistency across species groups suggests an underlying genetic mechanism associated with sex chromosomes. (*See* Haldane, John Burdon Sanderson.)

No single genetic mechanism appears capable of explaining all observations of Haldane's rule. At present, two hypotheses are most consistent with the data. The first and arguably the largest contributor to Haldane's rule is the dominance theory (Orr, 1997). This theory is actually a slightly modified reincarnation of a hypothesis proposed some 60 years ago (Muller, 1942). The dominance theory explains Haldane's rule for hybrid inviability and contributes to (but cannot fully explain) the sterility of heterogametic hybrids. The efficacy of the dominance theory relies on the recessive nature of most genetic incompatibility factors. Because hybrid males that are heterogametic have hemizygous X chromosomes, X-linked loci contributing to hybrid inviability will be expressed, whereas in homogametic females, the expression of these X-linked loci may be masked by dominance. Although this was the basic idea postulated by Muller, the dominance theory does point out an oversight by Muller: females possess twice as many X chromosomes and are therefore expected to possess twice as many inviability alleles. If these loci act additively, males and females will be affected equally. Only when these loci are on average partially recessive will their effects be preferentially expressed in males. It is also assumed that the effects of inviability loci are the same in males and females.

Although the dominance theory explains most attributes of Haldane's rule, one puzzling observation remains unresolved. When males are the heterogametic sex, Haldane's rule is about 10 times more likely to cause male sterility than male inviability. This trend all but disappears when females are heterogametic. Wu *et al.* (1996) have suggested a 'rapid male evolution' model to explain the high frequency of hybrid male sterility relative to inviability. They argue that loci for male sterility evolve faster owing to sexual selection or some physiological characteristic of spermatogenesis that makes it more vulnerable to mutation. Although this theory does not explain Haldane's rule in organisms that have heterogametic females, there is strong evidence that male sterility evolves faster than female sterility.

Hybrid Speciation

The preceding sections have discussed divergent modes of speciation in which single lineages split to form two or more descendent lineages. However, there are many well-characterized instances of hybrid speciation in which new species arise from the merger of the genomes of genetically differentiated species. These hybrid lineages subsequently form reproductive barriers with their progenitors and can found major organismal groups. Hybrid modes of speciation are important not only because they are the chief cause of reticulate evolution, but also because they demonstrate that new species can arise rapidly and in some cases in sympatry. Hybrid speciation appears to be most common in plants, and most detailed genetic studies of this mode focus on plant systems.

Polyploid speciation

Polyploid speciation represents a rapid, sympatric mode of speciation in which new species may arise in a single generation through genome duplication. Genome duplication also leads to instantaneous reproductive isolation between the new polyploid species and its parents. Polyploid formation usually proceeds via fusion of unreduced gametes derived from individuals of the same species (autopolyploidy) or from individuals of different species (allopolyploidy). Thus, interspecific hybridization is a requirement for allopolyploid speciation, but autopolyploid species may arise by crosses between individuals from the same population or by hybridization between geographic races or subspecies. (*See* Polyploid Origin of the Human Genome.)

Theoretical studies have identified a number of factors that favor the establishment of polyploids in nature (Roderiguez, 1996). These include differential niche preference, a selfing mating system, high fecundity and a perennial life history. Niche separation and selfing enhance the probability of successful matings during the early stages of establishment. Random mating would result in a preponderance of mating with the diploid parental species and subsequent production of sterile triploid seed. Stochastic events due to a small number of polyploid colonizers decrease the chance of establishment, but this barrier is minimized by high fecundity and a perennial life history. Roderiguez' concluding assertion that 'the establishment of polyploids in higher plants is not an unlikely event' is supported by the high frequency of polyploidy in plants and by recent evidence for the multiple origins of many polyploid species, even those originating in the twentieth century (see below).

Recent applications of molecular tools to studies of polyploid evolution have yielded insights into the extent and distribution of duplicated sequences in polyploid genomes, the evolutionary potential of auto- versus allopolyploids, the nature of genomic changes occurring after polyploid formation and the number of independent origins of polyploid species. Results from studies of the origins of polyploid species have been particularly striking, because they suggest that most polyploid taxa have arisen on multiple independent occasions. For example, early in the twentieth century three diploid goatsbeard species (*Tragopogon dubius*, *T. porrifolius* and *T. pratensis*) were introduced from Europe into the Palouse region of eastern Washington State, USA. Shortly thereafter, morphological and cytological studies revealed that interspecific hybridization between the diploids had led to the formation of two allotetraploid species, *Tragopogon mirus* and *T. miscellus*, which were parented by *T. dubius* × *T. porrifolius* and *T. dubius* × *T. pratensis* respectively. Molecular marker surveys in natural populations of the three diploid and two tetraploid species identified multiple distinct genotypes, suggesting 5–9 independent origins for *T. mirus* and two to 2–21 origins for *T. miscellus* (Soltis *et al.*, 1995).

Homoploid hybrid speciation

Hybridization can also lead to the formation of new species that are at the same ploidal level as their progenitors. In this mode, interspecific hybridization is followed by selection for the most viable and fertile hybrid segregants. One of these may become established if it becomes reproductively isolated from the parental species. Theory suggests that reproductive isolation may arise via the sorting of chromosomal or genic sterility factors that differentiate the parental species or via the fixation of novel chromosomal mutations induced by hybridization.

Although this mode of speciation is probably rare, it does appear to have occurred on multiple occasions in plants. The process is perhaps best illustrated by the origin of a wild sunflower species, *Helianthus anomalus*, which is a stabilized hybrid derivative of two widespread sunflowers, *H. annuus* and *H. petiolaris* (reviewed in Rieseberg, 1998). Comparative genetic mapping studies indicate that the two parental species differ by a minimum of 10 chromosomal rearrangements. Sorting of these rearrangements in the hybrid speciation process placed two rearrangements from each parental species into the *H. anomalus* genome. However, considerable chromosomal breakage occurred during the speciation process as well: three chromosomal breakages, three fusions and one duplication are required to derive the *H. anomalus* genome from its parents. As a result of this karyotypic divergence, artificial hybrids between *H. anomalus* and either parental species are almost completely sterile.

Comparison of the *H. anomalus* genome to three experimentally synthesized hybrid lineages of *H. annuus* × *H. petiolaris* provided further insights into the speciation process. Although these synthetic lineages were generated independently, all three converged onto near identical gene combinations, and these combinations were recognizably similar to those of *H. anomalus*. Similarity in genomic composition suggests that deterministic forces such as selection, rather than stochastic forces, largely govern the formation of hybrid species. Furthermore, the apparent repeatability of this mode of speciation suggests that homoploid hybrid species, like allopolyploids, may have multiple, independent origins.

Conclusions

It is now feasible to estimate the number and magnitude of genetic changes required for evolution of reproductive barriers. Initial studies exploiting this power (many of which are described here) indicate that sweeping generalizations about the genetic basis of species barriers are no longer tenable: reproductive barriers can result from the combined effects of many loci (as envisioned by the neo-Darwinian theory) but often require few genetic changes. This does not mean that speciation has no regularities. On the contrary, observed discrepancies almost surely reflect real differences among organismal groups, speciational modes and kinds of reproductive barriers. We expect, for example, that different genetic architectures will be associated with different speciational modes. In fact, the genetic signatures left by different speciational processes may provide the only real clues for estimating their frequency. Similarly, organismal groups that differ in genomic constitution and life history might be expected to vary with regard to the kinds of genetic changes required for species formation. On the other hand, consistent patterns such as Haldane's rule indicate that there are regularities in the process of speciation that can override differences among organismal groups or speciational modes. A major challenge is to identify and explain these regularities and account for differences that are observed among organismal groups or among kinds of reproductive barriers.

There has also been considerable progress in defining and answering specific questions about speciation. After years of fierce debate, Haldane's rule now appears to be a composite phenomenon; both the dominance theory and rapid male evolution are required to explain all observed cases. Specific predictions derived from the standard model for the evolution of postzygotic barriers have been confirmed. Polyploid formation is no longer considered to be rare or sporadic, and most polyploid taxa are now recognized to be of multiple, independent origin. Homoploid hybrid speciation has been confirmed and, like polyploid speciation, it appears to be a highly repeatable process. These successes have not only raised the profile of speciation research, but also portend future accomplishments in this area.

See also

Evolution: Views of Speciation

References

Barton NH and Hewitt GM (1981) The genetic basis of hybrid inviability in the grasshopper *Podisma pedestris*. *Heredity* **47**: 367–383.

Bradshaw Jr HD, Wilbert SM, Otto KG and Schemske DW (1995) Genetic mapping of floral traits associated with reproductive isolation in monkeyflowers (*Mimulus*). *Nature* **376**: 762–765.

Coyne JA, Crittenden AP and Mah K (1994) Genetics of a pheromonal difference contributing to reproductive isolation in *Drosophila*. *Science* **265**: 1461–1464.

Coyne JA and Orr HA (1998) The evolutionary genetics of speciation. *Philosophical Transactions of the Royal Society B* **353**: 287–305.

Dobzhansky T (1937) *Genetics and the Origin of Species.* New York, NY: Columbia University Press.

Gottlieb LD (1984) Genetics and morphological evolution in plants. *American Naturalist* **123**: 681–709.

Haldane JBS (1922) Sex ratio and unisexual sterility in animal hybrids. *Journal of Genetics* **12**: 101–109.

Mayr E (1942) *Systematics and the Origin of Species.* New York, NY: Columbia University Press.

Muller HJ (1942) Isolating mechanisms, evolution, and temperature. *Biological Symposia* **6**: 71–125.

Orr HA (1995) The population genetics of speciation: the evolution of hybrid incompatibilities. *Genetics* **139**: 1805–1813.

Orr HA (1997) Haldane's rule. *Annual Review of Ecology and Systematics* **28**: 195–218.

Rieseberg LH (1998) Genetic mapping as a tool for studying speciation. In: Soltis DE, Soltis PS and Doyle JJ (eds.) *Molecular Systematics of Plants*, 2nd edn, pp. 459–487. New York, NY: Chapman & Hall.

Roderiguez DJ (1996) A model for the establishment of polyploidy in plants: viable but infertile hybrids, iteroparity, and demographic stochasticity. *Journal of Theoretical Biology* **180**: 189–196.

Soltis PS, Plunkett GM, Novak SJ and Soltis DE (1995) Genetic variation in *Tragopogon* species: additional origins of the allotetraploids *T. mirus* and *T. miscellus* (Compositae). *American Journal of Botany* **82**(10): 1329–1341.

Szymura JM and Barton NH (1991) The genetic structure of the hybrid zone between the fire-bellied toads *Bombina bombina* and

B. variegata: comparisons between transects and between loci. *Evolution* **45**: 237–261.

Wittbrodt J, Adam D, Malitschek B, *et al.* (1989) Novel putative receptor tyrosine kinase encoded by the melanoma-inducing *Tu* locus in *Xiphophorus*. *Nature* **341**: 415–421.

Wu C-I and Beckenbach AT (1983) Evidence for extensive genetic differentiation between the sex-ratio and the standard arrangement of *Drosophila pseudoobscura* and *D. persimilis* and identification of hybrid sterility factors. *Genetics* **105**: 71–86.

Wu C-I, Johnson NA and Palopoli MF (1996) Haldane's rule and its legacy: why are there so many sterile males? *Trends in Ecology and Evolution* **11**: 281–284.

Further Reading

Barton NH and Hewitt GM (1989) Adaptations, speciation and hybrid zones. *Nature* **341**: 497–503.

Otte D and Endler J (eds.) (1989) *Speciation and its Consequences.* Sunderland, MA: Sinauer Associates.

Rice WR and Hostert EE (1993) Laboratory experiments in speciation: what have we learned in 40 years? *Evolution* **47**: 1637–1653.

Rieseberg LH (1997) Hybrid origins of plant species. *Annual Review of Ecology and Systematics* **28**: 359–389.

Soltis DE and Soltis PS (1993) Molecular data and the dynamic nature of polyploidy. *Critical Reviews in Plant Sciences* **12**: 243–273.

Templeton AR (1981) Mechanisms of speciation – a population genetic approach. *Annual Review of Ecology and Systematics* **12**: 23–48.

Wu C-I and Palopoli M (1994) Genetics of postmating reproductive isolation in animals. *Annual Review of Genetics* **27**: 283–308.

Species and Speciation

Daniel Otte, *The Academy of Natural Sciences, Philadelphia, Pennsylvania, USA*

Introductory article

Article contents

- Species Defined
- Speciation (Origin of Species)

A species is the fundamental or basic unit of organic or biological diversity, usually characterized as a set of individuals or populations of individuals that constitute an interbreeding or potentially interbreeding unit. Speciation is the formation or origin of species, usually the formation of several new daughter species from an existing ancestral species.

Species Defined

Biological diversity is an observational fact familiar to everyone. In their daily lives people encounter many different kinds of animal and plant that are often equivalent to what biologists call species: people, dogs, horses, fleas, carrots, apples. However, the word 'kind' is a more general term that can refer to a variety of classes or groups of animals and plants – including species. Thus, both dogs and greyhounds may be called kinds, but only dogs rank as species. The species in biology is a more precise term that is based on the

most inclusive group whose members are able to interbreed.

Taxonomists (the biologists who describe and classify the living world) classify individuals by first putting them into populations, these in turn are assigned to species, species are put into genera, genera into subfamilies and so on. Taxonomists must choose which populations belong in a particular species, which species belong to a genus and so on. Because in most cases there are no rules other than that the members of a group must share a common ancestry, the classification is somewhat arbitrary. For example, a case might be made for grouping five species into one genus or into two or even three related genera. Opinions on how to group organisms are therefore numerous and disagreements are common.

Only one classification category, the species, is not arbitrary (with certain exceptions; see below). Species therefore tend to remain stable in any classification system. The fundamental attribute that sets species apart from other categories is that the members of species – those with sexual reproduction – are part of an interbreeding entity. Because the species is definable and stable, it is treated as the most important basal unit in the classification of organisms. The species described by Linnaeus more than 200 years ago are much more likely to be recognized today than are the genera or families that he recognized.

A definition of species that is satisfactory to all biologists remains elusive. There are three main reasons for this:

1. The system of naming organisms – the Linnaean binomial system (genus and species, as in *Homo sapiens*) – was first applied to sexual forms. Later, systematists applied the system to forms with very different kinds of reproduction, for example, to asexual (uniparental) organisms where a definition based on interbreeding cannot be applied because interbreeding does not occur. To reduce confusion, it is usually convenient to refer to two kinds of species – biparental species and uniparental species.
2. Paleontologists use the species category somewhat differently from biologists concerned with living species. They find it convenient to subdivide a long lineage in which much evolutionary change has occurred into a set of related species, one sequentially replacing another.
3. Because the formation of species is a gradual process, it is often true that two forms have not yet reached a state in which they can be called fully formed species. Often such populations are geographically separated and their ability to interbreed can be tested only with difficulty or not at all. In such cases, the decision as to whether they should be considered separate species or not is arbitrary.

Biparental (sexual) species

The essential feature of biparental reproduction is that two individuals (male and female) are needed to produce the next generation. Half of an individual's genes come from the female parent and half from the male parent. Genes from any individual in the population can be combined with those of any other of the opposite sex. In uniparental forms, the genes in one parent cannot be combined with those in another parent.

The most widely (although not universally) accepted definition of biparental, currently living species is the following. Species are groups of actually or potentially interbreeding natural populations that are reproductively isolated from other such groups. Species consist of populations that allow genes to spread more or less freely among them, sometimes indirectly through a chain of populations (e.g. from Chihuahuas to Great Danes), but which prevent the flow of genetic information across their boundaries (e.g. between dogs and cats). Sexual species may be thought of as containment devices that prevent the flow of genetic information across their boundaries.

When two forms coexist and do not interbreed, their status as distinct species is clear. However when they are separated (geographically, by year or season of breeding, or by utilizing different hosts) and no interbreeding tests can be applied, their status must remain uncertain. Still, taxonomists usually take a stand and state whether they believe two forms have reached the status of species or not. This they do by making an intelligent guess as to whether interbreeding might occur on the basis of their experience with other cases and other situations. Related forms that are known to belong to different species where they occur together may be used as a reference point. Some taxonomists and systematists refuse to make a guess as to whether interbreeding might occur, and recognize all geographically isolated forms that are reliably distinguishable as distinct species (this is known as the phylogenetic species concept). Others simply apply some nontaxonomic, neutral label to such populations and refer to them as 'evolutionary units' or 'operational taxonomic units'.

Incomplete species

The naming of species and their ordering into a classification is made more difficult by the fact that many species are still being formed. The following kinds of relationship between diverging forms appear common in nature; some present no difficulties to the taxonomist, while others are more problematical. All are of interest to the ecologist and evolutionary biologist.

1. Two forms live together; they do not interbreed and they remain discrete. Such forms are universally treated as members of two different species.
2. Two forms live in different areas and interbreed where they meet. If they appear to be coalescing into a single interbreeding population, such forms may be called subspecies or races, but not species.
3. Two forms live mostly in different areas and they interbreed where they meet. Interbreeding continues along a narrow front; but away from the front

the two forms retain their discreteness. Such forms would normally be called species.

4. Two forms are separated by a small or moderate barrier and acquire small differences. They may be either treated as discrete populations of the same species or treated as distinct species. In the absence of breeding tests, the decision is arbitrary.

5. Two forms are separated by a major barrier that will forever keep the forms apart. They have acquired some differences by which they can be distinguished. These forms may be treated as varieties, races, subspecies or species, depending on how similar they are to examples of related species that overlap geographically.

The naming of units is often a largely arbitrary endeavor. It is for this reason that emphasis should be put on describing the situation, rather than naming the units.

Uniparental (asexual) species

When a single parent is capable of producing offspring without the genetic contribution of another individual, reproduction is called uniparental or asexual. Some organisms reproduce through simple fission, or by the production of new buds or new sprouts, each of which becomes a separate individual. In others, the eggs or ovules develop without fertilization (parthenogenesis); in still others, both gametes come from the same individual (self-fertilization). The genetic consequence is that the progeny acquires the same genotype as its ancestor – except for the occurrence of mutation. Clones of genetically similar individuals are formed and every clone is isolated from every other.

Genetic changes may occur in clones or in pure lines through gene mutations or chromosomal alterations. A new altered clone or pure line arises in this manner. If the effects of the mutation are favorable, the new clone may replace the existing one; otherwise, it is eliminated by natural selection. The formation of favorable gene combinations (as occurs in biparental speciation) is only possible with the occurrence of a series of mutations in the same line of descent.

For purposes of classification, it is fortunate that the variation in asexually reproducing groups is not absolutely continuous. Taxonomists find aggregations of more or less clearly distinct types, each of which is constant and reproduces its kind (biotype) if allowed to breed. These types are sometimes called species, but they are not united into integrated groups as in species of cross-fertilizing forms.

Clusters of asexual forms can be arranged in a hierarchical order, in a way that is analogous to that encountered in sexual forms. The different clusters may be given some rank – some as species, others as subgenera or genera and so on. Which rank is ascribed to a given cluster is based on convenience; the decision is in this sense purely arbitrary.

Species in paleontology

In living forms, different names are applied only to taxa that have undergone lineage splitting. However, in paleontology it is common practice, for the purposes of classification, to break up a single lineage into segments and to give them different names. In the case of lineages broken by fossil gaps the task is easy, but in other cases the splitting must be done arbitrarily because there is no nonarbitrary way to subdivide a continuous line. Usually some criterion is needed to indicate how large to make the segments. Successive species are ordinarily defined so as to make the differences between them as great as differences among contemporaneous species of an allied group.

Because use of the criterion of interbreeding to recognize species cannot apply in temporally separated populations, paleontologists employ the evolutionary species concept: a species is a single lineage of ancestral descendant populations of organisms that maintains its identity from other such lineages and which has its own evolutionary tendencies and historical fate.

Speciation (Origin of Species)

Origin of biparental species

As soon as two parts of a species are separated from one another, they begin to evolve differences through mutation, selection and sampling error (genetic drift). Hence the beginnings of speciation may be seen everywhere. If the isolation is ended (as is usually the case), the two parts may reamalgamate by interbreeding and their distinctiveness is lost. If they remain separated long enough, however, they come to differ from each other in many genes and chromosomal alterations so as to make the two populations reproductively incompatible and unable to produce a viable hybrid generation.

A daughter species may arise from one or a few individuals that by a rare accident reach a geographically isolated region, or two or more daughter species may arise when an entire population is subdivided by some event.

The first stages of species formation may be seen in most species. Species are frequently aggregates of populations, each one possessing its own complex of characteristics. Many plant species are split into numerous 'ecotypes' – groups of plants adapted to living in definite ecological situations. Coastal, dune, forest, swamp, alpine and other ecotypes may be recognized. The formation of geographical races is probably the most usual method of differentiation of species in both animal and the plant kingdoms. Geographical races may in turn be subdivided into smaller secondary races.

Two modes of gradual speciation have been postulated, those involving geographical separation of the diverging populations (allopatric speciation) and those without geographic separation (sympatric speciation). The formation of species without geographical separation (the sympatric model) has remained controversial ever since Darwin. Some biologists believe it occurs, but note that to bring it about an interruption of interbreeding is essential. Life cycle switches or host switches are possible mechanisms. A life cycle switch may occur in which a portion of the population gets accidentally switched to a new life cycle; for example, in species that take more than 1 year to emerge, a part of a population may emerge in odd years and so become separated from the part that breeds in even years. A host shift involves the movement of a part of a species onto a new host. In this case, genetic isolation is achieved if mating also occurs on the new host.

Reproductive isolation

Isolation of populations is essential for them to diverge genetically. Factors producing reproductive isolation are generally of two kinds: geographical and physiological. Physiological barriers to interbreeding can develop only after a period of spatial or temporal isolation. Variation in nature furnishes evidence to support this thesis. Ever since Darwin, it has been regarded as probable that the formation of geographical races is an important antecedent of species formation. The forms of reproductive isolation involved in speciation can be classified as follows.

Geographical and temporal isolation (extrinsic isolation)

Groups of individuals may be prevented from interbreeding by living in different geographical regions or by breeding at different times. A special case of extrinsic isolation (which might be termed microgeographic isolation) is one in which two forms are isolated on two different hosts and in which mating also occurs on the host (sympatric speciation; see above). Prolonged geographical and temporal isolation leads inevitably to physiological isolation.

Physiological isolation (intrinsic isolation)

This type of isolation evolves both as a result of geographical or other kinds of extrinsic separation but also as a result of selection against interbreeding. Physiological isolation is any isolation resulting from an incompatibility of the parental forms that prevents

mating or the production of hybrid zygotes, or which produces disturbances in the development such that no hybrids reach the reproductive stage. Physiological isolation can take several forms:

- *Hybrids are unfit*: Zygotes form but they do not develop or the offspring are sterile, or the offspring themselves produce inviable offspring. The life of a hybrid zygote may be cut short at any stage, beginning with the first cleave of the egg and up to the late embryonic or postembryonic development.
- *Parental gametes cannot join*: The parents mate but the gametes do not join to form a zygote. Copulation in animals with internal fertilization, or the release of the sexual products into the medium in forms with external fertilization, or the placing of the pollen on the stigma of the flower in plants, is followed by chains of reactions that bring about the actual union of the gametes or fertilization proper. These reactions may be out of balance in representatives of different species, with a consequent hindrance or a prevention of the formation of hybrid zygotes.
- *Parental forms cannot mate (mechanical isolation)*: The parents are physically prevented from mating by their physical attributes. According to the 'lock-and-key' theory, the female and the male genitalia of the same species (at least in insects) are so closely matched to each other that deviations in the structure of either make copulation physically impossible or restrictive. Such physical incompatibilities are important in some groups, but not in others. The differences in the flower structure in related species of plants may prevent cross-fertilization because the flowers are pollinated by different insects.
- *Parental forms will not mate (psychological isolation)*: Mating does not take place because the parents do not recognize one another as potential mates or they choose not to mate with one another. A lack of attraction may be due to differences in scents, courtship behavior, sexual recognition signs, etc.
- *Parental forms never meet (ecological isolation)*: Parents have different ecological stations – they live in different places or breed at different times.

The flow diagram presented in **Figure 1** illustrates possible causal relations. The horizontal arrow indicates that extrinsic (usually geographical) isolation must occur for physiological isolation to evolve. The first physiological difficulties to arise are developmental disorders – hybrids are unfit or gametes do not join. Such difficulties occur after mating and can be circumvented by the prevention of mating. Hence any tendency not to mate would be beneficial. The vertical arrow points to forms of isolation that could

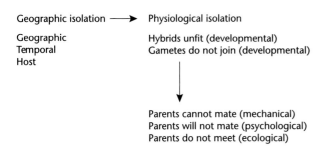

Geographic isolation ⟶ Physiological isolation

Geographic
Temporal
Host

Hybrids unfit (developmental)
Gametes do not join (developmental)

Parents cannot mate (mechanical)
Parents will not mate (psychological)
Parents do not meet (ecological)

Figure 1 Possible causal relations between different types of isolation.

evolve (be selected) as a consequence of postmating disorders.

The origins of various kinds of isolation are of great interest. Any or all of the physiological forms of isolation could be the result of evolution during geographical isolation. However, it has been proposed and supported with evidence that hybrid inferiority could act selectively on psychological, mechanical or ecological isolation. The occurrence of hybridization is often disadvantageous because it breaks down gene complexes that work well together in the developing organism. Genetic factors that tended to prevent the formation of hybrids (by improving mechanical, psychological and ecological isolation) may therefore be favored by natural selection.

Sexual selection in relation to speciation

The failure to recognize a potential mate or the absence of mutual attractions between males and females may be the result of differences in scents, courtship behavior, sexual recognition signs, etc. Darwin coined the term sexual selection to refer to the evolutionary effect that acts or features of one sex have on the other sex. The two most commonly held views on the causes of nonrecognition or discrimination are that (1) they are an accidental by-product of the geographical separation – as a result of adaptation to different environments during the separation, the isolated populations diverge to the point where they are no longer recognizable as potential mates; or (2) they result from selection that favors those parents who will not mate because of the negative consequences of mating. Organisms do not mate indiscriminately – they choose their mates. Because individuals vary in their quality as potential mates, we expect natural selection to favor traits that ensure that mating occurs with partners of the highest possible quality. The payoff from mating reliably with a member of the correct species may be much greater than that to be gained by discriminating between members of the same species simply because the consequences of such a mating are much more likely to be detrimental.

Sudden origin of species (polyploidy)

The sudden origin of a species by gene mutation is an impossibility. Races of a species, or species of a genus, differ from each other in many genes and often also in chromosome structure. A mutation that would bring a new species into being must, therefore, involve simultaneous changes in many gene loci. The known mutation rates show that the probability of such an event is negligible.

The only known mode of instantaneous speciation is through polyploidy – a multiplication of the normal chromosome number. If the normal diploid chromosome number ($2n$) of a species is 14, then multiples of 7 other than 14 are polyploids. Individuals with three chromosome sets are called triploids, individuals with four sets are tetraploids and so forth. Polyploidy is very widespread among plants and is one of the important mechanisms of speciation in the plant kingdom. Probably more than a third of all species of plants have arisen by polyploidy, but it is rare in conifers and fungi.

Speciation through polyploidy is of two kinds. Autopolyploid species arise when more than two haploid chromosome sets of a single species participate in the formation of the zygote. An allopolyploid species arises when the chromosome sets come from two species. Normally an allopolyploid arises through the doubling of the chromosomes in a hybrid.

It is believed that polyploidy among animals is rare. Nearly all proven cases occur in species that have abandoned sexual reproduction in favor of permanent parthenogenesis or self-fertilizing hermaphroditism.

Breeding structure and speciation

The kinds of barrier that isolate gene pools are extraordinary in their variety, and current knowledge of the minimal barriers needed to bring about speciation is exceedingly poor. Furthermore, there is good evidence that barriers need not be absolute; partial or porous boundaries may be sufficient to restrict the flow of genes so that the homogenizing effects of migration are overcome by the evolution occurring in the separated parts.

The magnitude of the natural barrier needed to isolate populations depends a great deal on the mobility of organisms; thus, a given barrier is greater for a flightless insect than for a flying one. Species of land snails, because of their limited means of locomotion, are particularly apt to be subdivided into colonies whose members seldom pass over the barriers that separate one colony from another.

Although geographical separation of populations is thought to be the most important event, cases have come to light in which portions of the same populations become isolated from one another ecologically, either through host shifts or through changes in time of breeding (sympatric speciation).

Partial isolation of local populations, even if only by distance, seems to be important not only as a possible precursor of the splitting of species, but also as leading to more rapid evolutionary change of the populations as a single system and thus more rapid differentiation from other populations from which it is completely isolated. Local differentiation within a species, based either on nonadaptive inbreeding and genetic drift effects or on local conditions of selection or both, permits trial and error both within series of multiple alleles and between gene combinations, and thus a more effective process of selection than is possible in a purely panmictic population (i.e. one with random interbreeding among members of the entire population).

Species as ecological entities

Some biologists treat species as ecological units and suggest that a classification based on their ecological roles is desirable. The ecological characteristics of a wide spectrum of species reveals some interesting patterns. At one end of the scale are species that are so ecologically variable or labile that their parts could be (and sometimes have been) treated as distinct species. Sometimes the environment of a single species provides multiple niches, leading to the evolution and coexistence of functionally diverse forms within a single gene pool. In some lakes, the Cladoceran zooplankton species *Daphnia pulex* exists as two genetically distinct forms. It is likely that many populations are faced with more than one possible strategy for survival and reproduction.

How easy is it for selection to produce two or more specialized forms in a population, each with its own unique configuration of traits? The principal problem is that of overcoming the effects of interbreeding. Models suggest that single species in multiple-niche environments should sometimes differentiate into specialized forms, even in single panmictic gene pools.

Differences between local populations, sometimes only a few meters apart from each other, can also have adaptive explanations. Natural selection is often far more intense and gene flow often more limited than previously envisioned. It is likely that ecologically plausible forms of selection can sometimes cause a randomly mating population to bifurcate into ecologically distinct forms whose intermediates are relatively unfit.

At the other end of the scale are whole groups of species that are so similar that they behave more like a single species than several. The similarity is so strong (in all respects) that one is forced to question the rule that different species can only coexist if they occupy different niches. In these cases an entire cluster of species may constitute an ecological unit.

Species as evolutionary units

Most species are broken up into more or less isolated colonies. The colonies, and not species as wholes, may be considered to be elementary interbreeding communities. Because populations respond to natural selection as a unit, evolution is defined as gene frequency change within populations. Therefore, evolutionary theory often focuses on the population and not the species as the place of evolutionary change. However, if a population is embedded in a network of populations – a species – its fate is dependent on those of the others. In such cases the evolving unit is the species.

But the interbreeding unit may be even larger and may extend beyond the species to any grouping within which there is any gene flow. In plants, taxonomists have defined species that exist in a larger unit known as the syngameon that are characterized by natural hybridization and limited gene exchange. The syngameon can be viewed as the most inclusive unit of interbreeding in a hybridizing species group.

Speciation and morphological change

Analysis of geographical variation in molecular features and in behavior can be used to identify the boundaries of species that are not perceptible using morphology alone; these boundaries are demonstrated by the maintenance of fixed genetic differences at the molecular level where reproductively incompatible populations geographically overlap, or they may be demonstrated by highly species-specific signaling systems. Molecular systems that demonstrate a variety of characteristic evolutionary rates show that speciation and morphological change are distinct and decoupled phenomena in many groups – this is especially evident in the 'sibling groups', which are groups of related species in which it is extremely difficult or even impossible to distinguish the species using their morphology. Clearly, speciation may occur at a higher rate than the origins of morphological novelties. This fact suggests that for groups in which morphology is the sole source of evidence, the number of species is seriously underestimated. In acoustical insects, for example, a taxonomy based on morphology alone would grossly underestimate the true number of species.

See also

Evolution: Views of
Gene Trees and Species Trees
Speciation
Speciation: Genetics

Further Reading

Dobzhansky T (1982) *Genetics and the Origin of Species*. New York, NY: Columbia University Press.

Ereshefsky M (ed.) (1992) *The Units of Evolution: Essays on the Nature of Species*. Boston, MA: MIT Press.

Grant V (1981) *Plant Speciation*, 2nd edn. New York, NY: Columbia University Press.

Mayr E (1963) *Animal Species and Evolution*. Cambridge, MA: Belknap Press of Harvard University Press.

Otte D and Endler JA (eds.) (1989) *Speciation and its Consequences*. Sunderland, MA: Sinauer Associates.

Provine WB (ed.) (1986) *Evolution: Selected Papers by Sewell Wright*. Chicago, IL: University of Chicago Press.

Stebbins GL (1950) *Variation and Evolution in Plants*. New York, NY: Columbia University Press.

Species Trees

See Gene Trees and Species Trees

Spinal Muscular Atrophy

Intermediate article

Henning Schmalbruch, *Department of Medical Physiology, University of Copenhagen, Denmark*

Spinal muscular atrophy is the name for denervating disorders of skeletal muscles caused by hereditary degeneration of lower motor neurons. This definition includes not only infantile and juvenile spinal muscular atrophy but also such diverse diseases such as Kennedy syndrome and neuronal Charcot–Marie–Tooth disease. These diseases differ in heredity and genetic basis; spinal muscular atrophy in the strict sense is a recessively inherited condition localized on chromosome 5.

Article contents

- Introduction
- Clinical Picture
- Genetics of Spinal Muscular Atrophy
- Histopathology
- Animal Models and Clues to the Pathogenesis
- Therapeutic Outlook
- Conclusion

Introduction

Spinal muscular atrophy (SMA) is the name for denervating disorders of skeletal muscles due to hereditary degeneration of the lower motor neurons. This definition includes not only infantile and juvenile spinal muscular atrophy but also such diverse diseases such as Kennedy syndrome and neuronal Charcot–Marie–Tooth disease. This article discusses only the recessively inherited forms of SMA that have been located on chromosome 5 (Brzustowicz *et al.*, 1990, Melki *et al.*, 1990). Since the beginning of the 1990s the common genetic basis of these seemingly different nosological entities has been clarified.

Clinical Picture

Infantile spinal muscular atrophy or SMA 1 is the most frequent form that affects 1 in 10 000 newborns; 2% of the general population are heterozygous carriers. Affected children are hypotonic ('floppy') and tend to lie in a frog-like position; proximal muscles are preferentially affected. Intellectual deficits are uncommon.

Most children die during the first year of life from respiratory complications; survival beyond 2 years is rare. The milder form, SMA 2, differs from SMA 1 in that children with SMA 1 never sit independently, while those with SMA 2 learn to sit but not to walk. The survival rates in large groups of SMA 2 patients were 99% and 69% after 5 and 25 years, respectively. Patients with the so-called juvenile form of Kugelberg–Welander (SMA 3) learn to walk independently and have a normal life expectancy. Despite the attribute 'juvenile', SMA 3 may start in infancy. Of a group of patients with clinical onset before the age of 3 years, 70% and 22% still walked after 10 and 40 years; the corresponding figures for clinical onset after that age were 97% and 59% (Zerres *et al.*, 1997). Some patients only complain that they cannot get out of a chair or climb stairs without using their hands. SMA 3 may be misdiagnosed clinically as limb-girdle muscular dystrophy; genetic analysis usually provides the correct diagnosis.

There is also a very severe form of SMA (SMA 0) that has recently been described (Dubowitz, 1999). Characteristics include reduced fetal movements, neonatal contractures that are rarely seen in SMA 1, asphyxia and neonatal death. Autopsy findings suggest a congenital axonal neuropathy, but genetic analysis demonstrates the typical SMA defect. These observations may be relevant for an understanding of the pathogenesis of SMA (see below).

The different forms of SMA were originally distinguished by the age of onset. This has, however, in individual patients led to erroneous prognostic predictions. Long-term observations of populations of patients have shown that the maximum achieved motor performance is a better prognostic indicator than the age at which the first symptoms occur.

Genetics of Spinal Muscular Atrophy

The SMA gene has been located on chromosome 5q11.2–13.3 (Brzustowicz *et al.*, 1990; Melki *et al.*, 1990). Four genes have been identified in this region, of which survival of motor neuron 1, telomeric (*SMN1*) and baculoviral IAP repeat-containing 1 (*BIRC1*) are the most interesting ones. The other two genes are Cockayne syndrome 1 (classical) (*CKN1*) and small EDRK-rich factor 1A (telomeric) (*SERF1A*). Owing to an inverted duplication of this region, each of the four genes exists in a telomeric and centromeric copy. It is now generally accepted that the telomeric *SMN1* gene (Lefebvre *et al.*, 1995) is the SMA causative gene. Patients, but also healthy subjects, may have deletions in any of the other genes. The *SMN1* gene contains nine exons and is 99% identical with its centromeric copy survival of motor neuron 2, centromeric (*SMN2*). Almost all SMA patients investigated have homozygous deletions in the *SMN1* gene, usually in exons 7 and 8. Less than 1% of the patients have missense or point mutations in any of the exons 1–7. Evidently, modifier genes must be responsible for the widely divergent clinical phenotypes. More than 90% of the patients have exon 7 deletions, but the size of the deletion and whether it also comprises *BIRC1* and *CKN1* apparently correlates with disease severity. Nevertheless, it is obscure whether *BIRC1* itself determines the severity of the disease.

The SMN protein expressed by *SMN1* has 294 amino acids and is a ubiquitous constituent of all cells. The protein expressed by *SMN2* resembles that expressed by *SMN1*, but expression is usually incomplete and the protein is less stable. SMN is localized in distinct nuclear organelles and has been thought to play a role in ribonucleic acid (RNA) metabolism. Other investigators suspect that SMN prevents neuronal apoptosis. The amount of SMN protein is reduced in the spinal cord of SMA fetuses as compared with normal controls. SMN is also reduced in skeletal muscles of patients, and quantitative analysis in

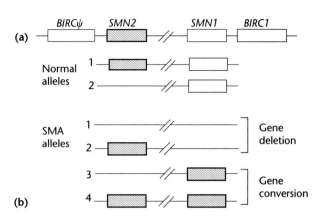

Figure 1 (a) Representation of the duplicated region of chromosome 5q13, containing the *SMN* and *BIRC* genes. *CKN1* and *SERF1A* are not represented as they are not considered candidate genes for SMA. *BIRCψ* is the *BIRC1* pseudogene copy. *SMN1* is the telomeric and *SMN2* the centromeric copy. (b) Illustration of normal and mutant *SMN* alleles. Thin-lined and thick-lined boxes represent *SMN1* and *SMN2* gene copies, respectively. Normal alleles contain one *SMN1* and one *SMN2* gene (1); in rare cases *SMN2* may be lacking (2). Deletion of *SMN1* does not alter the number of copies of *SMN2* and causes SMA 1 (or SMA 0) (1, 2). Conversion of *SMN1* into *SMN2* increases the number of copies of *SMN2* and is found in SMA 2 or 3 (3, 4). (Modified from Campbell *et al.* (1997) with the kind permission of Chicago University Press.)

lymphoblasts suggests a correlation between SMN content and disease severity. The *SMN1* gene may convert into *SMN2*, and SMA 2 or SMA 3 patients in contrast to SMA 1 patients with one (or rarely zero) copy of *SMN2* may have up to three *SMN2* copies (Campbell *et al.*, 1997) (**Figure 1**). It has been proposed that SMN protein derived from *SMN2* partly compensates for the lack of intact SMN protein.

Histopathology

Almost all workers who investigated autopsies of SMA 1 patients report a gross reduction of the number of anterior horn cells. Phrenic motor neurons innervating the diaphragm and motor neurons of external eye muscles are less affected, and cortical neurons and the pyramidal tracts are normal. Data for SMA 2 and SMA 3 patients are scarce. Surprisingly, a normal density of anterior horn cells was found in the few autopsied cases with early fatal SMA 0. Peripheral nerves in those patients showed axonal degeneration which confirms electrophysiological findings of reduced amplitude and conduction velocity of the compound motor nerve action potential. These findings suggest that the disease starts with axonal degeneration, and that children with SMA 0 died before retrograde neuronal loss was complete. Some observations in mice with non-SMA motor neuron disease (see below) suggest that motor neurons die by apoptosis. Nevertheless, reports of neuronal apoptosis in SMA patients are contradictory. This probably reflects the difficulty of working with autopsy material and that at any one time only a few cells may be undergoing apoptosis.

Neuronal degeneration is not restricted to motor neurons. Sensory ganglia and sural nerves of SMA 1 patients show neuronal and axonal degeneration although no sensory deficits have been reported. This is not surprising considering the domineering motor deficits and the age of the children.

Muscle biopsies from SMA patients differ from those from patients with other denervating disorders. Muscles of SMA 1 patients are characterized by pronounced fatty infiltration and the presence of numerous very small muscle fibers of less than 10 μm in diameter. Few fibers are of normal size or hypertrophic. These fibers are almost exclusively of type 1 (**Figure 2**), either because slow-twitch motor units selectively survive, or because overload has caused transformation of fiber type. The myofibrils of the large fibers are always normal, while those of the small fibers are either normal or degenerating. The small fibers have peripheral nuclei and do not resemble immature myotubes. Cellular remnants are found between small fibers, which indicates fiber breakdown. Structural and immunohistochemical signs of apoptosis of myonuclei of these small fibers have been described, but this has not been confirmed. All small fibers contain, in contrast to muscle fibers of age-matched controls, the intermediate-filament protein vimentin (**Figure 2**). This is a

useful diagnostic indicator. Fiber type grouping due to collateral reinnervation as in other neurogenic disorders does not occur. Correspondingly, electrophysiological motor unit counts reveal pronounced loss of motor units without any compensatory increase in the size of the remaining ones.

Interestingly, immature muscles of newborn rats after partial denervation develop the same pattern of few large and many small fibers without signs of collateral reinnervation. Also in these muscles, the denervated small fibers are eventually replaced by fat cells. This suggests that the characteristic appearance of SMA 1 muscles is not disease-specific but reflects the developmental stage at onset of disease (Schmalbruch, 1988).

Muscle biopsies from SMA 3 patients vary (**Figure 3**). Some consist of a majority of normal sized fibers of type 1 or type 2 and groups of small fibers resembling those in SMA 1. Fatty infiltration is restricted to parts of the muscle. Type grouping may be present or absent. Muscles from older patients probably representing late stages of the disease often show type grouping or even fiber type dominance. either of type 1 or type 2, and also myopathic changes such as random variability of fiber diameters, internal myonuclei and endomysial fibrosis (**Figure 3c**). The latter changes suggest secondary muscle fiber necrosis followed by regeneration. Group atrophy and angulated fibers indicating fresh denervation are uncommon. These findings agree with the fact that motor units are enlarged due to collateral sprouting and that modest elevations of muscle enzymes in the serum due to muscle fiber necrosis occur in SMA 3 but rarely in SMA 1.

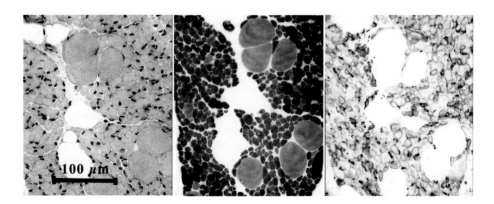

Figure 2 Three serial cross sections through a typical muscle biopsy of a 1-year-old girl with SMA 1. The left section stained by the Gomori trichrome method shows few hypertrophic muscle fibers and numerous small round fibers of less than 10 μm in diameter. The nuclei of all fibers are peripheral. The large unstained cells are fat cells. Staining for adenosine triphosphatase (pH 10.3) (middle) shows weak reaction in large fibers (presumably slow-twitch type 1 fibers) and more intense reaction in most small fibers. The right section was immunostained for the mesenchymal cytoskeletal protein vimentin, which is present in developing but not mature muscle fibers. The small but not the large fibers retain reactivity with antivimentin. (From Schmalbruch and Haase (2001), reproduced with the kind permission of *Brain Pathology*.)

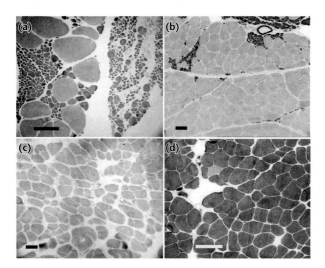

Figure 3 Examples of muscle biopsies that illustrate the histological variability of SMA 3. All cross-sections have been stained for adenosine triphosphatase (pH 10.3). Scale bars, 100 μm. The micrographs of this panel suggest that the small fibers may increasingly vanish. (a) Modestly affected 5-year-old boy who is still able to play football (soccer) in kindergarten. The picture resembles that of the SMA 1 biopsy shown in **Figure 2**. Numerous small fibers without evidence of reinnervation surround a few hypertrophic fibers. This biopsy sample illustrates that it is not possible to make prognostic predictions based on biopsy findings. (b) This 6-year-old girl is still mildly affected. She has difficulties in getting up from a sitting position and uses both hands when climbing staircases. The muscles consist of normal-sized type 1 fibers and interspersed small type 2 fibers. (c, d) This 29-year-old man had been suspected of having Becker muscular dystrophy. His main complaint is that he has difficulties in getting up from a sitting position. Biopsies from the anterior tibial and medial vastus muscles show almost complete fiber type homogeneity: the biopsy from the anterior tibial muscle consists of weakly stained type 1 fibers (c) and that from the vastus (d) consists of intensely stained type 2 fibers; only one lightly stained type 1 fiber is seen. (c) Distinct myopathic features such as increased endomysial connective tissue and random variability of fiber diameters. (From Schmalbruch and Haase (2001), reproduced with the kind permission of *Brain Pathology*.)

Animal Models and Clues to the Pathogenesis

Spontaneous hereditary motor neuron diseases have been observed in a variety of animals, but only two mouse mutants, Wobbler (*wr*) and progressive motoneuronopathy (*pmn*), and a mutation in dogs (hereditary canine spinal muscular atrophy, HCSMA) have been studied in detail. Based on defects in the gene coding for superoxide dismutase (SOD) that have been found in some patients with familial amyotrophic lateral sclerosis, several transgenic mouse strains (SOD mice) have been developed. None of these motor

neuron diseases is, however, homologous to human SMA. Two spontaneous mutations in mice (motor neuron disease, *mnd*, and motor neuron disease 2, *mnd2*) were wrongly classified and do not have motor neuron disease at all.

Heterozygous *Smn* knockout mice develop late-onset and mild motor neuron disease, whereas homozygous *Smn* knockout is lethal. Mice, in contrast to humans, do not have second gene copies, and embryos die at the blastocyst stage. However, homozygous *Smn* knockout mice may be born alive when a human *SMN2* transgene is introduced. These mice show a variable phenotype that has been only incompletely analyzed.

In a most elegant approach, Melki and colleagues (Frugier *et al.*, 2000) produced mice with a neuron-specific deletion of the *Smn* exon 7. Homozygotes were born normally, but after 2 weeks developed severe motor neuron disease and died within a few weeks. Skeletal muscles showed neurogenic atrophy, while histopathological changes in the anterior horn were subtle without obvious neuronal loss.

For decades, Hausmanova-Petrusewicz and her school had maintained that SMA 1 affects not only motor neurons but also skeletal muscles directly, and that the interplay between immature neurons and muscles is disturbed. This view was supported by the experimental observation that immature but not mature motor neurons eventually die when not connected to muscle. Furthermore, it has been shown that myoblasts from SMA patients cocultured with neurons impede the neurotrophic effect of normal myoblasts, and that these do not develop into innervated myotubes. The crucial role of the SMN protein for skeletal muscle is illustrated by the fact that mice with a muscle-specific deletion of *Smn* exon 7 develop a necrotizing myopathy with dystrophic changes (Cifuentes-Diaz *et al.*, 2001).

During normal development, about half of the motor neurons die and their number is fitted to the size of the muscular target. It has been speculated that this normal process is exaggerated in SMA. This would support the assumption that SMA is not a progressive neurodegenerative disorder, and that the number of surviving motor neurons determines the course of the disease. A small number in SMA 1 might suffice for a newborn child but not when the child grows. If many motor neurons survive, the patient may present with SMA 3. The decrease of the number of small fibers characteristic of SMA 1, the paucity of signs of fresh denervation and the long survival even of severely disabled SMA 3 patients may support this view. No loss of muscle strength has been found in a prospective study, although motor function declined (Iannacone *et al.*, 2000).

Therapeutic Outlook

Degeneration of motor neurons and death have been delayed in non-SMA homologous mouse models by various neurotrophic factors such as ciliary neurotrophic factor (CNTF), neurotrophin-3 (NT3), insulin-like growth factor 1 (IGF-1) or brain-derived neurotrophic factor (BDNF). Therapeutic trials with the same substances in patients with amyotrophic lateral sclerosis, a noninherited motor neuron disease, have, however, failed. Whether mice with SMA-homologous disease models or even patients with SMA respond to neurotrophic factors has not yet been investigated. Hypothetically, one might also consider replacing the lacking SMN protein either by gene transfer of *SMN1* or by activating *SMN2* expression. Finally, neuronal apoptosis might be blocked by antiapoptotic peptides. These options entail a plethora of technical difficulties and are so far speculative. Nevertheless, the most relevant question is whether there is continuous loss of motor neurons that might be therapeutically prevented or delayed, or whether SMA is basically a nonprogressive disease with loss of motor neurons completed when the disease becomes clinically manifest. In the latter case, a therapy can only try to preserve motor abilities once acquired.

Conclusion

The gene locus of SMA and the defective gene product have been identified since the 1990s. Clinically divergent hereditary motor neuron diseases (SMA 0–3) have become understood as different facets of the same nosological entity, and diagnostic accuracy has increased. The role of the gene product SMN in RNA metabolism is still not generally accepted, and it is unknown why lack of this central and universally active protein preferentially affects motor neurons. The pathogenesis of SMA and whether it is a progressive disease or not is still a matter of speculation. The study of early lethal cases of SMA 0, with their unusual histopathology, may be of importance in determining pathogenesis and potential forms of treatment.

See also
Charcot–Marie–Tooth Disease and Associated Peripheral Neuropathies

References

Brzustowicz LM, Lehner T, Castilla LH, *et al.* (1990) Genetic mapping of chronic childhood onset spinal muscular atrophy to chromosome 5q11.2–13.3. *Nature* **344**: 540–541.

Campbell L, Potter A, Ignatius J, Dubowitz V and Davies K (1997) Genomic variation and gene conversion in spinal muscular atrophy: implications for disease process and clinical phenotype. *American Journal of Human Genetics* **61**: 40–50.

Cifuentes-Diaz C, Frugier T, Tiziano FD, *et al.* (2001) Deletion of murine SMN exon 7 directed to skeletal muscle leads to severe muscular dystrophy. *Journal of Cell Biology* **152**: 1107–1114.

Dubowitz V (1999) Very severe spinal muscular atrophy (SMA type O): an expanding clinical phenotype. *European Journal of Paediatric Neurology* **3**: 49–51.

Frugier T, Tiziano FD, Cifuentes-Diaz C, *et al.* (2000) Nuclear targeting defect of SMN lacking the *C*-terminus in a mouse model of spinal muscular atrophy. *Human Molecular Genetics* **9**: 849–858.

Iannaccone ST, Russman BS, Browne RH, *et al.* (2000) Prospective analysis of strength in spinal muscular atrophy. DCN/Spinal Muscular Atrophy Group. *Journal of Child Neurology* **15**: 97–101.

Lefebvre S, Bürglen L, Reboullet S, *et al.* (1995) Identification and characterization of a spinal muscular atrophy-determining gene. *Cell* **80**: 155–165.

Melki J, Abdelhak S, Sheth P, *et al.* (1990) Gene for chronic proximal spinal muscular atrophies maps to chromosome 5q. *Nature* **344**: 767–768.

Schmalbruch H (1988) The effect of peripheral nerve injury on immature motor and sensory neurons and on muscle fibers. Possible relation to the histogenesis of Werdnig–Hoffmann disease. *Revue Neurologique (Paris)* **144**: 721–729.

Schmalbruch H and Haase G (2001) Spinal muscular atrophy: present state. *Brain Pathology* **11**: 231–247.

Zerres K, Rudnik-Schöneborn S, Forrest E, *et al.* (1997) A collaborative study on the natural history of childhood and juvenile onset proximal spinal muscular atrophy (type II and III SMA): 569 patients. *Journal of Neurological Sciences* **146**: 67–72.

Further Reading

Burek MJ and Oppenheim RW (1996) Programmed cell death in the developing nervous system. *Brain Pathology* **6**: 427–446.

Crawford TO and Pardo CA (1996) The neurobiology of childhood spinal muscular atrophy. *Neurobiology of Disease* **3**: 97–110.

Gomez MR (1994) Motor neuron diseases in children. In: Engel AG and Franzini-Armstrong C (eds.) *Myology*, 2nd edn, vol. 2, pp. 1837–1853. New York, NY: McGraw-Hill.

Jablonka S, Rossoll W, Schrank B and Sendtner M (2000) The role of SMN in spinal muscular atrophy. *Journal of Neurology* **247**(13): I37–I42.

Jessell TM and Sanes JR (2000) The generation and survival of nerve cells. In: Kandel ER, Schwartz JH and Jessell TM (eds.) *Principles of Neural Science*, pp. 1041–1062. New York, NY: McGraw-Hill.

Robertson GS, Crocker SJ, Nicholson DW and Schulz JB (2000) Neuroprotection by the inhibition of apoptosis. *Brain Pathology* **10**: 283–292.

Scheffer H, Cobben JM, Matthijs G and Wirth B (2001) Best practice guidelines of molecular analysis in spinal muscular atrophy. *European Journal of Human Genetics* **9**: 484–491.

Son YJ, Trachtenberg JT and Thompson WJ (1996) Schwann cells induce and guide sprouting and reinnervation of neuromuscular junctions. *Trends in Neurosciences* **19**: 280–285.

Talbot K and Davies KR (2001) Spinal muscular atrophy. *Seminars in Neurology* **21**: 189–197.

Web Links

Baculoviral IAP repeat-containing 1 (*BIRC1*); Locus ID: 4671. LocusLink:
http://www.ncbi.nlm.nih.gov/LocusLink/LocRpt.cgi?l = 4671

Cockayne syndrome 1 (classical) (*CKN1*); Locus ID: 1161.
  LocusLink:
    http://www.ncbi.nlm.nih.gov/LocusLink/LocRpt.cgi?l = 1161
Small EDRK-rich factor 1A (telomeric) (*SERF1A*); Locus ID: 8293.
  LocusLink:
    http://www.ncbi.nlm.nih.gov/LocusLink/LocRpt.cgi?l = 8293
Survival of motor neuron 1, telomeric (*SMN1*); Locus ID: 6606.
  LocusLink:
    http://www.ncbi.nlm.nih.gov/LocusLink/LocRpt.cgi?l = 6606
Survival of motor neuron 2, centromeric (*SMN2*); Locus ID: 6607.
  LocusLink:
    http://www.ncbi.nlm.nih.gov/LocusLink/LocRpt.cgi?l = 6607
Baculoviral IAP repeat-containing 1 (*BIRC1*); MIM number: 600355. OMIM:
    http://www.ncbi.nlm.nih.gov/htbin-post/Omim/dispmim?600355

Cockayne syndrome 1 (classical) (*CKN1*); MIM number: 216400. OMIM:
    http://www.ncbi.nlm.nih.gov/htbin-post/Omim/dispmim?216400
Small EDRK-rich factor 1A (telomeric) (*SERF1A*); MIM number: 603011. OMIM:
    http://www.ncbi.nlm.nih.gov/htbin-post/Omim/dispmim?603011
Survival of motor neuron 1, telomeric (*SMN1*); MIM number: 600354. OMIM:
    http://www.ncbi.nlm.nih.gov/htbin-post/Omim/dispmim?600354
Survival of motor neuron 2, centromeric (*SMN2*); MIM number: 601627. OMIM:
    http://www.ncbi.nlm.nih.gov/htbin-post/Omim/dispmim?601627

Spliceosome

Daniel A Pomeranz Krummel, *Medical Research Council Laboratory of Molecular Biology, Cambridge, UK*

Kiyoshi Nagai, *Medical Research Council Laboratory of Molecular Biology, Cambridge, UK*

Intermediate article

Article contents
• Intron/Exon Structure
• Spliceosome Components
• Spliceosome Complexes and Functional Interactions
• Spliceosome Protein Diversity

The removal of introns and the splicing together of the protein-coding regions or exons from precursor messenger RNA (pre-mRNA) transcripts is fundamental to the development and maintenance of human cells. The nuclear process of pre-mRNA splicing is a complex phenomenon catalyzed by the 'spliceosome', which comprises more than 100 hundred protein and RNA molecules that come on and off during its assembly and activity.

Intron/Exon Structure

The introns of protein-coding genes are defined by three short sequence elements: a 5′ splice or donor site, a 3′ splice or acceptor site, and the branchpoint region (**Figure 1a**). Each of these three sequence elements is important to the assembly and catalytic reactions of the spliceosome.

The junction of the 5′ exon and intron (5′ splice or donor site) in most metazoan precursor messenger ribonucleic acid (pre-mRNA) transcripts is defined by the sequence YRG/GURRGU (where R is A or G, Y is C or U, and the slash marks the 5′ exon/intron junction). The junction of the intron and 3′ exon (3′ splice or acceptor site) is defined most often by a smaller sequence, YAG/ (where the slash marks the intron/3′ exon junction). Close to the 3′ splice site (20–40 nucleotides upstream) is the branchpoint region, which commonly has the sequence YNRAY, containing a highly conserved adenosine. The 2′-OH group of the sugar of this adenosine acts as the nucleophile in the first of two *trans*-esterification (phosphodiester bond exchange) reactions that are catalyzed by the spliceosome to remove the intron. The branchpoint region precedes a pyrimidine-rich sequence (the pyrimidine tract). In addition to these primary sequence elements, there are less well-defined exonic elements that act as RNA splicing enhancers (Blencowe, 2000). (*See* Exonic Splicing Enhancers; Gene Structure and Organization; Splice Sites.)

It is essential that the spliceosome both excises the intron and splices together the 5′ and 3′ exons with single-nucleotide precision, otherwise a frameshift mutation or premature stop codon will be generated. But it is not unusual for a prospective codon triplet to exist at the 5′ exon/intron junction and the next triplet to start at the beginning of a downstream 3′ exon junction (a phase 0 intron; **Figure 1b**). The triplet may be discontinuous at the 5′ exon end, such that either the first and the second or only the first base of the triplet is encoded in the 5′ exon and the remainder is encoded in the 3′ exon (phase 1 or 2 introns respectively). The consensus sequence elements described above are not perturbed by the existence of these various types of intron (**Figure 1b**).

Spliceosome Components

Genetic studies in the yeast *Saccharomyces cerevisiae* indicate that more than 50 genes encode proteins that

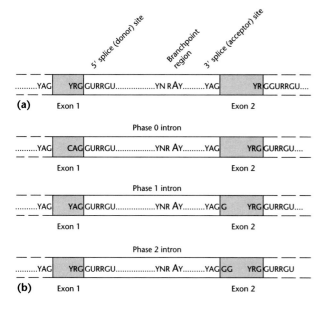

Figure 1 Intron/exon structure. (a) A metazoan pre-mRNA transcript showing the three common sequence elements and their sequence: 5′ splice (donor) site; branchpoint region containing the highly conserved adenosine (large letter); and 3′ splice (acceptor) site. Exonic regions are shaded. Y: pyrimidine; R: purine; N: pyrimidine or purine. (b) Introns of protein-coding genes may be one of three types: phase 0 (top; codon CAG for glutamine in exon 1); phase 1 (middle; codon AGG for arginine in exons 1 and 2); phase 2 (bottom; codon GGG for glycine in exons 1 and 2).

are essential for RNA splicing (Rymond and Rosbash, 1992). Less is known about the essentiality of these types of gene in the human cell. Whereas less than 5% of the protein-coding genes of *S. cerevisiae* contain an intron, most human protein-coding genes contain introns and alternative splicing is highly prevalent (it is estimated that on average each human gene encodes 2.6 mRNA transcripts). The number of protein-coding genes involved in RNA splicing is likely to be much greater in human than in yeast. (*See* Protein Coding; Splicing of pre-mRNA.)

Partly to deal with this much greater obligation in the human cell versus the yeast cell, the copy number of the particles that make up the human spliceosome are greater in number (10^5–10^6 splicing particles per human cell versus 50–100 particles per yeast cell; Burge *et al.*, 1999). The spliceosome is composed of five such splicing particles that contain both RNA and protein subunits. Each particle has a single RNA subunit, a set of seven similar proteins, and 3–12 proteins that are specific to the particle (**Table 1**). The RNA subunit of each of these five particles is rich in uridine residues, relatively small ($\sim$ 100–200 nucleotides) and primarily present in the nucleus. The RNAs have been therefore called U small nuclear RNAs (snRNAs).

Four of the five U snRNAs (U1, U2, U4, U5) are under the control of an RNA polymerase II promoter, whereas the U6 snRNA is under the control of a promoter for RNA polymerase III. As a consequence, the 5′ modification and subsequent biogenesis of the U6 snRNA differs from that of the other four U snRNAs. The U1, U2, U4 and U5 snRNAs first obtain a monomethylated guanosine cap at their 5′ end, which is later trimethylated in the cytoplasm. The 5′ monomethylated cap modification and protein factors including a specific cap-binding protein allow the U snRNA to move into the cytoplasm, where a set of seven Sm proteins, D1, D2, B/B′, D3, E, F and G (B can exist as two splice variants, the B′ isoform seems to be specific to neuronal cells), bind to a specific single-stranded sequence of the U1, U2, U4 and U5 snRNAs. The Sm proteins bind as two heterodimeric protein complexes (D1–D2 and D3–B/B′) and one heterotrimeric protein complex (E–F–G) to form a stable globular, doughnut-like structure (the proposed order of interaction is G-E-F-D2-D1-B-D3; Kambach *et al.*, 1999). (*See* RNA Polymerases and the Eukaryotic Transcription Machinery.)

The Sm proteins act as part of a nuclear bipartite transport signal in conjunction with the trimethylated guanosine at the 5′ end of the U snRNAs for the snRNA to re-enter the nucleus. In the nucleus, the U snRNAs can be fully assembled to form the small nuclear ribonucleoprotein particles (snRNPs) by the incorporation of the particle-specific proteins. For example, the U1 snRNP has three specific proteins: U1 A, U1 70K and U1 C (**Table 1**). The U1 A and 70K proteins bind at specific RNA loops of the U1 snRNA, recognizing an RNA sequence in a specific context, whereas the U1 C protein seems to require protein–protein interactions for its stable incorporation into the particle.

Notably, the Sm proteins function not only as part of the signal for nuclear import but also to maintain the integrity of the particle through interactions with the snRNP specific proteins. Encoded in the human genome are the genes for a separate set of seven Sm-like proteins (Lsm2–Lsm8) that bind to the U6 snRNA. These Sm homologous proteins also form a heptameric ring and seem to interact similarly with RNA.

Spliceosome Complexes and Functional Interactions

The five snRNPs in conjunction with non-snRNP splicing-associated proteins coordinate the assembly of the spliceosome onto the pre-mRNA transcript and the catalytic events of RNA splicing. The initial events in the assembly of the spliceosome are the recognition

Table 1 The protein subunits of the human U snRNPs[a]

Human protein	Molecular mass (kDa)	Primary sequence motif	U1 (12S)	U2 (12S)	U2 (17S)	U5 (20S)	U4/U6 (12S)	U4/U6 • U5 (25S)
B/B′	24/25	Sm	•	•	•	•	•	••
D1	13	Sm	•	•	•	•	•	••
D2	14	Sm	•	•	•	•	•	••
D3	14	Sm	•	•	•	•	•	••
E	11	Sm	•	•	•	•	•	••
F	10	Sm	•	•	•	•	•	••
G	9	Sm	•	•	•	•		••
Lsm2	11	Sm					•	•
Lsm3	12	Sm					•	•
Lsm4	15	Sm					•	•
Lsm5	10	Sm					•	•
Lsm6	9	Sm					•	•
Lsm7	12	Sm					•	•
Lsm8	10	Sm					•	•
U1A	32	2RRMs	•					
U170K	52	1RRM	•					
U1C	17	Zinc-finger	•					
U2B″	29	LRR		•	•			
U2A′	31	RRM		•	•			
	33				•			
	35				•			
SF353/SAP49	53	2RRMs			•			
SF3a60/SAP61	60	Zinc-finger			•			
SF3a66SAP62	66	Zinc-finger; Pro-rich			•			
	92				•			
SF3a120/SAP114	110	Surp 1 module			•			
SF3b120/SAP1130	120	UV-DDB			•			
SF3b150/SAP145	150	Pro-rich			•			
SF3b160/SAP155	160				•			
	15					•		•
	40					•		•
	52					•		•
	100	DEAD box				•		•
	102					•		•
	110					•		•
	116	G domain				•		•
HBRR2	200	DEIH, DxxH				•		•
HPRP8	220					•		•
HPRP3	77	PWI motif					•	•
hPRP4	58	WD40 repeats					•	•
	15.5							•
	20	*cis–trans* isomerase						•
	27	RS						•
	61							•
	63							•

G domain: regulatory GTPase-binding motif; LRR: leucine-rich repeat; PWI motif: sequence of roughly 80 residues that has successive proline, tryptophan and isoleucine residues near the *N*-terminus of the motif; RRM: RNA recognition motif; RS: arginine/serine-rich repeat where serine is commonly phosphorylated; S: sedimentation coefficient reported in Svedbergs; UV-DDB: ultraviolet-damaged DNA-binding domain; WD40 repeats, tryptophan/asparagine-rich repeats that often contain the dipeptide glycine-histidine near the *N*-terminus of the motif. DEAD, DEIH/H and DxxH are variants of an RNA helicase box.

[a]Dot indicates the presence of the protein in the U snRNP.

NATURE ENCYCLOPEDIA OF THE HUMAN GENOME / ©2003 Macmillan Publishers Ltd, Nature Publishing Group / www.ehgonline.net

of the 5′ splice site and branchpoint sequence by the 12S U1 and 17S U2 snRNPs, respectively. These snRNPs are assisted in their function by the association of other non-snRNP proteins. For example, binding of the U2 snRNP to the pre-mRNA transcript is assisted by the non-snRNP splicing factor heterodimeric complex comprising the U2 auxiliary factor small and large subunits (U2AF^{35} and U2AF^{65}, respectively), which recognize the pyrimidine tract and 3′ splice site (Gozani et al., 1998; Wu et al., 1999).

The U1 and U2 snRNPs are thought to interact, in conjunction with the U2AF complex and other factors, to 'orient' the correct scissile bonds at the 5′ and 3′ exon borders for the subsequent two trans-esterification reactions (Fu and Maniatis, 1992). Once this is accomplished, the U2AF complex dissociates. The interaction of the U1 and U2 snRNPs is mediated partly by complementary base-pairing between their respective snRNAs and the 5′ exon/intron sequence and branchpoint region, respectively (**Figure 2a**); however, the active-site nucleophilic adenosine in the pre-mRNA branchpoint sequence is not base-paired with the U2 snRNA. The stage is then set for the joining of a pre-assembled 25S U4/U6 • U5 tri-snRNP (**Figure 2a**).

The interaction between the U4 and U6 snRNPs is also mediated in part by base-pairing between their respective snRNAs. Commensurate with these assembly events, energy in the form of nucleoside triphosphate is expended to drive the structural rearrangements between RNA strands by the activity of RNA helicases of the DExD/H (Asp-Glu-X-Asp/His) box family of proteins (Staley and Guthrie, 1998). The U6 snRNA now forms base pairs with the U2 snRNA, as well as the 5′ splice site, which displaces the U1 snRNA (**Figure 2b**). The U1 snRNP dissociates as does the U4 snRNP, and then the first of the two trans-esterification reactions occurs, whereby the 2′-OH of the sugar of the branchpoint adenosine attacks the phosphate of the first nucleotide of the intron (guanosine) at the donor site (**Figure 2b**). Subsequently, the 3′-OH of the sugar at the 5′ exon acts as the nucleophile attacking the phosphodiester bond at the 3′ splice or acceptor site (**Figure 2c**). These steps are assisted by a loop in U5 snRNA, which seems to align the 5′ and 3′ exons (Sontheimer and Steitz, 1993). The 5′ and 3′ exons are finally ligated to form the canonical 3′–5′ phosphodiester linkage, while the intron is liberated as a lariat-like structure (**Figure 2d**).

At each phase of spliceosome assembly there are many non-snRNP splicing factors that associate/dissociate, as illustrated for the U2AF complex. There are also protein factors that act to chemically modify snRNP-associated proteins. For example, a protein kinase associates with the U1 snRNP and phosphorylates the U1 70K protein subunit – an event that seems

to alter the capacity of the U1 snRNP to recognize the 5′ splice site (Tazi et al., 1993). In addition, several of the seven Sm proteins have a carboxy (C)-terminal arginine/glycine-rich domain in which the arginine residues are subject to methylation by a methyltransferase (Friesen et al., 2001). (**See** SR Proteins.)

Spliceosome Protein Diversity

The third most common protein domain encoded in human genes is termed the RNA recognition motif (RRM; it is also known as the RNP-type RNA-binding domain (RBD) and the ribonucleoprotein (RNP) motif). The two most common protein domains are the zinc-finger and immunoglobulin (Ig) domains. The RRM is often present in 1–4 copies in a single protein and in conjunction with other motifs. For example, the U1 snRNP specific protein U1 70K has a single amino (N)-terminal RRM and a C-terminal arginine/serine-rich repeat. Although the RRM is the most frequently occurring recognition element for binding RNA, it may also be involved in mediating protein–protein interactions. Notably, this also seems to be true for the two other most prevalent RNA-binding domains, the KH and dsRBD domains. The most biochemically well-characterized RRM is found in the U1 snRNP specific protein U1 A, which has two RRMs.

Sequence alignment of RRMs from various proteins indicates that the RRM consists of about 80 amino acids within which are two highly conserved sequence motifs, the RNP1 octamer and RNP2 hexamer, that are separated by about 30 amino acids. The RNP1 octamer and RNP2 hexamer have the consensus sequences (Arg/Lys)-Gly-(Phe/Tyr)-(Gly/Ala)-(Phe/Tyr)-Val-X-(Phe/Tyr) and (Leu/Ile)-(Phe/Tyr)-(Val/Ile)-X-(Asn/Gly)-Leu, respectively. As one might predict, given the knowledge that the RRM mediates interaction with an RNA molecule, many of the conserved amino acid residues have positively charged or aromatic side chains – properties that are important for hydrogen bonding to and base-stacking with RNA bases. (**See** Protein Structure.)

The high-resolution X-ray crystallographic structure of the N-terminal RRM of the U1 A protein in complex with its cognate RNA-binding site in the U1 snRNA has shown that the RRM forms a four-stranded antiparallel β sheet flanked on one side by α helices. The RNP1 and RNP2 elements are present in two adjacent β strands and form the primary surface of interaction with the RNA bases (Kambach et al., 1999). Notably, the RRM is not always sufficient for a protein recognition of an RNA, and other proteins may modulate such binding. For the U2 snRNP specific protein U2 B″, interaction with RNA through

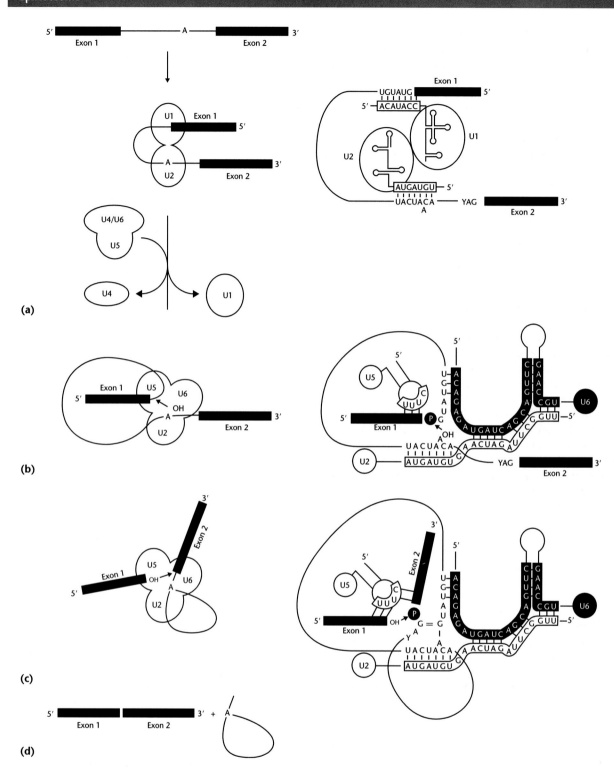

Figure 2 Stages of assembly of the U snRNPs onto the pre-mRNA transcript to form the spliceosome, which catalyzes intron removal and exon splicing. (a) Recognition of the 5′ splice site by U1 snRNP and branchpoint sequences by U2 snRNP, which is mediated by base-pairing of the respective snRNAs with these pre-mRNA sequence elements. (b) First *trans*-esterification reaction is catalyzed by nucleophilic attack of the 2′-OH of the sugar of the branchpoint adenosine at the 5′ splice site. (c) Second *trans*-esterification reaction is catalyzed by nucleophilic attack of the 3′-OH of the 5′ splice site at the 3′ splice site – a step that is assisted by a loop in U5 snRNA. (d) The 5′ and 3′ exons are ligated to form a canonical 3′–5′ phosphodiester linkage, and the intron is liberated as a lariat-like structure.

its single RRM requires interaction with the U2 snRNP specific protein U2 A′ (Kambach *et al.*, 1999). The U2 A′ has a leucine-rich repeat that mediates this protein–protein interaction.

The second most common motif present in snRNP proteins is that shared among the Sm proteins. The Sm proteins have a conserved sequence – which is unrelated to other known protein sequence motifs – that comprises two segments (Sm1 and Sm2) connected by a linker of variable length (Kambach *et al.*, 1999). The high-resolution structure of two Sm protein heterodimers has shown that the Sm proteins assume a common fold – an *N*-terminal α helix, followed by a five-stranded antiparallel β sheet. The segments Sm1 and Sm2 form the first three and remaining two β strands, and the last strand turns to interact with that of the first (Kambach *et al.*, 1999).

The high degree of protein–protein, RNA–protein and RNA–RNA structural rearrangements that occur in the spliceosome necessitate factors that are capable of initiating such changes, such as ATP-dependent helicases, GTPases and peptidyl/prolyl isomerases. The U5 snRNP specific proteins of 100 and 200 kDa are both members of the DExD/H box family of ATPase proteins, whereas the 116 kDa protein has a GTPase motif (Seraphin, 1995; Staley and Guthrie, 1998). In addition, the U4/U6 snRNP associated 20 kDa protein is a member of the peptidyl/propyl *cis–trans* isomerase class of proteins and thus may accelerate the rate of folding or assembly of proteins that are important to the function of this particle (Horowitz *et al.*, 1997).

See also
Posttranscriptional Processing
RNA Processing
Splicing of pre-mRNA
SR Proteins
Trans Splicing

References

Blencowe BJ (2000) Exonic splicing enhancers: mechanism of action, diversity and role in human genetic diseases. *Trends in Biochemical Science* **25**: 106–110.

Burge C, Tuschl T and Sharp PA (1999) Splicing of precursors to mRNAs by the spliceosomes. In: Gesteland R, Cech T and Atkins J (eds) *RNA World II*, pp. 525–560. Cold Spring Harbor, NY: Cold Spring Harbor Laboratory Press.

Friesen WJ, Massenet S, Paushkin S, Wyce A and Dreyfuss G (2001) SMN, the product of the spinal muscular atrophy gene, binds preferentially to dimethylarginine-containing protein targets. *Molecular Cell* **7**: 1111–1117.

Fu XD and Maniatis T (1992) The 35-kDa mammalian splicing factor SC35 mediates specific interactions between U1 and U2 small nuclear ribonucleoprotein particles at the 3′ splice site. *Proceedings of the National Academy of Sciences of the United States of America* **89**: 1725–1729.

Gozani O, Potashkin J and Reed R (1998) A potential role for U2AF-SAP 155 interactions in recruiting U2 snRNP to the branch site. *Molecular Cellular Biology* **18**: 4752–4760.

Horowitz DS, Kobayashi R and Krainer AR (1997) A new cyclophilin and the human homologues of yeast Prp3 and Prp4 form a complex associated with U4/U6 snRNPs. *RNA* **3**: 1374–1387.

Kambach C, Walke S and Nagai K (1999) Structure and assembly of the spliceosomal snRNPs. *Current Opinion in Structural Biology* **9**: 222–230.

Rymond BC and Rosbash M (1992) Yeast pre-mRNA splicing. In: Broach JR, Pringle J and Jones EW (eds.) *The Molecular and Cellular Biology of the Yeast* Saccharomyces, vol. 2, p. 143. Cold Spring Harbor, NY: Cold Spring Harbor Laboratory Press.

Seraphin B (1995) Sm and Sm-like proteins belong to a large family: identification of proteins of the U6 as well as the U1, U2, U4 and U5 snRNPs. *EMBO Journal* **14**: 2089–2098.

Sontheimer EJ and Steitz JA (1993) The U5 and U6 small nuclear RNAs as active site components of the spliceosome. *Science* **262**: 1989–1996.

Staley JP and Guthrie C (1998) Mechanical devices of the spliceosome: motors, clocks, springs, and things. *Cell* **92**: 315–326.

Tazi J, Kornstadt U, Rossi F, *et al.* (1993) Thiophosphorylation of U1-70K protein inhibits pre-mRNA splicing. *Nature* **363**: 283–286.

Wu S, Romfo CM, Nilsen TW and Green MR (1999) Functional recognition of the 3′ splice site AG by the splicing factor U2AF35. *Nature* **402**: 832–835.

Further Reading

Berget SM, Moore C and Sharp PA (1977) Spliced segments at the 5′ terminus of adenovirus 2 late mRNA. *Proceedings of the National Academy of Sciences of the United States of America* **74**: 3171–3175.

Burd CG and Dreyfuss G (1994) Conserved structures and diversity of functions of RNA-binding proteins. *Science* **265**: 615–621.

Chow LT, Gelinas RE, Broker TR and Roberts RJ (1977) An amazing sequence arrangement at the 5′ ends of adenovirus 2 messenger RNA. *Cell* **12**: 1–8.

Kramer A (1996) The structure and function of proteins involved in mammalian pre-mRNA splicing. *Annual Review of Biochemistry* **65**: 367–409.

Lamm GM and Lamond AI (1993) Non-snRNP protein splicing factors. *Biochimica Biophysica Acta* **1173**: 247–265.

Mermoud JE, Cohen PT and Lamond AI (1994) Regulation of mammalian spliceosome assembly by a protein phosphorylation mechanism. *EMBO Journal* **13**: 5679–5688.

Mount SM (2000) Genomic sequence, splicing, and gene annotation. *American Journal of Human Genetics* **67**: 788–792.

Oubridge C, Ito N, Evans PR, Teo CH and Nagai K (1994) Crystal structure at 1.92 Å resolution of the RNA-binding domain of the U1A spliceosomal protein complexed with an RNA hairpin. *Nature* **372**: 432–438.

Reed R (2000) Mechanisms of fidelity in pre-mRNA splicing. *Current Opinion in Cell Biology* **12**: 340–345.

Sharp PA (1994) Split genes and RNA splicing (Nobel lecture). *Cell* **77**: 805–815.

Sleeman JE and Lamond AI (1999) Nuclear organization of pre-mRNA splicing factors. *Current Opinion in Cell Biology* **11**: 372–377.

Varani G and Nagai K (1998) RNA processing by RNP proteins during RNA processing. *Annual Review of Biophysics and Biomolecular Structure* **27**: 407–445.

Web Links

Proteome Bioknowledge™ Library
http://www.proteome.com

Splice Sites

Richard A Padgett, *Cleveland Clinic Foundation, Cleveland, Ohio, USA*

Christopher B Burge, *Massachusetts Institute of Technology, Cambridge, Massachusetts, USA*

Splice sites are the sequences that define the junctions between introns and exons in eukaryotic genes.

Introduction

Most genes in higher eukaryotes contain intervening sequences or introns, which interrupt the coding sequence into two or more exons. As a correct translational reading frame must be maintained in the final messenger ribonucleic acid (mRNA), the splicing process that removes introns and connects exons must be accurate to the nucleotide. The signals within the pre-mRNA that direct the splicing process are the splice sites. This term refers to the actual sites of cleavage and ligation as well as to the conserved sequence elements or signals that direct the splicing process. These splice site signals are largely located within intron sequences immediately adjacent to the intron/exon junctions. Precursor mRNA (pre-mRNA) splicing is accomplished by the recognition of these splice site signal elements by the splicing machinery, which assembles around them a large multicomponent complex termed the spliceosome. This structure juxtaposes the two splice sites and carries out the cleavage and ligation reactions of splicing that join the exons and remove the intron (see Burge *et al.* (1999), Beggs (2001), Newman (2001) and Padgett (2001) for reviews). In this article, the splice site joining the upstream exon to the 5′ end of the intron is termed the 5′ splice site, while the splice site joining the 3′ end of the intron to the downstream exon is termed the 3′ splice site. These two sites are also often called the 'splice donor site' and the 'splice acceptor site', respectively. (*See* Spliceosome.)

Most metazoan organisms, including humans, contain two distinct classes of intron, which are removed by different spliceosomes (Burge *et al.*, 1998). In humans, the vast majority of introns belong to a single class, called the U2-dependent class, and most genes contain introns of only this class. The minor class of intron, called the U12-dependent class, comprises about 0.2% of introns that are scattered throughout the genome in genes that also contain multiple major class introns. Thus, while numerically small, the minor class of introns must be properly spliced for the expression of a wide array of genes. This interspersion of different classes of intron suggests that the two types of spliceosome must communicate with

each other to properly join exons adjacent to introns of different types. Minor class introns are present in both tissue-specific and ubiquitously expressed genes. Their inclusion in some genes of central importance to all cells (e.g. ribosomal protein genes) implies that the minor class splicing system is active in all cell types.

Interaction of Splice Sites with Small Nuclear Ribonucleic Acids

The central function of the splice site signals in the process of the removal of introns from primary transcripts is to interact with small nuclear ribonucleic acids (snRNAs) mainly through base-pairing (see Nilsen (1998) for a review). **Figure 1a** shows the two-step nature of the pre-mRNA splicing reaction and identifies the three functional sites: the 5′ splice site, branch site and 3′ splice site. **Figures 1b** and **1c** show the interactions of these sites with the snRNAs specific for the major splicing pathway (**Figure 1b**) and the minor splicing pathway (**Figure 1c**). The snRNAs function within complexes of RNA and protein known as small nuclear ribonucleoproteins (snRNPs).

Major class (U2-dependent) intron splicing

Spliceosome assembly begins with the binding of U1 snRNP to the 5′ splice site and U2 snRNP to the branch site near the 3′ splice site (**Figure 1b**). These interactions are made through both Watson–Crick base-pairing between the snRNAs and the pre-mRNA and protein–RNA interactions. These early recognition events serve to specify the sites of splicing and lead to the formation of the spliceosome. In the next step of spliceosome formation, the U4/U6•U5 tri-snRNP complex associates with the pre-mRNA and triggers a number of changes in the RNA–RNA interactions. First, the U4/U6 snRNA base-pairing interaction is destabilized, probably through the action of an adenosine 5′ triphosphate (ATP)-dependent RNA

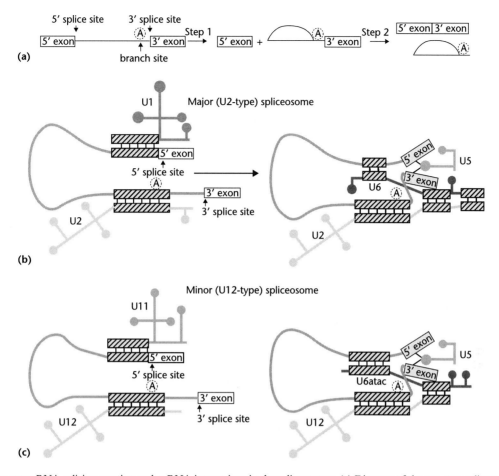

Figure 1 The pre-mRNA splicing reaction and snRNA interactions in the spliceosomes. (a) Diagram of the two-step spliceosomal splicing reaction described in the text. (b) RNA–RNA interactions in the major, U2-dependent spliceosome. (Left) The initial interactions of the pre-mRNA with U1 and U2 snRNPs. (Right) The interactions in the mature spliceosome. Hatched bars indicate regions that interact by base-pairing. The adenosine residue at the branch site is circled. (c) RNA–RNA interactions in the minor, U12-dependent spliceosome.

helicase or unwindase. This destabilizes the association of U4 snRNP, which can be now dissociated from the active spliceosome. The newly freed U6 snRNA sequences now form two major types of RNA–RNA interaction. First, a highly conserved region of U6 snRNA base-pairs to the 5′ splice site region of the intron, displacing U1 snRNA in the process. U1 snRNP is thus destabilized from the spliceosome and can be dissociated from the active complex. Second, U6 snRNA base-pairs with U2 snRNA and also forms an intramolecular stem–loop structure, both of which appear to be required for spliceosome function. At this stage, U5 snRNA also interacts with the exon nucleotides adjacent to the 5′ splice site.

Following these rearrangements, the spliceosome is competent to carry out the first step of splicing (**Figure 1a**). In this reaction, the 2′ hydroxyl group of an adenosine residue at the branch site to which U2 snRNA is bound (circled A in **Figure 1**) carries out a nucleophilic attack at the phosphodiester bond between the last nucleotide of the 5′ exon and the first

nucleotide of the intron. This produces two RNAs that are the intermediates in the complete splicing reaction: the 5′ exon fragment, which has a free 3′ hydroxyl group and is bound to U5 snRNA, and a lariat RNA consisting of the intron still attached to the downstream exon in which the 5′ nucleotide of the intron is now linked in a branch structure to the branch site adenosine. Before the second and final chemical step of splicing can take place, the spliceosome must undergo another series of rearrangements.

In contrast to the dramatic changes in snRNP content and RNA–RNA interactions that accompany the activation of the spliceosome for the first step of the reaction, the changes that occur between the two splicing steps are more subtle. One change that occurs at this point is the association of the exon nucleotides adjacent to the 3′ splice site with residues in U5 snRNA that are next to those associated with the upstream exon. This is thought to position the end of the upstream exon so that it can react with the 3′ splice site in the second step. Other internal rearrangements of the RNAs within the spliceosome have been detected that presumably serve to configure the spliceosome for the second catalytic step. In this step, the 3′ hydroxyl of the upstream exon, which was generated by the cleavage of the 5′ splice site in the first step, now attacks the 3′ splice site in another nucleophilic transesterification or phosphate transfer reaction, which leads to the ligation of the two exons through a standard 3′–5′ phosphodiester bond (**Figure 1a**). The intron is released as a lariat RNA with a terminal 3′ hydroxyl group.

Minor class (U12-dependent) intron splicing

Although the events involving the assembly and function of the minor U12-dependent spliceosome are less well studied than those of the major U2-dependent system, many of the interactions appear to be very similar (**Figure 1c**). The major difference between the two systems is the replacement of several snRNPs of the major pathway with analogous snRNPs in the minor pathway. Thus, U1 is replaced by U11, U2 by U12, U4 by U4atac and U6 by U6atac snRNPs. Only U5 snRNP appears to function in both spliceosomes (Tarn and Steitz, 1996). These analogous snRNPs interact with the intron splice sites and with each other in ways that very closely resemble the interactions that occur in the U2-dependent spliceosome. For example, U11 snRNP base-pairs to the 5′ splice site and U12 snRNP base-pairs to the branch site in an early step of the reaction. Prior to the first step of splicing, U6atac snRNA replaces U11 snRNA at the 5′ splice site, U4atac snRNP appears to be destabilized from the spliceosome and a U12–U6atac interaction can be detected. It has yet to be shown that U5 snRNP plays the same role in the minor spliceosome as in the major spliceosome; however, this seems quite likely.

Consensus Sequences for Splice Sites

A comparison between the complementary deoxyribonucleic acid (cDNA) sequence derived from the mature mRNA from a gene and the genomic sequence for that gene will locate the sites at which splicing has occurred in the processing of the pre-mRNA. Often the precise location of the splicing event in the nucleotide sequence is uncertain owing to repeating nucleotides at the junctions. This uncertainty can almost always be resolved by applying the 'GU–AG rule' which holds that introns normally begin with the dinucleotide GU and end in the dinucleotide AG. Exceptions observed in a small number of cases in major class introns are GC–AG and AU–AC terminal intron dinucleotides. The majority of minor class introns have either GU–AG or AU–AC terminal

dinucleotides. Splice sites that appear to violate these rules are often the result of sequencing or annotation errors and should be carefully reevaluated (e.g. see Jackson (1991)).

Using this terminal dinucleotide rule, the adjacent exon and intron sequences can then be aligned to produce consensus sequences for the 5′ and 3′ splice site signals. A pictorial representation of these sites in shown in **Figures 2** and **3** where the size of the letter at each position is proportional to its frequency in that position. Separate consensus sequences are shown for the major U2-dependent class (**Figure 2**) and the minor U12-dependent class of intron (**Figure 3**).

5′ Splice Site Function

At the 5′ splice site, the conserved sequence spans the site of splicing for both the major and minor classes of intron. As discussed above, this sequence is recognized by components of the splicing machinery mainly by base-pairing to U1 or U11 snRNAs. Recent evidence shows that the U4/U6 • U5 triple snRNP also interacts with the last few nucleotides of major class exons to facilitate U1 snRNP binding (Maroney *et al.*, 2000). A similar interaction of the U4atac/U6atac.U5 triple snRNPs of the minor splicing pathway has not been demonstrated.

The functional importance of the nucleotides within the splice site consensus is mirrored by their level of conservation among all splice sites of a given class. For example, in the major class intron 5′ splice site, the +1G, +2U and +5G are highly conserved and also critical for splice site function (**Figure 2a**). Many genetic diseases are caused by mutations in highly conserved residues at splice sites. These mutations lead to exon skipping or missplicing events such as cryptic splice site activation or intron inclusion, which produce aberrant or nonfunctional mRNAs from the affected gene. There are also correlations between positions in the consensus sequence. For example, the consensus G residues at positions −1 and +5 are not always present in 5′ splice sites. However, only a very few 5′ splice sites lack a G residue at both −1 and +5 (Burge and Karlin, 1997). (*See* RNA Processing and Human Disorders.)

For minor class GU 5′ splice sites, a negative correlation exists between a G at +1 and a G at −1 (**Figure 3a**). Almost no 5′ splice sites of this class have a G at both positions. In fact, one of the rare cases in which a G is found in both positions is in a 5′ splice site that seems to be able to alternatively splice to both a major and a minor class 3′ splice site. A distinctly different bias is seen in the AU class of minor 5′ splice site signals with T being rare at positions −1 and −3 and G being rare at position −2 (**Figure 3d**).

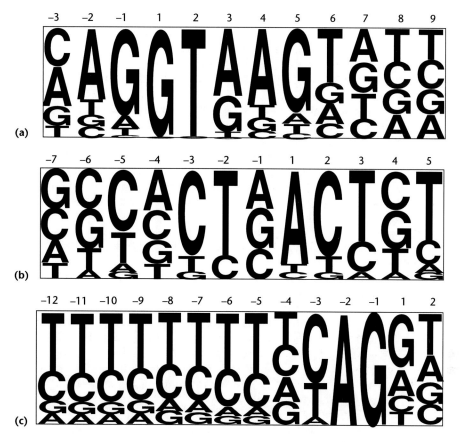

Figure 2 Pictograms of the major U2-dependent intron class consensus splice site signals. Approximately 20 000 5′ and 3′ splice sites from annotated GenBank files were extracted and aligned as described in Burge *et al.* (1999). In these pictograms, the size of a letter corresponds to the frequency with which that base is present at each position in a compilation of splice sites. (a) Major class 5′ splice site consensus sequence. The position labeled 1 is the first nucleotide of the intron and the position labeled −1 is the last nucleotide of the upstream exon. (b) Major class branch site consensus. A small database of experimentally confirmed branch sites (Nelson and Green, 1989) was used to generate this pictogram. The position labeled 1 is the branch site residue. (c) Major class 3′ splice site consensus. The position labeled −1 is the last nucleotide of the intron and the position labeled 1 is the first nucleotide of the downstream exon.

The other positions of the minor class 5′ splice site signals are all highly conserved. Mutations at any of these positions can lead to inactivation of the splice site or conversion of a minor class 5′ splice site to a major class 5′ splice site (Dietrich *et al.*, 1997).

3′ Splice Site Function

The organization of the 3′ splice site is more complex, with three distinguishable components in major class splice site signals and two components in minor class splice site signals. For both classes, the terminal AG (or occasionally AC) intron dinucleotide defines the actual site of splicing but elements further upstream in the intron determine which AG is used as the splice site. The upstream element common to both intron classes is the branch site consensus sequence. For the major class introns in humans, this site has a very weak

consensus (**Figure 2b**) and is typically located between 14 and 40 nucleotides from the 3′ splice site AG (counting from the branch site A residue to the 3′ splice site). Some unusual introns have branch sites located over 100 nucleotides upstream of the 3′ splice site. In contrast, the minor class intron branch site is highly conserved (**Figure 3b**) and is located between 10 and 20 nucleotides upstream of the 3′ splice site (Dietrich *et al.*, 2001). The branch site is recognized by a base-pairing interaction with U2 snRNA or U12 snRNA in major or minor class introns, respectively. This base-pairing is believed to specify and to help activate the specific adenosine residue that initiates the chemical steps of the splicing reaction. The high conservation of the branch site sequence in minor class introns suggests that this is the primary recognition element for the minor class 3′ splice site. In contrast, the weak conservation of the major class branch site element in mammalian introns suggests that other signals predominate.

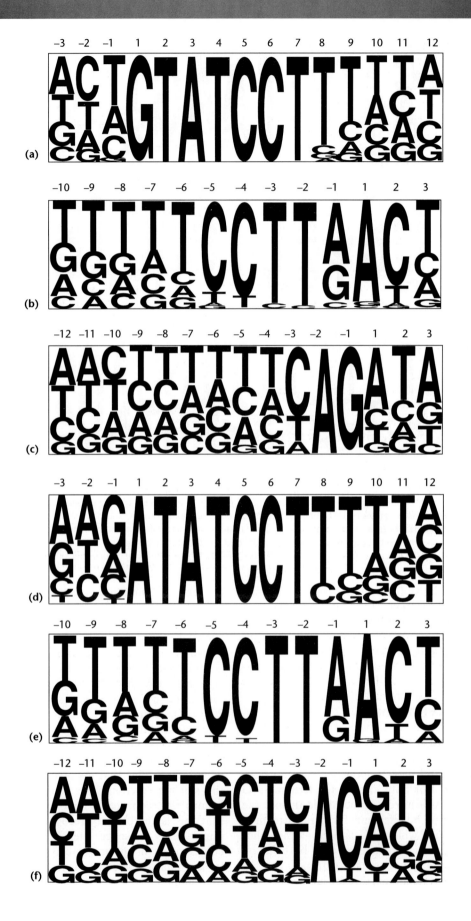

The primary recognition element for the major class 3′ splice site is a pyrimidine (C or U)-rich region located just upstream of the 3′ splice site. This polypyrimidine tract separates the branch site and 3′ splice site. The strength of a 3′ splice site has been associated with the length and purity of the polypyrimidine tract. An important function of the polypyrimidine tract is to bind protein factors that recruit the U2 snRNP particle to the branch site. In mammalian major class introns, the U2 auxiliary factor U2AF recognizes both the polypyrimidine tract and the 3′ splice site AG dinucleotide, while the factor SF1 recognizes the branch site. These proteins then recruit the U2 snRNP to the branch site (see Moore, 2000). Minor class introns do not have a polypyrimidine tract between the branch site and the 3′ splice site. Whether U2AF or SF1 plays a role in recruiting U12 snRNP to the branch site in these introns is not clear.

Additional Sequence Elements

These 5′ and 3′ splice site consensus sequences do not by themselves contain enough information to uniquely identify the intron/exon structure of a gene. Particularly in the major class of intron, additional information supplied by sequences surrounding the splice sites serves to define a region within which the consensus sequences function. These additional sequences include various types of splicing enhancer element, which bind proteins that help to recruit splicing components to adjacent splice sites. These enhancer elements are frequently poorly conserved at the sequence level and probably function as members of arrays of low-affinity binding sites which together activate particular splice sites for splicing (see Blencowe, 2000). (*See* Exonic Splicing Enhancers.)

An additional role of enhancer elements is to regulate the use of alternative splice sites. The abundance and/or activity of different enhancer binding proteins varies between cell types. This leads to differential activation of enhancer-dependent splice sites and differences in the alternative splicing patterns of specific genes. (*See* Alternative Processing: Neuronal Nitric Oxide Synthase; SR Proteins.)

Negative or suppressor elements are also known to exist near alternative splice sites. These elements are frequently recognized and bound by proteins with a tissue-specific distribution, which can regulate the alternative splicing pattern of a gene.

Conservation of Splice Sites between Species

The above discussion is focused on splice sites in human genes. Other vertebrates appear to have similar splice sites and most vertebrate genes will be properly spliced when transferred to another vertebrate. Other eukaryotic organisms have splice site sequences that are recognizably similar to those of vertebrates and, in particular, adhere to the 'GU–AG rule'. However, in many cases the importance of particular splice site residues is different. In addition, some classes of organism, such as many plants, require that additional signals be present within introns. As a result, most genes transferred between vertebrates and invertebrates, plants or fungi will not be properly processed.

See also
RNA Processing
Spliceosome
Splicing of pre-mRNA

References

Beggs JD (2001) Spliceosomal machinery. In: *Encyclopedia of Life Sciences*, http://www.els.net, London: Nature Publishing Group.

Blencowe BJ (2000) Exonic splicing enhancers: mechanism of action, diversity and role in human genetic diseases. *Trends in Biochemical Science* 25: 106–110.

Burge C and Karlin S (1997) Prediction of complete gene structures in human genomic DNA. *Journal of Molecular Biology* 268: 78–94.

Burge CB, Padgett RA and Sharp PA (1998) Evolutionary fates and origins of U12-type introns. *Molecular Cell* 2: 773–785.

Burge CB, Tuschl T and Sharp PA (1999) Splicing of precursors to mRNAs by the spliceosome. In: Gestland RF, Cech T and Atkins JF (eds.) *The RNA World II*, pp. 525–560. Cold Spring Harbor, NY: Cold Spring Harbor Laboratory Press.

Dietrich RC, Incoravia R and Padgett RA (1997) Terminal intron dinucleotide sequences do not distinguish between U2- and U12-dependent introns. *Molecular Cell* 1: 151–160.

Dietrich RC, Peris MJ, Seyboldt AS and Padgett RA (2001) Role of the 3′ splice site in U12-dependent intron splicing. *Molecular Cell Biology* 21: 1942–1952.

Jackson IJ (1991) A reappraisal of non-consensus mRNA splice sites. *Nucleic Acids Research* 19: 3795–3798.

Figure 3 Pictograms of the minor U12-dependent intron class consensus splice site signals. Because the consensus sequences differ between the GU–AG and AU–AC subclasses of U12-dependent introns, these are presented separately. The GU–AG subclass consensus was derived from 160 examples, while the AU–AG subclass consensus was derived from 30 examples. (a) Minor class GU 5′ splice site consensus sequence. The position labeled 1 is the first nucleotide of the intron and the position labeled −1 is the last nucleotide of the upstream exon. (b) Minor class GU–AG branch site consensus sequence. The position labeled 1 is the branch site residue. (c) Minor class AG 3′ splice site consensus. The position labeled −1 is the last nucleotide of the intron and the position labeled 1 is the first nucleotide of the downstream exon. (d) Minor class AU 5′ splice site consensus sequence. (e) Minor class AU–AC branch site consensus sequence. (f) Minor class AC 3′ splice site consensus sequence.

Maroney PA, Romfo CM and Nilsen TW (2000) Functional recognition of 5′ splice site by U4/U6 • U5 tri-snRNP defines a novel ATP-dependent step in early spliceosome assembly. *Molecular Cell* **6**: 317–328.

Moore MJ (2000) Intron recognition comes of AGe. *Nature Structural Biology* **7**: 14–16.

Newman AJ (2001) RNA interactions in mRNA splicing. In: *Encyclopedia of Life Sciences*. London, UK: Nature Publishing Group.

Nilsen TW (1998) RNA–RNA interactions in nuclear pre-mRNA splicing. In: Simons RW and Grunberg-Manago M (eds.) *RNA Structure and Function*, pp. 279–307. Cold Spring Harbor, NY: Cold Spring Harbor Laboratory Press.

Padgett RA (2001) mRNA splicing: role of snRNAs. In: *Encyclopedia of Life Sciences*. London, UK: Nature Publishing Group.

Tarn W-Y and Steitz JA (1996) Highly diverged U4 and U6 small nuclear RNAs required for splicing rare AT–AC introns. *Science* **273**: 1824–1832.

Further Reading

Lund M and Kjems J (2002) Defining a 5′ splice site by functional selection in the presence and absence of U1 snRNA 5′ end. *RNA* **8**: 166–179.

Tarn W-Y and Steitz JA (1997) Pre-mRNA splicing: the discovery of a new spliceosome doubles the challenge. *Trends in Biochemical Sciences* **22**: 132–137.

Splicing of pre-mRNA

Intermediate article

Article contents

- Introduction
- Spliceosome Formation
- SR Proteins
- Exonic Splicing Enhancers and Regulated Splicing
- Enhancers and Silencers of Splicing

Andrew Newman, *Medical Research Council Laboratory of Molecular Biology, Cambridge, UK*

Eukaryotic messenger ribonucleic acids (mRNAs) are transcribed as precursors containing introns, which are excised by splicing together the surrounding exon sequences to produce the mature mRNA. Alternative splicing of precursor mRNAs in higher eukaryotes contributes to proteomic diversity by allowing a single gene to direct the synthesis of functionally diverse proteins.

Introduction

Variations in the structure of messenger RNA (mRNA) arise through a wide range of alternative splicing events. Exons can be included or omitted, and introns that are usually removed can be retained in the mRNA. The use of alternative 5′ or 3′ splice sites can make specific exons longer or shorter. The encoded proteins can differ by inclusion or exclusion of functional domains or specific short peptide sequences or they can be truncated by inclusion of premature translational stop codons in the mRNA. Such changes can affect the biochemical activity, subcellular location or state of phosphorylation of the protein.

In some genes there are tandem arrays of potential alternative exons, which appear in the mRNA in a mutually exclusive manner. This mechanism can allow the synthesis of an immense number of different polypeptides from one gene. In *Drosophila melanogaster* the gene encoding the homolog of the Down syndrome cell adhesion molecule (DSCAM) yields mRNAs comprising 24 exons (Schmucker *et al.*, 2000). Exons 4, 6 and 9 are each encoded by a tandem array of multiple, alternative exons and these appear in mRNAs in a wide range of combinations. If all possible arrangements of the alternative exons were to be used, the Down syndrome cell adhesion molecule (*DSCAM*)

gene could produce more than 38 000 different proteins!

The ways in which cells control splice site selection are not yet thoroughly understood. The consensus sequences that define splice sites in metazoans (**Figure 1**) do not themselves carry enough information to ensure accurate splicing: many sequences that conform to the consensus are not actually used, while other functional sites match the consensus only poorly. Clearly choice of splice site must be strongly influenced by sequences outside the splice sites themselves.

Spliceosome Formation

Each intron in a precursor mRNA (pre-mRNA) is assembled into a spliceosome, which recognizes the splice sites and catalyzes the transesterification reactions that result in intron excision and exon ligation. Spliceosomes are complex ribonucleoprotein particles

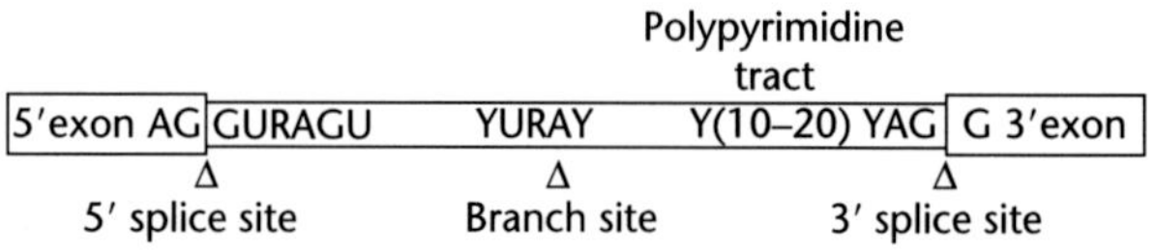

Figure 1 Consensus sequences that define a typical metazoan intron (R: purine; Y: pyrimidine).

built up from five small nuclear ribonucleoproteins (snRNPs), the U1, U2, U4, U5 and U6 snRNPs, together with many accessory splicing factors. U1 snRNP recognizes and binds to the 5′ splice site and splicing factor 1/branchpoint-binding protein (SF1/BBP) binds to the branchpoint. The 65 and 35 kDa subunits of U2 snRNP auxiliary factor (U2AF65 and U2AF35) bind to the polypyrimidine tract and the 3′ splice site sequence YAG, respectively (**Figure 2**). (*See* Splice Sites.)

At this stage, 'bridging interactions' (probably mediated by members of the SR family of splicing regulators, see below) between U1 snRNP and SF1/U2AF have paired the 5′ and 3′ ends of the intron, thus defining it for excision. Subsequently, U2 snRNP supplants SF1/BBP at the branchpoint and the remaining snRNPs (U4, U5 and U6) are incorporated into the spliceosome as a preformed tri-snRNP. The spliceosome is then reorganized to put the splice sites and branchpoint into the active site so that catalysis can proceed.

Strictly conserved short sequences in U2 and U6 snRNAs base-pair with the branchpoint and 5′ splice site, respectively, in active spliceosomes. Invariant nucleotides lying nearby in U2 and U6 are believed to play crucial roles in the catalysis of splicing. Intron excision proceeds via two transesterification reactions. In the first step, the 2′ OH of the branchpoint adenosine attacks the 5′ splice site to release the 5′ exon and form a branched intron–3′ exon intermediate. In the second step, the 3′ OH of the 5′ exon attacks the 3′ splice site to generate the free intron and spliced mRNA products (reviewed in Staley and Guthrie (1998)).

SR Proteins

Pre-mRNA splicing patterns are believed to be controlled by proteins that bind to target sites distinct from the splice sites and can locally enhance or inhibit splicing complex formation. Proteins of the SR family are the best understood of these splicing regulators. There are currently 10 characterized human SR proteins. They all have a modular structure consisting of an *N*-terminal RNA recognition motif (RRM) and a *C*-terminal RS domain (rich in arginine–serine residues) that mediates protein–protein interactions (reviewed in Gravely (2000)). RS domain-mediated interactions between SR proteins and the U2AF35 and U1 snRNP 70 kDa proteins (that also contain RS domains) may play important roles as early 'bridging factors' to commit introns to spliceosome assembly (**Figure 2**). (*See* SR Proteins.)

Functional assays and *in vitro* evolution (SELEX) experiments have been used to determine the RNA sequences that act as high-affinity binding sites for SR proteins, and the data show that a wide variety of sequences are recognized. Individual SR proteins have characteristic preferences, but the consensus sequences are quite degenerate (Tacke and Manley, 1995; Liu *et al.*, 1998). SR proteins interact with exonic splicing enhancers (ESEs), which have been characterized in both constitutive and regulated splicing substrates. ESEs are typically purine rich, and different ESEs are preferentially recognized by specific subsets of SR proteins. SR protein–ESE interactions may play a crucial role in splice site recognition by a process

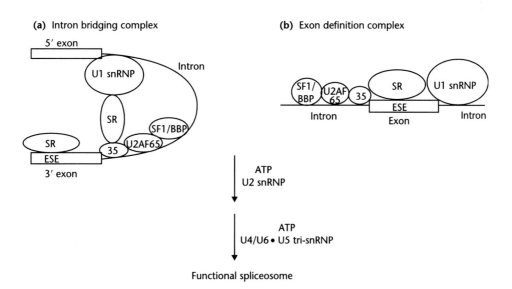

Figure 2 Early events in splicing complex formation are promoted by SR-protein-mediated bridging interactions. (a) SR proteins can stabilize U1 small nuclear ribonucleoprotein (snRNP) and U2AF interactions by bridging across the intron. (b) In exon definition complexes, SR proteins may function by binding to an exonic splicing enhancer (ESE) (SF1/BBP: splicing factor 1/branchpoint-binding protein; AF: auxiliary factor; ATP: adenosine 5′ triphosphate).

known as exon definition (Berget, 1995). Introns are often very long, whereas exons are typically short (100–300 nucleotides) so that one feature that might help to define an authentic 3′ splice site is the presence of a 5′ splice site downstream at the other end of the exon. Indeed, in exon definition the splice sites are thought to be recognized initially by bridging interactions across the exon (**Figure 2**). SR proteins bound to an ESE may play a crucial early role in the choice of splice site by promoting the interactions of U2AF and U1 snRNP with the 3′ and 5′ splice sites in exon definition complexes.

The RS domains of SR proteins are extensively phosphorylated on serine residues, and modulation of the extent of phosphorylation can affect protein–protein interactions and so change SR protein functions in splicing (Xiao and Manley, 1997; Wang et al., 1998). Three types of protein kinase have been shown to phosphorylate RS domains in vitro, including kinases of the Clk/Sty family. In D. melanogaster, mutations in the Clk/Sty homolog disrupt the regulated splicing of the doublesex gene, demonstrating that SR protein phosphorylation can control the output of this important developmentally regulated splicing pathway (Du et al., 1998). Similarly, during adenovirus infection, one of the viral proteins regulates an alternative splicing event by inducing dephosphorylation of bound SR proteins (Kanopka et al., 1998).

Exonic Splicing Enhancers and Regulated Splicing

Sex-specific alternative splicing of pre-mRNAs from the doublesex gene in D. melanogaster provides us with a well-characterized example of regulated ESE function: exon 4 is skipped in males but included in females. The female-specific splicing pattern requires an ESE in exon 4, which consists of six tandem repeats of a 13-nucleotide element. Activation of splicing of the upstream intron is achieved in females by the binding of a complex of three proteins to the ESE: Tra, Tra2 and a specific SR protein called RBP1 (Lynch and Maniatis, 1996). The crucial component of this complex is the Tra protein, which is expressed only in females and is essential for ESE function.

How does formation of a stable enhancer complex on the ESE activate splicing of the upstream intron? This point is still controversial, but one likely explanation is that the presence of the enhancer protein complex facilitates recruitment of splicing factors to the female-specific 3′ splice site, which has a relatively weak polypyrimidine tract. In support of this idea, substituting a stronger polypyrimidine tract for the natural doublesex sequence makes splicing

enhancer-independent. Conversely, reducing the polypyrimidine tract strength of a constitutively spliced intron can make splicing enhancer-dependent. These findings suggest that the enhancer-bound complex may work by recruiting U2AF65 to the pyrimidine tract of the regulated intron, presumably by RS domain-mediated protein–protein interactions.

Splicing enhancers can also promote the recognition of alternative 5′ splice sites. Pre-mRNAs from the fruitless gene in D. melanogaster are alternatively spliced in a sex-specific fashion. Analysis of the fruitless gene sequence showed that it contains three copies of the 13-nucleotide ESE repeats originally found in the doublesex gene, and the fruitless element is also the target of a similar Tra–Tra2–RBP1 trimeric enhancer binding complex (Ryner et al., 1996). In the case of fruitless pre-mRNA, the binding of the regulatory complex to the enhancer stimulates the use of a downstream 5′ splice site, probably by RS-domain-mediated interactions with the U1 snRNP component U1 70 kDa.

Enhancers and Silencers of Splicing

In addition to associating with splicing factors and SR proteins, pre-mRNAs are targeted by members of the heterogeneous nuclear ribonucleoprotein (hnRNP) family of RNA-binding proteins. In some cases, hnRNP proteins can influence the choice of splice site. In simian virus 40, alternative 5′ splice site selection controls the synthesis of large T and small t viral proteins. The SR protein ASF/SF2 (alternative splicng factor/splicing factor 2) activates use of the weaker 5′ splice site. In contrast, hnRNP A1 antagonizes the activity of ASF/SF2 (Caceres et al., 1994). There are several examples of pre-mRNAs that contain sequences that inhibit splicing, designated exonic silencers, and these may also operate via interactions with hnRNP proteins. The efficiency of exon recognition can therefore be controlled by the balance of silencer and enhancer functions, and the relative concentrations of hnRNP proteins and splicing factors.

Dedicated, gene-specific regulators operate in this environment of general antagonism between silencing and enhancement to achieve cell-type-specific splicing. In D. melanogaster, the P-element transposase is produced only in germline cells. In somatic cells, splicing of transposase intron 3 is blocked by the action of a silencer element in the upstream exon. Proteins present in somatic cell nuclear extracts have been shown to bind to the silencer and this prevents recruitment of U1 snRNP to the authentic intron 3 5′ splice site. In the presence of the silencer-binding complex, U1 snRNP instead binds nonproductively to a pseudo-5′ splice site in the silencer element (Adams et al., 1996). Protein purification and cloning have shown that the silencer-binding complex includes at least four proteins including a 97 kDa polypeptide called P-element somatic inhibitor (PSI), which is expressed only in somatic tissue. Ectopic expression of PSI in the *D. melanogaster* germ line inhibits splicing of P-element intron 3, and antibodies against PSI activate intron 3 splicing in somatic cell extracts. Recent biochemical experiments show that PSI interacts directly with the *C*-terminal arginine-serine-rich domain of the U1-snRNP-specific 70 kDa protein. This interaction serves to recruit U1 snRNP to the silencer element (Labourier *et al.*, 2001).

Interconnections between regulated splicing and other RNA processing events such as polyadenylation provide a further layer of complexity to the generation of proteomic diversity. Competition between splicing and export of RNA to the cytoplasm can also be modulated: the *human immunodeficiency virus* Rev protein enhances the expression of viral structural proteins by promoting export of unspliced RNA. The interplay of splicing regulatory sequences, general and gene-specific factors and other RNA processing events allows the cell to exert exquisite control over the mRNAs delivered to the cytoplasm for translation. (*See* RNA Processing.)

See also

Alternative Splicing: Evolution
Exonic Splicing Enhancers
Spliceosome
SR Proteins

References

Adams MD, Rudner DZ and Rio DC (1996) Biochemistry and regulation of pre-mRNA splicing. *Current Opinion in Cell Biology* 8: 331–339.

Berget SM (1995) Exon recognition in vertebrate splicing. *Journal of Biological Chemistry* 270: 2411–2414.

Caceres JF, Stamm S, Helfman DM and Krainer AR (1994) Regulation of alternative splicing *in vivo* by overexpression of antagonistic splicing factors. *Science* 265: 1706–1709.

Du C, McGuffin ME, Dauwalder B, Rabinow L and Mattox W (1998) Protein phosphorylation plays an essential role in the regulation of alternative splicing and sex determination in *Drosophila*. *Molecular Cell* 2: 741–750.

Graveley BR (2000) Sorting out the complexity of SR protein functions. *RNA* 6: 1197–1211.

Kanopka A, Muhlemann O, Petersen-Mahrt S, *et al.* (1998) Regulation of adenovirus alternative RNA splicing by dephosphorylation of SR proteins. *Nature* 393: 185–187.

Labourier E, Adams MD and Rio DC (2001) Modulation of P-element pre-mRNA splicing by a direct interaction between PSI and U1 snRNP 70 K protein. *Molecular Cell* 8: 363–373.

Liu H-X, Zhang M and Manley JL (1998) Identification of functional exonic splicing enhancer motifs recognized by individual SR proteins. *Genes and Development* 12: 1998–2012.

Lynch KW and Maniatis T (1996) Assembly of specific SR protein complexes on distinct regulatory elements of the *Drosophila* doublesex splicing enhancer. *Genes and Development* 10: 2089–2101.

Ryner LC, Goodwin SF, Castrillon DH, *et al.* (1996) Control of male sexual behaviour and sexual orientation in *Drosophila* by the fruitless gene. *Cell* 87: 1079–1089.

Schmucker D, Clemens JC, Shu H, *et al.* (2000) *Drosophila* DSCAM is an axon guidance receptor exhibiting extraordinary molecular diversity. *Cell* 101: 671–684.

Staley JP and Guthrie C (1998) Mechanical devices of the spliceosome: motors, clocks, springs and things. *Cell* 92: 315–326.

Tacke R and Manley JL (1995) The human splicing factors ASF/SF2 and SC35 possess distinct, functionally significant RNA binding specificities. *EMBO Journal* 14: 3540–3551.

Wang H-Y, Lin W and Dyck JA, *et al.* (1998) A differentially expressed SR protein-specific kinase involved in mediating the interaction and localization of pre-mRNA splicing factors in mammalian cells. *Journal of Cell Biology* 140: 737–750.

Xiao S-H and Manley JL (1997) Phosphorylation of the ASF/SF2 RS domain affects protein–protein interactions and is necessary for splicing. *Genes and Development* 11: 334–344.

Further Reading

Black DL (2000) Protein diversity from alternative splicing: a challenge for bio-informatics and post-genome biology. *Cell* 103: 367–370.

Blencowe BJ (2000) Exonic splicing enhancers: mechanism of action, diversity and role in human genetic diseases. *Trends in Biochemical Science* 25: 106–110.

Burge CB, Tuschl T and Sharp PA (1999) Splicing of precursors to mRNAs by the spliceosomes. In: Gesteland RF, Cech TR and Atkins JF (eds) *The RNA World*, 2nd edn, pp. 525–560. Cold Spring Harbor, NY: Cold Spring Harbor Press.

Collins CA and Guthrie C (2000) The question remains: is the spliceosome a ribozyme? *Nature Structural Biology* 7: 850–854.

Moore MJ (2000) Intron recognition comes of AGe. *Nature Structural Biology* 7: 14–16.

Misteli T (1999) RNA splicing: what has phosphorylation got to do with it? *Current Biology* 9: R198–R200.

Reed R (2000) Mechanisms of fidelity in pre-mRNA splicing. *Current Opinion in Cell Biology* 12: 340–345.

Smith CW and Valcarcel J (2000) Alternative pre-mRNA splicing: the logic of combinatorial control. *Trends in Biochemical Science* 25: 381–388.

Tacke R and Manley JL (1999) Determinants of SR protein specificity. *Current Opinion in Cell Biology* 11: 358–362.

Yu Y-T, Scharl EC, Smith CM and Steitz JA (1999) The growing world of small nuclear ribonucleoproteins. In: Gesteland RF, Cech TR and Atkins JF (eds) *The RNA World*, 2nd edn, pp. 487–524. Cold Spring Harbor, NY: Cold Spring Harbor Press.

Web Links

NCBI: The Human Genome. A Guide to online information resources
http://www.ncbi.nlm.nih.gov/genome/guide/human/
UCSC Genome Bioinformatics Site
http://genome.cse.ucsc.edu/
Down syndrome cell adhesion molecule (DSCAM); Locus ID: 1826. LocusLink:
http://www.ncbi.nlm.nih.gov/LocusLink/LocRpt.cgi?l = 1826
Down syndrome cell adhesion molecule (DSCAM); MIM number: 602523. OMIM:
http://www.ncbi.nlm.nih.gov/htbin-post/Omim/dispmim?602523

Spontaneous Function Correction of Pathogenic Alleles in Inherited Diseases Resulting in Somatic Mosaicism

Quinten Waisfisz, *VU University Medical Center, Amsterdam, The Netherlands*

Hans Joenje, *VU University Medical Center, Amsterdam, The Netherlands*

Inherited mutations can cause disease, but in monogenic disease correction of just one pathogenic mutation can lead to somatic mosaicism and revert the phenotype, which may ameliorate symptoms. The diseases for which this 'natural gene therapy' has been described include Fanconi anemia and Wiskott–Aldrich syndrome, among others.

Intermediate article

Introduction

The spontaneous mutation rate for individual genes in humans is of the order of 10^{-6} per gene per cell generation. This figure implies that during each cell generation, there is a fair chance of each daughter cell acquiring a genetically modified copy of any of its 40 000 genes. As it takes some 40–50 cell generations to develop from a zygote into a complete individual, it is clear that the human body is bound to be a mosaic of genetically nonidentical cells. Is this something to worry about? Acquired mutations in oncogenes or tumor suppressor genes are clear examples of an adverse effect on the individual. In some cases, however, genetically altered cells may be beneficial. In patients affected with certain inherited monogenic disorders (**Table 1**), secondary alterations at the disease locus can (partially) correct the pathogenic mutation. These alterations may result from homologous recombination between the two alleles in compound heterozygotes (with or without actual chromatid crossing over) or consist of alterations in *cis*, that is, changes within one and the same mutated allele. (*See* Genetic Disorders; Mosaicism; Retroviral Vectors in Gene Therapy.)

Correction by Intragenic Homologous Recombination with Crossover

Homologous recombination associated with intragenic crossover is an important mechanism involved in somatic reversion of pathogenic mutations, first described in leukocytes from patients with Bloom syndrome (BS) (MIM#210900; MIM number for identification on OMIM (see Web Links)) (Ellis *et al.*, 1995). Patients with BS are clinically characterized by a small body size, immunodeficiency and a sunsensitive facial skin. Cells from BS patients typically show chromosomal instability and a strikingly elevated frequency of spontaneous sister chromatid exchanges (SCEs). In approximately 20% of BS patients, however, a proportion of lymphocytes have a normal level of SCEs. The proportion of reverted (normal–SCE) lymphocytes varies from patient to patient but may be as high as 75%, suggesting that a reverted lymphoid progenitor tends to have an increased ability to expand. Molecular genetic analysis of reverted cell lines from 11 compound heterozygous patients with markers surrounding the disease locus revealed that somatic intragenic recombination between the two inherited mutations had resulted in the generation of a wild-type allele (Ellis *et al.*, 1995).

Figure 1 shows how intragenic recombination in somatic cells can result in daughter cells that have gained one wild-type gene. In 5 of the 11 patients, markers distal to the *BLM* locus showed reduction to homozygosity, which fits well with the proportion of 50% predicted by a model of random segregation of recombinant chromatids (**Figure 1**). Thus far, mosaicism in BS patients has only been observed in B and T lymphocytes, suggesting that intragenic recombination took place in the lymphoid lineage. Despite the

Table 1 Reported corrective mechanisms for inherited mutations in monogenic disorders

Disorder	Back mutation	Recombination and crossover	Recombination and gene conversion	Second-site *cis* mutation
Bloom syndrome	+	+		
Fanconi anemia	+	+	+	+
Epidermolysis bullosa			+	
Tyrosinemia type I		+		
Adenosine deaminase deficiency	+			
X-linked SCID	+			
Wiskott–Aldrich syndrome	+			
Lesch–Nyhan syndrome				+

SCID: severe combined immune deficiency.
(*See* Lesch–Nyhan Disease; Severe Combined Immune Deficiency (SCID): Genetics.)

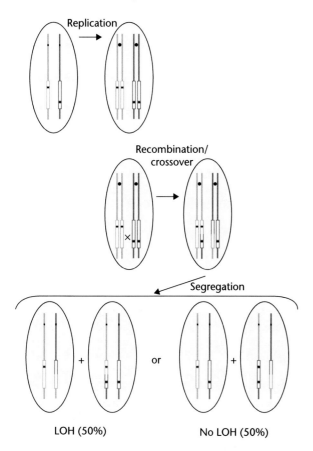

Figure 1 Generation of a wild-type allele in somatic cells through intragenic homologous recombination with crossover, as described in compound heterozygous patients with nonoverlapping mutations. After replication, intragenic recombination involving the homologous chromatids leads to a wild-type allele and a doubly mutated allele. Depending on the segregation of chromatids the genetically corrected daughter cells will (left) or will not (right) exhibit loss of heterozygosity for markers distal to the disease gene.

phenotypic reversion in part of the hematopoietic compartment of mosaic patients, no ameliorative effect, for example on immunodeficiency, has yet been observed.

Somatic mosaicism due to intragenic recombination has also been reported in lymphocytes from patients with Fanconi anemia (FA) (MIM#227650) (Lo Ten Foe *et al.*, 1997). Both BS and FA cells display chromosomal instability, but whether this has caused or facilitated the observed somatic intragenic recombination is unknown. (*See* DNA Recombination.)

Correction by Intragenic Homologous Recombination with Gene Conversion

A wild-type allele at the disease locus can be also generated by gene conversion, that is, homologous recombination without crossover. This has been shown in keratinocytes from a compound heterozygous patient with generalized atrophic benign epidermolysis bullosa (GABEB) (MIM#226650) (Jonkman *et al.*, 1997). GABEB is an autosomal recessive inherited disease characterized by blistering of all areas of the skin. The patient described by Jonkman *et al.* (1997), however, had a mosaic pattern of both affected skin patches and patches that never blistered (**Figure 2**). Whereas the inherited mutant genes did not express any COL17A1 protein product in affected skin, this protein was readily detectable in keratinocytes from the unaffected areas. Analysis of the haplotypes and collagen, type XVII, alpha 1 (*COL17A1*) gene sequences from both types of keratinocyte revealed that in the reverted keratinocytes, one of the alleles had lost the original mutation, a 1-base-pair (bp) deletion. In addition, a heterozygous intronic polymorphism situated nearby and downstream of the original mutation was lost from the reverted allele. As all other polymorphic markers were still present, however, the reversion was apparently not associated with mitotic crossover. Both genetic changes were best explained by an event of

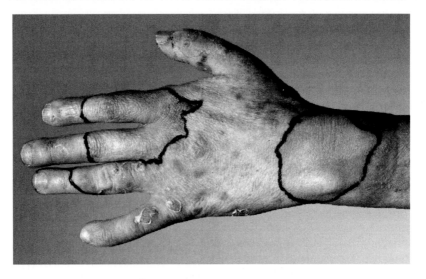

Figure 2 Mosaicism of the skin in a patient with generalized atrophic benign epidermolysis bullosa (GABEB). The picture shows the characteristic blistering of the skin next to unaffected patches of the skin where somatic recombination with gene conversion has generated a wild-type allele of the disease gene. (Reproduced from Jonkman *et al.* (1997).)

gene conversion in which the wild-type sequence from one allele was involved in strand exchange with the sequence from the mutant allele and the subsequent fixation of the wild-type sequence by mismatch repair. Reversion due to gene conversion is not restricted to keratinocytes, as this has also been observed in lymphoid cells from FA patients (Lo Ten Foe *et al.*, 1997).

Back Mutation

When a mutation changes back to the exact original wild-type sequence, without recombination, it is called a back mutation. Although back mutations seem infrequent events, they have been observed in mosaic patients affected with different autosomal and X-linked monogenic disorders. Clear examples are mosaic patients who are homozygous for their mutation but have reverted cells that have lost the mutation at one allele. This has been described for patients with hereditary tyrosinemia type I (FAH) (MIM#276700) (Kvittingen *et al.*, 1994). FAH patients have an enzymatic deficiency of fumarylacetoacetate hyroxylase (FAH) that is predominantly expressed in the liver and kidney. Patients display a variable degree of severity of the disease, even among siblings. More mildly affected patients appear to have residual enzyme activity that correlates with the presence of immunoreactive protein. Genetic analysis of immunoreactive liver nodules from four mosaic patients, two of whom carried homozygous mutations, has revealed back

mutations in one allele. Similarly, in a mildly affected patient with mutations in the adenosine deaminase gene, which normally results in adenosine deaminase deficiency with severe combined immunodeficiency (ADA/SCID) (MIM#102700), spontaneous *in vivo* reversion of lymphoid cells was caused by a back mutation of inherited single base pair substitutions (Hirschhorn *et al.*, 1996). In contrast to a severely affected sibling who had died before 3 years of age, this mosaic patient, aged 12 years, had a progressively mild course and had survived so far without any therapy, presumably owing to the back mutation. There are additional examples of back mutations in patients with BS (Ellis *et al.*, 2001), FA and a mildly affected patient with X-linked SCID (SCIDX1) (MIM#300400) (Stephan *et al.*, 1996).

In most cases, the molecular mechanism of back mutations has remained unknown. It is difficult to define whether specific sequences associated with (the surroundings of) the inherited mutation have facilitated the occurrence of back mutations. Some back mutations may be caused by deoxyribonucleic acid (DNA) polymerase slippage, as reported for a mosaic patient with the X-linked Wiskott–Aldrich syndrome (WAS) (MIM#301000) (Wada *et al.*, 2001). This male patient inherited a 6-bp insertion in the Wiskott–Aldrich syndrome (eczema-thrombocytopenia) (*WAS*) gene, which abrogates WAS protein (WASP) expression. Despite this, the majority of the patient's T cells were positive for WASP protein, which was associated with a clinical improvement over the years. Sequence analysis of the reverted T cells revealed that the 6-bp insertion was absent from the WASP-positive cells. The inherited mutation appeared to be a duplication of the exact 6 bp adjacent to the insertion, suggesting that DNA polymerase slippage had caused both the original germline insertion and the back mutation in the patient's T cells.

Correction by Intragenic Second-site Mutations in *cis*

Exact restoration of the wild-type sequence is not required for a mutated gene to regain its wild-type function. Somatic *de novo* insertions or deletions situated up- or downstream of an inherited frameshift mutation may restore the original reading frame, leaving only a short stretch of altered codons in the sequence. Also, inherited nonsense and missense mutations may be reverted by *de novo* mutations changing a triplet to encode an alternative amino acid that can still support the protein's function.

Examples of such second-site mutations are those observed in mosaic FA patients (Waisfisz *et al.*, 1999). FA is clinically characterized by pancytopenia due to bone marrow failure and predisposition to cancer. Cells from FA patients feature a characteristic chromosomal instability and hypersensitivity to bi-functional alkylating (cross-linking) agents. However, approximately 25% of FA patients have phenotypically reverted lymphocytes, that is, cells that have lost their hypersensitivity to cross-linking agents. This reversion to a wild-type cellular phenotype may occur through various mechanisms, and in some, but not all, mosaic patients, there is clear evidence for relatively mild hematological symptoms. The mechanism that results in correcting secondary/second-site mutations is often difficult to trace. However, in one patient who was homozygous for a 1-bp insertion, reverted cells were corrected by an insertion of 5 bp downstream that restored the open reading frame, suggesting DNA polymerase slippage as the underlying mechanism.

In some cases, second-site mutations can result in restoration of wild-type protein sequence without restoration of the genomic sequence (Yang *et al.*, 1988), as illustrated by a patient with Lesch–Nyhan syndrome (LNS) (MIM#300322). The primary (inherited) mutation in the X-linked hypoxanthine phosphoribosyltransferase 1 (Lesch–Nyhan syndrome) (*HPRT1*) gene was a large partial gene duplication. Reverted lymphoblastoid cell lines from the patient had a second major rearrangement that deleted most or all of the inherited mutation. Although the normal gene structure was not restored, the gene's reading frame had become indistinguishable from the wild type.

Conclusions

The list of inherited monogenic diseases with spontaneous correction of cellular phenotypes is increasing rapidly and appears not to be limited to disorders due to genomic instability. Reversion appears to be an event that occurs with a low probability, considering that mosaicism may discordantly occur in patients within the same family. On the other hand, different patients with the same mutation may show exactly the same correcting mutation, suggesting that in such cases a specific mechanism is operative, possibly evoked by the mutation itself.

So far, somatic mosaicism has been found predominantly in the hematopoietic compartment, probably owing to the relative ease of obtaining material in combination with its unique fluid nature allowing continuous mixing of its constituent cells. The ameliorative effect of corrective mutations on the disease in somatic mosaics varies from undetected, such as in BS, to clearly milder phenotypes, as illustrated by ADA/SCID and reverted skin areas in GABEB. The benefits for a patient with somatic mosaicism will depend strongly on the developmental stage in which the reversion took place, which cell types are reverted and whether the reverted cells have any selective advantage in a mutant background. Diseases where spontaneously reverted cells have a selective advantage and ameliorate symptoms may be considered promising targets for experimental gene therapeutic intervention. (*See* Gene Therapy.)

See also
Mosaicism

References

Ellis NA, Ciocci S and German J (2001) Back mutation can produce phenotype reversion in Bloom syndrome somatic cells. *Human Genetics* **108**: 167–173.

Ellis NA, Lennon DJ, Proytcheva M, *et al.* (1995) Somatic intragenic recombination within the mutated locus BLM can correct the high sister-chromatid exchange phenotype of Bloom syndrome cells. *American Journal of Human Genetics* **57**: 1019–1027.

Hirschhorn R, Yang DR, Puck JM, *et al.* (1996) Spontaneous *in vivo* reversion to normal of an inherited mutation in a patient with adenosine deaminase deficiency. *Nature Genetics* **13**: 290–295.

Jonkman MF, Scheffer H, Stulp R, *et al.* (1997) Revertant mosaicism in epidermolysis bullosa caused by mitotic gene conversion. *Cell* **21**: 543–551.

Kvittingen EA, Rootwelt H, Berger R and Brandtzaeg P (1994) Self-induced correction of the genetic defect in tyrosinemia type I. *Journal of Clinical Investigation* **94**: 1657–1661.

Lo Ten Foe JR, Kwee ML, Rooimans MA, *et al.* (1997) Somatic mosaicism in Fanconi anemia: molecular basis and clinical significance. *European Journal of Human Genetics* **5**: 137–148.

Stephan V, Wahn V, Le Deist F, *et al.* (1996) Atypical X-linked severe combined immunodeficiency due to possible spontaneous reversion of the genetic defect in T cells. *New England Journal of Medicine* **335**: 1563–1567.

Wada T, Schurman SH, Otsu M, *et al.* (2001) Somatic mosaicism in Wiskott–Aldrich syndrome suggests *in vivo* reversion by a DNA slippage mechanism. *Proceedings of the National Academy of Sciences of the United States of America* **98**: 8697–8702.

Waisfisz Q, Morgan NV, Savino M, *et al.* (1999) Spontaneous functional correction of homozygous Fanconi anaemia alleles reveals novel mechanistic basis for reverse mosaicism. *Nature Genetics* **22**: 379–383.

Yang TP, Stout JT, Konecki DS, *et al.* (1988) Spontaneous reversion of novel Lesch–Nyhan mutation by HPRT gene rearrangement. *Somatic Cell and Molecular Genetics* **14**: 293–303.

Further Reading

Ariga T, Kondoh T, Yamaguchi K, *et al.* (2001) Spontaneous *in vivo* reversion of an inherited mutation in the Wiskott–Aldrich syndrome. *Journal of Immunology* **166**: 5245–5249.

Ariga T, Oda N, Yamaguchi K, *et al.* (2001) T-cell lines from 2 patients with adenosine deaminase (ADA) deficiency showed the restoration of ADA activity resulted from the reversion of an inherited mutation. *Blood* **97**: 2896–2899.

Gregory Jr JJ, Wagner JE, Verlander PC, *et al.* (2001) Somatic mosaicism in Fanconi anemia: evidence of genotypic reversion in lymphohematopoietic stem cells. *Proceedings of the National Academy of Sciences of the United States of America* **98**: 2532–2537.

Jonkman MF (1999) Revertant mosaicism in human genetic disorders. *American Journal of Medical Genetics* **85**: 361–364.

Stephan V, Wahn V, Le Deist F, *et al.* (1996) Atypical X-linked severe combined immunodeficiency due to possible spontaneous reversion of the genetic defect in T cells. *New England Journal of Medicine* **335**: 1563–1567.

Youssoufian H and Pyeritz RE (2002) Mechanisms and consequences of somatic mosaicism in humans. *Nature Review Genetics* **10**: 748–758.

Web Links

NCBI Online Mendelian Inheritance in Man (OMIM). An online catalog of human genes and genetic disorders
http://www.ncbi.nlm.nih.gov/Omim/

Collagen, type XVII, alpha 1 (*COL17A1*); Locus ID: 1308. LocusLink:
http://www.ncbi.nlm.nih.gov/LocusLink/LocRpt.cgi?l = 1308

Hypoxanthine phosphoribosyltransferase 1 (Lesch–Nyhan syndrome) (*HPRT1*); Locus ID: 3251. LocusLink:
http://www.ncbi.nlm.nih.gov/LocusLink/LocRpt.cgi?l = 3251

Wiskott-Aldrich syndrome (eczema-thrombocytopenia) (*WAS*); Locus ID: 7454. LocusLink:
http://www.ncbi.nlm.nih.gov/LocusLink/LocRpt.cgi?l = 7454

Collagen, type XVII, alpha 1 (*COL17A1*); MIM number: 113811. OMIM:
http://www.ncbi.nlm.nih.gov/htbin-post/Omim/dispmim?113811

Hypoxanthine phosphoribosyltransferase 1 (Lesch–Nyhan syndrome) (*HPRT1*); MIM number: 308000. OMIM:
http://www.ncbi.nlm.nih.gov/htbin-post/Omim/dispmim?308000

Wiskott–Aldrich syndrome (eczema-thrombocytopenia) (*WAS*); MIM number: 300392. OMIM:
http://www.ncbi.nlm.nih.gov/htbin-post/Omim/dispmim?300392

SR Proteins

Brenton R Graveley, *University of Connecticut Health Center, Farmington, Connecticut, USA*

Klemens J Hertel, *University of California Irvine, Irvine, California, USA*

Advanced article

Article contents

- The SR Protein Family
- Roles in Pre-mRNA Processing
- Splicing Repression
- Role of Phosphorylation in Regulating SR Protein Activities
- SR Protein Localization and Roles in mRNA Export

Members of the serine/arginine-rich (SR) protein family have several important functions in mRNA biogenesis. SR proteins are essential splicing factors that also participate in the regulation of alternative splicing and the export of mRNAs from the nucleus to the cytoplasm.

The SR Protein Family

Members of the serine/arginine-rich (SR) protein family function in both constitutive and alternative splicing, and have a role in export of messenger ribonucleic acid (mRNA). In humans, the SR protein family is encoded by nine *splicing factor, arginine/ serine-rich* genes, designated *SFRS1*, *SFRS2*, *SFRS3*, *SFRS4*, *SFRS5*, *SFRS6*, *SFRS7*, *SFRS9* and *SFRS11* (**Table 1** and **Figure 1**). All nine members of the SR protein family, SF2/ASF, SC35, SRp20, SRp40, SRp55, SRp75, SRp30c, 9G8 and SRp54, have a common structural organization (**Figure 2**). Each SR protein contains either one or two amino (*N*)-terminal

Table 1 Human genes encoding SR proteins

Gene	SR protein	Chromosomal location
SFRS1	SF2/ASF/SRp30a	17q21.3–q22
SFRS2	SC35/SRp30b	17q25.1
SFRS3	SRp20	6p21.31
SFRS4	SRp75	1p35.3
SFRS5	SRp40	14q24.2
SFRS6	SRp55	20q13.11
SFRS7	9G8	2p22.1
SFRS9	SRp30c	12q24.23
SFRS11	SRp54	1p31.1

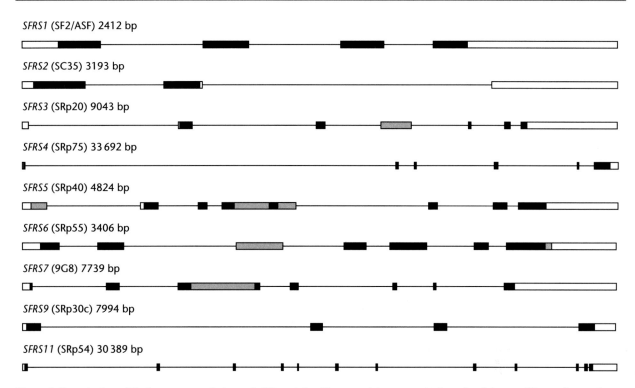

Figure 1 Organization of the human genes that encode SR proteins. The exon–intron organization of each human SR protein gene is shown. Noncoding portions of each exon are represented by white boxes, coding portions of exons are indicated by black boxes. Alternatively spliced exons are indicated in gray. The splicing of the alternative exons is autoregulated by the SR protein encoded by the gene. The exons and introns for each gene are drawn to scale and the size of each gene is indicated.

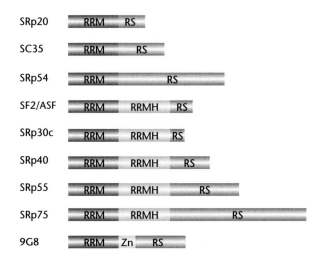

Figure 2 The human SR protein family. The structural organization of the nine human SR proteins is shown. RRM: RNA recognition motif; RRMH: RRM homology; Zn: zinc-knuckle; RS: arginine/serine-rich domain.

ribonucleic acid (RNA)-binding domains and a variable-length arginine/serine-rich (RS) domain at their carboxy (C)-terminus that functions as a protein interaction domain. In addition to their structural similarities, all members of the SR protein family share several biochemical properties. All SR proteins

contain a phosphoepitope within the RS domain that is recognized by the monoclonal antibody mAb104; individual SR proteins can complement HeLa cell S100 (the supernatant of a 100,000g spin of a cytoplasmic extract) extracts deficient in SR proteins; and all SR proteins can be precipitated in 20 mM $MgCl_2$. (*See* Alternative Processing: Neuronal Nitric Oxide Synthase; RNA Processing; Spliceosome; Splicing of pre-mRNA; *Trans* Splicing.)

SR proteins are highly conserved and exist in all metazoan species (Zahler *et al.*, 1992) as well as some lower eukaryotes such as *Schizosaccharomyces pombe*. But SR proteins are not present in all eukaryotes; they are apparently missing from *Saccharomyces cerevisiae*. As a general rule, the species-specific presence of SR proteins correlates with the presence of RS domains within other components of the general splicing machinery. (*See* Alternative Splicing: Evolution.)

All SR proteins contain one RNA-binding domain of the RNA recognition motif (RRM) type. For most SR proteins with two RNA-binding domains, the second is a poor match to the RRM consensus and is therefore referred to as an RRM homolog (RRMH). The only exception is 9G8, which, in addition to one RRM, contains a zinc-knuckle that is thought to contact RNA. In the cases where it has been determined, SR proteins have specific but rather degenerate RNA-binding specificities (Liu *et al.*, 1998); however, tight RNA binding does not necessarily correlate with maximal activity. (*See* RNA-binding Proteins: Regulation of mRNA Splicing, Export and Decay.)

The RS domains of SR proteins participate in protein interactions with several other RS-domain-containing splicing factors (Wu and Maniatis, 1993; Kohtz *et al.*, 1994). These include other SR proteins, a second class of proteins known as the SR-related proteins (SRrps) and, most importantly, constituents of the general splicing machinery such as the U1-70K (relative molecular mass, M_r, 70 000) component of the U1 small nuclear ribonuclear protein (snRNP); U2AF[35]; and possibly the RS-domain-containing 100K component of U5 snRNP and the 27K, 65K and 110K components of the U4/U6·U5 tri-snRNP. Although RS domains have been shown to be dispensable for the ability of SR proteins to promote splice-site switching in certain pre-mRNAs and for the splicing of some precursor mRNAs (pre-mRNAs) that contain strong splice sites, RS domains are required for most activities of SR proteins.

Roles in Pre-mRNA Processing

The best-characterized activities of SR proteins are those that involve their binding to exon sequences and are therefore referred to as exon-dependent. It is now well documented that SR proteins bind to sequences within regulated exons and thereby activate exon inclusion. These binding sites have been designated exonic splicing enhancers (ESEs). When bound to an ESE, SR proteins can enhance the recognition of suboptimal upstream 3′ splice sites, possibly by directly recruiting the essential splicing factor U2AF to the pyrimidine tract (Zuo and Maniatis, 1996; **Figure 3a**). However, SR proteins may also recruit components of the splicing machinery through interactions with other essential splicing factors or by relieving the inhibitory effect of splicing repressor elements (**Figure 3a**). (*See* Exonic Splicing Enhancers.)

In addition, SR proteins regulate choice of 5′ splice site when bound to ESEs that are positioned upstream of 5′ splice sites (Ryner *et al.*, 1996). Analogous to the U2AF recruitment model for activation of 3′ splice sites, SR proteins may recruit U1 snRNP to the regulated 5′ splice site by directly interacting with the U1-70K component of U1 snRNP (**Figure 3b**). In support of this view, the SR protein SF2/ASF has been shown to recruit U1 snRNP to a 5′ splice site containing RNA (Kohtz *et al.*, 1994). But further experiments are required to delineate decisively the mechanism of SR-protein-dependent 5′ splice-site activation. (*See* Splice Sites.)

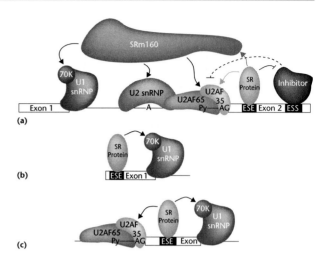

Figure 3 Exon-dependent functions of SR proteins. (a) ESE-bound SR proteins may function by activating upstream 3′ splice sites. This can be achieved through recruitment of the general splicing factor U2AF, through interactions with the splicing coactivator SRm160, or by antagonizing the activity of splicing inhibitors. (b) Alternative 5′ splice sites may be activated on the recruitment of U1 snRNP by ESE-bound SR proteins. (c) SR proteins may function in constitutive splicing by simultaneously interacting with U2AF bound to the upstream 3′ splice site and U1 snRNP bound to the downstream 5′ splice site. Py: pyrimidine tract.

Surprisingly, SR proteins promote constitutive splicing in an exon-dependent manner (Mayeda *et al.*, 1999). It is thought that SR proteins bound to constitutively spliced exons simultaneously interact with U2AF[35] that is bound to the upstream 3′ splice site and with U1-70K of U1 snRNP that is bound to the downstream 5′ splice site (Wu and Maniatis, 1993; **Figure 3c**). These cross-exon interactions most probably facilitate exon definition and thus promote incorporation of the exon into the resulting mRNA. These observations suggest that SR proteins are essential not only for the recognition of alternatively spliced exons but also for the definition of constitutively spliced exons. The degenerate RNA-binding specificities of SR proteins may ensure that at least one member of the SR protein family can bind to each constitutively spliced exon.

In addition to their exon-dependent functions, SR proteins have activities that do not require interactions with exon sequences. A potentially important exon-independent function is the pairing of 5′ and 3′ splice sites (**Figure 4**). Similar to the exon-bridging model, this activity requires simultaneous interactions with U1-70K and U2AF[35] (Wu and Maniatis, 1993). In this case, however, the interactions span across the intron and thus allow juxtaposition of the splice sites that are to be joined together. SR proteins have also been shown to facilitate the incorporation of the U4/U6·U5 tri-snRNP into the spliceosome (**Figure 4**). This activity probably involves interactions between

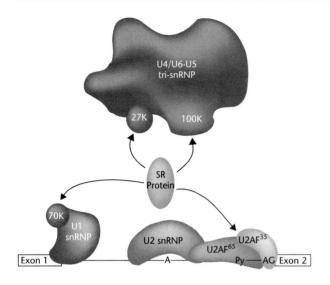

Figure 4 Exon-independent functions of SR proteins. SR proteins have two exon-independent functions. First, SR proteins facilitate splice-site pairing by simultaneously interacting with U1 snRNP and U2AF across the intron. Second, SR proteins in the partially assembled spliceosome are involved in recruiting the U4/U6·U5 tri-snRNP.

an SR protein within the partially assembled spliceosome and an RS-domain-containing component of the tri-snRNP such as U5-100K or the 27K, 65K or 110K proteins of the tri-snRNP complex.

Splicing Repression

In certain situations, SR proteins can also function as negative splicing regulators. The best-characterized example of this occurs during adenovirus infection (Kanopka *et al.*, 1996), where splicing is repressed by the binding of the SR protein SF2/ASF to an intronic repressor element located upstream of the branch site of a regulated 3′ splice site in the adenovirus pre-mRNA. When bound to the repressor element, SF2/ASF prevents the interaction of U2 snRNP with the branch site, which inactivates the 3′ splice site. Because SR proteins are thought to interact predominantly with exonic sequences, binding to intronic sequences may be a general mechanism of splicing repression.

Role of Phosphorylation in Regulating SR Protein Activities

All activities of SR proteins are modulated by phosphorylation within the RS domain. Whereas both hypo- and hyperphosphorylated SR proteins are inactive in splicing assays (Sanford and Bruzik, 1999), moderately phosphorylated SR proteins can

participate in the splicing reaction. The phosphorylation level of SR proteins, and thus their activity, is regulated throughout development (Sanford and Bruzik, 1999) and during the course of a single round of intron removal. These observations suggest that SR protein phosphorylation and dephosphorylation are highly dynamic and essential aspects of their biological activities.

Several kinases have been identified that phosphorylate SR proteins, including SR protein kinases 1 and 2 (SRPK1 and SRPK2), Clk/Sty and DNA topoisomerase I. As the RS domain provides a rather extensive platform for phosphorylation, the extent and specificity of modifications required for SR protein activity are only beginning to be understood.

SR Protein Localization and Roles in mRNA Export

Virtually all proteins involved in pre-mRNA splicing, including the SR proteins, are enriched in numerous nuclear compartments called 'speckles', which consist of two distinct structures: interchromatin granule clusters (IGCs) of 20–25 nm in diameter, which act as storage or reassembly sites for pre-mRNA splicing factors; and perichromatin fibrils (PFs) of ∼5 nm in diameter, which form the site of actively transcribing genes and cotranscriptional splicing. The SR proteins are one prominent component of nuclear speckles, and biochemical analyses have indicated that RS domains are responsible for targeting the SR proteins to speckles. As the nuclear organization of SR proteins is dynamic, SR proteins are recruited from the IGC storage clusters to the site of cotranscriptional splicing (PFs). Notably, both the RNA-binding domains and the RS domains are required for the recruitment of SR proteins from IGCs to PFs, as is phosphorylation of the RS domain.

Some SR proteins – SF2/ASF, SRp20 and 9G8 – shuttle continuously between the nucleus and the cytoplasm. The movement of these SR proteins requires an appropriately phosphorylated RS domain and the RNA-binding domain. These unique intracellular transport properties suggest that a subset of SR proteins function not only in pre-mRNA processing but also in mRNA export. In fact, the SR proteins 9G8 and SRp20 promote nuclear export of the intronless histone H2a mRNA in mammalian cells and *Xenopus* oocytes (Huang and Steitz, 2001) by binding to a 22-nucleotide sequence within the H2a mRNA. In addition, the *S. cerevisiae* protein Npl3p – a protein that is closely related to SR proteins – assists in mRNA export in yeast. Once again, phosphorylation of specific serine residues within the RS domain

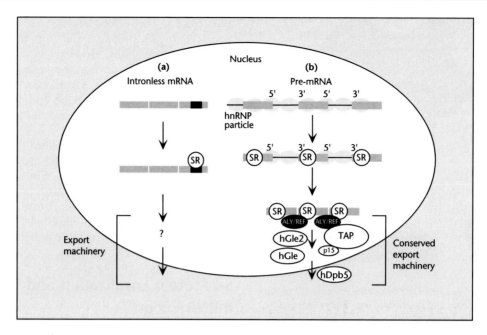

Figure 5 Models of mRNA export. (a) Intronless mRNAs such as H2a contain high-affinity binding sites (black box) for SR proteins that shuttle continuously between the nucleus and the cytoplasm. Association of shuttling SR proteins with the intronless mRNA leads to nuclear export via an unknown pathway. (b) Before splicing, nascent pre-mRNAs are coated by hnRNPs (heterogeneous nuclear ribonucleoproteins). During spliceosomal assembly, SR proteins bind to each exon with sufficient affinity to replace hnRNPs effectively. During intron removal, conserved export factors, in particular ALY/REF, are recruited to the exons. Because of their association with exonic sequences throughout the splicing cycle, SR proteins may assist in linking pre-mRNA splicing to mRNA export either directly by interacting with the export machinery or indirectly by inhibiting the association of nuclear retention factors such as hnRNPs with the mRNA. The spliced mRNA is then exported from the nucleus via a pathway involving TAP/p15, Gle, Gle2 or Dbp5 (nuclear export factors).

seems to control the functional efficiency of mRNA export mediated by Npl3p. It is not known yet whether SR proteins also promote the export of spliced mRNAs; however, given the fact that SR proteins are essential for splicing, remain associated with the spliced mRNA after intron removal, and shuttle between the nucleus and the cytoplasm, it seems highly likely that SR proteins have an important role in the export of spliced mRNAs. In the proposed model (**Figure 5**), SR proteins participate in the export of mRNAs not only through interactions with specific *cis*-acting export elements of intronless mRNAs, but also by assisting the functions of other export factors through stable associations with spliced mRNAs after intron removal. (*See* mRNA Export.)

See also

Exonic Splicing Enhancers
Posttranscriptional Processing
Splicing of pre-mRNA

References

Huang Y and Steitz JA (2001) Splicing factors SRp20 and 9G8 promote the nucleocytoplasmic export of mRNA. *Molecular Cell* **7**: 899–905.

Kanopka A, Muhlemann O and Akusjarvi G (1996) Inhibition by SR proteins of splicing of a regulated adenovirus pre-mRNA. *Nature* **381**: 535–538.

Kohtz JD, Jamison SF, Will CL, *et al.* (1994) Protein–protein interactions and 5′ splice site recognition in mammalian mRNA precursors. *Nature* **368**: 119–124.

Liu HX, Zhang M and Krainer AR (1998) Identification of functional exonic splicing enhancer motifs recognized by individual SR proteins. *Genes and Development* **12**: 1998–2012.

Mayeda A, Screaton GR, Chandler SD, Fu XD and Krainer AR (1999) Substrate specificities of SR proteins in constitutive splicing are determined by their RNA recognition motifs and composite pre-mRNA exonic elements. *Molecular and Cellular Biology* **19**: 1853–1863.

Ryner LC, Goodwin SF, Castrillon DH, *et al.* (1996) Control of male sexual behavior and sexual orientation in *Drosophila* by the *fruitless* gene. *Cell* **87**: 1079–1089.

Sanford JR and Bruzik JP (1999) Developmental regulation of SR protein phosphorylation and activity. *Genes and Development* **13**: 1513–1518.

Wu JY and Maniatis T (1993) Specific interactions between proteins implicated in splice site selection and regulated alternative splicing. *Cell* **75**: 1061–1070.

Zahler AM, Lane WS, Stolk JA and Roth MB (1992) SR proteins: a conserved family of pre-mRNA splicing factors. *Genes and Development* **6**: 837–847.

Zuo P and Maniatis T (1996) The splicing factor U2AF[35] mediates critical protein–protein interactions in constitutive and enhancer-dependent splicing. *Genes and Development* **10**: 1356–1368.

Further Reading

Blencowe BJ, Bowman JAL, McCracken S and Ronina E (1999) SR-related proteins and the processing of messenger RNA precursors. *Biochemistry and Cell Biology* **77**: 277–291.

Caceres JF, Screaton GR and Krainer AR (1998) A specific subset of SR proteins shuttles continuously between the nucleus and the cytoplasm. *Genes and Development* **12**: 55–66.

Graveley BR (2000) Sorting out the complexity of SR protein functions. *RNA* **6**: 1197–1211.

Kan JLC and Green MR (1999) Pre-mRNA splicing of IgM exons M1 and M2 is directed by a juxtaposed splicing enhancer and inhibitor. *Genes and Development* **13**: 462–471.

Li Y and Blencowe BJ (1999) Distinct factor requirements for exonic splicing enhancer function and binding of U2AF to the polypyrimidine tract. *Journal of Biological Chemistry* **274**: 35074–35079.

Prasad J, Colwill K, Pawson T and Manley JL (1999) The protein kinase Clk/Sty directly modulates SR protein activity: both hyper- and hypophosphorylation inhibit splicing. *Molecular and Cellular Biology* **19**: 6991–7000.

Reed R and Magni K (2001) A new view of mRNA export: separating the wheat from the chaff. *Nature Cell Biology* **3**: E201–E204.

Roscigno RF and Garcia-Blanco MA (1995) SR proteins escort the U4/U6•U5 tri-snRNP to the spliceosome. *RNA* **1**: 692–706.

Schaal TD and Maniatis T (1999) Multiple distinct splicing enhancers in the protein-coding sequences of a constitutively spliced pre-mRNA. *Molecular and Cellular Biology* **19**: 261–273.

Spector DL (1993) Macromolecular domains within the cell nucleus. *Annual Reviews in Cell Biology* **9**: 265–315.

Web Links

Splicing factor, arginine/serine-rich 1 (*SFRS1*); LocusID: 6426. Locus Link:
http://www.ncbi.nlm.nih.gov/LocusLink/LocRpt.cgi?l = 6426

Splicing factor, arginine/serine-rich 1 (*SFRS1*); MIM number: 600812. OMIM:
http://www.ncbi.nlm.nih.gov/htbin-post/Omim/dispmim?600812

Stationary Allele Frequency Distributions

Bruce Rannala, *University of Alberta, Edmonton, Canada*

Forces that determine the allele frequencies in natural populations include genetic drift, natural selection, migration and mutation. A balance of opposing forces can, in some cases, cause allele frequencies to approach a stationary distribution over time. The form of this distribution is not influenced by initial allele frequencies, but instead is determined by the relative magnitudes of different evolutionary forces.

Advanced article

Article contents

- Introduction
- Theoretical Background
- Stationary Allele Frequency Distributions
- Sampling Distributions of Alleles

Introduction

Frequencies of alleles at a genetic locus in a population are the outcome of a complex interplay among several forces including genetic drift, natural selection, gene flow and mutation. Forces such as selection, mutation and gene flow impose systematic pressures, causing allele frequencies to change in a particular direction. Genetic drift causes random (directionless) changes of allele frequency and contributes to the variation of allele frequencies observed within a single population over time, as well as the variation in allele frequencies among populations. (*See* Genetic Drift; Migration.)

Allele frequency change in finite populations is a stochastic process. Consequently, the allele frequency in a population at some future generation cannot be exactly predicted, even if current allele frequencies, and the magnitudes of evolutionary forces causing allele frequency changes, are known. At best, probabilities can be assigned to each possible future combination of allele frequencies on the basis of a stochastic model of allele frequency change. The probability distribution of future population allele frequencies, in general, depends on the number of generations of reproduction and the initial allele frequ-

encies, as well as the relative magnitudes of various evolutionary pressures influencing allele frequencies.

In special cases, when opposing forces of evolutionary change balance one another, a 'stationary' probability distribution of allele frequencies may exist. Over time, the allele frequencies in the population invariably approach this distribution, regardless of the initial frequencies of alleles in the population. Once this distribution is achieved, the population remains in this state so long as evolutionary pressures do not change. The stationary allele frequency distribution has long been of interest to population geneticists because it is informative about the long-term outcomes of evolutionary forces and is thought to be of relevance in interpreting gene frequencies in many (but not all) natural populations.

Theoretical Background

Fisher–Wright model

Mathematical models aimed at predicting population allele frequency changes over time were independently developed by Fisher (1930) and Wright (1931). A fundamental model, considered by both, was of a diploid population of N individuals ($2N$ alleles) with a random mating structure. The so-called Fisher–Wright model assumes that each allele produces a random number of offspring (with mean one) in each generation (subject to the constraint that population size remains constant). The generations are discrete and nonoverlapping. The process of genetic drift is intrinsic to the model. Other processes such as migration, mutation and natural selection have been incorporated. (*See* Fisher, Ronald Aylmer; Wright, Sewall.)

Many results obtained by studying a Fisher–Wright model also apply to organisms with more complex mating systems, including humans. Consider a single genetic locus with two alleles, A and a. If no mutations occur and all other factors, apart from genetic drift, can be neglected, the probability distribution of the number of copies of a at the next generation, given that there are x copies of a in the current generation, is a binomial (n, p) with parameters $p = x/2N$ and $n = 2N$. The probability distribution of allele frequencies at the next generation can be understood as the frequency with which populations with particular combinations of allele frequencies would be observed if one were to repeat a population mating experiment many times, starting each experiment with the same number, x, of copies of a.

Continuous approximation

The frequency of an allele in a diploid population of size N is an integer value that ranges from 0 to $2N$. To facilitate the study of allele frequency evolution over many generations under the Fisher–Wright model, Fisher (1930), Wright (1931) and Kimura (1955) all made use of an approximation that treats allele frequency as continuous (real valued) rather than discrete (integer valued). The relative frequency of a is $q = x/2N$, where x is the number of copies of a in the population. An increase, by one, of the number of copies of a in the population increases the relative frequency by $\delta q = 1/2N$. For large N (tending to infinity), δq becomes small (infinitesimal) and q takes on an approximately continuous range of values. (*See* Diffusion Theory.)

Probability density of allele frequency

Making use of a continuous approximation simplifies the study of the probability distribution of allele frequencies. If we define $f(q\,|\,t)$ to be the probability density function (pdf) of q after t generations of

reproduction, the probability that q is in the interval (a, b) is

$$\int_a^b f(q\,|\,t)\,\mathrm{d}q$$

Properties of the process of population allele frequency change over time, such as the expected (or average) population allele frequency at time t, the variance of allele frequencies across replicate populations, etc., can be studied by evaluating statistical moments of the pdf of population allele frequencies.

Stationary Allele Frequency Distributions

Stationary distributions

If a stationary allele frequency distribution exists, the pdf of the stationary allele frequency distribution can be obtained by taking the limit

$$\phi(q) = \lim_{t \to \infty} f(q\,|\,t)$$

Wright (1969) gives analytical expressions for the stationary distribution of allele frequencies under several models of mutation, migration and selection. Here we briefly outline results for the allele frequency distribution under mutation models with finite numbers of possible alleles, and models of subdivided populations with symmetrical migration patterns. The models presented here assume neutrality of alleles.

Two alleles: mutation or migration

Two-state mutation model

A simple model of mutation considered by Wright (1931) assumes that only two alleles are possible at a locus, a and A. The mutation process is reversible with a mutation rate υ from allele a to A and rate u from A to a. The stationary pdf of q is

$$\varphi(q) = \frac{\Gamma(4Nu + 4Nv)}{\Gamma(4Nu)\Gamma(4Nv)} q^{4Nv-1}(1-q)^{4Nu-1}$$

where $\Gamma(\cdots)$ denotes the gamma function. This is a beta distribution with parameters $\alpha = 4Nu$ and $\beta = 4Nv$ (see Johnson *et al.*, 1995). The expectation (mean) of the allele frequency at stationarity is $u/(u+v)$ and the variance (across populations that are at stationarity and experiencing identical evolutionary pressures) is $u\upsilon/[(u+v)^2(4N(u+v)+1)]$.

Wright island model

A simple model of migration, the so-called 'Wright island model', assumes that a fraction m of the alleles in a population are replaced by migrant alleles in each generation. The allele frequency among migrants is assumed to be constant, $\bar{q}$. The stationary pdf of allele frequency in the population is

$$\varphi(q) = \frac{\Gamma(4Nm)}{\Gamma(4Nm\bar{q})\Gamma[4Nm(1-\bar{q})]} q^{4Nm\bar{q}-1}(1-q)^{4Nm(1-\bar{q})-1}$$

This is a beta distribution with parameters

$$\alpha = 4Nm\bar{q} \quad \beta = 4Nm(1-\bar{q})$$

The mean allele frequency at stationarity is $\bar{q}$.

Multiple alleles: mutation or migration

Mutation model (k alleles)

A general formula is available for the stationary allele frequency distribution only under a (quite unrealistic) model of mutation in which all alleles mutate to allele j with the same rate v_j. In that case, the stationary joint pdf of allele frequencies is

$$\varphi(q_1,q_2,\ldots K,q_k) = \Gamma\left(4N\sum_{i=1}^{k} v_i \prod_{j=1}^{k} \frac{q_j^{4Nv_j-1}}{\Gamma(4Nv_j)}\right)$$

Wright island model (k alleles)

The Wright island model can be extended to k alleles (see Wright, 1969). If we define q_j to be the frequency of the jth allele on an island, and define $\bar{q}_j$ to be the frequency of the allele among migrants, the stationary joint pdf of allele frequencies is

$$\varphi(q_1,q_2,\ldots K,q_k) = \Gamma(4Nm) \prod_{j=1}^{k} \frac{q_j^{4Nm\bar{q}_j-1}}{\Gamma(4Nm\bar{q}_j)}$$

The above distributions are special cases of a general distribution known as the Dirichlet distribution (Kotz et al., 2000), which arises in many population genetic models.

Sampling Distributions of Alleles

The stationary distributions of allele frequencies outlined above apply to a population. Experimental studies of natural populations usually characterize the number of copies of each distinct allele observed in a sample of n diploid individuals ($2n$ copies in total). This is referred to as the sampling distribution of alleles and can be used to estimate parameters of population genetic models. For example, one can derive a maximum-likelihood estimator of $4Nm$ under the Wright island model (Rannala and Hartigan, 1996). Assuming that individuals are sampled at random (with respect to genotype), and that

genotypes are in Hardy–Weinberg equilibrium in a population, the probability distribution of allelic sample configurations (conditional on the allele frequencies) is either binomial (two alleles) or multinomial (more than two alleles)

$$\Pr(x_1,x_2,\ldots K,x_k \mid \mathbf{q}) = \binom{2n}{x_1,x_2,\ldots K,x_k} \prod_{j=1}^{k} q_j^{x_j}$$

where $\mathbf{q} = \{q_j\}$ is a vector of the population allele frequencies, q_j is the frequency of allele j, and x_j is the number of copies of allele j in the sample. If the population is at equilibrium, and the stationary allele frequency distribution is specified under a particular model, the sampling distribution is

$$\Pr(x_1,x_2,K,x_k) = \int \Pr(x_1,x_2,K,x_k \mid \mathbf{q})\varphi(\mathbf{q})\,d\mathbf{q}$$

where integration is over the multidimensional simplex of population allele frequencies. For most models of finite numbers of alleles that have been a subject of analysis, the sampling distribution is a multinomial Dirichlet distribution (Johnson et al., 1997). In the special case of two alleles, this simplifies to the beta binomial distribution.

Mutation model (k alleles)

For the k-allele symmetrical mutation model, the stationary sampling distribution of alleles is

$$\Pr(x_1,x_2,\ldots K,x_k) = \left[\Gamma(2n+1)\Gamma\left(4N\sum_{i=1}^{k}v_i\right)\right]\left[\Gamma\left(2n + 4N\sum_{i=1}^{k}v_i\right)\right]^{-1} \prod_{j=1}^{k} \frac{\Gamma(x_j+4Nv_j)}{\Gamma(x_j+1)\Gamma(4Nv_j)}$$

Wright island model (k alleles)

For the island model, the stationary sampling distribution of alleles is

$$\Pr(x_1,x_2,\ldots K,x_k) = \frac{\Gamma(2n+1)\Gamma(4Nm)}{\Gamma(2n+4Nm)} \prod_{j=1}^{k} \frac{\Gamma(x_j+4Nm\bar{q}_j)}{\Gamma(x_j+1)\Gamma(4Nm\bar{q}_j)}$$

See also
Evolution: Neutralist View
Genetic Drift
Migration
Mutational Change in Evolution
Population Genetics

References

Fisher RA (1930) *The Genetical Theory of Natural Selection*. Oxford, UK: Oxford University Press.
Johnson NL, Kotz S and Balakrishnan N (1995) *Continuous Univariate Distributions*, vol. 2. New York, NY: John Wiley.
Johnson NL, Kotz S and Balakrishnan N (1997) *Discrete Multivariate Distributions*. New York, NY: John Wiley.

Kimura M (1955) Solution of a process of random genetic drift with a continuous model. *Proceedings of the National Academy of Sciences of the United States of America* **41**: 144–150.

Kotz S, Balakrishnan N and Johnson NL (2000) *Continuous Multivariate Distributions*, vol. 1. New York, NY: John Wiley.

Rannala B and Hartigan JA (1996) Estimating gene flow in island populations. *Genetical Research* **67**: 147–158.

Wright S (1931) Evolution in Mendelian populations. *Genetics* **16**: 97–159.

Wright S (1969) Evolution and the genetics of populations. Volume 2. *The Theory of Gene Frequencies*. Chicago, IL: University of Chicago Press.

Further Reading

Ewens WJ (1979) *Mathematical Population Genetics*. New York, NY: Springer.

Gale JS (1990) *Theoretical Population Genetics*. London, UK: Unwin Hyman.

Medhi J (1994) *Stochastic Processes*, 2nd edn. New Delhi, India: Wiley Eastern.

Rice JA (1995) *Mathematical Statistics and Data Analysis*, 2nd edn. Belmont, CA: Duxbury Press.

Statistical Methods in Physical Mapping

Sophie Schbath, *Institut National de la Recherche Agronomique, Jouy-en-Josas, France*

Statistical methods are used to predict the progress of a physical mapping project in terms of the number and size of the contigs and the proportion of genome covered by contigs.

Advanced article

Article contents

- Introduction
- Probabilistic Model and Properties
- Preliminary Result
- Proportion of Genome Covered by Anchored Islands

Introduction

A physical mapping project usually aims to produce a physical map of each chromosome of an organism. Such a physical map is a set of ordered and overlapping genomic fragments (clones) spanning the entire chromosome. A genomic library of clones is first constructed, generally by partial restriction digest, and is supposed to represent the chromosome. Clones are then chosen at random and overlaps are inferred, these overlapping clones being linked into islands or contigs. Several approaches can be used to infer clone overlaps. For example, clones and clone ends are characterized by a fingerprint (sequence or restriction fragment information), and fingerprints are compared. Two clones are said to overlap if their fingerprints are sufficiently similar. For long clones, genomic mapping by anchoring random clones is more adapted. This requires a genomic library of anchors, very small genomic sequences assumed to uniquely identify a chromosome location. Thus, clones containing a common anchor overlap. (*See* Contig assembly; Genome Map; Genome Mapping; Yeast Artifical Chromosome (YAC) Clones.)

Statistical questions addressed in a physical mapping project concern the prediction of its progress. How many islands are obtained, what is their mean length, how many clones are the islands composed of, what is the proportion of chromosome covered by the islands, and so on? One is interested here in studying the evolution of the physical map with respect to the number of clones studied, their length and some parameters related to the mapping strategy. Lander and Waterman (1988) and Port *et al.* (1995) have presented such a mathematical analysis for physical mapping by fingerprinting or by end-characterized random clones. In this paper, we restrict ourself to the random anchoring strategy. The main reference considered is Arratia *et al.* (1991).

Solving these questions requires the modeling of clone locations along the genome, the problem falling under the general heading of coverage processes (Hall, 1988). Clone locations are usually modeled by a homogeneous Poisson process along the genome with rate $c = LN/G$, where G is the genome length in base pairs, L is the clone length in base pairs and N is the (mean) number of clones. This means that clones are uniformly distributed along the chromosome, and c represents the expected number of clones covering a random point (also called the redundancy of coverage). It is possible to relax the fixed length of the clones by considering random lengths as independent and identically distributed with a given probability density function. Homogeneity assumptions can be also relaxed by considering a clone length distribution and/or a clone rate c that depends on the location of the clones (Schbath, 1997; Schbath *et al.*, 2000). The aim here is to give an idea of the statistical methods used to determine the progress of a sequence tagged site (STS) mapping project. For convenience, we will use the homogeneous framework. (*See* Sequence-tagged Site (STS).)

In the next section, we outline the probabilistic model for the clones and anchors, before presenting how to calculate the mean proportion of genome covered by anchored islands. This choice is made for two reasons: biologically, the proportion of genome covered by islands is probably the most important area of interest; and mathematically, it gives a good insight into the techniques used.

Probabilistic Model and Properties

It is assumed that clone lengths L are independent and identically distributed random variables with a mean length $\mathbb{E}(L)$. We rescale the genome by $\mathbb{E}(L)$ so that the normalized genome corresponds to the interval $[0, g]$ on the real line, with $g = G/\mathbb{E}(L)$. Let f be the probability density function of the normalized clone lengths denoted by $Q = L/\mathbb{E}(L)$. We define the function $\mathcal{F}(u) = \mathbb{P}(Q > u) = \int_u^\infty f(q)\,\mathrm{d}q$. We assume that the right-hand ends of clones are distributed independently across the real line $(-\infty, +\infty)$ according to a homogeneous Poisson process with rate $c = N/g$. Recall that N is the mean number of clones used in the mapping project. The anchors are also randomly drawn along the real line according to a homogeneous Poisson process, independent of the clone process, and with rate $a = M/g$, M being the mean number of anchors used.

A general Poisson process is defined so that the number of points falling into two disjoint sets are independent and, moreover, the number of points falling in a given set is Poisson-distributed. More precisely, for a Poisson process on the real line with rate $\lambda(t)$, the number of points falling into a set A is distributed according to the Poisson distribution with mean $\int_A \lambda(t)\,\mathrm{d}t$.

The main property of a homogeneous Poisson process is that the distances between consecutive points are independent and identically distributed according to an exponential distribution. Let us prove the following related property that will be used later.

Property 1: *Let* V_t *be the distance forward from a given position* t *to the next anchor, and let* W_t *be the distance back from* t *to the previous anchor.*

1. V_t *is exponentially distributed with mean* a^{-1}*; its probability density function is defined by*

$$g(v) = \begin{cases} a\exp(-av), & v \geq 0 \\ 0, & v < 0 \end{cases}$$

2. W_t *is exponentially distributed with mean* a^{-1}*; its probability density function is* g.
3. V_t *and* W_t *are independent random variables.*

Proof:
1. Set $v \geq 0$; V_t is strictly greater than v if and only if no anchors fall into the interval $(t, t+v)$. It occurs with a probability equal to $\exp(-av)$ because the number of anchors falling into $(t, t+v)$ is a Poisson variable with mean av. Therefore, we have

$$\mathbb{P}(V_t \leq v) = 1 - \exp(-av)$$

By differentiating this equality, it gives $g(v) = a\exp(-av)$.
2. The proof is similar to that above.
3. Set $v \geq 0$ and $w \geq 0$; simultaneously, V_t is strictly greater than v and W_t is strictly greater than w, if and only if no anchors fall into $(t, t+v)$ and no anchors fall into $[t-w, t)$. As these two intervals are disjoint, the anchors falling into $(t, t+v)$ are independent of the ones falling into $[t-w, t)$. Therefore, $\mathbb{P}(V_t > v, W_t > w) = \mathbb{P}(V_t > v)\mathbb{P}(W_t > w)$ and independence is established.

Preliminary Result

Proposition 1: *The probability* J(x) *that two points separated by distance x are not covered by a common clone is*

$$J(x) = \exp\left(-c \int_x^\infty \mathcal{F}(u)\,\mathrm{d}u\right)$$

Proof: Set $t \in \mathbb{R}$. A clone of length Q ending at position y covers $[t, t+x]$ if and only if $t+x \leq y$ and $y - Q \leq t$. Averaging over the distribution of Q gives that the probability that a clone ending at y covers $[t, t+x]$ is $\mathcal{F}(y-t)\mathbb{I}\{y \geq t+x\}$. As the right-hand ends of clones are a Poisson process with rate c, the process of right-hand ends of clones covering $[t, t+x]$ is a Poisson process with inhomogeneous rate $c\mathcal{F}(y-t)\,\mathbb{I}\{y \geq t+x\}$ (marking theorem; Kingman, 1993). The total number of clones covering $[t, t+x]$ is then a Poisson variable with mean $\int_\mathbb{R} c\mathcal{F}(y-t)\,\mathbb{I}\{y \geq t+x\}\,\mathrm{d}y$ and $J(x)$ is the probability that this Poisson variable is equal to zero:

$$J(x) = \exp\left(-c \int_{t+x}^{+\infty} \mathcal{F}(y-t)\,\mathrm{d}y\right)$$

The change of variable $u = y-t$ leads to the result.

Remark 1: *The probability that a given location is covered by a clone is given by* $1 - J(0) = 1 - \exp(-c)$.

Proportion of Genome Covered by Anchored Islands

Because of the homogeneity assumption, the mean proportion of chromosome not covered by anchored islands is also the probability that a given location t on the chromosome is not covered by an anchored island. To calculate this probability, denoted by $r_0(t) = r_0$, we start working conditionally with V_t the distance forward from t to the next anchor, and W_t the distance back from t to the previous anchor. Given V_t and W_t, the location t is not covered by an anchored island if and only if no clones simultaneously cover the locations $t - W_t$ and t, and no clones simultaneously cover the locations t and $t + V_t$. Because clones covering $[t - W_t, t]$ are not independent of the ones covering $[t, t + V_t]$, the previous conditions are rewritten as follows. Location t is not covered by an anchored island if and only if no clones simultaneously cover t and $t + V_t$ (event E_1) and no clones ending in $[t, t + V_t)$ cover $t - W_t$ (event E_2). Events E_1 and E_2 are independent because E_1 is related to clones ending in $[t + V_t, +\infty)$, E_2 is related to clones ending in $[t, t + V_t)$ and the right-hand end of clones is a Poisson process. We then have

$$\mathbb{P}(t \text{ is not covered by an anchored island}$$
$$|V_t, W_t) = \mathbb{P}(E_1|V_t, W_t)\mathbb{P}(E_2|V_t, W_t)$$

Proposition 1 gives $\mathbb{P}(E_1|V_t, W_t) = J(V_t)$. To calculate the conditional probability of E_2, we consider a clone ending at position y. Given its length Q, this clone ends in $[t, t + V_t)$ and covers $t - W_t$ if and only if $t \le y < t + V_t$ and $y - Q < t - W_t$. Therefore, a clone ending at y ends in $[t, t + V_t)$ and covers $t - W_t$ with probability

$$p(y) = \mathcal{F}(y - t + W_t)\mathbb{I}\{t \le y < t + V_t\}$$

Because the right-hand end of the clones is a Poisson process with rate c, marking theorem ensures that the right-hand end of clones ending in $[t, t+V_t]$ and covering $t-W_t$ is an inhomogeneous Poisson process with rate $cp(\cdot)$. The total number of such clones is then a Poisson variable with mean $\int_{\mathbb{R}} cp(y)\,dy$. It

follows that

$$\mathbb{P}(E_2|V_t, W_t) = \exp\left(-\int_t^{t+V_t} c\mathcal{F}(y - t + W_t)\,dy\right)$$
$$= \exp\left(-c\int_{W_t}^{V_t+W_t} \mathcal{F}(u)\,du\right)$$

Using Proposition 1, we can write $\mathbb{P}(E_2|V_t, W_t) = J(W_t)/J(V_t + W_t)$ and

$$\mathbb{P}(t \text{ is not covered by an anchored island}|V_t, W_t)$$
$$= \frac{J(W_t)J(V_t)}{J(V_t + W_t)}$$

We then have to average the previous conditional probability according to the distribution of the couple (V_t, W_t). Using Property 1, this leads to

$$r_0 = r_0(t) = \int_{v=0}^{\infty}\int_{w=0}^{\infty} \frac{J(w)J(v)}{J(v+w)}a^2 \exp[-a(v+w)]\,dv\,dw$$

See also

Cytogenetic and Physical Chromosomal Maps: Integration
Genetic and Physical Map Correlation
Genome Mapping

References

Arratia R, Lander ES, Tavaré S and Waterman MS (1991) Genomic mapping by anchoring random clones: a mathematical analysis. *Genomics* **11**: 806–827.

Hall P (1988) *Introduction to the Theory of Coverage Processes*. New York, NY: Wiley.

Kingman JFC (1993) *Poisson Processes*. Oxford, UK: Oxford University Press.

Lander ES and Waterman MS (1988) Genomic mapping by fingerprinting random clones: a mathematical analysis. *Genomics* **2**: 231–239.

Port E, Sun F, Martin D and Waterman MS (1995) Genomic mapping by end-characterized random clones: a mathematical analysis. *Genomics* **26**: 84–100.

Schbath S (1997) Coverage processes in physical mapping by anchoring random clones. *Journal of Computational Biology* **4**: 61–82.

Schbath S, Bossard N and Tavaré S (2000) The effect of nonhomogeneous clone length distribution on the progress of an STS mapping project. *Journal of Computational Biology* **7**: 47–57.

Stem Cells

See Hematopoietic Stem Cells
Neural Stem Cells

Stored Genetic Material: Use in Research

Loane Skene, *University of Melbourne, Parkville, Australia*

The use of stored genetic material depends on how it was obtained and the consent that was given by the original person. If that person was fully informed and consented to the material being used in research, it is lawful to use it in research following guidelines for ethical conduct. If consent was not or cannot be obtained, the legal authority to use the material is less clear and research should be done only on anonymized and unlinked samples.

Intermediate article

Article contents

- Legality of Using Genetic Material in Research Depends on its Source
- Genetic Material Obtained with Donor's Prior Authority or by Direct Approach to Donor
- Stored Genetic Material for Which the 'Donor' is not Available

---

Legality of Using Genetic Material in Research Depends on its Source

Genetic material includes human bodies, body parts (limbs and organs) and tissue (such as blood and cells). If the genetic material was obtained with consent from the person from whom it was derived and that person was fully informed and consented to it being used in research, it is lawful to use the material in research following guidelines for ethical conduct of research. The consent may be given before the material is taken or after it has been stored.

If consent cannot be obtained, however, the legal authority to use the material is less clear and research should generally be done only on anonymized and unlinked samples, following approval by a research ethics committee. Generally, research should not be done at all on samples taken without a person's consent for forensic purposes or pursuant to a statutory requirement. These principles are explained more fully below.

Genetic Material Obtained with Donor's Prior Authority or by Direct Approach to Donor

Genetic material may be obtained after a person, or his or her representative, has given authority for it to be used in research; alternatively, the consent may be sought later, when the material has been removed already and is stored. Material obtained in this way includes human bodies, body parts and tissue either stored in hospital laboratories after therapeutic tests, or held for retesting, coronial investigation, medical audit or use in teaching and research. People may authorize the use of their bodies after death, or their body parts or tissue in research. Tissue may be removed specifically for research. The researcher, or someone else such as a doctor, may approach the

relevant person for consent to use a body, or surgically removed body parts or tissue, in research. Consent may relate to a specific project, or to more general research.

Legal authority to remove and retain genetic material

The legal authority for the initial removal of genetic material from a person and its subsequent retention and use may come from one of three sources:

1. *Consent*: the consent of the person concerned (express or implied), the consent of a parent if the person is a minor, or the consent of the next of kin or personal representative if the person has died.
2. *Statute*: a statute requiring the removal and retention of the material, such as coronial legislation requiring that postmortem examination be conducted, or road safety legislation requiring blood alcohol testing after a motor accident.
3. *Court order*: a court order, such as an order authorizing tests for paternity, identification of criminal suspects or evidence in a medical negligence case.

If the authority is a statute or a court order, material may be removed without the person's consent, but, in law, the material should be used only for the purpose for which it was acquired, in other words, postmortem investigation, blood alcohol testing, paternity testing and so on. It cannot be used for research without consent from the donor unless there is statutory authority allowing that to occur. Such authority exists under Australian human tissue legislation where a coroner performs a postmortem. For example, the Human Tissue Act 1982 (Vic) s 30(3) states that 'An order by a coroner under the Coroners Act 1985 (Vic) directing a postmortem examination is ... authority for the use, *for therapeutic, medical or scientific purposes*, of tissue removed from the body of the deceased person for the purpose of the postmortem examination'. There is no equivalent provision in the United Kingdom, and a coroner has no authority to retain tissue when an investigation is completed (Bristol Report, 2000, Annex B, paras 38, 55–58, 141, 145).

Consent forms and information

When a person (or a parent or personal representative) consents to the taking and use of genetic material in research, the person will usually sign a formal consent form (e.g. see the model consent form for research involving new samples of human biological material; MRC, 2001, pp. 34–36). But this is not legally necessary. The person should be fully informed of what is proposed and, if tissue is removed specifically for a clinical trial, will usually be given a printed description of the project. But, again, printed information is not legally necessary. Verbal information and verbal consent are equally valid. Provided that the person has consented to his or her genetic material being used in a research project, that consent is effective even if the person has not been fully informed about the nature of the proposed research. 'Blanket consent' – consent to the use of genetic material 'in research' without more detail – is less desirable than properly informed consent, however, because it undermines the person's autonomy and right to self-determination.

Ethical approval

When research projects are undertaken in a research institution or are publicly funded, the institution or funding body generally requires that the project must be approved in advance by a research ethics committee (e.g. see MRC, 2001). Again, ethical approval is not always legally required, although it may be made a legal requirement by a contract of employment or a funding agreement, or even be regarded as part of a health professional's common law duty of care for a patient. The matters to be considered by research ethics committees are suggested in professional guidelines (e.g. see MRC, 2001). Such guidelines state that the potential benefits of research should outweigh the potential risks to donors of samples (including the risk of discovery or wrongful use of genetic information). The body and its parts should be treated with respect, taking into account religious and cultural factors. Confidentiality must be preserved. And people are entitled to know the results of tests on their genetic material.

If ethical approval is not obtained or the research is not undertaken in accordance with that approval, research funding for the project can be withdrawn. The

researchers may also face disciplinary action. The law is not completely clear concerning the rights, if any, of the person from whom the genetic material was obtained in such circumstances. There may be a right to be compensated for genetic material being removed without the person's informed consent (see the United States court case, *Moore versus Regents of the University of California*, 1990); however, the person has no property right in the removed material and is not entitled to prevent its use or to share in the proceeds of commercialization. Many academic writers have argued that people should have a property interest in excised material removed from their bodies (Harris, 1996; Mason and Laurie, 2001; contra, Skene, 2002a,b), but that is not the law. People do not 'own' their bodies or anything removed from their bodies (*Doodeward versus Spence*, 1908).

Authority to use genetic material

The direct consent method is the least problematic way to obtain genetic material for research. Provided that people are properly informed about the use that will be made of their material and they then consent to its use for a specific project, or for research in general, there is sufficient authority for the researcher to take and/or use the tissue in the research. If the researcher later develops a commercial product from the material, he or she is legally entitled to patent and exploit it. The person from whom the material came has no legal right to prevent the commercialization or to share in the profits (see 'Ethical approval').

Change in project

If a person has consented to his or her genetic material being used in a particular project and the researcher later wants to use it for another project of a different kind, ethical guidelines recommend that the person should be recontacted and asked for consent to the new use (MRC, 2001, para 10.3). Note that this may not be legally necessary – the guidelines are perhaps unduly cautious in making this recommendation. Nevertheless, it protects the donor's right to autonomy. Consent to the new use cannot be presumed. People may be prepared to consent to their tissue being used in diabetes research for example, but not in research that involves contraception or cloning. The guidelines make an exception where later research is on samples anonymized and unlinked before use and it is not possible or practicable to contact donors (MRC, 2001, para 10.3).

Note that it is legally possible for people to place restrictions on the use of their tissue and, although this may be inconvenient for researchers, these conditions should be observed. In law, the person might be entitled to sue for compensation if the conditions were not observed and substantial damages might be awarded. However, the person would have no property rights and could not prevent the use or commercialization of the genetic material (*Moore versus Regents of the University of California*, 1990).

Stored Genetic Material for Which the 'Donor' is not Available

Examples of genetic material stored in hospitals and laboratories for which the 'donor' is not available

Hospitals and laboratories have large stores of organs and tissue for which it may not be possible to contact the people from whom the material was derived to seek their consent to use it in research. For example, there are bodies, organs and tissue retained after death for postmortem examination; there are organs, blood, serum, skin and other living body parts for use in medical care, and living cells that can be reproduced indefinitely for research, such as cell lines and pluripotent stem cells. Tissue is also preserved on slides for therapeutic pathology tests, including genetic tests with implications for relatives. Other tissue is stored for forensic tests, including testing for alcohol or drugs, identification of criminal suspects, paternity or evidence in medical negligence cases.

When this stored material can be used lawfully in research

If the material was initially acquired pursuant to a statutory authority or a court order, it should not be used in research without consent from the person from whom the tissue was derived (see 'Legal authority to remove and retain genetic material'). If it was acquired with the person's consent for a therapeutic purpose but the possibility of its use in research was not discussed, then it can probably be used lawfully in research, although there is little legal authority on this point.

Ethical guidelines, which are not law, support the use of stored tissue in such a case without consent when it is not possible to obtain consent and ethical approval is obtained from a research ethics committee. The Medical Research Council guidelines state, for example, that 'It is acceptable to use human material surplus to clinical requirements for research without consent if it is anonymous and unlinked' (MRC, 2001, para. 3.2), and that it is also acceptable to do research on anonymous, linked samples if the predictive value

of the genetic test is not known and 'there is strong scientific justification for not irreversibly anonymizing the samples' (MRC, 2001, para 10.2). The guidelines recommend, however, that even if consent is not needed for anonymized research, individuals should be informed of the possibility that their samples may be used in research and be given an opportunity to refuse (MRC, 2001, para. 3.1). Ethics committees should ensure that the tissue was not collected unethically or improperly and that there was valid consent to the taking (MRC, 2001, para 10.2).

Despite the Medical Research Council's guidelines permitting some use of stored genetic material without consent, researchers should bear in mind that there have been several inquiries into the use of stored human tissue in research in the United Kingdom. All have emphasized the need to obtain consent from donors or next of kin, even when that is not a legal requirement. They have said that it makes no difference if the tissue is used in an 'anonymized' way, even if identifiers are removed completely, rather than coded. The emphasis on the need to inform and to involve people in decision-making, especially in the wake of the Bristol (Bristol Report, 2000) and Alder Hey (Alder Hey Report, 2001) Inquiries, indicates a new direction in community attitudes. In the past, researchers have used stored body parts or tissue commonly without donors' consent, especially in an anonymous form that is monitored by ethics committee, but this practice is now fraught with difficulties, although many people still support it for promoting community interests.

Note also that where stored genetic material has come from people who have died, the executor or personal representative of a deceased person has a legal right to obtain possession of the body and body parts for burial or cremation. There is an argument that this right extends only to the body itself and to body parts needed for burial, and not to tissue held on slides or in other preserved form (see Alder Hey Report, 2001, chap. 11, paras 3.3, 3.4). This means that there is a gap in the law. The coroner does not have lawful authority to hold the tissue after the coronial investigation is completed (see 'Legal authority to remove and retain genetic material'), but the personal representative is not entitled to possession of it for disposal. Nevertheless, in view of the sensitivity concerning these issues, especially in the United Kingdom, researchers would be well advised not to undertake research on this tissue without consent from the person before death, or the executor or personal representative after death.

See also
Informed Consent in Human Genetic Research
Ownership of Genetic Material and Information

References

Alder Hey Report (2001) The Royal Liverpool Children's Inquiry Report, 30 January 2001. http://www.rlcinquiry.org.uk/

Australian Law Reform Commission, Discussion paper 66, Protection of Genetic Information, August 2002.

Bristol Report (2000) The Inquiry into the Management of Care of Children Receiving Complex Heart Surgery at The Bristol Royal Infirmary, Chair, Professor Ian Kennedy, Interim Report, May 2000; Final Report, July 2001. http://www.bristol-inquiry.org.uk/index.htm

Doodeward versus Spence (1908) 6 Commonwealth Law Reports 406, High Court of Australia.

Graeme Laurie (2002) Genetic privacy. A challenge to medico-legal norms, Cambridge University Press.

Harris JW (1996) Property and Justice, pp. 351–359. Oxford, UK: Clarendon Press.

Mason JK and Laurie GT (2001) Consent or property? Dealing with the body and its parts in the shadow of Bristol and Alder Hey. Med Law Review, pp. 710–729.

MRC (2001) Human Tissue and Biological Research Samples for Use in Research. London, UK: Medical Research Council.

Moore versus Regents of the University of California (1990) 793 P 2d 479 California Supreme Court.

Skene L (2002a) Proprietary rights in human bodies, body parts and tissue: regulatory contexts and proposals for new laws. Legal Studies 22(1): 102–127.

Skene L (2002b) Arguments against people 'owning' their own bodies, body parts and tissue. Macquaire Law Journal 2: 165–176.

Further Reading

Chief Medical Officer, UK (2001) Organ Retention: Interim Guidance on Post-Mortem Examination. London, UK.

Council of Europe (1997) Convention on Human Rights and Biomedicine.

Grubb A (1998) I, me, mine: bodies parts and property. Medical Law International 3: 299.

Human Genetics Commission, UK (October–December 2000) Public Attitudes to Human Genetic Information.

Human Genome Organisation Ethics Committee (1998) Statement on DNA Sampling: Control and Access. London, UK.

Kennedy I and Grubb A (2000) Medical Law, 3rd edn. London, UK: Butterworths.

Magnusson RS (1998) Proprietary rights in human tissue. In: Palmer N and McKendrick E (eds.) Interests in Goods, 2nd edn. London, UK: Lloyds of London Press.

MRC (2000) Public Perceptions of the Collection of Human Biological Samples. London, UK: Medical Research Council.

Morgan D (2001) Issues in Medical Law and Ethics, pp. 83–104. London, UK: Cavendish.

National Health and Medical Research Council (1998) National Statement on Ethical Conduct in Research involving Humans. Canberra, Australia.

Nuffield Council on Bioethics, UK (1995) Human Tissue, Ethical and Legal Issues. London, UK.

Skene L (1998) Patients' rights or family responsibilities? Two approaches to genetic testing. Medical Law Review 6: 1–41.

Structural Databases of Biological Macromolecules

Helen M Berman, Rutgers, The State University of New Jersey, New Brunswick, New Jersey, USA

The Protein Data Bank began as an archive of the structural data available about known biological macromolecules. The advances made in all technologies have been mirrored in further development of the Protein Data Bank and in the structural, speciality and structural characteristic databases that have also evolved.

Intermediate article

Article contents

- Historical Background
- The Protein Data Bank
- Structural Databases
- Speciality Databases
- Databases of Structural Characteristics
- Challenges

Historical Background

In 1957, the first structure of a biological macromolecule (myoglobin) was determined (Kendrew et al., 1958). This was followed by the determinations of several more key molecules, including hemoglobin (Perutz et al., 1960), lysozyme (Blake et al., 1965) and ribonuclease (Kartha et al., 1967; Wyckoff et al., 1967). In 1971, small-molecule and protein crystallographers from both sides of the Atlantic agreed to establish a data bank of the protein structures being determined. Its mission would be to collect, archive and disseminate data on the three-dimensional structures of biological macromolecules. Walter Hamilton of the Brookhaven National Laboratory and Olga Kennard of the Cambridge Structural Database (CSD) collaborated to manage the Protein Data Bank (PDB) resource (1971). Hamilton's interest was borne from his work on the high-resolution determination of amino acid crystal structures and from his visionary idea of setting up distributed computing resources whereby every crystallographer would have a graphics workstation on his/her desk with full network access to powerful high-speed computers. Kennard had founded the CSD in 1965 to create a database of organic and metal-organic compounds studied by X-ray and neutron diffraction, and was well experienced in managing structural data. (*See* Crystallization of Nucleic Acids; Protein Structure.)

The PDB contained less than a dozen structures at its inception, with a few more structures added each year. The structures themselves were relatively small. The PDB file format was simple, and it was relatively easy to extract the structures from magnetic tape to find out what you wanted to know about any particular molecule.

In the 1980s, the improvements in the technology required to do crystal structures began to evolve rapidly. Now, two decades later, modern molecular biology techniques have made it much more straightforward to obtain large quantities of proteins. Crystallization methods have emerged that allow investigators to screen many different conditions using exceedingly small amounts of material. Data collection methods have improved at all levels. The lifetimes of crystals are routinely extended by flash freezing. The radiation sources are much more intense, especially with the emergence of powerful synchrotron beam lines. Detectors are much more sensitive and allow the very rapid collection of arrays of reflections. Methods for phase determination and refinement have improved. Indeed, crystallography is now part of the armament of techniques that is readily accessible to biologists.

As crystallographic methods continue to improve, another method of structure determination has come of age: nuclear magnetic resonance (NMR). This method, which allows the determination of structures in solution, is currently responsible for approximately 15% of the structures released in the PDB.

The improvements in technology have also made it possible to determine the structures of very complex molecules. Several structures of ribosomal subunits (Moore, 2001), as well as the entire 70S ribosome structure (Yusupov *et al.*, 2001), have been deposited in the PDB. During this same period, the structural genomics initiative (2000) has begun with the goal of determining thousands of structures in a high-throughput mode. Thus, the PDB holdings will continue to grow (**Figure 1**).

The level of activity in structural biology has made it essential that the PDB use the most modern technologies to collect, archive and disseminate data. The PDB is an *archival database*, which contains coordinates of biological macromolecules determined using public funds as well as many from the private sector. It also contains information about the methods and materials used to determine those structures.

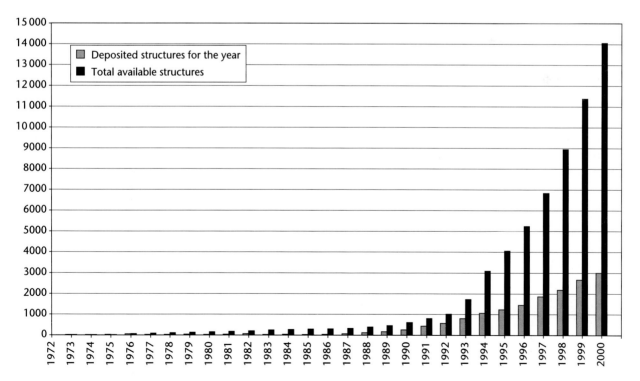

Figure 1 Growth of the contents of the Protein Data Bank. The number of structures deposited each year is shown in gray, the total number of structures available in black. This chart is regularly updated at http://www.rcsb.org/pdb/holdings.html.

Other databases have emerged (**Table 1**) that extract some of the information contained in the PDB and organize that information in different ways so as to enable different types of query. These are *value-added databases*, which serve the needs of particular users. In this article we describe the PDB and some of these other structural databases.

The Protein Data Bank

After 27 years at Brookhaven National Laboratory, the PDB is now managed by the Research Collaboratory of Structural Bioinformatics (RCSB) (Berman *et al.*, 2000). The RCSB is a consortium consisting of three member groups: Rutgers, the State University of New Jersey; the San Diego Supercomputer Center of the University of California, San Diego; and the National Institute of Standards and Technology. The PDB collects information about biological macromolecular structures and the methods used to determine those structures. Coordinates, primary experimental data, statistics about the structure determination and refinement, information about the source, sequence and chemistry of the molecule and the solution and/or crystallization conditions are collected and assembled using a software tool called the AutoDep Input Tool (ADIT) (see Web Links). Annotation, checking and validation of the data are carried out with a variety of programs whose output is reviewed by skilled

Table 1 Selected database resources for macromolecular structures

---

Archival database of biological macromolecules

Protein Data Bank (Berman *et al.*, 2000; Bernstein *et al.*, 1977)	http://www.pdb.org/

Structural databases

3D ALI (a database of aligned protein structures and related sequences) (Pascarella and Argos, 1992)	http://www-embl-heidelberg.de/argos/ali/ali_info.html
CAMPASS (Sowdhamini *et al.*, 1998)	http://www-cryst.bioc.cam.ac.uk/~campass/
CATH (Orengo *et al.*, 1997)	http://www.biochem.ucl.ac.uk/bsm/cath/
CSD (Allen *et al.*, 1979)	http://www.ccdc.cam.ac.uk/
FSSP (Holm and Sander, 1998)	http://www2.ebi.ac.uk/dali/fssp/
HSSP (Dodge *et al.*, 1998)	http://www.sander.embl-heidelberg.de/hssp/
ISSD (Adzhubei *et al.*, 1998)	http://www.protein.bio.msu.su/issd/
Library of Protein Family Cores (LPFC) (Schmidt *et al.*, 1997)	http://WWW-SMI.Stanford.EDU/projects/helix/LPFC/
Molecular Modeling Database (Holm and Sander, 1994)	http://www.ncbi.nlm.nih.gov/Structure/
SCOP (Murzin *et al.*, 1995)	http://scop.mrc-lmb.cam.ac.uk/scop/

Speciality databases

ENZYME database (Bairoch, 2000)	http://www.expasy.ch/enzyme/
Enzyme Structures Database	http://www.biochem.ucl.ac.uk/bsm/enzymes/
HIV Protease Database (Vondrasek *et al.*, 1997)	http://srdata.nist.gov/hivdb/
International Immunogenetics Database (IMGT) (Lefranc *et al.*, 1998)	http://imgt.cines.fr:8104/
Nucleic Acid Database (Berman *et al.*, 1992)	http://ndbserver.rutgers.edu/
Prolysis (protease and protease inhibitor web server)	http://delphi.phys.univ-tours.fr/Prolysis/
Protein Kinase Resource (Smith *et al.*, 1997)	http://pkr.sdsc.edu/html/index.shtml

Structural characteristic databases

Biological Macromolecule Crystallization Database (BMCD) (Gilliland, 1997)	http://wwwbmcd.nist.gov:8080/bmcd/bmcd.html
Dictionary of Interfaces in Proteins (DIP)	http://www.drug-redesign.de/
ISOSTAR (Bruno *et al.*, 1997)	http://www.ccdc.cam.ac.uk/prods/isostar/
Molecular Movements Database (Gerstein and Krebs, 1998)	http://bioinfo.mbb.yale.edu/MolMovDB/
OLDERADO (Kelley and Sutcliffe, 1997)	http://neon.chem.le.ac.uk/olderado/
PDBSum (Laskowski *et al.*, 1997)	http://www.biochem.ucl.ac.uk/bsm/pdbsum/
PROCAT (Wallace *et al.*, 1996)	http://www.biochem.ucl.ac.uk/bsm/PROCAT/PROCAT.html
PROMISE (Degtyarenko *et al.*, 1998)	http://metallo.scripps.edu/PROMISE/
Protein Quaternary Structures (PQS)	http://pqs.ebi.ac.uk/
ReLiBase (Receptor/ligand complexes database) (Hendlich *et al.*, 2003)	http://relibase.ccdc.cam.ac.uk/
TESS	

Table 2 Protein Data Bank holdings (as of 14 August 2001)

	Proteins, peptides and viruses	Protein–nucleic acid complexes	Nucleic acids	Carbohydrates	Total
X-ray diffraction and other	11 893	569	579	14	13 055
NMR	1964	73	390	4	2431
Theoretical modeling	293	20	23	0	336
Total	14 150	662	992	18	15 822

annotation staff working in close collaboration with the depositors of the data.

Once the data are fully checked and approved for release, they are loaded into a series of databases. Two different search engines can query the databases: *SearchLite* and *SearchFields*. A rich set of reporting options make it possible to access information about a single molecule, compare it with other molecules, and access other databases that contain information about that molecule. Particular groups of macromolecules can be selected according to their features so that a variety of reports can be created. The PDB maintains mirrors around the world, which provide the same capabilities as the main RCSB site.

As of this writing there are more than 15 800 molecules in the PDB. The distribution of these data is shown in **Table 2**.

Structural Databases

While the PDB focuses on individual structures, some databases organize their data according to tertiary structural characteristics. SCOP (a Structural Classification of Proteins) classifies each structure in the PDB according to *family, superfamily, common fold* and *class*. *Families* are classified according to their sequence similarities. Families with similar structure and function belong to the same *superfamily*. Families and superfamilies with the same arrangement of secondary structures, which are connected with one another in the same way, have the same *common fold*. *Class* refers to the types of secondary structures (all alpha, all beta, alpha–beta, etc.). SCOP was one of the earliest databases that attempted to integrate sequence, structure and function information; it continues to be a major resource in structural biology.

CATH provides another classification scheme based on class (C), architecture (A), topology (T) and homologous superfamilies (H). *Class* defines the secondary structure content as in SCOP. *Architecture* defines the description of the arrangement of these secondary structures without consideration of the connectivities. *Topology* is equivalent to fold in SCOP. Finally, *homologous superfamilies* contain all

folds with a similar function. CATH has a systematic classification system for all structures analogous to the EC classification for enzyme function. The type of research possible with this database is exemplified by an analysis of all enzymes in which it was shown that the topology of enzymes is more related to the ligands bound than the enzyme EC class (Martin *et al.*, 1998).

Speciality Databases

Another type of database that has proved invaluable in research has been the speciality database. These databases are curated by experts in the field and provide information beyond the structures themselves. These may include derived structural data, sequence information and other biochemical information. An example is the Protein Kinase Resource, which provides not only structural but also functional and pharmacological data about these key drug targets. The same is true of the HIV Protease Database, which has captured all the information about HIV protease structures to be included in one place with the goal of aiding drug development. The Nucleic Acid Database (NDB) has provided a searchable resource about nucleic acids. The Enzyme Structures Database organizes all the enzyme structures in the PDB according to the EC codes contained in the ENZYME Data Bank and provides information about them. (*See* DNA Structure.)

Databases of Structural Characteristics

Databases of structural features contained within macromolecules have also emerged. The Molecular Movements Database provides information about the possible motions of macromolecules by analyzing the various structures of particular molecules. OLDERADO (On Line Database of Ensemble Representatives And Domains) provides a database of structures for which there are several representatives, such as an ensemble of NMR structures. The TESS

(Template Search and Superimposition) algorithm has allowed for the creation of a database of active site templates called PROCAT. This type of database will become invaluable in the quest for relating structure to function. PROMISE is a database that provides information about the prosthetic centers and metal ions in the active sites. ISOSTAR provides an integrated view of the nonbonded interactions geometry around ligands in proteins. PDBsum gives a variety of carefully curated information about all the structures in the PDB. The Dictionary of Interfaces in Proteins (DIP) is a data bank of complementary molecular surface patches and is meant to enable molecular recognition research.

Challenges

The PDB is now much more than a repository of coordinate data. To make this resource even more useful, all the files need to be in a uniform format so that the many new databases of derived information can be easily constructed without having to first clean the files. A project at the PDB is underway to re-examine the archive to achieve this uniformity (Bhat *et al.*, 2001). The PDB will also integrate the validation criteria that have been developed by a variety of researchers (Wilson *et al.*, 1998).

The various methods that have been developed for classification (Gerstein and Levitt, 1998; Orengo and Taylor, 1996) and structure comparison (Alexandrov and Fischer, 1996; Gibrat *et al.*, 1996; Holm and Sander, 1996; Shindyalov and Bourne, 1998) will continue to improve and their results incorporated into the databases, as will methods to understand macromolecular interactions with one another (Jones and Thornton, 1997), with nucleic acids (Jones *et al.*, 1999) and with small molecule ligands (Wallace *et al.*, 1995).

The goal of being able to relate structure to function will be facilitated by different types of database efforts. Databases that assemble information about particular protein families will be one avenue that will provide this information. In these databases the coverage is very narrow and deep, so that a truly full understanding of a single class of proteins with known function is possible. The lessons learned from these types of resource will perhaps allow us to develop some general principles about the relationships of structure and function.

The structural genomics project is an outgrowth of the various genome projects. Its goal is to determine macromolecular structures on a genomic scale – the discovery, analysis and dissemination of three-dimensional structures of biological macromolecules representing the entire range of structural diversity found in nature (see Web Links). The sequences being targeted by many of these efforts are being stored in a database (see Web Links). Once the anticipated large volume of three-dimensional data is collected and assembled, it will be critical to coordinate and to relate the structural and sequence data in order to create a full picture of protein fold space.

While these efforts are ongoing, databases of information about chemical and biological properties of macromolecules and their complexes will provide yet another avenue to understanding function.

With the large number of databases that have been created, it is important to develop methods to query across all of these databases in a seamless way. To help in this effort, the RCSB has developed a standard application interface for macromolecular data based on the Common Object Request Broker Architecture (Corba). The proposal was adopted by the Object Management Group (OMB) in February 2001 (see Web Links). This specification opens the door to more seamless and specific access to PDB data. More specifically, it provides a standard application programming interface (API) that will allow direct access by remote programs to the binary data structures of the PDB. This and other similar initiatives will help to ensure that the world of biology *in silico* will be readily accessible.

Acknowledgements

Parts of this work have appeared in 'The past and future of structural databases' by Helen M. Berman (1999) *Current Opinion in Biotechnology* **10**: 76–80. The US National Science Foundation, the Department of Energy and the National Institutes of Health supported this work.

See also
Protein Databases

References

(1971) Protein Data Bank. *Nature New Biology* **233**: 223.

Adzhubei IA, Adzhubei AA and Neidle S (1998) An integrated sequence-structure database incorporating matching mRNA sequence and protein three-dimensional structure data. *Nucleic Acids Research* **26**: 327–331.

Alexandrov NN and Fischer D (1996) Analysis of topological and nontopological structural similarities in the PDB: new examples with old structures. *Proteins* **25**: 354–365.

Allen FH, Bellard S, Brice MD, et al. (1979) The Cambridge Crystallographic Data Centre: computer-based search, retrieval, analysis and display of information. *Acta Crystallographica B* **35**: 2331–2339.

Bairoch A (2000) The ENZYME database in 2000. *Nucleic Acids Research* **28**: 304–305.

Berman HM, Olson WK, Beveridge DL, et al. (1992) The Nucleic Acid Database – a comprehensive relational database of three-dimensional structures of nucleic acids. *Biophysical Journal* **63**: 751–759.

Berman HM, Westbrook J, Feng Z, et al. (2000) The Protein Data Bank. *Nucleic Acids Research* **28**: 235–242.

Bernstein FC, Koetzle TF, Williams GJB, et al. (1977) Protein Data Bank: a computer-based archival file for macromolecular structures. *Journal of Molecular Biology* **112**: 535–542.

Bhat TN, Bourne P, Feng Z, et al. (2001) The PDB data uniformity project. *Nucleic Acids Research* **29**: 214–218.

Blake CC, Koenig DF, Mair GA, et al. (1965) Structure of hen egg-white lysozyme. A three dimensional Fourier synthesis at 2 Å resolution. *Nature* **206**: 757–761.

Bruno IJ, Cole JC, Lommerse JP, et al. (1997) ISOSTAR: a library of information about nonbonded interactions. *Journal of Computer-Aided Molecular design* **11**: 525–537.

Degtyarenko KN, North ACT, Perkins DN and Findlay JBC (1998) PROMISE: a database of information on prosthetic centres and metal ions in protein active sites. *Nucleic Acids Research* **26**: 376–381.

Dodge C, Schneider R and Sander C (1998) The HSSP database of protein structure-sequence alignments and family profiles. *Nucleic Acids Research* **26**: 313–315.

Gerstein M and Krebs W (1998) A database of macromolecular motions. *Nucleic Acids Research* **26**: 4280–4290.

Gerstein M and Levitt M (1998) Comprehensive assessment of automatic structural alignment against a manual standard, the SCOP classification of proteins. *Protein Science* **7**: 445–456.

Gibrat J-F, Madej T and Bryant SH (1996) Surprising similarities in structure comparison. *Current Opinions in Structural Biology* **6**: 377–385.

Gilliland GL (1997) Biological Macromolecule Crystallization Database. *Methods in Enzymology* **277**: 546–556.

Hendlich M, Bergner A, Günther J and Klebe G (2003) ReLiBase: Design and development of a database for comprehensive analysis of protein-ligand interactions. *Journal of Molecular Biology* **326**: 607–620.

Holm L and Sander C (1994) Searching protein structure databases has come of age. *Proteins* **19**: 165–173.

Holm L and Sander C (1996) Mapping the protein universe. *Science* **273**: 595–603.

Holm L and Sander C (1998) Touring protein fold space with Dali/FSSP. *Nucleic Acids Research* **26**: 316–319.

Jones S, van Heyningen P, Berman HM and Thornton JM (1999) Protein–DNA interactions: a structural analysis. *Journal of Molecular Biology* **287**: 877–896.

Jones S and Thornton JM (1997) Analysis of protein–protein interaction sites using surface patches. *Journal of Molecular Biology* **272**: 121–132.

Kartha G, Bello J and Harker D (1967) Tertiary structure of ribonuclease. *Nature* **213**: 862–865.

Kelley LA and Sutcliffe MJ (1997) OLDERADO: On-line database of ensemble representatives and domains. *Protein Science* **6**: 2628–2630.

Kendrew JC, Bodo G, Dintzis HM, Parrish RG and Wyckoff H (1958) A three-dimensional model of the myoglobin molecule obtained by X-ray analysis. *Nature* **181**: 662–666.

Laskowski RA, Hutchinson EG, Michie AD, et al. (1997) PDBsum: a Web-based database of summaries and analyses of all PDB structures. *Trends in Biochemical Sciences* **22**: 488–490.

Lefranc MP, Giudicelli V, Busin C, et al. (1998) IMGT, the International ImMunoGeneTics database. *Nucleic Acids Research*, **26**: 297–303.

Martin ACR, Orengo CA, Hutchinson EG, et al. (1998) Protein folds and functions. *Structure* **6**: 875–884.

Moore P (2001) The ribosome at atomic resolution. *Biochemistry* **40**: 3243–3250.

Murzin AG, Brenner SE, Hubbard T and Chothia C (1995) SCOP: a structural classification of proteins database for the investigation of sequences and structures. *Journal of Molecular Biology* **247**: 536–540.

Orengo CA, Michie AD, Jones S, et al. (1997) CATH – a hierarchic classification of protein domain structures. *Structure* **5**: 1093–1108.

Orengo CA and Taylor WR (1996) SSAP: sequential structure alignment program for protein structure comparison. *Methods in Enzymology* **266**: 617–635.

Pascarella S and Argos P (1992) Analysis of insertions/deletions in protein structures. *Journal of Molecular Biology* **224**: 461–471.

Perutz MF, Rossmann MG, Cullis AF, Muirhead G and Will G (1960) Structure of haemoglobin: a three-dimensional Fourier synthesis at 5.5 Å resolution. *Nature* **185**: 416–422.

Schmidt R, Gerstein M and Altman R (1997) LPFC: An Internet library of protein family core structures. *Protein Science* **6**: 246–248.

Shindyalov IN and Bourne PE (1998) Protein structure alignment by incremental combinatory extension of the optimum path. *Protein Engineering* **11**: 739–747.

Smith C, Gribskov M, Shindyalov IN, et al. (1997) The Protein Kinase Resource. *Trends in Biochemical Sciences* **22**: 444–446.

Sowdhamini R, Burke DF, Huang J-f, et al. (1998) CAMPASS: a database of structurally aligned protein. *Structure* **6**: 1087–1094.

Vondrasek J, Buskirk C and Wlodawer A (1997) Database of three-dimensional structures of HIV proteinases. *Nature Structural Biology* **4**: 8.

Wallace A, Laskowski R and Thornton J (1996) Derivation of 3D coordinate templates for searching structural databases: application to the Ser-His-Asp catalytic triads of the serine proteinases and lipases. *Protein Science*, **5**: 1001–1013.

Wallace AC, Laskowski RA and Thornton JM (1995) LIGPLOT: a program to generate schematic diagrams of protein ligand interactions. *Protein Engineering* **8**: 127–134.

Wilson KS, Butterworth S, Dauter Z, et al. (1998) Who checks the checkers? Four validation tools applied to eight atomic resolution structures. *Journal of Molecular Biology* **276**: 417–436.

Wyckoff HW, Hardman KD, Allewell NM, et al. (1967) The structure of ribonuclease-S at 6 Å resolution. *Journal of Biological Chemistry* **242**: 3749–3753.

Yusupov MM, Yusupova GZ, Baucom A, et al. (2001) Crystal structure of the ribosome at 5.5 Å resolution. *Science* **282**: 883–896.

Further Reading

(2000) Structural genomics supplement. *Nature Structural Biology* **7**: 927–994 [entire issue].

(2001) Database Issue. *Nucleic Acids Research* **29**: 1–349.

(2001) *The Structures of Life*. NIH publication number 01-2778. http://www.nigms.nih.gov/news/science_ed/structlife.pdf.

Benton D (1996) Bioinformatics – principles and potential of a new multidisciplinary tool. *Trends in Biotechnology* **14**: 261–272.

Berman HM, Bhat TN, Bourne PE, et al. (2000) The Protein Data Bank and the challenge of structural genomics. *Nature Structural Biology* **7**: 957–959.

Berman HM, Gelbin A and Westbrook J (1996) Nucleic acid crystallography: a view from the Nucleic Acid Database. *Progress in Biophysics and Molecular Biology* **66**: 255–288.

Gaasterland T (1998) Structural genomics: bioinformatics in the driver's seat. *Nature Biotechnology* **16**: 625–627.

Holm L and Sander C (1994) Searching protein structure databases has come of age. *Proteins* **19**: 165–173.

Rost B (1998) Marrying structure and genomics. *Structure* **6**: 259–263.

Swindells MB, Orengo CA, Jones DT, Hutchinson EG and Thornton JM (1998) Contemporary approaches to protein structure classification. *BioEssays* **20**: 884–891.

Westbrook J and Bourne PE (2000) STAR/mmCIF: an extensive ontology for macromolecular structure and beyond. *Bioinformatics* **16**: 159–168.

Wilson KS, Butterworth S, Dauter Z, *et al.* (1998) Who checks the checkers? Four validation tools applied to eight atomic resolution structures. *Journal of Molecular Biology* **276**: 417–436.

Zou J-Y and Mowbray SL (1994) An evaluation of the use of databases in protein structure refinement. *Acta Crystallographica D* **50**: 237–249.

Web Links

PDB Deposition Information. Links to the AutoDep Input Tool (ADIT), AutoDep, and other deposition resources
http://www.pdb.org/

Second International Structural Genomics Meeting. NIGMS statement on coordinate deposition, highlights, agreed principles and procedures, roster, agenda, and Task Force Reports
http://www.nigms.nih.gov/news/meetings/airlie.html

TargetDB. Target Registration Database that contains sequences from the worldwide structural genomics centers, and the PDB
http://targetdb.pdb.org/

OMG/LSR Corba Standard for Macromolecular Structure Data (OMG specification formal/02-05-01). First formal version of the Macromolecular Structure specification
http://www.omg.org/technology/documents/formal/macro_molecular.htm

The OpenMMS Toolkit. Corba, Relation Database and XML Software for Macromolecular Structure
http://openmms.sdsc.edu/

Structural Predictions and Modeling

Martin Norin, *Biovitrum AB, Stockholm, Sweden*

Computer models form the basis for the predictions of the molecular three-dimensional structures of polypeptides – the proteins. The different models all use the polypeptide amino acid sequence as the starting point for the prediction.

Intermediate article

Article contents

- Introduction
- Comparative Modeling
- Fold Recognition and Threading Methods
- *Ab Initio* Methods
- Prediction of Protein Function from Structure

Introduction

The function of a protein depends on its three-dimensional structure, which, in turn, is determined by its amino acid sequence. The most precise and accurate information on the structure of a particular protein or a protein complex can be obtained using experimental methods, such as X-ray crystallography and nuclear magnetic resonance (NMR). However, the experimental methods require large resources, and it has been shown that many proteins are very difficult to study by means of experiments. The key limiting factors are difficulties in obtaining pure soluble protein material, growing protein crystals and the physical sizes of larger proteins. The human genome is believed to encode at least 30 000–40 000 individual proteins, with the highest estimates reaching numbers between 100 000 and 150 000. In the beginning of 2002 the collection of publicly available protein structures in the Research Collaboratory for Structural Bioinformatics (RCSB) protein data bank comprised approximately 1300 nonredundant structures, but as few as about 300 are of human origin. This gap in information has triggered a demand for higher-throughput theoretical methods to predict and to analyze protein

structures computationally. The use of structural analyses of protein structures, including those based on predicted structures, has expanded as a consequence of the exponential growth of genome research; for example, in the pharmaceutical industry, protein modeling is applied throughout the value chain from the discovery of target proteins through the generation of lead molecules to the prediction of pharmacological polymorphic effects in clinical trials.

Prediction methods of protein structures all use the primary amino acid sequence as the starting point for modeling the conformation of the target protein. Current methods can be categorized into three classes: (1) comparative modeling, (2) fold recognition and (3) *ab initio* methods. The comparative modeling methods are used to create detailed atomic models based on significant sequence identities (>30%) to experimentally determined template structures. Threading (fold recognition) methods are used for sequences that have distant sequence relationships to known structures,

Table 1 Protein structure prediction methods

Method	Required structural knowledge	Resolution
Comparative modeling/homology modeling	Experimentally determined sequence homolog	Amino acid backbone and side-chain conformations of core residues
Threading/fold recognition	Relationship to known protein fold	Overall protein fold. Amino acid intramolecular relationships
Ab initio	None, except primary amino acid sequence of target protein	Ideas of possible protein structural elements

undetectable using sequence information alone. Finally, *ab initio* methods can be also applied when attempting to predict conformation of proteins devoid of a structural template. Quite accurate models having less than 2 Å root mean square deviations of the Cα amino acid backbone atoms to experimental structures may be achieved when the model sequence has an overall amino acid identity of more than 30% to the structural template (Sternberg *et al.*, 1999). Below this threshold the quality of the models decreases rapidly. **Table 1** summarizes the overall features of the modeling methods.

The progress and performances of protein prediction methods are assessed on a regular basis at the Critical Assessment of Techniques for Protein Structure Prediction (CASP) meetings. The reviews from the 1998 CASP3 experiments serve as a good starting point for both learning about the state-of-the-art of protein structure prediction and the CASP experiments (Moult, 1999; Murzin, 2001). The CASP4 experiments are published in *Proteins*.

Comparative Modeling

Comparative modeling methods predict the conformation of a protein by using a closely related high-quality experimental structure as a template. Comparative modeling can be broken down into five steps:

1. Identification of an experimentally determined structural homolog to serve as a template for the subsequent modeling steps.
2. Production of an optimal sequence alignment to the template.
3. Creation of an initial model based on the alignment to the template structure.
4. Modeling of missing parts, loops (insertions and deletions), in the sequence alignment.
5. Optimization of side-chain conformation and final overall refinement of the model.

A number of commercial protein modeling packages are available. Many of these are rather expensive and include high-performance computer graphics capabilities to examine and to analyze the protein models. For academic scientists conducting noncommercial research, an alternative is to use free publicly available Web-based tools such as the SWISS-MODEL server. These servers also often provide easy-to-use computer graphics software for visualizing protein structures.

Structural templates

The search for structural templates can be performed using established sequence comparison methods, for example PSI-BLAST (Altschul *et al.*, 1997). The limited number of available three-dimensional protein structure templates restricts the number of proteins that can be modeled using comparative modeling. About 20–70% (depending on the organism) of the proteome can be modeled at the 30% sequence identity threshold level. In a recent estimate, Vitkup *et al.* (2001) suggest that 16 000 novel structures need to be experimentally determined to model 90% of the global proteome at >30% sequence identity, given that the selection of novel structures is optimally coordinated. It is important to note here that very few structures of membrane-bound proteins have been experimentally determined. Rough models of the G-protein-coupled receptor family (7TM proteins) can be obtained from the related structure of the rhodopsin protein. Thus, one of the biggest challenges in structural biology is the successful study of membrane proteins.

Often sequence comparison methods retrieve several template candidates. As a rule-of-thumb, the closest sequence similarity template is the best candidate. For regions with low local sequence similarity, however, fragments from other homologous structures may be considered.

Sequence alignment

The key step in modeling is to produce a correct sequence-to-structure alignment of the query sequence to the template. Multiple sequence alignments and superimposed multiple template structures may significantly improve the alignment. If the alignment is incorrect, however, there are no model refinement methods available to rescue the models from these errors. Automatic freely available alignment algorithms such as CLUSTAL W (Thompson *et al.*, 1994) provide excellent starting points. However, detailed knowledge of the particular protein family and in-depth experience of protein structure are still important manually to improve sequence alignments produced by automatic methods. It is especially important to examine the boundaries to insertions and deletions in the alignment. In general, one should avoid insertions and deletions within secondary structure elements. Cysteine amino acid residues often form disulfide bridges between structural elements and may therefore form distant constraints. Furthermore, important conserved residues may be detected by a careful examination of the structures within a protein family.

Modeling of the missing parts of the backbone (loops)

The modeling of the gaps and insertions to the template structure, often referred to as loop regions, constitutes a great challenge for protein modelers. These loops often appear at structurally diverse regions between protein secondary structure elements. Database methods are used to fit known structural fragments to the loop boundaries, while *ab initio* methods apply energy refinement schemes. The methods to model these loops have undergone significant improvement, but still need to be further developed to provide reliable predictions. Some recent work (Fiser *et al.*, 2000) in which energy potential functions are combined with statistical terms looks promising when these loops are of reasonable size.

Modeling of side-chain conformations and final model refinement

Accurate prediction of side-chain conformations can be very useful, not only in comparative modeling of novel structures, but also in rational re-engineering of proteins starting with well-defined structures. The combinatorial complexity of the number of rotamers that needs to be considered grows exponentially with the size of the protein. The combinatorial problem may successfully be reduced to a tractable size by limiting the choices using combinations of potential energy functions together with knowledge-based sidechain rotamer libraries (Ponder and Richards, 1987) and/or dead-end elimination algorithms (Desmet *et al.*, 1992).

The last step in a homology modeling scheme is usually a global energy refinement of the 'crude' model. Some improvements in the conformation of the amino acid side chains can be achieved here, but

in general these steps do not make any big improvements in bringing the model closer to the native conformation.

Examination and analysis of the protein model

The quality of the predicted protein model needs to be assessed. One should be careful when drawing functional conclusions derived from regions with low local similarities to the template structure. The degrees of freedom for rotational angles of the polypeptide backbone are limited by steric constraints. In so-called Ramachandran plots, the allowed conformations of the polypeptide backbone can be compared to the measured angles of the model. In general, a consistent model contains very few amino acid residues, except for glycine residues, outside the allowed regions. Algorithms such as PROCHECK are useful to check the model for internal consistency. In addition, energy calculations using potential energy functions may spot problematic parts in the model.

Fold Recognition and Threading Methods

Comparative modeling methods do not work if the query sequence lacks homologs to proteins of known structure. An alternative way to find template sequences is by using 'threading' algorithms. Here the fold is predicted by assessing the compatibility of a sequence to a particular fold by analyzing the intramolecular distances as the sequence is threaded through a collection of known folds (Sippl, 1990). The environments of the amino acid residues are scored using knowledge-based scoring functions. These methods can sometimes assign the correct fold even for very distant sequence relationships. However, experience from the blind-prediction CASP3 experiment is that the most sensitive sequence comparison methods, such as PSI-BLAST, perform at least equally well.

Ab Initio Methods

In *ab initio* modeling of protein conformation, only the primary amino acid sequence need be used as the specific input for predicting protein structure. These methods can be applied when other methods described above are not able to identify related proteins with known folds. Currently, these *ab initio* models are very crude, having in the very best cases 4–7 Å root mean square deviations from the native structure. However, some significant progress made in the CASP3 experiment indicates that these methods may play an important role in complementing future structural genomics efforts. *Ab initio* methods may already, together with sequence comparison and threading methods, synergistically contribute to the fold recognition of proteins that have weakly homologous proteins of known structures (Kolinsky *et al.*, 2001).

Prediction of Protein Function from Structure

Proteins sharing the same fold may have quite different functions. In most proteins the functional machinery is built up by a limited number of specific amino acid residues in precise three-dimensional relative positions ('hot-spots'). The PROSITE database (Hofmann *et al.*, 1999) contains sequence patterns of functional sites, but has no three-dimensional information. Kasuya and Thornton (1999) have analyzed the three-dimensional features of the PROSITE patterns in order to define structural templates for structure-based mining for functional sites. In another example, three-dimensional descriptors based on intramolecular distances between amino acid residues were able to identify a potential oxidoreductase active site, the serine/threonine phosphatase-1 subfamily (Fetrow *et al.*, 1999).

In a recent interesting example by Sowa *et al.* (2001), an important site for the regulation of G protein signaling was identified and confirmed through site-specific mutagenesis. They applied an 'evolutionary trace' method by combining structural and sequence data. The method detects three-dimensional clusters of amino acid residues that are specific for a related subset of sequences within a protein family. This method is especially interesting because it is able to propose novel sites without any prior knowledge of the specific amino acid features.

See also
Protein Structure
Protein Structure Prediction and Databases

References

Altschul S, Madden T, Schaffen A, *et al.* (1997) Gapped BLAST and PSI-BLAST: a new generation of protein database search programs. *Nucleic Acids Research* **25**: 3389–3402.
Desmet J, De Maeyer M, Hasez B and Lasters I (1992) The dead-end elimination theorem and its use in protein side-chain positioning. *Nature* **356**: 539–542.
Fetrow J, Siew N and Skolnick J (1999) Structure-based functional motif identifies a potential disulphide oxidoreductase active site in the serine/threonine protein phosphatase-1 subfamily. *Federation of American Societies for Experimental Biology* **13**: 1866–1874.
Fiser A, Do R and Sali A (2000) Modeling of loops in protein structures. *Protein Science* **9**: 1753–1773.
Hofmann K, Bucher P, Falquet L and Bairoch A (1999) The PROSITE database, its status in 1999. *Nucleic Acids Research* **27**: 215–219.
Kasuya A and Thornton J (1999) Three-dimensional structure analysis of PROSITE patterns. *Journal of Molecular Biology* **286**: 1673–1691.
Kolinsky A, Betancourt MR, Kihara D, Rotkiewicz R and Skolnick J (2001) Generalized comparative modeling (GENECOMP): a combination of sequence comparisons, threading, and lattice modeling for protein structure prediction and refinement. *Proteins* **44**: 133–149.
Moult J (1999) Predicting protein three-dimensional structure. *Current Opinion in Biotechnology* **10**: 583–588.
Murzin AG (2001) Progress in protein structure prediction. *Nature Structural Biology* **8**: 110–112.
Ponder J and Richards F (1987) Use of packing criteria in the enumeration of allowed sequences for different structural classes. *Journal of Molecular Biology* **193**: 775–791.
Sippl M (1990) The calculation of conformational ensembles from potentials of mean force, an approach to the prediction of local structures in globular proteins. *Journal of Molecular Biology* **213**: 659–883.
Sowa M, He W, Slep K, *et al.* (2001) Prediction and confirmation of a site critical for effector regulation of RGS domain activity. *Nature Structural Biology* **8**: 234–237.
Sternberg MJE, Bates PA, Kelley LA and MacCallum RM (1999) Progress in protein structure prediction: assessment of CASP3. *Current Opinion in Structural Biology* **9**: 368–373.
Thompson J, Higgins D and Gibson T (1994) CLUSTAL W: improving the sensitivity of progressive multiple sequence alignment through sequence weighting, position-specific gap penalties and weight matrix choice. *Nucleic Acids Research* **22**: 4673–4680.
Vitkup D, Melamud E, Moult J and Sander C (2001) Completeness in structural genomics. *Nature Structural Biology* **8**(6): 559–566.

Further Reading

Fetrow J, Godzik A and Skolnick J (1998) Functional analysis of the *Escherichia coli* genome using the sequence-to-structure-to-function paradigm: identification of proteins exhibiting the glutaredoxin/thioredoxin disulfide oxidoreductase activity. *Journal of Molecular Biology* **281**: 949–968.
Looger L and Helliga H (2000) Generalized dead-end elimination algorithms make large-scale protein side-chain prediction tractable: implications for protein design and structural genomics. *Journal of Molecular Biology* **307**: 429–445.
Madej T, Boguski M and Bryant S (1995) Threading analysis suggests that the obese gene product may be a helical cytokine. *FEBS Letters* **373**: 13–18.
Mendes J, Nagarajaram H, Soares C, Blundell T and Carrondo M (2001) Incorporating knowledge-based biases into an energy-based side-chain modeling method: application to comparative modeling of protein structure. *Biopolymers* **59**: 72–86.
Norin M and Sundström M (2002) Structural proteomics: developments in structure-to-function predictions. *Trends in Biotechnology* **20**: 79–84.
Panchenko A, Marchler-Bauer A and Bryant S (2000). Combination of threading potentials and sequence profiles improves fold recognition. *Journal of Molecular Biology* **296**: 1319–1331.

Sánches R, Pieper U, Melo F, *et al.* (2000) Protein structure modeling for structural genomics. *Nature Structural Biology* **7**(supplement): 986–990.

Todd A, Orengo C and Thornton J (2001) Evolution of function in protein superfamilies, from a structural perspective. *Journal of Molecular Biology* **307**: 1113–1143.

Wilson C, Kreychman J and Gerstein M (2000) Assessing annotation transfer for genomics: quantifying the relations between protein sequence, structure and function through traditional and probabilistic scores. *Journal of Molecular Biology* **297**: 233–249.

Web Links

ExPASy Molecular Biology Server. Provides easy-to-use computer graphics software for visualization of protein structures
www.expasy.ch

Protein Data Bank. A collection of publicly available protein structures
http://www.rcsb.org/pdb

Protein Structure Prediction Centre. Reviews of the CASP meetings.
http://PredictionCenter.llnl.gov

Structural Proteins: Genes

| | Intermediate article |

Mon-Li Chu, *Thomas Jefferson University, Philadelphia, Pennsylvania, USA*

Structural proteins are proteins required for the building and support of body tissues. These include collagens, which provide the integrity and tensile strength of connective tissues, bone and cartilage, and myosins, which are important for the elasticity and contractility of muscles. Genes for structural proteins are usually present as a single copy per haploid genome.

Article contents
• Introduction
• Collagen Gene Family
• Gene Sizes and Chromosomal Locations
• Genes for the Fibrillar Collagens
• Genes for the Nonfibrillar Collagens
• Conclusion

Introduction

Structural proteins are proteins required for the building and support of body tissues. Collagens and myosins are among the major structural proteins in the human body. Collagens provide the integrity and tensile strength of connective tissues, bone and cartilage, while myosins are important for the elasticity and contractility of muscles. In addition to their prominent role in muscle cells, it should be noted that myosins are also present in nonmuscle cells. It is becoming increasingly clear that structural proteins such as collagens and myosins also serve other biological functions, examples of which include an influence on cell growth and involvement in signal transduction.

Collagens and myosins each constitute a large superfamily of proteins that share common structural features. Myosins share a common domain, the motor domain, which has been shown to interact with actin and which binds adenosine 5′ triphosphate (ATP). They are grouped into 15 subclasses according to phylogenetic analyses, the majority of which contain more than one member (Sellers, 2000). Collagens are the most abundant structural proteins in the body, accounting for about 25% of the total protein mass. The collagen family, consisting of more than 20 distinct types, will be used as an example in this article for the discussion of genes encoding structural proteins. Genes for structural proteins are usually present as a single copy per haploid genome.

Collagen Gene Family

Collagens are homo- or heterotrimers, in which three collagen polypeptides, called α chains, align in parallel and fold into a unique triple-helical structure, which is based on the repeating Gly–X–Y amino acid sequence of the α chains (**Figure 1**). The presence of glycine, the smallest amino acid, at every third position is crucial for the tight packing of the α chains into the triple-helical conformation. The X and Y positions can be filled by any amino acid but are often proline and hydroxyproline respectively, which stabilize the triple helix. Because glycine and proline are encoded by GGN and CCN respectively, the coding sequences of the collagen genes have a high G+C content.

To date, 20 vertebrate collagen types, I–XX, encoded by at least 34 distinct genes, have been described in the literature, and at least four more collagen chains are currently under characterization (Myllyharju and Kivirikko, 2001). The α chains of different collagen types vary greatly in size, ranging from 600 to 3000 amino acids, and differ considerably in the number and size of their triple-helical (collagenous) domains (Prockop and Kivirikko, 1995; Brown and Timpl, 1995). The α chains also contain one or more globular (noncollagenous) domains, which again vary considerably in size and modular structure. The globular domains consist of modules that are frequently shared among different collagens and other

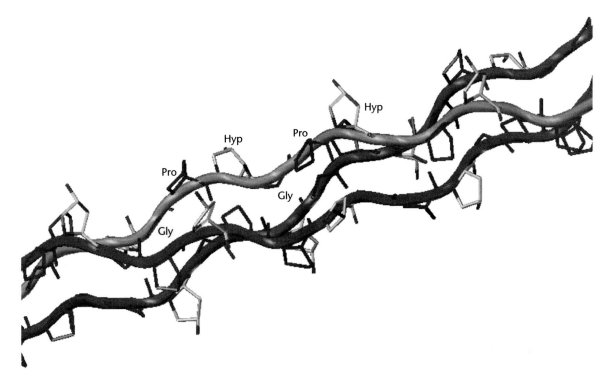

Figure 1 [*Figure is also reproduced in color section.*] A computer model of the collagen triple helix. Three collagen α chains intertwine to form a triple helix. The repeating sequence in the α chain shown is glycine (Gly), proline (Pro), hydroxyproline (Hyp). The presence of glycine in every third position is crucial for the tight packing of the α chains into the triple helix. Proline and hydroxyproline are frequent amino acid residues in collagen molecules as they stabilize the triple helix.

extracellular proteins, such as fibronectin, thrombospondin and von Willebrand factor. On the basis of the primary structure and supramolecular assembly, the collagens are generally divided into two major classes: fibrillar collagens that form characteristic cross-striated fibrils and nonfibrillar collagens that do not form cross-striated fibers. The structures of the collagen genes are consistent with this classification in that fibrillar collagen genes share similarities that distinguish them from nonfibrillar collagen genes.

Gene Sizes and Chromosomal Locations

In general, collagen genes are large and highly interrupted, except for the genes for types VIII and X collagen (**Table 1**). The type VII collagen gene consists of 118 exons and is apparently the most complex gene described to date (Christiano *et al.*, 1994). Most collagen genes are scattered in the human genome (**Table 1**). For example, the two genes encoding type I collagen, collagen, type I alpha 1 (*COL1A1*) and collagen, type I alpha 2 (*COL1A2*), are located on chromosomes 17 and 7, respectively. By contrast, some collagen genes are grouped in clusters. Among

the clustered collagen genes, the most remarkable is the head-to-head arrangement of the three type IV collagen gene pairs on chromosomes 13, 2 and X. The sequences of each gene pair are encoded on opposite strands of the deoxyribonucleic acid (DNA) with the first exons separated by only 5–442 base pairs (bp). The genes for type III collagen, collagen, type III alpha 1 (Ehlers–Danlos syndrome type IV, autosomal dominant) (*COL3A1*), and for the α_2 chain of type V collagen, *COL5A2*, are in the long arm of chromosome 2 with their 3′ ends within a region of 35 kb. The genes for the α_1 and α_2 chains of type VI collagen, *COL6A1* and *COL6A2*, and for type XVIII collagen, *COL18A1*, are situated in a head-to-tail orientation at the end of the long arm of chromosome 21, and the gene order is *COL18A1*–*COL6A1*–*COL6A2*–telomere. The genes for the α_1 chain of type IX collagen (*COL9A1*), type XII collagen (*COL12A1*) and type XIX collagen (*COL19A1*) are located in the long arm of chromosome 6.

Genes for the Fibrillar Collagens

The hallmark of fibrillar collagens is a long central triple-helical domain with about 1000 amino acids of uninterrupted repeating Gly–X–Y sequences flanked

Table 1 Sizes, number of exons and chromosomal locations of human collagen genes

Collagen type	Gene	Gene size (kb)	No. of exons	Chromosomal location
I	COL1A1	18	51	17q21.33
	COL1A2	38	52	7q21.3
II	COL2A1	30	54	12q13.11
III	COL3A1	44	52	2q32.2
IV	COL4A1	159	52	13q34
	COL4A2	207	47[a]	13q34
	COL4A3	151	60	2q36.3
	COL4A4	159	48	2q36.3
	COL4A5	258	51	Xq23
	COL4A6	285	46	Xq23
V	COL5A1	201	66	9q34.2–q34.3
	COL5A2			2q24.3–q31
	COL5A3	51	67	19p13.2
VI	COL6A1	23	35	21q22.3
	COL6A2	35	30	21q22.3
	COL6A3	90	44	2q37
VII	COL7A1	30.5	118	3p21.31
VIII	COL8A1	6	4[b]	3q12.1
	COL8A2			1p32.3–34.2
IX	COL9A1	90	38	6q13
	COL9A2	15	32	1p34.2
	COL9A3	23	32	20q13.33
X	COL10A1	7	3	6q22.1
XI	COL11A1	238	69	1p21.1
	COL11A2	31	66	6p21.2
	COL2A1[c]	30	53	12q13.11
XII	COL12A1	120	65	6q14.1
XIII	COL13A1	140	42	10q22.1
XIV	COL14A1	175	44	8q23
XV	COL15A1	145	42	9q22.33
XVI	COL16A1	51	71	1p35.1
XVII	COL17A1	55	56	10q25.1
XVIII	COL18A1	104[a]	43[a]	21q22.3
XIX	COL19A1	> 250	51	6q13

For references see text, Prockop and Kivirikko (1995) and the draft sequence of the human genome (Web Links).
[a]For the mouse gene.
[b]For the rabbit gene.
[c]The α_3(XI) chain is encoded by the COL2A1 gene.

by N- and C-propeptides, which are usually cleaved by proteinases before assembly into fibrils. The N-propeptides often contain one or more short triple-helical domains. Types I, II and III are the three major fibrillar collagens. Type I collagen, a heterotrimer of two α_1(I) chains and one α_2(I) chain, is most abundant and a major constituent of skin, bone and tendon, whereas types II and III, both homotrimers, are predominantly located in the cartilage and blood vessel walls, respectively. Types V and XI are minor fibrillar collagens, which are found in small amounts in fibrils composed of major fibrillar collagens.

The four major fibrillar collagen genes (COL1A1, COL1A2, COL2A1 (collagen type II, alpha 1 (primary osteoarthritis, spondyloepiphyseal dysplasia,

congenital)) and COL3A1) have strikingly similar gene structures, suggesting that they have evolved from a common ancestral gene. The basic structure consists of 52 exons, of which 42 exons encode the central triple-helical domain, six exons the N-propeptide and four exons the C-propeptide (Boedtker et al., 1985). The first and last exons encoding the triple-helical domain are 'junctional' exons, which contain sequences for both the collagenous and noncollagenous domains. In the COL1A1 gene, two triple-helical coding exons (exons 33 and 34) found in the other genes are fused (**Table 2**). In the COL2A1 gene, two additional exons code for the N-propeptide, one of which encodes a cysteine-rich domain that is alternatively spliced and developmentally regulated. In the COL3A1 gene, the triple-helical

Table 2 Exons encoding the triple-helical domains of the fibrillar collagens

Major fibrillar collagen genes		Minor fibrillar collagen genes	
Exon no.	Size (bp)	Exon no.	Size (bp)
7	**45**[a]		
8	54	15	54
9	54	16	54
10	54	17	54
11	54	18	54
12	54	19	54
13	45	20	45
14	54	21	54
15	45	22	45
16	54	23	54
17	**99**	24	45
		25	54
18	45	26	45
19	**99**	27	54
		28	45
20	54	29	54
21	108	30	108
22	54	31	54
23	**99**	**32**	**54**
		33	45
24	54	34	54
25	**99**	35	45
		36	54
26	54	37	54
27	54	38	54
28	**54**	**39**	**108**
29	**54**		
30	**45**	**40**	**90**
31	**99**	**41**	**54**
32	108	42	108
33[b]	**54**	43	108
34[b]	**54**		
35	54	44	54
36	54	45	54
37	108	46	108
38	54	47	54
39	**54**	**48**[c]	**108**
40	**162**	49	54
		50	108
41	108	51	54
42	108	52	54
		53	54
43	54	54	54
44	108	55	108
45	54	56	54
46	**108**	57	54
47	**54**	58	108
48	**108**	59	54
		60	36
		61	54

Data from the *COL1A2* and *COL5A1* genes are taken as the prototypic major and minor fibrillar collagen genes. Exons that differ in size between genes are in bold type.
[a] 54 bp in the *COL3A1* gene.
[b] Exons 33 and 34 are fused in the *COL1A1* gene.
[c] Split into two exons of 54 bp in the *COL11A2* gene.

domain is 15 amino acids longer than those of the related genes, but two exons coding for the short triple-helical domain in the *N*-propeptide are fused.

Exons coding for the triple-helical domains are very unusual in that they are short, multiples of 9 bp in size, and begin with complete codons for glycine, thereby specifying a discrete number of Gly–X–Y repeats (**Table 2**). Most exons are 54 bp and the others are twice 54, three times 54, or combinations of 45 and 54 bp. The exon size pattern has led to the hypothesis that an ancestral collagen gene might have evolved through duplication and recombination of a 54-bp primordial exon (Yamada *et al.*, 1980). Exons coding for the *C*-propeptides of the four genes also share striking similarities in size, except for the junctional exon, the size of which is somewhat variable among the four genes. The number and sizes of exons encoding the *N*-propeptides, in contrast to the *C*-propeptide exons, are more variable among the four genes (Chu and Prockop, 1993). This reflects the presence or absence of a cysteine-rich region and the sizes of the short triple-helical domains in the *N*-propeptides of the four chains.

The characteristic 54-bp exon pattern exhibited in the major fibrillar collagens is largely conserved in the minor fibrillar collagens, but there are some notable differences (Takahara *et al.*, 1995). For example, the triple-helical domains of minor fibrillar collagens are encoded by 47–48 exons, six to seven more than in the major fibrillar collagens (**Table 2**). This is because exons of 99 and 162 bp found in the major fibrillar collagens are split into two to three exons of 45 and 54 bp in the minor fibrillar collagens. Another difference is that minor fibrillar collagens contain exons of 36 and 90 bp, which are not found in the major fibrillar collagen genes. The *N*-propeptides of the minor fibrillar collagens, larger and more variable than those of the major fibrillar collagens, are encoded by 14 exons, several of which are alternatively spliced. The *C*-propeptides of the minor fibrillar collagens are encoded by either three or four exons of sizes similar to their counterparts in the major fibrillar collagen genes.

Genes for the Nonfibrillar Collagens

The nonfibrillar collagens represent a diverse group of proteins as they are classified together on the basis of one shared unique feature, the triple helix. In some cases, the noncollagenous domains represent the majority of the total protein mass. These collagens usually have one or more triple-helical domains, which are considerably shorter than those in the fibrillar collagens and contain imperfections in the repeating

Gly–X–Y sequence, such as Gly–X or X–Y. They form a wide range of supramolecular structures, including beaded microfibrils (type VI), short filaments (types VII and XVII) and sheet-like networks (types IV, VIII and X) (Myllyharju and Kivirikko, 2001). Types IX, XII and XIV collagens are found on the surface of cross-striated collagen fibrils and have common primary structural features, which are also shared by types XVI and XIX collagens. These five collagen types are classified as the FACIT (fibril-associated collagens with interrupted triple helices) subgroup. Types XIII and XVII are collagens that possess a transmembrane domain. Lastly, types XV and XVIII are homologous collagens that contain endostatins, the angiogenesis inhibitors, in their *C*-terminal globular domains.

As a consequence of the diverse modular structures, the exon number and sizes of the nonfibrillar collagen genes vary greatly within this group and also differ from the fibrillar collagen genes. However, genes encoding different α chains of the same collagen type or genes encoding homologous collagen types do share similar gene structures. For example, of the six type IV collagen genes, each odd-numbered gene (*COL4A1*, *COL4A3* (collagen, type IV, alpha 3 (Good pasture antigen)) and *COL4A5* (collagen, type IV, alpha 5 (Alport syndrome))) is paired head-to-head with an even-numbered gene (*COL4A2*, *COL4A4* and *COL4A6*). The exon structures of the odd-numbered genes are highly similar, as is the case for the even-numbered genes (Zhang *et al.*, 1996). Also, the odd- and even-numbered type IV collagen genes share some similarities. Likewise, homologies in gene structures are also found between the genes for α$_1$ and α$_2$ chains of type VI collagen and among the three genes of type IX collagen.

Exons coding for the triple-helical domains of the nonfibrillar collagen genes do not show the 54-bp size pattern of the fibrillar collagen genes, except for exons coding for certain triple-helical domains of types IX and XIX collagens. In the majority of nonfibrillar collagen genes, exons coding for the triple-helical domains without Gly–X–Y imperfections start with complete codons for glycine, and are multiples of 9 bp in size, like those in the fibrillar collagen genes. However, the predominant exon size is not 54 bp, and exon sizes not usually observed in the fibrillar collagen genes, such as 18, 27, 36, 63 and 72 bp, are found. For example, of the exons coding for the triple-helical domain of type VII collagen gene, 25 exons are 36 bp, 12 exons 45 bp, eight exons 63 bp, six exons 81 bp, five exons 72 bp, four exons 54 bp, three exons 27 bp and only one exon is 108 bp in length (Christiano *et al.*, 1994). In types VI and XIII collagen genes, the predominant exon sizes are 63 and 27 bp, respectively. A striking finding is that some exons coding for the

triple-helical domains begin with split codons for glycine. In particular, split glycine codons are found in all triple-helical coding exons of type XVII collagen gene and in the second halves of all type IV collagen genes. Another unexpected finding is that the entire triple-helical domains of types VIII and X collagen are each encoded by a single, large exon and the genes consist of only three to four exons, in sharp contrast to the other collagen genes.

Many collagen genes utilize alternative promoters and alternative splicing as mechanisms to further vary their primary structures. Variations in primary structure usually occur in the noncollagenous domains. An exception is the type XIII collagen gene, in which eight exons encoding the collagenous domains may be alternatively spliced in any combination. It is not known, however, whether those α chains with different lengths of the triple-helical domains are able to fold together into a triple helix.

Conclusion

The collagen gene family comprises more than 34 genes encoding 20 homo- or heterotrimeric collagen types. With some exceptions, most collagen genes are scattered on different chromosomes. All collagen genes, except for genes encoding types VIII and X, are highly interrupted, with exon number ranging from 30 to 118. This is primarily because exons coding for the collagenous domain are relatively small and usually encode a discrete number of Gly–X–Y repeated amino sequences, and hence are multiples of 9 bp long. The fibrillar collagen genes, consisting of approximately 50 exons each, have highly similar exon size patterns. Exons coding for the Gly–X–Y sequences are most frequently 54 bp or a variation thereof, and start with complete codons for glycine. This has led to the suggestion that fibrillar collagen genes have evolved from a primordial exon of 54 bp. The nonfibrillar collagen genes do not show the 54-bp exon size pattern. The most frequent exon sizes are 27, 36 or 63 bp, and some exons start with a split codon for glycine. The exon size patterns support the kinship of all collagen genes.

See also
Cell Adhesion Molecules and Human Disorders
Myosins
Skeletal Dysplasias: Genetics
Skeletogenesis: Genetics
Skin: Hereditary Disorders

References

Boedtker H, Finer M and Aho S (1985) The structure of the chicken α2 collagen gene. *Annals of the New York Academy of Sciences* **460**: 85–116.

Brown JC and Timpl R (1995) The collagen superfamily. *International Archives of Allergy and Immunology* **107**: 484–490.

Christiano AM, Hoffman GG, Chung-Honet LC, *et al.* (1994) Structural organization of the human type VII collagen gene (*COL7A1*), composed of more exons than any previously characterized gene. *Genomics* **21**: 169–179.

Chu ML and Prockop DJ (2002) Collagen: gene structure. In: Royce PM and Steinmann B (eds.) *Connective Tissue and its Heritable Disorders: Molecular, Genetic and Medical Aspects*, 2nd edition, pp. 223–248. New York, NY: Wiley-Liss.

Myllyharju J and Kivirikko KI (2001) Collagens and collagen-related diseases. *Annals of Medicine* **33**: 7–21.

Prockop DJ and Kivirikko KI (1995) Collagens: molecular biology, diseases, and potentials for therapy. *Annual Review of Biochemistry* **64**: 403–434.

Sellers JR (2000) Myosins: a diverse superfamily. *Biochimica Biophysica Acta* **1496**: 3–22.

Takahara K, Hoffman GG and Greenspan DS (1995) Complete structural organization of the human α1 (V) collagen gene (*COL5A1*): divergence from the conserved organization of other characterized fibrillar collagen genes. *Genomics* **29**: 588–597.

Yamada Y, Avvedimento VE, Mudryj M, *et al.* (1980) The collagen gene: evidence for its evolutionary assembly by amplification of a DNA segment containing an exon of 54 bp. *Cell* **22**: 887–892.

Zhang X, Zhou J, Reeders ST and Tryggvason K (1996) Structure of the human type IV collagen *COL4A6* gene, which is mutated in Alport syndrome-associated leiomyomatosis. *Genomics* **33**: 473–479.

In: Royce PM and Steinmann B (eds.) *Connective Tissue and its Heritable Disorders: Molecular, Genetic and Medical Aspects*, 2nd edition, pp. 159–221. New York, NY: Wiley-Liss.

Mayne R and Brewton RG (1993) New members of the collagen superfamily. *Current Biology* **5**: 883–890.

Mayne R and Burgeson RE (eds.) (1987) *Structure and Function of Collagen Types*. San Diego, CA: Academic Press.

van der Rest M and Garrone R (1991) Collagen family of proteins. *FASEB Journal* **5**: 2814–2823.

Vuorio E and de Crombrugghe B (1990) The family of collagen genes. *Annual Review of Biochemistry* **59**: 837–872.

Yurchenco PD, Birk De and Mecham RP (eds.) (1994) *Extracellular Matrix Assembly and Function*. San Diego, CA: Academic Press.

Web Links

NCBI: The Human Genome. A guide to online information resources.
http://www.ncbi.nlm.nih.gov/genome/guide/human/
UCSC Genome Bioinformatics Site
http://genome.cse.ucsc.edu/
Genew: Human Gene Nomenclature Database Search Engine. This site can be used to search for further information about the many collagen genes cited in this article. Enter the gene symbol in the search engine. An example is give below for *COL1A1*
http://www.gene.ucl.ac.uk/cgi-bin/nomenclature/searchgenes.pl
Collagen, type I, alpha 1 (*COL1A1*); Locus ID: 1277. LocusLink:
http://www.ncbi.nlm.nih.gov/LocusLink/LocRpt.cgi?l = 1277
Collagen, type I, alpha 1 (*COL1A1*); MIM number: 120150. OMIM:
http://www.ncbi.nlm.nih.gov/htbin-post/Omim/dispmim?120150

Further Reading

Kielty CM and Grant ME (2002) Collagen: the collagen family: structure, assembly, and organization in the extracellular matrix.

Structural Proteomics: Large-scale Studies

Martin Norin, *Biovitrum AB, Stockholm, Sweden*
Michael Sundström, *Biovitrum AB, Stockholm, Sweden*

Intermediate article

Article contents

- Introduction
- Collection of Protein Folds in the Proteome
- Protein Production
- Structure Determination Using NMR
- High-throughput Protein Crystallography
- Structure–Function Annotations Derived Directly from Experimental Structures
- Theoretical Prediction of Protein Function from Structure

Structural proteomics or structural genomics is a common term for systematic efforts to functionally annotate protein molecular structures of whole or selected parts of genomes and/or proteomes.

Introduction

The function of a protein depends on its three-dimensional (3D) structure, which in turn is determined by its amino acid sequence. The ultimate goal of structural proteomics (or genomics) is to provide the structural basis for functional annotations of the proteome. The uses of structural analyses of protein structures, including those based on predicted structures, have expanded as a consequence of the exponential growth of genome research. For example, in the pharmaceutical industry, protein modeling is applied throughout the value-chain from the discovery of target proteins to the generation of lead molecules to the prediction of pharmacological polymorphic effects in clinical trials.

The human genome is believed to encode at least 30 000–40 000 individual proteins, with the highest estimates reaching numbers of 100 000–150 000. The most precise and accurate information on the structure of a particular protein or a protein complex can be obtained from experimental methods, such as X-ray crystallography and nuclear magnetic resonance (NMR). However, the experimental methods require large resources, and many proteins have proved to be very difficult to approach experimentally. The collection of publicly available protein structures in the Research Collaboratory for Structural Bioinformatics (RCSB) protein data bank (see Web Links) consisted at the beginning of year 2002 of approximately 1300 nonredundant structures, but only as few as about 300 were of human origin. This gap of information has prompted large initiatives in structural genomics/ proteomics to experimentally determine and to analyze arrays of protein structures (**Table 1**).

The structural genomics projects all aim at systematically mapping the protein structural space, either targeting specific organisms (e.g. thermophilic bacteria, *Caenorhabditis elegans, Mycobacterium tuberculosis,* etc.), different protein classes (e.g. membrane proteins, metabolic enzymes, kinases, proteases, etc.), targets of specific diseases or biological function relevance, or targeting proteins that have the potential of providing examples of novel structure folds (note that novel experimental protein structures provide templates for structure predictions of homologous proteins). However, the technological challenges are common to any of these strategies. The key limiting factors are difficulties obtaining pure soluble protein material, growing protein crystals, the manual intervention and time required for X-ray crystallographic data collection and evaluation, and the time required for data collection and spectral interpretation using NMR approaches.

Specific recent technology developments include high-throughput (HT) parallel cloning and multivariate approaches for expression and purification, core domain identification using proteolysis methods and the use of expression and detection tags. Protein crystallography has undergone a dramatic series of improvements: freezing of crystals at liquid-nitrogen temperature (cryofreezing), multiple-wavelength anomalous dispersion (MAD) phasing, robotization, automated data collection and the use of synchrotron beamlines have been adopted as standard methodologies. The improvements in structure determination by biomolecular NMR using isotope-enriched protein samples include the use of high-field spectroscopy instrumentation, cryogenic probes as well as automated spectra assignment and structure determi-

nation. **Figure 1** summarizes the experimental flow in structural proteomics, the current bottlenecks and technology developments.

Collection of Protein Folds in the Proteome

Structural proteomics aims to generate useful 3D structures of entire proteomes by a combination of experimental structure determination and modeling. Here, the key question is obvious: how many structural templates are needed to model most proteins, or their domains?

Until recently, the number of true novel structures having novel folds added each year had been low. In 2001, approximately 3300 structures were deposited but less than 10% contained novel folds. The limited number of available 3D protein structure templates restricts attempts to make large-scale functional annotation using structural information. Depending on the organism, 20–70% of the proteome can be modeled from homologous templates at the 30% amino-acid sequence identity threshold level (reviewed by Sánches *et al.*, 2000).

In a recent estimate, Vitkup *et al.* (2001) suggested that 16 000 novel structures need to be experimentally determined to model 90% of the global proteome, at >30% amino-acid sequence identity, given that the selected novel structures are optimally coordinated. The remaining 10% would require substantial additional investments as it contains a high degree of singletons, that is, sequences without relatives, or families with few members. Here it is important to remark that very few structures of membrane-bound proteins have been experimentally determined. Rough models of the G-protein-coupled receptor family (7TM proteins) can be obtained from the related structure of the rhodopsin protein. Thus, one of the biggest challenges in structural biology is to successfully approach membrane proteins.

Protein Production

The success of HT structure determination and subsequent structural analysis is totally dependent on high-throughput protein production. Other critical factors involve the availability of methods for rapid and accurate analysis of purity, homogeneity and structural integrity.

For a research effort in structural proteomics one can pick the 'winners', that is, target proteins that with minimum amount of effort are easy to express with the appropriate characteristics and give good quality NMR spectra or form diffracting crystals. Thus, in

Table 1 Structural proteomics intitatives

Name and main affiliation	Strategic goals	Website
NIH-funded consortia		
The Berkeley Structural Genomics Center, University of California, Berkeley, USA (S.-H. Kim)	To obtain a near-complete structure of two minimal mycoplasma genomes	www.strgen.org
Center for Eukaryotic Structural Genomics. University of Wisconsin, Madison (J. L. Markley)	Method and technology development. HT structure determination with focus on *Arabidopsis thaliana*	www.uwstructuralgenomics.org
Joint Center for Structural Genomics, The Scripps Research Institute (I. Wilson)	Novel structures from *C. elegans* and human proteins involved in cell signaling	www.jcsg.org
Midwest Center for Structural Genomics, Argonne National Laboratory (A. Joachimiak)	Streamlined and cost-effective processes. Structures of targets of unknown fold and proteins from disease-causing organisms	www.mcsg.anl.gov
New York Structural Genomics Research Consortium, Rockefeller University (S. Burley)	Streamlined processes. Solving hundreds of protein structures from human and model organisms	www.nysgrc.org
Northeast Structural Genomics Consortium, Rutgers University (G. Montelione)	X-ray and NMR methodologies. Targets from model organisms and related human proteins	www.nesg.org
Southeast Collaboratory for Structural Genomics, University of Georgia, Athens (B. C. Wang)	Structures of targets from the human genome and two model organisms (*C. elegans, Pyrococcus furiosus*)	www.secsg.org
Structural Genomics of Pathogenic Protozoa, University of Washington (W. G. J. Hol)	Crystal structures of key proteins from tropical pathogens	www.sgpp.org
TB Structural Genomics Consortium, Los Alamos National Laboratory (T. Terwilliger)	Structure determination and analysis of about 400 proteins from *M. tuberculosis*	www.doe-mbi.ucla.edu/TB
Other academic initiatives		
Ontario Center for Structural Proteomics, University of Toronto, Canada (C. Arrowsmith, A. Edwards)	Genome-scale structural biology. Function from structure. Provides protein samples for various structural research groups worldwide	www.uhnres.utoronto.ca/proteomics
Montreal–Kingston Bacterial Structural Genomics Initiative, NRC Biotechnology Research Institute, Montreal, Canada (M. Cygler)	Bacterial structural genomics with *Escherichia coli* as model system. Proteins of unknown functions and metabolic pathways	euler.bri.nrc.ca/brimsg/bsgi.html
Protein Structure Factory, Berlin, Germany (P. Umbach)	Structural determination of proteins selected on grounds of predicted novelty and/or medical or biological relevance	www.proteinstrukturfabrik.de
RIKEN Structural Genomics Initiative, Yokohama, Japan (S. Yokoyama)	Large-scale structural biology of prokaryotes (replication, repair, transcription, translation) and eukaryotes (cell growth and differentiation genetic systems)	www.rsgi.riken.go.jp
Paris-Sud Yeast Structural Genomics, CNRS Université Paris-Sud, France (H. v. Tilbeurgh, J. Janin)	Initial 3 years are focused on obtaining 100 proteins from 200 yeast ORFs and the resolution of about 20 3D structures	genomics.eu.org
Oxford Protein Production Facility, MRC, Oxford, UK (D. Stuart)	HT production of proteins and protein crystals by automating and miniaturizing. Focus on human proteins and those of human pathogens	www.oppf.ox.ac.uk
North West Structural Genomics Centre, Daresbury Laboratory and University of Manchester (S. S. Hasnain and J. R. Heliwell)	Structure determination of targets from pathogens, especially regarding host–pathogen interactions. Specific focus on *M. tuberculosis*	www.nwsgc.ac.uk
Marseilles Structural Genomics Programs. AFMB, CNRS, Marseille, France (C. Cambillau)	Disease targets from human (cancer and central nervous system) and bacterial/viral pathogens. Structure determination of *E. coli* proteins of unknown function	afmb.cnrs-mrs.fr/stgen

Table 1 Continued

Name and main affiliation	Strategic goals	Website
Selected companies		
Structural GenomiX, San Diego	Disease targets from prokaryotic pathogens	www.stromix.com
SYRRX, San Diego	Technology and methodology development. Human disease targets	www.syrrx.com
Affinium Pharmaceuticals, Toronto, Canada	Technology and methodology development. Disease targets from prokaryotic pathogens	www.affinium.com

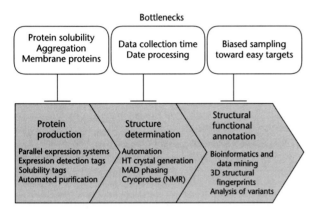

Figure 1 Experimental flow in structural proteomics, the current bottlenecks and important technology developments.

the initial phase of structural proteomics, one should probably streamline the expression efforts and use one or a few constructs us most proteins that will be structurally determined are likely to be of significant value. Obviously, the risk with this approach is that certain folds could become overrepresented in time and that other target types will not appear until a directed effort is attempted (e.g. causing a biased sampling of the structural space).

HT approaches, by necessity, most often utilize affinity and detection tags to allow rapid protein purification. Smaller tags, such as histidine clusters, can often be kept throughout the structure determination while larger tags need to be removed before NMR and X-ray studies. Structural studies by NMR require the tag to be small and to not interfere with the target protein. An example of such an approach was the identification of a solubility enhancement tag (SET) from the protein GB1 domain (Zhou *et al.*, 2001). In the test cases reported, the SET tag improved the characteristics of the expressed proteins in terms of solubility and stability and did not interact with the target proteins.

Regardless of the choice of fusion partners, either smaller tags or larger proteins such as green fluorescent protein (GFP) will to a certain extent give misleading data by solubilizing poorly behaving expression constructs or protein components lacking their natural interaction partner. Thus, 'blind' optimization for the best fusion tag using solubility screens needs to be accompanied by functional assays to assure that the constructs chosen for further studies are biologically relevant.

Structure Determination Using NMR

Solution state NMR is likely to play a key role in structural proteomics projects. Although current methodologies do not allow NMR to be applied for structural investigations on full proteomes, one can estimate that around 25% of the proteome of yeasts (likely to be somewhat lower for higher organisms) will fall within the size limit for NMR studies (around 300 residues for a monomeric protein). Because such a fraction of any given proteome constitutes a wealth of novel targets for structural studies for the foreseeable future, a potentially more serious limitation for NMR as a key structural proteomics methodology arises from the requirements for the sample's qualities, which include very high solubility in aqueous solutions and stability in solution over extended time periods during data collection.

HT structure characterization for selecting target proteins or constructs is likely to be a research area where NMR could make significant contributions as a complement to more traditional characterization methods. Recently, approaches toward the automated structure elucidation from NMR spectra have been published, such as the NOESY-Jigsaw, in which sparse and unassigned NMR data can be used to assess secondary structure reasonably accurately and to align it. The information thus retrieved could be useful in quick structural assays to assess folds prior to full structure determination and analysis as a complement to fold prediction approaches (Bailey-Kellog *et al.*, 2000).

High-throughput Protein Crystallography

HT crystallography has been facilitated by improved phasing and model-building methods, decreased sample requirements through miniaturization and also robotization and automation from the crystallization stage to structure determination. Companies such as SYRRX and Structural GenomiX, as well as academic initiatives such as the Protein Structure Factory (**Table 1**), are developing HT capabilities for protein expression, crystallization and image analysis for automatic crystal detection and structure determination. A late-stage rate-limiting step is the manual intervention needed for crystal mounting and alignment in the X-ray beam. Muchmore *et al.* (2000) reported the setup of a system to allow automatic mounting, optical crystal alignment and data collection with the same range of accuracy as the manual operations, allowing uninterrupted data collection using conventional X-ray systems.

Important technology developments that address effective methods to solve phase information in X-ray data have recently advanced the field. One of these developments utilizes seleno-methionine (Se-Met) labeled protein samples and data collection using single or multiwavelength anomalous diffraction (SAD, MAD) to retrieve phase information (Rice *et al.*, 2000). Yet another promising approach utilizes halides, such as bromine, bound to protein crystals. Dauter *et al.* (2001) reported the crystal structure of a pepstatin-insensitive carboxyl proteinase devoid of structural template that was solved using phase information derived from the bromine peak absorption. The phase information was compared to traditional MAD and found to give similar structural quality, indicating that halide phasing may be a complementary approach for structure determination.

Structure–Function Annotations Derived Directly from Experimental Structures

Proteins sharing the same folding may have quite different functions. Studies by Wilson *et al.* (2000) and Todd *et al.* (2001) conclude that precise function seems to be conserved down to 40% sequence identity, whereas a broader definition of a functional class is conserved down to 25–30% identity. In a limited but significant number of cases, direct electron density for 'native' ligands or cofactors bound to the protein could be observed in structures derived from X-ray crystallography. When such data are available at high resolution, hypothesis generation on the function of the protein often can be relatively straightforward.

Databases and database mining tools to store, organize and identify protein folds are becoming more and more important as the number of protein structures grows. A novel structure that has been determined may be scanned against databases of known structures. Such resources include DALI and CATH (see Web Links).

A good example of direct functional annotation from structure was reported by Zarembinski *et al.* (1998). In this study, the crystal structure of an unannotated protein, MJ0577, from *Methanococcus jannaschii* clearly revealed a bound ATP in the 1.7 Å electron density maps, suggesting that MJ0577 was an ATPase or an ATP-mediated molecular switch. The structure-based hypothesis could subsequently be confirmed by biochemical experiments. In addition, the structural analysis of the ATP binding motif could be used to suggest other putative ATP binding sequences among the many homologous, but previously unannotated, proteins in this family.

Although a few studies on structure-based assignment of single proteins from experimental structures have emerged, the structural proteomics effort on the archaeon *Methanobacterium thermoautotrophicum* represents the best published case study (Christendat *et al.*, 2000). Here, 424 out of 900 target proteins, predicted to be soluble and without a template in the Protein Data Bank, were chosen for structure determination and subsequent functional assignment. The selected proteins represented around 25% of the organism's proteome (1871 open reading frames). The targets were cloned, expressed and purified in a streamlined approach and attempts were made to solve the structures by both NMR (< 20 kDa) and crystallographic methods at various laboratories. Approximately 20% of the target proteins were found to be suitable candidates for structure determination.

Furthermore, the study revealed that poor expression and solubility of the proteins accounted for close to 60% of the failures. It was also observed that NMR data collection and crystallization were the two major time and resource consumers in the process. Ten structures (including MTH538 discussed above) by NMR and X-ray were simultaneously published. Five of the ten structures contained a bound ligand or a ligand-binding site that could be inferred from structural homology. Thus, many of the structures suggested a number of functional assays that could be used to provide insights of function.

Theoretical Prediction of Protein Function from Structure

The PROSITE database (Hofmann *et al.*, 1999) contains sequence patterns of functional sites, but have no 3D information. Thornton and colleagues (Kasuya and Thornton, 1999) have analyzed the 3D features of the PROSITE patterns in order to define structural templates for structure-based mining for functional sites. In another example, 3D descriptors based on intramolecular distances between amino acid residues were able to identify a potential oxidoreductase active site in the serine/threonine phospatase-1 subfamily (Fetrow et al., 1999).

In a recent interesting example by Sowa et al. (2001), an important site for the regulation of G protein signaling was identified and confirmed through site-specific mutagenesis. They applied a 'evolutionary trace' (ET) method by combining structural and sequence data. The method detects 3D clusters of amino acid residues that are specific for a related subset of sequences within a protein family. This method is especially interesting as it is able to propose novel sites without any prior knowledge of the specific amino acid features.

See also

Industrialization of Proteomics: Scaling-up Proteomics Processes
Macromolecular Structure Determination: Comparison of Crystallography and Nuclear Magnetic Resonance (NMR)
Mass Spectrometry in Protein Characterization
Molecular Entry Point: Strategies in Proteomics
Protein Characterization in Proteomics

References

Bailey-Kellog C, Widge A, Kelley JJ, et al. (2000) The NOESY Jigsaw: automated protein secondary structure and main-chain assignment from sparse, unassigned data. Journal of Computational Biology 7: 537–558.

Christendat D, Yee A, Dharamsi A, et al. (2000) Structural proteomics of an archaeon. Nature Structural Biology 7: 903–909.

Dauter Z, Li M and Wlodawer A (2001) Practical experience with the use of halides for phasing macromolecular structures: a powerful tool for structural genomics. Acta Crystallographica D 57: 239–249.

Fetrow J, Siew N and Skolnick J (1999) Structure-based functional motif identifies a potential disulphide oxidoreductase active site in the serine/threonine protein phosphatase-1 subfamily. FASEB Journal 13: 1866–1874.

Hofmann K, Bucher P, Falquet L and Bairoch A (1999) The PROSITE database, its status in 1999. Nucleic Acids Research 27: 215–219.

Kasuya A and Thornton J (1999) Three-dimensional structure analysis of PROSITE patterns. Journal of Molecular Biology 286: 1673–1691.

Muchmore SW, Olson J, Jones R, et al. (2000) Automated crystal mounting and data collection for protein crystallography. Structure 8: 243–246.

Rice LM, Earnest TN and Brunger AT (2000) Single-wavelength anomalous diffraction phasing revisited. Acta Crystallography D 56: 1413–1420.

Sánchez R, Pieper U, Melo F, et al. (2000) Protein structure modeling for structural genomics. Nature Structural Biology 7(supplement): 986–990.

Sowa ME, He W, Slep KC, et al. (2001) Prediction and confirmation of a site critical for effector regulation of RGS domain activity. Nature Structural Biology 8: 234–237.

Todd A, Orengo CA and Thornton JM (2001) Evolution of function in protein superfamilies, from a structural perspective. Journal of Molecular Biology 307: 1113–1143.

Vitkup D, Melamud E, Moult J and Sander C (2001) Completeness in structural genomics. Nature Structural Biology 8: 559–566.

Wilson CA, Kreychman J and Gerstein M (2000) Assessing annotation transfer for genomics: quantifying the relations between protein sequence, structure and function through traditional and probabilistic scores. Journal of Molecular Biology 297: 233–249.

Zarembinski TI, Hung LW, Mueller-Dieckmann HJ, et al. (1998) Structure-based assignment of the biochemical function of a hypothetical protein: a test case of structural genomics. Proceedings of the National Academy of Sciences of the United States of America. 95: 15 189–15 193.

Zhou P, Lugovskoy AA and Wagner G, et al. (2001) A solubility-enhancement tag (SET) for NMR studies of poorly behaving proteins. Journal of Biomolecular Nuclear Magnetic Resonance 20: 11–14.

Further Reading

Baker D and Sali A (2001) Protein structure prediction and structural genomics. Science 7: 93–96.

Nature Structural Biology (2000) Structural genomics supplement. Nature Structural Biology 7: 927–994. [whole issue].

Norin M and Sundström M (2002) Structural proteomics: developments in structure-to-function predictions. Trends in Biotechnology 20: 79–84.

Stevens RC, Yokoyama S and Wilson IA (2001) Global efforts in structural genomics. Science 7: 89–92.

Yee A, Chang X, Pineda-Lucena A, et al. (2002) An NMR approach to structural proteomics. Proceedings of the National Academy of Sciences of the United States of America 99: 1825–1830.

Web Links

CATH – Protein Structure Classification. CATH is a novel hierarchical classification of protein domain structures, which clusters proteins at four major levels: Class (C)
http://www.biochem.ucl.ac.uk/bsm/cath_new/cath_info.html# C_Level,
Architecture (A)
http://www.biochem.ucl.ac.uk/bsm/ cath_new/cath_info.html#A_Level,
Topology (T)
http://www.biochem.ucl.ac.uk/bsm/cath_new/cath_info.html#-T_Level
Homologous superfamily (H)
http://www.biochem.ucl.ac.uk/bsm/cath_new/cath_in-fo.html#H_Level

DALI. The Dali server is a network service for comparing protein structures in 3D
http://www2.ebi.ac.uk/dali/

The Protein Databank (PDB). A worldwide repository for the processing and distribution of 3-D biological macromolecular structure data
http://www.rcsb.org/

STS

See Sequence-tagged Site (STS)

Substitution Matrices

Stephen F Altschul, *National Center for Biotechnology Information, Bethesda, Maryland, USA*

A substitution matrix is a collection of scores for aligning nucleotides or amino acids with one another. These scores generally represent the relative ease with which one nucleotide or amino acid may mutate into or substitute for another, and they are used to measure similarity in sequence alignments.

Introduction

The relationship between a pair of deoxyribonucleic acid (DNA) or protein sequences is often represented by an *alignment*. A *global alignment* places into correspondence the whole of two sequences, while a *local alignment* does so only for a contiguous segment from each. Alignments are generally assigned scores, both to choose among the many possible ways of aligning a given pair of sequences, and to compare alignments of different pairs of sequences. Higher scores are usually considered superior, and the highest scoring alignment for a given pair of sequences is called an optimal alignment. The score of this alignment is often taken as a measure of the sequences' *similarity*. (*See* Similarity Search.)

An alignment's score is usually defined as the sum of substitution scores for the pairs of nucleotides or amino acids placed into correspondence by the alignment, and gap scores for each stretch of residues in one sequence aligned with null characters inserted into the other. A *substitution matrix* is simply the collection of scores specified for aligning all possible pairs of residues.

DNA and protein sequences frequently share only isolated regions of similarity, so the most widely used *similarity search* programs (e.g. Smith–Waterman, FASTA and basic local alignment search tool (BLAST)) seek optimal local alignments. Conveniently, much more is known about the properties of substitution matrices when used for local rather than global alignment, and we will consider such matrices primarily in the local alignment context. (*See* BLAST Algorithm; FASTA Algorithm; Similarity Search; Smith–Waterman Algorithm.)

Match/Mismatch Matrices

The earliest used and simplest type of substitution score form a 'match/mismatch' matrix, which assigns a fixed positive score to all matching pairs of nucleotides or amino acids, and a fixed negative score to all mismatching pairs. This sort of matrix is still frequently used for DNA sequence comparison, but it has a major disadvantage in the protein alignment context. Specifically, it ignores the fact that amino acids with similar physical/chemical properties, such as the hydrophobic isoleucine and valine, the negatively charged aspartic and glutamic acid, and the aromatic tyrosine and phenylalanine, are much more likely to substitute for one another over evolutionary time than are amino acids with dissimilar properties. Accordingly, matrices that assign better scores to similar than to dissimilar pairs of amino acids generally outperform match/mismatch matrices in distinguishing biological relationships from chance similarities.

Log-odds Substitution Matrices

A detailed analytic theory exists describing the behavior of substitution matrices in the context of local alignments that may not contain gaps (Karlin and Altschul, 1990; Altschul, 1991). This fairly restricted case, however, sheds much light on the general properties of scoring matrices. For proteins, the theory assumes a simplified 'random protein' model, in which the 20 amino acids occur independently at each protein position, with a 'background probability' p_i for amino acid i. Then, if biologically accurate alignments can be characterized by the 'target frequencies' q_{ij} with which they align amino acids i and j, the substitution matrix that best distinguishes real from chance sequence similarities can be shown to have scores s_{ij}, given by

$$s_{ij} = \log_b \frac{q_{ij}}{p_i p_j} = \left(\ln \frac{q_{ij}}{p_i p_j} \right) \Big/ \lambda \qquad (1)$$

The base b of the logarithm, or the parameter λ, simply establishes an arbitrary scale for the scoring system.

One may define substitution matrices using a broad range of rationales, but if one seeks optimal local alignments of unspecified length, it always makes sense for the expected score $\Sigma_{ij} p_i p_j s_{ij}$ for aligning two random amino acids to be negative. Were this not the case, long local alignments of unrelated proteins would tend to outscore short alignments representing true biologic relationships. When the expected score of a substitution matrix is negative, the matrix may always be written in the form of eqn [1], with a unique set of target frequencies and a unique scale parameter (Karlin and Altschul, 1990; Altschul, 1991). Thus, *every* sensible local alignment substitution matrix is implicitly a log-odds matrix, optimized for detecting ungapped alignments characterized by a specific set of target frequencies q_{ij}. Given this, most present-day attempts to construct substitution matrices aim at estimating appropriate target frequencies, from which the scores may be determined, up to an arbitrary scale, using eqn [1].

Point accepted mutation matrices for protein sequence comparison

The earliest explicit construction of substitution matrices using the log-odds formalism of eqn [1] was by Margaret Dayhoff and co-workers. She observed, however, that the target frequencies characterizing true alignments cannot be represented by a *unique* set of q_{ij}. The reason, most simply, is that biologically accurate alignments of closely related proteins primarily align identical residues, implying one set of q_{ij}, while equally accurate alignments of proteins that have diverged substantially imply a different set of q_{ij}.

By studying many families of closely related proteins, Dayhoff approximated the relative rates at which the various amino acids tend to replace one another over evolutionary time. She used these data to construct a stochastic model of protein evolution, from which she could calculate the q_{ij}, and thus an attendant substitution matrix, for any specified degree of evolutionary change (Dayhoff *et al.*, 1978; Schwartz and Dayhoff, 1978). In Dayhoff's model, 1 point accepted mutation (PAM) of evolution represents the substitution, on average, of 1% of all amino acids by another. Because the amino acids in some protein positions may remain fixed, while those in other positions may change multiple times, perhaps even returning to the original residue, an alignment of two proteins diverged by 100 PAMs of evolution using Dayhoff's model still has, on average, identical residues at over 43% of all positions.

For the global alignments Dayhoff studied, and the relatively small database of proteins available at the time, quite distant similarities could be distinguished from chance, and Dayhoff accordingly advocated the use of the PAM-250 substitution matrix, optimized for alignments with approximately 20% identical residues.

For current protein database similarity searches, however, this matrix focuses on relationships that are in general too weak to be distinguished from chance, and the PAM-180 matrix, tailored to alignments that are approximately 27% identical, is usually more appropriate. There is no uniform mapping between PAM distances and evolutionary time because different protein families tend to accumulate changes at greatly varying rates.

Since the original PAM model, the data upon which it was based have ballooned (Jones *et al.*, 1992), and the mathematical details of the model itself have been criticized and refined (Muller and Vingron, 2000). These developments have led to the construction of newer PAM-like matrices that may be better for detecting distant protein relationships.

BLOSUM matrices for protein sequence comparison

A more direct approach to estimating target frequencies has been taken by Henikoff and Henikoff (1992). They built a database of 'blocks', which consist of aligned protein segments that are relatively well conserved across whole protein families. The q_{ij} used to construct the BLOSUM (for 'block-sum') substitution matrices are derived simply by counting all aligned pairs of residues implied by the blocks database. BLOSUM matrices optimal for differing degrees of evolutionary divergence are constructed by clustering segments within blocks that are more than a given percentage identical and counting aligned pairs of residues only between but not within clusters. The BLOSUM-62 matrix is thus derived by first clustering all segments that are more than 62% identical. As the number associated with a BLOSUM matrix grows, a greater proportion of closely related protein segments is used in deriving the q_{ij}. Thus, somewhat confusingly, higher-numbered BLOSUM and lower-numbered PAM matrices are tailored for closely related sequences. Various tests have shown BLOSUM matrices generally to be more effective than the original PAM matrices at distinguishing true biological relationships from chance similarities (Henikoff and Henikoff, 1993; Pearson, 1995).

Gap Costs

The theory described above ignores the issue of gaps within local alignments. However, most local alignment algorithms use a substitution matrix in conjunction with gap scores to define optimal alignments. It is not known whether or how substitution matrices should be altered when they are used in the gapped alignment context, nor is there a good theory for selecting optimal gap costs. Therefore, the usual practice has been to employ substitution matrices unaltered when they are used with gap costs, and to choose optimal gap costs essentially by trial and error (Pearson, 1995).

Hidden Markov models, an alternative formalism for sequence comparison, do not seek the 'optimal alignment' of two sequences but instead integrate over all possible alignments. The analogs of substitution matrices and gap scores in this formalism are unified in a single probabilistic framework. (*See* Hidden Markov Models.)

DNA Alignment Matrices

DNA mutation models similar to the PAM model of protein evolution can be constructed. The simplest yield match/mismatch substitution matrices, but more sophisticated evolutionary models, yield differing mismatch scores for transitions and transversions (States *et al.*, 1991). It is also possible to base DNA substitution matrices directly upon alignment data. As with proteins, the optimal scoring system depends strongly upon the degree of conservation in the alignments sought (States *et al.*, 1991). Thus, one matrix would be appropriate for comparing human expressed sequence tag and genomic data, where almost all mismatches in the alignments of interest are due to rare sequencing errors, but quite a different matrix for comparing human and mouse genomic data in a search for homologous sequences.

In general, owing largely to the redundancy of the genetic code, homologies between protein-coding DNA regions are much easier to detect at the protein than at the DNA level (States *et al.*, 1991). Thus, when studying such regions, it often pays to translate them conceptually to protein, in multiple reading frames if the correct frame or strand is unknown, and compare the resulting sequences with other protein or DNA sequences using an amino acid substitution matrix.

Substitution Matrices for Global Alignment

Very little is known about the statistics of global as opposed to local sequence alignment, and theory describing optimal substitution matrices and gap costs for global alignment is correspondingly poorly developed. In practice, optimal global alignments are often constructed employing the same substitution matrices and gap costs used for local alignments. The only difference is that the alignments returned are required to involve the complete sequences, as opposed to only segments from them. In this context, it is always possible to add a constant $2a$ to each substitution score, and ka to the score associated with a gap involving k letters, without altering the relative scores of any alignments. Thus, global alignment substitution matrices and their associated gap scores may be rendered uniformly nonnegative without in any way changing their essence. Negative expected substitution scores are required in the local alignment context because the scores there determine which segments will be aligned, as well as the optimal alignment of these segments.

See also

BLAST Algorithm
FASTA Algorithm
Global Alignment
Sequence Similarity
Similarity Search

References

Altschul SF (1991) Amino acid substitution matrices from an information theoretic perspective. *Journal of Molecular Biology* **219**: 555–565.

Dayhoff MO, Schwartz RM and Orcutt BC (1978) A model of evolutionary change in proteins. In: Dayhoff MO (ed.) *Atlas of Protein Sequence and Structure*, vol. 5, supplement 3, pp. 345–352. Washington DC: National Biomedical Research Foundation.

Henikoff S and Henikoff JG (1992) Amino acid substitution matrices from protein blocks. *Proceedings of the National Academy of Sciences of the United States of America* **89**: 10915–10919.

Henikoff S and Henikoff JG (1993) Performance evaluation of amino acid substitution matrices. *Proteins* **17**: 49–61.

Jones DT, Taylor WR and Thornton JM (1992) The rapid generation of mutation data matrices from protein sequences. *Computer Applications in the Biosciences* **8**: 275–282.

Karlin S and Altschul SF (1990) Methods for assessing the statistical significance of molecular sequence features by using general scoring schemes. *Proceedings of the National Academy of Sciences of the United States of America* **87**: 2264–2268.

Muller T and Vingron M (2000) Modeling amino acid replacement. *Journal of Computational Biology* **7**: 761–776.

Pearson WR (1995) Comparison of methods for searching protein sequence databases. *Protein Science* **4**: 1145–1160.

Schwartz RM and Dayhoff MO (1978) Matrices for detecting distant relationships. In: Dayhoff MO (ed.) *Atlas of Protein Sequence and Structure*, vol. 5, supplement 3, pp. 353–358. Washington DC: National Biomedical Research Foundation.

States DJ, Gish W and Altschul SF (1991) Improved sensitivity of nucleic acid database searches using application-specific scoring matrices. *Methods* **3**: 66–70.

Further Reading

Altschul SF (1993) A protein alignment scoring system sensitive at all evolutionary distances. *Journal of Molecular Evolution* **36**: 290–300.

Chiaromonte F, Yap VB and Miller W (2002) Scoring pairwise genomic sequence alignments. In: Altman RB, Dunker AK, Hunter L, Lauderdale K and Klein TE (eds.) *Pacific Symposium on Biocomputing*. pp. 115–126. Singapore: World Scientific Publishing.

Claverie J-M (1993) Detecting frameshifts by amino acid sequence comparison. *Journal of Molecular Biology* **234**: 1140–1157.

Dembo A, Karlin S and Zeitouni O (1994) Limit distribution of maximal non-aligned two-sequence segmental score. *Annals of Probability* **22**: 2022–2039.

Durbin R, Eddy S, Krogh A and Mitchison G (1998) *Biological Sequence Analysis. Probabilistic Models of Proteins and Nucleic Acids.* Cambridge, UK: Cambridge University Press.

Ewens WJ and Grant GR (2001) *Statistical Methods in Bioinformatics.* New York, NY: Springer-Verlag.

Gonnet GH, Cohen MA and Benner SA (1992) Exhaustive matching of the entire protein sequence database. *Science* **256**: 1443–1445.

Henikoff S and Henikoff JG (2000) Amino acid substitution matrices. *Advances in Protein Chemistry* **54**: 73–97.

Kann M, Qian B and Goldstein RA (2000) Optimization of a new score function for the detection of remote homologs. *Proteins* **41**: 498–503.

Vingron M and Waterman MS (1994) Sequence alignment and penalty choice. Review of concepts, case studies and implications. *Journal of Molecular Biology* **235**: 1–12.

Superfamilies

See Gene Families: Multigene Families and Superfamilies

Surface Plasmon Resonance

Matthew J Fivash, *National Cancer Institute, Frederick, Maryland, USA*

Surface plasmon resonance is an optical resonance effect where the back side of a thin conductive mirror affects the angle at which there is a minimum of reflected light.

Advanced article

Article contents
• Introduction
• Outline of Methods
• Applications
• Future Developments

Introduction

Surface plasmon resonance (SPR) is an optical resonance effect. The well-studied Kretschmann configuration for exciting surface plasmons (the plasmon is to a plasma wave what the photon is to a light wave) is shown in **Figure 1** (Welford, 1991). The SPR angle is partly determined by the dielectric constant in the region adjacent to the back side of the thin mirror (about 50 nm for 760 nm light). Modifying the dielectric constant, for example, by adsorbing molecular material in this region, produces changes in the SPR angle, making this effect useful for studying biomolecular interactions.

Most biological applications use SPR with a fluidics system. In this way both static and dynamic binding reactions may be designed and followed in real time. For example, the Biacore® (Biacore AB, Uppsala, Sweden) instrument uses this approach (Karlsson and Fält, 1997). Binding reactions between the surface and the analyte have been studied for many years using the Langmuir model (Langmuir, 1918). This model describes a binary reaction where one analyte molecule interacts with one analyte molecule in one way. Signals from this type of instrument may show kinetic pathways that are more complex than the simple Langmuir model would suggest. Understanding and modeling these signals to illuminate the dynamics of binding reactions is an active area of research.

Figure 1 Some of the important components of a surface plasmon resonance (SPR) detector system. The stippled area on the back side of the thin mirror is the active area where changes in the dielectric constant change the angle of the dark band in the reflected light. The light source (monochromatic) is directed through a prism to a detector (here represented by an eye), where changes in the dielectric constant are recorded. When proteins are used to change the dielectric constant in the active area, the change in SPR angle is highly correlated with the protein's molecular weight.

Outline of Methods

An SPR instrument records dielectric changes on the back of a mirror. Over small ranges (up to about 3°) the change of the SPR angle is essentially linear with the dielectric constant, and is highly correlated with molecular weight. Signals from these systems, then, show directly the amount of binding in the reaction area as an analyte binds to a ligand.

Currently, most studies of biological molecules using SPR detection are made with the Biacore instrument. To affect the local dielectric constant, this instrument attaches one molecule to the surface (with one of several chemistries) and delivers the second biomolecule through a flow system. The additional mass of the binding molecules shifts the dielectric constant and thus the SPR angle. These signals are recorded in real time, and follow the progress of the binding reaction. This instrument can pump simple flow buffer, producing a signal that relaxes to a baseline as the complexes dissociate. A typical signal from a Biacore instrument is shown in **Figure 2**.

Applications

Most biological applications of SPR are studies of antigen–antibody and receptor–ligand reactions. Significant work is also done in studying protein–nucleic acid interactions, protein–small molecule and cell–cell adhesion studies. In addition, there are an increasing number of other uses, including analyses pairing an SPR instrument and a mass spectro-scopy instrument, using SPR for thermodynamic measurements, and monitoring purification of ligands. The surfaces used in the Biacore instrument may be built to mimic a cell membrane and so to observe membrane interactions.

Future Developments

The number of biological uses of SPR is increasing both in bimolecular and multimolecular reactions. Examining the kinetic pathway can shed light on molecular dynamics. SPR experiments require careful design and thorough mathematical modeling. This technique is currently providing unexpected insights into biological machinery, and has the potential to describe much of the basic kinetic action of biomolecular interactions.

Acknowledgement

This project has been funded in whole or in part by Federal funds from the US National Cancer Institute, DHHS, under contract with #NO1-CO-46002. The content of this publication does not necessarily reflect the views or policies of the US Department of Health and Human Services, nor does the mention of trade names, commercial products or organization imply endorsement by the US Government.

References

Karlsson R and Fält A (1997) Experimental design for kinetic analysis of protein–protein interactions with surface plasmon resonance biosensors. *Journal of Immunological Methods* **200**: 121–133.

Langmuir I (1918) The adsorption of gases on plane surfaces of glass, mica and platinum. *Journal of the American Chemical Society* **40**: 1361.

Welford K (1991) Surface plasmon-polaritons and their uses. *Optical and Quantum Electronics* **23**: 1–27.

Further Reading

Fisher RJ, Fivash M, Casas-Finet J, *et al*. (1994) Real-time BIAcore measurements of *Escherichia coli* single-stranded DNA binding (SSB) protein to polydeoxythymidylic acid reveal single-state kinetics with steric cooperativity. *Methods: A Companion to Methods in Enzymology* **6**: 121–133. [This paper details methods for developing kinetic models of more complex binding systems.]

Fivash M, Towler EM and Fisher RJ (1998) BIAcore for macromolecular interaction. *Current Opinion in Biotechnology* **9**(1): 97–101. [This review details the growth and use of SPR technology.]

Myszka DG, Morton TA, Doyle ML and Chaiken IM (1997) Kinetic analysis of protein antigen–antibody interaction limited by mass transport on an optical biosensor. *Biophysical Chemistry* **64**: 127–137. [This paper examines the issues of transport with flow systems like the Biacore instrument, and shows that a simple model for this effect is usually adequate.]

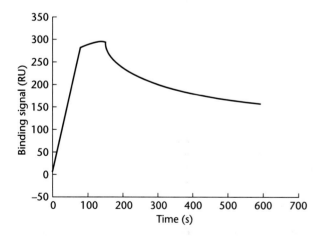

Figure 2 Binding signal (here ETS-1 DNA-binding domain and a peptide, R105V) showing initial binding followed by a stochastic steady state (loosely, an equilibrium) from a Biacore instrument. When the flow of peptide/analyte is replaced with a flow-on buffer, this reversible binding reaction relaxes toward a zero bound concentration. The signal is measured in response units (RU). Each RU represents about 1/3000 of a degree.

Surnames and Genetics

Mark A Jobling, *University of Leicester, Leicester, UK*

Heritable surnames contain information about the relatedness of individuals within and between populations and thus can be used to estimate inbreeding and population structure. These estimates are affected adversely by the failure of underlying assumptions: many surnames have several founders, and the link between surname and inheritance can be broken by illegitimacy and other factors.

Introduction

Most human societies use surnames, and in most societies these are heritable. People sharing a surname (whether patrilineal, matrilineal or clan-based) are likely to be more closely related than are people with different surnames, and thus the distribution of surnames might reflect the degree of inbreeding in a population. This departure from random mating (or panmixia) is an important property that is referred to as population structure. This principle was realized over 100 years ago by George Darwin (the son of Charles Darwin), who used the proportion of same-surname marriages to estimate the frequency of first-cousin marriages in different social spheres (Darwin, 1875). (*See* Darwin, Charles.)

Isonymy, Degrees of Relationship and Inbreeding

Extension of this idea led to the development of methods (Crow and Mange, 1965) to calculate the coefficient of inbreeding (for a population, the average probability that an individual inherits two copies of a gene from a common ancestor) from surname data. Confusingly, the term 'isonymy' (which simply means having the same surname) is used often in this field to refer to any one of several different and specific measures, some of which are described below. (*See* Inbreeding.)

Crow and Mange (1965) noticed that, for most relationships, an isonymous marriage indicates the same amount of inbreeding in a population regardless of how closely related the marriage partners are. For example, a brother and sister always share a surname, and will produce offspring with an inbreeding

coefficient of one-quarter. First cousins, on the other hand, will have a probability of sharing a surname of one-quarter, and their children will have an inbreeding coefficient of one-sixteenth. In both cases, the probability of surname sharing divided by the inbreeding coefficient gives the same value (four). It can be shown that this is true for most marriage relationships; there are some for which it does not hold but these are across-generation examples, which are unlikely to be common. Thus, the population inbreeding coefficient can be estimated by dividing the marital isonymy (frequency of isonymous marriages) by four. (*See* Kinship and Inbreeding.)

Two different factors contribute to the population inbreeding coefficient. One is the nonrandom contribution owing to mate choice; the second is the random contribution caused by genetic drift in the population. These can be separated by considering the frequencies of surnames. With random marriage, the frequency of isonymous marriages will be the sum of squares of the surname population frequencies, provided that the values in males and females are the same. (*See* Genetic Drift; Nonrandom Mating.)

A difficulty with estimates that are based on marital isonymy is that the number of isonymous marriages is usually small (typically 0–4%), and it is therefore difficult for the figures to reach statistical significance. This difficulty does not apply to measures that are based simply on the population frequencies of surnames, which are also easier to obtain. Here, the proportion of isonymy is given by

$$I_{ii} = \sum_k p_{ik}^2$$

where p_{ik} is the frequency of surname k in population i, and the sum is across all surnames; and the interpopulation isonymy is given by

$$I_{ij} = \sum_k p_{ik}p_{jk}$$

where p_{ik} and p_{jk} are the relative frequencies of surname k in populations i and j, respectively, and the sum is again across all surnames. From the interpopulation measure, a distance measure (Lasker's distance) can be derived as $-\log(I_{ij})$, which can be used in comparisons with geographical distance (see Barrai et al., 2001).

Although patterns of surname inheritance vary between populations, the most common system is patrilineal, in which the surname passes down from father to children. Analogy is often made to a Y-chromosome-linked gene with many alleles (well over a million worldwide), and examining this analogy will illustrate some of the advantages and disadvantages of surname studies.

Advantages of Surname Studies

The advantages of surname studies actually arise from failures of the analogy. First, to assay a million alleles in a large population sample would be enormously costly and complex, whereas accurate surname data on samples of millions of individuals can be obtained simply and cheaply from electoral rolls, from registers of births, marriages and deaths, and from telephone directories, many of which are available on CD-ROM for instant computer-based analysis. Notably, data are often available on past as well as present populations. Second, the frequencies of alleles of many genes are influenced by natural selection, whereas surname frequencies can be reasonably regarded as selectively neutral and thus understandable in terms of drift, mutation and migration alone. (See Fitness and Selection.)

A further departure from the genetic analogy is to do with time depth. Whereas gene genealogies coalesce over hundreds or thousands of generations and reliable pedigree information typically spans 3–5 generations, heritable surnames give information about an intermediate time span. The oldest hereditary surnames are those in China (about 5000 years old) and there are surname systems (such as those in Turkey or Wales) that have become established only in the past century or two; however, most surnames – and certainly those that have been studied most intensively – originated around 500–1000 years ago.

Although this time depth is too shallow to reflect the ancient migrations that established the current populations in their places, it is deep enough to give useful information about a period beyond the reach of most pedigree studies.

Problems of Surname Studies

The failure of genetic analogy is also the cause of the clear disadvantages of surname studies. First, although two individuals sharing alleles of a gene generally share them by descent from a common ancestor (monophyly), many surnames, particularly the common ones that derived from occupations, characteristics or first names, clearly have many independent origins (polyphyly). This is 'identity by state' as opposed to 'identity by descent'. Second, mutation generates new alleles through processes that conform to certain biological rules, but the relationship between patrilineal surnames and lineages can change arbitrarily and abruptly through illegitimacy, child adoption, instances of maternal surname inheritance, adoption of new surnames and fixation of spelling variants.

Critics of isonymy studies (Rogers, 1991) base their objections mostly on the departure from monophyly and breakages in the link between surname and inheritance. When inbreeding is estimated from surnames, and can be assessed independently using pedigree records, isonymy measures tend to over estimate inbreeding – in one example by a factor of 2.2 (Mathias et al., 2000). Inbreeding estimates can also differ substantially when either monophyletic or known polyphyletic surnames are used (Rojas-Alvarado and Garza-Chapa, 1994). There are further limitations: surname sharing is a binary indicator of a continuous variable, that is, the degree of inbreeding – for example, siblings share a surname whether or not they are themselves inbred.

Surnames as Indicators of Population Origins

Aside from the controversy that surrounds the use of isonymy, surnames can be used more simply to give information about the origins of a population or subpopulation. In new populations that arise from admixture of older populations with established surname systems, components with different origins can be identified readily from surname information, and this can be used to examine the relative risk of disorders in the subpopulations. Examples are the identification of Hispanic components in the United States in an epidemiological cancer study (Stewart *et al.*, 1999), and the demonstration that the incidence of noninsulin-dependent diabetes mellitus in Mexicans increases with increasing Amerindian ancestry, as estimated from surnames (Garza-Chapa *et al.*, 2000). (*See* Ethnicity and Disease.)

Testing Assumptions by Y-chromosomal Polymorphisms

Despite the objections to surname studies, publications continue to accrue on this subject: a spectacular example is the examination of the isonymy structure of the United States, which analyzed a sample of more than 18 million telephone users and 899 585 surnames (Barrai *et al.*, 2001). The triviality of data acquisition and ease of analysis clearly continue to hold a strong allure for human biologists, even when the information that emerges might seem to be of dubious value.

For patrilineal surnames at least, developments in molecular genetics now offer ways to test some of the assumptions that underlie surname studies, and either to place them on a firmer footing or to reveal their limitations in starker contrast.

We now have available a wealth of deoxyribonucleic acid (DNA) polymorphisms on the Y chromosome, including binary markers such as base substitutions, and multiallelic markers such as microsatellites. These markers define highly informative haplotypes that can distinguish between most males in a population. Under the assumptions that underlie surname studies, males sharing a patrilineal surname should also share a Y chromosome and this has been tested for one English surname, Sykes (Sykes and Irven, 2000). Out of 48 randomly ascertained Sykes males, 21 shared an identical Y-chromosomal haplotype, albeit at rather low resolution. The remaining Sykes men had diverse haplotypes and were ascribed to illegitimacy after a founding event by a single man. Under these assumptions, and apparently assuming a 'star-like' genealogy for descent from the founder, an average nonpaternity rate of 1.3% per generation was calculated. Although this study is striking, many questions remain. Is the observed pattern also compatible with a greater number of founders, followed by loss or depletion of most of the lineages by drift? Are some haplotypes the result of mutation, rather than illegitimacy? And is the assumption that underlies the

nonpaternity rate calculation reasonable? (*See* Chromosome Y: General and Special Features.)

These questions seem amenable both to investigation by the analysis of more surnames and to modeling experiments. Nonetheless, the apparent low rate of nonpaternity and the coherence of haplotypes in the surname group have offered solace to researchers attached to isonymy studies (Barrai *et al.*, 2001). In addition to this, the new techniques of high-resolution Y-chromosome analysis offer genealogists an alternative DNA-based way to their traditional methods of searching for their roots.

See also
Genetic Isolates and Behavioral Gene Searches
Inheritance and Society
Kinship and Inbreeding
Nonrandom Mating

References

Barrai I, Rodgriguez-Larralde A, Mamolini E, Manni F and Scapoli C (2001) Isonymy structure of USA population. *American Journal of Physical Anthropology* **114**: 109–123.

Crow JF and Mange JF (1965) Measurement of inbreeding from the frequency of marriages between persons of the same surname. *Eugenics Quarterly* **12**: 199–203.

Darwin GH (1875) Marriages between first cousins in England and their effects. *Journal of the Statistical Society* **38**: 153–184.

Garza-Chapa R, Rojas-Alvarado MA and Cerda-Flores RM (2000) Prevalence of NIDDM in Mexicans with paraphyletic and polyphyletic surnames. *American Journal of Human Biology* **12**: 721–728.

Mathias RA, Bickel CA, Beaty TH, *et al.* (2000) A study of contemporary levels and temporal trends in inbreeding in the Tangier Island, Virginia, population using pedigree data and isonymy. *American Journal of Physical Anthropology* **112**: 29–38.

Rogers AR (1991) Doubts about isonymy. *Human Biology* **63**: 663–668.

Rojas-Alvarado MA and Garza-Chapa R (1994) Relationships by isonymy between persons with monophyletic and polyphyletic surnames from the Monterrey metropolitan area, Mexico. *Human Biology* **66**: 1021–1036.

Stewart SL, Swallen KC, Glaser SL, Horn-Ross PL and West DW (1999) Comparison of methods for classifying Hispanic ethnicity in a population-based cancer registry. *American Journal of Epidemiology* **149**: 1063–1071.

Sykes B and Irven C (2000) Surnames and the Y chromosome. *American Journal of Human Genetics* **66**: 1417–1419.

Further Reading

Jobling MA (2001) In the name of the father: surnames and genetics. *Trends in Genetics* **17**: 353–357.

Lasker GW (1985) *Surnames and Genetic Structure*. Cambridge, UK: Cambridge University Press.

Relethford JH (1988) Estimation of kinship and genetic distance from surnames. *Human Biology* **60**: 475–492.

Susceptibility Genes: Detection

Nicola J White, *GlaxoSmithKline, Stevenage, UK*

John H Riley, *GlaxoSmithKline, Stevenage, UK*

Complex disorders result from the interaction of multiple susceptibility genes with diverse environmental factors. Identification of susceptibility genes is a challenging process; however, the publication of the human genome sequence and a wealth of data on single nucleotide polymorphisms will greatly facilitate this process.

Introduction

There are many reports identifying causative genes for Mendelian disorders. These include mutations in the cystic fibrosis transmembrane conductance regulator, ATP-binding cassette (sub-family C, member 7) (*CFTR*) gene on chromosome 7 that cause cystic fibrosis (CF) and in the dystrophin (muscular dystrophy, Duchenne and Becker types) (*DMD*) gene on the X chromosome that cause Duchenne muscular dystrophy (DMD). Mendelian disorders, also referred to as monogenic disorders, result from mutations in a single gene and these alone can cause the disease phenotype. These disorders tend to be very rare in the general population.

In contrast, more common disorders tend to show non-Mendelian inheritance. By definition, these disorders are more prevalent in the population. A few or many genes may be involved, and these disorders are thus referred to as oligogenic or polygenic, respectively. There are many examples of diseases that are thought to be caused by multiple genes including migraine, asthma and Alzheimer disease. The identification of the genes involved in polygenic disorders has proved to be challenging, and there are only a limited number of success stories. The reason for this is that rather than being causative, the genes involved in these disorders are merely susceptibility factors, and mutations in them alone are not usually sufficient to cause the disease. As represented in **Figure 1**, additional complicating factors, including variable environmental effects, may interact with these genes to produce the resulting complex or multifactorial disease phenotype.

Identification of susceptibility genes can be achieved via a multistage process known as positional cloning (**Figure 2**). The process initially requires the collection of blood samples from related family members or from unrelated individuals in populations. Deoxyribonucleic acid (DNA) is extracted from each

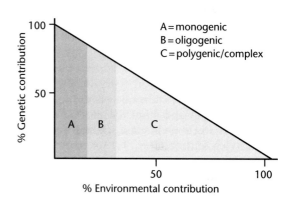

Figure 1 Genetic disorders and the environment.

of these blood samples. In addition, a comprehensive marker set is required in order to screen the DNA samples to identify genetic loci that cosegregate with the disorder of interest. This is done using either linkage or association analysis, or a combination of both. Subsequently, some locus refinement is usually necessary to narrow the genetic region of interest before gene identification and candidate gene analysis can begin.

Sample Collection

A aspect of effective analysis of complex disease is a well-characterized sample collection. DNA samples should be collected according to appropriate ethical and consent guidelines for the country of origin, and

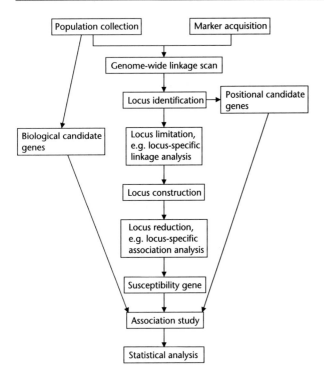

Figure 2 Flow chart for detection of susceptibility genes.

for each individual accurate age, sex, disease status (phenotype) and ethnicity information should be recorded. Phenotypic data should include any subphenotypic information, for example migraine with or without aura, as these subphenotypes may result from polymorphisms in different susceptibility genes.

Linkage analysis is family-based and requires the collection of DNA from affected and unaffected individuals from multiple generations within each family. Ideally, each analysis would use only a single large, multigeneration family for each disease of interest. In practice, however, most families are too small to provide sufficient statistical power, so multiple families are used and the data derived from each family are summed across all families during the analysis. Consistent phenotypic information is therefore critical. Ideally, linkage analysis is performed using multigeneration families. In late-onset diseases such as Alzheimer disease, however, it is often difficult to identify individuals from several generations; therefore affected sibling pairs may be used as this enables the analysis to be performed without any prior knowledge about the mode of inheritance of the disease.

In contrast to linkage analysis, association analysis is population-based. Unrelated, affected individuals

are collected alongside unrelated, unaffected individuals who are matched for age, sex and ethnicity with the affected population. Accurate phenotypic data are critical, as is accurate information about ethnicity for each individual so that multiple population subsets are not combined. These population subsets may have different allele frequencies at the loci of interest such that a combined analysis is inappropriate. This phenomenon is known as population stratification and is a common problem in association studies.

It has been argued that isolated or homogeneous populations should be used for complex disease analysis. This increases the probability that all affected individuals will have derived from a common ancestor and will therefore carry the same disease-associated polymorphisms. Within an isolated population, there is a limited pool of genetic recombinations. This results in a high degree of linkage disequilibrium (LD) between markers and means that the span of associated loci tends to be very large. This phenomenon can facilitate the initial identification of disease-associated loci because a greater number of polymorphic markers will be in LD with the disease-causing variant(s). Once association has been detected in this way, a heterogeneous or more outbred population may be used to refine the region of interest. This is because outbred populations provide a greater pool of recombination breakpoints such that LD is broken down and the span of associated loci is smaller.

There has been much debate regarding the optimal sample size for association analysis. It is widely agreed that larger population collections provide more power for statistical analysis; however, this must be considered in conjunction with the number of markers to be genotyped and the genotyping technology available so that the study does not become unmanageable. In general, sample sizes involving thousands to tens of thousands of individuals are considered necessary for thorough complex disease analysis.

Evolution of Markers

Restriction fragment length polymorphisms

Early genome scans for linkage were performed using restriction fragment length polymorphisms (RFLPs). RFLP assays test for the presence or absence of restriction sites that are created or deleted by naturally occurring nucleotide polymorphisms. An individual's genomic DNA is digested with the appropriate restriction endonuclease, the cleaved products are subjected to agarose gel electrophoresis and Southern blot, and the polymorphic fragments of interest are visualized using a radioactive probe. Although polymorphisms are easily scored using this method, the procedure is both time-consuming and costly and provides only limited information because only two alleles (indicated by the presence or absence of the restriction site) are available for each marker.

Tandem repeat markers

Minisatellite or variable number of tandem repeat (VNTR) markers facilitate linkage analysis by providing more alleles, but their use has been limited as they tend to be clustered around chromosomal telomeres. This has been overcome by the development of microsatellite or short tandem repeat (STR) markers. These markers consist of di-, tri- or tetranucleotide repeats, which are highly polymorphic for repeat number. In addition, they are widely dispersed throughout the genome, with an approximate frequency of one every megabase; they are easily amplified by the polymerase chain reaction (PCR). Detection of individual alleles is performed using polyacrylamide gel separation of radioactive or fluorescent PCR products. The use of these markers has greatly accelerated genome-wide scans for linkage.

Single nucleotide polymorphisms

Most recently, interest has focused on the use of single nucleotide polymorphisms (SNPs) in the search for disease susceptibility genes. This includes RFLP-based SNP markers and also encompasses polymorphisms that do not inherently change a restriction site. SNPs occur at random in the human genome and are found, on average, once every kilobase (kb). Although they provide only limited polymorphic information, they are easily assayed by PCR followed by fluorescent detection which is extremely cost-effective.

Association studies are primarily SNP-based and have been fueled by major SNP identification projects such as the Human Genome Project and The SNP Consortium (TSC). Through the efforts of these and other groups, the number of SNP markers available in the public domain exceeds 3 million. This gives an average frequency of one SNP every 1000 bases throughout the human genome. The SNPs are also annotated with mapping information and, in many cases, allele frequency information, thus providing a wealth of data for susceptibility gene mapping.

Locus- or genome-wide association analysis is thought to require a density of one SNP every 10–30 kb. This frequency is derived from studies that have examined the extent of LD 'blocks' within the human genome. An LD block describes a genomic region within which recombination rates are low. As a result, markers within this region will be in high LD with each other, but they will be in lower LD with markers in adjacent blocks. Individual blocks are separated from each other by regions of high recombination, such as hot spots. Blocks can vary between approximately 5 and 100 kb, with the average being 20–50 kb. The series of alleles within a block is known as a haplotype and, for most blocks, only two or three SNPs may be sufficient to capture more than 85% of the haplotype diversity within that block. It should be noted, therefore, that this type of association-based approach might only locate a potential disease-causing variant to within an LD block, rather than determine its location precisely.

Due to incomplete sequence coverage, it is likely that there will be gaps greater than 50 kb between published SNPs within any given locus of interest. In this case, some de novo identification of SNPs will be necessary to achieve the required SNP coverage. A random-based strategy is appropriate for this type of de novo SNP identification. Sequence information can be derived from genomic clones that map to the locus using PCR primer pairs designed at predetermined intervals throughout each clone. SNPs can be identified by resequencing these PCR products in a panel of unrelated individuals and then comparing the sequence of each PCR product in turn. SNP selection for genotyping is based on several factors, including physical location within the locus, LD information and frequency of the minor allele. A comparison of marker types is shown in **Table 1**.

Locus Identification

It is a straightforward process to determine the mode of inheritance of a monogenic disorder because the disease phenotype usually segregates clearly through the family pedigree. However, this is not the case for complex disorders, and preliminary evidence for the involvement of genetic factors in complex diseases often comes from the analysis of monozygotic (MZ) and dizygotic (DZ) twins. MZ twins are genetically identical, therefore for an inherited disorder it is expected that MZ twins will share the same phenotype more often than DZ twins. Consequently, twin studies have been used to identify a genetic component in many complex disorders including migraine and asthma.

Given that there is evidence implicating genetic factors, the first step in locus identification is to perform a genome-wide scan for linkage. Linkage analysis provides a measure of how often two or more

Table 1 Types of polymorphic marker

Marker type	Frequency in genome	Method of analysis	Advantages	Disadvantages
RFLPs	About one every 30 kb	Agarose gel electrophoresis of restriction endonuclease digests followed by Southern hybridization	Only two alleles; easy to score	Not very informative. Time and labor intensive
Minisatellites (VNTRs)	Clustered at telomeres	As above	Many alleles; highly informative	Clustered at telomeres
Microsatellites (di-, tri- or tetranucleotide repeats; STRs)	About one every 30 kb	PCR followed by radioactive or fluorescent gel detection	Many alleles; highly informative. Located throughout the genome. Multiplex analysis	More costly than SNP genotyping
SNPs	About one every 1 kb	PCR then as for RFLP or fluorescent detection	Only two alleles; easy to score. High frequency in the genome. Potential for high-throughput genotyping	Less informative than microsatellites

RFLPs: restriction fragment length polymorphisms; kb: kilobase (1000 bases); VNTRs: variable number of tandem repeats; STRs: short tandem repeats; PCR: polymerase chain reaction; SNP: single nucleotide polymorphism.

markers will be coinherited owing to their chromosomal proximity. In practice, family members are genotyped using a panel of microsatellite markers. Usually, 300–400 markers are used that fall at approximately 10 centiMorgan (cM) intervals throughout the genome. Linkage analysis relies on the fact that markers that are close to a disease gene are separated less often from that gene by recombination than markers that are further away. Statistical methods are used to detect the cosegregation of linked markers with the disease phenotype within family pedigrees. The degree of linkage is represented as a lod (log₁₀ of the odds) score which is determined for all markers in each family and then summed across all families of interest for a particular disease. In monogenic diseases, an lod score greater than 3 generally indicates linkage between disease and marker, while a lod score of less than −2 suggests that linkage can be excluded. For complex diseases, where multiple loci are likely to be involved, a lod score of more than 1 generally indicates a chromosomal region worthy of further investigation.

Since the late 1980s, linkage analysis has been undertaken using panels of microsatellite markers. In the future, it is likely that this approach will be superseded with linkage analysis using panels of SNP markers. Although bialleleic SNP markers are less informative than multiallelic microsatellite markers, high SNP density and easy genotypic interpretation makes them ideal for automated, high-throughput linkage analysis. This type of approach was being tested in 2002 using evenly spaced or clustered SNP marker sets, and future linkage studies are likely to use a combination of both.

Locus Limitation

Linkage analysis may identify susceptibility loci that span up to 20 cM, or approximately 20 megabases (Mb). Although this indicates regions worthy of further investigation, 20 cM is too large for immediate detailed investigation; therefore, some locus refinement is usually required to try to limit each linked region to around 1 cM or 1 Mb.

As indicated in **Figure 2**, this may be addressed by performing higher-density linkage analysis within the locus of interest. This requires the identification of additional polymorphic markers across the linked locus followed by genotyping of those markers in the original set of families. Subsequent statistical analysis may help to narrow the linkage peak and thus the span of the disease-linked locus.

Once individuals have been typed for a series of markers, genotypes are compiled for each individual at each marker position. Genotypes from family members are then compared to determine the ancestry of each chromosomal segment. This is termed 'haplotype analysis' and can be used to trace the transmission of the disease-associated chromosomal segment at the linked locus. Affected individuals within a particular family will share an ancestral haplotype at the linked locus, and a comparison of recombination breakpoints within these individuals may further delimit the span of the linked region.

However, in complex disease analysis there are several factors that may complicate locus refinement. The first of these is penetrance. Penetrance describes the probability that a person with a given, disease-associated genotype will manifest the disease phenotype. Where nonpenetrance or reduced penetrance is observed, that is, the phenotype in question is absent or limited in an individual despite their genotype, it is possible that other genetic loci or the environment may be involved in moderating the effect of the genotype. Linkage analysis can be adapted to account for reduced penetrance, but often only an estimate of the penetrance level can be applied.

A further complication to susceptibility gene mapping is locus heterogeneity, where several different genetic loci cause a single apparent phenotype. Migraine headache provides a clear example of genetic heterogeneity. A causative gene for familial hemiplegic migraine (FHM) has been located to the short arm of chromosome 19, but mutations in this gene, calcium channel, voltage-dependent, P/Q type, alpha 1A subunit (CACNA1A), account for only approximately 50% of cases of FHM. Two further loci have been mapped to the long arm of chromosome 1, but the genes in question are yet to be elucidated. Although FHM is a rare, autosomal dominant form of migraine, it would be expected that the more typical forms of migraine (migraine with and without aura) would also show heterogeneity. Indeed, several genetic loci have been cited in the literature, although only one study has identified a disease-associated gene.

Locus Construction

Having narrowed the locus of interest as far as possible, a traditional positional cloning approach would then aim to gather physical information about the locus to facilitate gene identification. The first step in locus construction involves the collation of marker and gene information between the two polymorphic markers which flank the linked region of interest. Marker information may include polymorphic markers such as microsatellites or SNPs and nonpolymorphic markers defined purely by a pair of PCR primers flanking an isolated DNA sequence. These unique marker sequences are referred to as sequence tagged sites or STSs. In silico databases can be a rich source of marker information, but marker order may vary between different databases and caution should be used when interpreting unverified physical data. The markers are then used to identify large genomic clones such as bacterial artificial chromosomes (BACs), P1-derived artificial chromosomes (PACs) and yeast artificial chromosomes (YACs) that map to the locus. Where successive clones overlap, a contig can be constructed to span the region between the two flanking markers. Where gaps exist in the clone coverage, de novo marker identification may be necessary. Several methods can be used to generate

new STS markers for clone isolation. These include the generation of sequence from genomic clone ends, from inter-Alu or random genomic clone libraries or from the ends of successive clone deletions. The ultimate aim of this approach is to define a complete contig of genomic clones between the two flanking markers. Sequence analysis of these genomic clones may then reveal the presence of additional or novel genes within the region. Ideally each gene would then be sequenced in a series of unrelated, affected individuals to determine the location of potentially disease-causing mutations that segregate with the disease phenotype.

Since the publication of the first draft sequence of the human genome in February 2001, emphasis has shifted from laboratory-based locus construction to an in-silico-based approach. By mid-2002 it was estimated that over 98% of the human genome had been sequenced. At that time, over 63% of the genome was covered with high-quality, finished sequence and the remaining 35% consisted of unordered, draft sequence. By early 2003, the amount of finished sequence had increased to 95% with a further 3% listed as draft. This provides an unsurpassed wealth of information regarding clones, markers and genes with detailed mapping information for each. Thus, the first stage in any locus construction project should now concentrate on the collation of electronic data prior to any de novo analysis. Thereafter, regular monitoring of in silico information is advisable as the databases are updated.

Association Studies

Locus-specific association studies

Several susceptibility gene identification studies have made use of the surge in human genome data by using a combination of linkage and association techniques to identify susceptibility genes. In this approach, linkage analysis is undertaken to identify broad chromosomal regions of interest, and association analysis is then used to refine the linked loci. Whereas linkage analysis measures the cosegregation of markers within pedigrees, association analysis compares allelic or genotypic frequencies in affected versus unaffected populations. Not only have association studies been facilitated by a dramatic increase in published polymorphic markers, but advances in SNP-based genotyping technologies mean that linked loci can also be rapidly and cost-effectively assessed for further interest.

A migraine susceptibility gene has been identified by this approach (**Figure 3**). Linkage analysis was performed on 16 families to identify a migraine locus of approximately 600 kb on chromosome 19. Subsequently, a combination of in silico data mining and

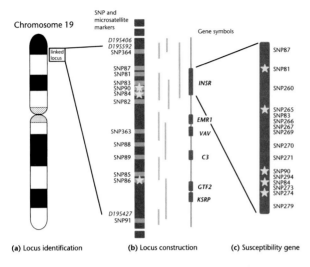

(a) Locus identification   **(b)** Locus construction   **(c)** Susceptibility gene

Figure 3 Identification of a susceptibility gene for migraine.
(a) Locus identification. Microsatellite markers were used to
identify a linked locus on chromosome 19. (b) Locus construction.
The vertical bar represents the linked region on chromosome 19.
Horizontal bars on this represent SNP markers. Vertical lines
represent large genomic clones. The thick vertical bars represent
positional candidate genes. The stars represent SNP markers
showing association with migraine. SNP markers are labeled in
roman type and microsatellite markers in italics. (c) Susceptibility
gene. The insulin receptor (*INSR*) is a potential susceptibility
gene for migraine with five associated SNPs.

laboratory-based practical work was used to construct
a contig of large genomic clones across the locus.
De novo SNP identification was undertaken and
association analysis performed in a population of
over 1500 migraineurs and nonmigraineurs. Significant association was observed with two SNPs lying
within the insulin receptor (*INSR*) gene in this locus. In
order to confirm this finding, additional SNPs were
identified throughout the gene and the association
analysis was repeated. In total, five SNPs within *INSR*
showed significant association, identifying this gene as
a potential susceptibility factor for migraine. However,
it is important to note that functional analysis will
be required to prove beyond doubt that *INSR* is a
susceptibility gene for migraine.

Candidate gene association studies

The final stage of the study above is an example of
positional candidate gene analysis. This type of study
generally requires no information about the underlying disease pathophysiology, but selects genes as
candidates purely on their physical location within a
genetic region of interest. The alternative to this
approach is to select genes that are of biological
interest. Biological candidate gene analysis involves
the dissection of disease-related biochemical pathways

in order to identify proteins, and hence genes, of
interest.

Where candidate gene analysis is required, the gene
of interest will be screened in depth for SNPs. This
usually involves resequencing each exon including at
least 100 base pairs (bp) of flanking intronic sequence,
all 5' and 3' untranslated regions (UTRs) and random
fragments of promoter sequence. SNPs are selected
for inclusion in the association study based on their
position within the gene, LD information and allelic
frequency as described previously. Once association
analysis has been performed, biological and positional
candidate genes must ultimately be screened for
mutations in affected individuals to provide proof of
their involvement in a disorder.

Statistical Analysis

There are several statistical tests for association
depending on the type of sample collection that is
available. Case–control analysis is widely used and
compares allelic or genotypic frequencies between
unrelated affected and unaffected populations. Statistical analysis is usually performed by means of the
chi-squared (χ^2) test with significance achieved at a
probability of less than 0.05. Although case–control
analyses are statistically straightforward to perform,
they rely heavily on accurate phenotypic annotation of
all study subjects. Where subjects have been collected
from multiple sites or where affected and unaffected
individuals are not appropriately matched, population
stratification may occur. If these factors are not
taken into account during statistical analysis, they
may lead to incorrect conclusions about LD or disease
association.

Alternative types of association analysis can be
performed to avoid the issue of population stratification. Several such tests exist, but the one most
commonly used is the transmission disequilibrium
test (TDT), which makes use of small nuclear families
(trios) consisting of one or more affected offspring plus
both parents. The parents must be heterozygous for
the marker of interest, and parental alleles are divided
into those that are transmitted to the affected offspring
versus those that are not transmitted. The nontransmitted alleles act as controls and the TDT evaluates
the disease risk conferred by the transmitted alleles
versus the nontransmitted alleles. Again, the results
are measured using the χ^2 test with associated *P* values.

Haplotype analysis measures the allelic association
of a series of linked markers with the disease
phenotype. This contrasts with case–control and
TDT analyses, which examine the association between
single markers and the disease of interest. Haplotype
analysis is, again, performed using the χ^2 test and can

be applied to case–control or TDT population collections. Simultaneous analysis of multiple markers can provide stronger evidence for disease association than single markers alone.

Genome-wide Association Studies

Based on the ever-increasing catalog of SNP marker information, whole-genome association studies are likely to soon become a possibility. The advantage of this approach over traditional linkage analysis is that associated loci will be better defined more quickly from the outset, thus hastening the process from disease to gene identification.

Estimates vary as to the number of SNPs required for a genome-wide scan for association. It is likely that at least 100 000 SNP markers will be necessary to detect association to within 30 kb of a disease locus. It is also predicted that association scans will be performed using case–control panels of 500 affected versus 500 unaffected individuals or TDT panels of 1000 trios. This represents in excess of 1×10^8 genotypes per scan, which is not only unfeasible but also too costly using the genotyping technology available in 2002.

For whole-genome association scans to become a reality, three major issues need to be addressed: throughput, cost and DNA sample size. Current genotyping technologies are PCR-based and are, therefore, relatively labor-intensive and costly. In order to enable whole-genome analysis, automation will be essential to enable high-throughput genotyping. In addition, individual reaction costs must be drastically reduced to make each scan affordable. This can be achieved by miniaturization of reactions in conjunction with multiplexing of several SNP markers at once. Finally, patient DNA will become a limiting factor given the number of markers to be screened. This, too, can be achieved by miniaturization and multiplex analysis. In addition, pooling of patient DNA samples prior to genotyping has been shown to be an effective method for increasing genotyping throughput while reducing the materials required. Therefore, many of these issues are already being addressed such that whole-genome linkage and association scans using SNPs will soon become routine.

See also
Complex Genetic Systems and Diseases
Complex Multifactorial Genetic Diseases
Genetic Susceptibility
Linkage Analysis
Linkage and Association Studies

Further Reading

Cardon LR and Bell JI (2001) Association study designs for complex diseases. *Nature Reviews Genetics* 2: 91–99.
Kwok P (2001) Methods for genotyping single nucleotide polymorphisms. *Annual Reviews of Genomics and Human Genetics* 2: 235–258.
Montagna P (2000) Molecular genetics of migraine headaches: a review. *Cephalalgia* 20: 3–14.
Risch NJ (2000) Searching for genetic determinants in the new millennium. *Nature* 405: 847–856.
Strachan T and Read AP (1999) *Human Molecular Genetics 2*. Oxford, UK: BIOS Scientific Publishers.
Stumpf MPH (2002) Haplotype diversity and the block structure of linkage disequilibrium. *Trends in Genetics* 18: 226–228.

Web Links

The SNP Consortium Ltd (TSC). Single nucleotide polymorphisms for biomedical research
http://snp.cshl.org/
National Center for Biotechnology Information (NCBI). Repository for databases of genomic information, related literature and data-mining tools for researchers
http://www.ncbi.nlm.nih.gov/
Ensembl. Automatically tracks, assembles and displays (human) genomic sequence including related feature such as genes
http://www.ensembl.org/
Golden Path. Displays current progress on human genome sequencing and enables detailed searching of human genome sequence information
http://genome.ucsc.edu/
Human Genome Project information. Provides information on the aims and progress of the Human Genome Project
http://www.ornl.gov/hgmis/
Calcium channel, voltage-dependent, P/Q type, alpha 1A subunit (CACNA1A); Locus ID: 773. LocusLink:
http://www.ncbi.nlm.nih.gov/LocusLink/LocRpt.cgi?l=773
Insulin receptor (INSR); Locus ID: 3643. LocusLink:
http://www.ncbi.nlm.nih.gov/LocusLink/LocRpt.cgi?l=3643
Calcium channel, voltage-dependent, P/Q type, alpha 1A subunit (CACNA1A); MIM number: 601011. OMIM:
http://www.ncbi.nlm.nih.gov/htbin-post/Omim/dispmim?601011
Insulin receptor (INSR); MIM number: 147670. OMIM:
http://www.ncbi.nlm.nih.gov/htbin-post/Omim/dispmim?147670

Sutton, Walter Stanborough

Introductory article

James F Crow, *University of Wisconsin, Madison, Wisconsin, USA*

Walter Stanborough Sutton (1877–1916) was an American scientist who united Mendelism and cytology to establish the chromosome theory of heredity.

Walter Stanborough Sutton was born in 1877 in Utica, New York. When he was 10, his family moved to a farm in Russell County, Kansas. The family farm was noted for its quality livestock, especially horses. Walter displayed inventive skills early. He enjoyed photography and made his own camera. He was also skilled in mechanics and was particularly adept at repairing farm machinery.

In 1896 he graduated from high school and enrolled at the University of Kansas in engineering, a subject he enjoyed greatly. When he returned home the next summer, the entire family fell ill with typhoid and his 17-year-old brother died. The tragedy affected Walter deeply and probably contributed to his decision to switch from engineering to medicine. Whatever the reason, when he returned to the university he started biology courses in preparation for a career in medicine.

He was an impressive figure, a 6-foot-tall 200-pounder. He was a popular student and a member of the basketball team. He had a distinguished academic record, being elected to both Sigma Xi and Phi Beta Kappa. He soon came in contact with C. E. McClung, who was known for the discovery of sex chromosomes, and they became fast friends. After graduation in 1900 he became McClung's first graduate student. Sutton was skilled in microscopy and in a summer collecting trip found that the lubber grasshopper (*Brachystola magna*) had unusually large chromosomes. The species became the basis of his master's thesis, a study of grasshopper spermatogenesis, completed in 1901.

On McClung's advice Sutton transferred to Columbia University to work with E. B. Wilson. In 1902, he heard William Bateson deliver a Mendelism lecture, which crystallized an idea of his that had been fermenting for some time. Noting that chromosome behavior during meiosis paralleled Mendel's rules of inheritance, he concluded that the chromosomes were the carriers of the genes. His paper, published in 1902, concludes:

> I may finally call attention to the probability that the association of paternal and maternal chromosomes in pairs and their subsequent separation during the reducing division … may constitute the physical basis of the Mendelian law of heredity.

In a more detailed article a year later he noted that as there are more genes than chromosomes, there must be several per chromosome – thus foreshadowing linkage. He also expressed the belief that chromosome pairs orient at random in meiosis, thereby giving rise to Mendelian-independent assortment. Ironically, Bateson was not persuaded; he did not accept the chromosome theory until much later. (*See* Mendel, Gregor Johann.)

Undoubtedly, several other biologists noted the parallelism, but Sutton was the first to publish a convincing argument. At about the same time, Theodor Boveri in Germany arrived at the same conclusion and published a paper soon after. The chromosome theory of heredity is sometimes called the Sutton–Boveri theory. (*See* Meiosis.)

For reasons that are not known, Sutton dropped out of graduate school in 1903 before completing his PhD thesis. The likely cause was financial, and he expected to return and complete the thesis. In the summer of 1903, he went to work as a foreman in the Kansas oil fields. There his creativity again became evident. He invented a method for starting large motors from the high-pressure gas emerging from the wells. He also invented a hoisting device for deep-well drilling.

After 2 years he had accumulated enough money to return to his education. Using his graduate courses as part of the requirement, he was able to complete his medical degree in 2 years. He was an outstanding medical student and received a coveted 2-year fellowship in surgery at Roosevelt Hospital in New York City. After that, he transferred to Kansas City, Kansas, and settled in the home of his parents, who had moved there meanwhile. He was appointed to the University of Kansas Medical School and 2 years later became Associate Professor.

As he had been in other areas, he was also inventive in surgery. He was especially interested in orthopedic and plastic surgery and used his skills in drawing and photography to document his surgical procedures. He developed methods for irrigating the abdominal cavity in the treatment of a ruptured appendix and for giving anesthesia by rectum. He invented a 'speedometer' for calibrating slow infusion of fluids into the body. He was a surgeon in France during the First World War and developed a clever surgical procedure, following localization of a foreign body (usually a bullet) by fluoroscopy. At the time of fluoroscopy, a thread was inserted, connecting the foreign body to the body surface, thus providing a guide for surgery. Ironically, Sutton's death came from a ruptured appendix, a subject on which he had made a major contribution. He was 39 when he died.

Although he was heavily involved in medicine, Sutton did not lose interest in cytology and genetics. That he expected to finish and publish his thesis is evident in a statement found in a paper by Eleanor Carothers in 1913, in which, incidentally, she proved for the first time the already-accepted fact of the independence of the different chromosomes in meiosis.

Sutton was respected by his colleagues. Many prominent physicians gave laudatory statements at his funeral. The University of Kansas Medical Center in Kansas City maintains a collection of Sutton archives, which were the source of much of the material in this article.

See also
Bateson, William
Wilson, Edmund Beecher

Further Reading

Crow EW and Crow JF (2002) 100 years ago: Walter Sutton and the chromosome theory of heredity. *Genetics* **160**: 1–4.

McKusick VA (1960) Walter S. Sutton and the physical basis of Mendelism. *Bulletin of the History of Medicine* **34**: 487–497.

Sturtevant AH (1965) *A History of Genetics*. New York, NY: Harper and Row.

Sutton WS (1902) On the morphology of the chromosome group in *Brachystola magna*. *Biological Bulletin* **4**: 24–39.

Sutton WS (1903) The chromosomes in heredity. *Biological Bulletin* **4**: 231–251.

Synchronous and Asynchronous Replication

Sara Selig, *Technion, Haifa, Israel*

Replication of nuclear DNA in animal cells is carried out according to a temporal program that is conserved from one cell division to the next. The order of replication of genomic regions can be determined by various methods. In most genomic regions both allelic copies replicate synchronously, with the exception of the X chromosomes in female mammals, regions containing imprinted genes or genes with one randomly inactivated allele.

Intermediate article

Article contents
• Methods for Determining the Timing of Replication
• Correlation between the Timing of Replication and Gene Activity
• Synchrony of Replication of Both Allelic Copies
• Regulation of Timing of Replication

Methods for Determining the Timing of Replication

The replication of deoxyribonucleic acid (DNA) in animal cells occurs during S phase, which takes approximately 8–10 h. The order of replication of different genomic regions is fixed according to a tight temporal program, and several methods exist for determining the replication timing of specific regions during S phase. Many of these methods are based on incorporation of a nucleotide analog into replicating DNA, the most commonly used is bromodeoxyuridine (BrdU), an analog of thymidine. Following labeling of cells with BrdU, the chromosomal or genomic regions that incorporated BrdU can be detected as described below. (*See* Chromosome Preparation; DNA Replication Origins.)

Replication banding

Visualization of replication bands on chromosomes is possible because clusters of adjacent replication origins fire simultaneously, and therefore BrdU is incorporated into large continuous genomic regions. After labeling with BrdU, chromosomes are prepared by conventional methods, and regions containing BrdU are visualized. This is achieved either by utilizing antibodies to BrdU or by various staining techniques (reviewed in Drouin *et al.* (1994)) as exemplified in **Figure 1a**. (*See* Chromosome Preparation; DNA Replication Origins.)

In contrast to other chromosomal bands, such as G bands, the replication banding pattern is dynamic and is determined by the BrdU labeling protocol. Two basic protocols exist – the continuous labeling protocol and the pulse labeling protocol. As the replication bands are only visualized on mitotic chromosomes, the duration of the continuous label before the harvest, or the duration of the chase after the BrdU pulse label, will produce different banding patterns. In the case of continuous labeling (**Figure 1b**), if the period of labeling is shorter than the duration of G$_2$, no

(a)

(b)

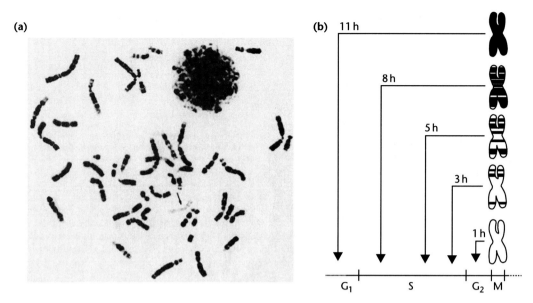

Figure 1 Replication bands. (a) Replication bands on human chromosomes. Human cells were incubated with BrdU for 5.5 h before harvesting, resulting in labeling of late-replicating regions. Chromosomes were prepared and stained for replication bands. Light bands indicate areas that incorporated BrdU. The inactive late-replicating X chromosome is marked by an arrow (see text) (From Verma RS and Babu A (1995) *Human Chromosomes: Principles and Techniques.* New York, NY: McGraw-Hill, reprinted by permission from McGraw-Hill, Inc.) (b) Replication bands after continuous labeling with BrdU. In this scheme, the regions that incorporated BrdU appear as dark bands. The pattern of replication bands varies according to the length of the incubation period with BrdU before harvest. The arrows point to the position in the cell cycle at which the chromosome was exposed to BrdU. The chromosome labeled for 1 h lacks replication bands because the BrdU incubation period initiated when the chromosome was in G_2, and had already completed DNA replication. However a 3 h incubation incorporated BrdU at the end of S phase, therefore the bands represent the latest replicating regions. As the incubation period lengthens, the chromosome is covered with more and more replication bands. For an incubation period that exceeds the sum of G_2 and S phases, the entire chromosome replicated in the presence of BrdU and is therefore stained uniformly.

replication bands will be visualized in mitotic chromosomes because introduction of the analog must have occurred following termination of DNA replication. Increasing incubation periods with BrdU before harvest will produce chromosomes with gradually larger regions staining as BrdU positive. At the point where the labeling extends over a period that covers the entire length of S phase plus G_2, the chromosome will be uniformly labeled with BrdU. Each incubation period produces a characteristic banding pattern, and by comparing one pattern with another it is possible to infer the order of replication of the different chromosomal regions. (*See* Banding Techniques; Chromosomes during Cell Division.)

The alternative pulse label protocol enables the visualization of the chromosomal regions that replicate only at a specific interval during S phase. This is achieved by a short period of BrdU labeling (pulse), for approximately 1 h, followed by growth in media lacking BrdU (chase) for different intervals before harvest. To visualize regions that replicate early in S phase, the length of the chase following the pulse should cover the duration of the remaining S phase plus G_2. If latest replicating regions are of interest, the

chase should extend for a period of time only slightly longer than G_2.

Overlap between replication bands and G bands

In studies that characterized the replication bands in mammalian chromosomes and their order of replication during S phase, a striking similarity was noted between replication banding patterns and G banding (reviewed in Drouin *et al.* (1994)). An almost complete overlap exists between light G bands and early S-phase replication bands, and between dark G bands and late S-phase replication bands. The dark and light G-band staining represents chromosomal domains that differ both at structural and functional levels; light G bands contain CG-rich DNA and most of the gene sequences, while dark G bands are AT-rich and gene-poor (reviewed in Drouin *et al.* (1994)). The overlap between the light/dark G bands and between early/late replication bands adds another view to the functional distinction between both classes of chromatin and suggests a correlation between replication timing and gene activity, as described below. (*See* Chromosomal Bands and Sequence Features; Gene Distribution on

Human Chromosomes; Genome Organization of Vertebrates; Isochores.)

Replication timing determination by fluorescent *in situ* hybridization

The replication banding technique enables the determination of replication timing of large chromosomal regions extending over hundreds of kilobases (kb) up to several megabases (Mb). In order to determine the replication timing of smaller genomic regions in the range of several kilobases to approximately 100 kb, one may utilize the fluorescent *in situ* hybridization (FISH) replication timing assay, which is based on the capacity to differentiate between nonreplicated and replicated regions (Selig *et al.*, 1992). Prior to replication, a FISH signal appears as one single hybridization dot (or 'singlet'). After replication and

sufficient separation of the two replication products, the FISH signal appears as two hybridization dots (a 'doublet'), each dot representing one of the sister chromatids. For an interphase nucleus in which the region of interest has not yet replicated, two singlets are visible, representing the two chromosomal copies, while for an interphase nucleus that has replicated the regions of interest, two doublets are usually detected (**Figure 2a**). (*See* Fluorescence *In Situ* Hybridization (FISH) Techniques.)

In order to determine the replication timing of a given region, a nonsynchronous culture that contains cells from all stages of S phase is labeled for 1 h with BrdU and harvested by conventional methods. FISH is performed on these cells simultaneously with detection of BrdU, and BrdU-positive cells, which represent S-phase cells, are scored for singlets and doublets. The earlier the region of interest replicates during S phase, the larger will be the percentage of

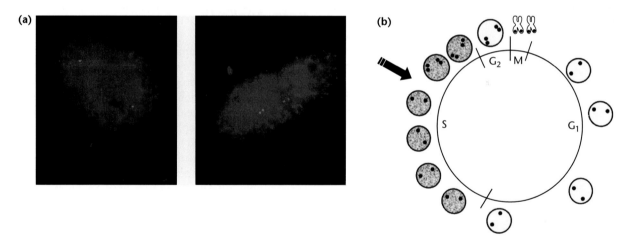

Figure 2 [*Figure is also reproduced in color section.*] Fluorescent *in situ* hybridization (FISH) replication timing assay. (a) Patterns of hybridization representing nonreplicated and replicated regions. The nucleus on the left represents a BrdU-positive cell (in S phase), indicated by its staining (by 7-amino-4-methylcoumarin-3-acetic acid (AMCA)), hybridizing with a singlet–singlet pattern. Each hybridization dot represents one chromosomal copy that has not yet replicated. The nucleus on the right represents a BrdU-positive cell with a doublet–doublet pattern, each doublet representing a pair of sister chromatids originating from one chromosome. The presence of a doublet indicates that the region of interest has replicated. (Figure courtesy of Rachel Ofir.) (b) A schematic explanation for the FISH replication timing assay. Distribution of the hybridization patterns relative to the position of the cells in the cell cycle is indicated in the figure. Nuclei from G_1 hybridize with a singlet–singlet pattern, while nuclei in G_2 hybridize with a doublet–doublet pattern. In metaphase, two signals emanate from each chromosome. In S-phase nuclei, the pattern of hybridization depends on the timing of replication of the region of interest. All nuclei from cells positioned in S prior to the point where replication takes place (shown by an arrow) will show the singlet–singlet pattern, while nuclei originating from cells positioned after the point of replication will display the doublet–doublet pattern. Thus the timing of replication will affect the percentage of nuclei showing either of the hybridization patterns. (Modified version of an illustration by Asaf Hellman.)

S-phase cells that display the doublet–doublet pattern of hybridization (**Figure 2b**). In contrast, if a region replicates very late in S phase, the majority of the S-phase cells will hybridize in the singlet–singlet pattern. By determining the percentages of cells displaying each of the patterns of hybridization, it is possible to determine whether the regions studied replicated in early, middle or late S phase and the order of replication of various genomic regions relative to each other (Selig *et al.*, 1992).

Timing of replication determined by separation of cells from different stages in S phase

A third set of methods for determination of the timing of replication is based on physical separation of cells from different stages of S phase and isolation of newly replicated DNA. Cells are pulsed for 1–2 h with BrdU, separated into four to six S-phase fractions, either by elutriation (based on the gradually increasing size of cells during S phase) or by cell sorting (based on DNA content), and BrdU–DNA is isolated from the individual cell fractions. The BrdU–DNA from each fraction represents the DNA that replicated at that specific interval of S phase. Therefore BrdU–DNA from the earliest fraction of S will contain the DNA regions that replicated early in S phase, and so forth. Southern blotting techniques or the polymerase chain reaction are then used to detect sequences in each fraction in order to determine the replication timing of the sequence of interest (Epner *et al.*, 1988; Selig *et al.*, 1992; Hansen *et al.*, 1997) (**Figure 3**).

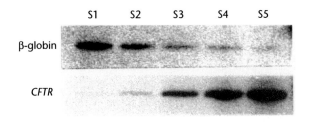

S1   S2   S3   S4   S5

β-globin

CFTR

Figure 3 Replication timing analysis by cell elutriation. Cells from the human erythroleukemia cell line K562 were labeled with BrdU and separated to five fractions in S phase (labeled S1–S5, from early to late in S phase) by elutriation. Southern blot hybridization was performed on BrdU–DNA isolated from the different fractions and digested with *Eco*RI. Different probes give different patterns of hybridization. A probe from the human β-globin gene, active in this cell line, hybridizes preferentially to BrdU–DNA that replicates early in S phase, while a probe from the region of the cystic fibrosis transmembrane conductance regulator (*CFTR*) gene, which is inactive in these cells, hybridizes to BrdU–DNA replicating late in S phase.

Correlation between the Timing of Replication and Gene Activity

The timing of replication of both housekeeping genes and tissue-specific genes has been determined by several methods and a general rule has emerged from the findings: active genes (including all housekeeping genes) replicate during the first half of S phase and inactive genes may replicate during all stages of S phase, but many replicate during the latter half of S phase (Goldman *et al.*, 1984). Many examples exist of tissue-specific genes that replicate early only in the tissue of expression and late in all other tissues (Goldman *et al.*, 1984; Epner *et al.*, 1988; Selig *et al.*, 1992). This differential timing of replication of tissue-specific genes indicates that during the organism's development, distinct patterns of timing of replication are established in different tissues. However, these differences must be limited to relatively small genomic regions, because clear differences between chromosomal replication bands across different tissues have not been observed (Drouin *et al.*, 1994).

Synchrony of Replication of Both Allelic Copies

The strict temporal replication program in animal cells is also evident in the synchronous replication of both parental alleles. At the gross level, replication banding patterns on the two homologous chromosomes are almost always identical, with some exceptions in the case of the latest replicating bands (Drouin *et al.*, 1994). Using the FISH replication timing assay, tight synchrony in replication of both alleles is found for most loci. However, several exceptions to the rule of synchronous allelic replication exist in animal cells, and these are described below.

Asynchronous replication

X chromosomes in female mammals

The most striking case of asynchronous replication is in the case of the two X chromosomes in female mammals. The inactive chromosome initiates replication much later in S phase in comparison with the active chromosome; therefore the two chromosomes may be distinguished from each other by replication banding (Drouin *et al.*, 1994; Heard *et al.*, 1997) (**Figure 1a**). The female X chromosomes provide a large-scale example of the correlation between late replication and gene inactivity, as mentioned above in relation to specific genes. It is notable that in the inactive X chromosome the specific genes that escape inactivation replicate earlier than the remaining inactive portion of the chromosome (Heard *et al.*, 1997). (*See* Chromosome X: General Features; X-chromosome Inactivation.)

Asynchronous replication of imprinted and allelically excluded genes

The differential replication timing of active and inactive alleles of a mammalian gene is not limited to genes on the X chromosome in females. At least another two classes of genes are known to exhibit asynchronous replication as well; imprinted genes (reviewed in Bartolomei and Tilghman (1997)) and allelically excluded genes, for example the olfactory receptor genes (Chess, 1998). In both cases, only one of the two parental alleles is expressed in a given individual cell. In the case of imprinted genes, the parental source of the active and inactive copies is fixed, while in the case of the allelically excluded genes it is stochastic. Such asynchrony was first demonstrated at the level of replication banding in the region of human chromosomes 15q11, where the genes for Angelman and Prader–Willi syndromes reside (Izumikawa *et al.*, 1991). Following this study, asynchronous replication of several imprinted regions, both in human and mouse cells, has been demonstrated by the FISH replication timing assay and by the molecular methods involving cell sorting (reviewed in Bartolomei and Tilghman (1997)). When using the FISH assay this asynchrony is demonstrated by the high percentage of nuclei displaying a singlet–doublet pattern, illustrating the replication of only one of the two alleles. The percentage of the asynchronous nuclei usually exceeds 20%, depending on the time gap between the replication timing of the two alleles. (*See* Epigenetic Factors and Chromosome Organization; Imprinting Disorders.)

In contrast to the general correlation found between activity and early replication, in the case of imprinted genes the paternal copy is usually the earlier replicating copy, regardless of whether it is the expressed allele. For example, in the case of the H19, imprinted maternally expressed untranslated mRNA *(H19)* and insulin-like growth factor 2 receptor *(IGF2R)* genes, the paternal copies are inactive, but earlier replicating than the maternal copies (Kitsberg *et al.*, 1993). In several cases where this inverse correlation is found, the imprinted gene is embedded in a region which includes other imprinted genes with an inverse pattern of expression. The replication marking of the active and inactive copies is already established in germ cells, and the asynchrony in timing of replication in the mouse has been demonstrated as early as in the preimplantation embryo soon after fertilization (Simon *et al.*, 1999).

Regulation of Timing of Replication

It is most likely that the interaction between *cis*- and *trans*-acting factors determines the timing of replication of each genomic segment during S phase. A replication-delaying position effect of *cis* elements has been shown in the case of the murine X-inactivation center region introduced artificially into transgenic mice (Lee and Jaenisch, 1997), the fragile X mental retardation 1 *(FMR1)* gene, with an expanded trinucleotide repeat (Hansen *et al.*, 1997), and human telomeric sequences (Ofir *et al.*, 1999; Smith and Higgs, 1999). An opposite position effect is evident at the region of the immunoglobulin constant genes where the juxtaposed variable regions advance their replication timing to earlier in S phase (Calza *et al.*, 1984). The existence of *trans*-acting factors that affect the timing of replication has also been shown in cases of cell fusions. For example the fusion between female mouse lymphocytes and teratocarcinoma cells advances the timing of replication of both the inactive X chromosome (Takagi *et al.*, 1983) and the satellite DNA sequences of the lymphocyte cells (Selig *et al.*, 1988). In addition, epigenetic modifications such as methylation and chromatin conformation have been shown to correlate with replication timing at a number of loci including the inactive X chromosome, mouse satellite DNA and imprinted regions (Selig *et al.*, 1988; Bartolomei and Tilghman, 1997; Heard *et al.*, 1997). (*See* CpG Islands and Methylation; Telomere.)

Additional studies are being carried out to define more precisely the genetic and epigenetic mechanisms that determine and regulate the replication timing program in mammalian cells. These studies will establish whether the timing of replication of specific genes is a by-product of their regulation or alternatively if the timing of replication in S phase plays a role in regulating gene expression.

See also
DNA Replication

References

Bartolomei MS and Tilghman SM (1997) Genomic imprinting in mammals. *Annual Review of Genetics* **31**: 493–525.

Calza RE, Eckhardt LA, DelGiudice T and Schildkraut CL (1984) Changes in gene position are accompanied by a change in the time of replication. *Cell* **26**: 689–696.

Chess A (1998) Expansion of the allelic exclusion principle? *Science* **279**: 2067–2068.

Drouin R, Holmquist GP and Richer C-L (1994) High-resolution replication bands compared with morphologic G- and R-bands. *Advances in Human Genetics* **22**: 47–115.

Epner E, Forrester WC and Groudine M (1988) Asynchronous DNA replication within the human β-globin gene locus.

Proceedings of the National Academy of Sciences of the United States of America **85**: 8081–8085.

Goldman MS, Holmquist GP, Gray MC, Caston LA and Nag A (1984) Replication timing of genes and middle repetitive sequences. *Science* **224**: 686–692.

Hansen RS, Canfield TK, Fjeld AD, *et al.* (1997) A variable domain of delayed replication in FRAXA fragile X chromosomes: X inactivation-like spread of late replication. *Proceedings of the National Academy of Sciences of the United States of America* **94**: 4587–4592.

Heard E, Clerc P and Avner P (1997) X-chromosome inactivation in mammals. *Annual Review of Genetics* **31**: 571–610.

Izumikawa Y, Naritomi K and Hirayama K (1991) Replication asynchrony between homologs 15q11.2: cytogenetic evidence for genomic imprinting. *Human Genetics* **87**: 1–5.

Kitsberg D, Selig S, Brandeis M, *et al.* (1993) Allele-specific replication timing of imprinted gene regions. *Nature* **364**: 459–463.

Lee JT and Jaenisch R (1997) Long-range cis effects of ectopic X-inactivation centres on a mouse autosome. *Nature* **386**: 275–279.

Ofir R, Wong ACC, McDermid HE, Skorecki KL and Selig S (1999) Position effect of human telomeric repeats on replication timing. *Proceedings of the National Academy of Sciences of the United States of America* **96**: 11434–11439.

Selig S, Ariel M, Goiten R, Marcus M and Cedar H (1988) Regulation of mouse satellite DNA replication timing. *EMBO Journal* **7**: 419–426.

Selig S, Okumura K, Ward DC and Cedar H (1992) Delineation of DNA replication time zones by fluorescence *in situ* hybridization. *EMBO Journal* **11**: 1217–1225.

Simon I, Tenzen T, Reubinoff BE, *et al.* (1999) Asynchronous replication of imprinted genes is established in the gametes and maintained during development. *Nature* **401**: 929–932.

Smith ZE and Higgs DR (1999) The pattern of replication at a human telomeric region (16p13.3): its relationship to chromosome structure and gene expression. *Human Molecular Genetics* **8**: 1373–1386.

Takagi N, Yoshida MA, Sugawara O and Sasaki M (1983) Reversal of X-inactivation in female mouse somatic cells hybridized with murine teratocarcinoma stem cells *in vitro*. *Cell* **34**: 1053–1062.

Further Reading

DePamphilis ML (2000) Review: nuclear structure and DNA replication. *Journal of Structural Biology* **129**: 186–197.

Holmquist GP (1987) Role of replication time in the control of tissue-specific gene expression. *American Journal of Human Genetics* **40**: 151–173.

Simon I and Cedar H (1996) Temporal order of DNA replication. In: DePamphilis ML (ed.) *DNA Replication in Eukaryotic Cells*, pp. 387–408. New York, NY: Cold Spring Harbor Laboratory Press.

Zannis-Hadjopoulos M and Price GB (1999) Eukaryotic DNA replication. *Journal of Cellular Biochemistry* **32**–**33**(supplement): 1–14.

Web Links

Cystic fibrosis transmembrane conductance regulator, ATP-binding cassette (sub-family C, member 7) (*CFTR*); Locus ID: 1080. LocusLink:
http://www.ncbi.nlm.nih.gov/LocusLink/LocRpt.cgi?l = 1080

Fragile X mental retardation 1 (*FMR1*); Locus ID: 2332. LocusLink:
http://www.ncbi.nlm.nih.gov/LocusLink/LocRpt.cgi?l = 2332

H19, imprinted maternally expressed untranslated mRNA (*H19*); Locus ID: 8043. LocusLink:
http://www.ncbi.nlm.nih.gov/LocusLink/LocRpt.cgi?l = 8043

Insulin-like growth factor 2 receptor (*IGF2R*); Locus ID: 3482. LocusLink:
http://www.ncbi.nlm.nih.gov/LocusLink/LocRpt.cgi?l = 3482

Cystic fibrosis transmembrane conductance regulator, ATP-binding cassette (sub-family C, member 7) (*CFTR*); MIM number: 602421. OMIM:
http://www.ncbi.nlm.nih.gov/htbin-post/Omim/dispmim = 602421

Fragile X mental retardation 1 (*FMR1*); MIM number: 309550. OMIM:
http://www.ncbi.nlm.nih.gov/htbin-post/Omim/dispmim = 309550

H19, imprinted maternally expressed untranslated mRNA (*H19*); MIM number: 103280. OMIM:
http://www.ncbi.nlm.nih.gov/htbin-post/Omim/dispmim = 103280

Insulin-like growth factor 2 receptor (IGF2R); MIM number: 147280. OMIM:
http://www.ncbi.nlm.nih.gov/htbin-post/Omim/dispmim = 147280

Synonymous and Nonsynonymous Rates

Soojin Yi, *University of Chicago, Chicago, Illinois, USA*

Intermediate article

Article contents

- Introduction
- Estimation of Synonymous and Nonsynonymous Substitutions
- Detecting Positive Selection from DNA Sequence Data

When comparing protein-coding DNA sequences between species, it is useful to distinguish between synonymous (silent) substitutions and nonsynonymous (amino-acid changing) substitutions. Because the two types of change are under different intensities of selective constraints, comparing their rates provides insights into the mechanisms of evolution of the gene under study.

Introduction

Protein sequences are encoded by strings of codons, each of which is a triplet of nucleotides and specifies an amino acid according to the genetic code. Because the genetic code is degenerate, an amino acid is often encoded by more than one codon. For example, proline is encoded by CCU, CCC, CCA and CCG.

Two codons are said to be synonymous if they code for the same amino acid, and nonsynonymous if they code for two different amino acids.

When comparing two homologous protein-coding sequences, it is conventional to distinguish between synonymous and nonsynonymous differences. For example, in the two sequences below, the T and C difference at the first position is nonsynonyomous, while the T and C difference at the third position of the second codon is synonymous.

Sequence 1 : TAT(Tyr) TAT(Tyr) GGT(Gly)

Sequence 2 : CAT(His) TAC(Tyr) GAG(Glu)

We make the above distinction because the two types of change are under different selective constraints. Because synonymous changes do not affect the protein product, they are often assumed to be selectively neutral and their rate of substitution may reflect the underlying mutation rate. On the other hand, nonsynonymous changes alter the protein product and so may affect the protein function. If an amino acid change is deleterious (as in most cases), it is likely to be eliminated from the population, while if the change is advantageous, it may have a chance to spread to the whole population. For this reason, the rate of nonsynonymous substitution is likely to be different from the mutation rate. Therefore, comparing the synonymous and nonsynonymous rates in the same gene can provide information on the strength of functional constraints on the protein product.

In view of the importance of distinguishing between synonymous and nonsynonymous substitutions, we shall first discuss methods for estimating the numbers of synonymous and nonsynonymous substitutions between two homologous sequences. Then we shall present a method that utilizes the ratio of nonsynonymous and synonymous rates to detect selection at the molecular level. Finally, an example of application of this method will be presented.

Estimation of Synonymous and Nonsynonymous Substitutions

Conventionally, the numbers of synonymous and nonsynonymous substitutions are defined as the number of synonymous substitutions per synonymous site ($K_S = S_S/L_S$, also called d_S) and the number of nonsynonymous substitutions per nonsynonymous site ($K_A = S_A/L_A$, also called d_N). The use of per site as the unit is to facilitate comparisons between different genes. If the divergence time (T) between the two species compared is known, then the rates of synonymous and nonsynonymous substitutions per unit time can be computed as $K_S/(2T)$ and $K_A/(2T)$, respectively.

Many methods have been devised to estimate K_S and K_A (Li, 1997; Nei and Kumar, 2000). These can be divided into two groups. The first group includes heuristic methods, each of which counts the numbers of synonymous and nonsynonymous sites and the numbers of synonymous and nonsynonymous differences by considering all possible evolutionary pathways. The second group includes codon-based maximum likelihood methods.

Heuristic methods

We shall first describe in detail the method of Li *et al.* (1985). In this method, nucleotide sites are classified into nondegenerate, twofold degenerate and fourfold degenerate sites. A site is nondegenerate if all possible changes are nonsynonymous, twofold degenerate if one of the three possible changes is synonymous and fourfold degenerate if all possible changes are synonymous. For example, in the codon CCG, which encodes proline (see above), the first and second positions are nondegenerate because any change at either of the two sites changes proline to another amino acid, while the third position is fourfold degenerate because it can change to any of the three other nucleotides without changing the amino acid. As another example, in the codon TTT (phenylalanine), the third position is twofold degenerate, while the first and second positions are nondegenerate.

Li *et al.* (1985) note that all substitutions at fourfold sites are synonymous, while all substitutions at nondegenerate sites are nonsynonymous. At twofold sites, transitional changes are mostly synonymous, while transversional changes are mostly nonsynonymous. They consider one-third of a twofold degenerate site as potentially synonymous and two-thirds of it as potentially nonsynonymous. Therefore, the number of synonymous sites in a sequence is approximately $L_4 + L_2/3$, and the number of nonsynonymous sites is approximately $L_0 + 2L_2/3$, where L_i is the number of i-fold degenerate sites.

To determine the numbers of synonymous and nonsynonymous differences between two sequences, the codons of the two sequences are compared one by one. Each difference is recorded as synonymous or nonsynonymous. This is straightforward for two codons differing by only one nucleotide. If two codons differ by more than one nucleotide, then all possible evolutionary pathways need to be considered. For example, for the differences in the third codons in the alignment of two sequences given in the Introduction, there are two possible parsimonious pathways:

GGT(Gly) ↔ GAT(Asp) ↔ GAG(Glu)

or

GGT(Gly) ↔ GGG(Gly) ↔ GAG(Glu)

The first pathway requires two nonsynonymous changes, while the second requires one synonymous and one nonsynonymous change. Li *et al.* (1985) used empirical data to estimate the relative weights for different pathways.

Each substitution is further classified into transitional and transversional changes and the proportions of transitional ($P_i = S_i/L_i$) and transversional ($Q_i = V_i/L_i$) differences are estimated. The numbers of transitional (A_i) and transversional (B_i) substitutions per *i*-fold degenerate site can then be estimated using Kimura's two-parameter method:

$$A_i = \tfrac{1}{2}\ln(a_i) - \tfrac{1}{4}\ln(b_i)$$

$$B_i = \tfrac{1}{2}\ln(b_i)$$

for $i = 0, 2$ and 4, where $a_i = 1/(1-2P_i-Q_i)$ and $b_i = 1/(1-2Q_i)$.

As most transitional changes at the twofold degenerate sites and all the changes at the fourfold degenerate sites are synonymous, the total number of synonymous substitutions can then be estimated by summing L_2A_2 and $L_4(A_4+B_4)$. Combining this with the total number of synonymous sites, the number of synonymous substitutions per synonymous site is computed as

$$K_S = [L_2A_2 + L_4(A_4 + B_4)]/(L_2/3 + L_4)$$

Likewise, the number of nonsynonymous substitutions per nonsynonymous site is computed as

$$K_A = [L_2B_2 + L_0(A_0 + B_0)]/(2L_2/3 + L_0)$$

The above method has been modified as follows. It is well known that transitional changes occur more frequently than transversional changes. At twofold degenerate sites, most transitional changes are synonymous changes. Therefore, counting one-third of a twofold degenerate site as synonymous is likely to overestimate the rate of synonymous substitution. For this reason, Li (1993) and Pamilo and Bianchi (1993) proposed the following modification. Assuming that the rates of transition are the same for twofold and fourfold degenerate sites, the rate of transitional synonymous substitution can be estimated as a weighted average of the transitional changes at twofold and fourfold degenerate sites. The total number of synonymous substitutions per synonymous site is then the sum of this value and the transversional synonymous substitutions at fourfold degenerate sites:

$$K_S = (L_2A_2 + L_4A_4)/(L_2 + L_4) + B_4$$

Likewise, the rate of nonsynonymous substitutions can now be estimated as the sum of the average transversional changes at nondegenerate and twofold degenerate sites and the transitional changes at nondegenerate sites:

$$K_A = A_0 + (L_0B_2 + L_2B_2)/(L_0 + L_2)$$

Nei and Gojobori (1986) proposed a method in which the numbers of potentially synonymous and potentially nonsynonymous sites are counted separately for each codon. The numbers of synonymous and nonsynonymous differences are computed considering all possible evolutionary pathways as done in the above method. They assume that all pathways occur with the same probability and use the one-parameter method to correct for multiple changes. (*See* Evolutionary Distance.)

Codon-based maximum likelihood method

This group includes the codon-based maximum-likelihood method developed by Goldman and Yang (1994). In this method a specific substitution from codon *i* to codon *j* is modeled to occur according to a unique probability. Each probability is independent of the substitution history before codon *i*. These are general properties of a class of stochastic models called Markov processes, and the method to obtain the probabilities of each change is a commonly used method in such models. Briefly, the substitution rate from codon *i* to codon *j* ($i \neq j$) for any of the 61 sense codons (excluding the nonsense codons) is assumed to be governed by the following three parameters:

π_j the equilibrium frequency of the *j*th codon
κ the transition/transversion ratio
ω the nonsynonymous/synonymous rate ratio

For example, the substitution rate for a synonymous transversion is π_j, and that for a synonymous transition is $\kappa\pi_j$. For nonsynonymous rates, the additional parameter ω is used, so that the substitution rate for a nonsynonymous transversion is $\omega\pi_j$ and that for a nonsynonymous transition is $\omega\kappa\pi_j$. The rates between two codons that differ by more than two sites are given as 0.

Each rate represents an element ('state') in a 61×61 rate matrix, and the probability of transition from one state to another can be computed by the following equation:

$$P(t) = \{p_{ij}(t)\} = e^{Qt}$$

where $p_{ij}(t)$ is the probability that codon *i* becomes codon *j* after time *t*. The probability of transition from any codon to any other codon is then represented as a different state in a 61×61 matrix, called the 'transition probability matrix'.

In the codon-based maximum likelihood model, the parameter π_j takes into account the codon usage bias in the data and is usually estimated from the observed codon frequencies (assuming they are in equilibrium). The other two parameters ω and κ are estimated from the data by the maximum likelihood method. Interested readers should refer to Goldman and Yang (1994) for the original implementation.

Detecting Positive Selection from DNA Sequence Data

We now describe a method for detecting positive selection during the evolution of a protein by comparing the nonsynonymous and synonymous rates. Following the above maximum likelihood method, let us denote the ratio of nonsynonymous to synonymous rates (K_A/K_S) by ω. In cases where the protein product is under selective constraint, ω is less than 1. If the protein product is under no selective constraint, then a nonsynonymous mutation has the same probability of fixation as does a synonymous mutation and ω is expected to be 1. If the protein product is under positive selection favoring its modification, then more nonsynonymous substitutions than synonymous substitutions will be fixed and ω will be greater than 1. Using this principle, we can conduct statistical tests to determine whether ω is significantly greater than 1. Thus, we can distinguish among the three types of functional constraint as follows:

$\frac{K_A}{K_S} < 1$ selective constraints on the protein (or 'purifying selection')

$\frac{K_A}{K_S} = 1$ no selective constraint (neutral)

$\frac{K_A}{K_S} > 1$ positive selection

Compared to morphological traits, it is often difficult to prove adaptive evolution at the molecular level. The test described above provides a very useful method to detect selection at DNA sequence level. Such a test statistic can be computed by the method described below.

Likelihood ratio test

The null model is $\omega = 1$. This means that there is no selective constraint on all nonsynonymous sites. The alternative model assumes a specific value of ω to be estimated from the data. The log maximum likelihood value from each model can then be obtained, say $\ln L_0$ for the null model and $\ln L_1$ for the alternative model. Then the log likelihood ratio is given by

$$LR = 2(\ln L_1 - \ln L_0)$$

which is approximately χ^2 distributed. Therefore, the significance of the LR value can be determined. If the

estimated ω value is greater than 1 and the LR is significant, it may be concluded that the rate of nonsynonymous substitution is greater than that of synonymous substitution, owing to positive selection.

The codon-based maximum likelihood model is further modified to be able to allow a different ω for different sites, hence allowing some sites to evolve under positive selection, while some other sites under purifying selection. For a review, see Yang and Bielawski (2000).

An example: positive selection in the evolution of primate lysozyme

Lysozyme is a ubiquitous bacteriolytic enzyme. It is secreted in many different kinds of body fluid (serum, saliva, tears, etc.) of almost all animals to act as the first line of defense against foreign bacteria.

Some herbivorous animals have a digestive system called the foregut fermentation. In this system, the anterior part of the stomach is modified and functions as a chamber for bacterial fermentation of ingested plant tissues. This type of digestive system has independently evolved twice in the placental mammals, once in the ruminants (cows) and once in leaf-eating colobine monkeys.

In such animals, lysozyme is secreted in the posterior parts of the digestive system as well as in its usual places. It is thought that this additional secretion helps freeing nutrients within bacterial cell wall. Interestingly, the lysozyme sequences from these two types of animals are found to be very similar. In fact, the lysozyme sequences from the langur and the cow are more similar than those from the horse and the cow, while the former species pair diverged much earlier than the latter species pair.

There are five uniquely shared amino acids between the lysozymes from the langur and the cow, as compared to only one uniquely shared amino acid between the cow and the horse. This observation suggests a series of adaptive parallel substitutions. Moreover, it is proposed that the five amino acids confer advantage for functioning at lower pH, such as the foregut environment. Therefore, the evolution of lysozymes is regarded as an example of convergent adaptive evolution at the molecular level.

Yang (1998) analyzed the DNA sequences of lysozyme from 24 primate species using the likelihood ratio test described above. Some results from an analysis of a subset of seven sequences are shown in **Figure 1**.

First, a likelihood ratio test between a model assigning different ω values for each branch and a model with a uniform ω value for all branches is performed. The log likelihood value for the first model

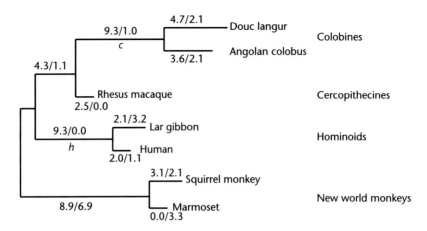

Figure 1 Phylogeny of lysozyme sequences from seven primate species, representing four major groups of species. The two numbers shown along each branch are the maximum likelihood estimates of the numbers of nonsynonymous and synonymous substitutions for the entire gene along that branch. A model of different ω for each branch of the tree, the so-called 'free-ratio' model, is used. Branches are drawn in proportion to estimates of their lengths. Adapted from Yang (1998) with premission of the society for Molecular Biology and Evolution.

is $\ln L_1 = -896.41$, and that of the second model is $\ln L_0 = -906.02$.

The log likelihood ratio is then computed as described above:

$$LR = 2(\ln L_1 - \ln L_0) = 19.21$$

This value can be compared with a χ^2 distribution with 10 degrees of freedom (as 11 parameters – different ω values for all branches – are estimated from the first model, while a single ω value is estimated from the second model, df $= 11-1 = 10$). The difference between the two models is significant ($0.01 < P < 0.05$), indicating that the former model (also called 'free-ratio model') performed significantly better.

In particular, the branch leading to the colobine monkeys (branch c) and the branch leading to hominoids (branch h) are long and have high ω values, both significantly greater than the background ratios. The ω estimate along branch h was significantly greater than 1, indicating that positive selection may have operated during the lysozyme evolution along this lineage. This is also in accord with a previous analysis of these sequences by heuristic methods (Messier and Stewart, 1997). The ω value along branch c was not significantly greater than one. This result is compatible with both relaxed selective constraints and presence of positive selection along lineage c. However, because the test has a low power and because lysozyme did not lose function along this branch, but rather acquired a new function, the hypothesis of positive selection appears more plausible than relaxed selective constraints.

See also
Mutation Rate
Mutation Rates: Evolution
Nucleotide Substitution: Rate of

References

Goldman N and Yang Z (1994) A codon-based model of nucleotide substitution for protein-coding DNA sequences. *Molecular Biology and Evolution* **11**: 725–736.

Li WH (1993) Unbiased estimation of the rates of synonymous and nonsynonymous substitution. *Journal of Molecular Evolution* **36**: 96–99.

Li WH (1997) *Molecular Evolution*. Sutherland, MA: Sinauer Associates.

Li WH, Wu CI and Luo CC (1985) A new method for estimating synonymous and nonsynonymous rates of nucleotide substitution considering the relative likelihood of nucleotide and codon changes. *Molecular Biology and Evolution* **2**: 150–174.

Messier W and Stewart C-B (1997) Episodic adaptive evolution of primate lysozymes. *Nature* **385**: 151–154.

Nei M and Gojobori T (1986) Simple methods for estimating the numbers of synonymous and nonsynonymous nucleotide substitutions. *Molecular Biology and Evolution* **3**: 418–426.

Nei M and Kumar S (2000) *Molecular Evolution and Phylogenetics*. New York, NY: Oxford University Press.

Pamilo P and Bianchi O (1993) Evolution of the Zfx and Zfy genes: rates and interdependence between the genes. *Molecular Biology and Evolution* **19**: 271–281.

Yang Z (1998) Likelihood ratio tests for detecting positive selection and application to primate lysozyme evolution. *Molecular Biology and Evolution* **15**: 568–573.

Yang Z and Bielawski JP (2000) Statistical methods for detecting molecular adaptation. *Trends in Ecology and Evolution* **15**: 496–503.

Synthetic Gene Delivery Systems in Gene Therapy

Andrew David Miller, *Imperial College of Science, Technology and Medicine, London, UK*

Synthetic gene delivery systems are agents prepared using synthetic organic chemistry and/or molecular biology expertise for the delivery of therapeutic genes to cells *in vitro, ex vivo* and *in vivo*. Liposomes/micelles and/or soluble polymers form the basis of most known synthetic systems.

Introduction

Gene therapy may be described as the use of genes as medicines to treat disease. A more precise definition would be the delivery of nucleic acids with a vector for some therapeutic purpose. The first description emphasizes the idea of gene therapy as a sophisticated somatic therapeutic modality able to direct genes to intervene in disease at the most basic level, resulting in more than just the treatment of symptoms, but also cure. That this state of affairs has not yet been realized (by 2002) is due in large part to a lack of access to clinically useful vector technology for the delivery of therapeutic genes to their required site of action in the body. Broadly speaking, available vectors or gene delivery systems are either viral or synthetic in character. Synthetic gene delivery systems have many potential advantages as compared with viral systems; for instance their much lower toxicity/immunogenicity and potential for oncogenicity, their size-independent delivery of nucleic acids (from oligonucleotides to artificial chromosomes), simpler quality control and less severe pharmaceutical/regulatory requirements. However, present (2002) synthetic gene delivery systems are unlikely to find routine clinical use because they are beset with too many basic problems (Miller, 1999, 2002). Therefore, new synthetic gene delivery platform technologies are required that have as background characteristics stability, reproducibility of formulation, consistency of transfection (i.e. gene delivery and expression) and ease of storage, but which are assembled in such a way as to be easily upgraded in a modular fashion for the application of interest. These characteristics need to be achieved in order to make synthetic gene delivery systems truly competitive with their viral counterparts. Nevertheless, the development of synthetic gene delivery systems has come a long way since the late 1990s suggesting that properly competitive synthetic gene delivery systems can be developed soon.

Simple Synthetic Gene Delivery Systems

Nucleic acids in the form of plasmid deoxyribonucleic acid (pDNA), oligodeoxynucleotides (ODNs) or ribonucleic acid (RNA), for example, possess a substantial negative charge at neutral pH owing to phosphodiester backbone charges. In order to traverse a biological membrane, itself typically of overall negative charge, nucleic acid charge must be neutralized, encouraging simultaneous compaction of the nucleic acid structure. In this form, nucleic acids may enter cells by mechanisms such as endocytosis, phagocytosis or even caveolar import. Nucleic acids may still face formidable intracellular barriers to access the nucleus or even the cytosol effectively, but the main problem of cellular entry is solved. Simple synthetic gene delivery systems were originally devised to neutralize and to compact nucleic acids for cell entry. They fall into two main classes: cationic liposome/micelles and cationic polymers.

Cationic liposome/micelle systems

Cationic liposome/micelle-based systems are formed either from a single synthetic cationic amphiphile (known as a cytofectin: 'cyto' for cell and 'fectin' for transfection) or more commonly from the combination of a synthetic cytofectin and a naturally available neutral lipid such as dioleoyl L-α-phosphatidylethanolamine (DOPE) or cholesterol (Chol) (**Figure 1**). Typically, cytofectin structures comprise a hydrophobic region, a polar linker and a cationic head group. The structures of a number of synthetic cytofectins are shown in **Figure 1**, illustrating how diverse polar linker and head group structures may

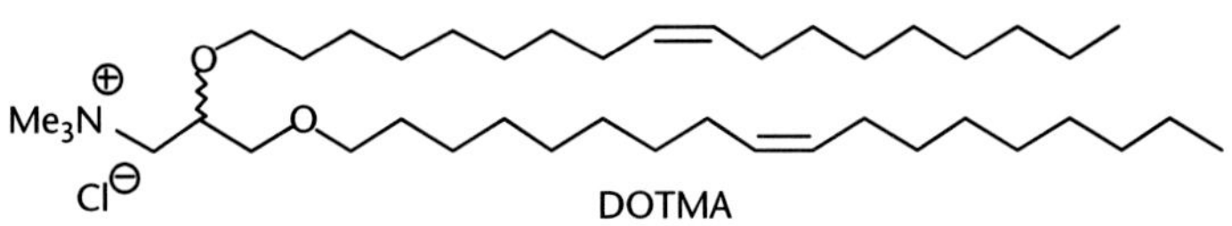

DOTMA

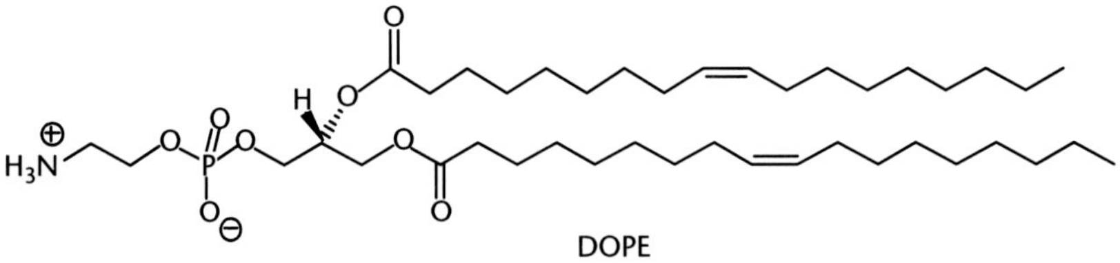

DOPE

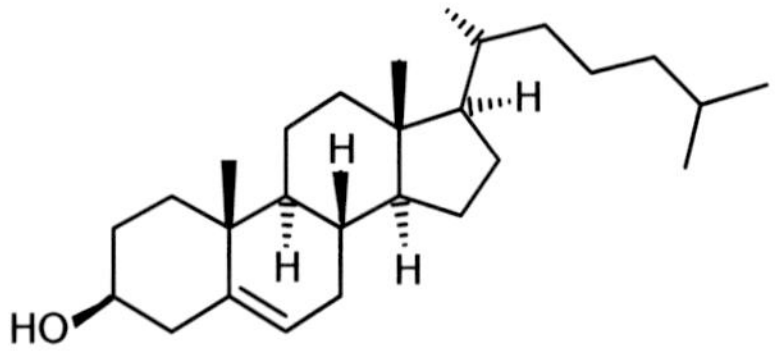

Chol

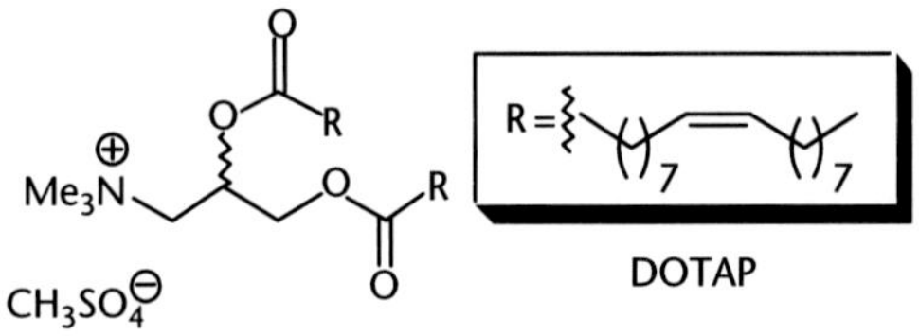

DOTAP

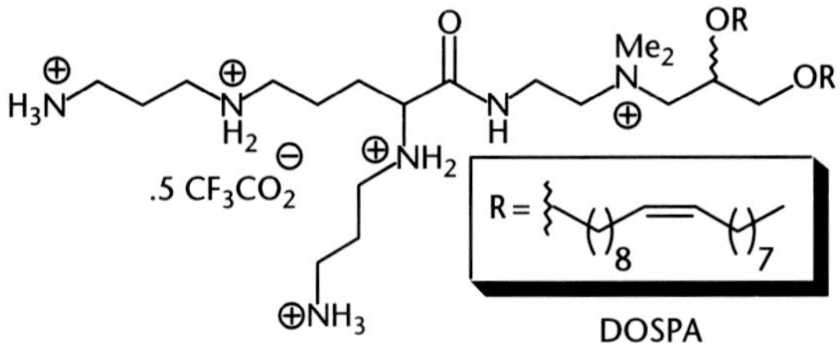

DOSPA

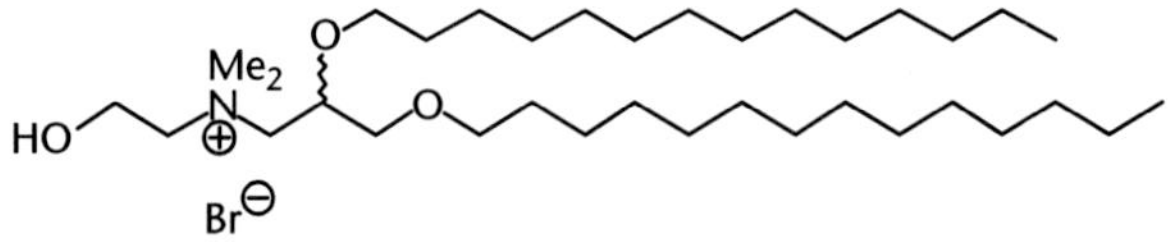

DMRIE

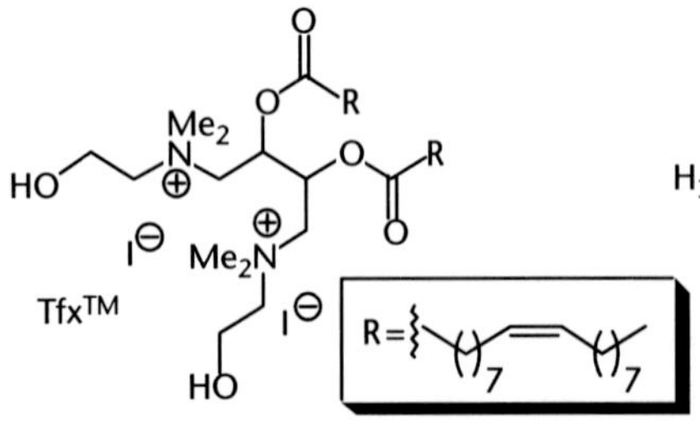

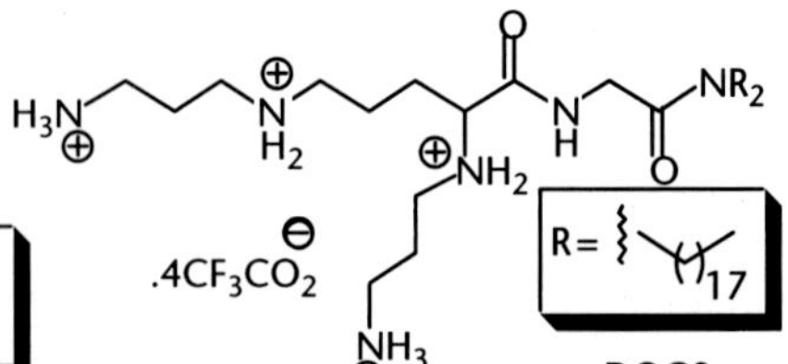

Tfx™

DOGS

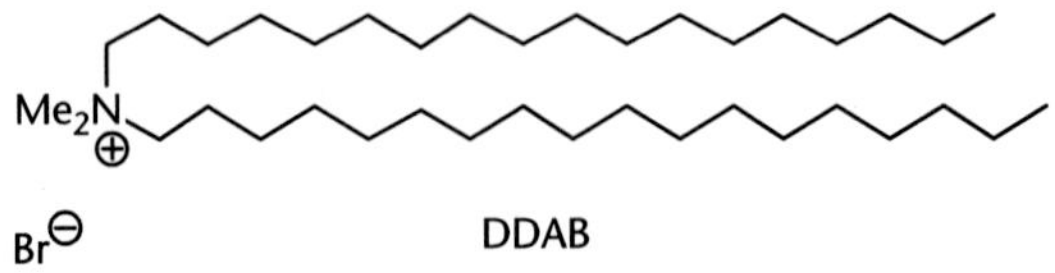

DDAB

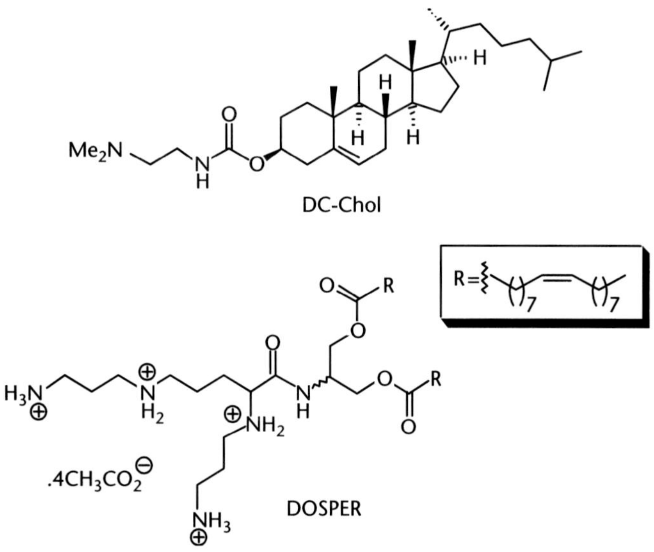

Figure 1 Structures of well-known cytofectins and neutral lipids used in the preparation of simple cationic liposome/micelle systems. DOTMA: *N*-[1-(2,3-dioleyloxy)propyl]-*N,N,N*-trimethyl ammonium chloride; DOPE: dioleoyl L-α-phosphatidylethanolamine; Chol: cholesterol; DOTAP: 1,2-dioleoyloxy-3-(trimethylammonio)propane; DOSPA: 2,3-dioleyloxy-*N*-[2-(sperminecarboxamido)ethyl]-*N,N*-dimethyl-1-propanaminium trifluoroacetate; DMRIE: 1,2-dimyristyloxypropyl-3-dimethylhydroxyethylammonium bromide; DOGS: dioctadecylamidoglycylspermine; DDAB: dimethyldioctadecylammonium bromide; DC-Chol: 3β-[*N*-(*N′,N′*-dimethylaminoethane) carbamoyl]cholesterol; DOSPER: 1,3-dioleoyloxy-2-(6-carboxyspermyl)propylamide.

Table 1 Selection of commercially available cationic liposome/micelle systems

Cytofectin	Formulation	Trade name/manufacturer
DOTMA	DOTMA/DOPE 1:1 (wt/wt)	Lipofectin™/GIBCO BRL
DOTAP	DOTAP	DOTAP/Roche Molecular
DOSPA	DOSPA/DOPE 3:1 (wt/wt)	LipofectAMINE™/GIBCO BRL
DMRIE	DMRIE/Chol 1:1 (m/m)	DMRIE-C/GIBCO BRL
Tfx	Tfx/DOPE (Tfx-10, -20, -50)	Tfx™/Promega
DOGS	Micelle	Transfectam®/Promega
DDAB	DDAB/DOPE 1:2.5 (wt/wt)	LipofectACE™/GIBCO BRL
DC-Chol	DC-Chol/DOPE 6:4 (m/m)	DC-Chol/Sigma
DOSPER	Micelle	DOSPER/Roche Molecular

DOTMA: *N*-[1-(2,3-dioleyloxy)propyl]-*N,N,N*-trimethyl ammonium chloride; DOPE: dioleoyl L-α-phosphatidylethanolamine; DOTAP: 1,2-dioleoyloxy-3-(trimethylammonio)propane; DOSPA: 2,3-dioleoyloxy-*N*-[2-(sperminecarboxamido)ethyl]-*N,N*-dimethyl-1-propanaminium trifluoroacetate; DMRIE: 1,2-dimyristyloxypropyl-3-dimethylhydroxyethylammonium bromide; DOGS: dioctadecylamidoglycylspermine; DDAB: dimethyldioctadecylammonium bromide; DC-Chol: 3β-[*N*-(*N′,N′*-dimethylaminoethane)carbamoyl]cholesterol; DOSPER: 1,3-dioleoyloxy-2-(6-carboxyspermyl)propylamide.

be. Usually, cytofectin and neutral lipid components are mixed together in an appropriate molar or weight ratio and then induced or formulated into vesicles. Alternatively, cytofectins may be assembled into micellar structures after being dispersed in water or aqueous organic solvents. There are a number of commercially available vesicular or micellar cationic systems available (**Table 1**).

Vesicles or micelles are combined with nucleic acids to form cationic liposome/micelle–nucleic acid complex (lipoplex; LD) mixtures consisting of nanometric particles that are able to deliver nucleic acids into cells (**Figure 2**). These nanometric particles are thought to enter cells largely by endocytosis triggered by nonspecific interactions between complexes and the cell surface proteoglycans of adherent cells (**Figure 2**). Once inside, a proportion of the bound nucleic acids either escape from early endosomes into the cytosol to perform a therapeutic function there, as in the case of RNA (path B), or else traffic from cytosol to the nucleus in order to perform a function there instead, as in the case of DNA (path C).

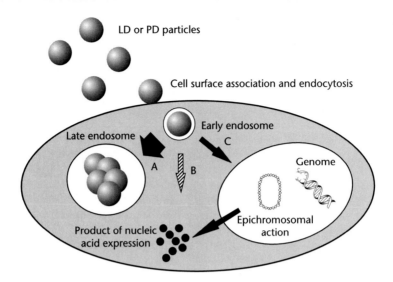

Figure 2 Diagram showing the process of cationic liposome/micelle–nucleic acid complex (lipoplex) (LD) cell entry. LD or cationic polymer–nucleic acid complex (polyplex) (PD) particles that have not succumbed to aggregation and/or serum inactivation associate with the cell surface and enter usually by endocytosis. The majority in early endosomes become trapped in late endosomes (path A) and the nucleic acids fail to reach the cytosol. A minority are able to release their bound nucleic acids into the cytosol. Path B is followed by RNA that acts directly in the cytosol. Path C is followed by DNA that enters the nucleus in order to act. The diagram is drawn on the assumption that plasmid DNA has been delivered and is expressed in an epichromosomal manner. Reproduced from Miller (1999) with the kind permission of Bios Scientific Publishers Ltd.

This process is seductively simple, but is in reality deeply inefficient. The problems of simple cationic liposome/micelle-mediated nucleic acid delivery are many. LD particles formed from cationic liposome/ micelles and nucleic acids are difficult to formulate reproducibly. They are susceptible to aggregation, are difficult to store long term and do not mediate reproducible transfections *in vivo, ex vivo* or even *in vitro*. Moreover, LD particles are not cell-type specific, they appear to be slow to enter cells (hours), are prone to endosome entrapment, and appear to be only weak facilitators of nucleic acid entry into the cell nucleus. DNA entry within the nuclear envelope appears impossible without the intervention of M phase in the cell cycle when the nuclear membrane is partially dismantled to allow mitosis and cell division to take place. If this were not enough, LD particles are very highly unstable in biological fluids (e.g. high salt and serum). *In vivo* topical lung delivery is beset by problems from mucus, intravenous (iv) and intraarterial (ia) delivery by serum components such as acidic serum albumin proteins, low-density lipoprotein, macroglobulins and other low-molecular-weight components. Hydrophobic, negatively charged proteins such as serum albumin associate with LD particles and inhibit direct cellular uptake, as well as opsonize complexes for scavenging by the reticuloendothelial system (RES). Low-molecular-weight lipids like oleic acid and large glycosides like heparin also disrupt LD structural integrity by displacing nucleic acids and lipid components leading to heavily

impaired transfection efficiency. LD particles even activate complement, and bacterially derived plasmid DNA, most usually used to prepare LD particles, now even appears to be immunogenic, eliciting immune responses from unmethylated CpG islets.

In spite of these problems, some simple cationic liposome systems have found favor in clinical trials, not the least for topical lung delivery in cystic fibrosis clinical trials (Alton *et al.*, 1999). Furthermore, there has been a steady stream of recent reports concerning the application of simple systems to deliver genes *in vivo*, for instance to brain, tumors, nasal passages and inflamed joints in animals. However, in general there has been a growing realization that nucleic acid delivery mediated by simple cationic liposome/micelle-based systems is unlikely to lead to routine clinical gene therapy applications without further adaptation.

Cationic polymer systems

Simpler in conception even than cationic liposome/ micelle-based systems, an impressive range of cationic polymers have been used to mediate gene delivery (**Table 2**).

Polymers are combined with nucleic acids to form cationic polymer–nucleic acid complex (polyplex; PD) mixtures also consisting of nanometric particles that are able to deliver nucleic acids into cells (**Figure 2**). Broadly speaking, transfection with PD systems prepared from simple cationic polymer systems appears to suffer from similar problems to transfection with

Table 2 Selection of polymer systems

Polymer	Typical composition
pLL	19–1116 L-lysine residues
PAMAM	Sixth-generation dendrimer; intact or fractured
PEI	580 ethylenimine units (branched); 510 units (linear)
APL PolyCat57	Glucaramide-based system
PEVP	P_W 500 ethyl vinylpyridine polymer
p(DMAEMA)	50±30 dimethylaminoethylmethacrylate units
PEG-pLL	Block copolymer of PEG and pLL
[NaeNpeNpe]$_{12}$	Aminoethyl glycine/phenylethyl glycine peptoid
TEG or GD5	Chimeric fusion proteins
KALA or [K]$_{16}$RGD	Bifunctional peptides

pLL: poly-L-lysine; PAMAM: polyamidoamine; PEI: polyethylenimine; PEVP: poly(N-ethyl-4-vinylpyridinium bromide); p(DMAEMA): poly(2-(dimethylamino)ethyl methacrylate); PEG: polyethylene glycol; Nae: N-(2-aminoethyl)glycine; Npe: N-(2-phenylethyl)glycine.

LD systems. The same patterns of irreproducibility, instability, poor targeting and even toxicity are frequently encountered giving rise to limited *in vivo* use in particular. Of these cationic polymer systems, polyethylenimine (PEI) is generally the most effective with a proven track record for gene delivery *in vitro* and *in vivo* in animals. Moreover, in its most recent incarnation, PEI is arguably the most effective simple synthetic gene delivery agent in mice for lung transfection following iv delivery of PEI-containing PD mixtures (Zou *et al.*, 2000). However, this is unlikely to be of immediate clinical use.

Complex Synthetic Gene Delivery Systems

Problems observed using simple LD and PD mixtures for transfection have led and will continue to lead to new innovations in the design of synthetic gene delivery systems in order to meet these problems head-on. It is unclear how many of these problems need to be solved before routine clinical utility is achieved. Partial or even complete solutions to all these problems will probably be necessary. Some approaches are as follows:

- Physical instability in biological fluids, immunogenicity and opsonization may be reduced using steric protection afforded by introducing polyethylene glycol (PEG) moieties to the surface of LD or PD particles. Such an approach has been well tested in the field of stealth liposomes. However, PEGylation frequently has an adverse effect on the efficiency of nucleic acid delivery. Other stealth moieties may be needed.

- Poor cell-type specificity and slow cell entry may be tackled using targeting moieties attached to the surface of LD or PD particles. Such an approach has been attempted using 'cell-surface-specific' small molecules, peptides, sugar and protein ligands. Convincing targeting *in vivo* is elusive. Cell surface receptor-specific antibodies appear to be promising ligands as are the protein transferrin and folate. Most applications with these ligands have been directed toward cancer cell targeting.

- Intracellular endosome entrapment appears to be easily solvable. Polymers such as PEI have an inbuilt endosmolytic capacity mediating endosome membrane disruption by an osmotic shock mechanism. Alternatively, LD or PD particles can be equipped with pH-sensitive lipid components and/or endosmolytic peptides, both triggered to disrupt endosome membranes as endosome pH drops below neutral after the process of endocytosis is complete.

- Entry of nucleic acids into the cell nucleus of resting cells is problematic. Details of the structure of the nuclear pore complex were still being elucidated in 2002. However, scrupulous use of nuclear localization signal (NLS) peptides may yet solve this problem, which represents a severe issue for synthetic gene delivery systems used for somatic gene delivery other than cancer (Zanta *et al.*, 1999).

No one synthetic gene delivery system has been shown to incorporate successfully all of the features described above, although combinations have been used. However, complex gene delivery systems comprising elements of all main points are certain to emerge. Fundamentally, synthetic gene delivery systems will be mostly likely effective *in vivo* if they can be made 'triggerable', that is stable and nonreactive in extracellular fluids but unstable once recognized and internalized by target cells in the target organ of choice. This paradox goes to the heart of the matter.

Platform Technologies

Research has amply demonstrated that attempts at systematic improvement of any synthetic gene delivery system are destined to be fruitless unless the most fundamental problems associated with achieving properly reproducible transfection outcomes, reproducible formulations and stable long-term storage of LD or PD mixtures are solved. Systematic improvements need to be based on properly defined, stable

LD or PD systems that can be easily upgraded in clearly defined, modular ways drawing from the list of possible improvements outlined above. Synthetic gene delivery platform technologies which answer some of these problems, for instance liposome : mu : DNA (LMD) and stabilized plasmid–lipid particle (SPLP) systems have begun to emerge since 1999.

LMD is a ternary LD system built around the μ (mu) peptide associated with the condensed core complex of the adenovirus (Tagawa *et al.*, 2002). Homogeneous LMD particles (120 ± 30 nm), that consist of a mu : DNA (MD) particle encapsulated within a cationic bilammellar liposome, may be formulated reproducibly, are amenable to long-term storage at $-80\,^{\circ}$C and are stable to aggregation at DNA concentrations appropriate for facile use *in vivo*. LMD transfections are notably time and dose efficient *in vitro*, and will also take place in the presence of biological fluids (e.g. up to 100% serum) as well as *in vivo*, suggesting that the LMD formulation exhibits an additional element of stability. LMD particles were designed to allow modular upgrading with relative ease for further specific applications of interest. SPLP systems are generated when plasmid DNA is trapped within very well defined vesicular particles (diameter approximately 100 ± 40 nm) stabilized by PEG moieties (PEG-CerC$_8$), using a detergent dialysis procedure that employs a dialysis medium dosed with citrate and NaCl designed to enhance the entrapment process (55–70% efficient) (Zhang *et al.*, 1999). Although free DNA needs to be removed before use, SPLP particles so obtained are reasonably transfection competent *in vitro* and *in vivo*, and show no changes in size or DNA encapsulation when stored at $4\,^{\circ}$C for at least 5 months. Good *in vivo* results have been reported using a variation of SPLP in the form of an immunocationic liposome system based around very low ratio cationic liposomes doped with PEG-lipid variants, one for stabilization and one for the covalent attachment of an antibody specific for the murine blood–brain barrier (BBB) transferrin receptor (Shi and Pardridge, 2000).

Other platform technologies may well emerge from the work of Huang and coworkers on liposome-entrapped polycation-condensed DNA systems (LPDII) (Lee and Huang, 1996). In these, DNA was first complexed to poly-L-lysine (pLL) in the ratio $1 : 0.75$ (wt/wt) and then entrapped into folate-targeted pH-sensitive anionic liposomes composed of DOPE/cholesteryl hemisuccinate/folate-PEG-DOPE $6 : 4 : 0.01$ (m/m/m) via charge interaction, giving small particles (74 ± 14 nm). A number of variations of LPDII have since been reported. Although there remain to be convincing demonstrations that LPDII systems will transfect *in vivo*, LPDII

systems appear robust and credible platform technologies.

Semisynthetic Gene Delivery Systems

The difficulties experienced in working with fully synthetic gene delivery systems have resulted in attention shifting to semisynthetic virosome systems. The term virosome, originally coined in reference to combinations of liposomes and various virus glycoproteins, is now used more generally to refer to viral/nonviral hybrid vector systems. Of these, the HVJ–liposome system is instructive. This semisynthetic system is prepared from a combination of ultraviolet-irradiated virions of the hemagglutinating virus of Japan (HVJ; *Sendai virus*) and liposomes in which are encapsulated nucleic acids complexed with the high-mobility group 1 (HMG-1) protein (Kaneda, 1999). Although negatively charged and approximately 350–500 nm in size, HVJ–liposomes themselves have been used in an impressive range of local systemic and anticancer applications *in vivo* involving both ODN and pDNA delivery. One major reason for the success of HVJ–liposomes is the presence of the hemagglutinin neuraminidase (H$_N$) and fusion (F) glycoproteins in the liposome bilayer. These are fusogenic proteins that allow HVJ–liposomes to interact with cell surface sialic residues, fuse with the cell membrane and then release encapsulated nucleic acids directly into the cytoplasm, bypassing endocytosis altogether. HVJ–cationic liposomes are also known and have proved able to mediate delivery of nucleic acids to tracheal and bronchiolar epithelial cells *in vivo* with reasonable efficiency. In addition, an HVJ–AVE liposome system has emerged assembled from anionic artificial viral envelope (AVE) liposomes with a lipid composition similar to that of *human immunodeficiency virus* (HIV) retroviral envelopes. These appear to be slightly more efficacious than conventional HVJ–liposomes.

Duration of Gene Expression

Seeking control of the duration of gene expression is an important avenue for research into synthetic gene delivery systems. Plasmid DNA delivery has dominated reported applications of synthetic delivery systems. Consequently gene expression is rarely persistent beyond 7–14 days after transfection. Applications of minigene constructs, artificial chromosomes and even viral genome constructs (for instance, adenoassociated virus constructs) could see widespread use, as the problems of working with fully synthetic gene delivery systems are resolved.

References

Alton EWFW, Stern M, Farley R, *et al.* (1999) Cationic lipid-mediated CFTR gene transfer to the lungs and nose of patients with cystic fibrosis: a double-blind placebo-controlled trial. *Lancet* **353**: 947–954.

Kaneda Y (1999) Development of a novel fusogenic viral liposome system (HVJ-liposomes) and its applications to the treatment of acquired diseases. *Molecular Membrane Biology* **16**: 119–122.

Lee RJ and Huang L (1996) Folate-targeted, anionic liposome-entrapped polylysine-condensed DNA for tumor cell-specific gene transfer. *Journal of Biological Chemistry* **271**: 8481–8487.

Miller AD (1999) Nonviral delivery systems for gene therapy. In: Lemoine NR (ed.) *Understanding Gene Therapy.* pp. 43–70. Oxford, UK: Bios Scientific Publishers.

Miller AD (2002) Non viral liposomes. In: Springer CJ (ed.) *Suicide Gene Therapy: Methods and Protocols for Cancer.* Totowa, NJ: Humana Press.

Shi NY and Pardridge WM (2000) Noninvasive gene targeting to the brain. *Proceedings of the National Academy of Sciences of the United States of America* **97**: 7567–7572.

Tagawa T, Manvell M, Brown N, *et al.* (2002) Characterization of LMD virus-like nanoparticles self-assembled from cationic liposomes, adenovirus core peptide μ (mu) and plasmid DNA. *Gene Therapy* **9**: 564–576.

Zanta MA, Belguise-Valladier P and Behr JP (1999) Gene delivery: a single nuclear localization signal peptide is sufficient to carry DNA to the cell nucleus. *Proceedings of the National Academy of Sciences of the United States of America* **96**: 91–96.

Zhang YP, Sekirov L, Saravolac EG, *et al.* (1999) Stabilized plasmid-lipid particles for regional gene therapy: formulation and transfection properties. *Gene Therapy* **6**: 1438–1447.

Zou SM, Erbacher P, Remy JS and Behr JP (2000) Systemic linear polyethylenimine (L-PEI)-mediated gene delivery in the mouse. *Journal of Gene Medicine* **2**: 128–134.

Further Reading

Chesnoy S and Huang L (2000) Structure and function of lipid–DNA complexes for gene delivery. *Annual Review of Biophysics and Biomolecular Structure* **29**: 27–47.

Cooper RG, Harbottle RP, Schneider H, Coutelle C and Miller AD (1999) Peptide mini-vectors for gene delivery. *Angewandte Chemie International Edition* **38**: 1949–1952.

Drummond DC, Zignani M and Leroux JC (2000) Current status of pH-sensitive liposomes in drug delivery. *Progress in Lipid Research* **39**: 409–460.

Gerasimov OV, Boomer JA, Qualls MM and Thompson DH (1999) Cytosolic drug delivery using pH- and light-sensitive liposomes. *Advanced Drug Delivery Reviews* **38**: 317–338.

Wente SR (2000) Gatekeepers of the nucleus. *Science* **288**: 1374–1377.

Wyman TB, Nicol F, Zelphati O, *et al.* (1997) Design, synthesis and characterization of a cationic peptide that binds to nucleic acids and permeabilizes bilayers. *Biochemistry* **36**: 3008–3017.

Systems Biology: Genomics Aspects

Andrzej K Konopka, *BioLingua Research Inc., Gaithersburg, Maryland, USA*

Intermediate article

Article contents

- Introduction
- Von Bertalanffy's General System Theory
- Systems Biology: The Legacy and the Future
- Reductionism versus Holism
- The Modeling Relation

Systems biology is concerned with functionally complex systems (such as an organism, immune system or ecosystem) that can be approximately described by several complementary models but cannot be adequately represented by a single model. It addresses foundational issues of 'entire' biology (such as system modeling, fractionation, integration and emergence) but it is also devoted to specific techniques for data integration and interpretation. This brings systems biology right in the middle of the old controversy between reductionism and holism.

Introduction

Systems biology – in the most general sense – is dedicated to studies of pairs consisting of observer and observed where the latter is assumed to be a living thing. For over two millennia the term 'system' has been a cliché used to signify a 'whole' as an opposite of an arbitrary collection of individual objects (parts). In biology the term means a complex conglomerate of objects and processes bound together in a functionally efficient unit such as an organism, immune system or ecosystem. Despite the existence of several technical meanings of 'system' in biology (such as, for instance, the Linnaeus classification system for plants and animals) the original general concept of system as a functionally robust whole has remained at the center of interest of biologists since at least Aristotle.

On the other hand an observer (often assumed to be a creature capable of rational thinking) observes the world in an active way. Observations are not only performed and memorized (registered), but the scope of every act of observing can be controlled. In idealized cases the control can be either internal (inside the observer's mind) or external (by manipulation of circumstances in which the observation is made), but in reality it is a convoluted combination of both of these extremes.

Most of the methodologies of science in use today attempt to minimize the effects of internal control (the observer's mind) on scientific observations by setting appropriate habits of experimentation and modeling. However, these efforts have only been partly effective in selected subfields of physics. Research in biology and chemistry continues to rely not only on external but also on internal controls that in some subcultures of physics could be considered 'subjective'.

One reason for this elevated role of the observer's (the knower's) mind in the process of observing (knowing) is the complexity of the observed (knowable) itself. Another reason is limitations of the (natural) object language in which observer's causal questions (what, how and why) are formulated and the answers are communicated. Yet another reason is constraints imposed by the meta-language in which the validity of questions and answers can be judged.

Von Bertalanffy's General System Theory

General system theory (also known as 'cybernetics', 'systems research' or 'systems thinking') constitutes a vaguely defined academic trend that spans several traditional fields of science. It promotes the idea that functional organization of the system as a whole can be described by concepts and principles that are independent of the properties of the components. In other words it is assumed that:

- there exist general laws (or rules) that govern the behavior of systems (functioning) as a whole and
- the rules do not need to be derivable from the properties of the components of the system.

Perhaps systems thinking in biology reached its maximum prominence between the 1930s and the 1970s. General system theory, while contributing useful language, is not specific enough to be usable for interpretation of biological data or formulation of theories about these data. Four more specific complementary paradigms have been tried in this respect, namely chaos theory, cellular automata, catastrophe theory and hierarchical systems.

The chart in **Figure 1** shows the methodological relationships contributing to a systems view of biology consistent with general system theory. It can be seen from this figure that hierarchical systems constitute a focal point for methods of systems biology. It should be noted that, as **Figure 1** also indicates, hierarchical

systems and the concept of hierarchy pertain strongly to a category of clear thinking, while three other incarnations of general system theory (chaos, cellular automata and catastrophes) are all variants of dynamic systems.

Systems Biology: The Legacy and the Future

The merits of systems biology today have almost nothing (direct) to do with either technology or instrumentation; nor are the developments in genomics and proteomics of particular importance for systems biology, although systems biology is a potential source of useful methods for these fields. Instead, systems biology is concerned with the methodological foundations of biology on the one hand and with techniques for approximate data integration on the other.

Perhaps the most prominent issue at the outset is establishing the material adequacy of definitions (or metaphors) of core concepts such as complexity, system, organism, machine and historic narrative. Another foundational task is establishing complementary methods and protocols for modeling complex systems. In this respect a general theory and practical principles of modeling, at the most general level, are currently at the core of research in systems biology as well as in other subfields of systems science.

As far as data integration is concerned, there are no reasons for even moderate optimism. Techniques for approximate integration of data from the molecular level with data and concepts from other levels of biological organization (such as cellular, organismal or population) are being discussed, but little, if any, progress in this area has been made since the 1960s. Even the simplest conceptual issues that involve data integration, such as the relationship between genotype and phenotype, are extremely difficult to model by means existing at present. As a matter of fact it is unclear if a derivation of epigenetic rules or properties of phenotypes from what is known about genes at the molecular level is even feasible in principle. Similar projects in other fields of research (such as for instance the complete derivation of macroscopic thermodynamics from the microscopic principles of statistical mechanics) have failed miserably in the past. Diverse techniques for combining qualitatively different databases are being proposed and tested along with new methods of data storage and acquisition. The outcomes of these proposals will certainly be useful for computer science and 'information' technology (IT) but such brute force database activities cannot guarantee proper addressing (not to mention a

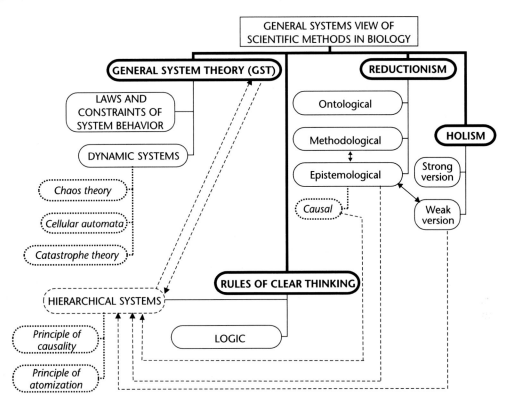

Figure 1 Systems view of biology. The borders (outlines surrounding boxes) and connecting lines indicate the level of derivation. Boxes with bold outlines (and bold connecting lines) are at the beginning of this particular order. Boxes with thin outlines (and thin connecting lines) indicate the second level, while boxes with dotted outlines and dotted connecting lines signify the third level (last in this diagram). Box with dashed outline and dashed arrows indicate a special role of hierarchical systems. This box should normally be at the second level (boxes with thin outlines and thin connecting lines), if it was not so special. The four main agencies, viz. (1) General system theory, (2) Rules of clear thinking, (3) Reductionism and (4) Holism, constitute the foundation for systems thinking in biology. Hierarchical systems appear to play a special role (indicated by the dashed arrows). Not only do they constitute one of two major aspects of clear thinking (the second aspect is logic) but also, at the same time, they are variants of general systems. In addition they can conform to both weak version of holism and an epistemological version of reductionism. Their conformity to causal reductionism (preference for upward causation and against downward causation) has also been prevalent in biological research to date.

solution) of the problem of data integration. Grand schemes of involving a vague metaphor of 'system thinking' into otherwise routine (and well-defined) computer simulations of complex processes do not have a very successful track record either.

In these circumstances one can safely guess that the real contribution of systems biology to science in the future will in fact be an efficient logicomethodological way of representing the complexity of the material world.

Reductionism versus Holism

All science (including biology) conforms to two universally accepted principles of clear thinking:

- The principle of causality – every scientifically explainable object, process or fact can be thought

of as materially or logically entailed by a cause (also a fact, object or process).
- The principle of logicolinguistic atomization – every scientifically explainable system can be represented either as a subsystem (or a part) of a plausible (or well defined) larger system or divided into subsystems whose properties are believed to be plausible or well defined.

To be precise, as indicated in **Figure 1**, on top of these two rules there is also an assumption of the existence of hierarchy of things (hierarchical system): a given object, process or fact belongs to a hierarchical organization and, at the same time, contains things organized hierarchically. This assumption appears to be as natural a habit for representing and naming as logic is believed to be for reasoning.(*See* Genetics, Reductionism and Autopoiesis.)

These two robust, common-sense, principles, plus the fact that concept of 'complexity' is difficult to adequately define other than metaphorically, constitute an excellent reason for rejuvenation of the millennia-old combat between reductionism and holism. According to the commonsense, naïve, interpretation of reductionism, life should be fully explained via exhaustive studies of systems of differential equations (dynamic systems) but very few, if any, biologists would see the point in following this research program. An extreme version of this paradigm adopts an assumption that phenotype can ultimately be modeled by dynamic systems (i.e. systems of differential equations in which rates and forces are time derivatives). In a slightly less extreme version of naïve reductionism, genes are assumed to prescribe (in an unknown way) the epigenetic rules (also unknown) that in turn control interactions (of unknown nature and number) of proteins (of unknown kind and function) with each other as well as with other (also unknown) ligands. Clearly the naïve reductionist agenda, that has been so successful in physics in the past three centuries, is standing on shaky ground in 'postgenome' molecular biology.

Some remarks on terminology

A closer look at the reasons why extreme reductionist strategies do not satisfy the rigors of biological explanation leads to the realization that reductionism is practiced in at least four variants:

- Ontological – a view that the whole is strictly 'nothing but' the sum of parts.
- Methodological – a research strategy that explores a concept of substituting a given system with a surrogate system.
- Epistemological – a view that the whole can (in principle) be a sum of parts subject to adequacy of the definition of 'parts' and 'sum'.
- Causal – a view that upward causation ('the whole is there because of the parts) should be legitimate but downward causation ('the parts are there because of the whole') should not.

The methodological version leads to a clear-cut distinction between the natural system and its theory. This distinction is in fact reflected in the Rosen–Hertz modeling relation discussed later (see also **Figure 2**). In the case of complex systems, the boundary between the modeled system and its model (the 'epistemic' cut) is not easy to find and, when found, often leads to paradoxes or inadequate theories.

Causal reductionism is technically derived from epistemological variety as a special case (see **Figure 1** for a fuller presentation of the systemic view of biology). Nonetheless this is the version of reductionism that fails in a most visible way in biology while being extraordinarily successful in physics. A probable

reason for this failure in the case of complex biological systems is the fact that final causation and the concept of function are in fact important ingredients of a plausible biological explanation. A programmatic rejection of downward causation must upset any methodology devoted to explanations in functional and not only structural terms. Complex systems (such as living things) are in fact convoluted in terms of their function (that is why the concept of functional complexity in biology needs to be separated from the mere complication due to a very large number of objects and the relations between them) and therefore causal reductionism often leads to inadequate or even irrelevant explanations of their behavior (functioning).

Holism is known in at least two variants:

- Strong – a view that putting together of their parts will not produce the wholes (such as living systems) or account for their properties and behaviors. Emergence is impossible even in principle.
- Weak – a view that the parts of a complex system (such as a living thing) have mutual relations that do not exist for the parts in isolation. Emergence is in principle possible but difficult to make sense of for pragmatic reasons.

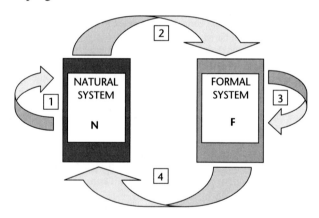

Figure 2 Rosen–Hertz modeling relation. A general diagram of a modeling relation (MR) between a natural system N and a surrogate formal system F (see Rosen's *Life Itself* (Further Reading) for more detailed explanations). Arrow (step 1) represents an act of descriptive observation of natural system N that generally reflects casual entailments within N. Arrow 2 symbolizes the process of abstraction via which the observer encodes observations into symbols (such as numbers or words). Arrow 3 represents all acts of formal inference (permitted in F) such as derivation of conclusions from premises, solving equations or generating grammatically correct sentences from words. Arrow 4 stands for a mapping of formal constructs, such as formulae or theorems, back into the natural system. In other words, arrow 4 signifies an act of prediction about N made on the basis of its formal description generated within F. Formal system F is defined as a model of the natural system N if, and only if, the outcome of observation 1 within N is the same as a final combined outcome of formalization 2 followed by inference 3 and then followed by prediction 4 (i.e. the MR commutes).

Most biologists appear to subscribe to a weak version of a holistic world view according to which properties of biological systems cannot be derived from the properties of their subsystems alone. The main reason why such derivation is questionable, even in principle, is the alleged fact that the observer cannot be naturally separated from the observed living system. (The analogy with complex numbers comes to mind here. A given complex number can be represented as a pair of real numbers but it cannot be derived within any set of individual real numbers (i.e. a complex number C can be interpreted as a point in R^2 but not in R^1).) On the other hand, a forced (i.e. not a natural) separation of the observer from the observed (the epistemic cut) can be done within every model of the system under consideration. The price to pay for this necessary separation is an increased risk of generating inadequate models. In these circumstances, a reasonable future for systems biology could be the invention of reliable methods of measuring the material adequacy of models and definitions. With such methods at hand, biologists would be able to effectively address, identify and perhaps even overcome the real misgivings about epistemological reductionism. In particular, an accurate evaluation of material adequacy could help to formulate research questions in answerable ways.

The Modeling Relation

Figure 2 shows a general scheme, called a modeling relation (MR), that summarizes a process of modeling a natural system (of causal entailments) N by substituting it with a surrogate formal system (of inferential entailments) F. This concept of a model not only clarifies fractionation – the methodological reduction needed for studies of simple systems (mechanisms) – but also gives rise to an adequate notion of biological complexity. In addition, its emphasis on the substitution of systems by their surrogates provides a universal tool (theory) for comparing entailment structures of any kind.

At least in biology there is a close relation between our ability to make controlled observations (experiments) and the language that we have at our disposal. In the case of complex systems, which biologists experiment with and theorize about, there is rarely a possibility to see the real natural system as it is. Instead we are making even initial observations on more or less adequate models of the inaccessible natural system (i.e. a living thing itself). For instance, we do not know what exactly makes the difference between a living animal and a dead one but we have good guesses (models) about the rate and nature of

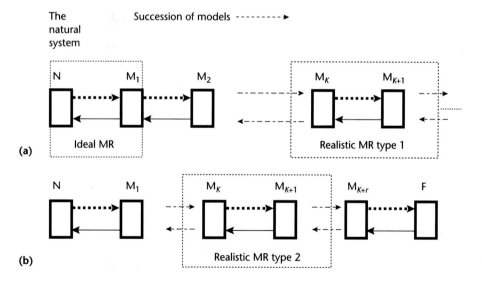

Figure 3 Selected aspects of modeling complex systems: cascades of models. For simplicity, observations within left sides and inferences within right sides of modeling relations (MRs) are not shown. Encoding 'lower-level' systems into their models ('higher level') is indicated by broken thick arrows, while decoding from models to modeled systems is symbolized by unbroken arrows. These arrows indicate real encodings that we know about, while the thin broken arrows indicate encodings (decodings) whose existence we surmise but do not know in detail. (a) The ideal MR could occur only between the original natural system and the first model in the cascade of successive models. The real MR of type 1 occurs between a model at the level k of the cascade and the following model $k + 1$. We generally do not know the value of k, but in the complex system theory it is assumed that there is no limit on how high this value could be (no ultimate formal model.). (b) A realistic MR occurs again between the model k and $k + 1$. However, the entire situation is different because there exists the ultimate (original) natural system N as well as an ultimate formal model F (the 'largest' model) at the end of the cascade of models. This situation holds only for simple (even if complicated) mechanisms but is believed not to be a case for functionally complex systems.

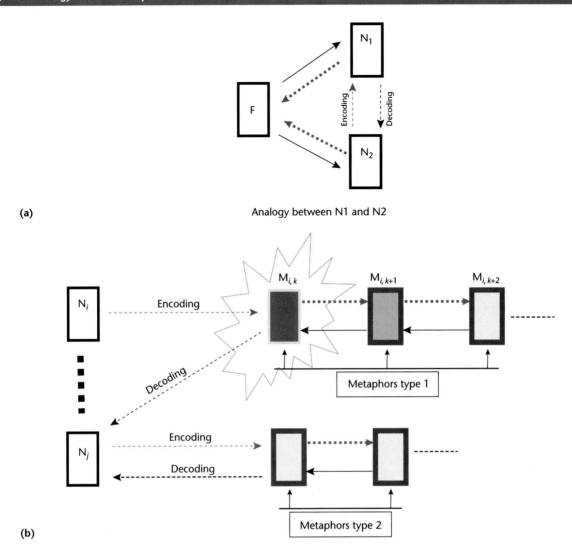

(a)

Analogy between N1 and N2

(b)

Figure 4 Systems (relational) view of analogies and metaphors. The rectangles signify systems with internal entailment structure. Both causal and inferential entailments are taken into account. Arrows between rectangles symbolize encodings (broken) and decodings (unbroken) taking place between modeled systems and their models. (a) Natural systems (or models) N_1 and N_2 are analogous if they both are in a commuting modeling relation with the same formalism F. Thin broken arrows indicate direct encodings and decodings of unknown intensity that may or may not take place between N_1 and N_2 in addition to their firm connection through a formalism (model) F.
(b) A kth model in the ith cascade does not decode into any previous model in the same cascade. Instead it decodes into a system that is modeled by a different cascade. In this sense model $M_{i,k}$ is a metaphor. Further encodings and decodings of it are metaphors as well.

metabolism, brain activity, mobility and so on. None of the individual models alone is expected to give us a reliable indication but a combination of the complementary models might. (*See* Biological Complexity: Beyond the Genome)

Cascades of models

Figure 3 shows a sequence of connected Rosen–Hertz MRs such that the right-hand side of each preceding MR is the left-hand ('natural') side of the next MR. We could also imagine that a first MR in this cascade,

say MR_0, exists, in which case its left-hand side would reflect the very natural system N that we intended to study. As we have already observed, the MR_0 is an ideal situation that is quite common for simple systems such as clockwork mechanisms but which does not take place at all in legitimate cases of complex systems. Some researchers even consider this 'indirect only' observability of functionally complex systems to be a good potential criterion for an operational definition of complexity.

The MR indicated in **Figure 3a** as type 1 has on its left-hand side a system that itself is a result of a certain (but generally unknown) number k of modeling steps (encodings). This system model at the level k is then the subject of further modeling into a model at the level $k+1$. Assuming that we feel comfortable with the model $k+1$, in other words we believe that it explains what we need to know about the system k, we may ask how much can we learn (from model $k+1$) about the original natural system N. The answer to this is a subject of intensive current research but cannot be provided quite yet. It depends on how strong the encodings of interest are and how we measure this strength. It also depends on how strong the decoding is all the way 'down' to the very natural system N (or at least its closest model). One thing is certain though: methods of measuring the strength of encoding (decoding) within MRs are in demand today and almost certainly will be invented in the future.

Analogies and metaphors

As illustrated in **Figure 4**, analogies can be understood as special kind of modeling relation in which two different systems are modeled by the same model. However, the concept of metaphors is more complicated. In order to create an elementary (initial) metaphor, we need to decode a model into a system that was not originally encoded into this model. Classic examples in biology include metaphors of machine, language and organic system from which a host of technical concepts of molecular biology are derived. (*See* Information Theories in Molecular Biology and Genomics)

Distant encoding and decoding that span an unknown number of modeling steps create an impression that metaphors are not precise enough to lead to a profound understanding of modeled systems. However, this lack of precision could be advantageous for model making if we could only measure its extent. Once again, quantitative measures of the strength of incompletely understood encodings and decodings appear to be in demand from a practical point of view.

See also

Biological Complexity: Beyond the Genome
Genetics, Reductionism and Autopoiesis

Further Reading

von Bertalanffy L (1969) *General System Theory: Foundations, Development, Applications*. New York, NY: George Braziller.

von Bertalanffy L (1975) *Perspectives in General System Theory*. New York, NY: George Braziller.

Konopka AK (2002) Grand metaphors of biology in the genome era. *Computers and Chemistry* **26**: 397–401.

Mayr E (1998) *This is Biology: The Science of the Living World*. Cambridge, MA: Harvard University Press.

Morange M (2001) *The Misunderstood Gene*. Cambridge, MA: Harvard University Press.

Pattee HH (2001) The physics of symbols: bridging the epistemic cut. *BioSystems* **60**: 5–21.

Peacocke A (1976) Reductionism: a review of the epistemological issues and their relevance to biology and the problem of consciousness. *Zygon* **11**: 307–336.

Rosen R (1991) *Life Itself*. New York, NY: Columbia University Press.

Rosen R (2000) *Essays on Life Itself*. New York, NY: Columbia University Press.

Smuts JC (1929) Holism. *Encyclopaedia Britannica*, p. 640. London, UK: Encyclopaedia Britannica.

Ulanowicz RE (1999) Life after Newton: an ecological metaphysics. *BioSystems* **50**: 127–142.

Tay–Sachs Disease

Don J Mahuran, *Hospital For Sick Children, Toronto, Ontario, Canada*

Tay–Sachs disease was first described in the late nineteenth century. Not only has it served as a model for other lysosomal storage diseases but, as its disease mechanisms have been elucidated, other biological processes have been identified and characterized. Interestingly, mouse models of the disease do not produce a phenotype until very late in life, demonstrating that data from mice cannot always be extrapolated to humans.

Introduction

Tay–Sachs disease (TSD) was independently described by Warren Tay, a British ophthalmologist, in 1881 (Tay, 1881) and Bernard Sachs, a physician at New York's Mount Sinai Hospital, in 1896 (Sachs, 1896). Tay was first to note the now classical clinical characteristic of TSD, a 'cherry red' spot in the retina of a 1-year-old child with both mental and physical retardation. Sachs described a fatal disease common to his Jewish patients that involved retardation and blindness which he called 'amaurotic family idiocy'. He also noted that neurons from such patients appeared to have distended cytoplasms. The dysmorphology of TSD neurons was later found to be caused by the presence of multiple 'membranous cytoplasmic bodies' (MCBs) (**Figure 1**). In turn, MCBs were found to be lysosomes storing large amounts of a member, GM2, of a novel class of sialic acid-containing glycolipids, called gangliosides (Klenk, 1939; Svennerholm,

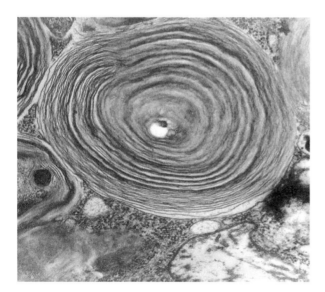

Figure 1 Membranous cytoplasmic body (MCB), as seen in electron micrographs of Tay–Sachs/Sandhoff disease neurons, which represents enlarged lysosomes containing stored ganglioside.

1962). Gangliosides are made up of a hydrophobic ceramide moiety that normally anchors them in the plasma membrane, and a negatively charged, hydrophilic oligosaccharide of varying length that faces the extracellular environment (**Figure 2**). Neuronal cells appear to be the primary producers of the 'higher' gangliosides, that is, those with longer oligosaccharide chains. Given the structure of GM2 ganglioside (**Figure 2**) and the fact that the lysosomal degradation of other oligosaccharides was known to take place through the sequential removal of each newly produced terminal, nonreducing sugar by a group of specific glycosidases, it was thought that TSD was caused by a missing β-*N*-acetyl-galactosaminidase or a more general β-hexosaminidase. In mammalian cells it was found that one enzyme utilized both β-linked *N*-acetylgalactosamine (GalNAc) and *N*-acetylglucosamine (GlcNAc); thus β-hexosaminidase (Hex) became the candidate enzyme to hydrolyze GM2 ganglioside (GM2). Because of the complexity of using membrane-bound GM2 as a substrate, artificial substrates were made to measure Hex activity, the most commonly used one today being 4-methylumbelliferyl-β-*N*-acetylglucosamine (MUG). On the cleavage of the β-GlcNAc the MU product fluoresces, which makes for a highly sensitive assay for Hex. Unfortunately when samples from Jewish Tay–Sachs patients were tested using this type of assay, they proved to have near normal levels of Hex.

A solution to this apparent dead-end came in 1968 when Robinson and Stirling found that there were two Hex isozymes in normal human tissue, an isozyme with an acidic isoelectric point (p*I*) of 4.8 that was heat labile, Hex A, and one with a more basic p*I* (6.9) that was heat stable, Hex B; and Okada and O'Brien (1969) found the A isozyme was missing in Jewish TSD patients. This comfortably simple model for TDS was challenged by Sandhoff, who examined the Hex isozyme profile from a group of TSD patients who were not all Jewish. He found that some patients did indeed lack Hex A, but a few others lacked both isozymes and still fewer had apparently normal levels of both Hex A and Hex B (Sandhoff, 1969). These data were viewed with some suspicion for several years until

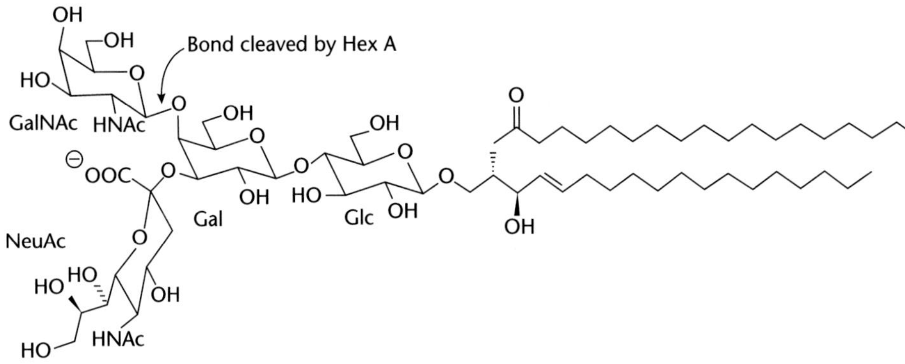

Figure 2 Chemical structure of GM2 ganglioside (NeuAc: sialic acid). The removal of the terminal GalNAc residue by Hex A and the activator produces GM3.

the structure and substrate specificity of each isozyme, and the physiological role of a small glycolipid transport protein in GM2 hydrolysis, the GM2 activator protein (Activator), was elucidated.

Biochemical Causes of GM2 Gangliosidosis

Structure of the Hex isozymes

An early immunological study of the partially purified Hex isozymes came to the now proven conclusion that each isozyme shared at least one subunit, called the β subunit, and that the A isozyme had at least one additional subunit, called the α subunit, with each subunit being encoded by a separate gene. Thus, the simplest structure for each isozyme was 2β for Hex B and αβ for Hex A (Srivastava and Beutler, 1973), with Jewish TSD being caused by the loss of the α subunit (encoded by the *hexosaminidase A (alpha polypeptide)* (*HEXA*) gene), which still allowed Hex B to be formed; and a new disease with the same clinical phenotype, ultimately called Sandhoff disease (SD), being caused by a missing β subunit (encoded by the *hexosaminidase B (beta polypeptide)* (*HEXB*) gene), which prevented the formation of either isozyme. This explained Sandhoff's observation of apparent TSD patients with neither Hex A nor Hex B (see above). However, on closer examination of samples from SD patients, a small amount, around 1–5%, of residual Hex-MUG activity was noted. This activity was associated with a new Hex isozyme with a very acidic pI (∼3.5), which was called Hex S (Beutler *et al.*, 1975; Ikonne *et al.*, 1975). Hex S was found to be composed of two α subunits (Geiger *et al.*, 1977). Thus any dimeric combination of α and/or β subunits produces an active Hex isozyme. These data also indicate that each subunit contains a potentially active site, but as no

active monomers have ever been observed, dimerization is required to make them functional.

Isolation of cDNA clones for the subunits of Hex A

Clones for the α and β subunits were isolated in the mid-1980s (Myerowitz and Proia, 1984; Myerowitz *et al.*, 1985; O'Dowd *et al.*, 1985; Korneluk *et al.*, 1986) and the gene structures were determined a few years later (Proia and Soravia, 1987; Neote *et al.*, 1988; Proia, 1988). These data demonstrated that the *HEXA* (15q23–q24 (Nakai *et al.*, 1991)) and *HEXB* (5q13 (Bikker *et al.*, 1988)) genes evolved from a common ancestral gene resulting in the placement of all but the first of the intron/exon junctions in identical positions in the genes and an approximately 60% identity in aligned, deduced amino acid sequences of the subunits (**Figure 3**). Methods for polymerase chain reaction (PCR) amplification of all exons and intron/exon junctions for purposes of mutational analyses of *HEXA* (Triggs-Raine *et al.*, 1991), *HEXB* (Wakamatsu *et al.*, 1992) and GM2 ganglioside activator protein (*GM2A*) (see below) (Chen *et al.*, 1999) have been published.

Substrate specificity of the Hex isozymes

By definition, all three isozymes can hydrolyze terminal nonreducing β-linked GalNAc and GlcNAc residues from naturally occurring macromolecules and artificial substrates, for example MUG. Despite their high degree of homology (**Figure 3**), however, genetic evidence demonstrates that *in vivo* only Hex A can hydrolyze the GalNAc residue from GM2 ganglioside (Kytzia *et al.*, 1984). There are two reasons for this restricted substrate specificity. Firstly, Kresse *et al.* (1981) noted that Hex S as well as Hex A, but not Hex B, could hydrolyze the terminal β-linked $GlcNAc \cdot 6SO_4$ from keratan sulfate (a

Chain		Amino acid no

α   MTSSRLWFSLLLAAAFAGRATA   22
                 *   *   *   *

β   MELCGLGLPRPPMLLALLLATLLAAMLALLTQVALVVQVAEA   42

α -------------[LWPWPQNFQTSDQRYVLYPNNFQFQYDVSSAAQPGCSVLDEAFQRY   68
                *** *   * * **   * * *   *   *** **

β ARAPSVS[AKPGPALWPLPLSVKMTPNLLHLAPENFYISHSPNSTAGPSCTLLEEAFRRY   101

α RDLLFG]SGSWPRPYLTGKRH{TLEKNVLVVSVVTPGCNQLPTLESVENYTLTINDDQCL   126
      **   *   *   **   *   *   * * ***

β HGYIFG]FYKWHHEPAEFQAK{TQVQQLLVSITLQSECDAFPNISSDESYTLLVKEPVAV   159

α LLSETVWGALRGLETFSQLVWKSAEGTFFINKTEIEDFPRFPHRGLLLDTSRHYLPLSS   185
  *   **************   *** **   * * *** *** * ***+***

β LKANRVWGALRGLETFSQLVYQDSYGTFTINESTIIDSPRFSHRGILIDTSRHYLPVKI   218

α ILDTLDVMAYNKLNVFHWHLVDDPSFPYESFTFPELMRKGSYNPVTHIYTAQDVKEVIEY   245
  ** *** ** ** ** *** **+** **** * ***** ****   * **   **   ****

β ILKTLDAMAFNKFNVLHWHIVDDQSFPYQSITFPELSNKGSYSLS-HVYTPNDVRMVIEY   277

α ARLRGIRVLAEFDTPGHTLSWGPGIPGLLTPCYS GSEP SGTFGPVNPSLNNTYEFMS   302
  ********* **+*** ****** *   *******   *** ** ** ** *

β ARLRGIRVLPEFDTPGHTLSWGKGQKDLLTPCYS}RQNK(LDSFGPINPTLNTTYSFLT   334

α TFFLEVSSVFPDFYLHLGGDEVDFTCWKSNPEIQDFMRKKGFGEDFKQLESFYIQTLLDI   362
  *** * * ****   ****+** * ** *** ****** **** *** ******* ***

β TFFKEISEVFPDQFIHLGGDEVEFKCWESNPKIQDFMRQKGFGTDFKKLESFYIQKVLDI   394

α VSSYGKGYVVWQEVFDNKVKIQPDTIIQVWREDIPVNYMKELELVTKAGFRALLSAPWYL   422
   **   *+****** *** ***   **   *+   *   **   ** **   ******+**

β IATINKGSIVWQEVFDDKVKLAPGTIVEVWK-DSA--YPEELSRVTASGFPVILSAPWYL   451

α NRISYGPDWKDFYVVEPLAFEGTPEQKALVIGGEACMWGEYVDNTNLVPRLWPRAGAVAE   482
    **** **   * **** * **   ** * ****** *+*+*** *** ******* ** *

β DLISYGQDWRKYYKVEPLDFGGTQKQKQLFIGGEACLWGEYVDATNLTPRLWPRASAVGE   511

α RLWSNKLTSDLTFAYERLSHFRCELLRRGVQAQPLNVGFCEQEFEQT}   529
  ****   *   ** **   **   ** **** * * *

β RLWSSKDVRDMDDAYDRLTRHRCRMVERGIAAQPLYAGYCNHENM)   556

Figure 3 The aligned primary structures of the α and β subunits of human Hex A. The *N*-terminal signal peptide of each prepropolypeptide is underlined. Sites (N–X–S/T) shown to contain N (Asn)-linked oligosaccharides are also underlined with the 'Ns' linked to those carbohydrates receiving a mannose-6-phosphate, lysosomal targeting moiety, double underlined. Sequences comprising the mature, lysosomal α_p and β_p chains are in square brackets [ ], those making up the α_m and β_b chains are in braces { }, and those associated with the β_a chain are in parentheses (). The residues between different bracket sets are presumably lost during lysosomal processing. Residues in bold, larger-sized fonts are predicted to be in the active site of the subunits based on molecular modeling from the crystal structure of bacterial chitobiase; the mutant residue associated with the B1-variant of TSD is underlined. The residues (NR) involved in binding the 6SO₄ group in MUG and the sialic acid group of GM2 are indicated by a wavy underline.

glycosaminoglycan). This observation led to the development of a Hex A and S 'specific' artificial substrate MUGS (MU-β-GlcNAc·6SO₄). In fact the MUG/MUGS ratio for each isozyme is Hex B, 200–300:1; Hex A, 3–4:1 and Hex S, 1–1.5:1 (Hou *et al.*, 1996). Thus Hex B has some residual activity toward MUGS. The mechanism behind this difference in specificity involves at least two nonconserved amino acid residues in the subunits, α-Asn423Arg424, which align with β-Asp452Leu453 (**Figure 3**). In the α-active site, the Arg424 actively binds the negatively charged 6SO₄, while in the β, the Asp453 actively repels it (Sharma

et al., 2001). Like GlcNAc·6SO₄, GM2 also contains a negatively charged group, sialic acid (**Figure 3**), which appears to interact at the same sites as the −6SO₄ (unpublished data). However, unlike GlcNAc·6SO₄, GM2 is membrane bound *in vivo* and forms micelles in water *in vitro*. Under these conditions, none of the Hex isozymes can utilize it as a substrate. The addition of detergent, for example sodium taurodeoxycholate, in an *in vitro* assay allows GM2 to be hydrolyzed by Hex S, as well as Hex A, and under these conditions, as with MUGS, Hex B has some residual activity (Geiger *et al.*, 1977; Sandhoff *et al.*, 1977).

GM2 activator protein

The second component that restricts the ability of both Hex S and Hex B to hydrolyze GM2 is the GM2 activator protein (Activator) (Hechtman and LeBlanc, 1977), which *in vivo* fulfills the role of the detergent *in vitro* (Li and Li, 1976), by extracting single molecules of GM2 from the membrane and presenting them to the Hex A for hydrolysis (Conzelmann and Sandhoff, 1979); thus the Activator–GM2 complex is the true *in vivo* substrate for Hex A, and the Activator can be considered a 'substrate specific cofactor' for Hex A (Meier *et al.*, 1991) (**Figure 4**). Whereas this complex can interact with both Hex A and Hex S, it appears that in the case of Hex S, the GM2 molecule cannot be properly orientated into either of its α-active sites

(Kytzia and Sandhoff, 1985). Therefore, elements of the β subunit are necessary for the constructive binding of the complex such that the oligosaccharide portion of GM2 is placed in Hex A's α-active site for hydrolysis (Hou *et al.*, 1998).

A patient with apparently normal levels of Hex A and Hex B, but with a TSD-like clinical phenotype was found to be lacking the Activator (Conzelmann and Sandhoff, 1978). This disease became known as the AB variant form of GM2 gangliosidosis and explained Sandhoff's most confusing observation of isozyme patterns in apparent TSD patients (see above).

The Activator has been 'cloned' (Schröder *et al.*, 1989; Xie *et al.*, 1991) and the gene structure elucidated (Chen *et al.*, 1999; Schepers *et al.*, 2000). The *GM2A* gene maps to 5q 31.3–33.1 (Heng *et al.*, 1993) with a

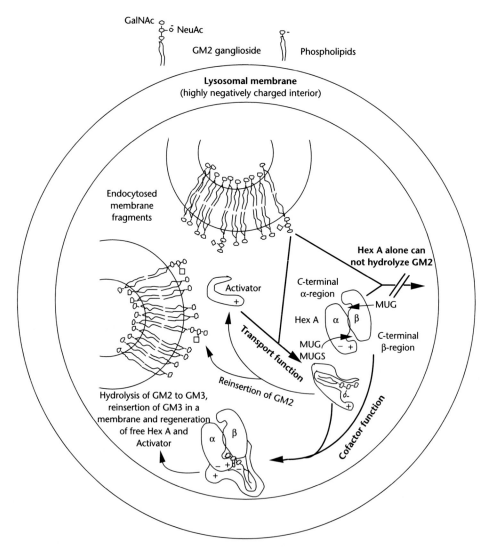

Figure 4 Diagram outlining the steps in the degradation of GM2 ganglioside by Hex A and the Activator.
GalNAc: *N*-acetylgalactosamine; NeuAc: *N*-acetylneuraminic acid; MUG: 4-methylumbelliferyl-β-*N*-acetylglucosamine; MUGS: 4-methylumbelliferyl-β-*N*-acetylglucosamine-6SO$_4$. (Adapted from Sandhoff *et al.* (1989).)

processed pseudogene on chromosome 3 (Xie *et al.*, 1992). The *GM2A* gene encodes a small, ~20 kDa, heat stable protein, ± one Asn-linked oligosaccharide (Glombitza *et al.*, 1997; Rigat *et al.*, 1997) with four disulfide bonds (Schütte *et al.*, 1998). The protein has been crystallized and demonstrated to contain a novel fold called a β cup (Wright *et al.*, 2000). The cup is composed of eight antiparallel β-pleated sheets. Thus, the protein possesses an accessible central hydrophobic cavity, $12\,\text{Å} \times 14\,\text{Å} \times 22\,\text{Å}$ (easily capable of holding the C_{18} acyl chain of GM2), rather than the standard buried hydrophobic core of most proteins.

Clinical Forms of the GM2 Gangliosidoses

The above data demonstrate that mutations in any one of three genes can produce a nearly identical clinical phenotype, ultimately caused by the storage of GM2 in the lysosomes of neuronal cells leading to apoptosis (Huang *et al.*, 1997). Mutations in the *HEXA* gene are associated with the most common Tay–Sachs disease (carrier frequency of 1 in 30–40 Jews of Central and Eastern European extraction, and 1 in 300 for non-Jews), the *HEXB* gene with the less common Sandhoff disease (carrier frequency estimated at 1 in 500 for Jews and 1 in 300 for non-Jews) and the *GM2A* gene with the very rare AB variant form (reviewed in Gravel *et al.* (1995)).

Although the most severe acute/infantile forms of GM2 gangliosidosis have a rather homogeneous clinical presentation, there are less severe forms that show extreme variability in expression (see below). Typically, the earlier the age of onset of clinical symptoms, the more severe the disease. In the past, patients have been classified strictly according to their age at onset, but because this system is so dependent on the initial timing of the diagnosis, a more general nomenclature based on the different clinical phenotypes and recognizing the dominance of the encephalopathy has been suggested (Gravel *et al.*, 1995): acute (infantile), subacute (late infantile and juvenile forms) and chronic (adult and chronic forms).

Acute (infantile)

Clinically classical infantile Tay–Sachs patients (or patients with other acute forms of GM2 gangliosidosis, see above) show normal development until 3–6 months of age. Parental concern is usually caused about then by apparent motor weakness, hypotonia, poor head control and decreasing attentiveness. Parents often initially note an exaggerated startle reaction to sharp (not necessarily loud) sounds,

characterized by a sudden extension of both arms and legs. By the age of 10–12 months, any motor skills that had been achieved earlier are lost, and over the next year vision diminishes. At this time, ophthalmosocopy reveals macular pallor with contrasted prominence of the macular fovea centralis (Tay's 'cherry red spot'). An enlargement of the head is also characteristic of this disorder, resulting from reactive cerebral gliosis. By 18 months, both upper and lower motor neuron deterioration become evident, accompanied by frequent seizures. By the age of 2, further deterioration leads to the disappearance of swallowing and gag reflexes, and finally to a completely vegetative state. Bronchopneumonia associated with aspiration and/or a diminished capacity to cough is frequently antecedent to death. Classical Tay–Sachs patients usually do not survive beyond 4 years of age (Sandhoff *et al.*, 1989; Gravel *et al.*, 1995). Infantile Tay–Sachs disease has its highest frequency in the French Canadian, Cajun and Ashkenazi and Moroccan Jewish populations (see **Table 1**). Fortunately, international carrier detection programs initially based on the isozyme profiles, that is, total Hex versus Hex A ratios (using differential heat denaturation and/or MUG versus MUGS ratios), of serum and/or leukocyte samples have reduced its incidence in these groups by 90%.

Subacute (late infantile and juvenile)

Patients with subacute GM2 gangliosidosis present the same symptoms as acute patients; however, onset is delayed, with motor ataxia first becoming evident at 2–6 years of age, and death occurring by the age of 10–15 years.

Chronic (adult)

Chronic TSD/SD is the most clinically variable GM2 gangliosidosis, with extreme variability being found even in members of the same family. Generally, chronic patients usually present with abnormalities of gait and posture between 2 and 5 years of age. Most of the patients with this condition are still living in their third to fourth decade of life. Symptoms of spinocerebellar and lower motor neuron dysfunction are most prominent. Psychosis, usually of the schizophrenia type with slow personality disintegration, often with episodes of depression, develops in a third of patients. Because of the diversity in clinical symptoms associated with the adult phenotype, patients have been previously misdiagnosed with an array of disorders including spinal muscular atrophy, atypical Friedreich ataxia and Kugelberg–Welander disease (Yaffe *et al.*, 1979; Jellinger *et al.*, 1982; Argov and Navon, 1984; Johnson and Wu, 1984).

Table 1 Tay–Sachs disease (TSD) mutations (for a complete list see the GM2 gangliosidosis mutations database in Web Links)

Mutation	Location	Result	Biochemical phenotype	Clinical	Heritage
Δ7.6 kb	5′ to IVS 1	No mRNA	Abnormal Southern blot	Acute	82% of French Canadian TSD
+TATC−1278	Exon 11	Frameshift	STOP at +4 codons, ND mRNA	Acute	20% non-Jews; 75% Ashkenazi and 2% Moroccan Jews; 92% Cajuns
c → g	IVS 12+1	Abnormal splicing	ND mRNA	Acute	15% of TSD in Ashkenazi, 2% in Moroccan Jews
G509 → A	Exon 5	Arg170 → Gln + abnormal splicing	mRNA splicing cryptic site −51 bp at 5′ end of exon	Acute	37% of TSD in Moroccan Jews; Japanese, Scottish
a → g	IVS 5−1	Abnormal splicing	ND mRNA	Acute	12% of TSD in Moroccan Jews
ΔTTC 910–912	Exon 8	ΔPhe304 or ΔPhe305	Transport-incompetent mutation	Acute	43% of TSD in Moroccan Jews; Irish, French
G805 → A	Exon 7	Gly269 → Ser + abnormal splicing	50% mRNA, 2–3% unstable Hex A; transport mutation	Chronic	Ashkenazi 3%, 5% in non-Jews with TSD
(1) C739 → T (2) C745 → T	Exon 7	(1) Arg247 or (2) Arg249 → Trp	Reduced activity and α-CRM, Hex A is formed but is unstable	Asymptomatic in compounds with null allele	(1) Diverse, ~5% of enzyme-based carriers in non-Jews, <3% in Jews (2) French Canadians
(1) C532 → T (2) G533 → A (3) G533 → T	Exon 5	(1) Arg178 → Cys (2) Arg178 → His (3) Arg178 → Leu	(1) Lower levels of Hex A, some transport problems (2, 3) Normal Hex A (α, β) levels with MUG, inactive with MUGS	Subacute in homozygotes (acute in compound with null allele)	B1-variants. (1) Diverse (1–3) Common in Portuguese

ND: not detectable; IVS: intervening sequence (intron); CRM: cross-reacting material; mRNA: messenger ribonucleic acid; MUG: 4-methylumbelliferyl-β-*N*-acetylglucosamine; MUGS: 4-methylumbelliferyl-β-*N*-acetylglucosamine-6SO$_4$; Δ: deletion.

Critical threshold hypothesis

Mutations that cause the complete loss of Hex A activity give rise to the acute form of GM2 gangliosidosis, while mutations leaving even very small levels of residual Hex A activity give rise to subacute or chronic diseases. A correlation between residual Hex A activity (using a natural GM2/Activator assay system) and the severity of the resulting disease has been made (Conzelmann et al., 1983). Activities found for acute, subacute and chronic patients were ≤0.1%, 0.5% and 2–4% of normal controls respectively. Two clinically healthy individuals with low Hex A activity were found to possess activities of 11% and 20% (Dreyfus et al., 1977). These data suggest a 'critical threshold', that is, the minimum amount of Hex A activity required to keep the rate of GM2 hydrolyzed greater than or equal to the rate of ganglioside transport and incorporation into the lysosome, for TSD and SD of between 5% and 10% of normal Hex A activity (Conzelmann and Sandhoff, 1984).

Genotype–Phenotype Correlations

Partial gene deletions

To date, there have been about 100 mutations in *HEXA*, 25 mutations in *HEXB* and five mutations in *GM2A* reported to cause GM2 gangliosidosis (see the gangliosidosis mutations database in Web Links). The first mutations to be identified were found using Southern blots, before the *HEXA* and *HEXB* gene structures were determined. These were partial 5′ gene deletions (see **Tables 1** and **2**). The *HEXA* deletion is a common mutation found in the French-Canadian populations. The *HEXB* deletion may account for as much as 25% of the SD alleles outside of South America. The mutations common to the Jewish population were found to be a splice junction mutation and a four base pair insertion leading to an early STOP codon (see **Table 1**). It turned out that while the splice junction mutation is only common in the Jewish population, the insertion mutation is also

Table 2 Sandhoff disease mutations (for a complete list see the GM2 gangliosidosis mutations database in Web Links)

Mutation	Location	Result	Biochemical phenotype	Clinical	Heritage
Δ16 kb	Promoter → exon 5	No mRNA	Abnormal Southern blot	Acute	Diverse ~25% of SD alleles (not S. American)
+18 bp duplication	IVS 13 last 16 bp + first 2 bp exon 14	~90% abnormal splicing	6 amino acid insertion, normal mRNA levels; transport incompetent, unstable elongated β chain, 10% Hex A, ND Hex B	Asymptomatic	French 'SD Paris'
g → a	IVS 12–26	~97% abnormal splicing	8 amino acid insertion, normal mRNA levels; transport incompetent unstable elongated β chain, 3% Hex A, ND Hex B	Subacute (2nd allele = Δ16 kb)	Caucasian, Japanese
C → T	IVS 10 + 8 (exon 11)	~92% abnormal splicing + Pro^{417}Leu	Pro417 Leu is silent, ~8% residual Hex A, ND Hex B, ND mRNA	Mild chronic to subacute	French-Canadian, Japanese, Italian

ND: not detectable (by Northern blots or isozyme assay); mRNA: messenger ribonucleic acid; IVS: intervening sequence (intron); Δ: deletion.

common to non-Jews including Cajuns. None of these mutations allows the formation of any active Hex A, and thus they are all associated with the acute form of TSD/SD.

Splice junction mutations

Splice junction mutations can be broken down into two groups: those that affect one of four invariant nucleotides in junctions, that is, the first and last two nucleotides of each intron (Krawczak et al., 1992), and those that do not. Patients homozygous for the former have no detectable messenger ribonucleic acid (mRNA) (improperly spliced mRNA or mRNA with early STOP codons are unstable (Muhlrad and Parker, 1994)) and present with the acute phenotype (reviewed in Gravel et al. (1995) and Mahuran (1999)). Those with the latter can have any of the three phenotypes or even be asymptomatic. Three *HEXB* gene mutations are the best examples of these wide variations. In all three cases some properly spliced mRNA is generated, resulting in detectable levels of Hex A ranging from 10% for asymptomatic individuals to 5% for two patients with subacute SD (**Table 2**). The third mutation is in a position not normally associated with splicing, + 8 nucleotides from the intron 10/exon 11 junction, and actually produces a βPro417Leu substitution which is silent (**Table 2**). It is also more common than the others having been found in Japanese, French Canadian and Italian patients. Furthermore it can produce either a subacute or very mild chronic disorder for reasons yet to be understood. The distribution of Hex isozymes in samples from any of these individuals is very similar; all have almost undetectable levels of Hex B. Thus the small amounts of β subunit being translated (with the wild-type

sequence) are preferentially formed into the less stable Hex A heterodimers. A mechanism to explain this unexpected observation is presented below.

Mutations producing early STOP codons

The types of mutation leading to GM2 gangliosidosis run the gamut of those described for other diseases. Insertions, deletions and nonsense mutations all produce the acute form of the disease. Interestingly all of these also result in no detectable protein in the lysosome and if the insertion/deletion does not retain the proper reading frame, that is, produces an early STOP, reduced or undetectable mRNA (reviewed in Mahuran (1999)).

Missense mutations

Missense mutations, as one might expect, can produce any of the three clinical phenotypes. Interestingly, in all but a group of mutations called the B1 variants, residual Hex A activity is proportional to the amount of Hex A protein in the lysosome. Several of these mutations, which lead to the chronic phenotype, also produce a heat-labile protein, most notably the αGly269Ser which is the most common mutation associated with adult TSD. It accounts for 3% of the mutant alleles in the Ashkenazi and 5% in the non-Jewish population (**Table 1**). Two missense mutations are associated with a so-called pseudodeficiency of Hex A, Arg247Trp and Arg249Trp (**Table 1**). These mutations again reduce the level of Hex A protein rather than lowering the specific activity of the protein, but not below the critical threshold needed to prevent GM2 accumulation.

Role of the endoplasmic reticulum in determining the clinical phenotype of mutations associated with the GM2 gangliosidoses

Like other lysosomal and secretory proteins, the α and β chains of Hex and the monomeric Activator contain cleavable signal peptides (**Figure 3**), which direct their synthesis to polysomes attached to the rough endoplasmic reticulum (ER). In the ER polypeptide chain folding, Asn-linked glycosylation, disulfide bond formation and dimerization occur. For proteins destined for the lysosome, a mannose-6-phosphate 'tag' on one or more Asn-linked oligosaccharide is specifically generated (**Figure 5**). It is believed that some three-dimensional structure common to lysosomal proteins is recognized; thus, each monomer has to have obtained its native folding pattern in order to be tagged (Lang et al., 1984). All proteins synthesized in the ER must fold into their near native conformation and in some cases form functional multimers in order to be passed out of the ER into the cis Golgi (reviewed in Lodish (1988), Hurtley and Helenius (1989) and Pelham (1989)), where further Golgi transport continues via bulk flow (Bergeron et al., 1994) (**Figure 4**). In the trans Golgi network, the tag is recognized by mannose-6-phosphate receptors, which divert the protein to the lysosome (**Figure 5**).

The retention, through recycling between the ER and the cis Golgi network, of unassembled subunit(s) of multimeric proteins, even though properly folded (Hsu et al., 1991), is achieved through interactions with resident proteins that are themselves normally recycled between these compartments (Munro and Pelham, 1987; Lewis and Pelham, 1990), that is,

chaperones (reviewed in Kim and Arvan (1998)). These proteins also assist normal monomers in their folding, while retaining those that are deemed to be 'misfolded', for example because of mutation, and accelerate their degradation (Knittler et al., 1995; Hayes and Dice, 1996). It appears that for Hex, dimer formation is necessary for ER to Golgi transport and that unlike β homodimers (Hex B), homodimers of the α subunits (Hex S) are not readily formed. Thus, the β subunit can be viewed as a 'transport subunit' (Kim and Arvan, 1998) necessary for transport of the α in Hex A out of the ER (Proia et al., 1984). This situation would also allow the cell to specifically retain α subunits, increasing their concentration in the ER, which in turn would promote the formation of heterodimeric Hex A by newly synthesized β subunits, rather than the more stable Hex B homodimers (**Figure 4**). This mechanism would explain why variable amounts of Hex A, but undetectable levels of Hex B, are found in patients with low levels of β-subunit translation due to inefficient mRNA splicing (see above). It also implies that mutant proteins may not be totally incapable of forming a functional enzyme, but are prevented from doing so by the ER's quality control system. Indeed overexpression of mutant β-glucuronidase (causing mucopolysaccharidosis VII) (Wu et al., 1994) or some α-subunit mutations, for example the αGly269Ser (Brown and Mahuran, 1993), in transfected COS cells has resulted in surprisingly high levels of enzyme activity, demonstrating that the quality control system can be saturated.

Once the Hex isozymes enter the lysosome, the single chain pro-α and pro-β polypeptides (∼68 kDa each) undergo posttranslational proteolytic as well as glycosidic processing to produce complex polypeptide

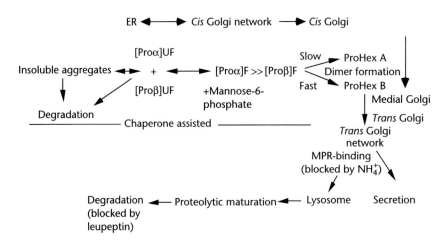

Figure 5 Flow chart depicting the biosynthesis, intracellular transport and assembly of the Hex isozymes. [Proα/β]UF indicates the concentration of unfolded α or β propolypeptides, whereas [Proα/β]F indicates the concentration of either folded subunit. Transport from the cis Golgi to the trans Golgi network is through bulk flow, with secretion being the cell's default pathway. ER: endoplasmic reticulum; MPR: mannose phosphate receptor.

structures, held together by disulfide bonds '{}', of the mature subunits (**Figure 3**), that is, $\{\alpha_p\ (\sim 14\,\text{kDa})\ \alpha_m\ (\sim 56\,\text{kDa})\}\ \{\beta_p\ (\sim 17\,\text{kDa})\ \beta_b\ (\sim 24\,\text{kDa})\ \beta_a\ (\sim 28\,\text{kDa})\}$ (reviewed in Gravel *et al.* (1995) and Mahuran (1999)) (**Figure 3**). The conversion of the propolypeptides to the mature subunits is easily detected by Western blot analysis (Brown *et al.*, 1989) and can be used as a marker for lysosomal incorporation (Hasilik and Neufeld, 1980).

B1 variant of TSD

Patients with the B1 variant were originally thought to have an Activator defect, because they express both Hex A and Hex B activities as assayed with MUG. However, unlike the normal Hex A found in the true AB variants, Kytzia *et al.* (1983) found that B1 variant Hex A was inactive toward an α-specific, MUGS-like substrate (as well as GM2 ganglioside even in the presence of Activator), and they suggested the presence of a mutation at or near the active site of the α subunit. Suzuki and colleagues identified the first specific mutation linked to the B1 phenotype in five of six patients examined, – a missense mutation substituting a His for αArg178. Two other mutations at the same B1 codon have been described, αArg178Cys and αArg178Leu (**Table 1**, last row).

All the members of a given hydrolase family (Henrissat and Bairoch, 1993) are believed to be evolutionarily related and have similar three-dimensional structures (e.g. Pons *et al.*, 1998). Thus, molecular modeling of human Hex has been carried out on the basis of the three-dimensional structure of bacterial chitobiase, as both are part of the glycosyl hydrolase 'family 20' (Tews *et al.*, 1996). From the chitobiase model it is known that the Arg aligning with αArg178, cArg379, sits at the base of a binding pocket for NAG-A, the β-1,4 linked 2-acetamido-2-deoxy-glucopyranosyl residue at the nonreducing end of chitin, and plays the most critical role of any residue in substrate binding and orientation. When human Hex B was mutated at its aligned residue, βArg211Lys, the K_m for MUG was increased 10-fold while the V_{max} was reduced 500-fold (Hou *et al.*, 2000). These data suggest that these active Arg residues are also involved in the stabilization of the transition state(s) of the Hex substrates, thus assisting catalysis.

Mouse Models for the GM2 Gangliosidoses

Mouse models have been made for TSD, SD and the AB variant. Interestingly, whereas the Sandhoff mouse presents with a clinical phenotype similar to that observed in human patients, the Tay–Sachs mouse does not develop a phenotype until very late in life. Based on radioactive GM1 feeding experiments with mutant mouse cells, it was concluded that in mice, unlike humans, the conversion of GM2 to GA2 (the asialo derivative of GM2) by a lysosomal sialidase is a functional alternative pathway. Because either Hex A or Hex B can hydrolyze GA2, although *in vivo* Hex B hydrolysis is very slow (Sango *et al.*, 1995; Phaneuf *et al.*, 1996), the normal levels of Hex B in Tay–Sachs mice are sufficient to dramatically slow down the storage of GM2 and GA2.

The Sandhoff mouse is fertile during the early part of its life, and was bred with the Tay–Sachs mouse to produce offspring with a total Hex deficiency. These mice display a severe clinical phenotype consistent with a combined gangliosidosis and mucopolysaccharidosis (Sango *et al.*, 1996). Thus glycosaminoglycans like gangliosides are critical substrates for Hex. It would also follow that the low residual level of Hex S ($\sim 1.5\%$ of normal in SD mice) in SD is sufficient to degrade MPS substrates in mice and human alike, as initially suggested by Kresse *et al.* (1981) and Ludolph *et al.* (1981).

The AB variant mouse produced a clinical phenotype intermediate to that of the other two. Storage of GM2 was only slightly increased over the Tay–Sachs mouse. However, in this case some GA2 storage was also evident, but still at a level around 10-fold less than that of the Sandhoff mouse (Liu *et al.*, 1997). The lower level of GA2 storage suggests that this glycolipid can be hydrolyzed in the absence of the Activator, but that the Activator is probably required for an optimal rate of degradation.

See also
Carrier Screening of Adolescents in Montreal

References

Argov Z and Navon R (1984) Clinical and genetic variations in the syndrome of adult GM2 gangliosidosis resulting from hexosaminidase A deficiency. *Annals of Neurology* **16**: 14–20.

Bergeron JJM, Brenner MB, Thomas DY and Williams DB (1994) Calnexin: a membrane-bound chaperone of the endoplasmic reticulum. *Trends in Biological Science* **19**: 124–128.

Beutler E, Kuhl W and Comings D (1975) Hexosaminidase isozyme in type O GM2 gangliosidosis (Sandhoff–Jatzkewitz disease). *American Journal of Human Genetics* **27**: 628–638.

Bikker H, Meyer MF, Merk AC, deVijlder JJ and Bolhuis PA (1988) XmnI RFLP at 5q13 detected by a 049 Xmn I fragment of human hexosaminidase (HEXB). *Nucleic Acids Research* **16**: 8198.

Brown CA and Mahuran DJ (1993) β-Hexosaminidase isozymes from cells co-transfected with α and β cDNA constructs: analysis of α subunit missense mutation associated with the adult form of Tay–Sachs disease. *American Journal of Human Genetics* **53**: 497–508.

Brown CA, Neote K, Leung A, Gravel RA and Mahuran DJ (1989) Introduction of the α subunit mutation associated with the B1 variant of Tay–Sachs disease into the β subunit produces a β-hexosaminidase B without catalytic activity. *Journal of Biological Chemistry* **264**: 21705–21710.

Chen B, Rigat B, Curry C and Mahuran DJ (1999) Structure of the GM2A gene: identification of an exon 2 nonsense mutation and a naturally occurring transcript with an in-frame deletion of exon 2. *American Journal of Human Genetics* **65**: 77–87.

Conzelmann E, Kytzia H-J, Navon R and Sandhoff K (1983) Ganglioside GM2 *N*-acetyl-beta-D-galactosaminidase activity in cultured fibroblasts of late-infantile and adult GM2 gangliosidosis patients and of healthy probands with low hexosaminidase level. *American Journal of Human Genetics* **35**: 900–913.

Conzelmann E and Sandhoff K (1978) AB variant of infantile GM2 gangliosidosis: deficiency of a factor necessary for stimulation of hexosaminidase A-catalyzed degradation of ganglioside GM2 and glycolipid GA2. *Proceedings of the National Academy of Sciences of the United States of America* **75**: 3979–3983.

Conzelmann E and Sandhoff K (1979) Purification and characterization of an activator protein for the degradation of glycolipids GM2 and GA2 by hexosaminidase A. *Hoppe-Seylers Zeitschrift für Physiologische Chemie* **360**: 1837–1849.

Conzelmann E and Sandhoff K (1984) Partial enzyme deficiencies: residual activities and the development of neurological disorders. *Developmental Neuroscience* **6**: 58–71.

Dreyfus JC, Poenaru L, Vibert M, Ravise N and Boue J (1977) Characterization of a variant of beta-hexosaminidase: 'Hexosaminidase Paris'. *American Journal of Human Genetics* **29**: 287–293.

Geiger B, Arnon R and Sandhoff K (1977) Immunochemical and biochemical investigation of hexosaminidase S. *American Journal of Human Genetics* **29**: 508–522.

Glombitza GJ, Becker E, Kaiser HW and Sandhoff K (1997) Biosynthesis, processing, and intracellular transport of GM2 activator protein in human epidermal keratinocytes – the lysosomal targeting of the GM2 activator is independent of a mannose-6-phosphate signal. *Journal of Biological Chemistry* **272**: 5199–5207.

Gravel RA, Clarke JTR, Kaback MM, *et al.* (1995) The GM2 gangliosidoses. In: Scriver CR, Beaudet AL, Sly WS and Valle D (eds.) *The Metabolic and Molecular Bases of Inherited Disease*, vol. 2, pp. 2839–2879. New York, NY: McGraw-Hill.

Hasilik A and Neufeld EF (1980) Biosynthesis of lysosomal enzymes in fibroblasts: synthesis as precursors of higher molecular weight. *Journal of Biological Chemistry* **255**: 4937–4945.

Hayes SA and Dice JF (1996) Roles of molecular chaperones in protein degradation. *Journal of Cell Biology* **132**: 255–258.

Hechtman P and LeBlanc D (1977) Purification and properties of the hexosaminidase A-activating protein from human liver. *Biochemical Journal* **167**: 693–701.

Heng HHQ, Xie B, Shi X-M, Tsui L-C and Mahuran DJ (1993) Refined mapping of the GM2 activator protein (GM2A) locus to 5q 31.3–33.1, distal to the spinal muscular atrophy locus. *Genomics* **18**: 429–431.

Henrissat B and Bairoch A (1993) New families in the classification of glycosyl hydrolases based on amino acid sequence similarities. *Biochemical Journal* **293**: 781–788.

Hou Y, McInnes B, Hinek A, Karpati G and Mahuran D (1998) A Pro504Ser substitution in the β-subunit of β-hexosaminidase A inhibits α-subunit hydrolysis of GM2 ganglioside, resulting in chronic Sandhoff disease. *Journal of Biological Chemistry* **273**: 21386–21392.

Hou Y, Tse R and Mahuran DJ (1996) The direct determination of the substrate specificity of the α-active site in heterodimeric β-hexosaminidase A. *Biochemistry* **35**: 3963–3969.

Hou Y, Vocadlo D, Withers S and Mahuran D (2000) The role of beta-Arg211 in the active site of human beta-hexosaminidase B. *Biochemistry* **39**: 6219–6227.

Hsu VW, Yuan LC, Nuchtern JG, *et al.* (1991) A recycling pathway between the endoplasmic reticulum and the Golgi apparatus for retention of unassembled MHC class I molecules. *Nature* **352**: 441–444.

Huang JQ, Trasler JM, Igdoura S, *et al.* (1997) Apoptotic cell death in mouse models of GM2 gangliosidosis and observations on human Tay–Sachs and Sandhoff diseases. *Human Molecular Genetics* **6**: 1879–1885.

Hurtley SM and Helenius A (1989) Protein oligomerization in the endoplasmic reticulum. *Annual Review of Cell Biology* **5**: 277–307.

Ikonne JV, Rattazzi MC and Desnick RJ (1975) Characterization of hex S, the major residual β-hexosaminidase activity in type O GM2 gangliosidosis (Sandhoff–Jatzkewitz disease). *American Journal of Human Genetics* **27**: 639–650.

Jellinger K, Anzil AP, Seemann D and Bernheimer H (1982) Adult GM2 gangliosidosis masquerading as slowly progressive muscular atrophy. *Clinical Neuropathology* **1**: 31.

Johnson W and Wu P (1984) Hexosaminidase deficiency with spinal muscular atrophy: biochemical characterization of the residual enzyme. *Neurology* **34**: 273.

Kim PS and Arvan P (1998) Endocrinopathies in the family of endoplasmic reticulum (ER) storage diseases: disorders of protein trafficking and the role of ER molecular chaperones. *Endocrine Reviews* **19**: 173–202.

Klenk E (1939) Beitrage zur Chemie der Lipidosen. I. Niemann–Pick'sche Krankheit und amaurotische Idiotie. *Hoppe-Seylers Zeitschrift für Physiologische Chemie* **262**: 128–134.

Knittler MR, Dirks S and Haas IG (1995) Molecular chaperones involved in protein degradation in the endoplasmic reticulum: quantitative interaction of the heat shock cognate protein BiP with partially folded immunoglobulin light chains that are degraded in the endoplasmic reticulum. *Proceedings of the National Academy of Sciences of the United States of America* **92**: 1764–1768.

Korneluk RG, Mahuran DJ, Neote K, *et al.* (1986) Isolation of cDNA clones coding for the alpha subunit of human β-hexosaminidase: extensive homology between the α and β subunits and studies on Tay–Sachs disease. *Journal of Biological Chemistry* **261**: 8407–8413.

Krawczak M, Reiss J and Cooper DN (1992) The mutational spectrum of single base-pair substitutions in mRNA splice junctions of human genes: causes and consequences. *Human Genetics* **90**: 41–54.

Kresse H, Fuchs W, Glossl J, Holtfrerich D and Gilberg W (1981) Liberation of *N*-acetylglucosamine-6-sulfate by human beta-*N*-acetylhexosaminidase A. *Journal of Biological Chemistry* **256**: 12926–12932.

Kytzia HJ, Hinrichs U, Maire I, Suzuki K and Sandhoff K (1983) Variant of GM2-gangliosidosis with hexosaminidase A having a severely changed substrate specificity. *EMBO Journal* **2**: 1201–1205.

Kytzia HJ, Hinrichs U and Sandhoff K (1984) Diagnosis of infantile and juvenile forms of GM2 gangliosidosis variant 0 Residual activities toward natural and different synthetic substrates. *Human Genetics* **67**: 414–418.

Kytzia HJ and Sandhoff K (1985) Evidence for two different active sites on human b-hexosaminidase A. *Journal of Biological Chemistry* **260**: 7568–7572.

Lang L, Reitman M, Tang J, Roberts RM and Kornfeld S (1984) Lysosomal enzyme phosphorylation: recognition of a protein-dependent determinant allows specific phosphorylation of oligosaccharides present on lysosomal enzymes. *Journal of Biological Chemistry* **259**: 14 663–14 671.

Lewis MJ and Pelham RB (1990) A human homologue of the yeast HDEL receptor. *Nature* **348**: 162–163.

Li S-C and Li Y-T (1976) An activator stimulating the enzymic hydrolysis of sphingoglycolipids. *Journal of Biological Chemistry* **251**: 1159–1163.

Liu YJ, Hoffmann A, Grinberg A, *et al.* (1997) Mouse model of GM2 activator deficiency manifests cerebellar pathology and motor impairment. *Proceedings of the National Academy of Sciences of the United States of America* **94**: 8138–8143.

Lodish HF (1988) Transport of secretory and membrane glycoproteins from the rough endoplasmic reticulum to the Golgi. *Journal of Biological Chemistry* **263**: 2107–2110.

Ludolph T, Paschke E, Glossl J and Kresse H (1981) Degradation of keratan sulphate by β-N-acetylhexosaminidases A and B. *Biochemical Journal* **193**: 811–818.

Mahuran DJ (1999) Biochemical consequences of mutations causing the GM2 Gangliosidoses. *Biochimica Biophysica Acta* **1455**: 105–138.

Meier EM, Schwarzmann G, Fürst W and Sandhoff K (1991) The human GM2 activator protein. A substrate specific cofactor of β-hexosaminidase A. *Journal of Biological Chemistry* **266**: 1879–1887.

Muhlrad D and Parker R (1994) Premature translational termination triggers mRNA decapping. *Nature* **370**: 578–581.

Munro S and Pelham HRB (1987) A C-terminal signal prevents secretion of luminal ER protein. *Cell* **48**: 899–907.

Myerowitz R, Piekarz R, Neufeld EF, Shows TB and Suzuki K (1985) Human β-hexosaminidase α chain: coding sequence and homology with the beta chain. *Proceedings of the National Academy of Sciences of the United States of America* **82**: 7830–7834.

Myerowitz R and Proia RL (1984) cDNA clone for the alpha-chain of human β-hexosaminidase: deficiency of α-chain mRNA in Ashkenazi Tay–Sachs fibroblasts. *Proceedings of the National Academy of Sciences of America* **81**: 5394–5398.

Nakai H, Byers MG, Nowak NJ and Shows TB (1991) Assignment of beta-hexosaminidase A alpha-subunit to human chromosomal region 15q23→q24. *Cytogenetics and Cell Genetics* **56**: 164.

Neote K, Bapat B, Dumbrille-Ross A, *et al.* (1988) Characterization of the human *HEXB* gene encoding lysosomal β-hexosaminidase. *Genomics* **3**: 279–286.

O'Dowd B, Quan F, Willard H, *et al.* (1985) Isolation of c-DNA clones coding for the β-subunit of human β-hexosaminidase. *Proceedings of the National Academy of Sciences of the United States of America* **82**: 1184–1188.

Okada S and O'Brien JS (1969) Tay–Sachs disease: generalized absence of a β-D-N-acetylhexosaminidase component. *Science* **165**: 698–700.

Pelham RB (1989) Control of protein exit from the endoplasmic reticulum. *Annual Review of Cell Biology* **5**: 1–23.

Phaneuf D, Wakamatsu N, Huang JQ, *et al.* (1996) Dramatically different phenotypes in mouse models of human Tay–Sachs and Sandhoff diseases. *Human Molecular Genetics* **5**: 1–14.

Pons T, Olmea O, Chinea G, *et al.* (1998) Structural model for family 32 of glycosyl-hydrolase enzymes. *Proteins* **33**: 383–395.

Proia RL (1988) Gene encoding the human β-hexosaminidase β-chain: extensive homology of intron placement in the α- and β-genes. *Proceedings of the National Academy of Sciences of the United States of America* **85**: 1883–1887.

Proia RL, d'Azzo A and Neufeld F (1984) Association of α- and β-subunits during the biosynthesis of β-hexosaminidase in cultured fibroblasts. *Journal of Biological Chemistry* **259**: 3350–3354.

Proia RL and Soravia E (1987) Organization of the gene encoding the human β-hexosaminidase α chain. *Journal of Biological Chemistry* **262**: 5677–5681.

Rigat B, Wang W, Leung A and Mahuran DJ (1997) Two mechanisms for the re-capture of extracellular GM2 activator protein: evidence for a major secretory form of the protein. *Biochemistry* **36**: 8325–8331.

Sachs B (1896) A family form of idiocy, generally fatal, associated with early blindness. *Journal of Nervous and Mental Diseases* **21**: 475–479.

Sandhoff K (1969) Variation of β-N-acetylhexosaminidase-pattern in Tay–Sachs disease. *FEBS Letters* **4**: 351–354.

Sandhoff K, Conzelmann E and Nehrkorn H (1977) Specificity of human liver hexosaminidase A and B against glycosphingolipids GM2 and GA2: purification of the enzymes by affinity chromatography employing specific elution. *Hoppe-Seylers Zeitschrift für Physiologische Chemie* **358**: 779–787.

Sandhoff K, Conzelmann E, Neufeld EF, Kaback MM and Suzuki K (1989) The GM2 gangliosidoses. In: Scriver CV, Beaudet AL, Sly WS and Valle D (eds.) *The Metabolic Basis of Inherited Disease*, vol. 2, pp. 1807–1839. New York, NY: McGraw-Hill.

Sango K, McDonald MP, Crawley JN, *et al.* (1996) Mice lacking both subunits of lysosomal β hexosaminidase display gangliosidosis and mucopolysaccharidosis. *Nature Genetics* **14**: 348–352.

Sango K, Yamanaka S, Hoffmann A, *et al.* (1995) Mouse models of Tay–Sachs and Sandhoff diseases differ in neurologic phenotype and ganglioside metabolism. *Nature Genetics* **11**: 170–176.

Schepers U, Lemm T, Herzog V and Sandhoff K (2000) Characterization of regulatory elements in the 5'-flanking region of the GM2 activator gene. *Biological Chemistry* **381**: 531–544.

Schröder M, Klima H, Nakano T, *et al.* (1989) Isolation of a cDNA encoding the human GM2 activator protein. *FEBS Letters* **251**: 197–200.

Schütte CG, Lemm T, Glombitza GJ and Sandhoff K (1998) Complete localization of disulfide bonds in GM2 activator protein. *Protein Science* **7**: 1039–1045.

Sharma R, Deng H, Leung A and Mahuran D (2001) Identification of the 6-sulfate binding site unique to α-subunit-containing isozymes of human β-hexosaminidase. *Biochemistry* **40**: 5440–5446.

Srivastava SK and Beutler E (1973) Hexosaminidase A and hexosaminidase B: studies in Tay–Sachs and Sandhoff's disease. *Nature* **241**: 463.

Svennerholm L (1962) The chemical structure of normal human brain and Tay–Sachs gangliosides. *Biochemical and Biophysical Research Communications* **9**: 436.

Tay W (1881) Symmetrical changes in the region of the yellow spot in each eye of an infant. *Transactions of the Ophthalmological Society of the United Kingdom* **1**: 155–157.

Tews I, Perrakis A, Oppenheim A, *et al.* (1996) Bacterial chitobiase structure provides insight into catalytic mechanism and the basis of Tay–Sachs disease. *Nature Structural Biology* **3**: 638–648.

Triggs-Raine BL, Akerman BR, Clarke JRT and Gravel RA (1991) Sequence of DNA flanking the exons of the *HEXA* gene, and identification of mutations in Tay–Sachs disease. *American Journal of Human Genetics* **49**: 1041–1054.

Wakamatsu N, Kobayashi H, Miyatake T and Tsuji S (1992) A novel exon mutation in human β-hexosaminidase β-subunit gene affecting the 3' splice site selection. *Journal of Biological Chemistry* **267**: 2406–2413.

Wright CS, Li SC and Rastinejad F (2000) Crystal structure of human GM2-activator protein with a novel beta-cup topology. *Journal of Molecular Biology* **304**: 411–422.

Wu BM, Tomatsu S, Fukuda S, *et al.* (1994) Overexpression rescues the mutant phenotype of L176F mutation causing beta-glucuronidase deficiency mucopolysaccharidosis in two Mennonite siblings. *Journal of Biological Chemistry* **269**: 23 681–23 688.

Xie B, Kennedy JL, McInnes B, Auger D and Mahuran D (1992) Identification of a processed pseudogene related to the functional gene encoding the GM2 activator protein: localization of the pseudogene to human chromosome 3 and the functional gene to human chromosome 5. *Genomics* **14**: 796–798.

Xie B, McInnes B, Neote K, Lamhonwah A-M and Mahuran D (1991) Isolation and expression of a full-length cDNA encoding the human GM2 activator protein. *Biochemical and Biophysical Research Communications* **177**: 1217–1223.

Yaffe M, Kaback MM, Goldberg M, *et al.* (1979) An amytrophic lateral sclerosis-like syndrome with hexosaminidase A deficiency. *Neurology* **29**: 611.

Web Links

GM2 ganglioside activator protein (*GM2A*); Locus ID: 2760. LocusLink:
http://www.ncbi.nlm.nih.gov/LocusLink/LocRpt.cgi?l = 2760

Hexosaminidase A (alpha polypeptide) (*HEXA*); Locus ID: 3073. LocusLink:
http://www.ncbi.nlm.nih.gov/LocusLink/LocRpt.cgi?l = 3073

Hexosaminidase B (beta polypeptide) (*HEXB*); Locus ID: 3074. LocusLink:
http://www.ncbi.nlm.nih.gov/LocusLink/LocRpt.cgi?l = 3074

GM2 ganglioside activator protein (*GM2A*); MIM number: 272750. OMIM:
http://www.ncbi.nlm.nih.gov/htbin-post/Omim/dispmim?272750

Hexosaminidase A (alpha polypeptide) (*HEXA*); MIM number: 606869. OMIM:
http://www3.ncbi.nlm.nih.gov/htbin-post/Omim/dispmim?606869

Hexosaminidase B (beta polypeptide) (*HEXB*); MIM number: 606873. OMIM:
http://www3.ncbi.nlm.nih.gov/htbin-post/Omim/dispmim?606873

TSD diseases features
http://www3.ncbi.nlm.nih.gov/htbin-post/Omim/dispmim?272800

AB-variant
http://www3.ncbi.nlm.nih.gov/htbin-post/Omim/dispmim?272800

Sandhoff disease
http://www3.ncbi.nlm.nih.gov/htbin-post/Omim/dispmim?268800

Telomerase: Structure and Function

Nathaniel J Hansen, *University of Alabama at Birmingham, Birmingham, Alabama, USA*

Joseph C Poole, *Medical College of Georgia, Augusta, Georgia, USA*

Lucy G Andrews, *University of Alabama at Birmingham, Birmingham, Alabama, USA*

Trygve O Tollefsbol, *University of Alabama at Birmingham, Birmingham, Alabama, USA*

Telomerase, the enzyme that completes DNA replication at the termini of eukaryotic chromosomes, is an important factor in chromosomal stability, cellular senescence and tumorigenesis.

Intermediate article

Article contents

- Telomere Synthesis, Chromosomal Stability and Replicative Potential
- Telomerase Holoenzyme
- Telomere Polymerization
- Transcriptional Regulation of the Human *TERT* Promoter
- Telomerase and Cancer

Telomere Synthesis, Chromosomal Stability and Replicative Potential

The eukaryotic enzymes involved in polymerizing the lagging strand of deoxyribonucleic acid (DNA) nucleotides fail to replicate completely the termini of linear chromosomes. Consequently telomerase, a specialized complex of ribonucleic acid (RNA) and protein, functions to recognize, synthesize and maintain the elements that are characteristic of eukaryotic chromosomal ends. These structures, known as 'telomeres', consist of a tandem series of guanosine-rich nucleic acid hexamers and various associated binding proteins, forming a structure that preserves chromosomal stability by preventing recombination, end-to-end fusion and degradation of unprotected DNA ends. (*See* Chromosome; DNA Structure; Telomere; Telomeres: Protection and Maintenance.)

Most human adult somatic cells do not express the telomerase ribonucleoprotein; however, highly proliferative hematopoietic, germline and endometrial cells, as well as about 90% of human tumor cells, show telomerase activity. In telomerase-negative somatic cells, each division is associated with attritional loss of telomeric DNA, which leads eventually to critically short telomeres, growth arrest and cellular senescence. By contrast, differentiated cells with ectopically expressed telomerase maintain elongated telomeres and are transformed to an immortalized state. These findings have shown that telomerase is an enzyme that is central to mechanisms of aging, specifically those in which interaction of cell regulatory factors with the telomere structure enables the detection of a minimal telomeric length necessary for further mitotic division. (*See* Aging: Genetics.)

Telomerase also has been found to be essential for the neoplastic transformation of differentiated human cells. The expression of telomerase in normal human fibroblast and epithelial cells in conjunction with that of two oncogenes results in a bypass of cellular replicative limits – that is, a direct conversion to cells with characteristic tumorigenic properties (Hahn *et al.*, 1999). Interaction among telomerase, components of cell-cycle regulatory pathways and oncogenic alleles that promote proliferative dysfunction is regarded as an early and critical step in the initiation of human cancers. Taken together, present evidence on its cellular effects indicates that telomerase has a complex range of actions beyond the simple maintenance of telomere length, bringing investigations into the function and regulation of its components to the forefront of research into aging and cancer. (*See* Cancer Cytogenetics; Telomeres and Telomerase in Aging and Cancer.)

Telomerase Holoenzyme

Telomerase elongates template DNA with telomeric repeats *in vitro* when its telomerase RNA (TER) subunit is added to cell lysates containing native or ectopic telomerase reverse transcriptase (TERT), the catalytically active subunit of the enzyme. Thus, these two components are regarded as the minimum requirement to reconstitute telomerase activity.

Analysis of telomerase-associated proteins in budding yeast, protozoan ciliates and vertebrate systems has identified homologous TER and TERT subunits, as well as many other protein factors that associate with the RNA–TERT ribonucleoprotein complex. These protein factors include telomerase-associated protein 1 (TEP1), the molecular chaperones p23 and heatshock protein 90 (Hsp90), the human dyskerin and Gar1 proteins, a human homolog of the *Drosophila* Staufen protein, L22 and the product encoded by the yeast *EST1p* gene. The established functions of these accessory proteins include accumulating small RNA molecules and binding to single-stranded DNA, suggesting that a complex variety of molecules are involved in assembly of telomerase and its recruitment to telomeric DNA *in vivo*. (*See* RNA-binding Proteins: Regulation of mRNA Splicing, Export and Decay.)

TERT subunit

The gene encoding the TERT subunit of telomerase shows interspecies variation in its nucleotide sequence, but the amino acid character and motifs known to be associated with reverse transcription activity are highly conserved (Lingner *et al.*, 1997). The human *telomerase reverse transcriptase* (*TERT*) gene product (hTERT) is a protein of roughly 125 kDa with seven reverse transcriptase motifs that are similar in structure to the reverse transcriptase motifs of non-long terminal repeats and group II introns. hTERT also contains an eighth, telomerase-specific motif. (*See* Gene Structure and Organization.)

TERT family reverse transcriptases contain large amino (*N*)-terminal and carboxy (*C*)-terminal domains with an internal catalytic site resembling that of the human immunodeficiency virus 1 (HIV-1) reverse transcriptase. This shared reverse transcriptase motif has a three-dimensional shape similar to a right hand, with the active site located in the palm (Lingner *et al.*, 1997), and its catalytic function is dependent on three completely conserved aspartic acid residues. TERT remains stably bound to its specific, associated TER and synthesizes DNA repeats from a small portion of its ribonucleotide template, which distinguishes it from related reverse transcriptases. (*See* Human Immunodeficiency Virus (HIV) Infection: Genetics.)

In addition to the reverse transcriptase active site, the TERT subunit contains other interactive domains implicated in telomerase function. The stable association with the TER subunit is mediated through interaction at one or both of two putative RNA-binding domains in its *N*-terminal region (Armbruster *et al.*, 2001). Deletion of other sequences in the *N*-terminal region of hTERT has been reported to interfere with overall enzymatic function (Beattie *et al.*, 2001). Moreover, additional *N*-terminal and *C*-terminal domains have been implicated in the oligomerization of hTERT subunits independent of the TER subunit. Recombinant hTERT with truncations of these regions has reduced telomerase activity *in vitro* (Arai *et al.*, 2002).

TER subunit

Like its catalytic associate, the TER subunit of telomerase shows interspecies homology and has a vital role in overall enzyme function. Although it primarily acts as the nucleotide template for the added telomeric repeats, evidence suggests that the TER subunit may also function as a structural backbone for binding other telomerase-associated proteins (Bachand *et al.*, 2001). Similar to hTERT with its pattern of multifunctional domains, the human telomerase RNA gene product (hTER or hTR) of 451 nt has been suggested to have functions in addition to its role as a template of the telomerase enzyme.

Several conserved sequences and secondary structural motifs presumably participate in oligomerization of the TERT–RNA complex, provide a binding substrate for additional enzyme assembly and recruitment factors, and define template 'reading' boundaries for polymerizing the correct telomeric repeat sequence by TERT. Among such regions analyzed in hTER, a helical 'pseudoknot' motif has been reported to be necessary for catalytic function but not for protein binding, and a domain known as conserved regions 4 and 5 (CR4–CR5) contains a stem–loop structure proposed to be a putative binding site for hTERT (Bachand *et al.*, 2001). (*See* RNA Secondary Structure Prediction.)

Also present in the 3′-terminal region of hTER is a helix–loop/5′-ACA-3′ (H/ACA) box motif that is similar to a domain conserved among a group of small nucleolar ribonucleoproteins (snoRNPs). Because these RNAs participate in the nucleolar splicing of ribosomal RNAs, the existence of an H/ACA box in hTER is consistent with nucleolar processing of the telomerase RNA. The H/ACA box region, although evolutionarily conserved, does not seem to participate actively in RNA–protein interactions or in the reconstitution of telomerase catalytic activity *in vitro*. Notably, different accessory proteins described as part of the telomerase complex have been implicated in the processing and stabilization of snoRNPs, suggesting that nucleolar processing mediated by H/ACA maturation factors may be essential for the *in vivo* generation of active telomerase. (*See* Nucleolus: Structure and Function; snoRNPs.)

Additional telomerase-associated factors

Much research has described the binding of additional protein factors to the two main subunits of telomerase. Although several of these factors have functional significance, the actual mechanisms by which they interact with telomerase are not understood completely.

TEP1 (telomerase-associated protein 1) has been shown to interact with mammalian telomerase RNA, but it seems to possess general RNA-binding properties that are not specific to telomerase and is dispensable for telomerase function. The well-characterized Hsp90 and p23 molecular chaperones are necessary to reconstitute telomerase activity *in vitro*, but their role in the catalytic action of the enzyme is unclear. Two snoRNP-related factors, Gar1 and dyskerin, associate with hTER, and the function of dyskerin seems to relate to the accumulation of hTER that precedes holoenzyme assembly.

Additional factors including several heterogeneous nuclear ribonucleoproteins (hnRNPs) and the yeast Est3p protein interact with telomerase and have been proposed as potential mediators of telomerase–telomere association. Intriguingly, Ku, a protein involved in the DNA damage repair cycle, reportedly binds to the yeast RNA template, *TLC1*. Because Ku is known to participate in the control of DNA

damage checkpoints, this potentially could provide a direct means of regulatory communication between telomerase and the cellular senescence pathways that are characteristic of critically short telomeres. Other related proteins, such as L22 and the human homolog of the *Drosophila* Staufen protein, remain uncharacterized with regard to telomerase function. (*See* DNA Damage Response; DNA Repair.)

Multimerization of functional telomerase

Purification of the human telomerase complex isolated a ribonucleoprotein that is several times larger than a single pair of hTERT–hTER subunits (Bachand *et al.*, 2001). Consequently, there is speculation that the telomerase holoenzyme might consist of an oligomer of several catalytic and/or template subunits *in vivo*, rather than the minimal composition of one TERT and one TER.

The hTER template region and a hairpin element of the H/ACA box domain have been identified as sites of direct interaction with the hTERT protein; such sites provide the potential for binding among several hTERT–TER monomers or homodimer pairs. Evidence from yeast and human systems supports the existence of one or more subunit dimers in endogenous telomerase. A nonfunctional mutant of the yeast *TLC1* RNA subunit can be complemented by a second, wild-type *TLC1*, implying that TER RNAs undergo a dimeric interaction in yeast and that the presence of the active template is sufficient to 'rescue' the function of mutant–wild-type dimers.

Similarly, gel filtration analysis of human telomerase shows that the hTER subunit may also exist as a dimer in the active enzyme (Wenz *et al.*, 2001). Human TER differs from that of yeast, however, because the combination of wild-type and function-deficient hTER monomers yields an enzyme with reduced telomerase activity. Thus, hTER dimers may consist of two strictly interdependent subunits. In addition, two inactive hTERT fragments from separate protein molecules have been shown to interact and to recreate telomerase function *in vitro* (Beattie *et al.*, 2001). Putative binding domains for this dimerization have been identified in the large *N*-terminal and *C*-terminal domains of the hTERT protein. (*See* Multidomain Proteins.)

Although it is clear that some oligomerization of the two principal subunits occurs in endogenous telomerase, the exact stoichiometry, as well as that of other accessory proteins in the 'true' telomerase holoenzyme, remains to be established. TEP1 binds to the active ribonucleoprotein complex but does not seem to be completely necessary for enzymatic function. Other factors, such as dyskerin, the molecular chaperone pair Hsp90 and p23, and the yeast

Est3p protein, probably interact with the core subunits in the assembly of the telomerase complex and its localization to the telomeres. However, there is little consensus regarding the accessory factors that are strictly required for telomere synthesis or are directly involved in the polymerization process. Further study will be necessary to elucidate the exact composition of telomerase as it exists *in vivo*.

Telomere Polymerization

Among the more significant features that a cell must maintain is genomic stability. Because telomere length decreases, telomerase-positive cells must have a method by which to localize the polymerization complex of this enzyme to chromosomal termini to prevent the DNA damage associated with cellular senescence. Although the precise processes have not been established, it is generally accepted that this localization must involve some form of molecular signaling between the higher order structure produced by telomeric DNA, and telomere-associated proteins, cell-cycle regulators, damage repair regulators and telomerase-associated factors. (*See* Apoptosis and the Cell Cycle in Human Disease; Cell Cycle Checkpoint Genes and Cancer.)

Functionally assembled telomerase must overcome the complex folding structures that protect chromosomes from end degradation. The extreme terminus of a telomere exists as a 'T-loop', which prevents the DNA from exposing its final 3'-OH at a linear end (Griffith *et al.*, 1999). Once telomerase recruitment factors have localized the holoenzyme to this structure, the complex gains access to the single-stranded 3' overhang at the telomere end and the template sequence of the TER subunit base-pairs with the existing telomeric repeats in a complementary manner (**Figure 1**).

Studies of telomerase polymerization in the ciliate genus *Tetrahymena* have suggested that telomerase adds repeats in a processive manner, with roughly 80–90 new repeats generated before 50% of the enzyme dissociates (Greider, 1991). Proper assembly of the telomerase holoenzyme and orientation on the DNA substrate determine the template boundaries on the TER subunit that correspond to the correct repeat sequence. Because synthesis on the substrate DNA occurs in the 5'→3' direction, blocking the RNA sequence through substrate binding effectively establishes the 3' boundary of the repeat template.

A domain of the *Tetrahymena* TER subunit contributes similarly to both multimerization of the holoenzyme and definition of the 5' template boundary through binding to the reverse transcriptase. Polymerization pauses at the nucleotide corresponding to the 5'

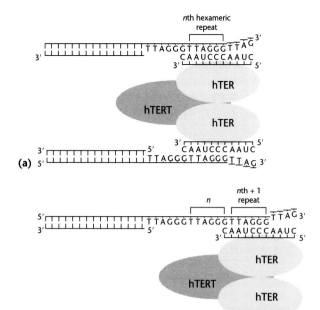

Figure 1 Simplified model of telomeric addition. After recruitment of the telomerase holoenzyme to telomeric DNA, the template domain of the telomerase RNA (hTER) allows complementary base-pairing to existing hexameric repeats (a). Reverse transcription by the telomerase reverse transcriptase (hTERT) continues until the RNA template sequence is fully replicated, at which time the enzyme complex must translocate to continue synthesis of the next repeat (b). In the ciliate *Tetrahymena*, this translocation is accompanied by a pause in polymerization, and telomerase can maintain processive synthesis for an average of roughly 520 nt before dissociating from the DNA substrate (Greider, 1991). Here, an oligomer of several RNA subunits is shown to illustrate how the enzyme might extend telomeric DNA using multiple active sites. (Modified with permission from Greider (1991).)

boundary of the RNA template (the final template nucleotide), while the enzyme complex translocates to the 3' end of the newly synthesized DNA to continue polymerization without complete dissociation (Greider, 1991; **Figure 1**). Interaction among the TER subunit, a protein 'anchor' and the substrate DNA may enable the translocation to occur without release of the telomere.

The likelihood that telomerase exists as a multimer of several TERT and/or TER subunits may add additional complexity to the mechanism of telomere synthesis. Because of the role that telomerase has in replicative potential, it has been suggested that subunit dimers or multiple active sites in the enzyme complex may enable telomeric addition to be coupled to the action of the DNA replication complex (Beattie *et al.*, 2001).

In a popular model, the DNA replication complex can associate with two unreplicated chromosomal ends: the parent template of the leading strand duplex; and that of the newly synthesized lagging strand, which is thought to bend to allow unidirectional synthesis of both daughter strands at the same replication fork. Alternatively, tandem translocation of telomere substrate DNA from one enzyme subunit to another has been proposed as a model that incorporates multiple active sites (Wenz et al., 2001). Given the lack of clarity regarding the precise composition of the telomerase holoenzyme complex, however, the actual manner of telomere synthesis may be more or less complex than is currently imagined. (*See* CpG Islands and Methylation; Promoters.)

Transcriptional Regulation of the Human *TERT* Promoter

Of the two main subunits of human telomerase, the RNA template hTER is far more widely expressed, and the hTERT subunit is found only in highly proliferative tissues such as the germline and adult endometrial and blood stem cells. Accordingly, transcriptional control of the human *TERT* gene has received considerable attention as a key target of telomerase regulation in humans. Structural analysis of the *TERT* 5′ control region led to the discovery of a putative core promoter region, transcriptional and translational initiation sites, and many consensus binding sites for potential transcriptional regulators (reviewed in Poole et al., 2001).

Significantly, a region surrounding the *TERT* minimal promoter contains two 5′-CACGTG-3′ (E-box) binding motifs for the transcription factors c-Myc and Mad1. These proteins act competitively in forming heterodimer complexes with a third protein, Max. Binding of c-Myc–Max dimers to the E-boxes in the *TERT* promoter activates gene transcription, whereas binding of Mad1–Max dimers leads to downregulation of *TERT* messenger RNA (see Poole et al., 2001). Treatment of a telomerase-positive tumor cell line with retinoic acid (a compound that stimulates cellular differentiation) causes a reduction in *TERT* mRNA levels, owing to an increase in the expression of Mad1. Given the importance of transcriptional control of *TERT* in telomerase regulation and the presence of regions that bind cMyc–Mad1 within the *TERT* gene promoter, Mad1-mediated displacement of c-Myc may be one of the main events underlying downregulation of *TERT* during the course of development. (*See* Transcription Factors.)

Additional mechanisms are necessary for stabilizing the inactive *TERT* promoter complex. Mad1 binding to the *TERT* regulatory region recruits histone deacetylases – enzymes that promote a heterochromatin-like, inactive transcriptional state. Binding to histone deacetylase enzymes also reduces the availability of Sp1, a known activator of telomerase that has five binding sites in the *TERT* core promoter. CpG dinucleotide methylation, another widespread repressor of gene expression, can repress the binding of c-Myc–Max; in addition, DNA methyltransferases, the enzymes that methylate cytosines, localize to heterochromatin through the action of histone deacetylases. Although the precise nature of their interactions in differentiating cells remains to be resolved, it is likely that several or all of these factors interact to repress telomerase stably in adult somatic tissues. (*See* Histone Acetylation: Long-range Patterns in the Genome; Methylation-mediated Transcriptional Silencing in Tumorigenesis.)

Telomerase and Cancer

The complex process by which normal cells become tumorigenic involves an accumulation of mutations or alterations in gene expression. Activation of telomerase has been shown to be one of these critical steps (Hahn et al., 1999). Cancer, which has at its core a loss of proliferative control, makes use of the stabilization of genetic material afforded by active telomere synthesis to bypass the growth regulatory signals imposed on ordinary tissues. Mechanisms of hTERT subunit repression are therefore essential for our understanding of the onset of cancer and serve as a promising target for future treatments. (*See* Tumor Formation: Number of Mutations Required.)

Unfortunately, an enzyme with such complex functionality as hTERT is unlikely to be maintained by one activator or inhibitor, and even a single event such as reactivation of telomerase transcription may require many different regulatory disruptions. Greater understanding of the factors that mediate interactions among the telomerase enzyme complex, telomeric structure and gene transcription will be needed to understand the tumorigenic process fully.

See also
Telomere
Telomeric and Subtelomeric Repeat Sequences
Telomeres: Protection and Maintenance
Telomeres and Telomerase in Aging and Cancer

References

Arai K, Masutomi K, Shilagardy K, et al. (2002) Two independent regions of human telomerase reverse transcriptase are important for its oligomerization and telomerase activity. *Journal of Biological Chemistry* **277**: 8538–8544.

Armbruster BN, Banik SS, Guo C, Smith AC and Counter CM (2001) *N*-terminal domains of the human telomerase catalytic subunit required for enzyme activity *in vivo*. *Molecular and Cellular Biology* **21**: 7775–7786.

Bachand F, Triki I and Autexier C (2001) Human telomerase RNA–protein interactions. *Nucleic Acids Research* **29**: 3385–3393.

Beattie TL, Zhou W, Robinson MO and Harrington L (2001) Functional multimerization of the human telomerase reverse transcriptase. *Molecular and Cellular Biology* **21**: 6151–6160.

Greider CW (1991) Telomerase is processive. *Molecular and Cellular Biology* **11**: 4572–4580.

Griffith JD, Comeau L, Rosenfield S, *et al.* (1999) Mammalian telomeres end in a large duplex loop. *Cell* **97**: 503–514.

Hahn WC, Counter CM, Lundberg AS, *et al.* (1999) Creation of human tumour cells with defined genetic elements. *Nature Medicine* **400**: 464–468.

Lingner J, Hughes TR, Shevchenko A, *et al.* (1997) Reverse transcriptase motifs in the catalytic subunit of telomerase. *Science* **276**: 561–567.

Poole JC, Andrews LG and Tollefsbol TO (2001) Activity, function, and gene regulation of the catalytic subunit of telomerase (hTERT). *Gene* **269**: 1–12.

Wenz C, Enenkel B, Amacker M, *et al.* (2001) Human telomerase contains two cooperating telomerase RNA molecules. *EMBO Journal* **20**: 3526–3534.

photo-cross-linking to single- and double-stranded DNA primers. *Molecular and Cellular Biology* **17**: 296–308.

Lai CK, Miller MC and Collins K (2002) Template boundary definition in *Tetrahymena* telomerase. *Genes and Development* **16**: 415–420.

Liu Y, Snow BE, Hande MP, *et al.* (2000) Telomerase-associated protein TEP1 is not essential for telomerase activity or telomere length maintenance *in vivo*. *Molecular and Cellular Biology* **20**: 8178–8184.

Nakamura TM, Morin GB, Chapman KB, *et al.* (1997) Telomerase catalytic subunit homologs from fission yeast and human. *Science* **277**: 955–959.

Peterson SE, Stellwagen AE, Diede SJ, *et al.* (2001) The function of a stem-loop in telomerase RNA is linked to the DNA repair protein Ku. *Nature Genetics* **27**: 64–67.

Xu D, Popov N, Hou M, *et al.* (2001) Switch from Myc/Max to Mad1/Max binding and decrease in histone acetylation at the telomerase reverse transcriptase promoter during differentiation of HL60 cells. *Proceedings of the National Academy of Sciences of the United States of America* **98**: 3826–3831.

Further Reading

Bodnar AG, Ouellette EH, Frolkis M, *et al.* (1998) Extension of life-span by introduction of telomerase into normal human cells. *Science* **279**: 349–352.

Chan SW and Blackburn EH (2002) New ways not to make ends meet: telomerase, DNA damage proteins and heterochromatin. *Oncogene* **21**: 553–563.

Doetzlhofer A, Rotheneder H, Lagger G, *et al.* (1999) Histone deacetylase 1 can repress transcription by binding to Sp1. *Molecular and Cellular Biology* **19**: 5504–5511.

Fuks F, Burgers WA, Brehm A, Hughes-Davies L and Kouzarides T (2000) DNA methyltransferase Dnmt1 associates with histone deacetylase activity. *Nature Genetics* **24**: 88–91.

Gunes C, Lichtsteiner S, Vasserot A and Englert C (2000) Expression of the hTERT gene is regulated at the level of transcriptional initiation and repressed by Mad1. *Cancer Research* **60**: 2116–2121.

Hammond PW, Lively TN and Cech TR (1997) The anchor site of telomerase from *Euplotes aediculatus* revealed by

Web Links

Telomerase-associated protein 1 (*TEP1*); LocusID: 7011. LocusLink:
http://www.ncbi.nlm.nih.gov/LocusLink/LocRpt.cgi?l = 7011

Telomerase RNA component (*TERC*); LocusID: 7012. LocusLink:
http://www.ncbi.nlm.nih.gov/LocusLink/LocRpt.cgi?l = 7012

Telomerase reverse transcriptase (*TERT*); LocusID: 7015. LocusLink:
http://www.ncbi.nlm.nih.gov/LocusLink/LocRpt.cgi?l = 7015

Telomerase-associated protein 1 (*TEP1*); MIM number: 601686. OMIM:
http://www.ncbi.nlm.nih.gov/htbin-post/Omim/dispmim?601686

Telomerase RNA component (*TERC*); MIM number: 602322. OMIM:
http://www.ncbi.nlm.nih.gov/htbin-post/Omim/dispmim?602322

Telomerase reverse transcriptase (*TERT*); MIM number: 187270. OMIM:
http://www.ncbi.nlm.nih.gov/htbin-post/Omim/dispmim?187270

Telomere

Christa Lese Martin, *University of Chicago, Chicago, Illinois, USA*

David H Ledbetter, *University of Chicago, Chicago, Illinois, USA*

Intermediate article

Article contents

- Introduction
- Simple Sequence Repeats Cap the Ends of Human Chromosomes
- Subtelomeric Region
- Properties of Unique Sequence DNA Adjacent to the Telomeres
- Other Important Features of Telomere Regions

Telomeres, the 'caps' at the ends of human chromosomes, consist of specialized DNA and protein complexes that help to protect chromosomes from degradation. Other unique characteristics of telomeric DNA include the sequence homology shared between subtelomeric DNA regions, a high gene density and an increased frequency of chromosomal imbalances.

Introduction

The telomeres of human chromosomes have many unique features that make them an essential part of understanding chromosome organization and function, and may make them disproportionately important in human pathology and disease. Here we address these specialized aspects of the telomeric region by focusing on three main subdivisions (**Figure 1**). First, we examine the ends, or 'caps', of

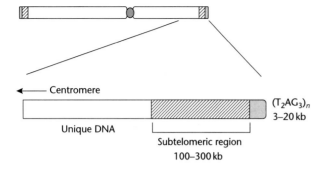

Figure 1 Diagram of the telomeric regions of human chromosomes. The telomeric regions of a human chromosome are depicted on the chromosome in the top of the figure. The bottom of the figure highlights a telomere region: the gray-shaded block at the end of the chromosome represents the TTAGGG repeat sequence, the diagonal striped region is the subtelomeric region, and the white area is unique DNA sequence.

human chromosomes, which are comprised of tandem repeats of a deoxyribonucleic acid (DNA) sequence of 6 base pairs (bp), TTAGGG. Next, we discuss the complex mosaic of repeats and shared sequence homologies of the subtelomeric region, which lies just proximal to the TTAGGG region. Finally, we explore the features of the unique DNA sequences that lie adjacent to these telomeric domains.

Simple Sequence Repeats Cap the Ends of Human Chromosomes

Telomeres are integral in understanding chromosome biology, structure and function. Acting as specialized DNA–protein complexes that cap the ends of eukaryotic chromosomes, telomeres are necessary for complete chromosomal end replication and provide chromosomal stability by protecting chromosome ends from degradation.

Human telomeric DNA consists of between 3 and 20 kilobases (kb) of tandemly repeated TTAGGG sequences, $(T_2AG_3)_n$, that have been shown to be evolutionarily conserved among vertebrate species. These telomeric sequences interact directly with two proteins: telomeric repeat binding factor 1 (TRF1) and 2 (TRF2). The identification and characterization of TRF1 and TRF2 were elucidated by the de Lange laboratory (Smogorzewska et al., 2000). Basically, the expression of these two proteins regulates telomere length either by their direct interactions with telomeric DNA or with interactions with other telomere-related proteins.

During each round of normal somatic cell division, telomeres shorten by about 50–200 bp. Because DNA

polymerase can synthesize DNA in the $5'$ to $3'$ direction only, a $3'$ overhang is created on the lagging strand during DNA synthesis, owing to an inability to replicate the extreme end of the DNA strand. This process, commonly referred to as the 'end replication problem', leads to the gradual loss of these repetitive telomeric sequences. As a result, telomere shortening serves as a 'mitotic clock' and is thought to control cellular aging and senescence.

Cells that proliferate actively, such as germline, embryonic or stem cells, maintain the length of telomeric sequences. In addition, some cancer cells do not lose telomeric sequences during each cell division and become immortalized. These processes occur as a result of the presence of telomerase, a ribonucleoprotein that consists of two main components: human telomerase reverse transcriptase (hTERT), the catalytic protein subunit; and human telomerase ribonucleic acid (RNA) (hTR), the RNA component. Telomerase catalyzes the addition of $(T_2AG_3)_n$ repeats to the telomeres of chromosomes by using its RNA component as a template for reverse transcription, thus helping to maintain telomeric length. Telomerase is not expressed in most normal somatic cells, accounting for the gradual loss of telomeric sequences observed in these cells.

Studies of telomere length in cells from individuals with accelerated aging syndromes have also added evidence of a correlation between telomere length and cellular senescence. For example, cells from individuals with Werner syndrome, Bloom syndrome, Hutchinson–Gilford progeria and Down syndrome all show accelerated telomere shortening as compared with age-matched controls.

Over the past several years, telomerase has gained increased attention in cancer research. In some cancer cells, telomerase is hypothesized to be reactivated, providing immortalized growth due to the stabilization of telomere length. Alterations of telomere length and telomerase activity have been reported in a number of malignancies. Telomere length is usually decreased in most cancer cells, because activation of telomerase is a late event in tumorigenesis; however, once telomerase is activated, telomere length is stabilized. Thus, the presence of telomerase activity could be an attractive target for cancer detection and treatment, although other mechanisms for telomere length maintenance may also exist, and the potential usefulness of telomerase as a cancer therapy agent has yet to be demonstrated.

Subtelomeric Region

Immediately proximal to the tandemly repeated $(T_2AG_3)_n$ sequences in humans is the subtelomeric region, commonly referred to as subtelomeric repeats, or telomere-associated DNA. These DNA sequences usually show segments of shared homology between subsets of different chromosomes, although a minority of subtelomeric sequences are specific to a single telomere region. Several subtelomeric sequences have been identified and characterized by their cross-hybridization to different subsets of human telomeres. A number of these subtelomeric DNA sequences have been shown to have a polymorphic chromosomal distribution in humans. They are present on a different group of chromosome ends in different individuals and create length polymorphisms for homologous chromosomes (Wilkie *et al.*, 1991).

The subtelomeric repeat regions of human chromosomes are basically mosaics of shared repetitive DNA and other shared sequence homologies. These regions are still relatively poorly characterized, owing to the complex nature of the shared DNA sequence homologies between multiple chromosome ends. However, emerging evidence indicates a complex organization and evolutionary history.

A model for the organization of subtelomeric DNA has been proposed from sequencing data from a number of telomeric regions. Studies by Flint *et al.* (1997) indicate the presence of two structurally distinct subdomains in human telomeres. The distal subdomain, closest to the $(T_2AG_3)_n$ repeat, contains a mosaic of very short sequences shared by many different chromosomes, whereas the more proximal subdomain is comprised of much longer sequences of shared homologies from fewer chromosomes. The shared homologies within the distal subdomain are less than 2 kb in length, suggesting the occurrence of frequent exchanges among all telomeres (Flint *et al.*, 1997). In contrast, the proximal subdomain is comprised of much longer segments (10–40 kb) of shared sequence homologies that arise from fewer chromosomes, indicating recent duplications of this domain (Flint *et al.*, 1997). These recent duplications could be the result of unbalanced translocations occurring between telomeres during primate evolution. The proximal and distal subdomains are separated by a block of degenerate $(T_2AG_3)_n$ repeats and sequences with homology to putative origins of replication. Surprisingly, this same organization of two subdomains separated by degenerate $(T_2AG_3)_n$ and origins of replication is also observed in yeast. This sequence conservation suggests an important functional role for the organization of telomere repeats.

Another interesting feature of the subtelomeric regions is the identification of DNA sequence-length polymorphisms. Wilkie *et al.* (1991) first described three different alleles for chromosome 16p with a length polymorphism of up to 260 kb. The three alleles differed not only in length, but one also had an entirely different subtelomeric repeat sequence. Thus, these three alleles were nonhomologous with each other, but homologous to Xq and other chromosome ends. This raises the interesting possibility that an individual heterozygous with respect to two different 16p telomere alleles could be at risk for an increased rate of nondisjunction for chromosome 16, or for unbalanced translocations involving this chromosome end. Another significant length polymorphism of 55 kb has also been identified for chromosome 2q (Macina *et al.*, 1994).

Recent genome-wide sequencing analysis is also adding to our understanding of the organization of subtelomeric sequences in the human genome. Duplications involving more than 1 kb of DNA have been shown to account for approximately 3.6% of the entire human genome sequence (Bailey *et al.*, 2001). Notably, these duplications are concentrated in the subtelomeric and pericentromeric regions up to 10-fold higher than in other regions of the genome, suggesting that these regions have a more dynamic evolutionary history.

Properties of Unique Sequence DNA Adjacent to the Telomeres

Unique sequence DNA is located proximal to the subtelomeric repeats, about 100–300 kb from the end of each chromosome (NIH Collaboration, 1996). This region of human chromosomes contains the highest density of genes in the genome, as initially described by Saccone *et al.* (1992), based on the identification of GC-rich isochores with a high frequency of CpG islands in 'T-bands' at human telomeres. This observation has subsequently been confirmed by DNA sequencing efforts for several telomeres (Flint *et al.*, 1997).

It has long been postulated that submicroscopic deletions or duplications in the genome may be responsible for a significant percentage of unexplained cases of mental retardation. This effect may be exaggerated in the telomeric regions as a result of the disproportionately high density of genes. However, the analysis of telomeric chromosome bands is a significant challenge using conventional cytogenetics methods (G-banding analysis at the 450–550 band level), as most terminal bands are similarly G-negative in appearance and subtle rearrangements (<3 Mb) are difficult to visualize. Recognizing the potential for such cryptic telomeric rearrangements to exist for every human telomere and contribute significantly to human pathology, a strategy to use fluorescence *in situ* hybridization (FISH) to screen for cryptic telomere rearrangements was proposed (Ledbetter, 1992). A unique telomere probe for every chromosome could be developed and combined in a multiplex telomere screening assay and used as a diagnostic tool for the efficient assessment of telomere integrity.

Toward this goal, a first-generation set of human telomere probes was developed. It consisted of 41 probes corresponding to the end of every human chromosome. The total number is 41 because the short arms of the acrocentric chromosomes were excluded, and the probes for Xp/Yp and Xq/Yq are contained within the pseudoautosomal regions (NIH Collaboration, 1996). The identification of a unique telomere probe for each human chromosome was significantly aided by the identification of half-yeast artificial chromosome (YAC) clones – that is, YACs that have only one yeast telomere and require a human insert that contains its own telomere sequences to be propagated. The vector–insert junction sequences from the half-YAC clones were used to screen genomic libraries. As most of the half-YAC clones are 150–300 kb, the resulting genomic clones, which extended proximal from this point, were more likely to contain only chromosome-specific sequences and therefore serve as probes for use in a FISH assay (NIH Collaboration, 1996).

This first telomere set provided a clone for each telomere. However, they were mostly cosmid clones that produce a hybridization signal suitable for single probe FISH, but not usually robust enough for multiplex formats. A second initiative was therefore undertaken to create a second-generation set of telomere probes consisting of bacterial artificial chromosome (BAC) or P1 artificial chromosome (PAC) clones, which contain larger insert sizes and work well when used in multiplex assays (Knight *et al.*, 2000).

Multiplex FISH telomere screening in various clinical populations has revealed that as much as 5–8% of unexplained mental retardation may be explained by unbalanced telomere rearrangements (Knight *et al.*, 1999). This significant percentage lends to rapid adoption of telomere screening in the evaluation of many children with unexplained mental retardation.

Other Important Features of Telomere Regions

In humans, genetic recombination rates increase in the telomeric regions for both sexes, but dramatically so in males. Generally speaking, female recombination rates are higher than those of males for most regions of the genome; however, this pattern is reversed at the majority of telomeres, where male recombination is higher than female. This increased rate of recombination at

telomeres could have a role in occasional unequal exchange and gene dosage imbalance.

Telomeres are also critical in chromosome pairing in meiosis. They are the first regions of the chromosome to pair and synapse, thereby mediating homologous chromosome pairing. The precise mechanisms underlying this process remain relatively obscure. However, recent FISH studies of humans, mouse and maize telomeres have demonstrated highly conserved features of telomere behavior in meiotic prophase. In all species studied, the telomeres move to the nuclear envelope and cluster into a 'bouquet formation'. The bouquet formation represents the first evidence of chromosome pairing, consistent with the notion that homology searches and pairing may initiate at the telomeres. This clustering of all chromosomal telomeres could provide an opportunity for promiscuous telomere pairing, which might occasionally lead to exchange events among nonhomologous chromosomes, causing telomeric rearrangements.

Although many of the aspects of telomere biology, such as those described here, have been elucidated over the years, the continued examination of human telomeres is sure to uncover many other interesting features of these specialized regions of human chromosomes.

See also
Chromosome
Telomeres: Protection and Maintenance
Telomerase: Structure and Function
Telomeric and Subtelomeric Repeat Sequences

References

Bailey JA, Yavor AM, Massa HF, Trask BJ and Eichler EE (2001) Segmental duplications: organization and impact within the current human genome project assembly. *Genome Research* **11**(6): 1005–1017.

Flint J, Bates GP, Clark K, *et al.* (1997) Sequence comparison of human and yeast telomeres identifies structurally distinct subtelomeric domains. *Human Molecular Genetics* **6**(8): 1305–1313.

Knight SJ, Regan R, Nicod A, *et al.* (1999) Subtle chromosomal rearrangements in children with unexplained mental retardation. *Lancet* **354**(9191): 1676–1681.

Knight SJL, Lese CM, Precht K, *et al.* (2000) An optimized set of human telomere clones for studying telomere integrity and architecture. *American Journal of Human Genetics* **67**(2): 320–332.

Ledbetter DH (1992) Cryptic translocations and telomere integrity. *American Journal of Human Genetics* **51**: 451–456.

Macina RA, Negorev DG, Spais C, *et al.* (1994) Sequence organization of the human chromosome 2q telomere. *Human Molecular Genetics* **3**: 1847–1853.

National Institutes of Health, Institute of Molecular Medicine Collaboration, *et al.* (1996) A complete set of human telomeric probes and their clinical application. *Nature Genetics* **14**: 86–89.

Saccone S, De Sario A, Della Valle G and Bernardi G (1992) The highest gene concentrations in the human genome are in telomeric bands of metaphase chromosomes. *Proceedings of the National Academy of Sciences of the United States of America* **89**: 4913–4917.

Smogorzewska A, van Steensel B, Bianchi A, *et al.* (2000) Control of human telomere length by TRF1 and TRF2. *Molecular and Cell Biology* **20**(5): 1659–1668.

Wilkie AO, Higgs DR, Rack KA, *et al.* (1991) Stable length polymorphism of up to 260 kb at the tip of the short arm of human chromosome 16. *Cell* **64**(3): 595–606.

Further Reading

Bass HW, Marshall WF, Sedat JW, Agard DA and Cande WZ (1997) Telomeres cluster *de novo* before the initiation of synapsis: a three-dimensional spatial analysis of telomere positions before and during meiotic prophase. *Cell Biology* **137**: 5–18.

Blackburn EH (1991) Structure and function of telomeres. *Nature* **350**: 569–573.

Cross SH, Allshire RC, McKay SJ, McGill NI and Cooke HJ (1987) Cloning of human telomeres by complementation in yeast. *Nature* **338**: 771–774.

Evans SK, Bertuch AA and Lundblad V (1999) Telomeres and telomerase: at the end, it all comes together. *Trends in Cell Biology* **9**(8): 329–331.

Klapper W, Parwaresch R and Krupp G (2001) Telomere biology in human aging and aging syndromes. *Mechanisms of Ageing and Development* **122**: 695–712.

Meyne J, Ratliff RL and Moyzis RK (1989) Conservation of the human telomere sequence (TTAGGG)*n* among vertebrates. *Proceedings of the National Academy of Sciences of the United States of America* **86**(18): 7049–7053.

Moyzis RK (1991) The human telomere. *Scientific American* **265**(2): 48–55.

Moyzis RK, Buckingham JM, Cram LS, *et al.* (1988) A highly conserved repetitive DNA sequence, (TTAGGG)*n*, present at the telomeres of human chromosomes. *Proceedings of the National Academy of Sciences of the United States of America* **85**: 6622–6626.

Riethman HC, Moyzis RK, Meyne J, Burke DT and Olson MV (1989) Cloning human telomeric DNA fragments into *Saccharomyces cerevesiae* using a yeast artificial chromosome vector. *Proceedings of the National Academy of Sciences of the United States of America* **86**: 6240–6244.

Scherthan H, Weich S, Schwegler H, Heyting C and Harle M (1996) Centromere and telomere movements during early meiotic prophase of mouse and man are associated with the onset of chromosome pairing. *Cell Biology* **134**: 1109–1125.

Shay JW, Zou Y, Hiyama E and Wright WE (2001) Telomerase and cancer. *Human Molecular Genetics* **10**(7): 677–685.

Telomeres and Telomerase in Aging and Cancer

Michel M Ouellette, *University of Nebraska Medical Center, Omaha, Nebraska, USA*

Intermediate article

Telomeres protect the ends of chromosomes and shorten each time somatic human cells divide. As a consequence, cells enter senescence after a limited number of divisions. The enzyme telomerase expressed in rare cells of the skin, blood and intestine can prevent the shortening of telomeres and extend cellular lifespan.

Article contents

- Telomeres
- Telomerase
- Telomere-controlled Senescence
- Role of Telomerase in Cancer Progression
- Role of Telomeres in Organismal Aging
- Conclusion

Telomeres

Telomeres are specialized structures that cap and protect the extremities of linear chromosomes (Blackburn, 1994). In humans, they are made of a simple double-stranded deoxyribonucleic acid (dsDNA) repeat, (TTAGGG)$_n$, reiterated over thousands of base pairs (3–20 kilobase pairs (kbp)). A single-stranded 3' overhang of 50–200 bases, made of the same (TTAGGG)$_n$ repeats, caps the very end of each telomere. A growing number of proteins have been identified that associate with either the single- or the double-stranded portion of each telomere. The resulting DNA/protein complex is often described as the 'telosome' (de Lange, 2002) (**Figure 1a**). A major function of this large complex is to protect the ends of chromosomes from degradation by nuclease, inappropriate recombination and interchromosomal fusions.

A second function, more clearly defined in yeast systems, is to assist in the alignment of homologous chromosomes during meiosis. A third function is to silence transcriptional activities over telomeres and nearby subtelomeric regions, producing a variegated pattern of expression described as the 'telomere position effect'. A fourth function is to prevent the recognition of the ends of chromosome by DNA damage checkpoints. Telomeres are natural interruptions in the collinearity of chromosomal DNA that could potentially be recognized as dsDNA breaks. (*See* Telomere; Telomeres: Protection and Maintenance.)

Recent studies have identified two features of the telosomes that play a role in hiding the telomeres from

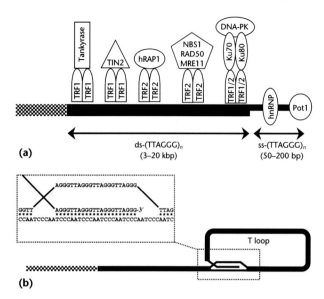

(a)

(b)

Figure 1 (a) Components of the human telosome. The protein components of the telosome include two sequence-specific DNA-binding proteins, TRF1/PIN2 and TRF2, which can interact directly with duplex telomeric DNA. Most of the other components of the telosome that are recruited through their interaction with either TRF1/PIN2, such as TIN2 and tankyrase, or else TRF2, such as hRAP1 and the MRE11/RAD50/NBS1 complex. The Ku/DNA-PK complex appears to be recruited by both TRF1 and TRF2. The 3′ telomeric overhang is also a binding site for a number of candidate proteins, including the protein Pot1 and the heterogeneous nuclear ribonucleoproteins (hnRNP) A1, A2–B1, D and E (ss: single-stranded; ds: double-stranded). (b) T-loop formation. The formation of this structure involves a looping of the telomeres, the displacement of the G-rich strand and the hybridization of the G-rich 3′ telomeric overhang with the C-rich strand of duplex telomeres.

these checkpoints. The first is a specialized T-loop structure. The formation of this structure involves a looping of the telomeres, the displacement of the G-rich strand and the hybridization of the G-rich 3′ telomeric overhang with the C-rich strand of duplex telomeres (**Figure 1b**). The second feature is TRF2, a sequence-specific DNA-binding protein that promotes T-loop formation. TRF2 forms a homodimer that recognizes duplex telomeric repeats and localizes to the telomeres. Loss of TRF2 function prevents the formation of T loops, results in interchromosomal fusions and leads to the activation of DNA damage checkpoints. Depending on the cell type considered, activation of DNA damage checkpoints by the unprotected telomeres will either produce a permanent state of cell cycle arrest or else lead to apoptosis.

The ends of linear chromosomes pose a challenge to the DNA replication machinery. While one DNA strand is replicated uninterruptedly (leading strand synthesis), the opposite strand has to be replicated through short successive stretches of DNA, termed

Okazaki fragments, each primed by a short ribonucleic acid (RNA) primer of 10–20 nt (lagging strand synthesis). When a replication fork reaches the end of a linear chromosome, the DNA replication machinery falls short of completing lagging strand synthesis because the last Okazaki fragment is not necessarily positioned at the very end of the template strand and/or because the removal of the RNA primer from the last Okazaki fragment creates a gap of unreplicated DNA. Recent evidence also points to the postreplication processing of telomeres as an additional source of telomere erosion. To compensate for this net loss of telomeric DNA during replication, an enzymatic activity exists that can synthesize and add new repeats to the very ends of chromosomes. This activity is that of the enzyme telomerase.

Telomerase

The enzyme telomerase is a DNA polymerase specialized in the synthesis of telomeric repeats. Its primary function is to compensate for the shortening of telomeres produced by the 'end replication problem'. The human enzyme contains two essential subunits: the protein hTERT (human telomerase reverse transcriptase) and the small nuclear RNA hTR (human telomerase RNA). The first provides catalytic activity and the second contains a short sequence (5′CUAACCCUAAC 3′) that the enzyme utilizes as an internal template for the synthesis of telomeric repeats (**Figure 2**). The enzyme associates with the 3′ telomeric overhang, allowing the template region of hTR to hybridize with the overhang. In the next step, the enzyme functions as reverse transcriptase as it creates a six-base telomeric repeat while using the overhangs as primer and the RNA hTR as a template. The enzyme is highly processive and can add many more repeats to the same telomere before falling off its substrate.

In humans, the telomerase complex has an estimated mass of over 1000 kDa and may function as a homodimer. In addition to its two essential subunits, hTR and hTERT, the telomerase complex contains numerous auxiliary factors. These include the heterogeneous nuclear ribonucleoproteins (hnRNP) A1, C1/C2 and D, the small nucleolar RNA (snoRNA) binding proteins dyskerin and hGAR1, the molecular chaperones p23/hsp90, the La autoantigen and the hStau, L22 and TEP1 proteins. The exact role of many of these subunits has yet to be defined but some probably participate in the recognition of native telomeres by telomerase. It is the interactions between these auxiliary subunits and components of the telosome that are thought to control the access of telomerase to native telomeres. One component of the telosome that plays such a role is sequence-specific

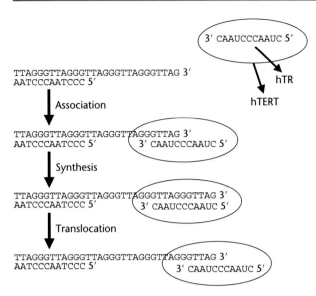

```
                                    3' CAAUCCCAAUC 5'

                                                    hTR
TTAGGGTTAGGGTTAGGGTTAGGGTTAG 3'
AATCCCAATCCC 5'                                     hTERT

      Association

TTAGGGTTAGGGTTAGGGTTAGGGTTAG 3'
AATCCCAATCCC 5'              3' CAAUCCCAAUC 5'

      Synthesis

TTAGGGTTAGGGTTAGGGTTAGGGTTAGGGTTAG 3'
AATCCCAATCCC 5'              3' CAAUCCCAAUC 5'

      Translocation

TTAGGGTTAGGGTTAGGGTTAGGGTTAGGGTTAG 3'
AATCCCAATCCC 5'                    3' CAAUCCCAAUC 5'
```

Figure 2 Biochemical activity of telomerase. The enzyme telomerase is made of two essential subunits, human telomerase reverse transcriptase (hTERT) and human telomerase RNA (hTR). The hTR RNA contains a short sequence, complementary to that of the telomeric repeats (5'-CUAACCCUAAC-3'), which the enzyme uses as a template. As the enzyme associates with the telomeres, this sequence hybridizes with the 3' overhang that caps all telomeres. The protein hTERT, acting as a reverse transcriptase, then uses the telomere as a primer and hTR as a template to synthesize a six-base telomeric repeat. The enzyme is highly processive and can add many repeats to the same DNA substrate through cycles of synthesis and translocation.

DNA-binding protein TRF1. Like TRF2, TRF1 forms homodimers, which can recognize duplex telomeric repeats and localize to telomeres. In human cells expressing active telomerase, the overexpression of TRF1 produces cells with shorter telomeres, while its inhibition has the opposite effect. It is believed that TRF1 serves as an anchor for a larger complex that limits the access of telomerase to individual telomere, such that the enzyme is preferentially recruited to the shortest telomeres. (*See* Telomerase: Structure and Function.)

The activity of telomerase can be measured in biological samples with the use of the telomeric repeat amplification protocol (TRAP) assay. The assay uses a polymerase chain reaction (PCR) to detect the elongation products formed by telomerase after exposure of a single-stranded DNA substrate to cell extracts. The availability of such a sensitive assay was quickly followed by the realization that most primary human cells lack telomerase. Telomerase activity is expressed early during human development but is later repressed. At birth, most somatic tissues lack the activity with the exception of rare cells of the skin, blood and digestive tract. Recent studies suggest that this pattern of activity is dictated by the expression of the hTERT. First, *telomerase reverse transcriptase*

(*TERT*) gene expression correlates with the presence of the activity, whereas the levels of many other components of the telomerase complex, such as hTR and TEP-1, do not. Second, the hTERT complementary DNA (cDNA) alone suffices to reconstitute active telomerase in primary human cells that would otherwise be negative for the activity.

Telomere-controlled Senescence

Because they lack telomerase activity, most human cells have a life span limited by the shortening of telomeres that they experience at each round of divisions. After a limited number of divisions, telomeres reach a threshold size, at which point cells cease to divide and enter senescence. At senescence, cells are viable but no longer divide. Similar to terminal differentiation, entry into senescence is accompanied by changes in both morphology (the flat and enlarged phenotype) and gene expression (affecting numerous genes). This limited life span of normal human cells has been most extensively studied *in vitro* with the use of primary cultures. In culture, human cells lose 50–200 bp of telomeric repeats per division and enter senescence after 15–100 divisions, depending on the cell type considered.

Because it limits the *in vitro* life span of primary human cells, telomere-controlled senescence is a major obstacle to the establishment of human cell lines. We and others have shown that the expression of an hTERT cDNA in primary human cells can reconstitute telomerase activity and prevent telomere shortening. In numerous cell types, the activity suffices to overcome senescence and provide cells with immortality. Cell types so far immortalized with hTERT alone include skin fibroblasts, keratinocytes, mammary epithelial cells, retinal pigmented epithelial cells, adrenocortical cells, osteoblasts and endothelial cells of the vasculature. A major advantage in using exogenous telomerase to establish lines of human cells is the observations that hTERT can overcome senescence without causing cancer-associated changes or altering phenotypic properties (Harley, 2002). By contrast, conventional approaches to establish cell lines, such as the cultivation of cancer samples or the introduction of viral oncogenes, invariably give rise to lines of cells that display cancer-associated changes.

However, not all types of human cell can readily be immortalized with hTERT alone. First, telomerase does not appear to alter phenotypic properties, such that postmitotic terminally differentiated cells are unlikely to be rescued by the enzyme. Second, certain cell types tend to experience other forms of senescence, independent of telomere shortening, which are not bypassed by hTERT alone. Quiescent states indistinguishable from telomere-controlled senescence can be induced by oxidative stress, oncogenic signals or inappropriate culture conditions (Shay and Wright, 2001).

Current data suggest that senescence is induced when the shortest telomere has become too short to prevent its recognition by DNA damage checkpoints. According to one model, the number of telomere-bound TRF2 molecules plummets as telomeres shorten, eventually causing a T loop to unfold and a single telomere to be revealed as a dsDNA break. In support of this model is the observation that the same pathways are mobilized during telomere-controlled senescence and following the introduction of dsDNA breaks by gamma irradiation, mainly the mobilization of both the p16INK4a/pRB and the p53/p21WAF pathways. Viral oncogenes that can block both pathways, such as the SV40 large T antigen, can greatly extend the life span of human cells beyond senescence. However, cells expressing the oncogene are not yet immortal, as telomeres continue to shorten with divisions. Terminal telomere shortening eventually leads to crisis, an antiproliferative state characterized by massive cell death. Rare clones sometimes emerge from crisis (at a frequency of 10^{-7}) that have gained the capacity to maintain telomere size, in most instances achieved through the induction of endogenous hTERT expression. A similar selection for hTERT-expressing cells is seen *in vivo* during cancer progression, which suggests a role for telomerase in cancer progression.

Role of Telomerase in Cancer Progression

Whereas most normal human cells lack telomerase activity, one notable exception is the presence of the activity in more than 85% of cancers, irrespective of tumor type (Shay and Bacchetti, 1997). In most cases, the activity is localized within the tumor with little activity in surrounding healthy tissues. A survey of the literature published in 1994 established telomerase activity as one of the best-known markers of cancer cells. Depending on tumor type, the marker has shown great promise for the early detection of cancer, for differentiating benign and malignant tumors, for refining prognosis and for following the efficacy of therapeutic regimens. The value of the TRAP assay for the detection of malignant cells and for the diagnosis of cancers has now been established for many of the major cancer sites, including the prostate, lung, pancreas, bladder and breast. The TRAP assay is versatile and highly sensitive. It can detect a single cancer cell mixed with thousands of normal human cells and can be used to reveal the presence of cancer

cells in almost any clinical specimen including biopsies, frozen sections, fine-needle aspirates, brushes, washes and biological fluids (urine, pancreatic juice, blood).

The observation that telomerase activity is abnormally expressed in most cancers implies that the enzyme has a role in cancer progression and that telomere-controlled senescence is an important obstacle to tumor progression. Carcinogenesis is a multistep process that necessitates successive alterations in a limited number of genes (four to seven). Each step generally involves the alteration of a gene, selection of mutant cells and the clonal expansion of these cells to favor the occurrence of the next mutation. By imposing limits on clonal expansion, the shortening of telomeres could halt the progression of individual clones at an early stage, much before their conversion to malignancy. According to this model, the role of active telomerase is to allow carcinogenesis to proceed without being blocked by the induction of senescence or crisis as well as to give cancer cells the extended life span needed for tumor growth, invasion and metastasis. While this hypothesis is supported by the observation that hTERT immortalizes without causing cancer-associated changes, emerging data suggest that the enzyme may have a more active role in the establishment of the malignant phenotype. First, active telomerase is an absolute requirement for the *in vitro* transformation of primary human cells, even for those strains that possess long telomeres. Second, the forced expression of exogenous telomerase in laboratory mice, known to possess excess telomere reserves, was found to be associated with the development of breast carcinomas (Artandi *et al.*, 2002). These observations suggest that active telomerase may have transforming properties beyond its capability to maintain telomeres size and function.

Role of Telomeres in Organismal Aging

Telomere-controlled senescence is believed to represent an effective tumor suppressive mechanism that long-lived species have acquired to better control the emergence of cancer cells. While it has the evolutionary advantage of increasing survival to the age of reproduction, this mechanism could, later in life, be a driving force behind the aging process. The absence of telomerase in most human tissues could lead, over a lifetime, to a net accumulation of senescent cells, which could then alter local physiology and hamper tissue regeneration. Current data clearly show that the shortening of telomeres is not limited to cultured human cells and that similar declines in telomere size are also seen *in vivo*. Such decreases with donor age have been observed in skin fibroblasts, adrenocortical cells, osteoblasts, lymphocytes, myoblasts and in the endothelial cells of blood vessels, in which case the rate of shortening was found to correlate with hemodynamic stress. Thus, even in renewal tissue in which active telomerase is expressed, telomeres were found to shorten with age, implying that this low-level telomerase activity is apparently insufficient to maintain telomeres. What is still unclear is whether these reductions in telomere size are sufficient to induce senescence and alter tissue physiology.

One approach that has been used to document the accumulation of senescent cells with age is the cultivation of biopsies followed by measurements of *in vitro* life span as a function of donor age. A major obstacle to these experiments is that some tissues contain sectors that differ in their telomere size and replicative history, some made of near-senescent cells and others of relatively young cells. This mosaicism can lead to extreme variations in life span, in which the contribution of senescent cells is almost always underestimated. Such mosaicism is most evident in the skin, where multiple biopsies derived from one single donor can yield cultures that show extreme differences in both telomere length and remaining life span. While this variation prevented the observation of more subtle age-related changes, perfect correlations were still noted between the size of telomeres and remaining life span, suggesting that telomeres do exist that are short enough to yield senescent cells. In other tissues, in which mosaicism is minimal, good correlations were seen between telomere length, remaining life span and donor age. In primary cultures of osteoblasts, myoblasts and adrenocortical cells, telomere size and remaining life span both decrease with donor age (Yang *et al.*, 2001). In primary cultures derived from the elderly, very few doublings were available, implying that sufficient telomere shortening had occurred to hamper tissue regeneration.

A role for telomere-controlled senescence in the aging process is also suggested from the use of transgenic mouse models (Goytisolo and Blasco, 2002). Laboratory mice express active telomerase in most of their adult tissues and have telomeres much longer than those of humans. Accordingly, the shortening of the telomeres is quite unlikely to be a cause of aging in these animals. However, the targeted deletion of the *Terc* gene, encoding the mouse telomerase RNA, does produce a phenotype with features reminiscent of normal human aging (Goytisolo and Blasco, 2002). While homozygous mice of the first generation lacked telomerase activity but displayed no phenotype, those of the sixth generation had telomeres comparable in size to those of humans. Most interestingly, these mice of the sixth generation displayed symptoms normally associated with the aging process, as they

are observed in the human species: infertility, graying and loss of hair, defects in wound healing, decreased body weight, loss of fat cells, immunological dysfunction and decreased resistance to stress and diseases. A related phenotype is seen in patients with dyskeratinosis congenita (Dokal, 2001). The disease is a human equivalent of the *Terc* knockout mouse. It is caused by mutations in either the human telomerase RNA gene or the gene encoding dyskerin, a component of the telomerase complex. The two mutations cause reduced levels of telomerase activity and affect those tissues that normally express the enzyme, namely the skin, blood and digestive system.

While telomere dysfunction is a probable cause of many age-related conditions, many of the most common afflictions of old age appear to be caused by other mechanisms. First, only a subset of all age-related conditions are reproduced in the knockout mice targeted for the deletion of the *Terc* gene. Second, in spite of their ample telomere reserves, rodents and other short-lived species are still experiencing the aging process.

Conclusion

The absence of active telomerase in most human somatic tissues is a double-edge sword. On one hand, it has the evolutionary advantage of adding senescence as an extra obstacle to the evolution of cancer cells. On the other hand, this lack of activity also imposes limits on the regeneration and repair of tissues that might be a cause of many age-related diseases.

See also
Telomere
Telomeres: Protection and Maintenance
Telomerase: Structure and Function

References

Artandi SE, Alson S, Tietze MK, *et al.* (2002) Constitutive telomerase expression promotes mammary carcinomas in aging mice. *Proceedings of the National Academy of Sciences of the United States of America* **99**: 8191–8196.

Blackburn EH (1994) Telomeres: no end in sight. *Cell* **77**: 621–623.

Dokal I (2001) Dyskeratosis congenita. A disease of premature ageing. *Lancet* **358**: S27.

Goytisolo FA and Blasco MA (2002) Many ways to telomere dysfunction: *in vivo* studies using mouse models. *Oncogene* **21**: 584–591.

Harley CB (2002) Telomerase is not an oncogene. *Oncogene* **21**: 494–502.

de Lange T (2002) Protection of mammalian telomeres. *Oncogene* **21**: 532–540.

Shay JW and Bacchetti S (1997) A survey of telomerase activity in human cancer. *European Journal of Cancer* **33**: 787–791.

Shay JW and Wright WE (2001) Aging: when do telomeres matter? *Science* **291**: 839–840.

Yang L, Suwa T, Wright WE, Shay JW and Hornsby PJ (2001) Telomere shortening and decline in replicative potential as a function of donor age in human adrenocortical cells. *Mechanisms of Ageing and Development* **122**: 1685–1694.

Wright WE and Shay JW (2000) Telomere dynamics in cancer progression and prevention: fundamental differences in human and mouse telomere biology. *Nature Medicine* **6**: 849–851.

Further Reading

Chang E and Harley CB (1995) Telomere length and replicative aging in human vascular tissues. *Proceedings of the National Academy of Sciences of the United States of America* **92**: 11 190–11 194.

Hahn WC, Counter CM, Lundberg AS, *et al.* (1999) Creation of human tumour cells with defined genetic elements. *Nature* **400**: 464–468.

Nakamura TM, Morin GB, Chapman KB, *et al.* (1997) Reverse transcriptase motifs in the catalytic subunit of telomerase. *Science* **276**: 561–567.

Vogelstein B and Kinzler KW (1993) The multistep nature of cancer. *Trends in Genetics* **9**: 138–141.

Wright WE, Piatyszek MA, Rainey WE, Byrd W and Shay JW (1996) Telomerase activity in human germline and embryonic tissues and cells. *Developmental Genetics* **18**: 173–179.

Web Links

TelDB: GenLink Multimedia Telomere Resource. TelDB is a telomere information center that provide access to a searchable database of more than 2100 telomere-related citations from 290 journals, covering over 120 species
http://www.genlink.wustl.edu/teldb/

Telomerase reverse transcriptase (TERT); Locus ID: 7015 at LocusLink:
http://www.ncbi.nlm.nih.gov/LocusLink/

Telomerase reverse transcriptase (TERT); OMIM number: 187270 at OMIM:
http://www.ncbi.nlm.nih.gov/entrez/query.fcgi?db = OMIM

Telomerase RNA component (TERC); Locus ID 7012 at LocusLink: OMIM:
http://www.ncbi.nlm.nih.gov/LocusLink/

Telomerase RNA component (TERC); OMIM number 602322 at OMIM:
http://www.ncbi.nlm.nih.gov/query.fcgi?db = OMIM

Telomeres: Protection and Maintenance

Titia de Lange, *The Rockefeller University, New York, USA*

Eukaryotic chromosomes terminate in telomeres, nucleoprotein complexes thought to mask chromosome ends from the machinery that detects and repairs damaged deoxyribonucleic acid (DNA). The maintenance of this protective telomeric cap requires a specialized mode of DNA synthesis, because chromosome duplication results in incomplete replication of DNA ends. Proteins associated with the telomeric DNA sequence at chromosome ends execute and regulate the maintenance and protection functions of telomeres.

Advanced article
Article contents
• DNA Component of Mammalian Telomeres
• Telomere Dynamics in Senescence and Cancer
• Proteins Associated with Mammalian Telomeres
• Telomere Protection
• Telomere Length Regulation
• T-loop-based Model of Telomere Protection and Length Regulation

DNA Component of Mammalian Telomeres

Like most eukaryotes, mammals maintain their telomeres with telomerase (Morin, 1989). The addition of short repeats to the 3′ end of the chromosome by this enzyme balances the loss of terminal sequences as it occurs during deoxyribonucleic acid (DNA) replication. Human telomerase is composed of a reverse transcriptase (TERT; Nakamura *et al.*, 1997) and a template ribonucleic acid (RNA; TERC) that directs telomerase to add arrays of TTAGGG repeats to chromosome ends (**Figure 1**). As a consequence of their maintenance by telomerase, human chromosomes terminate in long tandem arrays of TTAGGG repeats. The same sequence is found at chromosome ends in all other vertebrates. The 3′ end of vertebrate chromosomes has a 100–300 nucleotide protrusion of single-stranded TTAGGG repeats. Electron microscopy analysis of telomeric DNA from human cells reveals a higher-order structure referred to as the 'T-loop'. T-loops are large duplex DNA loops that frequently occur at the ends of chromosomes in mammals and in several unicellular eukaryotes. It is thought that T-loops are generated by invasion of the 3′ telomeric overhang into the duplex part of the telomeric repeat array, forming a D-loop at the loop–tail junction.

Telomere Dynamics in Senescence and Cancer

Many primary human cell types are known to have a finite replicative potential *in vitro*. Upon reaching this

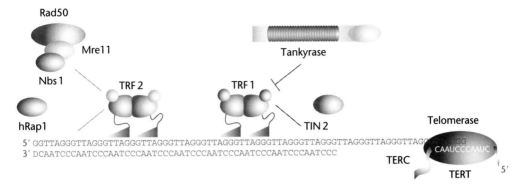

Figure 1 Proteins associated with human telomeres. Additional components (not shown) are Ku, DNA-PKcs and the RECQ helicases BLM and WRN (see text for details).

so-called Hayflick limit, the cells exit the cell cycle with a diploid (2n) or tetraploid (4n) DNA content and undergo a series of morphological and molecular changes, collectively referred to as senescence. Senescent cells are large and flat, express increased levels of the proteins p53, p21 and p16, and often have aberrant chromosomes (end-to-end fusions, dicentric and multicentric chromosomes, gaps, breaks and rings). The mitotic clock that allows cells to count the cell divisions preceding the Hayflick limit is programmed into telomere dynamics. Primary cells that lack telomerase lose telomeric sequences at a constant rate. Their exit from the cell cycle occurs when telomeres have shortened to a set critical length (**Figure 2**). Direct proof of this hypothesis is the observation that forced telomere maintenance by introduction of exogenous telomerase can circumvent senescence in several human cell types. Although other mechanisms (primarily acting on the retinoblastoma (Rb) tumor suppressor pathway) may also curb proliferation in some cell types, it is now generally believed that maintenance of telomeric DNA is required for cellular immortalization and that telomerase activation is a major pathway for the bypass of replicative senescence.

Most human somatic cells undergo programmed telomere shortening *in vivo* (**Figure 2**). In many cases, this is a result of the absence of TERT protein, although more complex regulatory systems have been identified as well. It is now established that activation of telomerase (through upregulation of TERT expression) is an important event in the etiology of human cancer, presumably because telomere maintenance is required for continued proliferation of transformed cells. Thus, telomere shortening may be viewed as a tumor-suppressor system that could curb the growth of malignant cells. Indeed, inhibition of telomerase in cultured tumor cells (using a dominant-negative allele of *TERT*) resulted in telomere shortening and, ultimately, death of the cells. More-

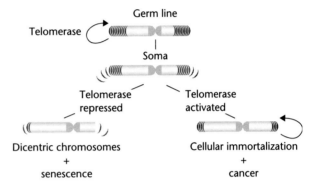

Figure 2 Dynamics of human telomeres.

over, targeted deletion of the ribonucleic acid (RNA) component of mouse telomerase (mTER) resulted in diminished tumor formation in the cancer-prone INK4a knockout mouse. In the context of this tumor-suppressor pathway, it is important to understand the difference between a fully functional telomere that allows cell growth and preserves chromosome integrity and the chromosome ends in senescent cells that signal cell cycle exit and fail to protect the termini from fusion.

Proteins Associated with Mammalian Telomeres

The two major telomeric DNA binding proteins in mammals are the related factors TRF1 and TRF2 (**Figure 1**). These paralogs have similar Myb-type helix–turn–helix domains in their carboxy (C)-termini with which they specifically bind to duplex TTAGGG repeats, both at telomeres and at internal sites of the chromosome. TRF1 and TRF2 associate with telomeric DNA throughout the cell cycle, and their messenger RNAs (mRNAs) are found in all human cell types. They share a second conserved domain, the

TRF homology domain (TRFH). This region is responsible for homotypic interactions that allow the proteins to form homodimers and higher-order homo-oligomers. In addition, the TRFH domain can mediate association with interacting partners. TRF1 and TRF2 do not form heteromeric complexes, and they are substantially different in their amino (N)-termini, where TRF1 is acidic and TRF2 is basic. TRF1 primarily controls the length of telomeres in telomerase-expressing cells, whereas TRF2 is essential for the protection of chromosome ends.

TRF1 has two types of interacting factor, the two closely related tankyrases (tankyrase 1 and tankyrase 2) and TIN2 (**Figure 1**). The tankyrases are newly discovered telomeric enzymes that associate specifically with the acidic N-terminus of TRF1. This interaction positions tankyrases at telomeres, although the majority of this enzyme has other subcellular localizations including nuclear pore complexes, centrosomes and the Golgi complex. Tankyrase 1 and 2 have been shown to be a poly(ADP-ribose) polymerase that can use TRF1 as a substrate. Poly(ADP-ribos)ylation of TRF1 *in vitro* results in inhibition of the DNA binding activity of TRF1. Both tankyrases appear to remove TRF1 from telomeres *in vivo*, as cells forced to overexpress these enzymes in the nucleus show a loss of the typical punctate staining pattern of TRF1 on chromosome ends. Moreover, such cells lose control over the length of their telomeres, showing rapid telomere elongation, a phenotype also seen in cells with diminished TRF1 activity. Thus, tankyrases act as an upstream negative regulators of TRF1-dependent control of telomere length (see below).

A second TRF1 interacting factor, TIN2, is a telomeric protein of 40 kDa with no obvious sequence motifs. TIN2 interacts with the TRFH domain of TRF1 (but not TRF2) and is associated with telomeres throughout the cell cycle. An integral component of the telomeric complex, TIN2 is important in regulation of telomere length. Overexpression of a truncated form of TIN2 results in dramatic telomere elongation in cells that express telomerase.

TRF2 recruits its own distinct set of proteins to the telomere (**Figure 1**). A two-hybrid screen identified RAP1, the human ortholog of the major telomeric binding protein of *Saccharomyces cerevisiae*, Rap1p. Like yeast Rap1p, hRap1 affects telomere length maintenance. Notably, hRap1 differs from its yeast counterpart in that it does not bind to telomeric DNA directly. Rather, it is recruited to telomeres by its interaction with TRF2, thus indicating that a major shift in the organization of the telomeric complex took place during evolution.

TRF2 also binds to the Mre11/Rad50/Nbs1 recombinational repair complex (**Figure 1**). Co-immunoprecipitation and immunofluorescence studies have shown that a small fraction of the cellular Mre11 complex is bound to the TRF2/Rap1 complex. This interaction tethers the Mre11 complex to telomeres throughout interphase. The interaction of Nbs1 with telomeres is predominantly observed in S phase. Further confirmation that the Mre11 complex functions at telomeres comes from the observation of these proteins at chromosome ends in meiosis. Recent data suggest that Nbs1 influences the regulation of telomere length.

In addition to the Mre11 complex, there are other factors involved in detection and processing of DNA damage that associate with telomeres. The Ku70/Ku86 heterodimer is an abundant DNA binding complex with a strong preference for DNA ends. The Ku protein binds to telomeric DNA ends *in vitro*, and cross-linking studies have indicated that Ku is detectable at telomeres *in vivo*. Evidence of the role of Ku at mammalian chromosome ends comes from the analysis of Ku knockout mice, which show an increase in chromosome end fusions, in particular when the cells are deficient for p53. Similarly, the PI3-related protein kinase DNA-PKcs, which is associated with Ku and telomeres, may contribute to telomere function, as DNA-PKcs deficiency results in an increased rate of chromosome end fusions and deregulation of telomere length. Finally, the RECQ helicases BLM and WRN have been observed in association with telomeric DNA. These helicases, notably, have been shown *in vitro* to have the unique ability to unwind the unusual G·G base-paired G4 DNA structures that can be formed by telomeric repeats.

Additional factors have been implicated in telomere metabolism on the basis of mutant phenotypes. Examples are the higher rate of telomere shortening in primary cells from individuals with ataxia-telangiectasia lacking a functional ATM kinase, and the effects of mutations in HNRPA1 and ADPRT on telomere length regulation in murine cells. However, it is currently unclear whether these effects are direct or reflect a secondary effect of these proteins, as so far they have not been demonstrated to bind chromosome ends.

Finally, human and fission yeast orthologs of a ciliate telomere terminus factor were recently identified. This factor, Pot1, can bind to single-stranded TTAGGG repeats *in vitro*. Fission yeast experiments suggest that Pot1 is involved in the protection of telomere ends (Baumann and Cech, 2001).

In summary the telomeric repeat array at the ends of mammalian chromosomes recruits two sequence-specific DNA binding proteins that each tether additional factors to the telomere. The overall complex is composed both of factors that are telomere-specific and a large number of factors known to function in DNA repair and recombination. Collectively, these

Plate 1 [Repetitive Elements and Human Disorders] Mapping recombination junctions in Alu/Alu unequal homologous recombination events. Alu/Alu recombination, as shown in **Figure 1**, can result in a crossover occurring at any point in the Alu element. Slight differences in the Alu elements can be used to map the site of crossover approximately. The dark bar at the top represents the approximately 300 bp Alu sequence. The white arrows within the bar mark the A-rich regions found at the downstream end of both halves of the dimer-like structure. The approximate locations of the recombination junctions for the Alu/Alu recombination events shown in **Table 3** are shown as lines with a dot to show the center position. Some of these are color-coded for the genes with the largest numbers of such events. The low-density lipoprotein receptor recombination events seem heavily biased to the left end, near the Alu promoter region.

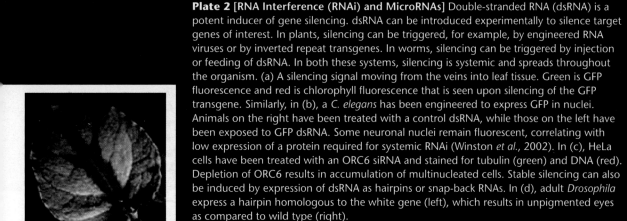

Alu/Alu recombination junctions

LDLR
MLL1
globin
assorted genes

Plate 2 [RNA Interference (RNAi) and MicroRNAs] Double-stranded RNA (dsRNA) is a potent inducer of gene silencing. dsRNA can be introduced experimentally to silence target genes of interest. In plants, silencing can be triggered, for example, by engineered RNA viruses or by inverted repeat transgenes. In worms, silencing can be triggered by injection or feeding of dsRNA. In both these systems, silencing is systemic and spreads throughout the organism. (a) A silencing signal moving from the veins into leaf tissue. Green is GFP fluorescence and red is chlorophyll fluorescence that is seen upon silencing of the GFP transgene. Similarly, in (b), a C. elegans has been engineered to express GFP in nuclei. Animals on the right have been treated with a control dsRNA, while those on the left have been exposed to GFP dsRNA. Some neuronal nuclei remain fluorescent, correlating with low expression of a protein required for systemic RNAi (Winston *et al.*, 2002). In (c), HeLa cells have been treated with an ORC6 siRNA and stained for tubulin (green) and DNA (red). Depletion of ORC6 results in accumulation of multinucleated cells. Stable silencing can also be induced by expression of dsRNA as hairpins or snap-back RNAs. In (d), adult *Drosophila* express a hairpin homologous to the white gene (left), which results in unpigmented eyes as compared to wild type (right).

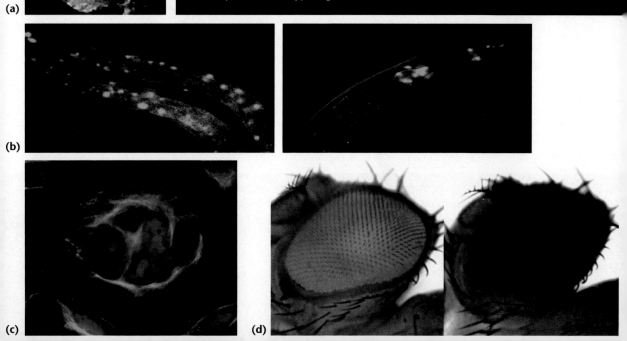

(a)

(b)

(c)

(d)

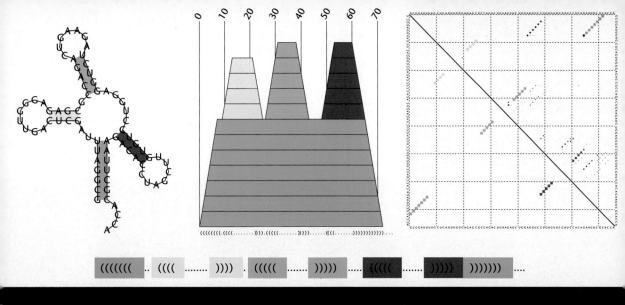

Plate 3 [RNA Secondary Structure Prediction] Transfer ribonucleic acid clover leaf structure as a secondary structure graph, mountain plot, dot plot and in bracket notation.

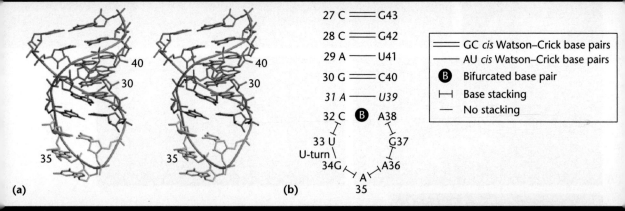

Plate 4 [RNA Tertiary Structure Prediction: Computational Techniques] Yeast tRNA^Phe anticodon stem–loop. Cylinders are drawn between covalently bonded atoms. Hydrogen atoms are not shown. (a) 3D X-ray crystal structure. The stem is shown in blue. The U-turn motif is shown in yellow. (b) Secondary structure using a newly proposed annotation by Leontis and Westhof. Regular font indicates C3′ endo puckers, italic font indicates C2′ exo pucker and all nucleotides have antiglycosyl torsions.

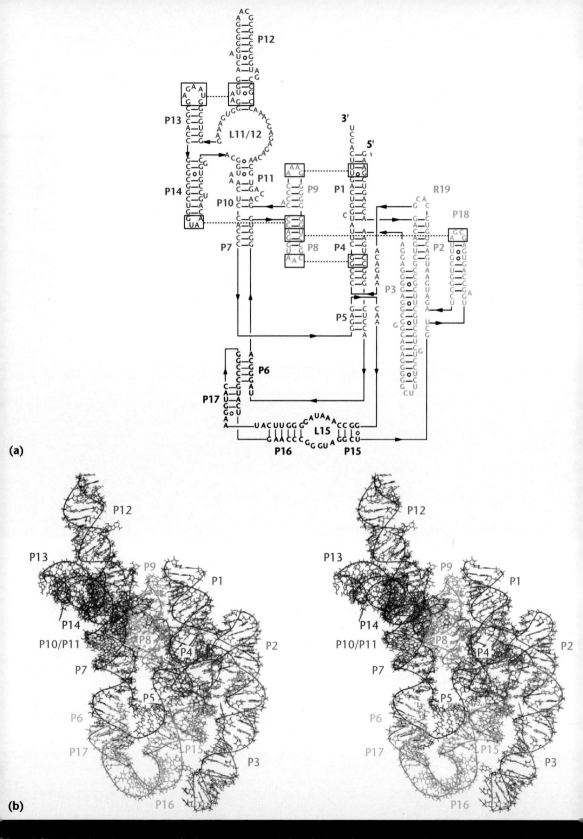

Plate 5 [RNA Tertiary Structure Prediction: Computational Techniques] RNase P RNA from Haas and co-workers. (a) Secondary structure. Straight lines indicate *cis* Watson–Crick base pairs. Dots represent GU wobble base pairs. Arrows indicate tertiary base pairs revealed by cross-linking data. (b) Stereoview of the tertiary structure.

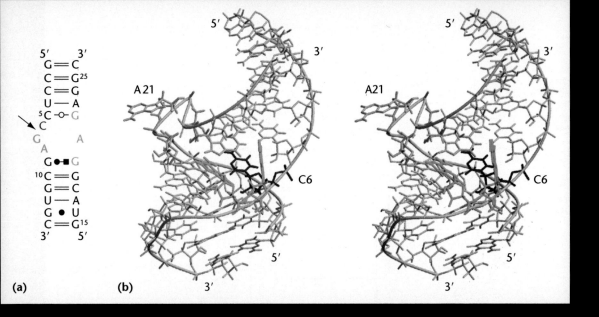

Plate 6 [RNA Tertiary Structure Prediction: Computational Techniques] Lead-activated ribozyme. (a) Secondary structure. The symbols defined in **Figure 1b** were used. The empty circle indicates a *trans* Watson–Crick base pair. The filled circle followed by the filled box indicates a *cis* base pair with Watson–Crick and Hoogsteen interacting edges. The single filled circle indicates a GU wobble base pair. The arrow indicates the cleavage site. (b) Stereoview of the tertiary structure. The stems flanking the internal loop are shown in gray. The nucleotides in the internal loop are colored: C6 in red, G7 and A8 in green, and G20, A21 and G22 in yellow.

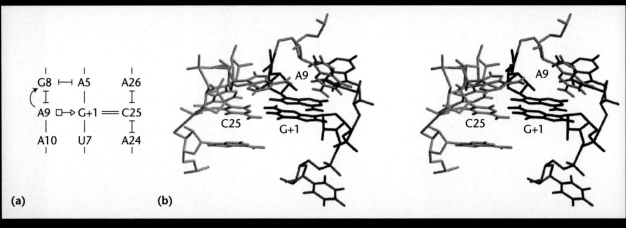

Plate 7 [RNA Tertiary Structure Prediction: Computational Techniques] Catalytic core of the hairpin ribozyme. (a) Secondary structure. The symbols defined in **Figure 1b** were used. The empty square followed by the arrow indicates a *trans* base pair with Hoogsteen and sugar interacting edges. The curved arrow indicates a direction change in the backbone. (b) Stereoview of the tertiary structure. A8–A10 are shown in blue, A24–A26 are shown in green and B5–B7 (substrate) are shown in red. G+1 (substrate), A9 and C25 form a base triple.

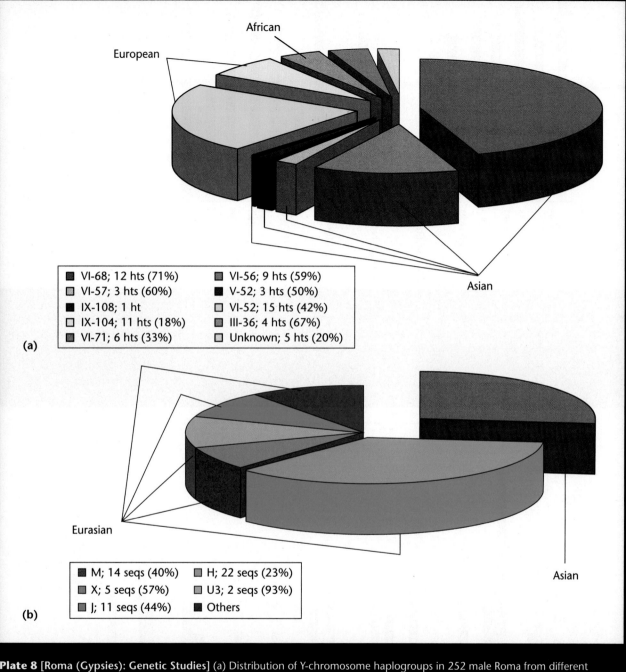

Plate 8 [**Roma (Gypsies): Genetic Studies**] (a) Distribution of Y-chromosome haplogroups in 252 male Roma from different endogamous groups. The haplogroups are defined by unique event polymorphisms as described by Underhill *et al.* (2000). Shaded in gray are haplogroups whose geographic association is not known. The number of haplotypes (defined by the alleles of seven STR loci) along each haplogroup is given next to the haplogroup. The figures in parentheses refer to the frequency of the most common STR haplotype within that haplogroup. (b) Distribution of mtDNA haplogroups in 275 Roma from different endogamous groups. The haplogroups are defined by restriction fragment length polymorphisms in the mitochondrial DNA. The number of different hypervariable segment 1 (HVS1) sequences within each haplogroup is given next to the haplogroup. The figures in parentheses refer to the frequency of the most common HVS1 sequence variant within that haplogroup. The category labeled 'Others' consists of haplogroups I, N, T, U1, U5, U(K) and W.

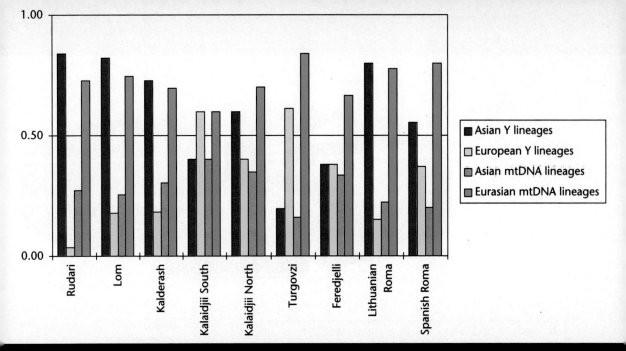

Plate 9 [Roma (Gypsies): Genetic Studies] Distribution of geographically localized Y-chromosome and mtDNA haplogroups in different endogamous Roma groups. Color coding: Asian Y chromosome haplogroups: red; Asian mtDNA haplogroups: orange; European Y chromosome haplogroups: yellow; Eurasian mtDNA haplogroups: green.

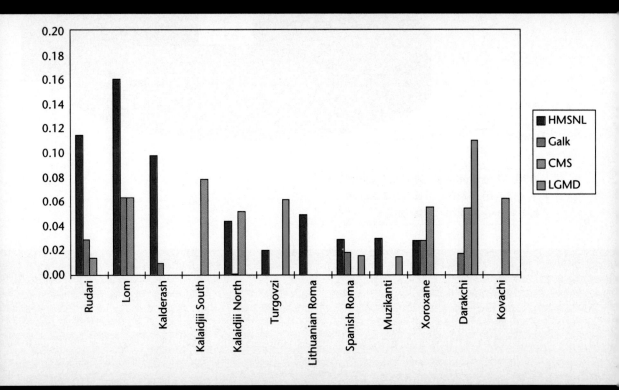

Plate 10 [Roma (Gypsies): Genetic Studies] Carrier rates of four founder mutations (R148X in *NDRG1*, 1267delG in *CHRNE*, 928T in *GK1* and C283Y in *SGCG*) among 800 Roma from different endogamous groups from Bulgaria, Lithuania and Spain. These mutations are responsible for the autosomal recessive disorders hereditary motor and sensory neuropathy type Lom (HMSNL),

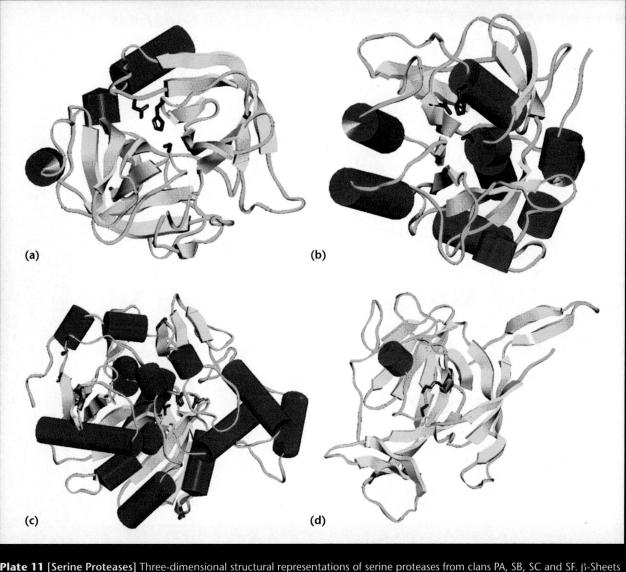

Plate 11 [Serine Proteases] Three-dimensional structural representations of serine proteases from clans PA, SB, SC and SF. β-Sheets are represented by broad ribbons and α-helices are represented by cylinders. Catalytic residues are shown in stick representation. Coordinates are from Protein Data Bank entries, indicated in parentheses after each enzyme name. (a) Human chymotrypsin (4CHA) of clan PA. (b) *Bacillus amyloliquefaciens* subtilisin BPN′ (2ST1) of clan SB. (c) Human lysosomal carboxypeptidase A (1IVY) of clan SC. (d) *Escherichia coli* signal peptidase (1B12) of clan SF. Coordinates for human representatives of clans SB and SF were not available at the time of writing.

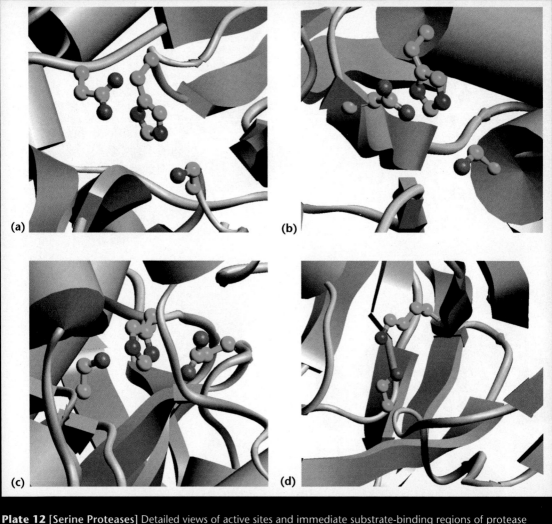

Plate 12 [Serine Proteases] Detailed views of active sites and immediate substrate-binding regions of protease structures in **Figure 1**. Catalytic residues are shown in ball-and-stick representation. It can be seen that the orientation of the catalytic triad for clan SC (c) is the mirror-image reverse of that seen for clans PA (a) and SB (b). The Lys seen in the *Escherichia coli* signal peptidase structure would be replaced by His in eukaryotic homologs, which would also have an Asp residue in the catalytic site.

Uninterrupted microsatellite	CTCTCTCTCTCTCTCTCTCT
Imperfect microsatellite	CTCTCTCTCTATCTCTCTCTC
Interrupted microsatellite	CTCTCTCTAAACTCTCTCTCT
Compound microsatellite	CTCTCTCTCCACACACACACAC

Plate 13 [Simple Repeats] Microsatellite nomenclature. The uninterrupted microsatellite consists of a perfect repetition of CT dinucleotides, the imperfect microsatellite has a C substituted by an A, the interrupted microsatellite has an insertion of three As, which interrupt the CT repeat, the composite repeat consists of two different juxtaposed repeats.

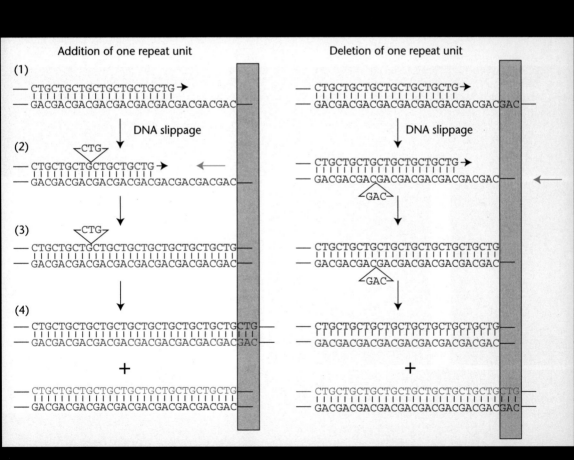

Plate 14 [Simple Repeats] Model of DNA replication slippage adding (left) or removing (right) one repeat unit. (1) First round of DNA replication. (2) DNA slippage, causing one repeat unit to loop out; the dashed arrow indicates the direction of DNA slippage. (3) DNA synthesis continues without repair of the loop. (4) Second round of DNA replication leads to the addition or deletion of one repeat unit in one of the strands.

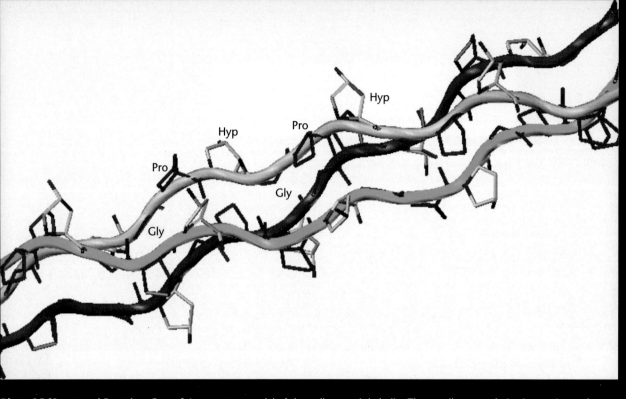

Plate 15 [Structural Proteins: Genes] A computer model of the collagen triple helix. Three collagen α chains intertwine to form a triple helix. The repeating sequence in the α chain shown is glycine (Gly), proline (Pro), hydroxyproline (Hyp). The presence of glycine in every third position is crucial for the tight packing of the α chains into the triple helix. Proline and hydroxyproline are frequent amino acid residues in collagen molecules as they stabilize the triple helix.

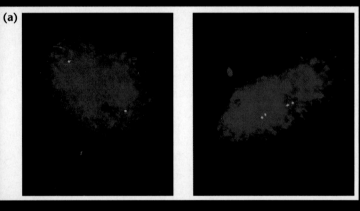

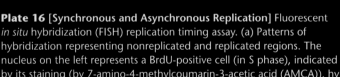

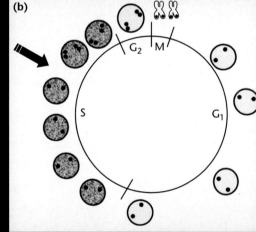

Plate 16 [Synchronous and Asynchronous Replication] Fluorescent *in situ* hybridization (FISH) replication timing assay. (a) Patterns of hybridization representing nonreplicated and replicated regions. The nucleus on the left represents a BrdU-positive cell (in S phase), indicated by its staining (by 7-amino-4-methylcoumarin-3-acetic acid (AMCA)), hybridizing with a singlet–singlet pattern. Each hybridization dot represents one chromosomal copy that has not yet replicated. The nucleus on the right represents a BrdU-positive cell with a doublet–doublet pattern, each doublet representing a pair of sister chromatids originating from one chromosome. The presence of a doublet indicates that the region of interest has replicated. (Figure courtesy of Rachel Ofir.) (b) A schematic explanation for the FISH replication timing assay. Distribution of the hybridization patterns relative to the position of the cells in the cell cycle is indicated in the figure. Nuclei from G_1 hybridize with a singlet–singlet pattern, while nuclei in G_2 hybridize with a doublet–doublet pattern. In metaphase, two signals emanate from each chromosome. In S-phase nuclei, the pattern of hybridization depends on the timing of replication of the region of interest. All nuclei from cells positioned in S prior to the point where replication takes place (shown by an arrow) will show the singlet–singlet pattern, while nuclei originating from cells positioned after the point of replication will display the doublet–doublet pattern. Thus the timing of replication will affect the percentage of nuclei showing either of the hybridization patterns. (Modified version of an illustration by Asaf Hellman.)

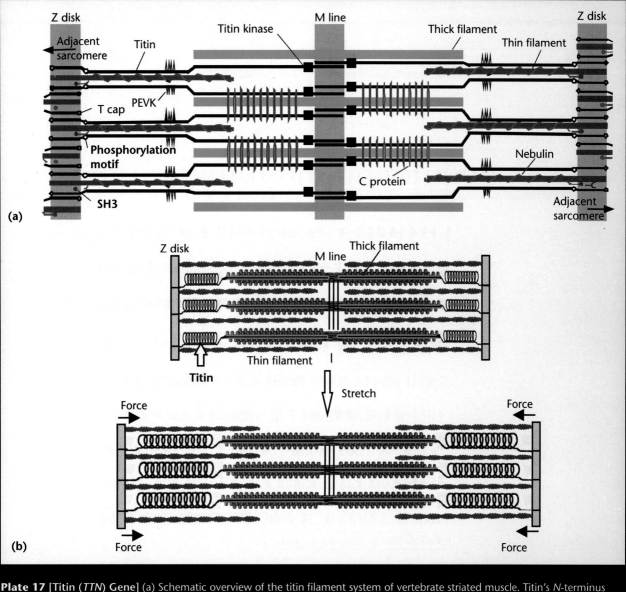

Plate 17 [Titin (*TTN*) Gene] (a) Schematic overview of the titin filament system of vertebrate striated muscle. Titin's *N*-terminus fully spans the Z disk while its *C*-terminus spans the M line. (b) Schematic diagram indicating titin's spring segment. The spring extends as the sarcomere is stretched and a restoring force ensues (for example, as occurs during filling of the heart). PEVK: spring region of titin rich in proline (P), glutamate (E), valine (V) and lysine (K) residues; SH3: nebulin's *C*-terminal region with homology to scr-3 sarc tyrosine kinase; T cap: brief name for the titin cap protein binding to titin's *N*-terminus.

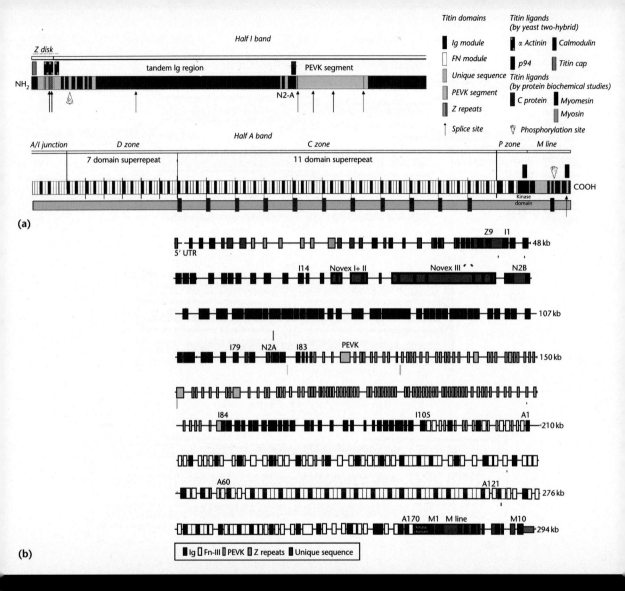

Plate 18 [Titin (*TTN*) Gene] (a) Domain architecture of soleus titin polypeptide. (Modified from Gregorio *et al.* (1998).) In addition to the 243 Ig/FN3 repeats with structural roles for Z disk and thick filament assembly, titin also contains nonrepetitive sequence elements. Since these elements include a serine/threonine kinase domain, phosphorylation motifs and calpain protease binding sites, titin appears to have also multiple roles in myofibrillar signal transduction. (b) Exon–intron structure and domain architecture of the human *titin* gene. (Modified from Bang *et al.* (2001).) *Titin* has a total coding mass of 4200 kDa, which is organized in 363 exons.

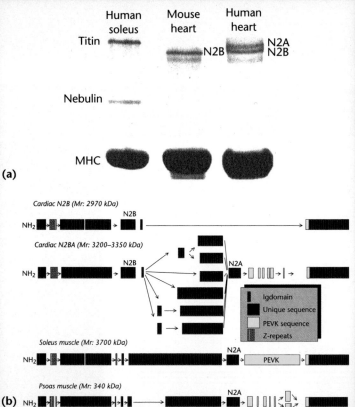

Plate 19 [Titin (*TTN*) Gene] (a) Identification of giant titin polypeptides in vertebrate striated muscle tissues. On denaturing 2% polyacrylamide gels, titin bands appear as low-mobility species far above the 220 kDa myosin heavy chain, and the about 800 kDa nebulin band. Note the mobility difference between the 3700 kDa titin from human soleus skeletal muscle, and the 2970 kDa titin from heart muscle (main band in mouse heart and lower titin band in human heart muscle). (b) The different titin polypeptide size classes are caused by the differential processing of titin transcripts by distinct splice pathways (Freiburg *et al.*, 2000). Titin cDNAs from cardiac muscle (top) and soleus and psoas skeletal muscle (bottom) predict titins that have very different I band regions. Identified splice routes are indicated by arrows. Predicted molecular weights of respective isoforms are given on the left.

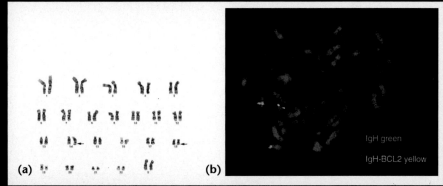

Plate 20 [Translocation Breakpoints in Cancer] (a) Follicular lymphoma showing the characteristic balanced translocation t(14;18) (q32;q21). Arrows indicate the translocation breakpoints on the rearranged copies of chromosomes 14 and 18. (b) Fluorescence *in situ* hybridization (FISH) image showing the same t(14;18)(q32;q21) translocation using probes that span the two loci involved (*IgH@* and *BCL2*). The translocation produces two red–green (yellow) fusion FISH signals, one each on the translocated copies of chromosomes 14 and 18.

```
                   A16V D22G K23R N29I/T
Cationic  MNPLLILTFVAAALAAPFDDDDKIVGGYNCEENSVPYQVSLNSGYHFCGGSLINEQWVVSAGHCY 65
Anionic   MNLLLILTFVAAAVAAPFDDDDKIVGGYICEENSVPYQVSLNSGYHFCGGSLISEQWVVSAGHCY 65
Meso      MNPFLILAFVGAAVAVPFDDDDKIVGGYTCEENSLPYQVSLNSGSHFCGGSLISEQWVVSAAHCY 65
T6        MNPLLILAFVGAAVAVPFDDDDKIVGGYTCEENSVPYQVSLNSGSHFCGGSLISEQWVVSAGHCY 65

                                                                R122H/C
Cationic  KSRIQVRLGEHNIEVLEGNEQFINAAKIIRHPQYDRKTLNNDIMLIKLSSRAVINARVSTISLPT 130
Anionic   KSRIQVRLGEHNIEVLEGNEQFINAAKIIRHPKYNSRTLDNDILLIKLSSPAVINSRVSAISLPT 130
Meso      KTRIQVRLGEHNIKVLEGNEQFINAAKIIRHPKYNRDTLDNDIMLIKLSSPAVINARVSTISLPT 130
T6        KPHIQVRLGEHNIEVLEGNEQFINAAKIIRHPKYNRITLNNDIMLIKLSTPAVINAHVSTISLPT 130

Cationic  APPATGTKCLISGWGNTASSGADYPDELQCLDAPVLSQAKCEASYPGKITSNMFCVGFLEGGKDS 195
Anionic   APPAAGTESLISGWGNTLSSGADYPDELQCLDAPVLSQAECEASYPGKITNNMFCVGFLEGGKDS 195
Meso      APPAAGTECLISGWGNTLSFGADYPDELKCLDAPVLTQAECKASYPGKITNSMFCVGFLEGGKDS 195
T6        APPAAGTECLISGWGNTLSSGADYPDELQCLDAPVLTQAKCKASYPLKITSKMFCVGFLEGGKDS 195

Cationic  CQGDSGGPVVCNGQLQGVVSWGDGCAQKNKPGVYTKVYNYVKWIKNTIAANS 247
Anionic   CQGDSGGPVVSNGELQGIVSWGYGCAQKNRPGVYTKVYNYVDWIKDTIAANS 247
Meso      CQRDSGGPVVCNGQLQGVVSWGHGCAWKNRPGVYTKVYNYVDWIKDTIAANS 247
T6        CQGDSGGPVVCNGQLQGIVSWGYGCAQKRRPGVYTKVYNYVDWIKDTIAANS 247
```

Plate 21 [Trypsinogen Genes: Evolution] Molecular properties of the human cationic, anionic and mesotrypsinogens and predicted amino acid sequence for the expressed pseudogene *T6*. The primary translation product is termed pretrypsinogen (underlined red). Removal of the signal peptide (highlighted in turquoise) results in trypsinogen (underlined blue). Removal of the activation peptide (highlighted in pink) by enteropeptidase or trypsin itself then converts trypsinogen into active trypsin (underlined black). The catalytic triad residues histidine (H), aspartic acid (D) and serine (S) are in red. The three residues determining trypsin specificity are in blue. Threonine (T) at residue 29 in mesotrypsinogen (highlighted in red) is conserved in orthologous sequences of frog, mouse, rat, dog, cow and pig. The primary autolysis site (arginine (R) at residue 122; highlighted in gray) is only present and stringently conserved in clearly functional mammalian trypsins; in *T6*, note the presence of H at this site (highlighted in green). The figure also shows mutations in cationic trypsinogen that are associated with pancreatitis. Aligned amino acid sequences are from GenBank accession numbers M22612, M27602 and D45417. G: glycine; K: lysine; N: asparagine; V: valine.

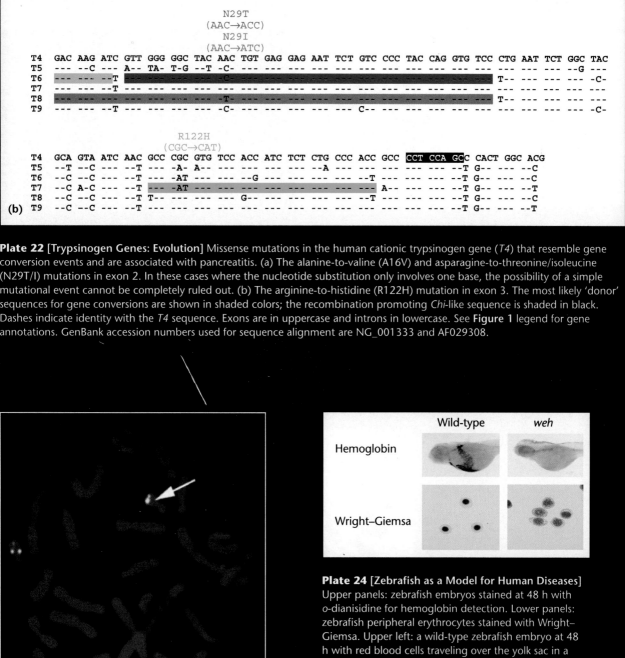

Plate 22 [Trypsinogen Genes: Evolution] Missense mutations in the human cationic trypsinogen gene (*T4*) that resemble gene conversion events and are associated with pancreatitis. (a) The alanine-to-valine (A16V) and asparagine-to-threonine/isoleucine (N29T/I) mutations in exon 2. In these cases where the nucleotide substitution only involves one base, the possibility of a simple mutational event cannot be completely ruled out. (b) The arginine-to-histidine (R122H) mutation in exon 3. The most likely 'donor' sequences for gene conversions are shown in shaded colors; the recombination promoting *Chi*-like sequence is shaded in black. Dashes indicate identity with the *T4* sequence. Exons are in uppercase and introns in lowercase. See **Figure 1** legend for gene annotations. GenBank accession numbers used for sequence alignment are NG_001333 and AF029308.

Plate 23 [Velocardiofacial Syndrome (VCFS) and Schizophrenia] Chromosome 22q11 microdeletion identified by fluorescence *in situ* hybridization (FISH).

Plate 24 [Zebrafish as a Model for Human Diseases] Upper panels: zebrafish embryos stained at 48 h with *o*-dianisidine for hemoglobin detection. Lower panels: zebrafish peripheral erythrocytes stained with Wright–Giemsa. Upper left: a wild-type zebrafish embryo at 48 h with red blood cells traveling over the yolk sac in a primitive circulation. Upper right: a *weissherbst* zebrafish embryo at 48 h with marked anemia. Lower left: wild-type zebrafish peripheral erythrocytes with normal nuclear size. Circulating erythrocytes are nucleated in zebrafish. Lower right: *weissherbst* peripheral erythrocytes with enlarged nuclei and immature morphology. (Reprinted by permission of Macmillan Magazines Ltd from Donovan A, Brownlie A, Zhou Y, *et al.* (2000) Positional cloning of zebra fish ferroportin 1 identifies a conserved vertebrate iron exporter. *Nature* **403**; 776–781.)

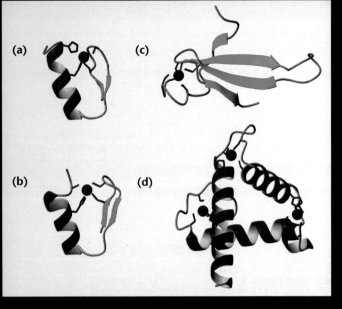

Plate 25 [Zinc-finger Genes] ZnF structures illustrating the variety
of small zinc-dependent protein folds. ZnFs are found as mixed αβ (a,
b), all-β (c) and all-α (d) structures. In each structure, only amino acid
side chains that ligate zinc atoms are shown. (a) Classical C2H2 ZnF
structure of finger 31 from *Xenopus laevis* XFIN-31 (Protein Data Bank
(PDB) accession code 1znf). (b) First CCHC finger from U-shaped
– an FOG-family protein from *Drosophila melanogaster* (PDB accession
code 1fv5). (c) Zinc-ribbon domain from human transcription factor
TFSII (PDB accession code 1tfi). (d) TAZ2 domain from mouse CBP
(PDB accession code 1f81).

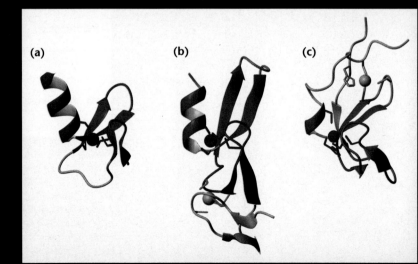

proteins are probably responsible for the two main functions of telomeres: the protection of chromosome ends and the maintenance of terminal sequences.

Telomere Protection

Cells have the ability to distinguish natural chromosome ends from damaged DNA. The telomeric complex is generally believed to be responsible for this distinction. TRF2 has a key role in the protection of chromosome ends. Inhibition of TRF2 function through the expression of a dominant-negative allele of the gene *TERF2* results in the immediate loss of protection of telomeres in a wide variety of human and mouse cell types. One of the first events detected in these cells is the (partial) loss of the 3' overhang from the telomere termini. In addition, the uncapped chromosomes undergo frequent end-to-end fusions resulting in dicentric and multicentric chromosomes. These fusions involve chromosome ends that have preserved most of the duplex part of the telomeric DNA tracts, demonstrating that telomeric DNA *per se* is not sufficient to cap chromosome ends. Binding of TRF2 to the TTAGGG repeats is apparently required for full telomere protection. The end-to-end fusions are created by covalent ligation of both strands of the telomere and result in stable telomere associations that persist through mitosis. As a consequence, cells lacking normal TRF2 function show frequent anaphase chromatin bridges and will often require either chromosome nondisjunction or rupture of dicentric chromosomes for progression to cytokinesis.

At the cellular level, the response to loss of TRF2 from telomeres is either apoptosis or growth arrest similar to replicative senescence. In cells that are known to respond to extensive DNA damage with the induction of apoptosis (such as T cells), telomere malfunction induced by TRF2 inhibition also induces this pathway. Within 24 h of removing TRF2 from the telomeres, cells upregulate p53 and BAX proteins and execute programmed cell death. In addition to p53, the ATM kinase is necessary for this induction. Passage through mitosis (or S phase) is not required for this apoptotic signal, suggesting that the signal can arise directly from defunct telomeres. In other words, the rupture of dicentric chromosomes or other secondary damage in mitosis is not required for the activation of the ATM/p53 pathway.

As noted above, the Ku heterodimer and DNA-PKcs have also been implicated in the capping function of telomeres. However, the effect of loss of Ku and DNA-PKcs seems relatively minor, as affected cells continue to proliferate and affected mice develop to adulthood.

Telomere Length Regulation

An intriguing aspect of telomere function is the regulation of telomere length in cells that express telomerase. In the germ line and in immortal human (tumor) cell lines, the maintenance of human telomeric DNA seems to be a highly regulated process. Telomeres in telomerase-positive cell lines are generally stable, suggesting a homeostasis mechanism that balances telomerase-mediated elongation with the forces that shorten telomeric tracts. How is the length of the telomeric tract measured, and how does this measurement translate into regulation of telomerase? In general, the regulation appears to play out in *cis*. It has been shown that a single short telomere in a cell can undergo rapid growth, allowing it to catch up with the length of the other telomeres. At the same time, the bulk of the telomeres in the cell remain stable. Thus, cells can monitor and modulate the length of a single telomere, indicating a *cis*-acting regulatory mechanism.

TRF1, TRF2 and their interacting factors – TIN2, tankyrase 1, hRap1 and possibly the Mre11 complex – all contribute to telomere length homeostasis. TRF1 and TIN2 have a negative influence on telomere length maintenance. These negative regulators appear to suppress the ability of telomerase to elongate the telomere. Tankyrase 1 has the opposite effect, behaving as a positive regulator of telomere length. It seems that Tankyrase 1 acts to remove the TRF1/TIN2 complex from telomeres, diminishing the presence of these negative regulators and promoting telomere elongation by telomerase. The TRF2/Rap1 complex also influences telomere length maintenance, but the details of this pathway are less clear. Telomere length regulation is influenced by many other factors. ATM kinase deficiency results in enhanced telomere shortening in primary human cells, and DNA-PKcs, hnRNPA1 and PARP1 affect telomere length setting in mouse cells.

As TRF1, TRF2, TIN2, Rap1 and tankyrase do not affect the expression of telomerase components or the activity of telomerase measured in cell extracts, it is more likely that their control is exerted in *cis*. Perhaps the telomere binding proteins alter the accessibility of the telomere terminus to telomerase or modulate the activity of the enzyme at the telomere terminus.

The current working hypothesis for telomere length regulation by TRF1 is shown in **Figure 3**. The model dictates that binding of TRF1 is governed by the length of the telomere. Longer telomeres are proposed to bind a larger mass of these proteins, whereas shorter telomeres bind less. Moreover, the model suggests that a critical mass of the TRF1 complex is required to block the access of telomerase to the telomere terminus. If telomerase is blocked, the telomere will

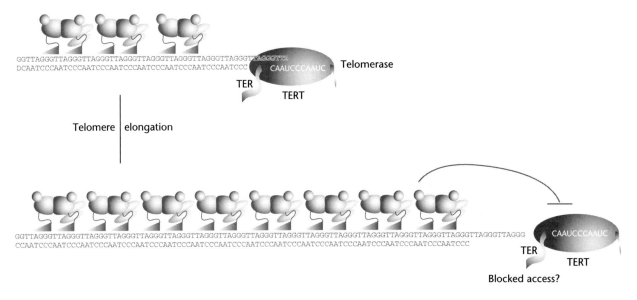

Figure 3 Model of telomere length regulation by TRF1.

shorten during proliferation until it no longer binds sufficient TRF1 to inhibit telomerase. Consequently, the telomere will enter an elongation phase. In this manner, telomeres will switch back and forth between elongation and shortening states, depending on their length and the level of TRF1/TIN2 in the cells. In addition, telomere length setting will depend on the activity of telomerase. Tankyrases may act as an upstream negative regulator of TRF1 that can remove this protein complex from the telomeres. Perhaps tankyrases are required at every cell division to 'erase' TRF1 from the telomere so that telomere length can be re-assessed.

T-loop-based Model of Telomere Protection and Length Regulation

The finding of T-loops has suggested a mechanism by which telomeres could protect chromosome ends and regulate telomerase-mediated telomere maintenance (**Figure 4**). This T-loop-based model dictates that telomerase is incapable of accessing the 3′ end of the chromosome when the telomere is in the T-loop configuration. Telomerase requires a 3′ overhang *in vitro*, arguing that T-loops would have to be opened (transiently) for telomerase to access the 3′ terminus. In this scheme, telomere length could be regulated by the duration of the open state, presumably during S phase. Telomere-binding proteins that promote T-loop formation would act as negative regulators of telomere length maintenance. As a given telomere

becomes inappropriately short, it might no longer bind enough of the proteins that are needed to form T-loops, and the telomeres thus elongate. Moreover, very short telomeres (shorter than the persistence length of roughly 400 base pairs) might not fold as efficiently into T-loops, owing to the rigidity of duplex DNA. Conversely, inappropriately long telomeres may readily recruit proteins that form T-loops, thus limiting the 'window of opportunity' for telomerase. In this manner, the length of the telomeric repeat array could dictate how telomerase interacts with the telomere terminus.

It has been also proposed that T-loops mask the chromosome ends from inappropriate activation of DNA damage checkpoints and repair factors. It is not known exactly how mammalian cells detect DNA damage. The mechanism by which T-loops might prevent detection by these pathways is therefore unclear. It may be that the invasion of the 3′ overhang is simply sufficient to mask the chromosome end. Alternatively, capping of the chromosome end could involve recruitment of a specific protein complex to the base of the T-loop.

The T-loop model predicts that inappropriate resolution of T-loops would result in activation of DNA damage checkpoints and attempts at repair of chromosome ends, leading to degradation and fusion. Furthermore, opened T-loops could lead to uncontrolled telomere elongation by telomerase. Testing this model will require manipulation of the T-loop state and dynamics *in vivo*. This can be achieved once it is known which proteins are involved in the resolution, formation and maintenance of T-loops. Current

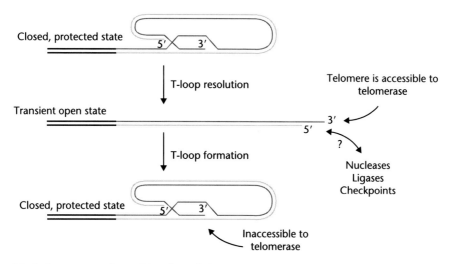

Figure 4 T-loop model of telomere protection and length regulation.

candidates for the factors that remodel the state of the telomere are TRF1 and TRF2, the Mre11 complex and the RECQ helicases. *In vitro* studies have shown that TRF1 can promote the pairing of telomeric DNA and that it has the ability to loop DNA. Both features are expected to facilitate T-loop formation. Notably, incubation of a telomeric DNA substrate with TRF2 can increase the frequency of *in vitro* T-loop formation. TRF2 accumulates at the base of the T-loop, perhaps recognizing and stabilizing a specific DNA configuration. Further studies are required to determine which of these factors is involved in T-loop formation *in vivo*.

See also

Telomerase: Structure and Function
Telomere
Telomeres and Telomerase in Aging and Cancer

References

Baumann P and Cech TR (2001) Pot1, the putative telomere end-binding protein in fission yeast and humans. *Science* **292**: 1171–1175.

Morin GB (1989) The human telomere terminal transferase enzyme is a ribonucleoprotein that synthesizes TTAGGG repeats. *Cell* **59**: 521–529.

Nakamura TM, Morin GB, Chapman KB, *et al.* (1997) Telomerase catalytic subunit homologs from fission yeast and human. *Science* **277**: 955–959.

Further Reading

Blackburn EH and Greider CW (eds.) (1995) *Telomeres.* Cold Spring Harbor, NY: Cold Spring Harbor Press.

Collins K (2000) Mammalian telomeres and telomerase. *Current Opinion in Cell Biology* **12**(3): 378–383.

de Lange T (2001) Telomere capping – one strand fits all. *Science* **292**: 1075–1076.

de Lange T and Jacks T (1999) For better or worse? Telomerase inhibition and cancer. *Cell* **98**: 273–275.

McEachern MJ, Krauskopf A and Blackburn EH (2000) Telomeres and their control. *Annual Review of Genetics* **34**: 331–358.

Nugent CI and Lundblad V (1998) The telomerase reverse transcriptase: components and regulation. *Genes and Development* **12**: 1073–1085.

Oulton R and Harrington L (2000) Telomeres, telomerase, and cancer: life on the edge of genomic stability. *Current Opinion in Oncology* **12**: 74–81.

Sherr CJ and DePinho RA (2000) Cellular senescence: mitotic clock or culture shock? *Cell* **102**: 407–410.

Web Links

TERF2 (telomeric repeat binding factor 2); LocusID: 7014. LocusLink:
  http://www.ncbi.nlm.nih.gov/LocusLink/LocRpt.cgi?l = 7014

TERT (telomerase reverse transcriptase); LocusID: 7015. LocusLink:
  http://www.ncbi.nlm.nih.gov/LocusLink/LocRpt.cgi?l = 7015

TERF2 (telomeric repeat binding factor 2); MIM number: 602027. OMIM:
  http://www.ncbi.nlm.nih.gov/htbin-post/Omim/dispmim?602027

TERT (telomerase reverse transcriptase); MIM number: 187270. OMIM:
  http://www.ncbi.nlm.nih.gov/htbin-post/Omim/dispmim?187270

Telomeric and Subtelomeric Repeat Sequences

Carolyn M Price, *University of Cincinnati College of Medicine, Cincinnati, Ohio, USA*

Telomeric repeats are the tandem arrays of a short G-rich sequence that are present at the ends of most eukaryotic chromosomes. Subtelomeric sequences lie adjacent to the telomeric repeats and are composed of complex, low-copy repeated sequences.

Introduction

Telomeres are the structures that are present at the ends of linear chromosomes. In humans, they are composed of tandem repeats of the sequence $T_2AG_3 \cdot A_2TC_3$ and specialized telomere proteins. Some of the telomere proteins (e.g. TRF1, TRF2 and Pot1) bind directly to the telomeric deoxyribonucleic acid (DNA), while others (e.g. Tin2, tankyrase and hRap1) associate with the telomere through the DNA-binding proteins. The resulting DNA–protein complexes protect chromosome ends from end-to-end joining and prevent the terminal DNA sequence from being recognized as a double-strand break. Telomeres also provide a way of preventing chromosome shortening due to incomplete DNA replication. Although DNA polymerase is unable to replicate the extreme 5′ end of a linear DNA molecule, the enzyme telomerase can add new T_2AG_3 repeats to telomeres, so a net loss of telomeric sequence is prevented. Telomerase is an unusual ribonucleic acid (RNA)-containing reverse transcriptase that uses its RNA component to template the synthesis of new T_2AG_3 sequence. Cells that lack telomerase usually have a limited life span because they stop dividing when their telomeres become critically short.

Subtelomeric repeats differ from the telomeric T_2AG_3 repeats in that no essential biological role has yet been identified for them. Chromosomes with terminal deletions that have been healed by the addition of telomeric repeats are found in some patients with α-thalassemia and can also be generated artificially. Although these healed chromosomes lack subtelomeric repeats, they propagate normally. This suggests that subtelomeric sequences may simply serve as a buffer between the telomeric tract and the gene-rich unique sequences located near the telomere.

Telomeric Repeat Sequences

The telomeric tract of a human chromosome usually extends for about 5–15 kilobases (kb) with the G-rich strand oriented toward the 3′ end of the chromosome

(de Lange *et al.*, 1990). The G-rich strand of some, and perhaps all, telomeres is about 150–200 nucleotides longer than the C-rich strand, so the telomeric DNA terminates in a long single-stranded overhang (Wright *et al.*, 1997). Electron microscopy studies indicate that these G-strand overhangs can tuck into the duplex region of the telomeric tract by strand invasion to generate a large loop structure on the end of chromosome (**Figure 1**) (Griffith *et al.*, 1999). These 'T loops' are an excellent way to hide the DNA terminus from DNA-modifying activities throughout much of the cell cycle. Their formation seems to be mediated by telomere proteins that bind to the duplex region of the telomeric tract.

Although human chromosomes contain many different types of repeated sequence DNA, the T_2AG_3 telomeric tracts are unusual because they are heterogeneous in length. This heterogeneity can be visualized by using Southern hybridization to identify the telomeric restriction fragments (Allsopp *et al.*, 1992). When genomic DNA is digested with enzymes that cut adjacent to the T_2AG_3 tract, the hybridization signal appears as a long smear instead of a series of discrete bands. This smear results from variation in the length of the telomeric tract on each chromosome

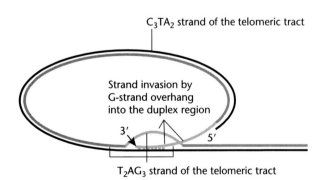

Figure 1 Structure of a T loop. The arrow points to the 3′ end of the G strand. 5′ refers to the end of the C strand.

within a cell, as well as between individual chromosomes from different cells. The cell-to-cell variation in telomere length can be measured by marking an individual telomere with unique sequence DNA and using that sequence to identify the telomeric restriction fragment (Sprung et al., 1999). For a telomere with an average length of 5 kb, there is ~2 kb difference in length between the longest and shortest fragments.

Telomeric restriction fragments do not normally contain sequences that are chromosome specific so Southern hybridization cannot be used to compare the lengths of the 46 different telomeres in a cell. However, individual telomere lengths can be compared by a technique known as quantitative fluorescent in situ hybridization (Q-FISH). In this procedure, a fluorescent-tagged peptide nucleic acid (PNA) probe is hybridized to the telomere sequences on metaphase chromosome spreads (Martens et al., 1998). The amount of fluorescence at each telomere is then quantified. Q-FISH analysis has revealed that the telomeres from each chromosome arm differ significantly in length, with the longest having two and a half to three times more T_2AG_3 sequence than the shortest. Chromosome 4p usually has the longest telomeres while chromosome 17p has some of the shortest.

The length heterogeneity of the telomeric tract is caused by the action of both lengthening and shortening activities on individual telomeres. The net length of a telomere is determined by the balance between these activities. Shortening of the telomeric tract can result from degradation by nucleases, removal of the RNA primer from the final Okazaki fragment during DNA replication and failure to place the final Okazaki fragment at the terminus of the parental strand. Most telomere lengthening results from telomerase adding new T_2AG_3 repeats. During this process, a short template region in the telomerase RNA subunit hybridizes to the complementary T_2AG_3 sequence at the terminus of the telomeric G strand. The catalytic subunit, TERT, then polymerizes addition of new G-strand DNA.

In humans, telomerase is expressed in the germ line, in some highly regenerative tissues (e.g. hematopoietic stem cells, intestinal cells and epidermis) and in many tumors, but not in most somatic tissues. The lack of telomerase in somatic cells leads to progressive telomere shortening with age so that the length of the telomeric tract declines by several kilobases during the course of a normal life span (Allsopp et al., 1992). Interestingly, the rate of telomere shortening differs between cell types (e.g. around 49 bp per population doubling in foreskin fibroblasts compared with 101 bp per population doubling in umbilical vein endothelial cells). This difference may result from variation in the levels of nuclease activity or positioning of the terminal Okazaki fragment.

In germ-line tissues, telomerase levels are sufficient to prevent telomere shortening, so the length of the T_2AG_3 tract is maintained from generation to generation. This maintenance of telomere length is essential because progressive telomere shortening eventually causes cells to stop dividing and become senescent (Bodnar et al., 1998). In regenerative tissue, the levels of expression are lower and/or expression is limited to certain progenitor cells, so telomere shortening does take place. However, the level of telomerase appears to be sufficient to ensure that this telomere shortening does not limit the proliferative potential of the tissue during a normal human life span. One of the hallmarks of human tumor cells is that they are immortal and hence can divide indefinitely. In more than 90% of human tumors, this capacity to divide indefinitely is achieved by turning on telomerase expression (Kim et al., 1994). The presence of telomerase results in the telomeres being maintained and hence stops the progressive telomere shortening that normally triggers replicative senescence.

While most of the T_2AG_3 sequence in the human genome is present at the telomere, a small amount is found elsewhere. Stretches of less than 2 kb are present within the subtelomeric region of many chromosomes (see below) and a tract of about 1 kb is found at chromosome 2q. The telomeric repeats on chromosome 2q are in a head-to-head orientation and probably reflect an ancient telomere–telomere fusion of two ancestral ape chromosomes. The sequence of the repeats present at interstitial sites tends to be quite degenerate (e.g. TGAGGG, TTGGGG, etc.). Interestingly, degenerate repeats are also found within the centromere proximal region of the telomeric tract but not in the distal region. Presumably this is because the distal region of the telomeric tract is maintained accurately by telomerase, while the proximal region is maintained by the cellular replication machinery that allows mutations to accumulate.

Subtelomeric Sequences

Subtelomeric sequences were first isolated by cloning restriction fragments that contained both T_2AG_3 repeats and some adjoining DNA (de Lange et al., 1990). Mapping and sequencing of these telomere-associated sequences revealed a family of low-copy repeat elements of more than 3 kb that were present on multiple chromosomes. Further studies revealed considerable polymorphism in the distribution of these elements between individuals. This polymorphism leads to large differences in the length of the terminal segments of some chromosomes; for example, three alleles of chromosome 16p have been identified where alternative arrangements of the subtelomeric sequences result in the α-globin genes being positioned either 170, 350 or 430 kb from the telomere.

In recent years, the entire telomeric region of a number of chromosomes has been mapped and sequenced (Flint *et al.*, 1997). Detailed comparison of the sequence from the ends of chromosomes 4p, 16p and 22q revealed that the subtelomeric region is divided into two distinct domains (**Figure 2**). The boundary between these domains is marked by a stretch of degenerate T_2AG_3 repeats. The distal domain lies immediately adjacent to the telomeric tract and spans 10–20 kb. It contains numerous short (< 2 kb) stretches of DNA that are found in the subtelomeric region of many other chromosomes. The more proximal subtelomeric domain starts just internal to the degenerate T_2AG_3 sequence and can stretch for several hundred kilobases toward the centromere. Although the sequences within this domain are found on only a few other chromosomes, the regions of

sequence identity are continuous for 10–40 kb. The homology among the subtelomeric regions of different chromosomes is thought to arise from relatively unrestrained recombination and gene conversion. However, the distinctive differences in both pattern and the frequency with which sequences from the proximal and distal subtelomeric domains are shared among chromosomes suggest that the two domains may interact with the equivalent domains of other chromosomes by quite different mechanisms.

One of the challenges in the Human Genome Project has been to obtain chromosome-specific telomeric probes. Because subtelomeric sequences stretch for many kilobases and are shared between chromosomes, these probes are of necessity located tens to hundreds of kilobases from the telomeric tract. Identification of chromosome-specific telomeric probes was essential to complete the map of the human genome. Since then the probes have proved to be medically important as they allow the mapping of small deletions that cannot be detected by standard karyotype analysis (Knight and Flint, 2000). The DNA adjacent to the subtelomeric sequences is more gene rich than other regions of the chromosome and is frequently subject to small deletions. Mapping studies with chromosome-specific telomere probes have demonstrated that these deletions are often correlated with severe mental retardation and hematological defects. It is possible that the high deletion frequency in the gene-rich regions results from their close proximity to the recombination-proficient subtelomeric sequences.

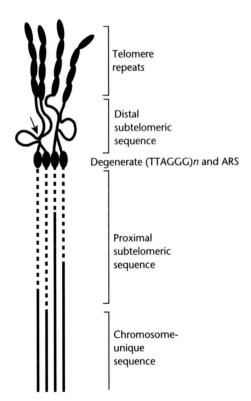

Figure 2 Organization of the subtelomeric domains. Four regions are shown with the telomeric repeats at the top of the figure. The distal subtelomeric sequence contains short sequences common to many chromosomes. The arrow marks a DNA loop from one chromosome interacting with another. The proximal subtelomeric sequence consists of long sequences common to a few chromosomes (marked by dotted lines). The degenerate TTAGGG repeats mark the boundary between the distal and proximal subtelomeric sequence (ARS = autonomous replication sequence). (Reproduced by permission of Oxford University Press from *Human Molecular Genetics* (1997) **6**: 1305–1314.)

Labels in figure:
- Telomere repeats
- Distal subtelomeric sequence
- Degenerate (TTAGGG)*n* and ARS
- Proximal subtelomeric sequence
- Chromosome-unique sequence

See also

Repetitive Elements: Detection
Telomerase: Structure and Function
Telomere
Telomeres and Telomerase in Aging and Cancer

References

Allsopp RC, Vaziri H, Patterson C, *et al.* (1992) Telomere length predicts replicative capacity of human fibroblasts. *Proceedings of the National Academy of Sciences of the United States of America* **89**: 10114–10118.

Bodnar AG, Ouellette M, Frolkis M, *et al.* (1998) Extension of life-span by introduction of telomerase into normal human cells. *Science* **279**: 349–352.

Flint J, Bates GP, Clark K, *et al.* (1997) Sequence comparison of human and yeast telomeres identifies structurally distinct subtelomeric domains. *Human Molecular Genetics* **6**: 1305–1314.

Griffith JD, Comeau L, Rosenfield S, *et al.* (1999) Mammalian telomeres end in a large duplex loop. *Cell* **97**: 503–514.

Kim NW, Piatyszek MA, Prowse KR, *et al.* (1994) Specific association of human telomerase activity with immortal cells and cancer. *Science* **266**: 2011–2015.

Knight SJL and Flint J (2000) Perfect endings: a review of subtelomeric probes and their use in clinical diagnosis. *Journal of Medical Genetics* **37**: 401–409.

de Lange T, Shiue L, Myers RM, *et al.* (1990) Structure and variability of human chromosome ends. *Molecular and Cellular Biology* **10**: 518–527.

Martens UM, Zijlmans JM, Poon SS, *et al.* (1998) Short telomeres on human chromosome 17p. *Nature Genetics* **18**: 76–80.

Sprung CN, Sabatier L and Murnane JP (1999) Telomere dynamics in a human cancer cell line. *Experimental Cell Research* **247**: 29–37.

Wright WE, Tesmer VM, Huffman KE, Levene SD and Shay JW (1997) Normal human chromosomes have long G-rich overhangs at one end. *Genes and Development* **11**: 2801–2809.

Further Reading

Blackburn E and Greider C (eds.) (1995) *Telomeres.* Cold Spring Harbor, NY: Cold Spring Harbor Laboratory Press.

Blackburn EH (2000) The end of the (DNA) line. *Nature Structural Biology* **7**: 847–850.

Evans SK and Lundblad V (2000) Positive and negative regulation of telomerase access to the telomere. *Journal of Cell Science* **113**: 3357–3364.

Greider CW (1996) Telomere length regulation. *Annual Reviews of Biochemistry* **65**: 337–365.

Holt S and Shay J (1999) Role of telomerase in cellular proliferation and cancer. *Journal of Cellular Physiology* **180**: 10–18.

Kippling D (1995) *The Telomere.* Oxford, UK: Oxford University Press.

de Lange T (2000) Cell biology. Telomere capping – one strand fits all. *Science* **292**: 1075–1076.

McEachern MJ, Krauskopf A and Blackburn EH (2000) Telomeres and their control. *Annual Review of Genetics* **34**: 331–358.

Oulton R and Harrington L (2000) Telomeres, telomerase and cancer: life on the edge of genomic stability. *Current Opinion in Oncology* **12**: 74–81.

Price CM (1999) Telomeres and telomerase: broad effects on cell growth. *Current Opinion in Genetics and Development* **9**: 218–224.

Testicular Cancer

RSK Chaganti, *Memorial Sloan Kettering Cancer Center, New York, USA*

Adult testicular germ cell tumors, although rare and curable, are an excellent model system for the study of mechanisms of germ cell transformation, chemotherapy sensitivity and resistance of tumors and regulation of embryonal lineage differentiation.

Intermediate article

Article contents
• Introduction
• Incidence and Clinical Presentation
• Pathobiology
• Predisposing Factors
• Treatment and Response
• Genetic Analysis
• Germ Cell Transformation
• Embryonal-like Differentiation
• Chemotherapy Resistance

Introduction

The testis mainly consists of seminiferous tubules, the site of spermatogenesis. The seminiferous tubules are lined by Sertoli cells and the developing germ cells. The space between the seminiferous tubules (the interstitium) is occupied by Leydig cells, blood vessels and elements of the lymphatic and nervous systems. Sertoli cells regulate many aspects of the developing germ cells, while the Leydig cells are secretory in function producing steroid hormones. The predominant tumor type in the testis is the germ cell tumor (GCT), comprising more than 95% of tumors that affect this organ. Tumors arising in the Sertoli, Leydig and lymphoid cells do occur, but very infrequently. Testicular cancer generally refers to GCT.

Incidence and Clinical Presentation

GCTs comprise only about 2% of all human malignancies; however, they are the most common cancers that affect young males aged 15–35 years. The incidence of GCTs has been on the increase in the Western world over the past several decades. GCTs present predominantly in the testis, with a small minority occurring in the so-called midline, namely the anterior mediastinum, retroperitoneum and the pineal.

Pathobiology

GCTs are divided into two broad groups, seminomas (SEs) and nonseminomas (NSEs). GCTs of all types are frequently associated with carcinoma *in situ* (CIS), which invariably progresses to invasive lesions. The tumor cells in SE GCTs morphologically resemble spermatogonial germ cells. NSE GCTs display differentiation patterns of embryonal and extraembryonal lineages similar to those seen in the developing embryo. These include a primitive zygotic type represented by embryonal carcinoma (EC), embryonal-like somatically differentiated teratoma (TE) and extraembryonally differentiated yolk-sac tumor (YST) and choriocarcinoma (CC). Among the NSE GCTs, mature TEs exhibit the most complete differentiation, presenting such tissue types as cartilage, neural tissue and mucinous and nonmucinous glands. Occasionally, somatically differentiated cells in the TE lesions undergo malignant transformation into lesions that exhibit histologies characteristic of *de novo* tumors in multiple normal somatic tissue types.

Predisposing Factors

Risk factors for the development of GCTs include cryptorchidism, spermatogenic or testicular dysgenesis, Klinefelter syndrome, prior history of a GCT and a positive family history. A 6- to 10-fold increased risk for development of GCT has been estimated for the first-degree relatives of an affected individual. Familial clustering of GCTs is rare, most reported cases comprising two affected individuals in a family, with rare families presenting three or four affected members, a pattern of inheritance that suggested the action of a low-penetrant dominant or recessive mode of inheritance. Linkage analysis of familial cases using polymorphic deoxyribonucleic acid (DNA) markers has identified a possible susceptibility locus at the chromosomal band Xq27.

Treatment and Response

SEs localized to the testis are often successfully treated by surgery alone (orchiectomy). SEs are highly sensitive to radiation, which is the choice of treatment in the presence of metastatic disease. NSEs are highly sensitive to cisplatin-based chemotherapy regimens, cure rates reaching 70–80% of cases with advanced disease. Patients failing chemotherapy treatment are given more intensive treatment regimens or autologous bone marrow transplantation. However, about 5% of tumors that are refractory to chemotherapy lead to patient death. The overall high cure rate and the presence of a small subset of lethally resistant tumors has led to greater attention being paid to morbidity and mortality of treatment and investigation of mechanisms of sensitivity and resistance to cisplatin therapy. During the 1980s, distinguishing between patients who were most likely to be cured (good risk) versus those least likely to be cured (poor risk) was the subject of careful clinical studies. Clinicopathological features that associate with poor risk include mediastinal NSE presentation and hepatic, osseous and/or brain metastases. Patients with high serum levels of lactate dehydrogenase, α-fetoprotein or human chorionic gonadotropin also have a lessened likelihood of complete remission. Another interesting aspect of resistance to cisplatin therapy relates to residual tumor following surgical resection. Frequently, such residual tumors are relatively resistant to cisplatin-based treatment. Finally, malignantly transformed lesions arising in TE respond to treatment appropriate for the transformed histology, rather than to cisplatin-based treatment.

Genetic Analysis

In common with most solid tumors, GCTs present complex karyotypes. Near-triploidy to near-tetraploidy is the rule. A unique cytogenetic feature of these tumors is the presence, in virtually all tumors, of increased copy number of the chromosomal arm 12p. This may be in the form of one or more copies of an isochromosome [i(12p)], tandem duplication of 12p into many copies, or transposition of 12p elsewhere in the genome followed by duplication or multiplication *in situ*. In addition to occurrence of multiple copies of 12p, the 12p11–12 subregion has been noted to be further amplified in a small proportion of tumors, most frequently SE. This subregional amplification has also been noted in many other solid tumor types and has been suggested to be associated with tumor progression. The constant 12p copy number increase in GCTs has been suggested to imply a critical role for this aberration in the origin of these tumors. Although no gene(s) have been conclusively identified to be the target of this genomic amplification, strong circumstantial evidence indicates cyclin D2 (*CCND2*), a gene mapped to 12p13, to be a potential deregulated gene. *CCND2* encodes cyclin D2, a member of the D-type family of cyclins that are involved in the regulation of cell cycle progression through the G_1 phase of the mitotic cycle. While cyclin D2 is not expressed in normal germ cells, CIS as well as GCTs of all histologies show varying levels of cyclin D2 expression. Thus, while CIS and seminomas express abundant cyclin D2, differentiated elements in EC and somatically differentiated lineages in TE appear to downregulate cyclin D2 expression. With regard to the 12p11–12 subregional amplification, recent studies using complementary DNA (cDNA) array technology have identified two novel genes of unknown function, *GCT1* and *GCT2*, to be amplified and overexpressed, suggesting them to be candidate targets for amplification. In addition to these aberrations affecting the 12p chromosome arm, multiple deletions have been described in the genomes of GCTs using the loss of heterozygosity (LOH) assay. These studies identified loss of known tumor suppressor genes (TSGs) such as deleted in colorectal carcinoma (*DCC*) and tumor protein p53 (Li–Fraumeni syndrome) (*TP53*), as well as novel sites unique to this tumor type. Notable among the latter is 12q22, which has been shown to present deletions in all histological subtypes as well as in CIS, suggesting an important role for one or more genes located in this chromosomal region in GCT development. Extensive finemapping studies narrowed the deletion to a region of less than 1 megabase (Mb) of DNA. Careful mutation analysis of all presumptive candidate genes in this region failed to identify a deleted gene with potential TSG function, suggesting that some genes in this region may be inactivated by nonmutational mechanisms of gene silencing such as methylation. Thus, chromosome 12 overall plays an important role in the genesis of GCTs and this role still needs to be fully understood.

Germ Cell Transformation

The pluripotentiality for differentiation into embryonal and extraembryonal lineages displayed by GCTs clearly indicates their germ cell origin. In normal development, the germ cell lineage is governed by a highly complex program of proliferation and differentiation, beginning from the first recognition of germ cells in the epiblasts to the fulfillment of their ultimate fate, namely the development of gametes (sperm). The point in development at which germ cells transform leading to development of a GCT is not clearly understood. Two possible mechanisms of origin of GCTs have been proposed, although neither of them has been experimentally verified. One suggests that fetal germ cells (gonocytes) that have escaped normal development into spermatogonia may undergo abnormal cell division mediated by a kit receptor/stem cell factor (SCF) paracrine loop, leading to the origin of CIS cells. Gonocytes so derailed in their normal development may be susceptible to subsequent invasive growth through mediation of postnatal and pubertal gonadotropin stimulation. The other hypothesis suggests that the target cell for transformation may be the spermatocyte at the zygotene–pachytene stage of meiosis. At this stage, wild-type p53 is expressed and DNA breaks associated with homologous chromosome recombination (crossing over) are generated. A recombination checkpoint operates that can provide an apoptotic trigger in the presence of unresolved DNA doublestrand breaks. According to this model, aberrant chromatid exchange events associated with chromosome recombination may lead to increased copy number of 12p and deregulated expression of *CCND2*, leading to an aberrant reinitiation of the cell cycle and genomic instability.

Embryonal-like Differentiation

As true experiments of nature, GCTs display, albeit in a spatially and temporally abnormal manner, differentiation patterns that mimic events in the developing zygote. They provide a unique opportunity to study embryonal (represented by TE) versus extraembryonal (represented by YST and CC) pathways of differentiation, as well as development of somatic lineages (in TE). In a recent study, expression of the *ID* family of development regulator genes by RNA *in situ* hybridization of tumor tissue sections was shown to parallel that in normal murine embryogenesis, thus validating the usefulness of these tumors as models for the study of normal developmental processes. Tumorderived pluripotent EC cells have been isolated and maintained as cell lines, which make powerful cellular tools, analogous to embryonal stem cells, for studying the coordination of molecular events associated with lineage development. One such cell line is NT2/D1, previous studies of which have established that its *de facto* as well as all-*trans* retinoic acid (ATRA)-induced developmental fate is the neuronal lineage. Recent attention has focused on the analysis of neuronal and other lineage differentiation pathways in this cell line using the powerful technology of expression profiling with oligonucleotide or cDNA arrays. In one study, analysis of time response to ATRA via clustering of more than 12 000 human transcripts revealed distinct stages in the transition from an EC cell to neuronal progenitor cells expressing patterning markers compatible with posterior hindbrain fates followed by the appearance of immature postmitotic neurons with an evolving synaptic apparatus. The analysis of neuronal lineage induction in NT2/D1 cells serves as a paradigm for the analysis of transcriptional regulation of other lineages in NT2/D1 as well as in other EC cells using multiple morphogens. The relevance of such studies to the understanding of human embryonal lineage development is obvious.

Chemotherapy Resistance

The small proportion of GCTs that display resistance to cisplatin-based treatment have served as good models for analyzing pathways involved in their sensitivity and resistance phenotypes. GCTs overall express abundant wild-type p53 and *TP53* gene mutations are rare. Predictably, the exquisite sensitivity of these tumors to DNA-damaging agents such as cisplatin has been shown to be rooted in a p53-mediated apoptotic response. Recent *in vitro* studies of GCT cell lines and tumors have identified several pathways to resistance that include functional inactivation of the *TP53* gene by mutation, gene amplification and possible perturbation in the DNA repair pathways, as well as involvement of apoptosis antagonists such as Bcl-X_L. The multiplicity of these pathways offers unusual opportunities to study the resistance phenomenon, of relevance not only to GCTs but also to many other tumor types.

The several unusual features that characterize GCTs, such as origin in totipotential germ cells, pluripotentiality of the tumors and defined pathways to chemotherapy sensitivity and resistance make them excellent model systems for studying germ cell biology, tumorigenesis, therapy resistance and embryonal lineage differentiation.

Further Reading

Bosl GJ and Motzer RJ (1997) Testicular germ-cell cancer. *New England Journal of Medicine* **337**: 242–253.

Bourdon V, Naef F, Rao PH, *et al.* (2002) Genomic and expression analysis of the 12p11–12p12 amplicon in germ cell tumors using EST arrays identifies two novel amplified and over-expressed genes. *Cancer Research* **62**: 6281–6233.

Chaganti RSK and Houldsworth J (2000) Genetics and biology of human male germ cell tumors. *Cancer Research* **60**: 1475–1482.

Heidenreich A, Srivastava S, Moul JW and Hofman R (2000) Molecular genetic parameters in pathogenesis and prognosis of testicular germ cell tumors. *European Urology* **37**: 121–135.

Horwich A (1996) *Testicular Cancer – Investigation and Management*. London, UK: Chapman & Hall.

Houldsworth J, Heath SC, Bosl GJ, Studer L and Chaganti RSK (2002) Expression profiling of lineage differentiation in pluripotential human embryonal carcinoma cells. *Cell Growth and Differentiation* **13**: 257–264.

Rajpert-DeMeyts E, Grigor KM and Skakkebaek NE (1998) *Neoplastic Transformation of Testicular Germ Cells*. Copenhagen: Munksgaard.

Ulbright TM (1993) Germ cell neoplasms of the testis. *American Journal of Surgical Pathology* **17**: 1075–1091.

Web Links

Cyclin D2 (*CCND2*); Locus ID: 894. LocusLink:
http://www.ncbi.nlm.nih.gov/LocusLink/LocRpt.cgi?l = 894

Cyclin D2 (*CCND2*); MIM number: 123833. OMIM:
http://www.ncbi.nlm.nih.gov/htbin-post/Omim/dispmim?123833

Thalassemias

David Weatherall, *Weatherall Institute of Molecular Medicine, Oxford, UK*

The thalassemias are the commonest genetic disorders in humans and present an increasing public health problem in the tropical countries in which they occur at a high frequency.

The thalassemias are a heterogeneous group of inherited disorders of hemoglobin. The word 'thalassemia' is derived from Greek roots for 'sea' and 'blood', reflecting the mistaken belief on the part of those who first described these diseases that they were confined to the Mediterranean region. It is now clear that they are the commonest human single-gene disorders and occur widely throughout many tropical countries of the world. Furthermore, because of large population movements, these conditions are now seen in almost every country.

Different Forms of Thalassemia

All the different forms of thalassemia result from inherited defects in the production of the globin chains of hemoglobin. Different hemoglobins are produced at various stages of human development. In adult life, there is a major hemoglobin (Hb) called Hb A, comprising about 90% of the total, and a minor hemoglobin, Hb A_2, which normally accounts for 2–3%. In fetal life, the main hemoglobin is Hb F, only traces of which are found in normal adults. There are three embryonic hemoglobins. All these hemoglobins are tetramers of two pairs of unlike globin chains; adult and fetal hemoglobins have α chains associated with β (Hb A, $\alpha_2\beta_2$), δ (Hb A_2, $\alpha_2\delta_2$) or γ chains (Hb F, $\alpha_2\gamma_2$). In the embryo, there are different α-like and β-like chains called ζ and ε chains respectively. Each globin chain has a heme molecule attached to it, to which oxygen is bound. (*See* Carrier Screening for Inherited Hemoglobin Disorders in Cyprus and the United Kingdom; Paroxysmal Nocturnal Hemoglobinuria.)

The structure and synthesis of the different globin chains is directed by two gene families: the α-like globin genes on chromosome 16 and the β-like genes on chromosome 11. These genes appear in their respective gene clusters in the order in which they are activated during development (**Figure 1**). The major

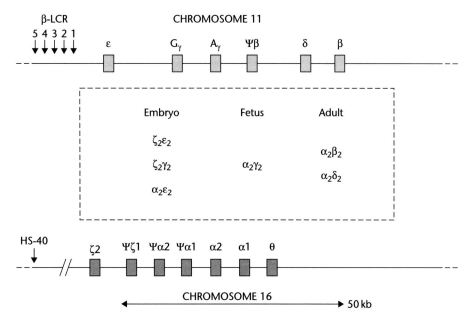

Figure 1 Human globin gene clusters on chromosomes 16 and 11. The different genes that are activated at various stages of development, together with the hemoglobins that are produced, are shown. The locus control region (LCR) and HS-40 are the major 'master' regulatory regions for the two clusters.

regulatory sequences that control the production of each of the individual globin chains, and 'master' sequences that control each of the clusters, have been defined. But although a great deal is known about the regulation of the globin genes at each stage of human development, the way in which they switch from embryonic to fetal to adult hemoglobin production is not yet understood.

There are two common forms of thalassemia, α and β thalassemia, which result from defective synthesis of either α or β chains. There are also rarer forms in which both δ- and β-chain synthesis is defective, the δβ thalassemias, or in which the ε-, γ-, δ- and β-chain genes are all lost by deletion.

The majority of the thalassemias are inherited in a Mendelian recessive fashion. In addition to their genetic classification into α and β thalassemias, they are also divided on clinical grounds into the severe, homozygous form of the disease, which is called thalassemia major, and the symptomless carrier state, in which only one defective gene is inherited, called thalassemia trait or thalassemia minor. However, because the disease has a very variable phenotype, many patients fall between these extremes; these conditions are called thalassemia intermedia.

Molecular Pathology

The thalassemias result from many different mutations that involve either the α- or β-globin genes.

Because normal persons receive two α genes from each parent, their genotype is written $\alpha\alpha/\alpha\alpha$. There are two main types of α thalassemia. First, there are the α^0 thalassemias, in which both α genes are deleted; the homozygous state is written $-/-$ and the heterozygous state $-/\alpha\alpha$. In the α^+ thalassemias, only one of the α genes is lost; the homozygous and heterozygous states are therefore designated $-\alpha/-\alpha$ and $-\alpha/\alpha\alpha$ respectively. Both α^0 and α^+ thalassemia are extremely heterogeneous at the molecular level, reflecting the fact that there are many differently sized deletions involved. In some cases, the α genes are intact and the disease results from a deletion of a major regulatory region upstream from the α-globin gene cluster, which completely inactivates all the genes in this region. Sometimes, α thalassemia results from a mutation that inactivates an individual α gene rather than deleting it. In this case, the heterozygous and homozygous states are represented as $\alpha^T\alpha/\alpha\alpha$ and $\alpha^T\alpha/\alpha^T\alpha$ respectively.

Over 200 different mutations of the β-globin genes have been found in patients with β thalassemia. They affect gene function at every level: transcription, processing of the primary messenger ribonucleic acid transcript, translation and posttranslational stability of the β-globin product. Less commonly, β thalassemia, like α thalassemia, may result from a partial deletion of the β-globin gene. Most of the δβ thalassemias result from deletions involving the β-globin gene cluster. Some of the mutations that cause β thalassemia cause a complete absence of β-globin gene production; the resulting condition is called β^0

thalassemia. Others cause a reduced output of β chains: β$^+$ thalassemia.

Cellular Pathology

As the result of these mutations, there is imbalanced globin chain production in all the thalassemias. In the β thalassemias, this leads to an excess of α chains, which precipitate in the red cell precursors, causing their damage in the bone marrow and shortening the survival of the red cells that reach the peripheral blood. Thus, the profound anemia of this condition is the result of both ineffective red cell production and a shortened survival of those that are produced. In α thalassemia, the excess γ chains that are produced in fetal life form γ$_4$ molecules called Hb Bart's, while in adults excess β chains form β$_4$ molecules called Hb H. These abnormal homotetramers do not release oxygen at normal physiologic tensions and, particularly in the case of Hb H, are unstable. This leads to a shortened red cell survival and anemia, these patients being further disadvantaged because of defective delivery of oxygen to the tissues.

Clinical Manifestations

The homozygous state for α^0 thalassemia, that is, the loss of all four α-globin genes, results in stillbirth late in pregnancy. These infants are anemic and edematous, and show all the features of severe intrauterine hypoxia. This reflects both anemia and a very high level of Hb Bart's, which further compromises the oxygenation of their tissues. Persons who have lost three out of their four α genes $(-\alpha/--)$ have a condition called Hb H, disease which is associated with moderate anemia. Those who have lost one or two of their α-globin genes are asymptomatic but may pass on the defective chromosomes to their children. In short, the clinical features of the α thalassemias reflect a gene dosage effect, with increasing severity mirroring the number of α genes that have been lost by deletion or point mutation.

Because there are so many different β thalassemia mutations, many affected children are compound heterozygotes for two different alleles. The homozygous or compound heterozygous states for severe forms of β thalassemia are characterized by severe anemia from the first year of life after the switch from γ- to β-globin production has occurred. If these children are not given regular blood transfusions, they usually die within a few months. If they are adequately transfused, they grow and develop normally but succumb to the effects of iron overload due to the iron received from transfusions. If they are inadequately transfused, they develop skeletal deformities, growth retardation and a variety of other complications consequent on over-expansion of the bone marrow.

The clinical picture of a child with inadequately treated β thalassemia is very characteristic. Because of chronic anemia and the accumulation of iron in the body from blood transfusion or increased absorption, these children do not grow and develop at a normal rate. They respond to severe anemia by producing the hormone erythropoietin, which causes massive proliferation of the bone marrow and hence deformities of the skull and other bones. The skull changes are characterized by frontal bossing and overgrowth of the upper jaws. These changes give these children a very characteristic facial appearance. Because of the profound degree of anemia, there is a reversion to fetal sites of erythropoiesis, particularly the liver and spleen. This leads to progressive and often extensive enlargement of these organs. Unless the iron that accumulates from transfusion and increased absorption from the bowel in response to anemia is removed, these children often die in their early 20s from the effects of iron loading. Iron has a particular predilection for the endocrine glands and pancreas but also for the myocardium. Indeed, the commonest form of death in this disease is cardiac damage, leading to either acute or chronic heart failure.

However, if children with β thalassemia are adequately transfused, and are able to tolerate nightly infusions of a drug to prevent excessive iron accumulation, they grow and develop normally and are able to live to adult life in good health.

Phenotypic Variability

There is remarkable variation in the clinical phenotype of β thalassemia. Part of this reflects the variable severity of the different β-thalassemia alleles; in some cases, they result in a very mild reduction in β-globin chain output, while in others there is a complete absence of β-globin chain production. Often, children are compound heterozygotes for mild and severe alleles, and hence these interactions cause a broad spectrum of clinical severity. However, children who inherit the same thalassemia alleles may also show wide differences in the severity of the disease. At least two genetic modifiers have been found to account for this observation. Firstly, if children with β thalassemia also inherit α thalassemia, there is less globin chain imbalance, and hence the disease is milder. Secondly, it appears that there are a number of genetic polymorphisms that may result in a greater ability to produce Hb F in adult life; if these are inherited together with β thalassemia, the increased propensity for γ-chain production also results in less globin chain imbalance and hence milder disease.

The phenotypic complexity of the β thalassemias also results from polymorphisms of genes that may be involved in modifying the complications of the disease, iron accumulation, bone deformity, overproduction of bilirubin in response to red cell breakdown and propensity to infection. The environment also plays a role in modifying the phenotype. In short, the clinical picture of β thalassemia reflects the action of different β-thalassemia alleles, modification by the rates of α- or γ-chain production, and the consequences of polymorphisms at a number of other loci unrelated to the globin genes.

Coinheritance of Thalassemia with Hemoglobin Variants

Although there are many structural hemoglobin variants, most of them are rare and only three, hemoglobins S, C and E, reach high frequencies. It is not uncommon therefore for a person with β thalassemia to coinherit a gene for one of these abnormal hemoglobins. (*See* Sickle Cell Disease as a Multifactorial Condition.).

The compound heterozygous state for β thalassemia and the sickle cell gene, sickle cell β thalassemia, results in a clinical picture that resembles sickle cell anemia. The inheritance of β thalassemia together with Hb E, the commonest structural variant in the world population, produces a relatively severe form of thalassemia, Hb E β thalassemia, which is usually, but not always, transfusion dependent. This reflects the fact that Hb E is synthesized at a reduced rate and therefore acts as a mild β^+-thalassemia allele.

World Distribution and Population Genetics

The thalassemias occur at a high frequency in the Mediterranean region, throughout the Middle East and the Indian subcontinent, and in Southeast Asia, where they are distributed in a line stretching from Southern China through the Malaysian peninsula into the island populations of Melanesia and Indonesia (**Figure 2**).

Remarkably, each population has its own particular varieties of α or β thalassemia. The gene frequencies for β thalassemia in these high-frequency regions range from 2% to 20%, while those for α^+ thalassemia range from 2% up to 70% in some parts of the world. The α^0 thalassemias are localized to parts of Southeast Asia and the Mediterranean islands, and hence it is only in these regions that stillborn babies homozygous for this condition are encountered.

It was originally suggested by J. B. S. Haldane that the thalassemias have reached their high frequencies

☐ α and β thalassemia

Figure 2 World distribution of the thalassemias.

owing to heterozygote advantage against severe malaria. In other words, they reflect a balanced polymorphism, that is a balance established between the relative protection of heterozygotes and the early death of severely affected homozygotes. While there is evidence suggest that this is the case for the β thalassemias, it has not yet been formally established. However, in the case of α^+ thalassemia, there is clear evidence that both heterozygotes ($-\alpha/\alpha\alpha$) and homozygotes ($-\alpha/-\alpha$) are protected against malaria and possibly against other childhood infections.

Overall, there is a fairly good correlation between the global distribution of the thalassemias and present or past malaria. There are a few exceptions, however. For example, α thalassemia is found quite commonly in some of the Pacific Island populations in which malaria has never occurred. Recent studies suggest that these islands were populated by settlers who took thalassemia with them and therefore acted as founders for particular island populations. In populations of middle and south America, where malaria has been quite common, thalassemia is not observed. It is currently believed that this is because malaria was only transported to these countries by invasions from Europe and therefore there has not been time for selection to increase the frequency of thalassemia mutations in this region. Similarly, it is presumed that thalassemia was not yet established on the Asian mainland when the first settlers of the Americas crossed the Bering Straits. (*See* Drug Metabolic Enzymes: Genetic Polymorphisms.)

Control and Management

All the severe forms of thalassemia can be easily identified in the carrier state and can also be diagnosed in fetal life; hence it is possible to offer counseling and prenatal diagnosis for parents to wish to terminate pregnancies carrying babies with these conditions. This approach has caused a marked reduction in the frequency of the birth of babies with the β thalassemias in some countries, notably Cyprus, Sardinia and Greece.

The only definitive form of treatment for thalassemia is bone marrow transplantation, which is possible only when there is a matching donor relative. Symptomatic treatment involves regular blood transfusion and the use of drugs to remove the excess iron that accumulates from transfused blood. If children receive regular transfusion and are able to comply with treatment with the drug desferrioxamine, which has to be given by intravenous infusion every night, they can grow and develop normally and reach adult life in good health. While many children who have been treated in this way are surviving to grow and have their own families in the developed world, the position is much less satisfactory in the developing countries, where the cost of regular transfusion and drugs to remove iron is such that many patients receive either no treatment or insufficient treatment to allow them to grow and develop normally.

Further research efforts are being directed at trying to stimulate the production of fetal hemoglobin, or at somatic gene therapy directed at replacing defective α- or β-globin genes.

See also

Carrier Screening for Inherited Hemoglobin Disorders in Cyprus and the United Kingdom
Globin Genes: Evolution
Globin Genes: Polymorphic Variants and Mutations
Hemoglobin Disorders: Gene Therapy
Sickle Cell Disease as a Multifactorial Condition

Further Reading

Cao A and Rosatelli MC (1993) Screening and prenatal diagnosis of the haemoglobinopathies. *Clinical Haematology* **6**: 263–286.

Olivieri NF, Nathan DG, MacMillan JH, *et al.* (1994) Survival of medically treated patients with homozygous β thalassemia. *New England Journal of Medicine* **331**: 574–578.

Weatherall DJ (2001) The thalassemias. In: Stamatoyannopoulos G, Perlmutter RM, Majerus PW and Varmus H (eds.) *Molecular Basis of Blood Diseases*, 3rd edn, pp. 183–227. Philadelphia, PA: WB Saunders.

Weatherall DJ (2001) Phenotype–genotype relationships in monogenic disease: lessons from the thalassaemias. *Nature Reviews* **2**: 245–255.

Weatherall DJ and Clegg JB (1996) Thalassaemia: a global public health problem. *Nature Medicine* **2**: 847–849.

Weatherall DJ and Clegg JB (2001) *The Thalassaemia Syndromes*, 4th edn. Oxford, UK: Blackwell Scientific Publications.

Weatherall DJ and Clegg JB (2001) Inherited hemoglobin disorders: an increasing global health problem. *Bulletin of the World Health Organization* **79**: 704–711.

Weatherall DJ, Clegg JB, Higgs DR and Wood WG (2001) The hemoglobinopathies. In: Scriver CR, Beaudet AL, Sly WS, *et al.* (eds.) *The Metabolic and Molecular Bases of Inherited Disease*, 8th edn, pp. 4571–4632. New York, NY: McGraw-Hill.

Thrifty Gene Hypothesis: Challenges

Robyn McDermott, *Queensland Health, Cairns, Queensland, Australia*

Type 2 diabetes has in the past been claimed to be genetically determined. However, nongenetic metabolic effects including complex social determinants better explain the current epidemic of obesity and diabetes worldwide.

Introductory article

Article contents

- 'Thrifty Gene' Hypothesis
- Barker Hypothesis: 'Thrifty Phenotype'
- Fetal Hyperinsulinemia and Intergenerational Risk (Freinkel's Hypothesis of Fuel-mediated Teratogenesis)
- Causes of Cases versus Causes of Incidence: Role of Energy Imbalance at a Population Level
- Nutrition, Risk Behaviors, Intergenerational Effects and the Social Gradient
- A 'Life-course' Approach to Type 2 Diabetes Mellitus and Other Chronic Diseases
- 'Genohype' and the Promise of Individualized Therapies

'Thrifty Gene' Hypothesis

In 1962, the population geneticist James Neel proposed the existence of a 'thrifty gene' to explain the apparent paradox of the emergence of diabetes (a disease with a 'well-defined genetic basis') in some populations despite its manifestly adverse effects on reproduction. This thrifty gene made its owners exceptionally efficient in the utilization of food, conferring a survival advantage during times of famine. As hunter–gatherer societies moved from subsistence through agriculture to urbanized Western lifestyles, those carrying the thrifty gene became more susceptible to obesity and diabetes (Neel, 1962).

Since then, numerous studies have tended to label affected populations as 'genetically susceptible', simply on the basis of higher prevalence compared with other ethnic or geographically separate groups. However, ecologic and migrant studies indicate that nongenetic factors acting at various stages in the life course account for a large part of the variation in chronic disease outcomes, including type 2 diabetes mellitus (T2DM), between genetically similar populations in different geographic areas (Cruickshank *et al.*, 2001).

Even in the face of the rapid appearance of T2DM in most populations in the space of one or two generations, including the doubling of diabetes prevalence in the USA and Australia in the last decade, some reviews and textbooks on the subject still claim that T2DM is strongly genetically influenced. What is the evidence that T2DM is a genetic condition?

In the absence of the culprit gene/s for T2DM, most evidence supporting the genetic theory comes from observed familial susceptibility, documented differences between certain ethnic groups, a high concordance among monozygotic twins and the discovery of rare maturity-onset diabetes of the young (MODY) genes.

Leaving aside the MODY pedigree (which is clearly genetic, accounts for fewer than 1% of diabetics, and is an entity distinct from T2DM), the high prevalence in certain societies, strong family history and observations in twins can be explained equally plausibly by other, nongenetic mechanisms. This is not to argue that genetic differences in susceptibility do not exist, but that the rapid evolution of the epidemic means that contributions from genotypes are probably insignificant compared with changes in phenotype.

Barker Hypothesis: 'Thrifty Phenotype'

In the 1980s the dominant model for the etiology of chronic conditions and T2DM concentrated on identifying adult risk factors. However, these explained only a fraction of observed disease. Barker and colleagues published a landmark series of large cohort studies in the early 1990s in the UK which demonstrated a clear link between low birth weight and poor nutrition in childhood and a range of adult chronic diseases, including abnormal glucose tolerance and T2DM. This effect was amplified among those who were born small and who subsequently gained excess weight as adults (Barker, 1998). These findings have since been confirmed in numerous cohort studies in different populations. Barker proposed a physiological mechanism for this phenomenon, the thrifty phenotype, where poor nutrition in fetal life and early infancy affects the development and function of pancreatic β cells which in turn predisposes the individual to the development

of the metabolic syndrome and T2DM (intrauterine programing).

Recent challenges to simple interpretations of Barker's hypothesis come from twins which, despite being lighter at birth than singletons, do not as a group have higher blood pressure or other manifestations of the metabolic syndrome. More recently, Hattersley has proposed the 'fetal insulin hypothesis' to explain the association of low birth weight with later diabetes and vascular disease. In this model, low birth weight and subsequent insulin resistance are merely phenotypes of the same insulin-resistant genotype which impairs insulin-mediated growth in the fetus. Thus, 'the predisposition to NIDDM (noninsulin-dependent diabetes mellitus) and vascular disease is likely to be the result of both genetic and fetal environmental factors' (Hattersley and Tooke, 1999).

Fetal Hyperinsulinemia and Intergenerational Risk (Freinkel's Hypothesis of Fuel-mediated Teratogenesis)

The observation that T2DM runs in families is used to support genetic inheritance theories. However, in 1980 Norbert Freinkel hypothesized that a mechanism for the heritability of diabetes might rest in a form of 'fuel-mediated teratogenesis' during pregnancy, where the fetus exposed to hyperglycemia *in utero* has long-term anthropometric and metabolic effects, including increased susceptibility to obesity and T2DM (Freinkel, 1980). Supporting this were data from the large Pima Indian cohort studies which showed that the offspring of women with T2DM were more obese and had greater prevalence of T2DM than offspring of nondiabetic women or of women who developed diabetes after the pregnancy. These effects persisted after controlling for age, fathers' diabetic status and the age of onset of the mothers' diabetes.

These studies also suggested that the very early age of onset of T2DM in the female offspring of these diabetic mothers set up a vicious circle where this generation was exposing the next to a diabetogenic intrauterine environment, thus amplifying the risk through the generations (Pettit *et al.*, 1993). Subsequent cohort studies in (mainly Caucasian) nonindigenous women and their offspring in Chicago showed a greatly increased risk of glucose intolerance in the offspring of diabetic mothers by 10 years of age (Silverman *et al.*, 1995). The risk of impaired glucose tolerance (IGT) in this cohort was linked to fetal hyperinsulinemia (correlated with maternal hyperglycemia) rather than to the mothers' type of diabetes, since most of these mothers had type 1 diabetes while their offspring were showing a T2DM syndrome (obesity and IGT).

Animal studies looking at artificially induced fetal hyperinsulinemia showed a similar phenomenon, where offspring developed abnormal glucose tolerance during subsequent pregnancies, and their offspring in turn were macrosomic (Susa *et al.*, 1993). This suggests a teratogenic effect which can persist into the third generation, but which is metabolic, not genetic, in origin.

Causes of Cases versus Causes of Incidence: Role of Energy Imbalance at a Population Level

Clinicians search for the cause of disease in an individual. Epidemiologists search for the causes of incidence in a population. In 1993, Rose proposed that populations with complex patterns of social history, current exposures and risk behaviors gave rise to 'sick' individuals (Rose, 1993). Complex determinants operating at the population level can shift the distribution of risk in a population, so that 'caseness' (the number of people reaching a diagnostic threshold, for example, for T2DM or hypertension, which is in fact based on probability of risk for certain outcomes, rather than strict 'objective' and immutable diagnostic criteria) can increase suddenly. This is largely what we are observing with the T2DM 'epidemic'.

Populations as a whole are getting fatter, by small increments per year, owing to changes in lifestyle affecting energy balance. Large cohort studies show that body mass index is the single most important predictor of incidence of T2DM, even in a 'low-risk' population of mostly Caucasian middle-class nurses in the USA (Colditz *et al.*, 1995). The epidemic of obesity and T2DM is appearing in populations (adults and now children) of every genetic type exposed to these social conditions (Visscher and Seidel, 2001). This is challenging for the genetic hypothesis: genotypes cannot change in two generations, so does that mean all populations carry the thrifty gene?

Nutrition, Risk Behaviors, Intergenerational Effects and the Social Gradient

Recent studies of the social epidemiology of chronic diseases in Western societies, including T2DM, demonstrate a clear social gradient in the distribution of risk, particularly obesity, poor micronutrient intake, smoking, low birth weight and poor outcomes.

'Systemic stress', neuroendocrine pathways and hyperglycemia

Another line of argument for the genetic hypothesis is the very high prevalence of T2DM among indigenous people, particularly those affected by 'coca-colonization' in the Pacific and North America. However, new lines of research have begun to address if and how social conditions can affect health outcomes by way of responses mediated by the central nervous system to the social environment. There is evidence that indigenous societies (and some other racial minorities) are subject to ongoing stresses ('systemic stress') in everyday social experience which are manifest in chronically elevated levels of catecholamines which, added to risks accumulated through the life course (low birth weight, adult obesity, poor nutrition) eventually lead to the metabolic syndrome and T2DM, as demonstrated by higher population-level distributions of glycated hemoglobin (Kelly *et al.*, 1997). The subjective experience of racism in daily living has been linked, among African-American men, to higher rates of raised blood pressure, when controlling for other known risk factors (Krieger, 1990).

A 'Life-course' Approach to Type 2 Diabetes Mellitus and Other Chronic Diseases

A review of the epidemiological, biological and sociological evidence from different settings over the last 100 years has led to a 'composite' model for explaining the incidence of chronic diseases in different populations, where risk accumulates over the life course: poor maternal nutrition and fetal development, poor childhood nutrition, obesity and inactivity in adulthood, smoking and other risk behaviors, adverse physical and social environments and low socioeconomic status all act to increase the likelihood of T2DM and other chronic diseases in adults (Kuh and Ben-Schlomo, 1997).

'Genohype' and the Promise of Individualized Therapies

The announcement of the deciphering of the human genome in February 2001 was accompanied by much fanfare and exaggerated claims that the genome will lead to the unraveling, not just of single-gene disorders, but also of the polygenic forms of common diseases like T2DM, and even 'will eventually tell us...what we are'. The Human Genome Project has become 'the new power base for big business and big money, as well as the platform for the launch of inflated hopes and monstrous hyperbole' (Radford, 2001). The truth will probably be much less exciting, and the spectacle so far resembles stage one of the 'hype cycle', a term coined by the information technology consulting firm Gartner following the 'dot.com' collapses.

The early phase of the cycle is characterized by a technological breakthrough that excites much media interest. This is followed by a marketing effort that inflates expectations with unrealistic forecasts, and the exaggerated publicizing of some successes, but then more shortfalls as the technology fails to deliver on promises. Much of the money made in this first stage is made by conference organizers and magazine publishers. This phase is followed by disillusion, and finally by a 'plateau of productivity' where the real, much more modest, benefits of the technology are demonstrated. The Human Genome Project has so far discovered over 1000 single-gene disorders that affect less than 4% of the world population. These discoveries may lead to better predictive tests and treatments for some of those individuals.

However, incidence and clinical outcomes in the metabolic syndrome and T2DM are determined by complex interactions throughout the life course. The simplistic reductionist paradigm of the thrifty gene hypothesis is wholly inadequate to describe the complexity of T2DM. The quest for the T2DM genotype has been compared to searching for the susceptibility genes for cholera in the middle of an outbreak.

The other promise, that the Human Genome Project will eventually benefit almost everyone in the world, is also improbable, as the Project proposes to deliver, at best, tailored drug treatments and gene therapy for affected individuals. However increasingly, sufferers of T2DM are poor people, and the roots of their health problems lie in the same poverty which will exclude them from individualized treatments. The logical imperative now is to address urgently the social and environmental determinants of the epidemic of T2DM at a population level, rather than being distracted by the chimera of genetic causes and costly 'cures' for individuals.

See also

Diabetes: Genetics
Obesity: Genetics
Thrifty Genotype Hypothesis and Complex Genetic Disease

References

Barker DJP (1998) *Mothers, Babies and Health in Later Life*, 2nd edn. Edinburgh, UK: Churchill Livingstone.

Colditz GA, Willett WC, Rotnitzky A and Manson JE (1995) Weight gain as a risk factor for clinical diabetes mellitus. *Annals of Internal Medicine* **122**: 481–486.

Cruickshank JK, Mbanya JC, Wilks R, *et al.* (2001) Sick genes, sick individuals or sick populations with chronic disease? The emergence of diabetes and high blood pressure in African-origin populations. *International Journal of Epidemiology* **30**: 111–117.

Freinkel N (1980) Of pregnancy and progeny. *Diabetes* **29**: 1023–1035.

Hattersley AT and Tooke JE (1999) The fetal insulin hypothesis: an alternative explanation of the association of low birthweight with diabetes and vascular disease. *Lancet* **353**: 1789–1792.

Kelly S, Hertzman C and Daniel M (1997) Searching for the biological pathways between stress and health. *Annual Reviews of Public Health* **18**: 437–462.

Kreiger N (1990) Racial and gender discrimination: risk factors for high blood pressure? *Social Science and Medicine* **30**: 1273–1278.

Kuh D and Ben-Schlomo B (eds.) (1997) *A Life Course Approach to Chronic Disease Epidemiology*. Oxford, UK: Oxford University Press.

Neel JV (1962) Diabetes mellitus: a 'thrifty' genotype rendered detrimental by 'progress'? *American Journal of Human Genetics* **14**: 353–362.

Pettit DJ, Nelson RG, Saad MF, *et al.* (1993) Diabetes and obesity in the offspring of Pima Indian women with diabetes during pregnancy. *Diabetes Care* **16**: 310–314.

Radford T (2001) Cracking the genome (Book review). *Lancet* **357**: 1537.

Rose G (1993) *Strategy of Preventive Medicine*. Oxford, UK: Oxford University Press.

Silverman BL, Metzger BE, Cho NH, *et al.* (1995) Impaired glucose tolerance in adolescent offspring of diabetic mothers. *Diabetes Care* **18**: 611–617.

Susa JB, Sehgal P and Schwartz R (1993) Rhesus monkeys made exogenously hyperinsulinemic *in utero* as fetuses display abnormal glucose homeostasis as pregnant adults and have macrosomic fetuses. *Diabetes* **42**(supplement): 86A.

Visscher TLS and Seidel JC (2001) The public health impact of obesity. *Annual Reviews of Public Health* **22**: 355–375.

Further Reading

Drever F and Whitehead M (1997) *Health Inequalities*. London, UK: HMSO.

Kuh D and Ben-Schlomo Y (1997) *A Life Course Approach to Chronic Disease Epidemiology*. Oxford, UK: Oxford University Press.

Marmot M and Wilkinson R (2000) *Social Determinants of Health*. Oxford, UK: Oxford University Press.

Pickup JC and Williams G (eds.) (1997) *Textbook of Diabetes*. Oxford, UK: Blackwell Science.

Web Links

Social Science and Medicine. McDermott R (1998) Ethics, Epidemiology and the Thrifty Gene
http://www.elsevier.com/locate/socscimed

Thrifty Genotype Hypothesis and Complex Genetic Disease

Pnina Vardi, *Felsenstein Medical Research Center, Tel Aviv University, Petah Tikva, Israel*

Konstantin Bloch, *Felsenstein Medical Research Center, Tel Aviv University, Petah Tikva, Israel*

Thrifty genes might be responsible for the appearance and increasing frequency of diseases that were uncommon before the emergence of the affluent Western lifestyle. Evolutionary selection of such genes enabled metabolic adaptation to disadvantageous conditions, but now thrifty gene expression is detrimental to health.

Advanced article

Article contents

Clinical and Epidemiological Application of the Thrifty Genes Hypothesis

According to the thrifty genotype hypothesis proposed by Neel (1962), the increasing incidence of age-related Western diseases such as diabetes mellitus (DM; in particular, type 2) obesity, hypertension and cardiovascular disease is due to a change in lifestyle from that of Paleolithic hunters, gatherers and subsistence agriculturists to a contemporary pattern characterized by sedentary occupations, high-energy fuel consumption and continuous urban stress. As a consequence, these diseases are emerging as major causes of morbidity and mortality, both in developing and in developed countries (Zimmet, 2000). The highest prevalence of some of these diseases is found in populations that have undergone rapid change in lifestyle, such as Australian Aborigines, Native Americans (particularly Pima Indians) and Pacific Islanders. Increasing prevalence of these diseases has also been reported among some migrant populations that have adopted, during a relatively short time, the modern Western lifestyle, such as Asian Indians, African and Mexican Americans, and Yemenite and Ethiopian Jews emigrating to Israel. In these latter two populations, DM was very uncommon before immigration, but reached the highest frequency afterwards, within 50 years in Yemenites and less than 10 years in Ethiopians. Such different rates of disease incidence are probably due to different levels of Westernization encountered by the new immigrants arriving in two different historical periods. (*See* Evolutionary History of the Human Genome; Humans: Demographic History; Human Genetic Diversity; Human Populations: Evolution; Modern Humans: Origin and Evolution; Thrifty Gene Hypothesis: Challenges.)

In addition, the thrifty genotype was also implicated in the process of physiological individual weight gain during adulthood (Vardi and Pinhas-Hamiel, 2000), an observation reported in both men and women of various Western communities. Diseases of thrifty gene background differ from those encoded by Mendelian inheritance. Unlike highly morbid or lethal syndromes of monogenic nature (e.g. thalassemia major, cystic fibrosis), characterized by relatively stable disease incidence and stormy clinical onset, usually during childhood, diseases of a polygenic nature develop gradually through adulthood, with no dramatic clinical onset. Such differences suggest a pure genetic effect of an abnormal monogene, unaffected by the surroundings, in the former, and morbid expression of normal thrifty polygenes, induced by continuous environmental influence, in the latter. In contrast to the thrifty genotype, the thrifty phenotype hypothesis attributes the pathological expression of adapted biological systems to postgenetic events, most commonly occurring during intrauterine life. (*See* Obesity: Genetics.)

Type 2 Diabetes Mellitus

The most common association of the thrifty genes hypothesis was made with type 2 DM reaching epidemic proportions, due to its growing incidence in the Western world. Type 2 DM is a heterogeneous condition that cannot be attributed to a single pathophysiological mechanism, with both defective insulin secretion and resistance to insulin action required for the disease to become manifest. The clustering of type 2 DM in certain families and ethnic populations suggests a role for a strong genetic background unmasked by the effect of the environment.

Table 1 Genes and diseases implicated in the thrifty-genotype hypothesis

Gene symbol	Gene name	Disease	MIM number
NIDDM2	non-insulin-dependent diabetes mellitus (common, type 2) 2	Type 2 DM	125853
PPARG	peroxisome proliferative activated receptor, gamma	Type 2 DM	601487
LEP	leptin (obesity homolog, mouse)	Obesity	164160
LEPR	leptin receptor	Type 2 DM, obesity	601007
MC4R	melanocortin 4 receptor	Obesity	155541
UCP3	uncoupling protein 3 (mitochondrial, proton carrier)	Obesity, Type 2 DM	602044
GCK	glucokinase	Maturity onset diabetes of the young (MODY)	138079
ACE	angiotensin I converging enzyme	Hypertension	106180

Type 2 DM: non-insulin-dependent diabetes mellitus (type 2).

Several genes have already been identified as candidates for the thrifty genotype in humans and animals, including those encoding energy pathways and insulin-signaling (Hasstedt et al., 2001; **Table 1**). (See Diabetes: Genetics.)

Obesity and the Metabolic Syndrome

In the last decades of the twentieth century, obesity was officially declared a disease, owing to the observation that obese patients are predisposed to a higher risk of morbidity and mortality caused by age-related Western diseases (Zimmet et al., 2001). In parallel with type 2 DM, the incidence of obesity reached epidemic proportions, affecting adults and children in affluent societies. A common genetic background and common metabolic pathways shared by these diseases is suggested by the frequent concurrent finding of obesity, glucose intolerance (or type 2 DM), hypertension and dyslipidemia, known as 'the metabolic syndrome', a disease believed to be initiated by oversecretion of insulin and resistance to its action. Studies of twins, adopted children and some adult populations indicate that body composition is significantly influenced by genetic factors. However, in no specific instance in either humans or animals is the precise etiology of obesity known at the molecular level. Attempts to identify the molecular basis of obesity have been hampered by the difficulty in accurately measuring food intake and energy expenditure, as well as the apparent polygenic control of body composition. Various thrifty genes have been implicated in the development of obesity (Comuzzie and Allison, 1998), such as those encoding leptin, a

hormone produced by adipocytes and involved in the control of food intake, its receptor (LEPR), the uncoupling proteins and the β_3-adrenergic receptor, involved in the regulation of temperature homeostasis and energy balance (Clement et al., 1995), as well as a few other genes found in human and animal models of obesity. (See Metabolism: Hereditary Errors.)

Individual Weight Gain

According to the set-point hypothesis, the brain continuously adjusts metabolism and subconsciously manipulates feeding behavior to maintain a target weight with an extraordinary degree of control. However, in most Western populations, a gradual and persistent increase in body weight is observed throughout the adult years, with a typical weight gain of 3–5 kg/decade beyond the age of 20 years. This phenomenon is associated with profound changes in body composition, with a loss of approximately 15% in skeletal muscle mass after the third decade of life, and with gradual development of abdominal adiposity accompanied by hyperinsulinemia and insulin resistance. According to the thrifty genotype hypothesis, weight gain during evolution evolved as a protective means of ensuring sufficient energy stores for basic metabolic needs during late adulthood, when energy-consuming activities such as hunting are curtailed owing to age-dependent muscle loss. Since the primary site of insulin resistance is the skeletal muscle, resistance to insulin activity during adulthood causes fuel previously directed to the muscle to be diverted and deposited as adipose tissue, in order to be used in times of shortage of external energy sources.

Hypertension

The ability to retain sodium by the kidney varies and is believed to be associated with development of hypertension through positive sodium balance, water retention and expansion of plasma volume. Insulin is known to affect sodium retention, and its relation to development of hypertension and cardiovascular diseases, in particular in association with diabetes and obesity, is described extensively in the literature. The variation in the ability to retain sodium is reflected by the finding that Blacks living in the Westernized nations retain sodium more than Whites in the West and more than Blacks in Africa. Such geographical and/or ethnic differences in sodium-handling capacity are believed to result from evolutionary selection of thrifty genes encoding greater efficiency of the kidney for conservation of sodium and water in subjects living in dry areas. Indeed, the kidney in Blacks differs from that of Whites; it contains larger glomeruli, but it is also subject to a higher frequency of nephrosclerosis with poorer outcome, when compared with White patients with renal disease. Larger glomeruli than those of African Americans were found in the Pima Indians, a community with the highest world prevalence of type 2 DM and obesity, suggesting a multisystemic effect of the thrifty genotype (Saad et al., 1991). Several thrifty genes have been implicated in the development of hypertension and cardiovascular diseases, including the angiotensin gene, which is involved in blood pressure control, and the *guanine nucleotide binding protein (G protein), beta polypeptide 3 (GNB3)* gene, encoding the ubiquitously expressed β_3-subunit of heterotrimeric G proteins.

Animal Models of Diabetes and Obesity

The concept of the thrifty genotype is strongly supported by the animal model of nutritionally induced type 2 DM-like syndrome in *Psammomys obesus*, a desert rodent (Ziv and Shafrir, 1995), which was described in the Nile delta as early as 1902. In the Dead Sea region of Israel, the *Psammomys* nutrition is based mainly on plants with very low energy value, such as the salt bush and other halophilic vegetation. Although it is extensively used as a model of type 2 DM by biologists, there is no evidence for spontaneous development of DM in *Psammomys* in its native habitat. However, when these animals are maintained on a high-energy diet, they gradually gain weight, develop hyperinsulinemia, and hyperglycemia with loss of beta cells, culminating in insulin deficiency, weight loss and ketosis. The different *Psammomys* phenotypes displayed in nature and in the passive experimental environment could be due to natural evolutionary selection of thrifty genes that enabled *Psammomys* to survive under extremely harsh desert conditions, but are detrimental in the sedentary animal-housing environment of experimental laboratories. The primary site for the phenotypic expression of the thrifty genes in the *Psammomys* could well be the insulin receptor. Having a very low number of insulin receptors, these desert rodents are found to display not only downregulation of the receptor number on a high-energy diet, but also impairment of the receptor function, resulting in peripheral insulin resistance.

Additional support for the thrifty gene hypothesis is found in the diabetes–obesity syndromes encoded by two single mutations, obese (*ob*) and diabetes (*db*) in mice (Coleman, 1979). Homozygosity for these mutations is expressed in the form of progressive obesity, hyperinsulinemia and hyperglycemia, with the clinical expression differing according to the genetic background; heterozygous mice do not progress to overt diabetes, and mutants of either ob^+ or db^+ have higher metabolic efficiency, as reflected by their ability to survive prolonged fasting when compared with homozygotes ($+/+$). Thus, in both humans and animals, conservation of thrifty genes during evolution most probably enabled longer survival, through adaptive metabolic advantages that were effective during prolonged periods of food shortage or when individuals were not able to reach nutrient sources.

See also

Metabolism: Hereditary Errors
Obesity: Genetics
Thrifty Gene Hypothesis: Challenges

References

Clement K, Vaisse C, Manning BS, et al. (1995) Genetic variation in the β_3-adrenergic-receptor and an increased capacity to gain weight in patients with morbid obesity. *New England Journal of Medicine* **333**(6): 352–354.

Coleman DL (1979) Obesity genes: beneficial effects in heterozygous mice. *Science* **203**(4381): 663–665.

Comuzzie AG and Allison DB (1998) The search for human obesity genes. *Science* **280**: 1374–1377.

Hasstedt SJ, Ren QF, Teng K and Elbein SCJ (2001) Effect of the peroxisome proliferator-activated receptor-gamma 2 Pro(12)Ala variant on obesity, glucose homeostasis, and blood pressure in members of familial type 2 diabetic kindred. *Journal of Clinical Endocrinology and Metabolism* **86**(2): 536–541.

Neel JV (1962) Diabetes mellitus, a thrifty genotype rendered detrimental by progress. *American Journal of Human Genetics* **324**: 353–362.

Saad MF, Lillioja S, Nyomba BL, et al. (1991) Racial differences in the relation between blood pressure and insulin resistance. *New England Journal of Medicine* **324**(11): 733–739.

Vardi P and Pinhas-Hamiel O (2000) The young hunter hypothesis: age-related weight gain – a tribute to the thrifty theories. *Medical Hypotheses* **55**(6): 521–523.

Zimmet P (2000) Globalization, coca-colonization and the chronic disease epidemic: can the Doomsday scenario be averted? *Journal of Internal Medicine* **247**(3): 301–310.

Zimmet P, Alberti KG and Shaw J (2001) Global and societal implications of the diabetes epidemic. *Nature* **414**(6865): 782–787.

Ziv E and Shafrir E (1995) *Psammomys obesus*: nutritionally induced NIDD-like syndrome on a 'thrifty gene' background. In: Shafrir E (ed.) *Lessons from Animal Diabetes*, vol. 5, pp. 285–300. London, UK/Japan: Smith-Gordon Company Limited/ Nishimura Company.

Further Reading

Dalgaard LT and Pedersen O (2001) Uncoupling proteins: functional characteristics and role in the pathogenesis of obesity and type II diabetes. *Diabetologia* **44**(8): 946–965.

Daniel M, Rowley KG, McDermott R and O'Dea K (2002) Diabetes and impaired glucose tolerance in Aboriginal Australians: prevalence and risk. *Diabetes Research and Clinical Practice* **57**(1): 23–33.

Hamman RF, Bennett PH and Miller M (1978) Incidence of diabetes among the Pima Indians. *Advances in Metabolic Disorders* **9**: 49–63.

Matsuda S, Ishikawa Y, Tsuchiya N, *et al.* (2002) Single nucleotide polymorphisms of thrifty genes for energy metabolism: evolutionary origins and prospects for intervention to prevent obesity-related diseases. *Biochemical and Biophysical Research Communications* **295**(2): 207–222.

Reaven GM (1991) Relationship between insulin resistance and hypertension. *Diabetes Care* **14**: 33–38.

Saltiel AR and Kahn CR (2001) Insulin signaling and the regulation of glucose and lipid metabolism. *Nature* **414**(6865): 799–806.

Sharma AM (1998) The thrifty-genotype hypothesis and its implications for the study of complex genetic disorders in man. *Journal of Molecular Medicine* **76**(8): 568–571.

Trostler N (1997) Health risks of immigration: the Yemenite and Ethiopian cases in Israel. *Biomedicine and Pharmacotherapy* **51**(8): 352–359.

Zimmet P, Boyko EJ, Collier GR and de Courten (1999) Etiology of the metabolic syndrome: potential role of insulin resistance, leptin resistance, and other players. *Annals of the New York Academy of Sciences* **892**: 25–44.

Zimmet P, Dowse G, Finch C, Serjeantson S and King H (1990) The epidemiology and natural history of NIDDM: lessons from the South Pacific. *Diabetes/Metabolism Research and Reviews* **6**(2): 91–124.

Thyroid Cancer: Molecular Genetics

Marian Ludgate, *University of Wales College of Medicine, Cardiff, UK*

Carol Evans, *Cardiff & Vale NHS Trust, Cardiff, UK*

Intermediate article

Article contents

- Introduction
- Proto-oncogenes and Growth Factors
- Tumor Suppressor Genes
- Concluding Comments

The thyroid comprises two specialized cell types, follicular thyrocytes and C cells. Papillary and follicular cancers are differentiated carcinomas of thyrocytes; medullary cancers arise from the C cells. A spectrum of thyroid-specific and nonspecific gene mutations and rearrangements underlines the molecular mechanisms resulting in thyroid cancer.

Introduction

Thyroid cancer affects 0.5–10 individuals per 100 000 per year worldwide. Although it accounts for only 1% of cancers, it results in more deaths than all other endocrine malignancies combined and is one of the commonest malignancies in those aged 20–40 years. The differentiated thyroid cancers (DTC), papillary and follicular, arise from follicular thyrocytes, the cells which produce thyroxine, and account for 90% of cases. Papillary thyroid cancer (PTC) is the commonest, but follicular cancer is the more aggressive form, often giving rise to metastases. The prognosis is generally good, although even after apparently successful initial therapy, the disease recurs and metastases develop in 15–30% of patients. Medullary thyroid cancer (MTC) is a differentiated cancer that arises from the calcitonin-producing parafollicular C cells. It accounts for 4–10% of cases of thyroid cancer and frequently metastasizes at an early stage. Approximately 75% of cases are sporadic and solitary in origin, 25% of cases are familial (FMTC) or associated with multiple endocrine neoplasia (MEN). MEN2A is defined as pheochromocytoma and parathyroid adenoma, and MEN2B as mucosal ganglioneuromas and marfanoid habitus. MEN2B is associated with the most aggressive form of MTC, and FMTC with the least. Anaplastic thyroid cancer is extremely aggressive, malignant and invariably fatal, representing the end-stage form of thyroid cancer, having lost all features of thyroid differentiation.

In common with other malignancies, the pathogenesis of thyroid cancer is thought to be a multistep process, in which tumors develop as the result of sequential accumulation of alterations in genes involved in the control of cell proliferation, differentiation or apoptosis. The genes implicated fall into two broad categories: proto-oncogenes (including hormones and/or growth factors) which become activated from mutation or overexpression to stimulate proliferation, and tumor suppressor genes which normally inhibit the neoplastic process, but if lost promote tumor growth.

Activation or loss of a variety of genes has been associated with the spectrum of thyroid neoplasia; some of these are thyroid specific, for example, activation of ret proto-oncogene (multiple endocrine neoplasia and medullary thyroid carcinoma 1, Hirschsprung disease) (*RET*) by point mutations in MTC or by PTC chromosomal rearrangement, while others such as the *RAS* proto-oncogenes and tumor protein p53 (Li–Fraumeni syndrome) (*TP53*) are not.

Proto-oncogenes and Growth Factors

RET

The *RET* proto-oncogene (chromosome 10q11.2) encodes a 170 kDa receptor tyrosine kinase with three major domains: extracellular, which includes ligand-binding and cysteine-rich regions; transmembrane; and an intracellular domain with tyrosine kinase activity. Ligand binding causes receptor dimerization and activation of the tyrosine kinase domain resulting in intracellular signaling. Chromosomal rearrangements that lead to the activation of the *RET* proto-oncogene without the need for ligand binding have been identified in papillary thyroid carcinomas (Grieco *et al.*, 1990). These typically involve fusion of the receptor tyrosine kinase to the *N*-terminal promoter region of an 'activating' gene either on the same (i.e. inversion) or a different chromosome (translocation). These activating genes are ubiquitously expressed and contain domains able to form dimers that activate the *RET* tyrosine kinase without ligand stimulation. As a consequence, the enzymatic activity of the tyrosine kinase is relocated from the membrane to the cytoplasm.

Eight *RET* rearrangements in PTC have been described to date, all of which result in constitutive expression of RET in thyroid follicular cells: *RET*/PTC1 is due to an inversion linking the RET tyrosine kinase domain to DNA segment on chromosome 10 (unique) 170 (*D10S170*), which has no known function (Pierotti *et al.*, 1992). *RET*/PTC2 is a translocation in which RET tyrosine kinase fuses with the regulatory subunit of R1 alpha of protein kinase A. *RET*/PTC3 and *RET*/PTC4 are chromosome inversions linking RET tyrosine kinase with nuclear receptor coactivator 4 (*NCOA4*), whose function is unknown.

Chromosomal rearrangements generating fusion oncoproteins are common in leukemia/lymphoma, but in solid carcinoma are almost exclusively restricted to DTC. The predisposition for *RET* gene rearrange-

ments in papillary cancer may be favored by the particular spatial arrangement of chromosome 10 in thyroid cells during cell division. *RET*/PTC rearrangements have been detected in papillary microcarcinomas, indicating that they may be an early event in thyroid carcinogenesis. In clinically detectable tumors, the presence of *RET*/PTC rearrangements varies widely, with frequencies between 0% and 59% reported in different series. The frequency is increased following exposure to radiation, and post-Chernobyl, the dramatic rise in pediatric papillary thyroid cancer was associated with *RET*/PTC rearrangements in 38–87% of cases. The mechanism is likely to be the increased incidence of chromosomal breaks following exposure of the thyroid gland to radiation.

RET in medullary thyroid cancer

MTC is caused by *RET* proto-oncogene activating mutations (Mulligan *et al.*, 1993; Hofstra *et al.*, 1994). Mutations that cause MEN2A and FMTC result in ligand-independent dimerization of receptor molecules and constitutive activation of the intracellular signaling pathways. These mainly affect the cysteine-rich extracellular domain, converting cysteine to another amino acid. Genomic mutations are identified in the index case in more than 97% of cases of MEN2A and 95% of FMTC. MEN2B is caused by mutations in the intracellular tyrosine kinase, 95% of which involve codon 918 in exon 16. Half of all MEN2B cases arise from new germ-line mutations. An activating mutation at codon 918 of *RET* is also found in tumor tissue of nearly half of the cases of sporadic MTC.

FMTC, MEN2A and MEN2B are autosomal dominantly inherited diseases; hence in an MEN2 kindred, 50% of family members are potentially affected, and almost all patients with *RET* mutations will develop MTC. MEN2 carrier detection forms the basis for recommending thyroidectomy to prevent or cure MTC in family members. *RET* codon mutations can be used to stratify family members of an affected individual into levels of risk that predict age at onset and aggressiveness of MTC. Although there is a clear genotype–phenotype relationship, aggressive disease can occur irrespective of the mutation involved; in these cases progression of the disease probably depends on additional somatic mutations in other proto-oncogenes and/or tumor suppressor genes. Diet, smoking and environmental factors may promote additional mutations.

Tyrosine kinase receptors

The neurotrophic tyrosine kinase, receptor, type 1 (*NTRK1*) gene encodes a tyrosine kinase (TK) receptor for nerve growth factor. Somatic rearrangements (both inversions and translocations) generate fusion transforming genes, which comprise an 'activating' gene with the tyrosine kinase enzymic activity of the receptor similar to the oncogene activation found with *RET*/PTC1 (Bongarzone *et al.*, 1989). These result in the expression of a chimeric protein with ectopic constitutive TK activity. *NTRK1* rearrangements have been found only in PTC, but with a lower prevalence than that reported for *RET*/PTC. More recently TRK receptors have been implicated in the onset and progression of MTC.

Peroxisome proliferator activated receptor

Translocations between chromosomes 2 and 3 have recently been detected in a series of follicular cancers. The genes involved are the thyroid transcription factor paired box gene 8 (*PAX8*) (chromosome 2) and a member of the nuclear hormone receptor superfamily peroxisome proliferative activated receptor, gamma (*PPARG*). The translocation causes an in-frame fusion of *PAX8* to *PPARG* (Kroll *et al.*, 2000). The predicted fusion protein has been detected in follicular cancer tissue by immunoprecipitation and has been postulated to have a role in the pathogenesis of thyroid follicular carcinogenesis. While *PAX8*, a paired box containing transcription factor, is central to the expression of several thyroid-specific genes including thyroglobulin and thyroperoxidase, the precise role of *PPARG* in this context has not been elucidated. *PAX8*/ *PPARG* rearrangements have not been reported in papillary thyroid cancer.

RAS genes

Mutations in the *RAS* proto-oncogenes, encoding a guanosine triphosphatase (GTPase) that cycles between the guanosine diphosphate (GDP)/guanosine triphosphate (GTP) bound states, play an important role in the pathogenesis of several human cancers. The three *RAS* genes, v-Ha-ras Harvey rat sarcoma viral oncogene homolog (*HRAS*), v-Ki-ras2 Kirsten rat sarcoma 2 viral oncogene homolog (*KRAS2*) and neuroblastoma RAS viral (v-ras) oncogene homolog (*NRAS*) code for 21 kDa Ras proteins which transduce signals from tyrosine kinase receptors in the membrane to a cascade of mitogen-activated phosphokinases which activate transcription of target genes. Mutations at amino acids 12, 13 and 61 cause constitutive activation of Ras proteins and convert them into active oncogenes, generating a continuous flow of growth signals. In contrast to other tumors in which one single *RAS* gene is mutated (*KRAS2* in pancreatic and colon cancer, *NRAS* in hematological malignancies and *HRAS* in bladder cancer), all three *RAS* genes are

activated in thyroid cancer. *RAS* oncogene activation has been found in a wide spectrum of benign and malignant thyroid pathologies, including multinodular goiters, follicular adenomas and follicular and papillary cancers, suggesting it is an early event in thyroid tumorigenesis. Overall, around 40% of benign and malignant thyroid neoplasms display *RAS* activation, and the frequency of *RAS* mutations is higher in follicular than in papillary cancers (Lemoine *et al.*, 1988). *RAS* gene amplifications and enhanced protein expression have also been described. *RAS* mutations are thought to cause genomic instability, predisposing to additional genetic alterations and tumor progression.

MET, c-myc and c-fos

Increased levels of the met proto-oncogene (hepatocyte growth factor receptor) (*MET*) gene product, a 190 kDa transmembrane protein with tyrosine kinase activity, have been detected in papillary thyroid carcinomas (Di Renzo *et al.*, 1992). Met is a potent thyrocyte mitogen whose expression is thought to be induced by *RAS* and *RET* oncogene activation. Increased concentrations of the transcription factors c-myc and c-fos have also been described, with the highest levels in poorly differentiated tumors (Terrier *et al.*, 1988).

Cyclic adenosine monophosphate signaling pathway

The growth and differentiated function of the follicular thyrocyte is positively regulated by cyclic adenosine monophosphate (cAMP). Gain-of-function mutations in the thyroid stimulating hormone receptor (*TSHR*) gene, resulting in constitutive activation of adenylate cyclase, have been found to be a major cause of benign hyperfunctioning thyroid adenomas, with mutations in GNAS complex locus (*GNAS*) accounting for most of the rest. Studies investigating the oncogenic potential of the cAMP pathway have reported mutations in *GNAS* in seven of 61 and activating *TSHR* mutations in five of 44 thyroid carcinomas. (*See* Thyroid Dysfunction: Molecular Genetics.)

Tumor Suppressor Genes

Inactivation of tumor suppressor genes such as *TP53* or retinoblastoma 1 (including osteosarcoma) (*RB1*) is an important mechanism in the development of most cancers. Inactivation can be caused by a point mutation in one allele in a cell in which the second allele has already been deleted, a process known as loss of heterozygosity (LOH). LOH affects follicular thyroid cancer more often than papillary cancer, and is frequently observed on chromosomes 3p, 3q, 7q, 10p, 11q, 13q and 22q.

TP53

Tumor protein p53 is a transcription factor that causes cell cycle arrest to allow repair of damaged deoxyribonucleic acid (DNA). In cases of severe irrecoverable damage, it directs the cell to the programmed cell death pathway resulting in apoptosis. Cells with impaired tumor protein p53 function are more likely to permit the accumulation of altered genes, and mutations in *TP53* are the commonest genetic lesion in human cancer. *TP53* mutations are associated with poorly differentiated and anaplastic thyroid tumors in the majority of cases (Ito *et al.*, 1992). In contrast to *RAS* and *RET*, *TP53* mutations are not targeted to a discrete region of the gene.

RB1

The *RB1* gene located on 13q14 produces a 105 kDa nuclear protein which has a key role in cell cycle control. *RB1* inactivation through biallelic *RB1* genetic mutations, in either a familial or sporadic setting, can result in retinoblastoma and other human tumors. Divergent results have been reported for levels of retinoblastoma protein expression in thyroid tumors with some authors finding moderate to strong positivity in the majority of thyroid carcinomas and others demonstrating the opposite.

Concluding Comments

In summary, thyroid cancer occurs when the normal equilibrium of regulatory pathways is disrupted, either through enhancement of stimulatory pathways or deficient inhibitory pathways. Furthermore, many mammalian cells carry an intrinsic, cell-autonomous program that limits their replicative potential and this must also be disrupted for a clone of cells to produce a tumor. The devices monitoring the cells' replicative clock are the telomeres, which shorten at each cell division until they are eventually unable to protect the ends of chromosomal DNA. Telomere length is maintained by upregulating expression of the telomerase enzyme, and it is estimated that 85–90% of human neoplasms express telomerase, including 79% of malignant thyroid neoplasms. In contrast, only 28% of benign thyroid lesions expressed telomerase and these were subsequently found to have lymphocytic thyroiditis, and all normal thyroids were telomerase negative.

References

Bongarzone I, Pierotti MA, Monzini N, *et al.* (1989) High frequency of activation of tyrosine kinase oncogenes in human papillary thyroid carcinoma. *Oncogene* **4**: 1457–1462.

Di Renzo MF, Olivero M, Ferro S, *et al.* (1992) Overexpression of the c-MET/HGF receptor gene in human thyroid carcinomas. *Oncogene* **7**: 2549–2553.

Grieco M, Santoro M, Berlingieri MG, *et al.* (1990) PTC is a novel rearranged form of the ret proto-oncogene and is frequently detected *in vivo* in human thyroid papillary carcinomas. *Cell* **60**: 557–563.

Hofstra RMW, Landsvater RM, Ceccherini I, *et al.* (1994) A mutation in the RET proto-oncogene associated with multiple endocrine neoplasia type 2b and sporadic MTC. *Nature* **367**: 375–376.

Ito T, Seyama T, Mizuno T, *et al.* (1992) Unique association of p53 mutations with undifferentiated but not with differentiated carcinomas of the thyroid gland. *Cancer Research* **52**: 1369–1371.

Kroll TG, Sarraf P, Pecciarini L, *et al.* (2000) PAX8-PPARgamma1 fusion oncogene in human thyroid carcinoma. *Science* **289**: 1357–1360.

Lemoine NR, Mayall ES, Wyllie FS, *et al.* (1988) Activated ras oncogenes in human thyroid cancers. *Cancer Research* **48**: 4459–4463.

Mulligan LM, Kwok JB, Healy CS and Elsdon MJ (1993) Germline mutations of the RET proto-oncogene in MEN 2A. *Nature* **363**: 458–460.

Pierotti MA, Santoro M, Jenkins RB, *et al.* (1992) Characterisation of an inversion on the long arm of chromosome 10 juxtaposing D10S170 and RET and creating the oncogenic sequence RET/PTC. *Proceedings of the National Academy of Sciences of the United States of America* **89**: 1616.

Terrier P, Sheng ZM, Schlumberger M, *et al.* (1988) Structure and expression of c-myc and c-fos proto-oncogene in thyroid carcinomas. *British Journal of Cancer* **57**: 43–47.

Further Reading

Baverstock K, Egloff B, Pinchera A, Ruchti C and Williams D (1992) Thyroid cancer after Chernobyl. *Nature* **359**: 21–22.

Bounacer A, Wicker R, Caillou B, *et al.* (1997) High prevalence of activating RET proto-oncogene rearrangements in thyroid tumors from patients who had received external radiation. *Oncogene* **15**: 1263–1273.

Brandi ML, Gagel RF, Angeli A, *et al.* (2001) Consensus: guidelines for diagnosis and therapy of MEN type 1 and type 2. *Journal of Clinical Endocrinology and Metabolism* **86**: 5658–5671.

Ito Y, Kobayashi T, Takeda T, *et al.* (1996) Expression of the retinoblastoma gene-product in clinical thyroid tissues, a reason for the slow growth of thyroid carcinoma. *Oncology Reports* **3**: 57–62.

Jhiang SM (2000) The RET proto-oncogene in human cancers. *Oncogene* **19**: 5590–5597.

Saji M, Xydas S, Westra WH, *et al.* (1999) Human telomerase reverse transcriptase (hTERT) gene expression in thyoid neoplasms. *Clinical Cancer Research* **5**: 1483–1489.

Schlumberger MJ (1998) Papillary and follicular thyroid carcinoma. *New England Journal of Medicine* **338**: 297–306.

Web Links

Met proto-oncogene (hepatocyte growth factor receptor) (*MET*); Locus ID: 4233. LocusLink:
http://www.ncbi.nlm.nih.gov/LocusLink/LocRpt.cgi?l = 4233

Neuroblastoma RAS viral (v-ras) oncogene homolog (*NRAS*); Locus ID: 4893. LocusLink:
    http://www.ncbi.nlm.nih.gov/LocusLink/LocRpt.cgi?l = 4893
Ret proto-oncogene (multiple endocrine neoplasia and medullary thyroid carcinoma 1, Hirschsprung disease) (*RET*); Locus ID: 5979. LocusLink:
    http://www.ncbi.nlm.nih.gov/LocusLink/LocRpt.cgi?l = 5979
Tumor protein p53 (Li–Fraumeni syndrome) (*TP53*); Locus ID: 7157. LocusLink:
    http://www.ncbi.nlm.nih.gov/LocusLink/LocRpt.cgi?l = 7157
v-Ha-ras Harvey rat sarcoma viral oncogene homolog (*HRAS*); Locus ID: 3265. LocusLink:
    http://www.ncbi.nlm.nih.gov/LocusLink/LocRpt.cgi?l = 3265
v-Ki-ras2 Kirsten rat sarcoma 2 viral oncogene homolog (*KRAS2*); Locus ID: 3845. Locus Link:
    http://www.ncbi.nlm.nih.gov/LocusLink/LocRpt.cgi?l = 3845
Met proto-oncogene (hepatocyte growth factor receptor) (*MET*); MIM number: 164860. OMIM:
    http://www.ncbi.nlm.nih.gov/htbin-post/Omim/dispmim?164860

Neuroblastoma RAS viral (v-ras) oncogene homolog (*NRAS*); MIM number: 164790. OMIM:
    http://www.ncbi.nlm.nih.gov/htbin-post/Omim/dispmim?164790
Ret proto-oncogene (multiple endocrine neoplasia and medullary thyroid carcinoma 1, Hirschsprung disease) (*RET*); MIM number: 164761. OMIM:
    http://www.ncbi.nlm.nih.gov/htbin-post/Omim/dispmim?164761
Tumor protein p53 (Li-Fraumeni syndrome) (*TP53*); MIM number: 191170. OMIM:
    http://www.ncbi.nlm.nih.gov/htbin-post/Omim/dispmim?191170
v-Ha-ras Harvey rat sarcoma viral oncogene homolog (*HRAS*); MIM number: 190020. OMIM:
    http://www.ncbi.nlm.nih.gov/htbin-post/Omim/dispmim?190020
v-Ki-ras2 Kirsten rat sarcoma 2 viral oncogene homolog (*KRAS2*); MIM number: 190070. OMIM:
    http://www.ncbi.nlm.nih.gov/htbin-post/Omim/dispmim?190070

Thyroid Dysfunction: Molecular Genetics

Marian Ludgate, *University of Wales College of Medicine, Cardiff, UK*

Dagmar Fuhrer, *Universität Leipzig, Leipzig, Germany*

Intermediate article

Article contents

- Introduction
- Pituitary Thyrotropin Receptor
- Sodium Iodide Symporter
- Thyroglobulin
- Thyroid Peroxidase
- Dual Oxidase 2 (DUOX2)
- Pendrin
- Thyroid Transcription Factor 1
- Forkhead Box E1 (Thyroid Transcription Factor 2)
- PAX8
- Conclusions

The thyroid produces thyroxine (T4) and triiodothyronine (T3). A surplus or a deficit in thyroid hormone production, due to thyroid dysfunction, results in hyper- or hypothyroidism respectively. A spectrum of thyroid-specific gene mutations, implicit in the control of thyroid function and growth, underline the molecular mechanisms leading to thyroid dysfunction.

Introduction

The follicular thyrocytes produce the thyroid hormones thyroxine (T4) and triiodothyronine (T3), under the positive regulation of pituitary thyrotropin (TSH) which is itself positively regulated by hypothalamic TSH releasing hormone (TRH). The production of TSH and TRH is under feedback control by T3/T4, to achieve euthyroidism. Excessive production of T3/T4 causes clinical hyperthyroidism and is accompanied by suppressed circulating TSH. Inadequate T3/T4 produces hypothyroidism and elevated circulating TSH that stimulates thyroid growth, manifest as a goiter, which is the hallmark of some forms of hypothyroidism.

Thyroid dysfunction (**Figure 1**) is common, with a prevalence in females for overt hyperthyroidism of 19/1000 and hypothyroidism of 10/1000. A major cause of hypothyroidism worldwide was once iodine deficiency, and despite the introduction of iodine supplementation programs in recent decades, endemic goiter remains a problem in Africa and other developing countries. Autoimmune disorders also account for a significant proportion of hyper- and hypothyroidism, but these polygenic disorders are outside the scope of this article. The syndrome of thyroid hormone resistance, in which there is adequate production of T3/T4 by the thyroid but single-gene defects in their nuclear hormone receptors prevent appropriate transcriptional activation or repression of target genes, will not be discussed.

This article is confined to somatic and germ-line mutations in genes central to the growth and function of the thyroid gland that have been found in sporadic cases and families exhibiting clinical signs and symptoms of hyper- or hypothyroidism (see **Table 1**).

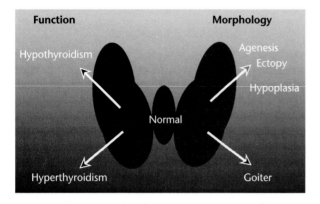

Figure 1 Thyroid dysfunction can involve abnormalities in thyroid hormone production and/or thyroid proliferation. The two conditions are not necessarily linked, for example, hyperthyroidism can be associated with a normal size thyroid gland or a goiter, while hypothyroidism may occur in the presence of a goiter, a normal size gland or thyroid hypoplasia/agenesis.

Pituitary Thyrotropin Receptor

TSH receptor (TSHR): synonym, thyrotropin receptor, thyroid-stimulating hormone receptor; GenBank ID M31774; MIM 603372 (see Web Links).

The *thyroid stimulating hormone receptor* (*TSHR*) gene is more than 60 kilobases (kb) long and maps to chromosome 14q31. It is a G-protein-coupled receptor, with major transcripts of 3.9 and 4.3 kb; exons 1–9 code for the extracellular domain (386 amino acids) while exon 10 encodes the entire transmembrane domain (349 amino acids). The native TSHR is a heterodimer and is activated by TSH, which stimulates both thyroid function and growth, predominantly via cyclic adenosine monophosphate (cAMP) signaling (**Figure 2**).

Gain-of-function mutations in *TSHR* cause constitutive receptor activation. Somatic activating *TSHR* mutations (Parma *et al.*, 1993; Duprez *et al.*, 1994) have been identified as the molecular cause of toxic thyroid nodules, subsets of toxic multinodular goiters and in rare thyroid cancers. Activating TSHR germ-line mutations (Duprez *et al.*, 1994) have been identified as the cause of autosomal dominantly inherited nonautoimmune hyperthyroidism, which occurs in familial ($n = 12$) and sporadic ($n = 7$) forms.

Table 1 Summary of selected genes altered in thyroid dysfunction, the nature of the molecular genetic defect, the functional characteristics and the clinical phenotype

Gene	Type of mutation	Mode of inheritance	Effect on gene function	Phenotype
TSHR	Activating: point mutation, two deletion mutations	Autosomal dominant	Constitutive activation of TSH receptor signaling	Hyperthyroidism and thyroid enlargement with variable disease onset
	Inactivating	Autosomal recessive: compound heterozygous, homozygous	Impairment of TSH-binding and or TSHR cell surface expression and/or TSHR signaling	Latent to overt congenital hypothyroidism, normal size thyroid gland or variable degree of thyroid hypoplasia
DUOX2	Inactivating	Autosomal recessive	Putatively impairment of thyroid peroxidase generation	Congenital hypothyroidism and goiter, iodide organification defect
TG	Inactivating	Autosomal recessive: one family with autosomal dominant inheritance	Abnormal TG folding → intracellular TG trapping, TG truncation, impairment of thyroid hormone binding	Euthyroid goiter, congenital hypothyroidism and goiter
TITF1	Inactivating	Autosomal dominant, haploinsufficiency	TITF1 truncation, TITF1 deletion, reduction of DNA-binding activity	Syndrome: congenital hypothyroidism and thyroid hypoplasia, pulmonary failure, movement disorders
PAX8	Inactivating	Autosomal dominant, haploinsufficiency	PAX8 truncation, reduction of DNA-binding activity	Congenital hypothyroidism and hypoplasia

Gene symbols: full names of genes and links to further information can be found on the Genew Human Gene Nomenclature Database (see Web Links).
TG: thyroglobulin; TSH: pituitary thyrotropin (thyroid-stimulating hormone); TSHR: pituitary thyrotropic receptor (thyroid-stimulating hormone receptor); TITF1: thyroid transcription factor 1.

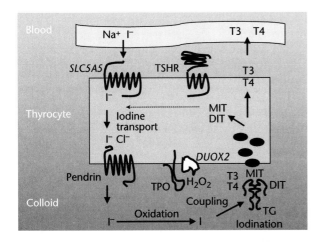

Figure 2 Diagram of a thyroid epithelial cell showing the steps of thyroid hormone production (thyroxine (T4) and triiodothyronine (T3)) comprising iodide (I⁻) uptake from the blood, iodide transport across the thyroid cell into the follicular lumen, iodide oxidation and coupling (organification), storage of thyroid hormones bound to thyroglobulin (TG) as the 'colloid' and lastly TG endocytosis and release of thyroid hormone into the blood. Disturbance of genes involved in thyroid hormone synthesis (*SLC5A5*, *SLC26A4* (pendrin), *TPO*, *TG*, *DUOX2*) usually results in a goiter with euthyroid or hypothyroid function. Depending on their nature, mutations in thyroid-stimulating hormone receptor (TSHR), which is the key regulator of both thyroid function and growth, cause either hyperthyroidism and goiter or hypothyroidism in a normal or hypoplastic gland. In contrast, defects in transcription factors regulating thyroid differentiation (TITF1, FOXE1, PAX8) result in hypothyroidism and a variable degree of thyroid hypoplasia. (Note putative function of pendrin as the basal iodide transporter.) MIT: mono-iodotyrosine; DIT: di-iodotyrosine.

Constitutively activating *TSHR* mutations have been found in 29 codons (with 41 different amino acid exchanges) almost exclusively within exon 10. Most mutations occur as heterozygous point mutations resulting in single amino acid exchanges; two activating deletion mutations have also been described. In addition, a mutation in exon 7 produces a TSHR with increased affinity for human choriogonadotropin, resulting in gestational thyrotoxicosis.

Germ-line loss-of-function *TSHR* mutations (Sunthornthepvarakul *et al.*, 1995) have been described in 23 individuals/families with variable phenotypic presentations ranging from euthyroid TSH resistance and a normal thyroid gland to severe congenital hypothyroidism (CH) with marked thyroid hypoplasia. Functional impairment of the mutant TSHR is due to reduced TSH-binding, TSHR cell surface expression and/or TSHR signaling, the degree of which determines the phenotype. Inactivating *TSHR* mutations occur as homozygotes and compound heterozygotes, are localized throughout the receptor and include missense ($n = 15$) and nonsense mutations ($n = 6$) as well as four deletion mutations.

A TSHR mutation database has been created (see Web Links).

Sodium Iodide Symporter

Sodium iodide symporter: synonym, sodium/iodide cotransporter; GenBank ID U66088; MIM 601843 (see Web Links).

The sodium iodide symporter gene *solute carrier family 5 (sodium iodide symporter), member 5 (SLC5A5)* is located on chromosome 19p13.2–p12 and has 15 exons. It has an open reading frame of 1929 nucleotides encoding a glycoprotein of 643 amino acids with a predicted molecular mass of 68.7 kDa. The protein has 13 putative transmembrane domains, is located on the basolateral surface of the thyroid and is responsible for transporting iodide into the cell using the Na⁺ gradient generated by an associated Na⁺, K⁺-ATPase.

SLC5A5 germ-line mutations (Fujiwara *et al.*, 1997) were first described in two unrelated Japanese displaying an iodide trapping defect, who were homozygous for Thr354Pro. This missense mutation has subsequently been shown to be highly prevalent in the Japanese population and also occurs in heterozygous combination with other loss-of-function *SLC5A5* mutations, for example, Gly93Arg. In common with other gene defects producing thyroid dysfunction, a wide variation in phenotype is observed, even in patients harboring the same *SLC5A5* mutation.

Thyroglobulin

Thyroglobulin (TG): GenBank ID NM003235; MIM 188450 (see Web Links).

The *thyroglobulin (TG)* gene is more than 300 kb long, maps to chromosome 8q24.2–q24.3, comprises at least 37 exons and yields a messenger ribonucleic acid (mRNA) transcript of ~8.3 kb. The mature 660 kDa TG protein is a glycosylated homodimer of two identical 300 kDa subunits and functions as the substrate and storage protein for thyroid hormones in the lumen (**Figure 2**).

Molecular alterations in TG (Ieri *et al.*, 1991) have been described in at least eight families with CH and goiter as well as three kindreds with euthyroid familial goiters. *TG* mutations have been identified as missense ($n = 5$) and nonsense mutations ($n = 2$), a 138 bp mRNA deletion and a 3′ splice site mutation. Functional impairment of TG through these mutations has been suggested to result from TG trapping due to altered protein folding, TG truncations or alteration of TG residues involved in thyroid hormone formation. Most *TG* mutations occur as homozygous mutations, and less frequently as compound hetero-zygosity. In addition, TG abnormalities (low or undetectable intrathyroidal TG levels) have long been described in a number of families with CH, even with an autosomal dominant trait of inheritance, and individuals with euthyroid goiters; the underlying genetic defects have yet to be clarified.

Thyroid Peroxidase

Thyroid peroxidase (TPO): synonym, thyroperoxi-dase, microsomal antigen; GenBank ID M25702; MIM 606765 (see Web Links).

The *thyroid peroxidase (TPO)* gene is 150 kb long, comprises 17 exons and is located on chromosome 2p25. *TPO* has two major transcripts of approximately 3 kb encoding a protein of 933 amino acids and a shorter inactive form. TPO is a glycosylated hemo-protein that catalyzes the iodination and coupling of tyrosyl residues in TG to produce T3/T4. The membrane-bound enzyme of 110 kDa is located mainly at the apex of the thyroid follicle with its catalytic site facing into the lumen.

A proportion of children detected with CH by newborn screening have an iodide organification defect which may be partial (PIOD) or total (TIOD) and caused by mutations in *TPO*, *TG*, pendrin or the H_2O_2 generating system.

TIOD has an autosomal recessive mode of inheri-tance, indicating the requirement for both *TPO* alleles to be affected, and usually results from homozygous mutations (Abramowicz et al., 1992), such as a GGCC duplication in exon 8 producing a frameshift and premature stop in *TPO*. Analysis of a large series of 35 families with TIOD identified 16 different mutations, predominantly in exons 8, 9 and 10, with the duplication being detected most frequently. Mutations in both *TPO* alleles were found in 29 families (13 homozygous and 16 compound heterozygous). In the remaining families mutations must be present in the introns or promoter regions of *TPO*.

Dual Oxidase 2 (DUOX2)

Dual oxidases 1 and 2: synonyms thyroid oxidases 1 and 2 (ThOX1, 2), LNOX 1 and 2; GenBank ID AF 230495 and AF 230496; MIM 606758 and 606759 (see Web Links).

The *dual oxidase 2 (DUOX2)* and the closely related *dual oxidase 1 (DUOX1)* genes have been mapped to chromosome 15q15. *DUOX2* is composed of 33 exons, spanning 20 kb and encodes a reduced nicotinamide–adenine dinucleotide phosphate (NADPH) oxidase of 1548 amino acids and ~177 kDa size. DUOX2

belongs to the family of flavoproteins and is part of the thyroid Ca^{2+}/NADPH-dependent peroxide gen-erating system, which is rate limiting to TPO-catalyzed iodide oxidation during thyroid hormone production (**Figure 2**).

In a group of 12 patients with CH and PIOD, a spectrum of 13 heterozygous point mutations (three intronic, three silent, seven missense and two nonsense mutations) has very recently been identified in the *DUOX2* gene (Moreno et al., 2000). However, the precise impact of these *DUOX2* mutations on thyroid disturbance remains to be clarified.

Pendrin

Pendrin: GenBank ID G36360 and G36379; MIM 274600 and 605646 (see Web Links).

Pendred syndrome (PDS) is a recessively inherited disorder in which sensorineural deafness is associated with goiter and PIOD. The gene responsible (*solute carrier family 26, member 4 (SLC26A4)*) maps to 7q31 and has a transcript of approximately 5 kb encoding a transmembrane protein related to a family of sulfate transport proteins. *In vitro* studies indicate that pendrin is a chloride–iodide transport protein and it is postulated to be the iodide efflux mechanism that would be located at the thyroid apex, in contact with the lumen.

Three deleterious mutations (Everett et al., 1997), each segregating with the disease in the respective families, have been identified in *SLC26A4*. Subsequent-ly, analysis of a cohort of 56 kindreds with PDS identified 47 of 60 mutant alleles in 31 families, including three homozygous consanguineous kindreds and four recurrent mutations accounting for 74% of PDS chromosomes.

Thyroid Transcription Factor 1

Thyroid transcription factor 1 (TITF1): synonyms, thyroid nuclear factor, NKX2-1, thyroid-specific enhancer-binding protein; GenBank ID NM003317; MIM 600635 (see Web Links).

The *thyroid transcription factor 1 (TITF1)* gene maps to chromosome 14q13, spans ~3.3 kb of genomic deoxyribonucleic acid (DNA) and is encoded by two exons with a 2.4 kb mRNA transcript, corresponding to a 371 amino acid protein of size 38 kDa. TITF1 belongs to the family of NK homeobox proteins which function as nuclear transcription factors. In the thyroid, TITF1 is elementary to organogenesis and regulates thyroid differentiation. In addition, TITF1 is highly expressed in lung tissue and also in the central nervous system.

TITF1 gene alterations (Devriendt *et al.*, 1998) have so far been described in seven patients, who showed combined features of hyperthyrotropinemia or CH, respiratory problems ($n = 7$) and movement disorders ($n = 6$). *TITF1* mutations occurred as heterozygous loss-of-function mutations in the form of nonsense ($n = 4$) and missense mutations ($n = 1$) or gene deletions ($n = 2$). Functional disturbance is believed to result from *TITF1* haploinsufficiency with the defective allele coding for a truncated protein or a TITF1 protein with reduced DNA-binding capacity.

Two animal models have been developed: heterozygous *Titf1* mice exhibit a phenotype of neurological symptoms and mild TSH resistance, while *Titf1* knock-out mice are born dead, lacking thyroid and lung parenchyma, the pituitary gland and exhibiting further severe structural defects of the central nervous system.

Forkhead Box E1 (Thyroid Transcription Factor 2)

Forkhead box E1 (FOXE1): synonyms, thyroid transcription factor 2, FKHL15; GenBank ID U89995; MIM 241850 and 602617 (see Web Links).

FOXE1 is a member of the forkhead/winged-helix domain transcription factor family. The *forkhead box E1 (thyroid transcription factor 2) (FOXE1)* gene is situated on chromosome 9q22. FOXE1 is encoded by a single-exon gene spanning 3.5 kb and producing a 376 amino acid residue protein with a predicted molecular mass of 42 kDa. FOXE1 is expressed in the developing thyroid, most of the foregut epithelium and in craniopharyngeal ectoderm. FOXE1 is a zinc-finger DNA-binding protein which functions as a transcriptional activator in adult thyroid regulating the expression of TG and TPO via FOXE1 binding sites in the promoters of both genes.

Foxe1-null mice have either a sublingual or a completely absent thyroid gland associated with cleft palate; Bamforth–Lazarus syndrome is the human equivalent comprising CH due to thyroid agenesis, cleft palate, choanal atresia and spiky hair. A homozygous missense mutation of the FOXE1 forkhead DNA-binding domain was found in two brothers having this phenotype (Clifton-Bligh *et al.*, 1998). Although a rare cause of CH, a second family, again of two brothers, harboring a different mutation in the DNA-binding site has been reported.

PAX8

PAX8 (paired box gene 8): synonym, paired domain gene 8; GenBank ID TM X69699 and S55490; MIM 167415 (see Web Links).

The *paired box gene 8 (PAX8)* gene has been mapped to chromosome 2q12–q14 and is encoded by at least 10 exons with an mRNA transcript size of 2.7 kb and an open reading frame of 450 amino acids. PAX8 belongs to the mammalian Pax family and functions as a nuclear transcription factor that recognizes DNA via its paired domain. PAX8 plays an important role in thyroid organogenesis and in the adult gland contributes to the activation of TG and TPO transcription. PAX8 is also expressed in other human tissues, including the kidneys.

PAX8 mutations (Macchia *et al.*, 1998) have been identified in three individuals and two families with variable penetrance of CH and thyroid hypoplasia and occur in the form of heterozygous missense ($n = 4$) or nonsense ($n = 1$) mutations affecting the paired box DNA-binding domain. The resulting thyroid abnormalities are most probably due to allelic haplo-insufficiency.

The *Pax8* mouse model exhibits hypothyroidism and thyroid hypoplasia with complete absence of thyroid follicular cells, when both *Pax8* alleles are knocked out.

Conclusions

Many aspects of thyroid dysfunction can now be explained by single-gene alterations causing disturbance of thyroid hormone production or thyroid growth or both. While this also extends to some forms of thyroid malignancy, which are reviewed elsewhere, it is important to emphasize that the molecular contribution to other thyroid disease, including nontoxic nodules, familial euthyroid goiter and the majority of cases of CH, remains to be clarified.

See also

Endocrine Disorders: Hereditary
Metabolism: Hereditary Errors
Pax Genes: Evolution and Function

References

Abramowicz MJ, Targovnik HM, Varela V, *et al.* (1992) Identification of a mutation in the coding sequence of the human thyroid peroxidase gene causing congenital goitre. *Journal of Clinical Investigation* **90**: 1200–1204.

Clifton-Bligh R, Wentworth J, Heinz P, *et al.* (1998) Mutation of the gene encoding human Ttf2 associated with thyroid agenesis, cleft palate and choanal atresia. *Nature Genetics* **19**: 399–401.

Devriendt K, Vanhole C, Matthijs G and Zegher F (1998) Deletion of thyroid transcription factor 1 gene in an infant with neonatal thyroid dysfunction and respiratory failure. *New England Journal of Medicine* **338**: 1317–1318.

Duprez L, Parma J, Van Sande J, *et al.* (1994) Germline mutations in the thyrotropin receptor gene cause nonautoimmune autosomal dominant hyperthyroidism. *Nature Genetics* **7**: 396–401.

Everett LA, Glaser B, Beck JC, *et al.* (1997) Pendred syndrome is caused by mutations in a putative sulphate transporter gene (PDS). *Nature Genetics* **17**: 411–422.

Fujiwara H, Tatsumi K, Miki K, *et al.* (1997) Congenital hypothyroidism caused by a mutation in the Na⁺/I⁻symporter. *Nature Genetics* **16**: 124–125.

Ieri T, Cochaux P, Targovnik HM, *et al.* (1991) A 3-prime splice site mutation in the thyroglobulin gene responsible for congenital goiter with hypothyroidism. *Journal of Clinical Investigation* **8**: 1901–1905.

Macchia PE, Lapi P, Krude H, *et al.* (1998) PAX8 mutations associated with congenital hypothyroidism caused by thyroid dysgenesis. *Nature Genetics* **19**: 83–86.

Moreno JC, Bikker H, De Randmie J, *et al.* (2000) Mutations in the ThOX2 gene in patients with congenital hypothyroidism due to iodine organification defects. *Endocrine Journal* **47**: 107.

Parma J, Duprez L, Van Sande J, *et al.* (1993) Somatic mutations in the thyrotropin receptor gene cause hyperfunctioning thyroid adenomas. *Nature* **365**: 649–651.

Sunthornthepvarakul T, Gottschalk ME, Hayashi Y, *et al.* (1995) Brief report: resistance to thyrotropin caused by mutations in the thyrotropin-receptor gene. *New England Journal of Medicine* **332**: 155–160.

Further Reading

Bakker B, Bikker H, Vusma T, *et al.* (2000) Two decades of screening for congenital hypothyroidism in the Netherlands: TPO gene mutations in total iodide organification defects (an update). *Journal of Clinical Endocrinology and Metabolism* **85**: 3708–3712.

Coyle B, Rearrdon W, Herbrick JA, *et al.* (1998) Molecular analysis of the PDS gene in Pendred syndrome (sensorineural hearing loss and goitre). *Human Molecular Genetics* **7**: 1105–1112.

DeFelice M, Ovitt C, Biffali E, *et al.* (1998) A mouse model for hereditary thyroid dysgenesis and cleft palate. *Nature Genetics* **19**: 395–398.

Kimura S, Hara Y, Pineau T, *et al.* (1996) The T/ebp null mouse: thyroid specific enhancer protein is essential for the organogenesis of the thyroid, lung, ventral forebrain and pituitary. *Genes and Development* **10**: 60–69.

Kosugi S, Inoue S, Matsuda A and Jhiang SM (1998) Novel, missense and loss of function mutations in the sodium/iodide symporter gene causing iodide transport defect in three Japanese patients. *Journal of Clinical Endocrinology and Metabolism* **83**: 3373–3376.

Mansouri A, Chowdhury K and Gruss P (1998) Follicular cells of the thyroid gland require PAX8 gene function. *Nature Genetics* **19**: 87–90.

Paschke R and Ludgate M (1997) The thyrotropin receptor in thyroid disease. *New England Journal of Medicine* **337**: 1675–1681.

Rodien P, Bremont C, RaffinSanson ML, *et al.* (1998) Familial gestational hyperthyroidism caused by a mutant thyrotropin receptor hypersensitive to human chorionic gonadotropin. *New England Journal of Medicine* **339**: 1823–1826.

Targovnik HM, MedeirosNeto G, Varela V, *et al.* (1993) A nonsense mutation causes human hereditary congenital goitre with preferential generation of a 171 nucleotide deleted thyroglobulin ribonucleic acid messenger. *Journal of Clinical Endocrinology and Metabolism* **77**: 210–215.

Web Links

Genew Human Gene Nomenclature Database. HUGO approved nomenclature and links to further information
http://www.gene.ucl.ac.uk/cgi-bin/nomenclature/searchgenes.pl

NCBI Entrez nucleotides database. Use GenBank ID number in search engine for further information
http://www.ncbi.nlm.nih.gov/entrez/query.fcgi?db = Nucleotide

NCBI OMIM: Online Mendelian Inheritance in Man. An online a catalog of human genes and genetic disorders (use OMIM number to search)
http://www.ncbi.nlm.nih.gov/Omim/

TSH Receptor Database II. The TSHR mutation database is aimed at scientists and clinicians. It comprises all TSHR mutations reported up to December 2002 and contains all available functional and clinical data on the different TSHR mutations including 55 pedigrees of patients with germline TSHR mutations with detailed information on molecular aspects, clinical courses and treatment options. In addition, a first compilation of site-directed mutagenesis studies has been included as well as special search tools and an administrator tool for submission of novel TSHR mutations. The TSHR mutation database is installed as one of the locus specific HUGO mutation databases (http://ariel.ucs.unimelb.edu.au/ ~cotton/dblist.htm). It is listed under index TSHR 603372 (http://ariel.ucs.unimelb.edu.au/ ~cotton/glsdbq.htm) and can be accessed via
www.uni-leipzig.de/innere/tshr.
http://www.uni-leipzig.de/~innere/tshrdb/index.php

Dual oxidase 2 (*DUOX2*); Locus ID: 50506. LocusLink:
http://www.ncbi.nlm.nih.gov/LocusLink/LocRpt.cgi?l = 50506

Thyroid peroxidase (*TPO*); Locus ID: 7173. LocusLink:
http://www.ncbi.nlm.nih.gov/LocusLink/LocRpt.cgi?l = 7173

Thyroglobulin (*TG*); Locus ID: 7038. LocusLink:
http://www.ncbi.nlm.nih.gov/LocusLink/LocRpt.cgi?l = 7038

Thyroid stimulating hormone receptor (*TSHR*); Locus ID: 7253. LocusLink:
http://www.ncbi.nlm.nih.gov/LocusLink/LocRpt.cgi?l = 7253

Thyroid transcription factor 1 (*TITF1*); Locus ID: 7080. LocusLink:
http://www.ncbi.nlm.nih.gov/LocusLink/LocRpt.cgi?l = 7080

Dual oxidase 2 (*DUOX2*); MIM number: 606759. OMIM:
http://www.ncbi.nlm.nih.gov/htbin-post/Omim/dispmim?606759

Thyroid peroxidase (*TPO*); MIM number: 606765. OMIM:
http://www.ncbi.nlm.nih.gov/htbin-post/Omim/dispmim?606765

Thyroglobulin (*TG*); MIM number: 188450. OMIM:
http://www.ncbi.nlm.nih.gov/htbin-post/Omim/dispmim?188450

Thyroid stimulating hormone receptor (*TSHR*); MIM number: 603372. OMIM:
http://www.ncbi.nlm.nih.gov/htbin-post/Omim/dispmim?603372

Thyroid transcription factor 1 (*TITF1*); MIM number: 600635. OMIM:
http://www.ncbi.nlm.nih.gov/htbin-post/Omim/dispmim?600635

Tissue-specific Locus Control: Structure and Function

Constanze Bonifer, *University of Leeds, Leeds, UK*

All developmental processes involve decisions in which the developmental fate of precursor cells is restricted to a single lineage by establishing a unique gene expression pattern. The activation at each of these gene loci involves restructuring chromatin as a necessary prerequisite for the assembly of the transcription apparatus. Tissue-specific locus activation is controlled by the interplay between sequence-specific transcription factors, chromatin-modifying activities and basic chromatin components.

> **Intermediate article**
>
> **Article contents**
> - Basic Mechanisms
> - Examples: α Globin Versus β Globin

Basic Mechanisms

The phenotypic changes in cells observed during development are based on consecutive alterations in the expression status of individual genes. Many so-called 'housekeeping' genes are active in all cells and stay active during cell differentiation. Other genes, however, encode proteins that are only required in specialized cell types. This presents the developing organism with two problems. First, specific genes need to be kept silent in inappropriate cell types but have to be activated in appropriate cell types, and second, their expression level has to be correctly regulated. Eukaryotes solved the first problem by organizing the genetic information into chromatin, which is tightly packed in transcriptionally inactive genes. Such chromatin structures are essentially nonpermissive for transcription. To activate a gene locus, chromatin has to be modified.

Higher-order chromatin structure involving several layers of specific multiprotein complexes represents an important level of control in the tissue-specific expression of eukaryotic genes. The activation of a eukaryotic gene is accompanied by the reorganization and modification of extensive regions of chromatin. Such chromatin reorganization and modification not only occurs at the coding region where transcription takes place, but extends into 5' and 3' flanking deoxyribonucleic acid (DNA). Different types of protein complexes are able to reorganize chromatin (Aalfs and Kingston, 2000). These include complexes that use energy to move nucleosomes, such as SWI/SNF-type complexes, as well as a number of different enzymes that modify nucleosomes by covalently adding small molecules to individual histones. These include histone acetyltransferases (HATs), kinases and histone methyltransferases (HMTs). Each core histone (H2A, H2B, H3, H4) can be modified at the N-terminal tails that protrude from the nucleosome core particle. The specific combinations of histone modifications ultimately determine the activation status of a gene

(Jenuwein and Allis, 2001). Acetylated and phosphorylated core histones are generally found on transcriptionally active genes. Consistent with this finding, HATs have a transcriptional coactivator function. In contrast, histone methylation correlates with both gene activity and inactivity, depending on which amino acid is modified in a specific histone. Histone acetylation and phosphorylation are dynamic processes. Enzymes removing histone modifications are histone deacetylases (HDACs) and phosphatases. Consequently, HDACs have a transcriptional co-repressor function (Laherty *et al.*, 1997). (*See* Chromatin in the Cell Nucleus: Higher-order Organization; Histone Acetylation: Long-range Patterns in the Genome.)

Eukaryotic genes are regulated by a number of different *cis*-regulatory elements, such as enhancers and promoters, that can be distributed over large distances. They are characterized by the presence of clusters of binding sites for sequence-specific transcription factors. The transcriptional activation of a gene locus in development is achieved through the cooperation of such elements. The binding of transcription factors and the formation of transcription factor complexes with cofactors results in distinct structural changes in the chromatin of regulatory regions. Active *cis*-regulatory elements often become hypersensitive to digestion with endonucleases, such as DNaseI (Gross and Garrard, 1988), indicating that chromatin structure and DNA topology have changed. Indeed, the enhanced accessibility of *cis*-regulatory elements can serve as means to determine their state of activity in different cell types. Many *cis*-regulatory elements exhibit highly organized chromatin structure even in their nonexpressed state, with nucleosomes accurately positioned over specific DNA sequences (Thoma, 1992). This very fact suggests that they contribute to the efficiency of transcription by generating a chromatin structure designed to interact with sequence-specific DNA-binding proteins in a highly specific fashion. Nucleosome positioning at *cis*-regulatory elements can change as a result of the binding of transcription factor complexes. (**See** Enhancers.)

The chromatin-modifying activities described above are found in every cell type, and in principle can interact with any nucleosomal substrate. The most important question in this context was, therefore, what directs chromatin-modifying activities to tissue-specifically expressed genes. It is already apparent that the way an extended chromatin region encompassing an active gene locus is reorganized is effectively defined by the distribution of regulatory elements. It is also clear that site-specific chromatin modifications are determined by sequence-specific transcription factors. These, in turn, can associate with architectural factors to form large protein assemblies on enhancers and promoters (Merika and Thanos, 2001). An ever-increasing number of different transcription factors has been shown to specifically interact with chromatin-modifying activities in a highly ordered fashion (Agalioti *et al.*, 2000). This is not only true for enzymes with transcriptional coactivator function like HATs and kinases, but also for repressive activities such as HDACs. Interestingly, depending on the context, the same transcription factors can associate with different enzymes. This leads to two conclusions. First, it is the transcription factor assemblies on the *cis*-regulatory element and not individual factors that determine which chromatin-modifying activity is recruited to specific sequences on their target genes. Second, it is the combination of sequence-specific transcription factors expressed in different cell types that ultimately determines whether a gene is activated or shut down. These factors either stay on the genome and establish gene expression patterns that are faithfully maintained throughout cell division, or change in development as a result of signaling from outside the cell. Thus, what drives the establishment of tissue-specific gene expression patterns and defines the identity of a given cell is the combination of functional DNA-binding and non-DNA-binding transcription factors together with their associated chromatin-modifying activities. (**See** Chromatin Structure and Domains.)

Examples: α Globin versus β Globin

Much research has been devoted to the elucidation of which type of sequence information is required to achieve correct cell-type- and cell-stage-specific gene expression in ontogeny. The answer to this question is not only of relevance for basic research, but also for all gene transfer experiments carried out in biotechnology and gene therapy applications. Two experimental strategies have been used to address this problem. First strategy is either by deletion or mutation of specific *cis*-regulatory elements at their endogenous

loci by gene targeting, generally in the mouse. In humans, the effects of naturally occurring mutations that lead to a specific disease phenotype have also been useful in this regard. The alternative strategy involves the analysis of gene constructs integrated at ectopic sites in transgenic animals. The most well-studied examples of genes looked at this way are the human α-globin and β-globin gene clusters (**Figure 1**). Because both globin gene clusters are prime candidates for gene therapy approaches, a large body of work has been carried out to examine what is needed for their correct regulation. (**See** Hemoglobin Disorders: Gene Therapy.)

Both globin gene clusters contain several genes encoding the α-globin and β-globin family of proteins that together make up the hemoglobin molecule and that are specifically expressed in red blood cell precursor cells. The expression of genes within both gene clusters is developmentally regulated as different types of hemoglobin are formed in the developing embryo and in the adult organism to accommodate their different oxygen environments. In both, the expression of the embryonic/fetal genes is switched off after birth and is replaced by expression of the adult-type genes. The developmental time course of expression follows the order of the genes on the chromosome, with the most 5′ embryonic gene being expressed first. In both cases, gene locus activation is accompanied by extensive chromatin remodeling events that result in a general increase in histone acetylation as well as in the appearance of DNaseI hypersensitive sites (DHSs) within and around the gene clusters (Forsberg and Bresnick, 2001; Schubeler *et al.*, 2000; Anguita *et al.*, 2001). From their similar genomic structure and expression patterns, it was originally assumed that the same type of control elements would be used to achieve a coordinated and balanced production of α-globin and β-globin proteins. However, experiments examining the minimal requirements of correctly regulated gene expression in transgenic mice uncovered significant regulatory differences between the two loci. (**See** Globin Genes: Evolution.)

The analysis of gene constructs in transgenic mice provided one of the most convincing arguments for a role for chromatin in the regulation of gene expression. Gene constructs inserted randomly into chromosomes often demonstrated an integration-site-dependent variation in expression level. In addition, the correct temporal and spatial pattern of transgene expression could be disturbed. This phenomenon is called the chromosomal position effect. For the β-globin locus, a high level of expression that is relatively independent of the chromosomal integration site requires the presence of an intact cluster of upstream *cis*-regulatory elements termed the locus control region (LCR) (**Figure 1**). Each of the five LCR elements binds many different ubiquitous

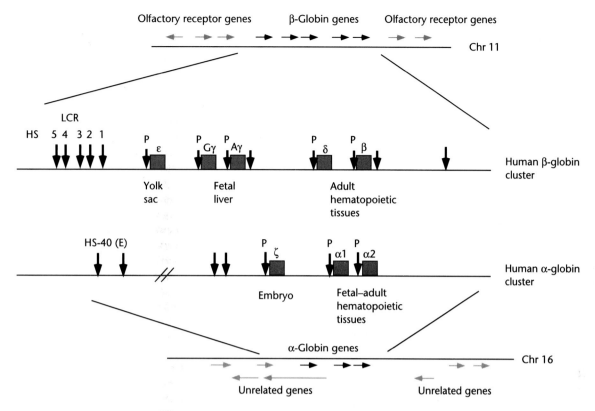

Figure 1 Organization of the human β-globin cluster on chromosome 11 (top) and the human α-globin cluster on chromosome 16 (bottom). Functional genes are indicated as boxes. DNaseI hypersensitive sites that mark the positions of *cis*-regulatory elements such as promoters (P) or enhancers (E) are indicated as vertical arrows. The top and bottom diagrams depict the chromosomal organization of globin and nonglobin genes.

and erythroid-specific transcription factors that have been shown to recruit chromatin remodeling activities. When coupled to a single β-globin gene, the LCR is sufficient for high-level integration-site-independent expression in erythroid cells (Grosveld *et al.*, 1987). However, correct developmental regulation requires that the locus is in its original conformation on the transgene, and the order of the genes with respect to the LCR is preserved (Hanscombe *et al.*, 1991). This was interpreted such that the promoters of the different β-globin genes compete with each other for the interaction with the LCR in a developmental-specific fashion. These principles also operate in the α-globin locus. High-level expression and correct developmental regulation in transgenic mice is observed when the gene locus is in its original conformation, and specific sets of *cis*-regulatory elements that bind a similar collection of factors as the β-globin genes are required for high-level expression (Sharpe *et al.*, 1993). However, integration-site-independent expression was not observed. Even with very large transgenes, no *cis*-regulatory elements could be identified that were able to confer this property. (*See* Locus Control Regions (LCRs); Transgenic Mice.)

The reason why an element which was able to confer integration-site-independent expression could be identified in the β-globin locus and not in the α-globin locus lay in the different chromosomal organization of each locus (Craddock *et al.*, 1995; Bulger *et al.*, 2000). The β-globin locus is surrounded by odorant receptor genes that are expressed in a different tissue and it does not contain any other genes (**Figure 1**). Its chromatin only becomes accessible in precursors of erythroid cells, and in all other cell types its chromatin is tightly packed and inaccessible. Hence, extended regions of chromatin need to be remodeled from the inactive state. In contrast, the α-globin locus lies within a region of constitutively accessible chromatin and partially overlaps with genes that are expressed in every cell (**Figure 1**). Here the opposite problem is presented: how is the α-globin locus kept silent in nonerythroid cells?

Additional evidence that the chromosomal environment can determine the developmental regulation of gene loci to a significant extent came from experiments in which the LCR of the endogenous β-globin locus was deleted by gene targeting. Mutations that at ectopic integration sites severely impaired β-globin expression had a much milder effect or no effect at all in the natural chromosomal environment (Bender *et al.*, 2000).

In summary, the comparison between these two gene loci has revealed general principles regarding tissue-specific gene locus control:

- Tissue specificity of gene expression is mediated by the combination of ubiquitous and tissue-specific transcription factors binding to diverse collections of *cis*-regulatory elements.
- These factors recruit chromatin remodeling activities that are required for the establishment and stable maintenance of a chromatin structure permissive for messenger ribonucleic acid expression.
- Every gene locus has evolved its own individual strategy to control appropriate tissue-specific activation in its own specific chromosomal environment.

See also

Locus Control Regions (LCRs)

References

Aalfs JD and Kingston RE (2000) What does 'chromatin remodeling' mean? *Trends in Biochemical Science* 25: 548–555.

Agalioti T, Lomvardas S, Parekh B, *et al.* (2000) Ordered recruitment of chromatin modifying and general transcription factors to the IFN-beta promoter. *Cell* 103: 667–678.

Anguita E, Johnson CA, Wood WG, Turner BM and Higgs DR (2001) Identification of a conserved erythroid specific domain of histone acetylation across the alpha-globin gene cluster. *Proceedings of the National Academy of Sciences of the United States of America* 98: 12114–12119.

Bender MA, Bulger M, Close J and Groudine M (2000) Beta-globin gene switching and DNase I sensitivity of the endogenous beta-globin locus in mice do not require the locus control region. *Molecular Cell* 5: 387–393.

Bulger M, Bender MA, van Doorninck JH, *et al.* (2000) Comparative structural and functional analysis of the olfactory receptor genes flanking the human and mouse beta-globin gene clusters. *Proceedings of the National Academy of Sciences of the United States of America* 97: 14560–14565.

Craddock CF, Vyas P, Sharpe JA, *et al.* (1995) Contrasting effects of alpha and beta globin regulatory elements on chromatin structure may be related to their different chromosomal environments. *EMBO Journal* 14: 1718–1726.

Forsberg EC and Bresnick EH (2001) Histone acetylation beyond promoters: long-range acetylation patterns in the chromatin world. *BioEssays* 23: 820–830.

Gross DS and Garrard WT (1988) Nuclease hypersensitive sites in chromatin. *Annual Review of Biochemistry* 57: 159–197.

Grosveld F, Bloom van Assendelft G, Greaves DR and Kollias G (1987) Position-independent, high-level expression of the human β-globin gene in transgenic mice. *Cell* 51: 975–985.

Hanscombe O, Whyatt D and Fraser P, *et al.* (1991) Importance of globin gene order for correct developmental expression. *Genes and Development* 5: 1387–1394.

Jenuwein T and Allis CD (2001) Translating the histone code. *Science* 293: 1074–1080.

Laherty CD, Yang WM, Sun JM, *et al.* (1997) Histone deacetylases associated with the mSin3 corepressor mediate mad transcriptional repression. *Cell* 89: 349–356.

Merika M and Thanos D (2001) Enhanceosomes. *Current Opinion in Genetics and Development* 11: 205–208.

Schubeler D, Francastel C, Cimbora DM, *et al.* (2000) Nuclear localization and histone acetylation: a pathway for chromatin opening and transcriptional activation of the human beta-globin locus. *Genes and Development* 14: 940–950.

Sharpe JA, Wells DJ, Whitelaw E, *et al.* (1993) Analysis of the human alpha-globin gene cluster in transgenic mice. *Proceedings of the National Academy of Sciences of the United States of America* 90: 11262–11266.

Thoma F (1992) Nucleosome positioning. *Biochimica et Biophysica Acta* 1130: 1–19.

Further Reading

Bonifer C (2000) Developmental regulation of eukaryotic gene loci: which *cis*-regulatory information is required? *Trends in Genetics* 16: 310–315.

Brown KE, Amoils S, Horn JM, *et al.* (2001) Expression of alpha- and beta-globin genes occurs within different nuclear domains in haemopoietic cells. *Nature Cell Biology* 3: 602–606.

Engel JD and Tanimoto K (2000) Looping, linking and chromatin activity: new insights into β-globin locus regulation. *Cell* 100: 499–502.

Flint J, Tufarelli C, Peden J, *et al.* (2001) Comparative genome analysis delimits a chromosomal domain and identifies key regulatory elements in the alpha globin cluster. *Human Molecular Genetics* 10: 371–382.

Fry CJ and Peterson L (2001) Chromatin remodelling enzymes: who's on first? *Current Biology* 11: R185–R197.

Gribnau J, Diderich K, Pruzina S, Calzolari R and Fraser P (2000) Intergenic transcription and developmental remodeling of chromatin subdomains in the human beta-globin locus. *Molecular Cell* 5: 377–386.

Higgs DR (1998) Do LCRs open chromatin domains? *Cell* 95: 299–302.

Higgs DR, Sharpe JA and Wood WG (1998) Understanding alpha globin gene expression: a step towards effective gene therapy. *Seminars in Hematology* 35: 93–104.

Li Q, Harju S and Peterson KR (1999) Locus control regions: coming of age at a decade plus. *Trends in Genetics* 15: 403–408.

Titin (*TTN*) Gene

Siegfried Labeit, *Universitätsklinikum, Mannheim, Germany*

Thomas Centner, *Universitätsklinikum, Mannheim, Germany*

Christian Witt, *Universitätsklinikum, Mannheim, Germany*

Dietmar Labeit, *Universitätsklinikum, Mannheim, Germany*

Henk Granzier, *Washington State University, Pullman, Washington, USA*

> **Intermediate article**
>
> **Article contents**
> - Introduction
> - History and Structure of the Titin Protein
> - Organization of the *Titin (TTN)* Gene
> - Genetic Diseases due to *Titin* Mutations

Muscle sarcomeres contain several thousand precisely assembled and tuned proteins that give rise to efficient muscle contraction. Titin is a multifunctional filamentous protein, in charge of sarcomeric assembly and turnover, and providing long-range elasticity.

Introduction

Titin is an unusually large protein that is expressed in all vertebrate striated muscle tissues. A single gene, *titin* (*TTN*), located on chromosome 2q24 in humans, encodes titin. The titin locus expresses up to 100 kilobases (kb) of full-length messenger ribonucleic acids (mRNAs), which are translated into 27 000–33 000 residue giant polypeptides. The titin *N*-terminus is anchored within the Z-disk region of the sarcomere, whereas its *C*-terminus is anchored within the sarcomere's M line lattice. Thus, *in situ* titin polypeptides are 1–2 µm in length and establish a sarcomeric filament system that is critical for myofibrillar integrity and elasticity.

History and Structure of the Titin Protein

Titin was discovered in the early 1980s as a protein of extremely low mobility on low-percentage-gradient gels (for reviews see Gregorio *et al.* (1998) and Granzier and Labeit (2002)). Titin is found in all vertebrate striated muscles and is, after myosin and actin, the third most abundant muscle protein. Development of titin-specific monoclonal antibodies and mapping of their epitopes by immunoelectron microscopy established that titin forms a myofilament system separate from the thick and thin filaments in both skeletal and cardiac muscle (Fürst *et al.*, 1988).

Different parts of titin perform different functions within the myofibril. Titin's *N*-terminal 80 kDa region spans the Z disk and contains multiple binding sites in charge of Z-disk assembly. Titin's I band region contains multiple distinct elastic spring elements that account for the elastic properties of the titin filament system (**Figure 1**). A ~2 MDa segment of titin is located within the A band. This segment is attached to the thick filaments via its multiple binding sites for myosin and C protein. Finally, the 200 kDa *C*-terminal region of the titin polypeptide is an integral component of the M line lattice. Similar to the titin overlap in the Z disk, titins from adjacent half-sarcomeres overlap in the M line region of the sarcomere. As a result, titin molecules form a filament system that is continuous along the full length of the myofibril (Granzier and Labeit, 2002, and references therein). This makes titin ideal for generating and transmitting passive tensions. During myofibrillogenesis, titin could perform a template function in sarcomere assembly.

About 90% of titin's polypeptide mass is contained in 90–100 residue repeats that belong to either the immunoglobulin (Ig) or fibronectin type 3 (FN3) domain superfamilies (Labeit and Kolmerer, 1995). The Ig and FN3 repeats are arranged in different patterns along the molecule (**Figure 2**). Most notably, within titin's A band section, the Ig/FN3 repeats are arranged in regular superrepeat patterns (Labeit and Kolmerer, 1995). About 10% of titin's mass is contained in 17 unique sequence insertions. These sequence insertions encode Z-disk phosphorylation motifs, α-actinin and calpain protease binding sites, an extensible region found exclusively in cardiac titin (within the N2B sequence element) and a binding site for the RING finger protein MURF-1 (Granzier and Labeit, 2002).

Organization of the *Titin (TTN)* Gene

Titin is encoded by a single gene that is located in both humans and mice on the long arm of chromosome 2 (Bang *et al.*, 2001). In humans, the titin gene is 300 kb in size (EMBL data library, see Web Links) and its coding sequence is contained within 363 exons (**Figure 2b**). Only one locus for the titin gene has been identified in vertebrate genomes, and differential

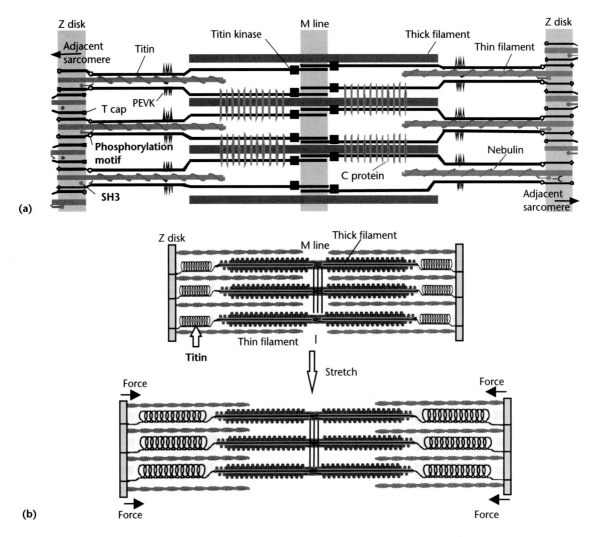

(a)

(b)

Figure 1 [*Figure is also reproduced in color section.*] (a) Schematic overview of the titin filament system of vertebrate striated muscle. Titin's *N*-terminus fully spans the Z disk while its *C*-terminus spans the M line. (b) Schematic diagram indicating titin's spring segment. The spring extends as the sarcomere is stretched and a restoring force ensues (for example, as occurs during filling of the heart). PEVK: spring region of titin rich in proline (P), glutamate (E), valine (V) and lysine (K) residues; SH3: nebulin's *C*-terminal region with homology to scr-3 sarc tyrosine kinase; T cap: brief name for the titin cap protein binding to titin's *N*-terminus.

splicing events control titin's primary structure in a tissue-specific fashion in order to adapt titin's functions to muscle-specific needs. For example, differential splicing cascades adjust the number and length of the expressed spring elements in heart and skeletal muscles. As a result, titin's I band segment varies from ~800 kDa in the stiff cardiac muscle N2B isoform, whereas a large I band titin up to ~1.5 MDa is expressed in the elastic soleus skeletal muscle (**Figure 3a**). On the genomic level, the region of differential splicing comprises a 100 kb segment which contains 161 differentially spliced exons (Freiburg *et al.*, 2000).

More recent studies have shown how different muscle types express distinct titin isoforms (**Figure 3**). In large

mammals, distinct titin isoforms are differentially expressed within different anatomical compartments of the heart. For example, the more compliant N2BA titins are expressed in larger quantities in the atria whereas the stiffer N2B titins dominate in the ventricle (Freiburg *et al.*, 2000; Cazorla *et al.*, 2000).

Genetic Diseases due to *Titin* Mutations

Linkage studies have identified familial cardiac and skeletal muscle diseases that are caused by mutations within the titin gene (Gerull *et al.*, 2002; Hackman

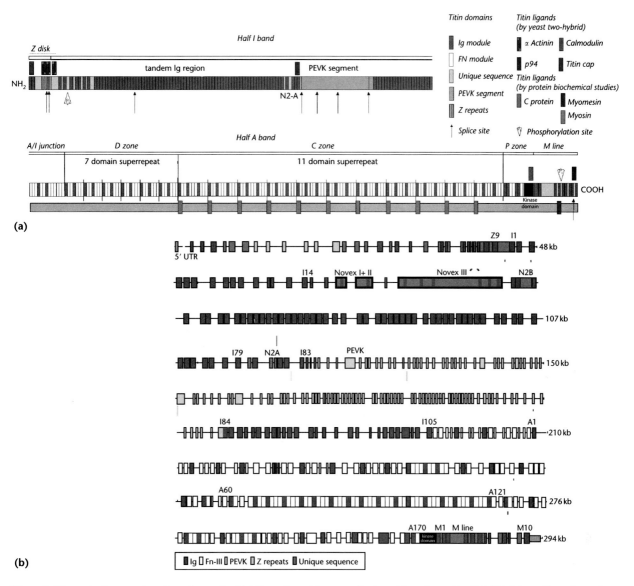

Figure 2 [*Figure is also reproduced in color section.*] (a) Domain architecture of soleus titin polypeptide. (Modified from Gregorio *et al.* (1998).) In addition to the 243 Ig/FN3 repeats with structural roles for Z disk and thick filament assembly, titin also contains nonrepetitive sequence elements. Since these elements include a serine/threonine kinase domain, phosphorylation motifs and calpain protease binding sites, titin appears to have also multiple roles in myofibrillar signal transduction. (b) Exon–intron structure and domain architecture of the human *titin* gene. (Modified from Bang *et al.* (2001).) *Titin* has a total coding mass of 4200 kDa, which is organized in 363 exons. Note novel exons 1–3. (Figure reproduced with the permission of Lippincott, Williams & Wilkins.)

et al., 2002). Titin mutations cause the skeletal muscular dystrophy TMD (tibialis anterior muscular dystrophy), which primarily affects distal leg muscles (Hackman *et al.*, 2002). A missense mutation in the Z-disk titin and a frameshift mutation in the A-band titin was identified in affected family members in two unrelated Caucasian families suffering from dilated cadiomyopathies, (Gerull *et al.*, 2002). Both the cardiac and the skeletal muscle diseases linked to titin are of dominant inheritance. Similarly,

Satoh-Itoh *et al.* (2002) identified two missense *titin* mutations in familial cases of dilated cardiomyopathy in a Japanese population. Interestingly, the Japanese study also identified two mutations in sporadic cases. Therefore, it is possible that *titin* mutations could be a more common cause for dilated cardiomyopathies. The availability of the *titin* gene sequence should facilitate future genetic studies and clarify how frequently mutations in *titin* contribute to human genetic diseases.

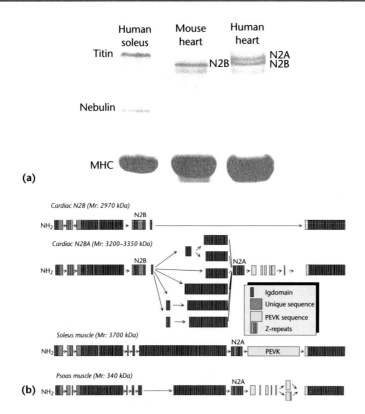

Figure 3 [*Figure is also reproduced in color section.*] (a) Identification of giant titin polypeptides in vertebrate striated muscle tissues. On denaturing 2% polyacrylamide gels, titin bands appear as low-mobility species far above the 220 kDa myosin heavy chain, and the about 800 kDa nebulin band. Note the mobility difference between the 3700 kDa titin from human soleus skeletal muscle, and the 2970 kDa titin from heart muscle (main band in mouse heart and lower titin band in human heart muscle). (b) The different titin polypeptide size classes are caused by the differential processing of titin transcripts by distinct splice pathways (Freiburg *et al.*, 2000). Titin cDNAs from cardiac muscle (top) and soleus and psoas skeletal muscle (bottom) predict titins that have very different I band regions. Identified splice routes are indicated by arrows. Predicted molecular weights of respective isoforms are given on the left.

See also

Muscular Dystrophies

References

Bang M, Centner T, Fornoff F, *et al.* (2001) The complete gene sequence of titin, expression of an unusual ~700 kDa titin isoform and its interaction with obscurin identify a novel Z-line to I band linking system. *Circulation Research* **89**: 1065–1072.

Cazorla O, Freiburg A, Helmes M, *et al.* (2000) Differential expression of cardiac titin isoforms and modulation of cellular stiffness. *Circulation Research* **86**: 59–67.

Freiburg A, Trombitas K, Hell W, *et al.* (2000) Series of exon-skipping events in titin's elastic spring region as the structural basis for myofibrillar elastic diversity. *Circulation Research* **86**: 1114–1121.

Fürst DO, Osborn M, Nave R and Weber K (1988) The organization of titin filaments in the half-sarcomere revealed by monoclonal antibodies in immunoelectron microscopy: a map of ten nonrepetitive epitopes starting at the Z line extends close to the M line. *Journal of Cell Biology* **106**: 1563–1572.

Gerull B, Gramlich M, Atherton J, *et al.* (2002) Mutations of TTN, encoding the giant muscle filament titin, cause familial dilated cardiomyopathy. *Nature Genetics* **30**: 201–204.

Granzier H and Labeit S (2002) Cardiac titin: an adjustable multi-functional spring. *Journal of Physiology* **541**: 335–342.

Gregorio C, Granzier H, Sorimachi H and Labeit S (1998) Muscle assembly: a titanic achievement? *Current Opinion in Cell Biology* **11**: 18–25.

Hackman P, Vihola A, Haravuori H, *et al.* (2002) Tibial muscular dystrophy is a titinopathy caused by mutations in TTN, the gene encoding the giant skeletal-muscle protein titin. *American Journal of Human Genetics* **71**: 492–500.

Labeit S and Kolmerer B (1995) Titins: giant proteins in charge of muscle ultrastructure and elasticity. *Science* **270**: 293–296.

Satoh-Itoh M, Hayashi T, Nishi H, *et al.* (2002) Titin mutations as the molecular basis for dilated cardiomyopathy. *Biochemical and Biophysical Research Communications* **291**: 385–393.

Further Reading

Gregorio CC and Antin PB (2000) To the heart of myofibril assembly. *Trends in Cell Biology* **10**: 355–362.

Maruyama K and Kimura S (2000) Connectin: from regular to giant sizes of sarcomeres. *Advances in Experimental Medicine and Biology* **481**: 25–33.

Trinick J and Tskhovrebova L (1999) Titin: a molecular control freak. *Trends in Cell Biology* **9**: 377–380.

Wang K (1996) Titin/connectin and nebulin: giant protein rulers of muscle structure and function. *Advances in Biophysics* **33**: 123–134.

Web Links

EMBL and NCBI data libraries, titin gene sequence under accession
  AJ277892:
  http://www.ncbi.nlm.nih.gov/
Synopsis of titin's structure and biological features:
  http://www.embl-heidelberg.de/ExternalInfo/Titin/

Transgenic mouse models of titin, link:
  http://titin.mdc-berlin.de/publicdata.htm)
Overview of titin-specific reagents:
  http://www.titin.de

Tjio, Joe-Hin

Introductory article

Maj A Hultén, *University of Warwick, Coventry, UK*

Joe-Hin Tjio (1919–2001) was a Java-born scientist who, in collaboration with Albert Levan, uncovered the correct number of human chromosomes.

Joe-Hin Tjio (**Figure 1**) is in human genetics best known for his contribution to the establishment that the number of chromosomes within a normal somatic human cell is 46 and not 48 as had been the dogma since the work by Hans von Winiwarter in 1912.

Figure 1 Joe-Hin Tjio (1919–2001), *ca.* 1960.

The revised, correct, estimate of 46 chromosomes resulted from a number of improvements in cytological techniques. T. C. Hsu, for example, had found that by using *in vitro* cultured cells that had been exposed to a hypotonic solution, he could make squash preparations in which the chromosomes were generally separate. Albert Levan had pioneered the use of mitotic inhibitors, such as colchicine, to arrest cells at metaphase of mitosis, the cell stage at which chromosomes are most easily counted. He had also experimented on tumor cells using the combination of colchicine treatment and hypotonic solutions. An additional factor in the correct enumeration was Tjio's acknowledged technical skill, including photography of microscopy images.

The crucial study was performed at the Institute of Genetics, University of Lund in Sweden. Fetal lung fibroblasts, cultured *in vitro* by Rune Grubb at the university's Department of Microbiology, were used, and they proved to be excellent source material for squash preparations following colchicine and mild hypotonic pretreatment. Counting chromosome numbers in metaphase plates of the material proved to be straightforward, and it was immediately apparent that human somatic cells had 46 chromosomes. The seminal paper by Tjio and Levan, titled 'The chromosome number of man', was published in 1956. The paper's publication marks the start point of the discipline of clinical cytogenetics, and it has had a major impact on human genetics, including the study of the initiation, progression and treatment of solid cancers, leukemia and allied conditions.

Joe-Hin Tjio was born on 11 February 1919 to Chinese parents and grew up in Java in Indonesia, which was then part of the Dutch West Indies. He was educated in Dutch colonial schools, which required him to learn French, German and English as well as Dutch. He came to cytogenetics indirectly: he first trained as an agronomist before deciding to take up a position as a cytogeneticist at the Botanical Institute in Bogor, Indonesia. However, in 1942 he was interned following the Japanese invasion of Indonesia, and remained imprisoned until 1945. After the war, he managed to get on a Red Cross boat for displaced people to the Netherlands, whose government provided him with a fellowship that allowed him to continue his studies in Europe. He worked as a research assistant in laboratories in both Denmark and Sweden, including that of Levan, before taking up a position as head of cytogenetics in Saragossa, Spain, a position he held from 1948 to 1959. His move to Spain did not, however, end his association with Levan. Tjio made regular study visits to the Institute of Genetics in Lund during the summers and holidays, and Levan encouraged him to extend his interests from plants and insect cytogenetics to the study of mammalian tissue.

Following the publication of the correct chromosome number and presentation of the observations at the First International Human Genetics Congress in Copenhagen in 1956, Tjio received a number of invitations to work in the United States. Although initially reluctant to move to the USA, as he was concerned about McCarthyism, he was eventually persuaded to emigrate by Herman Muller, Nobel laureate in genetics and Professor at Indiana University. In 1957, Tjio

joined Theodore Puck at the University of Colorado. While there, he continued to work on human material and in 1960 he completed his PhD with a paper titled *The Somatic Chromosomes of Man*. He published a number of papers with Puck and A. Robinson on chromosome abnormalities in constitutional genetic defects. He later became especially interested in the Philadelphia chromosome (a marker chromosome resulting from a reciprocal translocation involving chromosomes 9 and 22) in chronic myeloid leukemia.

In 1959, Tjio moved to the National Institutes of Health (NIH) in Bethesda, Maryland, as a visiting scientist, initially at the National Institute of Arthritis and Metabolic Diseases and subsequently at the National Institute of Diabetes and Digestive and Kidney Diseases. At the latter, he was head of the cytogenetics section, a position he held until retirement in 1992. He continued to work after retirement, having received the status of scientist emeritus and retaining his laboratory space and resources until 1997. For many years he applied his cytogenetic expertise in studies on autoimmunity in mouse model systems.

He received a number of awards and honors for his scientific endeavors, including honorary doctoral degrees from the University of Claude Bernard, France, in 1974 and the Science University in Saragossa in 1981. The Kennedy International Award from the Joseph P. Kennedy Jr Foundation that he received in 1962 (**Figure 2**) is probably the most prestigious. It recognized his contribution to our

Figure 2 Photograph taken at the ceremony of the Kennedy International Award in 1962. Tjio is standing to the left of Jerome Lejeune. (Photo courtesy of Tjio.)

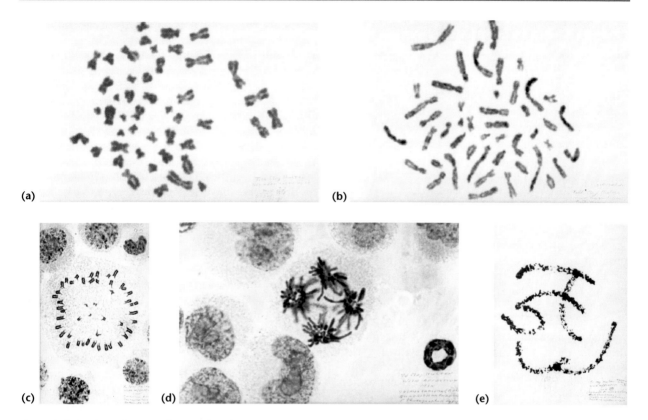

Figure 3 Examples of the microphotographs taken by Tjio. On each photograph he has written, in a combination of Chinese and Swedish, 'to Maj Hultén with affection' followed by the species, the time the photo was taken, and the time it was given to her. (a,b) Metaphase plates of human chromosomes; (c) Ehrlich ascites tumor metaphase; (d) Yoshida sarcoma (rat) quadripolar anaphase; (e) *Luzula purpurea* (2N = 6) late prophase, diffuse centromere.

understanding of genetically determined learning disability.

Tjio had many interests and hobbies. He was proficient at black-and-white photography, a skill taught to him by his father, who was a portrait photographer and had a darkroom where Tjio at a young age learned to develop film and make prints. Later in life Tjio made large copies of his photographs of chromosomes in plants, insects and mammals (**Figure 3**) as well as of his son – which he presented to his numerous friends, and also exhibited at the NIH. Tjio was proud of his skill in many languages, speaking Chinese, English, Spanish, German, Dutch, Danish and Swedish fluently or very nearly so. His wife, Inga, was originally from Iceland. They met during his visit to Denmark in the 1940s. They had one son, Ju-Hin, born in 1970.

See also

Chromosome
Chromosome Analysis and Identification
Levan, Albert
Muller, Herman Joseph

Further Reading

Hultén MA (2002) Numbers, bands and recombination of human chromosomes: historical anecdotes from a Swedish student. *Cytogenetics and Genome* **96**: 14–19.

Levan A (1978) The background of the determination of the human chromosome number. *American Journal of Obstetrics and Gynecology* **130**: 725–726.

Tjio J-H and Levan A (1956) The chromosome number of man. *Hereditas* **42**: 1–6.

Tjio JH (1978) The chromosome number of man. *American Journal of Obstetrics and Gynecology* **130**: 723–724.

Tooth Agenesis

Heleni Vastardis, *Tufts University School of Dental Medicine, Massachusetts, USA*

Evidence supporting the genetic basis of tooth agenesis is being established through identification of human mutations. Two genes, *MSX1* and *PAX9*, have been associated with two different types of tooth agenesis, indicating that the development of differently shaped teeth is determined by different genetic pathways.

Introduction

Tooth agenesis, or failure of development of certain teeth, is the most common anomaly of human dentition. Other terms such as hypodontia, partial anodontia or oligodontia are used interchangeably in the literature to describe numeric dental anomalies. Tooth agenesis occurs as part of a genetic syndrome or as an isolated sporadic or familial finding. Familial tooth agenesis (FTA) is transmitted as an autosomal dominant, autosomal recessive or X-linked trait, although there are cases of FTA with no clear segregation pattern. The condition presents inter-familial and intrafamilial variation in its expression. Maxillary lateral incisors, second premolars and third molars are the most frequently targeted teeth of the permanent dentition, with an incidence that varies by approximately 2.2%, 3.4% and 30% respectively. Deciduous dentition is less frequently affected (0.1%).

The tooth has long been used as a model for understanding organogenesis, as a typical example of an organ undergoing complex morphogenesis regulated by epithelial–mesenchymal interactions. Although considerable information has been accumulated – more than 200 genes have been identified to be expressed during mouse odontogenesis – and the elucidation of the molecular events of odontogenic induction is a work-in-progress, the molecular basis of human tooth development and patterning remains largely undefined. The Human Genome Project has given fresh impetus to the experimental investigation of odontogenesis, providing the art and science of locating genes on chromosomes. Naturally, gene mapping reflects gene-defect mapping via identification of human mutations that cause conditions such as tooth agenesis.

Pinpointing Human Defective Dental Genes – Two Genes for the 'Odontogenic Code'

Although the 200 genes identified in mouse odontogenesis could potentially be candidates for tooth agenesis in humans, only two of them have been associated with the condition. Genetic defects causing tooth agenesis have just recently started to emerge. Genetic linkage studies in a family with autosomal dominant agenesis of second premolars and third molars have identified a locus on chromosome 4p, where the homeobox gene *MSX1* (*msh homeobox homolog 1* (*Drosophila*)) resides (Vastardis *et al.*, 1996).

MSX1 is a transcription factor expressed in several embryonic structures, including the dental mesenchyme. Targeted inactivation of *Msx1* in transgenic mice leads to arrested tooth development at the bud stage. Homozygous *Msx1*-deficient mice exhibit multiple craniofacial abnormalities, including cleft palate, whereas the heterozygotes appear completely normal.

The dental phenotype of the *Msx1*-deficient mice, the expression assays in murine teeth and the availability of the genomic sequence of *MSX1* have prompted researchers to screen for *MSX1* mutations. Sequence analyses have detected a missense mutation (Arg196Pro) in the homeodomain of *MSX1* in all affected family members, providing evidence of involvement of the human *MSX1* in tooth development (Vastardis *et al.*, 1996). Although the pattern of agenesis in this family is bilateral, symmetric and involves second premolars and third molars, some affected members also present phenotypic variability by lacking flanking teeth such as maxillary first premolars or mandibular first permanent molars (**Figure 1**).

This mutation results in a protein that contains an arginine-to-proline substitution at position 31 of the homeodomain of the *MSX1*. To investigate the consequences of this substitution on the structure and function of the gene, biochemical and biological analyses have been performed. These analyses have shown that the single-point mutation causes structural perturbation and reduced thermostability of the MSX1 protein, rendering it inactive. Such data point to haploinsufficiency, that is, insufficient functional protein, as the mechanism causing tooth agenesis in this family (Hu *et al.*, 1998). The selective effect of a single amino acid change in a protein, presumably present in all teeth, on only certain teeth can be best explained in terms of a threshold active during odontogenesis. For a complete tooth to develop, this threshold needs to be overcome. Since tooth agenesis

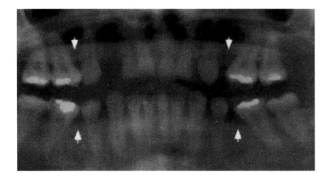

Figure 1 Oral orthopantomograph of a carrier of the Arg196Pro *MSX1* mutation. Note the typical pattern of agenesis, with all second premolars (arrows) and third molars missing. In addition, maxillary first premolars are absent and adjacent teeth are severely rearranged.

in this family segregates as an autosomal dominant trait, inactivation of one copy of *MSX1* deprives tooth primordium of its odontogenic potential and highlights the importance of dosage for mediating the biological actions of MSX1.

Apparently the reduced dosage of MSX1 in other teeth is tolerated, suggesting that morphogenesis of the affected teeth requires a greater amount of MSX1. The idea of an individualized need for MSX1 among different types of teeth is also supported by the clinical observation that, while individuals with tooth agenesis always fail to develop second premolars and third molars, flanking teeth are more variably affected. Alternatively, tooth morphogenesis in the affected family might be particularly susceptible to a reduced MSX1 dosage because of specific effects on genetic background. Interestingly, all affected members report a complete deciduous dentition, suggesting functional redundancy of homeoprotein signals and/or that other genetic mechanisms are involved in the development of the deciduous teeth. This can also be explained by a reduced need for MSX1 in the deciduous dentition or could even be in accordance with the lack of defects in heterozygous mice.

MSX1 has also been associated with syndromic forms of tooth agenesis. A nonsense mutation (Ser105Stop) in the *MSX1* gene, in a family with autosomal dominant tooth agenesis and combinations of cleft palate only and cleft lip and palate, has been identified, providing additional evidence for the importance of this gene in craniofacial development. The clinical image of the affected individuals in this family resembles the phenotype of the *Msx1* knockout mice. The *MSX1* effect on tooth, lip and palate development offers an appreciation of the associations between different organs in the developing body (van den Boorgaard *et al.*, 2000).

Another nonsense mutation (Ser202Stop) in the homeodomain of *MSX1* has been found to cosegregate

with the phenotype of Witkop disease (also called tooth and nail syndrome), suggesting that *MSX1* controls the fate of both nail and tooth buds (Jumlongras *et al.*, 2001).

The dental phenotypes associated with the Ser105-Stop and Ser202Stop mutations are typical for *MSX1* mutations, affecting second premolars and third molars. It has been suggested that tissue-specific differences in the expressivity of the *MSX1* mutations, resulting from a comparable loss of *MSX1* function, can be explained by modifier genes' actions (Jumlongras *et al.*, 2001).

Another mutation (Met61Lys), upstream of the homeodomain of *MSX1*, has been identified in a family presenting a tooth agenesis profile similar to that of the Arg196Pro mutation (Lidral and Reising, 2002).

Based on the clinical information provided in the previous reports, *MSX1* mutations appear responsible for a specific pattern of severe tooth agenesis involving second premolars and third molars. As far as the peripheral dental phenotype is concerned, both Met61Lys and Arg196Pro missense mutations do not produce maxillary molar agenesis. In contrast, the autosomal dominant orofacial clefting and tooth agenesis and the Witkop disease nonsense mutations produce a variable degree of maxillary molar agenesis. Whether the severity and/or extension of the phenotype depends on the location of the genetic defect (complete absence of the homeodomain for the Ser105Stop, clipping of the *C*-terminal of the homeodomain for the Ser202Stop mutation, etc.) or tissue-specific genes work in concert with *MSX1*, modifying its expression, remains to be elucidated.

Other cases of tooth agenesis have not been explained by *MSX1* defects. Five additional families presenting with different types of missing teeth (incisal, premolar or canine agenesis) were evaluated for linkage to *MSX1*. None of these families was found linked to *MSX1*, demonstrating that defects in different genes contribute to the clinical variation of this disorder (Vastardis, 2000). Between 1996 and 2002, two more independent groups failed to associate *MSX1* with familial or sporadic tooth agenesis (Nieminen *et al.*, 1995; Scarel *et al.*, 2000). In particular, studies in Finnish families have shown absence of linkage between *MSX1* and hypodontia (Nieminen *et al.*, 1995). The pattern of hypodontia in these families is a nonuniform one, ranging from numerical and morphological dental malformations, that is, incisor agenesis and premolar agenesis and conically shaped teeth. Additionally, gene carriers exhibit no dental phenotype, suggesting a model of reduced penetrance for these families. These results, however, agree with the genetic heterogeneity hypothesis, namely that more than one gene is responsible for the phenotypic variability of tooth agenesis.

PAX9 (*paired box gene 9*) is the only other defective gene found to cause human tooth agenesis. A family with agenesis of mainly permanent molars, and sporadic agenesis of second premolars and lower permanent lateral incisors is associated with a frameshift mutation of the *PAX9* gene via genome-wide linkage analyses and candidate-gene sequencing (Stockton *et al.*, 2000).

PAX9 is a member of the Pax family of homologous genes that code for the paired box-containing transcription factors. Pax9 is known to be expressed in neural-crest derived mesenchymal cells in the craniofacial region. The need for a functional *Pax9* gene has been shown by the creation of Pax9 null mouse. Homozygote knockouts have secondary cleft palate and other abnormalities in the craniofacial skeleton; they lack all teeth and derivatives of the third and fourth pharyngeal pouches. The heterozygotes with one functional allele appear completely normal.

In a small nuclear family, in which a father and his daughter are affected with agenesis of all primary and permanent molars, evidence has been provided that deletion of the entire *PAX9* gene is the cause of molar agenesis. These data support a model of haploinsufficiency for *PAX9* as the underlying mechanism for hypodontia (Das *et al.*, 2002).

Another nonsense mutation in the *PAX9* gene has been associated with autosomal dominant molar tooth agenesis in a Finnish family. The Arg340Thr transversion creates a stop codon at lysine 114, resulting in a truncated PAX9 protein at the end of the DNA-binding paired box. The tooth agenesis phenotype involves all permanent second and third molars, and the majority of the first molars and resembles the previously reported *PAX9* phenotype, for which haploinsufficiency of *PAX9* is the likely mechanism (Nieminen *et al.*, 2001). In another Finnish family with molar tooth agenesis, similar sequence changes in *PAX9* have not been found (Nieminen *et al.*, 2001).

Attempts to detect more participants in the human odontogenic code have given two more dental loci. Autosomal recessive hypodontia associated with various dental anomalies such as enamel hypoplasia and failure of eruption has been linked to a region on chromosome 16q in one family (Ahmad *et al.*, 1998). There is no report on the specific genetic defect responsible for this type of hypodontia. Another locus on chromosome 10q11.2 has been identified recently as causing agenesis of permanent teeth (Liu *et al.*, 2001).

Conclusions

MSX1 and *PAX9* mutations can explain a fraction of tooth agenesis cases and particularly the most severe ones, where multiple posterior teeth are missing. Gene defects responsible for anterior teeth, that is, incisor and canine agenesis, have not been discovered in humans.

Experimental animal models have given clear indications of patterning in the mouse dentition. To mention a few, Dlx1 and Dlx2 null mutant mice are missing maxillary molars, whereas maxillary and mandibular incisors and mandibular molars are normal. The opposite phenotype can be seen in the activin-βA, activin receptor IIA and IIB, and Smad2 mutant mice.

As knowledge of tooth agenesis genes grows, a deeper understanding of the condition will be obtained in terms of the relation between clinical characteristics – severity and expansion of the phenotype – and sequence features. Interpretation of the genomic data will help in reconstructing odontogenesis and relate protein function to new treatment modalities.

See also

Dental Anomalies: Genetics
Tooth Morphogenesis and Patterning: Molecular Genetics

References

Ahmad W, Brancolini V, Faiyaz ul Haque M, *et al.* (1998) A locus for autosomal recessive hypodontia with associated dental anomalies maps to chromosome 16q12.1. *American Journal of Human Genetics* **62**: 987–991 (letter).

van den Boorgaard M-JH, Dorland M, Beemer FA and Ploos van Amstel HK (2000) *MSX1* mutation is associated with orofacial clefting and tooth agenesis in humans. *Nature Genetics* **24**: 342–343 (letter).

Das P, Stockton DW, Bauer C, *et al.* (2002) Haploinsufficiency of *PAX9* is associated with autosomal dominant hypodontia. *Human Genetics* **110**(4): 371–376.

Hu G, Vastardis H, Bendall AJ, *et al.* (1998) Haploinsufficiency of *MSX1*: a mechanism for selective tooth agenesis. *Molecular and Cellular Biology* **18**(10): 6044–6051.

Jumlongras D, Bei M, Stimson JM, *et al.* (2001) A nonsense mutation in *MSX1* causes Witkop syndrome. *American Journal of Human Genetics* **69**(1): 67–74.

Lidral AC and Reising BC (2002) The role of *MSX1* mutation in human tooth agenesis. *Journal of Dental Research* **81**: 274–278.

Liu W, Wang H, Zhao S, *et al.* (2001) The novel gene locus for agenesis of permanent teeth (He-Zhao deficiency) maps to chromosome 10q11.2. *Journal of Dental Research* **80**(8): 1716–1720.

Nieminen P, Arte S, Pirinen S, Peltonen L and Thesleff I (1995) Gene defect in hypodontia: exclusion of *MSX1* and *MSX2* as candidate genes. *Human Genetics* **96**(3): 305–308.

Nieminen P, Arte S, Tanner D, *et al.* (2001) Identification of a nonsense mutation in the *PAX9* gene in molar oligodontia. *European Journal of Human Genetics* **9**(10): 743–746.

Scarel RM, Trevilatto PC, Di Hipolito Jr O, Camargo LE and Line SR (2000) Absence of mutations in the homeodomain of the *MSX1* gene in patients with hypodontia. *American Journal of Medical Genetics* **92**(5): 346–349.

Stockton DW, Das P, Goldenberg M, D'Souza RN and Patel PI (2000) Mutation of PAX9 is associated with oligodontia. *Nature Genetics* 24: 18–19.

Vastardis H (2000) The genetics of human tooth agenesis: new discoveries for understanding dental anomalies. *American Journal of Orthodontics and Dentofacial Orthopedics* 117(6): 650–656.

Vastardis H, Karimbux N, Guthua SW, Seidman JG and Seidman CE (1996) A human *MSX1* homeodomain missense mutation causes selective tooth agenesis. *Nature Genetics* 13: 417–421.

Further Reading

Ferguson CA, Tucker AS, Christensen L, *et al.* (1998) Activin is an essential early mesenchymal signal in tooth development that is required for patterning of the murine dentition. *Genes and Development* 12(16): 2636–2649.

Ferguson CA, Tucker AS, Heikinheimo K, *et al.* (2001) The role of effectors of the activin signalling pathway, activin receptors IIA and IIB, and Smad2, in patterning of tooth development. *Development* 128(22): 4605–4613.

Peck S, Peck L and Kataja M (2002) Concomitant occurrence of canine malposition and tooth agenesis: evidence of orofacial genetic fields. *American Journal of Orthodontics and Dentofacial Orthopedics* 122: 657–660.

Peters H, Neubuser A, Kratochwil K and Balling R (1998) Pax9-deficient mice lack pharyngeal pouch derivatives and teeth and exhibit craniofacial and limb abnormalities. *Genes and Development* 12(17): 2735–2747.

Satokata I and Maas R (1994) Msx1-deficient mice exhibit cleft palate and abnormalities of craniofacial and tooth development. *Nature Genetics* 6: 348–356.

Thesleff I (2000) Genetic basis of tooth development and dental defects. *Acta Odontologica Scandinavica* 58(5): 191–194.

Thomas BL, Tucker AS, Qui M, *et al.* (1997) Role of *Dlx-1* and *Dlx-2* genes in patterning of the murine dentition. *Development* 124(23): 4811–4818.

Web Links

Gene Expression in Tooth
   http://bite-it.helsinki.fi.

Online Mendelian Inheritance in Man (OMIM)
   http://www.ncbi.nlm.nih.gov/Omim.

MSX1 (*msh homeo box homolog 1 (Drosophila)*); Locus ID: 4487. LocusLink:
   http://www.ncbi.nlm.nih.gov/LocusLink/LocRpt.cgi?l = 4487.

PAX9 (*paired box gene 9*); LocusID: 5083. LocusLink:
   http://www.ncbi.nlm.nih.gov/LocusLink/LocRpt.cgi?l = 5083.

MSX1 (*msh homeo box homolog 1 (Drosophila)*); MIM number: 142983. OMIM:
   http://www.ncbi.nlm.nih.gov/htbin-post/Omim/dispmim?142983.

PAX9 (*paired box gene 9*); MIM number: 167416. OMIM:
   http://www.ncbi.nlm.nih.gov/htbin-post/Omim/dispmim?167416.

Tooth Morphogenesis and Patterning: Molecular Genetics

Abigail Saffron Tucker, *King's College London, London, UK*

Teeth develop by a series of epithelial–mesenchymal interactions that govern not only where the tooth will form within the developing jaw, but also what type of tooth will develop.

Intermediate article

Article contents
• Stages of Tooth Development
• Tooth Initiation
• Tooth Identity
• Bud Formation
• Role of the Enamel Knot

Stages of Tooth Development

Teeth develop on the oral surface of the developing mandible, maxilla and frontal nasal process. The first indication of a developing tooth bud is a thickening of the epithelium. The thickened epithelium invaginates into the underlying neural crest-derived mesenchyme that starts to condense around the forming bud. The epithelial tissue then extends downwards to enclose the condensed mesenchyme, first forming a cap shape, which develops into a bell-shaped tooth germ. The mesenchymal tissue within the tooth germ forms the dental papilla, with those cells in close contact with the inner enamel epithelium differentiating as odontoblasts, producing dentin. The adjacent epithelial cells differentiate as ameloblasts, which produce the enamel matrix. The fully differentiated tooth is then ready to erupt (see **Figure 1**).

Tooth Initiation

The genetic pathways involved in setting up this chain of events are starting to be elucidated using the mouse as the model organism. The first problem is where the tooth buds are to be initiated. The oral/aboral axis of the developing mandible appears to be controlled by the signaling molecule fibroblast growth factor 8 (Fgf8), which is expressed in the oral epithelium from embryonic day (E) 8.75. This induces the expression of genes necessary for tooth development, such as the homeobox genes *Lhx6* (*LIM homeobox protein 6*) and *Lhx8* (*LIM homeobox protein 8*), in the

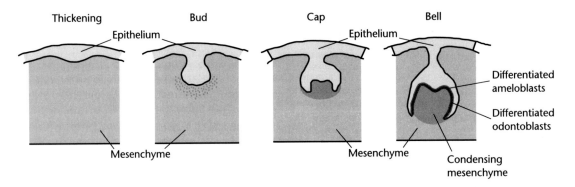

Figure 1 Stages of tooth development. Condensing mesenchyme will form part of the dental papilla.

oral mesenchyme. When Fgf8 is knocked out in the head from E 9 using the Cre/LoxP system, molar teeth are lost but incisors develop normally (Trumpp *et al.*, 1999). The presence of incisors, however, appears to be due to compensation by other fibroblast growth factors (Fgfs) such as Fgf9, as inhibition of Fgf signaling using a chemical block leads to loss of all teeth. Induction of odontogenic genes in the oral mesenchyme restricts the expression of genes necessary for correct skeletal development, such as the homeobox gene *goosecoid* (*Gsc*), to the aboral mesenchyme (Tucker *et al.*, 1998b). The oral–aboral axis of the developing jaw is thus set up to make sure that teeth only develop within the oral cavity. (*See* Transgenic Mice; Mouse Genetics as a Research Tool; Cre-*lox* Inducible Gene Targeting; Fibroblast Growth Factors: Evolution.)

On the oral surface of the mandible in the mouse, four thickenings form, which will develop into the two incisors and two primary molars. The mouse dentition thus represents a much more simplified version of tooth development than that found in humans, where there are four tooth types (incisors, canines, premolars, molars). The exact location of these thickenings in the mouse seems to involve coordinated signaling from four families of signaling factors: the fibroblast growth factors, bone morphogenetic proteins (Bmps), members of the Wingless family (Wnts) and Sonic hedgehog (*Shh*). The expression of these molecules is spatially restricted in the epithelium. *Fgf8* is expressed in the proximal region of the developing jaw, while *Bmp4* is expressed distally. Fgf8 induces the expression of *Pax9*, a paired-box transcription factor, while *Bmp2* and *Bmp4* inhibit its expression. This antagonistic signaling results in *Pax9* expression in mandibular mesenchyme underlying the sites of tooth bud invagination at E 10.5 (Neubüser *et al.*, 1997). In mouse knockouts of *Pax9*, teeth do initiate correctly, however, and arrest does not occur until the bud stage. Thus *Pax9* cannot be responsible for determining the sites of tooth initiation. The spatially restricted expression of Bmps and Fgfs appears to be controlled by *Pitx2* (*paired-like homeodomain transcription factor 2*). Mutations in this gene are responsible for Rieger syndrome in humans, which is characterized by missing teeth. *Pitx2* induces the expression of Fgf8 and represses the expression of *Bmp4* in the oral epithelium, so that in mutant mice the expression of Fgf8 is lost and that of *Bmp4* extends proximally (Lu *et al.*, 1999). (*See* Development: Disorders; Gene Targeting by Homologous Recombination; L Isochore Map: Gene-poor Isochores; Tooth Agenesis.)

Tooth Identity

The early expression of BMPs and FGFs in the oral epithelium appears to have a role in determining not only where the tooth germs will develop, but also what type of tooth will form. Proximally expressed FGFs induce the expression of the homeobox genes *Barx1*, *Dlx1* and *Dlx2* in the underlying proximal mesenchyme, whereas distally expressed BMPs induce the expression of the homeobox genes *Msx1* and *Msx2* in the underlying distal mesenchyme, while inhibiting expression of the proximally expressed homeobox genes. This leads to the establishment of a spatially restricted set of homeobox genes in the developing jaw, which can be thought of as a homeobox code that specifies tooth type. Molars develop from tooth germs that form from mesenchyme expressing *Dlx1/2* and *Barx1*, while incisors develop from tooth germs that form from mesenchyme expressing *Msx1/2*. Evidence for this theory comes from gene knockout and misexpression studies. For example, in *Dlx1* and *Dlx2* double-mutant mice, the incisors develop as normal, but development of maxillary molars is arrested at the epithelial-thickening stage (Thomas *et al.*, 1997). Development of the mandibular molars is normal, but this may be due to compensation for loss of *Dlx1* and

Dlx2 by other members of the Dlx family. *Dlx5* and *Dlx6*, for example, are expressed in mandibular molar mesenchyme, but are absent from maxillary mesenchyme. Loss of *Dlx1/2* leads to loss of *Barx1* in the mesenchyme above the thickening epithelium, and a switch from an odontogenic to a chondrogenic line of differentiation. Double knockouts of *Msx1* and *Msx2* mice result in an arrest of tooth development at the epithelial thickening stage in all teeth, rather than just affecting incisors as would be predicted by the model. This, however, may be due to the fact that, after the initial expression of *Msx1* in the distal mesenchyme of the jaw, it becomes upregulated in the condensing mesenchyme surrounding all teeth. The Msx genes therefore have a later role in general tooth-bud formation as well as a potential earlier role in specifying incisors. The homeobox genes can be misexpressed by interfering with the signals from the epithelium at E 10. If noggin protein-soaked beads are placed in early distal mesenchyme, BMP signaling is blocked in this region. This leads not only to loss of *Msx1*, but also leads to upregulation of *Barx1*, owing to removal of the usual repression in this region. Expansion of *Barx1* into the distal region results in a conversion of incisor tooth germs to a molar fate (Tucker *et al.*, 1998a). Interestingly, in some cases the transformation is not complete and teeth with both incisor and molar characteristics develop. In humans, the formation of premolars and canines in addition to molars and incisors may represent a more complex homeobox code, with overlaps between homeobox genes producing teeth of intermediate character. The initiation of a tooth and the type of tooth formed are thus intrinsically linked. The expression pattern of some of the homeobox genes expressed in the developing mandible is outlined in **Figure 2**. (*See* Gene Families.)

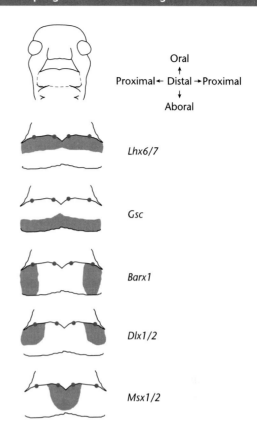

Figure 2 Expression of homeobox genes in the developing mandible at E 10. Frontal view of the head of an E 10 embryo showing the mandible (dashed line). The expression domains of five homeobox genes are represented as gray blocks within a schematic of the developing mandible. The spots indicate the sites where tooth germs will initiate. Tooth germs develop within *Lhx*-expressing mesenchyme and are excluded from *Gsc*-expressing mesenchyme. Tooth germs that develop within *Barx1* and *Dlx* domains develop as molars, while tooth germs that develop within *Msx* domains develop as incisors.

Bud Formation

After the initial role of FGFs and BMPs in coordinating tooth initiation, the Wnts and *Shh* signaling pathways come into operation. *Shh* is first expressed in the oral epithelium at the thickening stage at approximately E 11. Expression then becomes restricted to the tooth germ epithelium and as such, in the mouse, can be seen as four spots in the developing mandible. In a complementary pattern to *Shh*, *Wnt7b* is expressed in the nontooth epithelium and is excluded from the tooth epithelium. Overexpression of Wnt7b in tooth epithelium results in a downregulation of Shh and loss of the tooth germ (Sarkar *et al.*, 2000). This effect can be rescued by addition of *Shh* protein to treated cultures, confirming that the repressive action of

Wnt7b acts specifically on *Shh*. Shh signaling induces proliferation of the epithelium and is thought to thus aid invagination of the epithelium and development of the tooth bud. Addition of ectopic Shh protein next to tooth germs causes the development of additional bud-like invaginations. However, these extra buds do not continue on to form teeth (Hardcastle *et al.*, 1998). Use of an antibody to block Shh signaling at this time leads to arrest of tooth development at the thickening stage. It thus appears that *Shh* is required for the transition from thickening to bud. The transition from thickening to bud is preceded by a shift in the expression of Bmp4 from the epithelium to the mesenchyme at E 11–12. It is at this stage that most of the mesenchymal expression domains become fixed, such that removal of the epithelium has little effect on their expression patterns, and it is the mesenchyme rather than the epithelium that now retains the odontogenic capability.

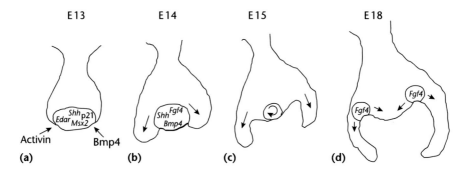

Figure 3 Schematic to show signaling involved in setting up the enamel knot and its action within the tooth germ. (a) Induction of the primary enamel knot (EK) at the bud stage by signals from the mesenchyme. (b, c) Stimulation of proliferation outside of the EK and apoptosis within the EK lead to changes in the shape of the tooth germ epithelium and eventual loss of the primary EK by the end of the cap stage. (d) Formation of secondary EKs in molar tooth germs act to control cusp morphogenesis.

Bmp4 induces its own expression via *Msx1*, and the expression of Bmp4 is intimately linked with that of *Msx1* throughout tooth development. Both genes are expressed in the condensing mesenchyme that surrounds the developing epithelial bud, and loss of *Msx1* leads to loss of *Bmp4* and arrest of tooth development at the bud stage.

Role of the Enamel Knot

The enamel knot, the putative tooth signaling center, forms at the tip of the tooth bud at E 13. This transient structure expresses a large number of signaling molecules, transcription factors and receptors that together define the shape of the developing tooth germ (see **Figure 3**). The enamel knot is thought to be induced by Bmp4 in the surrounding dental mesenchyme. In *Msx1* knockout mice, teeth arrest before enamel knot formation. These arrested teeth can be partially rescued to the cap stage by the addition of Bmp4 protein to cultured tooth germs (Zhao *et al.*, 2000). Bmp4 can also induce the expression of two early markers of the enamel knot, *p21* and *Msx2*, when it is added to isolated dental epithelium (Jernvall *et al.*, 1998). The enamel knot itself represents a group of nonproliferating cells within the invaginated epithelium. These enamel knot cells express high levels of *Shh*, *Fgf4* and *Fgf9*, but low levels of any Fgf receptors, thus allowing the knot to remain in a nonproliferative state, while the epithelial cells surrounding the knot have high proliferation. This results in the distinctive folding of the epithelium seen as the tooth germs enter the cap stage of development at E 14. The role of *Shh* in the enamel knot has been investigated using the Cre/LoxP system to generate a conditional knockout of *Shh*. Loss of *Shh* in the oral epithelium at E 12.5 leads to a reduction in tooth size and defects in cusp morphology, although expression of other enamel knot genes (*Fgf4*, *Bmp2*) is unaffected. *Shh* would therefore appear to play an important role in controlling proliferation at this stage of development.

Defects in enamel knot formation have been shown to be directly responsible for defects in molar cusp morphology. Mutations in *ectodysplasin-A* (*Eda*) result in tooth defects in both humans (hypohidrotic ectodermal dysplasia) and mice (Tabby mutants). Loss of *Eda* in the tooth epithelium leads to a small enamel knot, which in turn results in molar teeth with flattened cusps (Pispa *et al.*, 1999). An almost identical phenotype is observed after mutation of the *Eda* receptor (*Edar*). In this case the cusp defect is due to the enamel knot failing to retain its compact shape and instead forming a long, rope-like structure. This distorted enamel knot still expresses all the usual signaling factors (*Shh*, *Fgf4*), but the change in shape is enough to affect cusp morphology. *Edar* appears to be induced in the enamel knot by activin, a transforming growth factor-beta (TGFβ) family member. *Activin-βA* has a very similar expression pattern to *Pax9* in the mesenchyme underlying the developing tooth germs, and in mutant mice tooth development arrests at the bud stage, except in maxillary molars, where development appears normal. (*See* Craniofacial Abnormalities: Molecular Basis; Dental Anomalies: Genetics; Genetic Disorders; Skin: Hereditary Disorders; Tooth Agenesis.)

In addition to being a region of nonproliferation, the knot represents a region of high apoptosis. This is possibly controlled by *Bmp4*, which, in addition to its expression in the condensing mesenchyme, becomes upregulated in the enamel knot from E 14. The expression of *Bmp4* is associated with sites of apoptosis throughout the embryo. Apoptosis within the enamel knot results in loss of this signaling center by E 15–16, as indicated by loss of expression of *Fgf4*. In addition to the primary enamel knot that is established at the late bud stage, molar tooth germs form secondary enamel knots during the bell stage. These secondary enamel knots are associated with formation of the multiple cusps characteristic of molars and are not seen in incisors. The secondary enamel knots are positioned above where additional cusps will develop and presumably again act to control proliferation and folding of the tooth germ epithelium. The degree and timing of apoptosis within the enamel knot could lead to differences in cusp shape and size. For example, apoptosis in the second molar starts later and is less intense than that of the first molar at a similar stage, which could account for the different sizes of these two teeth.

See also

Dental Anomalies: Genetics

Tooth Agenesis

References

Hardcastle Z, Mo R, Hui C-C and Sharpe PT (1998) The *Shh* signaling pathway in tooth development: defects in Gli2 and Gli3 mutants. *Development* **125**: 2803–2811.

Jernvall J, Åberg T, Kettunen P, Keränen S and Thesleff I (1998) The life history of an embryonic signaling center: *Bmp4* induces p21 and is associated with apoptosis in the mouse tooth enamel knot. *Development* **125**: 161–169.

Lu M-F, Pressman C, Dyer R, Johnson RL and Martin JF (1999) Function of Rieger syndrome gene in left–right asymmetry and craniofacial development. *Nature* **401**: 276–278.

Neubüser A, Peters H, Balling R and Martin GR (1997) Antagonistic interactions between FGF and BMP signaling pathways: a mechanism for positioning the sites of tooth formation. *Cell* **90**: 247–255.

Pispa J, Jung H-S, Jernvall J, *et al.* (1999) Cusp patterning defect in Tabby mouse teeth and its partial rescue by FGF. *Developmental Biology* **216**: 521–534.

Sarkar, L, Cobourne M, Naylor S, *et al.* (2000) Wnt/*Shh* interactions regulate ectodermal boundary formation during mammalian tooth development. *Proceedings of the National Academy of Sciences of the United States of America* **97**: 4520–4524.

Thomas BT, Tucker AS, Qui M, *et al.* (1997) Role of Dlx-1 and Dlx-2 genes in patterning of the murine dentition. *Development* **124**: 4811–4818.

Trumpp A, Depew MJ, Rubenstein JL, Bishop JM and Martin GR (1999) Cre-mediated gene inactivation demonstrates that FGF8 is required for cell survival and patterning of the first branchial arch. *Genes and Development* **13**: 3136–3148.

Tucker AS, Matthews KL and Sharpe PT (1998a) Transformation of tooth type induced by inhibition of BMP signaling. *Science* **82**: 1136–1138.

Tucker AS, Yamada G, Grigoriou M, Pachnis V and Sharpe PT (1998b) Fgf8 determines rostral–caudal polarity in the first branchial arch. *Development* **126**: 51–61.

Zhao X, Zhang Z, Song Y, *et al.* (2000) Transgenically ectopic expression of *Bmp4* to the *Msx1* mutant dental mesenchyme restores downstream gene expression but represses *Shh* and *Bmp2* in the enamel knot of wild type tooth germs. *Mechanisms of Development* **99**: 29–38.

Further Reading

Chen Y, Bei M, Woo I, Satokata I and Maas R (1996) *Msx1* controls inductive signaling in mammalian tooth morphogenesis. *Development* **122**: 3035–3044.

Dassule M, Lewis P, Bei M, Maas R and McMahon A (2000) Sonic hedgehog regulates growth and morphogenesis of the tooth. *Development* **127**: 4775–4785.

Depew M, Tucker AS and Sharpe PT (2002) Craniofacial development. In: Janet R and Patrick T (eds.) *Mouse Development: Patterning, Morphogenesis and Organogenesis*, pp. 436–441, 454–456. New York, NY: Academic Press.

Ferguson CA, Tucker AS and Sharpe PT (2000) Temporo-spatial cell interactions regulating mandibular and maxillary arch patterning. *Development* **127**: 403–412.

Jernvall J and Thesleff I (2000) Reiterative signaling and patterning during mammalian tooth morphogenesis. *Mechanisms of Development* **92**: 19–29.

Jernvall J, Kettunen P, Karavanova I, Martin LB and Thesleff I (1994) Evidence for the role of the enamel knot as a control center in mammalian tooth cusp formation: non-dividing cells express growth stimulating *Fgf4* gene. *International Journal of Developmental Biology* **38**: 463–469.

Ruch JV (1995) Tooth crown morphogenesis and cytodifferentiations: candid questions and critical comments. *Connective Tissue Research* **32**: 17–25.

Thomas BT and Sharpe PT (1998) Patterning of the murine dentition by homeobox genes. *European Journal of Oral Science* **106**: 48–54.

Tucker AS, Headon DJ, Schneider P, *et al.* (2000) Edar/Eda interactions regulate enamel knot formation in tooth morphogenesis. *Development* **127**: 4691–4700.

Vaahtokari A, Åberg T, Jernvall J, Keränen S and Thesleff I (1996) The enamel knot as a signaling center in the developing mouse tooth. *Mechanisms of Development* **54**: 39–43.

Web Links

Gene Expression in Tooth. Gene expression patterns of a wide variety of signaling molecules, transcription factors etc. involved in tooth development from the epithelial thickening stage to the bell stage of development
http://bite-it.helsinki.fi/

Topoisomerases

Intermediate article

Article contents

- Introduction
- Classification of Topoisomerases
- Human Topoisomerases: their Genes and Function

Andrzej Stasiak, *University of Lausanne, Lausanne-Dorigny, Switzerland*

Topoisomerases are specialized enzymes that can change the topological state of the DNA by transiently cleaving one or two DNA strands and allowing a passage of single- or double-stranded DNA segments through the cleavage site. After the passage the cleavage site is resealed.

Introduction

Topoisomerases, as their name suggests, catalyze interconversions between different deoxyribonucleic acid (DNA) topoisomers (covalently closed circular DNA molecules of identical length and sequence but with different number of turns of their double helix). Such interconversion reactions require the temporary cutting of one or both strands of the DNA. This cutting then permits a complex series of movements of the DNA strands, which leads ultimately to the re-establishment of the original connectivity within the transiently cut strands.

The ability to change the topological state of DNA is necessary for many fundamental processes, such as DNA replication, transcription and recombination, chromosome division and chromosome condensation and decondensation. Because each of these processes has different topological requirements, several different topoisomerases have evolved to fulfill best their function. (*See* Chromosomes during Cell Division; DNA Coiling and Unwinding; DNA Replication; DNA Recombination; Transcriptional Regulation: Coordination.)

Classification of Topoisomerases

Currently known topoisomerases are divided into two general types, I and II, which are each additionally divided into two subtypes, A and B, that strongly differ in sequence and structure (Champoux, 2001). It is easy to remember that type I topoisomerases cleave only one strand of DNA, whereas type II topoisomerases cleave both strands.

All topoisomerases use the O4 oxygen of a tyrosine side chain for a nucleophilic displacement attack on a phosphorus atom in the DNA chain. Such an attack results in the covalent attachment of the involved polypeptide chain to the 5′ end of the cut DNA strand in the case of subtypes IA, IIA and IIB. For subtype IB, the nucleophilic displacement results in DNA cleavage with the phosphate positioned at the 3′ end, to which the topoisomerase is then attached.

For all topoisomerases, the energy of the cleaved phosphodiester bond in the DNA is maintained in the phosphodiester bond to the tyrosine, and each cycle of reaction finishes with the closure of the DNA cleavage and the elimination of the covalent bond between the topoisomerase and DNA. The cleavage and closure are therefore isoenergetic reactions.

Topoisomerases I

Subtypes IA and IB differ significantly from each other in the actual mechanism that introduces the topological change (Keck and Berger, 1999).

Subtype IA

In reactions mediated by subtype IA topoisomerases, the uncut strand traverses the opening in the transiently opened strand, after which the cut site recloses. The resulting change of the number of helical turns in DNA is one per catalytic cycle. With the exception of a specific class of reverse gyrases, all other topoisomerases of subtype IA do not require adenosine triphosphate (ATP) but need Mg^{2+} ions for their activity, and relax only DNA molecules that are significantly negatively supercoiled and therefore have a tendency to separate their strands.

The relaxation reaction is thought to proceed according to the enzyme-bridging model (Champoux, 2001). On cleavage, the 5′-ended strand of the cut site is bound covalently to the active tyrosine and the second end is firmly nested in a nucleotide-binding site in a distal domain of the protein. A conformational change in the enzyme then moves the cut ends away from each other, creating an enzyme-bridged gap in which the topoisomerase has the form of an arc that spans the gap. At this point of the reaction, the intact strand is attracted towards the interior of the topoisomerase arc, which provokes a new conformational change that brings the cleaved ends together and results in religation of the cut strand. The directionality and driving force for the reaction is provided by the relaxation of negative supercoils (**Figure 1a**).

For reverse gyrases, which are specialized topoisomerases of subtype IA that are present only in thermophilic bacteria and archaea, ATP is required for activity, which introduces positive supercoiling into the DNA. Positive supercoiling stabilizes DNA molecules by significantly increasing their melting

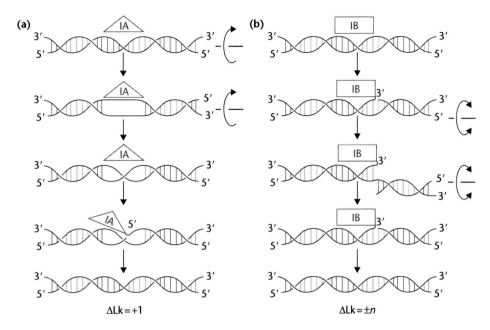

Figure 1 Principal mechanistic differences between topoisomerases of subtypes IA and IB. To visualize DNA rotation in static drawings, the bases visible from a major or minor groove side are shown in different shades. For easier presentation, the left side of each shown helix is immobilized and the rotation concerns the right side. Only small regions of a long covalently closed circular DNA molecule are shown. (a) The negative supercoiling and helix destabilizing properties of subtype IA topoisomerases contribute together to local unwinding and rotation of the helix. This rotation leads to the relaxation of negative supercoiling in the remaining portion of the molecule (not shown). A destabilized helix under the torsional tension of negative supercoiling has a tendency to coil in a left-handed direction and this further relaxes negative supercoiling in the molecule. Such an arrangement of DNA strands is a good substrate for the actual strand-passage reaction that is driven by the formation of Watson–Crick base pairs in previously unpaired DNA regions. During the strand-passage reaction, the topoisomerase is attached covalently to the 5′ end of the cleaved strand. The resulting change in the number of helical turns, which is more precisely defined as the linking number (Lk) in the DNA, is +1. A DNA molecule that is negatively supercoiled and has a deficit of Watson–Crick helical turns gains one turn and decreases its torsional stress. The model shown explains why subtype IA topoisomerases can relax only DNA with a significant level of negative supercoiling and why the relaxation reaction does not proceed to completion. (b) A subtype IB topoisomerase cleaves one strand and attaches itself to the 3′ end of the cleaved strand. The topoisomerase does not firmly bridge both cleaved ends and therefore a torsional stress of positive or negative supercoiling can lead to one or more rotations around the uncut strand. The model shown explains why topoisomerases of subtype IB relax to completion both positively and negatively supercoiled DNA molecules and why the change in the linking number can be more than one per catalytic cycle (cleavage and closure).

temperature, which is essential for organisms that live at temperatures higher than the melting point of negatively supercoiled or relaxed DNA molecules. (*See* DNA Coiling and Unwinding; DNA Structure.)

Subtype IB

For topoisomerases of subtype IB, the reaction proceeds by a swiveling mechanism (Stewart *et al.*, 1998) in which the uncut strand serves as an axis of rotation. DNA molecules under significant stress can rotate several times before the cut site is closed again. The relaxation reaction proceeds according to the controlled rotation mechanism (**Figure 1b**). During DNA cleavage, the 3′-ended strand of the cut site is bound covalently to the active tyrosine and the other end remains relatively free, although the whole cleavage region is enclosed in the grip of the topoisomerase. After the cleavage, however, the grip

of topoisomerase weakens and the unbound end can start to rotate if there is positive or negative torsional stress in the DNA molecule. After a while, the topoisomerase grip becomes tighter again and this stops the rotation when the cleaved ends are in proper apposition for their religation. At this point the topoisomerase can either dissociate or initiate another round of reaction, which leads eventually to complete relaxation of the torsional stress in the molecule. The smallest number of rotations per relaxation cycle (from cleavage to religation) is one, but frequently more rotations can occur. For example, for subtype IB topoisomerases of vaccinia virus, on average five rotations occur during each cleavage and closure cycle (Champoux, 2001). Subtype IB topoisomerases do not require ATP and Mg^{2+} ions for their activity and can relax equally well torsional stress caused by under- or overwinding of the DNA helix.

Topoisomerases II

For type II topoisomerases, the reaction proceeds by the creation of a transient double-strand break that opens a gate for the transport of a double-stranded DNA segment that lies proximal or distal to the cut site. Eukaryotic type II topoisomerases assume a dyad symmetry structure through a homodimeric arrangement, and prokaryotic A_2B_2 tetramers adopt a very similar structure. This structural similarity is explained by the fact that eukaryotic topoisomerase II proteins evolved in such a way that they correspond to a proper-order fusion of A and B subunits of prokaryotic topoisomerases of subtype IIA.

The functional enzyme with the dyad symmetry arrangement binds the DNA that will be cleaved in the reaction. This bound region of DNA is called a G segment because it serves as a passage 'gate' for the 'transported' T segment. After initial binding, the topoisomerase captures the T segment (**Figure 2**). Subsequently, both strands of the G segment are cleaved with a four-base stagger by subtype IIA enzymes (subtype IIB topoisomerases can create a two-base stagger). The cleavage reaction leads to formation of covalent bonds involving both of the 5′ ends of the cleaved strands and two active tyrosines in dyad symmetry positions. A conformational change of the enzyme then moves the cut ends apart from each other to form the gate, which is supported by a partially open symmetric arrangement of the protein subunits (**Figure 2a**). The T segment then moves across the opened gate.

When the transfer reaction is intramolecular, it can lead to knotting or unknotting, but most frequently it results simply in a change in the number of DNA helical turns by exactly two. When the transferred T segment is from another DNA molecule, the reaction can lead to catenation or decatenation. After the transfer, the gate closes and both cut strands religate. Although the cleavage and religation do not require energy input, DNA transport in reactions by type II topoisomerases is unidirectional and this necessitates energy input; therefore, type II topoisomerases require ATP in addition to Mg^{2+} ions for their activity (Vologodskii et al., 2001).

Most known type II topoisomerases are subtype IIA; subtype IIB topoisomerases are only found in archaea, as exemplified by DNA topoisomerase VI from *Sulfolobus shibatae*. With the exception of bacterial DNA gyrases, all remaining topoisomerases that are subtype IIA show only relaxation activity and can relax positive and negative supercoils. Bacterial gyrases are specialized topoisomerases that actively introduce negative supercoiling into DNA because this facilitates all processes that require strand separation of the DNA, such as replication, transcription and

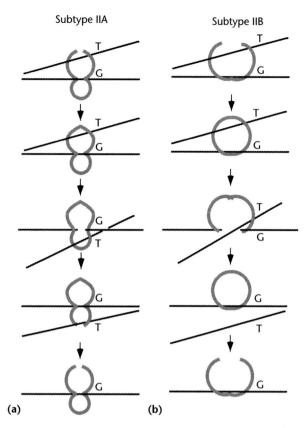

Subtype IIA   Subtype IIB

(a)   **(b)**

Figure 2 Principal mechanistic differences between topoisomerases of subtypes IIA and IIB. Both subtypes have a similar mechanism of double-stranded DNA cleavage but differ significantly in their structure. (a) Subtype IIA topoisomerases contain a cavity that accommodates the transported double-stranded DNA segment after the passage. (b) Subtype IIB topoisomerases lack such a cavity. The gate segment and the transported segment are denoted by G and T respectively and are both double-stranded.

recombination. In subtype IIA topoisomerases, the transported segment traverses not only the gate in the DNA but also a special exit gate placed between the symmetry-related subunits of the topoisomerase (**Figure 2a**). DNA topoisomerases of subtype IIB lack the region that corresponds to the exit gate in topoisomerases of subtype IIA (**Figure 2b**).

Human Topoisomerases: their Genes and Function

As the sequencing of the human genome nears its completion, five human topoisomerases have been identified. One belonging to subtype IB is known as topoisomerase I and is encoded by the gene *topoisomerase (DNA) I* having the gene symbol *TOP1*. Two topoisomerases of subtype IA, known as topoisomerase IIIα and topoisomerase IIIβ, are encoded by the genes *topoisomerase (DNA) III alpha* (*TOP3A*) and *topoisomerase III (DNA) beta* (*TOP3B*) respectively. There are also two topoisomerases of subtype IIA, topoisomerase IIα and topoisomerase IIβ, which are encoded by the genes *topoisomerase (DNA) II alpha* (*TOP2A*) and *topoisomerase (DNA) II beta* (*TOP2B*) respectively.

TOP1

Human topoisomerase I (subtype IB) creates swivel sites for the controlled rotation of DNA during DNA replication and transcription, and chromosome segregation and condensation (Li and Liu, 2001). The protein acts as a monomer of 765 amino acids (91 kDa) and its gene *TOP1* is located on human chromosome 20q12–13.1; the coding sequence is split into 21 exons and extends over at least 85 kilobases (kb).

TOP1 is essential for an organism's survival, as shown in studies of knockout mice (Morham *et al.*, 1996). Abnormalities of human topoisomerase I have been connected with genetic disorders such as ataxia telangiectasia and Fanconi anemia. Human topoisomerase I has been shown to interact with various nuclear proteins such as the Werner syndrome gene product (WRN), p53, T-antigen, TBP, HMG1, HMG2 and UBC9 (Li and Liu, 2001). (*See* Chromatin in the Cell Nucleus: Higher-order Organization; DNA Helicase-deficiency Disorders.)

TOP3A, TOP3B

Human topoisomerase IIIα (subtype IA) acts as a monomer of 1001 amino acids, and its gene *TOP3A* maps to chromosome 17p11.2–12 (Li and Liu, 2001). Studies in mice have shown that topoisomerase IIIα is required in early embryogenesis (Li and Wang, 1998), and it has been proposed that the enzyme is required to unlink precatenanes that are formed during DNA replication (Champoux, 2001). Topoisomerase IIIα interacts with the Bloom syndrome gene product (BLM), which is a RecQ family helicase (Li and Liu, 2001). A complex of topoisomerase IIIα with BLM has been proposed to be involved in the regulation of recombination in somatic cells.

Human topoisomerase IIIβ (subtype IA) comprises 862 amino acids, and its gene *TOP3B* maps to chromosome 22q11. Studies in mouse have shown that, in contrast to the embryonic lethality of mutants lacking topoisomerase IIIα, mutants that lack topoisomerase IIIβ are viable and reach maturity without visible deficiencies other than a significantly shorter life span. As the mutant mice lacking topoisomerase IIIβ grow older, however, they show lesions in several organs including hypertrophy of the spleen and submandib-

ular lymph nodes, glomerulonephritis and perivascular infiltrates in various organs.

Topoisomerase IIIα and IIIβ interact with human RecQ DNA helicase (Li and Liu, 2001), and defects in these helicases are implicated in human progeroid syndromes such as Bloom syndrome, Werner syndrome and Rothmund–Thomson syndrome. Topoisomerase IIIβ presumably cooperates with RecQ helicases during DNA repair and recombination; therefore, the lack of this topoisomerase may lead to premature aging owing to defects in DNA repair that frequently involve DNA recombination. The involvement of topoisomerase IIIβ in DNA recombination is supported by the observation that its gene is highly expressed in the testis and shows a considerable increase in expression 17 days after birth, when cells in the pachytene phase start to appear (Seki *et al.*, 1998). (*See* DNA Helicase-deficiency Disorders; DNA Repair; DNA Repair: Disorders; Meiosis; Transcription-coupled DNA Repair.)

TOP2A, TOP2B

Human topoisomerase IIα and topoisomerase IIβ (subtype IIA) are two isozymes that share 72% amino acid identity. Topoisomerase IIα is encoded by the gene *TOP2A*, which is located on chromosome 17q21–22. *TOP2A* spans about 30 kb and contains 35 exons. The protein has 1531 amino acids and acts as a dimer. Topoisomerase IIα functions in DNA replication and transcription, and chromosome condensation and segregation, and is essential for cell growth (Li and Liu, 2001). The concentration of topoisomerase IIα is highest in the late S/G2 phase of the cell cycle and it is the principal component of mitotic chromosome scaffolds, as is expected for a protein that functions in chromosome condensation and segregation (Woessner *et al.*, 1991). Topoisomerase IIα associates with the ribonucleic acid (RNA) polymerase II holoenzyme, and *in vitro* studies have shown that topoisomerase IIα activity is required for transcription on chromatin templates but not for transcription on protein-free DNA templates.

Human topoisomerase IIβ has 1626 amino acids and is coded by the gene *TOP2B*, which is located on chromosome 3p24. The gene spans more than 49 kb and contains 36 exons. In contrast to topoisomerase IIα, constant amounts of topoisomerase IIβ are expressed during the cell cycle (Woessner *et al.*, 1991). The precise function of topoisomerase IIβ is not known as yet. Unexpectedly, topoisomerase IIβ knockout mice are born alive but show defects in development and growth of motor axons. The pups also have a strong breathing impairment that leads to their death shortly after birth (Yang *et al.*, 2000). Cell lines from the knockout mice can be established, and these show that in the presence of topoisomerase IIα, the function of topoisomerase IIβ is not essential for cell growth. (*See* Cell Cycle: Chromosomal Organization; Mitosis: Chromosomal Rearrangements.)

References

Champoux JJ (2001) DNA topoisomerases: structure, function and mechanism. *Annual Reviews in Biochemistry* **70**: 369–413.

Keck JL and Berger JM (1999) Enzymes that push DNA around. *Nature Structural Biology* **6**: 900–902.

Li T-K and Liu LF (2001) Tumor cell death induced by topoisomeraes-targeting drugs. *Annual Reviews in Pharmacology and Toxicology* **41**: 53–77.

Li W and Wang JC (1998) Mammalian DNA topoisomerase IIIα is essential in early embryogenesis. *Proceedings of the National Academy of Sciences of the United States of America* **95**: 1010–1013.

Morham SG, Kluckman KD, Voulomanos N and Smithies O (1996) Targeted disruption of the mouse topoisomerase I gene by camptothecin selection. *Molecular and Cellular Biology* **16**: 6804–6809.

Seki T, Seki M, Onodera R, Katada T and Enomoto T (1998) Cloning of cDNA encoding a novel mouse DNA topoisomerase III (Topo III-β) possessing negatively supercoiled DNA relaxing activity, whose message is highly expressed in the testis. *Journal of Biological Chemistry* **273**: 28 553–28 556.

Stewart L, Redinbo MR, Qiu X, Hol WG and Champoux JJ (1998) A model for the mechanism of human topoisomerase I. *Science* **279**: 1534–1541.

Vologodskii AV, Zhang W, Rybenkov VV, *et al.* (2001) Mechanism of topology simplification by type II DNA topoisomerases. *Proceedings of the National Academy of Sciences of the United States of America* **98**: 3045–3049.

Woessner RD, Mattern MR, Mirabelli CK, Johnson RK and Drake FH (1991) Proliferation- and cell cycle-dependent differences in expression of the 170 kilodalton and 180 kilodalton forms of topoisomerase II in NIH-3T3 cells. *Cell Growth and Differentiation* **2**: 209–214.

Yang X, Li W, Prescott ED, Burden SJ and Wang JC (2000) DNA topoisomerase IIβ and neural development. *Science* **287**: 131–134.

Further Reading

Berger JM (1998) Type II DNA topoisomerases. *Current Opinion in Structural Biology* **8**: 26–32.

Collins I, Weber A and Levens D (2001) Transcriptional consequences of topoisomerase inhibition. *Molecular and Cellular Biology* **21**: 8437–8451.

Durrieu F, Samejima K, Fortune JM, *et al.* (2000) DNA topoisomerase IIα interacts with CAD nuclease and is involved in chromatin condensation during apoptotic execution. *Current Biology* **10**: 923–936.

Fass D, Bogden CE and Berger JM (1999) Quaternary changes in topoisomerase II may direct orthogonal movement of two DNA strands. *Nature Structural Biology* **6**: 322–326.

Hu P, Beresten SF, van Brabant AJ, *et al.* (2001) Evidence for BLM and topoisomerase III-alpha interaction in genomic stability. *Human Molecular Genetics* **10**: 1287–1298.

Mao Y, Sun M, Desai SD and Liu LF (2000) SUMO-1 conjugation to topoisomerase I: a possible repair response to topoisomerase-mediated DNA damage. *Proceedings of the National Academy of Sciences of the United States of America* **97**: 4046–4051.

Mondal N and Parvin JD (2001) DNA topoisomerase IIα is required for RNA polymerase II transcription on chromatin templates. *Nature* **413**: 435–438.

Stasiak A (2000) Feeling the pulse of a DNA topoisomerase. *Current Biology* **10**: R526–R528.

Strick TR, Croquette V and Bensimon D (2000) Single-molecule analysis of DNA uncoiling by a type II topoisomerase. *Nature* **404**: 901–904.

Wu L, Davies SL, North PS, *et al.* (2000) The Bloom's syndrome gene product interacts with topoisomerase III. *Journal of Biological Chemistry* **275**: 9636–9644.

Web Links

Topoisomerase (DNA) I (*TOP1*); LocusID: 7150. LocusLink:
http://www.ncbi.nlm.nih.gov/LocusLink/LocRpt.cgi?l = 7150

Topoisomerase (DNA) III alpha (*TOP3A*); LocusID: 7156. LocusLink:
http://www.ncbi.nlm.nih.gov/LocusLink/LocRpt.cgi?l = 7156

Topoisomerase (DNA) III beta (*TOP3B*); LocusID: 8940. LocusLink:
http://www.ncbi.nlm.nih.gov/LocusLink/LocRpt.cgi?l = 8940

Topoisomerase (DNA) II alpha (*TOP2A*); LocusID: 7153. LocusLink:
http://www.ncbi.nlm.nih.gov/LocusLink/LocRpt.cgi?l = 7153

Topoisomerase (DNA) II beta (*TOP2B*); LocusID: 7155. LocusLink:
http://www.ncbi.nlm.nih.gov/LocusLink/LocRpt.cgi?l = 7155

Topoisomerase (DNA) I (*TOP1*); MIM number: 126420. OMIM:
http://www.ncbi.nlm.nih.gov/htbin-post/Omim/
dispmim = 126420

Topoisomerase (DNA) III alpha (*TOP3A*); MIM number: 601243. OMIM:
http://www.ncbi.nlm.nih.gov/htbin-post/Omim/
dispmim?601243

Topoisomerase (DNA) III beta (*TOP3B*); MIM number: 603582. OMIM:
http://www.ncbi.nlm.nih.gov/htbin-post/Omim/
dispmim?603582

Topoisomerase (DNA) II alpha (*TOP2A*); MIM number: 126430. OMIM:
http://www.ncbi.nlm.nih.gov/htbin-post/Omim/
dispmim?126430

Topoisomerase (DNA) II beta (*TOP2B*); MIM number: 126431. OMIM:
http://www.ncbi.nlm.nih.gov/htbin-post/Omim/
dispmim?126431

Tp53, Tumor Protein p53 Gene

See Apoptosis and the Cell Cycle in Human Disease
Apoptosis: Regulatory Genes and Disease
Li–Fraumeni Syndrome
Tumor Suppressor Genes

Transcriptional Channelopathies of the Nervous System

Stephen G Waxman, *Yale School of Medicine, New Haven, Connecticut, and VA Hospital, West Haven, Connecticut, USA*

Recent studies have provided evidence for the existence of transcriptional channelopathies, which result from dysregulated expression of nonmutated channel genes. Examples of this new class of disorders are provided by peripheral nerve injury which triggers spinal sensory neurons to turn off some previously active sodium channel genes and turn on other previously silent sodium channel genes, a set of changes that can produce hyperexcitability of these cells, and by changes in sodium channel gene expression that may perturb cerebellar function in multiple sclerosis.

Introduction

Channelopathies are defined as disorders in which abnormal ion channel function results in clinical signs and symptoms (Rose and Griggs, 2001). In the genetic channelopathies, channel protein structure is altered so that channels function abnormally or fail to function as a result of mutations of ion channel genes. An example is provided by several forms of epilepsy in animal models and in humans. The acquired channelopathies include autoimmune channelopathies in which antibodies perturb channel function. Examples include the Lambert–Eaton myasthenic syndrome in which antibodies against presynaptic calcium channels impair transmitter release at the neuromuscular junction, producing weakness. By contrast, the production of an aberrant repertoire of channels whose protein structure is not abnormal, as a result of dysregulated transcription of (nonmutated) channel genes, occurs in the transcriptional channelopathies (**Figure 1**), which can affect the function of the nervous system in important ways (Waxman, 2001).

Sodium Channel Gene Transcription: Dynamic and Complex

Sodium channels play crucial roles in the generation and transmission of electrical signals by neurons. Molecular analysis has now shown that 10 genes encode distinct voltage-gated sodium channels, all sharing a common overall structural motif but with different amino acid sequences. Different sodium channels can exhibit different physiological properties. The selective expression of different ensembles of sodium channels within distinct types of neurons endows them with different functional properties such as threshold, refractory period and firing patterns (Waxman, 2000). It is not surprising, therefore, that changes in sodium channel transcription in disease states can have significant effects on neuronal function.

The transcription of sodium channel genes within neurons is a dynamic process. During development, the levels of transcription for some sodium channels (e.g. sodium channel, voltage-gated, type I, alpha

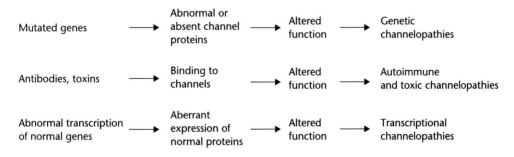

Figure 1 Channelopathies can occur as a result of several types of molecular pathology. Genetic channelopathies are the result of mutations in the genes encoding channel proteins. In the autoimmune and toxic channelopathies, the binding of autoantibodies or toxins to channels alters their function. Transcriptional channelopathies are due to dysregulated expression of nonmutated genes which results in the production of an abnormal repertoire of channels whose protein structure is not abnormal.

polypeptide (*SCN1A*) (Na$_v$1.1), *SCN2A1* (Na$_v$1.2) and *SCN8A* (Na$_v$1.6)) rise while transcription of others (e.g. *SCN3A* (Na$_v$1.3)) falls. Many factors, including trophic factors such as nerve growth factor (NGF) and glial cell line-derived neurotrophic factor (GDNF), participate in regulating the transcription of channel genes. Electrical activity may also modulate the expression of sodium channels within neurons. In addition, there is evidence that in some types of neurons, such as the magnocellular neurosecretory neurons within the hypothalamic supraoptic nucleus, the transition from a quiescent to a bursting state is associated with a change in sodium channel expression (Tanaka *et al.*, 1999).

Nerve Injury

Since ion channel expression is dynamic in normal neurons, it is not surprising that neuronal injury can trigger changes in channel gene expression. Nerve injury and multiple sclerosis (MS) provide two examples of neurological disorders associated with changes in sodium channel gene transcription (Waxman, 2001).

Evidence from animal models and humans has revealed a transcriptional channelopathy that is associated with nerve injury and suggests that this channelopathy can contribute to neuropathic pain. Neuropathic pain arises, at least in part, from spontaneous, sustained firing of injured axons and the spinal sensory neurons from which they arise (Devor and Seltzer, 1999; Zhang *et al.*, 1997). Early electrophysiological studies suggested that altered sodium channel activity can contribute to hyperexcitability of injured nerves. For example, **Figure 2** shows an intracellular recording from a previously injured axon in rat sciatic nerve that had regenerated for 1 year following injury, and shows aberrant repetitive action potential activity which does not occur in uninjured axons (Kocsis and Waxman, 1983). A prolonged or 'slow' depolarization lasts for more than 10 ms, and underlies the abnormal repetitive activity. The long-lasting duration of the depolarization suggested that it might be produced by the opening of a slow or 'persistent' sodium channel, but recordings of the slow depolarization could not prove that it was a result of the activity of a newly expressed channel. These early studies could not provide information about the pathological mechanisms (altered gene transcription, altered translation or abnormal channel insertion into the membrane?) or the molecular identity of the channels that were involved.

Molecular studies have demonstrated that nerve injury triggers dysregulation of sodium channel gene expression. Much of this work has utilized spinal

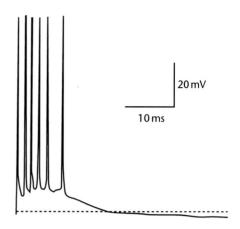

Figure 2 Spinal sensory neurons and their axons can become hyperexcitable after nerve injury. This intracellular recording shows repetitive action potential activity in a previously transected, regenerating axon from rat sciatic nerve (1 year postcrush) following blockade of potassium channels with 4-aminopyridine. The repetitive impulses arise from a prolonged depolarization that follows the first action potential. This bursting and the slow depolarization are not seen in uninjured axons. (Modified from Kocsis and Waxman (1983).)

sensory neurons (dorsal root ganglion (DRG) neurons), which express multiple sodium channels, as a model system (Dib-Hajj *et al.*, 1996). These studies show that following transection of the axons of spinal sensory neurons by nerve injury, gene expression for sodium channel, voltage-gated, type X, alpha polypeptide (*SCN10A*) and sodium channel, voltage-gated, type XI, alpha polypeptide (*SCN11A*) is downregulated (**Figure 3**, top and middle). In parallel, the ionic currents produced by Na$_v$1.8 (*SCN10A*) and Na$_v$1.9 (*SCN11A*) channels are attenuated (**Figure 4**) (Sleeper *et al.*, 2000).

In tandem with downregulation of Na$_v$1.8 and Na$_v$1.9, expression of the Na$_v$1.3 (formerly called type III) sodium channel gene (sodium channel, voltage-gated, type III, alpha polypeptide (*SCN3A*)) is upregulated (**Figure 3**, bottom) so that SCN3A sodium channel protein (which is not detectable in uninjured spinal sensory neurons or their axons) accumulates at distal ends of transected axons, which are sites of abnormal impulse generation within neuromas. Concomitant with the upregulation of SCN3A expression, the neurons produce a new sodium current which recovers (reprimes) rapidly from inactivation (Cummins and Waxman, 1997) (**Figure 4c–4e**). Rapid repriming can produce a shorter refractory period which predisposes the neurons to fire in high- frequency bursts, thus contributing to hyperexcitability following nerve injury.

The changes in sodium channel expression after nerve injury are due, at least in part, to interrupted access to peripheral pools of neurotrophic factors

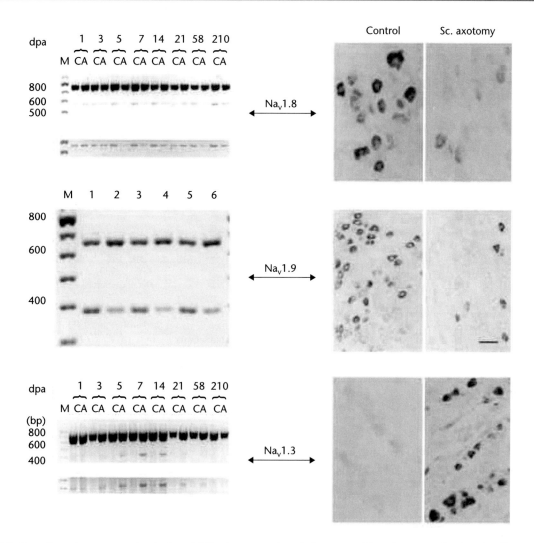

Figure 3 Sodium channel gene expression is altered following transection of the axons of dorsal root ganglion (DRG) neurons. *SCN10A* (Na$_v$1.8) (top row) and *SCN11A* (Na$_v$1.9) (middle row) sodium channel genes turn off, while the *SCN3A* (Na$_v$1.3) sodium channel gene turns on in DRG neurons (lower row) following axonal transection within the sciatic nerve. Micrographs (right column) show *in situ* hybridizations in control DRG and 5–7 days following axotomy. Gels (left) show reverse transcriptase polymerase chain reaction (RT-PCR) products following coamplification of Na$_v$1.8 (top left) and Na$_v$1.3 mRNA (bottom left) together with β-actin transcripts in control (C) and axotomized DRG (A) (days postaxotomy indicated above gels), with computer-enhanced images of amplification products shown below the gels. Coamplification of Na$_v$1.9 and glyceraldehyde-3-phosphate dehydrogenase (GAPDH) (middle row, left) shows decreased expression of Na$_v$1.9 mRNA at 7 days' postaxotomy (lanes 2, 4, 6) compared with uninjured controls (lanes 1, 3, 5). (Upper and lower panels modified from Dib-Hajj *et al.* (1996) and middle panels from Dib-Hajj *et al.* (1998).)

including NGF and GDNF (Black *et al.*, 1997; Cummins *et al.*, 2000). Delivery of NGF and GDNF to injured axons upregulates *SCN10A* (Na$_v$1.8) and *SCN11A* (Na$_v$1.9) messenger ribonucleic acid (mRNA) and protein and the sodium currents that are produced by these channels in spinal sensory neurons. These neurotrophic factors also downregulate SCN3A (Na$_v$1.3) expression in spinal sensory neurons. There is evidence for behavioral consequences of the changes in channel expression, since similar changes in the activity of the *SCN10A*, *SCN11A* and *SCN3A* sodium channel genes and in the SCN10A

(Na$_v$1.8), SCN11A (Na$_v$1.9) and SCN3A (Na$_v$1.3) channel currents are associated with lowered pain threshold in a neuropathic pain model in rats. Delivery of GDNF (which, as described above, participates in the regulation of sodium channel expression in spinal sensory neurons) can prevent and reverse sensory abnormalities in some rodent neuropathic pain models (Boucher *et al.*, 2000).

A similar transcriptional channelopathy also appears to occur as a consequence of nerve injury in humans. As in the experimental nerves described above, axons in injured human nerves are

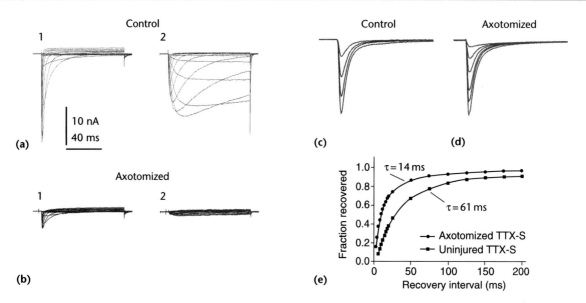

Figure 4 Amplitudes of the currents produced by the SCN10A (Na$_V$1.8), SCN11A (Na$_V$1.9) and SCN3A (Na$_V$1.3) channels are altered following transection of the axons of spinal sensory neurons within the sciatic nerve. (a) Panels 1 and 2 show patch clamp records of currents produced by Na$_V$1.8 and Na$_V$1.9 channels in control (uninjured) spinal sensory neurons. (b) Panels 1 and 2 show the currents from similar neurons 7 days after transection of their axons within the sciatic nerve. (Reproduced from Cummins *et al.* (2000).) (c–e) A rapidly repriming tetrodotoxin-sensitive (TTX-S) current emerges in peripherally axotomized spinal sensory neurons. (c) Family of tetrodotoxin-sensitive sodium current traces (with recovery times indicated) showing the time course of recovery from inactivation (repriming) at −80 mV, from a control spinal sensory neuron. (d) Similar family of traces showing accelerated repriming in a spinal sensory neuron whose axon had been transected 7 days previously. (e) Single exponential fits showing accelerated recovery from inactivation in spinal sensory neurons following axonal transection. (Reproduced from Black *et al.* (1999a).)

hyperexcitable and fire spontaneously, with repetitive impulses superimposed on a slow depolarization that is not present in normal nerve fibers. Abnormal expression patterns of SCN10A (Na$_V$1.8) and SCN11A (Na$_V$1.9) sodium channel proteins, consistent with the altered transcription observed in animal models, have been reported in human dorsal root ganglion neurons from patients with nerve injuries and neuropathic pain. Thus, studies on human tissue as well as animal models indicate that axonal injury can cause a transcriptional channelopathy.

Multiple Sclerosis

Multiple sclerosis (MS) has classically been described as a 'demyelinating disease'. Demyelination, the pathological hallmark of MS, produces axonal conduction slowing and conduction block which underlies some of the clinical abnormalities that occur in MS. Attention has also focused, over the past decade, on axonal degeneration, which may cause irreversible (nonremitting) deficits in MS. In addition to these two types of pathology, recent studies suggest that some symptoms in MS (e.g. cerebellar ataxia) may be caused by a transcriptional channelopathy.

Demyelination of an axon might be predicted to trigger altered channel expression in the neuronal cell body from which it arises, since sodium channels in healthy myelinated axons are clustered at a high density in the axon membrane at the node of Ranvier, but are present at a much lower density (too low to support secure impulse conduction) in the internodal axon membrane under the myelin. Conduction block occurs in demyelinated axons, in part because loss of myelin uncovers sodium channel-poor membrane which drains current from the axon without generating an action potential. Action potential conduction is restored in some chronically demyelinated axons by the acquisition of a higher than normal density of sodium channels within the demyelinated axon membrane. Little is known about the molecular mechanisms that derepress or promote sodium channel gene expression in demyelinated neurons, and the sodium channel genes that are involved have not been identified.

To determine whether sodium channel gene expression is altered in neurons in the demyelinating diseases and, if so, which sodium channel genes are upregulated or downregulated, the Taiep mutant rat model, in which myelin develops normally and subsequently degenerates due to an abnormality of oligodendrocytes, was studied (Black *et al.*, 1999b). Expression of the SCN10A (Na$_V$1.8) sodium channel protein, which is normally detectable only in spinal sensory neurons and trigeminal neurons, was examined because earlier studies had shown that expression of $Na_v1.8$ is dynamic, changing markedly after axon transection. Black *et al.* (1999b) showed that, following loss of myelin in Taiep rats, there was enhanced expression of $Na_v1.8$ mRNA within cerebellar Purkinje cells. Immunocytochemical experiments showed upregulation of $Na_v1.8$ protein in these neurons.

A more recent study has examined the expression of SCN10A ($Na_v1.8$) in the brains of mice with chronic relapsing experimental allergic encephalomyelitis (CR-EAE), an inflammatory model of MS, and in humans with MS (Black *et al.*, 2000). This study revealed increased expression of $Na_v1.8$ mRNA within Purkinje cells in mice with CR-EAE. $Na_v1.8$ protein levels were also upregulated in these cells. Upregulated $Na_v1.8$ expression was not part of a global increase in expression of all sodium channels, because expression of SCN11A ($Na_v1.9$), another sodium channel that is normally preferentially expressed, like $Na_v1.8$, in

spinal sensory and trigeminal ganglion neurons and not in brain, was not upregulated. These changes in sodium channel gene transcription are not limited to animal models. Examination of postmortem brain tissue from patients with a history of cerebellar deficits due to disabling progressive MS has demonstrated upregulation of $Na_v1.8$ mRNA and protein within Purkinje cells (**Figure 5**).

The functional consequences of upregulated SCN10A ($Na_v1.8$) transcription and translation in Purkinje cells in MS have not yet been directly studied. One possibility is that upregulation of $Na_v1.8$ is an adaptive change, contributing to the restoration of action potential conduction along demyelinated Purkinje cell axons. Consistent with this idea, $Na_v1.8$ channels contribute substantially to inward current flow during the depolarizing phase of the action potential in the cells (spinal sensory neurons) in which they are normally present. If $Na_v1.8$ expression

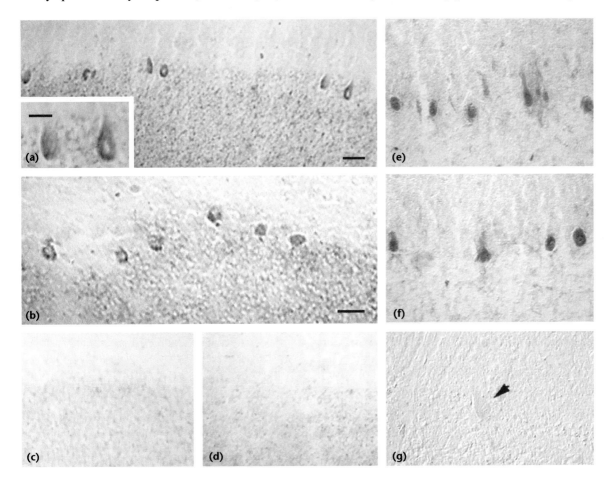

Figure 5 Expression of the sensory neuron-specific (SNS) sodium channel SCN10A ($Na_v1.8$) is upregulated within cerebellar Purkinje cells in patients with MS. *In situ* hybridization with $Na_v1.8$-specific antisense riboprobes demonstrates increased $Na_v1.8$ mRNA in Purkinje cells from MS patients obtained at post-mortem (a), (b), compared with control without neurological disease (c). No signal is present following hybridization with sense riboprobe (d). Immunocytochemistry with SNS-specific antibodies demonstrates upregulation of $Na_v1.8$ channel protein in Purkinje cells from MS patients (e, f) compared with controls (g) (arrowhead indicates Purkinje cell). Magnifications: (a) ×120, inset ×280; (b–d) ×165; (e–g) ×175. (Reproduced from Black *et al.* (2000).)

restores conduction in demyelinated axons, it would not produce a disease phenotype and should not be classified as a channelopathy.

More likely, however, is the possibility that abnormal SCN10A (Na$_v$1.8) expression in Purkinje cells perturbs the pattern of electrical activity in the cerebellum. Na$_v$1.8 displays a unique physiological signature which includes depolarized voltage dependence of inactivation, slow development of inactivation and rapid recovery from inactivation. Sodium channels play a major role in electrogenesis within Purkinje cells, and mutations of the sodium channels that are normally expressed in Purkinje cells can change the pattern of impulse generation in these cells, thereby producing ataxia (Raman et al., 1977). A study on transgenic Na$_v$1.8 knockout spinal sensory neurons showed that the presence of Na$_v$1.8 channels affects both the configuration of action potentials and the pattern of action potential firing in response to depolarizing stimuli (Renganathan et al., 2001).

More recent studies have begun to focus on Purkinje cells and suggest that, when Na$_v$1.8 sodium channels are experimentally expressed within these cells, there is a striking change in their firing pattern. Although the mechanisms responsible for these changes in electrical activity need to be examined, the results of these initial studies suggest that altered patterns of activity within the cerebellum, due to abnormal Na$_v$1.8 expression in Purkinje cells, may disrupt cerebellar coding, a change that could lead to clinical dysfunction in MS. If this speculation is correct, a next step would be to determine whether subtype-specific blockade of Na$_v$1.8 channels can reduce the degree of ataxia in animal models of MS.

Extent of the Transcriptional Channelopathies

Do the transcriptional channelopathies affect other classes of channels and other types of neurons? Upregulated SCN2A1 (Na$_v$1.2) and KCNA1 (K$_v$1.1) and KCNA2 (K$_v$1.2) potassium channel expression has been observed along dysmyelinated axons within the brain of the Shiverer mouse, in which myelin fails to compact owing to a mutation of myelin basic protein. A recent study has suggested the ectopic expression of the α_{1B} pore-forming subunit of N-type calcium channels within dystrophic (presumably degenerating) axons in EAE and MS. There is also immunocytochemical and electrophysiological evidence for downregulation of potassium channels and upregulation of some calcium channels in dorsal root ganglion (DRG) cells after their axons are injured, and γ-aminobutyric acid (GABA) A receptor expression is also altered in axotomized DRG neurons.

It is possible that channel transcription may be perturbed in other disorders of the nervous system. Reduced SCN10A (Na$_v$1.8) mRNA levels have been reported within spinal sensory neurons in a rat model of diabetic neuropathy and there is also evidence for an upregulation of SCN3A (Na$_v$1.3) mRNA. Transcriptional channelopathies may also contribute to some types of epilepsy. Increased expression of SCN2A1 (Na$_v$1.2) and Na$_v$1.3 sodium channel mRNA has been reported in rat hippocampus following kainate-induced seizures. Whether these transcriptional changes, or a change in properties of preexisting channels, for example, by phosphorylation, produce the increased sodium current density and shifted voltage dependence is not known.

An increased understanding of transcriptional channelopathies of neurons may point to new therapeutic strategies, since the different amino acid sequences of various channels may endow them with differential sensitivities to activators or blockers. The preferential expression of some channels within specific groups of neurons may facilitate the targeting of these cells with minimal side effects. Transcriptional channelopathies of neurons, and the molecular targets they present, may thus provide new inroads for treatment of diseases of the nervous system.

See also
Ion Channels and Human Disorders
Multiple Sclerosis (MS): Genetics
Potassium Channels

References

Black JA, Cummins TR, Plumpton C, et al. (1999a) Upregulation of a silent sodium channel after peripheral, but not central, nerve injury in DRG neurons. Journal of Neurophysiology 82: 2776–2785.

Black JA, Dib-Hajj SD, Baker D, et al. (2000) Sensory neuron specific sodium channel SNS is abnormally expressed in the brains of mice with experimental allergic encephalomyelitis and humans with multiple sclerosis. Proceedings of the National Academy of Sciences of the United States of America 97: 11 598–11 602.

Black JA, Fjell J, Dib-Hajj S, et al. (1999b) Abnormal expression of SNS/PN3 sodium channel in cerebellar Purkinje cells following loss of myelin in the taiep rat. NeuroReport 10: 913–918.

Black JA, Langworthy K, Hinson AW, Dib-Hajj SD and Waxman SG (1997) NGF has opposing effects on Na$^+$ channel III and SNS gene expression in spinal sensory neurons. NeuroReport 8: 2331–2335.

Boucher TJ, Okuse K, Bennett DL, et al. (2000) Potent analgesic effects of GDNF in neuropathic pain states. Science 290: 124–127.

Cummins TR and Waxman SG (1997) Down-regulation of tetrodotoxin-resistant sodium currents and up-regulation of a rapidly repriming tetrodotoxin-sensitive sodium current in small spinal sensory neurons following nerve injury. Journal of Neuroscience 17: 3503–3514.

Cummins TR, Black JA, Dib-Hajj SD and Waxman SG (2000) Glial-derived neurotrophic factor upregulates expression of functional SNS and NaN sodium channels and their currents in axotomized dorsal root ganglion neurons. Journal of Neuroscience 20: 8754–8761.

Devor M and Seltzer Z (1999) The pathophysiology of damaged peripheral nerves in relation to chronic pain. In: Wall PD and Melzack R (eds.) *Textbook of Pain*, 4th edn, pp. 129–164. Edinburgh UK: Churchill Livingstone.

Dib-Hajj S, Black JA, Felts P and Waxman SG (1996) Down-regulation of transcripts for Na channel α-SNS in spinal sensory neurons following axotomy. *Proceedings of the National Academy of Sciences of the United States of America* **93**: 14 950–14 954.

Dib-Hajj SD, Tyrrell L, Black JA and Waxman SG (1998) NaN, a novel voltage-gated Na channel preferentially expressed in peripheral sensory neurons and down-regulated following axotomy. *Proceedings of the National Academy of Sciences of the United States of America* **95**: 8963–8968.

Kocsis JD and Waxman SG (1983) Long-term regenerated nerve fibres retain sensitivity to potassium channel blocking agents. *Nature* **304**: 640–642.

Raman IM, Sprunger LK, Meisler MH and Bean BP (1997) Altered subthreshold sodium currents and disrupted firing patterns in Purkinje neurons in Scn8a mutant mice. *Neuron* **19**: 881–891.

Renganathan M, Cummins TR and Waxman SG (2001) The contribution of $Na_v1.8$ sodium channels to action potential electrogenesis in DRG neurons. *Journal of Neurophysiology* **86**: 629–640.

Rose M and Griggs RE (eds.) (2001) *Channelopathies of the Nervous System*. Oxford, UK: Butterworth-Heinemann.

Sleeper AA, Cummins TR, Dib-Hajj SD, *et al.* (2000) Changes in expression of two tetrodotoxin-resistant sodium channels and their currents in dorsal root ganglion neurons after sciatic nerve injury but not rhizotomy. *Journal of Neuroscience* **20**: 7279–7289.

Tanaka M, Cummins TR, Ishikawa K, *et al.* (1999) Molecular and functional remodeling of electrogenic membrane of hypothalamic neurons in response to changes in their input. *Proceedings of the National Academy of Sciences of the United States of America* **96**: 1088–1093.

Waxman SG (2000) The neuron as a dynamic electrogenic machine: modulation of sodium channel expression as a basis for functional plasticity in neurons. *Philosophical Transactions of the Royal Society of London, Series B: Biological Sciences* **355**: 199–213.

Waxman SG (2001) Transcriptional channelopathies: an emerging class of disorders. *Nature Reviews Neuroscience* **2**: 652–659.

Zhang J-M, Donnelly DF, Song X-J and LaMotte RH (1997) Axotomy increases the excitability of dorsal root ganglion cells with unmyelinated axons. *Journal of Neurophysiology* **78**: 2790–2794.

Further Reading

Waxman, SG (2001) Transcriptional channelopathies: an emerging class of disorders. *Nature Reviews Neuroscience* **2**: 654–659.

Web Links

Sodium channel, voltage-gated, type I, alpha polypeptide (*SCN1A*); Locus ID: 6323. LocusLink:
http://www.ncbi.nlm.nih.gov/LocusLink/LocRpt.cgi?l = 6323

Sodium channel, voltage-gated, type X, alpha polypeptide (*SCN10A*); Locus ID: 6336. LocusLink:
http://www.ncbi.nlm.nih.gov/LocusLink/LocRpt.cgi?l = 6336

Sodium channel, voltage-gated, type XI, alpha polypeptide (*SCN11A*); Locus ID: 11280. LocusLink:
http://www.ncbi.nlm.nih.gov/LocusLink/LocRpt.cgi?l = 11280

Sodium channel, voltage-gated, type III, alpha polypeptide (*SCN3A*); Locus ID: 6328. LocusLink:
http://www.ncbi.nlm.nih.gov/LocusLink/LocRpt.cgi?l = 6328

Sodium channel, voltage-gated, type I, alpha polypeptide (*SCN1A*); MIM number: 182389. OMIM:
http://www.ncbi.nlm.nih.gov/entrez/dispomim.cgi?id = 182389

Sodium channel, voltage-gated, type X, alpha polypeptide (*SCN10A*); MIM number: 604427. OMIM:
http://www.ncbi.nlm.nih.gov/entrez/dispomim.cgi?id = 604427

Sodium channel, voltage-gated, type XI, alpha polypeptide (*SCN11A*); MIM number: 604385. OMIM:
http://www.ncbi.nlm.nih.gov/entrez/dispomim.cgi?id = 604385

Sodium channel, voltage-gated, type III, alpha polypeptide (*SCN3A*); MIM number: 182391. OMIM:
http://www.ncbi.nlm.nih.gov/entrez/dispomim.cgi?id = 182391

Transcriptional Regulation: Coordination

Jean-Marc Egly, INSERM/CNRS/ULP, Illkirch, France
Cécile Rochette-Egly, INSERM/CNRS/ULP, Illkirch, France

Intermediate article

Article contents

- Introduction
- Coordination of Transcription Initiation and Pre-messenger RNA Processing
- Activators
- Chromatin Decompaction and Recruitment of the Transcription Machinery
- Conclusion

In higher eukaryotes, gene expression is tightly coordinated at multiple stages with other events that orchestrate the life of the cell. Prompted by an initial signal, regulatory transcription factors are activated, assemble at specific enhancer sequences and trigger a cascade of reactions, resulting in an appropriately remodeled chromatin template, with functional transcription machinery on the promoter. Phosphorylation, acetylation and methylation orchestrate the interconnection of the different steps.

Introduction

The identification of most of the components of transcription machinery, followed by the discovery of their primary enzymatic functions, has led to advances in our understanding of the different steps in gene expression. It then became important to investigate the order of these steps: how and when can a gene, initially packed in a dense chromatin structure, be transcribed? What sort of signals induce chromatin remodeling and allow the recruitment of the deoxyribonucleic acid (DNA)-processing enzymes to initiate transcription? Is it sufficient to explain that activation of transcription increases ribonucleic acid (RNA) synthesis without considering the posttranscriptional events that result in protein synthesis and the function of that protein in the cell? Clearly transcription has to be coordinated with other processes and must be ready to adopt a standby or 'hold' position if unexpected events occur. For example, in the case of genotoxic attack, cell-cycle progression, as well as transcription, has to stop, to allow DNA repair. Although not fully understood, it appears that these various reactions are coordinated by common signals and/or factors. Indeed, the TATA-binding protein (TBP) can associate with the three RNA polymerases, the proliferating cell nuclear antigen (PCNA) coordinates both DNA replication and repair and the transcription factor IIH (TFIIH) complex is involved in both transcription and DNA repair.

Mechanisms have been developed by higher eukaryotes for the tight control of transcription. The different steps of these mechanisms are interconnected and they are coordinated with other events that orchestrate the life of the cell. The expression of most genes is regulated by 'activators' that bind specific sequences located upstream of the transcription initiation site and whose efficiency depends on physiological signals (levels of insulin, hormones, etc.) or environmental stimuli (genotoxic attacks or oxidative stresses). These activators are targeted by a battery of intermediary proteins, also called coregulators, including chromatin remodelers and modifiers, which act in a coordinated and/or combinatorial manner to decompact chromatin and direct RNA polymerase II (RNA pol II) and the general transcription factors (GTFs) to the promoter. Each factor and coregulator undergoes sequential and coordinated covalent modifications that regulate its activity and/or promotes the following step, to ensure the correct transcription of a specific gene.

Coordination of Transcription Initiation and Pre-messenger RNA Processing

RNA is synthesized by RNA pol II, whose accurate recruitment to the promoter depends on the six GTFs, two of which, TFIID and TFIIH, have enzymatic activities (Woychik and Hampsey, 2002). The large, multisubunit TFIID binds to the promoter through its TBP, and possesses associated factors or TAFIIs, which develop kinase, acetylase and ubiquitination activities, and/or bridge the complex to activators.

TFIIH also exhibits a cyclin-dependent kinase and two DNA helicase activities. The transcription process can be divided into several steps: (i) promoter recognition; (ii) initiation, including promoter opening and the formation of the first phosphodiester bond that initiates RNA synthesis; (iii) promoter clearance and elongation, in which RNA pol II escapes from the promoter and (iv) termination before recycling of part if not all of the components of the transcription apparatus. A number of recent results have popularized the view that RNA pol II and GTFs are recruited to the promoter as a preformed complex, also called the holoenzyme. However, the fact that GTFs were identified and purified independently argues in favor of a step-wise assembly model in which the preinitiation complex is nucleated by the recognition of the TATA box by TBP, followed by the sequential recruitment of TFIIA, TFIIB, the unphosphorylated RNA pol II associated with TFIIF, TFIIE and TFIIH. Thereafter, the helicases of TFIIH wrap RNA pol II and associated factors around the promoter, and, upon adenosine triphosphate (ATP) hydrolysis, the preinitiation complex becomes activated. The xeroderma pigmentosum group B (XPB) helicase of TFIIH unwinds and opens the DNA around the initiation site (promoter melting), leading to the formation of the 'open and active' initiation complex, a prerequisite for RNA synthesis. Finally, the transcription initiation complex is disrupted (promoter clearance) and elongation can proceed. (*See* DNA Polymerases: Eukaryotic; Transcription Factors; Promoters.)

In addition to reading the coding strand, RNA pol II also has to provide a 'ready-to-use' template for the translation machinery by coordinating a number of posttranscriptional events such as capping, splicing and polyadenylation (Proudfoot *et al.*, 2002). This function is partially assumed by the serine-rich *C*-terminal domain (CTD) of the largest subunit of RNA pol II, which is subjected to phosphorylation–dephosphorylation processes (Orphanides and Reinberg, 2002). CTD phosphorylation is governed essentially by cyclin-dependent kinases (CDKs), including the Cdk7 and Cdk9 subunits (also involved in cell-cycle regulation) of TFIIH and the elongation factor positive transcription elongation factor b (P-TEFb), as well as by Fcp1 phosphatase. It is noteworthy that the CTD can be used as a substrate by other kinases such as the mitogen-activated protein kinases (MAPKs). (*See* RNA Processing; Splicing of pre-mRNA.)

When in the preinitiation complex, the CTD of RNA pol II is unphosphorylated. Following, or in parallel to, the formation of the first phosphodiester bond, the CTD becomes phosphorylated, most probably by the Cdk7 kinase. It has been suggested that such a phosphorylation disturbs the interactions between RNA pol II and the initiation factors and therefore favors the transition from the initiation to the elongation phase. Thereafter RNA pol II pauses to recruit, via its phosphorylated CTD, the capping enzyme, which adds a cap structure to the 5′ end of the nascent transcript, to protect the new messenger RNA (mRNA) from attack by nucleases. In a further step, the stalled RNA pol II recruits elongation factors, one of them being the elongation complex P-TEFb that, through its Cdk9 subunit, phosphorylates other serine residues of the CTD, thus favoring the recruitment and/or the selection of the next processing factors, involved in splicing, cleavage and/or polyadenylation. During these mRNA-processing steps, it seems that the additional elongation factors (TFIIS, TFIIF, elongins, the Elongator complex and others) are recruited and/or activated to allow RNA pol II to resume elongation.

Finally, having read the gene, RNA pol II must be either dephosphorylated by the Fcp1 CTP phosphatase for further reinitiation or ubiquitinated and degraded by the proteasome. It thus appears that changes in the phosphorylation state of the CTD define a 'CTD code' that coordinates the recruitment and/or release of factors associated not only with transcription but also with mRNA processing.

Activators

The events described in the previous section occur at core promoter elements and are therefore involved in the control of transcription of any gene. However, for the transcription of a given gene at the right time, the cell must control the activators which bind to specific DNA sequences, also called enhancers (Brivanlou and Darnell, 2002). Once bound to the enhancer through their DNA-binding domain, activators act on the transcription machinery through contacts with mediators, coregulators and general factors. This activation potential can reside in one specific activator or result from the combined and sophisticated action of several effectors that respond to multiple signaling pathways, thus leading to a unique and specific regulation of the transcription of a given gene. Transcription regulation may also involve the action of several activators recognizing distinct binding sites within the same enhancer. Among the activators, several are constitutively nuclear and active, with acidic, glutamine-, proline- or leucine-rich activation domains, and as such participate in the transcription of housekeeping genes. However, a broad class of other activators require, to become active, multiple modes of regulation, most of them involving phosphorylation processes. (*See* Enhancers; Posttranscriptional Processing.)

Indeed, phosphorylation, triggered by kinase cascades (MAPKs, CDKs, protein kinase A (PKA),

protein kinase C (PKC), Akt) in response to environmental or physiological signals (Hunter, 2000), can act as an on–off switch, promoting the translocation of several 'latent' cytoplasmic activators into the nucleus and/or the proteolysis of inhibitors. In contrast, for other activators already present in the nucleus, covalent modifications such as phosphorylations would simply increase their transcriptional activity, through enhancing the recruitment of coactivators and components of the transcription machinery and/or by helping the release of corepressors.

As part of the preinitiation complex and/or when recruited for transactivation, the phosphorylation and subsequent activation of 'activators' can occur upon contact with the basal transcription machinery. This is the case of the nuclear hormone receptors which become able to stimulate ligand-dependent transcription when phosphorylated by the Cdk7 subunit of TFIIH, most probably during the formation of the preinitiation complex, once the activator and the GTF are bound to the response element and the promoter respectively (Keriel *et al.*, 2002). Although the significance of this process remains undetermined, it is hypothesized that this might increase the efficiency of preinitiation complex assembly and/or the activator–mediator interaction.

Once activated and bound to the enhancer, activators can be also acetylated or methylated by the coregulators, thus enhancing their DNA-binding activity or their association with coregulators. Finally, after transcription has been initiated, activators may be subject to other modifications that decrease their activity. Indeed, certain phosphorylation processes can promote the dissociation of activators from the transcription initiation complex through either their rapid exclusion from the nucleus or the disruption of their interaction with DNA. Phosphorylations can also trigger their degradation by the ubiquitin–proteasome pathway (Gianni *et al.*, 2002), allowing the cell to continuously monitor its environment.

Chromatin Decompaction and Recruitment of the Transcription Machinery

DNA is packaged into a highly organized and compact nucleoprotein structure known as chromatin, which consists of a tight assembly of nucleosomes, in which DNA is wrapped around a protein core containing the four histone proteins. Protruding from the nucleosomes are the *N*-terminal 'tails' of the core histones which can be subjected to covalent modifications. To facilitate the recruitment of RNA pol II and the general transcriptional machinery to the transcriptional start site, activators must decompact repressive chromatin, which creates barriers for all of the DNA-processing enzymes. Activators recruit diverse families of protein complexes, also called coregulators (Narlikar et al., 2002), that either reposition nucleosomes or change histone–DNA or histone–histone contacts in order to facilitate the positioning of the transcription apparatus to the promoter. The ATP-dependent remodeling coregulators use the energy of ATP hydrolysis to reposition nucleosomes at the promoter through sliding them *in cis* or displacing them *in trans*, allowing the formation of nucleosome-free or nucleosome-spaced regions. The histone-modifying complexes which include histone acetyltransferases (HATs) and histone methyltransferases (HMTs), acetylate and methylate lysine or arginine residues located at the N-terminal tails of histones, respectively. In some cases, the efficiency of histone acetylation and methylation can be regulated upon phosphorylation of the nearby serines by associated kinases. (**See** Chromatin Structure and Domains; Histones.)

Such histone modifications create tags or binding sites that form a 'code' read by a specialized bromo-domain, present in chromatin remodelers and modifiers and in some transcription factors. It has been hypothesized that this 'code' would coordinate the recruitment of additional histone ATP-dependent remodelers and/or acetyltransferases for further chromatin decompaction and transcription apparatus recruitment. Finally, HATs, HMTs and the associated kinases target not only histones but also themselves and/or the activators, promoting their dissociation from activators, and the subsequent recruitment of other coregulators and/or additional transcription factors for initiation as well as for elongation and termination.

As ATP-dependent remodelers and covalent modifiers work together to facilitate access of the general transcription machinery to DNA, their recruitment should be temporally coordinated. However, there is *a priori*, no obligate temporal order for the function of ATP-dependent remodelers and covalent modifiers that would be general for all promoters. This is due to the fact that all promoter regions differ in regard to the arrangement of nucleosomes in the repressed state.

As an example, in a context of repressive chromatin where nucleosomes do not impede the binding of activators to their DNA recognition sequences, HAT complexes can be recruited first. Then, subsequent to histone acetylation, ATP-dependent remodeling complexes are recruited, leading to the displacement of impeding nucleosomes within the proximal promoter region, thus facilitating access of the general transcription machinery to DNA. Such an order of complex recruitment has been observed on the interferon-beta (IFNβ) promoter and for transcription directed by

nuclear retinoid receptors. However, the reverse, that is, the recruitment of ATP-dependent remodelers before HATs, has been observed for genes that are transcribed during mitosis when the genome is highly condensed. Finally, for a particular gene, whose expression can be regulated during different phases of the cell cycle, the order of recruitment of chromatin remodelers and modifiers may vary: activation during interphase will require only HATs, while activation of the same gene during late mitosis would require the prior recruitment of ATP-dependent remodelers to assist HATs. The requirement is that an appropriately decondensed chromatin state with a functional preinitiation complex positioned for transcription must be attained efficiently by each gene.

Finally, for efficient transcription, chromatin disruption has to be spread from the promoter to the remainder of the gene to be read. This process is coupled to elongation and involves chromatin-remodeling and -modifying factors that track with elongating RNA pol II (Orphanides and Reinberg, 2002).

Once repressive chromatin has been decondensed, enhancer-bound activators can participate in the entry of RNA pol II (and GTFs) into the preinitiation complex. They do this through the recruitment of a mediator complex (Woychik and Hampsey, 2002). Although sometimes isolated by fractionation in association with the RNA pol II and some GTFs, the mediator complex does not seem to be recruited at the promoter in a single step as a preassembled 'RNA pol II holoenzyme complex' (Malik and Roeder, 2000). Rather, once recruited by activators, the mediator complex helps to expedite entry of RNA pol II (and GTFs) into the preinitiation complex.

A multiplicity of mediator complexes have been characterized, including, in addition to conserved modules interacting with the CTD of Pol II, specialized units interacting with activators. This heterogeneity led to the proposal that the mediator would physically bridge specific activators to RNA pol II in response to specific transduction signals. Therefore, it would transduce both positive and negative information from gene-specific activators to the core transcriptional machinery. Being able to interact simultaneously with multiple activators bound to different sequence motifs within the same promoter, the mediator would also promote synergy between diverse activators.

Finally, a mediator complex can also regulate activated transcription. This function is fulfilled by two of its subunits carrying kinase activity, Srb10 (also called Cdk8) and a RING-3 subunit. Importantly, depending on the integrated signal and the targeted kinase, the mediator complex regulates activated transcription either positively or negatively. Activation of the RING-3-like kinase would potentiate transcription activation through stimulating the

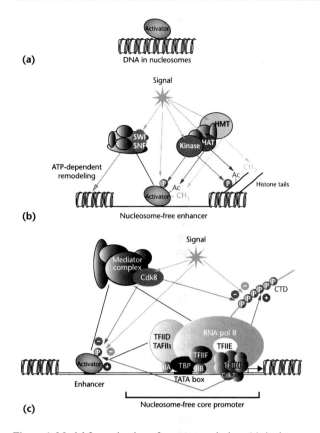

(a)

(b)

(c)

Figure 1 Model for activation of gene transcription. (a) Activators bind to their DNA-specific recognition sequences in a context of repressive chromatin; (b) once turned on, activators recruit a battery of coregulators acting in a coordinated manner to decompact chromatin: (i) ATP-dependent remodeling complexes (SWI/SNF) which displace the impeding nucleosomes; (ii) histone acetyltransferase (HAT) complexes associated with histone methyltransferases (HMTs) and histone kinases which acetylate, methylate and phosphorylate histones, thus changing histone–DNA and histone–histone contacts. Each coregulator can also modify activators as well as the other coregulators, enhancing and/or disrupting the interactions between activators and coregulators. The efficiency of these events is increased by phosphorylation processes in response to physiological or environmental signals; (c) activators recruit the mediator complex which will expedite entry of RNA polymerase II (RNA pol II) and general transcription factors into the preinitiation complex. At this step, the Cdk7 subunit of transcription factor IIH (TFIIH) can phosphorylate certain activators, increasing the efficiency of the transcription initiation complex assembly. Cdk7 also phosphorylates the RNA pol II (PRNA pol II) C-terminal domain (CTD), favoring the transition from the initiation to the elongation phase. At this step, phosphorylation processes in response to specific signals would negatively regulate activated transcription by inactivating activators or targeting the CTD. Ac: acetylation; P: phosphate; TAFIIs: transcription associated factors; TBP: TATA-binding protein; SWI: mating type switching; SNF: sucrose nonfermenting.

CTD kinase activity of TFIIH (Jiang *et al.*, 1998). In contrast, activation of Cdk8 negatively regulates activated transcription (Akoulitchev *et al.*, 2000). If the stimulus occurs early in the transcription initiation process, before RNA pol II assembly, Cdk8 halts the formation of the preinitiation complex through phosphorylation of the CTD at specific serine residues. However, if RNA pol II has already associated with the initiation complex, Cdk8 stops transcription through inhibiting the CTD kinase activity of TFIIH.

Conclusion

Transcription of the right gene at the right time is finely tuned by multiple players acting in a spatially and temporally coordinated manner (see **Figure 1**). Dictated by an initial signal (**Figure 1**), regulatory transcription factors assembled at distal specific enhancer sites become 'activated' and trigger a cascade of reactions, resulting in an appropriately remodeled chromatin template with a functional preinitiation complex poised ready for transcription at the promoter. The order of the different steps is orchestrated mainly by phosphorylation, acetylation and methylation events that target either activators, coregulators, chromatin histones, a mediator complex or the CTD of RNA pol II. Such events also determine whether transcription stops through excluding activators and coactivators from the transcription complex or through their degradation. Considering the diversity of physiological signals, transcription factors may be subjected to multiple combinatorial regulations, resulting in positive or negative regulation of transcription. Finally, given the functional importance of the regulatory networks governing the normal transcription of a given gene, it is evident that the deregulation of any factor or of the signal transduction pathways may contribute to diseases (Egly, 2001) or to tumoral processes (Hunter, 2000; Klochendler-Yeivin *et al.*, 2002). (*See* Histone Acetylation and Disease.)

References

Akoulitchev S, Chuikov S and Reinberg D (2000) TFIIH is negatively regulated by Cdk8-containing mediator complexes. *Nature* **407**: 102–106.

Brivanlou AH and Darnell Jr JE (2002) Signal transduction and the control of gene expression. *Science* **295**: 813–818.

Egly JM (2001) The 14th Datta Lecture. TFIIH: from transcription to clinic. *FEBS Letters* **498**: 124–128.

Gianni M, Bauer A, Garattini E, Chambon P and Rochette-Egly C (2002) Phosphorylation by p38MAPK and recruitment of SUG-1 are required for RA-induced RARγ degradation and transactivation. *EMBO Journal* **21**: 3760–3769.

Hunter T (2000) Signaling – 2000 and beyond. *Cell* **100**: 113–127.

Jiang YW, Veschambre P, Erdjument-Bromage H, *et al.* (1998) Mammalian mediator of transcriptional regulation and its possible role as an end-point of signal transduction pathways. *Proceedings of the National Academy of Sciences of the United States of America* **95**: 8538–8543.

Keriel A, Stary A, Sarasin A, Rochette-Egly C and Egly JM (2002) XPD mutations prevent TFIIH-dependent transactivation by nuclear receptors and phosphorylation of RARalpha. *Cell* **109**: 125–135.

Klochendler-Yeivin A, Muchardt C and Yaniv M (2002) SWI/SNF chromatin remodeling and cancer. *Current Opinion in Genetics and Development* **12**: 73–79.

Malik S and Roeder RG (2000) Transcriptional regulation through mediator-like coactivators in yeast and metazoan cells. *Trends in Biochemical Sciences* **25**: 277–283.

Narlikar GJ, Fan HY and Kingston RE (2002) Cooperation between complexes that regulate chromatin structure and transcription. *Cell* **108**: 475–487.

Orphanides G and Reinberg D (2002) A unified theory of gene expression. *Cell* **108**: 439–451.

Proudfoot NJ, Furger A and Dye MJ (2002) Integrating mRNA processing with transcription. *Cell* **108**: 501–512.

Woychik NA and Hampsey M (2002) The RNA polymerase II machinery: structure illuminates function. *Cell* **108**: 453–463.

Further Reading

Agalioti T, Lomvardas S, Parekh B, *et al.* (2000) Ordered recruitment of chromatin modifying and general transcription factors to the IFN-beta promoter. *Cell* **103**: 667–678.

Cheung P, Allis CD and Sassone-Corsi P (2000) Signaling to chromatin through histone modifications. *Cell* **103**: 263–271.

Cosma MP (2002) Ordered recruitment: gene specific mechanism of transcription activation. *Molecular Cell* **10**: 227–236.

Dilworth FJ and Chambon P (2001) Nuclear receptors coordinate the activities of chromatin remodeling complexes and coactivators to facilitate initiation of transcription. *Oncogene* **20**: 3047–3054.

Featherstone M (2002) Coactivators in transcription initiation: here are your orders. *Current Opinion in Genetics and Development* **12**: 149–155.

Kouzarides T (2002) Histone methylation in transcriptional control. *Current Opinion in Genetics and Development* **12**: 198–209.

Krebs JE, Kuo MH, Allis CD and Peterson CL (1999) Cell cycle-regulated histone acetylation required for expression of the yeast HO gene. *Genes and Development* **13**: 1412–1421.

Riedl T and Egly JM (2000) Phosphorylation in transcription: the CTD and more. *Gene Expression* **9**: 3–13.

Shang Y, Myers M and Brown M (2002) Formation of the androgen receptor transcription complex. *Molecular Cell* **9**: 601–610.

Thomas D and Tyers M (2000) Transcriptional regulation: kamikaze activators. *Current Biology* **10**: 341–343.

Transcriptional Regulation: Evolution

Gregory A Wray, *Duke University, Durham, North Carolina, USA*

Changes in transcriptional regulation can result from mutations in *cis* (within regulatory sequences) or *trans* (in the structure or expression of transcription factors). Modifications in transcription can have a profound impact on gene function and constitute an important class of evolutionary change within genomes.

Intermediate article

Article contents

- Evolutionary Significance of Transcriptional Regulation
- Evolution of *cis*-regulatory versus Coding Sequences
- Polymorphisms in *cis*-regulatory Sequences
- Interspecies Differences in Transcriptional Regulation
- Evolutionary Mechanisms Operating on Transcription
- Macroevolutionary Changes in Phenotype

Evolutionary Significance of Transcriptional Regulation

Regulatory sequences surround every gene, determining where, when, at what level and under what conditions it is transcribed (Davidson, 2001; White, 2001). Because the context in which a gene product is produced is often closely linked with its function, these *cis*-regulatory sequences have been considered important sites of evolutionary change in the genome since their discovery more than 30 years ago. Empirical evidence of such a role remained limited, however, until the experimental tools were developed that allowed functional changes in transcriptional regulation among species to be detected.

During the 1990s, the finding that most animals share a common set of developmental regulatory genes (Carroll *et al.*, 2001) drew considerable attention to the evolutionary significance of changes in transcriptional regulation. The ability of these conserved regulatory genes to pattern the diverse body plans and morphological features of animals has resulted from evolutionary changes in their expression profiles and their influence on the expression of downstream genes (Carroll *et al.*, 2001; Davidson, 2001). These

differences, along with several other lines of evidence, make it clear that changes in transcriptional regulation have been crucial in the evolution of humans and other organisms (Wilkins, 2002; Wray *et al.*, 2003).

Evolution of *cis*-regulatory versus Coding Sequences

Most of what we know about the evolution of genes comes from coding sequences and the introns embedded within them. Because *cis*-regulatory sequences differ in structure and function from coding sequences, however, their evolutionary dynamics are quite different. The functionally important components of *cis*-regulatory sequences are binding sites for transcription factors, proteins that associate with deoxyribonucleic acid (DNA) in a sequence-specific manner. Transcription factor binding sites are short (generally, 4–9 base pairs) and imprecise, in the sense that transcription factors are typically able to bind several variants on a consensus sequence rather than a single, exact sequence. Binding sites do not occupy any consistent position relative to the start site of transcription or other features of a gene. Furthermore, whether a transcription factor actually binds often depends upon the presence of other proteins in the nucleus and upon the state of chromatin condensation.

These features of binding site organization and function have important consequences for understanding how transcriptional regulation evolves (Wray et al., 2003). One consequence is that many potential binding sites lack a biological function. Identifying functional binding sites requires experimental verification, which makes evolutionary comparisons of cis-regulatory sequences considerably more difficult than comparisons of coding sequences. Another important consequence is that the transcription profile of a gene is determined by both cis and trans sequences. An evolutionary difference in transcription could thus result from changes in nearby binding sites or from differences in the activity or expression of any of its regulators.

Once bound to DNA, transcription factors exert their influence by altering the state of chromatin condensation, bending or looping DNA so as to modulate binding by other proteins, recruiting ribonucleic acid (RNA) polymerase to the transcription initiation site, or interfering with these processes (White, 2001). The consequences of a particular transcription factor binding to DNA often depend upon what cofactors or other transcription factors are present. It is not unusual for the same transcription factor to act as an activator of transcription in some contexts and as a repressor in others. Thus, inspection of evolutionary differences in cis-regulatory sequences cannot alone reveal what changes might have evolved in the transcription profile. Additional information, including the identity and expression profiles of the proteins that bind to it, and experimental tests of cis-regulatory sequence function in various contexts, are needed.

A substantial fraction of a eukaryotic genome is devoted to the task of regulating transcription. According to one estimate, approximately the same number of functionally constrained nucleotides fall in coding and noncoding regions in humans (Shabalina et al., 2001).

Polymorphisms in *cis*-regulatory Sequences

All evolutionary differences begin as genetic polymorphisms within populations. The nature and level of genetic diversity within cis-regulatory sequences therefore provides insights into how evolutionary differences in transcriptional regulation between species first arose. Most of what we know about genetic diversity in cis-regulatory sequences comes from humans, where a large number of polymorphisms have been identified that have clinical consequences.

Loci containing cis-regulatory polymorphisms are distributed throughout the human genome and encode a wide variety of protein products (Cooper, 1999). The phenotypic consequences of cis-regulatory mutations range in magnitude from relatively subtle propensities to outright lethality, and affect diverse aspects of anatomy, physiology and behavior. As many as 40% of all loci in a typical individual may be heterozygous with respect to a functional cis-regulatory polymorphism (Rockman and Wray, 2002), as compared with about 20% of loci that are likely to be polymorphic with respect to an amino acid substitution.

Most known functional polymorphisms in human cis-regulatory DNA are the result of small-scale mutations (Rockman and Wray, 2002). Approximately 73% are due to single-base substitutions, with insertion/deletion and tandem repeat variation comprising roughly 7% and 20% respectively. Of these mutational classes, only tandem repeat variations are overrepresented in functional cis-regulatory polymorphisms relative to the genome as a whole. $(AC)_n$ microsatellites are particularly abundant and comprise half the known cis-regulatory polymorphisms that involve tandem repeat variation. Transposon insertions and chromosomal rearrangements, by contrast, are relatively rare as human cis-regulatory polymorphisms with few well-documented cases. Comparable information does not exist for any other species, so the generality of these results is unclear.

As expected, the biochemical consequences of a mutation in cis-regulatory DNA vary considerably (Cooper, 1999). Some mutations have no detectable impact on gene expression, presumably because they do not alter protein–DNA interaction or because any such alteration does not affect the transcription profile. Of the cis-regulatory alleles that are known to have a biochemical consequence in humans, approximately two-thirds have at least a twofold impact on transcription rates as measured in cell culture assays. Several alleles are known that eliminate transcription and others that boost transcription rates by an order of magnitude or more. Much less is known about the impact that cis-regulatory mutations have on spatial and temporal aspects of transcription, nor about consequences for transcription during prenatal development, all of which are difficult to assay in humans.

Interspecies Differences in Transcriptional Regulation

Comparisons among species reveal that transcription factor binding sites can be gained or lost over relatively short evolutionary timescales (Wray *et al.*, 2003) Polymorphisms in human *cis*-regulatory sequences can be polarized by comparison with homologous sequences in the great apes (Rockman and Wray, 2002). For those loci where such data are available, known losses of transcription factor binding sites only slightly outnumber gains. In a few cases, the new allele in humans binds a different transcription factor with higher affinity, a situation known as transcription factor switching. The functional consequence of the new allele in humans is approximately as likely to involve a decrease as it is an increase in transcription rate. Although these comparisons involve only polymorphisms and exclude binding sites that have become fixed in humans since they diverged from the other great apes, they demonstrate how rapidly functional changes in *cis*-regulatory regions can evolve.

More distant evolutionary comparisons among mammalian orders reveal a correspondingly greater degree of binding site turnover. Alignments of functionally verified *cis*-regulatory sequences from human and rodent genomes often reveal a combination of conserved and lineage-specific transcription factor binding sites (Dermitzakis and Clark, 2002). Approximately one-third of the functional binding sites in humans apparently do not exist in rodents. Comparisons among species of *Drosophila* also reveal differences in functional binding sites over relatively short evolutionary timescales (Ludwig and Kreitman, 1995; Shaw *et al.*, 2002). In some cases, these gains and losses of binding sites do not have a measurable impact on transcription profiles, suggesting that functionally compensatory changes have evolved.

Despite these examples of rapid evolutionary change, other *cis*-regulatory regions have apparently persisted for much longer intervals of time. For example, blocks of highly similar sequence are present in *cis*-regulatory regions of the *Hox* complex in mammals, birds and teleosts (Tümpel *et al.*, 2002). In at least one case, the fish *cis*-regulatory sequence can direct-correct spatial expression in transgenic mouse embryos (Aparicio *et al.*, 1995). These results imply that selection can maintain *cis*-regulatory sequence and function for hundreds of millions of years.

Evolutionary Mechanisms Operating on Transcription

The evolutionary mechanisms that shape *cis*-regulatory sequences are not particularly well understood, primarily because the functional consequences of mutation in *cis*-regulatory DNA are difficult to assess. In a few cases, natural selection is probably acting directly on a particular *cis*-regulatory allele in

humans. For instance, some *cis*-regulatory alleles at the *IL-10* locus alter expression levels of the cell-surface protein it encodes. Since the IL-10 protein is used by the human immunodeficiency virus (HIV) to infect lymphocytes, allelic differences in levels of expression have a direct impact on life expectancy in infected individuals (Bamshad *et al.*, 2002). Positive selection on promoter sequences is also apparent in the genomes of some pathogens of humans, including *Mycobacterium tuberculanum* (Rinder *et al.*, 1998) and the hepatitis C virus (Buckwold *et al.*, 1997).

Examples of balancing selection on *cis*-regulatory alleles have also been found. For instance, a *cis*-regulatory allele at the Duffy blood group locus (*FY*) of humans increases blood pressure but also provides protection against malaria infection. This allele has gone to fixation in West Africa, but remains at low frequencies in most other locations (Tournamille *et al.*, 1995; Hamblin and Di Rienzo, 2000). Similarly, one *cis*-regulatory allele at the *LDH* locus in the teleost *Fundulus* is favored in colder water, and another in warm water. The distribution of these alleles across the range of the fish is the result of balancing selection on different local optima (Segal *et al.*, 1999).

In general, however, we know much less about the mechanisms that sort variation in *cis*-regulatory sequences than in coding sequences. It is not clear, for instance, why many human *cis*-regulatory alleles are present at intermediate frequencies (Rockman and Wray, 2002). This is strikingly different from polymorphisms in coding regions, where a single allele nearly always predominates. Nor is it clear why some binding sites differ between humans and their close relatives. Such binding site turnover may involve functional compensation or may be associated with changes in transcription profiles.

Macroevolutionary Changes in Phenotype

Much of the original motivation for understanding the evolution of *cis*-regulatory sequences and the transcription factors that interact with them was based on the expectation that mutations in these regions of the genome are particularly likely to produce interesting phenotypic consequences (Raff and Kauffman, 1983; Wilkins, 2002). Many cases are now known where changes in transcriptional regulation are phylogenetically correlated with macroevolutionary (e.g. interspecific) differences in morphology. In several cases, differences in the expression of *Hox* transcription factors along the anteroposterior body axis are correlated with changes in body proportions in arthropods and vertebrates (Burke *et al.*, 1995; Averof and Patel, 1997). Other examples include a change in the expression of the *tb* transcription factor, which was a key component in the domestication of maize from teosinte (Wang *et al.*, 1999), and differences in the expression of *bric-a-brac* among species of *Drosophila* that are correlated with abdominal pigmentation patterns (Kopp *et al.*, 2001).

Despite a growing list of such cases, much remains to be learned about what kinds of changes in anatomy are likely to result from mutations in transcriptional regulation, and which evolutionary mechanisms operate on *cis*-regulatory sequences. The role of transcriptional regulation in the evolution of new structures presents a particular challenge. In many cases, transcription factors have been recruited to regulate expression within new structures (Lowe and Wray, 1997; Wilkins, 2002). Although this process of recruitment seems to be pervasive, few cases have been examined in any detail.

The clear challenge for the future is to understand the evolution of *cis*-regulatory sequences as well as we currently understand the evolution of coding sequences. With this understanding will come important insights into the genetic basis for phenotypic divergence among species and the evolutionary origin of humans.

See also

Gene Structure and Organization
Transcription Factors
Transcription Factors and Human Disorders
Transcriptional Regulation: Coordination

References

Aparicio S, Morrison A, Gould A, *et al.* (1995) Detecting conserved regulatory elements with the model genome of the Japanese pufferfish, *Fugu rubipes*. *Proceedings of the National Academy of Sciences of the United States of America* **92**: 1684–1688.

Averof M and Patel NH (1997) Crustacean appendage evolution associated with changes in *Hox* gene expression. *Nature* **388**: 682–686.

Bamshad MJ, Mummidi S, Gonzles E, *et al.* (2002) A strong signature of balancing selection in the 5′ cis-regulatory region of CCR5. *Proceedings of the National Academy of Sciences of the United States of America* **99**: 10 539–10 544.

Buckwold VE, Xu ZC, Yen TSB and Ou JH (1997) Effects of a frequent double-nucleotide basal core promoter mutation and its putative single-nucleotide precursor mutations on hepatitis B virus gene expression and replication. *Journal of General Virology* **78**: 2055–2065.

Burke AC, Nelson CE, Morgan BA and Tabin C (1995) *Hox* genes and the evolution of vertebrate axial morphology. *Development* **121**: 333–346.

Carroll SB, Grenier JK and Weatherbee SD (2001) *From DNA to Diversity: Molecular Genetics and the Evolution of Animal Design*. Malden: Blackwell Science.

Cooper DN (1999) *Human Gene Evolution*. Oxford, UK: BIOS Scientific Publishers.

Davidson EH (2001) *Genomic Regulatory Systems: Development and Evolution*. San Diego, CA: Academic Press.

Dermitzakis ET and Clark AG (2002) Evolution of transcription factor binding sites in mammalian gene regulatory regions: conservation and turnover. *Molecular Biology and Evolution* **19**: 1114–1121.

Hamblin MT and Di Rienzo A (2000) Detection of the signature of natural selection in humans: evidence from the Duffy blood group locus. *American Journal of Human Genetics* **66**: 1669–1679.

Kopp A, Duncan I, Godt D and Carroll SB (2001) Genetic control and evolution of sexually dimorphic characters. *Nature* **408**: 553–559.

Lowe CJ and Wray GA (1997) Radical alterations in the roles of homeobox genes during echinoderm evolution. *Nature* **389**: 718–721.

Ludwig M and Krteiman M (1995) Evolutionary dynamics of the enhancer region of *even-skipped* in Drosophila. *Molecular Biology and Evolution* **12**: 1002–1011.

Raff RA and Kauffman TC (1983) *Embryos, Genes, and Evolution*. Bloomington, IN: Indiana University Press.

Rinder H, Thomschke A, Rusch-Gerdes S, *et al.* (1998) Significance of *ahpC* promoter mutations for the prediction of isoniazid resistance in *Mycobacterium tuberculosis*. *European Journal of Clinical Microbiology and Infectious Diseases* **17**: 508–511.

Rockman MV and Wray GA (2002) Abundant raw material for cis-regulatory evolution in humans. *Molecular Biology and Evolution* **19**: 1991–2004.

Segal JA, Lynn Barnett J and Crawford DL (1999) Functional analysis of natural variation in Sp1 binding sites of a TATA-less promoter. *Journal of Molecular Evolution* **49**: 736–749.

Shabalina SA, Ogurtsov AY, Kondrashov VA and Kondrashov AS (2001) Selective constraint in intergenic regions of human and mouse genomes. *Trends in Genetics* **17**: 373–376.

Shaw PJ, Wratten NS, McGregor AP and Dover GA (2002) Coevolution in bicoid-dependent promoters and the inception of regulatory incompatibilities among species of higher Diptera. *Evolution and Development* **4**: 265–277.

Tournamille C, Colin Y, Cartron JP and Le Van Kim C (1995) Disruption of a GATA motif in the *Duffy* gene promoter abolishes erythroid gene expression in Duffy-negative individuals. *Nature Genetics* **10**: 224–228.

Tümpel S, Maconochie M, Wiedeman LM and Krumlauf R (2002) Conservation and diversity in the *cis*-regulatory networks that integrate information controlling expression of *Hoxa2* in hindbrain and cranial neural crest cells in vertebrates. *Developmental Biology* **246**: 45–56.

Wang RL, Stec A, Hey J, Lukens L and Doebley J (1999) The limits of selection during maize domestication. *Nature* **398**: 236–239.

White RJ (2001) *Gene Transcription: Mechanisms and Control*. Oxford, UK: Blackwell Science.

Wilkins AS (2002) *The Evolution of Developmental Pathways*. Sunderland, MA: Sinauer Associates.

Wray GA, Hahn MW, Abouheif E, *et al.* (2003) The evolution of transcriptional regulation in eukaryotes. *Molecular Biology and Evolution* (in press).

Transcriptional Silencing in Tumorigenesis

See Methylation-mediated Transcriptional Silencing in Tumorigenesis

Transcription-coupled DNA Repair

Isabel Mellon, *University of Kentucky, Lexington, KY, USA*

Excision repair pathways remove a wide variety of lesions formed by endogenous and environmental agents, and protect humans from the mutagenic and carcinogenic consequences of persisting damage. Transcription-coupled repair is a subpathway of excision repair that selectively removes lesions from the transcribed strands of expressed genes.

Introduction

Cells are continuously exposed to a multitude of deoxyribonucleic acid (DNA)-damaging agents. Damage can be produced by exposure to agents present in the environment, such as ultraviolet (UV) light and chemical carcinogens. In addition, by-products of normal cellular metabolism can damage DNA. Left unrepaired, DNA damage contributes to increased genomic instability and neoplasia. To protect from the potentially deleterious consequences of persisting damage, cells possess a wide array of DNA repair pathways that recognize and remove damaged DNA. (*See* DNA Repair.)

Nucleotide excision repair (NER) is the major pathway for the removal of bulky lesions and it operates on a variety of structurally unrelated lesions. Hence, it is likely that the distortion of the DNA helix produced by the lesion is recognized and not the lesion itself. Substrates include cyclobutane pyrimidine dimers (CPDs) and (6–4) photoproducts produced by UV light, and many lesions formed by chemical agents. Transcription-coupled repair (TCR) has been clearly demonstrated to be a subpathway of NER and is often defined as more rapid or more efficient removal of lesions from the transcribed strand of an expressed gene compared with the nontranscribed strand. This was first demonstrated when studying the removal of UV-induced CPDs from each strand of the dihydrofolate reductase (*DHFR*) gene in hamster and human cell lines (Mellon *et al.*, 1987). Subsequent studies have demonstrated TCR of CPDs in *Escherichia coli* and yeast. In addition, certain bulky chemical lesions are substrates for TCR. Hence, this subpathway of NER has been conserved from bacteria to humans and operates on many different adducts. While the precise mechanism of TCR has not yet been completely elucidated, an early event probably involves blockage of the ribonucleic acid (RNA) polymerase (RNA pol) complex at lesions present in the transcribed strands of expressed genes. Most lesions that are substrates of TCR have been found to block transcription elongation.

Base excision repair (BER) represents a collection of different glycosylase-initiated repair pathways that, in general, recognize and remove smaller, less-distorting lesions. These lesions can be produced by endogenous metabolic activities that generate reactive oxygen species, deamination and abasic sites. In addition, they can be formed by exposure to environmental agents such as ionizing radiation and alkylating agents. Several studies that focused on the repair of oxidative damage have reported that TCR is also a subpathway of BER (Cooper *et al.*, 1997). Many studies have quantified repair by measuring the incorporation of bromodeoxyuridine into repair patches using an antibody to bromodeoxyuridine (BrdUrd). However, this method does not quantify any specific lesion. Ionizing radiation and other forms of oxidative agents produce a wide spectrum of different lesions, including some that are substrates for NER. Hence it is unclear whether the selective repair of oxidative damage from the transcribed strands of active genes represents BER, NER or some other repair pathway. Moreover, there has been no demonstration (either genetic or biochemical) of a direct role of BER in TCR of oxidative damage. In fact, while it has been reported that 8-oxoguanine is a substrate for TCR, the glycosylase encoded by *Ogg1*, which is required for the removal of 8-oxoguanine from nontranscribed sequences, is not required for its removal from transcribed sequences (Le Page *et al.*, 2000). Furthermore, 8-oxoguanine (Viswanathan and Doetsch, 1998) and thymine glycol (Tornaletti *et al.*, 2001) do not appear to block RNA polymerase elongation when examined in cell-free systems. In contrast to oxidative damage, TCR is not found in the repair of alkylation-induced lesions that are also substrates for BER pathways (Plosky *et al.*, 2002). Hence, it is unclear whether TCR is a bona fide subpathway of a glycosylase-mediated BER pathway.

Mechanisms of TCR in *E. coli*

TCR in *E. coli* was first alluded to by studies of mutation frequency decline in transfer RNA (tRNA) operons. It was documented as more rapid removal of CPDs from the transcribed strand of the *lac* operon.

Based on observations in mammalian cells and bacteria, it was proposed that blockage of elongating RNA polymerase complexes at CPDs is an initial step in TCR. CPDs in the transcribed strands of expressed genes pose blocks to RNA polymerase elongation, while those in the nontranscribed strand are generally bypassed. Also, the RNA polymerase complex and/or some feature of the transcription bubble probably play important roles in TCR by providing signals for the recruitment of NER proteins.

NER in *E. coli* is well understood and has served as a paradigm for the investigation of other organisms. Damage recognition and processing is carried out by the UvrABC system. UvrA dimerizes (UvrA$_2$) and binds UvrB and the UvrA$_2$B complex binds DNA. The helicase activity of the complex may enable scanning for damage by translocation along the DNA and it unwinds DNA at the site of the lesion. UvrA$_2$ dissociates from the damaged site, leaving an unwound preincision complex containing UvrB that is recognized and bound by UvrC. UvrBC produces an incision on each side of the lesion, both made by UvrC. UvrD unwinds and displaces the damaged oligonucleotide produced by the bracketed incisions. The gap is filled in by DNA polymerase I and the repair patch is sealed by DNA ligase.

It remains somewhat unclear what precise mechanism facilitates NER in the transcribed strands of active genes. The mutation frequency decline protein Mfd is required for TCR in a cell-free system and the *lac* operon in intact cells, but is not required for repair of nontranscribed DNA. Furthermore, Selby and Sancar (1993) have found that Mfd promotes the release of RNA polymerase complexes stalled at lesions in the transcribed strand of a gene expressed in a cell-free system. However, this interesting observation has provided a conundrum, in that, if the stalled polymerase complex or the transcription bubble is an important signal for TCR, presumably this signal is lost when the polymerase complex becomes displaced from the lesion. Attempts to reconcile this have proposed that Mfd also recruits UvrA to damaged sites using its UvrB homology sequence. However, there is no evidence that Mfd remains bound to a lesion once the RNA polymerase complex is displaced. Moreover, if UvrA exists as a dimer already bound to UvrB, then it is difficult to imagine how this 'hand-off' with Mfd takes place. More recent studies have provided additional insights into possible mechanisms. First, it is now clear that certain lesions are bound more efficiently when present in 'bubble' substrates and incision can occur in the absence of certain NER factors. For example, bubble substrates containing a chemical adduct are efficiently incised in the absence of UvrA (Zou *et al.*, 2001). In addition, bubble-like structures trigger the 3' and 5' endonuclease activities

of UvrBC. Hence, it is likely that some aspect of the transcription bubble plays a key role in TCR. Second, Park *et al.* (2002) have recently identified a novel function for Mfd; it has the ability to reverse 'backtracked' RNA polymerase complexes. Backtracking involves translocation of the RNA polymerase complex and the transcription bubble backward from the site of blockage. In fact, Hanawalt and colleagues propose that RNA polymerase complexes backtrack at CPDs, and studies performed *in vitro* demonstrate that RNA polymerase II (pol II) complexes can translocate backwards from CPDs for up to 25 nt. While similar studies have yet to be performed using *E. coli* RNA polymerase and CPDs, a model for TCR in *E. coli* is described in **Figure 1**, which incorporates backtracking and loading of NER factors onto the transcription bubble.

The model is as follows: after UV irradiation, RNA polymerase complex elongates until it encounters a CPD on the transcribed strand. It stalls because the CPD is a noncoding lesion and lacks information for template-directed addition of the ribonucleotide into the nascent strand of RNA. The polymerase then translocates backwards. Mfd recognizes the backtracked complex and binds to DNA upstream of the RNA polymerase complex and to the RNA polymerase complex (Park *et al.*, 2002). Mfd then induces forward translocation of the polymerase until it reencounters and perhaps even bypasses the lesion for a short distance. UvrA$_2$B or perhaps UvrB alone loads 5' to the lesion (relative to the damaged strand). The loading of UvrB is facilitated by features of the transcription bubble brought about by the forward translocation induced by Mfd. At this point the polymerase may backtrack again or may be completely released by Mfd. UvrC then binds to the lesion-bound UvrB complex, resulting in a stable preincision complex, and this and subsequent downstream NER events continue as they would in nontranscribed DNA. The salient point of this model is that the 'coupling' of NER to transcription is mediated by the correct positioning of the transcription bubble at the lesion rather than direct physical interactions between NER proteins and transcription factors. During NER in nontranscribed DNA, the loading of UvrB is an asymmetric process that initiates on the 5' side of the damage and involves denaturing the DNA near the site of the lesion (Moolenaar *et al.*, 2000). In a transcribed strand, RNA polymerase denatures DNA at the lesion when it forms a bubble around it. The damage-containing bubble substrate may be recognized by UvrA$_2$B or UvrB alone when it encounters the stalled polymerase complex from the 5' side (relative to the damaged strand). Mfd may serve two functions: one is to maintain the transcription bubble at the site of the lesion by reversing backtracked complexes; the other

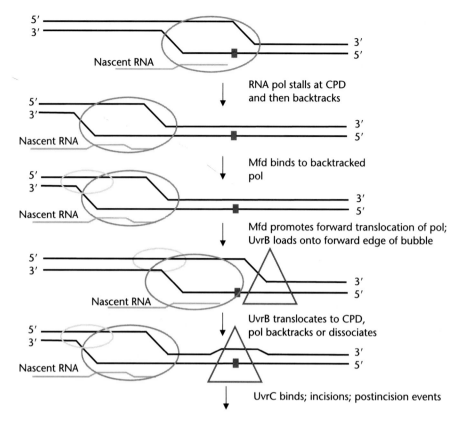

Figure 1 Model for transcription-coupled repair in *E. coli*. Elongating RNA polymerase complex (RNA pol; large oval) stalls at a cyclobutane pyrimidine dimer (CPD; small shaded square) in the transcribed strand. The polymerase complex, transcription bubble and nascent RNA translocate backwards. The mutation frequency decline protein Mfd (small oval) binds backtracked polymerase and DNA upstream of the bubble. Mfd promotes the forward translocation of the polymerase complex. The protein UvrB (triangle) binds 5′ (relative to the damaged strand), loads onto the forward edge of the bubble and translocates to the lesion. The polymerase complex backtracks or is dissociated by Mfd. Subsequent nucleotide excision repair (NER) processing events continue as they would in nontranscribed DNA. (After Park *et al.*, 2002.)

may be to ultimately displace the complex from the damaged site to allow incision and DNA synthesis.

Mechanisms of TCR in Mammalian Systems

The general strategy of NER in mammalian systems closely parallels that of *E. coli*. However the repertoire of proteins required for mammalian NER is significantly more complex. While somewhat controversial, there is considerable evidence that the XPC/hHR23B complex is involved in an early step of damage recognition at least in the repair of (6–4) photoproducts. General transcription factor TFIIH is then recruited, which results in unwinding near the lesion by virtue of the helicase activities of XPB and XPD, components of TFIIH. XPG, XPA/RPA and ERCC1/XPF are assembled to form a stable preincision complex. Dual incisions are carried out: the 3′

incision by XPG and the 5′ incision by ERCC1/XPF, followed by postincision events. With the exception of XPC (and perhaps hHR23B) the same repertoire of proteins described above are required for the repair of nontranscribed DNA and for TCR. It is likely that in TCR, the RNA polymerase complex replaces the function of XPC/hHR23B in damage recognition. Genetic studies have indicated a requirement for additional genes in TCR. These include Cockayne syndrome group A and B (*CKN1* and *ERCC6*) genes, genes involved in mismatch repair, UV-sensitive syndrome (*ERCC5*) and the XPA-binding protein 2 gene (*HCNP*). Mutations in these genes result in a selective loss of TCR, while repair in nontranscribed DNA is not affected or less affected. (*See* DNA Repair: Disorders; Mismatch Repair Genes.)

For the purposes of a description of a putative mechanism of TCR in mammalian systems (**Figure 2**), CSA, CSB and XAB2 are included, because biochemical studies have implicated some mechanism for their direct involvement. CSB interacts with pol II but

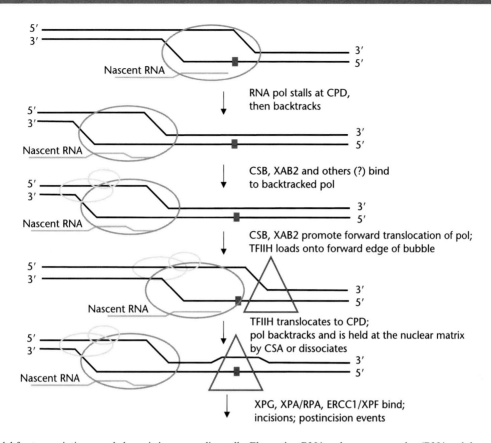

5′
3′
Nascent RNA

RNA pol stalls at CPD, then backtracks

5′
3′
Nascent RNA
3′
5′

CSB, XAB2 and others (?) bind to backtracked pol

5′
3′
Nascent RNA
3′
5′

CSB, XAB2 promote forward translocation of pol; TFIIH loads onto forward edge of bubble

5′
3′
Nascent RNA
3′
5′

TFIIH translocates to CPD; pol backtracks and is held at the nuclear matrix by CSA or dissociates

5′
3′
Nascent RNA
3′
5′

XPG, XPA/RPA, ERCC1/XPF bind; incisions; postincision events

Figure 2 Model for transcription-coupled repair in mammalian cells. Elongating RNA polymerase complex (RNA pol; large oval) stalls at a cyclobutane pyrimidine dimer (CPD; small shaded square) in the transcribed strand. The polymerase complex, transcription bubble and nascent RNA translocate backwards. Proteins CSB and XAB2 (small ovals) bind the backtracked polymerase complex and promote forward translocation. General transcription factor TFIIH (triangle) binds 5′ to the damage and loads onto the forward edge of the bubble. The polymerase backtracks and is held at the nuclear matrix by CSA or dissociates. Subsequent NER events continue as they would in nontranscribed DNA.

unlike Mfd, CSB does not dissociate pol II complexes stalled at lesions. CSA and XAB2 interact with each other and with CSB and pol II. In addition, CSA has been found to translocate to the nuclear matrix after UV irradiation. In addition to having roles in TCR, CSB and XAB2 appear to have direct roles in transcription. A model for TCR in mammalian cells (**Figure 2**) makes use of the same key features proposed for *E. coli*. 'Coupling' of NER to transcription is dependent upon the positioning of the transcription bubble at the lesion and not necessarily on direct interactions between NER proteins and transcription factors. CSB and XAB2 serve roles that are equivalent to Mfd. They bind pol II, play a direct role in elongation and help maintain the transcription bubble at the lesion. CSB can bind double-stranded DNA and, perhaps like Mfd, binds the polymerase complex and DNA upstream of the polymerase complex. TFIIH serves a role that is equivalent to UvrB; it encounters the transcription bubble on the 5′ side (relative to the damaged strand) and translocates to

the lesion. The polymerase complex may backtrack and be held upstream while repair takes place by virtue of CSA-mediated association with the nuclear matrix. Subsequent NER events occur as they would in the repair of nontranscribed DNA.

The importance of TCR in protecting humans from disease is clearly illustrated by the severe symptoms associated with inheritable genetic diseases such as Cockayne syndrome and UV-sensitive syndrome. Continued investigation of the mechanisms of TCR in *E. coli* and eukaryotic systems should provide a better understanding of how alterations in TCR cause human disease.

See also
DNA Repair

References

Cooper PK, Nouspikel T, Clarkson SG and Leadon SA (1997) Defective transcription-coupled repair of oxidative base damage in Cockayne syndrome patients form XP group G. *Science* **275**: 990–993.

Le Page F, Klungland A, Barnes DE, Sarasin A and Boiteux S (2000) Transcription coupled repair of 8-oxoguanine in murine cells: The *Ogg1* protein is required for repair in nontranscribed sequences but not in transcribed sequences. *Proceedings of the National Academy of Sciences of the United States of America* **97**: 8397–8402.

Mellon I, Spivak G and Hanawalt PC (1987) Selective removal of transcription-blocking DNA damage from the transcribed strand of the mammalian *DHFR* gene. *Cell* **51**: 241–249.

Moolenaar GF, Monaco V, van der Marel GA, *et al.* (2000) The effect of the DNA flanking the lesion on formation of the UvrB–DNA preincision complex. *Journal of Biological Chemistry* **275**: 8038–8043.

Park J-S, Marr MT and Roberts JW (2002) *E. coli* transcription repair coupling factor (Mfd protein) rescues arrested complexes by promoting forward translocation. *Cell* **109**: 757–767.

Plosky B, Samson L, Engelward B, *et al.* (2002) Base excision repair and nucleotide excision repair contribute to the removal of *N*-methylpurines from active genes. *DNA Repair* **1**: 683–696.

Selby CP and Sancar A (1993) Molecular mechanism of transcription-repair coupling. *Science* **260**: 53–58.

Tornaletti S, Maeda LS, Lloyd DR, Reines D and Hanawalt PC (2001) Effect of thymine glycol on transcription elongation by T7 RNA polymerase and mammalian RNA polymerase II. *Journal of Biological Chemistry* **276**: 45376–45371.

Viswanathan A and Doetsch PW (1998) Effects of nonbulky DNA base damages on *Escherichia coli* RNA polymerase-mediated elongation and promoter clearance. *Journal of Biological Chemistry* **272**: 21276–21281.

Zou Y, Luo C and Geacintov NE (2001) Hierarchy of DNA damage recognition in *Escherichia coli* nucleotide excision repair. *Biochemistry* **40**: 2923–2931.

Further Reading

Batty DP and Wood RD (2000) Damage recognition in nucleotide excision repair of DNA. *Gene* **241**: 193–204.

Engstrom J, Larsen S, Rogers S and Bockrath R (1984) UV-mutagenesis at a cloned target sequence: converted suppressor mutation is insensitive to mutation frequency decline regardless of the gene orientation. *Mutation Research* **132**: 143–152.

Friedberg EC, Walker GC and Siede W (1995) *DNA Repair and Mutagenesis*. Washington, DC: ASM Press.

Kamiuchi S, Saijo M, Citterio E, *et al.* (2002) Translocation of Cockayne syndrome group A protein to the nuclear matrix: possible relevance to transcription-coupled repair. *Proceedings of the National Academy of Sciences of the United States of America* **99**: 201–206.

Mellon I and Hanawalt PC (1989) Induction of the *Escherichia coli* lactose operon selectively increases repair of its transcribed DNA strand. *Nature* **342**: 95–98.

Scicchitano DA and Mellon I (1997) Transcription and DNA damage: a link to a kink. *Environmental Health Perspectives* **105**: 145–153.

Spivak G, Itoh T, Matsunaga T, *et al.* (2002) Ultraviolet-sensitive syndrome cells are defective in transcription-coupled repair of cyclobutane pyrimidine dimers. *DNA Repair* **50**: 1–15.

Sugasawa K, Ng JMY, Masutani CSI, *et al.* (1998) Xeroderma pigmentosum group C protein complex is the initiator of global genome nucleotide excision repair. *Molecular Cell* **2**: 223–232.

Svejstrup JQ (2002) Mechanisms of transcription coupled DNA repair. *Nature Reviews* **3**: 21–29.

Tornaletti S, Reines D and Hanawalt PC (1999) Structural characterization of RNA polymerase II complexes arrested by a cyclobutane pyrimidine dimer in the transcribed strand of template DNA. *Journal of Biological Chemistry* **274**: 24124–24130.

Transcription Factors

Intermediate article

David S Latchman, *Birkbeck College, University of London, London, UK*

Transcription factors are regulatory proteins that can increase or decrease the transcription of a particular gene from deoxyribonucleic acid into the corresponding ribonucleic acid. They play a key role in embryonic development, the creation and maintenance of cell type- and tissue-specific patterns of protein synthesis and the response to cellular signaling pathways.

Article contents

Transcriptional Control and Transcription Factors

The regulation of gene transcription is central both to tissue-specific gene expression and to the regulation of gene activity in response to specific stimuli. Thus, while some cases of regulation after transcription do exist, in most cases regulation occurs at the level of transcription by deciding which genes will be transcribed into the primary ribonucleic acid (RNA) transcript.

Inspection of the regulatory regions of genes that show similar patterns of transcription reveals the presence of short deoxyribonucleic acid (DNA) sequences that are held in common by genes with a particular pattern of regulation but are absent from other genes that do not show this pattern of regulation. Thus, for example, genes whose transcription is induced in response to exposure to elevated temperature contain a common regulatory element known as the heat shock element (HSE), which is absent from genes that do not show heat-inducible transcription. Transfer of the HSE to a nonheat-inducible gene is sufficient to render it heat-inducible (Pelham, 1982). Similarly, genes induced by glucocorticoid hormone or by cyclic adenosine monophosphate (cAMP) show other specific elements, known as the glucocorticoid response element (GRE) and cAMP response element (CRE), respectively, which mediate their response to these stimuli.

It is now clear that these short DNA sequences act by binding specific regulatory proteins, known as transcription factors, which in turn regulate the transcription of the gene either positively or negatively to produce the observed effect on transcription.

Direct versus Indirect Interactions of Transcription Factors with DNA

Many transcription factors exert their effects on transcription by binding directly to DNA sequences such as the HSE, GRE or CRE. An increasing number of factors are being characterized, however, that exert their effects by binding via a protein–protein interaction to another factor, which is itself bound to DNA.

DNA binding

Detailed analysis of a number of different transcription factors has indicated that they have a modular structure in which specific regions of the molecule are responsible for binding to the DNA, while other regions produce a stimulatory or inhibitory effect on transcription (**Figure 1**). Studies on the DNA-binding regions of different transcription factors have revealed several

distinct structural elements that can produce DNA binding (for review, see Travers, 1993). Transcription factors are frequently classified on the basis of their DNA-binding domains, and a selection of these binding domains is listed in **Table 1**. Thus a wide variety of DNA-binding domains allow transcription factors to bind to their appropriate DNA sequences within target genes.

Non-DNA-binding transcription factors

Although many transcription factors exert their effect by binding directly to DNA, not all do so. This is seen in the case of the herpes simplex virus (HSV) transactivator VP16, which interacts with the cellular transcription factor Oct-1. Thus, Oct-1 can bind directly to an 8-base sequence in the DNA but activates transcription only weakly. However, in HSV-infected cells, VP16 binds to DNA-bound Oct-1 via a

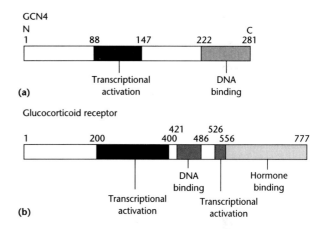

Figure 1 Structure of (a) the yeast GCN4 factor and (b) the mammalian glucocorticoid receptor, indicating the distinct regions that mediate DNA binding or transcription activation.

Table 1 Transcription factor families classified by their DNA-binding domains

Domain	Factor-containing domain	Comments
Homeobox	Numerous *Drosophila* homeotic genes, related genes in other organisms	DNA binding mediated via helix–turn–helix motif
POU	Oct-1, Oct-2, Pit-1, Unc-86	Consists of POU-specific domain and POU-homeobox
Paired box	Various *Drosophila* segmentation genes, PAX factors	Often found together with a homeobox in PAX factors
Cysteine–histidine zinc-finger	TFIIIA, Kruppel, SP1, etc.	Multiple copies of finger motif
Cysteine–cysteine zinc-finger	Steroid-thyroid hormone receptor family	Single pairs of fingers, related motifs in adenovirus E1A and yeast GAL4, etc.
Basic element	C/EBP, c-Fos, c-Jun, GCN4	Often found in association with leucine zipper or helix–loop–helix dimerization motifs
Ets domain	Ets-1, Elk-1, SAP	Contain helix–turn–helix motif

protein–protein interaction and strongly stimulates transcription. Hence VP16 acts by interacting with another factor bound to DNA and then stimulates transcription.

Since the discovery of VP16, many cellular transcription factors, which act without binding to DNA, have been discovered. Factors of this type can either activate or repress transcription (see the section Activation and Repression of Transcription) and are called, respectively, coactivators or corepressors. A coactivator of particular interest is the cAMP response element binding protein (CREB) CBP (CREB binding protein). CBP was originally identified on the basis of its binding to the CREB transcription factor only following phosphorylation of CREB on serine 133 (Ser133). This effect is of vital importance in gene regulation by cAMP.

Thus CREB acts by binding to the CRE. After treatment with cAMP, the CREB factor becomes phosphorylated on Ser133. However, such phosphorylation does not result in enhanced ability of CREB to bind to the CRE. Indeed, CREB is already bound to the CRE before exposure of cells to cAMP but is unable to activate transcription. Such phosphorylation results, however, in CREB being able to bind CBP.

In turn, CBP is able to bind to specific components of the basal transcriptional complex, thereby linking CREB to this complex and allowing stimulation of transcription after cAMP treatment. In addition, however, CBP has been shown to possess histone acetyl transferase activity. Such enhanced acetylation of histones has been shown to occur in regions of DNA, which are active or potentially active in transcription (for review, see Grunstein, 1997) and to be involved in the open chromatin structure characteristic of such regions.

Hence CBP plays a key role as a coactivator binding to DNA-bound CREB via a protein–protein interaction and allowing transcriptional activation to occur. CBP has been shown to interact with a variety of other transcription factors and to play a key role in a number of different aspects of gene regulation (for review, see Shikama et al., 1997).

Activation and Repression of Transcription

After direct DNA binding or protein–protein interaction, transcription factors must act either to enhance or reduce the rate of gene transcription.

Activation of transcription

As shown in **Figure 1**, many transcription factors contain, in addition to the DNA-binding domain, specific regions that are necessary for the activation of transcription and are known as activation domains (for review, see Mitchell and Tjian, 1989). As with DNA-binding domains, a number of distinct types of activation domains have been identified, which are defined on the basis that they are rich in acidic amino acids, glutamine residues or proline residues.

These activation domains appear to function by interacting with components of the basal transcriptional complex. This is a complex of RNA polymerase II and various transcription factors such as TFIIB and TFIID which assembles at the gene promoter and is essential for transcription to occur (for review, see Nikolev and Burley, 1997). Activation domains have been shown to interact either directly with specific components of this complex or indirectly by interacting with coactivator molecules, which then interact with the basal complex itself. Whatever the case, such interactions appear to result in enhanced transcription either by stimulating the rate of transcription factor complex assembly or by stimulating the level of its activity.

Hence, following binding to their appropriate DNA-binding site mediated via the DNA-binding domain, the activation domains of specific activating transcription factors can interact with the basal transcriptional complex so as to stimulate transcription.

Repression of transcription

Although it was originally thought that most eukaryotic transcription factors acted by stimulating transcription, it has now become clear that a wide variety of factors act by inhibiting the transcription of specific genes and that such inhibitory transcription factors may be at least as important as stimulatory factors (for review, see Hanna-Rose and Hansen, 1996; Latchman, 1996b).

The earliest examples of such inhibitory transcription factors were shown to act by interfering with the activity of a positively acting factor, thereby blocking its stimulatory effect on transcription (**Figure 2a–2d**). This could be achieved, for example, by preventing the positively acting factor from binding to DNA, either via the negatively acting factor organizing an inactive chromatin structure (**Figure 2a**) or by binding to the DNA-binding site of the activator (**Figure 2b**) or by the formation of a non-DNA-binding protein–protein complex between the positively acting factor and the negatively acting factor (**Figure 2c**). Alternatively, the negatively acting factor could act by interacting with the positively acting factor to block the activity of its activation domain in a phenomenon known as 'quenching' (**Figure 2d**).

It later became clear, however, that a class of inhibitory transcription factors exists that can directly inhibit transcription even in the absence of a positively acting factor (**Figure 2e**). These factors can thus reduce the basal level of transcription below that observed even in the absence of any activating molecule, and they appear to function by interacting with the basal transcriptional complex either directly or indirectly via a corepressor so as to reduce its activity. They thus constitute the antithesis of activating molecules and possess defined inhibitory domains, which are responsible for their effects.

Hence the balance between binding of transcriptional activators and transcriptional repressors to the regulatory region of a particular gene will determine the rate of its transcription in any particular situation. Clearly, however, in order for a particular gene to respond to specific signals or to be regulated in a cell type-specific manner, the balance between these activating and repressing molecules must change in different situations. The mechanisms that are used to regulate transcription factor activity are discussed in the next section.

Regulation of Transcription Factors by Cellular Differentiation and Signaling Pathways

Transcription factors can be regulated at two levels, namely by the regulation of transcription factor synthesis and the regulation of transcription factor activity (**Figure 3**).

Regulation of synthesis

In a number of different situations, a transcription factor is regulated by being synthesized in one particular tissue or cell type and not in other tissues. The most dramatic example of this concerns the MyoD transcription factor, which is synthesized only in skeletal muscle cells. Thus, in this case, the over-expression of the MyoD factor in undifferentiated fibroblast cells is sufficient to convert them to skeletal muscle cells, indicating the critical role for this factor in the induction of muscle-specific gene expression.

Regulation of transcription factor activity

Although the regulation of transcription factor synthesis is an important control point, it cannot be the only regulatory mechanism that controls transcription factor activity. If this were the case, the enhanced synthesis of a transcription factor in response to a particular stimulus would be controlled by enhanced transcription of its corresponding gene, which in turn would require the *de novo* synthesis of further transcription factors, so resulting in the need for new transcription of these genes and so on.

Therefore, it is necessary to have an additional mechanism that allows *de novo* gene transcription by the activation of preexisting transcription factors (**Figure 3b**), and this mechanism is used extensively

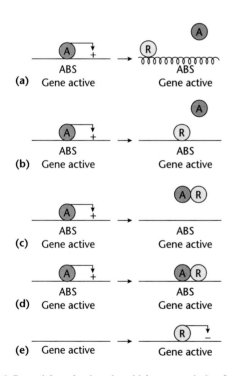

Figure 2 Potential mechanisms by which a transcription factor can repress gene expression. This can occur: (a) by the repressor (R) producing a tightly packed chromatin structure which prevents an activator (A) from binding; (b) by the repressor binding to the DNA-binding site of the activator and preventing it from binding and activating gene expression; (c) by the repressor interacting with the activator in solution and preventing its DNA binding; (d) by the repressor binding to DNA with the activator and neutralizing its ability to activate gene expression; or (e) by direct repression by an inhibitory transcription factor. ABS: activator-binding site.

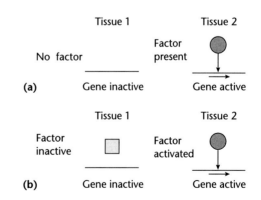

Figure 3 Gene activation mediated by (a) the synthesis of a transcription factor only in a specific tissue; or (b) activation of the transcription factor only in a specific tissue.

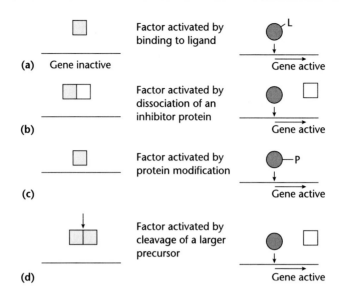

Figure 4 Mechanisms by which transcription factors can be activated by posttranslational changes. The circle represents an active transcription factor, while the square represents a non-active factor. A shaded box represents an inactive form of a transcription factor, while the open box represents either an inhibitory protein or a portion of the factor which has been cleaved off in the process of activation. L: ligand; P: protein.

to modulate transcription factor function in response to specific cellular signaling pathways. Such activation of preexisting transcription factors can occur via a number of different mechanisms (**Figure 4**), which can involve ligand binding, alterations in protein–protein interaction, transcription factor phosphorylation or cleavage of an inactive factor to generate a smaller active factor. Thus, for example, in the case of the steroid receptors, the inactive receptor is associated with an inhibitory protein hsp90. Following binding of the steroid hormone ligand, hsp90 dissociates and the steroid receptor moves to the nucleus, where it can bind to its appropriate response element and switch on transcription.

Similar dissociation of an inhibitory protein allowing DNA binding by the active transcription factor also occurs for the nuclear factor-κB (NF-κB). However, in this latter case, the dissociation of the inhibitory protein IκB from the NF-κB is mediated via the phosphorylation of the IκB protein, which results in its dissociation from NF-κB and targets it for rapid degradation. Hence, in this case, the mechanisms of regulatory protein–protein interaction and transcription factor phosphorylation are combined. A similar regulation of transcription factor activity by phosphorylation is also seen in the case of the CREB factor, which plays a critical role in the regulation of transcription in response to cAMP (see above).

Hence, in a specific cell type or in response to a specific stimulus, specific transcription factors are either synthesized or become activated following posttranslational modification. The binding of these

transcription factors to their appropriate recognition sequences thus produces specific patterns of gene transcription and is responsible for the observed dependence of particular patterns of gene activity on specific DNA-binding sequences, as discussed in the section Transcriptional Control and Transcription Factors.

Transcription Factors and Disease

Given the vital role of transcription factors in a wide variety of cellular processes, it is not surprising that alterations in these factors can result in human disease (for review, see Latchman, 1996a). Thus, for example, a number of developmental abnormalities have been shown to result from mutations, which result in the inactivation of specific transcription factors. Mutations in the gene encoding the POU family transcription factor Pit-1 have been identified in patients with combined pituitary hormone deficiency in whom there is no production of growth hormone, prolactin or thyrotropin, resulting in mental retardation and growth deficiency. Similarly, several different diseases have been shown to result from mutations in the genes encoding members of the PAX family of transcription factors. Mutations in *PAX3* (*paired box gene 3 (Waardenburg syndrome 1)*) result in Waardenburg syndrome, while mutations in *PAX6* (*paired box gene 6 (aniridia, keratitis)*) are associated with eye defects.

Such mutations resulting in developmental defects can also affect non-DNA-binding cofactors, as well as DNA-binding transcription factors. This is seen in the case of the CBP factor, discussed in the section Non-DNA-binding transcription factors. Thus inactivation of the gene encoding CBP results in Rubinstein–Taybi syndrome, involving mental retardation and various physical abnormalities (for review, see D'Arcangelo and Curran, 1995).

Hence, a variety of developmental disorders can arise from inactivation of transcription factors by mutation. It is likely, however, that such mutations that allow live individuals to be born represent the tip of the iceberg of transcription factor mutation, with the inactivation of many genes encoding transcription factors producing such a severe defect that it is not compatible with the individual's survival.

Alterations in transcription factors have also been observed in a number of human cancers such as leukemias, in which genes encoding specific transcription factors have been involved in chromosomal rearrangements so that they are either expressed at a higher level or become fused to part of another protein so that a novel protein with oncogenic activity is produced. Such chromosomal rearrangements are of particular importance in the case of acute and chronic myeloid leukemia but also occur in solid tumors.

In addition to such rearrangements of oncogenic transcription factors, mutations inactivating specific antioncogenic transcription factors have also been observed in human cancers. Mutations in the gene encoding the transcription factor p53 are particularly common in cancers, and it has been estimated that the majority of human cancers contain mutations in the p53 gene (*TP53; tumor protein p53 (Li–Fraumeni syndrome)*). When taken together with the existence of other antioncogenes, encoding transcription factors such as the retinoblastoma gene and the Wilms tumor gene, and the existence of many oncogenic transcription factors, it is clear that alterations in such transcription factors are likely to be involved in virtually all human cancers.

See also

Transcriptional Regulation: Coordination
Transcriptional Regulation: Evolution
Transcription Factors and Human Disorders

Conclusion

It is clear that transcription factors are vital to the process of transcriptional control of gene expression, which in turn underlies normal embryonic development, the creation and maintenance of cell type- and tissue-specific protein synthesis, and the response to specific cellular signaling pathways.

References

D'Arcangelo G and Curran T (1995) Smart transcription factors. *Nature* **376**: 292–293.

Grunstein M (1997) Histone acetylation in chromatin structure and transcription. *Nature* **389**: 349–352.

Hanna-Rose W and Hansen U (1996) Active repression mechanisms of eukaryotic transcription repressors. *Trends in Genetics* **12**: 229–234.

Latchman DS (1996a) Transcription factor mutations and disease. *New England Journal of Medicine* **334**: 28–33.

Latchman DS (1996b) Inhibitory transcription factors. *International Journal of Biochemistry and Cell Biology* **28**: 965–974.

Mitchell PJ and Tjian R (1989) Transcriptional regulation in mammalian cells by sequence specific DNA-binding proteins. *Science* **245**: 371–378.

Nikolev DB and Burley SK (1997) RNA polymerase II transcription initiation: a structural view. *Proceedings of the National Academy of Sciences of the United States of America* **94**: 15–22.

Pelham HRB (1982) A regulatory upstream promoter element in the *Drosophila* hsp70 heat shock gene. *Cell* **30**: 517–528.

Shikama N, Lyon J and La Thangue NB (1997) The p300/CBP family: integrating signals with transcription factors and chromatin. *Trends in Cell Biology* **7**: 230–236.

Travers A (1993) *DNA–Protein Interactions*. London, UK: Chapman & Hall.

Further Reading

Latchman DS (ed.) (1997) *Landmarks in Gene Regulation*. Colchester, UK: Portland Press Limited.

Latchman DS (ed.) (2002) *Gene Regulation: A Eukaryotic Perspective*, 4th edn. Cheltenham, UK: Nelson Thornes Publishers.

Latchman DS (1998) *Eukaryotic Transcription Factors*, 3rd edn. London, UK: Academic Press.

Latchman DS (ed.) (1998) *Transcription Factors: A Practical Approach*, 2nd edn. Oxford, UK: IRL Press.

Transcription Factors and Human Disorders

Gregg L Semenza, *Johns Hopkins University School of Medicine, Baltimore, Maryland, USA*

Transcription factors play a major role in the regulation of gene expression which underlies human development, physiology and pathophysiology. Germ-line mutations in genes encoding transcription factors are responsible for a large number of congenital malformation syndromes. Somatic mutations in these genes also play important roles in the pathogenesis of cancer.

Introduction

Underlying human development, physiology and pathophysiology is the concept of differential gene expression. The transcription of deoxyribonucleic acid (DNA) into ribonucleic acid (RNA) is the first step in the process of gene expression and is thus a major site of regulation. Transcriptional regulation is based upon the binding of transcription factors to specific DNA sequences within target genes. These bound factors influence the rate of initiation and/or elongation of the primary RNA transcript by RNA polymerase II (RNAP II), the enzyme responsible for the synthesis of RNAs encoding most proteins. Activators and repressors are transcription factors that have positive and negative effects respectively on the rate of gene transcription. Transcriptional activators can stabilize the formation of the transcription initiation complex, a megadalton macromolecular assembly comparable in size to the ribosome that contains RNAP II, general transcription factors and other regulatory proteins. Activators can stabilize the transcription initiation complex via direct protein–protein interactions or indirectly by binding to coactivators, which in turn interact with one or more components of the transcription initiation complex. The recruitment of coactivators is also essential because these proteins can be considered not only as structural proteins (see above) but also as enzymes that have histone acetyltransferase activity. Histones are highly basic molecules that bind tightly to the negatively charged DNA (Kornberg and Lorch, 1999). Acetylation of histones alters their ability to bind DNA tightly, which is believed to be essential to allow transcription of the DNA into RNA. Whereas transcriptional activators recruit coactivators with histone acetyltransferase activity, repressors recruit corepressors with histone deacetylase activity (Lemon and Tjian, 2000).

Transcription factors are categorized according to the protein motif utilized for DNA binding. Examples of such motifs are the zinc finger, basic helix–loop–helix, basic-leucine zipper, high-mobility group (HMG) domain and homeodomain proteins. The human genome contains over 2000 genes encoding transcription factors. Because transcription factors are essential in regulating development and postnatal physiology, germ-line mutations in genes encoding transcription factors are responsible for a large number (over 100) of congenital malformation syndromes. Somatic mutations in these genes also play important roles in the pathogenesis of cancer. To illustrate these principles, mutations in selected genes encoding zinc finger, paired box (PAX) domain, homeodomain, and HMG domain transcription factors are defined and described in the text and tables below. For a more comprehensive summary, see Further Reading.

Zinc-finger Transcription Factors

The zinc finger is a motif containing four amino acids, often four cysteine or two cysteine and two histidine residues, that form a tetrahedral coordination complex with a $Zn(II)$ ion such that the resulting tertiary structure has DNA-binding properties (Wolfe *et al.*, 2000). Transcription factors that utilize this motif usually contain two or more zinc fingers per molecule. Zinc-finger proteins probably constitute the largest group of transcription factors, which in humans may consist of several thousand members. The zinc-finger proteins include members of the nuclear receptor superfamily which are ligand-activated transcription factors that include the steroid (androgen, estrogen, glucocorticoid, mineralocorticoid and progesterone) and nonsteroid (retinoic acid, thyroid hormone, vitamin D_3) hormone receptors. Since the biological effects of these hormones are mediated via activation of their cognate receptors, mutations in the genes encoding these nuclear receptor transcription factors result in hormone insensitivity syndromes (**Table 1**). In addition, somatic mutations in the genes encoding the androgen and estrogen receptors play important roles in the pathogenesis of prostate and breast cancer respectively. In addition to the nuclear receptor superfamily, mutations in genes encoding other zinc-finger transcription factors also result in human disease. Mutations involving the Wilms tumor 1 (*WT1*) gene result in the autosomal dominant WAGR (Wilms tumor, Aniridia, Genitourinary malformation, Retardation) and Denys–Drash syndromes, which are characterized by genitourinary malformations, renal dysgenesis and predisposition to Wilms tumor of the kidney (**Table 2**). This provides an example of how a transcription factor can affect both prenatal organ development and postnatal cancer susceptibility within the same organ.

Table 1 Germ-line mutations in genes encoding nuclear receptor zinc-finger factors

Gene	Syndrome
AR	Androgen insensitivity syndromes; spinal/bulbar muscular atrophy
NR0B1	X-linked adrenal hypoplasia congenita; dosage-sensitive sex reversal
ESR1	Estrogen resistance
NR3C1	Glucocorticoid resistance
HNF4A	Maturity onset diabetes of the young
NR3C2	Pseudohypoaldosteronism; pregnancy-induced hypertension
NR2E3	Enhanced S cone syndrome
SF1	Adrenal insufficiency
THRB	Thyroid hormone resistance
VDR	Hereditary vitamin D resistant rickets type II

The full names of the genes mentioned in this table and details of links to further information can be found at http://www.gene.ucl.ac.uk/cgi-bin/nomenclature/searchgenes.pl

Table 2 Germ-line mutations in genes encoding other zinc-finger factors

Gene	Syndrome
EGR2	Congenital hypomyelinating neuropathy; Charcot–Marie–Tooth type 1
GATA1	Familial dyserythropoietic anemia and thrombocytopenia
GATA3	Hypoparathyroidism, sensorineural deafness and renal anomaly syndrome
GLI3	Greig syndrome; Pallister–Hall syndrome; postaxial polydactyly
HR	Alopecia universalis/congenital atrichia
MID1	Opitz G/BBB syndrome
SALL1	Townes–Brocks syndrome
TRPS1	Tricho-rhino-phalangeal syndrome type I
WT1	WAGR syndrome; Denys–Drash syndrome; Frasier syndrome
ZIC2	Holoprosencephaly
ZIC3	Situs ambiguus

The full names of the genes mentioned in this table and details of links to further information can be found at http://www.gene.ucl.ac.uk/cgi-bin/nomenclature/searchgenes.pl

PAX Proteins

The human genome contains nine genes encoding PAX proteins, which all share in common the presence of a DNA-binding motif that was first identified in the paired protein of *Drosophila melanogaster*. The PAX proteins play key roles in the development of both invertebrates and vertebrates. Perhaps the most striking example is the conserved role of PAX6 in eye development in flies, mice and humans (Gehring and Ikeo, 1999). The paired box gene 6 (aniridia, keratitis) (*PAX6*) gene is located adjacent to the *WT1* gene on human chromosome 11p13. Both genes are deleted in the WAGR syndrome, with aniridia resulting from loss of PAX6 expression, which is required for proper development of all ocular structures. Heterozygosity for missense mutations in the *PAX6* gene result in other ocular conditions including autosomal dominant keratitis, isolated foveal hypoplasia and Peter's anomaly. Heterozygosity for loss of function mutations in the paired box gene 3 (Waardenburg syndrome 1) (*PAX3*) gene result in Waardenburg syndromes type I and III, and the craniofacial–deafness–hand syndrome, both of which involved craniofacial, hearing and limb defects (**Table 3**). Homozygosity for loss of function mutations in *PAX3* or *PAX6* result in severe malformation syndromes associated with perinatal lethality. This is evidence of a gene dosage effect, indicating that the expression of transcription factors is tightly controlled during development such that a 50% reduction in expression or functional activity results in a disease phenotype, and a complete absence of functional protein has even more devastating effects on development. Involvement of *PAX* genes in cancer has also been demonstrated (Mansouri, 1998).

Homeodomain Proteins

Homeodomain transcription factors are characterized by the presence of a 60-amino-acid DNA-binding

Table 3 Germ-line mutations in genes encoding PAX proteins

Gene	Syndrome
PAX2	Optic nerve coloboma and renal hypoplasia
PAX3	Waardenburg syndrome types I/III; craniofacial–deafness–hand syndrome
PAX6	Aniridia; Peter's anomaly; isolated foveal hypoplasia; keratitis
PAX8	Thyroid dysgenesis
PAX9	Oligodontia

See footnote to **Table 2**.

domain. In humans and other mammals, the HOX proteins represent a family of 39 homeodomain proteins that play key roles in embryonic development and are encoded by genes that are tightly clustered at four loci (Veraksa *et al.*, 2000). In humans, there are 11 *HOXA* genes on chromosome 7p, 10 *HOXB* genes on chromosome 17q, nine *HOXC* genes on chromosome 12q, and nine *HOXD* genes on chromosome 2q. Mutations in homeo box A13 (*HOXA13*) and homeo box D13 (*HOXD13*) are associated with the autosomal dominant limb malformation syndromes known as the hand–foot–genital syndrome and synpolydactyly respectively (see **Table 4**). In addition to the *HOX* genes, there are a large number of additional human genes that encode homeodomain transcription factors, which when mutated resulted in malformation syndromes. Mutations in insulin promoter factor 1, homeodomain transcription factor (*IPF1*) are particularly interesting, as heterozygosity for loss of function alleles is associated with maturity onset diabetes whereas homozygosity is associated with pancreatic

agenesis, providing another illustration of gene dosage and a demonstration of how transcription factors control all aspects of organ structure and function from embryonic development to adult physiology. Involvement of *HOX* genes in cancer has also been demonstrated (Cillo *et al.*, 1999).

HMG Domain Proteins

The HMG domain is an 80-amino-acid motif that was first identified in the high-mobility group nuclear proteins HMG1 and HMG2. This domain defines a superfamily of DNA-binding proteins containing over 100 members expressed in yeast, plants and animals (Grosschedl *et al.*, 1994). These proteins appear to function as architectural factors whose primary function is to bend DNA. Sex determining region Y (*SRY*) is a gene localized to the sex determining region of the Y chromosome, which encodes an HMG domain protein that is essential for normal testicular determination and differentiation such that loss of function results in sex reversal, that is, genotypic (XY) males develop as phenotypic females (reviewed by Capel (1998)). A large family of genes with at least 20 members encoding transcription factors with an SRY-type DNA-binding domain has been identified (Kamachi *et al.*, 2000). The part of these *SOX* genes encoding the DNA-binding domain is referred to as the SRY box. Loss of function mutations in SRY (sex determining region Y)-box 9 (campomelic dysplasia, autosomal sex reversal) (*SOX9*) are also associated with sex reversal (**Table 5**), indicating that two HMG domain proteins (SRY and SOX9) are required for male sexual differentiation.

Other Transcription Factor Families

In addition to the families described above, additional groups of transcription factors have been identified based upon their DNA-binding domain structure. Two of the first such groups to be identified are the basic helix–loop–helix (bHLH) and basic-leukine zipper (bZIP) families. Remarkably, there are no known disease-causing germ-line mutations in genes

Table 4 Germ-line mutations in genes encoding homeodomain proteins

Gene	Syndrome
ALX4	Symmetric parietal foramina
CHX10	Microphthalmia, cataracts and iris abnormalities
CRX	Cone–rod dystrophy; Leber congenital amaurosis; retinitis pigmentosa
DLX3	Trichodento-osseous syndrome
HESX1	Septo-optic dysplasia
HLXB9	Hereditary sacral agenesis
TCF2	Maturity onset diabetes of the young
HOXA11	Amegakaryocytic thrombocytopenia and radioulnar synostosis
HOXA13	Hand–foot–genital syndrome
HOXD13	Synpolydactyly
IPF1	Pancreatic agenesis; maturity onset diabetes
LHX3	Combined pituitary hormone deficiency
LMX1B	Nail–patella syndrome
MSX1	Autosomal dominant tooth agenesis
MSX2	Boston-type craniosynostosis; foramina parietalia permagna
CSX	Congenital heart disease
PITX2	Rieger syndrome
PITX3	Anterior segment mesenchymal dysgenesis; autosomal dominant cataracts
PROP1	Combined pituitary hormone deficiency
SHOX	Short stature
SIX3	Holoprosencephaly
SIX6	Anophthalmia and pituitary anomalies
TGIF	Holoprosencephaly

See footnote to **Table 2**.

Table 5 Germ-line mutations in genes encoding HMG domain proteins

Gene	Syndrome
SRY	Sex reversal
SOX9	Campomelic dysplasia; sex reversal
SOX10	Waardenburg–Hirschsprung disease

See footnote to **Table 2**.

encoding bZIP proteins, and only two genes encoding bHLH proteins are targets for such mutations (**Table 6**). Three other families encode developmental regulators, loss of function for which is associated with congenital malformation syndromes: the POU domain (**Table 7**), the T domain (**Table 8**) and the forkhead domain (**Table 9**). Each of these are large gene families, and it is likely that in the future other family members will be identified as targets of mutation in other syndromes. There are other DNA-binding transcription factors that fall into other groups that for the purposes of this article will be labeled as other (**Table 10**).

Table 6 Germ-line mutations in genes encoding bHLH domain proteins

Gene	Syndrome
MITF	Waardenburg syndrome type II
TWIST	Saethre–Chotzen syndrome

See footnote to **Table 2**.

Table 7 Germ-line mutations in genes encoding POU domain proteins

Gene	Syndrome
POU1F1	Hypopituitary dwarfism
POU3F4	X-linked deafness type 3
POU4F3	Inherited progressive hearing loss

See footnote to **Table 2**.

Table 8 Germ-line mutations in genes encoding T domain proteins

Gene	Syndrome
TBX1	Velocardiofacial syndrome
TBX3	Ulnar mammary syndrome
TBX5	Holt–Oram syndrome
TBX22	X-linked cleft palate and ankyloglossia

See footnote to **Table 2**.

Table 9 Germ-line mutations in genes encoding forkhead domain proteins

Gene	Syndrome
FOXC1	Axenfeld–Rieger syndrome
FOXC2	Hereditary lymphedema distichiasis syndrome
FOXE3	Anterior segment ocular dysgenesis and cataracts
FOXP2	Speech and language disorder
FOXP3	X-linked immune dysregulation, polyendocrinopathy, enteropathy syndrome

See footnote to **Table 2**.

Finally, germ-line disease-causing mutations have been identified in genes encoding coactivators, co-repressors, chromatin remodeling factors and general transcription factors that do not bind to DNA in a sequence-specific manner (**Table 11**). Among the latter are proteins that are required for both transcription

Table 10 Germ-line mutations in genes encoding other DNA-binding proteins

Gene	Syndrome
RUNX2	Cleidocranial dysplasia
RUNX1	Thrombocytopenia; acute myelogenous leukemia
MHC2TA	Bare lymphocyte syndrome
EYA1	Branchio-oto-renal syndrome; congenital cataracts/anterior ocular syndrome
EYA4	Late onset deafness
TCF1	Maturity onset diabetes of the young
LHX4[a]	Syndromic short stature
NRL	Retinitis pigmentosa
TP53	Li–Fraumeni syndrome
TP63	Ectrodactyly, ectodermal dysplasia, cleft lip/palate; Hays–Wells syndrome; split hand foot malformation
RFX5	Bare lymphocyte syndrome
RFXAP	Bare lymphocyte syndrome
RFXANK	Bare lymphocyte syndrome
STAT1	Susceptibility to mycobacterial disease
TFAP2B	Char syndrome
TULP1	Retinitis pigmentosa type 14
TCF2	Glomerulocystic kidney disease

See footnote to **Table 2**.
[a]This is an interim gene symbol. The Locus ID in Locus Link is 89884.

Table 11 Germ-line mutations in genes encoding other non-DNA-binding factors

Gene	Syndrome
BRCA1	Familial breast cancer
CREBBP	Rubenstein–Taybi syndrome
CKN1	Cockayne syndrome
PCBD	Hyperphenylalaninemia
ERCC3	Xeroderma pigmentosa; trichothiodystrophy
ERCC2	Xeroderma pigmentosa; trichothiodystrophy
ERCC6	Cockayne syndrome
PER2	Familial advanced sleep phase syndrome
MECP2	Rett syndrome, X-linked mental retardation
RB1	Familial retinoblastoma
SMARCAL1	Schimke immuno-osseous dysplasia
SMARCB1	Rhabdoid predisposition syndrome
ATRX	X-linked alpha-thalassemia and mental retardation

See footnote to **Table 2**.

and DNA repair. The reader can find detailed descriptions of all the diseases listed in the tables by consulting the Online Mendelian Inheritance in Man website (see Web Links).

Understanding Disease Pathophysiology at the Molecular Level

The vast majority of the disease phenotypes listed in the tables are inherited as autosomal dominant traits. There are three pathophysiological mechanisms that provide an explanation as to why a single mutant allele can be sufficient to result in disease despite the presence of a normally functioning wild-type allele. The least common cause is a gain of function mechanism in which the mutation imparts either increased or novel activity or expression of the protein product. A missense mutation in msh homeo box homolog 2 (Drosophila) (*MSX2*) results in a protein with increased DNA-binding affinity at the molecular level, resulting in a craniosynostosis syndrome characterized by excessive growth of the skull bones during cranial development. A second mechanism is the generation of a dominant negative protein that interferes with the functioning of the wild-type protein. As a result, the activity of the protein is reduced to much less than 50% despite the presence of one mutant and one wild-type allele. In the case of transcription factors, however, the third potential mechanism, gene dosage or haploinsufficiency most commonly underlies the observed disease pathophysiology. In contrast to metabolic enzymes, for which 10% or less of normal activity may be sufficient for normal development and physiology, the expression and activity of transcription factors is precisely regulated and a 50% reduction is usually sufficient to result in a disease phenotype. As described above in the discussion of PAX domain proteins, homozygosity for a mutant allele results in an even more severe phenotype, which is usually not compatible with prolonged survival *ex utero*.

See also
Transcription Factors

References

Capel B (1998) Sex in the 90s: SRY and the switch to the male pathway. *Annual Reviews of Physiology* 60: 497–523.
Cillo C, Faiella A, Cantile M and Boncinelli E (1999) Homeobox genes and cancer. *Experimental Cell Research* 248: 1–9.
Gehring WJ and Ikeo K (1999) Pax 6: mastering eye morphogenesis and eye evolution. *Trends in Genetics* 15: 371–377.
Grosschedl R, Giese K and Pagel J (1994) HMG domain proteins: architectural elements in the assembly of nucleoprotein structures. *Trends in Genetics* 10: 94–100.
Kamachi Y, Uchikawa M and Kondoh H (2000) Pairing SOX off: with partners in the regulation of embryonic development. *Trends in Genetics* 16: 182–187.
Kornberg RD and Lorch Y (1999) Twenty five years of the nucleosome, fundamental particle of the eukaryote chromosome. *Cell* 98: 285–294.
Lemon B and Tjian R (2000) Orchestrated response: a symphony of transcription factors for gene control. *Genes in Development* 14: 2551–2569.
Mansouri A (1998) The role of Pax3 and Pax7 in development and cancer. *Critical Reviews in Oncology* 9: 141–149.
Veraksa A, Del Campo M and McGinnis W (2000) Developmental patterning genes and their conserved functions: from model organisms to humans. *Molecular Genetics and Metabolism* 69: 85–100.
Wolfe SA, Nekludova L and Pabo CO (2000) DNA recognition by Cys2His2 zinc finger proteins. *Annual Reviews of Biophysics and Biomolecular Structure* 29: 183–212.

Further Reading

Latchman DS (ed.) (1999) *Transcription Factors: A Practical Approach*, 2nd edn. New York: Oxford University Press.
Papavassiliou AG (1997) *Transcription Factors in Eukaryotes*. Austin, TX: Landes Bioscience.
Semenza GL (1998) *Transcription Factors and Human Disease*. New York: Oxford University Press.

Web Links

Online Mendelian Inheritance in Man. A detailed description of all the diseases and genes listed in the tables can be found on this website. Links for some important genes mentioned in the text are specified below
http://www.ncbi.nlm.nih.gov
Homeo box A13 (*HOXA13*); Locus ID: 3209. LocusLink:
http://www.ncbi.nlm.nih.gov/LocusLink/LocRpt.cgi?l = 3209
Homeo box D13 (*HOXD13*); Locus ID: 3239. LocusLink:
http://www.ncbi.nlm.nih.gov/LocusLink/LocRpt.cgi?l = 3239
Insulin promoter factor 1, homeodomain transcription factor (*IPF1*); Locus ID: 3651. LocusLink:
http://www.ncbi.nlm.nih.gov/LocusLink/LocRpt.cgi?l = 3651
msh homeo box homolog 2 (Drosophila) (*MSX2*); Locus ID: 4488. LocusLink:
http://www.ncbi.nlm.nih.gov/LocusLink/LocRpt.cgi?l = 4488
Paired box gene 3 (Waardenburg syndrome 1) (*PAX3*); Locus ID: 5077. LocusLink:
http://www.ncbi.nlm.nih.gov/LocusLink/LocRpt.cgi?l = 5077
Paired box gene 6 (aniridia, keratitis) (*PAX6*); Locus ID: 5080. LocusLink:
http://www.ncbi.nlm.nih.gov/LocusLink/LocRpt.cgi?l = 5080
Sex determining region Y (*SRY*); Locus ID: 6736. LocusLink:
http://www.ncbi.nlm.nih.gov/LocusLink/LocRpt.cgi?l = 6736
SRY (sex determining region Y)-box 9 (campomelic dysplasia, autosomal sex-reversal) (*SOX9*); Locus ID: 6662. LocusLink:
http://www.ncbi.nlm.nih.gov/LocusLink/LocRpt.cgi?l = 6662
Wilms tumor 1(*WT1*); Locus ID: 7490. LocusLink:
http://www.ncbi.nlm.nih.gov/LocusLink/LocRpt.cgi?l = 7490
Homeo box A13 (*HOXA13*); MIM number: 142959. OMIM:
http://www.ncbi.nlm.nih.gov/htbin-post/Omim/dispmim?142959

Homeo box D13 (*HOXD13*); MIM number: 142989. OMIM:
  http://www.ncbi.nlm.nih.gov/htbin-post/Omim/
  dispmim?142989
Insulin promoter factor 1, homeodomain transcription factor
  (*IPF1*); MIM number: 600733. OMIM:
  http://www.ncbi.nlm.nih.gov/htbin-post/Omim/
  dispmim?600733
msh homeo box homolog 2 (Drosophila) (*MSX2*); MIM number:
  123101. OMIM:
  http://www.ncbi.nlm.nih.gov/htbin-post/Omim/
  dispmim?123101
Paired box gene 3 (Waardenburg syndrome 1) (*PAX3*); MIM
  number: 606597. OMIM:
  http://www.ncbi.nlm.nih.gov/htbin-post/Omim/
  dispmim?606597

Paired box gene 6 (aniridia, keratitis) (*PAX6*); MIM number 106210.
  OMIM:
  http://www.ncbi.nlm.nih.gov/htbin-post/Omim/
  dispmim?106210
Sex determining region Y (*SRY*); MIM number 480000. OMIM:
  http://www.ncbi.nlm.nih.gov/htbin-post/Omim/
  dispmim?480000
SRY (sex determining region Y)-box 9 (campomelic dysplasia,
  autosomal sex-reversal) (*SOX9*); MIM number: 114290. OMIM:
  http://www.ncbi.nlm.nih.gov/htbin-post/Omim/
  dispmim?114290
Wilms tumor 1(*WT1*); MIM number: 194070. OMIM:
  http://www.ncbi.nlm.nih.gov/htbin-post/Omim/
  dispmim?194070

Transcriptomics and Proteomics: Integration?

Bent Honoré, *University of Aarhus, Aarhus, Denmark*

Morten Østergaard, *University of Aarhus, Aarhus, Denmark*

Intermediate article

Article contents

- Flow of Genetic Information
- Transcriptome – Transcriptomics
- Proteome – Proteomics
- Integration and Complementarity of Transcriptomic and Proteomic Data

The genes are transcribed to pre-messenger RNA, further processed to messenger RNA (transcripts), and finally translated into protein. The transcriptome constitutes the complete set of different transcripts that is synthesized in the lifetime of a cell or a tissue. The proteome refers to the complete set of proteins that is expressed, and modified following expression in the lifetime of a cell or a given tissue. Transcriptomics refers to the study of the transcriptome, while proteomics refers to the study of the proteome using large-scale technologies for transcript or protein analyses. The two disciplines each provide unique information and supplement each other.

Flow of Genetic Information

The human genome consists of about 30 000–40 000 genes (Lander *et al.*, 2001; Venter *et al.*, 2001). The sequence information gathered in the genes of the chromosomes are brought from the nucleus to the cytosol of the cell through various regulated steps (Day and Tuite, 1998; Kornberg, 1999) (**Figure 1**). First, the genes are transcribed into pre-messenger ribonucleic acid (pre-mRNA) (step 1; **Figure 1**). The transcriptional process simply copies the coding strand of the gene consisting of the deoxyribonucleotides A, G, C and T, into pre-mRNA consisting of the ribonucleotides A, G, C and U. The RNA may in some occasions undergo an editing process (step 2; **Figure 1**), which results in the exchange of nucleotides so that the coding part of the messenger RNA (mRNA) is altered. The pre-mRNA contains the exons as well as the introns of the coding strand of

the gene and undergoes a further processing control (step 3; **Figure 1**) where the exons are spliced together leaving the introns in the nucleus. (*See* Human Genome: Draft Sequence; mRNA Editing; Sequence Finishing.)

The RNA processing may lead to several different mRNAs by the process known as alternative or differential splicing where several different mRNAs may be produced by combining different exons in the same pre-mRNA and ultimately give rise to several different protein products. The number of different transcripts present in humans is unknown. It is estimated that 40–60% of the human genes have multiple splicing variants (Lander *et al.*, 2001). The spliced mRNAs are then transported out of the nucleus to the cytosol (step 4; **Figure 1**). In the cytosol, the mRNA level is controlled by several mechanisms. Some mRNAs may undergo degradation (step 5; **Figure 1**). Translation of the mRNA into protein, the

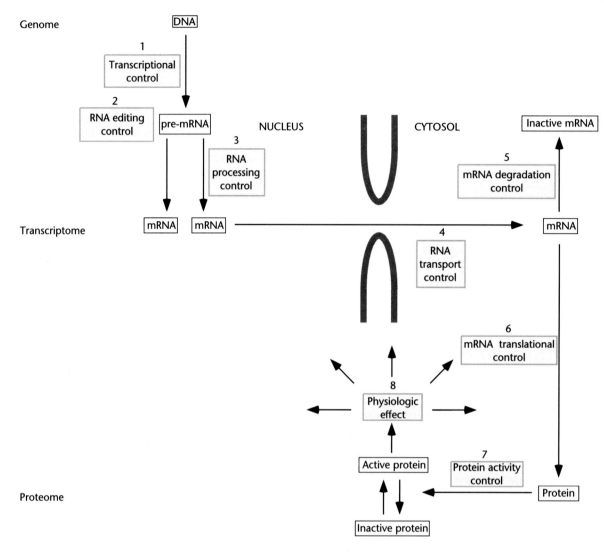

Figure 1 Flow of information from DNA via mRNA to protein. A gene (DNA) is transcribed (step 1) to the various forms of RNA, first to pre-mRNA that may be edited (step 2) and then processed (step 3) to one or by alternative splicing to several forms of mRNAs. The mRNAs are then transported (step 4) out of the nucleus to the cytosol. In the cytosol, the mRNA may be degraded (step 5) or translated (step 6) into protein. The activities of the proteins are controlled (step 7). They may be synthesized as inactive proteins that later are reversibly or irreversibly activated, or alternatively synthesized as active proteins that later are inactivated. Proteins are the ultimate effecting molecules producing the physiologic effect (step 8) in virtually every mechanism in the cell.

molecules that directly operate in the cell, is controlled in step 6 (**Figure 1**). The activity of the resulting protein is controlled in step 7 (**Figure 1**) as it may be synthesized in an active form that later can be reversibly or irreversibly inactivated. Alternatively, proteins may be synthesized first in their inactive forms and later activated by processes such as proteolysis, phosphorylation or otherwise. (*See* Alternative Splicing: Evolution; mRNA Export; mRNA Stability and the Control of Gene Expression; Posttranscriptional Processing; RNA Processing.)

Thus, there are several regulated steps involved in the expression of a gene through the different forms of mRNA constituting the transcriptome to the final protein products constituting the proteome. It is the active proteins that serve as the operating molecules producing the physiologic effects in virtually every mechanism in the cell, be it in the cytosol, in an organelle, as membrane attached or outside the cell in the extracellular space or in the bloodstream.

Transcriptome – Transcriptomics

The transcriptome constitutes the complete set of different transcripts that is synthesized in the lifetime of a cell or a tissue. Transcriptomics refers to the study of the transcriptome using large-scale technologies for transcript analyses. Transcript analyses may give information on the primary structure (nucleotide sequence) of the transcript or about the level of transcript present. In general, these techniques are rather fast and highly suitable for large-scale analyses due to the chemical nature of the nucleotides (Honoré, 2001).

Transcript sequence analysis

Transcriptome analysis includes the purification of mRNA from cells or tissues, reverse transcription of the mRNA to complementary deoxyribonucleic acid (cDNA) and finally sequencing of the cDNA. The cDNAs ideally contain a 5′ untranslated region, a start codon, an open reading frame with the complete information about the primary structure of the encoded protein and finally a 3′ untranslated region ending with the poly-A tail. Until recently, transcript sequence analysis was mostly performed by selectively cloning and analyzing specific cDNAs by using small protein sequences reverse translated into degenerated nucleotide sequences (oligonucleotides). By using these labeled oligonucleotides as probes in hybridization techniques, the specific cDNAs are selectively isolated from cDNA libraries.

As sequencing techniques have improved, many cDNAs have been sequenced randomly from similar cDNA libraries giving expressed-sequence tags (ESTs) (Adams et al., 1991). These results give mostly partial sequence information about the 5′ and 3′ ends of the cDNAs while others can be selected for complete sequencing. The partial or complete cDNA sequences are deposited in the DNA databanks from where researchers may search and retrieve the data via the internet (GenBANK, EMBL and DDBJ; see Web Links). The sequencing techniques give information on the sequence of nucleotides in the mRNA from which the primary structure of the encoded proteins may be deduced partially or completely. (*See* Expressed-sequence Tag (EST); Nucleotide Sequence Databases; Protein Databases.)

Transcript abundancy analysis

Transcript analyses may also give information about the levels present of various transcripts. Northern blotting is the classical technique for measuring abundancies of transcripts. Transcripts or total RNA is purified and size selected on a gel with subsequent blotting onto a membrane. A specific nucleotide probe is hybridized to the filter and the signal intensity is a measure of the abundancy of the transcript. Techniques that are based on polymerase chain reaction (PCR) amplification of mRNAs, quantitative PCR, may also be used to measure transcript levels (Ginzinger, 2002). Both techniques require that the

sequence is known and are generally not especially suitable for high-throughput analyses. (*See* Genetic Distance and Mapping Functions.)

Large-scale transcript abundancy analyses can be performed with serial analysis of gene expression (SAGE), a technique that uses specific restriction enzymes that generally have 4 bp recognition sequences thereby statistically cleaving every 256 bp. By a series of steps, the 3′-most fragments of up to 20 bp are constructed as di-tags that are concatenized, cloned and sequenced. The frequency of the presence of the di-tags will thus be a measure of the abundance of the transcript present (Velculescu et al., 1995).

Transcript abundancy analyses are especially suited to be performed with the DNA chip technology (DNA array). With this technique DNA that represent selected parts of known genes are immobilized on a chip. mRNA is then extracted from tissues or cells and reverse transcribed into cDNA. The cDNA may then be directly, or after transcription to complementary RNA (cRNA), probed to the immobilized DNA giving a fluorescence signal which is proportional to the level of transcript present (Schena et al., 1995). (*See* DNA Chips and Microarrays; Microarrays in Disease Diagnosis and Prognosis; Microarrays and Expression Profiling in Cancer.)

Proteome – Proteomics

The proteome may be defined as the complete set of proteins that is expressed and posttranslationally modified in the lifetime of a cell or a tissue. The posttranslational modifications include proteolytic processing, phosphorylations, glycosylations, etc. Proteomics is the discipline that uses large-scale techniques to analyze the proteins present. In general, proteomics is not equally suited for large-scale or high-throughput analyses as the techniques that measure transcripts, simply because of the highly variable chemical nature of proteins as opposed to the uniform chemical nature of nucleotides (Honoré, 2001). However, proteins represent the molecules that directly operate in the cells and are therefore closer to the functional level.

Protein separation

Proteomics analyses often involve techniques for separation of proteins. Today, the one technique that possesses the highest resolvability and may separate the largest number of proteins in a relatively short time is two-dimensional polyacrylamide gel electrophoresis (2D-PAGE), which separates proteins in one dimension according to the isoelectric point of the protein and in the other dimension with respect to the molecular mass of the protein (Görg *et al.*, 2000). By this procedure, thousands of proteins may be separated from cells or tissues in a few steps. The separated proteins in the gel may subsequently be visualized by various staining techniques. Each staining technique possesses advantages and disadvantages. Coomassie brilliant blue staining is applicable for very abundant proteins, while less abundant proteins may be visualized with silver staining, fluorescent staining or radioactive labeling.

The protein range where linearity is observed is limited, and different proteins may bind the staining substances with different affinity. When proteins are analyzed on a large-scale basis, the precise amount present of a specific protein may thus be estimated. However, relative comparisons of protein levels may be performed between normal and diseased cells or tissues by differential protein expression where the levels of upregulation or downregulation of various proteins are determined. (*See* Image Analysis Tools in Proteomics; Two-dimensional Gel Electrophoresis.)

Identification and partial sequencing of proteins

Mass spectrometry (MS) is the method of choice for the identification of proteins on a large-scale basis (Mann *et al.*, 2001). Techniques are available where single protein spots (silver-stained levels) may be cut out from 2D-PAGE gels. The proteins in the spot are then subjected to enzymatic digestion, mostly with trypsin cutting *C*-terminal to lysine and arginine. The mass divided by the charge (m/z) of the peptides may then be determined in a mass spectrometer giving a peptide fingerprint of the protein. If the protein is pure enough and not contaminated with other proteins, this peptide fingerprint can be sufficient to identify the protein. This is done by searching databases containing information about the known proteins or cDNAs translated to proteins with the possibility of deducing m/z values for putative peptides generated by tryptic digestion.

If this is not sufficient to identify the protein, each of the peptides may be further analyzed by a second round of mass spectrometry, tandem mass spectrometry (MS/MS). Here, each peptide is further partially fragmented, mostly in the peptide bonds. By such a procedure, partial amino acid sequence information can be obtained for the successfully analyzed peptides. The latter procedure is referred to as *de novo* protein sequencing. (*See* Mass Spectrometry Instrumentation in Proteomics; Mass Spectrometry in Protein Characterization.)

It is not only the protein level that determines the biological activity of a protein. Several proteins are subjected to modification processes posttranslationally (step 7; **Figure 1**). These may be proteolytic cleavage, covalent modifications such as phosphorylations, glycosylations, acylations, etc. or noncovalent addition of prosthetic groups. Such modifications may be essential for the biological activity of a protein, for example phosphorylations. At present, some of these posttranslational modifications may be analyzed by proteomics techniques such as MS. Such information is impossible to obtain by analyses of the transcripts alone. It can only be obtained by detailed analyses of the proteins. (*See* Posttranslational Modification and Human Disorders; Protein Glycosylation.)

Proteins in their physiologic context

Proteins do function inside or outside the cell by interacting with other proteins. Therefore, some techniques focus on the identification of other proteins that a given protein interacts with. Techniques that focus on protein–protein interactions include the yeast two-hybrid technique where protein interactions occur inside a yeast cell nucleus. The technique is based on the interaction of a 'bait' protein with a 'prey' protein from a cDNA library. The bait protein consists of a target protein fused to the DNA-binding domain of a transcription factor. The prey protein consists of a binding protein fused to the transcriptional activator domain of the transcription factor. Interaction of the bait with the prey creates a functional transcription factor that turns on the reporter gene. The cDNA coding for the binding protein can then be characterized by sequencing. (*See* Yeast Two-hybrid System and Related Methodology.)

Another method is the phage display technique where the cDNAs are expressed as peptides on the surface of the phages. By isolating the phage that interacts with a certain protein, the cDNA can be sequenced providing information about the primary structure of the interacting peptide. Various protein array techniques are also under way where proteins are immobilized to a chip which subsequently is used for screening of binding proteins (Emili and Cagney, 2000; MacBeath and Schreiber, 2000; Pandey and Mann, 2000). At the moment, the proteomics techniques are not yet at such high-throughput levels as the DNA array techniques. (*See* Array-based Proteomics.)

Integration and Complementarity of Transcriptomic and Proteomic Data

The genome is, with few exceptions, present at the same concentration in each cell and is thereby static in nature. Since the recent completion of its sequencing, we are close to having all the information available about the 30 000–40 000 human genes present although it is not straightforward to locate genes in the rough sequence unless supported by other type of information coming from separate analysis by cDNA sequencing or EST sequencing (Lander *et al.*, 2001). (*See* Gene Structure and Organization.)

As opposed to the genome, the transcriptome as well as the proteome is dynamic in nature. Each different type of cell in a given tissue may produce different sets of transcripts originating from different sets of genes selected to be transcribed at a particular time. Furthermore, different RNA processing events may take place including editing, splicing and alternative splicing. Information about the detailed structure of the transcripts does not come from the human genome sequencing but has to be obtained by separate cDNA sequencing. The nucleotide sequences may be computer translated to protein sequence and used for further analyses by being deposited in the databases.

Transcriptomics data mainly produce information about the sequence and the levels of various transcripts present. The techniques for these analyses are generally highly suitable for large-scale analyses; huge amounts of data may be accumulated owing to the chemical nature of the molecules where it is relatively easy to produce probes and to hybridize the probes to the target.

Only few studies have been performed on the analysis of the relation between transcript levels and protein levels. In human liver tissue (Anderson and Seilhamer, 1997) as well as in yeast cells (Futcher *et al.*, 1999; Gygi *et al.*, 1999), it has been shown that there may be quite large variations in transcript levels that are not reflected in similar changes in protein levels. Thus, translation of transcripts may be regulated at step 6 (**Figure 1**) as for example seen for the translation of the ferritin mRNA (Day and Tuite, 1998). Initiatives that serve to circumvent such shortcomings with the transcript analyses are being tried at the moment by purification of transcripts that are polysome-bound, that is, being actively translated. By analyzing such transcripts it may thus be possible to estimate the corresponding protein levels more reliably (Pradet-Balade *et al.*, 2001), bringing transcriptomics data closer to proteomics data. However, transcriptome analyses alone will still not be able to monitor posttranslational modifications.

The techniques used in proteomics are currently not at the level where the data amount produced is as easily obtained as for transcriptome data. However, because the protein molecules are closer to the functional level in the cell, they will be the focus of intense research in future.

The data obtained by transcriptomics thus give information that may supplement proteomics data and vice versa. Data obtained by transcriptomics produce information about transcript sequences and levels while data obtained from proteomics can give information about proteins, their identity and levels and with the possibility of obtaining information about posttranslational modifications. (*See* 3′ UTRs and Regulation; 5′ UTRs and Regulation.)

References

Adams MD, Kelley JM, Gocayne JD, *et al.* (1991) Complementary DNA sequencing: expressed sequence tags and human genome project. *Science* **252**: 1651–1656.

Anderson L and Seilhamer J (1997) A comparison of selected mRNA and protein abundances in human liver. *Electrophoresis* **18**: 533–537.

Day DA and Tuite MF (1998) Post-transcriptional gene regulatory mechanisms in eukaryotes: an overview. *Journal of Endocrinology* **157**: 361–371.

Emili AQ and Cagney G (2000) Large-scale functional analysis using peptide or protein arrays. *Nature Biotechnology* **18**: 393–397.

Futcher B, Latter GI, Monardo P, McLaughlin CS and Garrels JI (1999) A sampling of the yeast proteome. *Molecular and Cellular Biology* **19**: 7357–7368.

Ginzinger DG (2002) Gene quantification using real-time quantitative PCR: an emerging technology hits the mainstream. *Experimental Hematology* **30**: 503–512.

Görg A, Obermaier C, Boguth G, *et al.* (2000) The current state of two-dimensional electrophoresis with immobilized pH gradients. *Electrophoresis* **21**: 1037–1053.

Gygi SP, Rochon Y, Franza BR and Aebersold R (1999) Correlation between protein and mRNA abundance in yeast. *Molecular and Cellular Biology* **19**: 1720–1730.

Honoré B (2001) Genome- and proteome-based technologies: status and applications in the postgenomic era. *Expert Review of Molecular Diagnostics* **1**: 265–274.

Kornberg RD (1999) Eukaryotic transcriptional control. *Trends in Cell Biology* **9**: M46–M49.

Lander ES, Linton LM, Birren B, *et al.* (2001) Initial sequencing and analysis of the human genome. International Human Genome Sequencing Consortium. *Nature* **409**: 860–921.

MacBeath G and Schreiber SL (2000) Printing proteins as microarrays for high-throughput function determination. *Science* **289**: 1760–1763.

Mann M, Hendrickson RC and Pandey A (2001) Analysis of proteins and proteomes by mass spectrometry. *Annual Review of Biochemistry* **70**: 437–473.

Pandey A and Mann M (2000) Proteomics to study genes and genomes. *Nature* **405**: 837–846.

Pradet-Balade B, Bolumé F, Beug H, Müllner EW and Garcia-Sanz JA (2001) Translation control: bridging the gap between genomics and proteomics? *Trends in Biochemical Sciences* **26**: 225–229.

Schena M, Shalon D, Davis RW and Brown PO (1995) Quantitative monitoring of gene expression patterns with a complementary DNA microarray. *Science* **270**: 467–470.

Velculescu VE, Zhang L, Vogelstein B and Kinzler KW (1995) Serial analysis of gene expression. *Science* **270**: 484–487.

Venter JC, Adams MD, Myers EW, *et al.* (2001) The sequence of the human genome. *Science* **291**: 1304–1351.

Further Reading

Alberts B, Bray D, Johnson A, *et al.* (1998) *Essential Cell Biology. An Introduction to the Molecular Biology of the Cell.* New York, NY: Garland.

Alberts B, Johnson A, Lewis J, *et al.* (2002) *Molecular Biology of the Cell.* New York, NY: Garland.

Cooper GK (2000) *The Cell. A Molecular Approach.* Washington, DC: ASM Press.

Lodish H, Berk A, Zipursky SL, *et al.* (2000) *Molecular Cell Biology.* New York, NY: WH Freeman & Co.

Web Links

DDBJ. The DNA Data Bank of Japan is officially certified to collect DNA sequences from researchers and to issue the internationally recognized accession number to data submitters. They collect data mainly from Japanese researchers, but accept data from researchers in any other country
http://www.ddbj.nig.ac.jp/

EMBL. The EMBL (European Molecular Biology Laboratory) Nucleotide Sequence Database constitutes Europe's primary nucleotide sequence resource. Main sources for DNA and RNA sequences are direct submissions from individual researchers, genome sequencing projects and patent applications
http://www.ebi.ac.uk/embl/

GenBANK. GenBANK is the NIH (National Institute of Health) genetic sequence database, an annotated collection of all publicly available DNA sequences. GenBANK is part of the International Nucleotide Sequence Database Collaboration, which comprises the DNA DataBank of Japan (DDBJ), the European Molecular Biology Laboratory (EMBL), and GenBank at NCBI. These three organizations exchange data on a daily basis
http://www.ncbi.nih.gov/Genbank/

Transfer RNA

See tRNA

Transgenic Mice

Advanced article

Article contents

- Introduction
- Generation of Transgenic Mice
- Transgenic Mice: A Unique Tool for the Study of Mammalian Biology
- Conclusion

Charles Babinet, *Institut Pasteur, Paris, France*

Transgenic mice carry exogenous genetic material introduced by the experimenter. Homologous recombination is used to introduce programmed modifications of the mouse genome.

Introduction

Over the last few decades of the twentieth century, the mouse became the animal model of choice for the study of the various aspects of mammalian development, physiology and physiopathology. Apart from being the easiest and cheapest laboratory mammal to maintain, its genetic map is very well known, and many inbred strains (the genetic background of which is defined and identical for all the individuals) as well as mutant strains are available. Furthermore, sequencing of the whole murine genome is almost complete, giving access in principle to all the genes present in the genome. It has been possible since the early 1980s to introduce experimentally new genetic information in the mouse germ line, giving rise to 'transgenic' mice. Finally, and this is unique to the mouse among mammals, the methods for creating transgenic mice have been refined in such a way that it has become possible to create mice bearing a whole range of programmed genetic modifications, including null

mutations as well as other types such as not only point mutations, but also alterations at the chromosome level, for example large deletions or translocations. Taken as a whole, the transgenic approach has revolutionized the study of all aspects of mammalian genetics and biology.

Generation of Transgenic Mice

Transgenic mice by DNA microinjection into the pronucleus of the zygote

The first route to generate transgenic mice consists in directly injecting a cloned deoxyribonucleic acid (DNA) of choice (the transgene) into one pronucleus of a fertilized egg. This procedure (**Figure 1**) gives rise, after reimplantation into a foster mother, to mice in which the transgene has integrated randomly in the mouse genome, generally as tandem head-to-tail

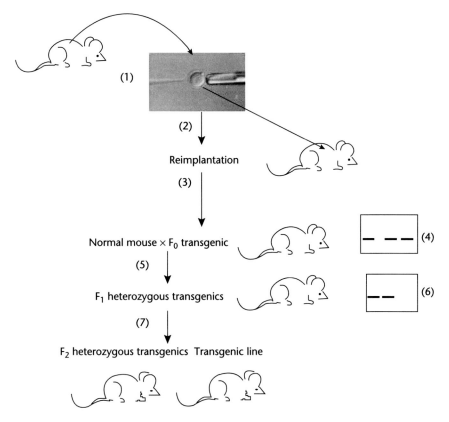

Figure 1 Generation of transgenic mice by DNA microinjection into the pronucleus of the zygote. A DNA solution is injected (1) into the pronucleus of a zygote. Injected eggs are then reimplanted into a foster mother (2, 3). In a proportion of cases, the injected DNA integrates into the chromosomes of the zygote. The integrated exogenous DNA (the transgene) is transmitted through cell division to all the cells of the mouse born from the injected zygote, giving rise to a transgenic mouse. The presence of the transgene in the host DNA is monitored by Southern blot, using a radioactive probe that specifically recognizes the injected DNA (4, 6). Crossing of transgenic founders (F_0) with a nontransgenic mouse will give rise to F_1 progeny, half of which are heterozygous for the transgene (5, 6). F_1 intercrosses (7) allow the experimenter to obtain mice homozygous for the transgene, therefore generating a line of transgenic mice.

arrays of variable length. Thus neither the number of copies integrated nor the site of integration is controlled. However, this route of transgenesis is instrumental in the study of gene regulation *in vivo*; furthermore, it offers a means of targeting expression of a given gene product into a chosen cell or tissue type, thereby opening the way for the study of its function in physiological or pathological situations.

Transgenic mice by the use of embryonic stem cells

Transgenic mice may be obtained by a second route, using embryonic stem (ES) cells (**Figure 2**). These cells, which are derived directly from the culture of blastocysts, retain the remarkable ability to colonize a host embryo, including its germ line; thus previous introduction of exogenous DNA into the cultured ES cells allows one to generate transgenic mice via the production of germ-line chimeras. However, the most important virtue of the use of ES cells to make

transgenic mice, in contrast to the transgene pronuclear injection, is that desirable and low-frequency genetic alterations may be selected and verified in the ES clones maintained in culture and then reintroduced in the animal. Thus, for any mutation introduced in ES cells *in vitro*, the corresponding mutated mouse may be obtained and the phenotypic effects of the mutation studied in detail in the context of the living embryo or animal. Such a scenario has been extremely fruitful for the study of gene function in mammals (Capecchi, 1989).

Transgenic Mice: A Unique Tool for the Study of Mammalian Biology

Pronuclear microinjection

The normal development of complex organisms such as the mouse implies the tightly regulated expression, both spatially and temporally, of genes or groups of

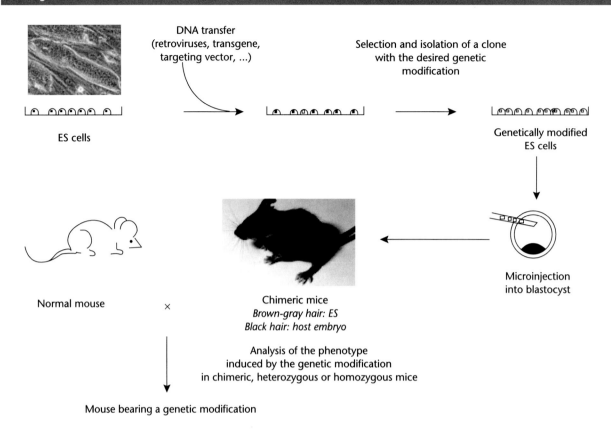

Figure 2 Different stages in the creation of genetically modified mice using embryonic stem (ES) cells.

genes. Thus, understanding the mechanisms that control gene expression in the context of the developing embryo or animal is a key issue. Transgenic mice generated by pronuclear microinjection offer a unique opportunity to address this question. Indeed, transgene expression can be monitored in any type of cell, at any time of embryonic and postnatal development. Thus, different versions of a given gene may be used to generate transgenic mice, thereby allowing one to map the functional regions of this gene necessary for its correct regulation. This approach was applied for the first time to elastase gene expression in the cells of the exocrine pancreas, and it was shown that a 134-bp-long sequence was necessary and sufficient for proper temporal and spatial regulation of the gene (Hammer et al., 1987). Since then, such a strategy has been applied successfully to many other genes (reviewed by Macdonald and Swift, 1998). Despite these achievements, it should be noted that analyses of transgene regulation may be obscured by the random character of the integration site; thus the genomic sequences flanking the transgene may result in its enhanced, repressed or even inappropriate expression. This is referred to as a 'position effect' (Wilson et al., 1990). Finally, it is generally observed that transgene

expression is not proportional to copy number. These observations point to the necessity of analyzing several independent transgenic lines to draw conclusions about the in vivo regulation regions of any gene (Macdonald and Swift, 1998). Interestingly, specific DNA sequences called locus control regions (LCRs) have been identified in the vicinity of some genes that render transgene expression proportional to the number of copies and independent of the insertion site (Grosveld, 1999). Finally, it should be noted that it is possible to buffer the position effect by using large transgenes such as yeast or bacterial artificial chromosomes (YACs and BACs) which encompass from 100 kb up to 2 Mb (reviewed by Giraldo and Montoliu, 2001).

Having defined the regulatory sequences that control specific gene expression, it is then possible to fuse them to the coding sequences of any gene of interest, resulting in a 'hybrid transgene'. Depending on the regulatory sequences used and the coding sequence fused to it, enhanced or ectopic expression of either the normal gene or a modified one can be obtained in transgenic mice. Such a strategy is extremely versatile (**Figure 3**) and has been widely used, for example to address the biological function of

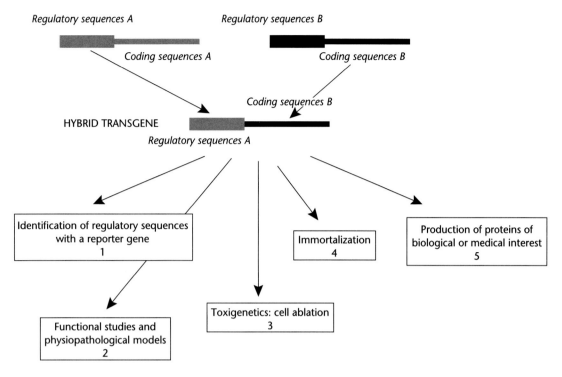

Figure 3 Hybrid transgenes and their use. A hybrid transgene is made of the regulatory sequences of a gene A bound to the coding sequences of a gene B. In a mouse carrying such a transgene, a gene of interest may be expressed in the cells or tissues where gene A is normally expressed. The gene of interest could code for: (1) A reporter gene (e.g. β-galactosidase or green fluorescent protein). This will help in determining precisely the pattern of expression of gene A. (2) Any protein. The phenotypic consequences of the ectopic expression of this protein will give insight into its function. This approach may also allow study of the function of gene products in various pathological situations, for example the role of oncogenes in malignancy. (3) A toxin, which will result in the ablation of a given type of cells and could illuminate the physiological function of the ablated cells. (4) An immortalizing oncogene, in which case cells expressing the transgene in the mouse may serve to derive cell lines in culture, from cell types that otherwise could not be maintained *in vitro*. (5) A protein of biological or medical interest. The hybrid transgene is constructed in such a way that the protein of interest will be synthesized in a tissue from which it could be extracted and purified, for example resulting in the extraction of milk or blood. Owing to its small size, the mouse serves only as a model system for bigger animals, for example farm animals.

a given protein. It may also allow the creation of murine models of human diseases, for example, the role of oncogenes in malignant transformation.

Knockout and knockin

The availability of ES cells and the design of methods using homologous recombination (HR) to alter a given gene has allowed scientists to extend considerably the use and interest of transgenesis in the mouse. Indeed, it has become possible to create mice with a mutation specifically in a given gene: the phenotypic analysis of the mutant embryo or mouse will then give insight into the function of this gene. The general scenario of targeted mutagenesis by HR in ES cells (outlined in **Figure 2**) was first used either to introduce or to correct null mutations in the *Hprt* gene; indeed, in this particular case, the mutated or corrected ES cells could be directly selected by the use of appropriate drugs.

HR between an incoming DNA and the homologous sequence in the genome is a rare event, as

compared with random insertion. Thus, appropriate targeting vectors and methods of selection were devised to achieve and select HR events and produce ES cells with an altered gene. The most frequently produced alteration is gene disruption, commonly called knockout (KO; **Figure 4a**). Once the mutated ES clones have been generated, mice bearing the mutation can be obtained via the generation of chimeras (**Figure 2**). The ability to generate knockout mice has been widely used and more than 1000 genes have been disrupted by HR in ES cells and the corresponding mutated mice generated, illuminating the function of genes in embryonic development as well as in the various (muscular, hematopoietic, nervous, immunological, etc.) biological systems of the mouse. A very interesting extension of the KO approach, the 'knockin', consists of introducing in the targeting vector the coding sequences of a gene of interest in-frame with those of the endogenous gene. After HR, the modified gene will express the gene of interest, precisely under the control of the promoter

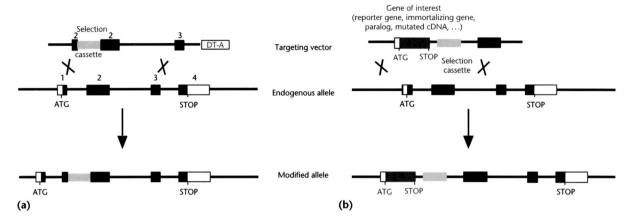

Figure 4 General principle of homologous recombination (HR): (a) Knockout. The targeting vector includes a selection cassette inserted in an exon (exon 2; black rectangles: coding region; white rectangles: noncoding regions) and surrounded by regions of homology with the target gene. In addition, a cassette may be added at one end of the targeting vector, in order to counterselect the cells in which the integration occurs outside of the targeted gene; a cassette expressing the A subunit of the diphtheria toxin (DT-A) is shown. Upon random integration, the DT-A-expressing cassette is retained and the cell is killed by the toxin. In contrast, upon HR, it is excised and therefore the cell survives. Recombination with the endogenous gene occurs within the homologous sequences and results in the creation of a null allele in which disruption of the gene is induced by insertion of the selection cassette into an exon. (b) Knockin: Besides invalidation of the target gene, a gene of interest is introduced in the locus. Following homologous recombination, the gene of interest is placed under the control of the promoter and regulatory sequences of the target gene, and is therefore expressed in place of the target gene. cDNA: complementary deoxyribonucleic acid (DNA); STOP: termination codon.

and regulatory sequences of the targeted gene, thus avoiding the position effect encountered in transgenesis by pronuclear microinjection (**Figure 4b**). The gene of interest may be, for example, a reporter gene, coding for proteins such as *E. coli* beta galactosidase or green fluorescent protein, whose activity is easily visualized and therefore enables determination of the pattern of a given gene even at the cellular level. Another interesting use of knockin deals with genes that have several homologs in the mouse genome but have overlapping or different patterns of expression and the disruption of which generates different pheno-

types. Knocking in a complementary DNA (cDNA) of a given homolog into another homolog will make it possible to determine if the proteins encoded by the two genes can replace each other and therefore have a similar biological function.

Gene targeting: subtle mutations and chromosomal rearrangements

While gene disruption is a valuable tool to address gene function, more subtle alterations are needed to refine the genetic analysis and also to create models of human

Figure 6 (opposite) The Cre–*lox*P system and its applications. (a) The *lox*P site (triangle) is a sequence of 34 bp composed of palindromic sequences of 13 bp separated by a sequence of 8 bp. Cre recombinase specifically recognizes this sequence, provokes the cleavage in DNA (vertical arrows) and (b) induces the recombination of DNA between the two *lox*P sites. This reaction is reversible. Several types of recombination events can be produced depending on whether the two *lox*P sites are carried by the same DNA molecule (recombination in *cis*) or by two different DNA molecules (recombination in *trans*) and depending on the respective orientation of the two *lox*P sites (the orientation of a *lox*P is given by the nonpalindromic 8-bp sequence). (c) Recombination in *cis*. If the two *lox*P sites have the same orientation, the DNA region situated between these sites is deleted during recombination. This type of configuration is used to create mutations devoid of the selection cassette (see **Figure 5**), deletions and conditional mutations (see **Figure 7**). If the orientation of the two *lox*P sites is opposed, recombination leads to the inversion of the region comprised between the two sites. (d) Recombination in *trans*. If one *lox*P site is integrated in the genome and the other is carried by a circular plasmid, there may be an insertion of sequences carried by the plasmid in the integrated *lox*P site. However, since the insertion is a rare event compared with deletion (i.e. the reverse reaction), this type of event requires the use of mutant *lox*P sites. When the *lox*P sites are both integrated in the genome, recombination in *trans* induces chromosomal rearrangements: deletions, duplications or translocations. Such recombination events are rare and have to be selected to be revealed. To do so, one can use truncated and nonfunctional *hp-lox*P and *lox*P-*rt* selection cassettes. After recombination between the *lox*P sites, and only in this case, a functional *hp*-loxP-*rt* cassette (the remaining *lox*P site is situated in an intron) is reconstituted, thus allowing selection of the chromosomal rearrangement desired (see legend to **Figure 5**). Furthermore, the relative orientation of *lox*P sites compared with the centromeric–telomeric axis of the chromosomes is important. Indeed, in the case of wrong relative orientation, recombination will result in the formation of acentric or dicentric chromosomes, which, in view of their great instability, will be eliminated and induce cell death.

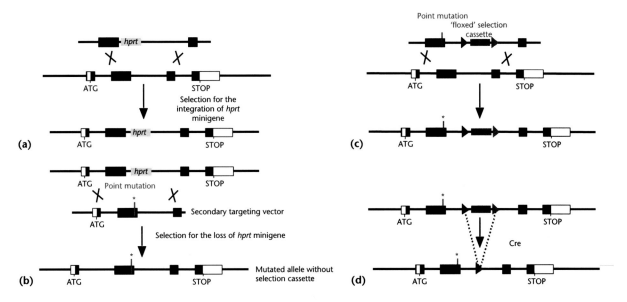

Figure 5 Subtle mutations. The persistence in a modified allele of a selection cassette with its own promoter and regulatory sequences may affect the target locus and surrounding loci. Creation of subtle mutations (point mutations, small deletions and insertions, etc.) therefore requires elimination of the selection cassette. The two strategies that may be used to create this type of modification are shown. Left: the double replacement strategy. This approach requires the use of embryonic stem cells (ES) bearing a null mutation in the endogenous *hprt* gene. The first step (a) consists of introducing a cassette expressing the *hprt* gene in the target gene. The recombinant cells (*hprt+*) are selected in the presence of hypoxanthine, aminopterin, thymidine (HAT). Homologous recombinants are identified using Southern blot. In the second step (b), targeted ES cells are transfected with a replacement vector presenting a subtle mutation (asterisk) and devoid of a selection cassette. The homologous recombination event results in the loss of the *hprt* expression cassette. Targeted cells therefore revert to an *hprt−* phenotype, an event selected in the presence of 6-thioguanine (6-TG) (in principle, all the *hprt−* clones could result only from HR). The use of other replacement vectors carrying different modifications permits the rapid creation of several alleles for the same target gene.
Right: Use of the Cre–*lox*P system (see **Figure 6**). In the first step (c), the target gene is modified by a target vector with a subtle mutation and a 'floxed' selection cassette, that is, surrounded by two *lox*P sites in the same orientation. Then (d) the transient expression of Cre recombinase in the recombinant cells induces deletion of the selection cassette. Apart from the desired subtle modification, only one *lox*P site of 34 bp persists in the final modified allele. The position of this *lox*P site is chosen such that it does not interfere with the expression of the target gene (generally in an intron).

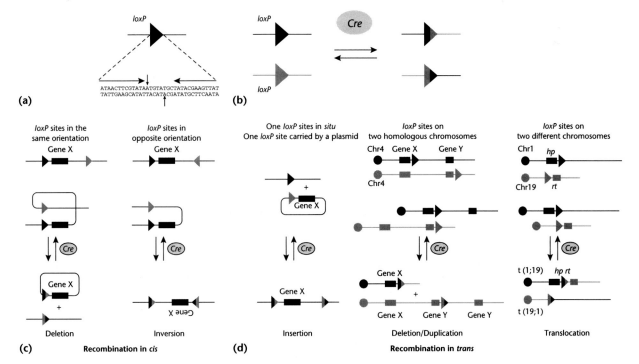

Figure 6 (caption opposite)

genetic disease, which are frequently caused by subtle mutations resulting in the synthesis of an abnormal protein. However, in this case the presence of the selection cassette in the mutated allele may interfere with the normal regulation and expression of the targeted gene. Methods have therefore been devised to generate a mutated allele devoid of foreign sequences (see **Figure 5**). One method relies on a hypoxanthine phosphoribosyl transferase (HPRT)-based selection system (**Figure 5a, 5b**) and has allowed, for example, the generation of mice bearing various point mutations in the prion protein *Prnp* gene (Barron *et al.*, 2001). The other method (**Figure 5b**) takes advantage of the properties of an enzyme, the Cre recombinase, isolated from the bacteriophage P1 (**Figure 6a, 6b**). The end result is a modified allele carrying the mutation and one *lox*P site as the only foreign sequence (e.g. the generation of a point mutation in the fibroblast growth factor receptor 3; Wang *et al.*, 1999).

Use of the Cre–*lox*P system has been extended, in combination with gene targeting, to the generation of chromosomal rearrangements (large deletions, inversions, duplications and translocations; **Figure 6c, 6d**). Chromosomal abnormalities are frequently involved in human fetal loss and in various types of tumors. Creating those types of chromosomal rearrangements in the mouse is therefore a very valuable tool to dissect the molecular mechanisms underlying the

defects they induce. Furthermore, large deletions could generate segmental haploidy in the diploid mouse genome, facilitating the detection of recessive mutations in the deleted regions (reviewed by Yu and Bradley, 2001).

The Cre–*lox*P system and conditional mutagenesis

As emphasized in the previous section, the generation of KO mice is extremely important in the study of gene function. However, this approach has some limitations: first, a mutation that induces lethality at a given stage of development precludes the study of gene function later in development; second, when a gene has a complex pattern of expression, namely in several cell types or tissues, analysis of its function, based on the effects of a mutation present in all the cells of the organism, might become extremely difficult. To overcome these problems, strategies have been developed that are based on the Cre–*lox*P system (**Figure 7**; reviewed by Sauer, 1998). Two requirements need to be fulfilled for conditional targeting with the Cre–*lox*P system: (a) a 'floxed' allele must be created in such a way that two *lox*P sites flank an essential region of the gene, without altering its normal activity; (b) targeting of Cre expression must be tightly controlled: to that end, either transgenic mice by pronuclear

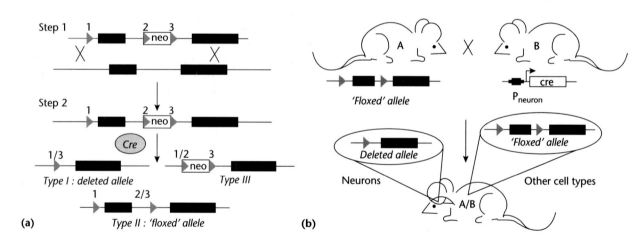

Figure 7 Conditional gene targeting. (a) Creation of a 'floxed allele' for conditional gene targeting. Step 1: the targeting construct contains three lox P sites in the same orientation, sites 1 and 2 flanking an essential region of the target gene (here an exon), and sites 2 and 3 flanking the neo selection cassette. Step 2: transient expression of the Cre recombinase in the targeted cell results in three types of alleles: (1) type I contains the deletion, the phenotype of which can be assessed *in vivo* after transferring the mutation back into the animal; (2) type II corresponds to the 'floxed allele'. *In vivo* deletion can be obtained by crossing mice carrying this allele with Cre-expressing transgenic mice; shown is a scenario that results in the disruption of the floxed gene, specifically in neurons; (3) the third allele can also be recovered but usually has no applications. (b) Conditional gene targeting is obtained by crossing two transgenic mice. The first one (mouse A) carries two 'floxed' alleles of a given gene (type II, see (a)) and exhibits no phenotype (only one allele is shown). The second one (mouse B) is a transgenic mouse for a hybrid transgene corresponding to the Cre recombinase coding sequence under the control of the *cis*-acting regulatory elements of a tissue-specific promoter (here a neuron-specific gene, P_{neuron}). In the progeny, only neurons express the Cre recombinase and consequently harbor deleted (type I, see (a)) allele; all other cells retain the active 'floxed' allele. Consequences of the absence of the target gene in neurons can therefore be assessed. Function of the target gene in other cell types could be addressed by crossing the first mouse with another transgenic mouse expressing the Cre under the control of an appropriate promoter.

microinjection or, better, knockin mice may be used. This scenario allows spatial control of the occurrence of a mutation and therefore insight into the function of a gene in a particular cell type or tissue. It has also proved to be instrumental in the creation of animal models of human disease. For example, transgenic mice were created in which disruption of the tumor suppressor gene *Brca1* was specifically induced in the epithelial cells of the mammary gland, therefore furnishing a model for the study of *BRCA1* (*breast cancer 1, early onset*) in breast cancer (Xu *et al.*, 1999). This could not have been possible without the use of a conditional approach, since homozygous null mutations in this gene result in embryonic lethality. A further refinement of the floxed gene approach relies on the production of fusion protein containing Cre and the ligand-binding domain (LBD) of a steroid receptor. In such chimeric proteins, the activity of Cre becomes hormone-dependent. Therefore, recombination between *lox*P sites could be induced in cells expressing the chimeric Cre by injection of the appropriate ligand into mice carrying floxed alleles and a fusion transgene expressing the Cre–LBD fusion protein. In this way, the occurrence of a mutation can be controlled, not only in a spatial but also in a temporal way (reviewed by Metzger and Feil, 1999).

Conclusion

There has been extraordinary development and refinement of the approaches allowing germ-line modification in the mouse. Indeed, almost any genetic alterations (null and point mutations, as well as chromosome rearrangements) may be introduced precisely and deliberately into the mouse genome. More recently, new approaches have been devised that allow control of the occurrence of a particular genome alteration, both spatially and temporally. No doubt the sophistication of the strategies of genomic engineering will increase in the next years, further establishing the mouse as a unique model for the study of mammalian development, physiology and physiopathology.

See also
Cre–*lox* Inducible Gene Targeting
Mouse Genetics as a Research Tool
Mouse as a Model for Human Diseases

References

Barron RM, Thomson V, Jamieson E, *et al.* (2001) Changing a single amino acid in the *N*-terminus of murine PrP alters TSE incubation time across three species barriers. *EMBO Journal* **20**: 5070–5078.

Capecchi MR (1989) Altering the genome by homologous recombination. *Science* **244**: 1288–1292.

Giraldo P and Montoliu L (2001) Size matters: use of YACs, BACs and PACs in transgenic animals. *Transgenic Research* **10**: 83–103.

Grosveld F (1999) Activation by locus control regions? *Current Opinion in Genetics and Development* **9**: 152–157.

Hammer RE, Swift GH, Ornitz DM, *et al.* (1987) The rat elastase I regulatory element is an enhancer that directs correct cells specificity and developmental onset of expression in transgenic mice. *Molecular and Cellular Biology* **7**: 2956–2967.

Macdonald RJ and Swift GH (1998) Analysis of transcriptional regulatory regions *in vivo*. *International Journal of Developmental Biology* **42**: 983–994.

Metzger D and Feil R (1999) Engineering the mouse genome by site-specific recombination. *Current Opinion in Biotechnology* **10**: 470–476.

Sauer B (1998) Inducible gene targeting in mice using the Cre/lox system. *Methods* **14**: 381–392.

Wang Y, Spatz MK, Kannan K, *et al.* (1999) A mouse model for achondroplasia produced by targeting fibroblast growth factor receptor 3. *Proceedings of the National Academy of Sciences of the United States of America* **96**: 4455–4460.

Wilson C, Bellen HJ and Gehring WJ (1990) Position effects on eukaryotic gene expression. *Annual Review of Cell Biology* **6**: 679–714.

Xu X, Wagner R-U, Larson D, *et al.* (1999) Conditional mutation of *Brca1* in mammary epithelial cells results in blunted ductal morphogenesis and tumor formation. *Nature Genetics* **22**: 37–43.

Yu Y and Bradley A (2001) Engineering chromosomal rearrangements in mice. *Nature Reviews. Genetics* **2**: 780–790.

Further Reading

Cid-Arregui A and Garcia-Carranca A (1998) *Microinjection and Transgenesis – Strategies and Protocols*. New York, NY: Springer-Verlag.

Cohen-Tannoudji M and Babinet C (1998) Beyond 'Knock-out' mice: new perspectives for the programmed modification of the mammalian genome. *Human Molecular Reproduction* **4**: 929–938.

Houdebine L-M (1997) *Transgenic Animals – Generation and Use*. Harwood Academic Publishers.

Leighton PA, Mitchell KJ, Goodrich LV, *et al.* (2001) Defining brain wiring patterns and mechanisms through gene trapping in mice. *Nature* **410**: 174–179.

Lewandoski M (2001) Conditional control of gene expression in the mouse. *Nature Reviews Genetics* **2**: 743–755.

Robertson EJ (1987) *Teratocarcinomas and Embryonic Stem Cells: A Practical Approach*. Washington, DC: IRL Press.

Rodriguez CI, Buchholz F, Galloway J, *et al.* (2000) High-efficiency deleter mice show that FLPe is an alternative to Cre-loxP. *Nature Genetics* **25**: 139–140.

Special issue (1998) Stem cells and transgenesis. *International Journal of Developmental Biology* **42**.

Stanford WL, Cohn JB and Cordes SP (2001) Gene-trap mutagenesis: past, present and beyond. *Nature Reviews. Genetics* **2**: 756–768.

Web Links

Nagy Lab. Cre-expressing mice. Samuel Lenenfeld Research Institute, Mount Sinai
http://www.mshri.on.ca/nagy/

TBASE (The Transgenic/Targeted Mutation Database). Knockout mice. The Jackson Laboratory
http://tbase.jax.org/

Translation Initiation: Molecular Mechanisms in Eukaryotes

Christopher UT Hellen, *State University of New York, Brooklyn, New York, USA*

Tatyana V Pestova, *Moscow State University, Moscow, Russia and State University of New York, Brooklyn, New York, USA*

Initiation is the first phase in protein synthesis, during which eukaryotic initiation factors assemble an 80S ribosome at an initiation codon of a messenger RNA (mRNA) such that this triplet is base-paired to the CAU anticodon of aminoacylated initiator methionyl-transfer RNA in the ribosomal peptidyl site. Initiation is followed by the elongation phase, in which the 80S ribosome moves along the mRNA, translating the open reading frame into protein.

Intermediate article

Article contents

Introduction

The initiation process involves large (60S) and small (40S) ribosomal subunits, messenger ribonucleic acid (mRNA), aminoacylated initiator methionyl-transfer RNA (Met-tRNA$^{Met}_i$), adenosine triphosphate (ATP) and guanosine triphosphate (GTP) and can be considered as a series of partial reactions mediated by at least 11 different eukaryotic initiation factors (eIFs; Hershey and Merrick, 2000; **Figures 1** and **2**). Initiation on most of eukaryotic mRNAs is 5′-end dependent and can be described by the ribosomal scanning model. Initiation on a small number of mRNAs is 5′-end independent and instead occurs by internal ribosomal entry (Hellen and Sarnow, 2001).

Structure of Cytoplasmic Eukaryotic mRNAs

Almost all eukaryotic cytoplasmic mRNAs have a 5′-terminal 7-methylguanosine (m^{7}G) cap and a 3′ poly(A) tail that contributes synergistically with the cap to initiation efficiency. The initiation codon is the AUG triplet closest to the 5′ end on most mRNAs and is usually separated from the 5′ end of the mRNA by a 50- to 70-nt-long 5′ untranslated region (UTR; Kozak, 1991). Conventional eukaryotic mRNAs are therefore functionally monocistronic. The initiation codon is usually an AUG triplet, but in rare instances initiation occurs at triplets that base-pair with only two of the three bases in the anticodon of initiator tRNA. (*See* mRNA Untranslated Regions (UTRs).)

Formation of 43S Preinitiation Complexes

The initiation process starts from individual ribosomal subunits (**Figure 1**). However, 80S ribosomes predominate under physiological conditions in the cytoplasm and they must be separated into 40S and 60S subunits for initiation to proceed. eIF3 and eIF1A have been implicated in the dissociation of 80S ribosomes. Both these factors interact directly with 40S subunits, but it is not yet clear whether they mediate dissociation allosterically (by inducing conformational changes in the 40S subunit) or by steric hindrance (by binding to its subunit interface). eIF1A is a homolog of the small prokaryotic initiation factor IF1 and, like it, may bind the small ribosomal subunit near the aminoacyl (A) site. eIF3 is a large (more than 600 kDa) factor that contains at least 11 nonidentical subunits. Its binding site on the 40S subunit has not definitively been established. (*See* Ribosomes and Ribosomal Proteins.)

The first step in ribosomal recruitment of initiator tRNA is formation of a ternary complex of eIF2 with GTP and Met-tRNA$^{Met}_i$ (Hinnebusch, 2000). eIF2 is a heterotrimeric protein that consists of α, β and γ subunits. The structure of eIF2γ is typical of GTP-binding proteins, and biochemical and genetic data indicate that this subunit plays the principal role in binding GTP and Met-tRNA$^{Met}_i$. Determinants in Met-tRNA$^{Met}_i$, such as the methionyl residue and the A–U base pair at the end of the acceptor stem, enable eIF2 to discriminate between it and all aminoacyl elongator tRNAs, thereby excluding them from the initiation process. The affinity of eIF2 for Met-tRNA$^{Met}_i$ is increased by binding GTP and is reduced by hydrolysis of GTP to GDP. The eIF2/GTP/

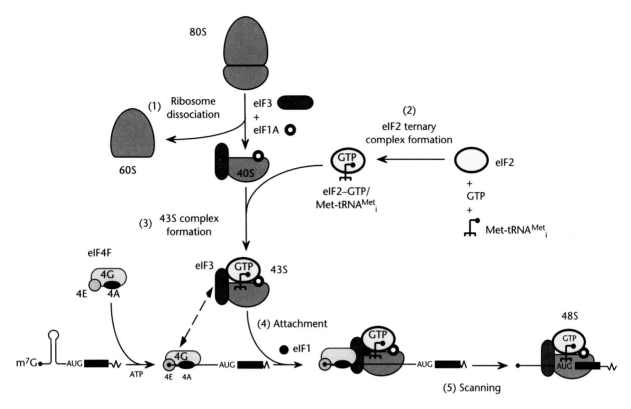

Figure 1 Schematic model of the pathway of 48S initiation complex formation on a capped eukaryotic messenger ribonucleic acid (mRNA). Dissociation of the 80S ribosome into 40S and 60S subunits is effected by eukaryotic translation initiation factor-1A (eIF1A) and -3 (eIF3; step 1). Aminoacylated initiator methionyl-transfer RNA (Met-tRNA$^{Met}_i$), guanosine triphosphate (GTP) and eIF2 form a ternary complex (step 2) that binds with eIF1A and eIF3 to a 40S subunit to form a 43S preinitiation complex (step 3). eIF4F and associated cofactors cooperate in adenosine triphosphate (ATP)-dependent binding of the 43S complex to mRNA by creating an unstructured cap-proximal ribosomal binding site (step 4). The interaction of the eIF4G subunit of cap-bound eIF4F with the eIF3 component of the 43S complex (indicated by a dashed, double-headed arrow) promotes attachment of the 43S complex to mRNA. The bound 43S complex requires eIF1 to scan downstream on the 5′ untranslated region (UTR) until it recognizes the initiation codon (step 5). The anticodon of Met-tRNA$^{Met}_i$ is base-paired to the initiation codon in the resulting 48S complex.

Met-tRNA$^{Met}_i$ ternary complex is an obligatory intermediate in the initiation process. Its binding to the 40S subunit is stabilized by eIF1A and eIF3 and results in the formation of a 43S ribosomal preinitiation complex that contains initiator tRNA in the P site. (*See* tRNA.)

Attachment of 43S Preinitiation Complexes to mRNA

Attachment of 43S complexes to most mRNAs is strictly end-dependent and is strongly enhanced by the m^{7}G cap. The initiation factors eIF4A, -4B and -4F mediate this process (Gingras *et al.*, 1999). eIF4F consists of eIF4E, eIF4A and eIF4G subunits: eIF4E has a specific affinity for the m^{7}G cap at the 5′ end of mRNA; eIF4A is a DEAD-box RNA-dependent ATPase/ATP-dependent RNA helicase that exists *in vivo* both in a free form and

as a part of eIF4F; eIF4G is a large bridging polypeptide that binds eIF4A, eIF4E, eIF3 and RNA. The helicase activity of eIF4F is greater than that of eIF4A alone and is further enhanced by two cofactors, eIF4B and eIF4H. The accessibility of the cap to eIF4F is a major determinant of an mRNA's efficiency of translation. These factors bind cooperatively to the cap-proximal region of mRNA through the m^{7}G cap–eIF4E interaction and re-structure it in a manner that facilitates ribosomal attachment. The interaction of the eIF4G subunit of eIF4F with ribosome-bound eIF3 probably coordinates ATP-dependent restructuring of mRNA with ribosomal attachment to the unwound region of the 5′ UTR.

The 3′ poly(A) tail together with bound poly(A)-binding protein (PABP) enhances ribosomal attachment to capped mRNA, acting synergistically with the m^{7}G cap and the eIF4F cap-binding complex (Sachs, 2000). PABP binds to eIF4G and, in yeast, synergism

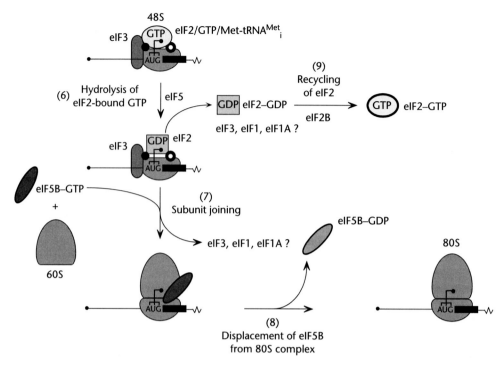

Figure 2 Schematic model of the pathway of 80S complex formation. Hydrolysis of eIF2-bound GTP in the 48S complex is triggered by eIF5, possibly leading to release of eIF2–guanosine diphosphate (GDP; step 6). The stage at which eIF1, eIF1A and eIF3 are released is not known. eIF5B–GTP mediates joining of the resulting complex to a 60S ribosomal subunit (step 7). Ribosome-activated hydrolysis of eIF5B-bound GTP leads to release of eIF5B–GDP to form an 80S ribosome that is competent to begin protein synthesis (step 8). Inactive eIF2–GDP is recycled by eIF2B to eIF2–GTP, which is again competent to bind Met-tRNA$^{Met}_i$ (step 9).

depends on the bridging interaction of eIF4G with the eIF4E–m^{7}G cap and 3′ poly(A)–PABP complexes, and may be due to their resulting enhanced interactions with mRNA and/or to allosteric changes in eIF4G that activate other bound factors.

Ribosomal Scanning and AUG Codon Selection

Binding to the 5′ end of mRNA does not position the 43S complex at the initiation codon. Indirect evidence suggests that after attachment the 43S complex scans the 5′ UTR in the 3′ direction to the initiation codon (Kozak, 1991). This scanning model accounts for the observations that initiation on most mRNAs is strictly end-dependent and occurs at the first AUG triplet from the 5′ end. Biochemical reconstitution experiments have revealed important roles for eIF1 and eIF1A in the scanning process. eIF1A enhances ribosomal attachment to mRNA and may enhance the process of scanning. eIF1 is required for initiation complexes to bind mRNA in a state in which they are competent to scan downstream from the initial binding site. It also monitors the fidelity of initiation codon

selection and destabilizes incorrectly assembled ribosomal complexes, for example on non-AUG triplets. However, many details of the molecular mechanism of scanning remain poorly understood. For example, scanning has been reported to be ATP-dependent. Scanning ribosomal complexes are able to penetrate in the 5′ UTR that have a free energy of less than -50 kcal mol^{-1}, and ATP is probably required for them to unwind such structures. However, it is not known whether ATP is also required to power ribosomal movement or whether it involves helicase-mediated translocation, and, if so, which factors mediate this process. The complete set of factors associated with the scanning complex is unknown, as is the stage at which eIF4F dissociates from the cap and from the 43S complex.

Scanning is arrested when the ribosomal complex encounters an AUG triplet and recognizes it as an initiation codon (Donahue, 2000). The dominant interaction that determines initiation codon recognition by the scanning complex is the codon/anticodon interaction, so that recognition of a mutated initiation codon can be restored by compensatory mutations in the anticodon of Met-tRNA$^{Met}_i$. Mutational analyses have shown that the sequences flanking an AUG triplet also influence its selection as the initiation codon by scanning ribosomes (Kozak, 1991). Ribosomes may bypass an AUG triplet and initiate further downstream if either the AUG triplet occurs very close to the 5' end of an mRNA or the flanking nucleotides deviate from the consensus sequence gcc(A/G)ccAUGG. Leaky scanning can be suppressed by downstream secondary structures that stall the scanning ribosome at an AUG triplet with suboptimal context. Genetic suppressor analyses of mutations have determined that eIF1 and all three subunits of eIF2 influence start site selection in *Saccharomyces cerevisiae* (Donahue, 2000). Mutations in these polypeptides enable the scanning ribosomal complex to begin translation at a UUG codon.

Ribosomal Subunit Joining

Ribosomal subunit joining is the last step of translation initiation (Pestova *et al.*, 2000; **Figure 2**). The 40S subunit in a 48S initiation complex assembled at the initiation AUG codon is still bound to initiation factors eIF2, -3, -1 and -1A and therefore is not ready to join a 60S subunit. Until recently, the only factor implicated in subunit joining was eIF5, a monomeric guanosine triphosphatase (GTPase)-activating protein specific for ribosome-associated eIF2 (Das and Maitra, 2001). eIF5 binds the β-subunit of eIF2. Binding of eIF5 to 40S-bound eIF2 activates the GTPase activity of the γ-subunit of eIF2. GDP-bound eIF2 can no longer bind initiator tRNA, and eIF2–GDP is released, leaving Met-tRNA$^{Met}_i$ in the ribosomal P site base-paired to the initiation codon. Initiation codon recognition by the scanning ribosomal complex may act as a trigger for this reaction, which acts as a molecular switch that converts the complex from a transiently bound form to a form that is stably bound to mRNA at the initiation codon and, as a result, mutations in eIF5 also influence initiation codon selection (Donahue, 2000). This model is supported by biochemical data that indicate that mutant forms of eIF1, eIF2 and eIF5 that permit initiation at non-AUG codons do so by prematurely releasing Met-tRNA$^{Met}_i$ into the P site. eIF2–GDP is recycled to the active eIF2–GTP moiety by eIF2B, a five-subunit guanine-nucleotide exchange factor specific for eIF2. However, hydrolysis of eIF2-bound GTP in 48S complexes is not sufficient to prepare 48S complexes for subunit joining, and this process requires a second factor, eIF5B, which is a homolog of the prokaryotic initiation factor IF2 (Pestova *et al.*, 2000). The exact role of eIF5B in subunit joining is not yet established but may be the release of other initiation factors (eIFs 1, 1A and 3) from the 48S complex. eIF5B has a ribosome-dependent GTPase activity and although GTP hydrolysis by eIF5B is not required for subunit joining, it is necessary for the release of eIF5B from the assembled ribosome. Bound eIF5B blocks the accessibility of the ribosomal A site and displacement of eIF5B–GDP is necessary for the ribosome to be able to begin translation. The canonical eukaryotic initiation process therefore involves two successive GTP hydrolysis steps.

Internal Initiation of Translation

The RNA genomes of several viruses are not translated by the canonical scanning mechanism. Instead they contain large, highly structured internal ribosomal entry sites (IRES) that promote cap-independent binding of the 40S subunit to an internal site in the 5' UTR without scanning from the 5' end (Hellen and Sarnow, 2001). IRESs also occur in a few cellular mRNAs, including several that encode growth factors, transcription factors and oncogenes. The mechanisms that they use have not been elucidated, but many function in circumstances in which factors such as eIF4F, which are required for cap-mediated initiation, are inactivated. (*See* 5' UTRs and Regulation.)

The sequences and structures of IRESs in the same virus family are conserved, whereas those from unrelated viruses are very different. Recent biochemical experiments have shown that unrelated IRESs mediate initiation by distinct mechanisms. The approximately 450-nt encephalomyocarditis virus (EMCV) IRES promotes direct binding of the 43S complex to the initiation codon through a specific interaction of the IRES and the central third of eIF4G. 48S complex formation on this IRES requires only eIF2, -3, -4A and a central fragment of eIF4G, and does not require eIFs 4E, 1 and 1A. Initiation on the hepatitis C virus (HCV) IRES is based on its unprecedented direct interaction with the 40S subunit. This IRES also specifically interacts with eIF3. Direct binding of the 43S complex to the HCV IRES positions the initiation codon precisely in the ribosomal P site so that initiation does not involve scanning or local unwinding of mRNA and therefore does not require eIFs 1, 1A, 4A, 4B and 4F. 48S complex formation on the cricket paralysis virus intergenomic region IRES also involves direct interaction with the 40S subunit, but does not require any initiation factors. Instead, a highly structured portion of the IRES probably occupies the ribosomal P site and mimics the codon–initiator tRNA interaction.

An important observation is that even related IRESs may function in a cell-specific manner, because they require different cell-specific IRES *trans*-acting factors (ITAFs). Initiation on the EMCV-like

Theiler's murine encephalitis virus (TMEV) IRES requires the pyrimidine tract-binding protein (PTB) in addition to eIF2, -3, -4A and -4G, whereas the related foot-and-mouth disease virus IRES requires a second *trans*-acting factor (ITAF45) in addition to PTB. However, the principal functional interaction of these two IRESs is with eIF4G, just as for EMCV. The role of PTB and ITAF45 is to alter the structure of these IRESs to enhance their binding to eIF4G.

In conclusion, current data show that unrelated viral IRESs use very different mechanisms of initiation. Despite different requirements for noncanonical factors, these mechanisms are all based on specific interactions of the IRESs with canonical components of the cellular translation apparatus.

Regulation of Translation Initiation

The efficiency of initiation on an mRNA is a function of its structural properties and this accounts for many of the differences in rates of expression of different proteins. In addition, initiation on some mRNAs may be regulated by specific RNA-binding proteins; for example, iron-regulatory proteins bind to ferritin mRNA and prevent it from binding to the 43S complex. More commonly, translation initiation is regulated by alteration of factor interactions or activities, either as a result of phosphorylation or by regulatory factor-binding proteins (Dever, 2002). Phosphorylation of eIF2α causes eIF2 to bind to and inhibit eIF2B, thereby reducing the intracellular pool of ternary complex. The activity of eIF4F in cap-mediated initiation is regulated by proteolysis of eIF4G during apoptosis and some viral infections, and by the eIF4E-binding proteins, which compete with eIF4G for binding to eIF4E (Gingras *et al.*, 1999). (*See* 5′ UTRs and Regulation; Iron Metabolism: Disorders; RNA-binding Proteins: Regulation of mRNA Splicing, Export and Decay.)

See also

Ribosomes and Ribosomal Proteins
tRNA
5′ UTRs and Regulation

References

Das S and Maitra U (2001) Functional significance and mechanism of eIF5-promoted GTP hydrolysis in eukaryotic translation initiation. *Progress in Nucleic Acids Research and Molecular Biology* 70: 207–231.

Dever TE (2002) Gene-specific regulation by general translation factors. *Cell* 108: 545–556.

Donahue TF (2000) Genetic approaches to translation initiation in *Saccharomyces cerevisiae*. In: Sonenberg N, Hershey JWB and Mathews MB (eds.) *Translational Control of Gene Expression*,

pp. 487–502. Cold Spring Harbor, NY: Cold Spring Harbor Laboratory Press.

Gingras A-C, Raught B and Sonenberg N (1999) eIF4 initiation factors: effectors of mRNA recruitment to ribosomes and regulators of translation. *Annual Reviews in Biochemistry* 68: 913–963.

Hellen CUT and Sarnow P (2001) Internal ribosomal entry sites in eukaryotic mRNA molecules. *Genes and Development* 15: 1593–1612.

Hershey JWB and Merrick WC (2000) Pathway and mechanism of initiation in protein synthesis. In: Sonenberg N, Hershey JWB and Mathews MB (eds.) *Translational Control of Gene Expression*, pp. 33–88. Cold Spring Harbor, NY: Cold Spring Harbor Laboratory Press.

Hinnebusch A (2000) Mechanisms and regulation of initiator methionyl-tRNA binding to ribosomes. In: Sonenberg N, Hershey JWB and Mathews MB (eds.) *Translational Control of Gene Expression*, pp. 185–243. Cold Spring Harbor, NY: Cold Spring Harbor Laboratory Press.

Kozak M (1991) Structural features in eukaryotic mRNAs that modulate the initiation of translation. *Journal of Biological Chemistry* 266: 19 867–19 870.

Pestova TV, Dever TE and Hellen CUT (2000) Ribosomal subunit joining. In: Sonenberg N, Hershey JWB and Mathews MB (eds.) *Translational Control of Gene Expression*, pp. 425–444. Cold Spring Harbor, NY: Cold Spring Harbor Laboratory Press.

Sachs A (2000) Physical and functional interactions between the mRNA cap structure and the poly(A) tail. In: Sonenberg N, Hershey JWB and Mathews MB (eds.) *Translational Control of Gene Expression*, pp. 447–465. Cold Spring Harbor, NY: Cold Spring Harbor Laboratory Press.

Further Reading

Asano K, Clayton J, Shalev A and Hinnebusch AG (2000) A multifactor complex of eukaryotic initiation factors, eIF1, eIF2, eIF3, eIF5, and initiator tRNA(Met) is an important translation initiation intermediate *in vivo*. *Genes and Development* 14: 2534–2546.

Cigan AM, Feng L and Donahue TF (1988) Met-tRNA$^{\text{Met}}_{i}$ functions in directing the scanning ribosome to the start site of translation. *Science* 242: 93–97.

Hentze MW and Kuhn LC (1996) Molecular control of vertebrate iron metabolism: mRNA-based regulatory circuits operated by iron, nitric oxide, and oxidative stress. *Proceedings of the National Academy of Sciences of the United States of America* 93: 8175–8182.

Huang HK, Yoon H, Hannig EM and Donahue TF (1997) GTP hydrolysis controls stringent selection of the AUG start codon during translation initiation in *Saccharomyces cerevisiae*. *Genes and Development* 11: 2396–2413.

Jackson RJ (2000) A comparative view of initiation site selection mechanisms. In: Sonenberg N, Hershey JWB and Mathews MB (eds.) *Translational Control of Gene Expression*, pp 127–183. Cold Spring Harbor, NY: Cold Spring Harbor Laboratory Press.

Kozak M (1989) The scanning model for translation: an update. *Journal of Cellular Biology* 108: 229–241.

Merrick WC (1992) Mechanism and regulation of eukaryotic protein synthesis. *Microbiological Reviews* 56: 291–315.

Pause A, Belsham GJ, Gingras AC, *et al.* (1994) Insulin-dependent stimulation of protein synthesis by phosphorylation of a regulator of 5′-cap function. *Nature* 371: 762–767.

Pestova TV, Borukhov SI and Hellen CUT (1998) Eukaryotic ribosomes require initiation factors 1 and 1A to locate initiation codons. *Nature* 394: 854–859.

Spahn CMT, Beckmann R, Eswar N, *et al.* (2001) Structure of the 80S ribosome from *Saccharomyces cerevisiae* – tRNA–ribosome and subunit–subunit interactions. *Cell* 107: 373–386.

Translocation Breakpoints in Cancer

Andre Mascarenhas Oliveira, *Brigham and Women's Hospital, Boston, Massachusetts, USA*

Jonathan Alfred Fletcher, *Brigham and Women's Hospital, Boston, Massachusetts, USA*

Chromosomal translocations are structural abnormalities generated by double-strand DNA breakages in two or more chromosomes, followed by reciprocal exchange of the segments between the chromosomes. In cancer cells, these translocations often juxtapose regulatory or coding sequences from two different genes, resulting in highly activated oncogenes that regulate proliferation, apoptosis and other aspects of the neoplastic phenotype. Many such translocation-associated oncogenes have been described in human cancers.

Advanced article

Article contents

- Introduction
- Etiology and Predisposing Factors for Chromosomal Translocations
- Tumor Specificity of Chromosomal Translocations
- Translocations Involving Transcription Factor Genes
- Translocations Involving Protein Tyrosine Kinase Genes
- Translocations Associated with Transcriptional Upregulation of Nonfusion Oncogenes
- Diagnosis of Chromosomal Translocations
- Clinical Relevance of Cancer Translocations

Introduction

Chromosomal translocations result from deoxyribonucleic acid (DNA) double-strand breakages in which segments of two or more chromosomes are reciprocally exchanged. Various chromosomal translocations are associated with specific types of human cancer, and such translocations can juxtapose regulatory or coding sequences from the two genes at the respective chromosome breakpoints. In general, it is believed that chromosomal translocations are formed more frequently in cancer cells than in normal cells, presumably as a manifestation of the genetic instability that is characteristic of many cancers. However, it appears that only a small fraction of the overall translocations result in activated oncogenes. Such activating translocations provide a survival advantage to the neoplastic cells and are selected for and retained within the ensuing neoplastic proliferation. Only a limited number of translocations can play this activating role in any given type of cancer, and therefore the recurring translocations can serve as diagnostic markers and therapeutic targets.

Translocation breakpoints that are associated with oncogene activation can either interrupt the coding sequences of the component genes or can fall outside the coding sequences and alter the transcriptional regulation of the gene. In the case of coding sequence fusions, one generally finds that important functional domains from the two component genes have been brought together. For example, many fusion kinase oncogenes result from fusion of a protein–protein association domain (oligomerization domain), for example from a transcriptional regulator gene, to the kinase gene catalytic domain. Such fusions result in constitutive activation of the kinase. Sometimes the intact coding sequence of a particular gene, usually one expressed at weak-to-undetectable levels in the normal cell, is overexpressed by its translocational juxtaposition to highly active promoter and enhancer sequences.

More than 100 recurring chromosomal translocations associated with activation of various oncogenes have been described in human cancers. Chromosomal translocations are particularly frequent in hematological and mesenchymal neoplasms, but are also found in some epithelial neoplasms (carcinomas). Here we briefly review the mechanisms for the generation of translocation-associated oncogenes, and provide key examples of their structural and functional features in human cancers.

Etiology and Predisposing Factors for Chromosomal Translocations

The underlying mechanisms responsible for the genesis of chromosomal translocations are still poorly understood, but DNA double-strand breakage is considered a key step in this process. DNA double-strand breakage may be induced by intracellular (endogenous) or extracellular (exogenous) agents. Some of the endogenous processes associated with DNA double-strand breakage include intrachromosomal rearrangements at the immunoglobulin or T-cell receptor (TCR) loci, meiotic recombination between homologous chromatids and the production of DNA-damaging agents (e.g. oxygen free radicals). Among the many exogenous insults that cause DNA double-stranded breakage, ionizing radiation is the most extensively studied. Ionizing radiation can cause a large variety of chemical injuries in the structure of DNA, including disruption of hydrogen bonds and sugar–phosphate backbones, damage to purine and pyrimidine bases and induction of cross-links between DNA strands. These may result in dosage-dependent single- or double-strand DNA breakage. Notably, the location and frequency of chromosomal breakage induced by ionizing radiation does not appear to be stochastically distributed. There are reports of an inverse relationship between chromosomal size and the frequency of chromosomal breakage, and it seems that sequences rich in GC repeats are particularly susceptible to breakage. Other environmental or artificial inductors of chromosomal breakage include UVA-activated psoralens, chemotherapy agents (e.g. nitrogen mustards, mitomycin C, nitrosureas and platinum compounds) and DNA endonucleases.

Studies on genomic instability mechanisms in the yeast *Saccharomyces cerevisiae* have implicated malfunctions of double-strand DNA breakage repair, namely homologous recombination repair and nonhomologous end joining, as important factors in the generation of chromosomal translocations. Homologous recombination repair requires the presence of the two homologous sister chromatids to guide the DNA repair machinery, and is therefore more active in the G_2 and S phases of the cell cycle. Nonhomologous end joining is intrinsically mutagenic and does not require extensive homology (usually less than 10 base pairs (bp)) between the two sequences to be joined. Nonhomologous end joining is more active in the G_1 phase of the cell cycle because of its independence of the presence of a guiding sister chromatid. Malfunction of these DNA repair systems is responsible for the chromosomal instability in some cancer and premature aging syndromes. Chromosomal instability in these syndromes is characterized by increased DNA double-strand breakage and chromosomal rearrangements. Bloom syndrome is an autosomal recessive disorder caused by inactivating mutations of the Bloom syndrome (*BLM*) gene. *BLM* encodes a nuclear protein related to the RecQ family of helicases, which are DNA-unwinding proteins involved in homologous recombination. Bloom syndrome patients show increased chromosomal breakage, deletions and rearrangements, and exhibit high frequencies of sister chromatid exchanges in the form of quadriradials, four-armed structures comprising two chromosomes intersecting at regions of chromatin homology. Patients with Bloom syndrome are predisposed to a large variety of benign and malignant neoplasms, which tend to occur at a much earlier age than in the normal population. Other chromosomal instability syndromes result from defects in RecQ helicase proteins. For example, Werner syndrome and Rothmund–Thomson syndrome are autosomal recessive disorders with germ-line mutations of *Werner syndrome* (*WRN*) and *RecQ protein-like 4* (*RECQL4*) respectively. Affected individuals are predisposed to many tumors, with osteosarcomas being particularly

characteristic of Rothmund–Thomson patients. Ataxia telangiectasia is an autosomal recessive disorder characterized by an exquisite sensitivity to ionizing radiation secondary to mutations of the *ataxia telangiectasia mutated* (*includes complementation groups A, C and D*) (*ATM*) gene. *ATM* encodes a protein kinase that participates in surveillance for double-strand DNA breakage. Lymphocytes from ataxia telangiectasia patients show increased levels of chromosomal rearrangement. Nijmegen breakage syndrome shares several features with ataxia telangiectasia but is caused by mutations of the *Nijmegen breakage syndrome 1* (*nibrin*) (*NBS1*) gene. The NBS1 protein participates in a complex with apparent roles in homologous recombination repair and nonhomologous end joining. Patients with ataxia telangiectasia and Nijmegen breakage syndrome are particularly susceptible to the development of leukemias and lymphomas. Fanconi anemia is a heterogeneous group of autosomal recessive disorders with predisposition to cancer, particularly leukemias and squamous cell carcinomas. At least eight distinct genes involved in DNA double-strand breakage repair have been implicated in the genesis of Fanconi anemia. The chromosomes of Fanconi anemia patients exhibit increased sensitivity to DNA cross-linking agents such as diepoxybutane, and this feature can be useful in diagnosis of the disease. Fanconi anemia DNA repair defects can also result from germ-line mutations of the *breast cancer 2, early onset* (*BRCA2*) gene.

DNA sequences near the chromosomal breakpoints may predispose to certain chromosomal rearrangements. Among those are repetitive DNA sequences, such as long interspersed nuclear elements (LINEs), Alu repeats and topoisomerase II DNA-consensus binding sites. Topoisomerase II DNA sequences are believed to be of particular relevance in the rearrangements of the *myeloid/lymphoid or mixed-lineage leukemia* (*trithorax homolog, Drosophila*) (*MLL*) transcription factor gene in various types of leukemia (Rowley, 1998). *MLL* is also rearranged in secondary leukemias induced by the use of topoisomerase II inhibitors, including epipodophyllotoxins. *MLL* rearrangements cluster to an 8.3 kilobase (kb) interval rich in topoisomerase II DNA consensus binding sites. Topoisomerase II catalyzes the sequential reaction of double-strand DNA breakage, DNA strand exchange and DNA strand rejoining. Because topoisomerase II inhibitors block the DNA strand rejoining step, they can encourage the formation of chromosome translocations (nonhomologous chromosomal recombination) after double-strand DNA breakage. Other examples of recombination-promoting sequences are the heptamer/nonamer sequences that are adjacent to translocation breakpoints in many B-cell lymphomas. These sequences serve as sites for V(D)J recombination, and are therefore recombination signals in the B-cell context.

Physical proximity of the rearranged genes in the interphase nuclei has also been implicated in the genesis of fusion genes. Radiation-induced thyroid carcinomas frequently exhibit rearrangements of the protein tyrosine kinase gene *ret proto-oncogene (multiple endocrine neoplasia and medullary thyroid carcinoma 1, Hirschsprung disease) (RET)* on chromosome 10q11. In many such thyroid carcinomas, *RET* is fused with the *DNA segment on chromosome 10 (unique) 170 (D10S170)* locus, which is telomeric (band 10q21) to *RET* on the same chromosome arm. Although on different chromosomes, *RET* and *D10S170* are located near each other in 35% of normal human thyroid cells during interphase, therefore providing a possible explanation as to why these two particular genes might be disproportionately 'fused' after radiation-induced double-strand DNA breakage.

An interesting aspect of cancer chromosome translocations is that they are sometimes amplified, providing additional copies of the associated fusion oncogenes. After their formation, the translocation breakpoint regions can be amplified as tandem repeats within the original chromosome, or can be amplified as extrachromosomal structures, such as double minute chromosomes. An example of intra-chromosomal amplification is the typical low-level amplification of *platelet-derived growth factor beta polypeptide (simian sarcoma viral (v-sis) oncogene homolog) (PDGFB)* fusion genes in some mesenchymal tumors (sarcomas). An example of extrachromosomal amplification is the high-level amplification, within double minute chromosomes, of *paired box gene 7 (PAX7)* fusion oncogenes in other sarcomas. In either of these scenarios, the presence of the amplification suggests that cancer cells may require multiple copies of the fusion genes to accomplish oncogenic cell transformation.

Tumor Specificity of Chromosomal Translocations

Various chromosomal translocations are found almost universally in particular types of human cancer. For example, virtually all chronic myelogenous leukemias have the translocation t(9;22) resulting in oncogenic *breakpoint cluster region (BCR)–v-abl Abelson murine leukemia viral oncogene homolog 1 (ABL1)* (commonly known as *BCR–ABL*) fusion. Such observations beg the question as to whether these characteristic translocations are 'initiating' tumorigenic events or, alternately, whether they are later events, responsible for tumor progression. For most cancer translocations, this question has not been answered. However, we do know that many cancer translocations create fusion oncogenes that can transform nonneoplastic cells, as is often observed when such genes are expressed in nonneoplastic cells *in vitro* or in mice. These observations do suggest that the translocation oncogenes are critical to the transformed properties of their associated human tumors, but do not prove that the translocations are the primary oncogenic events in those same tumors.

There is ample evidence that most translocations are not sufficient to create cancer. Rather, tumors require many oncogenic mutations, perturbing different aspects of the cell's biology (including apoptosis, proliferation, differentiation and adhesion), which collectively create the final neoplastic phenotype. These notions are supported by the fact that many of the characteristic chromosomal translocations observed in cancer can be detected at very low levels, for example one in 10 000 cells, among normal cell populations. The rare cells containing these translocations are individually at low risk of progression to cancer because it is unlikely they will acquire the other mutations that are required to create a clinically evident tumor.

It has been empirically observed that most recurring chromosome translocations are associated with one or a few types of cancer, and are therefore useful as diagnostic markers. Such diagnostic specificity has a biological basis in that the neoplastic phenotype is a symbiotic interaction between the cell environment and the translocation gene, each supporting the function of the other (Barr, 1999). An example of this concept is provided by B-cell lymphomas, where many of the diagnostic translocations juxtapose various oncogenes to the immunoglobulin loci. The immunoglobulin genes have extremely active promoters in B cells, and therefore the several described B-cell lymphoma translocations result in striking overexpression of the juxtaposed oncogenes. This translocation mechanism is not relevant in other cell types, where the immunoglobulin genes are transcriptionally silent. Indeed, B-cell lymphoma translocations have not been found in other types of human cancer. Another factor accounting for diagnostic specificity is that the transforming properties of translocation-associated oncoproteins are generally restricted to specific cell lineages. Fusion oncoproteins require interactions with specific cell proteins that enhance the function of the oncoprotein. Cell lineages lacking the relevant interacting proteins will not be transformed by a given fusion oncoprotein. These ideas are nicely illustrated by the observation that the BCR–ABL fusion oncoprotein, which is highly transforming in hematopoietic progenitor cells, fails to transform fibroblast cell lines (Daley *et al.*, 1987).

Translocations Involving Transcription Factor Genes

Transcription factors are a heterogeneous group of DNA-binding proteins that control the expression of genes, including those regulating cell proliferation and differentiation. Most transcription factors have domains involved in DNA binding, protein dimerization (for interaction with homologous proteins) and gene transactivation (for activation of gene transcription). Several classes of transcription factors exist, and they are classified primarily according to the structure of their DNA-binding domains. Some of the major categories include homeodomain proteins, zinc-finger proteins, leukine zipper proteins, forkhead proteins and helix–loop–helix proteins. Transcription factor genes are targeted by many chromosomal translocations, resulting in their aberrant expression and function. Very often these transcription factor genes control lineage-specific developmental pathways, and their abnormal activation can induce expression of corresponding lineage-specific antigenic markers, which then become a defining aspect of the transformed phenotype.

Hematological neoplasms

A large number of chromosomal translocations involving transcription factors have been described in leukemias and lymphomas. Chromosomal translocations in leukemias seem to interfere, in many instances, with the normal differentiation program of myeloid and lymphoid lineages (Lock, 1997). In acute leukemias, the targets for a variety of chromosomal translocations are genes encoding hematopoiesis-related transcription factors, including *runt-related transcription factor 1 (acute myeloid leukemia 1; aml1 oncogene) (RUNX1)* and *core-binding factor, beta subunit (CBFB)* (**Figure 1**). Runt-related transcription factor 1 (Runx1) is a DNA-binding protein that shows significant homology with the *Drosophila melanogaster* developmental protein Runt. Runx1 binds to the DNA enhancer sequence TGTGGT and regulates the expression of several genes involved in hematopoiesis. The DNA-binding capability of Runx1 is enhanced by its binding to CBFB. The most common chromosomal translocation observed in acute myeloid leukemias (approximately 12% of cases) is the t(8;21)(q22;q22), which fuses *RUNX1* on chromosome 21q to *core-binding factor, runt domain, alpha subunit 2; translocated to, 1; cyclin D-related (CBFA2T1)* on chromosome 8. *CBFA2T1* encodes a nuclear phosphoprotein normally expressed in the nervous system and in CD34 + hematopoietic progenitors. The CBFA2T1 protein normally participates in a multiprotein complex

involved in chromatin remodeling and transcriptional repression, and can transform NIH3T3 cells. The fusion protein Runx1–CBFA2T1 retains the Runt domain of Runx1 and almost the entire sequence of CBFA2T1 (**Figure 2**). It seems that CBFA2T1 domains can dominantly repress transcription of certain genes whose expression is normally activated by Runx1. As an example, the tumor suppressor protein p14ARF is transcriptionally activated by Runx1, but transcriptionally repressed by Runx1–CBFA2T1.

RUNX1 is also involved in several other chromosomal translocations in leukemias and myeloproliferative disorders, including *RUNX1–myelodysplasia syndrome 1 (MDS1)* in myelodysplastic (preleukemic) syndromes, *RUNX1–ecotropic viral integration site 1 (EVI1)* in chronic myelogenous leukemia in blast crisis, and *ets variant gene 6 (TEL oncogene) (ETV6)–RUNX1*

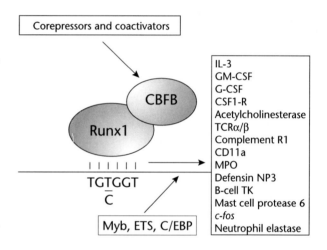

Figure 1 Runx1–CBFB transcription factor complex. Runx1 binds to the consensus DNA enhancer sequence TG(T/C)GGT and activates the transcription of several genes involved in hematopoiesis (right). Runx1 also cooperates with other transcription factors, such as Myb, ETS and C/EBP (bottom). Interactions with CBFB increase the DNA-binding activity of Runx1, and several other coactivators and corepressors further regulate the Runx1–CBFB transcription machinery.

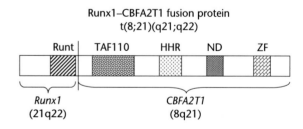

Figure 2 Fusion protein Runx1–CBFA2T1. Runx1–CBFA2T1 retains the Runx1 Runt domain and almost the entire CBFA2T1 sequence. The CBFA2T1 component recruits a transcriptional repressor complex, including the NcoR and mSin3A proteins and several histone deacetylases, which collectively create a myeloid differentiation block.

in pre-B acute lymphoblastic leukemia. All of these fusion genes retain the Runx1 Runt domain.

Another example of a fusion gene involving the *RUNX1/CBFB* complex is *CBFB–myosin, heavy polypeptide 11, smooth muscle (MYH11)*, which results from rearrangements of chromosome 16 in a subtype of acute myelogenous leukemias (Alcalay *et al.*, 2001). *CBFB–MYH11* can result from translocation or pericentric inversion of chromosome 16, thereby fusing 5′ end sequences of *CBFB* to 3′ end sequences of *MYH11*. The resultant fusion protein brings together the CBFB–Runx1-binding heterodimerization domain to MYH11 coiled-coil domains. The fusion protein CBFB–MYH11 binds to the Runx1 Runt domain more effectively than does the normal CBFB, thereby explaining its dominant negative effect on the Runx1–CBFB complex.

The possibility of developing therapeutic strategies targeting tumor-specific translocations is exemplified in acute promyelocytic leukemia (APL). APL is a clonal proliferative disorder of myeloid cells arrested at the promyelocytic stage of differentiation, and is highly responsive to all-*trans* retinoic acid (ATRA). At least five chromosomal translocations involving the *retinoic acid receptor, alpha (RARA)* gene have been found in APL (Zelent *et al.*, 2001). The chromosomal translocation t(15;17)(q22;q21) is the most common and generates the fusion gene *promyelocytic leukemia (PML)–RARA* (**Figure 3**). *PML* encodes a transcription factor with the RING zinc-finger motif, and contributes to cellular senescence and apoptosis. *RARA* is a member of the steroid-thyroid receptor superfamily of nuclear hormone receptors that heterodimerize with retinoic-X receptors (RXR). In the presence of physiological concentrations of retinoic acid, the RARA–RXR complex recruits coactivator proteins with histone acetyltransferase activity and functions as a transcriptional activator. In contrast, PML–RARA seems to function as a transcriptional repressor, through recruitment of corepressors with histone deacetylase activity. ATRA displaces corepressors

from the complex and recruits coactivators, restoring the granulocytic differentiation program (Zelent *et al.*, 2001).

Soft tissue neoplasms

Both benign and malignant soft tissue tumors can contain transcription factor fusion oncogenes. A much-studied example is the fusion oncogene *Ewing sarcoma breakpoint region 1 (EWSR1)–Friend leukemia virus integration 1 (FLI1)* resulting from translocation t(11;22)(q24;q12) (Arvant and Denny, 2001) which is found in up to 85–90% of the Ewing sarcoma family of tumors. These are highly malignant, small round cell tumors that occur predominantly in adolescents and young adults. The translocation fuses *EWSR1* gene 5′ sequences to *FLI1* gene 3′ sequences (**Figure 4**). *EWSR1* encodes a ubiquitously expressed protein involved in DNA transcription, and FLI1 is a member of a large family of DNA transcription factors that contain the highly conserved ETS domain. The *EWSR1–FLI1* fusion oncogene structure varies, depending on which *EWSR1* and *FLI1* exons are retained in the chimeric gene, but the *EWSR1–FLI1* oncogene resulting from fusion of *EWSR1* exon 7 to *FLI1* exon 6 is the most common form and is referred to as 'type 1'. Type 1 fusions have been associated with a better survival when compared with the other *EWSR1–FLI1* fusion types.

EWSR1 is targeted by several chromosomal translocations in mesenchymal neoplasias, always resulting in fusion oncoproteins with EWSR1 at the *N*-terminal end of the fusion gene. One example is desmoplastic small round cell tumor (DSRCT), which is an exceptionally malignant cancer composed of nests of small round tumor cells immersed in an intense reactive fibroblastic proliferation. DSRCT is cytogenetically characterized by the translocation t(11;22)(p13;q12), which fuses *EWSR1* to the *Wilms tumor 1 (WT1)* gene (Gerald *et al.*, 1995). *WT1* is a tumor suppressor gene located on chromosome 11p13 that encodes a zinc-finger transcription factor with crucial roles in the development of the genitourinary tract. Germ-line

PML–RARA fusion protein
t(15;17)(q22;q21)

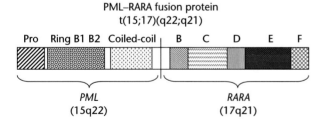

Figure 3 Fusion protein PML–RARA. PML–RARA proteins retain the RARA DNA-binding domain (region C) and the ligand-binding domain (region E). Most PML–RARA fusions retain the PML proline-rich region (Pro), the zinc-finger motif RING, the cysteine-rich B1 and B2 boxes and the α-helical coiled-coil domain.

EWSR1–FLI1 fusion protein
t(11;22)(q24;q12)

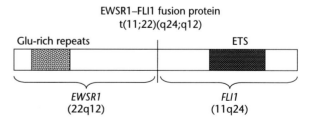

Figure 4 Fusion protein EWSR1–FLI1. The EWSR1–FLI1 fusion protein invariably retains the ETS DNA-binding domain of FLI1 at the *C*-terminal region.

WT1 mutations cause Wilms tumor syndromes, such as Denys–Drash and WAGR (Wilms tumor–aniridia–genitourinary abnormalities and mental retardation). The EWSR1–WT1 fusion oncoprotein retains the EWSR1 transactivation domain and the WT1 DNA-binding domain.

Epithelial neoplasms

Tumors of epithelial differentiation (carcinomas) often have complex karyotypes which can obscure the presence of recurring translocations. However, several translocations have been identified in thyroid carcinomas, where one finds fusion kinase oncoproteins in papillary thyroid carcinoma, and *paired box gene 8 (PAX8)–peroxisome proliferative activated receptor, gamma (PPARG)* fusion transcription factor genes resulting from translocation t(2;3)(q13;p15) in follicular thyroid carcinoma (Kroll *et al.*, 2000) (**Figure 5**). Another example is the translocation t(15;19)(q13;p13)-associated fusion oncogene seen in a very aggressive form of carcinoma in young adults (French *et al.*, 2001). Initial studies have localized the translocation breakpoint within the coding region of the transcriptional regulator *bromodomain containing 4 (BRD4)*.

Translocations Involving Protein Tyrosine Kinase Genes

Chromosomal translocations in various neoplasms produce fusion forms of protein tyrosine kinase genes, and such oncogenes are of considerable clinical importance. Protein tyrosine kinases comprise a large family of proteins, which are primarily involved in signal transduction. All tyrosine kinase proteins contain a highly conserved kinase (catalytic) domain which mediates the phosphorylation of tyrosine residues in protein substrates. Phosphorylated tyrosines serve to stabilize various protein–protein interactions, and also to enhance kinase activity (for those substrates with intrinsic kinase function). Tyrosine kinase

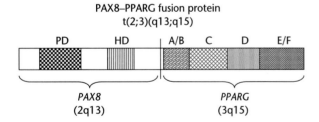

PAX8–PPARG fusion protein
t(2;3)(q13;q15)

Figure 5 PAX8–PPARG fusion protein. PAX8 paired and partial homeobox DNA-binding domains are represented by PD and HD respectively; PPARG DNA-binding, ligand-binding, RXR dimerization and transactivation nuclear receptor domains are represented by A/B, C, D and E/F respectively.

fusion oncoproteins resulting from chromosomal translocations in cancer have a very consistent structure. The *C*-terminal end of these fusion oncoproteins typically contains the entire kinase domain from the tyrosine kinase protein, whereas the *N*-terminal end contains an oligomerization domain from the non-kinase fusion partner. The oligomerization domain enables spontaneous interactions between neighboring kinase fusion oncoproteins, and the complexed oncoproteins are then able to phosphorylate each other, resulting in further upregulation of the kinase activity.

Hematological neoplasms

The Philadelphia (Ph) chromosome – the cytogenetic hallmark of chronic myelogenous leukemia (CML) – was the first diagnostic translocation identified in a human cancer, and also provides the classic example of a fusion tyrosine kinase oncogene (Deininger *et al.*, 2000). The Ph chromosome results from chromosomal translocation t(9;22)(q34;q11.2), leading to fusion of the *BCR* gene on chromosome 22 to the *ABL1* kinase gene on chromosome 9. In addition to CML, this translocation is also observed in 25% of cases of acute lymphoblastic leukemia in adults, in a smaller number of acute lymphoblastic leukemias in children and in the rare CML variant chronic neutrophilic leukemia. *ABL1* encodes a 145 kilodaltons (kDa) nonreceptor protein tyrosine kinase that shuttles between the nucleus and the cytoplasm and is involved in variety of cellular processes, such as cell cycle regulation and apoptosis. *BCR* encodes a 160 kDa protein with dimerization, serine/threonine kinase, Rho-GEF (guanine nucleotide exchange factor) and Rac-GTPase domains. The BCR–ABL fusion oncoprotein is expressed as three structural variants, depending on the location of the breakpoint in the *BCR* gene (**Figure 6**). The BCR–ABL type correlates with specific neoplasms such that the 190, 210 and 230 kDa BCR–ABL proteins are typically expressed in acute lymphoblastic leukemia, CML and chronic neutrophilic leukemia respectively. The BCR–ABL fusion oncoproteins feature constitutive activation of the ABL kinase domain, resulting in autophosphorylation and tyrosine phosphorylation in various substrates. The phosphotyrosine residues serve as binding sites for various signaling proteins, resulting in activation of signaling pathways such as Ras/MAPK (implicated in cell proliferation), and PI3K/AKT (implicated in cell survival), among others.

Soft tissue neoplasms

Translocation-associated fusion kinase oncogenes are found in various soft tissue neoplasms. Congenital fibrosarcoma is a pediatric spindle cell neoplasm with

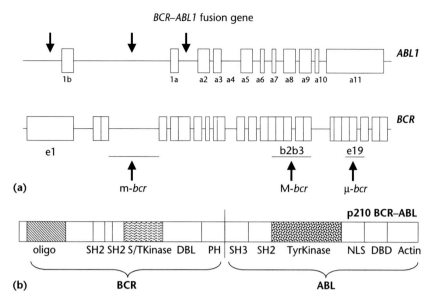

Figure 6 *BCR–ABL1* fusion gene. The overall structures of the *BCR* and *ABL1* genes are shown in (a). The arrows indicate the location of the breakpoints in the *ABL1* gene on chromosome 9. In the *BCR* gene on chromosome 22, the breakpoints localize to three areas: m-*bcr*, M-*bcr* and μ-*bcr*. In most patients with CML, the BCR–ABL fusion oncoprotein is 210 kDa and includes the BCR oligomerization (oligo) and serine threonine kinase (S/TKinase) domains and the ABL tyrosine kinase domain.

an excellent prognosis. Cytogenetically, congenital fibrosarcoma is characterized by the balanced translocation t(12;15)(p13;q25), which fuses the 5′ end of the transcription factor gene *ETV6* to the 3′ end of the *neurotrophic tyrosine kinase, receptor, type 3* (*NTRK3*) protein tyrosine kinase gene (Knezevich *et al.*, 1998). The ETV6–NTRK3 fusion oncoprotein retains the ETV6 helix–loop–helix dimerization domain and the NTRK3 kinase domain. *ETV6* is also rearranged with other genes, including protein tyrosine kinase genes, in a variety of leukemias. Interestingly, the *ETV6–NTRK3* fusion gene is seen in a subset of pediatric renal tumors known as cellular nephroblastic nephroma, which are histologically similar to congenital fibrosarcoma (Rubin *et al.*, 1998). This finding suggests that these two pathological entities belong to the same spectrum of tumors. Interestingly, *ETV6–NTRK3* is also found occasionally in myeloid leukemia (Eguchi *et al.*, 1999), and indeed this is one of several oncogenes that are known to play transforming roles in both soft tissue and hematopoietic neoplasms.

Inflammatory myofibroblastic tumor (IMT) is an unusual entity that arises predominantly in the abdominal cavity and thoracic/pulmonary region of young patients. Similar to congenital fibrosarcoma, IMT is associated with an excellent prognosis, and metastases are rare. Some IMTs have chromosomal translocations that rearrange the protein tyrosine kinase gene *anaplastic lymphoma kinase (Ki-1)* (*ALK*) on the chromosomal band 2p23 (Lawrence *et al.*, 2000). *ALK* rearrangements were initially described in the chromosomal translocation t(2;5)(p23;q35), which occurs in a subset

of anaplastic large cell lymphomas (ALCL) (Falini, 2001). At least nine *ALK* fusion genes have been described in IMT and ALCL, and in all of them the resultant fusion oncoproteins contain the ALK kinase domain fused to an oligomerization domain of another protein. Interestingly, two of the *ALK* fusion genes, *tropomyosin 3* (*TPM3*)–*ALK* and *clathrin, heavy polypeptide* (*Hc*) (*CLTC*)–*ALK*, have been found in both IMT and ALCL, indicating that similar oncogenic mechanisms can contribute to these different tumors.

Another mechanism of protein tyrosine kinase activation generated by a chromosomal translocation is observed in a rare mesenchymal tumor known as dermatofibrosarcoma protuberans (DFSP) (Simon *et al.*, 1997) . DFSP is an infiltrative subcutaneous sarcoma that exhibits high local recurrence rates but rarely metastasizes. Cytogenetically, this tumor is characterized by the chromosomal translocation t(17;22) (q21;q13), often amplified within a circular 'ring' chromosome. The t(17;22) creates a *collagen, type I, alpha 1* (*COL1A1*)–*PDGFB* fusion gene, in which the entire *PDGFB* coding sequence is placed under the transcriptional control of the highly active *COL1A1* promoter. This mechanism results in *PDGFB* overexpression with resultant autocrine activation of the *PDGFB* receptor.

Epithelial neoplasms

Chromosomal translocations targeting the *RET* or *neurotrophic tyrosine kinase, receptor, type 1* (*NTRK1*) receptor tyrosine kinase genes are found in up to 45%

of papillary thyroid carcinomas (Pierotti *et al.*, 1998). In all of these fusion proteins, the kinase domain of RET or NTRK1 is retained.

Translocations Associated with Transcriptional Upregulation of Nonfusion Oncogenes

Hematological neoplasms

Many lymphomas feature juxtaposition of intact proto-oncogenes to transcriptionally active loci, for example the immunoglobulin (Ig) locus in B-cell lymphomas or the T-cell receptor (TCR) locus in T-cell lymphomas. V(D)J recombination and class switch recombination, which normally occur within the immunoglobulin or TCR loci in lymphocytes, have been implicated in the genesis of these nonfusion genes (Kuppers and Dalla-Favera, 2001).

Follicular lymphoma is an indolent neoplasm that accounts for 25–35% of non-Hodgkin lymphomas. In approximately 70–80% of cases, follicular lymphomas carry the chromosomal translocation t(14;18)(q32;q21), which juxtaposes the antiapoptotic gene *B-cell CLL/lymphoma 2* (*BCL2*) to the *immunoglobulin heavy chain locus* (*IGH*). This rearrangement involves the *IGH* V(D)J recombination process, with the breakpoint regions on exon 3 of *BCL2* fused to the 5' end of either J_H or $D–J_H$ heavy chain segments. Therefore, transcription of the *BCL2* gene is placed under the control of the *IGH* enhancer. Current evidence suggests that the translocation process is necessary but not sufficient for lymphomagenesis. Deregulated BCL2 expression confers survival advantage, but additional cellular events are needed for full neoplastic transformation.

Rearrangements of the *v-myc myelocytomatosis viral oncogene homolog* (*avian*) (*MYC*) proto-oncogene in Burkitt lymphoma exemplify V(D)J recombination and class switch recombination (Boxer and Dang, 2001). Burkitt lymphomas are a neoplasm of B cells, and therefore V(D)J recombination of the three immunoglobulin loci (heavy chain, kappa light chain and lambda light chain) is involved in the chromosomal translocation t(8;14)(q24;q32) in 'endemic' Burkitt lymphomas found in patients from equatorial Africa. These translocations juxtapose *MYC* to the *IGH* intron enhancer, leading to MYC overexpression. In contrast, nonendemic forms of Burkitt lymphoma have translocation breakpoints downstream of the *IGH* intron enhancer, juxtaposing *MYC* to other *IGH* enhancers. In other chromosomal translocations, *MYC* is juxtaposed to the Ig κ and Ig λ light chains. Since MYC is involved in a variety of cell processes (e.g. cell proliferation, growth and apoptosis), deregulation of MYC expression is expected to contribute in different ways to the generation and/or maintenance of the transformed phenotype.

Rearrangements of the TCR locus in leukemias with chromosome 14q11 rearrangements, including the juxtaposition of *MYC*, the basic helix–loop–helix transcription factor *T-cell acute lymphocytic leukemia 1* (*TAL1*), or the transcription regulator *LIM domain only 2* (*rhombotin-like 1*) (*LMO2*) to the TCR locus in T-cell acute lymphoblastic leukemia are observed. These rearrangements, similar to those described for B-cell lymphomas, result in the abnormal expression of these genes when juxtaposed to TCR locus.

Soft tissue tumors and epithelial neoplasms

Promoter exchange has also been described in both soft tissue and epithelial tumors. Rearrangements of the developmental zinc-finger transcription factor *pleiomorphic adenoma gene 1* (*PLAG1*) are frequently observed in the salivary gland tumor pleiomorphic adenoma and the pediatric adipose tissue tumor lipoblastoma (Hibbard *et al.*, 2000). In both tumors chromosomal breakpoints occur in the 5' noncoding region of the involved genes, resulting in overexpression of PLAG1.

Diagnosis of Chromosomal Translocations

Cancer chromosomal translocations can be diagnosed by various methods, including assays directed to DNA, ribonucleic acid (RNA) or encoded oncoproteins. Conventional karyotyping, in which metaphase cells are collected from tissue cultures and then stained to produce 'banding' patterns, is a time-honored method (**Figure 7a**). However, karyotyping is fallible in that the requisite metaphase cells are not always obtained. Molecular cytogenetic methods, such as fluorescence *in situ* hybridization (FISH), are more commonly used because such assays can be performed in archival paraffin-embedded or frozen specimens and do not require cell culture (**Figure 7b**). However, FISH probes for many of the diagnostic cancer translocations are not yet available commercially. Polymerase chain reaction (PCR) methods are also widely used, typically with RNA templates but also with applicability to genomic targets, and have the advantage of much higher sensitivity than conventional karyotyping or FISH. Therefore, PCR assays have been essential in evaluations of minimal residual disease after a chemotherapy course for hematological neoplasms. The major disadvantage of PCR is the

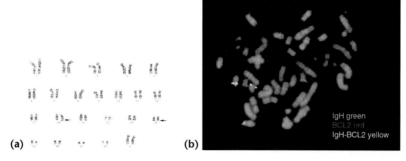

Figure 7 [*Figure is also reproduced in color section.*] (a) Follicular lymphoma showing the characteristic balanced translocation t(14;18)
(q32;q21). Arrows indicate the translocation breakpoints on the rearranged copies of chromosomes 14 and 18. (b) Fluorescence *in situ*
hybridization (FISH) image showing the same t(14;18)(q32;q21) translocation using probes that span the two loci involved (*IgH@* and *BCL2*).
The translocation produces two red–green (yellow) fusion FISH signals, one each on the translocated copies of chromosomes 14 and 18.

risk of false-positive results due to cross-contamination in these exquisitely sensitive reactions. Cross-
contamination is a particular issue when the PCR
target is present in very small amounts, for example
when evaluating fusion oncogene transcripts from
paraffin-embedded specimens, and where many cycles
of PCR are needed to demonstrate the target.
Southern blotting is another popular method for
demonstrating translocation breakpoints but is less
widely used today than in previous years because it is
more labor intensive and requires larger amounts of
tumor material than FISH and PCR methods.

Clinical Relevance of Cancer Translocations

The clinical relevance of cancer translocations is
discussed at various points in the sections above.
However, the major clinical roles of chromosomal
translocations can be summarized as: (1) diagnostic
markers, (2) response markers and (3) therapeutic
targets. The diagnostic role is clear, since most cancer
translocations are only associated with one or a few
types of cancer. Their detection can be very useful in
pinpointing a diagnosis, particularly in those tumors
where the conventional histology is atypical or in
tumors that have become undifferentiated and have
lost their defining morphological and immunopheno-
typical features. The use of cancer translocations as
'response markers' is exemplified by the increasing role
of FISH and PCR surveillance in patients with
leukemia, and where such methods can reveal small
amounts of residual or recurrent leukemia cells that
are not evident otherwise. The role of cancer translocations as 'therapeutic targets' is also shown in various
leukemias, but most notably in the example of the
BCR–ABL translocation product in chronic myelogenous leukemia. The small molecule kinase inhibitor
imatinib mesylate is the prototype of a new class of

drugs that have been designed to inhibit the kinase
function (Savage and Antman, 2002). Imatinib
mesylate inhibits ABL and several other tyrosine
kinase proteins by occupying the ATP-binding pocket
of the kinase domain, therefore preventing substrate
phosphorylation. Trials with imatinib mesylate in
chronic myelogenous leukemia have been highly successful, with many patients having loss of all detectable
leukemia cells. Similarly, the ETV6–PDGFRB kinase
oncoprotein in chronic myelomonocytic leukemia is
exquisitely sensitive to imatinib kinase inhibition.

See also
Cancer Cytogenetics
Fusion Proteins and Diseases
Mitosis: Chromosomal Rearrangements
Oncogenes

References

Alcalay M, Orleth A, Sebastiani C, *et al.* (2001) Common themes
in the pathogenesis of acute myeloid leukemia. *Oncogene* **20**:
5680–5694.
Arvant A and Denny C (2001) Biology of EWS/ETS fusions in
Ewing's family of tumors. *Oncogene* **20**: 5747–5754.
Barr F (1999) Translocations, cancer and the puzzle of specificity.
Nature Genetics **19**: 121–124.
Boxer L and Dang C (2001) Translocations involving c-myc and
c-myc function. *Oncogene* **20**: 5595–5610.
Daley G, McLaughlin J, Witte O and Baltimore D (1987) The
CML-specific P210 bcr/abl protein, unlike v-abl, does not
transform NIH/3T3 fibroblasts. *Science* **237**: 532–535.
Deininger MW, Goldman JM and Melo JV (2000) The molecular
biology of chronic myeloid leukemia. *Blood* **96**(10): 3343–3356.
Eguchi M, Eguchi-Ishimae M, Tojo A, *et al.* (1999) Fusion of ETV6
to neurotrophin-3 receptor TRKC in acute myeloid leukemia
with t(12;15)(p13;q25). *Blood* **93**(4): 1355–1363.
Falini B (2001) Anaplastic large cell lymphoma: pathological,
molecular and clinical features. *British Journal of Haematology*
114: 741–760.
French C, Miyoshi I, Aster J, *et al.* (2001) BRD4 bromodomain gene
rearrangements in aggressive carcinoma with translocation
t(15;19). *American Journal of Pathology* **159**: 1987–1992.
Gerald W, Rosai J and Ladanyi M (1995) Characterization of the
genomic breakpoint and chimeric transcripts in the EWS-WT1
gene fusion in desmoplastic small round cell tumor. *Proceedings*

of the National Academy of Sciences of the United States of America
92: 1028–1032.

Hibbard M, Kozakewich H, Dal Cin P, *et al.* (2000) PLAG1 fusion
oncogene in lipoblastoma. *Cancer Research* **60**: 4869–4872.

Knezevich SR, McFadden DE, Tao W, Lim JF and Sorensen PH
(1998) A novel ETV6–NTRK3 gene fusion in congenital
fibrosarcoma. *Nature Genetics* **18**(2): 184–187.

Kroll TG, Sarraf P, Pecciarini L, *et al.* (2000) PAX8-PPAR gamma1
fusion oncogene in human thyroid carcinoma [corrected]. *Science*
289(5483): 1357–1360.

Kuppers R and Dalla-Favera R (2001) Mechanisms of chromosomal
translocations in B-cell lymphomas. *Oncogene* **20**: 5580–5594.

Lawrence B, Perez-Atayde A, Hibbard M, *et al.* (2000) TPM3-ALK
and TPM4-ALK oncogenes in inflammatory myofibroblastic
tumors. *American Journal of Pathology* **157**(2): 377–384.

Lock A (1997) Oncogenic transcription factors in the human acute
leukemias. *Science* **278**: 1059–1064.

Pierotti MA, Vigneri P and Bongarzone I (1998) Rearrangements of
RET and NTRK1 tyrosine kinase receptors in papillary thyroid
carcinomas. *Recent Results in Cancer Research* **154**: 237–247.

Rowley J (1998) The critical role of chromosome translocations in
human leukemias. *Annual Review of Genetics* **32**: 495–519.

Rubin BP, Chen CJ, Morgan TW, *et al.* (1998) Congenital meso-
blastic nephroma t(12;15) is associated with ETV6–NTRK3 gene
fusion: cytogenetic and molecular relationship to congenital
(infantile) fibrosarcoma. *American Journal of Pathology* **153**(5):
1451–1458.

Savage DG and Antman KH (2002) Imatinib mesylate – a new
oral targeted therapy. *New England Journal of Medicine* **346**(9):
683–693.

Simon MP, Pedeutour F, Sirvent N, *et al.* (1997) Deregulation of the
platelet-derived growth factor B-chain gene via fusion with
collagen gene COL1A1 in dermatofibrosarcoma protuberans and
giant-cell fibroblastoma. *Nature Genetics* **15**(1): 95–98.

Zelent A, Guidez F, Melnick A, Waxman S and Licht J (2001)
Translocations of the RAR alpha gene in acute promyelocytic
leukemia. *Oncogene* **20**: 7186–7203.

Web Links

Ataxia telangiectasia mutated (includes complementation groups A,
C and D) (*ATM*); Locus ID: 472. LocusLink:
http://www.ncbi.nlm.nih.gov/LocusLink/LocRpt.cgi?l = 472

Breakpoint cluster region (*BCR*); Locus ID: 613. LocusLink:
http://www.ncbi.nlm.nih.gov/LocusLink/LocRpt.cgi?l = 613

ets variant gene 6 (TEL oncogene) (*ETV6*); Locus ID: 2120.
LocusLink:
http://www.ncbi.nlm.nih.gov/LocusLink/LocRpt.cgi?l = 2120

Runt-related transcription factor 1 (acute myeloid leukemia 1; aml1
oncogene) (*RUNX1*); Locus ID: 861. LocusLink:
http://www.ncbi.nlm.nih.gov/LocusLink/LocRpt.cgi?l = 861

v-abl Abelson murine leukemia viral oncogene homolog 1 (*ABL1*);
Locus ID: 25. LocusLink:
http://www.ncbi.nlm.nih.gov/LocusLink/LocRpt.cgi?l = 25

Ataxia telangiectasia mutated (includes complementation groups A,
C and D) (*ATM*); MIM number: 208900. OMIM:
http://www.ncbi.nlm.nih.gov/htbin-post/Omim
/dispmim?208900

Breakpoint cluster region (*BCR*); MIM number: 151410. OMIM:
http://www.ncbi.nlm.nih.gov/htbin-post/Omim/
dispmim?151410

ets variant gene 6 (TEL oncogene) (*ETV6*); MIM number: 600618.
OMIM:
http://www.ncbi.nlm.nih.gov/htbin-post/Omim/
dispmim?600618

Runt-related transcription factor 1 (acute myeloid leukemia 1; aml1
oncogene) (*RUNX1*); MIM number: 151385. OMIM:
http://www.ncbi.nlm.nih.gov/htbin-post/Omim/
dispmim?151385

v-abl Abelson murine leukemia viral oncogene homolog 1 (*ABL1*);
MIM number: 189980. OMIM:
http://www.ncbi.nlm.nih.gov/htbin-post/
Omim/dispmim?189980

Atlas of Genetics and Cytogenetics in Oncology and Hematology
http://wwww.infobiogen.fr/services/chromcancer

Transmembrane Domains

Intermediate article

T Ramasarma, *Indian Institute of Science, Bangalore, India*

NV Joshi, *Indian Institute of Science, Bangalore, India*

Article contents

- Distinctive Architecture of Membrane Proteins
- Membrane-spanning Proteins
- The Hydropathic Analysis
- Exon Analysis of the Nucleotide Sequences of
  Complementary Deoxyribonucleic Acid
  Corresponding to Transmembrane Domains

One or more passes of polypeptide chains of the lipid bilayer, each consisting of about
25 hydrophobic residues with the occasional presence of a polar residue, anchor integral
proteins in the cell membrane. These are known as transmembrane domains and
participate in the functions of these proteins in some unspecified way. Encoded by a
random exon make-up, sequences of these short stretches are highly variable with hardly
any repetition.

Distinctive Architecture of Membrane Proteins

The human genome must be distinctive in some way.
The 'evolution of novel extracellular and transmembrane architecture' is the greatest innovation in the
human lineage, according to the analysis of the
International Human Genome Sequence Consortium
(2001). Of the 32 000 identified genes in the human
genome, transmembrane proteins account for about
20%, a relatively high proportion compared with
other species. Acquiring new potentials depends on
the arrangement of the polypeptide chain around the
membrane and its relation to the rest of the protein,

implied by architecture. This placing of the protein in the membrane is programmed in the gene by nucleotide sequences encoding hydrophobic domains. Occupying a high or low proportion of the total protein, these domains are not mere anchors but participate decisively in the actions of some proteins. It is all the more interesting that they possess uncomplicated helical structures of about 30 Å in length with little or no help from the side chains. Sequences, the exon make-up and contributions to membrane activities of some examples of these little, versatile structures are briefly described here.

Membrane-spanning Proteins

Transmembrane (TM) proteins, also known as integral membrane proteins, are embedded in the lipid bilayer with the polypeptide chain crossing the membrane. **Figure 1** is a stylized diagram showing possible membrane passes. An extended polypeptide chain (**Figure 1a**), even with hydrophobic residues, cannot normally pass the hydrophobic environment of the membrane because of its polar peptide units. But its segment can do so after forming hydrogen bonds between them, thus attaining a stable structure (α-helix or β-sheet) (**Figure 1b, 1h**). Also a pass of the membrane always seems to be completed, leaving no loose end within the lipid bilayer. Examples of a β-sheet with a multipass barrel-type arrangement are known in some channel proteins with about 10–12 residues for each hydrophobic stretch. A short sheet–turn–sheet loop structure halfway into the lipid bilayer (**Figure 1g**) along with other helices is considered typical of the pore region of many channels. Proteins with single, four and seven TM passes are common. Other examples of up to 14 and 17 spans also occur. It is appropriate to refer to these as spanins, appending the number of times the chain crosses the membrane.

The Hydropathic Analysis

A stretch of about 25 hydrophobic residues in a protein ideally fits a single α-helical pass of the

membrane lipid bilayer. The hydropathic analysis of Kyte and Doolittle (1982) is a convenient method of determining such stretches showing as positive peaks when the value of hydropathy of residues is plotted against the sequence number of the polypeptide (see **Figure 2** for an example). This method is widely used for predicting the existence, location and number of TM domains. The following authors have developed other algorithms: Klein *et al.* (1985); von Heijne (1992); Peron and Argos (1994) and Casadio *et al.* (1995). Differences in the number of domains are encountered using these methods. Words such as 'putative', 'predicted', 'potential' or 'purported' are used to qualify sequences thus identified.

The presence of a couple of hydrophilic residues is not uncommon in several sequences claimed to be TM domains. It is difficult sometimes to understand why such sequences are included while other cases with long enough stretches of hydrophobic residues are ignored. The presence of glycation sites in the vicinity and the release of expected peptides on protease action from the connecting loops provide confirmatory evidence. Some helical sequences as short as 15 residue and others as long as 40 residues are known and are referred to as negative and positive 'hydrophobic mismatch' respectively (Monne and von Heijne, 2001). Such domains are likely to be accommodated in the membrane by thinning or thickening of its lipid layers and by tilting and bending of the helices (**Figure 1e, 1f**). Indeed, more will be learnt about membrane placement of these domains by studying the lipids in the vicinity. Attachment of these proteins to the membrane is beyond doubt. The helical nature of folding is also probably correct as supported by data on crystal structure, circular dichroism and two-dimensional nuclear magnetic resonance (NMR) spectroscopy of some membrane proteins. Placing variable amounts of protein exposed in the two aqueous phases must therefore serve some purpose.

Monospanins

One span across will suffice to firmly place a protein in the membrane. This simple design is used by families

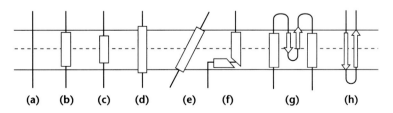

Figure 1 Arrangement of polypeptide chain in membrane spans. The membrane bilayer is represented as two lines with a middle broken line. The possible membrane passes are shown: (a) extended polypeptide chain, normally not found; (b) α-helix shown as a box; (c) short helix, negative mismatch; (d) long helix, positive mismatch; (e) tilted helix; (f) bent helix; (g) half occupied sheet characteristic of channel proteins; and (h) β-sheet shown as parallel arrows.

of proteins such as enzyme-linked receptors for signal transduction, transport proteins for moving compounds and ions in and out of the cell, and cell-surface proteins employed as recognition, linker and adhesion molecules. **Figure 3** shows the distribution of some examples of the polypeptide chain across the membrane.

Signals received by extracellular domains from a variety of ligands such as growth factors and peptide hormones are transferred into the cell where the protein kinase activity of intracellular domains becomes active (**Figure 3a, 3b**). How does the bit of membrane-locked α-helix transfer the signal into the cell with hardly any help from its hydrophobic residues? Such stimulation of ligand-sensitive kinase activity is retained in the absence of the lipid in a purified preparation of the receptor. The only link between the two bulk portions of the polypeptide on

the two sides is the tiny TM domain. Signal transfer function must reside in its helical polypeptide backbone. Mobilization and transfer of electrons over the α-helix across its intrinsic supramolecular structure of helical sequences of alternating peptide group and hydrogen bond (..HN–C = O..HN–C = O..) has been surmised as a possibility (Ramasarma, 2000).

Only short intracellular tails, sufficient to recognize the coated pits, are employed in transport proteins, both for internalization (low-density lipoprotein (LDL) receptor) and externalization (immunoglobulin A (IgA) receptor) (**Figure 3c, 3d**). The strategy of noncovalent dimerization of the membrane-bound polypeptides is found in receptors for insulin and transferrin, for example (**Figure 3b, 3e**).

A large number of monospanin proteins that recognize and bind to cell surface structures (e.g. CD4, integrins and selectins) and form functional supramolecular complexes are known. These polypeptide chains (*N*-terminal) are extended outside the cell where action occurs by way of recognition of other molecules, substrates and signals. Extracellular placement of the *C*-terminal of the polypeptide is also found, e.g. in the transferrin receptor, membrane-type frizzled-related protein (MFRP) and corin (**Figure 3e–3g**).

Dispanins

The two-membrane span occurs in a few membrane-bound proteins. They have no common type of action. Some proteins are ion channels: a chloride channel (e.g. CFTCR), inward rectifier potassium channel, calcium homeostasis endoplasmic reticulum protein and sodium channel (DEG-ENaC) (**Figure 4a**). Others have unrelated enzyme activities such as

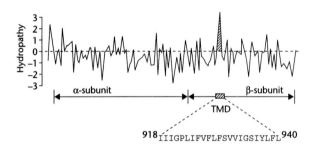

918 IIIGPLIFVFLFSVVIGSIYLFL 940

Figure 2 Hydropathic plot according to Kyte and Doolittle (1982). The hatched peak corresponds to the hydrophobic residues given below, the purported membrane span of the insulin precursor protein: TMD, transmembrane domain; I: isoleucine; G: glycine; P: proline; L: leucine; F: phenylalanine; V: valine; S: serine; Y: tyrosine.

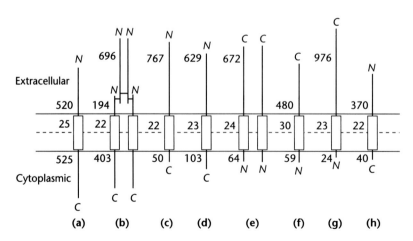

Figure 3 Distribution of the polypeptide chain in some typical monospanin proteins. The membrane span is shown as a box with the polypeptide chain extended into the extracellular and cytoplasmic sides. The number of residues of each domain is given: (a) α-platelet-derived growth factor receptor; (b) insulin receptor; (c) low-density lipoprotein receptor; (d) polyIg receptor; (e) transferrin receptor; (f) membrane-type frizzled-related protein; (g) corin; and (h) CD4 protein.

acylCoA : cholesterol acyl transferase (ACAT), tyrosine phosphatase and an ecto adenosine triphosphatase (ATPase), uridine diphosphatase (**Figure 4b**). Both their *N*- and *C*-terminals are in the cytoplasm with the connecting loop as the extracellular domain.

Used in the small subunit c in the F_0 part of mitochondrial adenosine 5' triphosphate (ATP) synthase, this architecture plays a pivotal role in the process of energy transduction. This enzyme complex is considered as a tiny molecular motor with a concentric ring of 10 molecules of subunit c (**Figure 4c**) surrounding the $F_1\gamma$ subunit (**Figure 4d**) together constituting the rotor. The subunit c acts as the link between electron transport and ATP synthase, both membrane based. Thus this dispanin complex undergoes an extraordinary mechanical rotation as part of the process of transferring energy. Nothing stands out as being different from other TM domains in the sequences of these two parts, and they are coded by two separate exons.

Trispanins

The occurrence of three spans in a membrane protein is infrequent. The example of leukotriene C4 synthase, involved in the pathogenesis of asthma, has rather a long middle span and is shown as tilted (**Figure 5a**). It is encoded by exon 3 up to the two polar residues in its middle and then changes to exon 4. The TM domains are near the *N*-terminal in a long chain in another example (muscle popeye gene product). How the three membrane domains contribute to making these proteins active besides anchoring is not known.

Tetraspanins

Proteins that span the membrane four times forming two extracellular loops, the second one usually large (e.g. CD9 antigen) (**Figure 5b**), are referred to as the transmembrane 4 superfamily (TM4SF), simplified as tetraspanins. Known for their action as molecular

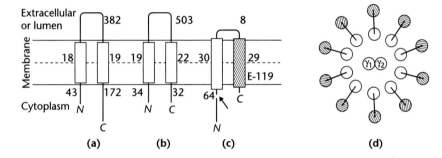

Figure 4 Architecture of the multimeric subunit c of F_0 adenosine triphosphatase (ATPase) and some dispanin proteins: (a) DEC/ENaC, a sodium channel; (b) UDPase, an ecto ATPase; (c) c subunit of F_0 ATPase (P1 form, *Escherichia coli*) (where the signal peptide is clipped shown by an arrow); the hatched helix forming the outer ring has the conserved residue E (glutamic acid)); and (d) arrangement of the 10 subunits around the two helices of subunit γ of F_0 ATPase which is a part of the rotary unit, as viewed from one side of the membrane. The number of residues of each domain is given.

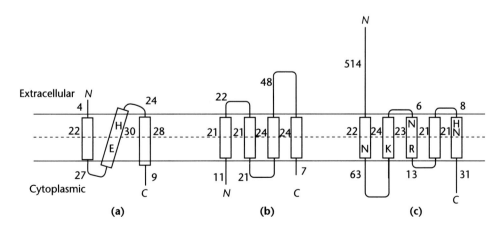

Figure 5 Distribution of the polypeptides of tri-, tetra- and pentaspanin proteins: (a) leukotriene C4 synthase (long middle helix is shown tilted); (b) CD9 antigen; and (c) M83 protein. Hydrophilic residues occurring within the helices are shown by letter code (H: histidine; E: glutamic acid; N: asparagine; K: lysine; R: arginine). The number of residues of each domain is given.

facilitators in signaling, adhesion, differentiation and proliferation, they bring together large molecular complexes by interacting with proteins such as integrins and other receptors. They are unfairly labeled as 'promiscuous' because their liaison with other proteins is widespread, albeit not entirely random. The tetraspanins include a large number of cluster differentiation (CD) proteins (9, 37, 53, 63, 81 and 82), receptors of γ-aminobutyric acid (GABA) and glycine and also some ion channels. In the voltage-dependent potassium channel, the center of the associated four helices forms a gate.

Pentaspanins

Proteins known to span the membrane five times are few. The examples of CD47, AC133, hcLcA2, M83 and Cig30 have extended *N*-terminal extracellular domains (**Figure 5c**). They bind to integrins, selectins and other adhesion molecules. Their functions appear to be similar to tetraspanins. The need for five spans, however, remains unclear.

Hexaspanins

Proteins with six spans have a variety of actions as enzymes (phosphatidate phosphatase, type III adenyl cyclase), as channel proteins (HCN2), as transporters (ZnT-3, PI transfer protein) and as growth factor activators (LMP1). The last two spans of LMP1 are necessary for anchoring the protein but are not sufficient for its action on nFκB. This strategy of assignment of protein functions to different spans seems to be used in this and other multispan proteins.

A growing superfamily of ATP-dependent proteins that translocate amphiphilic and lipophilic substrates belong to this category. Adenylyl cyclase (type III), multidrug resistance-ATPase and cystic fibrosis TM regulator belong to this group. A half-size transporter containing a single nucleotide-binding domain, ABCG1, has six TM helices coded by five exons 11–15 with two exons contributing partly to the sequences of domains II and IV (Langmann *et al.*, 2000).

The seven-transmembrane proteins, heptaspanins

By far the best-known TM proteins are the G-protein-coupled receptors on the cell surface, characterized by seven membrane spans with the *N*-terminal outside and the *C*-terminal inside the cell (**Figure 6a**). An arrangement of clockwise connectivity of the helices believed to be oriented perpendicular to the membrane has been proposed. This provides a membrane-embedded surface for the receptor protein (**Figure 6b**). There are several receptor families for the ligands such as noradrenaline, acetylcholine, serotonin, peptides, glycoprotein hormones, adenosine, prostaglandin E2 and thromboxane A2. A good proportion of the polypeptide in these is conserved, in contrast to the monospanins. It is also utilized to build the seven TM domains and the loop between domains V and VI for the β-adrenergic receptor (Emorine *et al.*, 1989), and all these are coded in one unusually long exon (**Table 1** (7)). Binding a ligand on the surface of helices outside the cell leads to dissociating a subunit of G protein acting as the second messenger system inside the cell. Here lies the enigma. How is the information carried through the simple architecture provided by TM helices? In the case of adrenergic receptors, it was

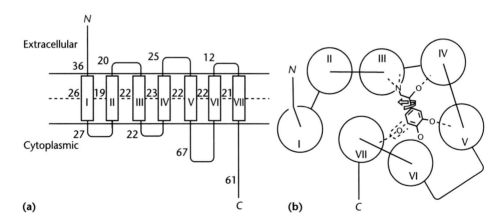

Figure 6 β-Adrenergic receptor: (a) distribution of the polypeptide; and (b) proposed architecture with clockwise connectivity of the helices viewed from the extracellular side. Critical for activity are the short loops on the extracellular side, and the loop between helices V and VI and the *C*-terminal chain. Noradrenaline is proposed to bind the helices as shown in (b), and its *m*-OH can bind to either helix V or VII by rotation of the molecule indicated by the arrow.

found that the essential amide group of noradrenaline binds to helix III. Taking advantage of rotation of its C1–C7 bond (see **Figure 6b**), noradrenaline can bind its other essential group, *m*-OH, to either helix V or VII. It was proposed that such a choice of helix pairs may be used by different ligands of the multiple subtypes of these receptors as in the adrenergic system.

Table 1 Sequences of amino acids of TM domains and of nucleotides of the corresponding cDNA segments with their distribution in exons.

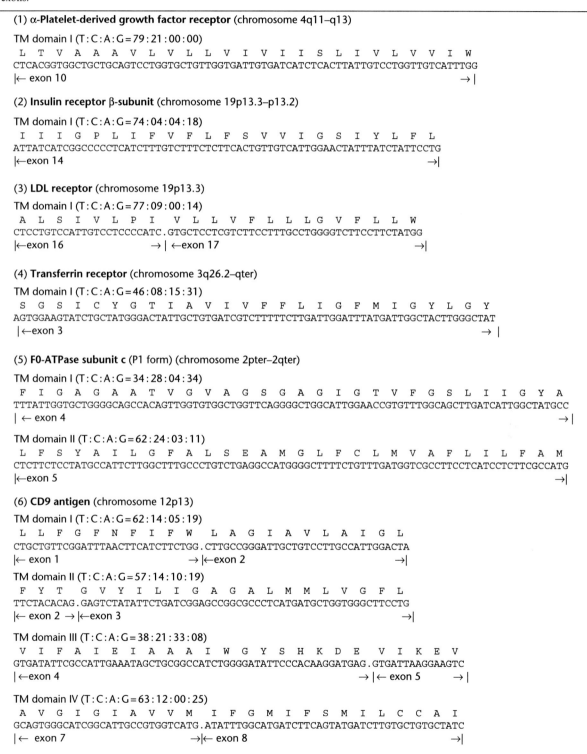

(1) α-**Platelet-derived growth factor receptor** (chromosome 4q11–q13)

TM domain I (T:C:A:G = 79:21:00:00)

```
  L   T   V   A   A   A   V   L   V   L   L   L   V   I   V   I   I   S   L   I   V   L   V   V   I   W
CTCACGGTGGCTGCTGCAGTCCTGGTGCTGTTGGTGATTGTGATCATCTCACTTATTGTCCTGGTTGTCATTTGG
|← exon 10                                                                  →|
```

(2) **Insulin receptor β-subunit** (chromosome 19p13.3–p13.2)

TM domain I (T:C:A:G = 74:04:04:18)

```
  I   I   I   G   P   L   I   F   V   F   L   F   S   V   V   I   G   S   I   Y   L   F   L
ATTATCATCGGCCCCCTCATCTTTGTCTTTCTCTTCACTGTTGTCATTGGAACTATTTATCTATTCCTG
|←exon 14                                                            →|
```

(3) **LDL receptor** (chromosome 19p13.3)

TM domain I (T:C:A:G = 77:09:00:14)

```
  A   L   S   I   V   L   P   I   V   L   L   V   F   L   L   L   G   V   F   L   L   W
CTCCTGTCCATTGTCCTCCCCATC.GTGCTCCTCGTCTTCCTTTGCCTGGGGTCTTCCTTCTATGG
|←exon 16              → | ←exon 17                                 →|
```

(4) **Transferrin receptor** (chromosome 3q26.2–qter)

TM domain I (T:C:A:G = 46:08:15:31)

```
  S   G   S   I   C   Y   G   T   I   A   V   I   V   F   F   L   I   G   F   M   I   G   Y   L   G   Y
AGTGGAAGTATCTGCTATGGGACTATTGCTGTGATCGTCTTTTTCTTGATTGGATTTATGATTGGCTACTTGGGCTAT
 |←exon 3                                                                       → |
```

(5) **F0-ATPase subunit c** (P1 form) (chromosome 2pter–2qter)

TM domain I (T:C:A:G = 34:28:04:34)

```
  F   I   G   A   G   A   A   T   V   G   V   A   G   S   G   A   G   I   G   T   V   F   G   S   L   I   I   G   Y   A
TTTATTGGTGCTGGGGCAGCCACAGTTGGTGTGGCTGGTTCAGGGGCTGGCATTGGAACCGTGTTTGGCAGCTTGATCATTGGCTATGCC
| ← exon 4                                                                                  →|
```

TM domain II (T:C:A:G = 62:24:03:11)

```
  L   F   S   Y   A   I   L   G   F   A   L   S   E   A   M   G   L   F   C   L   M   V   A   F   L   I   L   F   A   M
CTCTTCTCCTATGCCATTCTTGGCTTTGCCCTGTCTGAGGCCATGGGGCTTTTCTGTTTGATGGTCGCCTTCCTCATCCTCTTCGCCATG
|←exon 5                                                                                    →|
```

(6) **CD9 antigen** (chromosome 12p13)

TM domain I (T:C:A:G = 62:14:05:19)

```
  L   L   F   G   F   N   F   I   F   W   L   A   G   I   A   V   L   A   I   G   L
CTGCTGTTCGGATTTAACTTCATCTTCTGG.CTTGCCGGGATTGCTGTCCTTGCCATTGGACTA
|← exon 1                     → |←exon 2                         →|
```

TM domain II (T:C:A:G = 57:14:10:19)

```
  F   Y   T   G   V   Y   I   L   I   G   A   G   A   L   M   M   L   V   G   F   L
TTCTACACAG.GAGTCTATATTCTGATCGGAGCCGGCGCCCTCATGATGCTGGTGGGCTTCCTG
|← exon 2 → |←exon 3                                             →|
```

TM domain III (T:C:A:G = 38:21:33:08)

```
  V   I   F   A   I   E   I   A   A   A   I   W   G   Y   S   H   K   D   E   V   I   K   E   V
GTGATATTCGCCATTGAAATAGCTGCGGCCATCTGGGGATATTCCCACAAGGATGAG.GTGATTAAGGAAGTC
|←exon 4                                              → |← exon 5        → |
```

TM domain IV (T:C:A:G = 63:12:00:25)

```
  A   V   G   I   G   I   A   V   V   M   I   F   G   M   I   F   S   M   I   L   C   C   A   I
GCAGTGGGCATCGGCATTGCCGTGGTCATG.ATATTTGGCATGATCTTCAGTATGATCTTGTGCTGTGCTATC
|← exon 7                      →|← exon 8                                   →|
```

(7) Adrenergic receptor β3 (chromosome 8p11.1–8p12)

TM domain I (T:C:A:G=48:37:04:11)

```
  A   A   L   A   G   A   L   L   A   L   A   V   L   A   T   V   G   G   N   L   L   L   V   I   V   A   I   A
GCGGCCCTAGCCGGGGCCCTGCTGGCGCTGGCGGTGCTGGCCACCGTGGGAGGCAACCTGCTGGTCATCGTGGCCATCGCC
| exon 1                                                                      →|
```

TM domain II (T:C:A:G=56:26:11:05)

```
  N   V   F   V   T   S   L   A   A   A   D   L   V   M   G   L   L   V   V
AACGTGTTCGTGACTTCGCTGGCCGCAGCCGACCTGGTGATGGACTCCTGGTGGTG
|←exon 1                                          →|
```

TM domain III (T:C:A:G=41:32:09:18)

```
  L   W   T   S   V   D   V   L   C   V   T   A   S   I   E   T   L   C   A   L   A   V
CTGTGGACCTCGGTGGACGTGCTGTGTGTGACCGCCAGCATCGAAACCCTGTGCGCCCTGGCCGTG
|←exon 1                                                   → |
```

TM domain IV (T:C:A:G=43:36:04:17)

```
   T   A   V   V   L   V   W   V   V   S   A   A   V   S   F   A   P   I   M   S   Q   W   W
ACAGCTGTGGTCCTGGTGTGGGTCGTGTCGGCCGCGGTGTCGTTTGCGCCCATCATGAGCCAGTGGTGG
|←exon 1                                                        →|
```

TM domain V (T:C:A:G=59:27:14:00)

```
   Y   V   L   L   S   S   S   V   S   F   Y   L   P   L   L   V   M   L   F   V   Y   A
TACGTGCTGCTGTCCTCCTCCGTCTCCTTCTACCTTCCTCTTCTCGTGATGCTCTTCGTCTACGCG
|←exon 1                                                    →|
```

TM domain VI (T:C:A:G=54:23:05:18)

```
   T   L   G   L   I   M   G   T   F   T   L   C   W   L   P   F   F   L   A   N   V   L
ACCTTGGGTCTCATCATGGGCACCTTCACTCTCTGCTGGTTGCCCTTCTTTCTGGCCAACGTGCTG
|←exon 1                                                    →|
```

TM domain VII (T:C:A:G=38:24:19:19)

```
   A   F   L   A   L   N   W   L   G   Y   A   N   S   A   F   N   P   L   I   Y   C
GCTTTCCTTGCCCTGAACTGGCTAGGTTATGCCAATTCTGCCTTCAACCCGCTCATCTACTGC
|←exon 1                                                 →|
```

The chromosomal location of the gene and the per cent values of nucleotides T, C, A and G occurring in the second position of the code are given in parentheses. T and C represent hydrophobic residues. Notice that the exon split occurs between residues and also within the triplet (example 6).

Amino acids: L: leucine (Leu); T: threonine (Thr); V: value (Val); A: alanine (Ala); I: isoleucine (Ile); S: serine (Ser); W: tryptophan (Trp); G: glycine (Gly); P: proline (Pro); F: phenylalanine (Phe); Y: tyrosine (Tyr); E: glutamic acid (Glu); M: methionine (Met); H: histidine (His); K: lysine (Lys); D: aspartic acid (Asp); C: cysteine (Cys); N: asparagine (Asn); Q: glutamine (Gen).

Multispanins

Multiple spans beyond seven are known in some proteins. Generally these are channel proteins. Reports of eight and nine spans are rare. Several 10-span proteins are known as transporters of amino acids and also as cotransporters of chloride and bicarbonate. With even-number spans both *N*- and *C*-terminals have to face the same side of the membrane, and they are more commonly inside the cell. The surface of the helices of the TM domains is connected by short loops and these few residues are therefore important in the action of the receptor. For example, glucose-6-phosphate transporter utilizes the polypeptide to build its 10 spans (I–X) of residues ranging from 18 to 30 with short connecting loops (Pan *et al.*, 1999). These are encoded by eight exons (1–8) thus: I, exon 1; II, exon 2; III and IV, exon 3; V, exon 4; VI, exon 5, VII, exon 6, VIII, exon 7; IX,

exons 7 and 8 (G/TG); X, exon 8. Larger numbers of spans (11, 12, 13, 14 and 17) are known to exist in some transporter/channel proteins. Any arrangement of such a large number of spans is expected to provide the multiple sites needed for action of these proteins.

Exon Analysis of the Nucleotide Sequences of Complementary Deoxyribonucleic Acid Corresponding to Transmembrane Domains

Dominating the small stretches of TM domains are the hydrophobic amino acids Ile, Val, Leu, Phe, Cys, Met, Ala, Gly and Trp. Hydroxy amino acids Ser, Thr and

Tyr, and also the helix-breaker, Pro, occur frequently in these helices. With this many residues, these can have innumerable sequences; and they do, conserving only hydrophobicity. No repetition of a sequence or a part of it was found in the examples of TM domains studied (**Table 1**). These are coded by exons necessarily differing in sequences in many ways: one exon coding for one or more domains; one domain coded by two exons with the split occurring between residues and in some cases between the nucleotides of a triplet. No doubt, desired sequences are fused thus, but hardly any repeating units are noticed. In the example of CD9 antigen, each of the four domains (I–IV) is coded by two exons (1–8) as follows: I, exons 1 and 2; II, exons 2 and 3 (triplet G/GA); III, exons 4 and 5; IV, exons 7 and 8 (Boucheix et al., 1991) (**Table 1** (6)). No generalization is possible with the choice of the sequences being so random in these domains.

The hydrophobic residues Phe, Leu, Ile, Met and Val are coded by the second letter T, and the second letter C codes for Ser, Pro, Thr and Ala in triplets in cDNA. An analysis of per cent nucleotide present in the second position showed the expected abundance of the pyrimidines, T followed by C, in the domain sequences (**Table 1**). Indeed values combined for T and C account for about 70%. Thus these stretches in the exons do show a repeating pattern of XTX or XCX. Animals also share the architecture of multiple TM domains and thus the feature that makes the human genome distinct is something beyond these domains.

References

Boucheix, Benoit P, Bachet P, et al. (1991) Molecular cloning of CD9 antigen, a new superfamily of cell surface proteins. *Journal of Biological Chemistry* **266**: 117–122.

Casadio RB, Fariselli R and Sander C (1995) Transmembrane helices predicted at 95% accuracy. *Protein Science* **4**: 521–533.

Emorine LJ, Marullo S, Briend-Sutren MM, et al. (1989) Molecular characterization of the human beta 3-adrenergic receptor. *Science* **245**: 1118–1121.

von Heijne G (1992) Membrane protein structure prediction: hydrophobic analysis and the positive-inside rule. *Journal of Molecular Biology* **225**: 487–494.

Klein P, Kanehisa M and Di Lisis C (1985) The detection and classification of membrane-spanning proteins. *Biochimica et Biophysica Acta* **815**: 468–476.

Kyte J and Doolittle RF (1982) A simple method for displaying the hydropathic character of a protein. *Journal of Molecular Biology* **157**: 105–132.

Langmann T, Porsch-Orzcurumez M, Unkelbach U, Kulcken J and Schmitz G (2000) Genomic organization and characterization of the promoter of the human ATP-binding cassette transporter-G1 (ABCG1) gene. *Biochimica et Biophysica Acta* **1494**: 175–180.

Monne M and von Heijne G (2001) Effects of 'hydrophobic mismatch' on the location of transmembrane helices in the ER membrane. *FEBS Letters* **496**: 96–100.

Pan CJ, Lin B and Chou JY (1999) Transmembrane topology of human glucose 6-phosphate transporter. *Journal of Biological Chemistry* **274**: 13 865–13 869.

Peron B and Argos P (1994) Prediction of transmembrane segments in proteins utilizing multiple sequence alignments. *Journal of Molecular Biology* **237**: 182–192.

Ramasarma T (2000) In praise of the hydrogen bond. In: Lal M, Lillford PJ, Naik VM and Prakash V (eds.) *Supramolecular and Colloidal Structures in Biomaterials and Biosubstrates*, pp. 450–462. London, UK: Imperial College Press of the Royal Society.

Further Reading

Baldwin JM (1993) The probable arrangement of the helices in G protein-coupled receptors. *EMBO Journal* **12**: 1693–1703.

Benovic JL, Bovier M, Caron MG and Lefkowitz RL (1980) Regulation of adenyl cyclase-coupled β-adrenergic receptors. *Annual Reviews of Cell Biology* **4**: 405–428.

Maecker HT, Todd SC and Levy S (1997) The tetraspanin superfamily: molecular facilitators. *FASEB Journal* **11**: 428–442.

Ramasarma T (1996) Transmembrane domains participate in functions of integral membrane proteins. *Indian Journal of Biochemistry and Biophysics* **33**: 20–29.

Sakharkar M, Long M, Tan TW and de Souza SJ (2000) Prediction tool for transmembrane segment in proteins. *Nucleic Acids Research* **28**: 191–192. See also Web Links.

Savonov S, Daizadeh I, Fedorov A and Gilbert W (2000) The exon-intron database: an exhaustive database of protein-coding intron-containing genes. *Nucleic Acids Research* **28**: 185–190. See also the Exon–Intron Database in Web Links.

Stock D, Leslie AGW and Walker JE (1999) Molecular architecture of the rotary motor in ATP synthase. *Science* **286**: 1700–1705.

Yardley Y and Ulrich A (1998) Growth factor receptor tyrosine kinases. *Annual Reviews of Biochemistry* **57**: 473–478.

Web Links

The Exon–Intron Database. An exhaustive database of protein-coding intron-containing genes
http://golgi.harvard.edu/gilbert/eid/

Genome Net. GenomeNet is a Japanese network of database and computational services for genome research and related research areas in molecular and cellular biology
http://www.genome.ad.jp

TSEG, the prediction tool for Transmembrance SEGment in proteins
http://www.genome.ad.jp/sit/tseg.html

Transplantation

See Gene Transfer in Transplantation

Transposable Elements: Evolution

Thomas H Eickbush, *University of Rochester, Rochester, New York, USA*

Danna G Eickbush, *University of Rochester, Rochester, New York, USA*

Transposable elements are DNA segments that only encode the enzymatic activity necessary to make more copies of themselves. Most of the human genome is composed of inactive copies of three distinct classes of elements that have accumulated over millions of years. These classes differ in their mechanisms of integration, origins and modes of evolution.

Introduction

The human genome is a model of inefficiency. Its genes are usually separated by tens to hundreds of thousands of base pairs. These long intergenic stretches contain short regions that serve as transcriptional regulatory sequences but for the most part have no known function. The genes within the human genome are themselves divided into small coding regions (exons), typically less than 300 bp in size, and larger noncoding regions (introns), typically many thousands of base pairs in size. These introns are spliced from the original transcript prior to protein translation and simply destroyed. The inefficiency of the human genome is further underscored by comparing it with that of yeast: more than 60% of the yeast genome encodes the 6000 or so proteins that are needed to build and regulate a yeast cell. The estimated 35 000 genes required in human development are scattered through a genome nearly 300 times larger than that of yeast, with protein-encoding regions comprising approximately 1% of the deoxyribonucleic acid (DNA).

What makes up the remaining 99% of the DNA? It has long been known that a high percentage of large eukaryotic genomes are composed of interspersed, repetitive sequences. Sequence analysis from the human genome project has convincingly revealed that most of this repetitive DNA corresponds to transposable elements. Transposable elements, sometimes simply called mobile elements, have evolved the remarkable ability to insert copies of themselves at other locations in the genome. Four million elements have been identified in the human genome, indicating that there are over 100 transposable element insertions for every gene. These elements constitute an estimated 45% of the regions between genes as well most introns. Because there is no selective pressure to eliminate the spontaneous mutations that arise in the individual copies of a transposable element, after 200 million years or more these mutations accumulate to the point where the individual copies are no longer recognized as such. Thus, while 45% of the human genome can

readily be attributed to transposable elements, a much larger fraction is probably derived from these elements. (*See* Repetitive Elements: Detection; Repetitive Elements and Human Disorders.)

While the human genome is enormous as a result of transposable element insertions, it is by no means the largest. Various plants, reptiles and amphibians have genomes that are 10 or even 100 times larger than the human genome. Because there is no reason to believe that these organisms have more genes than humans, these genomes appear to have suffered even more rampant accumulations of transposable elements. In some cases the accumulation is rapid: it has been estimated that the maize (corn) genome has more than doubled in the past 3 million years as a result of new transposable element insertions.

The impact of these elements on the genome goes beyond a simple increase in size. When transposable elements insert into or near a gene, they can alter or disrupt its function. No doubt such insertions have given rise to countless mutations in our human ancestry. It has been estimated that one out of every 100 human births contains a new transposable element insertion. Paradoxically, because transposable elements already compose a large portion of the genome and consequently are the most likely sites of new insertions, very few characterized human spontaneous mutations have been associated with transposable element insertions.

Why do Humans Have Transposable Elements?

While transposable elements currently cause few mutations, their sheer number affects the amount of DNA that must be synthesized at each cell division, as well as the amount of ribonucleic acid (RNA) that is synthesized for every functional messenger RNA (mRNA) produced. Surely our genomes would be more efficient without these insertions. Why over the past hundreds of millions of years has our genome permitted these elements to expand to such high numbers?

It has been speculated that transposable elements are maintained by organisms to create variation that allows the genome to evolve more rapidly in response to environmental stresses. In support of this 'function' model, there are many examples in which new genes have been formed, modified or differentially expressed as a direct result of transposable element insertions. Even simple recombination between the repeated transposable element sequences can lead to gene duplications and modifications. While such changes can occasionally be viewed as a benefit to the host, it is more likely that these effects on host fitness are a natural consequence of the abundance of transposable elements rather than their function.

An alternative explanation for the presence of transposable elements is that eukaryotic genomes are simply not able to prevent them from spreading. As beautifully explained by Doolittle and Sapienza in 1980, whenever a segment of DNA has an advantage in replicating or being transferred to the next generation, it can expand in a species even if it serves no useful function. How widely the DNA expands will be influenced by its efficiency in spreading and the extent of its negative effect on the genome. This 'functionless' model suggests that transposable elements exist simply because they can make more copies of themselves. They will continue to do so until the host stops them or all active copies fall victim to mutation. Thus transposable elements have been compared to parasites, selfishly appropriating normal cellular functions to maintain their existence. For simplicity, transposable elements can be viewed as viruses, living and replicating inside a cell even though they provide no direct benefit to the host. However, unlike viruses, transposable elements cannot readily leave one cell to infect another (though some do at a low frequency). Thus in order for transposable elements to survive they cannot irreparably harm the cell. While transposable elements increase their numbers at a glacial pace compared with a virus, persistence pays off, because over millions of years they have expanded to impressive numbers in most eukaryotic genomes.

Classes of Transposable Elements

Full-length transposable elements vary in size from 2 kb to more than 10 kb. While many different families, or evolutionary lineages, can be found in eukaryotes, the vast majority fall into one of three classes. These classes differ in their mechanism of movement, as outlined in **Figure 1**. The human genome contains representatives of each of these classes.

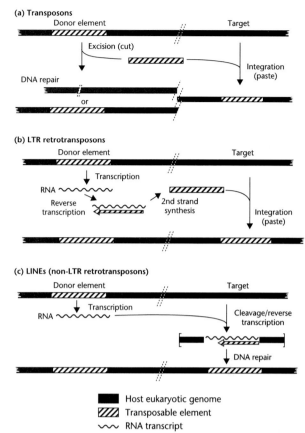

Figure 1 Mechanism of expansion of the three classes of eukaryotic transposable elements. (a) Excision and integration of the transposon is carried out by the element-encoded transposase. The cleaved donor site can be religated by DNA repair, which frequently generates small deletions. Alternatively the homologous chromosome can serve as template to fill in the gap, which results in resynthesis of the element at the donor site. (b) Long terminal repeat (LTR) retrotransposons are transcribed and the RNA used as template by the element-encoded reverse transcriptase to generate a double-stranded DNA intermediate. Integration of the element is carried out by the transposase-like enzyme integrase. (c) Reverse transcription of the RNA transcript of a long interspersed nuclear element (LINE; non-LTR retrotransposon) occurs directly onto the cleaved site of the chromosome. Most subsequent steps are assumed to be carried out by the host cellular repair machinery.

The first class of elements moves through the excision of a genomic copy from its original donor site and the reinsertion of that copy to a target site elsewhere in the genome (**Figure 1a**). These elements usually encode a single protein, termed a transposase, that recognizes 10- to 30-bp inverted repeat sequences at the ends of the element and cleaves the DNA at the junction of the element and flanking non-element DNA. While remaining associated with the released DNA copy, the transposase cleaves a distant 'target' DNA and integrates the excised element.

This 'cut and paste' mechanism, as it is frequently termed, appears at first glance to merely maintain the number of genomic copies. However, such elements have been shown to excise immediately after DNA replication and insert into target sites that have not yet been replicated. This insures an increase in at least one of the newly replicated chromosomes. Furthermore, the DNA break that is generated by the excised element at the donor site must be repaired. Often the cellular DNA repair machinery does not simply religate the two free ends but rather uses the sister chromatid, or the homologous chromosome, as a template to fill in the gap generated by the excision event. The result is re-formation of the element at the excision site. Elements utilizing only a transposase are classified as the DNA-mediated transposable elements and are usually referred to simply as transposons. There are approximately 0.4 million transposon insertions in the human genome, constituting nearly 3% of the DNA. (*See* Transposons.)

The remaining two classes of transposable elements encode an unusual enzyme, termed a reverse transcriptase. This enzyme generates a DNA copy (complementary DNA, cDNA) from the element's RNA transcript. Because of this capability, these elements are classified as the RNA-mediated transposable elements and are usually referred to as the retrotransposons ('retro' from the Latin meaning 'backwards') to emphasize the unusual backwards flow of information from RNA to DNA.

The two classes of retrotransposons differ in where and how the cDNA copy is made. In one class (**Figure 1b**) reverse transcription occurs in the cytoplasm of the cell and is primed by a host transfer RNA (tRNA) molecule. After the cDNA strand is generated and the RNA template destroyed, synthesis of the second DNA strand returns the element to a double-stranded DNA state. At this point another element encoded enzyme, a transposase-like enzyme called integrase, binds inverted repeats at the ends of the newly synthesized DNA, cleaves a target site and inserts (pastes) the new copy into the chromosome. This class of retrotransposons is called the 'long terminal repeat (LTR) retrotransposons', after the 300- to 500-bp long terminal repeats that flank the elements and are involved in many aspects of the elements' synthesis and insertion. They are also referred to as the viral-like retrotransposons because of the similarity of their structure and means of replication to that of the retroviruses. There are about 0.7 million LTR retrotransposon insertions in the human genome, constituting 8% of the DNA.

The second class of retrotransposons utilizes a much simpler mechanism to insert new copies (**Figure 1c**). In addition to the reverse transcriptase, these elements encode a simple endonuclease which cleaves chromosomal target sites. The 3′ hydroxyl group of the sugar–phosphate backbone which is exposed by this cleavage is used to prime the reverse transcription reaction such that the cDNA is polymerized directly onto the chromosomal target site. The means by which this cDNA is attached to the upstream sequences of the target site as well as the means by which the second strand of DNA is synthesized are not understood. One possibility is the recruitment to the target site of the cellular repair machinery involved in the healing of chromosomal breaks. This class of element was not originally recognized as mobile and the elements were simply referred to as long interspersed nuclear elements (LINEs). These elements are also referred to as non-LTR retrotransposons, emphasizing the absence of the terminal repeats associated with the previous group, and retroposons, emphasizing the absence of the transposase step. There are about 1.3 million LINE insertions in the human genome, constituting 20% of the DNA. (*See* Long Interspersed Nuclear Elements (LINEs); Long Interspersed Nuclear Elements (LINEs): Evolution.)

A remarkable aspect of the LINE insertion mechanism is that the reverse transcriptase encoded by these elements shows little sequence specificity in recognizing its own RNA template. As a result, unrelated RNA transcripts can be used as template in the formation of cDNA. The human genome contains abundant examples in which reverse-transcribed cellular RNA sequences have been integrated by the LINE machinery. For example, processed pseudogenes are insertions derived from mature mRNAs after their introns have been removed. The 5′ promoters are absent from these pseudogenes and hence they are transcriptionally inactive. A much larger number of insertions in the human genome have resulted from small, stable cellular RNAs encoded by genes that contain promoter sequences that lie within the transcribed region. When such an RNA transcript is reverse-transcribed, the new insertion contains its own promoter, which can lead to multiple rounds of transcription and insertion. These insertions are called short interspersed nuclear elements (SINEs). There are over 1.8 million SINE insertions in the human genome, constituting 13% of the DNA. While most eukaryotic SINEs are derived from tRNA genes, the majority of human SINEs are derived from 7SL RNA, an RNA involved in cellular protein secretion. These latter insertions are about 300 bp in length and have been called Alu sequences because their initial characterization revealed a common cleavage site for the restriction enzyme, *Alu*I. (*See* Promoters: Evolution; Pseudogenes and their Evolution; Retrosequences and Evolution of Alu Elements.)

Origin of Transposable Elements

The ubiquitous nature of transposable elements raises interesting questions concerning their origins. Phylogenetic analyses using the protein sequences encoded by the elements suggest that each of the three classes is very old. Are transposable elements derivations from an early eukaryotic cell's recombinational machinery, or are they derived from more primitive replication machinery? The two enzymes that are the cornerstones of mobility, transposase and reverse transcriptase, are not highly similar in structure or function to any eukaryotic protein. Indeed, transposases are most related to the transposases found in prokaryotic mobile elements as well as to certain prokaryotic proteins involved in site-directed recombination. These prokaryotic mobile elements utilize mechanisms to increase in number which are essentially identical to the eukaryotic transposons. Likewise, the reverse transcriptases of LINEs are most related to reverse transcriptases found in other prokaryotic mobile elements, group II introns. While group II introns utilize a different mechanism to cleave the chromosomal target site, like the LINEs, they use the cleaved chromosomal DNA to prime reverse transcription.

Thus there is good reason to believe that at least two of the major classes of eukaryotic transposable elements were inherited from our prokaryotic ancestors (see **Figure 2**). These two classes, the transposons and LINEs (non-LTR retrotransposons), have essentially a one-step mechanism for the insertion

of new copies into the genome (**Figure 1**). LTR retrotransposons, on the other hand, use components of both methods. In this class, a reverse transcriptase generates a complete cytoplasmic double-stranded DNA copy of the element from its RNA transcript. This DNA copy is then bound by the transposase (integrase) and inserted into the chromosome by a mechanism similar to that of the DNA-mediated transposons. There are no prokaryotic elements identified to date that utilize this combined mechanism. The model in **Figure 2** explains the appearance of LTR retrotransposons and their two-step mechanism as the fusion of a transposon and an LINE sometime early in the formation of eukaryotes. The success of this hybrid element was probably due to the increased stability provided for its intermediates in eukaryotic cells where transcription and translation are no longer coupled as they were in prokaryotes.

Modes of Evolution

There are two means by which eukaryotic transposable elements survive over long periods: the stable inheritance of active copies much like host cellular genes, and the transfer of active copies from one species to another. The movement of elements between species is called horizontal or lateral transmission. Inheritance of a transposable element along with the rest of the genome is called vertical transmission. Evidence for strict vertical descent is obtained if the sequence divergence among the transposable elements from different species agrees with the length of separation of the species (i.e. the species phylogeny). Unfortunately, in those cases where disagreements arise, one cannot automatically attribute the disagreement to horizontal transfers. In any organism, the different active copies of a transposable element can diverge into multiple lineages, because there is no means to maintain uniformity among all copies. Without selective pressure to maintain any particular lineage, the diverging lineages will be differentially retained in the organism's descendants. As a result, it is often difficult to determine whether an inconsistent sequence relationship among elements is the result of a comparison between different lineages or a true horizontal event. Because of this uncertainty, the best evidence for a horizontal event is the presence of nearly identical elements in species that diverged many millions of years ago.

In the case of the DNA transposons, most eukaryotes contain numerous distinct families of elements that bear no consistent relationship to the transposons in related species. Furthermore, in most of these families, all copies of the element have accumulated mutations, which eliminate their ability

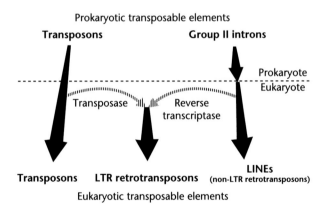

Prokaryotic transposable elements
Transposons   Group II introns
                                              Prokaryote
                                              Eukaryote
Transposase   Reverse
                   transcriptase

Transposons   LTR retrotransposons   LINEs
                                     (non-LTR retrotransposons)
Eukaryotic transposable elements

Figure 2 Possible origin of eukaryotic transposable elements. Eukaryotic transposons are similar in sequence and mechanism of insertion to the transposons of bacteria and presumably evolved with little change directly from these prokaryotic elements. LINE elements are most similar in sequence and mechanism of insertion to prokaryotic group II introns. Group II introns are speculated to have been lost in eukaryotes, with aspects of their integration machinery surviving as LINEs. No equivalent of an LTR retrotransposon has been found in prokaryotes. Their mechanism of expansion suggests that they resulted from the fusion of a transposon and an LINE.

to produce functional transposase. The loss of this essential activity is believed to be directly related to how these elements move about. Because protein synthesis occurs in the cytoplasm, the newly made transposase must migrate into the nucleus to find an element for transposition. Once there, the transposase relies solely on the 10- to 30-bp inverted DNA repeats at the ends of a potential donor element to carry out the cut-and-paste mechanism. Therefore, as long as these short sequences at the ends of the element are intact, defective copies will be transposed as readily as active copies. The inevitable buildup of mutations in all copies of transposable elements means the ever-increasing population of inactive copies will eventually overwhelm the process.

The continued survival of DNA transposons therefore requires that an active copy be transferred to a new species prior to the total inactivation of a transposon lineage within the old species. In the new species, the element is free to start a new cycle of expansion unencumbered by defective copies. There are many documented instances of horizontal transfers of DNA transposons between species. In the best-studied cases, a widespread lineage of elements called 'mariner' was shown to have undergone horizontal transfers between different groups of insects and even between insects, hydra and flatworms. The mechanism by which these elements are transferred from organism to organism is not known, but viruses or parasites that transfer cellular components between hosts are likely candidates.

The transposons of the human genome have no doubt been derived from horizontal events. There are at least seven different families in the human genome, none of which at present contains active members. Indeed, based on the divergence among the copies within each family, there has not been an active DNA transposon in the human genome for the past 50 million years. The extreme age of these elements will probably make it impossible to determine from what species we obtained these elements.

In contrast to the transposons, most data suggest that LINEs can be stably maintained in eukaryotic genomes by vertical descent. Why are LINEs for the most part stable? Again the answer lies in their mechanism of expansion. The reverse transcriptases of the LINEs associate with their mRNAs in the cytoplasm. While there is little RNA sequence require-ment in this association, as is illustrated by the generation of processed pseudogenes and SINEs, mounting evidence suggests that the newly synthesized proteins preferentially bind to the RNA from which they were translated. This 'cis preference' provides an effective mechanism to insure that transcripts from active elements are predominately used in the reverse transcription process. As a result, the accumulation of

defective copies over time does not as readily overwhelm the transposition machinery.

Despite their stability, the continued transmission of active elements is not a certainty. Phylogenetic analyses suggest that 15–20 ancient lineages of LINEs exist in eukaryotes. The human genome has representatives of three of these lineages. Of these three, two appear to have long ago lost all active copies and consequently stopped expanding. Each of these LINEs had propagated a predominate SINE family, which was incapable of expanding once the corresponding LINE machinery disappeared. Only one lineage, L1, is still active today. Indeed, this lineage represents the most abundant transposable element in all mammalian genomes. While the L1 lineage remains very active in some mammals (e.g. mice), it has become much less active in the human lineage. It has been estimated that only a few dozen human L1 copies are still capable of making new copies. Of the millions of transposable element insertions, these few L1s are probably the only active transposable elements in the entire human genome.

As in the case of the LINEs, the LTR retrotransposons clearly exhibit long-term stability in a species lineage. This stability is again likely to be provided by the utilization of RNA templates in the integration machinery, as only those RNAs of actively transcribed copies are present in the cytoplasm. In addition, the more complex sequence requirements at the 5′ and 3′ ends as well as internal regions of the RNA make it unlikely that a transcribed defective copy will be used for the formation of a double-stranded DNA inter-mediate. While perhaps not essential to their long-term survival, LTR retrotransposons have also been shown to undergo rare horizontal transfers. As in the case of the DNA transposons, the mechanism of this transfer is unknown.

One of the more interesting aspects of the evolution of the LTR retrotransposons is that they have given rise to a variety of viruses. These include the retroviruses and hepadnaviruses of vertebrates, and the caulimoviruses of plants. While the origin of some of these viruses appears rather complex, that of the vertebrate retroviruses is easy to postulate. Retroviral structure and mechanism of integration is like that of LTR retrotransposons, with one key difference. All retroviruses contain an extra gene, the envelope gene, which codes for a transmembrane receptor-binding protein which allows the RNA intermediate to bud off as a membrane-bound virion. What is the origin of this new gene? Unfortunately, the retroviral lineage is quite old and the envelope proteins lack sequence similarity to other known proteins. Thus it has not been possible to determine whether the envelope gene was acquired from a host gene or a receptor-binding or membrane fusion gene of another virus. However, surveys of LTR retrotransposons in plants, insects and nematodes have revealed a number of additional lineages that have acquired envelope-like genes. In a few of the most recent acquisitions, the envelope proteins appear to have originated from other DNA or RNA viruses. The propensity of the LTR retrotransposable elements to become viruses is most probably linked to their ability to package their RNA template in stable, 'virus-like' particles in the cytoplasm prior to reverse transcription. (*See* Retroviral Repeat Sequences.)

As in the case of the DNA transposons, no LTR retrotransposon lineage is currently active in the human genome. The inactive copies that remain are from the same lineage as the retroviruses, but it remains unclear as to whether they are remnants of LTR retrotransposons or retroviruses that subsequently lost their envelope gene. While the LTR retrotransposons themselves are not active, their retroviral descendants continue to play a devastating role in human health.

Conclusion

The enormous impact of transposable elements on the structure of the human genome cannot be over-emphasized. Their amazing ability to increase in number has left few genes untouched. In fact, the human genome is composed largely of transposable elements and the by-products of their transposition machinery. The vast majority of these elements are now inactive, overwhelmed in many cases by their own high numbers. Surprisingly, the human genome is doing little to rid itself of this excess baggage. Millions of insertions remain as ancient relics of the genome's evolutionary past and are surely, albeit slowly, becoming unrecognizable through mutations.

See also

Long Interspersed Nuclear Elements (LINEs): Evolution
Retrosequences and Evolution of Alu Elements
Retroviral Repeat Sequences
Repetitive Elements and Human Disorders
Transposons

Further Reading

Boeke JD and Stoye JP (1997) Retrotransposons, endogenous retroviruses, and the evolution of retroelements. In: Coffin JM, Hughes SH and Varmus HE (eds.) *Retroviruses*, pp. 343–435. Cold Spring Harbor, NY: Cold Spring Harbor Laboratory Press.

Craig N, Craigie R, Gellert M and Lambowitz A (eds.) (2001) *Mobile DNA II*. Washington, DC: American Society of Microbiology Press.

Doolittle WF and Sapienza C (1980) Selfish genes, the phenotype paradigm and genome evolution. *Nature* **284**: 601–603.

International Human Genome Sequencing Consortium (2001) Initial sequencing and analysis of the human genome. *Nature* **409**: 860–921.

Kazazian HH and Moran JV (1998) The impact of L1 retrotransposition on the human genome. *Nature Genetics* **19**: 19–24.

Li W-H, Gu Z, Wang H and Nekrutenko A (2001) Evolutionary analysis of the human genome. *Nature* **409**: 847–849.

Malik HS and Eickbush TH (2001) Phylogenetic analysis of ribonuclease H domains suggests a late, chimeric origin of LTR retrotransposable elements and retroviruses. *Genome Research* **11**: 1187–1197.

Malik HS, Henikoff S and Eickbush TH (2000) Poised for contagion: evolutionary origins of the infectious abilities of insect errantiviruses and nematode retroviruses. *Genome Research* **10**: 1307–1318.

Transposons

Pierre Capy, *Centre National de la Recherche Scientifique, Gif-sur-Yvette, France*

Jean-Marc Deragon, *Centre National de la Recherche Scientifique, Clermont-Ferrand, France*

Genomes are highly plastic entities submitted to several types of mutations in somatic and germinal lines. These mutations can alter the general structure of the genome and also its regulation. A large proportion of these events is due to repeated and mobile sequences – the transposable elements.

Advanced article

Article contents

- Introduction
- Classes and Sizes of Transposable Elements
- Most Abundant Elements in the Human Genome
- Functional and Structural Impacts on the Host Genome
- Evolutionary Relationships between Transposable Elements and Retroviruses

Introduction

Genomes are highly plastic entities and are submitted to several types of mutations in somatic and germinal lines. These mutations can alter both the general structure of the genome and its regulation. Many mutation events are caused by repeated and mobile sequences – the transposable elements (TEs). These elements are defined as repeated sequences that have an intrinsic capacity to move around in the host genome. The presence of these elements in genomes was suspected in the 1940s from the work of Barbara McClintock on chromosomal instability in maize. Their existence was shown molecularly at the end of the 1970s, and she won the Nobel prize in 1983.

Transposable elements can be characterized by several structural features such as duplication of the target site on insertion and, for some of them, the presence of repeats (inverted or direct) at both ends of the element. They are present in all living organisms: bacteria, archea and eukaryotes. In eukaryotes, TEs represent a variable part of the total genome, from about 3% (e.g. *Saccharomyces cerevisiae*) to more than 50% (e.g. higher plants such as *Zea mays*). In humans, TEs correspond to 40–45% of the genome (3000 megabases) and they constitute the middle repetitive fraction of the deoxyribonucleic acid (DNA). The copy number of TEs varies greatly from one family of elements to another. For example, in the human genome the average copy number of class II elements is generally low, sometimes limited to a few elements per family; by contrast, there are about 100 000 and 1 million copies of long interspersed nuclear elements (LINEs) or short interspersed nuclear elements (SINEs) respectively. TEs can also jump from one species to another in a process called 'horizontal transfer'. In prokaryotes, most of these transfers probably occur by plasmid intermediates, but no clear mechanism has been shown in eukaryotes.

Transposable elements are considered as selfish genes because they 'spread by forming additional copies of themselves within the host genome' (Dawkins, 1976, 1982). Their mobility can be deleterious for the host, but they can also be at the origin of new advantageous genetic variants and help the genome to face environmental changes (McClintock, 1984; see the paragraphs about their impact). Therefore, they probably represent a useful genetic load.

Classes and Sizes of Transposable Elements

In the human genome, two main classes of TEs can be found: class I elements (or retrotransposons) use a 'copy-and-paste' mechanism with a ribonucleic acid (RNA) intermediate, whereas class II elements (or transposon) use a 'cut-and-paste' mechanism with a DNA intermediate (**Figure 1**). Transposons encode a single protein – the transposase – that is required for excision and insertion of the elements. Most human class II elements belong to the mariner/Tc1 superfamily. The majority of them are dead, such as the miniature inverted-repeats TEs (MITEs; Smit and Riggs, 1996).

Retrotransposons can be split into two main subclasses: long terminal repeat (LTR) retrotransposons, and retroposons. LTR retrotransposons are flanked by two LTRs in direct orientation, between which are at least two open reading frames (ORFs). These ORFs, named *gag* and *pol*, are similar to those found in retroviruses. The *gag* gene encodes the proteins involved in forming the capsid and the *pol* gene encodes the four proteins required for copying and integrating the elements. The retroposons are made up of two superfamilies: the LINEs and the SINEs.

The size of TEs varies widely according to the class. Class II elements are generally short sequences of about 2–3 kilobases (kb), whereas retroposons and LTR retrotransposons are about 5–7 and 5–10 kb respectively. SINE elements that belong to class I are relatively short (less than 0.5 kb). Not all elements in a genome are active. Only a few of them are able to encode the functional proteins required for their mobility, but defective elements can move by using the proteins provided by functional elements (*trans*-mobilization). This occurs for many class II elements such as the MITEs (Feschotte and Mouches, 2000).

Most Abundant Elements in the Human Genome

As mentioned above, LINE and SINE elements are the most abundant TEs in the human genome. LINEs are autonomous elements with two ORFs: ORF1 encodes a nucleic-acid-binding protein, and ORF2 encodes a reverse transcriptase and an endonuclease activity. For the human LINE-1 elements, the ORF1 product is associated with a large cytoplasmic ribonucleoprotein complex (Hohjoh and Singer, 1997). LINEs can be classified into 13 distinct clades on the basis of the phylogenetic analysis of their reverse transcriptase domain. Four of these clades are represented in the human genome: the L1 clade (with L1 elements), the Jockey clade (with L2 elements), the retrotransposable element (RTE) clade (with BovB elements) and the chicken repeat element (CR1) clade (with L3 and CR1HS elements). SINEs are short nonautonomous (and noncoding) elements that are transcribed by the RNA polymerase III complex. Human SINEs are ancestrally related to 7SL RNA (the Alu elements) or to transfer RNA (the mammalian interspersed (MIR) elements). (*See* Long Interspersed Nuclear Elements (LINEs): Evolution; Retrosequences and Evolution of Alu Elements.)

The 'master gene' model of evolution (Deininger *et al.*, 1992) is particularly adapted to these types of elements, and mammalian LINEs and SINEs can be separated into subfamilies of different evolutionary

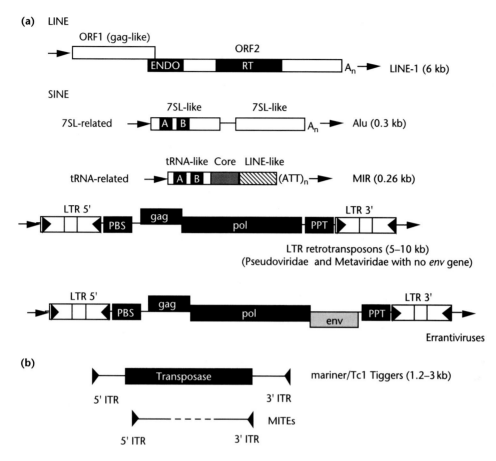

Figure 1 Structure of the main classes of transposable elements found in human genome. (a) Class I elements: A and B: RNA polymerase III promoters; Core: conserved region of unknown function; ENDO: endonuclease; gag: capsid gene; pol: polyprotein including reverse transcriptase, integrase, ribonuclease H and protease domains; LTR: long terminal repeat; PBS: primer-binding site; PTT: polypurine tract; RT: reverse transcriptase; transposase: protein required for the excision and insertion of class II elements. The gray shading for the *envelope* (*env*) gene indicates that a functional gene is not always found. (b) Class II elements: ITR: inverted terminal repeat; MITEs: miniature inverted-repeated transposable elements.

age on the basis of the activity of a few founding sequences. The retroposition of an SINE family has been suggested to depend on the presence of a virus-like particle (VLP) that is encoded by an LINE partner. Thus, SINEs may have evolved as parasites of the LINE's VLP. This SINE/LINE partnership probably relies on a shared poly(A) tail (Alu/L1; Boeke, 1997) or a shared 3′ region (MIR/L2; Okada *et al.*, 1997).

Functional and Structural Impacts on the Host Genome

Transposable elements can be assimilated to endogenous mutagenic agents. For example, TEs can affect the host genome by inserting themselves into a gene or into its regulatory region. It is obvious that insertion of an element into an exon is likely to inactivate the gene, but insertion into an intron can also affect gene

expression by modifying the positions of donor and acceptor sites.

In humans, several diseases are caused by such events (Deragon and Capy, 2000, and references therein). For example, insertions in gene-coding regions of Alu and L1 elements are responsible for breast cancer and hemophilia A and B. An insertion in the regulatory region of a gene can change its tissue-specificity and timing of expression, usually by providing new binding sites for transcription factors. TEs can also act as new promoters, as in the case of the salivary amylase gene, or can provide new polyadenylation signals. Transcription initiated in TEs can also perturb the expression of neighboring genes (transcription interference). Transcription of a TE is silenced during early embryogenesis, but this epigenetic process (i.e. based on DNA methylation) is imperfect and produces a mosaic pattern of TE expression in somatic cells. Thus, TEs can perturb the expression of neighboring genes in a mosaic pattern that corresponds to the activity of each element. The stochastic nature of TE activity, and the very large number of genes that may be affected, can produce subtle phenotypic variations, which may affect, for example, disease risk. TEs could therefore act as epigenetic mediators of phenotypic variation in humans (Whitelaw and Martin, 2001). (*See* Repetitive Elements and Human Disorders; Retrotransposition and Human Disorders.)

Transposable elements can also be responsible for chromosomal rearrangements. For example, elements in different genomic positions can be involved in homologous or nonhomologous recombination, leading to translocation (when elements are on different chromosomes), inversion (when they are on the same chromosome and in inverted orientations) or deletion (same chromosome and in direct orientations). This last process is also at the origin of solo-LTRs, where ectopic recombination occurs between the two LTRs of a retrotransposon (**Figure 2**). In humans, several diseases result from recombination between TEs (Deragon and Capy, 2000). Most of them are due to deletions after recombination between Alu elements.

Transposable elements can also be recruited by the host genome for new functions. The most striking example of TE domestication is the probable DNA transposon origin of our immune system's recombination mechanism. The V(D)J antibody recombination system is dependent on *recombinase activating gene 1* (*RAG1*) and *RAG2*. *RAG1* encodes a protein that is involved in cleavage site recognition, whereas *RAG2* encodes an endonuclease. These genes originate from a single class II element (probably a Tc1-like element) that was immobilized by the loss of its two inverted terminal repeats (ITRs). The repeated sequences flanking the V(D)J system (the recombinase-binding site) are assumed to be the old ITRs of the element (Agrawal *et al.*, 1998). Other examples of domestication include the centromere-binding protein CENP-B, which is an ancient descendant of a Pogo-like transposase (Kipling and Warburton, 1997), and the possible recruitment of SINE RNA to regulate the interferon response through interaction with the protein kinase (PKR) (Schmid, 1998).

Evolutionary Relationships between Transposable Elements and Retroviruses

Evolutionary relationships exist between the main classes of TEs. Phylogenetic analyses indicate that all LTR retrotransposons arose well after the origin of LINE retroposons (Malik and Eickbush, 2001) and that

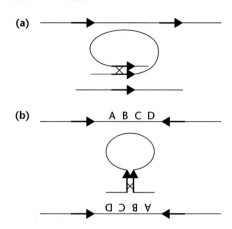

Figure 2 Structural rearrangements caused by ectopic recombination of TEs. (a) Ectopic recombination involving copies in direct orientation leads to the deletion of the chromosomal portion between the two copies. This mechanism is at the origin of solo-LTRs, where recombination occurs between the two LTRs of LTR retrotransposons. (b) Ectopic recombination involving copies in inverted orientation leads to the inversion of the chromosomal portion between the two copies.

they are closely related to retroviruses. Structurally, the main difference between LTR retrotransposons and retroviruses is the presence of an *envelope* (*env*) gene in the latter. This gene is partly responsible for the capacity of retroviruses to move from one cell to another. The phylogenetic relationships between TEs and viruses are now well established. There is no clear-cut separation between these sequences, and for several TEs a virus nomenclature has been adopted. For instance, the few LTR retrotransposons containing a functional *env* gene, like the *gypsy* element of *Drosophila*, or containing a defective *env* gene, are now classified into the errantiviruses (Boeke *et al.*, 1999). (*See* Retroviral Repeat Sequences.)

On the basis of phylogenetic analyses, several models of TEs evolution have been proposed (Lerat *et al.*, 1999; Xiong and Eickbush, 1990). These analyses use domains that are present in several types of elements, such as the reverse transcriptase found in all retroelements or the integrase/transposase detected in LTR retrotransposons and in several class II elements. Today, it is clear that TEs show a modular evolution. They are like a construction set in which each piece has its own history. Discussion about their origin remains highly controversial. Because they are found in all living organisms, it has been proposed that they were present in the last common ancestor. In this scheme, TEs evolution should go from the simplest elements to the more complex ones – in other words, from the element found in bacteria to LTR retrotransposons and retroviruses. But an alternative hypothesis that suggests that they derive from pre-existing (RNA) viruses cannot be ruled out. In this scheme, the simplest elements could result from the degradation of complex elements. These two hypotheses are not totally exclusive, and it is likely that there is 'to-ing and fro-ing' between different classes of elements. This is probably the case between LTR retrotransposons and retroviruses. (*See* Transposable Elements: Evolution.)

See also

Long Interspersed Nuclear Elements (LINEs)
Long Interspersed Nuclear Elements (LINEs): Evolution
Retroviral Repeat Sequences
Short Interspersed Elements (SINEs)
Transposable Elements: Evolution

References

Agrawal A, Eastman QM and Schatz DG (1998) Transposition mediated by RAG1 and RAG2 and its implications for the evolution of the immune system. *Nature* **394**: 744–751.

Boeke JD (1997) LINEs and Alus – the polyA connection. *Nature Genetics* **16**: 6–7.

Boeke JD, Eickbush TH, Sandmeyer SB and Voytas OF (1999) Metaviridae. In: Murphy FA (ed.) *Virus Taxonomy: ICTV VIIth Report.* New York, NY: Springer-Verlag.

Dawkins R (1976) *The Selfish Gene.* Oxford, UK: Oxford University Press.

Dawkins R (1982) *The Extended Phenotype.* Oxford, UK: Oxford University Press.

Deininger PL, Batzer MA, Hutchison III CA and Edgell MH (1992) Master genes in mammalian repetitive DNA amplification. *Trends in Genetics* **8**: 307–311.

Deragon JM and Capy P (2000) Impact of transposable element on the human genome. *Annals of Medicine* **32**: 264–273.

Feschotte C and Mouches C (2000) Evidence that a family of miniature inverted-repeat transposable elements (MITEs) from the *Arabidopsis thaliana* genome has arisen from a pogo-like DNA transposon. *Molecular Biology of Evolution* **17**: 730–737.

Hohjoh H and Singer MF (1997) Ribonuclease and high salt sensitivity of the ribonucleoprotein complex formed by the human LINE-1 retrotransposon. *Journal of Molecular Biology* **271**: 7–12.

Kipling D and Warburton PE (1997) Centromeres, CENP-B and Tigger too. *Trends in Genetics* **13**: 141–145.

Lerat E, Brunet F, Bazin C and Capy P (1999) Is the evolution of transposable elements modular? *Genetica* **107**: 15–25.

Malik HS and Eickbush TH (2001) Phylogenetic analysis of ribonuclease H domains suggests a late, chimeric origin of LTR retrotransposable elements and retroviruses. *Genome Research* **11**: 1187–1197.

McClintock B (1984) The significance of responses of the genome to challenge. *Science* **226**: 792–801.

Okada N, Hamada M, Ogiwara I and Ohshima K (1997) SINEs and LINEs share common 3′ sequences: a review. *Gene* **205**: 229–243.

Schmid CW (1998) Does SINE evolution preclude Alu function? *Nucleic Acids Research* **26**: 4541–4550.

Smit AF and Riggs AD (1996) Tiggers and other DNA transposon fossils in the human genome. *Proceedings of the National Academy of Sciences of the United States of America* **93**: 1443–1448.

Whitelaw E and Martin DI (2001) Retrotransposons as epigenetic mediators of phenotypic variation in mammals. *Nature Genetics* **27**: 361–365.

Xiong Y and Eickbush TH (1990) Origin and evolution of retroelements based upon their reverse transcriptase sequences. *EMBO Journal* **9**: 3353–3362.

Further Reading

Capy P, Bazin C, Higuet D and Langin T (1997) *Dynamics and Evolution of Transposable elements.* Landes Company, Heidelberg, Germany: Springer-Verlag.

Capy P (1996) *Genetica.* vol. 100 [Special issue on transposable elements.]

Maraia RJ (1995) *The Impact of Short Interspersed Elements (SINEs) on the Host Genome.* MBIU, RG Landes Company, Heidelberg, Germany: Springer-Verlag.

McDonald JF (1992) *Genetica* **86**. [Special issue on transposable elements.]

McDonald JF (1999) *Genetica* **107**. [Special issue on transposable elements.]

Sherrat DJ (1995) *Mobile Genetic Elements.* Oxford, UK: IRL Press.

Trans Splicing

Suzanne Furuyama, *Case Western Reserve University, Cleveland, Ohio, USA*

James P Bruzik, *Case Western Reserve University, Cleveland, Ohio, USA*

Intermediate article

Article contents

- Introduction
- Discovery of SL-dependent *Trans*-splicing
- Structure of the SL RNA
- Structure/Function Analysis of the Spliced Leader RNA/RNP
- Required Components of a *Trans*-spliceosome
- Assembly of the *Trans*-spliceosome
- Biological Function of SL-dependent *Trans*-splicing
- Potential for *Trans*-splicing in Higher Eukaryotes

Exon sequences are normally tethered through introns, which are then spliced out to generate mRNA. In *trans*-splicing, the exons are not linked together, which raises issues concerning how the two splice sites are juxtaposed for splicing to occur.

Introduction

Trans-splicing involves the joining of exons that originate on separate transcripts. Initially, split group II introns in the organelles of plants and algae, and spliced-leader (SL)-dependent *trans*-splicing in many lower eukaryotes, were the only documented examples of *trans*-splicing in nature. Since the early 1990s, evidence has been obtained to suggest that *trans*-splicing may also occur in higher eukaryotes. With the surprisingly low number of identified genes in humans, there has been an interest in determining whether *trans*-splicing occurs in mammals because this process could potentially increase the coding capacity of a genome.

Discovery of SL-dependent *Trans*-splicing

The term '*trans*-splicing' has been used to describe several reactions in the literature. These include split group II introns, split group I introns, split precursor messenger ribonucleic acid (pre-mRNA) transcripts and reactions involving an SL RNA. *Trans*-splicing dependent on an SL RNA is the best-characterized type of *trans*-splicing (Nilsen, 1997). In this reaction, the 5′ exon is donated from a small RNA polymerase II transcript – the SL RNA. *Trans*-splicing occurs at a 3′ splice site located on an RNA molecule that is transcribed separately. Addition of SL RNA to a 3′ *trans* acceptor molecule is not dependent on the formation of base pairs between the two transcripts. The reaction itself is chemically analogous to *cis*-splicing, which involves two sequential transesterification reactions (**Figure 1**).

Trans-splicing through an SL RNA was first identified in trypanosomes as a result of experiments aimed at elucidating the origin of a common 39-nucleotide sequence present at the 5′ end of the variant surface glycoprotein mRNAs. Since its discovery in these organisms, SL-dependent *trans*-splicing has been reported in several lower eukaryotes, including nematodes, trematodes, protists, ascidians and cnidarians (Nilsen, 2001). Although most SL-dependent *trans*-splicing organisms use a single SL exon, there are two distinct and functional SL exons in the free-living nematode *Caenorhabditis elegans* (SL1 and SL2) and the cnidarian *Hydra vulgaris* (SL-A and SL-B).

Structure of the SL RNA

Spliced leader RNA is composed of the SL exon and intron sequences separated by a splice donor site. This structural organization results in consumption of the SL RNA during the *trans*-splicing reaction. Cross-species comparison of SL RNAs reveals an overall lack of conservation of the nucleotide sequence. But two elements that are common to all SL RNAs are a GU dinucleotide at the 5′ splice site and a binding site for Sm proteins, both of which are essential components of most U-rich small nuclear RNAs (snRNAs) involved in pre-mRNA splicing.

Analysis of computer-generated secondary structures (**Figure 2**) suggests that the conserved GU dinucleotide is present in a stem–loop structure that is generated by a folding back of the nucleotides present at the 5′ end of the SL RNA, whereas the intron binding site for Sm proteins is single-stranded. A direct comparison of SL RNAs present in any one phylum has revealed more extensive sequence conservation. For example, the 22-nucleotide SL exon derived from the SL RNAs in the nematodes *C. elegans* (SL1) and *Ascaris lumbricoides* are identical, even though they diverged about 500 million years ago. In addition, like the U snRNAs that are required for *cis*-splicing, all SL RNAs except those found in trypanosomes have a trimethylguanosine cap.

Structure/Function Analysis of the Spliced Leader RNA/RNP

Owing to the lack of experimentally tractable systems in many of the known *trans*-splicing organisms, structure/function analysis of the conserved features of the SL RNA has been achieved primarily in studies in trypanosomes and nematodes. The stem–loop structure that contains the GU dinucleotide at the exon–intron junction of the SL RNA is conserved throughout phylogeny; however, its presence is not essential for the formation of *trans*-spliced products.

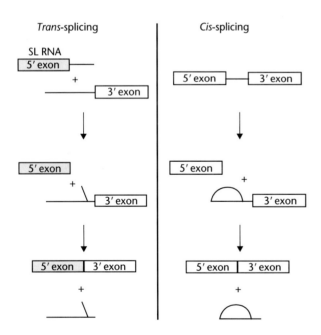

Figure 1 *Trans*- and *cis*-splicing pathways. Comparable steps in the two reactions are drawn next to each other for direct comparison. The 5′ exon in the *trans*-splicing reaction is contained in the SL RNA.

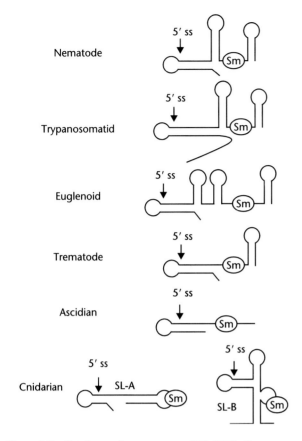

Figure 2 Predicted secondary structures of SL RNAs from different organisms. Each contains a 5′ terminal stem loop that includes the SL exon and the 5′ splice site (ss, indicated by an arrow). In the 3′ intronic region of the molecule, each SL RNA contains a binding site for the common Sm proteins. Secondary structures are not drawn to scale, and subtle features such as bulges are not noted.

Studies with the SL RNA of *A. lumbricoides* and with the SL1 and SL2 RNAs of *C. elegans* have determined that neither the overall length nor the sequence of the donated SL exon is required for *trans*-splicing. In addition, SL RNAs capable of folding into three stem loops require either the primary sequence or structure of stem loop II for *trans*-splicing. This fundamental requirement for stem loop II may be related to protein binding. Unlike the necessity for stem loop II, the necessity for stem loop III varies and it is difficult to make gross generalizations about sequence or structure.

The snRNAs bind to proteins and function in the form of small nuclear ribonucleoprotein particles (snRNPs). In addition to the common Sm proteins and the proposed protein that binds stem loop II, the SL RNP contains at least two specific proteins, as indicated by biochemical experiments in *A. lumbricoides* (Denker *et al.*, 1996). The association of these proteins with the SL RNP depends on prior Sm protein interactions and also correlates with

trans-splicing activity. Another common feature of most SL RNPs is the presence of a trimethylguanosine cap. In *A. lumbricoides*, this hypermethylated cap structure is not necessary for the formation of *trans*-spliced products.

Required Components of a *Trans*-spliceosome

The *trans*-spliceosome, the multiprotein–RNA complex that is responsible for exon joining in *trans*-splicing, requires the function of U2, U4, U5 and U6 snRNPs – the essential components of a *cis*-spliceosome. But in contrast to *cis*-splicing, the 5′ end of U1 snRNA, which normally recognizes the 5′ splice site, is not required for *trans*-splicing. In addition, RNA–RNA interactions that are distinct from those in a *cis*-spliceosome occur during assembly of the *trans*-spliceosome. These include specific SL–U6 and SL–U5 interactions, both of which occur in the absence of a 3′ splice site containing *trans*-acceptor. These data, coupled with the ability to affinity purify U4/U6 · U5 triple snRNP with the SL RNP (Maroney *et al.*, 1996), indicate that the SL RNP is a member of a four-component snRNP (SL · U4/U6 · U5).

The role of protein factors in *trans*-splicing has been less thoroughly studied than that of the RNP components of the *trans*-spliceosome. In addition to the proteins known to be associated with the above-mentioned SL and U snRNPs, the conserved, non-snRNP-associated serine/arginine-rich (SR) proteins are required for *trans*-splicing (Sanford and Bruzik, 1999). Although distinct roles for these splicing factors in *trans*-splicing are beginning to be elucidated, modification of the phosphorylation state of SR proteins has been identified as a mechanism by which the cell can regulate both *cis*- and *trans*-splicing activity. Documented roles for SR proteins in splice-site selection, snRNP recruitment and splice-site communication during *cis*-splicing suggest possible interactions leading to the juxtaposition of splice sites during *trans*-splicing.

Assembly of the *Trans*-spliceosome

All of the required components for *trans*-splicing must be assembled accurately into a catalytically active *trans*-spliceosome. The current model (**Figure 3**) suggests that assembly of the *trans*-spliceosome is initiated with the recognition of the 3′ splice site through the recruitment of U2 snRNP to the branch-point sequence. This recognition is dependent on both the sequence of the 3′ splice site and interactions

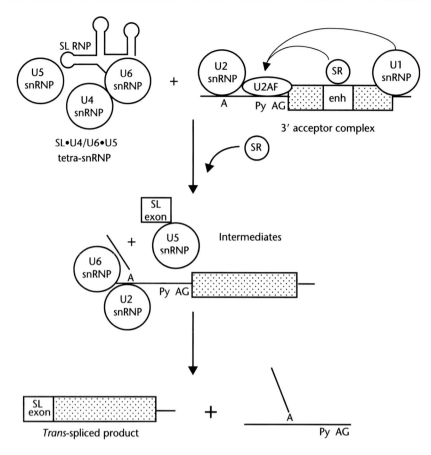

Figure 3 *Trans*-spliceosome assembly. The RNA containing the 3′ splice site (3′ acceptor) interacts with U2 snRNP, assisted by interactions with SR proteins bound to splicing enhancers (enh) or by interactions with U1 snRNP bound downstream (exon definition). The SL RNP interacts with the U4/U6·U5 tri-snRNP to form the SL·U4/U6·U5 tetra-snRNP. In a manner dependent on SR proteins, these initial complexes associate functionally, leading to the first catalytic step of the reaction – generation of the free SL exon and the Y-branched intermediate. In the second step of the reaction, the *trans*-spliced product is formed.

between an enhancer and SR proteins or exon definition interactions that promote use of the *trans*-acceptor.

Trans-spliceosome assembly progresses to a catalytically competent complex with recruitment of the SL RNP and the U4/U6·U5 triple snRNP in the context of a preformed tetra-snRNP (SL·U4/U6·U5); this complex allows simultaneous recognition of the 5′ splice site and splice-site juxtaposition. The transition from a complex containing U2 snRNP to a complete *trans*-spliceosome is also dependent on SR protein function.

Biological Function of SL-dependent *Trans*-splicing

A longstanding aim of the field of *trans*-splicing is to determine the biological function of SL addition.

With its discovery in trypanosomes and *C. elegans*, SL-dependent *trans*-splicing was presumed to aid in the maturation of monocistronic transcripts from polycistronic genes. In *C. elegans*, SL2 *trans*-splicing, which occurs predominately on pre-mRNAs derived from internal genes in an operon, is coupled with 3′-end maturation of the upstream pre-mRNA through a complex containing SL2 RNP and the polyadenylation factor CstF64 (Evans *et al.*, 2001).

Because *trans*-splicing occurs at a 3′ splice site positioned upstream from the translational start codon of the gene, it has been suggested that adding both the SL sequence and the trimethylguanosine cap increases the translational efficiency of a *trans*-spliced mRNA. In addition, work in other systems has shown that on splicing, the mRNA is 'marked' by the deposition of specific proteins upstream from the splice junction. The interaction of these proteins with the mRNA promotes export to the cytoplasm (Maquat and Carmichael, 2001). *Trans*-splicing might serve as a mechanism for mRNAs to acquire these protein factors. Whether *trans*-splicing affects other aspects of RNA metabolism remains unclear; however, the extent of *trans*-splicing in trypanosomes and nematodes indicates that SL addition is an important step in gene expression in these organisms.

Potential for *Trans*-splicing in Higher Eukaryotes

Split pre-mRNA transcripts have been used in assays to understand the mechanism of pre-mRNA splicing. More specifically, exons originating from a gene that is normally *cis*-spliced could be spliced in *trans* when located on separate transcripts. These bimolecular splicing events required either the formation of base pairs between the two transcripts or recognition of the 3′ splice site by enhancers or exon definition (Bruzik, 1996). Thus, splicing could occur even if the 5′ and 3′ splice sites were not linked by a continuous phosphodiester backbone.

Several endogenous *trans*-splicing events have been proposed to occur in higher eukaryotes. Examples include bimolecular splicing of the thymus-specific first exon of the c-*myb* proto-oncogene, carnitine octanolytransferase pre-mRNA, human neural cadherin-like (protocadherin) exons and the human acyl-coenzyme A/cholesterol acyltransferase pre-mRNA. Perhaps the strongest evidence for *trans*-splicing in complex eukaryotes is based on studies of the *mod*(*mdg4*) locus in *Drosophila melanogaster*. Initial reports showed that the first four exons were transcribed from one deoxyribonucleic acid (DNA) strand, whereas the final two exons were transcribed from the opposite strand. It was subsequently shown that at least seven isoforms of *mod*(*mdg4*) exist, each containing the first four exons in common. In fact, transgene analysis has shown that complete *mod*(*mdg4*) transcripts can be generated even when the two gene fragments are located at distinct chromosomal locations. It has been proposed that *trans*-splicing may in fact allow interallelic complementation (Mongelard *et al.*, 2002). Mod(mdg4) protein was produced only when distinct mutations in sequences encoding each of the transcripts were present on separate alleles.

Assuming that these events do indeed occur in RNA, they would most probably use the basic splicing apparatus. But interactions between the separate transcripts or novel interactions between already characterized factors and/or unique factors might be required for these bimolecular splicing events. If *trans*-splicing does occur in higher eukaryotes, then it raises important issues about how such events are regulated in the cell. The level of control would need to be exquisite, as aberrant *trans*-splicing would lead to a gross disruption of gene expression.

Trans-splicing has been reported as a possible tool for gene therapy. One approach, referred to as 'spliceosome-mediated RNA *trans*-splicing', requires an intron of one pre-mRNA to interact with the intron of another, resulting in the use of splice sites located on separate transcripts. The idea is to use this technique to repair defects in RNA with the hope of producing functional protein. An example that demonstrates the potential utility of this approach is the partial correction of the mutated ΔF508 *cystic fibrosis transmembrane conductance regulator* (*CFTR*) transcript both *in vitro* and *in vivo* (Liu *et al.*, 2002).

In conclusion, with the completion of the human genome project came the unexpectedly low estimate of 35 000 human genes. With about 18 000 genes in nematodes and 26 000 genes in plants, we are left wondering how organismal complexity at the level of the proteome arises. Attention has focused on alternative pre-mRNA splicing, as roughly 40–60% of human genes are expected to have at least two alternatively spliced isoforms. If *trans*-splicing occurs in more complex eukaryotes, then this nonlinear exon pairing reaction might serve as another mechanism by which to increase the coding capacity of the genome – the functional significance of which awaits further analysis.

See also
Caenorhabditis elegans Genome Project
Gene Clustering in Eukaryotes
mRNA Editing

References

Bruzik JP (1996) Splicing glue: a role for SR proteins in *trans* splicing? *Microbial Pathogenesis* **21**: 149–155.

Denker JA, Maroney PA, Yu Y-T, Kanost RA and Nilsen TW (1996) Multiple requirements for nematode spliced leader RNP function in *trans*-splicing. *RNA* **2**: 746–755.

Evans D, Perez I, MacMorris M, *et al.* (2001) A complex containing CstF-64 and the SL2 snRNP connects mRNA 3′ end formation and *trans*-splicing in C. elegans operons. *Genes and Development* **15**: 2562–2571.

Liu X, Jiang Q, Mansfield SG, *et al.* (2002) Partial correction of endogenous ΔF508 CFTR in human cystic fibrosis airway epithelia by spliceosome-mediated RNA *trans*-splicing. *Nature Biotechnology* **20**: 47–52.

Maquat LE and Carmichael GG (2001) Quality control of mRNA function. *Cell* **104**: 173–176.

Maroney PA, Yu Y-T, Jankowska M and Nilsen TW (1996) Direct analysis of nematode *cis*- and *trans*-spliceosomes: a functional role for U5 snRNA in spliced leader addition *trans*-splicing and the identification of novel Sm snRNPs. *RNA* **2**: 735–745.

Mongelard F, Labrador M, Baxter EM, Gerasimova TI and Corces VG (2002) *Trans*-splicing as a novel mechanism to explain interallelic complementation in *Drosophila*. *Genetics* **160**: 1481–1487.

Nilsen TW (1997) *Trans*-splicing. In: Krainer AR (ed.) *Eukaryotic mRNA Processing*, pp. 310–334. Oxford, UK: IRL Press.

Nilsen TW (2001) Evolutionary origin of SL-addition *trans*-splicing: still an enigma. *Trends in Genetics* **17**: 678–680.

Sanford JR and Bruzik JP (1999) SR proteins are required for nematode *trans*-splicing *in vitro*. *RNA* **5**: 918–928.

Further Reading

Blumenthal T and Steward K (1997) RNA processing and gene structure. In: Riddle DL, Blumenthal T, Meyer BJ and Priess JR (eds.) *C. elegans*, vol. II, pp. 117–146. Cold Spring Harbor, NY: Cold Spring Harbor Laboratory Press.

Bruzik JP, Van Doren K, Hirsh D and Steitz JA (1988) *Trans* splicing involves a novel form of small nuclear ribonucleoprotein particles. *Nature* **335**: 559–562.

Evans D and Blumenthal T (2000) *Trans* splicing of polycistronic *Caenorhabditis elegans* pre-mRNAs: analysis of the SL2 RNA. *Molecular and Cellular Biology* **20**: 6659–6667.

Kikumori T, Cote GJ and Gagel RF (2001) Promiscuity of pre-mRNA spliceosome-mediated *trans* splicing: a problem for gene therapy? *Human Gene Therapy* **12**: 1429–1441.

Krause M and Hirsh D (1987) A *trans*-spliced leader sequence on actin mRNA in *C. elegans. Cell* **49**: 753–761.

Murphy WJ, Watkins KP and Agabian N (1986) Identification of a novel Y branch structure as an intermediate in trypanosome mRNA processing: evidence for *trans* splicing. *Cell* **47**: 517–525.

Reed R (1996) Initial splice-site recognition and pairing during pre-mRNA splicing. *Current Opinion in Genetics and Development* **6**: 215–220.

Sutton RE and Boothroyd JC (1986) Evidence for *trans* splicing in trypanosomes. *Cell* **47**: 527–535.

Wu Q and Maniatis T (1999) A striking organization of a large family of human neural cadherin-like cell adhesion genes. *Cell* **97**: 779–790.

Zorio DA, Cheng NN, Blumenthal T and Spieth J (1994) Operons as a common form of chromosomal organization in *C. elegans. Nature* **372**: 270–272.

Trinucleotide Repeat Expansions: Disorders

David Nelson, *Baylor College of Medicine, Houston, Texas, USA*

Huda Zoghbi, *Baylor College of Medicine, Houston, Texas, USA*

Trinucleotide repeat mutations, also known as dynamic mutations, were discovered only in the early 1990s, but are responsible for many important neurological disorders, such as Huntington disease, fragile X mental retardation and myotonic dystrophy.

Intermediate article

Article contents

- Introduction to Dynamic Mutations
- Diseases Resulting from Expansions of Noncoding Repeats
- Diseases Resulting from Expansions of Coding Repeats
- Pathogenesis of Polyglutamine Disorders

Introduction to Dynamic Mutations

Since the early 1990s, many unusual clinical phenomena have become less mysterious, a precise diagnosis of clinically indistinguishable phenotypes has become possible, and the pathogeneses of several disorders have begun to unfold. The rapidity with which all this happened is due, in part, to the discovery of 'dynamic mutations', that is, the expansion of unstable trinucleotide repeats. So far, 14 neurological disorders have been found to result from dynamic mutations, and the list will probably continue to grow. Our understanding of the pathogenic mechanism underlying these disorders is improving, owing to the power of model organisms and genetic studies.

Triplet repeat mutations are 'dynamic' in that they are unstable and prone to expand as alleles are passed from one generation to the next, although contractions can also occur. Repeats at different loci differ markedly in their rates and extents of expansion, and the instability of repeats at each locus often depends on the sex of the transmitting parent. Somatic instability is observed at several of the loci; this ranges from small alterations in some tissues to widespread

variation in most organs. Polyglutamine expansions in coding sequences, which are inherited in a dominant manner, cause the resulting protein to gain some toxic function. In contrast, expansions in noncoding repeats can result in both dominant and recessive alleles, with both loss and gain of function effects. It is likely that additional types of repeats and effects will be discovered, but for now we can divide the triplet repeat diseases according to the location of the expansion.

Diseases Resulting from Expansions of Noncoding Repeats

Several types of repeats have been found to affect gene function and result in disease even though they are located outside the coding sequence of the gene. A brief description of each is provided below (see **Table 1** for a list of the disorders and their prevalences).

Table 1 Prevalence of triplet repeat diseases

Disease	Prevalence
Fragile X syndrome	1 in 3500
FRAXE	1 in 100 000
Myotonic dystrophy	1 in 8000
Friedreich ataxia	2–4 in 100 000
SBMA (Kennedy disease)	1 in 40 000 men
Huntington disease	3–7 in 100 000
All autosomal dominant cerebellar ataxias, including DRPLA[a]	1–2 in 100 000[a]

[a]Altogether the known spinocerebellar ataxias (SCAs) have a prevalence of 1–2 affected individuals in 100 000 people in most countries, except in the case of populations with probable founder effects. For example, SCA3 has a prevalence of 1 in 4000 in the Azores, and DRPLA accounts for about 20% of all autosomal dominant ataxias in Japan. This wide geographical variation in prevalence rates can even occur within the same country: in southern Italy, for example, SCA2 accounts for roughly two-thirds of all cases of autosomal dominant ataxia, but SCA1 is responsible for three-quarters of individuals with autosomal dominant cerebellar ataxia (ADCA) in northern Italy. SBMA: spinobulbar muscular atrophy.

Fragile X syndrome

The X-linked dominant fragile X (FRAX) syndrome was the first disease discovered to result from a triplet repeat mutation. Its unusual genetic features, including reduced penetrance and variable expression, were thought to be a function of ascertainment bias (the ability of family members and health care providers to recognize a clinical syndrome based on having seen it before in a family member) until the dynamic nature of the mutation was appreciated. Attention to the locus at Xq27.3 was initially focused by the presence of a folate-sensitive fragile site (FRAXA) that results from an expansion mutation of CGG repeats. The CGG repeat is found in the 5′ untranslated region of the *fragile X mental retardation 1* (*FMR1*) gene and varies from five to 50 triplets in the general population. Individuals affected with fragile X carry more than 230 and up to several thousand repeats. This large expansion of the CGG repeat causes a loss of function of *FMR1*, which is mediated through changes in chromatin (histone deacetylation and methylation) that abolish transcription of *FMR1*.

Tracts of 45–200 CGG repeats are called 'premutations', because they are not long enough to result in mental retardation but are prerequisite for further expansion to the full mutation. An escalating risk of affected offspring is associated with increasing size of premutations in female carriers. Curiously, offspring of male carriers of premutations are not at risk for fragile X syndrome, as their alleles have not been observed to expand to full mutations on male transmission. The normal function of *FMR1* remains

to be fully determined, but the protein seems to be involved in metabolism of ribonucleic acid, probably in the areas of transport, stability and/or translation.

FRAXE mental retardation

The mental retardation disorder FRAXE results from mutations in a second gene with triplet repeat amplification associated with the fragile site that is located only 600 000 bp from *FMR1* at Xq27.3. This gene is called *fragile X mental retardation 2* (*FMR2*), and numerous families that have a fragile site and X-linked mental retardation that was previously thought to result from fragile X syndrome have been found to be affected instead by the repeat expansion of FRAXE. A GCC repeat is located in the 5′ untranslated region of the *FMR2* gene and behaves similarly to the CGG repeat in *FMR1*. As with *FMR1*, large repeat expansions lead to chromatin alteration and repression of the *FMR2* gene. *FMR2* differs from *FMR1* in derived amino acid sequence, and functionally seems to be a nuclear protein with the characteristics of a transcriptional activator. The highest levels of *FMR2* expression are found in embryonic development, and loss-of-function mutations in FRAXE syndrome result in a mild mental impairment.

Myotonic dystrophy

Myotonic dystrophy remains a particularly enigmatic disorder of triplet repeats. It is autosomal dominant and affects several systems, causing myotonia, muscle weakness, cardiac conduction defects, diabetes, cataracts, premature balding and sometimes dysmorphic features and mild mental retardation. The tendency of these features to appear in more severe form in subsequent generations of a family brought this disorder to the early attention of geneticists, who described this worsening of symptoms as 'genetic anticipation'. Considerable debate ensued because this phenomenon had no clear explanation. With the discovery of a CTG repeat in the 3′ untranslated region of the *dystrophia myotonica-protein kinase* (*DMPK*) gene and its subsequent characterization, a molecular basis for genetic anticipation is now well described. This CTG tract is small in the general population (5–37 triplets), but can expand markedly to thousands of repeats. Slightly longer repeats (70–100) can result in mild effects such as late-onset cataracts, and the classic symptoms of early adulthood myotonia are found in individuals with hundreds of repeats. Very large expansions can cause severe disease in newborns. Genetic anticipation is thus almost certainly a direct manifestation of the expansion of repeats from one generation to the next.

The CTG repeat expansion in this chromosome-19-linked dominant disorder was identified in 1992, but the mechanism through which the mutation leads to the various clinical symptoms remains uncertain. Several effects have been described and the repeat expansions clearly affect both the *DMPK* gene and nearby genes. It is also likely that messenger RNAs carrying the expanded CTG repeat may have diminished amounts of nuclear proteins binding to them, resulting in the mis-regulation of genes that are not linked to *DMPK*. Mutations in equivalent mouse genes have not provided a complete model of the human disease. No other mutations have been found in *DMPK* in humans affected with myotonic dystrophy, suggesting that the phenotype in this disease results from several effects of the repeat expansion, rather than from a simple loss or gain of function mechanism.

Friedreich ataxia

Friedreich ataxia causes both limb and gait ataxia, loss of position and vibration sense, decreased tendon reflexes, cardiomyopathy, diabetes and in some people optic atrophy and deafness. It is the only triplet repeat disorder known to involve a GAA sequence. Expansion of an intronic stretch of GAA repeats in the *Friedreich ataxia* (*FRDA*) gene, which encodes frataxin, located at 9q13 is the most common mutation found in this recessive disease. Carriers of Friedreich ataxia are rather common (1/85), and consequently so is the GAA expansion. Members of the general population carry between seven and 34 repeats, whereas pathogenic alleles can contain hundreds of repeats. Some large increases have been identified in transmissions of intermediate length alleles, resulting in the designation of premutations as in fragile X syndrome.

The *FRDA* gene product seems to be involved in mitochondrial iron homeostasis, and its absence affects postmitotic cells rich in mitochondria. The mechanism by which the GAA repeat affects gene function seems to be the inhibition of primary ribonucleic acid (RNA) transcripts, probably owing to the formation of unusual structures by deoxyribonucleic acid (DNA), or by DNA and RNA in combination. GAA repeats can form stable triplex structures, which have been suggested to cause reduction in RNA levels.

Spinocerebellar ataxia type 8

Spinocerebellar ataxia type 8 (SCA8) is a dominantly inherited ataxia caused by an expanded noncoding CAG/CTG repeat that is found on antisense transcripts in the *kelch-like 1* (*KLHL1*) locus at 13q21. Repeat lengths can be large; more than 100 repeats is usually pathogenic, although there is considerable variation between affected individuals, leading to some perplexity regarding this mutation. For example, one unaffected individual has been found to carry 800 repeats at this locus. It is possible that pathology is restricted to a range of lengths, with very long repeats having a reduced effect. Significant meiotic instability occurs with expanded repeats, and large contractions may account for the reduced penetrance found in this dominant disorder. The basis for pathology in this disorder remains unclear.

Spinocerebellar ataxia type 12

Spinocerebellar ataxia type 12 (SCA12) is caused by a CAG repeat expansion in the 5′ untranslated region of the *protein phosphatase 2 regulatory subunit B, beta isoform* (*PPP2R2B*) gene, which encodes a brain-specific regulatory subunit of protein phosphatase 2A (PP2A). The gene is located at 5q31–q33. Expanded tracts were found to contain roughly 100 repeats in the single large pedigree studied. No indication of long repeats could be found in a large sample of normal individuals. Although this expansion probably affects regulation of the *PPP2R2B* gene, demonstration of this hypothesis is pending.

Diseases Resulting from Expansions of Coding Repeats

The second class of triplet repeat disorders includes those caused by expansion of translated CAG repeats, which encode a polyglutamine tract in each of the respective proteins. These 'polyglutamine' disorders share many features, which suggests that there is an underlying common pathogenetic mechanism even though the mutated genes share no homology other than the CAG repeats. For example, all of the diseases are progressive neurological diseases with onset of symptoms in young-to-mid adulthood; neuronal dysfunction and eventual death occur 10–20 years after the onset of symptoms and lead to failure of many vital functions such as swallowing and breathing; in all of the diseases, symptoms develop when the number of uninterrupted repeats exceeds about 35 glutamines, except for *CACNA1A* in which repeats of only 21 or more can cause disease (see below); intergenerational instability of translated CAG repeats is common and is more pronounced in paternal transmissions; for all of the diseases, the phenotype is caused not by loss of function of the relevant protein but rather by a gain of function conferred by the expanded polyglutamine tract; only a specific group of neurons is vulnerable in each disease, despite the wide and overlapping expression pattern of all eight genes. But the specificity of the phenotypes and patterns of neuronal degeneration are lost when the expansions are very large, and this results in severe, juvenile-onset disease. The overlapping and broader phenotypes observed with very large CAG expansions suggest that sufficiently long polyglutamine tracts can render the proteins toxic to many more cell types and might explain why expansions of coding repeats are typically small (one would predict very large expansions to cause embryonic lethality).

Spinobulbar muscular atrophy

Spinobulbar muscular atrophy (SBMA or Kennedy disease), which is the only polyglutamine disease with an X-linked recessive pattern of inheritance, is characterized by the selective degeneration of anterior horn cells, bulbar brain stem neurons and dorsal root ganglia. Affected males develop proximal muscle weakness and atrophy, and difficulties in speech and swallowing. Mild androgen insensitivity that manifests as gynecomastia (enlarged breasts) and testicular atrophy becomes apparent during adolescence.

SBMA is caused by expansion of a CAG repeat in the first coding exon of the androgen receptor (*AR*) gene of Xq13–q21. Normal alleles contain up to 36 repeats, whereas expanded alleles contain more than 38 repeats. The AR is a steroid hormone-activated transcription factor, which on binding to androgen in the cytoplasm translocates into the nucleus and transactivates androgen-responsive genes. Expansion of the glutamine repeat has no effect on hormone binding and only slightly reduces the ability of the receptor to transactivate, but it may cause a partial loss of function. This would account for symptoms of androgen insensitivity in affected males.

Nuclear inclusions containing mutant AR have been reported in motor neurons within the pons, medulla and spinal cord. Mouse models expressing truncated *AR* genes with an expanded polyglutamine tract develop progressive motor weakness.

Huntington disease

Huntington disease (HD) causes sporadic involuntary movements (chorea), which become progressively worse and interfere with motor function. Dystonia, rigidity and epilepsy can also accompany the chorea, particularly in juvenile-onset cases. The motor disorder is often preceded or accompanied by cognitive decline, mood disturbance and/or changes in personality; in the terminal phase, affected individuals are bedridden and suffer from severe dementia and failure of swallowing and breathing control. The pathology of HD is characterized by atrophy of medium spiny neurons in two brain structures known as the caudate and putamen. The disease is caused by expansion of a highly variable CAG repeat in the *huntingtin* (*HD*) gene, which maps to 4p16.3. Normal alleles contain 6–35 repeats, whereas disease alleles contain 36–121 repeats. Individuals with 70 or more repeats develop the more severe, juvenile form of the disease. Genetic studies in mice have shown that HD is not caused by loss of function of the huntingtin protein. Mice with one copy of the *Hdh* gene do not develop HD-like symptoms, and mice lacking both copies die in gestation between embryonic days 8.5 and 10.5. This suggests that huntingtin is essential for embryonic development and neurogenesis.

The *HD* gene is expressed throughout the brain, and the protein localizes to the cytoplasm of neurons as well as to the dendrites and axons. It is associated with microtubules in dendrites and with synaptic vesicles in axon terminals, perhaps having a function in synaptic transmission. Abnormal aggregation of mutant huntingtin has been found in postmortem brains from people with HD, especially in the dendrites and dendritic spines. Huntingtin aggregates are recognized only with antibodies specific for the *N*-terminus of the protein, suggesting that proteolytic cleavage and release of a toxic expanded polyglutamine protein may be an important step in pathogenesis.

Many proteins that interact with huntingtin have been identified, but their role in pathogenesis remains unknown. The finding that huntingtin localizes and/or interacts with transcription activators raises the possibility that alterations in gene expression may be one consequence of the expanded glutamine tract in HD.

Spinocerebellar ataxias

The spinocerebellar ataxias (SCAs) are distinct genetically but overlap in clinical presentation between subtypes and even within families affected by the same disease. It is nearly impossible to distinguish one subtype from another on the basis of clinical signs alone, but here we highlight features that sometimes point toward one SCA or another.

SCA type 1 (SCA1) was the first of the inherited ataxias to be mapped, hence its name. Individuals affected with SCA1 suffer from progressive ataxia (loss of balance and coordination), dysarthria and eventual respiratory failure. SCA1 is characterized pathologically by cerebellar atrophy with severe loss of Purkinje cells and brain stem neurons. People with SCA1 have an expanded 'perfect' CAG repeat tract, usually 39–82 repeats, in the *spinocerebellar ataxia 1* (*SCA1*) gene, which maps to 6p23 and encodes a polyglutamine tract in the protein ataxin-1. Normal alleles contain 6–44 repeats; larger normal alleles (> 20 repeats) are interrupted by 1–4 CAT repeat units that encode histidine, which seem to stabilize the tract.

The function of the ataxin-1 protein is unknown. It may be involved in neuronal functions that are important for learning, because mice lacking the *Sca1* gene have impaired spatial and motor learning. Ataxin-1 is expressed predominantly in the nucleus in neurons and in the cytoplasm in peripheral cells such as muscles and blood cells. In individuals with SCA1, mutant ataxin-1 localizes to a single nuclear inclusion in brain stem neurons, and this nuclear inclusion contains ubiquitin and proteasome components of the cellular machinery that degrades proteins.

SCA type 2 (SCA2) is another inherited ataxia that is characterized by degeneration of neurons in the cerebellum and brain stem. Individuals with SCA2 have 36–63 CAG repeats in the *SCA2* gene, whereas unaffected individuals have 15–31 repeats. The *spinocerebellar ataxia 2* (*SCA2*) gene, which maps to 12q24.1, encodes ataxin-2 – a cytoplasmic protein whose function is unknown. *SCA2* is expressed widely throughout the body and brain, with high levels in cerebellar Purkinje cells. Ataxin-2 tends to accumulate in the cytoplasm of neurons from people affected with SCA2, suggesting that its catabolism is altered.

Machado–Joseph disease or spinocerebellar ataxia type 3 (SCA3) causes progressive ataxia, bulging eyes, rippling movements of facial and lingual muscles and rigidity. Degeneration is most prominent in the basal ganglia, brain stem, spinal cord and dentate neurons of the cerebellum. The disease is caused by the expansion of CAG repeats (55–84) in the *Machado–Joseph disease* (*MJD*) gene at 14q32.1. Normal alleles contain 12–40 repeats. The *MJD* gene encodes a widely expressed protein (ataxin-3) that has no known function. Ataxin-3 is predominantly a cytoplasmic protein but also aggregates in the nucleus of affected neurons in people with SCA3.

Spinocerebellar ataxia type 6 (SCA6) causes affected individuals to develop either episodic or very slowly progressive ataxia. Neuropathological findings in SCA6 include marked cerebellar atrophy with loss of cerebellar Purkinje cells. The CAG repeat in individuals with SCA6 is relatively stable and small, ranging from 21 to 33 repeats. Normal alleles contain fewer than 18 repeats. The *calcium channel, voltage-dependent P/Q type alpha 1A subunit* (*CACNA1A*) gene, which is located at 19p13, encodes the P/Q type calcium channel α1A subunit. This is a voltage-sensitive calcium channel that modulates the entry of calcium into cerebellar Purkinje cells. The mutant α1A calcium channels aggregate in the cytoplasm of SCA6-affected Purkinje cells. The repeat expansion is likely to alter channel activity, thus disturbing calcium balance within vulnerable neurons. Notably, the missense and splicing mutations in the α1A voltage-dependent calcium channel have been identified in other neurological disorders, including hereditary paroxysmal cerebellar ataxia (or episodic ataxia type 2) and familial hemiplegic migraine.

Individuals with spinocerebellar ataxia type 7 (SCA7) develop cerebellar ataxia and/or visual problems, including pigmentary macular degeneration. Childhood-onset cases have a more severe and rapid course that affects both neuronal and non-neuronal (heart) tissues. Primary neuronal loss is in the cerebellum, inferior olive and photoreceptors. The SCA7 mutation is the most unstable in the polyglutamine disease family. It can expand by as many as 200–263 repeats as it is passed from one generation to the next. Normal alleles contain 4–35 repeats and disease alleles have 37–306 repeats. The *spinocerebellar ataxia 7* (*SCA7*) gene maps to 3p12–13, and its product, ataxin-7, is expressed widely but its function is unknown. In people with SCA7, ataxin-7-positive nuclear aggregates appear in several brain regions.

Dentatorubro-pallidoluysian atrophy

Dentatorubro-pallidoluysian atrophy (DRPLA) leads to choreoathetosis, myoclonic epilepsy, balance disorder and dementia; the epilepsy and intellectual impairment are more common in cases of juvenile onset. The neuropathology of DRPLA includes marked neuronal loss, massive gliosis and severe demyelination in the cerebral and cerebellar cortices, the globus pallidus, the striatum, and in dentate, subthalamic and red nuclei. In some families, calcifications are noted in the basal ganglia. The *dentatorubral-pallidoluysian atrophy* (*DRPLA*) gene maps to 12q13.31. Its product, atrophin-1, contains 6–35 repeats on normal alleles and 49–88 on expanded alleles. The atrophin-1 protein has an unknown function, but is expressed widely. It is localized primarily in the cytoplasm in neurons and peripheral cells.

Pathogenesis of Polyglutamine Disorders

Studies of humans with deletions of the *AR*, *HD* or *SCA1* genes (and of mice that lack these genes) confirm that each disease is not caused by the loss of function of the respective protein. More importantly, studies of mouse and fruit fly models that overexpress full-length or truncated forms of either ataxin-1, ataxin-3, huntingtin, AR, atrophin-1 or ataxin-7 show that there is progressive neuronal dysfunction consistent with a gain-of-function mechanism. Several illuminating findings have emerged from studying affected humans, cell culture systems and various animal models.

First, the protein context of the polyglutamine tract clearly influences its toxicity. In humans, a polyglutamine tract of 36 repeats in ataxin-1 or ataxin-3 does not cause disease, but the same length repeat tract in ataxin-2 will result in ataxia. In transgenic mice, expression of truncated polypeptides containing expanded glutamine tracts caused widespread dysfunction that extends beyond the specific groups of neurons affected by the full-length protein. Second, there is a relationship between protein concentrations and repeat lengths. Either high concentrations (increases > 5–10-fold) of expanded proteins or very long glutamine tracts are required to see a phenotype within the short life span of the mouse. This is consistent with the inverse relationship between age of onset and repeat length and suggests that the toxicity that takes decades to show in humans can be expedited in model organisms by boosting protein concentrations or repeat lengths.

Several studies of polyglutamine toxicity have been carried out in cell culture, mouse models and, more recently, invertebrate animal models. For some of the proteins (huntingtin, AR and ataxin-3), polyglutamine toxicity is more readily detectable when a fragment rather than the whole protein is used. This raises the possibility that protein cleavage, perhaps by caspases, might promote toxicity in vulnerable neurons. Most of the expanded proteins that are involved in polyglutamine disorders tend to accumulate in the nucleus (huntingtin, ataxin-1, ataxin-3, ataxin-7, atrophin-1 and AR); only two so far are predominantly cytoplasmic (ataxin-2 and CACNA1A). The redistribution of components of the ubiquitin–proteasome pathway and protein folding machinery (chaperones) to the site of the nuclear aggregates suggests that the expanded polyglutamine tracts may cause the respective proteins to misfold or resist degradation. Studies in cell culture and in animal models have provided evidence in support of this hypothesis. The data suggest that misfolded proteins alter gene transcription either through aberrant protein–protein interactions or by sequestering key components of the transcriptional mechanism. Alterations in the levels of proteins essential for the normal functioning of neurons (such as proteins that regulate concentrations of Ca^{2+} and neurotransmitters) might, in turn, exacerbate neuronal dysfunction and lead to neuronal death.

The finding that chaperone overexpression mitigates the phenotype in several invertebrate models of polyglutamine disorders and the identification of several additional modifiers of the neuronal phenotype provide a platform for identifying key pathways of pathogenesis and potential therapeutic targets.

See also
Anticipation
Friedreich Ataxia
Huntington Disease

Protein Aggregation and Human Disorders
Trinucleotide Repeat Expansions: Mechanisms and Disease Associations

Further Reading

Abel A, Walcott J, Woods J, Duda J and Merry DE (2001) Expression of expanded repeat androgen receptor produces neurologic disease in transgenic mice. *Human Molecular Genetics* **10**(2): 107–116.

Chan HY and Bonini NM (2000) *Drosophila* models of human neurodegenerative disease. *Cell Death and Differentiation* **11**: 1075–1080.

Cummings CJ and Zoghbi HY (2000) Trinucleotide repeats: mechanisms and pathophysiology. *Annual Review of Genomics and Human Genetics* **1**: 281–328.

Cummings CJ and Zoghbi HY (2000) Trinucleotide repeats: mechanisms and pathophysiology. *Annual Review of Genomics and Human Genetics* **1**: 281–328.

Fernandez-Funez P, Nino-Rosales ML, de Gouyon B, *et al.* (2000) Identification of genes that modify ataxin-1-induced neurodegeneration. *Nature* **408**(6808): 101–106.

Gecz J (2000) The FMR2 gene, FRAXE and non-specific X-linked mental retardation: clinical and molecular aspects. *Annals of Human Genetics* **64**(2): 95–106.

Groenen P and Wieringa B (1998) Expanding complexity in myotonic dystrophy. *BioEssays* **20**(11): 901–912.

Gusella JF and MacDonald ME (2000) Molecular genetics: unmasking polyglutamine triggers in neurodegenerative disease. *Nature Reviews Neuroscience* **1**(2): 109–115.

Jin P and Warren ST (2000). Understanding the molecular basis of fragile X syndrome. *Human Molecular Genetics* **9**(6): 901–908.

McCampbell A, Taylor JP, Taye AA, *et al.* (2000) CREB-binding protein sequestration by expanded polyglutamine. *Human Molecular Genetics* **9**(14): 2197–2202.

Osborne LR (2000) Polyglutamine stretches suppress transcription. *Molecular Medicine Today* **6**(12): 457–458.

Puccio H and Koenig M (2000). Recent advances in the molecular pathogenesis of Friedreich ataxia. *Human Molecular Genetics* **9**(6): 887–892.

Wells RD and Warren ST (eds.) (1998) *Genetic Instabilities and Hereditary Neurological Diseases*. San Diego, CA: Academic Press.

Web Links

Androgen receptor (*AR*); LocusID: 367. LocusLink
http://www.ncbi.nlm.nih.gov/LocusLink/LocRpt.cgi?l = 367

Fragile X mental retardation 1 (*FMR1*); LocusID: 2332. LocusLink
http://www.ncbi.nlm.nih.gov/LocusLink/LocRpt.cgi?l = 2332

Friedreich ataxia (*FRDA*); LocusID: 2395. LocusLink
http://www.ncbi.nlm.nih.gov/LocusLink/LocRpt.cgi?l = 2395

Machado–Joseph disease (*MJD*); LocusID: 4287. LocusLink
http://www.ncbi.nlm.nih.gov/LocusLink/LocRpt.cgi?l = 4287

Spinocerebellar ataxia 1 (*SCA1*); LocusID: 6310. LocusLink
http://www.ncbi.nlm.nih.gov/LocusLink/LocRpt.cgi?l = 6310

Androgen receptor (*AR*); MIM number: 313700. OMIM
http://www.ncbi.nlm.nih.gov/htbin-post/Omim/dispmim?313700

Fragile X mental retardation 1 (*FMR1*); MIM number: 309550. OMIM
http://www.ncbi.nlm.nih.gov/htbin-post/Omim/dispmim?309550

Friedriech ataxia (*FRDA*); MIM number: 606829. OMIM
    http://www.ncbi.nlm.nih.gov/htbin-post/Omim/
    dispmim?606829
Machado–Joseph disease (*MJD*); MIM number: 607047. OMIM
    http://www.ncbi.nlm.nih.gov/htbin-post/Omim/
    dispmim?607047

Spinocerebellar ataxia 1 (*SCA1*); MIM number: 601556. OMIM
    http://www.ncbi.nlm.nih.gov/htbin-post/Omim/
    dispmim?601556

Trinucleotide Repeat Expansions: Mechanisms and Disease Associations

Jörg T Epplen, *Ruhr-University, Bochum, Germany*

Martin Gencik, *Ruhr-University, Bochum, Germany*

Trinucleotide (and selected other) repeat expansions may cause severe neurodegenerative disorders with adult onset. Either elongated polyglutamine blocks or reduced gene expression are responsible for the differential pathogenesis of this novel disease category.

Intermediate article

Article contents

- Definitions of Dynamic Mutations and Disease
- Trinucleotide Repeat Polymorphisms and Expansions
- Repeat Instability and Pathogenesis
- Anticipation

Definitions of Dynamic Mutations and Disease

Trinucleotide repeat or more general repeat block expansion diseases are caused by expansions of naturally occurring polymorphic microsatellites within or in the vicinity of human genes. The major characteristic of these mutations is their instability in successive generations, as well as between cells of the same organisms ('dynamic' mutation). A common clinical phenomenon in trinucleotide repeat disorders is 'anticipation'. Until now, more than a dozen human diseases with Mendelian inheritance were shown to be caused by expansions of polymorphic tandem repeats. Natural animal or plant model systems have not been observed, but valuable transgenic models may be available in the near future.

Short tandem repeats (microsatellites) are a broadly represented class of repetitive sequences found in virtually all genomes investigated. In the human genome, short tandem repeats occur on average every 2000 base pairs (bp) and localize in up to 8% in coding gene regions. The high mutation rate in short tandem repeats makes them most polymorphic and therefore suitable for genetic linkage or association studies because of their high degree of informativeness. However, there is a significant difference between the mutational rate of single short tandem repeats, the major factors being the locus type itself and the mean allele size. Several disease-causing short tandem repeats have been shown to have the highest size variation and heterozygosity compared with anonymous loci.

A further property is the allele size and gender-dependent segregation deviation of short tandem repeats.

There are two main groups of repeat block repeat diseases: (1) $(CAG)_n$ and $(GCG)_n$ expansions in coding regions, translated into polyglutamine and polyalanine stretches; and (2) very large and untranslated expansions of $(GAA)_n$, $(CTG)_n$, $(CGG)_n$, $(GGC)_n$, $(CCTG)_n$, $(ATTCT)_n$ or $(C_4GC_4GCG)_n$ blocks (**Table 1**). The pathophysiological characteristics of both disease categories differ, because the polyglutamine diseases are caused by distorted protein degradation, accumulation and/or other mechanisms interfering with cellular hemostasis, whereas the causal pathogenesis of the group characterized by noncoding repeat expansions is due to reduced gene expression at the transcriptional or the translational levels respectively (see Cummings and Zoghbi, 2000).

Trinucleotide Repeat Polymorphisms and Expansions

Polyglutamine diseases

Polyglutamine diseases are due to translated $(CAG)_n$ repeats in the coding regions of different genes. Polyglutamine diseases probably share similar pathophysiological mechanisms, although expansion

Table 1 Repeat block expansion diseases

Disease	Gene designation	Protein	Chromosome locus	Repeat motif	Localization in respective gene	Physiological repeat length	Expanded repeat range
DRPLA	*DRPLA*	Atrophin-1	13p13.31	CAG	Coding	6–35	49–88
Fragile XA syndrome	*FMR1*	FMR-1	Xq27.3	CGG	5′ UTR	6–53	>230
Fragile XE syndrome	*FMR2*	FMR-2	Xq28	GCC	5′ UTR	6–35	>200
Friedreich ataxia	*FRDA*	Frataxin	9q13–21.1	GAA	Intronic	7–34	>100
Huntington disease	*HD*	Huntingtin	4p16.3	CAG	Coding	4–35	36–250
Kennedy disease	*AR*	Androgen rec.	Xq13–21	CAG	Coding	9–36	38–62
Myotonic dystrophy 1	*DM1*	DMPK?	19q3	CTG	3′ UTR?	5–37	>50
Myotonic dystrophy 2	*DM2*	ZNF9	3q13.3–q24	CCTG	Intronic	104–176 bp	75–11 000
SCA1	*SCA1*	Ataxin-1	6p63	CAG	Coding	6–44	39–82
SCA2	*SCA2*	Ataxin-2	12q24.1	CAG	Coding	15–31	36–63
SCA3/Machado–Joseph disease	*SCA3*	Ataxin-3	14q32.1	CAG	Coding	12–40	55–84
SCA6	*SCA6*	CACNA1A	19p13	CAG	Coding	4–18	21–33
SCA7	*SCA7*	Ataxin-7	13p12–13	CAG	Coding	4–35	37–306
SCA8	*SCA8*	None	13q21	CTG	3′ UTR	16–37	110–250
SCA10	*SCA10*	Ataxin-10	22q13	ATTCT	Intronic	10–22	800–4600
SCA12	*SCA12*	PP2A-PR55B	5q31–33	CAG	Promoter	7–28	66–78
SCA17	*TBP*	TBP	6q27	CAG	Coding	25–42	45–63
PME1	*CSTB*	Cystatin B	21q22.3	C_4GC_4GCG	Promoter	2–3	30–75
Oculopharyngeal muscular dystrophy	*PABPN1*	PABPN1	14q11.2–13	GCG	Coding	6	7–13
Synpolydactyly	*HOXD13*	HOXD13	2q31–32	GCN	Coding	15	22–24

thresholds and localizations of the pathogenetic principle are disease specific (**Table 1**). In addition, polyglutamine diseases represent mostly adult-onset neurological traits, with progressive neuronal dysfunction leading to neurodegeneration in specific brain regions. Until now, this group has consisted of nine diseases, among which Huntington disease (HD) is most frequent. HD is characterized by neurological and/or psychiatric symptoms with onset in the fourth to fifth decade of life. On average, HD patients die after 10–20 years of disease duration. Furthermore, the genetically heterogeneous spinocerebellar ataxias (SCAs) represent a group of dominantly inherited neurodegenerative diseases with cerebellar atrophy. SCAs have a broad range of overlapping clinical symptoms with typical cerebellar signs. The disease entities SCA1-3, SCA6-7 and SCA17 are based on critical $(CAG)_n$ expansions, whereas SCA8 and SCA12 are caused by noncoding repeats and SCA10 is associated with respective expansions. Machado–Joseph disease is the Japanese variant of SCA3, with slightly different clinical symptoms. Dentatorubral-pallidoluysian atrophy (DRPLA) is more common in Japan and shares symptoms of HD and the SCAs. Finally, Kennedy disease (spinobulbar muscular atrophy) is an X-chromosomal recessive trait in which the patient presents with neuromuscular symptoms and endocrine deficiency due to androgen insensitivity. The causative $(CAG)_n$ expansion is localized within the *androgen receptor* gene (*AR*) on chromosome X.

Autosomal dominantly inherited spondylodactyly is caused by expansion of an imperfect repeat translated into a polyalanine tract of the HOXD13 protein (Goodman *et al.*, 1997). Usually 15 alanine residues are expressed, whereas affected individuals harbor 7–9 extra alanine residues in the respective mutated allele. In contrast to other trinucleotide diseases, the polyalanine tract in HOXD13 protein is stable. Thus these mutations appear as products of nonreciprocal recombination or misalignment during recombination.

Diseases with repeat expansions not expressed at the protein level

Expansions of 'nonprotein' coding deoxyribonucleic acid (DNA) may also cause disease (see **Figure 1**). Such expansions show large variability concerning repeat positions in the genes, sizes and sequence motifs. Similarly variable are the known pathophysiological mechanisms involved. The fragile XA and the rare fragile XE mental retardation syndromes are X-chromosomal recessively inherited. The $(CGG)_n$ and

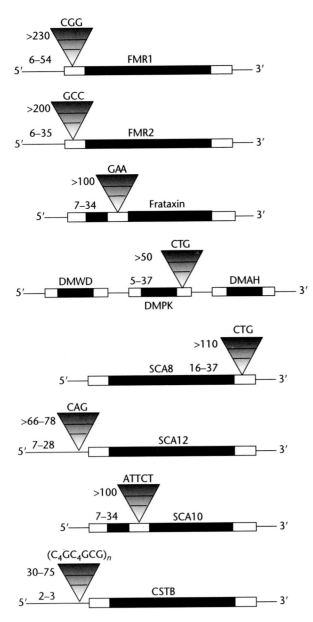

Figure 1 Schematic representation of genes and their trinucleotide blocks involved in selected repeat expansion diseases other than polyglutamine disorders (black bars, protein coding parts). Detailed exon–intron structures of the genes are not shown or have been minimized to demonstrate intronic localization of the repeat blocks. The physiological length ranges of the polymorphic simple repeat tracts are enumerated at the tips of the inverted triangles, whereas the pathological range is annotated at the top portion of each triangle.

$(GCC)_n$ repeats are each localized in the $5'$ untranslated regions of *FMR1* (*fragile X mental retardation 1*) and *FMR2* genes respectively. Repeat expansions entail increased methylation of neighboring DNA sequences and lead to inactivation of gene expression, resulting in loss of function.

An intronic expansion of the $(GAA)_n$ repeat in the autosomal-recessive *Friedreich ataxia* (*FRDA*) gene prevents expression of frataxin, a protein important for iron homeostasis and mitochondrial function (Campuzano *et al.*, 1996). Myotonic dystrophy is a disorder characterized mainly by muscle wasting and myotonia and is caused by either a $(CTG)_n$ expansion in the $3'$ untranslated region of the *DMPK* gene in myotonic dystrophy 1 or by the tetranucleotide repeat $(CCTG)_n$ expansions in the first intron of the *ZNF9* gene in myotonic dystrophy 2. In myotonic dystrophy 1, the dominant inheritance suggests a negatively dominant effect of the expansion on gene transcription, eventually influencing the expression of neighboring genes. In analogy, *SCA8*, *SCA10* and *SCA12* have dominant expansions of a $(CTG)_n$ repeat in the $3'$ untranslated region, an intronic $(ATTCT)_n$ block or a $(CAG)_n$ tract in the promoter region of the respective disease genes. Finally, in the autosomal recessively inherited progressive myoclonus epilepsy of Unverricht–Lundborg type (EPM1), the expansion of the dodecamer repeat $(C_4GC_4GCG)_n$ in the promoter region of the *CSTB* (cystatin B (stefin B)) gene reduces the transcriptional activity, resulting in low levels of cystatin B. The $(ATTCT)_n$, $(CCTG)_n$ and $(C_4GC_4GCG)_n$ expansions in the *SCA10*, *TBP* and *CSTB* genes are the sole pathologic repeats that are not trinucleotides. Debilitating influences of other simple tandem repeats and/or minisatellites are covered elsewhere in this encyclopedia. The multifactorial nature of common diseases such as diabetes mellitus and certain malignancies includes genetic contributions of e.g. minisatellite loci, but these latter disease entities characteristically lack Mendelian inheritance patterns. (*See* Diabetes: Genetics; Minisatellites.)

Repeat Instability and Pathogenesis

Over the years, a number of large-scale repeat expansions have been shown to cause cytogenetically detectable 'fragile' sites, demonstrable in metaphase chromosomes. Yet this phenomenon is not inevitably pathogenic, except for critical expansions in the *FMR1* gene (Fu *et al.*, 1991). Here expansion interferes with gene expression, unlike in polyglutamine disorders, when the tract is translated into a glutamine block. Approximately 8% of human genes bear a repeat polymorphism and are potential candidates for pathogenic expansions (Wren *et al.*, 2000). Two general mechanisms account for repeat instability. The first is 'slippage' of DNA polymerase during mitotic and meiotic DNA replication (Sinden, 2001). The latter mechanism explains exclusively the low-degree length variations; it is thought to be due to imprecise DNA replication during error-prone cell divisions, especially during gametogenesis. Second, disease-associated triplet repeats form stable, 'alternative' DNA conformations such as triplex and quadruplex helices as well as hairpins, and promote intermediate- to large-range expansions during replication. Finally, ineffective DNA repair mechanisms may account for some additional instability.

Depending on repeat class, different mechanisms are responsible for disease development:

1. $(CAG)_n$ repeats are translated into polyglutamine domains within proteins. Depending on individual thresholds, the resulting proteins form peptides (containing β-sheets), accumulate, and are deposited as inclusion bodies in the cytoplasm and nucleus. Furthermore, these misfolded proteins interfere with the proteasome and metabolic pathways of the cell and lead finally to neuronal cell death (Sherman and Goldberg, 2001).
2. $(CGG)_n$ and $(GCC)_n$ expansions in fragile X syndromes and the intronic $(GAA)_n$ expansion in FRDA downregulate or abolish gene expression on the transcriptional level via loss of function (Sinden, 1999; Jin and Warren, 2000).
3. The pathomechanisms behind $(CTG)_n$ and $(CCTG)_n$ expansions in myotonic dystrophy 1, SCA8, and myotonic dystrophy 2; $(CAG)_n$ expansion in SCA12; as well as $(ATCTT)_n$ expansion in SCA10 are not sufficiently understood at present, especially the apparent dominant negative effect.
4. $(GCG)_n$ expansions in the coding region of the *PABP2* gene are translated into polyalanine tracts that may cause mutated PABP2 peptide oligomers to accumulate as filamentary inclusions in the nuclei of certain cell populations (Brais *et al.*, 1998). A similar mechanism has been proposed in synpolydactyly, where the polyglutamine tract in the HOXD13 protein promotes enhanced di(poly)merization and binding to additional HOX proteins (Goodman *et al.*, 1997). This dominant negative effect causes the proteins to be sequestered and degraded at an increased rate.

Anticipation

The repeats involved in human repeat expansion diseases harbor intrinsic properties that are responsible for pathogenic expansions beyond thresholds in terms of a new mutation on the one hand, and for additional intergenerational expansion of already pathogenic expansions on the other hand. Anticipation is the clinical correlate of prolongation of the repeat expansion in the successive generation: earlier age of onset and, in some cases, aggravation of the

Table 2 Intergenerational instabilities in repeat expansion diseases

Locus	Intergenerational instability
SCA1, SCA2, SCA(3), SCA7, DRPLA, HD, AR	P > M
FMR1	P < M
FRDA	P < M
DM1, SCA8	P < M
DM2	P < M

P: paternal; M: maternal.

clinical course in the offspring generation. In $(CAG)_n$ expansion diseases, the intergenerational repeat instability is a result of male gonadic mosaicism and is thought to be the result of slippage or to be mediated by hairpin formation during DNA replication. In contrast, diseases with maternal bias toward expansion often produce dramatic repeat elongations in the range of several thousand units. These two types of expansion have in common age and size dependency; that is, the longer the parental repeat and the greater the age of the parent, the greater the probability of repeat size increase in the subsequent generation. The bias of intergenerational instabilities in repeat expansion diseases is shown in **Table 2**.

In summary, molecular biological investigations into trinucleotide expansion diseases have revealed unprecedented insights into the genetics and pathogenesis of these disease models in humans. The ramifications for other diseases cannot be emphasized enough; yet it is also to be expected that more simple repeat expansion diseases will be discovered in the future for rare Mendelian traits.

See also
Anticipation
Friedreich Ataxia
Huntington Disease
Microsatellites
Trinucleotide Repeat Expansions: Disorders

References

Brais B, Bouchard J-P, Xie Y-G, *et al.* (1998) Short GCG expansions in the *PABP2* gene cause oculopharyngeal muscular dystrophy. *Nature Genetics* **18**: 164–167.

Campuzano V, Montermini L, Molto MD, *et al.* (1996) Friedreich's ataxia: autosomal recessive disease caused by an intronic GAA triplet repeat expansion. *Science* **271**: 1423–1427.

Cummings CJ and Zoghbi HY (2000) Fourteen and counting: unraveling trinucleotide repeat diseases. *Human Molecular Genetics* **9**: 909–916.

Fu YH, Kuhl DP, Pizzuti A, *et al.* (1991) Variation of the CGG repeat at the fragile X site results in genetic instability: resolution of the Sherman paradox. *Cell* **67**: 1047–1058.

Goodman FR, Mundlos S, Muragaki Y, *et al.* (1997) Synpolydactyly phenotypes correlate with size of expansions in HOXD13 polyalanine tract. *Proceedings of the National Academy of Sciences of the United States of America* **94**: 7458–7463.

Jin P and Warren ST (2000) Understanding the molecular basis of fragile X syndrome. *Human Molecular Genetics* **9**: 901–908.

Sherman MY and Goldberg AL (2001) Cellular defenses against unfolded proteins: a cell biologist thinks about neurodegenerative diseases. *Neuron* **29**: 15–32.

Sinden RR (1999) Human genetics '99: trinucleotide repeats: biological implications of the DNA structures associated with disease-causing triplet repeats. *American Journal of Human Genetics* **64**: 346–353.

Sinden RR (2001) Origins of instability. *Nature* **411**: 757–758.

Wren JD, Forgacs E, Fondon III JW, *et al.* (2000) Repeat polymorphisms within gene regions: phenotypic and evolutionary implications. *American Journal of Human Genetics* **67**: 345–356.

Further Reading

Ashley CT and Warren ST (1995) Trinucleotide repeat expansion and human disease. *Annual Review of Genetics* **29**: 703–728.

Gusella JF and MacDonald ME (2000) Molecular genetics: unmasking polyglutamine triggers in neurodegenerative disease. *Nature Review* **1**: 109–115.

Web Links

Androgen receptor (dihydrotestosterone receptor; testicular feminization; spinal and bulbar muscular atrophy; Kennedy disease)

(*AR*); LocusID: 367. LocusLink:
http://www.ncbi.nlm.nih.gov/LocusLink/LocRpt.cgi?l = 367

Cystatin B (stefin B) (*CSTB*); LocusID: 1476. LocusLink:
http://www.ncbi.nlm.nih.gov/LocusLink/LocRpt.cgi?l = 1476

Fragile X mental retardation 1 (*FMR1*); LocusID: 2332. LocusLink:
http://www.ncbi.nlm.nih.gov/LocusLink/LocRpt.cgi?l = 2332

Friedreich ataxia (*FRDA*); LocusID: 2395. LocusLink:
http://www.ncbi.nlm.nih.gov/LocusLink/LocRpt.cgi?l = 2395

Spinocerebellar ataxia 8 (*SCA8*); LocusID: 6315. LocusLink:
http://www.ncbi.nlm.nih.gov/LocusLink/LocRpt.cgi?l = 6315

Androgen receptor (dihydrotestosterone receptor; testicular feminization; spinal and bulbar muscular atrophy; Kennedy disease) (*AR*); MIM number: 313700. OMIM:
http://www3.ncbi.nlm.nih.gov/htbin-post/Omim/dispmim?313700

Cystatin B (stefin B) (*CSTB*); MIM number: 601145. OMIM:
http://www3.ncbi.nlm.nih.gov/htbin-post/Omim/dispmim?601145

Fragile X mental retardation 1 (*FMR1*); MIM number: 309550. OMIM:
http://www3.ncbi.nlm.nih.gov/htbin-post/Omim/dispmim?309550

Friedreich ataxia (*FRDA*); MIM number: 606829. OMIM:
http://www3.ncbi.nlm.nih.gov/htbin-post/Omim/dispmim?606829

Spinocerebellar ataxia 8 (*SCA8*); MIM number: 603680. OMIM:
http://www3.ncbi.nlm.nih.gov/htbin-post/Omim/dispmim?603680

Trisomy

Michel Vekemans, *Hôpital Necker–Enfants Malades, Paris, France*

Human trisomy is a leading cause of fetal wastage, perinatal and infant mortality, congenital malformation and mental retardation.

Intermediate article

Article contents

- Introduction
- Incidence
- Origin
- Mechanisms
- Etiology
- Phenotypic Consequences

Introduction

Human trisomy is a leading cause of fetal wastage, perinatal and infant mortality, congenital malformation and mental retardation. This includes only the impact of trisomy where the individual's constitutional karyotype is determined during fertilization or the first few cell divisions thereafter. If this restriction were to be relaxed, then a large proportion, if not the majority, of malignancies would also be included as well as an uncertain proportion of disorders associated with aging and/or autoimmune processes. In the material that follows, only human trisomy as a cause of constitutional genetic disease will be addressed.

Incidence

Trisomy is extremely common in humans. Among newborns, 0.3% of live births are trisomic with the most common anomalies being trisomy 21 and sex chromosome polysomies (47,XXX, 47,XYY and 47,XXY). The incidence increases to 4% among stillbirths, with the autosomal trisomies being similar to those observed among newborns. Among clinically recognized spontaneous abortions, 26.1% are trisomic, but unlike stillbirths or live births, different trisomies are found. The most common anomaly is trisomy 16, which constitutes about one-third of all trisomies identified in spontaneous abortions. Using these figures and assuming that 15% of clinically recognized pregnancies spontaneously abort whereas 1% are stillborn and the rest are liveborn, it has been estimated that at least 4.1% of all human conceptions are trisomic. Of course this figure underestimates the real incidence of trisomy in humans because it does not include

Table 1 Origins of human trisomy

| | Meiotic | | | | |
| | Paternal | | Maternal | | |
Trisomy	MI (%)	MII (%)	MI (%)	MII (%)	Mitotic (%)
15		15	76	9	
16			100		
18			33	56	11
21	3	5	65	23	4
XXY	46		38	14	2
XXX		6	60	16	18

Adapted from Hassold and Hunt (2001).

conceptions that spontaneously abort during the first month of gestation (Hassold and Jacobs, 1984).

Origin

Over the past decade, DNA polymorphisms have been used to track down the parental and the meiotic/mitotic stage of origin of different trisomic conditions. The results from such studies have been reviewed by Hassold and Hunt (2001). They are summarized in **Table 1**. Although maternal meiosis I (MI) errors predominate, the parent and meiotic stage of origin of the additional chromosome do vary from one trisomic condition to another. For example, paternal errors account for 46% of the 47,XXY condition but only 8% of trisomy 21. They are rarely the cause of trisomy 16 or trisomy 18. Similarly, the importance of MI versus meiosis II (MII) errors varies among chromosomes. For example, among trisomies of maternal origin, all cases of trisomy 16 seem to be due to MI errors but for trisomy 18 about 60% of cases involve MII nondisjunction.

It is important to stress, however, that studies tracking DNA polymorphisms between parents and offspring is an indirect way of looking at meiotic processes. Investigations using fluorescence *in situ* hybridization (FISH) provide conflicting results regarding both stage and type of segregation errors (Mahmood *et al.*, 2000). Therefore, it is fair to say that we are still ignorant about the mechanisms of origin of constitutional trisomies, comprising the most common type of genetic disease in humans.

Mechanisms

Recently, several studies comparing the frequency and distribution of meiotic exchanges in meioses generating a trisomy with those observed in normal meioses have shown that a reduction in the number of chromosome exchanges is a feature of all trisomies derived from MI error. This includes trisomies 15, 16, 18, 21, sex chromosome polysomies derived from a maternal error and cases of trisomy 21 derived from a paternal error. The most pronounced reduction is observed in the XXY condition resulting from a paternal error. In this instance, the genetic map of the XY pseudoautosomal pairing region is decreased from 50 cM in normal meioses to 10 cM in meioses generating the polysomy. In some instances, an absence of exchange is responsible for this overall decrease in the map length. For example, about 40% of cases of trisomy 21 derived from a maternal MI error result from an achiasmate bivalent. Similar findings have been observed for trisomy 21 and the 47,XXY condition resulting from a paternal error, as well as trisomies 15, 18 and sex chromosome polysomies resulting from an MI maternal error. In addition, when a single exchange is observed in maternal MI-derived cases of trisomy 21, the exchange is displaced toward the telomere. In contrast, an increase in pericentromeric exchanges is observed in cases of trisomy 21 resulting from an MII maternal error. It has been suggested that these pericentromeric exchanges might produce chromosome entanglement. The subsequent segregation at MII would then result in a disomic gamete having identical centromeres. This would be scored as an MII nondisjunctional event. Although this might be the case for trisomy 21, more recent studies have shown that chromosome 18 and sex chromosomes behave differently (Hassold and Hunt, 2001).

Etiology

After many years of study, apart from maternal age no single parental factor, whether environmental or genetic in nature, has been consistently implicated as a determinant of human trisomy. The association between increasing maternal age and Down syndrome was recognized long before the syndrome was known to result from trisomy 21 (Penrose, 1933). A similar effect is also observed for other viable and nonviable trisomies. The rapid increase begins around 35 years of age. For example, among woman aged 20 years, 2% of all clinically recognized pregnancies are trisomic, but among women over 40 years this value approaches 35% (Hassold and Chiu, 1985). Also, studies of double and mosaic trisomies have shown that they are affected by increasing maternal age, the effect being more important in the case of double trisomies. In addition, the relationship is specific to the chromosome involved in the trisomy, that is, the maternal age effect is greatest for trisomies involving the smallest chromosomes and absent for most larger chromosomes (Risch *et al.*, 1986). Interestingly, trisomy 16 is a clear outlier in this relationship as its frequency increases linearly and not exponentially with maternal age. Nevertheless, much remains to be learned about the biological basis of the maternal age effect. One model has suggested that abnormal recombination and nondisjunction associated with maternal age are causally related (Lamb *et al.*, 1996). Indeed, this model proposes that at least two hits are required for age-dependent trisomy. The first hit is age independent as it occurs in the fetal ovary and involves the establishment of bivalents liable to nondisjunction. The second hit is age dependent as it occurs at metaphase I in the adult ovary and involves an abnormal processing of these liable bivalents. The abnormal processing might result from a defect in oocyte maturation, a change in the hormonal milieu or a change in the level of proteins that influence proper chromosome segregation.

Phenotypic Consequences

From classical studies in plant genetics, it was known that trisomy produced easily recognizable phenotypes (Blakeslee, 1932). Similarly, the adverse effects of trisomy on the phenotype in humans are now well established. Among live births which represent the least affected of all trisomic conceptions, the presence of an additional autosome is often associated with a severe mental and physical handicap and early lethality (**Table 2**). In contrast, the gain of an additional sex chromosome is relatively well tolerated and is associated with much variation in phenotype (**Table 3**). In addition, mosaicism including a normal cell line is common and accounts for much variation between individual cases of autosomal trisomies as well as sex chromosome aneuploidy. How the inheritance of three 'normal' copies of a gene on the additional chromosome results in disruption of normal development remains unknown (Reeves *et al.*, 2001). The developmental instability hypothesis suggests that the correct balance of gene expression in pathways regulating development is upset by the dosage imbalance of genes on the additional chromosome. This leads to an increased frequency and increased variability of phenodeviant characters among trisomic individuals when compared with euploid individuals. Another hypothesis holds that dosage imbalance of a specific gene or group of genes on the additional chromosome is responsible for specific individual traits. Although the first proposal has been considered as an untestable hypothesis, the second hypothesis has been rather difficult to demonstrate (Korenberg *et al.*, 1994). Presumably, a comprehensive explanation for the phenotypes observed in trisomic conditions should consider the consequences of trisomy on the embryo's

Table 2 Autosomal trisomy syndromes

	Trisomy 8	Trisomy 13	Trisomy 18	Trisomy 21
Birth incidence	1 : 25 000, most often mosaicism	1 : 15 000, early childhood lethality	1 : 8000, early childhood lethality	1 : 700
Sex ratio	M : F = 2 : 1	M : F = 1 : 1.2	M : F = 1 : 4	M : F = 3 : 2
Neurology	Moderate retardation	Mental retardation, forebrain defects	Mental retardation, hypertonia	Mental retardation, hypotonia
Head		Sloping forehead, scalp defects	Prominent occiput	Flat facies, flat occiput
Eyes		Microphthalmia, colobomata	Small palpebral fissures	Upslanting epicanthal folds
Nose				Low nasal bridge
Mouth	Prominent lower lip	Cleft lip and/or palate	Micrognathia	Protruding tongue
Heart	Heart defects (< 50%), VSD and ASD	Heart defects (90%), VSD and PDA	Heart defects (99%), VSD and PDA	Heart defects (40%), AV canal and VSD
Abdomen	Kyphoscoliosis, vertebral anomalies	Absent spleen	Short sternum, Meckel's diverticulum	Duodenal atresia
Hands	'Plis capitonnés', camptodactyly	Postaxial polydactyly	Clenched hands with overlapping fingers. Excess of arches	Short hands and fingers, clinodactyly of fifth finger, simian crease, distal axial triradius
Feet		Rocker-bottom feet	Rocker-bottom feet	Sandal gap
Urogenital	Hypogonadism	Polycystic kidneys, cryptorchidism, bicornuate uterus	Renal anomalies	

VSD: ventricular septal defect; ASD: atrial septal defect; PDA: patent ductus arteriosus; AV: atrioventricular.
Adapted from Nora JJ and Fraser FC (1994) *Medical Genetics Principles and Practice*, 4th edn. Philadelphia, PA: Lea & Febiger.

Table 3 Sex chromosome polysomy syndromes

Syndrome	Incidence	Phenotype
47,XXY	1 : 1000 males	Male usually with long arms and legs. Small genitalia. Educational problems frequent. Mostly infertile
47,XYY	1 : 1000 males	Male with tall stature. Normal to reduced intelligence. Behavioral problems. Normal to reduced fertility
47,XXX	1 : 1000 females	Female with tall stature. Educational problems frequent. Normal fertility

Adapted from Thompson MW, McInnes RR and Willard HF (1991) *Thompson and Thompson: Genetics in Medicine*, 5th edn. Philadelphia, PA: WB Saunders.

normal developmental pattern (i.e. its norm of reaction) rather than the direct consequences of the overexpression of the triplicated genes (Vekemans and Trasler, 1986).

See also

Down Syndrome
Karyotype Interpretation
Monosomies
Uniparental Disomy
XYY Syndrome

References

Blakeslee A (1932) *Proceedings of the Sixth International Congress of Genetics*, vol. 1, pp. 104–120. Philadelphia, PA: Jones.

Hassold T and Chiu D (1985) Maternal age-specific rates of numerical chromosome abnormalities with special reference to trisomy. *Human Genetics* **70**: 11–17.

Hassold T and Hunt P (2001) To err (meiotically) is human: the genesis of human aneuploidy. *Nature Reviews Genetics* **2**: 280–291.

Hassold TJ and Jacobs PA (1984) Trisomy in man. *Annual Review of Genetics* **18**: 69–97.

Korenberg JR, Chen XN, Schipper R, *et al.* (1994) Down syndrome phenotypes: the consequences of chromosomal imbalance.

Proceedings of the National Academy of Sciences of the United States of America **91**: 4997–5001.

Lamb NE, Freeman SB, Savage-Austin A, *et al.* (1996) Susceptible chiasmate configurations of chromosome 21 predispose to non-disjunction in both maternal meiosis I and meiosis II. *Nature Genetics* **14**: 374–376.

Mahmood R, Brierley CH, Faed MJ, Mills JA and Delhanty JD (2000) Mechanisms of maternal aneuploidy: FISH analysis of oocytes and polar bodies in patients undergoing assisted conception. *Human Genetics* **106**(6): 620–626.

Penrose L (1933) The relative effects of paternal and maternal age in mongolism. *Journal of Genetics* **27**: 219–224.

Reeves RH, Baxter LL and Richtsmeier JT (2001) Too much of a good thing: mechanisms of gene action in Down syndrome. *Trends in Genetics* **17**: 83–88.

Risch N, Stein Z, Kline J and Warburton D (1986) The relationship between maternal age and chromosome size in autosomal trisomy. *American Journal of Human Genetics* **39**: 68–78.

Vekemans M and Trasler T (1986) Liability to cleft palate in trisomy 19 mouse embryos. *Journal of Craniofacial Genetics and Developmental Biology* **2** (supplement): 235–240.

Further Reading

De Grouchy J and Turleau C (1984) *Clinical Atlas of Human Chromosomes*, 2nd edn. New York, NY: John Wiley.

Dellarco VL, Voyteck PE and Hollaender A (1985) *Aneuploidy: Etiology and Mechanisms*. New York, NY: Plenum Press.

Epstein CJ (1986) *The Consequences of Chromosome Imbalance – Principles, Mechanisms, and Models*. Cambridge, UK: Cambridge University Press.

Hook EB (1992) Chromosome abnormalities: prevalence, risks and recurrence. In: Brock DJH, Rodeck CH and Ferguson-Smith MA (eds.) *Prenatal Diagnosis and Screening*, pp. 351–392. Edinburg/London, UK: Churchill Livingstone.

Schinzel A (2001) *Catalogue of Unbalanced Chromosome Aberrations in Man*. Berlin: Walter de Gruyter.

Shapiro BL (1983) Down syndrome – a disruption of homeostasis. *American Journal of Medical Genetics* **14**: 241–269.

Warburton D (1989) The effect of maternal age on the frequency of trisomy: change in meiosis or *in utero* selection? In: Terry J. Hassold and Charles J. Epstein (eds.) *Molecular and Cytogenetics Studies of Non-Disjunction*, pp. 165–186. New York, NY: Alan R Liss.

tRNA

Emanuel Goldman, *University of Medicine and Dentistry of New Jersey, Newark, New Jersey, USA*

Transfer RNAs decode the genetic code and carry attached amino acids to the growing protein chain on the ribosome.

Intermediate article

Article contents
• General Features
• Transfer RNA Structure
• Transfer RNA Function
• Diversity of tRNA Genes

General Features

Transfer ribonucleic acid (tRNA) comprises the crucial class of adapter molecule that decodes the information in messenger RNA (mRNA) and delivers the requisite amino acids to growing protein chains on ribosomes. (*See* Protein Coding.)

tRNA constitutes about 10% of the total RNA in human cells. It sediments at 4 S, has an average relative molecular weight of 25 000 and in general a chain length of 75–95 nucleotides. Hundreds of individual tRNA molecules have been sequenced (Sprinzl *et al.*,

1998) and, with the completion of sequencing of the human genome, there is of course an immense database of genes encoding human tRNAs (International Human Genome Sequencing Consortium, 2001, pp. 893–895). (*See* Nonprotein-coding Genes.)

tRNAs are subdivided into families of isoacceptors. Each family is defined by the corresponding or 'cognate' amino acid that it carries. There are 20 families of isoacceptors in cells, one family for each cognate amino acid. Each individual tRNA species in a family carries the same amino acid. Individual tRNA species can be separated from one another by sophisticated chromatographic or gel electrophoresis techniques. Some clever innovations have taken advantage of functional characteristics of tRNA for use in purification. For example, the binding of aminoacyl-tRNA to elongation factor 1A (EF-1A; EF-Tu in bacteria) was used as a basis for purification (Derwenskus *et al.*, 1984), and in 2002 an assay was developed to measure the degree of aminoacylation of tRNA by amino acid attachment to radiolabeled tRNA (Wolfson and Uhlenbeck, 2002). (*See* Gel Electrophoresis.)

Members of a family of isoacceptors are designated by a superscript abbreviation for the amino acid; thus, members of the valine family are written as 'tRNA^{Val}'. Before comprehensive primary tRNA sequence data became available, individual isoacceptor species were identified by a numerical subscript; for example, the first species isolated in a chromatographic separation of the valine isoacceptors would be called tRNA$_1^{Val}$. Because of occasional confusion about the species that was being specified by this nomenclature, many workers have taken to using triplet sequences related to the decoding function of the individual isoacceptor species as the subscript.

Because the role of tRNAs is to carry amino acids, these RNAs can exist in two states, either with or without an attached amino acid. When the amino acid is attached, it is abbreviated as a prefix to the tRNA. Thus, when valine is attached to tRNA$_1^{Val}$, the whole molecule is designated as Val-tRNA$_1^{Val}$.

Each family of tRNA isoacceptors is recognized by a single cognate enzyme, called an aminoacyl-tRNA synthetase. For example, the valyl-tRNA synthetase recognizes all of the individual tRNA^{Val} isoacceptor species and attaches valine to them. Similarly, seryl-tRNA synthetase does the same for all tRNA^{Ser} species, and so on for all 20 amino acids.

Transfer RNA Structure

Inspection of the primary sequences of many tRNAs led to the identification of a generalized secondary structure, called a 'cloverleaf' model (e.g. see Rich and Kim, 1978). This structure, which is stabilized by hydrogen bonding in the various stems, consists of an acceptor stem, three stem–loops (or 'arms') and a variable loop. A numbering system of the tRNA bases runs from position 1 to position 76, with additional bases, when present, designated by letter suffixes (e.g. A, B, C) to the numbered base at the insertion point.

The 3′ end of the molecule is always terminated with the sequence CCA, and all cells contain enzymes that can add CCA ends posttranscriptionally to tRNAs. In the folded structure, the 5′ and 3′ parts of the molecule form a stem with seven base pairs (bp) followed by an overhanging unpaired nucleotide at position 73 and the CCA. This stem is called the 'acceptor stem' because the amino acid that is used in protein synthesis is attached to the ribose of the A residue at the 3′ terminus.

Unlike eubacteria, eukaryotes do not have the 3′ CCA encoded in tRNA genes. CCA-adding enzymes are therefore essential for life in eukaryotes. Because the CCA end of tRNA is not base-paired and 'sticks out' from the tRNA structure, it is subject to nuclease degradation; however, the CCA-adding enzyme can regenerate complete tRNA from damaged molecules.

On the left side of the cloverleaf, following a stem of 3–4 bp, is the 8–11-base 'D loop', which derives its name from the near universal-presence of one or more unusual dihydrouridine (D) residues in this portion of the tRNA (in dihydrouridine, one of the double bonds in the pyrimidine ring is hydrogenated). On the right side of the cloverleaf, after a stem of 5 bp, is the 7-base TψC loop (often called simply the 'T loop'), which derives its name from the near-universal presence of this sequence, containing the unusual nucleotides ribothymidine (T; in deoxyribonucleic acid (DNA) thymine is attached to deoxyribose) and pseudouridine (ψ; in pseudouridine uracil is turned upside-down in its connection to ribose). Just below the TψC arm is the 'variable loop' – the main source of variability in tRNA size from one tRNA species to another. This may contain as few as 4 to as many as 21 (or even more) bases.

At the bottom of the cloverleaf, following another stem of 5 bp, is the 7-base 'anticodon loop'. The anticodon is a three-base sequence that is complementary to the codon that it decodes. It pairs with the codon in an antiparallel fashion (as is the case in all double-stranded nucleic acids). This means that the 5′ anticodon base in the tRNA (nucleotide 34) pairs with the 3′ codon base in the mRNA, and the 3′ anticodon base (nucleotide 36) pairs with the 5′ codon base. There is some flexibility in base-pairing between the 3′ codon base in the mRNA and the 5′ anticodon base, known as 'wobble': G and U can pair with each other in these positions, and the rare base I (inosinic acid) in the first position of the anticodon (nucleotide

34) can pair with U, C or A in the codon. In some tRNAs, especially in mitochondria, wobble is even more extensive than this. Some mitochondrial tRNAs from several organisms show varying degrees of deviation from the standard cloverleaf structure, including in extreme cases the absence of either the TψC or the D arms altogether. (See Genetic Code: Evolution; Mitochondrial Genome: Evolution.)

The tertiary three-dimensional structure of tRNA (see Rich and Kim, 1978), solved by X-ray crystallography, is essentially an 'L-shaped' molecule that retains the secondary structure interactions of the cloverleaf model: the D arm of the cloverleaf is folded over on top of the TψC arm, with a bend near the topographical center of the molecule. This leaves the anticodon and the CCA acceptor at opposite ends of the molecule. The proper three-dimensional conformation of tRNA is essential for normal function; for example, there are temperature-sensitive mutant tRNAs that function in protein synthesis at low but not high temperature (ordinarily, one thinks of temperature-sensitive mutations only in proteins). This indicates that a conformational change in the tRNA molecule controls its biological activity.

tRNA contains numerous nucleotide modifications (such as methylation) and unusual bases, which are added posttranscriptionally at many positions in the molecule. Some of these are involved in the specificity of tRNA aminoacylation, some are involved in codon recognition and others seem to mediate the efficiency of translation and influence the three-dimensional structure of the molecule. Modification status of tRNAs has been correlated with malignancy in some cancers (Dirheimer et al., 1995). Modifications increase the surface exposure of the molecule. Without modifications, it is estimated that only 10% of the nucleotides are accessible to solvent, whereas modifications alter the solvent-accessible fraction to 20%. Thus, modifications allow twice as much access to proteins for interaction with tRNA. (See RNA-binding Proteins: Regulation of mRNA Splicing, Export and Decay.)

Transfer RNA Function

A specific initiator tRNA, which is aminoacylated with methionine, generally begins all rounds of protein synthesis in response to an AUG initiation codon. Aminoacylated tRNA molecules form complexes with elongation factor EF-1A and GTP. This is generally referred to as the 'ternary complex'. EF-1A–GTP escorts the aminoacylated tRNA to the 40S ribosomal subunit 'A site'. GTP is hydrolyzed and EF-1A–GDP is released as the aminoacylated tRNA is then locked into the 60S ribosomal subunit A site (this sequence of events is extrapolated from work with bacterial ribosomes). Peptidyl-tRNA, in the 'P site' of the ribosome, then transfers the peptide to the amino acid attached to the tRNA in the A site. The energy for this 'peptidyl transferase' reaction, which is thought to be catalyzed by the 28S ribosomal RNA of the 60S subunit (again, extrapolating from data obtained in bacteria), is derived from the high-energy bond between the amino acid and the tRNA. (See Translation Initiation: Molecular Mechanisms in Eukaryotes.)

The ribosome is thought to assume one of two conformations, pretranslocational or posttranslocational, corresponding to the status of tRNA binding. In the pretranslocational state, the A (aminoacyl) and P (peptidyl) sites are occupied, and the E (exit) site is vacant. In the posttranslocational state, the P and E sites are occupied and the vacant A site is available for an incoming aminoacyl-tRNA to start the next cycle. The P site tRNA always contains the growing peptide chain except for the transient period during peptide bond formation (where the peptide is transferred to the tRNA in the A site) and translocation. After translocation, the discharged tRNA that had been in the P site is found in the E site, and the tRNA carrying the growing peptide chain has moved from the A site to the P site. The E site tRNA is released on binding of the next aminoacyl-tRNA in the A site, switching the ribosome to the pretranslocational conformation (models for the translation elongation cycle are discussed in Burkhardt et al., 1998). (See Ribosomes and Ribosomal Proteins.)

Uncharged tRNA is also capable of entering the A site of the ribosome. There is evidence in bacteria that uncharged cognate tRNA in large excess can compete with ternary complex for entry into the small subunit portion of the ribosomal A site, thereby slowing the rate of elongation of protein synthesis (Rojiani et al., 1990). (See Noncoding RNAs: A Regulatory Role?)

Transfer RNA abundance is also thought to affect efficiency of translation (Kanaya et al., 2001). For example, proteins that are needed in large amounts in several model organisms are almost exclusively encoded by codons corresponding to abundant tRNAs, whereas tandem codons corresponding to rare tRNAs have been shown, in some examples, to be inhibitory to protein synthesis; this inhibition is reversed when a plasmid carrying a gene for the rare tRNA is introduced. This can become a practical issue when trying to express some human proteins in heterologous systems such as bacteria, which have a very different codon bias; frequently in such situations, a quick fix has been to resynthesize the gene using synonymous codons corresponding to the codons that are favored in the expression organism. tRNA abundance can also determine the extent of certain programmed translational frameshifts. (See Codon Usage; Expression Studies.)

Especially notable is the translation of the gene for silk fibroin protein, which comprises mostly codons for glycine, alanine and serine. To accommodate expression of this highly specialized message, the tRNA composition in differentiated worm silk glands consists of primarily cognate tRNAs to codons for these three amino acids (Chevallier and Garel, 1982).

Diversity of tRNA Genes

With the publication of the complete sequence of the human genome (International Human Genome Sequencing Consortium, 2001), there is now a wealth of data on the numbers of tRNA genes that are cognate to each codon. In the published draft of the human genome, there are 497 identified tRNA genes and 324 putative tRNA pseudogenes; this analysis allows for the possibility that a few more tRNA genes will be identified as the sequence is refined, but the overall number is not expected to change by very much. (*See* Genetic Redundancy; RNA Gene Prediction.)

Several of these genes represent gene duplications that encode identical tRNA isoacceptors. **Figure 1**, taken from the International Human Genome Sequencing Consortium (2001) article, delineates the number of tRNA genes in the human genome with anticodons corresponding to each codon in the universal genetic code. Apparently omitted from this figure is the specialized seryl-tRNA that is used to decode the UGA stop codon under specialized circumstances for incorporating selenocysteine (the '21st amino acid'). The absence of any other tRNAs to decode the stop codons settles the long-standing nagging issue of whether human cells have the capacity to suppress nonsense codons; we now know that if there is any such suppression it is not tRNA-based. The vast database is only beginning to be mined for information, and this is an exciting area for future bioinformatics research. (*See* Bioinformatics; Genetic Databases: Mining; Gene Duplication: Evolution; Gene Mapping and Positional Cloning.)

```
Phe  [ 171 UUU \ AAA 0        Ser [ 147 UCU 7 AGA 10      Tyr [ 124 UAU \ AUA 1      Cys [ 99 UGU \ ACA 0
     [ 203 UUC - GAA 14           [ 172 UCC / GGA 0            [ 158 UAC - GUA 11         [ 119 UGC - GCA 30
                                  [ 118 UCA - UGA 5       Stop -  0 UAA - UUA 0     Stop -  0 UGA - UCA 0
Leu  [ 73 UUA - UAA 8             [ 45 UCG - CGA 4        Stop -  0 UAG - CUA 0     Trp - 122 UGG - CCA 7
     [ 125 UUG - CAA 5

Leu  [ 127 CUU 7 AAG 13      Pro [ 175 CCU 7 AGG 11       His [ 104 CAU \ AUG 0     Arg [ 47 CGU 7 ACG 9
     [ 187 CUC / GAG 0           [ 197 CCC / GGG 0            [ 147 CAC - GUG 12        [ 107 CGC / GCG 0
     [ 69 CUA - UAG 2            [ 170 CCA - UGG 10      Gln  [ 121 CAA - UUG 11        [ 63 CGA - UCG 7
     [ 392 CUG - CAG 6           [ 69 CCG - CGG 4            [ 343 CAG - CUG 21         [ 115 CGG - CCG 5

Ile  [ 165 AUU 7 AAU 13      Thr [ 131 ACU 7 AGU 8        Asn [ 174 AAU \ AUU 1     His [ 121 AGU \ ACU 0
     [ 218 AUC / GAU 1           [ 192 ACC / GGU 0            [ 199 AAC - GUU 33        [ 191 AGC - GCU 7
     [ 71 AUA - UAU 5            [ 150 ACA - UGU 10      Lys  [ 248 AAA - UUU 16    Arg  [ 113 AGA - UCU 5
Met -  221 AUG - CAU 17          [ 63 ACG - CGU 4            [ 331 AAG - CUU 22         [ 110 AGG - CCU 4

Val  [ 111 GUU 7 AAC 20      Ala [ 185 GCU 7 AGC 25       Asp [ 230 GAU \ AUC 0     Gly [ 112 GGU \ ACC 0
     [ 146 GUC / GAC 0           [ 282 GCC / GGC 0            [ 262 GAC - GUC 10        [ 230 GGC - GCC 11
     [ 72 GUA - UAC 5            [ 150 GCA - UGC 10      Glu  [ 301 GAA - UUC 14        [ 168 GGA - UCC 5
     [ 288 GUG - CAC 19          [ 74 GCG - CGC 5            [ 404 GAG - CUC 8          [ 160 GGG - CCC 8
```

Figure 1 The human genetic code and associated tRNA genes. For each of the 64 codons, we show the corresponding amino acid, the observed frequency of the codon per 10 000 codons, the codon, the predicted wobble-pairing to a tRNA anticodon (diagonal lines), an unmodified tRNA sequence and the number of tRNA genes found with this anticodon. For example, phenylalanine is encoded by UUU or UUC: UUC is seen more frequently, 203 versus 171 occurrences per 10 000 total codons; both codons are expected to be decoded by a single tRNA anticodon type, GAA, using a G·U wobble pair; and there are 14 tRNA genes found with this anticodon. The modified anticodon sequence in the mature tRNA is not shown, even where posttranscriptional modifications can be predicted confidently (e.g. when an A is used to decode a U or C third position, the A is almost certainly an inosine in the mature tRNA). The number of distinct tRNA species (such as distinct sequence families) for each anticodon is also not shown; often there is more than one species for each anticodon. (Reproduced with permission from International Human Genome Sequencing Consortium (2001), pp. 894–895.)

See also
Codon Usage
Translation Initiation: Molecular Mechanisms in Eukaryotes

References

Burkhardt N, Junemann R, Spahn CM and Nierhaus KH (1998) Ribosomal tRNA binding sites: three-site models of translation. *Critical Reviews in Biochemistry and Molecular Biology* **33**: 95–149.

Chevallier A and Garel JP (1982) Differential synthesis rates of tRNA species in the silk gland of *Bombyx mori* are required to promote tRNA adaptation to silk messages. *European Journal of Biochemistry* **124**: 477–482.

Derwenskus KH, Fischer W and Sprinzl M (1984) Isolation of tRNA isoacceptors by affinity chromatography on immobilized bacterial elongation factor Tu. *Analytical Biochemistry* **136**: 161–167.

Dirheimer G, Baranowski W and Keith G (1995) Variations in tRNA modifications, particularly of their queuine content in higher eukaryotes. Its relation to malignancy grading. *Biochimie* **77**: 99–103.

International Human Genome Sequencing Consortium (2001) Initial sequencing and analysis of the human genome. *Nature* **409**: 860–921.

Kanaya S, Yamada Y, Kinouchi M, Kudo Y and Ikemura T (2001) Codon usage and tRNA genes in eukaryotes: correlation of codon usage diversity with translation efficiency and with CG-dinucleotide usage as assessed by multivariate analysis. *Journal of Molecular Evolution* **53**: 290–298.

Rich A and Kim SH (1978) The three-dimensional structure of transfer RNA. *Scientific American* **238**: 52–62.

Rojiani MV, Jakubowski H and Goldman E (1990) Relationship between protein synthesis and concentrations of charged and uncharged tRNA^Trp in *Escherichia coli*. *Proceedings of the National Academy of Sciences of the United States of America* **87**: 1511–1515.

Sprinzl M, Horn C, Brown M, Ioudovitch A and Steinberg S (1998) Compilation of tRNA sequences and sequences of tRNA genes. *Nucleic Acids Research* **26**: 148–153.

Wolfson AD and Uhlenbeck OC (2002) Modulation of tRNA^Ala identity by inorganic pyrophosphatase. *Proceedings of the National Academy of Sciences of the United States of America* **99**: 5965–5970.

Further Reading

Bjork GR, Ericson JU, Gustafsson CE, *et al.* (1987) Transfer RNA modification. *Annual Reviews in Biochemistry* **56**: 263–287.

Cold Spring Harbor Laboratory (2001) The ribosome. *Cold Spring Harbor Symposia on Quantitative Biology* **66**. [This volume contains several articles relevant to transfer RNA.]

Giegé R, Sissler M and Florentz C (1998) Universal rules and idiosyncratic features in tRNA identity. *Nucleic Acids Research* **26**: 5017–5035.

Goddard JP (1977) The structures and functions of transfer RNA. *Progress in Biophysics and Molecular Biology* **32**: 233–308.

Kisselev L (ed.) (2002) *FEBS Letters* **514**(1). [Special issue devoted to a conference on protein biosynthesis held in 2001; it contains several articles relevant to transfer RNA.]

Rich A and RajBhandary UL (1976) Transfer RNA: molecular structure, sequence and properties. *Annual Reviews in Biochemistry* **45**: 805–860.

Schimmel PR, Söll D and Abelson JN (eds.) (1979) *Transfer RNA: Structure, Properties and Recognition*. Cold Spring Harbor, NY: Cold Spring Harbor Laboratory Press.

Sharp SJ, Schaack J, Cooley L, Burke DJ and Söll D (1985) Structure and transcription of eukaryotic tRNA genes. *CRC Critical Reviews in Biochemistry* **19**: 107–144.

Söll D, Abelson JN and Schimmel PR (eds.) (1980) *Transfer RNA: Biological Aspects*. Cold Spring Harbor, NY: Cold Spring Harbor Laboratory Press.

Söll D and RajBhandary UL (eds.) (1995) *tRNA*. Washington, DC: American Society for Microbiology.

Trypsinogen Genes: Evolution

Intermediate article

Jian-Min Chen, *Etablissement Français du Sang, Brest, France*

Claude Férec, *Université de Bretagne Occidentale, Brest, France*

An evolutionary approach was used to study disease-associated genes, to elucidate gene conversion events occurring between the highly homologous trypsinogen genes and to understand the nature of natural selection acting on the gene products. Genomic, transcriptomic and proteomic data were brought together to understand the formation of pseudogenes and the creation of new genes at this locus.

Article contents

- Introduction
- Disease-associated Genes Provide Insights into Molecular Evolution
- Coupling Evolutionary Theory to Current Genomic, Transcriptomic and Proteomic Data
- Conclusions and Perspective

Introduction

Trypsinogen is the precursor of trypsin. In humans, it accounts for about 30% of protein secreted by the pancreas. Trypsinogen is converted into trypsin upon cleavage of the activation peptide by enteropeptidase (enterokinase), when it enters the duodenum. The newly formed trypsin has a central role in digestion, by activating all other pancreatic digestive zymogens as well as trypsinogen itself.

Historically, trypsinogen has been among the most extensively studied enzyme models of protein structure and function. Recently, the molecular evolution of trypsinogen genes has also been studied. A comprehensive analysis of vertebrate trypsinogen sequences suggested that trypsinogen did not evolve as a classical single-locus gene accumulating mutations at a constant rate, but rather that its evolution has been dynamic and multimodal (Roach *et al.*, 1997). A comparison of the clusters of five trypsin genes in *Drosophila melanogaster* and *Drosophila erecta* demonstrated that while two genes are evolving in concert, a third gene appears to be evolving independently, and the remaining two genes show an intermediate pattern of evolution (Wang *et al.*, 1999). In short, both studies suggested that in the trypsinogen gene family, the patterns of molecular evolution are complex.

In this article, we focus on the molecular evolution of the human trypsinogen gene family. We first demonstrate that mutations in cationic trypsinogen that are associated with pancreatitis can help us understand how the trypsinogen genes might have evolved. We then bring together data on deoxyribonucleic acid (DNA), ribonucleic acid (RNA) and proteins, to assess the formation of pseudogenes and the generation of new genes. To this end, **Figure 1** shows a diagram of the genomic organization of the human trypsinogen gene family, and **Figure 2** shows the molecular properties of the three human trypsinogen isoforms. Note that in fact, all the three human trypsinogen isoforms are anionic, but to avoid confusion, the least anionic isoform was termed 'cationic' and the most anionic isoform, 'anionic'; mesotrypsinogen is characterized by its intermediate electrophoretic mobility and isoelectric point.

Disease-associated Genes Provide Insights into Molecular Evolution

Evolutionary information has often been used to interpret the functional consequences of mutations. Conversely, understanding the effects of mutations that are associated with disease can improve our understanding of the evolutionary processes that have shaped those genes.

Traces of gene conversion events

Since the trypsinogen genes are tandemly repeated, gene conversion is likely to have played a part in their evolution, although 'molecular traces of such events, if present, have been largely obliterated by subsequent mutation' (Roach *et al.*, 1997). Luckily, clues to these molecular events have recently been gleaned by the identification of gene conversion-like mutations (**Figure 3**). Evidence that gene conversion may be involved includes (1) high overall sequence homology ($\sim$91%) between the group I trypsinogen genes; (2) the presence of possible 'donor' sequences among *T5–T9* for conversion of the gene *T4* and (3) the presence of recombination-promoting factors such as *Chi*-like sequence. These findings, together with the indication that other genetic mechanisms resulting in sequence convergence, such as unequal crossover, do not affect the *TRB@* (*TCRB*)/trypsinogen locus, strongly suggest that gene conversion has contributed to the concerted evolution of human trypsinogen genes (Chen and Ferec, 2000a). (*See* Concerted Evolution; Gene Conversion in Health and Disease.)

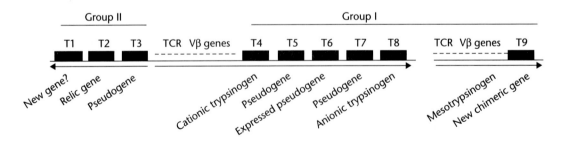

Figure 1 Schematic representation (not to scale) of the human trypsinogen gene family. The family was inadvertently sequenced, owing to its intercalation within the T-cell receptor (TCR) β locus (*TRB @*), which was among the earliest targets of the human genome project (Rowen *et al.*, 1996). The nine trypsinogen genes are separated physically into two groups. Group I (*T4–T8*) is located toward the 3′ end of the *TRB@* locus and consists of five tandem 10-kilobase (kb) repeats in 70 kb between *TRBV29-1* and *TRBD1*. A sixth member (*T9*) has been translocated from chromosome 7q34 to chromosome 9p11.2. Group II (*T1–T3*) is located toward the 5′ end of the *TRB@* locus. The assignment of *T4*, *T8* and *T9* to the three well-characterized isoforms of trypsinogen (cationic, anionic and mesotrypsinogens), and the annotation of *T2*, *T3*, *T5* and *T7* as relic genes or pseudogenes, follow Rowen *et al.* (1996). See the text in detail for the annotation of T1, T6 and T9. Interim gene symbols for T1–T3 and T5–T7 are *TRY1–TRY3* and *TRY5–TRY7*, respectively; official gene symbols for T4, T8 and T9 are *PRSS1* (PRSS denotes protease, serine), *PRSS2* and *PRSS3* respectively. Horizontal arrows indicate the transcriptional direction.

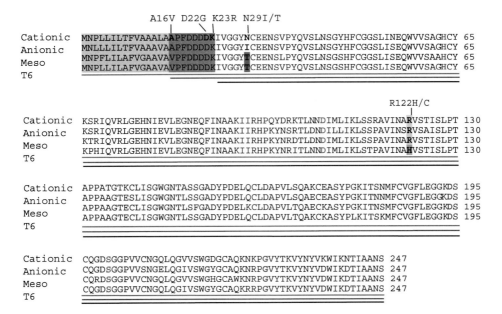

Figure 2 [*Figure is also reproduced in color section.*] Molecular properties of the human cationic, anionic and mesotrypsinogens and predicted amino acid sequence for the expressed pseudogene *T6*. The primary translation product is termed pretrypsinogen (underlined red). Removal of the signal peptide (highlighted in turquoise) results in trypsinogen (underlined blue). Removal of the activation peptide (highlighted in pink) by enteropeptidase or trypsin itself then converts trypsinogen into active trypsin (underlined black). The catalytic triad residues histidine (H), aspartic acid (D) and serine (S) are in red. The three residues determining trypsin specificity are in blue. Threonine (T) at residue 29 in mesotrypsinogen (highlighted in red) is conserved in orthologous sequences of frog, mouse, rat, dog, cow and pig. The primary autolysis site (arginine (R) at residue 122; highlighted in gray) is only present and stringently conserved in clearly functional mammalian trypsins; in *T6*, note the presence of H at this site (highlighted in green). The figure also shows mutations in cationic trypsinogen that are associated with pancreatitis. Aligned amino acid sequences are from GenBank accession numbers M22612, M27602 and D45417. G: glycine; K: lysine; N: asparagine; V: valine.

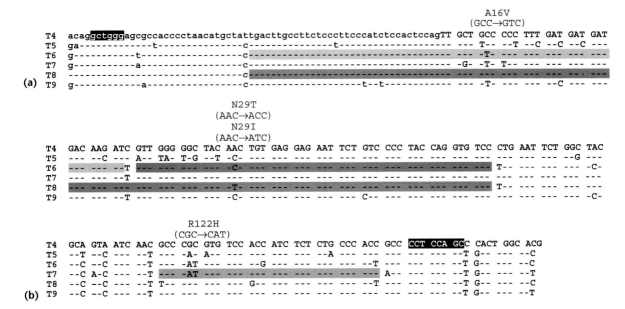

Figure 3 [*Figure is also reproduced in color section.*] Missense mutations in the human cationic trypsinogen gene (*T4*) that resemble gene conversion events and are associated with pancreatitis. (a) The alanine-to-valine (A16V) and asparagine-to-threonine/isoleucine (N29T/I) mutations in exon 2. In these cases where the nucleotide substitution only involves one base, the possibility of a simple mutational event cannot be completely ruled out. (b) The arginine-to-histidine (R122H) mutation in exon 3. The most likely 'donor' sequences for gene conversions are shown in shaded colors; the recombination promoting *Chi*-like sequence is shaded in black. Dashes indicate identity with the *T4* sequence. Exons are in uppercase and introns in lowercase. See **Figure 1** legend for gene annotations. GenBank accession numbers used for sequence alignment are NG_001333 and AF029308.

Details of natural selection acting on trypsinogen gene products

The basic elements of evolution are random variation and natural selection. Understanding evolution involves understanding how these processes affect gene products, and how those gene products affect the viability of the organism. The examination of mutations associated with diseases provides a reliable and efficient way to understand these evolutionary processes. (*See* Fitness and Selection.)

A positively selected residue in human cationic trypsinogen

Evolutionary change results from differences in the fitness and thus reproductive success of individuals with different genotypes. The downside of this process is easy to grasp: selection against disease constantly removes deleterious mutations from the gene pool. However, we know remarkably little about evolution's upside, that is, about the types of mutations that lead to increased fitness (Olson, 1999).

The identification of a hereditary pancreatitis-causing mutation – Asn29Ile (replacement of asparagine by isoleukine at amino acid residue 29; originally termed Asn21Ile in the chymotrypsin numbering system) – in the human cationic trypsinogen led us to propose that the sequence divergences from threonine (Thr) to Asn in cationic trypsinogen, and Thr to Ile in human anionic trypsinogen respectively (**Figure 2**),

> may represent rare mutations of the two functional
> human trypsinogen genes, positively selected by evolution
> to somehow endow an as-yet unknown advantageous
> effect on their respective protein's structure and function.
> (Chen and Ferec, 2000b)

This theory was given support by later developments. A back mutation from Asn to Thr at residue 29, Asn29Thr, in the human cationic trypsinogen was found to cause hereditary pancreatitis (Pfutzer *et al.*, 2002; **Figure 2**). In addition, *in vitro* analysis showed that the wild-type Asn29 human cationic trypsin degraded more rapidly than the mutant Thr29 molecule (Sahin-Toth, 2000). Thus, the Asn29 of the human cationic trypsinogen indeed confers a selective advantage, that is, it renders cationic trypsin that has been activated prematurely, within the pancreas, more susceptible to autolysis. Whether the evolutionary divergence from Thr to Ile in the human anionic trypsinogen has the same advantageous effect, although highly likely, remains to be functionally evaluated. Nevertheless, substitution of Asn29 in the cationic trypsinogen gene by an Ile, which is used in the corresponding position of the wild-type anionic trypsinogen gene (**Figure 2**) is disease-causing reveals

the subtlety of natural selection acting on the different functional genes.

A rare view of stepwise selective pressures acting on trypsinogen gene products

In addition to the above, two mutations – Asp22Gly (aspartic acid to glycine) and Lys23Arg (lysine to arginine) – occurring in the activation peptide sequence and two mutations – Arg122His (arginine to histidine; originally termed Arg117His) and Arg122Cys (arginine to cysteine) – occurring in the Arg122 primary autolysis site (**Figure 2**) have been identified and functionally analyzed. An evaluation of these in the alignments of more than 30 vertebrate trypsinogen sequences yields a rare view of how natural selection has progressively acted on trypsinogen gene products. Briefly, this can be seen in three stages. First, the activation peptide sequence is subject to strong selection pressure in higher vertebrates to keep trypsinogen autoactivation at the lowest level. Second, the Arg122 autolysis site as an important 'self-destruct' mechanism has further evolved in mammals. Finally, evolutionary divergence from Thr to Asn at residue 29 provides additional advantage in the human cationic trypsinogen. These findings unravel the complex, multilevel, 'built-in' mechanisms that the body has evolved to prevent intrapancreatic trypsin activation (Chen *et al.*, 2001).

Coupling Evolutionary Theory to Current Genomic, Transcriptomic and Proteomic Data

A primary driving force in genome evolution is duplication. Provided that one copy of a gene remains functional, the second copy is redundant and thus is free to accumulate mutations without deleterious effects on an organism's phenotype. This is important in the generation of new genes. Here we integrate different sources of information about the human trypsinogen gene family to address this important issue. (*See* Gene Duplication: Evolution.)

T1: a likely new gene generated from the highly diverged group II trypsinogen genes

The group I and group II trypsinogen genes tended not to exchange genetic information with each other, and acted largely as separate gene families after duplication (Roach *et al.*, 1997; **Figure 1**). At the nucleotide level, there is no significant homology between these two groups.

The three group II genes were annotated as relic or pseudogenes by Rowen *et al.* (1996). However, unlike the relic gene *T2* and the pseudogene *T3*, *T1* has an inframe coding sequence, as does its orthologous gene in the mouse (GenBank accession numbers U66059 and AE000663). Notably, although they have no significant homology at the nucleotide level, the amino acids are 66% identical and 85% similar, and align with no insertions and deletions. This evolutionary conservation suggests that the gene may have an important function in cells. There are transcripts corresponding to the human and mouse T1 genes in the expressed-sequence tags (ESTs) database. Nevertheless, these two genes are unlikely to still form a functional trypsinogen protein, owing to their highly divergent evolution. They presumably represent a new gene, possibly with a new function, that has evolved from an ancient duplicated copy of trypsinogen. (*See* Pseudogenes and their Evolution; Homologous, Orthologous, and Paralogous Genes; Expressed-sequence Tag (EST).)

T9: gradual loss of function as a trypsinogen, and possible gain of a new function as a chimeric protein

There is no doubt that the *T9* gene encodes mesotrypsinogen. However, mesotrypsinogen is quite different from cationic and anionic trypsinogens, at both the DNA and protein levels. As illustrated in **Figure 1**, the mesotrypsinogen gene (*T9*) has been translocated from 7q34 to 9p11.2. This may liberate it from shared functional or regulatory constraints exerted at the original locus. It has also presumably escaped the gene conversion events that result in sequence converge between the highly homologous group I trypsinogen genes. There is evidence that gene conversion occurs less frequently between homologous genes in different chromosomes, than between homologous genes in the same chromosome. Consistent with this, none of the most likely donor sequences for gene conversion in cationic trypsinogen is from *T9* (**Figure 3**).

More importantly, at the protein level, cationic trypsinogen and anionic trypsinogen account for 18.1% and 9.3% of the secretory proteins in the human pancreatic juice, respectively, but mesotrypsinogen is nearly undetectable. Furthermore, mesotrypsin is almost totally resistant to biological trypsin inhibitors, such as the pancreatic secretory trypsin inhibitor. Thus, in evolutionary terms, mesotrypsinogen may be gradually losing its role as a digestive enzyme.

Interestingly, an alternatively spliced transcript of the mesotrypsinogen gene, known as trypsinogen IV, is expressed specifically in the brain. Trypsinogen IV shares exons 2–5 with the pancreatic mesotrypsinogen messenger RNA (mRNA), and they are in the same reading frame, but its exon 1 is about 45 kilobases (kb) further upstream. This alternative exon 1 does not encode a typical signal peptide sequence, suggesting that the protein is not secreted. In a transgenic mouse model, trypsinogen IV is expressed in the cytoplasm of neurons, and there is no direct evidence that the protein has trypsin activity (Minn *et al.*, 1998). Thus, it is clearly inappropriate to label trypsinogen IV as a 'human brain-specific trypsinogen', although this term is unanimously accepted both in peer-reviewed publications and in public sequence databases. Instead, it appears to be a new chimeric gene product, derived from the mesotrypsinogen gene, but perhaps with a different biological function. There is a precedent for such divergence: a blood protein that keeps certain Antarctic fish from freezing arose from a pancreatic trypsinogen-like protease. (*See* Signal Peptides; Alternative Splicing: Evolution; Exons: Shuffling.)

T6: an expressed pseudogene

The genomic sequence of *T6* predicted it to be a functional gene (Rowen *et al.*, 1996). Its mRNA has been found in certain tissues, but in minute amounts, and its protein has never been detected. If translated, the protein's surface charge and shape would differ significantly from those of the cationic and anionic trypsinogens. Notably, the presence of histidine instead of arginine at residue 122 indicates relaxation of the selection pressure acting on this evolutionarily conserved autolysis site (remember that the Arg122His mutation usually causes disease; **Figure 2**). In addition, the *T6* gene is often deleted, since it is located in a common insertion/deletion polymorphism. Taking all these observations together, T6 may represent a transition state between a functioning duplicated gene and a nonfunctional pseudogene, that is, an expressed pseudogene (Chen *et al.*, 2001).

Conclusions and Perspective

The integration of genomic, transcriptomic, proteomic and disease-association data provides a broad view of the evolution of the trypsinogen gene family and, at the same time, yields a more reliable annotation of this locus. In particular, the awareness that two genes may have developed new functions is significant, although one has to keep in mind that their proteins have yet to be fully characterized. The ever-increasing availability of genome sequences from other species, and the enormous efforts of large-scale transcript identification and protein analysis, may eventually provide further information on these issues.

References

Chen JM and Ferec C (2000a) Gene conversion-like missense mutations in the human cationic trypsinogen gene and insights into the molecular evolution of the human trypsinogen family. *Molecular Genetics and Metabolism* **71**: 463–469.

Chen JM and Ferec C (2000b) Origin and implication of the hereditary pancreatitis-associated N21I mutation in the cationic trypsinogen gene. *Human Genetics* **106**: 125–126.

Chen JM, Montier T and Ferec C (2001) Molecular pathology and evolutionary and physiological implications of pancreatitis-associated cationic trypsinogen mutations. *Human Genetics* **109**: 245–252.

Minn A, Schubert M, Neiss WF and Muller-Hill B (1998) Enhanced GFAP expression in astrocytes of transgenic mice expressing the human brain-specific trypsinogen IV. *Glia* **22**: 338–347.

Olson MV (1999) When less is more: gene loss as an engine of evolutionary change. *American Journal of Human Genetics* **64**: 18–23.

Pfutzer R, Myers E, Applebaum-Shapiro S, *et al.* (2002) Novel cationic trypsinogen (PRSS1) N29T and R122C mutations cause autosomal dominant hereditary pancreatitis. *Gut* **50**: 271–272.

Roach JC, Wang K, Gan L and Hood L (1997) The molecular evolution of the vertebrate trypsinogens. *Journal of Molecular Evolution* **45**: 640–652.

Rowen L, Koop BF and Hood L (1996) The complete 685-kilobase DNA sequence of the human β T cell receptor locus. *Science* **272**: 1755–1762.

Sahin-Toth M (2000) Human cationic trypsinogen. Role of Asn-21 in zymogen activation and implications in hereditary pancreatitis. *Journal of Biological Chemistry* **275**: 22 750–22 755.

Wang S, Magoulas C and Hickey D (1999) Concerted evolution within a trypsin gene cluster in *Drosophila*. *Molecular Biology and Evolution* **16**: 1117–1124.

Further Reading

Bork P and Copley R (2001) The draft sequences. Filling in the gaps. *Nature* **409**: 818–820.

Cheng CH and Chen L (1999) Evolution of an antifreeze glycoprotein. *Nature* **401**: 443–444.

Chen JM and Ferec C (2000) Wanted: a consensus nomenclature for cationic trypsinogen mutations. *Gastroenterology* **119**: 277–278.

Chen JM and Ferec C (2003) Human trypsins. In: Barrett AJ, Rawlings ND and Woessner JF (eds.) *Handbook of Proteolytic Enzymes*. London, UK: Academic.

Glusman G, Rowen L, Lee I, *et al.* (2001) Comparative genomics of the human and mouse T cell receptor loci. *Immunity* **15**: 337–349.

Li WH, Gu Z, Wang H and Nekrutenko A (2001) Evolutionary analyses of the human genome. *Nature* **409**: 847–849.

Roach JC (2001) A clade of trypsins found in cold-adapted fish. *Proteins* **47**: 31–44.

Strachan T and Read AP (1999) *Human Molecular Genetics*. New York, NY: John Wiley & Sons.

Whitcomb DC, Gorry MC, Preston RA, *et al.* (1996) Hereditary pancreatitis is caused by a mutation in the cationic trypsinogen gene. *Nature Genetics* **14**: 141–145.

Wiegand U, Corbach S, Minn A, Kang J and Muller-Hill B (1993) Cloning of the cDNA encoding human brain trypsinogen and characterization of its product. *Gene* **136**: 167–175.

Tuberous Sclerosis: Genetics

Intermediate article

Jeremy P Cheadle, *University of Wales College of Medicine, Heath Park, Cardiff, UK*

Julian R Sampson, *University of Wales College of Medicine, Heath Park, Cardiff, UK*

Tuberous sclerosis is an autosomal dominant disorder characterized by the development of hamartomatous growths in a wide variety of tissues and organs. It is caused by mutations in either the *TSC1* gene on 9q34 or the *TSC2* gene on 16p13.3. Hamartin and tuberin, the gene products, interact and play a role in the control of cell size.

Article contents

- Introduction
- Cloning and Characterization of *TSC1*
- Cloning and Characterization of *TSC2*
- Mutation Spectrum and Distribution
- Genotype–Phenotype Correlations
- Animal Models of TSC
- Functions of the TSC Genes

Introduction

Tuberous sclerosis complex (TSC) is an autosomal dominant disorder characterized by the development of hamartomatous growths in a variety of tissues and organs (Gomez *et al.*, 1999). Involvement of the brain is associated with intellectual handicap, epilepsy and abnormal behavioral phenotypes. Other organs commonly involved include the skin, kidneys and heart, where the associated features include facial angiofibromas, ungual fibromas, forehead plaques, shagreen patches, renal angiomyolipomas and cysts, and cardiac rhabdomyomas. About 60% of TSC cases are sporadic, representing new mutations, and TSC occurs in at least one in 10 000 live births, without apparent ethnic clustering (Osborne *et al.*, 1991).

Linkage of TSC to 9q34 (the tuberous sclerosis 1 (*TSC1*) locus) was first reported in 1987 (Fryer *et al.*, 1987); however, subsequent analyses suggested locus heterogeneity, and in 1992, linkage to 16p13.3 (the tuberous sclerosis 2 (*TSC2*) locus) was identified (Kandt *et al.*, 1992). Approximately half of multiplex families were linked to 9q34 and half to 16p13 (Povey *et al.*, 1994). (*See* Genetic Linkage Mapping.)

Cloning and Characterization of *TSC1*

A 1.5 megabase (Mb) candidate region on 9q34 was defined by key meiotic recombination events in large *TSC1* families (Haines *et al.*, 1991; Nellist *et al.*, 1993). Microsatellite markers from this region were then identified and a cosmid, P1-derived artificial chromosome (PAC) and bacterial artificial chromosome (BAC) contig assembled (van Slegtenhorst *et al.*, 1995; Hornigold *et al.*, 1997). Two recombinants identified in unaffected individuals narrowed the candidate region to 900 kilobases (kb) between the markers D9S2127 and DBH. Over 30 genes were identified in this region by a variety of techniques, many of which were assessed as candidates for *TSC1* without success (van Slegtenhorst *et al.*, 1997). (*See* Gene Mapping and Positional Cloning.)

Complete genomic sequencing of the region was initiated and computer software was employed to predict further putative exons and genes. Mutations in predicted exons were then sought in patients with TSC by heteroduplex and sequence analysis. Mobility shifts corresponding to small truncating mutations were identified in an exon that corresponded to a previously identified complementary deoxyribonucleic acid (cDNA) clone, and a combination of 5′ RACE (rapid amplification of cDNA ends), reverse transcriptase polymerase chain reaction (RT-PCR) and isolation of other cDNA clones defined the remainder of the open reading frame (van Slegtenhorst *et al.*, 1997). Comparison of cDNA and genomic sequences revealed 23 exons, the first two of which were untranslated. The 8.6 kb transcript had a 4.5 kb 3′ untranslated region (UTR) and was predicted to encode a novel 1164 amino acid/130 kDa protein that was called hamartin. Hamartin is predicted to be generally hydrophilic with a single potential transmembrane domain at amino acids 127–144 and with regions of coiled coils between residues 730 and 996. These regions are conserved in orthologs of *TSC1* from mouse, rat and *Drosophila*.

Cloning and Characterization of *TSC2*

Linkage studies identified an approximately 1.5 Mb candidate region of 16 p as likely to contain the *TSC2* gene (Kandt *et al.*, 1992; Kwiatkowski *et al.*, 1993). A child with TSC and autosomal dominant polycystic kidney disease was identified who had inherited an unbalanced karyotype and was deleted for 16p13.3→16pter. The translocation breakpoint was shown to disrupt the previously unidentified polycystic kidney disease 1 (*PKD1*) gene (European Polycystic Kidney Disease Consortium, 1994) and it was reasoned that

the child manifested TSC because of loss of one copy of the more telomerically located *TSC2* gene. The breakpoint was mapped by fluorescence *in situ* hybridization and pulsed field and conventional gel electrophoresis to 150 kb telomeric to 16AC2.5, the most centromeric flanking marker then identified for *TSC2*. The telomeric limit of the candidate region was reduced by a second breakpoint in a previously reported patient who had a *de novo* truncation of 16p but no clinical evidence of TSC; this deletion excluded approximately 1.1 Mb of the remaining 1.4 Mb candidate region.

A cosmid contig of the remaining 300 kb candidate region was constructed and probes generated from it were used to assay a panel of 255 unrelated patients with TSC for rearrangements by pulsed field gel electrophoresis and Southern blotting. Five patients had genomic deletions of 30–100 kb, which involved the same 120 kb interval. cDNA clones were isolated corresponding to four genes in the interval, and one was found to be disrupted by all five deletions. Four smaller intragenic deletions were subsequently identified in patients with TSC, including a *de novo* deletion that was associated with a shortened transcript. These findings confirmed the identity of the *TSC2* gene. A novel protein of approximately 198 kDa was predicted from the sequence of the 5.5 kb transcript and named tuberin (European Chromosome 16 Tuberous Sclerosis Consortium, 1993).

TSC2 comprises 41 coding exons and a noncoding leader exon. The sequence predicts a protein of 1807 amino acids (Maheshwar *et al.*, 1996). Centromeric to *TSC2* is the *PKD1* gene; the genes are oriented 3′ to 3′ and their polyadenylation signals are separated by only 60 base pairs (bp). Distal to *TSC2* and oriented 5′ to 5′ is the *n*th endonuclease III-like 1 (*E. coli*) (*NTHL1*) gene, a component of the base excision repair pathway.

Tuberin shows homology with rap1GAP or GAP3, extending over approximately 160 amino acid residues encoded by exons 34–38 (Maheshwar *et al.*, 1997). Other potentially important domains include a leukine zipper consensus sequence, four potential tyrosine kinase phosphorylation sites, several regions rich in hydrophobic residues and two small potential coiled-coil domains. *TSC2* orthologs have been characterized in mouse, rat, the Japanese pufferfish, *Fugu rubripes*, and *Drosophila*. Comparative analysis of the *TSC2* gene in humans and *F. rubripes* revealed an overall amino acid identity of 60%, but four regions of the gene, including the guanosine-triphosphatase activating protein (GAP) related domain, were more highly conserved and may be functionally significant (Maheshwar *et al.*, 1996). Alternative splicing involving exon 25, the first 3 bp of exon 26 and exon 31 has been documented in humans, mouse, rat and *F. rubripes*.

Mutation Spectrum and Distribution

Three hundred and thirty-seven different constitutional mutations (105 in *TSC1* and 232 in *TSC2*) in 446 unrelated cases have been reviewed (Cheadle *et al.*, 2000). Although truncating mutations make up 98% of the reported mutations in *TSC1* and 77% of those in *TSC2*, and single base substitutions account for approximately half of all *TSC1* and *TSC2* mutations, the mutation spectra in these two genes differ significantly. In *TSC1*, the majority of pathogenic single base substitutions are nonsense changes, with missense changes being rare or nonexistent. In *TSC2*, both nonsense and missense mutations are equally prevalent and contribute substantially to this class of mutation. The *TSC1* mutation spectrum also includes many small (less than 30 bp) insertions and deletions, but apparently very few large deletions. In contrast, *TSC2* contains both small insertions and deletions, and in approximately 18% of cases, large deletions and rearrangements involving one or more exons.

The heterogeneous mutation spectra of *TSC1* and *TSC2* present a clear challenge in the design of a simple strategy for the identification of mutations. Denaturing high-performance liquid chromatography (dHPLC) analysis has been found to be more sensitive than single-strand conformation polymorphism and heteroduplex analysis for mutation detection in TSC, and detects mutations in about 68% of cases (Jones *et al.*, 2000). (*See* Mutation Detection.)

Genotype–Phenotype Correlations

TSC1 mutations have been identified in about 10–15% of patients with sporadic TSC, while *TSC2* mutations account for about 70%, with the remainder as yet unclassified (Jones *et al.*, 1999). In contrast, the two genes appear to be equally frequently mutated in large TSC families suitable for linkage analysis. These observations suggest that patients with *TSC1* mutations may be less severely affected than those with *TSC2* mutations, so that each new sporadic *TSC1* patient has a better chance of founding a family than a new *TSC2* patient. (*See* Genotype–Phenotype Relationships.)

Intellectual disability

Jones *et al.* (1999) assessed the frequency of intellectual disability in sporadic cases as a reliable and important aspect of disease severity. Unlike many other components of the TSC phenotype which show age-dependent penetrance, intellectual disability is almost invariably present from early childhood if it develops at all, and rarely escapes detection. Moderate to severe intellectual disability also clearly limits reproductive

capacity. Intellectual disability was significantly more frequent among sporadic cases with *TSC2* than *TSC1* mutations.

Clinical severity

A comprehensive survey of 224 TSC patients found that a range of clinical features occurred with greater severity in *TSC2* than in *TSC1* sporadic cases (Dabora *et al.*, 2001). These included seizures, mental retardation, number of subependymal nodules, tuber count, kidney angiomyolipomas, grade of facial angiofibromas, forehead plaques and retinal hamartomas.

Polycystic kidney disease in TSC

Renal cysts are common in TSC patients, but only a small minority develop severe renal cystic disease similar to autosomal dominant polycystic kidney disease (ADPKD). In these cases, the renal cystic disease is a serious complication accounting for significant morbidity and mortality. Analysis of TSC patients with severe polycystic kidney disease has shown that contiguous deletions of *TSC2* and *PKD1* are present in the majority while approximately 10% have other mutations at the *TSC2*/*PKD1* locus (Sampson *et al.*, 1997). (*See* Autosomal Dominant Polycystic Kidney Disease.)

Lymphangioleiomyomatosis

Lymphangioleiomyomatosis (LAM) is seen almost exclusively in females and is characterized by bronchiolar smooth muscle infiltration and cystic changes in the lung parenchyma. LAM patients often have angiomyolipoma of the kidneys and/or abdominal and hilar lymph nodes. Symptomatic LAM is estimated to occur in approximately one per million of the population without other evidence of TSC but in up to 5% of females with TSC, implicating the TSC genes in the etiology of LAM. Somatic biallelic inactivating mutations of *TSC2* have been demonstrated in affected tissues from five sporadic LAM patients (Carsillo *et al.*, 2000).

Mosaicism

Mosaicism has been sought in several collections of patients who have been screened for mutations in the TSC genes. The highest level of mosaicism was reported in a series of patients with the *TSC2* and *PKD1* contiguous gene deletion syndrome (Sampson *et al.*, 1997). In that series of 27 unrelated families, mosaicism was found in the first affected member in seven families (26%). This may reflect the dramatic renal involvement in these patients making it easier to detect disease in mosaic cases. Verhoef *et al.* (1999)

presented evidence for mosaicism in founders of six TSC families in a series of 62 unrelated families with a mutation in either *TSC1* or *TSC2*. In general, patients with mosaicism have less severe disease than those with nonmosaic mutations.

At least 10 sets of siblings affected by TSC but with apparently normal parents have been reported (Rose *et al.*, 1999). In several instances, identical *TSC2* mutations have been found in DNA from the affected siblings but not in blood-derived DNA from their parents, implicating gonadal mosaicism in one of the parents. (*See* Mosaicism.)

Animal Models of TSC

Drosophila models

Mutations in the *Drosophila* mutants *tsc1* and 'gigas' (*tsc2*) have been found in the orthologs of *TSC1* and *TSC2* respectively (Ito and Rubin, 1999; Potter *et al.*, 2001; Tapon *et al.*, 2001). Homozygous mutant *tsc1* and 'gigas' *Drosophila* die during larval development. Using a tissue-specific recombination system, it has been shown that homozygous mutant cells develop to an abnormally large size in the eye and wing. In the eye, all null cells are enlarged by about two to three times in area, but the ploidy, structure and organization are normal. (*See Drosophila* as a Model for Human Diseases.)

Mouse models

Tsc1 (*Tsc1+/−*) and *Tsc2* (*Tsc2+/−*) knockout mice have been engineered (Onda *et al.*, 1999; Kobayashi *et al.*, 1999, 2001; Kwiatkowski *et al.*, 2002). Consistent with the phenotype severity difference seen in humans with mutations in *TSC1* and *TSC2*, *Tsc1+/−* mice develop renal cystadenomas by 15–18 months whereas *Tsc2+/−* mice develop such lesions by 6 months. Liver hemangiomas were more common, more severe and caused higher mortality in female than in male *Tsc1+/−* mice (Kwiatkowski *et al.*, 2002). Other tumors seen infrequently in the knockout mice include uterine leiomyomas, lung adenomas and tail, paw and oral hemangiosarcomas. Brain lesions similar to those found in patients with TSC have not yet been described. Loss of heterozygosity (LOH) or somatic intragenic mutations have been reported in renal and extrarenal tumors consistent with inactivation of the wild-type allele. (*See* Mouse as a Model for Human Diseases; Transgenic Mice.)

Tsc1−/− and *Tsc2−/−* embryos died at midgestation (embryonic age 10.5–12.5 days) and were less developed than their heterozygous and wild-type littermates. Null embryos were paler, edematous and had pericardial effusions, and a proportion had open neural tubes. Upon histological analysis, liver hypoplasia was most striking in null embryos and appeared to be the primary cause of fetal demise, with secondary growth retardation and circulatory failure from anemia.

The Eker rat

The Eker rat is a naturally occurring autosomal dominant hereditary model of predisposition to renal adenoma and carcinoma. Kidney lesions vary in morphology and include pure cysts, cysts with papillary projections and solid adenomas (Eker *et al.*, 1981) which can be seen as early as 4 months; a small minority of these tumors become malignant. Eker rats also develop pituitary adenomas, uterine leiomyomas and leiomyosarcomas, splenic hemangiomas (Everitt *et al.*, 1992), and at low frequency a variety of brain lesions including lesions resembling human TSC subependymal nodules and cortical tubers (Mizuguchi *et al.*, 2000).

Molecular analyses showed that Eker rats carry an insertion of a 6.3 kb intracisternal A particle in one allele of the *Tsc2* gene (*Tsc2+/Ek*) (Yeung *et al.*, 1994; Kobayashi *et al.*, 1995). LOH studies in Eker tumors have provided strong evidence for the two-hit model of tumorigenesis and revealed organ-specific variation in the extent of LOH (Yeung *et al.*, 1995). Renal tumors from Eker rats treated with *N*-ethyl-*N*-nitrosourea (ENU) show a high frequency of somatic point mutations and such treatment greatly accelerates the formation of renal tumors.

Tsc2Ek/Ek embryos die at an embryonic age of 10–12 days of uncertain causes. Null embryos in the Long–Evans strain display an abnormal head morphology with exencephaly in some cases. However, brain and skull development of null embryos in the Fisher 344 strain is normal. Since null embryos in each strain die at the same age, other anatomical or physiological processes are likely to be the cause of death.

Functions of the TSC Genes

Tumor suppressor properties

Knudson proposed that inherited predisposition to some tumors might reflect the germ-line mutation of tumor suppressor genes with tumor development resulting from somatic 'second-hit' mutation of the corresponding wild-type allele. The detection of somatic *TSC1* or *TSC2* mutations in a variety of TSC hamartomas supports classification of the TSC genes as tumor suppressor genes (Au *et al.*, 1999).

Expression and localization

Northern blot analysis of *TSC1* and *TSC2* expression revealed major transcripts of 8.6 and 5.5 kb respectively that are present in a wide variety of human tissues. In cultured cells, hamartin has been localized to the cytoplasm, with a punctate pattern of immunofluorescence (Plank *et al.*, 1998), while tuberin has shown perinuclear staining (Wienecke *et al.*, 1996). An association of tuberin with the *cis*-medial Golgi was suggested by colocalization with mannosidase-II. Hamartin has been shown to augment the expression of tuberin by inhibiting its ubiquitination (Benvenuto *et al.*, 2000). Both hamartin and tuberin pellet from cytoplasmic extracts and the failure of either protein to be solubilized from the pellet fraction when treated with anionic detergents suggests that the proteins are not associated with cell membranes (Plank *et al.*, 1998; Nellist *et al.*, 1999).

Cellular roles of hamartin and tuberin

Hamartin and tuberin interact

An interaction between hamartin and tuberin has been demonstrated in the yeast two-hybrid (Y2H) system and by coimmunoprecipitation (van Slegtenhorst *et al.*, 1998; Plank *et al.*, 1998). Tuberin phosphorylation appears to regulate its interaction with hamartin (Aicher *et al.*, 2001) and TSC-associated mutations in the tuberin–hamartin binding domains abolish the interaction (Hodges *et al.*, 2001; Nellist *et al.*, 2001). The association between tuberin and hamartin appears to be stoichiometric, and fractionation and gel filtration studies indicate that they are found in a cytosolic complex of approximately 450 kDa (Nellist *et al.*, 1999).

Control of cell size

The finding that *Drosophila tsc1* and *tsc2* null cells develop to an abnormally large size suggests that hamartin and tuberin may play a role in the control of cell size (Potter *et al.*, 2001; Tapon *et al.*, 2001). Goncharova *et al.* (2002) showed that the ribosomal protein S6, which exerts translational control of protein synthesis and is required for cell growth, is hyperphosphorylated in *TSC2*−/− smooth muscle and epithelial cells. These cells also displayed constitutive activation of p70 S6 kinase (p70S6K) and increased basal DNA synthesis. Rapamycin, an immunosuppressant, inhibited hyperphosphorylation of S6, p70S6K activation and DNA synthesis in mutant cells. Analogous results were obtained in hamartin-deficient murine fibroblast cells (Kwiatkowski *et al.*, 2002). These data suggest that tuberin and hamartin negatively regulate the activity of S6 and p70S6K, and suggest a potential mechanism for abnormal cell growth.

Role in cell cycle control

Tuberin-deficient Eker rat-derived fibroblasts have been shown to transit faster through the cell cycle (Soucek *et al.*, 1998b), and overexpression of either tuberin or hamartin inhibits cell proliferation in many cell types (Miloloza *et al.*, 2000). However, the apparently cytoplasmic localization of the proteins would argue against a direct role in the cell cycle events in the nucleus.

Hamartin and cell adhesion

Using the Y2H system, Lamb *et al.* (2000) provided evidence that hamartin binds to ezrin and participates in cell adhesion events and rho signaling. Inactivation of hamartin led to rapid endothelial cell retraction, accompanied by loss of focal adhesions, cell rounding and progressive detachment from the substrate. Overexpression of both full-length and the *N*-terminal domain of hamartin (not containing the ezrin binding domain) induced actin stress fiber formation through activation of the rho guanosine triphosphatase (GTPase).

Guanosine triphosphatase activating protein activity of tuberin

Tuberin has homology with rap1GAP, a GTPase activating protein (GAP) that stimulates the hydrolysis of active GTP-bound rap1a and rap1b to their inactive guanosine diphosphate (GDP) bound forms. rap1a and b are members of the ras superfamily of small GTP binding proteins whose functions include transduction of mitogenic signals from plasma membrane receptors to the nucleus. By catalyzing the conversion of GTP binding proteins to their inactive state, GAPs can function as negative regulators of cellular processes, including proliferation. Modest GAP activity toward rap1a, but not rap2, ras, rho or rac, has been reported for native tuberin immunoprecipitated from K-562 cells (Wienecke *et al.*, 1995). Glutathione *S*-transferase (GST) fusion proteins incorporating the GAP containing the *C*-terminal part of tuberin and expressed in *Escherichia coli* and in Sf9 insect cells were reported to show similar activity.

Immunoprecipitates of native tuberin and the recombinant protein have also been reported to have modest GAP activity for the GTPase Rab5, which serves a role in regulating endosome fusion (Xiao *et al.*, 1997). Increased fluid-phase endocytosis has been reported in a single tuberin-deficient cell line and reexpression of tuberin appeared to normalize this. The relevance of these observations to the etiology of TSC remains speculative.

Other putative functions

Transcriptional activation by tuberin

Tsuchiya *et al.* (1996) found that a construct expressing the *C*-terminus of tuberin autoactivated expression of the LacZ reporter in the Y2H system. Tuberin has also been demonstrated to bind to members of the steroid receptor superfamily of genes and selectively modulate transcription; however, the importance of these observations to the biological function of tuberin remains uncertain.

Neuronal differentiation

Soucek *et al.* (1998a) demonstrated that tuberin expression was upregulated upon induction of neuronal differentiation in two neuroblastoma cell lines and that this upregulation occurred at the posttranscriptional level. Antisense inhibition of tuberin expression was shown to inhibit neuronal differentiation. Further studies on the role of tuberin in neuronal differentiation are therefore warranted.

Tuberin determines polycystin-1 localization

Intracellular trafficking of polycystin-1, the product of the *PKD1* gene, may be disrupted in tuberin-deficient cells (Kleymenova *et al.*, 2001). Reexpression of tuberin was found to restore localization of polycystin-1 to the plasma membrane. A direct role for tuberin in relation to polycystin may contribute to the severity of the polycystic kidney phenotype in patients with deletion of both *TSC2* and *PKD1*.

References

Aicher LD, Campbell JS and Yeung RS (2001) Tuberin phosphorylation regulates its interaction with hamartin: two proteins involved in tuberous sclerosis. *Journal of Biological Chemistry* **276**: 21017–21021.

Au KS, Hebert AA, Roach ES and Northrup H (1999) Complete inactivation of the *TSC2* gene leads to formation of hamartomas. *American Journal of Human Genetics* **65**: 1790–1795.

Benvenuto G, Li S, Brown SJ, *et al.* (2000) The tuberous sclerosis-1 (*TSC1*) gene product hamartin suppresses cell growth and augments the expression of the *TSC2* product tuberin by inhibiting its ubiquitination. *Oncogene* **19**: 6306–6316.

Carsillo T, Astrinidis A and Henske EP (2000) Mutations in the tuberous sclerosis complex gene *TSC2* are the cause of sporadic pulmonary lymphangioleiomyomatosis. *Proceedings of the National Academy of Sciences of the United States of America* **97**: 6085–6090.

Cheadle JP, Reeve MP, Sampson JR and Kwiatkowski DJ (2000) Molecular genetic advances in tuberous sclerosis. *Human Genetics* **107**: 97–114.

Dabora SL, Jozwiak S, Franz DN, *et al.* (2001) Mutational analysis in a cohort of 224 tuberous sclerosis patients indicates increased severity of *TSC2* compared with *TSC1* disease in multiple organs. *American Journal of Human Genetics* **68**: 64–80.

Eker R, Mossige J, Johannessen JV and Aars H (1981) Hereditary renal adenomas and adenocarcinomas in rats. *Diagnostic Histopathology* **4**: 99–110.

European Chromosome 16 Tuberous Sclerosis Consortium (1993) Identification and characterisation of the tuberous sclerosis gene on chromosome 16. *Cell* **75**: 1305–1315.

European Polycystic Kidney Disease Consortium (1994) The polycystic kidney disease 1 gene encodes a 14 kb transcript and lies within a duplicated region on chromosome 16. *Cell* **77**: 881–894.

Everitt JI, Goldsworthy TL, Wolf DC and Walker CL (1992) Hereditary renal cell carcinoma in the Eker rat: a rodent familial cancer syndrome. *Journal of Urology* **148**: 1932–1936.

Fryer AE, Chalmers A, Connor JM, *et al.* (1987) Evidence that the gene for tuberous sclerosis is on chromosome 9. *Lancet* **i**: 659–661.

Gomez MR, Sampson JR and Holets-Whittemore V (1999) *Tuberous Sclerosis Complex*, 3rd edn. New York, NY: Oxford University Press.

Goncharova EA, Goncharov DA, Eszterhas A, *et al.* (2002) Tuberin regulates p70 S6 kinase activation and ribosomal protein S6 phosphorylation: a role for the *TSC2* tumor suppressor gene in pulmonary lymphangioleiomyomatosis (LAM). *Journal of Biological Chemistry* **277**: 30958–30967.

Haines JL, Short MP, Kwiatkowski DJ, *et al.* (1991) Localisation of one gene for tuberous sclerosis within 9q32–9q34, and further evidence for heterogeneity. *American Journal of Human Genetics* **49**: 764–772.

Hodges AK, Li S, Maynard J, *et al.* (2001) Pathological mutations in *TSC1* and *TSC2* disrupt the interaction between hamartin and tuberin. *Human Molecular Genetics* **10**: 2899–2905.

Hornigold N, van Slegtenhorst M, Nahmias J, *et al.* (1997) A 1.7-megabase sequence-ready cosmid contig covering the *TSC1* candidate region in 9q34. *Genomics* **41**: 385–389.

Ito N and Rubin G (1999) gigas, a *Drosophila* homolog of tuberous sclerosis gene product-2, regulates the cell cycle. *Cell* **96**: 529–539.

Jones AC, Sampson JR, Hoogendoorn B, Cohen D and Cheadle JP (2000) Application and evaluation of denaturing HPLC for molecular genetic analysis in tuberous sclerosis. *Human Genetics* **106**: 663–668.

Jones AC, Shyamsundar MM, Thomas MW, *et al.* (1999) Comprehensive mutation analysis of *TSC1* and *TSC2* and phenotypic correlations in 150 families with tuberous sclerosis. *American Journal of Human Genetics* **64**: 1305–1315.

Kandt RS, Haines JL, Smith M, *et al.* (1992) Linkage of an important gene locus for tuberous sclerosis to a chromosome 16 marker for polycystic kidney disease. *Nature Genetics* **2**: 37–41.

Kleymenova E, Ibraghimov-Beskrovnaya O, Kugoh H, *et al.* (2001) Tuberin-dependent membrane localization of polycystin-1: A functional link between polycystic kidney disease and the *TSC2* tumor suppressor gene. *Molecular Cell* **7**: 823–832.

Kobayashi T, Hirayama Y, Kobayashi E, Kubo Y and Hino O (1995) A germ line insertion in the tuberous sclerosis (*Tsc2*) gene gives rise to the Eker rat model of dominantly inherited cancer. *Nature Genetics* **9**: 70–74.

Kobayashi T, Minowa O, Kuno J, *et al.* (1999) Renal carcinogenesis, hepatic hemangiomatosis, and embryonic lethality caused by a germ-line *Tsc2* mutation in mice. *Cancer Research* **59**: 1206–1211.

Kobayashi T, Minowa O, Sugitani Y, *et al.* (2001) A germ-line *Tsc1* mutation causes tumor development and embryonic lethality that are similar, but not identical, to those caused by *Tsc2* mutation in mice. *Proceedings of the National Academy of Sciences of the United States of America* **98**: 8762–8767.

Kwiatkowski DJ, Armour J, Bale AE, *et al.* (1993) Report on the second international workshop on human chromosome 9. *Cytogenetics and Cell Genetics* **64**: 94–106.

Kwiatkowski DJ, Zhang H, Bandura JL, *et al.* (2002) A mouse model of *TSC1* reveals sex-dependent lethality from liver hemangiomas, and up-regulation of p70S6 kinase activity in *Tsc1* null cells. *Human Molecular Genetics* **11**: 525–534.

Lamb RF, Roy C, Diefenbach TJ, *et al.* (2000) The *TSC1* tumour suppressor hamartin regulates cell adhesion through ERM proteins and the GTPase Rho. *Nature Cell Biology* **2**: 281–287.

Maheshwar MM, Cheadle JP, Jones AC, *et al.* (1997) The GAP-related domain of tuberin, the product of the *TSC2* gene, is a target for missense mutations in tuberous sclerosis. *Human Molecular Genetics* **6**: 1991–1996.

Maheshwar MM, Sandford R, Nellist M, *et al.* (1996) Comparative analysis and genomic structure of the tuberous sclerosis 2 (*TSC2*) gene in human and pufferfish. *Human Molecular Genetics* **5**: 131–137.

Miloloza A, Rosner M, Nellist M, *et al.* (2000) The TSC1 gene product, hamartin, negatively regulates cell proliferation. *Human Molecular Genetics* **9**: 1721–1727.

Mizuguchi M, Takashima S, Yamanouchi H, *et al.* (2000) Novel cerebral lesions in the Eker rat model of tuberous sclerosis: cortical tuber and anaplastic ganglioglioma. *Journal of Neuropathology and Experimental Neurology* **59**: 188–196.

Nellist M, Brook-Carter PT, Connor JM, *et al.* (1993) Identification of markers flanking the tuberous sclerosis locus on chromosome 9 (*TSC1*). *Journal of Medical Genetics* **30**: 224–227.

Nellist M, van Slegtenhorst MA, Goedbloed M, *et al.* (1999) Characterization of the cytosolic tuberin–hamartin complex: tuberin is a cytosolic chaperone for hamartin. *Journal of Biological Chemistry* **274**: 35647–35652.

Nellist M, Verhaaf B, Goedbloed MA, *et al.* (2001) TSC2 missense mutations inhibit tuberin phosphorylation and prevent formation of the tuberin–hamartin complex. *Human Molecular Genetics* **10**: 2889–2898.

Onda H, Lueck A, Marks PW, Warren HB and Kwiatkowski DJ (1999) $Tsc2(+/-)$ mice develop tumours in multiple sites that express gelsolin and are influenced by genetic background. *Journal of Clinical Investigation* **104**: 687–695.

Osborne JP, Fryer A and Webb D (1991) Epidemiology of tuberous sclerosis. *Annals of the New York Academy of Science* **615**: 125–127.

Plank TL, Yeung RS and Henske EP (1998) Hamartin, the product of the tuberous sclerosis 1 (*TSC1*) gene, interacts with tuberin and appears to be localised to cytoplasmic vesicles. *Cancer Research* **58**: 4766–4770.

Potter CJ, Huang H and Xu T (2001) Drosophila *Tsc1* functions with *Tsc2* to antagonize insulin signaling in regulating cell growth, cell proliferation, and organ size. *Cell* **105**: 357–368.

Povey S, Burley MW, Attwood J, *et al.* (1994) Two loci for tuberous sclerosis: one on 9q34 and one on 16p13. *Annals of Human Genetics* **58**: 107–127.

Rose VM, Au KS, Pollom G, *et al.* (1999) Germ-line mosaicism in tuberous sclerosis: how common? *American Journal of Human Genetics* **64**: 986–992.

Sampson JR, Maheshwar MM, Aspinwall R, *et al.* (1997) Renal cystic disease in tuberous sclerosis: role of the polycystic kidney disease 1 gene. *American Journal of Human Genetics* **61**: 843–851.

van Slegtenhorst M, deHoogt R, Hermans C, *et al.* (1997) Identification of the tuberous sclerosis gene *TSC1* on chromosome 9q34. *Science* **77**: 805–808.

van Slegtenhorst M, Janssen B, Nellist M, *et al.* (1995) Cosmid contigs from the tuberous sclerosis candidate region on chromosome 9q34. *European Journal of Human Genetics* **3**: 78–86.

van Slegtenhorst M, Nellist M, Nagelkerken B, *et al.* (1998) Interaction between hamartin and tuberin, the *TSC1* and *TSC2* gene products. *Human Molecular Genetics* **7**: 1053–1057.

Soucek T, Holzl G, Bernaschek G and Hengstschlager M (1998a) A role of the tuberous sclerosis gene-2 product during neuronal differentiation. *Oncogene* **16**: 2197–2204.

Soucek T, Yeung RS and Hengstschlager M (1998b) Inactivation of the cyclin-dependent kinase inhibitor p27 upon loss of the tuberous sclerosis complex gene-2. *Proceedings of the National Academy of Sciences of the United States of America* **95**: 15653–15658.

Tapon N, Ito N, Dickson BJ, Treisman JE and Hariharan IK (2001) The *Drosophila* tuberous sclerosis complex gene homologues restrict cell growth and cell proliferation. *Cell* **105**: 345–355.

Tsuchiya H, Orimoto K, Kobayashi T and Hino O (1996) Presence of potent transcriptional activation domains in the predisposing tuberous sclerosis (*Tsc2*) gene product of the Eker rat model. *Cancer Research* **56**: 429–433.

Verhoef S, Bakker L, Tempelaars AMP, *et al.* (1999) High rate of mosaicism in tuberous sclerosis complex. *American Journal of Human Genetics* **64**: 1632–1637.

Wienecke R, Konig A and DeClue JE (1995) Identification of tuberin, the tuberous sclerosis-2 product – tuberin possesses specific rap1GAP activity. *Journal of Biological Chemistry* **270**: 16409–16414.

Wienecke R, Maize JC, Shoarinejad F, *et al.* (1996) Co-localization of the *TSC2* product tuberin with its target rap1 in the Golgi apparatus. *Oncogene* **13**: 913–923.

Xiao G-H, Shoarinejad F, Jin F, Golemis EA and Yeung RS (1997) The tuberous sclerosis 2 gene product, tuberin, functions as a rab5 GTPase activating protein (GAP) in modulating endocytosis. *Journal of Biological Chemistry* **272**: 6097–6100.

Yeung RS, Xiao GH, Everitt JI, Jin F and Walker CL (1995) Allelic loss at the tuberous sclerosis 2 locus in spontaneous tumours in the Eker rat. *Molecular Carcinogenesis* **14**: 28–36.

Yeung RS, Xiao GH, Jin F, *et al.* (1994) Predisposition to renal carcinoma in the Eker rat is determined by germ-line mutation of the tuberous sclerosis 2 (*Tsc2*) gene. *Proceedings of the National Academy of Sciences of the United States of America* **91**: 11413–11416.

Web Links

nth endonuclease III-like 1 (*E. coli*) (*NTHL1*); Locus ID: 4913. LocusLink:
http://www.ncbi.nlm.nih.gov/LocusLink/LocRpt.cgi?l = 4913

Polycystic kidney disease 1 (autosomal dominant) (*PKD1*); Locus ID: 5310. LocusLink:
http://www.ncbi.nlm.nih.gov/LocusLink/LocRpt.cgi?l = 5310

Tuberous sclerosis 1 (*TSC1*); Locus ID: 7248. LocusLink:
http://www.ncbi.nlm.nih.gov/LocusLink/LocRpt.cgi?l = 7248

Tuberous sclerosis 2 (*TSC2*); Locus ID: 7249. LocusLink:
http://www.ncbi.nlm.nih.gov/LocusLink/LocRpt.cgi?l = 7249

nth endonuclease III-like 1 (*E. coli*) (*NTHL1*); MIM number: 602656. OMIM:
http://www.ncbi.nlm.nih.gov/htbin-post/Omim/dispmim?602656

Polycystic kidney disease 1 (autosomal dominant) (*PKD1*); MIM number: 601313. OMIM:
http://www.ncbi.nlm.nih.gov/htbin-post/Omim/dispmim?601313

Tuberous sclerosis 1 (*TSC1*); MIM number: 605284. OMIM:
http://www.ncbi.nlm.nih.gov/htbin-post/Omim/dispmim?605284

Tuberous sclerosis 2 (*TSC2*); MIM number: 191092. OMIM:
http://www.ncbi.nlm.nih.gov/htbin-post/Omim/dispmim?191092

Tumor Formation: Number of Mutations Required

C Richard Boland, *University of California, San Diego, La Jolla, California, USA*

Tumors develop as a consequence of mutations in critical growth-regulating genes. A large number of mutations can be found in most tumors, frequently in functionally irrelevant DNA sequences. This suggests that in most instances, a form of widespread genomic instability is an antecedent process that generates mutations randomly, some of which provide growth or survival advantages, which lead to clonal expansion and the emergence of a tumor.

Introduction

Tumors develop as a consequence of the accumulation of a critical number of mutations in genes that regulate cellular behaviors, including the rate of proliferation, programmed cell death, the ability to invade adjacent tissue boundaries and the ability to grow in an inappropriate environment, to name but a few. Mutations in deoxyribonucleic acid (DNA) sequences ordinarily occur relatively infrequently, but given the number of cells in the adult human (10^{13}–10^{14}) and the relative likelihood for a mutation to occur at each nucleotide during each division (perhaps of the order of 10^{-10}), given enough time and cell divisions, it is likely that multiple mutations could eventually accumulate in an occasional cell. Since individual cells express only a minority of their encoded genes in the differentiated state, the likelihood that a single mutation will alter the behavior of that cell is quite low. It is therefore apparent why cancer occurs more frequently with aging.

Growth Advantage, Clonal Expansion and Natural Selection

A mutation in a gene which is critical for regulating the behavior of a cell might provide it with a growth or survival advantage; alternatively, the mutation could lead to cell death. Mutations that lead to cell death are self-resolving, since the mutation is eliminated. Mutations that provide a growth advantage will lead to clonal expansion and increase representation of that new DNA sequence in the cell population. Furthermore, if the proliferation rate of a cell is increased, this expands the opportunity for more errors to occur. The occurrence of additional mutations in the expanding clone again permits the opportunity for cumulative growth advantages to occur, and further clonal expansion. This process could occur over and over

again, and a simplistic view of tumor development is similar to an evolutionary process where mutations occur randomly, but natural selection favors the expansion of certain cells.

Studies of naturally occurring human tumors suggest that the truth is more complicated, because of a process known as genomic instability, which can either increase the appearance of mutations or decrease the rate of DNA repair processes. The presence of 'signature mutations' in certain tumors suggests that tumorigenesis is not always explained by randomly occurring mutations followed by a simple evolutionary process. In fact, some investigators suggest that most cancers are characterized by some form of genomic instability that dramatically increases the rate of mutation.

Which Types of Mutation are Found in Cancers?

Very broadly defined, a mutation is an alteration in DNA that changes the nature of gene expression from that sequence. Narrowly defined, a mutation most often refers to an alteration in the genetic sequence. For this discussion, a broad definition will include 'epigenetic' changes such as promoter methylation that silences gene expression without altering the DNA sequence, since this is one mechanism for altering gene activity in cancer (Toyota *et al.*, 2000). Genetic alterations that occur in germ cells give rise to germ-line mutations, which are then found in every cell of the offspring of that germ cell. Tumors, however, are formed by the acquisition of somatic mutations that are only found in the tumor cells.

Point mutations

The simplest mutation to understand is a 'point mutation' in which one nucleotide is converted to another in the DNA sequence. This can occur through a variety of mechanisms. First, there are many biochemical processes that result in the decay of DNA structure, some of which result in miscoding during the subsequent round of DNA replication. Second, DNA polymerase occasionally incorporates the wrong nucleotide base in the newly synthesized strand, creating a mismatch. The proofreading subunit of DNA polymerase will remove approximately 99% of these errors; an additional 99.9% of proofreading occurs by virtue of the DNA mismatch repair (MMR) system. Chemical carcinogens can form adducts with DNA by chemical interaction (typically with a purine), which can lead to errors during the next round of replication. Insertion or deletion mutations, in which a nucleotide is dropped or added, will result in a frameshift, which will alter the amino acids encoded downstream of the mutation, and frequently leads to the generation of a premature stop codon.

Frameshifts

Frameshift mutations are particularly likely to occur at simple repetitive sequences called microsatellites (such as A_n or $[CA]_n$). During replication, microsatellite sequences are prone to slippage errors, and one of the progeny will have one more or one less element in the microsatellite sequence. Preventing the propagation of this is a principal function of the DNA MMR system.

Nucleotide excision repair

Ultraviolet irradiation gives rise to pyrimidine dimers, which interfere with both transcription and replication of the DNA strand. The nucleotide excision repair (NER) system is responsible for removal of these sunlight-induced lesions. Loss of NER activity will result in an increase in mutational load. Therefore, one will observe an increase in the number of mutations either by acceleration of the rate of mutation formation, or by reduction in the activity of the systems that repair these forms of DNA damage.

Chromosomal instability

It was recognized in the early part of the twentieth century that cancers frequently have an inappropriate number of chromosomes, including duplications,

deletions and rearrangements. In fact, the variety of mutations and chromosomal rearrangements, and their apparent randomness, served to confuse early investigators in the field. It initially appeared that each tumor was unique, and that there would never be unifying principles. To some degree, this may be true.

Which Genes are Mutated in Cancers?

Oncogenes

Some of the first insights in the genetic basis of cancer were due to the diligence of Bishop and Varmus, who studied the ability of the Rous sarcoma virus to induce tumors in chickens (Bishop, 1991). This virus contained only four genes, three of which were required for viral replication. The fourth of them, *src*, was a mutant version of a normal cellular gene which, when reintroduced with the viral genome, led to cancer. This work led to the discovery of a large number of oncogenes which are normal cellular genes that can participate in carcinogenesis either by point mutations which alter their activity or ability to be regulated, or through amplification of the gene. Generally, oncogenes participate in carcinogenesis by their excessive activity. In this specific instance, it appears that activation of one potent oncogene in one cell can lead to a tumor, but this model is not sufficient to explain most tumorigenesis in humans.

Tumor suppressor genes

It soon became apparent that another group of genes normally functioned to restrain cell growth, and their inactivation was found in cancers. Tumor suppressor genes (TSGs) can be inactivated by nonsense mutations, insertion–deletion mutations that create frameshifts and premature stop codons, interactions at the protein level that inhibit or destroy the protein and (frequently) loss of the gene from the genome. Genes that are essential for halting proliferation and regulators of cell cycle checkpoints or programmed cell death (apoptosis) are prime examples of TSGs, as their inactivation permits inappropriate survival of cells.

Genomic Instability in Cancer

Chromosomal instability and microsatellite instability

While there are often many mutations in a cancer, the oncogenes and tumor suppressor genes are the mutational targets in cancer. The types of mutation found in a tumor are not always random; one type of signature mutation sometimes predominates in an individual tumor, providing some clue to its pathogenesis. In aneuploid tumors, for example, one finds a large number of chromosomal deletions, amplifications and rearrangements; this is called 'chromosomal instability' (CIN) (Lengauer et al., 1997). The mechanism responsible for this is controversial, and there may be multiple causes. Some tumors are characterized by a very large number of mutations at microsatellite sequences, called 'microsatellite instability' (MSI) (Ionov et al., 1993). MSI in a cancer is caused by loss of DNA MMR activity.

Models of Carcinogenesis

The single mutation model

The initial work with *src* in chickens focused attention on the possibility that single genetic alterations might cause cancer. In humans, the Philadelphia chromosome was recognized in patients with chronic myelogenous leukemia (CML) by its morphological features which revealed that there was a translocation between chromosomes 9 and 22. It was subsequently recognized that this translocation results in a Bcr/Abl chimeric tyrosine kinase which functions as a human oncogene (De Klein et al., 1982). This tyrosine kinase becomes dysregulated in its translocated position, which leads to the clonal expansion of those lymphocytes. The discovery that inhibition of this kinase by a drug leads to the resolution of CML raises the possibility that this tumor is caused, or at least supported, by a single mutation. Again, this simple model does not apply to most cancers, which appear to require multiple alterations.

A simple *in vitro* model of carcinogenesis

There have been multiple approaches to developing experimental models of carcinogenesis, both *in vitro* and *in vivo*, using carcinogens and newer types of transgenic or gene knockout techniques in mice. One *in vitro* model required the addition of only three genetic elements; the expression of the telomerase catalytic subunit (hTERT), a mutant H-*ras* gene, and the SV40 T antigen in cultured human fibroblasts or epithelial cells resulted in tumorigenic cells (Hahn et al., 1999). This may be a biologically unique model, or possibly the T antigen may have induced genomic instability and additional mutations. However, in certain settings, a small and defined set of mutations may be sufficient for tumorigenesis.

Multistep carcinogenesis

In the late 1980s, the concept of tumor suppressor genes was explored, and a group of colon tumors was examined for chromosomal losses at randomly selected locations. Over 20% of the loci tested had experienced allelic loss or loss of heterozygosity (LOH) (Vogelstein et al., 1988). Certain chromosomal locations were more frequently deleted than others, and a few loci were found in which LOH was found in most of the tumors. The lost sites were presumed to be the locations of tumor suppressor genes involved in the genesis of colon cancer (Vogelstein et al., 1989).

Microsatellite instability and the mutator phenotype

Some colorectal cancers have no LOH events. Loss of DNA MMR activity and MSI were subsequently linked to this subset of tumors. The technique used permitted the estimate that loss of DNA MMR activity leads to the generation of about 100 000 microsatellite mutations in each tumor (Ionov et al., 1993). This type of cancer was referred to as the mutator phenotype, and is responsible for about 10–15% of colorectal cancers.

How Many Mutations in a Tumor?

Simple sequence alterations

The number of mutations one detects in a tumor depends upon the method used to find them. Direct sequencing of the DNA from a tumor would reveal the number of point mutations, and the number of relatively small insertion–deletion mutations present in the tumor. However, it would not indicate how many of these mutations are necessary for tumor development. One attempt to estimate the number of point mutations and small deletions in colon cancers involved sequencing around 3.2 megabases (Mb) of coding exons from 12 human tumors that did not have MSI. This project revealed only three distinct mutations, which predicted a total of about 6000 of these types of mutations in coding genes in the neoplastic genome (Wang et al., 2002).

Deletions and loss of heterozygosity

Genetic deletions are quite another matter. The human genome is diploid. DNA sequencing is performed in individual bites that are less than 1 kilobase (kb) long. Deletions occurring as a consequence of CIN in tumors can involve multiple megabases, or whole chromosomes. If one sequences cellular DNA in which a whole chromosome has been deleted, the sequence will appear to be normal because of the presence of the remaining chromosome. One only recognizes the deletion if the 'breakpoint' of an interstitial deletion is encountered during the sequencing. Sequencing through the genome will not provide an accurate estimate of the magnitude of CIN. With very precise karyotype analysis, one might be able to estimate the number of deletions, duplications and rearrangements. Accurately estimating the number of mutations in a tumor is therefore a daunting task, and would require multiple approaches.

Are tumors like snowflakes?

It may be the case that most tumors are, more or less, unique. Each has its own series of point mutations, insertion–deletion mutations, small interstitial chromosomal deletions, large-scale rearrangements, chromosomal duplications and chromosomal losses. There is also a growing awareness of the role of epigenetic changes that are important in carcinogenesis. Specifically, a gene may be present in the genome and have no alterations in the primary nucleotide sequence, yet the gene may be silent because of promoter hypermethylation. There may be other types of epigenetic changes that alter the pattern of gene expression even in the presence of normal sequences.

Which Mutations are Essential for Tumor Formation?

Ordinary DNA decay, accidents occurring during cell division and environmental carcinogenesis all conspire to create a background of mutational damage as each person matures and ages. Somatic mutations are passed on to the progeny of the somatic cell until that lineage is extinguished. In some instances, perhaps commonly, a form of genomic instability may be superimposed upon this background process, which accelerates genomic alterations and would be associated with a large number of signature mutations. This is particularly notable in diseases such as xeroderma pigmentosum, and hereditary nonpolyposis colorectal cancer (HNPCC) which occur due to loss of NER or DNA MMR activities respectively.

Genetic targets in colorectal carcinogenesis

Only certain genes are actually required for tumor formation. The full complement of genes required is not known, but the critical mutations required for the formation of a cancer are probably better known for colon tumors than most others. In the setting of CIN, it appears that loss of the adenomatosis polyposis coli

(*APC*) gene is sufficient for the appearance of an early, small, benign tumor (the adenoma). The growth of the adenoma may be accelerated by a Ras mutation. The benign tumor is converted into a malignant tumor when there is biallelic inactivation of the tumor protein p53 (Li–Fraumeni syndrome) (*TP53*) gene.

In colon cancers with MSI, the target genes are different. Interestingly, the sequence of events may also begin with an inactivating event in the *APC* gene or in another gene in a growth regulatory pathway mediated by Wnt signaling. The APC protein participates in the middle of the Wnt signaling pathway, and when the *APC* gene is expressed, it extinguishes the growth stimulating response to the Wnt ligand. Loss of the *APC* gene leaves the Wnt signaling pathway unregulatable. However, one can achieve the same result through mutational inactivation of one of the downstream effectors of Wnt signaling. In tumors with MSI, one can find point mutations in the *APC* gene, or mutations in microsatellite sequences present in either the catenin (cadherin-associated protein), beta 1, 88 kDa (*CTNNB1*) or WNT1 inducible signaling pathway protein 3 (*WISP3*) genes. Any of these events can theoretically lead to the same result, that is, unregulated proliferation of the colonic epithelial cell.

A similar phenomenon is seen for events that occur later in the evolution of a colon cancer. Biallelic inactivation of *TP53* occurs as the benign tumor is converted into a malignant tumor in colon cancers with CIN. By contrast, mutations in the transforming growth factor-β (TGF-β) signaling pathway occur near the benign-to-malignant transition in tumors with MSI. At least two genes in this pathway, including the transforming growth factor, beta receptor II (70/80 kDa) (*TGFBR2*) and the insulin-like growth factor 2 receptor (*IGF2R*), can be inactivated by insertion–deletion mutations in the A_{10} or G_8 sequences respectively of these two genes. Interestingly, in tumors with CIN, in which one is very unlikely to see mutations at these repetitive sequences, it is not unusual to find allelic losses of one of the *MADH* genes, which are downstream effectors of TGF-β regulation of epithelial cell growth. Therefore, it is the inactivation of growth regulatory pathways that is essential for tumor development, and the specific genes involved will vary from tumor to tumor, depending in part upon the type of genomic instability involved and in part on chance.

Pathways to Cancer

Thus there are several pathways to cancer, and the number and type of mutations will be unique for each tumor. Colorectal cancers with CIN are characterized by the accumulation of mutations that activate oncogenes and inactivate tumor suppressor genes, compounding the appearance of a process that leads to abnormal chromosomal segregation during mitosis. Some tumors will have allelic losses at a majority of chromosomal locations studied, whereas others will have a relatively smaller number of these. As mentioned, the number of point mutations in the coding exons of these tumors may be relatively small, perhaps in the range of 6000 per tumor.

Tumors with MSI are typically diploid, and have exactly 46 chromosomes without rearrangements. However, they may have in excess of 10^5 mutations at microsatellite sequences, most of which occur in noncoding areas of the genome, and these will be functionally meaningless. However, it is now recognized that there are critical genes with microsatellite sequences in exons. In the absence of DNA MMR activity, mutations will accumulate at these sequences at a very rapid rate. Only mutations at these critical growth-regulatory genes will result in clonal expansion. However, any microsatellite sequence that has already undergone mutation in that cell will be present in the tumor as a 'passenger'.

There are multiple opinions on the issue of how many mutations there are in a cancer. One group has offered the opinion that the ordinary mutation rate that occurs during a cell division over the lifetime of an organism is sufficient to account for the mutations found in a tumor (Tomlinson et al., 2002). These estimates are challenged by the fact that cells have homeostatic mechanisms to censor the propagation of mutated genomes, which lead to programmed cell death in the presence of an excessive number of mutations. In fact, what is most remarkable is the stability of human DNA. Searching for mutations in individual genes in a panel of tumors typically results in either no mutations or a relatively small number of them.

The fundamental dynamic of tumor formation is essentially Darwinian with progressive waves of clonal expansion mediated by mutations that alter the growth characteristics of a cell. However, the presence of very large number of irrelevant 'signature mutations' in some tumors indicates the advent of a destabilizing process that accelerates this process.

Acknowledgements

Supported in part by a grant from the National Cancer Institute CA72851.

See also

Chromosomal Instability (CIN) in Cancer
Gene Amplification and Cancer
Methylation-mediated Transcriptional Silencing in Tumorigenesis
Oncogenes
Tumor Suppressor Genes

References

Bishop JM (1991) Molecular themes in oncogenesis. Cell 64: 235–242.

De Klein A, Van Kessel AG, Grosveld G, et al. (1982) A cellular oncogene is translocated to the Philadelphia chromosome in chronic myelocytic leukaemia. Nature 300: 765–767.

Hahn WC, Counter CM, Lundberg AS, et al. (1999) Creation of human tumour cells with defined genetic elements. Nature 400: 464–468.

Ionov Y, Peinado MA, Malkhosyan S, Shibata D and Perucho M (1993) Ubiquitous somatic mutations in simple repeated sequences reveal a new mechanism for colonic carcinogenesis. Nature 363: 558–561.

Lengauer C, Kinzler KW and Vogelstein B (1997) Genetic instability in colorectal cancers. Nature 386: 623–627.

Tomlinson I, Sasieni P and Bodmer W (2002) How many mutations in a cancer? American Journal of Pathology 160: 755–758.

Toyota M, Ohe-Toyota M, Ahuja N and Issa JP (2000) Distinct genetic profiles in colorectal tumors with or without the CpG island methylator phenotype. Proceedings of the National Academy of Sciences of the United States of America 97: 710–715.

Vogelstein B, Fearon ER, Hamilton SR, et al. (1988) Genetic alterations during colorectal-tumor development. New England Journal of Medicine 319: 525–532.

Vogelstein B, Fearon ER, Kern SE, et al. (1989) Allelotype of colorectal carcinomas. Science 244: 207–211.

Wang TL, Rago C, Silliman N, et al. (2002) Prevalence of somatic alterations in the colorectal cancer cell genome. Proceedings of the National Academy of Sciences of the United States of America 99: 3076–3080.

Further Reading

Vogelstein B and Kinzler KW (eds.) (2001) The Genetic Basis of Human Cancer, 2nd edn. McGraw-Hill.

Web Links

Adenomatosis polyposis coli (APC); Locus ID: 324. LocusLink:
http://www.ncbi.nlm.nih.gov/LocusLink/LocRpt.cgi?l = 324

Catenin (cadherin-associated protein), beta 1, 88 kDa (CTNNB1); Locus ID: 1499. LocusLink:
http://www.ncbi.nlm.nih.gov/LocusLink/LocRpt.cgi?l = 1499

Tumor protein p53 (Li-Fraumeni syndrome) (TP53); Locus ID: 7157. LocusLink:
http://www.ncbi.nlm.nih.gov/LocusLink/LocRpt.cgi?l = 7157

WNT1 inducible signaling pathway protein 3 (WISP3); Locus ID: 8838. LocusLink:
http://www.ncbi.nlm.nih.gov/LocusLink/LocRpt.cgi?l = 8838

Adenomatosis polyposis coli (APC); MIM number: 175100. OMIM:
http://www.ncbi.nlm.nih.gov/htbin-post/Omim/dispmim?175100

Catenin (cadherin-associated protein), beta 1, 88 kDa (CTNNB1); MIM number: 116806. OMIM:
http://www.ncbi.nlm.nih.gov/htbin-post/Omim/dispmim?116806

Tumor protein p53 (Li-Fraumeni syndrome) (TP53); MIM number: 191170. OMIM:
http://www.ncbi.nlm.nih.gov/htbin-post/Omim/dispmim?191170

WNT1 inducible signaling pathway protein 3 (WISP3); MIM number: 603400. OMIM:
http://www.ncbi.nlm.nih.gov/htbin-post/Omim/dispmim?603400

Tumor Suppressor Genes

Introductory article

Martin L Hooper, *University of Edinburgh, Edinburgh, UK*

When functioning normally, a tumor suppressor gene prevents the formation of one or more types of cancer. Mutations in tumor suppressor genes that interfere with their function can be inherited in the germ line or can occur in somatic cells; the accumulation of such mutations can allow cancer to develop.

Cancer as a Multistep Process

For cancer to develop, defects must arise in the regulatory processes that normally ensure that cells in our bodies divide, differentiate and die at exactly the right time and place. The number of regulatory circuits that must be disrupted for a cell to become cancerous depends on the cell type. But in all situations the change from a normal cell to a cancerous cell – 'carcinogenesis' – is a multistep process with each step reflecting a genetic alteration. Although the genes that must be mutated for cancer to develop differ between cell types, Hanahan and Weinberg have suggested that the catalog of cancer cell genotypes reflects six essential perturbations of cell physiology, some or all of which are necessary:

- *Acquisition of self-sufficiency in growth signals.* Normal cells only move from a quiescent state into an actively proliferating state in the presence of mitogenic growth signals, which are transmitted into the cell when signaling molecules bind to transmembrane receptors. Cancer cells acquire genetic alterations that enable them to generate their own growth signals.
- *Loss of sensitivity to antiproliferative signals.* Antiproliferative signals can prevent normal cell growth, either by driving cells reversibly into a quiescent state or by inducing them to undergo an irreversible process of differentiation in which they lose their capacity to divide. Cancer cells acquire changes that block these signals or their effects.
- *Evasion of apoptosis.* Programmed cell death by the process known as apoptosis is a mechanism that protects against uncontrolled cell growth in most if not all cell types. Mutations that disrupt this process commonly occur in cancer.
- *Acquisition of limitless replicative potential.* Once normal cells have undergone a certain number of divisions they stop growing – a process that is known as 'senescence'. Mutations in cancer cells can disrupt this inbuilt program, conferring unlimited replicative potential or 'immortalization'.
- *Acquisition of capacity for sustained angiogenesis.* The oxygen and nutrient supply necessary for normal cell function is ensured by precise regulation of angiogenesis – the formation and growth of blood vessels. When tumors begin to grow they are at first limited in size by the availability of already existing blood vessels and to continue to grow they must develop the ability to induce new blood vessel growth.
- *Acquisition of invasive and metastatic capacity.* Most deaths from cancer in humans are not due directly to the growth of the primary tumor but to the formation of secondary tumors (metastases) at distant sites in the body. The process of metastasis requires that the cells become capable of invading into the tissue that surrounds the primary tumor, reaching the circulation and initiating new growth in sites where their normal parent cells cannot grow. (*See* Tumor Formation: Number of Mutations Required.)

Oncogenes and Tumor Suppressor Genes

The types of genetic change listed in the previous section fall into two broad classes. The first class results in altered genes that have acquired new functions that the normal version of the gene does not possess: these are gain-of-function mutations. The second class results in genes that have become inactivated: these are loss-of-function mutations. When gain-of-function mutations are involved in carcinogenesis, the activated mutant gene is termed an oncogene and the normal version of the gene a proto-oncogene. When loss-of-function mutations are involved in carcinogenesis, the gene is termed a tumor suppressor gene.

Thus, when a tumor suppressor gene is functioning normally it acts to prevent cancer formation. Of the six types of perturbation listed in the previous section, tumor suppressor gene inactivation is involved most commonly in loss of sensitivity to antiproliferative signals and in evasion of apoptosis (see below). (*See* Cancer Cytogenetics.)

Germ Line and Somatic Events

In general, activation of one allele of a proto-oncogene is sufficient to confer altered properties on a cell, whereas it is necessary to inactivate both alleles of a tumor suppressor gene to alter cellular properties. Therefore, when oncogenes are involved in cancer formation, they are usually activated in somatic cells rather than inherited through the germ line, because the presence of the activated oncogene in the germ line normally affects embryonic development so severely that it causes embryonic lethality.

By contrast, an inactivating mutation in one allele of a tumor suppressor gene may be inherited through the germ line, although mutations can also occur in somatic cells. Individuals heterozygous for such a mutation are at increased risk of developing specific types of cancer because they require fewer somatic mutations for this to happen, and they transmit this risk to offspring that inherit the mutant allele. The same gene can be affected by both germ line and somatic mutations. This explains why similar cancers run in families in some cases and occur sporadically in others, owing completely to somatic mutation.

Gatekeeper and Caretaker Functions

Inactivation of a tumor suppressor gene can directly affect the processes of cell proliferation and apoptosis if the gene has a direct role in antiproliferative signaling or in the cell death program. But there is another mechanism by which inactivation of tumor suppressor genes can function, that is, by making the genome less stable such that mutations or altered numbers of copies of other, directly acting genes occur more frequently. Kinzler and Vogelstein introduced the term 'caretaker' to describe genes that act in this more indirect way, distinguishing them from the 'gatekeeper' genes that directly affect cell proliferation or apoptosis.

The normal function of caretaker genes is to ensure the stable propagation of the genome when cells divide. Two types of genomic instability that arise from caretaker gene inactivation are important in carcinogenesis. The first, chromosome instability, affects the accuracy of chromosome disjunction and results in an increased incidence of aneuploid cells. The second, microsatellite instability, leads to alterations in the number of repeats of microsatellite sequences and therefore has a selective effect on genes that contain microsatellites at sites that are crucial to their function. Microsatellite instability results from mutations in deoxyribonucleic acid (DNA) repair genes. (*See* Chromosomal Instability (CIN) in Cancer; Mismatch Repair Genes.)

Four subdivisions of gatekeeper genes can be made: first, genes encoding intracellular proteins that regulate progression through a specific stage of the cell cycle; second, genes encoding receptors for secreted molecules that function to inhibit cell proliferation; third, genes encoding proteins that regulate cell differentiation; and last, genes encoding proteins that signal or are otherwise involved in apoptosis. The classification of tumor suppressor genes into caretakers and gatekeepers is useful but should not be applied too rigidly as some genes can function in both capacities. The *p53* (*tumor protein p53*; *TP53*) gene, for example, functions as a caretaker in ensuring chromosome stability and as a gatekeeper in signaling apoptosis in response to DNA damage. The concept of caretaker and gatekeeper functions is perhaps more useful than that of caretaker and gatekeeper genes. (*See* Caretakers and Gatekeepers.)

RB1 Gene and Retinoblastoma

As is often the case in science, our understanding of tumor suppressor genes was advanced by finding an example that was particularly straightforward to study and then applying the lessons to progressively more complex situations. The straightforward example emerged from a study of a relatively rare cancer called childhood retinoblastoma. Almost half of the cases of this cancer are familial and the remainder are sporadic. In familial retinoblastoma, tumors develop in the retina, usually before the individual is 5 years old, and in general affect both eyes. If diagnosed early enough, the tumors respond well to radiation or other treatments that preserve useful vision; if they have progressed beyond this point, however, surgical removal of the eye is necessary to prevent further spread. In sporadic retinoblastoma, tumor formation occurs later and there is usually only a single tumor in one eye. (*See* Retinoblastoma.)

In the early 1970s, Knudson found that the age distribution at diagnosis in individuals with retinoblastoma in both eyes (bilateral retinoblastoma) corresponded to what was predicted if each tumor arose as a result of a single event that affected a retinal cell. But for unilateral retinoblastoma, in which only one eye is affected, the data corresponded to what was predicted if two events affecting a single retinal cell were necessary for tumor development. As the resulting tumors are similar in both groups of individuals, he drew the inspired conclusion that the 'events' correspond to genetic mutations and that one mutation is already present in the germ line of individuals with bilateral retinoblastoma.

Comings suggested shortly thereafter that the two mutations involved the two alleles of the same gene.

But it was another 20 years before the technology was developed that allowed these predictions to be confirmed at the molecular level. The gene responsible was identified and named *retinoblastoma 1* (*RB1*), and mapped to the long arm of chromosome 13.

Individuals with familial retinoblastoma are therefore heterozygous for a mutation that inactivates *RB1*. They have an increased risk (more than 90%) of developing retinoblastoma because only one allele remains to be mutated in each retinal cell and the chance of this happening in at least one cell is very high. By contrast, in the rest of the population, this chance is low because two mutations have to happen in the same cell. Although both *RB1* alleles have to be inactivated for a cell to grow into a tumor, it is heterozygous individuals, who inherit only one mutant gene, that are at risk. Another way of saying this is that susceptibility to retinoblastoma is inherited as a dominant trait (to be precise, an autosomal dominant trait). This is an example of the more general phenomenon that the same mutation can have both a recessive and a dominant effect depending on the phenotype under study. In cells, where the phenotype is the growth of a cell into a tumor, the mutation is recessive. In the individual, where the phenotype is susceptibility to retinoblastoma, the mutation is dominant, and of course this is what is clinically significant.

Loss of heterozygosity

The loss of the remaining normal allele from heterozygous cells can occur in individuals with familial retinoblastoma by one of many mechanisms (**Figure 1**). Whereas germ-line mutations are usually either point mutations or small deletions, somatic events that lead to the production of cells with no remaining wild-type allele often involve whole chromosomes or large chromosome segments. These more extensive genome changes are not compatible with normal embryonic development, which explains their absence from the spectrum of germ-line mutations, whereas the lesser requirements for uncontrolled growth as a tumor cell place fewer restrictions on the types of somatic event that are commonly seen.

Function of the *RB1* gene product

The *RB1* gene encodes a protein with a relative molecular mass of about 105 000 that regulates the passage of the cell through different phases of the cell cycle. This protein, retinoblastoma 1 (RB1), undergoes changes in phosphorylation mediated by cyclin-dependent kinases whose activity is itself dependent on cell-cycle phase. The unphosphorylated

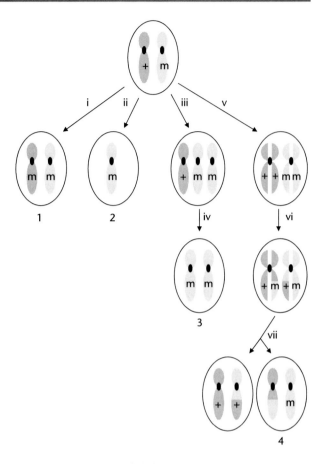

Figure 1 Mechanisms of loss of heterozygosity. A cell containing two copies of chromosome 13, one with the normal or wild-type *RB1* allele (+) and one with the mutant *RB1* allele (m) is shown at the top (the remaining 44 chromosomes are not shown for clarity). Independent mutation of the wild-type allele (i) is possible but relatively rare. Aberrant cell division in which segregation of chromosomes into daughter cells does not occur faithfully can lead to loss of the whole chromosome carrying the wild-type allele (ii), and this may be combined with duplication of its homolog carrying the mutant allele (iii, iv). Alternatively, after DNA replication (v) exchange of material between chromatid arms can occur by mitotic recombination (vi). As the two copies of the mutant allele are now on different copies of the chromosome, they will segregate independently at the following cell division (vii), and some daughter cells will acquire two copies of the mutant allele and no wild-type allele. These pathways lead to four possible genotypes (1–4), all of which have no remaining wild-type allele. Not all of these pathways are important for all tumor suppressor genes; in particular, for some chromosomes genotype 2 may be inconsistent with cell survival because essential gene products are not produced in sufficient amounts.

protein binds to and inhibits the function of other proteins that are required for progression through the G1 phase of the cell cycle and entry into S phase. During passage through G1, different residues of RB1 are phosphorylated at different time points and this releases the bound proteins in a progressive manner, allowing the cell cycle to proceed. This process is regulated in response to extracellular growth-promoting and growth-inhibiting molecules that transmit signals into the cell by binding to receptors on the cell surface. (*See* Evolution: Selectionist View.)

Mouse *Rb* mutants

The mouse gene equivalent to *RB1* is designated *Rb* (sometimes *Rb1*). Mice heterozygous for an inactive allele of *Rb* do not develop retinoblastomas but instead develop tumors in the pituitary. The latter do not occur in human *RB1* heterozygotes, probably because the structure of the mouse and human pituitaries is different. The reason for the absence of retinoblastomas in the mice is still under study and may help us to understand in more detail how retinoblastomas develop. Where the mice provide new information, however, is in allowing us to study homozygotes.

No human *RB1* mutant homozygotes have been reported, because families in which both parents are heterozygotes are extremely rare. In mice, however, planned matings of heterozygotes can be carried out. These matings have shown that homozygous embryos die before birth and have abnormal development of the blood and nervous system, which arises from increased apoptosis and overabundant or inappropriately located cell division. The tissues affected are those where cells normally differentiate and stop dividing earliest in embryonic development, which indicates that the embryos die because this process cannot occur normally. This provides evidence for a role for *Rb* in the process of maturation of dividing precursor cells to nondividing differentiated cells. This ties in with what we know about the role of RB1 in regulating the cell cycle (see above).

Generalization of Knudson's Hypothesis

Although Knudson's studies are consistent with a requirement for only two mutations for retinoblastoma to occur, they do not exclude the possibility that other mutations might be needed. Most retinoblastomas do have some additional genetic abnormalities, but it is not clear whether these are required for the tumor to begin growing or whether they occur after tumor growth has begun and produce faster-growing cells that eventually dominate the population of cells in the tumor. In other types of cancer, it is well established that the situation is more complex and that inactivation of more than one tumor suppressor gene is required for cancer formation.

Several tumor suppressor genes are now known and some examples are listed in **Table 1**. In some cancers, tumor formation proceeds through well-defined intermediate stages and the requirement for mutation of a particular gene can be identified with a particular stage in the process. The process of developing colon cancer, for example, proceeds through the formation of a benign tumor known as a polyp or adenoma. Mutation of the *adenomatous polyposis coli* (*APC*) gene is a requirement for adenoma formation, whereas mutation of either *p53* or one of a group of mismatch repair

Table 1 Some tumor suppressor genes and their associated familial cancer syndromes

Gene	Syndrome	Affected tissues or organs
RB1	Familial retinoblastoma	Retina, bone
Wilms tumor 1 (*WT1*)	Syndromes associated with Wilms' tumor, e.g. WAGR syndrome, Denys–Drash syndrome	Kidney
p53 (*TP53*)	Li–Fraumeni syndrome	Many, most commonly breast and brain
neurofibromin 1 (*NF1*)	Neurofibromatosis type 1	Peripheral nervous system, skin, iris
neurofibromin 2 (*NF2*)	Neurofibromatosis type 2	Central nervous system
breast cancer 1, early onset (*BRCA1*)	Breast cancer	Breast
breast cancer 2, early onset (*BRCA2*)	Breast cancer	Breast
adenomatous polyposis coli (*APC*)	Familial adenomatous polyposis	Colon
MSH2	HNPCC (hereditary nonpolyposis colorectal cancer)	Colon
MSH6	HNPCC	Colon
MLH1	HNPCC	Colon
PMS2	HNPCC	Colon
DPC4 (*MADH4*)	Juvenile polyposis	Gastrointestinal tract
VHL	von Hippel–Lindau syndrome	Central nervous system, adrenal, kidney, pancreas
E-CAD (*CDH18*)	Familial gastric cancer	Stomach
p16 (*CDKN2A*)	Familial melanoma	Skin

genes (*mutS homolog 2*; *MSH2*, *mutS homolog 6*, *MSH6*; *mutL homolog 1*, *MLH1*, *postmeiotic segregation increased 2*, *PMS2*) is associated with the transition from adenoma to full-blown cancer.

Notably, inactivation of *RB1* is required in many cancers whose incidence is not noticeably elevated in individuals with familial retinoblastoma. This apparent anomaly occurs because in these cancers *RB1* is merely one of a group of tumor suppressor genes whose inactivation is necessary; therefore, several somatic events still need to accumulate even when an *RB1* germ-line mutation is already present, and this results in an increase in incidence that is too small to be distinguished from a chance occurrence.

Although it is a general rule that mutations in both alleles of a tumor suppressor gene are required for cancer to develop, there are exceptions to this rule. There are three common reasons for this:

- *Haploinsufficiency.* Usually the amount of gene product produced by one functioning allele is sufficient for a cell to have the normal properties of a cell with two functioning alleles. But this is not always so, and the reduction in the amount of gene product in a cell with only one functioning allele can have a role in cancer development. This phenomenon is termed haploinsufficiency.
- *Dominant-negative mutations.* Some mutations lead to the production of an altered protein that is not only inactive itself but can interact with the protein produced by the normal allele in such a way as to inactivate it. These mutations are described as dominant-negative and can reduce the functioning protein to very low amounts without the need for a mutation in the normal allele.
- *Imprinting.* For some genes, the allele inherited from one parent is inactivated without mutation, in some or all of the tissues in which it is usually expressed, by a normal developmental process called imprinting. For such genes, it is necessary to mutate only the copy of the gene inherited from the other parent to reduce the amount of functioning protein in the cell to zero. (*See* Genomic Imprinting at the Transcriptional Level.)

Tumor Suppressor Gene Pathways

In some cell types, cancer development requires the disruption of a regulatory pathway that involves more than one tumor suppressor gene. For example, the phosphorylation of RB1 is carried out by a cyclin-dependent kinase that can be inhibited by the protein produced by the *p16* (*cyclin-dependent kinase inhibitor 2A*; *CDKN2A*) gene. If *p16* is inactivated by mutation, RB1 is permanently inactivated by phosphorylation in cell types where there is no alternative mechanism to

regulate the activity of the kinase. In these cell types, the same result can be achieved by inactivating either *p16* or *RB1*. For example, this happens in pancreatic carcinomas, in which the abnormalities usually include mutations in either *p16* or *RB1*, but not both. But individuals with germ-line mutations in *p16* develop melanomas and not retinoblastomas, which emphasizes the fact that differences in regulatory circuits are present between different cell types because they express a different gene repertoire. (*See* Cell Cycle Checkpoint Genes and Cancer; Pancreatic Cancer.)

See also
Caretakers and Gatekeepers
Oncogenes
Retinoblastoma
Tumor Formation: Number of Mutations Required

Further Reading

DiCiommo D, Gallie BL and Bremner R (2000) Retinoblastoma: the disease, gene and protein provide critical leads to understand cancer. *Seminars in Cancer Biology* 10: 255–269.

Fishel R (2001) The selection for mismatch repair defects in hereditary nonpolyposis colorectal cancer: revising the mutator hypothesis. *Cancer Research* 61: 7369–7374.

Hanahan D and Weinberg RA (2000) The hallmarks of cancer. *Cell* 100: 55–70.

Heinrichs A, Hodges M and Eccleston A (2001) A *Trends* guide to cancer biology. *Trends in Cell Biology* 11: S1.

Hooper ML (1998) Tumour suppressor gene mutations in humans and mice: parallels and contrasts. *EMBO Journal* 17: 6783–6789.

Knudson AG (1971) Mutation and cancer: statistical study of retinoblastoma. *Proceedings of the National Academy of Sciences of the United States of America* 68: 820–823.

Knudson AG (1997) Hereditary predisposition to cancer. *Annals of the New York Academy of Sciences* 833: 58–67.

Lodish H, Berk A, Zipursky SL, *et al.* (2000) Cancer. *Molecular Cell Biology*, 4th edn, pp. 1055–1084. New York: WH Freeman.

Ponder BAJ (2001) Cancer genetics. *Nature* 411: 336–341.

Quon KC and Berns A (2001) Haplo-insufficiency? Let me count the ways. *Genes and Development* 15: 2917–2921.

Vogelstein B and Kinzler KW (1998) *The Genetic Basis of Human Cancer*. New York: McGraw-Hill.

Web Links

Cyclin-dependent kinase 2A (*CDKN2A*); LocusID: 1029. LocusLink:
http://www.ncbi.nlm.nih.gov/LocusLink/LocRpt.cgi?l = 1029

Retinoblastoma 1 (*RB1*); LocusID: 5925. LocusLink:
http://www.ncbi.nlm.nih.gov/LocusLink/LocRpt.cgi?l = 5925

Tumor protein p53 (*TP53*); LocusID: 7157. LocusLink:
http://www.ncbi.nlm.nih.gov/LocusLink/LocRpt.cgi?l = 7157

Cyclin-dependent kinase 2A (*CDKN2A*); MIM number: 600160. OMIM:
http://www.ncbi.nlm.nih.gov/htbin-post/Omim/dispmim?600160

Retinoblastoma 1 (*RB1*); MIM number: 180200. OMIM:
http://www.ncbi.nlm.nih.gov/htbin-post/Omim/dispmim?180200

Tumor protein p53 (*TP53*); MIM number: 191170. OMIM:
http://www.ncbi.nlm.nih.gov/htbin-post/Omim/dispmim?191170

Turner Syndrome

Laura L Hall, *Annapolis, Maryland, USA*

Turner syndrome (TS) results from the absence of some or all of the X chromosome, with a karyotype of 45,X in the fully developed syndrome. Because not all genes on the X chromosome are inactivated normally, the diminished expression of these genes presumably leads to the TS phenotype, including short stature, incomplete sexual development and reproductive function, and other structural and cognitive abnormalities. Research has begun to identify genes that are potentially linked to TS, including the short stature homeobox (*SHOX*) gene and *DFRX*, the human homolog of a fruitfly gene involved in oogenesis.

Introduction

The disorder Turner syndrome (TS) was named after Henry Turner, who in 1938 was the first person to identify its phenotype in adolescent girls – 8 years after Ullrich characterized the condition in younger girls. Although an estimated 99% of embryos and fetuses with TS are spontaneously miscarried in the early stages of pregnancy, about 1 in every 2000 live female newborns is affected. Many physical malformations are expressed to various degrees in TS, the most common and recognizable features being short stature and stalled reproductive maturation and function. The brain is not spared, as impairments in visuospatial abilities and nonverbal problem solving are often present.

The genetic cause of TS is the complete or partial loss of one of the X chromosomes in all or just a portion of cells, leaving a karyotype of 45,X or 45,X/46,XX or various X-chromosome deletions. Although X-chromosome inactivation (X-inactivation) is a normal epigenetic phenomenon that ensures that males and females experience the same dose of genes located on the X chromosome, not all X-chromosome genes are suppressed in this process; thus, it may be insufficient gene dose and expression, haploinsufficiency, and the escape of some genes from X-inactivation that underpin the pathology of TS. Scientists are probing both the molecular mechanisms of normal X-inactivation and the genes linked to the TS phenotype, including the cognitive traits. (*See* Chromosome X; Chromosome X: General Features.)

X-inactivation

The presence of two X chromosomes in females is not correlated with double doses of most X-linked gene products when compared with the presence of a single X chromosome in males. To avoid such genetic overdosing, much of one of the X chromosomes is inactivated with, as Mary Lyon first discovered in 1961, its genes silenced from very early in development. The unique molecular mechanisms of X-inactivation used by mammals are slowly coming into focus. Several regions of the genome, referred to as *cis*-acting factors when they are on the X chromosome itself and *trans*-acting factors when they are elsewhere, are involved.

For example, one gene discovered in 1991, called *X (inactive)-specific transcript (XIST)*, is located in a portion of the X chromosome called the X-inactivation center or XIC – a region of the chromosome that is essential for inactivation. *XIST* does not encode a protein but rather a large untranslated ribonucleic acid (RNA) molecule that functions in silencing the inactivated X chromosome. Other genes on the X chromosome, including *X (inactive)-specific transcript, antisense (TSIX)*, and a *trans*-acting factor, *CCCTC-binding factor (zinc-finger protein) (CTCF)*, have also been shown to participate in selecting and silencing the inactivated X chromosome in mammals. The elaborate process for X-inactivation shows that normal development and function hinge not only on the right set of genes, but also on the proper dosage of gene product. Relevant to TS is the fact that in normal females some regions of the X chromosome are not inactivated; that is, expression of genes on both copies of the X chromosome occurs and is, in fact, requisite. (*See* X-chromosome Inactivation; X-chromosome Inactivation and Disease.)

Turner Syndrome Karyotype

In the late 1950s, the genetic malformation underpinning TS – the absence of one X chromosome in females – was realized. The complete loss of one X chromosome, karyotype 45,X, results in the fully developed syndrome. As often as not, however, the cells of an affected individual vary in their karyotype, showing both 45,X and 46,XX. The latter karyotype is referred to as mosaicism and it actually encompasses an assortment of karyotypes. The percentage of cells and tissues that show the 45,X karyotype in mosaicism can range widely, and different tissues can have different karyotypes. Another karyotype in this genetic family has the deletion of one arm of one X chromosome. In this case, the normal X chromosome will see its arm duplicated. Sometimes the ends of the chromosome may stick together, giving the appearance of a ring (a defect that, unlike TS, is associated with mental retardation). More rarely, individuals with TS may also have a partial Y chromosome present.

Scientists are just beginning to understand specific genes that contribute to TS and the resulting phenotype. It is well established that individuals with the 45,X karyotype generally show most or all of the abnormalities associated with TS, whereas mosaicism gives rise to many fewer and less severe problems. In fact, for mosaicism, correlations between the results of prenatal karyotyping and the degree to which TS pathologies are expressed have not been well documented.

Overview of the Turner Syndrome Phenotype

The two prototypic problems associated with TS are short stature and failure to mature and function reproductively.

Girls with TS fall behind in their growth: by the age of 5 or 6 years they generally fall below the third percentile, and ultimately they fail to realize growth spurts that normally occur in adolescence. By adulthood, TS is associated with a 21-cm difference in height, on average, when compared with unaffected women. Women with TS rarely achieve a height above 150 cm (or 5 feet). Researchers are working to identify both the genetic basis for short stature in TS (see below) and its pathophysiology, which seems to relate to bone development. On the treatment side, studies show that high doses of growth hormone over the course of a few years do stimulate growth, especially if started at a relatively young age (8–9 years) and if the artificial prompting of menstruation is delayed.

Typically, TS results in incomplete sexual development and reproductive function. In essence, ovarian senescence is accelerated. Ovarian follicles begin involuting while the female affected with TS is still a fetus, and ovarian failure is completed during their infancy. Menstruation hardly ever commences in individuals with the 45,X karyotype (and only does so in 20–30% of young women who have mosaicism).

Even if menarche does occur, an ongoing menstrual cycle is rarely maintained in the 45,X karyotype. Naturally occurring ovulation and fertility are rare. Adult secondary sexual characteristics do not develop and the uterus remains preadolescent in structure.

The sexual and reproductive impairments that are prototypical of TS have increasingly retreated with modern treatments. Currently, the reproductive problems associated with TS are redressed largely with hormone replacement therapy, which also increases height and, over the long term, enhances cardiovascular, bone and other organs' health. *In vitro* fertilization with donor oocytes has also been successful in women with TS. Replacement estrogen undoubtedly also enhances brain maturation and function.

Other abnormalities associated with TS include congenital lymphedema that leads, for example, to a webbed neck and swollen hands and feet, coarctation of the aorta, abnormal structure and/or location of the kidney. Morbidity in adulthood includes increased risk of various disorders of the bone (e.g. osteoporosis, scoliosis and spinal fractures), endocrine systems (e.g. diabetes and thyroid disease), cardiovascular system (e.g. hypertension), and partial or complete deafness. As a result, life expectancy is reduced in adult women with TS.

Impact on Brain and Behavior

The brain is not spared the impact of TS, although the specific impairments and their root causes have proved to be subtle and elusive. Although early studies and lore implied that mental retardation was a correlate of TS, evidence clearly shows that overall intelligence quotient (IQ) is normal or potentially above average in females with TS. But affected individuals often have impairment of specific cognitive abilities, especially visuospatial skills. So whereas verbal and language skills are normal, or even better than normal in some individuals, performance IQ is decreased as are visuomotor skills and attention. Money and Alexander called this deficit 'space–form blindness', because identifying positions in space, mentally rotating geometric shapes, drawing geometric shapes and human figures, orienting to left–right directions, and even handwriting are all impaired. Girls with TS tend to have difficulty with mathematics and geography in school.

Beyond these cognitive skills, reports suggest that girls and women with TS may experience greater degrees of hyperactivity and may be more prone to depression and anorexia nervosa. Social and emotional maturity is often delayed and, as a group, women with TS report increased social isolation and lower self-esteem; they are less likely to be married and their professional success tends to be lower than that of their peers. It must be noted, however, that many women with TS go on to achieve personal, social and professional success and fulfillment.

Neuroimaging studies are beginning to identify structural correlates of the neurocognitive impairments in TS. A well-designed study using psychological testing and imaging technology (mass resonance imaging; MRI) did not reveal overall cerebral or subcortical volume differences, but the distribution of gray and white matter was significantly different in both the right and the left parietal regions. Specifically, both the right and the left parietal areas showed smaller total tissue volume ratios, but only the right parietal and occipital regions had a larger proportion of gray and white matter as compared with controls. Functional MRI and positron emission tomography – approaches that measure brain function – have shown that there is overall hypermetabolism in most brain areas but reduced functioning in the right parietal lobe.

The cause of these cognitive, emotional, behavioral and social problems is not clear, nor is it likely to be singular. Undoubtedly nature and nurture are interlocked in their contributions. Girls with TS are small and their sexual maturation does not occur on its own. The impact of social responses to these features (on the part of parents, educators, peers and others in the community) and the influence of sexual steroids (or the lack thereof) on the development and function of the brain are likely to contribute significantly. There might be also direct genetic effects on the structure and function of the brain, possibilities of which are discussed below.

Potential Genetic Mechanisms

Because of the various phenotypic expressions of TS, several genes are thought to be involved. Short stature has been associated with haploinsufficiency of an X-chromosome region that escapes inactivation – the location of the *short stature homeobox* (*SHOX*) gene. The *SHOX* gene is expressed primarily in the limbs during development, and defects of this gene have been linked to short stature in various conditions. This gene is therefore a strong candidate for involvement in TS, although it is unlikely to explain fully the reduced growth.

Another candidate gene has been identified for the reproductive failure seen in TS. The gene *D1APH2*, a human homolog of a fruitfly gene involved in oogenesis, is known to escape X-inactivation and its

location has been implicated in ovarian failure. But as with *SHOX*, it seems likely that other genes will be involved.

Insofar as the cognitive deficits are concerned, deletions on the X chromosome, where genes escaping inactivation are presumably located, have been linked to neurocognitive deficits similar to those seen in TS. Specifically, individuals missing only a small interval at the distal tip of Xp manifest TS-like impairments in neurocognition. Data from another study, as yet unreplicated, suggest that genomic imprinting may be involved in TS. In that study, girls with TS who had an X chromosome of paternal origin showed normal aspects of cognitive function as compared with those with an X chromosome of maternal origin. But, as noted above, this study has not been replicated and therefore has caused some controversy in the field. Much remains to be learned about the specific genetic and molecular mechanisms that underpin TS, as well as the kinds of interventions that can help girls and women with this genetic disorder to lead as healthy a life as possible.

See also
Monosomies

Further Reading

Heard E, Clerc P and Avner P (1997) X-chromosome inactivation in mammals. *Annual Reviews in Genetics* 31: 571–610.

Percec I and Bartolomei MS (2002) Do X chromosomes set boundaries? *Science* 295: 287–288.

Ranke MB and Saenger P (2001) Turner's syndrome. *Lancet* 358: 309–314.

Reiss A, Mazzocco M, Greenlaw R, Freund L and Ross J (1995) Neurodevelopmental effects of X monosomy and volumetric imaging study. *Annals of Neurology* 38: 731–738.

Ross JL, Roeltgen D, Kushner H, Wei F and Zinn AR (2000) The Turner syndrome-associated neuroscognitive phenotype maps to distal Xp. *American Journal of Human Genetics* 67: 672–681.

Saenger P (1996) Turner's syndrome. *New England Journal of Medicine* 335(23): 1749–1754.

Zinn AR and Ross JL (1998) Turner syndrome and haploinsufficiency. *Current Opinion in Genetics and Development* 8(3): 322–327.

Web Links

Short stature homeobox (*SHOX*); LocusID: 6473. LocusLink: http://www.ncbi.nlm.nih.gov/LocusLink/LocRpt.cgi?l = 6473

X (inactive)-specific transcript (*XIST*); LocusID: 7503. LocusLink: http://www.ncbi.nlm.nih.gov/LocusLink/LocRpt.cgi?l = 7503

Short stature homeobox (*SHOX*); MIM number: 400020. OMIM: http://www.ncbi.nlm.nih.gov/htbin-post/Omim/dispmim?400020

X (inactive)-specific transcript (*XIST*); MIM number: 314670. OMIM http://www.ncbi.nlm.nih.gov/htbin-post/Omim/dispmim?314670

Twin Methodology

Harold Snieder, *Georgia Prevention Institute, Augusta, Georgia, USA*

Alex J MacGregor, *St Thomas's Hospital, London, UK*

The classic twin study is established as the ideal study design with which to investigate the relative importance of genetic and environmental factors to traits and diseases in human populations. Twin methodology is concerned with analysis techniques that have been developed to estimate and quantify the relative contribution of genes and environment to the disease or trait of interest.

The Classic Twin Study

The principal objective of collecting twin data is to assess whether familial aggregation of a disease or trait can be explained by common genetic or environmental factors. The fact that a trait runs in families is not sufficient evidence that its etiology is genetic, because families may share predisposing environments as well as genes. One design that successfully disentangles genetic and environmental causes of familial resemblance is the adoption study, which studies genetically unrelated subjects living together (to assess shared environmental influences) and genetically related individuals living apart (to test genetic influences). Another more practical solution is offered by the natural experiment of twinning. The classic twin study builds on the biological fact that there are, genetically, two kinds of twins that provide contrasting degrees of genetic relationship in siblings of the same age and family circumstances. Monozygotic (MZ) twins share identical genotypes, so any differences between them are due to their environments. Dizygotic (DZ) twins, in contrast, are no more alike genetically than siblings, sharing on average 50% of their segregating genes. If it is assumed that both types of twins share environmental influences to the same extent, any greater similarity in attributes among MZ twins when compared with DZ twins reflects the importance of genetic influences. The assumption of equal environmental sharing in MZ and DZ twins has been frequently criticized as a potential weakness of the twin design. However, studies specifically carried out to test it (e.g. studies conducted among twins where zygosities had been misassigned) have shown no instances where violation of this assumption leads to important bias in interpretation of the results of classic twin studies (Kyvik, 2000).

Concordance Measures

For dichotomous traits such as the presence or absence of a disease, the similarity of twins can be expressed in terms of their concordance. Two measures of concordance are in common use: pairwise concordance and casewise (sometimes referred to as probandwise) concordance. Both provide different information. Pairwise concordance is the proportion of affected pairs in which both twins have the disease out of all pairs in which at least one of the co-twins is affected. Casewise concordance measures the proportion of co-twins of affected twin individuals ('cases') that express the disease or trait themselves. Pairwise concordance is of value in predicting disease status of the pair when it is known that only one of the twins is affected. Casewise concordance provides a measure of risk to an individual with an affected relative and may be of use, for example, in genetic counseling.

The statistical estimators for both concordance measures need to take into account the mode of ascertainment of the twins in the sample. For example, the statistics used to estimate pairwise and casewise concordance will differ in studies where data are available from a twin registry (in which there is complete ascertainment) when compared with studies in which twin pairs have been ascertained nonrandomly through an affected proband (in which ascertainment may be complete or incomplete). When ascertainment is complete, pairwise concordance is estimated by the ratio of concordant (C) to the sum of concordant and discordant (D) pairs ($C/(C + D)$) and casewise concordance from the ratio $2C/(2C + D)$. Where ascertainment is incomplete, concordance estimators need to take into account the probability that an affected twin was identified from the original population (the ascertainment probability). This probability is reflected in the number of concordant pairs that have been ascertained 'doubly' (i.e. both twins were identified independently in the original sampling) and 'singly' (i.e. only one of the twins was identified in the original sampling and the second affected twin was identified through subsequent examination). Formulae for estimating casewise and pairwise concordance under incomplete ascertainment (which incorporate the numbers of singly and doubly ascertained pairs) are derived by Witte *et al.* (1999).

An excess concordance for disease in MZ twins compared with DZ twins suggests a contribution from genetic factors, and the statistical significance of the difference in their concordance can be tested (Witte et al., 1999). It should be noted, however, that the magnitude of the difference in concordance between MZ and DZ twins cannot be equated to the size of the genetic contribution to a trait in a simple manner. The absolute size of the MZ–DZ concordance difference is influenced by the underlying prevalence of the trait. For example, for diseases with an equal underlying genetic contribution, a greater absolute MZ–DZ concordance difference would be expected for diseases that are common in the population when compared with those that are rare. Where the cumulative prevalence of disease increases with age, the underlying genetic contribution may also be hard to assess directly from concordance measures, which will also tend to increase. In these circumstances, estimating heritability provides a more useful and readily interpretable measure of the genetic contribution.

Variance Components Approaches

Most phenotypes show large individual differences (i.e. variance), the source of which may be genetic or environmental. For continuous traits, the contribution of genetic and environmental factors can be estimated directly in quantitative terms by comparing MZ and DZ covariances (or correlations). For discontinuous traits (such as the presence or absence of disease), their degree of correlation can be inferred by considering the quantitative association of an underlying (unobserved) continuous and normally distributed 'liability' variable that gives rise to disease when its value exceeds a certain threshold. The level of the threshold is determined by the population prevalence of the disease (Falconer, 1989).

One aim of studying the components of variation among relatives is to assess heritability, which can be defined as the proportion of phenotypic variation that can be attributed to genetic variation. Heritability provides an index of the importance of genetic variation in explaining the distribution of disease in a population. Heritability does not represent percentage of disease attributable to genetic factors (Hopper, 1998), as it is sometimes naively interpreted. A number of approaches have been used in studies of twins to estimate the genetic and environmental components underlying trait variation in the normal population. (See Heritability; Heritability Wars.)

Analysis of variance (ANOVA)

The analysis of variance approach to studying continuous data in twins is based on a comparison of the expected value of mean-square trait scores within (WMS) and among (AMS) MZ and DZ twin pairs. These values contain different contributions from genetic and environmental factors. For example, in MZ twins (who share the same genotype within pairs), the WMS contains no genetic variation – all of which is included in the AMS. By contrast, in DZ twins, the WMS is expected to contain half of the additive genetic population variance. Factors in the shared family environment in twins contribute only to AMS in both MZ and DZ twin pairs.

Various hypotheses concerning the contribution of individual variance components can be tested within this framework (see Christian and Williams (2000) for a comprehensive overview of the approach). One commonly reported measure from ANOVA is the intraclass correlation coefficient (r) between twins. This is calculated from $(AMS - WMS)/(AMS + WMS)$. In the absence of genetic dominance, an estimate of heritability is provided by $2(r_{MZ} - r_{DZ})$. The size of the common environmental variance contribution is estimated from $2(r_{DZ}) - r_{MZ}$.

DeFries–Fulker regression method

DeFries and Fulker (1985) have developed a multiple regression method to test for genetic etiology specifically for selected samples of MZ and DZ twins, in which the study design involves ascertainment of individuals with extreme trait scores. The idea behind this procedure is that the mean trait score of co-twins of selected probands should differentially regress back toward the population mean as a function of the heritability of the trait. That is, for DZ twins, who share, on average, 50% of their genetic material, co-twins of selected probands should regress further back to the population mean than should co-twins of MZ twins, who share 100% of their genes.

The basic multiple regression model for the analysis of selected twin data is

$$C = B_0 + B_1 P + B_2 R \qquad (1)$$

where C is the co-twin's predicted score, P is the proband's score, R is the coefficient of relationship (1 for MZ, 0.5 for DZ) and B_0 is the regression constant. B_1 is a measure of twin resemblance that is independent of zygosity. B_2 reflects the differential regression toward the mean in MZ and DZ co-twins and provides a test of genetic etiology.

DeFries and Fulker (1985, 1988) have also developed an augmented model for use in both unselected and selected samples:

$$C = B_0 + B_3 P + B_4 R + B_5 PR \qquad (2)$$

where PR is the product of the proband's score with the coefficient of relationship. The advantage of this parameterization is that B_5 is a direct estimate of the heritability and B_3 a direct estimate of the shared environmental variance.

Covariates such as gender and age can easily be included in the regression models. Recent extensions of this method have involved bivariate regressions, thereby estimating the correlation between genetic factors that influence two traits.

Structural equation modeling

Structural equation modeling (also known as path analysis or variance components modeling) has become a standard tool in twin research in recent years (Neale and Cardon, 1992). It involves solving a series of simultaneous linear structural equations in order to estimate genetic and environmental parameters that best fit the observed twin variances and covariances (**Figure 1**). Model-fitting analysis of twin data has some major advantages over the classic twin methodology: (1) models make assumptions explicit; (2) a test of the goodness-of-fit of the model is provided; (3) estimates of genetic and environmental variance components and their confidence intervals are given; (4) the fit of alternative models can be compared;

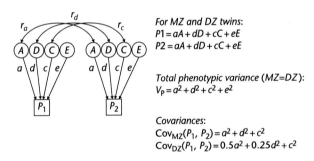

For MZ and DZ twins:
$P1 = aA + dD + cC + eE$
$P2 = aA + dD + cC + eE$

Total phenotypic variance (MZ=DZ):
$V_P = a^2 + d^2 + c^2 + e^2$

Covariances:
$Cov_{MZ}(P_1, P_2) = a^2 + d^2 + c^2$
$Cov_{DZ}(P_1, P_2) = 0.5a^2 + 0.25d^2 + c^2$

Figure 1 On the left the path diagram for the simple twin model is shown. This path diagram can be translated into linear structural equations (right), i.e. they are equivalent representations of the same twin model. A number of conventions are used in path analysis and the representation of the path diagram. Observed variables for twin 1 and twin 2 are shown in squares; latent variables (or factors) are shown in circles. A single-headed arrow indicates a direct influence of one variable on another, its value represented by a path coefficient (comparable to a factor loading). Double-headed arrows between two variables indicate a correlation without any assumed direct relationship. A: additive genetic latent factor; D: dominance genetic latent factor; C: shared (or common) environmental latent factor; E: unique environmental latent factor; a: additive genetic factor loading; d: dominance genetic factor loading; c: shared (or common) environmental factor loading; e: unique environmental factor loading; P_1, P_2: phenotypic value of twin 1, twin 2; r_a: additive genetic correlation (1 for MZ and 0.5 for DZ twins); r_d: dominance genetic correlation (1 for MZ and 0.25 for DZ twins); r_c: shared environmental correlation (1 for both MZ and DZ twins); V_P: total phenotypic variance; a^2: additive genetic variance; d^2: dominance genetic variance; c^2: shared environmental variance; e^2: unique environmental variance.

(5) more than two groups of twins can be analyzed simultaneously; and (6) generalization from the analysis of one variable (univariate) to multiple variables (multivariate) and from one time point (cross-sectional) to multiple time points (longitudinal) is relatively easy. (*See* Path Analysis in Genetic Epidemiology; Quantitative Genetics.)

Genetic model fitting of twin data allows the decomposition of the observed phenotypic variance into two genetic and two environmental components. Additive genetic variance (V_A) is the variance that results from the additive effects of alleles at each contributing locus. Dominance genetic variance (V_D) is the variance that results from the nonadditive effects of two alleles at the same locus summed over all loci that contribute to the variance of the trait. Shared (common) environmental variance (V_C) is the variance that results from environmental events shared by both members of a twin pair (e.g. rearing, school, neighborhood, diet). Specific (unique) environmental variance (V_E) is the variance that results from environmental effects that are not shared by members of a twin pair and also includes measurement error (**Figure 1**). Age can be incorporated in the model if the trait of interest is related to age. Failure to allow for the association between age and the disease trait may introduce a spurious shared environmental effect (Snieder, 2000).

A, D, C and E can be conceived of as uncorrelated latent factors with zero mean and unit variance. a, d, c and e are regression coefficients (or factor loadings) of the observed variable on the latent factors; they indicate the degree of relationship between latent factors and the phenotype. V_P is the phenotypic variance. Squaring the regression coefficients yields the (unstandardized) variance components ($V_A = a^2$, $V_D = d^2$, $V_C = c^2$, $V_E = e^2$), whose sum is equal to the total phenotypic variance. The contributions of genes and environment to the total variance are often reported in their standardized form. This standardization is achieved by dividing the specific variance component by the total phenotypic variance (e.g. narrow-sense heritability = V_A/V_P).

For MZ twins, correlations between the additive and dominance genetic factors between twin and co-twin are unity. For DZ twins, these values are 0.5 and 0.25 respectively. By definition, in both MZ and DZ same-sex pairs, correlations are unity between common environmental factors and zero between specific environmental factors. The model further assumes random mating and absence of gene–environment interaction and gene–environment correlation.

In twin studies, the effects of D and C are confounded, which means that they cannot both be included in the same univariate model. However, D and C have opposite effects on the patterns of MZ and

DZ twin correlations. *D* tends to produce DZ twin correlations less than one-half the MZ twin correlations, and *C* inflates the DZ correlation to be greater than one-half the MZ correlation. Models constraining all genetic effects to be nonadditive are considered unlikely, because they lack a plausible interpretation (Neale and Cardon, 1992).

Future Role of Twin Studies

In applying the methodology described in this overview, twin studies have often provided the first crucial proof of the importance of genes in the etiology of a disease or trait, justifying further search for the underlying genes. Twin studies will continue to play an important role in the postgenomic era, especially in the study of complex traits and diseases. (*See* Identical Twins Reared Apart; Twin Studies; Twin Studies: Software and Algorithms.)

References

Christian JC and Williams CJ (2000) Comparison of analysis of variance and likelihood models of twin data analysis. In: Spector TD, Snieder H and MacGregor AJ (eds.) *Advances in Twin and Sib-pair Analysis*, pp. 103–118. London: Greenwich Medical Media.

DeFries JC and Fulker DW (1985) Multiple regression analysis of twin data. *Behavior Genetics* 15: 467–473.

DeFries JC and Fulker DW (1988) Multiple regression analysis of twin data: etiology of deviant scores versus individual differences. *Acta Geneticae Medicae et Gemellologiae (Roma)* 37: 205–216.

Falconer DS (1989) *Introduction to Quantitative Genetics,* 3rd edn. Harlow, UK: Longman.

Hopper JL (1998) Heritability. In: Armitage P and Colton T (eds.) *Encyclopaedia of Biostatistics*, pp. 1905–1906. New York: John Wiley & Sons.

Kyvik KO (2000) Generalizability and assumptions of twin studies. In: Spector TD, Snieder H and MacGregor AJ (eds.) *Advances in Twin and Sib-pair Analysis*, pp. 67–77. London: Greenwich Medical Media.

Neale MC and Cardon LR (1992) *Methodology for Genetic Studies of Twins and Families.* Dordrecht, The Netherlands: Kluwer Academic Publishers.

Snieder H (2000) Path analysis of age related disease traits. In: Spector TD, Snieder H and MacGregor AJ (eds.) *Advances in Twin and Sib-pair Analysis*, pp. 119–130. London: Greenwich Medical Media.

Witte JS, Carlin JB and Hopper JL (1999) Likelihood-based approach to estimating twin concordance for dichotomous traits. *Genetic Epidemiology* 16: 290–304.

Further Reading

MacGregor AJ, Snieder H, Schork N and Spector TD (2000) Twins; novel uses to study complex traits and genetic diseases. *Trends in Genetics* 16: 131–134.

Martin N, Boomsma D and Machin G (1997) A twin-pronged attack on complex traits. *Nature Genetics* 17: 387–392.

Plomin R, DeFries JC, McClearn GE and McGuffin P (2001) *Behavioral Genetics*, 4th edn. New York: Worth Publishers.

Snieder H, Boomsma DI and van Doornen LJP (1999) Dissecting the genetic architecture of lipids, lipoproteins and apolipoproteins: lessons from twin studies. *Arteriosclerosis, Thrombosis and Vascular Biology* 19: 2826–2834.

Spector TD, Snieder H and MacGregor AJ (eds.) (2000) *Advances in Twin and Sib-pair Analysis*. London: Greenwich Medical Media.

Twin Studies

Susan L Trumbetta, *Vassar College, Poughkeepsie, New York, USA*

Introductory article

Article contents

- Introduction
- The Classic Twin Study
- Assumptions of the Classic Twin Study
- Calculations of Parameters in More Complex Twin Models
- Unique Applications of Classic Twin Study Design
- Beyond Classic Twin Design
- Twin Studies Provide a Genetically Informative Control for Environment-only Family Studies
- Conclusion

Twin studies that compare monozygotic with dizygotic twin pairs for their level of phenotypic similarity have long provided useful estimates of the relative influences of genetic and environmental factors on phenotypic variability in the population. Variations on the classic twin study, such as twins reared apart, co-twin control and children-of-twins designs, may provide additional clues to the effects of specific environments and of gene-by-environment interactions in the expression of certain phenotypes.

Introduction

In the early fifth century, Augustine's *The City of God* provided perhaps the first use of twins as evidence for evaluating causes of human behavior. Augustine refuted popular astrological theories by noting that the constellations shared by the biblical twins Jacob and Esau did not prevent their emergence as dramatically different personalities. However, it was not until the late nineteenth century that Francis Galton reasoned that the degree of resemblance among relatives phenotypes varies as a function of genetic similarity, and it was at the beginning of the twentieth century that R. A. Fisher, Merriman, Siemens and others laid the mathematical foundations necessary for the systematic study of twins as tools for genetic research.

The Biology of twinning

No two individuals are more closely related than monozygotic (MZ) twins. The division of a single zygote into two separate organisms creates MZ twins. Because the nuclear deoxyribonucleic acid (DNA) in all their cells replicates that of the original zygote, we assume that MZ twins share all their segregating genes, those genes whose alleles differ across individuals. MZ twins usually bear a striking physical resemblance, hence the term 'identical twins'. Dizygotic (DZ) twins, on the other hand, result from two zygotes, created by the fertilization of two separate ova by two separate spermatozoa. Because each sperm and egg contains one-half of the father's and mother's genotypes, respectively, the exact percentage of genetic similarity between DZ twins varies, in theory, from 0% to 100%. On average, however, DZ twins share half of their segregating genes. Because the similarity of DZ twin pairs is equivalent to that of nontwin siblings, we also call them 'fraternal' twins.

Determining zygosity by sex, chorionicity and phenotype

There are several simple ways, short of genotyping, to determine zygosity. Virtually all opposite-sex twin pairs are dizygotic: discordance for anomalies of sex chromosomes is possible, but extremely rare, among MZ twins. Opposite-sex twins account for about one-third of all twin births.

Because they develop from separate zygotes, all DZ twins are dichorionic, born with their own placenta or chorion. MZ pairs may be either monochorionic or dichorionic, depending upon the timing of twinning relative to placental development. All twin pairs born with a single placenta (about 20%) can be identified reliably as MZ. Dichorionic, same-sex twin pairs account for slightly less than half of all twin births and may be either MZ or DZ. Within this group, DZ twinship can often be established easily. Obvious physical dissimilarities or less obvious ones such as different blood types will establish dizygosity. In cases of greater intrapair similarity, however, zygosity may be less clear. Nevertheless, twins' self-reports of similarity are fairly accurate. Twin pair responses to the question of whether or not, through their childhood, they were as similar as 'two peas in a pod' provide a reliable index of zygosity, with approximately 95–97% accuracy.

The Classic Twin Study

A classic, univariate (one variable) twin study begins with phenotypic data obtained from sizeable, representative samples of MZ and DZ twins. The classic twin method decomposes phenotypic variation into its genetic and environmental components by fitting the observed covariation between twins to models of their expected covariation. The proportion of phenotypic variance attributable to genetic factors is also called the 'heritability' (h^2) of the phenotype. The proportion of variance attributable to environmental factors can be subdivided into two parts: the environmental variance expected to be shared by twins, or variance attributable to 'common environment' (c^2); and variance expected to differ between twins, or variance attributable to 'unique environment' (e^2). In any model that contains only these three factors, $h^2 + c^2 + e^2 = 1.00$, because the sum of the proportions must equal the whole.

The terms 'common environment' and 'unique environment' are a bit misleading, as common environmental factors are not simply shared family environments, but instead encompass all nongenetic sources of phenotypic variance associated with twin similarity. Common environmental factors include not only shared fetal and family environments, but also shared, extrafamilial environments such as birth cohort or military service. Similarly, unique (nonshared) environment is not limited to extrafamilial environments, for it includes all nongenetic sources of phenotypic variance associated with twin differences. Unique environmental factors include not only extrafamilial environments such as different childhood classrooms, different spouses or different lifetime differences in exposure to illness or injury, but also any different and differentiating experiences that occur in the context of a common family environment such as a parent's favoritism of one twin over the other. Unique environmental factors also include measurement error, so that any estimates we derive of unique environmental contributions to phenotypic variability are, by definition, overestimates.

Decomposing phenotypic variability into its genetic and environmental components

We can estimate heritability (h^2) by comparing MZ with DZ correlations. By definition, MZ and DZ twins alike share all of their common environments. MZ twins are genetically identical, whereas DZ twins, on average, share only half of their genes. Twin similarity in MZ pairs, therefore, results from full sharing of both genes and common environments, whereas twin similarity in DZ pairs results from half-sharing of genes and full-sharing of common environments. These concepts are restated in the following equations:

$$r_{MZ} = h^2 + c^2 \qquad (1)$$

$$r_{DZ} = 0.5h^2 + c^2 \qquad (2)$$

By subtraction:

$$r_{MZ} - r_{DZ} = 0.5h^2 \qquad (3)$$

Equation [1] states that the phenotypic correlation between MZ twins equals the sum of the variability attributable to genetics (h^2) plus the variability attributable to common environments. Equation [2] states that the phenotypic correlation between DZ twins equals the sum of half of the heritability plus the common environmental variability. By subtracting the second equation from the first, we see that the difference between MZ and DZ correlations equals one-half of the phenotype's heritability. If the difference between MZ and DZ correlations is doubled, an estimate of heritability is obtained. In mathematical terms, $2(r_{MZ} - r_{DZ}) = h^2$. Now suppose that we want to estimate the heritability of a trait such as extraversion from twin data. The correlations we obtain for extraversion are as follows: $r_{MZ} = 0.52$ and $r_{DZ} = 0.28$. The difference between them is 0.24. By doubling this difference, we estimate that the heritability of extraversion is 0.48.

Estimates of nonshared, nongenetic contributions to phenotypic variability (unique environments, e^2) can be calculated using only the MZ twin correlation. Because MZ twins share all of their segregating genes and are perfectly correlated for common environmental factors, any differences between MZ twins must derive solely from nonshared, nongenetic factors (e^2). Therefore, for any phenotype, the degree to which MZ twins are less than perfectly correlated provides an estimate of the e^2. In mathematical terms, $e^2 = 1.00 - r_{MZ}$. In the extraversion example, where MZ twins are correlated 0.52 for extraversion, 48% of the variance in extraversion is attributable to unique, nongenetic factors: $1.00 - 0.52 = 0.48 = e^2$.

There are at least two ways to estimate c^2 in a classic twin study. DZ twins share, on average, half of their genes. Therefore, when twin similarity is due to additive genetic factors alone, with no contributions of the common environment, the expected value of the DZ twin correlation will equal half the value of h^2. Whenever similarity between DZ twins (r_{DZ}) exceeds $h^2/2$, any additional similarity must be attributable to common environmental influences. In the case of social extraversion, where $h^2 = 0.48$, the expected DZ correlation in the presence of genetic effects alone would be 0.24 (half of h^2). However, because DZ twins are actually correlated 0.28 for extraversion, there is

evidence of common environmental influence. To estimate its magnitude, 0.24 (the amount of the DZ correlation accounted for by genetic factors) is subtracted from 0.28 (the full DZ correlation), leaving 0.04 as the proportion of phenotypic variance attributable to common environmental factors (c^2). To estimate c^2, half of h^2 is subtracted from the DZ twin correlation (r_{DZ}). Another way to derive c^2 is to subtract the sum of h^2 and e^2 from 1.0, as 1.0 represents 100% of the phenotypic variance. In this case, $h^2 = 0.48$ and $e^2 = 0.48$, and, by subtracting their sum of 0.96 from 1.00, we also find that $c^2 = 0.04$.

Assumptions of the Classic Twin Study

These estimations of h^2, c^2 and e^2 from univariate twin data make several assumptions. The first is that genes will have additive effects. That is, for any polygenic trait, the number of contributory alleles at relevant loci will determine the strength of the phenotype. Because DZ twins share half of their genotype, they will show half of MZ twins' similarity plus additional similarity due to effects of common environment. However, not all genetic effects are additive.

Some phenotypes are determined, in part, by interactions between genes. This gene–gene interaction is called 'epistasis' and it has multiplicative effects. Epistasis causes MZ twins to resemble each other more than twice as much as DZ twins do. Genetic dominance also causes MZ twins to be more than twice as similar as DZ twins, because DZ twins are correlated, on average, 0.25 for the genetic factors that are perfectly correlated (1.00) between MZ twins. Whenever r_{MZ} exceeds $2r_{DZ}$, nonadditive genetic effects (whether epistatic or dominant, or both) are evident and there is no evidence of common environmental effects. In such cases, we generally substitute for the common environmental factor, a factor that estimates either dominance or epistasis.

Second, simple twin models also assume that common environments are no more similar for MZ than for DZ twins. There are ways to test and control for violations of the common environments assumption. For example, we can measure social environments directly to determine whether they are more similar for MZ than for DZ twins and can include the effects of specific environments in our models. In adult twins, we can measure and control for frequency of contact between twin pairs, where increased contact may increase common environmental experiences.

A third assumption of the simple twin model is that genetic and environmental factors are uncorrelated. However, to the extent that we passively receive environmental influences associated with our parents' genotypes, elicit gene-correlated, environmental feedback for our phenotypic behaviors, or actively select environments amenable to our genotypes, we experience gene–environment (GE) correlations. With the advent of genotyping, it has become possible to identify specific allelic loci and to measure their associations with particular environments. Short of genotyping, however, a 'children-of-twins design' can help us to estimate specific GE correlations. Insofar as parental genes and childhood rearing environments are correlated, a greater similarity of rearing environments between the children of MZ twins than between the children of DZ twins is expected, because of their parents' differing levels of genetic relatedness.

A fourth assumption of the classic twin model is that there are no gene-by-environment (G × E) interactions. One need only consider how different individuals thrive in dissimilar environments to realize that G × E interactions exist. One way to test for the presence of G × E interactions is to study MZ twin pairs discordant for a specific environmental exposure.

A fifth assumption of the simple twin model is that parents of twins are randomly mated. That is, the expected phenotypic correlation between parents is zero. One need only look at heights of spouses in the general population to see that this assumption is not always correct. When spouses resemble each other more than would be expected by chance, we can assume that they are also correlated for genetic factors contributing to their phenotypic similarity. If mother and father carry some of the same segregating genes, then DZ twins will share, on average, more than 50% of their segregating genes. Twin models that do not control for parental similarity are therefore likely to yield underestimates of h^2. As the difference in the genetic similarity of MZ and DZ twins will be less than 50%, doubling the difference between MZ and DZ correlations will amount to less than 100% of the actual heritability. Twin-family designs including parents can test, measure and control for nonrandom mating and its effects on heritability estimates.

A sixth assumption is that any differences between MZ twins are due to nongenetic, nonshared factors. In fact, phenotypic differences between MZ twins sometimes occur because of genetic factors, particularly where there is discordant inactivation of maternally and paternally inherited X chromosomes in female twins. Human male sex chromosomes are XY and human female sex chromosomes are XX. For those loci of the X chromosome with a counterpart on the Y chromosome, females' X chromosomes contain no redundancy and both alleles are active. However, there are some loci on the X chromosome without Y counterparts, so that two alleles are redundant. It appears that many of females' redundant alleles are inactivated randomly during embryonic development across all of the organism's cells. If the inactivation of maternal and paternal X chromosomes is not symmetric when the organism divides in MZ twinning, it is possible that one twin's cells will have predominantly activated the maternal X alleles, while her co-twin's cells will have predominantly activated the paternal X alleles.

This kind of asymmetry in X-inactivation before twinning may lead to identical twin sisters discordant for red–green color blindness. When a father has normal color vision and a mother carries the allele for color blindness on one X chromosome, all of their sons who inherit the maternal X chromosome carrying the putative allele will be affected, as there is no corresponding allele on the Y chromosome to dilute its effects. In most cases, however, daughters who inherit the putative allele will not themselves develop color blindness, because approximately half of these alleles will be inactivated and their counterparts for normal color vision on the paternal X chromosome will be activated. However, in the case of MZ twinning, if cells with activated alleles for normal color vision line up, early in development, on one side of the blastomere, while cells with activated alleles for color blindness line up on the other side, the blastomere's division into MZ twins may leave one twin with the majority of alleles for normal color vision and the other with a majority of alleles for color blindness.

MZ discordance for mitochondrial disorder provides another example of genetically based discordance, as these disorders occur when mitochondrial genetic material in the cytoplasm is distributed asymmetrically between twins. Finally, postzygotic genetic mutations may result in MZ twin discordance. Although these genetically based phenotypic differences between MZ twins have been noted in the literature, they occur with such rarity that they make little difference when we estimate genetic and environmental effects from large twin samples.

Calculations of Parameters in More Complex Twin Models

The relative contributions of genetic and environmental factors to a single phenotype are estimated easily from twin data by solving a series of regression equations. However, because univariate methods are insufficient for more complex questions, multivariate methods are often used. When, for example, we suspect that assortative mating has biased our estimates of heritability, parents of twins may be added to our analyses. In a study of psychological disorders that co-occur more often than by chance, a twin model can be applied to both disorders simultaneously to find out whether the same genes contribute to the comorbid disorders. In developmental studies, to determine

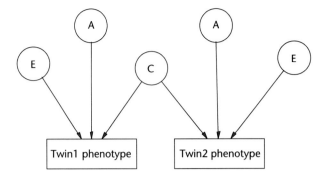

Figure 1 Path diagram of the simple twin model in which additive genetic factors (A), common environmental factors (C) and unique environmental factors (E) are latent variables, all contributing to observed or manifest variables of twin phenotype.

whether or not different genetic factors contribute to the organism's phenotypic expression at different ages, longitudinal, repeated measures of the same phenotype for each twin are added to the twin model. Whether we include additional family members, additional phenotypes, or additional occasions, our calculations become increasingly unwieldy as the number of unknowns and the number of equations to solve for them grow. Fortunately, computer software is available to handle these more complex twin models.

Because the equations used describe the structure of relationships among multiple variables and because they provide models of potential genetic and environmental influences, they are called structural equation models (SEM). Computerized structural equation modeling uses routines for matrix algebra and model fitting to estimate genetic and environmental influences on phenotypes. These programs identify those parameter values that best minimize the difference between predicted and observed phenotypes for each individual. In addition to their efficiency, computerized SEM software also produces path diagrams of the relationships among variables (**Figure 1**).

Unique Applications of Classic Twin Study Design

Until recently, most twin studies in the behavioral sciences focused on observable behavior for either continuous, quantitative traits such as social extraversion or for discrete phenotypes such as schizophrenia. Investigators reasoned that multiple, unseen phenotypes mediated the effects of genes on observed behavioral phenotypes. Many investigators hypothesized that genes influenced brain structure and function and that these, in turn, affected the probability of schizophrenia. Insofar as these mediating phenotypes

could not be observed, they were known as 'endophenotypes'.

With neuroimaging techniques, aspects of brain structure and activity previously considered endophenotypic for observable behavior are now observable phenotypes themselves. Thompson and colleagues conducted a twin study in 2001 using neuroimaging and found that genetically identical twins were virtually identical in their distribution of their frontal gray matter, more similar than were DZ pairs and far more similar than were genetically unrelated control pairs. This, of course, suggests strong heritability. Previous twin studies of intelligence have shown strong heritability for general intellectual function, and previous studies of brain morphology have shown an association between general intelligence and activity in the frontal gray matter. The study of Thompson and colleagues suggests one path by which genes may influence general intellectual function.

Beyond Classic Twin Design

In addition to the classic twin design, there are several other ways to use twin data to examine genetic and environmental contributions to particular phenotypes. General categories of other twin designs include studies of twins reared apart, co-twin control studies and children-of-twins studies.

Twins reared apart

The study of twins reared apart allows one to combine the strengths of twin design with those of adoption design. Twins reared apart provide a control group for twins reared together by which to estimate any effects of being reared together with a twin. MZ twins reared together (MZTs) are more similar to each other than are MZ twins reared apart (MZAs) for many personality traits, including positive emotionality, suggesting the influence of shared rearing environment. Perhaps surprisingly, however, MZAs show greater similarity for other personality traits than to MZTs, suggesting a disidentifying effect of shared rearing environment for these traits, including negative emotionality. It appears that twins reared together may experience pressures to differentiate from each other in ways that prevent them from expressing their genotype as fully as they might in the absence of a co-twin reared in the same family.

Co-twin control studies

Phenotypically discordant twin studies

Often, co-twins of twins affected by a particular trait or disorder may serve as a natural experimental control. For example, twins discordant for schizophrenia can help us in several ways to understand the disorder better. First, discordant MZ pairs may provide evidence of any brain anomalies associated with schizophrenia, because the well twin presumably has a brain like the ill twin would have, if not for the disorder. Magnetic resonance imaging (MRI) of the brain of a schizophrenic patient may be compared with the MRI of the nonaffected co-twin. Any consistent differences between the brains of schizophrenic twins and those of their well co-twins may indicate changes in brain morphology associated with the disorder. This is an example of one type of co-twin control study, in which well co-twins serve as controls, or as a baseline, from which their ill twins' differences are measured and evaluated.

Environmentally discordant MZ twin studies

Twins may be used to identify genotypes particularly vulnerable or resilient to specific environments. Among MZ twins discordant for exposure to an environmental risk or protective factor, those pairs whose target phenotypes differ most may be considered most vulnerable to the effects of that environmental factor. For example, one might test the effects of sleep deprivation on the performance of a memory task across many genotypes using a sample of MZ twin pairs. One MZ twin would be sleep-deprived, while the co-twin control would receive sufficient sleep before completing the task. Those twin pairs whose performance scores diverged most on the performance task would represent genotypes most vulnerable to memory deficits from sleep deprivation. Those pairs with the least discrepant scores would likely represent genotypes most resilient to the effects of sleep deprivation on memory.

Children-of-twins studies

A third form of co-twin control study can be seen in certain children-of-twins designs. An important, early application of the children-of-twins design examined how schizophrenia is transmitted from generation to generation. Despite evidence from twin studies of a high heritability for schizophrenia, some researchers continued to argue that family environment was the most important risk for schizophrenia. Gottesman and Bertelsen, in 1989, used a unique twin design that considered the children of MZ and DZ twin pairs, in which one or both twins had schizophrenia. Contrary to environment-oriented predictions that maternal schizophrenia would increase offspring risk for schizophrenia, the level of schizophrenia risk to the offspring of *well* MZ twins was equal to the level of risk experienced by their schizophrenic co-twins' offspring, and also equal to the risk for schizophrenia among offspring of schizophrenic DZ twins. This children-of-twins design provided evidence that genetic

vulnerability to schizophrenia existed independent of parental expression of the schizophrenic phenotype.

In 1993, Segal proposed another use for children of twin designs, as a test of evolutionary hypotheses about kin preference in altruism. For example, if altruism increases with genetic relatedness, then MZ twins will show greater altruism toward their co-twin's children than DZ twins would toward their co-twin's children. Similarly, the children of MZ twins, genetic half-siblings, should show greater altruism toward each other than the children of DZ twins, genetic cousins, under conditions of kin preference.

Twin Studies Provide a Genetically Informative Control for Environment-only Family Studies

Perhaps owing to the environmental interests of Freud and Skinner, decades of research have probed family environments for causes of behaviors that show familial resemblance. Divorce, for example, tends to run in families: offspring of divorced parents are at higher statistical risk for divorce than are offspring of nondivorced parents. Decades of research on the intergenerational transmission of divorce has investigated how the pre- and postdivorce family environment might affect children in ways that would make them more vulnerable to divorce in adulthood. Genetically uninformative data have suggested that the lack of a successful marital role model, greater psychological insecurity associated with parental departure from the home, more parental conflict, less parental monitoring, earlier sexual activity and lower household socioeconomic status all contribute to offspring's risk of divorce. Researchers usually attribute the intergenerational transmission divorce to the direct or indirect effects of an adverse family environment. In 1993, however, McGue and Lykken examined the phenotype of divorce using data from the Minnesota Twin Registry. Their analyses show that genetic factors and nonshared, nongenetic factors both contribute significantly to divorce risk. Furthermore, the data show that c^2 for divorce is zero: shared family environment apparently contributes nothing to divorce risk. Familial resemblance for divorce derives solely from genetic factors. (Of course, genetic factors may include gene-correlated environments.)

In a follow-up study using the same sample, Jockin and colleagues in 1996 found that heritable personality traits, associated with divorce risk, accounted for a sizeable portion of genetic influences on divorce. These data suggest that, as genes are passed along from parent to child that increase the probability of certain personality traits, the probability of divorce is also passed along. The heritable personality traits positively associated with divorce risk included negative emotionality (similar to neuroticism) and positive emotionality (similar to extraversion) for both sexes, whereas constraint showed a negative association with divorce risk, stronger in women than in men. One can easily imagine how these personality traits might be associated with the environmental risk factors for offspring of divorce observed in genetically noninformative studies.

Conclusion

Twins provide a natural experiment through which we can begin to discern the relative importance of genes and environments to the development of specific traits, behaviors and disorders. Although twin studies also have practical and methodological difficulties, they provide benefits not available in adoption studies, where atypical biological and adoptive parents, selective placement in adoptive homes, additional confidentiality issues, and diminishing numbers in the general population create methodological and practical challenges.

Twin studies may be particularly helpful in developmental studies or studies of phenotypes with variable age of onset, as they control for age effects. Twin studies provide one way to begin disentangling the separate and interactive influences so often confounded in genetically uninformative studies of 'environmental' effects. Until we understand the specific effects of genes both independently and in combination, we cannot estimate accurately the true effects of the environment on phenotypic development.

Further Reading

Augustine (2000) *The City of God*, translated by M Dods. New York: Modern Library Paperback Classics.

Bouchard TJ and Propping P (1993) *Twins as a Tool of Behavioral Genetics*. New York: John Wiley.

Bulmer MG (1970) *The Biology of Twinning in Man*. Oxford, UK: Clarendon Press.

Gottesman II (1991) *Schizophrenia Genesis: The Origins of Madness*. New York: WH Freeman.

Gottesman II and Bertelsen A (1989) Confirming unexpressed genotypes for schizophrenia: risks in the offspring of Fischer's Danish identical and fraternal discordant twins. *Archives of General Psychiatry* **46**: 867–872.

Heath AC, Eaves LJ and Martin NG (1998) Interaction of marital status and genetic risk for symptoms of depression. *Twin Research* **1**: 119–122.

Jockin V, McGue M and Lykken DT (1996) Personality and divorce: a genetic analysis. *Journal of Personality and Social Psychology* **71**(2): 288–299.

Kendler KS (1993) Twin studies of psychiatric illness: current status and future directions. *Archives of General Psychiatry* **50**: 905–915.

Kendler KS (2001) Twins studies of psychiatric illness: an update. *Archives of General Psychiatry* **58**: 1005–1014.

MacGilivray I, Campbell DM and Thompson B (1988) *Twinning and Twins*. New York: John Wiley.

McGue M and Lykken DT (1993) Genetic influence on risk of divorce. *Psychological Science* **3**(6): 368–373.

Neale MC and Cardon LR (1992) *Methodology for Genetic Studies of Twins and Families*. Boston, MA: Kluwer Academic Publishers.

Rende RD, Plomin R and Vandenberg SG (1990) Who discovered the twin method? *Behavior Genetics* **20**(2): 277–285.

Segal N (1993) Twin, sibling, and adoption methods: tests of evolutionary hypotheses. *American Psychologist* **9**: 943–956.

Tellegen A, Lykken DT, Bouchard TJ, *et al.* (1988) Personality similarity in twins reared apart and together. *Journal of Personality and Social Psychology* **54**: 1031–1039.

Thompson PM, Cannon TD, Narr KL, *et al.* (2001) Genetic influences on brain structure. *Nature Neuroscience* **4**(12): 1253–1258.

Twin Studies: Software and Algorithms

Michael C Neale, *Virginia Commonwealth University, Richmond, Virginia, USA*

The implementation and advantages of likelihood-based model fitting approaches to the analysis of data collected from MZ and DZ twins. Estimation of heritability and environmental components of variance that contribute to individual differences.

Advanced article

Article contents

- Historical Background
- Measuring Similarity
- Estimating Heritability
- Structural Equation Modeling of Twin Data
- Model Fitting to Covariance Matrices
- Model Fitting to Raw Data
- Binary and Ordinal Data

Historical Background

In 1865, Sir Francis Galton (1822–1911) formally described the differences between identical (monozygotic, MZ) and fraternal (dizygotic, DZ) twins, and suggested that MZ twins were genetically identical (Galton, 1865). This distinction yields a natural experiment, because comparison of the similarity of these two types yields information about the relative importance of heredity versus environment

to individual differences. To quantify these sources of variation, it is necessary to have (1) a way to assess similarity and (2) an algorithm to use this information to estimate the parameters of interest. The optimal choices for (1) and (2) vary according to the type of measurement that is taken from the twins.

Measuring Similarity

For continuous, normally distributed measures the Pearson correlation coefficient is a useful summary statistic. It is a very simple function of the variances and covariance of a pair of variables. The variance is calculated as the average of the squared deviations of scores from the mean, which is in turn calculated as the sum of the scores divided by the number of scores in the list. The variance can be computed for both members of a twin pair, where one member is arbitrarily assigned as twin 1 and the other as twin 2. Covariance is calculated as the average of the products of the two scores from a twin pair. The correlation coefficient is the covariance divided by the square root of the product of the two variances. This standardized statistic varies from -1 indicating complete disagreement to $+1$ (perfect agreement) with 0 indicating that they are independent.

Estimating Heritability

The classical twin study yields three basic statistics: the total phenotypic variance V_P, and the covariances of MZ and DZ twins cov_{MZ} and cov_{DZ}. To estimate heritability, measures of similarity of MZ and of DZ twin pairs are used in combination with a model for variation in the population that makes predictions about the resemblance between relatives. Under the polygenic model described in the quantitative genetics entry in this Encyclopedia, variation is caused by additive genetic factors (A) and environmental factors which may be divided into those that are shared in common (C) by members of a twin pair and those that are unique to each twin (E), and the amounts of variance caused by these components are a^2, c^2 and e^2 respectively. The total variance of an individual under this 'ACE' model is predicted to be $V_P = a^2 + c^2 + e^2$. MZ twins sharing all their genes in common have a predicted covariance of $cov_{MZ} = a^2 + c^2$, while that of DZ twins is $cov_{DZ} = 0.5a^2 + c^2$. Given observed values of V_P, cov_{MZ} and cov_{DZ}, these simultaneous equations may be solved. (*See* Quantitative Genetics.)

Twice the difference between the MZ and DZ covariance correlations gives an estimate of the additive genetic variance: $2(cov_{MZ} - cov_{DZ}) = a^2$. This a^2 estimate may be subtracted from the MZ covariance to find $cov_{MZ} - a^2 = c^2$, and the difference between the MZ covariance and the phenotypic variance yields $V_P - cov_{MZ} = e^2$. All three estimates may be divided by the phenotypic variance to yield the proportions of variance associated with each source. This algorithm, along with some alternative statistics, such as the Holzinger (1929) ratio $H = (r_{MZ} - r_{DZ})/(1 - r_{DZ})$, was popular until the 1970s.

Although it is useful as a rule of thumb, there are at least eight problems with this algorithm for estimating components of variance:

1. It is possible to obtain nonsensical estimates of the heritability, both greater than 1.0 and less than zero.
2. It takes no account of the relative precision of the cov_{MZ} and cov_{DZ} statistics, which may be unequal if the sample sizes differ or if the magnitude of the correlation is different, or (as is usually the case) both.
3. There is no assessment of whether the correlations are consistent with the additive genetic model.
4. It does not easily generalize to the multivariate case to permit the examination of why variables correlate with each other.
5. There is no easy way to correct estimates for the effects of covariates such as age and sex.
6. It does not generalize to extended twin studies that involve other relatives.
7. It is inefficient when there are missing data.
8. It is not suitable for selected samples of twins.

Modern methods overcome these limitations by using more sophisticated computer software and algorithms.

Structural Equation Modeling of Twin Data

Structural equation modeling is very popular in the social sciences, and applications are common in econometrics, sociology, psychology and many other fields. A structural equation model (SEM) is essentially a linear regression model, in which two forms of statistical relationship are imposed between variables: correlational and causal. Two types of variables may be specified: observed, the measures being analyzed; and latent, those factors that have not been observed but that are hypothesized to exist. There is a mathematically complete description of a structural equation model, called a path diagram, in which the observed variables are shown in boxes, latent variables are shown in circles, causal relations are shown as single-headed arrows, and correlational ones are shown as double-headed arrows. The classical twin study is analyzed as a multiple group SEM because MZ twins have a different predicted covariance from DZ twins. Therefore, almost any SEM software that has multiple group capabilities may be used to fit the simple univariate twin model. Examples include LISREL (Jöreskog and Sörbom, 1989), EQS (Bentler, 1989), Amos (Arbuckle, 1994) and Mx (Neale *et al.*, 1999). Of these programs, only Mx is freely available on the Internet (see Web Links), and it is used most widely for twin studies. It also has features for the analysis of raw data and correction for ascertainment not found in the other programs.

Model Fitting to Covariance Matrices

The simple ACE model described above yields the following predicted covariance matrices for MZ and DZ twin pairs:

$$\sum = \begin{bmatrix} a^2 + c^2 + e^2 & \alpha a^2 + c^2 \\ \alpha a^2 + c^2 & a^2 + c^2 + e^2 \end{bmatrix} \quad (1)$$

where *a*, *c* and *e* are path coefficents for additive genetic, common environment, random environment and genetic dominance effects respectively; and $\alpha = 1.0$ for MZ and $\alpha = 0.5$ for DZ twins.

The parameters of this structural equation model for twin data may be estimated by several approaches, of which the most popular is maximum likelihood (ML). ML estimation is usually accomplished by numerical optimization, which uses information on the slope and curvature of the likelihood at specific trial values of parameters to find the set of parameter values that maximizes the likelihood. When model fitting to covariance matrices, the goodness-of-fit function for each group *g* is

$$\text{ML}_g = N_g \left\{ \ln |\Sigma_g| - \ln |S_g| + \text{tr}(S_g \Sigma_g^{-1}) - 2m \right\} \quad (2)$$

where N_g is the sample size minus one for group *g*, S_g is the observed covariance matrix, Σ_g is the covariance matrix predicted by the model for the trial values of the parameters (eqn [1]); *m* is the number of phenotypes measured on each twin; and Σ and Σ^{-1} denote the determinant and inverse of the matrix Σ respectively. This function is actually the difference between the likelihood of a saturated model in which all the covariances are estimated freely, and the model being tested. Under certain regularity conditions, it is asymptotically distributed as χ^2 with degrees of freedom equal to the difference between the number of free parameters in the model and the number of observed statistics. That is, the function value gives a measure of the fit of the model to the data.

Typically, several reduced models (setting either $a^2 = 0$ or $c^2 = 0$ or both) are fitted to the covariances to test specific hypotheses about influence of genes and environment on the phenotype. The difference between the fit statistic of a model and that of a submodel with *w* parameters fixed at prespecified values may be interpreted as χ^2 with *w* degrees of freedom. Therefore, likelihood-ratio tests of alternative hypotheses may be conducted (e.g. that *a* or *c* is zero), and, given sufficient statistical power, parsimony-based fit indices such as Akaike's Information Criterion may be used to find the simplest model that explains the data adequately. At the same time, confidence intervals on the parameter estimates may be computed (Neale and Miller, 1997).

The model-fitting approach solves the first three problems identified above: heritability estimates can never exceed 1.0; it incorporates sample sizes and the precision of the data; and the overall fit of the model is assessed. Problem (4), generalization to the multivariate case, is also handled easily. An elaboration of the predicted covariance matrix Σ is to substitute $m \times m$ genetic and environmental covariance matrices *A*, *C* and *E* for the single values a^2, c^2 and d^2. This $2m \times 2m$ predicted covariance matrix is then fitted to the observed covariance matrix organized with twin 1 variables $T_1V_1, \ldots, T_1V_m$ followed by twin 2 variables $T_2V_1, \ldots, T_2V_m$. The structure of the predicted genetic and environmental covariance matrices may be specified according to a variety of models.

One popular approach is to specify a small number of factors that influence all traits plus some variance specific to each trait, so that $A = FF' + G$ where *F* is an $m \times f$ matrix of genetic factor loadings from the *f* factors to the *m* phenotypes, and *G* is a diagonal matrix of phenotype-specific genetic effects. Another is to freely estimate all the genetic variances and covariances with $A = LL'$ where matrix *L* is lower triangular. Comparisons between these models can be used to assess how well simple factor models account for the genetic variance and covariance pattern of multiple traits. Originally implemented by Martin and Eaves (1977), these methods were described in detail by Neale and Cardon (1992), and Mx scripts for them may be downloaded from the website (see Web Links).

Several approaches may be used to solve problem (5), the addition of covariates such as age. One is to subclassify twin pairs according to age, and to test for heterogeneity between age groups. The second is to regress the phenotypes on age and analyze the residual scores. The third approach is to model the individual means of the twins as well as the covariances, which is accomplished by fitting models to raw data.

Model Fitting to Raw Data

Significant flexibility is added when models are fitted using the raw data instead of summary statistics like covariance matrices. All that is required is to directly evaluate the normal theory likelihood of each raw data vector, which in the case of twins consists of the data on both twin 1 and twin 2. If there are *m* observed variables on each twin, the normal probability density function of a column vector of twins observed scores x_i is given by

$$|2\pi\Sigma|^{-2m/2} \exp\left[-\tfrac{1}{2}(x_i - \mu_i)'\Sigma^{-1}(x_i - \mu_i)\right] \qquad (3)$$

where Σ is the predicted covariance matrix and μ_i is the (column) vector of predicted means of the variables. The joint likelihood of the N independent pairs in the sample is computed as the product of the likelihoods of all pairs. Typically, the likelihood of any pair is less than 1.0, which causes computational problems because computers have limited precision representation of real numbers. In practice, therefore, software algorithms compute the logarithm of the likelihood and sum the log-likelihoods over the sample.

If some of the variables are not observed on either one or both twins, the likelihood function in eqn [3] is revised. The predicted mean vector is reduced so that it contains only those predicted means corresponding to those variables actually measured. The covariance matrix Σ is also filtered so that it only contains the rows and columns corresponding to the data vector. The likelihood is then computed as usual. This method solves the problem of missing data, as long as the mechanism of missingness is either completely random or random (MCAR or MAR) (Little and Rubin, 1987). It also provides a treatment for extended twin studies that include other relatives, since family sizes and participation may be expected to vary.

An extension of the method is to weight the likelihood in eqn [3] by some relevant quantity. Nonrandomly ascertained pairs of twins, for example, those in which at least one has an extreme value of a trait, may be analyzed by using a weight equal to the reciprocal of the probability that the pair is ascertained. Fitting models to raw data therefore addresses problems (5)–(8), albeit at the expense of increased computer time. Yet further extensions of the method allow fitting of mixture distributions, where the predicted covariances or means differ between classes. The likelihood is computed as the sum of the weighted likelihoods of each class. Such extensions prove useful in the analysis of genetic linkage and in the analysis of data collected from twin pairs in which zygosity diagnosis is not perfect.

Binary and Ordinal Data

In many substantive areas, phenotypic measures are not made on a continuous scale with a normal distribution, but are limited to simple binary affected versus unaffected or ordinal scales of the type zero/few/many. One approach to this type of data is to assume that underlying these distinctions there is a normal distribution of liability on which there are abrupt thresholds, so that those with a liability greater than t_j will have an observed score of j. This threshold model, which has been popular in genetics for many years (Wright, 1934; Falconer, 1960), provides a way to apply multifactorial quantitative genetic theory to the analysis of binary and ordinal data.

One analytic approach is to compute summary statistics consisting of tetrachoric or polychoric correlations and a weight matrix based on the variance–covariance matrix of these correlations and then use a weighted least-squares fit function to estimate parameters. As in the analysis of continuous data, this approach does not handle problems of missing data very effectively, so it may be preferable to compute directly the likelihood of the raw observed data vectors via numerical integration of the likelihood function of eqn [3]. A problem with this latter approach is that it becomes computationally intensive, especially in the multivariate case. The cost to integrate over p dimensions is an exponential function of the form K^p which dramatically increases with p.

Furthermore, the number of unique response sets that the data may contain increases rapidly according to r^{2m} for r responses on m variables measured in each twin. At the time of writing, the direct ML method is practical for up to seven ordinal variables per twin. Future improvements in computer hardware performance, software algorithms for optimization and integration, and evaluation of the likelihood should make the method viable even for quite large numbers of variables.

See also

Quantitative Genetics
Twin Methodology
Twin Studies

References

Arbuckle J (1994) *Amos 3.5 Documentation Package*. Chicago, IL: Smallwaters.

Bentler PM (1989) *EQS: Structural Equations Program Manual*. Los Angeles, CA: BMDP Statistical Software.

Falconer DS (1960) *Quantitative Genetics*. Edinburgh, UK: Oliver & Boyd.

Galton F (1865) Hereditary talent and character. *Macmillan's Magazine* **12**: 157–166, 318–327.

Holzinger KJ (1929) The relative effect of nature and nurture influences on twin differences. *Journal of Educational Psychology* **20**: 245–248.

Jöreskog KG and Sörbom D (1989) *LISREL 7: A Guide to the Program and Applications*, 2nd edn. Chicago, IL: SPSS, Inc.

Little RJA and Rubin DB (1987) *Statistical Analysis with Missing Data*. New York: John Wiley.

Martin NG and Eaves LJ (1977) The genetical analysis of covariance structure. *Heredity* **38**: 79–95.

Neale MC and Cardon LR (1992) *Methodology for Genetic Studies of Twins and Families.* New York: Kluwer.

Neale MC and Miller MM (1997) The use of likelihood-based confidence intervals in genetic models. *Behavior Genetics* **27**: 113–120.

Neale MC, Boker S, Xie G and Maes H (1999) *Mx: Statistical Modeling*, 5th edn. Richmond, VA: Virginia Commonwealth University.

Wright S (1934) The method of path coefficients. *Annals of Mathematical Statistics* **5**: 161–215.

Web Links

Mx. A matrix algebra interpreter and numerical optimizer for structural equation modeling
http://www.vcu.edu/mx

Twin Study Contributions to Understanding Ontogeny

Kathryn S Lemery, *Arizona State University, Tempe, Arizona, USA*

The genetic contribution to individual differences in development during infancy and childhood can be estimated with a twin study.

Introduction

The twin study is a potent tool for advancing the understanding of neurodevelopmental and ontogenetic aspects of behavior. By comparing the similarity of identical co-twins to the similarity of fraternal co-twins on a given trait, the relative contributions of genetic and environmental factors can be estimated. The power of molecular genetic designs can be increased by identifying highly heritable phenotypes, and by determining the proportion of the total genetic effect due to the candidate chromosomal region within a twin design. Opposite-sex twins provide a unique method of studying sex differences, and important environmental influences on development can be identified after controlling for genetic effects. Further, in a multivariate design, the twin study can identify relevant biological substrates, elucidate the etiology of the co-occurrence of two or more phenotypes and assist in forming more homogenous categories of deviation and disorder. (*See* Twin Studies.)

Contributions of Twin Studies to Understanding Ontogeny

The twin study is a powerful design for elucidating neurodevelopmental and ontogenetic aspects of behavior. The classic twins-reared-together design includes both identical, monozygotic (MZ) twins, who share 100% of their genes, and fraternal, dizygotic (DZ) twins, who share on average 50% of their segregating genes. MZ twins are more similar to each other than DZ twins for a heritable trait. The heritability coefficient is the proportion of the differences among individuals within a population for a particular trait that are due to genetic differences. Thus, heritability does not apply to the development of single individuals nor to differences between populations. Results from twin studies have indicated that individual differences in both cognitive and emotional behaviors during infancy and childhood are heritable, with approximately 50% of the variance due to genetic influences (commonly ranging from 30% to 70%). Twin studies are equally informative for studying environmental effects. Environmental influences are parsed into shared environmental influences that create similarities among individuals, and nonshared environmental influences that create differences among individuals. Evidence for shared environmental effects exists for some traits such as positive emotionality in infancy and early childhood; yet the majority of environmental influences are not shared. (*See* Identical Twins Reared Apart; Twin Methodology.)

This article outlines seven ways in which twin studies can advance the study of human development. Specifically, twin studies:

- estimate the relative contributions of genetic and environmental factors;
- increase the power of molecular genetic designs;
- provide a unique method for studying sex differences;
- help to form better definitions of deviation and disorder;
- elucidate the etiology of the co-occurrence of two or more phenotypes;
- evaluate environmental risk factors; and
- identify biological substrates and correlates.

In these ways, the study of twins provides a powerful method for understanding the influence of genes and environments throughout ontogeny.

Twin Studies Estimate the Relative Contributions of Genetic and Environmental Factors

The phenotypic variance of a trait at a particular age is parsed into its genetic and environmental underpinnings. A heritable trait is not necessarily present at birth or unmodifiable. Results from twin studies suggest that the influence of genes changes over the lifetime. For example, variation in neonatal temperament (e.g. irritability, activity) is due to the environment (Riese, 1990), whereas temperamental variation later in infancy is heritable (Goldsmith *et al*., 1999). The heritability of intelligence increases with age (Pederson *et al*., 1992; Boomsma, 1993), and the environment begins to have a larger impact on antisocial behavior after the transition to adolescence. Genetic factors are not as important in predicting antisocial behavior in adolescence because it becomes the norm for adolescents to engage in antisocial acts (Moffitt, 1993), and many more individuals began to express these behaviors for sociocultural reasons. (*See* Personality and Temperament.)

Genetic and environmental effects on the onset of behaviors can also be studied within the context of a twin design. Influences on the timing of developmental milestones of infancy can be examined – such as the onset of crawling, stranger anxiety and understanding that objects are permanent and exist even when out of sight. Developmental theorists are split on whether infant milestones are biologically driven or experientially driven. The twin design can inform this debate. For developmental deviations or anomalies, whether or not a higher genetic liability (i.e. predisposition) corresponds to an earlier onset can be considered.

When extended to the multivariate case, genetic and environmental effects on the structure of continuity and change across age can be examined. For early temperament and cognition, genes contribute to change as well as continuity (see Emde and Hewitt, 2001). A phenotype may be highly heritable at both early and later ages, but different genetic factors may be influencing the phenotype at the different ages. In this case, the genetic contribution to continuity would be low. Thus, significant progress in understanding development can be made by documenting genetic and environmental effects on individual differences in behavior. Influences on the timing of developmental onsets, and continuity and change across age can be explored.

Twin Studies Increase the Power of Molecular Genetic Designs

Molecular genetic designs, such as linkage and association, require well-defined phenotypes that are highly heritable for good statistical power. The twin design is useful in identifying highly heritable traits that can then be subjected to molecular methods. Even more striking, the power of molecular genetic techniques is increased by using them within the context of a twin design including both MZ and DZ twins. The DZ twins are used to determine whether or not genetic similarity at a particular chromosomal region is related to phenotypic similarity. By also including MZ twins, the phenotypic variance can be parsed into genetic effects due to variation at the candidate chromosomal region, other unidentified genetic effects and environmental effects. In this way, the proportion of the overall genetic effect due to the candidate chromosomal region can be estimated, indicating the importance of this region for overall phenotypic variation.

Twin studies directly address the limitations of the power of current molecular genetic techniques. Further, they facilitate the study of age and sex differences, which have a long history of being misunderstood or ignored in the psychological literature.

Twin Studies Provide a Unique Method of Studying Sex Differences

Opposite-sex twins provide a unique opportunity to study the development of sex differences. Twin studies offer advantages over the traditional comparison of single-born girls and boys, because opposite-sex twins are precisely the same age, they simultaneously experience a range of family variables (such as socioeconomic status, family constellation and parental attitudes), and they share on average half of their genes. Most sex differences are not present at birth or during infancy, suggesting that differences seen later in childhood may be largely due to socialization factors. Longitudinal twin studies that span infancy and childhood can determine whether estimates of heritability become significantly different for boys and girls across age, suggesting that genes may also contribute to later-appearing sex differences. In addition to facilitating the study of sex differences, twin studies help to distinguish between disordered and nondisordered complex behaviors.

Twin Studies Assist in Forming Better Definitions of Deviation and Disorder to Inform the Taxonomy of Psychiatric Disorders

Diagnostic categories are constellations of features that are sometimes weakly related. These categories are helpful in clinical decision-making, but their heterogeneous nature can make it difficult for research. One problematic area is identifying where the boundary between disordered and nondisordered behavior should be placed. This is unclear for many psychiatric phenotypes, such as major depression. A twin study can decompose the relationship between subthreshold levels of symptoms and the diagnosed disorder into common or separate genetic and environmental effects. For example, the phenotype may appear continuous across individuals but have different underlying genetic influences. In addition, high levels of a trait may be associated with different frequencies of an influential gene, or may be linked with different alleles than lower levels of the trait. A phenotypic threshold may be placed based on genetic information gained from the twin method.

In addition to identifying meaningful thresholds between normal range variation and disordered extremes, the relationship among components of a single clinical disorder may be examined. Many disorders have multiple components, and these components may have different genetic roots. Genetic heterogeneity may be examined within the twin design by considering the genetic correlations among these features of a disorder. For example, attention deficit hyperactivity disorder (ADHD) is characterized by disinhibition, impulsivity and inattention (American Psychiatric Association, 1994). There is some evidence

both from twin studies and molecular genetic studies that these features may have different genetic underpinnings (e.g. Waldman et al., 1998), which suggests the utility of dividing ADHD into more homogeneous subgroups. In addition to informing the taxonomy of psychiatric disorders, the twin design can clarify the association between disorders or other phenotypes. (See Autism: Genetics; Developmental Psychopathology.)

Twin Studies Elucidate the Etiology of the Co-occurrence of Two or More Phenotypes

Rates of the co-occurrence or comorbidity of childhood psychiatric disorders exceed chance. Those with one disorder are more likely to have a second disorder. These disorders may really be components of the same overarching disorder, or they may share a genetic liability. The co-occurrence of anxiety and depression in both childhood and adulthood is high. These disorders may have common genetic underpinnings, and the high rates of comorbidity reported within families support this hypothesis. The twin design can elucidate comorbidity by indicating whether the basis of the observed association is common genetic effects, common environmental effects or both. A twin study illustrated that Tourette syndrome, ADHD and conduct disorder share a common genetic influence in childhood (Comings, 1995). Twin studies are equally powerful for identifying environmental effects. (See Reading and Dyslexias.)

Twin Studies Evaluate Environmental Risk Factors Free from Genetic Contamination

Twin designs offer a research context for studying unambiguous environmental influences by controlling for genetic effects on environmental measures. Measures of the environment that have been developed in traditional developmental psychology research can be incorporated into a twin study. For example, the oft-cited relationship between parenting style and child behavior has been found to be partially due to shared genes (Plomin and Bergeman, 1991). By controlling for the genetic component of the covariance within a twin design, true environmental influences can be uncovered. Of course, identifying these environmental influences is important for designing effective interventions.

Evidence stemming from twin studies underscores the important role of the nonshared environment on behaviors in childhood. One way to identify nonshared environmental influences is to study MZ co-twin differences. Because MZ twins are genetically identical, the environment is at the root of their differences. Importantly, the availability of magnetic resonance imaging (MRI) and positron emission tomography (PET) methodologies for research has dramatically increased since the early 1990s, and these tools can be used to identify brain differences between MZ twins discordant for a developmental anomaly. Documenting these brain differences adds another piece to the puzzle of the mechanisms involved in both normal and abnormal development. Case studies of discordant MZ co-twins have elucidated several etiologies: postzygotic unequal division of the inner cell mass, unequal sharing of the blood supply from a shared placenta, mutation of a modifying gene, maladaptive development of the central nervous system and mosaicism. For example, X inactivation is thought to be the cause of MZ discordance of Duchenne muscular dystrophy, red–green color blindness and Hunter disease (Shotelersuk et al., 1999).

An experimental approach to studying environmental effects is the MZ co-twin control design. One twin receives the treatment while the other twin is studied as the control. This design was originally used by Gesell and colleagues to study practice effects on physical development in infancy (Gesell and Thompson, 1941). They demonstrated that physical training can cause physical skills to appear sooner. However, MZ co-twins who were then trained later performed equally well after a relatively shorter period of training. The MZ co-twin control design is also used to consider drug or treatment effectiveness, although ethical considerations surrounding treating one twin but not the other hinder the use of this methodology.

In summary, the twin design provides a powerful way to investigate environmental influences by controlling for genetic contamination of environmental measures and by studying MZ twin differences. In addition, the twin study can advance the investigation of development by identifying biological systems that share genetic underpinnings with complex behavior.

Twin Studies Identify Biological Substrates and Correlates of Behavior

The seventh way in which the twin design informs our understanding of development is by elucidating biological substrates of complex behavior. Genes are expressed through physiological pathways, so a genetic etiology suggests a neurobiological diathesis. By including biological measures within a twin design, phenotypic composites can be formed that are defined

across the physiological and individual behavioral levels. It follows that these composites would display a stronger association with genetic influences, and could be used to pinpoint gene action.

Activation of the hypothalamic–pituitary axis (HPA) and resulting increased levels of the hormone cortisol have been linked to both temperament and disorder in infancy and childhood. For example, newborns who display more behavioral distress during examinations also have higher salivary cortisol levels (Gunnar et al., 1989). Later in infancy at 9 and 13 months, infants with higher basal cortisol levels were better copers and had fewer negative reactions during a maternal separation. Thus, cortisol is a promising biological marker of infant emotion-related behavior and can be incorporated into a twin design to identify genetic and environmental links to complex behaviors. (See Personality and Temperament.)

Identifying biological substrates and correlates of complex behavior can help individualize pharmacological therapy that can maximize the probability of response and successful treatment. The relatively new field of psychopharmacogenetics utilizes drug responsiveness to help dissect complex genetic traits or disorders. In addition, identifying genetic variation through twin designs can individualize pharmacological therapy and maximize the probability of response. Reducing genetic heterogeneity and delineating subtypes of disorder will increase the probability of identifying associations between behavioral anomalies and genes involved in etiology, targets of drug effects and metabolism of drugs. For example, African American children diagnosed with ADHD and having two copies of the long form of the dopamine transporter gene solute carrier family 6 (neurotransmitter transporter, dopamine), member 3 (SLC6A3) were more likely to be nonresponders to methylphenidate than those with one or no copies (Winsberg and Comings, 1999). Thus, twin studies can isolate common genetic underpinnings of biological and behavioral phenotypes to inform drug therapy and treatment. (See Psychopharmacogenetics.)

Examining both biological and behavioral phenotypes within the twin design helps inform etiology by examining the basis of the association between the biological and behavioral levels of analysis, pinpointing the level of analysis where development is occurring and exploring how change at one level affects processes at other levels.

Conclusion

Elucidating the genetic and environmental underpinnings of the associations among alleles, physiological pathways and complex behaviors using the twin method has tremendous power for creating links between divergent fields (e.g. genetics, biology, psychology). Even at the behavioral level, it provides a unique method for studying sex differences, defining true environmental influences and identifying thresholds between normal and abnormal behavior. In addition, the assumptions of the twin design can be empirically tested. For example, whether or not the results from twin studies generalize to other nontwin groups can be determined by comparing them with the results from other behavioral genetics designs, such as adoption and family studies.

This article focuses on the many ways in which twin studies can inform the study of development; however, twin studies do not address several ontogenic issues. The focus on individual differences does not generalize to the complex interaction of genetic and environmental influences that determine an individual's behavior. These influences differ from person to person. Individuals with a high genetic risk within a protective environment may display the same phenotype as an individual with low genetic risk and hazardous environment, for example. Also, a behavioral anomaly may have a genetic origin, but the mechanism of effect may be environmental. Although smoking is heritable, the carcinogenic effect is environmental. The disease poliomyelitis is heritable, yet it is caused by a virus.

Coupled with advances in molecular genetic techniques and the completed map of the human genome, the twin study is a potent tool for advancing the understanding of neurodevelopmental and ontogenetic aspects of behavior.

See also

Identical Twins Reared Apart
Twin Methodology
Twin Studies

References

American Psychiatric Association (1994) *Diagnostic and Statistical Manual of Mental Disorders*, 4th edn. Washington, DC: American Psychiatric Association.

Boomsma DI (1993) Current status and future prospects in twin studies of the development of cognitive abilities: infancy to old age. In: Bouchard Jr TJ and Propping P (eds.) *Twins as a Tool of Behavioral Genetics*, pp. 67–82. New York: John Wiley.

Comings DE (1995) The role of genetic factors in conduct disorder based on studies of Tourette syndrome and attention deficit hyperactivity disorder probands and their relatives. *Journal of Developmental and Behavioral Pediatrics* 16: 142–157.

Emde RN and Hewitt JK (2001) *Infancy to Early Childhood: Genetic and Environmental Influences on Developmental Change*. New York: Oxford University Press.

Gesell A and Thompson H (1941) Twins T and C from infancy to adolescence: a biogenetic study of individual differences by the method of cotwin control. *Genetic Psychology Monographs* 24(1): 3–122.

Goldsmith HH, Lemery KS, Buss KA and Campos J (1999) Genetic analyses of focal aspects of infant temperament. *Developmental Psychology* 35: 972–985.

Gunnar MR, Mangelsdorf S, Larson M and Hertsgaard L (1989) Attachment, temperament, and adrenocortical activity in infancy: a study of psychoendocrine regulation. *Developmental Psychology* 25: 355–363.

Moffitt TE (1993) Adolecence-limited and life-course-persistent antisocial behavior: a developmental taxonomy. *Psychological Review* 100: 674–701.

Pedersen NL, Plomin R, Nesselroade JR and McClearn GE (1992) A quantitative genetic analysis of cognitive abilities during the second half of the life span. *Psychological Science* 3: 346–353.

Plomin R and Bergeman CS (1991) The nature of nurture: genetic influence on 'environmental' measures. *Behavioral and Brain Sciences* 14: 373–427.

Riese ML (1990) Neonatal temperament in monozygotic and dizygotic twin pairs. *Child Development* 61: 1230–1237.

Shotelersuk V, Tifft CJ, Vacha S, Peters KF and Biesecker LG (1999) Discordance of oral–facial–digital syndrome type 1 in monozygotic twin girls. *American Journal of Medical Genetics* 86: 269–273.

Waldman ID, Rowe DC, Abramowitz A, *et al.* (1998) Association and linkage of the dopamine transporter gene and attention-deficit hyperactivity disorder in children: heterogeneity owing to diagnostic subtypes and severity. *American Journal of Human Genetics* 63: 1767–1776.

Winsberg BG and Comings DE (1999) Association of the dopamine transporter gene (DAT1) with poor methylphenidate response. *Journal of the American Academy of Child and Adolescent Psychiatry* 38: 1474–1477.

Further Reading

Benjamin J, Ebstein RP and Belmaker RH (2002) *Molecular Genetics and the Human Personality*. Washington, DC: American Psychiatric Publishing.

Plomin R, DeFries JC, Craig IW and McGuffin P (2003) *Behavioral Genetics in the Postgenomic Era*. Washington, DC: American Psychological Association.

Plomin R, DeFries JC, McClearn GE and McGuffin P (2000) *Behavioral Genetics*. New York: WH Freeman.

LaBuda MC, Gottesman II and Pauls DL (1993) Usefulness of twin studies for exploring the etiology of childhood and adolescent psychiatric disorders. *American Journal of Medical Genetics* 48: 47–59.

Web Links

Solute carrier family 6 (neurotransmitter transporter, dopamine), member 3 (*SLC6A3*); Locus ID: 6531. LocusLink:
http://www.ncbi.nlm.nih.gov/LocusLink/LocRpt.cgi?l = 6531

Solute carrier family 6 (neurotransmitter transporter, dopamine), member 3 (*SLC6A3*); MIM number: 126455. OMIM:
http://www.ncbi.nlm.nih.gov/htbin-post/Omim/dispmim?126455

Two-dimensional Gel Electrophoresis

Shao-En Ong, *University of Southern Denmark, Odense, Denmark*

Cynthia Mui Yee Rosa Liang, *National University of Singapore, Singapore, Malaysia*

Two-dimensional gel electrophoresis separates proteins according to their isoelectric points and molecular weight. Recent developments allowing improved resolution, detection and reproducibility, coupled with sensitive protein identification by mass spectrometry has made this methodology one of the most common tools for the study of proteomes in the postgenomic age.

Introduction

A timely convergence of several technological advances has made the study of proteins easier than ever before. These include the tremendous increase in the size of primary sequence databases and the development of biological mass spectrometry that complement protein separation methods like two-dimensional gel electrophoresis (2DE). These are the cornerstones of the new science of proteomics, a term first coined by Marc Wilkins in 1994. The high degree of sophistication in microarray technology now allows the simultaneous study of gene expression for the whole genome on the messenger ribonucleic acid (mRNA) level. Why then is it so important to study protein expression? After all, proteins are notoriously difficult to work with in comparison to deoxyribonucleic acid (DNA). There is no technique in protein science that is analogous to the polymerase chain reaction (PCR) for amplifying DNA. Proteins may be present in several alternatively spliced forms and are often posttranslationally modified, with the end result that several active forms may exist for each gene. Several studies show that the level of mRNA expression correlates poorly with protein abundance. This is largely due to the various mechanisms by which gene expression and protein levels can be regulated. These factors combined make the study of proteins a challenging endeavor and also crucial for a more comprehensive understanding of cellular mechanisms.

In its early days, proteomics was largely defined by the large-scale identification and cataloging of proteins. This focus has grown to include functional studies and studies of protein–protein interaction, but all with a view to acquiring data sets on a scale that was unheard of in the days of single protein characterization. 2DE has had a considerable part to play in the development of the field, and thanks perhaps to its old-fashioned familiarity, as well as some of its unique advantages as a separation technique, it has firmly entrenched itself as one of the major proteomics tools in many laboratories today.

Little has changed in the basic principles that define 2DE since its description in 1975 by Klose and, independently, by O'Farrell. Proteins are first separated in the first dimension according to their electrical charge through a process called isoelectric focusing (IEF). As zwitterionic molecules, proteins can carry varying charges, even different polarities, depending on the pH of their surrounding environment. The isoelectric point (pI) of a protein can be approximated by the number of acidic and basic amino acids in its sequence, while the real pI depends on the microenvironment surrounding the protein's ionic groups. A pH gradient is critical to the process of IEF as the net charge of proteins changes at different points in the pH gradient. For example, under acidic conditions all acidic groups such as carboxylic groups and phenols are neutralized, while all basic groups are protonated. This leaves the protein with a net positive charge. In this case, the protein will migrate in the gel toward the cathode until it reaches the point where the pH in the gradient equals the protein's pI. With a zero net charge, the protein reaches its equilibrium position in the electric field (see also **Figure 1**). In this manner, all proteins in a sample will migrate and 'focus' to their pIs in the first-dimension gel. Subsequently, the IEF gel is transferred to the second dimension, for separation by sodium dodecyl sulfate–polyacrylamide gel electrophoresis (SDS-PAGE). The proteins, now coated with the ionic detergent SDS (1.4 g SDS per g protein) and bearing a net negative charge, behave similarly in the electrical field and migrate at rates proportional to their molecular weights.

Separating proteins by 2DE results in characteristic patterns or 'maps' of proteins. In the early days of 2DE, it was not possible to couple the separation

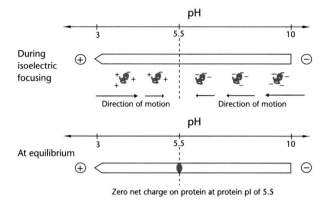

Figure 1 Isoelectric focusing – the first dimension in 2DE. Sample introduction by rehydrating the immobilized pH gradient (IPG) strip in the protein sample results in a distribution of protein along the entire length of the strip. A protein of p*I* 5.5 has different charge states at various positions along the pH gradient. After a sufficiently long isoelectric focusing (IEF) step, proteins in the IPG reach their equilibrium positions at their respective p*I*s.

step with subsequent identification of the protein. Researchers had to be content with the generation of these two-dimensional (2D) maps and the rudimentary comparisons of protein mixtures based on the patterns generated after 2DE separation. Protein identification required the use of antibodies in immunoblotting, and it was only in the late 1980s that the transfer of protein to polyvinylidene fluoride (PVDF) membranes for Edman sequencing allowed direct identification of proteins from 2D gels. Despite this limitation, the method was still attractive because separation is based on the physicochemical properties of proteins rather than some functional characteristics inherent to the sample. Therefore, it is relatively easy to start with a completely unknown mixture of proteins, separate them by 2DE and proceed to study the patterns. This unbiased assessment of a protein sample is perhaps the most attractive feature of 2DE.

Furthermore, the capability of 2DE as a protein separation technique is still largely unparalleled. With the most sensitive method of detection (proteins labeled metabolically with ^{35}S-methionine), dense 2DE maps containing more than 5000 distinct 'spots' are not uncommon. Having said that, it is worth remembering that the complexity at the protein level in the cell far exceeds the resolving capacity of any method. The protein expression profile can change dramatically in different tissue types and conditional stimuli often result in differentially modified proteins as well as expression of splice variants. The dynamic range of protein expression can be up to 12 orders of magnitude. These temporal variations at the cellular level compound the difficulties in any global studies of protein expression.

Recent Technological Advances in Two-dimensional Gel Electrophoresis

Immobilized pH gradients for isoelectric focusing

Before the introduction of immobilized pH gradients (IPG) strips, the first-dimension gel was cast in rods/ tubes with carrier ampholytes (CAs), small zwitterionic molecules that possess a high buffering capacity about their isoelectric points. Typically available as a mixture of different polymeric species with varying p*I*s, they specify a pH range according to their varying compositions. It is well known that the composition of these CA mixtures can vary from batch to batch, adversely affecting reproducibility.

IPGs were first introduced in 1982 by Bjellqvist and colleagues, and much of the subsequent development for use in 2DE was continued by Gorg and coworkers. Rather than using CAs for generation of the pH gradient, buffering species are covalently linked to the polyacrylamide gel matrix when the IEF gel is cast. This greatly improves reproducibility of the 2DE technique as cathodic drift due to the breakdown of CAs over extended focusing periods is eliminated. Tube gels for IEF are delicate to handle, often tending to swell in water and adding to the variability of the original protocol. To address this, IPG gels are cast in large sheets with a plastic backing and then cut into small strips, which are used for the first-dimension focusing. They are far more convenient to use, as they may be stored frozen and rehydrated just before use. After the focusing step, IPG strips can even be frozen and stored for later separation in the second-dimension SDS-PAGE. This is not possible with tube gels, which have to be transferred to the second dimension directly.

Despite the good resolving abilities of 2DE, mass spectrometric analyses often reveal that a single gel spot can contain several proteins. Proteins may not be sufficiently resolved if a broad pH range is used. To alleviate this problem, IPG strips with a narrow pH range are available, spanning as little as a single pH unit (**Figure 2**). Proteins that are not within the pH range of the IPG simply migrate to the edge of the strip and can be removed during IEF with electrode wicks. Assuming that sample amounts are not a limiting factor, one could potentially run 'zoom gels' with narrow-range IPG strips to cover a much broader pH range, thus producing a 'virtual' 2DE gel that has a far greater resolving power than a single gel.

Improved detection methods

Several options exist for the detection and visualization of proteins in 2DE. The choice of method relies

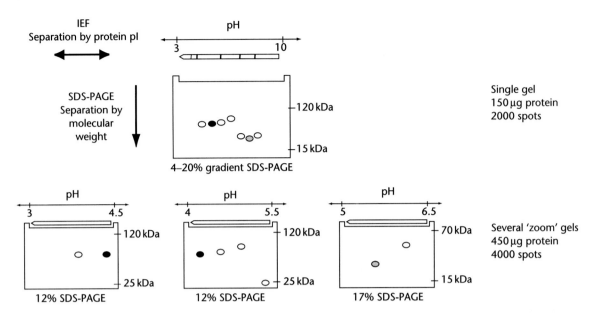

Figure 2 Zooming in on target proteins with broad- and narrow-range immobilized pH gradients (IPGs) in 2DE. A broad-range IPG and a gradient sodium dodecyl sulfate–polyacrylamide gel electrophoresis (SDS-PAGE) is used to get an overview of the 2DE map and to survey for proteins of interest. A more focused analysis with overlapping narrow-range IPGs spanning a p*I* range from 3 to 6.5 can improve resolution of proteins. Choices of IPG, pH range and density of SDS-PAGE gels can be used to isolate a region of interest. The numbers cited illustrate the greater number of proteins visualized in a targeted approach. The full and shaded ovals are landmarks, representing the same proteins in the various gels.

most heavily on the eventual goal of the experiment. For instance, radiolabeled samples incorporating either ^{35}S, ^{125}I, ^{32}P or other radioactive isotopes generally provide the highest degree of sensitivity and visualization is straightforward using well-established methods for radiolabel detection. The standard safety precautions involved in the use of radioactivity also apply here, limiting its use to specialized applications.

The most commonly used protein stains are either the Coomassie-based dyes or silver-based stains. With the increased interest in 2DE proteomics applications that rely on the mass spectrometric (MS) identification as the final readout, greater emphasis has been placed on MS compatibility in the development of protein stains. The limit of detection for Coomassie-based dyes is in the low- to mid-picomole range and sensitivity about three orders higher is possible with silver staining. Although generally useful in many applications, silver stains have a rather narrow linear dynamic range, that is, an increased intensity of a stained spot is only linearly proportional to the increase in protein abundance in a 10-fold range. Silver staining also exhibits different staining characteristics for different proteins, which may lead to misleading results in quantitative comparisons of 2DE gels.

Fluorescent dyes which are at least as sensitive as silver stains, with a broader linear dynamic range and which are MS compatible, have recently been introduced. These stains are best visualized with scanners

and ultraviolet (UV) light-boxes with excitation at 300 nm, making gel spot excision inconvenient unless specialized gel cutter robots are employed. The high cost of these stains also puts them beyond the reach of most academic laboratories.

Quantitation of gene expression with two-dimensional gel electrophoresis

Comparison of gel patterns in the study of protein expression is one of the most important applications of 2DE. For quantitations based on image analyses, the volume of the spots is determined based on the area of the spot as well as the intensity of staining; hence the importance of a broad linear dynamic range for the staining method used. The recent introduction of fluorescent dyes and techniques that utilize such dyes for quantitative analysis has revitalized the field. In the difference gel electrophoresis (DIGE) approach, a fluorescent dye is covalently linked to a small proportion of each of the proteins in one population; a second fluorescent dye (with different excitation/ emission properties) is used to label the sample for comparison. Fluorescently labeled populations are mixed before resolution by 2DE on a single gel. Visualization of proteins with different excitation and detection wavelengths produces a gel image of each sample. False color analyses with overlaid gel images reveal gel spots unique to each sample as well as spots that are common to both. This effectively allows direct quantitative comparisons from a single gel, reducing the amount of work required as well as reducing errors arising from the comparison of two separate gels.

MS-based methods for quantitation are based on the comparison of peptide signals from two populations that can be discriminated through the incorporation of a characteristic mass tag. The most established approach for this is the isotope coded affinity tag (ICAT) approach introduced by Gygi and colleagues in 1999. The chemical modification targets cysteine-containing proteins, and is typically used to enrich for cysteine-containing peptides which are used to identify proteins and quantitate their relative abundance in liquid chromatography–mass spectrometry (LC-MS) analyses. A recent report describes the use of ICAT to label two protein populations before mixing and separation by 2DE. Relative protein quantitation is performed by obtaining a ratio of peptide signals from MS analyses of each excised gel spot. A parallel experiment is required to identify differentially expressed proteins and decide which gel spots should be analyzed. In an alternative approach to MS-based quantitation, mammalian cell cultures labeled by stable isotope labeling with amino acids in cell culture (the SILAC approach) incorporate a labeled amino acid in every peptide chain synthesized. Labeling does not involve chemical manipulation and every single protein would contain several labeled peptides and is ideal for 2DE-MS quantitation. These MS-based approaches have primarily been developed for use with gel-free analyses, but it is clear that they are complementary approaches that can be used in tandem with 2DE to provide quantitative information.

An Overview of Two-dimensional Gel Electrophoresis in Proteomic Applications

Many of the large-scale approaches in genomics and proteomics have altered the way we plan experiments. The ability to monitor simultaneous changes in gene or protein expression with relatively little effort has made it possible to adopt a discovery-based approach to experimental design. A common starting point in a 2DE experiment is the comparison of two complex protein mixtures. We first present a general overview of the 2DE-based approach in proteomics (see also **Figure 3**) in order to provide a background for the discussion of the merits of the 2DE approach as well as its limitations.

Any soluble protein mixture should be analyzable by 2DE. Proteins can be extracted from mammalian

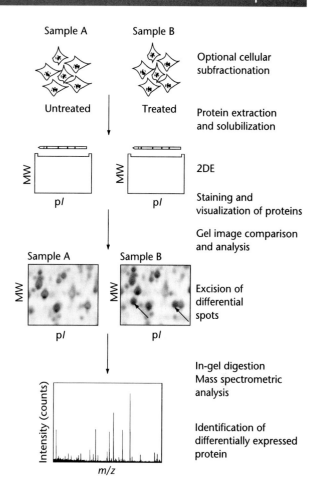

Figure 3 Identifying differentially expressed proteins with 2DE mass spectrometry (2DE-MS). The workflow for a 2DE-MS approach to studying differential protein expression in cells with or without drug treatment is illustrated here. Proteins from an equivalent number of cells are extracted and solubilized in a buffer suitable for 2DE. After 2DE separation, proteins are stained and visualized with a protein stain. Comparison of gel images is performed and spots that are found differentially expressed (spots marked by arrows) are excised. Proteins are digested in the gel and peptides extracted and identified following matrix-assisted laser desorption ionization time-of-flight (MALDI-TOF) MS analyses. The combination of mass peaks in the mass spectrum identifies a protein in peptide mass fingerprinting (MW: molecular weight; m/z: mass-to-charge ratio).

tissue, plant extracts, cell culture and bodily fluids (urine, saliva and even tears).

Preparation of the sample is especially critical in 2DE as contaminants like lipids, nucleic acids, polysaccharides, high-salt ionic substances, etc. can lead to smearing and poor protein resolution. A typical solubilization solution contains protease inhibitors, high concentrations of chaotropic agents, reducing agent and nonionic detergents, and is kept at high pH. Insoluble proteins do not enter the gel and hence will not be resolved. Protein solubilization cocktails capable of solubilizing a broad range of proteins are highly desirable, although no single protocol exists that is universally applicable as the diversity in samples requires optimization on a case-by-case basis. For an accurate assessment of quantitative relationships when comparing 2D maps, proteins must not be modified or degraded in the process of sample preparation. Any artifacts generated through sample preparation would result in a biased analysis.

The first-dimension IEF step is now greatly simplified with the use of IPG strips and specialized equipment capable of running multiple IPG strips. After IEF, IPG strips can be stored frozen or run directly in second-dimension SDS-PAGE. Upon completion of the SDS-PAGE step, proteins are typically fixed within the gel after treatment with dilute acetic acid and methanol and stained for protein visualization. Gel images are analyzed and protein spots of interest are excised from the gel manually or with gel excision robots. These gel-immobilized proteins are digested with a protease (typically trypsin) to produce peptides.

Several different types of MS analyses are available for protein identification. Broadly defined, each method is most suitable for a particular sample type. For instance, nanoscale LC-MS employs a peptide separation step before MS and is very useful for the analysis of complex mixtures. Because proteins are already separated in 2DE and gel spots typically do not contain more than a few overlapping proteins, matrix-assisted laser desorption ionization time-of-flight (MALDI-TOF) instruments and peptide mass fingerprinting (PMF) can be employed. PMF identifies a protein by matching theoretical peptide masses from a sequence in the database with observed peptide masses from the mass spectrometer. This approach is complicated when a mixture of proteins (more than five) is analyzed, as the large number of data points leads to decreased statistical confidence in identifications, even if iterative algorithms are applied. Putative identifications have a statistical probability score based typically on the size of the database searched, the specificity of the search parameters as well as mass error tolerances. In 2DE-MS protein identification, an approximate gauge of the protein's molecular weight and pI is available from the 2DE map and hence can be useful as an additional constraint for excluding false positives.

Known Limitations of Two-dimensional Gel Electrophoresis

It is well known that extremely large or small proteins and some classes of proteins that are difficult to solubilize in buffers conducive for IEF do not enter the first-dimension gel and are therefore refractory to 2DE analysis. Very basic proteins are also difficult to resolve in IEF. These limitations should be borne in mind if whole-proteome analysis is the primary goal.

Furthermore, the dynamic range of protein expression in a cell exceeds the dynamic range detectable on a single 2D gel by several orders of magnitude. If a particular protein is expressed at low levels in the cell and has to be resolved alongside a protein present at high amounts, the simultaneous visualization of both these proteins may be difficult. Dynamic range limitations arise due to the finite loading capacity of the IPG strip as well as the method of detection. For example, a gel may be silver stained for an extended period of time to visualize less abundant proteins, but this cannot be the general practice as overstaining will occur for abundant proteins and gel background staining would also increase, ultimately diminishing returns. Some of this may be appropriately addressed by running more gels across narrower pH ranges, increasing the effective resolution of proteins than what would normally be achievable with a single gel.

Although 2DE is a relatively straightforward procedure to perform for a single gel, reproducibility of the entire process from sample preparation, separation to staining for two parallel samples is not trivial. The difficulty would be compounded if it were necessary to run several zoom gels for each sample. The 2DE workflow is also labor intensive, requiring a degree of manual dexterity and attention to detail that may make it seem more art than science. An important component in 2DE analysis is the analysis of gel images after the staining and visualization step. There exists several software packages designed specifically for this purpose, yet these still require human intervention and interpretation, and the process can be time consuming. As 2DE image analyses hinge on the reproducibility of individual gels that are averaged to make up a sample set, the importance of standardization and reproducible gel runs cannot be overstated. These issues are not strictly limitations of the method itself, as robotics, automation, improved software development and strict quality control of reagents could potentially improve the situation.

Sample procurement is a problem common to all proteomics methods but may be especially important with 2DE as more sample material is required for the analysis. Often, comparison of proteins obtained from tissue can be difficult as the heterogeneity of the cell types within the sample is indeterminate. Cell populations should be closely matched in order to draw valid conclusions from such studies. This is the case for all quantitative comparisons of gene expression, whether they are mRNA- or protein-based studies. Laser capture microdissection can harvest proteins from single cells, making the selective analysis of specific cell types possible. Unfortunately, it still takes considerable amounts of time (half a day) to harvest sufficient protein material for a single 2DE gel.

Where Does Two-dimensional Gel Electrophoresis Fit in the Postgenomic Era?

2DE now has a renewed importance in many protein analysis laboratories. Its ability to separate proteins in a complex mixture is still unmatched. Reproducibility, sensitivity and ease of use of the 2DE approach is better than it ever was. However, advances in parallel technologies such as protein chips, complementary DNA (cDNA) microarrays and protein analysis with LC-MS techniques provide new tools for the study of gene expression. What biological questions are most appropriately addressed based on the strengths of the 2DE approach?

A quick search of the primary literature is telling. 2DE is a very popular choice for studying gene expression in species that have not been studied extensively at the genomic level. It is not a trivial task to prepare the cDNA microarrays or high-density oligonucleotide chips required for such studies. In contrast, differential protein expression can easily be investigated by the direct comparison of 2DE gel patterns. Although protein identification still assumes the availability of sequence databases, peptides can be sequenced *de novo* and used to design primers for the cloning of novel genes.

2DE is also extremely useful for analysis of protein modification. The separation of proteins on the basis of charge and molecular weight can effectively separate proteins modified by phosphorylation or glycosylation. One report describes the identification of 10 different phosphorylated isoforms of the eukaryotic initiation factor eIF4E-binding protein through a combination of ^{32}P radiolabeling and 2DE separation. Such an analysis would be extremely difficult to perform with other methods and clearly demonstrates where the 2DE approach is superior. In another example, Western blotting of 2DE gels with modification-specific antibodies (antiphosphotyrosine) was used for the identification of phosphorylated substrates after platelet-derived growth factor (PDGF) stimulation of mouse fibroblasts.

The limitations of dynamic range previously discussed could in part be improved by prefractionation of the sample, which helps reduce sample complexity and enriches less abundant proteins. A popular concept in proteomics is that of organellar proteomics – the isolation of cellular organelles like the nuclei or mitochondrion and the specific analysis of these fractions. Targeted protein analysis with 2DE can be achieved with a combination of narrow-range IPG strips and SDS-PAGE for resolving the appropriate molecular weight range. With selective analysis of a specific subset of the protein population, 2DE performed in this manner could potentially lead to very high resolution of protein species within the area of interest. A less focused analysis would be far more time consuming and could even be impossible, as some proteins are not analyzable by 2DE. LC-MS approaches that involve a 2D (strong cation exchange and reversed phase separation (MudPIT, for multidimensional protein identification technology)) chromatographic separation show promise as an alternative approach for protein identification, identifying an impressive 1484 yeast proteins from a single experiment, including some proteins that would not have been detectable by 2DE approaches.

The 2DE-MS approach to the study of differential protein expression has been widely used in the study of human disease such as cancer of the bladder, liver and prostate. Some groups have developed 2DE databases as a means to communicate their experimental findings and to offer an extra dimension to the annotation of the human genome. There is an overwhelming need for standardization in 2DE protocols in order for widespread data sharing to be successful. It is a situation reminiscent of that of microarrays, where the need for standardization in data formats was long recognized but has only recently been more widely adopted.

Conclusion

2DE has some unique analytical advantages that can be exploited to the fullest with judicious experimental planning as well as a good working knowledge of the potential difficulties involved. Continued development on several technological fronts promises to bring added functionality to users of 2DE as well as to improve existing standards of reproducibility and sensitivity.

See also
Gel Electrophoresis
Image Analysis Tools in Proteomics
Protein Characterization in Proteomics
Proteomics: A Shotgun Approach without Two-dimensional Gels
Quantitative Proteomics (ICAT™)

Further Reading

Celis JE and Gromov P (1999) 2D protein electrophoresis: can it be perfected? *Current Opinion in Biotechnology* **10**: 16–21.

Gorg A, Obermaier C, Boguth G, *et al.* (2000) The current state of two-dimensional electrophoresis with immobilized pH gradients. *Electrophoresis* **21**: 1037–1053.

Pandey A and Mann M (2000) Proteomics to study genes and genomes. *Nature* **405**: 837–846.

Patton W (2002) Detection technologies in proteome analysis. *Journal of Chromatography B* **771**: 3–31.

Patton W, Schulenberg B and Steinberg T (2002) Two-dimensional gel electrophoresis; better than a poke in the ICAT? *Current Opinion in Biotechnology* **13**: 321–328.

Peng J and Gygi SP (2001) Proteomics: the move to mixtures. *Journal of Mass Spectrometry* **36**: 1083–1091.

Rabilloud T (1999) Solubilization of proteins in 2-D electrophoresis: an outline. In: Link AJ (ed.) *2-D Proteome Analysis Protocols*, pp. 9–19. Totowa, NJ: Humana Press.

Rabilloud T (2002) Two-dimensional gel electrophoresis in proteomics: old, old fashioned, but it still climbs up the mountains. *Proteomics* **2**: 3–10.

Westermeier R and Naven T (2002) *Proteomics in Practice*. Weinheim: Wiley-VCH.

Yarmush ML and Jayaraman A (2002) Advances in proteomic technologies. *Annual Reviews in Biomedical Engineering* **4**: 349–373.

Uniparental Disomy

Albert Schinzel, *Institute of Medical Genetics, University of Zurich, Zurich, Switzerland*

Alessandra Baumer, *Institute of Medical Genetics, University of Zurich, Zurich, Switzerland*

The inheritance of both homologs of one pair of chromosomes from the same parent causes abnormal development if it leads to loss of the active allele of an imprinted (parent of origin specific) gene or genes or reduction to homozygosity of segments containing mutated recessive genes.

Introduction

Uniparental disomy (UPD) is the inheritance of both homologs of one chromosome from only one parent, either the father or the mother (instead of inheriting one from the father and the other from the mother) (Engel, 1980; Engel and Antonarakis, 2001).

Genomic imprinting is the differential modification of the maternal and paternal contributions to the zygote, resulting in the differential expression of parental alleles during development and in the adult. In the following, the imprinted homolog is defined as the one which is switched off (thus, paternal imprinting is when only the maternal allele is active).

Mechanisms of Formation of Uniparental Disomy

There are three ways by which UPD can arise:

1. Gamete complementation: a gamete with disomy of one chromosome fuses with a gamete nullisomic for the same chromosome.
2. Trisomy correction: a gamete with disomy for one chromosome fuses with a normal gamete, and subsequently the chromosome from the normal gamete, for which two homologs exist in the other gamete, is lost.
3. Monosomy duplication: a gamete nullisomic for one chromosome fuses with a normal gamete, and subsequently the single chromosome is duplicated.

Incidence of Uniparental Disomy

The incidence of UPD is poorly known. If the incidence of all trisomies at birth is about $1:500$, and if the loss through prenatal diagnosis is disregarded, the incidence of UPD must be distinctly lower, may be $1:20\,000$ to $1:30\,000$. This estimation is derived from the following observations: the incidence at birth of Prader–Willi syndrome patients is about $1:25\,000$; about one-third of Prader–Willi patients have maternal UPD 15; maternal UPD 7 leading to Silver–Russell syndrome might be equally frequent, and all other instances of UPD are distinctly less frequent. However, the incidences of UPD without abnormal phenotype are generally unknown, and thus the true figures, not considering phenotypic abnormalities or disease, might be much higher.

Stages of Formation of Uniparental Disomy

For clinical consequences, it does not matter whether the chromosome lost from one parent was lost before, during or after meiosis unless (in the latter case) the loss occurred only in mosaic form. The gain of a chromosome can occur before, at and after meiosis. Meiotic nondisjunction leading to a hyperhaploid gamete with disomy for one chromosome leads to heterodisomy. In case of meiosis I nondisjunction, markers close to the centromere are expected to be present in the heterozygous state while, due to crossover, more peripheral markers may or may not be reduced to homozygosity. The opposite situation prevails in the case of meiosis II nondisjunction: reduction to homozygosity at the centromere, maintainance of heterozygosity or reduction to homozygosity peripherally (closer to the telomere) of the chromosome. Therefore, in most heterodisomic UPD cases one does not find complete heterodisomy throughout the length of the chromosome. In contrast, mitotic nondisjunction will always lead to complete isodisomy (reduction to homozygosity throughout the length of the chromosome).

Consequences of Uniparental Disomy on the Phenotype of the Carrier

Loss of the active homolog of an imprinted gene or imprinted genes

If a gene is paternally imprinted, i.e. the paternal homolog is not active (functional monosomy), then paternal UPD will lead to functional nullisomy and thus to loss of the gene product; if this gene product is necessary to maintain a physiological function, it would lead to disease. (See later for a discussion of the syndromes cause by paternal UPD 6 and 15 and maternal UPD 7, 14 and 15.) If, on the other hand, increased dosage of a gene product leads to congenital developmental defects, and the active gene is, for example, the paternal one, paternal UPD would lead to an abnormal phenotype (see below for mosaic segmental paternal UPD 11p15).

Exposure of a recessive condition due to uniparental disomy

This is due to reduction to homozygosity of a mutated recessive gene and has been described for different diseases, most often in patients who did not display other abnormalities. It occurs predominantly in paternal UPD because the latter is almost always isodisomy; only one parent needs to carry a mutation, cases always occur sporadically and the family does not face the 25% risk of recurrence for further affected sibs which is normally predicted after occurrence of a recessive condition in a family.

Coexistence of the phenotype due to uniparental disomy and that of a recessive disease

This has been described for cystic fibrosis and Silver–Russell syndrome due to maternal UPD 7 and for Bloom syndrome and Prader–Willi syndrome due to maternal UPD 15.

Characteristics of and Differences between Maternal and Paternal Uniparental Disomy

The proportion of maternal versus paternal, meiotic versus mitotic nondisjunction in UPD parallels that of trisomy: predominance of maternal origin and increased mean maternal age. Thus, the vast majority of maternal UPD is heterodisomy, mostly stemming from maternal meiosis I nondisjunction (maternal UPD 18 has so far never been observed, it is most likely lethal, and it would, were it viable, be expected to occur predominantly as meiosis II nondisjunction similar to trisomy 18).

Mean maternal age in maternal UPD equals that usually found in maternal trisomy, i.e. an elevation of about 6 years over euploid controls. As the risk of reduction to homozygosity of mutated recessive alleles is confined to isodisomy, recessive diseases predominantly occur in paternal and not in maternal UPD cases. Paternal UPD, in contrast to maternal, is almost always isodisomy, thus occurring after zygote formation. The only study investigating a series of paternal UPD cases (with Angelman syndrome) for parental ages unexpectedly showed an increase in the mean maternal age. This finding lends some evidence to the hypothesis that, in these cases, the zygote has already lost the maternal contribution, and thus increased mean maternal age may lead both to gain and to loss of a chromosome, the latter, however, always with fatal outcome unless there is early correction through disomy in the sperm (gamete complementation) or postzygotic mitotic gain of the same chromosome which is lost in the female gamete (monosomy duplication) (Ginsburg et al., 2000).

Uniparental disomy and confined placental mosaicism

After the introduction of chorionic villus sampling (CVS), it soon became evident that placental trisomy with a diploid fetus is not uncommon (Kalousek et al., 1993). These cases predominantly occur in pregnancies of 'older' mothers, and it can be concluded that these cases start as trisomies of meiotic origin (as evidenced by microsatellite markers showing three distinct alleles), and the additional chromosome is lost giving the fetus the chance to survive. Indeed, it has repeatedly been observed that newborns with Prader–Willi or Silver–Russell syndrome were born after detection of a trisomy 15 and 7 respectively at CVS examination while amniocyte analysis disclosed normal diploid karyotypes. Thus, trisomy correction is not uncommon, and is probably the most frequent mechanism of formation of maternal UPD. The obvious consequence of this observation is that a UPD examination should be performed before a couple are told that their fetus is most likely healthy and normal if (mosaic) trisomy 7, 14 or 15 is found at CVS while a normal karyotype is ascertained at amniocentesis. For other chromosomes, predictions are not possible unless the normal outcome is documented from a sufficient number of cases.

Uniparental disomy and hidden mosaic trisomy

If the phenotypes of cases with UPD of the same origin and chromosome are very different, and if mosaic trisomy of that chromosome is known to exist, then the most likely explanation of these differences is additional mosaic trisomy with different proportions of trisomic cells in different organs, hence leading to different phenotypes. This is the likely reason why both maternal and paternal UPD 14 cases have varying phenotypes. In only one case has the mosaic trisomy (mat) been documented: a case of mosaicism between mosaic trisomy 9 and maternal UPD 9 only showed the trisomy phenotype.

Cytogenetic Findings in Uniparental Disomy

The majority of UPD cases show a normal diploid karyotype. However, a minority display structural chromosomal aberrations. If a translocation between two homologous chromosomes or (in fact much more often) an isochromosome is found, at least half of the cases show UPD, often isodisomy (Robinson et al., 1994). The most frequent finding is isochromosomes of acrocentrics, found in both maternal and paternal UPD 14 and 15 (Robinson et al., 1994), or two isochromosomes replacing two normal homologs, e.g. 7p and 7q in cases of Silver–Russell syndrome (Eggerding et al., 1994). Other cases with interhomologous translocations are detected due to repeated spontaneous abortions (e.g. UPD 2, 13, 21 or 22 with isochromosome formation (Robinson et al., 1994), rarely with occasional transmission from mother to offspring). Furthermore, UPD 15 in Prader–Willi syndrome was first detected in a case with a familial t(13;15) translocation (Nicholls et al., 1989). Both trisomy correction and monosomy complementation can theoretically occur, although the latter is less likely to be detected since the karyotype would be normal and the rearrangement only found in one of the parents.

Most recently, several cases with segmental UPD other than paternal 11p15 have been found. In addition, mosaic duplication of a chromosome segment has been shown to be sometimes connected with 'segmental UPD' (i.e. UPD not for an entire chromosome, but only for a chromosomal segment) (Schinzel et al., 1997).

Specific Phenotypes due to Uniparental Disomy

Paternal UPD 6

The clinical phenotype is benign neonatal diabetes mellitus usually resolving by the age of 3 years. Segmental UPD confined to the segment 6q24 was found in one patient, narrowing down the candidate region for the responsible, maternally imprinted, gene and allowing the discovery of the candidate gene pleiomorphic adenoma gene-like 1 (*PLAGL1*) (Kamiya et al., 2000).

Maternal UPD 7

The clinical correlate is the Silver–Russell syndrome which is characterized by pre- and postnatal growth retardation, no microcephaly, but a skull with prominent occiput and high forehead, a delicate triangular face, clinodactyly with brachymesophalangy V, frequent occurrence of body asymmetry and areas of altered skin pigmentation (hyper- or hypopigmentation) with normal or near normal intelligence (Kotzot et al., 1995).

Paternal segmental mosaic UPD 11p15

The clinical consequence of this type of UPD is either the full Beckwith–Wiedemann syndrome (also called EMG for exomphthalos–macroglossia–gigantism syndrome) or an incomplete version of the latter. The full syndrome is characterized by large placenta, pre- and postnatal overgrowth, macroglossia, umbilical hernia, organomegaly, earlobe creases, a coarse facies with midface hypoplasia, exophthalmos, suborbital creases and upturned nares, not infrequent body asymmetry and facultatively mild mental retardation. The reason for the abnormal phenotype is not absence of the normal homolog of a maternally imprinted gene, but overexpression of a maternally imprinted gene (insulin-like growth factor 2 (somatomedin A) (*IGF2*)) (Henry et al., 1991; Dutly et al., 1998). Apart from UPD, there are cases with paternal duplication of a segment including 11p15 (always with more distinct mental retardation), maternally inherited translocation with a common breakpoint at 11p15 disrupting the 'control locus', and abnormal methylation of KCNQ1 overlapping transcript 1 (*KCNQ1OT1*). A minority of patients show a diminished phenotype, e.g. only macrosomia and body asymmetry (personal observation). An increased (approximately 15%) risk for Wilms tumor and occasionally other tumors is mostly observed in cases with structural mutations involving the *IGF2* region.

Maternal UPD 14

The minimal phenotype is pre- and postnatal growth retardation (as in maternal UPD 7) and early onset of puberty. Many of the cases discovered so far have a *de novo* isochromosome 14. (For patients with additional anomalies see hidden mosaic trisomy.)

Maternal UPD 15

This type of UPD is found in about 25–30% of Prader–Willi syndrome patients, the remaining almost all showing a paternal 15q11.2–q12 deletion. Characteristic features of this syndrome include diminished fetal activity, mildly reduced birthweight, severe neonatal (and postnatal) hypotonia, feeding difficulties, onset of obesity due to inappropriate hunger at about 9 months of age, delicate hands and feet, male genital hypoplasia with cryptorchidism and small penis, a hypotonic facies with almond-shaped eyes, moderate mental retardation with behavioral disturbances, and others. The mean maternal age of UPD cases is significantly increased (Robinson *et al.*, 1991; Ginsburg *et al.*, 2000).

Paternal UPD 15

Paternal UPD 15 is present in 2–5% of patients with Angelman ('happy puppet') syndrome. Characteristic features include mildly diminished birth measurements, postnatal microcephaly, narrow forehead, broad mouth and prominent mandible, severe to profound mental deficiency with epilepsy and a characteristic electroencephalogram (EEG), ataxia, unmotivated outbursts of laughter and happy, pleasant disposition. Several different molecular and cytogenetic findings can be found, including maternal deletion of the 15q11.2–q12 segment, imprinting mutations and mutations in the ubiquitin protein ligase E3A (human papilloma virus E6-associated protein, Angelman syndrome) (*UBE3A*) gene.

Other types of uniparental disomy

Other types of UPD are covered in detail in the review by Kotzot (1999). Intrauterine growth retardation has been observed in cases of maternal UPD 2, 16 (Kalousek *et al.*, 1993) and 22. However, as for example in confined placental trisomy 16 intrauterine growth retardation is found with normal biparental contribution of chromosomes 16, it seems more likely

(in fact in all of these UPDs) that the growth deficit is due to dysfunction of the trisomic placenta and not to the UPD proper. Normal phenotypes or just recessive conditions have been described in UPD of some further chromosomes including 1, 9, 13, 21 and 22, mostly with a very limited number of cases. The phenotype of paternal UPD 14 is very variable, and it is yet unclear how much undetected mosaicism for trisomy contributes to this variability and whether UPD alone goes along with an abnormal phenotype.

See also

Imprinting Disorders
Prader–Willi Syndrome and Angelman Syndrome

References

Dutly F, Baumer A, Kayserili H, *et al.* (1998) Seven cases of Wiedemann–Beckwith syndrome, including the first reported case of mosaic paternal isodisomy along the whole chromosome 11. *American Journal of Medical Genetics* **79**: 347–353.

Eggerding FA, Schonberg SA, Chehab FF, *et al.* (1994) Uniparental isodisomy for paternal 7 p and maternal 7 q in a child with growth retardation. *American Journal of Human Genetics* **55**: 253–265.

Engel E (1980) A new genetic concept: uniparental disomy and its potential effect, isodisomy. *American Journal of Medical Genetics* **6**: 137–143.

Engel E and Antonarakis SE (2001) *Genomic Imprinting and Uniparental Disomy in Medicine. Clinical and Molecular Aspects.* New York: John Wiley.

Ginsburg C, Fokstuen S and Schinzel A (2000) The contribution of uniparental disomy to congenital developmental defects in children born to mothers at advanced childbearing age. *American Journal of Medical Genetics* **95**: 454–460.

Henry I, Bonaiti-Pellié C, Chehensse V, *et al.* (1991) Uniparental paternal disomy in a genetic cancer-predisposing syndrome. *Nature* **351**: 665–667.

Kalousek DK, Langlois S, Barrett I, *et al.* (1993) Uniparental disomy for chromosome 16 in humans. *American Journal of Human Genetics* **52**: 8–16.

Kamiya M, Judson H, Okazaki Y, *et al.* (2000) The cell cycle control gene ZAC/PLAGL1 is imprinted: a strong candidate gene for transient neonatal diabetes. *Human Molecular Genetics* **9**: 453–460.

Kotzot D (1999) Abnormal phenotypes in uniparental disomy (UPD): fundamental aspects and a critical review with bibliography of UPD other than 15. *American Journal of Medical Genetics* **82**: 265–274.

Kotzot D, Schmitt S, Bernasconi F, *et al.* (1995) Uniparental disomy 7 in Silver–Russell syndrome and primordial growth retardation. *Human Molecular Genetics* **4**: 583–587.

Nicholls RD, Knoll JHM, Butler MG, Karam S and LaLande M (1989) Genetic imprinting suggested by maternal heterodisomy in non-deletion Prader–Willi syndrome. *Nature* **342**: 281–285.

Robinson WP, Bernasconi F, Basaran S, *et al.* (1994) A somatic origin of homologous Robertsonian translocations and isochromosomes. *American Journal of Human Genetics* **54**: 290–302.

Robinson WP, Bottani A, Yagang X, *et al.* (1991) Molecular, cytogenetic, and clinical investigations of Prader–Willi syndrome patients. *American Journal of Human Genetics* **49**: 1219–1234.

Schinzel A, Kotzot D, Brecevic L, *et al.* (1997) Trisomy first, translocation second, uniparental disomy and partial trisomy third: a new mechanism for complex chromosomal aneuploidy. *European Journal of Human Genetics* **5**: 308–314.

Further Reading

Baumer A, Balmer D and Schinzel A (1999) Screening for UBE3A gene mutations in a group of Angelman syndrome patients selected according to non-stringent clinical criteria. *Human Genetics* **105**: 598–602.

Bernasconi F, Karaguzel A, Celep F, *et al.* (1996) A normal phenotype with maternal isodisomy in a female with two isochromosomes: i(2p) and i(2q). *American Journal of Human Genetics* **59**: 1114–1118.

Bottani A, Robinson WP, DeLozier-Blanchet CD, *et al.* (1994) Angelman syndrome due to paternal uniparental disomy of chromosome 15: a milder phenotype? *American Journal of Medical Genetics* **51**: 35–40.

Kirkels VGHJ, Hustinx TWJ and Scheres JMJC (1980) Habitual abortion and translocation (22q;22q): unexpected transmission from a mother to her phenotypically normal daughter. *Clinical Genetics* **18**: 456–461.

Malcolm S, Clayton-Smith J, Nichols M, *et al.* (1991) Uniparental paternal disomy in Angelman's syndrome. *Lancet* **337**: 694–697.

Mannens M, Hoovers JMN, Redeker E, *et al.* (1994) Parental imprinting of human chromosome region 11p15.3-pter involved in the Beckwith–Wiedemann syndrome and various neoplasia. *European Journal of Human Genetics* **2**: 3–23.

Pentao L, Lewis RA, Ledbetter DH, Patel PI and Lupski JR (1992) Maternal uniparental isodisomy of chromosome 14: association with autosomal recessive rod monochromacy. *American Journal of Human Genetics* **50**: 690–699.

Purvis-Smith SG, Saville T, Manass MY, *et al.* (1992) Uniparental disomy 15 resulting from 'correction' of an initial trisomy 15. *American Journal of Human Genetics* **50**: 1348–1349.

Wolstenholme J (1995) An audit of trisomy 16 in man. *Prenatal Diagnosis* **15**: 109–121.

Web Links

Insulin-like growth factor 2 (somatomedin A) (*IGF2*); Locus ID: 3481. LocusLink:
http://www.ncbi.nlm.nih.gov/LocusLink/LocRpt.cgi?l=3481
KCNQ1 overlapping transcript 1 (*KCNQ1OT1*); Locus ID: 10984. Locus Link:
http://www.ncbi.nlm.nih.gov/LocusLink/LocRpt.cgi?l=10984
Pleiomorphic adenoma gene-like 1 (*PLAGL1*); Locus ID: 5325. LocusLink:
http://www.ncbi.nlm.nih.gov/LocusLink/LocRpt.cgi?l=5325
Ubiquitin protein ligase E3A (human papilloma virus E6-associated protein, Angelman syndrome) (*UBE3A*); Locus ID: 7337. LocusLink:
http://www.ncbi.nlm.nih.gov/LocusLink/LocRpt.cgi?l=7337
Insulin-like growth factor 2 (somatomedin A) (*IGF2*); MIM number: 147470. OMIM:
http://www.ncbi.nlm.nih.gov/htbin-post/Omim/dispmim?147470
KCNQ1 overlapping transcript 1 (*KCNQ1OT1*); MIM number: 604115. OMIM:
http://www.ncbi.nlm.nih.gov/htbin-post/Omim/dispmim?604115
Pleiomorphic adenoma gene-like 1 (*PLAGL1*); MIM number: 603044. OMIM:
http://www.ncbi.nlm.nih.gov/htbin-post/Omim/dispmim?603044
Ubiquitin protein ligase E3A (human papilloma virus E6-associated protein, Angelman syndrome) (*UBE3A*); MIM number: 601623. OMIM:
http://www.ncbi.nlm.nih.gov/htbin-post/Omim/dispmim?601623

Untranslated Regions

See mRNA Untranslated Regions (UTRs)
3′ UTR Mutations and Human Disorders
3′ UTRs and Regulation
5′ UTRs and Regulation

3' UTR Mutations and Human Disorders

Béatrice Conne, *University of Geneva, Geneva, Switzerland*

André Stutz, *University of Geneva, Geneva, Switzerland*

Jean-Dominique Vassalli, *University of Geneva, Geneva, Switzerland*

The 3' untranslated region (3' UTR) of a number of messenger ribonucleic acids (mRNAs) is involved in the regulation of the processing, localization, translation or degradation of the transcript. Perturbations in such 3' UTR-mediated functions are implicated in a variety of human diseases.

Intermediate article

Article contents

- Introduction
- A Defect in mRNA Processing and Export
- Functional Disturbances of the ARE Machinery
- Abnormally Destabilized mRNAs
- *trans*-Acting Factors as Regulators of Translation
- A Systematic Search for '3' UTR-mediated Diseases'

Introduction

As a function of the cell type in which it is expressed and of the physiological state of the cell, the processing, localization, translation or degradation of a given messenger ribonucleic acid (mRNA) may vary dramatically. Evidence is accumulating on the role of the 3' untranslated region (3' UTR) of mRNAs in the regulation of gene expression. The 3' UTR is not under the same rigid structural constraints as the coding or the 5' untranslated regions that have to accommodate the translational machinery. Evolutionary pressure may have taken advantage of the greater degree of freedom of 3' UTRs in order to modulate the fate of mRNA molecules. Regulated translation allows mRNAs to be used at different times and in specific subcellular compartments, and such an uncoupling between transcription and translation is required in gametogenesis, embryogenesis and for the targeting of specific mRNAs into neuronal dendrites. *cis*-Acting determinants are being identified in the 3' UTR together with the *trans*-acting factors with which they interact. Overall, it is now clear that the 3' UTR can specifically control the nuclear export, polyadenylation status, subcellular targeting and rates of translation and degradation of mRNAs. Modifications in 3' UTR-mediated functions may affect the expression of one (e.g. a gene carrying a mutation in its 3' UTR) or more genes (e.g. by changes in a *trans*-acting factor affecting the fate of different mRNAs). Moreover, the 3' UTR transcribed from one mutated allele can have a dominant-negative effect by sequestering *trans*-acting transport or regulatory proteins. Thus, the entire range of molecular perturbations can be played, with the 3' UTR as a primary target. We review here some examples of human diseases whose expression may depend on *cis* and/or *trans* factors acting at the mRNA 3' UTR level. (**See** mRNA Untranslated Regions (UTRs).)

A Defect in mRNA Processing and Export

Myotonic dystrophy (DM1), a dominantly inherited, multisystem disorder, shows a wide range of presentation and progression. DM1 is caused by an expanded number of trinucleotide (CTG) repeats in the 3' UTR of a serine threonine protein kinase gene (*DMPK; dystrophia myotonica-protein kinase*). The most frequent allele in the normal population contains five repeats. In the most severe, congenital form of the disease, characterized by hypotonia, respiratory distress, cataracts, endocrine dysfunction and mental retardation, the number of repeats reaches 1000 or more (Timchenko, 1999). (**See** Trinucleotide Repeat Expansions: Disorders; Trinucleotide Repeat Expansions: Mechanisms and Disease Associations.)

How does the expanded CTG repeats region affect the function of the *DMPK* gene, and how does this account for the phenotype? In healthy individuals, alternatively initiated translation products as well as multiple spliced forms of *DMPK* mRNA suggest several functions for this kinase, including the phosphorylation of ion channels. An impairment of this function may well account for the altered muscle excitability. Homozygous loss of *DMPK* in mice leads to a myopathy, sharing some but not all the pathological features of DM1. By contrast, mice that express expanded CUG repeats develop a DM1 phenotype (Mankodi *et al.*, 2000). Mutated *DMPK* transcripts are retained within DM1 myoblast nuclei, leading to an impaired kinase synthesis. This phenomenon involves an RNA-binding protein (CUG-BP), highly homologous to several 'embryonic lethal abnormal vision' (ELAV)-type ribonucleoproteins that recognize CUG repeats containing

 NATURE ENCYCLOPEDIA OF THE HUMAN GENOME / ©2003 Macmillan Publishers Ltd, Nature Publishing Group / www.ehgonline.net

mRNAs. The function of the CUG-BP is not yet clear. At least two isoforms have been identified, which seem to differ by their phosphorylation status. One of these, the hyperphosphorylated form, CUG-BP1, predominates in the cytoplasm, whereas the second, hypophosphorylated form, CUG-BP2, is present in the nucleus of normal cells (Roberts *et al.*, 1997). These CUG-BP molecules bind to the 3' UTR of *DMPK* RNA and may regulate its splicing and transport through a self-regulating loop, whereby DMPK phosphorylates the CUG-BP itself. The decrease in DMPK itself may contribute to the accumulation of abnormally hypophosphorylated CUG-BP in the nucleus, leading to subsequent alteration in the fate of other RNAs containing CUG triplets. Indeed, the nuclear sequestration of mutant *DMPK* transcripts is compatible with a loss-of-function model for DM1. CUG repeats are also found within muscle-specific splicing enhancers of the human *troponin T2, cardiac* (*TNNT2*) gene. In DM1 patients, the regulation of the alternative *TNNT2* mRNA splicing is altered (Philips *et al.*, 1998). Moreover, CUG-BP is also able to bind U/G elements on pre-mRNA targets such as the muscle specific chloride channel (CIC-1). An overexpression of CUG-BP may be responsible for the disruption of CIC-1 alternative splicing regulation, and may thereby cause a predominant pathologic feature of DM1 (Charlet *et al.*, 2002). The nuclear accumulation of hypophosphorylated CUG-BP in DM1 cells may thus affect the expression not only of the mutated but also of the wild-type *DMPK* allele, as well as that of a number of genes that require this factor for posttranscriptional processing. It should also be pointed out that the close proximity of the CTG expansion may result in transcriptional repression of *DMPK* downstream genes. Indeed, the triplet expansion leads to the suppression of the *Six5* homeobox transcription factor gene as well as its target gene products. The silencing of direct six5 targets such as *Igfbp5* and *Igf2*, encoding components of insulin growth factor signaling, may contribute to metabolic abnormalities frequently seen in DM1 patients (Sato *et al.*, 2002). (*See* mRNA Export; Posttranscriptional Processing.)

Functional Disturbances of the ARE Machinery

The cytoplasmic half-life of a variety of mRNAs is controlled by *cis*-acting AU-rich elements (AREs). A major class of such elements consists of pentanucleotide sequences (AUUUA) mapping in the 3' UTR of transcripts encoding oncoproteins, cytokines and growth and transcription factors. Many RNA-binding proteins, mostly members of ELAV family, recognize

AREs (Peng *et al.*, 1998). Defective function of AREs leading to the abnormal stabilization of an mRNA may form the basis of several human diseases. (*See* RNA-binding Proteins: Regulation of mRNA Splicing, Export and Decay.)

Most malignancies of hematopoietic lineages are associated with nonrandom chromosomal alterations, and genes located at the chromosomal breakpoints are directly involved in tumorigenesis. The t(11;14)(q13;32) translocation and its molecular counterpart, a rearrangement of the *cyclin D1* (*PRAD1: parathyroid adenomatosis 1*) (*CCND1*) oncogene, are consistent features of lymphomas called mantle-cell lymphomas (MCL). As a result of this translocation, the *CCND1* oncogene is juxtaposed to an IgH-enhancer sequence leading to the overexpression of a gene product (CCND1) belonging to the cyclin G1 family. In addition to this transcriptional effect, the chromosomal lesion may also contribute through posttranscriptional mechanisms to CCND1 overexpression. Indeed, in some primary tumors and in cell lines, 1.5-, 2- and 3-kb mRNAs can be detected with the normal 4.5-kb mRNA. These smaller transcripts result from deletions of part of the 3' end of the gene and lack therefore several AREs responsible for the short half-life of the wild-type mRNA. Thus, increased CCND1 protein, and the ensuing neoplastic pathology, can result from the additive contributions of two independent genetic events, one acting at the transcriptional level and the other affecting the structure of the mRNA's 3' UTR, hence its half-life (Rimokh *et al.*, 1994). Similarly, because of chromosomal breaks, aberrant HGMA transcripts encoding for a family of architectural transcription factors have been shown to be truncated in their 3'UTR. As a consequence, the loss of ARE motifs alters the stability and expression of these genes whose overexpression is related to several mesenchymal tumors (Borrmann *et al.*, 2001). (*See* Translocation Breakpoints in Cancer.)

Neuroblastoma, a common neoplasm in children, arises from neural crest-derived cells. These tumors comprise different cell types, and heterogeneous cellular subpopulations are frequently observed in cell cultures derived from human neuroblastoma; subclones consisting of neuroblast-like cells (N) and larger substrate-adherent cells (S) have been established. The steady-state levels of N-*myc* mRNA and protein and of c-*fos* mRNA are much higher in N than in S cells: this correlates with the tumorigenicity of the N cells and results, at least in part, from differences in the rate of degradation of the mRNAs. The 3' UTR of N-*myc* mRNA contains two distinct AREs that can bind a member of the ELAV-like family (p40). This protein also interacts with elements in the 3' UTR of c-*fos* mRNA. A large amount of p40 has been demonstrated in N cells, whereas it is barely detectable in S cells. This phenomenon may play a role in the stabilization of N-*myc* and c-*fos* mRNAs in N cells and seems to correlate with the aggressive clinical behavior of neuroblastomas (Chagnovich and Cohn, 1997). (*See* Neuroblastoma.)

Tumor necrosis factor (TNF) is central to various immune and inflammatory phenomena through effects ranging from cellular activation and proliferation to toxicity and apoptosis. Given the pleiotropism of TNF action, it is not unexpected that its biosynthesis is under the control of multiple and complex regulatory mechanisms. Some of these may be mediated by ARE motifs. Experiments in transgenic mice, aimed at analyzing the tissue-specific effects of ARE mutations on TNF mRNA, have shown that the absence of regulatory AREs leads to profound temporal and spatial deregulation of TNF production (Kontoyiannis *et al.*, 1999). This is characterized by the persistent accumulation and decreased rate of decay of the mutant TNF mRNA and by its unresponsiveness to translational inhibition in hematopoietic and stromal tissues. The binding of a member of a class of Cys-Cys-Cys-His zinc-finger proteins (tristetraprolin) to the AREs is directly responsible for the destabilization of the mRNA. This protein is thus likely to play a role both in normal and in pathologic conditions. A deregulation of the rate of decay of TNF mRNA results in the development of chronic inflammatory arthritis and inflammatory bowel disease, suggesting a possible link between a defective function of the ARE regulatory machinery and the development of analogous abnormalities in humans.

Cancer cells maintain a high glycolytic rate, a phenomenon known as the Warburg effect. Fructose-2-6-biphosphate, a regulator of glycolysis, depends on the activity of the 6-phosphofructo-2 kinase. An isoform of this enzyme, induced by proinflammatory stimuli, has been shown to contain multiple AREs in the 3' UTR of its mRNA. This kinase is constitutively expressed in several human cancer cell lines and is required for tumor cell growth. Its inhibition decreases the pentose pathway products required for nucleic acid synthesis. Increased production of this kinase may provide an explanation for the apparent coupling of enhanced glycolysis and cell proliferation (Chesney *et al.*, 1999). The continuous expansion of knowledge on the functions of the ARE machinery will likely lead to the understanding of a variety of different diseases. As a last example, we can mention the characterization of a cluster of three single nucleotide polymorphisms in the 3'UTR of the human *PC-1* gene. This haplotype stabilizes *PC-1* mRNA and is associated with PC-1 overexpression and insulin resistance (Frittitta *et al.*, 2001). (*See* Thalassemias.)

Abnormally Destabilized mRNAs

The disease α-thalassemia, caused by the presence of an antitermination mutation, represents the best-studied example of abnormal mRNA half-life not involving ARE-related mechanisms. The expression of globin genes is regulated at both transcriptional and posttranscriptional levels. One salient feature of globin mRNA is its exceptional stability. An understanding of the structural basis of this stability has been facilitated by the study of an α-globin gene (*HBA1*) variant (α Constant Spring or α^{CS}). This allele contains a UAA to CAA antitermination mutation that allows translating ribosomes to proceed into the 3' UTR; this is in turn associated with a dramatic decrease in the α^{CS} mRNA half-life, suggesting that the ribosomes may somehow mask stability determinants within the 3' UTR. Such determinants have been mapped to three C-rich regions, and mutations in any of these tend to destabilize the mRNA, independent of its translation. These C-rich regions interact with a ribonucleoprotein complex, the α-complex, containing a poly(C) binding protein as well as several other proteins. Evidence from transgenic mice carrying the human α^{CS} gene indicates a function for the α-complex, with the poly(A)-binding protein, in controlling the rate of deadenylation of globin mRNA. The α-complex may protect the poly(A) tail, through a combined interaction with the C-rich regions of the 3' UTR and the poly(A) binding protein, and stabilize α-globin mRNA. If this interaction is prevented, as in the case of α^{CS} globin mRNA, the poly(A) tail undergoes an accelerated shortening, the mRNA is prematurely degraded, and thalassemia ensues (Morales *et al.*, 1997). (*See* mRNA Stability and the Control of Gene Expression; Repetitive Elements and Human Disorders.)

Fukuyama-type congenital muscular dystrophy (FCMD), one of the most common autosomal recessive disorders in Japan, is accompanied by a neuronal migration defect. It has recently been shown that FCMD is associated with a retrotransposed 3-kb insertion of tandemly repeated sequences within the 3' UTR of a gene encoding a protein named fukutin. The protein contains an *N*-terminal signal sequence, suggesting that it is an extracellular component. The fukutin gene (*FCMD; Fukuyama type congenital muscular dystrophy* (*fukutin*)) is normally expressed in various tissues from unaffected individuals, but transcripts are virtually undetectable in FCMD patients, suggesting that the 3' UTR insertion may affect the *FCMD* mRNA transcription rate and/or stability (Toda *et al.*, 2000).

trans-Acting Factors as Regulators of Translation

Human acute myelogenous leukemia (AML) is a clonal disease arising in early hematopoietic progenitor cells through a multistep process. Mutations of the p53 tumor suppressor gene (*TP53; tumor protein p53 (Li−Fraumeni syndrome)*) are infrequent in AML blast cells, and *TP53* mRNA is in general abundant. The p53 protein, however, is often undetectable, raising the possibility that its half-life is altered in AML cells and/or that p53 expression is translationally regulated. Indeed, in two AML blast cell lines with similar levels of protein expression but with different levels of mRNA, there is a preferential association of *TP53* mRNA, with large polysomes in the cell containing less *p53* mRNA, indicating increased translational activity. *In vivo* and *in vitro* experiments have demonstrated that translation of human *TP53* was inhibited by *TP53* 3′ UTR. More recently, a U-rich sequence that mediates translational repression has been characterized; an RNA-binding protein interacts specifically with this sequence (Fu *et al.*, 1999). This *trans*-acting factor may either override this negative regulatory element in normal cells or suppress *TP53* mRNA translation in leukemic blasts.

A Systematic Search for '3′ UTR-mediated Diseases'

The information summarized so far represents compelling evidence for involvement of alterations in 3′ UTR-mediated functions in the pathogenesis of different diseases. It is likely that a more systematic study will demonstrate other such '3′ UTR-mediated diseases'. Given the well-documented involvement of the 3′ UTR in controlling mRNA translation during both male and female gametogenesis and in early embryonic development, and as such posttranscriptional controls are essential to these processes, it is a reasonable assumption that certain reproductive disorders will be found to belong to this class of diseases. Recent findings have identified neuronal mRNAs that also appear to depend on their 3′ UTR for appropriate subcellular targeting or translational control. The possibility that certain disorders of neuronal plasticity and learning are due to perturbations in 3′ UTR-mediated functions thus deserves attention.

The identification of putative additional 3′ UTR-mediated diseases will require a specific set of cellular and molecular approaches. Careful attention must be paid to the localization of the transcript, its functional half-life as well as its translational efficiency. The identification of *trans*-acting effectors that recognize specific 3′ UTR determinants is an additional avenue that should be useful.

The few 3′ UTR-mediated diseases identified to date remind us that between genome and proteome lies a still-mysterious world, with its own set of rules and disorders, the 'ribonucleome'. The 3′ UTR appears as a particularly intriguing regulatory region, which may be well worth a more systematic exploration.

See also
mRNA Stability and the Control of Gene Expression
mRNA Untranslated Regions (UTRs)
Posttranscriptional Processing
RNA Processing

References

Borrmann L, Wilkening S and Bullerdiek J (2001) The expression of HMGA genes is regulated by their 3′ UTR. *Oncogene* **20**: 4537–4541.

Chagnovich D and Cohn SL (1997) Activity of a 40 kDa RNA-binding protein correlates with MYCN and c-*fos* mRNA stability in human neuroblastoma. *European Journal of Cancer* **33**: 2064–2067.

Charlet BN, Savkur RS, Singh G, *et al.* (2002) Loss of the muscle-specific chloride channel in type 1 myotonic dystrophy due to misregulated alternative splicing. *Molecular Cell* **10**: 45–53.

Chesney J, Mitchell R, Benigni F, *et al.* (1999) An inducible gene product for 6-phosphofructo-2-kinase with an AU-rich instability element: role in tumor cell glycolysis and the Warburg effect. *Proceedings of the National Academy of Sciences of the United States of America* **96**: 3047–3052.

Frittitta L, Ercolino T, Bozzali M, *et al.* (2001). A cluster of three nucleotide polymorphisms in the 3′untranslated region of human glycoprotein PC-1 gene stabilizes PC-1 mRNA and is associated with increased PC-1 protein content and insulin resistance-related abnormalities. *Diabetes* **50**: 1952–1955.

Fu L, Ma W and Benchimol S (1999) A translation repressor element resides in the 3′ untranslated region of human p53 mRNA. *Oncogene* **18**: 6419–6424.

Kontoyiannis D, Pasparakis M, Pizarro TT, Cominelli F and Kollias G (1999) Impaired on/off regulation of TNF biosynthesis in mice lacking TNF AU-rich elements: implications for joint and gut-associated immunopathologies. *Immunity* **10**: 387–398.

Mankodi A, Logigian E, Callahan L, *et al.* (2000) Myotonic dystrophy in transgenic mice expressing an expanded CUG repeat. *Science* **289**: 1769–1773.

Morales J, Russell JE and Liebhaber SA (1997) Destabilization of human α-globin mRNA by translation anti-termination is controlled during erythroid differentiation and is paralleled by phased shortening of the poly(A) tail. *Journal of Biological Chemistry* **272**: 6607–6613.

Peng SS, Chen CY, Xu N and Shyu AB (1998) RNA stabilization by the AU-rich element binding protein, HuR, an ELAV protein. *EMBO Journal* **17**: 3461–3470.

Philips AV, Timchenko LT and Cooper TA (1998) Disruption of splicing regulated by a CUG-binding protein in myotonic dystrophy. *Science* **280**: 737–741.

Rimokh R, Berger F, Bastard C, *et al.* (1994) Rearrangement of CCND1 (BCL1/PRAD1) 3′ untranslated region in mantle-cell lymphomas and t(11q13)-associated leukemias. *Blood* **83**: 3689–3696.

Roberts R, Timchenko NA, Miller JW, *et al.* (1997) Altered phosphorylation and intracellular distribution of a (CUG)n triplet repeat RNA-binding protein in patients with myotonic dystrophy and in myotonin protein kinase knockout mice. *Proceedings of the National Academy of Sciences of the United States of America* **94**: 13 221–13 226.

Sato S, Nakamura M, Cho DH, *et al.* (2001) Identification of transcriptional targets for Six5: implication for the pathogenesis of myotonic dystrophy type 1. *Human Molecular Genetics* **11**: 1045–1058.

Timchenko LT (1999) Myotonic dystrophy: the role of RNA CUG triplet repeats. *American Journal of Human Genetics* **64**: 360–364.

Toda T, Kobayashi K, Kondo-Iida E, Sasaki J and Nakamura Y (2000) The Fukuyama congenital muscular dystrophy story. *Neuromuscular Disorders* **10**: 153–159.

Further Reading

Cazzola M and Radeck C (2000) Translational pathophysiology: a novel molecular mechanism of human disease. *Blood* **95**: 3280–3288.

De Moor CH and Richter JD (2001) Translational control in vertebrate development. *International Review of Cytology* **203**: 567–608.

Macdonald P (2001) Diversity in translational regulation. *Current Opinion in Cell Biology* **13**: 326–331.

Mendez R and Richter JD (2001) Translational control by CPEB: a means to the end. *Nature Review* **2**: 521–529.

Sonenberg N, Hershey WB and Matthews MB (2000) Translational control of gene expression. Plainview, NY: Cold Spring Harbor Laboratory Press.

Web Links

CCND1 (cyclin D1 (PRAD1: parathyroid adenomatosis 1)); LocusID: 595. LocusLink:
http://www.ncbi.nlm.nih.gov/LocusLink/LocRpt.cgi?l = 595

DMPK (dystrophia myotonica-protein kinase); LocusID: 1760. LocusLink:
http://www.ncbi.nlm.nih.gov/LocusLink/LocRpt.cgi?l = 1760

FCMD (*Fukuyama type congenital muscular dystrophy (fukutin)*); LocusID: 2218. LocusLink:
http://www.ncbi.nlm.nih.gov/LocusLink/LocRpt.cgi?l = 2218

TP53 (*tumor protein p53 (Li-Fraumeni syndrome)*); LocusID: 7157. LocusLink:
http://www.ncbi.nlm.nih.gov/LocusLink/LocRpt.cgi?l = 7157

CCND1 (*cyclin D1 (PRAD1: parathyroid adenomatosis 1)*); MIM number: 168461. OMIM:
http://www3.ncbi.nlm.nih.gov/htbin-post/Omim/dispmim?168461

DMPK (*dystrophia myotonica-protein kinase*); MIM number: 605377. OMIM:
http://www3.ncbi.nlm.nih.gov/htbin-post/Omim/dispmim?605377

FCMD (*Fukuyama type congenital muscular dystrophy (fukutin)*); MIM number: 253800. OMIM:
http://www3.ncbi.nlm.nih.gov/htbin-post/Omim/dispmim?253800

TP53 (*tumor protein p53 (Li–Fraumeni syndrome)*); MIM number: 191170. OMIM:
http://www3.ncbi.nlm.nih.gov/htbin-post/Omim/dispmim?191170

3′ UTRs and Regulation

Advanced article

John Hesketh, *University of Newcastle, Newcastle upon Tyne, UK*

Messenger RNAs contain sequences that do not encode proteins and are not translated. These occur at the 5′ and 3′ ends of the coding region and are referred to as untranslated regions (UTRs). 3′ UTRs have various roles in the posttranscriptional control of gene expression.

Article contents
- Structure and Composition of 3′ UTRs
- Regulatory Functions of 3′ UTRs
- Specific 3′ UTR Regulatory Motifs
- Regulation by RNA–RNA Interactions
- Conclusions

Structure and Composition of 3′ UTRs

In the mammalian genome, the length of 3′ untranslated regions (UTRs) shows considerable variation (**Table 1**), ranging from 60 to 80 nucleotides (nt) to about 4000. The average length in human genes is roughly 740 nt, and this is over three times the average length of the 5′ UTR (210 nt). Notably, the average length of the 3′ UTR in human genes is greater than that in other mammalian and vertebrate genes (420–500 nt) and in plants and fungi (241–274 nt);

this is paralleled by a reduction in the relative length of the 5′ UTR as compared with the 3′ UTR (Pesole *et al.*, 1997). The relatively long length of 3′ UTRs in human genes suggests that the human genome has evolved to make increased use of posttranscriptional control mechanisms to regulate gene expression. (*See* mRNA Untranslated Regions (UTRs).)

Among taxa and species including humans, analysis of gene sequences shows that the 3′ UTR has a lower G+C content than has the corresponding 5′ UTR. In

Table 1 Lengths of 3′ UTRs of mRNAs encoded by human genes

mRNA	Length of 3′ UTR (no. of nucleotides)	Functions
c-*fos*	785	Stability, localization
metallothionein IG	137	Localization
lipoprotein lipase	1946	Translation
glutathione peroxidase 4	221	Se incorporation
glutathione peroxidase 3	876	Se incorporation
glutathione peroxidase 2	363	Se incorporation
c-*myc*	224	Localization, stability
dystrophin	2690	?
15-lipoxygenase	678	Translation
transferrin receptor 2α	443	Stability

Available database information shows that there is wide variation in the lengths of 3′ UTRs of mRNAs encoded by the human genome. A selection is shown in the above table. Data are from the UTRdb.

humans, the average G+C content is about 44% in the 3′ UTR and about 61% in both the 5′ UTR and in the third silent codon position in the coding region. Indeed, a high AU content is typical of 3′ UTRs (Pesole *et al.*, 1997).

The gene regions corresponding to the UTRs in messenger ribonucleic acid (mRNA) can contain introns, although these occur infrequently in the 3′ UTR. In the few examples identified so far, alternative splicing has been found to give rise to mRNAs with identical coding regions but different 3′ UTRs. For example, alternative splicing produces two forms (3.6 and 3.2 kb) of lipoprotein lipase mRNA with alternative polyadenylation signals: the 3.2-kb form has a truncated 3′ UTR. Both forms are expressed in adipose tissue, but only the longer, more efficiently translated form of the mRNA is found in the heart; this illustrates the basis of tissue-specific differences in translation.

Regulatory Functions of 3′ UTRs

Although 3′ UTRs were regarded originally as being unimportant, it is now realized that in many genes the 3′ UTR is an important regulatory region that has a principal role in the posttranscriptional control of gene expression. With the use of recombinant deoxyribonucleic acid (DNA) technology, 3′ UTR sequences can be exchanged between genes or attached to reporter sequences, the chimeric constructs can be introduced by transfection, and their function can be investigated. Using such approaches, it has been shown that 3′ UTRs can contain elements that regulate mRNA polyadenylation and translation, that confer instabil-

ity, or that determine specific localization of the mRNA in the cytoplasm. In all examples that have been examined in detail, only relatively short regions of the 3′ UTR (10−200 nt), rather than the full 3′ UTR, are necessary for any of these different regulatory functions. Detailed analysis has often been carried out by *in situ* mutagenesis.

All mRNAs except those encoding histones are polyadenylated, and this is primarily brought about by nuclear polyadenylation through the canonical nuclear polyadenylation element (AAUAAA) present toward the end of the 3′ UTR. In addition, cytoplasmic polyadenylation can occur. This process has been observed in organisms as diverse as clam and mouse, and it regulates the translational activation of maternal mRNAs in early development. Such cytoplasmic adenylation requires a specific regulatory element (CPE) that is AU-rich and located in the 3′ UTR near to the classical nuclear polyadenylation signal. The CPE is not identical in all mRNAs, but it has the general structure UUUUUUAU with a minimum of UUU-UAU; its precise position relative to the AAUAAA varies, but the two signals are usually within 100 nt.

Many mRNAs encoding proto-oncogenes and cytokines are highly unstable. Attaching the 3′ UTR from one of these genes (e.g. granulocyte-macrophage colony-stimulating factor (*GM-CSF*) or c-*fos*) to a reporter (e.g. the β-*globin* coding region) has been found to cause mRNA instability and rapid degradation. The whole 3′ UTR is not required for these effects, and the instability has been shown to be due to specific AU-rich elements (AREs) present in the 3′ UTR of many proto-oncogene, lymphokine and cytokine mRNAs (Chen and Shyu, 1995; Xu *et al.*, 1997). AREs vary in size from 50 to 150 nt, often contain several reiterations of the AUUUA sequence, and have a high U (and sometimes A) content. Each ARE contains a combination of distinct motifs such as AUUUA, UUAUUUA(U/A)(U/A) nonamers, U stretches or a U-rich domain, and it is the combination of these domains that determines instability. AREs have been shown to direct mRNA degradation together with cytoplasmic deadenylation of the poly(A) tail. The decay kinetics are always biphasic, and deadenylation precedes degradation of the main part of the mRNA. (*See* mRNA Stability and the Control of Gene Expression.)

AREs have been classified into groups according to their sequence features and functional properties (Chen and Shyu, 1995).

- AREI elements have 1−3 separate AUUUA sequences close to a U-rich region and promote deadenylation that is synchronous with degradation. They are found in early response gene mRNAs that encode nuclear transcription factors and some cytokine mRNAs.

- AREII elements (e.g. *GM-CSF*) have copies of the AUUUA sequence arranged so that they form at least two overlapping copies of the UUAUUUA(U/A)(U/A) nonamers in a U-rich region. They promote deadenylation that is asynchronous with degradation. So far, all of the mRNAs in this class encode cytokines.
- AREIII elements (e.g. *c-jun*) do not contain a AUUUA but have a U-rich region. They mediate deadenylation that is synchronous with degradation.

In addition to AREs, other motifs have been identified that are important in determining mRNA stability. For example, a 10-nt motif regulates stability of *glucose transporter-1* mRNA and a conserved, 29-nt, U-rich element controls stability of *amyloid precursor protein* mRNA. The 3′ UTR of erythroid *15-lipoxygenase* mRNA contains tandem repeat sequences that bind specific regulatory proteins, thereby causing inhibition of translation (Ostareck-Lederer *et al.*, 1994). In rabbits, the *15-lipoxygenase* 3′ UTR contains 10 tandem repeats of the following 19-nt pyrimidine-rich motif: CCCCA/GCCCU-CUUCCCCAAG, where A/G indicates that the A and G are alternatives. In the corresponding human mRNA, there is a homologous shorter motif (CCCC/UA/GCCCUCCCAAG) in only four tandem repeats.

Single, highly conserved AUUUA and AUAUUUA motifs also have been identified in specific 3′ UTRs. Originally thought to be part of elements regulating mRNA stability, they have now been also implicated in the elements that control mRNA localization and translation. For example, in the *tumor-necrosis factor-α* 3′ UTR, there are two regulatory elements that regulate mRNA stability and translational efficiency: one is an AREII; the other, located 150 nt downstream, contains a conserved AUAUUUA. In the *c-myc* 3′ UTR, a conserved AUUUA is essential for determining perinuclear mRNA localization (Veyrune *et al.*, 1996): mutation of the AUUUA causes a marked reduction in mRNA localization and association with the cytoskeleton.

Some mRNAs are localized in different regions of the cytoplasm. The phenomenon is seen in many different mammalian cells from the highly polarized neuron to the hepatocyte and Chinese hamster ovary cell. In all examples examined so far, the 3′ UTR is essential for localization of the mRNA, and attachment of the 3′ UTR to a reporter is sufficient to cause targeting of the reporter (Jansen, 2001). The 3′ UTRs from β-actin and myelin basic protein mRNAs are needed for transport of the mRNAs to the cell periphery, whereas those from *c-myc*, *c-fos* and

metallothionein mRNAs are essential for localization in the perinuclear cytoplasm and association with the cytoskeleton. Although it is still unclear whether the signal elements involved in such mRNA targeting are based on sequence or structural motifs, the parts of the 3′ UTRs involved have been defined to greater and lesser extents: the *β-actin* and *myelin basic protein* signals are located in 45 and 21-nt regions, respectively, and the perinuclear localization signal is in an 86-nt region of the *c-myc* 3′ UTR and a 145-nt region of the *c-fos* 3′ UTR. (*See* mRNA Localization: Mechanisms.)

Specific 3′ UTR Regulatory Motifs

3′ UTR sequences have a well-defined role in the translation of selenoprotein mRNAs. The essential micronutrient selenium is incorporated into this small family of proteins (roughly 30 members in humans) as the amino acid selenocysteine during translation. Selenocysteine is encoded by UGA codons in the coding regions of these mRNAs. The translation of UGA, which is usually a stop codon, as selenocysteine requires a conserved stem−loop structure (called the SECIS element), which in eukaryotic mRNAs is found in 3′ UTRs. Consensus SECIS structures have been formulated (**Figure 1**) and binding proteins are beginning to be identified; notably, the stem−loop

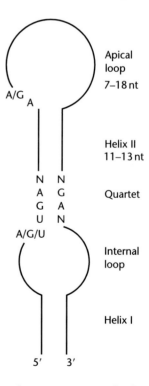

Figure 1 A proposed consensus structure for the selenocysteine insertion sequence found in the 3′ UTRs of selenoprotein genes (see Web Links).

includes a non-Watson–Crick base-pairing (Walczak *et al.*, 1998). (*See* RNA Secondary Structure Prediction.)

Histone mRNAs are not polyadenylated but contain a highly conserved stem–loop structure with a six-base stem and a four-base loop (Williams and Marzluff, 1995). Unusually for conserved structures, both the stem bases and the single-stranded loop region are conserved, and the flanking sequences are also essential. The loop has dual roles: in the nucleus, it is essential for RNA processing and transport; in the cytoplasm, it regulates mRNA stability and translational efficiency. Processing also requires protein binding to a purine-rich element located 10–20 nt downstream of the stem–loop.

The iron-responsive element (IRE) is a stem–loop structure located in either the 3′ UTR or the 5′ UTR of various mRNAs that encode proteins whose expression is regulated by concentrations of cellular iron (Hentze and Kuhn, 1996). Five IREs are found in the 3′ UTR of *transferrin receptor* mRNA. The binding of specific proteins to this IRE protects the mRNA from degradation; under conditions of high intracellular iron, protein binding does not occur and the mRNA is degraded. Two alternative consensus structures have been found: one with a single cytosine 'bulge' in the stem, and one with a cytosine and two additional bases. (*See* 5′ UTRs and Regulation.)

Regulation by RNA–RNA Interactions

In *Caenorhabditis elegans*, the expression of certain genes important in determining gene expression during cell fate determination is regulated by RNA–RNA interactions involving 3′ UTR sequences (Reinhart *et al.*, 2000). For example the genes *let-7* and *lin-4* encode small RNAs (the *let-7* gene product is 21 nt) that are complementary to elements in the 3′ UTR of several developmentally regulated genes. The interaction between these small RNAs and 3′ UTR sequences is thought to regulate gene expression at the stage of translation.

Conclusions

3′ UTRs should be considered as principal regulatory regions in control of gene expression. This regulation is brought about by several different types of motif, mainly RNA secondary structures that bind specific proteins. It is these specific RNA–protein interactions that determine the fate of the mRNAs in the cytoplasm. A 3′ UTR may contain several elements that are functionally distinct. For example, *utrophin* and c-*fos* 3′ UTRs contain regions that are required for

mRNA instability and separate regions that are required for mRNA localization. Thus, 3′ UTRs should be regarded as potentially multifunctional control regions. Information on 3′ UTRs and their regulatory motifs is increasing rapidly and is available in specific databases.

As with other genomic regions, sequences corresponding to 3′ UTRs can contain genetic variation. Indeed, it has been estimated that there are 300 000 single nucleotide polymorphisms in the sections of the human genome corresponding to 3′ UTRs. Variation in 3′ UTR sequences has the potential to give rise to variation in regulation of gene expression and thus may have phenotypic consequences. The functional significance of polymorphisms in these regions is an important area of future research. Mutations in 3′ UTR sequences can also have phenotypic impact, as in the disease myotonic dystrophy, which has been associated with the occurrence of extended CTG repeats. (*See* 3′ UTR Mutations and Human Disorders.)

See also
mRNA Untranslated Regions (UTRs)
5′ UTRs and Regulation

References

Chen C-YA and Shyu A-B (1995) AU-rich elements: characterization and importance in mRNA degradation. *Trends in Biochemical Sciences* **20**: 465–470.

Hentze MW and Kuhn LC (1996) Molecular control of vertebrate iron metabolism: mRNA based regulatory circuits operated by iron, nitric oxide and oxidative stress. *Proceedings of the National Academy of Sciences of the United States of America* **93**: 8175–8182.

Jansen RP (2001) mRNA localization: message on the move. *Nature Reviews in Molecular Cell Biology* **2**: 247–256.

Ostareck-Lederer A, Ostareck DH, Standart N and Thiele BJ (1994) Translation of 15-lipoxygenase mRNA is inhibited by a protein that binds to a repeated sequence in the 3′ untranslated region. *EMBO Journal* **13**: 1476–1481.

Pesole G, Luini S, Grillo G and Saccone C (1997) Structural and compositional features of untranslated regions of eukaryotic mRNAs. *Gene* **205**: 95–102.

Reinhart BJ, Slack FJ, Basson M, *et al.* (2000) The 21-nucleotide *let-7* RNA regulates developmental timing on *Caenorhabditis elegans*. *Nature* **403**: 901–906.

Veyrune J-L, Campbell GP, Wiseman J, Blanchard J-L and Hesketh JE (1996) A localization signal in the 3′ untranslated region of c-*myc* mRNA targets c-*myc* mRNA and β-globin reporter sequences to the perinuclear cytoplasm and cytoskeletal-bound polysomes. *Journal of Cell Science* **109**: 1185–1194.

Walczak R, Carbon P and Krol A (1998) An essential non-Watson–Crick base pair motif in 3′ UTR to mediate selenoprotein translation. *RNA* **4**: 74–84.

Williams AS and Marzluff WF (1995) The sequence of the stem and flanking sequences at the 3′ end of histone mRNA are critical determinants for the binding of the stem–loop binding protein. *Nucleic Acids Research* **23**: 654–662.

Xu N, Chen C-YA and Shyu A-B (1997) Modulation of the fate of cytoplasmic mRNA by AU-rich elelments; key sequence features controlling mRNA deadenylation and decay. *Molecular and Cellular Biology* **17**: 4611–4621.

Further Reading

Bergsten SE, Huang T, Chatterjee S and Gavis ER (2001) Recognition and long-range interactions of a minimal nanos RNA localization signal element. *Development* **128**: 427–435.

Dalgleish G, Veyrune J-L, Blanchard J-M and Hesketh JE (2000) mRNA localization by a 145 nt region of the c-*fos* 3' untranslated region: links to translation but not stability. *Journal of Biology Chemistry* **276**: 13593–13599.

Gramolini AO, Belanger G and Jasmin BJ (2001) Distinct regions in the 3' untranslated region are responsible for targeting and stabilizing utrophin transcripts in skeletal muscle cells. *Journal of Cell Biology* **154**: 1173–1183.

Grzybowska EA, Wilczynska A and Siedlecki JA (2001) Regulatory functions of 3' UTRs. *Biochemical and Biophysical Research Communications* **288**: 291–295.

Kislauskis EH, Li Z, Taneja KL and Singer RH (1993) Isoform-specific 3' untranslated sequences sort α-cardiac and β-cytoplasmic actin messenger RNAs to different cytoplasmic compartments. *Journal of Cell Biology* **123**: 165–172.

Kwon S, Barbarese E and Carson JH (1999) The *cis*-acting RNA trafficking signal from myelin basic protein mRNA and its cognate *trans*-acting ligand hnRNP A2 enhance cap-dependent translation. *Journal of Cell Biology* **147**: 247–256.

Pesole G, Liuni S, Grillo G, et al. (1999) UTRdB: a specialized database of 5' and 3' untranslated regions of eukaryotic mRNAs. *Nucleic Acids Research* **27**: 188–191.

Stoecklin G, Soeeckle P, Lu M, Muehlemann O and Moroni C (2001) Cellular mutants define a common mRNA degradation pathway targeting cytokine AU-rich elements. *RNA* **7**: 1578–1588.

Wagner C, Palacios I, Jeager L, et al. (2001) Dimerization of the 3' UTR of bicoid mRNA involves a two-step mechanism. *Journal of Molecular Biology* **313**: 511–524.

Wilkinson MF and Shyu AB (2001) Multifunctional regulatory proteins that control gene expression in both the nucleus and the cytoplasm. *BioEssays* **23**: 775–787.

Web Links

UTR Database (UTRdb). 5' and 3' UTRs of eukaryotic mRNAs
http://bighost.area.ba.cnr.it/BIG/UTRHome

Gladyshev website. Prediction of selenocysteine insertion motifs (SECIS)
http://bio-selenium.unl.edu

SECI Search website. Identification of SECIS elements.
http://genome.unl.edu/SECISearch.html

5' UTRs and Regulation

Advanced article

Flavio Mignone, *Università Milano, Milan, Italy*

Graziano Pesole, *Università Milano, Milan, Italy*

Article contents

- Structural and Compositional Features
- Regulation of Translation Efficiency
- Regulation of mRNA Stability

The 5' untranslated region consists of the leading part of a cellular mRNA from the 5' end to the translation start codon AUG. The 5' end of the precursor mRNA is posttranscriptionally modified by the removal of introns and the addition of a 7-methyl-guanilate cap structure, which has a crucial role in translation initiation and in regulating transcript stability.

Structural and Compositional Features

The average length of 5' untranslated regions (UTRs) is roughly constant over diverse taxonomic classes of eukaryote and ranges from 100 to 200 nt. Within a species, however, there can be remarkable length heterogeneity, ranging from a few to a few thousand nucleotides (Mignone *et al.*, 2002). In fact, it has been shown using a mammalian *in vitro* system that even a single nucleotide is a sufficient 5' UTR for initiating translation. The length of 5' UTRs correlates with the translation efficiency of the coded protein: messenger ribonucleic acid (mRNAs) encoding housekeeping proteins usually have shorter 5' UTRs than do mRNAs encoding regulatory proteins, and they are translated more efficiently.

Diverse 5' UTRs may be encoded by a single gene, owing to the use of alternative promoters that result in different transcription start sites (TSSs) and to alternative splicing (**Figure 1**). Indeed, about 30% of genes in Metazoa have fully untranslated exons, that is, the translation start codon (sAUG) occurs in the second or following exons. The expression pattern of mRNAs with alternative 5' UTRs has been shown to vary considerably with tissue type, developmental stage and disease state. Cancer-specific alternative splicing forms, differing in 5' UTRs, have been shown for several genes, including, for example, the ataxia-telangiectasia mutated (*ATM*) gene and the glioma-associated oncogene homolog (*GLI*). (*See* Alternative Promoter Usage; Promoters.)

The average G+C content of 5' UTRs, about 60% in warm-blooded vertebrates, has been shown to be

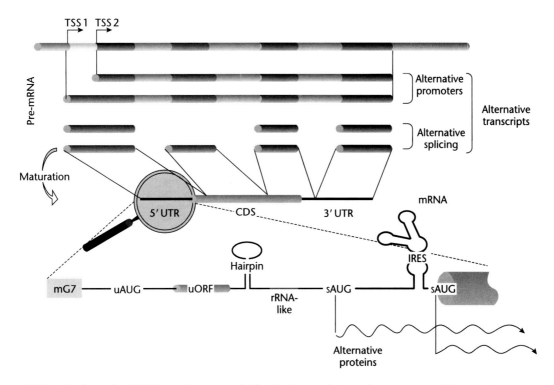

Figure 1 mRNAs with alternative 5′ UTRs can be generated either by the use of alternative promoters, which results in different transcription start sites (TSS1 and TSS2), or by alternative spicing events. In this way, alternative 5′ UTRs corresponding to the same gene may contain different combinations of regulatory elements, as shown here. CDS: coding sequence; mG7: 7-methyl-guanilate; IRES: internal ribosome entry site; sAUG: start AUG; uAUG: upstream AUG; uORF: upstream open reading frame.

inversely correlated with their length; that is, the 5′ UTRs of genes localized in GC-rich chromosome regions tend to be shorter than 5′ UTRs localized in GC-poor chromosome regions. The high G+C content of 5′ UTRs in warm-blooded vertebrates is also due to the presence of CpG islands at the 5′ end of a large fraction of genes. Accordingly, the CpG dinucleotide is much less depleted in the 5′ UTRs than in the remaining portion of the genes. There is a significant difference in the G+C content of the 5′ UTR between housekeeping and regulatory mRNAs, with a higher G+C content in regulatory mRNAs. In addition, housekeeping mRNAs show higher G/C and A/T asymmetry, disfavoring the formation of stable hairpins that can reduce translation efficiency (Kochetov *et al.*, 1998). (*See* CpG Islands and Methylation; mRNA Untranslated Regions (UTRs).)

Regulation of Translation Efficiency

Upstream AUGs and ORFs

According to the scanning model, translation initiates at the first AUG encountered by the 40S ribosomal subunit after it binds at the 5′ 7-methyl-guanilate (m7G) cap site and starts scanning the 5′ UTR (**Figure 2a**). However, many mRNAs contain AUG codons and open reading frames (ORFs) upstream of the sAUG, denoted respectively as uAUG and uORFs, which can control expression of the encoded protein (**Figure 2b–h**). uAUGs and uORFs can induce the formation of translation-competent ribosomes that (1) may terminate and reinitiate; (2) may terminate and leave the mRNA, thereby reducing the translation efficiency of the main ORF (leaky scanning); (3) may synthesize an *N*-terminally extended protein that is either in or out of frame with respect to the main ORF. The uAUG and uORFs may also regulate initiation at a different sAUG, resulting in the synthesis of different proteins and also determining their isoform ratio. A crucial factor for modulating translation is the nucleotide context around the uAUGs and the sAUG. Indeed, the ideal context in eukaryotic mRNAs is RCCAUGG, where the most conserved nucleotides are the purine (R) at -3 and the guanine at $+4$ (the A of the AUG codon is designated $+1$). (*See* Ribosomes and Ribosomal Proteins; Translation Initiation: Molecular Mechanisms in Eukaryotes.)

When the ribosome terminates the translation of an uORF, the reloading of several initiation factors released at the beginning of the uORF translation is

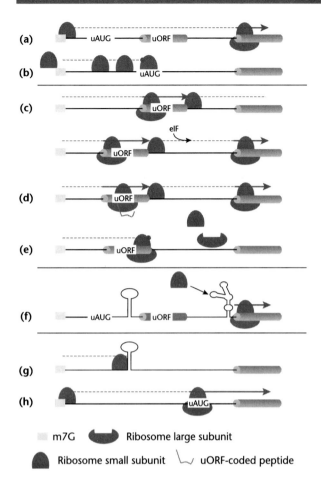

(a)

(b)

(c)

(d)

(e)

(f)

(g)

(h)

elF

uAUG | uORF

uAUG

uORF

uORF

uORF

uORF

uAUG | uORF

uAUG

☐ m7G ⬤ Ribosome large subunit

⬤ Ribosome small subunit ⌣ uORF-coded peptide

Figure 2 Translation initiation regulation. (a) Canonical scanning model; (b) ribosome stalling; (c) inefficient (top) and efficient (bottom) reinitiation, depending on the distance between the uORF and the sAUG; (d) regulation by uORF-encoded peptide; (e) detachment of the ribosome mediated by an uORF stop codon; (f) IRES-mediated initiation; (g) inhibition of secondary structure; (h) in-frame or out-of-frame translation starting from an uAUG.

required to reinitiate translation. This makes reinitiation an intrinsically inefficient process that depends on the concentration of the various factors and on the distance between the uORF and the main AUG. In addition, the context surrounding the stop codon of the uORF is important because it can modulate the efficiency of the termination: depending on the stop codon context, the ribosome can efficiently terminate translation and remain bound to the mRNA, or it can stall at the stop codon and block following ribosomes. Nucleotides downstream of the stop codon can also promote detachment of the ribosome, thus impairing the reinitiation process. The best-known example of translational control through reinitiation is that of the *GCN4* mRNA of *Saccharomyces cerevisiae*.

Increasing evidence seems to suggest that the peptide sequence coded by the uORF might have

regulatory function. The best-studied example is the *AdoMetDC* mRNA, which contains an uORF in the 5′ UTR. The scanning ribosome stalls during translation of the uORF, thereby decreasing the efficiency of translating the main ORF. The mechanism is not completely clear, but the experiments are consistent with a model in which the stalling of the ribosome is dependent on both the coding sequence of the uORF and the content of cellular polyamines (Shantz and Pegg, 1999).

Internal ribosome entry sites

Internal mRNA ribosome binding is an alternative mechanism of translation initiation to conventional cap-dependent ribosome scanning and was first discovered for some uncapped viral mRNAs. This mechanism, which is activated in specific cell states (particularly when general protein synthesis is inhibited), involves ribosome recruitment inside the 5′ UTR in close proximity to the sAUG by a complex structural element termed the internal ribosome entry site (IRES). IRES-mediated translation occurs for many cellular transcripts, such as c-*myc*, *apaf-1* and *ODC*, which are generally involved in growth control, cell cycle and apoptosis.

Comparative analysis of a set of cellular mRNAs containing known IRESs identified a common structural motif that forms a Y-type stem−loop structure either followed by the AUG triplet or followed by additional stem−loop structures and the AUG triplet. However, alternative structures, as well as regions of sequence complementarity to 18S ribosomal RNA (rRNA), have been observed to exert IRES-mediated initiation.

rRNA-like sequences

Several computational and experimental analyses have shown that many mRNAs contain rRNA-like sequences. An increasing body of evidence seems to indicate that specific rRNA-like sequences may mediate precise interactions between mRNAs and the ribosome, thereby modulating translation efficiency. In some cases, such as the mouse GTX homeodomain mRNA, a correlation between the extent of complementarity and translation inhibition has been observed (Hu *et al.*, 1999).

Repeats

5′ UTRs are known to contain several types of repeat, including short interspersed elements (SINEs), long interspersed elements (LINEs), minisatellites and microsatellites, although their functional meaning has not been elucidated as yet. In particular, trinucleotide repeats, such as $(CUG)n$ and $(CAG)n$, located in the 5' UTR of a reporter gene have been shown to depress translation as a function of the repeat length. A similar effect has been shown for retroelements such as Alu and L1. (*See* Long Interspersed Nuclear Elements (LINEs); Simple Repeats; Short Interspersed Elements (SINEs).)

Secondary structures

Secondary structures in 5' UTRs may also be important in regulating translation. Experimental data suggest that moderately stable secondary structures ($\Delta G > -30\,\mathrm{kcal\,mol^{-1}}$) do not significantly reduce translation efficiency. A reduction has been observed to occur only when very stable structures ($\Delta G < -50\,\mathrm{kcal\,mol^{-1}}$) are formed. In the latter case, translation efficiency can be rescued by an increase in the initiation factors (e.g. eIF4F) involved in RNA structure unwinding during ribosome scanning (Svitkin *et al.*, 2001). (*See* RNA Secondary Structure Prediction.)

Specific RNA structure may regulate translation by interacting with RNA-binding proteins. This occurs during the translational control of some mRNAs involved in iron homeostasis (e.g. ferritin, erythroid δ-aminolevulinate synthase) that contain in their 5' UTR a conserved hairpin structure, denoted as an 'iron-responsive element' (IRE). This element is the target of specific RNA-binding proteins, known as iron regulatory proteins (IRPs), that function as sensors of iron concentration. A low intracellular concentration of iron enables IRPs to bind to IREs, which prevents ribosome scanning and inhibits translation. (*See* RNA-binding Proteins: Regulation of mRNA Splicing, Export and Decay.)

Regulation of mRNA Stability

The cap structure is a crucial stability determinant for eukaryotic mRNAs because it blocks the action of $5' \rightarrow 3'$ exonucleases. Indeed, the activation of decapping enzymes triggered by shortening or removing the mRNA poly(A) tail induces transcript degradation. (*See* mRNA Stability and the Control of Gene Expression.)

Although most *cis*-acting elements that modulate mRNA stability are located in the 3' UTR, they also can be found in the 5' UTR (e.g. the JRE element that stabilizes *interleukin 2* mRNA in response to specific activating signals). uORFs in the 5' UTR may also be involved in determining mRNA stability through the

nonsense-mediated decay (NMD) mechanism. The signal that triggers NMD is a nonsense codon followed by a splicing junction. If the ribosome encounters a stop codon that is either premature or due to the presence of an upstream ORF, it disassembles and marker proteins deposited at the splice junction during splicing activate mRNA decay. In some mRNAs (e.g. *GCN4* and *YAP1* mRNAs in yeast), specific stabilizer elements located between the uORF and the sAUG have been observed to block the NMD pathway. (*See* 3' UTRs and Regulation; mRNA Turnover; Nonsense-mediated mRNA Decay.)

See also
mRNA Untranslated Regions (UTRs)
3' UTRs and Regulation

References

Hu MC, Tranque P, Edelman GM and Mauro VP (1999) rRNA-complementarity in the 5' untranslated region of mRNA specifying the Gtx homeodomain protein: evidence that base-pairing to 18S rRNA affects translational efficiency. *Proceedings of the National Academy of Sciences of the United States of America* **96**: 1339–1344.

Kochetov AV, Ischenko IV, Vorobiev DG, *et al.* (1998) Eukaryotic mRNAs encoding abundant and scarce proteins are statistically dissimilar in many structural features. *FEBS Letters* **440**: 351–355.

Mignone F, Gissi C, Liuni S and Pesole G (2002) Untranslated regions of mRNAs. *Genome Biology* **3**: reviews0004.1–0004.10.

Shantz LM and Pegg AE (1999) Translational regulation of ornithine decarboxylase and other enzymes of the polyamine pathway. *International Journal of Biochemistry and Cell Biology* **31**: 107–122.

Svitkin YV, Pause A, Haghighat A, *et al.* (2001) The requirement for eukaryotic initiation factor 4A (eIF4A) in translation is in direct proportion to the degree of mRNA 5' secondary structure. *RNA* **7**: 382–394.

Further Reading

Hellen CU and Sarnow P (2001) Internal ribosome entry sites in eukaryotic mRNA molecules. *Genes and Development* **15**: 1593–1612.

Kozak M (1999) Initiation of translation in prokaryotes and eukaryotes. *Gene* **234**: 187–208.

Meijer HA and Thomas AA (2002) Control of eukaryotic protein synthesis by upstream open reading frames in the 5'-untranslated region of an mRNA. *Biochemical Journal* **367**: 1–11.

van der Velden AW and Thomas AA (1999) The role of the 5' untranslated region of an mRNA in translation regulation during development. *International Journal of Biochemistry and Cell Biology* **31**: 87–106.

Vilela C, Ramirez CV, Linz B, Rodrigues-Pousada C and McCarthy JE (1999) Post-termination ribosome interactions with the 5' UTR modulate yeast mRNA stability. *EMBO Journal* **18**: 3139–3152.

Vandenberg, Steven G

Alan B Zonderman, *National Institute on Aging, National Institutes of Health, Bethesda, MD, USA*

Steven G. Vandenberg (1915–1992) was a pioneer in the study of twins.

In a career spanning four decades, Steven G. Vandenberg did pioneering work in twin studies, family studies and individual differences. Vandenberg's Hereditary Abilities Study, whose first results were published in 1956, initiated the contemporary twin studies, particularly heritability studies on specific cognitive abilities. (*See* Twin Studies.)

Born in 1915 in The Netherlands, he served from 1960 to 1967 as Director of the Louisville Twin Study, the first such study to investigate the influence of heredity on rates of growth and age-related behavioral changes. Using multivariate techniques, Vandenberg demonstrated significant heritable influences on the rate and acceleration of growth in height from birth to age four. Other findings showed that neonatal Apgar ratings were unrelated to subsequent cognitive abilities or personality; there were larger behavioral differences in fraternal twins than in identical twins; and there were significant heritabilities for cognitive abilities. These latter findings stimulated his work in multivariate analyses and the publication of two collections he edited, *Methods and Goals in Human Behavior Genetics* and *Progress in Human Behavior Genetics*. In several papers in the two volumes, Vandenberg described methods using Bartlett's F ratio for testing the differences between two variances. In an extension to canonical form, univariate monozygous and dizygous within-pair variances were replaced by within-pair covariance matrices for seeking linear combinations of variables best discriminating monozygous from dizygous twins. These methods presaged more sophisticated partitioning of genetic and environmental covariance matrices. (*See* Twin Methodology.)

In 1967, Vandenberg moved to the University of Colorado as Professor of Psychology and Fellow of the Institute for Behavioral Genetics. His interests turned toward family studies represented by two separate projects examining familialities (familial resemblance due to genetic or environmental influences) in cognitive abilities. In the Hawaii Family Study of Cognition, he was coinvestigator in a cooperative study between the Institute for Behavioral Genetics and the Behavioral Biology Laboratory in Hawaii examining familialities in two ethnic groups, Americans of European ancestry and Americans of Japanese ancestry. In the Boulder Family Study in Colorado, he was principal investigator in a similar study of familiality with a focus on parental child-rearing attitudes and behaviors, and assortative marriage (spouse resemblance). Data from the Hawaii Family Study showed identical structures of intellect in the two ethnic groups represented by four cognitive factors: verbal ability, spatial visualization, perceptual speed and accuracy, and visual memory. Data from both family studies showed significant assortative marriage for cognitive abilities and for physical attractiveness. Data from the Hawaii Family Study showed moderately high upper-bound heritability estimates for verbal and spatial abilities (approximately 0.6) and somewhat smaller values for memory and perceptual speed (0.4) with no evidence for sex linkage based on correlations between parents and children. (*See* Adoption Studies; Intelligence and Cognition.)

Vandenberg's legacy in behavior genetics is represented by his prolific publications; his international recognition, culminating in election as President of the Behavior Genetics Association; and his founding influences on the journal *Behavior Genetics* as one of the first outlets for papers in this area. For those who met or knew him, Vandenberg's legacy is his open-minded scientific outlook and his personal humility despite all his accomplishments.

See also
Twin Studies

Further Reading

DeFries JC, Ashton GC, Johnson RC, *et al.* (1976) Parent–offspring resemblance for specific cognitive abilities in two ethnic groups. *Nature* **261**: 131–133.

DeFries J, Plomin R, Vandenberg S and Kuse A (1981) Parent–offspring resemblance for cognitive abilities in the Colorado Adoption Project: biological, adoptive, and control parents and 1-year-old children. *Intelligence* **5**(3): 245–277.

DeFries J, Vandenberg SG, McClearn GE, *et al.* (1974) Near identity of cognitive structure in two ethnic groups. *Science* **183**: 338–339.

Johnson RC, DeFries J, Wilson JR, *et al.* (1976) Assortative marriage for specific cognitive abilities in two ethnic groups. *Human Biology* **48**: 343–352.

Spuhler KP and Vandenberg SG (1980) Comparison of parent–offspring resemblance for specific cognitive abilities. *Behavior Genetics* **10**(4): 413–418.

Vandenberg SG (1962) The Hereditary Abilities Study: hereditary components in a psychological test battery. *American Journal of Human Genetics* **14**(2): 220–237.

Vandenberg SG (ed.) (1965) *Methods and Goals in Human Behavior Genetics*. New York: Academic Press.

Vandenberg SG (ed.) (1968) *Progress in Human Behavior Genetics: Recent Reports on Genetic Syndromes, Twin Studies, and Statistical Advances*. Baltimore, MD: Johns Hopkins University Press.

Zonderman AB, Vandenberg SG, Spuhler KP and Fain PR (1977) Assortative marriage for cognitive abilities. *Behavior Genetics* **7**(3): 261–271.

Variable Drug Response: Genetic Evaluation

Michael PH Stumpf, *University College London, London, UK*

David B Goldstein, *University College London, London, UK*

Intermediate article

Article contents

- Introduction
- Population Structure
- Association Mapping in Pharmacogenetics
- Road to Individualized Medicine

Pharmacogenetics is the study of genetic variation in response to drugs. Although there is some evidence that drug response genotypes cluster in ethnic groups, this effect is less important than individual variation.

Introduction

Pharmacogenetics emerged in the late 1950s and early 1960s when variation in the metabolism of the antimalarial agents primaquine and isoniazid and the antihypertensive debrisoquine was shown to follow a simple Mendelian inheritance pattern. Following such discoveries, Motulsky set the conceptual foundation for pharmacogenetics by pointing out that common nonclinical or subclinical genetic variation may underlie genetic variation contributing to pharmacokinetic differences among individuals. The classical debrisoquine polymorphism also led the way into molecular work with the cloning of the gene responsible, cytochrome P450, subfamily IID (debrisoquine, sparteine, etc. -metabolizing), polypeptide 6 (*CYP2D6*), in the 1980s. Following this the genes responsible for most drug metabolism were rapidly discovered, leading to a concentrated research program devoted to the identification and frequency characterization of variants that might be expected to influence drug response. Because of their obvious importance, drug metabolizing enzymes (DMEs) figured prominently in this work, but more recently other classes of genes such as drug targets have been studied in the same framework.

Despite this relatively long history, pharmacogenetics is none the less viewed by some as a new discipline or as in an editorial in *Science* 'the fledgling science of pharmacogenetics' (Marx, 2002). This and similar pronouncements reflect optimism that advances in genomics will allow systematic identification of variants that influence drug response. Practiced in this way, it is certainly a new discipline that is beginning to attract extraordinary interest.

Here we consider two aspects of how the genetic basis of variable drug response (VDR) may be addressed. First, we consider how differences among individuals from different parts of the world might influence how they respond to drug administration, and we review approaches for representing this source of variation in the context of drug trials. Ultimately, however, the aim is to identify the precise genetic basis of VDR in order to allow the personalization of medicines. We therefore turn next to a description of genetic association studies in VDR, widely seen as the most promising approach for identifying the individual genetic basis of VDR.

Population Structure

Variants that influence drug responses often show sharp geographical localization, with DMEs being the most prominent examples (Xie *et al.*, 2001). Functional variants have been identified in most genes encoding DMEs and these often show significant frequency differences among at least some pairs of populations. For example, functional variants in genes belonging to the *CYP2C* family often show significant frequency differences among Caucasians, Asians and people of African descent. (**See** Human Genetic Diversity.)

Increasingly, other genes such as drug targets, transporters and channels, are being studied for their effects on drug response and many of their genetic variants also show frequency differences among populations. For example variants of the gene sodium channel, voltage-gated, type V, alpha polypeptide (long (electrocardiographic) QT syndrome 3) (*SCN5A*) have been implicated in a rare familial form of arrhythmia, including long QT intervals (Splawski *et al.*, 2002). Recently one such long QT associated variant, Y1102, was shown to be confined to individuals with African ancestry where it appears at moderate frequencies (8–15%). It is thought that this variant may predispose individuals to drug-induced problems with long QT intervals, which are a leading cause of drugs either failing in phase III clinical trials or being removed from the market.

In addition to between-population frequency differences of drug response gene (DRG) variants, there is also more direct evidence of differences among populations. For example, population differences have been reported for various drugs both for recommended doses (e.g. therapeutic dosage of the anticoagulant warfarin is lower in Japanese than in Americans and nifedipine requires lower dosage in South Asians than Caucasians) and for efficacies (the drug BiDIL has recently been approved for clinical trials specifically in African Americans).

Although between-population genetic differences clearly exist, it is important to note that work with multiple genetic marker systems has always confirmed the observation that most of human genetic variation is due to differences between individuals, not average differences between groups (Cavalli-Sforza *et al.*, 1996). Nevertheless observations such as those reported above and others have raised the question of how population genetic structure is best represented in drug evaluation. Two strategies have been proposed: racial labeling and explicit genetic inference. (*See* Human Populations: Evolution.)

Ethnic or racial labeling

This strategy is based on the notion that racial labels, properly defined, provide a reasonable guide to the pattern of geographical structuring of human genetic variation. The reasoning behind this approach is implicitly grounded in the view of modern human origins that postulates a moderately high level of geographical isolation among the continents of the world, which, according to Risch *et al.* (2002), leads to a racial classification that generally follows continental boundaries: in all, Risch counts five racial groups (Caucasians, Africans, Asians, Pacific Islanders and Native Americans) but acknowledges the existence of hybrid populations; for example, Ethiopians are generally considered to result from an admixed Caucasian/African population. Similarly Hispanics show admixture from Caucasian, African and Native American ancestry.

Explicit genetic inference

The alternative approach ignores all racial, ethnic and geographical labels and explicitly infers groups of more or less related individuals from genetic data. Statistical (cluster) analysis of the observed correlation of alleles across many unlinked loci allows the separation of individuals into distinct groups. A promising inferential procedure, implemented in the program STRUCTURE (Pritchard *et al.*, 2000), is based on a simulation approach that estimates the number of genetically separable populations. It then determines the probability that an individual belongs to each of the clusters, or under an admixture model it estimates the proportion of each individual's ancestry that is derived from each postulated cluster. (*See* Population Genetics: Historical Aspects.)

The use of explicit genetic inference in the evaluation of drug response was recently investigated by Wilson *et al.* (2001) who used 354 individuals sampled from eight global populations, including African, European, Asian and Papua New Guinean derived groups. Ignoring all ethnic labels they used STRUCTURE to determine population groupings based on 39 unlinked microsatellites. Four genetically distinct clusters were found, the majority of Ethiopians being found to cluster with the bulk of the Europeans, despite their African geographical origin; Papua New Guineans were well separated from Asians. These clusters were then studied with respect to common genetic variation in DMEs. Interindividual variation was found to exhibit geographical structuring and the inferred subpopulations often showed significant differences in the frequencies of DME mutations among these clusters. This demonstrates that there is room for considerable differences in VDR between the groups and that routine genotyping of populations in drug trials may be necessary if racial grouping schemes are shown to be insufficient to represent geographical ancestry.

Explicit genetic inference of population structure is in its infancy and there are several pressing problems: the number of subgroups/clusters into which the individuals are divided is not estimated rigorously in a statistical sense; there is no reliable and stable way to decide at which level of population resolution one should stop; and the more markers used the finer is the subdivision that could be recognized but it is unclear how fine a level is needed with respect to assessing VDR.

It is also unknown how many markers and how many individuals are required to yield stable clustering schemes. Different cluster criteria (e.g. model-based approaches as outlined above or methods that group individuals together based on summary statistics of measures of genetic diversity) do not result in identical partitions of a population, and the same method applied to different data sets has also been shown to yield contradictory results (Romualdi et al., 2002). This is, however, not surprising and we would expect some of these problems to evaporate as more markers and more individuals are included in future studies. It is also unknown if the samples used in present studies provide a realistic sample of global human diversity. Only when these questions have been addressed and when reliable genetic population structures have been inferred can we make detailed comparisons between genetically inferred clusters and ethnic labels. This will require considerably more work.

Explicit genetic inference versus ethnic labels

In comparing racial labels with explicit genetic inference the central question is whether human genetic history really is as simple as postulated by those advocating a simple scheme of five major population groupings. In the simplest case it was a straightforward process of successive colonization of the continents followed by isolated evolution within the resulting isolated populations and small levels of admixture between them (as e.g. observed in Ethiopians) and a few outliers (e.g. Papua New Guineans). Our own view, however, is that it is not prudent to assume that our genetic history is simple, and we are not at all convinced that current evidence warrants this view. In particular, the set of samples of human diversity upon which most current analyses are based is by no means exhaustive. Those who study human evolution have tended to focus on a small set of population samples that routinely reappear in nearly all analyses (e.g. Mbuti and Central African Republic pygmies, Papua New Guinea highlanders, etc.), which have been studied because of exceptional anthropological interest. The pattern that would result from alternative sampling schemes remains extremely poorly known. Apart from the already known hybrid and outlying populations, clinically important population differences could appear in other cases too. Principal component analyses of Asian populations (Cavalli-Sforza et al., 1996) demonstrated that Japanese, South Chinese and North Chinese individuals are well separated. There is thus considerable room for functionally important variation at DRGs between these three large Asian populations. Similarly,

Southeast Asian populations cluster with Melanesians; which populations cluster with other eastern populations is entirely unclear.

The distribution of variants of cytochrome P450, subfamily IIC (mephenytoin 4-hydroxylase), polypeptide 19 (CYP2C19) that is involved in the metabolism of diazepam, for example, confirms this. A significant increase in the prevalence of a poor metabolizing genotype was observed among Han Chinese compared with the Dai Chinese populations (He et al., 2002). Such cases are not easily reconciled with simple notions of race but are supported by explicit genetic analysis, and it is important to continue this line of research.

A stable and reliable clustering scheme is crucial if explicit structural inferences are to be used routinely in the evaluation of VDR. The advantages of such a potential standard inferential framework will be that comparisons across drug trials will be possible, allowing increasingly improved assessments of variable drug response as more data become available from other populations. It will also allow an assessment of whether a drug trial population is representative of global human diversity or of the genetic diversity in a given area (e.g. East Asia). The necessary complexity and sophistication of schemes to represent the pattern of human genetic diversity in phase III trials will only become clear when explicit genetic inference has been compared with racial labels in a sufficiently broad and diverse sample set. This is an area of ongoing work, and until this question is resolved, it is better to speak about the role of geographical ancestry as opposed to racial groups, which just assume a simple pattern.

What is clear, however, is that geographical ancestry or ethnic labels will rarely if ever be a sufficient basis for the personalization of medicines. The analyses of Wilson et al. (2001) and others, while demonstrating that individuals with different geographical ancestry may show average differences in drug response that would influence the decision of whether to bring a drug to market, will rarely result in absolute differences. Any adverse reaction present in one group will be present in others, although often at different frequencies. That is, drugs will never work perfectly, for example, in Britain, and terribly in Japan. This reflects the high degree of similarity among population groups. Most of the genetic variation in our species traces back to the common ancestral population in Africa, and is common to all human populations. For this reason, the road to the personalization of medicines depends on individual genotype, not racial labels or explicit genetic inference; even differences at this level may be relevant to the decision about whether to bring a drug to market.

Association Mapping in Pharmacogenetics

Drug response is a complex trait that is influenced by multiple genes and environmental dependences. This makes family-based approaches, which have been so successful in the study of Mendelian traits, unsuitable for most studies of drug reaction. Association studies are widely seen as the most promising approach for dealing with this complexity. However unlike in the case of diseases (presence or absence of disease/clinical phenotype) the reaction to a drug is generally a quantitative effect, and one that is not always readily measured. (*See* Linkage and Association Studies; Quantitative Genetics.)

In a standard association setting, statistically significant association between a marker and a phenotype is taken as evidence for the marker being either causal or linked, and therefore physically close, to a causal variant. The level of significance is, however, confounded by population and genomic processes, such as population structure and mutation, recombination and selection. In particular the level at which physically close markers tend to be coinherited, that is the level of linkage disequilibrium (LD), determines the statistical power and feasibility of association studies. Traditionally, LD was thought to decay in a simple relationship with physical distance. As experimental data became more reliable, this simple scenario could not generally be confirmed and levels of LD showed considerable variation between population and genomic regions.

Recently, however, several studies have provided good evidence for a much more encouraging scenario (Daly *et al.*, 2001). It appears that the extent of LD and haplotype structure in large parts of the human genome may be simpler than previously thought: extended blocks of high average LD and low haplotype diversity seem to prevail in many genomic regions. It is often possible to identify sets of representative single nucleotide polymorphisms (SNPs), so-called haplotype tagging SNPs (htSNPs), that capture, in principle, any desired level of the observed variation. This strongly suggests a haplotype-based strategy for the design of association studies. First, the haplotype structure of candidate regions is determined in the populations of interest. Initially only a few SNPs may be typed to detect large-scale blocks. From these SNPs a set of htSNPs are then selected that explain a given amount of the observed diversity, for example 95% of chromosomes. These are then typed in samples of cases and controls respectively and tested for association with the trait of interest. In this framework, variable drug reactions are, in many ways, a tractable problem in terms of the genetics. For example, the entire set of DMEs acting on commonly used drugs is a very manageable set of genes, and could be fairly easily screened for common variants influencing drug reactions. (*See* Blocks of Limited Haplotype Diversity.)

The definition of the phenotype will, however, pose a problem in association studies for VDR. Drug efficacy is continuously variable and can range from no effect to a profound physiological change in the patient, and it will be necessary to reconsider how medical data are processed in the course of clinical studies. Success will depend on the close integration of clinical data, such as various physiological parameters, with genetic data. Such integration is likely to be a major immediate challenge for medical informatics. Lessons can be learned from epidemiology and the population genetics of quantitative traits: for example, rather than dividing particular observed phenotypes into strict but potentially arbitrary case and control categories, logistic regression over physiological data can be used to detect correlation between quantitative drug effects and particular genetic variants.

Road to Individualized Medicine

It is important to realize that there are important differences between clinical trials and day-to-day clinical practice. In the former we can evaluate the genetic makeup of the drug recipients through direct genotyping (whether of candidate genes or, in the future, of a whole-genome marker set). For the clinical administration of a drug this will not be the case and it is crucial to find convenient and safe ways to translate knowledge about VDR gained in clinical trials into workable guidelines for doctors. Geographical ancestry, especially if validated by genetic data, will be one important guideline in the evaluation of whether to bring drugs to market. But the role of genetics in the clinic will almost certainly focus largely on individual genotype.

See also
Drug Metabolic Enzymes: Genetic Polymorphisms
Genomics, Pharmacogenetics, and Proteomics: An Integration
Pharmacogenetics and Pharmacogenomics

References

Cavalli-Sforza LL, Mennazzi P and Piazza A (1996) *The History and Geography of Human Genes.* Princeton, NJ: Priceton University Press.

Daly MJ, Rioux JD, Schaffner SF, Hudson TJ and Lander ES (2001) High-resolution haplotype structure in the human genome. *Nature Genetics* **29**: 229–232.

He N, Yan FX, Huang SL, *et al.* (2002) CYP2C19 genotype and *S*-mephenytoin 4′-hydroxylation phenotype in a Chinese Dai population. *European Journal of Clinical Pharmacology* **58**: 15–18.

Marx J (2002) Gene mutation may boost risk of heart arrhythmias. *Science* **297**: 1252.

Pritchard JK, Stephens M and Donnelly P (2000) Inference of population structure using multilocus genotype data. *Genetics* **155**: 945–959.

Risch N, Burchard EG, Ziv E and Tang H (2002) Categorization of humans in biomedical research: genes, race and disease. *Genome Biology* **3**: 1–12.

Romualdi C, Balding D, Nasidze IS, *et al.* (2002) Patterns of human diversity, within and among continents, inferred from biallelic DNA polymorphisms. *Genome Research* **12**: 602–612.

Splawski I, Timothy KW, Tateyama M, *et al.* (2002) Variant of SCN5A sodium channel implicated in risk of cardiac arrhythmia. *Science* **297**: 1333–1336.

Wilson JF, Weale ME, Smith AC, *et al.* (2001) Population genetic structure of variable drug response. *Nature Genetics* **29**: 265–269.

Xie HG, Kim RB, Wood AJ and Stein CM (2001) Molecular basis of ethnic differences in drug disposition and response. *Annual Review of Pharmacology and Toxicology* **41**: 815–850.

Further Reading

Elston RC, Olson JM and Palmer LJ (eds.) (2002) *Biostatistical Genetics and Genetic Epidemiology*. Chichester, UK: John Wiley.

Goldstein DB and Chiki L (2002) Human migrations and population structure: what we know and why it matters. *Annual Review of Genomics and Human Genetics* **3**: 129–152.

Lewontin RC (1972) The appointment of human diversity. *Evolutionary Biology* **6**: 381–398.

Licino J and Wong ML (eds.) (2002) *Pharmacogenomics*. Weinheim: Wiley-VCH.

McLeod HL and Ameyaw MM (2002) Ethnicity and pharmacogenomics. In: Licino J and Wong ML (eds.) *Pharmacogenomics*, pp. 489–514. Weinheim: Wiley-VCH.

Meyer UA and Zanger UM (1997) Molecular mechanisms of genetic polymorphisms of drug metabolism. *Annual Review of Pharmacology and Toxicology* **37**: 269–296.

Nebert DW (1999) Pharmacogenetics and pharmacogenomics: why is this relevant to the clinical geneticist? *Clinical Genetics* **56**: 247–258.

Weber WW (1997) *Pharmacogenetics*. New York: Oxford University Press.

Xie HG, Kim RB, Wood AJ and Stein CM (2001) Molecular basis of ethnic differences in drug disposition and response. *Annual Review of Pharmacology and Toxicology* **41**: 815–850.

Web Links

Cytochrome P450, subfamily IIC (mephenytoin 4-hydroxylase), polypeptide 19 (*CYP2C19*); Locus ID: 1557. LocusLink:
http://www.ncbi.nlm.nih.gov/LocusLink/LocRpt.cgi?l = 1557

Cytochrome P450, subfamily IID (debrisoquine, sparteine, etc., -metabolizing), polypeptide 6 (*CYP2D6*); Locus ID: 1565. LocusLink:
http://www.ncbi.nlm.nih.gov/LocusLink/LocRpt.cgi?l = 1565

Sodium channel, voltage-gated, type V, alpha polypeptide (long (electrocardiographic) QT syndrome 3) (*SCN5A*); Locus ID: 6331. LocusLink:
http://www.ncbi.nlm.nih.gov/LocusLink/LocRpt.cgi?l = 6331

Cytochrome P450, subfamily IIC (mephenytoin 4-hydroxylase), polypeptide 19 (*CYP2C19*); MIM number: 124020. OMIM:
http://www.ncbi.nlm.nih.gov/htbin-post/Omim/dispmim?124020

Cytochrome P450, subfamily IID (debrisoquine, sparteine, etc., -metabolizing), polypeptide 6 (*CYP2D6*); MIM number: 124030. OMIM:
http://www.ncbi.nlm.nih.gov/htbin-post/Omim/dispmim?124030

Sodium channel, voltage-gated, type V, alpha polypeptide (long (electrocardiographic) QT syndrome 3) (*SCN5A*); MIM number: 600163. OMIM:
http://www.ncbi.nlm.nih.gov/htbin-post/Omim/dispmim?600163

Velocardiofacial Syndrome (VCFS) and Schizophrenia

Kieran C Murphy, *Royal College of Surgeons in Ireland, Dublin, Ireland*

Velocardiofacial syndrome, the most frequent known interstitial deletion found in humans, is associated with high rates of psychiatric disorder, particularly schizophrenia.

Intermediate article

Article contents

- Introduction
- Elevated Rates of Psychiatric Disorder in VCFS
- Chromosome 22 and Schizophrenia
- Neurodevelopmental or Neurochemical Schizophrenia in VCFS?
- Animal Models
- Conclusions

Introduction

Velocardiofacial syndrome (VCFS) is the most frequent known interstitial deletion found in humans and occurs with an estimated prevalence of 1 in 4000–5000 live births. In around 85% of individuals, VCFS is associated with chromosomal microdeletions in the q11 band of chromosome 22 (**Figure 1**). The VCFS phenotype is complex, with multiple congenital abnormalities affecting a number of tissues and organs, many of which are embryologically derived from neural crest cells. Although considerable phenotypic variability occurs, common clinical features

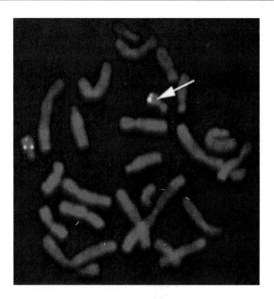

Figure 1 [*Figure is also reproduced in color section.*] Chromosome 22q11 microdeletion identified by fluorescence *in situ* hybridization (FISH).

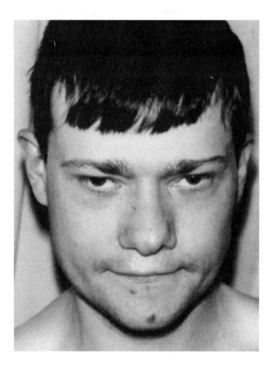

Figure 2 Characteristic facial dysmorphism in velocardiofacial syndrome.

include characteristic dysmorphology (**Figure 2**), cleft palate, congenital heart defects, borderline learning disability and high rates of psychiatric disorder, particularly schizophrenia. (*See* DiGeorge Syndrome and Velocardiofacial Syndrome (VCFS).)

Elevated Rates of Psychiatric Disorder in VCFS

Children with VCFS are reported to have poor social interaction skills, a bland affect with minimal facial expression, and high rates of attention-deficit hyperactivity disorder (ADHD; 36%), bipolar disorder (64%) and psychosis (16%) (Papolos *et al.*, 1996). As the first recognized cohort of children and adolescents with VCFS was followed up into adulthood, high rates of major psychiatric disorder were observed. Shprintzen *et al.* (1992) suggested that more than 10% had developed psychiatric disorders that mostly resembled chronic schizophrenia with paranoid delusions, although operational criteria were not used. Subsequently, in a small study of VCFS adults ($n = 14$), 11 (79%) were found to have a psychiatric diagnosis, of which four (29%) had schizophrenia or schizoaffective disorder (Pulver *et al.*, 1994). In a further study of a group of VCFS children and adults ($n = 25$), Papolos *et al.* (1996) reported that four (16%) of their sample had psychotic symptoms, while 16 (64%) met DSM-III-R criteria for a spectrum of bipolar affective disorders. Interestingly, although no individual had schizophrenia, the two oldest members of this cohort (aged 29 and 34 years) both had schizoaffective disorder.

Murphy *et al.* (1999) have suggested that the psychiatric phenotype observed in VCFS children and adolescents (containing prominent affective symptoms) may in some cases evolve into schizophrenia or schizoaffective disorder as the children age. In the largest study of its kind yet performed, they found that 18 (42%) of a sample of 50 VCFS adults had a major psychiatric disorder; 15 (30%) had a psychotic disorder with 12 (24%) fulfilling DSM-IV criteria for schizophrenia, while a further six (12%) had major depression without psychotic features. The individuals with schizophrenia had fewer negative symptoms and a relatively later age of onset (mean age 26 years) compared with nondeleted controls. Using different ascertainment strategies, however, Bassett *et al.* (1998) have reported a relatively early age of onset (mean age 19 years) in their sample of 10 individuals with VCFS and schizophrenia. This discrepancy is likely to be the result of small sample sizes and differences in ascertainment between these studies.

Chromosome 22 and Schizophrenia

Evidence from family, twin and adoption studies demonstrates a major genetic contribution to the etiology of psychotic disorders such as schizophrenia. The high rates of psychosis (especially schizophrenia) in VCFS suggest that, with the exception of being the offspring of a dual mating or the monozygotic (MZ) co-twin of an affected individual, deletion of chromosome 22q11 represents the highest known risk factor for the development of schizophrenia identified to date. (*See* Schizophrenia: Molecular Genetics.)

There are also several other lines of evidence to suggest that a locus conferring susceptibility to schizophrenia resides on chromosome 22q. Karayiorgou *et al.* (1995) have reported that two of 100 individuals with schizophrenia recruited from a large-scale epidemiological sample were found to have a previously undetected 22q11 deletion. In addition, when subjects with schizophrenia have been selected for the presence of clinical features consistent with VCFS, 22q11 deletions have been identified in 20–59% of cases (Bassett *et al.*, 1998). Furthermore, results from linkage studies of individuals with schizophrenia provide supportive evidence for a susceptibility locus on 22q. Although markers telomeric to the VCFS region have been implicated in some of these studies (Schizophrenia Collaborative Linkage Group, 1996), evidence for linkage to markers close to the VCFS region both for schizophrenia itself and for an inhibitory neurophysiological phenotype associated with schizophrenia has also been reported.

Neurodevelopmental or Neurochemical Schizophrenia in VCFS?

The strong association between schizophrenia and VCFS suggests that a gene or genes mapping to chromosome 22q11 may play a role in the etiology of both disorders. If this is so, what is the common pathogenetic mechanism? There is compelling evidence that a defect in early embryonic development is the cause of many of the abnormalities present in VCFS individuals. The importance of cephalic neural crest-derived cells in the development of the conotruncal region of the heart, the thymus, parathyroid glands and the palate, all structures that are affected in VCFS, has been demonstrated by microablation and transplantation studies in avian embryos. Based on these observations, it is therefore reasonable to hypothesize that a gene or gene located within the 22q11 deleted region is involved in the process of neural cell migration or differentiation in the pharyngeal arches and that haploinsufficiency of such a gene(s) disrupts proper development of these systems, leading to multiple organ and tissue abnormalities.

The perceptual, cognitive and behavioral difficulties that characterize schizophrenia are thought to arise from alterations of neuronal chemistry and function in the forebrain. There is increasing evidence that the etiology of schizophrenia reflects aberrations in forebrain development rather than a degenerative process that begins in maturity. One of the most promising lines of evidence that schizophrenia is associated with abnormal early brain development comes from reports suggesting that the laminar distribution of cortical neurons in people with schizophrenia may be displaced inwards in some individuals. As the normal migration of neurons destined for the cerebral cortex proceeds outwards from the periventricular zone toward the pial cortical surface during the second trimester of gestation, a defect in the 'inside-out' organization of cortical layers suggests a defect in the process of neuronal migration during this developmental stage.

There has been a recent resurgence of interest in a possible association between what are now termed 'minor physical anomalies' (MPAs) and schizophrenia. MPAs are slight anatomical defects of the head, hair, eyes, mouth, hands and feet and are usually attributed to disturbed neurodevelopment during the first or second trimester of fetal life. Several studies have indicated that MPAs occur in excess in patients with schizophrenia, particularly subtle dysmorphogenesis of the craniofacial area. While a high arched palate is one of the most consistent findings, the topography of MPAs in schizophrenia is poorly understood. Using an anthropometric approach, several studies have reported multiple quantitative and qualitative abnormalities of craniofacial and other structures in individuals with schizophrenia. Lane *et al.* (1997) have reported a core topography of dysmorphology characterized primarily by an overall narrowing and elongation of the middle and lower anterior facial region, in terms of heightening of the palate and reduced mouth width, with widening of the skull base and extensive abnormalities of the mouth, ears and eyes. It is interesting to observe that these abnormalities are reminiscent of the craniofacial abnormalities characterizing individuals with VCFS. This provides further supportive evidence that VCFS and schizophrenia might both be associated with similar mechanisms that disrupt neural crest cell migration. As the entire sequence of chromosome 22 has now been determined and as increasing numbers of neurodevelopmental candidate genes mapping to chromosome 22q11 are identified, future work will be required to determine the possible relevance of such genes to the development of schizophrenia in VCFS individuals.

Although schizophrenia is increasingly seen as a neurodevelopmental disorder, disturbances in catecholamine neurotransmission have also long been postulated to play a key etiological role. Consequently, the gene coding for catechol-*O*-methyltransferase

(COMT), an enzyme catalyzing the *O*-methylation of catecholamine neurotransmitters (dopamine, adrenaline and noradrenaline), has been considered a candidate gene in the etiology of schizophrenia. The gene coding for COMT maps to the region of chromosome 22q11 frequently deleted in VCFS. As an amino acid polymorphism (Val-108-Met) of this gene determines high and low activity of COMT, Dunham *et al.* (1992) have postulated that individuals hemizygous for COMT and carrying a low-activity allele on their nondeleted chromosome may be predisposed to the development of schizophrenia by a resulting increase in brain dopamine levels. Recently, Murphy *et al.* (1999) have found no evidence that the possession of the low-activity COMT allele is associated with the presence of schizophrenia in a series of VCFS individuals. However, as this study was limited by a relatively small sample size, a minor effect for the genetic variation in COMT in the development of schizophrenia could not be excluded.

Animal Models

Historically, animal models for psychiatric disorders have been difficult to construct as such disorders are still predominantly defined in terms of subjective experiences described by affected individuals. Recently, however, more objective measures, including subtle abnormalities of cell migration and sensorimotor gating abnormalities, which include defects in prepulse inhibition, have also been described in individuals with schizophrenia. Such objectively measured abnormalities hold great promise as they can be measured in both humans and animals. A mouse model deleted for the syntenic region of mouse chromosome 16 that corresponds to human chromosome 22q11 has now been produced and congenital cardiac abnormalies similar to those in VCFS observed. Such animals should be investigated closely for neuroanatomical and behavioral phenotypes of possible relevance to schizophrenia. Recently, a mutation of the proline dehydrogenase gene, which maps to the syntenic region of mouse chromosome 16 and is a candidate gene for schizophrenia, resulted in sensorimotor gating deficits in mice (Gogos *et al.*, 1999).

Conclusions

Apart from the offspring of a dual mating or the MZ co-twin of an affected individual, the presence of a chromosome 22q11 deletion represents the highest known risk factor for the development of schizophrenia identified to date. It is likely that haploinsufficiency of a neurodevelopmental gene or genes mapping to

chromosome 22q11, leading to disturbed neural cell migration, underlies susceptibility to psychosis in VCFS. (*See* Schizophrenia: Molecular Genetics.)

While deletion of chromosome 22q11 may only account for a small proportion of risk to the development of schizophrenia in the general population, nondeletion mutations or polymorphisms in genes within the VCFS region may make a more general and widespread contribution to susceptibility to schizophrenia in the wider population. Experience with other complex diseases (e.g. Alzheimer disease, diabetes and breast cancer) suggests that understanding the molecular basis for uncommon subtypes with high penetrance has been shown to be the most successful approach to understanding the genetics and underlying pathophysiology of complex diseases. As the entire sequence of chromosome 22 has now been determined, the future identification of the genetic determinants of the psychosis in VCFS individuals will have profound implications for our understanding of the molecular genetics and pathogenesis of psychosis in the wider nondeleted population. (*See* DiGeorge Syndrome and Velocardiofacial Syndrome (VCFS).)

See also
DiGeorge Syndrome and Velocardiofacial Syndrome (VCFS)
Schizophrenia: Molecular Genetics

References

Bassett AS, Hodgkinson K, Chow EWC, *et al.* (1998) 22q11 deletion syndrome in adults with schizophrenia. *American Journal of Medical Genetics* **81**: 328–337.
Dunham I, Collins J, Wadey R and Scambler P (1992) Possible role for COMT in psychosis associated with velo-cardio-facial syndrome. *Lancet* **340**: 1361–1362.
Gogos JA, Santha M, Takacs Z, *et al.* (1999) The gene encoding proline dehydrogenase modulates sensorimotor gating in mice. *Nature Genetics* **21**: 434–439.
Karayiorgou M, Morris MA, Morrow B, *et al.* (1995) Schizophrenia susceptibility associated with interstitial deletions of chromosome 22q11. *Proceedings of the National Academy of Sciences of the United States of America* **92**: 7612–7616.
Lane A, Kinsella A, Murphy P, *et al.* (1997) The anthropometric assessment of dysmorphic features in schizophrenia as an index of its developmental origins. *Psychological Medicine* **27**: 1155–1164.
Murphy KC, Jones LA and Owen MJ (1999) High rates of schizophrenia in adults with velocardiofacial syndrome. *Archives of General Psychiatry* **56**: 940–945.
Papolos DF, Faedda GL, Veit S, *et al.* (1996) Bipolar spectrum disorders in patients diagnosed with velocardiofacial syndrome: does a hemizygous deletion of chromosome 22q11 result in bipolar affective disorder? *American Journal of Psychiatry* **153**: 1541–1547.
Pulver AE, Nestadt G, Goldberg R, *et al.* (1994) Psychotic illness in patients diagnosed with velocardiofacial syndrome and their relatives. *Journal of Nervous and Mental Disease* **182**: 476–478.
Schizophrenia Collaborative Linkage Group (1996) A combined analysis of D22S278 marker alleles in affected sib-pairs: support

for a susceptibility locus for schizophrenia at chromosome 22q12. *American Journal of Medical Genetics* **67**: 40–45.

Shprintzen RJ, Goldberg R, Goldingkushner KJ and Marion RW (1992) Late-onset psychosis in the velo-cardio-facial syndrome. *American Journal of Medical Genetics* **42**: 141–142.

Further Reading

Akbarian S, Bunney Jr WE, Potkin SG, *et al.* (1993) Altered distribution of nicotamide-adenine dinucleotide phosphate-diaphorase cells in frontal lobe of schizophrenics implies disturbances of cortical development. *Archives of General Psychiatry* **50**: 169–177.

Blouin JL, Dombroski BA, Nath SK, *et al.* (1998) Schizophrenia susceptibility loci on chromosomes 13q32 and 8p21. *Nature Genetics* **20**: 70–73.

Lindsay EA, Botta A, Jurecic V, *et al.* (1999) Congenital heart disease in mice deficient for the DiGeorge syndrome region. *Nature* **401**: 379–83.

Murphy KC (2002) Schizophrenia and velocardiofacial syndrome. *Lancet* **359**: 426–430.

Murphy KC and Owen MJ (1996) Minor physical anomalies and their relationship to the aetiology of schizophrenia. *British Journal of Psychiatry* **168**: 139–142.

Murphy KC and Owen MJ (2001) Velocardio facial syndrome (VCFS): a model for understanding the genetics and pathogenesis of schizophrenia. *British Journal of Psychiatry* **179**: 397–402.

Shaw SH, Kelly M, Smith AB, *et al.* (1998) A genome-wide search for schizophrenia susceptibility genes. *American Journal of Medical Genetics* **81**: 364–376.

Swillen A, Devriendt K, Legius E, *et al.* (1997) Intelligence and psychological adjustment in velocardiofacial syndrome: a study of 37 children and adolescents with VCFS. *Journal of Medical Genetics* **34**: 453–458.

Weinberger DR (1995) From neuropathology to neurodevelopment. *Lancet* **346**: 552–557.

Venous Thrombosis: Genetics

Introductory article

Pieter H Reitsma, *Laboratory for Experimental Internal Medicine, Academic Medical Center, University of Amsterdam, Amsterdam, The Netherlands*

Article contents

- Introduction
- Coagulation Inhibitors
- Procoagulant Proteins

The genetics of venous thrombosis provides a well worked out example of a multifactorial disease. Several genetic risk factors have been characterized in considerable detail, and evaluation of these risk factors is helpful in assessing individual risk profiles.

Introduction

When blood vessels are injured, the coagulation system becomes activated in order to help stem the bleeding. Blood clot formation is accomplished by a network of stepwise reactions between procoagulant proteins, blood platelets and blood vessel lining cells. These procoagulant reactions are offset by complex anticoagulant reactions that protect from unnecessary and excessive clot formation.

Under normal circumstances, the procoagulant and anticoagulant systems are quiescent and in balance, and profuse bleeding or unprovoked coagulation are therefore rare events. Only in the case of unusual circumstances (vascular injury, treatment with blood coagulation inhibitors, septic shock, hereditary coagulation defects, etc.) do major problems occur. In addition, events of apparently unprovoked thrombus formation may occur, particularly in older persons. Most commonly, these thrombotic events occur in the venous system of the legs (deep venous thrombosis) or in the vasculature of the lungs (lung embolism).

Venous thrombosis is diagnosed frequently among people from Western populations. An estimated 10% of the population will experience venous thrombosis in their lifetime.

Venous thrombosis does not occur randomly, and both acquired and genetic risk factors play a role. Important nongenetic risk factors include advancing age, cancer, prolonged bed rest and surgery. In approximately 25% of patients, known genetic risk factors can be identified. These genetic risk factors consist of variations in genes encoding proteins that inhibit clot formation (coagulation inhibitors) or in genes encoding proteins that promote clotting (procoagulant proteins). In isolation, the risk associated with each of the inherited or noninherited factors may be low, but when several risk factors are simultaneously present, the risks multiply. Venous thrombosis is therefore considered to be a multifactorial disease that results from complex overlaps of hereditary and acquired pathophysiological mechanisms.

For five thrombosis genes, there is extensive and convincing genetic evidence: mutations and polymorphisms in protein C, protein S, antithrombin, factor V and prothrombin genes. In particular, the mutations in the latter two genes have become very popular targets for testing in individuals with venous thrombosis. This is not surprising in view of the ease with which genetic testing can be performed for factor V Leiden or prothrombin 20210A, and the relatively high prevalence of these two mutations. Genetic testing for the whole spectrum of rare mutations in coagulation inhibitors remains expensive and time consuming.

Coagulation Inhibitors

Antithrombin

Antithrombin is the most important direct inhibitor of the coagulation cascade in the blood. Loss-of-function mutations in the antithrombin gene cause antithrombin deficiency. Homozygosity for antithrombin deficiency is extremely rare, and nil antithrombin levels are probably not compatible with life. Heterozygous antithrombin deficiency is associated with a predisposition to venous thrombosis and has an estimated prevalence of 0.2–11.0 per 1000.

On the basis of blood tests, two principal types of deficiency are recognized. Type I is a quantitative deficiency, with plasma levels of antithrombin around 50% of normal. In type II deficiency, levels are roughly normal, but functional activity is impaired. Type II can be further subdivided on the basis of the nature of the functional defect. Mutations occur in the reactive site (RS) of antithrombin, and also in the heparin-binding site (HBS). The remaining functional mutations are classified as pleiotropic (PE).

Numerous mutations have been discovered in families with antithrombin deficiency (more than 130 unique mutations to date). Type I deficiency is associated with a wide variety of mutations, including large deletions, frameshift mutations, premature stop codons and splice-junction mutations. In type II RS deficiency, mutations are in (or influencing) the RS, whereas type II HBS mutations are in the heparin-binding region of antithrombin. The missense mutations that underlie type II PE deficiency tend to cluster in the *C*-terminal region of the protein.

Heparin cofactor II

This circulating anticoagulant protein is very similar to antithrombin. Heterozygous deficiency appears to predispose patients to arterial and venous thrombotic disease, but solid evidence to support these claims is still lacking. Two frameshift mutations and Arg189His and Pro433Leu missense mutations have been described in heterozygous deficiency.

Protein C

Protein C plays a central role in limiting the activity of the coagulation cascade. Thrombin formed during coagulation is responsible for the conversion of protein C to activated protein C (APC). This activation takes place on the surface of endothelial cells and monocytes by thrombin in complex with thrombomodulin. Activated protein C restrains blood coagulation by cleaving the activated cofactors Va and VIIIa into inactive fragments. This cleavage requires protein S (see below) as a cofactor in order to proceed efficiently.

Loss-of-function mutations in the protein C gene lead to protein C deficiency, which has an estimated prevalence of 3/1000. In 1981, the first family with heterozygous deficiency was discovered and soon thereafter individuals with homozygous deficiency. This condition is much more severe than heterozygous deficiency, and life-threatening thrombotic complications start immediately after birth. This differs sharply from the late-onset episodes of deep venous thrombosis or pulmonary embolism that arise in heterozygous carriers.

Around 200 different mutations in protein C deficiency are known. This fact serves to illustrate that loss-of-function mutations have a propensity to be very heterogeneous. It is as if every family possesses its private mutation, and the mutations that have been discovered cover the whole spectrum of imaginable abnormalities. These include missense mutations, promoter mutations, ribonucleic acid (RNA)-processing mutations and frameshift mutations.

Protein S

Protein S is the cofactor of protein C in the inactivation of factors Va and VIIIa. The human genome contains two copies of the protein S gene on chromosome 3. Only one of these copies is intact and active.

Clinical symptoms in heterozygous deficiency are similar to those in heterozygous protein C deficiency, thus being of late onset and relatively mild. The prevalence of heterozygous protein S deficiency is not known with certainty. Only a few cases of homozygous deficiency have been thoroughly documented, and these were associated with extremely severe thrombotic disease in newborns.

Around 100 different mutations have been documented in protein S deficiency, and the spectrum is wide – as expected for loss-of-function mutations – and comparable to that observed for protein C deficiency or type I antithrombin deficiency. The mutations are spread across the gene and include splice-junction abnormalities, premature stop codons, frameshift mutations, missense mutations and large deletions.

Several polymorphisms are known for the protein S gene. The Heerlen polymorphism is a notable example. This polymorphism was originally considered to be neutral and of little clinical importance because plasmatic tests failed to show abnormal activity or levels. However, the neutrality of the Heerlen polymorphism has been recently challenged by independent studies, which showed that the Heerlen polymorphism may be associated with low levels of free protein S. Therefore, it cannot be excluded that this polymorphism is a relevant risk factor for thrombotic disease.

Thrombomodulin

Thrombomodulin is an integral membrane protein of endothelial cells and monocytes. When thrombin binds to thrombomodulin, it loses its procoagulant activity but becomes capable of activating protein C. Given this crucial function in the protein C pathway, a hereditary deficiency of thrombomodulin might very well play a role as a risk factor for thrombotic disease.

At present, a handful of missense mutations are known in the thrombomodulin gene of patients with venous thrombosis. It remains difficult, however, to judge the relationship between these mutations and disease. One reason for this is that simple functional assays are not in place to prove that these mutations are detrimental to thrombomodulin gene function. Indeed, *in vitro* data obtained with the first mutation ever reported in the thrombomodulin gene, Asp468Tyr, did not reveal any abnormality in production levels or functional activity.

Endothelial-cell protein C receptor

The recently discovered endothelial-cell protein C receptor (EPCR) on endothelial cells is an important regulator of the protein C anticoagulant pathway. Recently, the EPCR gene has been characterized, which opens up the possibility of searching for mutations in this gene. Indeed, a 23-bp insertion in exon 3 has been identified, which may predispose patients to venous thrombosis. Further studies are clearly needed in order to evaluate the importance of these findings.

Tissue factor pathway inhibitor

Tissue factor pathway inhibitor (TFPI) is a so-called Kunitz-type inhibitor that plays a major role in the inhibition of coagulation. Inhibition takes place by the formation of a quaternary inactive complex between tissue factor, factor VIIa, factor Xa and TFPI. Systematic sequencing of the TFPI gene has yielded four different polymorphisms (Pro151Leu, Val264-Met, T384C exon 4 and C-33T intron 7). It is not known whether these sequence variations affect plasma levels of antigen and/or activity, with the possible exception of the Val264Met mutation, for which lower TFPI levels have been claimed. It is also uncertain whether these polymorphisms influence thrombotic risk.

Procoagulant Proteins

Factor V

In the coagulation cascade, factor V acts as cofactor for factor Xa. Homozygosity for loss-of-function mutations in the factor V gene leads to a bleeding disorder. The most common defect in the factor V gene, however, is a gain-of-function mutation that is not associated with bleeding but with venous thrombosis. This mutation is often referred to as factor V Leiden, and is responsible for a plasmatic abnormality called APC resistance. The genetic basis of factor V Leiden is mutation of the Arg506 inactivation site in factor Va. As a result, factor Va Leiden is less easily inactivated by APC, whereby the procoagulant properties of factor V are enhanced. The factor V Leiden abnormality has a prevalence of 3–6% in Caucasians, and is the most common risk factor for venous thrombosis in Caucasians. (*See* Gain-of-function Mutations in Human Genetic Disorders; Factor V Leiden.)

Most individual carriers of factor V Leiden are heterozygotes, but homozygous individuals are also common. Homozygotes have a higher risk of thrombosis than heterozygotes. Please note, however, that the symptoms in these homozygotes are much milder than in those with homozygous protein C or protein S deficiency. In this respect, there is a critical difference between mutations in procoagulant proteins versus coagulation inhibitors.

At least in Caucasian populations, the prevalence of factor V Leiden is much higher (3–6%) than the prevalence of protein C, protein S or antithrombin deficiency. In part, this high prevalence can be understood because factor V Leiden homozygotes are mildly affected, so that there is less selective pressure on this abnormality than on mutations responsible for anticoagulant abnormalities. On the other hand, there is evidence that being a factor V Leiden carrier has a negative effect on pregnancy outcome, and therefore it is reasonable to assume that there must also be positive selective pressures acting on the prevalence of factor V Leiden. One possibility is that factor V Leiden protects against intrapartum bleeding.

Two other major cleavage sites for APC exist in factor Va, at Arg306 and Arg679. It is logical to assume, therefore, that mutations at these sites also have occurred and that these too play a key role as risk factors for venous thrombosis. For unknown reasons, this does not seem to be the case. Although two mutations have been found at the 306 position (Arg306Thr and Arg306Gly), only replacement by Thr (factor V Cambridge) appears to lead to APC resistance in a plasmatic assay. However, this mutation appears to be rare in all populations tested, and its significance as a thrombotic risk factor remains disputable at best. The Arg306Gly mutation (factor V Hong Kong) is prevalent among Hong Kong Chinese, is not associated with plasmatic APC resistance, and, based on currently available data, does not predispose to thrombosis.

Several gene polymorphisms are known for the factor V gene. A particular haplotype (HR2) seems to be associated with partial APC resistance. Several studies have claimed that this haplotype predisposes to venous thrombosis, but others could not confirm this relationship. Consequently, the importance of this haplotype remains controversial.

Prothrombin

Activation of the coagulation system ultimately results in the conversion of prothrombin to thrombin. The enzyme thrombin in turn activates blood platelets and produces fibrin from the precursor fibrinogen, and combines the two in a clot. Deficiency of prothrombin therefore leads to decreased clot formation and a bleeding disorder. (*See* Gain-of-function Mutations in Human Genetic Disorders; 3′ UTR Mutations and Human Disorders.)

Systematic screening of the prothrombin gene identified an unusual mutation that affects RNA metabolism and that turned out to be a common risk factor for venous thrombosis. The mutation is a G-to-A transition at nucleotide position 20210, which is the last nucleotide of the prothrombin mRNA. This mutation represents a gain-of-function mutation causing increased mRNA accumulation and protein synthesis. Indeed, prothrombin 20210A is associated with increased prothrombin levels in blood (~120% of normal), and is linked to a two- to fivefold increased risk for venous thrombosis.

The prothrombin mutation is prevalent only in Caucasian populations (range 1–8%). Also, in accordance with the findings with factor V Leiden, the relatively high prevalence in the population goes with mild symptoms in homozygous subjects.

Factor XIII

Factor XIII has an important role in stabilizing fibrin clots. This is accomplished by creating bonds between lysine and glutamine side-chain residues of fibrin.

A common mutation in the factor XIII gene, Val34Leu, has been implicated as a risk factor for thrombotic disease. It seems protective against myocardial infarction, and predisposes to intracerebral hemorrhage. Two studies suggest that the Leu allele may also protect against venous thrombotic disease, but other studies negate this relationship to venous thrombotic disease.

Tissue factor

The major initiator of the blood coagulation cascade is tissue factor (TF). A recent study raised the question as to whether individual differences in TF gene expression would predispose to thrombosis. Sequencing of the promoter region of the TF gene yielded six novel polymorphisms that were distributed over two haplotypes with equal frequencies (designated 1208 D and 1208 I).

In a case–control study of venous thromboembolism, the odds ratio associated with the presence of at least one D allele was 0.72. Interestingly, the protective D allele was associated with lower levels of circulating tissue factor than the I allele. Therefore, polymorphisms in the TF gene might well be weak but nevertheless significant genetic risk factors for venous thrombosis.

Homocysteine

High plasma levels of homocysteine are important risk indicators for venous and arterial thrombosis. The level of homocysteine in blood is dependent on the activity of cystathione β-synthetase and methylene tetrahydrofolate reductase. Therefore, genetic abnormalities that increase homocysteine levels are candidate genetic risk factors for venous thrombotic disease. A 68 bp insertion in the cysthatione β-synthetase gene and a C677T mutation in the methylene tetrahydrofolate reductase gene have been studied in considerable detail, but there is controversy with respect to the robustness of the findings.

Emerging risk factors

Recent studies have yielded potentially relevant novel associations between the levels of coagulation factors and thrombotic risk. In particular, levels of factor VIII, factor XI and thrombin activatable fibrinolysis inhibitor were positively associated with risk. Genetic markers in these genes, which rival the importance of factor V and prothrombin markers, have not been found. It may be that, in the near future, such genetic markers will be identified that further extend the collection of useful inherited predictors of venous thrombosis.

See also
Factor V Leiden

Further Reading

Bertina RM (2001) Genetic approach to thrombophilia. *Thrombosis and Haemostasis* **86**: 92–103.

Bertina RM, Koeleman BP, Koster T, *et al.* (1994) Mutation in blood coagulation factor V associated with resistance to activated protein C. *Nature* **369**: 64–67.

Cooper DN and Krawczak M (1997) *Venous Thrombosis: From Gene to Clinical Medicine*. Oxford, UK: BIOS Scientific Publishers.

Dahlbäck B, Carlsson M and Svensson PJ (1993) Familial thrombophilia due to a previously unrecognized mechanism characterized by poor anticoagulant response to activated

protein C: prediction of a cofactor to activated protein C. *Proceedings of the National Academy of Sciences of the United States of America* **90**: 1004–1008.

Franco RF and Reitsma PH (2001) Genetic risk factors of venous thrombosis. *Human Genetics* **109**: 369–384.

Lane DA, Mannucci PM, Bauer KA, *et al.* (1996) Inherited thrombophilia. 1 [review]. *Thrombosis and Haemostasis* **76**(5): 651–662.

Lane DA, Mannucci PM, Bauer KA, *et al.* (1996) Inherited thrombophilia. 2 [review]. *Thrombosis and Haemostasis* **76**(6): 824–834.

Poort SR, Rosendaal FR, Reitsma PH and Bertina RM (1996) A common genetic variation in the 3′-untranslated region of the prothrombin gene is associated with elevated plasma prothrombin levels and an increase in venous thrombosis. *Blood* **88**(10): 3698–3703.

Vertebrate Evolution: Genes and Development

Anthony Graham, *King's College, London, UK*

The most significant differences between the vertebrates and the other chordates lie in the head. These novel features probably emerged as a result of modifications to the developmental program, and they are thought to have aided the vertebrates in assuming a predacious lifestyle. More specifically, the evolution of these characteristics must have involved extensive regionalization of the central nervous system, the evolution of neural crest and ectodermal placodes and alterations to the pharynx.

Intermediate article

Article contents

- Introduction
- Alterations to the Developmental Program
- Genetics of Vertebrate Evolution
- Conclusions

Introduction

Vertebrates are chordates, and like the other members of this phylum they possess a notochord, a dorsal nerve cord, lateral muscle blocks and pharyngeal gill slits (**Figures 1** and **2**). The vertebrates, however, differ significantly from the others, collectively termed the protochordates, in their possession of a number of unique features: a variety of special sense organs; the presence of a muscularized pharynx, with skeletal support; an enhanced respiratory system, with gills and aortic arches, and a massively complex nervous system (**Figure 3**). An important point about these features is that they are primarily found in the head, and it has been suggested that their development aided the vertebrates in their transition from being sessile filter feeders to assuming a more active predacious lifestyle (Gans and Northcutt, 1983). The more complex brain and the accumulation of sense organs at the anterior end of the animal would undoubtedly aid the detection and capture of prey; the enhanced respiratory apparatus would enable them to be more

active; and the more robust pharynx would assist in the ingestion and breaking-up of food.

The evolution of the vertebrates undoubtedly involved alterations to the developmental program, and comparative studies between vertebrate and protochordate species have pinpointed those developments that are most likely to be relevant. More specifically, the appearance of the key vertebrate characteristics listed above must have involved extensive regionalization of the dorsal nerve cord, the emergence of neural crest and ectodermal placodes and modification to the development of the pharyngeal segments.

Alterations to the Developmental Program

Central nervous system

The vertebrate central nervous system (CNS) can be seen to be organized along the dorsoventral and

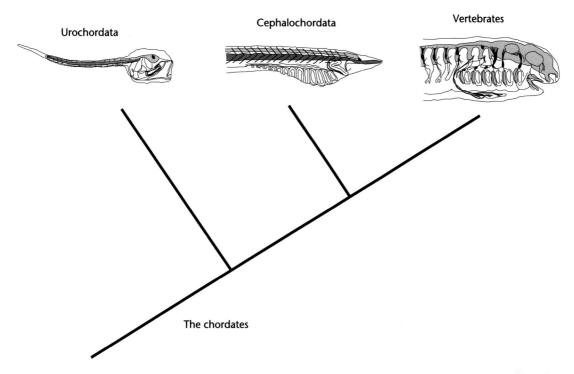

Figure 1 The chordate phylum consists of three subphyla: the vertebrates, the cephalochordates and the urochordates, which are united by their possession of a number of common features at the embryonic stage. These three subphyla do, however, exhibit differing degrees of anatomical complexity, with the vertebrates being the most complex.

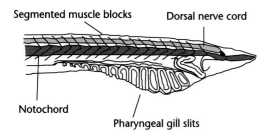

Figure 2 Schematic representation of a cephalochordate, *Amphioxus*, highlighting the typical shared characteristic chordate features: notochord, dorsal nerve cord, segmented muscle blocks and a segmented pharyngeal region.

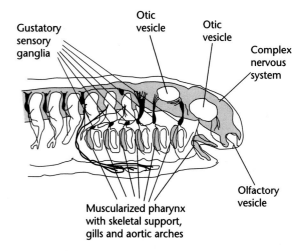

Figure 3 Unique features of the vertebrate head. The vertebrates differ from the other chordates in having a number of features concentrated in the head: they have an elaborate brain, numerous special sense organs, a muscularized pharynx with skeletal support, a backbone and an extensive peripheral nervous system.

anteroposterior axes into distinct functional territories. The ventral portion of the vertebrate CNS is primarily devoted to motor function and the dorsal portion to sensory function, while the long axis of the neural tube is subdivided into forebrain, midbrain, hindbrain and spinal cord territories. Interestingly, it would seem that these general subdivisions of the nervous system preceded the evolution of the vertebrates themselves, as they are evident in protochordate species (Williams and Holland, 1998; Holland and Holland, 1999). *Amphioxus*, for example, a cephalochordate which is the nearest living relative of the vertebrates, has dorsally located sensory neurons, ventral motor neurons,

and at its ventral midline a floor plate. Furthermore, the development of these different dorsoventral cell types in amphioxus seems to be under the control of the same molecules as those that are involved in organizing the vertebrates. Thus, in both *Amphioxus* and vertebrates, the dorsal regions are marked through their expression of specific *Pax* genes, the motor neurons via *neurogenin* and *islet* gene expression, and the floor plate through its expression of *hedgehog* genes, *netrins* and *HNF-3β* orthologs. Similarly, other studies have shown that all chordates have an anterior *Otx* expressing region, and a posterior *Hox* domain, territories that correspond to forebrain/anterior midbrain and hindbrain/spinal cord regions in vertebrates. It may also be, given that these regions do not abut in *Amphioxus*, that there is also a posterior midbrain homolog in the protochordates. Indeed, support for this comes from the fact that the posterior midbrain region in vertebrates is marked by its expression of the *Pax-2*, *Pax-5* and *Pax-8* genes and in the urochordate ciona, lying between the *Otx* and *Hox* domains, there is a territory that expresses the *Pax2/5/8* gene.

Thus it would seem that the vertebrate central nervous system was built upon a nervous system that was already regionalized along both the dorsoventral and anteroposterior axes, and that the basic genetic mechanisms for achieving this were already in place. The CNS of the vertebrates is, however, vastly more complex than that of any protochordate. It is anatomically sophisticated, and it contains many more neurons and a greater number of cell types. However, as a result of a lack of extant species that link the cephalochordates with the vertebrates, and a paucity of information on the intricacies of vertebrate nervous system development, we are some way from understanding the sequence of events that led to the emergence of a definitively vertebrate nervous system.

Neural crest

The neural crest is a transient embryonic cell type that arises at the dorsal aspect of the vertebrate neural tube and then migrates into the embryo, and it is of fundamental importance, as it gives rise to a range of derivatives that are definitively vertebrate. These include the majority of neurons, and all of the glia of the peripheral nervous system, melanocytes, and additionally, in the head, skeletal tissues (Le Douarin and Kalcheim, 1999). Indeed, the neural crest itself is thought to be a defining vertebrate characteristic. Consequently, an understanding of the evolutionary origins of the neural crest is central to an understanding of the origin of the vertebrates themselves.

During early development, the epidermal ectoderm signals, through the action of *Bmp-4* and *Bmp-7*, to the adjacent neural plate to induce the formation of the neural crest at the interface between these two tissues (Liem *et al.*, 1995). This process occurs as the neural plate is folding, and with time the crest cells come to lie at the dorsal aspect of the neural tube, and their location can be readily visualized through their

expression of a number of transcription factors, including members of the *msx* and *slug/snail* gene families. Interestingly, studies of orthologs of these genes in protochordates have shown that the components of this pathway for the induction of the crest exist and are expressed in analogous manner in these animals. In both *Amphioxus* and *Ciona*, a single *Bmp2/4* gene has been isolated and found to be expressed in the non-neural ectoderm, and correspondingly, in both of these species, *msx* and *slug/snail* orthologs have been found to be expressed at the lateral neural plate.

However, these results do not indicate that protochordate species have neural crest cells *per se*, migratory cells that emerge from the dorsal aspect of the neural tube. Indeed, there is no evidence of such a group of cells in either *Amphioxus* or *Ciona*. Rather, as these genes are also involved in specifying other aspects of the dorsal neural tube, the results reinforce the point that the mechanisms underlying the dorsoventral patterning of the neural tube are conserved within the chordates. These results would, however, be consistent with the neural crest evolving as part of an elaboration of the genetic program that patterns the dorsal neural tube.

Ectodermal placodes

The ectodermal placodes are a diverse group of structures that are only linked by their appearance in the developing embryo as a focal thickening of the cranial ectoderm (Graham and Begbie, 2000). They are important, however, as they give rise to many of the components of the special sense organs of the vertebrates: the olfactory receptor neurons and olfactory epithelia arise from the olfactory placode; the otic placode generates the inner ear and its associated ganglion; the epibranchial placodes produce the gustatory sensory neurons that innervate the taste buds; the lateral line placodes give rise to the lateral line system of aquatic anamniotes. Ectodermal placodes were thus felt to be central to the evolution of vertebrates, and to have evolved concomitantly. However, as we gain a greater, molecular, understanding of how the placodes develop, and then analyze orthologs of key genes in the protochordates, it is apparent that some of the placodes have clear homologs in these species.

There is good evidence for the existence of an olfactory placode-like structure in both *Amphioxus* and a urochordate species, *Botryllus schlosseri*. The olfactory placode in vertebrates, which is located proximal to the anterior extremity of the neural plate, presses the *msx-1* gene, and in *Amphioxus* a very anterior group of *msx*-expressing cells has been identified. In *Botryllus*, the neurohypophyseal duct shows features indicative of an olfactory placode homolog, in that it is found at the most anterior end of the embryo, and, like the olfactory placode of vertebrates, it generates migratory neuronal cells. There is also some evidence in another urochordate, *Halocynthia roretzi*, of a possible otic placode-like structure. In vertebrates this placode expresses the *Pax-2* gene, and in *Halocynthia* the atrial primordia expresses the *HrPax2/5/8* gene, and similarly both the vertebrate otic placode and the atrial primordia develop as a localized thickening that generates ciliated sensory cells.

Contrastingly, there are no indications of any lateral line structures in any protochordates, nor are there any suggestions of epibranchial placodes, and certainly these two groups of placodes seem to be vertebrate-specific and to have evolved with the vertebrates. We know very little about the early development of the lateral line placodes, and correspondingly it is unclear how they evolved. However, it is known that the epibranchial placodes are induced to form in the ectoderm by the pharyngeal pouch endoderm, and it is likely that the evolution of these structures involved the acquisition of localized expression of the inductive signal, *Bmp-7*, at these sites.

Pharynx

The vertebrate pharyngeal apparatus differs significantly from that of the protochordates in being muscularized and having skeletal support. During embryogenesis, these structures arise from a series of bulges that are found on the lateral surface of the head, the pharyngeal arches, the development of which is complex as it involves coordinated interactions between a number of disparate embryonic cell types: ectoderm, endoderm, mesoderm and neural crest (Graham, 2001). Of these it would seem that the neural crest and the endoderm play pivotal roles. It is the neural crest that gives rise to the skeletal elements, and it is also thought to play a role in patterning these structures, in assigning them their individual identities. Contrastingly, the endoderm forms the pouches that define the anterior and posterior boundaries of the arches, and helps maintain the segregation of the neural crest of each arch. Importantly, interactions between these two tissues are also central to the development of the arches, and an example of this is the fact that it is the endoderm that induces the neural crest cells to become skeletogenic.

Clearly, a key event in the evolution of the vertebrate pharynx was the generation of the neural crest cells and their differentiation into skeletogenic cell types within the pharyngeal arches. Interestingly, however, a number of studies have shown that endodermal segmentation and regionalization preceded the evolution of the neural crest. As such, the evolution of the vertebrate pharynx involved an integration between these two tissues. However, modifications to the pharyngeal endoderm also occurred with the emergence of the vertebrates. One notable change is the ability of this tissue to generate taste buds, which are defining features of the vertebrates.

Genetics of Vertebrate Evolution

Obviously, underpinning any changes to the developmental program are genetic alterations: the evolution of new genes or new gene functions, and, importantly, vertebrates have considerably more genes than protochordates. It was long thought that the expansion in gene number that occurred with the evolution of the vertebrates was a result of two whole-genome duplications, and support for this was forthcoming from the analysis of a number of important gene families; for example, *Amphioxus* has one *Hox* gene cluster, while most vertebrates have four (Ferrier and Holland, 2001). However, the recent analyses of the human genome have found no evidence for this. Rather, it now seems likely that there has been selective expansion in the number of genes of particular gene families. There is a long way to go before we begin to understand the genetic alterations that were associated with the evolution of the vertebrates, and, in this regard, the most important pieces of information that are still missing are the genome sequences of protochordates, and particularly that of *Amphioxus*, the nearest extant relative to the vertebrates, and the genome sequences of a number of other vertebrates, such as lamprey and shark species. These genome sequences are important as they would allow us to begin to relate the evolution of novel genes and the expansion of particular gene families to the morphological evolution.

Conclusions

Comparative molecular and developmental studies between vertebrate and protochordate species have given us real insights into the sequence of events that led to the evolution of the vertebrates, and they suggest that this was not an abrupt transition. Rather, we can now see how many features that were long thought to be exclusive to the vertebrates were built upon preexisting programs.

See also
Brain: Neurodevelopmental Genetics
Developmental Evolution
Hox Genes: Embryonic Development
Human Developmental Molecular Genetics
Pax Genes: Evolution and Function

References

Ferrier DE and Holland PW (2001) Ancient origin of the *Hox* cluster. *Nature Reviews of Genetics* **2**: 33–38.

Gans C and Northcutt RG (1983) Neural crest and the origins of vertebrates: a new head. *Science* **220**: 268–274.

Graham A (2001) The development and evolution of the pharyngeal arches. *Journal of Anatomy* **199**: 133–141.

Graham A and Begbie J (2000) Neurogenic placodes: a common front. *Trends in Neurosciences* **23**: 309–312.

Holland LZ and Holland N (1999) Chordate origins of the vertebrate nervous system. *Current Opinion in Neurobiology* **9**: 596–602.

Le Douarin NM and Kalcheim C (1999). *The Neural Crest.* New York: Cambridge University Press.

Liem Jr KF, Tremml G, Roelink H and Jessell TM (1995). Dorsal differentiation of neural plate cells induced by BMP-mediated signals from epidermal ectoderm. *Cell* **82**: 969–979.

Williams NA and Holland P (1998) Molecular evolution of the brain of chordates. *Brain Behaviour and Evolution* **52**: 177–185.

Further Reading

Butler AB (2000) Sensory system evolution at the origin of craniates. *Philosophical Transactions of the Royal Society of London B* **355**: 1309–1313.

Kardong KV (2002) *Vertebrates: Comparative Anatomy, Function, Evolution.* New York: McGraw-Hill.

Lacalli TC (2001) New perspectives on the evolution of proto-chordate sensory and locomotory systems, and the origin of brain and heads. *Philosophical Transactions of the Royal Society of London B* **356**: 1565–1572.

Schaeffer B (1987) Deuterostome monophyly and phylogeny. *Evolutionary Biology* **21**: 179–234.

Web Links

The Tree of Life. A collaborative internet project containing information about phylogeny and biodiversity
http://tolweb.org/tree/phylogeny.html

Vertebrate Immune System: Evolution

Austin L Hughes, *University of South Carolina, Columbia, South Carolina, USA*

Intermediate article

The vertebrates possess unique adaptations for immune defense that have apparently evolved gradually over the long evolutionary history of the vertebrate lineage. Molecular diversity is one of the hallmarks of vertebrate immune mechanisms, and molecular analyses have clearly implicated natural selection as the major factor in promoting molecular diversity and a consequent enhanced immune surveillance.

Article contents

- Vertebrate Immune System
- Innate Immunity: Continuity or Independent Evolution?
- Role of Natural Selection
- Conclusion

Vertebrate Immune System

The immune system of vertebrates is an extraordinarily fertile area for evolutionary study. Parasites (understood in the broad sense to include any organism – whether virus, prokaryote or eukaryote – living at the expense of a host) are expected to be an important source of natural selection on their hosts. Selection is expected to favor mechanisms that enable hosts to detect and eliminate parasites; there will be corresponding selection on parasites to evade recognition and elimination. The availability of molecular data has made it possible to answer some questions regarding the evolution of immune mechanisms that were previously mysterious. Yet there are still many important questions that remain unsolved.

Immunologists typically classify immune mechanisms of vertebrates into two categories: (1) innate immune mechanisms, which do not involve highly specific recognition of foreign antigens, and (2) specific immune mechanisms, which involve the production of receptors that are highly specific for individual antigens. (The latter are sometimes called 'adaptive immunity', but in the evolutionary sense all immune mechanisms are presumably adaptive in that they increase the host's fitness.) While invertebrate animals and even plants possess defense mechanisms that show certain analogies with vertebrate mechanisms of innate immunity, nothing remotely like the vertebrate-specific immune system is known to occur outside the vertebrates.

Vertebrate-specific immune mechanisms depend on three families of receptor proteins, all of which are members of the immunoglobulin superfamily: (1) the class I and class II molecules of the major histocompatibility complex (MHC), which bind peptides and present them to T cells; (2) the T-cell receptors, which recognize the complex of MHC molecule and bound peptide; and (3) the immunoglobulins, which are receptors for soluble antigens. T-cell receptors and immunoglobulins are characterized by the ability to create extensive receptor diversity by somatic rearrangement of gene segments and, in the case of immunoglobulins, by somatic mutation. The origin of vertebrate-specific immunity is thus tied to the origin of the mechanisms involved in rearrangement and somatic mutation.

MHC, T-cell receptors and immunoglobulins have not been reported outside the jawed vertebrates. The jawless vertebrates (lampreys and hagfish) are not known to possess any of these molecules, nor are any of the invertebrate chordates or other invertebrates (**Figure 1**). On the other hand, the complement system is found in invertebrate deuterostomes as well as vertebrates (**Figure 1**). Thus, one of the major problems posed for evolutionary biology is the problem of the origin of the MHC, T-cell receptors and immunoglobulins in the jawed vertebrates.

Origin of vertebrate-specific immunity

The problem of the origin of vertebrate-specific immune mechanisms was particularly acute when biologists accepted the hypothesis of a so-called 'Cambrian explosion'. This hypothesis was based on the sudden appearance in the fossil record in the

Cambrian period of animals clearly belonging to the major phyla seen on earth today. Thus, it was believed that these phyla originated 'explosively' in a comparatively short evolutionary time. However, molecular data do not support the hypothesis of a Cambrian explosion. Rather, they indicate that the origin of major animal phyla occurred deep in the Precambrian (**Figure 1**). For example, the last common ancestor of the protostome phyla (including insects) and the deuterostome phyla (including chordates) has been dated at between 800 and 900 million years ago, well before the Cambrian period (Wang *et al.*, 1999). Evidence of an ancient origin of vertebrates is important because it implies that vertebrates may have had a long time to evolve unique immune defenses (Hughes, 1999a).

Perhaps because of the perceived requirement for rapid evolution of vertebrate-specific immunity, biologists have been prone to invoke highly unlikely evolutionary scenarios to explain their origin. For example, it was suggested that the recombination activator genes, *RAG1* and *RAG2*, which are involved in segmental recombination of immunoglobulins and T-cell receptors, originated through horizontal gene transfer of bacterial genes. This hypothesis was based on sequence similarity between portions of RAG1 and RAG2 proteins and certain bacterial proteins (Bernstein *et al.*, 1996). On the other hand, RAG1 shows far greater sequence similarity to a yeast protein than it does to any bacterial protein, suggesting that these recombination proteins may have evolved from precursors present in ancestral eukaryotes (Hughes, 1999a).

Similarly, evidence that RAG1 and RAG2 can act together as a transposase *in vitro* has been taken as evidence that these genes arose from an ancestral transposable element that was captured by vertebrates and 'tamed' to carry out an immune system function (Agrawal *et al.*, 1998). While this scenario is theoretically possible, the *in vitro* transposase activity of RAG1 and RAG2 by no means proves it. We do not know the evolutionary origin of transposable elements. It is just as plausible that certain transposable elements have originated from eukaryotic genes that have 'escaped' as it is that transposable elements have been 'captured' by eukaryotic genomes.

It has recently been suggested that duplication of the vertebrate genome by polyploidization had a role in the origin of vertebrate-specific immunity (Kasahara *et al.*, 1997). A widely popular hypothesis states that two rounds of polyploidization occurred early in vertebrate history. Certainly polyploidization, if it occurred, may have increased the number of genes in certain gene families of the immune system, but it cannot explain the origin of the unique gene families essential to vertebrate-specific immunity. In any event, now that the complete human genome is available, it is obvious that

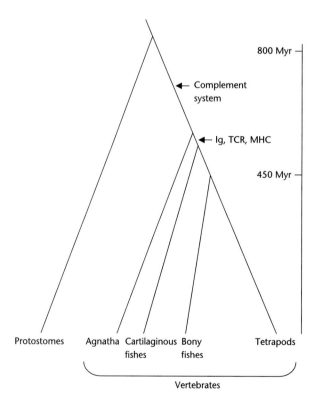

Figure 1 Schematic phylogeny of the chordates, indicating the time of origin of molecules involved in specific immunity in vertebrates. Ig: immunoglobulin; MHC: major histocompatibility complex; Myr: million years; TCR: T-cell receptors.

the hypothesis of polyploidization early in vertebrate history can explain little if any of the vertebrate genome organization (Friedman and Hughes, 2001).

Innate Immunity: Continuity or Independent Evolution?

Although we still know relatively little about immune mechanisms in invertebrates, we know enough to have observed certain similarities between invertebrate immunity and vertebrate innate immunity. According to the continuity hypothesis (Medzhitov and Janeway, 1998), there is evolutionary continuity between invertebrate immune mechanisms and vertebrate innate immune mechanisms; in other words, these mechanisms have been inherited from the common ancestor of vertebrates and invertebrates. Alternatively, the independent evolution hypothesis (Friedman and Hughes,

2002) holds that vertebrate innate immune mechanisms have largely evolved independently in the vertebrate lineage and have coevolved with vertebrate-specific immune mechanisms.

The Toll family of transmembrane receptors is involved in immune system signaling in both vertebrates and invertebrates. The existence of such a shared family might seem to favor the continuity hypothesis. In both vertebrates and invertebrates, these proteins are involved in stimulating immune responses, through the NF-κB signaling system, in response to the presence of molecules such as lipopolysaccharides in bacterial cell walls. However, the mechanisms by which toll proteins interact with such molecules are very different in insects and in vertebrates. In addition, although all insect-derived members of the Toll family are closely related to each other, some are involved in development rather than in immune reactions (**Figure 2**). Thus, it seems most likely that Toll represents a generalized type of signal-transduction mechanism that has independently been

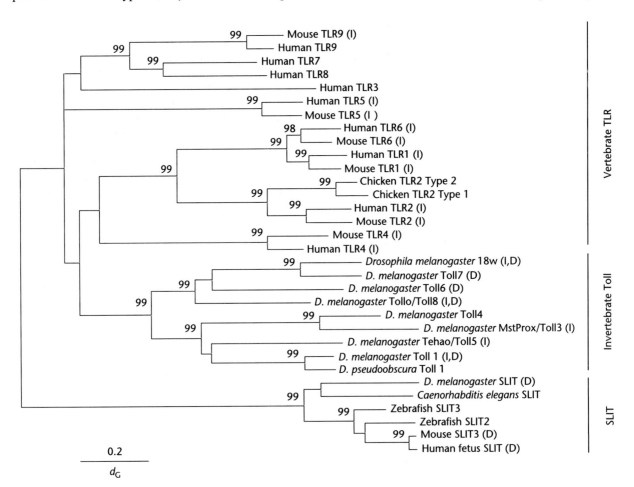

Figure 2 Phylogeny of the Toll family of receptors from insects and vertebrates TLR: Toll-like receptor; d_G = amino acid distance (corrected for multiple hits on the assumption that rates vary among sites following of gamma distribution). (Reproduced with permission from Friedman and Hughes, 2002.)

co-opted for an immune function in both insects and vertebrates (Friedman and Hughes, 2002).

The independent evolution hypothesis is consistent with the molecular evidence against the hypothesis of a Cambrian explosion. If the vertebrate lineage has evolved on its own for over 800 million years, it is easy to explain the independent evolution of both innate and specific immune mechanisms. However, there are a few immune mechanisms that do show evidence of evolutionary continuity. One example is the lysozymes, a family of antibacterial proteins found in both insects and vertebrates. Another family of antibacterial compounds, the defensins, may provide another example, although it is not certain whether defensins of vertebrates are actually evolutionarily related to those of insects (Hughes, 1999b).

Role of Natural Selection

Because the immune system is involved in the defense against parasites, it is not surprising that immune system genes have provided some of the best-documented evidence of natural selection at the molecular level. The immune system includes many highly diversified multigene families, and there is evidence that natural selection has had a role in diversifying the products of these genes at the level of amino acids. For example, immunoglobulin gene clusters include numerous variable (V) gene segments. Comparison of the patterns of nucleotide substitution between recently duplicated members of these clusters has revealed an excess of nonsynonymous (amino acid-altering) nucleotide substitutions over synonymous substitutions in the portion of the gene segment encoding the complementarity-determining (CDR) region of the V segment (Tanaka and Nei, 1989). The CDR region of the V segment is the antigen-binding portion of the molecule. These results thus imply that V gene segments have been

diversified as a result of the selective advantage conferred on a host that can bind a diverse array of antigens.

Interestingly, there is evidence of similarly diversifying selection in two families of mammalian genes whose products function in innate defense – the α and β defensins (Hughes, 1999b). These two families of antimicrobial peptides have undergone recent gene duplication in different mammalian lineages, and natural selection has acted to diversify them at the amino acid level. Thus, this aspect of innate defense is clearly anything but an evolutionary relic.

The genes of the MHC differ from those of other immune system families in that they are not encoded by an extensive multigene family. Instead, MHC molecules are encoded by a small number of loci. Several of these loci are extraordinarily polymorphic, and there is evidence that this polymorphism is selectively favored. MHC molecules bind foreign peptides and present them to T cells. In the codons encoding the peptide-binding region of the MHC molecule, the rate of nonsynonymous substitution is significantly greater than that of synonymous substitution (Hughes and Nei, 1988). This pattern of nucleotide substitution implies that natural selection has favored diversity in the peptide-binding region and thus enhanced immune surveillance.

There is evidence that proteins having immune system functions in general evolve at a faster rate than other proteins. Murphy (1993) compared a large set of human and rodent (mouse or rat) orthologous proteins, which were categorized according to function. The fastest evolving proteins were in the category of 'host defense ligands and receptors'. Similarly, Hughes (1997) compared orthologous immunoglobulin superfamily C2-set domains between human and rodent and found a higher rate of nonsynonymous (amino acid-altering) nucleotide substitution in those with the immune system expressed (**Figure 3**). The results showed little difference in the rate of

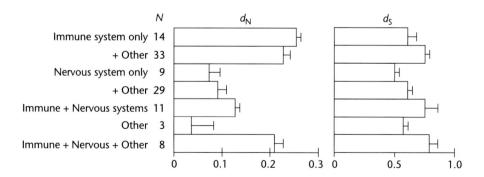

Figure 3 Mean numbers of synonymous (d_S) and nonsynonymous (d_N) nucleotide substitutions per site in comparisons between orthologous human and rodent immunoglobulin superfamily C2-set domains. N: number of genes compared. (Adapted with permission from Hughes, 1997.)

synonymous substitution among different expression categories, but the rate of nonsynonymous substitution was highest in molecules with immune system expression (**Figure 3**). Rapid evolution of immune system proteins is apparently a reflection of the ongoing coevolutionary race between host and parasite.

Conclusion

The origins of vertebrate immune mechanisms appear to stretch back for a long time in the vertebrate lineage. So far, it has been difficult to understand the origin of the unique vertebrate molecules (immunoglobulins, T-cell receptors and MHC) involved in specific immunity. Analysis of sequence data from immune system genes shows that they are characterized by rapid evolution, and there is evidence that natural selection arising from parasite pressure is the source of this selection. If such selection has been a constant feature of immune system evolution, it may be very difficult to trace the origins of vertebrate immune adaptations, because rapid evolution over millions of years may have obscured the sequence of changes that has taken place.

See also
Defensins: Evolution
Immunoglobulin Genes
Major Histocompatibility Complex (MHC) Genes: Evolution

References

Agrawal A, Eastmann QE and Schatz DG (1998) Transposition mediated by RAG1 and RAG2 and its implications for the evolution of the immune system. *Nature* **394**: 744–751.

Bernstein RM, Schluter SF, Bernstein H and Marchalonis J (1996) Primordial emergence of the recombination activating gene 1 (*RAG1*): sequence of the complete shark gene indicates homology to microbial integrases. *Proceedings of the National Academy of Sciences of the United States of America* **93**: 9454–9459.

Friedman R and Hughes AL (2001) Pattern and timing of gene duplication in animal genomes. *Genome Research* **11**: 1842–1847.

Friedman R and Hughes AL (2002) Molecular evolution of the NF-κB signaling system. *Immunogenetics* **53**: 964–974.

Hughes AL (1997) Rapid evolution of immunoglobulin superfamily C2 domains expressed in immune system cells. *Molecular Biology and Evolution* **14**: 1–5.

Hughes AL (1999a) Genomic catastrophism and the origin of vertebrate immunity. *Archivum Immunologiae et Therapiae Experimentalis* **47**: 347–353.

Hughes AL (1999b) Evolutionary diversification of the mammalian defensins. *Cellular and Molecular Life Sciences* **56**: 94–103.

Hughes AL and Nei M (1988) Pattern of nucleotide substitution at class I MHC loci reveals overdominant selection. *Nature* **335**: 167–170.

Kasahara M, Nayaka J, Satta Y and Takahata N (1997) Chromosomal duplication and the emergence of the adaptive immune system. *Trends in Genetics* **13**: 90–92.

Medzhitov R and Janeway Jr C (1998) An ancient system of host defense. *Current Opinion in Immunology* **10**: 12–15.

Murphy PM (1993) Molecular mimicry and the generation of host defense protein diversity. *Cell* **72**: 823–826.

Tanaka T and Nei M (1989) Positive Darwinian selection observed at the variable-region genes of immunoglobulin. *Molecular Biology and Evolution* **6**: 447–459.

Wang DY-C, Kumar S and Hedges SB (1999) Divergence time estimates for the early history of animal phyla and the origin of plants, animals and fungi. *Proceedings of the Royal Society of London, Series B. Biological Sciences* **266**: 163–171.

Further Reading

Hughes AL (1998) Protein phylogenies provide evidence of a radical discontinuity between arthropod and vertebrate immune systems. *Immunogenetics* **47**: 283–296.

Hughes AL (1999) *Adaptive Evolution of Genes and Genomes*. New York: Oxford University Press.

Klein J and Hořejší V (1998) *Immunology*, 2nd edn. Oxford, UK: Blackwell Scientific.

Web Links

recombination activating gene 1 (*RAG1*); LocusID: 5896. LocusLink:
http://www.ncbi.nlm.nih.gov/LocusLink/LocRpt.cgi?l = 5896

recombination activating gene 2 (*RAG2*); LocusID: 5897. LocusLink:
http://www.ncbi.nlm.nih.gov/LocusLink/LocRpt.cgi?l = 5897

recombination activating gene 1 (*RAG1*); MIM number: 179615. OMIM:
http://www.ncbi.nlm.nih.gov/htbin-post/Omim/dispmim?179615

recombination activating gene 2 (*RAG2*); MIM number: 179616. OMIM:
http://www.ncbi.nlm.nih.gov/htbin-post/Omim/dispmim?179616

Viral Vectors in Gene Therapy

See Adeno-associated Viral Vectors in Gene Therapy
Adenoviral Vectors in Gene Therapy
DNA Viral Vectors in Gene Therapy
Herpes Simplex Viral Vectors in Gene Therapy
Lentiviral Vectors in Gene Therapy
Retroviral Vectors in Gene Therapy
RNA Viral Vectors in Gene Therapy

Visual Pigment Genes: Evolution

Shozo Yokoyama, *Syracuse University, Syracuse, New York, USA*

More than 100 visual pigment genes have been cloned from a diverse range of vertebrates. Comparative sequence analyses of these genes and *in vitro* assays of engineered visual pigments have been used to elucidate not only the molecular bases of color vision but also the processes of adaptive evolution at the molecular level.

Visual Pigment Genes and Visual Pigments

Vision begins when photons are absorbed by photosensitive molecules, visual pigments. The visual pigments in rod cells are referred to as rhodopsin, whereas those in cone cells are often called cone pigments. Each visual pigment consists of an apoprotein, opsin and the chromophore, usually 11-*cis*-retinal, whose spectral sensitivity is characterized by the wavelength of maximal absorption (λ_{max}). The two molecules are bound to each other by a Schiff base linkage to the rhodopsin lysine residue, K296, or equivalent lysine of the cone pigments (Palczewski *et al.*, 2000). The Schiff base of 11-*cis*-retinal is usually protonated by the glutamate counterion, E113, of the opsin. The protonated Schiff base has a λ_{max} of 440 nm in solution. Interacting with an opsin, however, the Schiff base-linked chromophore in a visual pigment can have a λ_{max} from 360 to 600 nm. This phenomenon is known as the spectral tuning of visual pigments.

The opsin is encoded by a specific visual pigment gene (or an opsin gene). The cone pigment genes were first isolated from humans (Nathans *et al.*, 1986). Using the corresponding opsin and rhodopsin complementary deoxyribonucleic acids (cDNAs), more than 100 complete opsin genes and cDNA clones have been isolated and sequenced. Based on their nucleotide and deduced amino acid sequences, the opsin genes (and visual pigments) in vertebrate retinas are classified into five evolutionary groups: (1) RH1 (rhodopsins); (2) RH2 (RH1-like); (3) short wavelength-sensitive type 1 (SWS1); (4) SWS type 2 (SWS2); and (5) long wavelength- and middle wavelength-sensitive (LWS/MWS) groups. The RH1 genes are usually expressed in rods and the other four groups of opsin genes usually in cones. The gene size ranges from approximately 1 kb of the fish RH1 genes to approximately 20 kb of the rat MWS gene (**Figure 1**). In the early stage of fish evolution, all introns of the RH1 genes were lost. Otherwise, the introns 1, 2, 3 and 4 of the RH1, RH2, SWS1 and SWS2 genes and

introns 2, 3, 4 and 5 of the LWS/MWS genes interrupt their coding sequences at exactly the same corresponding sites.

The RH1 genes (and pigments) are most closely related to RH2, and then to the SWS2, SWS1 and LWS/MWS groups, in that order, which are strongly supported by high bootstrap values (**Figure 2**). In **Figure 2**, however, the phylogenetic positions of chameleon (P491) pigment in the RH1 group and gecko (P521) pigment in the LWS/MWS group do not agree with the phylogenetic relationships of organisms. The diurnal chameleon (*Anolis carolinensis*) has only cones, whereas the nocturnal gecko (*Gekko gekko*) has only rods. Thus, the incorrect phylogenetic positions of the two pigments seem to reflect their rapid mutant substitutions associated with the switch in the photoreceptor cell-specificity. Here two additional comments are in order. First, the five groups of genes have arisen through four gene-duplication events. The RH1 group contains visual pigments from a wide variety of organisms, ranging from lampreys to mammals. As the most recent gene duplication event of the four occurred prior to the divergence of various vertebrates, the vertebrate ancestor must have possessed all five groups of opsin genes. Second, the ability of humans to see light ranging in wavelength from 400 to 650 nm is controlled by the RH1 (human; P497), SWS1 (human; P414), MWS (human; P530) and LWS (human; P560) pigments (**Figure 2**). So far, neither RH2 nor SWS2 genes has been found in the human and other mammalian genomes. These genes must have become nonfunctional and been lost in an early stage of mammalian evolution. From **Figure 2**, we can also see that RH1, RH2, SWS1, SWS2 and LWS/MWS pigments have a λ_{max} of 480–510, 470–510, 360–430, 410–460 and 510–560 nm. In the following discussion, the amino acid site numbers are those of the bovine (P500) pigment in the RH1 group.

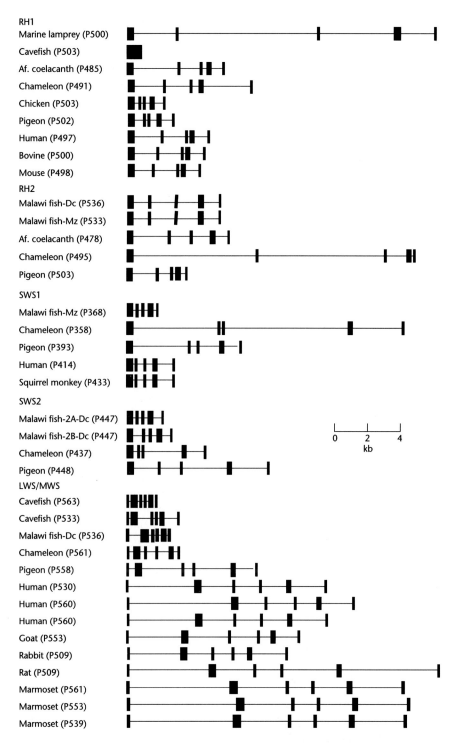

Figure 1 Structures of visual pigment genes, where exons and introns are represented by black boxes and horizontal lines respectively. The numbers after P refer to λ_{max}. For Malawi fish pigments, Dc and Mz denote *Dimidiochromis compressiceps* and *Metriaclima zebra* respectively. Malawi fish-Dc (P536), Malawi fish-Mz (P533), Malawi fish-2A-Dc (P447), Malawi fish-2B-Dc (P488), Malawi fish-Dc (P368), Malawi fish-Dc (P569), marmoset (P561), marmoset (P553) and marmoset (P539) pigment genes are from GenBank (accession nos. AF247121, AF247122, AF247113, AF247118, AF191220, AF247125, AB046549s1–s6, AB046555s1–s6 and AB046561s1–s6 respectively). For other genes, see Yokoyama (2000). The gene duplication of the human P530 and P560 genes occurred some 30 million years (MY) ago (Nathans *et al.*, 1986). Two human (P560) genes have intron 1 length polymorphism, one of them being 2 kb longer than the other. Af: African; LWS/MWS: long wavelength- and middle wavelength-sensitive; RH1: rhodopsins; RH2: RH1-like; SWS1: short wavelength-sensitive type 1; SWS2: SWS type 2.

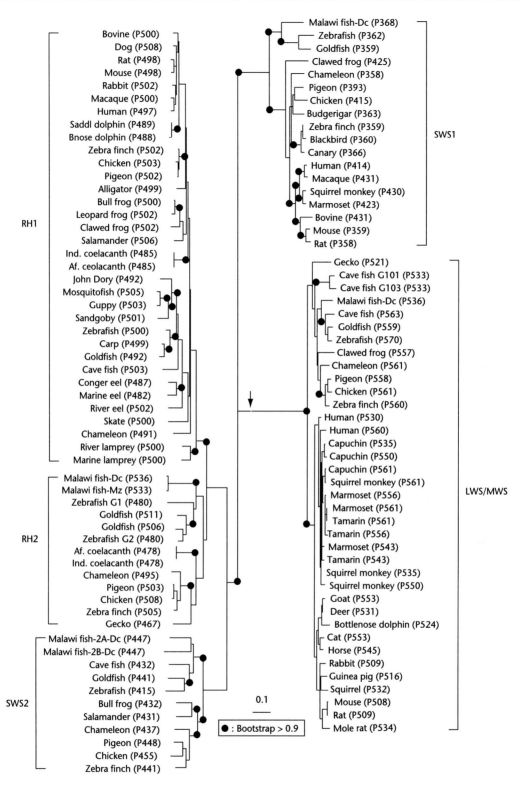

Figure 2 Phylogenetic tree for the vertebrate visual pigments by applying the neighbor-joining method (Saitou and Nei, 1987) to their amino acid sequences. Ind. coelacanth (P485) and Ind. coelacanth (P478) are from Indonesian coelacanth (*Latimeria menadoensis*. Salamander (P431), bull frog (P432) and mole rat (P534) pigments are from GenBank (accession nos. AF038946, AB010085 and AF139726 respectively). Blackbird (P360) is from red-winged blackbird (*Agelaius pheniceus*). For other sequences, see Yokoyama (2000). The arrow indicates the root of the phylogenetic tree. The bar at the bottom indicates evolutionary distance measured as the number of amino acid replacements per site.

Coelacanths and RH1 and RH2 Pigments

How did organisms modify their color vision to adapt to various environments? This evolutionary question is closely related to a central question in phototransduction: How do visual pigments detect a wide range of wavelengths using the same 11-*cis*-retinal? Thus, evolutionary biology and vision science have an important common goal. Shortly after the cloning of the human cone pigment genes, the functional assay of visual pigments was developed, where virtually any opsin cDNAs can be expressed in cultured cells, reconstituted with 11-*cis*-retinal, and the absorption spectra of the resulting visual pigments can be measured. These advances in vision research also provide a rare opportunity for the study of adaptive evolution at the molecular level.

The coelacanths (*Latimeria chalumnae*) live at a depth of 200 m near the coast of the Comoros Islands in the western Indian Ocean. The ocean floor at the depth of approximately 200 m receives only a narrow range of sunlight at approximately 480 nm. Out of the five groups of visual pigments, the coelacanths have retained only RH1 (African coelacanth; P485) and RH2 (African coelacanth; P478) pigments. Note that, compared with those of most orthologous pigments, the λ_{max} of these two pigments are reduced by approximately 10–20 nm (**Figure 2**) and their absorption spectra have been devised to visualize the entire spectrum of color available to the coelacanths (Yokoyama *et al.*, 1999). How did the coelacanths achieve these exquisitely coordinated blue shifts in the λ_{max} of the two pigments? Comparative amino acid sequence analyses suggest that E122Q/A292S (amino acid changes E → Q and A → S at sites 122 and 292 respectively, and E122Q/M207L occurred along the branches leading to the coelacanth RH1 and RH2 pigments respectively. Indeed, amino acid changes Q122E/S292A in the RH1 pigment and Q122E/L207M in the RH2 pigment increase the λ_{max} by 26 and 21 nm respectively. Thus, the blue shift in the λ_{max} in the RH1 pigment has been explained well by E122Q/A292S and that of the RH2 pigment by E122Q/M207L (Yokoyama *et al.*, 1999). These amino acid sites are located in the transmembrane segments (**Figure 3**).

LWS and MWS Pigments

Many MWS and LWS pigments have λ_{max} of approximately 530 and 560 nm respectively (**Figure 2**). It can be shown experimentally that this 30 nm difference in the λ_{max} is caused by amino acid

differences at three sites: A164/F261/A269 in the MWS pigments and S164/Y261/T269 in the LWS pigments. Having LWS pigment-like amino acids A164/Y261/T269, however, the orthologous pigments in mouse, rat and rabbit have λ_{max} of approximately 510 nm (**Figure 2**). These extremely blue-shifted λ_{max} are shown to be achieved by H181Y/A292S. Thus, to explore the spectral tuning in the LWS/MWS pigments in all vertebrates, we should consider amino acid replacements at five sites: 164, 181, 261, 269 and 292. Multiple regression analyses based on these five sites of 26 currently known LWS/MWS pigments strongly suggest that the pigment in the vertebrate ancestor had an amino acid composition of S164/H181/Y261/T269/A292 with a λ_{max} of 559 nm and that mutations S164A, H181Y, Y261F, T269A, A292S and S164A/H181Y in this ancestral pigment shift the λ_{max} by −7, −28, −8, −15, −27 and 11 nm respectively (Yokoyama and Radlwimmer, 2001). These sites are located either in the transmembrane segments or very close to them (**Figure 3**). Importantly, extensive mutagenesis experiments also reveal that amino acid changes at the background sites do not cause any λ_{max} shift, showing that the spectral tuning in the LWS/MWS pigments is explained fully by the 'five-sites' rule (Yokoyama and Radlwimmer, 2001).

In higher primates, the LWS/MWS pigments evolved in two different ways. Hominoid and Old World monkeys use LWS and MWS opsins, which are encoded by two separate X-linked loci. Most New

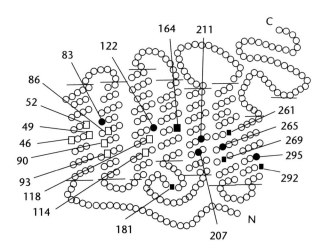

Figure 3 Secondary structure of bovine RH1 opsin, showing naturally occurring amino acid mutations that cause more than 5 nm of λ_{max} shift. The model is based on Palczewski *et al.* (2000). Open square, filled square and filled circles indicate the amino acid sites that are involved mainly in the spectral tuning of SWS1, LWS/MWS and RH1/RH2 pigments respectively (see Yokoyama *et al.*, 2000; Shi *et al.*, 2001).

World monkeys, however, have one corresponding X-linked locus with three alleles (**Figure 1**). In these species, therefore, all males are so-called red–green color blind, whereas females are either 'color blind' or have complete red–green color vision depending on the genotype.

In human, red–green color vision is controlled by LWS and MWS pigments. On the human X chromosome, one LWS and one or multiple MWS genes are located. Intragenic recombination between LWS and MWS genes produces both 5′ MWS–LWS 3′ and 5′ LWS–MWS 3′ hybrid genes (Nathans *et al.*, 1986). The addition of the 5′ MWS–LWS 3′ gene to an otherwise normal gene array containing wild-type LWS and MWS genes causes the most common inherited color vision anomaly, deuteranomaly (Nathans *et al.*, 1986). Many deuteranomalous men have wild-type MWS gene(s) that do not contribute to correct the anomalous color vision (Neitz *et al.*, 1996). It should be noted that, as long as the exon 5 of the hybrid gene encodes Y261 and T269, the red vision is restored. Thus, if we define any LWS/MWS genes that encode Y261 and T269 as LWS genes, many human X chromosomes have more than one LWS gene, and sometimes even up to four (Neitz and Neitz, 1995).

In human populations, it has been observed that approximately 60% of the LWS pigments have S164 and approximately 40% of the allelic LWS pigments have A164 (Winderickx *et al.*, 1992). Among MWS genes, variation at site 164 is less common, as at least 90% encode A164 (Nathans, 1999). The amino acid dimorphism can cause 7 nm difference in the light-sensitivities of the two types of LWS pigments.

SWS1 Pigments

Many fishes, amphibians, reptiles, birds and some mammals use ultraviolet (UV) vision for such basic activities as foraging and mate choice. These species detect light maximally at 360–370 nm by using UV pigments. These UV pigments and violet (or blue) pigments with λ_{max} of 390–430 nm belong to the same SWS1 group (**Figure 2**). The spectral tuning in the UV pigments has been studied first by considering avian pigments, and then mammalian pigments. The zebra finch, blackbird, canary and budgerigar SWS1 pigments have λ_{max} of 358–366 nm, whereas the orthologous violet pigments of pigeon and chicken have λ_{max} of more than 390 nm (**Figure 2**). It has been shown that the avian UV pigments evolved from the violet pigment by one amino acid replacement, S90C (Wilkie *et al.*, 2000; Yokoyama *et al.*, 2000). On the other hand, the mouse SWS1 pigment has a λ_{max} of 359 nm, whereas the closely related human blue pigment has a λ_{max} of 414 nm. The mouse UV pigment can be made

blue-sensitive ($\lambda_{max} = 411$ nm) by introducing seven amino acid changes F46T/F49L/T52F/F86L/T93P/A114G/S118T, whereas the human blue pigment can be made into UV pigment with a λ_{max} of 360 nm by introducing the seven reverse mutations (Shi *et al.*, 2001). These analyses show that the violet pigments evolved from the UV pigment by accumulating at least two of the eight amino acid replacements. However, F86Y had the major impact in shifting the λ_{max} by more than 50 nm (Fasick *et al.*, 2002).

These results suggest that the difference between the UV and violet pigments in vertebrates is based on a total of eight amino acid sites 46, 49, 52, 86, 90, 93, 114 and 118, which are all located in the transmembrane segments (**Figure 3**). Comparative amino acid sequence analyses suggest that the common ancestral SWS1 pigment in vertebrates had amino acids F46/F49/T52/F86/S90/T93/A114/S118. This amino acid composition is identical to those of the contemporary salamander, chameleon, mouse and rat UV pigments with λ_{max} of approximately 360 nm (**Figure 2**), but T93Q occurred in the ancestral fish UV pigment. Using the goldfish UV pigment, it has been shown that Q93T does not shift the λ_{max} from that of the wild-type pigment. Thus, the ancestral pigment in vertebrates must have had a λ_{max} of approximately 360 nm, and the fish, salamander, chameleon, mouse and rat pigments have maintained their UV sensitivities through purifying selection (Shi *et al.*, 2001). In the avian lineage, the ancestral pigment lost UV sensitivity, but some descendants regained it by S90C. Because of the nonadditive effects of amino acid changes on the λ_{max} shift, the evolutionary processes of the functional differentiation of various violet pigments remain to be elucidated.

Perspectives

The comparative sequence analyses followed by the mutagenesis experiments demonstrate that the evolutionary approach is a powerful method in enhancing our understanding of the functional differentiations of a wide variety of visual pigment genes. As more amino acid sequences and absorption spectra of visual pigments accumulate, the prediction of potentially important amino acid changes in the spectral tuning in visual pigments will become more accurate. Sampling of visual pigments from various photic environments or those associated with different behavioral characteristics would be of particular interest, because we may also uncover previously unknown amino acid sites that are involved in the spectral tuning in visual pigments. Visual pigments associated with specific photic environments or unique behaviors also provide an excellent opportunity to analyze adaptive evolution at the molecular level.

See also

Color Vision Defects
Eye: Proteomics

References

Fasick JI, Applebury ML and Oprian DD (2002) Spectral tuning in the mammalian short-wavelength sensitive cone pigments. *Biochemistry* **41**: 6860–6865.

Nathans J (1999) The evolution and physiology of human color vision: insights from molecular genetic studies of visual pigments. *Neuron* **24**: 299–312.

Nathans J, Thomas D and Hogness DS (1986) Molecular genetics of human color vision: the genes encoding blue, green, and red pigments. *Science* **232**: 193–201.

Neitz M and Neitz J (1995) Numbers and ratios of visual pigment genes for normal red–green color vision. *Science* **267**: 1013–1016.

Neitz J, Neitz M and Kainz PM (1996) Visual pigment gene structure and the severity of color vision defects. *Science* **274**: 801–803.

Palczewski K, Kumasaka T, Hori T, *et al.* (2000) Crystal structure of rhodopsin: a G protein-coupled receptor. *Science* **289**: 739–745.

Saitou N and Nei M (1987) The neighbor-joining method: a new method for estimating phylogenetic trees. *Molecular Biology and Evolution* **4**: 406–425.

Shi Y, Radlwimmer FB and Yokoyama S (2001) Molecular genetics and the evolution of ultraviolet vision in vertebrates. *Proceedings of the National Academy of Sciences of the United States of America* **98**: 11 731–11 736.

Wilkie SE, Robinson PR, Cronin TW, *et al.* (2000) Spectral tuning of avian violet- and ultraviolet-sensitive visual pigments. *Biochemistry* **39**: 7895–7901.

Winderickx J, Lindsey DT, Sanocki E, *et al.* (1992) Polymorphism in red photopigment underlies variation in colour matching. *Nature* **356**: 431–433.

Yokoyama S (2000) Molecular evolution of vertebrate visual pigments. *Progress in Retinal and Eye Research* **19**: 385–419.

Yokoyama S and Radlwimmer FB (2001) The molecular genetics and evolution of red and green color vision in vertebrates. *Genetics* **158**: 1697–1710.

Yokoyama S, Radlwimmer FB and Blow NS (2000) Ultraviolet pigments in birds evolved from violet pigments by a single amino acid change. *Proceedings of National Academy of Sciences of the United States of America* **97**: 7366–7371.

Yokoyama S, Zhang H, Radlwimmer FB and Blow NS (1999) Adaptive evolution of color vision of the Comoran coelacanth (*Latimeria chalumnae*). *Proceedings of National Academy of Sciences of the United States of America* **96**: 6279–6284.

Further Reading

Ebrey T and Koutalos Y (2001) Vertebrate photoreceptors. *Progress in Retinal and Eye Research* **20**: 49–94.

Kochendoerfer GG, Lin SW, Sakmar TP and Mathies RA (1999) How color visual pigments are tuned. *Trends in Biochemical Sciences* **24**: 300–305.

Nathans J (1990) Determinants of visual pigment absorbance: role of changed amino acids in the putative transmembrane segments. *Biochemistry* **29**: 937–942.

Sharpe LT, Stockman A, Jagle H, *et al.* (1998) Red, green and red–green hybrid pigments in the human retina: correlations between deduced protein sequences and psychophysically measured spectral sensitivities. *Journal of Neuroscience* **18**: 10 053–10 069.

Yokoyama S (2002) Molecular evolution of color vision in vertebartes. *Gene* **300**: 69–78.

Watson, James Dewey

JA Witkowski, *Cold Spring Harbor Laboratory, New York, USA*

James Dewey Watson (1928–) is an American scientist who discovered, with Francis Crick, the double-helical structure of DNA and went on to establish the Human Genome Project in the United States.

James Watson was born on 6 April 1928 in Chicago, Illinois. His house was filled by books and his father inspired him with a love of both learning and bird watching. Although a radio quiz kid contestant, Watson claims not to have been a childhood genius. His early entry to the University of Chicago (at age 15) was due to the university's enlightened policy and the encouragement of his mother.

Watson studied zoology at Chicago and was awarded a Bachelor of Science degree in 1947. During his undergraduate work, he read and was greatly impressed by Erwin Schrödinger's *What Is Life?*, with its discussion of the physical nature of the gene. Nevertheless, still interested in ornithology, he went to Indiana University after graduating. Hermann Muller was in the Department of Zoology and Watson took both his course on genetics and that of Salvador Luria on viruses. He realized that Muller's *Drosophila* was unlikely to yield new insights and that the cutting edge of genetics was in the analysis of microorganisms. He began his thesis research on X-ray inactivation of bacteriophage with Luria.

At that time, Luria, Max Delbrück and Alfred Hershey were using bacteriophage to study genetics and, although all three were at other institutions, phage research became inextricably linked with Cold Spring Harbor where Delbrück and Luria had established the 'Phage Course' in 1945. (Hershey moved to Cold Spring Harbor in 1950.) This course trained many of the future leaders of molecular genetics. In 1948, Luria brought his two graduate students, Watson and Renato Dulbecco, to Cold Spring Harbor for a summer of phage experiments.

Watson completed his doctorate in 1950; his thesis research had not led to new insights into the nature of the gene, but he was determined to pursue research on its chemical basis. Inspired by his contacts with European scientists such as Luria and Delbrück, Watson decided to carry out research in Europe and Luria arranged for him to take up a National Research Fellowship in Herman Kalckar's laboratory in Copenhagen. He discovered, however, that his and Kalckar's interests were not the same. Watson was fortunate to find that Kalckar's friend Ole Maaløe was interested in phage replication, and he began work on

phage but had to move when Maaløe went to the California Institute of Technology (Caltech). In the spring of 1951, Watson had met Maurice Wilkins, from King's College London, at a meeting in Naples. Wilkins had shown him an X-ray diffraction picture of DNA that suggested that DNA must have a regular structure that might be determined. Watson was convinced that knowing the structure of DNA was the key to understanding the gene and that he had to go to the European center for X-ray crystallography – the Cavendish Laboratory in Cambridge, England.

He arrived there in September 1951 and began a collaboration with Francis Crick. Using information from Rosalind Franklin of King's College, together with data on the chemical composition and properties of DNA and model building, he and Crick arrived eventually at the double-helical structure of DNA. Published in *Nature* on 25 April 1953, it established Watson as an intellectual leader in the new discipline of molecular genetics. But DNA did not occupy Watson's energies completely: he also studied the X-ray study structure of tobacco mosaic virus.

Between 1953 and 1955, Watson was a Senior Research Fellow at Caltech. There he began studies aimed at understanding the role of ribonucleic acid, again by determining first its structure. Unfortunately, the X-ray diffraction patterns of RNA were not distinct and Watson and Alex Rich had to admit defeat. Watson did not find Caltech a congenial environment and spent 1955 back at the Cavendish Laboratory. During that year, he and Francis Crick wrote an important paper on the structure of small viruses. In 1956, Watson became an assistant professor of biology at Harvard University, becoming a professor in 1961. He was noted at this time for his strenuous efforts to introduce molecular studies into the Harvard department and curriculum.

Watson continued his research on RNA but used molecular biology techniques rather than X-ray crystallography to try to understand its role. With Alfred Tissieres, he published an important paper on the ribosome and later one of the two papers that reported the discovery of messenger RNA. Watson's insistence in not putting his name on the papers of his graduate students gives a false impression of the extent of his research interests at this time. His students included R. W. Risebrough, J. Steitz, N. Hopkins, R. Burgess and M. Capecchi – all of whom have gone on to achieve great distinction as molecular biologists.

As a consequence of teaching at Harvard, Watson wrote the textbook *The Molecular Biology of the Gene*, which set a new style and standard in biology textbooks. (Subsequently, he was a coauthor of two other textbooks, *The Molecular Biology of the Cell* and *Recombinant DNA: A Short Course*.) It was while he was at Harvard that Watson wrote an account of the discovery of the structure of DNA – *The Double Helix*. This controversial account (Harvard's President Nathan Pusey barred the Harvard University Press from publishing it) instantly became an international bestseller, was translated into over 20 languages and has remained in print ever since it was published in 1968.

Watson had been the Harvard University representative on the Board of Trustees of Cold Spring Harbor Laboratory on Long Island, New York, and in 1968 he became director of the Laboratory, while retaining his Harvard position. In 1972, he moved to Cold Spring Harbor to work full-time as director. Watson steered the laboratory into the field of tumor virology and other research on cancer, as well as plant molecular biology, cell biology and neuroscience. He expanded the program of meetings and courses held at the laboratory and the activities of Cold Spring Harbor Laboratory Press.

In 1974, Watson was one of the signatories of the 'Berg' letter – which was written to the journal *Science* – that warned of the possible dangers of recombinant DNA technology. This letter led to the Asilomar Conference, where scientists decided that voluntary restrictions on certain types of experiment were appropriate. Watson argued strongly against such restrictions, fearing that the government would impose regulations, as indeed it did. The restrictions were later relaxed and there is no evidence that recombinant DNA techniques have caused any hazard to scientists or the public. He and John Tooze subsequently published a history of the recombinant DNA controversy – *The DNA Story*.

Between 1988 and 1992, Watson was responsible for directing the National Institutes of Health (NIH) Human Genome Project, while maintaining his activities as director of Cold Spring Harbor. It was through his efforts that the scientific community came to embrace the Project and that the Project received the necessary funding from Congress. Watson, worried about the societal applications of human molecular

genetics, established the Ethical, Legal and Social Issues program, allocating it about 5% of the Project's budget. Watson resigned in 1992 after a disagreement with Bernadine Healey, director of the NIH, over the patenting of DNA sequences.

Watson returned full-time to Cold Spring Harbor Laboratory, where he became its first President in 1994. He continues to promote the Human Genome Project and its applications, his fear being that overemphasis on possible abuses will delay the beneficial use of genetic knowledge. In 2000, he published a collection of his essays that deal with this and other controversial topics in genetics.

Watson has received many honors, including the Eli Lilly Award in Biochemistry (1960), the Albert Lasker Prize (1960), the Nobel Prize for Physiology or Medicine (1962), the John J. Carty Gold Medal of the National Academy of Sciences (1971), the Presidential Medal of Freedom (1977), the Copley Medal of the Royal Society (1993), the Charles A. Dana Distinguished Achievement Award in Health (1994), the National Medal of Science (1997) and the Liberty Medal (2000). He was elected a member of the National Academy of Sciences (1962) and the Academy of Sciences, Russia (1989), and a Foreign Member of the Royal Society (1981). He is an Honorary Fellow of Clare College, Cambridge (1968).

See also
Crick, Francis
DNA Structure
Franklin, Rosalind Elsie
Muller, Herman Joseph
Wilkins, Maurice Hugh Frederick

Further Reading

Watson JD (1965) *The Molecular Biology of the Gene*. Menlo Park, CA: Benjamin/Cummings.
Watson JD (1968) *The Double Helix*. London: Weidenfeld & Nicolson.
Watson JD, Gilman M, Witkowski JA and Zoller M (1992) *Recombinant DNA*. New York: WH Freeman.
Watson JD (2000) *A Passion for DNA: Genes, Genomes, and Society*. Plainview, NY: Cold Spring Harbor Laboratory Press.
Watson JD and Tooze J (1981) *The DNA Story: A Documentary History of Gene Cloning*. San Francisco, CA: WH Freeman.
Watson JD (2001) *Genes, Girls and Gamow*. Oxford, UK: Oxford University Press.
Alberts B, Bray D, Lewis J, *et al.* (1994) *Molecular Biology of the Cell*. New York: Garland Publishing.
Judson HF (1996) *The Eighth Day of Creation: Makers of the Revolution in Biology*, expanded edition. Plainview, NY: Cold Spring Harbor Laboratory Press.
Olby R (1974) *The Path to the Double Helix*. Seattle, WA: University of Washington Press.

Weapons of Mass Destruction: Genotoxicity

Christine Gosden, *University of Liverpool, Liverpool, UK*

Derek Gardener, *University of Liverpool, Liverpool, UK*

Intermediate article

Article contents

- Introduction
- Exposures of Human Populations
- Susceptibility and Detoxification
- Protection, Treatment and Aid for Survivors
- Conclusions

Many chemical, biological and radiological weapons of mass destruction are genotoxic and damage DNA so that survivors of exposures are at increased risk of cancers, infertility, medical disorders and having children with birth defects. Each agent has specific effects – for example, aflatoxin causes liver cancer and nuclear weapons cause leukemia – indicating that each type of weapon has precise genomic targets and suggesting that therapies might be developed to aid victims and to counteract the threats posed by such weapons.

Introduction

The twentieth century has seen many millions of deaths from wars, conflicts, terrorist attacks and genocidal policies of ethnic cleansing. There are no humane weapons. There are, however, fundamental differences between conventional weapons and weapons that, in addition to their immediate ability to kill or incapacitate, have the potential to cause long-term genetic damage among survivors. The use of deoxyribonucleic acid (DNA)-damaging weapons can lead to increased death from cancer and can cause malformations and cancers in the children of exposed people. These weapons act as genetic time bombs by exerting harmful effects on victims years after exposure. Genotoxic weapons include nuclear and radiological weapons, chemical weapons such as mustard gas (a powerful alkylating agent) and biological toxins such as aflatoxin. (*See* Eugenics; Eugenics: Contemporary Echoes; Gene–Environment Interaction; Mutation Rate; Protein Families: Evolution.)

The term 'weapons of mass destruction' (WMD) conveys the concept of highly lethal weapons, such as nuclear bombs or nerve gases, that have been scientifically targeted to kill opposing military forces rapidly. In most major incidents involving WMD, including mustard gas attacks in the First World War, Hiroshima and Nagasaki, the Tokyo subway attack, the Iran–Iraq war and the Kurdish Anfal campaign, however, less than 10% of the victims died and over 90% of the exposed populations survived the immediate attacks. Many died later as a result of medical conditions such as cancers and birth defects. Deployment of genotoxic WMD allows perpetrators to escape censure for the full scale of their crime, because deaths (through cancers and birth defects) that occur years after exposure are not usually counted as direct casualties. (*See* DNA Damage Response; Genetic Variation: Polymorphisms and Mutations; Mutational Change in Evolution.)

There are stark contrasts in the vulnerability of military and civilian victims. Military victims of WMD tend to be fit young men with protective clothing, gas masks and battlefield training, who are armed with detectors, antidotes and decontaminants. By contrast, civilian victims are unprepared, lack knowledge about WMD agents, how to protect themselves or detoxify their food, water and environment, and are thus vulnerable. Each weapon has specific ways in which it damages DNA: the risks depend on dose, route of exposure and the types of cell that are exposed and vulnerable to damage (**Figure 1**). One of the principal achievements of sequencing the human genome has been the identification of mutations that have adverse effects on human health. Genomic approaches to examining the types of genetic damage arising from WMD use (including cancers, birth defects, genetic susceptibility and genetically encoded detoxification mechanisms) will aid the detection of WMD use and facilitate treatments for victims at risk. (*See* Cancer Genome Anatomy Project; Disability, Human Rights and Contemporary Genetics; Genome Organization of Vertebrates; Tumor Formation: Number of Mutations Required.)

Exposures of Human Populations

Many powerful chemical and biological WMD kill rapidly but do not induce DNA damage (although they may cause other long-term medical effects such as neurological damage). These include most nerve agents (sarin, tabun, VX (*S*-(diisopropylaminomethyl) methylphosphonothiolate *O*-ethyl ester), BZ (3-quinuclidinyl benzilate)) and infectious bioagents (anthrax, plague, smallpox, Ebola). **Table 1** shows details of WMD and their genotoxicity. **Tables 1–3** are compiled from detailed information taken from key references (Patura and Rall, 1993; Hayes, 1994; Schull, 1995; US Institute of Medicine, 1996;

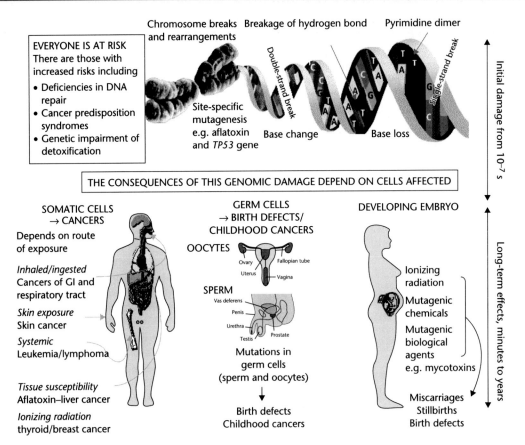

Figure 1 Human genomic damage from weapons of mass destruction. The figure illustrates the way in which genotoxic agents can damage DNA and the consequences of DNA damage according to the cell types affected; somatic cells (cancers), male and female germ cells (transmission to future generations) and cells of developing embryo (birth defects). Also under genetic control are the mechanisms for DNA repair and detoxification that affect susceptibility and contribute to the severity of the effects. The time course of the response indicates that while DNA damage may occur in microseconds (emphasising the need for early detection and protection), clinical manifestations of this damage may occur hours to years after exposure. Because damage occurs at random within the genome and in different cell types, the same agent may cause a spectrum of effects that depend also on the agent, route of exposure, genetic susceptibility and dose in those affected. GI: gastrointestinal.

Table 1 Chemical, biological and radiological WMD

Type	Mutations caused	Agent	Genotoxic?
Chemical agents		Mustard gas	Yes
		Agent orange	Yes
		Nerve agents (sarin, tabun, VX, BZ)	Some
			Most nongenotoxic, some weakly mutagenic
Biological toxins	Carcinogenic, mutagenic and embryotoxic	Aflatoxin	Yes
		Mycotoxins	Yes
Biological infectious agents	Highly lethal agents, but do not damage DNA	Viruses, e.g. smallpox, Ebola	No
		Bacteria, e.g. anthrax, plague	No
Radiological weapons	Cause extensive damage to chromosomes, cells and DNA	Atom bombs	Yes
		Irradiated zirconium bombs	Yes
		Neutron bombs	Yes
		Enriched uranium pellets	Yes
		Radioactive waste	Yes

Table 2 Radiological, chemical and biological WMD: deployment and affected populations

Conflict or region	Details
Radiological WMD: atomic, nuclear, irradiated zirconium bombs and radioactive waste	
Second World War 1945	Hiroshima and Nagasaki atomic bombs, thousands of immediate civilian deaths, long-term fatalities caused by cancers, birth defects and medical disorders
Development, manufacture, testing and disposal	
Pacific	Bikini, Eniwetok, Johnston Atoll, Mororua, Fangataufa Atolls, Christmas and Malden Islands; testing by the United States, United Kingdom and France
Australia	Monte Bello Island, Emu Fields and Maralinga; testing by the United Kingdom
United States	Nevada, New Mexico, Colorado and Alaska; testing by the United States
Soviet Union	Artic Islands, Kazakhstan, Ukriane, Uzbekistan and Turkmenistan; testing by the Soviet Union
Algeria	Testing by France
China	Xingiiang Province; testing by China
India	Pokhran; testing by India
1987 Iraq	Irradiated zirconium bombs tested (UNSCOM reports)
1945–2002	More than 2000 nuclear weapons tests worldwide since July 1945
Genotoxic chemical or biological WMD	
Mustard gas	
First World War 1917–1918	Battlefield use in France and Belgium with more than a million casualties in the British Empire, US, French, Belgian, German, Russian troops
1919	British use against Afghan rebels
1935–1936	Italian use against Abyssinians
1937	Japanese use against Chinese
Second World War 1943	Air raid hits American vessel carrying mustard weapons at Bari Harbour
1980–1988	Iran–Iraq War: extensive use of nerve agents and mustard gas – many thousands of military and civilian casualties
1987–1988	Iraq use against population of Iraqi Kurdistan/Anfal Campaign. Over 200 separate attacks, over 5000 fatalities in Halabja attack
Twentieth century	Many regional conflicts have allegedly used chemical or biological agents but this is unconfirmed
Agent orange	
Vietnam War	Mass spraying of dioxin-contaminated agent orange defoliant from high altitude; acute lymphocytic leukemia in children of exposed veterans
Development, manufacture and disposal	
Mustard gas	
First World War to Second World War	Mustard gas factory workers; munitions fillers; test-chamber volunteers
Disposal after Second World War	British sea-dumping Baltic, Beaufort's Dyke, Hebrides and Land's End; United States sea-dumping Operation CHASE ('cut holes, and sink'em')
Weapons disposal	Convention on the prohibition of development, production, stockpiling and use of chemical weapons and on their destruction, 29 April 1997, OPCW
Aflatoxin, mycotoxins	
Iraq	UNSCOM reports aflatoxin weaponized by Iraq

Ellenhorn, 1997). Infectious bioweapon agents are highly lethal but do not cause DNA damage and are therefore beyond the scope of this article.

Strategically, WMD are immensely powerful; they are cheap to produce and can be deployed in novel ways, including as landmines, hand grenades and weaponized radioactive waste. Analyses of the major incidents (the First World War, Hiroshima and Nagasaki, the Iran–Iraq war and the Kurdish Anfal campaign), and of exposures from conflicts, manufacture, weapons testing and disposal, indicate that the immediate destructive power of WMD is subsequently amplified by lethal effects in survivors (**Table 2**).

Genotoxic WMD

What are the long-term consequences of genomic damage from WMD used in wars, conflicts and ethnic cleansing programs, and in weapons manufacture, testing and disposal? Genotoxic agents may damage

Table 3 Long-term medical and genomic effects of WMD

Condition	Agents	Details
Solid tumors and hematological malignancies in adults and children		
Damage in somatic cells	Radiation	Leukemia, lymphoma and solid tumors according to isotopes
	Mustard gas	Laryngeal, pharyngeal, lung, leukemia, skin and other cancers
	Agent Orange	
	Aflatoxin	Liver and other cancers
	Mycotoxins	Some are potent carcinogens
Birth defects: teratogenic and embryonic effects		
Damage to germ cells and developing embryo	Radiation	Cranial neural crest, cleft palate, cardiac, skeletal and other
	Mustard gas	Cranial neural crest, cleft palate, cardiac, skeletal and other
	Agent Orange	Leukemia in children of exposed vets.
	Mycotoxins	Embryonic and fetal death
Disturbed sex ratios: X-linked lethal mutations		
Damage in germ cells	Radiation	Mutations, especially on X chromosome, causing
	Mustard gas	lethality in males
	Agent Orange	Data being collected
	Mycotoxins	Embryonic deaths
Infertility, testicular, ovarian and germ cell failure		
Germ cell effects, transmissible genetic damage and mutations	Radiation	Sterility from radiation exposures. Human and animal studies have given quantitative data for males and females
	Mustard gas	Iran–Iraq war: testicular damage
	Alkylating agents	Known to sterilize male and female individuals undergoing chemotherapy
Medical disorders		
Organ and tissue damage	Nerve agents	Long-term neuropathy
	Mustard gas	Heart failure, corneal scarring, cataracts, skin burns, lung damage
	Radiation	Heart failure, renal and/or liver damage, cataracts
Neuropsychiatric disorders		
Organ and tissue damage	Radiation	Depression and impaired neurological function. Microcephaly in fetuses exposed *in utero*
	Nerve agents	Long-term neuropathy (paraoxonase alleles susceptible?)
	Mustard gas	Psychosis and depression (accumulates in myelin sheaths, fatty tissue of brain and nervous system)
Immunosuppression		
Organ, tissue and immune damage	Radiation	Depressed immune system function
	Nerve agents	
	Mustard gas	
Animal experiments		
Somatic, germ cell, tissue and immune damage	Radiation	Animal studies have confirmed all of the effects found in humans, and allow dose–response curves to be constructed to quantify effects
	Nerve agents	
	Mustard gas	
	Mycotoxins	

both somatic cells and germ cells (**Table 3**). Somatic cell damage leads to cancers, immunological impairment, cell death and tissue damage. Germ cell damage gives rise to infertility, miscarriage, stillbirth and congenital malformations (**Figure 1**). It can even give rise to cancers in the children born to survivors, for example, the children of Vietnam veterans who were exposed to agent orange have increased risks of acute lymphocytic leukemia. The most extreme consequence of genotoxicity is death; this removes a whole genome from the gene pool and ensures that the individual no longer reproduces. (*See* DNA Damage Response;

Gene Structure and Organization; Genetic Disorders; Genetic Load; Infertility: Genetic Disorders.)

The Geneva Convention proscribes using weapons either against civilians, women and children, or for ethnic cleansing, but innumerable human genomes have been damaged by genotoxic WMD (**Table 3**). The ability of WMD to cause DNA and chromosomal damage was recognized as a consequence of the atomic bombs dropped on Hiroshima and Nagasaki. Radiation kills and impairs life quality by causing conditions such as cancer and heart failure 5–50 years after exposure (Schull, 1995). The Human Genome Project has conferred the ability to measure genomic damage directly. This should enable effective testing to be carried out, so that it should no longer be necessary to demand proof of where people were or to force them to have intimate medical examinations to try to assess their possible exposure. Dose–response curves show that risk is proportional to dose, making it important to develop effective ways of providing counseling, support and treatment for victims of exposure (**Table 3**). (*See* Gene–Environment Interaction; Infertility: Genetic Disorders; Path Analysis in Genetic Epidemiology; Tumor Formation: Number of Mutations Required.)

Agents, constituencies and deployment

In the First World War, chemical weapons were used to gain military advantage by killing, incapacitating and terrifying troops in the trenches, and in this respect mustard gas, phosgene and chorine were used to great effect. There were over 1 000 000 victims of mustard gas (a powerful carcinogenic agent) among the British, French, German, US and Russian troops. There were also biological threats, with typhus and other diseases being carried by ticks, rodents and sick troops. The long-term effects of these weapons were largely overlooked because there was no systematic follow-up of people who had been exposed to WMD (**Table 2**). Many thousands of people subsequently died or suffered from conditions such as lung cancer, pulmonary disease, heart failure or infertility (Patura and Rall, 1993). (*See* Gene–Environment Interaction; Infertility: Genetic Disorders; Path Analysis in Genetic Epidemiology; Tumor Formation: Number of Mutations Required.)

In the Second World War, there was large-scale manufacture of chemical weapons, such as mustard gas and nerve gases including sarin and tabun, but these were never used in battle, largely through the efforts of commanders who had experienced the horrors of these weapons in the First World War. There were, however, many civilian casualties of manufacture, distribution (the Bari Harbour disaster), testing and subsequent disposal in countries such as the United States, Great Britain, Germany and Japan. In addition, there were thousands of military 'volunteers' on whom these agents were tested. The civilian men and women who worked in the poison-gas factories in the United States, Great Britain, Germany and Japan during the Second World War became extremely ill as a result of their exposure and have been found to show an eightfold increase in head, neck and lung cancers, major pulmonary problems and other major disorders (Easton *et al.*, 1988; Azizi *et al.*, 1995). (*See* Gene–Environment Interaction; Genetic Counseling: Psychological Issues; Tumor Suppressor Genes.)

The military imperative to develop awesome weapons led to the advent of radiological weapons in the Second World War, culminating in deaths and over 750 000 survivors from the atomic bombs at Hiroshima and Nagasaki in 1945. Conflicts using different WMD have resulted in many casualties. These include victims of the 1980–1988 Iran–Iraq war, over 500 000, Kurds as victims of ethnic cleansing and of WMD used in the Anfal campaign of 1987–1988 (Middle East Watch, 1993), and 750 000 US troops fearful of the consequences of Agent Orange because of the risks of acute myeloid leukemia in the children of exposed veterans.

Immediate effects and long-term genotoxicity

Among the important facts to emerge from the atomic bomb casualties are the direct relationships between risks for cancer, leukemia, birth defects and medical disorders and the radiological dose. The main risk factor for cancers is dose; associated risks include increased susceptibility owing to factors such as age and genetic ability to repair DNA. For breast cancer, for example, the highest risks occur in individuals who are younger than 20 years of age at the time of exposure, suggesting that the developing breast is especially at high risk. (*See* Alcoholism: Collaborative Study on the Genetics of Alcoholism (COGA); Genetic Disorders; Genetic Load; Limb Development Anomalies: Genetics; Orofacial Clefting.)

Some individuals and families prove to be much more prone to cancer than others, consistent with our knowledge that certain inherited genes (such as genes for DNA repair and tumor suppressor genes) carry specific mutations and that variations in some families can result in increased susceptibility to cancer. (*See* Alcoholism: Collaborative Study on the Genetics of Alcoholism (COGA); Cancer Genome Anatomy Project; Environmental Mutagenesis; Tumor Formation: Number of Mutations Required; Tumor Suppressor Genes.)

Mechanisms and consequences of genomic damage

The assessment of effects of genotoxic WMD involves four different scientific approaches. The first approach uses dosimetry to estimate doses and risks. There are several ways of estimating doses and assessing the total chromosomal and DNA damage, including quantitative cytogenetic analysis (measuring the chromosomal breaks and characteristic radiation-induced rearrangements), DNA adducts and sister-chromatid exchanges (Black *et al.*, 1997). These exchanges provide significant information about recent exposures to mutagens but are of limited use for long-term assessments. The second approach uses detailed studies of the medical effects (cancers, birth defects, miscarriage, stillbirths and medical disorders) to establish risk figures. Assessment indicates that cancers, cataracts and heart failure are all dose-dependent. Reproductive consequences (miscarriage, stillbirth, birth defects and infertility) require careful approaches, particularly in cultures where infertility or abnormal babies are regarded as shameful. (*See* Development: Disorders; Disability, Human Rights and Contemporary Genetics; Genetic Disorders; Genetic Variation: Polymorphisms and Mutations; Informed Consent: Ethical and Legal Issues.)

The third approach uses environmental monitoring to establish which agents were used and to ensure no further significant exposures. The fourth approach, which uses advanced genome investigations, is of fundamental importance for future studies to identify the genomic damage induced by WMD, establish weapons use and enable the development of targeted individual therapeutic interventions. (*See* Microarrays and Expression Profiling in Cancer; Microarrays in Disease Diagnosis and Prognosis; Microarrays in Toxicological Research.)

Protection of vulnerable genomes in conflicts

To use WMD with the deliberate intent of causing infertility, damaging offspring or causing genetic damage that results in cancer contravenes the Geneva Conventions. Sadly, this has not protected civilian populations from attack. Despite international conventions designed to protect them, civilian populations have been increasingly targeted in conflicts. In addition to exposures resulting from conflicts, genotoxic damage has also resulted from weapons development, manufacture, testing and disposal as well as major accidents including Chernobyl, Seveso and Bhopal. Exposed populations may have to exist for weeks with contaminated environment, food and water.

In the hands of terrorists or rogue states with a ruthless drive to terrorize or to subdue, WMD can effect cruel destructiveness. Effective protection and countermeasures are dependent on knowledge about WMD, the ways in which they can be deployed, their effects on the body and effective medical treatment. The largest civilian population ever exposed to WMD – the Kurdish population in northern Iraq, which was attacked using WMD between 1987 and 1988 – is isolated, both politically and geographically. They suffered hundreds of attacks with weapons including mustard gas and nerve agents and need humanitarian and medical relief. The United Nations Investigation of the Iran–Iraq war proved that there was large-scale use of WMD including mustard and nerve agents. Given the scale of the attacks and the genotoxicity of the WMD used, it seems difficult to explain why this Kurdish population of 4 million people has not received significant help.

Susceptibility and Detoxification

The question of who dies and who survives exposures to WMD, both in the short term and the long term, is important. For example, who is at risk? Is it the elderly, the very young, pregnant women, their fetuses, or people with specific genotypes? There are important lessons from long-term survivors. People may be at risk because of their genetic vulnerability in detoxification and cancer susceptibility. Some families are known to be more cancer-prone than others because they carry mutations in genes that confer susceptibility to cancer. (*See* Cancer Cytogenetics; DNA Repair.)

In addition to genetic predispositions, there are other questions about susceptibility, survival and injury. Do activities such as sheltering in cellars, protecting the eyes, head or body with wet sheets, or undergoing immediate decontamination offer protection? Which age groups are most susceptible? Concerns about domestic preparedness indicate that these issues are important.

Susceptibility and detoxification are genetically determined

Animal experiments and studies of human populations have shown that there are genetic variations in the ability to detoxify WMD and to repair damaged DNA and in susceptibility to cancer at given doses of radiation, with subgroups of people showing specific susceptibilities to radiation and radiomimetic chemicals. Carriers of some of these genes show chromosomal instability and have high numbers of micronuclei, which indicate that even without exposure to damaging agents they have higher risks of malignancy. (*See* Cancer: Chromosomal Abnormalities.)

Our knowledge about the molecular basis of genes predisposing to cancer is expanding. We know of many cancer-related genes, including tumor suppressor genes such as *tumor protein p53 (Li–Fraumeni syndrome) (TP53)*, breast and ovarian cancer genes (*breast cancer, early onset 1 (BRCA1)* and *breast cancer, early onset 2 (BRCA2)*), the *retinoblastoma 1 (RB1)* gene and DNA repair genes associated with Fanconi anemia and xeroderma pigmentosa. Despite these advances, there are few studies of susceptibilities to cancer and birth defects after exposure to genotoxic WMD, which makes it important to undertake such studies as they will facilitate diagnosis, treatment, screening and approaches to prevention and chemoprevention. (*See* DNA Repair: Disorders;

Leukemias and Lymphomas: Genetics; Tumor Suppressor Genes.)

Genetic basis of detoxification and implications for victims

The Ayum Shienko attack on the Tokyo subway, in which the nerve gas sarin was used, killed and injured several hundred people. It also showed that, in addition to the subway victims, the first responders on site and medical staff at the receiving hospitals are also vulnerable. Sarin is colorless and odorless. Confronted by unconscious or severely ill victims, dedicated but unprepared medical attendants leant over the sick and, without protection of gas masks or protective clothing, breathed fumes of the nerve agent carried by the bodies and clothing of victims. Twenty per cent of the victims with long-term neurological effects are the medical and emergency response personnel who aided victims (Ohbu et al., 1997). Those people with severe neurological effects are those homozygous for the paraoxonase alleles that are least able to detoxify sarin.

There are several genetically determined systems responsible for the detoxification of WMD. For nerve agents and mustard gas, these include cytochrome P450 and paraoxonase. Paraoxonase activity occurs when it is coupled to albumin. Fetuses in utero and neonates lack albumin, having instead α-fetoprotein, the fetal form of the albumin. Fetuses and neonates lack paraoxonase and therefore cannot detoxify nerve agents including sarin and VX, making them susceptible to severe neurological damage. Paraoxonase is polymorphic for glycine and arginine at base 192. The Tokyo subway victims and Gulf war victims with severe long-term neurological damage have been shown to have paraoxonase alleles that confer an inability to detoxify nerve gases. The susceptibility of genetic subgroups to bioweapons is also genetically determined. For people with ABO blood groups, for example, those with blood group A are at a selective advantage and those with blood group O are at a selective disadvantage against challenges with plague bacilli (Yersinia pestis). (See Brain: Neurodevelopmental Genetics; Infectious Diseases: Predisposition; Infectomics: Study of Response to Infection using Microarrays; Twin Study Contributions to Understanding Ontogeny.)

Protection, Treatment and Aid for Survivors

The use of WMD to target vulnerable genomes in civilian and military populations will inevitably affect the population structures of exposed groups. In some cases, the primary purpose of using WMD may be ethnic cleansing. The gas chambers of the Nazi Holocaust decimated the previously large Jewish and Gypsy populations of Europe; 55 years after the Holocaust, European Jewish populations are fewer than 2 million in contrast to their prewar populations of nearly 10 million. In such circumstances, sterilization, infertility, infant deaths and increased cancers contribute not just to the immediate decline of a population, but also to its inability to recover. (See Eugenics: Contemporary Echoes; Nazi Movement and Eugenics; Roma (Gypsies): Genetic Studies.)

Effects on large populations: assessing genomic damage

Given that there were over 1 million victims of mustard gas in the First World War in German, French, British, US and Russian troops, it seems extraordinary that there was no long-term follow-up of the victims, despite the high risks of cancer and reproductive effects (Patura and Rall, 1993). Studies of exposed populations need to be sensitive to problems of assessing reproductive failure (infertility, fetal and infant deaths, severe malformations) and cancer risks. (See Genetic Disease: Prevalence; Infertility: Genetic Disorders; Patenting of Genes: Discoveries or Inventions?)

The use of WMD as a silent form of long-term genocide and ethnic cleansing is an ingenious, sophisticated and diabolical form of genetic terrorism. In northern Iraq, for example, people have been exposed to WMD either directly or through exposures from food, water and environment, but no systematic forensic testing has been undertaken by UN organizations, including the Organization for the Prohibition of Chemical Weapons (OPCW) and the UN Special Commission (UNSCOM), since 1987–1988 because UN organizations respect Government sovereignty and must await an invitation from the Government of Iraq.

Treating and aiding exposed populations and survivors

In conflict-torn areas where WMD are used, who should protect, treat and aid civilian victims? Emergency humanitarian aid organizations and nongovernmental organizations do admirable work in challenging and dangerous catastrophes and emergencies, but have neither the primary objective nor the resources for long-term treatment, research and follow-up. As studies of the atomic bomb exemplify, analyses of exposures and research on effects of WMD in conflict zones are demanding, time-consuming and expensive, and cover difficult and sensitive issues including reproductive failure. UN organizations (including OPCW and the World Health Organization) can only operate at the express invitation of sovereign governments. It is unlikely that governments using WMD against their own people or adjacent populations and in contravention of treaties would invite UN agencies to investigate their wrongdoing. Thus, despite the long and significant lists of people exposed to WMD (**Table 2**), there have been remarkably few detailed studies of long-term effects.

Conclusions

Rogue states and terrorist groups exploit WMD to put lives under threat. In many instances of alleged WMD use, detailed investigations have not been undertaken. Whether this is due to military or political expediency, simple neglect or the feeling that the task is too difficult or too challenging is unclear. To counteract threats, initiatives are needed to implement weapons treaties and to advance disarmament and conflict resolution. Victims of WMD attacks need help and reassurance that they will not be subjected to further experimentation. Medical treatment should be evidence-based and not harmful; however, there is the paradox that without research little is known about how to treat people exposed to WMD. It is analogous to the process of trying to trap a ghost: how do you design a ghost trap if you do not have a ghost on which to practice? Research on WMD faces several problems. Exposed populations are often displaced, dispersed or seeking asylum from torture and death in any country that allows them entry and therefore elude systematic follow-up. Deaths or serious health consequences of WMD are thus lost to follow-up in the flotsam of war and conflict, being diluted among the populations of host countries.

The Human Genome Project has facilitated studies of disease genes and potential therapeutic interventions. New microchip technology permits the identification of single base mutations and alterations in gene expression. Although this technology has not been used as yet to help populations exposed to WMD, it should enable genomic damage to be characterized and treatments, including advanced pharmacogenetics and gene therapy, to be developed for WMD victims. Information about genomic damage could be related to information about susceptibility (cancer predisposition, tumor suppressor and DNA repair genes) and could be used as the basis for providing cancer therapy and targeted treatments for victims specifically tailored to the genetic damage. (*See* DNA Damage Response; Gene–Environment Interaction; Genetic Disorders; Microarrays and Expression Profiling in Cancer.)

As beneficiaries of the knowledge from the Human Genome Project, we should be custodians to protect the genome from malevolent damage. If we apply our knowledge of the genome wisely, it might be possible to negate the damaging effects of WMD and overcome the threats to human populations by detecting genomic damage and developing specific therapies for damaged genomes. (*See* Gene Therapy.)

See also
Environmental Mutagenesis

References

Azizi F, Keshavarz A, Roshanzamir F and Nafarabadi M (1995) Reproductive function in men following exposure to chemical warfare with sulphur mustard. *Medicine and War* **11**: 34–44.

Black R, Clarke R, Harrison J and Read R (1997) Biological fate of sulphur mustard: identification of valine and histidine adducts in haemoglobin from casualties of sulphur mustard poisoning. *Xenobiotica* **27**: 499–512.

Easton D, Peto J and Doll R (1988) Cancers of the respiratory tract in mustard gas workers. *British Journal of Industrial Medicine* **45**: 652–659.

Ellenhorn MJ (1997) *Medical Toxicology: Diagnosis and Treatment of Human Poisoning*, 2nd edn. Baltimore, MD: Williams & Wilkins.

Hayes AW (1994) *Principals and Methods of Toxicology*, 3rd edn. New York: Raven Press.

Ohbu S, Yamashina A, Takasu N, *et al.* (1997) Sarin poisoning on Tokyo subway. *Southern Medical Journal* **90**: 587–593.

Patura CM and Rall DP (eds.) (1993) *Veterans at Risk: The Health Effects of Mustard Gas and Lewisite*. Washington, DC: Institute of Medicine, National Academy Press.

Schull WJ (1995) *Effects of Atomic Radiation: A Half Century of Studies from Hiroshima and Nagasaki*. New York: Wiley-Liss.

US Institute of Medicine (1996) *Veterans and Agent Orange: Update 1996. Committee to Review the Health Effects in Vietnam Veterans of Exposure to Herbicides*. Washington, DC: National Academy Press.

Further Reading

Emad A and Rezaian GR (1997) The diversity of the effects of sulfur mustard gas inhalation on respiratory system 10 years after a single, heavy exposure: analysis of 197 cases. *Chest* **112**: 734–738.

Furlong CE, Li WF, Brophy VH, *et al.* (2000) The *PON* gene and detoxification. *Neurotoxicology* **21**: 581–587.

Middle East Watch Report (1993) *Genocide in Iraq: The Anfal Campaign Against the Kurds*. New York: Human Rights Watch.

Pour-Jafari H (1994) Congenital malformations in the progenies of Iranian chemical victims. *Veterinary and Human Toxicology* **36**: 562–563.

Puga A and Wallace KB (eds.) (1998) *Molecular Biology of the Toxic Response*. London: Taylor & Francis.

Sasser L, Cushing J and Dacre J (1993) Dominant lethal study of sulfur mustard in male and female rats. *Journal of Applied Toxicology* **13**: 359–368.

Somani SM and Romano JA (eds.) (2001) *Chemical Warfare Agents: Toxicity at Low Levels*. Washington, DC: CRC Press.

Taher AA (1992) Cleft lip and cleft palate in Tehran. *Cleft Palate Craniofacial Journal* **29**: 15–16.

Web Links

Greenpeace.Evidence about Weapons testing
http://www.greenpeace.org/homepage
United Nations Special Commission on Iraq. Contains reports on Iraqi WMD, including mustard gas, sarin, VX, aflatoxin, botulinum toxin, anthrax and irradiated zirconium bombs
http://www.un.org/Depts/unscom/

21st Century complete guide to bioterrorism, biological and chemical weapons, germs and germ warfare, nuclear and radiation terrorism – military manuals and federal documents with practical emergency plans, protective measures, medical treatment and survival information
http://forum.nomi.med.navy.mil/resources.htm

Whole-genome Amplification

Dagan Wells, *University College London, London, UK*

Whole-genome amplification (WGA) techniques utilize the principles of the polymerase chain reaction (PCR) to copy enzymatically the deoxyribonucleic acid (DNA) contained within a sample. However, unlike most PCR applications, the amplification is nonspecific and theoretically provides an increase in the copy number of all genomic sequences. WGA is used in areas of research where minute DNA samples are routinely encountered. Applications include preimplantation genetic diagnosis, the creation of probes from microdissected or flow-sorted chromosomes, forensics and the study of ancient DNA samples.

Advanced article

Article contents

- Introduction
- Primer Extension Preamplification
- Degenerate Oligonucleotide Primed PCR
- Tagged-PCR
- Linker Adapter PCR
- Alu-PCR
- Pitfalls and Problems Using WGA
- Future Applications

Introduction

In some fields of genetic research, extremely small deoxyribonucleic acid (DNA) samples are routinely encountered. In such cases relatively few tests can be conducted before the DNA sample is expended, restricting the number of investigations that can be performed. Furthermore, any analytical techniques that require large quantities of DNA cannot be applied. Minute DNA samples may be obtained from microdissected pieces of tumor, from forensic or archaeological specimens, and, in the most extreme cases, may consist of a single cell. This last category presents the greatest challenge for genetic analysis and is encountered during preimplantation genetic diagnosis and also in certain prenatal tests, which involve the sampling of fetal cells from the maternal bloodstream. Additionally, irreplaceable DNA samples, which have been obtained from deceased individuals or from small biopsies taken during surgery, can also prove to be insufficient for extensive investigation. A solution to these limitations is to perform whole-genome amplification (WGA). WGA techniques employ the principles of the polymerase chain reaction (PCR) to copy enzymatically the DNA contained within a sample. However, unlike most PCR applications, the amplification is not targeted at a specific fragment of DNA, rather the amplification is nonspecific and, theoretically, provides an increase in the copy number of all genomic sequences. Aliquots of the WGA reaction can be taken and used for subsequent amplification of specific DNA fragments. Performing WGA generates a resource of DNA that can be stored or subjected to additional genetic testing. Several WGA methods have been described and a short discussion of these follows. (*See* Ancient DNA: Recovery and Analysis; Fetal Diagnosis; Polymerase Chain Reaction (PCR); Preimplantation Diagnosis; Preimplantation Genetic Diagnosis: Ethical Aspects.)

Primer Extension Preamplification

The WGA method that has most often been applied to the analysis of single cells is primer extension preamplification (PEP) (Zhang *et al.*, 1992). PEP has been used for single cell analysis of diseases such as Tay–Sachs, cystic fibrosis, hemophilia A and Duchenne muscular dystrophy as well as for sex determination (Snabes *et al.*, 1994), and is the only WGA technique to have been applied clinically to preimplantation genetic diagnosis (the detection of genetic disorders in single cells taken at biopsy from a human embryo) (Ao *et al.*, 1998). (*See* Preimplantation Diagnosis.)

The reagents used for PEP are essentially identical to those found in routine PCR experiments, with the exception of the primers. Instead of a pair of oligo-nucleotides of defined base sequence, specific to a single locus, a heterogenous mixture of degenerate oligo-nucleotides is used. In most cases the random primer mixture is composed of oligonucleotides 15 nucleotides in length. Cycling conditions involve denaturation followed by a temperature ramp from 37° to 72°C, during which primers can anneal throughout the genome; finally a few minutes at 72°C are allowed to ensure that large fragments are amplified. Between 40 and 50 cycles of this type are usually employed.

The use of 'forward' and 'reverse' (or 'sense' and 'antisense') primers in traditional PCR procedures provides exponential amplification by allowing ampli-fication to be initiated at each end of a DNA fragment. PEP primers, on the other hand, anneal at random and do not necessarily initiate DNA synthesis from the ends of fragments. As a result the increase in DNA afforded by PEP occurs in a linear fashion and is relatively modest. It has been predicted that PEP grants at least a 30-fold increase of 70–90% of genomic sequences if applied to a single cell (Zhang et al., 1992; Wells et al., 1999).

Degenerate Oligonucleotide Primed PCR

One of the most widely applied forms of WGA is degenerate oligonucleotide primed PCR (DOP-PCR) (Telenius et al., 1992). This method utilizes oligonu-cleotide primers with specified 5′ and 3′ ends separated by a short stretch of random bases. A number of PCR cycles conducted at low annealing temperatures allow the semidegenerate primers to anneal at numerous sites genome wide. Later, a change in reaction conditions allows the products of earlier cycles to be amplified in an exponential fashion (**Figure 1**). Consequently DOP-PCR generates a much larger quantity of DNA than PEP. It has been shown that DOP-PCR performed on a single cell can provide enough DNA for over 90 subsequent PCR amplifications (Wells et al., 1999).

Although, in the vast majority of cases, DOP-PCR has been used for the generation of DNA probes, usually chromosome paints made from microdissected or flow-sorted chromosomes, there is potential for this technique to be used in other molecular genetic applications. Experimental data suggest that >85% of genomic sequences can be amplified from single-cell DOP-PCR products and this figure approaches 100% for larger tissue samples (Wells et al., 1999). (**See** Chromosome Analysis and Identification; Flow-sorted Chromosomes; Fluorescence *In Situ* Hybridization (FISH) Techniques.)

Not only do minute tissue samples constrain the number of genetic loci that can be analyzed, they also place limits on the amount of cytogenetic data that can be obtained. Problems arise because techniques for the generation of metaphase chromosomes are inefficient when applied to small samples and require living cells. Fluorescence *in situ* hybridization (FISH) analysis has been applied to interphase nuclei and has allowed the analysis of several chromosomes per single cell, however technical limitations prevent a more extensive analysis. An alternative is to employ comparative genomic hybridization (CGH), which, in a single experiment, provides data regarding the copy number of every chromosomal region greater than 10 Mb. The CGH procedure requires ~100 ng of DNA, so for it to be performed on a single cell, the DNA must therefore undergo a 10 000-fold amplification. This has been achieved using DOP-PCR. Thus, following DOP-PCR, both cytogenetic and molecular genetic analyses can be performed on the same isolated cell (Wells et al., 1999). (**See** Chromosome Analysis and Identification; Comparative Genomic Hybridization in the Study of Human Disease; Fluorescence *In Situ* Hybridization (FISH) Techniques; Preimplantation Diagnosis.)

Tagged-PCR

Essentially a variant of DOP-PCR, tagged-PCR (T-PCR) achieves WGA by employing a mixture of primers that contain a specified 5′ sequence ('tag') followed by several random nucleotides (Grothues et al., 1993). In some cases, the last few bases at the 3′ end are also specified, such that the random sequence is flanked by defined sequences and thus resembles the primers used for DOP-PCR. Fragments that have sequences complementary to the tag at both ends are generated after several PCR cycles at low annealing temperatures (**Figure 1a–g**). Excess degenerate primers are then removed and a second amplification reaction is initiated using a single primer of identical sequence to the tag. This provides the opportunity for expo-nential amplification of all DNA fragments containing tag sequences at each end. T-PCR has been success-fully applied to the generation of DNA probes from microdissected or flow-sorted chromosomes.

Linker Adapter PCR

In contrast to the methods discussed above, linker adapter PCR (LA-PCR) does not employ degenerate primers. Instead the sample DNA is digested using a frequent cutting restriction endonuclease (e.g. *Sau*3AI, a four base cutter). This produces a large number of fragments of varying sizes. The restriction enzyme cleaves the DNA in an asymmetric fashion producing

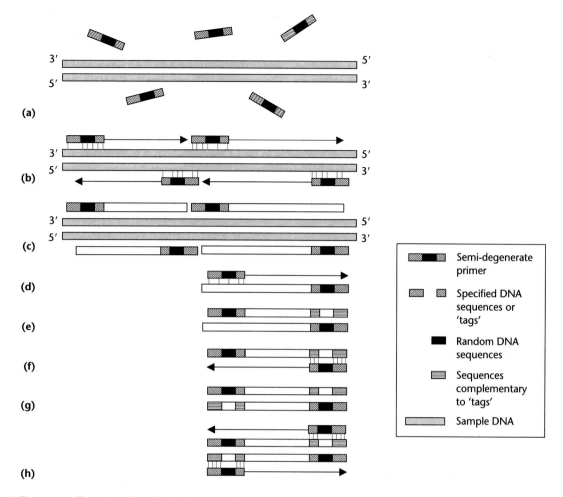

Figure 1 Degenerate oligonucleotide primed polymerase chain reaction (DOP-PCR). (a) Sample deoxyribonucleic acid (DNA) is added to a reaction mixture containing a semidegenerate primer. The primer consists of two specified 'tag' sequences separated by several random nucleotides. (b) Low annealing temperatures allow the primers to anneal at many sites throughout the genome. (c) Initiation of DNA synthesis from these sites permits the entire genome to be copied. (d) The fragments generated in this way can also serve as templates during later cycles. (e–g) After several rounds of amplification most fragments contain regions complementary to the tag sequences at both ends. Next the PCR annealing temperatures are increased. At these higher temperatures primers can only successfully anneal to sequences complementary to their tags. (h) Fragments that contain these sequences at each end are amplified in a highly efficient exponential fashion.

fragments that have single-stranded DNA overhangs at both ends. An adapter oligonucleotide that will specifically anneal to the single-stranded overhangs can then be ligated, producing fragments composed of genomic DNA flanked by adapters of defined sequence (**Figure 2**). After removal of surplus adapter oligonucleotides, a primer, complementary to the adapter, is used to amplify all ligated fragments. DNA synthesis is initiated from adapters at both ends of each fragment and, consequently, there is an exponential increase of amplified DNA. LA-PCR has been successfully applied to the generation of chromosome-specific libraries and paint probes (Miyashita et al., 1994). The average fragment size produced by LA-PCR can be adjusted by using restriction enzymes that cut more or less frequently (shorter or longer recognition site). The choice of restriction enzyme will

also influence the efficiency with which different genomic regions are amplified. For example, an enzyme with an A : T rich recognition sequence will cleave G : C rich regions of the genome less frequently, leading to large fragments that are amplified less efficiently or not at all.

Alu-PCR

Unlike other methods of whole-genome amplification, Alu-PCR targets specific genomic sequences for amplification. This is achieved by using primers of complementary sequence to Alu repeats. Alu repeats are DNA elements composed of a core sequence ~280 bp in length, flanked by short (16–18 bp) direct repeats. Approximately 900 000 Alu repeats are

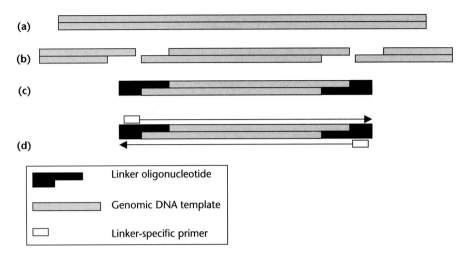

Figure 2 Linker adapter polymerase chain reaction (LA-PCR). (a) Genomic deoxyribonucleic acid (DNA) template. (b) Template is digested with a restriction enzyme. (c) Adapter oligonucleotides are ligated and (d) serve as sites for the annealing of PCR primers.

interspersed throughout the human genome and thus, by initiating DNA synthesis from each of these points, an amplification of the entire genome is achieved (Brooks-Wilson *et al.*, 1990; Lichter *et al.*, 1990). Furthermore, many of the fragments generated will extend from one Alu repeat to another. The second repeat can also serve as a site for primer annealing, and, if it is in the correct orientation, the intervening DNA can be amplified from both ends of the fragment. The exponential amplification that this provides is highly efficient, leading to a dramatic increase in fragments amplified in this way.

Alu-PCR has been effectively employed for the generation of probes from cloned DNA fragments and flow-sorted chromosomes. An advantage of this approach is that Alu repeats are particular to humans and consequently only human DNA is amplified. This feature of Alu-PCR has allowed chromosome-specific probes to be produced from somatic cell hybrids that contain single human chromosomes (or chromosome fragments) on a rodent background (Lichter *et al.*, 1990) and it has enabled amplification of human DNA prior to cloning.

It should be noted that Alu repeats are not evenly distributed in the human genome and, consequently, areas that are Alu dense are amplified with greater efficiency than others. This may be particularly significant when amplifying extremely small DNA samples or single cells (Wells *et al.*, 1999).

Pitfalls and Problems Using WGA

Clearly, any PCR-based technique carries with it an inherent risk of contamination by extraneous DNA. For very low concentrations of DNA, this problem is exacerbated by the large number of amplification cycles that are required. Consequently, for many WGA applications, significant precautions to avoid contamination must be undertaken (Wells and Sherlock, 1998).

Direct amplification of DNA from single cells is also associated with problems related to preferential amplification. Overrepresentation of one of the two alleles in a heterozygous cell is frequently observed and interferes with attempts to employ quantitative PCR. Preferential amplification in its most extreme form manifests as the phenomenon known as allele dropout (Wells and Sherlock, 1998, and references therein). In this case only one of the two alleles is amplified, causing the cell to appear homozygous. This phenomenon, which affects approximately 10% of amplifications, has caused significant problems for single-cell testing (e.g. preimplantation genetic diagnosis). Allele dropout occurs at similar frequencies, regardless of whether a locus is amplified directly from a single cell or from a single-cell WGA product. (*See* Preimplantation Diagnosis.)

As is the case with most PCR applications, WGA amplifies short DNA sequences more efficiently than long sequences. However, for WGA techniques that involve the use of degenerate primers, this effect is somewhat more significant, because during each random-priming cycle existing fragments are subject to internal annealing of primers. Despite the reduction in average size that results, fragments over 2 kb in length can still be amplified from WGA products.

The accuracy of amplification using WGA techniques is generally very good. DNA sequencing of genes amplified by WGA has revealed that unique sequences are amplified with high fidelity. However, the amplification of highly repetitive stretches of DNA, such as the microsatellites, can produce artifacts. These errors take the form of DNA fragments of unexpected length, caused by insertion or deletion of a number of repeat units. This type of artifact is not seen when microsatellites are amplified directly from single cells and thus it seems to be a peculiarity of WGA. However, the artifact has only been reported in extremely small DNA samples (< 10 cells) and does not seem to affect WGA performed on larger quantities of DNA.

Future Applications

Areas of research in which small DNA samples are encountered are becoming increasingly common and, consequently, the use of WGA is likely to increase. It is likely that WGA techniques applied to DNA or complementary DNA (cDNA) will be an essential part of attempts to apply emerging micro-array technology to small cell numbers.

See also

Ancient DNA: Recovery and Analysis
Polymerase Chain Reaction (PCR)
Polymerase Chain Reaction (PCR): Design and Optimization of Reactions
Preimplantation Diagnosis

References

Ao A, Wells D, Handyside AH, Winston RM and Delhanty JDA (1998) Preimplantation genetic diagnosis of inherited cancer: familial adenomatous polyposis coli. *Journal of Assisted Reproduction and Genetics* **15**(3): 140–144.

Brooks-Wilson AR, Goodfellow PN, Povey S, *et al.* (1990) Rapid cloning and characterization of new chromosome 10 DNA markers by Alu element-mediated PCR. *Genomics* **7**(4): 614–620.

Grothues D, Cantor CR and Smith CL (1993) PCR amplification of megabase DNA with tagged random primers (T-PCR). *Nucleic Acids Research* **11**(5): 1321–1322.

Lichter P, Ledbetter SA, Ledbetter DH and Ward DC (1990) Fluorescence *in situ* hybridization with Alu and L1 polymerase chain reaction probes for rapid characterization of human chromosomes in hybrid cell lines. *Proceedings of the National Academy of Sciences of the United States of America* **87**(17): 6634–6638.

Miyashita K, Vooijs MA, Tucker JD, *et al.* (1994) A mouse chromosome 11 library generated from sorted chromosomes using linker-adapter polymerase chain reaction. *Cytogenetics and Cell Genetics* **66**(1): 54–57.

Snabes MC, Chong SS, Subramanian SB, *et al.* (1994) Preimplantation single-cell analysis of multiple genetic loci by whole-genome amplification. *Proceedings of the National Academy of Sciences of the United States of America* **91**(13): 6181–6185.

Telenius H, Carter NP, Bebb CE, *et al.* (1992) Degenerate oligonucleotide-primed PCR: general amplification of target DNA by a single degenerate primer. *Genomics* **13**(3): 718–725.

Wells D and Sherlock JK (1998) Strategies for pre-implantation genetic diagnosis of single gene disorders by DNA amplification. *Prenatal Diagnosis* **18**(13): 1389–1401.

Wells D, Sherlock JK, Handyside AH and Delhanty JDA (1999) Detailed chromosomal and molecular genetic analysis of single cells by whole genome amplification and comparative genomic hybridization (CGH). *Nucleic Acids Research* **27**(4): 1214–1218.

Zhang L, Cui X, Schmitt K, *et al.* (1992) Whole genome amplification from a single cell: implications for genetic analyses. *Proceedings of the National Academy of Sciences of the United States of America* **89**: 5847–5851.

Further Reading

Cotter FE, Das S, Douek E, Carter NP and Young BD (1991) The generation of DNA probes to chromosome 11q23 by Alu PCR on small numbers of flow-sorted 22q-derivative chromosomes. *Genomics* **9**(3): 473–480.

Kristjansson K, Chong SS, Van den Veyver IB, *et al.* (1994) Preimplantation single cell analyses of dystrophin gene deletions using whole genome amplification. *Nature Genetics* **6**(1): 19–23.

Lengauer C, Green ED and Cremer T (1992) Fluorescence *in situ* hybridization of YAC clones after Alu-PCR amplification. *Genomics* **13**(3): 826–828.

Mao YW, Liang SY, Song WQ and Li XL (1998) Construction of a DNA library from chromosome 4 of rice (*Oryza sativa*) by microdissection. *Cell Research* **8**(4): 285–293.

Montgomery KD, Tedford KL and McDougall JK (1995) Genetic instability of chromosome 3 in HPV-immortalized and tumorigenic human keratinocytes. *Genes Chromosomes and Cancer* **14**(2): 97–105.

Telenius H, Pelmear AH, Tunnacliffe A, *et al.* (1992) Cytogenetic analysis by chromosome painting using DOP-PCR amplified flow-sorted chromosomes. *Genes Chromosomes and Cancer* **4**(3): 257–263.

Voullaire L, Slater H, Williamson R and Wilton L (2000) Chromosome analysis of blastomeres from human embryos by using comparative genomic hybridization. *Human Genetics* **106**(2): 210–217.

Weier HU, Polikoff D, Fawcett JJ, *et al.* (1994) Generation of five high-complexity painting probe libraries from flow-sorted mouse chromosomes. *Genomics* **21**(3): 641–644.

Wells D and Delhanty JDA (2000) Comprehensive chromosomal analysis of human pre-implantation embryos using whole genome amplification and single cell comparative genomic hybridization. *Molecular Human Reproduction* **6**(11): 1055–1062.

Wells D and Delhanty JDA (2001) Preimplantation genetic diagnosis: applications for molecular medicine. *Trends in Molecular Medicine* **7**: 23–30.

Wilkins, Maurice Hugh Frederick

Watson Fuller, *Keele University, Keele, UK*

Introductory article

Maurice Hugh Frederick Wilkins (1916–) is a British biophysicist (born in New Zealand) noted for his work on DNA.

The discovery of the double-helical structure of deoxyribonucleic acid (DNA) in 1953 is widely regarded as the most significant discovery in biology of the twentieth century. The 1962 Nobel Prize for Physiology or Medicine was awarded to Francis Crick and James Watson, who proposed the model, and Maurice Wilkins who, together with Rosalind Franklin and colleagues at King's College, London, had provided the X-ray diffraction data on which the model was based. The claim by Oswald Avery in 1944 that DNA carried the genetic information had been received with scepticism. It is a mark of Wilkins's insight that by 1950 he was persuaded that not only was DNA the genetic material, but that X-ray fiber diffraction offered the most promising way of obtaining clues as to how it functioned, by allowing the determination of its three-dimensional structure. As early as 1950, Wilkins assembled a parallel array of uniform thin fibers (each a fraction of a millimeter in diameter) drawn from a DNA gel. The diffraction pattern from this specimen indicated a high degree of regularity, demonstrating, in Wilkins's words, that 'now it was really obvious – genes had a crystalline structure' (**Figure 1**). This regularity in the coding of the genetic information in chemical structure is a crucial feature of the double helix and is central to the biochemical processes that have evolved for copying and translating this information into the structure of the proteins which give organisms their distinguishing characteristics. The fact that these processes are independent of the particular message being read, with essentially the same mechanisms for processing information being employed in all organisms, is inextricably linked to an understanding of Darwinian evolution. (*See* Brenner, Sydney; Chromosome Analysis and Identification; Crick, Francis; DNA Repair; Watson, James Dewey.)

Maurice Wilkins was born in 1916 in Pongaroa, New Zealand, to where his father had recently moved from Dublin to work as a doctor. His Anglo-Irish family had engaged, over many generations, in a wide range of activities, which can be seen as stimulating his enthusiasm for science and engineering, and a wide range of social and political concerns. Back in England his father made him a workshop where, as a schoolboy, he developed skills in designing and making equipment, particularly telescopes. At Cambridge, where he studied natural sciences, he was much

Figure 1 Maurice Wilkins adjusting an X-ray camera to record the diffraction pattern from a fiber of DNA (about 1953).

influenced by the Cambridge Scientists Anti-War Group and by J. D. Bernal's use of X-ray diffraction to study the structure of proteins and viruses. He obtained his PhD at Birmingham, working with J. T. Randall, for pioneering studies on luminescence of solids. This was followed by work on the separation of uranium isotopes with the Manhattan Project at the University of California.

After the Second World War, Wilkins decided that his future research would be in biology and he rejoined Randall, first at St Andrews and then at King's College, London, where, within the Physics Department, a unique Biophysics Research Unit employing biologists, physicists and chemists was established with major Medical Research Council support. In addition to his work on DNA and the more general development of teaching and research in biophysics, Maurice Wilkins maintained his early concerns on the wider impact of science. He played a major role in establishing the British Society for Social Responsibility in Science (President, 1969–1991) and pioneered undergraduate courses on the social impact of science, characterized by the same intellectual openness and rigor displayed in his scientific research.

See also

Avery, Oswald Theodore
Crick, Francis
DNA Structure
Franklin, Rosalind Elsie
Watson, James Dewey

Further Reading

Wilkins MHF (1964) The molecular configuration of nucleic acids (Nobel Lecture, December 11, 1962) *Nobel Laureates in Physiology or Medicine 1942–1962*, pp. 754–782. Amsterdam: Elsevier.

Olby R (1974) *The Path to the Double Helix*. London: Macmillan.

Maddox B (2002) *Rosalind Franklin: The Dark Lady of DNA*. London: HarperCollins.

Williams Syndrome: A Neurogenetic Model of Human Behavior

Julie R Korenberg, *University of California, Los Angeles, California, USA*

Ursula Bellugi, *The Salk Institute for Biological Studies, La Jolla, California, USA*

Lora S Salandanan, *University of California, Los Angeles, California, USA*

Debra L Mills, *Emory University, Atlanta, Georgia, USA*

Allan L Reiss, *Stanford University School of Medicine, Stanford, California, USA*

Introductory article

Article contents

- Introduction
- Clinical Features
- Molecular-genetic Profile
- Cognitive Profile
- Neurophysiological Profile
- Neuroanatomical Characteristics
- Development of a Phenotypic Map of WMS Cognition

Williams syndrome (WMS) is a rare neurogenetic disorder, usually caused by a deletion at 7q11.23. Integrating the genetic with the distinctive clinical, cognitive, neurophysiological and neuroanatomical profiles of WMS provides an opportunity to understand the molecular basis of human cognition and behavior.

Introduction

Even when times are bad we still try to shine a light upon other people, and to give that sense of glow to our friends... You have to accept things you cannot change. (Twin 1 with Williams syndrome)

First of all I am a human being, and I'm a man... And I want people to realize that I do have feelings, I'm not a freak... So respectability is a main factor in learning how to deal with people with Williams syndrome or any other syndrome. (Twin 2 with Williams syndrome)

These are quotes from patients who knew the impact that the diagnosis of Williams syndrome (WMS) had on their lives. However, researchers and doctors have only recently begun to realize the potential neurogenetic impact of WMS, as was first hinted at in 1952 by Fanconi *et al.*, who noticed patients with hypercalcemia and 'elfin' facial features. Later, in 1961, Williams, Barratt-Boyes and Lowe independently reported clinical findings of supravalvular stenosis and the same elfin facial features in patients. For a time, it was considered a genetic disorder of unknown

origin, until 1993, when it was discovered that WMS is caused by the deletion of one copy of a small set of genes on chromosome 7 (7q11.23).

WMS is a special syndrome for many reasons. It is a rare and complex genetic disorder, occurring in one in 20 000 live births. Due to a deletion in chromosome band 7q11.23, the patients have cardiovascular, connective tissue and neurodevelopmental deficits **(Table 1)**. In addition, WMS patients have an unusual cognitive profile, with strengths and weaknesses in cognitive abilities, such as engaging language ability being masked by low IQ. Using WMS as a model, one of the greatest challenges faced in understanding the relationships between genetics, brain anatomy and cognition is the need to link investigations across disciplines within the neurosciences. Here we present studies on WMS that reveal distinctive profiles of clinical, genetic, cognitive, neurophysiological and neuroanatomical characteristics, which allow us to begin to infer anatomical and physiological interconnections and to understand the molecular basis of human behavior.

Table 1 Summary of major phenotypic features of Williams syndrome

Neurological	Cardiovascular	Genitourinary	Neurocognitive	Musculoskeletal	Endocrine	Facies
Average IQ 55 (range 40–80)	Supravalvular aortic stenosis	Nephrocalcinosis	Friendly, loquacious personality	Joint limitations	Transient infantile hypercalcemia	Medial eyebrow flare
Mild neurological dysfunction	Pulmonic valvular stenosis	Small, solitary and/or pelvic kidneys	Enhanced musical ability	Kyphoscoliosis		Short nasal palpebral fissures; epicanthal folds
• Tight heel cords	Peripheral pulmonary artery stenosis	Vesicoureteral reflux	General anxiety	Hallux valgus		Flat nasal bridge, small upturned nose
• Poor coordination	Ventricular or atrial septal defects		Delayed but relatively spared language development, enhanced vocabulary and social use of language versus significant visual–spatial processing	Hypoplastic nails		Stellate iris
Hyperacusis	Defects of arterial walls			Adult stature slightly smaller than average		Long philtrum
Harsh, brassy or hoarse voice						Prominent lips with open mouth
						Puffiness around the eyes
						Small, widely spaced teeth

NATURE ENCYCLOPEDIA OF THE HUMAN GENOME / ©2003 Macmillan Publishers Ltd, Nature Publishing Group / www.ehgonline.net

Clinical Features

Diagnostic characteristics of WMS embody a set of specific physical and facial features that include a constellation of cardiovascular difficulties such as supravalvular aortic stenosis (SVAS), a narrowing of the aorta; failure to thrive in infancy; transient neonatal hypercalcemia; delayed language and motor milestones; and abnormal sensitivities to classes of sound (hyperacusis). The facial features of WMS patients are quite distinctive and have been described as 'pixie-like' and 'elfin'. Persons with WMS often look more like each other than like their own family members. **Table 1** lists other physical and cognitive features of WMS.

Molecular-genetic Profile

What are the genetic origins of the observed features in WMS? It is caused by a microdeletion on chromosome 7 that involves the gene encoding elastin (ELN) and approximately 20 other genes, some of which are described in **Table 2**. To investigate the size and extent of deletions in individuals with WMS, a physical map of the deleted region of chromosome 7 band 7q11.23 has been constructed, using multicolor fluorescence *in situ* hybridization (FISH) of bacterial artificial chromosomes (BACs) on metaphase and interphase chromosomes, large-fragment library screening, genomic Southern blot and pulse-field gel analyses, sequence tagged site (STS) and polymorphic-marker analyses.

A working model of the genome organization was developed to characterize chromosome band 7q11.2, and it suggests that this region includes highly homologous chromosomal duplications that are also characterized by a number of repeat-sequence families, genes and pseudogenes. The overall region is comprised of inner and outer duplicated regions organized as nested repeated structures that surround the largely unique region occupied by elastin and the

Table 2 Common deletions in genes in subjects with Williams syndrome

Symbol	Gene name	Possible functions
FKBP6	FK506 binding protein 6	Structural homolog to FKBP immunophilins, cellular receptors for the immunosuppressive drugs FK506 and rapamycin
FZD9	Frizzled homolog 9 (*Drosophila*)	Transmembrane receptor for WNT. Highly expressed in all areas of the brain, eye, testis, skeletal muscle and kidney
BAZ1B	bromodomain adjacent to zinc-finger domain, 1B	Structural motifs involved in chromatin-dependent regulation of transcription
BCL7B	B-cell CLL/lymphoma 7B	Gene encodes serine-rich protein product, unknown function
TBL2	Transducin (beta)-like 2	Member of beta transducin protein family, with four putative WD 40 repeats, high expression in heart, brain, placenta, skeletal muscle and pancreas
WBSCR14	Williams–Beuren syndrome chromosome region 14	Member of the family of transcription factors, with nuclear localization signal
STX1A	Syntaxin 1A (brain)	Involved in neurotransmitter release from synaptic vesicles
CLDN4	Claudin 4	Functional high-affinity receptor for CPE. Specific expression in lung, kidney and intestine. Absence in brain and spleen
CLDN3	Claudin 3	Putative apoptosis-associated gene. Specific expression in lung, kidney and intestine. Low levels in brain, heart, muscle and spleen
ELN	Elastin (supravalvular aortic stenosis, Williams–Beuren syndrome)	Major structural protein in the development of arterial walls and connective tissue. Mutation causes SVAS
LIMK1	LIM domain kinase 1	Plays a signaling role in actin reorganization and cell movement
WBSCR1	Williams–Beuren syndrome chromosome region 1	A positive regulator of protein synthesis during translation initiation
WBSCR5	Williams–Beuren syndrome chromosome region 5	Unknown function
RFC2	Replication factor C (activator 1) 2 (40 kDa)	A subunit of a five protein complex involved in DNA strand elongation during replication
CYLN2	Cytoplasmic linker 2	Mediates interaction between specific organelles to microtubules
GTF2IRD1	GTF2I repeat domain containing 1	Set of five GTF2I-like repeats that possess the ability to interact with other HLH proteins and function as a transcription factor
GTF2I	General transcription factor II, i	Plays a role in transcription and signal transduction, a potential phosphorylation target

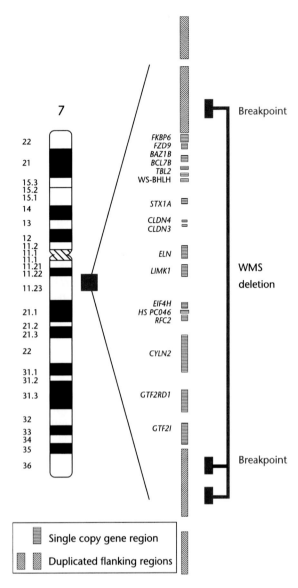

7

22	FKBP6
21	FZD9
	BAZ1B
	BCL7B
15.3	TBL2
15.2	WS-BHLH
15.1	
14	STX1A
13	CLDN4
	CLDN3
12	
11.2	
11.1	ELN
11.1	
11.21	LIMK1
11.22	
11.23	
	EIF4H
21.1	HS PC046
21.2	RFC2
21.3	
22	CYLN2
31.1	
31.2	
31.3	GTF2RD1
32	
33	GTF2I
34	
35	
36	

Breakpoint

WMS deletion

Breakpoint

▦ Single copy gene region

▨ ▩ Duplicated flanking regions

Figure 1 Region of chromosome 7, band 7q11.23, that is commonly deleted in Williams syndrome (WMS) is represented by the solid square. This region is expanded to the right to illustrate its genomic organization, a region of largely single copy genes flanked by a series of genomic duplications.

other deleted genes (**Figure 1**). This implies that the region of deoxyribonucleic acid (DNA) deleted in WMS individuals is located within an apparently single copy region of chromosome 7 that appears to be surrounded by a series of genomic duplications, some of which must be recent and others which have been duplicated earlier in primate evolution. The outer duplicated regions appear asymmetrically located with respect to the gene *ELN* (*elastin* (*supravalvular aortic stenosis, Williams–Beuren syndrome*)).

The common deletion breakpoints in WMS are located in the inner duplicated regions, clustered at one centromeric and two telomeric sites on chromosome 7.

Why do deletions occur? It is important to note that the causes of deletions involve a number of factors that may predispose to genomic instability. BAC and marker analyses have illustrated localization of the breakpoints, which suggest that particular sequences (i.e. frequency of Alu elements in the region, high degree of sequence conservation in the duplicated regions (99% identical), inversions) or chromatin structures may also contribute to regions of chromosome band borders that may themselves be unstable and may promote the tendency to delete. In any event, it is likely that the meiotic mispairing of subsets of the numerous repeated families, combined with crossing over, results in a series of different chromosomal aberrations of the deletion/duplication types. Because the deletion breakpoints in WMS occur largely in common regions, most individuals with WMS have the same genes deleted. In the following sections, unusual dissociations in higher cortical functioning and brain anatomy in WMS provide opportunities to explore some of the central issues of cognitive neuroscience that tie cognitive functions to brain organization and, ultimately, to the human genome.

Cognitive Profile

General cognitive functioning

From studies across different populations, a characteristic WMS cognitive profile is emerging. Although characterized with general anxiety, individuals with WMS have an excessively social personality and an increased appreciation of music. Moreover, WMS patients display characteristic 'peaks' and 'valleys' in specific cognitive abilities. Across an array of standardized conceptual and problem-solving tasks (some verbal, some nonverbal), WMS patients demonstrate a consistent, serious impairment in general cognitive functioning. In general cognitive tasks such as IQ probes, most individuals with WMS rank in the 'mild-to-moderate mentally retarded' range, with global standard scores on IQ test ranging from 40 to 80, with a mean of 55. A hallmark of WMS is the dissociation between language (which is a strength in adolescents and adults) and spatial cognition at global levels (which is profoundly impaired), as shown in **Figure 2**. Complex expressive language abilities and face-processing abilities are remarkably strong.

Expressive language abilities and linguistic affect

One striking aspect of WMS is the strong language ability in adolescents and adults, in contrast to the

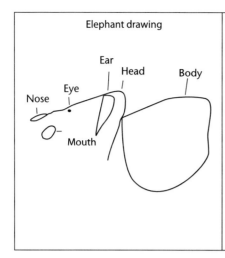

Elephant drawing	Elephant description
	And what an elephant, it is one of the animals. And what the elephant does, it lives in the jungle. It can also live in the zoo. And what it has, it has long gray ears, fan ears, ears that can blow in the wind. It has a long trunk that can pick up grass, or pick up hay… If they're in a bad mood it can be terrible… if the elephant gets mad it could stamp; it could charge. Sometimes elephants can charge. They have big long tusks. They can damage a car… It could be dangerous. When they're in a pinch. when they're in a bad mood it can be terrible. You don't want an elephant as a pet. You want a cat or a dog or a bird…

Figure 2 Contrasts of drawing and description of an elephant by a teenager with WMS. The dissociation between language and spatial cognition in WMS is evident (full-scale IQ of 49, verbal IQ of 52 and performance IQ of 54).

WMS, age 17, IQ 50:
Once upon a time when it was dark at night… the boy had a frog. The boy was looking at the frog… sitting on the chair, on the table, and the dog was looking through… looking up to the frog in a jar. That night he slept and slept for a long time, the dog did. But, the frog was not gonna go to sleep. The frog went out from the jar. And when the frog went out… the boy and the dog were still sleeping. Next morning it was beautiful in the morning. It was bright and the sun was nice and warm. Then suddenly when he opened his eyes… he looked at the jar and then suddenly the frog was not there. The jar was empty. There was no frog to be found.

DNS, age 18, IQ 55:
The frog is in the jar. The jar is on the floor. The jar on the floor. That's it. The stool is broke. The clothes is laying there.

Figure 3 WMS individuals use affective devices in storytelling, compared with subjects with Down syndrome (DNS). Examples from narratives of the *Frog, Where Are You?* story show the excessive use of narrative evaluative devices in adolescents with WMS.

overall impairment seen in cognitive abilities. In the earliest stages of language development, children with WMS show significant delay. However, once language is acquired, this acquisition tends to become a relative strength. Compared with age-matched and full-scale IQ-matched subjects with Down syndrome (DNS), subjects with WMS perform far better on a wide variety of grammar probes.

Another distinctive facet of the language abilities of WMS patients is their ability to use their heightened linguistic skills to engage others socially. Many individuals with WMS display a strong impulse toward social contact and affective expression. The intersection of language behavior and social engagement in individuals with WMS has been investigated and compared with DNS patients through a series of

narrative tasks in which subjects are asked to tell a story from a wordless book (**Figure 3**). The most obvious distinction between patients with WMS, DNS and age-matched normal controls is in the use of narrative enrichment devices (affective qualities) during the story-telling task by WMS patients. Individuals with WMS, therefore, exhibit a striking contrast to the social and language profiles of individuals with other disorders.

Visual–spatial cognition in WMS

In studies that examined global spatial cognition, in contrast to language, patients with WMS were significantly more impaired than subjects with DNS across all age ranges examined. For example, a WMS

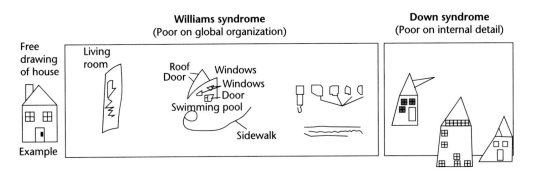

Figure 4 Free drawings of houses by age- and IQ-matched adolescents with WMS and DNS show different spatial deficits. The drawings by subjects with WMS contain many parts of houses but the parts are not organized coherently. In contrast, the DNS subjects' drawings are simplified but have the correct overall configuration of houses.

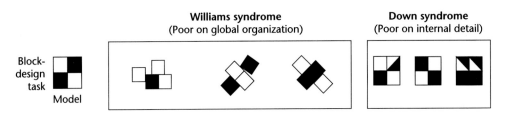

Figure 5 Block-design task showing spatial deficits in WMS and DNS. WMS subjects typically show disjointed and fragmented designs, while age- and IQ-matched DNS subjects tend to make errors in internal details while maintaining the overall configuration.

patient may draw different parts of a house as separate entities from the house itself, thus lacking global organization (**Figure 4**). On block-design tasks, both WMS and DNS subjects performed poorly, but in different ways: those with WMS were typically unable to organize the blocks into a global pattern, and patients with DNS instead made errors of internal detail (**Figure 5**). However, across difficult face-processing and recognition tasks, individuals with WMS showed strong performance in contrast to spatial cognition (**Figure 6**).

Neurophysiological Profile

Connections between cognition and neurophysiological characteristics of WMS are being revealed through studies of brain function (event-related potentials, ERPs). ERP techniques are used to assess the timing and organization of the neural systems that are active during sensory, cognitive and linguistic processing in patients with WMS and provide information about the timing and temporal sequence of neural activity.

Neurophysiological markers for auditory language and face processing

Are the spared abilities such as language and face processing normally organized in WMS brains or do

they show a different, perhaps unique pattern of functional organization? The ERP technique is ideally suited to examine this question. ERPs are recorded from electrodes placed on the scalp and measure the electrical activity of the brain time-locked to particular events. They are characterized by a series of positive and negative fluctuations in voltage (measured in microvolts) and reflect changes in brain activity over time on a millisecond by millisecond basis.

ERPs to auditory words are markedly different between WMS and controls. Individuals with WMS show larger amounts of neural activity within the first 200 ms after hearing a word than do same-age controls. This increased activity is reflected by larger ERP amplitudes for two positive ERP components peaking at 50 and 200 ms (P50 and P200; **Figure 7a**). These components, generated in primary auditory cortex, may index increased sensitivity to sound in WMS. Whether the hyperexcitability of the auditory system is related to the sparing of language abilities in WMS is unknown.

ERPs to faces presented one at a time in a face-recognition task show that the first negative component, peaking at 100 ms (N100), is abnormally small in WMS, whereas the second negative component (N200) is abnormally large (about 5 times the amount of activity in controls; **Figure 7b**). The abnormally small N100 amplitude reflects decreased neural activity in primary visual cortex and is likely to be linked to structural curtailment of the occipital lobes. In

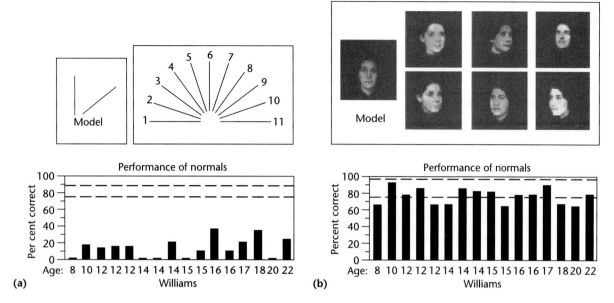

Figure 6 (a) Line- and (b) face-processing in WMS. The results are shown from two tasks that are both visuoperceptual tasks, where the correct answer requires only pointing to a picture without any constructional component. The contrast in performance on (a) line orientation (Benton judgment of line orientation) and (b) face discrimination (Benton face recognition) is shown for 16 individuals with WMS. On the line-orientation task, several individuals with WMS could not even pass the warm-up items. In great contrast, another 16 subjects with WMS perform remarkably well on a very difficult face-discrimination task that involves recognizing the same individual under different conditions of lighting, shadow and orientation. In both tasks, performance of normal individuals is indicated by the broken lines.

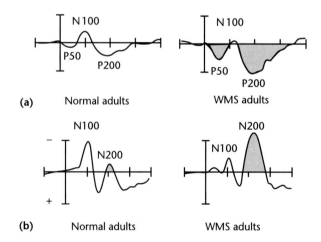

Figure 7 Neurophysiological markers for language and face processing: (a) ERPs to auditory words and (b) ERPs to upright faces.

contrast the abnormally large N200 is thought to be linked to increased attention to faces in WMS.

The ERP patterns observed for both auditory language and faces are characteristic of almost all WMS subjects, but not observed in normal adults, children or in other atypical populations that have been studied. Individuals with WMS show equal amounts of activity over the left and the right hemispheres to both spoken words and faces. That is, they do not show the lateral asymmetries (i.e. left greater than right to words, and right greater than left to faces) observed in normal adults. Thus, the unique phenotypic markers for language and face processing may serve as neurophysiological indices that relate brain and behavior characteristic in WMS and, taken together, these findings suggest that, in individuals with WMS, the neural systems that subserve higher cognitive functions such as language and face processing are different from normal individuals.

Neuroanatomical Characteristics

With the presence of a predictable neurobehavioral phenotype, coupled with the availability of increasingly sophisticated technology for assessing brain structure and function, initial neuroimaging studies have been carried out to investigate the neural systems that mediate the cognitive profile of individuals with WMS. The results to date point to specific morphological differences between WMS and comparison groups and also provide clues as to the neuroanatomical basis of the uneven cognitive profile observed in individuals with WMS.

Two groups, WMS subjects and normal controls, were studied using magnetic resonance imaging

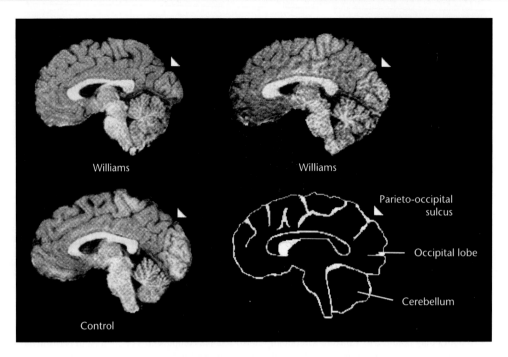

Figure 8 Occipital lobe reduction and preservation of cerebellum. Comparable sagittal MRI brain images from two subjects with WMS and a normal control. The images demonstrate that the occipital lobe, separated from the parietal lobe by the parietal-occipital sulcus (triangle), is greatly reduced in the subjects with WMS. Relative preservation of cerebellar size in WMS relative to controls is also shown.

(MRI). Initial findings from preliminary imaging studies suggested that the brains of individuals with WMS were reduced (15%) in volume overall compared with normal controls, with relative preservation of cerebellar volume. In addition, occipital lobe gray matter volume, particularly on the right, was noted to be decreased in the WMS group compared with controls (**Figure 8**). Other brain areas found to be preserved (or disproportionately enlarged) in subjects with WMS were the superior temporal region and midline cerebellum (vermis). The superior temporal region of the brain is thought to be important in the perception and processing of music, as well as in auditory and language processing, while the cerebellar vermis is involved in sensory information processing and social behavior.

Planned future studies will focus on limbic, mesial temporal and subcortical regions, as well as subregions of the cerebellum. Shape-based analyses of the cerebral lobes and their relation to the posterior fossa also are underway to more fully elucidate possible patterns of altered brain development in individuals with WMS. Eventual correlation of neurocognitive and structural imaging findings with results obtained from functional imaging studies will be particularly important in elucidating the topography of neuroanatomical function and dysfunction underlying cognition and behavior in individuals affected with this enigmatic genetic condition.

Development of a Phenotypic Map of WMS Cognition

To understand the pathways that bridge genes and behavior, the genes deleted in different individuals with WMS are correlated across levels. This encompasses neural information on gene expression, neuroanatomy, physiology and neurocognition. Since the common WMS deletion involves essentially the same genes deleted, accompanied by variable phenotypes observed in individuals with WMS, assigning specific genes to certain features has become an arduous task. Ideally, to link genes to features, some variability in deletions and variations in cognitive behavior are essential to determine the relationships. Based on that, a database of individuals who carry a partial deletion and show only some features of WMS has been started. It is by combining current molecular and cognitive data from these subjects with other individuals carrying atypical deletions that variability in specific neurocognitive features of WMS has been assigned to different regions with known genes. The current phenotypic map of WMS is shown in **Figure 9**.

The construction of the phenotypic map allows us to ask not only which genes are likely to affect cognition, but also which regions and their genes have been demonstrated in humans to be associated with changes in cognition when deleted. For example,

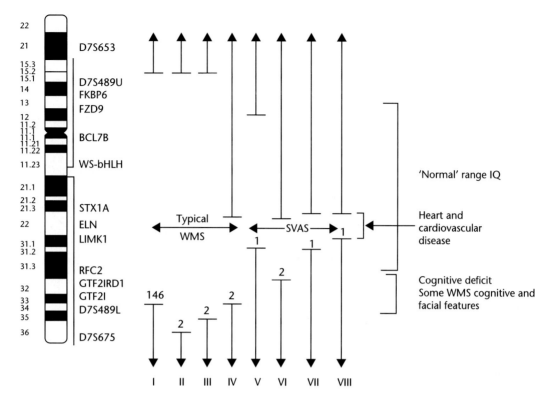

Figure 9 Phenotypic map of WMS. Spaces between vertical lines indicate the regions deleted, and the number of subjects carrying the deletion (i.e. typical deletion, $n=146$). Square brackets indicate regions that are likely to contain a gene or genes that when deleted contribute in some measure to the characteristic features of WMS. The significance of these data is that deletion of *FZD9* through *RFC2* does not appear to be associated with the characteristic facial features or cognitive deficits seen in WMS, although it could contribute.

from previous studies of isolated cases with small deletions and mutations of *ELN*, it appears that the absence of one copy of the gene is probably responsible for the cardiovascular defects in WMS. The lack of characteristic features (i.e. facies, hoarse voice) in individuals with *ELN* single-base mutations also suggests that none of the other physical features are largely due to deletion of *ELN*. In contrast, although the absence of one copy of *LIMK1* has been implicated in the spatial deficit characteristic of WMS, a study of three isolated subjects whose deletions included *LIMK1* reveals that the absence of this gene and others in the region is compatible with normal range function. Moreover, analyses of two individuals who carry a smaller deletion of the region from *FZD9* through *RFC2* indicate that the genes in this region are unlikely to be the major cause of variation characteristic of WMS cognitive and visual–spatial deficits or facial features. This is because both individuals have heart disease with essentially normal range or mildly impaired cognitive and physical features. Therefore, one can infer that the genes responsible for a significant part of mental retardation and other features are located in the region of *RFC2* through *GTF2I*, although genes located outside these regions

may also be involved. On the whole, these findings are important in setting out the approach for defining the consequences of specific deletions. Greater understanding of the range of such consequences thus requires further studies of further individuals and model systems.

Besides studying individuals with atypical deletions, what other molecular perspectives are considered necessary for understanding the factors underlying WMS cognition? To begin, it is important to note that the variable expression of genes that are not deleted may also contribute to phenotype. Sources of their variation may involve the effects of transcribed but not translated pseudogenes, including those for *GTF2I*. Further sources of phenotypic variability include interactions of genes in the deleted region, with others mapping both inside and outside of the WMS region, as well as possible regional disturbances of replication and transcription due to deletion and chromatin rearrangement. Such larger rearrangements that do not result in deletion may therefore be a cause of the WMS phenotype. Furthermore, some rearrangements may lead to the decreased expression of nondeleted genes located in the region close to the deletion. Other potential sources of genetic effects on cognition include imprinting, or the phenomenon by which the expression of a gene is differentially modified by passing through the maternal versus the paternal germ line.

In summary, to increase the understanding of the genes and their role in the development of WMS features, future studies of rare WMS subjects with atypical deletions are necessary to identify gene candidates for human cognitive features and characteristic WMS phenotype. The resulting multidimensional data from well-studied individuals with atypical deletions will help elucidate the many levels requiring intersection in the neurosciences; the role of cellular processes and neuroanatomical structures in cognition and behavior; and the role of gene expression in determining structure, cognition and disease.

Acknowledgements

The research was supported by grants from the National Institute of Child Health and Development (HD33113) and the James S. McDonnell Foundation to U. B. and J. R. K., as well as the Oak Tree Philanthropic Foundation to U. B. J. R. K. holds the Geri and Richard Brawerman Chair in Molecular Genetics.

See also

Microdeletion Syndromes

Microdeletions and Microduplications: Mechanism

Further Reading

Bellugi U, Klima ES and Wang PP (1996) Cognitive and neural development: clues from genetically based syndromes. *The Lifespan Development of Individuals: Behavioral, Neurobiological, and Psychosocial Perspectives*, pp. 223–243. New York: Cambridge University Press.

Bellugi U, Lichtenberger L, Jones W, *et al.* (2000) The neurocognitive profile of Williams syndrome: a complex pattern of strengths and weaknesses. *Journal of Cognitive Neuroscience* **12**(supplement 1): 7–29.

Bellugi U, Lichtenberger U, Mills D, *et al.* (1999) Bridging cognition, the brain and molecular genetics: evidence from Williams syndrome. *Trends in Neuroscience* **22**: 197–207.

Botta A, Novelli G, Mari A, *et al.* (1999) Detection of an atypical 7q11.23 deletion in Williams syndrome patients which does not include the *STX1A* and *FZD9* genes. *Journal of Medical Genetics* **36**: 478–480.

Ewart AK, Morris CA, Atkinson D, *et al.* (1993) Hemizygosity at the elastin locus in the developmental disorder, Williams syndrome. *Nature Genetics* **5**: 11–18.

Francke U (1999) Williams–Beuren syndrome: genes and mechanisms. *Human Molecular Genetics* **8**(10): 1947–1954.

Jones W, Bellugi U, Lai Z, *et al.* (2000) Hypersociability in Williams syndrome. *Journal of Cognitive Neuroscience* **12**(supplement 1): 30–46.

Korenberg JR, Chen XN, Hamao H, *et al.* (2000) Genome structure and cognitive map of Williams syndrome. *Journal of Cognitive Neuroscience* **12**(supplement 1): 89–107.

Lenoff HM, Wang PP, Greenberg F and Bellugi U (1997) Williams syndrome and the brain. *Scientific American* **27**(7): 68–73.

Li DY, Toland AE, Boak BB, *et al.* (1997) Elastin point mutations cause an obstructive vascular disease, supravalvular aortic stenosis. *Human Molecular Genetics* **6**: 1021–1028.

Mervis CB, Morris CA, Bertrand J and Robinson BF (1999) Williams syndrome: findings from an integrated program of research. *Neurodevelopmental Disorders: Contributions to a New Framework from the Cognitive Neurosciences*, pp. 65–110. Cambridge, MA: MIT Press.

Mills DL, Alvarez TD, St George M, *et al.* (2000) Electrophysiological studies of face processing in Williams syndrome. *Journal of Cognitive Neuroscience* **12**(supplement 1): 47–64.

Morris CA, Leonard CO and Dilates C (1988) Natural history of Williams syndrome: physical characteristics. *Journal of Pediatrics* **113**: 318–325.

Morris CA, Loker J, Ensing G and Stock AD (1993) Supravalvular aortic stenosis cosegregates with familial 6;7 translocation which disrupts the elastin gene. *American Journal of Medical Genetics* **46**: 737–744.

Osborne LR, Li M, Pober B, *et al.* (2001) A 1.5-million base pair inversion polymorphism in families with Williams–Beuren syndrome. *Nature Genetics* **29**: 321–325.

Perez Jurado LA, Peoples R, Kaplan P, *et al.* (1996) Molecular definition of the chromosome 7 deletion in Williams syndrome and parent-of-origin effects on growth. *American Journal of Human Genetics* **59**(4): 781–792.

Reiss AL, Eliez S, Schmitt JE, *et al.* (2000) Neuroanatomy of Williams syndrome: a high-resolution MRI Study. *Journal of Cognitive Neuroscience* **12**(supplement 1): 65–73.

Robinson WP, Waslynka J, Bernasconi F, *et al.* (1996) Delineation of 7q11.2 deletions associated with Williams–Beuren syndrome and mapping of a repetitive sequence to within and to either side of the common deletion. *Genomics* **34**: 17–23.

Tassabehji M, Metcalfe K, Karmiloff-Smith A, *et al.* (1999) Williams syndrome: use of chromosomal microdeletions as a tool to dissect cognitive and physical phenotypes. *American Journal of Human Genetics* **64**(1): 118–125.

Wu YQ, Sutton VR, Nickerson E, *et al.* (1998) Delineation of the common critical region in Williams syndrome and clinical correlation of growth, heart defects, ethnicity, and parental origin. *American Journal of Medical Genetics* **78**(1): 82–89.

Web Links

elastin (supravalvular aortic stenosis, Williams-Beuren syndrome) (ELN); LocusID: 2006. LocusLink:
http://www.ncbi.nlm.nih.gov/LocusLink/LocRpt.cgi?l = 2006

elastin (supravalvular aortic stenosis, Williams-Beuren syndrome) (ELN); MIM number: 130160. OMIM:
http://www3.ncbi.nlm.nih.gov/htbin-post/Omim/dispmim?130160

Wilms Tumor

Kathy Pritchard-Jones, *Institute of Cancer Research and Royal Marsden Hospital, Sutton, UK*

Wilms tumor, otherwise known as nephroblastoma, is a childhood embryonal kidney cancer. This potentially heritable tumor involves several different genes, many of which illustrate the close relationship between cancer and maldevelopment. The majority of children are cured of their cancer but new molecular markers to predict tumor recurrence risk are needed to allow better stratification of treatment intensity.

Introduction

Wilms tumor belongs to the category of embryonal tumors of childhood, so-called because of the close resemblance of tumor cells to their presumed fetal tissue of origin. For Wilms tumor, this is the embryonic kidney and tumors show variable degrees of differentiation toward cell types and structures seen during normal kidney development (nephrogenesis) (**Figure 1**). The presumed precursor cell population for Wilms tumor are nephrogenic rests, islands of residual poorly differentiated embryonic kidney found in ~1% of neonatal autopsies. Since the overall incidence of Wilms tumor is 1 in 10 000 children, this suggests that only 1% of nephrogenic rests undergo malignant conversion. Nephrogenic rests can be subdivided into two types, intralobar and perilobar, which are thought to represent mutations occurring at an early or late

stage of nephrogenesis respectively (**Figures 2** and **3**). These changes have been correlated with defects in different Wilms tumor genes (Beckwith, 1998).

Epidemiology of Wilms Tumor

Wilms tumor is by far the commonest type of kidney tumor occurring in childhood. The median age of presentation is 3–4 years and the incidence declines rapidly thereafter (**Figure 4**) (Breslow *et al.*, 1993). For as yet unknown reasons, the age at onset is slightly but significantly earlier in males than in females. There is also a degree of ethnic variation: Wilms tumor is commoner in Blacks and relatively rare in Asians. The age at onset distributions and the tendency for incidence rates to vary more strongly with ethnicity

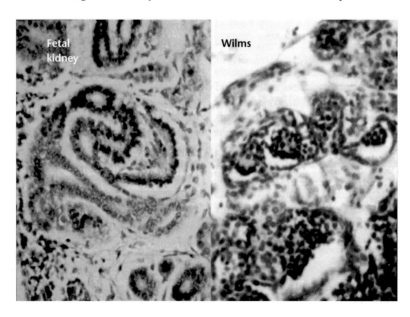

Figure 1 Wilms tumor shows remarkable mimicry of the nephrogenic process. The left panel shows the nephrogenic zone of an 18-week gestation human fetal kidney. There is orderly differentiation of the primitive blastema into tubular (epithelial) components in response to reciprocal inductive signals from the branching ureteric bud. The right panel shows a classic triphasic Wilms tumor in which all three cell types (primitive blastema, epithelia and stroma) are seen. In this particular tumor, epithelial differentiation is proceeding as far as an attempt to make pseudoglomeruli.

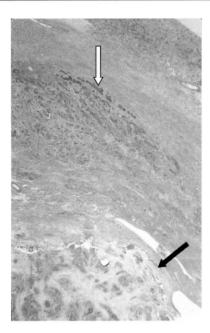

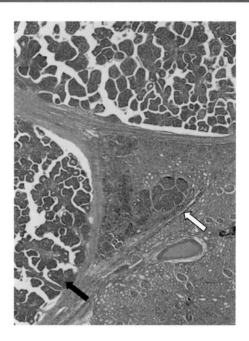

Figure 2 Intralobar nephrogenic rest (ILNR). ILNRs are regions of residual primitive nephrogenic tissue thought to arise at an early stage of nephrogenesis. As in this case (white arrow), they usually lie immediately adjacent to the associated Wilms tumor (black arrow), the latter often showing prominent stromal differentiation. ILNRs usually mimic the entire spectrum of renal histogenesis, are usually single and are thought to have a high tumorigenic potential. (Photograph courtesy of Dr Gordan Vujanic, University Hospital of Wales, Cardiff.)

Figure 3 Perilobar nephrogenic rest (PLNR). A PLNR (white arrow) is a well-circumscribed area of incompletely differentiated fetal renal tissue. These are usually found at the periphery of the renal lobule, are often multiple and show restricted differentiation potential, mimicking late stages of nephrogenesis with mainly tubular differentiation. The associated Wilms tumors (black arrow) usually have restricted, mainly epithelial differentiation. (Photograph courtesy of Dr Gordan Vujanic, University Hospital of Wales, Cardiff.)

than with geography or over time suggest that genetic factors play an important role in etiology. Usually only one kidney is affected, but in 5–8% of cases, there are tumors in both kidneys (bilateral disease). Based on mathematical modeling of the age at onset of tumor development in heritable and sporadic cases Knudson proposed that his 'two-hit' hypothesis could be applied to the development of Wilms tumor. According to this model, the two 'hits' represent inactivating mutations in both alleles of a tumor suppressor gene. The first mutation is present in all cells in familial or bilateral cases, but both mutations have to occur in the same somatic cell in sporadic cases, explaining their tendency to occur later and in only a single kidney. With the subsequent discovery of genetic heterogeneity in this disease and the complexity of the biology of the *WT1* gene, it is now clear that this model holds true in only a minority of cases (Hastie, 1994).

Genetics of Wilms Tumor

Heritable genes

The majority of cases of Wilms tumor are 'sporadic' with no obvious cause. In total, 1–2% have a family

history of Wilms tumor and 2–3% occur in children with congenital malformation syndromes that carry a greatly increased risk of Wilms tumor (**Table 1**). A further 5% occur in children with an isolated congenital malformation, either hemihypertrophy (asymmetrical overgrowth) or genital abnormalities such as undescended testis (cryptorchidism) or abnormal placement of the penile urethral opening (hypospadias).

The first of these associations to be worked out at a molecular level was the association of Wilms tumor with aniridia, a lack of development of the iris in the eye. Such children usually have other defects including variable mental retardation and abnormal genital development. (Pelletier *et al.*, 1991) hence the acronym, WAGR syndrome. This defect is due to a chromosomal deletion that removes one copy of the Wilms tumor gene, *WT1*, and the adjacent *PAX6* gene at 11p13. Loss of one allele of *PAX6* is dominant, but development of the tumor requires loss or mutation of the remaining *WT1* allele in one or more kidney cells (i.e. tumor development is recessive, hence *WT1* belongs to the class of tumor suppressor genes in this context). Germ-line mutations in *WT1* are also responsible for the predisposition to Wilms tumor associated with abnormal kidney development seen in

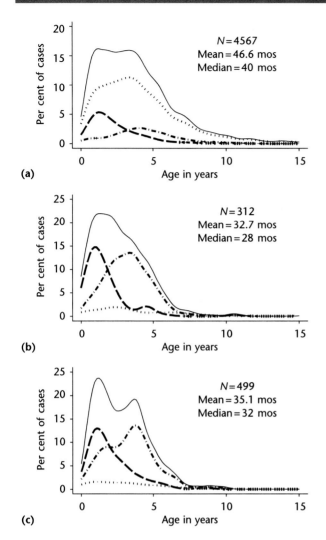

Figure 4 Age distribution of Wilms tumor in relation to bilaterality and type of associated nephrogenic rest (a) unilateral unicentric tumors; (b) bilateral tumors; (c) unilateral, multicentric tumors. (Data courtesy of Professor Norman Breslow and the National Wilms Tumor Study Group, North America.)

Denys–Drash syndrome, where children lose large quantities of protein in their urine (nephrotic syndrome) and often have abnormal genital development (Pelletier *et al.*, 1991). Thus, Wilms tumor provides a fascinating example of how normal development of an organ can be intimately linked to cancer predisposition in that organ. The biology of the *WT1* gene is discussed in more detail later.

A second category of genetic predisposition to Wilms tumor occurs in the overgrowth syndromes of childhood, of which Beckwith–Wiedemann syndrome (BWS) is the best recognized. The genetics of BWS are complex, involving several different genes within the 11p15.5 chromosomal locus and the phenomenon of 'imprinting' whereby expression of a gene depends on its parental origin (DeBaun *et al.*, 2002). The overall tumor risk is ~10%, of which half are Wilms tumors. It appears that children with early nephromegaly (i.e. overgrowth of the kidneys) or asymmetrical overgrowth (hemihypertrophy) are at greatest risk (DeBaun *et al.*, 1998). Recently, the gene underlying a second syndrome with phenotypic overlap with BWS, the Simpson–Golabi–Behmel syndrome, has been defined as the glypican 3 gene on the X chromosome. However, its somatic mutation rate in sporadic Wilms tumor appears to be low or negligible.

Familial Wilms tumor usually occurs in small pedigrees without associated congenital abnormalities, making genetic linkage studies difficult. Two genes, *FWT1* and *FWT2*, have been mapped to chromosomes 17q and 19q respectively, but there are other pedigrees unlinked to any currently known Wilms tumor gene locus (Rahman *et al.*, 1996; Rapley *et al.*, 2000). The penetrance of these genes appears to be low, meaning that a proportion of patients with apparently sporadic Wilms tumor may be gene carriers. However, the absolute risk to offspring must be low, as only four cases among 362 offspring of 462 survivors of unilateral Wilms tumor have been reported (Hawkins *et al.*, 1995).

Table 1 Congenital malformation syndromes predisposing to Wilms tumor

Syndrome	Locus	Gene(s)	Mutation type	Risk of WT
WAGR	11p13	*WT1*	Complete deletion of *WT1*	30–50%
Denys–Drash/Frasier	11p13	*WT1*	Missense mutation/aberrant splicing of *WT1*	30–50%
Overgrowth syndromes				
Beckwith–Weidemann syndrome	11p15.5	*P57/IGF2/H19*	Abnormal imprinting	<10%
Hemihypertrophy	?	?		<1%
Simpson–Golabi–Behmel	Xq26	*GPC3*	Loss of function	~1%
Perlman	?	?		?

Maldevelopment is dominant, tumorigenesis is recessive. WAGR: Wilms–aniridia genitourinary malformation and mental retardation.

Somatic gene mutations in Wilms tumor

The *WT1* gene, isolated in 1990, does not explain the majority of sporadic Wilms tumors. Among over 600 published cases analyzed for intragenic mutations, *WT1* mutation occurs in only ~10%. In some tumors, mutation of both *WT1* alleles appears to be sufficient for tumorigenesis, in accordance with Knudson's two-hit hypothesis. However, in other tumors, either only one *WT1* allele is mutated or other genetic events are clearly interacting, for example, Beta-catenin mutations occur at a similar frequency to *WT1* mutation in Wilms tumor and often coexist (see Web Links).

The molecular biology of the WT1 protein is fascinating. It is expressed in a tissue-specific fashion, mainly in the embryonic gonads and kidney, and complete loss of function in a murine 'knockout' model leads to complete absence or severe abnormalities of these organs. The protein contains four zinc-finger motifs and an *N*-terminal effector domain, and is believed to function as a transcription factor whose targets are influenced by the presence or absence of the second splice site, the three amino acids KTS (**Figure 5**). In humans and in mice, mutations that disrupt this splice site cause severe genitourinary abnormalities.

Allele loss (loss of heterozygosity) studies suggest that several different genetic loci exist for other Wilms tumor genes at 1p, 11p15, 16q and 22q. Some of these may be involved in tumor progression rather than initiation. Although allele loss spanning the BWS locus at 11p15 occurs in 30–50% of Wilms tumors, none of the several genes implicated in BWS are mutated in sporadic Wilms tumor. There does, however, appear to be a common cellular phenotype of loss of imprinting (LOI) of the insulin-like growth factor II gene in both BWS and sporadic Wilms tumor. There is evidence that the mechanism for LOI of IGFII may involve disruption of other imprinted genes within the region.

Clinical Characteristics

For the last 30 years, Wilms tumor has been one of the success stories of pediatric oncology with long-term disease-free survival approaching 90% in localized disease and over 70% even for metastatic disease. The commonest site of such metastases is the lung, followed by lymph nodes and liver. Wilms tumor rarely metastasizes to bone, bone marrow or brain. The treatment consists of chemotherapy with one to three different drugs (usually vincristine, actinomycin D ± adriamycin) together with surgical excision of the affected kidney. Radiotherapy is also used where there is residual or spilt tumor in the abdomen or metastases.

There is a different philosophical approach to the organization of treatment between different national and international childhood cancer study groups. The National Wilms Tumor Study Group (NWTSG) of North America favors immediate surgical excision of the affected kidney, followed by chemotherapy ± radiotherapy according to the tumor extent found at the time of surgery (Green *et al.*, 1998). The approach of the International Society of Paediatric Oncology (SIOP) is to use preoperative chemotherapy to shrink the tumor prior to surgery (Tournade *et al.*, 1993). Their studies show that this reduces the risk of tumor rupture during operation and also reduces the tumor stage, hence allowing less intensive postoperative treatment. The two groups have comparable cure rates, but the NWTSG approach uses slightly more radiotherapy whereas the SIOP approach uses more anthracyclines (adriamycin). Both these treatments have the potential for long-term side effects, on growth and fertility and on the heart muscle respectively. The majority of children with Wilms tumor are cured

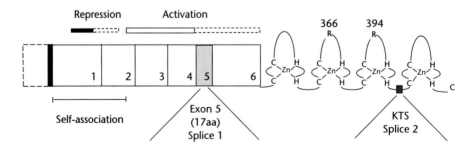

Figure 5 Structure of the *WT1* gene, showing the position of the two alternative splice sites, comprising the whole of exon 5 and the additional amino acids, KTS, in the linker between the third and fourth zinc-fingers. The position of two commonly mutated arginine residues in Denys–Drash syndrome is indicated (R^{366} and R^{394}) – these residues are critical for DNA binding by these zinc-fingers. The *N*-terminal transcriptional regulatory domains defined by *in vitro* studies are indicated, the dotted line indicates the possible usage of an alternative initiation codon.

without the need for either of these agents and are unlikely to suffer any long-term sequelae.

Pathological Subgroups, Etiology and Prognosis

As with most cancers, various 'prognostic factors' can be recognized in Wilms tumor. The most obvious adverse factor is increasing tumor stage. However, a distinct histological subtype called anaplasia carries a poor prognosis, especially when associated with advanced stage disease. Anaplasia is associated with mutations in the *p53* gene, which can occur focally as part of clonal evolution of a tumor. Other molecular characteristics that may be associated with worse outcome are allele loss on chromosomes 16q, 1p and possibly 22q and 11q. These have been tested prospectively in the current NWTS 5 clinical trial, which aims to use molecular characteristics of a tumor to better define risk groups (Grundy *et al.*, 1996).

There has been limited analysis of nephrogenic rests and their associated Wilms tumors to provide molecular evidence for tumor progression from a precursor lesion. In the few cases tested that involve the *WT1* gene, the mutation was often identical in both the rest and the tumor although allele loss did show evidence for clonal evolution (Charles *et al.*, 1998). The age distribution of Wilms tumors associated with the two types of nephrogenic rests does suggest that these are genetically distinct and that the ILNR-associated tumors arise from earlier defects in nephrogenesis (**Figure 4**).

Treatment of Relapsed Wilms Tumor and Future Therapeutic Possibilities

Although Wilms tumor is one of the most curable of childhood cancers at initial diagnosis, those cases that relapse carry a much worse prognosis, even with intensive retreatment. Less than a third of relapses are due to the anaplastic variant. Hence, one of the goals of current Wilms tumor clinical studies is to identify factors present at diagnosis that are predictive of outcome. Future treatment intensity could then be stratified according to predicted risk of relapse, using molecular characteristics of individual tumors. Another potential avenue is to use knowledge of the biology of the various 'Wilms tumor genes' to devise novel therapeutic approaches. In the future, this might also lead to preventative strategies for children at increased genetic risk of Wilms tumor.

Screening

Early detection of Wilms tumor has the potential to increase survival and reduce treatment morbidity. Children known to be genetically predisposed to Wilms tumor, such as those with WAGR, Beckwith–Wiedemann or Denys–Drash syndromes, are candidates for screening. This is usually done by regular abdominal ultrasound scanning, although teaching the parents to perform regular palpation of the abdomen is an acceptable alternative. Since Wilms tumors can grow very rapidly, ultrasound screening is recommended at intervals of no greater than 3–4 months. However, there are no definitive clinical trials to determine which screening method or interval is superior for detecting tumors at a low stage. In the future, if a larger proportion of children are shown to be carriers of one of the familial Wilms tumor genes, then more information should become available on the efficacy of screening. A further benefit of the application of molecular genetics is that children with germline *WT1* mutations appear to be at risk of late renal failure and hence require appropriate monitoring. (*See* Cancer Cytogenetics; Kidney: Hereditary Disorders; Microdeletion Syndromes.)

References

Beckwith JB (1998) Nephrogenic rests and the pathogenesis of Wilms tumor: developmental and clinical considerations. *American Journal of Medical Genetics* **79**: 268–273.

Breslow N, Olshan A, Beckwith JB and Green DM (1993) Epidemiology of Wilms tumor. *Medical and Pediatric Oncology* **21**: 172–181.

Charles AK, Brown KW and Berry PJ (1998) Microdissecting the genetic events in nephrogenic rests and Wilms' tumor development. *American Journal of Pathology*. **153**: 991–1000.

DeBaun MR, Niemitz EL, McNeil DE, *et al.* (2002) Epigenetic alterations of H19 and LIT1 distinguish patients with Beckwith–Wiedemann syndrome with cancer and birth defects. *American Journal of Human Genetics* **70**: 604–611.

DeBaun MR, Siegel MJ, Choyke PL, *et al.* (1998) Nephromegaly in infancy and early childhood: a risk factor for Wilms tumor in Beckwith–Wiedemann syndrome. *Journal of Pediatrics* **132**: 401–404.

Green DM, Breslow NE, Beckwith JB, *et al.* (1998) Effect of duration of treatment on treatment outcome and cost of treatment for Wilms tumour: a report from the National Wilms Tumour Study Group. *Journal of Clinical Oncology* **16**: 3744–3751.

Grundy P, Telzerow P, Moksness J and Breslow NE (1996) Clinicopathologic correlates of loss of heterozygosity in Wilms tumor: a preliminary analysis. *Medical and Pediatric Oncology*. **27**: 429–433.

Hastie ND (1994) The genetics of Wilms' tumor: a case of disrupted development. *Annual Review of Genetics* **28**: 523–558.

Hawkins MM, Winter DL, Burton HS, *et al.* (1995) Heritability of Wilms' tumour. *Journal of the National Cancer Institute* **87**: 1323–1324.

Pelletier J, Bruening W, Kashtac CE, *et al.* (1991) Germline mutations in the Wilms' tumor suppressor gene are associated with abnormal urogential development in Denys–Drash syndrome. *Cell* **67**: 437–447.

Rahman N, Arbour L, Tonin P, *et al.* (1996) Evidence for a familial Wilms' tumour gene (*FWT1*) on chromosome 17q12–q211. *Nature Genetics* **13**: 461–463.

Rapley EA, Barfoot R, Bonaiti-Pellie C, *et al.* (2000) Evidence for susceptibility genes to familial Wilms tumor in addition to *WT1*, *FWT1* and *FWT2*. *British Journal of Cancer* **83**: 177–183.

Tournade MF, Com-Nougue C, Voute PA, *et al.* (1993) Results of the sixth International Society of Paediatric Oncology Wilms Tumour Trial and Study: a risk-adapted therapeutic approach in Wilms tumour. *Journal of Clinical Oncology* **11**: 1014–1023.

Further Reading

Choyke PL, Siegel MJ, Craft AW, *et al.* (1999) Screening for Wilms tumor in children with Beckwith–Wiedemann syndrome or idiopathic hemihypertrophy. *Medical and Pediatric Oncology* **32**: 196–200.

Hammes A, Guo JK, Lutsch G, *et al.* (2001) Two splice variants of the Wilms' tumor 1 gene have distinct functions during sex determination and nephron formation. *Cell* **106**: 319–329.

Koziell A, Charmandari E, Hindmarsh PC, *et al.* (2000) Frasier syndrome, part of the Denys Drash continuum or simply a *WT1*

gene associated disorder of intersex and nephropathy? *Clinical Endocrinology* **52**: 519–524.

Kreidberg JA, Sariola H, Loring JM, *et al.* (1993) WT-1 is required for early kidney development. *Cell* **74**: 679–691.

Little M, Holmes G and Walsh P (1999). WT1: what has the last decade told us? *BioEssays* **21**: 191–202.

Pritchard-Jones K (1997) Molecular genetic pathways to Wilms tumour. *Critical Reviews in Oncogenesis* **8**: 1–27.

Schumacher V, Schneider S, Figge A, *et al.* (1997) Correlation of germ-line mutations and two-hit inactivation *WT1* gene with Wilms tumors of stromal-predominant histology. *Proceedings of the National Academy of Sciences of the United States of America* **94**: 3972–3977.

Veugelers M, De Cat B, Muyldermans SY, *et al.* (2000) Mutational analysis of the GPC3/GPC4 glypican gene cluster on Xq26 in patients with Simpson–Golabi–Behmel syndrome: identification of loss-of-function mutations in the GPC3 gene. *Human Molecular Genetics* **9**: 1321–1328.

Web Links

Database of *WT1* mutations.
  http://archive.uwcm.ac.uk/uwcm/mg/hgmd0.html

Wilson, Edmund Beecher

Introductory article

WF Bynum, *Wellcome Trust Center for the History of Medicine at University College London, London, UK*

Edmund Beecher Wilson (1856–1939) was a biologist who made fundamental contributions to our understanding of the cell in heredity and development.

Edmund Beecher Wilson, the son of a successful lawyer, spent a year teaching in a small country school in Illinois when he was 16 years old. His experience encouraged him to continue his own education, first at Antioch College in Ohio, then at the University of Chicago and finally at Yale University, where his passion for zoology, embryology and evolutionary biology was awakened. Following graduation in 1878, he was introduced to the joys of original research at Johns Hopkins University, where he received his PhD in 1881. His PhD research, under the physiologist H. Newell Martin and the morphologist William Keith Brooks, used the cnidarian *Renilla* to investigate the cellular changes during early embryological development. He then spent a year studying in Europe, where William Bateson, Rudolf Leuckart and Anton Dohrn influenced him. His months with the latter at the Zoological Station in Naples were especially rewarding.

Returning to the United States in 1883, Wilson taught at Williams College and the Massachusetts Institute of Technology before becoming head of the Biology Department at Bryn Mawr College in

Pennsylvania. He did little research during these years, but he wrote a textbook, *General Biology* (1886), with William T. Sedgwick. It was notable for its emphasis on ecological and experimental aspects of biology. In 1891, he was appointed head of the Department of Zoology at Columbia University, New York City, but before he took up his duties he returned to Europe to work with Theodor Boveri in Munich and with Hans Driesch in Naples. Although he remained at Columbia for the rest of his career, he still worked regularly at marine research stations, such as those at Wood's Hole, Massachusetts and the Chesapeake Zoological Station of Johns Hopkins University. An expert sailor, he often combined pleasure and zoological collecting during summer expeditions.

Wilson was an outstanding experimental zoologist, but he also presided over a golden era at Columbia, where his students and colleagues included H. J. Muller, C. E. McClung, T. H. Morgan, C. B. Bridges and A. H. Sturtevant. His cytological investigations were fundamental for the genetic work in the famous 'fly room', where *Drosophila* studies revolutionized modern genetics. Wilson was also an inspirational teacher at both the undergraduate and postgraduate level.

Wilson's early research was primarily concerned with descriptive embryology and morphology. He used two annelids – the earthworm, *Lumbricus*, and a marine polychaete, *Nereis* – to develop techniques of following cell lineage in early development. In addition to its intrinsic interest, this work opened another way of investigating phylogenetic relationships. For example, he differentiated two basic developmental patterns of mesodermal formation among triploblastic animals, the spiral or mosaic pattern he had observed in earthworms, and a radial pattern in echinoderms, primitive chordates and vertebrates.

Following his sabbatical with Boveri, Wilson turned to questions of cell division, such as the relationship between chromosomal pairing and events in the cell's cytoplasm. He carried out, partially with McClung, a series of experiments and observations that established the roles of the X and Y chromosomes in sex determination. In other studies he investigated the relationship between nucleus and cytoplasm during cell division, and chromosomal phenomena such as nondisjunction (a word he and Bridges coined) and crossing-over. He argued forcefully that the chromosome contains the hereditary material, and that the events during cell division are consistent with Mendelian patterns of inheritance. This vision of the cell provided the conceptual framework for the *Drosophila* research carried out in his department and under his watchful eye.

Wilson discussed those and many other issues of cell biology in his influential monograph, *The Cell in Development and Inheritance*, first published in 1896. Its third edition, by then titled *The Cell in Development and Heredity* (1925), was a fitting summary of Wilson's life work. Encyclopedic in scope, its 1200 pages synthesized the whole of cell biology. It was awarded medals from the National Academy of Sciences and the Linnean Society of London.

Wilson was also a skilled linguist and an accomplished cellist, who was described as the 'foremost nonprofessional player in New York' (Muller, 1943). He saw beauty in music as well as biology, and once compared artistic and scientific creativity.

See also
Morgan, Thomas Hunt
Muller, Herman Joseph

Further Reading

Gillispie CC (ed.) (1970–1980) *Dictionary of Scientific Biography.* New York: Charles Scribner's Sons.
Morgan TH (1941) Edmund Beecher Wilson (1856–1939). *Biographical Memoirs, National Academy of Sciences* 21: 315–342.
Muller HJ (1943) Edmund B. Wilson, an appreciation. *American Naturalist* 77: 5–37; 142–172.
Wilson EB (1896) *The Cell in Development and Inheritance.* New York: Macmillan.

Wilson, Edward Osborne
Introductory article

Michael Ruse, *Florida State University, Tallahassee, Florida, USA*

Edward Osborne Wilson (1929–) is an American biologist, the world's leading authority on ants, who has explored biogeography.

Edward Osborne Wilson was born in Birmingham, Alabama, on 10 June 1929. He was educated at the University of Alabama and then moved to Harvard University for graduate work. He stayed at that institution until he retired in 2000, rising from doctoral student and Junior Fellow to Professor. Wilson, who is married with one child, has received many honors, including the Pulitzer Prize for nonfiction (twice), the Swedish Academy of Science's Crafoord Prize, and Membership in the National Academy of Sciences, USA and Fellow of the Royal Society.

Wilson is the world's living authority on ants. He has done major biogeographical studies of ants in the Far East, in Brazil, and in his homeland. He has written on their nature, their behavior and their systematics, culminating in *The Ants* (1990), written in collaboration with Bert Holldöbler, for which he won one of his Pulitzer Prizes.

Wilson has also been interested in the theory of biogeography, putting forward in the 1960s, with the late Robert H. MacArthur, a seminal theory of island flora and fauna, showing how immigration and emigration and extinction eventually reach equilibrium. Wilson's focus on ants led to an interest in chemical communication, and to an important study of the use of pheromones for information transmission, and that then led to broader studies of social behavior. In 1975, he published his magisterial

Sociobiology: The New Synthesis, in which he discussed social behavior from an evolutionary perspective, going all the way from slime molds to humans. In a shorter, later work, *On Human Nature* (1978), for which Wilson won the other of his Pulitzer Prizes, he extended and developed his thinking about human nature.

His excursions into human sociobiology brought on much criticism, especially from Marxists (who disliked any systematic attempt to relate human nature to biology), social scientists (who felt threatened in their own work) and feminists (who said Wilson was justifying patriarchy in the name of the genes). Undeterred, Wilson defended and extended his thinking, including thinking in a formal mode, gaining much from a collaboration with a young Canadian physicist, Charles Lumsden. He stressed then, as he stresses now, that taking a biological perspective does not at once commit one to a hard-line deterministic position, whereby the genes are the sole causal factor behind human nature. It is just that one can and should no more deny biology than one can or should deny the environment and culture.

At the same time as he was working on the science of human nature, Wilson turned to more philosophical issues. Epistemologically speaking (i.e. in the realm of the theory of knowledge), Wilson sees all knowledge as interconnected. When his critics accuse him of being unduly 'reductionistic' (meaning that he wants to explain everything in terms of just a few basic principles), he takes it as praise. He waxes eloquently on such themes in his recent book *Consilience*, the title of which is borrowed from William Whewell, the Victorian historian and philosopher of science who saw a 'consilience of inductions' as the supreme unifying move in all good science.

Wilson is also interested in the philosophical ideas and challenges of morality. Much in a tradition going back to the English evolutionist Herbert Spencer, a thinker whom he admires, Wilson argues that morality is and must be based in human nature, and that means human nature as created and preserved by evolution. For Wilson, ultimately one must refer back to natural selection, although Wilson has always interpreted selection in a catholic sense, being particularly open to the idea of selection working directly on groups rather than on groups only indirectly as it works on individuals. One should nevertheless understand that for Wilson the evolutionary process does not depend on the traditional Social Darwinian notion of *laissez faire*, whereby the struggle for existence is given total freedom. Wilson sees cooperation as being a major key

to evolutionary success, and with that he would link the need to be sensitive to the environment around us. It is not too much of an exaggeration to say he believes that in a world of plastic, humans would literally die. We have need physically and psychologically of other organisms. Therefore the supreme Wilsonian moral imperative centers on biodiversity. Without many different species on the earth, humans are doomed. In recent years, Wilson has been involved practically with the preservation of the Brazilian rain forests, and at a more general level with documentation of still-extant species.

Wilson is a traditionalist in that he justifies his evolutionary ethics in terms of progress. He sees the evolutionary process as having a significant direction, from lesser to greater, with humans at the pinnacle. That is why humans are worth cherishing – we are the best – and why a legitimate appeal can be made to the forces of nature and our obligation to abide by them. This vision for Wilson is much bound up with his feelings about religion. An intensely religious man, he lost his childhood faith in Christianity in his teens when he found Darwinism. He nevertheless sees religion as an essential part of human culture, binding the tribe together. For Wilson therefore we must keep the content but replace the foundations: We must put atheistic materialism in place of a good deity and a son who died on the cross. Hence for Wilson, his science, his ethics and his religion are a seamless whole. It is the consilience beyond measure.

See also

Sociobiology, Evolutionary Psychology and Genetics

Further Reading

MacArthur RH and Wilson EO (1967) *The Theory of Island Biogeography*. Princeton, NJ: Princeton University Press.

Ruse M (1996) *Monad to Man: The Concept of Progress in Evolutionary Biology*. Cambridge, MA: Harvard University Press.

Ruse M (1998) *Taking Darwin Seriously: A Naturalistic Approach to Philosophy*, 2nd edn. Buffalo, NY: Prometheus Books.

Ruse M (1999) *Mystery of Mysteries: Is Evolution a Social Construction?* Cambridge, MA: Harvard University Press.

Wilson EO (1971) *The Insect Societies*. Cambridge, MA: Harvard University Press.

Wilson EO (1975) *Sociobiology: The New Synthesis*. Cambridge, MA: Harvard University Press.

Wilson EO (1978) *On Human Nature*. Cambridge, MA: Harvard University Press.

Wright R (1987) *Three Scientists and their Gods*. New York: Times Books.

Wilson EO (1992) *The Diversity of Life*. Cambridge, MA: Harvard University Press.

Wilson EO (1998) *Consilience*. New York: Knopf.

Wright, Sewall

Newton Morton, *University of Southampton, Southampton, UK*

Sewall Wright (1889–1988) was an American biologist who was one of the three founders of population genetics.

Sewall Wright was born on 21 December 1889 in Melrose, Massachusetts and died in Madison, Wisconsin on 3 March 1988 at the age of 98. After graduating from Lombard College, he was committed to biology, confirmed by graduate school at the University of Illinois and summers at Cold Spring Harbor. As assistant to William Ernest Castle at Harvard, he specialized in genetics, which he pursued in appointments at the US Department of Agriculture, the University of Chicago and (in official retirement) the University of Wisconsin. With R. A. Fisher and J. B. S. Haldane, he developed evolutionary theory in which selection was the primary force, but he uniquely emphasized the role of stochastic factors. He derived solutions of the Kolmogorov forward equation, which describes changes of gene frequency with time, and prepared the way for Kimura's work on neutral mutations when the extent of nucleotide diversity became known. Whereas Haldane and Fisher considered stochastic factors to be negligible in evolution, Wright's contributions were guided by his 'shifting balance' theory in which population subdivision and contraction caused genetic drift and facilitated intergroup selection in a changing environment, thereby leading to a higher fitness than would be likely under panmixia in the species as a whole. The shifting balance theory has not been formulated mathematically nor tested by experiment or observation. Although controversial, it has influenced many biologists, either directly or through the advocacy of Th. Dobzhansky.

Wright revolutionized animal breeding through his papers on systems of mating, inbreeding theory, effective population size and prediction of response to selection. Many of his results were obtained through path analysis, which he developed to distinguish correlation between observed variables from causation by latent variables (such as genes in the premolecular era). After extensive use by animal breeders from 1920 to 1960, path analysis became influential in the social sciences and in human genetics to test hypotheses about family resemblance. Today, the principles of path analysis are usually expressed as equivalent variance component models. His inbreeding theory was subsequently derived in terms of probability by Malecot and his successors, gaining clarity but not altering the equations. Wright introduced the hierarchical model for population structure and the concept of isolation by distance, which have been widely applied in genetics and anthropology, especially in the study of human diversity. Extensions include forensic use of deoxyribonucleic acid identification and positional cloning by linkage disequilibrium.

Wright also contributed to mammalian genetics through his studies on guinea pigs. Much of his theoretical work was stimulated by these experiments in physiological and developmental genetics, to which he applied models of pigment chemistry, enzyme kinetics and gene homology that were in advance of his time. His greatest achievement remains his impact on genetics and evolutionary biology, for which he was awarded 10 honorary degrees. There is little in population genetics that does not derive in part from his seminal work, which spanned most of the twentieth century and continues to be applied and extended by his academic descendants.

See also

Evolution: Views of
Fisher, Ronald Aylmer
Haldane, John Burdon Sanderson
Malecot, Gustave
Population Genetics: Historical Aspects

Further Reading

Crow JF (1994) Sewall Wright. *Biographical Memoirs*, vol. 64, pp. 439–455. Washington, DC: National Academy Press.

Hill WG (1996) Sewall Wright's 'Systems of mating'. *Genetics* **143**: 499–506. [Reprinted in Crow JF and Dove WF (2000) *Perspectives on Genetics*. Madison, WI: University of Wisconsin Press.]

Provine WB (1986) *Sewall Wright and Evolutionary Biology*. Chicago, IL: University of Chicago Press.

Wrongful Birth, Wrongful Life

See Genetic Disability and Legal Action: Wrongful Birth, Wrongful Life

X and Y Chromosomes: Homologous Regions

Nisrine El-Mogharbel, *Australian National University, Canberra, Australia*

Jennifer AM Graves, *Australian National University, Canberra, Australia*

The human X and Y chromosomes have quite different DNA and gene contents, and seem to be completely cytologically nonhomologous. However, they share many homologous regions, the relics of a process of gradual evolution from a homologous autosomal pair into the differentiated pair they are today.

Intermediate article

Article contents

Introduction

Sex determination in humans relies on an X/Y sex chromosome mechanism. Normal females have an XX sex chromosome complement; males are XY. The presence of the Y determines maleness via the testis-determining factor, identified as the gene *SRY* on the distal short arm of the Y at Yp11.32 (Sinclair, 1990). The human X and Y chromosomes have quite different deoxyribonucleic acid (DNA) and gene contents, and seem to be completely cytologically nonhomologous. However, they share many homologous regions, the relics of a process of gradual evolution from a homologous autosomal pair into the differentiated pair they are today.

The human X chromosome is a large, submetacentric chromosome with about 165 megabases (Mb) of DNA. It contains some 1500 genes that are highly conserved between species. Although the X is present in two copies in females and a single copy in males, X inactivation in females ensures that only one X is active in both sexes. In marked contrast, the human Y chromosome is a small, gene-depleted, submetacentric chromosome containing about 60 Mb of DNA. The distal 30 Mb of Yq consists of repetitive noncoding DNA, cytologically distinguished as differentially stained heterochromatin. The rest of the Y is euchromatic, although it too has a high content of repeats. It spans roughly 35 Mb and bears only about 30 functional genes, some of which are part of gene families and a number of others that are putative (see the map of Y genes at the National Center for Biotechnology Information website). Many genes on the Y have male-specific functions, and several genes have homologs on the X chromosome.

Four types of shared regions are found on the sex chromosomes. Two homologous pseudoautosomal regions at either end of the X and Y are identical, and pair and recombine at meiosis. Within the nonrecombining differential region (YDR), there are also genes and pseudogenes (nonfunctional copies) that have copies on the X, as well as several blocks of

repeated sequences shared by the X and Y. These regions have separate evolutionary histories.

Pseudoautosomal Regions and Meiosis

In the testis and ovary of mammals, chromosome pairing at meiosis between structurally identical pairs of homologous autosomes is facilitated by their sequence near-identity. Homologs pair and exchange their genetic information by homologous recombination (crossing-over), wherein they break and rejoin their DNA molecules without loss of genetic material. Pairing and recombination together are essential for random segregation of chromosomes to the poles of the dividing germ cells.

Although they are morphologically distinct, the X and Y chromosomes pair during male meiosis. They exchange sequence information within small regions of homology known as pseudoautosomal regions (PARs). These regions lie at two ends of the sex chromosomes; PAR1 is on the short arms and PAR2 is on the long arms (Rappold, 1993) (**Figure 1**).

PAR1

PAR1 extends over 2.6 Mb at the tips of the short arms of the X and Y. It is one of the longest X/Y homologous regions. PAR1 is the site of an obligatory cross-over during male meiosis. It is required for correct meiotic segregation of the sex chromosomes, as deletion of PAR1 results in failure of pairing and male sterility.

Genes within the Y have been mapped by sequencing bacterial artificial chromosome (BAC) contigs, and by linkage analysis. The physical map exactly corresponds to the linkage map. There is a gradient of sex linkage whereby genes close to the telomere act as if they are autosomal (hence the name), but genes closer to the boundary that separates PAR1 and the sex-specific region behave as if they are sex-linked. The

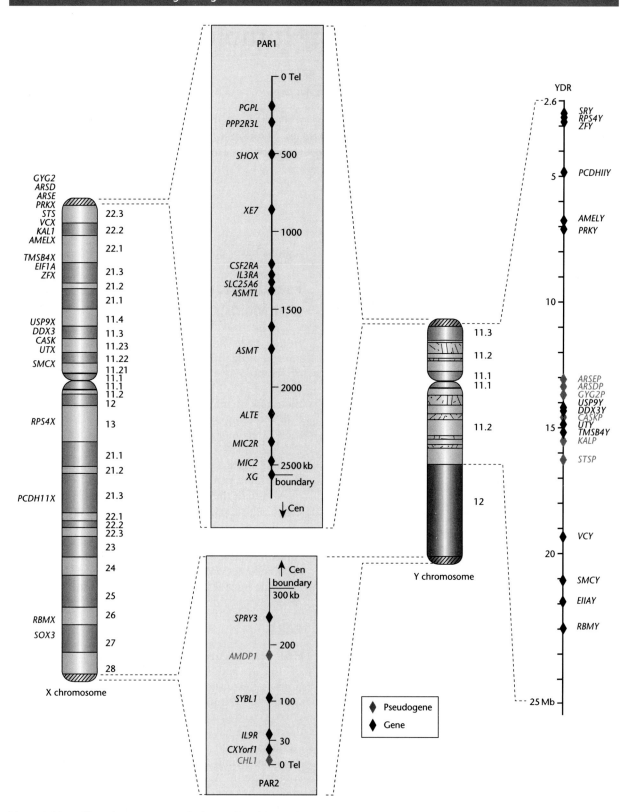

Figure 1 Map of homologous genes on the human X and Y sex chromosomes. Genes (black) and pseudogenes (gray) of PAR1, PAR2 and the YDR are shown as diamonds. Pseudogene positioning is tentative. The hashed regions represent X/Y homologous blocks. The PAR scale is in kb; the Y scale is in Mb and represents the euchromatic YDR. Cen: centromere; Tel: telomere. (Modified from Ried *et al.* (1998), Lahn and Page (1999), Ciccodicola *et al.* (2000), Tilford *et al.* (2001) and the Human Genome Resources website of NCBI.)

recombination rate within PAR1 is normal in females; however, it is very high (approaching 50%) in males, reflecting the obligatory crossover in PAR1. The boundary that separates PAR1 from the differentiated (nonrecombining) region of the human Y has an Alu element (Rappold, 1993).

To date, 13 genes have been found to lie in PAR1 (**Table 1**). Since these genes are expressed from both sex chromosomes, it makes sense that they are not subject to X-inactivation in females. Many PAR1 genes fall into groups that share sequence fragments (*ASMTL/ASMT/XE7*, *MIC2/MIC2R/XG* and *IL3RA/CSF2RA*). Others (including *SLC25A6*, *PPP2R3L* and *PGPL*) bear sequence similarity to genes elsewhere in the genome.

PAR1 genes are implicated in several human diseases. *PPP2R3L* is suggested to be involved in breast and kidney cancer, and possibly with a common lymph node cancer, Hodgkin disease. Some of the phenotypic abnormalities in XO females with Turner syndrome probably result from haploinsufficiency (single dosage) of a PAR1 gene. Their short stature may be ascribed to monosomy for *SHOX*, which is highly expressed in osteogenic tissue. Single mutations or

deletions of *SHOX* can lead to idiopathic short stature, as well as Leri–Weill dyschondrosteosis (Blaschke and Rappold, 2000).

PAR2

The pseudoautosomal region on the tip of the long arms of the X and Y is only 320 kilobases (kb) long. Unlike PAR1, crossing-over between the X and Y is rare (only 2%) in PAR2, although this still represents a frequency higher than that for female meiosis in this region (Freije *et al.*, 1992). Recombination in PAR2 represents reciprocal exchange of information rather than exchange of flanking sequences (gene conversion), as evidenced by the involvement of a single sequence of DNA (Li and Hamer, 1995). The boundary between PAR2 and the nonrecombining differential region occurs within repetitive sequences of long interspersed nuclear elements (LINE elements).

Complete sequencing of PAR2 reveals four genes and two fragmentary pseudogenes (nonfunctional copies of exons) (**Table 1** and **Figure 1**). On the basis of GC content of repetitive sequences (such as Alu and Long interspersed nuclear elements (LINEs)) and mode of expression, PAR2 was subdivided into two characteristic subregions. Two PAR genes, *SPRY3* and *SYBL1* (**Table 1**), are inactivated on both the inactive X and the Y chromosomes, a newly discovered mechanism that maintains equal dosage between the sexes. The distal zone harbors the other two genes, *IL9R* and *CXYorf1*, which behave like autosomal genes, as they escape X-inactivation and are expressed from both the X and Y (Ciccodicola *et al.*, 2000).

PAR2 pairing seems to be neither necessary nor sufficient for successful male meiosis. This is evidenced by the absence of PAR2 crossover in most males and by the normal segregation and phenotype of male carriers of Y chromosomes with PAR2 deleted (Kuhl *et al.*, 2001).

Table 1 Known genes (and pseudogenes) on the human pseudoautosomal regions

Region	Gene symbol	Name
PAR1	*PGPL*	Pseudoautosomal GTP-binding protein-like
	PPP2R3L	Protein phosphatase 2 regulatory subunit B
	SHOX	Short stature homeobox
	XE7	Pseudoautosomal gene XE7
	CSF2RA	Colony-stimulating factor 2 receptor
	IL3RA	Interleukin 3 receptor
	SLC25A6	Solute carrier family 25, member 6
	ASMTL	Acetylserotonin *O*-methyltransferase-like
	ASMT	Acetylserotonin *O*-methyltransferase
	ALTE	Ac-like transposable element
	MIC2R	MIC2-related
	MIC2	Antigen identified by monoclonal antibodies 12E7, F21 and O13
	XG	XG blood group
PAR2	*SPRY3*	Sprouty homolog 3 (*Drosophila*)
	AMDP1	*S*-adenosylmethionine decarboxylase pseudogene 1
	SYBL1	Synaptobrevin-like 1
	IL9R	Interleukin 9 receptor
	CXYorf1	CXYorf1 pseudoautosomal region gene
	CHL1	Cell adhesion molecule with homology to L1CAM

Differential (Nonrecombining) Region

X/Y shared genes in the YDR

It is impossible to make a recombination map of this region, and it has been difficult to construct a physical map because of the high concentration of repetitive elements. A deletion map and an early yeast artificial chromosome (YAC) map have been constructed. Recently, a high-resolution physical map of the Y was made using genomic subtraction and sequence family variants (Tilford *et al.*, 2001). The 35-Mb euchromatic region of the Y contains 30 genes in the

YDR that have been recovered by positional cloning and by screening a testis complementary DNA (cDNA) library with Y-specific YACs.

Many single-copy, Y-borne genes in this non-recombining region have an obvious X homolog. Most are ubiquitously expressed. These characteristics were used to distinguish these genes from others that are multicopy and have testis-specific expression (Lahn and Page, 1997). However, the distinction between these class I and class II genes breaks down if other species are considered; for instance, ZFY is considered class I in humans and class II in mouse. Also, two proto-type class II genes, RBMY1A1 (candidate spermatogenesis gene) and SRY, have both been found to have X-linked homologs (Foster and Graves, 1994; Delbridge et al., 1999). Class I genes are likely to be implicated in the etiology of Turner syndrome and in growth defects and gonadoblastoma, whereas class II genes could control gonadal sex reversal and spermatogenic failure. X/Y shared and Y-specific genes are intermingled within YDR (Lahn and Page, 1997) (**Table 2** and **Figure 1**).

X/Y shared repetitive blocks

Several regions of extensive homology occur within repetitive sequence blocks in the nonrecombining

Table 2 Known X/Y shared genes in the nonrecombining region

Gene symbol	Name
GYG2/P	Glycogenin 2
ARSD/P	Arylsulfatase D
ARSE/P	Arylsulfatase E
PRKX/Y	Protein kinase
STS/P	Steroid sulfatase
KAL1/P	Kallmann syndrome
AMELX/Y	Amelogenin
TMSB4X/Y	Thymosin, β 4
EIF1A/EI1AY	Eukaryotic translation initiation factor 1 A
ZFX/Y	Zinc-finger protein
USP9X/Y	Ubiquitin-specific protease 9
DDX3/Y	DEAD/H (Asp-Glu-Ala-Asp/His) box polypeptide 3
CASK/CASKP	Calcium/calmodulin-dependent serine protein kinase
UTX/Y	Ubiquitously transcribed tetratricopeptide repeat
SMCX/Y	Smcy homolog, X/Y chromosome (mouse)
VCX/Y	Variable charge
RPS4X/Y	Ribosomal protein S4
PCDH11X/Y	Protocadherin 11
RBMX/Y	RNA-binding motif protein
SOX3/SRY	SRY-box 3/sex-determining region Y

The X homolog is listed first; the Y homolog may be a pseudogene, denoted by P.

region of the human X and Y chromosomes (**Figure 1**). Regions that include highly conserved sequence are shared by Xp/Yq, Xq/Yp and Xq/Yq and are located at Yp11.1/Xp21.2–22.1 (ZFX/Y); Yp11.1/Xq13 (RPS4X/Y); Yp11.1/Xq21.1–21.3; Yp11.1/Xq22.3–ter; Yp11.1/Xp21.2–22.1 (AMELX/Y); Yp11.1/Xq13; Yp11.1/Xq21.1–21.3; Ycen–q11.21/Xp22.3–pter; Yq11.22–q11.3/Xq28. A few of these regions have been studied extensively, while little is known about the size, internal structure and gene content of others. This complex pattern of sequence similarity between the X and Y is the product of transpositions, inversions and other rearrangements during evolution (Affara and Ferguson-Smith, 1994).

The Xp22.3/Yq11.2 homologous region is implicated in rare X–Y translocations. The order of markers such as STS and KAL1 within this segment of the X lie in the reverse order to their pseudogene counterparts on the Yq. The order of other loci within this region is also reversed, indicating a pericentric inversion (Rappold, 1993). A hot spot for ectopic recombination was identified on Xp22.3/Yp11. Sequence analysis of the X–Y junctions in rearranged X chromosomes of XX males revealed high sequence homology in this region. X–Y pairing and interchange promoted by this high homology results in XX males whose X contains SRY from the translocated Yp11.1 segment (Weil et al., 1994).

The most extensive (4 Mb) region of homology in Xq21.3/Yp11.1 was revealed by making YAC contigs of this region and comparing them with Y deletion panels. Overall, sequence analysis reveals 99% sequence identity between the X and Y sequence. Marker order on the Y shows an inversion of this homologous block. The only gene described so far on Xq21.1/Yp11.1 is PCDH11X/Y. Exon trapping experiments found several pseudogenes (Sargent et al., 2001). In addition, a small segment within the 4-Mb segment was found to be X-specific (Mumm et al., 1997). A more recent detailed analysis of this X-specific region showed a complex structure with loss of particular small subregions that produce X-specific markers (Sargent et al., 2001). Individuals who are monosomic with respect to this homologous region show minor features of Turner syndrome, suggesting that a gene with a minor role in Turner syndrome is located here, but more critical genes lie elsewhere on the X and Y (Schwartz et al., 1998).

Evolution of X/Y shared regions

Y chromosome degradation and addition

X/Y homology provides evidence supporting the hypothesis that the X and Y chromosomes differentiated from an ancestral autosome pair when one member of this pair acquired a male-determining locus (presumed to be *SRY* in mammals). Other alleles with a male-specific function then accumulated nearby, and to prevent disruption of this male-specific package, recombination was suppressed. This allowed sexually antagonistic genes to evolve. Genes such as the candidate spermatogenesis gene *RBMY1A1* benefit one sex and are favored by selection in that sex, but may be disadvantageous in the other (Rice, 1992). Suppression of recombination then leads to the accumulation of mutations, insertions and deletions in the genetically isolated region. The mutant alleles either 'hitchhiked' along with a favorable allele in the nonrecombining region or drifted to fixation as Y chromosomes with fewer mutations were lost, by chance, in a process known as Muller's ratchet (Charlesworth, 1991). Loss of active genes and insertion of repetitive elements into the Y chromosome rendered parts of it essentially transcriptionally inert and composed largely of repetitive DNA sequence, leaving intact only the most terminal segments (PAR1) and genes with a selectable function in males.

Localization of human X-linked genes in distantly related mammals (marsupials and monotremes) revealed that genes on the long arm and pericentromeric region of the human X are conserved on the X in all mammals. Chromosome painting, using DNA from a flow-sorted marsupial X to hybridize *in situ* to human chromosomes, confirms this division. This conserved region of the X represents the original mammalian X before the divergence of extant mammalian groups 170 million years ago.

However, genes from the short arm of the human X (including the PAR1) cluster on an autosome in both marsupials and monotremes, and therefore represent a recently added autosomal region. Similarly, the human Y comprises a conserved and added region (Graves, 1995, 2000), which accounts for all but about 10 Mb of the human Y (Waters *et al.*, 2001). The paucity of genes within the added region of the Y, as well as the conserved region, implies that the added region has also been subjected to degradation on the Y and compensatory inactivation on the X. Addition of autosomal material onto the X and Y probably occurred by translocation onto the PAR of an ancient X and Y that was only partially differentiated. It greatly augmented the XY homologous region.

Evolution of PAR1

PAR1 therefore represents the last vestige of homology of the Y added region. At the interface between sex-specific and PAR1 sequences on the Y are Alu repeats that were inserted and shortened on an earlier PAR in a ratchet-like mode (Kvaloy *et al.*, 1994).

The location of the PAR1 genes within a region of high recombination frequency facilitated gene duplication events. This allowed the creation of new gene fragments and combinations created by fusions, as observed in such groups as *MIC2/MIC2R/XG*, driven by unequal crossing-over in the ancestral PAR1 (Ried *et al.*, 1998). The PAR1 is completely nonconserved with the pseudoautosomal region shared by the mouse X and Y, suggesting that the PAR is subject to high variability, for reasons that are quite obscure.

Evolution of X/Y shared genes in the differential region

X/Y shared genes may have diverged at different times according to their origin in the ancient sex chromosomes or the added region and the course of differentiation. For example, *SOX3/SRY* is located in the ancient conserved region and shared between marsupials and eutherians, whereas *ZFX/Y* is autosomal in marsupials and was recently added to the human X and Y. These ancient and added regions were further subdivided by the sequence divergence between the X and Y homologs (Lahn and Page, 1999). Gene pairs that were most divergent represent the first region to differentiate. The most ancient evolutionary stratum 1 includes *SOX3/SRY*, *RBMX/Y* and *RPS4X/Y*. Genes with the lowest divergence (and therefore the most recently separated) include the genes in the recently added region *PRKX/Y*, *STS/P* and *KAL1/P*. The PAR represents a fifth nondivergent stratum. The oldest two strata correspond to the conserved regions of the X and Y, defined by comparative mapping, and the three youngest ones to the added region.

Genes on the ancient region of the Y chromosome have presumably been exempt from degradation on the Y because they confer a selective advantage. Since they are present only in males, the selectable function is likely to be either male-specific or to maintain dosage. The Y homolog of many of these genes has acquired a male-specific function (e.g. sex determination by *SRY* and spermatogenesis by *RBMY*). Ubiquitously expressed genes on the Y such as *RPS4Y* may be required in two doses in both sexes. Despite some divergence at the nucleotide level, the X and Y homologs of this gene have been demonstrated to be functionally interchangeable (Lahn and Page, 1997)

The evolution of male-specific genes such as *SRY* has been suggested to be driven by a selective pressure to differentiate species. Whitfield *et al.* (1993) measured the frequency of synonymous (Ks) and nonsynonymous (Ka) changes (i.e. those that affect the amino acid) and concluded from the Ka/Ks ratio that directional selection did occur in primates. However, O'Neill *et al.* (1997) found no evidence of directional selection in the large species complex represented by rock wallabies (*Petrogale*).

Genes located within the nonrecombining region have a high probability of accumulating mutations and undergoing loss of function. Gene conversion has been proposed to explain how gene function can be maintained in this region without recombination. Gene conversion is a nonreciprocal transfer of sequence information between a pair of allelic or nonallelic sequences, perhaps resulting from homology searches prior to and fundamental to synapsis (pairing) of homologous chromosomes (Slattery *et al.*, 2000).

Evolution of X/Y shared sequence blocks

X/Y homology outside of the PARs could be the relics of an ancestral PAR. Mapping in primates reveals that the pericentric inversion on Xp22.3/Yp11.1 occurred about 40 million years ago, when higher primates diverged from prosimians. The homology of genes (*STS*, *KAL*) to pseudogenes on the Y within this region is evidence of divergence of the Y copy after the inversion on the Y and exclusion from the PAR. However, the Xp22.3/Yq11.3 region was found to be sex-specific in Old World monkeys, and so it must have been excluded from the PAR before divergence of the higher primates and then homology maintained by gene conversion (Rappold, 1993; Weil *et al.*, 1994).

The Yp inversion of the 4-Mb region shared with Xp22.3/Yp11.1 is present in all males from diverse human populations, but not in chimps. Thus, X/Y transposition and the subsequent inversion must have occurred after the divergence of hominid and chimp lineages. Most of the pseudogenes identified within this region contain no exons, so were probably retroposed (Sargent *et al.*, 2001). The X-specific segment, found within this region, originated as an insertion onto the X via an L1-mediated mechanism. It remains X-specific throughout primate evolution and so was lost from the Y in humans subsequent to the transposition event from the X to the Y (Schwartz *et al.*, 1998; Vacca *et al.*, 1999).

Similarly, comparisons between primates showed that other homologous blocks within Xq/Yp and Xq/Yq resulted from duplication and transposition during primate evolution (Affara and Ferguson-Smith, 1994).

Evolution of PAR2

PAR2 at the terminus of Xq and Yq has quite a different origin from PAR1. The four genes in human PAR2 are present on the X in great apes, but have no Y homologs. This points to a recent event in the hominid lineage that probably represents a transfer of the 320 kb from the X to the Y in a recent illegitimate recombination exchange event facilitated by LINE repetitive elements late in the primate mammalian lineage. Fragmentary elements in PAR2 are related to subtelomere sequences at the end of other human chromosomes, consistent with frequent genetic exchange (Ciccodicola *et al.*, 2000).

Conclusion

Despite having diverged almost entirely in gene content and sequence, the X and Y chromosomes still show vestiges of homology that point to a common origin from an autosomal pair. Homology varies between near identity in the PARs and some repetitive blocks to isolated and divergent gene pairs in the YDR. These homologous regions have different evolutionary histories that explain the extent of divergence.

See also

Chromosome X: General Features
Chromosome Y: General and Special Features
Mammalian Sex Chromosome Evolution
Sex Chromosomes

References

Affara NA and Ferguson-Smith MA (1994) DNA sequence homology between the human sex chromosomes. In: Wachtel SS (ed.) *Molecular Genetics of Sex Determination*, pp. 225–266. San Diego, CA: Academic Press.

Blaschke RJ and Rappold GA (2000) SHOX: growth, Leri–Weill and Turner syndromes. *Trends in Endocrinology and Metabolism* **11**: 227–230.

Charlesworth B (1991) The evolution of sex chromosomes. *Science* **251**: 1030–1033.

Ciccodicola A, D'Esposito M, Esposito T, *et al.* (2000) Differentially regulated and evolved genes in the fully sequenced Xq/Yq pseudoautosomal region. *Human Molecular Genetics* **9**: 395–401.

Delbridge ML, Lingenfelter PA, Disteche C and Graves JAM (1999) The candidate spermatogenesis gene *RBMY* has a homologue on the human X chromosome. *Nature Genetics* **22**: 223–224.

Foster JW and Graves JAM (1994) An SRY-related sequence on the marsupial X chromosome: implications for the evolution of the mammalian testis-determining gene. *Proceedings of the National Academy of Sciences of the United States of America* **91**: 1927–1931.

Freije D, Helms C, Watson MS and Donis-Keller H (1992) Identification of a second pseudoautosomal region near the Xq and Yq telomeres. *Science* **258**: 1784–1787.

Graves JAM (1995) The origin and function of the mammalian Y chromosome and Y-borne genes: an evolving understanding. *BioEssays* **17**: 311–320.

Graves JAM (2000) Human Y chromosome, sex determination, and spermatogenesis: a feminist view. *Biology of Reproduction* **63**: 667–676.

Kuhl H, Rottger S, Heilbronner H, Enders H and Schempp W (2001) Loss of the Y chromosomal PAR2-region in four familial cases of satellited Y chromosomes (Yqs). *Chromosome Research* **9**: 215–222.

Kvaloy K, Galvagni F and Brown WR (1994) The sequence organization of the long arm pseudoautosomal region of the human sex chromosomes. *Human Molecular Genetics* **3**: 771–778.

Lahn BT and Page DC (1997) Functional coherence of the human Y chromosome. *Science* **278**: 675–680.

Lahn BT and Page DC (1999) Four evolutionary strata on the human X chromosome. *Science* **286**: 964–967.

Li L and Hamer DH (1995) Recombination and allelic association in the Xq/Yq homology region. *Human Molecular Genetics* **4**: 2013–2016.

Mumm S, Molini B, Terrell J, Srivastava A and Schlessinger D (1997) Evolutionary features of the 4-Mb Xq21.3 XY homology region revealed by a map at 60-kb resolution. *Genome Research* **7**: 307–314.

O'Neill RJ, Eldridge MD, Crozier RH and Graves JA (1997) Low levels of sequence divergence in rock wallabies (*Petrogale*) suggest a lack of positive directional selection in *SRY*. *Molecular Biology and Evolution* **14**: 350–353.

Pecon Slattery J, Sanner-Wachter L and O'Brien SJ (2000) Novel gene conversion between X–Y homologues located in the nonrecombining region of the Y chromosome in Felidae (Mammalia). *Proceedings of the National Academy of Sciences of the United States of America* **97**: 5307–5312.

Rappold GA (1993) The pseudoautosomal regions of the human sex chromosomes. *Human Genetics* **92**: 315–324.

Rice WR (1992) Sexually antagonistic genes: experimental evidence. *Science* **256**: 1436–1439.

Ried K, Rao E, Schiebel K and Rappold GA (1998) Gene duplications as a recurrent theme in the evolution of the human pseudoautosomal region 1: isolation of the gene *ASMTL*. *Human Molecular Genetics* **7**: 1771–1778.

Sargent CA, Boucher CA, Blanco P, *et al.* (2001) Characterization of the human Xq21.3/Yp11 homology block and conservation of organization in primates. *Genomics* **73**: 77–85.

Schwartz A, Chan DC, Brown LG, *et al.* (1998) Reconstructing hominid Y evolution: X-homologous block, created by X–Y transposition, was disrupted by Yp inversion through LINE–LINE recombination. *Human Molecular Genetics* **7**: 1–11.

Sinclair AH, Berta P, Palmer MS, *et al.* (1990) A gene from the human sex-determining region encodes a protein with homology to a conserved DNA-binding motif. *Nature* **346**: 240–244.

Tilford CA, Kuroda-Kawaguchi T, Skaletsky H, *et al.* (2001) A physical map of the human Y chromosome. *Nature* **409**: 943–945.

Vacca M, Matarazzo MR, Jones J, *et al.* (1999) Evolution of the X-specific block embedded in the human Xq21.3/Yp11.1 homology region. *Genomics* **62**: 293–296.

Waters PD, Duffy B, Frost CJ, Delbridge ML and Graves JAM (2001) The human Y chromosome derives largely from a single autosomal region added to the sex chromosomes 80–130 million years ago. *Cytogenetics and Cell Genetics* **92**: 74–79.

Weil D, Wang I, Dietrich A, *et al.* (1994) Highly homologous loci on the X and Y chromosomes are hot-spots for ectopic recombinations leading to XX maleness. *Nature Genetics* **7**: 414–419.

Whitfield LS, Lovell-Badge R and Goodfellow PN (1993) Rapid sequence evolution of the mammalian sex-determining gene *SRY*. *Nature* **364**: 713–715.

Web Links

The Human Genome. National Center for Biotechnology Information's guide to online information resources.
http://www.ncbi.nlm.nih.gov/genome/guide/human/

SRY (sex determining region Y)-box 3 (*SOX3*); LocusID: 6658. LocusLink:
http://www.ncbi.nlm.nih.gov/LocusLink/LocRpt.cgi?l = 6658

sex determining region Y (*SRY*); LocusID: 6736. LocusLink:
http://www.ncbi.nlm.nih.gov/LocusLink/LocRpt.cgi?l = 6736

zinc-finger protein, X-linked (*ZFX*); LocusID: 7543. LocusLink:
http://www.ncbi.nlm.nih.gov/LocusLink/LocRpt.cgi?l = 7543

zinc-finger protein, Y-linked (*ZFY*); LocusID: 7544. LocusLink:
http://www.ncbi.nlm.nih.gov/LocusLink/LocRpt.cgi?l = 7544

SRY (sex determining region Y)-box 3 (*SOX3*); MIM number: 313430. OMIM:
http://www.ncbi.nlm.nih.gov/htbin-post/Omim/dispmim?313430

sex determining region Y (*SRY*); MIM number: 480000. OMIM:
http://www.ncbi.nlm.nih.gov/htbin-post/Omim/dispmim?480000

zinc-finger protein, X-linked (*ZFX*); MIM number: 314980. OMIM:
http://www.ncbi.nlm.nih.gov/htbin-post/Omim/dispmim?314980

zinc-finger protein, Y-linked (*ZFY*); MIM number: 490000. OMIM:
http://www.ncbi.nlm.nih.gov/htbin-post/Omim/dispmim?490000

X Chromosome

See Chromosome X

X-chromosome Inactivation

Intermediate article

Matthew J Wakefield, *Australian National University, Canberra, Australia*

Jennifer AM Graves, *Australian National University, Canberra, Australia*

X-chromosome inactivation is the epigenetic silencing of one of the two X chromosomes in female mammals that equalizes dosage of X-linked genes between males and females.

Article contents

- X-chromosome Inactivation
- Lyon Hypothesis
- Molecular Basis of X-chromosome Inactivation
- Nonrandom X-chromosome Inactivation
- Genes that Escape X-chromosome Inactivation
- Conclusion

X-chromosome Inactivation

X-chromosome inactivation (XCI) turns off the majority of the genes on one of the two X chromosomes in females, a unique example of large-scale gene regulation. Mammals have an XX female : XY male chromosome sex determination system that is based on a male-determining Y chromosome. The X is large and contains approximately 1000 genes. By contrast, the Y chromosome is a small, mainly heterochromatic chromosome that contains few genes other than the testis-determining factor and spermatogenesis genes. The X and Y pair over only a small region of homology and share about 14 genes in the non-recombining region. Thus, most genes on the X have no partner on the Y and are present in only one copy in males. These heteromorphic sex chromosomes present an obvious difference in gene dosage. It is apparent from the phenotype of trisomies such as Down syndrome that the dosage of at least some genes is finely balanced and is critical for normal development and function. It is therefore not surprising that mechanisms have evolved to compensate for the differences in genetic dosage caused by sex chromosome heteromorphy. The importance of this dosage compensation system is underlined by the observation that mutations that disrupt the inactivation system are lethal early in female development.

Lyon Hypothesis

X chromosome inactivation was first proposed by Mary Lyon (1961). Her hypothesis was based on the observations that several X-linked, mutated, coat color genes result in a mosaic phenotype consisting of patches of mutant color and wild-type color in heterozygous female mice. Lyon proposed that one of the X chromosomes was randomly inactivated in each somatic cell of a female embryo early in development. This inactive state is then stably inherited, resulting in patches of mutant and wild-type cells (**Figure 1**). The sizes and shapes of the patches depend on the migration of cells and the proportion of cells in which the normal and mutant alleles are active, and vary between individuals by chance. Since mosaicism is also apparent for coat color in tortoiseshell cats and in the eyes of human patients with sex-linked ocular albinism, as well as other genes with localized action, Lyon correctly predicted that X-chromosome inactivation is a general phenomenon occurring in all eutherian mammals.

The Lyon hypothesis was supported by observations that one X chromosome was condensed into a heteropyknotic sex chromatin body. The inactive X chromosome was found to replicate later than the active X and, combined with the heterochromatinization of the X, indicated that inactivation is a whole-chromosome effect (Lyon, 1962).

Molecular Basis of X-chromosome Inactivation

The inactivation of the X chromosome in mammals results from transcriptional repression. It is a complex,

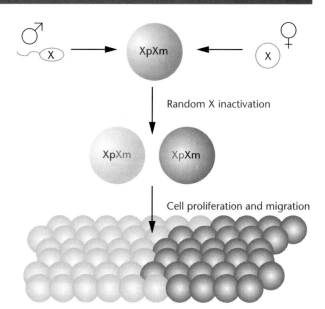

Figure 1 In female cells one X chromosome is inherited from each parent. Random inactivation means that in some cells the paternally inherited X chromosome is inactivated, while in other cells the maternal X is inactivated. Cell proliferation and migration continue after inactivation and results in clumps of cells with either the paternal or maternal X inactivated. When the X chromosomes have different alleles this is apparent as clumps of cells with a different phenotype, such as the coat pattern in tortoise shell cats.

multistep process that involves changes in chromosome conformation, late deoxyribonucleic acid (DNA) replication, DNA methylation, nonprotein-coding ribonucleic acids (RNAs), histone modification and accumulation and exclusion of specific histone isoforms.

Initiation

Initiation of X-chromosome inactivation involves two critical mechanisms. An X-chromosome counting mechanism determines whether inactivation is required, and a mechanism chooses at random the X chromosome to inactivate. In cells that contain more than two X chromosomes, a single X chromosome is maintained in the active state, and all other X chromosomes are inactivated (Lyon, 1962). This X-chromosome counting mechanism may act by preventing only one X chromosome from inactivating; however, the molecular basis of counting and choice remain unknown.

The most likely mechanism for counting and choice is modulation of the expression of the gene *X-inactive specific transcript* (*XIST*). *XIST* expression is the first molecular change to occur in X inactivation. *XIST* is expressed only from the inactive X chromosome and produces a long (17-kilobase (kb)) RNA molecule that does not encode a protein (Brown *et al.*, 1991, 1992).

XIST is essential for X-chromosome inactivation. Knockout mice that lack a functional *XIST* gene cannot inactivate the X chromosome carrying the mutant allele, causing female embryonic lethality. Several mechanisms have been suggested for the control of *XIST* expression, including differential RNA stability, transcription from different promotors, differential promotor methylation and acetylation, limiting transcription factors and control of an antisense transcript (*TSIX*). Dissecting the mechanism of these initial switches in X inactivation is an area of active research and scientific debate (Willard and Carrel, 2001).

Spreading

A critical region that must be present for inactivation to occur was defined as the X inactivation center by both genetic mapping of allelic variants and by X–autosome translocations. The gene *XIST* was cloned from this region. It has the unique property of being expressed only from the inactive X chromosome. X inactivation spreads out from this region along the entire X chromosome. *XIST* RNA is untranslated but is believed to have a major role in this spreading, as it coats the entire inactive X chromosome (Clemson *et al.*, 1996). The mechanism of this spreading, and the means by which *XIST* RNA stays associated with the X chromosome, is not yet understood. It has been speculated that 'waystation' elements on the X chromosome may associate with *XIST* RNA (Lyon, 2000). This theory is supported by the limited spread of *XIST* RNA into autosome regions that are translocated onto the X chromosome.

Although the mechanism of action of *XIST* is unknown, it probably controls X inactivation by modulating the composition and structure of chromatin. This is supported by the observation that in cells in which *XIST* has been knocked out, there is not the normal coalescence of a histone isoform macroH2A, which is usually associated with the initiation of X inactivation (Csankovszki *et al.*, 1999).

Maintenance

In eutherian mammals X-chromosome inactivation is extraordinarily stable. Once established, inactivation is faithfully maintained through numerous somatic cell divisions. This epigenetic stability is provided by a multistep maintenance mechanism.

Chromatin modification

Chromatin on the inactive X chromosome is highly modified. There is an accumulation of histone isoforms that are associated with inactive chromatin, such as macroH2A, and an exclusion of isoforms that are associated with highly active chromatin, such as H2AZ (Costanzi and Pehrson, 1998; Chadwick and Willard, 2001). Core histones on the inactive X chromosome are also modified by the removal of acetyl groups from the amino (*N*)-terminal end of the protein, reducing the transcriptional potential of the chromatin (Jeppesen and Turner, 1993). At least one of these modifications is conserved in marsupials and is likely to represent the ancient X inactivation system in the ancestor of all mammals (Wakefield *et al.*, 1997).

Chromosome conformation

The major changes in the structure of the inactive X chromatin result in a visible change in the X chromosome, which forms a distinct, tightly packed heteropyknotic body at interphase. Sex chromatin, often observed on the periphery of the nucleus, may represent a distinct nuclear compartment in which the inactive X is physically isolated from the transcriptional processes in the rest of the nucleus. The inactive X does not change in volume, but is more spherical than the active X chromosome. The reduction in the surface area of the inactive X chromosome may be involved in the repression of transcription (Eils *et al.*, 1996).

Methylation

In eutherian mammals, X-chromosome inactivation is stabilized by the methylation of cytosine residues at clusters of CpG sites in the promotor regions of genes. Methylation was originally proposed as the major mechanism by which X inactivation asserts its silencing effect. The first experimental evidence of DNA modification in X inactivation was the ability of purified DNA from the active, but not the inactive, X chromosome to transform a mutant cell line (Liskay and Evans, 1980). Treatment with a demethylating agent was shown to reactivate the inactive X in rodent–human somatic cell hybrids and interspecific mouse hybrid embryo cells, and purified DNA from these reactivated hybrids was shown to transform mutant cells. Analysis of the promoter regions of X-linked genes identified GC-rich regions that were hypermethylated on the inactive X and hypomethylated on the active X. Extraembryonic tissues show no promotor methylation of inactivated genes, and its DNA from the inactive X in the tissues is capable of transforming cells. The failure of demethylating agents to reactivate the inactive X chromosome in cells other than hybrids, and the onset of methylation in X inactivation well after other observable signs of X inactivation differentiating in embryonic stem (ES) cells, implies that methylation is not the primary mechanism of X chromosome inactivation. Methylation is therefore involved in X inactivation, but as a secondary stabilizing system in a multistep system (Gartler *et al.*, 1985). The absence of differential methylation of CpG islands in marsupial X inactivation indicates that the differential methylation evolved recently in the eutherian linage and was not a part of X-chromosome inactivation in the common therian ancestor (Loebel and Johnston, 1996).

There is a strong interaction between DNA methylation and histone de-acetylation. Enzyme complexes that bind hypoacetylated histones methylate DNA, and enzyme complexes that bind methylated DNA deacetylate histones. This bidirectional feedback between the molecular components of X inactivation probably contributes to the extremely robust silencing of X-inactivated genes.

Late replication

One of the most consistent features of the inactive X chromosome is that it replicates later in the cell cycle than the active X chromosome and the autosomes. The mechanism of this late replication and its role in the X inactivation process are unknown. Late replication may be merely a consequence of a less-open chromatin structure or the spacial isolation of the inactive X chromosome in the nucleus; however, it may be important in maintaining the differential features of the inactive X chromosome.

Nonrandom X-chromosome Inactivation

Although most X-chromosome inactivation is random, there are several cases where inactivation is nonrandom.

Paternal inactivation

In the extraembryonic tissues of rodents, X-chromosome inactivation is imprinted so that the paternal X chromosome is always inactivated. This may reflect differences in timing, as extraembryonic tissues are the first to differentiate in the developing embryo, possibly before the random choice mechanism for determining X-chromosome inactivation is active. The commonality of paternal X inactivation in rodent extra-embryonic tissues and in all tissues in marsupials suggests imprinted X inactivation is ancestral, but its apparent absence in humans leaves the question of its evolutionary origins open. Ohlsson et al. (2001) have argued that random X inactivation evolved from stochastic monoallelic expression, which would make random inactivation the ancestral state.

Skewing

The ratio of cells with paternal and maternal inactive X chromosomes is not always 1:1 in random X inactivation. This can be caused by variation at the gene locus that controls X inactivation biasing the choice of the chromosome to be inactivated. The controlling element was initially discovered and genetically mapped using interstrain variation in the extent of coat color variegation (Cattanach and Isaacson, 1967). The XIST locus and its antisense transcript TSIX are probably responsible for this effect. In humans, a mutation in the promotor of XIST has been identified that causes a complete biasing of inactivation (Plenge et al., 1997).

Apparent nonrandom X inactivation commonly results from selection against cells heterozygous for a recessive mutation when the only functional allele is inactivated. When the mutation is in a tissue-specific gene, skewing may be seen in one tissue and not others. Skewing of X inactivation also occurs when autosomes are translocated onto the X chromosome, as spreading of inactivation into the autosomal segment is nearly always deleterious (Migeon, 1998).

Genes that Escape X-chromosome Inactivation

Although X-chromosome inactivation is a whole-chromosome phenomenon in eutherians, it is not complete. Several genes on the human X show partial or complete escape from X inactivation. One class of these exempt genes are located in the pseudoautosomal region of the X chromosome, which pairs and recombines with the Y at meiosis. This region is homologous between the X and Y chromosomes, so it is present in two copies in males as well as females, and dosage compensation is unnecessary. The necessity for dosage equivalence may have provided the pressure for the evolution of boundary elements or other mechanisms to prevent the spread of X inactivation into this region.

Several genes from outside the pseudoautosomal region also escape X chromosome inactivation on the human X. These genes retain active homologs within the male-specific portion of the Y chromosome. The Y-borne copies of many of these genes are not identical, but at least some encode a product similar to the X-linked copy. It is likely that incorporation into the X inactivation system followed degradation of the Y-borne allele. A few exceptional genes that escape X inactivation though they have no Y homologs have lost their Y partner only recently, so that the selective pressure has not yet been strong enough to induce compensatory changes to its dosage by inclusion in the X inactivation system (Graves et al., 1998). This is supported by the clustering of most human genes that escape inactivation to a region on the short arm of the

X (Carrel *et al.*, 1999), which has been shown to have been added recently to the X chromosome (Graves, 1995).

Conclusion

X-chromosome inactivation is a complex, multistep process involving noncoding RNAs, chromatin remodeling and DNA modification in a way that is still poorly understood. Ongoing research into this process promises to illuminate the mechanisms and control of X chromosome inactivation and its implications for basic mechanisms of transcriptional control across the genome.

See also

Chromosome X: General Features
Dosage Compensation Mechanisms: Evolution
Epigenetic Factors and Chromosome Organization
Sex Chromosomes
X and Y Chromosomes: Homologous Regions
X-chromosome Inactivation and Disease

References

Brown CJ, Ballabio A, Rupert JL, *et al.* (1991) A gene from the region of the human X inactivation centre is expressed exclusively from the inactive X chromosome. *Nature* **349**: 38–44.

Brown CJ, Hendrich BD, Rupert JL, *et al.* (1992) The human *XIST* gene: analysis of a 17 kb inactive X-specific RNA that contains conserved repeats and is highly localized within the nucleus. *Cell* **71**: 527–542.

Carrel L, Cottle AA, Goglin KC and Willard HF (1999) A first-generation X-inactivation profile of the human X chromosome. *Proceedings of the National Academy of Sciences of the United States of America* **96**: 14 440–14 444.

Cattanach BM and Isaacson JH (1967) Controlling elements in the mouse X chromosome. *Genetics* **57**: 331–346.

Chadwick BP and Willard HF (2001) Histone H2A variants and the inactive X chromosome: identification of a second macroH2A variant. *Human Molecular Genetics* **10**: 1101–1113.

Clemson CM, McNeil JA, Willard HF and Lawrence JB (1996) *XIST* RNA paints the inactive X chromosome at interphase: evidence for a novel RNA involved in nuclear/chromosome structure. *Journal of Cell Biology* **132**: 259–275.

Costanzi C and Pehrson JR (1998) Histone macroH2A1 is concentrated in the inactive X chromosome of female mammals. *Nature* **393**: 599–601.

Csankovszki G, Panning B, Bates B, Pehrson JR and Jaenisch R (1999) Conditional deletion of *Xist* disrupts histone macroH2A localization but not maintenance of X inactivation. *Nature Genetics* **22**: 323–324.

Eils R, Dietzel S, Bertin E, *et al.* (1996) Three-dimensional reconstruction of painted human interphase chromosomes: active and inactive X chromosome territories have similar volumes but differ in shape and surface structure. *Journal of Cell Biology* **135**: 1427–1440.

Gartler SM, Dyer KA, Graves JA and Rocchi M (1985) A two-step model for mammalian X-chromosome inactivation. *Progress in Clinical and Biological Research* **198**: 223–235.

Graves JA (1995) The evolution of mammalian sex chromosomes and the origin of sex determining genes. *Philosophical Transactions of the Royal Society of London B* **350**: 305–311.

Graves JA, Disteche CM and Toder R (1998) Gene dosage in the evolution and function of mammalian sex chromosomes. *Cytogenetics and Cell Genetics* **80**: 94–103.

Jeppesen P and Turner BM (1993) The inactive X chromosome in female mammals is distinguished by a lack of histone H4 acetylation: a cytogenetic marker for gene expression. *Cell* **74**: 281–289.

Liskay RM and Evans RJ (1980) Inactive X chromosome DNA does not function in DNA mediated cell transformation for the hypoxanthine phosphoribotransferase gene. *Proceedings of the National Academy of Sciences of the United States of America* **77**: 4895–4898.

Loebel DA and Johnston PG (1996) Methylation analysis of a marsupial X-linked CpG island by bisulfite genomic sequencing. *Genome Research* **6**: 114–123.

Lyon MF (1961) Gene action in the X-chromosome of the mouse. *Nature* **190**: 372–373.

Lyon MF (1962) Sex chromatin and gene action in the mammalian X-chromosome. *American Journal of Human Genetics* **14**: 135–148.

Lyon MF (2000) LINE-1 elements and X chromosome inactivation: a function for 'junk' DNA? *Proceedings of the National Academy of Sciences of the United States of America* **97**: 6248–6249.

Migeon BR (1998) Non-random X chromosome inactivation in mammalian cells. *Cytogenetics and Cell Genetics* **80**: 142–148.

Ohlsson R, Paldi A and Graves JA (2001) Did genomic imprinting and X chromosome inactivation arise from stochastic expression? *Trends in Genetics* **17**: 136–141.

Wakefield MJ, Keohane AM, Turner BM and Graves JA (1997) Histone underacetylation is an ancient component of mammalian X chromosome inactivation. *Proceedings of the National Academy of Sciences of the United States of America* **94**: 9665–9668.

Willard HF and Carrel L (2001) Making sense (and antisense) of the X inactivation center. *Proceedings of the National Academy of Sciences of the United States of America* **98**: 10 025–10 027.

Further Reading

Avner P and Heard E (2001) X-chromosome inactivation: counting, choice and initiation. *Nature Reviews Genetics* **2**: 59–67.

Brockdorff N (1998) The role of Xist in X-inactivation. *Current Opinion in Genetics and Development* **8**: 328–333.

Brown CJ and Robinson WP (2000) The causes and consequences of random and non-random X chromosome inactivation in humans. *Clinical Genetics* **58**: 353–363.

Disteche CM (1999) Escapees on the X chromosome. *Proceedings of the National Academy of Sciences of the United States of America* **96**: 14180–14182.

Gartler SM and Goldman MA (1994) Reactivation of inactive X-linked genes. *Developmental Genetics* **15**: 504–514.

Goto T and Monk M (1998) Regulation of X-chromosome inactivation in development in mice and humans. *Microbiology and Molecular Biology Reviews* **62**: 362–378.

Heard E, Clerc P and Avner P (1997) X-chromosome inactivation in mammals. *Annual Review of Genetics* **31**: 571–610.

Keohane AM, Lavender JS, O'Neill LP and Turner BM (1998) Histone acetylation and X inactivation. *Developmental Genetics* **22**: 65–73.

Lyon MF (1999) X-chromosome inactivation. *Current Biology* **9**: R235–R237.

Mlynarczyk SK and Panning B (2000) X inactivation: *Tsix* and *Xist* as yin and yang. *Current Biology* **10**: R899–R903.

Tsuchiya KD and Willard HF (2000) Chromosomal domains and escape from X inactivation: comparative X inactivation analysis in mouse and human. *Mammalian Genome* **11**: 849–854.

Web Links

X(inactive)-specific transcript (*XIST*); LocusID: 7503. Locus Link:
http://www.ncbi.nlm.nih.gov/LocusLink/LocRpt.cgi?l7503
X(inactive)-specific transcript, antisense (*TSIX*); LocusID: 9383.
Locus Link:
http://www.ncbi.nlm.nih.gov/LocusLink/LocRpt.cgi?l9383

X(inactive)-specific transcript (*XIST*); MIM number: 314670.
OMIM:
http://www.ncbi.nlm.nih.gov/htbin-post/Omim/
dispmim?314670
X(inactive)-specific transcript, antisense (*TSIX*); MIM number:
300181. OMIM:
http://www.ncbi.nlm.nih.gov/htbin-post/Omim/
dispmim?300181

X-chromosome Inactivation and Disease

Mary F Lyon, *Medical Research Council Mammalian Genetics Unit, Harwell, UK*

Intermediate article

X-chromosome inactivation is the phenomenon in which one of the two X chromosomes in every somatic cell of female mammals becomes transcriptionally inactive early in embryonic development. This has the result of equalizing the effective gene dosage of X-linked genes in chromosomally XX females and XY males, and is hence known as a dosage compensation mechanism. In humans, X-chromosome inactivation has implications for the effects seen in diseases due either to X-linked genes or to numerical or structural anomalies of the X chromosome.

Article contents

- Mechanism of X-chromosome Inactivation
- X Inactivation and X-linked Genes
- X-chromosome Numerical and Structural Anomalies

Mechanism of X-chromosome Inactivation

In order to understand the implications of X-chromosome inactivation (XCI) in diseases, one must first know something of its mechanism. It occurs in all mammals but in no other group, and has been studied mainly in the mouse and the human. The onset of XCI occurs early in development when the embryo has relatively few cells; before this both X chromosomes are active. At the onset of XCI, one or the other X chromosome is inactivated in each cell of the embryo, and remains inactive, throughout all further cell divisions in that individual. In female germ cells the inactive X chromosome (Xi) is reactivated as the cells enter meiosis, and in male germ cells the single X chromosome becomes inactive. When the X chromosome becomes inactive, it takes on a set of unusual properties, mainly characteristic of heterochromatin. It replicates its deoxyribonucleic acid (DNA) late in the synthesis (S) phase of the cell cycle. It remains condensed during interphase and forms the sex chromatin body lying against the nuclear membrane. Its associated histone proteins, particularly H4, are underacetylated and the cytosines in its CpG islands are methylated, whereas in the rest of the genome these islands are unmethylated. In addition, an unusual histone, macro H2A1, is associated with the Xi.

The initiation of XCI requires the presence of a specific site on the X chromosome, located in band Xq13 in humans, and known as the X-inactivation center. Altered X chromosomes with deletions or translocations that remove the center cannot be inactivated. In individuals with supernumerary or missing X chromosomes (e.g. XXY, XXX, XO), a single X chromosome remains active (Xa) no matter how many are present. However, in triploids there may be two Xa. There is thus a counting mechanism, which protects a single X chromosome from inactivation per two autosome sets. Although in the embryo proper either X chromosome can undergo XCI, in the extraembryonic membranes of mice and rats, and probably also humans, the paternally derived X chromosome is preferentially inactivated. This is known as the imprinted type of XCI. The X-inactivation center is thought to be involved in counting and choice of X chromosome for inactivation, as well as in initiating the inactive state. It encodes (among others) a gene termed *X inactive-specific transcript* (*XIST*) (or *Xist* in the mouse), which is unique in being transcribed from the Xi only. The *XIST* gene has no open reading frame and codes for a 17-kilobase (kb) ribonucleic acid (RNA), which appears to coat or paint the Xi. In the very early mouse embryo, before XCI, *Xist* RNA is present at a low level and appears by fluorescence *in situ* hybridization (FISH) as a dot over the *Xist* locus. At the onset of XCI, the *Xist* RNA expands to coat the entire chromosome. Knockouts of *Xist* have shown that it is essential for XCI. Knockout of the 5′ end of the gene prevents the mutated X chromosome from being inactive, whereas deletion of the 3′ end affects the counting mechanism so that the mutated X chromosome cannot be selected to remain active and becomes inactive in all cells. Thus, different sequences are needed for initiation of inactivity and for counting. In the region 3′ to *Xist*, there is the origin of an antisense transcript, termed *Tsix*, which is transcribed across the *Xist* locus. Knockout of *Tsix* results in the mutated X chromosome becoming inactive in all cells (Lee and Lu, 1999). *Tsix* thus appears to have a regulatory role in XCI, by switching off *Xist*. Work with transgenes has shown that *Xist* RNA is capable of silencing transcription in autosomes, and can travel long distances along autosomal DNA, as it can on the X chromosome. However, *Xist* RNA alone is not sufficient to bring about the characteristic heterochromatic properties of the Xi. Some unknown developmental factor is also needed (Wutz and Jaenisch, 2000). These other properties are thought to be involved in locking in the inactive state. Thus, *Xist* alone can initiate transcriptional silencing, but is not sufficient to make this silencing permanent. (*See* X-chromosome Inactivation.)

X Inactivation and X-linked Genes

Ohno's law

As a consequence of XCI, transfer of genes between the X chromosome and autosomes during evolution would lead to genetic imbalance, and hence there would be selection against such transfers. This led Ohno to postulate Ohno's law, that a gene X-linked in one mammalian species is X-linked in all. This law is very widely obeyed, except for X-linked genes that have homologs on the Y chromosome (pseudoautosomal genes), which do not require dosage compensation and are not inactivated. The law is valuable since it implies that genes found on the X chromosome in the mouse or other mammals will have orthologs on the human X chromosome, suggesting candidate genes for X-linked diseases or providing animal models of disease. (*See* Chromosome X: General Features; X and Y Chromosomes: Homologous Regions.)

Effects in heterozygotes for X-linked genes

In considering diseases due to X-linked genes, an important consequence of XCI is that heterozygous females are mosaics, with two types of cells, having one or the other X chromosome, and hence a normal or a disease gene, active. Since XCI happens very early in development, the shapes and sizes of patches seen in the adult depend on the manner of cell growth and

mingling in the tissue concerned. In the skin an example of such patches is given by Blaschko's lines. Some structures, such as intestinal crypts, are monoclonal (i.e. they arise from a single cell) and hence each individual crypt has all cells either mutant or normal. Conversely, intestinal villi are polyclonal and each villus shows cells of both types. Muscle fibers have a polyclonal origin, and hence in heterozygotes for Duchenne muscular dystrophy one does not see distinct patches of normal and mutant fibers. Rather there are gradations of effect among the fibers.

Typically in heterozygotes for X-linked genes, the ratio of the two types of cell is near 50 : 50; however, for various reasons the ratio may deviate from this. Differences between the Xi and Xa in methylation of certain polymorphic sequences on the two X chromosomes may be used as a marker of per cent of cells of each type. An example is provided by a polymorphism in the androgen receptor (dihydrotestosterone receptor; testicular feminization; spinal and bulbar muscular atrophy; Kennedy disease) (*AR*) gene, in which the Xi is differentially methylated (Naumova *et al.*, 1996). Departures from a 50 : 50 ratio of cells, known as skewing of XCI, may arise by chance due to there being only a small pool of cells in the embryo at the time of initiation of XCI, or by deviation from random choice of X chromosome at initiation, or by random choice followed by cell selection.

Skewing of inactivation

Skewing due to cell selection usually arises when cells with the mutant gene active, divide more slowly or have low viability and may affect only those cell lines in which the gene is expressed, typically rapidly dividing cells such as blood cells. Examples are provided by hypoxanthine phosphoribosyl transferase (HPRT) deficiency in Lesch–Nyhan syndrome and by the immunodeficiency disease Wiskott–Aldrich syndrome. In both these cases all the leukocytes of heterozygotes have the normal X chromosome active, whereas in skin fibroblasts there is a mixture of both cell types. In heterozygotes for other diseases, cell selection may affect all cells rather than one tissue. This can occur when a female is a carrier of a recessive lethal X-linked gene, and may be associated with increased spontaneous abortion (Pegoraro *et al.*, 1997).

Primary nonrandom XCI, due to biased choice of X chromosome at initiation of XCI, appears to occur more rarely in humans. Examples have been found in two families with a mutation in the promoter region of the *XIST* gene (Plenge *et al.*, 1997) and in a family in which the causative factor was apparently not at the *XIST* locus (Naumova *et al.*, 1996).

Skewing of observed patterns of XCI due to chance occurs in a sizable minority of human females. In a large survey Naumova *et al.* (1996) found that 22% of women had $\geq 80\%$ of cells with the same X chromosome active. This can lead to expression of an X-linked gene for which the female is heterozygous as though she was in fact homozygous. Examples of this have been found with various X-linked diseases. This aberrant expression appears to occur more commonly in monozygotic twins, one member of the pair showing aberrant expression and the other not. Chitnis *et al.* (1999) found that discordance between members of a twin pair was greater in those pairs in which the splitting of the embryo occurred earliest, suggesting that the presence of only a few cells in each twin at splitting was the cause of the skewing.

X-chromosome Numerical and Structural Anomalies

Aneuploids

The occurrence of XCI has implications for the effects seen in both numerical and structural abnormalities of the X chromosome. Numerical abnormalities are the various forms of X-chromosome aneuploidy. The most common types are the absence of one sex chromosome, giving an XO karyotype and resulting in females with Turner syndrome, and the presence of an additional X chromosome, in XXY males with Klinefelter syndrome. Less common types include XXX, XXXX and XXXY. The first point of note is that all these types are viable, although in the case of XO poorly so. Autosomal aneuploidy has much more severe effects. All autosomal monosomies (corresponding to XO) in humans are lethal early in development, and among autosomal trisomies (corresponding to XXY, XXX) only three (for chromosomes 15, 18 and 21) are compatible with live birth, and these involve severe malformation. Clearly the inactivation of all save one of the X chromosomes present explains the less severe phenotypes of individuals with X-chromosome aneuploidy. The fact that there is some abnormality is thought to be due to escape of some genes from XCI. As XCI is a dosage compensation mechanism, and genes with homologs on the Y chromosome do not require compensation, it was expected that such genes would escape XCI, and this in fact is so. However, in the human, but less so in the mouse, some other X-linked genes escape XCI. Carrel *et al.* (1999) found that 34 out of a sample of 224 human X-linked genes and expressed sequence tags (ESTs) escaped XCI. Thus, individuals with X-chromosome aneuploidy would have wrong effective dosage of these genes. Most of the escaping genes were located on the short arm, Xp. Hence, among cases of partial aneuploidy, involving deletions or duplications of parts of the X chromosome, those affecting the short arm would be expected to have a more severe effect. In addition to their somatic abnormalities, patients with XO or XXY karyotypes are sterile, with the death of germ cells, and this is thought to be due to incorrect X-chromosome dosage in the germ cells, or to abnormal X-chromosome pairing at meiosis. (*See* Turner Syndrome.)

Deletions

Structural abnormalities of the X chromosome include deletions or duplications of parts of the chromosome, and also translocations between the X chromosome and an autosome. One of the most consistent features of these structural abnormalities is the occurrence of marked skewing of XCI. This skewing is thought to result from cell selection. In the case of deletions, in cells with the deleted X chromosome as the Xa and the normal X chromosome inactive, there is a deficiency of gene products from the deleted region. Consequently, there is selection against such cells and the result is that, in females heterozygous for X-chromosome deletions, all surviving cells have the deleted X chromosome as the Xi.

Translocations

In heterozygotes for X-autosome translocations, the picture is more complex. Only one of the two segments into which the X chromosome is broken will carry the X-inactivation center. In cells with the translocated X chromosome as the Xi, inactivation will spread from this segment into the attached autosomal material and inappropriately inactivate genes there. The other X-chromosome segment will remain active. Thus, in such cells there is incorrect gene dosage of both autosomal and X-linked genes, and there is strong selection against them. Hence in females heterozygous for X-autosome translocations, the translocated X chromosome typically forms the Xa in all cells. Most such females show no somatic abnormalities, but may be sterile (Schmidt and Du Sart, 1992). However, a consequence of the extreme skewing of XCI is that any mutant genes on the translocated X chromosome will be fully expressed, as in a male. This may happen if a gene at the chromosomal breakpoint has been disrupted by the translocation. Examples of this are provided by certain girls with Duchenne muscular dystrophy. In a small proportion of balanced X-autosome translocations, typically those in which the breakpoint is near to either end of the X chromosome, some cells with the translocated X chromosome as the Xi persist. These patients usually show congenital malformations, which result mainly from excess X-chromosome dosage, rather than from a deficiency of autosomal gene products (Schmidt and Du Sart, 1992). Some patients have unbalanced X-autosome translocations, and have only one of the translocated products. In such cases, inactivation of the translocated X chromosome may reduce the potential genetic imbalance and selection may favor such cells. However, although XCI travels into attached autosomal segments, the extent of travel is variable, and some autosomal genes resist inactivation. Thus, such patients often show congenital malformation resulting from genetic imbalance. On the other hand, White *et al.* (1998) described a patient who was normal except for reduced fertility, and who had an unbalanced translocation involving the X chromosome and chromosome 4. XCI had spread extensively into chromosome 4, and had thus reduced the potential genetic imbalance due to additional genes in this segment. However, about one-third of the genes there had escaped inactivation, a higher proportion than occurs in a normal X chromosome.

Tiny rings

A further type of X-chromosome structural anomaly in which XCI is relevant is that of tiny ring X chromosomes. Such rings constitute a specific type of X-chromosome deletion. Females bearing such tiny rings are very severely affected, with mental retardation and multiple congenital malformations, and it is not immediately clear why the effects are so severe. Studies showed that in the tiny rings the *XIST* locus is either deleted or very poorly expressed. Hence, the ring cannot undergo XCI and the female consequently has excess products of the genes in the ring, accounting for the phenotypic abnormalities (Migeon *et al.*, 1994).

See also
Chromosome X: General Features
Sex Chromosomes
X-chromosome Inactivation

References

Carrel L, Cottle AA, Goglin KC and Willard HF (1999) A first-generation X-inactivation profile of the human X chromosome. *Proceedings of the National Academy of Sciences of the United States of America* **96**: 14 440–14 444.
Chitnis S, Derom C, Vlietinck R, *et al.* (1999) X chromosome-inactivation patterns confirm the late timing of monoamniotic-MZ twinning. *American Journal of Human Genetics* **65**: 570–571.
Lee JT and Lu N (1999) Targeted mutagenesis of *Tsix* leads to nonrandom X inactivation. *Cell* **99**: 47–57.
Migeon BR, Luo S, Jani M and Jeppesen P (1994) The severe phenotype of females with tiny ring X chromosomes is associated with inability of these chromosomes to undergo X inactivation. *American Journal of Human Genetics* **55**: 497–504.
Naumova AK, Plenge RM, Bird LM, *et al.* (1996) Heritability of X chromosome inactivation phenotype in a large family. *American Journal of Human Genetics* **58**: 1111–1119.
Pegoraro E, Whitaker J, Mowery-Rushton P, *et al.* (1997) Familial skewed X inactivation: a molecular trait associated with high spontaneous abortion rate maps to Xq28. *American Journal of Human Genetics* **61**: 695–705.
Plenge RM, Hendrich BD, Schwartz C, *et al.* (1997) A promoter mutation in the *XIST* gene in two unrelated families with skewed X-chromosome inactivation. *Nature Genetics* **17**: 353–356.
Schmidt M and Du Sart D (1992) Functional disomies of the X chromosome influence the cell selection and hence the X inactivation pattern in females with balanced X–autosome translocations: a review of 122 cases. *American Journal of Medical Genetics* **42**: 161–169.
White WM, Willard HF, Van Dyke DL and Wolfe DJ (1998) The spreading of X inactivation into autosome material of an X-autosome translocation: evidence for a difference between autosomal and X-chromosomal DNA. *American Journal of Human Genetics* **63**: 20–28.
Wutz A and Jaenisch R (2000) A shift from reversible to irreversible X inactivation is triggered during ES cell differentiation. *Molecular Cell* **5**: 695–795.

Further Reading

Avner P and Heard E (2001) X-chromosome inactivation: counting, choice and initiation. *Nature Reviews Genetics* **2**: 58–67.
Belmont JW (1996) Genetic control of X inactivation and processes leading to X inactivation skewing. *American Journal of Human Genetics* **58**: 1101–1108.
Brockdorff N (1998) The role of *Xist* in X-inactivation. *Current Opinion in Genetics and Development* **8**: 328–333.
Disteche CM (1995) Escape from X inactivation in human and mouse. *Trends in Genetics* **11**: 17–22.
Heard E, Clerc P and Avner P (1997) X-chromosome inactivation in mammals. *Annual Review of Genetics* **31**: 571–610.
Lyon MF (1988) Clones and X-chromosomes. *Journal of Pathology* **155**: 97–99.
Migeon BR (1994) X-chromosome inactivation: molecular mechanisms and genetic consequences. *Trends in Genetics* **10**: 230–235.

Web Links

Androgen receptor (dihydrotestosterone receptor; testicular feminization; spinal and bulbar muscular atrophy; Kennedy disease) (*AR*); Locus ID: 367. LocusLink:
http://www.ncbi.nlm.nih.gov/LocusLink/LocRpt.cgi?l = 367
X inactive-specific transcript (*XIST*); Locus ID: 7503. LocusLink:
http://www.ncbi.nlm.nih.gov/LocusLink/LocRpt.cgi?l = 7503
Androgen receptor (dihydrotestosterone receptor; testicular feminization; spinal and bulbar muscular atrophy; Kennedy disease) (*AR*); MIM number: 313700. OMIM:
http://www.ncbi.nlm.nih.gov/htbin-post/Omim/dispmim?313700
X inactive-specific transcript (*XIST*); MIM number: 314670. OMIM:
http://www.ncbi.nlm.nih.gov/htbin-post/Omim/dispmim?314670

Xenologs

See Orthologs, Paralogs and Xenologs in Human and Other Genomes

X-linked Genes for General Cognitive Abilities

Hildegard Kehrer-Sawatzki, *University of Ulm, Ulm, Germany*

Peter Steinbach, *University of Ulm, Ulm, Germany*

Horst Hameister, *University of Ulm, Ulm, Germany*

More than 200 syndromic and nonsyndromic disorders associated with X-linked mental retardation are known, and the gene responsible for the condition has been identified in many cases. The phenotypic effects of known MRX genes, and their obvious overrepresentation on the X chromosome, suggest that they played an important role during human speciation.

Accumulation of Genes for Cognitive Abilities on the X Chromosome

Genes that contribute to general cognitive abilities can be recognized in their default state when mutations cause mental retardation (MR). More than 100 years ago it was realized that more males than females are found among the mentally retarded. Since then, X-linked inheritance of MR has been observed in many families, and it thus became clear that the excess of mentally retarded males is caused by mutations in X-linked genes. Interest in X-linked MR (XLMR) genes was further stimulated by the discovery of the *FMR1* gene. Expansion of a CGG-trinucleotide repeat in the 5′ UTR of the *FMR1* gene results in fragile X syndrome, the most frequent inherited form of MR. The latest available update on XLMR disorders lists 136 syndromic (MRXS) and 66 nonsyndromic (MRX) entities. At present, 42 genes mutated in XLMR probands have been identified (**Figure 1**). Compared to autosomal loci associated with MR, genes that interfere with general cognitive abilities if mutant are overrepresented on the X chromosome: According to the OMIM database, a 3.1-fold higher density of MR genes is observed on the X chromosome after correction of ascertainment bias for X-chromosomal linkage (Zechner *et al.*, 2001). (*See* Intelligence and Cognition; Intelligence, Genetic Basis: The IQ Quantitative Trait Loci Project; Intelligence, Heredity and Genes: A Historical Perspective.)

Functional Spectrum of X-chromosomal Genes for Cognitive Abilities

The syndromes associated with mutation of XLMR genes can be categorized as metabolic disorders, as disorders of brain development with a more or less widespread pattern of somatic abnormalities or as neuromuscular diseases (**Figure 1**). In these syndromic disorders, a causative link between gene function and cognitive impairment is relatively obvious. In a fourth category of XLMR phenotypes, designated MRX, mental retardation is the only symptom. This last category of genes is particularly interesting for understanding the evolution of general cognitive abilities. The following discussion focuses on genes of this type. (*See* Rett Syndrome.)

Nonsyndromic X-linked Mental Retardation Genes

The genes involved in seven MRX disorders have been identified (**Table 1**). Strikingly, at least three of them (*OPHN1*, *PAK3*, *ARHGEF6*) are involved in signal transduction mediated by small GTPases of the Rho family (**Figure 2a**). Rho proteins are involved in the modulation of cytoskeletal organization, which is instrumental to migration and polarity of neuronal cells, axon outgrowth and guidance, as well as dendrite development and synapse formation.

When bound to GDP, Rho proteins are inactive, but become active upon binding of GTP. Switching between the two states is accomplished by guanine nucleotide exchange factors (GEFs), which mediate the replacement of GDP by GTP, resulting in activation, and GTPase-activating proteins (GAPs), which increase the intrinsic GTPase activity of the Rho proteins and thus reduce the duration of the active phase. In response to extracellular stimuli, Rho family members forward signals to the actin cytoskeleton and the nucleus.

A mutation in *OPHN1* has been observed in an MRX family. The protein encoded by this gene,

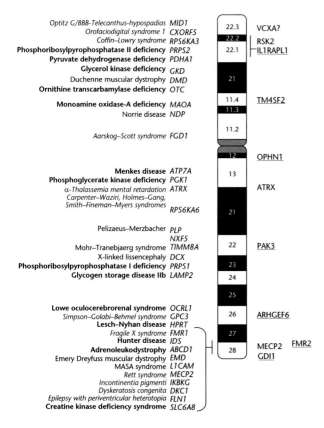

Optitz G/BBB-Telecanthus-hypospadias	MID1	VCXA?
Orofaciodigital syndrome 1	CXORF5	
Coffin–Lowry syndrome	RPS6KA3	RSK2
Phosphoribosylpyrophosphatase II deficiency	PRPS2	IL1RAPL1
Pyruvate dehydrogenase deficiency	PDHA1	
Glycerol kinase deficiency	GKD	
Duchenne muscular dystrophy	DMD	
Ornithine transcarbamylase deficiency	OTC	
Monoamine oxidase-A deficiency	MAOA	TM4SF2
Norrie disease	NDP	
Aarskog–Scott syndrome	FGD1	
		OPHN1
Menkes disease	ATP7A	
Phosphoglycerate kinase deficiency	PGK1	
α-*Thalassaemia mental retardation*	ATRX	ATRX
Carpenter–Waziri, Holmes–Gang,		
Smith–Fineman–Myers syndromes	RPS6KA6	
Pelizaeus–Merzbacher	PLP	
	NXF5	
Mohr–Tranebjaerg syndrome	TIMM8A	PAK3
X-linked lissencephaly	DCX	
Phosphoribosylpyrophosphatase I deficiency	PRPS1	
Glycogen storage disease IIb	LAMP2	
Lowe oculocerebrorenal syndrome	OCRL1	
Simpson–Golabi–Behmel syndrome	GPC3	ARHGEF6
Lesch–Nyhan disease	HPRT	
Fragile X syndrome	FMR1	
Hunter disease	IDS	
Adrenoleukodystrophy	ABCD1	MECP2 FMR2
Emery Dreyfuss muscular dystrophy	EMD	GDI1
MASA syndrome	L1CAM	
Rett syndrome	MECP2	
Incontinentia pigmenti	IKBKG	
Dyskeratosis congenita	DKC1	
Epilepsy with periventricular heterotopia	FLN1	
Creatine kinase deficiency syndrome	SLC6A8	

Figure 1 Positions of the genes associated with syndromic forms of X-linked mental retardation (MRXS) are indicated on the left. Mutant phenotypes are classified as metabolic disorders (bold type), disorders of morphogenesis (italic) or neuromuscular disorders (roman). Genes involved in nonsyndromic X-linked mental retardation (MRX) are shown on the right. The question marks indicate that these genes are strong candidates to cause mental retardation, although mutations in these genes have not yet been identified. Genes associated only with nonsyndromic MR are underlined. *ABCD1*: ATP-binding cassette, subfamily D, member 1; *ATP7A*: ATPase, Cu²⁺-transporting α-polypetide; *ATRX*: α thalassemia/mental retardation syndrome, X-linked; *ARHGEF6*: Rac/Cdc42 guanine nucleotide exchange factor (GEF) 6; *CXORF5*: chromosome X open reading frame; *DCX*: doublecortex, lissencephaly, X-linked (doublecortin); *FGD1*: faciogenital dysplasia; *FLN1*: filamin A, α-lactin-binding-protein-280; *FMR1*: fragile X mental retardation; *GPC3*: glypican 3; *HPRT*: hypoxanthine phosphoribosyltransferase 1; *IL1RAPL1*: IL-1 receptor accessory protein-like 1; *GDI1*: (αGDI), Rab GDP dissociation inhibitor 1; *IDS*: iduronate 2-sulfatase; *LAMP2*: lysosome-associated membrane protein 2; *L1CAM*: L1 cell adhesion molecule; *MECP2*: methyl CpG-binding protein 2; *NXF5*: nuclear export factor 5; *MID1*: midline 1; *OPHN1*: oligophrenin 1; *PAK3*: p21 activated kinase 3; *PLP*: proteolipid protein 1; *RSK2* (*RPS6KA3*): ribosomal protein S6 kinase polypeptide 3; *RSK4* (*RPS6KA6*): ribosomal protein S6 kinase polypeptide 6; *SLC6A8*: solute carrier family 6, member 8; *TIMM8A*: translocase of inner mitochondrial membrane 8; *TM4SF4*: Xp11.4 tetraspanin, transmembrane 4 superfamily member 2; *VCXA*: variable charged, X chromosome mRNA on CRI-S232A.

oligophrenin 1, acts as a GAP for RhoA and is highly expressed in fetal and adult mouse brain (Billuart *et al.*, 1998). Another link between MRX and disturbance of Rho-regulated signaling is provided by the identification of a splice-site mutation in the *ARHGEF6* gene in an MRX kindred (Kutsche *et al.*, 2000). The ARHGEF6 protein is one of at least 20 different GEFs for Rho proteins. Mutations in the *PAK3* gene have been identified in two families with MRX (Allen *et al.*, 1998; Bienvenu *et al.*, 2000). The PAK3 protein (p21 (Cdc42/Rac)-activated kinase) acts as a downstream effector of Rac and Cdc42, is highly expressed in fetal and adult brain, and is involved in cytoskeleton organization and regulation of gene expression. Activated PAK3 interacts with paxillin α in focal adhesions, the sites of cell contact with the extracellular matrix.

Mammalian PAKs consist of three isoforms (PAK1–3) that show high sequence similarity to each other and probably have overlapping functions. One of the substrates of PAK1 is the LIM kinase, which regulates the activity of cofilin, an actin-depolymerizing factor. Overexpression of cofilin results in increased neurite outgrowth. Loss-of-function mutations in the *PAK* gene of *Drosophila melanogaster* result in defects in photoreceptor axon guidance, suggesting a role for Rho/PAK-mediated pathways in dendrite development and plasticity.

In addition to small GTPases of the Rho family, RabGTPases have also been implicated in brain function. Proteins of this class are involved in membrane trafficking. Mutations in the gene for RabGDIα have been identified in three MRX probands (Bienvenu *et al.*, 1998; D'Adamo *et al.*, 1998). As illustrated in **Figure 2b**, RabGDIα mediates the association of Rab3a with the membrane. Rab3a is a key regulator of vesicular transport and neurotransmitter release. This highly dynamic process involves transport, docking and fusion of vesicles at presynaptic membranes, and is thus critical for synaptic plasticity. RabGDIs retrieve Rab proteins from the target membrane and keep them in the GDP-bound inactive state as a cytosolic reservoir. RabGDI delivers RabGDP to the vesicle membrane, where GDI dissociates from Rab and enables specific GEFs to exchange GTP for GDP, thus recruiting Rab for a further catalytic cycle. RabGDIα is one of three identified RabGDI isoforms, and is highly expressed in brain, where it interacts with the locally abundant Rab3a. In RabGDIα-deficient mice, neuronal hyperexcitability and epileptic seizures are observed. Consistent with an additional function of these proteins during brain development, RabGDIα and Rab3a are already highly expressed in the embryonic mouse brain prior to the onset of synaptic activity.

Another MRX gene associated with nonsyndromic XLMR is *TM4SF2* (Zemni *et al.*, 2000). The function

Table 1 Genes associated with nonsyndromic forms of X-linked mental retardation[a]

Gene	Mutations identified	Sites of highest expression in adult mouse or rat brain
OPHN1 (oligophrenin 1)	Disrupted by t(X;12) in a sporadic case; 1578delA in MRX60 family	Unknown
PAK3 (p21 activated kinase)	R419X in the MRX30 family; R67C in the MRX47 family	Cortex, limbic regions of cortex, piriform cortex, hippocampus, thalamus, pontine nucleus, reticulotegmental, external cuneate and lateral reticular nuclei
ARHGEF6 (α-PIX)	Disrupted by t(X;21) in a sporadic case; IVS1–11 T → C in a family	Unknown
GDI1 (αGDI) (GDP dissocation inhibitor)	R423P in family R; R70X in family MRX48; L92P in family MRX41	Hippocampus, Purkinje cells of cerebellum, ganglion cell layer of the retina, cortex, olfactory bulb, nasal epithelium
TM4SF4 (transmembrane 4 superfamily 2)	Disrupted by t(X;2) in a patient; G218X in family L28; P172H in family T15	Hippocampus, cortex, primary olfactory cortex
IL1RAPL1 (IL-1 receptor accessory protein-like)	Intragenic deletion in MRX family 43; Y459X in an MRX family	Hippocampus, primary olfactory cortex, ento- and perirhinal cortex, mammillary bodies and supramammillary nucleus
FMR2	Deletions; CGG expansions	Cingulate gyrus, hippocampus, neocortical neurons, Purkinje cells

[a]Some mutations in the *RSK2*, *ATRX* and *MECP2* genes are also observed in nonsyndromic MR, but other mutations in these genes are involved in syndromic forms of XLMR.

of this gene is not well defined. The predicted product belongs to the tetraspanin family of transmembrane proteins comprising at least 28 members in mammals. Tetraspanins contain four membrane-spanning domains separated by two extracellular loops. In the membrane they form mobile complexes either with each other, or with β1-integrins, proteoglycans, transmembrane immunoglobulins and with other signaling enzymes. The involvement of tetraspanins in homotypic and heterotypic cell–cell adhesion has been repeatedly described. The contribution of tetraspanins to β-integrin-driven cell migration is also well established. Little is known about the function of TM4SF2, except that it interacts with phosphoinositide 4-kinase, thereby affecting the activity of the kinase, which has an essential role in neurotransmitter release. In *D. melanogaster*, a member of the tetraspanin family is involved in neuromuscular synapse formation.

A truncating mutation in the *IL1RAPL1* gene has been observed to segregate in a family with nonsyndromic XLMR (Carrie *et al.*, 1999). This gene encodes a homolog of the interleukin 1 receptor accessory protein (IL-1RACP) and of type I and type II interleukin-1 receptors. However, IL1RAPL1 has a specific 150-residue *C*-terminal extension that shows no similarity to known proteins. The interaction of IL-1RACP with the IL-1β/IL-1R(I) complex has been confirmed whereas the participation of IL1RAPL1 in a similar complex remains to be demonstrated. With respect to the high-level expression of IL1RAPL1 in fetal and adult brain, it is worth

noting that interleukin receptors and their ligands regulate glial activities in neural tissues after trauma of the perinatal brain. IL-1β induces early healing responses that restore the integrity of damaged structures and create conditions that induce the dendrite sprouting necessary for neuronal plasticity.

Another form of MRX with an estimated incidence of one in 50 000–100 000 males is associated with expansion of a GCC trinucleotide repeat in the 5' UTR of the *FMR2* gene at the *FRAXE* locus 600 kb distal to *FMR1* (Gécz *et al.*, 1996). The retardation phenotype associated with repeat expansion is mild to moderate mental impairment. Loss of *FMR2* expression owing to repeat expansion and subsequent hypermethylation is the molecular basis of the disease. The FMR2 protein (FMR2P) belongs to the AF4-like family of transcription factors, and is localized in brain structures involved in learning and memory. In *D. melanogaster*, a maternally provided transcriptional regulator, termed Lilliputian, with homology to FMR2P has been identified. Lilliputian deficiency results in severe defects in cytoskeleton function during cellularization and segmentation. The function of FMR2P in mammalian cells remains to be investigated.

In some genes associated with well-established entities of syndromic X-linked mental retardation (MRXS), rare allelic variants have been linked to further nonsyndromic forms of mental retardation. One such gene, *RPS6KA3*, also known as *RSK2* (ribosomal S6 kinase 2) is mutated in patients with Coffin–Lowry syndrome. The Rsk2 protein is a

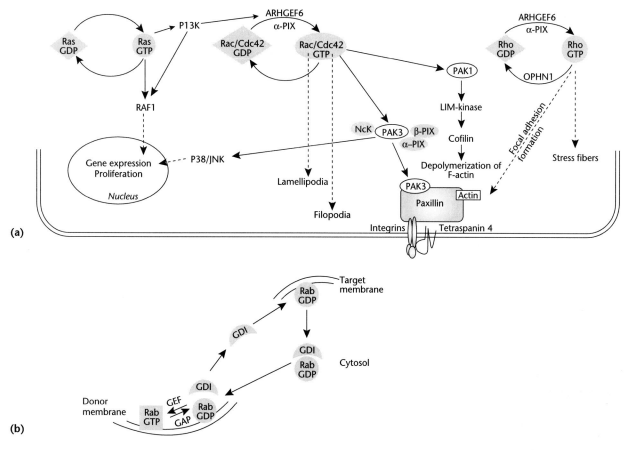

Figure 2 (a) Genes/proteins (ARHGEF6, OPHN1, PAK3, TM4SF4) involved in the pathogenesis of X-linked mental retardation and their functional context in signal transduction pathways. The activation of the small GTPases Ras, Rac, Cdc42 is dependent on diverse signals mediated by lipids or growth factors and their receptors. Activated Rho, Rac and Cdc42 GTPases are involved in the formation of lamellipodia, filopodia, stress fibers and in signal transmission via focal adhesions. In neuronal cells, these processes influence axon growth and dendrite plasticity. JNK: c-jun *N*-terminal kinase; PIX: PAK-interacting exchange factor; arrows indicate activating influences; dotted arrows represent simplification of downstream pathways leading to the indicated effects. (b) Involvement of RabGDI (GDP dissociation inhibitor) in Rab cycling. RabGTPase bound to GDP is retrieved from the target membrane of an organelle or vesicle by GDI, and maintained in this inactive state in the cytosol. Subsequently, GDI is able to deliver RapGDP back to the donor membrane, where GDI dissociates from the complex and a GEF regulates GDP replacement by GTP. In its active, GTP-bound, state, Rab acts on effector proteins. Following conversion of the GTP-bound state to the GDP-bound form, during or after vesicle fusion and neurotransmitter release, GDI retrieves RabGDP to the cytosolic pool.

serine/threonine kinase and a target of extracellular regulated kinases (ERKs). Upon activation, Rsk2 phosphorylates the transcription factor CREB and histone H3. A second gene that has been linked with both syndromic and nonsyndromic XLMR is *MECP2* (methyl CpG binding protein 2). Heterozygous mutations in the *MECP2* gene are observed in individuals with Rett syndrome, a severe neurologic disorder that occurs almost exclusively in females. The mutations were thought to be lethal in males, but rarely, boys with severe neonatal encephalopathy have been born. (*See* Rett Syndrome; X-chromosome Inactivation.)

Recently, *MECP2* mutations have been also found in males with different degrees of MRX. These findings suggest that *MECP2* mutations are the second most frequent cause of XLMR in males, with *FMR1* mutations being the most common. Also in case of mutations in the *ATRX* gene (α-thalassemia and X-linked mental retardation), a broad spectrum of allelic heterogeneity has been observed. Carriers of *ATRX* mutations have been identified who showed MR as the only symptom. A common feature of these MRXS/MRX genes with allelic and phenotypic heterogeneity is the involvement of their products in chromatin remodeling. In this respect, these genes are functionally different from the MRX genes discussed in the previous paragraphs. (*See* Chromatin Structure and Domains; Chromatin Structure and Modification: Defects.)

Conclusion

The identification of loci associated with MR has revealed an overrepresentation of genes for cognitive abilities on the X chromosome. This is reflected by the higher frequency of males among mentally retarded individuals. Furthermore, intelligence quotient (IQ) scores have repeatedly shown that males present with a higher variance. Males are overrepresented among both low-scoring and high-scoring individuals. This observation suggests that there are specific combinations of X-linked alleles associated with low cognitive abilities, whereas other combinations endow the carrier with extraordinarily high cognitive abilities. (*See* Intelligence, Heredity, and Genes: A Historical Perspective.)

This prompts to the question why so many genes involved in the development of cognitive abilities reside on the X chromosome? The high density of genes for cognitive abilities on the X chromosome is reminiscent of the recently described excess of sex- and reproduction-related genes on this chromosome. Sex- and reproduction-related genes are specifically involved in the speciation process. Indeed, for speciation genes, a large X chromosome effect, that is, a high concentration of these genes on the X chromosome, is a common finding. The most striking species-specific trait that developed during human evolution is high cognitive ability. It is this trait that most obviously distinguishes us from all other species. Therefore, it is no surprise to find a high density of genes for general cognitive abilities on the human X chromosome. (*See* Forkhead Domains.)

Most interesting in this respect are the genes of nonsyndromic XLMR. Many of them are highly expressed in adult mammalian brain, and regulate processes essential for neuronal function, such as axon growth and guidance, dendritic spine formation, and synaptic plasticity. In the adult brain, high-level expression of some of the MRX genes is observed in regions involved in learning and long-term memory. Dendrite sprouting and synaptic plasticity became highly developed during human speciation. Thus, the MRX genes are the most attractive candidates for true human speciation genes. (*See* Speciation: Genetics.)

See also

Chromosome X: General Features
Intellectual Disability: Genetics

References

Allen KM, Gleeson JG, Bagrodia S, *et al.* (1998) PAK3 mutation in nonsyndromic X-linked mental retardation. *Nature Genetics* **20**: 25–30.

Bienvenu T, Portes des V, McDonell N, *et al.* (2000) Missense mutation in PAK3, R67C, causes X-linked nonspecific mental retardation. *American Journal of Medical Genetics* **93**: 294–298.

Bienvenu T, Portes des V, Saint Martin A, *et al.* (1998) Non-specific X-linked semidominant mental retardation by mutations in a Rab GDP-dissociation inhibitor. *Human Molecular Genetics* **7**: 1311–1315.

Billuart P, Bienvenu T, Ronce N, *et al.* (1998) Oligophrenin-1 encodes a rhoGAP protein involved in X-linked mental retardation. *Nature* **392**: 923–926.

Carrie A, Jun L, Bienvenu T, *et al.* (1999) A new member of the IL-1 receptor family highly expressed in hippocampus and involved in X-linked mental retardation. *Nature Genetics* **23**: 25–31.

D'Adamo P, Menegon A, Lo Nigro C, *et al.* (1998) Mutations in GDI1 are responsible for X-linked non-specific mental retardation. *Nature Genetics* **19**: 134–139.

Gécz J, Gedeon AK, Sutherland GR and Mulley JC (1996) Identification of the gene FMR2, associated with FRAXE mental retardation. *Nature Genetics* **13**: 105–108.

Kutsche K, Yntema H, Brandt A, *et al.* (2000) Mutations in ARHGEF6, encoding a guanine nucleotide exchange factor for Rho GTPases, in patients with X-linked mental retardation. *Nature Genetics* **26**: 247–250.

Zechner U, Wilda M, Kehrer-Sawatzki H, *et al.* (2001) A high density of X-linked genes for general cognitive ability: a runaway process shaping human evolution? *Trends in Genetics* **17**: 697–701.

Zemni R, Bienvenu T, Vinet MC, *et al.* (2000) A new gene involved in X-linked mental retardation identified by analysis of an X-2 balanced translocation. *Nature Genetics* **24**: 167–170.

Further Reading

Bagrodia S and Cerione RA (1999) Pak to the future. *Trends in Cell Biology* **9**: 350–355.

Berditchevski F (2001) Complexes of tetraspanins with integrins: more than meets the eye. *Journal of Cell Science* **114**: 4143–4151.

Chelly J and Mandel JL (2001) Monogenic causes of X-linked mental retardation. *Nature Review of Genetics* **2**: 669–680.

Chiurazzi P, Hamel BC and Neri G (2001) XLMR genes: update 2000. *European Journal of Human Genetics* **9**: 71–81.

Hedges LV and Nowell A (1995) Sex differences in mental test scores, variability, and numbers of high-scoring individuals. *Science* **269**: 41–45.

Ide CF, Scripter JL, Coltman BW, *et al.* (1996) Cellular and molecular correlates to plasticity during recovery from injury in the developing mammalian brain. *Progress in Brain Research* **108**: 365–377.

Ishizaki H, Miyoshi J, Kamiya H, *et al.* (2000) Role of rab GDP dissociation inhibitor alpha in regulating plasticity of hippocampal neurotransmission. *Proceedings of the National Academy of Sciences of the United States of America* **97**: 11 587–11 592.

Luo L (2000) Rho GTPases in neuronal morphogenesis. *Nature Review of Neuroscience* **1**: 173–180.

Plomin R (1999) Genetics and general cognitive ability. *Nature* **402**: C25–C29.

Saifi GM and Chandra HS (1999) An apparent excess of sex- and reproduction-related genes on the human X chromosome. *Proceedings of the Royal Society of London, Series B* **266**: 203–209.

Web Links

Coffin–Lowry syndrome (CLS) LocusID: 1210. LocusLink:
http://www.ncbi.nlm.nih.gov/LocusLink/LocRpt.cgi?l = 1210

X-ray Diffraction: Principles

Jan Drenth, *University of Groningen, Groningen, The Netherlands*

The properties of biological macromolecules cannot be fully understood without knowledge of their three-dimensional structure. X-ray diffraction is one technique for structure determination.

Advanced article

Article contents
• Crystals
• Bragg Law
• Summation of Waves
• Structure Factor
• Reciprocal Space and the Ewald Sphere
• Fourier Summation
• Summary

Crystals

There is a large gap between the practice of crystal growth of macromolecules and the theory behind it. The crystallographer uses trial-and-error methods in a more or less organized form and relies on personal experience. The size of the crystals should be 0.5 mm or less, down to 0.1 mm or even somewhat smaller. Very small crystals and crystals of very large molecules, or complexes of molecules (for instance viruses), require the extremely intense X-ray radiation from a synchrotron. In the home laboratory, a rotating anode tube is the preferred source of radiation. The faceted shapes of crystals express the internal three-dimensional arrangement of the molecules. Each molecule or group of molecules repeats itself through the entire crystal lattice. In this lattice, three axes can be assigned: x, y and z with repeating distances a along x, b along y and c along z. The three axes make angles α, β and γ with each other (**Figure 1**). The six parameters a, b, c, α, β and γ determine one unit cell and the crystal lattice consists of an enormous number of unit cells all having exactly the same content.

In the lattice, planes can be constructed through the lattice points. These planes are distinguished by three indices, h, k and l, called Miller indices. If, for a given set of parallel planes, the a edge is divided into two parts, h is equal to 2; similarly, k and l are equal to the number of parts into which b and c are divided. The boundary planes of the unit cell are therefore $(1\,0\,0)$, $(0\,1\,0)$ and $(0\,0\,1)$. Note that h, k and l are always integers. X-ray reflections against a set of lattice planes are given the same indices: $(h\,k\,l)$ for the first-order and $(n \times h\ n \times k\ n \times l)$ for the nth-order reflection.

Translation of the crystal structure by the repeating distances a, b and c results in exactly the same structure: this is translational symmetry. Most crystals have additional symmetry elements: axes, mirror planes or a center of symmetry. Application of these symmetry operations also results in a crystal structure indistinguishable from the starting structure. Because biological macromolecules are enantiomeric, their crystals have neither mirror planes nor a center of

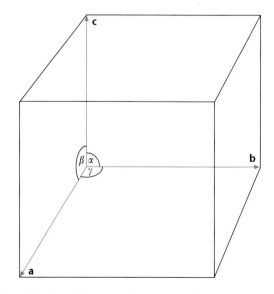

Figure 1 Unit cell in a crystal. The vectors **a**, **b**, and **c** indicate the repeating distances in the crystal structure. Depending on the relationships between, and the values of, the six parameters **a**, **b**, **c**, α, β and γ, the crystal belongs to one of the seven crystal systems (see **Table 1**). (Reproduced from *Encyclopedia of Life Sciences*.)

Table 1 Seven crystal systems

Crystal system	Conditions imposed on cell geometry
Triclinic	None
Monoclinic	$\alpha = \gamma = 90°$ (b is the unique axis; for proteins this is a twofold axis or screw axis); or $\alpha = \beta = 90°$ (c is the unique axis; for proteins this is a twofold axis or screw axis)
Orthorhombic	$\alpha = \beta = \gamma = 90°$
Tetragonal	$a = b$; $\alpha = \beta = \gamma = 90°$
Trigonal	$a = b$; $\alpha = \beta = 90°$; $\gamma = 120°$ (hexagonal axes); or $a = b = c$; $\alpha = \beta = \gamma$ (rhombohedral axes)
Hexagonal	$a = b$; $\alpha = \beta = 90°$; $\gamma = 120°$
Cubic	$a = b = c$; $\alpha = \beta = \gamma = 90°$

symmetry, since these operations would change the chirality. They do have symmetry axes, which can be two-, three-, four- or sixfold, either as normal axes or combined with a translation (screw axes). Crystallographic symmetry operators multiply the number of molecules in the unit cell. For instance, with a twofold axis, at least two identical molecules related by this axis are present. The unit cell then consists of two asymmetric units.

On the basis of the parameters a, b, c, α, β and γ, and on symmetry operations in the crystal, seven crystal systems can be distinguished (**Table 1**).

Bragg Law

A crystal scatters an incident X-ray beam in a huge number of different directions. It was Bragg (1913) who pointed out that each scattered beam can be regarded as the result of a reflection against a set of parallel lattice planes with the indices h, k and l. From this idea it is easy to derive the Bragg law (eqn [1]), where d is the distance between the parallel planes, θ is the reflection angle, λ is the X-ray wavelength and n is an integer (**Figure 2**):

$$2d \sin\theta = n \times \lambda \qquad (1)$$

The incident beam is reflected, but only in the directions for which the Bragg law is obeyed.

Summation of Waves

In an X-ray diffraction experiment, the electrons in the material absorb the radiation and emit it again. Part of the emitted radiation will have a longer wavelength than the incident radiation, but for diffraction we can neglect this part and consider only the radiation with the same wavelength. The waves are electromagnetic waves, but only the electric component contributes to the scattering. The waves from all electrons in the specimen must be combined. Waves are characterized

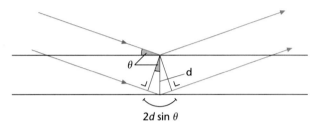

Figure 2 Beams reflected by a set of parallel planes amplify each other when the path difference between reflections from successive planes is equal to λ, 2λ, etc. Because the path difference is also $2d \sin\theta$, the Bragg law is obtained: $2d \sin\theta = n\lambda$. (Reproduced, in modified form, from Drenth (1999), with the permission of Springer-Verlag.)

by an amplitude (E_0), a phase angle (α), the speed at which they travel through a medium and their wavelength. X-rays travel as cosine waves at the speed of light, c, through vacuum and with approximately the same speed through air. At a chosen time $t = 0$, the electromagnetic field strength at position z is given by

$$E(t = 0; z) = E_0 \cos 2\pi \frac{z}{\lambda} \qquad (2)$$

At the origin of the system, where $z = 0$, $E = E_0$. A little later, at time t, the electromagnetic field strength at z is equal to what it was at $t = 0$ at position $z - (t \times c)$:

$$E(t; z) = E_0 \cos 2\pi \frac{z - (t \times c)}{\lambda} \qquad (3)$$

The field strength at the origin is now given by

$$E(t; z = 0) = E_0 \cos \frac{2\pi}{\lambda}(-t \times c) = E_0 \cos 2\pi v t = E_0 \cos \omega t \qquad (4)$$

where $v = c/\lambda$, the X-ray frequency, and ω is short for $2\pi v$.

Suppose there is a second wave with the same amplitude E_0 but displaced over a distance Z with respect to the first one. This displacement corresponds

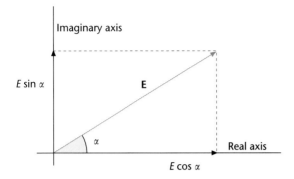

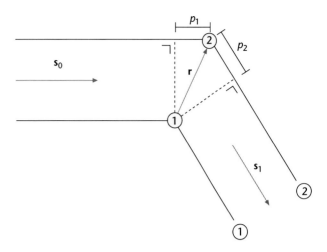

Figure 3 A wave can be represented in the Argand diagram as a vector **E** in a plane with horizontal and vertical axes. The wave **E** can be regarded as the sum of two waves, one along the horizontal axis with amplitude **E** cos α, and the other along the vertical axis with amplitude **E** sin α. The mathematical expression for **E** is $(E \cos \alpha) + i (E \sin \alpha) = E \exp(i\alpha)$. (Reproduced, in modified form, from Drenth (1999) with the permission of Springer-Verlag.)

Figure 4 A simple system consisting of two electrons, 1 and 2. The position of electron 2 with respect to electron 1 is given by vector **r**. Vectors $\mathbf{s}_0$ and $\mathbf{s}_1$ indicate the direction of the incident and scattered beams. The path difference between the wave involving electron 1 and the wave involving electron 2 is $p_1 + p_2$. (Reproduced, in modified form, from Drenth (1999) with the permission of Springer-Verlag.)

to a phase difference of $2\pi Z/\lambda = \alpha$. Hence for the second wave we have

$$E(t; z = 0) = E_0 \cos(\omega t + \alpha) \qquad (5)$$

More generally, any wave that has a phase difference α with respect to a chosen origin can be expressed as $E_0 \cos(\omega t + \alpha)$:

$$\begin{aligned} E_0 \cos(\omega t + \alpha) &= E_0 \cos \alpha \cos \omega t - E_0 \sin \alpha \sin \omega t \\ &= E_0 \cos \alpha \cos \omega t - E_0 \sin \alpha \cos(\omega t + 90°) \end{aligned}$$
$$(6)$$

Hence any wave can be regarded as consisting of two waves with a phase difference of 90°. Their amplitudes are $E_0 \cos \alpha$ and $E_0 \sin \alpha$. This splitting of the waves can also be represented in graphical form. Each wave can be expressed as a vector in a plane with a horizontal and a vertical axis (**Figure 3**). The plane is called the complex plane and the set of axes forms an Argand diagram. If many waves must be added, it is easy to sum their horizontal components because they are all in phase. The same is true for the vertical components, and the result is that summation of waves can be visualized as a vector summation in the Argand diagram. To facilitate mathematical processing, the wave is written as

$$(E \cos \alpha) + i(E \sin \alpha) \qquad (7)$$

where i means that the $E \sin \alpha$ component is along the vertical axis, or more generally rotated 90° counterclockwise. For easy manipulation, $(E \cos \alpha) + i (E \sin \alpha)$ is replaced by its equivalent $E \exp(i\alpha)$.

Structure Factor

The Bragg law provides a visualization of X-ray reflection but it leaves open the question of why

diffraction is indeed reflection against lattice planes, and it does not give an expression for the diffraction intensity. This requires one to look at diffraction from a different point of view.

When an X-ray beam impinges on the sample, the electrons in the sample begin to oscillate; they act as the 'antenna' of a transmitter and emit X-ray radiation themselves. The amplitude of the emitted radiation in a specific direction is the same for each electron. The radiation from all electrons in the sample must be summed to give the diffracted beam in a certain direction. Although the amplitude of the beams radiated by the individual electrons is the same, their phase angles are not, because of path differences.

For a simple system of only two electrons (**Figure 4**), the beam involving electron 2 follows a longer path than the beam involving electron 1. The total path difference is $p_1 + p_2$ and hence beam 2 is retarded in phase with respect to beam 1 by $2\pi(p_1 + p_2)/\lambda$. The directions of the incident and scattered beam can be expressed by assigning to them a vector: $\mathbf{s}_0$ for the incident beam and $\mathbf{s}_1$ for the scattered beam. The lengths of $\mathbf{s}_0$ and $\mathbf{s}_1$ are not important; for convenience, both vectors will be given the length $1/\lambda$. The path differences p_1 and p_2 can be expressed as a scalar product:

$$p_1 = \lambda \mathbf{r} \cdot \mathbf{s}_0 \quad \text{and} \quad p_2 = -\lambda \mathbf{r} \cdot \mathbf{s}_1 \qquad (8)$$

The phase difference is then $2\pi \mathbf{r} \cdot (\mathbf{s}_0 - \mathbf{s}_1)$. Because beam 2 lags behind in phase, it has a phase angle $[-2\pi \mathbf{r} \cdot (\mathbf{s}_0 - \mathbf{s}_1)]$ with respect to beam 1, or with respect

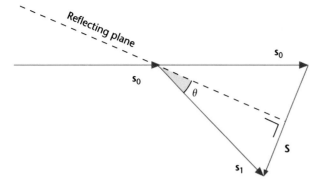

Figure 5 Vectors $\mathbf{s}_0$ and $\mathbf{s}_1$ indicate the directions of the primary and diffracted beams respectively. Vector $\mathbf{S} = \mathbf{s}_1 - \mathbf{s}_0$ is perpendicular to a plane that can be regarded as a plane reflecting the primary beam. (Reproduced from *Encyclopedia of Life Sciences*.)

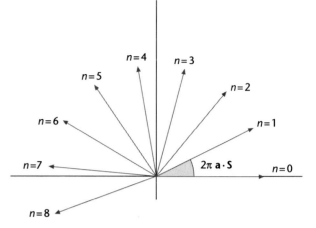

Figure 6 Each arrow in the Argand diagram represents the scattering by one unit cell. In the figure the scattering by nine unit cells ($n = 0$–8) is shown. In fact n is a very large number and the vectors point in all directions; the scattering by a crystal would be zero. However, in the special case that $\mathbf{a} \cdot \mathbf{S}$ is an integer, all vectors point to the right and the crystal does scatter. (Reproduced, in modified form, from Drenth (1999) with the permission of Springer-Verlag.)

to the origin of the system in electron 1. If $[-(\mathbf{s}_0 - \mathbf{s}_1)]$ is replaced by $\mathbf{S}$, the phase of beam 2 is $(2\pi\mathbf{r} \cdot \mathbf{S})$ and its scattering is $1 \times \exp[2\pi i \mathbf{r} \cdot \mathbf{S}]$. Its amplitude is 1, because the amplitude of scattering by one electron is taken as 1. The vector $\mathbf{S}$ can be regarded as being perpendicular to a plane that reflects $\mathbf{s}_0$ to $\mathbf{s}_1$ (**Figure 5**). The length of $\mathbf{S}$ is $|\mathbf{S}| = (2\sin\theta)/\lambda$, where θ is the reflection angle and 2θ the scattering angle.

If this procedure is applied to the electrons in a unit cell, the total scattering by all electrons in the unit cell with an electron density distribution $\rho(\mathbf{r})$ is

$$\mathbf{F}(\mathbf{S}) = \int \rho(\mathbf{r}) \exp(2\pi i \mathbf{r} \cdot \mathbf{S})\, d\mathbf{r} \qquad (9)$$
$$\mathbf{r} \text{ in the cell}$$

$\mathbf{F}(\mathbf{S})$ is a vector in the Argand diagram with amplitude $|\mathbf{F}(\mathbf{S})|$ and a phase angle α determined by the position of the many electrons in the unit cell. $\mathbf{F}(\mathbf{S}) = |F|\exp(i\alpha)$ is called the structure factor, because it depends on the structure (the position of the electrons) in the unit cell. A crystal consists of a large number of unit cells, and to obtain the scattering by the crystal the contributions from all unit cells must be added, taking into account their position with respect to a chosen origin. Take this origin in the origin of an arbitrary unit cell. The unit cells are shifted with respect to each other over distances equal to an integral number times a, times b and times c. For simplicity, choose a second unit cell shifted by $n \times a$, where n is an integer. As for the two-electron system, the path difference for the X-ray beam is $n \times a$ and the phase difference is $2\pi na \cdot \mathbf{S}$. Hence the scattering by the second unit cell is given by

$$\mathbf{F}(\mathbf{S}) \exp(2\pi i n\mathbf{a} \cdot \mathbf{S}) \qquad (10)$$

The summation over n results in zero scattering unless the scalar product $\mathbf{a} \cdot \mathbf{S}$ is an integer (**Figure 6**). In that case, the scattering by the crystal is proportional to $\mathbf{F}(\mathbf{S})$. Thus, the conditions for scattering are

as follows:

$$\mathbf{a} \cdot \mathbf{S} = h, \quad \mathbf{b} \cdot \mathbf{S} = k, \quad \mathbf{c} \cdot \mathbf{S} = l \qquad (11)$$

These are called the Laue conditions; h, k and l are integers, and it can be shown that they are equal to the indices of the reflecting lattice planes (Drenth, 1999).

Note that scattering by a crystal occurs only in those directions for which $\mathbf{S}$ obeys the Laue conditions. These directions correspond to the directions of the Bragg reflections against lattice planes.

Reciprocal Space and the Ewald Sphere

The scattering of X-rays by a crystal can be mimicked in a very simplified way with visible light and a grating; for instance, by a laser beam pointing perpendicularly onto an electron microscope grid or any fine net. If the grid spacing is smaller, the diffraction angle is larger. This can be understood from the Bragg law: if d is smaller, θ must be larger for a constant λ. Hence, the angle of reflection changes reciprocally to the unit distance in the grating. This reciprocity led Ewald (1913) to construct a convenient tool in crystallography: the reciprocal lattice. This lattice has unit cells determined by the six parameters a^*, b^*, c^*, α^*, β^* and γ^*, which are obtained in the following way:

a^* is drawn perpendicular to plane $(1\,0\,0)$ and has a length $[d(1\,0\,0)]^{-1}$.

b^* is drawn perpendicular to plane $(0\,1\,0)$ and has a length $[d(0\,1\,0)]^{-1}$.

c^* is drawn perpendicular to plane $(0\,0\,1)$ and has a length $[d(0\,0\,1)]^{-1}$.

α^* is the angle between b^* and c^*; β^* is the angle between a^* and c^*; and γ^* is the angle between a^* and b^*.

The following properties of the reciprocal lattice should be appreciated:

- The crystal lattice is a real lattice, but the reciprocal lattice is an imaginary lattice.
- If the crystal is reoriented, the reciprocal lattice rotates in exactly the same way.
- A set of parallel lattice planes in the crystal with indices $(h\,k\,l)$ corresponds with one grid point in the reciprocal lattice. The vector from the origin of the reciprocal lattice to this grid point is equal to $\mathbf{S}(h\,k\,l)$.
- The length of vector $\mathbf{S}(h\,k\,l)$ is equal to $|\mathbf{S}| = (2\sin\theta)/\lambda$ (see the previous section); applying the Bragg law, this is equal to $n/d(h\,k\,l)$ where $d(h\,k\,l)$ is the lattice plane distance in the real ($=$crystal) lattice.

The reciprocal lattice is a convenient tool for constructing the directions of the beams diffracted by the crystal. The procedure is as follows (**Figure 7**).

Step 1. Draw a sphere with radius $1/\lambda$. This sphere is called the Ewald sphere.

Step 2. Direct the incident X-ray beam toward the center M of the sphere.

Step 3. The point O where the beam leaves the sphere is taken as the origin of the reciprocal lattice.

Step 4. Construct the reciprocal lattice in an orientation corresponding to the actual orientation of the crystal.

Step 5. Grid points of the reciprocal lattice that are located on the surface of the Ewald sphere are in reflecting positions, because for those points $2d\sin\theta = n\times\lambda$ and the Bragg law is obeyed. The other points in reciprocal space can be put into a reflecting position by rotating the crystal and thus the reciprocal lattice.

Fourier Summation

A crystal scatters an X-ray beam only in specific directions determined by the Laue conditions. The scattering is proportional to $\mathbf{F}(\mathbf{S})$, given by

$$\mathbf{F}(\mathbf{S}) = \int \rho(\mathbf{r}) \exp(2\pi i \mathbf{r}.\mathbf{S})\,d\mathbf{r} \tag{12}$$

$$\mathbf{r} \text{ in the cell}$$

It is common to replace the position vector $\mathbf{r}$ in the unit cell by $\mathbf{a}x + \mathbf{b}y + \mathbf{c}z$, where x, y and z are fractional coordinates with a value between 0 and 1. The Laue conditions give the following:

$$\mathbf{r}\cdot\mathbf{S} = (\mathbf{a}x + \mathbf{b}y + \mathbf{c}z)\cdot\mathbf{S} = \mathbf{a}\cdot\mathbf{S}x + \mathbf{b}\cdot\mathbf{S}y + \mathbf{c}\cdot\mathbf{S}z$$
$$= hx + ky + lz \tag{13}$$

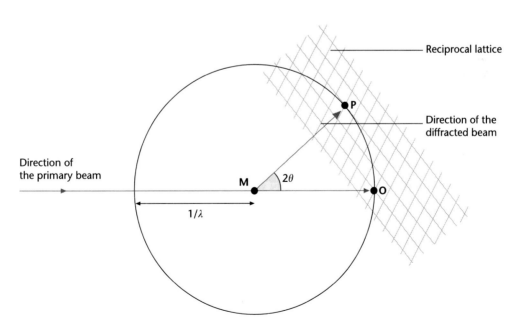

Figure 7 Construction of the diffracted beam by means of the Ewald sphere and the reciprocal lattice. M is the center of the sphere and O is the origin of the reciprocal lattice. The primary beam passes through M and is directed toward O. If the crystal is regarded as being situated at M, a diffracted beam leaves the crystal in the direction of MP where P is any reciprocal lattice point on the surface of the Ewald sphere. (Reproduced from *Encyclopedia of Life Sciences*.)

Hence $\mathbf{F}(\mathbf{S})$ can also be written as

$$\mathbf{F}(h\,k\,l)$$
$$= V \int\limits_{x=0}^{1} \int\limits_{y=0}^{1} \int\limits_{z=0}^{1} \rho(x,y,z)\exp[2\pi i(hx+ky+lz)]\mathrm{d}x\,\mathrm{d}y\,\mathrm{d}z$$

$$(14)$$

where V is the volume of the unit cell. With this equation, the scattering by a crystal can be calculated if the electron density distribution is known. This process has to be reversed if the electron density is to be calculated from the scattering information. The reversal is mathematically known as Fourier inversion. It results in

$$\rho(x,y,z) = \frac{1}{V}\sum_{h}\sum_{k}\sum_{\ell}\mathbf{F}(h\,k\,l)\exp[-2\pi i(hx+ky+lz)]$$

$$(15)$$

Because diffraction occurs only in specific directions, the integration is replaced by summation. Since $\mathbf{F} = |\mathbf{F}|\exp(i\alpha)$, we can also write

$$\rho(x,y,z) = \frac{1}{V}\sum_{h}\sum_{k}\sum_{l}|\mathbf{F}(h\,k\,l)|\exp[-2\pi i(hx$$
$$+ ky + lz) + i\alpha(h\,k\,l)]$$

$$(16)$$

The amplitudes $|F(h\,k\,l)|$ can be derived from the scattered intensity because, apart from correction factors, $|F(h\,k\,l)| = [I(h\,k\,l)]^{1/2}$ where $I(h\,k\,l)$ is the intensity of the reflection $(h\,k\,l)$. Unfortunately, the phase angles $\alpha(h\,k\,l)$ cannot be derived straightforwardly from the diffraction pattern. Several techniques are available to find them: multiple isomorphous replacement, multiple wavelength anomalous dispersion, molecular replacement and (less common) direct methods. (*See* Perutz, Max Ferdinand.)

Summary

The internal structure of crystals can be determined with X-ray diffraction. When an X-ray beam impinges on a crystal, a large number of scattered beams are observed in specific directions. It is easy to find these directions with the reciprocal lattice and the Ewald sphere. The relation between the scattering angle 2θ and the lattice plane distance d in the crystal is given by Bragg's law: $2d\sin\theta = n \times \lambda$. Alternatively, the scattering directions can be found with the Laue conditions: $\mathbf{a}\cdot\mathbf{S} = h$; $\mathbf{b}\cdot\mathbf{S} = k$; $\mathbf{c}\cdot\mathbf{S} = l$, where $\mathbf{S}$ is a reciprocal lattice vector perpendicular to the Bragg planes with a length n/d. The structure factor $\mathbf{F}(h\,k\,l)$ expresses the wave with amplitude $|F(h\,k\,l)|$ and phase angle $\alpha(h\,k\,l)$. This wave is the sum of the waves scattered by the individual electrons in the crystals. The electron density distribution $\rho(x,y,z)$ in the unit cell can be calculated from

$$\rho(x,y,z) = \frac{1}{V}\sum_{h}\sum_{k}\sum_{\ell}|\mathbf{F}(h\,k\,l)|\exp[-2\pi i(hx+ky+lz) + i\alpha(h\,k\,l)]$$

$$(17)$$

The major problem to solve here is the determination of the phase angles $\alpha(h\,k\,l)$ for each scattered beam.

See also
Crystallization of Nucleic Acids
Crystallization of Proteins and Protein–Ligand Complexes
Macromolecular Structure Determination: Comparison of Crystallography and Nuclear Magnetic Resonance (NMR)

References

Bragg WL (1913) The diffraction of short electromagnetic waves by a crystal. *Proceedings of the Cambridge Philosophical Society* **17**: 43–57.
Drenth J (1999) The theory of X-ray diffraction by a crystal. *Principles of Protein X-ray Crystallography*, 2nd edn, pp. 83–84. New York: Springer-Verlag.
Ewald PP (1913) Zur Theorie der Interferenzen der Röntgenstrahlen in Kristallen. *Physikalische Zeitschrift* **14**: 465–472.

Further Reading

Blundell TL and Johnson LN (1976) *Protein Crystallography*. London: Academic Press.
Drenth J (1999) *Principles of Protein X-ray Crystallography*, 2nd edn. New York: Springer-Verlag.
Glusker JP, Lewis M and Rossi M (1994) *Crystal Structure Analysis for Chemists and Biologists*. New York: VCH.
International Union for Crystallography (1995) *International Tables for Crystallography*. Dordrecht, The Netherlands: Kluwer Academic Publishers.
McRee DE (1993) *Practical Protein Crystallography*. San Diego, CA: Academic Press.
Rhodes G (1993) *Crystallography Made Crystal Clear*. San Diego, CA: Academic Press.

XYY Syndrome

Kerry-Jane Galenzoski, *Health Sciences Centre, Winnipeg, Manitoba, Canada*
Chitra Prasad, *Health Sciences Centre, Winnipeg, Manitoba, Canada*

XYY is a condition associated with tall stature and occasional learning difficulties in males with an extra Y chromosome.

Genetics of XYY

In XYY the karyotype (chromosome count) is 47,XYY (**Figure 1**). This means that each of the somatic cells contains 47 chromosomes, 22 normal pairs of autosomes, and three sex chromosomes, one X chromosome and two Y chromosomes. XYY results from an egg containing an X chromosome joining with a sperm cell that contains two Y chromosomes, instead of the normal single Y chromosome. In some cases the sperm cell may contain multiple extra copies of the Y chromosome, leading to an embryo with a 48,XYYY or 49,XYYYY chromosome complement. These are generally more severe disorders than XYY and may be associated with mental retardation, tall stature and infertility. Fortunately, they are very rare.

Technology for studying large populations of XYY patients has only been available since the development of fluorescence techniques beginning in the early 1970s. Therefore, either prospective or retrospective studies of very large groups of XYY patients have been somewhat limited. However, there have been some key studies that have greatly improved understanding of XYY. These include the prospective sex-chromosome study conducted in Denver, Colorado, as well as a more recent study in which 39 males with XYY were examined, and their body habitus, medical history and psychosocial development were noted.

Physical Features of XYY

Most individuals with XYY have no unusual characteristics. Their weight and length at birth are usually normal. Some boys with XYY may be taller than their peers, and the average adult height is more than 6 feet tall. Extremely tall boys may be self-conscious about their height.

In some cases, XYY is associated with a prominent glabella, which refers to the area on the face between

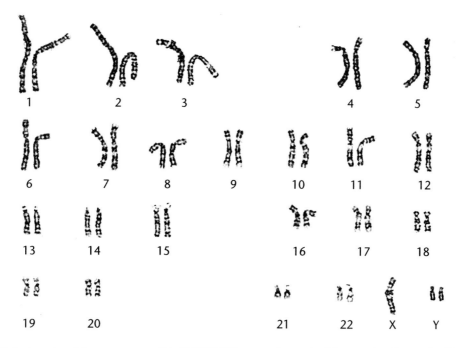

Figure 1 In XYY, the karyotype (chromosome count) is 47,XYY. This means that each of the somatic cells contains 47 chromosomes, 22 normal pairs of autosomes, and three sex chromosomes, one X chromosome and two Y chromosomes. (Figure Courtesy of A. J. Dawson.)

the eyebrows. They may have clinodactyly, a subtle angulation of their fifth finger. Some boys with XYY have a pectus carinatum, which is a condition where the breastbone curves outwards. This is not usually a serious condition, but may cause some self-consciousness. XYY has also been associated with hernias, particularly in the groin. However, this is the exception rather than the rule. Severe acne has also been described.

Boys with XYY begin puberty on average 6 months later than their peers; however, adult testosterone levels are normal. Sexual development is normal, and most XYY men have healthy children. Some reports have described a slightly increased risk of miscarriage, stillbirth and aneuploidy in pregnancies fathered by XYY men. One study of spermatogenesis in a 47,XYY male examined the karyotype of 75 spermatozoa and found that 4% of cells had a numerical chromosome abnormality; 10% of sperm were structurally abnormal. However, when these numbers were compared with 46,XY controls, they were found not to be significantly different.

In some instances, males with XYY may have very small testes or absent sperm production, but infertility is not a common feature of XYY.

XYY and Learning Difficulties

Most males with XYY are of normal intelligence. However, studies have shown an increased frequency of learning disabilities in XYY males compared with the general population. In particular, XYY boys appear to have difficulties with speech and language acquisition. While mental retardation is not a usual feature of XYY, boys on average have been found to have an IQ 10–15 points below their siblings. Performance IQ is generally higher than verbal IQ and, while many XYY males have difficulty processing sentences, they often perform well at nonverbal problem solving and are average or slightly below average in tests of verbal-conceptual skills.

In some cases, XYY has been associated with pervasive developmental disorders (PDD) such as autism. PDD refers to a group of disorders characterized by lack of communication, poor social skills and characteristic repetitive behaviors, called stereotypical behaviors. PDD may be mild and easily treated with social-skills training, or it may be more severe and significantly impact the ability to establish relationships and communicate. While most boys with XYY do not suffer from PDD, an association between these two disorders has been found. One study in a PDD population revealed that of 92 patients presenting with PDD features, one had XYY syndrome. Neuromotor deficits such as incoordination and reduced speed of gross motor activities have been found in approximately 50% of XYY males.

Up to 25% of XYY males display hyperactive behavior or have emotional issues that may contribute to learning problems. Mild depression is common. Distractibility and an increased tendency toward temper tantrums may also be seen, even in boys with a normal IQ. These boys may benefit from special assistance in the classroom to enable them to concentrate and optimize their learning. Professional counseling has also been found to be useful in helping parents and teachers learn how best to manage difficult behavior.

XYY has been associated with difficulty in managing stressful situations. XYY boys with unhappy family lives have been found to be less well adjusted than their siblings in the same environment. However, in a positive environment there was no difference in coping skills between XYY boys and their siblings.

XYY males struggling with hyperactivity, learning disabilities and behavioral issues respond to the same techniques as XY males. A supportive teaching environment and structured teaching program are often instrumental in enabling children and adolescents to optimize their learning.

XYY and Violence

In the 1960s, XYY was a topic of great debate in the popular press. An increased frequency of XYY among inmates at a high-security Scottish prison led to the designation of the extra Y chromosome as the 'criminal chromosome'. Parents of XYY males were counseled that there was a significant risk of violence in their sons. Unfortunately, this misinformation was due to an ascertainment bias in the original study. This meant that the data showing violence in XYY males had been gathered in a very specific population of incarcerated males and could not be generalized to the population as a whole. While there were XYY males in the prison, there were many more XYY males in the general population who had never committed a crime. However, the study did not take these XYY males into account. Many careful studies since have shown no increase in violence among males with XYY. Among these studies is a review of patients diagnosed with various aneuploid conditions antenatally, and followed during childhood and adolescence. While speech delay and distractibility were frequent findings among XYY males, researchers noted that 'aggression was not frequently observed in children and adolescents'. It should be noted that, while prenatal diagnosis has allowed for the study of an unbiased population of XYY patients, many of these individuals would probably never otherwise have been ascertained, owing to lack of symptoms.

See also
Trisomy

Further Reading

Benet J and Martin R (1988) Chromosome complements in a 47,XYY male. *Human Genetics* **78**(4): 313–315.

Cowen P and Mullen P (1979) An XYY man. *British Journal of Psychiatry* **135**: 79–81.

Gaylin W (1980) The XYY controversy: researching violence and genetics. *The Hastings Center Report* (supplement): August 1980.

Kopelman L (1978) Ethical controversies in medical research: the case of XYY screening. *Perspectives in Biology and Medicine* 196–204.

Linden M, Bender B and Robinson A (1996) Intrauterine diagnosis of sex chromosome aneuploidy. *Obstetrics and Gynecology* **87**: 468–476.

Marcus AM and Richmond G (1970) The XYY syndrome: review with a case study. *Canadian Psychiatric Association Journal* **15**: 389–396.

Money J (1994) *Sex Errors of the Body and Related Syndromes.* Baltimore, MD: Paul H Brooks.

Rimoin DL, Connor JM and Pyeritz RE (1996) *Emery and Rimoin's Principles and Practice of Medical Genetics*, 3rd edn. New York: Churchill Livingstone.

Robinson A, Bender B and Linden MG (1990) Summary of clinical findings in children and young adults with sex chromosome anomalies. *Birth Defects Original Article Series* **26**(4): 225–228.

Weidmer-Mikhail E, Sheldon S and Ghaziuddin M (1998) Chromosomes in autism and related pervasive developmental disorders: a cytogenetic study. *Journal of Intellectual Disability Research* **42**(1): 8–12.

Wiedeman HR, Grosse KR and Dibbern H (1985) *An Atlas of Characteristic Syndromes.* London: Wolfe Medical.

Y Chromosome

See Chromosome Y

Yeast Artificial Chromosome (YAC) Clones

Ramaiah Nagaraja, *National Institute on Aging, Baltimore, Maryland, USA*

David Schlessinger, *National Institute on Aging, Baltimore, Maryland, USA*

Artificial chromosome vector systems have facilitated mapping and sequencing of complex genomes at an increasingly rapid pace. They include cloned DNA fragments ranging from 50 kb to more than 1 million base pairs, as well as sequences that render them capable of growth in yeast or bacteria. Traditional bacterial cloning systems have remained important for the study of relatively short clones, but for the cloning of very large DNA segments artificial chromosomes have completely replaced earlier bacterial systems, including lambda phage-based cosmids.

Advanced article

Article contents

- Yeast Artificial Chromosomes
- Bacterial Artificial Chromosomes
- P1-based Artificial Chromosomes

Yeast Artificial Chromosomes

The discovery that linear plasmids can be constructed in yeast using heterologous sequences from *Tetrahymena pyriformis*, and the demonstration that linear artificial chromosomes can be constructed and faithfully maintained in the yeast *Saccharomyces cerevisiae*, was based on the identification of sequences that provide the function of centromeres (CEN), telomeres (TEL) and autonomously replicating sequences (ARS) (see Murray and Szostak, 1983, and references therein). This led to the notion that such artificial chromosomes could be used to clone heterologous sequences from other organisms. Following this line of thought, Burke *et al.* (1987) constructed the first artificial chromosome vector, and developed a collection ('library') of yeast artificial chromosomes (YACs) containing inserts of fragments of human deoxyribonucleic acid (DNA), which were soon shown to carry and faithfully maintain full copies of human genes. After this initial demonstration, the construction of comparable libraries from many complex genomes rapidly became commonplace. (*See* Artificial Chromosomes; DNA Cloning.)

Most studies have used the vector and preparative protocols developed by Burke *et al.* (1987). The standard YAC vector pYAC4 (Genbank acc. no. GI 401849) has the following features:

- A circular plasmid, it has a pBR322 origin of replication to sustain propagation in bacteria and provide large amounts of DNA for cloning purposes.

- A marker selectable in bacteria, *Amp*^{r}, ensures its retention in cultures during bacterial growth.
- Digestion of the vector with *Bam*HI and *Eco*RI releases three fragments. One 'stuffer' fragment, between the *T. pyriformis* telomere sequences, is discarded during the cloning procedure; the others become the 'arms' of the linear YAC (**Figure 1**). One arm contains CEN4 and ARS sequences required for replication and segregation in a yeast host, along with a yeast selectable marker *TRP1*; the other contains another yeast selectable marker, *URA3*. Both fragments terminate with a cloning site at one end and TEL sequences at the other. The digestion with *Eco*RI also disrupts an *SUP4* gene that is required to suppress an *ADE4* mutation, which otherwise renders the colonies red. Clones that incorporate a DNA insert will thus be red, whereas those in which the arms religate without an insert will be white.

During cloning in a standard protocol, source DNA is first partially digested with *Eco*RI and size-fractionated by pulsed field gel electrophoresis or by zonal sedimentation in a sucrose gradient. Fragments larger than about 200 kilobases (kb) are recovered and ligated to the YAC vector arms, transformed into the host strain AB1380, and plated on medium lacking uracil and tryptophan. Clones with an insert (red) that are *URA*$^+$*TRP*$^+$ (that is, have incorporated the selective markers from the YAC vector arms) are isolated from the primary transformation plates, purified by regrowth on the selective medium and tested to verify the presence of an insert – for

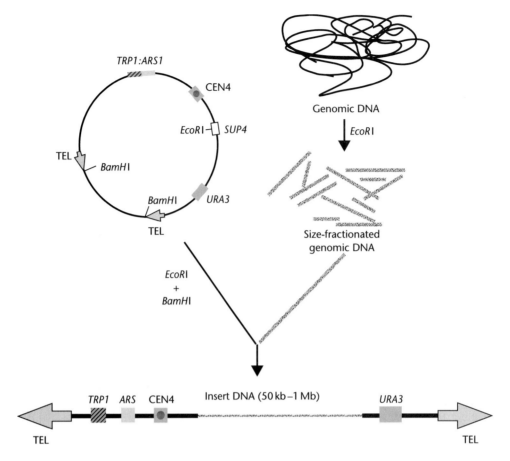

Figure 1 Generalized features of a typical YAC cloning vector and cloning steps. The YAC vector has yeast selectable auxotrophic markers for tryptophan (*TRP1*) and uracil (*URA3*) biosysnthesis. The centromere (CEN4) sequence is derived from yeast centromere 4. Cleavage with *BamH*I and *EcoR*I releases two arms of the YAC vector terminating in *Tetrahymena pyriformis* telomere (TEL) sequences associated with one of the selectable markers and disrupts the *SUP4* gene that suppresses the *ADE4* mutation. The stuffer fragment released by digestion with *BamH*I between the telomeres is removed during purification of the arms. The genomic DNA is digested with *EcoR*I in the presence of *EcoR*I methylase to prevent excessive digestion, fractionated by pulsed field gel electrophoresis (or zonal sedimentation through a sucrose gradient) and ligated to the arms and transformed into *Saccharomyces cerevisiae* strain AB1380. Mutation in the *ADE4* gene in the host strain yields transformed colonies that are CEN4 in the absence of the *SUP4* gene on a selection medium lacking uracil and tryptophan. Such colonies are further purified and the presence of insert verified.

example, by blot hybridization with a radioactive probe generated from source DNA. Many modifications of the vector and host have been developed to favor higher transformation rates, permit the use of other selectable markers or increase the stability of YACs (e.g. Shero *et al.*, 1991).

Although YAC cloning has been of great utility, many clones are 'chimeric' or 'cocloned', bringing together catenated segments from two or more locations in genomic DNA. They are observed more frequently (in up to 50% or more of clones) among larger YACs. Such clones may arise by spurious ligation or recombination between partially homologous repeat sequences of different segments during the outgrowth of transformed strains. They are probably favored during the outgrowth of strains transformed with multiple segments of source DNA, especially

because many abortive clones may contain two identical vector arms (with two CEN sequences or none).

One approach to alleviate the problem of chimeric YACs uses targeted recombination-based cloning. Targeted cloning of specific chromosomal loci has been based on transformation-associated recombination (TAR) (Larionov *et al.*, 1996). This approach capitalizes on recombination in yeast between the free ends of defined DNA and longer tracts of source DNA (**Figure 2**). In one version, segments of source DNA are cloned between a unique sequence, ligated to one YAC vector arm, and a common repeat sequence (such as an Alu element) ligated to the other arm. Because such repeat elements are located at varying distances from the unique tract, a series of YACs of increasing size is generated, with inserts all

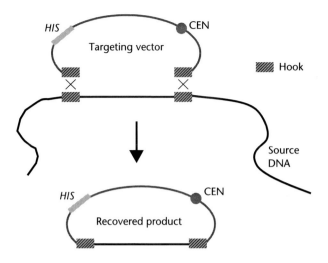

Figure 2 Principles of transformation-associated recombination (TAR) cloning. A BAC-based vector containing yeast centromere (CEN) and selectable marker (*HIS3*) sequences is cleaved with a restriction enzyme at a unique site placed between the 'hooks' (DNA sequences of at least 200 bp at the edges of the region to be cloned). The digested vector is transformed into yeast along with high molecular weight genomic DNA. The 'hooks' pair with the corresponding sequences in genomic DNA, and recombination produces the TAR clone. The cells are plated and recombinants that grow depend on the presence of replication origins in the cloned DNA, thus ensuring that the DNA between the 'hooks' is rescued. Note that this vector system rescues genomic DNA in circular form; unlike YAC construction it does not require the presence of telomere sequences, and discourages cocloning events; in fact, no cocloned TAR-derived YAC has been observed.

starting from the unique sequence and terminating in Alu. Alternatively, each YAC arm is first ligated to one of two unique 'hook' sequences that flanks a region to be cloned at distances of 50–300 kb in the genome. Resultant YACs then incorporate the intervening genomic DNA by homologous recombination during the cloning procedure.

This system tends to avoid chimerism. For example, 3′ and 5′ flanking regions of the breast cancer 1 gene (*BRCA1*) have been used to rescue the intact gene (Annab *et al.*, 2000). The resulting clones are circular, avoiding the need for telomeres for the stable maintenance of the YACs. A comparable principle can be applied to circularize existing YAC clones and fit them with markers to permit their mobilization and growth in bacteria (Cocchia *et al.*, 2000).

Bacterial Artificial Chromosomes

A second way to avoid cocloning and facilitate recovery of purified large-insert clones is provided by bacterial artificial chromosomes (BACs). The BAC vector has an origin of replication from the F episome that is functional in bacteria. It also incorporates

elements involved in partition during cell division, which permits its maintenance at one copy per cell. This avoids deletions and rearrangements that are very frequent during growth of cosmid-based high copy number cloning systems (Shizuya *et al.*, 1992).

BAC vectors can accommodate inserts of up to 300 kb, with an average size of about 150 kb. Recent vectors have incorporated screening aids including blue–white selection based on *LacZ* gene expression (to recognize clones with inserts), selection against the *SacBII* gene during growth on sucrose (to facilitate clone recovery), and addition of Epstein–Barr virus (EBV) oriP sequences and the blasticidin deaminase gene (to permit growth and selection of clones transferred into mammalian cells) (see Frengen *et al.*, 2000, and references therein).

P1-based Artificial Chromosomes

A variant of BACs, P1-based artificial chromosomes (PACs) use the DNA packaging system of P1 bacteriophage which incorporates DNA by 'headfuls'. Packaging begins at the *Pac* site by digestion with the phage-encoded Pacase, with no sequence specificity. DNA is stuffed into the phage head starting at the *Pac* site and continuing until the head is full. Tail structures are then added to produce infectious phage particles. For a first-generation vector, size-fractionated DNA inserts were cloned between two P1 loxP sites. Upon delivery of linear DNA molecules into cells, a kanamycin resistance gene in the vector aided in the selection of cells infected with DNA, and phage P1 Cre recombinase catalyzed recombination between the two loxP sites to generate a circular molecule.

Because the phage head can accommodate no more than 113 kb of DNA including the vector (Pierce and Sternberg, 1992, and references therein), the packaging step limited the size of cloned DNA. The limitation of insert size has since been bypassed in a vector that overcomes the need to package DNA into viral heads (pCYPAC1) (Ioannou *et al.*, 1994). Modifications included the removal of one of the loxP sites and insertion of a pUC19 plasmid between its *Escherichia coli* promoter and the *SacBII* gene. Toxicity of the *SacB* gene, which ferments sucrose in the medium to generate toxic products, is thereby reduced, and large quantities of the vector can be prepared as 'pUC19 plasmids'. Later, during the preparation of the vector for cloning, the pUC19 sequences are removed and replaced with insert DNA. Thus, the features of the P1 replicon system are retained but the vector can be handled like a typical plasmid. After the insert is ligated into a *BamHI* site, these plasmids can be introduced into cells by electroporation. Kanamycin is retained as a selectable marker, and clones lacking inserts are again selected against by growth on sucrose. Further modifications to vectors, similar to BACs, include addition of EBV oriP sequences and blasticidin deaminase genes. Such constructs with inserts are thereby selectable in mammalian cells (Frengen *et al.*, 2000).

The sequences and the structures for the current generation BAC and PAC vectors can be found on the Vector Information website (see Web Links).

In summary, yeast and bacteria are both viable hosts for artificial chromosomes with inserts of hundreds of kilobases of DNA, and are providing the materials and protocols to be adapted to mammalian artificial chromosomes.

See also

Artificial Chromosomes
Contig assembly
DNA Cloning

References

Annab LA, Kouprina N, Solomon G, *et al.* (2000) Isolation of a functional copy of the human BRCA1 gene by transformation-associated recombination in yeast. *Gene* **250**: 201–208.

Burke DT, Carle GF and Olson MV (1987) Cloning of large segments of exogenous DNA into yeast by means of artificial chromosome vectors. *Science* **236**: 806–812.

Cocchia M, Kouprina N, Kim SJ, *et al.* (2000) Recovery and potential utility of YACs as circular YACs/BACs. *Nucleic Acids Research* **28**: E81.

Frengen E, Zhao B, Howe S, *et al.* (2000) Modular bacterial artificial chromosome vectors for transfer of large inserts into mammalian cells. *Genomics* **68**: 118–126.

Ioannou PA, Amemiya CT, Garnes J, *et al.* (1994) A new bacteriophage P1-derived vector for the propagation of large human DNA fragments. *Nature Genetics* **6**: 84–89.

Larionov V, Kouprina N, Graves J, *et al.* (1996) Specific cloning of human DNA as yeast artificial chromosomes by transformation-associated recombination. *Proceedings of the National Academy of Sciences of the United States of America* **93**: 491–496.

Murray AW and Szostak JW (1983) Construction of artificial chromosomes in yeast. *Nature* **305**: 189–193.

Pierce JC and Sternberg NL (1992) Using bacteriophage P1 system to clone high molecular weight genomic DNA. *Methods in Enzymology* **216**: 549–574.

Shero JH, McCormick MK, Antonarakis SE and Hieter P (1991) Yeast artificial chromosome vectors for efficient clone manipulation and mapping. *Genomics* **10**: 505–508.

Shizuya H, Birren B, Kim UJ, *et al.* (1992) Cloning and stable maintenance of 300-kilobase-pair fragments of human DNA in *Escherichia coli* using an F-factor-based vector. *Proceedings of the National Academy of Sciences of the United States of America* **89**: 8794–8797.

Further Reading

Sambrook J, Fritsch EF and Maniatis T (1989) *Molecular Cloning, a Laboratory Manual*, 2nd edn, vol. 1, pp. 2.2–3.58. Plainview, NY: Cold Spring Harbor Laboratory Press.

Web Links

Vector Information. For a description of PAC/BAC vectors, thier sequences, and library resources.
http://www.chori.org/bacpac/vectors.htm

Yeast as a Model for Human Diseases

Judith L Fridovich-Keil, *Emory University School of Medicine, Atlanta, Georgia, USA*

Intermediate article

Article contents

- Introduction
- Cell Cycle Control
- Genetic Instability and Colon Cancer
- Aging
- Present and Predicted Contributions of Yeast to Drug Discovery and Production

Yeasts are single-celled eukaryotes that provide a genetically and biochemically amenable model system for biomedical research. Over the past decades, studies in yeast have established a foundation for much of our current knowledge on such fundamental processes as eukaryotic cell cycle control, genetic instability and colon cancer, metabolism and metabolic disease and even aging. Through basic science and pharmaceutical applications, studies in yeast hold promise for even greater contributions in the future.

Introduction

If we are truly to understand any human disease, we must first understand humans. As any medical or biomedical graduate student can tell you, however, the human body is enormously complex, so that understanding the normal structure and function of any given organ or tissue or even any given cell, is a Herculean task. Indeed, despite many decades to centuries of intense research, and even given recent access to a nearly completed database of the human genome, what we do not know about our own bodies, tissues and cells still far outweighs what we do know.

 NATURE ENCYCLOPEDIA OF THE HUMAN GENOME / ©2003 Macmillan Publishers Ltd, Nature Publishing Group / www.ehgonline.net

Model genetic organisms, including mice, zebrafish, flies, nematodes, the plant *Arabidopsis thaliana* and bacteria, among others, have all proved extremely useful in helping scientists to probe the mysteries of living systems. Indeed, some of the most powerful model organisms have been those single-celled eukaryotes – the yeasts (Guthrie and Fink, 1991). A variety of different species of yeast have been used, the two most common being *Saccharomyces cerevisiae* or baker's yeast and *Schizosaccharomyces pombe* or fission yeast. A number of specific features render these yeasts particularly amenable to genetic and biochemical manipulation. First, both are small, single-celled eukaryotes that proliferate rapidly in inexpensive media and that can survive indefinitely as frozen glycerol stocks. Second, both can be maintained for generations either as haploids or as diploids, thereby enabling genetic selection and screening for rare mutations, either dominant or recessive. Third, when diploids are induced to undergo meiosis, all four resultant haploid spores are retained within a single shell, or ascus. Tetrad dissection followed by spore germination and characterization thereby enables classic genetic analysis. Fourth, both yeasts are adept at maintaining and expressing exogenous genes either as chromosomal integrants or on episomal plasmids; homologous recombination is extremely efficient in yeast, so that relatively simple procedures that enable researchers to insert, delete or change virtually any base or sequence within the yeast genome are now commonplace. Finally, since 1996, the entire *Sa. cerevisiae* genome sequence has been available as an annotated, searchable public database (see Web Links). A genome database for *Sc. pombe* is also nearly complete (see Web Links).

Perhaps most important, however, is the observation that despite the vast evolutionary distance that separates yeasts from humans, many fundamental genetic and biochemical processes have been conserved between the two. Furthermore, a comparative homology search of the predicted amino acid sequences of recognized genes in *Sa. cerevisiae* with mammalian sequences in GenBank revealed that more than 30% of the predicted yeast proteins (1914 proteins) had clear mammalian homologs (Botstein *et al.*, 1997). This number surely represents an underestimate, as no mammalian genomes were fully sequenced and available at the time of the study; no doubt an update of this estimate will be forthcoming. Studies begun in yeast have therefore provided a foundation for much of our current biomedical knowledge in such diverse areas as cell cycle control, genetic instability and colon cancer, metabolism and metabolic disease and even aging. The remainder of this article will describe a handful of selected examples of seminal contributions from yeast studies to our current knowledge, as well as discuss some of the present and potential future applications of yeast research to drug discovery and pharmaceutical production.

Cell Cycle Control

The progression of cells through the various stages and checkpoints of the eukaryotic cell cycle is a carefully orchestrated process, involving the activities and interplay of a multitude of gene products. Key experiments conducted using both *Sa. cerevisiae* and *Sc. pombe* laid much of the ground work for our current understanding of cell cycle progression and regulation (for review see Nurse, 2000), by demonstrating that conditional mutations in single genes could result in the synchronous arrest of a culture of cells. The cloning of these mutant alleles and the wild-type genes from which they were derived established a foundation upon which subsequent biochemical experiments were then able to define the functions, interactions and regulation of these genes and their encoded proteins. One of the most informative of the yeast genes identified by this strategy was called *CDC2* ((schpo) *Sc. pombe* nomenclature) or *CDC28* ((yeast) *Sa. cerevisiae* nomenclature). The protein product of this gene turned out to be a protein kinase that functioned as part of a complex, the composition of which changed through the cell cycle. The mammalian homologs of this gene are now known as the cyclin-dependent kinases (CDKs), and encode a family of protein kinases believed to regulate both the G_1–S transition and the G_2–M transition of the cell cycle.

With regard to *CDC2*, yeast played an important role not only in the identification of this first CDK gene in yeast, but also in the identification of the first human homolog of this gene. Once it became clear how central *CDC2* function was to cell cycle regulation in yeast, a number of research groups quickly attempted the next step, namely screening mammalian complementary deoxyribonucleic acid (cDNA) libraries by nucleic acid hybridization with a *CDC2* probe looking for homologs. None was found. Lee and Nurse (1987) then took another approach, namely complementation cloning. They screened a human cDNA library for clones that could restore *CDC2* function to a host strain of yeast carrying a conditional-null allele of *cdc2*, and hit the jackpot. From this first human CDK clone, scientists were then able to identify both other CDKs and their interacting proteins in humans and other mammalian species.

Genetic Instability and Colon Cancer

Genetic instability, especially of dinucleotide repeat tracts, has long been recognized as a characteristic of aggressive human cancer, in particular colon cancer. Nonetheless, until the mid-1990s, the molecular basis for this instability was unknown. A breakthrough came when researchers made the connection that the genetic instability observed in many colon cancers, and particularly in a familial form known as hereditary nonpolyposis colorectal cancer (HNPCC), was reminiscent of the genetic instability seen in yeast with mutations in their mismatch repair genes, especially (yeast)*MSH2* and (yeast)*MLH1* (Fishel *et al.*, 1993; Leach *et al.*, 1993) (for review see Kolodner and Alani (1994)). This was an intriguing observation, because it immediately led to the testable hypothesis that inherited mutations in the human homologs of these yeast genes might underlie the cancer predisposition seen in HNPCC families. Indeed, gene mapping studies demonstrated that the human homolog of (yeast)*MSH2* (mutS homolog 2, colon cancer, nonpolyposis type 1 (*E. coli*) ((human)*MSH2*)) is located within one of the HNPCC loci that had previously been assigned by family linkage studies (chromosome 2p). What is more, sequence analyses of the (yeast)*MSH2* alleles in these families demonstrated that although different families often carried different specific mutations, most if not all indeed carried germ-line mutations in (human)*MSH2*, and these mutations cosegregated through each family with the disease.

A similar story unfolded for the human homolog of (yeast)*MLH1*, called mutL homolog 1, colon cancer, nonpolyposis type 2 (*E. coli*) ((human)*MLH1*), which was mapped to a small region of chromosome 3 previously assigned as an alternate HNPCC locus (for review see Kolodner and Alani, 1994). As with (human)*MSH2*, germ-line mutations in (human)*MLH1* were identified in these HNPCC families that cosegregated with the disease.

Finally, sequence analyses of the mismatch repair gene loci in matched samples representing normal (nontumor) versus tumor tissue derived from patients with HNPCC demonstrated that one (familial) mutation was identified in the nontumor tissue while two mutations (one familial, one novel) were found in the tumor sample (Leach *et al.*, 1993). These data clearly supported a modified Knudson two-hit model (Hutchinson, 2001) to explain the genesis of cancer in HNPCC. The two-hit hypothesis of inherited cancer predisposition was first proposed by Alfred Knudson as a result of his work with retinoblastoma. According to this hypothesis, patients with a familial risk for retinoblastoma inherit one functional copy, and one mutant copy of the *retinoblastoma 1* (including osteosarcoma) (*RB1*) tumor suppressor gene. Although cells with even one functional copy of *RB1* do not develop into cancer, statistically speaking, individuals who inherit one mutant copy of *RB1* in each of their cells

are extremely likely to experience at least one somatic *RB1* mutation in at least one of their retinal cells, thereby leaving that cell with no functional copies of *RB1*. It is these cells that go on to become cancerous. With regard to HNPCC, therefore, according to this hypothesis, individuals who inherit a single mutation in a mismatch repair gene (e.g. in (human)*MSH2*) are at risk to experience a 'second hit', namely a second mutation, in one or more of their somatic cells. When, by chance, this second mutation strikes the only remaining 'good allele' of the mismatch repair gene in question, the host cell, often a colonic cell, is left with no functional copies of that gene, and the downward spiral toward further mutation and genetic instability ensues, eventually leading to cancer.

While providing an explanation for a given human disease is important, offering tools for prevention, treatment or cure is better. Now that scientists and physicians understand the basis of HNPCC, and have clear targets for study (e.g. (human)*MSH2* and (human)*MLH1*), they are better armed to search out effective therapies. In the meanwhile, knowing which genes to check for inherited mutations provides an option for presymptomatic diagnosis for a majority of HNPCC families. Once the inherited mutation in an affected individual has been identified, family members can be screened for the presence or absence of that mutation. Those individuals determined to carry the mutation are encouraged, despite their current good health, to undergo frequent colonoscopy, a procedure that inspects the colon for evidence of polyps or tumor growth. Given the age-related penetrance of the disease, for most of these 'at-risk' individuals, it is not a question of if a tumor will be found, but a question of when. At this point, many patients choose to undergo surgical removal of the colon at the first sign of tumor appearance, thereby markedly improving their long-term outcome. For the case of HNPCC, there is no doubt that presymptomatic diagnosis, coupled with increased surveillance and early intervention, is saving lives.

Aging

Although the effects of aging that are so well known in humans and other multicellular animals were for decades if not centuries believed to spare the single-celled species, studies in yeast since the mid-1980s (Jazwinski, 2001) have demonstrated that this is not the case. Any given population of yeast may be immortal – proliferating indefinitely under good conditions – nonetheless, individual yeast cells exhibit a finite lifespan; they age and die, just as we do. This simple observation opened the door to a flood of genetic and biochemical studies that have helped to identify some of the key players and interactions that underlie the eukaryotic processes of cellular senescence, or aging. Indeed, at least 30 different yeast genes have been implicated in aging (Jazwinski, 2001), many with apparent human homologs (see the Saccharomyces Genome Database in Web Links). Although the individual functions of the proteins encoded by these genes are diverse, they tend to fall into four categories: metabolism, stress resistance, gene dysregulation and genetic instability.

One particularly intriguing feature of aging in both human cells and yeast is a form of genetic instability that results from replication-dependent telomere shortening (Johnson *et al.*, 2001). Indeed, primary human fibroblasts derived from healthy individuals have been known for decades to demonstrate a finite 'Hayflick limit' of replicative potential, dependent upon the age of the cell donor (Hayflick, 1980); it is now known that this limit reflects progressive telomere shortening with repeated cell divisions resulting from the natural absence of telomerase activity in most somatic cells. Consistent with this conclusion, genetic restoration of telomerase activity to many normal human cells allows them to bypass their Hayflick limit, and achieve apparent immortality (Bodnar *et al.*, 1998). Furthermore, many cancer cells have been found to express telomerase activity aberrantly, thereby ostensibly attaining cellular immortality (Keith *et al.*, 2001). Yeasts genetically depleted of telomerase, like their normal human counterparts, lose telomeric repeats with successive rounds of cell division, and age (Lundblad and Szostak, 1989).

Normal human fibroblasts, like normal humans, age at a fairly uniform rate. However, a small number of devastating genetic disorders, such as Werner syndrome (WS), have been noted in which patients 'age' at a markedly increased rate. As predicted, cells derived from patients with WS also shorten their telomeres and senesce at a faster rate than do their normal counterparts, and this cellular senescence can be blocked by the genetic restoration of telomerase activity (Oullette *et al.*, 2000). WS is now known to result from a loss of function of the RecQ family DNA helicase encoded by the Werner syndrome (*WRN*) gene. Yeasts also have a RecQ family helicase, encoded by the yeast gene *SGS1*. Recent studies have demonstrated that in yeast lacking telomerase, Sgs1p, the protein product of *SGS1*, plays a role in telomere maintenance; telomerase-null *sgs1* mutants demonstrate unusually rapid telomere shortening, and senesce prematurely (Johnson *et al.*, 2001). These data support the hypothesis that *WRN* may also function in telomere maintenance in normal human cells.

Present and Predicted Contributions of Yeast to Drug Discovery and Production

Because of their biochemical and genetic facility, yeasts have contributed to the biomedical field not only as model systems for research but also as tools for more applied purposes, such as the production of recombinant vaccines. Indeed, the recombinant hepatitis B vaccine currently administered in the United States and around the world is produced using yeast as a host organism to support the production of the recombinant viral protein.

More complex applications of yeast relate to the field of drug discovery rather than drug production, and serve both research and pharmaceutical objectives (Ma, 2001). For example, 'proteomic' applications are under way at a number of institutions using yeast as a host for large-scale definition of the protein–protein interactions that underlie the higher-order organization of functional networks, metabolic and otherwise, in living cells (Kumar and Snyder, 2002). The value of this work to drug discovery is clear: if key proteins associated with human disease exist *in vivo* as components of complexes, a knowledge of these complexes and their constituents is essential to efforts to design and test drugs that have an impact upon the relevant function.

References

Bodnar AG, Ouellette M, Frolkis M, *et al.* (1998) Extension of lifespan by introduction of telomerase into normal human cells. *Science* **279**: 349–352.

Botstein D, Chervitz SA and Cherry JM (1997) Genetics: yeast as a model organism. *Science* **277**: 1259–1260.

Fishel RA, Lescoe MK, Rao MRS, *et al.* (1993) The human mutator gene homolog MSH2 and its association with hereditary nonpolyposis colon cancer. *Cell* **75**: 1027–1038.

Guthrie C and Fink G (eds.) (1991) *Guide to Yeast Genetics and Molecular Biology.* San Diego, CA: Academic Press.

Hayflick L (1980) Cell aging. *Annual Review of Gerontology and Geriatrics* **1**: 26–67.

Hutchinson E (2001) Alfred Knudson and his two-hit hypothesis (interview of Alfred Knudson by Ezzie Hutchinson). *Lancet Oncology* **2**: 642–645.

Jazwinski SM (2001) New clues to old yeast. *Mechanisms of Ageing and Development* **122**: 865–882.

Johnson FB, Marciniak RA, McVey M, *et al.* (2001) The *Saccharomyces cerevisiae* WRN homolog Sgs1p participates in telomere maintenance in cells lacking telomerase. *EMBO Journal* **20**: 905–913.

Keith NW, Evans JTR and Glasspool RM (2001) Telomerase and cancer: time to move from a promising target to a clinical reality. *Journal of Pathology* **195**: 404–414.

Kolodner RD and Alani E (1994) Mismatch repair and cancer susceptibility. *Current Opinion in Biotechnology* **5**: 585–594.

Kumar A and Snyder M (2002) Protein complexes take the bait. *Nature* **415**: 123–124.

Leach FS, Nicolaides NC, Papadopoulos N, *et al.* (1993) Mutations of a *mutS* homolog in hereditary nonpolyposis colorectal cancer. *Cell* **75**: 1215–1225.

Lee MG and Nurse P (1987) Complementation used to clone a human homolog of the fission yeast cell cycle control gene *cdc2*. *Nature* **327**: 31–35.

Lundblad V and Szostak JW (1989) A mutant with a defect in telomere elongation leads to senescence in yeast. *Cell* **57**: 633–643.

Ma D (2001) Applications of yeast in drug discovery. *Progress in Drug Research* **57**: 117–162.

Nurse P (2000) A long twentieth century of the cell cycle and beyond. *Cell* **100**: 71–78.

Oullette MM, McDaniel LD, Wright WE, Shay JW and Schultz RA (2000) The establishment of telomerase-immortalized cell lines representing human chromosome instability syndromes. *Human Molecular Genetics* **9**: 403–411.

Further Reading

Adams A, Gottschling DE, Kaiser CA and Stearns T (1997) *Methods in Yeast Genetics*. Plainview, NY: Cold Spring Harbor Laboratory Press.

Broach JR, Pringle JR and Jones EW (1991) *The Molecular and Cellular Biology of the Yeast* Saccharomyces, vol. 1, *Genome Dynamics, Protein Synthesis and Energetics*. Plainview, NY: Cold Spring Harbor Laboratory Press.

Jones EW, Pringle JR and Broach JR (1992) *The Molecular and Cellular Biology of the Yeast* Saccharomyces, vol. 2, *Gene Expression*. Plainview, NY: Cold Spring Harbor Laboratory Press.

Pringle JR, Broach JR and Jones EW (1997) *The Molecular and Cellular Biology of the Yeast* Saccharomyces, vol. 3, *Cell Cycle and Cell Biology*. Plainview, NY: Cold Spring Harbor Laboratory Press.

Web Links

Saccharomyces Genome Database (SGD). This is a scientific database of the molecular biology and genetics of the yeast *Saccharomyces cerevisiae*
http://genome-www.stanford.edu/Saccharomyces/

The Wellcome Trust Sanger Institute: the *S. pombe* Genome Project. A useful resource for *S. pombe* genome project data as well as other tools and links
http://www.sanger.ac.uk/Projects/S_pombe/

mutL homolog 1, colon cancer, nonpolyposis type 2 (*E. coli*) (*MLH1*). A resource for molecular biology and clinical information concerning *MLH1*; Locus ID: 4292. LocusLink:
http://www.ncbi.nlm.nih.gov/LocusLink/LocRpt.cgi?l = 4292

mutS homolog 2, colon cancer, nonpolyposis type 1 (*E. coli*) (*MSH2*). A resource for molecular biology and clinical information concerning *MSH2*; Locus ID: 4436. LocusLink:
http://www.ncbi.nlm.nih.gov/LocusLink/LocRpt.cgi?l = 4436

Werner syndrome (*WRN*); A resource for molecular biology and clinical information concerning *WRN*; Locus ID: 7486. LocusLink:
http://www.ncbi.nlm.nih.gov/LocusLink/LocRpt.cgi?l = 7486

mutL homolog 1, colon cancer, nonpolyposis type 2 (*E. coli*) (*MLH1*). Information on hereditary nonpolyposis colorectal cancer (HNPCC type 2); MIM number: 120436. OMIM:
http://www.ncbi.nlm.nih.gov/htbin-post/Omim/dispmim?120436

mutS homolog 2, colon cancer, nonpolyposis type 1 (*E. coli*) (*MSH2*). Information on hereditary nonpolyposis colorectal cancer (HNPCC type 1); MIM number: 120435. OMIM:
http://www.ncbi.nlm.nih.gov/htbin-post/Omim/dispmim?120435

Werner syndrome (*WRN*). Information on the *WRN* gene and encoded gene product; MIM number: 604611. OMIM:
http://www.ncbi.nlm.nih.gov/htbin-post/Omim/dispmim?604611

Yeast Two-hybrid System and Related Methodology

Russell L Finley Jr, *Wayne State University School of Medicine, Detroit, Michigan, USA*

Advanced article

Article contents

- Introduction
- Two-hybrid Assays
- Yeast Two-hybrid System
- Limitations
- Two-hybrid Systems in Other Organisms
- Variations on a Theme – Other Two-hybrid Assays
- Summary

The identification and detailed characterization of the binary interactions between gene products in a cell is an essential step toward understanding the functions of genes and how they control cellular systems. Two-hybrid assays have become powerful tools for this purpose.

Introduction

Genetic control is mediated by precise interactions between gene products. The ability of one protein to specifically recognize and bind to another protein, for example, determines the role the protein plays in a cell. Such binary interactions mediate the flow of energy and information within a cell. Signal transduction pathways, for example, consist of networks of interacting proteins that convert an extracellular signal into changes in a cellular process such as transcription or replication. Specific binary interactions are also fundamental to the assembly of important complexes, including the multiprotein complexes that carry out deoxyribonucleic acid (DNA) replication and transcription, or the ribonucleic acid (RNA)–protein complexes that mediate RNA splicing and protein translation.

Two-hybrid Assays

Two-hybrid assays detect the binary interaction between two molecules. The two-hybrid assays discussed here detect protein–protein interactions, though a variety of related systems have been described to detect other macromolecular interactions. Two-hybrid assays take the general form shown in **Figure 1**. To detect an interaction between two macromolecules, X and Y, they are synthesized so that each is covalently attached to a different tag, A or B. This results in two hybrid molecules, B–X and A–Y. The tags are designed such that when they come into proximity with each other they activate a detection system. In the original yeast two-hybrid system, the tags are two domains of a transcription factor. An interaction between the two hybrid proteins reconstitutes the transcription factor, which results in transcriptional activation of reporter genes in yeast. In other variations, the two tags are two halves of an enzyme, which is reconstituted through the interaction between X and Y.

There is an important and defining distinction between two-hybrid assays and other assays that use a tag to uniquely identify and isolate a protein. A 'pull-down assay', for example, uses an affinity matrix to isolate a tagged protein from a cell lysate; one commonly used tag, for example, is glutathione-*S*-transferase (GST), which binds to a glutathione matrix. This approach can be used as an interaction assay by identifying and quantifying the associated

proteins that are isolated along with the tagged protein. In contrast, the tags in a two-hybrid assay are used to enable a detection system, not to separate the proteins from a mixture. Importantly, this means that two-hybrid assays can be performed within a milieu, including inside a living cell, and they do not require the ability to purify the proteins being studied. Moreover, the two-hybrid assay provides a means for detecting interactions in a setting that may mimic physiological conditions. (*See* Fusion Proteins as Research Tools.)

Yeast Two-hybrid System

The yeast two-hybrid system (Fields and Song, 1989) is one of the most widely used assays for protein–protein interactions. The assay takes place inside a yeast nucleus and uses two tags that together function as a transcription factor (**Figure 2**). One tag is a DNA-binding domain (DBD), which mediates binding to specific DNA sequences located upstream of a reporter gene. The other tag is a transcription activation domain (AD). In a cell expressing two hybrid proteins, DBD-X and AD-Y, an interaction between X and Y results in recruitment of the AD to the reporter gene, where it activates transcription. A number of different versions of the two-hybrid system have been described, each using a different combination of reporter genes and tags. Most versions use at least two reporters, including the *Escherichia coli lacZ* and a yeast marker gene required for growth; a protein interaction leads to growth of the yeast and β-galactosidase (β-gal) activity. Various DBD and ADs have also been used. The most common DBDs are derived from the yeast Gal4 protein or from *E. coli* LexA. ADs that have been used include the AD from yeast Gal4, from the herpes simplex virus protein VP16, and from an *E. coli* open reading frame (ORF), B42. The assay works even when most of the components are not derived from yeast; for example, systems that use the LexA DBD, the B42 AD and the *lacZ* reporter to detect interaction between nonyeast proteins. In essence, the yeast cell is being used as a

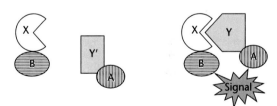

Figure 1 Two-hybrid assays for protein–protein interactions. In a two-hybrid assay, each protein to be tested is expressed fused to a tag, A or B. In this example, X interacts with Y (right) but not Y′ (left). When A and B are brought near each other through the X–Y interaction, they activate a signal that can be detected (right). The signal may result from direct interaction between A and B, for example, if together A and B form an active enzyme. Alternatively, the signal may be generated when one tag becomes localized to a particular subcellular location as a result of the interaction, as in transcription-based two-hybrid systems.

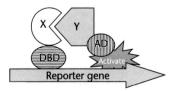

Figure 2 In the transcription-based yeast two-hybrid system, one tag is a DNA-binding domain (DBD) which binds to specific sites in a reporter gene. The other tag is a transcription-activation domain (AD). An interaction between X and Y localizes the AD to the reporter, where it activates transcription.

living test tube in which to conduct interaction assays. (*See* Transcription Factors.)

The simplicity of the yeast two-hybrid system has made it a routine tool for identifying new protein interactions and for characterizing known interactions. The assay has been used successfully with proteins from many different organisms, with a variety of protein types, and with proteins normally found in various subcellular compartments. Once established for a particular pair of proteins, the two-hybrid assay can be used to characterize an interaction in detail. For example, deletion and point mutant variants for each protein can be generated and tested to map interacting domains and amino acid contacts. The assay can also be used to follow the ability of small molecules or other proteins to enhance or inhibit the interaction. One of the most powerful uses has been to discover new protein interactions. This is usually accomplished by screening libraries that express AD-tagged proteins to identify partners for a DBD-tagged protein of interest. New, interacting partners can provide clues about a protein's function. The success of this approach has inspired efforts to use the two-hybrid system on a genome-wide scale. High-throughput two-hybrid screens have been developed using a mating assay in which the AD- and DBD-tagged proteins are expressed in haploid yeast with opposite mating types (Bendixen *et al.*, 1994; Finley and Brent, 1994). The interaction assay is conducted by mating the haploid yeast and measuring reporter activity in the diploid cells. Large libraries or arrays of yeast strains expressing different AD- or DBD-tagged proteins can be collected and systematically mated to sample all possible interactions. High-throughput two-hybrid screens can be used to construct protein interaction maps for yeast and several other organisms. The interaction maps from these efforts include hundreds of new protein–protein interactions. (*See* Protein–Protein Interaction Maps.)

Limitations

Yeast two-hybrid interaction data provide clues about protein and network function by defining possible binary contacts among functionally linked sets of proteins. Interpretation of two-hybrid data, however, requires an understanding of its limitations. One important limitation to consider is that two-hybrid experiments can generate false-positives. These arise when two proteins appear to interact in the assay but have no biologically relevant interaction *in vivo*. One class of false-positives emerges when the reporter genes are activated in the absence of a true interaction between the DBD- and AD-tagged proteins. Most two-hybrid systems provide methods to identify and eliminate this class of false-positives. Another class of false-positives,

which is more difficult to identify, includes the true two-hybrid interactions between proteins that do not normally interact in their native environments. For example, in some cases the two proteins may normally be found in different organelles or different tissues and only interact when forced into a yeast nucleus. The insidious nature of these false-positives demands that yeast two-hybrid data be confirmed by other means. Ideally, this involves showing that an interaction takes place in cells in which the proteins are normally found, often by coimmunoprecipitation. However, such assays may be prohibitively inefficient for the thousands of new interactions being identified by high-throughput yeast two-hybrid projects. For these, it will be necessary to use computational approaches and to combine the interaction data with other functional genomics-scale data to derive testable hypotheses about protein and network functions.

The second major limitation of the yeast two-hybrid system is that it cannot detect all protein–protein interactions. The possible causes of false-negatives are worth considering. First, a particular protein may be difficult to express in yeast, for example, because it is degraded by specific yeast proteases or is toxic. Second, some proteins may fail to enter the yeast nucleus; though a nuclear localization signal is included in the hybrid tag, other signals in a protein may dominate. Third, a protein outside its native environment may fail to fold properly or may lack posttranslational modifications that are needed for interactions. Fourth, the presence of the AD or DBD tag may in some cases block proper folding or obscure surfaces needed for interactions. Finally, some binary protein interactions are too weak to be detected by the yeast two-hybrid assay. Studies with protein pairs having a known equilibrium dissociation constant (K_d) have suggested that yeast two-hybrid systems may miss interactions with $K_d > 10^{-6}$ M (Estojak *et al.*, 1995).

In conclusion, interactions detected in yeast two-hybrid experiments often require verification by other assays. Moreover, some interactions will require another assay to be detected. While no assay is immune to the problem, different types of interaction assay technologies are generally subject to different false-positives and false-negatives. Thus, efforts to define and characterize complete maps of the protein interactions in a cell will benefit from the development of several other two-hybrid assays.

Two-hybrid Systems in Other Organisms

Transcription-based assays similar to the yeast two-hybrid system have been developed in other organisms. A version that is conducted in cultured mammalian cells, for example, has been used to detect interactions between mammalian proteins (Dang *et al.*, 1991). Assays in mammalian cells may detect interactions that cannot be detected in yeast, such as those that may depend on specific posttranslational modifications. However, mammalian two-hybrid systems have been used primarily with individual known proteins, because robust library screening approaches have not been developed. In contrast, bacterial two-hybrid systems have been described that may facilitate easier library screening (Dove *et al.*, 1997). In most bacterial systems, one tag is a DBD and the other is a subunit of the RNA polymerase (RNAP). An interaction between the hybrids results in increased recruitment of the RNAP to the promoter and transcription of the reporter. As in the yeast two-hybrid assay, the level of reporter activation can correlate with the strength of the protein interaction, and interactions with K_d of at least 10^{-6} M can be detected (Dove *et al.*, 1997). Although eukaryotic proteins are less likely to be properly folded or modified in bacteria than in yeast, bacterial two-hybrid systems have been used to detect interactions with yeast and mammalian proteins.

Variations on a Theme – Other Two-hybrid Assays

A number of two-hybrid systems have been developed that do not rely directly on transcription factors. Although many systems exist, the following examples are intended to provide a glimpse of the possibilities. In the Sos and Ras recruitment systems (Aronheim *et al.*, 1997), the interaction assay takes place at the cytoplasmic membrane of a yeast cell rather than in the nucleus. These assays take advantage of a temperature-sensitive mutation in the yeast Ras gene *CDC25*. At the nonpermissive temperature, the Cdc25 protein is trapped in the inactive guanosine diphosphate (GDP)-bound form and the yeast fails to grow. The mutant can be complemented either by human Ras (hRas) or human Sos (hSos), a guanyl nucleotide exchange factor, but only if one of these proteins is brought to the membrane. In the interaction assay, one protein is tagged with either hSos or hRas, and the other is tagged with a myristoylation signal sequence, which leads to its attachment to the plasma membrane. An interaction between the tagged proteins recruits hSos or hRas to the membrane and allows growth at the nonpermissive temperature. In another system, which is particularly suited for proteins naturally targeted to the plasma membrane, the two proteins are tagged with complementary halves of ubiquitin

(Johnsson and Varshavsky, 1994). One protein has an additional *N*-terminal tag, constituting a transcription factor, but, since the protein is anchored in the membrane, the reporter genes are not activated. When the ubiquitin halves are brought together as a result of a protein interaction, the ubiquitin-dependent proteolysis machinery liberates the *N*-terminal transcription factor, which then goes to the nucleus and activates reporter genes.

Several two-hybrid assays are independent of a subcellular compartment and do not depend on activation of reporter genes. One assay is based on the ability of two different, inactive mutant versions of β-gal to form an active enzyme by coming together and sharing their active domains (Rossi *et al.*, 1997). If two weakly complementing mutants are used as tags in a two-hybrid assay, they can be forced together by the interacting hybrid proteins, resulting in active β-gal. A key advantage to this assay is that it allows detection of protein interactions in a variety of cell types and in various intracellular compartments. A related approach, called protein fragment complementation (PCA), uses tags derived from two halves of an enzyme such as dihydrofolate reductase (DHFR) (Pelletier *et al.*, 1998). An interaction between the two hybrids leads to reconstitution of the enzyme. In principle, the PCA assay could be adapted to any protein, provided the protein can be split into fragments that are not capable of spontaneously assembling unless they are brought together through interacting proteins.

Finally, naturally fluorescent proteins such as green fluorescent protein (GFP) make ideal detection systems for protein interactions. GFP and related proteins can be used for PCA or in fluorescence resonance energy transfer (FRET) (Day *et al.*, 2001). In an FRET-based two-hybrid assay, the two proteins are tagged with two fluorescent proteins with different excitation and emission spectra. If the proteins interact, the excitation energy from one fluorescent protein (the donor) will be transferred to the other (acceptor), resulting in fluorescence at the emission wavelength of the acceptor protein. Combined with sophisticated imaging techniques, FRET can be used to assay the kinetics of interactions in live cells and whole organisms.

Summary

Over the past 10 years, the yeast two-hybrid system has been used to characterize thousands of proteins. It has also evolved into a high-throughput assay that is being used to generate protein-interaction maps for a number of organisms. The maps are rich with clues about protein and network function, and also with data that require validation. Moreover, the maps are static, lacking information about the dynamics of how network interactions dissolve and re-form as cells develop and respond to their environment. Thus, hypotheses generated with yeast two-hybrid data will require more careful analysis of selected interactions by testing them in other assays, preferably in more native environments and in real time. New two-hybrid assays may provide methods for validating high-throughput data and for detailed *in vivo* studies of interaction networks.

See also

Array-based Proteomics
Disease-related Genes: Functional Analysis
Fusion Proteins as Research Tools
Protein–Protein Interaction Maps

References

Aronheim A, Zandi E, Hennemann H, Elledge SJ and Karin M (1997) Isolation of an AP-1 repressor by a novel method for detecting protein–protein interactions. *Molecular Cell Biology* **17**: 3094–3102.

Bendixen C, Gangloff S and Rothstein R (1994) A yeast mating-selection scheme for detection of protein–protein interactions. *Nucleic Acids Research* **22**: 1778–1779.

Dang CV, Barrett J, Villa-Garcia M, *et al.* (1991) Intracellular leucine zipper interactions suggest c-*myc* hetero-oligomerization. *Molecular Cell Biology* **11**: 954–962.

Day RN, Periasamy A and Schaufele F (2001) Fluorescence resonance energy transfer microscopy of localized protein interactions in the living cell nucleus. *Methods (San Diego, CA)* **25**: 4–18.

Dove SL, Joung JK and Hochschild A (1997) Activation of prokaryotic transcription through arbitrary protein–protein contacts. *Nature* **386**: 627–630.

Estojak J, Brent R and Golemis EA (1995) Correlation of two-hybrid affinity data with *in vitro* measurements. *Molecular Cell Biology* **15**: 5820–5829.

Fields S and Song O (1989) A novel genetic system to detect protein–protein interactions. *Nature* **340**: 245–246.

Finley Jr RL and Brent R (1994) Interaction mating reveals binary and ternary connections between *Drosophila* cell cycle regulators. *Proceedings of the National Academy of Sciences of the United States of America* **91**: 12980–12984.

Johnsson N and Varshavsky A (1994) Split ubiquitin as a sensor of protein interactions *in vivo*. *Proceedings of the National Academy of Sciences of the United States of America* **91**: 10340–10344.

Pelletier JN, Campbell-Valois FX and Michnick SW (1998) Oligomerization domain-directed reassembly of active dihydrofolate reductase from rationally designed fragments. *Proceedings of the National Academy of Sciences of the United States of America* **95**: 12141–12146.

Rossi F, Charlton CA and Blau HM (1997) Monitoring protein–protein interactions in intact eukaryotic cells by beta-galactosidase complementation. *Proceedings of the National Academy of Sciences of the United States of America* **94**: 8405–8410.

Further Reading

Bartel PL and Fields S (1997) The two-hybrid system: a personal view. In: Bartel PL and Fields S (eds.) *The Yeast Two-hybrid System*, pp. 3–7. Oxford, UK: Oxford University Press.

Brent R and Finley Jr RL (1997) Understanding gene and allele function with two-hybrid methods. *Annual Review of Genetics* **31**: 663–704.

Golemis EA (2002) *Protein–Protein Interactions*. Plainview, NY: Cold Spring Harbor Laboratory Press.

Hazbun TR and Fields S (2001) Networking proteins in yeast. *Proceedings of the National Academy of Sciences of the United States of America* **98**: 4277–4278.

Hu JC, Kornacker MG and Hochschild A (2000) *Escherichia coli* one- and two-hybrid systems for the analysis and identification of protein–protein interactions. *Methods (San Diego, CA)* **20**: 80–94.

Ito T, Chiba T, Ozawa R, *et al.* (2001) A comprehensive two-hybrid analysis to explore the yeast protein interactome. *Proceedings of the National Academy of Sciences of the United States of America* **98**: 4569–4574.

Johnston M and Fields S (2000) Grass-roots genomics. *Nature Genetics* **24**: 5–6.

Mendelsohn AR and Brent R (1999) Protein interaction methods: toward an endgame. *Science* **284**: 1948–1950.

Rain JC, Selig L, De Reuse H, *et al.* (2001) The protein–protein interaction map of *Helicobacter pylori*. *Nature* **409**: 211–215.

Remy I and Michnick SW (2001) Visualization of biochemical networks in living cells. *Proceedings of the National Academy of Sciences of the United States of America* **98**: 7678–7683.

Uetz P, Giot L, Cagney G, *et al.* (2000) A comprehensive analysis of protein–protein interactions in *Saccharomyces cerevisiae*. *Nature* **403**: 623–627.

Walhout AJ and Vidal M (2001) Protein interaction maps for model organisms. *Nature Reviews Molecular Cell Biology* **2**: 55–62.

Zebrafish as a Model for Human Diseases

Paula Goodman Fraenkel, *Children's Hospital and Beth Israel Deaconess Medical Center, Department of Medicine, Harvard Medical School, Boston, Massachusetts, USA*

Leonard I Zon, *Howard Hughes Medical Institute, Children's Hospital, Department of Pediatrics, Harvard Medical School, Boston, Massachusetts, USA*

Intermediate article

Article contents
- Zebrafish Genomics
- Zebrafish Models of Hematologic Diseases
- Zebrafish Models of Cardiovascular Diseases
- Zebrafish Models of Renal Disorders
- Zebrafish Models of Neurologic Disorders
- Future Directions of Zebrafish Research

Zebrafish (*Danio rerio*) are small freshwater fish, the only vertebrate model in which a large-scale mutagenesis strategy has been undertaken as an approach to the study of the development of blood and vital organs. The development of techniques for gene mapping in zebra fish and the anticipated completion of the zebra fish genome sequencing project in 2002 will enhance the use of zebra fish as models for human diseases.

Zebrafish Genomics

Random mutagenesis of the 1.7 billion basepair zebrafish (**Figure 1**) genome has resulted in the identification of more than 500 mutant phenotypes in large-scale screens for defects in specific aspects of early development. Mutations have been introduced in zebrafish by treatment with ethylnitrosourea (ENU), gamma irradiation or insertion of genetic material. This article will focus on a few of the many characterized mutants with direct relevance to human diseases affecting blood, cardiovascular, renal or neurologic development (**Table 1**).

Recent progress in zebrafish gene mapping has revealed extensive synteny with the human genome, and sequence homology of many genes. Zebrafish genes have been identified by positional cloning, by identification of a mutated gene by its chromosomal map position, and by using the candidate gene approach, identification of a mutated gene based on its suspected function. Phenotypic rescue of an embryo following microinjection with complementary deoxy-

ribonucleic acid (cDNA) or a genomic clone assists in identifying the gene of interest. The creation of transgenic zebrafish that express the green fluorescent protein gene linked to tissue-specific promoters has facilitated the study of cell lineage and migration in the developing embryo.

Searching the zebrafish genome and expressed sequence tag (EST) databases has already revealed analogs to genes affected in many human diseases, such as those for amyotrophic lateral sclerosis, fragile X syndrome and Huntington disease (Dodd *et al.*, 2000). Although techniques for targeted gene deletion have not been developed in the zebrafish, recent innovations suggest that it will be possible to generate human disease models by lowering the expression of specific genes. This can be accomplished by injecting embryos with morpholinos, chemically modified oligonucleotides that inhibit translation of messenger ribonucleic acid (mRNA) (Nasevicius and Ekker, 2000).

Figure 1 The zebra fish, *Danio rerio*, is a small freshwater fish that has been used in large-scale mutagenesis studies to generate models of human diseases and disorders of development. (Photograph used by permission of ZFIN, http://zfin.org/)

Zebrafish Models of Hematologic Diseases

Analogous to higher vertebrates, zebrafish exhibit multilineage hematopoiesis, resulting in erythroid, monocyte, granulocyte and thrombocyte lineages, and undergo hemoglobin switching. Mouse and human homologs of blood-specific genes including *scl*, *lmo-2*, *gata-1* and *c-myb* have been cloned in the zebrafish. In a large-scale mutagenesis screen, more than 50 mutants with defects in different stages of hematopoiesis were identified by screening zebrafish embryos (**Figure 2**) for absent or hypochromic blood. Several of the mutant phenotypes, named after varieties of wine, are analogous to human blood

Table 1 Selected zebra fish mutants with relevance to human diseases

Mutant	Phenotype	Analogous human disease	Mutated zebra fish gene
Zinfandel	Defective early globin expression with recovery in the adult	Thalassemia	Regulator of globin expression
Sauternes	Hypochromic anemia with declining blood counts	X-linked sideroblastic anemia	δ-Aminolevulinate synthase
Yquem	Photosensitive, autofluorescent erythrocytes. Absent blood when constantly exposed to light	Hepatoerythropoietic porphyria	Uroporphyrinogen decarboxylase
Riesling	Arrest of erythrocyte differentiation with declining numbers of erythrocytes after 72 h	Hereditary spherocytosis	β-Spectrin
Gridlock	Absent circulation to the tail. Inability of the lateral dorsal aortas to fuse	Coarctation of the aorta	Gridlock/basic helix–loop–helix protein
Double bubble	Glomerular cysts. Alterations in the glomerular basement membrane and altered localization of Na^+/K^+ ATPase	Autosomal dominant polycystic kidney disease	Unknown
Fleer	Glomerular cysts. Alterations in the glomerular basement membrane and altered localization of Na^+/K^+ ATPase. Eye degeneration	Autosomal dominant polycystic kidney disease. Senior–Loken syndrome (combined retinal and renal dysplasia)	Unknown
Elipsa	Glomerular cysts. Eye degeneration	Autosomal dominant polycystic kidney disease. Senior–Loken syndrome (combined retinal and renal dysplasia)	Unknown
One-eyed pinhead	Cyclopia. Notochord present, but lacks prechordal plate and ventral neuroectoderm	Holoprosencephaly	One-eyed pinhead/EGF-related ligand

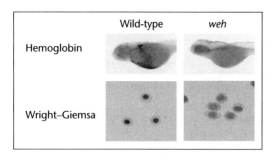

Figure 2 [*Figure is also reproduced in color section.*] Upper panels: zebra fish embryos stained at 48 h with o-dianisidine for hemoglobin detection. Lower panels: zebrafish peripheral erythrocytes stained with Wright–Giemsa. Upper left: a wild-type zebrafish embryo at 48 h with red blood cells traveling over the yolk sac in a primitive circulation. Upper right: a *weissherbst* zebrafish embryo at 48 h with marked anemia. Lower left: wild-type zebrafish peripheral erythrocytes with normal nuclear size. Circulating erythrocytes are nucleated in zebrafish. Lower right: *weissherbst* peripheral erythrocytes with enlarged nuclei and immature morphology. (Reprinted by permission of Macmillan Magazines Ltd from Donovan A, Brownlie A, Zhou Y, *et al.* (2000) Positional cloning of zebrafish ferroportin 1 identifies a conserved vertebrate iron exporter. *Nature* **403**: 776–781.)

diseases. *zinfandel* lacks early globin expression and is a model for thalassemia. *sauternes*, with hypochromic anemia, delayed erythroid maturation and abnormal globin expression, encodes a mutation in δ-aminolevulinate synthase, a heme biosynthetic enzyme. *sauternes* is the first animal model of X-linked sideroblastic anemia, an inherited defect in δ-amino-levulinate synthase (Brownlie *et al.*, 1998). This defect results in decreased heme synthesis and ineffective erythropoiesis, i.e. low numbers of mature red blood cells in the circulation despite active production of immature erythroid cells in the bone marrow.

The zebrafish mutant *yquem* has a loss-of-function mutation in another enzyme in the heme synthetic pathway, uroporphyrinogen decarboxylase (Wang *et al.*, 1998). Patients with the disease hepato-erythropoietic porphyria (HEP) also have decreased uroporphyrinogen decarboxylase activity. The human disease presents at a young age with blistering skin lesions in areas of sun exposure, liver damage and red urine due to the accumulation of photosensitive porphyrins. 'Yquem' zebrafish also have complications of porphyrin buildup including fluorescent blood that hemolyzes in the presence of light. Ablation of uroporphyrinogen decarboxylase expression in the wild-type zebrafish embryo, by injecting morpholinos targeted against this gene, reproduces the *yquem* phenotype (Nascevicius and Ekker, 2000).

In addition to generating models for human blood diseases, study of zebrafish mutants has resulted in the discovery of novel human genes. Positional cloning of the mutation responsible for the hypochromic zebrafish mutant *weissherbst* resulted in identification of the novel iron transporter, ferroportin 1. Efforts to clone the affected genes in *cloche*, *bloodless* and *moonshine* mutants defective in early hematopoiesis are likely to result in the identification of novel genes that are important in hematopoietic stem cell development and may improve our understanding of diseases such as aplastic anemia and leukemia. The recent discovery of zebrafish immunoglobulin light chain and T-cell antigen receptor genes (Haire *et al.*, 2000) and observations of zebrafish thymic development (Trede and Zon, 1998) will aid in studying mutants affecting the development of lymphocytes. The identification of zebrafish mutants with coagulation pathway defects may lead to zebrafish models for the study of hemostasis and thrombosis.

Zebrafish Models of Cardiovascular Diseases

The zebrafish embryonic heart consists of atria, ventricles and valves, and resembles the human fetal heart at 3 weeks' gestation. It is lined by endocardium and begins to beat at 22 h postfertilization, with sequential contraction of the atrium and ventricle by 36 h postfertilization. Unlike mouse or human embryos, zebrafish embryos do not depend on cardiac circulation for survival; thus, many cardiovascular mutant embryos can survive to be studied for several days. Cardiovascular mutations in zebrafish (Stainier *et al.*, 1996) include defects in contractility, heart rhythm or cardiovascular structure. *gridlock* homozygotes have obstructed blood flow to the tail due to a failure of fusion of the lateral dorsal aortas. This trait is similar to the human anomaly, coarctation of the aorta, in which the aorta is narrowed below the origin of the subclavian artery, limiting blood flow to the trunk and limbs. *gridlock* is the first naturally occurring animal model for coarctation of the aorta and the subsequent development of collateral blood flow, which can also occur in humans. The *gridlock* gene was recently shown to encode a novel protein in the *hairy/enhancer of split family* (Zhong *et al.*, 2000), basic helix–loop–helix proteins that act as transcriptional regulators downstream of notch. *gridlock* is expressed in the developing dorsal aorta, but not in the adjacent vein, suggesting that notch signaling could regulate the specification of blood vessels.

Zebrafish Models of Renal Disorders

The pronephric kidney is the first kidney to form in both human and zebrafish embryos. In both humans and zebrafish, the pronephric kidney is subsequently replaced by a more complex mesonephric kidney; however, the human mesonephric kidney gives rise to a metanephric kidney while the zebrafish kidney remains in the mesonephric stage. In recessive zebrafish mutants such as *double bubble*, *fleer* and *elipsa*, fluid-filled cysts replace the normal pronephros. In general, the cysts expand, gross edema develops, and the embryos die 5–6 days postfertilization. These zebrafish mutants are being studied as models of glomerular cyst formation, a process that occurs in many human inherited diseases that result in kidney failure, such as autosomal dominant polycystic kidney disease (ADPKD). In the zebrafish mutants *double bubble* and *fleer*, as in the human disease ADPKD, the localization of the Na^+/K^+ ATPase is altered from the basolateral to the apical surface of cells forming the renal tubule (Drummond *et al.*, 1998). The genes causing cystic degeneration of the zebrafish pronephros have not yet been identified. There is no evidence, at present, that any of the mutants is caused by mutations in homologs of the human polycystic kidney gene.

Zebrafish Models of Neurologic Disorders

Defects in brain, nerve and retinal development are being studied extensively via zebrafish genetic models. In several cases, the mutated genes have been identified and reviewed (Talbot and Hopkins, 2000). Many of the signaling proteins involved are similar to those in humans, mice and *Drosophila*. *one-eyed pinhead*, a zebrafish mutant with fusion of the eyes and diminished ventral brain, has been compared with the human congenital anomaly, holoprosencephaly, which, in its most severe form, is characterized by a small brain with a single ventricle, failure to separate the right and left cerebral hemispheres and a single eye. Holoprosencephaly is the most common human developmental defect affecting the forebrain and midface. In humans, mutations in at least 12 different genes (Wallis and Muenke, 2000) have been associated with holoprosencephaly, including *TG-interacting factor* and *sonic hedgehog*. Positional cloning of the *one-eyed pinhead gene* in zebrafish resulted in the identification of a novel epidermal growth factor (EGF)-related ligand (Zhang *et al.*, 1998) with similarity to human CFC1. Recently, loss-of-function mutations in *CFC1* were identified in patients with heterotaxia (randomized organ positioning) and, in one case, holoprosencephaly. Microinjection of wild-type human *CFC1* RNA rescued *one-eyed pinhead* mutant zebrafish embryos, while injection of *CFC1* mutant RNAs failed to ameliorate the *one-eyed pinhead* phenotype (Bamford *et al.*, 2000). This experiment illustrates the versatility of the zebrafish system for assessing the function of human genes.

Future Directions of Zebrafish Research

A new large-scale mutagenesis screen is currently under way at the Max-Planck Institute in Tübingen, Germany. In addition to screening for defects in early embryonic patterning, blood and organ development, embryos will now be fixed and stained to search for abnormalities in the development of blood vessels, bone and cartilage (Wixon, 2000).

Most of the known zebrafish genes have been identified since the late 1990s following advances in zebrafish mapping techniques. The pace of gene identification in the zebrafish will likely accelerate with the completion of the zebrafish genome sequencing project, begun in 2000 by the Sanger Centre in Cambridge, UK. In addition, use of the internet has facilitated the sharing of zebrafish sequence information and genetic maps (see Web Links). Genetic manipulation of zebrafish via random mutagenesis strategies, as well as more targeted approaches to knock down gene expression, will continue to be employed to develop human disease models and to test the function of novel genes identified in the Human Genome Project. In the future, zebrafish disease models may be used in high-throughput screens for therapeutic agents and in preclinical gene therapy trials.

See also
Animal Models

References

Bamford RN, Roessler E, Burdine RD, *et al.* (2000) Loss-of-function mutations in the EGF-CFC gene *CFC1* are associated with human left–right laterality defects. *Nature Genetics* **26**(3): 365–369.

Brownlie A, Donovan A, Pratt SJ, *et al.* (1998) Positional cloning of the zebrafish sauternes gene: a model for congenital sideroblastic anaemia. *Nature Genetics* **20**(3): 244–250.

Dodd A, Curtis PM, Williams LC and Love DR (2000) Zebrafish: bridging the gap between development and disease. *Human Molecular Genetics* **9**(16): 2443–2449.

Drummond IA, Majumdar A, Hentschel H, *et al.* (1998) Early development of the zebrafish pronephros and analysis of mutations affecting pronephric function. *Development* **125**(23): 4655–4667.

Haire RN, Rast JP, Litman RT and Litman GW (2000) Characterization of three isotypes of immunoglobulin light chains and T-cell antigen receptor alpha in zebrafish. *Immunogenetics* **51**(11): 915–923.

Nasevicius A and Ekker SC (2000) Effective targeted gene 'knock-down' in zebrafish. *Nature Genetics* **26**(2): 216–220.

Stainier DY, Fouquet B, Chen JN, *et al.* (1996) Mutations affecting the formation and function of the cardiovascular system in the zebrafish embryo. *Development* **123**: 285–292.

Talbot WS and Hopkins N (2000) Zebrafish mutations and functional analysis of the vertebrate genome. *Genes and Development* **14**(7): 755–762.

Trede NS and Zon LI (1998) Development of T-cells during fish embryogenesis. *Developmental and Comparative Immunology* **22**(3): 253–263.

Wallis D and Muenke M (2000) Mutations in holoprosencephaly. *Human Mutation* **16**(2): 99–108.

Wang H, Long Q, Marty SD, Sassa S and Lin S (1998) A zebrafish model for hepatoerythropoietic porphyria. *Nature Genetics* **20**(3): 239–243.

Wixon J (2000) Interview with Stefan Schulte-Merker: zebrafish functional genomics. *Yeast* **17**(3): 232–234.

Zhang J, Talbot WS and Schier AF (1998) Positional cloning identifies zebrafish one-eyed pinhead as a permissive EGF-related ligand required during gastrulation. *Cell* **92**(2): 241–251.

Zhong TP, Rosenberg M, Mohideen MA, Weinstein B and Fishman MC (2000). Gridlock, an HLH gene required for assembly of the aorta in zebrafish. *Science* **287**(5459): 1820–1824.

Further Reading

Amsterdam A and Hopkins N (1999) Retrovirus-mediated insertional mutagenesis in zebrafish. *Methods in Cell Biology* **60**: 87–98.

Barbazuk WB, Korf I, Kadavi C, *et al.* (2000) The syntenic relationship of the zebrafish and human genomes. *Genome Research* **10**(9): 1351–1358.

Chan FY, Robinson J, Brownlie A, *et al.* (1997) Characterization of adult α- and β-globin genes in the zebrafish. *Blood* **89**(2): 688–700.

Donovan A, Brownlie A, Zhou Y, *et al.* (2000) Positional cloning of zebrafish ferroportin1 identifies a conserved vertebrate iron exporter. *Nature* **403**(6771): 776–781.

Dooley K and Zon LI (2000) Zebrafish: a model system for the study of human disease. *Current Opinion in Genetics and Development* **10**(3): 252–256.

Neumann CJ and Nuesslein-Volhard C (2000) Patterning of the zebrafish retina by a wave of sonic hedgehog activity. *Science* **289**(5487): 2137–2139.

Postlethwait JH, Yan YL, Gates MA, *et al.* (1998) Vertebrate genome evolution and the zebrafish gene map. *Nature Genetics* **18**(4): 345–349.

Ransom DG, Haffter P, Odenthal J, *et al.* (1996) Characterization of zebrafish mutants with defects in embryonic hematopoiesis. *Development* **123**: 311–319.

Weinstein BM, Stemple DL, Driever W and Fishman MC (1995) Gridlock, a localized heritable vascular patterning defect in the zebrafish. *Nature Medicine* **1**(11): 1143–1147.

Web Links

WashU-Zebrafish Genome Resources Project
http://zfish.wustl.edu/
Stanford Zebrafish Genome Project
http://zebrafish.stanford.edu/
The Zebrafish Information Network
http://zfin.org/

Zinc-finger Genes

David Gell, *University of Sydney, Sydney, Australia*

Merlin Crossley, *University of Sydney, Sydney, Australia*

Joel Mackay, *University of Sydney, Sydney, Australia*

Intermediate article

The term zinc-finger is used to describe several small protein domains in which one or more zinc ions stabilize the folded structure. Many zinc-finger genes contain more than one zinc-finger domain and these domains act as interaction motifs, contacting such diverse partners as other proteins, DNA, RNA or lipids.

Introduction

The class of zinc-finger (ZnF)-encoding genes seems to be larger than any other comparable set in the human genome. The term ZnF was coined after the discovery of a repeated sequence of about 30 amino acids in the transcription factor TFIIIA; it was proposed that the conserved cysteine and histidine residues in each repeat allowed each domain to coordinate a zinc ion (Miller *et al.*, 1985). This proved to be correct, and now over 8200 such motifs have been identified in around 550 genes in the human genome. Since this initial discovery, the term ZnF has been extended to include other classes of small protein domains (less than 100 amino acids) that contain one or more zinc ions in a structural (rather than a catalytic) capacity. ZnFs are generally found in intracellular proteins. The requirement for a metal ion is probably related to the difficulty of creating a stable hydrophobic core in a small domain in the reducing environment in a cell, where disulfide formation is not an option. (*See* Domain Duplication and Gene Elongation; Evolutionary History of the Human Genome; Gene Duplication: Evolution.)

At least 14 distinct classes of ZnF domain have been defined by functional and structural analysis (**Table 1**). In addition to these verified ZnFs, several putative zinc-binding sequence motifs have been identified. The cellular functions of the different classes of ZnFs vary greatly, but in general ZnFs can be considered to be 'contact domains'. The 'contact partner' may be single- or double-stranded deoxyribonucleic acid (DNA), ribonucleic acid (RNA), another protein or a lipid. In addition, many proteins contain many ZnF domains (often from distinct classes) and so a single ZnF protein may mediate several interactions. (*See* Multidomain Proteins.)

Genes Encoding Classical Zinc-fingers

Classical ZnF domains, or C2H2 ZnFs, are the most numerous type of ZnF (**Table 1**). They are often implicated in the regulation of gene expression, and many act as sequence-specific DNA-binding motifs that recognize sequences in the promoter or enhancer regions of target genes. (*See* Transcription Factors.)

Several three-dimensional structures of classical ZnFs have been determined, both alone (**Figure 1a**) and in complex with DNA. The basic and hydrophobic residues located in the α helix of the ZnF make specific contacts with 2–4 bases in the major groove of the DNA helix. In general, these interactions are not sufficient to define a DNA sequence that occurs infrequently enough to function as a binding site for transcription factors. To allow recognition of longer DNA sequences, combinations of three or more C2H2 ZnFs in series are generally used, often separated by a conserved five-residue linker (Thr-Gly-Glu-Lys-Pro). Certain ZnF transcription factors also form oligomers with themselves or other ZnF factors to generate a binding site for a particular DNA sequence. Around 70% of classical ZnF proteins contain more than three ZnF domains, with some proteins possessing more than 30. In these proteins, the function of the full set of ZnF domains is currently unknown. (*See* Crystallization of Nucleic Acids; Protein Structure; Transcriptional Regulation: Evolution.)

Much interest has been focused on the challenge of designing ZnF structures that can recognize chosen DNA sequences by combining engineered C2H2 modules. In this way, novel transcription factors can be designed to bind specific DNA sequences. Two complementary approaches have been used. The first is a rational design approach in which a knowledge of known interactions between C2H2 ZnFs and DNA is used to deduce the determinants of DNA recognition. The second approach involves the large-scale screening of degenerate libraries of ZnF peptides to select for binding to the required DNA sequence.

The function of C2H2 fingers is not limited to DNA recognition; they also bind proteins and often other ZnFs. For example, the Ikaros family of transcription

Table 1 Zinc-finger (ZnF) families in the human genome[a]

ZnF motif	No. of genes[b]	Order of zinc ligands in sequence[c]	Structure class[d]	Example protein	Sequence reference	PDB code[d]
ARF-GAP	23	C2-C2	Treble clef (+)	PYK2-associated protein β	6730310	1dcq
GATA	10	C2-C2	Treble clef	GATA-1 transcription factor	P17679	1gnf
LIM	65	$C2_A$-[H/C]2_A-$C2_B$-C[C/H/D]$_B$	Treble clef	CRP1 cysteine-rich protein	Q05158	1a7i
Nuclear hormone receptor	46	$C2_A$-$C2_A$-$C2_B$-$C2_B$	Treble clef	Glucocorticoid receptor	P06536	1glu
PKC	45	H_A-$C2_B$-$C2_A$-[H/D]C_B-C_A	Treble clef (x)	PKC-δ	S35704	1ptq
XPA	1	C2-C2	Treble clef	XPA DNA repair factor	P23025	1d4u
PHD	62	$C2_A$-$C2_B$-HC_A-C[H/C]$_B$	Treble clef (x)	WSTF transcription factor	AAC97879	1f62
RING	189	$C2_A$-CH_B-[C/H]C_A-$C2_B$	Treble clef (x)	Not-4 transcriptional repressor	O95628	1e4u
FYVE	19	$C2_A$-$C2_B$-$C2_A$-$C2_B$	Treble clef (x)	Eaa1 lipid-binding protein	AAA79121	1hyi
C2H2 classical ZnF	556	C2-H[C/H]	Classical ββα	TFIIIA transcription factor	P03001	1tf6
BIR	10	C2-HC	Classical ββα (+)	c-AIP-1 inhibitor of apoptosis	U37547	1qbh
TAZ	2	HC_A-CC_A-HC_B-CC_B-HC_C-CC_C	TAZ	CREB-binding protein CBP	P45481	1f81
Zinc-ribbon	7	C2-C2	Zinc-knuckle	TFIIB transcription factor	Q00403	1dl6
C2HC, nucleocapsid-like	21	C2-HC	Zinc-knuckle	HIV nucleocapsid p55	CAC05361	2znf
B-box	39	C[C/H]-CH	B-box	Xnf7 nuclear protein	A43906	1fre

[a] Each distinct ZnF fold described in the Structural Classification of Proteins (SCOP) database is included plus two newly published structures (PHD and TAZ domains).

[b] The number of genes containing each finger domain is taken from the annotated Ensembl database, which is estimated to include 90% of human genes.

[c] Zinc ligands appear in pairs (separated by 2–6 residues) that interact with the same metal ion. These pairs are often separated by variable gaps in different examples of each finger motif. In structures in which more than one zinc ion is bound, subscript letters are used to indicate the ion bound by each pair of ligands. One exception is the PKC ZnF, in which the first and last zinc ligands in the sequence form a pair in three-dimensional space.

[d] Each ZnF motif is assigned to a structural class on the basis of SCOP classification. Some SCOP classes have been unified to form a single treble-clef family (**Figure 2**). (+) denotes that the finger exists as part of a larger conserved domain; (x) denotes motifs that bind two zinc ions in a cross-brace fashion (**Figure 2b, c**).

factors contain a cluster of four ZnFs at their amino (*N*)-termini that mediate DNA binding, and two ZnFs at their carboxy (*C*)-termini that mediate protein dimerization. ZnF interactions can occur between all family members, which results in different transcriptional activities. Ikaros family proteins regulate the development of lymphoid cells, and mutations in Ikaros have been implicated in specific childhood leukemias (Cortes *et al.*, 1999). (**See** Gene Families: Formation and Evolution; Leukemias and Lymphomas: Genetics.)

In contrast to DNA interactions, protein–protein interactions are often mediated by a single classical finger. Several individual ZnFs from the protein Friend of GATA-1 (FOG-1), which has a variant CCHC zinc coordination pattern but a conserved structure (**Figure 1b**) (Liew *et al.*, 2000), bind the *N*-terminal ZnF of the transcription factor GATA-1. Notably, about 50 human proteins contain a single classical ZnF. Given that a single C2H2 finger cannot generally mediate DNA binding, it is possible that these ZnFs act as

protein–protein interfaces. Individual C2H2 fingers may also contact several partners. For example, the three-ZnF domain of Sp1 family transcription factors recognizes both DNA and the ZnFs of GATA-1 (Merika and Orkin, 1995).

Mutations in several genes encoding classical ZnFs have been associated with human diseases. For instance, deletions and mutations in the Wilms tumor suppressor gene *WT1* are associated with a kidney tumor that occurs in children. (**See** Cancer Cytogenetics; Wilms Tumor.)

A close structural relative of the classical ZnF, the baculoviral inhibitor of apoptosis repeat (BIR) domain, comprises an additional BIR-specific fold that is *N*-terminal to the classical ZnF fold seen in **Figure 1a** and **b**. These domains occur in 'inhibitor of apoptosis' proteins, and the mechanism by which they inhibit caspases (the executioner proteases of apoptosis) has been elucidated (Silke and Vaux, 2001). (**See** Apoptosis and the Cell Cycle in Human Disease; Apoptosis: Disorders; Apoptosis: Regulatory Genes and Disease.)

Genes Containing Treble-clef Fingers

Several classes of ZnFs with apparently different three-dimensional structures actually share a common core fold (**Figure 2**). This has been termed the 'treble-clef motif' because of its resemblance to the musical symbol. Thus, the ZnF domains in GATA, Really interesting new gene (RING), protein kinase C (PKC) cysteine-rich region, nuclear hormone receptor, LIM (named from the three proteins it was first found in, Lin-11, Isl-1, and Mec-3), FYVE (named from the four proteins it was first found in, Fab1p, YOTB, Vac1p and EEA1), plant homeodomain (PHD) – note this displays no homology to the DNA-binding homeodomains, ADP-ribosylation factor-GTPase-activating protein (ARF-GAP) and xeroderma pigmentosum group A (XPA) all share a zinc-knuckle (a β hairpin containing the sequence Cys-X-X-Cys), followed by a β hairpin and an α helix. In all treble-clef structures, one zinc ion is coordinated by two cysteine residues in the zinc-knuckle, together with two cysteine or histidine residues at the junction of the β hairpin and the α helix. Additional zinc ions are bound in many other positions in several of the treble-clef structures (**Figure 2**). (*See* DNA Repair.)

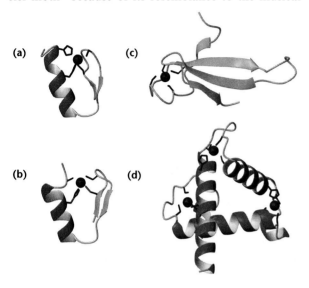

Figure 1 [*Figure is also reproduced in color section.*] ZnF structures illustrating the variety of small zinc-dependent protein folds. ZnFs are found as mixed αβ (a, b), all-β (c) and all-α (d) structures. In each structure, only amino acid side chains that ligate zinc atoms are shown. (a) Classical C2H2 ZnF structure of finger 31 from *Xenopus laevis* XFIN-31 (Protein Data Bank (PDB) accession code 1znf). (b) First CCHC finger from U-shaped – an FOG-family protein from *Drosophila melanogaster* (PDB accession code 1fv5). (c) Zinc-ribbon domain from human transcription factor TFSII (PDB accession code 1tfi). (d) TAZ2 domain from mouse CBP (PDB accession code 1f81).

GATA-type zinc-fingers

The GATA-type ZnFs coordinate a single zinc ion through four cysteine residues (**Figure 2a**). There are 17 such ZnFs in the human genome. Twelve of these are found in tandem pairs in the GATA family of transcription factors (GATA-1 to GATA-6), one is found in trichorhinophalangeal syndrome protein 1 (TRPS1), and the other five are located in proteins about which comparatively little is known. GATA-type ZnFs recognize DNA containing the GATA sequence, hence their name. In contrast to the classical ZnF, a single GATA-type ZnF (together with an ancillary basic domain) is sufficient for DNA binding, so the fact that most human GATA proteins contain

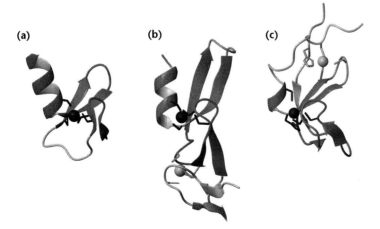

Figure 2 [*Figure is also reproduced in color section.*] The treble-clef fold is conserved in many ZnF structures. Representative structures from three ZnF families (see **Table 1**) are shown with the central treble-clef fold (red) and its conserved zinc ion (blue) highlighted. Note that additional zinc atoms can be accommodated at several positions around the treble-clef in the different fingers (gray spheres). (a) Carboxy-terminal GATA ZnF from chicken GATA-1 (PDB accession code 1gat). (b) FYVE domain from human EEA1 (PDB accession code 1hyi). (c) PKC-type ZnF from human RAF-1 (PDB accession code 1far).

two tandem fingers is intriguing. For human GATA-1, the *N*-terminal finger binds to the classical ZnFs of FOG-1 to regulate blood cell development (Fox *et al.*, 1999).

Haploinsufficiency of GATA-3 has been associated with human hypoparathyroidism, sensorineural deafness and renal anomalies (HDR) syndrome, at least four separate mutations in the *N*-terminal ZnF of GATA-1 seem to cause inherited blood disorders, and several mutations in *TRPS1* cause trichorhinophalangeal syndrome. (*See* Hemophilia and other Bleeding Disorders: Genetics; Transcription Factors and Human Disorders.)

Hormone receptor zinc-finger genes

The nuclear hormone receptor proteins interact (through a non-ZnF domain) with hormones such as vitamin D, 9-*cis* retinoic acid and thyroid hormone. The loaded receptors dimerize and bind to adjacent DNA sequences of AGGTCA (or a variant). Each ZnF domain binds two zinc atoms through Cys-Cys-Cys-Cys motifs, and the α helix of the *N*-terminal motif makes most of the sequence-specific DNA contacts (in the major groove). Both homo- and heterodimers of different receptor molecules bind to DNA, and different dimers bind to sites with different spacings (between one and five intervening bases). Mutations in the vitamin D receptor protein are associated with familial rickets. (*See* Nuclear Receptor Genes.)

Cross-brace zinc-fingers

The RING, PHD and FYVE domains adopt a 'cross-brace' topology in which two zinc atoms are bound by alternate pairs of cysteine or histidine ligands (**Figure 2b**). PKC-type ZnFs also exhibit a modified form of this arrangement (**Figure 2c**).

RING-fingers (Barlow *et al.*, 1994) comprise the second largest group of ZnFs in humans, are found in around 200 proteins and are therefore one of the most common of all protein domains. RING-fingers have been implicated as regulators of protein ubiquitination – the process by which the small protein ubiquitin is covalently linked to the target protein. Ubiquitination acts as a marker for the degradation of misfolded proteins but is also thought to regulate other cellular processes. (*See* Protein Degradation and Turnover.)

RING-fingers have also been proposed to have a general role in assembling multiprotein complexes. For example, RING domains are frequently found in a tripartite motif that also contains a B-box and a coiled-coil domain and is known as an 'RBCC' domain. The RBCC domain of a transcriptional corepressor (KAP-1) has been shown to mediate

both homodimerization and an interaction with the Kruppel associated box (KRAB) domains of C2H2 ZnF transcription factors. Several diseases have been associated with defects in RING-finger proteins, including promyelocytic leukemia, which can arise from a chromosomal translocation involving the RBCC domain of promyelocytic leukaemia (PML) and the retinoic acid receptor. (*See* Fusion Proteins and Diseases; Translocation Breakpoints in Cancer.)

The FYVE, PHD (or LAP) and PKC domains have structures that resemble the RING domain. FYVE domains are specific recognition domains for the lipid phosphatidylinositol-3-phosphate (PtdIns3P), a component of membranes in eukaryotic cells. Thus, proteins with an FYVE domain are localized to cell membranes, and many FYVE proteins have functions in vesicular trafficking but may also be involved in other PtdIns3P-mediated events such as signal transduction and phagocytosis. PKC-type ZnFs also bind small ligands such as diacylglycerol, which is important in signal transduction. (*See* G-protein-mediated Signal Transduction and Human Disorders.)

The PHD domain (**Table 1**) is thought to mediate protein–protein interactions in multiprotein complexes, specifically chromatin remodeling complexes (Asland *et al.*, 1995). Defects in PHD-containing proteins have been associated with human disorders including a form of α-thalassemia associated with mental retardation (caused by mutations in ATRX), autoimmune polyglandular disease (APECED; caused by mutations in AIRE) and several types of myeloid leukemia. (*See* Chromatin in the Cell Nucleus: Higher-order Organization; Chromatin Structure and Domains; Thalassemias.)

Other treble-clef zinc-finger genes

LIM domains bind two atoms of zinc and resemble two contiguous GATA-type fingers with the α helix of the first module absent (Perez-Alvarado *et al.*, 1994). These domains are found in several nuclear and cytoplasmic proteins and are thought to mediate protein–protein interactions. A subclass of LIM-containing proteins (so-called LIM-only (LMO) proteins) consist of little more than two to four LIM domains and function as molecular bridges. Abnormal expression of the LIM-only protein LMO2 causes acute myeloid leukemia in children. (*See* Leukemias and Lymphomas: Genetics.)

Two other treble-clef fingers have been defined. The ARF-GAP family of GTPase proteins contain a ZnF domain that may contribute to their phospholipid-binding function. The XPA ZnF domain is present in a single protein – the DNA repair protein XPA – in which it acts as the binding interface for replication protein A. Mutations in this protein cause the auto-somal recessive disorder xeroderma pigmentosum. Affected individuals are susceptible to skin cancers and their skin is hypersensitive to ultraviolet radiation.

Zinc-ribbons

Zinc-ribbons are all-β structures in which a three-stranded β sheet incorporating two zinc-knuckles binds a zinc ion (**Figure 1c**). Several proteins associated with RNA polymerase (RNA pol) from both eukryotic and archaeal organisms contain this motif, making it one of the most highly conserved ZnF motifs. These proteins include the RNA pol-II-associated transcription factors TFIIS and TFIIB, as well as the RNA pol III homolog of TFIIB, BRF. (*See* RNA Polymerases and the Eukaryotic Transcription Machinery.)

Zinc-ribbons are known to function as protein-binding domains (e.g. in interactions between TFIIB and RNA pol II) but may also recognize nucleic acids (TFIIS is required for pol II to recognize premature termination sites during transcription). Many ribosomal proteins also contain zinc-ribbon motifs.

Other Zinc-finger-encoding Genes

Other types of ZnF domain have been characterized. The transcriptional regulator CBP/p300 contains two such domains, transcriptional adaptor TAZ1 and TAZ2. Each of these domains binds three zinc atoms and forms a triangular domain, in which each zinc atom is found in a loop region between two antiparallel α helices (De Guzman *et al.*, 2000) (**Figure 1d**). This domain seems to be found only in CBP/p300, where it functions as a protein recognition motif that is able to contact a range of partner proteins. The chaparone DnaJ also contains a novel zinc-binding motif that is involved in recognizing unfolded proteins. The domain uses two zinc ions to stabilize its structure and form an extended V-shaped structure. (*See* Protein Folding and Chaperones.)

In summary, nature has produced many different protein folds that use one or more zinc atoms to help maintain their structures. The zinc ion is usually ligated by cysteine or histidine side chains, but aspartate side chains are also found in LIM domains. The structure is stabilized by a small fraction of the residues in the finger and the remaining residues can vary considerably. Thus, different fingers with different binding surfaces can be generated and, by linking distinct fingers, proteins with highly specialized binding properties can be potentially generated.

See also
Transcription Factors
Transcription Factors and Human Disorders

References

Asland R, Gibson T and Stewart F (1995) The PHD finger: implications for chromatin mediated transcriptional regulation. *Trends in Biochemical Sciences* **20**: 56–59.

Barlow PN, Luisi B, Milner A, Elliott M and Everett R (1994) Structure of the C3HC4 domain by [1]H-nuclear magnetic resonance spectroscopy: a new structural class of zinc-finger. *Journal of Molecular Biology* **237**: 201–211.

Cortes M, Wong E, Koipally J and Georgopoulos K (1999) Control of lymphocyte development by the Ikaros gene family. *Current Opinion in Immunology* **11**: 167–171.

De Guzman RN, Liu HY, Martinez-Yamout M, Dyson HJ and Wright PE (2000) Solution structure of the TAZ2 (CH3) domain of the transcriptional adaptor protein CBP. *Journal of Molecular Biology* **303**: 243–253.

Fox AH, Liew CK, Holmes M, *et al.* (1999) Transcriptional cofactors of the FOG family interact with GATA proteins by means of multiple zinc fingers. *EMBO Journal* **18**: 2812–2822.

Liew CK, Kowalski K, Fox AH, *et al.* (2000) Solution structures of two CCHC zinc fingers from the FOG family protein U-shaped that mediate protein–protein interactions. *Structure with Folding and Design* **8**: 1157–1166.

Merika M and Orkin SH (1995) Functional synergy and physical interactions of the erythroid transcription factor GATA-1 with the Kruppel family proteins Sp1 and EKLF. *Molecular and Cellular Biology* **15**: 2437–2447.

Miller J, McLachlan AD and Klug A (1985) Repetitive zinc-binding domains in the protein transcription factor IIIA from *Xenopus* oocytes. *EMBO Journal* **4**: 1609–1614.

Perez-Alvarado GC, Miles C, Michelsen JW, *et al.* (1994) Structure of the carboxy-terminal LIM domain from the cysteine-rich protein CRP. *Nature Structural Biology* **1**: 388–398.

Silke J and Vaux DL (2001) Two kinds of BIR-containing protein: inhibitors of apoptosis, or required for mitosis. *Journal of Cell Science* **114**: 1821–1827.

Further Reading

Chen H, Legault P, Glushka J, Omichinski JG and Scott RA (2000) Structure of a (Cys3His) zinc ribbon, a ubiquitous motif in archael and eukaryal transcription. *Protein Science* **9**: 1743–1752.

Choo Y and Klug A (1997) Physical basis of a protein–DNA recognition code. *Current Opinion in Structural Biology* **7**: 117–125.

Dawid I, Breen JJ and Toyama R (1998) LIM domains: multiple roles as adapters and functional modifiers in protein interactions. *Trends in Genetics* **14**: 156–162.

Freemont PS (2000) RING for destruction? *Current Biology* **10**: R84–87.

Grishin NV (2001) Treble clef finger: a functionally diverse zinc-binding structural motif. *Nucleic Acids Research* **29**: 1703–1714.

Hahn S and Roberts S (2000) The zinc ribbon domains of the general transcription factors TFIIB and Brf: conserved functional surfaces but different roles in transcription initiation. *Genes and Development* **14**: 719–730.

Joazeiro CAP and Weissman AM (2000) RING finger proteins: mediators of ubiquitin ligase activity. *Cell* **102**: 549–552.

Laity JH, Lee BM and Wright PE (2001) Zinc finger proteins: new insights into structural and functional diversity. *Current Opinion in Structural Biology* **11**: 39–46.

Lee MS, Gippert GP, Soman KV, Case DA and Wright PE (1989) Three-dimensional solution structure of a single zinc finger DNA-binding domain. *Science* **245**: 635–637.

Rastinejad F (2001) Retinoid X receptor and its partners in the nuclear receptor family. *Current Opinion in Structural Biology* **11**: 33–38.

Web Links

Ensembl Genome Browser
    http://www.ensembl.org
Structural Classification of Proteins (SCOP)
    http://scop.mrc-lmb.cam.ac.uk/scop/

Zuckerkandl, Emile

Introductory article

Gregory J Morgan, *Johns Hopkins University, Baltimore, Maryland, USA*

Emile Zuckerkandl (1922–) discovered the molecular evolutionary clock and founded the *Journal of Molecular Evolution*.

Emile Zuckerkandl was born into a prominent Viennese family on 4 July 1922. He grew up in Vienna until the threat of Nazi persecution prompted him and his family to flee Austria for Paris and later Algiers. After a delay due to the Second World War, Zuckerkandl gained a master's degree in physiology at the University of Illinois under C. Ladd Prosser in 1947 and then a doctorate from the Sorbonne. Following graduation, he took a position with the Centre National de la Recherche Scientifique (CNRS) and worked at the marine biology laboratory in Roscoff, Brittany, where he studied hemocyanin and the molting cycle of crabs.

Although the position with CNRS was secure, Zuckerkandl and his wife, Jane, wanted to return to the United States. While Linus Pauling was visiting Europe in 1957, Zuckerkandl arranged to meet him. Zuckerkandl made such an impression that he secured a postdoctoral position under Pauling at the California Institute of Technology. When Zuckerkandl arrived at Cal Tech in 1959, Pauling, with some foresight, shepherded him into beginning a project on the evolution of hemoglobin. Zuckerkandl and Richard T. Jones used the electrophoretic/chromatographic technique of 'fingerprinting' to analyze hemoglobin from a variety of species. They discovered that the primate hemoglobins were almost identical. Later, Zuckerkandl determined that gorilla and human hemoglobin chains differ by only one or two amino acid residues. (*See* Great Apes and Humans: Genetic Differences; Pauling, Linus Carl; Primates: Phylogenetics.)

In 1961, using the first data from the amino acid sequence of hemoglobin, Zuckerkandl proposed the influential notion of a molecular evolutionary clock. The hypothesis was that molecules evolve at a roughly constant rate. Using the clock, biologists could date the branching points in evolutionary trees. Zuckerkandl and Pauling published the clock hypothesis in a *Festschrift* volume for Albert Szent-Györgyi in 1962 and later in a widely cited article of 1965. The molecular evolutionary clock later became intertwined with the neutralism/selectionism debate of the 1970s, but as originally conceived, it was intended to be consistent with selection at the molecular level. Ernst Mayr, George Gaylord Simpson and other traditional biologists objected to the idea that molecular evolution proceeded at a constant rate. (*See* Evolution: Neutralist View; Evolution: Selectionist View; Kimura, Motoo; Molecular Clocks.)

In the 1960s, Zuckerkandl made a number of predictions that have largely been borne out: He predicted that pseudogenes would be found, that molecular biology would revolutionize anthropology and that morphological evolution is largely due to changes in gene regulation rather than changes in structural genes. (*See* Developmental Evolution; Pseudoexons.)

In 1965, Zuckerkandl returned to France and founded the innovative CNRS laboratory Centre de Recherches en Biochimie Macromoléculaire in Montpellier.

Zuckerkandl founded the *Journal of Molecular Evolution* in 1971 and served as its editor-in-chief. During the next three decades, he published numerous articles on molecular evolution. For example, in 1975 he argued that new protein classes could evolve only from old proteins and that most protein classes evolved before the major elements of the translation apparatus of modern cells were fully evolved. In 1981, he proposed a novel function for introns: that they help stabilize high-order chromatin structures. Since 1997, he has argued that random drift is needed to explain the evolution of complex gene regulation in higher organisms. In the late 1970s he returned to the United States to direct the Linus Pauling Institute, and in 1992 he became President of the Institute of Molecular Medical Sciences in Palo Alto, California.

See also

Pauling, Linus Carl

Further Reading

Morgan GJ (1998) Emile Zuckerkandl, Linus Pauling, and the molecular evolutionary clock, 1959–1965. *Journal of the History of Biology* **31**: 155–178.

Zuckerkandl E (1975) The appearance of new structures and functions in proteins during evolution. *Journal of Molecular Evolution* **7**(1): 1–57.

Zuckerkandl E (1981) A general function of noncoding polynucleotide sequences: mass binding of transconformational proteins. *Molecular Biology Reports* **7**(1–3): 149–158.

Zuckerkandl E (1987) On the molecular evolutionary clock. *Journal of Molecular Evolution* **26**: 34–46.

Zuckerkandl E (1997) Neutral and nonneutral mutations: the creative mix – evolution of complexity in gene interaction systems. *Journal of Molecular Evolution* **44**: S2–S8.

Zuckerkandl E (2001) Intrinsically driven changes in gene interaction complexity. I.C3 Growth of regulatory complexes and increase in number of genes. *Journal of Molecular Evolution* **53**: 539–554.

Zuckerkandl E and Pauling L (1962) Molecular disease, evolution and genic heterogeneity. In: Kasha M and Pullman B (eds.) *Horizons in Biochemistry: Albert Szent-Györgyi Dedicatory Volume*, pp. 189–225. New York: Academic Press.

Zuckerkandl E and Pauling L (1965) Evolutionary divergence and convergence in proteins. In: Bryson V and Vogel H (eds.) *Evolving Genes and Proteins*, pp. 97–166. New York: Academic Press.

Zuckerkandl E and Schroeder W (1960) Amino acid sequences of gorilla hemoglobin. *Nature* **192**: 984–985.

Contributors

Roger Abseher Boehringer Ingelheim Austria, Vienna, Austria
Microarrays in Disease Diagnosis and Prognosis

John C Achermann University College London, London, UK
j.achermann@ich.ucl.ac.uk
Gonadotropin Hormones: Disorders

Ronald M Adkins University of Tennessee, Memphis, Tennessee, USA
radkins1@utmem.edu
Coevolution: Molecular

Gwen Adshead Traumatic Stress Clinic, London, UK; Broadmoor Hospital, Crowthorne
rak@wlmht.nhs.uk
Criminal Responsibility and Genetics

Ruedi Aebersold Institute for Systems Biology, Seattle, Washington, USA
raebersold@systemsbiology.org
Quantitative Proteomics (ICAT^{TM})

Nabeel A Affara University of Cambridge, Cambridge, UK
na@mole.bio.cam.ac.uk
Chromosome Y

Adriano Aguzzi University Hospital Zurich, Zurich, Switzerland
adriano@pathol.unizh.ch
Prion Disorders

Slimane Ait-Si-Ali Institut Andre Lwoff, Villejuif, France
aitsiali@vjf.cnrs.fr
Histone Acetylation and Disease

H Akashi Pennsylvania State University, University Park, Pennsylvania, USA
akashi@psu.edu
Purifying Selection: Action on Silent Sites

Mirit I Aladjem National Institutes of Health, Bethesda, Maryland, USA
Cell Cycle Control: Molecular Interaction Map

Donna Albertson University of California San Francisco, San Francisco, California, USA
albertson@cc.ucsf.edu
Comparative Genomic Hybridization in the Study of Human Disease

Werner Albig Georg-August-Universität Göttingen, Göttingen, Germany
Histones; Intronless Genes

Maher Albitar The University of Texas MD Anderson Cancer Center, Houston, Texas, USA
malbitar@mdanderson.org
Myeloproliferative Diseases: Molecular Genetics

Gary L Albrecht University of Illinois, Chicago, Illinois, USA
garya@uic.edu
Quality of Life: Human Worth Reduced to Measures of Ability

Priscilla Alderson University of London, London, UK
Children in Genetic Research

Micheala A Aldred University of Leicester, Leicester, UK
maldred@hgmp.mrc.ac.uk
Activating and Inactivating Mutations in the GNAS1 Gene

Layla Al-Jader University of Wales College of Medicine, Cardiff, UK
al-jaderla@cf.ac.uk
Reproductive Genetic Screening: A Public Health Perspective from the United Kingdom

Wynand Alkema Karoliniska Institutet, Stockholm, Sweden
Phylogenetic Footprinting

Garland E Allen Washington University, St Louis, Missouri, USA
allen@biology.wustl.edu
Intelligence Tests and Immigration to the United States, 1900–1940

David C Allison Medical College of Ohio, Toledo, Ohio, USA
dallison@mco.edu
Chromosomes during Cell Division

Joseph S Alper University of Massachusetts, Boston, Massachusetts, USA
j.alper@umb.edu
'Race', IQ, and Genes

Michael R Altherr Los Alamos National Laboratory, Los Alamos, New Mexico, USA
Chromosome 4

Stephen F Altschul National Center for Biotechnology Information, Bethesda, Maryland, USA
altschul@ncbi.nlm.nih.gov
BLAST Algorithm; Substitution Matrices

Jan O Andersson Dalhousie University, Halifax, Canada
jan.andersson@icm.uu.se
Bacterial DNA in the Human Genome

Siv Andersson Uppsala University, Sweden
siv.andersson@ebc.uu.se
Mitochondrial Proteome: Origin

Brage S Andresen Aarhus University Hospital, Aarhus, Denmark
Protein Misfolding and Degradation in Genetic Disease

Lucy G Andrews University of Alabama at Birmingham, Birmingham, Alabama, USA
Telomerase: Structure and Function

Nancy C Andrews Harvard Medical School, Boston, Massachusetts, USA
nandrews@enders.tch.harvard.edu
Iron Metabolism: Disorders

Ignacio Anegon INSERM U437, Nantes, France
ianegon@nantes.inserm.fr
Gene Transfer in Transplantation

Francisco Antequera CSIC/Universidad de Salamanca, Salamanca, Spain
cpg@gugu.usal.es
CpG Islands and Methylation

Stylianos E Antonarakis University of Geneva Medical School, Geneva, Switzerland
stylianos.antonarakis@medecine.unige.ch
CpG Dinucleotides and Human Disorders; Mutations in Human Genetic Disease

Ron D Appel Swiss Institute of Bioinformatics, Geneva, Switzerland
sonja.voordijk@isb-sib.ch
Image Analysis Tools in Proteomics

M Amin Arnaout Massachusetts General Hospital, Charlestown, Massachusetts, USA
arnaout@receptor.mgh.harvard.edu
Autosomal Dominant Polycystic Kidney Disease

Einar Árnason University of Iceland, Reykjavik, Iceland
einar@lif.hi.is
deCODE and Iceland: A Critique

Ulfur Arnason University of Lund, Lund, Sweden
ulfur.arnason@cob.lu.se
Hominids: Molecular Phylogenetics

Bruce J Aronow Cincinnati Children's Hospital Medical Center, Cincinnati, Ohio, USA
bruce.aronow@cchmc.org
Evolutionarily Conserved Noncoding DNA

Adrienne Asch Wellesley College, Wellesley, Massachusetts, USA
aasch@wellesley.edu
Disability and Genetics: A Disability Rights Perspective

David J Askew Children's Hospital, Harvard Medical School, Boston, Massachusetts, USA
Serpins: Evolution

Paul Atkinson Cardiff University, Cardiff, UK
atkinsonpa@cardiff.ac.uk
Inheritance and Society

Henri Atlan Hadassah University Hospital, Jerusalem, Israel and Ecole des Hautes Etudes en Sciences Sociales, Paris, France
atlan@ehess.fr
Genetics as Explanation: Limits to the Human Genome Project

Teresa K Attwood University of Manchester, Manchester, UK
attwood@bioinf.man.ac.uk
Genetic Databases

Denise Avard   University of Montreal, Montreal, Canada
denise.avard@umontreal.ca
Informed Consent; Informed Consent and Multiplex Screening

Neil D Avent   University of the West of England, Bristol, UK
neil.avent@uwe.ac.uk
Blood Groups: Molecular Genetic Basis

Torik Ayoubi   Flanders Interuniversity Institute for Biotechnology,
Zwijnaarde, Belgium
torik.ayoubi@vib.be
Alternative Promoter Usage

Tatyana Azhikina   Shemyakin-Ovchinnikov Institute of Bioorganic
Chemistry, Russian Academy of Sciences, Russia
tanya@humgen.siobc.ras.ru
Primer Walking

Paul Badham   University of Wales Lampeter, Ceredigion, UK
p.badham@lamp.ac.uk
Christianity and Genetics

J Michael Bailey   Northwestern University, Evanston, Illinois, USA
jm-bailey@northwestern.edu
Sexual Orientation: Genetics

Joan E Bailey-Wilson   National Institutes of Health, Baltimore,
Maryland, USA
jebw@nhgri.nih.gov
Parametric and Nonparametric Linkage Analysis

Vivienne Baillie Gerritsen   Swiss Institute of Bioinformatics, Geneva,
Switzerland
Protein Databases

Amos Bairoch   Swiss Institute of Bioinformatics, Geneva, Switzerland
amos.bairoch@isb-sib.ch
Protein Databases

Kate D Baker   Institute of Child Health, University College London, UK
d.skuse@ich.ucl.ac.uk
Brain: Neurodevelopmental Genetics

Richard Baldock   MRC Human Genetics Unit, Edinburgh, UK
richard.baldock@hgu.mrc.ac.uk
Mapping Gene Function in the Embryo

Andrea Ballabio   Telethon Institute of Genetics and Medicine (TIGEM),
Naples, Italy
ballabio@tigem.it
*Disease-related Genes: Functional Analysis; Disease-related Genes:
Identification*

Esteban Ballestar   Centro Nacional de Investigaciones Oncológicas,
Madrid, Spain
eballestar@cnio.es
DNA Methylation and Histone Acetylation; Methylated DNA-binding Proteins

Rudi Balling   Gesellschaft für Biotechnologische Forschung (GBF),
Braunschweig, Germany
balling@gbf.de
Expression Analysis In Vivo

Hans-Jürgen Bandelt   University of Hamburg, Hamburg, Germany
bandelt@math.uni-hamburg.de
Evolutionary Networks

Sandro Banfi   Telethon Institute of Genetics and Medicine (TIGEM),
Naples, Italy
Disease-related Genes: Identification

Rosamonde E Banks   Cancer Research UK, Leeds, UK
r.banks@leeds.ac.uk
Laser Capture Microdissection in Proteomics

Aruna Bansal   University of Utah, Utah, USA
Complex Multifactorial Genetic Diseases

Guido Barbujani   University of Ferrara, Ferrara, Italy
g.barbujani@unife.it
Migration; Population History of Europe: Genetics

David L Barker   Illumina Inc., San Diego, California, USA
Microarrays: Use in Gene Identification

Winona C Barker   National Biomedical Research Foundation,
Washington, DC, USA
Protein Sequence Databases

Colin Barnes   University of Leeds, Leeds, UK
Disability: Western Theories

John A Barranger   University of Pittsburgh, Pittsburgh, Pennsylvania, USA
john.barranger@mail.hgen.pitt.edu
Lysosomal Storage Disorders: Gene Therapy

Ara Barsam   Oxford University, Oxford, UK
ara_barsam@yahoo.com
Cloning of Animals in Genetic Research

Adrian H Batchelor   Royal Melbourne Hospital, Melbourne, Australia
abatchel@rx.umaryland.edu
Crystallization of Protein–DNA Complexes

Alex Bateman   The Wellcome Trust Sanger Institute, Cambridge, UK
agb@sanger.ac.uk
Gene Families

Patrick Bateson   University of Cambridge, Cambridge, UK
ppgb@cam.ac.uk
Sociobiology, Evolutionary Psychology, and Genetics

Mark A Batzer   Louisiana State University, Baton Rouge, Louisiana, USA
Retrosequences and Evolution of Alu Elements

Matthias F Bauer   Academic Hospital, Munich-Schwabing, Germany
bauer@bio.med.uni-muenchen.de
Mitochondrial Disorders

Marc Baumann   University of Helsinki, Helsinki, Finland
marc.baumann@helsinki.fi
Protein Characterization in Proteomics

Alessandra Baumer   University of Zurich, Zurich, Switzerland
baumer@medgen.unizh.ch
Microdeletion Syndromes; Uniparental Disomy

Robert C Baumiller   Xavier University, Cincinnati, Ohio, USA
baumillr@xavier.edu
Code of Ethical Principles for Genetics Professionals

Kurt Bayertz   Westfälische Wilhelms-Universität, Muenster, Germany
bayertz@uni-muenster.de
Genetic Enhancement

Viviana Bazan   University of Palermo, Palermo, Italy
lab-oncobiologia@usa.net
Caretakers and Gatekeepers

Christine Beahler   University of Washington, Seattle, Washington, USA
Databases in Genetics Clinics

Stephan Beck   The Sanger Centre, Cambridge, UK
beck@sanger.ac.uk
Sequence Accuracy and Verification

Karl-Friedrich Becker   Technische Universität München, München
Germany
kf.becker@lrz.tum.de
Gastric Cancer

Jon Beckwith   Harvard Medical School, Boston, Massachusetts, USA
jbeckwith@hms.harvard.edu
'Race', IQ, and Genes

Henri Begleiter   SUNY HSC, Brooklyn, New York, USA
hb@cns.hscbklyn.edu
Alcoholism: Collaborative Study on the Genetics of Alcoholism (COGA)

Soshana Behrstock   University of Wisconsin Madison, Madison,
Wisconsin, USA
Neural Stem Cells

David R Beier   Brigham and Women's Hospital, Harvard Medical School,
Boston, Massachusetts, USA
beier@rascal.med.harvard.edu
Animal Models

Hilary L Bekker   University of Leeds, Leeds, UK
h.l.bekker@leeds.ac.uk
Genetic Screening: Facilitating Informed Choices

Ursula Bellugi   Salk Institute for Biological Studies, La Jolla, California,
USA
bellugi@salk.edu
Williams Syndrome: A Neurogenetic Model of Human Behavior

Andrew S Belmont University of Illinois, Urbana, Illinois, Illinois, USA
asbel@uiuc.edu
Chromosomes and Chromatin

Juan Carlos Izpisúa Belmonte Salk Institute for Biological Studies, La Jolla, California, USA
belmonte@salk.edu
Left–Right Asymmetry in Humans

Robert Benezra Memorial Sloan-Kettering Cancer Center, New York, New York, USA
r-benezra@ski.mskcc.org
Mitosis: Chromosome Segregation and Stability

Jonathan Benjamin Ben Gurion University of the Negev, Beersheba, Israel
Anxiety Disorders

Jacqueline M Benson Centocor, Inc., Malvern, Pennsylvania, USA
jbenson4@cntus.jnj.com
Autoimmune Diseases: Gene Therapy

Lionel Bently King's College, London, UK
lionel.bently@kcl.ac.uk
Patent Issues in Biotechnology

Philip L Bereano University of Washington, Seattle, Washington, USA
pbereano@u.washington.edu
Human Genome Project as a Social Enterprise

Lutz-Peter Berg Science and Technology Office, Embassy of Switzerland, London, UK
lutz-peter.berg@lon.rep.admin.ch
Ectopic Transcription; Genes: Types

Helen M Berman Rutgers, The State University of New Jersey, New Brunswick, New Jersey, USA
berman@rcsb.rutgers.edu
Structural Databases of Biological Macromolecules

Giorgio Bernardi Stazione Zoologica 'A. Dohrn', Napoli, Italy
bernardi@szn.it
Chromosomes 21 and 22: Gene Density; Evolutionary History of the Human Genome; GC-rich Isochores: Origin; Gene Distribution on Human Chromosomes; Genome Organization of Vertebrates; Isochores

Giorgio Bertorelle University of Ferrara, Ferrara, Italy
ggb@unife.it
Population Differentiation: Measures

Daniel Bertrand University of Geneva, Medical School, Geneva, Switzerland
daniel.bertrand@medecine.unige.ch
Nicotinic Receptors; Receptors and Human Disorders

Jaume Bertranpetit Universitat Pompeu Fabra, Barcelona, Spain
Pseudogenes: Patterns of Mutation

Esther Betrán University of Chicago, Chicago, Illinois, USA
ebetran@uchicago.edu
Gene Fusion

Ernest Beutler The Scripps Research Institute, La Jolla, California, USA
beutler@scripps.edu
Hemochromatosis

Aditya Bharadwaj Cardiff University, Cardiff, UK
bharadwaja@cardiff.ac.uk
Inheritance and Society

Wendy A Bickmore MRC Human Genetics Unit, Edinburgh, UK
w.bickmore@hgu.mrc.ac.uk
Chromatin Structure and Domains; Chromosome Structures: Visualization

Barbara B Biesecker National Human Genome Research Institute, Bethesda, Maryland, USA
barbarab@nhgri.nih.gov
Genetic Counseling: Psychological Issues

Darell D Bigner Duke University Medical Center, Durham, North Carolina, USA
Brain Tumors

Phillip I Bird Monash University, Victoria, Australia
phil.bird@med.monash.edu.au
Ovalbumin Serpins; Serpins: Evolution

C William Birky Jr University of Arizona, Tucson, Arizona, USA
birky@u.arizona.edu
Mitochondrial Non-Mendelian Inheritance: Evolutionary Origin and Consequences

Mercedes Blázquez Instituto de Ciencias del Mar (CSIC), Barcelona Spain
blazquez@icm.csic.es
Protein Secretory Pathways

Elisabeth Blennow Karolinska Institutet, Stockholm, Sweden
elisabeth.blennow@cmm.ki.se
Banding Techniques; Chromosome Preparation

Konstantin Bloch Tel Aviv University, Petah Tikva, Israel
kbloch@post.tau.ac.il
Thrifty Genotype Hypothesis and Complex Genetic Disease

Daniel F Bogenhagen State University of New York at Stony Brook, Stony Brook, New York, USA
dan@pharm.sunysb.edu
Mitochondrial DNA Repair in Mammals

Andrew J Bohonak San Diego State University, San Diego, California, USA
bohonak@sciences.sdsu.edu
Genetic Drift

Stefan Böhringer Ruhr-University, Bochum, Germany
Microsatellites

Stéphane Boissinot National Institutes of Health, Bethesda, Maryland, USA
Long Interspersed Nuclear Elements (LINEs): Evolution

Jacky Boivin Cardiff University, Cardiff, UK
boivin@cardiff.ac.uk
In Vitro *Fertilization*

C Richard Boland University of California, San Diego, La Jolla, California, USA
rickbo@baylorhealth.edu
Tumor Formation: Number of Mutations Required

Edoardo Boncinelli Università Vita-Salute San Raffaele, Milan, Italy
Development: Disorders

Constanze Bonifer University of Leeds, Leeds, UK
medcb@leeds.ac.uk
Tissue-specific Locus Control: Structure and Function

Ingrid B Borecki Washington University School of Medicine, St Louis, Missouri, USA
ingrid@wubios.wustl.edu
Linkage and Association Studies

Jennifer Bostock Kings College London, London, UK
jl.bostock@virgin.net
Criminal Responsibility and Genetics

Maxime Bouchard Research Institute of Molecular Pathology, Vienna, Austria
Pax *Genes: Evolution and Function*

Thomas J Bouchard Jr University of Minnesota, Minneapolis, Minnesota, USA
bouch001@umn.edu
Identical Twins Reared Apart

Gérard Bouchet Swiss Institute of Bioinformatics, Geneva, Switzerland
Image Analysis Tools in Proteomics

Déborah Bourc'his Institut Jacques Monod, Paris, France
Epigenetic Factors and Chromosome Organization

Judith VMG Bovée Leiden University Medical Centre, Leiden, The Netherlands
Skeletogenesis: Genetics

Richard P Bowater University of East Anglia, Norwich, UK
r.bowater@uea.ac.uk
DNA Structure

Derrick John Bowen University of Wales College of Medicine, Cardiff, UK
bowendj1@cf.ac.uk
In Vitro *Mutagenesis*

Peter J Bowler   Queen's University, Belfast UK
Lamarckian Inheritance; Malthus, Darwin, and Social Darwinism

Dana Boyd   Harvard Medical School, Boston, Massachusetts, USA
Art and Genetics

Joy T Boyer   National Human Genome Research Institute, Bethesda,
Maryland, USA
boyerj@exchange.nih.gov
*ELSI Research Program of the National Human Genome Research Institute
(NHGRI)*

Hannah Bradby   University of Warwick, Coventry, UK
h.bradby@warwick.ac.uk
Racism, Ethnicity, Biology, and Society

Don M Bradley   University Hospital of Wales, Cardiff, UK
Genetic Risk: Social Construction; Newborn Screening Programs

Alvis Brazma   European Molecular Biology Laboratory, Cambridge, UK
brazma@ebi.ac.uk
Gene Expression Databases

Matthew Breen   The Animal Health Trust, Newmarket, UK
Comparative Cytogenetics

Charles H Brenner   Oakland, California, USA
cbrenner@uclink.berkeley.edu
Forensic Genetics: Mathematics

Emery H Bresnick   University of Wisconsin Medical School, Madison,
Wisconsin, USA
ehbresni@facstaff.wisc.edu
Histone Acetylation: Long-range Patterns in the Genome

Albert O Brinkmann   University Medical Center Rotterdam, Rotterdam,
The Netherlands
a.brinkmann@erasmusmc.nl
Androgen Insensitivity

Stephen C Brock   Medical Education Consultant, Bexley, Ohio, USA
stevebrock@columbus.rr.com
Narrative Ethics

Anthony J Brookes   Karolinska Institute, Stockholm, Sweden
anthony.brookes@cgb.ki.se
Single Nucleotide Polymorphism (SNP)

Doug A Brooks   Women's and Children's Hospital, North Adelaide,
Australia
douglas.brooks@adelaide.edu.au
Protein: Cotranslational and Posttranslational Modification in Organelles

Peter Bross   Aarhus University Hospital, Aarhus, Denmark
peter.bross@mmf.au.dk
Protein Misfolding and Degradation in Genetic Disease

Amy M Brower   Third Wave Technologies, Madison, Wisconsin, USA
Reading and Dyslexias

Dennis Brown   Massachusetts General Hospital, Boston, Massachusetts,
USA
brown@receptor.mgh.harvard.edu
Protein Targeting

Steve David Macleod Brown   MRC Mammalian Genetics Unit and
Mouse Genome Centre, Harwell, UK
s.brown@har.mrc.ac.uk
Mouse as a Model for Human Diseases

Janet Browne   Wellcome Trust Centre for the History of Medicine at UCL,
London, UK
Darwin, Charles Robert

James P Bruzik   Case Western Reserve University, Cleveland, Ohio, USA
jxb83@po.cwru.edu
Trans Splicing

Graham Budd   Uppsala University, Uppsala, Sweden
graham.budd@pal.uu.se
Morphological Diversity: Evolution

Michael Bulger   Fred Hutchinson Cancer Research Center, Seattle,
Washington, USA
mbulger@fred.fhcrc.org
Enhancers; Locus Control Regions (LCRs)

Christopher B Burge   Massachusetts Institute of Technology, Cambridge,
Massachusetts, USA
cburge@mit.edu
Splice Sites

David Burgner   University of Western Australia, Perth, Australia
dburgner@paed.uwa.edu.au
Infectious Diseases: Predisposition

Kate Bushby   International Centre for Life, Newcastle upon Tyne, UK
kate.bushby@newcastle.ac.uk
Muscular Dystrophies

Meinrad Busslinger   Research Institute of Molecular Pathology, Vienna,
Austria
Pax Genes: Evolution and Function

Konrad Büssow   Max Planck Institute of Molecular Genetics, Berlin,
Germany
Array-based Proteomics

Terry D Butters   University of Oxford, Oxford, UK
terry@glycob.ox.ac.uk
Glycoproteins

Helen J Bynum   Wellcome Trust Centre for the History of Medicine at
UCL, London, UK
almo43@uk.uumail.com
Garrod, Archibald Edward; Hardy, Godfrey Harold

WF Bynum Wellcome Trust Center for the History of Medicine at UCL,
London, UK
w.bynum@ucl.ac.uk
Wilson, Edmund Beecher

Mario A Cabrera-Salazar   University of Pittsburgh, Pittsburgh,
Pennsylvania, USA
Lysosomal Storage Disorders: Gene Therapy

Francesc Calafell   Universitat Pompeu Fabra, Barcelona, Spain
francesc.calafell@cexs.upf.es
Pseudogenes: Patterns of Mutation

Aleth Callé   INSERM, Faculté de Médecine Lyon-R.T.H. Laennec, Lyon,
France
Nucleolar Proteomics

David F Callen   Bionomics Ltd, Thebarton, South Australia
david.callen@imvs.sa.gov.au
Chromosome 16

Nicola J Camp   University of Utah, Utah, USA
nicki@genepi.med.utah.edu
Complex Multifactorial Genetic Diseases

Rebecca L Cann   University of Hawaii-Manoa, Honolulu, Hawaii, USA
rcann@hawaii.edu
DNA Markers and Human Evolution; Human Populations: Evolution

Chris Cannings   University of Sheffield, Sheffield, UK
c.cannings@shef.ac.uk
Evolutionary Conflicts

Pierre Capy   Centre National de la Recherche Scientifique, Gif-sur-Yvette,
France
pierre.capy@pge.cnrs-gif.fr
Transposons

Rebecca Caraway   Southern Illinois University School of Medicine,
Carbondale, Illinois, USA
ldilalla@siu.edu
Developmental Psychopathology

Romeo Carrozzo   IRCCS Ospedale San Raffaele, Milan, Italy
Development: Disorders

Nigel P Carter   The Sanger Centre, Cambridge, UK
npc@sanger.ac.uk
Flow-sorted Chromosomes

Richard J Carter   Lawerence Berkeley National Laboratory, Berkeley,
California, USA
RNA Gene Prediction

Ronald Frederick Carter   McMaster University, Hamilton, Canada
rcarter@mcmaster.ca
Human Genetics: Ethical Issues and Social Impact

MG Castro Gene Therapeutics Research Institute, Cedars-Sinai Medical Center and Department of Medicine, David Geffer School of Medicine, University of California, Los Angeles, California USA
castromg@cshs.org
Neurological Diseases: Gene Therapy

Timothy Caulfield University of Alberta, Edmonton, Canada
tcaulfld@law.ualberta.ca
Commercialization of Human Genetic Research

Thomas Centner Universitätsklinikum, Mannheim, Germany
Titin (TTN) Gene

Rüdiger Cerff Technische Universität Braunschweig, Braunschweig, Germany
r.cerff@tu-bs.de
Introns: Movements

Ruth Chadwick Lancaster University, Lancaster, UK
r.chadwick@lancaster.ac.uk
Personal Identity: Genetics and Determinism

R S K Chaganti Memorial Sloan Kettering Cancer Center, New York, New York, USA
chagantr@mskcc.org
Testicular Cancer

Ranajit Chakraborty University of Cincinnati, Cincinnati, Ohio, USA
ranajit.chakraborty@uc.edu
Population Genetics: Historical Aspects

Jeffrey S Chamberlain University of Washington, Seattle, Washington, USA
jsc5@u.washington.edu
Minigenes

Lawrence Chan Baylor College of Medicine, Houston, Texas, USA
lchan@bcm.tmc.edu
Apolipoprotein Gene Structure and Function

Sarah WL Chan Murdoch Childrens Research Institute, Parkville, Australia
Nucleic Acid Hybridization

Srinivasan Chandrasegaran Johns Hopkins University, Baltimore, Maryland, USA
schandra@jhsph.edu
Restriction Enzymes

Brian Charlesworth University of Edinburgh, Edinburgh, UK
brian.charlesworth@ed.ac.uk
Fitness and Selection

Deborah Charlesworth University of Edinburgh, Edinburgh UK
deborah.charlesworth@ed.ac.uk
Inbreeding

Inês Chaves Erasmus Medical Center, Rotterdam, The Netherlands
i.machadoferreirachaves@erasmusmc.nl
Circadian Rhythm Genetics

Jeremy P Cheadle University of Wales College of Medicine, Cardiff UK
cheadlejp@cardiff.ac.uk
Tuberous Sclerosis: Genetics

Jian-Min Chen Etablissement Français du Sang, Brest, France
jian-min.chen@univ-brest.fr
Cystic Fibrosis Transmembrane Conductance Regulator Sequences: Comparative Analysis; Trypsinogen Genes: Evolution

Qiuyun Chen Cleveland Clinic Foundation, Cleveland, Ohio, USA
chenq2@ccf.org
Cardiovascular Disease and Congenital Heart Defects; Retinitis Pigmentosa

Ting Chen University of Southern California, Los Angeles, California, USA
tingchen@usc.edu
Dynamic Programming

Wensheng Chen Aeomica Inc., Sunnyvale, California, USA
Microarrays: Use in Gene Identification

Jan-Fang Cheng Lawrence Berkeley National Laboratory, Berkeley, California, USA
jfcheng@lbl.gov
Chromosome 5

Roberta Chiaraluce University of Rome, Rome, Italy
Chaperones, Chaperonins, and Heat Shock Proteins (HSPs)

Lounès Chikhi Université Paul Sabatier, Toulouse, France
chikhi@cict.fr
Genetics of Large Populations and Association Studies

Andrew Chisholm University of California, Santa Cruz, California, USA
chisholm@biology.ucsc.edu
Caenorhabditis elegans as an Experimental Organism

Andy KH Choo Murdoch Childrens Research Institute, Parkville, Australia
choo@cryptic.rch.unimelb.edu.au
Nucleic Acid Hybridization

Mon-Li Chu Thomas Jefferson University, Philadelphia, Pennsylvania, USA
mon-li.chu@mail.tju.edu
Structural Proteins: Genes

Man-Wei Chung Centenary Institute, Newtown, Australia
Hypertrophic Cardiomyopathy

Erna Claes Center for Human Genetics, University of Leuven, Belgium
erna.claes@med.kuleuven.ac.be
Predictive Genetic Testing: Psychological Impact

Abbot F Clark Alcon Research Ltd, Fort Worth, Texas, USA
abe.clark@alconlabs.com
Eye: Proteomics

Anne-Marie Cleton-Jansen Leiden University Medical Centre, Leiden, The Netherlands
Skeletogenesis: Genetics

Natalie Coe The Jackson Laboratory, Bar Harbor, Maine, USA
Obesity: Genetics

William A Coetzee New York University School of Medicine, New York, New York, USA
william.coetzee@med.nyu.edu
Potassium Channels

Irun R Cohen The Weizmann Institute of Science, Rehovot, Israel
irun.cohen@weizmann.ac.il
Genetics as Explanation: Limits to the Human Genome Project

Noel G Coley The Open University, Milton Keynes, UK
n.g.coley@surrey28.freeserve.co.uk
Perutz, Max Ferdinand; Sanger, Frederick

Greg R Collier Deakin University, Geelong, Australia
Diabetes: Genetics

Andrew Collins University of Southampton, Southampton, UK
arc@soton.ac.uk
Genetic Maps: Integration

Roberto Colombo Catholic University of the Sacred Heart, Milan, Italy
Mutations: Dating

Alfredo Colosimo University of Rome 'La Sapienza', Rome, Italy
a.colosimo@caspur.it
Expression Analysis In Vitro

David E Comings City of Hope Medical Center, Duarte, California, USA
dcomings@coh.org
Polygenic Disorders

Béatrice Conne University of Geneva, Geneva, Switzerland
beatrice.conne@medcli.unige.ch
3' UTR Mutations and Human Disorders

Christian Conrad German Cancer Research Center, Heidelberg, Germany
Digital Image Analysis

Thomas P Conrads National Cancer Institute, Frederick, Maryland, USA
conrads@ncifcrf.gov
Phosphoproteomics

Valerio Consalvi University of Rome, Rome, Italy
consalvi@caspur.it
Chaperones, Chaperonins, and Heat Shock Proteins (HSPs)

Miguel Constância The Babraham Institute, Cambridge, UK
miguel.constancia@bbsrc.ac.uk
Genomic Imprinting at the Transcriptional Level

Edwin H Cook Jr, University of Chicago, Chicago, Illinois, USA
ed@yoda.bsd.uchicago.edu
http://psychiatry.uchicago.edu/ldn
Autism: Genetics

William OCM Cookson   University of Oxford, Oxford, UK
william.cookson@ndm.ox.ac.uk
Asthma: Genetics

David N Cooper   University of Wales College of Medicine, Cardiff, UK
cooperdn@cardiff.ac.uk
Exons: Insertion and Deletion during Evolution; Gene Deletions in Evolution; Genes: Types; Gross Insertions and Microinsertions in Evolution; Mutations in Human Genetic Disease; Primate Evolution: Gene Loss and Inactivation; Pseudogenes and their Evolution

Richard R Copley   Wellcome Trust Centre for Human Genetics, Oxford, UK
copley@well.ox.ac.uk
Protein Families: Evolution

Laura D Corden   University of Dundee, Dundee, UK
lcorden@hgmp.mrc.ac.uk
Skin: Hereditary Disorders

Robin P Corley   University of Colorado, Boulder, Colorado, USA
robin.corley@colorado.edu
Adoption Strategies

Amy Corrigan   Aeomica Inc., Sunnyvale, California, USA
Microarrays: Use in Gene Identification

Thomas J Corydon   Aarhus University, Aarhus, Denmark
Protein Misfolding and Degradation in Genetic Disease

Gina L Costa   SurroMed, Inc., Mountain View, California, USA
gcosta@454.com
Autoimmune Diseases: Gene Therapy

RGH Cotton   Genomic Disorders Research Centre, Melbourne, Australia
cotton@medstv.unimelb.edu.au
Human Genome Variation Society; Mutation Detection

Alan Coulson   Wellcome Trust Sanger Institute, Cambridge, UK
arc@mrc-lmb.cam.ac.uk
Caenorhabditis elegans Genome Project

Amandine Cournil   University of Montpellier, France
Longevity: Genetics

Yohann Couté   INSERM, Faculté de Médecine Lyon-R.T.H. Laennec, Lyon, France
Nucleolar Proteomics

Diane W Cox   University of Alberta, Alberta, Canada
diane.cox@ualberta.ca
α₁-Antitrypsin (AAT) Deficiency; Chromosome 14

Nick Craddock   University of Wales College of Medicine, Cardiff, UK
craddockn@cardiff.ac.uk
Mood Disorders: Molecular; Psychiatric Disorders: The Search for Genes

David Craufurd   St Mary's Hospital, Manchester, UK
Huntington Disease: Predictive Genetic Testing

Rachel A Craven   Cancer Research UK, Leeds, UK
r.craven@cancermed.leeds.ac.uk
Laser Capture Microdissection in Proteomics

Christoph G Cremer   Ruprecht Karls University, Heidelberg, Germany
cremer@kip.uni-heidelberg.de
Chromatin in the Cell Nucleus: Higher-order Organization

Marion Cremer   University of Munich, Munich, Germany
marion.cremer@lrz.uni-muenchen.de
Cell Cycle: Chromosomal Organization

Thomas Cremer   University of Munich, Munich, Germany
thomas.cremer@lrz.uni-muenchen.de
Cell Cycle: Chromosomal Organization; Chromatin in the Cell Nucleus: Higher-order Organization

Gay M Crooks   Childrens Hospital Los Angeles, Los Angeles, California, USA
Hematopoietic Stem Cells

Merlin Crossley   University of Sydney, Sydney, Australia
Zinc-finger Genes

Peter J P Croucher   Christian-Albrechts-University, Kiel, Germany
p.croucher@mucosa.de
Linkage Disequilibrium

James F Crow   University of Wisconsin, Madison, USA
jfcrow@wisc.edu
Evolution: Views of; Sutton, Walter Stanborough

Raymond R Crowe   University of Iowa College of Medicine, Iowa City, Iowa, USA
Alcoholism: Collaborative Study on the Genetics of Alcoholism (COGA)

Michael Crowe   University of Southern California, Los Angeles, California USA
Alzheimer Disease

Marina Cuchel   University of Pennsylvania, Philadelphia, Pennsylvania, USA
Familial Hypercholesterolemia: Gene Therapy

Thomas J Cummings   Duke University Medical Center, Durham, North Carolina, USA
cummi008@mc.duke.edu
Brain Tumors

Jim Cummins   Murdoch University, Perth, Australia
cummins@central.murdoch.edu.au
Mitochondrial DNA: Fate of the Paternal Mitochondrial Genome

Sarah Cunningham-Burley   University of Edinburgh, Edinburgh, UK
sarah.c.burley@ed.ac.uk
Public and Professional Understandings of Genetics

David T Curiel   University of Alabama at Birmingham, Birmingham, Alabama, USA
david.curiel@ccc.uab.edu
Delivery Targeting in Gene Therapy

Catherine Déon   Geneva University Hospital, Switzerland
Nucleolar Proteomics

Mohamed R Daha   Leiden University Medical Centre, Leiden, The Netherlands
Complement System and Fc Receptors: Genetics

Björn Dahlbäck   University of Lund, Malmö, Sweden
bjorn.dahlback@klkemi.mas.lu.se
Factor V Leiden

Nicole Datson   Leiden/Amsterdam Center for Drug Research, Leiden, The Netherlands
datson_n@lacdr.leidenuniv.nl
Expression Studies

David Tollervey   University of Edinburg, Edinburg, UK
d.tollervey@ed.ac.uk
Nonprotein-coding Genes

Duncan Davidson   MRC Human Genetics Unit, Edinburgh, UK
duncan.davidson@hgu.mrc.acd.uk
Mapping Gene Function in the Embryo

John F Davidson   University of Washington School of Medicine, Seattle, Washington, USA
davidsoj@u.washington.edu
Genomic and Chromosomal Instability

Kay Davies   University of Oxford, Oxford, UK
kay.davies@human-anatomy.oxford.ac.uk
Duchenne Muscular Dystrophy (DMD) Gene

Nicholas P Davies   Institute of Neurology, London, UK
n.davies@ion.ucl.ac.uk
Ion Channels and Human Disorders

Joe Davis   Massachusetts Institute of Technology, Cambridge, Massachusetts, USA
joedavis@mit.edu
Art and Genetics

Dena S Davis   Cleveland State University, Cleveland, Ohio, USA
dena.davis@law.csuohio.edu
Sex Selection; Genetic Disability and Legal Action: Wrongful Birth, Wrongful Life

Lennard J Davis   University of Illinois, Chicago, Illinois, USA
lendavis@uic.edu
Heredity and the Novel

Lisa M Davis   eXagen Corporation, Albuquerque, New Mexico, USA
Comparative Chromosome Mapping: Rodent Models

Ian N M Day   University of Southampton, Southampton, UK
inmd@soton.ac.uk
Monogenic Hypercholesterolemia: Genetics

Michael Dean   National Cancer Institute, Frederick, Maryland, USA
dean@ncifcrf.gov
ATP-Binding Cassette (ABC) Transporter Supergene Family: Genetics and Evolution

Paul H Dear   Medical Research Council Laboratory of Molecular Biology, Cambridge, UK
phd@mrc-lmb.cam.ac.uk
Basepair (bp); Expressed-sequence Tag (EST); Framework Map; Genome Map; Genome Mapping; Genome Map: Resolution; HAPPY Mapping; Kilobase Pair (kbp); Molecular Genetics; Megabase Pair (Mbp); OD Unit; Recombinant DNA; Restriction Fragment Length Polymorphism (RFLP); Sequence-tagged Site (STS)

Giovanna De Benedictis   University of Calabria, Rende, Italy
g.debenedictis@unical.it
Mitochondrial DNA Polymorphisms

Paul G Debenham   LGC, Teddington, UK
paul.debenham@lgc.co.uk
DNA Fingerprinting, Paternity Testing, and Relationship (Immigration) Analysis

Christine Debouck   GlaxoSmithKline, King of Prussia, Pennsylvania, USA
christine.m.debouck@gsk.com
Microarrays in Drug Discovery and Development

Marleen Decruyenaere   Center for Human Genetics, University of Leuven, Belgium
marleen.decruyenaere@med.kuleuven.ac.be
Predictive Genetic Testing: Psychological Impact

Karen E Deffenbacher   University of Nebraska Medical Center, Omaha, Nebraska, USA
Reading and Dyslexias

Prescott Deininger   Tulane University Health Sciences Center, New Orleans, Louisiana, USA
pdeinin@tulane.edu
Repetitive Elements and Human Disorders; Retrosequences and Evolution of Alu Elements

Jean Maurice Delabar   Université Paris 7, Paris, France
delabar@paris7.jussieu.fr
Chromosome 21

Claire Delahunty   The Scripps Research Institute, La Jolla, California, USA
Proteomics: A Shotgun Approach without Two-dimensional Gels

Panos Deloukas   The Wellcome Trust Sanger Institute, Hinxton, UK
panos@sanger.ac.uk
Radiation Hybrid Mapping

Erick Denamur   Hôpital Robert Debré, Paris, France
denamur@bichat.inserm.fr
Cystic Fibrosis Transmembrane Conductance Regulator Sequences: Comparative Analysis

Lieve Denayer   Center for Human Genetics, University of Leuven, Belgium
lieve.denayer@med.kuleuven.ac.be
Predictive Genetic Testing: Psychological Impact

Jean-Marc Deragon   Centre National de la Recherche Scientifique, Clermont-Ferrand, France
Transposons

Enrico Di Cera   Washington University School of Medicine, St Louis, Missouri, USA
enrico@biochem.wustl.edu
Serine Proteases

Marilyn Diaz   National Institute of Environmental Health Sciences, Research Triangle Park, North Carolina, USA
Somatic Hypermutation of Antigen Receptor Genes: Evolution

Jean-Jacques Diaz   INSERM, Faculté de Médecine Lyon-R.T.H. Laennec, Lyon, France
diaz@cgmc.univ-lyon1.fr
Nucleolar Proteomics

Donna L Dickenson   University of Birmingham, Birmingham, UK
d.l.dickenson@bham.ac.uk
Feminist Perspectives on Human Genetics and Reproductive Technologies

Régis Dieckmann   Geneva University Hospital, Geneva, Switzerland
Nucleolar Proteomics

Alexander Diemand   Glaxo Wellcome Experimental Research and Swiss Institute of Bioinformatics, Geneva, Switzerland
azd93529@glaxowellcome.co.uk
Protein Homology Modeling

Harry C Dietz   Johns Hopkins University School of Medicine and Howard Hughes Medical Institute, Baltimore, Maryland, USA
hdietz@jhmi.edu
Nonsense-mediated mRNA Decay

Lisabeth F DiLalla   Southern Illinois University School of Medicine, Carbondale, Illinois, USA
Developmental Psychopathology

Roeland W Dirks   Leiden University Medical Center, Leiden, The Netherlands
r.w.dirks@lumc.nl
Expression Analysis In Vivo: Cell Systems

Todd R Disotell   New York University, New York, New York, USA
todd.disotell@nyu.edu
Humans: Archaic and Modern; Modern Humans: Origin and Evolution; Primates: Phylogenetics

David J Dix   United States Environmental Protection Agency, Research Triangle Park, North Carolina, USA
Gene Expression Networks

Michael J Dixon   University of Manchester, Manchester, UK
mike.dixon@man.ac.uk
Orofacial Clefting

Detlef Doenecke   Georg-August-Universität Göttingen, Göttingen, Germany
ddoenec@gwdg.de
Histones; Intronless Genes

Francisco S Domingues   Max-Planck Institute for Informatics, Saarbrücken, Germany
doming@mpi-sb.mpg.de
Protein Structure Prediction and Databases

Luisa Doneda   Milan University, Milan, Italy
Heterochromatin: Constitutive

Nijsje Dorman   Harvard Medical School, Boston, Massachusetts, USA
mRNA Export

Josephine C Dorsman   University of Leiden, Leiden, The Netherlands
jc. dorsman@vumc.nl
Fusion Proteins as Research Tools

Steve Dorus   University of Chicago, Chicago, Illinois, USA
sdorus@midway.uchicago.edu
Mammalian Sex Chromosome Evolution

Gabriel Dover   University of Leicester, Leicester, UK
gabrieldover2000@yahoo.co.uk
Darwin and the Idea of Natural Selection

Jan Drenth   University of Groningen, Groningen, The Netherlands
j.drenth@chem.rug.nl
X-ray Diffraction: Principles

Lee Alan Dugatkin   University of Louisville, Louisville, Kentucky, USA
lee.dugatkin@Louisville.edu
Group Selection

Michel Duguet   Université Paris-Sud, Orsay, France
duguet@igmors.u-psud.fr
DNA Coiling and Unwinding

Cornelia M van Duijn   Erasmus Medical Center, Rotterdam, The Netherlands
c.vanduijn@erasmusmc.nl
Disease Associations: Human Leukocyte Antigen (HLA) and Apolipoprotein E

Andrew Dunham   The Wellcome Trust Sanger Institute, Hinxton, UK
ad1@sanger.ac.uk
Chromosome 13

Contributors

Ian Dunham The Sanger Institute, Wellcome Trust Genome Campus, Hinxton, UK
id1@sanger.ac.uk
Chromosomes 21 and 22: Comparisons; Chromosome 22; Genome Sequencing

Johan T den Dunnen Leiden University Medical Center, Leiden, The Netherlands
j.t.den_dunnen@lumc.nl
Mutation Nomenclature

Troy Duster New York University, New York, USA
troy.duster@nyu.edu
Eugenics: Contemporary Echoes

D A Dyment University of Oxford, Oxford, UK
Multiple Sclerosis (MS): Genetics

G C Ebers Radcliffe Infirmary, Oxford, UK
george.ebers@clinical-neurology.oxford.ac.uk
Multiple Sclerosis (MS): Genetics

Richard P Ebstein S. Herzog Hospital, Jerusalem, Israel
ebstein@netvision.net.il
Aggression and Criminal Behavior; Dopamine D4 Receptor (DRD4) Box Score

Edward Mitchell Eddy National Institute of Environmental Health Science, Research Triangle Park, North Carolina, USA
eddy@niehs.nih.gov
Germ Plasm and the Molecular Determinants of Germ Cell Fate

Howard J Edenberg Indiana University School of Medicine, Indianapolis, Indiana, USA
Alcoholism: Collaborative Study on the Genetics of Alcoholism (COGA)

Suni M Edson Armed Forces DNA Identification Laboratory, Rockville, Maryland, USA
edsonsm@afip.osd.mil
Romanov Family

Steven D Edwards Centre for Philosophy and Health Care, University of Wales Swansea, Swansea, UK
s.d.edwards@swansea.ac.uk
Conceptualization of the Body in 'Disability'; Disability: Philosophical Issues

Edward H Egelman University of Virginia, Charottesville, Virginia, USA
DNA Helicases

Jean-Marc Egly INSERM/CNRS/ULP, Illkirch, France
egly@igbmc.u-strasbg.fr
Transcription Regulation: Coordination

E E Eichler Case Western University School of Medicine, Cleveland, Ohio, USA
eee@po.cwru.edu
Chromosome-specific Repeats (Low-copy reeats)

Danna G Eickbush University of Rochester, Rochester, New York, USA
Transposable Elements: Evolution

Thomas H Eickbush University of Rochester, Rochester, New York, USA
eick@mail.rochester.edu
Transposable Elements: Evolution

Roland Eils German Cancer Research Center, Heidelberg, Germany
r.eils@dkfz.de
Digital Image Analysis

Leon Eisenberg Harvard Medical School, Boston, Massachusetts, USA
leon_eisenberg@hms.harvard.edu
Human Cloning: Arguments Against

Frank Eisenhaber Research Institute of Molecular Pathology, Vienna, Austria
Pax Genes: Evolution and Function

George H Elder University of Wales College of Medicine, Cardiff, UK
elder@cardiff.ac.uk
Porphyrias: Genetics

Nisrine El-Mogharbel Australian National University, Canberra, Australia
el-mogharbel@rsbs.anu.edu.au
X and Y Chromosomes: Homologous Regions

Mohammed El-Sawy Tulane University Health Sciences Center, New Orleans, Louisiana, USA
Repetitive Elements and Human Disorders

Chris Elkin US DOE Joint Genome Institute, Walnut Creek, California, USA
Robotics and Automation in Molecular Genetics

Hans Ellegren Uppsala University, Uppsala, Sweden
hans.ellegren@ebc.uu.s
Male-driven Evolution

Lora Hedrick Ellenson Weill Medical College of Cornell University, New York, USA
lhellens@med.cornell.edu
Endometrial Cancer

Carl Elliott Center for Bioethics, University of Minnesota, Minneapolis, Minnesota, USA
ellio023@umn.edu
Bioethics: Practice

R John Ellis University of Warwick, Coventry, UK
jellis@bio.warwick.ac.uk
Protein Folding and Chaperones

Lynne W Elmore Virginia Commonwealth University, Richmond, Virginia, USA
Hepatocellular Carcinoma

Joanna L Elson University of Newcastle, Newcastle upon Tyne, UK
J.L.Elson@ncl.ac.uk
Mitochondrial Heteroplasmy and Disease

Beverly S Emanuel The Children's Hospital of Philadelphia, Philadelphia, Pennsylvania, USA
beverly@mail.med.upenn.edu
Chromosome 22; Segmental Duplications and Genetic Disease

Ronald B Emeson Vanderbilt University, Nashville, Tennessee, USA
ron.emeson@vanderbilt.edu
mRNA Editing

Steve Emmott Green Group in the European Parliament, Brussels, Belgium
semmott@europarl.eu.int
DNA Technology: A Critical European Perspective; Far-field Light Microscopy

Carol Emslie MRC Social & Public Health Sciences Unit, Glasgow, UK
c.emslie@msoc.mrc.gla.ac.uk
Genetic Susceptibility

H Tristram Engelhardt Jr Rice University, Houston, Texas, USA
Bioethics: Institutionalization of; Secular Humanism

Veronica English British Medical Association, London, UK
venglish@bma.org.uk
Privacy and Genetic Information

Bryan K Epperson Michigan State University, East Lansing, Michigan, USA
epperson@msu.edu
Malécot, Gustave

Jörg T Epplen Ruhr-University, Bochum, Germany
joerg.t.epplen@ruhr-uni-bochum.de
Microsatellites; Minisatellites; Trinucleotide Repeat Expansions: Mechanisms and Disease Associations

Charles J Epstein University of California, San Francisco, California, USA
cepst@itsa.ucsf.edu
Down Syndrome

Ernie L Esquivel Yale University School of Medicine, New Haven, Connecticut, USA
Kidney: Hereditary Disorders

Manel Esteller Centro Nacional de Investigaciones Oncológicas, Madrid, Spain
mesteller@cnio.es
DNA Methylation and Histone Acetylation; Methylated DNA-binding Proteins

Louise H Eunson Institute of Neurology, London, UK
Ion Channels and Human Disorders

Carol Evans Cardiff & Vale NHS Trust, Cardiff, UK
Thyroid Cancer: Molecular Genetics

Christine Evans University of Wales College of Medicine, Cardiff UK
Confronting Genetic Disease: Psychological Issues; Genetic Counseling: Impact on the Family System

Gerry Evers-Kiebooms University of Leuven, Leuven, Belgium
gerry.kiebooms@med.kuleuven.ac.be
Predictive Genetic Testing: Psychological Impact

Warren J Ewens University of Pennsylvania, Philadelphia, Pennsylvania, USA
wewens@sas.upenn.edu
Population Genetics; Relatives-based Test for Linkage Disequilibrium: The Transmission/Disequilibrium Test (TDT)

M Daniele Fallin Johns Hopkins Bloomberg School of Public Health, Baltimore, Maryland, USA
dfallin@jhsph.edu
Linkage and Association Studies: Replication

Liam J Fanning National University of Ireland, Cork, Ireland
l.fanning@ucc.ie
Immunoglobulin Genes

Bernardino Fantini University of Geneva, Geneva, Switzerland
bernardino.fantini@medecine.unige.ch
Monod, Jacques Lucien

Christine J Farr University of Cambridge, Cambridge, UK
c_farr@mole.bio.cam.ac.uk
Mammalian Artificial Chromosomes (MACs)

Chantal Farra Brigham and Women's Hospital, Boston, Massachusetts, USA
Chromosome 15

Colin Farrelly University of Manchester, Manchester, UK
colin.farrelly@man.ac.uk
Distributive Justice and Genetics

C Garrison Fathman Stanford University School of Medicine, Stanford, California, USA
cfathman@stanford.edu
Autoimmune Diseases: Gene Therapy

Isabelle Favre University of Geneva Medical School, Geneva, Switzerland
Receptors and Human Disorders

Katie Featherstone Cardiff University, Cardiff, UK
FeatherstoneK@cardiff.ac.uk
Inheritance and Society

Alexei Fedorov Harvard University, Cambridge, Massachusetts, USA
Exonic Splicing Enhancers

Larisa Fedorova Tufts University, Boston, Massachusetts, USA
Exonic Splicing Enhancers

Andrew P Feinberg Johns Hopkins University School of Medicine, Baltimore, Maryland, USA
afeinberg@jhmi.edu
Imprinting Disorders

Josué Feingold Hôpital Necker-Enfants Malades, Paris, France
Genotype–Phenotype Relationships

Claude Férec Université de Bretagne Occidentale, Brest, France
claude.ferec@univ-brest.fr
Cystic Fibrosis Transmembrane Conductance Regulator; Sequences: Comparative Analysis; Trypsinogen Genes: Evolution

Rohan L Fernando Iowa State University, Ames, Iowa, USA
rohan@iastate.edu
*Kinship and Inbreeding
Segregation Analysis Software*

Janet Fernihough University of Oxford, Oxford, UK
Duchenne Muscular Dystrophy (DMD) Gene

Adolfo A Ferrando Dana Farber Cancer Institute, Boston, Massachusetts, USA
Leukemias and Lymphomas: Genetics

S Fidler Imperial College School of Medicine, London, UK
s.fidler@imperial.ac.uk
Infectious Diseases: Gene Therapy

Witold Filipowicz Friedrich Miescher Institut, Basel, Switzerland
witold.filipowicz@fmi.ch
snoRNPs

Jo-Anne Finegan The Hospital for Sick Children, Toronto, Canada
Behavioral Phenotypes: Goals and Methods

David J Fink University of Pittsburgh, School of Medicine, Pittsburgh, Pennsylvania, USA
Herpes Simplex Viral Vectors in Gene Therapy

Russell L Finley Jr Wayne State University School of Medicine, Detroit, Michigan, USA
rfinley@genetics.wayne.edu
Yeast Two-hybrid System and Related Methodology

A Fischer Hôpital Necker, Paris, France
fischer@necker.fr
Immunodeficiency Syndromes: Gene Therapy; Severe Combined Immune Deficiency (SCID): Genetics

Paula L Fischhaber UT Southwestern Medical Center, Dallas, Texas, USA
DNA Replication Fidelity

Kevin FitzGerald Georgetown University, Washington, DC, USA
Code of Ethical Principles for Genetics Professionals

Matthew J Fivash National Cancer Institute, Frederick, Maryland, USA
fivash@ncifcrf.gov
Surface Plasmon Resonance

Jonathan Alfred Fletcher Brigham and Women's Hospital, Boston, Massachusetts, USA
jfletcher@partners.org
Translocation Breakpoints in Cancer

Jonathan Flint University of Oxford, Oxford, UK
jf@molbiol.ox.ac.uk
Intellectual Disability: Genetics; Rett Syndrome; Telomere; Williams Syndrome: A Neurogenetic Model of Human Behavior

Terence R Flotte Department of Pediatrics and Powell Gene Therapy Center at the University of Florida Genetics Institute, Gainesville, Florida, USA
flotttr@peds.ufl.edu
Cystic Fibrosis: Gene Therapy

Mary Ford Cardiff University, Cardiff, UK
Human Cloning: Legal Aspects; In Vitro Fertilization: Regulation

Tatiana Foroud Indiana University School of Medicine, Indianapolis, Indiana, USA
Alcoholism: Collaborative Study on the Genetics of Alcoholism (COGA)

Donald R Forsdyke Queen's University, Kingston, Canada
forsdyke@post.queensu.ca
Bateson, William; Haldane, John Burdon Sanderson; Mendel, Gregor Johann; Muller, Hermann Joseph

Michael A Fortun Rensselaer Polytechnic Institute, Troy, New York, USA
fortum@rpi.edu
Celera Genomics: The Race for the Human Genome Sequence

Morris W Foster University of Oklahoma, Norman, Oklahoma, USA
morris.w.foster-1@ou.edu
Human Genome Diversity Project (HGDP)

Paula Goodman Fraenkel Harvard Medical School, Boston, Massachusetts, USA
Zebrafish: as a Model for Human Diseases

Peter Francis University College London, London, UK
peterjamesfrancis@ukonline.co.uk
Lens Disorders

Brunella Franco TIGEM, Naples, Italy
franco@tigem.it
Chromosome X

Patricia Françon Institute of Human Genetics, Montpellier, France
DNA Replication Origins

Ian M Frayling Addenbrooke's Hospital, Cambridge, UK
ifraylin@hgmp.mrc.ac.uk
ian.frayling@nhs.net
Colorectal Cancer: Genetics; Familial Adenomatous Polyposis; Hereditary Nonpolyposis Colorectal Cancer

Karl Fredga Uppsala University, Sweden
karl.fredga@ebc.uu.se
Levan, Albert

Jean Frézal Hôpital Necker-Enfants Malades, Paris, France
Genotype–Phenotype Relationships

Judith L Fridovich-Keil Emory University School of Medicine, Atlanta, Georgia, USA
jfridov@emory.edu
Yeast as a Model for Human Diseases

Errol C Friedberg UT Southwestern Medical Center, Dallas, Texas, USA
errol.friedberg@utsouthwestern.edu
DNA Replication Fidelity

Felix Friedberg Howard University College of Medicine, Washington, DC, USA
Pseudogenes: Age

Lauriane Fritsch Institut Andre Lwoff, Villejuif, France
Histone Acetylation and Disease

Alan Fryer Royal Liverpool Children's Hospital, Liverpool, UK
alan.fryer@rlch-tr.nwest.nhs.uk
Clinical Genetic Services in the United Kingdom

Jean-Pierre Fryns University Hospital of Leuven, Leuven, Belgium
jean-pierre.fryns@med.kuleuven.ac.be
Dysmorphic Syndromes; Fetal Wastage: Molecular Basis; Monosomies

Dagmar Fuhrer Universität Leipzig, Leipzig, Germany
Thyroid Dysfunction: Molecular Genetics

Yukio Fujiki Kyushu University, Fukuoka, Japan
yfujiscb@mbox.nc.kyushu-u.ac.jp
Functional Complementation; Peroxisome Biogenesis Disorders

Watson Fuller Keele University, Keele, UK
Wilkins, Maurice Hugh Frederick

Eric T Fung Ciphergen Biosystems, Inc., Fremont, California, USA
efung@ciphergen.com
Molecular Entry Point: Strategies in Proteomics

Jens Oliver Funk University of Erlangen-Nuremberg, Erlangen, Germany and Merck KGaA, Darmstadt, Germany
jens-oliver.funk@merck.de
Cell Cycle Checkpoint Genes and Cancer

Anthony V Furano National Institutes of Health, Bethesda, Maryland, USA
avf@helix.nih.gov
Long Interspersed Nuclear Elements (LINEs): Evolution

Suzanne Furuyama Case Western Reserve University, Cleveland, Ohio, USA
Trans Splicing

Giuliana Fuscaldo University of Melbourne, Victoria, Australia
g.fuscaldo@pgrad.unimelb.edu.au
Gamete Donation and 'Race'

Pascal Gagneux Glycobiology Research and Training Center, La Jolla, California, USA
gagneux@biomail.ucsd.edu
Great Apes and Humans: Genetic Differences

Isabelle Gaillard Institut de Génétique Humaine, CNRS, Montpellier, France
Olfactory Receptors

Kerry-Jane Galenzoski Health Sciences Centre, Winnipeg, Canada
XYY Syndrome

Brenda L Gallie Princess Margaret University, Toronto, Canada
gallie@attglobal.net
Retinoblastoma

Michael J Galsworthy King's College, London, UK
Intelligence and Cognition

Dawn K Garcia University of Texas Health Science Center, San Antonio, Texas, USA
Chromosome 3

Derek Gardener University of Liverpool, Liverpool, UK
Weapons of Mass Destruction: Genotoxicity

Mark Gardiner University College London, UK
mark.gardiner@ucl.ac.uk
Epilepsy: Genetics

Jessica Gardner Bognor Regis, UK
Genetic Testing of Children: Parental Requests

Pauline Hélène Garnier-Géré INRA Recherches Forestières, Gazinet, France
Genetics of Large Populations and Association Studies

Nicola Gebbia University of Palermo, Palermo, Italy
Caretakers and Gatekeepers

Bruce D Gelb Mount Sinai School of Medicine, New York, New York, USA
bruce.gelb@mssm.edu
Noonan Syndrome

David Gell University of Sydney, Sydney, Australia
Zinc-finger Genes

Martin Gencik Ruhr-University, Bochum, Germany
Trinucleotide Repeat Expansions: Mechanisms and Disease Associations

Patrick Gendron Université de Montréal, Montréal, Canada
RNA Tertiary Structure Prediction: Computational Techniques

Emanuelle Génin INSERM U155, Paris, France
Homozygosity Mapping

Daniela S Gerhard Office of Cancer Genomics, Bethesda, Maryland, USA
gerhardd@mail.nih.gov
Chromosome 11

Kris Gevaert Ghent University, Ghent, Belgium
kris.gevaert@rug.ac.be
Protein Characterization: Analytical Approaches and Applications to Proteomics

Pamela K Geyer University of Iowa, Iowa City, Iowa, USA
pamela-geyer@uiowa.edu
Chromosomes: Higher-order Organization; Position Effect Variegation in Human Genetic Disease

Ian Giddings Institute of Cancer Research, Sutton, UK
Microarrays and Expression Profiling in Cancer

Susan Gilchrist MRC Human Genetics Unit, Edinburgh, UK
Chromosome Structures: Visualization

Raanan Gillon Imperial College of Science Technology and Medicine, London, UK
raanan.gillon:@ic.ac.uk
Health Care Ethics: The Four Principles

Dominique Giorgi Institut de Génétique Humaine, Montpellier, France
giorgi@igh.cnrs.fr
Olfactory Receptors

Joel N Glasgow University of Alabama at Birmingham, Birmingham, Alabama, USA
Delivery Targeting in Gene Therapy

Jörn Glökler Max Planck Institute of Molecular Genetics, Berlin, Germany
Array-based Proteomics

Joseph C Glorioso University of Pittsburgh, School of Medicine, Pittsburgh, Pennsylvania, USA
glorioso@pitt.edu
Herpes Simplex Viral Vectors in Gene Therapy

Alison Goate Washington University School of Medicine, St Louis, Missouri, USA
Alcoholism: Collaborative Study on the Genetics of Alcoholism (COGA)

Hans H Goebel Johannes Gutenberg-University Medical Center, Mainz, Germany
Desmin-related Myopathies

Peter Gogarten University of Connecticut, Storrs, Connecticut, USA
gogarten@uconn.edu
Orthologs, Paralogs, and Xenologs in Human and Other Genomes

E Richard Gold McGill University, Montreal, Canada
richard.gold2@mcgill.ca
Patenting of Genes: Discoveries or Inventions?

Alfred L Goldberg Harvard Medical School, Boston, Massachusetts, USA
alfred_goldberg@hms.harvard.edu
Protein Degradation and Turnover

Lev G Goldfarb National Institute of Neurological Disorders and Stroke, National Institutes of Health, Bethesda, Maryland, USA
goldfarbl@ninds.nih.gov
Desmin-related Myopathies; Genotype–Phenotype Relationships: Fatal Familial Insomnia and Creutzfeldt–Jakob Disease

Emanuel Goldman   University of Medicine and Dentistry of New Jersey, Newark, New Jersey, USA
egoldman@umdnj.edu
tRNA

H Hill Goldsmith   University of Wisconsin–Madison, Madison, Wisconsin, USA
hhgoldsm@wiscmail.wisc.edu
Gene-to-behavior Pathways; Personality and Temperament

David B Goldstein   University College London, London, UK
d.goldstein@ucl.ac.uk
Variable Drug Response: Genetic Evaluation

G Golfier   CNRS UMR 7637, Paris, France
Chromosome-specific Repeats (Low-copy repeats)

Susanne M Gollin   University of Pittsburgh, Pittsburgh, Pennsylvania, USA
Mitosis: Chromosomal Rearrangements

Sarah E Gollust   National Human Genome Research Institute, Maryland USA
Population Carrier Screening: Psychological Impact

Yoichi Gondo   RIKEN Genomic Sciences Center, Yokohama, Japan
gondo@gsc.riken.go.jp
Megasatellite DNA

Iris L Gonzalez   Al duPont Hospital for Children, Wilmington, Delaware, USA
igonzale@nemours.org
rRNA Genes: Evolution

CF Goodey   The Open University, London, UK
cgoodey@aol.com
Intelligence, Heredity, and Genes: A Historical Perspective

John Goodier   University of Pennsylvania School of Medicine, Philadelphia, Pennsylvania, USA
Long Interspersed Nuclear Elements (LINEs)

William Goodwin   University of Central Lancashire, Preston, UK
Neanderthal DNA

Robert J Gorlin   University of Minnesota School of Dentistry, Minneapolis, Minnesota, USA
gorli002@tc.umn.edu
Nevoid Basal Cell Carcinoma (Gorlin) Syndrome

Christine M Gosden   University of Liverpool, Liverpool, UK
cgosden@liverpool.ac.uk
Genetics and the Control of Human Reproduction; Weapons of Mass Destruction: Genotoxicity

Jonatha M Gott   Case Western Reserve University, Cleveland, Ohio, USA
jmg13@po.cwru.edu
RNA Editing and Human Disorders

Irving I Gottesman   University of Minnesota, Minneapolis, Minnesota, USA
gotte003@umn.edu
Slater, Eliot Trevor Oakeshott

Stephanie K Grade   University of Iowa, Iowa City, Iowa, USA
Position Effect Variegation in Human Genetic Disease

Darren Grafham   The Sanger Centre, Wellcome Trust Genome Campus, Hinxton, UK
dg1@sanger.ac.uk
Sequence Finishing

Anthony Graham   Kings College, London, UK
anthony.graham@kcl.ac.uk
Vertebrate Evolution: Genes and Development

FL Graham   McMaster University, Hamilton, Canada
graham@mcmaster.ca
Adenoviral Vectors in Gene Therapy

Henk Granzier   Washington State University, Pullman, Washington, USA
Titin (TTN) Gene

Dan Graur   Tel Aviv University, Tel Aviv, Israel
gruar@post.tau.ac.il
Mutational Change in Evolution; Single-base Mutation

Brenton R Graveley   University of Connecticut Health Center, Farmington, Connecticut, USA
graveley@neuron.uchc.edu
SR Proteins

Jennifer AM Graves   The Australian National University, Canberra, Australia
Chromosome X; X-chromosome Inactivation; X and Y Chromosomes: Homologous Regions

Courtney Gray-McGuire   Case Western Reserve University, Cleveland, Ohio, USA
Relatives-based Nonparametric Analysis of Linkage

Anna Greco   INSERM, Faculté de Médecine Lyon-R.T.H. Laennec, Lyon, France
Nucleolar Proteomics

Niels Gregersen   Aarhus University Hospital, Aarhus, Denmark
Protein Misfolding and Degradation in Genetic Disease

Peter K Gregersen   North Shore Long Island Jewish Research Institute, Manhasset, New York, USA
peterg@nshs.edu
Absolute Pitch: Genetics; Rheumatoid Arthritis: Genetics

Simon G Gregory   The Wellcome Trust Sanger Institute, Cambridge, UK
sgregory@chg.duhs.duke.edu
Contig Assembly

Timothy J Griffin   Institute for Systems Biology, Seattle, Washington, USA
tgriffin@systemsbiology.org
Quantitative Proteomics (ICAT^{TM})

Tiemo Grimm   Department of Medical Genetics, University of Würzburg, Germany
tgrimm@biozentrum.uni-wuerzburg.de
Genetic Risk: Computation; Mutation Rate

Leif Groop   Lund University, Malmö Sweden
leif.groop@endo.mas.lu.se
Metabolism: Hereditary Errors

Peter C Groot   Utrecht University, Utrecht, The Netherlands
Representational Difference Analysis

Frank G Grosveld   Erasmus University Medical Center, Rotterdam, The Netherlands
Gene Conversion in Health and Disease

Mark Groudine   Fred Hutchinson Cancer Research Center and University of Washington School of Medicine, Seattle, Washington, USA
markg@fhcrc.org
Enhancers; Locus Control Regions (LCRs)

Yizhong Gu   Aeomica Inc., Sunnyvale, California, USA
Microarrays: Use in Gene Identification

Valentina Guasconi   Institut Andre Lwoff, Villejuif, France
Histone Acetylation and Disease

Jayanthi P Gudikote   University of Texas MD Anderson Cancer Center, Houston, Texas, USA
RNA-binding Proteins: Regulation of mRNA Splicing, Export, and Decay

Nicolas Guex   Glaxo Wellcome Experimental Research and Swiss Institute of Bioinformatics, Geneva, Switzerland
Protein Homology Modeling

Jeffrey R Gulcher   deCODE Genetics, Reykjavik, Iceland
jgulcher@decode.is
deCODE: A Genealogical Approach to Human Genetics in Iceland

Per Guldberg   John F. Kennedy Institute, Glostrup, Denmark
Phenylketonuria

Deborah L Gumucio   University of Michigan, Ann Arbor, Michigan, USA
dgumucio@umich.edu
Promoters: Evolution

Peter William Gunning   The Children's Hospital at Westmead, Sydney, Australia
Protein Isoforms and Isozymes

Haiwei H Guo   University of Washington School of Medicine, Seattle, Washington, USA
Genomic and Chromosomal Instability

Jinjiao Guo   Aeomica Inc., Sunnyvale, California, USA
Microarrays: Use in Gene Identification

Sun-Wei Guo   Medical College of Wisconsin, Milwaukee, Wisconsin, USA
swguo@mcw.edu
China: The Maternal and Infant Health Care Law

J Mitchell Guss   University of Sydney, Sydney, Australia
Macromolecular Structure Determination: Comparison of Crystallography and Nuclear Magnetic Resonance (NMR)

Flemming Güttler   John F. Kennedy Institute, Glostrup, Denmark
Phenylketonuria

Marta Gwinn   Centers for Disease Control, Atlanta, Georgia, USA
genetics@cdc.gov
Epidemiological Tools

Thomas Haaf   Mainz University School of Medicine, Mainz, Germany
haaf@humgen.klinik.uni-mainz.de
Centromeres

Rebecca P Haberman   University of North Carolina at Chapel Hill, Chapel Hill, North Carolina, USA
rahabs@med.unc.edu
Gene Therapy: Technology

Guy Haegeman   University of Gent, Gent, Belgium
guy.haegeman@dmb.rug.ac.be
DNA Viral Vectors in Gene Therapy

Arndt von Haeseler   Heinrich-Heine-Univesitaet 1, Duesseldorf and John von Neumann-Institut für Computing, Juelich, Germany
haeseler@cs.uni-duesseldorf.de
Population History and Linkage Disequilibrium

Sinuhe Hahn   University of Basel, Basel, Switzerland
Fetal Diagnosis

Toshio Hakoshima   Nara Institute of Science and Technology, Nara, Japan
hakosima@bs.aist-nara.ac.jp
Leucine Zippers

Judith G Hall   University of British Columbia, Vancouver, Canada
jhall@cw.bc.ca
Mosaicism

Laura L Hall   Annapolis, Maryland, USA
lauraleehall@onebox.com
Turner Syndrome

Mark A Hall   Wake Forest University, Winston-Salem, North Carolina, USA
mhall@wfubmc.edu
Discrimination in Insurance: Experience in the United States

Finn Hallböök   Biomedical Centre, Uppsala, Sweden
finn.hallbook@neuro.uu.se
Paralogous Genes and Gene Duplications

Nina Hallowell   The University of Edinburgh Medical School, Edinburgh, UK
nina.hallowell@ed.ac.uk
Ethics and Evidence

Mahmoud Hamdan   University of Verona, Verona, Italy
Preparation Artefacts in Proteomics: Technological Issues

Horst Hameister   University of Ulm, Germany
horst.hameister@medizin.uni-ulm.de
X-linked Genes for General Cognitive Abilities

Michael Hampsey   Robert Wood Johnson Medical School, Piscataway, New Jersey, USA
michael.hampsey@umdnj.edu
Negative Regulatory Elements (NREs)

Clemens Oliver Hanemann   Ulm University, Ulm, Germany
oliver.hanemann@medizin.uni-ulm.de
Demyelinating Neuropathies: Hereditary

Neil Anthony Hanley   Southampton University, Southampton, UK
n.a.hanley@soton.ac.uk
Human Developmental Molecular Genetics

Michael G Hanna   Institute of Neurology, London, UK
Ion Channels and Human Disorders

Gregory J Hannon   Cold Spring Harbor Laboratory, Cold Spring Harbor, New York, USA
RNA Interference (RNAi) and MicroRNAs

Nathaniel J Hansen   University of Alabama at Birmingham, Birmingham, Alabama, USA
Telomerase: Structure and Function

David K Hanzel   Aeomica Inc., Sunnyvale, California, USA
Microarrays: Use in Gene Identification

Jean-Pierre Hardelin   Pasteur Institute, Paris, France
hardelin@pasteur.fr
Deafness: Hereditary

Ross C Hardison   Pennsylvania State University, University Park, Pennsylvania, USA
rch8@psu.edu
Globin Genes: Evolution

Annick Harel-Bellan   Institut Andre Lwoff, Villejuif, France
annick.harel-bellan@vjf.cnrs.fr
Histone Acetylation and Disease

Eric Hugh Harley   University of Cape Town, Rondebosch, South Africa
harley@chempath.uct.ac.za
DNA Fingerprinting in Ecology

Henry Harpending   University of Utah, Salt Lake City, Utah, USA
henry.harpending@anthro.utah.edu
Humans: Demographic History

Peter S Harper   University of Wales College of Medicine, Cardiff, UK
harperps@cardiff.ac.uk
Huntington Disease

Curtis C Harris   National Institute of Health, Bethesda, Maryland, USA
harrisc@intra.nci.nih.go
Hepatocellular Carcinoma

John Harris   University of Manchester, Manchester, UK
john.m.harris@man.ac.uk
Reproductive Choice

Rodney Harris   St Mary's Hospital, Manchester, UK
rodney.harris@man.ac.uk
Genetic Services: Access

Michael T Harrison   Strand Chambers, London, UK
Gene Therapy: Ethics and Regulation

F-Ulrich Hartl   Max Planck Institute for Biochemistry, Martinsried, Germany
Protein Folding and Chaperones; Protein Folding In Vivo

Ian M Hastings   Liverpool School of Tropical Medicine, Liverpool, UK
hastings@liverpool.ac.uk
Genetic Load

Nobutaka Hattori   Juntendo University School of Medicine, Tokyo, Japan
nhattori@med.juntendo.ac.jp
Parkinson Disease

Silke Hauf   Research Institute of Molecular Pathology, Vienna, Austria
hauf@nt.imp.univie.ac.at
Meiosis and Mitosis: Molecular Control of Chromosome Separation

Henk AMJ ten Have   University Medical Center Nijmegen, Nijmegen, The Netherlands
h.tenhave@efg.umcn.nl
Geneticization: Concept

Trevor Hawkins   US DOE Joint Genome Institute, Walnut Creek, California, USA
trevor.hawkins@am.amershambiosciences.com
Robotics and Automation in Molecular Genetics

Adam M Hedgecoe   University College London, London UK
a.m.hedgecoe@sussex.ac.uk
Geneticization: Debates and Controversies

Gunnar von Heijne   Stockholm University, Stockholm, Sweden
gunnar@biokenu.su.se
Signal Peptides

Christopher UT Hellen   State University of New York, Brooklyn, New York, USA
Translation Initiation: Molecular Mechanisms in Eucaryotes

Sirkku K Hellsten   University of Dar es Salaam, Dar es Salaam, Tanzania and The University of Helsinki, Finland
skhellsten@yahoo.com
Autonomy and Responsibility in Reproductive Genetics

Brian Hendrich   University of Edinburgh, Edinburgh, UK
brian.hendrich@ed.ac.uk
Chromatin Structure and Modification: Defects

Lidewij Henneman   VU University Medical Center, Amsterdam, The Netherlands
l.henneman.emgo@med.vu.nl
Genetic Carrier Testing

Jaap Heringa   Free University, Amsterdam, The Netherlands
heringa@cs.vu.nl
Sequence Similarity

Klemens J Hertel   University of California Irvine, Irvine, California, USA
SR Proteins

John Hesketh   University of Newcastle, Newcastle upon Tyne, UK
j.e.hesketh@ncl.ac.uk
3′ UTRs and Regulation

Victor Hesselbrock   University of Connecticut Health Center, Farmington, Connecticut, USA
Alcoholism: Collaborative Study on the Genetics of Alcoholism (COGA)

Lambert van den Heuvel   Nijmegen University Medical Centre, Nijmegen, The Netherlands
Oxidative Phosphorylation (OXPHOS) System: Nuclear Genes and Genetic Disease

Ian D Hickson   University of Oxford, Oxford, UK
ian.hickson@cancer.org.uk
DNA Helicase-deficiency Disorders

Winston Hide   South African National Bioinformatics Institute and University of the Western Cape, Bellville, South Africa
winhide@sanbi.ac.za
Genome Databases

Friedhelm Hildebrandt   University of Michigan, Freiburg, Michigan, USA
Chromosome 2

Matthew J Hilton   University of Houston, Houston, Texas, USA
Chromosome 8

Tetsuro Hirose   Yale University School of Medicine, New Haven, Connecticut, USA
th88@email.med.yale.edu
GAS5 Gene

Denis Hochstrasser   Geneva University Hospital, Geneva, Switzerland
denis.hochstrasser@dim.hcuge.ch
Nucleolar Proteomics

Angela Kaye Hodges   University of Wales College of Medicine, Cardiff, UK
hodgesak@cardiff.ac.uk
Posttranscriptional Processing

Shirley V Hodgson   University of London, London, UK
s.hodgson@sghms.ac.uk
Penrose, Lionel Sharples

Ivo L Hofacker   University of Vienna, Vienna, Austria
ivo@tbi.univie.ac.at
RNA Secondary Structure Prediction

Kristen C Hoffbuhr   Children's National Medical Center, Washington, DC, USA
Rett Syndrome

Eric P Hoffman   Children's National Medical Center, Washington, DC, USA
Rett Syndrome

Marten Hofker   University of Maastricht, Maastricht, The Netherlands
m.hofker@gen.unimaas.nl
Mouse Genetics as a Research Tool

Kay Hofmann   MEMOREC Stoffel GmbH, Cologne, Germany
kay.hofmann@memorec.com
Profile Searching

Sabine Hofmann   Academic Hospital, Munich-Schwabing, Germany
sabine.hofmann@lrz.uni-muenchen.de
Mitochondrial Disorders

Michael Hofreiter   Max Planck Institute for Evolutionary Anthropology, Leipzig, Germany
hofreiter@eva.mpg.de
Ancient DNA: Phylogenetic Applications

Pancras CW Hogendoorn   Leiden University Medical Centre, Leiden, The Netherlands
p.c.w.Hogendoorn@lumc.nl
Skeletogenesis: Genetics

Josephine Hoh   Rockefeller University, New York, New York, USA
Genetic Linkage Mapping

Elizabeth L Holbrook   Lawrence Berkeley National Laboratory, Berkeley, California, USA
Crystallization of Nucleic Acids

Stephen R Holbrook   Lawrence Berkeley National Laboratory, Berkeley, California, USA
srholbrook@lbl.gov
Crystallization of Nucleic Acids; RNA Gene Prediction

Sheila Hollins   St George's Hospital Medical School, London, UK
s.hollins@sghms.ac.uk
Mentally Handicapped in Britain: Sexuality and Procreation

Søren Holm   University of Manchester, Manchester, UK
soren.holm@man.ac.uk
Data Protection Legislation; Informed Consent: Ethical and Legal Issues

Peter Holmans   Medical Research Council Biostatistics Unit, Cambridge UK
peter.holmans@mrc-bsu.cam.ac.uk
Linkage Analysis Software

Gerald P. Holmquist   City of Hope, Duarte, California, USA
gholm@coh.org
Chromosomal Bands and Sequence Features

Nils Holtug   University of Copenhagen, Copenhagen Denmark
nhol@hum.ku.dk
Genetic Harm

Neil A Holtzman   Johns Hopkins University, School of Medicine and School of Public Health, Baltimore, Maryland, USA
nholtzma@jhsph.edu
Neonatal Screening

Wolfgang Holzgreve   University of Basel, Basel, Switzerland
wolfgang.holzgreve@unibas.ch
Fetal Diagnosis

Bent Honoré   University of Aarhus, Denmark
bh@biokemi.au.dk
Transcriptomics and Proteomics: Integration?

Rob Hooft van Huijsduijnen   Serono Pharmaceutical Research Institute, Geneva, Switzerland
rob.hooft@serono.com
Kinases and Phosphatases and Human Disorders

Christine Hoogland   Swiss Institute of Bioinformatics, Geneva, Switzerland
christine.hoogland@isb-sib.ch
Nucleolar Proteomics

Martin L Hooper   University of Edinburgh, Edinburgh, UK
Tumor Suppressor Genes

O Horaitis   Genomic Disorders Research Centre, Melbourne, Australia
Human Genome Variation Society

Martin P Horan   University of Wales College of Medicine, Cardiff, UK
horanmp@cardiff.ac.uk
Methylation-mediated Transcriptional Silencing in Tumorigenesis

Gijsbertus TJ van der Horst   Erasmus Medical Center, Rotterdam, The Netherlands
g.vanderhorst@erasmusmc.nl
Circadian Rhythm Genetics

Bernhard Horsthemke   Universität Essen, Essen, Germany
b.horsthemke@uni-essen.de
Prader–Willi Syndrome and Angelman Syndrome

Richard Houlston   Institute of Cancer Research, Sutton, UK
r.houlston@icr.ac.uk
Mutations: Penetrance

Charles HV Hoyle University College London, London, UK
c.hoyle@ucl.ac.uk
Neuropeptides and their Receptors: Evolution

John C Hozier University of New Mexico, Albuquerque, New Mexico, USA
jhozier@salud.unm.edu
Comparative Chromosome Mapping: Rodent Models

Tianhua Hu Aeomica Inc., Sunnyvale, California, USA
Microarrays: Use in Gene Identification

Sheng-He Huang University of Southern California, Los Angeles, California, USA
shuang@chla.usc.edu
Infectomics: Study of Response to Infection using Microarrays

Tim JP Hubbard Wellcome Trust Sanger Institute, Hinxton, UK
th@sanger.ac.uk
Human Genome: Draft Sequence

Ulrich Hübscher University of Zürich, Zürich, Switzerland
hubscher@vetbio.unizh.ch
DNA Polymerases: Eukaryotic

Austin L Hughes University of South Carolina, Columbia, South Carolina, USA
austin@biol.sc.edu
Defensins: Evolution; Genomics, Pharmacogenetics and Proteomics: An Integration; Interferons; Major Histocompatibility Complex (MHC) Genes: Evolution; Selective and Structural Constraints; Vertebrate Immune System: Evolution

Yannick Hugodot Université Paris-Sud, Orsay, France
DNA Coiling and Unwinding

Maj Hultén University of Warwick, Coventry, UK
maj.hulten@warwick.ac.uk
Genetic Mapping: Comparison of Direct and Indirect Approaches; Genetic Maps: Direct Meiotic Analysis; Levan, Albert; Meiosis; Tjio, Joe-Hin

Susanne Hummel University of Göttingen, Göttingen, Germany
shummel1@gwdg.de
Ancient DNA: Recovery and Analysis

Kate Hunt MRC Social & Public Health Sciences Unit, Glasgow, UK
k.hunt@msoc.mrc.gla.ac.uk
Genetic Susceptibility

Laurence D Hurst University of Bath, Bath, UK
bssldh@bath.ac.uk
Mutation Rate: Sex Biases

Joh-E Ikeda Tokai University, Isehara, Japan
joh-e@nga.med.u-tokai.ac.jp
Megasatellite DNA

Ana Smith Iltis Center for Health Care Ethics, St Louis, Missouri, USA
Bioethics: Institutionalization of

Hogune Im University of Wisconsin Medical School, Madison, Wisconsin, USA
Histone Acetylation: Long-range Patterns in the Genome

Magnus Ingelman-Sundberg Karolinska Institutet, Stockholm, Sweden
magnus.ingelman-sundberg@imm.ki.se
Drug Metabolic Enzymes: Genetic Polymorphisms

Hidetoshi Inoko Tokai University School of Medicine, Kanagawa, Japan
hinoko@is.icc.u-tokai.ac.jp
Major Histocompatibility Complex (MHC) Genes

Ken Inoue Baylor College of Medicine, Houston, Texas, USA
kinoue@bcm.tmc.edu
Charcot–Marie–Tooth Disease and Associated Peripheral Neuropathies

Andreea Ioan-Facsinay Leiden University Medical Centre, Leiden, The Netherlands
Complement System and Fc Receptors: Genetics

Alan Irvine Our Lady's Hospital for Sick Children, Dublin, Ireland
alan.irvine@olhsc.ie
Hair Loss: Genetics

Alan D Irvine University of Dundee, Dundee, UK
Skin: Hereditary Disorders

James A Irving Monash University, Victoria, Australia
Serpins: Evolution

Takahiro Isaka Teikyo University, Tokyo, Japan
Chromosomes during Cell Division

Toshihisa Ishikawa Tokyo Institute of Technology, Yokohama, Japan
tishikaw@bio.titech.ac.jp
Multidrug Resistance in Cancer: Genetics of ABC Transporters

Valérie Itier University of Geneva, Medical School, Geneva, Switzerland
Nicotinic Receptors

J Larry Jameson Northwestern University, Chicago, Illinois, USA
Gonadotropin Hormones: Disorders

Axel Janke University of Lund, Lund, Sweden
axel.janke@cob.lu.se
Hominids: Molecular Phylogenetics

Timo Järvilehto University of Oulu, Oulu, Finland
timo.jarvilehto@oulu.fi
Behavior: Role of Genes

Anil G Jegga Cincinnati Children's Hospital Medical Center, Cincinnati, Ohio, USA
Evolutionarily Conserved Noncoding DNA

David Jenkins Aeomica Inc., Sunnyvale, California, USA
Microarrays: Use in Gene Identification

Yonggang Ji Aeomica Inc., Sunnyvale, California, USA
Microarrays: Use in Gene Identification

Yishi Jin University of California, Santa Cruz, California, USA
Caenorhabditis elegans as an Experimental Organism

HA Jinnah Johns Hopkins Hospital, Baltimore, Maryland, USA
hjinnah@jhmi.edu
Lesch–Nyhan Disease

Mark A Jobling University of Leicester, Leicester, UK
maj4@leicester.ac.uk
Surnames and Genetics

Hans Joenje VU University Medical Center, Amsterdam, The Netherlands
h.joenje.humgen@med.vu.nl
Spontaneous Function Correction of Pathogenic Alleles in Inherited Diseases resulting in Somatic Mosaicism

Kirby D Johnson University of Wisconsin Medical School, Madison, Wisconsin, USA
Histone Acetylation: Long-range Patterns in the Genome

Josephine Johnston University of Minnesota, Minneapolis, Minnesota, USA
johnstonj@thehastingscenter.org
Bioethics: Practice

Jukka Jokela Swiss Federal Institute of Technology (ETH), Zürich, Switzerland
Sex: Evolutionary Advantages

Gregory V Jones University of Warwick, Coventry, UK
G.V.Jones@warwick.ac.uk
Handedness, Left/Right: Genetics

Ian Jones University of Birmingham, Queen Elizabeth Psychiatric Hospital, Birmingham, UK
Mood Disorders: Molecular

Louise Kathleen Jones Institute of Cancer Research, London, UK
Fusion Proteins and Diseases

Wilfried W de Jong University of Nijmegen, Nijmegen, The Netherlands
w.dejong@ncmls.kun.nl
Crystallins

C Victor Jongeneel Ludwig Institute for Cancer Research and Swiss Institute of Bioinformatics, Epalinges, Switzerland
victor.jongeneel@licr.org
Bioinformatics: Technical Aspects

Paulus HLJ Joosten University of Nijmegen, Nijmegen, The Netherlands
p.joosten@sci.kun.nl
Promoter Haplotypes and Gene Expression

I King Jordan National Institutes of Health, Bethesda, Maryland, USA
jordan@ncbi.nlm.nih.gov
Comparative Genomics

Lynn B Jorde University of Utah Health Sciences Center, Salt Lake City, Utah, USA
Human Genetic Diversity

NV Joshi Indian Institute of Science, Bangalore India
Transmembrane Domains

Fabrice Jotterand Rice University, Houston, Texas, USA
Bioethics: Institutionalization of

Peter R Jungblut Max-Planck Institute for Infection Biology, Berlin, Germany
jungblut@mpiib-berlin.mpg.de
Bacterial Proteomes

Jerzy Jurka Genetic Information Research Institute, Mountain View, California, USA
jurka@girinst.org
Repetitive Elements: Detection

Monica J Justice Baylor College of Medicine, Houston, Texas, USA
mjustice@bcm.tmc.edu
Mouse N-ethyl-N-nitrosourea (ENU) Mutagenesis

Henrik Kaessmann Max-Planck Institute for Evolutionary Anthropology, Leipzig, Germany
kaessmann@eva.mpg.de
Human and Chimpanzee Nucleotide Diversity

Dion Kaiserman Monash University, Victoria, Australia
Serpins: Evolution

Luba Kalaydjieva Western Australian Institute for Medical Research, Australia
l.kalaydjieva@ecu.edu.au
Roma (Gypsies): Genetic Studies

Deepak Kamnasaran University of Alberta, Edmonton, Canada
Chromosome 14

Karthikeyan Kandavelou Johns Hopkins University, Baltimore, Maryland, USA
Restriction Enzymes

Vladimir V Kapitonov Genetic Information Research Institute, Mountain View, California USA
Repetitive Elements: Detection

Josseline Kaplan Hôpital Necker–Enfants Malades, Paris, France
Eye Disorders: Hereditary

Gulshan A Karbani University of Leeds, Leeds, UK
Genetic Counseling: Consanguinity and Cultural Expectations

Sharon LR Kardia University of Michigan, Ann Arbor, Michigan, USA
skardia@umich.edu
Gene–Environment Interaction

Olof Karlberg Uppsala University, Sweden
olof.karlberg@ebc.uu.se
Mitochondrial Proteome: Origin

Marcel Karperien Leiden University Medical Centre, Leiden, The Netherlands
Skeletogenesis: Genetics

Masanori Kasahara The Graduate University for Advanced Studies, Hayama, Japan
kasaharams@soken.ac.jp
Polyploid Origin of the Human Genome

Mohammed Kashani-Sabet University of California at San Francisco, San Francisco, California, USA
kashanisabet@orca.ucsf.edu
Antisense and Ribozymes

Richard Kaslow Birmingham University of Alabama, Birmingham, Alabama, USA
Human Immunodeficiency Virus (HIV) Infection: Genetics

Gregory J Kato Johns Hopkins University School of Medicine, Baltimore, Maryland, USA
gkato@jhmi.edu
Proteases and Human Disorders

Randal J Kaufman University of Michigan Medical Center, Ann Arbor, Michigan, USA
kaufmanr@umich.edu
Hemophilias: Gene Therapy

Eckhard Kaufmann University of Ulm, Ulm, Germany
eckhard.kaufmann@medizin.uni-ulm.de
Forkhead Domains

Haig H Kazazian Jr University of Pennsylvania School of Medicine, Philadelphia, Pennsylvania, USA
kazazian@mail.med.upenn.edu
Long Interspersed Nuclear Elements (LINEs); Retrotransposition and Human Disorders

Hildegard Kehrer-Sawatzki University of Ulm, Germany
hildegard.kehrer-sawatzki@medizin.uni-ulm.de
X-linked Genes for General Cognitive Abilities

Uddhav P Kelavkar University of Pittsburgh, Pittsburgh, Pennsylvania, USA
kelavkarup@msx.upmc.edu
Human Genome Project

Naoya Kenmochi Miyazaki Medical College, Miyazaki, Japan
kenmochi@post.miyazaki-med.ac.jp
Ribosomes and Ribosomal Proteins

James Kennedy University of Toronto, Toronto, Canada
james_kennedy@camh.net
Anxiety Disorders

Lindsey Kent University of Birmingham, Birmingham, UK
l.s.kent@bham.ac.uk
Mood Disorders: Molecular

Daniel Keppler Louisiana State University Health Sciences Center, Shreveport, Louisiana, USA
dkeppl@lsuhsc.edu
Cystatins

Lauren Kerzin-Storrar Central Manchester Hospitals Healthcare Trust and University of Manchester, Manchester, UK
Genetic Registers

Ad Geurts van Kessel University Medical Center Nijmegen, Nijmegen, The Netherlands
a.geurtsvankessel@antrg.azn.nl
Chromosome 18

Saeed A Khan JD Institute for Social Policy and Understanding, Rochester Hills, Michigan, USA
skhan@ispu.us
'Race' and Difference: Orientalism and Western Concepts

Kum Kum Khanna The Queensland Institute of Medical Research, Brisbane, Australia
kumkumk@qimr.edu.au
DNA Damage Response

Linda Kharaboyan University of Montreal, Québec, Canada
linda.kharaboyan@umontreal.ca
Informed Consent and Multiplex Screening

Muin J Khoury Centers for Disease Control, Atlanta, Georgia, USA
mlg1@cdc.gov
Epidemiological Tools

Karine Kindbeiter INSERM, Faculté de Médecine Lyon-R.T.H. Laennec, Lyon, France
Nucleolar Proteomics

David King Human Genetics Alert, London, UK
davidking22@blueyonder.co.uk
Intelligence: Ethical Debates about the Search for IQ Quantitative Trait Loci

Glenn F King University of Connecticut Health Center, Farmington, Connecticut, USA
glenn@psel.uchc.edu
Macromolecular Structure Determination: Comparison of Crystallography and Nuclear Magnetic Resonance (NMR)

Taroh Kinoshita Osaka University, Osaka, Japan
tkinoshi@biken.osaka-u.ac.jp
Paroxysmal Nocturnal Hemoglobinuria

Maggie Kirk University of Glamorgan, Pontypridd, UK
mkirk@glam.ac.uk
Genetic Education of Primary Care Health Professionals in Britain

Timothy J Kirkpatrick The University of Texas Medical School at Houston, Houston, Texas, USA
Neural Tube Defects: Genetics

Ilan R Kirsch National Cancer Institute, Bethesda, Maryland, USA
kirschi@exchange.nih.gov
Cytogenetic and Physical Chromosomal Maps: Integration

Katsumi Kitagawa St Jude Children's Research Hospital, Memphis, Tennessee, USA
katsumi.kitagawa@stjude.org
Kinetochore: Structure, Function and Evolution

Aaron Klug Medical Research Council, Cambridge, UK
Franklin, Rosalind Elsie

Avraham N Kluger The Hebrew University of Jerusalem, Jerusalem, Israel
avraham.kluger@huji.ac.il
Dopamine D4 Receptor (DRD4) Box Score

Eric B Kmiec University of Delaware, Newark, Delaware, USA
ekmiec@udel.edu
Chimera-directed Gene Repair

Joseph E Knapp University of Pittsburgh School of Pharmacy, Pittsburgh, Pennsylvania, USA
Naked DNA in Gene Therapy

Michael Knapp University of Bonn, Bonn Germany
knapp@uni-bonn.de
Sample Size Requirements

Walter Knöchel University of Ulm, Ulm, Germany
walter.knoechel@medizin.uni-ulm.de
Forkhead Domains

Bartha M Knoppers University of Montreal, Montreal, Canada
knoppers@droit.umontreal.ca
Informed Consent

Jorn Koch Institute of Pathology, Aarhus Kommunehospital, Aarhus, Denmark
jokoc@akh.aaa.dk
Centromeric Sequences and Sequence Structures

Michel Koenig Institut de Génétique et de Biologie Moleculaire et Cellulaire, Illkirch, France
Friedreich Ataxia

Donald B Kohn Childrens Hospital, Los Angeles, California, USA
dkohn@chla.usc.edu
Hematopoietic Stem Cells

Kurt W Kohn National Cancer Institute, National Institutes of Health, Bethesda, Maryland, USA
kohnk@dc37a.nci.nih.gov
Cell Cycle Control: Molecular Interaction Map

Regine Kollek University of Hamburg, Germany
kollek@uni-hamburg.de
Reprogenetics: Visions of the Future

Inke R König Medical University of Lübeck, Lübeck, Germany
Genetic Distance and Mapping Functions

Andrzej K Konopka BioLingua Research Inc., Gaithersburg, Maryland, USA
akk@blingua.org
Biomolecular Sequence Analysis: Pattern Acquisition and Frequency Counts; Sequence Complexity and Composition; Systems Biology: Genomics Aspects

Zoltan Konthur Max Planck Institute of Molecular Genetics, Berlin, Germany
Array-based Proteomics

Eugene V Koonin National Institutes of Health, Bethesda, Maryland, USA
koonin@ncbi.nlm.nih.gov
Comparative Genomics; DNA Repair: Evolution

Julie R Korenberg University of California, Los Angeles, California, USA
Williams Syndrome: A Neurogenetic Model of Human Behavior

Moshe Kotler Ben-Gurion University of the Negev, Beersheva, Israel
Aggression and Criminal Behavior

Viktor Kozich Charles University-1st Faculty of Medicine, Prague, Czech Republic
vkozich@lf1.cuni.cz
Cystathionineβ-Synthase (CBS) Deficiency: Genetics

David C Krakauer Santa Fe Institute, Sante Fe, New Mexico, USA
krakauer@santafe.edu
Genetic Redundancy

Jan Peter Kraus University of Colorado School of Medicine, Denver, Colorado, USA
jan.kraus@uchsc.edu
Cystathionineβ-Synthase (CBS) Deficiency: Genetics

Michael Krawczak Christian-Albrechts-Universität Kiel, Germany
krawczak@medinfo.uni-kiel.de
Kinship Testing; Multiple Significance Testing in Genetic Epidemiology

Thane Kreiner Affymetrix Inc., Santa Clara, California, USA
thane_kreiner@affymetrix.com
Genetic Age: A Vision

Maxwell M Krem Washington University School of Medicine, St Louis, Missouri, USA
kremm@medicine.wustl.edu
Serine Proteases

Raju Kucherlapati Harvard Partners Genome Center, Cambridge, Massachusetts, USA
rkucherlapati@partners.org
Chromosome 12

Anver Kuliev Reproductive Genetics Institute, Chicago, Illinois, USA
rgi@flash.net
Preimplantation Diagnosis

Jerzy K Kulski Tokai University School of Medicine, Kanagawa, Japan and CBBC, Information Technology, Murdoch University, Murdoch, Australia
jkulski@cbbc.murdoch.edu.au
Major Histocompatibility Complex (MHC) Genes

Erdmute M Kunstmann Ruhr University, Bochum, Germany
Minisatellites

Pui-Yan Kwok University of California, San Francisco, California, USA
kwok@cvrimail.ucsf.edu
Single Nucleotide Polymorphisms (SNPs): Identification and Scoring

Dietmar Labeit Universitätsklinikum, Mannheim, Germany
Titin (TTN) Gene

Siegfried Labeit Universitätsklinikum, Mannheim, Germany
labeit@embl.de
Titin (TTN) Gene

Bruce T Lahn University of Chicago, Chicago, Illinois, USA
blahn@genetics.uchicago.edu
Mammalian Sex Chromosome Evolution

Renu B Lal Centers for Disease Control and Prevention, Atlanta, Georgia, USA
Human Immunodeficiency Virus (HIV) Infection: Genetics

Stephen C-T Lam University of Illinois at Chicago, Chicago, Illinois, USA
sclam@uic.edu
Integrins

Titia de Lange The Rockefeller University, New York, New York, USA
delange@rockefeller.edu
Telomeres: Protection and Maintenance

Ulrich Langenbeck University Hospital Frankfurt, Frankfurt/Main, Germany
u.langenbeck@em.uni-frankfurt.de
Genetic Disorders

Cordelia F Langford The Sanger Institute, Cambridge UK
cfl@sanger.ac.uk
Comparative Cytogenetics

Dan Larhammar Biomedical Centre, Uppsala, Sweden
dan.larhammar@neuro.uu.se
Paralogous Genes and Gene Duplications

Lidia Larizza Milan University, Milan, Italy
lidia.larizza@unimi.it
Heterochromatin: Constitutive

David S Latchman University College London, London, UK
d.latchman@bbk.ac.uk
Transcription Factors

Vincent Laudet Ecole Normale Supérieure de Lyon, Lyon, France
vincent.laudet@ens-lyon.fr
Nuclear Receptor Genes

Graeme Laurie University of Edinburgh, Edinburgh, UK
graeme.laurie@ed.ac.uk
Ownership of Genetic Material and Information

CE Lawrence Wadsworth Center, Albany, New York, USA
Gibbs Sampling and Bayesian Inference

Jeffrey G Lawrence University of Pittsburgh, Pittsburgh, Pennsylvania, USA
jlawrenc@pitt.edu
Gene Clustering in Eukaryotes

Daniel Lawson Wellcome Trust Sanger Institute, Cambridge, UK
Caenorhabditis elegans *Genome Project*

Guillaume Lecointre Muséum National d'Histoire Naturelle, Paris, France
lecointr@mnhn.fr
Cystic Fibrosis Transmembrane Conductance Regulator Sequences: Comparative Analysis

David H Ledbetter University of Chicago, Chicago, Illinois, USA
dledbetter@genetics.emory.edu
Telomere

Pierre Legrain Hybrigenics, Paris, France
plegrain@hybrigenics.fr
Protein–Protein Interaction Maps

Alan R Lehmann University of Sussex, Brighton, UK
a.r.lehmann@sussex.ac.uk
DNA Repair: Disorders

Heike Lehrmann Institut Andre Lwoff, Villejuif, France
Histone Acetylation and Disease

Fabrice Lejeune University of Rochester, Rochester, New York, USA
RNA Processing and Human Disorders

Caryn E Lerman University of Pennsylvania, Philadelphia, Pennsylvania, USA
Familial Breast Cancer: Genetic Testing

Kathryn S Lemery Arizona State University, Tempe, Arizona, USA
klemery@asu.edu
Twin Study Contributions to Understanding Ontogeny

Trudo Lemmens University of Toronto, Toronto, Canada
trudo.lemmens@utoronto.ca
Insurance and Human Genetics: Approaches to Regulation

Thomas Lengauer Max-Planck Institute for Informatics, Saarbrücken, Germany
Protein Structure Prediction and Databases

Bonnie S LeRoy University of Minnesota, Minneapolis, Minnesota, USA
leroy001@umn.edu
Nondirectiveness

Arthur M Lesk University of Cambridge, Cambridge UK
aml2@mrc-lmb.cam.ac.uk
Proteins: Mutational Effects in; Protein Structure

AML Lever Addenbrooke's Hospital, Cambridge, UK
tdb21@medschl.cam.ac.uk
Lentiviral Vectors in Gene Therapy

Kate E Leverton Imperial College at Hammersmith Hospital, London, UK
Expression Targeting in Gene Therapy

Jacqueline Levilliers Pasteur Institute, Paris, France
jaclev@pasteur.fr
Deafness: Hereditary

Michael Levin Imperial College School of Science, Technology and Medicine, London, UK
Infectious Diseases: Predisposition

Nicola C Levitt University of Oxford, Oxford, UK
DNA Helicase-deficiency Disorders

Fran Lewitter Whitehead Institute for Biomedical Research, Cambridge, Massachusetts, USA
lewitter@wi.mit.edu
Nucleotide Sequence Databases

Wen-Hsiung Li University of Chicago, Chicago, Illinois, USA
whli@uchicago.edu
Domain Duplication and Gene Elongation; Evolutionary Distance; Gene Duplication: Evolution; Gene Trees and Species Trees; Heterozygosity; Homologous, Orthologous, and Paralogous Genes; Molecular Clocks; Nucleotide Substitution: Rate of; Phylogenetics

Cynthia Mui Yee Rosa Liang National University of Singapore, Singapore
Two-Dimensional Gel Electrophoresis

Daiqing Liao University of Florida, Gainesville, Florida, USA
dliao@ufl.edu
Concerted Evolution

Stephen A Liebhaber University of Pennsylvania School of Medicine, Philadelphia, Pennsylvania, USA
liebhabe@mail.med.upenn.edu
mRNA Stability and the Control of Gene Expression

Melissa C Liechty Applied Genetics Laboratories, Inc, Melbourne, Florida, USA
Comparative Chromosome Mapping: Rodent Models

Robert N Lightowlers University of Newcastle, Newcastle upon Tyne, UK
r.n.lightowlers@ncl.ac.uk
Mitochondrial Heteroplasmy and Disease

Susan Lindsay University of Newcastle upon Tyne, Newcastle-upon-Tyne UK
s.lindsay@newcastle.ac.uk
Developmental Evolution

Andrew Linzey Oxford University, Oxford, UK
andrewlinzey@aol.com
Cloning of Animals in Genetic Research: Ethical and Religious Issues

Frédérique Lisacek Geneva Bioinformatics SA, Geneva, Switzerland
frederique.lisacek@genebio.com
Pattern Searches

Elwira Lisowska Ludwik Hirszfeld Institute of Immunology and Experimental Therapy, Wrocaw, Poland
lisowska@iitd.pan.wroc.pl
Protein Glycosylation

Peter FR Little University of New South Wales, Sydney, Australia
p.little@unsw.edu.au
Gene Mapping and Positional Cloning

Dexi Liu University of Pittsburgh School of Pharmacy, Pittsburgh, Pennsylvania, USA
dliu@pitt.edu
Naked DNA in Gene Therapy

David J Lockhart Ambit Biosciences, San Diego, California, USA
dlockhart@ambitbio.com
DNA Chip Revolution

Lawrence A Loeb University of Washington School of Medicine, Seattle, Washington, USA
laloeb@u.washington.edu
Genomic and Chromosomal Instability

Joseph C Loftus Mayo Clinic Scottsdale, Scottsdale, Arizona, Arizona, USA
Integrins

Dietmar Rudolf Lohmann Universität Essen, Essen, Germany
dr.lohmann@uni-essen.de
Retinoblastoma

David A Lomas Cambridge Institute for Medical Research, Cambridge, UK
dal16@cam.ac.uk
Pulmonary Disorders: Hereditary

Manyuan Long The University of Chicago, Chicago, Illinois, USA
mlong@midway.uchicago.edu
Gene Fusion; Introns: Movements; Pseudoexons

Marcus J Longley University of Glamorgan, Pontypridd UK
mlongley@glam.ac.uk
Citizens' Jury on Genetic Testing for Common Disorders

A Thomas Look Dana Farber Cancer Institute, Boston, Massachusetts, USA
thomas_look@dfci.harvard.edu
Leukemias and Lymphomas: Genetics

Alexandra Louis Infobiogen, Evry, France
Genetic Databases: Mining

James Daniel Love Medical Research Council Laboratory of Molecular Biology, Cambridge, UK
jdl@mrc-lmb.cam.ac.uk
Nuclear Receptors and Disease

PR Lowenstein Cedars-Sinai Medical Center and University of California, Los Angeles, California USA
lowensteinp@cshs.org
Neurological Diseases: Gene Therapy

Nicolette H Lubsen University of Nijmegen, Nijmegen, The Netherlands
Crystallins

John C Lucchesi Emory University, Atlanta, Georgia, USA
lucchesi@biology.emory.edu
Sex Chromosomes

Marian Ludgate University of Wales College of Medicine, Cardiff, UK
ludgate@cf.ac.uk
Thyroid Cancer: Molecular Genetics; Thyroid Dysfunction: Molecular Genetics

Stefan Müller Ludwig Maximilians University, Müchen, Germany
Chromosomes in Mammals: Diversity and Evolution; Chromosome Rearrangement Patterns in Mammalian Evolution

Angelika Lueking Max Planck Institute of Molecular Genetics, Berlin, Germany
Array-based Proteomics

Thomas Lufkin Mount Sinai School of Medicine, New York, New York, USA
thomas.lufkin@mssm.edu
Hox *Genes: Embryonic Development*

Cliff J Luke Children's Hospital, Harvard Medical School, Boston, Massachusetts, USA
Serpins: Evolution

Tshilobo Prosper Lukusa Centre for Human Genetics, Leuven, Belgium
prosper.lukusa@med.kuleuven.ac.be
Monosomies

Lars-Gustav Lundin Biomedical Centre, Uppsala, Sweden
lg.lundin@neuro.uu.se
Paralogous Genes and Gene Duplications

Ming-Juan Luo Harvard Medical School, Boston, Massachusetts, USA
mRNA Export

James R Lupski Baylor College of Medicine, Houston, Texas, USA
jlupski@bcm.tmc.edu
Charcot–Marie–Tooth Disease and Associated Peripheral Neuropathies

Mary F Lyon MRC Mammalian Genetics Unit, Harwell, UK
m.lyon@har.mrc.ac.uk
X-chromosome Inactivation and Disease

Stanislas Lyonnet Hôpital Necker-Enfants Malades, Paris, France
lyonnet@necker.fr
Genotype–Phenotype Relationships

Marcella Macaluso Temple University, Philadelphia, Pennsylvania, USA
Caretakers and Gatekeepers

Angus S MacDonald Herriot-Watt University, Edinburgh, UK
Genetic Factors in Life Insurance: Actuarial Basis

Darryl RJ Macer Institute of Biological Sciences, University of Tsukuba, Tsukuba Science City, Japan
macer@biol.tsukuba.ac.jp
Bioethics in Asia; Genetic Information and the Family in Japan

Alex J MacGregor St Thomas's Hospital, London, UK
Twin Methodology

Steven J Mack Children's Hospital Oakland Research Institute, Oakland, California, USA and Roche Molecular Systems, Berkeley, California, USA
Major Histocompatibility Complex (MHC) Genes: Polymorphism

Joel Mackay University of Sydney, Sydney, Australia
j.mackay@mmb.usyd.edu.au
Zinc-finger Genes

Dixie L Mager British Columbia Cancer Agency, Vancouver, Canada
dmager@bccrc.ca
Retroviral Repeat Sequences

Eileen Magnello Wellcome Trust Centre for the History of Medicine at UCL, London, UK
meileenmagnello@aol.com
Pearson, Karl

Keith Magni Harvard Medical School, Boston, Massachusetts, USA
mRNA Export

Don J Mahuran Hospital For Sick Children, Toronto, Canada
hex@sickkids.ca
Tay–Sachs Disease

Christiane Maier University of Ulm, Ulm, Germany
christiane.maier@medizin.uni-ulm.de
Prostate Cancer

François Major Université de Montréal, Montréal, Canada
francois.major@umontreal.ca
RNA Tertiary Structure Prediction: Computational Techniques

Partha P Majumder Indian Statistical Institute, Calcutta, India
ppm@isical.ac.in
Peopling of India: Insights from Genetics

Izabela Makalowska Pennsylvania State University, University Park, Pennsylvania, USA
izabelam@psu.edu
Gene Feature Identification

Wojciech Makalowski Pennsylvania State University, University Park, Pennsylvania, USA
wojtek@psu.edu
Bioinformatics

Tomi Mäkelä University of Helsinki and Helsinki University Hospital, Helsinki Finland
tomi.makela@helsinki.fi
Apoptosis and the Cell Cycle in Human Disease

Kateryna D Makova University of Chicago, Chicago, Illinois, USA
Domain Duplication and Gene Elongation; Molecular Clocks

Carlo C Maley Fred Hutchinson Cancer Research Center, Seattle, Washington, USA
cmaley@alum.mit.edu
Barrett Esophagus

David Malkin The Hospital for Sick Children, Toronto, Canada
david.malkin@sickkids.ca
Li–Fraumeni Syndrome

Grazia MS Mancini Erasmus University Medical Centre, Rotterdam, The Netherlands
g.mancini@erasmusmc.nl
Lysosomal Transport Disorders

Mala Mani Johns Hopkins University, Baltimore, Maryland, USA
Restriction Enzymes

Sylvie Manouvrier-Hanu Hôpital Jeanne de Flandre, Lille, France
smanouvrier@chru-lille.fr
Limb Development Anomalies: Genetics

Lynne E Maquat University of Rochester, Rochester, New York, USA
lynne_maquat@urmc.rochester.edu
RNA Processing and Human Disorders

Ignacio Marín Universidad de Valencia, Valencia, Spain
ignacio.marin@uv.es
Dosage Compensation Mechanisms: Evolution

John Matthew Maris Children's Hospital of Philadelphia and University of Pennsylvania School of Medicine, Philadelphia, Pennsylvania, USA
maris@email.chop.edu
Neuroblastoma

Jonathan Marks University of North Carolina at Charlotte, Charlotte, North Carolina, USA
jmarks@email.uncc.edu
Human Genome Diversity Project (HGDP): Impact on Indigenous Communities

Contributors

Philip A Marsden   St Michael's Hospital and University of Toronto, Toronto, Canada
p.marsden@utoronto.ca
Alternative Processing: Neuronal Nitric Oxide Synthase

Wallace F Marshall   Yale University, New Haven, Connecticut, USA
wfm5@pantheon.yale.edu
Mitosis

Christa Lese Martin   University of Chicago, Chicago, Illinois, USA
clmartin@genetics.uchicago.edu
Telomere

Maryanne Martin   University of Oxford, Oxford, UK
Handedness, Left/Right: Genetics

Paul A Martin   University of Nottingham, Nottingham, UK
paul.martin@nottingham.ac.uk
Gene Therapy: Expectations and Results

William F Martin   University of Düsseldorf, Düsseldorf, Germany
w.martin@uni-duesseldorf.de
Mitochondrial Origins of Human Nuclear Genes and DNA Sequences

Rosa Martínez-Arias   Georg-August-University, Göttingen, Germany
Pseudogenes: Patterns of Mutation

Philip J Mason   The Hammersmith Hospital, London, UK
p.mason1@imperial.ac.uk
Glucose-6-Phosphate Dehydrogenase (G6PD) Deficiency: Genetics

Tara C Matise   Rutgers University, Piscataway, New Jersey, USA
matise@biology.rutgers.edu
Chromosome 1

John Stanley Mattick   University of Queensland, Brisbane, Australia
j.mattick@imb.uq.edu.au
Noncoding RNAs: A Regulatory Role?

Stephen C Maxson   The University of Connecticut, Storrs, Connecticut, USA
stephen.maxson@uconn.edu
Animal Models of Human Behavior

Patrick H Maxwell   Imperial College of Science, Technology and Medicine, London, UK
p.maxwell@imperial.ac.uk
Renal Carcinoma and von Hippel–Lindau Disease

Michel Maziade   Laval University, Sainte-Foy, Quebec, Canada
michel.maziade@psa.ulaval.ca
Schizophrenia and Bipolar Disorder: Linkage on Chromosomes 5 and 11

Pauline MH Mazumdar   University of Toronto, Toronto, Canada
pmazumda@chass.utoronto.ca
Eugenics Society

David McCarthy   University of Bristol, Bristol, UK
Human Cloning: Arguments for

Jane McCarthy   St George's Hospital Medical School, London, UK
Mentally Handicapped in Britain: Sexuality and Procreation

Tommie V McCarthy   University College Cork, Cork, Ireland
t.mccarthy@ucc.ie
Malignant Hyperthermia

Maclyn McCarty   Rockefeller University, New York, New York, USA
mccartm@rockvax.rockefeller.edu
Avery, Oswald Theodore

Robyn McDermott   Queensland Health, Cairns, Queensland, Australia
robyn_mcdermott@health.qld.gov.au
Thrifty Gene Hypothesis: Challenges

Jean E McEwen   National Human Genome Research Institute, Bethesda, Maryland, USA
ELSI Research Program of the National Human Genome Research Institute (NHGRI)

Glenn McGee   University of Pennsylvania Center for Bioethics, Philadelphia, Pennsylvania, USA
mcgee@mail.med.upenn.edu
Genetic Enhancement: The Role of Parents

William McGinnis   University of California, San Diego, California, USA
wmcginnis@ucsd.edu
Hox Genes and Body Plan: Evolution

Tony McGleenan   Queen's University, Belfast, UK
t.mcgleenan@ulster.ac.uk
Genetic Information, Genetic Testing, and Employment; Insurance and Genetic Information

Anne McLaren   Wellcome/CRC Institute, Cambridge, UK
Human Cloning

Jerry W McLarty,   Louisiana State University Health Sciences Center, Shreveport, Louisiana, USA
jmclar@lsuhsc.edu
Neural Networks

Sheila AM McLean   University of Glasgow School of Law, Glasgow, UK
s.mclean@law.gla.ac.uk
Genetic Screening Programs; Interests of the Future Child

WH Irwin McLean   Our Lady's Hospital for Sick Children, Dublin, Ireland
wmclean@hgmp.mrc.ac.uk
Skin: Hereditary Disorders

Roger E McLendon   Duke University Medical Center, Durham, North Carolina, USA
mclendon@duke.edu
Brain Tumors

Janet M McNicholl   Centers for Disease Control and Prevention, Atlanta, Georgia, USA
jkm7@cdc.gov
Human Immunodeficiency Virus (HIV) Infection: Genetics

Alexander McPherson   University of California, Irvine, California, USA
amcphers@uci.edu
Crystallization of Proteins and Protein–Ligand Complexes

Gil McVean   University of Oxford, Oxford UK
mcvean@stats.ox.ac.uk
Diffusion Theory

John H McVey   MRC Clinical Sciences Centre, London, UK
john.mcvey@csc.mrc.ac.uk
Hemophilia and Other Bleeding Disorders: Genetics

Marcel Méchali   Institute of Human Genetics, Montpellier, France
marcel.mechali@igh.cnrs.fr
DNA Replication Origins

Patrik Medstrand   Lund University, Lund, Sweden
Retroviral Repeat Sequences

GJ te Meerman   University of Groningen, Groningen, The Netherlands
g.j.te.meerman@med.rug.nl
Linkage Analysis

Maxwell J Mehlman   Case Western Reserve University, Cleveland, Ohio, USA
mjm10@po.cwru.edu
Health Care and Health Insurance in the United States

Joao Meidanis   University of Campinas, Campinas, Brazil
meidanis@ic.unicamp.br
Global Alignment; Sequence Assembly

Timothy J Meier   Stanford University, Cincinnati, Ohio, USA
Code of Ethical Principles for Genetics Professionals

Frauke Mekus   Medizinische Hochschule Hannover, Hannover, Germany
Genomic DNA: Purification; Restriction Mapping

Jaroslaw Meller   Children's Hospital Research Foundation, Cincinnati, Ohio, USA and Copernicus University, Torun Poland
jmeller@chmcc.org
Molecular Dynamics

Isabel Mellon   University of Kentucky, Lexington, Kentucky, USA
mellon@uky.edu
Transcription-coupled DNA Repair

Joshua T Mendell   Johns Hopkins University School of Medicine and Howard Hughes Medical Institute, Baltimore, Maryland, USA
Nonsense-mediated mRNA Decay

Richard F Meraz   Lawerence Berkeley National Laboratory, Berkeley, California, USA
RNA Gene Prediction

Chantal Mérette   Laval University, Sainte-Foy, Quebec, Canada
Schizophrenia and Bipolar Disorder: Linkage on Chromosomes 5 and 11

NATURE ENCYCLOPEDIA OF THE HUMAN GENOME / ©2003 Macmillan Publishers Ltd, Nature Publishing Group / www.ehgonline.net   **849**

Seppo Meri University of Helsinki, Helsinki, Finland
seppo.meri@helsinki.fi
Protein Characterization in Proteomics

Andres Metspalu University of Tartu, Tartu, Estonia
andres@ebc.ee
Microarrays and Single Nucleotide Polymorphism (SNP) Genotyping

David Metzgar The Scripps Research Institute, La Jolla, California, USA
dmetzgar@hermes.scripps.edu
Mutation Rates: Evolution

Diogo Meyer University of California, Berkeley, California, USA
diogo@allele5.biol.berkeley.edu
Major Histocompatibility Complex (MHC) Genes: Polymorphism

Loren Michel Memorial Sloan-Kettering Cancer Center, New York, New York, USA
Mitosis: Chromosome Segregation and Stability

Anna Middleton Addenbrooke's Hospital, Cambridge, UK
anna.middleton@addenbrookes.nhs.uk
Deaf Community and Genetics

William Mifsud University of Malta Medical School, Guardamangia, Malta
Gene Families

Manuela Migliavacca University of Palermo, Palermo, Italy
Caretakers and Gatekeepers

Flavio Mignone Università Milano, Milano, Italy
5' UTRs and Regulation

A Dusty Miller Fred Hutchinson Cancer Research Center, Seattle, Washington, USA
dmiller@fhcrc.org
Retroviral Vectors in Gene Therapy

Andrew David Miller Imperial College of Science, Technology and Medicine, London, UK
a.miller@imperial.ac.uk
Synthetic Gene Delivery Systems in Gene Therapy

Debra L Mills Emory University, Atlanta, Georgia, USA
Williams Syndrome: A Neurogenetic Model of Human Behavior

Bud Mishra Courant Institute, New York University and Cold Spring Harbor Laboratory, New York, New York, USA
mishra@nyu.edu
Optical Mapping

John J Mitchell McGill University, Montreal, Canada
jojames66@hotmail.com
Carrier Screening of Adolescents in Montreal

Philip Mitchell University of Edinburgh, Edinburgh, UK
pmitch@holyrood.ed.ac.uk
mRNA Turnover

Felix Mitelman University of Lund, Lund, Sweden
felix.mitelman@klingen.lu.se
Cancer: Chromosomal Abnormalities

Yoshikuni Mizuno Juntendo University School of Medicine, Tokyo, Japan
Parkinson Disease

Mirella Mochi University of Bologna Medical School, Bologna, Italy
Migraine: Genetics

Bernadette Modell University College London, London, UK
modell@pcps.ucl.ac.uk
Carrier Screening for Inherited Hemoglobin Disorders in Cyprus and the United Kingdom

Corina Moen Leiden University Medical Center, Leiden, The Netherlands
cmoen@lumc.nl
Mouse Genetics as a Research Tool

Evita Mohr University of Hamburg, Hamburg, Germany
emohr@uke.uni-hamburg.de
mRNA Localization: Mechanisms

Robert D Moir Massachusetts General Hospital and Harvard Medical School, Charlestown, Massachusetts, USA
Protein Aggregation and Human Disorders

Hans W Moises University of Kiel, Kiel, Germany
hmoises@web.de
Psychoses

Chris Molenaar Leiden University Medical Center, Leiden, The Netherlands
Expression Analysis In Vivo: Cell Systems

Pasquale Montagna University of Bologna Medical School, Italy
pmontagn@kaiser.alma.unibo.it
Migraine: Genetics

Jonathan Montgomery University of Southampton, Southampton, UK
j.r.montgomery@soton.ac.uk
Genetic Testing of Children: Capacity of Children to Consent

Kate Montgomery Harvard Partners Genome Center, Cambridge, Massachusetts, USA
kmontgomery@rics.bwh.harvard.edu
Chromosome 12

Tom Moore University College Cork, Cork, Ireland
t.moore@ucc.ie
Genetic Conflict and Imprinting

Bharti Morar Edith Cowan University, Perth, Australia
bmorar@cyllene.uwa.edu.au
Roma (Gypsies): Genetic Studies

Gregory J Morgan Johns Hopkins University, Baltimore, Maryland, USA
morgan@jhu.edu
Zuckerkandl, Emile

Derek Morgan Cardiff University, Cardiff, UK
morgandm1@cardiff.ac.uk
Human Cloning: Legal Aspects; In Vitro *Fertilization: Regulation*

Ryuichi Morishita Graduate School of Medicine, Osaka University, Osaka, Japan
morishit@cgt.med.osaka-u.ac.jp
Cardiovascular Disease: Gene Therapy

Etsuko N Moriyama University of Nebraska, Lincoln, Nebraska, USA
emoriyama2@unl.edu
Codon Usage

Bernice E Morrow Albert Einstein College of Medicine, Bronx, New York, USA
morrow@aecom.yu.edu
Microdeletions and Microduplications: Mechanism

Cynthia C Morton Brigham and Women's Hospital and Harvard Medical School, Boston, Massachusetts, USA
cmorton@partners.org
Chromosome 15

Newton Morton University of Southampton, Southampton, UK
nem@soton.ac.uk
Blocks of Limited Haplotype Diversity; Wright, Sewall

Nicholas K Moschonas Department of Biology, University of Crete & Institute of Molecular Biology and Biotechnology (IMBB-FORTH), Crete, Greece
moschon@imbb.forth.gr
Chromosome 10

Linda Moses Research Center for Genetic Medicine, Children's National Medical Center, Washington, DC, USA
Rett Syndrome

Richard Mott University of Oxford, Oxford, UK
richard.mott@well.ox.ac.uk
Alignment: Statistical Significance; Quantitative Trait Loci (QTL) Mapping Methods; Smith–Waterman Algorithm

Margaretha CA van Mourik Institute of Medical Genetics, Glasgow, UK
Pregnancy Termination for Fetal Abnormality: Psychosocial Consequences

RF Mueller University of Leeds, Leeds, UK
Genetic Counseling: Consanguinity

Leonhard Müllauer University of Vienna, Vienna, Austria
leonhard.muellauer@akh-wien.ac.at
Apoptosis: Regulatory Genes and Disease

Stefan Müller University of Munich, Munich, Germany
s.mueller@lrz.uni-muenchen.de
Comparative Cytogenetics Technologies

Ulrich Müller Institut für Humangenetik, Justus-Liebig-Universität, Giessen, Germany
ulrich.mueller@humangenetik.med.uni-giessen.de
Craniofacial Abnormalities: Molecular Basis

Ulrich Müller The Scripps Research Institute, La Jolla, California, USA
umuller@scripps.edu
Cell Adhesion Molecules and Human Disorders

Werner Müller Mascheroder Weg 1, Braunschweig, Germany
wmueller@gbf.de
Expression Analysis In Vivo

Benno Müller-Hill Institute of Genetics, Cologne, Germany
muellerhill@uni-koeln.de
Chromosome 6; Nazi Scientists

Bertram Müller-Myhsok Max-Planck-Institute of Psychiatry, Munich, Germany
bmm@mpipsykl.mpg.de
Genetic Risk: Computation

Karl Munger Harvard Medical School, Boston, Massachusetts, USA
Oncogenes

Christian Munthe Gothenburg University, Gothenburg, Sweden
christian.munthe@phil.gu.se
Preimplantation Genetic Diagnosis: Ethical Aspects

Kieran C Murphy Royal College of Surgeons in Ireland, Dublin, Ireland
kmurphy@rcsi.ie
Velocardiofacial Syndrome (VCFS) and Schizophrenia

Jeffrey C Murray University of Iowa, Iowa City, Iowa, USA
Orofacial Clefting

Adele Murrell The Babraham Institute, Cambridge UK
Genomic Imprinting at the Transcriptional Level

Kiyoshi Nagai MRC Laboratory of Molecular Biology, Cambridge, UK
kn@mrc-lmb.cam.ac.uk
Spliceosome

Ramaiah Nagaraja National Institute on Aging, Baltimore, Maryland, USA
nagarajar@grc.nia.nih.gov
Artificial Chromosomes; Yeast Artificial Chromosome (YAC) Clones

Jürgen Naggert The Jackson Laboratory, Bar Harbor, Maine, USA
jkn@jax.org
Obesity: Genetics

Susan L Naylor University of Texas Health Science Center, San Antonio, Texas, USA
naylor@uthscsa.edu
Chromosome 3

James Y Nazroo University College London, UK
j.nazroo@public-health.ucl.ac.uk
Ethnic Inequalities in Health

Michael C Neale Virginia Commonwealth University, Richmond, Virginia, USA
neal@hydro.psi.vcu.edu
Quantitative Genetics; Twin Studies: Software and Algorithms

Daniel W Nebert University of Cincinnati Medical Center, Cincinnati, Ohio, USA
dan.nebert@uc.edu
Cytochrome P450 (CYP) Gene Superfamily; Drug Metabolism: Evolution

Masatoshi Nei Pennsylvania State University, University Park, Pennsylvania, USA
nxm2@psu.edu
Evolutionary Distance: Estimation

Jay Neitz Medical College of Wisconsin, Milwaukee, Wisconsin, USA
Color Vision Defects

Maureen Neitz Medical College of Wisconsin, Milwaukee, Wisconsin, USA
mneitz@mcw.edu
Color Vision Defects

Dorothy Nelkin Department of Sociology and School of Law, New York University, New York, USA
dorothy.nelkin@nyu.edu
Gene as a Cultural Icon

David Nelson Baylor College of Medicine, Houston, Texas, USA
nelson@bcm.tmc.ed
Trinucleotide Repeat Expansions: Disorders

David R Nelson University of Tennessee, Memphis, Tennessee, USA
dnelson@utmem.edu
Cytochrome P450 (CYP) Gene Superfamily

Andrea L Nestor Medical College of Ohio, Toledo, Ohio, USA
Chromosomes during Cell Division

Andrew Newman MRC Laboratory of Molecular Biology, Cambridge, UK
newman@mrc-lmb.cam.ac.uk
Splicing of pre-mRNA

Derek C Newton St Michael's Hospital and University of Toronto, Toronto, Canada
Alternative Processing: Neuronal Nitric Oxide Synthase

Richard H Nicholson Bioethics Publications Limited, London, UK
Helsinki Declaration

Henrik Nielsen University of Copenhagen, Copenhagen Denmark
hamra@imbg.ku.dk
Genetic Disorders in History and Prehistory

R Nielsen Cornell University, Ithaca, New York, USA
rn28@cornell.edu
Purifying Selection: Action on Silent Sites

Emily L Niemitz Johns Hopkins University School of Medicine, Baltimore, Massachusetts, USA
Imprinting Disorders

Arthur W Nienhuis St Jude Children's Research Hospital, Memphis, Tennessee, USA
arthur.nienhuis@stjude.org
Hemoglobin Disorders: Gene Therapy

Joel T Nigg Michigan State University, East Lansing, Michigan, USA
nigg@msu.edu
Personality and Temperament

Leo Nijtmans Nijmegen University Medical Centre, Nijmegen, The Netherlands
Oxidative Phosphorylation (OXPHOS) System: Nuclear Genes and Genetic Disease

Irmgard Nippert Westfälische Wilhelms-Universität, Münster Germany
nippert@uni-muenster.de
Genetics in Contemporary Germany

Jan A Nolta Childrens Hospital Los Angeles, California, USA
Hematopoietic Stem Cells

Masaru Nonaka University of Tokyo, Tokyo, Japan
mnonaka@biol.s.u-tokyo.ac.jp
Complement System: Evolution

Martin Norin Biovitrum AB, Stockholm, Sweden
martin.norin@biovitrum.com
Protein Structure: Prediction and Modeling; Structural Proteomics: Large-scale Studies

Jennifer L Northrop Baylor College of Medicine, Houston, Texas, USA
Mouse N-ethyl-N-nitrosourea (ENU) Mutagenesis

Hope Northrup University of Texas Medical School at Houston, Houston, Texas, USA
hope.northup@utn.tmc.edu
Neural Tube Defects: Genetics

Neil J Nosworthy University of Sydney, Sydney, Australia
neiln@anatomy.usyd.edu.au
Proteomics and Genomics Technology: Comparison in Heart Failure and Apoptosis

Uri Nudel Weizmann Institute of Science, Rehovot, Israel
uri.nudel@weizmann.ac.il
Alternative Promoters: Duchenne Muscular Dystrophy (DMD) Gene

John I Nurnberger Jr Indiana University School of Medicine, Indianapolis, Indiana, USA
Alcoholism: Collaborative Study on the Genetics of Alcoholism (COGA)

Anne C Nye University of Illinois, Urbana, Illinois, USA
Chromosomes and Chromatin

Minna Nyström-Lahti University of Helsinki, Helsinki, Finland
Mismatch Repair Genes

Michael O'Donovan   University of Wales College of Medicine, Cardiff, UK
odonovanmc@cardiff.ac.uk
Anticipation

William S Oetting   University of Minnesota, Minneapolis, Minnesota, USA
oetti001@umn.edu
Albinism: Genetics

Tomoko Ohta   National Institute of Genetics, Mishima, Japan
tohta@lab.nig.ac.jp
Gene Families: Formation and Evolution; Gene Families: Multigene Families and Superfamilies; Kimura, Motoo

John Michael Old   Oxford Haemophilia Centre, The Churchill Hospital, Oxford, UK
john.old@orh.nhs.uk
Globin Genes: Polymorphic Variants and Mutations

Lorraine Olendzenski   University of Connecticut, Storrs, Connecticut, USA
Orthologs, Paralogs, and Xenologs in Human and Other Genomes

Andre Mascarenhas Oliveira   Brigham and Women's Hospital, Boston, Massachusetts, USA
aoliveira@rics.bwh.harvard.edu
Translocation Breakpoints in Cancer

Anne S Olsen   DOE Joint Genome Institute, Walnut Creek, California, USA
olsen2@llnl.gov
Chromosome 19

Mark OJ Olson   University of Mississippi Medical Center, Jackson, Mississippi, USA
molson@biochem.umsmed.edu
Nucleolus: Structure and Function

Gert-Jan B van Ommen   Leiden University Medical Center, Leiden, The Netherlands
gjvo@lumc.nl
Expression Studies; Human Genome Project, HUGO, and Future Health Care

Jane M Olson   Case Western Reserve University, Cleveland, Ohio, USA
olson@darwin.cwru.edu
Relatives-based Nonparametric Analysis of Linkage

Dennis H O'Neil   Palomar College, San Marcos, California, USA
doneil@palomar.edu
Nonrandom Mating

Shao-En Ong   University of Southern Denmark, Odense Denmark
shaoen@bmb.sdu.dk
Two-Dimensional Gel Electrophoresis

Fumio Oosawa   Aichi Institute of Technology, Aichi, Japan
Molecular Machines and Human Disorders

Bernard A van Oost   Utrecht University, Utrecht, The Netherlands
Representational Difference Analysis

Hunter O'Reilly   University of Wisconsin Parkside, Kenosha, Wisconsin, USA
Art and Genetics

David M Ornitz   Washington University Medical School, St Louis, Missouri, USA
dornitz@molecool.wustl.edu
Fibroblast Growth Factors: Evolution

Torben F Orntoft   Aarhus University Hospital, Skejby, Denmark
orntoft@kba.sks.au.dk
Bladder Cancer; DNA Chips and Microarrays

Morten Østergaard   University of Aarhus, Denmark
moej@biokemi.au.dk
Transcriptomics and Proteomics: Integration

Eric M Ostertag   University of Pennsylvania School of Medicine, Philadelphia, Pennsylvania, USA
ostertag@mail.med.upenn.edu
Retrotransposition and Human Disorders

Harry Ostrer   New York University School of Medicine, New York, New York, USA
harry.ostrer@med.nyu.edu
Male Sex Determination: Genetics

Jurg Ott   Rockefeller University, New York, New York, USA
ott@rockefeller.edu
Genetic Linkage Mapping

Daniel Otte   The Academy of Natural Sciences, Philadelphia, Pennsylvania, USA
otte@acnatsci.org
Species and Speciation

Sarah P Otto   University of British Columbia, Vancouver, Canada
otto@zoology.ubc.ca
Fixation Probabilities and Times

Michel M Ouellette   University of Nebraska Medical Center, Omaha, Nebraska, USA
mouellet@unmc.edu
Telomeres and Telomerase in Aging and Cancer

Igor Ovchinnikov   Columbia University, New York, New York, USA
Neanderthal DNA

Gareth I Owen   Centro de Genómica y Bioinformática, Pontificia Universidad Catolica de Chile, Santiago, Chile
gowen@bio.puc.cl
Nuclear Receptor Genes: Evolution

Michael J Owen   University of Wales College of Medicine, Cardiff, UK
owenmj@cardiff.ac.uk
Schizophrenia: Molecular Genetics

Norio Ozaki   Fujita Health University School of Medicine, Toyoake, Japan
nozaki@fujita-hu.ac.jp
Psychopharmacogenetics

Seiji Ozawa   Gunma University School of Medicine, Gunma, Japan
Glutamate Receptors

Laurie J Ozelius   Albert Einstein College of Medicine, Bronx, New York, USA
Chromosome 9

Richard A Padgett   Cleveland Clinic Foundation, Cleveland, Ohio, USA
padgetr@ccf.org
Splice Sites

Roberta A Pagon   University of Washington, Seattle, Washington, USA
bpagon@u.washington.edu
Databases in Genetics Clinics

Thomas Paiss   University of Ulm, Ulm, Germany
Prostate Cancer

Roberta Palmour   McGill University, Montreal, Canada
Schizophrenia and Bipolar Disorder: Linkage on Chromosomes 5 and 11

Akhilesh Pandey   Johns Hopkins University School of Medicine, Baltimore, Maryland, USA
pandey@jhmi.edu
Mass Spectrometry in Protein Characterization; Proteomics: A Molecular Scanner in Medicine

Fernando Pardo-Manuel de Villena   University of North Carolina at Chapel Hill, Chapel Hill, North Carolina, USA
fernando@med.unc.edu
Karyotype Evolution

Timothy J Parnell   University of Iowa, Iowa City, Iowa, USA
Chromosomes: Higher-order Organization; Position Effect Variegation in Human Genetic Disease

Silvio Parodi   University of Genoa and National Institute for Cancer Research, Genoa, Italy
Cell Cycle Control: Molecular Interaction Map

Elizabeth M Parry   University of Wales, Swansea, UK
Environmental Mutagenesis

James M Parry   University of Wales, Swansea, UK
j.m.parry@swansea.ac.uk
Environmental Mutagenesis

Evelyn P Parsons   University of Wales College of Medicine, Cardiff, UK
parsonsep@cf.ac.uk
Genetic Risk: Social Construction; Newborn Screening Programs

Terence Partridge   Imperial College School of Medicine, London, UK
terence.partridge@csc.mrc.ac.uk
Muscle Disease: Gene Therapy

Stefania Pasa   University of Genoa, Genoa, Italy
Cell Cycle Control: Molecular Interaction Map

Eberhard Passarge   Universitätsklinikum Essen, Essen, Germany
eberhard.passarge@uni-essen.de
Gastrointestinal Tract: Molecular Genetics of Hirschsprung Disease

Giuseppe Passarino   University of Calabria, Rende, Italy
g.passarino@unical.it
Mitochondrial DNA Polymorphisms

Annalisa Pastore   National Institute for Medical Research, London, UK
apastor@nimr.mrc.ac.uk
Protein Structure

Carlos N Pato   State University of New York, Syracuse, New York, USA
cmpato@aol.com
Genetic Isolates and Behavioral Gene Searches

Michele T Pato   State University of New York, Syracuse, New York, USA
cmpato@aol.com
Genetic Isolates and Behavioral Gene Searches

George P Patrinos   Erasmus University Medical Center, Rotterdam, The Netherlands
g.patrinos@erasmusmc.nl
Gene Conversion in Health and Disease

Cam Patterson   University of North Carolina, Chapel Hill, North Carolina, USA
cpatters@med.unc.edu
Atherosclerosis

László Patthy   Hungarian Academy of Sciences, Budapest, Hungary
patthy@enzim.hu
Alternative Splicing: Evolution; Exons and Protein Modules; Exons: Shuffling; Introns: Phase Compatibility

Margie M Paz   Iowa State University, Ames, Iowa, USA
mmpaz@iastate.edu
Genetic and Physical Map Correlation

William R Pearson   University of Virginia, Charlottesville, Virginia, USA
wrp@virginia.edu
FASTA Algorithm; Similarity Search

Thoru Pederson   University of Massachusetts Medical School, Worcester, Massachusetts, USA
mRNA: Intranuclear Transport

Manuel C Peitsch   Novartis Pharma AG, Basel & Swiss Institute of Bioinformatics, Basel, Switzerland
manuel.peitsch@pharma.novartis.com
Protein Homology Modeling

Päivi Peltomäki   University of Helsinki, Helsinki, Finland
paivi.peltomaki@helsinki.fi
Mismatch Repair Genes

Leena Peltonen   University of California at Los Angeles, Los Angeles, California, USA and Department of Medical Genetics, University of Helsinki, Finland
lpeltonen@mednet.ucla.edu
Ethnicity and Disease

Sharron G Penn   Aeomica Inc., Sunnyvale, California, USA
sharron.penn@amersham.com
Microarrays: Use in Gene Identification

William D Pennie   Pfizer Drug Safety Evaluation, Groton, Connecticut, USA
william_d_pennie@groton.pfizer.com
Microarrays in Toxicological Research

David Penny   Massey University, Palmerston North, New Zealand
d.penny@massey.ac.nz
Mammalian Phylogeny

Derek A Persons   St Jude Children's Research Hospital, Memphis, Tennessee, USA
derek.persons@stjude.org
Hemoglobin Disorders: Gene Therapy

Beth N Peshkin   Georgetown University, Washington, DC, USA
peshkinb@georgetown.edu
Familial Breast Cancer: Genetic Testing

Graziano Pesole   Università Milano, Milano, Italy
graziano.pesole@unimi.it
mRNA Untranslated Regions (UTRs); 5' UTRs and Regulation

Tatyana V Pestova   Moscow State University, Moscow, Russia and State University of New York, Brooklyn, New York, USA
tatyana.pestova@downstate.edu
Translation Initiation: Molecular Mechanisms in Eukaryotes

Jan-Michael Peters   Research Institute of Molecular Pathology, Vienna, Austria
peters@imp.univie.ac.at
Meiosis and Mitosis: Molecular Control of Chromosome Separation

Alan Petersen   University of Plymouth, Plymouth, UK
a.petersen@plymouth.ac.uk
Dolly and Polly

Christine Petit   Pasteur Institute, Paris, France
cpetit@pasteur.fr
Deafness: Hereditary

Arturas Petronis   Centre for Addiction and Mental Health, Toronto, Canada
arturas_petronis@camh.net
Epigenetics: Influence on Behavioral Disorders

Gerd P Pfeifer   Beckman Research Institute of the City of Hope, Duarte, California, USA
gpfeifer@coh.org
DNA Methylation and Mutation

Daniel Phaneuf   Laval University, Sainte-Foy, Canada
Schizophrenia and Bipolar Disorder: Linkage on Chromosomes 5 and 11

John A Phillips III   Vanderbilt University Medical Center, Nashville, Tennessee, USA
john.phillips@vanderbilt.edu
Endocrine Disorders: Hereditary; Growth Disorders: Hereditary

Matthew Phillips   Massey University, Palmerston North, New Zealand
Mammalian Phylogeny

Tamara J Phillips   Oregon Health & Science University and Veterans Affairs Medical Center, Portland, Oregon, USA
phillipt@ohsu.edu
Addiction and Genes: Animal Models

Craig S Pikaard   Washington University, St Louis, Missouri, USA
pikaard@biology.wustl.edu
Nucleolar Dominance

Daniel Pinkel   University of California San Francisco, San Francisco, California, USA
pinkel@cc.ucsf.edu
Comparative Genomic Hybridization in the Study of Human Disease

Derek E Piper   Howard Hughes Medical Institute and the Johns Hopkins University School of Medicine, Baltimore, Maryland, USA
Crystallization of Protein–DNA Complexes

Sinikka Pirinen   University of Helsinki, Helsinki, Finland
Dental Anomalies: Genetics

Lila Pirkkala   Hormos Medical Corp., Turku, Finland
lila.pirkkala@hormos-med.com
Heat Shock Proteins (HSPs): Structure, Function and Genetics

Robert Plomin   King's College, London, UK
r.plomin@iop.kcl.ac.uk
Intelligence and Cognition

Olivier Poch   Institut de Génétique et de Biologie Moléculaire et Cellulaire, Strasbourg, France
poch@igbmc.u-strasbg.fr
Multiple Alignment; Sequence Alignment

Flemming Pociot   Steno Diabetes Center, Gentofte, Denmark
Insulin-dependent Diabetes Mellitus (IDDM): Identifying the Disease-causing Gene at the IDDM11 Locus

Vanda Pogačić   Friedrich Miescher Institut, Basel, Switzerland
snoRNPs

Michael F Pogue-Geile   University of Pittsburgh, Pittsburgh, Pennsylvania, USA
mfpg@pitt.edu
Schizophrenia Spectrum Disorders

Joan C Politz University of Massachusetts Medical School, Worcester, Massachusetts, USA
joan.politz@umassmed.edu
mRNA: Intranuclear Transport

Martin Pollard US DOE Joint Genome Institute, Walnut Creek, California, USA
mjpollard@lbl.gov
Robotics and Automation in Molecular Genetics

Nicolas Pollet CNRS and Université Paris-Sud, Orsay, France
nicolas.pollet@ibaic.u-psud.fr
Chromosome 20

Pamela M Pollock National Institutes of Health, Bethesda, Maryland, USA
ppollock@tgen.org
Melanoma: Genetics

Daniel A Pomeranz Krummel MRC Laboratory of Molecular Biology, Cambridge, UK
dapk@mrc-lmb.cam.ac.uk
Spliceosome

Yves Pommier National Cancer Institute, Bethesda, Maryland, USA
nicolas.pollet@ibaic.u-psud.frl
Cell Cycle Control: Molecular Interaction Map

Joseph C Poole Medical College of Georgia, Augusta, Georgia, USA
Telomerase: Structure and Function

Bernice Porjesz SUNY HSC, Brooklyn, New York, USA
Alcoholism: Collaborative Study on the Genetics of Alcoholism (COGA)

Carol Portwine The Hospital for Sick Children, Toronto, Canada
Li–Fraumeni Syndrome

M-C Potier CNRS UMR 7637, Paris, France
marie-claude.potier@espci.fr
Chromosome-specific Repeats (Low-copy repeats)

Sue Povey University College London, London, UK
sue@galton.ucl.ac.uk
Human Gene Nomenclature

Chitra Prasad Health Sciences Centre, Winnipeg, Canada
XYY Syndrome

Carol A Prescott Virginia Commonwealth University, Richmond, Virginia, USA
cprescot@hsc.vcu.edu
Alcoholism and Drug Addictions

Carolyn M Price University of Cincinnati College of Medicine, Cincinnati, Ohio, USA
Telomeric and Subtelomeric Repeat Sequences

Gerald B Price McGill University, Montreal, Canada
DNA Replication

Lindsay Prior Cardiff University, Cardiff, UK
priorl@cardiff.ac.uk
Genetic Risk

Kathy Pritchard-Jones Institute of Cancer Research and Royal Marsden Hospital, Sutton, UK
kathy.pritchard-jones@icr.ac.uk
Wilms Tumor

Annie Procter Institute of Medical Genetics, Cardiff, UK
procteram@cardiff.ac.uk
Genetic Testing of Children

Teresa M Przytycka National Institute of Health, Bethesda, Maryland, USA
przytyck@ncbi.nlm.nih.gov
Hidden Markov Models

Jennifer M Puck National Institutes of Health, Bethesda, Maryland, USA
jpuck@mail.nih.gov
Apoptosis: Disorders; Immunological Disorders: Hereditary

Ann E Pulver Johns Hopkins University School of Medicine, Baltimore, Maryland, USA
aepulver@jhmi.edu
Linkage and Association Studies: Replication

Michael Andrew Quail The Wellcome Trust Sanger Institute, Cambridge, UK
mq1@sanger.ac.uk
DNA Cloning; DNA: Mechanical Breakage

Michael Quante Westfälische Wilhelms-Universität, Muenster, Germany
quante@uni-muenster.de
Genetic Enhancement

Thierry Rabilloud DRDC/BECP, CEA-Grenoble, Grenoble, France
thierry.rabilloud@cea.fr
Limits of Proteomics: Protein Solubilization Issues

Daniel J Rader University of Pennsylvania, Philadelphia, Pennsylvania, USA
rader@mail.med.upenn.edu
Familial Hypercholesterolemia: Gene Therapy

Ramji R Rajendran University of Illinois, Urbana, USA
rajendra@uiuc.edu
Chromosomes and Chromatin

T Ramasarma Indian Institute of Science, Bangalore, India
trs@biochem.iisc.ernet.in
Transmembrane Domains

Brinda K Rana University of California at San Diego, La Jolla, California, USA
bkrana@ucsd.edunschork@ucsd.edu
Genetic Epidemiology of Complex Traits

David R Rank Aeomica Inc., Sunnyvale, California, USA
Microarrays: Use in Gene Identification

Bruce Rannala University of Alberta, Edmonton, Canada
brannala@ualberta.ca
Stationary Allele Frequency Distributions

DC Rao Washington University School of Medicine, St Louis, Missouri, USA
rao@wubios.wustl.edu
Path Analysis in Genetic Epidemiology

Lisa M Rasmussen Rice University, Houston, Texas, USA
and University of Alabama, Birmingham, Alabama, USA
Secular Humanism

Angel Raya Salk Institute for Biological Studies, La Jolla, California, USA
Left–Right Asymmetry in Humans

Aharon Razin Hebrew University Medical School, Jerusalem, Israel
DNA Methylation: Evolution

Sergey V Razin Institute of Gene Biology, Moscow, Russia
sergey.v.razin@usa.net
Matrix-associated Regions (MARs) and Scaffold Attachment Regions (SARs)

Andrew P Read University of Manchester, Manchester, UK
Andrew.Read@man.ac.uk
Haploinsufficiency

Robin Reed Harvard Medical School, Boston, Massachusetts, USA
robin_reed@hms.harvard.edu
mRNA Export

Simon Huw Reed University of Wales College of Medicine, Cardiff, UK
DNA Repair

Jonathan L Rees University of Edinburgh, Edinburgh, UK
jrees@staffmail.ed.ac.uk
Skin Pigmentation: Genetics

Theodore Reich Washington University School of Medicine, St Louis, Missouri, USA
Alcoholism: Collaborative Study on the Genetics of Alcoholism (COGA)

Brian J Reid Fred Hutchinson Cancer Research Center, Seattle, Washington, USA
bjr@fhcrc.org
Barrett Esophagus

Wolf Reik The Babraham Institute, Cambridge, UK
wolf.reik@bbsrc.ac.uk
Genomic Imprinting at the Transcriptional Level

Philip R Reilly Interleukin Genetics Inc., Waltham, Massachusetts, USA
preilly@ilgenetics.com
Informed Consent in Human Genetic Research

Allan L Reiss Stanford University School of Medicine, Stanford, California, USA
reiss@stanford.edu
Williams Syndrome: A Neurogenetic Model of Human Behavior

Lawrence T Reiter University of California, San Diego, California, USA
lreiter@biomail.ucsd.edu
Drosophila *as a Model for Human Diseases*

Pieter H Reitsma University of Amsterdam, Amsterdam, The Netherlands
p.h.reitsma@amc.uva.nl
Venous Thrombosis: Genetics

Cristobal G dos Remedios University of Sydney, New South Wales, Australia
Proteomics and Genomics Technology: Comparison in Heart Failure and Apoptosis

Chandra A Reynolds University of California, Riverside, California, USA
chandra.reynolds@ucr.edu
Alzheimer Disease

Tim Reynolds Queen's Hospital, Burton-on-Trent, UK
Down Syndrome: Antenatal Screening

Allen R Rhoads Howard University College of Medicine, Washington, DC, USA
arhoads@howard.edu
Pseudogenes: Age

David A Rhodes Cambridge Institute for Medical Research, Cambridge, UK
dar32@mole.bio.cam.ac.uk
Human Leukocyte Antigen (HLA) System and Human Disorders

Jonathan Rhodes University of Liverpool, Liverpool, UK
j.m.rhodes@liverpool.ac.uk
Glycosylation and Disease

John P Rice Washington University School of Medicine, St Louis, Missouri, USA
john@zork.wustl.edu
Alcoholism: Collaborative Study on the Genetics of Alcoholism (COGA)

Treva Rice Washington University School of Medicine, St Louis, Missouri, USA
treva@wubios.wustl.edu
Path Analysis in Genetic Epidemiology

Martin PM Richards University of Cambridge, Cambridge, UK
mpmr@cam.ac.uk
Heredity: Lay Understanding

Dietmar Richter University of Hamburg, Hamburg, Germany
richter@uke.uni-hamburg.de
mRNA Localization: Mechanisms

Thomas Ried National Cancer Institute, Bethesda, Maryland, USA
Cytogenetic and Physical Chromosomal Maps: Integration

Mariluce Riegel Institute of Medical Genetics, University of Zürich, Switzerland
Microdeletion Syndromes

Loren H Rieseberg Indiana University, Bloomington, Indiana, USA
lriesebe@bio.indiana.edu
Speciation: Genetics

Jeremy Rifkin The Foundation on Economic Trends, Washington, DC, USA
jrifkin@foet.org
Patenting of Genes: A Personal View

Pier Giorgio Righetti University of Verona, Verona, Italy
righetti@sci.univr.it
Capillary Electrophoresis; Preparation Artefacts in Proteomics: Technological Issues

John H Riley GlaxoSmithKline, Stevenage, UK
jhr24535@gsk.com
Susceptibility Genes: Detection

Jean-Loup Risler Génome et Informatique, Evry, France
risler@genopole.cnrs.fr
Genetic Databases: Mining

Paul Roberts St James's University Hospital, Leeds, UK
paul.roberts@leedsth.nhs.uk
Cancer Cytogenetics

Peter N Robinson Charité University Hospital, Berlin, Germany
peter.robinson@charite.de
Marfan Syndrome

Fiona M Robson St James's University Hospital, Leeds, UK
fiona.robson@leedsth.nhs.uk
Privacy: Confidentiality and Responsibility

Cécile Rochette-Egly INSERM/CNRS/ULP, Illkirch, France
cegly@igbmc.u-strasbg.fr
Transcription Regulation: Coordination

John C Rockett United States Environmental Protection Agency, Research Triangle Park, North Carolina, USA
rockett.john@epa.gov
Gene Expression Networks

Peter Roepstorff University of Southern Denmark, Odense, Denmark
roe@bmb.sdu.dk
Mass Spectrometry Instrumentation in Proteomics

MA Rogers German Cancer Research Center, Heidelberg, Germany
m.rogers@dkfz.de
Keratins and Keratin Diseases

Serge P Romana Hôpital Necker-Enfants Malades, Paris, France
serge.romana@nck.ap-hop-paris.fr
Clinical Molecular Cytogenetics

Johanna M Rommens The Hospital for Sick Children and University of Toronto, Toronto, Canada
johanna@genet.sickkids.on.ca
Cystic Fibrosis Gene: Identification

Anja Roos Leiden University Medical Centre, Leiden, The Netherlands
Complement System and Fc Receptors: Genetics

Keith Rose GeneProt Inc., Geneva, Switzerland
keith.rose@geneprot.com
Industrialization of Proteomics: Scaling Up Proteomics Processes

Steven PR Rose The Open University, Milton Keynes, UK
s.p.r.rose@open.ac.uk
Genetics, Reductionism, and Autopoiesis

Allen D Roses GlaxoSmithKline, Research Triangle Park, North Carolina, USA
christine_m_debouck@gsk.com
Microarrays in Drug Discovery and Development

Mark T Ross Wellcome Trust Sanger Institute, Hinxton, UK
mtr@sanger.ac.uk
L Isochore Map: Gene-poor Isochores

Fabio Rossi University of British Columbia, Vancouver, Canada
fabio@brc.ubc.ca
Regulatable Gene Expression in Gene Therapy

Sylvie Rouquier Institut de Génétique Humaine, CNRS, Montpellier, France
rouquier@igh.cnrs.fr
Olfactory Receptors

Lee Rowen Institute for Systems Biology, Seattle, Washington, USA
leerowen@systemsbiology.org
Gene Structure and Organization

Marc-André Roy Laval University, Sainte-Foy, Canada
Schizophrenia and Bipolar Disorder: Linkage on Chromosomes 5 and 11

Astrid M Roy-Engel Tulane University Health Sciences Center, New Orleans, Lousiana, USA
Retrosequences and Evolution of Alu Elements

Jean-Michel Rozet Hôpital Necker–Enfants Malades, Paris, France
Eye Disorders: Hereditary

Carol M Rubin University of California, Davis, California, USA
cmrubin@ucdavis.edu
Short Interspersed Elements (SINEs)

Bernardo Rudy New York University School of Medicine, New York, New York, USA
Potassium Channels

Elena I Rugarli Telethon Institute of Genetics and Medicine (TIGEM), Naples, Italy
Disease-related Genes: Functional Analysis

Michael Ruse   Florida State University, Tallahassee, Florida, USA
mruse@mailer.fsu.edu
Creationism; Darlington, Cyril Dean; Fisher, Ronald Aylmer; Hamilton, William Donald; Malthus,Thomas Robert; Wilson, Edward Osborne

Shelley J Russek   Boston University School of Medicine, Boston, Massachusetts, USA
srussek@bu.edu
γ-Aminobutyric Acid (GABA) Receptors

Stephen J Russell   Mayo Clinic, Rochester, Minnesota, USA
sjr@mayo.edu
RNA Viral Vectors in Gene Therapy

Antonio Russo   University of Palermo, Palermo, Italy
lab-oncobiologia@usa.net
Caretakers and Gatekeepers

Cecilia Saccone   University of Bari, Bari, Italy
saccone@area.ba.cnr.it
Mitochondrial Genome

Nancy L Saccone   Washington University School of Medicine, St Louis, Missouri, USA
nlims@vodka.wustl.edu
Alcoholism: Collaborative Study on the Genetics of Alcoholism (COGA)

Salvatore Saccone   University of Catania, Catania, Italy
saccosal@unict.it
Chromosomes 21 and 22: Gene Density; Gene Distribution on Human Chromosomes; GC-rich Isochores: Origin

Vaskar Saha   Institute of Cancer Research, London, UK
vaskar.saha@cancer.org.uk
Fusion Proteins and Diseases

George H Sakorafas   251 Hellenic Air Force Hospital, Athens, Greece
georgesakorafas@yahoo.com
Pancreatic Cancer

Lora S Salandanan   University of California, Los Angeles, California, USA
Williams Syndrome: A Neurogenetic Model of Human Behavior

Stanley N Salthe   City University of New York, New York, New York, USA and Binghamton University, Binghamton, New York, USA
ssalthe@binghamton.edu
Biological Complexity: Beyond the Genome; Health Care and Health Insurance in the United States

Julian R Sampson   University of Wales College of Medicine, Cardiff, UK
Tuberous Sclerosis: Genetics

R Jude Samulski   University of North Carolina at Chapel Hill, Chapel Hill, North Carolina, USA
rjs@med.unc.edu
Gene Therapy: Technology

Jean-Charles Sanchez   Geneva University Hospital, Geneva, Switzerland
Nucleolar Proteomics

Christopher L Sansam   Vanderbilt University, Nashville, Tennessee, USA
mRNA Editing

Kyriakie Sarafoglou   New York Presbyterian Hospital and Weill Medical College of Cornell University, New York, New York, USA
kys2001@med.cornell.edu
Intersex Disorders

Srikant Sarangi   Cardiff University, Cardiff, UK
sarangi@cardiff.ac.uk
Genetic Counseling Communication: A Discourse-analytical Approach

Carole A Sargent   University of Cambridge, Cambridge, UK
cas1001@hermes.cam.ac.uk
Chromosome Y

Tomo Saric   ATABIS GmbH, Cologne, Germany
tomosaric@msn.com
Protein Degradation and Turnover

Ugis Sarkans   European Molecular Biology Laboratory, Cambridge, UK
Gene Expression Databases

Sahotra Sarkar   University of Texas at Austin, Austin, Texas, USA
sarkar@mail.utexas.edu
Complex Genetic Systems and Diseases

Etsuko Satoh   University of Pittsburgh School of Pharmacy, Pittsburgh, Pennsylvania, USA
Naked DNA in Gene Therapy

Brian Sauer   Stowers Institute for Medical Research, Kansas City, Missouri, USA
bls@stowers-institute.org
Cre–lox Inducible Gene Targeting

Julian Savulescu   University of Oxford, Oxford, UK
julian.savulescu@stx.ox.ac.uk
Bioethics: Utilitarianism

Cheryl AG Scacheri   Children's National Medical Center, Washington, DC, USA
Rett Syndrome

ADJ Scadden   University of Cambridge, Cambridge, UK
RNA Processing

Peter Scambler   Institute of Child Health, London, UK
pscamble@hgmp.msc.ac.uk
DiGeorge Syndrome and Velocardiofacial Syndrome (VCFS)

Kevin Scanlon   Keck Graduate Institute, Claremont, California, USA
Antisense and Ribozymes

Sophie Schbath   Institut National de la Recherche Agronomique, Jouy-en-Josas, France
sophie.schbath@jouy.inra.fr
Statistical Methods in Physical Mapping

Harold A Scheraga   Cornell University, Ithaca, New York, USA
has5@cornell.edu
Protein Folding Pathways

Stephen W Scherer   The Hospital for Sick Children, Toronto, Canada
steve@genet.sickkids.on.ca
Chromosome 7

Alexander Scherl   Geneva University Hospital, Geneva, Switzerland
Nucleolar Proteomics

Lothar Schermelleh   University of Munich, Munich, Germany
Cell Cycle: Chromosomal Organization

Harry Scherthan   University of Kaiserslautern, Kaiserslautern, Germany and MPI for Molecular Genetics, Berlin, Germany
scherth@uni.de
Chromosome Numbers in Mammals

Brigitte Scheucher   University of Vienna, Vienna, Austria
Apoptosis: Regulatory Genes and Disease

Kim M Schindler   State University of New York at Buffalo, Buffalo, New York, USA
Genetic Isolates and Behavioral Gene Searches

Albert Schinzel   University of Zurich, Zurich, Switzerland
schinzel@medgen.unizh.ch
Microdeletion Syndromes; Uniparental Disomy

Alexander Schleiffer   Research Institute of Molecular Pathology, Vienna, Austria
Pax Genes: Evolution and Function

David Schlessinger   National Institute on Aging, Baltimore, Maryland, USA
schlessingerd@grc.nia.nih.gov
Artificial Chromosomes; Yeast Artificial Chromosome (YAC) Clones

Christian Schlötterer   Veterinärmedizinische Universität Wien, Vienna, Austria
christian.schloetterer@vu-wien.ac.at
Simple Repeats

Henning Schmalbruch   University of Copenhagen, Copenhagen, Denmark
h.schmalbruch@mfi.ku.dk
Spinal Muscular Atrophy

Carl W Schmid   University of California, Davis, California, USA
cwschmid@ucdavis.edu
Short Interspersed Elements (SINEs)

Jörg Schmidtke   Hannover Medical School, Hannover, Germany
schmidtke.joerg@mh-hannover.de
Presymptomatic Diagnosis

Chris Schmutte Thomas Jefferson University, Philadelphia, Pennsylvania, USA
cschmutte@mail.jci.tju.edu
Chromosome Instability (CIN) in Cancer

Ulrich Schneider Scienion AG, Berlin, Germany
Array-based Proteomics

Wolfgang Johann Schneider University of Vienna, Vienna, Austria
wjs@mol.univie.ac.at
Low-density Lipoprotein Receptor (LDLR) Family: Genetics and Evolution

Peter H Schonemann Purdue University, West Lafayette, Indiana, USA
phs@psych.purdue.edu
Heritability

Nicholas J Schork University of California at San Diego, La Jolla, California, USA
Genetic Epidemiology of Complex Traits

Marc Schuckit University of California at San Diego, San Diego, California, USA
mschuckit@ucsd.edu
Alcoholism: Collaborative Study on the Genetics of Alcoholism (COGA)

Torsten Schwede Biozentrum, Universität Basel & Swiss Institute of Bioinformatics, Basel, Switzerland
ts80960@glaxowellcome.co.uk
Protein Homology Modeling

Charles R Scriver McGill University, Montreal, Canada
charles.scriver@mcgill.ca
Allelic and Locus Heterogeneity; Carrier Screening of Adolescents in Montreal

Jackie Leach Scully University of Basel, Basel, Switzerland
scully@bluewin.ch
Disability: Stigma and Discrimination

Redha Sekhri Institut Andre Lwoff, Villejuif, France
Histone Acetylation and Disease

Peter J Selby Cancer Research UK, Leeds, UK
Laser Capture Microdissection in Proteomics

Sara Selig Technion, Haifa, Israel
seligs@technunix.technion.ac.il
Synchronous and Asynchronous Replication

James R Sellers National Heart, Lung and Blood Institute, Bethesda, Maryland, USA
jsellers@helix.nih.gov
Myosins

Gregg L Semenza The Johns Hopkins University School of Medicine, Baltimore, Maryland, USA
gsemenza@jhmi.edu
Transcription Factors and Human Disorders

Christopher Semsarian Centenary Institute, Newtown, Australia
c.semsarian@centenary.usyd.edu.au
Hypertrophic Cardiomyopathy

Lisa G Shaffer Washington State University, Spokane, Washington, USA
lshaffer@wsu.edu
Karyotype Interpretation

Tamim H Shaikh The Children's Hospital of Philadelphia, Philadelphia, Pennsylvania, USA
shaikh@email.chop.edu
Chromosome 22; Segmental Duplications and Genetic Disease

Tom Shakespeare University of Newcastle upon Tyne, Newcastle upon Tyne, UK
t.w.shakespeare@ncl.ac.uk
Disability, Human Rights, and Contemporary Genetics

Pak Sham Institute of Psychiatry, London, UK
p.sham@iop.kcl.ac.uk
Heritability Wars

Mark Shannon Aeomica Inc., Sunnyvale, California, USA
Microarrays: Use in Gene Identification

Neil F Sharpe Genetic Testing Research Group, Hamilton, Canada
Human Genetics: Ethical Issues and Social Impact

Alison Shaw Brunel University, Uxbridge UK
alison.shaw@brunel.ac.uk
Genetic Counseling for Muslim Families of Pakistani and Bangladeshi Origin in Britain

Ruth Shemer Hebrew University Medical School, Jerusalem, Israel
DNA Methylation: Evolution

Kathleen IJ Shennan University of Aberdeen, Aberdeen, UK
k.i.shennan@abdn.ac.uk
Protein Secretory Pathways

Brad Sherman Griffith University, Brisbane and Australian National University, Canberra, Australia
Patent Issues in Biotechnology

Jamie Sherman University of Washington, Seattle, Washington, USA
Quantitative Proteomics (ICAT™)

Stephen T Sherry National Center for Biotechnology Information, Bethesda, Maryland, USA
steve_sherry@nih.gov
Human Variation Databases

Tsuyoshi Shiga Imperial College, London, UK
Genomics, Pharmacogenetics, and Proteomics: An Integration

Shoshana Shiloh Tel Aviv University, Tel Aviv, Israel
shoshi@freud.tau.ac.il
Genetic Counseling: Psychological Models in Research and Practice

Randy C Shoemaker Iowa State University, Ames, Iowa, USA
Genetic and Physical Map Correlation

Ravi Shridhar Wayne State University, Detroit, Michigan, USA
rs0421@hotmail.com
Cystatins

Zahava Siegfried The Hebrew University of Jerusalem, Jerusalem, Israel
Dopamine D4 Receptor (DRD4) Box Score

Karol Sikora Hammersmith Hospital, London, UK
karol.sikora@astrazeneca.com
Gene Therapy

Gary A Silverman Harvard Medical School, Boston, Massachusetts, USA
gary.silverman@tch.harvard.edu
Chromosome 18; Ovalbumin Serpins; Serpins: Evolution

Valerie Sinason St George's Hospital Medical School, London, UK
Mentally Handicapped in Britain: Sexuality and Procreation

Lea Sistonen Åbo Akademi University, Turku, Finland
Heat Shock Proteins (HSPs): Structure, Function, and Genetics

Johanna Sjöström Helsinki University Hospital, Helsinki, Finland
johanna.sjostrom@hus.fi
Apoptosis and the Cell Cycle in Human Disease

Claus Skaanning Dezide, Aalborg, Denmark
claus.skaanning@dezide.com
Markov Chain Monte Carlo Methods

Shalini Reshmi Skarja University of Pittsburgh, Pittsburgh, Pennsylvania, USA
Mitosis: Chromosomal Rearrangements

Loane Skene University of Melbourne, Parkville, Australia
l.skene@unimelb.edu.au
Stored Genetic Material: Use in Research

Heather Skirton University of Wales College of Medicine, Cardiff, UK
heather.skirton@tst.nhs.uk
Genetic Counseling Profession in Europe; Genetic Counseling Services: Outcomes

David H Skuse Institute of Child Health, London, UK
Brain: Neurodevelopmental Genetics

Susan A Slaugenhaupt Harvard Institute of Human Genetics, Boston, Massachusetts, USA
slaugenhaupt@helix.mgh.harvard.edu
Chromosome 9

Bonnie F Sloane Wayne State University, Detroit, Michigan, USA
bsloane@med.wayne.edu
Cystatins

Stephen T Smale University of California, Los Angeles, California, USA
smale@mednet.ucla.edu
Promoters

Jan AM Smeitink   Nijmegen University Medical Centre, Nijmegen, The Netherlands
j.smeitink@cukz.umcn.nl
Oxidative Phosphorylation (OXPHOS) System: Nuclear Genes and Genetic Disease

Arian FA Smit   The Institute for Systems Biology, Seattle, Washington, USA
Repetitive Elements: Detection

Christopher WJ Smith   University of Cambridge, Cambridge, UK
cwjs1@mole.bio.cam.ac.uk
RNA Processing

Shelley D Smith   University of Nebraska Medical Center, Omaha, Nebraska, USA
ssmith@unmc.edu
Reading and Dyslexias

Harold Snieder   Georgia Prevention Institute, Augusta, Georgia, USA
hsnieder@mcg.edu
Twin Methodology

Irina Solovei   University of Munich, Munich, Germany
irina.solovei@lrz.uni-muenchen.de
Cell Cycle: Chromosomal Organization

Stefan Somlo   Yale University School of Medicine, New Haven, Connecticut, USA
stefan.somlo@yale.edu
Kidney: Hereditary Disorders

Ingolf Sommer   Max-Planck Institute for Informatics, Saarbrücken, Germany
Protein Structure Prediction and Databases

Ann Sommerville   British Medical Association, London, UK
Privacy and Genetic Information

Christopher Southan   Oxford GlycoSciences Ltd, Abingdon, UK
chris.southan@ogs.co.uk
Proteases: Evolution

Demetrios A Spandidos   University of Crete, Crete, Greece
spandidos@spandidos.gr
Oncogenic Kinases in Cancer

Michael R Speicher   Technical University Munich and GSF-Neuherberg Munich, Germany
speicher@humangenetik.med.tu-muenchen.de
Chromosome; Fluorescence In Situ Hybridization (FISH) Techniques

Peter J van der Spek   Johnson & Johnson Pharmaceutical Research & Development, Beerse, Belgium pvdspek@prdbe.jnj.com
Microarray Bioinformatics

Jennifer M Spence   University of Cambridge, Cambridge, UK
Mammalian Artificial Chromosomes (MACs)

Hamish G Spencer   University of Otago, Dunedin, New Zealand
h.spencer@otago.ac.nz
Imprinting

Allen M Spiegel   National Institutes of Health, Bethesda, Maryland, USA
G-protein-mediated Signal Transduction and Human Disorders

Richard S Spielman   University of Pennsylvania School of Medicine, Philadelphia, Pennsylvania, USA
spielman@pobox.upenn.edu
Relatives-based Test for Linkage Disequilibrium: The Transmission/Disequilibrium Test (TDT)

Arun Srivastava   Indiana University School of Medicine, Indianapolis, Indiana, USA
asrivast@iupui.edu
Adeno-associated Viral Vectors in Gene Therapy

Roscoe Stanyon   National Cancer Institute, Frederick, Maryland, USA
stanyonr@ncifcrf.gov
Chromosomal Rearrangements in Primates

Andrzej Stasiak   University of Lausanne, Lausanne-Dorigny, Switzerland
andrzej.stasiak@lau.unil.ch
Topoisomerases

Hans H Stassen   Psychiatric University Hospital Zürich, Zürich, Switzerland
k454910@bli.unizh.ch
Electroencephalogram (EEG) and Evoked Potentials

Stephen C Stearns   Yale University, New Haven, Connecticut USA
stephen.stearns@yale.edu
Evolutionary Thinking in the Medical Sciences

H Thomas Steely Jr   Alcon Research Ltd, Fort Worth, Texas, USA
thomas.steely@alconlabs.com
Eye: Proteomics

Hanno Steen   Harvard Medical School, Boston, Massachusetts, USA
hanno_steen@hms.harvard.edu
Mass Spectrometry in Protein Characterization; Proteomics: A Molecular Scanner in Medicine

Kari Stefansson   deCODE Genetics, Reykjavik, Iceland
kstefans@decode.is
deCODE: A Genealogical Approach to Human Genetics in Iceland

Peter Steinbach   University of Ulm, Ulm, Germany
peter.steinbach@medizin.uni-ulm.de
X-linked Genes for General Cognitive Abilities

Martin H Steinberg   Boston University School of Medicine, Boston, Massachusetts, USA
msteinberg@medicine.bu.edu
Sickle Cell Disease as a Multifactorial Condition

Joan A Steitz   Yale University School of Medicine, New Haven, Connecticut, USA
js228@email.med.yale.edu
Gas5 Gene

Gunther S Stent   University of California, Berkeley, California, USA
stent@uclink4.berkeley.edu
Beyond the Genome

W Stephan   University of Munich, Munich, Germany
Multigene Families: Evolution

Richard von Sternberg   National Museum of Natural History, Washington, DC, USA
and National Center for Biotechnology Information (GenBank), Bethesda, Maryland, USA
sternber@ncbi.nlm.nih.gov
Biological Complexity: Beyond the Genome

Caro-Beth Stewart   State University of New York at Albany, Albany, New York, USA
cstewart@albany.edu
Evolution: Convergent and Parallel

Alan Stockdale   Center for Applied Ethics and Professional Practice, Newton, Massachusetts, USA
astockdale@edc.org
Gene Therapy: Motivations for Research

Mark Stoneking   Max Planck Institute for Evolutionary Anthropology, Leipzig, Germany
stoneking@eva.mpg.de
Mitochondrial Genome: Evolution

George A Stouffer   University of North Carolina, Chapel Hill, North Carolina, USA
Atherosclerosis

Marilyn Strathern   University of Cambridge, Cambridge, UK
ms10026@cam.ac.uk
Kinship in Flux

Christian Stratowa   Boehringer Ingelheim Austria, Vienna, Austria
christian.stratowa@vie.boehringer-ingelheim.com
Microarrays in Disease Diagnosis and Prognosis

Konstantin Strauch   University of Bonn, Bonn Germany
strauch@uni-bonn.de
Multilocus Linkage Analysis

Robert L Strausberg   National Cancer Institute, Bethesda, Maryland, USA
rls@nih.gov
Cancer Genome Anatomy Project

Eddy Street   University of Wales College of Medicine, Cardiff, UK
streetec@cf.ac.uk
Confronting Genetic Disease: Psychological Issues; Genetic Counseling: Impact on the Family System

Christian Stricker   Applied Genetics Network, Altendorf, Switzerland
stricker@genetics-network.ch
Segregation Analysis Software

Carson Strong   University of Tennessee College of Medicine, Memphis, Tennessee, USA
cstrong@utmem.edu
Eugenics

Andrew Stubbs   Johnson & Johnson Pharmaceutical Research & Development, Beerse, Belgium
Microarray Bioinformatics

Michael PH Stumpf   University College London, London, UK
Variable Drug Response: Genetic Evaluation

André Stutz   University of Geneva, Geneva, Switzerland
3′ UTR Mutations and Human Disorders

Kathleen E Sullivan   University of Pennsylvania School of Medicine, Philadelphia, Pennsylvania, USA
Complement: Deficiency Diseases

Michael Sundström   Biovitrum AB, Stockholm, Sweden
Structural Proteomics: Large-scale Studies

Andrea Superti-Furga   University of Lausanne, Lausanne, Switzerland
asuperti@chuv.unil.ch
Skeletal Dysplasias: Genetics

Grant R Sutherland   Women's and Children's Hospital, Adelaide, Australia
grant.sutherland@adelaide.edu.au
Fragile Sites

Clive Svendsen   University of Wisconsin Madison, Madison, Wisconsin, USA
svendsen@waisman.wisc.edu
Neural Stem Cells

Eugene Sverdlov   Shemyakin–Ovchinnikov Institute of Bioorganic Chemistry, Moscow, Russia
edsver@freemail.ru
Primer Walking

James E Sylvester   Nemours Children's Clinic, Jacksonville, Florida, USA
rRNA Genes: Evolution

Ann-Christine Syvänen   Uppsala University, Uppsala, Sweden
ann-christine.syvanen@medsci.uu.se
Microarrays: Use in Mutation Detection

Moshe Szyf   McGill University, Montreal, Canada
mszyf@pharma.mcgill.ca
DNA Methylation: Enzymology

Shunichi Takeda   Kyoto University, Kyoto, Japan
stakeda@rg.med.kyoto-u.ac.jp
DNA Recombination

Hans J Tanke   Leiden University Medical Center, Leiden, The Netherlands
h.j.tanke@lumc.nl
Expression Analysis In Vivo*: Cell Systems; Fluorescence Microscopy*

Rudolph E Tanzi   Massachusetts General Hospital and Harvard Medical School, Charlestown, Massachusetts, USA
tanzi@helix.mgh.harvard.edu
Protein Aggregation and Human Disorders

Peter Tarczy-Hornoch   University of Washington, Seattle, Washington, USA
Databases in Genetics Clinics

Marco Tartaglia   Istituto Superiore di Sanità, Rome, Italy
Noonan Syndrome

Simon Tavaré   University of Southern California, Los Angeles, California, USA
stavare@usc.edu
Coalescent Theory

Marion Dawn Teare   University of Sheffield, Sheffield, UK
m.d.teare@sheffield.ac.uk
Model-based Linkage Analysis

Charles Tease   University of Warwick, Coventry, UK
ctease@bio.warwick.ac.uk
Genetic Mapping: Comparison of Direct and Indirect Approaches; Genetic Maps: Direct Meiotic Analysis; Meiosis

Sara Teter   University of California at Davis, Davis, California, USA
Protein Folding In Vivo

Irma Thesleff   University of Helsinki, Helsinki, Finland
irma.thesleff@helsinki. fi
Dental Anomalies: Genetics

Russell S Thomas   Aeomica Inc., Sunnyvale, California, USA
Microarrays: Use in Gene Identification

Elizabeth A Thompson   University of Washington, Seattle, Washington, USA
thompson@stat.washington.edu
Lod Score

Julie D Thompson   Institut de Génétique et de Biologie Moléculaire et Cellulaire, Strasbourg, France
julie@igbmc.u-strasbg.fr
Multiple Alignment; Sequence Alignment

Elizabeth J Thomson   National Human Genome Research Institute, Bethesda, Maryland, USA
ELSI Research Program of the National Human Genome Research Institute (NHGRI)

Aad Tibben   Leiden University Medical Centre, Leiden, The Netherlands
a.tibben@lumc.nl
Predictive Genetic Testing

Randal S Tibbetts   The University of Wisconsin, Madison, Wisconsin, USA
rstibbetts@wisc.edu
DNA Damage Response

Randy Todd   Massachusetts General Hospital/Harvard School of Dental Medicine, Boston, Massachusetts, USA
randy_todd@hms.harvard.edu
Oncogenes

Richard D Todd   Washington University School of Medicine, St Louis, Missouri, USA
toddr@psychiatry.wustl.edu
Attention Deficit–Hyperactivity Disorder (ADHD)

Alexandre A Todorov   Washington University School of Medicine, St Louis, Missouri, USA
todorov@matlock.wustl.edu
Homozygosity Mapping

Trygve O Tollefsbol   University of Alabama at Birmingham, Birmingham, Alabama, USA
trygve@uab.edu
Telomerase: Structure and Function

David Tollervey   University of Edinburgh, Edinburgh, UK
d.tollervey@ed.ac.uk
Nonprotein-coding Genes

J Bruce Tomblin   University of Iowa, Iowa City, Iowa, USA
j-tomblin@uiowa.edu
Language and Genes

Ian PM Tomlinson   Cancer Research UK, London, UK
ian.tomlinson@cancer.org.uk
Hamartomatous Polyposis Syndromes: Peutz–Jeghers Syndrome and Familial Juvenile Polyposis

Ella Tour   University of California, San Diego, California, USA
etour@biomail.ucsd.edu
Hox *Genes and Body Plan: Evolution*

Margaret Town   Imperial College, London, UK
Genomics, Pharmacogenetics, and Proteomics: An Integration

Thomas Traut   University of North Carolina School of Medicine, Chapel Hill, North Carolina, USA
tom_traut@unc.edu
Multidomain Proteins

Richard C Trembath   University of Leicester, Leicester, UK
rtrembat@hgmp.mrc.ac.uk
Activating and Inactivating Mutations in the GNAS1 *Gene*

Edward N Trifonov   University of Haifa, Haifa, Israel
trifonov@research.haifa.ac.il
Genetic Code: Evolution

Leendert A Trouw   Leiden University Medical Centre, Leiden, The Netherlands
Complement System and Fc Receptors: Genetics

Susan L Trumbetta   Vassar College, Poughkeepsie, New York, USA
trumbetta@vassar.edu
Twins Studies

Christos Tsatsanis   University of Crete, Crete, Greece
tsatsani@med.uoc.gr
Oncogenic Kinases in Cancer

Willian T Tse   Children's Hospital, Harvard Medical School, Boston, Massachusetts, USA
william.tse@tch.harvard.edu
Erythrocyte Membrane Disorders

Masako Tsubakihara   The University of Sydney, Sydney, Australia
Proteomics and Genomics Technology: Comparison in Heart Failure and Apoptosis

Lap-Chee Tsui   The University of Hong Kong, Hong Kong SAR, China
tsuilc@hkucc.hku.hk
Cystic Fibrosis (CF)

Keisuke Tsuzuki   Gunma University School of Medicine, Gunma, Japan
tsuzuki@med.gunma-u.ac.jp
Glutamate Receptors

Abigail Saffron Tucker   King's College London, London, UK
abigail.tucker@kcl.ac.uk
Tooth Morphogenesis and Patterning: Molecular Genetics

Edward GD Tuddenham   MRC Clinical Sciences Centre, London, UK
edward.tuddenham@csc.mrc.ac.uk
Hemophilia and Other Bleeding Disorders: Genetics

Burkhard Tümmler   Medizinische Hochschule Hannover, Hannover, Germany
tuemmler.burkhard@mh_hannover.de
Genomic DNA: Purification; Restriction Mapping

Mitchell S Turker   Oregon Health and Science University, Portland, Oregon, USA
turkerm@ohsu.edu
DNA Methylation in Development

Jon Turney   University College London, London, UK
j.turney@ucl.ac.uk
Public Understanding of Genetics: The Deficit Model

Damian Twerenbold   Université Neuchâtel, Neuchâtel, Switzerland and GenSpec SA, Boudry, Switzerland
twerenbold@genspec.ch
Cryogenic Detectors: Detection of Single Molecules

Chris Tyler-Smith   University of Oxford, Oxford, UK
chris@bioch.ox.ac.uk
Chromosome Y: General and Special Features

Kazuhiko Uchida   University of Tsukuba, Tsukuba Ibaraki, Japan
kzuchida@md.tsukuba.ac.jp
Gene Amplification and Cancer

Patricia Uelmen Huey   Emory University School of Medicine, Decatur, Georgia, USA
Apolipoprotein Gene Structure and Function

Sheila Unger   The Hospital for Sick Children, Toronto, Canada
Skeletal Dysplasias: Genetics

Mark C Ungerer   North Carolina State University, Raleigh, North Carolina, USA
Speciation: Genetics

Meena Upadhyaya   Institute of Medical Genetics, University of Wales College of Medicine, Cardiff, UK
upadhyaya@cardiff.ac.uk
Facioscapulohumeral Muscular Dystrophy (FSHD): Genetics

Kathleen Van Craenenbroeck   University of Gent, Gent, Belgium
DNA Viral Vectors in Gene Therapy

Joël Vandekerckhove   University of Gent, Gent, Belgium
jjoel.vandekerckhove@ugent.be
Protein Characterization: Analytical Approaches and Applications to Proteomics

Hilde Van Esch   University Hospital of Leuven, Leuven, Belgium
hilde.vanesch@med.kuleuven.ac.be
Dysmorphic Syndromes

Peter Vanhoenacker   University of Gent, Gent, Belgium
petervh@dmb.rug.ac.be
DNA Viral Vectors in Gene Therapy

Pnina Vardi   Tel Aviv University, Petah Tikva, Israel
pvardi@post.tau.ac.il
Thrifty Genotype Hypothesis and Complex Genetic Disease

Karen M Vasquez   University of Texas MD Anderson Cancer Center, Smithville, Texas, USA
kvasquez@sprd1.mdacc.tmc.edu
Gene Targeting by Homologous Recombination

Jean-Dominique Vassalli   University of Geneva, Geneva, Switzerland
jean-dominique.vassalli@medecine.unige.ch
3′ UTR Mutations and Human Disorders

George Vassaux   Imperial College at Hammersmith Hospital, London, UK
georges.vassaux@cancer.org.uk
Expression Targeting in Gene Therapy

Heleni Vastardis   Tufts University School of Dental Medicine, Boston, Massachusetts, USA
orthogr@yahoo.com
Tooth Agenesis

Timothy D Veenstra   National Cancer Institute, Frederick, Maryland, USA
Phosphoproteomics

Michel Vekemans   Hôpital Necker–Enfants Malades, Paris, France
vekemans@necker.fr
Clinical Molecular Cytogenetics; Trisomy

B Venkatesh   Institute of Molecular and Cell Biology, Singapore
mcbbv@imcb.nus.edu.sg
Fugu: *The Pufferfish Model Genome; Genome Size*

J Craig Venter   The Center for the Advancement of Genomics, Rockville, Maryland, USA
jcventer@tcag.org
Shotgunning the Human Genome: A Personal View

J Sjef Verbeek   Leiden University Medical Centre, Leiden, The Netherlands
j.s.verbeek@lumc.nl
Complement System and Fc Receptors: Genetics

Frans W Verheijen   Erasmus University Medical Centre, Rotterdam, The Netherlands
f.verheijen@erasmusmc.nl
Lysosomal Transport Disorders

Laurent K Verkoczy   The Scripps Research Institute, La Jolla, California, USA
Somatic Hypermutation of Antigen Receptor Genes: Evolution

Yury Verlinsky   Reproductive Genetics Institute, Chicago, Illinois, USA
Preimplantation Diagnosis

Rogier Versteeg   University of Amsterdam, Amsterdam, The Netherlands
r.versteeg@amc.uva.nl
Clustering of Highly Expressed Genes in the Human Genome

Patrick Vicart   Faculté de Médecine Pitié-Salpétrière, Paris, France
Desmin-related Myopathies

Evani Viegas-Péquignot   Institut Jacques Monod, Paris, France
viegas@ijm.jussieu.fr
Epigenetic Factors and Chromosome Organization

Linda Vigilant   Max Planck Institute for Evolutionary Anthropology, Leipzig, Germany
Ancient DNA: Phylogenetic Applications

Angela J Villar   University of California, San Francisco, California, USA
avillar@itsa.ucsf.edu
Down Syndrome

Isabel Virella-Lowell   University of Florida, Gainesville, Florida, USA
virelil@peds.ufl.edu
Cystic Fibrosis: Gene Therapy

David Viskochil   University of Utah, Salt Lake City, Utah, USA
dave.viskochil@hsc.utah.edu
Neurofibromatosis: Type I (NF1)

JE Visser University Medical Center St Radboud, Nijmegen, The Netherlands
j.visser@neuro.umcn.nl
Lesch–Nyhan Disease

Jerry Vockley Mayo Clinic, Rochester, Minnesota, USA
vockley@mayo.edu
Digenic Inheritance

Friedrich Vogel University of Heidelberg, Heidelberg, Germany
Medical Genetics: History

Walther Vogel University of Ulm, Ulm, Germany
walther.vogel@medizin.uni-ulm.de
Prostate Cancer

Peter H Vogt University of Heidelberg, Heidelberg, Germany
peter_vogt@med.uni-heidelberg.de
Infertility: Genetic Disorders

Sonja Voordijk Swiss Institute of Bioinformatics, Geneva, Switzerland
Image Analysis Tools in Proteomics

Erno Vreugdenhil Leiden University, Leiden, The Netherlands
Expression Studies

Thomas J Vulliamy The Hammersmith Hospital, London, UK
t.vulliamy@imperial.ac.uk
Glucose-6-Phosphate Dehydrogenase (G6PD) Deficiency: Genetics

Paul A Wade Emory University School of Medicine, Atlanta, Georgia, USA
pwade@emory.edu
DNA Demethylation

Claes Wadelius Rudbeck Laboratory, Uppsala, Sweden
claes.wadelius@genpat.uu.se
Chromosome 17

Andreas Wagner University of New Mexico, Albuquerque, New Mexico, USA
wagnera@unm.edu
Evolution: Neutralist View; Evolution: Selectionist View

Hester M Wain University College London, London, UK
Human Gene Nomenclature

Quinten Waisfisz VU University Medical Center, Amsterdam, The Netherlands
q.waisfaisz.humgen@med.vu.nl
Spontaneous Function Correction of Pathogenic Alleles in Inherited Diseases Resulting in Somatic Mosaicism

Matthew J Wakefield The Australian National University, Canberra, Australia
matthew.wakefield@anu.edu.au
Chromosome X; X-chromosome Inactivation

Ken Walder Deakin University, Geelong, Australia
Diabetes: Genetics

Ann P Walker University of California, Irvine, California, USA
awalker@uci.edu
Genetic Counseling

John C Waller University College London, London, UK
ucgajcw@ucl.ac.uk
Galton, Francis

Lori L Wallrath University of Iowa, Iowa City, Iowa, USA
lori-wallrath@uiowa.edu
Position Effect Variegation in Human Genetic Disease

JB Walsh University of Arizona, Tucson, Arizona, USA
jbwalsh@u.arizona.edu
Multigene Families: Evolution

Gerald Walter Biorchard Limited, London, UK
gerald@biorchard.com
Array-based Proteomics

Daniel Walther Swiss Institute of Bioinformatics, Geneva, Switzerland
Image Analysis Tools in Proteomics

Qing Wang The Cleveland Clinic Foundation, Cleveland, Ohio, USA
wangq2@ccf.org
Retinitis Pigmentosa

Xin Wei Wang NIH, NCI, CCR, Bethesda, Maryland, USA
Hepatocellular Carcinoma

Dorothy Warburton Columbia University, New York, New York, USA
Chromosome Analysis and Identification

Laura E Warner University of Washington, Seattle, Washington, USA
lauraewarner@hotmail.com
Minigenes

Wyeth W Wasserman University of British Columbia, Vancouver, Canada
wyeth@cmmt.ubc.ca
Phylogenetic Footprinting

Raymond Waters University of Wales College of Medicine, Cardiff, UK
watersr1@cf.ac.uk
DNA Repair

Katherine D Watson University of Oxford, Wolfson College, Oxford, UK
wolf8005@mail.wolfson.ox.ac.uk
Pauling, Linus Carl

Nick Watson University of Edinburgh, Edinburgh, UK
n.watson@ed.ac.uk.
Disability: Diagnostic Labeling

Stephen G Waxman Yale School of Medicine, New Haven, Connecticut, USA and VA Hospital, West Haven, Connecticut, USA
Transcriptional Channelopathies of the Nervous System

David Weatherall Weatherall Institute of Molecular Medicine, Oxford, UK
liz@gwmail.jr2.ox.ac.uk
Thalassemias

J Weber Imperial College School of Medicine, London, UK
Infectious Diseases: Gene Therapy

William J Wedemeyer Cornell University, Ithaca, New York, USA
Protein Folding Pathways

Charles Weijer Dalhousie University, Halifax, Canada
charles.weijer@dal.ca
Community Consent for Genetic Research

Paul Weindling Oxford Brookes University, Oxford, UK
pjweindling@brookes.ac.uk
Nazi Movement and Eugenics

Lee S Weinstein National Institutes of Health, Bethesda, Maryland, USA
leew@amb.niddk.nih.gov
Gain-of-function Mutations in Human Genetic Disorders

Bruce Weir North Carolina State University, Raleigh, North Carolina, USA
weir@stat.ncsu.edu
DNA Evidence: Inferring Identity

Kenneth M Weiss Penn State University, University Park, Pennsylvania, USA
kenweiss@psu.edu
Allele Spectrum of Human Genetic Disease; Polymorphisms: Origins and Maintenance

Myriam Welkenhuysen University Hospital Leuven, Leuven, Belgium
Genetic Carrier Testing

Dagan Wells University College London, London, UK
dagan.wells@embryos.net
Whole-genome Amplification

Dan E Wells University of Houston, Houston, Texas, USA
dwells@uh.edu
Chromosome 8

Frank Wells Faculty of Pharmaceutical Medicine, London, UK
fow5851@aol.com
deCODE and Iceland: A Critique

Nils Welsh Uppsala University, Uppsala, Sweden
nils.welsh@medcellbiol.uu.se
Diabetes Mellitus: Gene Therapy

Dorothy C Wertz University of Massachusetts Medical School, Waltham, Massachusetts, USA
dorothy.wertz@umassmed.edu
Clinical Genetics and Genetic Counseling Professionals: Attitudes to Contentious Issues

Reiner Westermeier Amersham Biosciences Europe GmbH, Freiburg, Germany
reiner.westermeier@amersham.com
Gel Electrophoresis

James C Whisstock Monash University, Victoria, Australia
Serpins: Evolution

Bruce A White University of Connecticut Health Center, Farmington, Connecticut, USA
bwhite@nso2.uchc.edu
Polymerase Chain Reaction (PCR): Design and Optimization of Reactions

Nicola J White GlaxoSmithKline, Stevenage, UK
Susceptibility Genes: Detection

Peter S White University of Pennsylvania School of Medicine, Philadelphia, Pennsylvania, USA
white@genome.chop.edu
Chromosome 1

Michael C Whitlock University of British Columbia, Vancouver, Canada
Effective Population Size; Fixation Probabilities and Times

Harvey Wickham Institute of Psychiatry, London, UK
h.wickham@iop.kcl.ac.uk
Heritability Wars

Marek Wieczorek University of Washington, Seattle, Washington, USA
Art and Genetics

Johannes Wienberg National Research Center for Environment and Health (GSF), München, Germany
j.wienberg@lrz.uni-muenchen.de
Chromosome Rearrangement Patterns in Mammalian Evolution; Chromosomes in Mammals: Diversity and Evolution; Comparative Cytogenetics Technologies

Benjamin S Wilfond National Human Genome Research Institute, Bethesda, Maryland, USA
wilfond@nih. gov
Population Carrier Screening: Psychological Impact

Andrew OM Wilkie University of Oxford, Oxford, UK
awilkie@molbiol.ox.ac.uk
Dominance and Recessivity; Polygenic Inheritance and Genetic Susceptibility Screening

Tom Wilkie Cambridge Publishers Ltd, Cambridge, UK
tom@campublishers.com
Genetic Futures and the Media

Martin R Wilkins Imperial College, London, UK
m.wilkins@imperial.ac.uk
Genomics, Pharmacogenetics, and Proteomics: An Integration

Miles F Wilkinson University of Texas MD Anderson Cancer Center, Houston, Texas, USA
mwilkins@mail.mdanderson.org
RNA-binding Proteins: Regulation of mRNA Splicing, Export, and Decay

David Willey The Sanger Centre, Wellcome Trust Genome Campus, Hinxton, UK
dw1@sanger.ac.uk
Sequence Finishing

David Ian Wilson Southampton University, Southampton, UK
d.i.wilson@soton.ac.uk
Human Developmental Molecular Genetics

John H Wilson Baylor College of Medicine, Houston, Texas, USA
Gene Targeting by Homologous Recombination

Susan R Wilson Australian National University, Canberra, Australia
sue.wilson@anu.edu.au
Epistasis

Guil Winchester Midhurst, West Sussex, UK
guil.winchester@virgin.net
Dobzhansky, Theodosius; Morgan, Thomas Hunt

Jerry A Winkelstein The Johns Hopkins University School of Medicine, Baltimore, Maryland, USA
jwinkels@jhmi.edu
Complement: Deficiency Diseases

Paul C Winter Belfast City Hospital, Belfast, UK
Polymerase Chain Reaction (PCR)

G Brian Wisdom The Queen's University, Belfast UK
b.wisdom@qub.ac.uk
Posttranslational Modification and Human Disorders

Steven M Wise Center for the Expansion of Fundamental Rights, Boston, Massachusetts, USA
cefr1@aol.com
Animal Rights: Animals in Genetics Research

JA Witkowski Cold Spring Harbor Laboratory, Cold Spring Harbor, New York, USA
banbury@cshl.edu
Brenner, Sydney; Brown, Michael Stuart; Crick, Francis; Goldstein, Joseph Leonard; Watson, James Dewey

Christian Witt Universitätsklinikum, Mannheim, Germany
Titin (TTN) Gene

Ingrid Witters University Hospital Leuven, Leuven, Belgium
ingrid.witters@ uz.kuleuven.ac.be
Fetal Wastage: Molecular Basis

Sabine Wohlfart University of Vienna, Vienna, Austria
Apoptosis: Regulatory Genes and Disease

Cynthia Wolberger Howard Hughes Medical Institute and the Johns Hopkins University School of Medicine, Baltimore, Maryland, USA
Crystallization of Protein–DNA Complexes

Ulrich Wolf University of Freiburg, Freiburg, Germany
Ohno, Susumu

Gerhard Wolff Institute of Human Genetics, Freiburg, Germany
wolff@ukl.uni-freiburg.de
Genetic Counseling: Nondirectiveness

Gane Ka-Shu Wong University of Washington, Seattle, Washington, USA and Beijing Institute of Genomics, Beijing, China
gksw@u.washington.edu
Protein Coding

Bernard Wood George Washington University, Washington, DC, USA
Hominids

Troy E Wood Indiana University, Bloomington, Indiana, USA
trowood@indiana.edu
Speciation

Richard Wooster Institute of Cancer Research, Sutton, UK
Microarrays and Expression Profiling in Cancer

Nancy A Woychik University of Medicine and Dentistry of New Jersey, Piscataway, New Jersey, USA
nancy.woychik@umdnj.edu
RNA Polymerases and the Eukaryotic Transcription Machinery

Gregory A Wray Duke University, Durham, North Carolina, USA
gwray@duke.edu
Transcriptional Regulation: Evolution

Alan F Wright MRC Human Genetics Unit, Edinburgh, UK
alan.wright@hgu.mrc.ac.uk
Genetic Variation: Polymorphisms and Mutations

Susan Wright University of Michigan, Ann Arbor, Michigan, USA
spwright@umich.edu
DNA Technology: Asilomar Conference and 'Moratorium' on Use

Cathy H Wu Georgetown University Medical Center, Washington, DC, USA
wuc@georgetown.edu
Neural Networks

Joy Wu Johns Hopkins University, Baltimore, Maryland, USA
Restriction Enzymes

Gerald J Wyckoff University of Chicago, Chicago, Illinois, USA
wyckoffg@umkc.edu
Mammalian Sex Chromosome Evolution

Mitsuyoshi Yamazoe Kyoto University, Kyoto, Japan
yamazoe@rg.med.kyoto-u.ac.jp
DNA Recombination

John R Yates III The Scripps Research Institute, La Jolla, California, USA
jyates@scripps.edu
Proteomics: A Shotgun Approach without Two-dimensional Gels

William R Yates   University of Oklahoma College of Medicine-Tulsa, Tulsa, Oklahoma, USA
william-yates@ouhsc.edu
Adoption Studies

Soojin Yi   University of Chicago, Chicago, Illinois, USA
soojinyi@midway.uchicago.edu
Synonymous and Nonsynonymous Rates

Seppo Ylä-Herttuala   University of Kuopio, Kuopio, Finland
seppo.ylaherttuala@uku.fi
Atherosclerosis: Gene Therapy

Shozo Yokoyama   Syracuse University, Syracuse, New York, USA
syokoyam@syr.edu
Visual Pigment Genes: Evolution

Yoshihiro Yoneda   Osaka University, Japan
yyoneda@anat3.med.osaka-u.ac.jp
Protein Transport

Lu-Gang Yu   University of Liverpool, Liverpool, UK
Glycosylation and Disease

Alexandros Zafiropoulos   University of Crete, Crete, Greece
zafeiros@med.uoc.gr
Oncogenic Kinases in Cancer

Doris T Zallen   Virginia Polytechnic Institute and State University, Blacksburg, Virginia, USA
dtzallen@vt.edu
Medical Genetics in Britain: Development Since the 1940s

Ines Zanna   University of Palermo, Palermo, Italy
Caretakers and Gatekeepers

Maria Zannis-Hadjopoulos   McGill Cancer Center, Montreal, Canada
maria.zannis@mcgill.ca
DNA Replication

Zhao-Bang Zeng   North Carolina State University, Raleigh, North Carolina, USA
zeng@sun01pt2-1523.statgen.ncsu.edu
Quantitative Trait Loci (QTL) Mapping

Massimo Zeviani   National Neurological Institute 'Carlo Besta', Milan, Italy
zeviani@tin.it
Mitochondrial Disorders: Nuclear Gene Mutations

Jian Zhang   Aeomica Inc., Sunnyvale, California, USA
Microarrays: Use in Gene Identification

Jianzhi Zhang   University of Michigan, Ann Arbor, Michigan, USA
Evolutionary Distance: Estimation

Olga Zhaxybayeva   University of Connecticut, Storrs, Connecticut, USA
Orthologs, Paralogs, and Xenologs in Human and Other Genomes

Andreas Ziegler   University of Lübeck, Lübeck, Germany
ziegler@imbs.uni-luebeck.de
Chromosome 6; Genetic Distance and Mapping Functions

Ronald L Zimmern   Public Health Genetics Unit, Cambridge, UK
phgu@srl.cam.ac.uk
Public Health Genetics

Paul Zimmet   International Diabetes Institute, Melbourne, Australia
pzimmet@idi.org.au
Diabetes: Genetics

Joel Zlotogora   Ministry of Health, Jerusalem, Israel
joelz@cc.huji.ac.il
Genetic Disease: Prevalence

Huda Zoghbi   Baylor College of Medicine, Houston, Texas, USA
hzoghbi@bcm.tmc.edu
Trinucleotide Repeat Expansions: Disorders

Sebastian Zöllner   University of Chicago, Chicago, Illinois, USA
Population History and Linkage Disequilibrium

Massimo Zollo   Telethon Institute of Genetics and Medicine (TIGEM), Naples, Italy
zollo@tigem.it
Insulin-dependent Diabetes Mellitus (IDDM): Identifying the Disease-causing Gene at the IDDM11 Locus

Leonard I Zon   Harvard Medical School, Boston, Massachusetts, USA
zon@hhmi.tchlab.org
Zebrafish: as a Model for Human Diseases

Alan B Zonderman   National Institute on Aging, National Institutes of Health, Bethesda, Maryland, USA
zonderman@nih.gov
Vandenberg, Steven G

Florence J van Zuuren   University of Amsterdam, Amsterdam, The Netherlands
fvanzuuren@fmg.uva.nl
Genetic Counseling Consultations: Uncertainty

Bastiaan Johannes Zwaan   Leiden University, Leiden, The Netherlands
zwaan@rulsfb.leidenuniv.nl
Aging: Genetics

Glossary

ab initio gene prediction Prediction of a gene model using the DNA sequence alone. Uses measures of coding bias based on DNA composition and consensus sequences for donor and acceptor splice sites.

ab initio prediction Method of prediction that has no input or bias from experimental data.

aberration The deviation from perfect imaging in an optical system, which is caused by the properties of the material of the lenses or by the geometric forms of the refracting or reflecting surfaces.

acceptor site See 3' splice site.

acceptor splice site See 3' splice site.

accommodation John Calvin's belief that scripture is written (by God) in a simple or metaphorical form to make it understandable to simple and uneducated people.

acetylation Covalent attachment of acetyl groups to residues, mainly lysines, in histone proteins.

acetyltransferase An enzyme that catalyzes the transfer of an acetyl group from one molecule to another. The acetylation of histones by acetyltransferases is associated with gene activation.

aCGH See array comparative genomic hybridization.

acquired character Phenotypic character acquired by the adult organism through exposure to a changed environment or adoption of new habits.

acrocentric chromosome A chromosome with the centromere very close to one end. In humans, chromosomes 13, 14, 15, 21 and 22 are acrocentric.

acute leukemia Malignant disease that consists of the uncontrolled proliferation of immature malignant hematopoietic precursor cells that invade the bone marrow and displace normal blood-forming cells, resulting in immunodeficiency, anemia and increased bleeding as a result of the failure to produce granulocytes, red cells and platelets respectively.

acute lymphoblastic leukemia Acute leukemia resulting from malignant cells morphologically resembling the bone-marrow progenitors that normally produce B and T lymphocytes.

acute myelogenous leukemia Acute leukemia resulting from malignant cells morphologically resembling the bone-marrow progenitors that normally produce red cells, granulocytes, monocytes or platelets.

activated protein C resistance Resistance shown by some individuals in in vitro tests to the anticoagulant activity of activated protein C (APC). It is manifest as a poor anticoagulant response to added APC in clotting-based assays and is caused by the factor V Leiden mutation.

activation The trigger that induces the egg to develop as an embryo. Experimentally, various physical or chemical treatments can substitute for sperm entry.

activator A protein that binds to a positive regulatory element (enhancer or upstream activating sequence) to stimulate initiation of transcription.

actuarial fairness (1) A perspective that judges the fairness of discrimination based solely on the accuracy of the classifications made, rather than on their social or moral implications.
(2) When a proposer for an insurance contract is allocated to a risk group on the basis of the risk that he or she brings to the insurance pool and is charged a premium which matches that risk as closely as possible.

additive allele An allele whose effects on phenotype increase linearly with the number of copies of the allele. In terms of fitness, if allele A is additive and if the average fitness of individuals that carry one copy of A is $(1 + s)$ relative to a fitness of 1 for individuals that carry no copies of A, then we would expect the fitness of individuals carrying two copies of allele A to have fitness $(1 + 2s)$.

additive effects See additive allele; additive gene effects.

additive gene effects The situation when each gene in a given set contributes its own share to variation in the population, independently of the other genes in the mix. cf. nonadditive gene effects, multiplicative gene effects.

additive genetic variance Genetic variance due to alleles that sum in their effects, in contrast to variance due to interacting alleles.

additive inheritance Several loci contribute to risk, but do not necessarily all have to be present, but variants at each locus add to risk.

additive model A statistical model in which the predicted outcome is the sum of the effects of the predictors and does not include either nonlinear effects or interactions between predictors.

adverse selection In insurance terms, the situation when an individual, having a better knowledge of their own level of risk than they need to disclose to the insurer, chooses to apply for insurance when they would not otherwise have done so, or to apply for a greater amount of insurance than they would have applied for in the absence of such information about their own risk.

agonist A molecule that binds to and activates a receptor.

aldosterone Steroid hormone produced by the zona glomerulosa cells of the adrenal cortex. It is synthesized in response to activation of the renin-angiotensin system. Aldosterone binds to the mineralocorticoid receptor and regulates sodium reabsorption and potassium secretion in the principal cells of the distal nephron, as well as proton secretion in the intercalated cells.

alignment The matching of two or more nucleic acid or amino acid sequences in such a way that maximizes any sequence similarity between them. Gaps are usually inserted in either sequence to help maximize similarity. A pairwise alignment contains two sequences, a multiple alignment more than two. Alignment is the basis of most methods of determining the evolutionary relatedness or homology of sequences and of the computational algorithms designed to search databases for related sequences.

alignment of expert identity The ways in which experts signal their neutral identities and/or roles in interaction, for example, by not showing explicit agreement or disagreement with the viewpoints of clients.

allele One of two or more alternative forms of a gene that can occupy a given genetic locus, that is a particular place on a chromosome. An individual possesses two alleles at each of the genetic loci on their autosomes (all chromosomes except the sex chromosomes), one on the chromosome inherited from the father and the other on the homologous chromosome inherited from the mother. Many genes in the human genome have multiple alleles in the population as a whole.

allele fixation See fixation.

allele loss The situation where no members of the population carry a particular allele that was once present in that population.

allelic association See linkage disequilibrium.

allelic exclusion The situation where only one allele at a given locus on a pair of homologous chromosomes is expressed, while the other allele is repressed.

allelic heterogeneity The existence of more than one disease-related mutation (or allele) for the same gene.

allotransplantation Transplantation of a tissue or organ from one individual to another, genetically non-identical, individual of the same species.

alternative splicing The mechanism by which potential coding regions (exons) in a primary RNA transcript can be combined in different ways by the pairing of different splice sites to generate different mRNAs from the same transcript.

Alu A family of short (approximately 300 base pair) repetitive DNA sequences that is the most abundant short interspersed nuclear element (SINE) in the human genome. Alu repeats are so named because they generally contain recognition site(s) for the restriction enzyme AluI. The Alu family consists of about a million members distributed throughout the genome. It is thought to be derived from a 7SL RNA that has been reverse transcribed and integrated into the genome.

amino-acid chronology Temporal order of appearance of the different amino acids in proteins in the course of early evolution.

amino-terminal At or towards the amino terminus of a protein.

amino terminus The end of a polypeptide chain with an unreacted amino group. Proteins are synthesized from the amino terminus.

amniocentesis A procedure used to collect a sample of the fluid surrounding the fetus (amniotic fluid) in the second trimester in order to obtain fetal cells and fluid for genetic or biochemical analysis.

amphiphilic α helix An α-helix element of secondary protein structure with positively charged amino acids on one side, facing the surface of the protein, while the other side is hydrophobic and is buried against the hydrophobic core.

amphipathic Having both hydrophobic and hydrophilic regions. The description can apply equally to small molecules, such as phospholipids, and macromolecules such as proteins.

ampholyte A molecule bearing both acidic and basic groups, and thus able to generate positive and negative charges (for example, amino acids, peptides and proteins).

amplicon A DNA sequence capable of replication.

amyloid Insoluble proteinaceous material that accumulates in tissues in certain diseases.

amyloidogenesis Production of amyloid.

amyloidosis Pathological condition caused by the accumulation of amyloid.

anaphase The fourth stage of mitosis, in which all the pairs of sister chromatids separate, the members of each pair migrating on the spindle to opposite poles of the cell.

anaphase lag Abnormally slow movement, during mitosis and cell division, of a chromosome toward one of the poles of the cell, resulting in this chromosome being excluded from one of the daughter cells.

anaphase-promoting complex (APC) A multiprotein enzyme complex, which acts as a ubiquitin ligase, catalysing the attachment of ubiquitin to a protein substrate, thus tagging it for proteolytic degradation by a proteasome. Its name comes from its role in mediating the degradation of specific proteins during anaphase of mitosis.

anchored primer A primer for a polymerase chain reaction (PCR) that includes both a sequence for hybdridizing to a target template and a sequence that encodes additional useful information such as a restriction site, the T7 polymerase promoter, and so on.

androgen insensitivity syndrome *See* testicular feminization.

aneuploid Having one or more extra or missing chromosomes compared to the normal diploid number (which in humans is 46). This abnormal condition is known as aneuploidy.

aneuploidy *See* aneuploid.

angiogenesis The development of blood vessels in the embryo or during tissue repair.

angiogenic Stimulating the growth of new blood vessels.

Angstrom (Å) A unit of length equal to 108 cm and used in describing atomic and molecular dimensions.

anionic channels Protein-lined pores in biological membranes that are permeable to anions.

ankyrin An erythrocyte membrane protein that links the membrane skeleton to the lipid bilayer of the membrane by binding to the spectrin protein and to the cytoplasmic domain of the band 3 protein. Mutations involving ankyrin can give rise to hereditary spherocytosis.

anomalous trichromacy Defective color vision mediated by three cone types in the retina, where the two cone types that absorb in the medium- to long-wavelength region of the spectrum are much more similar in spectral sensitivity than are the cones underlying normal trichromatic color vision.

anosmia The absence of a sense of smell.

antagonistic pleiotropy The situation when the action of a gene on one trait is beneficial, but its action on another trait is detrimental.

anticipation In genetic disease, the increased severity and earlier onset of a genetic disorder with successive generations of a family.

anticoagulant Any substance that inhibits the clotting of blood.

antigen Any substance that can be recognized by an antibody or a T-cell receptor and is thus, in principle, able to induce a specific immune response. *See also* immunogen.

antimicrobial Compound that inhibits the growth of or kills microorganisms.

antigen-presenting cell Cell that is presenting antigen on its surface in the form of a peptide fragment bound to a major histocompatibility complex (MHC) molecule, in which form the antigen can be recognized by a T lymphocyte. The term usually refers to one of a small number of specialized immune-system cells that display both antigen bound to MHC molecules and the co-stimulatory molecules required for the initial activation of a naive T cell. Dendritic cells are the most efficient of these 'professional' antigen-presenting cells but B cells and macrophages also have this function.

antisense RNA A transcript from the antisense strand of a gene. Naturally occurring antisense RNAs may negatively regulate gene expression.

antisense Describes a nucleotide sequence complementary to the DNA strand that carries the coding information for a given gene or other element.

antisocial behavior Actions that contravene accepted societal practices, whether actually illegal or not.

APC (1) Anaphase-promoting complex.
(2) Activated protein C.

APC resistance *See* activated protein C resistance.

aphakia Absence of the lens of the eye.

apolipoprotein The protein component of a lipoprotein (a specific complex of protein and lipid). Apolipoproteins function as ligands for specific lipoprotein receptors on cells, as structural components of lipoproteins, and as activators for plasma enzymes.

apoptosis The controlled self-destruction of a cell by an orderly and programmed sequence of cellular events in response to an internal or external stimulus. It is a normal and essential occurrence in biological processes such as embryonic development and immune-system function and is quite distinct morphologically from necrotic cell death caused by tissue damage.

application layer The top-most layer of the Internet, and the layer most familiar to users, in which users write emails, surf the web, and analyze data.

aquaporin-2 A transmembrane protein that facilitates movement of water across a membrane by forming an aqueous pore across the lipid bilayer. In the kidney, it is inserted into the plasma membrane of collecting duct epithelial cells, in response to vasopressin, and is necessary for urine concentration.

aqueous humor Clear, thin, aqueous fluid produced by the ciliary epithelium, which provides nutritional support for the avascular tissues of the anterior segment of the eye.

ARE *See* AU-rich element.

areflexia An absence of the normal reflex muscle jerks.

arranged marriage Marriage in which the main decisions about the choice of spouse are made by the couple's parents, other relatives or guardians.

array comparative genomic hybridization (aCGH) A version of comparative genomic hybridization in which the targets of the simultaneous hybridization are not metaphase chromosome spreads but arrays of BAC clones or cDNAs with a known coverage and location in the genome affixed to a solid support (such as a glass slide). Depending on the genomic spacing of the clones used in the array, a greater resolution is possible than that offered by the metaphase chromosome spread.

arthropods A phylum of invertebrate animals with multijointed appendages, which includes crustaceans, insects and spiders.

artificial chromosome A DNA segment constructed to contain all the elements required for self-replication (telomere, centromere, autonomously replicating sequence, and so on) so that when introduced into a cell it functions as a supernumerary chromosome, replicating and distributing daughter chromatids into cells at each cell division. When used as a cloning vector, it is constructed to include genes as well.

Aryan In Nazi terminology, Christian Caucasians, especially those of Nordic ancestry.

ARS *See* autonomously replicating sequence.

ART *See* assisted reproductive technology.

assisted reproductive technology (ART) General term for all forms of assisted conception and the methods used to achieve this, such as *in vitro* fertilization (IVF), donor insemination (DI), intracytoplasmic sperm injection (ICSI).

association study Population genetic study designed to seek a link between particular alleles at a polymorphic gene (or genes) and a particular phenotype (commonly a disease phenotype). The method typically uses case-control samples of unrelated individuals, or sets of families in which the 'controls' are provided by unaffected members within each family. The approach is particularly useful for fine localization of a disease susceptibility gene and direct testing of 'candidate' genes.

association A tendency for two or more factors (for example, particular alleles of a gene or genes) to occur together.

assortative mating Mating that takes little or no account of any similarity or dissimilarity in the phenotypes of the two parents.

astral microtubules Microtubules nucleated by the spindle poles that extend out to the cell cortex and position the spindle during mitosis.

asymmetrical exon An exon flanked by introns of different phase.

asymptomatic Showing no obvious signs of disease.

ataxia An inability to coordinate movement.

ataxia telangiectasia Hereditary disorder characterized by problems with muscle coordination, immunodeficiency, inadequate DNA repair and an increased risk of developing cancer.

atherosclerosis The accumulation of lipids, macrophages and matrix deposits on the interior walls of medium- and large-calibre blood vessels.

attributable fraction The proportion of cases that would not occur within a specified population and time period in the absence of a particular risk factor. Also called population attributable risk.

attributable risk Proportion of disease in a population that can be attributed to a particular susceptibility-conferring allele.

AU-rich element (ARE) Nucleotide sequence typically located in the 3′ untranslated region of an mRNA, and which can influence its half-life.

autism spectrum disorders Autism and related disorders.

auto-focus In microscopy, a function that automatically finds the focus plane of the object contour.

autoimmunity Inappropriate and abnormal adaptive immune response against self-antigens.

automation The use of programmable robots and similar computer-controlled or mechanical devices for repetitive laboratory tasks and procedures.

autosomal recessive disease A pattern of inheritance of a genetic disease that is primarily characterized by affected children being born to unaffected parents who each carry one copy of the mutation. The child is affected only if the child inherits a mutation from both parents, and on average, this will occur in one out of four conceptions for such couples.

autonomous element Transposable genetic element that encodes all the functions required for its transposition, for example a transposase that both excises a copy of the element from the DNA and mediates its insertion elsewhere.

autonomously replicating sequence (ARS) A specialized DNA sequence in eukaryotic DNA that confers the ability to be replicated on any DNA that contains it (including prokaryotic DNA). It is the site at which the proteins that mediate DNA replication in eukaryotic cells are first assembled to initiate replication.

autonomy In ethics, the governing of oneself according to one's own system of morals and beliefs. *See also* principle of autonomy.

autoradiography Sensitive detection method for proteins or nucleic acids that have been radioactively labelled by the incorporation of a radioactive isotope, and which thus cause whitening of X-ray film. It is used, for example, in gel electrophoresis to detect distinct bands of separated proteins on the gel and to detect the distribution of particular proteins and mRNAs in tissues.

autosomal Pertaining to any chromosome that is not a sex chromosome.

autosomal dominant A mutation of a gene on an autosome, not a sex chromosome, and which affects the phenotype even when the other copy of the gene is normal.

autosome Any chromosome except a sex chromosome. In humans these are chromosomes 1–22.

autozygous The state of having the identical allele at the equivalent position of a pair of homologous chromosomes.

B cell B lymphocyte, an immune-system cell that produces antibodies (immunoglobulins).

B lymphocyte *See* B cell.

BAC *See* bacterial artificial chromosome.

BAC clone Genomic DNA inserted and maintained in a bacterial artificial chromosome vector, which contains sequences that enable it to replicate and be maintained in a bacterial cell.

bacterial artificial chromosome (BAC) A cloning vector based on the F-factor of *Escherichia coli* that serves as the vehicle for the incorporation and propagation of relatively large (100–200 kilobase) pieces of genomic DNA. BACs have provided the backbone for the sequencing of the human genome and also provide a resource for the refined linkage of sequence, restriction map and cytogenetic location.

bacterial artificial chromosome clone (BAC clone) Genomic DNA inserted and maintained in a bacterial artificial chromosome vector, which contains sequences that enable it to replicate and be maintained in a bacterial cell.

BAC library A human (or other) genomic DNA library of bacterial clones containing large human (or other) DNA inserts, typically 100–200 kilobases in size in vector DNA (a bacterial artificial chromosome or BAC) that contains all the necessary elements to maintain a single copy per bacterial cell.

background selection Effect of purifying selection (acting on a particular locus) over the adjacent regions of the genome. It usually diminishes variability at adjacent unselected loci and thus can be modeled as a reduction in the effective population size in population genetic studies.

bacteriophage A virus that infects bacteria.

balancing selection Any form of natural selection that maintains a polymorphism at a genetic locus. A typical selection pattern is where a heterozygote has a higher fitness than either homozygote and thus the frequency of the alleles in the favored heterozygote tend to increase.

band 3 protein The major erythrocyte integral membrane protein in red blood cells that functions both as an anion exchanger and as a major site of attachment for the membrane skeleton. Mutations involving band 3 protein can give rise to hereditary spherocytosis and ovalocytosis.

bands (1) Specific areas on a metaphase chromosome that appear dark or light when subjected to one or more of the chromosome staining techniques. Regions and bands are numbered consecutively from the centromere outward along each chromosome arm. A particular band is designated by four parameters: the chromosome number within that region; whether it is on the long arm (q) or short arm (p); the region number; and the band number. These items are given in consecutive order without spacing or punctuation. For example, 9q34 means chromosome 9, the long arm, region 3, band 4.
(2) Sharply delimited regions of identical DNA fragments or proteins that form when a mixture of DNA or of proteins is subjected to gel electrophoresis.

banding techniques The use of particular dyes to reveal a characteristic pattern of light and dark regions, or bands, on metaphase chromosomes.

β barrel Protein secondary structure motif in which the strands of a β sheet are arranged in a configuration resembling the walls of a cylinder or barrel.

basal ganglia The deep portions of the brain controlling movement and other important functions.

base In molecular biology and genetics, usually refers to a purine or pyrimidine. These are nitrogen-containing ring compounds (heterocycles), which are termed bases because they can combine with H^+ in acidic solutions. Four different types of base are usually present in DNA: adenine and guanine, which are purines; and cytosine and thymine (replaced by uracil in RNA), which are pyrimidines.

base-excision repair Form of DNA repair that involves removing a damaged or misincorporated base, typically by the action of glycosylases.

base pair (bp) Two complementary nitrogenous bases held together by hydrogen bonding in nucleic acids. Adenine (A) always pairs with thymine (T) in DNA, and with uracil (U) in RNA, and guanine (G) pairs with cytosine (C). Two strands of DNA are held together in the shape of a double helix by the bonds between successive base pairs. The number of base pairs is commonly used as a measure of the size of a gene or DNA fragment.

base quality In DNA sequencing, a statistical measure of how likely it is that the base identified in the sequence is correct, encapsulated in the base-quality score. A high-quality base in a raw sequence read has been defined as one with an accuracy of at least 99%.

base-calling The process of identifying each base in the sequence read from the output of the DNA sequence.

basic region A region in a polypeptide containing positively charged amino-acid residues such as lysine, arginine and histidine. It facilitates interactions with negatively charged molecules such as DNA.

Bayes' formula Mathematical formula that relates the prior odds in favor of a finite number of mutually exclusive hypotheses, and the conditional probabilitiy of some observed data, given each of these hypotheses, to their posterior odds. Named after the English non-conformist minister Thomas Bayes (1702-1761).

Bayes' theorem The rule of probabilities by which $P(A|B)$ is obtained from $P(B|A)$.

beneficence In biomedical ethics, a moral principle that requires that investigators (1) do no harm and (2) maximize potential benefits to study participants while minimizing risks, that is, being of benefit to the patient and putting clinical knowledge and experience at the service of the patient's well-being. (3) The state of doing or producing good.

benign tumor A tumor that grows locally but does not invade surrounding tissues or produce secondary tumors (metastases) at distant sites in the body.

benzodiazepines A class of drugs that enhance the response of GABA receptors to their ligand gamma-aminobutyric acid (GABA). They are widely used for the treatment of anxiety and sleep disorders.

beta-amyloid (Aβ) A protein fragment normally made up of 40 amino acids, or a toxic form made up of 42 amino acids that is present in amyloid plaques between neurons.

betaine *N,N,N*-trimethylglycine, a compound naturally occurring in beetroots. It is used as an alternative source of methyl groups in treatment of cystathionine β-synthase (CBS) deficiency by stimulating the remethylation of homocysteine.

between-group component of hierarchical selection A measure of how group productivity is influenced by the frequency of altruists within a group. Between-group selection favors groups with many altruists.

B-form DNA The DNA structure solved by Crick, Watson and collaborators in 1953. In cells, B-form DNA is the most 'typical' form of DNA – a double-stranded molecule of right-handed helix.

bias A systematic deviation from truth as a result of errors in measurement, analysis, or inference.

bifurcation point The point of greatest instability within a dynamic cellular gene-expression network that is sensitive to an infinitesimal change, either quantitative or qualitative, that will lead to an altered phenotype and new stationary state.

bilateral descent The situation where group membership or kinship ties are traced equally through the male and female lines, as found in Euro-American cultures. Also called bilateral reckoning or cognatic descent.

bilateral reckoning *See* bilateral descent.

binary trait A genetic trait that exists in one of two distinct states, for example affected (by a given disease) or unaffected.

biocentrism The attribution of equality to all forms of life, which considers the view of the individual organism, and may ascribe equal rights to all forms of life.

biocolonialism Political relationship in which economically developed nations or corporations acquire the environmental knowledge or biological resources of indigenous peoples cheaply and transform them for profit, without distributing the gains equitably among those peoples.

bioethics The discipline of analyzing, clarifying and resolving moral problems in relation to life and life sciences. It includes the ethics of biological and biomedical experimentation, and its legal, philosophical, theological, medical and technological implications, especially as bearing on experimentation on animals and humans and the implications of research such as genetic modification, which have given rise to rules and regulation of the conduct of aspects of biological research.

bioinformatics The study and application of computational and statistical methods to biological information, especially genetic and genomic information.

biological determinism A theory that human characteristics are biologically caused and consequently unchangeable.

biomarker A biochemical, genetic or molecular indicator that can be used to screen for disease or toxicity.

biopiracy A term, used by critics of biotechnology, to describe a form of neocolonialism wherein the developed world is seen to exploit the underdeveloped world for its genetic heritage and its traditional knowledge for commercial gain, primarily through the enforcement of patent rights.

biopolitics The use of modern technologies to discipline not only the individual body but also the social body, a process begun in the eighteenth century.

biopsychosocial model The view that biological, psychological and social factors are all involved in any state of health and illness.

biosynthetic labeling The incorporation of radioactively-labeled amino acids into newly synthesized proteins in cell culture. The rate at which new protein is synthesized is calculated from the total amount of protein and the fraction of labeled protein. The half-life of a protein can also be determined with this technique.

biotechnology The application of science and engineering in the direct or indirect use of living organisms in their natural or modified forms in an innovative manner in the production of goods and services.

bipolar disorder Severe psychiatric disorder with onset usually in young adulthood, characterized by pathological episodic mood disturbance ranging from elation (manic mood) to depression and associated changes in cognition, behavior and functioning. Delusions and hallucinations may occur. Also known as manic–depressive illness.

bisatellited marker chromosome The reciprocal product of a Robertsonian translocation, consisting of the fused short arms (p arms) of two acrocentric chromosomes.

bivalent A pair of homologous chromosomes, each composed of two sister chromatids, that are paired in the pachytene phase of the first division of meiosis.

BLAST Basic local alignment search tool. A computer program for detecting similarity between two DNA sequences. It is most widely used for comparing, in successive pairwise comparisons, a DNA or protein sequence against all other sequences in a database in order to identify the sequence or find related sequences.

blastocyst An early stage in mammalian embryogenesis, consisting of an inner cell mass surrounded by a layer of cells that will form the placenta.

blastula An embryo in an early stage of development consisting of a single layer of cells (the blastoderm) that forms a hollow sphere.

blood spots The several drops of blood collected from newborn babies shortly after birth for screening. The blood is blotted and dried onto personally identified cards. Also known as Guthrie cards.

blotting Transfer in their *in situ* position of bands of protein or nucleic acids after separation by gel electrophoresis onto an immobilizing membrane where they can be hybridized with nucleic-acid probes or treated with specific antibodies to detect particular molecules. Transfer can be made by diffusion, capillary forces, negative pressure (vacuum blotting) or electrophoretic migration.

body as an instrument The body as something to be 'used' by the self which resides within it.

body as object The body conceived of as a wholly material thing. *cf.* body as subject.

body as subject The body considered as identical to the self, and directly engaged and immersed in the world. *cf.* body as object.

Boolean query A query composed of several words or items that are linked by operators such as 'AND', 'OR' or 'NOT'.

bottleneck A severe reduction in population size of a species or population, causing an alteration in gene frequencies, which can reduce the genetic diversity of a population.

bottom-up strategy An approach to genome sequencing that starts with the highest possible level of resolution (the DNA sequence in small pieces) and attempts to reassemble the overall structure from the pieces. Shotgun sequencing is a bottom-up strategy, although in all its applications to genome sequencing some element of lower-resolution information is used to guide assembly.

box C/D snoRNA A subset of small nucleolar RNAs possessing characteristic box elements (boxes C, D, C′ and D′) and a terminal stem structure.

box H/ACA snoRNA A subset of small nucleolar RNAs possessing characteristic box elements (boxes H and ACA) and two stem–loop structures.

bp *See* base pair.

BPS *See* branchpoint sequence.

brachycephaly Reduction in the length of the skull relative to its width.

brachyclinodactyly Shortened and bent or deflected fingers.

branch migration In DNA recombination, the movement of the recombination joint along the stretch of hybrid DNA, which can be inhibited by DNA secondary structure.

branch site The nucleotide (usually an adenosine) near the 3′ end of an intron, the 2′ hydroxyl group of which will attack the 5′ splice site in the first step of RNA splicing.

branchial arch One of a series of bony or cartilaginous arches that develop in the walls of the mouth cavity and pharynx of a vertebrate embryo.

branchpoint sequence (BPS) In mRNA processing, a consensus sequence CTRAY (R = purine, Y = pyrimidine), which is located 10–50 bases upstream of the 3′ end of the intron, and which contains the adenosine at which the lariat-splicing intermediate is formed.

Brushfield spots Small white spots surrounding the pupil resulting from aggregation of the normal connective tissue of the iris.

BRCA1/BRCA2 **genetic testing** Mutation searching or predictive DNA testing (mutation searching) is offered to individuals who have a family history of breast and/or ovarian cancer. The criteria for offering mutation searching to individuals affected with breast or ovarian cancer normally include several cases of disease in the family with an earlier than average age of onset. Predictive testing is normally offered to at-risk individuals after the identification of a family-specific mutation in a relative affected by breast or ovarian cancer.

broader phenotype Social or communication impairment or restricted and repetitive behavior insufficient to lead to a formal diagnosis of an autism spectrum disorder. The phenotype may include increased visuospatial ability for some relatives of probands with autism spectrum disorders.

Bushman model The model of pre-agricultural demography in which hunter–gatherers lived peaceful lives with low birth rates, long-term zero population growth and in ecological balance with their environments.

caged compound A compound whose usual chemical activity or physical properties are attenuated by the presence of an additional chemical group. Caged fluorescent compounds used in cell biology research are often covalently linked to nitrobenzyl groups, which lock the fluorochrome in a nonfluorescent conformation. When the linkage of such caging groups to the fluorochrome is photolabile, the masking group can be released using a pulse of light of appropriate wavelength.

Cajal bodies Nuclear organelles implicated in the assembly, recycling and possibly also the function of snoRNPs and U snRNPs. Also called coiled bodies.

calibration In phylogenetics, validating mutation rates by correlating with time points corresponding to fossil evidence for the last possible common ancestor between evolutionary lineages.

calnexin Protein found in the endoplasmic reticulum that acts as a molecular chaperone for the assembly of folding of proteins (such as the MHC molecules).

calreticulin Protein found in the endoplasmic reticulum that acts as a molecular chaperone for the assembly of folding of proteins (such as the MHC molecules).

calpain A calcium-activated, cytoplasmic cysteine protease.

Cambrian period Geological period 544–505 million years ago when a great diversification of vertebrate life appeared.

cAMP response element binding protein *See* CREB.

candidate gene A gene that is suspected, but not proven, to be involved in the genetic aetiology of a disorder or condition. For example, a gene that resides within a chromosomal interval to which a genetic disease has been mapped.

5′ cap A chemical modification made to the 5′ end of precursor mRNA during RNA processing. It consists of addition of GMP to the first nucleotide of the RNA in a 5′5′ triphosphate linkage, methylation of the guanosine to form 7-methylguanosine, and sometimes methylation of the C2′ of the ribose of the adjacent nucleotide and its neighbor.

capitated Describes an insurance system that pays healthcare providers a fixed amount per patient regardless of the services the patient receives.

carboxybetaines Class of chemicals containing both a quaternary ammonium and a carboxylic acid group, separated by a hydrocarbon linker. This linker is generally one carbon long, but can be longer.

carboxy-terminal At or towards the carboxyl terminus of a protein.

carboxyl terminus The end of a polypeptide chain with an unreacted carboxyl group.

carcinoma A cancer of epithelial cells, that is, cells that line external or internal surfaces of the body.

carrier test A test that indicates the likelihood of an individual being at risk of passing a genetic disorder to their offspring. A carrier of a disorder may be completely unaffected by the condition.

carrier testing Identifies individuals with a gene or a chromosome abnormality that may affect the health of the offspring. Although carriers are not affected by the disease, two carriers may produce a child who has the disease.

carrier screening *See* carrier testing.

carrier (1) An individual who carries a recessive allele and a normal allele for an inherited disease, and who is not themselves affected by the disease. In the case of disease genes carried on autosomes, two carriers can produce a child with the disease; in the case of X-linked genes, a mother can pass the faulty gene, and thus the disease, on to a son.
(2) Any integral membrane protein that is involved in the transport of solutes or ions across the membrane. Carriers include passive transporters, which carry a solute or ion down its concentration gradient, and active transporters, which use energy to transport material against a concentration gradient. In bacteria carriers are often called permeases.

Carter effect That in a disease with multigenic (or multifactorial) inheritance first degree relatives of patients with the gender that is less frequently affected in the general population, show a higher proportion of affected individuals.

cascade screening Testing that is actively offered to the closest family members of a newly diagnosed individual (the index case) as they are at increased risk of having a mutant gene themselves. Testing is subsequently offered to the relatives of those who test positive, and so on.

case-control study A study comparing the frequency of a risk factor in persons with disease and a suitable control (reference) group.

cataract Opacification of the lens of the eye, which is normally clear.

catechol-O-methyl transferase (COMT) One of two major enzymes involved in the breakdown of catecholamine neurotransmitters.

catenins Group of proteins involved in cell adhesion. They constitute the intracellular anchorage of cadherins, key proteins in cell junctions.

cathepsins A class of proteases found in lysosomes, whose enzymatic activity is optimal in acidic environments (for example, cathepsins B, D, L and G).

cation channel Protein-lined pore in a biological membrane that is permeable to cations and causes the depolarization of the membrane.

CBP A human cruciform-binding protein, identified as a member of the 14-3-3 protein family, which binds to DNA replication origins in a structure-specific manner.

CCD Charge-coupled device, a light-sensitive imaging silicon chip used in cameras.

cDNA See complementary DNA

cDNA library A collection of cloned cDNAs, usually representing the total mRNA expressed by a particular cell type, tissue or developmental stage.

cDNA selection A method for identifying genomic clones that hybridize to a cDNA library.

CDS See coding sequence.

cell adhesion molecules Proteins in the cell membrane or immediate extracellular environment that promote adhesion between cells or between cells and the extracellular matrix.

cell cycle The fixed sequence of events by which a cell grows, replicates its DNA and divides to form two daughter cells.

cell-cycle checkpoint A point in the eukaryotic cell cycle at which the cell arrests and checks its status before progressing to the next stage of the cycle.

cell fusion Process in which the plasma membranes of two cells fuse down at the point of contact between them, allowing the two cytoplasms to mingle.

cell line A population of genetically identical cells, of a particular cell type, that are being permanently maintained in culture. They all derive from a small initial sample of cells from a particular individual.

cell nuclear replacement Method of cloning by placing the nucleus of a diploid cell into an enucleated egg to produce a viable embryo genetically identical to the donor of the diploid cell.

cell nuclear transfer Cloning technique that involves removing the nucleus of an egg cell and substituting the nucleus of a cell taken from one of the body's organs or tissues. Cloned sheep, cows, goats, pigs and mice have been produced by this technique. Also known as reproductive cloning.

cell polarity The segregation of different plasma membrane protein and lipid components into distinct apical and basolateral domains on the surface of an epithelial cell.

cell separation A necessary step before analyzing gene expression *ex vivo*, so that the starting material consists of a homogeneous cell population. The most commonly used techniques are flow cytometry and magnetic cell separation. Alternatively, laser-capture microdissection is used to isolate cells from tissue sections.

CENP See centromeric protein.

CENP-B An abundant centromere-associated protein that binds a specific DNA motif (the CENP-B box). The function of CENP-B at the centromere, if any, is unknown.

centimorgan (cM) A genetic distance between loci with a (calculated or expected) recombinant frequency of 0.01 (1%). Very roughly, 1 cM is similar to 1 000 000 base pairs.

central database A collection of mutations in all genes; the information contained is usually little more than the mutation itself and a reference, but is nevertheless useful. Central databases are also known as general databases.

central element Thin proteinaceous band present in the central region of synapsed chromosomes at meiosis, bounded by the lateral elements, along the synaptonemal complex.

centromere A constricted area visible on metaphase chromosomes. It is made up of densely-packed DNA (satellite repeats), which is visible as darkly stained heterochromatin, and an associated protein complex, the kinetochore. It is the site at which two sister chromatids are pulled apart during mitosis or meiosis by microtubules of the spindle that attach to the kinetochores of the individual chromatids. Chromosomes are classified according to the position of the centromere: metacentric chromosomes have the centromere in the middle and acrocentric chromosomes have it (almost) at one end of the chromosome.

centromeric protein (CENP) A protein that is constitutively or transiently associated with the centromere/kinetochore and involved in centromere structure and function.

centromerization The process in which a specific chromosome region is marked for centromere formation.

centrosome A cellular organelle that consists primarily of a pair of centrioles, cylindrical arrays of mictrotubules. During mitosis, the centrosomes form the spindle poles from which mitotic spindles emanate.

CEPH families A collection of 60 families, each comprising at least six children, their parents and their four grandparents, from which samples have been made available by the Centre d'Études du Polymorphisme Humain (CEPH), Paris, France, for genetic mapping by the scientific community.

cephalochordates Sub-phylum of the chordates which includes the lancelet (amphioxus), which have a persistent notochord extending the length of the animal, muscle blocks organized into somites, a complex nervous system and an intricate pharynx.

CFR See Code of Federal Regulations.

CGH See comparative genomic hybridization.

α chain In collagen, the polypeptide chain. Each collagen molecule consists of three identical α chains, or two to three different α chains. For example, the constituent chains for type I collagen are called α1(I) and α2(I) collagen chains. Note that the designation α chain is also used for many other chains in many other types of proteins.

chain extension Synthesis of a new DNA strand from a primer hybridized to a template strand.

channel An ion channel, an integral membrane protein that forms an aqueous pore in the membrane through which ions can pass. For most channels, opening is regulated by a chemical ligand (ligand-gated channels), or membrane voltage (voltage-gated channels), or a mechanical deformation of the membrane.

chaotrope Chemical able to disrupt, at least in part, the hydrogen-bonded structure of water when dissolved in it. Although many water-miscible solvents fall into this class (for example, acetone, pyridine) this term is usually reserved for solid chemicals. The most commonly used chaotropes in biochemistry are urea and guanidine.

character In phylogenetic studies, a feature that, when compared between organisms, shows at least two states (ancestral or derived). Characters include behavioral traits, morphological entities, enzymatic pathways and the DNA sequences of genes.

chelicerates The class of arthropods that includes spiders, scorpions and horseshoe crabs.

chemokine A general name for any protein secreted by leukocytes and other cell types that attracts other cells and causes them to change their behavior through interaction with specific cell-surface receptors.

chemoprevention The use of medication or dietary supplements to reduce cancer risk.

chiasma Crossover between paired homologous chromosomes in meiosis that is visible under the microscope; plural **chiasmata**.

children-of-twins design A study that compares similarities between children of monozygotic twins (genetic half-siblings) with children of dizygotic twins (genetic cousins).

chimeric gene A gene constructed from parts of different genes.

chimeric intron Intron constructed from parts of different introns.

chimeric protein A protein encoded by a gene constructed from parts of different genes.

chimeric Describes an organism, a protein or a DNA that is composed of parts from two or more different sources. In the case of an animal, a chimera derives from an early embryo in which cells from two genetically different embryos have been mixed.

chimerism When a single individual is made up of two genetically distinct cell lineages that have come together early in development.

CHIP assay See chromatin immunoprecipitation assay.

chromatin immunoprecipitation assay (CHIP assay) Assay for detecting proteins bound to DNA *in vivo*, in which such proteins are first covalently cross-linked to the DNA and then the chromatin complex is immunoprecipitated with specific antibodies.

chordates The animal phylum that includes the vertebrates, cephalochordates and urochordates. All chordates possess a number of common features, including a notochord (embryonic only in vertebrates), a dorsal nerve cord, muscle blocks (somites) and a segmented pharynx.

chorionic villus sampling Method of obtaining embryonic tissue for diagnostic purposes by retrieval of tissue from the trophoblast around the embryonic sac.

chromatid One of the longitudinal halves (sister chromatids) of a replicated chromosome, while still held together at the centromere.

chromatin The complex of DNA wound around octamers of histones to form a chain of nucleosomes, which is the basic form in which DNA exists in the eukaryotic cell nucleus. The chain of nucleosomes is packaged into higher-order structures that include numerous other proteins.

chromatin domain A distinct structural region in chromatin. The term has been used to denote both large (around 1 megabase) higher-order chromatin structures and smaller (around 100 kilobases) chromatin loops. Chromatin domains of around 1 megabase have diameters of some 300–800 nm and are maintained throughout the cell cycle as higher-order structures.

chromatin fiber The basic material of which eukaryotic chromosomes are made, comprising a single molecule of DNA complexed with histones. Numerous other DNA-binding proteins become part of the complex either permanently or at various times.

chromatin insulator A DNA sequence element that can block the repressive effects of neighboring heterochromatin when flanking a stably integrated transgene. The term insulator is also applied to elements that prevent enhancer-mediated gene activation when placed between the enhancer and promoter. The two activities are not identical, and a specific element may possess one or both functions.

chromatin structure The higher-order organization of the complex of DNA and histones in a chromosome.

chromatography A technique for separating the components of complex mixtures by means of their different absorption rates on solid media such as a column of inert material or absorbent paper, which is called the stationary phase. The mixture is applied to the start of the paper or column and an inert solvent or gas – the mobile phase – is run through to separate the components of different adsorptive properties which can then be collected separately for analysis or further separation.

chromosomal aberration Any abnormality of chromosome number or structure visible under the light microscope.

chromosome A self-replicating structure consisting of a single strand of DNA complexed with proteins, that carries the genetic information in the form of genes along its length and transmits it from a cell to its descendants and from generation to generation. In eukaryotes, chromosomes are linear structures and are contained within the cell nucleus; most prokaryotes have a single circular 'chromosome'. In interphase nuclei, chromosomes are present as fine extended strands and are not visible in the light microscope. Before cell division, chromosomes become more compact and become visible as rod-like structures in both the light and electron microscope. Human cells contain 22 autosomes, shared by men and women, and two sex chromosomes, the X and the Y, of which men have one copy of each, and women two copies of the X and no Y chromosome.

chromosome abnormality A change in chromosome number or in the structure of a chromosome. Numerical aberrations are designated by a plus or minus sign before the chromosome to indicate additional or missing copies, for example, +21 for a trisomy of chromosome 21. Structural aberrations are designated by single- and three-letter abbreviations, for example t for translocation and inv for inversion. The chromosomes involved in the rearrangement are specified within parentheses immediately following the symbol indicating the type of rearrangement. A semicolon (;) is used to separate the chromosome designations. The breakpoints, given within parentheses, are specified in the same order as the chromosomes involved, and a semicolon is again used to separate the breakpoints. For example, t(9;22)(q34;q11) describes a translocation between chromosomes 9 and 22 at bands q34 and q11 respectively.

chromosome degeneration Loss of genetic functions associated with accumulation of mutations in a sex-limited chromosome.

chromosome paint Fluorescently-labeled probes consisting of DNA complementary to the sequence of an entire chromosome. They are obtained by universal DNA amplification of microdissected or flow-sorted chromosomes and are used to identify chromosomes in the technique of fluorescence *in situ* hybridization (FISH). Probes are either attached directly to a fluorescent hapten or detected after hybridization by tagging with fluorescently labeled antibodies.

chromosome painting probe See chromosome paint.

chromosome territory (CT) A distinct subregion occupied by a particular chromosome in the interphase nucleus of cells of multicellular organisms such as mammals, birds and plants.

chromosome walking Construction of a physical map of a piece of DNA by moving on one cloned segment at a time. Before the completion of the human genome sequence it was the only way of getting from a known point on the genome to a nearby gene suspected, for example, of causing a disease, in order to isolate it.

chylomicron A large triglyceride-rich lipoprotein made by the intestine in response to a meal.

cirrhosis Scarring of the liver, associated with severe liver dysfunction, caused by chronic liver infection or chronic exposure to liver-damaging toxins.

***cis*-acting element** A stretch of DNA that affects the activity of DNA sequences on that same molecule of DNA, usually by binding a transcriptional regulator.

***cis* element** See *cis*-acting element.

***cis*-acting region** See *cis*-acting element.

***cis*-regulatory element** See *cis*-acting element.

clade A group that shares a common ancestor that is exclusive to the group in question. In phylogenetic terms a clade is monophyletic.

cladistic analysis A rigorous method for reconstructing the evolutionary history and relationships of species which aims to reflect the actual sequence of events and avoid artificial groupings.

class switch recombination Switching of immunoglobulin isotype by changing the heavy-chain C gene that is expressed from Cμ to Cγ, Cε or Cα.

classic twin study A study that decomposes the variance of a phenotype into its genetic and environmental components, by comparing the patterns of phenotypic similarity between monozygotic and dizygotic twin pairs.

classical conditioning Type of associative learning in which a conditioned stimulus on its own (for example a bell) elicits a reflex (for example, salivation in dogs) after training by pairing it with an unconditioned stimulus (presentation of food).

classical satellite In human chromosomes, tandemly repeated sequences found mainly on regions around the centromeres of chromosomes 1, 9, 16, on the short arms of acrocentric chromosomes (chomosomes 13, 14, 15, 21 and 22) and on the long arm of the Y chromosome. Three families have been defined: classical satellite 1 includes AT-rich families and one chromosome Y-specific sequence; classical satellites 2 and 3 define two related families containing a 5-base pair ATTCC repeat. Classical satellite 2 is mainly located on chromosomes 1 and 16 and classical satellite 3 on chromosome 9, Y and acrocentric short arms.

classification The process of deciding which of several groups or classes an object (such as a protein sequence) is most similar to.

client-centered counseling A counseling style in which the client, and not a specific problem, is at the center of the counseling process. Originally developed by Carl Rogers, the founder of humanistic psychology.

cline A gradient of allele frequencies, or of genetic diversity, in geographical space.

clinical genetics A medical specialty involved in the diagnosis of genetic disorders and the genetic counseling of affected persons and their families.

clinical trial A biomedical experiment involving humans that follows very precise rules so that a particular drug or treatment can be evaluated for safety and therapeutic potential.

clone library A collection of random fragments of DNA representing part or the whole of a genome, each fragment (a clone) isolated, maintained and amplified as part of a carrier (vector) DNA in a bacterial host.

clone map A representation of a region of human (or other) DNA in a series of overlapping cloned DNA fragments. Clone maps are now mainly of historical interest.

clone (1) The asexually produced, genetically identical, offspring of a single individual, whether produced naturally, as in many organisms, or artificially by embryo splitting or transplantation of a somatic nucleus into an egg.
(2) A cell population derived from a single progenitor.
(3) An individual fragment of human (or other) genomic or cDNA that is being maintained and amplified in a bacterial host.

cloned DNA Multiple exact copies of a single DNA sequence isolated, maintained and amplified by incorporation into a carrier DNA (a vector) and maintained and amplified in a bacterial host.

cloning (1) In general terms, the production of an organism with identical genetic material to another organism. The creation of a population of genetically identical organisms or objects. See also reproductive cloning; therapeutic cloning.
(2) A recombinant DNA technique used to isolate a single fragment of DNA from a genome and produce millions of copies of it. A DNA fragment is spliced into a cloning vector (such as a modified plasmid, artificial chromosome or phage genome). The resulting DNA construct is introduced into a bacterial cell, which then replicates the introduced DNA as it grows and multiplies.

closing a gap Finding the relative positions of enough clones on the physical map of the genome so that the whole genome is covered.

CLSM *See* confocal laser scanning microscopy.

clustering In microarray analysis, a statistical technique for correlating information from individual genes to show those that respond similarly (or differently) in the particular conditions being studied. Dividing data into smaller subclusters helps to identify genes linked to particular responses or patient subpopulations.

cluster analysis A method of selecting genes that have a similar behavior in different samples, from genes that behave in an independent and random way. The most similar genes cluster tightly together; those that are very different in behavior cluster far apart.

cM *See* centimorgan.

Cnidaria Metazoan diploblastic phylum that arose early in animal evolution and which contains jellyfish, corals, sea anemones and hydra.

coactivator A protein that activates the transcription of one or more genes without directly binding to DNA, usually acting by binding to another gene regulatory protein or transcription factor.

coalescence In the reconstruction of a phylogenetic tree or a pedigree, the event whereby two sequences have the same ancestor in the previous generation.

coancestry The sharing between two individuals, or a group of individuals, of one or more common ancestors in the not too distant past. In genetics, the term coancestry (or kinship) coefficient is also used to denote the probability that two copies of a gene, each one drawn at random from one of two individuals, have been inherited from a common ancestor.

Code of Federal Regulations (CFR) The corpus of regulations issued by departments of the executive branch of the United States government. The federal regulations pertaining to the oversight of federally funded research that involves human subjects are found in Title 45, Part 46 of the CFR.

coding sequence (CDS) (1) Sequence of nucleotides that corresponds to the sequence of amino acids in a protein.
(2) Region(s) of deoxyribonucleic acid (DNA) or ribonucleic acid (RNA) whose sequence determines the sequence of amino acids in a protein.

codon capture In the evolution of the genetic code, the reassignment of some codons from repertoires of earlier amino acids to encode new amino acids acquired later.

codon repertoire For a given amino acid, a list of all the codons that encode it. The number varies from one (for example, for methionine) to six (for example, for serine).

codon usage bias The nonrandom usage of different codons for the same amino acid in a gene or genome.

codon A set of three consecutive nucleotides in a strand of RNA (the corresponding set in DNA is formally known as a triplet) that represents the instruction for incorporation of a specific amino acid into a growing polypeptide chain or serves as a termination signal for translation.

coefficient of relatedness (r) The probability that a random gene sampled in one individual is identical by descent to a gene present in another individual. For full siblings, r = 0.5; for half-siblings, r = 0.25.

coefficient of selection (s) Proportional reduction, or increase, in contribution to the gene pool made by individuals of a particular genotype. If there is selection against, then the value of s may vary between zero and one.

coevolution (1) The parallel evolution of two species, or two genes, in which changes in one tend to result in changes in the other.
(2) Concerted evolution of amino acid and codon composition based on similarity of codons for biosynthetically related similar amino acids.

cognition Awareness with perception, reasoning, judgment, intuition and memory.

cognitive appraisal The mental process that people use in assessing whether a demand is threatening and what resources are available to meet the demand.

cohesins Proteins that hold the sister chromatids together following DNA replication.

cohesion The holding of replicated sister chromatids together from S phase until the metaphase-to-anaphase transition.

cohort (1) A group of individuals of the same age in a population.
(2) In the context of genetic counseling, these are a consecutive series of individuals applying for genetic counseling at a clinical genetics center over a particular time.

cohort study A study comparing the incidence of disease in two or more groups of people with different risk factors followed over time.

coiled coil A stable rod-like protein structure formed by two α helices coiled around each other.

coinsurance A percentage of the healthcare provider's fee for which an insured individual is responsible.

collagenous domain The protein domain containing repeating Gly–X–Y amino-acid sequences, found in collagen proteins. Also referred to as the 'triple-helical' domain, as it enables the formation of a triple helix of protein chains.

collision-activated dissociation Fragmentation of peptides induced by collision onto inert gas molecules such as argon.

coloboma Absence of part of the lens of the eye.

combinatorial Of, or relating to, the arrangement of, operation on, and selection of discrete mathematical elements belonging to finite sets or making up geometric configurations.

commissioning parent Person claiming to be mother or father of offspring through arranging for its conception. Also known as intending parent. *See also* genetic parent.

commodification The transformation of values and objects previously not marketed into objects for purchase or sale.

common complex diseases Diseases that are frequent in a population and have a heterogeneous set of causes such as coronary heart disease, hypertension, diabetes and cancer.

common environmental factors Nongenetic sources of phenotypic similarity that are due to people living in a similar environment. Also known as shared environment.

communitarian ethics Seeking to promote the common good, community values, social goals and cooperative virtues.

communitarianism A philosophy that stresses social participation and service to the common good, in contrast to liberal individualism. Individuals have responsibilities as well as rights, and their freedom has to be balanced with the good of the community.

community consent (1) Formal and binding authorization by the legitimate authority of a community on behalf of its members.
(2) An emerging doctrine in the moral literature on informed consent that asserts that among certain self-defined groups in the human population, especially those that have historically been the target of sustained prejudice from other groups, efforts to enroll members as research subjects ought to be preceded by review and approval by an entity that represents the interests of the body politic of that group.

community consultation Dialogue between investigators and community representatives as to the planned research.

community A complex form of human association, in which the group is bound together by common history, traditions, values or interests.

comparative genomic hybridization (CGH) A method of determining differences between genomes by hybridizing different complete genomes to each other. In its simplest application each of two genomes is converted into a set of differentially labeled probes and co-hybridized to a metaphase spread containing a normal complement of chromosomes of the reference genome. The ratio of differential label intensity across each chromosome is a function of the relative contribution of each of the distinct genomes.

comparative genomics Comparing genomic DNA sequences across different species to identify regions of conservation that may indicate genes, regulatory elements or splice variants. This type of analysis can also be used to determine evolutionary relationships.

compensasome Complex of five proteins (encoded by the male-specific lethal genes) and two noncoding RNAs (encoded by the genes *roX1* and *roX2*) that mediates dosage compensation in *Drosophila* flies.

competent cells Cells (usually bacterial) which have been treated so that they readily take up DNA molecules from solution.

complement system A set of plasma proteins that are part of the innate immune response against foreign antigens and are also activated by antibodies made against pathogens in an adaptive immune response. Their activation in the process of complement fixation facilitates the removal of the pathogen by phagocytes or its direct lysis.

complementary DNA (cDNA) DNA molecule made as a copy of an mRNA and thus lacking the introns present in genomic DNA. As they can be easily cloned and replicated, cDNAs are widely used as hybridization probes in place of less stable mRNA and can also be incorporated into an expression vector to make protein.

complex allele Allele resulting from illegitimate recombination or gene conversion rather than from point mutation.

complex disease A disease (such as cancer, heart disease, Alzheimer disease) that is attributed to a combination of several different genes interacting with environmental and random factors.

complex disorder *See* complex disease.

complex genetic disease *See* complex disease.

complex system A type of system that contains a perplexingly large number of diverse parts that interact as a whole and can generate emergent properties. A living cell, for example, is a complex system because life emerges from the interactions of the cell's many component molecules, despite the fact that none of the cell's molecules is alive.

complex trait A heritable trait influenced by multiple genetic and environmental factors.

computational genomics Computational methods for identifying and characterizing genes from DNA and protein sequence data through sequence database searches, statistical analysis, and multiple sequence alignment, gene recognition (exon/intron prediction), identifying signals in unaligned sequences, and integration of genetic and sequence information in biological databases.

concatemer DNA structure consisting of a number of genes connected end-to-end in a chain.

concerted evolution Sequence homology that is maintained among members of a gene family within a species, despite their divergence when compared with family members in other species.

concordance In twins, the extent to which a particular phenotype, for example, height, is correlated between twin pairs. It is expressed as the proportion of second twins sharing the same character as the proband twins.

concurrence Agreement between different investigators in assigning individual items to a set of categories.

cone Type of retinal photoreceptor cell that provide sharp visual acuity and color discrimination. There are several different types of cone cell in the retina, sensitive to different wavelengths of light.

confined placental mosaicism The condition in which mosaicism, mostly mosaic trisomy, is present in the placenta but not in the fetus proper.

conflict theory of imprinting An explanation, derived from kin selection, for the evolution of imprinting in terms of parental allelic conflict over the level of maternal investment in offspring.

conflict An interaction between individuals that results in payoffs and losses to the players according to the behaviors they adopt.

confocal imaging A technique in which a point source is imaged into a pinhole, which is placed in front of the photodetector of a scanning optical microscope, resulting in a strong optical sectioning property that can be used to give three-dimensional images. *See also* confocal laser scanning microscopy.

confocal laser scanning microscopy (CLSM) Technique of microscopy used for generating an image of a section through a three-dimensional structure. It uses a laser as the source of light, and records the image by scanning across a plane through the object, recording the returned fluorescence point-by-point. Images of successive planes can be assembled by computation to give a three-dimensional image. *See also* confocal imaging.

conformational change Alteration in the three-dimensional shape of a protein as a result of reorganization of structural elements without breakage of the protein chain.

consanguineous marriage Marriage between people who are related by descent from a fairly recent common ancestor, and the preferred form of marriage in about 20% of the world's population. The term is usually limited in meaning to marriages between second cousins or individuals who are more closely related.

consanguinity Relation by descent from a common ancestor; blood relationship.

consensus sequence A DNA or protein sequence that represents the most commonly occurring nucleotide or amino acid at each position in, for example, a regulatory DNA element or a protein motif, when all known examples are taken into consideration.

consent A common ethical imperative requiring assent from individuals to authorize any touching of their body or interference with their integrity. Consent must be informed, that is, given in the full knowledge of what will be undertaken, and voluntary in order to be valid. Agreement can be verbal, written, or implied by acquiescence.

consequentialism The ethical theory stating that the moral evaluation of an act exclusively depends upon the consequences of the act.

conserved sequence A sequence of amino acids in a polypeptide or of nucleotides in DNA or RNA that is similar in many different species.

constitutive Not subject to normal processes of regulation.

constitutive heterochromatin Highly repeated, late-replicating and transcriptionally inert chromosomal DNA sequences that remain as heterochromatin throughout most of the cell cycle. In mammals, they are mainly located in centromeric regions and stain after C banding. In humans, the major components of constitutive heterochromatin are classical and α satellites.

constitutive promoter An unregulated promoter that allows for continuous transcription of its associated gene.

construct Plasmid, phage or DNA product of the polymerase chain reaction that have been modified in a specific way.

contextual fear conditioning Classical conditioning in which some aspect of the test chamber is paired with a mild shock.

contig A continuous segment of DNA sequence assembled from overlapping clones.

contiguous gene syndrome A set of clinical symptoms caused by deletion of one chromosome of two or more genes that map close to each other.

control element A region of DNA, such as a promoter or enhancer adjacent to (or within) a gene, that regulates expression of that gene by the binding of transcription factors and other gene regulatory proteins.

convergent evolution Independent evolution of similar traits from different genetic backgrounds.

co-payment A fixed amount per healthcare service that an insured individual must pay.

coping The process by which people try to manage the stress that they experience.

copyright A right to prevent unauthorized use or copying of original literary, artistic or graphic works, which lasts for the life of the creator plus 70 years after his death.

corepressor A protein that represses the transcription of one or more genes without directly binding to DNA. It usually acts by binding to transcription factors or gene regulatory proteins such as repressors.

corporate liberalism Corporate liberalism is a political and/or social philosophy based on: a constitutionally limited government that protects individual (including corporate) rights; a materialistic emphasis on property as the basis for society, especially a free-market system; and a belief in social progress.

correlation A measure of the dependency of one variable on another.

cosegregation (1) The statistical tendency of closely adjacent DNA sequences to remain connected when chromosomes or DNA are broken randomly, and hence to be found together when the fragments are dispersed by meiosis, dilution or other means.
(2) The co-inheritance of two genetic markers from a parent.

cosmid Bacterial cloning vector for use in *Escherichia coli* and which utilizes the lambda phage packaging system and is able to accommodate inserts of genomic DNA between 35 and 45 kilobases.

coverage In DNA sequencing, the mean number of times a nucleotide is represented in a collection of random raw sequence. In practice, only high-quality bases will be counted.

CpG dinucleotide A dinucleotide of adjacent cytosine and guanine (therefore linked by a phosphate group) in DNA. In the mammalian genome, most methylated cytosines occur in CpG residues. CpG dinucleotides occur at about a quarter of the expected frequency in vertebrate DNA, presumably as a result of the deamination of 5-methylcytosine to thymine.

CpG island A GC-rich region of DNA, usually located upstream of a gene, that has more CpG than GpC dinucleotides. CpG islands can be detected by the restriction enzyme *HpaII* and account for 1–2% of the genome. They are located at the 5′ end of all housekeeping genes and on some tissue-specific genes, and in mammals they are generally unmethylated except for certain genes on the mammalian inactive X chromosome.

craniology The science, now regarded as unfounded, of studying the mind and human behavior by measuring the shape and dimensions of the skull.

craniosynostosis Premature fusion of the calvarial sutures in the skull.

creation science A form of biblical literalism couched in scientific language that first appeared in the 1960s and 1970s in the United States.

creationism A generic term for the belief that all life was produced in six days, miraculously, as described in the early chapters of Genesis.

CREB The cAMP response element binding protein, a gene regulatory protein that recognizes the palindromic cAMP regulatory sequence (CRE) in many genes with a role in development and in memory processes in humans. The activity of CREB is regulated by phosphorylation at multiple sites and by its interaction with a variety of other proteins.

Cre-loxP A site-specific recombinase (Cre) and its recognition site (*loxP*), originally isolated from bacteriophage P1.

criminal behavior Actions that contravene accepted legal standards, societal practices or both.

criminal responsibility Legal accountability for one's actions.

criteria of chronology Various single- and multifactor considerations and theories that suggest specific amino-acid chronologies during the evolution of the genetic code.

cross-fostering Transfer of the biological offspring of one set of parents to the care and nurture of a different set of rearing parents.

cross-genomic Describes the analysis of a gene or genes from one organism in a different organism either by bioinformatics methods or through mis-expression experiments.

crossing over Reciprocal exchange between sister chromatids during DNA recombination. The crossover point corresponds to the site of chromatid breakage and rejoining.

crossover *See* crossing over.

cross-validation A method used to validate a selected set of genes or samples, for example, when testing a classifier. One sample or gene from the set is left out and the rest used to classify this gene or sample. This is repeated with all the genes or samples in the set. If a high fraction can be correctly classified by the remaining (*n*−1) genes or samples, it is a good validation.

crown group A taxonomic grouping that includes a set of extant (living) taxa, their last common ancestor and all intermediates.

cruciform DNA Secondary DNA structure, not in the B-DNA conformation, forming at inverted repeat sequences in topologically constrained linear or circular DNA. It acts as a regulatory signal at replication origins.

crustaceans The class of anthropods that includes crabs, shrimps and lobsters. Insects branched off from crustaceans approximately 400 million years ago.

crystallins Several different classes of highly soluble, stable proteins (α, β, γ, and δ crystallins) that give the lens of the eye its transparency.

C terminus *See* carboxyl terminus.

cultural heritability The proportion of the phenotypic variance explained by the familial environment.

culture The central value system of a particular society.

cycle-sequencing reaction A process in which DNA is copied to produce a population of DNA whose elements differ by one base pair and in which the terminal base pair is fluorescently labeled. Subsequent electrophoretic separation of the fragments and detection of the fluorescent labels identifies the sequence of the original DNA strand.

cyclin-dependent kinases A family of enzymes that phosphorylate other proteins and drive progression from one phase of the eukaryotic cell cycle to another. They become active after binding to a partner cyclin protein, whose concentration oscillates during the cell cycle.

cyclobutane pyrimidine dimer The product of the ultraviolet radiation-induced chemical reaction between two thymine or cytosine bases, linking them at the C5 and C6 carbon centers by a cyclobutane ring. It is readily formed in duplex DNA upon exposure to ultraviolet light.

cyclosome *See* anaphase-promoting complex.

cystatin Endogenous protein inhibitor of papain-like cysteine proteases.

cysteine-rich motif A protein motif that has areas with regularly spaced cysteine residues and which usually corresponds to a zinc-finger domain.

cytogenetic band Region on a metaphase chromosome that shows up after particular staining procedures. Each chromosome shows a specific pattern of bands and they are used for low-resolution genetic mapping.

cytogenetic band Differentially stained regions in metaphase chromosomes that are useful for low-resolution mapping.

cytogenetic mapping A type of genetic mapping that relates positions of genes to chromosomal banding patterns. Cytogenetic maps are built from relating the positions of genes to cytogenetic markers or by *in situ* hybridization.

cytogenetics The study of the chromosome constitution of a cell using optical microscopy, especially to detect chromosomal aberrations that may indicate disease.

cytokine A general name for any protein secreted by leukocytes and other cell types that affects the function of another cell through interaction with specific cell-surface receptors. They are distinguished from protein hormones by acting at relatively short range on nearby cells. The term is used particularly for the numerous proteins produced by lymphocytes and other immune-system cells that help generate and coordinate the immune response.

cytoskeleton A complex network of intracellular protein filaments characteristic of eukaryotic cells, but not present in prokaryotes. It consists of actin filaments, intermediate filaments and microtubules. These structures provide internal structure and are variously involved in intracellular movement of organelles, cell division and cell motility.

Darwinism The belief that the main cause of evolution is the mechanism of natural selection proposed by Charles Darwin.

data encoding The process of manipulating or extracting information from data such as a protein sequence to make it suitable for input to a neural network.

data protection legislation Legislation aimed at regulating the acquisition, storage, transfer and processing of personal data.

deadenylation The sequential loss of adenylate residues from the polyadenylate 'tail' present at the 3′ end of mRNAs. This is the initial and rate-limiting step in the mRNA degradation pathway of most transcripts, and the time required for deadenylation to occur therefore defines their functional life span.

decision-making The process of making choices among competing courses of action.

degenerate primers A complex mix of primers for the polymerase chain reaction (PCR) whose sequences have been designed to anneal to multiple potential combinations of DNA sequences within an incompletely characterized DNA template.

deleterious mutation A mutation that is known to be associated with an increased risk of a disease or other undesirable condition.

deletion (1) A mutation that entails the removal of a contiguous segment from a DNA, RNA or protein sequence.
(2) A chromosomal change involving the loss of chromosomal material. Deletions result from one or more breaks in the chromosome and may occur at an end or in the interior of the chromosome.

denaturing high-performance liquid chromatography (DHPLC) A scanning method for detecting single-nucleotide polymorphism (SNP) which is based on the observation that mismatched deoxyribonucleic acid (DNA) duplexes migrate through a special resin at slower speeds than their perfectly matching counterparts during high-performance liquid chromatography at elevated temperatures.

dendrites The branching structures of a neuron that radiate out from the cell body to form synaptic contacts with other neurons and receive signals from them.

dendrogram Bifurcating, tree-like diagram indicating relationships among things.

dense fibrillar component A densely staining region in the nucleolus where the newly synthesized preribosomal RNA accumulates.

deontological ethics An ethical perspective that the moral worth of an individual's actions is not based on their consequences, but on whether the actions are carried out in accordance with (universally applicable) moral rules or principles.

deontology The ethical theory stating that the moral evaluation of an act depends on features of the act itself, not on its consequences.

deoxynucleotide The monomer unit in the DNA polymer chain, consisting of a base (adenine, cytosine, guanine or thymine), the sugar deoxyribose and either a 3′ or 5′ monophosphate.

deoxyribonucleic acid *See* DNA.

depth of field In microscopy, the axial depth of the space on both sides of the object plane within which the object can be moved without detectable loss of sharpness of the image, and within which features of the object appear acceptably sharp in the image while the position of the image plane is maintained.

dermatoglyphics The study of skin ridges on the fingers, palms and soles, which may have a characteristic appearance in some genetic conditions.

dermatoglyphic patterns *See* dermatoglyphics.

descent Culturally defined principle of transmission of group membership. Principles can include patrilineal, matrilineal or bilateral descent.

desmosome A type of cell junction between epithelial cells, characterized by thickening of the apposed cell membranes separated by an extracellular space filled with filamentous material, and attachment of keratin filaments at the intercellular faces.

determinism A philosophical concept arguing that all events, including human behavior, are predetermined. *See also* genetic determinism.

deutaneranopia Relatively common type of X-linked inherited defective color vision in which one of the three cone pigments is lacking. The ability to see blue and yellow is retained, but red and green are not seen.

deuteranomaly Common type of X-linked inherited anomalous color vision in which all three cone pigments are present but one is abnormal or deficient leading to anomalous perception of certain colors.

deuterostomes Animals in which the anus develops from the primary invagination of the blastopore, with the mouth opening developing separately. They include the chordates, echinoderms and a few other phyla.

deviance Any trait or behavior that makes an individual differ from the 'norm' in some way.

DHPLC *See* denaturing high-performance liquid chromatography.

DI *See* donor insemination.

diabetes insipidus A clinical condition characterized by polyuria (urine output greater than three liters/day) involving the excretion of dilute urine. It may be from central causes (lack of antidiuretic hormone production by the pituitary gland) or nephrogenic causes (failure of cells of the distal nephron to respond to antidiuretic hormone).

dichromacy Defective color vision in which one of the three cone pigments is missing.

differential gene loss Random independent loss of a gene in different evolutionary lineages, which creates a gene distribution pattern where the gene is shared between distantly related organisms but absent in more closely related organisms. In phylogenetic studies it can be mistaken for horizontal gene transfer.

differentiated cell A specialized cell that has developed from a single (undifferentiated) cell, the zygote.

digital imaging microscopy A method used in cell biology to quantitatively record the fluorescent signal present in various regions of a suitably labeled cell. The fluorescent light coming from the cells is directed through a camera port to the face of silicon chip in a charge-coupled device (CCD) camera, where the distribution and intensity of light is transformed into electronic signals that can be interpreted and manipulated using computer software.

dimerization The binding together of two molecules; usually applied to the binding of two protein subunits to form a functional protein molecule.

diploblast Animal with a body plan composed of two cellular layers, an ectoderm and an endoderm.

diploid Having two sets of homologous chromosomes, which is the situation in the nuclei of the somatic cells of most eukaryotes. The diploid number is denoted by 2*N*, where *N* is the haploid number of chromosomes.

diplotype A multilocus genotype with unknown linkage phase.

direct damage reversal Form of DNA repair that disrupts abnormal structures in DNA formed as a result of mutagenic factors, for example, the cleavage of thymine dimers by photolyases.

direct repeats Serial repetition of a sequence, for example, ACGT....ACGT.

directed sequencing Sequencing a part of a genome (or other DNA source) near a region of known sequence by designing primers that are based on the known region.

directive European legislation harmonizing law throughout the European Union.

directiveness A counseling style in which the counselor defines the problem and the goals, and openly or covertly tries to exert a direct influence on the feeling, thinking, and decision of the person being counseled.

disability rights The campaign for the extension of civil, social, political and legal rights to all disabled people.

disability Factors or forces that limit an individual's ability to function effectively, defined legally in the US as 'a physical or mental impairment that substantially limits one or more of the major life activities of such individual; a record of such an impairment; or being regarded as having such an impairment.'

disablement The process of becoming disabled.

discourse The use of language to form our concepts and understanding of people or things.

discovery The identification of something that exists in nature in the form that nature would provide.

discrete generations A reproductive strategy in which all parents are the same age, they produce their offspring at the same point in time, and then die.

discriminative Describes selection between alternatives based on some characteristics. Usually implies a negative selection that harms the interests of a person or group.

disease gene A gene which, when altered by mutation, causes a particular disease.

disease mutation A change in DNA that results in a lacking or dysfunctional protein, causing the symptoms of the particular disease.

disease odds ratio The odds of disease among persons with a given factor divided by odds of disease among persons without the factor.

disjunction Segregation of chromosomes at anaphase I to spindle poles or of chromatids at anaphase II.

disomy The condition of having two copies of a given chromosome or genetic locus.

dissociability The ability of particular features or anatomical parts to evolve independently of each other.

distance A measure of the difference between two DNA or protein sequences, which is equal to zero for two identical sequences.

distance geometry A method of analyzing genetic linkage that is tolerant of the fact that estimates of the genetic distance between pairs of markers are prone to error. The markers are first arranged in multidimensional space to accommodate these errors, and best projection of this arrangement onto a one-dimensional line is then calculated.

distributive DNA polymerase A DNA polymerase that synthesizes just a few nucleotides before dissociating from its template.

district general hospital A general hospital providing secondary medical care to a defined health district population, usually between 2 000 000 and 3 000 000 people.

divergence Accumulation of genetic changes after speciation.

dizygotic Originating from two zygotes, or fertilized eggs.

dizygotic twins (DZ twins) 'Fraternal' or non-identical twins, originating from two eggs that are fertilized by two different sperm.

DM *See* double minute.

DNA Deoxyribonucleic acid, the chemical form in which genetic information is stored in all living cells and transmitted from generation to generation. It is a long double-stranded helical molecule (the double helix) consisting of two backbones of alternating sugar (deoxyribose) and phosphate residues with four different kinds of nitrogenous bases (adenine (A), thymine (T), cytosine (C) and guanine (G)) attached to the sugars. The bases along each strand are an exact complement of those along the other strand (A pairs with T and C with G), and the order of bases along the strand encodes genetic information in the form of instructions for making RNA and proteins.

DNA array *See* microarray.

DNA microarray *See* microarray.

DNA demethylation Removal of a methyl group from 5-methylcytosine in DNA to give cytosine.

DNA fiber FISH A method for visualizing the location of DNA fragments on the genome by direct microscopic observation of the positions of hybridization of fluorescently labeled DNA on extended DNA fibers.

DNA fingerprinting Demonstration of genetic individuality by digestion of genomic DNA of an individual by restriction enzymes, separation by agarose gel electrophoresis and hybridization to one or more minisatellite probes.

DNA glycosylase An enzyme that catalyzes the hydrolysis of the covalent bond between N1 of a base in DNA and C1 of the deoxyribose.

DNA hybridization (1) Generally, the ability of a single strand of DNA to form a duplex with a strand of exactly complementary sequence, a property used in numerous techniques in molecular biology to identify DNA sequences.
(2) Technique for comparing sequence similarities between total genomic DNA by observing the annealing of heat-denatured DNA from different species.

DNA library A collection of cloned DNA fragments that represents at least part of the genome of an organism or RNA complement of a particular cell type.

DNA ligase An enzyme which can join (ligate) the ends of two DNA molecules together.

DNA marker Single-nucleotide positions or stretches of DNA of different lengths that are present in at least two different variants in a population and which can be used to track the associated DNA through different generations or in an experiment.

DNA methylation A chemical modification of DNA by the addition of a methyl (CH_3) group, usually to the C5 of cytosine bases in CpG pairs to form 5-methylcytosine, or to N4 of cytosine to give N^4-methylcytosine or to N5 of adenine to form N^6-methyladenine.

DNA methyltransferases Enzymes that catalyze the transfer of a methyl (CH_3) group from *S*-adenosylmethionine to the cytosine or adenine in DNA.

DNA polymerase An enzyme that replicates DNA with reference to a pre-existing DNA or RNA template, using the four deoxyribonucleoside 5'-triphosphates as substrates.

DNA pooling The combination of DNA from individuals within a particular study or disease group, making it possible to screen numerous DNA markers for a systematic scan of the genome for associations between DNA markers and a disease or disease trait.

DNA primase Specialized RNA polymerase that can synthesize an RNA primer at which DNA polymerases can initiate DNA replication.

DNA profiling Demonstration of genetic individuality by typing variable or hypervariable markers at one or more genetic loci simultaneously.

DNA repair The general term for a variety of biochemical reactions that repair DNA after it has been damaged.

DNA replication The process by which DNA is duplicated.

DNA sequencing Determination of the exact order of the base pairs in a segment of DNA.

DNA transposon A mobile segment of DNA that can insert itself into, and excise itself from, genomic DNA.

DNase I-sensitive genomic domains Extended regions of chromosomes (up to 100 kilobases or more) that are preferentially digested by exogenous DNase I introduced into permeabilized cells.

dolichol A polyisoprenoid lipid that acts as a carbohydrate carrier and donor in protein glycosylation in the endoplasmic reticulum.

domain (1) In proteins, a discrete portion of the protein with a distinct compact three-dimensional structure and usually a specific function. Some proteins are composed of a single domain, others of two or more domains. (2) In the nucleus, a section of the nuclear volume occupied by a given genomic region such as a whole chromosome, individual chromosome arm, or single genetic locus.

dominant Describes an allele that is expressed in the phenotype when only one copy is present, and is capable of masking the effect of a recessive allele.

donor insemination (DI) The current term for assisted reproduction by donation of sperm by a 'third party', namely a man other than the intending parent.

donor site *See* 5' splice site.

donor splice site The junction between the 3' end of an intron and the 5' end of the adjacent exon. Also called the 3' splice site.

dopamine Brain neurotransmitter involved in reward response and behavioral control, among other functions.

dosage compensation Differential regulation in the two sexes of genes carried on sex chromosomes, such that the expression levels of these genes are comparable between males and females despite the fact that they are present in different numbers of copies in the two sexes.

double minute (DM) Gene amplification in the form of small, paired extrachromosomal elements that lack centromeres, which can be observed cytogenetically by fluorescent *in situ* hybridization.

double-strand break A break in both strands of the DNA helix at the same place or very close to each other.

DP *See* dynamic programming.

draft sequence An unfinished version of genomic sequence that is produced by assembling overlapping sequence reads from individual clones or from a whole genome. The quality of a draft sequence is usually defined by the redundancy of the high-quality sequenced nucleotides representing the target.

dual specificity Kinases or phosphatases that act on tyrosine in addition to serine and threonine residues in a substrate protein.

duodenal atresia An interruption of the duodenal lumen.

duplication (1) Doubling of a segment of DNA, RNA, or protein sequence so that it is present in two copies in the sequence. Also known as endoduplication.
(2) In cytogenetics, a segment of a chromosome that is present in three instead of two copies in a diploid cell. If the duplication involves an entire chromosome, it is called a trisomy. If a whole genome becomes duplicated this is known as polyploidization.

duplicon One copy of a gene that has been duplicated.

dynamic programming (DP) A type of alogrithm used for making DNA and protein sequence alignments in which a recursive procedure that combines the solutions to subproblems is used to find optimal pathways (optimal alignment) among a series of weighted steps.

dynamics The specification of the changes in frequencies as a function of the current values.

dysgenesis Incorrect or abnormal growth of a tissue outside the normal developmental pattern.

dystonia Sustained simultaneous contraction of agonist and antagonist muscles, resulting in unusual postures that are transiently maintained.

DZ twins *See* dizygotic twins.

ecocentric Viewpoint that gives importance to ecological processes, living in tune with nature, with emphasis on ecobalance, recycling and conservation of natural resources.

ectodermal placodes Localized thickenings of the cranial ectoderm of vertebrate embryos that contribute extensively to the sensory systems of the head.

ectopia lentis Dislocation of the eye lens.

ectopic recombination Recombination, either homologous or non-homologous, between nonallelic regions of the genome.

ectopic transcripts Transcripts that are produced at an extremely low level, potentially from any gene in any tissue.

ectopic Not where normally found.

Edman degradation Sequential analysis of amino acids starting from the free *N*-terminal amino group of proteins and peptides using a series of chemical reactions that splits off and identifies one amino acid at a time.

effective population size A theoretical estimate of the average number of breeding individuals in a population over time that is based on the genetic variation of a population sample. It is the size of an idealized population in which all alleles have the same probability of being passed between generations.

egg donation In assisted reproduction, the donation of ova by a 'third party', namely, a woman other than the intending parent.

EIAV *See* equine infectious anemia virus.

electric field strength Voltage per centimeter separation distance applied on a separation medium, for example a gel. The higher the electric field strength, the faster the migration of a charged molecule.

electroelution Removal of a separated nucleic acid fraction from a gel by applying an electric field.

electroencephalogram A graphical record of the electrical activity of the brain, recorded by an electroencephalograph machine that detects weak currents generated by the brain via electrodes placed on the head.

electroendosmosis Water flow in gels caused by fixed charges exposed to an electric field. Because the usually negative charges cannot migrate, there is a counterflow of protonated water molecules toward the cathode. This results in blurred bands of the material being separated.

electrophoresis A method of separating large molecules (such as DNA fragments or proteins) from a mixture of similar molecules. An electric current is passed through a medium containing the mixture, and each kind of molecule travels through the medium at a different rate, depending on its electric charge and size. Agarose and acrylamide gels are the media commonly used for electrophoresis of proteins and nucleic acids.

electroporation The use of a high-voltage pulse of electrical current to generate transient pores in the cell membrane and facilitate the uptake of exogenous DNA molecules by cells.

electroretinogram (ERG) Record of the electrical changes in the retina after stimulation by light. It is used as a test for evaluating the electrical function of the eye.

electrospray ionization (ESI) The generation of charged species from an aqueous or organic solution.

elliptocytosis A condition in which red blood cells lose their normal biconcave, disk-like appearance and become oblong in shape. It can be seen in inherited and acquired forms and is usually asymptomatic or associated only with mild hemolysis.

ELSI *See* Ethical, Legal, Social Implications.

embryo The early stages of development of a multicellular organism before it is fully formed and capable of independent life. In mammals, the term is applied to the developing zygote from fertilization up to the end of the implantation period. In the earlier literature, the term 'embryo' was sometimes used up until the time (8–10 weeks after fertilization) that the conceptus begins to look like a baby.

embryogenesis The development of an embryo from a fertilized egg.

ELAV protein Any member of a family of RNA-binding proteins (characterized by the embryonic lethal abnormal vision protein from *Drosophila*). The HuR protein from this family appears to bind simultaneously to the AU-rich element and the poly(A) tail of certain mRNAs.

embryonic stem cells (ES cells) Totipotent cells derived from the inner cell mass of early mammalian embryos. Mutated cultured embryonic stem cells are injected into blastocysts to produce transgenic embryos or knockouts.

emergence The process by which a system generates properties beyond those which could be anticipated through analysis of its component parts. Emergence is the product of complexity.

emphysema Lung disease characterized by distension of lung air sacs, with loss of elastic fibers and rupture of their walls, leading to difficulty in breathing.

end repair To convert damaged or incompatible DNA strand ends to a form in which the DNAs can then be used for cloning.

endogamy Laws, customs or rules that specify that couples must marry within defined boundaries (such as a religion or language group).

endogenous retrovirus A retroviral provirus that has integrated into a germline cell and is inherited in mendelian fashion.

endonuclease Enzymatic activity that cleaves DNA or RNA at an internal phosphodiester bond.

endophenotype (1) A phenotype (trait) associated with a disease diagnosis, but which potentially has a more direct biological or biochemical basis and is measurable in both affected and unaffected individuals. An example would be cholesterol level as an endophenotype for coronary heart disease.
(2) Unobserved phenotypes that mediate the effects of genes on observable phenotypes.

endoplasmic reticulum Extensive intracellular organelle into which newly synthesized proteins destined for the plasma membrane, Golgi apparatus, and some other organelles, and for secretion from the cell, are transported. Protein glycosylation also begins within the endoplasmic reticulum.

endoribonuclease An enzyme that cuts RNA at an internal phosphodiester bond.

endosymbiosis A type of symbiosis in which one cell (the endosymbiont) lives inside another cell to their mutual advantage.

endothelium The single layer of cells that lines all blood vessels and acts as an interface between the blood compartment and the vessel wall.

end-stage renal disease (ESRD) Progressive decline in renal function leading to clinical symptoms and necessitating either kidney transplantation or dialysis.

enhancer A regulatory element in DNA that binds an activator protein or protein complex to stimulate gene expression in eukaryotes. Enhancers are able to function at variable distance and variable orientation relative to the promoter of the gene they affect.

enhancer-blocking element *See* insulator.

enucleation Removal of the nucleus from a cell.

environmental architecture of disease An overarching view of the environmental causes of disease. It typically involves knowledge of the different environmental factors, the frequency distribution of those factors, the correlation between factors, and the influence of these environmental factors on disease initiation, progression and severity, as well as the risk factors for the disease.

environmentalism A school of thought holding that cognitive abilities and other psychological traits are almost exclusively determined by environmental (non-genetic) factors.

environmentality A statistic that indicates the proportion of variation in the disease risk or liability in a population that is accounted for by environmental influences.

enzyme A protein that catalyzes a chemical reaction in a living system.

epicanthic folds Vertical folds of skin over the inner corners of eyes.

epigenetic (1) Changes that can affect the phenotype without altering the genotype.
(2) Aspects of development that are not programmed directly by gene action.

epigenetic code Information encoded by modifications of chromatin rather than irreversible changes to the DNA.

epigenetic mechanism Any heritable influence on chromosome or gene function that is not accompanied by a change in DNA sequence.

epigenetic modification Change that can affect phenotype without altering the genotype.

epimutation Inherited and/or acquired alteration in epigenetic modification that causes or predisposes to a disease.

epiphyses Parts of long bones developed from a center of ossification distinct from that of the shaft and separated at first from the latter by a layer of cartilage.

episome A piece of DNA within a cell that is separate from, and much smaller than, the cell's chromosomes, and which can exist as free autonomously replicating DNA or be attached to and integrated into a chromosome.

epistasis Interaction between two genetic loci that has an effect on phenotype.

epistemic modality The use of auxiliary verbs, such as must, may, should or ought, to indicate the speaker's claims to knowledge and his or her level of certainty and uncertainty about the stated claim.

epsilon-modified fluorescent lysine Fluorescently labeled lysine binding a fluorophore at the epsilon amino group. This modified amino acid can be incorporated into nascent proteins during translation through the use of epsilon-modified charged lysine tRNAs.

equine infectious anemia virus (EIAV) A lentivirus that infects horses.

ERG *See* electroretinogram.

ERP *See* event-related potential.

error function In neural networks, a measure of performance of a neural network during training. Typically the error function is the squared difference between actual output values and desired (target values), summed over all input patterns. The goal of training is to reduce the value of the error function to an acceptably small number.

erythema Redness of the skin as a result of increased blood circulation.

erythrocyte membrane skeleton A structural scaffold consisting of interconnected proteins located on the inner surface of the plasma membrane of the red blood cell, which gives the membrane its strength and flexibility. Its main components include α and β spectrin, ankyrin, protein 4.1, protein 4.2 and actin.

ES cells *See* embryonic stem cells.

ESI *See* electrospray ionization.

ESRD *See* end-stage renal disease.

essential gene A gene (or a set of genes) without which an organism cannot exist.

EST *See* expressed sequence tag.

Ethical, Legal, Social Implications (ELSI) A program of the Human Genome Program that was intended to investigate the consequences of human genetics research.

ethnic affiliation estimation A strategy to provide a probability that a subject can be determined by DNA marker analysis to have a particular probability from coming from a select population group.

ethnic minority An ethnic group that forms a minority of a population that does not share their religion, food preferences, family patterns, or language.

ethnicity The real, or perceived, common origins of a people with visions of a shared culture and destiny. For minority ethnic groups in a population it is often associated with stereotyping and social exclusion.

ethnocultural diversity Population diversity arising from a variety of ethnic, racial, and cultural backgrounds.

euchromatin Chromatin that is less compact and stains less intensely than heterochromatin, and which corresponds to the genetically active or potentially active part of the genome, comprising genes, pseudogenes and slightly or mdoerately repeated elements.

eugenics Originally, the 'science' of improving the quality of a human population, which attempted to use principles of mendelian genetics to control reproduction. A popular policy throughout Europe and North America in the early and mid-twentieth century, it fell out of favour partly as a result of revulsion against the eugenic ethos of the Nazi Third Reich, and also as the limitations of using selective breeding to modify population characteristics became apparent. Debate over the ethics of eugenics has revived with the advent of more effective methods of genetic manipulation.

eukaryote An organism whose cells (or cell in single-celled eukaryotes) are divided into a distinct membrane-bounded nucleus, in which the DNA is located, and cytoplasm. Eukaryotes constitute one of the three domains of life along with bacteria and archaea.

Euler tour An ordering of all the arcs in a graph that contains each arc exactly once, with the end point of each arc being the starting point of the next.

euploid The normal state of possessing a complete set of chromosomes characteristic of the species.

Euro-American Anthropologists' designation of discourse (for example, on life, society, nature) originating in self-defined 'Western' societies but not confined to them.

eutherian A mammal with an allantoic placenta and which gives birth to live young after a long period of gestation, that is all mammals except the marsupials and monotremes.

E-value *See* expectation value.

event-related potential (ERP) An electroencephalogram (EEG) activity pattern that is evoked by a specific stimulus such as a light or a sound.

evo-devo Evolutionary developmental genetics, the study of the macroevolution of genes with major effects on development.

evolution The development of new species of organism from pre-existing types by the accumulation of genetic differences over long periods of time.

evolutionary tree A graphic representation of the evolutionary relationships among alleles at a given locus (gene tree) or of the genealogical relationships among individuals in a population (population tree). A gene tree and a population tree are not the same thing.

exocrine pancreatic insufficiency Inadequate supply of enzymes from the pancreas to digest food.

exon definition Interaction between U1 small nuclear (sn) RNP at a downstream 5′ splice site and U2 snRNP at an upstream 3′ splice site, which promotes splicing of the upstream intron.

exon shuffling Intron-mediated genomic rearrangement leading to new combinations of exons. Exon shuffling may result in the formation of chimeric genes consisting of exons of different genes (gene fusion, exon insertion) or in intragenic exon duplication.

exon skipping Alternative splicing of a transcript in which splice junction sites that are normally used in RNA splicing are not used, resulting in deletions of exons from the final product.

exon trapping A technique for identifying putative exon sequences within a cloned genomic DNA by taking advantage of their capability of becoming spliced to exons within a specialized vector.

exon A coding portion of part of a eukaryotic gene, separated in the genomic DNA from other coding regions of the same gene by stretches of noncoding DNA called introns. After transcription of the gene, introns are spliced out and the exons joined together to form a mature mRNA (or other functional RNA).

exonic splicing enhancer A particular sequence in an exon that interacts with SR proteins and promotes constitutive and regulated RNA splicing.

exonuclease An enzyme that degrades nucleic acid molecules by progressively shortening them from one end. Exonucleases typically exhibit a specific directionality of action and are referred to as 5′ → 3′ or 3′ → 5′ exonucleases.

ex vivo gene therapy Gene therapy in which cells are obtained from an individual, cultivated in the laboratory and incubated with vectors carrying a corrective or therapeutic gene. Cells containing the new genetic information are then transplanted back into the person from whom they were derived.

ex vivo transfection Technique in which cells are removed from the body, transfected *in vitro* and then reintroduced into the body.

ex vivo Outside the body.

expectation value (E-value) The number of times an event (e.g. a score greater than 100) is expected to occur based on a random model. Used as a means of assessing similarity scores between two nucleotide or amino-acid sequences when searching databases. Sequence similarity score E-values (e) can range from zero to the size of the database being examined. E-values less than 0.001 generally indicate homology (common biological ancestry).

expected frequency Average, weighted by probability of occurrence, of possible frequencies.

expected value The expected value of a variable is the long-run average value of that variable.

experiential counseling A development of the client-centered counseling approach in which the counselor is led through the counseling process by the 'experience', in other words the feeling and thinking of the person being counseled.

exposure odds ratio In a case-control study, the odds of a given factor (exposure or attribute) among cases divided by the odds of the factor among controls.

expressed sequence tag (EST) A short strand of DNA (200–400 bases) that is part of a cDNA molecule and can act as an identifier for a gene. Used in locating and mapping genes.

expression vector A plasmid that carries a DNA sequence into a suitable host cell and there directs the synthesis of the protein encoded by the sequence.

extension The property of being physically located, of occupying space.

externalities Effects of an action that are not considered in deciding whether or not to go ahead with that action.

extracellular Outside the cell.

extracellular matrix An insoluble network of proteins secreted from cells that provides structural support for tissues and plays an important role in cellular development and function.

factor V (FV) A blood coagulation protein that after activation functions as a cofactor to factor X in the activation of prothrombin.

factor V Leiden A mutant form of factor V which has glutamine at position 506 compared with the normal arginine.

factor IX A blood coagulation protein, a deficiency of which leads to the bleeding disorder hemophilia B (Christmas disease).

factor VIII A blood coagulation protein, a deficiency of which leads to the bleeding disorder hemophilia A.

facultative heterochromatin Chromosomes or parts of chromosomes that are found as heterochromatin only under certain conditions, such as the X-inactive chromosome in the female genome. This chromosome forms a condensed and globally late-replicating structure in human nuclei named the Barr body.

familial Describes any trait that occurs more often in the relatives of an affected person than in the general population, that is traits that tend to 'run in families'.

familial defective apolipoprotein B (FDB) A hypercholesterolemia inherited in autosomal dominant fashion, often incompletely penetrant, and attributable to defects in the *APOB* gene which encodes apolipoprotein B, the ligand that enables clearance of low-density lipoprotein from plasma.

familial hypercholesterolemia (FH) The historic description of severe autosomal dominant hypercholesterolemia leading to early coronary disease, attributable to defects in the *LDLR* gene.

familial disease Disease transmitted through and expressed by members of a family.

14-3-3 family Protein family whose members all contain the 14-3-3 protein interaction domain.

family classification Classification of biomolecules into large classes or families of molecules with similar structure and/or function.

family history information Information about the illnesses suffered by parents or other close relatives, and in particular, where applicable, the cause of their death, usually in the context of disclosures required by an insurer of a proposer for insurance, in order to inform the underwriting process.

family systems Interactive effects in relationships among relatives.

Far-Western Describes a modification of the Western blot technique for identifying proteins, in which a protein–protein interaction is identified using a labeled protein as probe.

FCS *See* fluorescence correlation spectroscopy.

FDB *See* familial defective apolipoprotein B.

feature extraction The process of manipulating input sequences to create or discover units of biologically relevant information that may provide insight into their structure or function.

feedback loop The mechanism by which a gene product stimulates or inhibits its own production.

fee-for-service A payment system that pays healthcare providers for each additional service they render.

fetus The developing mammal from the end of implantation, once the primitive streak has formed, to birth.

FH *See* familial hypercholesterolemia.

fibrillar center A lightly staining region of the nucleolus that is surrounded by the dense fibrillar component.

fibrillar collagen Collagen that assembles into cross-striated fibrils. These types of collagen contain a long, uninterrupted triple-helical domain of approximately 1000 amino acids flanked by propeptides at the amino and carboxy termini.

fibroblast Connective tissue cell that secretes extracellular matrix and which is the source of chondroblasts (cartilage-producing cell), collagenoblasts (collagen-producing cell) and osteoblasts (bone-producing cells).

finished sequence Complete sequence of a cloned DNA or a genome, with an accuracy of at least 99.99% and few or no gaps. In practice some sequences remain unclonable by current techniques and so cannot be determined. Finishing is also used to describe the process of obtaining the finished sequence after initial shotgun, by using a variety of directed sequencing approaches.

first-degree relative Parent, sibling or child of the person in question.

FISH *See* fluorescence *In situ* hybridization.

fitness Difference in reproductive success of an individual or genotype relative to another.

FIV Feline immunodeficiency virus. A pathogenic lentivirus of cats.

fixation The complete replacement of one allele (or other DNA element) by another in a given population.

FLP/FRT A technique for making deletions in the DNA of somatic cells using the flip recombinase and sequence-specific recombinase target (FRT) sites.

fluorescence correlation spectroscopy (FCS) A method used to measure the rate of movement of fluorescent molecules in a very small volume. A beam of light is directed into a sample, and the total intensity of light is measured repetitively at very short intervals. The more fluctuation in the intensity, the more rapidly the fluorescent molecules are moving in and out of the sampled volume. Using various statistical algorithms, data can be analyzed to estimate an apparent diffusion coefficient and fraction of each component present.

fluorescence *in situ* hybridization (FISH) Visualization of specific DNA targets in metaphase chromosomes or fixed interphase nuclei by hybridizing a single-stranded fluorochrome-labeled DNA probe to its complementary DNA sequence on chromosomal DNA.

fluorescence recovery after photobleaching (FRAP) A method used to measure a diffusion coefficient for fluorescently labeled molecules in cells. A beam of high-intensity light is directed into a cell and the fluorescence in the targeted spot is completely bleached. The recovery, or influx, of fluorescent molecules into the bleached area, is monitored using a photomultiplier tube, and the rate of increase of intensity is proportional to the diffusion coefficient.

fluorochrome A molecule that, when excited by a certain wavelength of light, emits fluorescent light of a particular color. Different fluorochromes emit different colors. They can be attached to proteins, nucleotides or other chemicals and observed using a fluorescence microscope.

fluorophore A chemical group that fluoresces strongly when suitably illuminated, and which can be observed using a fluorescence microscope.

foam cell A macrophage that has accumulated esterified cholesterol in an atherosclerotic plaque.

focal adhesion A cell junction consisting of clustered integrins for cell attachment to the extracellular matrix. Focal adhesions are attached to the actin filaments of the cytoskeleton.

fold The three-dimensional pathway inn space of the polypeptide backbone.

forensics The attempt to use scientific methods, usually at the crime scene, to find information that can be used in a court of law.

fosmid Bacterial cloning vector that contains the lambda phage packaging system and the *Escherichia coli* F factor replication origin so that it is maintained at a single copy per cell.

founder effect Increased prevalence of a specific allele in a population owing to its existence in one of the original founders of the population.

founder population The individuals from whom the population derives. The alleles carried by these individuals form the gene pool of this population.

Fourier transform ion cyclotron resonance (FTICR) A form of mass spectroscopy in which all ionic analyte species are forced into circular trajectories by means of high magnetic fields. As the frequency of the ions' circular movements, which induce a so-called image current between two detector plates, is determined by the mass-to-charge ratio of ions, a mass spectrum can be obtained by Fourier transforming the image current.

fragile X syndrome A genetic condition, due to an aberration of the X chromosome, which is characterized by enlarged jaw, ears and testes, a high forehead and mental retardation in most men, and mild mental retardation in women.

fragment A sequenced piece of DNA.

frameshift Changes in the sequence that cause a shift of the normal reading frame used during translation of the RNA sequence into a protein.

FRAP *See* fluorescence recovery after photobleaching.

free will A philosophical concept arguing that human behavior is the result of individual free choice.

free-enterprise liberalism Ideology that seeks to minimize the restrictions of the state on individual activity, especially economic activity.

frequency Proportion of successes in a particular sample.

FTICR *See* Fourier transform ion cyclotron resonance.

function In mathematics, a mapping (or rule of assignment) of one or more variables into another.

functional genomics The scientific field of understanding how cell function is directed and modulated by genetic variation and control of gene expression, using information from complete genome sequences.

functional psychoses A broad (and somewhat poorly defined) term for the set of severe psychiatric illnesses that include delusions, hallucinations, other psychotic symptoms or severe mood disturbance that historically were not known to have a clear organic cause. They include schizophrenia, bipolar disorder and illnesses that share features of both of these.

functional redundancy The situation when complete elimination of a particular isoform or isozyme has no functional impact.

fundamentalism An early twentieth century form of Creationism, which led eventually to the prosecution (and conviction) of the teacher John Thomas Scopes, in 1925, in Dayton, Tennessee.

fundus The interior posterior surface of the eyeball that includes retina, optic disc, macula, and posterior pole and which is visible through the ophthalmoscope.

fusion protein A protein composed of two different proteins (or parts of two proteins) joined end to end.

G+C content *See* GC content.

G banding A staining technique using the dye Giemsa that generates a banded pattern on metaphase chromosomes which distinguishes the different chromosomes.

G bands Specific chromosome regions in metaphase chromosomes that show up on Giemsa staining. They are late-replicating at metaphase, relatively AT-rich and have relatively few genes. They nearly coincide with the Q bands generated by staining with the fluorochrome quinacrine.

G1 phase The phase of the cell cycle (first gap or growth phase) that follows cell division and precedes DNA replication.

G protein Any of a large family of proteins with intrinsic GTPase activity that undergo large conformational changes on binding GTP or GDP. Many act as molecular switches in signal transduction pathways.

gain of function Describes a mutant allele that causes an increase in normal function or the gain of a new phenotype, rather than loss of the normal function.

GAL4/UAS Mis-expression system developed for targeted gene expression in *Drosophila*. Requires yeast GAL4 protein *in trans* and a yeast upstream activator sequence (UAS) in *cis*.

GAL4 Yeast transcription factor with discrete DNA-binding and transcriptional activation domains which has been used in the yeast two-hybrid system.

gamete A haploid reproductive cell that will fuse with another gamete to form a diploid zygote. In mammals they are the mature sperm and the egg cells.

gametic phase Used to indicate a given combination of specific alleles on the same chromosome in the gamete.

gap penalty function Formula that computes the penalty associated with a gap in a sequence alignment given its length.

gap In a sequence alignment, a group of consecutive spaces inserted in a sequence. For instance, the sequence TCA---GAC---ATCC has two gaps.

GC content Percentage of guanine plus cytosine in a given piece of DNA.

GC$_3$ The average GC level at third codon positions in a coding sequence.

GC-rich A region of DNA that contains a higher proportion of guanine and cytosine bases than the rest of the DNA.

gender The socially-constructed behavioral and psychological traits typically associated with one or the other sex.

gene A functional unit of inheritance composed of a specific sequence of bases along the DNA molecule. Most genes encode instructions for making proteins in the sequence of their bases, but some produce functional RNAs such as rRNAs and tRNAs.

genealogy A lineage of descent from an ancestor; the study of lineages and of descent.

gene cluster A genomic region that contains a group of tightly linked genes of similar sequence. Gene clusters have arisen by duplication and divergence of an ancestral gene.

gene clustering The close physical arrangement on a chromosome of genes whose products contribute to a single function.

gene conversion (1) Nonreciprocal transfer of genetic information between highly homologous loci.
(2) The modification of one of two alleles of the same gene by the other. It involves a nonreciprocal recombination event that replaces an 'acceptor' gene or DNA sequence by a 'donor' sequence which itself remains physically unchanged.

gene desert Region of the genome that is devoid of or has very few genes.

gene dosage effect The effect of the quantity (dosage) of a gene product (protein) produced by one of a pair of alleles of a gene compared with the amount produced by both.

gene duplication The generation of an additional copy of a gene in a genome. A process by which paralogous genes can be created in a chromosome, whereby a single ancestral gene may give rise to multiple descendant genes with distinct functions.

gene–environment interaction A nonadditive statistical relationship between genotype variation and environmental variation in the explanation of phenotype variation. It indicates that genotype and environment are not separate, independent contributors to difference in phenotypes but act synergistically to create more or less than the sum of their parts.

gene expression network A set of genes that interact, often in a hierarchical manner, to perform a specific cellular function(s).

gene expression The initiation of transcription and the subsequent events (transcription, RNA splicing, RNA editing (if applicable) and translation of mRNA) that result in production of the protein product of a gene.

gene fission An evolutionary change in which one gene is partitioned into two or more separate genes.

gene flow The exchange of genes between different populations of the same species as a result of migration of individuals.

gene genealogy A phylogeny that relates how different alleles found for a given genetic locus have mutated over time.

gene knockout The procedure of permanently deleting all or part of a gene in the genome by homologous DNA recombination, resulting in loss of the function of the gene.

gene map A depiction of the genome or genomic region that shows the relative positions of particular genes.

gene penetrance That proportion of individuals with a particular allele who actually show its effect.

gene pool The total genetic information present in the reproductive members of a population of sexually reproducing organisms.

gene prediction Finding genes in a genome sequence by sequence similarity searches (homology methods) or by using programs that rely only on the statistical qualities of exons (*ab initio* methods).

gene signature A set of genes linked to a particular event or process in a cell.

gene targeting The procedure of replacing a gene in the genome with a mutant copy via homologous recombination, which is used to produce transgenic or gene knockout animals.

gene therapy The correction of a genetic defect by the introduction of normal copies of the gene into bone marrow or other cell types.

gene transfer The deliberate transfer of DNA sequences containing relevant genetic information for experimental or therapeutic purposes into somatic cells, or into germ cells for the generation of transgenic animals.

gene tree Graphical depiction of the history of descent of a particular stretch of DNA.

generic consent An all-purpose consent to multiplex screening.

genetic anticipation Earlier onset and increased severity of a disease in later generations.

genetic architecture An overarching view of the genetic causes of disease. It typically involves knowledge of the number of genes, the number of functional alleles in each gene, the relative frequency of those alleles or haplotypes, and the influence of these allelic variations on disease initiation, progression, severity, as well as its risk factors.

genetic carrier testing Testing the carrier status of a healthy individual or a couple known to be at high risk because of a positive family history, so that they can make reproductive decisions.

genetic counseling A process, involving an individual or family, comprising evaluation to confirm, diagnose or exclude a genetic condition, malformation syndrome or isolated birth defect; discussion of natural history and the role of heredity; identification of medical management issues; calculation and communication of genetic risks; and provision of or referral for psychosocial support.

genetic defense The use of behavioral genetic evidence in a criminal's defense.

genetic determinism (1) The belief that genes are the only or chief determinant of the future development of an organism, that genes provide the essence of the organism's nature.
(2) The claim that all the significant characters of the mature organism are predetermined by the genes.

genetic discrimination Discrimination, especially in regard to socioeconomic matters, against an otherwise healthy individual on the basis of genetic information that suggests an unusually high predisposition to future illness or to reproductive risks.

genetic distance The average amount of DNA sequence difference between two populations.

genetic drift A process by which the number of copies of an allele in a population changes randomly over time due to variation in the numbers of offspring of individuals.

genetic enhancement The amplification of 'normal' genes in order to amplify a normal trait.

genetic epidemiology Study of the prevalence of a disorder in a genetically informative sample, such as families, twin pairs or adoptees.

genetic essentialism The belief that human beings in all their complexity are products of a molecular text.

genetic harm Damage done to an individual by the possession of a particular gene (or set of genes).

genetic heritability The proportion of the phenotypic variance explained by genetic factors.

genetic heterogeneity The phenomenon that a similar phenotype can be caused by different allelic or nonallelic mutations.

genetic hitchhiking The phenomenon that, as a selectively favored allele at a locus is driven to high frequency by positive selection, alleles at nearby loci on the chromosome also become more frequent as a result of physical linkage.

genetic–environmental complexity Interactions among genetic and environmental influences on a trait or behavior that make it impossible to separate the genetic from the environmental causes of the variation of that trait between individuals.

genetic interference The phenomenon that formation of one chiasma (crossover) on a chromosome in meiosis inhibits the formation of others in its vicinity.

geneticization The social–cultural process of interpreting and explaining human beings within the terminology and concepts of genetics.

genetic linkage (1) The inheritance together of particular alleles at two or more genetic loci, which results from their physical proximity on the DNA. (2) The correlation between a particular chromosomal region or genetic locus and a particular phenotype, for example a disease phenotype.

genetic linkage mapping Genetic linkage depends on the identification of co-inheritance of a particular chromosomal region with disease. A genome screen typically involves the typing of polymorphisms spaced 5–20 megabases (Mb) apart in members of families with disease. In a complex disorder such as asthma, disease genes are usually mapped to within 10–30 Mb of DNA. This interval is too large for simple gene identification.

genetic locus A specific location on a chromosome which is occupied by an allele of a gene.

genetic map The linear arrangement of genes and DNA markers on a chromosome, as determined by recombination frequencies between pairs of genes.

genetic mapping Ascertaining the relative positions of genes or markers on a chromosome by means of the frequency of recombination between pairs of genes. This mapping method is based on the fact that the frequency of crossover between two genes on the same chromosome during meiosis is directly related to the physical distance between them.

genetic marker A DNA sequence on a chromosome that is easily detectable and which can be used to map the chromosome. It may be part of coding or noncoding DNA.

genetic parent Person claiming to be mother or father of offspring through genetic endowment via sperm or ova.

genetic polymorphism The case where there are two or more variant alleles of a gene each occurring at a frequency greater than 1% in the population.

genetic recombination The reciprocal exchange of DNA between maternal and paternal homologous chromosomes at meiosis, leading to new combinations of alleles on the resulting recombinant chromosomes in the gametes.

genetic resistance to disease Absolute or relative resistance to disease conferred by particular genetic variants.

genetic screen A search for genetic mutants with particular phenotypes.

genetic screening A search in a population of apparently healthy individuals for genotypes that may have adverse effects. Tests such as analysis of DNA, detection of chromosome abnormalities or of biochemical correlates of a particular condition are the usual way in which the search is conducted.

genetic test The analysis of DNA, RNA, chromosomes, proteins or certain metabolites with the aim of detecting inherited disease-related genotypes, mutations, phenotypes or karyotypes to predict risk of disease, identify carriers and establish prenatal and postnatal diagnosis or prognosis.

genetic variation Differences in the sequence of chromosomal DNA among individuals of the same species.

genetics The science that deals with the structure, functions and inheritance of genes and of inherited variation.

genodermatosis Any hereditary disorder of epithelium.

genome The total genetic information contained in a single chromosome set of a diploid cell. The human nuclear genome is composed of approximately three billion nucleotide pairs.

genome informatics The systematic development and application of computing and mathematical techniques for analyzing and making predictions based on genomic DNA sequences and related data.

genome integrity Maintenance of the genomic DNA sequence without alterations.

genome project A group of investigators cooperating to produce a complete sequence of the genome of a particular species or set of species.

genome scan An analysis of the whole genome of an individual against a set of markers whose positions on the chromosomes are well known, in order to determine common patterns of inheritance between these markers and disease characteristics.

genome survey sequences (GSS) Short genomic DNA sequences that are representative of a genome.

genomic cluster A region on the human chromosome that contains multiple related genes whose gene products contribute to either the expression of related proteins or subunits that contribute to the formation of a distinct protein complex.

genomic disorder A human disease resulting from a recurrent DNA rearrangement due to nonallelic homologous recombination between region-specific, low-copy repeats. This rearrangement leads to the loss or gain of dosage-sensitive genes.

genomic DNA microarray Slides on which cloned genomic DNA segments have been spotted at high density. *See also* microarray.

genomic imprinting The expression of only the paternal or the maternal copy of some genes in the offspring, thus leading to functional monosomy. Paternal imprinting means that an allele inherited from the father is not expressed in offspring. Maternal imprinting means that an allele inherited from the mother is not expressed in offspring.

genomic instability The tendency for a mutated genome to accumulate an increasing number of further mutations or aberrations over time as compared with a wild-type genome.

genomic junk Apparently superfluous genetic material that is not known to code for any RNA and/or protein and that serves no other function in the nucleus of the cell and the organism as a whole.

genomics Area of research that applies both computational and experimental techniques to the description, cataloguing and analysis of whole genomes and the application of that information. It is based on the determination of the DNA sequences of genes, chromosomes and whole genomes and the large-scale analysis of gene expression.

genotoxic Describes substances that are harmful to the genetic material, for example drugs that cause damage to DNA.

genotoxins Agents which are capable of causing damage to DNA.

genotype The particular alleles present at a given genetic locus, or set of specified loci, in an individual.

genotype–environment interaction *See* gene–environment interaction.

genotype relative risk (GRR) The relative disease risk associated with two different genotypes (for example, *Aa* and *aa*) at a given locus.

genotyping Analysis of the genetic constitution (genotype) of an individual by detecting the alleles present at particular loci, or by the detection of particular DNA polymorphisms, mutations and marker sequences.

germ cell Sperm or egg cell.

germ-cell fate determinants Maternal proteins and RNAs that cause cells of the embryo to become germ-cell precursors.

germ-cell fate induction The process by which signals within the embryo cause pluripotent embryonic cells to become germ-cell precursors.

germ line The lineage of cells that produces the reproductive cells (gametes) that transmit the genetic information to the next generation of offspring.

germ-line gene therapy The transfer of genetic material into the gametes with the aim of correcting a genetic defect, resulting in permanent change to the germ line.

germ-line mosaicism Multiple germ cells (egg and sperm) from one parent with a change in the genetic makeup.

germ-line genetic modification A modification resulting in a change in the genetic make-up of sperm or ova.

germ-line mutation A mutation in a germ cell that will thus be present in all cells of the organism that develops from that germ cell and which can be inherited by future generations.

germ plasm (1) August Weismann's term for the material substance which transmits characters from parent to offspring, located in the cell nucleus. Equivalent to the genome in modern terminology. (2) Fibrogranular structures in the cytoplasm of oocytes and germline cells of embryos in some species that are associated with germ-cell fate determinants.

GFP *See* green fluorescent protein.

Giemsa Stain used to produce characteristic dark and light banding patterns on metaphase chromosomes.

Glanzmann's thrombasthenia An autosomal hereditary disorder caused by defects in integrin $\alpha_{IIb}\beta_3$ which result in an abnormal susceptibility to bleeding.

glomerulonephritis A disease involving inflammation predominantly of the renal glomerulus, most likely of immunological origin. It manifests clinically with acute renal failure, hypertension, proteinuria (appearance of protein in the urine) and hematuria (red blood cells in the urine).

glomerulosclerosis A pathologic term describing scarring or fibrosis involving the glomerulus. Focal or diffuse refers to the number of glomeruli involved, whereas segmental or global refers to involvement of only portions or of the entire glomerulus.

glomerulus A highly specialized structure in the kidney consisting of a capillary tuft supported on a basement membrane lined by podocytes, specialized renal epithelial cells. Its primary function is selective ultrafiltration of plasma so that essential proteins are retained in the blood.

glycomics Study of the glycome, the whole set of glycans (oligosaccharide chains) produced in an organism.

glycosidase An enzyme that hydrolyzes glycosidic bonds between adjacent monosaccharides.

glycosylation Covalent addition of oligosaccharide, usually to a protein, but also to lipids or other oligosaccharides.

glycosyltransferase An enzyme that catalyzes the transfer of monosaccharide from an activated donor to protein, lipid or oligosaccharide acceptors.

golden path assembly One of the publicly available databases of assembled sequence from the human genome. The sequence is arranged by chromosome and features such as genes, polymorphisms, genetic markers and repeats are displayed.

Golgi complex Network of intracellular membrane vesicles in which the oligosarrachride chains of newly synthesized glycoproteins are further modified, and through which glycoproteins are transported from the endoplasmic reticulum to the plasma membrane.

gonadotropes Anterior pituitary cells that synthesize gonadotropin hormones (FSH, LH).

GPCR *See* G-protein-coupled receptor.

G-protein-coupled receptor (GPCR) A cell-surface receptor that acts via coupling to an intracellular heterotrimeric G protein. G-protein-coupled receptors comprise a very large superfamily of transmembrane proteins with seven transmembrane regions.

G-protein-regulated effectors Enzymes and ion channels whose activity is regulated by activated subunits of heterotrimeric G proteins.

graft tolerance Absence of a destructive anti-graft immune response in the short and long term in recipients not receiving immunosuppressors and with normal immune responses against other antigens.

graft-versus-host disease Disease caused when mature T cells introduced as part of a tissue graft into a non-identical immunoincompetent recipient (for example, in bone marrow transplantation) attack the recipient. Tissues affected typically are the skin, the intestinal tract and the liver.

granular component A region in the nucleolus where nearly mature preribosomal complexes accumulate and which have a punctate or granular appearance under electron microscopy.

granulocyte White blood cell containing granules filled with oxidative enzymes that can kill ingested microbes. They comprise the neutrophils, eosinophils and basophils.

graph An abstract model consisting of points (vertices) and directed lines (arcs) joining them.

graphical model A model describing the dependencies (causal and noncausal) among a set of variables. Dependencies may be quantified with joint or conditional probabilities.

gravity center Gravity of gray levels over the image coordinates.

green fluorescent protein (GFP) A fluorescent protein isolated from jellyfish that absorbs blue light and emits green light. By fusing the gene for this protein to that of another protein a recombinant protein is expressed that can be visualized in a living cell.

group II intron Any of a class of self-splicing introns found in bacteria, mitochondria and chloroplast genes. Some group II introns encode a reverse transcriptase and are mobile.

group insurance Insurance that charges all individuals the same premium regardless of health risks, and that an individual is eligible for by virtue of being a member of a group, for example an employee.

GRR *See* genotype relative risk.

GSS *See* genome survey sequences.

GTPase A protein that hydrolyzes GTP to GDP, releasing inorganic phosphate. GTPases are, for example, part of certain switch proteins in signal transduction pathways and of the tubulin proteins that assemble to form microtubules.

HA *See* heteroduplex analysis.

β hairpin Element of protein secondary structure in which a segment of the polypeptide chain bends back sharply on itself.

haploid Having one set of chromosomes, which is the situation in a normal gamete. In humans, the haploid number (N) is 23.

haploinsufficiency Insufficiency of gene product, resulting in a distinct phenotype, caused by the presence of only one functional copy of a gene instead of the usual two.

haplotype Set of alleles found together at neighboring loci in the same chromosome of a homologous pair and which is usually inherited as a unit from one parent.

haplotype map A resource that identifies and describes the location of blocks of SNPs on a chromosome that are inherited together, including a subset of specific SNPs that identify these blocks. This resource will be used to do efficient association studies that relate genetic variation to disease or response to drugs.

HapMap Haplotype map; the initiative of the Human Genome Project to identify blocks of sequences maintained by linkage disequilibrium and the common haplotypes in the population.

hapten A chemical group that can be tightly bound by a specific antibody or other protein.

Hardy–Weinberg equilibrium Ideal population state in which the proportions of alernative genotypes are as predicted by the product rule.

harm avoidance A measure of anxiety proneness that is one of the dimensions of personality assessed by the Tri-dimensional Personality Questionnaire (TPQ).

harmful genetic information Information about an individual's (or group of individuals') genes the disclosure of which harms, typically by rendering this individual (or group) or his or her relatives worse off.

HBOC *See* hereditary breast–ovarian cancer.

HDL *See* high-density lipoprotein.

health behavior Behavioral patterns, actions and habits that affect health, either positively or negatively.

Hebb–Williams maze A cognitive task for rodents in which the animal must navigate through a classic maze design in order to get to a goal box.

hegemony Political dominance of an authority supported by power.

helical repeat The number of base pairs present in a complete turn of the DNA helix.

helicase Any of a family of ATP-driven proteins responsible for unwinding nucleic acid duplexes in RNA and DNA during recombination, replication, repair, transcription and RNA splicing.

α helix Element of protein secondary structure in which the polypeptide chain forms a regular helical coil.

helper CD4 T lymphocyte Type of T cell bearing the CD4 cell-surface protein and which helps stimulate B cells to produce antibodies in an immune response. A broader definition also includes other CD4$^+$ T cells that stimulate macrophages.

helper virus Virus that is capable of replication. This term originally referred to replication-competent viruses that 'helped' replication-defective oncogenic retroviruses to replicate.

hematopoiesis The generation of blood cells.

hematopoietic growth factors Cytokines that affect the activities of hematopoietic cells such as proliferation or quiescence, survival, patterns of gene expression, movement or activation, etc.

hematopoietic stem cell Self-renewing cell that can give rise to all blood cell types. Hematopoietic stem cells are obtained from bone marrow but can also be isolated from the blood following mobilization from the bone marrow by administration of cytokines.

heme Red pigment formed by insertion of ferrous iron into the center of the protoporphyrin IX macrocycle and which forms the reactive center of proteins involved in the transport and metabolism of oxygen.

hemidesmosome Specialized cell junction that joins epithelial cells to their basement membranes, and which are attached internally to intermediate filaments..

hemimethylated DNA DNA that is methylated on the parental DNA strand and unmethylated on the newly synthesized strand.

hemizygous Having only one copy of a gene or group of genes normally present in two copies in one sex or in normal diploid cells.

hemophilia A disease caused by failure of the blood to clot quickly, or at all, which is controlled by a sex-linked, recessive Mendelian gene.

hemostasis Arrest of hemorrhage.

heptad repeat A repetitive amino-acid sequence with a repeat length of seven residues.

hereditary breast–ovarian cancer (HBOC) Cases of breast and ovarian cancer thought to be caused by an inherited susceptibility (comprising between 5–10% of all cases). Of these, 50% are thought to be caused by mutations in the genes *BRCA1* or *BRCA2*.

heritability In population genetics, the proportion of the variance (the statistical measure of differences) in a trait within a group of people that is due to genetic variance.

hermaphrodite An organism that has both male and female organs and produces both male and female gametes.

heterochromatin A state of chromatin which remains condensed during interphase, stains darkly, and either contains few genes or is transcriptionally inactive.

heterochrony A change in timing of the development of an organ or feature relative to the same structure in an ancestor.

heterodimer A protein made up of two different peptide chains (protein subunits).

heterodimerization The binding together of two distinct proteins to form a complex.

heterodisomy Uniparental disomy (UPD) with the presence of two copies of one of the two homologs from one parent.

heteroduplex Double-stranded DNA molecule containing one or more positions at which nucleotides on opposing strands are mismatches.

heteroduplex analysis (HA) A method of detecting single-nucleotide polymorphisms that relies on separation of perfectly matching DNA duplexes from mismatched molecules by the use of a gel that promotes melting of DNA duplexes during electrophoresis.

heterogametic The sex of a given species whose gametes can contain different sex chromosomes (for example the male in mammals).

heterogametic sex The sex that produces gametes that contain one of two different types of sex chromosomes.

heterogeneous nuclear RNA *See* pre-messenger RNA.

heteronuclear ribonucleoproteins (hnRNPs) A large family of chromatin-associated RNA-binding proteins that are associated with mRNA trafficking within the nucleus and the cytoplasm and with mRNA splicing, transcript packaging, mRNA stability and turnover, and have some effects on translation.

heteroplasmy Coexistence of wild-type and mutated mitochondrial DNA within a cell.

heterotrimeric G proteins A subfamily of G proteins in which a guanine nucleotide-binding α subunit is reversibly associated with a dimer composed of β and γ subunits.

heterotypic continuity The tendency of a behavior pattern to persist over time but to take on the form of different, age-appropriate problems or behaviors at different ages.

heterozygosity In a diploid organism, the state of having two different alleles at a given genetic locus.

heterozygote Individual with two different alleles at a given genetic locus.

heterozygote advantage A form of balancing selection that maintains a polymorphism at a genetic locus because individuals that are heterozygous at that locus have higher fitness than homozygotes. Also known as overdominant selection.

heterozygote carrier An individual that carries one defective copy of a gene (that causes disease when present in two copies) and can transmit that allele to their offspring, but are typically unaffected by the disease as a result of compensaation by the normal copy of the gene.

heterozygous The condition of having a pair of dissimilar alleles at one or more genetic loci.

heuristics Rules of thumb to judge uncertainties by simple cognitive operations that are useful but sometimes lead to systematic errors.

HFEA *See* Human Fertilization and Embryology Authority.

H-form DNA *See* triplex DNA.

hidden layers The inner layers of a neural network, not the input or output layers.

hidden Markov model (HMM) A widely used probabilistic model for data that are observed in a sequential fashion. A HMM makes two primary assumptions. The first is that the observed data arise from a mixture of K probability distributions. The second assumption is that there is a discrete-time Markov chain with K states, which is generating the observed data by visiting the K distributions in Markov fashion.

hierarchical selection A framework in which natural selection may act on a nested hierarchy of units, in which the unit of selection can range from the subcellular to the ecosystem.

high-density lipoprotein (HDL) Plasma lipoproteins of density 1.063–1.210 g/ml that can scavenge excess cholesterol from other lipoproteins or from cells. High levels of HDL are generally associated with a decreased risk of cardiovascular disease.

high-fidelity DNA polymerase Thermostable DNA polymerases used in the polymerase chain reaction (PCR) that possess a $3'$ to $5'$ exonuclease proofreading activity.

high-throughput genomic sequence (HTG) Unfinished DNA sequences generated by the genome sequencing centers.

hippocampus A forebrain structure found in all vertebrates which is believed to have a role in explicit learning and memory such as that for spatial relations.

histocompatibility The genetically determined properties of a tissue that determine whether it will be accepted or rejected on grafting. Histocompatibility is specified by cell-surface proteins encoded by genes of the major histocompatibility complex (MHC).

histocompatibility gene Any of the genes responsible for causing a grafted tissue to be recognized as self or as non-self (foreign).

histone fold motif Highly conserved three-dimensional structure that is found in the globular domain of all core histones and also in some proteins associated with regulation of transcription.

histone gene clusters Groups of histone genes that are organized as tandem repeats of gene quintets (encoding histones H1, H2A, H2B, H3 and H4) in lower eukaryotes or as apparently random clusters of multiple copies of genes encoding the five histone classes. In the human genome, two major histone gene clusters are located on the short arm of chromosome 6 and a small cluster on the long arm of chromosome 1.

histones Highly conserved basic chromosomal proteins that together with DNA form the basic subunit of eukaryotic chromatin, the nucleosome. Five histone classes are known: H1 (lysine rich), H2A and H2B (moderately lysine rich), and H3 and H4 (arginine rich). Most histones are synthesized in coordination with DNA replication. Postsynthetic modifications of histone proteins locally affect the chromatin structure.

hitchhiking effect *See* genetic hitchhiking.

HIV Human immunodeficiency virus, of which there are two types, 1 and 2.

HLA *See* human leukocyte antigens.

HLA genes The name given to the highly polymorphic genes in the major histocompatibility complex in humans that encode the MHC class I and class II proteins which are involved in antigen presentation in an immune response and which determine an individual's tissue type.

HMM *See* Hidden Markov model.

hMORF Human microarrayed open reading frames.

hnRNA Heterogeneous nuclear RNA. See pre-messenger RNA.

hnRNP *See* heteronuclear ribonucleoproteins.

Holliday junction A hybrid DNA molecule that forms during recombination between two homologous double-stranded DNA molecules.

holoenzyme A generic term referring to a multiprotein enzyme complex containing the complete complement of protein subunits.

homeobox A 180-base pair sequence of DNA that encodes the 60-amino-acid homeodomain in the homeodomain family of DNA-binding proteins.

homeodomain A highly conserved helix-turn-helix type of sequence-specific DNA-binding domain of about 60 amino acids.

homeoprotein Any member of the large superfamily of proteins that contain a homeodomain. The developmentally important Hox proteins encoded by homeotic genes are a small subclass of this superfamily.

homeotic gene Gene that determines the developmental fate of groups of cells in discrete regions of a developing organism and that when mutant can result in one part of the body being transformed into another.

homeotic mutation A mutation that transforms one body part on the anterior–posterior axis into another structure that normally develops elsewhere.

hominoids Superfamily of Primates including humans, chimpanzees, bonobos, gorillas and orangutans.

Hominidae Family within the order Primates that includes the living taxa *Pongo*, *Pan*, *Gorilla* and *Homo*, and all the extinct taxa that are more closely related to them than to any other living primate genus.

Hominini Tribe within the order Primates that includes modern humans and all extinct taxa that are more closely related to modern humans than they are to any other living primate taxon.

homocysteine A metabolite of methionine with pro-oxidative properties that is normally converted to cystathionine in a vitamin-B6-dependent reaction.

homocystinuria A class of rare genetic disorders characterized by grossly elevated plasma total homocysteine (usually about 10 times the normal reference range) with subsequent elimination of homocystine in urine (homocystinuria).

homodimer A protein made up of two identical peptide chains (protein subunits).

homodimerization The binding together of two identical proteins.

homoduplex Double-stranded DNA molecule in which all nucleotide positions are complementary (no mismatches are present).

homogametic The sex of a given species whose gametes all bear the same sex chromosome (for example the female in mammals).

homogeneously staining region (HSR) Long unbanded regions on chromosomes created by gene amplification, which can be observed by fluorescent *in situ* hybridization.

homologous (1) Any biological structures (anatomical features, DNA sequences, and so on) that are similar due to a common evolutionary origin.
(2) When applied to chromosomes in the diploid genome, it usually refers to the maternal and paternal chromosomes of each type.

homologous characters Shared characters that are related by descent from a common ancestral character.

homologous DNA recombination DNA recombination between two highly similar DNAs. It often refers to the technique of gene targeting in ES cells in which such recombination occurs between introduced DNA and the homologous target region in the genome.

homologous genes Genes that have similar DNA sequences, and therefore encode proteins of similar amino-acid sequence, due to descent from a common ancestral gene.

homologous loci Genetic loci that are descended from a common ancestral locus.

homologous recombination Genetic recombination between identical (or nearly identical) DNA sequences. In this context homologous means only that the recombining sequences are (near) identical and implies nothing about their phylogenetic relationship.

homologous unequal recombination Recombination involving the cleavage and rejoining of nonsister chromatids at identical (or near identical) but nonallelic sequences.

homology Similarities or identities in a biological feature that are due to descent from a common ancestor.

homoplasic characters Shared characters that are not related by common descent.

homoplasy Independent evolution of similar or identical traits.

homozygosity The condition of possessing two identical alleles at a given gene locus on a pair of homologous chromosomes.

homozygote Individual with identical alleles at a given genetic locus.

homozygous Having identical alleles at a given gene locus.

horizontal gene transfer Transfer of genetic material between distantly related species. Also known as lateral gene transfer.

hot-start PCR Any protocol for polymerase chain reaction (PCR) that inhibits the ability of the thermostable DNA polymerase to commence polymerization from a primed double-stranded DNA segment before the initial melting temperature is reached.

housekeeping gene A gene whose expression is essential for the function of all cell types and is thought to be involved in routine cellular metabolism.

Hox **genes** A group of homeobox-containing genes that have roles in specifying regional identity along the anterior-posterior axis in animal development. In mammals, the 39 genes *Hox* genes are divided into 13 paralogy groups and separated into four different complexes on different chromosomes.

HS *See* hypersensitive site.

hSERT *See* serotonin transporter.

HSR *See* homogeneously staining region.

HTG *See* high-throughput genomic sequence.

5-HTT *See* serotonin transporter.

Human Fertilization and Embryology Authority (HFEA) The UK Government appointed regulatory body which oversees fertility treatments in the UK involving fertilization outside the body, storage and use of human gametes and embryos and the use of embryos for research.

Human Genome Project An international research project to map each human gene and to determine the complete sequence of human DNA.

human genome sequence The complete string of genetic information, the sequence of nucleotide bases, in the human chromosomes.

human leukocyte antigen (HLA) A set of proteins on the surface of cells, encoded by the MHC class I and class II genes, whose normal role is in antigen presentation in immune responses, but which are also the major target for the immune reactions between cells of different people that lead to graft rejection.

humanistic psychology Founded by Carl Rogers in the 1940s, a school of psychology that starts from the hypothesis that the personality of each individual has the potential to grow, provided that he/she experiences a sensitive, nonjudging understanding.

Huntington disease An inherited genetic disorder that results in damage to the nervous system and ultimately death.

hybridization The formation of double-stranded DNA or RNA molecules from complementary single strands, which is used in numerous methods to detect and identify nucleic acids of similar sequence.

hydrodynamic DNA transfer Transfer of DNA into cells by means of hydrostatic pressure.

hydropathy The proportion of nonpolar or hydrophobic amino-acid residues in a protein.

hydrophobic core The hydrophobic interior of a globular protein formed by amino-acid residues with nonpolar side chains which cannot form favorable interactions with polar water molecules and thus form favorable contacts with other nonpolar residues during protein folding to become buried within the interior.

hypercholesterolemia A high concentration of cholesterol in the bloodstream.

hyperlink A portion of an electronic document that points to another document on the Web that can be reached directly by a single mouse click.

Hyperoartia A chordate order of jawless fish that includes the lampreys. This order is an example of the earliest living vertebrate. Also referred to as Petromyzontiforma.

Hyperotreti A chordate order of jawless fish that includes the hagfishes. This order is classified under craniates but not vertebrates. Also referred to as Myxiniforma.

hypomorphic mutations Mutations that reduce but do not completely abolish the activity of the gene product.

hypoparathyroidism Clinical problem resulting from low circulating parathyroid hormone levels. This reduces circulating blood calcium levels and may result in fitting.

hypotonia Decreased muscle tone.

hypoxia-inducible factor-1 (HIF-1) A mammalian transcription factor involved in responses to low levels of oxygen and in the response to xenobiotic compounds.

IBD *See* identical by descent.

IC *See* interchromatin compartment.

ICSI *See* intracytoplasmic sperm injection.

identical by descent (IBD) Two alleles in two individuals that have been inherited from the same common ancestor.

identical by state Two identical alleles in two individuals.

identity by descent The inheritance of DNA from common ancestors without mutation and recombination.

identity-constituting Characteristics that determine the identity of that person.

ideal population A population in which each offspring has an equal and independent change of having any parent.

IEF *See* isoelectric focusing.

IF *See* immunofluorescence.

iid sample Independent and identically distributed sample.

imitation experiment The simulation of the chemical and environmental conditions of the presumed primordial atmosphere in which more complex substances are synthesized.

immunofluorescence (IF) Experimental method that uses antibodies with attached fluorescent tags to visualize proteins and modified DNA within a cell.

immunogen Any substance that is capable of inducing an immune reponse.

immunoglobulin E (IgE) The major class of immunoglobulin involved in arming mast cells and basophils. IgE binds to high-affinity receptors on these cells, and triggers the release of acute inflammatory mediators in the presence of antigens, causing an allergic reaction.

immunoprecipitation A technique that uses antigen-specific antibodies to selectively recognize a specific protein and remove it from a complex mixture of proteins in solution.

immunotherapy The correction of an immunological state by the introduction of gene(s) into leukocytes.

impairment An accredited biological abnormality.

imprinted expression *See* genomic imprinting.

imprinted genes Genes that are expressed differentially depending upon whether they come from the mother or the father (for example, genes involved in Prader–Willi and Angelman syndromes on 15q11-q13).

imprinting *See* genomic imprinting.

in situ In the natural location.

in situ **hybridization** After both the probe and the target DNA are denatured they are placed together and reanneal according to the specific sequences present. The location of the reannealing can be detected and mapped in relationship to cellular or subcellular structures such as chromosomes.

in vitro **fertilization (IVF)** The creation of an embryo outside the human body, literally 'in glass', in the laboratory. This technique is the one most widely used in assisted reproduction and also enables the screening of embryos before they are implanted.

in vivo In the body.

in vivo **gene therapy** Type of gene therapy in which a vector carrying the therapeutic gene or genes is administered directly to the person.

inborn error of metabolism A genetic defect that disrupts any of the many chemical pathways within the body, often leading to serious health problems.

inbreeding Mating between relatives that occurs more frequently than if mates were chosen at random from a population, which can lead to a reduction in the genetic diversity of the population.

incapacity A person lacks capacity if some impairment or disturbance of mental functions renders them unable to make a decision whether to consent to or to refuse treatment. Under English law no one else can consent for an adult who does not have capacity.

incidence The rate at which new cases of a disorder occur.

inclusion body Proteinaceous structure in the cytoplasm or the nucleus which is demonstrable histologically and can fucntion as repositories for biomolecules.

indel A DNA variation in which the alleles differ by the presence or absence of one or more bases.

infectome Collective word for global genotypic and phenotypic changes in microbial pathogens and their host that contribute to microbial infection. These changes occur at the levels of genomes, proteomes and glycomes.

informatics The study and application of computational and statistical methods to the management of information.

information asymmetry The situation which arises when one party to an insurance contract has more information relative to the risk propensity than does the other party. This typically arises when the proposed does not disclose all material information to the insurer.

information content A standard method of measuring the degree of conservation at a nucleotide or amino-acid position in DNA or protein sequences, which was designed by using the information theory of communication. It has an advantage over goodness-of-fit methods, such as the χ^2-test, in that it is additive across sites and free from the influence of sample size.

informed consent A process or act by which – on the basis of comprehensive information – a legally competent person makes a decision on a medical intervention or participation in a research project or clinical trial.

informed decision Decision based on all available information.

inheritance The transmission of goods, names, membership, and personal characteristics from generation to generation. What is inherited, or believed to be inherited, and their patterns of transmission are highly variable from one society to another.

innate immunity A wide range of intrinsic defenses against infection, such as the physical barriers of skin and mucosal membranes, the complement system, and germline-encoded receptors for molecular features common to different types of pathogen, that are deployed in the initial stages of an infection.

insertion A change that adds a new segment to a DNA, RNA or protein sequence.

institutional review board A group of individuals that act as a duly constituted body, as described in relevant federal regulations, to evaluate and monitor proposals to conduct research that involves human subjects.

insulator (1) A DNA sequence that blocks enhancer-mediated activation when placed between an enhancer and promoter. This activity is distinguished from silencing activity in that the insulator has no such function when located elsewhere. Also known as an enhancer-blocking element.
(2) DNA sequences that are capable of protecting stably integrated transgenes from position effects.

intake session First counseling session aimed at exploring and defining the genetic question at issue.

integrase Enzyme encoded by a virus or phage (or retroelements derived from viruses) that catalyzes the integration of the viral DNA into another DNA molecule.

integration The recombinational process by which a piece of DNA becomes inserted into and becomes part of another DNA molecule. Retroviral DNA integrated into host-cell chromosomes is replicated along with the chromosome and is considered to be a permanent part of the cell.

integrins A family of heterodimeric cell-surface receptors mediating cell–cell and cell–matrix interactions.

intellectual property That area of the law dealing with patents, copyright, trademarks and trade secrets wherein the property is a novel, intangible product of someone's intellect.

intelligence quotient (IQ) A measure of performance on a certain type of test that is calculated by dividing a person's mental age as determined by the IQ test, by their chronological age. It is usualy expressed as a multiple of 100.

intelligent design A contemporary belief that organisms show the evidence of divine planning and direct manufacture.

interallelic complementation The restoration of gene function when different mutations in the same gene are present in separate alleles.

interchromatin compartment (IC) A space between chromosome domains which is of variable width (from nanometers to micrometers) and starts at te nuclear pores and runs throughout the nucleus between the chromosome territories.

interference In DNA recombination, the tendency of one crossover to influence further crossing over within the same region of the paired chromosomes.

interferon A group of antiviral cytokines that are synthesized and secreted by virus-infected cells. Interferons-α and -β inhibit virus replication and cell proliferation, interferon-γ acts primarily by modulating the immune response.

interleukins (IL) A family of glycoproteins secreted by a variety of leukocytes that have effects on other leukocytes. The effects are mediated through specific receptors on the surface of target cells, which are coupled to intracellular signal transduction and second-messenger pathways to activate cell growth and/or differentiation. The name 'interleukin' came from 'between leukocytes'.

interphase The stage when the cell is resting and not dividing.

interstitial For chromosomes, a segment neither at the ends nor at the centromere.

intima The layers of cells between the internal and external elastic laminae that are the most common site for developing atherosclerotic lesions.

intracellular Inside the cell (projecting into the cytoplasm).

intracytoplasmic sperm injection (ICSI) A widely used technique for assisted reproduction that allows the egg to be fertilized *in vitro* by the insertion of a single sperm, thereby enabling men who produce few sperm to have children.

intron A noncoding segment of eukaryotic DNA that forms part of a gene, is transcribed, but is removed from the primary transcript by splicing together the remaining portions of the transcript, termed exons. Introns are present not only in protein-coding genes, where the exons represent the protein-coding sequences, but in genes for ribosomal RNA or transfer RNA species. Also called intervening sequences.

invention Something that is not a discovery – that is, it involves human intervention – and is new, has an inventive step and has an industrial application.

inventive step The creative element to an invention.

inversion Recombination of DNA that inverts gene order with respect to flanking markers.

inverted repeats Closely related or identical DNA sequences oriented in the opposite direction near to each other in the same DNA strand. As they are complementary to each other, they can allow the strand to fold back on itself to form a hairpin loop.

ion channel A membrane-spanning protein that mediates the flow of ions through a cell membrane.

ionophoresis A modification of electrophoresis which can be applied to proteins and other charged molecules. The method facilitates the separation of components of different mobility into distinct zones and has been applied to the separation of partial cleavage products of proteins and nucleic acids. Also called zone electrophoresis.

ionotropic receptor A transmembrane receptor consisting of four to five subunits that includes both a ligand-binding site and an ion channel.

IQ *See* intelligence quotient.

irreducible complexity The intricate organization of living things, supposedly making impossible a natural (that is, evolutionary) origin.

isochore A long stretch of DNA, in the range of hundreds of kilobases, that is homogeneous in base composition. Both GC- and AT-rich isochores have been found. The best studied isochores are those in the genomes of warm-blooded vertebrates, but they also exist in other genomes.

isochromosome A type of chromosomal aberration that arises as a result of transverse division of the centromere during cell division giving two daughter chromosomes, each lacking one chromosome arm but comprising the other arm doubled.

isodisomy Uniparental disomy with the presence of one copy of each of the two homologs from one parent.

isoelectric focusing (IEF) A separation technique in which the analytes (almost exclusively peptides and proteins) are separated according to their isoelectric point, using an electric field as the driving force and a supporting medium in which a pH gradient is established.

isoelectric point The pH at which the net charge of a molecule is zero. A characteristic of molecules, such as proteins, containing both weakly ionizable acidic and basic groups.

isoenzyme One of a set of alternative allelic forms of the same enzyme.

isoform (1) Member of any family of functionally related proteins that differ slightly in their amino-acid sequences.
(2) One of a set of forms of a protein with slightly different amino-acid sequences and often differences in function and distribution.

isolate Any geographically defined group of individuals who are more likely to reproduce with one another than with individuals of different origin.

isonymy Possession of the same surname.

isozyme *See* isoenzyme.

IVF *See* in vitro fertilization.

jawed vertebrates All extant vertebrates except for cyclostomes.

joint configuration A set of stochastic variables all being in one of their states: $\{X_1 = x_1, \ldots, X_K = x_K\}$.

justice Fair, just conduct, self-authority in maintenance of equity; in bioethics, love of others. A moral principle that requires that investigators treat research subjects equitably.

karyotype A pictorial or photographic representation of the complete chromosomal complement of a species. The autosomes are normally ordered according to size and numbered, and in mammals the sex chromosomes are labeled as X and Y.

kb Kilobase, one thousand bases (or base pairs) of DNA.

kbp *See* kilobase.

kilobase (kb) One thousand bases (or base pairs) of DNA.

kinase An enzyme that transfers a phosphoryl group from a simple precursor such as ATP to another molecule, often a protein.

kinetochore (1) The protein structure assembled at, and embedded in, the centromere which is responsible for microtubule binding. The kinetochore also contains the molecular motors providing force for chromosomal movement and tension-sensing mechanism(s) to monitor alignment of the sister chromatid pairs at the metaphase plate.
(2) Specialized multiprotein complex located at the constricted region of a mitotic or meiotic chromosome that interacts with spindle microtubules and contains motor molecules for moving chromosomes toward the spindle poles.

kin selection Selection involving trade-offs in individual fitness between family members.

kinship The social recognition of relatedness based on ties of procreation and succession. Kinship systems are highly variable across cultures and do not follow the patterns of inheritance identified by biomedical science.

knobs-into-holes packing A tight side-chain packing at the interface between two α helices. Side chains from one helix pack as 'knobs' into 'holes' formed by side chains from the opposite helix.

knockout Experimental deletion of a gene or part of a gene *in vivo* using recombinant DNA methods, leading to its permanent inactivation and loss of function.

knockout mouse A genetically engineered mouse in which a gene of interest has been specifically removed or inactivated by targeted integration of a selectable marker gene into the chromosomal locus.

knowledge base *See* locus-specific database.

***k*-tuples** Groups of elements in a sequence of size *k*. Counts of *k*-tuples, such as consecutive amino-acid pairs (for example, AA, AC, AD), are often used as features in a training input pattern for search algorithms.

LAD *See* leukocyte adhesion deficiency.

Lamarckism The theory that characters acquired by organisms in the course of their lifetime (for example, as a result of adopting changed habits) can be inherited by their offspring.

lamellipodium The fan-shaped flattened leading edge of a cell such as a moving fibrobalst, corresponding to a highly motile region.

latency The state of a nonlytic, dormant viral genome residing in a neuronal cell nucleus.

latent class analysis (LCA) A statistical method that introduces 'latent variables' to model relationships between categorical variables. The key assumption of 'local independence' supposes that the relationship between two variables of interest can be explained by a latent variable having the property that for a given level of this latent variable, the original two variables are independent of each other.

lateral gene transfer *See* horizontal gene transfer.

LCA *See* latent class analysis.

LCAT *See* lecithin:cholesterol acyltransferase.

LCR *See* locus control region.

LD *See* linkage disequilibrium.

long-term depression (LTD) A long-lasting decrease of synaptic transmission following a low-frequency repetitive stimulation.

learning In neural netowrks, the process by which the network is adjusted to perform its approximation, classification or prediction tasks. Learning is usually accomplishing by adjusting the connections between units in a network to minimize an error function.

learning disability Current UK term for what used to be called 'mental handicap' and in other countries is called 'intellectual disability', or 'mental retardation'. Learning disability includes the presence of a significantly reduced ability to understand new or complex information, to learn new skills (impaired intelligence) with a reduced ability to cope independently (impaired social functioning) which started before adulthood, with a lasting effect on development.

lecithin:cholesterol acyltransferase (LCAT) A plasma enzyme that catalyzes the esterification of cholesterol scavenged by the HDL particle from extrahepatic tissues.

legumain A lysosomal cysteine protease initially found in legumes but also recently identified in mammals (porcine and human).

lentivirus A family of retroviruses named 'lenti' after the slow onset of disease caused by the prototype virus *Maedi-Visna virus* (MV V) which infects sheep. Lentiviruses are found in primates and some domesticated mammals and are characterized by the ability to infect and integrate into the chromosomes of nondividing cells.

leukocyte adhesion deficiency (LAD) A genetic disease resulting from the absence or greatly reduced levels of integrin β_2 resulting in recurrent life-threatening bacterial infections and impaired wound healing.

liability Propensity, genetic risk or vulnerability to develop a disease. Liability can have genetic and/or environmental sources.

library Collection of cloned DNA fragments generated by a restriction enzyme that cuts DNA at a particular site.

life insurance Insurance payable on the survival of humans for particular periods or on death within certain periods, including whole of life insurance, endowment insurance and temporary life insurance (also known as term insurance or term life insurance).

life-narrative The story we tell of our life in which we are its main character.

ligand Any molecule that binds to another molecule.

ligand-gated channel Transmembrane proteins that form at the same time the recognition site for a neurotransmitter (the ligand) and the effector site, an ion channel. Binding of the neurotransmitter causes within a few microseconds the opening of the ion channel.

ligate To link nucleotides together by the formation of a phosphodiester bond.

likelihood Numerical measure of the appeal of a hypothesis or theoretical model in the light of empirically observed data. Mathematically, the likelihood equals the conditional probability of observing the data, given that the hypothesis or model in question is correct.

LINE *See* long interspersed nuclear element.

lineage Evolutionary descent, from the oldest (ancestral) sequences to the youngest.

linkage Co-inheritance of particular alleles of two or more different genes because their loci are close together on the same chromosome, such that after meiosis they remain associated more often than the 50% expected for unlinked genes.

linkage analysis General term for any of several statistical methods of genetic analysis applied to family-based samples to detect cosegregation of a trait and alleles at a genetic locus, and which can be used to infer the order and spacing of DNA sequences.

linkage disequilibrium (LD) The situation where particular alleles at two or more neighboring loci (typically up to around 50 kilobases apart) occur together in the same chromosome with frequencies significantly higher than those predicted from the individual allele frequencies. Also known as allelic association.

linkage disequilibrium mapping The use of a dense map of polymorphisms, spaced at 1–10 kilobase intervals, across the genome or a region of known genetic linkage to detect association between a disease and a particular polymorphism, which indicates that a disease-causing gene is located approximately 50 kilobases either side of the polymorphism.

linkage map A chromosome map showing the relative positions of genetic markers of a given species, as determined by linkage analysis. This is not the same as a physical map or a gene map, which are generated using linkage analysis, cytogenetic examination and physical techniques.

linkage study Study design for locating the gene for a given phenotype, such as a disease, by looking at the cosegregation of the phenotype (usually within families) with alleles of polymorphic genetic markers of known location.

linker/adapter Synthetic DNA sequence that can be ligated to double-stranded cDNA to provide restriction sites or annealing sites for polymerase chain reaciton (PCR) primers.

linker DNA DNA segment connecting the nucleosomal cores. The length of linker DNA varies between 20 and 80 nucleotides. Binding of histone H1 (the linker histone) seals two rounds of DNA at the DNA exit/entry site of the nucleosomal core.

lipoprotein (1) A water-soluble complex of lipids and protein (the apoprotein) that circulates in the bloodstream. Lipoproteins are commonly classified according to their flotation in an ultracentrifugal density gradient.
(2) Particle circulating in the blood that is a complex of a protein (the apoprotein) and numerous lipid molecules (triglycerides, cholesterol). There are several different types.

liposome Small vesicular structure that forms when phospholipids are placed in water, and the hydrophilic portions of the molecules face outward and the hydrophobic portions of the molecules face inward. Can be utilized to deliver target DNA to cells.

liquid chromatography MS/MS The coupling of (mainly) reverse-phase liquid chromatography separations with electro-spray ionization-based tandem mass spectrometry (MS/MS) for the high-throughput and sensitive sequencing of peptides.

local sequence similarity A measure of protein or DNA sequence similarity that focuses on the most similar regions, ignoring surrounding dissimilar sequences.

locus *See* genetic locus.

locus control region (LCR) A sequence element that confers high-level expression onto linked transgenes independent of the site of transgene insertion in the genome.

locus heterogeneity The existence of more than one genetic locus that contains a disease-related mutation (or allele).

locus-specific database (LSDB) A listing of mutations in individual genes, together with phenotype, clinical information, assay data, and so on, with up to 80 fields. Also known as a 'knowledge base'.

lod score Logarithm (base 10) of the odds for linkage divided by the odds for no linkage. By convention, for a single Mendelian disorder, a lod score of +3 (1,000:1 odds) is taken as proof of linkage; a score of −2 (100:1 odds against) indicates no linkage.

logo graph The graphic representation of the degree of conservation of nucleotides or amino acids in a consensus sequence motif. The relative size of each letter represents its proportionate use in the set of sequences from which this consensus is derived. The total height of the stacked letters represents total information content and corresponds to the degree of conservation at a nucleotide or amino-acid site in the sequences.

long interspersed nuclear element (LINE) A class of repetitive DNA sequences of heterogeneous size ranging up to about 6–7 kilobases and constituting about 20% of the human genome. LINE sequences are generally flanked by short direct repeats and often contain a sequence similar to the reverse transcriptase of retroviruses.

long terminal repeat (LTRs) Repeated sequence found at the 5' and 3' ends of an integrated retrovirus genome. It contains various elements that control transcription of the integrated viral DNA.

loss-of-function mutation A mutation associated with the loss of the normal function of a protein.

low-density lipoprotein (LDL) A plasma lipoprotein (density 1.019–1.063 g/ml) made up of one molecule of the apoB protein and molecules of cholesterol. Oxidized LDL particles are major contributors to atherosclerotic plaques.

low-density lipoprotein receptor (LDLR) A receptor that promotes uptake of low-density lipoprotein (LDL) into cells, particularly in the liver but also in other tissues, by binding to apoB. Liver uptake and clearance of LDL cholesterol is particularly important in protecting the vascular system from hypercholesterolemia.

LSDB *See* locus-specific database.

LTD *See* long-term depression.

LTP *See* long-term potentiation.

LTR *See* long terminal repeat.

lymphedema An accumulation of lymphatic fluid that leads to abnormal development of nearby structures when it occurs in the fetus.

lymphoma A general term used to describe various abnormal proliferative diseases of the lymphoid tissues (lymph nodes, spleen, thymus, tonsils, and so on).

lysosomal integral membrane protein (LAMP) Transmembrane protein found in lysosomes. Type I LAMPs have a single transmembrane domain, a short carboxy-terminal tail and a heavily glycosylated intraluminal amino terminus.

magnetic resonance imaging (MRI) A neuroimaging technique involving a powerful magnet to induce nuclear magnetic resonance and which produces pictures of the brain that localize metabolic activity.

major gene The primary gene responsible for producing a particular phenotype.

major histocompatibility complex (MHC) A tightly linked cluster of genes on chromosome 6 (in humans) in which are located the genes for the MHC class I and II cell-surface glycoproteins (human leukocyte antigens, HLAs) whose primary immunological function is to bind and 'present' antigenic peptides on the surfaces of cells for recognition by the antigen receptors of T lymphocytes. The MHC region also contains many other genes involved in immune function. Sometimes called the HLA region in humans.

MALDI *See* matrix-assisted laser desorption ionization.

Malthusian prediction The prediction of the 18th-century economist and clergyman Thomas Malthus that the 'natural' unchecked geometric increase in population growth would produce a crisis in food supply. Only war, famine, disease or other natural disasters would slow this crisis, which he theorized was inevitable.

mammalian artificial chromosome (MAC) A piece of DNA containing sequences that enable it to replicate, segregate and be expressed alongside the endogenous chromosomes in mammalian cells.

managed care plan An organization that bears health insurance risk and provides healthcare services to its insureds, and that typically strives to hold down spending on healthcare by employing a number of techniques to discourage utilization.

manic–depressive illness *See* bipolar disorder.

MAO *See* monoamine oxidase.

map unit The measure of distance between two loci on a chromosome as inferred from the percentage frequency of crossing over (recombination) between them. Expressed in units called centimorgans (cM).

MAR *See* matrix-associated region; scaffold-attachment region.

MAR-binding proteins Proteins that bind *in vitro* in a specific fashion to matrix-associated regions (MARs) in nuclear DNA.

Marfan syndrome A group of connective tissue disorders characterized by excessive growth of long bones, numerous skeletal abnormalities including arachnodactyly and joint laxity, lens dislocation and dilatation and aneurysma aortae.

marker Any feature (such as a particular sequence or a restriction site) whose position in a genome can be found.

marker chromosome A structurally abnormal chromosome of unknown origin.

marker genes Genes, such as those for green fluorescent protein (GFP) or luciferase, used in analysis of gene expression to give an easily detectable phenotype to cells carrying another gene of interest.

matrix-associated region (MAR) DNA sequence element that can interact *in vitro* in a specific fashion with isolated nuclear matrix.

mass spectrometry (MS) Analytical technique that measures the mass-to-charge ratio of a charged molecule such as a small peptide.

mast cell Type of bone-marrow-derived cell that is the primary effector cell of allergic reactions. Found at cutaneous and mucosal surfaces and in deeper tissues about venules, they contain dense cytoplasmic granules full of mediators of acute inflammation, which are released when the mast cell is stimulated.

master gene hypothesis Proposition that few copies of a given transposable element are active in a genome.

matchmaker An enzyme (for example, replication factor C) that can bring another protein (for example, proliferating cell nuclear antigen) to the place where it acts.

materialism Explanatory model that focuses on the physical, and often the economic and productive, properties of a society.

materialist Describes the viewpoint that prioritizes material and economic forces when explaining human development or action.

maternal inheritance Inheritance of genes in mitochondrial DNA, which is exclusively transmitted from a mother to her offspring.

matrilineal descent Situation where group membership is traced through females only, so that children are affiliated to the group of their mother and their mother's brother, but not their father.

matrix-assisted laser desorption ionization (MALDI) The generation of charged molecules from a solid phase consisting of matrix crystals (mainly small organic molecules) using infrared or ultraviolet lasers.

matrix-attachment region *See* scaffold-attachment region.

maximal heritability The proportion of the phenotype variance explained by both genetic and familial environments.

maximum-likelihood estimator A point estimate of a parameter obtained by maximizing the probability of the observed sample data as a function of the parameter under a specified statistical model.

Mb *See* megabase.

MDS *See* myelodysplasia.

meaning The impact of information.

meconium Ileus Inability of a newborn infant to pass meconium through the rectum, resulting in an acute, life-threatening intestinal obstruction requiring emergency surgery.

MeCP2 Methyl-CpG-binding protein, which specifically interacts with methylated DNA to mediate transcriptional repression.

median network Network obtained from a set of sampled binary haplotypes by successively taking consensus sequences of triplets of haplotypes and finally linking all resulting types by the minimum spanning network.

medical confidentiality The right to control identifiable personal health information; also, an ethical duty owed to all patients by health professionals.

medical underwriting An assessment of insurance risk-based health status or medical indicators.

medicalization The social process of the increasing power of physicians over laymen resulting from increasing use of the terms 'ill' or 'healthy' to judge things.

megabase (Mb) A million bases or base pairs (Mbp) of nucleotide sequence.

meiosis The specialized type of cell division by which gametes containing the haploid number of chromosomes are produced from diploid cells.

meiotic arrest A block in the differentiation process of the male or female germ cells at their first meiotic division cycle in the germ line.

meiotic drive Significant deviation from the expected Mendelian transmission ratios due to preferential segregation of chromosomes during meiosis or gametic selection.

meiotic recombination Reciprocal exchange of genetic material between homologous chromosomes, which occurs during meiosis.

Mendelian disorder A disease resulting from a mutation in a single gene, with recessive or dominant action, that segregates in families.

Mendelian genetics *See* Mendelian inheritance.

Mendelian inheritance Pattern of inheritance that obeys the laws of segregation, first described by Gregor Mendel (1823–1884), whereby members of each pair of alleles of a gene separate during the production of germ cells.

Mendelism *See* Mendelian inheritance.

messenger RNA (mRNA) An RNA molecule that specifies the sequence of the amino acids comprising a protein.

metabolic activation Enzymatic conversion of a chemical from one state to another, for example by hydroxylation, epoxidation or conjugation.

metabotropic receptor A receptor consisting of a subunit that has seven membrane-spanning domains and that couples to GTP-binding proteins which activate intracellular second messengers.

metacentric chromosome Chromosome in which the centromere is around the middle of the chromosome and the two arms are of roughly equal length.

metaphase The stage of mitotic or meiotic cell division (mitosis or meiosis) when chromosomes are highly compacted and are lined up at the equatorial plane of the spindle ready for chromatid separation and segregation to daughter nuclei.

metaphor The representation of one idea by another idea based on some similarity apparently shared by the ideas.

methylation (1) The addition of a methyl group to a molecule.
(2) *See* DNA methylation.

5-methylcytosine A modified form of cytosine bearing a methyl group on C5.

7mG cap *See* 5′ cap.

MHC *See* major histocompatibility complex.

micelle A large, ellipsoidal complex of noncovalently bound monomers of amphiphilic molecules (such as surfactants or detergents) in water, with the polar parts of the molecules all facing outwards and the hydrophobic parts into the interior.

microarray An ordered grid-like array of molecular probes, immobilized on a solid substrate such as nylon or glass, that can be used to simultaneously detect the presence of large numbers of different biomolecules, such as RNA, DNA or protein, in a biological sample. The two main approaches to making DNA microarrays are to spot cDNA molecules onto the solid surface or to synthesize oligonucleotide probes directly on the surface. Spot sizes can be as small as $100\,\mu m$ in diameter, while synthesized arrays usually have squares of DNA molecules with dimensions as small as $50\,\mu m$.

microbiome The collective genome of non-pathogenic microorganisms normally colonizing the body surface of adult animals. The components of a microbiome have important effects on immune function, nutrient processing and a broad range of other host activities.

microcephaly Abnormal smallness of the head.

microdeletion A small deletion of a region of a chromosome, demonstration of which in general requires application of either high-resolution chromosome banding or molecular cytogenetic techniques.

microinjection Injection of genetic material into an embryo or oocyte (egg).

microphylogeny Recent biological history of closely related populations.

microsatellite A region of DNA, usually no more than 300 base pairs in length, comprising numerous repeats of short sequences of 2–6 base pairs.

microsatellite marker A polymorphic site comprising numerous repeats of short sequences of 2–6 basepairs, normally extending less than a total of 300 basepairs in length. The best known are CA (alternatively GT) dinucleotide repeats.

microsatellite typing Demonstration of microsatellite length variation between different individuals by any of a variety of molecular genetic methods.

midline The median line or median plane of the body or some part of the body.

midzone An antiparallel array of microtubules that forms during anaphase in the center of the spindle and directs progression of the cleavage furrow during cytokinesis.

migration The movement of individuals from one location to another. If the migrating individuals reproduce in the new area, migration results in gene flow.

minimal risk The probability and magnitude of the harm or discomfort anticipated for participation of a subject in research is no greater than that encountered in daily life or during psychological tests or routine physical tests.

minisatellite Region of repetitive DNA with a repeat unit length of 15–100 basepairs, extending up to a total of 20 kilobases in length.

minisatellite typing *See* DNA fingerprinting.

misexpression The expression of a transgene in a novel expression pattern.

mismatch distribution Distribution of the number of sequence differences between pairs of individuals. For neutral genes, its shape depends on episodes in previous demographic history, such as population expansions and contractions.

mismatch repair Form of DNA repair that involves removing mismatches, such as A-G pairs, that result from nucleotide misincorporation.

misogyny Hatred of women.

missense mutation Change in a DNA sequence that causes the substitution of one amino acid for another during translation of the RNA transcript into a protein.

mitochondria *See* mitochondrion.

mitochondrial disorder A clinical syndrome caused by faulty oxidative phophorylation (OXPHOS). Mitochondrial disorders are usually genetically determined disorders caused by mutations of mtDNA, or by mutations in nuclear genes related to OXPHOS.

mitochondrial DNA (mtDNA) DNA carried by mitochondria which contains the genetic information for the synthesis of some essential protein components of the respiratory-chain complexes. These genes are transcribed and translated in the mitochondrion. Most mitochondrial proteins are encoded by the nuclear DNA.

mitochondrion Rod-shaped or oval cytoplasmic containing organelle with dimensions of approximately $2\,\mu lt\ 0.5\ \mu m$ and surrounded by a double membrane. Mitochondria are the main energy-generating organelles of eukaryotic cells and contain the enzymes of the citric acid cycle, the molecular machinery for oxidative phosphorylation (the manufacture of ATP from ADP and phosphate) and the oxidation of fatty acids. Mitochondria also carry their own distinct DNA, which contains around 13 protein-coding genes. Most mitochondrial proteins are encoded by the nuclear DNA.

mitogen Any agent that stimulates cell division.

mitosis The process which eukaryotic cells go through after DNA replication, to complete chromosome division and segregation into new daughter cells. It consists of five distinct stages: prophase, prometaphase, metaphase, anaphase, and telophase.

mitosis promoting factor (MPF) A protein kinase made up (in humans) of the cyclin B regulatory subunit and the catalytic subunit Cdc2, which governs the transition from G_2 to M phase.

mitotic checkpoint A mechanism in mitosis that comprises several proteins that monitor the attachment of kinetochores to microtubules and delay cells at metaphase until all the chromosomes have a bipolar attachment to the mitotic spindle.

mitotic recombination The exchange of corresponding segments between chromosomes by recombination during cell division in somatic tissues.

mitotic segregation The random distribution of mitochondria between daughter cells during mitosis.

mobile element Mobile elements are segments of genomic DNA that have specific mechanisms that allow the preferential amplification of those specific segments into multiple chromosomal locations in the genome. This includes retroelements, as well as elements that use a DNA-replication-based amplification mechanism.

model organism An organism selected to study particular, frequently universal, phenomena for which the model organism is particularly appropriate.

model-free methods Methods for testing linkage that do not require specification of the disease model.

modifier gene A gene that changes or augments the phenotypic expression of a major gene.

module (1) Any structural unit that is found in a variety of different contexts in different molecules.
(2) A protein domain that is used as a versatile building block in different types of multidomain proteins.
(3) A potentially definable element in a developmental process.

molecular chaperone Protein that associates temporarily with other molecules and facilitates their transport or promotes their structural stability.

molecular clock The idea of a constant, or near-constant, rate of change of proteins or DNA sequences over evolutionary time.

molecular genetics Tests on DNA or RNA to ascertain abnormalities or alterations of sequence.

molecular genetic testing Testing that involves the analysis of deoxyribonucleic acid (DNA), either through linkage analysis, sequencing or one of several methods of mutation detection.

molecular marker Detectable DNA landmark on a chromosome, which can be used to identify a linked gene.

monoamine oxidase (MAO) An enzyme that breaks down a class of neurotransmitters that includes serotonin, dopamine and norepinephrine. It occurs in two forms, MAO-A and MAO-B, which are encoded by genes on the X chromosome.

monoamine A class of chemical messengers (neurotransmitters) that includes dopamine, norepinephrine and serotonin, which convey signals between nerve cells.

monochromacy Rare defective color vision mediated by a single type of cone cell.

monoclonal Arising from a single founder cell.

monogenic Caused by a mutation in only one gene.

monogenic genetic disease A hereditary disorder caused by a mutation in a single gene.

monophyletic Describes a group of organisms that share a common ancestor.

monophyletic group No more and no less than all the taxa descended from a common ancestor.

monophyly Descent from a single common ancestor.

monosomic Having only a single copy of one particular chromosome.

monosomy Possession by an embryo or adult of only one copy of a given chromosome instead of two.

monozygotic Originating from one zygote, or fertilized egg.

monozygotic twins (MZ twins) Twins genetically identical at conception, arising from a single egg that splits after fertilization.

moral responsibility Moral accountability for one's actions.

morphogenesis The precise and orderly mechanism by which particular three-dimensional cell–cell arrangements are achieved in the development of tissues and organs.

morpholino A chemically modified oligonucleotide that is resistant to enzymatic degradation and reduces the function of a particular gene when introduced into a cell by inhibiting translation of the corresponding mRNA.

morphology The study of external and internal form and structure of organisms.

Morris water maze A cognitive task for rodents in which the animal must learn and swim to the location of a platform hidden under the surface of the water in a large tank.

mosaic A tissue or an individual deriving from a single zygote but made up of genetically different cells.

mosaicism The presence in an individual or tissue derived from a single zygote of at least two cell lines differing in genotype or karyotype.

most recent common ancestor In phylogenetics, a single ancestral allele at a genomic locus from which all alleles in a present-day population sample are descended.

motif A recurring, recognizable arrangement of protein secondary structure elements or amino-acid sequence.

moving platform Ring-shaped proliferating cell nuclear antigen molecule that encircles the DNA and can move along the DNA to find a junction from double-strand to single-strand DNA.

MPF See mitosis-promoting factor.

MRI See magnetic resonance imaging.

mRNA 5′ cap See 5′ cap.

MS/MS See tandem mass spectrometry.

mtDNA See mitochondrial DNA.

Muller's ratchet A theory proposed by Herman Muller which postulates that, in the face of mainly deleterious mutations, only recombination can effectively regenerate fit haplotypes. Hence, in asexual genomes or regions of suppressed recombination, deleterious alleles would inevitably accumulate.

multidomain protein Protein consisting of more than one discrete structural domain.

multifactorial Caused by the interaction of genetic and environmental factors.

multifactorial etiology Describes the origins of diseases that are caused or influenced by several different factors.

multifactorial inheritance A pattern of inheritance (such as for height, intelligence) which results from the interaction of several different types of genes with environmental factors.

multifactorial trait Any feature to which both genetic and environmental factors contribute.

multigene family A group or cluster of related genes originating from an ancestral gene by the process of duplication.

multilocus probe A minisatellite sequence that can be used to identify partial matches with a group of related minisatellite loci.

multiple alignment An alignment of more than two protein or DNA sequences.

multiplex fluorescence *in situ* **hybridization (M-FISH)** A multicolor karyotyping technique where 24 painting probes specific for the autosomes and the sex chromosomes are labeled with a combination of five dyes.

multiplex screening The screening procedure in which tests for completely different genetic conditions are offered in a single panel.

multiplicative inheritance The pattern of inheritance when susceptibility variants at several loci must all be present for the disorder to occur.

multiplicity Ratio of infectious virus particles to cells.

multivariate distribution A probability distribution over multiple variables.

mutagen A substance or agent which can induce changes in nucleic acid.

mutagenesis Deliberate alteration of DNA sequence by any of a variety of means to induce changes in gene function.

mutation A change in a DNA sequence and any consequent change in the RNA or protein sequence derived from it.

mutation rate The frequency with which mutations occur at a given locus per generation.

mutation search The use of small DNA probes to find mutated sequences in a particular gene in an individual manifesting the effects of mutation in that gene. In many conditions there are unrecognized disease-causing mutations.

mutational hot spot Region of the genome with a mutation rate much higher than the average.

mutator phenotype Exhibiting a tendency to accumulate changes in the information encoded by DNA.

myelin Sheath of compacted glial cell processes around nerve axons. It is made up of protein and lipids. The myelin sheath around nerves speeds the transmission of impulses along the nerve cells. Myelin-free areas are called Ranvier nodes.

myelodysplasia (MDS) A group of disorders in which the bone marrow overproduces cells, but they do not mature normally.

myenteric plexus A network of neural cells (ganglion cells) within the muscular wall of the intestines.

MZ twins See monozygotic twins.

N terminus See amino terminus.

naked DNA DNA that has not formed a complex with synthetic or naturally occurring compounds.

natural selection Charles Darwin's theory in which the best-adapted individuals survive and reproduce, thereby transmitting their characters to the next generation. The process by which the number of copies of an allele in a population changes over time due to variation in fitness (as determined by the number of viable offspring) among individuals with different genotypes.

N_e See variance effective population size.

Neanderthals An archaic species of humans (*Homo neanderthalensis*) or subspecies (*H. sapiens neanderthalensis*) different from anatomically modern humans (*H. sapiens*), classically known from the early Wurm glacial period of Europe (about 75 000 years ago).

needful falsehoods In Plato's Republic, Socrates argues that the rules of the states should have the 'privilege of lying... for the public good'.

Nematoda Invertebrate phylum of smooth-skinned unsegmented worms with a long cylindrical body tapered at the ends, for example, roundworms and threadworms.

neocentromere Ectopic centromere formed on a chromosome at a location other than normal centromere. It does not contain typical repetitive centromeric DNA, but does acquire centromeric chromatin and shows patterns of kinetochore-associated proteins identical to those of normal centromeres.

neonatal screening Usually the collection, on filter paper, of a small amount of the newborn's blood, obtained by heel prick, on which genetic tests can be performed.

neoteny Species-specific delay in the developmental schedule, that leads to evolutionary novelty.

nested primer A polymerase chain reaction (PCR) primer that corresponds to a sequence on an amplified DNA fragment that lies just interior to the original primers used to generate that fragment.

network Many-to-many relationships showing involvement of a protein or compound in multiple processes. This is in contrast to hierarchical data, such as phylogenetic relationships within a gene family.

neural crest An anatomical structure, localized longitudinally along the spinal cord during embryonic development, containing pluripotent precursor cells which migrate to various locations during embryonic development and give rise to a wide range of different types of cells and tissues.

neurobiological diathesis A biological predisposition toward a particular phenotype.

neurocristopathy Heterogeneous group of genetic disorders resulting from various malfunctions of cells originating from the neural crest.

neurodegeneration The process of degeneration of brain cells leading to many serious neurological diseases.

neurofibrillary tangle Abnormally entwined tau protein filaments or threads found in a neuron's interior that compromise its microtubule structure.

neuron (1) Nerve cell.
(2) In a neural network, the basic unit of a layer. Typically artificial neurons perform a weighted sum of input signals and output a signal, which is constrained between limits such as 0 and 1. This is analogous to biological neurons that are turned on or off depending upon the strength of incoming signals from connecting neurons.

neuropathy Disease of the peripheral nerves.

neurotransmission The mechanism by which one neuron transmits an excitatory or inhibitory signal to another neuron.

neurotransmitter Molecule involved in the transmission of signals from one neuron to another neuron or to a final target cell such as muscle. Examples include noradrenaline, serotonin and acetylcholine.

neurotrophic Supporting the survival and/or growth of neurons. Several families of naturally occurring neurotrophic factors have been identified.

neurotropic Possessing a natural affinity for neurons. Neurotropic viruses target neurons and invade the nervous system.

neutral evolution Theoretical model of evolution in which most mutations have no effects on fitness, and thus allele frequencies change at random by genetic drift.

neutral polymorphism The presence in a population of two or more alleles at a given genetic locus that is not maintained by natural selection but by genetic drift.

neutral theory of evolution *See* neutral evolution.

niche The range of physical and biological conditions and interactions among which an organism lives and reproduces.

nicotine Alkaloid extracted from tobacco leaves.

nicotinic receptor Type of acetylcholine receptor that is characterized experimentally by its ability to be activated by nicotine. Its normal ligand is acetylcholine.

NMR *See* nuclear magnetic resonance.

nonadditive gene effects The situation when the contribution of a given set of genes to behavioral or other variation in the population is due to interaction between them.

nonautonomous element Transposable genetic element that does not encode all the functions required for its transposition, but which can be mobilized by a transposase provided by an autonomous element present in the same genome.

noncoding RNA RNA that does not encode protein.

noncoding sequences General name for all parts of the genome that do not encode a protein or a functional RNA, such as intergenic sequences and introns.

noncollagenous domain A domain in collagens that does not contain repeating Gly-X-Y amino-acid sequences. Also referred to as the globular domain.

nondirective counseling The official ethos of genetic counselors who do not impose their value judgments or recommendations for action on clients, but support individuals and families to exercise autonomy.

nonhomologous end-joining One of the double-strand break repair pathways by which two DNA ends are simply ligated together.

nondisjunction Condition in which two homologous chromosomes fail to separate during cell division and migrate together into the same daughter cell.

nonfibrillar collagens Collagens that do not form cross-striated fibrils. They assemble into other supramolecular structures, examples of which include fine filaments and hexagonal networks.

non-genomic intergenerational transfer Transmission of traits from parents to offspring through the enduring effects of parental behavior on the biology of the offspring.

noninvasive diagnosis Attempts to achieve a prenatal diagnosis without invasive procedures, for example, from fetal cells or cell-free DNA from the maternal circulation.

noninvasive *in vivo* imaging Methods in which gene expression is observed over time in the intact organism, by using light-emitting enzymes as markers.

noninvasive sample Sample that can be collected from wild animals without disturbance, such as feces, shed hair, shed feathers.

nonmaleficence The state of not doing harm or evil. An obligation to not inflict pain, harm, misinformation or risk on others.

nonmalfeasance *See* nonmaleficence.

nonpenetrance The absence of the characteristic phenotype in an individual with a disease-causing mutation. Such a phenotype may be expressed as a proportion of the 'at-risk' cases.

nonprescriptive counseling *See* nondirective counseling.

nonsense-mediated decay Pathway of mRNA degradation that is triggered by the presence of a premature translation termination codon.

nonsense mutation A mutation that replaces the codon for an amino acid by a stop codon (UAA, UAG or UGA).

nonshared environment Experiences that children have uniquely that tend to make them less similar to one another in their observed trait (phenotype).

nonsynonymous substitution Nucleotide substitution that causes an amino-acid change in the encoded protein.

nonvertebrate eukaryotes All eukaryotes except vertebrates. Not to be confused with invertebrates. For example, *Plasmodium falciparum*, the malaria parasite, is a nonvertebrate but not an invertebrate.

nonviral vector A carrier DNA, not derived from a virus, that can transfer a gene into cells.

norm of reaction The range of phenotypes (reactions) associated with varying an environmental factor. Different genotypes or strains of an organism typically have different norms of reaction.

normal color vision Trichromatic color vision, the most common form of color vision experienced by humans, mediated by S, M and L cones.

normality What is considered usual, regular and typical.

normalization The process of adjusting the signal from two different reporter channels on a single microarray or a single-reporter channel on multiple microarrays, to a common scale. Current methods involve either normalization to some representative statistic of all of the data, or to a statistic of some subset of the data, such as a set of house-keeping genes.

not-for-profit Enterprises that are exempt from certain taxes by virtue of their structure and public service function.

notochord A stiff contractile rod of tissue running the length of the animal body in embryonic chordates (and in some larval forms). In vertebrates the backbone forms around the notochord and nerve cord.

novelty An invention that has never been previously described.

nuclear export pathway A series of events that leads to the movement of a protein or RNA out of the nucleus into the cytoplasm.

nuclear magnetic resonance (NMR) A technique that examines the structure of molecules by detecting nuclear spin reorientation (radiofrequency transitions) in an applied magnetic field.

nuclear matrix Residual nuclear structure that can be obtained by extraction of isolated nuclei with 2 M NaCl solution and extensive digestion of nucleic acids.

nuclear scaffold Residual nuclear structure that can be obtained by extraction of isolated nuclei with 25 mM lithium diiodosalicylate solution and extensive digestion of nucleic acids.

nuclear transfer The process of transferring the nucleus of an adult diploid somatic cell into an egg cell that has its nucleus removed, as used in the cloning of animals.

nuclease hypersensitive site (HS) A short (100–400 basepairs) genomic region that is highly sensitive to nuclease digestion, compared with surrounding regions. Sensitivity is thought to arise from the binding of transcription factors to the DNA, with the concomitant disruption of chromatin structure, although DNA structure may also contribute to the formation of such a site.

nucleic acids Biological polymers consisting of nucleotide subunits linked together by phosphodiester bonds. See DNA, RNA.

nucleolar organizer regions Regions on metaphase chromosomes that contain the genes for preribosomal RNA.

nucleolus A highly organized organelle within the nucleus that forms the site of ribosomal RNA synthesis, processing and ribosomal subunit assembly.

nucleolus-derived foci Large particles (1–3 μm in diameter) that contain nucleolar material and are seen in the cytoplasm during anaphase and telophase of mitosis.

nucleoside A small molecule composed of a nitrogenous base and a sugar (in cells these are deoxyribose or ribose). It corresponds to a nucleotide lacking its phosphate group or groups.

nucleosome Repeated basic subunit of eukaryotic chromosomes. It consists of about 200 basepairs of DNA and histones. DNA (146 basepairs) is wrapped (in one and three-quarter turns) around a histone octamer consisting of two molecules of each of the core histones H2A, H2B, H3 and H4 to form the core of the nucleosome. The linker histone H1 is bound to the DNA linking the nucleosomal cores. Hence, a nucleosome is formed by the nucleosomal core, one H1 histone and linker DNA.

nucleosome phasing The binding of nucleosomes to a specific region of DNA rather than the more normal random binding of nucleosomes to DNA. It occurs because either the binding is influenced by sequence-specific topology or the first nucleosome in a region is preferentially assembled at a particular site.

nucleotide Phosphate ester of a pentose sugar in which a nitrogenous purine or pyrimidine base is linked to the sugar. In ribonucleotides (sugar ribose) the pentose residue is D-ribose, in deoxyribonucleotides it is 2-deoxy-D-ribose. The phosphate group may be bonded to the C3 or C5 carbon atom of the pentose forming $3'$ or $5'$ nucleotides. They are the basic subunits of nucleic acids and various nucleotides also act as cofactors and energy carriers in cells.

nucleotide diversity The average number of nucleotide differences per site between two DNA sequences at a genomic locus randomly chosen from a population.

nucleotide excision repair Form of DNA repair that involves removing a damaged or misincorporated nucleotide, typically together with a surrounding stretch of DNA, and then filling in the gap through a concerted action of nucleases, helicases, polymerases and ligases.

nucleus An organelle present in all eukaryotic cells (and defines them as such), which contains the genetic material (DNA) packaged into chromosomes.

null mutant A mutant organism that has lost complete expression of a particular protein through mutation of the corresponding gene.

OBA/Ku A mammalian protein, identified as mouse Ku86, that binds to replication origins in DNA in a sequence-specific manner.

ODN *See* oligonucleotide.

odor A smell or fragrance. The conscious perception of an odor is the result of the stimulation of a particular panel of olfactory receptors by chemical(s) composing that precise odor and which generate multiple signals that are integrated in the higher brain structures. An odor is therefore encoded by a combinatorial code.

odorant A chemical that binds olfactory receptor(s) and initiates a stimulus resulting in the perception of an odor.

odds   Ratio between the probability and the counterprobability of an event or hypothesis or model.

odds ratio   In a case-control study, the odds of a given factor (exposure or attribute) among cases divided by the odds of the factor among controls is termed the exposure-odds ratio. It is equivalent to the cohort study disease-odds ratio: the odds of disease among persons with a given factor divided by odds of disease among persons without the factor.

Okazaki fragments   Short fragment of DNA synthesized on the template DNA at the lagging strand of the replication fork. Numbers of such fragments are formed discontinuously because all known DNA polymerases have a $5' \rightarrow 3'$ directionality in synthesizing DNA. Fragments are ligated together after synthesis to form a continuous strand.

olfactory receptor   Any member of a large family of seven-span G-protein-coupled receptors in the nasal sensory neuroepithelium that bind odorants. The receptor initiates a signaling cascade that results in transmission of an electrical signal to the brain.

oligo(dT)   Stretch of thymidine $5'$ triphosphate (dT TP) complementary to the poly(A) tail of mRNA used for cDNA synthesis.

oligogenic inheritance   Phenotype reflects combined effects of genetic variation in several (less than 10) genes. Each risk variant has a modest effect.

oligonucleotide   A short single-stranded DNA, up to around 50 nucleotides long. Synthetic oligonucleotides are used as molecular probes to detect or quantify the presence of specific DNA or RNA molecules, and as primers for polymerase chain reaction (PCR).

oligosaccharides   A series of monosaccharides linked together by their hydroxyl groups (glycosidic bonds) as linear or branched structures.

omics   The study of entities in aggregate, such as the DNA (genomics), RNA (transcriptomics), protein (proteomics) or other molecular complement of a cell, tissue or organism.

oncogene   A gene which, when turned on inappropriately or expressed at higher than normal levels, participates in the evolution of a normal cell to a cancer cell.

oncogenesis   The process by which normal cells become cancerous.

ontogeny   The development of the individual organism.

ontological   Related to the identity or being of an individual.

ontology   A set of concepts and relations between them for some particular domain of knowledge. The simplest form of ontology is a controlled vocabulary, that is, a list of terms.

oocyte   Egg cell.

oophorectomy   Removal of the ovaries and related anatomical structures, such as the fallopian tubes, which may be carried out to reduce the risk of developing ovarian cancer.

open reading frame (ORF)   Corresponds to a stretch of DNA that could potentially be translated into a polypeptide.

operator   A prokaryotic negative regulatory element that binds a repressor protein to block initiation of transcription at the adjacent gene.

operon   Type of gene organization in which a cluster of separate genes are transcribed together into a single RNA transcript from which the individual proteins are translated.

oppression   The process of subjugation.

opsonic complement system   Complement system that enhances phagocytosis of pathogens by macrophages and neutrophils.

optimal codon   Synonymous codons that correspond to the most abundant tRNAs and that have optimal interaction with anticodons. Such codons are translated most efficiently.

optimal score   The maximum among the scores of all alignments for a set of sequences.

opt-out   The right not to participate via presumed consent and to be entered on an opt-out list instead.

ORC   *See* origin recognition complex.

ORE   *See* origin recognition element.

ORF   *See* open reading frame.

organelle   Membrane-bounded compartment in a eukaryotic cell that has a distinct structure, macromolecular composition, and function. Examples are nucleus, mitochondrion, peroxisome, endoplasmic reticulum and Golgi apparatus.

origin of replication   A point or region on the chromosome where DNA replication initiates.

origin recognition complex (ORC)   A complex of six polypeptides isolated from *Saccharomyces cerevisiae* from the origin of replication.

origin recognition element (ORE)   The binding site in DNA of proteins that initiate DNA replication.

origin usage   The selective activation of replication origins during the S phase of the cell cycle.

origin-binding proteins   Proteins that bind to the origin recognition element and are involved in the initiation of DNA replication.

orthologs   Genes that share a common ancestor but have diversified during their evolution in different species.

orthologous   Describes genes in different species that are similar at both the DNA and protein level as a result of descent from a common ancestral gene, and which may have similar functions.

orthologous loci   Genetic loci descended by divergent evolution from a common ancestral locus without gene duplication.

outreach clinic   A clinic held usually in a district general hospital where specialists from the regional center visit, so that families can be seen nearer to their home.

overdominant selection   *See* heterozygote advantage.

overlapping generations   Reproductive system in which offspring are produced by parents with a range of ages and parents can reproduce multiple times.

oxidative phosphorylation (OXPHOS)   The process by which the condensation of adenosine $5'$-diphosphate (ADP) and inorganic phosphate to adenosine $5'$-triphosphate (ATP), the common energy currency of the cell, is powered by the energy derived from respiration. It relies on a proton gradient set up across the inner mitochondrial membrane (or cell membrane in bacteria) as a result of cellular oxidative respiration.

OXPHOS-related nuclear genes   Nuclear genes encoding structural components of the respiratory chain, or controlling its formation, function and turnover.

p arm   The short arm of a chromosome.

P1-based artificial chromosome (PAC)   *See* P1-derived artificial chromosome.

P1-derived artificial chromosome (PAC)   Bacterial cloning vector able to carry a large genomic DNA insert, typically 100–200 kilobases, and which utilizes the P1 phage replication origin.

PAC   *See* P1-derived artificial chromosome.

2D-PAGE   *See* two-dimensional polyacrylamide electrophoresis.

packaging cells   Eukaryotic cells that express viral proteins needed to package retroviral RNA genomes into infectious virions but that do not produce replication-competent virus. These cells can be used to produce retroviral vectors following transient transfection of vector DNA or to generate stable vector-producing cell lines following stable introduction of vector DNA.

packaging signal   One or a number of RNA structures, usually in the form of complex helices and loops formed by folding of the viral RNA, that provide a recognition motif for the viral structural proteins to capture the RNA genome and select it specifically against the background of cellular RNAs.

paired domain   A sequence-specific DNA-binding protein domain, encoded by the paired box, found in a number of developmentally important proteins.

Paneth cell   Cells located in intestinal glands that function in immune defense.

panin   Vernacular term for the tribe Panini.

Panini   Tribe within the order Primates that includes modern chimpanzees and all extinct taxa that are more closely related to modern chimpanzees than they are to any other living primate taxon.

panmictic population   A population in which mating is random with respect to the genes under consideration.

panmixia   Complete interbreeding in a population.

papain   A cysteine protease found in the latex of papayas. All proteases with an active-site cysteine and folding similar to papain are classified as papain-like cysteine proteases (for example cathepsins B, H, L and S).

PAR   *See* pseudoautosomal region.

paracentric inversion   Inversion of a chromosomal segment that does not include the centromere.

paradigm   According to the philosopher of science, Thomas Kuhn, a piece of scientific work that founds a new theoretical concept in a science by a special example of scientific analysis.

parallel evolution   Independent evolution of identical traits from similar genetic backgrounds.

paralogous   Describes genes belonging to the same gene family within the same species. They have arisen by gene duplication within the same genome.

paralogous group   Clusters of related genes that have resulted from successive duplication events within a single genome.

paralogous region   Chromosomal segment that contains a closely linked set of paralogous genes.

paralogs   Related genes that have arisen by duplication within the same ancestral species. Compare orthologs.

parametric methods   Methods that require parameters governing the disease model (gene frequencies, penetrances) to be specified.

paranodal   Region next to a node of Ranvier in peripheral nerves.

parental investment   Investment by a parent in an individual offspring that increases the offspring's chance of survival at a cost to the parent's ability to invest in other offspring.

parietal lobe A part of the cerebral cortex that is located, in humans, on the two sides of the brain. It is involved in integrating sensory information to form a single percept and in constructing a spatial coordinate system that represents the outside world.

parsing Automatic extraction of relevant words in a text.

patent A legal right to prevent all others in the country in which it was granted from making, selling, using or importing an invention registered with the appropriate authority for a period typically up to 20 years. It includes the right to license others to make, use or sell it.

parthenogenesis Literally 'virgin birth', a culturally specific belief that sexual intercourse has no direct bearing on conception and procreation.

path diagram Pictorial representation of a structural equations model.

pathology A manifestation of disease or abnormality.

pathogen A microbe capable of causing host damage.

pathway Molecular connectivity diagram reflecting protein, enzyme, and ligand interactions, and also metabolic and signal transduction cascades. They provide insight into the mechanism of action of novel compounds or existing drugs.

patrilineal Passing from father to children.

patrilineal descent This situation when group membership is traced exclusively through the male line.

patterning The developmental mechanisms that direct appropriate differentiation and morphogenesis during embryogenesis.

Pax genes Genes encoding developmentally important proteins containing the DNA-binding paired domain.

PBSC *See* peripheral blood stem cells.

PCR *See* polymerase chain reaction.

PDZ proteins A diverse group of proteins that are involved in cross-linking other proteins, the so-called PDZ-binding proteins, to the actin cytoskeleton.

pedigree A family tree diagram showing inheritance patterns of a particular trait or disease.

peeling Calculation of pedigree probabilities by solving nuclear pedigrees and incorporating results in subsequent nuclear pedigrees.

P-element Transposable element used in *Drosophila* for transgenic expression studies and creation of loss-of-function alleles.

penetrance (1) Degree to which the features of a disease are apparent in a person carrying the disease gene.
(2) The proportion of individuals with a specific genotype who express that character in the phenotype.

PEP *See* primer extension preamplification (PEP).

peptide aptamers Antibody-like recognition agents consisting of conformationally constrained peptides displayed on the surface of scaffold proteins.

peptide mass fingerprinting Analysis of the molecular masses of the peptides obtained by chemical or enzymatic digestion of a protein.

peptide nucleic acid (PNA) A sequence of nucleosides that are linked together 5′ to 3′ by a peptide linkage instead of the normal phosphodiester backbone.

perceived personal control The subjective belief that one has at one's disposal a response that can reduce, minimize, eliminate or offset the adverse effects of an unpleasant event.

pericentric inversion A structural chromosomal abnormality resulting from reversal of a chromosomal segment that includes the centromere.

peripheral blood stem cells (PBSC) Hematopoietic stem and progenitor cells that can be mobilized to leave the bone marrow and circulate in the blood by administration of high doses of certain hematopoietic growth factors. After mobilization they can be collected by leukophoresis (isolation of the cells from large volumes of the blood) and used for transplantation instead of bone marrow.

peroxisome biogenesis disorders Human fatal, autosomal recessive disorders caused by defects of matrix protein import and/or impaired membrane assembly. This disease includes Zellweger (hepato-cerebro-renal) syndrome, neonatal adrenoleukodystrophy, infantile Refsum disease, and rhizomelic chondrodysplasia punctata.

personal control Perceived likelihood of changing the situation or managing one's emotional reactions.

personal data Any information relating to an identified or identifiable person.

personal obligation The expectations, rights and duties of individuals who have special emotional, social or genetic ties with each other.

Pfam A large database of multiple sequence alignments and hidden Markov models covering many common protein domains, based on Swiss-Prot and SP-TrEMBL protein sequence databases.

PFGE *See* pulsed-field gel electrophoresis.

PGD *See* preimplantation genetic diagnosis.

pharmacogenetic test A test that determines or predicts a genetically determined individual response to a medicine.

pharming The genetic alteration of animals to produce therapeutically valuable materials, usually proteins.

phase 0 intron The intron that lies between the third nucleotide of the upstream codon and the first nucleotide of the downstream codons.

phase 1 intron The intron that lies between the first and second nucleotides of a codon.

phase 2 intron The intron that lies between the second and third nucleotides of a codon.

phase class The position of an intron (or splice site) relative to the reading frame of adjacent translated exons.

phenomenology A form of philosophical inquiry that starts from the subjective description of lived experience, rather than from objective observation or theory.

phenotype The biochemical, physiological and physical characteristics of an organism. The phenotype collectively is considered to result from the interaction of genotype with enviornment, but a given individual phenotypic trait need not necessarily be genetically determined.

phenotype rescue The restoration of the wild-type phenotype in an animal or cellular model following an experimental modification, for example, the introduction of a gene.

phenotypic plasticity A characteristic of the norm of reaction. It denotes the type of sensitivity or insensitivity to environmental change that a particular genotype exhibits.

phenylketonuria (PKU) Hereditary metabolic disorder resulting in reduced ability to metabolize phenylalanine, which unless treated before 1 month of age causes mental retardation. It is inherited in autosomal recessive fashion and is due to a defect in the gene for phenylalanine hydroxylase, which catalyzes the conversion of phenylalanine to tyrosine.

pheochromocytoma Chromaffin cell tumors that usually develop in the adrenal medulla.

Philadelphia chromosome Abnormal chromosome 22 that results from a reciprocal translocation between the distal portion of chromosome 22q and the distal portion of chromosome 9q. This translocation is found in most patients with chronic myelogenous leukemia.

phlebotomy Large-scale drainage of blood for research purposes. Originally, the therapeutic practice of opening a vein for the purpose of bloodletting. Also called venesection.

phosphatase Enzyme that removes a phosphate group from a substrate.

phosphodiester bond Either of the two bonds covalently linking oxygens of a single phosphate and carbon. In nucleic acid polymers, these are the bonds connecting the individual nucleotide monomers, and the carbons are the 3′ and 5′ carbons of deoxyribose (DNA) or ribose (RNA).

phosphorylation The addition of one or more phosphate groups to a protein (or other molecules).

photolithography A chemical process by which a pattern is imprinted on a silicon or glass wafer by shining light through a specially designed mask. Multiple masks can be used to create a complex, layered pattern.

photoreceptor (1) Any cell or protein that is specifically responsive to light.
(2) Light-sensitive cell in the eye that mediates vision.

phototransduction The process by which photons cause a photochemical conformational change in membrane-bound rhodopsin in the retina. This signal is converted into an electric impulse, amplified and integrated before being carried to the brain by the optic nerve.

phylogenetic Dealing with the evolutionary and taxonomic relationships between species, or other related objects, such as protein sequences.

phylogenetic tree A tree-like representation of the evolutionary relationship between objects with a common origin. The objects can be species, nucleic acid or protein sequences or any other objects that evolve through time. In molecular evolutionary studies, phylogenetic trees are calculated from a multiple sequence alignment of the sequences under consideration, which provides an estimate of evolutionary distance between sequences.

phylogeny The evolutionary history of a group of species (or sequences), often presented as a phylogenetic tree.

physical map A map of the physical location of genes (and other sequence features) on the chromosomes. Can be constructed by analysis of restriction enzyme digestions or by contig (clone) mapping. Such a map shows distances between and within genes or specified markers measured in base pairs. The genome sequence is the highest-resolution physical map.

physical mapping Any process by which any nucleotide sequence is positioned within a longer nucleotide sequence.

PIC *See* polymorphism information content.

PKU *See* phenylketonuria.

plasmid A small circular DNA molecule that replicates independently of the main genome. Bacterial and eukaryotic (yeast) plasmids are used extensively as vectors for DNA cloning and expression.

pleiotropy The control or determination of more than one characteristic or function by a single gene.

Pleistocene Climatic period between 2 million and 10 000 years before present. The Pleistocene was characterized by extreme climatic oscillations, with cold periods when large parts of Europe and North America were covered by glaciers, interspersed with interglacials, when it was often warmer than today.

PLP *See* pyridoxal 5′-phosphate.

pluripotent cells Cells that are not committed to a single pathway of differentiation.

PNA *See* peptide nucleic acid.

point spread function (PSF) In fluorescence microscopy, the normalized spatial fluorescence intensity distribution in the image space obtained for a 'point-like' fluorescent object. The width of this distribution at half the maximum intensity (full-width at half-maximum, FWHM) is a measure of the optical resolution.

Poisson distribution A discrete probability distribution whose mean and variance are equal to one another.

polar ejection force Force generated by astral microtubules that pushes the chromosomes away from the poles, toward the center of the spindle.

polarity (1) Electrical polarity due to + and − charges. (2) The directional nature of either a nucleotide or polynucleotide.

poly(A) tail (1) A string of multiple adenosine nucleotides that is added to the 3′ end of nascent mRNA molecules following site-specific cleavage, and which is important in stabilizing the mRNA. (2) In transposable elements, refers to the presence of several adenine nucleotides in the 3′ end region.

polyadenylation Addition of a string of adenine nucleotide residues (typically around 200) to the 3′ end of mRNA.

polyadenylation site A specific sequence in the 3′ untranslated region of an RNA transcript that determines where the poly(A) tail is added to the 3′ end of the mRNA.

polycistronic messages mRNAs that encode multiple protein products.

polyclonal Arising from many founder cells.

polygenetic *See* polygenic.

polygenic Describes a trait determined by a group of nonallelic genes that interact such that the effect is cumulative.

polygenic disease Disease that results from the interaction of many different genes, each of which has a relatively minor effect.

polygenic inheritance Describes a pattern of inheritance in which the phenotype reflects the interaction of many different genes.

polyglutamine diseases A class of neurodegenerative diseases caused by expansion of a trinucleotide (CAG) that encodes the amino acid glutamine.

polyglutamine repeat A protein sequence that contains a long continuous series of the amino acid glutamine.

polymerase chain reaction (PCR) An enzymatic method of selectively copying a segment of DNA *in vitro*, using a thermostable DNA polymerase and a pair of oligonucleotide 'primers' complementary to either end of the segment to be copied. Only a small number of copies of the starting sequence are required and many millions of copies of the exact sequence can be made within hours.

polymorphic *See* polymorphism.

polymorphism A phenotypic trait (including sites of variation in DNA and proteins) that occurs in at least two forms in the population with a frequency of not less than 1% for the rarest form.

polymorphism information content (PIC) A measure of the variation in a region of DNA between different individuals.

polynomial time algorithm An algorithm whose running time is bounded by a polynomial function of the input size.

polyphyly Descent from more than one ancestor.

polyploid Having more than two sets of chromosomes as the natural genome.

polyploidization The acquisition, by various means, of more than two haploid sets of chromosomes in a cell or species.

population A group of organisms of the same species.

population substructure Existence in a population of subpopulations that tend to mate endogenously.

porphyrin Any of a group of substances containing four pyrrole rings linked together to form a macrocycle in which one or more of the eight peripheral hydrogenations is replaced by a side chain. Position isomers in which the order of the side chains around the macrocycle differs are indicated by roman numerals (for example, I, III, IX).

porphyrinogen Unstable reduced (hexahydro-) form of a porphyrin.

position effect The effect of the position of a gene in the chromosomes on its expression.

position-effect variegation The effects of large-scale chromatin regulation on the transcription of genes that are placed at a new genomic locus, either artificially or because of genome instability. The term derives from the fact that the natural phenomenon was originally observed as color variegation of plant seeds and leaves.

positional cloning Strategy for identifying and cloning a gene on the basis of its location in the genome rather than the biological function of its product. The method usually involves using linkage studies to link the genetic locus of interest to one that has already been mapped and to map the gene to a small chromosomal interval containing few genes. The later stages of positional cloning have been greatly helped by the availability of the sequence of the human genome.

positive Darwinian selection Natural selection that favors nucleotide substitutions.

postmodernist A term used to denote economically, technologically and socially advanced societies with a diverse array of ever-changing cultural and social forms.

postnatal genetic screen Analysis of tissue or blood samples from an asymptomatic population, usually newborn infants, to detect a particular chromosomal abnormality or a genetic condition.

postsource decay (PSD) Analysis of peptide fragments of selected peptides generated by MALDI.

posttranslational modification Reversible or irreversible covalent chemical modification of a protein after it has been synthesized, such as glycosylation on serine or asparagine residues and phosphorylation on serine, threonine or tyrosine residues.

PP *See* protein phosphatase.

predictive genetic testing The use of a genetic test in an asymptomatic person to predict the future risk of a particular disease. Commonly used to provide an estimate of an individual's risk of developing a particular multifactorial or polygenic disorder, or to detect the presence of a mutation that causes a specific late-onset disease.

predictive test *See* predictive genetic testing; predisposition testing.

predictive test protocol Guidelines for a systematic pre- and post-test approach regarding predictive genetic test requests.

predisposition testing Testing that is done to look for a mutation or a biochemical marker (such as elevated blood cholesterol or iron) that significantly increases the chance of, but does not guarantee, the development of a genetically influenced disease later in life.

pre-embryo The product of gametic union from fertilization to the formation of the primitive streak at roughly 14 days after fertilization, which marks the beginning of the embryonic period.

preimplantation genetic diagnosis (PGD) A technique used when conception occurs through *in vitro* fertilization in which DNA testing is carried out to ascertain the presence of disabling traits in the embryo before it is transferred to the woman's uterus. Typically, embryos with diagnosed abnormalities are not transferred.

pre-messenger RNA (pre-mRNA, hnRNA) The RNA transcript of a eukaryotic gene before it is processed to form mRNA. It gives rise to mRNA by processing steps such as splicing to remove introns, and addition of a poly(A) tail.

pre-mRNA *See* pre-messenger RNA.

pre-mRNA intron An intron in pre-messenger RNA.

premutation A change in DNA that does not quite cause a different phenotype (such as a disease) but is prone to become a disease-causing mutation in successive generations.

prenatal diagnosis *See* prenatal testing; pre-implantation genetic diagnosis.

prenatal testing Any of the various methods to ascertain fetal characteristics, usually 'abnormalities' or disabling conditions. Some tests will also determine the sex of the fetus. These methods include, but are not limited to, amniocentesis, chorionic villus sampling, determination of α-fetoprotein, ultrasound, fetoscopy, percutaneous umbilical blood sampling and DNA and cytogenetic testing.

prenucleolar bodies Particles visible inside the nuclei of telophase cells and which contain nucleolar material.

pre-ribosomal RNA processing The maturation process of ribosomal RNAs, which takes place in the nucleolus. During pre-rRNA processing, the initial transcripts of the rRNA genes are cleaved both internally and from the ends and many nucleotides undergo modifications.

pre-rRNA processing *See* pre-ribosomal RNA processing.

presequence A sequence in a mitochondria-derived nuclear gene that encodes a mitochondrion-targeting function in the protein.

presumed consent Implied consent when a patient has not expressed any unwillingness for their medical records to be used for research, information in writing on the possibility of such use of the records having been generally available to them.

presymptomatic diagnosis Discovery of indications of future disease in an asymptomatic person.

presymptomatic test *See* predictive genetic testing.

presymptomatic testing *See* predictive genetic testing.

prevalence At a given point in time, the proportion of persons in a given population with a disorder or trait.

primary care The providers of general medical care to the population. In the United Kingdom, the primary care providers are general practitioners and community doctors, nurses and therapists. They will refer to specialists (secondary care) for specific advice, diagnosis and management when appropriate.

primary RNA transcript *See* primary transcript.

primary transcript An RNA molecule when it is first transcribed from a eukaryotic gene before it is processed into a mature mRNA or other functional RNA.

primates Mammalian order comprising the tree-shrews, lemurs, monkeys, apes and humans.

primer In the polymerase chain reaction (PCR), a short (10–25 bases) single-stranded DNA or RNA (in reverse transcriptase PCR) of specified sequence that hybridizes with the target DNA and forms the starting point for DNA replication.

primer extension preamplification (PEP) A variant of the polymerase chain reaction (PCR) that uses a nonspecific oligonucleotide primer to amplify all (or most) DNA sequences in a mixture.

primitive streak An aggregation of undifferentiated cells that forms at the posterior end of the mammalian embryonic plate just before implantation, and which begins to develop into the fetus.

primordial germ cells First generation of embryonic cells committed to becoming germ cells.

principle In medical ethics, an abstract moral statement, such as 'take seriously the choices of autonomous persons', from which more specific moral rules such as 'seek the informed consent of study participants', are derived.

principle of autonomy The moral obligation to respect the autonomy (literally self-rule, as a special attribute of all moral agents) of others.

principle of population The eighteenth-century economist and clergyman Thomas Malthus's assumption that, in theory, a population can increase at a geometrical rate (although, in practice, the population size is limited by the available food supply).

prion Infectious proteinaceous particle considered to be the cause of transmissible spongiform encephalopathies such as bovine spongiform encephalopathy (BSE) in cattle, scrapie in sheep and Creutzfeld–Jacob disease (CJD) in humans.

privacy A general human right to control and limit information about oneself.

probability Commonly defined as the frequency of an event's occurrence in a long series of 'trials', such as the frequency of getting 'heads' in 100 tosses of a coin.

proband The first person in the family who undergoes genetic testing.

probe A DNA molecule carrying a fluorophore or other identifiable label. It will tend to hybridize (bind) selectively to the part of another DNA molecule of complementary sequence.

processed genes The genetic units created by transposition of reverse transcribed mRNAs (cDNAs). Such units, if nonfunctional, are processed pseudogenes.

processed pseudogene Inactive, intronless copy of a gene derived from the insertion of a reverse-transcribed mRNA into the genome.

processivity The capacity of an enzyme such as DNA polymerase to move along the DNA template without disengaging after each nucleotide addition, incorporating hundreds or thousands of nucleotides before dropping off the template.

procoagulant Describes any substance (such as the clotting factor proteins in plasma) or condition that promotes the clotting of blood.

products of nature Discoveries, not inventions.

profiling Determination of levels of expression of a set of genes or proteins in a tissue and relating this to, for example, disease status.

progenitor cell In hematopoiesis, a cell with a relatively limited lifespan (weeks or months) that can give rise to a subset of blood cell types.

program A detailed and comprehensive plan for a sequence of events.

progressive ophthalmoplegia Progressive paralysis of the muscles that move the eye in the orbit and elevate the eye lids.

prokaryote Any of a large group of organisms (the bacteria and archaea) in which the DNA is not enclosed in a nucleus and whose cells lack a cytoskeleton and other eukaryotic organelles. They are generally unicellular.

prometaphase The second stage of mitosis during which the nuclear envelope dissolves, the mitotic spindle apparatus is assembled and attaches to the chromosomes, followed by chromosome migration toward the metaphase plate where they form compact, ring-like structures called prometaphase rosettes.

promoter Region within a gene, usually (but not always) located immediately to the 5′ side of the transcription start site, which contains the binding site for RNA polymerase and other sequence elements to which regulatory proteins bind to control the initiation of gene transcription.

propeptide (1) Polypeptide sequence in an inactive protease precursor the cleavage of which results in activation of the protease.
(2) The amino- or carboxy-terminal noncollagenous domain of fibrillar collagens. The amino- and carboxy-propeptides are cleaved off by proteinases before assembly of the collagen molecules into collagen fibers.

property A legal regime of recognized rights of exclusive possession and control over specified subject matter.

prophase The first stage of mitosis and meiosis, during which the chromosomes condense into worm-like structures and the centrosome divides, with the new centrosomes migrating around the still-intact nuclear envelope to form the nuclear poles.

prophylactic surgery Surgery, such as removal of the breasts or ovaries, in carriers of a mutation predisposing to a hereditary cancer, to reduce the risk of a cancer developing.

proposal density A distribution that is used to sample (propose) from.

protanomaly Common inherited anomalous trichromatic color vision mediated by S cones and at least two cones of the M class.

protanopia Common dichromatic inherited color vision mediated by S cones and a single type of M cone.

protease Any of a large group of enzymes that cleave proteins by hydrolyzing peptide bonds. Also known as proteinase. The terms cysteine protease or serine protease indicate what the active-site residue is and the molecular mechanisms of cleavage.

protease clan Group of proteases whose homology is established by structural similarity or common order of catalytic residues in the primary sequence. It may include one or more protease families.

protease family Group of proteases whose homology is established by similarity of amino-acid sequence.

proteasome An intracellular multisubunit protein complex, cylindrical in shape with the internal chamber lined with proteases in which specifically targeted proteins are degraded into small peptides. It is found in both prokaryotic and eukaryotic cells. In eukaryotes, proteins destined for degration are often marked by covalent coupling of multiple copies of the small protein ubiquitin to lysines on the protein.

protein degradation Partial or complete breakdown of a protein, usually by enzymes, with consequent loss of its function.

protein domain Discrete and compact region of structure within a protein, with its own characteristic fold. Some proteins are formed of a single domain, others of two or more, often with different biochemical functions.

protein family Group of evolutionarily related proteins that share similar amino-acid sequences and three-dimensional structures, and which consequently generally perform closely related functions.

protein kinase An enzyme that transfers a phosphoryl group from ATP to tyrosine, threonine or serine side chains on a protein.

protein phosphatase (PP) Any of a large group of enzymes that remove phosphate groups from phosphorylated sites on proteins.

protein tyrosine phosphatase (PTP) Any of a large family of enzymes that dephosphorylate phosphotyrosine-containing proteins.

protein site Region of a protein having a specific function.

proteolysis Cleavage of the covalent chemical bond coupling two adjacent amino acids in a target protein.

proteome The full complement of proteins produced in a cell or an organism.

proteomics The isolation, purification and study of protein structure and function, particularly with reference to the complements of proteins produced in a cell or organism.

prothrombinase complex A complex of clotting factors Va and Xa which is formed on the surface of negatively charged phospholipid membranes and catalyzes the activation of prothrombin to thrombin.

protochordates Those chordates that are not vertebrates, consisting of the urochordates and the cephalochordates.

protofibril Small highly ordered protein assemblies that provide nuclei for the growth of larger amyloid fibrils.

proto-oncogene Gene that has the potential to initiate cancer if it becomes dysregulated through mutation, chromosomal rearrangement, gene amplification or transcriptional activation after provirus integration.

protostomes Animals in which the mouth is derived from the first involution of the blastopore and the anus is formed by the second opening. They include the arthropods, nematodes and molluscs, among other phyla.

provirus A viral genome integrated into the genomic DNA of the host cell. In mammals it usually refers to the integrated DNA form of a retrovirus.

PSD *See* postsource decay.

pseudoautosomal region (PAR) A small region of homology between X and Y chromosomes at which there may be synapsis and recombination during meiosis.

pseudoexons Gene regions that are functional in a gene from one species and that have been conserved in nonfunctional form in another species.

pseudogene A nonfunctional DNA sequence in the genome that resembles a given gene but which has lost its function through the presence of one or more inactivating mutations incurred during evolution. Pseudogenes can also be formed by the insertion of a DNA copy of an RNA followed by mutation.

pseudosplice site A short sequence in an intron highly similar to the consensus sequence of a splice site, but which is not used in an RNA splicing reaction.

psychiatry Branch of medicine that deals with mental disorders, that is pathological disturbances of the higher functions of the brain, including mood (affect), cognition and complex behaviors.

psychological well-being of testees Aspects of pre- and post-test psychological functioning assessed by a battery of psychometric tests and questionnaires.

psychosis Mental disturbance, usually in clear consciousness, characterized by loss of contact with reality and typically manifested by delusions and/or hallucinations.

PTP *See* protein tyrosine phosphatase.

public interest A common justification for breaching privacy and medical confidentiality.

puerperal psychosis Episodes of severe mental disorder precipitated by childbirth, usually severe and/or psychotic episodes of mood disorder occurring within 2 weeks of delivery.

pull-down experiment Isolation of associating proteins that can bind to a fusion protein of interest by centrifugation.

pulsed-field gel electrophoresis (PFGE) A modification of agarose gel electrophoresis which can separate DNA fragments in the size range 50 kilobases to 10 megabases.

pump Informal name for a membrane transport protein that can transport a solute against its electrochemical gradient using energy usually derived from ATP hydrolysis.

purifying selection Selection that tends to remove detrimental alleles from the genome.

pyridoxal 5′-phosphate (PLP) A vitamin B6-derived cofactor required by numerous enzymes.

pyrimidine dimer *See* cyclobutane pyrimidine dimer.

q arm The long arm of a chromosome.

QTL *See* quantitative trait locus.

quadrupole mass analyzer Instrument for analyzing ion species by their mass-to-charge value. It comprises four parallel rods (either cylindrical or hyperbolic profile) that are connected pairwise. A direct current (DC) superimposed on an alternating current (AC) is applied to the rod pairs. Depending on the amplitude of the AC and DC voltages, only ion species with a particular mass-to-charge value have a stable trajectory and can pass the analyzer.

quantitative trait locus (QTL) The location on a chromosome of a gene that affects a phenotypic trait that is measured on a quantitative scale, such as blood pressure or height. These traits are typically affected by more than one gene, and sometimes by the environment.

quartet Group of codons with common first and second bases. The genetic code table consists of 16 such quartets.

r *See* coefficient of relatedness.

R bands Type of chromosome bands that are the reverse of the G bands defined by Giemsa staining. They are specific chromosome domains, which are early replicating at metaphase, relatively GC rich and enriched in genes. A subset of the R bands, the T bands, are located in the telomeric regions and contain more than 60% of the mapped genes.

RACE *See* rapid amplification of cDNA ends (RACE).

race Formerly, a large natural division of the human species. Now, since such groups are recognized as illusory, the term is generally used to designate a politically meaningful category into which people may be allocated on the basis of some aspect of ancestry.

racemization The conversion of a L amino acid into a D form and vice versa. In living eukaryotes, only the L form exists. After the death of the organism, the L form is converted to the D form until both exist in equal amounts in dynamic equilibrium.

racial hygiene The movement in Germany in the early to mid-twentieth century designed to 'cleanse' the German gene pool of genes from non-'Aryan' people.

racialist (1) A synonym for racist (especially in the UK).
(2) A person who believes that races can be distinguished, that racial differences are important, but that the differences do not imply that one race is superior to another.

racialization The situation where meaning is attributed to particular biological features of human beings, as a result of which individuals may be assigned to a general category of people that reproduces itself biologically. The way in which a particular group is constructed as a race or as a racialized group is a matter for historical investigation.

racialized minority When a culture racializes those people who share a cultural heritage that is not seen as part of the majority, then varying degrees of discrimination are part of the experience of being identified with an ethnic minority. In addition to adhering to each other because of a mixture of self-identified commonalties, based on a language or religion or place of origin that differs from that of the majority, the shared experience of discrimination at the hands of the majority population may create commonalties that did not exist before, and force an ethnic minority to dwell on them.

racism A discourse of the signification of somatic features and attributions of negatively evaluated characteristics to groups.

racist A person who believes that races can be distinguished, that some races are superior to others, and that his/her race is one of the superior ones.

radial arrangement The position of chromosome territories, chromosome subregions or other nuclear structures in relation to their distance from the nuclear center.

radiation hybrid map (RH map) A genome map in which sequence-tagged sites are positioned on the basis of the frequency at which they are separated by radiation-induced breaks, assayed by analyzing a panel of hamster hybrid cell lines containing a collection of radiation-generated chromosomal fragments from the target genome.

radiation hybrid mapping (RH mapping) A physical mapping method in which a lethally irradiated human cell is fused with a rodent host cell to produce a hybrid cell that contains numerous small fragments of human chromosomes. The human DNA fragments are analyzed statistically to form a map.

radiation hybrids A panel of somatic cell hybrids, with each cell line containing a random set of fragments of, for example, irradiated human genomic DNA in a hamster background.

random mating The probability of a mating is independent of the genotypes of the partners.

random shearing Method of breaking DNA molecules in random positions, usually done by nebulization or by sonication.

rapid amplification of cDNA ends (RACE) A modified version of reverse transcriptase polymerase chain reaction. This technique allows the cloning and subsequent identification of the 5′ and/or 3′ regions of an mRNA molecule.

rapid cycling bipolar disorder Bipolar disorder in which the course is characterized by rapidly recurrent episodes of illness, typically cycling between low and elevated mood states, with at least four distinct episodes of mood disturbance in a 12-month period.

rare variant A variant of a phenotypic trait (including protein and DNA markers) that occurs at a frequency of less than 1% in the population.

rate of mutation The frequency with which gametes carrying a new mutant allele arise per generation.

rater contrast effects Tendency of a rater (typically a parent) to emphasize differences between siblings in their ratings and to sometimes do so differentially for identical than fraternal twins.

rational–emotional therapy A form of psychotherapy that focuses on the use of cognitive and emotional restructuring to foster adaptive behavior.

Rawls's two principles of justice The first principle states that each person has an equal right to a fully adequate scheme of equal basic liberties which is compatible with a similar scheme of liberties for all. The second principle states that social and economic inequalities are to satisfy two conditions: first, they must be attached to offices and positions open to all under conditions of fair equality of opportunity; and second, they must be to the greatest benefit of the least-advantaged members of society.

reaction surface A three-dimensional representation of phenotypic values, usually graphed on axes representing low to high genotypic values and restrictive to facilitative environmental values.

read-through transcription The transcription of a gene that is extended beyond its terminating signal.

real person Those characteristics of a person which determine their identity.

rearrangement A change in chromosomal structure.

recapitulation theory Theory which supposes that the development of the embryo (ontogeny) recapitulates the course of the species' evolution (phylogeny).

receptor Integral membrane protein that transduces an extracellular signal (usually the binding of another molecule) into an intracellular signal.

receptor kinase A membrane receptor that contains an intrinsic protein kinase activity in its cytoplasmic portion and signals through phosphorylation.

recessive Describes an allele whose effect is masked by a dominant allele when the latter is present.

recessive gene An allele that shows no effect in a carrier but can be passed on to the next generation.

recessive inheritance Pattern of inheritance of a recessive allele. For example, if both parents are carriers, each child has a 25% chance of inheriting a recessive characteristic, a 25% chance of not inheriting it and a 50% chance of being a carrier.

recessive mutation Mutation that gives rise to a recessive gene.

reciprocal translocation A balanced chromosome rearrangement of material between two chromosomes that usually has no effect on health. In some cases, infertility may result because of the chromosomes' inability to line up during their meiotic pairing process.

recombinant A gene, protein, virus or organismal genome that contains elements derived from two sources, either derived by natural recombination during meiosis, or by genetic engineering.

recombination The physical exchange of genetic information between corresponding DNA molecules.

recombination fraction (θ) In linkage analysis, the fraction of meiotic events that show recombination between two loci.

recombinational repair Form of DNA repair that involves homologous or nonhomologous recombination between two DNA molecules.

recurrence risk Chance that a hereditary disorder will be manifest in one's offspring.

recurrent spontaneous abortion A situation where there have been three consecutive losses of the embryo before the twentieth week of gestation.

reducing conditions Conditions without free oxygen.

reduction A program for explaining a scientific subject that is based on dissecting the subject into its component parts or underlying principles.

regional genetic center The staff that provide specialist genetic services for a defined population of usually between 2 000 000 and 5 000 000 people.

regulatory element A DNA region that regulates expression patterns of genes by binding to transcription factors.

regulatory sequence *See* regulatory element.

reinsurance Insurance that the insurer will purchase to offer protection against extremely large claims.

relational database A database in which the information is stored in tables that are linked in a coherent and logical manner consistent with their relations. The most common current form of database architecture.

relative recurrence risk to siblings (λ$_s$) Relative risk to siblings, defined in autism, for examples, as the risk of recurrence of autism in siblings born to parents after the first child with autism divided by the population prevalence (given low mortality, the prevalence and incidence are roughly similar).

relative risk The probability of disease in persons with a given factor (exposure or attribute), divided by the probability in persons without the factor.

remnant The product of lipoprotein catabolism by lipases, which hydrolyze the triglyceride moiety of very-low density lipoprotein and chylomicrons.

repetitive DNA DNA segments present in multiple copies in the genome, without a clearly assigned biological function.

replacement histones Histones that can be synthesized in the absence of DNA replication. They are particularly found in nondividing, differentiated cells. Genes encoding replacement histones are solitary genes outside the clusters of replication-dependent genes.

replication labeling Incorporation of hapten or fluorochrome labeled nucleotides during a given time window in S phase allows the visualization of replication foci that are active within this time.

replication origin A region of DNA at which replication is initiated.

repressor A protein that binds a repressor protein or protein complex to block transcription initiation.

reproductive advantage A genetically determined biological feature of an individual to survive in their environment long enough to be able to produce offspring.

reproductive cloning The production of offspring, by one of the artificial methods of cloning, that are genetically identical to another individual (the donor of the diploid cell used for nuclear replacement). The cloned embryo is implanted into the uterus of a surrogate mother who carries it to term and gives birth to it.

reproductive technology A general term for practices of assisted conception, usually but not always involving biomedical techniques.

research grant peer review A process by which research proposals are reviewed by other researchers with relevant expertise to determine their quality.

research The systematic collection of data that aims at producing generalizable knowledge.

respect for autonomy Respect for the patient's wishes, which assumes a rational agent making informed and voluntary decisions.

respect for communities A new moral principle requiring that investigators take seriously the choices and values of the community and, where possible, protect the community from harm.

respect for persons A moral principle that requires that investigators take seriously the choices of autonomous persons and protect those incapable of autonomous choice.

respiratory chain The metabolic pathway that carries out oxidative phosphorylation. The respiratory chain is composed of five multiheteromeric enzymes, the respiratory chain complexes I, II, III, IV and V, cytochrome *c* and coenzyme Q, all embedded in the inner membrane of mitochondria. Four of the five respiratory chain complexes (complexes I, III, IV and V) are composed of protein subunits that are encoded by nuclear genes as well as by subunits encoded by genes contained in the mitochondrial genome (mtDNA).

restricted and repetitive behavior One of the three domains that must be qualitatively impaired for a diagnosis of autism. Examples of such behavior are obsessions, compulsions, preoccupations, insistence that routines be followed in exact order, and complex stereotyped motor behaviors such as flapping of the hands while jumping up and down.

restriction endonuclease An enzyme that recognises a particular short sequence of bases in double-stranded DNA and cleaves the DNA at that site.

restriction enzyme *See* restriction endonuclease.

restriction enzyme fingerprinting A method for deriving contigs of clones by matching the patterns of DNA fragments obtained from each clone after treatment with specified restriction endonucleases.

restriction fragment length polymorphism (RFLP) A variation in the DNA of different individuals that changes the properties of a restriction site and thus can be detected by the different lengths of different variants after treatment of the DNA with a given restriction endonuclease.

restriction fragment length polymorphism (RFLP) analysis A method that detects variation among individuals by looking at the length of DNA fragments created by a given restriction endonuclease.

restriction site A short DNA sequence (usually 4–8 basepairs long) that is recognized and cleaved by a restriction enzyme.

restriction site polymorphism DNA sequence variation that is detected by the presence or absence of a recognition sequence for a specific restriction enzyme. *See also* restriction fragment length polymorphism.

retina A thin layer of photosensitive epithelium lining the rear two-thirds of the eye on which the image is formed. It contains the photoreceptor rod and cone cells, and nerve cells that serve to relay signals from these photoreceptors to the optic nerve.

retroelement A DNA element in a genome that has been derived through an RNA intermediate by reverse transcription.

retroposition *See* retrotransposition.

retroposon *See* retrotransposon.

retropseudogene DNA sequence that has arisen from the fortuitous reverse transcription of a gene-derived RNA and its subsequent integration into the genome. They are considered retrosequences, but not retroelements.

retrosequence Any DNA sequence that has been either specifically or fortuitously amplified in a genome by the integration of a reverse transcribed copy of its RNA.

retrotransposition Process by which an RNA transcript derived from a transposable DNA element or a retrovirus is reverse transcribed to create a cDNA copy which is then integrated into chromosomal DNA.

retrotransposon A transposable element that originates from reverse transcription of RNA to cDNA that becomes inserted into the genome and which transposes through an RNA intermediate.

retrovirus Member of a class of enveloped single-stranded RNA viruses that replicate via a DNA intermediate that integrates into the host-cell genome. The viral RNA genome is converted to DNA by the viral enzyme reverse transcriptase when it enters the infected cell. The integrated DNA is transcribed to produce new viral RNAs.

reversal A 'back mutation' to an ancestral state.

reverse transcriptase Enzyme that synthesizes single-stranded DNA using an RNA template. Also called RNA-dependent DNA polymerase.

reverse transcriptase polymerase chain reaction (RT-PCR) The synthesis of cDNA from an RNA template by a reverse transcriptase and the use of the cDNA as the template for amplification by the polymerase chain reaction.

RFLP *See* restriction fragment length polymorphism.

RH map *See* radiation hybrid map.

RH mapping *See* radiation hybrid mapping.

ribonucleic acid *See* RNA.

ribosomal active center A domain of the ribosome that associates with tRNA during translation, such as an aminoacyl-tRNA binding site (A site), a peptidyl-tRNA binding site (P site), and a deacylated-tRNA binding or exit site (E site).

ribozyme An RNA molecule that has catalytic activity.

risk assessment Calculation of an individual's risk, employing appropriate mathematical equations, to have inherited a certain gene mutation, to develop a particular disorder, or to have a child with a certain disorder based on analysis of multiple factors, including family medical history and ethnic background.

risk factor An attribute that is associated with an increased probability of a disease occurring. A risk factor is not to be understood as a cause of disease.

RNA Ribonucleic acid, a single-stranded linear molecule consisting of a backbone of alternating sugar (ribose) and phosphate molecules to which four different kinds of nitrogenous bases (adenine (A), uracil (U), cytosine (C), and guanine (G)) are attached to the sugars. RNA molecules are synthesized in cells by RNA polymerases that use segments of DNA as templates. The sequence of an RNA is therefore complementary to the sequence of the DNA strand from which it was transcribed, with U substituting for the T in DNA. There are three main types of RNA in cells: messenger RNAs (mRNA), which are transcribed from protein-coding genes and translated into protein; transfer RNAs (tRNA), which pick up amino acids and match them to the mRNA at the ribosome; and ribosomal RNA (rRNA), which is a structural component of the ribosome and carries the peptidyl synthetase activity. Numerous small RNAs with a variety of roles are also encoded by the genome.

RNA-binding protein Any protein that interacts specifically with the sequence and/or secondary structure of an RNA molecule. RNA-binding proteins often perform structural or functional roles.

RNA-dependent DNA polymerase *See* reverse transcriptase.

RNA editing Informational change that occurs at the level of mRNA, in which the coding sequence in an RNA has been chemically modified after transcription so that it differs from the sequence of the DNA from which it was transcribed.

RNA helicase Proteins that unwind local secondary structural elements in RNAs, thereby promoting an alteration in ribonucleoprotein structure. Also known as RNA resolvases.

RNAi *See* RNA interference.

RNA interference Technique by which a particular type of antisense DNA construct is made that will silence a gene of interest by binding to and inactivating the endogenous RNA.

RNA polymerase II RNA polymerase that transcribes protein-coding genes.

RNA polymerase III RNA polymerase that transcribes an RNA from the antisense strand of a DNA. It is involved in the synthesis of tRNA and other RNAs such as 7SL RNA.

RNA resolvase *See* RNA helicase.

RNA secondary structure Three-dimensional structure formed by intramolecular pairing between residues of the same RNA. It is usually characterized by the presence of one or more stems (paired double-stranded regions) and loops (unpaired regions within secondary structure elements) that can result in complex structures. RNA secondary structures often mediate interactions with RNA-binding proteins.

RNA splicing The removal of introns from a primary RNA transcript and the joining of the exons to produce a functional RNA. In the production of mRNA, the exons represent the coding sequences.

Robertsonian fusion *See* Robertsonian translocation.

Robertsonian translocation Translocation in which the centromeres of two acrocentric chromosomes (chromosomes with the centromere near a chromosome end) appear to have fused, forming an abnormal chromosome consisting of the long arms (q arms) of two different chromosomes. Also called a Robertsonian fusion.

robotics *See* automation

rod Type of photoreceptor cell in the retina that functions at low levels of light.

root In phylogeny, the point on an evolutionary tree that defines the last common ancestor of the taxa being considered.

royalty A payment made to the patent holder by someone using an invention.

RPA *See* ribonuclease protection assay.

ribonuclease protection assay (RPA) A technique for detecting specific mRNAs on the basis of their hybridization to a labeled sequence of complementary RNA and subsequent treatment with ribonuclease. Among other uses, the technique can be used to quantify and/or size a particular mRNA species.

RS domain Protein domain rich in arginine and serine dipeptides, found in SR proteins and other proteins involved in RNA splicing.

RT-PCR *See* reverse transcriptase polymerase chain reaction.

ruffle border In osteoclasts, the apical portion of the plasma membrane, which is specialized to aid bone resorption.

s *See* coefficient of selection.

S phase The phase of the eukaryotic cell cycle in which DNA is replicated.

safe harbor A private-sector privacy scheme outside the European Economic Area which fulfills the requirements of European data protection legislation.

SAGE *See* serial analysis of gene expression.

salvage anthropology Research on indigenous peoples requiring immediate action due to the impending destruction of those peoples or their ways of life.

sampling The process of drawing a state x of a stochastic variable X from its probability distribution.

SAR *See* scaffold-associated region; scaffold-attachment region.

sarcolemma The membrane surrounding the multinucleated muscle cells within a muscle fiber.

α-satellite DNA Predominant class of human centromeric DNA based on 169–172 basepair repeats that are organized into chromosome-specific arrays. Classically, the term satellite DNA referred to DNA that separated from the bulk of the genomic DNA upon gradient centrifugation. Gradually the meaning of the term has shifted to imply more-or-less any tandemly repeated noncoding DNA element.

satellite DNA DNA in eukaryotic chromosomes that contains many copies (from hundreds to more than 10 000) of identical or related tandem repeats of 100–300 basepairs, and which forms a separate band in density gradient centrifugation because of its different nucleotide composition and lower buoyant density compared with the rest of the genomic DNA. It is generally late replicating and constitutively heterochromatic and occurs mostly at centromeres and telomeres.

satellite sequence *See* satellite DNA.

scaffold Complex framework observed in both metaphase and interphase chromosomes that may aid chromatin fiber condensation. DNA and protein interact to produce a highly structured skeleton of DNA polymer loops radiating out from a central backbone.

scaffold-associated region (SAR) DNA sequence element that possesses an ability to interact *in vitro* in a specific fashion with isolated nuclear scaffold.

scaffold-attachment region (SAR) DNA sequence that is thought to bind to chromatin-domain-structuring proteins. These regions potentially keep large loops of chromatin organized in such a way as to facilitate transcriptional activation or repression.

schizophrenia Severe psychiatric disorder with onset usually in young adulthood, characterized by psychotic symptoms and deterioration in functioning in a variety of domains of activity including intellectual, occupational and social.

Schmidt–Lantermann incisures Clefts of noncompacted cytoplasm that run from the ab- to the adaxonal side of the myelin sheath.

SCID *See* severe combined immunodeficiency.

scissile bond In proteins, an amide bond that is cleaved by a protease.

scoring matrix A table used in scoring an alignment to provide it with a quantitative measure. For protein sequences, the table is set out as a 20 by 20 matrix of the amino acids in proteins with each cell given a predetermined score (derived from the study of numerous multiple alignments) that provides a value for a match between a particular pair of amino acids.

scoring scheme A method of giving an alignment a score to assess its accuracy. It usually consists of the use of a scoring matrix combined with a penalty value for gaps that are introduced to increase the degree of matching.

screening (1) Testing of a population for a disease state, which may already be present or might develop in the future.
(2) Analysis that aims to prevent a disease from developing or to minimize symptoms and improve the health of a person. In general, the focus is on whole populations of healthy, asymptomatic people or on particular groups coming from an identified geographic, racial or ethnic background.

seasonal affective disorder (SAD) Recurrent mood disorder that follows a clear-cut seasonal pattern, usually depressions during winter months.

secondary structure Types of local structure that occur in many different proteins (for example α helices, β turns, β strands) or within a nucleotide sequence (for example hairpin loops).

segregating genes Those genes that show allelic variation. They are so-called because when present in a parent in two distinguishable forms (alleles), only one will be found in each gamete after meiosis.

segregation analysis Formal analysis of distribution of disease or other phenotypic trait in a large pedigree to seek Mendelian pattern of inheritance, that is, autosomal dominant or recessive, or sex-chromosome linked.

segregation of chromosomes The separation of chromosomes during meiosis so that each gamete contains only one member of each pair of chromosomes.

selection That force by which alleles that contribute to survival and production of offspring are retained in the population whereas those that are deleterious are lost.

selective constraint The degree of intolerance that is characteristic for a DNA region in evolution. It is approximately equal to the proportion of mutations unaccepted by natural selection.

selective placement Similarity or dissimilarity between the measurable characteristics of biological and rearing parents, or the processes that produce the resemblance.

selenocysteine The twenty-first amino acid in proteins, which is formed when a selenium atom replaces the sulfur atom in cysteine.

selfish DNA DNA that is self-propagating without benefit to the organism.

self-MHC The major histocompatibility complex molecules expressed on the surface of an individual's own cells, recognized as non-foreign by the individual's immune system.

self-organization The process by which a system generates new information within itself.

self-selection of predictive test applicants The phenomenon that a relatively higher number of persons with better coping resources apply for predictive testing.

sensitization (1) Nonassociative learning in which an unpaired stimulus increases the strength of a reflex.
(2) Increased response to a drug following repeated exposure.

sequence alignment *See* alignment.

sequence assembly The process of merging overlapping sequence reads obtained from a shotgun sequencing project. The assembly results in a set of contiguous consensus sequences, referred to as 'contigs'.

sequence conversion A process, usually involving some type of non-homologous recombination, in which two nonidentical repeated sequences become identical.

sequence read An individual segment of sequence produced by performing a single sequencing reaction on a DNA sample and detecting (reading) the reaction products in the order of size as separated by gel electrophoresis. These days the reaction output is read by computer software integrated in an automated DNA sequencer. A sequence read will be only a few hundred basepairs in length, will probably not encompass all of the DNA substrate that was sequenced, and will contain errors and regions of high- and low-quality sequence.

sequence-tagged site (STS) A short (typically 50–500 basepairs) segment of known DNA sequence that permits placement of the sequence in the map of the human genome.

serial analysis of gene expression (SAGE) A method for comprehensive analysis of gene expression that employs short 10- to 14-basepair sequence tags linked together to identify and measure the abundance of RNA sequences in a sample.

serotonin An amine neurotransmitter involved in emotion and behavior regulation, among other functions.

serotonin transporter (5-HTT, hSERT) Membrane-bound protein involved in the reuptake of serotonin into neurons. It is the site of action of a major class of antidepressants, the selective serotonin reuptake inhibitors.

serum marker screening A technique for identifying pregnancies in which there may be a higher likelihood of a fetal birth defect (such as spina bifida) or chromosome abnormality (such as Down syndrome). The screening involves comparing the amounts of various proteins in the mother's blood against established norms.

severe combined immunodeficiency (SCID) Severe immunodeficiency due to an absence of T and B lymphocytes, which is fatal in early childhood if not treated.

second-degree relative Grandparent, grandchild, aunt, uncle, niece or nephew of the person in question.

segmental aneusomy Deviation from the normal chromosome copy number (two in the case of humans) for part of a chromosome.

segmentation In microscopy, the decomposition of an image into objects and background.

segregation The 'splitting' of pairs of alleles at a locus, by the distribution of the homologous chromosomes to different gametes at meiosis, so that individual alleles are transmitted to the offspring.

selection coefficient A term that is used to describe the difference in fitness between two genotypes. The fitness of a genotype is measured relative to a reference genotype; the difference between the relative fitness of the genotype and one is then the selection coefficient acting on that genotype.

self-determination Motivated self-direction.

sex chromatin A small, darkly staining body seen lying against the nuclear membrane of cells of female mammals but not males. It represents the inactivated X chromosome.

sex chromosome A chromosome or group of chromosomes which are responsible for the genetic determination of sex.

sex determination The process by which the sex of an organism is fixed. In humans, sex is established at the time of fertilization by the presence or absence of the Y chromosome.

sex-determining factor A gene (or its protein product) whose presence triggers the determination of one sex, but whose absence results in the development of the other sex.

sex linked Describes a trait determined by a gene on the X chromosome and which therefore shows a sex-linked pattern of inheritance.

sex ratio distortion Deviation of the sex ratio from unity, or some other expected or predicted value.

sex reversal A process whereby the sexual identity of an individual is changed from one sex to the other by a combination of surgical, pharmacological and psychiatric procedures or as part of the life history of pseudohermaphroditic individuals.

sexual determination *See* sex determination.

sexual differentiation Differentiation of male and female tissues and organs during embryogenesis but after sex determination.

shared environment (1) Experiences shared by children that tend to make them more similar in their observed trait (phenotype).
(2) *See* common environmental factors.

shearing A method of breaking DNA molecules in solution by mechanical agitation, for example using intense high-frequency sound waves (sonication).

β sheet Element of protein secondary structure in which straight segments of polypeptide chain (β strands) are arranged side by side.

short interspersed element (SINE) A class of short (100–400 basepairs) interspersed repeated DNA retroelements that are incapable of autonomous retrotransposition and are derived from either 7S RNA (Alu elements in primates and B1 elements in rodents) or various tRNAs or short interspersed nuclear elements in rodents and other organisms.

short tandem repeats (STR) Short repeats of 2–5 basepairs. *See* microsatellite DNA.

shotgun sequencing A method for reconstructing the sequence of a genome by randomly shearing it into many relatiely small fragments and then finding overlaps in the sequences of these small fragments.

sibling-pair analysis Analysis of sibships with one or more affected siblings, on the basis that if affected siblings inherit a particular allele more or less often than would be expected by chance (50%), this indicates that the allele is linked to the disease susceptibility.

sibship All the brothers and sisters in a family.

side-by-side arrangement Positions of chromosome territories or other nuclear structures with respect to their neighborhoods.

side chain That part of an amino acid that differs between amino acids and which extends from the polypeptide backbone when amino acids are incorporated into a protein.

signal transduction A series of steps by which external stimuli are converted into intracellular chemical or molecular signals and then into cellular responses.

silencer A regulatory element in DNA that binds a protein or protein complex that shuts down gene expression.

silencing Repression of a gene's transcription that is dependent on the gene's position on the chromosome. Silencing involves formation of heterochromatin over large regions of the chromosome.

silent codon position *See* silent site.

silent site Site within a gene or genome which, if mutated, may be supposed to have no significant effect on fitness and hence may evolve neutrally.

similarity Degree of closeness between two sequences. In a technical sense, the same as optimal score.

simple sequence repeat (SSR) Short DNA sequence repeated in a tandem array or other distinguishable pattern.

simple sequence repeat polymorphism (SSRP) Variation in the sequence or number of repeats in sequences such as (CA)n, (GATA)n and (ATA)n.

single locus probe (SLP) DNA probe that contains a sequence that will detect variations at a single minisatellite locus.

SINE *See* short interspersed element.

single-gene disorder A disorder caused by a mutation at a single genetic locus.

single-nucleotide polymorphism (SNP) (1) A DNA variant involving only a single base change, usually a single-base substitution, or an insertion or deletion, with the minor allele found in at least 1% of the population. SNPs are used in linkage, common disorder and pharmacogenomic studies.
(2) Common single-base variations that occur in human DNA at a frequency of one in every 1000 bases. These variations can be used to track inheritance in families. SNP is pronounced 'snip'.

single-strand chain polymorphism (SSCP) When a DNA strand carrying a mutation is denatured and run electrophoretically in a nondenaturing gel, it can refold into different secondary structures and thus appear polymorphic.

single-stranded DNA conformation analysis (SSCA) A method for scanning DNA for single-nucleotide polymorphisms based on the fact that single-stranded DNA will assume a particular tertiary structure determined by its sequence. Even a one-base difference in sequence will change this structure and cause the different DNA molecules to travel at different speeds during native gel electrophoresis.

sister chromatid The two copies of a deoxyribonucleic acid (DNA) molecule that are present after replication. They only become microscopically visible as 'sister chromatids' once chromatin condenses to form chromosomes during prophase of mitosis.

sister chromatid pair Identical copies of replicated chromosomes held together at specific sites by the cohesion proteins. In late S and G2 phases, the replicated DNA is held together at multiple sites along the entire lengths of the interphase chromosomes. During prophase chromosome condensation, the cohesion proteins holding the chromosome arms together are removed while the centromeres remain joined, producing mitotic sister chromatid pairs with free arms joined at the centromere. The sister chromatid pairs separate into individual chromosomes at the beginning of anaphase.

simian immunodeficiency virus (SIV) Family of lentiviruses that infect nonhuman primates. Subscripts identify the species, for example, SIV_{mac} infects rhesus macaque, SIV_{agm} infects African green monkey.

SIV *See* simian immunodeficiency virus.

skew Asymmetric distribution of the complementary nucleotides (A/T and G/C) between the two strands of DNA.

skewness A measure of the tendency of values in a distribution to have deviations from the mean that are larger in one direction than the other, so that the distribution is nonsymmetric about the mean. More formally, skewness for a population of observations is defined to be the expected value of the cubed difference between the values and the population mean, divided by the cube of the population standard deviation.

SKY *See* spectral karyotyping.

slipped mispairing Mechanism of gene deletion involving the misalignment of short direct repeat sequences.

SLP *See* single locus probe.

Sm protein Any of a group of related small proteins (8–28 kDa) that share the Sm domain recognized by autoantibodies from patients with systemic lupus erythematosus. Sm proteins are found associated with many snRNPs.

small nuclear RNA (snRNA) Small RNA molecule, typically 100–300 nucleotides long, which associates with specific proteins to form an snRNP.

small nucleolar ribonucleoprotein (snoRNP) One of a family of small nucleolar ribonucleoproteins that function in pre-rRNA processing in the nucleolus of eukaryotic cells.

small nuclear RNA (snRNA) Small RNA molecule, typically 100–300 nucleotides long, which associates with specific proteins to form an snRNP.

small ribonucleoprotein particle A part of the RNA splicing mechanism, where such particles form a complex with hnRNP proteins.

smiling effect The electrophoretic migration front in a gel and the distribution of bands is round shaped because of faster zone migration in the center than at the lateral sides of the gel, resulting from generation of heat in the buffer. It can be prevented by reducing the electric field strength, removal of heat by cooling or by applying a heated plate on the back of the gel.

SNARE proteins A group of transmembrane and cytosolic (soluble) proteins that work together to determine the specificity of membrane–membrane fusion events in cells.

snoRNA *See* small nucleolar RNA.

snoRNA host gene Gene that encodes snoRNA(s) within its intron(s).

snoRNP *See* small nucleolar ribonucleoprotein.

SNP *See* single-nucleotide polymorphism.

snRNA *See* small nuclear RNA.

snRNP *See* small nuclear ribonucleoprotein.

social constructionism Rooted in a social interactionist approach to society. Constructionism accepts that individuals have the capacity to 'construct' their own patterns of social action and each unique piece of social action emerges as a result of how a situation is defined and understood by an individual.

social Darwinism Any social policy that sees struggle or competition as inevitable, usually by analogy with natural selection.

social interactionism An approach within sociology which argues that social structures are the product, or sum, of individual social interaction. This approach is very different from that of the structuralists, who support the view that social structures largely determine social action.

social relations model of technology A philosophy that holds that technical phenomena are an outgrowth of social dynamics in a society, in particular the distribution of power in the society, so that the technologies produced reflect and manifest that power.

sodium dodecyl sulfate-polyacrylamide gel electrophoresis (SDS-PAGE) Polyacrylamide gel electrophoresis using a stacking and separating gel containing the detergent sodium dodecyl sulfate, which separates proteins by their size.

soft heredity Transmission of characters in such a way that modifications resulting from environmental effects can be incorporated into the germ plasm.

soma The region of a nerve cell that surrounds the nucleus, also termed the cell body.

somatic cell Cell that forms part of the body of an individual but does not contribute to the germ line.

somatic cell hybrid Cultured cell line obtained by the fusion of cells from two species, in which most of the chromosomes from one of the parent species have been lost.

somatic cell mosaicism The presence of two or more genetically different cell lineages within the somatic tissues.

somatic gene therapy The transfer of genetic material into tissues, but not into the germ cells, to correct a gene defect.

somatic hypermutation A step in the process of immunoglobulin gene maturation during the humoral response to antigen challenge. Point mutations are introduced into the variable region of an assembled immunoglobulin gene at a rate a million times greater than normal, thus increasing the diversity of antibodies that can be produced.

somatic mutation A mutation that is restricted to the tissue in question.

sorting signals Short amino-acid sequences (usually 2–20 amino acids), found within many proteins, that interact with a variety of other intracellular proteins to determine the intracellular fate and membrane insertion specificity of the proteins in which they occur.

source gene Gene from which a pseudogene has originated, either by duplication or retrotransposition.

Southern hybridization Identification of a specific DNA fragment by hybridizing a tagged piece of complementary DNA to a series of DNA fragments that have been separated by a technique such as gel electrophoresis.

spanin A protein with short stretches of polypeptide passing through the membrane lipid bilayer.

species extinction The disappearance of a plant or animal species when its environmental niche is so disturbed that it cannot reproduce itself.

species tree Graphic depiction of the history of descent of populations belonging to different species.

specific DNA binding Each gene regulatory protein binds a specific regulatory sequence by recognizing the DNA sequence through direct interactions with the DNA bases.

specificity Suitability of a substrate for undergoing catalysis by an enzyme.

spectral karyotyping (SKY) A chromosomal identification and categorization technique based on the differential binding of 24 probes specific for the autosomes and the sex chromosomes and labeled with combinations of five dyes. After the treatment, each chromosome can be detected as a unique color.

spindle assembly checkpoint A mechanism that ensures that the cell cycle is stopped at metaphase in cases where spindle damage occurs or not all chromosomes are properly attached to the mitotic spindle.

3′ splice site The junction between the 3′ end of an intron and the 5′ end of an exon, which is one of the sites involved in excision of the intron and joining of exons. Also called the acceptor splice site or acceptor site.

5′ splice site The junction between the 5′ end of an intron and the 3′ end of an exon, which is one of the sites involved in excision of the intron and joining of exons. Also called the donor splice site or donor site.

splice variant A variation in the combination of exons used during splicing of an RNA that leads to production of a variant of the complete protein. Splice variants increase protein diversity by allowing multiple, sometimes functionally distinct, proteins to be encoded by the same gene.

spliceosomal intron The class of introns in nuclear genes that are spliced using spliceosomes.

spliceosome A large complex of small nuclear RNAs and proteins that carries out the reactions of RNA splicing.

spliceosome assembly The binding of a large particle, composed of several small RNA–protein complexes, to the intron-exon borders of pre-messenger RNAs. This particle facilitates the removal of the intron and splicing of the adjacent exons.

splicing *See* RNA splicing.

splicing enhancer *Cis*-acting RNA sequence that interacts with splicing factors (SR proteins) to promote RNA splicing.

splicing factor *See* SR protein.

spongiform encephalopathy Non-inflammatory brain disorder characterized by vacuolation within neurons and through to be caused by prions. This group includes five human disorders: Creutzfeldt–Jakob disease, kuru, Gerstmann–Straussler–Scheinker disease, fatal familial insomnia, and variant CJD associated with consumption of contaminated meat from animals incubating bovine spongiform encephalopathy (BSE; mad cow disease). The spongiform encephalopathies also include six animal conditions, the best known of which are scrapie in sheep and BSE. Also known as transmissible spongiform encephalopathy.

sporadic Occurring apparently at random with no obvious familial pattern.

SR protein Any of a family of nuclear RNA-binding proteins involved in RNA splicing.

SSCP *See* single-strand chain polymorphism.

SSR *See* simple sequence repeat.

stabilizing selection A kind of natural selection that keeps a quantitative character at an optimum.

stable equilibrium An equilibrium to which the population converges from other points in the space of genotype frequencies.

state of the art The sum total of human knowledge currently available in the public domain.

stefin Type 1 family cystatins. In the early literature, type 1 cystatins were called cystatin A or B, for example. These proteins were isolated at the Jozef Stefan Institute in Slovenia. To distinguish family 1 cystatins from other cystatins, these proteins were termed stefins.

stem cell *See* embryonic stem cell; hematopoietic stem cell; peripheral blood stem cells.

stem taxa Organisms not included within a crown group, but arising from the lineage (stem) that extends from the crown group to its last common ancestor with its next closest living relatives.

stigma In the sociological context, social disapproval or exclusion associated with a recognizable characteristic, such as impairment.

stigmatization Refers to the process of excluding individuals or groups because of their perceived abnormality.

stochastic (1) Changing randomly in time or space.
(2) Refers to a mathematical approach based on random variables dependent upon a time parameter.

stoppage rules The reduction in family size after the birth of an affected offspring (with any phenotype).

STR *See* short tandem repeat.

strategy The specification of the behavior to be adopted in a conflict. It comprises a repertoire and a set of probabilities with which the behaviors are selected.

stressor Internal or external stimulus that cannot be responded to automatically, but that necessitates a coping process in the individual.

structural equations Algebraic equations that specify the structural interrelationships between a given set of variables.

structuralism An approach within sociology that sees social action as the product of structures within society rather than the result of individual social action.

struggle for existence Darwin's assumption that, as a result of population pressure, the organisms making up a population must compete for scarce resources.

STS *See* sequence-tagged site.

subcloned DNA A part of an already cloned DNA which is recloned and then maintained as a separate clone.

subdivided population A population in which mating is not random, because it is composed of distinct subpopulations or isolates.

submetacentric A chromosome with arms of different lengths.

subpopulation A group of individuals within a population that more frequently breed with each other than with members from other groups.

substitute cells Cells modified genetically to achieve replacement of another cell type that has lost its function.

substitution A mutation in which one nucleotide is relaced by another. When used in reference to mutant protein sequences it means the replacement of one amino acid by another.

substrate A substance on which an enzyme acts.

sulfobetaine Class of chemicals containing both a quaternary ammonium and a sulfonic acid group, separated by a hydrocarbon linker. This linker is generally two to four carbons long, most often three carbons long.

surname analysis Useful in Western cultures to estimate the relative homogeneity of a study population, where the surnames act as polymorphic markers for the Y chromosome. The hypothesis is that the higher the degree of homogeneity of the population, the smaller the number of surnames.

surrogacy The act of carrying and giving birth to a child intended for another to rear as its parent, whether the surrogate mother so-called offers both ova and uterus ('traditional surrogacy') or her uterus alone ('gestational surrogacy').

surrogate end point A substitute, usually biochemical, end point of ultimate clinical effectiveness.

surrogate mother *See* surrogacy.

survival of the fittest Term coined by Herbert Spencer to denote natural selection. Strictly speaking, 'fittest' means only best adapted to the local environment.

survivor guilt in noncarriers Feelings of guilt toward affected family members and problems with adapting to the new identity experienced by individuals receiving a favorable predictive test result.

susceptibility allele An allele that is associated with an altered risk (usually an increased risk) of developing a disease, but which is usually neither necessary nor sufficient for the disease to occur.

susceptibility gene *See* susceptibility allele.

susceptibility testing Genetic tests applied to healthy people to identify a mutation(s) causing an increased risk for future onset of a disease. For example, this type of 'predictive' testing can be used to identify people at increased risk for future cancers due to inheritance of a mutation in a cancer gene, or onset of an adult neurological disorder such as Huntington disease.

symmetric element Sequence that possesses an axis of internal symmetry, for example CTGAAGTC.

symmetrical exon An exon flanked by introns of identical phase.

synapse The specialized site of communication between two neurons or between a neuron and its effector cell (muscle or secretory cell).

synapsis (1) The pairing of two homologous chromosomes during meiosis.
(2) The bringing together of two DNA molecules by Cre–Cre interactions.

synaptic plasticity Increase or decrease in the efficacy of synaptic transmission following activity. This is thought to be a cellular basis for learning and memory.

synaptonemal complex Tripartite, proteinaceous structure found only in meiotic cells. It is present between the axes of paired homologous chromosomes in zygotene to diplotene.

syndactyly Fused or webbed fingers or toes.

syndrome (1) A characteristic association of several anomalies in the same individual implying that they are causally related.
(2) A group of symptoms and/or signs that occur together and tend to indicate the presence of a particular disease condition.

synonymous codons Codons that encode the same amino acid.

synonymous nucleotide divergence Nucleotide differences between two homologous genes in their protein coding sequences that do not result in amino acid replacements (i.e. nucleotide divergence between homologous genes at degenerate sites of codons).

synonymous substitution Nucleotide substitution that does not cause an amino-acid change in the encoded protein.

syntenic *See* synteny.

synteny (1) Location of genes or DNA sequences on the same chromosome.
(2) The maintenance of the same order of genes along a chromosome in two species.

tagged primer *See* anchored primer

tandem array Two or more copies of a gene arranged in series in the same orientation.

tandem mass spectrometry (MS/MS) A mass spectrometric technique in which a peptide is first selected after mass spectrometry and then fragmented by collision with an inert gas and the fragments analyzed by mass spectrometry. The fragmentation pattern allows one to deduce the sequence of the peptide.

tandem repeats *See* tandem array.

TATA box A regulatory region present in a large number of genes that binds the TATA-box binding factor and is required for the initiation of transcription.

taxa Plural of taxon.

taxon In phylogenetic analysis, a taxon is a biological entity such as a gene or species.

taxonomy The science of classifying animal and plant species.

TBR *See* temperance board registration.

technological fix A public policy approach that holds that because modern social problems are too messy and complex to solve by direct social interventions, the use of technical factors should be relied on to produce changes that are easier to have accepted socially and politically.

telomerase A ribonucleoprotein enzyme that extends chromosome 3′ ends by adding one strand of tandem DNA repeats. Two evolutionarily conserved key components of telomerase are the telomerase reverse transcriptase protein and the telomerase RNA, which acts as a template during the maintenance of chromosome ends.

telomere A repeated DNA sequence at the ends of linear chromosomes that protects the internal chromosomal structure and ensures complete replication of DNA content.

telomeric repeat Hexamer tandem repeats of TTAGGG of 4–15 kilobases that are found in the telomeres at the ends of chromosomes.

telophase The fifth and last stage of mitosis, during which the daughter chromosomes arrive at the nuclear poles, followed by breakdown of the mitotic spindle, reformation of the nuclear envelope and cell division.

telangiectasia A small permanently dilated blood vessel.

teleosts The largest group of ray-finned fishes. It does not include primitive lineages such as sturgeons, gar, and bowfin fishes.

temperance board registration (TBR) A registry of alcohol offenses (such as public intoxication and disorderly conduct) maintained in the past by governments in some Scandinavian countries.

template The strand of DNA being copied by a polymerase.

tenase A complex between blood coagulation factors VIIIa and IXa that is formed on the surface of negatively charged phospholipid membranes. This complex efficiently catalyzes the activation of factor X to factor Xa.

teratogen An agent (usually chemical or infectious) that has been shown to induce one or more birth defects or other developmental abnormalities in an exposed embryo or fetus. Examples of human teratogens include alcohol, thalidomide and the rubella virus (German measles).

tertiary structure Overall three-dimensional fold of a complete protein or nucleotide sequence, formed by the packing of its secondary structure elements.

testicular feminization Condition in which the tissues supposed to develop into male sex organs (testes) do not respond to the male hormones testosterone and dihydrotestosterone. These individuals are therefore phenotypically female but genotypically male (46XY).

tetralogy of Fallot Heart defect comprising an overriding aorta, pulmonary stenosis, ventriculoseptal defect, and right ventricular hypertrophy.

TF (1) *See* tissue factor.
(2) *See* transcription factor.

TGN *See* trans-Golgi network.

α-thalassemia A genetic defect resulting from chromosome breakage and loss of a portion of the α-globin gene.

theoretical plates A concept related to resolution in chromatography. A high number of theoretical plates means a column that has high resolution.

therapeutic cloning The use of a cloned embryo to produce stem cells that have the potential to be used for therapy for conditions such as degenerative disorders, such as Parkinson disease.

thin-layer chromatography Analytical technique involving distribution of samples in a mixture between a stationary phase (thin layer of silica gel or alumina) and a mobile (liquid) phase. The latter travels up the stationary phase, carrying the samples at different rates depending on their physico-chemical properties.

three-dimensional reconstruction of microscopic images Information about the three-dimensional arrangement of fluorescent structures obtained by the analysis of 'image stacks' that were obtained from a series of light-optical sections taken at different positions along the structure.

tight junction Type of cell–cell junction that seals the membranes of adjacent epithelial cells together locally, forming a physical barrier between the apical and basolateral faces of the epithelium.

time of flight (TOF) *See* time-of-flight mass spectrometry.

time-of-flight mass spectrometry Analysis of proteins using a mass spectrometer equipped with a detector that measures the time of flight of ionized molecules from the sample source to the detector. Time of flight is directly proportional to the mass-to-charge ratio and a profile of the peptides or proteins analyzed is generated.

tissue factor (TF) A membrane protein on cells surrounding blood vessels, and which triggers blood coagulation when it comes in contact with blood.

tissue-specific gene A gene that has an expression pattern restricted to one or a few tissues, and is usually involved in a specialized tissue- or cell-specific function.

tissue-specific promoter A regulated promoter that restricts gene expression to a particular tissue.

tolerance Reduced response to a drug following repeated exposure to it.

tonic-clonic Describes a generalized seizure in which the muscles stiffen in the 'tonic' phase and the extremities begin to jerk in the 'clonic' phase.

5′ TOP An oligopyrimidine tract found in the 5′ terminus of the RNA transcripts encoding proteins involved in translation.

top-down strategy An approach to genome sequencing that starts with maps of low resolution and gradually works through levels of increasing resolution to eventually obtain the DNA sequence itself. The advantage of this is that each sequence is ultimately linked to an overall map of the genome.

topoisomerase Any of a class of DNA-modifying enzymes that decatenate DNA strands and release topological stress, for example during DNA replication.

touchdown PCR A protocol for the polymerase chain reaction in which the annealing temperature is set at several degrees above optimal in the first few cycles, decreased incrementally, and subsequently maintained in the final, lowest annealing temperature for the remaining cycles.

trade-off Situation when an evolutionary change in one trait that increases fitness is connected to an evolutionary change in another trait that decreases fitness.

trait group The set of all individuals that affect each other's fitness.

trans factor *See* trans-acting factor.

trans-acting factor A protein or RNA that binds to a regulatory region of a gene and affects its expression.

transcript An RNA copy of a gene or other DNA sequence.

transcription The synthesis of RNA from nucleotide monomers by the enzyme RNA polymerase using a DNA strand as a template to specify the RNA sequence. The RNA produced is complementary in sequence to the template DNA strand.

transcription factor A general term for any protein, other than RNA polymerase, that participates in the initiation or regulation of transcription of genes. More specifically it refers to those proteins that initiate gene expression through binding to specific DNA sequences within the promoter region and interacting with RNA polymerase.

transcription initiation site The site in a DNA sequence at which transcription begins.

transcription unit A stretch of DNA that is transcribed to produce a single primary RNA transcript.

transcriptional adaptor Any molecule that plays a part in altering the transcriptional activity of a gene but does not necessarily bind to a transcription-factor-binding site in DNA. An example is histone deacetylase, which opens up the chromatin to make a promoter region more accessible to transcription factors.

transcriptome All the messenger RNAs that can be produced by a living organism, that is, the RNA product of the genome.

transduce (1) *See* transduction.
(2) *See* signal transduction.

transduction (1) The genetic modification of cells using DNA carried by viral vectors.
(2) *See* signal transduction.

transesterification Chemical reaction involving the transfer of phosphate ester bonds that occurs at each of the two steps of the RNA splicing reaction.

transfection The genetic modification of cultured cells by uptake of DNA from the culture medium. The DNA transfected is usually in the form of recombinant plasmids or other types of vector containing genes of interest.

transfer RNA (tRNA) Any of a number of small RNA molecules that specifically bind different amino acids and are involved in the addition of the correct amino acid to a growing polypeptide chain during protein synthesis.

transformation (1) The introduction of new genetic properties into bacteria by the incorporation of foreign DNA.
(2) The alteration in properties that mammalian cells undergo as they become cancerous.

transgene A gene that has been introduced into the genome of a plant or animal by genetic manipulation.

transgenic Describes an organism containing deliberately introduced foreign DNA in its germ line.

trans-Golgi network (TGN) The most distal portion of the Golgi processing pathway for transmembrane and secreted proteins, in which apical and basolateral membrane proteins, as well as other proteins such as lysosomal hydrolases, are sorted and packaged into distinct sets of transport vesicles that are directed toward different intracellular destinations.

transition A mutation in which A is replaced by G (or vice versa) or T by C (or vice versa).

translation The synthesis of a protein from amino-acid subunits, which is directed by a messenger RNA template that specifies the order of the amino acids.

translesion DNA synthesis DNA synthesis on a DNA molecule that contains altered bases or missed bases.

translocation (1) Rearrangement of a chromosome in which a segment is moved from one location to another, either within the same chromosome or to another chromosome.
(2) Chromosome rearrangements that can be caused by the mobility of transposable elements or by nonhomologous recombination between copies of a transposable element. A translocation is a change in the chromosomal position of a DNA segment. An inversion is the modification of the orientation of a DNA segment. A deletion is the elimination of a DNA segment.
(3) A chromosome, or a portion thereof, shifts to another chromosome, possibly disrupting the regulation of one or more genes.
(4) Transfer of chromosomal regions between nonhomologous chromosomes.
(5) Change in location of a chromosomal segment.

transmembrane Across the membrane, spanning the membrane lipid bilayer.

transposable element (TE) A DNA sequence that has the ability to move around the genome, whether by excision and reinsertion or by making copies of itself that are then inserted elsewhere. They can occur in up to hundreds of thousands of copies in eukaryotic genomes, although many of these copies have lost their ability to jump into new sites in the genome. Examples include LINES and SINES (long interspersed nuclear elements and short interspersed nuclear elements) found in mammals, Alu sequences (in primates), P elements in *Drosophila*, Ty elements in yeast and IS sequences and transposons in bacteria. *See also* retrotransposon.

transposase Enzyme encoded by a transposable element that mediates the movement of the sequence to another site in the genome.

transposon Transposable element, especially in bacteria, that transposes directly from a DNA form to a DNA form, without an RNA intermediate.

trans-**splicing** Splicing of two mRNA segments which are transcribed separately.

transsulfuration A series of essentially irreversible enzymatic reactions that convert methionine to cysteine, taurine and ultimately into inorganic sulfate.

transvection A homology-dependent physical interaction between alleles that allows transcriptional regulation in *trans*.

transversion A mutation in which a purine base (A or G) is replaced by a pyrimidine (T or C).

trimethylguanosine The name of the inverted cap structure on snRNA molecules. The base is modified by the addition of two methyl groups at position 2 and one at position 7.

trinucleotide repeat A sequence of three base-pairs repeated in tandem.

trinucleotide repeat disorders A group of important genetic disorders whose underlying mutation is an increase in the number of repeats of a sequence of three base-pairs at a particular locus, in comparison with the repeat number seen in the general population.

triple helix The characteristic structure of a collagen molecule, comprising three polypeptide chains intertwined in a helical fashion.

triplex DNA DNA formed by pairing between three strands of DNA, as opposed to the normal double-stranded DNA. It can be formed by homogenous runs of purines (cytosine or guanine) or pyrimidines (adenine or thymidine). The third strand binds to the first two via an alternative form of base-pairing known as Hoogsteen base-pairing. Also known as H-form DNA.

triploblast Animal with a body plan made of three main layers: ectoderm, mesoderm and endoderm.

trisomic Having three copies of a particular chromosome.

trisomy Possession of three copies of a chromosome instead of two. Commonly encountered human trisomies include trisomy 21 (Down syndrome), trisomy 18 (Edwards syndrome), trisomy 13 (Patau syndrome), XXY (Klinefelter syndrome) and two other sex-chromosome abnormalities, XXX and XYY.

tritanopia Rare dichromatic inherited color vision defect mediated by L and M cones.

trithorax Group of proteins that act as transcriptional activators and are involved in maintaining the correct expression of several key developmental regulator genes.

tropism In the context of viruses, used to describe the cell type which the virus can infect and in which it can replicate. Tropism can be regulated either by the ability to bind to specific cell surface receptors or to use intracellular processes such as cell-specific transcription factors.

true negative A test result indicating that a mutation identified in a relative has been ruled out in the tested individual.

true positive Test result indicating that a deleterious mutation known to be associated with a given disease has been identified.

truncus arteriosus Common arterial vessel present during early embryogenesis that septates to form the aortic and pulmonary trunks.

tumor suppressor gene A gene whose loss is associated with the development of cancer. Their normal function, directly or indirectly, is to inhibit cell growth and cell division.

tumorigenesis The development of tumors caused by the unregulated growth of cells.

twin registry A large database (often national or regional) of individuals from multiple births, used to identify potential research subjects or twins treated for a disorder.

twin studies Comparison of the resemblances of identical and fraternal twins, to estimate genetic and environmental components of variance.

two-dimensional polyacrylamide electrophoresis (2D-PAGE) The electrophoretic separation of proteins on the basis of isoelectric point (charge) in the first dimension and molecular mass in the second dimension.

two-hybrid screening *See* two-hybrid system.

two-hybrid system Method that detects proteins capable of interacting with a known yeast protein and that results in the immediate availability of the cloned genes for these interacting proteins.

U12-dependent Denotes the minor class of pre-mRNA introns that require the function of the U12 snRNA for splicing.

U2-dependent Denotes the major class of pre-mRNA introns that require the function of the U2 snRNA for splicing.

ubiquitin A small protein found in all eukaryotic organisms in a nearly identical form, covalently coupled to lysine residues on specific proteins by a complex series of enzymes, in a process called ubiquitinylation. It targets proteins for degradation in proteasomes.

ubiquitin-protein ligase (E3) Enzyme required for the assembly of multiubiquitin chains on proteins. Ubiquitin is first activated in an adenosine triphosphate-dependent step by thioester linkage to an ubiquitin-activating enzyme (E1). Ubiquitin is then passed on to a ubiquitin-conjugating enzyme (E2), and finally transferred to the substrate with the help of a ubiquitin-protein ligase (E3).

ultrasound A noninvasive diagnostic technique for visualizing, for example, a developing fetus in the uterus.

umbilical cord blood cells Cells isolated from the umbilical cord and placenta of newborns that contain hematopoietic stem cells and thus can be used for transplantation instead of bone marrow.

uncertain significance Describes the situation when a change is identified in a disease gene, which may or may not be associated with heightened disease risk. Further evaluation in the laboratory or clinic (via tracking within families) may be indicated to make this determination.

underdominance Lower fitness of heterozygotes relative to either homozygote.

unequal crossing-over A crossover event that results in duplication or deletion of a DNA region as a result of shifted chromosome pairing.

unequal homologous recombination Mispairing of duplicated sequences on two homologous sister chromatids followed by crossing over. The result is one chromatid containing a duplication and the other containing a deletion.

unequal recombination *See* unequal homologous recombination.

unfair genetic discrimination Discrimination when people are regarded as elevated risks or unhealthy because of information about an illness they may acquire in the future, rather than information about illness they currently have or previously had.

uninformative result General interpretation rendered when no mutations or clinically significant alterations in a known disease gene have been identified in a genetic test.

uniparental disomy The presence of two copies of a chromosome (or part of a chromosome) from one parent and none from the other.

unique environmental factors Also known as 'nonshared environment': the proportion of phenotypic variability due to nongenetic sources of phenotypic differences.

unrealistic optimism The tendency of people, on average, to think that they are less likely than the average person to experience negative events.

untranslated region (UTR) A sequence of nucleotides that are present in the mRNA strand, but do not code for amino acids because they are located either upstream (5′) of the translational start codon or downstream (3′) of the translational termination codon. UTRs can contain regulatory sequences.

3′ untranslated region The region at the 3′ end of an mRNA that is not translated as it lies beyond the termination codon.

uORF *See* upstream open reading frame.

UPD *See* uniparental disomy.

upstream open reading frame (uORF) Open reading frame found upstream of a gene. Its expression often controls expression of downstream genes in a polycistronic mRNA.

urochordates A subphylum of chordates, also known as the tunicates. They are generally the least complex of the chordates. The notochord is confined to the tail region, as are the muscle blocks, and the nervous system contains relatively few neurons.

use inheritance Inheritance of characters acquired through the body being exercised in a new way through the adoption of new habits.

use/abuse model Under the corporate liberal model, a theory of the interactions between technical phenomena and social forces that holds that technologies are neutral, value-free and capable of being used for good or abused for evil. Other than externalities, bad consequences from technologies are caused by misuse.

U snRNA Small nuclear RNAs that function in pre-mRNA splicing in association with specific proteins in the form of ribonucleoprotein particles.

utilitarian analysis An ethical position that seeks to maximize happiness by offsetting potential benefit against possible risks.

utility The usefulness of a statistic.

UTR *See* untranslated region.

3′ UTR *See* 3′ untranslated region.

validity Applied to screening tests, the probability that the test will detect a person who will develop the disease in question, and the probability that a person with a positive test result will develop the disease.

variable number of tandem repeats (VNTR) A highly polymorphic piece of DNA composed of tandem repeats.

variance effective population size (N_e) The size of an unstructured, randomly mating population that would exhibit the same amount of genetic drift as that found in the population of interest.

variance The dispersion of a random variable in a given population.

variation A statistical measure of the degree of differences among individuals in a population.

variegation Expression of a gene in only some of a population of otherwise identical cells, presumably by stochastically determined silencing or activation mechanisms.

VCFS *See* velocardiofacial syndrome.

V(D)J recombination Recombination process by which the exons encoding immunoglobulin or T-cell receptor variable regions are assembled together to form a functional gene.

vector (1) A collection of related numbers, usually in a row or column. In pattern recognition, an input vector is the set of features (a pattern) to be presented to a neural network for classification.
(2) Any agent used to transfer a transgene from the outside to the inside of a cell. Vectors are often modified viruses or plasmids into which the reequired gene has been inserted, but can also be lipid (liposome) or polymer vehicles in which the the DNA is contained.

velocardiofacial syndrome (VCFS) A genetic disorder associated with a microdeletion of chromosome 22q11.

very low-density lipoprotein (VLDL) Large, triglyceride-rich lipoprotein produced by the liver and which circulates in the blood. It is the major carrier of endogenously synthesized (that is, nondietary) lipids to peripheral tissues. The protein portion is apoB.

virus An infectious particle consisting of a nucleic acid molecule enclosed by a protective, mainly protein, coat. It is biologically inert until it enters a susceptible cell in which the viral nucleic acid is replicated and uses the cell's biochemical machinery to produce new viruses. Many viruses cause disease.

virus-like particle A particle composed of several proteins that protect the genetic material of a virus or in some cases the genetic material of a transposable element.

visual field The full area visible to the eye when fixating straight ahead.

vitalism Belief that the living body is governed by nonphysical forces.

VLDL *See* very low-density lipoprotein.

V_{max} The maximal activity of an enzyme.

VNTR *See* variable number of tandem repeats.

wait anaphase checkpoint The mechanism controlling the delay before anaphase of mitosis that ensures all sister-chromatid pairs are properly aligned on the metaphase plate.

weight The strength of a connection between units in a neural network. Incoming signals are multiplied by the weight of the connection before summation and subsequent output by a threshold function.

Western blotting The electrophoretic transfer of proteins onto membranes (usually nitrocellulose, nylon or polyvinylidene fluoride) for subsequent detection by probes such as antibodies or lectins.

wild type The most frequently observed, normal allele.

within-group component of hierarchical selection A measure of the fitness of different behavioral strategies within a trait group. Within-group selection usually favors 'cheating' strategies.

wobble Theory whereby some anticodons can recognize a wider range of codons than those strictly specified by the genetic code, because of a relaxation of the base-pairing rules between the third base of the codon and the first base of the anticodon.

working draft Of a genome sequence. A sequence generated by combining information from individually sequenced clones and positioning this sequence along a physical map. In the case of the public-domain human genome sequence, it is formed from a series of large-insert clones with at least a half-shotgun sequence, that is half the amount required to reach the finishing stage (an average coverage of fourfold).

X chromosome One of the mammalian sex chromosomes. Females have two X chromosomes and males have one X chromosome and one Y chromosome.

X-chromosome inactivation The inactivation of one of the two X chromosomes in somatic cells of female mammals, occurring early in embryonic life.

X inactivation *See* X-chromosome inactivation.

X-linked Describes a gene located on the X chromosome (sex chromosome). In males, a recessive mutation in such a gene may cause disease since they have only one allele of (almost) every gene on the X chromosome.

xenobiotic A synthetic chemical (e.g. a synthetic pesticide) that is toxic or damaging to a biological system.

xenotransplantation Transplantation from a donor to a recipient of a different species.

X-inactivation center The region on the X chromosome which is responsible for the initiation of the X-chromosome inactivation process.

X-ray crystallography X-ray diffraction studies of protein crystals to determine the three-dimensional structure of the protein.

XYY syndrome The condition of men who possess one extra Y chromosome.

YAC *See* yeast artificial chromosome.

yeast artificial chromosome (YAC) An artificial chromosome constructed from yeast (*Saccharomyces cerevisiae*) telomeres and centromeres and replication origins that can be maintained and replicate in yeast cells. It is used as a recombinant DNA vector for large fragments (up to 1 megabase) of DNA.

yeast mating-type switching A genetic rearrangement in yeast cells by which the cells switch their mating type from a to α or vice versa.

zinc-finger domain A protein motif present in a large number of DNA-binding proteins. It is characterized by two closely spaced cysteine and two histidine residues that serve as ligands for a single Zn^{2+}. They form a module from which protrude amino-acid side chains that interact with the bases partially exposed in the DNA major groove. The name zinc finger originated because a diagram of the structure of such a region resembles a finger.

zone electrophoresis *See* ionophoresis.

zymogen An enzyme produced in an inactive form, which can be processed into an active form.

Notes on Use of Subject Index

Note

Abbreviations

Abbreviations used as subentries are included as main entries (with appropriate cross references e.g. PAGE *see* Polyacrylamide gel electrophoresis) and hence are not listed in this note.

Cross-references

Cross-references in *italics* are either general cross-references e.g. *see also specific disorders* or refer to subentry terms within the main entry e.g. see also *diseases (above)*. Although cross-references are usually from the general to the specific, due to the length of this encyclopedia some references have been included which also go back from the specific to the general.

Entry wording

The wording of index entries relating to specific article titles has where possible been kept to the original wording.

Order

This index is presented in a letter-by-letter order, Greek letters are, where necessary, included in the sort e.g. α-fetoprotein is listed under α (alpha).

Page numbers

Page numbers in *italics* refer to figures/tables, whilst page number strings have been avoided as much as possible, due to the large and comprehensive nature of this encyclopedia some page number strings have had to be included.

Specific issues

As the subject of this encyclopedia is the human genome, please note that in general all entries refer to humans unless otherwise stated. Genes are indexed under their current HUGO nomenclature and also (if relevant) under their protein product. Proteins are indexed under their full names e.g. protein kinase C (PKC).

Subentries

Subentries (or subsubentries) to a specific index entry having the same page number have been included in order to indicate the breadth of the subject under discussion (as opposed to merely location) as an additional assistance to the reader.

Subject Index

A

A2M (α-2 macroglobulin), Alzheimer disease (AD) **1**:104
Abbe's law **2**:508
ABC genes see ATP-binding cassette (ABC) transporters
Abciximab **3**:502
ABC transporters *see* ATP-binding cassette (ABC) transporters
Ab initio protein modeling **4**:865, **5**:412, **5**:414
ABI PRISM 3700 **5**:271
ABL1 (v-abl Abelson murine leukemia viral oncogene homolog 1) **1**:379
  drug target **3**:622
  leukemia **2**:*546*
  translocation (Philadelphia chromosome) **1**:376, **1**:379–380, **1**:*380*, **1**:615, **1**:703, **3**:39, **3**:684, **4**:245, **4**:436, **5**:*603*
    receptor tyosine kinase mutation **4**:443, **5**:602
    tumor specificity **5**:599
    see also BCR (B-cell receptor) gene
  see also Leukemia
ABO blood group
  biological weapon susceptibility **5**:749
  genetics **1**:333, **1**:*334*, **1**:*335*, **3**:139
Aboriginal communities
  consanguinity **4**:360
  genetic research
    consent **1**:872
    data/sample access **1**:872
    dissemination **1**:872
    involvement in conduct **1**:872
    publication **1**:872
Abortion
  China **1**:506
  disabled fetus **2**:15
  Down syndrome, opinion surveys **2**:231
  ethics **3**:182
  feminist perspectives **2**:476
  genetic counseling **2**:761
    direction **1**:819
    Down syndrome **1**:819
  Muslim families **2**:764
  opinion surveys **2**:231
  'pro-life/pro-choice' debate **2**:185, **3**:182
  psychosocial consequences of **4**:677–681
  Russia **2**:231
  selective **3**:528–529
    cystic fibrosis **2**:854
    disability and **2**:6, **2**:12, **2**:13, **2**:15
    sex selection **5**:262
  spontaneous
    chromosomal anomalies **2**:480–481
    molecular anomalies **2**:482–484
    occurrence **2**:480
    see also Fetal wastage
'Abortion overshoot,' **3**:431
Absence epilepsy **1**:610, **2**:315
Absolute pitch (AP) **1**:3
  associated traits **1**:2–3
  definition **1**:1
  evolutionary considerations **1**:3
  genetic component **1**:2
  human ear **1**:1
  musical exposure **1**:2
  musicians **1**:1
  perception **1**:1
  relative pitch **1**:1–2
  tests **1**:2
Acanthocytes **2**:322
AceDB database **1**:368, **1**:*369*, **3**:8
  annotation **1**:370
  genome analysis **1**:371
  sequence **1**:369
'ACE' model, twin studies **5**:680, **5**:681
Aceruloplasminemia **3**:585
Acetaldehyde dehydrogenase (ALDH), drug metabolism (functionalization) **2**:240

Acetylcholine esterase (AChE) **5**:240
  blood group genetics **1**:340–341
Acetylcholine receptors
  muscarinic **4**:339
  nicotinic *see* Nicotinic acetylcholine receptors
N-Acetylgalactosaminyl transferase, blood group A **1**:333
Acheiropodia **3**:*697*
Achondrogenesis type 2 **5**:302
Achondroplasia **2**:*276*, **3**:*147*, **3**:150
  FGFR3 mutations **1**:79, **3**:170, **5**:302
    genotype-phenotype relations **3**:57
    mutation 'hot spot,' **1**:79, **4**:93, **5**:302
  genetic counseling **1**:819
  haploinsufficiency **3**:170
  heterogeneity **1**:79
  incidence **5**:298
  multiplex screening **3**:476
  prevalence **2**:807
Achromatopsia **1**:860
Ackroyd, Heather, phenotypic art **1**:189
Acquired immunodeficiency syndrome *see* AIDS
Acral lentiginous melanoma **3**:881
Acrocentric chromosomes **1**:480, **1**:740, **1**:779
Acrocephalosyndactyly **1**:961
Acrodental dysostoses **3**:*697*
Acro-dermato-ungual-lacrimal-tooth syndrome (ADULT) **3**:696, **3**:*697*
Acromegaly **2**:*276*, **2**:280, **3**:*147*, **3**:149
Acromelic dysplasia **3**:*697*
Acrylamide, alkylation of proteins **4**:691
Actin/actin filaments
  dystrophin interactions **2**:248
  erythrocyte membrane **2**:320, **2**:*321*
  isoforms **4**:838
  protein targeting **4**:871
α-Actinin
  α-actinin-4 and glomerulosclerosis **3**:614
  dystrophin homology **2**:250, **2**:*251*
Action potentials **4**:993, **4**:*994*
  cardiac **1**:396, **1**:404
    see also Cardiac arrhythmias
  sodium channel expression **5**:549
Activation function (AF) domains **4**:386–387
Activation-induced cytidine deaminase (AID), RNA editing role **4**:112
Activators *see* Transcriptional activators
'Activity type'
  doctor-patient relationship **2**:748
  genetic counselor-patient relationship **2**:748
ACTN3, redundancy **2**:893
Actuarial fairness, life insurance discrimination **2**:28
Acute intermittent porphyria (AIP)
  clinical management **4**:639
  genetics **4**:639
  penetrance **4**:639
Acute lymphoblastic leukemia (ALL) *see* Leukemia
Acute myelogenous leukemia (AML) *see* Leukemia
Acute promyelocytic leukemia (APL) *see* Leukemia
ADA (adenosine deaminase) gene
  gene therapy **2**:710, **2**:714, **3**:411
    approaches **3**:411
    hemopoietic progenitors **3**:411, **3**:*412*
    trial **3**:411
  mutations **1**:681, **3**:411, **4**:240, **5**:249
    back mutation **5**:390
ADAM1, loss in primate evolution **4**:702
Adams, Joseph **3**:853
Adaptation
  evolution and medicine **2**:371
  karyotype evolution **3**:596
  speciation **3**:356
  see also Natural selection
Adapter protein complex-1, DNA-binding (AP-1 site) **3**:683

Adaptive immune system **5**:730
  B-cell receptors *see* B-cell receptors (BCR)
  complement **3**:422
  immunoglobulins *see* Immunoglobulin(s)
  lymphocytes **3**:421
    see also B-cells; T-cells
  MHC *see* Major histocompatibility complex (MHC)
  recognition molecules **5**:344
  T-cell receptors *see* T-cell receptors (TCRs)
  see also Complement pathway
ADARs (adenosine deaminases that act on RNA)
  hyperediting **5**:126
  RNA editing **4**:113, **5**:91, **5**:120, **5**:*120*
    physiological impact **5**:91
ADATs (adenosine deaminases that act on tRNA) **5**:123
Addiction **1**:7, **1**:12, **1**:13, **1**:50–58
  alcoholism *see* Alcoholism
  animal models **1**:7–13, **1**:*10*, **1**:57–58
    knockout mice **1**:9
    mouse genetic study **1**:8, **1**:*10*
    random mutagenesis **1**:9
  disorders associated **1**:54
  drugs of abuse
    disorders associated **1**:54
    family studies **1**:54
    genetic epidemiology **1**:54
    heritability **1**:54
    twin studies **1**:54, **1**:*56*
  gene expression studies **1**:12
  gene isolation **1**:9, **1**:*12*
  genetic epidemiology **1**:54
  genetic influences **1**:58
    limitations **1**:58
    specific *versus* general **1**:54
    variation and risk **1**:57
  genetic models **1**:8
  gene transfer **1**:12
  heritability **1**:54
  human studies **1**:7
  nicotine
    genetic epidemiology **1**:54
    heritability **1**:54
    twin studies **1**:*55*
  psychiatric disorders and **1**:54
Adenine **1**:257
Adenine nucleotide translocator, mitochondrial disorders **3**:997
Adeno-associated virus (AAV)
  gene delivery vector **1**:16–20, **1**:1021, **2**:716, **2**:719
    animal models of human disease **1**:*18*
    clinical applications **1**:18, **1**:19, **1**:1022
      cystic fibrosis **1**:19
      familial hypercholesterolemia **2**:461
      hemophilia **1**:19, **3**:213, **3**:214
      lysosomal storage diseases **3**:762
      muscular dystrophy **4**:185
      phase I/II trials **1**:19
    combination approaches **1**:17
    episomal existence **1**:17
    first generation systems **1**:17
    high capacity systems **1**:17
    model systems **1**:1022
    recombinant vector construction **1**:17
    safety issues **1**:18
      advantages **1**:18, **1**:19
      immune aspects **1**:18
      replication-competent AAV contamination **1**:17
      tumorigenic potential **1**:18
    self-complementary (scAAV) **1**:17
    serotypes used **1**:19
    targeting **4**:319
    transgene expression levels **1**:18
    in vivo organ targets **1**:*18*
  genome **1**:16
  life cycle **1**:16

Adeno-associated virus (AAV) (*continued*)
  chromosome 19-specific integration **1**:17
  helper virus coinfection **1**:16
  Rep proteins **1**:16, **1**:17
Adenoma-carcinoma sequence **2**:453
Adenomas, bowel cancer arising from **1**:864
Adenomatous polyposis coli (APC) **1**:865
  Gardner syndrome **1**:1096
  gene *see APC* (adenomatous polyposis coli) gene
  medulloblastoma **1**:354
  mitosis role **4**:24, **4**:32
  see also Colorectal cancer
Adenosine deaminases (ADAs)
  deficiency **5**:249
    ADA gene therapy **2**:710, **2**:714, **3**:411
      approaches **3**:411
      hemopoietic progenitors **3**:411, **3**:*412*
      trial **3**:411
    mutations **1**:681, **3**:411, **4**:240, **5**:249
      back mutation **5**:390
  that act on RNA (ADARs)
    hyperediting **5**:126
    mRNA editing **4**:113, **5**:91, **5**:120, **5**:*120*
    physiological impact **5**:91
  that act on tRNA (ADATs), tRNA editing **5**:123
Adenosine triphosphatases *see* ATPase(s)
Adenoviral vectors **1**:21–26
  animal models **1**:24
  attenuation of vectors **1**:25
  expression levels **1**:24
  first-generation **1**:22
    design strategy **1**:22, **1**:*24*
  gene therapy delivery **1**:1021
    clinical applications **1**:1022
      cancer immunotherapy **1**:26
      familial hypercholesterolemia **2**:461
      hemophilia **3**:213, **3**:214
      muscular dystrophy **4**:185
    efficiency **1**:1083
    model systems **1**:1021
    targeting **1**:1083–1086
      adapter-based **1**:1084–1085, **1**:*1084*
      genetic modification **1**:1085–1086
      structural constraints **1**:1085, **1**:1086
    transient immunosuppression **2**:461
    see also Gene therapy
  gene transfer **1**:22–24
  high-capacity systems **1**:24
  longevity **1**:24
  pathogenic potential **1**:25
  safety features **1**:25
  second generation **1**:23
    design strategy **1**:*25*
Adenovirus (Ads) **1**:21–22, **1**:1083
  life cycle **1**:21–22
    cell entry **1**:1083, **1**:1084, **1**:*1084*
      receptor (CAR) tropism **1**:1083
    DNA replication **1**:22, **1**:*23*
  as vectors *see* Adenoviral vectors
  viral genome **1**:21–22, **1**:*22*, **1**:1083
  virion structure **1**:21, **1**:*22*, **1**:1083
    capsid **1**:*1083*
    fibers/knobs **1**:1083, **1**:*1084*
      'fiber pseudotyping,' **1**:1085
      structural analysis **1**:1085
    genetic modification **1**:1085–1086
Adenylyl cyclase, G proteins **2**:*280*
ADH1B (alcohol dehydrogenase IB beta polypeptide), alcoholism **1**:57
Adiabatic (Born-Oppenheimer) approximation **4**:50
Admixture **3**:964
  admixture rate **3**:964
  genetic epidemiology **2**:840, **2**:842

Cancer (*continued*)
 precancerous changes **3**:117
 as multistep process **5**:662
 phosphatases **3**:623
 protein kinases **3**:622, **4**:443–447
 see also Cancer genetics;
  Carcinogenesis
 skin disorders **5**:319–321, **5**:*321*
 treatment *see* Cancer therapy
 see also Tumors; *individual types*
Cancer chemotherapy *see* Cancer
 therapy
Cancer Chromosome Aberration Project
 (CCAP) **1**:1037
 commercial clone sets **1**:1037
 cytogenetic-physical map integration
  1:1037–1042
  Bac clone localization **1**:1037,
   1:*1038*
  Bac clone sequencing **1**:1040
  database integration **1**:1041–1042
  karyotype refinement **1**:1039–1040
  see also Human genome sequencing
 Mitelman database **1**:1037, **1**:*1041*
 NCBI website **1**:1037, **1**:*1039*, **1**:*1040*,
  1:*1041*, **1**:*1042*
 see also Cancer Genome Anatomy
  project (CGAP)
Cancer cytogenetics **1**:373–378, **1**:379–
 386, **1**:741, **1**:742, **1**:743, **1**:829,
 3:38–39, **5**:604, **5**:*605*
 chromosome abnormalities **1**:828,
  3:605
  aneuploidy **1**:254, **3**:39
  translocations *see under*
   Chromosomal translocations
 comparative genome hybridization
  (CGH) **1**:385–386, **1**:*385*
 cytogenetic-physical map integration
  1:1037–1042
 diagnostic tool **1**:377, **1**:743, **5**:599
 karyotyping *see* Karyotype analysis
 leukemias **1**:381, **1**:667, **1**:699, **1**:701,
  3:38, **3**:39
  Philadelphia chromosome *see*
   Philadelphia chromosome
 lymphomas **1**:381–382, **1**:*382*, **1**:701
 problems **1**:375, **1**:383
 prognostic tool **1**:377–378
 solid tumors **1**:383–385, **1**:*384*
  Ewing sarcoma family **1**:384
  neuroblastoma **1**:384–385
  rhabdomyosarcoma **1**:384
  see also specific tumors
 websites/databases **1**:374, **1**:375
 see also Chromosomal abnormalities;
  Chromosomal rearrangements;
  specific cancers
Cancer genetics **1**:376–377, **1**:379,
 1:639, **1**:1059, **3**:38–42, **5**:748
 apoptotic pathways **1**:166–168,
  1:174, **1**:379, **4**:446
  see also Apoptosis
 CCAP *see* Cancer Chromosome
  Aberration Project (CCAP)
 cell cycle role **1**:166–168, **4**:446
  checkpoints *see* Cell cycle control
 CGAP *see* Cancer Genome Anatomy
  project (CGAP)
 chromosomal level **4**:390
  bladder cancer **1**:323, **1**:*325*
  chromosome instability *see*
   Chromosome instability (CIN)
  loss of heterozygosity *see* Loss of
   heterozygosity (LOH)
  rearrangements *see* Chromosomal
   rearrangements
  X-Y homology regions **5**:779
  see also Cancer cytogenetics;
   individual chromosomes
 DNA repair defects **1**:534, **3**:40
  DSB repair *see* Double-strand break
   repair (DSBR)
  mismatch repair *see* Mismatch repair
  NER repair *see* Nucleotide excision
   repair (NER)
  see also DNA repair
 epigenetic level **3**:41
  CpG mutation **2**:119, **2**:121
  DNA methylation **2**:116–117,
   2:*116*, **3**:41
  loss of imprinting (LOI) **3**:41, **3**:427

Cancer genetics (*continued*)
 see also DNA methylation; Genomic
  imprinting
 familial *see* Hereditary cancer
 fusion proteins *see* Fusion proteins
 gene level **1**:679
  amplification **2**:593–598, **2**:*595*,
   3:40, **5**:599
  deregulation **1**:*375*, **1**:376–377,
   1:*383*
  expression analysis **1**:895, **3**:931–
   932
   see also Gene expression analysis;
    Microarray(s)
  fusions **1**:376–377, **1**:*376*, **1**:379
   see also Chromosomal
    translocations; Fusion proteins;
    Gene fusion
  gene conversion **2**:608
  Hox genes **3**:291
  'mutator' genes/phenotypes **1**:379,
   3:41, **3**:*42*, **3**:980
  nuclear receptor genes **1**:656,
   2:596, **4**:388, **4**:390
  oncogenes *see* Oncogenes
  phosphatases **3**:623
  protein kinases **3**:622, **4**:443–447
  telomerase expression **5**:490, **5**:499
  tumor suppressor genes *see* Tumor
   suppressor genes
 genetic redundancy **2**:893
 genomic instability *see* Genomic
  instability
 hereditary *see* Hereditary cancer
 methods **3**:42
  PCR **4**:605
  see also specific techniques
 sequence level **3**:40–41
 see also Carcinogenesis; *specific
  cancers; specific genes*
Cancer Genome Anatomy Project
 (CGAP) **1**:386–391
 anticancer drug design **1**:390–391
 applications **1**:390
 goals **1**:387
 informatics tools **1**:387, **1**:388–389,
  1:*389*
  cDNA xProfiler **1**:388
  Differential Gene Expression
   Displayer (DGED) **1**:389, **1**:*389*
  Gene Library Summerizer (GLS)
   1:388
  Virtual Northerns **1**:389, **1**:*390*
 methodologies
  EST analysis **1**:387
  FISH **1**:387
  SAGE analysis **1**:387
  overview **1**:387
  SNP identification **5**:295
 web site **1**:387, **1**:*388*, **1**:1037,
  1:*1038*
  'chromosomes' section **1**:388
  'genes' section **1**:387
  'pathways' section **1**:388
  see also Cancer Chromosome
   Aberration Project (CCAP)
Cancer therapy
 antisense therapy **1**:149–151, **1**:*150*,
  1:*151*, **1**:168
 apoptosis induction **1**:168, **1**:181
 DNA damage and **3**:694
 drug design, CGAP contribution
  1:390–391
 drug resistance **4**:154
  ABC transporters **4**:154–159
  apoptosis mutations and **1**:180
  gene amplification **2**:596, **2**:597,
   3:40
  p53 role **1**:167, **1**:180, **3**:694
  prediction **1**:168
  ribozyme therapy **1**:150
  testicular cancer **5**:503–504
  see also ATP-binding cassette
   transporters; Pharmacogenetics
 gene therapy **1**:167, **2**:699–700,
  2:718, **4**:321
  drug activation systems **2**:700
  genetic tagging **2**:699–700
  immunogenicity enhancement
   2:700
  naked DNA **4**:270
  oncogene downregulation **2**:700

Cancer therapy (*continued*)
 tumor suppressor gene replacement
  2:700
 see also Gene therapy
 hemopoietic stem cell transplant
  following **3**:197
 host glycosylation effects **3**:117
 Li-Fraumeni syndrome **3**:694
 mitotic spindle microtubules **4**:22
 oncogenes as targets **4**:441–442
 ribozyme therapy **1**:150, **1**:*150*, **1**:151
 selective targeting **2**:420–421
  endothelium vasculature **2**:421
  promoters responsive to treatment
   2:421
  tissue-specific promoters **2**:421
  tumor-selective promoters **2**:421
  see also individual cancers; *specific
   treatments*
Candida infections, host glycosylation
 and
 3:118
Candidate gene approach *see*
 Association studies
Canine models, spinal muscular atrophy
 (HCSMA) **5**:370
CAP2/3/4 sequence assemblers **5**:216,
 5:*216*
Capillary electrophoresis **1**:392–394
 advantages **1**:392
 applications **1**:393–394
 buffers **1**:394
 detection methods **1**:392
 electroendosmotic flow (EOF) **1**:392
 future developments **1**:394
 methods **1**:392
 modes **1**:392
 nucleic acid analysis **1**:393, **1**:*394*
 peptide/protein analysis **1**:393–394,
  1:*395*
 sample introduction **1**:392
 single-strand chain polymorphism
  (SSCP) **1**:393
 system **1**:*393*
 thermal gradient (TGCE) **1**:393
Capillary sequencers **5**:157
 see also DNA sequencing
Capitalism
 Darwinism and **3**:807–808
 disability theories **2**:23
 see also Commercialization (of
  research)
Carbohydrate-deficiency glycoprotein
 syndromes **3**:118–119, **4**:653
Carbohydrates
 blood group antigens **1**:333–334,
  1:*334*, **1**:*335*
 metabolism
  inherited disorders **3**:894–895
  mitochondrial **3**:983, **3**:*984*
Carboxypeptidase N **1**:*916*
 deficiency **1**:921
 genes **1**:*918*
Carboxypeptidase R **1**:*916*
 genes **1**:*918*
Carcinoembryonic antigen (CEA)
 in cancer **3**:115
 selective targeting **2**:421
Carcinogenesis **1**:374, **1**:611, **4**:435,
 4:439–441, **4**:443
 cellular dysregulation
  apoptosis role **1**:174
  cell cycle role **1**:447–449, **4**:446
  cell growth **4**:441
  cell proliferation **4**:440
  cell survival **4**:440–441
  DNA repair defects **1**:534, **3**:40,
   5:658
  see also DNA repair; *specific
   pathways*
  see also Apoptosis; Cell cycle control;
   DNA damage
 gene conversion **2**:608
 genes mutated **5**:658
  oncogenes *see* Oncogenes
  protein kinases **4**:443–447
   antiapoptotic kinases **4**:446
   cell cycle kinases **4**:446
   cytoplasmic/nuclear kinases
    4:445–446
   receptor tyrosine kinases **4**:443–
    445

Carcinogenesis (*continued*)
 see also individual protein kinases
 tumor suppressors *see* Tumor
  suppressor genes
 see also Cancer genetics; *specific
  genes*
 genetic instability **5**:658, **5**:659
  chromosomal *see* Chromosome
   instability (CIN)
  microsatellites *see* Microsatellite
   instability (MSI)
  see also Genomic instability
 heterozygosity loss *see* Loss of
  heterozygosity (LOH)
 methylation-mediated *see* DNA
  methylation
 models/theories **5**:659
  'cooperative oncogene' hypothesis
   4:440
  Darwinian evolution view **1**:533,
   1:534, **1**:535
   Barrett esophagus model **1**:253–
    256
  Knudson two-hit hypothesis **1**:383,
   3:691, **5**:665–666, **5**:812
  single mutation model **5**:659
  somatic mutation theory **1**:373
  in vitro model **5**:659
 multistep process **5**:659, **5**:662
  multiple mutations **3**:41, **3**:*42*
 mutation(s) **5**:657–661
  advantageous mutation **5**:657
  CpG mutation **2**:119, **2**:121
   see also CpG dinucleotides; DNA
    methylation
  deletion mutations **5**:659
  essential mutations **5**:660
  frameshift mutations **5**:658
  number **5**:659–660
  point mutation **5**:658
  signature mutation **5**:657
  see also Mutation(s)
 pathways **5**:660–661
  telomerase role **5**:490, **5**:499
 unique mutations **5**:660
 in vitro model **5**:659
 see also Cell cycle; Cell senescence;
  individual cancers/tumors
Cardiac arrhythmias **1**:404–407
 atrial fibrillation **1**:407
  gene loci **1**:*405*
 heart (AV) block **1**:407
 idiopathic ventricular fibrillation **1**:406
 long QT syndrome (LQT) **1**:404–406,
  4:661
  genes **1**:404–406, **1**:*405*
  genotype-phenotype correlation
   1:406
 right ventricular outflow tract
  tachycardia **1**:406
Cardiac biopsy, hypertrophic
 cardiomyopathy **3**:398
Cardiac fibroma, nevoid basal cell
 carcinoma syndrome **4**:332–333
Cardiac pacemaker **1**:396
Cardiac precursors **1**:398
Cardiomyopathy **1**:397, **4**:884
 arrhythmogenic right ventricular
  dysplasia (ARVD) **1**:*400*
 dilated *see* Dilated cardiomyopathy
  (DCM)
 hypertrophic *see* Hypertrophic
  cardiomyopathy (HCM)
 muscular dystrophies **1**:397, **2**:250,
  4:189
 mutations causing **5**:530
 proteomics *vs.* genomics **4**:884–888
Cardiovascular disease
 APOE*4 association **1**:409, **2**:32
 coronary artery occlusion *see* Coronary
  artery disease (CAD)
 gene therapy **1**:411–414, **2**:700–701
  angiogenic growth factors **1**:411–
   412, **1**:*412*
  antisense oligonucleotides **1**:413
  coronary disease **1**:412–413
  future directions **1**:414
  ischemic heart disease **4**:271
  myocardial disease **1**:414
  peripheral disease **1**:411, **1**:*412*
  problems **1**:414
  see also Gene therapy

NATURE ENCYCLOPEDIA OF THE HUMAN GENOME / ©2003 Macmillan Publishers Ltd, Nature Publishing Group / www.ehgonline.net

HIV-1 infection (*continued*)
  multi-drug resistance gene, ATP-
    binding cassette 3:368
  transporter associated with antigen
    processing (TAP) 3:365
  vitamin D receptor 3:368
  genetic vaccination 3:441
  health insurance discrimination 2:27
  host glycosylation importance 3:118
  life insurance discrimination 2:28
  pathogenesis 3:363
  see also AIDS
HIV Protease Database 5:409
HLA *see* Human leukocyte antigens (HLA)
HMG-CoA reductase (3-hydroxyl-3-
  methylglutaryl coenzyme A reductase)
  1:358
HMG proteins 5:576
  germ-line mutations 5:576, 5:*576*
  SOX proteins *see* SOX proteins
  SRY proteins *see* SRY proteins
HMMER2 searches 5:413
HMMer, *C. elegans* genome analysis
  1:370
HMMR, Smith-Waterman algorithm
  5:*334*
HNF4A (hepatocyte nuclear factor 4,
  alpha)
  mapping 1:684
  mutation 1:681
hnRNPs 4:647, 4:*647*
  classes 4:647
  location 4:647
  mRNA degradation, hnRNP D (AUF1)
    5:88
  mRNA export 5:87
  pre-mRNA splicing 2:387, 5:83, 5:386
    hnRNP1 (NOVA-1) 5:85
  telomerase holoenzyme 5:480
Holism, systems biology 5:461–462
Holliday junctions 2:152
  branch migration 2:99, 2:*99*
  DNA repair 2:152
  gene conversion 2:608
  meiotic recombination 2:727, 2:*727*,
    3:*870*
Holm's method 4:179, 4:*180*
Holocaust 4:977
  race inferiority 4:980
Holoprosencephaly (HPE) 1:672, 1:963–
  964, 1:1114, 2:256–257, 4:303
  environmental factors 1:1114
  genetic basis 1:1114
  mutations 1:963–964, 2:256–257
  phenotype 1:963, 2:256
  prevalence 1:1114
  zebrafish models 5:821
Holt-Oram syndrome 3:*697*, 3:701
  genetic basis 1:638
  septal defects 1:401
Homeobox *see* Homeodomain
Homeobox genes 3:287
  development role 1:1112
    brain development 1:1113
    limb development 3:702–703
    tooth development 5:539
  disease role 3:291
    cancer 3:291
    craniosynostosis syndromes 1:961
    dysregulation in leukemia 3:685
    germ-line mutations 1:961
    holoprosencephaly 1:964
    parietal foramina 1:963
    Turner syndrome 4:77
  Drosophila homeotic genes 1:1104
  gene family evolution 3:287, 3:*288*,
    4:325–326
    evolution of teleosts 2:537
  Hox genes *see* Hox gene family
  see also individual genes
Homeodomain 3:287, 3:290, 4:325
  functions 3:283
  genes encoding *see* Homeobox genes
  helix-turn-helix (HTH) motif 3:681
  proteins *see* Homeodomain proteins
Homeodomain-leucine zipper (HD-ZIP)
  motif 3:681
Homeodomain proteins 3:287
  cofactors for DNA-binding 3:290,
    3:*290*
  genes encoding *see* Homeobox genes
  helix-turn-helix (HTH) motif 3:681

Homeodomain proteins (*continued*)
  Hox proteins *see* Hox proteins
  leucine zipper motifs (HD-ZIP) 3:681
  mutations 5:575, 5:*575*
    morphological evolution 3:286
  transcription factors 5:575
  see also HMG proteins
Homeoproteins *see* Homeodomain
  proteins
Homeotic genes *see* Homeobox genes
Hominid evolution
  analysis
    ancient DNA analysis 1:118, 4:282–
      286
    codon bias 1:851, 1:*852*
    genetic markers 2:107–110
    mtDNA analysis *see* Mitochondrial
      DNA (mtDNA) analysis
    Y chromosome analysis *see* Y
      chromosome analysis
    see also Ancient DNA (aDNA)
  chromosome rearrangements 1:759–
    761
    centromeres 1:759–760
    chromosome 2 1:559, 1:759, 1:760,
      1:*760*, 1:761, 1:884, 1:890,
      1:*890*
      centromeric repeats 1:480
      telomeric repeats 1:499
    inversions 1:760–761
    sequence 1:761, 1:*762*
    telomeres 1:759, 1:760
    translocations 1:792
  gene expression alterations 4:731,
    4:733
  genome variation 4:731, 5:292
    see also Genetic variation;
      Polymorphism
  hominoid divergence times 3:272–
    273
  Homo sapiens 3:274–276, 4:38–40
    archaic humans 3:382–383
    codon usage bias 1:851, 1:*852*
    Darwin's views 1:1052, 1:1053–
      1054
    genetic diversity *see* Human genetic
      diversity
    hominoid divergence times 3:272–
      273
    karyotype 3:593–598
    'mitochondrial Eve,' 3:274, 4:39
    multiregional hypothesis 3:324–
      325, 4:38, 4:*39*, 4:41–42
    peopling of India 4:538–540
    populations *see* Human populations
    hybridization and assimilation
      hypothesis 4:282, 4:285
    last common maternal ancestor 2:109,
      3:322, 3:376, 3:377
    last common paternal ancestor 2:109,
      3:322, 3:377
    'mitochondrial Eve,' 3:274, 4:39
    multiregional hypothesis 4:38, 4:*39*,
      4:41–42, *map* 4:282
    Neanderthals *vs.* modern humans
      3:274, 4:285, 4:*285*
      mtDNA 4:283–286, 4:*284*
    Orronin tugenensis 3:374
    Out of Africa theory *see* 'Out of Africa'
      theory
    phylogeny/phylogenetics 3:271–276,
      4:284, 4:*285*, 4:586–588, 4:*705*,
      4:707, 4:708
      genetic disorders 2:814
      mtDNA polymorphisms 3:1004–
        1006, 3:*1006*, 3:*1007*
    see also Human genetic diversity;
      Human genome evolution; *individual
      species*
  Hominids 3:270–271
    evolution *see* Hominid evolution
    taxonomy 3:270, 3:*271*
    see also individual species
  Homo (genus)
    emergence 3:374
    phylogeny/phylogenetics 3:271–276,
      4:586–588
  Homocysteine
    accumulation 1:1006
    venous thrombosis 5:725
  Homocystinuria
    CBS deficiency 1:1002–1003

Homocystinuria (*continued*)
  clinical/biochemical features
    1:1002, 1:*1002*
  coronary artery disease (CAD) 1:409
  genetic basis 1:688
Homo erectus 3:374
  human demographic history 3:383
  Europe 4:632
  multiregional theory 4:38
  regional continuity model 3:375
Homo ergaster 3:374
Homogeneity, genetic isolates 2:862–
  863
HOMOG, multiple disease loci 3:715
Homo habilis 3:374
Homologous genes 1:897, 3:277,
  4:458–461
  detection of remote 4:729
  identical by descent (ibd) 3:632
  orthologs *see* Orthologous genes
  paralogs *see* Paralogous genes
  sex chromosomes 5:779–780
    see also X chromosome; Y
      chromosome
  xenologs 4:458, 4:461
  see also Comparative genomics; Gene
    families
Homologous recombination (HR) 2:142,
  3:*946*
  applications 2:694–695
  consequences of
    chromosomal translocations 2:544,
      4:26, 5:597
    chromosome instability 2:147, 4:26
    gene conversion *see* Gene
      conversion
    somatic mosaicism 5:388–389,
      5:*389*
  defects 3:945
  double strand break repair (DSBR)
    1:492, 2:78, 2:143–144, 2:*143*,
    2:152–153, 2:157, 4:26
  frequency of events 2:694
  gene targeting 2:693–695
    conditional alleles 2:694
    goals 2:694
    human disease models 2:694
      treatment strategy testing 2:695
    knock-out/knock-in strategies
      2:694
      improving targeting 2:694
      limitations 2:694
    proof of principle experiments
      2:695
    recombination-based cloning
      5:808–809, 5:*809*
    transgenic mice 5:587, 5:*588*
    triplex technology 2:695
    see also Gene targeting; Knockin
      mice; Knockout mice; Transgenic
      animals
  mechanisms 2:143–144, 2:*143*
    Brac1/2 2:144
    Holliday junctions *see* Holliday
      junctions
    low-copy repeats 1:492, 1:*493*,
      1:805, 1:*805*
    Rad52 proteins *see* Rad52 epistasis
      group
    RecQ proteins *see* RecQ family DNA
      helicases
  meiosis *see* Meiotic recombination
  nonallelic (NAHR) 1:492
    minimal efficient processing
      segment (MEPS) 1:492
  unequal 4:*233*
    concerted evolution 2:680
    gene duplication 4:835
    gene family evolution 4:164
    in genetic disease 4:233–235,
      4:*234*, 4:236
    low-copy repeats 1:492, 1:*493*
    meiosis 2:647
    mitosis 2:647
    opsin genes 1:861, 1:*862*
    transposable elements 5:5, 5:*5*,
      5:55–56
      SINE elements 5:*5*, 5:6–8, 5:56,
        5:269
  see also Chromosomal
    rearrangements; Meiosis; Mitosis
Homology evolution 2:373

Homology search
  bioinformatics 1:297
  protein structure prediction 4:865–
    866
  significance value 4:865
  see also Sequence homology;
    Similarity search
Homophila database 2:236
Homoplasmy 2:373, 2:375
  DNA/protein sequences 2:377
  mitochondrial DNA 3:985, 4:40
Homoplastic similarity 2:375
Homoploid hybrid speciation 5:360
Homo sapiens neandertalensis see
  Neanderthal man
Homo sapiens, origins and evolution *see*
  Hominid evolution
Homo sapiens sapiens 4:633, 4:635
Homosexuality
  prenatal screening 2:5, 2:6
  see also Sexual orientation
Homozygosity
  definition 1:77
  dominant disorders 2:209–210, 2:*209*
  population studies 2:820
  recombinant inbred mouse strains
    (RIS) 4:102
  see also Homozygosity mapping
Homozygosity mapping 2:842, 3:278–
  281
  applications 3:279
    Bardet-Biedl syndrome 3:279
    Bloom syndrome 3:279
    congenital hypothyroidism 3:280
    Hurler syndrome 3:279
    metachromic leukodystrophy 3:279
  false positives 3:279
  genotyping 3:280
    DNA pooling 3:278, 3:280
    LOD scores 3:278
    screening stage 3:278, 3:280
  history 3:278–279
  homogeneous population assumption
    3:279
  methodology 3:279
  statistical significance 3:279, 3:280–
    281
    Binomial Law 3:280
    thresholds 3:280
  see also Consanguinity; Inbreeding
Hoogsteen basepairs 2:*177*, 2:178
Hopeful monster 4:89
Horizontal (lateral) gene transfer
  bacterial DNA in humans 1:241–243
  comparative genomics 1:900
  detection 1:242–243
    BLAST 1:242
    phylogenetic analyses 1:243
  mitochondrial origins of nuclear genes
    4:12, 4:13
  transposable elements 5:617, 5:620
    LTRs 5:618
    transposons 5:617–618
  xenologous genes 4:458, 4:461
Hormone response elements (HREs)
  4:386
Horse, as model organism 1:135
Host defence, complement role 1:904
Hostile parenting 4:553
House dust mite, asthma 1:200
Housekeeping genes 2:261
  CpG islands 2:687
  GC-rich promoters 4:738
  see also R banding (reverse banding)
House of Commons Science and
  Technology Committee 4:721
Howler monkeys, taxonomy 1:763
HOXA gene complex 3:287, 3:*288*,
  5:575
  HOXA1 autism link 1:222
  HOXA9 cancer association 3:291
  HOXA10 limb development role 3:703
  HOXA11 limb development role
    3:703
  HOXA13
    hand-foot-genital syndrome 3:291
    limb development 3:703
  limb development 3:702
HOXB gene complex 3:287, 3:*288*,
  5:575
  chromosome 17 location 1:663
  HOXB4, limb development 3:696

Major histocompatibility complex (MHC)
(*continued*)
  interchromatin compartments
    1:515
  linkage advantages 3:371
    MIC 3:371
    see also Chromosome 4
  polymorphism 3:371–372, 3:783,
    3:787–788, 3:789–791
    fluctuating selection 3:790–791
    frequency-dependent models
      3:790–791
    generation 3:790
    heterozygosity 3:778
    heterozygote advantage 3:787–
      788, 3:790–791
    nature 3:789
    selective processes 3:790–791
    study prospects 3:791
  gene therapy in transplantation 2:973
  as genetic markers 2:109
  genomic map 3:779
  HLA system *see* Human leukocyte
    antigens (HLA)
  recombination 'hotspots,' 1:331,
    2:966
  MHC Sequencing Consortium 3:778
  'virtual,' 3:778
  see also Immunoglobulin superfamily
Malaria 3:447
  genetic variation and protection 1:73,
    3:449, 3:450
    genetic disease 2:331, 2:808
    see also Sickle cell disease (SCD);
      Thalassemia(s)
  resistance 3:93
  susceptibility in ethnic groups 3:447
Mal de Meleda 5:319, 5:319
MALDI 3:841, 3:843, 4:890, 4:*890*
  glycoprotein analysis 3:111
  high pressure 3:842
  high-throughput screening 3:843
  MALDI-MS/MS 3:842, 3:843
  MALDI-MS, protein analysis 4:766–
    767
  MALDI-TOF 2:405, 3:842, 4:766–767,
    4:*767*, 4:890, 5:692
    applications 3:848
    cryogenic detection *see* Cryogenic
      detectors
    microarrays 1:184
    protein-protein interactions 4:847
    protein sequencing 2:438, 4:770–
      771
  peptide mass fingerprinting 4:766
  peptide sequencing 2:438, 3:842,
    4:770–771
  variations 3:842
Male bias, DNA mutation rate 4:218–
  221
Malécot, G. 3:792–793
Male-driven evolution 3:794–796
  autosomal substitution rate 3:795
  human genetic disease 3:794
  male-to-female mutation rate ratio
    (α_m) 3:794
    ancestral polymorphism effects
      3:795–796
  molecular support 3:794–795
Malignancy *see* Cancer
Malignant fibrosis histiocytoma (MFH)
  5:*311*
Malignant hyperthermia (MH) 3:802–
  806
  central core disease (CCD) and 3:804
  clinical features 3:802–803
  diagnosis
    genetic 3:805–806
    susceptibility testing 3:803
  genetics 3:804–805
    DHP receptor 3:805
    locus heterogeneity 3:805, 3:*806*
    ryanodine receptor 1 (*RyR1*) 3:803
      functional analysis 3:805
      mutations 3:804–805, 3:*804*
  incidence 3:803, 3:805–806
  porcine 3:803
Malignant liposarcoma 1:660
Malignant melanoma *see* Melanoma
MaLR (mammalian apparent LTR
  retrotransposon) 5:49, 5:60, 5:61,
  5:62

Malthus, Thomas Robert 1:1051, 3:810–
  811
  An Essay on the Principle of Population
    1:1051
  influence on Darwin 3:807, 3:808,
    3:810
  population explosion prediction
    2:341, 3:807, 3:810
    consequence 3:810
Mammal(s)
  artificial chromosomes *see* Mammalian
    artificial chromosomes (MACs)
  circadian rhythm genetics 1:811
  development 1:1111–1112
  DNA repair 5:566–567, 5:*567*
  genomes *see* Mammalian genome(s)
  imprinting *see* Genomic imprinting
  methyltransferases (MTases) *see* DNA
    methyltransferases
  N-linked glycoproteins 3:108
  phylogeny 3:816–821, 3:*817*
    comparative genomics 1:889
    primates 1:759, 4:705–708
    rodent/primate divergence 1:875,
      1:*876*
  primates *see* Primate(s)
  reproduction 3:46
  see also specific groups/orders/species
Mammalian apparent LTR
  retrotransposon (MaLR) 5:49, 5:60,
  5:61, 5:62
Mammalian artificial chromosomes
  (MACs) 1:198, 3:811–815
  advantages 3:812
  applications 3:811–812
  biotechnology 3:812
  composition 3:812
  delivery problems 1:198
  development approaches 3:812–815,
    3:*813*
    bottom-up approach 3:813–814,
      3:*813*
    centromeric DNA 3:813–814
    obstacles 1:198, 3:813, 3:815
    satellite-DNA-based chromosomes
      3:814–815
    top-down approach 3:*813*, 3:814
  ethical considerations 1:198
  functional elements 3:812
  gene therapy 3:812
  human 3:627
    minichromosome generation
      3:812–814
    naturally arising 3:814
  minimal size 3:812
  mitotic stability 3:812
  requirements 3:812
  research 3:811–812
  structural stability 3:812
Mammalian genome(s)
  chromosome number 3:36, 3:*36*
  evolution 3:789–795, 2:363–365
    changes over time 1:795
    chromosome rearrangements 1:758–
      764, 1:789–795, 1:790, 1:884
      complex 1:794–795
      duplications 1:793–794, 2:623
      fission/fusion 1:789–792, 1:*790*,
        1:*791*, 1:884
      inversions 1:793, 1:*794*, 1:884
      translocations 1:792–793, 1:884
      see also Chromosomal
        rearrangements
    conservation 1:874, 2:731
    evolutionarily conserved
      chromosome segments (ECCS)
      1:884, 1:*886*
    gross changes 1:789
    karyotypes 1:747–754, 1:*749*,
      1:884
    primates *see* Primate evolution
    X chromosome conservation 1:763
    see also Comparative genomics;
      Genome evolution
  human *see* Human genome
  imprinting *see* Genomic imprinting
  marine mammals 1:790
  organization 1:883, 3:22
    isochores *see* Isochores
  sizes 3:20, 3:*36*
  synteny 1:763, 1:789, 1:*790*
  see also specific genomes

Mammalian interspersed repeats (MIR
  elements) 2:352; 5:49
  structure 5:*621*
Mammalian protein-protein interaction
  trap (MAPPIT) 4:772
Manhattan Project 4:525
Mania 4:81
Manic depression *see* Bipolar disorder
MANIP 5:143, 5:*144*
Mannan-binding lectin (MBL) 1:915
  genes 1:*918*
Mannosidases, glycoprotein biosynthesis
  3:106, 3:107
Mantle-cell lymphomas (MCL), gene
  mutation 502:5
Many-body problem 4:50
MAOs *see* Monoamine oxidases (MAO)
MAPaint program 5:*309*
Maple syrup disease 3:900
Mapmaker 2:873
MAPMAKER program 4:167
MAPMAKER/SIBS program 3:716, 4:168
Mapping functions *see* Genetic distance
  and mapping
Map-specific lod scores 3:738
MAPT (microtubule associated protein
  tau), mutation 4:240
Mapviewer 1:1041
Marble bone disease 3:769–770
Marfan syndrome 1:403–404, 1:655,
  3:832–836, 5:323
  clinical manifestations 3:832, 3:*834*
  diagnosis 3:832–833, 3:*834*
  ectopia lentis 3:665
  genetic basis 3:832, 3:833–835,
    3:836, 4:240
  genetic counseling 2:760
  genotype-phenotype correlations
    3:835–836
  inheritance, complications 3:940
  life expectancy 3:832, 3:833
  management 3:832–833
  neonatal 3:836
  nonsense-mediated RNA decay 4:365
  pathological mechanism 3:832, 3:836,
    3:*837*
  prevalence 3:832
Margarita syndrome 1:1095
Marine mammals, chromosome fusions
  1:790
Mariner/Tc1 elements 2:352, 5:5, 5:620
Marker chromosomes, karyotype analysis
  3:604
Marketing *see* Commercialization (of
  research)
Markov chain Monte Carlo methods
  (MCMC) 1:842, 3:838–840
  advanced topics 3:840
  advantages 3:72, 3:838
  calculation 3:838
  communication 3:838
  'detailed balance,' 3:838
  ergodic distribution 3:839
  Gibbs sampler *see* Gibbs sampler
  independent and identically
    distributed samples 3:839
  invariant distribution 3:838
  irreducible chains 3:839
  linkage analysis 3:838
  Metropolis-Hastings algorithm 3:840
  pedigree analysis 3:838, 3:839
  rate of mixing 3:839
  reducible chains 3:839
  slow mixing 3:839
  'warm-up' phase 3:839
Markov transition matrix, mutation
  dating 4:224
MARs *see* Matrix (scaffold) attachment
  regions (MARs; SARs)
Marseilles Structural Genomics Programs
  5:*423*
Marshall syndrome 5:302
Marshfield Maps 2:889
  chromosome 2 1:559
  chromosome 3 1:567
  chromosome 6 1:592
  chromosome 8 1:606
  chromosome 9 1:613
Marsupials
  evolutionary tree 3:*817*
  phylogeny 3:820–821

Marx, Karl 4:973
MASA syndrome 1:443
Maskless photolithography 3:923
MASP 1:915
  genes 1:*918*
Maspin, tumor progression 4:465
Mass spectrometry (MS) 3:841–844,
  3:*846*, 4:881, 4:890, 4:891, 4:959,
  4:961–962
  2D-PAGE and 3:405, 3:843, 4:770–
    771, 5:690, 5:691, 5:*691*, 5:692
    applications 4:891
    bacterial proteomics 1:244
    proteomics 4:959
    see also Polyacrylamide gel
      electrophoresis (PAGE)
  cryogenic detectors *see* Cryogenic
    detectors
  database search 3:845
  electrospray ionization *see* Electrospray
    ionization (ESI)
  elements 3:*846*
  fourier transform ion cyclotron
    resonance (FT-ICR) 3:842, 3:848
  ionization modes 3:847
  liquid chromatography and 3:843
    LC-MS 4:58
    LC-MS/MS 5:692
  MALDI *see* MALDI
  parameters to consider 3:847
  principles 1:979
  protein analysis 2:405, 3:841–844,
    4:770, 4:890, 4:891
    bacterial proteomics 1:244
    de novo sequencing 3:842
    glycoprotein analysis 3:111
    higher-order structure 3:843
    high-throughput 3:843
    nucleolus studies 4:404
    peptide mass mapping 4:890
    phosphoproteomics 4:573–574
    posttranslational modifications
      3:842, 4:772
    protein identification 3:842, 5:582
    protein-protein interactions 4:847
    in proteomics 4:884
    quantification 3:843
  proteomics instrumentation 3:845–
    849
    choice 3:848–849
  sensitivity 1:978
  separation procedures coupled 3:848
  tandem *see* Tandem mass
    spectrometry (MS-MS)
  tandem (MS/MS) 3:842
  techniques 3:847
  types 4:58–59
Mastectomy, bilateral on BCRA1/2
  carriers 2:458
'Master gene' evolution model 5:620
Matchmaker, DNA synthesis 2:138
Maternal age effects
  monosomies 4:76
  trisomies 5:639, 5:*639*
    Down syndrome *see* Down
      syndrome
  uniparental disomy 5:696, 5:698
Maternal and Child Health Care Act
  1:821
Maternal diabetes, holoprosencephaly
  (HPE) 1:1114
Maternal effects, asthma 1:201–202
Maternally inherited diabetes and
  deafness (MIDD) 3:988
Maternity, legal issues 2:564
Mathematical genetics 3:433–434
  diffusion methods *see* Diffusion Theory
  mouse *see* Inbred mouse strains
  see also Inbreeding
Matrilineal descent
  inheritance 3:485–486
  Trobriand Islanders 3:485
Matrix-assisted laser desorption/
  ionization *see* MALDI
Matrix (scaffold) attachment regions
  (MARs; SARs) 1:769, 1:798–800,
  3:850–851
  evolutionary conservation 2:353
  functional role 3:851
  genomic distribution 3:851
  insulator activity 1:800
  sequence features 3:850

Mobile genetic elements *see* Transposable elements
Mobius, Gregor 1:192
Möbius syndrome 2:*276*, 3:*147*
Mo cell line, property issues 4:468
MODBASE project 4:867
Model genetic organisms 3:357, 5:811
  aging research 1:41–42
    yeast 5:812–813
  developmental biology 1:1104
  gene expression analysis 2:406
  position-effect variegation 4:643
  see also Animal models; *specific organisms*
Modeling relations, systems biology 5:*462*, 5:*463*–465
Modifier genes 2:811, 3:61–62, 4:101, 4:731
  gene expression networks 2:637–638
  levels of action 2:638
  mechanisms 2:638
  modifiers at same locus 3:61
  modifiers of methylation 2:638
    see also DNA methylation
  mouse models 1:134
  nonallelic strongly-linked modifiers 3:62
  nonallelic unlinked modifiers 3:61–62
  screens, ENU mice mutations 4:109
  trans-acting 2:638
  see also Epistasis
Modularity, evolutionary psychology 5:341
Mohr-Tranebjaerg syndrome 3:990
  mutations 3:998
Molecular beacons 2:408
Molecular chaperones *see* Chaperones
Molecular clocks 4:43–48, 4:223, 4:224
  calibration points 3:273
  definitions 4:43, 4:44
  evolutionary studies 2:386
  fixation probabilities and times 2:494
  generation-time effect 4:47
  global clocks 4:45, 4:47
  hominid divergence 3:272–273
    Neanderthal/modern humans 4:285, 4:*285*
  humans 3:321, 3:375, 4:45
    Neanderthal/modern human divergence 4:285, 4:*285*
  local clocks 4:44–45, 4:48
  mutation rate 4:47
  natural selection 4:48
  population size 4:48
  primates 4:45
    divergence date assumptions 4:45
    substitution rates 4:*46*
  protein function 4:47
  relative-rate test 4:44, 4:*44*
    calculation 4:44
  rodents 4:44–45, 4:47
    GC homogenization 4:47
    generation-time effect 4:47
    marsupials *vs.* 4:47
    rate ratio 4:47
    substitution rates 4:44, 4:*45*, 4:47
      calculation 4:44
      comparison 4:45
  Zuckerland, E. 5:828
  see also Mutation dating
Molecular coevolution *see* Coevolution
Molecular cytogenetics 1:741, 1:757, 1:827–829
  applications 1:827–828
  cancer cytogenetics *see* Cancer cytogenetics
  chromosomal painting *see* Chromosome painting
  clinical cytogenetics 1:829
  clinical genetics 1:827–828
  comparative genomic hybridization (CGH) 1:828
  diagnostic algorithms 1:829, 1:*829*
  genome-wide study 1:828
  microdeletion syndromes 1:827, 1:*829*
  targeted studies 1:827–828
  see also Comparative genome hybridization (CGH); Fluorescence *in situ* hybridization (FISH)
Molecular drive 1:1049

Molecular dynamics (MD) 4:49–56
  algorithm 4:51
  atomic force field model 4:50–51
  'classical' mechanics 4:49
  force calculation 4:51
  generalizations 4:53
  ligand-binding studies 4:*52*, 4:55–56, 4:*56*
  limitations 4:53–54
    interatomic potential reliability 4:53–54
    quantum effects 4:53
    size limitations 4:54
    time limitations 4:54
  long-range interactions 4:51
  molecular modeller kit 4:54–55
  Newton's equations of motion 4:51
    numerical integration 4:51
  protein conformational changes 4:55
    ligand-binding/diffusion 4:*52*
    protein folding 4:*52*
  'quantum' simulations 4:49
  as research tool 4:49–50
  as statistical dynamics method 4:53
Molecular evolution
  codon usage *see* Codon usage
  coevolution *see* Coevolution
  DNA repair 2:159–161
  gene structure 2:679–683
  genomic congruity within phylogenies 2:729–731
  historical aspects 1:897
    Kimura's contribution 3:620–621
      see also Neutral theory
  measures 1:897
    codon usage bias 1:849
    genetic markers 2:108
    see also Codon bias
  mechanisms 4:324–327
    deletions 2:613–616
      microdeletions 2:616
      see also Deletions; Microdeletion(s)
    exon deletion/insertion 2:394–397
    exon duplication 2:395, 4:325, 4:326
    exon shuffling *see* Exon shuffling
    gene conversion 2:607–608, 2:609–611
    gene duplication *see* Gene duplication
    gene fixation *see* Fixation
    genetic drift *see* Genetic drift
    genome duplication (polyploidization) 4:324, 4:325–326
    mutational changes 4:194–203
    point mutations 4:325, 4:327
    see also Exons; Genome evolution; Introns; Mutation(s)
  neuropeptides and their receptors 4:324–329
  neutralist hypotheses 2:379, 2:384–385, 4:202–203
    neutral theory 3:620
    neutral theory 3:620–621
  orthologous genes/proteins 4:804
  phylogenetic analysis *see* Phylogenetic analysis
  polymorphism 1:71, 4:609–613
    nucleotide substitution *see* Base substitution
    SNPs 5:292
    see also Genetic variation; Mutation(s); Polymorphism
    protein isoforms 4:836–837
    pseudoexons 4:896–897
    pseudogenes *see* Pseudogenes
    rRNA genes 5:168–170
    selectionist view 2:380
      see also Natural selection
    selective constraints *see* Selective constraints
    structural constraints *see* Structural constraints
    see also Comparative genomics; Genetic variation; Mutation(s); Polymorphism; *specific genes/proteins*
Molecular genetics 4:62
  classical genetics *versus* 4:62
  genomics *versus* 4:62

Molecular genetics (*continued*)
  Hirschsprung disease (HSCR) 2:579–580
  schizophrenia 5:182–184
Molecular interaction maps 1:458
  cell cycle control 1:459–471
  computer simulation and 1:459
  interaction coordinate 1:458
  interaction number (#) 1:458
  lines 1:458
  nodes 1:458, 1:*459*
  protein modifications 1:*459*
  symbol defininitions/conventions 1:458–459, 1:*458*, 1:*460*
Molecular machines 4:62–63
  associated disease 4:63–66
    mutations 4:63–64
    pathophysiology 4:65–66
    in vivo function 4:64–65
  ATPases *see* ATPase(s)
  chaperonins *see* Chaperonins
  examples 4:62–63
  motors *see* Motor proteins
  see also Myosin(s); *specific molecules*
Molecular misreading, in genetic disease 4:237
Molecular modeling 4:54–55
  free energy methods 4:55
  molecular dynamics *see* Molecular dynamics (MD)
  Monte Carlo method 4:54
Molecular motors *see* Motor proteins
Molecular Movements Database 5:408–410
Molecular phenotyping *see* DNA fingerprinting/profiling
Molecular Sciences Institute 1:357
Molecular systematics 2:379
  atomic force field model 4:50–51
Moloney murine leukemia virus (MoMLV)
  autoimmune disease animal models 1:226
  gene delivery systems 5:65, 5:66
    hemophilia 3:214
    immunodeficiency syndromes 3:410–414
    lysosomal storage diseases 3:761
    safety issues 1:226
'Molten globule,' protein folding 4:781
Moment of conception 3:634
Mondrian 3:*937*, 3:*938*
Monilethrix 3:153
  genetic basis 1:635
Monitoring and blunting hypothesis, genetic counseling 2:784
Monkeyflowers (*Mimulus* spp.), prezygotic barriers 5:356–357
Monoallelic expression, discovery 2:738
Monoamine oxidases (MAO)
  alcoholism 1:63
  MAO-A (*MAOA*) gene
    aggression 1:36
    bipolar disorder 4:83, 4:915
    mutation 3:517
    OCD 1:158
    panic disorder 1:158
Monocyte colony-forming unit (CFU-M) 3:195
Monocyte neutrophil elastase inhibitor (MNEI) 4:466
Monod, Jacques Lucien 4:67–69, 4:287
  Chance and Necessity 1:512, 4:68
  collaboration with Jacob 4:68
  enzymatic adaptation 4:67
  Nobel Prize 4:68
  operon model 4:68
  Pasteur Institute 4:67, 4:68
  see also Transcriptional regulation
Monogenetic procreation 3:486
Monogenic traits *see* Mendelian inheritance
Monogenism 4:973
Monopoly rights, intellectual property 4:471
Monosomies 4:74–80
  autosomal 4:75, 4:77–80, 4:*78*
    chromosome 7 1:604
    chromosome 21 4:80
    partial 4:80
    see also Chromosome deletions; Microdeletion(s)

Monosomies (*continued*)
  clinical features 4:75, 4:77
  definition 4:74
  full 4:74, 4:80
  functional 5:696
  karyotype analysis 3:600
  maternal age and 4:76
  mechanisms 4:75–76
    de novo rearrangement 4:76
    malsegregation 4:76, 4:*77*
    nondisjunction 3:873, 4:75–76, 4:*75*
  mosaicism 4:74, 4:76
  sex chromosomes 4:74, 4:77
    X-inactivation effects 5:790
  uniparental disomy (UPD) 5:695
  see also individual chromosomes
Monotremes
  evolutionary tree 3:*817*
  phylogeny 3:821
Monozygotic twins (MZTs) 5:670
  biology 5:674
  concordance rates
    Alzheimer disease 1:102
    criminal behavior 1:36
    multiple sclerosis 4:175
    psychiatric disorders 4:913
      anxiety 1:157
  environmentally discordant 5:678
  heritability estimates 5:674
    twins raised apart 5:677
    twins raised together 3:232
    see also Heritability
  mosaicism 4:92, 4:94
  phenotypic variation in 5:675–676
    co-twin control studies 5:677
    epistasis effects 5:675
    gene-environment interactions 5:675–676
    X-inactivation/imprinting 5:676
  zygosity determination 5:674
  see also Dizygotic twins (DZTs); Twin studies
Monte Carlo methods
  Markov chain methods *see* Markov chain Monte Carlo methods (MCMC)
  molecular modeling 4:54
Montreal-Kingston Bacterial Structural Genomics Initiative 5:*423*
Mood disorders 1:1108, 4:81–85
  age of onset 1:1108
  association studies 4:83–85, 4:915, 5:179
    errors 4:83
  bipolar *see* Bipolar disorder
  characteristics 1:1108
  children at risk 1:1110
  classical genetics 4:81
  clinical features 4:81
  definition 4:81
  diagnosis 4:81
    DSM-IV criteria 4:*82*
  epidemiology 4:81
  inheritance 4:82
  linkage analysis 4:83, 4:*84*, 4:915, 5:178–181
    see also Linkage analysis
  longitudinal family studies 1:1110
  relative risk 1:1108, 1:*1108*, 4:81, 4:*83*
  seasonal affective disorder 4:85
  subtypes 4:85
  unipolar *see* Depression (major)
Mood stabilizers, pharmacodynamics 4:921
Moore case, property rights 4:468
Moore, G.E., ideal utilitarianism 1:289
Morality/moral theories 1:288
  patents 4:504
  secular humanism *see* Secular humanism
  utilitarianism *see* Utilitarianism
  see also Bioethics; Ethical issues
Morgan, genetic distance (M) 2:878, 2:888
  centiMorgan (cM) 2:728, 2:872, 2:878, 2:883, 2:*884*, 2:888, 3:18
Morgan, Thomas Hunt 2:234, 3:644, 4:86–87, 4:144–145
  '*Drosophila* group,' 4:67
  Pauling and 4:524
  publications 4:86

Mutation rate (*continued*)
mutation with age 4:219
mitochondrial DNA (mtDNA) 3:985,
3:1004, 3:1009, 4:1, 5:289
neutral, L1 element measures 3:750
second-order selection 4:215
Mutation-selection balance 2:961, 2:964
'Mutator' genes/phenotypes 1:379,
3:41, 3:42, 3:980
MUTYH, familial adenomatous polyposis
(FAP) 2:455
Myasthenic syndromes, congenital
4:994–996
see also Nicotinic acetylcholine
receptors (nAChRs)
MYB, antisense therapy 1:149, 1:413
Myc box 4:309
MYCL1, bladder cancer LOH 1:323
MYCN 4:309
medulloblastoma 1:354
MYC homology 4:309
neuroblastoma 1:385
gene amplification 2:594, 2:596,
3:40, 4:308, 4:309, 2:597
MycN protein 4:309
gene *see MYCN*
Mycobacteria
microarray 3:456
predisposition to infection 3:448,
3:449
Mycobacterium bovis bacillus Calmette-
Guérin (BCG), microarray 3:456
Mycobacterium tuberculosis
gene therapy 3:445
proteomics 1:246
Mycoplasma capricolum, codon usage
bias 1:851
Mycoplasma genitalium, genome size
3:33
Mycoplasma pneumoniae
genome size 3:33
proteomics 1:247
c-Myc protein 1:465
apoptosis 4:441
cell cycle control 1:460, 1:463, 1:465–
467, 1:466
see also Max protein
gene *see MYC* proto-oncogene
leucine zipper domain 3:679
negative regulation by GSK3β 1:465
telomerase transcriptional regulation
5:482
MYC proto-oncogene 1:609
alternative promoter 1:91
antisense therapy, coronary disease
1:413
array CGH 1:895
cancer associations 1:376, 1:606,
1:609, 4:440–441
Burkitt lymphoma 1:381–382,
1:382, 4:435, 4:436, 5:604
gene amplification 2:593
gene transcription 4:439, 4:440
immunocytomas 3:418
medulloblastoma 1:354
MYCN homology 4:309
structure 1:382, 1:382
Myelin
genes 1:491, 1:1087, 1:1088
PMP22 1:493–494, 1:1087
myelinopathies 1:494–495
multiple sclerosis *see* Multiple
sclerosis (MS)
sodium channels 5:552–554
peripheral components 1:494–495,
1:1087
Myelin-associated glycoprotein (MAG)
3:414
Myelodysplasia
chromosome 7 1:604
chromosome 16 1:660
Myelodysplastic syndrome (MDS) 4:256
Myelofibrosis (MF), chronic idiopathic
4:259
Myeloid cells 3:194
see also specific cells
Myeloid, lymphoid mixed leukemia
(MLL) 1:525
Myeloid progenitor cells 3:195
Myelomonocytic leukemia with
eosinophilia 1:660

8p11 myeloproliferative disease (MPD)
4:260
Myeloproliferative disorders (MPDs)
4:256–260
classification 4:256, 4:256
cytogenetic/molecular abnormalities
4:258
gene mapping 1:684
genes linked 1:681
see also Leukemia; *specific disorders*
MYH11, fusion 3:685
MYO18B (myosin-like heart gene)
discovery 3:939
expression data 3:938
structure 3:939
Myoblast transfer, muscular dystrophy
gene therapy 4:184
Myocardial disease, gene therapy 1:414
Myocardial infarction
atherosclerosis 1:204
genes associated 1:206
Myocilin, glaucoma 2:437, 2:437
Myoclonic epilepsy with ragged-red
fibers (MERRF) 2:314, 3:988
Myoclonus-dystonia syndrome (MDS),
imprinting 1:604
MyoD, gene therapy 1:414
Myogenic factors, gene therapy 1:414
Myoglobin 3:79
gene 3:80
Myomesin, expression in
cardiomyopathy 4:885
Myopathy 4:182
CAM related 1:443
cardiac *see* Cardiomyopathy
muscular dystrophies *see* Muscular
dystrophies (MD)
skeletal 5:530
see also specific myopathies
Myosin(s) 4:261–267, 5:416
classes 4:264–266, 4:265
composed 4:262–264
conventional 4:264
domains 4:262–264, 4:263
functions 4:261–264
gene superfamily 5:416
developmental regulation 4:836
mutations 4:63, 4:64
deafness 4:64, 4:65, 4:96,
5:435
familial hypertrophic
cardiomyopathy 4:64
genotype-phenotype correlations
3:398, 3:398
hypertrophic cardiomyopathy
(βMHC) 3:396, 3:397, 3:398,
3:398
myosin-actin interactions 4:63
see also Actin/actin filaments
nomenclature 4:264
phylogenetic analysis 4:261, 4:262
tail region 4:264
see also individual molecules
Myosin binding protein C (*MyBP-C*),
hypertrophic cardiomyopathy 3:396,
3:397, 3:398, 3:398
Myosin I 4:264
Myosin II 4:264
Myosin IIA syndrome 4:63, 4:64
Myosin III 4:264
Myosin IX 4:266
Myosin light chain type 2, expression in
cardiomyopathy 4:886
Myosin V 4:266
Myosin VA (MyoSa) gene 2:399
Myosin VI 4:266
Myosin VII 4:266
Myosin X 4:266
Myosin XV 4:266
Myosin XVI 4:266
Myosin XVIII 4:266
Myotilin mutation, muscular dystrophies
4:192
Myotonia congenita 3:571–574
CLCN1 chloride channel mutations
3:571–572
clinical features 3:571
dominant mutations 3:573
genetic counseling 3:574
recessive mutations 3:573
Myotonic dystrophy 4:244
anticipation 1:143

Myotonic dystrophy (*continued*)
repeat expansion 1:575, 4:193, 5:635,
5:636
instability 5:637
type 1 (DM1)502:2-3, 5:629–630
gene mutation502:2
genetic basis 5:629–630
mechanism502:3
type 2 (MD2) 4:233, 4:233
Myotubularin-related phosphatase family
1:495
Myriad Genetics, patent dispute 3:328
effects 3:329
enforcement 3:329
opposition filing claims 3:329
supporters 3:329
susceptibility testing 3:328
Myxopapillary ependymoma 1:353

N

N4 ERP component, alcoholism, linkage
analysis 1:64
NAB 5:143
N-acetyltransferase (NAT),
polymorphism 2:240, 2:240
NADH-ubiquinone oxidoreductase
(complex I) 3:993–994, 4:475, 4:476,
4:477
Nagasaki mice, prion diseases 4:715
Nail-patella syndrome (NPS) 3:615,
3:697, 3:701
modifying genes 3:61
Naming practices 3:485
Euro-American societies 3:485
Nanoelectrospray ionization 4:768
Nanopore technology, DNA sequencing
5:157, 5:158–159
Nanos, germ cell development 3:69,
3:70
NARP syndrome, ATP synthase mutation
4:63
Narrative ethics 1:287, 4:272–274
clinical application 5:341
complement 5:340
modest viewpoint 5:340
moral methodology 5:340–341
personal identity and 5:341
postmodernist viewpoint 5:340
propositions 5:340
Nascent chain-associated complex
(NAC) 4:814
Nasse, C. F. 3:853
National Agricultural Library (NAL)
3:353
National Association of Black Social
Workers 2:565
National Bioethics Advisory Commission
2:197
National Cancer Institute (NCI)
CGAP *see* Cancer Genome Anatomy
project (CGAP)
NC160 database 3:913
National Center for Human Genome
Research (NCHGR) 3:344
National Center for Biotechnology
Information (NCBI) 1:1060, 2:677,
2:800, 3:5, 3:6
cancer genomics 1:1037, 1:1039,
1:1040, 1:1041, 1:1042
CGAP informatics tools 1:388–389,
1:389
GenBank *see* GenBank
genome resources 3:5
Human Genome Project 2:671,
3:343–344
National Flow Libraries 2:498
The National Human Genome Research
Institute 3:934
National Institute of Genetics (Japan)
4:106
National Institutes of Health (NIH)
3:343–344
advisory committee on DNA
technology 2:188
commercialization of research 1:867
ELSI program *see* ELSI (Ethical, Legal
and Social Implications) program
mutant mouse resources 4:106

National Library of Medicine, MedLine
1:1058
National Organization for Rare Disorders
(NORD) 1:1060
National Society of Genetic Counselors
(NSGC) 2:746
Code of ethics 1:846
URL 1:1060
National Wilms Tumour Study Group
(NWTSG) 5:770
Natural killer (NK) cells
gene cluster 1:635
HLA class I interaction 3:371
Natural selection 1:76, 2:380
adaptation 1:1047
advantageous mutations 1:254
aging and 1:39–44
pre- *versus* post-reproductive genes
1:39, 1:40
balancing selection 1:73–74, 4:198
codominance 4:196, 4:197
competition 1:255
concept 1:1046–1047
defensins 1:1080–1081
definitions 1:1046–1047, 4:195
dominance 4:196–197, 4:197
see also Dominance
ecology 1:255
environment 1:255
evolutionary medicine 2:370, 2:371
factor V mutations 2:452, 5:724
Fisher's Fundamental Theorem of
Natural Selection 2:383
fixation 1:254
probabilities and times 2:492
see also Fixation
gene family evolution 2:650, 2:650,
2:654
gene structure 2:682
genetic variation 1:71, 4:611–612
genotypic fitness 2:491, 4:214
see also Fitness
global molecular clocks 4:47
hard 2:875
historical aspects
Bateson's views 1:259–260
Darwin's theory 1:1046–1049,
1:1051–1052, 3:808
see also Darwin, Charles Robert;
Evolutionary theory
Fisher's Fundamental Theorem
2:383
Mendel and 1:1047, 2:382, 3:888
Wallace, Alfred Russell 1:1052
immune system evolution 5:733
karyotype evolution 3:595, 3:596
levels of selection 1:1049
mutational change in evolution
4:195–198
mutation elimination 2:875
neoplastic evolution 1:254–255
overdominance 4:197–198, 4:198
see also Heterozygote advantage
phenotype/genotype relations 1:1048
prezygotic barriers, 0461.3
protein evolution 4:802
purifying selection 4:195
selective sweeps 1:254–255
'selfish gene' hypothesis 1:1048
sexual, Darwin's views 1:1053
SNPs 5:292, 5:293
soft 2:876
tests 2:380
trypsinogen 5:648
tumorigenesis 5:657
underdominance 4:197–198,
4:198
see also Genotype-environment
relationships; Genotype-phenotype
correlations; Mutation(s)
Natural substances, patents 4:503
Nature, genetic enhancement 2:825
Nature *versus* nurture debate
absolute pitch (AP) 1:2
Democritas 3:853
multiple sclerosis (MS) 4:175–176
see also Environmental factors; Genetic
risk
Navashin, M., nucleolar dominance
4:399
Naxos disease 5:315
genetic basis 3:156

SRY proteins (*continued*)
  identification 1:736
  SOX proteins *see* SOX proteins
  SRY genes see SRY genes
SSEARCH 5:*334*
SSX, in sarcoma 2:*546*
Stabilized plasmid-lipid particle (SPLP), gene therapy delivery systems 5:457–458
Staden Package 5:216, 5:*216*
Staining, PAGE 2:591, 2:*592*
Stalin, Lysenko and 3:645
Stanford-Binet test
  development 3:524
  US army applications 3:525
Stanford Microarray Database (SMD) 2:629, 3:934
Stargadt macular dystrophy 2:438
Stargardt disease (STGD) 1:215, 2:428–430
Startle disease 4:996, 4:*997*
STAT1
  interferon signaling 3:531–532, 3:*531*
  viral interference 3:533
STAT2
  interferon signaling 3:531, 3:*531*
  viral interference 3:533
Statism 1:231
Statistical methods
  in genetic epidemiology 4:179–182
  hypothesis testings 4:179–180
  linkage analysis *see* Linkage analysis
  molecular coevolution detection 1:858
  multiple significance testing *see* Multiple significance testing
  multivariate analysis 3:*509*, 3:510, 3:*512*, 4:956–957
  physical mapping 5:400–402
  sequence alignment *see* Sequence alignment
  see also specific methods
Statistical modelling 5:180
  Lander and Kruglyak's criteria 5:180
  susceptibility genes 5:440
Statistical theory of signal transmission *see* Information theory
Steatococcus tuberculatia, chromosome number 3:34
Stefins 1:1008
  pathology 1:1012
  structure and distribution 1:1008
Steichen, Edward, phenotypic art 1:188
Steinert's muscle dystrophia, genetic counseling 2:760
Stem cell research
  embryo 'harvesting,' 2:185
  ethical aspects 2:918–919
  European perspectives 2:185
  human development 3:314
  regulation 2:917
  US regulation 2:915, 2:918
  see also Embryonic stem cells (ESs)
Stem cells 3:314, 4:298
  ATP-binding cassette transporters 4:159
  cell lines, racial diversity of 4:975
  embryonic (pleuripotent) *see* Embryonic stem cells (ESs)
  hemopoietic *see* Hemopoietic stem cells (HSCs)
  neural *see* Neural stem cells (NSCs)
  research *see* Stem cell research
Stem-loop binding protein (SLBP) 5:120
  histone mRNA 4:133
Stepping stone model, migration 3:964, 3:*964*
Stereocilia, hair cells 1:1066, 1:*1067*
Sterile transcripts, immunoglobulins 3:417
Sterilization, compulsory 2:336, 2:337, 5:329–330
  aims/targets
    hereditary disease prevention 5:330
    Nazi Germany 4:276
    psychoses prevention 4:924–925
    schizophrenia 5:329, 5:330
  Canada 2:346
    British Colombia 4:276
  China 1:505
  deaths 2:336
  Denmark 4:276
  Finland 4:276
  Germany 2:342, 2:346, 4:278, 5:329, 5:330

Sterilization, compulsory (*continued*)
  Nazi Germany 2:337, 4:275, 4:276
  Norway 4:276
  Sweden 3:890, 4:276
  Switzerland 4:276
  United Kingdom 2:346
  United States 2:336, 2:346, 3:890
    California 2:341, 2:342
Steroid 5-α reductase 2 (SRD5A2), deficiency 3:540
Steroidogenesis, CYPs 1:1034
Steroidogenic factor-1 (*SF1*), hypogonadism 3:126
Steroids
  gene therapy expression regulation 4:1001, 4:1002
  hormone receptors
    evolution 4:384, 4:*384*
    see also Nuclear receptor superfamily
  metabolism in prostate cancer 4:750
  steroidogenesis 1:1034
Sterol regulatory element binding factor 2 (SREBF2) gene 2:399
Sterol synthesis, inherited defects 3:898–899, 3:*898*
Stickler syndrome 5:302
  type I, genetic basis 1:635
Stigmatization
  genetic counseling 2:779
  see also Discrimination
STIMULATE 5:174
Stimulated emission depletion microscopy (STED) 2:512
STK11 (serine/threonine kinase 11)
  disease link 1:679
  inactivation 3:162
Stochastic context-free grammars (SCFGs) 1:297
  RNA analysis 2:659
Stomatocytes 3:322
Stomatocytosis, hereditary 2:322–323
Stone, D.A., disability theories 2:22
Stored genetic material 3:482
  data protection legislation *see* Data protection legislation
  use in research *see* Genetic research
  see also Tissue samples
StrepII affinity tag 1:184
Stress, psychological models 2:783–784
Stress response
  kinases 3:622
  SINE elements 5:270
Stria vascularis 1:1066, 1:*1066*
Stroke
  apoptosis 1:169
  atherosclerosis 1:203, 1:204
  gene mapping 1:1073
  gene therapy 4:321
  stem cell therapy 4:301
Structural constraints
  amino acid sequence evolution 5:201–202
  evolutionary conserved positions 5:201
  functional constraints 5:201
  gene therapy targeting 1:1085, 1:1086
  mutation and molecular evolution 5:200–201
  RNA tertiary structure 5:135
  synonymous and nonsynonymous substitution 5:201
Structural equation modelling 5:672, 5:*672*
  advantages 5:672
Structural genomics 4:831, 4:863
  multiple alignment 4:170
Structural Genomics of Pathogenic Protozoa 5:*423*
Structural GenomiX 5:*423*
Structural graph, RNA structure prediction 5:142
Structural proteins
  collagen *see* Collagen
  genes 5:416–420
  myosin *see* Myosin
STRUCTURE program 5:715
Struggle for existence 3:808
STSs *see* Sequence-tagged sites (STSs)
Sturtevant, Alfred H. 4:145
Subependymoma 1:353
Subfunctionalization, of duplicate genes 2:648

Submetacentric chromosomes 1:740, 1:779
Subsignature network analysis 3:915
Substitution matrices 5:427–429
  BLOSUM matrices 5:428
  DNA alignment 5:429
  gap costs 5:428–429
  global alignment 5:429
  log-odds 5:427–428
  match/mismatch 5:427
  point accepted mutation (PAM) matrices 5:428
  Smith-Waterman algorithm 5:332, 5:*333*
Subtilisin *see* Serine proteases
Subtractive hybridization 2:414–415, 2:*415*
  method 2:414
Succinate dehydrogenase flavoprotein 4:885
Succinate-ubiquinone oxidoreductase (complex II) 3:995, 4:475, 4:477, 4:*477*
Sudden cardiac death
  hypertrophic cardiomyopathy (HCM) 3:396
  idiopathic ventricular fibrillation 1:406
Sugars, nucleic acid structure 1:257
Sulfate transporter 5:303
Sulfur amino acid metabolism 1:1000, 1:*1001*
  CBS mutation 1:1006
Sulfur-containing proteins, protein folding 1:485–486
Sundaland (land bridge) 3:376
Supercoiling *see* DNA supercoiling
Superconducting tunnel junctions (STJ) *see* Cryogenic detectors
Superficial spreading melanoma (SSM) 3:881, 3:*882*
Supernumerary teeth 1:1094, 1:1096
Superoxide dismutase (*SOD*)
  aging role 1:42, 1:*42*
  ALS 4:763
  Down syndrome 2:224
  SOD1 disease link 1:688
Suprachiasmatic nuclei (SCN), transplantation 1:808
Supravalvular aortic stenosis (SVAS) 1:403
Surface enhanced laser desorption/ionization (SELDI)
  applications 4:59
  microarrays 1:184
  ocular proteomics 2:438
Surface plasmon resonance (SPR) 4:59, 5:430–431
  biological applications 5:430, 5:431
  detector system components 5:*430*
  future developments 5:431
  instruments 5:430
  Langmuir model 5:430
  methods 5:431
  microarrays 1:184
  protein characterization 4:771
  protein-protein interactions 1:184, 4:772
Surfeit genes
  chromosome 9 location 1:615
  SURF1, mitochondrial disorders 3:989, 3:996, 4:479
Surname studies 5:432–434
  advantages 5:433
  degree of relationship 5:432–433
  population origins 5:434
  problems 5:433
  Y-chromosome polymorphisms 5:434
Surrogacy
  biological parents 3:635
  commerce 3:635
  donor anonymity 3:635
  gamete donation *see* Gamete donation
  gestational 3:635
  intending parents 3:635
  nonscientific issues 3:635
  selection criteria 2:564
  traditional 3:635
Surveys, public understanding 4:936
'Survival of the fittest,' 1:1053, 3:808
Survivor guilt 2:459
Survivor syndrome 3:394
Susceptibility genes 4:596–600, 4:*598*
  Alzheimer disease (AD) 1:103–104

Susceptibility genes (*continued*)
  Crohn disease 1:662
  epistatic interactions 4:599, 4:*600*
  genotype relative risk (GRR) 4:597, 4:*597*
  identification 1:841, 1:928, 2:30, 3:718–721, 5:435–441, 5:*436*
    association studies *see* Association studies
    genome-wide 1:1118, 3:720, 4:597, 5:441
    linkage disequilibrium 1:841
    linkage studies *see* Linkage analysis
    locus construction 5:439
    locus identification 5:437–438
    locus refinement 5:438–439
    markers 5:436–437, 5:*438*
    sample collection 5:436
    statistics 5:440
  locus heterogeneity 5:439
  migraine 5:439, 5:*440*
  penetrance and 3:723, 5:438
  rheumatoid arthritis 5:75
  schizophrenia 5:183
  see also Complex multifactorial traits/diseases; Quantitative trait loci (QTL)
Sutton-Boveri theory 5:442–443
Sutton, W.S. 5:442–443
  chromosome theory of heredity 5:442–443
SV40
  biology 2:191
  gene delivery vector 2:191–192
    safety 2:192
  large T antigen 2:103, 2:191, 2:192
    mechanism 2:105, 2:*106*
  origin of replication 2:*172*, 2:*173*, 2:*173*
Sweat chloride test 1:1015
Sweden, compulsory sterilization 3:890, 4:276
Swedish Adoption Twin Study of Aging (SATSA) 3:402
  extroversion 3:403
  intelligence 3:402
  sources of variance 3:404
Swiss-2Dpage 4:404, 4:788
SWISS-MODEL 4:866
SWISS-MODEL Repository 4:866
SWISS-MODEL server 4:413
SWISS-PROT 2:798, 2:800, 2:802, 4:*153*, 4:771, 4:786, 4:831, 4:856
  annotations 4:856
  comprehensiveness 4:855
  development 2:791
  entry 2:*793*
  foundation 4:785
  redundancy 4:855
  structure 2:798
Switzerland, compulsory sterilization 4:276
SYBR Green
  agarose gel electrophoresis 2:589
  PAGE 2:591
Syk, integrin signalling 3:501
Symphalanganism, proximal 3:697
Symplesiomorphies 2:374
Synapomorphies 2:374
Synapses
  chemical 4:993, 4:*994*
  plasticity 3:100
Synapsis 3:865–867, 3:*866*
  errors 3:*870*, 3:873
Synaptonemal complex (SC) 3:867, 3:*868*
Synesthesia, absolute pitch (AP) 1:2
Synonymous-nonsynonymous rates 5:287–288, 5:*287*, 5:*288*, 5:448–452
  estimation 5:449–450
    codon-based maximum likelihood method 5:450
    heuristic methods 5:449–450
  occurrence 5:288
  selection detection from DNA sequence data 5:451–452
    example 5:451–452
    likelihood ratio test 5:451
  selective constraints 5:449
Synonymous substitutions 2:361–362, 5:201
  major codon preference 4:945
    gene expression in mammals 4:946
    species comparisons 4:946–947
  in molecular coevolution 1:858
  purifying selection 4:945–947